MICROBIOLOGY

BERNARD D. DAVIS, M.D.

Adele Lehman Professor of Bacterial Physiology,
Harvard Medical School, Boston, Massachusetts

RENATO DULBECCO, M.D.

Assistant Director of Research,
Imperial Cancer Research Fund Laboratories,
London, England; Fellow of The Salk Institute,
La Jolla, California

HERMAN N. EISEN, M.D.

Professor of Immunology,
Massachusetts Institute of Technology,
Cambridge, Massachusetts; formerly
Professor and Head, Department of Microbiology,
Washington University School of Medicine,
St. Louis, Missouri

HAROLD S. GINSBERG, M.D.

Professor and Chairman, Department of Microbiology,
College of Physicians and Surgeons,
Columbia University, New York, New York; formerly
Professor and Chairman,
Department of Microbiology, School of Medicine,
University of Pennsylvania, Philadelphia, Pennsylvania

W. BARRY WOOD, JR., M.D.

(Deceased)
Boury Professor and Director, Department of Microbiology,
Johns Hopkins University School of Medicine,
Baltimore, Maryland

With special editorial help from

MACLYN McCARTY, M.D.

Vice-President and Physician-in-Chief,
The Rockefeller University, New York, New York

MEDICAL DEPARTMENT

Harper & Row, Publishers
Hagerstown, Maryland
New York, Evanston, San Francisco, London

microbiology

INCLUDING IMMUNOLOGY AND MOLECULAR GENETICS
SECOND EDITION

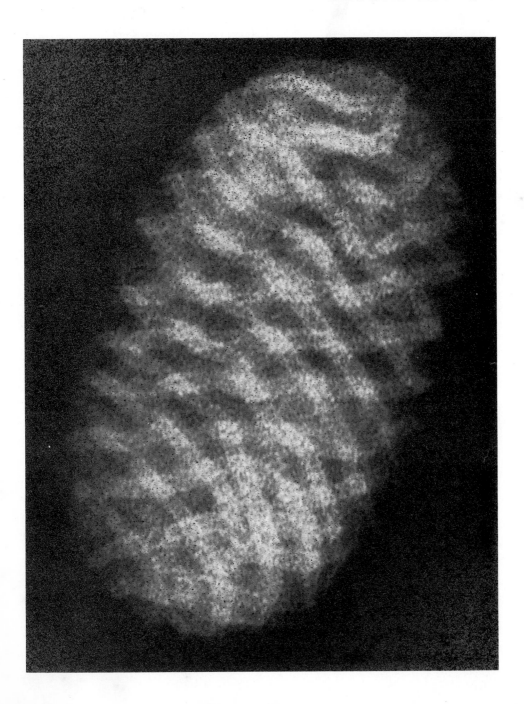

To the memory of Oswald T. Avery,
whose lifelong study of a single pathogenic bacterium
culminated in the discovery that DNA is the prime carrier
of genetic information.

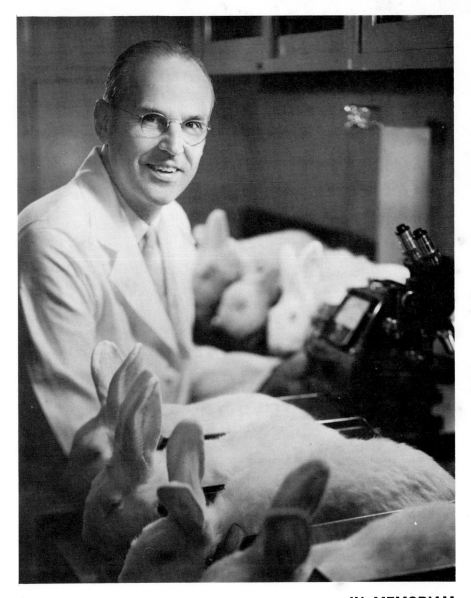

IN MEMORIAM

W. Barry Wood, Jr., was one of those rare individuals who grace all the people and institutions with whom they come in contact. His qualities of leadership, manifest in his undergraduate performance as an outstanding scholar–athlete, led rapidly to positions of major responsibility in clinical medicine and administration, but his lifelong interest in bacterial infections eventually drew him back into microbiology. This interest also led to the inception of this book, for he was primarily responsible for gathering its original team of authors. We were all strongly influenced by his devotion to microbiology and by his conviction that its advance as a pure science is inseparable from its role as one of the foundation stones of modern medicine. With his death, his coauthors and the community of biomedical scientists have lost an exceptional source of wisdom and judgment.

CONTENTS

PREFACE TO THE SECOND EDITION   xi

PREFACE TO THE FIRST EDITION   xiii

ACKNOWLEDGMENTS   xv

1. EVOLUTION OF MICROBES AND OF MICROBIOLOGY   1

I. BACTERIAL PHYSIOLOGY   Bernard D. Davis   19

2. BACTERIAL STRUCTURE   21
3. ENERGY PRODUCTION   39
4. BIOSYNTHESIS   59
5. BACTERIAL NUTRITION AND GROWTH   89
6. THE CELL ENVELOPE; SPORES   105
7. CHEMOTHERAPY   147

II. BACTERIAL AND MOLECULAR GENETICS
Bernard D. Davis/Renato Dulbecco   167

8. BACTERIAL VARIATION AND POPULATION DYNAMICS   169
9. GENE TRANSFER IN BACTERIA   181
10. STRUCTURE AND REPLICATION OF NUCLEIC ACIDS   201
11. ORGANIZATION, ALTERATION, AND EXPRESSION OF THE
    GENETIC INFORMATION   231
12. PROTEIN SYNTHESIS   271
13. METABOLIC REGULATION   311

III. IMMUNOLOGY   Herman N. Eisen   349

14. INTRODUCTION TO IMMUNE RESPONSES   351
15. ANTIBODY–ANTIGEN REACTIONS   359
16. ANTIBODY STRUCTURE: THE IMMUNOGLOBULINS   405
17. ANTIBODY FORMATION   449
18. COMPLEMENT   511
19. ANTIBODY-MEDIATED (IMMEDIATE-TYPE) HYPERSENSITIVITY   527
20. CELL-MEDIATED HYPERSENSITIVITY AND IMMUNITY   557
21. ALLOANTIGENS ON CELL SURFACES: BLOOD GROUP
    SUBSTANCES AND HISTOCOMPATIBILITY ANTIGENS   597

IV. BACTERIAL AND MYCOTIC INFECTIONS   Edited by Maclyn McCarty   625

22. HOST–PARASITE RELATIONS IN BACTERIAL INFECTIONS   627
23. CHEMOTHERAPY OF BACTERIAL DISEASES   667
24. CORYNEBACTERIA   681
25. PNEUMOCOCCI   693
26. STREPTOCOCCI   707
27. STAPHYLOCOCCI   727
28. THE NEISSERIAE   741
29. THE ENTERIC BACILLI AND SIMILAR GRAM-NEGATIVE
    BACTERIA   753

30. THE HEMOPHILUS-BORDETELLA GROUP   791
31. THE YERSINIAE, FRANCISELLA, AND PASTEURELLA   801
32. THE BRUCELLAE   811
33. AEROBIC SPORE-FORMING BACILLI   819
34. ANAEROBIC SPORE-FORMING BACILLI   829
35. MYCOBACTERIA   843
36. ACTINOMYCETES: THE FUNGUS-LIKE BACTERIA   871
37. THE SPIROCHETES   881
38. RICKETTSIAE   897
39. CHLAMYDIAE   915
40. MYCOPLASMAS AND L FORMS   929
41. OTHER PATHOGENIC BACTERIA   945
42. BACTERIA INDIGENOUS TO MAN   952
43. FUNGI   964

V. VIROLOGY   Renato Dulbecco/Harold S. Ginsberg   1007
44. THE NATURE OF VIRUSES   1009
45. MULTIPLICATION AND GENETICS OF BACTERIOPHAGES   1046
46. LYSOGENY, EPISOMES, AND TRANSDUCING BACTERIOPHAGES   1087
47. ANIMAL CELL CULTURES   1121
48. MULTIPLICATION AND GENETICS OF ANIMAL VIRUSES   1139
49. INTERFERENCE WITH VIRAL MULTIPLICATION   1171
50. VIRAL IMMUNOLOGY   1189
51. PATHOGENESIS OF VIRAL INFECTIONS   1205
52. ADENOVIRUSES   1221
53. HERPESVIRUSES   1239
54. POXVIRUSES   1257
55. PICORNAVIRUSES   1279
56. MYXOVIRUSES   1307
57. PARAMYXOVIRUSES AND RUBELLA VIRUS   1329
58. CORONAVIRUSES   1361
59. RHABDOVIRUSES   1367
60. TOGAVIRUSES AND OTHER ARTHROPOD-BORNE VIRUSES;
    ARENAVIRUSES   1377
61. REOVIRUSES   1399
62. HEPATITIS VIRUSES   1409
63. TUMOR VIRUSES   1417

64. STERILIZATION AND DISINFECTION   1451

INDEX   1467

PREFACE TO THE SECOND EDITION

The second edition of this text, like the first, is designed for the student seeking to understand modern biology in depth, including not only what we know but also how we came to know it. In order to keep the book from expanding, despite the introduction of much new material, nearly every page has been rewritten.

The blurring of the boundaries between microbiology and related disciplines has led to appreciable changes in the scope and organization of the book. We have introduced a major section on molecular genetics, including a new chapter on protein synthesis and expanded chapters on nucleic acids and on metabolic regulation. Moreover, because animal cells are now commonly cultivated in a manner similar to bacteria, and because such cultures are now the usual hosts for the investigation of viruses, a chapter on animal cell culture has been added. New chapters have also been added on the bacterial cell envelope and on several viruses, and the extensive advances in cellular immunology have led to an expanded chapter on antibody formation and a new one on cell-mediated immunity.

The authors responsible for the various sections are indicated in the table of contents, and the various experts who revised individual chapters in Bacterial and Mycotic Infections are indicated on the title pages of those chapters. In the section on virology, Renato Dulbecco was responsible for general virology and the tumor viruses and Harold S. Ginsberg for viral pathogenesis and the specific agents. Overall editing was provided by Bernard D. Davis.

B. D. DAVIS
R. DULBECCO
H. N. EISEN
H. S. GINSBERG

PREFACE TO THE FIRST EDITION

"What is new and significant must always be connected with old roots, the truly vital roots that are chosen with great care from the ones that merely survive."

This principle, professed by the composer Bela Bartok, is as applicable to science as it is to music. Indeed, it highlights the most difficult aspect of writing a modern textbook of microbiology, for few branches of natural science have been so rapidly altered by recent advances. Only a few years ago microbiology was largely an applied field, concerned with controlling those microbes that affect man's health or his economic welfare, but with the recent development of molecular genetics, stemming largely from the study of microbial mutants, microbiology has rapidly been drawn to the center of the biological stage.

As a result, infectious disease no longer constitutes the sole bridge between microbiology and medicine. An additional, rapidly broadening span is provided by the use of microbes as model cells in the study of molecular genetics and cell physiology, for the principles and the successful approaches developed in such studies will surely prove widely applicable to human cells, which can now be cultured much like bacteria. In addition, studies at a molecular level are also rapidly providing a deeper insight into problems directly related to infectious disease, including the action of chemotherapeutic drugs, the structure of antibodies and cellular antigens, and the nature of viruses. Hence to prepare the student for the scientific medicine of his future it has seemed to us desirable to increase emphasis on the molecular and genetic aspects of microbiology. At the same time, the authors, having all had clinical experience, are vividly aware of the importance of providing a thorough understanding of host–parasite relationships and mechanisms of pathogenicity, even though many aspects cannot yet be explained in molecular terms.

In short, we have tried to identify the "truly vital roots" of classical bacteriology, immunology, and virology, and to engraft upon them the recent molecular advances. To keep the volume to a reasonable size we have eliminated much traditional information that did not seem to have either theoretical or practical importance for the student of medicine. Moreover, the clinical and epidemiological aspects of infectious diseases have been largely left for later courses in the medical curriculum, and we have provided only a small number of selected references, primarily for access to the original literature and not for documentation. In the hope of making the book more useful and versatile, we have included in smaller type a good deal of material that seemed not essential for an introduction to the subject, but still likely to interest many readers.

The demands of the medical curriculum frequently lead to a condensed memorizing of conclusions; yet courses in the basic medical sciences should surely illustrate the scientific method as well as transmit a body of information. We have therefore briefly reviewed the history of many major discoveries in order to show how scientific advances may depend on new concepts or technics, or on ingenious experiments, or on alertness to the significance of unexpected observations. Moreover, we have endeavored throughout to indicate the nature of the evidence underlying the conclusions presented—for otherwise the student sees only the shadow and not the substance of science.

This book is designed primarily as a text for students and investigators of medicine and the allied professions: hence the exposition proceeds from general principles to specific pathogenic microorganisms. However, we hope that the discussion of general principles will also prove useful to graduate students and investigators in the biological sciences.

The preparation of this volume has been a truly cooperative effort: the chapters drafted by each author have been critically reviewed by most or all of the others. We are deeply grateful for the education and for the warm friendships that have resulted.

A new book of this size will inevitably contain errors and weaknesses. We shall welcome corrections and suggestions for future editions.

B. D. DAVIS
R. DULBECCO
H. N. EISEN
H. S. GINSBERG
W. B. WOOD, JR.

ACKNOWLEDGMENTS

Because of the untimely death of W. Barry Wood shortly after this revision was undertaken, most of the chapters on pathogenic bacteria were revised by various experts, with Maclyn McCarty as special editor; we are immensely grateful to these colleagues who responded to the emergency so generously on such short notice. We also wish to note that these authors should not be held accountable for any infelicitous expressions since their contributions were modified to fit the style of the rest of the book. These contributors include: Robert Austrian, William A. Blyth, Zanvil A. Cohn, Ronald Gibbons, Emil C. Gotschlich, Paul H. Hardy, George S. Kobayashi, Stephen I. Morse, James W. Moulder, Roger L. Nichols, A. M. Pappenheimer, Sigmund Socransky, Alex C. Sonnenwirth, Morton N. Swartz, Lewis Thomas, and Emanuel Wolinsky. In addition, Loretta Leive was most helpful as coauthor of Chapter 6.

We are deeply indebted to the many professional colleagues who have critically reviewed various parts of the manuscript. Those who have made particularly extensive suggestions include: K. F. Austen, L. K. Bailey, B. Bloom, J. M. Davie, E. P. Geiduschek, R. J. Graff, E. A. Kabat, A. D. Kaiser, F. Karush, E. D. Kilbourne, S. Kinsky, W. K. Maas, M. M. Mayer, D. E. Morse, H. M. Paulus, R. W. Schlesinger, D. H. Smith, L. A. Steiner, J. L. Strominger, P.-C. Tai, R. A. Weiss, and T. J. Wiktor.

We are grateful to the publisher's staff for skillful help, pleasant cooperation, and patience during the three years of preparation. In addition, we are indebted to the Marine Biological Laboratory at Woods Hole, Massachusetts, for the hospitality of its library, where much of this work was written.

Many investigators and publishers have provided illustrations or permisson to utilize previously published material. We are most grateful for all generous new contributions and for material repeated from the first edition. Sources are acknowledged in the legends to the figures and tables.

MICROBIOLOGY

chapter 1

EVOLUTION OF MICROBES AND OF MICROBIOLOGY

EVOLUTION OF MICROBIOLOGY   2
   The First Microscopic Observations   2
   Spontaneous Generation   3
   The Role of Microbes in Fermentations   5
  MICROBIOLOGY AND MEDICINE   6
   The Germ Theory of Disease   6
   Recognition of Agents of Infection   8
   Viruses   9
   The Host Response: Immunology   10
   Control of Infectious Diseases   10
   The Impact of Microbiology on the Concept of Disease   11
  THE DEVELOPMENT OF MICROBIAL PHYSIOLOGY AND
     MOLECULAR GENETICS   11

MICROBIOLOGY AND BIOLOGICAL EVOLUTION   12
   Role of Microbes in Human Evolution   12
   Evolution and Teleonomy   13
  MICROBIAL TAXONOMY   13
   Plants, Animals, and Protists   13
   Eukaryotes and Prokaryotes   14
   Precellular Evolution and the Origin of Life   15

There are similarities between the diseases of animals or man and the diseases of beer and wine. . . . If fermentations were diseases one could speak of epidemics of fermentation. L. PASTEUR

EVOLUTION OF MICROBIOLOGY

THE FIRST MICROSCOPIC OBSERVATIONS

The spread of certain diseases from one person to another long ago suggested the existence of invisible, transmissible agents of infection. Thus in the poem *De rerum natura* Lucretius (96?–55 B.C.) presciently recognized not only the atomistic nature of matter but also the existence of "seeds" of disease. Microscopic organisms (microbes) were not seen, however, until Antony van Leeuwenhoek (1632–1723) made microscopes with sufficient magnification: the science of microbiology began with his letter in the *Philosophical Transactions of the Royal Society of London* in 1677.

Even in an age when science was still in the hands of gifted amateurs, Leeuwenhoek was unusual for his isolation from the learned world and his lack of formal education. A cloth merchant in the town of Delft, Holland, with a political sinecure as custodian of the Town Hall, he spent much of his leisure time in grinding tiny lenses of high magnification (probably up to 300×), with which he made simple (one-lens) microscopes (Fig. 1-1). With these instruments this patient and curious man discovered an incredible variety of hitherto unseen structures, including the major morphological classes of bacteria (spheres, rods, and spirals; Fig. 1-2), as well as the larger microbes (protozoa, algae, yeasts), erythrocytes, spermatozoa, and the capillary circulation. Leeuwenhoek's discoveries were described in a flow of letters to the Royal Society of London, whose distinguished members he apparently never met.* Moreover, by keeping his methods se-

cret he remained throughout his long lifetime the sole occupant of the field he had created.

No other observers succeeded in using single lenses as effectively as Leeuwenhoek, and further advances depended on the perfection of the compound microscope, which had already been invented but suffered from serious optical aberrations. Following its improvement further descriptive observations on microbes accumulated; and though Linnaeus in 1767 distinguished only 6 species in assigning microbes to the class "Chaos," 600 types were figured in Ehrenberg's *Atlas* in 1838.

An experimental science of microbiology, however, emerged only slowly, and required the development of a special methodology. The key was the use of sterilized materials and aseptic technics; for while the chemist defines purity in terms of percentage of contaminating material, in microbiology a single contaminating cell can ruin an experiment. Only after learning to avoid such contamination could investigators recognize the existence of a great variety of microbes and

---

* Leeuwenhoek's letters have a charming colloquial style, no longer seen in scientific communications. For example, he describes the following observations on a decayed tooth, which emphasize that motility was then the only available criterion for considering a microscopic object alive. "I took

this stuff out of the hollows in the roots, and mixed it with clean rainwater, and set it before the magnifying-glass so as to see if there were as many living creatures in it as I had aforetime discovered in such material: and I must confess that the whole stuff seemed to me to be alive. But notwithstanding the number of these animalcules was so extraordinarily great (though they were so little withal, that 'twould take a thousand million of some of 'em to make up the bulk of a coarse sand-grain, and though several thousands were a-swimming in a quantity of water that was no bigger than a coarse sand-grain is), yet their number appeared even greater than it really was: because the animalcules, with their strong swimming through the water, put many little particles which had no life in them into a like motion, so that many people might well have taken these particles for living creatures too."

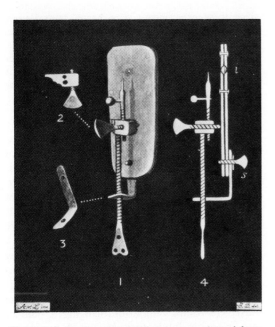

Fig. 1-1. A Leeuwenhoek microscope, viewed from the back (1) and in diagrammatic section (4). The specimen, on a movable pin, is examined through a minute biconvex lens (l), held between two metal plates. (From Dobell, C. *Antony van Leeuwenhoek and His "Little Animals."* Dover, New York, 1960.)

determine their distribution and their activities. The development of reliable methods was very much stimulated by a prolonged controversy over an issue with religious overtones: the spontaneous generation of life. This topic therefore occupies a prominent place in the early history of microbiology.

SPONTANEOUS GENERATION

Until recent centuries it was widely believed that living organisms can arise spontaneously in decomposing organic matter. For visible organisms this notion was dispelled in the seventeenth century, when Redi demonstrated that the appearance of maggots in decomposing meat depended on the deposition of eggs by flies. However, the idea of spontaneous generation persisted for the new world of microbes and delayed recognition of their relation to the rest of biology.

The question would appear to have been settled in the eighteenth century by Spallanzani (1729–1799), who introduced the use of sterile culture media: he showed that a

"putrescible fluid," such as an infusion of meat, would remain clear indefinitely if boiled and properly sealed.* Moreover, in 1837 Schwann elegantly showed that similar results could be obtained even when air was allowed to reenter the cooling flask before sealing, provided the air passed through a heated tube. Skeptics could claim, however, that the absence of decomposition in these sealed vessels was due to limitations in the supply of air rather than to the exclusion of dustborne living contaminants. To answer this objection Schroeder and von Dusch introduced the use of the cotton plug, which is still used today to exclude airborne contaminants.

Nevertheless, the controversy continued, for some investigators were unable to reproduce the alleged stability of dust-free sterilized organic infusions. Louis Pasteur (1822–1895) then entered the lists. He showed that boiled medium could remain clear in a "swan-neck flask open to the air through a sinuous horizontal tube, in which dust particles would

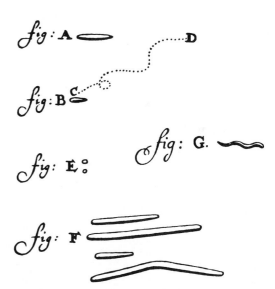

Fig. 1-2. Leeuwenhoek's figures of bacteria from the human mouth, from letter of 17 September, 1683. Dotted line between B and D indicates motility.

---

* Indeed, the soundness of this discovery was confirmed in the early nineteenth century, when a Parisian confectioner, Appert, competing for a prize offered by Napoleon, developed the art of preserving food by canning.

settle as air reentered the cooling vessel (Fig. 1–3). Pasteur also demonstrated that in the relatively dust-free atmosphere of a quiet cellar, or of a mountain top, sealed flasks could be opened and then resealed with a good chance of escaping contamination.

Pasteur's experiments were a public sensation. Though his contributions were in principle no more decisive than those of his predecessors who had also achieved stable sterility, his zeal and skill as a polemicist were largely responsible for laying the ghost of spontaneous generation. Because of his crusading spirit, as well as his experimental skill and intuitive genius, Pasteur was for the nineteenth century, in Dubos' words, "not only the arm but also the voice, and finally the symbol, of triumphant science." His style is illustrated by the following, from a lecture delivered at the Sorbonne in 1864.

"I have taken my drop of water from the immensity of creation, and I have taken it full of the elements appropriate to the development of microscopic oragnisms. And I wait, I watch, I question it!—begging it to recommence for me the beautiful spectacle of the first creation. But it is dumb, dumb since these experiments were begun several years ago; it is dumb because I have kept it sheltered from the only thing man does not know how to produce, from the germs which float in the air, from Life, for Life is a germ and a germ is Life. Never will the doctrine of spontaneous generation recover from the mortal blow of this simple experiment!"

The Problem of Spores. Despite Pasteur's dramatic success it turned out that his accusations of technical incompetence on the part of his opponents did not really explain why their infusions stubbornly refused to remain clear. The key fact was that they used infusions of hay, and Pasteur preferred other materials (sugar plus yeast extract) for his culture media.

The decisive experiments were provided by the British physicist John Tyndall, who became engaged in the problem through his interest in the optical effects of atmospheric dust. He found that after bringing a bale of hay into his laboratory he could no longer achieve sterility, in the same room, by boiling. Tyndall concluded that the hay had contaminated his laboratory with **an incredible kind of living organism: one that could survive**

Fig. 1-3. Pasteur's "swan-neck" flasks. After use in his studies on spontaneous generation these flasks were sealed, and they have since been preserved, with their original contents, in the Pasteur Museum. (Courtesy of Institut Pasteur, Paris.)

boiling. In the same year (1877) Ferdinand Cohn demonstrated the resistant forms as small, refractile **endospores** (Ch. 5) and showed that they were stages in the life cycle of the hay bacillus (*Bacillus subtilis*). Even the most resistant bacterial spores, however, are readily sterilized in the presence of moisture at 120°; hence the autoclave, which reaches this temperature through the use of steam under pressure, became the hallmark of the bacteriology laboratory.

Once the biological continuity of microbes was recognized, it soon became apparent that there are different kinds, which occupy different ecological niches and have different actions. Microbiology then developed largely through interest in three different groups of microbes: those responsible for fermentations, for much of the cycle of organic matter in nature, and for diseases of man, lower animals, and plants. These developments gave rise to the corresponding applied fields of industrial, agricultural, and medical microbiology. The study of fermentations came earliest and provided much of the impetus for the development of other areas in microbiology, as well as for the development of biochemistry.

THE ROLE OF MICROBES IN FERMENTATIONS

We have seen that practical men proceeded to preserve foods while the savants continued to dispute spontaneous generation. Useful fermentations* have an even longer history of achievement without a theoretical foundation: lost in antiquity are the origins of leavened bread, wine, or the fermentations that preserve food through the accumulation of lactic acid (soured milk, cheese, sauerkraut, ensilage).

In the 1830s, with the development of microscopes of sufficient resolving power, Schwann, Cagniard-Latour, and Kutzing concluded independently that the sediment of microscopic globules accumu-

lating in an alcoholic fermentation consists of tiny, growing plants, whose metabolic activities are responsible for the fermentation. However, the leading chemists of the day considered that fermentation was a chemical process, due to a self-perpetuating instability of the grape juice initiated by its exposure to the air. The amorphous sediment would thus be a byproduct of the fermentation, analogous to the frequent crystalline precipitates of tartaric acid. In particular, the distinguished Liebig, father of biochemistry, advanced this view with such vehemence that Schwann's excellent evidence was essentially discarded for two decades.† Liebig's authority was eventually refuted by Pasteur.

Educated as a chemist, Pasteur became interested in fermentations through discovering optical isomerism as a property of certain fermentation products (isoamyl alcohol, tartaric acid). In 1857, in his first paper on fermentation, Pasteur showed that **different kinds of microbes are associated with different kinds of fermentation:** spheres of variable size (now known as yeast cells) in the alcoholic fermentation, and smaller rods (lactobacilli) in the lactic fermentation.‡

In the course of establishing the nature of fermentation Pasteur founded the study of **microbial metabolism** and developed profound insight into many of its problems. Indeed, our knowledge of microbial physiology hardly advanced any further until the development of sophisticated biochemical technics many decades later. In particular, Pasteur showed that **life is possible without air,** that some organisms (obligatory anaerobes) are even inhibited by air, and that fermentation is much less efficient than respiration in terms of the growth yield per unit of substrate consumed.

---

* In general usage the term **fermentation** refers to the microbial decomposition of vegetable matter, which contains mostly carbohydrates; while **putrefaction** refers to the formation of more unpleasant products by the decomposition of high-protein materials, such as meat or eggs. (For a more rigorous definition of fermentation see Ch. 3.) The etymology of the word ferment (L. *fervere* = to boil), and its figurative use today, reflect the heating and bubbling that are generated in a vat of fermenting grape juice.

† Indeed, in 1839 Liebig and Wohler published an anonymous, scatological hoax in which they reported the microscopic visualization of an animal, shaped somewhat like a distilling apparatus, that took in sugar at one end and excreted alcohol at the other, while its large gonads released bubbles of CO_2.

‡ As in many scientific disputes, the losing argument possessed at least a grain of truth. The view of fermentation as a chemical rather than a vital process was ultimately vindicated in 1897, though in a profoundly modified form, when Edouard Buchner accidentally discovered that a cell-free extract, made by grinding yeast cells with sand, fermented a concentrated sugar solution added to preserve the extract.

Selective Cultivation. Since the nature of a specific fermentation depends on the organism responsible, how can a given kind of substrate, when not deliberately inoculated with a particular microbe, regularly undergo a given kind of fermentation? Pasteur recognized that the explanation lay in the principle of selective cultivation: various organisms are ubiquitous, and the one best adapted to a given environment eventually predominates. For example, in an initially mixed fermentation in grape juice the high sugar concentration and the low protein content (i.e., low buffering power) lead to a condition, now recognized as low pH, that favors the outgrowth of acid-resistant yeasts and thus yields an alcoholic fermentation. In milk, in contrast, neutralization of the acid by the high protein content favors the outgrowth of faster-growing but more acid-sensitive bacteria, which cause a lactic fermentation.

An excess of the wrong organism in the starting inoculum, however, can grow out sufficiently to cause formation of a product with a poor flavor; hence specific microbes play a role in "diseases" of beer and wine, as well as in different normal fermentations. This finding led to the fruitful suggestion that specific microbes might also be causes of diseases of man.

With an eye always to practical as well as to theoretical problems, Pasteur developed the procedure of gentle heating (pasteurization) to prevent the spoilage of beer and wine by undesired contaminating microbes. Many years later this process was extended to prevent milkborne diseases of man.

Of great economic importance was the extension of industrial fermentations from the production of foods and beverages to that of valuable chemicals, such as glycerol, acetone, and later vitamins, antibiotics, and alkaloids. In these applications, just as in the study of cell physiology, successful developments have depended largely on the possibility of selecting microbial mutants with desired metabolic properties.

Soil Microbiology: Geochemical Cycles. The types of carbohydrate metabolism studied by Pasteur subsequently proved very similar to those of mammalian "physiological" chemistry, thus leading eventually to the concept of the **unity** of biology at a molecular level.

In contrast, toward the end of the Pasteurian era other investigators, notably Winogradsky in Russia and Beijerinck in Holland, began exploring the microbiology of the soil, and discovered an astonishing **variety** of metabolic patterns by which different kinds of bacteria, adapted to different ecological niches, make their living. These organisms were isolated by the use of a systematic extension of Pasteur's principle of selective cultivation: **enrichment cultures,** in which only a particular energy source is provided, and growth is thus restricted to those organisms that can use that source. Some of the unusual patterns of bacterial energy production, ranging from the oxidation of sulfur to the formation of CH_4 from CO_2 and H_2, are described in Chapter 3.

With the development of soil microbiology it became clear that the major role of microbes in nature is **geochemical:** mineralization of organic carbon, nitrogen, and sulfur (i.e., conversion to CO_2, NH_3 or NO_3^-, $SO_4^=$ or $S^=$), so that these elements can be used cyclically for generation after generation of growth of higher plants and animals, rather than being tied up in dead organic matter. Moreover, soil fertility is enhanced by the bacteria that convert atmospheric nitrogen or the ammonia from organic decay into the nonvolatile form of nitrate. In addition, marine algae, and to a much smaller extent photosynthetic bacteria, are responsible for a considerable fraction of the other half of the geochemical cycle: the formation of organic matter from CO_2 by photosynthesis.

The beneficent geochemical role of microbes is worth emphasizing, since the prominent historical connection between infectious disease and microbiology has given rise to the popular image of a malignant and hostile microbial world. In fact, human pathogens constitute only a small fraction of the number of recognized bacterial species and an infinitesimal fraction of the total mass of microbes on the earth: most microbes attack organic matter only after it is dead and buried.

MICROBIOLOGY AND MEDICINE

THE GERM THEORY OF DISEASE

Among the major classes of disease infections have undoubtedly presented the greatest burden to mankind: not only were they a leading cause of death, but these deaths were often especially heartbreaking because they were so frequent among the young. Moreover, by their epidemic nature infections have often disabled and terrorized communities and determined the fate of armies and nations. Clearly the discovery of the causes of infectious diseases, and the development of methods for their control, have been the greatest achievement of medical science. Today it is easy to take these advances for granted while focusing on problems created by technology; but an earlier public enshrined Pasteur in France and Robert Koch (1843–1910) in Germany as national heroes for rescuing mankind from the then greatest menace to its environment: microbial pollution.

Epidemiological Evidence. The discovery of infectious agents was long preceded by the concept of **contagious** disease, i.e., one initiated by **contact** with a diseased person or with objects contaminated by him. Thus even though the ancient Hebrews viewed pestilences as punishments visited on peoples by the Lord, the Mosaic code also contains numerous public health regulations, including the isolation of lepers, the discarding of various unclean materials, and the avoidance of shellfish and pork as foods. Later shrewd observers of epidemics, such as Lucretius and Boccaccio, explicitly recognized their contagious nature.

In 1546 Fracastorius of Verona presented an impressive body of evidence in *De Contagione*. This book founded the science of **epidemiology,** which analyzes the distribution in human populations of events affecting health. After carefully studying epidemics of several diseases, including plague and syphilis, Fracastorius concluded that these diseases were spread by **seminaria** ("seeds"), transmitted from one person to another either directly or via inanimate objects. But like later writers, for some time to come, he presented a curious mixture of common sense and superstition: though it seemed prudent to avoid exposure to patients, or to plagued communities, as sources of transmissible seeds, older views were maintained through the proposal that the initial seeds in an epidemic were generated by supernatural or telluric forces.

The epidemiological evidence for the germ theory of disease was not generally taken seriously, and a century after Fracastorius leading physicians, such as William Harvey, rejected his conclusions. Most physicians continued to follow the views of Hippocrates and Galen, who ascribed epidemics to **miasmas,** i.e., poisonous vapors, created by the influence of planetary conjunctions or by disturbances arising within the earth.

Part of the difficulty arose from the existence of many communicable diseases that are not contagious in a strict sense. We now know that these are transmitted by less obvious routes, such as air, water, food, and insects; and it is easy to see how an airborne disease could logically be ascribed to poisoned air until the particulate nature of the agent was demonstrated. Moreover, before the development of chemistry the idea that living organisms too small to be seen could exist, and could mortally harm large animals, was clearly contrary to common sense. Thus even the recognition of transmissible "seeds of disease" did not quite hit the mark: the notion of seed was taken literally, the actual *contagium vivum* within the diseased body being considered something more complex derived from the seed. Experimental evidence was required, and it accumulated slowly, from several directions: transmission of infection, its prevention, and finally identification of the agents.

Transmission of infection was demonstrated boldly in the eighteenth century by the renowned surgeon John Hunter, who inoculated himself with purulent material from a patient with gonorrhea; unfortunately, both for him and for the understanding of etiology, he transmitted at the same time a much more serious disease, syphilis. The use of experimental animals in such studies was introduced later: for example, in 1865 Villemin so transmitted tuberculosis, though the nature of the responsible agent was not established until 20 years later.

The role of **indirect** transmission from one person to another was recognized later. In the 1840s Ignaz Semmelweis in Vienna, and Oliver Wendell Holmes (perhaps better known as a poet than as a physician) in Boston, shockingly (but unsuccessfully) asserted that obstetricians, moving with unwashed hands from one patient to the next, were responsible for the prevalence in hospitals of puerperal sepsis, a frequent cause of maternal death. Communicability was demonstrated in more detail for cholera in 1854 by John Snow, who traced a localized epidemic to the Broad Street pump in London and deduced that the disease was spread by a water supply contaminated with fecal material.

Preventive measures also lent support to the germ theory. In 1796 Edward Jenner, observing that milkmaids rarely came down with smallpox, introduced **vaccination** (L. *vacca* = cow) against this disease by inoculating material from lesions of a similar disease of cattle, cowpox; yet the theoretical implications of this finding with respect to the infectious origin of the disease were not appreciated. In the 1860s, however, Joseph Lister introduced **antiseptic surgery** on a much firmer theoretical foundation. Impressed by Pasteur's evidence for the ubiquity of airborne microbes, he reasoned that they might contaminate wounds as well as sterile culture media and hence might be responsible for the frequent development of pus. He found that direct application of a disinfectant, phenol (carbolic acid), markedly reduced the incidence of serious infections. It is noteworthy that this major advance was achieved a decade before any specific agent of human infection was identified. Emphasis later shifted from antiseptic to **aseptic surgery.**

RECOGNITION OF AGENTS OF INFECTION

The epidemiological evidence for communicability, though logically convincing, did not carry the weight of a direct demonstration of the agents of infection. The first to be recognized were **fungi,** which are larger than bacteria: in 1836 Agostino Bassi demonstrated experimentally that a fungus was the cause of a disease (of silkworms), and 3 years later Schönlein discovered the association of a fungus with a human skin disease (favus). In 1865 Pasteur entered the field of pathogenic microbiology with the discovery of a **protozoon**

that was threatening to ruin the European silkworm industry.

The etiological role of **bacteria** in anthrax was unequivocally established by Robert Koch in 1876, and was confirmed by Pasteur and his medical colleague, Joubert. This disease offered the investigator several advantages: the organism is unusually large and is readily identified morphologically; the disease, primarily one of cattle and sheep, may be conveniently transmitted to small animals; and unusually dense bacterial populations may appear in the blood. Indeed, Davaine in 1850 had already seen rod-shaped bodies in the blood of sheep dying of anthrax, and had later transmitted the disease by inoculating as little as 10^{-6} ml of blood. However, this evidence, though highly suggestive, did not prove that these bodies were the cause rather than a result of the disease, especially since the rods could not always been seen in infectious blood.

Koch, then a rural physician, solved the problem by isolating the anthrax bacillus in pure culture and showing that such cultures could transmit the disease to mice. In addition, though only vegetative rods were found in the blood, he found that cultures developed spores, recognized by virtue of their refractility, which were highly resistant to sterilization (Fig. 1–4). This finding accounted for the observation that fields once inhabited by anthrax-infected animals could infect fresh herds years later.

Pure Cultures. The key to the identification of bacteria as pathogens was the isolation of pure cultures. The theoretical necessity of this step had been pointed out in 1840 by Henle, following Schwann's identification of the agents of fermentation as living organisms. Lister showed that pure cultures could be obtained by the method of **limiting dilutions,** in which the source material is diluted until the individual inocula each contain either one infectious particle or none; but this method is awkward. Koch meticulously perfected the technics of identification that are used today; these include the use of **solid media,** on which individual cells give rise to separate colonies, and the use of **stains.** Koch's genius is perhaps best reflected in his patient modifications of his own methods that led to the identification of the tubercle bacillus in 1882: because this

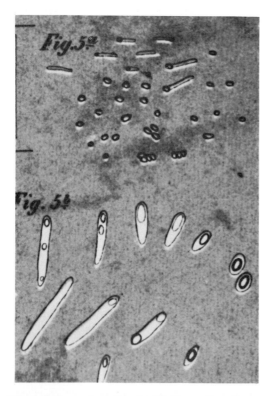

Fig. 1-4. Spore formation in *Bacillus anthracis,* as independently described and simultaneously published by Robert Koch ("Fig. 5a") and Ferdinand Cohn ("Fig. 5b"). (Courtesy of Koch Museum, Berlin.)

organism grows very slowly the usual 1 to 2 days of cultivation had to be extended to several weeks, and because it is so impervious the usual few minutes of staining had to be extended to 12 hours.

After identifying the tubercle bacillus Koch formalized the criteria, introduced by Henle but known as **Koch's postulates,** for distinguishing a pathogenic from an adventitious microbe: 1) the organism is regularly found in the lesions of the disease, 2) it can be isolated in pure culture on artificial media, 3) inoculation of this culture produces a similar disease in experimental animals, and 4) the organism can be recovered from the lesions in these animals. These criteria have proved invaluable in identifying pathogens, but they cannot always be met: some organisms (including all viruses) cannot be grown on artificial media and some are pathogenic only for man.

The powerful methodology developed by Koch introduced the "Golden Era" of medical bacteriology. Between 1879 and 1889 various members of the German school isolated (in addition to the tubercle bacillus) the cholera vibrio, typhoid bacillus, diphtheria bacillus, pneumococcus, staphylococcus, streptococcus, meningococcus, gonococcus, and tetanus bacillus. Studies naturally followed on the mechanisms of pathogenicity of these organisms, the host responses, and the methods of prevention and treatment.

Curiously, Pasteur, despite his early start with anthrax, did not enter the race to identify pathogens; as a chemist he was uninterested in the problems of isolating and classifying organisms. Instead, he devoted his later years to another aspect of infectious disease: the development of **vaccines.** By accident he found that chickens injected with an old culture of bacterium of chicken cholera were subsequently resistant to a fresh, virulent culture. Having earlier emphasized the constancy of the organisms responsible for different fermentations, he quickly recognized that this observation suggested an important **variability** as well. Within the incredible space of 4 years, and before the development of any understanding of the nature of the immune response, Pasteur discovered four methods of "attenuating" organisms and thus converting them to useful vaccines: aging of the culture (chicken cholera), cultivation at high temperature (anthrax), passage through another host species (swine erysipelas), and drying (rabies). As we shall see in later chapters, this "attenuation" comprises two distinct processes:

selection of less virulent mutants, and killing of virulent organisms with retention of their immunizing activity.

Koch's development of solid media had important consequences for general bacteriology. The isolation of pure cultures finally dispelled the claim of the "pleomorphists" (based on the fluctuating properties of impure cultures) that the various bacterial cell types were merely different stages in the life cycle of a single organism. With this misconception out of the way, a classification of bacteria proposed by Ferdinand Cohn was generally accepted and has persisted, in its main features, to the present day.

VIRUSES

The term virus (L., "poison"), long used as a synonym for infectious agent, is now reserved for the true viruses. These were originally defined as infectious agents small enough to pass filters that retain bacteria, and were later characterized also by the requirement of a living host for their multiplication. Despite this requirement other means have been developed for obtaining pure cultures and thus returning to the spirit of Koch's criteria.

The first virus to be recognized as filtrable was a plant pathogen, tobacco mosaic virus, discovered independently by Iwanowski in Russia in 1892 and by Beijerinck in Holland in 1899. Filtrable animal viruses* were first demonstrated for foot-and-mouth disease of cattle by Löffler and Frosch in 1898, and for a human disease, yellow fever, by the U.S. Army Commission under Walter Reed in 1900. Viruses that infect bacteria (bacteriophages) were discovered by Twort in England and by d'Herelle in France in 1916–1917.

For the first third of this century viruses could be detected only by their pathogenic effects on living hosts, and progress was slow. Eventually sophisticated physical and chemical methods were developed for purifying and characterizing viruses, while the development of the electron microscope and the advance of molecular genetics made it possible to analyze their mechanism of reproduction.

---

* It is curious that though the most dramatic part of Pasteur's work in the 1880s on the development of vaccines was performed with tissues containing rabies virus, he did not recognize the filtrability of the agent.

Precise quantitative studies became possible with the development of monolayer cultures of host cells, in which viruses can form discrete plaques analogous to the bacterial colonies formed on solid media. Virology has accordingly flourished in recent years, and viruses are now known to be a special class of particles with a unique method of reproduction.

With the development of rigorous criteria for distinguishing viruses from cells, certain filtrable agents long considered to be viruses have turned out to have a cellular organization; these groups (mycoplasmas, rickettsiae, chlamydiae) are therefore now classified as bacteria. At the same time, the viral kingdom has been greatly enlarged, with the recent recognition that acute viral diseases reveal only its most conspicuous members: with tissue cultures we are now identifying more and more additional animal viruses which cause chronic, "slow," mild, or inapparent disease. Moreover, it seems probable that viruses also play an important role in evolution by transmitting blocks of genetic material from one organism to another.

THE HOST RESPONSE: IMMUNOLOGY

Vertebrates infected by a microbial parasite exhibit a specific **"immune" response** which contributes to recovery and also to protection against reinfection. Analysis of this response has given rise to the major field of **immunology,** which is concerned with learning how foreign materials (antigens) elicit the appearance of specific cells and specific circulating proteins (antibodies) with a high affinity for the antigens, and with studying the properties and the uses of these components.

As with many other practical problems, the search for protection against infection resulted in successes that ran ahead of theory. Jenner's vaccination against smallpox in 1798, using the less virulent cowpox virus, was a remarkable achievement; and we have already noted that Pasteur developed several vaccines in the early 1880s before the immune response and the organisms were at all understood.

The Scope of Immunology. Immunity or **resistance** to infectious disease involves not only specific immune responses but also nonspecific factors, such as **enzymes** that attack the parasites and specialized **phagocytic cells** that engulf them. The science of immunology is thus in a sense more restricted than immunity, being concerned with only its specific, induced aspects. But in another sense immunology is a broader branch of pathophysiology, extending into areas distinct from infectious disease; for the same responses are equally brought into play when foreign substances of nonmicrobial origin gain access to the tissues (e.g., pollens, insect venom, drugs, foreign serum or other proteins, transplanted tissues). In addition, immunological methods are being used increasingly in biochemistry as delicate tools for the study of protein structure and interactions and for assays; and the immune response is of increasing interest to biologists as a model for differentiation.

Immunology can thus no longer be considered a branch of microbiology. Nevertheless, the two areas are still inextricably connected: not only are immune reactions of vital importance in host-parasite interactions, but immunological methods are indispensable for identifying and classifying various pathogenic microbes and for identifying those individuals in a host population who have previously been infected with a given organism. Finally, infectious disease probably constitutes the principal selective force in the evolution of the immune response; and, conversely, this response has had great influence on the evolution of those microbes that are pathogenic for man and other vertebrates.

CONTROL OF INFECTIOUS DISEASES

Identification of the agents of various infectious diseases soon led to several remarkably effective methods of control. 1) in technologically advanced countries environmental **sanitation** and improved personal **hygiene** have strikingly reduced in incidence, and sometimes even eliminated, certain diseases, particularly those spread by water or food (e.g., typhoid, cholera) or by insects (e.g., typhus, yellow fever). However, a knowledge of these diseases is still essential for the physician, since they may at any time be reintroduced by travelers. 2) **Vaccination** has drastically reduced the incidence of several serious epidemic diseases (e.g., smallpox, diphtheria, whooping cough, poliomyelitis). It has been especially valuable for diseases transmitted by respiratory droplets, whose distribution is difficult to control; but for many organisms vaccination is not effective or feasible. 3) The development of effective antibacterial **chemotherapy** has dramatically reduced the seriousness of many infectious diseases and the incidence of some. Indeed, this development represents the most striking advance in medical bacteriology since the 1880s.

In principle it should ultimately be possible to **eradicate,** by one or more of these methods, those organisms that are **obligatory** human pathogens. However, this hope does not exist for those pathogens that can also be widely carried by man without causing disease, or those that have **reservoirs** in lower animals or in the soil.

THE IMPACT OF MICROBIOLOGY ON THE CONCEPT OF DISEASE

Against a background of vague speculations on the etiology of disease, the discovery of specific agents of infection represented a tremendous theoretical advance for medicine. But as often happens with a new principle, its limits were not promptly recognized. The success of the Pasteurian approach thus led to unwarranted confidence that a single cause was waiting to be discovered for every disease. Such an oversimplified view still appears today: some will argue, for example, with total irrelevance, that tobacco-smoking cannot be **the** cause of bronchogenic cancer since this disease occasionally arises in nonsmokers.

It has become apparent that the principle of multifactorial causation is applicable even to many infectious diseases. Thus even though the concept of the tubercle bacillus as the etiological agent of tuberculosis proved much more fruitful than the preceding concept of the "phthisical diathesis" (i.e., a constitutional tendency to develop this disease), tuberculin testing has shown that many more people are **infected** with tubercle bacilli than have the **disease.** Hence, the presence of the tubercle bacillus is a necessary but not a sufficient condition for the disease tuberculosis: other factors, involving genetic constitution and physiological state of the host, can be decisive. Studies on infectious disease are therefore focusing increasingly on these subtle host factors, i.e., on the development of a molecular pathology.

THE DEVELOPMENT OF MICROBIAL PHYSIOLOGY AND MOLECULAR GENETICS

Until quite recently microbiology was largely an applied field, concerned primarily with controlling the activities, whether beneficial or harmful, of various microorganisms. Microbiology therefore long remained largely separated from the rest of biology, in its goals as well as in its technics. Indeed, it was not until the 1940s that heredity in bacteria was found to have any relation to the science of genetics. However, microbes eventually proved to be especially suitable for studying many basic problems of cell physiology and genetics, and so these organisms have now come to occupy a central position in biology.

The course of this spectacular development will be outlined in later chapters. Here we may simply note that it was first necessary to recognize the **unity of biology at a molecular level,** as shown by the many close biochemical resemblances between microbial cells and those of higher organisms—in their building blocks, enzymes, and metabolic pathways, and in the structure and function of their genetic material. Their advantages, in studying problems common to all cells, include their relatively simple structure, homogeneous cell populations, and extremely rapid growth. But by far the greatest advantage lies in the possibility of easily cultivating billions of individuals and then selecting, from these huge populations, rare mutants and rare genetic recombinants between these mutants.

Mutants, each with a single biochemical defect, have proved to be exceptionally sharp tools for dissecting complex intracellular processes. At first they were used to advance biochemistry, particularly in the analysis of biosynthetic pathways. Increasingly, however, attention was focused on analyzing the biochemical basis of gene action and its regulation. Here the use of bacterial and phage mutants converged with the biochemistry of nucleic acids and proteins, and with electron microscopy, in the direct exploration of the fine structure and the biosynthesis of macromolecules. These developments have given rise to the vigorous interdisciplinary activity called **molecular genetics.** We shall consider this area in some detail, not only because it is crucial for understanding the action of viruses and that of many antimicrobial agents, but even more because it is rapidly being extended to human cells and hence to aspects of medicine distinct from infectious disease.

Since the study of genetics is intimately linked with that of evolution it seems appropriate to consider here some evolutionary aspects of microbiology.

MICROBIOLOGY AND BIOLOGICAL EVOLUTION

Over the past century biological thought has flowed in two major streams, **evolutionary** (genetic) and **mechanistic** (physiological), concerned respectively with the origin of various organisms and with the mechanisms by which they function. These streams have often diverged widely. However, with the development of molecular genetics the two streams have converged again; the study of a biochemical process often includes consideration of its genetic variation and regulation; and the study of genes is increasingly concerned with their functions in the living organism and not simply with their distribution among progeny.

Microbial Evolution. The rapid population changes that can be readily seen with microbes remind us that evolution is not a finished historical process but is still going on. For example, because bacteria grow so rapidly, and because the presence of a growth inhibitor can apply extreme selective pressures, one can demonstrate the natural selection of fitter (i.e., drug-resistant) mutants overnight in a laboratory culture, or in a few months or years in the flora of a drug-treated human population. We shall therefore frequently invoke evolutionary principles in this volume in considering the genetic adaptation of various microbes to various ecological niches.

In medical bacteriology, in particular, the existence of a large number of variants of certain pathogens, differing from each other only in the structure of their surface macromolecules (antigens), may at first appear chaotic. These variations become understandable, however, in evolutionary and immunological terms: in a host that has developed specific immunological means of attacking the original surface a mutation of the parasite to the formation of a different surface represents increased fitness. Conversely, widespread in-

fection by a given agent will tend to select for increased resistance of the host species to that agent.

For example, the virus of myxoma, a highly lethal disease of rabbits, was introduced into Australia to reduce crop destruction by a plague of these animals; and at first this biological warfare was dramatically effective. Within a few years, however, the rabbit population recovered its initial density through the outgrowth of strains with increased resistance to myxoma. In addition, a less virulent mutant strain of the virus outgrew the original form, evidently because it did not kill off its hosts too rapidly.

Since hosts and parasites thus have a marked reciprocal effect on each other's evolution, it seems appropriate to dwell briefly on the impact of microbes (and of microbiology) on the human gene pool.

ROLE OF MICROBES IN HUMAN EVOLUTION

It is generally considered that human evolution has moved continuously in the direction of increasing intelligence and manual dexterity—the features most responsible for man's dominating position. However, as Haldane has pointed out, when *Homo sapiens* shifted his social pattern some 10 thousand years ago, from isolated family units to larger and more freely communicating groups, epidemic diseases must have become an increasingly prominent cause of death; and the resulting selection for increased resistance to these diseases would automatically conflict with the selection for more interesting traits.*

---

* It is a well-established principle that selection for one trait will interfere with the efficiency of selection for an uncorrelated trait. For example, a school may select its student body on the basis of both ability to pay tuition and personal qualities; but if scholarships are used to eliminate the first selection the second will become more effective because of the enlarged pool of applicants.

This selective pressure was strong: a century ago in the United States (and still today in some parts of the world) 25 to 50% of children died of infectious diseases before reaching puberty. Today this figure is below 2%. But our fantastic success in freeing our environment of this particularly dangerous form of pollution has, like many other technological advances, created a major new problem: the imbalance between a suddenly decreased death rate and persistence of the old birth rate.

Molecular Individuality in Man. Evolution selects for survival not of individuals but of a species (i.e., an interbreeding group); and since a species occupies varied, and often fluctuating, environments, it will be selected for a balance of fitness for various environments, rather than for the single type best fitted for a single environment. One important source of variation in the environment is the encounter with various infectious agents; and no one human genotype is maximally resistant to all possible agents. Selective pressures from a **variety** of infectious agents will therefore promote **genetic heterogeneity** (polymorphism) in the genes that influence resistance to these agents. The results of such selection can be clearly seen with sickle-cell hemoglobin; this variant, which protects against falciparum malaria, is prevalent in African tribes constantly exposed to this disease, even though the gene is strongly selected against in other environments because its homozygous state causes a serious anemia. Other infectious agents must exert similar selective pressures, though the specific gene products that they select for are as yet obscure.

This consideration may explain why the cells of a vertebrate species possess an extraordinary variety of specific surface antigens, differing from one individual to another. We recognize these antigens because they prevent successful organ transplantation between individuals (Ch. 21), but they can hardly have evolved for this function. An alternative explanation would be that a specific antigen promotes resistance to a specific disease, just as different surfaces on microbial cells influence their virulence for different hosts. For instance, different allelic forms of the best-studied human antigens, the ABO blood group substances (Ch. 21), may be associated with differences in resistance to major killers of the past, such as smallpox. Moreover, the ethnic distribution of these antigens is very uneven and shows some correlation with the historical distribution of certain major infectious diseases.

EVOLUTION AND TELEONOMY

The convergence of the evolutionary and the mechanistic approaches to biology has had another important consequence: a wider acceptance of the relevance of **teleological** considerations. The essential feature of teleology is the principle that any structure or mechanism found in a living organism is likely to have value or purpose for that organism. Though Aristotle formulated this concept in terms of a "final cause"—an agency that had foreseen this value—biologists after Darwin could profitably employ the same concept in a modified form, substituting the hindsight of natural selection for divine providence. Nevertheless, the use of teleology continued to be plagued by its earlier supernaturalistic connotations. The term **teleonomy** has therefore recently been introduced to eliminate these connotations: it simply implies that an organism's genetic characters reflect evolutionary adaptation to its environment.

Teleonomic reasoning has long been held suspect in biochemistry, in part because the fragmentary nature of our knowledge made inferences concerning the "purpose" of a reaction quite unreliable, in contrast to the explicit characterization of the components of the reaction. However, the focus of biochemistry has shifted: the role of an enzyme can be unequivocally established by its position in a detailed sequence, by the consequences of a mutation that causes its deletion, and by identification of the metabolites that regulate its formation and activity. Hence criteria are now available for distinguishing, for example, whether an enzyme has a biosynthetic or a degradative function. Indeed, the chapter on regulatory mechanisms (Ch. 13) would be nearly meaningless unless one were willing to infer a purpose for these mechanisms, i.e., the increased efficiency of growth for which natural selection inexorably presses.

Hence at the level of the individual cell and its molecular components we can now accept the physiological concept implicit in the etymology of the word "organism": an entity with unique properties derived from the interactions of its component parts. As we review, in the next few chapters, the intricate pattern of organization that has been revealed within a 1-μ cell, we may well feel a sense of awe at the marvels of evolution, perhaps akin to Leeuwenhoek's feelings on discovering that the world was teeming with such organisms.

MICROBIAL TAXONOMY

PLANTS, ANIMALS, AND PROTISTS

The living world has long been divided into two kingdoms: the nonmotile, photosynthetic plants, and the motile, nonphotosynthetic animals. Hence when the microbial world began to be systematically explored, in the early nineteenth century, the newly discovered organisms were fitted into this familiar pattern. Those single-celled organisms that have a flexible cell integument, as do animal cells, were considered the most primitive animals, or **Protozoa.** The **Algae,** on the other hand, were considered plants, since they are photosynthetic and have a rigid cell wall, like plant cells. On vaguer grounds the **Fungi** or **Eumycetes** (molds and yeasts) were also lumped with the plants, as were the **Bacteria** (also called **Schizomycetes,** or fission-fungi, because they divide by transverse fission).

Nevertheless, this Procrustean classification into plants and animals presented too many inconsistencies. Fungi and most bacteria are nonphotosynthetic; many bacteria are motile; some fungi and algae have motile spores (zoospores) which by themselves could be taken for protozoa; and the small group of slime molds were claimed by both zoologists and botanists. Moreover, Darwin's publication of the monumental *Origin of Species,* in 1859, implied that living organisms do not **have** to be either plants or animals; transition forms are expected. Indeed, microbes strongly support the theory of evolution by providing the required connection between the very disparate kingdoms of plants and animals. The microbes of today have apparently descended, perhaps with little change, from the primitive common ancestors of these kingdoms. The most logical solution therefore appears to be that proposed in 1866 by Haeckel, but long ignored: the establishment of a **third** biological kingdom, the **Protista** (Fig. 1–5), which are distinguished from animals and plants by their <u>relatively simple organization.</u>*

Most protists are unicellular or coenocytic (multinucleate) throughout their life cycle, though some form large, multicellular, superficially plant-like structures, such as seaweeds (many marine algae) and mushrooms (the basidiomycete group of fungi). However, the tissues of these organisms are aggregates of similar cells, with only very primitive differentiation, whereas true plants and animals have

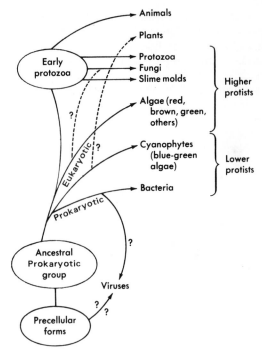

Fig. 1-5. Probable evolutionary relations of the major groups. Postulated ancestral, extinct groups are encircled.

highly differentiated multicellular adult forms, alternating with transient unicellular gametes.

EUKARYOTES AND PROKARYOTES

The electron microscope revealed a fundamental division among the protists, based on complexity of organization. The cells of higher protists (protozoa, fungi, and most algae) are **eukaryotic** (Gr., "true nucleus"), like those of plants and animals: they contain <u>a nucleus with a nuclear membrane,</u> <u>multiple chromosomes,</u> and <u>a mitotic apparatus to ensure equipartition of the products of chromosomal replication among the daughter nuclei.</u> In contrast, the lower protists, which include all bacteria and the small group of blue-green algae **(Cyanophyceae),**† are charac-

---

* The earlier classification into two kingdoms is still perpetuated in many introductory biology books. Indeed, it is reflected in the present volume by the omission of the protozoa, which are traditionally taken up in the medical curriculum along with the higher animal (metazoan) parasites.

† Algae carry out photosynthesis by the same mechanism as higher plants, while the few photosynthesizing bacteria employ a more primitive mechanism (Ch. 3). The "blue-greens" were originally classified as algae because of their plant-like photosynthesis, but they are better grouped with the bacteria in the light of their typically prokaryotic structure.

terized by smaller, **prokaryotic** cells, in which the "nucleus" is in fact a single naked chromosome, without a nuclear membrane. Prokaryotic cells also differ from eukaryotes in lacking other membrane-bound organelles (such as mitochondria).

Several biochemical peculiarities of prokaryotes will be described in later chapters. General properties include the absence (with few exceptions) of steroids and the presence of unique components (muramate and often diaminopimelate) in the cell wall. Specialized properties of certain groups include carbon storage as poly-β-hydroxybutyrate, N_2 fixation, obligate anaerobiosis, a mode of photosynthesis that does not release O_2, and derivation of energy from the oxidation of inorganic compounds.

The lower protists thus appear to be a distinct evolutionary group, stabilized at a primitive stage in the evolution of the cell. A wide evolutionary gap separates them from the eukaryotic protists. Since the further evolution of higher plants and animals depended on aggregation, differentiation, and specialization of cells, but not on any radical change in cell design, yeasts are of growing interest as a simple organism for studying the molecular genetics of eukaryotic cells. They offer many of the experimental advantages of bacteria, and at the same time they possess multiple chromosomes (15–18 in *Saccharomyces*); moreover, some can be grown in the haploid as well as the diploid state, thus permitting study of gene function without the complication of dominance.

The classification of bacteria is considered in Chapter 2. We shall also discuss fungi (Ch. 43), which include some pathogens. Algae, however, do not cause infectious disease, although some are extremely poisonous when ingested. For a broader introductory survey of bacterial and other microbial groups we recommend *The Microbial World* by Stanier *et al.* (see Selected References).

PRECELLULAR EVOLUTION AND THE ORIGIN OF LIFE

As we have seen, the work of Pasteur and of Tyndall dispelled earlier claims of spontaneous generation of life from nonliving matter under experimental conditions. However, this result does not exclude the possibility that it could arise under other conditions and given eons of time. Indeed, Darwin's theory of evolution (published within 2 years of Pasteur's first paper on spontaneous generation) logically required such an initial evolution of life, preceding the evolution of the contemporary living world. The problem is of particular interest to microbiologists since it concerns the development of the simplest living forms.

A satisfactory general explanation, now widely accepted, was proposed independently in 1924 by Oparin in the Soviet Union and in 1929 by Haldane in England. They suggested that the development of obviously living organisms was preceded by a period of **chemical, prebiotic evolution,** occupying perhaps the first 2 billion of the earth's 4 to 5 billion years. During this period bodies of water accumulated an increasing variety of organic compounds, formed with the aid of such agencies as ultraviolet light, lightning, volcanic heat, inorganic surface catalysis, and concentration by freezing. These substances could accumulate because the primitive earth lacked the two agents that make them so unstable under present terrestrial conditions: microbial cells and molecular oxygen. The former would be absent by definition; and there is geological evidence that free oxygen appeared in the earth's atmosphere rather late, arising as a consequence of the biological evolution of photosynthesis and/or the inorganic splitting of water followed by loss of the light H_2 molecules from the gravitational field.

The thin "soups" resulting from this organic accumulation are believed to have developed systems that slowly catalyzed their own formation from simpler substrates. With the selection of improved catalysts, which permitted an increasingly complex system to be condensed in a smaller volume, chemical evolution would merge into precellular biological evolution, and would eventually yield the minimal unit of life that we can recognize: a genome-containing, membrane-bounded cell, within which a concentrated and efficient set of catalysts bring about both replication of the genetic material and synthesis of further catalysts.

This hypothesis eliminates the concept of a sharp division between the living and nonliving: in early evolution there would be no moment at which the first living being suddenly began to stir, or at least to grow. Viewed in these terms, "life" cannot be defined in terms of a cellular pattern of organization, or in terms of a given kind of molecule, but is defined in terms of self-replication from simpler substrates. And though nucleic acids provide the basis for this genetic continuity in the organisms that we know today, the same function may originally have been provided, in a primitive form, in a system of catalytic molecules whose over-all capacity for self-replication was not concentrated in any special informational macromolecules.

The postulated precellular living systems have not been demonstrated. They would be difficult to recognize; and though multiple primitive systems may well have arisen independently, in various localities, we would expect them to be displaced by spread of the first efficient cellular system. This development would account for the apparently monophyletic origin of the present biological kingdom—an origin suggested by the construction of proteins only from L-amino acids, and not from the equally likely D-amino acids.

Viruses have been suggested as possible representatives of a precellular stage in evolution, since their organization is simpler than that of cells. Indeed, the first crystallization of a virus (tobacco mosaic virus), by Stanley in 1935, was widely hailed as having philosophical implications, because it apparently bridged the gap between the living and the nonliving. In reality, however, crystallinity simply reflects structural uniformity and surface complementarity of the particles, leading to orderly aggregation. Moreover, subsequent discoveries showed a special, intimate relation between viruses and their host cells rather than a simple nutritional dependence: viruses depend entirely upon the cell for biosynthetic machinery (Ch. 44); and some viruses can even exchange genes with the host cell (Ch. 46). It therefore seems quite possible that the viruses we know have evolved from host cell components, rather than from early precursors of cells.

Experimental Approaches. The Haldane-Oparin hypothesis received considerable support when Miller and Urey showed that various amino acids can be formed, in detectable amounts, by the action of an electric spark on a gas designed to simulate the atmosphere of the primitive earth (a mixture of water vapor, ammonia, methane, and hydrogen). Similar nonspecific reactions have yielded products as complex as nucleic acid bases, and have formed polynucleotides and polypeptides from their monomers. The problem of the origin of life has thus become a respectable area of experimentation, spurred on by the prospect of exploring for extraterrestrial life, in perhaps quite unfamiliar form. In this search, however, a liquid in which molecules can diffuse still seems indispensable.

SELECTED REFERENCES

BARGHOORN, E. S. The oldest fossils. *Sci Amer,* May, 1971, p. 30.

BERNAL, J. D. *The Origin of Life.* World Publishing, Cleveland, 1967.

BROCK, T. D. (ed. and trans.). *Milestones in Microbiology.* Prentice-Hall, Englewood Cliffs, N.J., 1961. Paperback. An excellent selection of classic papers, with helpful annotations; probably the best introduction to the history of the field.

BULLOCH, W. *The History of Bacteriology.* Oxford Univ. Press, London, 1960. A detailed account, with emphasis on medical bacteriology.

CAIRNS, J., STENT, G. S., and WATSON, J. D., eds. Phage and the Origins of Molecular Biology. Cold Spring Harbor Lab., N.Y., 1966.

CALVIN, M. *Chemical Evolution.* Oxford Univ. Press, London, 1969.

Chemical and Bacterial (Biological) Weapons and the Effects of Their Possible Use. United Nations, New York, 1969.

COHEN, S. S. Are/were mitochondria and chloroplasts microorganisms? Am Scientist *58*:281 (1970).

DOBELL, C. *Antony van Leeuwenhoek and His "Little Animals."* Constable, London, 1932. Reprinted in paperback by Dover, New York, 1960.

DUBOS, R. J. *Louis Pasteur: Free Lance of Science.* Little, Brown, Boston, 1950.

KENYON, D. H., and STEINMAN, G. *Biochemical Predestination.* McGraw-Hill, New York, 1969. Emphasizes experimental work on the origin of life.

LARGE, E. C. *Advance of the Fungi.* Cope, London, 1940. Reprinted in paperback by Dover, New York, 1962. An entertaining account of the development of the infectious microbiology of higher plants.

LECHEVALIER, H. A., and SOLOTOROVSKY, M. *Three Centuries of Microbiology.* McGraw-Hill, New York, 1965.

MARGULIS, L. *Origin of Eukaryotic Cells.* Yale Univ. Press, New Haven, 1970.

MESELSON, M. Chemical and biological weapons. *Sci Amer,* May, 1970, p. 15.

Microbial Classification (12th Symposium, Society for General Microbiology). Cambridge Univ. Press, London, 1962.

OPARIN, A. I. *The Origin of Life on Earth,* 3rd ed. Academic Press, New York, 1957; or 2nd ed. (1938), reprinted in paperback by Dover, New York, 1953.

SCHOPF, J. W. Precambrian micro-organisms and evolutionary events prior to the origin of vascular plants. *Biol Rev 45*:319 (1970).

STANIER, R. Y. Toward a definition of the bacteria. In *The Bacteria,* vol. V, p. 445. (I. C. Gunsalus and R. Y. Stanier, eds.) Academic Press, New York, 1964.

STANIER, R. Y., DOUDOROFF, M., and ADELBERG, E. A. *The Microbial World,* 3rd ed. Prentice-Hall,

Englewood Cliffs, N.J., 1970. An excellent introduction to general microbiology.

VALLERY-RADOT, R. *The Life of Pasteur*. Constable, London, 1901. Reprinted as paperback by Dover, New York, 1960. This biography is detailed and chronological; that of Dubos (see above) is more interpretive.

VAN NIEL, C. B. Natural selection in the microbial world. *J Gen Microbiol 13*:201 (1955).

WILSON, G. S., and MILES, A. A., *Topley and Wilson's Principles of Bacteriology and Immunology,* 3rd ed., 2 vols. Williams & Wilkins, Baltimore, 1964. An excellent reference work on medical bacteriology, including its history.

part **I**

BACTERIAL PHYSIOLOGY

BERNARD D. DAVIS

chapter 2

BACTERIAL STRUCTURE

METHODOLOGY   22

GROSS FORMS OF BACTERIA   24

FINE STRUCTURE OF BACTERIA   24
Cell Wall   24
Capsules   25
Flagella   26
Pili (Fimbriae)   30
Cytoplasmic Contents   31
Nuclear body (Nucleoid)   31

CELL SAP AND CYTOPLASMIC ORGANIZATION   32

BACTERIAL CLASSIFICATION   33
The Present Classification   33
The Arbitrary Definition of a Species   35
Taxonomy Based on DNA Homology   36

Bacteria are the smallest organisms that contain all the machinery required for growth and self-replication at the expense of foodstuffs, and they are morphologically simpler than the cells of higher organisms. The diameter is usually about 1 μm. Bacteria are designated as **prokaryotic,** because they lack the organized nucleus (with nuclear membrane and mitotic apparatus) of higher, **eukaryotic** cells. Other differences between the two groups have been noted in Chapter 1.

Though they are simpler than animal cells inside, bacteria have a more complex surface structure, with a **rigid cell wall** surrounding the **cytoplasmic membrane** (cell membrane, plasma membrane). The wall protects the cell against mechanical damage. Moreover, since cell function requires a high intracellular concentration of specific salts and metabolites the wall protects bacteria against osmotic rupture in dilute media. In addition, the wall is responsible for many of the taxonomically significant features of bacteria: their shapes, their major division into gram-positive and gram-negative organisms (see below), and antigenic specificities that are important in classification and in the interactions of pathogens with the host.

The development of bacterial cytology as a substantial discipline was long delayed, since the limited resolving power of the light microscope (0.2 μm) can reveal little detail in such small cells; hence bacteria were long regarded as essentially bags of enzymes, lacking any interesting organization. The electron microscope revealed the distinctive architecture of the prokaryotic cell, which is represented schematically in Figure 2-1. The cell wall and membrane are often referred to together as the **cell envelope,** while the more optional capsule, flagella, and pili are considered **appendages.**

This chapter will introduce bacterial structure in general and will consider the surface appendages; the wall and membrane will be discussed in detail in Chapter 6, and intracellular structures in various other chapters.

METHODOLOGY

Staining. Since classification has depended heavily on size, shape, and staining properties, the light microscope was long the hallmark of the bacteriologist. Unstained preparations may be used, especially with the phase-contrast microscope, but medical bacteriologists have generally studied heat-fixed, stained preparations.

Special stains used for certain organisms (e.g., mycobacteria, corynebacteria) will be described in the chapters on these groups. For most organisms the **Gram stain** is preferred. This method was developed empirically by the Danish bacteriologist Christian Gram in 1884. The cells are first fixed to the slide by heat and stained with a basic dye (e.g., crystal violet), which is taken up in similar amounts by all bacteria. The slides are then treated with an I_2-KI mixture to fix (mordant) the stain, washed with acetone or alcohol, and finally counterstained with a paler dye of different color (e.g., safranin). Gram-positive organisms retain the initial violet stain, while gram-negative organisms are decolorized by the organic solvent and hence show the counterstain.

Nearly 75 years after the introduction of this very useful procedure the discovery of a consistent difference in wall composition of gram-positive and gram-negative bacteria (see below) focused attention on the wall as the key to the gram stain. Salton then showed that the gram-positive wall is not stained itself but presents a **permeability barrier** to elution of the dye-I_2 complex by alcohol. This mechanism accounts for the old observation that gram-positive cells often become negative in aging cultures (in which autolytic enzymes attack the wall).

The dehydrating effect of the organic solvent seems to contribute to dye retention, since with gram-positive (but not with gram-negative) cells 95% ethanol is less effective than 50% ethanol in eluting metabolites.

Darkfield microscopy is a special procedure that recognizes objects through their reflection of light admitted from the side of the field of vision, rather than by virtue of their opacity or refractility. It is <u>useful for identifying objects that are too thin for resolution in the light microscope but long enough to have a characteristic shape</u> (e.g., certain spirochetes, Ch. 37).

Electron microscopy has increased the available resolving power at least 200-fold (i.e., ca. 1 nm). It is used in bacteriology in several ways. 1) **Shadowcasting** with metal vapor has been valuable for studying surface appendages of intact bacteria (Figs. 2-3 and 2-5), as well as shapes of viruses (Ch. 44, Fig. 44-2). 2) **Negative staining** is accomplished by drying in the presence of an electron-opaque solution (e.g., phosphotungstate), which forms thicker deposits in crevices (Fig. 44-3). This procedure has made it possible to resolve subunits that are obscured in metal-shadowed preparations, e.g., the fine structure of flagella or the surface of the cell wall. 3) The most important method is the study of **thin sections** (ca. 0.02 μm). 4) In **freeze-etching** the cells are frozen at very low temperature ($-150°$) in a block of ice, which is then cleaved with a knife. Some of the ice is sublimed from the fresh surface at low temperature ("etched") to reveal underlying structures, and a replica of the surface, with shadowing, is prepared. The fracture often passes through cells, and the new surfaces thus formed may separate natural lines of cleavage, e.g., between wall and membrane, and also between the inner and outer faces of a membrane. This method eliminates artefacts due to the fixation and drying ordinarily required in preparing specimens; and both the inner and outer surfaces

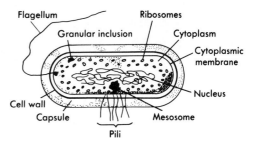

Fig. 2-1. Diagram of prototype eubacterial cell.

of a layer (concave and convex) can be studied.

Successful electron microscopy of thin sections of bacteria required modification of the methods of fixation and embedding developed for animal cells. The method of Ryter and Kellenberger, which is one of the most widely used, employs osmium tetroxide fixation at a low pH (6.0), in the presence of polyvalent cations (e.g., Ca^{++}) plus amino acids to preserve nuclear structure, followed by embedding in a nonshrinking epoxy resin (e.g., Araldite) or polyester resin (e.g., Vestopal). Contrast is enhanced by treating the sections with heavy ions (lead salts, uranyl acetate, lanthanum nitrate). Different procedures are optimal for demonstrating different structures.

Fractionation of bacteria (by mechanical disintegration, differential centrifugation, and selective digestion or solubilization) permits study of the chemical composition and molecular organization of specific components; electron microscopy has been invaluable in establishing the identity and homogeneity of the fractions. Because the cells are small and have tough walls it has been necessary to develop special methods of lysis.

Thus by violent agitation (e.g., in a Waring Blendor) it is possible to separate flagella and pili (Fig. 2–1) from bacteria. Shaking with glass beads disrupts the envelope as well, releasing the cell contents; the envelope fraction can be recovered by differential centrifugation. Bacteria can also be disrupted by grinding as a wet paste with abrasives (e.g., alumina), by sonic or supersonic oscillation (sonication), or by explosive release of high pressure (e.g., release past a needle valve in a French press); the products are often more useful for enzymatic than for morphological studies. For gentle disruption, with minimal fragmentation of long DNA or RNA chains, enzymatic digestion of the wall is followed by lysis of the resulting fragile spheres (see below, Protoplasts).

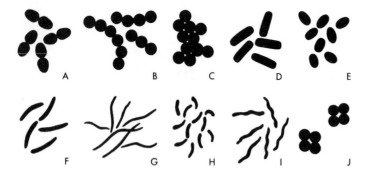

Fig. 2-2. Bacterial forms: **A,** diplococci; **B,** streptococci; **C,** staphylococci; **D,** bacilli; **E,** coccobacilli; **F,** fusiform bacilli; **G,** filamentous bacillary forms; **H,** vibrios; **I,** spirilla; and **J,** sarcinae.

Osmotically sensitive spheres may be lysed by transfer to a hypotonic medium, or by adding reagents that dissolve the membrane (lipase, deoxycholate). Because the DNA threads are relatively unfragmented such lysates are very viscous, in contrast to those produced by grinding or sonication; hence treatment with DNase is helpful in further fractionation.

Other Methods. Radioautographic studies have occasionally been valuable, e.g., for localizing DNA by means of radioactive thymidine, or for following wall synthesis. Cytochemical procedures have had limited use because of the small dimensions involved, but they have been of great value in animal virology, as will be seen in later chapters: they include the use of specific antibodies tagged with a fluorescent dye (Ch. 28, Fig. 28–4) or coupled to a substance, such as ferritin, that forms a sharp image in electron micrographs (Ch. 25, Fig. 25–1). Electron transport systems can be revealed by the oxidants tellurite or tetrazolium, whose reduction yields an unsoluble, opaque product (formazan).

We shall consider here the structures of only the typical or "true" bacteria (eubacteria); later chapters will deal with the less typical bacteria (including actinomycetes, spirochetes, mycoplasmas, rickettsiae, and chlamydiae), as well as with fungi and viruses.

GROSS FORMS OF BACTERIA

The light microscope reveals two principal forms of eubacteria: more or less spherical organisms known as **cocci** (Gr. and L., "berry") and cylindrical ones called **bacilli** (L., "stick").* Incompletely separated cocci may appear in a number of different patterns, depending upon the planes in which they

_____

* The term "bacillus" is unfortunately used both as a general name for rod-shaped bacteria and as the name of a particular genus (capitalized and italicized).

divide: when predominantly in pairs they are known as **diplococci;** in chains as **streptococci** (Gr. *streptos* = twisted); and in clusters as **staphylococci** (Gr. *staphyle* = bunch of grapes). Cocci that remain adherent after splitting successively in two or three perpendicular directions, yielding square tetrads or cubical packets, are known as **sarcinae** (L., "bundles"). Bacilli when unusually short are referred to as **coccobacilli;** when tapered at both ends as **fusiform bacilli;** when growing in long threads as **filamentous forms;** and when curved as **vibros** or **spirilla** (Fig. 2-2).

FINE STRUCTURE OF BACTERIA

CELL WALL

The presence of a rigid cell wall, outside the cytoplasmic membrane, may be demonstrated by plasmolysis, i.e., exposure of bacterial cells (especially of gram-negative species) to a hypertonic solution, which causes the protoplast to contract and shrink away from the wall (Fig. 2-3). Moreover, in bacterial preparations subjected to mechanical damage (e.g., crushing) many of the cells lose refractility, from loss of their intracellular contents, but leave a pale, empty ghost which retains the characteristic shape of the original cell.

Electron microscopy of thin sections reveals the walls of gram-positive organisms as a relatively thick (150 to 800 A), uniform, dense layer, with the cytoplasmic membrane closely apposed to its inner surface (Fig. 2-4). This dense material was eventually found to be a **peptidoglycan:** a network of polysaccharide chains cross-linked by an oligopeptide (Ch. 6). Gram-negative cells have a similar but much thinner peptidoglycan layer, covered by an outer layer of lipopoly-saccharide and lipoprotein; the several layers are not all distinct in electron micrographs (Fig. 6-3). The details of wall and membrane structure, function, and biosynthesis will be presented in Chapter 6.

Surface Topography. Though sections of fixed preparations give the impression of a uniform wall thickness, surface photographs sometimes reveal considerable structural detail. Negative staining of some gram-positive cells, or of their isolated wall preparations, reveals a surface layer of regular, hexagonally packed spherical subunits of 80 to 120 Å (Fig. 2-5), analogous to a tile wall. The subunits have been identified as proteins, and in some pathogens an outer protein layer is important in virulence and in classification (e.g., the M protein of streptococci). The walls of some other bacteria and of fungi, in contrast, have a matted, irregular fibrous texture, analogous to a thatched wall, but this appearance can also be produced by the drying of a slime layer.

With unfixed gram-negative bacilli negative staining has revealed a highly convoluted surface (Fig. 2–6). This organized structure is presumably the outermost layer of gram-negative cells (Ch. 6). However, the convolutions may be an artefact of drying, since they are not seen in freeze-etched preparations.

Protoplasts and Spheroplasts. When the rigid "basal" layer of the cell wall is digested by lysozyme (Ch. 6) the cell ordinarily lyses, but Weibull showed that in a hypertonic medium (e.g., 20% sucrose or 0.5 M KCl) such treatment releases the membrane with its contents as an osmotically sensitive sphere. With gram-positive organisms this product appears to be free of wall constituents, by both microscopic and chemical criteria, and is called a **protoplast.** Gram-negative organisms similarly yield osmotically sensitive spheres, but these retain an outer wall layer; they are called **spheroplasts,** to distinguish them from the presumably wall-free protoplasts. Spheroplasts can also be formed by specifically preventing basal wall synthesis in growing cells, either by adding an inhibitor or by withholding, in a mutant, a compound specifically required for that synthesis (e.g., diaminopimelate).

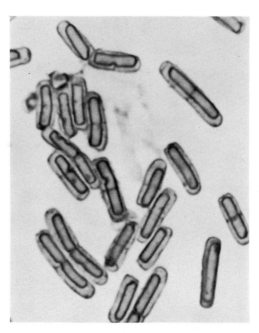

Fig. 2-3. Plasmolysis of *Bacillus megaterium.* The cells on a slide were successively treated with ether vapor (which loosens the attachment of membrane to wall), air-dried, post-fixed with Bouin's fluid (which cause contraction of the protoplast), and stained with Victoria blue (which enhances the visibility of the membrane enclosing the protoplast). (Courtesy of C. Robinow.)

CAPSULES

Capsules often protect pathogenic bacteria from phagocytosis and thus play a major role in determining virulence (Ch. 22, Antiphagocytic capsules). These loose, gel-like structures are most easily demonstrated by **negative staining** in Indian ink suspension, where they form a clear zone between the opaque medium and the more refractile (or stained) cell body (Ch. 27, Fig. 27-2). They can also be stained by special stains, and exposure to specific anticapsular antibodies increases their size and refractility ("quellung" reaction; Ch. 25, Fig. 25-1).

The capsular layer usually does not reveal any structural detail in electron micrographs. True capsules have a well-defined border (Ch. 22, Fig. 22-2A) and may require enzymatic attack for removal from the cell, although soluble capsular substance can often be detected immunologically in culture filtrates. Some "capsules" or slime layers, how-

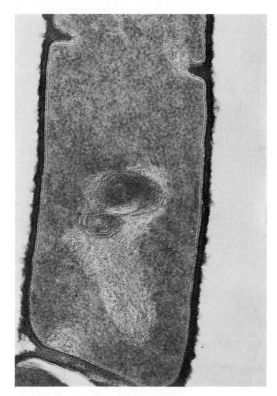

Fig. 2-4. Section of portion of gram-positive *Bacillus subtilis,* including a completed but not yet cleaved septum at one end of the cell and a beginning septum forming as equatorial ring in the midzone of the cell. Note well-defined plasma membrane ("double track" lining inner surface of dense wall), continuous with small vesicular mesosome at one portion of growing septum. Also note two large, concentric lamellar mesosomes in higher nuclear region; connection to plasma membrane not seen in this section. (Courtesy of A. Ryter.)

ever, have no definite border and can be seen to diffuse into the surrounding medium. These properties blur the distinction between a highly viscous excretion product and a cellular structure. The thickness of the capsule varies with growth conditions. In addition, some mutants that have lost the ability to make a visible capsule nevertheless make an immunologically detectable **microcapsule,** with unchanged specificity.

A given bacterial species may include strains of different antigenic types, with immunologically distinct capsules. Most capsules consist of relatively simple polysaccharides, containing repeating sequences of two or three sugars and often uronic acids. The synthesis of these polysaccharides, and their transfer across the cell membrane by a carrier lipid, involve the same mechanisms described in Chapter 6 for the polysaccharide portion of lipopolysaccharides of the gram-negative cell wall. The structures of some representative capsules of the pneumococcus are described in Chapter 25, and others are noted briefly in chapters on other organisms.

FLAGELLA

Flagella (*L. flagellum,* sing. = whip), when present, are responsible for the motility of eubacteria. Cells from which flagella are mechanically removed regain motility when the flagella regenerate. Motility can be recognized under the microscope in liquid medium (in a hanging drop or under a cover slip), where it must be carefully distinguished from brownian movement; it can also be revealed by the spread of visible growth in semisolid medium (e.g., 0.3% agar).

Flagella are long (3 to 12 μm), fine, wavy, filamentous appendages. Though readily visualized by darkfield microscopy, they are too thin (120 to 250 A) to be seen by ordinary microscopy unless heavily coated with special flagellar stains containing a precipitating agent (mordant), such as tannic acid (Fig. 2-7A). The wave length of the flagella, whether in live or in fixed preparations, is characteristic for a given bacterial strain.

Some species exhibit **peritrichous flagellation** (Gr. *trichos* = hair) with flagella distributed at random over the cell surface (Fig. 2-7), while in others one or a few flagella are found only at one or both poles **(polar flagellation).** The pattern is a genetically stable characteristic: strains that mutate to loss of flagellation revert to the original pattern. On this basis the common bacteria are separated into two orders, Eubacteriales (with peritrichous flagellation) and Pseudomonadales (with polar flagellation).

The vigor of bacterial movement depends on the number of flagella, which varies widely. Certain species (e.g., *Proteus*) may produce a huge number (Fig. 2-7C), especially when cell division is slowed; such cells spread in a thin film ("swarm") on the surface of the usual moist agar media. There is evidence that the flagella of a cell assume parallel di-

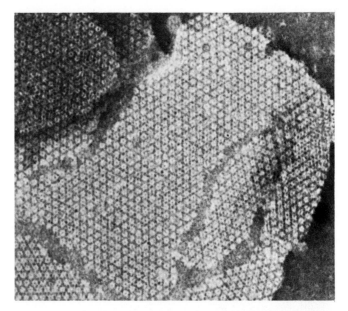

Fig. 2-5. Regular hexagonal array of granules on outer surface of isolated cell wall of a spirillum. Similarly regular rectangular arrays are found on many gram-positive bacteria, but the layer bearing the pattern is easily lost in preparation. Negatively stained. ×204,000, reduced. (Courtesy of R. G. E. Murray.)

Fig. 2-6. Surface of gram-negative organism, *Veillonella* (which closely resembles *Neisseria*); one-half a diplococcus is shown. Preparation freeze-dried without fixation and then negatively stained with silicotungstate, ×104,000. [From Bladen, H. A., and Mergenhagen, S. E. *J Bacteriol 88*:1482 (1964).]

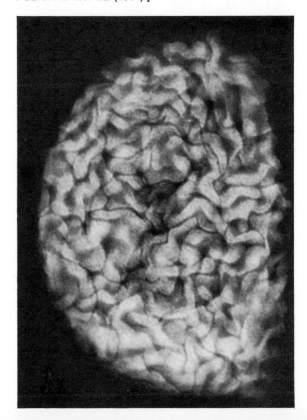

Fig. 2-7. Flagella. A. Flagellated bacillus stained with tannic acid-basic fuchsin, a flagellar stain. **B.** Electron micrograph of palladium-shadowed bacillus showing peritrichous flagellation. ×13,000. **C.** Highly motile form of *Proteus mirabilis* ("swarmer"), with innumerable peritrichous flagella. [**A** from Leifson, E., Carhart, S. R., and Fulton, M. *J Bacteriol* 69:73 (1955). **B** from Labaw, L. W., and Mosley, V. M. *Biochim Biophys Acta* 17:322 (1955). **C** from Hoeniger, J. F. M. *J Gen Microbiol* 40:29, (1965).]

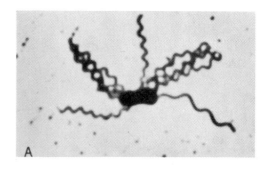

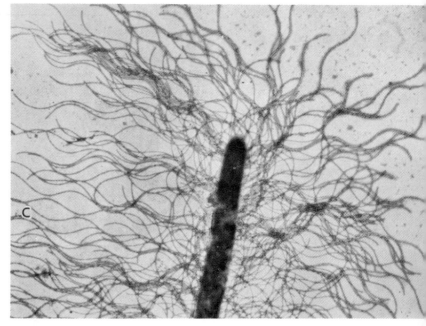

rections, and even move as adherent bundles, which would account for the linear motion observed (see Chemotaxis, Ch. 6).

Fine Structure. Being much thinner than the cilia of vertebrate cells or the flagella of protozoa, bacterial flagella do not contain the microtubules found in these structures. Rather, they consist of tightly wound chains, usually in a triple helix (Fig. 2-8A); some species have a different number of strands. Purified flagella dissociate at relatively mild acidity to yield a soluble globular protein, called **flagellin** (MW 40,000 in some species).* This material can reaggregate spon-

taneously in neutral solution to form the filament of flagella; the reaction is promoted by fragments of flagella, or membrane fragments containing the basal body, which serves as primer.

The **basal body** of the flagellum is attached to the plasma membrane, for the whole flagellum is retained by protoplasts when the wall is digested away. With purified flagella, obtained by dissolving away the membrane as well, the filament can be seen to pass through

---

* Some flagellins contain a novel amino acid, ε-N-methyl lysine. Since the genetic code (Ch. 12)

permits incorporation of only the 20 standard amino acids into protein, other amino acids are formed only by modification of residues after incorporation (e.g., conversion of proline to hydroxyproline in collagen). The capacity for such modification evidently evolved in prokaryotes.

a short, hook-like sheath at its base. In *E. coli* the cellular end of this sheath carries four very thin parallel rings (Fig. 2-8), while in a gram-positive organism, which has fewer envelope layers (Ch. 6), it has only two rings. The rings evidently provide areas of contact with the several layers of the surrounding cell envelope. One gene codes for the protein of the filament in *E. coli* and 11 for the other proteins making up the basal body; the proteins of the various components are also antigenically distinct.

Motility. The mechanism responsible for the whip-like **flagellar movement** is not known but may involve either contraction of the

rings or rotation of the hook-like sheath relative to the rigid layer of the wall: flagella are no longer motile when the wall is removed. The motility of the cell is not entirely random but exhibits some orientation in a gradient **(chemotaxis),** as will be discussed in Chapter 6.

Certain bacteria use other means of locomotion. In **spirochetes** the cell proper forms a helix around a relatively rigid axial rod (Ch. 37), and its contraction relative to the rod causes the cell to bend and thus to move. Certain **myxobacteria** (slime bacteria), and some blue-green algae, exhibit a slow **gliding** movement on solid surfaces; its mechanism is not understood.

The **mechanism of growth** of the flagellar filament may involve flow of subunits in a hollow core,

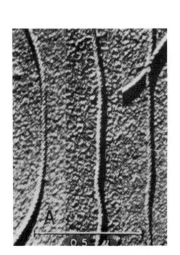

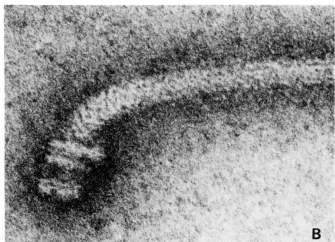

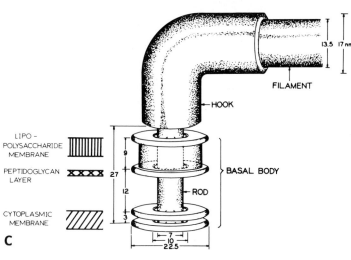

Fig. 2-8. Fine structure of flagella. **A.** Metal-shadowed filaments of bacillus in Figure 2-7B, at higher magnification showing helical structure. **B.** Negatively stained basal body of flagellum of *E. coli,* freed of attached wall and membrane. Note four rings and thick, hook-like sheath. **C.** Diagrammatic structure of basal body, inferred from preparations like B. [**B** from DePamphilis, M. L., and Adler, J. *J Bacteriol* *105*:384 (1971). **C** courtesy of M. L. DePamphilis and J. Adler.]

since labeling experiments have demonstrated growth at the tip; the alternative of recovery from free solution after excretion would seem too inefficient.

PILI (FIMBRIAE)

With many gram-negative bacilli the electron microscope revealed a group of still finer filamentous appendages, called pili (L., "hairs") or fimbriae (L., "fringe"). They are shorter, thinner, and straighter than flagella (Fig. 2-9); and on the same cell they vary in thick-ness (ca. 75 to 100 A) and length (up to several μm). Like flagella, they arise from basal bodies in the cytoplasmic membrane and they consist of a protein (pilin).

As we shall see in Chapter 9, in bacterial conjugation the male cell carries one or two special hollow sex pili that can form a bridge with the female cell. The same cell may carry hundreds of other pili whose function is unknown. In mutants that can no longer form pili the only significant change observed has been a decreased tendency to adhere to red blood cells. Cells mechanically de-

Fig. 2-9. Pili. Piliated strain of *E. coli,* grown in liquid medium without aeration. Each cell possesses hundreds of pili (diameter 70 A), and their presence promotes aggregation. Many isolated, broken pili are also seen. A few flagella, much longer and of larger diameter (140 A), extend from the cells to the edge of the photograph. Platinum shadowed, ×45,000, reduced. (Courtesy of Charles C. Brinton, Jr.)

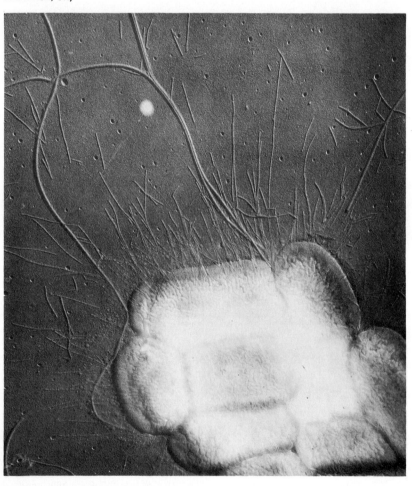

prived of pili rapidly form them again, which suggests that pili may be constantly shed and replaced.

CYTOPLASMIC CONTENTS

Mesosomes. Thin sections of bacteria often reveal one or more large, irregular invaginations of the plasma membrane called mesosomes; additional membrane within these pockets may be lamellar (Fig. 2-7A) or vesicular. Their function will be discussed in Chapter 6.

Ribosomes. The cytoplasm of bacteria is thickly populated with ribosomes: roughly spherical, dense objects, ca. 180 A in diameter (Fig. 2-4). Their structure and function will be considered in Chapter 12. In cells engaged in protein synthesis the ribosomes are mostly grouped in chains called **polysomes,** but these cannot be recognized in sections of cells because of the close packing. In gently lysed protoplasts ribosomes and polysomes are found not only free but also in the membrane fraction, but it is not certain whether they are physiologically attached (as in some animal cells) or simply trapped in the membrane fragments.

Granular Inclusions. The cytoplasm of many kinds of bacteria often contains relatively large granules, composed of cellular storage materials, in amounts that vary widely with nutritional conditions. **Polymetaphosphate** granules $[(PO_3{}^-)_n]$ are stained metachromatically (i.e., with a change in color) by certain dyes (methylene blue or toluidine blue). Iodine reveals the characteristic red of **glycogen** or the blue of a starch-like **granulose,** whose high molecular weights permit nutrients to be stored without increasing the osmolarity of the cytoplasm.

NUCLEAR BODY (NUCLEOID)

Though basic dyes stain all nucleic acids they can be used in mammalian cells to reveal the nucleus and the chromosomes because the DNA is ordinarily more densely aggregated than the RNA of the cytoplasm. In bacteria, in contrast, the cytoplasm is densely packed with RNA and hence is at least as basophilic as the nucleus. However, discrete nuclear bodies in bacteria can be recognized under the light microscope through special procedures, such as the DNA-specific Feulgen

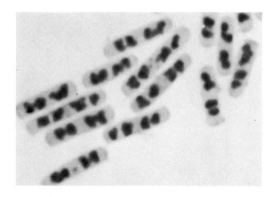

Fig. 2-10. Bacterial nuclear bodies. Demonstration of "chromatin bodies" in *Bacillus cereus* by a basic stain (Giemsa) after hydrolysis of the RNA by treatment with HCl. ×3600. (Courtesy of C. Robinow).

(fuchsin sulfite) stain, or selective hydrolysis of the RNA (in fixed cells) by HCl or RNase, followed by staining with a basic dye (Fig. 2-10). Most bacilli contain two or more such bodies per cell, since cell division lags behind nuclear division (Ch. 13).

More detailed study of these bodies with the electron microscope established the concept of <u>bacteria as prokaryotic cells, lacking the discrete chromosomes, mitotic apparatus, nucleolus, histones,</u> and <u>nuclear membrane of the nuclei of eukaryotic cells.</u> Hence the nuclear region in bacteria is referred to as the nuclear body or nucleoid. As we shall see in Chapter 10, the bacterial chromosome, which presumably constitutes the nuclear body, is a continuous cyclic molecule of double-stranded DNA, about 1000 times as long as the bacterium.

Even after DNA is stained with uranyl or lanthanum ion the nuclear region in bacteria is less dense to the electron beam than the cytoplasm. Different methods of fixation yield strikingly different pictures: one type reveals a mass of relatively fine fibers, often arranged in parallel, curved bundles (Fig. 2–11).

The fine detail resolved in such preparations generates confidence in the significance of the pattern observed. Moreover, in live cells growing in a medium with a refractive index close to that of the cell a compact nuclear region, with a higher refractive index, can be seen by phase-contrast microscopy; and compact nucleoids, held together by RNA, can be isolated by gently lysing bacteria in

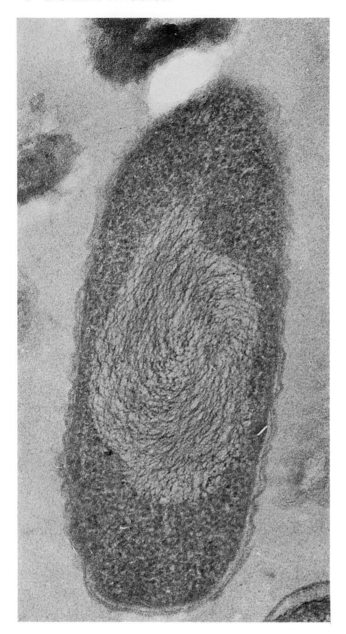

Fig. 2-11. Section of *Agrobacterium tumefaciens.* Special fixation to bring out fibrillar structure of nucleoid. (Courtesy of A. Ryter.)

media of high ionic strength. Nevertheless, the total exclusion of ribosomes from a large nucleoid (Fig. 2–11) is difficult to reconcile with evidence (Chs. 11 and 13) that much of the DNA in a growing cell carries growing chains of messenger RNA, each of which carries a closely packed train of ribosomes.

The compact nucleoid is apparently quite loosely packed: for though its area in sections corresponds to at least 10% of the cell's volume, DNA represents only 2 to 3% of the cell's dry weight. Moreover, in virus particles, in which the DNA is meta-

bolically inactive, the same methods reveal a much denser DNA-containing region.

CELL SAP AND CYTOPLASMIC ORGANIZATION

Bacterial cell volumes can be determined from measurements of the volume of packed wet cells, corrected for intercellular space (ca. 20%) by the use of a macromolecule that does not penetrate or adsorb to the bacteria (e.g., hemoglobin, or isotopically labeled dextrin or albumin). Dry weight determinations then show that an average bacterium (e.g., *E. coli*) contains about 75% water. Since the wall is dense, and contributes a substantial fraction of the cell weight, the cytoplasm (including the nuclear region) averages about 80% water. It contains a great variety of small molecules (the metabolite pool) and inorganic ions, as well as ribosomes and enzymes.

The metabolite pool can be recovered for chemical estimation by transferring cells to boiling water or cold 5% trichloroacetic acid, which disrupts the cytoplasmic membrane and precipitates the macromolecules. The cells must first be washed free of extracellular fluid. To carry out this step rapidly, in order to minimize metabolic alteration of the pool, nitrocellulose membrane filters have proved invaluable, since they retain bacteria quantitatively but allow fluid to pass quickly.

The high osmotic pressure of the bacterial cell shows that the components of the metabolite pool are largely in a free state. This osmotic pressure can be estimated by determining at what point increasing external osmolarity causes an abrupt decrease in cell volume (i.e., plasmolysis). The values observed are about 20 atmospheres in some gram-positive organisms and about 5 atmospheres in thinner-walled gram-negative organisms. (It will be recalled that a 1-M solution of a nonelectrolyte ideally has an osmotic pressure of 22.4 atmospheres.)

BACTERIAL CLASSIFICATION

Since we cannot communicate meaningfully about any objects of study unless we can identify them it is necessary to develop a nomenclature, classifying entities according to their similarities and differences. In biology, however, the ordering goes beyond a determinative, purely descriptive classification and aims at "natural" taxonomic classification of groups of organisms, reflecting their lines of evolutionary descent.

With higher animals and plants morphological, physiological, and developmental homologies, as well as the fossil record, yield taxonomic schemes that are generally accepted as reflecting natural, phylogenetic relations. A keystone is the definition of a **species** as a group of individuals capable of continued fertile interbreeding, for the reproductive separation between species places some limits on the range of gene flow and thus minimizes the production of unworkable combinations of genes. Indeed, despite the conti-

nuity of evolution it has proved feasible to divide the world of plants and animals, with few exceptions, into hundreds of thousands of such discontinuous groups. In contrast, the higher levels of classification (genus, tribe, family, order, class) lack such a clean, operational definition as the species, and so their assignments have been more arbitrary.

With bacteria the situation is quite different: the number of available characteristics is small; there is no ontogeny to recapitulate phylogeny; the fossil record is scanty; and reproduction is ordinarily not sexual. Hence we cannot reliably determine which characters of an organism were acquired earlier and which later.

For example, the morphological subdivision of bacteria into spheres, rods, and spirals would appear to reflect a stable trait, settled early in evolution; and the same might be thought of the major energy-yielding patterns, such as various specific fermentations, oxidative metabolism, or photosynthesis. Yet each of these metabolic patterns is found in each of the morphological groups, and the decision to subordinate one of these criteria to the other is arbitrary. Accordingly, in the most extensive cooperative effort at classification (*Bergey's Manual of Determinative Bacteriology*) the shape of the taxonomic tree has changed substantially with each edition.

THE PRESENT CLASSIFICATION

In the absence of reliable criteria for phylogenetic relations the classification that has developed is primarily useful as a determinative key, i.e., for identifying an unknown organism and relating it to previously described organisms. Many properties are used for this purpose: visible features (including shape, size, color, staining, motility, flagellar pattern, capsule, and colonial morphology); energy-yielding pattern; formation of characteristic chemical products; nutrition (including both growth requirements and ability to utilize various sugars and other foods); presence of characteristic surface macromolecules (usually detected immunologically, but also chemically); and ecological relations (including the ability to parasitize higher organisms and to cause disease). Biosynthetic pathways vary little; but differences in their regulatory mechanisms (Ch. 13) have recently furnished another variable.

By consulting published descriptions of

TABLE 2-1.   Main Groups of Bacteria

I. GRAM-POSITIVE EUBACTERIA

Cell shape	Motility	Other distinguishing characteristics		Genera	Families
Cocci	Nearly all permanently immotile	Cells in cubical packets Cells irregularly arranged		*Sarcina* *Micrococcus* *Staphylococcus*	Micrococcaceae
		Cells in chains, lactic fermentation of sugars		*Streptococcus* *Diplococcus* *Leuconostoc*	Lactobacillaceae
Straight rods	Nearly all permanently immotile	Lactic fermentation of sugars Propionic fermentation of sugars Oxidative, weakly fermentative		*Lactobacillus* *Propionibacterium* *Corynebacterium* *Listeria* *Erysipelothrix*	Lactobacillaceae Propionibacteriaceae Corynebacteriaceae
	Motile with peritrichous flagella, and related immotile forms	Endospores produced	Aerobic	*Bacillus*	Bacillaceae
			Anaerobic	*Clostridium*	

II. GRAM-NEGATIVE BACTERIA, EXCLUDING PHOTOSYNTHETIC FORMS

Cell shape	Motility	Other distinguishing characteristics		Genera	Families
Cocci	Permanently immotile	Aerobic Anaerobic		*Neisseria* *Veillonella*	Neisseriaceae
				Brucella *Pasteurella* *Hemophilus* *Bordetella*	Brucellaceae
Straight rods	Motile with peritrichous flagella, and related immotile forms	Facultative anaerobic	Mixed acid fermentation of sugars	*Escherichia* *Erwinia* *Shigella* *Salmonella* *Proteus*	Enterobacteriaceae
			Butylene glycol fermentation of sugars	*Enterobacter* *Serratia*	
		Aerobic	Free-living nitrogen fixers	*Azotobacter*	Azotobacteriaceae
			Symbiotic nitrogen fixers	*Rhizobium*	Rhizobiaceae
	Motile with polar flagella	Aerobic	Oxidize inorganic compounds Oxidize organic compounds	*Nitrosomonas* *Nitrobacter* *Thiobacillus* *Pseudomonas* *Acetobacter*	Nitrobacteriaceae Thiobacteriaceae Pseudomonadaceae
		Facultative anaerobic		*Photobacterium* *Zymomonas* *Aeromonas*	
Curved rods	Motile with polar flagella	Comma-shaped Spiral	Aerobic Anaerobic	*Vibrio* *Desulfovibrio* *Spirillum*	Spirillaceae

Table 2-1. Main Groups of Bacteria (*cont'd*)

III. OTHER MAJOR GROUPS

Characteristics	Genera	Orders (-ales) or families (-aceae)
Acid-fast rods	*Mycobacterium*	Actinomycetales
Ray-forming rods (actinomycetes)	*Actinomyces*	
Ray-forming rods (actinomycetes)	*Nocardia*	
Ray-forming rods (actinomycetes)	*Streptomyces*	
Spiral organisms, motile	*Treponema*	Spirochetales
	Borrelia	
	Leptospira	
	Spirocheta	
Small, pleomorphic; lack rigid wall	*Mycoplasma*	Mycoplasmataceae
Small intracellular parasites	*Rickettsia*	Rickettsiaceae
	Coxiella	
Small intracellular parasites, readily filtrable	*Chlamydia*	Chlamydiae
Intracellular parasites; borderline with protozoa	*Bartonella*	Bartonellaceae

Modified from Stanier, R. Y., Doudoroff, M., and Adelberg, E. A. The Microbial World. Prentice-Hall, Englewood Cliffs, N.J., 1963.

such features, and by reference to collections of "type cultures" (i.e., standard strains) that are maintained in several countries, bacteriologists can communicate meaningfully with each other. Following the Linnean* tradition of zoology and botany, each "species" of bacterium is assigned an official Latin binomial, with a capitalized genus followed by an uncapitalized species designation: this name is printed in italics to indicate the foreign language. Unitalicized vernacular or colloquial names may be derived from, or may be identical with, the official name (e.g., pneumococcus = *Diplococcus pneumoniae;* salmonella = *Salmonella* spp.) Some of the major "families" of bacteria are described in Table 2-1.

Unfortunately, international codes of nomenclature were developed late, have not always been adhered to, and have undergone frequent change. Hence the same organism can be encountered in the literature under several names: e.g., *Bacillus typhosus, Bacterium typhosum, Eberthella typhosa,* and *Salmonella typhi.* In virology, which developed later and emphasized molecular properties, the connection with systematic botany and zoology has been weak, and a proposed binomial system has not been generally adopted.

THE ARBITRARY DEFINITION OF A SPECIES

Even if we settle for the limited goal of a descriptive classification, bacteria still present problems. First, the grouping of individual strains into the same "kind" or "species" had to be settled without benefit of the interfertility test used in higher organisms; and even with the more recent development of bacterial genetics (Ch. 8), this test has not proved very useful (see below). Moreover, the mutability and the rapid growth of bacteria, combined with strong selection pressures from changes in the environment (Ch. 8), often lead to striking changes in the heritable properties of a strain during its cultivation in the laboratory. The "art" of the systematic bacteriologist therefore lies in distinguishing secondary details from the more stable properties, which presumably depend on large numbers of genes, or on genes whose mutations are lethal or are strongly selected against in nature. Surprisingly, some properties that may be lost

---

* Carolus Linnaeus (Carl von Linné) (1707–1778), Swedish botanist and professor of medicine.

by a single mutation, such as the ability to utilize lactose or other sugars, have proved very useful.

We thus find that a bacterial species usually includes a continuum of organisms with a relatively wide range of properties. In the Enterobacteriaceae in particular (Ch. 23), which appear to undergo much genetic recombination in nature, a large fraction of the strains encountered are intermediate in properties rather than identical with one or another type culture; each species thus represents a **cluster of biotypes,** more or less resembling a strain that is maintained as the type culture of the species, but without the sharp boundaries that we are accustomed to in higher organisms.

Specific molecules on the surfaces of bacteria are of particular importance for medicine, since they strongly influence host-parasite interactions. Hence the immunological characterization of surface macromolecules plays a large role in medical bacteriology (and virology), though not in general microbiology.* The deeper layers of the wall, and the cell membrane, vary much less among related organisms, for they are subject to less selection pressure from agents that attack the surface (antibodies, enzymes, bacteriophages). Hence immunological or chemical characterization of these deeper components provides a useful basis for recognizing natural relations. With the advance of bacterial cytology and chemistry we may expect increasing reliance on this approach, which has been extensively used with viruses.

Numerical Taxonomy. To avoid the arbitrary weighting of characters some investigators, with the aid of computers, have revived an approach to taxonomy introduced in the eighteenth century by the French biologist Adanson. In this system a large number of characters are determined for each strain, and strains are grouped on the basis of the proportion of characters shared, without giving any characters more weight than others. The approach is laborious, but it may lead to significant regrouping of certain classes of organisms.†

TAXONOMY BASED ON DNA HOMOLOGY

The approaches described above classify bacteria in terms of phenotypic traits. In recent years the development of molecular genetics has revolutionized the field by providing a precise, quantitative basis for defining biological relatedness. For as organisms drift apart in evolution, through the accumulation of mutations, their genes not only code for different structures and functions but also differ increasingly in their base sequence (Ch. 10). Indeed, the interfertility test for a species in higher organisms depends in part on just this homology: the ability of essentially any region in any chromosome of one parent to pair and recombine with the corresponding region of the other parent. **The chemical study of DNA** now offers a direct approach to the measurement of relatedness, through observations both on composition and on hybridization.

Thus, the **base composition** of DNA, though essentially the same in all vertebrates (ca. 40 mole % guanine + cytosine [=GC] and 60% adenine + thymine [=AT]), varies remarkably among bacteria, from about 30 to 70 mole % GC (Table 2-2).‡ Moreover, the chromosome of a bacterium is strikingly homogeneous in this respect: in a given strain the small DNA segments produced by fragmentation have a narrow range of GC ratios. Thus the DNA base composition of an organism is evidently a stable characteristic,

---

* However, since only a finite number of repeating polysaccharides are possible (in contrast to the virtually unlimited variety of proteins), the same antigenic determinant occasionally appears in widely separated organisms. For example, the surface polysaccharide of *E. coli* type O-86 cross-reacts with blood group substance B of human red cells.

---

† The attitude of many bacterial physiologists toward problems of classification is reflected in Duclaux' story of the microscopist who pointed out to Pasteur that an organism which he had taken for a coccus was in reality a small bacillus. The reply was, "If you only knew how little difference that makes to me!" However, as the study of bacterial variation has developed into bacterial genetics and then into molecular genetics, and as the study of visible structure has merged with that of chemical structure, the gap between the taxonomist and the physiologist has narrowed a good deal.

‡ Chapter 12 will show how such variation can be reconciled with a universal genetic code and a relatively uniform protein composition.

TABLE 2-2.   DNA Base Compositions of Representative Bacteria

Mole % guanine plus cytosine	Organism
28–30	*Spirillum linum*
30–32	*Clostridium perfringens, C. tetani*
32–34	*C. bifermentans, Leptospira pomona, Staphylococcus aureus*
34–36	*Bacillus anthracis,* other bacilli, *Clostridium kluyveri, Leptospira pomona, Mycoplasma gallisepticum, Pasteurella aviseptica, Staphylococcus albus, Streptococcus faecalis, Treponema pallidum*
38–40	*Bacillus megaterium, Hemophilus influenzae, Diplococcus pneumoniae, Lactobacillus acidophilus, Leuconostoc mesenteroides, Streptococcus pyogenes,* other streptococci, *Proteus vulgaris, Sporosarcina ureae*
40–42	*Bacillus laterosporus, Leptospira biflexa, Neisseria catarrhalis*
42–44	*Bacillus subtilis, B. stearothermophilus, Bacteroides insolitus, Coxiella burneti*
46–48	*Bacillus licheniformis, Clostridium nigrificans, Corynebacterium acnes, Pasteurella pestis, Vibrio cholerae*
48–50	*Neisseria gonorrheae,* other neisseriae
50–52	*Bacillus macerans, Escherichia coli,* other escherichiae, *Erwinia* spp., *Neisseria meningitidis, Proteus morgani, Salmonella* spp., *Shigella* spp.
52–54	*Aerobacter-Klebsiella* spp., *Corynebacterium diphtheriae, Erwinia* spp.
54–56	*Aerobacter-Klebsiella* spp., *Alcaligenes faecalis, Azotobacter agile, Brucella abortus*
56–58	*Corynebacterium* spp., *Lactobacillus bifidus*
58–60	*Agrobacterium tumefaciens, Corynebacterium* spp., *Serratia marcescens*
60–62	*Azotobacter vinelandii, Pseudomonas fluorescens, Rhodospirillum rubrum, Vibrio* spp.
64–66	*Desulfovibrio desulfuricans, Pseudomonas* spp.
66–68	*Pseudomonas aeruginosa, Mycobacterium tuberculosis*
68–70	*Pseudomonas saccharophila, Sarcina flava*
70–80	*Micrococcus lysoideikticus, Mycobacterium smegmatis, Nocardia* spp., *Sarcina lutea, Streptomyces* spp.

Modified from Marmur, J., Falkow, S., and Mandel, M. *Annu Rev Microbiol* 17:329 (1963).

settled over a long period of evolution. The base compositions presented in Table 2-2 confirm most previously accepted ideas concerning taxonomic relations, but have also revealed some interlopers.

For example, a similar composition (38 to 40% GC) is observed for various streptococci, pneumococci, and lactobacilli, which have traditionally been grouped together as lactic acid bacteria because of their characteristic fermentation. On the other hand, *Lactobacillus bifidus* is far removed (56% GC).

Similarly, though *Proteus* is traditionally grouped with the Enterobacteriaceae, and some species have the same GC content (50 to 52%) as the majority of this "family," others are far removed (38 to 40%). We also see considerable spread within other large groups that have been considered a single genus primarily on morphological grounds, e.g., *Bacillus, Corynebacterium, Pseudomonas.* Finally, the weakness of gross morphology as a major taxonomic criterion (cf. whales and fishes) is well illustrated by the anomalous position of *Sporosarcina ureae,* the only coccus that sporulates: this organism has a base composition (38 to 40% GC)

close to that of the sporulating rods (bacilli and clostridia), and very far from that of other sarcinae (70 to 80% GC).

Similarity of base composition represents only a minimal basis for close genetic relatedness, since even distant organisms can by chance have a similar composition. Moreover, as groups separate in evolution the composition changes only long after the **DNA sequence** has differentiated extensively. Homology of detailed base sequence can be measured quantitatively in terms of the ability of DNA strands from two different sources to **form molecular hybrids with each other in vitro** (Ch. 10, Hybridization of DNA). Surveys have revealed striking parallelism between the relatedness of organisms (bacterial and higher) by various criteria and the extent of such hybridization between their DNAs.

For example, the Enterobacteriaceae *Escherichia, Shigella,* and *Salmonella* have a similar DNA composition, as well as extensive **macrohomology** (i.e., map order of homologous genes); yet when DNA is transferred from one to another there is almost no recombination between *Salmonella* and the other two, though recombination is almost as frequent between *Escherichia* and *Shigella* DNA as within the same genus. The inferred differences between these groups in **microhomology** (i.e., homology of base sequence) has been verified by studies of DNA hybridization.

Since base sequences are ultimately translated into amino acid sequences in proteins, evolutionary relations among microbes will surely also be illuminated by analysis of the **amino acid sequences of homologous proteins** (e.g., cytochromes, various enzymes), which have already proved fruitful with higher organisms (Ch. 12, Fig. 12-35). However, gene flow can occur between even fairly distant bacterial "species," causing replacement of individual proteins. Moreover, proteins vary widely in their rates of evolution. Hence overall DNA homology appears to provide the most reliable basis for a true bacterial taxonomy, which now seems likely to develop quite quickly.

SELECTED REFERENCES: STRUCTURE

Books and Review Articles

GUNSALUS, I. C., and STANIER, R. Y. (eds). *The Bacteria,* vol. I, *Structure.* Academic Press, New York, 1960.

REMSEN, C. C., and WATSON, S. W. Freeze-etching of bacteria. *Internat Rev Cytol 33*:253 (1972).

RYTER, A. Association of the nucleus and the membrane of bacteria: A morphological study. *Bacteriol Rev 32*:39 (1968).

SMITH, R. W., and KOFFLER, H. Bacterial flagella. *Adv Microbial Physiol 6*:219 (1971).

Specific Articles

BRINTON, C. C., JR. Contributions of Pili to the Specificity of the Bacterial Surface. In *The Specificity of Cell Surfaces.* (B. D. Davis and L. Warren, eds.) Prentice-Hall, Englewood Cliffs, N.J., 1966.

DUGUID, J. P., and WILKINSON, J. F. Environmentally Induced Changes in Bacterial Morphology. In *Microbial Reaction to Environment* (11th Symposium, Society for General Microbiology), p. 69. Cambridge Univ. Press, London, 1961.

RYTER, A., and LANDMAN, O. E. Electron microscope study of the relationships between mesosome loss and the stable L state (or protoplast state) in *Bacillus subtilis. J Bacteriol 8*:457 (1964).

SELECTED REFERENCES: CLASSIFICATION

Books and Review Articles

AINSWORTH, G. G., and SNEATH, P. H. A. (eds.). *Microbial Classification* (12th Symposium, Society for General Microbiology). Cambridge Univ. Press, London, 1962.

BREED, R. S., MURRAY, E. D. G., and SMITH, N. R. (eds.). *Bergey's Manual of Determinative Bacteriology,* 7th ed. Wiliams & Wilkins, Baltimore, 1957. 1094 pp.

MANDEL, M. New approaches to bacterial taxonomy: Perspective and prospects. *Annu Rev Microbiol 23*:239 (1969).

SKERMAN, U. B. D. *A Guide to the Identification of the Genera of Bacteria.* Williams & Wilkins, Baltimore, 1959. 217 pp. A useful set of diagnostic keys, intended as a companion to *Bergey's Manual.*

Symposium: La notion d'espèce bactérienne à la lumière des découvertes récentes. *Ann Inst Pasteur 94*:137 (1958).

chapter

ENERGY PRODUCTION

FERMENTATIONS 40

SOURCE OF ENERGY IN FERMENTATION 41
Couplings of Fermentation with Phosphorylation 41
COENZYMES 42
Ferredoxins and Flavodoxins 43
ALTERNATIVE FATES OF PYRUVATE 43
Phosphogluconate Pathways 46
Alternative Pathways Within An Organism 46
ADAPTIVE VALUE OF DIFFERENT FERMENTATIONS 47

RESPIRATION 48

AEROBIC AND ANAEROBIC METABOLISM 48
Respiratory Components 49
Anaerobic Respiration 49
Incomplete Oxidations 49
ELECTRON TRANSPORT 50
Oxidative Phosphorylation 50
Flavoproteins 51
ENERGETICS AND GROWTH 52

AUTOTROPHIC METABOLISM 53

CHEMOAUTOTROPHY AND PHOTOSYNTHESIS 53
Photosynthesis 54
Autotrophic Assimilation of CO_2 55
NITROGEN CYCLE; NITROGEN FIXATION 55

ALTERNATIVE SUBSTRATES 56

GENOTYPIC ADAPTATION (SELECTION OF ORGANISMS) 56
PHENOTYPIC ADAPTATION (SELECTION OF ENZYMES) 56

The immediate principles of living bodies would be, to a degree, indestructible if, of all the organisms created by God, the smallest and apparently most useless were to be suppressed. And because the return to the atmosphere and to the mineral kingdom of everything which had ceased to live would be suddenly suspended, life would become impossible. LOUIS PASTEUR

In filling all possible ecological niches the microbial world has evolved members that thrive on the most remarkably different diets: conversely, their attack on all manner of compounds plays an essential geochemical role in the cycle of organic matter. The energy from these varied fuels becomes useful when converted into adenosine triphosphate (ATP): this mediator of most metabolic energy transfer links **catabolism** (energy-yielding, degradative processes) and **anabolism** (biosynthetic reactions).

There are three main modes of ATP production: fermentation, respiration, and photosynthesis.* As we noted in Chapter 1, fermentation evolved first. Respiration yields several times as much ATP per mole of substrate, but it could evolve only after the appearance of oxygen in the terrestrial atmosphere. The less efficient fermentative way of life has been widely retained in microbes because in general they grow only in aqueous solutions, of which only a thin surface layer can receive a rapid supply of oxygen from the atmosphere. Hence vigorous fermentations, occurring in the depths of sugar-rich

fluids such as grape juice, were recognized long before the development of microbiology as a discipline.

Complete respiration yields only CO_2 and H_2O but fermentation yield various organic compounds in large amounts. Their identification, starting even before Pasteur, was the first stage in the study of microbial metabolism. Only much later (after 1930) were methods developed for dissecting the pathways to those compounds and eventually isolating their enzymes.

In this chapter we shall focus on the variety of energy-yielding patterns that have evolved, and on the key reactions that provide metabolically useful energy; for further details the reader is referred to textbooks of biochemistry. We shall find that organisms of medical interest, which grow in the same environments as mammalian cells, also have the same central pathways (glycolysis and terminal respiration). To broaden acquaintanceship with the microbial world we shall also consider briefly certain other energy-yielding pathways, essential for maintaining the biosphere, that are seen only in nonpathogens (e.g., photosynthesis, chemosynthesis).

FERMENTATIONS

Pasteur defined fermentation as **life without air;** and he recognized that metabolic energy is derived, in this remarkable process, by the organism's "property of performing its respiratory functions, somehow or other, with the oxygen existing combined in sugar." This chapter will be largely concerned with the

"somehow or other." But first we should note that a more sophisticated definition of fermentation is now required because of the later discovery of **anaerobic respiration,** a fundamentally respiratory mechanism in which O_2 is replaced by another inorganic electron acceptor such as nitrate or sulfate. Fermentation is therefore now better defined as **metabolism in which organic compounds serve as both the electron donors and the electron acceptors.**

A fermentation must balance: since there is no net oxidation, the number of moles of

---

* The term fermentation here refers to an energy-yielding rearrangement of the atoms of organic substrates, without net oxidation, while respiration refers to the net oxidation of substrates at the expense of molecular oxygen. Modified uses of the terms will be noted below.

C, H, and O must be the same in the products as in the substrates (including water, and corrected for any assimilation into cell constituents). For example:

$$C_6H_{12}O_6 \longrightarrow 2\,C_3H_6O_3 \quad (CH_3CHOHCOOH)$$
Glucose　　　　　　　　　Lactic acid

or

$$C_6H_{12}O_6 \longrightarrow 2\,CH_3CH_2OH + 2\,CO_2$$
Ethanol

The anaerobic conditions required for fermentation are easily achieved with liquid media. In the laboratory it is convenient to use completely filled bottles provided with loose glass stoppers, to allow venting of any gas that may be produced. In industry deep vats, incubated without agitation, provide sufficient anaerobiosis; oxygen at the surface does not penetrate far, and in the gas phase above the surface it may be displaced by the heavier CO_2 evolved. (In industrial jargon the term fermentation is loosely extended to any large-scale process catalyzed by microbes, even when the process involves strong aeration, as in antibiotic production.)

SOURCE OF ENERGY IN FERMENTATION

The energetics of fermentation may be considered from several points of view. In **thermodynamic** terms fermentations proceed because the product(s) have a lower energy content than the substrates. Thus the molar free energy difference (ΔF) between a mole of glucose and 2 moles of lactate* (at neutral pH) is 58 kcal. This difference arises essentially from the much lower energy level of the carboxyl (-COOH) group, its decarboxylation product CO_2, and H_2O, compared with the same atoms in carbonyl or hydroxyl groups. Fermentation may thus be viewed, in terms of **bond energy**, as a regrouping of H and O atoms in organic molecules to yield a carboxyl group (as in the lactic fermentation above) or CO_2 (as in the ethanol fermentation). (Similar considerations explain the much higher energy yield (688 kcal) obtained on oxidizing all the C and H of glucose into CO_2 and H_2O.)

---

* For organic acids terms such as lactic acid and lactate will generally be used interchangeably in this volume; and for the sake of simplicity the formula will be generally written as the free acid, though at the usual PHs organic acids are largely ionized.

COUPLINGS OF FERMENTATION WITH PHOSPHORYLATION

In terms of **intermediate metabolism** the problem is to evolve sequences in which a spontaneous reaction is linked with the generation of ATP. For this purpose an organic phosphate must be formed whose energy of hydrolysis ($-\Delta F$) exceeds the -8 kcal released on conversion of ATP to ADP $+$ P_i; an enzyme can then exergonically transfer the phosphate group from the donor to ADP (i.e., reverse of a phosphokinase reaction).

Three classes of reactions yield such "substrate-level" phosphorylation, which provides all the useful energy from fermentation (and a small part of the energy of respiratory metabolism; see below). The simplest reaction is the conversion of a low-energy phosphate ester to an **enol phosphate,** whose hydrolysis to a free carbonyl group releases enough energy to permit transfer of the phosphate to ADP. Thus in the glycolytic pathway:

$$CH_2\text{-}CH\text{-}COOH \xrightarrow{-H_2O} CH_2 = C\text{-}COOH \longrightarrow$$
$$\underset{OH\ \ O℗}{|\ \ \ |} \qquad\qquad \underset{O \sim ℗\ \ ADP}{|}$$
2-P-glycerate † 　　　　　P-enolpyruvate

$$ATP + CH_3\text{-}C\text{-}COOH$$
$$\overset{||}{}$$
$$O$$
Pyruvate

In a second class, such as the oxidative step of the glycolytic pathway, an **aldehyde** is oxidized to a carboxyl, with the intermediate formation of a high-energy carboxyphosphate anhydride; its phosphate is then transferred to form ATP, whose P-P bond has a slightly lower level.

$$℗\ OCH_2\text{-}CHOH\text{-}CHO + P_i + DPN^+$$
3 - P - glyceraldehyde

$$\downarrow$$

$$℗\ OCH_2\text{-}CHOH\text{-}COO\sim℗ + DPNH + H^+$$
1, 3 - DiP - glyceric acid

$$\downarrow + ADP$$

$$℗\ OCH_2\text{-}CHOH\text{-}COOH + ATP$$
3 - P - glyceric acid

---

† Abbreviations: The phosphate group ($-PO_3H_2$) is symbolized by the letter P in the names of compounds, and by a circled P in structural formulas; P_i, inorganic phosphate; $\sim P$, high-energy phosphate; $DPN+$ and DPNH, the oxidized and the reduced forms of diphosphopyridine nucleotide (nicotinamide adenine dinucleotide, NAD).

The key to the formation of a high-energy group is the reaction of the aldehyde with an —SH of the enzyme (E—SH) to form a low-energy thiohemiacetal, which is then oxidized to a **high-energy thioester**. A phosphate replaces the thio group, and the resulting carboxyphosphate anhydride is released from the enzyme for further reaction.

The third class of reactions for creating ~P involves conversion of a **ketone** to a carboxyl, which requires cleavage of the ketone from an adjacent carbon. α-Keto acids are activated and decarboxylated for this purpose by the coenzyme thiamine pyrophosphate (TPP), in which C-2 of the thiazole forms a highly reactive carbanion. For example, its reaction with pyruvate as substrate yields enzyme-bound hydroxyethyl TPP.

This product of pyruvate cleavage can then undergo several different reactions, depending on the enzyme to which the TPP is bound and the available further reactants. Thus in the ethanol fermentation free acetaldehyde is released for further reduction, while in the acetoin fermentation the active acetaldehyde is transferred to another mole of pyruvate, yielding acetolactate (see below).

In the fermentations that yield formate, in a reaction formerly called "phosphoroclastic," hydroxyethyl TPP is oxidized to an enzyme-bound acetyl group, at the expense of reduction of the remainder of the pyruvate to formic acid (HCOOH). The acetyl group is then converted to acetyl phosphate, via acetyl CoA (which has a high-energy thioester bond).

A related reaction of **oxidative metabolism**, catalyzed by the pyruvate oxidase enzyme complex, releases CO_2 while the S-S compound lipoate oxidizes the 2-C fragment of hydroxyethyl TPP to form a high-energy thioester, acetyl dihydrolipoate. The acetyl group is transferred to form acetyl CoA,

while the reduced lipoate is reoxidized by DPN+ (generated by respiration). A precisely analogous reaction in the tricarboxylate cycle (Ch. 4) oxidizes α-ketoglutarate to succinyl CoA and CO_2.

COENZYMES

Bacterial fermentations and respiration employ a variety of cofactors (coenzymes) identical with those found in mammalian tissues. These substances, of relatively low molecular weight and high thermostability, were initially distinguished from enzymes because they could often be removed from extracts by dialysis, with consequent loss of enzymatic activity; and activity could be restored to the "resolved" enzymes by supplying the coenzyme-containing supernatant from an extract in which the enzymes had been denatured by boiling. Such familiar cofactors include coenzyme A for acyl transfer; thiamine pyrophosphate for transferring groups derived from a ketone (e.g., decarboxylation of α-keto acids, or the transketolase reaction of ketoses described in Ch. 4); biotin for CO_2 transfer; lipoic acid, DPN (NAD), TPN (NADP), riboflavin derivatives, ubiquinone, naphthoquinone, and cytochromes for hydrogen or electron transfer; and, of course, ATP for phosphate and energy transfer. In addition, various biosynthetic reactions (Ch. 4) employ pyridoxal phosphate for amino acid transamination, decarboxylation, and racemization; tetrahydrofolate for 1-carbon group transfer and reduction; and cobamide (from vitamin B_{12}, cobalamin) in methyl group transfers and certain reductions.

FERREDOXINS AND FLAVODOXINS

These cofactors could not account for those reactions that occur at a very low redox (oxidation-reduction) potential: for example, the release of H_2 gas by certain anaerobes (clostridia) or the utilization of H_2 as a fuel for respiration by other bacteria (Autotrophic metabolism, below). These processes evidently require biological catalysis of the reaction characteristic of the hydrogen electrode:

$$2H^+ + 2\textcircled{e} \rightleftharpoons H_2$$

The mechanism was revealed when Mortenson, in 1962, discovered a new iron-containing cofactor, **ferredoxin,** with a remarkably low standard redox potential (-417 mV), similar to that of the H_2 electrode. This compound made possible the utilization and the formation of molecular H_2 by appropriate extracts. The role of ferredoxin turned out to be much wider, for similar coenzymes have now been found in a variety of strictly anaerobic bacteria, in which they provide the strong reducing power (high-energy electrons) required for photosynthesis and for nitrogen fixation (see below). Ferredoxins are also the immediately reduced products in the photosynthesis of higher plants. Finally, anaerobic bacteria possessing ferredoxin can directly reduce acetate + CO_2 to pyruvate (see Ch. 4, Glyoxylate cycle).

The ferredoxins of various bacteria are small proteins of MW 6000 to 10,000, containing 8 atoms of $Fe^{++(+)}$ complexed with an equivalent amount of acid-labile S that readily yields H_2S. This strongly reducing sulfide group evidently contributes to the low redox potential, which is far below the range seen with $Fe^{++(+)}$ in the heme of cytochromes, or with other nonheme iron proteins.

Flavodoxins have a similar low potential and can replace ferredoxins in certain reactions. They lack Fe; the redox cofactor is flavin mononucleotide (FMN), shifted to an unusually low potential by the properties of its protein apoenzyme. Flavodoxins are synthesized by many bacteria (including *E. coli*), especially in low-Fe media.

ALTERNATIVE FATES OF PYRUVATE

A variety of fermentations, yielding quite different products, are based on the glycolytic pathway, which is presented, up to pyruvate, in Figure 3-1. ATP generation in this pathway depends on the oxidation of triose phosphate to an acid, at the expense of the reduction of DPN+ to DPNH + H+. Since the total DPN in the cell is very limited, fermentation would cease very rapidly if the DPNH were not reoxidized in the further metabolism of pyruvate. Microbes have evolved a variety of pathways for this purpose, illustrated in Figure 3-2; some of these pathways also yield additional ATP.

Lactic Fermentation. This is the simplest fermentation: a one-step reaction, catalyzed by DPN-linked lactic dehydrogenase (really a pyruvate reductase), reduces pyruvate to lactate; no gas is formed. Since two ATP are consumed in the formation of hexose diphosphate from glucose, and since four ATP are subsequently produced, the net yield is two ATP per hexose. This fermentation is identical with the glycolysis in mammalian cells.

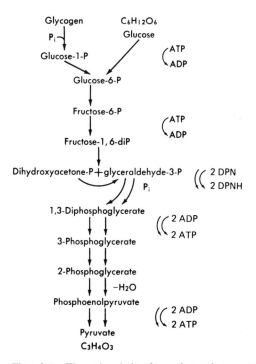

Fig. 3-1. The glycolytic formation of pyruvate (Embden-Meyerhof pathway). Sum: Glucose + 2 ADP + 2 P_i + 2 DPN+ $\longrightarrow$ 2 Pyruvate + 2 ATP + 2 DPNH + 2H+. Double arrows signify two moles reacting per mole of glucose.

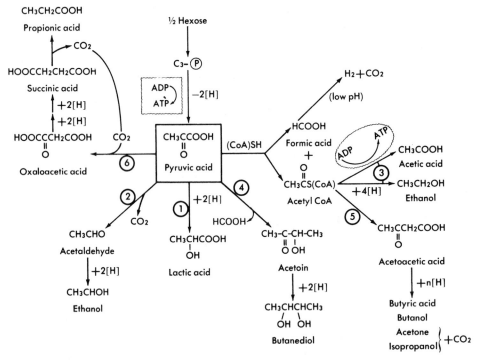

Fig. 3-2. Key role of pyruvate in principal fermentations. **1.** Lactic (*Streptococcus, Lactobacillus*). **2.** Alcoholic (many yeasts, few bacteria). **3.** Mixed acid (most Enterobacteriaceae). **4.** Butanediol (*Aerobacter*). **5.** Butyric (*Clostridium*). **6.** Propionic (*Propionibacterium*).

The **homolactic** fermentation, which forms only lactate, is characteristic of many of the lactic bacteria (e.g., rods such as *Lactobacillus casei;* cocci such as *Streptococcus cremoris*). Other members of this group carry out a rather different **heterolactic** fermentation (Phosphogluconate pathways, below), which converts only half of each glucose molecule to lactate. Both these lactic fermentations are responsible for the souring of milk and certain other foods; and such acidification, achieved at the expense of some carbohydrate energy, preserves food from further decomposition as long as it is kept anaerobic. The accompanying changes in flavor also add to the economic importance of these fermentations. Examples include cheeses, sauerkraut, and ensilage. The homolactic fermentation occurs in certain pathogens (the streptococci and pneumococci) which are included by taxonomists among the lactic bacteria.

Alcoholic Fermentation. Pyruvate (activated by TPP) is converted to CO_2 (retained in some beverages) plus acetaldehyde, which is then reduced to ethanol in a DPN-linked reaction. This fermentation is characteristic of yeasts; as a major pathway it is uncommon in

bacteria. Its value in the leavening of bread, as well as in the formation of beverages, has been recognized for millennia.

Propionic Fermentation. This pathway extracts additional energy from the substrate. Pyruvate is carboxylated to yield oxaloacetate, which is reduced to yield succinate and then is decarboxylated (after activation by CoA) to yield propionate.* The overall process takes up 4[H] per pyruvate. Since only 2 are required to balance the oxidation of triose to pyruvate, 2 are left over to balance a further oxidation. Hence for each 2 pyruvates reduced to propionate a third pyruvate can be formed from triose (−2[H]) and then oxidized to acetate and CO_2 (−2[H] more), with the generation (via acetyl CoA, as noted above) of an ATP.

Organisms possessing this pathway can eke out a living by fermenting lactate, the endproduct of an-

_____

* The decarboxylation is accomplished via a cobamide-linked reaction in which succinyl CoA is rearranged to form the less stable methylmalonyl CoA [CH₃CH(COOH)CO—CoA]. This compound is then split to yield propionyl CoA and CO_2.

other fermentation. The lactate is first oxidized to pyruvate; part is then reduced to propionate and the rest oxidized to acetate and CO_2.

$$3 \text{ CH}_3\text{CHOHCOOH} \xrightarrow{-6[H]} 3 \text{ CH}_3\text{COCOOH} \xrightarrow{+6[H]}$$
$$\text{Lactate} \qquad\qquad \text{Pyruvate}$$

$$2 \text{ CH}_3\text{CH}_2\text{COOH} + \text{CH}_3\text{COOH} + \text{CO}_2 + \text{H}_2\text{O}$$
$$\text{Propionate} \qquad\quad \text{Acetate}$$

This arduous process of extracting energy from lactate, by the over-all conversion of 3 hydroxyls to 1 carboxyl, yields only 1 ATP per 9 carbons fermented. Hence propionic acid bacteria grow slowly. In Swiss cheese their late formation of CO_2, after the completion of the lactic fermentation of milk by other organisms, is responsible for the formation of the holes, while the propionic acid contributes to the flavor.

The mixed acid fermentation of glucose noted immediately below also involves reduction of pyruvate to succinate, but not the subsequent decarboxylation to propionate. To ferment lactate, however, decarboxylation is essential in order to preserve carbon balance, since no free CO_2 is provided.

The mixed acid (formic) fermentation is characteristic of most Enterobacteriaceae. These organisms dispose of part of their substrate through a lactic fermentation, but most of it goes through a fermentation characterized by the splitting of pyruvate, without net oxidation or reduction, to an acetyl group and formate. The accompanying generation of $\sim$P (via acetyl CoA and acetyl-P) was described above.

Since this reaction does not absorb the 2[H] released in forming pyruvate, fermentation balance requires that an equal amount of pyruvate be reduced by pathways absorbing more [H]: a) formation of ethanol and formate, and b) reduction (after carboxylation) to succinate (as in the propionic fermentation).

$$(a) \quad \text{Glucose} \xrightarrow{-[4H]} 2 \text{ CH}_3\text{COCOOH} \xrightarrow{\text{(CoA)SH}} 2 \text{ HCOOH} +$$

$$2 \text{ CH}_3\text{CO-S(CoA)} \xrightarrow{+[4H]} \text{CH}_3\text{CH}_2\text{OH} + \text{(CoA)-SH}$$
$$\downarrow P_i$$
$$\text{CH}_3\text{COO} \textcircled{P} + \text{(CoA)-SH}$$

$$(b) \quad \text{CH}_3\text{COCOOH} \xrightarrow{\text{CO}_2} \text{HOOCCH}_2\text{CH}_2\text{COOH} + \text{H}_2\text{O}$$
$$\xrightarrow{+[4H]}$$
$$\text{Pyruvate} \qquad\qquad\qquad \text{Succinate}$$

Either process consumes 4[H] per pyruvate reduced. The formic fermentation thus yields 3 ATP per glucose fermented (compared with 2 in the lactic fermentation).

Gas Formation. The formate from this fermentation may remain as such if the pH is kept alkaline. However, most fermentations become acidic; and at a pH of 6 or less "gas-formers" (e.g., *E. coli*) form an enzyme, formic hydrogenlyase, that converts formic acid (HCOOH) to CO_2 and H_2. Some Enterobacteriaceae (e.g., *Shigella*) do not form this enzyme; in diagnostic tests (Ch. 29) they are found, like homolactic fermenters, to produce acid but no gas.

CO_2, which is formed in many fermentations, is readily observed in simple laboratory tests for gas formation. However, it is highly soluble in water, and so in the body is readily absorbed into the circulation. H_2, in contrast, has low solubility in water and forms free gas under some circumstances, such as the "gas gangrene" caused by certain clostridia (Ch. 34).

Accumulation of **intestinal gas** is promoted by several factors. Impaired peristalsis or pancreatic or biliary secretion prevents intestinal absorption from keeping pace with gas formation; and fermentation in the lower gut, where absorption is poor, is promoted by the ingestion of foods that are slowly hydrolyzed or foods that yield poorly absorbed but readily fermented products. Ruminants form a poorly soluble gas, methane (CH_4), in the gut by a fermentation (the methane fermentation) that oxidizes fatty acids at the expense of the reduction of CO_2 to CH_4, but this fermentation is not prominent in man.

Butylene Glycol (Acetoin) Fermentation. This pattern, observed in *Aerobacter,* certain

$$\text{CH}_3\text{-CO-COOH} + \text{TPP} \xrightarrow{[2H]} [\text{CH}_3\text{-CHO}] \cdot \text{TPP} + \text{HCOOH} \nearrow \text{H}_2 + \text{CO}_2$$
$$\text{"Active acetaldehyde"}$$

$$[\text{CH}_3\text{-CHO}] \cdot \text{TPP} + \text{CH}_3\text{-CO-COOH} \xrightarrow{\text{TPP}} \text{CH}_3\text{-CO}-\overset{\overset{\textstyle \text{COOH}}{|}}{\underset{\underset{\textstyle \text{OH}}{|}}{\text{C}}}-\text{CH}_3$$
$$\text{Acetolactate}$$

$$\xrightarrow{-\text{CO}_2} \text{CH}_3\text{-CO-CHOH-CH}_3 \xrightarrow{[2H]} \text{CH}_3\text{-CHOH-CHOH-CH}_3$$
$$\text{Acetoin} \qquad\qquad\qquad 2,3\text{-Butylene glycol}$$
$$\text{(acetylmethyl carbinol)}$$

other Enterobacteriaceae, and some species of *Bacillus,* also releases formate. However, the remaining "active acetaldehyde" is not oxidized but condenses with a second pyruvate. The product is converted to butylene glycol (butanediol) in the following reactions, in which the total of 4 H absorbed balances that created in forming the 2 pyruvates.

This fermentation, like the alcoholic fermentation, yields only neutral products and produces 2 ATP per glucose. It is often called the **acetoin fermentation** because exposure to air oxidizes some of the butylene glycol to acetoin, which is readily recognized by a specific color test (Voges-Proskauer). In sanitary engineering this test is of considerable diagnostic value (Ch. 29) in discriminating between *E. coli,* which reaches bodies of water primarily from the mammalian gut, and *Aerobacter,* originating primarily from vegetation.

The formation of neutral rather than acidic products permits the fermentation of larger amounts of carbohydrate, without self-inhibition, and therefore the production of more gas; hence the name *Aerobacter aerogenes.*

Butyric-Butylic Fermentation. This pattern of pyruvate reduction is seen in certain strict anaerobes that can activate molecular H_2 by the use of ferredoxin. The initial scission yields H_2, CO_2, and 2-carbon fragments at the acetate level of oxidation. Two such fragments are then condensed, not head-to-head as in acetoin but head-to-tail as in fatty acid synthesis. The resulting acetoacetyl CoA is decarboxylated and/or reduced, yielding acetone, isopropanol, butyric acid, and *n*-butanol, in varying proportions.*

Mixed Amino Acid Fermentations. These fermentations occur where there is considerable proteolysis; they are prominent in putrefactive processes, including the gangrene associated with anaerobic wound infections. Certain amino acids (or their deamination products) serve as electron donors and others as acceptors. For example:

---

* The use of this fermentation for the industrial production of acetone was developed by Chaim Weizmann in England in 1915. This scientific contribution, which solved an urgent problem in explosives manufacture in World War I, is said to have promoted the Balfour Declaration, and thus contributed eventually to the founding of the state of Israel, with Weizmann as its first president.

$$CH_3\text{-}CHNH_2\text{-}COOH + 2\,CH_2NH_2\text{-}COOH + 2\,H_2O \longrightarrow$$
Alanine   Glycine

$$3\,CH_3\text{-}COOH + CO_2 + 3\,NH_3$$
Acetate

In addition, decarboxylation of various amino acids, and further reactions, yield products that are pharmacologically active (e.g., histamine) or malodorous (e.g., indole from the breakdown of tryptophan, and even more mephitic —SH compounds derived from cysteine and methionine). More pleasant, empirically selected fermentations of minor constituents are responsible for the characteristic flavors of various wines and cheeses; on the other hand the toxic effect of poorly fermented wines is largely due to longer-chain aldehydes (fusel oil) derived from amino acids.

PHOSPHOGLUCONATE PATHWAYS

Not all metabolism of glucose proceeds via the Embden-Meyerhof pathway. For example, the **heterolactic fermenters** (e.g., *Leuconostoc mesenteroides*) yield approximately equimolar quantities of lactate, ethanol, and CO_2, via P-gluconate decarboxylation (pathway B, Fig. 3-3). In pseudomonads the major route of hexose metabolism is the **ketodeoxygluconate (Entner-Doudoroff)** pathway (C, Fig. 3-3). It is also used in *E. coli* to metabolize gluconate but not hexose.

These pathways were first suspected when the distribution of specific labeled C atoms from glucose was found to be very different from that observed in organisms employing the Embden-Meyerhof pathway. The enzymes were then isolated.

ALTERNATIVE PATHWAYS WITHIN AN ORGANISM

Glucose is metabolized in *E. coli* mainly via the Embden-Meyerhof pathway, but it can also pass through the so-called **hexose monophosphate shunt,** which converts it, via P-gluconate, to pentose-P (Fig. 3-3). In respiration the pentose-P is oxidized via acetyl-P and triose-P, and in fermentation it is converted to fructose-P by a complex series of 2-carbon and 3-carbon transfers (the reverse of those shown in Fig. 4-20).

The shunt is not essential: its block by mutation does not slow growth on glucose. On the other hand, a mutant blocked in the Embden-Meyerhof pathway can grow at ⅓ the normal rate by using the shunt to capacity; and isotopic studies show that in wild-type *E. coli* it functions simultaneously

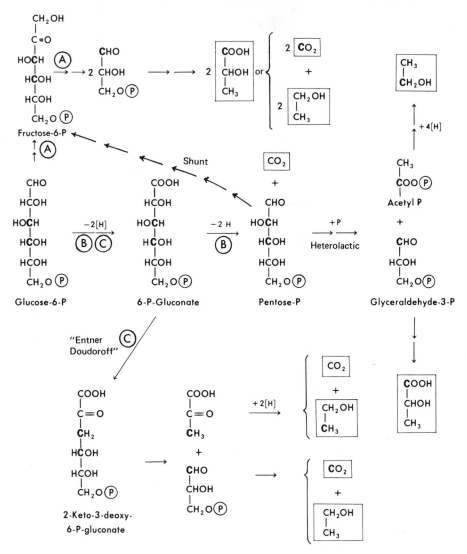

Fig. 3-3. Alternative ways of metabolizing glucose to lactate and/or ethanol. Glucose-3,4-14C yields alcohol labeled **(A)** in neither C atom (Embden-Meyerhof pathway), **(B)** in the CH_2OH (heterolactic pathway), or **(C)** in the CH_3 (ketodeoxygluconate pathway).

with the glycolytic pathway. The explanation is that while both pathways yield triose-P they also have different functions: glycolysis generates DPNH (which functions primarily in energy metabolism), while the shunt generates pentose-P and also TPNH (which is required for reductive steps in biosynthesis; Ch. 4). The alternative pathways allow reducing power to be distributed as needed between DPNH and TPNH; and a block in the shunt pathway has no perceptible effect because the cell has alternative routes to pentose-P and to TPNH (Ch. 4).

E. coli can also make the enzymes of the ketodeoxygluconate pathway, but does so only when the cells are grown on gluconate (which cannot enter the Embden-Meyerhof pathway). A mutant blocked in converting P-gluconate to ketodeoxygluconate

(Fig. 3-3) still grows on gluconate, again at ⅓ the normal rate, by using the shunt pathway.

It is clear that cells do not carry alternative pathways simply as options or reserves: apparent alternatives have different functions, though they may replace each other when necessary.

ADAPTIVE VALUE OF DIFFERENT FERMENTATIONS

The central role of the glycolytic pathway in many different fermentations illustrates the unity of biology at a molecular level, which presumably reflects evolutionary selection for the most effective mechanism. The diversity

47

of final products, on the other hand, reflects the variety of ecological opportunities available, which select for different variants of the basic mechanism. Thus as we have already noted in Chapter 1, Pasteur recognized that grape juice is much richer than milk in sugar but poorer in protein (i.e., buffering power). Hence in the fermentation of grape juice lactic acid bacteria soon produce a self-inhibitory pH, while the yeasts, though slower-growing, produce a neutral product and outgrow the bacteria. (Grape juice also fails to meet the requirement of these bacteria for various amino acids.) Milk, in contrast, supports the growth of either organism, and the faster-growing bacteria soon predominate.

Enteric organisms show a similar adaptation to their environment. The gut is largely anaerobic; and since its absorptive properties leave only a dilute solution of amino acids and sugar in its lower regions, it should select for an organism that has no specific growth requirements and that extracts as much energy as possible from its fermentation. These are the properties of *E. coli*. In contrast *Aerobacter,* which grows primarily on vegetation, can ferment larger concentrations of carbohydrate than *E. coli* before reaching a limiting pH, but with a less efficient energy yield.

It is evident that nature selects for a variety of fermentative talents, which might be compared to sprinters and long-distance runners.

Substrate-specific Pathways. Glucose is considered a typical substrate for fermentation, for it is much the most widely distributed sugar (being the monomer of starch and cellulose), and most (though not all) bacteria can utilize it. The first steps in the metabolism of most other substrates convert them to some intermediate in a central pathway (glycolysis or TCA cycle). The specific connecting pathway is often quite short. For example, a single step will convert fructose to fructose-6-P; a one-step hydrolysis will yield glucose from sucrose or from lactose; three steps are necessary to convert galactose to glucose-1-P (see Fig. 4-24); and two will convert glycerol to glyceraldehyde-P.

Under aerobic conditions facultative organisms will eventually oxidize, and not simply ferment, a substrate that they can use; but this response is still referred to as "fermentation-positive," since it is most conveniently recognized through the initial accumulation of acid. Highly reduced compounds (e.g., fatty acids), which cannot be fermented, generally are converted to acetate or members of the TCA cycle.

Though a facultative organism can make both fermentative and respiratory enzymes the existence of obligate aerobes and obligate anaerobes suggests that there is selective advantage in dispensing with the genes for one set when the customary habitat employs only the other. For example, yeasts, which tend to grow both on the surfaces of living fruits and within the juices of crushed or decaying fruit, are generally facultative, while the closely related molds, which form fluffy aerial growths (mycelia) on surfaces (Ch. 43), tend to be strictly aerobes.

RESPIRATION

AEROBIC AND ANAEROBIC METABOLISM

Bacteria fall into several groups with respect to the effect of oxygen on their growth and metabolism.

1) **Obligate aerobes** (e.g., the tubercle bacillus and some spore-forming bacilli) require oxygen and lack the capacity for substantial fermentation.

2) **Obligate anaerobes** (e.g., clostridia, propionibacter) can grow only in the absence of oxygen. A subgroup, called microaero-

philic, can tolerate or even prefer O_2 at low tension, but not at that of air.

3) **Facultative organisms** (e.g., many yeasts, enterobacteria) can grow with or without air, and they shift in its presence to a respiratory metabolism.

4) **Aerotolerant anaerobes** (e.g., most lactic acid bacteria) resemble facultative organisms in growing either with or without oxygen, but their metabolism remains fermentative.

While it is not certain why the addition of oxygen prevents obligate anaerobes from continuing to ferment and grow, the probable mechanism is the maintenance of certain enzymes in an oxidized state that prevents them from carrying out an essential reductive reaction. For example, succinate, which is required for biosynthetic purposes in *E. coli,* is formed by an oxidative pathway under aerobic conditions and by reduction under anaerobic conditions (Ch. 4, Fig. 4-5). O_2 prevents the reductive flavoproteins from functioning, for a mutant that has lost the oxidative pathway grows normally without air but stops growing as soon as O_2 is present, unless succinate is provided.

Some obligate anaerobes (especially clostridia) not only are inhibited but are killed by the presence of oxygen. It was long thought that the reason might be poisoning by accumulated hydrogen peroxide, formed by their flavoproteins in the presence of air; for catalase, which destroys hydrogen peroxide, is present only in aerobes. However, aerotolerant organisms also lack catalase and yet are not poisoned. Recent findings suggest that the poison may be a highly reactive free-radical form of O_2, **superoxide** (O_2^-), formed by flavoenzymes: superoxide dismutase destroys this product (in the reaction $2O_2^- + 2H^+ \rightarrow H_2O_2 + O_2$), and this enzyme is present in aerobes and aerotolerant organisms but not in strict anaerobes.

RESPIRATORY COMPONENTS

Obligate aerobes, and facultative organisms adapted to a respiratory metabolism, contain a complete electron transport chain in which electrons flow from DPNH (or directly from a few substrates, such as succinate or lactate) to a flavoprotein, and thence via several cytochromes to oxygen; ATP is generated in the accompanying process of oxidative phosphorylation. Most cytochromes are absent, however, from strictly fermentative organisms.

Some of the latter organisms, nevertheless can carry out a limited respiration. *Lactobacillus delbrueckii,* for example, can oxidize glucose via glucose-6-phosphate to 6-phosphogluconate through a flavoprotein, which in turn is directly reoxidized by air. This short respiratory chain yields little or no ATP, but it may serve to scavenge dissolved oxygen and thus to make anerobic growth possible.

Oxygenases carry out a small class of oxidative reactions in which molecular oxygen is transferred directly to the substrate or used to remove H. These reactions do not yield ATP, but they convert a refractory compound to one that is useful, either for structural purposes (e.g., saturated → unsaturated fatty acids) or for further metabolism. For example, pseudomonads utilize various benzenoid compounds via the reaction sequence:

Catechol · cis, cis-Muconic acid · α-Ketoglutarate

ANAEROBIC RESPIRATION

This paradoxical term refers to a process that is a respiration in the sophisticated sense of carrying out electron transport, though to an ultimate electron acceptor other than O_2. Thus an electron transport system in many organisms, including *E. coli,* can reduce nitrate to nitrite under anaerobic conditions, obtaining half as much free energy as in the reduction of O_2; the metabolic pattern is otherwise the same as in aerobiosis.

More specialized organisms can reduce nitrate to N_2 (denitrifiers), or can utilize sulfate as electron acceptor (yielding the H_2S characteristic of some polluted streams), or can use H_2 to reduce CO_2 to CH_4 (marsh gas) (see Table 3-2, below).

INCOMPLETE OXIDATIONS

Just as men discovered how to make wine and cheese long before the role of microbes was known, so they learned that wine trickled repeatedly over wood shavings (to provide a large surface) became vinegar. (The aerobic oxidation of ethanol to acetic acid has been repeatedly rediscovered by impecunious students who kept an opened bottle of wine too long!) The process is due to a group of organisms called *Acetobacter.* Some of these adapt to oxidizing the acetate after all the ethanol is consumed, while others (e.g., *A. suboxydans*) lack part of the tricarboxylate cycle and hence are genetically incapable of oxidizing acetate. Indeed, many organisms carry out the early stages of respiration faster than the late ones. *E. coli,* for example, growing aerobically on glucose, will at first oxidize this substance only as far as acetic acid and CO_2, yielding about 40% of the possible ATP (Table 3-1, below); only after acetic acid is heavily accumulated will it begin to be metabolized via the tricarboxylate cycle.

Many products of aerobic microbial metabolism are commercially valuable. For example, the citric acid used widely in carbonated beverages is derived more cheaply from *Aspergillus* cultures than from citrus fruits or from chemical synthesis. The reduction product of glucose, sorbitol, is oxidized by an *Acetobacter* to the previously rare sugar sorbose, which has made possible the inexpensive synthesis of ascorbic acid; and microbial oxidation of C-11 of the steroid nucleus has similarly facilitated the synthesis of steroid hormones and their analogs. Antibiotics and certain vitamins, obtained from well-aerated cultures, are the most valuable products of microbial technology today. Finally, bacterial mutants are increasingly used for the commercial production of a few amino acids, including glutamate and lysine. When we learn how to make amino acids, or palatable protein, cheaply from petroleum or wood, the contribution of microbial technology to the world food supply may rival that of agriculture.

ELECTRON TRANSPORT

In bacteria, as in higher organisms, the combustion of glucose depends on the oxidation of pyruvate to acetyl CoA $+$ CO_2 by pyruvate oxidase; the acetate is then oxidized to CO_2 and H_2O (terminal respiration) in one turn of the tricarboxylate cycle (Ch. 4, Fig. 4-6). Each oxidative reaction in these several processes, and in the preceding formation of pyruvate, is accomplished at the expense of the reduction of DPN to DPNH (except that

succinate oxidation is coupled directly to reduction of a flavoprotein). The DPNH is regenerated to DPN at the expense of the reduction of O_2 to H_2O. The large amount of energy thus released (effectively, that of burning H_2 gas) is used to form 3 molecules of ATP per 2 H oxidized. This oxidative phosphorylation occurs at three stages in electron transport that are each associated with a sufficient energy drop (Fig. 3-4). About ⅓ of the energy available in glucose is obtained by oxidation to the level of acetate and ⅔ by the oxidation of acetate (see Table 3-1, below).

The Cytochromes. Many bacterial cytochromes closely resemble those of mammalian mitochondria. Cytochromes c are defined as those with a protoheme group covalently linked to the proteins; they are therefore more stable and easier to isolate. Cytochromes b have the same heme more loosely bound, while in cytochromes a a different heme (a) is present. These three groups differ markedly in redox potential and also in the position of the peaks in the absorption spectrum of the reduced form.

OXIDATIVE PHOSPHORYLATION

Experiments with mitochondria from mammalian cells have demonstrated the incorporation of 3

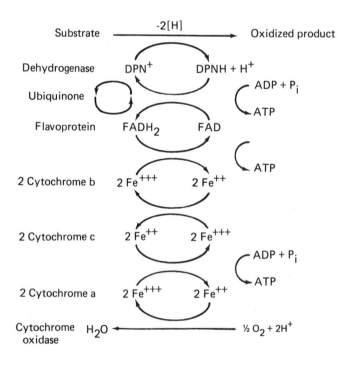

Fig. 3-4. Electron transport and associated oxidative phosphorylation (P/O = 3). The bacterial system also contains nonheme iron proteins, whose position in the scheme is uncertain. Some bacteria appear to use naphthoquinones rather than a benzoquinone (ubiquinone); they presumably go through a reversible cycle of reduction to hydroquinones.

moles of P_i into ATP per $\frac{1}{2}$ mole of O_2 consumed (P/O ratio = 3; Fig. 3-4). Since intact bacteria or protoplasts (unlike mitochondria) are impermeable to ATP cell disruption is required for studies on oxidative phosphorylation, and it evidently damages the respiratory apparatus since the observed P/O ratio, as with damaged mitochondria, rarely exceeds 1.0. For bacteria the value of 3.0 is inferred from the resemblance of individual components of the electron transport system to those of mammalian cells and from the energetics of growth of intact cells (i.e., calculation of the ATP required for the amount of biosynthesis observed).

Cytochrome a is identified with cytochrome oxidase (i.e., the last member of the chain, which interacts directly with O_2), since its spectrum is immediately altered on addition of O_2 and its catalytic activity is inhibited by CO, competitor of O_2. The electron transport sequence flavin—b—c—a—O_2 was established by measurements of the redox potential of the separated units. Moreover, on addition of inhibitors that selectively interrupt the chain at various points (e.g., antimycin, which blocks between b and c), spectral studies show that the components of the chain shift from their steady-state level of oxidation; those on one side of the block become fully reduced and those on the other side become fully oxidized.

The cytochromes with a given heme in different organisms generally have different proteins, just as do the mitochondrial cytochromes of different eukaryotes. The resulting differences in absorption spectrum have given rise to numerical subscripts, which may be arbitrary (a_1, a_2, etc.) or may be based on an absorption peak (e.g., c_{551}). However, these are different compounds only in the sense in which any homologous enzymes, with the same function, differ from one organism to another. Many staphylococci lack cytochrome c and hence are "oxidase-negative" in a diagnostically useful color test with N,N-diMe-p-phenylenediamine. Some aerobic bacteria lack a cytochrome a, and their terminal oxidase is called cytochrome o.

The successive members of a chain must have not only suitable potentials but also a suitable enzymatic (stereospecific) relation of their surfaces: in vitro the cytochrome oxidase of certain bacteria has been found to oxidize the cytochrome c of that species much more rapidly than the cytochrome c of beef heart or even that of other bacterial species.

The cytochromes and respiratory enzymes of disrupted bacteria are found in the particulate fraction, whereas most other enzymes are found in the soluble fraction. The particles with respiratory activity appear to be artificial fragments of the cell membrane, for they vary widely and continuously in size and contain much phospholipid. The bacterial plasma membrane carries attached stalked particles on its inner surface, and these probably contain the machinery of electron transport.

Since phosphorylation depends on the spatial relation of components organized within the membrane, including membrane-bound particles, the resulting difficulties in fractionation and reconstitution have slowed the analysis of the mechanism linking electron transport and oxidative phosphorylation. The basic mechanism, however, is probably rather like that observed in the "substrate-level" phosphorylation of fermentation or of α-keto acid oxidation as noted above: spontaneous phosphorylation of a molecule, then elevation of the phosphate energy level by an increase in oxidation level of the compound, then phosphate transfer to yield ATP. The reversibly reducible cofactors in this kind of reaction include a variety of benzoquinones called **ubiquinones** (coenzyme Q), and possibly **naphthoquinones** (e.g., vitamin K_2= menaquinone). Various bacteria contain mostly one group or the other; E. coli contains both. A long alkyl side chain presumably contributes to the localization of these compounds in the lipid-rich membrane.

Progress in this refractory field should be accelerated by the recent isolation of mutants blocked in ubiquinone synthesis or blocked at various stages in oxidative phosphorylation.

FLAVOPROTEINS

This group of cofactors of oxidation was discovered by Warburg on the basis of simple observations on lactobacilli, which lack the red cytochromes: on exposure to air the intact cells become yellow. Pursuit of the yellow substance led to the isolation of the first flavoprotein (L. *flavus*-yellow), glucose-6-P dehydrogenase. Some flavoproteins contain riboflavin phosphate (FMP, flavin monophosphate) and others flavin adenine dinucleotide (FAD). Unlike DPN and TPN, these redox factors function as prosthetic groups tightly bound to their enzymes rather than as diffusible hydrogen carriers.

Some flavoproteins are directly oxidizable by air, though more slowly than via the cytochrome chain. We do not know whether the energy-wasting direct bypass to oxygen is an in vitro artefact or whether it contributes to the metabolism of the cell under some circumstances. This short-circuited electron transport usually results in the reduction of O_2 to H_2O_2 (hydrogen peroxide) rather than to H_2O. While there are some enzymes (peroxidases) that can use H_2O_2 for valuable oxidations, in most biological systems this toxic material is immediately destroyed by catalase. The lactic acid bacteria characteristically lack catalase.

Bioluminescence is a curious accompaniment of respiration in certain bacteria, as in some higher forms. It is caused by the oxidation of a flavoprotein (luciferin) by O_2 in the presence of the enzyme luciferase. Though its value in the mating of a fish or a firefly is not hard to imagine, in bacteria it may be an adventitious concomitant of a primitive respiratory system. Luminescent bacteria have proved useful as delicate detectors of O_2 in solution, because their light production responds to even very low concentrations of O_2.

ENERGETICS AND GROWTH

Table 3-1 shows that respiration is considerably more efficient than glycolysis: it yields about 10 times as much free energy (ΔF) and 19 times as much ATP per mole of glucose metabolized. The energy not used in ATP formation, plus part of the ATP energy subsequently used in biosynthesis, appears as heat. Indeed, dissipation of heat is one of the major problems of large-scale fermentation; if inadequate, a fermentation may sterilize itself.

Assimilation. Because of this difference in $\sim$P yield, a facultative organism grown on a limited amount of glucose will exhibit a larger growth yield (dry weight of bacteria/weight of substrate metabolized) under aerobic than under anaerobic conditions; but the increase will be close to 3- to 5-fold, rather than the 19-fold that might superficially be expected from the data of Table 3-1. The major reason is that part of the substrate taken up is assimilated rather than used to yield energy: in fermentation this growth yield represents an almost negligible fraction ($< 10\%$) of the substrate consumed, whereas in respiration it may reach as much as ⅔. Among organisms that metabolize glucose in various ways, yielding different amounts of ATP, the **growth yield** is quite proportional to the ATP yield.

With cells that are not growing (for lack of a N source) oxidative assimilation may convert much substrate into **storage materials.** These are predominantly "fat" or carbohydrates, depending on the species and conditions; such storage is more prominent in yeasts than in bacteria. Many bacteria form a unique storage product, **poly-β-hydroxybutyric acid,** which is an incompletely reduced, highly polymerized equivalent of fat. This substance is intermediate in composition ($(C_2H_3O_2)_n$), and in energy content, between fat and carbohydrate ($(CH_2O)_n$); it is presumably more readily available than fat as a reserve material.

$$2n \; CH_3\text{-}\overset{\overset{\text{O}}{\|}}{C}\text{-S (CoA)} \longrightarrow (-\overset{\overset{\text{CH}_3}{|}}{O}CH\text{--}CH_2\text{--}\overset{\overset{\text{O}}{\|}}{C}-)_n$$

Acetyl CoA   Poly-β-hydroxybutyrate

The Pasteur Effect. In the presence of air a facultative anaerobe, such as the yeast of alcoholic fermentation, not only forms more CO_2 per mole of glucose consumed but also yields more growth and grows faster. Yet as Pasteur discovered, the **rate** of CO_2 evolution and of glucose utilization decreases. A reasonable explanation for this old paradox is that respiration, producing much more ATP per glucose molecule, soon saturates the cell's capacity to utilize the ATP; i.e., the ADP/ATP ratio ("energy ratio") is critical.

In more specific terms, glycolysis requires a supply of ADP for accepting phosphate from 1,3-di-P-glycerate (Fig. 3-1), and respiration will keep the ADP level low and the ATP high. This explanation is supported by the early observation

---

TABLE 3-1.   Energetics of Metabolism

Reaction	No. of ATP generated	$-\Delta F$ (cal)	Efficiency* (%)
Glucose → 2 lactic acid	2	58,000	28
Glucose → 2 ethanol + 2 CO_2	2	57,000	28
Glucose + 6 O_2 → 6 CO_2 + H_2O Stages:			
Glucose $\xrightarrow{\;-\,[4H]\;}$ 2 pyruvate (substrate) (2 DPN)	2 6		
$\xrightarrow{\;-\,[4H]\;}$ 2 AcCoA + 2 CO_2 (2 DPN)	6		
$\xrightarrow{\;-\,[16H]\;}$ 4 CO_2 + 4 H_2O (8 DPN or equivalent)	24		
Total	38	688,000	44

* Assuming ΔF of ATP formation = 8000 cal.

that the uncoupling of electron transport from phosphorylation, by dinitrophenol, increases the aerobic utilization of glucose. However, the Pasteur effect may not depend entirely on substrate limitations, since numerous interactions between metabolic pathways are now known to involve specific regulatory effects of metabolites on the activity of allosteric enzymes (Ch. 13).

Growth and Energy Production. The ADP/ATP ratio is influenced not only by the rate of formation of ATP by catabolism but also by its rate of utilization in anabolism. Hence the rate of fuel consumption is increased in a growing culture compared with a "resting" culture, deprived of a material required for growth such as a source of N. This effect is observed in respiratory as well as in fermentative metabolism, and so the rapid oxidation of fuel evidently also requires a supply of ADP. The fact that nongrowing cells ferment or respire at all probably reflects several processes: use of ATP for maintenance and for synthesis of storage materials, a background level of ATPase activity in the cell, and imperfect coupling of electron transport with phosphorylation.

AUTOTROPHIC METABOLISM

CHEMOAUTOTROPHY AND PHOTOSYNTHESIS

In the metabolic patterns described above the foodstuffs are organic compounds, which are formed in nature by animals or plants; hence this mode of metabolism is called **hetero-** **trophic** (feeding on others). Some microbes, however, **do not depend on organic compounds** but tap quite different sources of energy and reducing power, which they use to reduce CO_2 to the organic compounds needed for growth. Table 3-2 presents examples of such **autotrophic** (self-feeding)

TABLE 3-2. Autotrophic Modes of Metabolism

	Organism or group	Source of energy	Remarks
I.	Aerobic lithotrophs (chemoautotrophs)		Use inorganic (litho-) electron donors
	Hydrogen bacteria	$H_2 + 1/2\ O_2 \rightarrow H_2O$	
	Sulfur bacteria	$H_2S + 1/2\ O_2 \rightarrow H_2O + S$	Can produce H_2SO_4 to pH as low as O
	(colorless)	$S + 1.5\ O_2 + H_2O \rightarrow H_2SO_4$	
	Iron bacteria	$2\ Fe^{++} + 1/2\ O_2 + H_2O \rightarrow 2\ Fe^{3+} + 2\ OH^-$	
	Nitrifying bacteria		
	Nitrosomonas	$NH_3 + 1.5\ O_2 \rightarrow HNO_2 + H_2O$	Convert soil N to non-volatile form, used by plants
	Nitrobacter	$HNO_2 + 1/2\ O_2 \rightarrow HNO_3$	
II.	Anaerobic respirers	Use inorganic electron acceptors	Most can also use organic electron donors
	Denitrifiers	$NH_2 + NO_3^- \rightarrow N_2O,\ N_2,\ or\ NH_3$	Cause N loss from anaerobic soil
	Desulfovibrio	$NH_2 + SO_4^= \rightarrow S\ or\ H_2S$	Odor of polluted streams, mud flats
	Methane bacteria	$4\ H_2 + CO_2 \rightarrow CH_4 + 2\ H_2O$	Sewage disposal plants; prevent accumulation of free H_2 in nature
	Clostridium aceticum	$4\ H_2 + 2\ CO_2 \rightarrow CH_3COOH + 2\ H_2O$	
III.	Photosynthesizers	light	"Bacterial" photosynthesis; $H_2(A) =$ various electron donors
	Purple sulfur bacteria	$4\ CO_2 + 2\ H_2S + 4\ H_2O \rightarrow 4\ (CH_2O) + 2\ H_2SO_4$	
	Nonsulfur purple bacteria	light $CO_2 + 2\ H_2(A) \rightarrow (CH_2O) + H_2O + 2\ (A)$	
	Algae; higher plants	light $CO_2 + 2\ H_2O \rightarrow (CH_2O) + 1/2\ O_2$	"Plant" photosynthesis

metabolism. A number of these processes were discovered before 1900 by the Russian soil microbiologist Winogradsky.*

Autotrophic processes can be divided into two classes. 1) **Photosynthesis** derives energy and reducing power from the absorption of visible light, and is found in algae and a few bacteria, as well as in higher plants. Microbial photosynthesis in the upper layers of the ocean and lakes, mostly by algae, accounts for about half the total terrestrial photosynthesis. 2) **Chemoautotrophy** (chemosynthesis) derives energy from the respiration of **inorganic** electron donors; it is found only in certain bacteria, which are also known as **lithotrophs** (Gr. *lithos*=stone). In their electron transport some of these organisms use O_2 (class I, Table 3-2), while others use other electron acceptors (class II).

Those autotrophic bacteria that oxidize H_2 can often similarly utilize simple carbon compounds, such as CO, CH_4, HCHO, HCOOH, or CH_3OH. Since these compounds are organic their utilization does not conform to the original nutritional definition of autotrophy. Hence autotrophy must be redefined in metabolic terms: the utilization of CO_2 as a major source of organic compounds. (Minor incorporation of CO_2 occurs in all organisms, in biosynthetic reactions described in the next chapter.)

PHOTOSYNTHESIS

In photosynthesis a quantum of visible light is absorbed by a molecule of chlorophyll (a Mg^{++}-tetrapyrrole) or of a carotenoid. The energy is transferred to another chlorophyll in a special **reaction center,** causing it to eject an electron: this electron is accepted by ferredoxin, while the oxidized chlorophyll oxidizes the terminal cytochrome of an electron transport system. The resulting charged system, with a large potential difference between its termini, is used to create both ATP and biosynthetic reducing power in the form of TPNH: both are needed for reduction of CO_2 (see Fig. 3-6, below).

The system functions in two different ways, which permit the proportions of ATP and TPNH formed to be varied as needed. In **cyclic photophosphory-**

---

lation (Fig. 3-5) the electron is transported from the reduced ferredoxin to the oxidized cytochrome through a chain of quinones and cytochromes, much as in respiratory electron transport (Fig. 3-4). The electrons thus flow through a closed circuit (hence cyclic) which converts part of the absorbed light energy into ~P bond energy. In the companion process of **noncyclic photophosphorylation** (Fig. 3-5) the electron at the reducing end of the chain is used to reduce TPN. To complete the electron transport, therefore, an electron must be supplied from another source. Van Niel showed that in plants and algae (including the prokaryotic blue-green algae) this electron is derived from the oxidation of water, releasing O_2; bacteria oxidize other substrates at a lower potential, and do not release O_2.

The large potential difference required for the release of O_2 in plant photosynthesis is

Fig. 3-5. Electron flow in photosynthesis. In photosynthetic bacteria H_2A is H_2, H_2S, or various organic compounds; two electrons must be derived by photosynthesis per molecule oxidized. In "plant photosynthesis" H_2A is H_2O; a second reaction center chlorophyll (not shown), with a higher oxidized potential, must be coupled to the first.

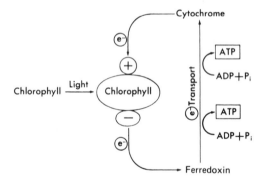

Cyclic photophosphorylation

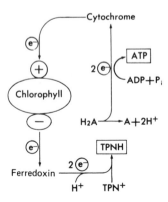

Noncyclic photophosphorylation

achieved by photoactivation of two kinds of reaction centers, connected in series:

I) reaches a low enough reduced potential to reduce ferredoxin (-0.42 V)

II) reaches a high enough oxidized potential to oxidize water ($+0.82$ V)

The bacterial system is more primitive and contains only reaction center I: hence its oxidized potential is not so high and requires an electron donor with a lower potential than that of water, e.g., H_2 or H_2S:

$$2Fe^{+++} \quad\quad 2Fe^{++}$$
$$H_2S \longrightarrow 2H^+ + S$$

Bacterial photosynthesis thus never releases O_2, and photosynthesizing bacteria are generally obligate anaerobes, whereas algae and plants must be at least aerotolerant. Bacterial photosynthesis is also more primitive in having the chlorophyll located in particles attached to lamellar extensions of the plasma membrane, rather than in separate organelles (chloroplasts) as in high plants.

The bacterial type of photosynthesis probably evolved first, creating the first electron transport system and converting much of the atmospheric CO_2 to organic compounds. Blue-green algae are obviously descended from later, intermediate forms: they are also prokaryotes, but they carry out plant-type photosynthesis. Their ancestors presumably generated the atmospheric O_2 that then permitted the electron transport system of photosynthesis to evolve into the powerful energy-yielding system of respiration. Blue-green algae may also have given rise in evolution, by symbiosis, to chloroplasts in eukaryotes.

In plants noncyclic photophosphorylation is essential, since no other source of reducing power is generally available. In bacteria, however, some of the reducing compounds used (e.g., H_2) can also reduce TPN directly, and so most photophosphorylation would then be cyclic.

Some photosynthetic bacteria show the property of **phototaxis**: a gradient of light intensity elicits differential flagellar activity in different parts of the cell, resulting in motion toward the light.

AUTOTROPHIC ASSIMILATION OF CO_2

With the use of the energy and the reducing power supplied by photosynthesis, or by chemosynthesis, an autotrophic cell derives its carbon from CO_2. Although this process is fundamentally biosynthetic, it is so intimately linked with autotrophic energy metabolism that it will be presented here rather than in the next chapter.

The elucidation of this pathway, largely by Calvin and Horecker, has been one of the triumphs of modern biochemistry. It turns out to involve surprisingly few novel enzymes, in addition to enzymes already present in universal pathways of carbohydrate metabolism (including pentose synthesis). The basic mechanism for the initial fixation of CO_2 is:

$$
\begin{array}{ccccc}
\begin{array}{l} CH_2O\,\text{\textcircled{P}} \\ | \\ C=O \\ | \\ CHOH \\ | \\ CHOH \\ | \\ CH_2O\,\text{\textcircled{P}} \end{array}
& \longrightarrow &
\left[\begin{array}{l} CH_2O\,\text{\textcircled{P}} \\ | \\ C\text{-}OH \\ || \\ C\text{-}OH \\ | \\ CHOH \\ | \\ CH_2O\,\text{\textcircled{P}} \end{array}\right]
& \xrightarrow[\text{Carboxy-dismutase}]{CO_2} &
\begin{array}{l} CH_2O\,\text{\textcircled{P}} \\ | \\ CHOH \\ | \\ COOH \\ + \\ COOH \\ | \\ CHOH \\ | \\ CH_2O\,\text{\textcircled{P}} \end{array}
\end{array}
$$

Ribulose di-P $\qquad\qquad\qquad\qquad$ 3-P-glycerate

At this stage the assimilated carbon is still at the fully oxidized, carboxy level. To serve as a general source of carbon this group is reduced by TPNH, with energy supplied by ATP (Fig. 3-6). In this process 3-P-glycerate is reduced to triose-P; of 6 such trioses, 5 are used to regenerate the pentoses of the cycle, while 1 represents the net gain. (The over-all process by which 5 trioses can become 3 pentoses is presented in Ch. 4, Pentose biosynthesis.)

NITROGEN CYCLE; NITROGEN FIXATION

Microbes play an essential role in the geochemical **nitrogen cycle**, in which some of the chemoautotrophic bacteria listed in Table 3-2 participate. The N in decomposing organic matter is at first largely converted to NH_3. This volatile compound is then stabilized in the soil by oxidation, through the action of **nitrifying** bacteria, to nonvolatile nitrate, which can be reduced by plants to organic amino com-

Fig. 3-6. The ribulose diP cycle requires 3 moles of ATP and 2 moles of TPNH per mole of CO_2 converted to carbohydrate.

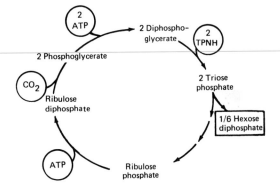

pounds. However, some NH_3 is lost immediately to the atmosphere, and additional amounts are lost, from anaerobic regions of the soil, through the action of denitrifying bacteria, which reduce nitrate to N_2 and NH_3. Maintenance of the biosphere therefore requires constant **fixation of atmospheric N_2.**

Biological N_2 fixation is accomplished only by bacteria and algae. The quantitatively most important group appears to be the bacterial genus *Rhizobium* (Gr. *rhizo*=root), which infects the roots of leguminous plants, leading to formation of symbiotic nitrogen-fixing nodules. Nitrogen can also be fixed, however, by many other bacteria, including

the photosynthetic purple sulfur bacteria (e.g., *Rhodospirillum rubrum*). These remarkably self-sufficient organisms can, therefore, derive all their major atoms (C, H, N, and O) from the atmosphere and water.

N_2 fixation, nitrification, and denitrification involve several different valence levels of N. Enzymes in two of these processes have been shown to contain molybdenum, which can also occupy several valence levels. The reduction of N_2 to NH_3 involves the powerful reductant ferredoxin and can now be accomplished in vitro.

ALTERNATIVE SUBSTRATES

GENOTYPIC ADAPTATION (SELECTION OF ORGANISMS)

Microbes exhibit not only a variety of fermentation patterns, adapted to different environments, but even greater variation in the ability to utilize different foodstuffs. It is axiomatic that every compound in the organic world can be metabolized by some microbe, for in the grand cycle of organic matter in nature the conversion of decaying organic matter to atmospheric CO_2 (**"mineralization"**) is just as essential as is the fixation of CO_2 by photosynthesis (see Pasteur quotation at head of chapter).* Indeed, the atmospheric reservoir of CO_2 would last only about 20 years if not replenished.

A small sample of soil may contain many different kinds of organisms, which play different roles in the process of mineralization. The initial stages of decay of animal and plant tissues generally occur in anaerobic environments; hence most organic constituents are first hydrolyzed and fermented and the products then diffuse to locations where they are ultimately oxidized, by a different group of organisms. Fatty compounds are too highly reduced to be readily fermented, and so their insoluble globules tend to be attacked by aerobes with a lipophilic surface, generally mycobacteria. Petroleum represents the ultimate in the natural reduction of organic

matter. It is formed when large amounts of organic matter decay below water, which ensures perpetual anaerobiosis; the hydrocarbons remain after all the oxygen has been eliminated from the molecules.

Many organisms have only narrow range of possible foodstuffs, but a single strain of *Pseudomonas,* a particularly versatile soil scavenger, has been found to use as many as 80 carbon sources.

Laboratory Selection. As noted in Chapter 1, a general microbiologist will sort out the organisms in a soil sample by using **enrichment culture** or **elective enrichment,** which exaggerates the selective pressures present in nature (e.g., by incubating the sample in medium containing a given compound as sole carbon source). Selective media were thus originally introduced to separate the organisms present in a natural mixed population, and were rapidly extended to medical microbiology. Very similar procedures are also used to isolate mutants arising in the laboratory (Ch. 8).

PHENOTYPIC ADAPTATION (SELECTION OF ENZYMES)

Not only do organisms differ genetically in their capacity to metabolize a given substrate, but the cells of a competent strain may or may not possess the enzymes required, depending on their immediate past history. This **adaptive enzyme formation** was first recognized by Karström in Finland in 1931; when a strain of lactic bacteria was grown on various sugars and then tested as "resting" cells (i.e., without a N source, to prevent further protein synthesis) most sugars could be fermented

---

* It is perhaps not surprising that the present microbial population does not always attack readily those synthetic products that have never been present in the soil until recently. Indeed toxic pesticides, which can be concentrated by plants and animals, are presenting an increasing danger, and synthetic detergents create problems in sewage disposal. Biodegradable substitutes are therefore being sought.

only by cells that had grown in their presence. However, the enzymes of glucose metabolism were **constitutive,** for glucose could be fermented regardless of the C source present during growth.

We now know that such adaptation involves changes in the ability of a sugar to penetrate, as well as the appearance of enzymes that attack it. Moreover, more refined studies have blurred the distinction between adaptive and constitutive enzymes. Nevertheless, these early observations were germinal.

The mechanisms responsible for regulating enzyme adaptation will be discussed in detail in Chapter 13. Meanwhile, as a background for the next few chapters, we should note that this response not only is stimulated by the presence of the substrate but also may be blocked by the presence of an alternative food that supports faster growth. Thus with *Micrococcus denitrificans,* which can use either O_2 or nitrate as electron acceptor, Kluyver's group in Holland discovered in 1940 that nitrate induces, and O_2 blocks, formation of the enzyme connecting nitrate with electron transport. In 1942 Monod in France found a similar interference between alternative carbon sources: *E. coli* growing on glucose plus almost any other carbon source consumes all the glucose first and then, after a lag, resumes growth at the expense of the second source (Fig. 3-7). This diphasic growth curve was called **diauxie.** Subsequent intensive study of one example, the formation of β-galactosidase induced by lactose, has provided much of our knowledge of the regulation of gene action (Ch. 13) and has established *E. coli* as the prototype species for studying this and related problems.

Though sugars have played a key role in the study of enzyme regulation, the pathways are now also known for the degradation of the various amino acids, purines, pyrimidines, and many other organic compounds. These adaptive exergonic pathways generally have little relation to the routes of biosynthesis.

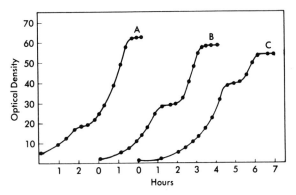

Fig. 3-7. Diauxic growth of *E. coli* on glucose and sorbitol in the proportions 1/3 **(A)**, 2/2 **(B)**, and 3/1 **(C)**. Minimal medium; inoculum grown on glucose. (From Monod, J. *La croissance des cultures bactériennes.* Hermann, Paris, 1942.)

SELECTED REFERENCES

Books and Review Articles

BARKER, H. A. *Bacterial Fermentations.* Wiley, New York, 1957.

BARTSCH, R. G. Bacterial cytochromes. *Ann Rev Microbiol 22*:181 (1968).

BENEMANN, J. R., and VALENTINE, R. C. High-energy electrons in bacteria. *Adv Microbiol Physiol 5*:135 (1971).

BUCHANAN, B. B., and ARNON, D. I. Ferredoxins: Chemistry and function in photosynthesis, nitrogen fixation, and fermentative metabolism. *Adv Enzymol 33*:119 (1970).

DOELLE, H. W. *Bacterial Metabolism.* Academic Press, New York, 1969.

GUNSALUS, I. C., and STANIER, R. Y., eds. *The Bacteria,* vol. II. Academic Press, New York, 1961. Espe-

cially Cyclic Mechanisms of Terminal Oxidation (Krampitz), Survey of Microbial Electron Transport Mechanisms (Dolin), Cytochrome Systems in Aerobic Electron Transport (Smith), and Fermentation of Carbohydrates and Related Compounds (Wood).

KELLY, D. P. Autotrophy: Concepts of lithotrophic bacteria and their organic metabolism. *Ann Rev Microbiol 25*:177 (1971).

KLUYER, A. J., and VAN NIEL, C. B. *The Microbe's Contribution to Biology.* Harvard Univ. Press, Cambridge, Mass. 1956. A thought-provoking set of lectures, providing perspective on the unity of biology and the evolutionary aspects of microbiology.

LEHNINGER, A. L. *Bioenergetics.* Benjamin, New York, 1956.

MORTENSON, L. E. Nitrogen fixation: Role of ferredoxin in anaerobic metabolism. *Annu Rev Microbiol 17*:115 (1963).

ORNSTON, L. N. Regulation of catabolic pathways in *Pseudomonas. Bacteriol Rev 35*:87 (1971).

POSTGATE, J. R., ed. *The Chemistry and Biochemistry of Nitrogen Fixation.* Plenum Press, New York, 1971.

SMITH, L. The Respiratory Chain System of Bacteria. In *Biological Oxidations.* (T. P. Singer, ed.) Interscience, New York, 1968.

VERNON, L. P. Photochemical and electron transport reactions of bacterial photosynthesis. *Bacteriol Rev 32*:243 (1968).

YOCH, D. C., and VALENTINE, R. C. Ferredoxins and flavodoxins of bacteria. *Annu Rev Microbiol 26*:139 (1972).

Specific Articles

ARNON, D. I. The light reactions of photosynthesis. *Proc Natl Acad Sci USA, 68*:2883 (1971).

MCCORD, J. M., KEELE, B. B., JR., and FRIDOVICH, I. An enzyme-based theory of obligate anaerobiosis: The physiological function of superoxide dismutase. *Proc Nat Acad Sci USA 68*:1024 (1971).

chapter 4

BIOSYNTHESIS

MICROBIAL COMPOSITION AND THE UNITY OF
    BIOCHEMISTRY 60
METHODS OF ANALYZING BIOSYNTHETIC PATHWAYS 61
CRITERIA FOR A BIOSYNTHETIC INTERMEDIATE 62
CATABOLIC, ANABOLIC, AND AMPHIBOLIC PATHWAYS 63
    Conversion of Amphibolic to Biosynthetic Pathways 63
    Biosynthesis from 2-C Compounds: The Glyoxylate Cycle 64
    Flow Between Glycolytic and TCA Pathways 67
PROMOTION OF UNIDIRECTIONAL FLOW 67
    Reversible and Irreversible Reactions 67
    Redox Cofactors: The Roles of DPN and TPN 68

AMINO ACIDS 69
    The Glutamate Family 69
    The Aspartate Family 71
    The Pyruvate Family: Aliphatic Amino Acids 72
    The Serine Family: 1-C Fragments 73
    1-C Transfer 74
    Histidine 74
    Aromatic Compounds 75
    D-Amino Acids 79
    Polyamines 79

NUCLEOTIDES 80
    Pentoses 81
    Purines 81
    Pyrimidines 82

OTHER PATHWAYS 83
SUGARS 83
POLYSACCHARIDES 83
POLYPEPTIDE SYNTHESIS 85
    Homopolymeric Polypeptides 85
    Peptide Antibiotics 85
SELECTED FEATURES OF SMALL-MOLECULE BIOSYNTHESIS 86
    Heterotrophic CO_2 Fixation 86
    The Importance of Being Ionized 86
    Economy in Mammalian Biosynthesis 87

The use of selective inhibitors and enzyme fractionation revealed the broad, energy-yielding pathways reviewed in Chapter 3, but not the more numerous, narrower pathways that use this energy for biosynthesis. Most biosynthetic intermediates were revealed later by the use of two new tools: microbial mutants blocked in various biosynthetic reactions, and radioactively labeled precursors; and once some of the intermediates were known the enzymes of a pathway could be identified and characterized. Between 1945 and 1965 these approaches revealed all the steps in the synthesis of most universal building blocks and of many special constituents of the cell envelope (Ch. 6); the paths to many cofactors are still incompletely explored.

Since the reader is assumed to be familiar with biochemistry this chapter will review the biosynthesis of small molecules selectively, with emphasis on fundamental principles and methodology, on the significance of these pathways for cell function, and on sequences unique to microbes.* Later chapters will consider the synthesis of lipids and wall polysaccharides (6), the polymerization of building blocks into nucleic acids (10, 11) and proteins (12), and the regulation of biosynthesis (13).

---

* In this chapter the following abbreviations are used: ATP, adenosine triphosphate; TCA, tricarboxylic acid; DPN, diphosphopyridine nucleotide (= NAD, nicotinamide adenine dinucleotide); TPN, triphosphopyridine nucleotide (= NADP); DPNH, TPNH, the reduced forms of DPN, TPN; CoA, coenzyme A; PRPP, 5-P-ribose-1-PP; DAP, diaminopimelate; THF, tetrahydrofolate; PEP, phosphoenolpyruvate; UMP, CMP, GMP, AMP, the ribonucleotide of uracil, cytosine, guanine, and adenine, respectively; dUMP, etc., deoxyuridine monoP, etc.; UDP, UTP, etc., uridine diP and uridine triP, etc.

MICROBIAL COMPOSITION AND THE UNITY OF BIOCHEMISTRY

Though many microorganisms can synthesize all their organic constitutents from a single carbon source (even, in autotrophs, from CO_2) others require a rich medium for growth. When these nutritional requirements were analyzed in terms of specific compounds (Ch. 5) a bewildering variety of patterns were observed, and it seemed obvious that these reflected differences in metabolic complexity (i.e., growth on a simple medium would reflect a simple metabolism). In the 1930s, however, Lwoff in France, and Knight in England, suggested the opposite interpretation, which stressed a unity underlying this diversity: cells all might utilize much the same set of building blocks and cofactors, but they would differ in their ability to make these compounds. The **essential nutrients** of any organism would then simply be those **essential metabolites** that it cannot make endogenously.

A general concept of the **unity of biochemistry** had already been formulated by the Dutch microbiologist Kluyver, in 1926, on the basis of the similarity of the classes of enzymatic reactions (hydrolyses, dehydrogenations, etc.) found throughout the biological kingdom; and the glycolytic pathway had been shown to have precisely the same intermediates and reactions in yeast cells and in mammalian muscle. However, the notion of a similar uniformity in biosynthesis, as suggested by Lwoff and Knight, did not receive strong support until suitable analytical methods were developed for determining the detailed composition of cells. It was then found that the proteins, nucleic acids, and enzyme cofactors of the most varied cells are made of the same **building blocks.** For example, such a metabolic eccentric as *Thiobacillus,* deriving its energy by converting sulfur to H_2SO_4, is not made of brimstone: it contains the same constituents as other organisms, with no excess of sulfur except in the pool of metabolites.

The final proof was the demonstration that biosynthetic pathways, as well as their products, are also remarkably uniform. Hence the elucidation of these pathways, often first in bacteria, has very general interest for biology. Moreover, the presence of the same pathways in the most diverse organisms probably reflects more than common ancestry: it also suggests that they are the most efficient pathways possible.

METHODS OF ANALYZING BIOSYNTHETIC PATHWAYS

Auxotrophic Mutants. Microbes offer several advantages over animal tissues for biosynthetic studies: more rapid growth and biosynthesis, homogeneity of the cell population, and the presence of a full complement of biosynthetic pathways (in the nutritionally simple strains). An even greater advantage is the possibility of isolating auxotrophic mutants, which require a particular metabolite for growth because they have acquired a genetic defect in its synthesis. Beadle and Tatum first systematically isolated such mutants, in 1943, from the red bread mold, *Neurospora crassa*. With bacteria similar mutants have proved even easier to isolate and to study quantitatively; their selection will be described in Chapter 8.

Two properties of auxotrophic mutants have been particularly useful in revealing biosynthetic intermediates: most strains accumulate and excrete the substrate of the blocked reaction in large amounts, often exceeding the weight of the cells; and many strains can grow not only on the endproduct of the blocked sequence but also on intermediates between the block and the endproduct. Hence strains blocked in different reactions in the same sequence often exhibit cross-feeding (**syntrophism**), the strain with the later block accumulating an intermediate to which the other can respond (Fig. 4-1). Moreover, the growth response can be used for bioassay (Ch. 5) of the compound, which is helpful in its isolation. Many intermediates, however,

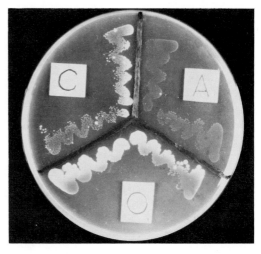

Fig. 4-1. Cross-feeding between different bacterial mutants blocked in the same pathway.

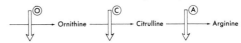

Note that a streak of mutant A, growing slightly on medium containing a trace of arginine, stimulates growth of adjacent streaks of mutants C and O, while C similarly stimulates O but not A. [From Davis, B. D. *Experientia* 6:41 (1950).]

cannot be detected by nutritional tests because they cannot penetrate into cells.

Other Methods. Once some intermediates in a pathway are available the enzymes that connect them can be characterized; additional enzymes and intermediates can then be identified by **enzyme fractionation**. **Radioactive isotopes** have also been extremely useful, in two general approaches: **precursor** competition and **specific atom distribution.**

In precursor competition, for example, a labeled possible precursor, glycine, is supplied along with an unlabeled general carbon source (e.g., glucose); since the purines subsequently recovered from the cells by hydrolysis are heavily labeled, glycine has served as a direct precursor. In the other approach the organism is supplied with a single carbon source (e.g., glucose) **selectively labeled in a specific atom**, and various endproducts are then isolated and are degraded in a way that permits determination of the isotope concentration (specific activity) in individual atoms. The results indicate where a given biosynthetic pathway branches off from a well-established central route (Fig. 4-2). The coherence of the mutant, enzymatic, and isotopic evidence has firmly established the pathways shown below.

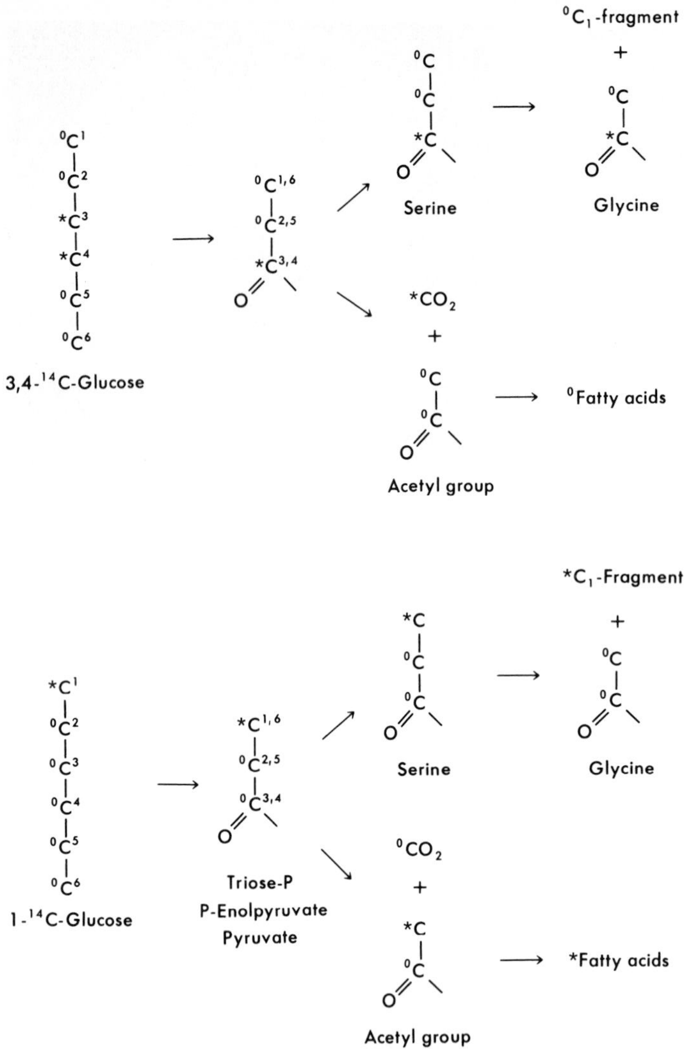

Fig. 4-2. Example of use of selectively labeled glucose in analysis of biosynthetic pathways. *C (radioactive); ^{0}C $= ^{12}$C (nonradioactive).

In the extension of the study of biosynthesis to complex macromolecules the amounts formed in vitro have generally been too small to detect by ordinary analytical methods. To measure these reactions **radioactive precursors are incorporated into polymers,** which are then precipitated by reagents (e.g., trichloracetic acid) that do not precipitate the substrates.

CRITERIA FOR A BIOSYNTHETIC INTERMEDIATE

By definition, a substance that is incorporated serves as a precursor. But it does not neces-

sarily follow that the compound serves in the cell as an intermediate between a general nutrient and a cell component. The distinction is significant for cell physiology: while most precursors have also turned out to be intermediates, there are exceptions.

For example, histidine provided in the medium induces *Aerobacter* to form degradative enzymes that convert its 5-C chain to α-ketoglutarate; hence it can serve as a source of glutamate. However, the level of histidine formed endogenously in the cell is not sufficient to induce its own degradation, and histidine biosynthesis has no connection with normal glutamate biosynthesis.

These considerations have led to the formulation of the following criteria: A and B are obligatory intermediates in the biosynthesis of product P

if 1) compounds A and B can each give rise to P (in the cell or in extracts), 2) a single enzyme converts A to B, and 3) loss of that enzyme (e.g., by mutation) results in a requirement for P.

These criteria exclude, for example, free bases as intermediates, even though they may be excreted by certain mutants and used by others; their biosynthetic pathway proceeds at the level of nucleotides (see below). Precursors that are not true intermediates are introduced by **salvage pathways,** which may not be active under some circumstances. Neglect of this consideration generated much confusion in study of the regulation of RNA synthesis (Ch. 13).

CATABOLIC, ANABOLIC, AND AMPHIBOLIC PATHWAYS

When the major energy-yielding pathways of glycolysis, pyruvate oxidation, and the tricarboxylate (TCA) cycle were discovered they were considered purely catabolic (degradative). However, it turned out that catabolic and anabolic processes are intimately connected: as Figure 4-3 shows, the various specific pathways of biosynthesis originate from various "catabolic" intermediates.

Since the glycolytic and the TCA pathways thus function just as directly and indispensably in biosynthesis as in energy production they are now designated as **amphibolic** (Gr. *amphi* = either) pathways. The anabolic pathways would then be those that branch from amphibolic intermediates to endproducts, while the term catabolic would be restricted to the short sequences that convert amphibolic intermediates only to degradative endproducts.*

Nature has obviously achieved a substantial sparing of genes by selecting biosynthetic and catabolic pathways that share intermediates. This evolutionary pressure for metabolic economy will receive additional support when we consider regulatory mechanisms in Chapter 13.

CONVERSION OF AMPHIBOLIC TO BIOSYNTHETIC PATHWAYS

Under various conditions, which are no less "normal" than aerobic growth on glucose, parts of the central, amphibolic pathways become purely biosynthetic. In particular, with any single C source, entering either the glycolytic or the TCA pathway, growth will require net flow into the other pathway, to keep pace with the biosynthetic drains. The reactions that connect the two pathways will be discussed below (Fig. 4-7). In mammals, of course, metabolic flow patterns are similarly affected by the use of different fuels (e.g., gluconeogenesis).

For example, a cell growing aerobically on succinate oxidizes it to oxaloacetate, part of which is used, directly or via the cycle, for biosynthesis. The rest is converted into pyruvate, P-enolpyruvate, and acetate, which are used for biosynthesis and for terminal respiration (Fig. 4-4). Reverse flow of carbon through the glycolytic pathway is essential to provide various cell components (Fig. 4-3). Similarly, when cells grow on glycerol, which is readily converted to triose-P, the lower part of the glycolytic pathway is used amphibolically and the upper part, in reverse, biosynthetically.

With facultative organisms terminal respiration ceases under anaerobic conditions, and the TCA cycle no longer functions as a cycle. Nevertheless, oxaloacetate, α-ketoglutarate, and succinate are still required as biosynthetic intermediates (Fig. 4-3), and they are formed by converting the TCA cycle into a pair of biosynthetic pathways branching from oxaloacetate: a portion of the original cycle remains as an oxidative branch forming α-ketoglutarate, while the usual pathway from succinate to oxaloacetate is reversed (by the induction of new enzymes) to form a reductive branch (Fig. 4-5). α-Ketoglutarate oxidase, which would be a useless enzyme, disappears.

_____

* It has been suggested that pathways difficult to classify (or perhaps those difficult to memorize) be designated as **diabolic.**

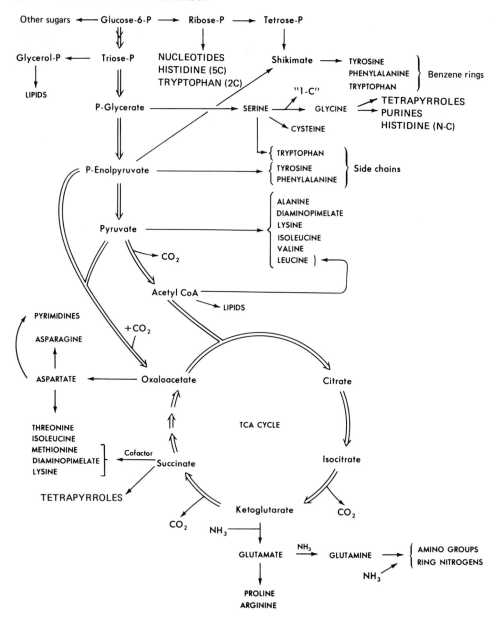

Fig. 4-3. Relation of amphibolic pathways (heavy arrows) to main anabolic pathways.

BIOSYNTHESIS FROM 2-C COMPOUNDS: THE GLYOXYLATE CYCLE

The central pathways reviewed above can account for the ability of organisms to grow on many compounds: sugars (or substances convertible to them), 3-C monocarboxylic acids, 4- to 6-C di- or tricarboxylic acids, and even CO_2 in autotrophs (Ch. 3). These pathways, however, do not account for the ability of many bacteria to grow on acetate, and hence on fat, as the sole C source; for though pyruvate is readily decarboxylated to acetate, the reverse reaction does not occur except in ferredoxin-containing anaerobes (see below). This inability of acetate to replace the biosynthetic drain on the TCA cycle is largely responsible for the acidosis of a diabetic mammal forced to consume fat (via acetyl

CoA) without carbohydrate. How bacteria can thrive on such a diet is the problem.

The answer, discovered in 1957 by Hans Kornberg, is a bypass or epicycle on the TCA cycle, involving reactions 1 and 2:

$$\underset{\substack{| \quad\;| \quad\;\; | \\ \text{COOH COOH COOH} \\ \text{Isocitrate}}}{CH_2-CH-CHOH} \xrightarrow[\text{(=Isocitrate lyase)}]{\text{Isocitratase}} \underset{\substack{| \quad\;\; | \\ \text{COOH COOH} \\ \text{Succinate}}}{CH_2-CH_2} + \underset{\substack{| \\ \text{COOH} \\ \text{Glyoxylate}}}{CHO} \quad (1)$$

$$\underset{\substack{| \quad\quad | \\ \text{COOH} \quad \text{COOH} \\ \text{Glyoxylate} \;\text{Acetate}}}{CHO \;+\; CH_3} \xrightarrow[\text{synthetase}]{\text{Malic}} \underset{\substack{| \quad\quad\; | \\ \text{COOH} \quad \text{COOH} \\ \text{Malate}}}{HOCH-CH_2} \quad (2)$$

$$\text{Succinate + Acetate} \xrightarrow[\substack{\text{(TCA cycle} \\ \text{reactions)}}]{-4[H]} \text{Isocitrate} \quad (3)$$

Net: 2 Acetate $\xrightarrow{-4[H]}$ Malate

The succinate and malate formed in reactions 1 and 2 can be used to regenerate isocitrate through familiar reactions of the TCA cycle (summed as reaction 3), yielding a glyoxylate cycle. Acetate is thus used for net synthesis of 4-C compounds rather than burned to CO_2 (Fig. 4-6).

The two enzymes of the glyoxylate bypass have been found in a variety of organisms grown on acetate as sole C source, but their formation is repressed by the simultaneous supply of a more rapidly used substrate, such as glucose or succinate.

Ferredoxin and Acetate Reduction. We have seen that the need for the glyoxylate cycle arises from the irreversibility, in most organisms, of the oxidative decarboxylation of pyruvate to acetyl CoA and CO_2. However, ferredoxin (Ch. 3) has a low

Fig. 4-4. Pathway of metabolic flow in aerobic growth on succinate. The glycolytic intermediates would be included, but are not specified, in the box.

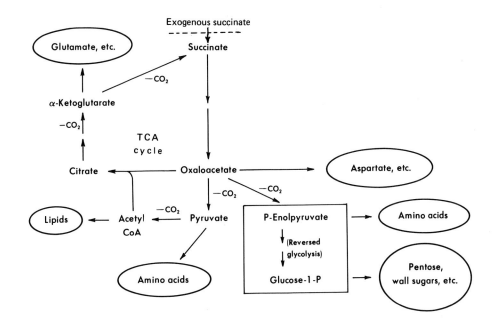

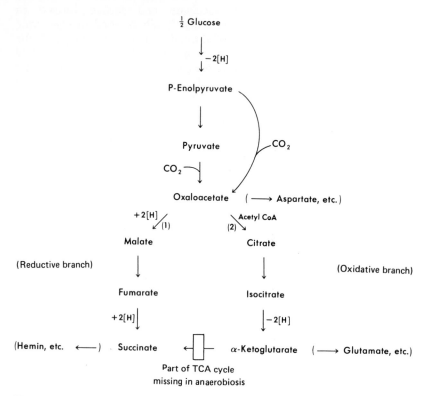

Fig. 4-5. Biosynthesis of TCA intermediates in anaerobic growth on glucose in *E. coli.* The TCA cycle becomes split into a reversed reductive branch and a normal oxidative branch.

Fig. 4-6. Krebs tricarboxylic acid (TCA) cycle, and within it the "glyoxylate bypass" (heavy arrows). The glyoxylate cycle substitutes these reactions, which conserve carbon, for the two decarboxylative reactions of the TCA cycle (dashed arrows) which release carbon as CO_2. In an organism growing on acetate alone the glyoxylate cycle provides net C-4 (and thus C-3) synthesis to replenish the biosynthetic drain (circled compounds). Since the glyoxylate cycle involves two oxidative steps, linked to electron transport, it also provides some energy; but most of the cell's energy is derived from the simultaneous oxidation of other acetate molecules via the TCA cycle.

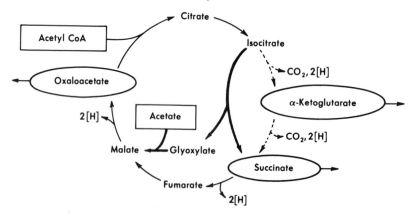

enough redox potential to permit photosynthetic bacteria and certain other obligate anaerobes to form pyruvate by reductive carboxylation.

FLOW BETWEEN GLYCOLYTIC AND TCA PATHWAYS

Several different carboxylation and decarboxylation reactions can mediate flow between the glycolytic and the TCA pathways (Fig. 4-7). The carboxylations replenish (= anaplerotic, Gr.) material drawn off from the TCA cycle for biosynthesis. The apparently alternative pathways are adaptable, like others described in Chapter 3, to different functions.

For example, malate can be oxidatively decarboxylated to pyruvate, generating TPNH; but when the need for TPNH is met (see below) a cell having a source of malate will convert it to oxaloacetate, generating the DPNH that is the substrate for electron transport. Similarly, oxaloacetate can be converted either to P-enolpyruvate to initiate reversed glycolysis, or to pyruvate, leading to terminal respiration.

Some of these reactions are reversible, and their

normal direction is inferred from their feedback regulatory responses (Ch. 13). For example, the malic enzyme (reaction 3) is inhibited by TPNH and acetyl CoA.

PROMOTION OF UNIDIRECTIONAL FLOW

REVERSIBLE AND IRREVERSIBLE REACTIONS

As we have noted, in the cell some over-all sequences (e.g., glycolysis) may exhibit net flow in either direction, depending on circumstances. For some steps in this reversal the same enzyme appears to be used reversibly. For example, aldolase catalyzes either the conversion of triose P to hexose diP or the reverse, depending on whether the C source is glycerol or glucose. However, for other steps, which are also reversible in vitro, cells have developed a somewhat different reverse reaction, presumably because the equilibrium constant of the first reaction is too unfavorable for its physiological reversal.

Fig. 4-7. Carboxylation and decarboxylation reactions connecting the glycolytic and the TCA pathways under various conditions and in different organisms. *Carbon source*
Glucose: Carboxylation by reaction 1 in *E. coli* (using energy of PEP), or by reaction 2 in *Pseudomonas* (using energy of ATP).
Lactate: Carboxylation by reaction 2.
TCA component (such as succinate): Decarboxylation to high-energy PEP by reaction 4 (using energy of GTP), for biosynthetic needs, and also by reactions 3 and 5 to pyruvate, for amphibolic purposes and to generate TPNH and DPNH.

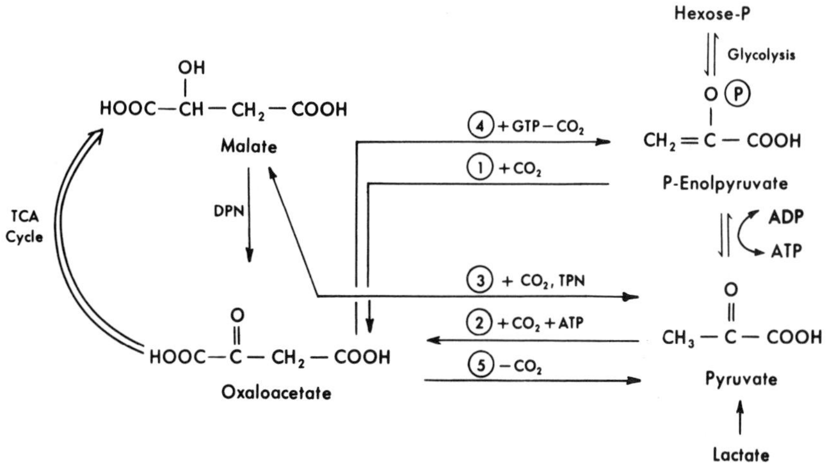

For example, the "clockwise" flow of the TCA cycle is favored by the release of acyl CoA energy in the reaction of the citrate-condensing enzyme:

Acetyl CoA + Oxaloacetate $\rightleftharpoons$ Citrate + CoA

In contrast, without the participation of CoA the equilibrium strongly favors cleavage; and organisms fermenting citrate start with this reaction, employing an inducible lyase:

$$
\underset{\text{Citrate}}{\begin{array}{c} CH_2-COOH \\ | \\ HO-C-COOH \\ | \\ CH_2-COOH \end{array}} \xrightarrow{\text{Citratase}} \underset{\text{Acetate}}{CH_3-COOH} \\ + \\ \underset{\text{Oxaloacetate}}{\begin{array}{c} O=C-COOH \\ | \\ CH_2-COOH \end{array}}
$$

The oxaloacetate can then be fermented by dismutation, some molecules being oxidized via pyruvate and others reduced to fumarate.

Many biosynthetic routes have mostly reversible reactions but flow in the desired direction because of a virtually irreversible reaction (usually ATP-linked) early in the sequence. This reaction functions rather like the initial, power-driven climb in a rollercoaster, followed by spontaneous fall past several peaks of decreasing energy level. It is therefore not surprising that pathways of degradation, which must flow downhill, overlap little with pathways of biosynthesis.

Flow in the desired direction is also promoted by release of pyrophosphate (PP) rather than P_i from ATP. The standard free-energy drop (i.e., at unit concentration of all components) is essentially the same for either of these hydrolyses of a P-O-P bond; but since the cell has a substantial P concentration, whereas PP is rapidly removed by a pyrophosphatase, the **actual** free-energy drop in the over-all reaction is greater for **PP** release. Since the PP is lost the cost is two high-energy bonds rather than one.

For example, in the biosynthetic use of ribose-5-P ATP donates PP to create an intermediate, 5-P-ribose-1-PP (PRPP), whose subsequent condensation is the reverse of a pyrophosphorolysis:*

$$PRPP + XNH_2 \rightleftharpoons XNH \cdot RP + PP_i$$

$$PP_i \longrightarrow 2\,P_i$$

---

* Unfortunately many enzymes that catalyze reversible reaction (e.g., glutamic dehydrogenase,

Though the R-1-PP is not a high-energy bond, the subsequent hydrolysis of the PP_i yields a large over-all ΔF.

REDOX COFACTORS: THE ROLES OF DPN AND TPN

Redox cofactors play a major role in directing metabolic flow: in order to ensure a smooth flow of energy and to promote economy, the cell contains several cofactors of different redox potential. In increasing order these are ferredoxin, lipoic acid, DPN and TPN, flavins, and cytochromes. (Flavins and hemes are firmly bound to proteins which influence their potentials.)

An example is the reduction of pyruvate to lactate and its reversal. The DPN-linked reduction, seen in the lactic fermentation (Ch. 3), is reversible in vitro; but the equilibrium strongly favors lactate.† To oxidize lactate cells employ a quite different, flavin-linked lactic oxidase, thus ensuring flow in the desired direction.

The necessity of ensuring a given direction of flow also explains the presence in the same cell of two pyridine nucleotides that are very similar in structure and redox potential. In a well-aerated culture in the steady state TPN is present predominantly as TPNH, generated primarily by the pentose phosphate pathway (Ch. 3) and by isocitrate dehydrogenase, and oxidized only by certain biosynthetic reactions. Though DPNH is reduced by many more reactions (glycolysis, pyruvate oxidation, and most TCA cycle dehydrogenations), it is present predominantly in the oxidized form because it is directly linked to the powerful oxidizing system of electron transport. Hence the cell can simultaneously carry out, in the same pool, DPN-linked oxidative biosynthetic reactions (e.g., histidinol $\rightarrow$ histidine) and

---

pyrophosphorylases) are named for what we now know to be the reverse of their physiological reactions.

† At 37° and pH 7.0 the reaction pyruvate + DPNH $\rightleftharpoons$ lactate + DPN has a $-\Delta F$ of 5000 cal, corresponding ($\Delta F = RT\ln K$) to an equilibrium constant, K, of about 10^4 [K = (lac) (DPN)/(pyr) (DPNH)]. Hence at a DPN/DPNH ratio of 1.0 the equilibrium lac/pyr ratio would be about 10,000. Though the DPN/DPNH ratio in aerobic cells exceeds 1.0, it is not high enough to ensure a good bow over this energy barrier.

TPN-linked reductive reactions (e.g., α-keto-glutarate $\rightarrow$ glutamate; acetate $\rightarrow$ fatty acid).

It may be noted that plant photosynthesis, which releases O_2, uses TPNH for CO_2 assimilation, but bacteria, whose photosynthesis is anaerobic (Ch. 3), use DPNH.

The reactions that reduce TPN have a larger energy drop than most other dehydrogenations; hence they can establish a very low TPN/TPNH ratio, which would support a reductive biosynthetic

function. Furthermore, the flow through these reactions is not rigidly coupled to over-all energy metabolism but can be adjusted precisely to meet biosynthetic requirements; for though the cell requires the product of each of these reactions for other purposes, it can also make these products by alternative reactions. Thus in addition to the TPN-linked reactions pentose P can be synthesized (from hexose P) by a nonoxidative reaction employing transketolase (Fig. 4-20), and α-ketoglutarate by a DPN-linked isocitrate dehydrogenase.

AMINO ACIDS

We shall note selected reactions in amino acid biosynthesis to illustrate the variety of patterns that have evolved.

The **carbon skeletons** are derived from a variety of amphibolic intermediates. Nitrogen is usually provided in the medium as NH_3 or NO_3, or is derived by breakdown of nitrogenous compounds. Many bacteria can reduce NO_3 to NH_3 for biosynthetic purposes by a TPNH-linked sequence, which is entirely different from the larger-scale cytochrome-linked reduction of NO_3 in anaerobic respiration (Ch. 3).

Sulfate Reduction. The usual inorganic S source in bacteriological media is sulfate rather than sulfide. Most bacteria can form sulfide from sulfate on a small scale for biosynthetic purposes, apparently by the same pathway that is used on a large scale for energy production by *Desulfovibrio* (Table 3-2). In this pathway sulfate is activated by the use of two ATP molecules to yield 3'-phosphoadeno-sine-5'-phosphosulfate:

$$\text{Adenine-ribose-3'-} \textcircled{P} \text{-5'-} \textcircled{P} \text{-OSO}_3\text{H}$$

Enzyme and Pool Distribution. The enzymes of amino acid biosynthesis are found in the soluble fraction of the cell (in contrast to certain membrane-bound enzymes: Ch. 6). Nevertheless, their distribution in the cytoplasm is not necessarily uniform, for though exogenous and endogenous amino acids mix in a single metabolite pool it is not always homogeneous in the growing cell. Thus diamino-pimelate (DAP) is incorporated into a cytoplasmic precursor of the cell wall (Ch. 6) and is also converted into lysine (Fig. 4-14), and when differentially labeled exogenous and endogenous DAP are competing the former is preferentially used as a source of wall DAP and the latter as a source of lysine. It thus appears that one set of enzymes is more peripherally located than the other, resulting in a gradient of labeling of the pool from the periphery to the center of the cell.

THE GLUTAMATE FAMILY

At an adequate NH_3 concentration α-keto-glutarate is converted to **glutamate** and to **glutamine** by the following reactions:*

However, at low NH_3 concentration (i.e., that provided by N_2 fixation) energy from an extra ATP is used to help take up NH_3:

$$\text{Glutamate} + \text{ATP} \xrightarrow{NH_3} \text{Glutamine}$$

$$\text{Glutamine} + \alpha\text{-Ketoglutarate} \xrightarrow{TPNH} 2 \text{ Glutamate}$$

In most bacteria, as in higher organisms, glutamate serves as the source, by transamination via pyridoxal-P, of the α-amino group of all other amino acids and of N atoms in certain other cell components. The amide group of glutamine also provides N, at a higher energy level, in the biosynthesis of various compounds (purines, pyrimidines, carbamyl-P, histidine, tryptophan, aminosugars). Finally, glutamate provides the C skeleton as well as an N atom in **proline** and in **ornithine,** a precursor of **arginine.** Ornithine, directly or

* Curiously, in many bacilli glutamate is converted to glutamine, by transamination from asparagine, after attachment to the tRNA (Ch. 13) specific for glutamine, whereas in Enterobacteriaceae glutamine is formed first and then attached to the tRNA.

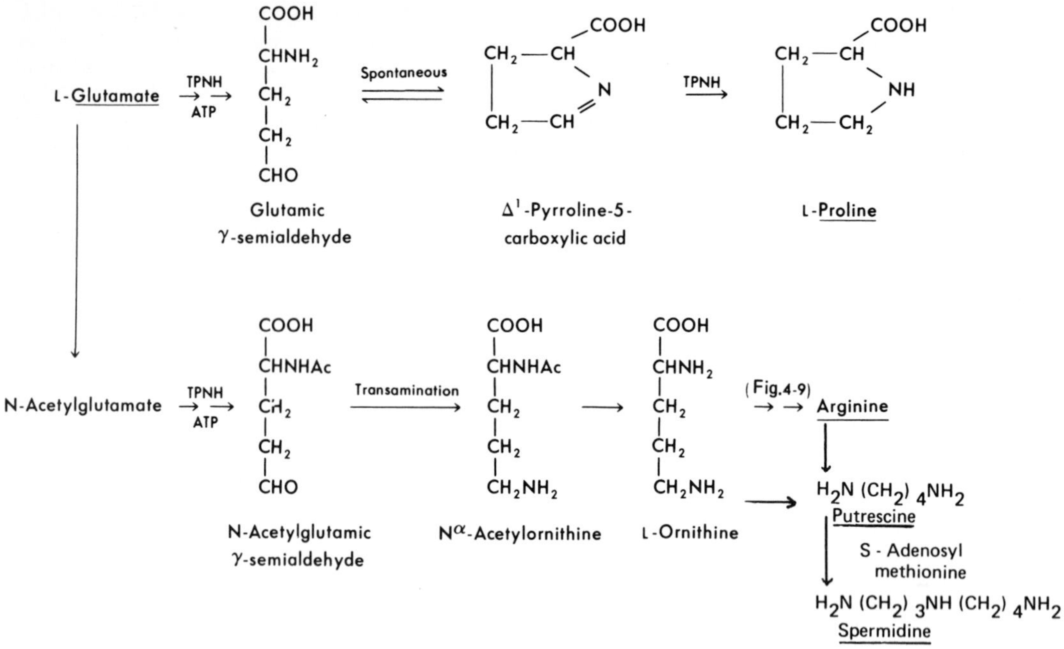

L-Glutamate $\xrightarrow[\text{ATP}]{\text{TPNH}}$

COOH
|
CHNH₂
|
CH₂
|
CH₂
|
CHO

Glutamic
γ-semialdehyde

$\underset{\text{Spontaneous}}{\rightleftharpoons}$

Δ¹-Pyrroline-5-
carboxylic acid

$\xrightarrow{\text{TPNH}}$

L-Proline

N-Acetylglutamate $\xrightarrow[\text{ATP}]{\text{TPNH}}$

COOH
|
CHNHAc
|
CH₂
|
CH₂
|
CHO

N-Acetylglutamic
γ-semialdehyde

$\xrightarrow{\text{Transamination}}$

COOH
|
CHNHAc
|
CH₂
|
CH₂
|
CH₂NH₂

Nᵅ-Acetylornithine

$\longrightarrow$

COOH
|
CHNH₂
|
CH₂
|
CH₂
|
CH₂NH₂

L-Ornithine

(Fig.4-9) → → Arginine

$\downarrow$

H₂N (CH₂)₄NH₂
Putrescine

S - Adenosyl
methionine

H₂N (CH₂)₃NH (CH₂)₄NH₂
Spermidine

Fig. 4-8. Biosynthesis of glutamate derivatives.

via arginine, also gives rise to part of the **polyamines** (see below). These pathways are outlined in Figure 4-8.

In **proline biosynthesis** a γ-semialdehyde spontaneously cyclizes with the NH₂ group, as in the formation of the O-containing furanose ring in sugars. In **arginine biosynthesis** this cyclization would interfere with the reduction of the open chain, and it is prevented by first blocking the α-NH₂ group of glutamate with an acetyl group, just

as one would do in a laboratory synthesis. The later reactions of arginine synthesis (Fig. 4-9) involve incorporation of CO_2 and NH_3 via carbamyl-P, followed by condensation with aspartate as a means of transferring its amino group.

The arginine pathway has been extensively used to study regulatory mechanisms. In mammalian liver the additional presence of arginase, which hydrolyzes arginine to ornithine and urea, yields the cycle for urea formation, which was discovered long before the arginine pathway.

Fig. 4-9. Biosynthesis of arginine from ornithine.

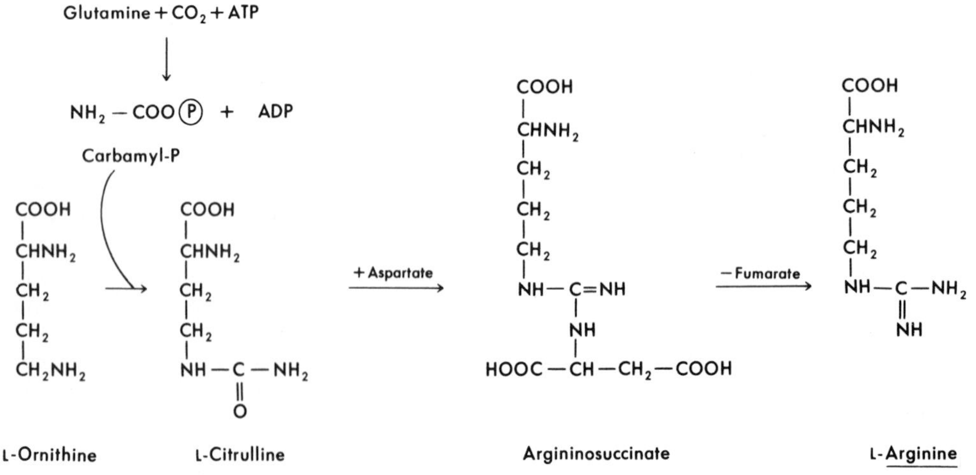

Glutamine + CO₂ + ATP

$\downarrow$

NH₂ — COO Ⓟ + ADP

Carbamyl-P

COOH
|
CHNH₂
|
CH₂
|
CH₂
|
CH₂NH₂

L-Ornithine

$\longrightarrow$

COOH
|
CHNH₂
|
CH₂
|
CH₂
|
NH — C — NH₂
‖
O

L-Citrulline

$\xrightarrow{+ \text{Aspartate}}$

COOH
|
CHNH₂
|
CH₂
|
CH₂
|
CH₂
|
NH — C=NH
|
NH
|
HOOC — CH — CH₂ — COOH

Argininosuccinate

$\xrightarrow{- \text{Fumarate}}$

COOH
|
CHNH₂
|
CH₂
|
CH₂
|
CH₂
|
NH — C — NH₂
‖
NH

L-Arginine

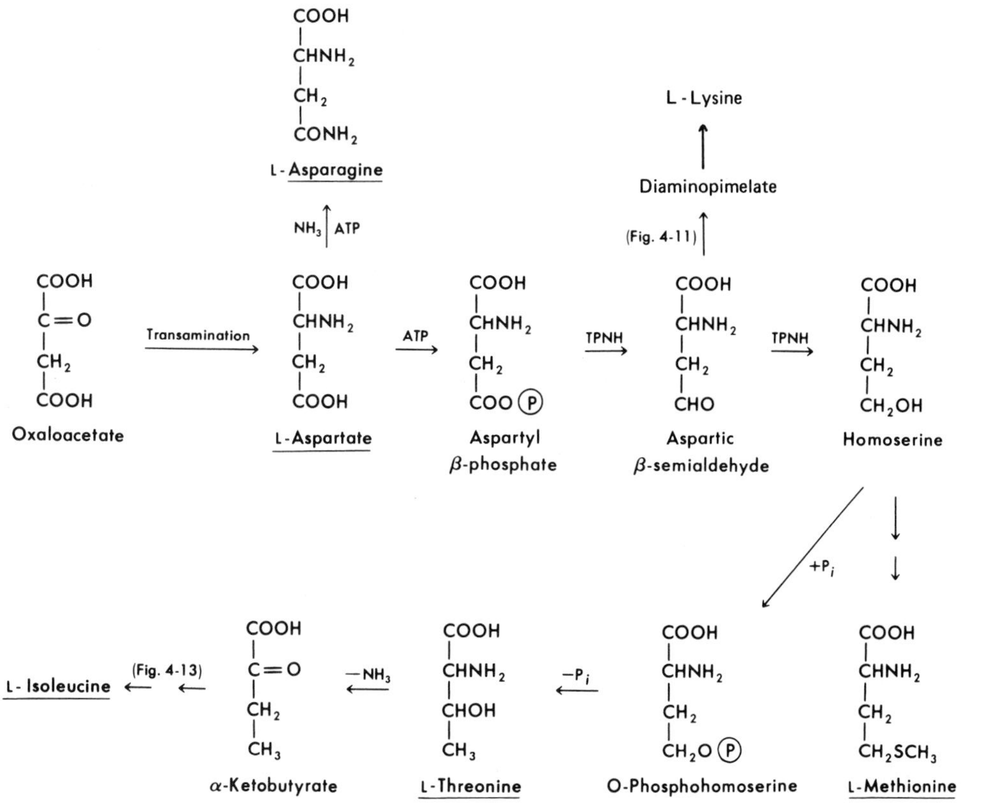

Fig. 4-10. Biosynthesis of the aspartate family.

THE ASPARTATE FAMILY

Aspartate is synthesized by transamination of oxaloacetate. Like glutamate, it is converted (via aspartyl adenylate) to an amide, **asparagine,** and reduced (via aspartyl-P) to a semialdehyde, whose further conversions are presented in Figure 4-10. Aspartate is also a direct precursor of the pyrimidine ring (see below).

Phosphorylation of the carboxyl, to make its reduction energetically possible, resembles the reduction of P-glycerate to P-glyceraldehyde in the reversal of glycolysis. An unusual feature of this pathway is the use of one natural amino acid (threonine) as a direct precursor of another (isoleucine).

Aspartate is also a precursor, together with a glycolytic intermediate, of nicotinamide (a component of DPN) in bacteria. Moreover, its amino group is the source of N atoms in the purine ring, the amino group of adenine, and the guanidino group of arginine.

Methionine derives its 4-C chain from aspartate, and the S from cysteine (see below), via the following reactions:

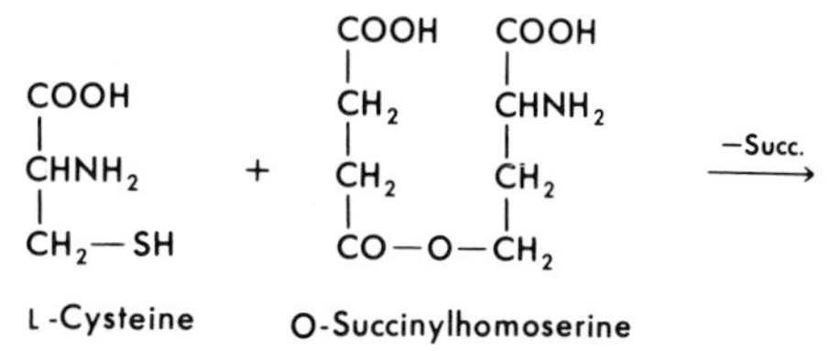

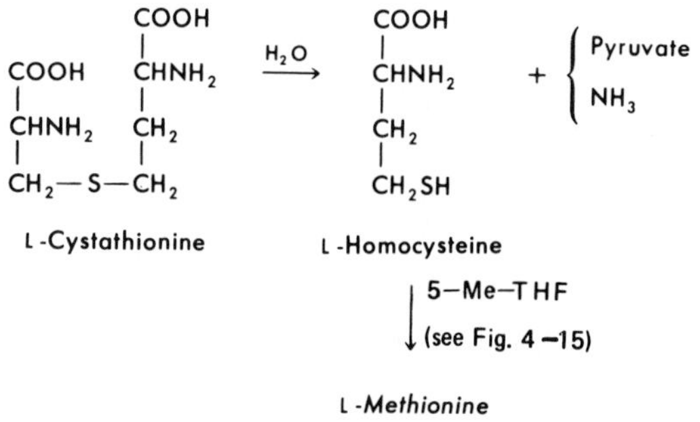

The intermediate preceding the thioether cystathionine is O-succinyl homoserine in Enterobacteriaceae, as depicted, but O-acetyl homoserine in bacilli. The methyl group is derived from 5-Me-THF (Fig. 4-15) in *E. coli* but via cobamide (a derivative of vitamin B_{12}) in the closely related *Aerobacter*. Remarkably, the former organism does not make or require B_{12} but has the enzymes for using the B_{12}-linked reaction as an alternative: a mutant blocked in the normal methylation reaction can grow when given either methionine or B_{12}.

Methionine links many processes in metabolism. In **protein synthesis** (Ch. 12) formyl-methionyl-tRNA participates in the initiation step, which regulates the rate. **S-adenosyl methionine**, a derivative in which the energy of the $S-CH_3$ bond is increased by conversion to the sulfonium (R_3S+) ion, provides the methyl group for a variety of **methylations**, including minor bases in DNA and RNA, the cyclopropane ring in some fatty acids, and cobamide. S-adenosyl methionine also provides the $-(CH_2)_3NH_2$ portion of **spermidine** (Fig. 4-8).

Diaminopimelate (DAP) synthesis (Fig. 4-11) begins with condensation of aspartic semialdehyde with pyruvate; a succinyl or acetyl group is temporarily attached to prevent cyclization. After suitable transformations removal of the acyl group yields L,L-DAP. A few bacterial species incorporate this compound into wall, but more use its derivative, the symmetrical, optically inactive *meso*-DAP, which contains a D-amino group (Ch. 6).

L-**Lysine** is synthesized by decarboxylation of *meso*-DAP. A mutant blocked before DAP cannot grow, for lack of both DAP and lysine; but if given lysine it will lyse, because it continues to synthesize protein without adequate wall. (Lysine was named, however, 60 years before the discovery of this property.)

The DAP pathway also yields **dipicolinate** (Fig. 4-11), a prominent constituent of bacterial spores (Ch. 6).

Lysine is the only amino acid for which **quite different biosynthetic routes** have been found in different organisms. In fungi, whose walls do not contain DAP, lysine is synthesized by a route involving the 6-C α-aminoadipate (Fig. 4-12), rather than the 7-C DAP (Fig. 4-11). Curiously, the DAP pathway is found in higher plants.

THE PYRUVATE FAMILY: ALIPHATIC AMINO ACIDS

Alanine is usually derived, like aspartate, by transamination of the corresponding α-keto

Fig. 4-11. Biosynthesis of diaminopimelate and lysine in *E. coli*.

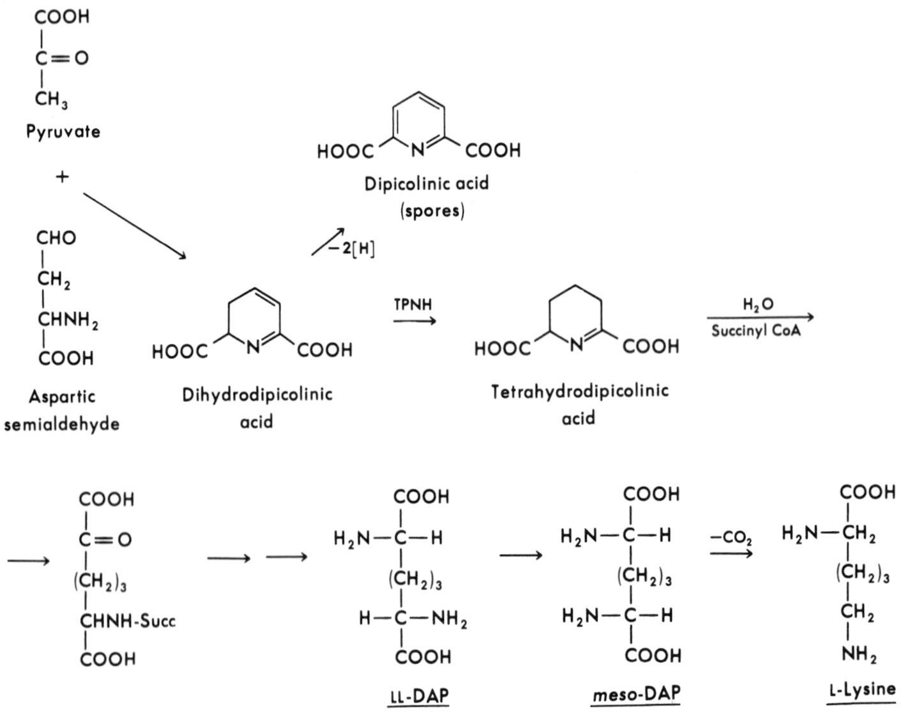

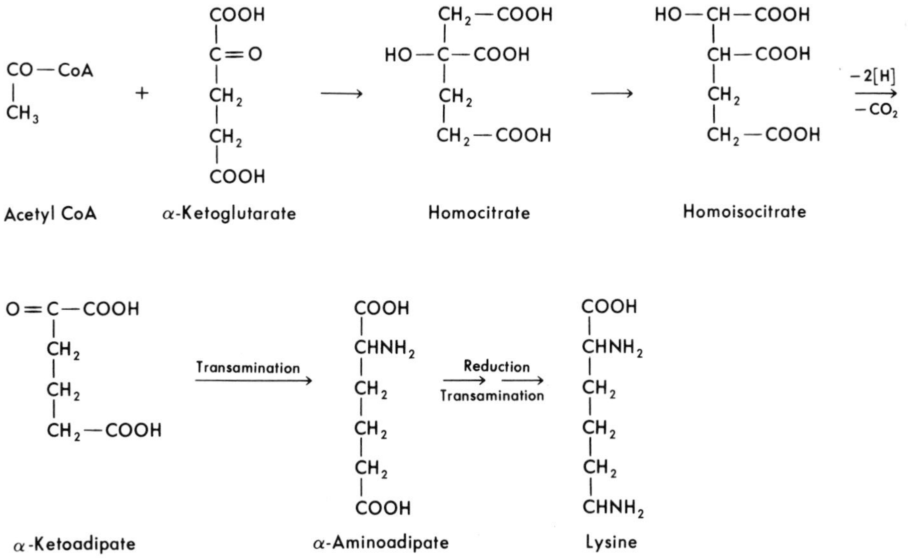

$$
\begin{array}{c}
\underset{\text{Acetyl CoA}}{\begin{array}{c}\text{CO}-\text{CoA}\\ |\\ \text{CH}_3\end{array}}
\;+\;
\underset{\alpha\text{-Ketoglutarate}}{\begin{array}{c}\text{COOH}\\ |\\ \text{C}=\text{O}\\ |\\ \text{CH}_2\\ |\\ \text{CH}_2\\ |\\ \text{COOH}\end{array}}
\;\longrightarrow\;
\underset{\text{Homocitrate}}{\begin{array}{c}\text{CH}_2-\text{COOH}\\ |\\ \text{HO}-\text{C}-\text{COOH}\\ |\\ \text{CH}_2\\ |\\ \text{CH}_2-\text{COOH}\end{array}}
\;\longrightarrow\;
\underset{\text{Homoisocitrate}}{\begin{array}{c}\text{HO}-\text{CH}-\text{COOH}\\ |\\ \text{CH}-\text{COOH}\\ |\\ \text{CH}_2\\ |\\ \text{CH}_2-\text{COOH}\end{array}}
\;\xrightarrow[-\text{CO}_2]{-2[\text{H}]}
$$

$$
\underset{\alpha\text{-Ketoadipate}}{\begin{array}{c}\text{O}=\text{C}-\text{COOH}\\ |\\ \text{CH}_2\\ |\\ \text{CH}_2\\ |\\ \text{CH}_2-\text{COOH}\end{array}}
\;\xrightarrow{\text{Transamination}}\;
\underset{\alpha\text{-Aminoadipate}}{\begin{array}{c}\text{COOH}\\ |\\ \text{CHNH}_2\\ |\\ \text{CH}_2\\ |\\ \text{CH}_2\\ |\\ \text{CH}_2\\ |\\ \text{COOH}\end{array}}
\;\xrightarrow[\text{Transamination}]{\text{Reduction}}\;
\underset{\text{Lysine}}{\begin{array}{c}\text{COOH}\\ |\\ \text{CHNH}_2\\ |\\ \text{CH}_2\\ |\\ \text{CH}_2\\ |\\ \text{CH}_2\\ |\\ \text{CHNH}_2\end{array}}
$$

Fig. 4-12. Biosynthesis of lysine in fungi. Note the precise homology between this pathway and that to glutamate (Fig. 4-3), the 5-C, lower homolog of α-aminoadipate.

acid, pyruvate.* Pyruvate also serves as the starting point for formation of the longer-chain aliphatic amino acids: **isoleucine, valine,** and **leucine.** These pathways are unusual in having a single set of enzymes catalyze two parallel sequences of reactions, whose substrates differ by a -CH$_2$- group (Fig. 4-13A).

Acetolacetate, a precursor of valine, is also an intermediate in the butylene glycol fermentation of *Aerobacter* (Ch. 3, Fig. 3-2). The regulatory roles of the biosynthetic and the degradative acetolactate synthetase will be discussed in Chapter 13 (Isozymes).

Leucine formation involves elongation of the chain of valine by 1 C, through addition of an acetyl group to "ketovaline," followed by oxidative decarboxylation (Fig. 4-13B). This series of reactions, like the formation of lysine in plants (Fig. 4-12), is precisely analogous to a familiar series in the TCA cycle, which converts oxalo-acetate to its higher homolog, α-ketoglutarate (Fig. 4-6).

The C skeleton of valine also takes up a hydroxymethyl group, via tetrahydrofolate, to yield **pantoate** (Fig. 4-13B). This compound combines with derivatives of other amino acids to form the coenzyme of acylation, **coenzyme A** (CoA) (Fig. 4-14).

_____

*Some members of the genus *Bacillus* form **alanine,** instead of glutamate, by reductive amination; it then serves as the source of the amino group in **transamination.**

THE SERINE FAMILY: 1-C FRAGMENTS

Serine is derived from 3-P-glycerate as follows:

$$
\underset{\text{3-P-Glycerate}}{\begin{array}{c}\text{COOH}\\ |\\ \text{CH}_2\text{OH}\\ |\\ \text{CH}_2\text{O}\;\circled{P}\end{array}}
\;\xrightarrow{\text{DPN}}\;
\underset{\text{3-P-Hydroxypyruvate}}{\begin{array}{c}\text{COOH}\\ |\\ \text{C}=\text{O}\\ |\\ \text{CH}_2\text{O}\;\circled{P}\end{array}}
\;\xrightarrow{\text{Transamination}}
$$

$$
\underset{\text{3-P-Serine}}{\begin{array}{c}\text{COOH}\\ |\\ \text{CHNH}_2\\ |\\ \text{CH}_2\text{O}\;\circled{P}\end{array}}
\;\xrightarrow{-\circled{P}}\;
\underset{\text{L-Serine}}{\begin{array}{c}\text{COOH}\\ |\\ \text{CHNH}_2\\ |\\ \text{CH}_2\text{OH}\end{array}}
$$

Serine can then give rise to **glycine** plus a "1-C" fragment, and the glycine can be further converted to a second such fragment, by the following two reactions:

$$
\underset{\text{L-Serine}}{\begin{array}{c}\text{COOH}\\ |\\ \text{CHNH}_2\\ |\\ \text{CH}_2\text{OH}\end{array}}
\;\underset{\substack{\text{(Tetrahydro-}\\ \text{folate)}}}{\overset{\text{THF}}{\rightleftharpoons}}\;
\begin{array}{c}\underset{\text{Glycine}}{\begin{array}{c}\text{COOH}\\ |\\ \text{CH}_2\text{NH}_2\end{array}}\\ +\\ \underset{\text{Hydroxymethyl THF}}{\text{HOCH}_2\text{-THF}}\end{array}
\qquad (1)
$$

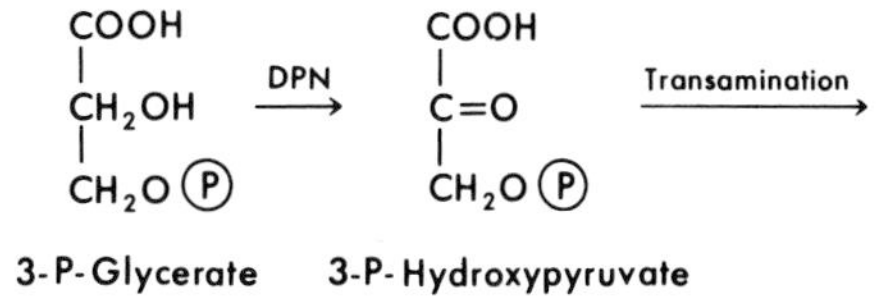

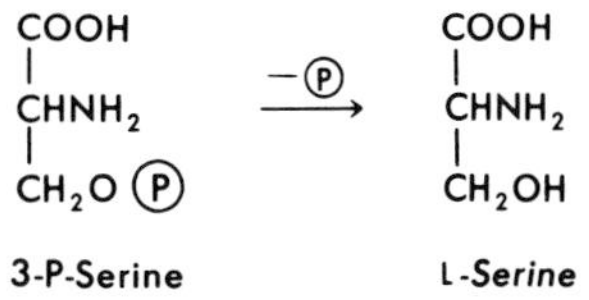

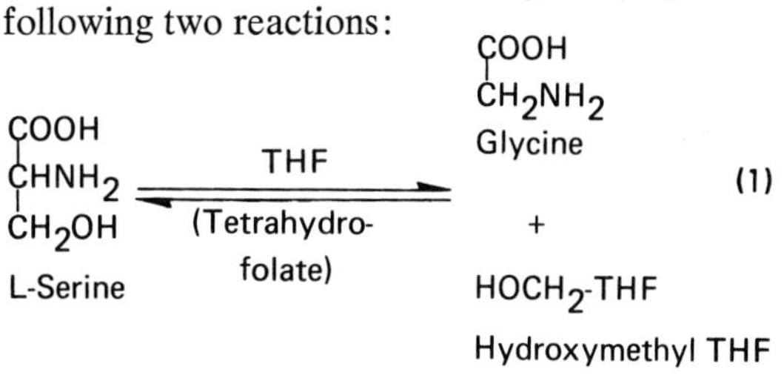

$$
\text{Glycine} + \text{THF} \rightleftharpoons \text{HOCH}_2\text{-THF} + \text{CO}_2 + \text{NH}_3 \qquad (2)
$$

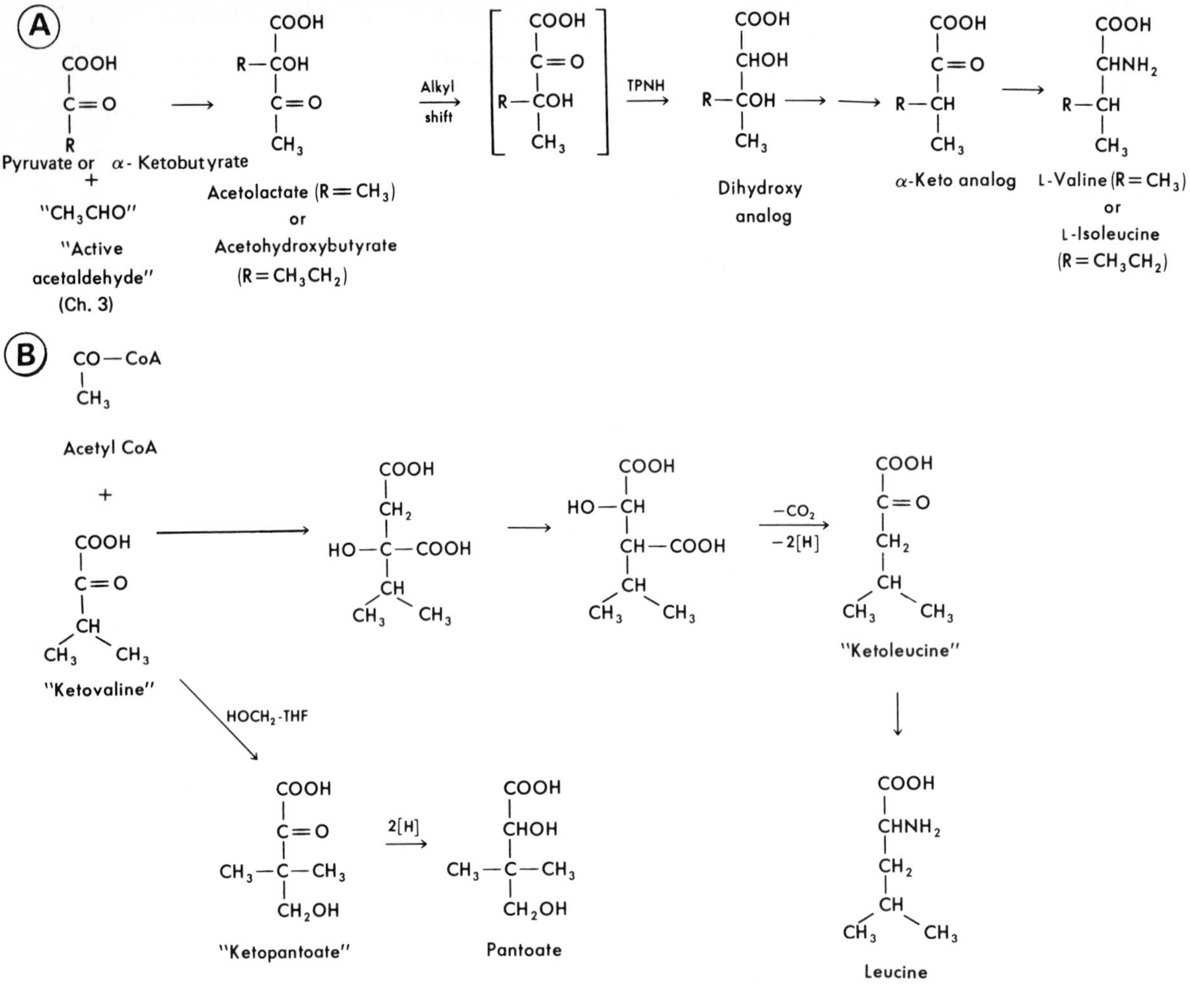

COOH
|
C=O
|
R
Pyruvate or
+
"CH$_3$CHO"
"Active
acetaldehyde"
(Ch. 3)

α-Ketobutyrate

COOH
|
R—COH
|
C=O
|
CH$_3$

Acetolactate (R=CH$_3$)
or
Acetohydroxybutyrate
(R=CH$_3$CH$_2$)

Alkyl shift →

[COOH
|
C=O
|
R—COH
|
CH$_3$]

TPNH →

COOH
|
CHOH
|
R—COH
|
CH$_3$

Dihydroxy analog

→

COOH
|
C=O
|
R—CH
|
CH$_3$

α-Keto analog

→

COOH
|
CHNH$_2$
|
R—CH
|
CH$_3$

L-Valine (R=CH$_3$)
or
L-Isoleucine
(R=CH$_3$CH$_2$)

B

CO—CoA
|
CH$_3$

Acetyl CoA

+

COOH
|
C=O
|
CH
/ \
CH$_3$ CH$_3$

"Ketovaline"

→

COOH
|
CH$_2$
|
HO—C—COOH
|
CH
/ \
CH$_3$ CH$_3$

→

COOH
|
HO—CH
|
CH—COOH
|
CH
/ \
CH$_3$ CH$_3$

$\xrightarrow[-2[H]]{-CO_2}$

COOH
|
C=O
|
CH$_2$
|
CH
/ \
CH$_3$ CH$_3$

"Ketoleucine"

↓

COOH
|
CHNH$_2$
|
CH$_2$
|
CH
/ \
CH$_3$ CH$_3$

Leucine

HOCH$_2$-THF ↘

COOH
|
C=O
|
CH$_3$—C—CH$_3$
|
CH$_2$OH

"Ketopantoate"

$\xrightarrow{2[H]}$

COOH
|
CHOH
|
CH$_3$—C—CH$_3$
|
CH$_2$OH

Pantoate

Fig. 4-13. Biosynthesis of **(A)** isoleucine and valine and **(B)** leucine and pantoate.

Serine thus yields different proportions of glycine and 1-C fragments, depending on the cell's needs. Moreover, since these reactions are reversible a cell can use them to form serine from exogenous glycine. Serine also provides the 3-C side chain of **tryptophan** (see below).

Cysteine is formed from serine, by the following reactions:

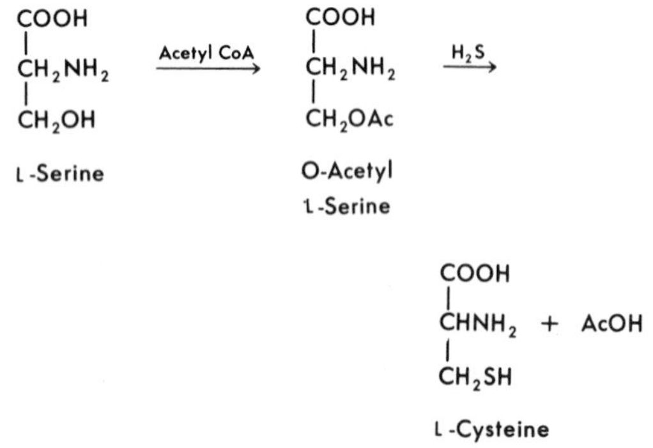

COOH
|
CH$_2$NH$_2$
|
CH$_2$OH

L-Serine

$\xrightarrow{Acetyl\ CoA}$

COOH
|
CH$_2$NH$_2$
|
CH$_2$OAc

O-Acetyl
L-Serine

$\xrightarrow{H_2S}$

COOH
|
CHNH$_2$ + AcOH
|
CH$_2$SH

L-Cysteine

1-C TRANSFER

The l-C fragment derived from serine or glycine, attached to THF, is used in a variety of biosynthetic reactions. The initial hydroxymethyl THF further condenses to form 5,10-methylene THF, in which the fragment remains at the aldehyde level of oxidation. Conversions to the methyl or the formyl level are outlined in Figure 4-15, which also summarizes the biosynthetic reactions that take up the fragment. Inhibition of THF synthesis (by sulfonamides) or function (by trimethoprim) are important mechanisms of antimetabolite action.

HISTIDINE

The biosynthesis of histidine (Fig. 4-16) from the 5-C chain of ribose-5-P has unique features. Whereas the carboxyl of all other amino acids is inherited from an amphibolic precursor, that of histidine is formed by the oxidation of an intermediate aminoalcohol (histidinol). Moreover, in this sequence ATP donates not only energy but also its N^2-C^2 fragment to form (with the addition of NH_3) the imidazole ring.

The residue of the ATP is the nucleotide of 5-aminoimidazole carboxamide, which is a normal intermediate in purine biosynthesis and hence is reconverted to ATP. Thus a cycle, grafted onto the pathway of purine biosynthesis, serves for the generation of an N-C portion of histidine.

AROMATIC COMPOUNDS

The key to this pathway was the finding that certain mutants of *E. coli* could use **shikimic acid** (Fig. 4-17), a hydroaromatic compound, to replace a mixture of aromatic compounds **(tyrosine, phenylalanine, tryptophan,** and cer-

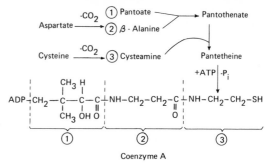

Fig. 4-14. Biosynthesis of pantetheine and coenzyme A.

tain trace substances) as a growth factor. The pathway is shown in Figure 4-17.

The same pathway is used in higher plants.* Since it yields the enormous amount of aromatic

---

* Many universal intermediates have first been isolated as compounds that accumulate, for obscure reasons, in various plants: e.g., shikimate in the shikimi tree, malate (L. malus = apple) in apples, citrate in citrus fruits.

Fig. 4-15. Reactions of tetrahydrofolate (THF) and its derivatives (DHF = dihydrofolate). In purine biosynthesis (Fig. 4-22) the formyl group from 5,10-methenyl THF closes the 5-membered ring (GAR → FGAR), while that from 10-formyl THF closes the 6-membered ring (AICAR → IMP). Formyl THF also contributes to the initiating tRNA in protein synthesis, formyl-Met-tRNA (Ch. 12). The more reduced derivatives of THF contribute the methyl groups of thymidylate (TMP), methionine, and a host of methylated derivatives (see text). THF also contributes 1-C fragments to pantoate (Fig. 4-14) and to thiamine.

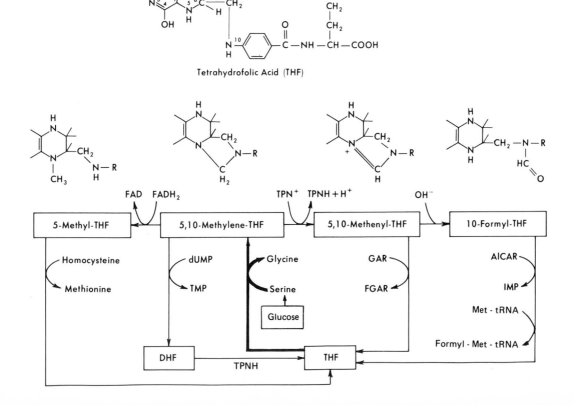

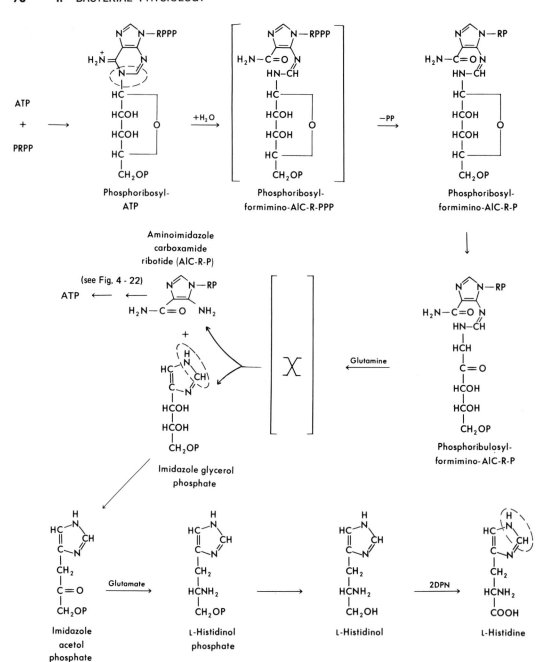

Fig. 4-16. Biosynthesis of histidine. The N-C fragment donated by ATP is circled by a dashed line. [After Ames, B. N., and Hartman, P. E. *Cold Spring Harbor Symp Quant Biol 28*:349 (1963).]

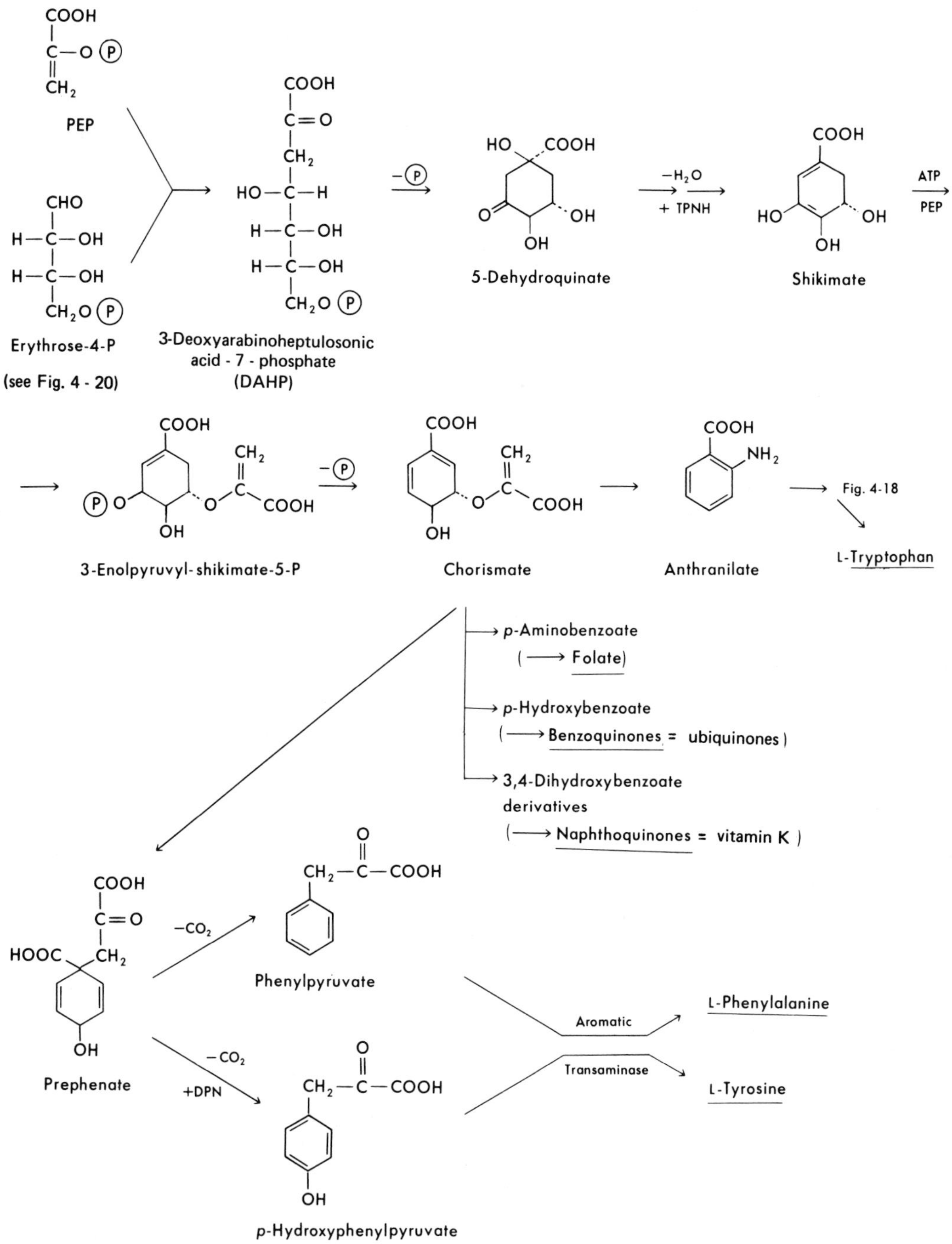

Fig. 4-17. The aromatic pathway.

material polymerized in lignin in woody plants, it has probably been the major source of the aromatic components of petroleum, and hence of the aromatic products of the chemical industry.

In this pathway aldol concentration of erythrose-P with P-enolpyruvate (PEP) yields a 7-C straight-chain compound that is readily **cyclized.** Further reactions (via shikimate), and addition of another PEP, yield a highly reactive, versatile compound, **chorismic acid** (Gr. = fork), which is the major branch point. In one branch a biochemically novel shift of the 3-C side chain, from an enol-ether to a C-C attachment, yields a compound that readily

aromatizes, with loss of the COOH group, to yield the C skeleton of tyrosine and phenylalanine. In other aromatization reactions the COOH is retained and the enolpyruvic chain is discarded; these reactions yield anthranilate (o-aminobenzoate) and other benzoic acid derivatives, including various quinones (Fig. 4-17).

Tryptophan synthesis from anthranilate is shown in Figure 4-18. The intermediate indoleglycerol-P exchanges its 3-C side chain for the 3-C chain of serine, to yield tryptophan in a single step. In histidine biosynthesis the same side chain, in imidazoleglycerol-P (Fig. 4-16), is retained and

Fig. 4-18. Biosynthesis of tryptophan, branching from the common aromatic pathway.

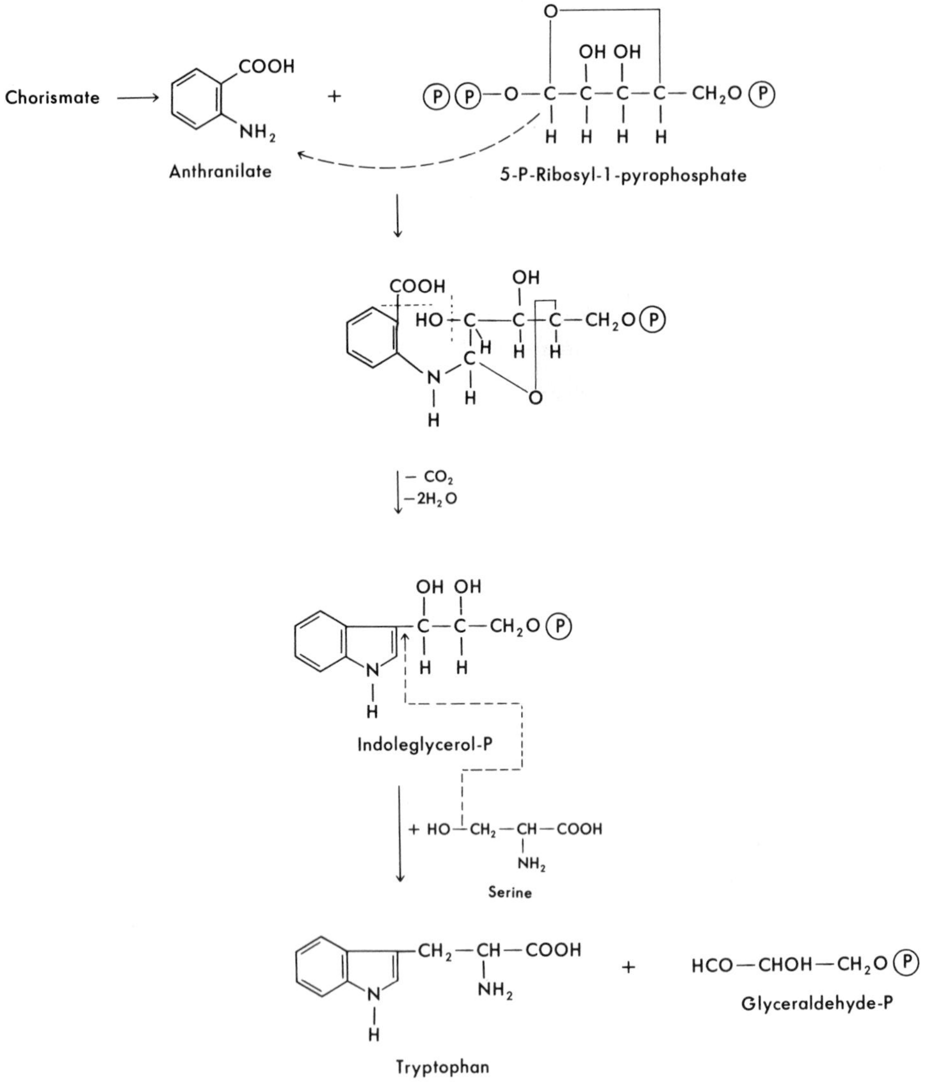

undergoes several further reactions. This short-cut in tryptophan biosynthesis is possible because the indole ring labilizes the side chain to electrophilic attack much more than does the imidazole ring. Mutations in the enzyme of this last step, investigated by Yanofsky, have contributed heavily to knowledge of gene-enzyme relations (Chs. 11 to 13).

In a rare exception to the unity of biochemistry, **nicotinamide** is made from tryptophan in fungi but from aspartate in bacteria.

D–AMINO ACIDS

D-Amino acids are present in bacteria in cell wall polypeptides, and sometimes in capsules and antibiotic products as well. Their synthesis involves a racemase that acts on alanine, plus a special transaminase that can transfer the D-amino group to various keto acids:

$$L\text{ - Alanine} \xrightleftharpoons[]{\text{Racemase}} D\text{ - Alanine}$$

$$D\text{ - Alanine} + \alpha\text{ - Ketoglutarate, etc.} \xrightleftharpoons[]{D\text{ - Transaminase}} D\text{ - Glutamate, etc.} + \text{Pyruvate}$$

POLYAMINES

Polyamines are not amino acids, but they are formed directly from ornithine and methionine (Fig. 4-8). Indeed, in *E. coli* the diamine **putrescine** and the triamine **spermidine** (Fig. 4-8), in about a 4:1 ratio, constitute about half the total product of the arginine pathway. **Spermine** (a tetramine), which is present in eukaryotes, has not been found in bacteria. However, exogenous spermine (which is present in yeast extract and meat extract) can be taken up by *E. coli* and then replaces the endogenous spermidine.

These polyamines bind readily to various nucleic acids in vitro, not only neutralizing the charge on these highly anionic polymers but also influencing their conformation through the formation of reversible, ionic cross-links. However, the functions of polyamines in the cell are as yet not clearly defined: they are readily displaced by other cations, and many of the same effects can be exerted on nucleic acids by an inorganic polyvalent cation, Mg^{++}, although greater concentrations are required.

The dimensions of putrescine are suitable for forming an ionic cross-link between opposite phosphates in a DNA or RNA double helix (Ch. 10), and the additional $-(CH_2)_3NH_2$ of spermidine can reach an adjacent phosphate and thereby increase the affinity. Hence these compounds increase the resistance of double-stranded nucleic acids to strand-separating agents and to enzymatic attack; they presumably also influence DNA functions in the cell. The bulk of the polyamine in cells, however, is probably bound to ribosomes, for the total content is almost proportional to the ribosome content of cells grown under various conditions. Moreover, ribosomes extracted from bacteria contain considerable spermidine; and though most kinds of ribosomes can function in vitro with adequate Mg^{++} without polyamine, much of the Mg^{++} may be replaced by a polyamine, with a resulting increase in activity.

Polyamines also bind to tRNA, where they probably strengthen the short double-stranded regions (Ch. 12, Fig. 12-1); they appear to exert a specific effect on conformation, for polyamines, unlike Mg^{++}, promote the correct pattern of methylation of tRNA precursors in vitro.

The best-defined function of polyamines is not in cells but in viruses: nearly half the charges on DNA in the T4 phage particle (Ch. 45) are neutralized by polyamines, whose cross-linking evidently promotes the required tight packing of the DNA in its coat.

Polyamines are required for, or promote, the growth of some bacteria (*Pasteurella, Mycoplasma*). However, this effect has not thrown light on their function in the cell, for they can be replaced by other, inorganic polycations, which evidently function from the outside to preserve the integrity of the cell membrane. Nevertheless, there is no doubt that polyamines play one or more essential roles in metabolism, for mutants of *E. coli* that require a polyamine have recently been isolated;* they should help to define the role of these compounds.

---

* Putrescine can be synthesized in *E. coli* via decarboxylation either of arginine or of its precursor ornithine (Fig. 4-8). The latter pathway is more economical; the former has evidently evolved so that the presence of arginine, which inhibits its own synthesis via ornithine (Ch. 13), will not inhibit polyamine synthesis. To isolate polyamine auxotrophs the presence of arginine was used to block the ornithine pathway functionally, so that a mutant blocked in arginine decarboxylase could be recognized by its requirement for putrescine in the presence of arginine. A second-step mutation, blocking ornithine decarboxylase, yielded a double mutant dependent on polyamines in any medium.

NUCLEOTIDES

Purine and pyrimidine nucleotides serve several functions: 1) building blocks of nucleic acids; 2) components of many coenzymes; 3) activators for the transfer and transformations of sugars, wall peptides, and complex lipids; 4) covalently added modifiers of enzymes; and 5) constituents of some antibiotics. The biosynthesis of the **purines** involves a sequence of ribose P derivatives that branches at hypoxanthine ribonucleotide to yield adenylate and guanylate; free bases are not formed. **Pyrimidine** nucleotides are similarly synthesized via a single pathway that adds ribose P, after completing the ring, and then branches at uridylate (UMP). These pathways were largely elucidated in animal tissues by the use of isotopic and enzymatic methods.

Many bacteria, like mammalian cells, can use exogenously supplied free purines and pyrimidines, or their nucleosides, as sources of nucleotides. The wide distribution of these "salvage" enzymes probably reflects the presence of nucleosides and free bases wherever cells are degraded, and their ready penetration (unlike the corresponding nucleotides) into cells. Two classes of reactions are known: 1) B + R − P → B − R (nucleoside), which is then phosphorylated; 2) B + PRPP → BRP (nucleotide).

The kinases that increase the level of phosphorylation of various nucleoside phosphates will not be discussed, except to note that the deoxyribonucleotide kinases in *E. coli,* which convert other deoxyribonucleotides to the triphosphates, do not act on dUMP. This specificity is important, for if dUTP were made it could be used by DNA polymerase (see below) in place of dTTP, forming a DNA of abnormal composition.

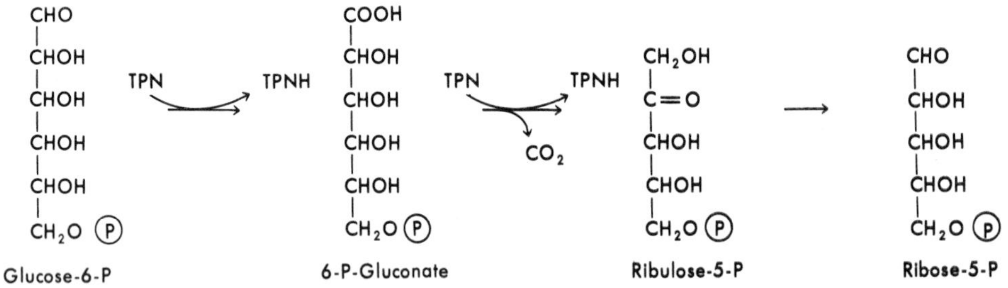

Glucose-6-P		6-P-Gluconate		Ribulose-5-P	Ribose-5-P

CHO   → TPN / TPNH →   COOH   → TPN / TPNH, CO_2 → CH_2OH → CHO

Fig. 4-19. Oxidative biosynthesis of ribose-5-P.

Fig. 4-20. Nonoxidative biosynthesis of ribose-5-P. Transketolase transfers the first two C atoms from 2-keto sugars, as a CH_2OH-CO group, and transaldolase transfers the first three C atoms from the same donors, as a CH_2OH-CO-CHOH group. The recipient of either is the aldehyde of an aldo sugar, and the condensation product is a 2-keto sugar. The C-C cleavage adjacent to a ketone (by transketolase) involves thiamine pyrophosphate, just as does α-keto acid decarboxylation (Ch. 3).

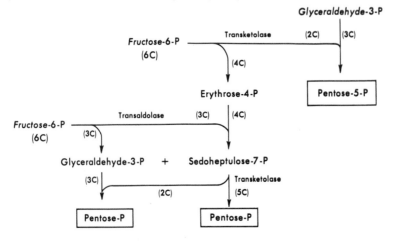

PENTOSES

Ribose-5-P can be formed in *E. coli* both oxidatively, from glucose-6-P (Fig. 4-19), and nonoxidatively, by the transfer of 2-C and 3-C groups (Fig. 4-20). Since the first pathway generates TPNH while the second does not, the presence of both permits the supply of this biosynthetic reductant (see above, Role of DPN and TPN) to be adjusted to the cell's needs under various conditions. Ribose-5-P enters nucleotide biosynthesis (and other biosyntheses) from its 1-pyrophosphate derivative, PRPP. **Deoxyribose** residues are formed by TPNH-linked reduction of the ribose of any ribonucleotide.

The substrates for the reduction are the nucleoside **triphosphates** in *Lactobacillus* and the **diphosphates** in *E. coli*. The reduction involves a derivative of vitamin B_{12} in the former organism, whose normal habitat (milk) contains that vitamin, while *E. coli*, which often grows in a simple environment, has evolved quite a different pathway that uses instead a novel small-protein factor, **thioredoxin**. In this compound a pair of -SH groups on adjacent cysteines undergo reversible oxidation to an -S-S group, as in the function of lipoate (Ch. 3).

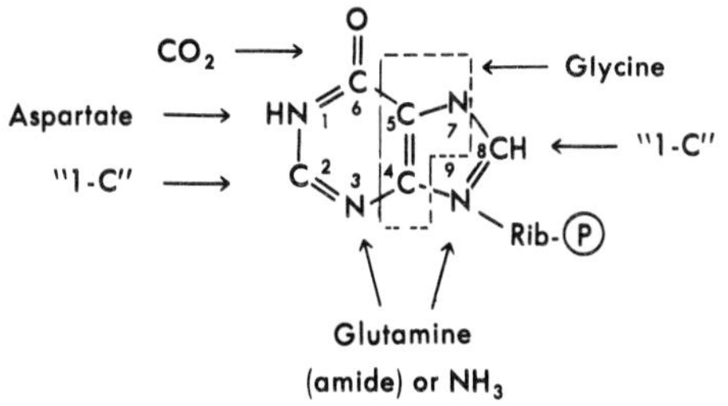

Fig. 4-21. Origin of purine ring atoms. The origin of C-2 and C-8 from 1-C fragments was first indicated through the incorporation of isotopically labeled formate, in a mammalian system. However, the condensation of formate with tetrahydrofolate is not a physiological process; even in those bacteria whose fermentations yield formate, the biosynthetic formyl group is derived from serine.

PURINES

The purine ring is built up from PRPP, glycine, and single C and N additions; the origin of its atoms is summarized in Figure 4-21, and the major reactions are presented in Figure 4-22. This sequence is highly endergonic, requiring at least $3 \sim P$.

Fig. 4-22. Biosynthesis of purine nucleotides.

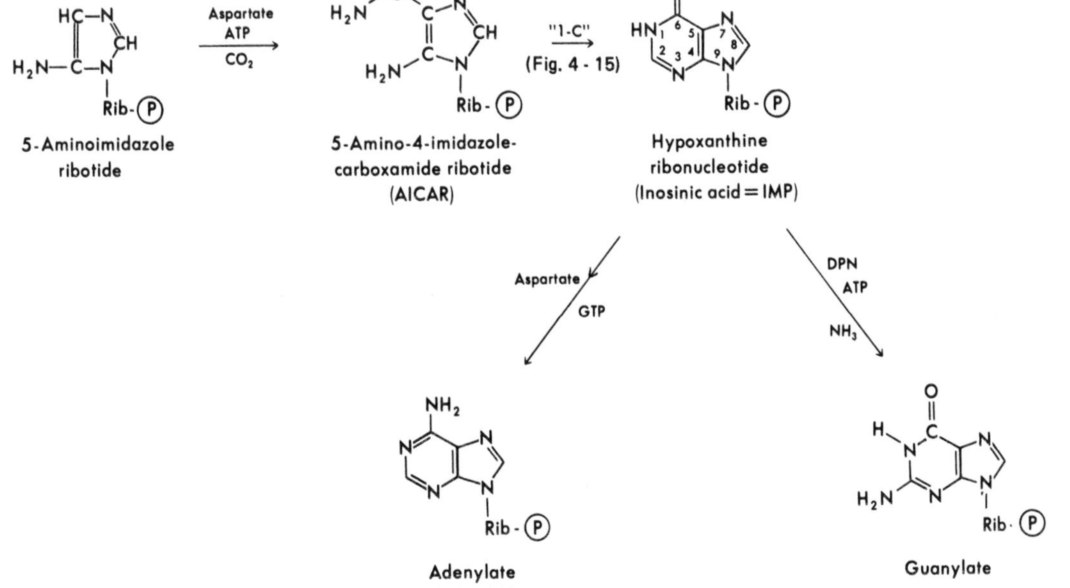

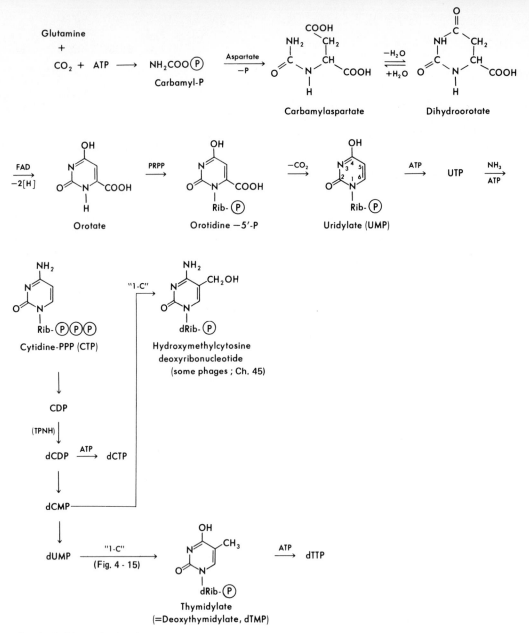

Fig. 4-23. Biosynthesis of pyrimidine nucleotides.

Adenylate (6-aminopurine nucleotide) is formed from IMP; the amino group is transferred from aspartate via a stable condensation product, just as in the formation of argininosuccinate in arginine biosynthesis (Fig. 4-9). Furthermore, the energy source for the condensation is GTP rather than ATP, which cleverly prevents any deficiency of adenylate from interfering with its own replenishment. The same condensing enzyme performs a similar reaction in the formation of an earlier intermediate (the carboxamide) in the same pathway. **Guanylate** is also formed from IMP: after oxidation by DPN to xanthylate (2,6-diketopurine nucleotide)

the 2-keto group is replaced by an amino group.

As was noted above, histidine biosynthesis involves a purine cycle: ATP donates its N^1-C^2 fragment (Fig. 4-16), forming as a byproduct a normal intermediate in purine biosynthesis (AICAR, Fig. 4-22).

PYRIMIDINES

The pathway to the pyrimidine ring (Fig. 4-23) is simpler than that to the purines. The transfer of a carbamyl group from carbamyl phosphate (see citrulline formation, Fig. 4-9) to aspartate leads to formation of the pyrimidine ring, in orotate. Addition of ribose-P from PRPP and decarboxylation yields UMP, which then gives rise to the other pyrimidine nucleotides.

Bacteria use NH_3 (and $\sim P$ energy) in the amination reactions that convert UTP to CTP and xanthylate to GMP, but animals use the amide of glutamine instead. This difference evidently reflects the prevalence of ammonia in the habitats of bacteria, contrasted with its toxicity to animals cells.

OTHER PATHWAYS

Since **lipids** are present only in membranes in bacteria their synthesis will be presented in correlation with their structure and function, in Chapter 6.

SUGARS

Though sugars are fuels, they are **not** metabolic intermediates within the bacterial (or the mammalian) cell: they are phosphorylated during (Ch. 6) or immediately after entry. Conversion into other sugars sometimes takes place at this phosphorylated level (cf. pentoses, Fig. 4-20; fructose, Fig. 3-1). However, most conversions of sugars occur at the level of sugar nucleotides (nucleoside diphosphate sugars), which were discovered by Leloir. These are formed by specific pyrophosphorylases:*

$$\text{Nucleoside-P-P-P} + \text{Hexose-1-P} \rightleftharpoons$$

$$\text{Nucleoside-P-P-hexose} + \text{P-P}$$

The utilization of such a nucleoside disphosphate "handle" in the conversion of glucose to galactose (and vice versa) is depicted in Figure 4-24; more extensive conversions include amination, reduction to deoxysugars, or oxidation to uronic acids. Nucleotide sugars are also the donors in transglycosylation reactions in the formation of glycogen in animals, starch in plants, and wall polysaccharides in bacteria (Ch. 6).

Intermediates in these pathways, as in amino acid biosynthesis, have often been discovered

through their accumulation by mutants blocked in the synthesis or transfer of a particular sugar. Indeed, such mutants often accumulate not only the precursor of the blocked reaction but also nucleotides of various other sugars, for the absence of one component prevents incorporation of any others into a repeating polysaccharide. Figure 4-25 outlines some pathways that have been worked out in this way.

Over 60 nucleotide sugars have now been isolated. Some of the known wall precursors are the following:

UDP-D-glucose (G)
UDP-D-acetyl glucosamine (GNAc)
UDP-D-galactose (Gal)
UDP-D-galacturonate
GDP-D-mannose
GDP-L-fucose (6-deoxy-L-galactose)
GDP-D-rhamnose (6-deoxy-D-mannose)
GDP-colitose (3,6-dideoxy-L-galactose)
GDP-D-glycero-D-mannoheptose
dTDP-D-galactose
dTDP-L-rhamnose
dTDP-D-fucose
dTDP-6-deoxy-D-glucose
dTDP-GNAc
dTDP-GalNAc
CDP-tyvelose (3,6-dideoxy-D-mannose)
CDP-abequose (3,6-dideoxy-D-galactose)
CDP-glycerol ⎫ Reduced sugar (polyol) pre-
CDP-ribitol ⎭ cursors of teichoic acids

The use of different nucleotide carriers may improve the specificity of the enzymes that incorporate sugars into polymers. It may also decrease the interference between pathways to different sugars. Such interference is suggested by the presence of D-mannose or L-fucose, but never both, in various capsular polysaccharides of *Aerobacter*. Since L-fucose is derived (as GDP-fucose) from GDP-mannose, the presence of the enzymes for its formation may depress the level of GDP-mannose in the cell enough to prevent mannose incorporation.

POLYSACCHARIDES

Capsules. Most bacterial capsules are repeating **heteropolymers** (i.e., contain more than

---

* The reaction, though reversible, normally flows in the direction opposite to pyrophosphorolysis; pyrophosphate is released from the triphosphate, whose remaining phosphate condenses with that of the hexose-P.

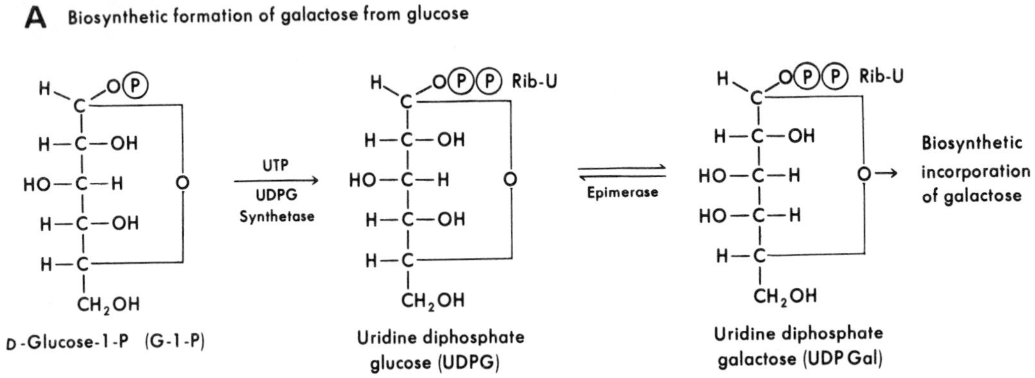

A   Biosynthetic formation of galactose from glucose

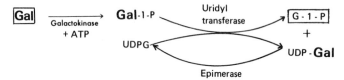

B   Utilization of galactose (Gal) as carbon source

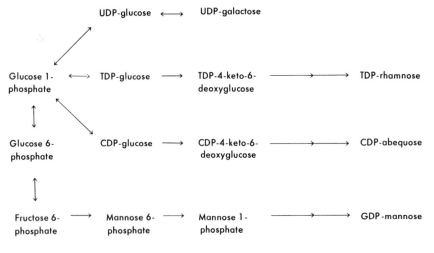

Fig. 4-24. Epimerization of galactose and glucose. The anabolic sequence from glucose to galactose (A) evidently employs the same isomerase as the pathway for the amphibolic utilization of galactose (B), for a gal⁻ mutant that lacks this enzyme has lost the ability both to use galactose as a carbon source and to make a galactose-containing wall polysaccharide from glucose. (These two sequences are also found in mammalian cells.)

Fig. 4-25. Biosynthesis of the precursors of a specific wall polysaccharide ("O" antigen) of group B *Salmonella*. [From Nikaido, H., Nikaido, K., and Mäkelä, P. M. *J Bacteriol 91*:1126 (1966).]

one kind of monomer), formed from sugar nucleotides by transglycosylation reactions. Examples in pneumococci will be given in Chapter 25. Each repeating subunit is built upon a membrane lipid, which then transfers it across the cytoplasmic membrane, as will be described in Chapter 6 for **wall polysaccharides.**

Certain lactic bacteria (e.g., *Leuconostoc*) form a homopolymer from exogenous sucrose (glucose-1,2-fructoside) by a simple reaction in which the energy of the glycoside bond is used to polymerize one of the two residues. Some species release and absorb the fructose moiety and polymerize the glucose (= dextrose) to a high-molecular-weight **dextran;** others free the glucose and polymerize the fructose (levulose) to a **levan.**

$$\text{n Sucrose} \longrightarrow (\text{Glucose})_n + \text{n Fructose}$$
or
$$\text{n Glucose} + (\text{Fructose})_n$$

In these polymers the predominant linkage is α-1,6, in contrast to the α-1,4 of starch or glycogen.

The polymerization appears to be catalyzed by an enzyme on the outer surface of the cell. The polymer makes the colonies highly mucoid when grown in the presence of sucrose, and it promotes adhesion of bacteria to surfaces; the role of sucrose in promoting dental caries involves this property (Ch. 42). Dextrans have been used in medicine as plasma extenders, for they are not especially antigenic. Preventing formation of these polymers is of great economic importance in the sugar industry.

Cellulose (poly-β-1,4-glucose), indistinguishable from that of higher plants, is formed by some aerobacters, obligate aerobes that oxidize ethanol to acetate. It appears to be excreted as separate macromolecules to the exterior, where it crystallizes into a mat. By trapping both cells and bubbles of CO_2 the mat floats the cells to the surface, which is obviously advantageous for obligate aerobes.

Glycogen (poly-α-1,4-glucose) is stored by many bacteria. It is formed, as in mammals, from UDPG (Fig. 4-24), and is utilized via phosphorolysis:

$$\text{Glycogen} + n\text{P}_i \longrightarrow \text{n Glucose-1-P}$$

The highly varied **wall polysaccharides** will be considered in Chapter 6.

POLYPEPTIDE SYNTHESIS

HOMOPOLYMERIC POLYPEPTIDES

For the synthesis of an unlimited variety of specific sequences in proteins an elaborate template mechanism has evolved (Ch. 12). In contrast, the synthesis of amino acid homopolymers, like that of short peptides (e.g., the tripeptide glutathione; wall peptides, see Ch. 6), employs a more classic mechanism, in which each successive monomer is recognized by the appropriate enzyme.

Thus some species of *Bacillus* form a capsule of poly-D-glutamic acid, or a mixture of poly-D- and poly-L-glutamic acid; in the anthrax bacillus this capsule is essential for virulence (Ch. 33). The residues are linked by a γ-peptide bond (in contrast to the α-peptide of proteins), and the MW may reach 250,000. The polymerization is accomplished by an enzyme plus ATP, without RNA.

PEPTIDE ANTIBIOTICS

Peptide antibiotics (of 10 to 20 residues) are synthesized by a mechanism intermediate in complexity between protein synthesis and homopolymer synthesis. The sequence of additions is determined neither by a sequence of separate enzymes nor by a template, but by a sequence of sites on a complex, multiheaded enzyme. An example is the synthesis by *Bacillus brevis* of **gramicidin S,** a head-to-tail ring of two pentapeptides:

L-Leu-D-Phe-L-Pro-L-Val-L-Orn
L-Orn-L-Val-L-Pro-D-Phe-L-Leu

In this process, the amino acids are activated as aminoacyladenylates (as in protein synthesis, Ch. 12), and then transferred to a high-energy thioester link with an -SH on one of two enzymes. Enzyme I racemizes and transfers the phenylalanine, to provide the D-Phe residue. Enzyme II, a multienzyme complex, can form thioester bonds with the other four amino acids of the pentapeptide, each in a specific site, but it does not polymerize them until **initiated** by the transfer of D-Phe from enzyme I. Moreover, if any of the four amino acid residues is absent enzyme II retains multiple growing chains, up to the empty site. The multienzyme complex (MW 280,000) has a single prosthetic group, pantetheine phosphate (Fig. 4-14), covalently attached in a position so that its free -SH end can reach the more peripheral sites of attachment of the four amino acids. The successive peptidyl transfers are achieved by transthiolation of the growing chain to the pantetheine, followed by peptide bond formation with the next, -SH-attached amino acid (Fig. 4-26).

This model is based on a very similar succession of transthiolation reactions by pantetheine in fatty acid biosynthesis (Ch. 6). The multiple sites in the latter process specify the sequence of reaction undergone by each acetate residue, while in peptide syn-

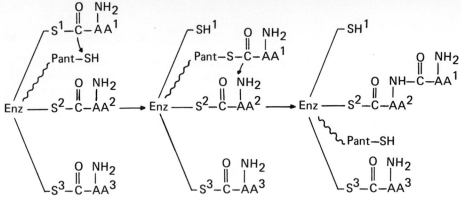

Fig. 4-26. Peptidyl transfer by transthiolation in synthesis of a peptide antibiotic. The pantetheine (Pant-SH) is attached at one end to some central position in the protein. Its free, reactive, -SH end can accept an amino acid (A) or peptide from a thioester and transfer it to the amino group of the next amino acid in the sequence.

thesis the multiple sites serve as a kind of template for specifying amino acid sequence.

As Lipman pointed out, before the genetic code could be evolved it was probably necessary that the primitive precursors of cells (Ch. 1, Prebiotic evolution) discover the extraordinary functional pliability of proteins. The current mechanism of synthesis of peptide antibiotics may be a metabolic "fossil" of such a process. On the other hand, this process may be an evolutionary dead end, suitable only for short sequences; some have speculated that nucleic acids, with their capacity for self-replication (Ch. 10) as well as for tertiary structure (Chs. 11 and 12), preceded proteins in evolution.

SELECTED FEATURES OF SMALL-MOLECULE BIOSYNTHESIS

Certain interesting generalizations emerge from the pathways we have just reviewed.

HETEROTROPHIC CO_2 FIXATION

In deriving their carbon and energy from organic matter heterotrophs produce CO_2, which long masked the fact that they also assimilate CO_2 (though on a much smaller scale than autotrophs).* However, in 1935 Wood

*Despite this masking, as early as 1916 Lebedev in Russia concluded that heterotrophs must fix CO_2, since the mold *Aspergillus* failed to grow in a current of CO_2-free air. This evidence does not constitute proof by contemporary standards; but the interpretation was bold and reasonable. Unfortunately, this important discovery was buried by publication in the *Journal of the Don Academy of Sciences* in 1921. In a moving introduction Lebedev apologized for the lack of specific data; his notebooks had been destroyed in the Revolution, and

and Werkman discovered that the propionic acid fermentation (Ch. 3) of glycerol could fix CO_2 (when provided in excess) into succinate, yielding more organic carbon than was supplied. Later studies with radioactive CO_2 showed that all **heterotrophs** (including mammals) **assimilate CO_2.** When biosynthetic pathways became known this CO_2 was found to be assimilated into purines, pyrimidines, and several amino acids.

The following reactions have been demonstrated:
1) Carboxylation of pyruvate or PEP (Fig. 4-7). In this reaction CO_2 yields one of the carbons of oxaloacetate (or malate), and thus of aspartate, glutamate, and their numerous derivatives. This is the largest route of heterotrophic CO_2 fixation.
2) Formation of carbamyl phosphate, which contributes both the guanido C of arginine (Fig. 4-19) and C_2 of the pyrimidine ring (Fig. 4-23).
3) The addition of CO_2 to 5-aminoimidazole ribotide, contributing C_6 of the purine ring (Fig. 4-22).
In addition, CO_2 participates cyclically, without being assimilated, in fatty acid biosynthesis (Ch. 6).

In contrast to autotrophic CO_2 fixation (Fig. 3-5), the limited heterotrophic fixation does not provide net synthesis of an amphibolic intermediate; hence it cannot be used for general biosynthesis.

THE IMPORTANCE OF BEING IONIZED

Although nonionized compounds may be excreted by cells as endproducts (e.g., ethanol), or may circulate between cells in higher organisms (e.g., glucose), all known intermediates in bacteria contain one or more groups

he was publishing a general account because he "did not know what the morrow will bring."

TABLE 4-1. The Number of Enzymes Involved in the Biosynthesis of the Amino Acids

Synthesized by mammals		Required by mammals		
Amino acid	No. of enzymes	Amino acid	No. of enzymes	Plus shared enzymes
Alanine	1	Arginine	8	1
Aspartate	1	Histidine	9	
Asparagine (f. aspartate)	1	Threonine	5	1
Glutamate	1	Methionine	5	9
Glutamine (f. glutamate)	1	Lysine	7	
Proline (f. glutamate)	2	Isoleucine	4	6
Serine	3	Valine	1	9
Glycine (f. serine)	1	Leucine	3	9
Cysteine (f. serine and S^{--})	2	Tyrosine	10	
		Phenylalanine	2	8
Total	13	Tryptophan	6	7
		Total	60	

The enzymes counted are those of the anabolic pathways, after they have branched off from amphibolic pathways. For branched anabolic pathways the enzymes of a shared portion are counted for the first endproduct cited, and are listed as "shared" for subsequently cited products. Modified from Davis, B. D. *Cold Spring Harbor Symp Quant Biol 26*:1 (1961).

that are largely ionized at physiological pH: generally a phosphate (as in glycolysis) or a carboxyl (as in the TCA cycle). The regular presence of such a group is illustrated by comparison of the purine and the pyrimidine pathways. Thus, from the first reaction of purine biosynthesis the intermediates are attached to ribose P; otherwise several of them would be un-ionized (Fig. 4-22). In contrast, in the first reaction of pyrimidine biosynthesis (Fig. 4-23) aspartate provides a carboxyl, which is retained until a phosphate is added, after ring formation. Similarly in glycolysis, the 2 Ps that are added to hexose from ATP, and are then recovered from 2 PEPs, contribute no net energy; but without them many intermediates would be un-ionized.

The function of these ubiquitous dissociable groups is not certain; they may promote the retention of a compound by a cell, and perhaps the efficiency and specificity of enzyme action.

ECONOMY IN MAMMALIAN BIOSYNTHESIS

Man can synthesize about half his amino acids but is an auxotroph for the other half. Analysis of the pathways in *E. coli* has shed light on this curious evolution: the 9 amino acids synthesized in man arise by pathways of 1 to 3 enzymes each, while the 11 required amino acids have individual path lengths of 6 to 13 enzymes (Table 4-1). The human species has thus spared itself genes for about 60 enzymes by losing the pathways to 11 amino acids, while the retained pathways to the other 9 amino acids are relatively inexpensive, involving only 13 enzymes in all.

SELECTED REFERENCES

Books and Review Articles

COHEN, S. S. *Introduction to the Polyamines.* Prentice-Hall, Englewood Cliffs, N.J., 1971.

DAVIES, D. A. L. Polysaccharides of gram-negative bacteria. *Adv Carbohyd Chem Biochem 15*:271, (1960).

GINSBURG, V. Sugar nucleotides and the synthesis of carbohydrates. *Adv Enzymol 26*:35 (1964).

GUNSALUS, I. C., and STANIER, R. Y. (eds). *The Bacteria,* vol. III, *Biosynthesis.* Academic Press, New York, 1962. Especially Wood and Stjernholm, CO$_2$ Fixation by Heterotrophs (Ch. 2); Umbarger and Davis, Amino Acid Biosynthesis (Ch. 4); Magasanik, Nucleotide Biosynthesis (Ch. 6).

KORNBERG, H. L. Anaplerotic sequences in microbial metabolism. *Angew Chem (Eng) 4*:558 (1965).

LARSON, A., and REICHARD, P. Enzymatic reduction of ribonucleotides. *Progr Nucleic Acid Res Mol Biol 7*:303 (1967).

LIPMANN, F. Attempts to map a process evolution of peptide biosynthesis. *Science 173*:875, 1971.

SHAPIRO, S. K., and SCHLENK, F. (eds.). *Transmethylation and Methionine Biosynthesis.* Univ Chicago Press, Chicago, 1965.

TABOR, H., and TABOR, C. W. Biosynthesis and metabolism of 1,4-diaminobutane, spermidine, spermine, and related amines. *Adv Enzymol 36*:203 (1972).

VOGEL, N. J., THOMPSON, J. S., and SCHOCKMAN, G. D. Characteristic metabolic patterns of prokaryotes and eukaryotes. In Organization and Control in Prokaryotic and Eukaryotic Cells. *Symp Soc Gen Microbiol,* Cambridge Univ Press, 1970, p. 107.

chapter 5

BACTERIAL NUTRITION AND GROWTH

NUTRITION   90

   Organic Growth Factors   90
   Inorganic Requirements   90
   Physical and Ionic Requirements   92
   Practical Bacterial Nutrition   94
   Attack on Nonpenetrating Nutrients; Exoenzymes   94

GROWTH IN LIQUID MEDIUM   95

   Methods of Measurement   95
   Growth Cycle   96
   Exponential Kinetics   96
   Relation of Growth to Substrate Concentration   98
   The Chemostat   99
   Synchronized Growth   100

GROWTH ON SOLID MEDIUM   101

   Solidifying Agents   101
   Uses of Solid Media   101
   Colonial Morphology   102
   Selective Media   104

When it is possible to catalogue the substances required by pathogenic bacteria for growth, it will probably be found that most of them are . . . important in animal metabolism, and . . . it is equally probable that some will be new. J. H. MUELLER, J BACTERIOL 7:309 (1922)

NUTRITION

ORGANIC GROWTH FACTORS

Microbiologists learned early that they could grow a wide variety of bacteria in "broths," obtained by cooking animal or vegetable tissues. In 1923 Mueller, undertaking to redefine this vague requirement in terms of specific compounds (see quotation at start of this chapter), discovered the previously unknown amino acid methionine. Bacterial nutrition became a lively field a dozen years later; but with recognition of the unity of biochemistry (Ch. 3) the many different nutritional patterns turned out to be simply minor variations on a central theme, and the study of microbial nutrition lost much of its theoretical interest. Nevertheless, this field retains its practical importance and still presents challenges; for example, the leprosy bacillus, the treponeme of syphilis, and rickettsiae cannot be cultivated in artificial media.

An important consequence of the study of growth factors was the development of **quantitative microbial (bio)assays** for amino acids and vitamins, as described in the preceding chapter for auxotrophic mutants. The simplicity of these assays, compared with those in animals, greatly facilitated the isolation of novel factors; hence a number of vitamins were first identified in this way, and were only later found to be also essential for mammals. A list is presented in Table 5-1.

Nutritional investigations with animals and with bacteria have sometimes converged unexpectedly on the same compound. Thus vitamin B_{12} was independently isolated through two programs, one based on arduous assays of the hematopoietic response of patients with pernicious anemia, and the other based on the purification of a growth factor for *Lactobacillus leichmannii*.

Bacteria that are adapted to growth in animal tissues, mucous surfaces, or milk often require various groups of amino acids and nucleic acid bases, as well as vitamins. Yeasts and molds, in contrast, usually grow on higher plants, and many strains exhibit requirements only for vitamins.

Other compounds required by various microbes include inositol and choline (as components of phospholipids or wall), vitamin K, hemin (or occasionally porphyrin), unsaturated fatty acids, mevalonic acid (a precursor of isoprenoid compounds), polyamines, and iron-chelating compounds (e.g., **mycobactin**, excreted by some mycobacteria and required by *M. paratuberculosis*).

INORGANIC REQUIREMENTS

Oxygen. The metabolic reasons for aerobic, anaerobic, and facultative requirements have been discussed in Chapter 3. O_2 has a low solubility in water: a solution in equilibrium with air at 34° contains ca. 5 $\mu g/ml$, which would be consumed in < 10 seconds by a fully grown culture of an aerobe. The diffusion of O_2 across the air-water interface therefore limits the density attained by an aerated, well-nourished culture: for example, with aeration by swirling in a flask growth is often limited to 1 to 2 mg dry weight per milliliter. Moreover, during the last portion of such growth the culture becomes anaerobic (often changing the composition of the organisms).

Methods that increase the area of the liquid-air interface, such as rapid bubbling of air through a porous sparger, or recycled dripping, will support heavier growth. Since a shift from small- to large-scale cultivation may markedly alter the adequacy of aeration it may lead to striking changes in composition of the cells and the culture fluid. The problem of adequately aerating dense cultures also makes large-scale production difficult with bacteria compared with yeasts, which respire more slowly (surface/volume ratio ca. 1/100 as great).

TABLE 5-1. Discovery of Vitamins

Vitamin	Microbe*	Animal
Discovered first in animals		
Riboflavin	Lactic bacteria	Rats
Thiamine (B₁)	Yeasts; *Staphylococcus aureus*	Rats, man
Pyridoxine (B₆)	Yeasts; lactic bacteria	Rats
Discovered independently or first in microbes		
p-Aminobenzoic acid	Sulfonamide inhibition; *Clostridium acetobutylicum*	
Folic acid	*Lactobacillus casei; Streptococcus faecalis*	Chicks
Biotin	Yeast; *Rhizobium*	Chicks
Nicotinic acid	*S. aureus; Corynebacterium diphtheriae*	Canine blacktongue, pellagra in man
DPN	*Hemophilus*	None
Pantothenic acid	Yeasts	Various; difficult to demonstrate
Lipoic acid	Lactic bacteria; *Tetrahymena geleii* (protozoon)	
Vitamin B₁₂	*Lactobacillus leichmannii; Lactobacillus lactis*	Pernicious anemia in man
Pyridoxal, pyridoxamine (physiological forms of pyridoxine)	Yeasts; lactic bacteria	

* The principles underlying microbiological assays are discussed below.

Anaerobiosis. The establishment of a strictly anaerobic atmosphere for the cultivation of obligate anaerobes (Ch. 3) on plates is difficult, since oxygen tensions as low as 10^{-5} atm can be inhibitory. However, supplementation of the medium with a sulfhydryl compound, such as sodium **thioglycollate** ($HSCH_2COONa$), permits even the strictest anaerobes, such as *Clostridium tetani,* to be grown in tubes exposed to air. It is helpful to add a layer of oil or paraffin to slow the diffusion of oxygen, plus semisolid agar (0.2 to 0.3%) to prevent convection. In nature mixed cultures are the rule, and the strict anaerobes may depend on neighboring facultative organisms to scavenge oxygen.

Some organisms are **microaerophilic,** initiating growth well at reduced but not at fully aerobic O_2 tensions. Some facultative organisms (e.g., *E. coli*) require cystine to permit initiation of anaerobic growth in a minimal medium; cysteine is ineffective. Presumably strict anaerobiosis reduces, and cystine restores, some S-S bonds required for growth.

Anaerobic conditions are associated with a low **oxidation-reduction (redox) potential** (E_h), measured with a platinum electrode immersed in the culture. Extensive observations on this variable have been reported. However, the data are difficult to interpret, for unlike acids and bases, whose states of dissociation are instantly determined by their pKs and the ambient pH, most redox systems do not

come to equilibrium with each other. Hence an observed redox potential in a complex mixture is the resultant of many competing reactions, rather than an equilibrium state.

Carbon Dioxide. The role of CO_2 as a universally essential nutrient has been discussed in Chapter 4 (Heterotrophic CO_2 fixation). Some organisms (e.g., meningococci, gonococci), especially when first isolated, initiate growth better at a pCO_2 higher than that found in air (ca. 0.03% outdoors); they presumably have some enzyme with a low affinity for CO_2. Elevated pCO_2 is conveniently provided in a **candle jar,** a closed vessel in which a candle is allowed to burn until it extinguishes itself. The accompanying lowering of pO_2, however, does not provide strict anaerobiosis.

All metabolizing aerobes (and many fermenters) elevate the pCO_2 in their environment. However, with a small inoculum this contribution is small, and there may be a long lag in the initiation of growth in a minimal medium because the low pCO_2 limits the required flow of carbon from glucose into the TCA cycle. Added TCA cycle intermediates may then exert a "sparking" effect in overcoming this lag. However, even with these additions growth can be prevented by measures that further reduce the CO_2 tension in the vessel. The explanation was provided by the discovery that fatty acid synthesis requires CO_2 for the conversion of acetyl CoA to malonyl CoA (Ch. 6). Hence even though this CO_2 is recycled rather than assimilated, **all growing cells have an absolute requirement for an adequate pCO_2,** for their lipids cannot all be supplied from without.

The inorganic ions required in substantial quantity are PO_4^{3-}, K^+, and Mg^{++}, NH_3 and SO_4^{--} (or a reduced product) are also required in the absence of organic sources of N and S. Unlike mammalian cells most bacteria can thrive in a broad range of concentration of the required ions. This flexibility long prevented recognition of their need for a constant *milieu interne,* just as simple nutritional requirements had hidden their complex metabolism. The constant internal K^+, Mg^{++}, and PO_4^{3-} concentrations depend on active membrane transport systems.

K^+ is required by many enzymes, and is especially important for ribosomal function: when the K^+ content of bacterial cells is progressively lowered (in a mutant with a defective K^+ transport system) protein synthesis ceases, while glycolysis continues. Mg^{++} is also essential for the integrity of ribosomes (Ch. 12) and for the function of many enzymes. The function of K^+ in ribosomes is antagonized by Na^+.

Trace Elements. $Fe^{++(+)}$ is required not only for heme proteins in aerobes but also for certain non-heme enzymes in anaerobes. Its concentration markedly influences diphtheria toxin formation (Ch. 24). Other requirements include Zn^{++} and Mn^{++} for certain enzymes, and Mo^{++} for N_2 fixation and nitrate reduction. Co^{++} is required by those bacteria that make vitamin B_{12}; since plants do not contain this vitamin bacteria may be its ultimate source for man. Cu^{++} has not been found in any bacterial enzymes, though it is widely used in higher organisms.

Ca^{++} does not appear to be required by gram-negative organisms. However, it is a major constituent of the wall of gram-positive bacilli and of their spores, and it appears to function in the excretion of proteases by these bacilli.

Siderochromes and Fe Transport. Iron transport presents a special problem, because at neutral pH Fe^{3+} forms very insoluble colloidal hydroxides. Accordingly, many bacteria and fungi form and excrete compounds that chelate Fe (i.e., form soluble coordination complexes), promoting its absorption; and their formation is sometimes induced by media with a low Fe content. Two main classes of such natural Fe-binding compounds (siderochromes) have been recognized: hydroxamic acids ($—CONH_2OH$) known as *sideramines,* and

catechols (2,3-dihydroxybenzene derivatives). Citrate also chelates Fe, and may either promote or impair its uptake in various organisms.

Enteric organisms (including *E. coli*) form a cyclic trimer of 2,3-dihydroxybenzoylserine known as *enterobactin* or enterochelin, while *Enterobacter* (*Aerobacter*) also forms a hydroxamic acid (aerobactin). One species of Mycobacterium requires for growth a sideramine, mycobactin, formed by other mycobacteria; conversely, some organisms form **sideromycins** (e.g., albomycin): non-chelating, **antibiotic** analogs that interfere with the uptake of sideramines.

PHYSICAL AND IONIC REQUIREMENTS

Pathogenic bacteria naturally thrive under physical conditions similar to those of the mammalian body, and can withstand a rather wide range of temperature, osmotic pressure, and pH. However, in filling all possible ecological niches the bacterial world has evolved members that can grow under conditions too extreme for any other group of organisms: temperatures up to 90°, pH below 1.0, or salinity up to 30% NaCl. These organisms are of considerable interest as experiments of nature that can help to correlate macromolecular structure and function.

The external temperature, of course, determines that within the cell; and the intracellular osmotic pressure cannot be less than that outside (though the specific composition is different). Hence in the cells that resist extremes of these parameters the enzymes and the ribosomes must also be resistant. A low external pH, in contrast, need not correspond to the internal pH; in the absence of accurate measurements of intracellular pH we can only be sure that the cell wall and membrane are exposed to the extreme conditions.

The existence of thermophiles shows that nature can evolve proteins, for all essential cellular functions, with stability far beyond the usual range. Conversely, many temperature-sensitive mutants isolated in the laboratory (Ch. 8) form specific altered enzymes that denature at ordinary temperatures.

Temperature. Most bacteria can grow over a **temperature range** at 30° or more but have quite a narrow range for optimal growth. Below the optimum the growth rate first falls off with a Q_{10} of ca. 2 to 3 (i.e., that typical of enzyme reactions) but then declines more rapidly, giving rise to a fairly well-defined **minimal growth temperature** (Fig. 5-1).

Above the optimum the growth rate decreases steeply with increasing temperature, giving rise to a sharply defined **maximum growth temperature.**

The lower temperature limit of growth may depend on solidification of membrane lipids (Ch. 6), or on the marked sensitivity of the initiation process in protein synthesis (Ch. 12) to cooling. Slightly above the upper temperature limit, in contrast, many enzymes are denatured and the cell dies: i.e., in general a cell evidently does not build enzymes with more stability than is useful. However, in the range where increasing temperature slows growth in *E. coli* the cells do not die but immediately shift to the new growth rate, which is limited by a reversible decrease in the activity of a particular biosynthetic enzyme. The teleonomic value of this adjustment is not obvious.

The **temperature range** for growth of an organism is a stable characteristic, of considerable taxonomic value. It is customary to divide bacteria into mesophiles, psychrophiles (or cryophiles), and thermophiles. Most bacteria are **mesophiles.** Those found in the mammalian body have a temperature optimum of 37 to 44°, but many others found in nature (e.g., *Bacillus megaterium*) grow better at 30°.

Psychrophiles (predominantly pseudomonads) tolerate rather than prefer very low temperatures: their optima are rarely below 29° (Fig. 5-1), but they multiply at a substantial rate even at 0°. These organisms are important in spoilage of refrigerated foods and are also found in naturally cold waters and soils. **Thermophiles** (predominantly bacilli), in contrast, may have temperature optima as high as 50 to 55°, with tolerance to 90°. They are found especially in hot springs and compost heaps.

Cold Shock. Though bacteria are often preserved successfully in the refrigerator, the sudden chilling of exponentially growing cells of some species (*E. coli, Pseudomonas*) results in substantial killing (> 90%). This curious phenomenon is not observed with gradual cooling or with stationary phase cells; it may be related to the changing composition of the cell membrane lipids with changes in temperature and in growth phase (Ch. 6).

pH. The pH range tolerated by most microorganisms extends over 3 to 4 units, but rapid growth may be confined to 1 unit or less. *E. coli* cannot withstand a pH much above 8 or below 4.5, while pathogens adapted to tissues (*Pneumococcus, Neisseria, Brucella*) have a narrower range. Vinegar-forming *Acetobacter* and sulfur-oxidizing bacteria can tol-

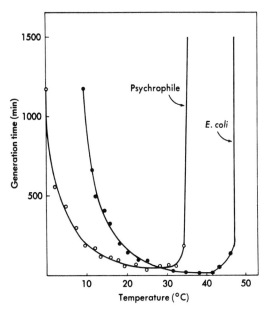

Fig. 5-1. Effect of temperature on the generation time of a typical mesophile (*E. coli*) and a psychrophilic pseudomonad. [After Ingraham, J. L. *J Bacteriol 76*:75 (1958); modified according to data of Ron, E. Z., and Davis, B. D. *J Bacteriol 107*:391 (1971).]

erate the acid that they produce up to 1 N (pH ~ 0 for sulfuric acid). In contrast, a few bacterial species (urea splitters, *Alcaligenes faecalis,* the cholera vibrio) thrive at a pH of 9.0 or more. Most yeasts and molds are highly acid-tolerant, and this feature is exploited in selective media for their cultivation.

The lower pH cutoff point depends in part on the concentrations of organic acids in the medium; a lower pH increases the proportion of an acid in the undissociated (and hence more permeable) form, thus making it more inhibitory. Hence a lactate-producing fermenter inhibits itself when it reaches a certain concentration of free lactic acid rather than a given pH.

In a culture growing aerobically on a limiting amount of sugar the pH often falls and then rises as acid accumulates and then is utilized. To restrict pH changes during growth, media are often heavily buffered; and for fine control automatic continual titration is sometimes employed. Incidentally, the concept of pH, which first clearly distinguished **extent** and **intensity** of acidity, was originally formulated by Sorensen in the course of determining what limits the growth of microbes in various media.

Halophiles. As was noted above, Na+ and Cl− are not widely required by bacteria, though moderate concentrations are generally tolerated. Most bacteria isolated from the ocean, however, are slight halophiles, requiring NaCl in a concentration approaching that of their natural habitat (3.5%). In addition, moderate and extreme halophiles, with NaCl requirements up to 20%, and with optima approaching saturation (slightly above 30%), are found in flats and lakes where salt water is evaporated, and in pickling fluids.

Although halophiles may contain a substantial concentration of Na+, the bulk of their unusually high salt content is K+. Furthermore, many of their enzymes, and their ribosomes, are stimulated in extracts by high K+ but not by Na+. Hence a major function of the external salt is probably purely osmotic, permitting the accumulation of a high intracellular K+ concentration, to which the enzymes of the cell are adapted. In addition, however, halophiles have an unusual cell wall whose integrity requires a high Na+ concentration.

Water. In contrast to higher organisms, whose specialized integument provides for retention of water, the growth and the metabolism of microbes are immediately dependent on ambient water. Thus in regions of high humidity textiles and leather rapidly become moldy, reminding us that when man wishes to withhold these organic materials from the geochemical cycle he must preserve them from moisture and hence from microbial attack. Indeed, armies bringing elaborate equipment into the tropics have had to provide fungicides to protect even glass lenses from attack by molds.

PRACTICAL BACTERIAL NUTRITION

Though chemically defined media are valuable for special purposes, the traditional rich "soups" are still generally employed in diagnostic bacteriology: they are less expensive, and initiation of growth by small inocula is often more reliable. These media are based primarily on **meat digest** (tryptic digest, peptone, nutrient broth), the soluble product of enzymatic hydrolysis of meat or fish. Many types are marketed, differing in source material or in method of preparation, and often in suitability for cultivating specific organisms.*

---

* The superiority of various digests is sometimes ascribed to their content of desirable peptides. Indeed, certain streptococci are stimulated by, though they do not require, various peptides; and some auxotrophic mutants of *E. coli* respond better to peptides than to the corresponding amino acids.

To provide vitamins and coenzymes media are often further enriched with **meat extract (meat infusion)** or **yeast extract.** In contrast to the extracts used by the enzymologist, in which denaturation is minimized, these nutritional extracts are composed of stable small molecules that have been released from the cells and concentrated by boiling. In **blood agar** the blood provides not only further nutrients but also a diagnostically useful index of hemolysis.

The genus *Hemophilus* requires hemin, DPN, or both (Ch. 30). These factors are readily supplied in heated blood agar (called **chocolate agar** because of its color): heat denaturation releases the heme from hemoglobin and inactivates an enzyme that hydrolyzes DPN.

Some organisms thrive best in media containing a high concentration of **serum** (e.g., 20%), in which the **albumin** serves as **protective, nonnutrient growth factor.** In the animal the versatile affinity of this protein for binding various small molecules serves to protect cells from such toxic compounds as fatty acids, cationic and anionic detergents, and dyes; it similarly protects bacteria when added to a culture.

In simple media small inocula of many bacteria (especially *Mycoplasma, Mycobacterium,* and *Neisseria*) are very sensitive to traces of detergents and heavy metal ions contaminating the glassware, and serum albumin (0.2 to 1.0%) or starch are used to bind the inhibitors.

Since individual amino acids are relatively expensive, **casein hydrolysate** is often used in preparing chemically defined, relatively rich media. In the usual acid hydrolysate (e.g., Casamino acids) the amino acids are all free, but tryptophan and glutamine have been destroyed. Enzymatic hydrolysates contain all the amino acids, but mostly as small peptides.

The compositions of some representative media are given in Table 5-2.

ATTACK ON NONPENETRATING NUTRIENTS: EXOENZYMES

Microbes can take up foods of low molecular weight, including oligopeptides, nucleosides, and small organic phosphates, e.g., glycerol phosphate (Ch. 6); nucleotides generally cannot penetrate at a substantial rate. But the organic matter initially returned to the soil

---

However, these responses reflect better penetration of the peptides via specific transport systems (Ch. 6), rather than direct incorporation.

TABLE 5-2. Compositions of Representative Bacteriological Media

Minimal medium for *E. coli*	
	Gm /L
K_2HPO_4	7.0
KH_2PO_4	3.0
Na_3citrate—$3H_2O$	0.5
$MgSO_4$—$7H_2O$	0.1
$FeSO_4$	0.01
$(NH_4)_2SO_4$	1.0
Glucose*	2.0

Penassay broth (typical rich medium)	
Peptone	5.0
Beef extract	1.5
Yeast extract	1.5
NaCl	3.5
Dipotassium phosphate	3.7
Monopotassium phosphate	1.3
Glucose	1.0

B_{12} Assay medium for *L. leichmannii*	
Vitamin-free casein hydrolysate	15.0
Tomato juice	10.0
Glucose	40.0
Asparagine	0.2
Sodium acetate	20.0
Ascorbic acid	4.0
Monopotassium phosphate	1.0
Dipotassium phosphate	1.0
Sorbitan monooleate	2.0
$MgSO_4$	0.4
NaCl	0.02
$FeSO_4$	0.02
$MnSO_4$	0.02
L-Cystine	0.4
DL-Tryptophan	0.4
Adenine sulfate	0.02
Guanine hydrochloride	0.02
Xanthine	0.02
Uracil	0.02
	Mg /L
Riboflavin	1.0
Thiamine	1.0
Niacin	2.0
p-Aminobenzoate	2.0
Ca pantothenate	1.0
Pyridoxine	4.0
Folic acid	0.2
Biotin	0.008

* Glucose is autoclaved separately, for when autoclaved in the presence of phosphate it produces discoloration and some toxicity.

in dead plants and animals is predominantly macromolecular. Preliminary extracellular hydrolysis is therefore necessary, just as in higher animals. For this purpose various bacteria and fungi elaborate a variety of **exoenzymes.** Many of these are secreted into the medium (especially by gram-positive rods) as **extracellular enzymes;** in pathogens these often play an important role by attacking tissue constituents. Other exoenzymes remain attached to the cell and at least some are **periplasmic** enzymes, held between the plasma membrane and the wall. The mechanism of enzyme secretion will be discussed in Chapter 6.

Extracellular enzymes include proteases and peptidases; polysaccharidases (amylase, cellulase, pectinase); mucopolysaccharidases (hyaluronidase, chitinase, lysozyme, neuraminidase); nucleases; lipases; and phospholipases. Some proteases are released from the cells as inactive zymogens, which catalyze their own activation. Protease formation can be repressed by a high concentration of amino acids in the medium.

Some Macromolecular Substrates. Recognition of various hydrolases is useful in diagnostic work: extracellular protease is generally detected by the liquefaction of gelatin (denatured collagen) around a colony; lecithinase is recognized by the formation of an opaque product from egg yolk (Ch. 34); clotting of plasma is affected by coagulase (Ch. 27) and by streptokinase and fibrinolysin (Ch. 26).

Microbial hydrolases offer considerable commercial promise, as in the tenderizing of meat. Moreover, some of these enzymes cleave only a particular site and hence have been helpful in the analysis of protein structure. (For example, subtilisin, from *Bacillus subtilis,* converts ovalbumin to crystalline plakalbumin plus a specific peptide.)

Starch is hydrolyzed by many bacteria and molds but by only few yeasts. Hence in the fermentation of grain to yield beer germinating barley is used to convert the starch to the dissaccharide maltose, which is then fermented by brewer's yeast (*Saccharomyces cerevisiae*). The mold *Aspergillus oryzae* is used in Japan in a one-step process that both splits and ferments starch.

GROWTH IN LIQUID MEDIUM

METHODS OF MEASUREMENT

For biochemical studies bacterial growth is usually defined in terms of **mass** of cellular material, while for studies of genetics or of infection **cell number** is more pertinent. The two remain proportional under conditions of steady-state growth, but their ratio can vary with growth conditions.

Cell mass can be determined in terms of **dry weight** or indirectly in terms of packed cell volume or nitrogen content. A much more convenient index is **turbidity,** whose rapid measurement in a photoelectric colorimeter or spectrophotometer allows the density of a culture to be followed while it is still growing. At wave lengths in the visible range the absorption of light by colored cell constituents is negligible; most of the turbidity is due to light scattering, dependent on the high refractive index of the bacteria (ca. 25% solid compared with 1 to 2% in the medium).

Wave lengths between 490 and 550 nm are generally used: the lower the wave length the greater the light scattering. However, below 490 nm absorption by yellow products of autoclaving may become significant. Turbidity is linear with bacterial density between 0.01 mg dry weight (ca. 10^7 cells) and 0.5 mg/ml. The addition of a salt or sugar will cause shrinkage of bacteria; and though their cross-section is decreased, their increased refractive index causes an increase in light scattering.

Cell Number. To determine the **viable number** of a culture a series of 10-fold or 100-fold dilutions is plated in, or on, a solid medium; the number of colonies is counted in those Petri plates that are not too crowded ($<$ 400 colonies). This method is useful down to extremely low bacterial densities. Precision is limited by the statistical sampling error: the **standard deviation** (SD) is the square root of the number counted (e.g., in a count of 400 colonies SD = 20, or 5%). It is therefore customary in precise work to make several replicate plates.

It is sometimes desirable, especially in studying antimicrobial action, to measure the **total cell count,** which is ordinarily identical with the viable count. Cells can be counted under the microscope in specially designed chambers, but it is more convenient to use an electronic particle analyzer (e.g., the Coulter counter), in which a precise volume (ca. 50 μl) of a dilute suspension flows past a pair of electrodes; these detect the effect of a passing particle on the resistance.

Cell number cannot be determined accurately when cells adhere to each other after division (e.g., streptococci) or when they aggregate. Indeed, if changing conditions in a culture lead to aggregation the measurements of cell number by turbidity yield grossly misleading results, simulating lysis. Freedom from aggregation is one of the properties that makes *E. coli* especially convenient for physiological investigation.

GROWTH CYCLE

Growth of bacteria in an adequate medium is characteristically **exponential** (see next section). However, an exponentially growing culture eventually slows down and ceases growth, either because a required nutrient (often O_2) becomes limiting or because inhibitory metabolic products (often organic acids or alcohol) accumulate. In this transition from the **log phase** to the **stationary phase** (Fig. 5-2) **the cells become smaller,** as a result of dividing faster than they grow; moreover, they develop major changes in their macromolecular composition (Ch. 13) and they may cease to maintain their high intracellular K^+ content.

Stationary phase cells transferred to fresh medium exhibit a **lag phase** which varies in length with the organisms and the medium; it is much more pronounced in minimal than in rich medium. Cell number lags more than cell mass (Fig. 5-2), for the small stationary phase cells increase in size before they begin to divide.

When cells from an exponentially growing culture in a rich medium are transferred to fresh, identical medium there is no lag in resuming growth. In transfer from minimal medium to fresh minimal medium there may or may not be a lag; it is more likely to occur with small inocula (e.g., below visible turbidity). One factor promoting this lag is inhibition of growth by trace contaminants in the medium (e.g., soap or heavy metal ions). Another is the need to accumulate the CO_2 required for biosynthesis.

Cells in the stationary phase develop adaptive chemical changes (Chs. 6 and 13) that increase their stability. However, on prolonged incubation cells do die, and the resulting membrane damage results in activation of autolytic enzymes. The released products support **cryptic growth** of surviving cells; hence more resistant mutants may accumulate even though there is no net growth.

EXPONENTIAL KINETICS

In the exponential phase of growth the rate of synthesis of bacterial substance at any time is proportional to the amount of substance present:

$$dB/dt = \alpha B \tag{1}$$

where B is bacterial mass, t is time, and α is the **instantaneous growth rate constant** for

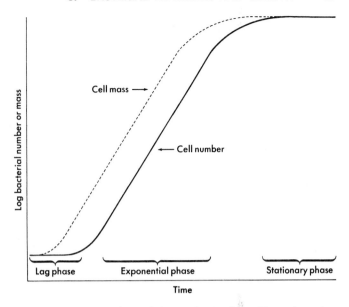

Log bacterial number or mass

Cell mass →

← Cell number

Lag phase   Exponential phase   Stationary phase

Time

Fig. 5-2. Phases of bacterial growth, starting with an inoculum of stationary phase cells. Note that the classic phases, defined in terms of cell number, do not precisely coincide with the phases of changing growth in terms of protoplasmic mass.

that culture (i.e., the relative increase per unit time). Hence

$$dB/B = \alpha dt \text{ (or } d\ln B = \alpha dt) \tag{2}$$

Integrating,

$$B_t = B_o e^{\alpha t} \text{ and} \tag{3}$$

$$\ln B_t/B_o = \alpha t, \text{ or } \ln B_t = \ln B_o + \alpha t \tag{4}$$

Hence in this phase a plot of the logarithm of B against time gives a straight line (Fig. 5-2). This semilogarithmic plot is generally used for bacterial growth curves. The exponential phase is also often called the **log phase,** because the logarithm of the mass increases linearly with time.

It is sometimes convenient to convert the instantaneous growth rate constant α, which has the dimensions of time^{-1}, into the more familiar dimensions of time: $\tau = 1/\alpha$. This **instantaneous generation time,** τ (Gr. tau), represents the time that would be required for a doubling of mass **if** the growth rate at zero time, αB_o, continued unchanged. In exponential growth, however, the value of B, and hence of αB, is constantly increasing, and at the end of a doubling the rate of cell synthesis is twice what it was at the beginning, hence the actual **doubling time,** t_D, is shorter

than τ (Fig. 5-3). The relation between the two is derivable by setting B_t at $2B_o$ (i.e., one doubling) in equation 4:

$$\ln B_t/B_o = \ln 2 = \alpha t_D$$
$$t_D = (1/\alpha) \ln 2 = 0.69 (1/\alpha) = 0.69 \tau$$
$$\text{or } \tau = 1.45 t_D$$

The doubling time is also called the **mean generation time,** or MGT.* Growth rate is usually expressed in terms of t_D or its reciprocal, μ ($= 1/t_D$), which is the exponential growth rate constant, expressed as generations per hour.

Figure 5-3 demonstrates the curvature of exponential growth when plotted linearly, rather than logarithmically, against time. Precisely the same curve would be obtained if cell number were measured instead of cell mass, because ordinary bacterial cultures are **asynchronous:** since the cells at any moment are randomly distributed with respect to stage in the division cycle the rate of cell formation

---

* The doubling time is often also called the generation time, which can be confusing since this term, as noted above, has been applied to another useful function, τ.

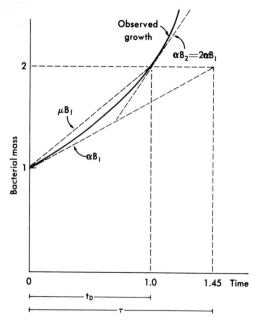

Fig. 5-3. Relation between exponential doubling time (t_D), growth rate constant (α), and linear (instantaneous) generation time ($\tau = 1/\alpha$).

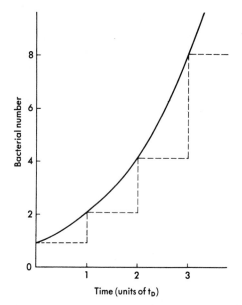

Fig. 5-4. Arithmetical plot of increase in **cell number** in asynchronous (solid line) and hypothetical synchronous (broken line) exponential growth. In either type of growth **mass** would follow the solid line.

rises continuously rather than discontinuously. The growth of a hypothetical perfectly synchronized culture is compared in Figure 5-4.

The linear relation between logarithm of number (or mass) and time is obtained regardless of the base of the logarithm. Logarithms to the base 10 are conventional, but the base 2 is more pertinent since the unit of growth then represents a doubling (one generation). The conversion is made through the relation:

$$\log_2 x = \log_{10} x \,/\, \log_{10} 2 = 3.3 \log_{10} x$$

It is convenient to remember that $2^{10} = 1024$: i.e., 10 generations equal a thousandfold increase and 20 generations a millionfold. In exponential growth plotted in terms of the conventional $\log_{10}$, an increase of 0.3 units = 1 generation.

From these considerations it is evident that exponential growth must be more the exception than the rule in the life of bacteria. A bacterium that doubles in 20 minutes will yield 10 generations (10^3 cells) in 3.3 hours, and 10^9 cells in 10 hours; while a bacterium doubling every 60 minutes would require three times as long. (These are approximately the values for many bacteria on a rich and on a minimal medium.) Hence single cells yield grossly visible colonies (ca. 10^6 to 10^7 cells) in overnight cultures. Furthermore, since the volume of an average bacterium is 1 μ^3, or 10^{-12} cm^3, the volume

of the earth (ca. 4×10^{27} cm^3) is equivalent to 4×10^{39} bacteria; and the progeny of our rapidly growing cell would reach this volume in only 45 hours if growth remained exponential. Fortunately, something becomes limiting earlier. What will limit human population growth has not been decided.

Linear Growth. The molecular basis of exponential growth is illustrated by the effect of adding certain amino acid analogs, whose incorporation instead of a normal amino acid results in formation of nonfunctional proteins (Ch. 7). Growth continues at first at a normal rate but is linear with time; and after a generation or so it ceases. Evidently in exponentially growing cultures the growth rate is limited by the content of at least one kind (and probably very many) of catalytic unit, and when this content is no longer being increased the growth rate is fixed, even though the cell mass increases.

RELATION OF GROWTH TO SUBSTRATE CONCENTRATION

Bioassays with microbes are performed by incubation with limited amounts of a required factor in an otherwise adequate medium; the amount of growth is measured after it levels off. In a satisfactory assay the response is strictly proportional to the amount of the factor provided (Fig. 5-5).

The **amounts** of various substances required to support a good growth yield involve **concentrations** that far exceed those required to saturate the corresponding uptake systems; hence only in the late stages of such a growth curve are the concentrations rate-limiting (Fig. 5-5).

Specifically, the initial concentrations required for maximal growth yield of various auxotrophs of *E. coli* are of the order of magnitude of 5 to 50 μg/ml of various amino acids, purines, or pyrimidines; 0.001 to 0.1 μg/ml of various vitamins; and 0.2 to 0.5% of a carbon source (for growth in shaken flasks). Maximal growth **rate,** in contrast, is observed with limiting amino acids down to ca. 0.01 to 1.0 μg/ml, and with glucose down to ca. 40 μg/ml (ca. 2×10^{-4} M). Hence with **building blocks,** or with general sources of cell material, **the transition from exponential growth to plateau is rapid** (ca. 1% of the range of visible growth; Fig. 5-5). With limiting amounts of required **cofactors,** however, the transition is more **gradual** (Fig. 5-5); as the concentration of a required cofactor in the cell drops the reaction in which it participates slows, but the cell can still make all its macromolecules.

THE CHEMOSTAT

In diagnostic bacteriology overnight cultures have been traditional, but in physiological studies more reproducible conditions are achieved by harvesting the cells in exponential growth. Even in this phase, however, the cells are growing in an ever-changing environment.

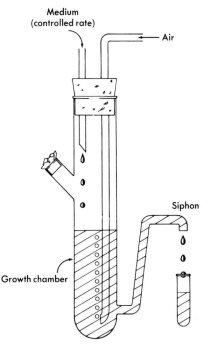

Fig. 5-6. Simplified diagram of the chemostat. [For a detailed description see Novick, A., and Szilard, L. *Proc Natl Acad Sci USA* 36:708 (1950).]

To obtain bacteria that have grown in a medium of truly constant composition Novick and Szilard, and Monod, devised the chemostat, which permits steady-state growth in a continuous-flow culture (Fig. 5-6).

In this instrument fresh medium flows into a filled, stirred growth chamber at a constant, carefully controlled rate; each drop causes a drop of culture to overflow, and so the volume of growing culture is constant. A specific required growth factor is provided in the medium at a **concentration** low enough to limit growth, and thus to determine the **cell density** in the steady-state culture. Moreover, the rate of flow of the medium is set low, so that the cell population can expand fast enough to utilize fully the added medium; the **rate of growth** is thus determined by the **rate of flow.**

The chemostat thus permits indefinite growth of bacteria in a constant medium, with independent control of growth rate and population density. It has made possible very precise analysis of certain regulatory mechanisms (Ch. 8, Mutation rates, and Ch. 13, Regulation of enzyme formation).

Since the mass of growing bacteria remains constant in the steady state in a chemostat, growth of the culture is **linear** rather than exponential. The observed doubling time is therefore identical with the instantaneous generation time (τ), as defined

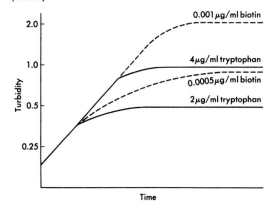

Fig. 5-5. Growth curves (semilogarithmic plot) of auxotrophs given limiting amounts of a required building block (tryptophan) or a required cofactor (biotin).

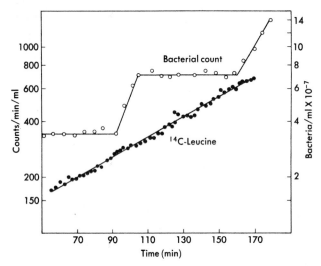

Fig. 5-7. Exponential synthesis of protein in an *E. coli* culture synchronized by filtration. Protein synthesis was measured as incorporation of radioactive leucine from the medium into cell components precipitated with trichloracetic acid. [Adapted from Abbo, F. E., and Pardee, A. B. *Biochim Biophys Acta 39:* 478 (1960).]

above. Growth of the individual cell, however, is exponential, for as the cell gets larger it takes up a larger fraction of the constantly limiting nutrient.

In a specific example, a 50-ml chemostat is inoculated with a tryptophan auxotroph, and medium containing 0.2 μg/ml of tryptophan flows in at 25 ml/hour. The generation time is then 2 hours; and as a first approximation the cell density is the same as that produced by the same organism and medium in an ordinary culture. The steady-state culture fluid will contain free tryptophan at a low concentration, which just permits the transport units to take up tryptophan at the rate corresponding to the growth rate.

SYNCHRONIZED GROWTH

In a growing bacterial culture the cells are distributed among all stages in their division cycle.* Hence chemical analyses yield only average values. But there are a number of things we would like to know about the properties of the individual cell in the course of its division cycle: the kinetics of any

---

* Bacteriologists frequently speak of "old" cells in referring to cells from old cultures, which may mean either cultures in the stationary phase or those stored for a long time. However, the concept of age in a physiological sense is not useful for bacteria, since they undergo binary fission rather than a life cycle that includes senescence and death.

cyclic change (or lack of change) in the synthesis of its major components (particularly DNA synthesis in relation to cell division), variations in mutability or stability of the genetic material, susceptibility to lethal agents, and competence for transfer of genetic material. Since it is not possible to study such problems with isolated cells, technics have been sought for synchronizing the growth of bacteria, i.e., for producing cultures in which all the cells would be in approximately the same part of the division cycle.

Synchronization of bacterial growth was first achieved by various procedures that delayed initiation of DNA replication and then restored the condition required (Ch. 13, Regulation of DNA synthesis). Methods used included repeated alternation between short periods of incubation at 37° and at 25°, or depriving an auxotroph of thymine followed by its restoration. These methods, however, yield cells whose composition and state of balanced growth have been drastically altered; and such cells do not appear to provide reliable information on the regulatory mechanisms responsible for balanced growth.

Physiological synchronization has been achieved by **mechanical separation** of the smaller, recent products of normal division from cells in other stages of the cycle. One approach consists in filtering cultures through a number of layers of filter paper, which act as sieves: the smallest cells are found in the effluent, while the larger cells, on the verge of dividing, may be recovered from the top layers. Under favorable circumstances either fraction gives synchronization on subsequent incubation, and it seems probable that balanced growth is resumed with little disturbance. In a simpler procedure cells are adsorbed or wedged in the **pores of a membrane filter;** subsequent reverse flow of fresh medium through the filter provides a continuous supply of newborn cells, derived by fission from the permanently trapped cells.

During a synchronized doubling of cell size the synthesis of the major constituents proceeds exponentially (Fig. 5-7), except for DNA (see Ch. 10). This finding leads to the important conclusion that **exponential growth,** $dB/dt = \alpha B$, **is a property of the growing cell,** and not simply a statistical property of the culture: a cell on the verge of dividing has twice as many enzyme molecules and ribosomes as a cell just formed, and it grows twice as fast.

The **persistence of synchronization** varies, but little ordinarily remains after two to four generations. This finding is in harmony with early microscopic observations on the division time of single cells on solid media, which revealed a rather large standard deviation from the mean. This deviation may reflect the absence of a mechanism for ensuring precise equipartition of cytoplasm in cell division.

GROWTH ON SOLID MEDIUM

SOLIDIFYING AGENTS

The earliest method available for enumerating bacteria or for isolating pure cultures (clones)* was **extinction dilution,** i.e., dilution up to loss of infectivity. An important advance was the introduction by Koch of **solid media.** The initial materials used were unsatisfactory: the cut surface of potato might permit confluent growth, while gelatin was solid only at relatively low temperatures and was liquefied by many organisms. The world is therefore much indebted to Frau Hesse, the wife of a German physician and amateur bacteriologist, for introducing agar, a Japanese product with which she was familiar for thickening soups.

Agar (Malay *agar-agar*) is an acidic polysaccharide derived from certain seaweeds (various red algae); it consists primarily of galactose, with 1 sulfate derivative per 10 to 50 residues. Agar proved to be an ideal substance for making solid media. It is nontoxic to bacteria, and very few attack it. At 1.5 to 2.0% agar the surface is wet enough to support growth but dry enough to keep colonies separate. (Exceptionally motile organisms, such as *Proteus*, require ca. 5% agar to prevent swarming.) After melting at 80 to 100° agar solutions remain fluid at temperatures down to 45 to 50°, which most bacteria can withstand briefly; and after solidifying at room temperature they remain solid far above 37°. The fibrous structure of the gel is fine enough to prevent motility of bacteria within it, but coarse enough to permit diffusion even of macromolecular nutrients.

Being a material of natural origin, agar contains traces of various metabolites; hence many auxotrophs will yield microscopic colonies on minimal agar. Being an anionic macromolecule, agar binds cations such as $Mg++$ and $Ca++$. A sulfate-free fraction, **agarose,** is available for special purposes. Purer solid media can be made, though inconveniently, through the use of **silica gel,** formed by neutralizing dilute Na_4SiO_4 with HCl.

USES OF SOLID MEDIA

For **enumerating viable cells** dispersed bacteria are spread on the surface of a prepared agar plate or are added to melted agar medium at 45°, which is then immediately poured.

In using a selective solid medium to **isolate pure cultures** from a naturally occurring mixture an important precaution must not be overlooked: if the selected cells are only a minor proportion of the population other organisms may be present in the area of a colony, and may even be fed by it. These contaminants may then become predominant if the colony is transferred to a nonselective medium for storage. Indeed, the literature is replete with results obtained with allegedly pure cultures that were actually mixtures iso-

---

* A clone is defined as a group of cells, derived by vegetative reproduction from a single parental cell. Thus any pure culture of bacteria is a clone, and the progeny of a mutant arising in it are a subclone.

lated from a single plating. To isolate clones reliably a colony from the first plate should be streaked again to yield single colonies on **a second plate of the same medium;** one of these colonies can then be used to furnish the stock culture.*

Solid media are used most widely for the **identification** of microorganisms, including pathogens. A practiced eye can recognize a variety of species in a single throat culture, simply on the basis of size, color, shape, opacity, hemolytic activity, and surface texture of the colonies (see below); the presumptive diagnosis can then be confirmed by microscopic examination and by subculture onto special diagnostic media.

Because molecules can diffuse through agar while bacteria cannot, agar media have lent themselves to a variety of special applications.

1) **Nutritional activity** of spots of added substances, known and unknown, can be detected on plates of inadequate medium containing a test organism (10^2 to 10^4 cells/ml); moreover, with known amounts of material added on a disc of thick filter paper the size of the zone of response yields reasonably accurate bioassays. Similarly, paper chromatograms can be incubated on large seeded plates to detect the positions of growth factors.

2) Conversely, with nutritionally adequate, heavily seeded plates a similar procedure provides a qualitative test or a quantitative bioassay for **antimicrobial agents** (Ch. 7, Fig. 7-2).

3) Streaking two strains close to each other on a medium that supports poor growth can reveal **cross-feeding** of one by a growth factor released by the other **(syntrophism).** This phenomenon, originally recognized as **satellite** growth on diagnostic plates, has proved invaluable in the recognition of biosynthetic intermediates excreted by auxotrophic mutants (Ch. 4, Fig. 4-1).

At the edge of a zone of inhibition growth may be heavier than elsewhere, owing to extra nutrient diffusing from the zone of inhibition.

COLONIAL MORPHOLOGY

Surface Texture. One of the most important diagnostic features of a colony is the texture

of its surface, ranging from rough (R) to smooth (S) to mucoid (M) (Fig. 5-8). **Smooth** colonies reflect the presence of a capsule or other surface component promoting a compact cellular orientation. **Rough** colonies have a dry and sometimes wrinkled surface; they are formed by cells that lack such a component or by cells growing in a filamentous manner. Different degrees of roughness are possible within a species, and are often correlated with differences in virulence.

Fig. 5-8. Variations in colonial morphology of *Pneumococcus* strains. **A.** Smooth colonies of capsulated, nonfilamentous cells. **B.** Rougher colonies of capsulated but filamentous variant. **C.** Nonfilamentous noncapsulated variant. **D.** Roughest variant, filamentous and noncapsulated. All photographs ×18, after 24 hours' incubation at 36° on blood agar. [From Austrian, R. *J Exp Med* **98:**21 (1953).]

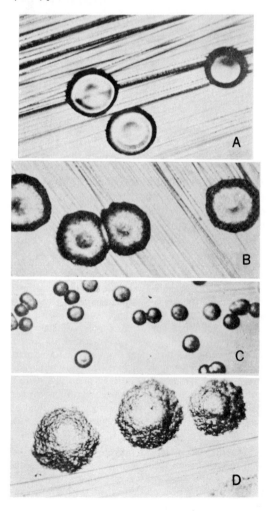

Thus with tubercle bacillus **virulence** is associated with a lipophilic surface component that causes the cells to adhere to each other in serpentine **cords,** while mutations to decreased virulence are often associated with a decreased tendency to form cords and hence with a smoother surface of the colonies (Ch. 35). With many organisms, in contrast, rough strains are avirulent, and **virulent strains** form a surface component associated with a **smooth** colony: polysaccharide side chains on the outer membrane in gram-negative organisms (e.g., Enterobacteriaceae) and a capsule in gram-positive organisms (e.g., pneumococci Fig. 5-8). In the former a capsule may also be formed, yielding a **mucoid** or **glossy** colony, but it is not very important for virulence. The structure of these surface molecules will be further considered in Chapter 6 and their role in virulence in Chapter 22.

Capsule formation, and therefore colonial morphology, sometimes depends on environmental as well as on genetic factors. Some nonmucoid Enterobacteriaceae become mucoid if grown at low temperature or in a medium with an excess of carbon source and a limitation of nitrogen or phosphorus. Mucoid colonies may be huge because of the volume of capsular material produced; some are liquid enough to run when the plate is tilted.

Size. Colony size provides a more sensitive comparison of growth rate than does the most careful direct measurement of growth rate in liquid medium. Thus, it is not possible to detect directly a 1% difference in growth rate of two organisms. However, growth from 1 cell to the 10^7 cells of a colony involves about 23 generations; and if the differential of 1% is maintained through this period the faster grower should multiply through $(1.01/1)^{23} = 1.3$ times as many generations, and hence should yield a colony larger in volume by the same ratio.

Differential media may reveal specific characteristics without being selective, and are very useful in bacterial identification. **Blood agar,** containing 5% sheep or horse blood, can reveal production of a hemolysin. In **fermentation plates** utilization of the particular sugar provided is revealed by indicator dyes, such as an eosin-methylene blue (EMB) mixture, which not only changes color but precipitates in the presence of acid; hence the colony itself is stained and precipitation is sufficiently localized to demarcate stained **sectors** in a colony that contains a mutant subclone (Fig. 5-9). Fermentation plates are used largely for facultative organisms; they are effective even when incubated in air, because such organisms convert sugars to organic acids faster than they can burn the latter to carbon dioxide.

A fermentation plate must contain other nutrients besides the test sugar, to permit growth of fermentation-negative organisms. When such negative colonies are fully grown they may give rise, on pro-

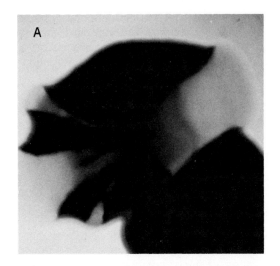

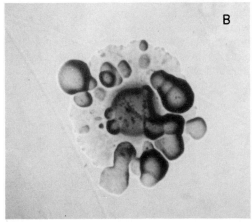

Fig. 5-9. A. Lac− (unstained) sectors in colony derived from an ultraviolet-irradiated lac+ *E. coli* cell, plated on EMB-lactose medium (which stains only lac+ cells). Note sharp demarcation of sectors, and adjacent lac+ colony without lac− mutants. **B.** Lac+ papillae, arising late in lac− colony incubated for several days on EMB-lactose medium. (**A** courtesy of H. B. Newcombe. **B** courtesy of V. Bryson.)

longed incubation, to positive **papillae** (Fig. 5-9) derived from mutant cells arising late in the growth of the colony. Indeed, even on nondifferential media prolonged cultivation frequently gives rise to papillae that can grow on nutrients not adequate for the mother colony; hence colonies on old plates are frequently warty and irregular.

Crowding. In diagnostic bacteriology it is important to deal with well-separated colonies: crowded colonies are too small to reveal characteristic morphology, and their extensive production of acid can obscure the staining of fermenting colonies and can cause diffuse nonspecific hemolysis. It is also important, of course, to inspect plates after a rather standard period of incubation (usually 18 to 24 hours).

The cells in different parts of a colony will differ in access of oxygen from above and of nutrients from below. The resulting differences in growth rate and metabolic state cause cells from solid media to vary in size more than those obtained from an exponentially growing liquid culture.

SELECTIVE MEDIA

The cell types that predominate in a population can be readily cloned on a plate. For the isolation of minor components, however, **selective media** are necessary. Selective liquid media provide **enrichment** of the population with respect to the desired organisms, while solid media permit direct **isolation.** General microbiologists use extreme conditions to "force" the outgrowth, from highly mixed

inocula, of organisms that can carry out autotrophic metabolism, fix N_2, etc., while medical microbiologists employ related procedures to recognize the known pathogens.

Enrichment is based on several principles. Organisms that can utilize a given sugar are easily screened for by making that compound the only carbon source. Selection of nonutilizers (nonfermenters), however, is more difficult. Similarly, the use of a minimal medium, or one containing few growth factors, will exclude fastidious organisms; selection for such organisms is harder and generally involves **selective inhibition** of the nutritionally simpler species (e.g., by unfavorable pH, salts, specific inhibitors).

A variety of useful selective inhibitors have been empirically discovered, such as tellurite for growth of diphtheria bacilli in throat cultures, bismuth for pathogenic *Salmonella* and *Shigella* in stool cultures, and various dyes (e.g., malachite green), as well as preliminary treatment with strong acid or alkali, to permit recovery of the slow-growing but hardy tubercle bacillus.

SELECTED REFERENCES

Books and Review Articles

ALEXANDER, M. *Microbial Ecology.* Wiley, New York, 1971.

EMERY, T. Hydroxamic acids of natural origin. *Adv Enzymol 35*:135 (1971).

FARRELL, J., and CAMPBELL, L. L. Thermophilic bacteria and bacteriophages. *Adv Microbial Physiol 3*:83 (1969).

GUIRARD, B. M., and SNELL, E. E. Nutritional Requirements of Microorganisms. In *The Bacteria,* vol. IV, p. 33. (I. C. Gunsalus and R. Y. Stanier, eds.) Academic Press, New York, 1962.

HUGHES, D. E., and WIMPENNY, J. W. T. Oxygen metabolism by microorganisms. *Adv Microbial Physiol 3*:197 (1969).

INGRAHAM, J. L. Temperature Relationships. In *The Bacteria,* vol. IV, p. 265. (I. C. Gunsalus and R. Y. Stanier, eds.) Academic Press, New York, 1962.

KNIGHT, B. C. J. G. Growth factors in microbiology. *Vitam Horm 3*:108 (1945).

LARSEN, H. Biochemical aspects of extreme halophilism. *Adv Microbial Physiol 1*:97 (1967).

MAALØE, O. Synchronous Growth. In *The Bacteria,* vol. IV, p. 1. (I. C. Gunsalus and R. Y. Stanier, eds.) Academic Press, New York, 1962.

SNOW, G. A. Myobactins: Iron-chelating growth factors from bacteria. *Bacteriol Rev 34*:99 (1970).

Symposium: Continuous Culture Methods and Their Application. In *Recent Progress in Microbiology* (7th International Congress of Microbiology, Stockholm, 1958). Thomas, Springfield, Ill., 1959.

TEMPEST, D. W. Quantitative relationships between inorganic cations and anionic polymers in growing bacteria. In *"Microbial Growth,"* 19th *Symp Soc Gen Microbiol,* Cambridge Univ. Press, 1969, p. 87.

TEMPEST, D. W. The place of continuous culture in microbiological research. *Adv Microbial Physiol 4*:223 (1970).

Specific Articles

BROCK, T. D. Life at high temperatures. *Science 158*:1012 (1967).

HITCHENS, A. P., and LEIKIND, M. C. The introduction of agar-agar into bacteriology. *J Bacteriol 37*:485 (1939).

MONOD, J. La technique de culture continue: Théorie et applications. *Ann Inst Pasteur 79*:390 (1950).

chapter 6

THE CELL ENVELOPE; SPORES

Revised in collaboration with LORETTA LEIVE, Ph.D.

CELL WALL   107
   Gram-positive and Gram-negative Walls   107
   PEPTIDOGLYCAN   110
   Structure   110
   Biosynthesis   111
   Enzymes Attacking Peptidoglycan   113
   Teichoic Acids   114
   OUTER LAYER OF GRAM-NEGATIVE ORGANISMS   116
   Structure   116
   Function   117
   Lipopolysaccharide Structures   117
   Lipopolysaccharide Synthesis   119
   ANTIBIOTICS ACTING ON WALL SYNTHESIS   120

CYTOPLASMIC MEMBRANE   121
   Composition   121
   Organization and Functions   121
   Mesosomes   122
   PERIPLASMIC PROTEINS   123
   STRUCTURE AND BIOSYNTHESIS OF LIPIDS   124
   Fatty Acids   124
   Complex Lipids   125
   Variations in Composition   125
   ANTIBIOTICS ACTING ON MEMBRANE   126
   Damage to Membrane   126
   Ionophores   127

THE ENVELOPE IN CELL DIVISION   128
   Wall Growth and Separation   128
   Sites of Growth of Wall   128
   Mutations in Wall Synthesis   129

MEMBRANE TRANSPORT   130
   Existence of Transport Systems in Bacteria   130
   β-Galactoside Transport   132
   The Exit Process   132
   Other Transport Systems   133
   Nonspecific Entry   134

MECHANISM OF ACTIVE TRANSPORT   134
   Isolation of Binding Proteins   134
   Carrier Model for Active Transport   135
THE PHOSPHOTRANSFERASE SYSTEM   136
CHEMOTAXIS AND CHEMORECEPTORS   137

SPORES   137

   Formation and Structure   138
   Composition of Core   141
   Regulatory Changes in Sporulation   142
   Germination   143
SUMMARY   145

CELL WALL

As we have seen in Chapter 2, the cell envelope of eubacteria contains a **cytoplasmic membrane** surrounded by a **rigid wall.** This chapter will discuss the molecular organization, biosynthesis, and functional properties of these structures and their cleavage during cell division. We shall also consider sporulation, which involves drastic changes in the cell envelope.

The unique structures of the cell wall are of particular interest for medicine. Thus the surface macromolecules (antigens) largely determine the interactions of bacteria with host defense mechanisms; the serological identification of these antigens is a major tool in diagnostic bacteriology; and wall biosynthesis is the site of action of many antibiotics. In contrast, the membrane is of interest more because of its universal properties, which may be readily manipulated in bacteria both by mutations and by environmental influences.

GRAM-POSITIVE AND GRAM-NEGATIVE WALLS

The Gram stain (Ch. 2) divides bacteria into two classes, which differ in their ability to retain a basic dye after fixation by I_2. This difference is based on major differences in the structure of the cell wall, demonstrated by electron microscopy and by chemical analysis.

Morphology. Sections of **gram-positive** cells (fixed by OsO_4) exhibit a rather thick (200 to 800 A) electron-dense outer layer, surrounding a cytoplasmic (plasma) membrane (75 A) (Fig. 2-4). The membrane has the trilaminar ("railroad track") cross-section typical of biological membranes, in which the charged groups are located on the two surfaces and become stained while the lipid interior forms a clear band. (A structure with this appearance is sometimes called a "unit membrane.") The typical envelope of **gram-negative** cells, in contrast, has three major layers (see Fig. 6-3A below): an innermost cytoplasmic membrane, a thin (20 to 30 A) electron-dense layer (corresponding to the thicker wall of gram-positive cells), and an outer layer (60 to 180 A). The latter two are not visibly separated in sections. The outer layer is also called the **outer membrane** or the **L** layer; it resembles typical membranes in its composition and in its trilaminar appearance. Freeze-etching, which in effect peels apart the layers (Ch. 2), confirms the presence of only two layers in gram-positive organisms and distinguishes three in gram-negative organisms (Fig. 6-1).

Composition. The **cytoplasmic membrane** of gram-positive cells is readily recovered, in fragments, as a particulate fraction after osmotic lysis of protoplasts. This material is largely composed, like other membranes, of lipid and protein.

The walls of **gram-positive** bacteria are easy to purify. If the cells are mechanically disrupted (e.g., by vigorous shaking with glass beads) the cell contents are released, and the insoluble envelope is readily separated from the soluble contents. The lipoprotein membrane is then removed from the wall by detergents. The residual insoluble wall retains the shape of the cell (Fig. 6-2) and is obviously the layer responsible for the rigidity of the cell and for its resistance to osmotic lysis;* it corresponds to the thick electron-

---

* Though the peptidoglycan is rigid enough to give various bacteria characteristic shapes, it is not entirely inflexible: in spirochetes (which are sensitive to penicillin and therefore must contain a peptidoglycan) motility involves extensive bending of the long, helical cells (Ch. 37).

Fig. 6-1. Surface of *E. coli* revealed by freeze-etching. W = outer surface of cell wall. At the arrow the outermost layer has been peeled away, revealing an underlying shelf; it is not certain whether this is peptidoglycan or lipoprotein. Both these layers have been removed over most of the surface, revealing the pebbly outer surface of the plasma membrane (PM). [From Bayer, M. E., and Remsen, C. C. *J Bacteriol 101*:304 (1970).]

dense layer of Figure 2-4A. It was identified chemically as **peptidoglycan** (also called **glycopeptide, mucopeptide,** or murein); and in many gram-positive cells accessory chains called **teichoic acids** are covalently attached on the outer surface. These structures will be discussed below.

In **gram-negative** bacteria the total wall cannot be similarly isolated free of membranes because its outer layer resembles the underlying membrane too closely. However, its rigid layer (basal wall) can be purified by use of detergents and lipid solvents (after cell disruption); it is fundamentally the same kind of peptidoglycan as that in gram-positive cells, but the amount is much smaller and there is no teichoic acid. This material can be identified with the middle electron-dense layer of the gram-negative wall, since it is eliminated by lysozyme (Fig. 6-3B), which specifically attacks the peptidoglycan (see below).

As was noted in Chapter 2, in a hypertonic medium dissolution of the peptidoglycan by lysozyme converts bacteria into osmotically sensitive spheres: **protoplasts** (bounded by the cytoplasmic membrane) from gram-positive cells, or **spheroplasts** (which also retain the outer membrane) from gram-negative cells.

Fig. 6-2. Shadow-cast electron micrograph of purified wall preparation from *B. megaterium.* Note flattened structure compared with intact cell. Latex balls: diam. 0.25 μ. [From Salton, M. R. J., and Williams, R. C. *Biochim Biophys Acta 14*:455 (1954).]

Fig. 6-3. A. Separation of cytoplasmic membrane (CM) from cell wall (CW) in purified envelope of *E. coli.* **B.** Partial dissolution of peptidoglycan layer of cell wall of *E. coli* on brief treatment with lysozyme. The upper half of the section of wall has a thick, dense band outside the plasma membrane and separated by a clear layer of constant thickness from an outer, thin, dense band; untreated cells have the same structure. In the lower half lysozyme has removed much material along the thick layer but a thinner residue remains. Thus in this gram-negative wall the peptidoglycan is fused with the outer layer, which is left after its removal [**A** from Schnaitman, C. A. *J Bacteriol 108*:545, 1971. **B** courtesy of R. G. E. Murray. See *Can J Microbiol 11*:547 (1965).]

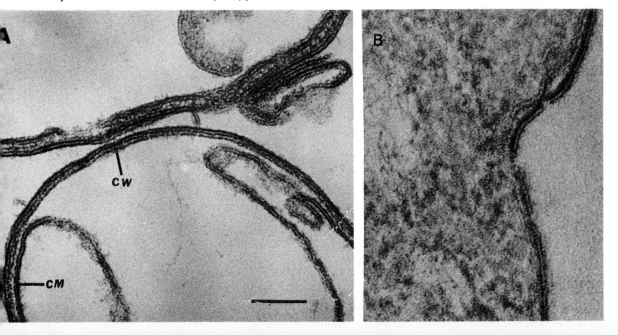

PEPTIDOGLYCAN

STRUCTURE

Acid hydrolysis of isolated membrane-free walls yields a small number of amino acids (some in the D configuration) and two sugars, N-acetylglucosamine (shown as GlcNAc in the figures) and its 3-O-D-lactyl ether, N-acetylmuramate (L. murus = wall; shown as MurNAc). Additional covalently attached residues vary from one organism to another. The enzyme lysozyme (from egg white) was found to hydrolyze a specific glycoside bond in the polymer, yielding a substituted disaccharide, GlcNAc-MurNAc. Later the use of some 20 additional bacteriolytic enzymes from various sources, each specific for a different linkage within the wall, yielded a variety of fragments. Their identfication, together with that of accumulated biosyn-

thetic intermediates (see below), revealed the linkages between the various monomers, establishing for *Staphylococcus aureus* the structure shown in Figure 6-4.

The essential features of this structure, found in virtually all bacteria,* are a **backbone** of alternating MurNAc and GlcNAc residues in β-1,4 linkage, a **tetrapeptide** substituent with alternating L and D residues, and a **peptide bridge** from the terminal carboxyl of one tetrapeptide (position 4) to an available NH₂ or COOH group of a neighboring tetrapeptide (Fig. 6-4). Within this frame-

---

* Exceptions are 1) the mycoplasmas, which lack a rigid wall and characteristic shape, and 2) certain halophilic bacteria (Ch. 5, whose evolutionary adaptation to a high external salt concentration has evidently eliminated the need for a peptidoglycan layer to prevent osmotic lysis.

Fig. 6-4. Structure of peptidoglycan of *S. aureus*. The polysaccharide chains (backbone) are β-1,4-linked polymers of alternating residues of N-acetylglucosamine (GlcNAc) and its 3-D-lactic ether (N-acetylmuramic acid = MurNAc; structure in Fig. 6-5). (The chain can also be considered a substituted homopolymer of GlcNAc, but because the lactic ether bond is very stable the muramic acid is ordinarily recovered as such in hydrolysates.) The COOH of the lactic group is attached to a tetrapeptide, which is in turn linked by a peptide cross-bridge to a tetrapeptide on a neighboring polysaccharide chain (which may be above, below, or in the depicted plane). Teichoic acid chains are attached to occasional MurNAc residues.

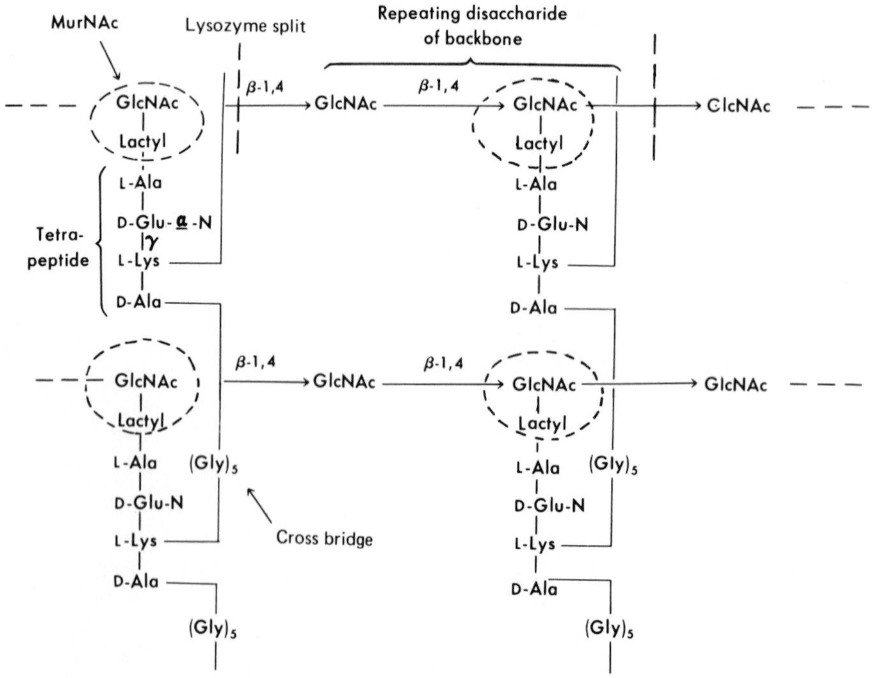

work many variations are found: in amino acids 2 and 3 of the tetrapeptide, in the structure of the cross-bridge, and in its frequency (Table 6-1).

Thus many gram-negative bacteria have a direct peptide bond between two tetrapeptides; *S. aureus* has a pentaglycine bridge (as noted above); and the frequency of peptide substitution on MurNAc varies from nearly 100% in *S. aureus* to ca. 30% in *E. coli. Micrococcus lysodeikticus* (which is named for its ready lysis by lysozyme) has a particularly loose structure: ca. 60% of the MurNAc residues have lost their tetrapeptides, which have been transferred to form long cross-bridges (consisting of several tetrapeptides linked end to end) between the re-

maining tetrapeptides. The wide variations in the peptidoglycans of various gram-positive organisms are of value for taxonomy.

The cross-linked structure not only joins the "backbone" polysaccharide chains into indefinite two-dimensional sheets but evidently also forms bridges **between** sheets, since the concentric layers fail to peel apart in isolated walls. The entire peptidoglycan of a cell would then be **one giant, bag-shaped, covalently linked molecule** (murein sacculus).

Additional features of the peptidoglycan contribute to its toughness. The β-1,4 provides a particularly compact, strong polysaccharide chain: it is also found in chitin (a poly-N-acetylglucosamine in fungal cell walls and crustacean exoskeletons) and in cellulose (a polyglucose in plants). Moreover, synthetic polymers of alternating D and L amino acids (as in the tetrapeptide) are stronger than those of either kind of monomer alone. Finally, extensive hydrogen bonding might be expected between the peptide groups, as in proteins.

BIOSYNTHESIS

Peptidoglycan biosynthesis occurs in four stages: 1) synthesis of water-soluble complex precursors; 2) attachment to a membrane lipid, followed by further additions; 3) formation of linear polymers outside the membrane; and 4) cross-linking of these polymers. It is the most complex known sequence of enzymatic reactions.

Soluble Intermediates. The key to peptidoglycan biosynthesis came not from a direct attack on this problem but from Park's observation that penicillin-treated staphylococci accumulate a novel type of compound: a short peptide attached to a nucleotide. Its function was a mystery until the basal wall was later isolated and was found to contain the same few amino acids; thus began a long mutual interaction of studies on wall synthesis and on penicillin action.

The accumulated compounds were found to be uridine nucleotides of MurNAc, with the carboxyl of the lactic acid moiety attached to a pentapeptide (Fig. 6-5) or to shorter peptides (presumably precursors). The biosynthetic sequence (Fig. 6-6) was then established by in vitro enzymatic studies of Strominger. The synthesis of the complete UDP-MurNAc-pentapeptide precursor, from

TABLE 6-1. Variations in Peptidoglycan in Gram-positive Bacteria

Type structure

```
                                    Bridge
      ①    ②   γ③    ④      |
MurNAc-L-Ala-D-Glu-L-Lys-D-Ala-C=O
             |α      |ε
            COOH  NH-bridge
```

Variations in tetrapeptide

Position 2: D-Glu-COOH (α)

D-Glu-CONH$_2$ (α)

D-Glu-Gly (α)

Position 3: L-Lys
meso-DAP
L,L-DAP
L-Ornithine
L-α,γ-Diaminobutyric acid
L-Homoserine
L-Glu
L-Ala

Variations in cross-bridge from D-Ala (position 4) of one tetrapeptide to NH$_2$ or COOH in another tetrapeptide

To NH$_2$ at position 3 via:
  Direct peptide bond (i.e., COOH of D-Ala to NH$_2$ of Lys)
  (Gly)$_5$
  (L-Ala)$_4$-L-Thr
  (Gly)$_3$-(L-Ser)$_2$
  L-Ser-L-Ala
  L-Ala-L-Ala
  D-Asp-NH$_2$
  D-Asp-L-Ala
  Standard tetrapeptide
To COOH at position 2 (requiring a diamino acid in the cross-bridge):
  D-Ornithine
  D-Diaminobutyric acid
  Gly-L-Lys

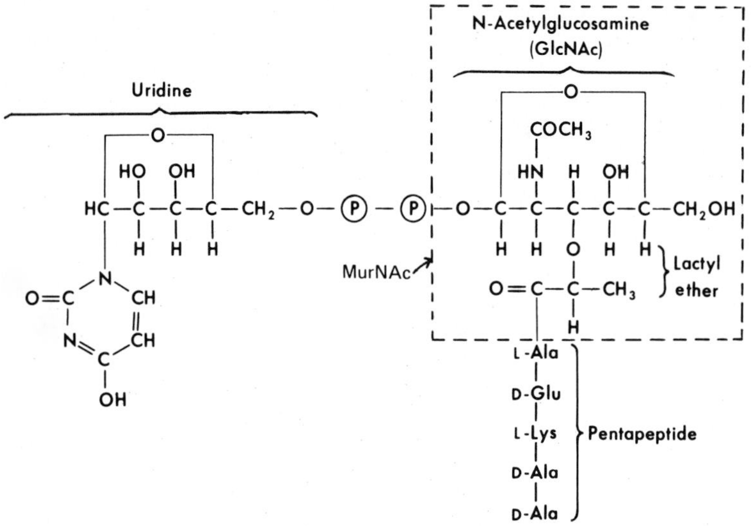

Fig. 6-5. Structure of UDP-Mur-NAc-pentapeptide, a uridine nucleotide from penicillin-treated *S. aureus.*

ordinary metabolites, requires 15 enzymes, present in the cytoplasm.

Unlike the synthesis of protein (Ch. 12) and of peptide antibiotics (Ch. 4), synthesis of this wall precursor utilizes a separate enzyme to catalyze each successive addition of an amino acid (except that the terminal D-ala-D-ala is made as a dipeptide and then added). Moreover, the amino acids are activated by the split of P_i from ATP, rather than by the more expensive split of PP used in these other processes.

Lipid-attached Intermediates. Further studies of Strominger and of Park with radioactive precursors showed that particulate, lipid-containing enzyme preparations, presumably derived from the cell membrane, can form a high-molecular-weight, acid-precipitable peptidoglycan from appropriate precursors. When the components were fractionated it was found that P-MurNAc-pentapeptide is first transferred from its nucleotide to the phosphate of a **membrane carrier lipid,** identified as **undecaprenol-P**

$$CH_3$$
$$H-(CH_2-C=CH-CH_2)_{11}-OP.$$

(This carrier is also employed in lipopolysaccharide biosynthesis; see below.) GlcNAc

Fig. 6-6. Pathway of biosynthesis of the UDP-muramic-pentapeptide wall precursor in *S. aureus.* The individual amino acids are added with the use of energy from ATP. The sites of action of various inhibitors, which have been useful in working out this pathway, are indicated. [After Strominger, J. L., and Tipper, D. J. *Am J Med* **39:**708 (1965).]

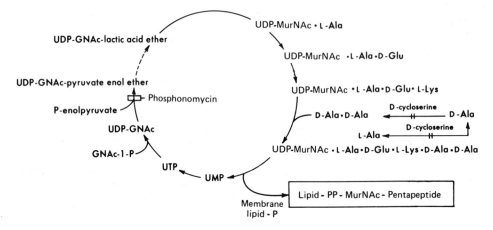

is then added from its UDP derivative, and those organisms that form a polypeptide bridge add it at this stage to the pentapeptide (Fig. 6-7).

The activated amino acids required for synthesis of bridge peptides are provided by the corresponding aminoacyl-tRNA (Ch. 12). In some cases (e.g., glycyl-tRNA of *S. aureus*) the cell has a special tRNA that functions only in peptidoglycan synthesis, though the tRNAs of protein synthesis can also serve in vitro.

Polymerization. In the polymerization reaction the completed disaccharide subunit is transferred from the lipid carrier to the growing end of a polysaccharide chain in the peptidoglycan. The undecaprenol-PP thus generated in the membrane then loses a P (Fig. 6-7), regenerating the form that can again accept a precursor, and possibly providing energy that promotes reorientation toward the inner face of the membrane.

Cross-linking. The terminal step, cross-linking of the polypeptide side chains, occurs **outside** the cell membrane and hence cannot draw directly on intracellular ATP to provide energy for creating a peptide bond. Instead, the energy source is built into the pentapeptide, which contains one more amino acid than is finally retained. The cross-link can then be formed by an energetically neutral **transpeptidation** reaction (Fig. 6-8), in which a free NH_2 group displaces the terminal D-alanine of a nearby sidechain, releasing that residue and forming a bridge to the sub-terminal D-alanine. This reaction was defined through its specific block by penicillin (see below, Antibiotics), in cells and subsequently in extracts: precursors are still polymerized but the products are not cross-linked.

The interference with cross-linking in penicillin-treated cells was demonstrated in several ways. 1) NH_2-terminal glycine accumulates. 2) An excess of labeled D-alanine is retained in the polymer. 3) The uncross-linked product is much more soluble than the cross-linked polymer formed in uninhibited cells. 4) Cells growing in the presence of penicillin accumulate large amounts of amorphous material between the membrane and the wall, presumably in zones of growth (Fig. 6-9).

ENZYMES ATTACKING PEPTIDOGLYCAN

Bacteria contain several kinds of **autolysins** (peptidoglycan hydrolases): **glycosidases**

Fig. 6-7. Lipid cycle in peptidoglycan biosynthesis in *S. aureus*. The acceptor is presumed to be the free end of a polysaccharide. The carrier lipid is a polyisoprene chain with a terminal phosphate (undecaprenol phosphate). [After Matsuhashi, M., Dietrich, C. P., and Strominger, J. L. *Proc Nat Acad Sci USA* 54:587 (1965).]

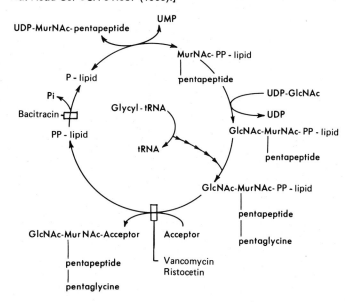

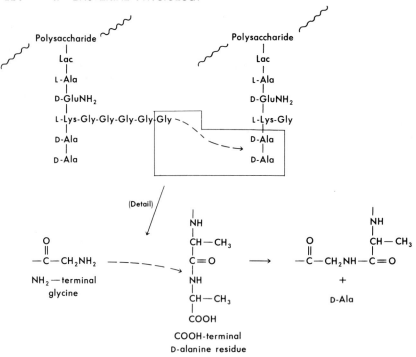

Fig. 6-8. The transpeptidation reaction in *S. aureus*, which completes the cross-link between different peptide side chains: the D-ala ⟶ D-ala peptide bond (CO ⟶ NH) is replaced by a similar D-ala ⟶ gly bond. Some species use other cross-links than the pentaglycine bridge.

(specific for one or the other glycoside bond between the alternating sugars in the backbone), an **amidase** (which releases the tetrapeptide from MurNAc), and **endopeptidases** (which attack various bonds in the bridge). The autolysis of dead cells first revealed the existence of these enzymes, but their main function is probably morphogenetic (i.e., splitting and reshaping the wall in the region of a dividing septum). Thus mutants lacking one or another autolysin may be unable to separate daughter cells or may have aberrant shapes.

Enzymes that attack various bonds in peptidoglycans are also widely distributed outside bacteria and have several functions: protection against infections (in plants as well as animals); food supply for animals (e.g., many protozoa) that ingest bacteria; and breakdown of organic matter in the soil. Animal lysozyme (which splits the MurNAc-GlcNAc bond) is probably very important in resistance to infection; the role of other enzymes of this general class has not been investigated.

The possibility of chemotherapeutic application of such enzymes initially seemed promising. However, as with other foreign enzymes injected into tissues, immune reactions interfere.

TEICHOIC ACIDS

These polyol-phosphate chains (Gr. *teichos* = wall), discovered by Baddiley, are present in the walls of many gram-positive bacteria, attached to occasional Mur-NAc residues of the peptidoglycan backbone. The polymer may be a chain of either **ribitol** residues or **glycerol** residues with phosphodiester links. Teichoic acids are major surface antigens, comparable to the lipopolysaccharides of gram-negative organisms (see below), but with a more limited range of constituents.

The chains vary in length, up to 30 residues. (However, the preparations studied may be partly degraded, since teichoic acids are readily hydrolyzed.) Various substituents (sugars, aminosugars, choline, D-alanine) may be attached, providing specific antigenic determinants. Teichoic acid is evidently located on the surface since in intact cells it

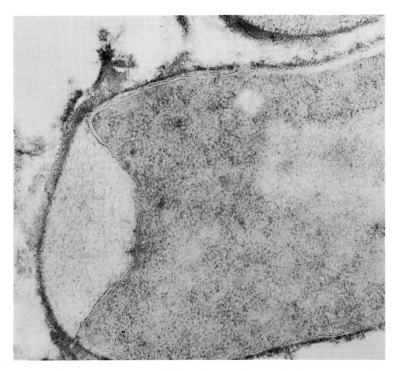

Fig. 6-9. Disorganized cell wall formation in *B. megaterium* grown in the presence of penicillin. Note the pocket of fibrous material between cell membrane and cell wall. Such accumulations are usually seen earliest in the region of the growing septum. [From Fitz-James, P., and Hancock, R. *J Cell Biol 26*:657 (1965).]

reacts with antibodies; it may amount to as much as 40% of the dry weight of the wall. The varied patterns of substitution can provide a variety of antigen types even within a species (e.g., *S. aureus*).

Ribitol teichoic acid is formed as follows, by a membrane-bound enzyme:

$$n[\text{Cytidine—P—P—ribitol}] \longrightarrow H-\left[-O-CH_2-CH-CH-CH-CH_2-O-\overset{\overset{\displaystyle O}{\|}}{\underset{\underset{\displaystyle O^-}{|}}{P}}- \right]_n -OH + n(\text{CMP})$$

OH OH OH

**Teichoic acid backbone
(polyribitol phosphate)**

The chain appears to be assembled, like peptidoglycan precursors, on a polyisoprenol intermediate, from which it is transferred to the peptidoglycan backbone. Glycerol teichoic acid is similarly formed from CDP-glycerol.

Environmental conditions can influence the amount and nature of the teichoic acid, which may have significance for pathogenicity. Thus when the normal choline of the teichoic acid in pneumococci is replaced by the analog ethanolamine the peptidoglycan becomes resistant to autolysis, the daughter cells remain linked in long chains, and the cells lose competence to take up transforming DNA (Ch.

9). Moreover, when lack of phosphate in the medium prevents synthesis of teichoic acid *B. subtilis* produces instead another acidic polymer, **teichuronic acid,** containing glycosidically linked uronic acid residues (with COOH replacing a sugar CH_2OH).

Lipoprotein. In *E. coli* as much as 10% of the peptides in the peptidoglycan are covalently attached (via the carboxyl of diaminopimelate) to a lipoprotein. Its relation to the outer membrane will be discussed below.

OUTER LAYER OF GRAM-NEGATIVE ORGANISMS

STRUCTURE

As noted above, removal of the peptidoglycan from gram-negative cell envelopes by lysozyme leaves fragments of both the outer and the inner membrane, containing phospholipid and protein (like all biological membranes). However, only the outer membrane also contains lipopolysaccharide (LPS), which is denser than the other components; hence two bands can be separated by density equilibrium centrifugation (Ch. 10) in a sucrose gradient. The heavier band, containing LPS, is derived from the outer membrane, while the lighter band contains the cytochromes and various biosynthetic enzymes characteristic of the inner membrane. In an alternative method of separation a nonionic detergent dissolves the inner membrane from the envelope fraction, leaving the outer membrane and peptidoglycan essentially intact.

LPS is accessible to antibodies on the surface of the cell. Its attachment to the underlying membrane is not uniform, for about half the LPS of *E. coli* can be released by treatment with a metal-chelating compound, ethylenediamine tetraacetic acid (= EDTA = Versene), while the entire LPS can be solubilized by phenol. These methods of extraction suggest that the attachment depends both on hydrophobic bonds and on ionic bonds mediated by polyvalent cations (e.g., Mg^{++}).

Anchoring Lipoprotein. As is indicated in Figure 6-10, the fusion of the peptidoglycan and outer membrane layers in *E. coli* depends on a virtually continuous layer of lipoprotein molecules, covalently attached to the former and embedded in the latter.

Thus cleavage of the covalent attachment by trypsin or pronase, in intact cells, permits separate middle and outer layers to be seen in the electron microscope. Moreover, with the purified basal wall fraction such treatment removes a layer of granules covering the outer surface, and in the residual pepti-

Fig. 6-10. Diagram of gram-negative cell envelope. Components are listed on the right, and probable sites of attack of destructive agents are indicated on the left. In the inner membrane many (and perhaps all) proteins have a dynamic role in transport or catalysis, but in the outer membrane few enzymes have been demonstrated and gel electrophoresis reveals a much smaller variety of proteins [From Schnaitman, C. A. *J Bacteriol 108*:553 (1971).]

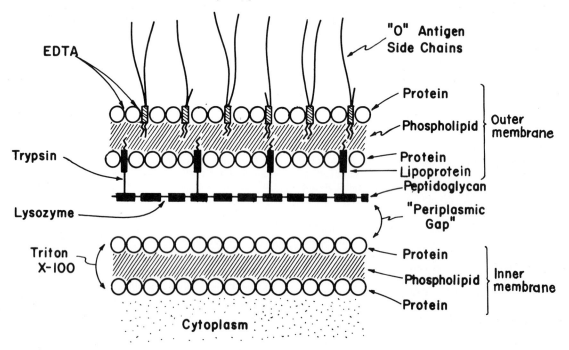

doglycan 1/10 of the DAP residues are found to carry, on their COOH, peptide remnants of the original connection. Conversely, on lysozyme digestion of a whole-envelope preparation most of the peptidoglycan fragments are freed but some (identified by radioactive DAP) remain attached to the outer membrane.

A similarly linked lipoprotein is found in various closely related organisms, and this linkage suggests an explanation for the presence of DAP and not lysine in the peptidoglycan of all gram-negative organisms (Table 6-1). However, some gram-negative bacteria (e.g., *Pseudomonas*) lack a covalently linked lipoprotein. These organisms evidently retain their outer membrane by a weaker, unknown mechanism, for under some conditions of fixation the outer membrane is seen under the electron microscope as a separate, often wrinkled layer.

FUNCTION

The outer membrane of gram-negative bacteria dominates their "social" activities, providing or covering receptors for phages, bacteriocins (Ch. 46), and defense mechanisms of infected hosts (including specific antibodies). In addition, the outer membrane influences permeability, for though it does not appear to have specific transport systems it does impair or exclude entry of many moderate-sized or large molecules. Thus in many gram-negative species (unlike gram-positive species) the peptidoglycan is not accessible to attack by lysozyme, and certain large drug molecules (such as actinomycin, Ch. 11) cannot reach their sites of action, unless the outer membrane is damaged (e.g., by treatment with EDTA).

Gram-negative bacteria also tend to be quantitatively more resistant than gram-positive cells to ionic detergents (which lyse the cell by attacking the cell membrane; Ch. 64) and to most antibiotics; and mutations that alter the LPS in *E. coli* can alter sensitivity to, for example, ampicillin. In addition, the outer membrane prevents escape of periplasmic proteins (see below).

LIPOPOLYSACCHARIDE STRUCTURES

LPS is also known as **endotoxin** because it is toxic and is firmly bound to the cells (Ch. 22). It can be extracted from intact cells (e.g., by incubation with 45% phenol at 90° for several hours) and then split by mild acid hydrolysis into **lipid A** (which retains the toxicity) and a **polysaccharide** (which is

responsible for the antigenic specificity).* Lipid A evidently plays an essential role in the cell envelope, for no mutants lacking it have been isolated.

Among enterobacteria lipid A is constant but the polysaccharide varies widely, providing the hundreds of **O antigens** of different strains; These antigens are of great diagnostic importance (Ch. 29). Moreover, their biosynthesis provides a model for the formation of polysaccharide surface antigens of mammalian cells. Indeed, some O antigens cross-react immunologically with certain human red cell antigens (Ch. 21).

Lipid A. The lipid moiety of LPS from *Salmonella* is a glycophospholipid, consisting of a chain of β-1,6-D-glucosamine disaccharide units, connected by 1,4′-pyrophosphate bridges and with all the hydroxyl groups substituted (Fig. 6-11). One of the sub-

---

* When the phenol-water mixture separates into two phases, on cooling, LPS is found in the aqueous phase. In contrast, lipid A contains a higher proportion of lipid to carbohydrate and is soluble in lipid solvents rather than in water; hence it is called a glycolipid.

Fig. 6-11. Tentative structure of subunit of lipid A from *Salmonella* lipopdysaccharide; the length of the chain of pyrophosphate-linked disaccharides is probably quite variable. The brackets enclose the ketose-linked ketodeoxyoctonate (KDO = 3-deoxy-D-mannooctulosonic acid), which links the lipid to a variable polysaccharide and may be considered part of lipid A. HM = β-hydroxymyristic acid (a C_{14} saturated fatty acid characteristic of lipid A); FA = other long-chain fatty acids. The fatty acids are very similar in all enterobacteria studied but are quite different in some other gram-negative organisms (e.g., *Pseudomonas*, *Brucella*). [After Rietschel, T., Gottert, H., Lüderitz, O., and Westphal, O. *Eur J Biochem* 28:166 (1972).]

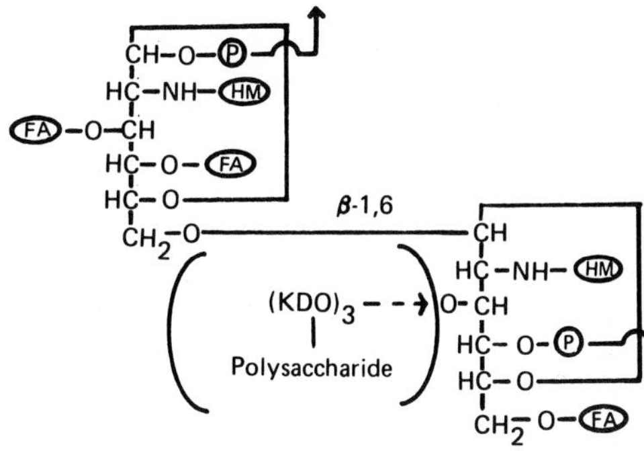

stituents is the polysaccharide of LPS; the others are long-chain fatty acids. Because there are five fatty acids and two phosphates per disaccharide this material behaves like a phospholipid.

The Core. The first step in analyzing the isolated polysaccharides of various strains consisted of identifying the monomers released by complete acid hydrolysis. Westphal discovered that the various O antigens contain a wide variety of sugars, often novel (e.g., deoxyhexoses, dideoxyhexoses; Ch. 29). In addition, five sugars are common to all wild-type, smooth (S) *Salmonella* LPS, and these are also present in certain rough (R) mutants which have lost their specific O antigen (Ch. 5). The common sugars evidently form a constant **core** polysaccharide, to which the **O-specific** residues of any given S strain are added.

Various R strains were found to be blocked either in the formation (Ch. 4, Fig. 4-25) or in the specific transfer of various individual components of the polysaccharide. Since the core polysaccharide chain is built up sequentially (from right to left in Fig. 6-12) the various R mutants yield polysaccharides with different degrees of incompleteness, up to the missing residue. These polysaccharides were very helpful in determining the sequence of the *Salmonella* core (Fig. 6-12): an outer pentasaccharide composed of three familiar sugars, followed by two heptose residues and then three residues of an 8-C sugar acid, KDO (Fig. 6-11). Other major groups of enterobacteria (e.g., *Shigella*) have a similar but not identical core.

Specific Side Chains. In a given LPS every side chain contains the same repeating sequence, in which the subunit is a linear trisaccharide (see Fig. 6-14) or a branched tetra- or pentasaccharide. The chains vary in length (even in the same organism), ranging up to 25 repeat units. The O antigen is thus a mat of "whiskers" of variable length, each attached, via a core chain, to lipid A embedded in the outer membrane.

Since the polysaccharide chains are uncharged they are not stained in the usual electron microscope preparations. Their length and location explain the presence of an

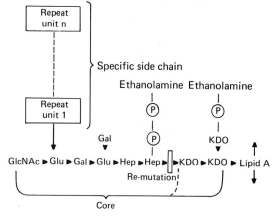

Fig. 6-12. Structure of core of *Salmonella* lipopolysaccharide (Hep = L-glycero-D-mannoheptose). The sugars are attached with their reducing group toward the lipid A (1→4, etc.); hence there are no free reducing groups. The Hep-KDO region has been called the backbone region of the core because it was thought to be the site of cross-link between parallel chains, but these links are now believed to be located between subunits of lipid A (Fig. 6-11).

In the biosynthesis of the core, from right to left, each residue is added by a specific enzyme. The genes for the core transferases map in a cluster (the *Rfa* "locus"), while those for the synthesis or transfer of a component of a specific, repeating chain (Fig. 6-13) map in another cluster (*Rfb* "locus;" see Ch. 11 for definition of locus). Re ("extreme rough") mutants have the most incomplete LPS compatible with viability.

electron-transparent zone of 20 to 40 A between adjacent smooth cells; this zone is absent with R cells, or with S cells from which the LPS has been extracted.

The structures of various O antigens were identified not only by the usual chemical methods (analysis of partial hydrolysis products and derivatives), but also by immunological tests. As later chapters will show (Hapten inhibition, Ch. 15), a particular antibody recognizes not only a particular sugar but also its immediate neighbor and the position and configuration of the linkage between them. Hence an oligosaccharide or glycoside of known structure competitively inhibits the reaction of specific antibodies with the corresponding group in the macromolecule being analyzed.

LPS structures are further discussed in Chapter 29 (Chemistry of the O Antigenic Determinants).

LIPOPOLYSACCHARIDE SYNTHESIS

All component chains of the LPS are synthesized on the cytoplasmic membrane, which contains the required enzymes. Moreover, the growing chains are also found there (in radioactive pulse-chase experiments) before being incorporated into the outer membrane. Both the core and the repeating chains derive each successive sugar residue, via a specific transferase, from a corresponding nucleoside-diP-sugar (Ch. 4). However, the two chains grow by quite different mechanisms. (R mutants have been useful in analyzing the biosynthesis of the core: the accumulated incomplete chains are substrates to which the next residue can be added.)

Lipid A is synthesized by unknown mechanisms. It also serves as the primer, and the membrane carrier, on which the **core polysaccharide** is built up. Because of this biosynthetic continuity the line between lipid A and core is arbitrary. Indeed, there is evidence that the core begins to be built onto lipid A while the latter is still growing.*

With the **repeating side chains,** in contrast, Robbins showed that each repeat subunit is formed by successive additions of its residues to a **glycosyl carrier lipid** in the membrane, identified as undecaprenol-P (Fig. 6-13A).† The carrier then evidently moves to the outer surface of the membrane, where the subunit is added to a growing chain. When the chain finally reaches an appropriate length, or

-----

* Though cleavage between the polysaccharide and lipid A by mild hydrolysis yields KDO as part of the former, KDO seems functionally to be part of lipid A, for both are essential for cell viability. Thus a mutant with a conditional block in KDO synthesis dies under conditions preventing that synthesis.

† It is remarkable that the same carrier lipid is employed as in peptidoglycan synthesis (see above), and that investigators of the two pathways identified the compound at the same time.

Fig. 6-13. Biosynthesis of repeating chain of LPS of *Salmonella newington.* Und = undecaprenol (a C_{55} polyisoprenol) carrier lipid, as in the synthesis of peptidoglycan (Fig. 6-7) and capsules. The nucleoside diP sugars (UDP-galactose, TDP-rhamnose, GDP-mannose) are synthesized as described in Chapter 4. The chemically reactive reducing group (C-1) of a sugar, by which it is linked to the nucleoside diP, is involved in its transfer to a growing repeat unit (A), in transfer of the "head" end of the growing chain to each additional repeat unit (B), and in final transfer (not shown) of the chain to the free end of a core chain (Fig. 6-12). [After Robbins, P. W., Bray, D., Dankert, M., and Wright, A. *Science 158*:1536 (1967).]

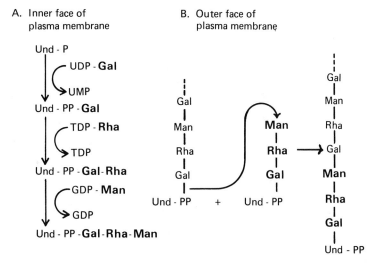

A. Inner face of plasma membrane

B. Outer face of plasma membrane

happens to encounter the transfer enzyme, it is transferred to the free end of a **core** side chain (built up on lipid A). The completed LPS molecule is eventually moved to the outer membrane, on the far side of the peptidoglycan; possible mechanisms will be discussed below (Sites of growth of wall).

During growth of the repeating side chain a brief radioactive pulse will label its attached end rather than the free end; hence the chain must grow by repeated transfers from its carrier to the free end of the next, carrier-linked subunit (Fig. 6-13B). This growth by **head** condensation resembles fatty acid synthesis (see below) or protein synthesis (Ch. 12); it permits synthesis of a long chain, whose free end may wander far from the site of growth while the growing end is fixed near the enzyme in the organized membrane. The long repeating chains of capsular polysaccharides are also synthesized in this way, but synthesis of the shorter core chain of LPS evidently does not require so elaborate a mechanism.

ANTIBIOTICS ACTING ON WALL SYNTHESIS

Cell wall biosynthesis is a frequent target of antibiotic action; several specific sites of inhibition have been identified.

Penicillin was recognized early as a lytic agent (Fig. 7-1) and was eventually found to block cross-linking of the peptidoglycan, as described above (see Fig. 6-9). Various features of this antibiotic are described in Chapter 7.

Atomic models show that part of penicillin is a close analog of D-ala-D-ala, with the exceptionally reactive peptide bond in the four-membered ring of the antibiotic (Fig. 7-6) corresponding to the bond that is cleaved in the transpeptidation reaction. As a result of this reactivity penicillin **irreversibly** blocks the cross-linking reaction in extracts of *E. coli,* probably forming a stable penicilloyl enzyme (in place of the normal transient peptidyl enzyme).

As noted above, in gram-negative cells the outer membrane may limit the access of penicillin to its sites of action. Thus ampicillin and penicillin G are equally effective in extracts of *E. coli* but the former is more effective with intact cells, presumably because its additional positive charge promotes transfer across the negatively charged LPS. Moreover, mutations in the LPS affect sensitivity to

penicillins. The sites of growth of the septum are evidently more accessible in the cell (or more sensitive) than the sites of cell elongation, for at borderline penicillin concentrations growing cells form long, nonseptate filaments.

Lysis by penicillin, or spheroplast formation in a hypertonic medium, does not result simply from explosion of the growing protoplast through an intact peptidoglycan layer; rather, it appears to be promoted by the action of **autolytic enzymes** in the regions of wall growth. Thus mutants defective in these enzymes are more resistant; and in the stages of penicillin action preceding lysis cells bulge in the septal region, where major growth is known to occur (see below, Sites of growth of wall).

Cycloserine, formed by a streptomycete, is a structural analog of D-alanine (Fig. 6-14). It blocks two successive reactions that involve this metabolite (Fig. 6-6). D-alanine competitively reverses inhibition by the analog, both in vitro and in cells.

The affinity of the analog for the dipeptide synthetase in *S. aureus* is about 20-fold greater than that of the natural substrate. Evidently the cyclic structure of the analog fixes it in the conformation required for binding, whereas the freely rotating bonds of D-alanine permit many conformations.

Phosphonomycin (Fig. 6-15) is an antibiotic analog of P-enolpyruvate. This metabolite provides the ether-linked lactyl group of UDP-MurNAc (Fig. 6-6), the enolpyruvyl group attaching to an SH of the enzyme and subsequently forming an enol-ether with UDP-GlcNAc. The analog, through its highly reactive epoxide group, forms a similar but stable derivative, irreversibly blocking the active site of the enzyme (just as in the action of penicillin on another enzyme).

Fig. 6-14. Structural analogy between D-cycloserine and D-alanine.

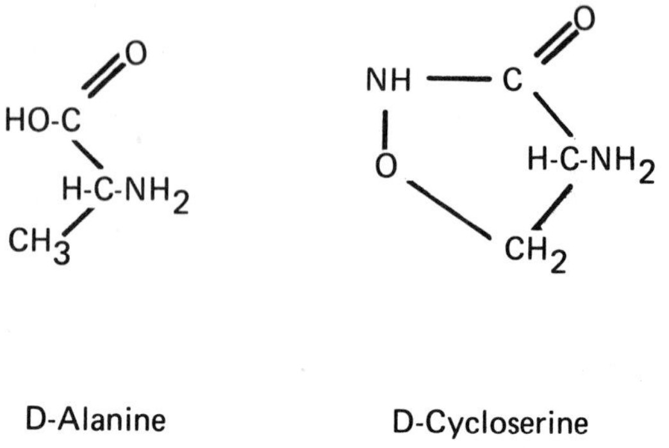

D-Alanine   D-Cycloserine

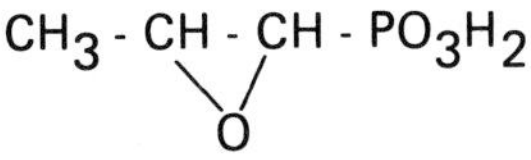

$$CH_3 - CH - CH - PO_3H_2$$
$$\diagdown O \diagup$$

Phosphonomycin

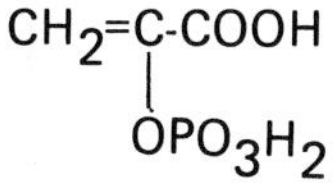

$$CH_2 = C - COOH$$
$$| \\ OPO_3H_2$$

Phosphoenolpyruvate

Fig. 6-15. Phosphoenolpyruvate and its analog phosphonomycin.

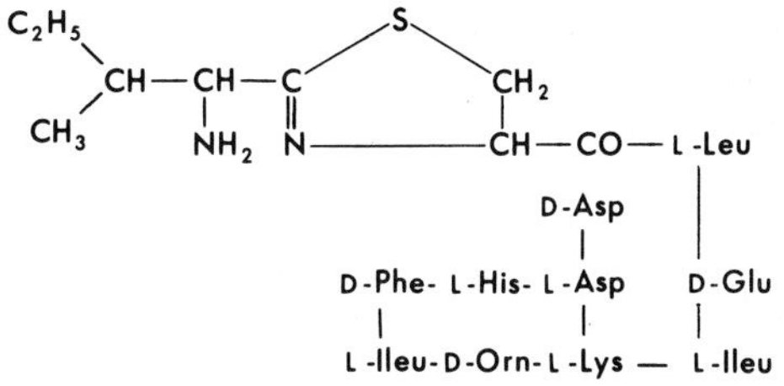

Vancomycin and the very similar but more toxic **ristocetin** (MW ca. 3500) are glycopeptides from an actinomycete of the genus *Nocardia*. They block subunit transfer from the carrier lipid to the growing peptidoglycan (Fig. 6-7), in cells or in extracts.

Bacitracin, a cyclic peptide (Fig. 6-16) from a strain of *B. subtilis* (from a patient named Tracy), blocks **dephosphorylation** of

Fig. 6-16. Bacitracin A. The portion presented in detail consists of an isoleucine and a cysteine residue, condensed to form a thiazoline ring rather than the usual peptide linkage of a polypeptide.

the undecaprenol-PP formed in the course of subunit transfer from carrier to wall (Fig. 6-7). (In fact, the resulting accumulation revealed the existence of the undecaprenol-PP.) In solution bacitracin forms a 1:1 complex with undecaprenol-PP in the presence of Zn^{++} or Mn^{++}, which coordinates the pyrophosphate with carboxyls of the antibiotic.

All these blocks in wall synthesis are bactericidal. Moreover, vancomycin and bacitracin (unlike penicillin) inhibit growth of spheroplasts, probably because they block the lipid carrier and thus disorganize the membrane.

CYTOPLASMIC MEMBRANE

COMPOSITION

Reasonably pure preparations of fragmented cytoplasmic membrane can be obtained from gram-positive cells by lysozyme digestion of isolated envelopes, or by lysis of protoplasts; with gram-negative cells separation from fragments of outer membrane can be accomplished by density gradient centrifugation (see above). The protein content is slightly higher than in mammalian cell membranes: about 60 to 70%, with 20 to 30% lipid and small amounts of carbohydrate. The membrane may contain as much as 20% of the total protein of a bacterium. Many proteins with specific functions in transport or catalysis (see below) have been identified; whether there are also purely structural proteins, as seems probable for the outer membrane, is uncertain.

The membrane lipids of bacteria, like those of eukaryotes, consist mainly of phosphatides.

Virtually no bacteria (except for certain mycoplasmas) contain sterols.

ORGANIZATION AND FUNCTIONS

The cytoplasmic membrane provides an **osmotic barrier,** transversed at intervals by **specific transport systems** (though they cannot be seen in the deceptively uniform cross-section of the membrane in electron micrographs). The barrier is very efficient in retaining metabolites and excluding external compounds: all ions and nonionized molecules larger than glycerol penetrate very slowly except by specific transport.

Because phospholipids are amphipathic, with a hydrophilic polar region and long, hydrophobic lipid chains, they will tend to aggregate from dispersions in aqueous solution, in vitro as well as in the cell, to form a highly oriented, thin bilayer of molecules

(Fig. 6-17). This film is the basis for the osmotic barrier. The pattern of organization of the protein in natural membranes is not certain, but since membrane proteins are generally not water-soluble unless coated by detergents they must possess large lipophilic regions, which can account for their stable spontaneous incorporation into the membrane.

Because the mutual attractions of lipids (unlike those of proteins and nucleic acids) are nonspecific the molecules readily exchange their contacts, and so biological membranes in general are **two-dimensional fluids:** physical studies and radioautography show that newly inserted molecules flow rapidly from one site to another. However, the bridges between peptidoglycan and membrane (see Fig. 6-24 below) may limit this mobility. Moreover, cytochemical studies provide evidence for clustering of certain membrane enzymes in bacteria, as in eukaryotic cells; and proteins appear to be excreted in specialized regions (see Mesosomes,

below). Since the lipids are fluid this regional differentiation of the membrane presumably depends on protein-protein interaction.

Proteins of transport systems are assumed to cross the membrane. Moreover, with red blood cells attachment of radioactive substituents to both surfaces of the membrane, compared with similar treatment of the outer surface only (in intact cells), has shown that some proteins extend across the membrane, with different parts of the polypeptide chain exposed on the two surfaces.

Enzymes. The cytoplasmic membrane contains a wide variety of enzymes. These include 1) the cytochromes and other proteins of **electron transport** and **oxidative phosphorylation;** 2) the enzymes of **complex lipid biosynthesis;** 3) the enzymes involved in the later stages of **wall biosynthesis**—i.e., assembly of subunits on the carrier molecules and subsequent polymerization; 4) **DNA replicase** (Ch. 10); and 5) **exoenzymes** that have not yet been released to the periplasmic space or the exterior (see below). Many of these enzymes **require** a lipophilic environment in order to have an active conformation: their activity is lost on isolation of the protein but is recovered on addition of a particular phospholipid.

Membrane enzymes vary widely in the ease with which they may be solubilized: increasingly drastic reagents proceed from various buffers to EDTA, osmotic shock, nonionic detergents, deoxycholate, dodecylsulfate, and finally phenol; and as with other detergent actions, solubilization increases rapidly with rising temperature. Many enzymes appear to be loosely bound, for in the usual gentle preparations they are found partly in the "particulate" fraction and partly in the supernate.

In higher organisms the membrane plays a central role in conduction and transmission of **nerve** impulses. Bacteria exhibit a primitive precursor of this process in the form of **chemotaxis** (see below), in which specific proteins on the outer surface of the membrane play a role. In eukaryotes **ribosomes** are often attached to membranes of the endoplasmic reticulum, but there is little evidence for such attachments in bacteria.

Fig. 6-17. Diagrammatic structure of membrane. The zigzag lines represent the fatty acid chains, and the solid circles represent the polar groups, of phospholipids. The shaded bodies are proteins, the thickened border representing regions with mostly hydrophobic side chains. Some proteins penetrate from one face of the membrane to the other, while others are embedded in the membrane and expose a hydrophilic region on only one face. Whether any proteins are bound only to the hydrophilic surface, as depicted, is uncertain. (After Fox, C. F. *Sci Amer* Feb. 1972, p. 30.)

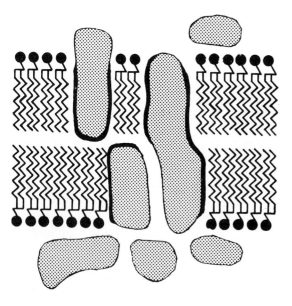

MESOSOMES

Gram-positive bacteria often contain one or more large, irregular, convoluted invaginations of the cytoplasmic membrane; in gram-negative cells they are smaller. These bodies, called mesosomes, often appear to be isolated in the cytoplasm, but serial sections reveal a connection with the cytoplasmic membrane, and electron-dense stains show that mesosomes are connected to the periplasmic space (Fig. 6-18). Some mesosomes are **vesicular** (Fig. 6-18) while others appear as concentric, **lamellar** whorls (Fig. 2-4). Vesicles (apparently pinched-off portions of membrane) are also released to the periplasmic space (Fig. 6-8); and mesosomes disappear on protoplast formation, evidently being drawn into the stretched cytoplasmic membrane.

Several functions of mesosomes have been identified. **Septal** mesosomes are involved in **cell division** (and often remain as polar mesosomes after division). These mesosomes often also provide the site of attachment of the bacterial chromosome to the cytoplasmic membrane. Since the attachment site must be split when replication is complete the extra membrane of the mesosome may facilitate separation of the segregating chromosomes, as well as the associated initiation of a cross-wall between them.

Lateral mesosomes, attached to nonseptal regions, function in **secretion.** Thus during induced formation of penicillinase in a *Bacillus* such mesosomes increase in number; in

addition, they appear to give rise to periplasmic vesicles (Fig. 6-18).

These vesicles are a specialized apparatus for penicillinase excretion in such cells: they may be recovered by conversion of the cells to protoplasts, and Lampen has shown that they contain the bulk of the cell-bound penicillinase, with little of other membrane-bound enzymes. The remainder of the bound penicillinase is found in the cell membrane, presumably as a precursor of the vesicles. Release to the exterior involves cleavage of a short terminal peptide "leash."

A third function of mesosomes is concerned with **electron transport.** Thus photosynthetic bacteria and blue-green algae, which carry out extensive reversed electron transport (Ch. 3), have an elaborate membranous reticulum, homologous to the more elaborate chloroplasts of photosynthetic eukaryotes (algae and higher plants). This structure provides a further indication that the machinery of electron transport does not simply attach to preexisting membrane as a matter of convenience but requires separation of charges across an insulating membrane.

PERIPLASMIC PROTEINS

Gram-negative bacteria release certain proteins when converted to **spheroplasts** or

Fig. 6-18. Mesosomes and periplasmic vesicles in *Bacillus licheniformis.* Two parallel cells, with negative stain penetrating into large septal mesosomes (S1) and into smaller lateral mesosomes; arrow shows vesicles being released from a lateral mesosome into periplasmic space. These cells are excreting penicillinase; in comparable cells not engaged in enzyme secretion lateral mesosomes and periplasmic vesicles are rare or absent. [From Ghosh, B. K., Lampen, J. O., and Remsen, C. C. *J Bacteriol 100*:1002 (1969).]

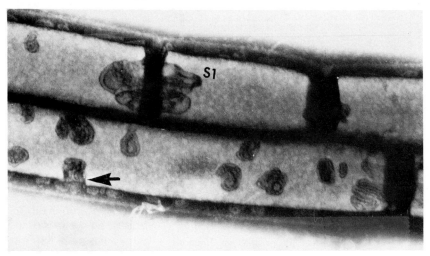

subjected to **osmotic shock.*** These proteins (mostly hydrolytic enzymes that attack impermeable foods, Ch. 5) are thought to lie in the **periplasmic space,** between the inner membrane and the peptidoglycan.† However, their restriction to gram-negative cells suggests that the outer membrane may be responsible for their retention. Moreover, mutants with certain wall defects excrete the same enzymes into the medium. Gram-positive bacteria usually excrete such digestive enzymes (e.g., proteases, nucleases, penicillinase), which may be present in periplasmic vesicles before release (see preceding section).

The mechanism of enzyme secretion across the cytoplasmic membrane is not known, but the problem is much the same as that of the origin of the lipoproteins of the gram-negative outer membrane. In animal cells excreted proteins are formed on ribosomes bound to the endoplasmic reticulum, but there is little evidence for a special class of membrane-bound ribosomes in bacteria. Flexibility and size of the protein may be important: exoenzymes generally lack cystine (and hence disulfide cross-links); and alkaline phosphatase has been shown to cross the membrane of *E. coli* as an inactive monomer, which then dimerizes in the presence of $Zn++$ (Ch. 12, Quaternary structure) in the periplasmic space to yield active enzyme. A membrane carrier may also be important, as in the proteolytic release of penicillinase noted in the preceding section.

Binding Proteins. Osmotic shock releases not only hydrolytic enzymes but also various proteins that can bind, tightly and specifically, an amino acid, ion, or sugar. These proteins are evidently involved in membrane transport: their elimination by mutation, or their release by osmotic shock, inactivates the corresponding membrane transport system (see below), and both systems are made temperature-sensitive by the same mutation. It is not clear, however, how a true periplasmic protein could be useful in transport. Moreover, with some sugar transport systems (e.g., β-galactoside in *E. coli;* see below) a similar specific binding protein is found in the membrane but not in the periplasmic space. Hence the apparently free binding proteins may well have been artificially detached, by osmotic shock, from transport systems embedded in the membrane.

STRUCTURE AND BIOSYNTHESIS OF LIPIDS

These topics are intimately related to membrane function and will be discussed here.

FATTY ACIDS

The apolar components of bacterial lipids are predominantly long (C_{14} to C_{18}) fatty acid chains, which may be saturated or mono-unsaturated, but not polyunsaturated. Unusual fatty acids are also found, containing methyl branches, cyclopropane rings, or extremely long chains (e.g., mycolic acids with two long branches totaling C_{83} in tubercle bacilli; see Ch. 35).

The **biosynthesis of fatty acids** in bacteria (like that in higher organisms) proceeds by successive addition of acetyl residues to the chain at its carboxyl end, which is attached, as Vagelos showed, to a small **acyl carrier protein** (ACP). (This "head addition" resembles the growth of specific wall polysaccharides; Fig. 6-13.) The reactivity of the acetyl group (initially on CoA) is first increased by adding CO_2 to form malonyl CoA (Fig. 6-19), and the malonyl group is then transferred to the -SH of a long-chain prosthetic group (phosphopantetheine, Fig. 4-14) on ACP.

The growing chain, on an -SH of the condensing enzyme, is transferred to the active CH_2 of the malonyl group, displacing its free carboxyl and creating a β-keto acid attached

---

* In the preparation of spheroplasts, by treatment with lysozyme, damage to the outer membrane (e.g., by EDTA) is an essential preliminary step. In osmotic shock, in which cells are exposed to EDTA in a hypertonic medium and then are rapidly diluted in a chilled solution of low osmotic strength, all layers may be damaged. Thus the EDTA releases a large fraction of the LPS and makes the outer membrane more permeable; cold makes the membrane more brittle; and a sudden outflow of osmolytes evidently places stress on the peptidoglycan.

† RNase I of *E. coli* illustrates the problem of defining the location of an enzyme in the cell. This basic protein, long considered a ribosomal enzyme, is evidently adsorbed to ribosomes during lysis, for if the cells are converted to spheroplasts before lysis the enzyme ís found in the periplasmic fraction instead.

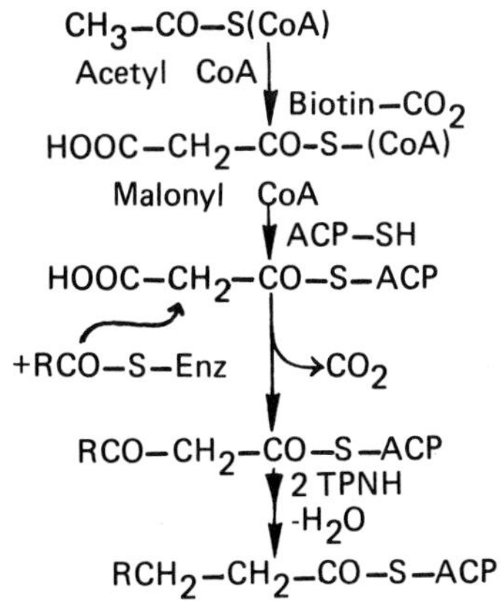

$$CH_3-CO-S(CoA)$$

Acetyl CoA

Biotin—CO_2

$$HOOC-CH_2-CO-S-(CoA)$$

Malonyl CoA

ACP—SH

$$HOOC-CH_2-CO-S-ACP$$

+RCO—S—Enz ⟶ CO_2

$$RCO-CH_2-CO-S-ACP$$

2 TPNH

-H_2O

$$RCH_2-CH_2-CO-S-ACP$$

Fig. 6-19. Chain elongation in fatty acid biosynthesis. ACP-SH = acyl carrier protein; RCO- = growing chain.

to ACP. (The added CO_2 is thus not incorporated, though it is essential for growth; Ch. 5.) The ACP is presumably in the center of the fatty acid synthetase complex,* in which rotation of the flexible pantetheine "leash" (as in peptide antibiotic synthesis; Ch. 4, Fig. 4-16) exposes the attached chain to a series of enzymatic centers. These reduce the β-keto group to a -CH_2 group, by means of two TPNH-linked reductions and a dehydration. When this process is completed the resulting longer fatty acid is again "parked" on an -SH group of the condensing enzyme, and the pantetheine-SH of the ACP is free to accept another malonyl residue; the cycle is repeated until the fatty acid reaches the proper length for transfer to a complex lipid.

Fatty acid synthesis is usually initiated by an acetyl group, leading to the formation of even-numbered, straight-chain fatty acids. However, fatty acids with an odd-numbered straight chain, or with a subterminal methyl branch, may be formed by initiation with propionyl CoA, isovaleryl CoA, etc.

Unsaturation. Aerobic bacteria, like animals and plants, make unsaturated fatty acids by O_2-linked oxidation of a saturated fatty acid. Anaerobes, however, cannot use this route. These organisms (and

* In yeasts and other eukaryotic cells the enzymes of fatty acid chain elongation form a tight complex, but *E. coli* yields only separate soluble components. However, radioautography suggests that these are present, presumably as a loose complex, in the membrane.

facultative organisms) make unsaturated fatty acids by preserving a double bond that appears in the dehydration step in the cycle of chain elongation. Thus this step usually produces a *trans* double bond, which is then reduced; but at a certain chain length molecules are occasionally dehydrated by a special enzyme that produces a *cis* double bond, which is not a substrate for reduction. In *E. coli* the main product is *cis*-vaccenate (*cis*-11,12-octadecenoate), rather than the familiar oleate (*cis*-9, 10-octadecenoate).

COMPLEX LIPIDS

The complex lipids of bacteria (other than lipid A, see above) consist largely of glycerol with two of the hydroxyls attached to hydrocarbon chains, through ester, vinyl ether (plasmalogen), or saturated ether linkages; the third hydroxyl is attached to a more polar moiety: either phosphate (in phosphatides) or a sugar (in some glycolipids). The phosphate in turn is linked either to a nitrogenous compound (serine, ethanolamine) or to a polyol (glycerol). Figure 6-20 outlines the **biosynthesis** of phosphatides.

In enterobacteria most of the phosphatide in the outer membrane is the neutral phosphatidyl ethanolamine, whereas the cytoplasmic membrane contains in addition the acidic phosphatidyl glycerol, diphosphatidyl glycerol (cardiolipin), and traces of phosphatidyl serine. Phosphatidyl choline, though widespread in higher organisms, is rare or absent in bacteria. Some bacteria form an additional class of complex lipids, **lipoamino acids,** with an amino acid (derived from its transfer RNA, Ch. 12) on one of the free hydroxyls of phosphatidyl glycerol.

Mevalonic Acid. As noted above, bacteria do not contain steroids, which are synthesized in yeast and higher organisms from acetyl CoA via mevalonic acid (β,γ-dihydroxy-β-methylvaleric acid). However, this intermediate also gives rise to the polyisoprenoid membrane carrier lipid described above, to carotenoids (present especially in photosynthetic bacteria), and to side chains on other constituents (e.g., quinones). Indeed, mevalonic acid was initially discovered as a growth factor for *Lactobacillus acidophilus*. This requirement is now explained in large part by the universal presence of polyisoprenols as vital, though quantitatively minor (0.1%), constituents of the membrane.

VARIATIONS IN COMPOSITION

The composition of bacterial lipids varies markedly with conditions of growth, including

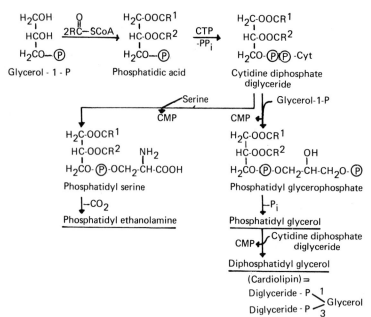

Fig. 6-20. Synthesis of phosphatides in *E. coli*. Only the underlined compounds are significant endproducts; the other phosphatides are short-lived intermediates (though they accumulate in other organisms). The fatty acids (R[1], R[2]) vary with growth conditions. The acyl group may initially be derived directly from acyl-ACP, as well as from acyl-CoA.

temperature, pH, and composition of the medium. At lower temperatures, for example, cells synthesize a larger proportion of unsaturated fatty acids (which have a lower melting point) and thus maintain the proper fluidity and permeability of the membrane.

The shift with temperature appears to be regulated by a very simple device: in extracts the enzyme that transfers a long-chain acyl residue to glycerol-P determines the proportion of saturated to unsaturated residues selected from a mixture, and the value is directly influenced by temperature.

As cells approach the **stationary phase** of growth (Ch. 5) their lipids undergo several changes. Most of the unsaturated fatty acids are converted to **cyclopropane** fatty acids, by addition of a methylene group (from S-adenosyl methionine) across the double bond. Moreover, phosphatidyl glycerol frequently adds an amino acid or a second phosphatidyl group to the CH₂OH of the glycerol moiety (Fig. 6-20). These changes may well "toughen" the membrane and thus increase the chances of cell survival through a period of starvation.

Auxotrophic Mutants. Mutants of *E. coli* deprived of the glycerol that they require provide a general block in complex lipid synthesis. Such cells remain alive, suggesting that in the absence of net membrane synthesis some regulatory mechanism pre-

vents the cell from growing and bursting. However, a complete block in phosphatide synthesis (in temperature-sensitive mutants) is coupled, in an unknown manner, with rapid cessation of DNA, RNA, and protein synthesis, and with early death. Mutants deprived of a required unsaturated fatty acid not only die but lyse rapidly, suggesting that membranes containing only saturated fatty acids are too "brittle" for survival.

The latter mutants can utilize not only the native *cis*-vaccenate but various other unsaturated fatty acids. The fatty acid composition of the membrane can thus be manipulated, and the results can be correlated with changes in its functional and physical properties, such as the transition from liquid to solid. Moreover, studies with mutants blocked in new membrane formation (through deprival of glycerol) suggest that transport systems cannot be effectively inserted into old membrane. Thus for some time after the onset of deprival such cells can be induced to form cytoplasmic β-galactosidase, and they undoubtedly make the corresponding specific transport protein (see below); yet they do not make the corresponding transport system in functional form.

ANTIBIOTICS ACTING ON MEMBRANE

DAMAGE TO MEMBRANE

Polymyxins, a group of cyclic polypeptides produced by *Bacillus polymyxa,* resemble cationic detergents in having basic groups (of the novel amino acid α,γ-diaminobutyrate) plus a fatty acid side chain (Fig. 6-21). They cause direct membrane damage by a detergent-like action (Ch. 64).

The membrane damage can be recognized by the decrease of turbidity, leakage of soluble constituents (including nucleotides and inorganic ions), penetration of normally excluded substrates into the cell, and staining of the cell by a dye that fluoresces when bound to proteins. Furthermore, a fluorescent derivative of polymyxin becomes concentrated at the cell membrane (Fig. 6-22). The lysis caused by polymyxin differs from that caused by penicillin or lysozyme: it leaves the intact wall as a relatively nonrefractile ghost, and it is not prevented by hypertonic media.

Polymyxin is unique among useful chemotherapeutic agents (Ch. 7) in being bactericidal even in the absence of cell growth. It is also unusual in being more effective against gram-negative than positive organisms, probably by attacking the outer membrane. Synthetic vesicles are not sensitive unless they contain considerable **phosphatidyl ethanolamine.** The large proportion of this component in bacterial membranes, compared with those of animal cells, may explain the selectivity that makes polymyxin useful.

Polyene antibiotics similarly impair membrane integrity, but by complexing with

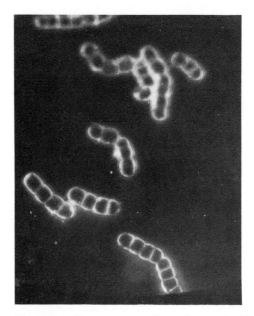

Fig. 6-22. Fluorescence photomicrograph of *B. megaterium* treated with a fluorescent derivative of polymyxin. [From Newton, B. A. *J Gen Microbiol 12*:226 (1955).]

sterols. They are therefore effective against certain fungi but not against bacteria (except those mycoplasmas that have incorporated sterols in their membrane, Ch. 40). These antibiotics are further described in Chapter 43.

Nalidixic acid and **novobiocin** (Ch. 7), and the synthetic **phenethyl alcohol** ($C_6H_5CH_2CH_2OH$), interfere with both DNA replication and membrane integrity. The biosynthetic impairment may well be secondary to an effect on the membrane, to which the DNA-synthesizing machinery is attached.

IONOPHORES

These molecules greatly increase the permeability of membranes to specific ions: they form rings that have a nonpolar periphery and therefore spontaneously insert themselves ("dissolve") in membranes, while the interior of the ring closely fits a particular inorganic cation and also provides carbonyl groups that coordinate with it, replacing its normal hydration shell. (Such caged ions are called clathrates.) These antibiotics are bactericidal but they are not selective enough to be useful

Fig. 6-21. Polymyxin B. DAB is α,γ-diaminobutyrate [NH₂CH₂CH₂CH(NH₂) COOH]. Aliphatic residue is 6-methyloctanoic acid.

$$
\begin{array}{c}
NH_3^+ \\
| \\
DAB \\
\diagup \quad \diagdown \\
\text{L-Leu} \qquad DAB\text{-}NH_3^+ \\
| \qquad\qquad | \\
\text{D-Phe} \qquad \text{L-Thr} \\
| \qquad\qquad | \\
{}^+NH_3\text{-}DAB \qquad DAB\text{-}NH_3^+ \\
\diagdown \quad \diagup \\
DAB \\
| \\
\text{L-Thr} \\
| \\
{}^+NH_3\text{-}DAB
\end{array}
$$

TABLE 6-2.   Classes of Antibiotic Ionophores

Enniatins	Cyclic depsipeptides with ring of 18 atoms
Gramicidins	Cyclic (Gramicidin S) and open-chain (A,B,C) peptides
Actins (e.g., Nonactin)	Tetralactone, ring of 32 atoms, 4 methyl or ethyl substituents
Nigericin, monensin	Open chain, form ring by hydrogen-bonding the terminal COOH to 2 OH groups at other end

in therapy and so they are of interest chiefly as tools for studying membrane physiology.

Several chemical classes of antibiotics have been found to act as ionophores. Some have a covalently closed ring, while others are linear but fold into a ring; some are neutral and others have a carboxyl group (Table 6-2). For example, **valinomycin,** which is highly specific for transporting K+, is a cyclic depsipeptide (i.e., a chain of alternating α-amino and α-hydroxy acids connected by peptide and ester bonds).

[D—hydroxyvalerate—D—valine—L—lactate—L—valine]$_3$

Valinomycin

Atomic models show that the aliphatic side chains readily face outward from the ring and the carboxyl groups inward.

The carrier function depends on diffusion or conformational change of the ionophore within its lipid matrix; the function is lost at temperatures low enough to solidify the lipids. Neutral ionophores permit the charged ion to move across the membrane in the direction of the electrical potential, thus eliminating that potential. (The simultaneous block in oxidative phosphorylation supports the view that energy is transferred in electron transport in the form of a membrane potential.) Acidic ionophores can exchange their metal ion for H+; hence they eliminate pH differences rather than potential differences across the membrane.

Oxidative phosphorylation is also the site of action of some other antibiotics, which are useful in its analysis. For example, **antimycin** blocks electron transport between cytochromes b and c, while **oligomycin** permits electron transport but uncouples the associated phosphorylation. Antimycin is fungicidal, but in intact bacteria it cannot reach its receptor.

THE ENVELOPE IN CELL DIVISION

WALL GROWTH AND SEPARATION

The division of bacteria (by binary fission) produces two daughter cells of approximately equal size, though the equality may be quite rough, since synchronization (Ch. 5) persists for only a few generations. Yeasts generally divide by a process of budding, resulting in the separation of a small daughter cell from the larger mother cell (Ch. 43).

Division of bacteria starts with ingrowth of cytoplasmic membrane, often in association with a mesosome (Fig. 2-7). The accompanying ingrowth of wall eventually forms a complete transverse septum (cross-wall), which is thicker than the peripheral cell wall. Septal cleavage, progressing inward from the periphery, leads to cell separation.

Differences in the details of septum formation and separation are responsible for characteristic differences in bacterial shape and arrangement. For example, linked cocci are formed by incomplete cleavage of septa: streptococci form successive septa in parallel orientation, yielding linear chains, while staphylococci begin a new septum perpendic-

ular to a still unseparated old one, yielding three-dimensional clumps. Prolonged delay in the cleavage of bacilli, often seen under conditions of impaired growth, yields long filaments.

The importance of the wall in orderly cell division is illustrated by the effect of its absence. Lysozyme-induced protoplasts cannot divide, though they can increase their protoplasm severalfold. Mycoplasmas (Ch. 40) and spheroplasts can divide slowly (especially in the depths of agar, whose fibers may help to pinch off daughter cells); but since there is no regulated septum formation the daughter cells vary widely in size.

SITES OF GROWTH OF WALL

The **equatorial zone** around a growing septum is a major (and perhaps the sole) region of growth of the peripheral wall. Such a region might be expected to be weak because of an abundance of "loose ends" of the growing peptidoglycan chain; and indeed, autolysis of staphylococci yields the wall as hemispheres.

In more direct tests the bacterial surface is labeled by staining with fluorescent antibodies or by incorporation of a tritiated component, and the cells are then allowed to continue growth without further labeling. The results showed that with **gram-positive** streptococci the old wall retains its label and the newly synthesized, unlabeled wall is laid down in an equatorial zone, where cell division finally takes place (Fig. 6-23). **Gram-negative** enterobacteria also have a middle, equatorial zone of rapid peptidoglycan synthesis, as shown by the initial bulging of this region in cells growing in the presence of penicillin. However, radioautography shows that incorporated ³H-DAP is redistributed during subsequent growth, suggesting diffuse growth of the peptidoglycan.*

---

* A similar conclusion had been suggested by the diffuse redistribution of fluorescent antibody on the cell surface during further growth, but this result could be accounted for by movement of the fluid surface LPS relative to the underlying peptidoglycan.

The mechanism of peptidoglycan synthesis is evidently modified in gram-negative cells to allow formation of the outer membrane, which requires transfer of components from the inner membrane. The required gaps in the peptidoglycan layer may not be restricted to the central equatorial zone, for radioautography after short pulses of labeling reveals synthesis of LPS at multiple sites. Moreover, when gram-negative cells are fixed in a hypertonic medium the retracted cytoplasmic membrane adheres to the outer membrane at several hundred scattered sites (Fig. 6-24). Conversely, osmotic shock creates multiple protrusions, with a diameter at the base of ca. 100 A. Since such protrusions are not formed in stationary-phase cells it appears that the holes in the peptidoglycan can be closed; and periodic formation and closure of such holes might account for the gradual diffuse redistribution of labeled DAP.

MUTATIONS IN WALL SYNTHESIS

Mutants with a straightforward genetic block in cell division cannot be preserved, since the

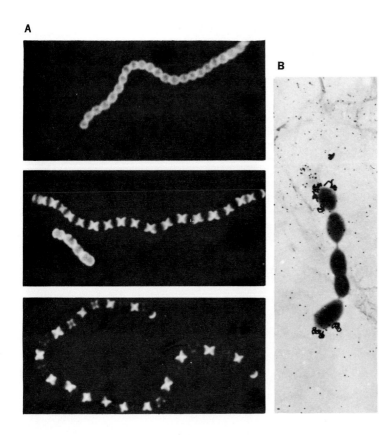

A

B

Fig. 6-23. Equatorial growth of gram-positive cell wall. **A.** Top: Streptococci, with surface antigen stained with fluorescent antibody (Ch. 15). The next two samples were taken after 15 and 30 minutes of growth without antibody: fresh, unstained wall has been deposited in each cell in an equatorial ring, which separates the older, stained parts of the cell. Each stained portion remains attached to the similar portion of the adjacent cell, forming an "X." **B.** Pneumococcus with wall of original cell labeled in its teichoic acid with ³H-choline, and then grown for several generations with ethanolamine (which promotes formation of chains of cells rather than separation into diplococci). Electron microscope radioautographs reveal localization of silver grains (from disintegration of ³H) only at ends of chain, indicating formation of new wall between conserved halves of original, labeled wall. [**A** from Cole, R. M. *Bacteriol Rev* 29:326 (1965). **B** from Briles, E. B., and Tomasz, A. *J Cell Biol* 47:786 (1970).]

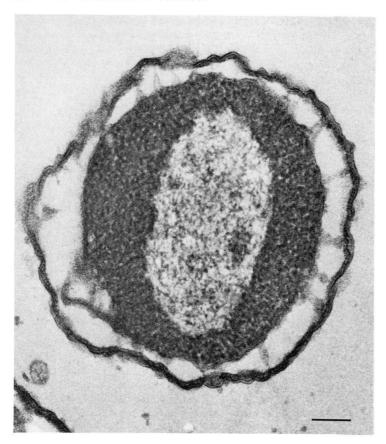

Fig. 6-24. Areas of adhesion between wall and inner membrane of *E. coli*. After plasmolysis (shrinkage in 20% sucrose for 2 minutes) the cells were fixed in formaldehyde and OsO₄, embedded,. and sectioned. Numerous duct-like extensions from membrane to wall are seen; other experiments showed that these are the sites of attachment of various phages to the outer surface of the wall. Bar = 1000 A. [From Bayer, M. E. *J Gen Microbiol 53*:395 (1968).]

defect is lethal. However, mutants with various **temperature-sensitive** (ts: Ch. 8) alterations in cell division have been isolated. These organisms can be maintained by growth at a low temperature, while growth at a higher temperature causes phenotypic expression of the defect. A variety of abnormalities are seen, suggesting that many genes contribute to the orderly process of cell division. The relation to chromosome replication will be discussed in Chapter 13.

For example, mutants that cannot make cross-walls yield **filaments,** while others frequently make a cross-wall near one end of the cell and thus form **minicells** (Fig. 6-25), without a chromosome. Some mutants with impaired peptidoglycan synthesis form **fragile** cells, and still others develop highly **aberrant shapes.**

Cells that multiply as pleomorphic L forms, without a rigid wall (Ch. 43), may have a genetic defect in wall synthesis, though some phenotypic impairments (e.g., by penicillin) can also be passed on to subsequent generations (Ch. 8).

MEMBRANE TRANSPORT

EXISTENCE OF TRANSPORT SYSTEMS IN BACTERIA

The presence of a **permeability barrier** in bacteria was suggested early by the finding that certain enzymes are **cryptic** (i.e., can attack added substrate only in a cell lysate and not in intact cells). Moreover, amino acids and mineral ions seemed to cross this barrier via systems capable of **active transport,** since equilibration with dilute external solutions of these compounds led to the accumulation of much higher concentrations in the cells.

Nevertheless, for many years there was widespread resistance to these interpretations. Thus bacteria seemed too small for such complexity; innumerable metabolic studies proceeded successfully

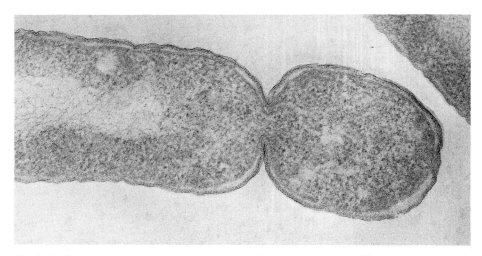

Fig. 6-25. *E. coli* mutant producing a minicell, without chromosome. [Courtesy of A. Jacobson and D. Allison. From Adler, H. I., Fisher, W. D., Cohen, A., and Hardigree, A. A. *Proc Natl Acad Sci USA 57*:321 (1967).]

on the assumption that bacteria are semipermeable bags of enzymes; and alternative explanations could be found for the apparent exceptions. (For example, an enzyme might be cryptic because it existed in an inactive state in the cell; and a permeant might attain a huge intracellular concentration by being bound to cellular constituents.)

Inducibility. The first unequivocal evidence for specific transport systems in bacteria was the finding that glucose-grown cells of *Aerobacter* cannot utilize exogenous citrate but cells grown in the presence of citrate can. Since citrate is a major obligatory intermediate the enzymes for its metabolism must be present in active form even in glucose-grown cells, and so the adaptive response has evidently increased the access of exogenous substrate. Cohen and Monod similarly showed that the adaptive response of *E. coli* to β-galactosides (Ch. 3, Diauxie) involves formation not only of the enzyme β-galactosidase (Ch. 13) but also of machinery permitting entry of its substrate into the cell.*

In both these systems the capacity for metabolizing the normal substrate exceeds the capacity for its uptake, and so the transport is "downhill" (with respect to the concentration gradient) and does not require energy. However, the induced cells also exhibit **active transport**† of various nonhydrolyzable analogs (e.g., methyl β-thiogalactoside = MTG: see Ch. 13, Table 13-1), against a concentration gradient. The "downhill" transport of substrate and the "uphill" transport of the analog evidently employ the same system, for the 2 kinds of compounds compete for entry; moreover, the 2 activities respond in parallel to induction (and to mutation).

Facilitated Transport. Active transport requires expenditure of energy. However, under

---

* Specific inducible transport systems in bacteria were initially called **permeases**. However, this term no longer seems useful: these systems are indistinguishable from the transport systems of eukaryotic cells; their mechanism of action need not involve an enzymatic ("-ase") reaction (i.e., one altering a covalent bond); and the term has been used ambiguously to refer to a specific binding protein or to the entire transport system.

† Active transport (or active concentration) can be defined as transport from a given external concentration (or, more precisely, chemical potential) to a higher level within the cell. It is usually studied by incubating cells with a radioactively labeled permeant and measuring the amount retained after rapid filtration. Two precautions may be noted: the radioactive compound recovered from the cells must be the same as that supplied, and not a derivative; and the washing procedure must not significantly remove accumulated material. (Chilled medium, which solidifies the membrane lipids, sometimes causes leakage.)

conditions of "downhill flow" (i.e., entry followed by rapid metabolism), or in the absence of an energy supply (e.g., cells poisoned by azide), systems capable of active transport become **uncoupled** from the energy supply but continue to shuttle back and forth; they then function as "facilitated" or "passive" transport systems. This kind of transport is also often called **facilitated diffusion,** though it has saturable, Michaelis kinetics; but true diffusion increases in rate indefinitely with increasing concentration of permeant.

Some systems are capable only of facilitated transport of specific sugars; they are found in red blood cells and many yeasts, which are genetically adapted to an environment providing a high concentration of the sugars. The contrasting active transport of sugars (and other metabolites) in bacteria evidently reflects their evolutionary adaptation to less opulent environments. To demonstrate active transport of a sugar in bacteria it is usually necessary to block subsequent metabolism (e.g., by a mutation) or to use a nonmetabolizable analog (as noted above for methyl β-thiogalactoside).

β-GALACTOSIDE TRANSPORT

The kinetics of active transport, as observed in the β-galactoside system in E. coli, may be summarized as follows. In the presence of an energy source cells possessing this system take up a corresponding radioactive nonmetabolized permeant (e.g., MTG) at an **initially constant** velocity (V_{in} = **influx** rate). (The rate of **net** uptake soon slows because the efflux begins to balance the influx.) V_{in} varies with external concentration of permeant (C_{ex}) according to the mass-law kinetics of a saturable system, with a characteristic dissociation constant (K_m). This constant, like the Michaelis constant of an enzyme, represents the concentration yielding half the maximal uptake rate, the latter value (V_{in}^{max}) being obtained at a saturating concentration of permeant.
Thus

$$V_{in} = V_{in}^{max} \cdot C_{ex} / (K_m + C_{ex}).$$

In contrast to the Michaelis kinetics of influx, the **efflux** rate at any time (in the presence of influx) is simply proportional to the internal concentration (C_{in}) of permeant; the exit rate constant is denoted as K. The

overall kinetics is thus specified by the equation:

$$\text{Net uptake} = \text{Influx} - \text{Efflux}$$
$$dC_{in}/dt = V_{in}^{max} C_{ex} / (K_m + C_{ex}) - KC_{in}$$

The cells come to equilibrium with the permeant after 2 to 3 minutes at 37°; the influx and the efflux rates are then equal, by definition, and there is no net uptake.

Some quantitative characteristics of the β-galactoside transport system are presented in Table 6-3. It is seen that different permeants using the same system differ from each other independently in **affinity for the entry system**, in **maximal rate of entry**, and in **exit constant**. These permeants therefore also differ in the degree to which they can be concentrated **(maximal concentration ratio)**: ratios of 100 or more are not uncommon.

THE EXIT PROCESS

The rate of efflux of a permeant is proportional to its intracellular concentration.* Such linear kinetics is compatible with two mechanisms of exit: nonspecific diffusion (through pores or by solution in the membrane), and transport on homologous or heterologous specific systems for which the permeant has a low affinity (since the early part of the concentration/rate curve of such systems, well below saturation, is also linear). Probably both mechanisms are used.

Thus with many permeants the rate constant for exit (as well as that for entry) increases with increasing induction, suggesting that the induced transport system participates in both processes. At the same time, a low osmolarity of the medium seems to "stretch" the membrane, since it increases exit rate constants (i.e., in general it lowers the maximal intracellular concentration of permeant, observed in cells at equilibrium with a saturating extracellular concentration).

Such "stretching" of the membrane should also be produced by increased internal osmolarity. This consideration suggests a general mechanism for

---

* Efflux is measured by the kinetics of exchange of labeled permeant during influx, for when the external supply of permeant is removed, in a preloaded cell still supplied with energy, the apparent efflux may be very slow because of the efficient **recapture** of molecules released to the periplasmic space.

TABLE 6-3. Parameters of β-Galactoside Transport System

Parameter	Thiomethyl-galactoside	Thiodiga-lactoside	Thiophenyl-galactoside	Lactose	o-Nitrophenyl-galactoside
Michaelis constant, K_m (mole/liter)	5×10^{-4}	2×10^{-5}	2.5×10^{-4}	7×10^{-5}	10^{-3}
Capacity Y (μmole/gm) at saturation	300 (14°) 160 (26°) 52 (34°)	40	32	550 (4°) — 125 (34°)	29 (14°) — 9 (34°)
Time for half-equilibrium, $t\frac{1}{2}$ (min) at 25°	0.75	1.35	< 0.25	2.4	—
Maximal rate of uptake, V_{1n}^{max} (μmole/gm/min)	148	20.4	> 86	158	—
Exit rate constant, K_{ex} (ml/gm/min)	0.82	0.59	> 2.7	—	—
Maximal concentration ratio, C_{1n}/C_{ex}	65	400	26	1950	—

From Kepes, A., and Cohen, G. N. In *The Bacteria*, vol. IV, p. 179. Academic Press, New York, 1962.

regulating internal osmolarity: uptake of a permeant leads to osmotic uptake of water, and the resulting swelling of the cell increases nonspecific exit. This mechanism may well function as a safety valve at high internal concentrations, limiting the difference in hydrostatic pressure between the inside and the outside of a cell. Moreover, this mechanism can explain how a nonpermeable intermediate, which cannot be utilized by mutants with earlier blocks (Ch. 4), can nevertheless be heavily excreted by mutants with a later block: the excretion is preceded by the accumulation of very high intracellular concentrations, which presumably cause swelling. (With a few intermediates, e.g., galactose-1-P, such accumulations are toxic and impair growth.) In the commercial production of metabolites (e.g., amino acids) by mutants the yield may be increased by controlled damage to the membrane (e.g., by limiting the biotin required for fatty acid production).

OTHER TRANSPORT SYSTEMS

Other Sugars. Similar specific transport systems have been demonstrated for several sugars. Others, however, enter by another mechanism, discussed below (Phosphotransferase system).

Amino acids are actively concentrated by transport systems with similar properties but with lower K_m values (which correlates with the smaller amounts of substrate generally encountered and utilized).

These systems are constitutive rather than inducible, and they have broader specificity: for example, leucine, isoleucine, and valine are transported by the same system. Unlike sugars, amino acids are generally transported faster than they can be utilized: hence their active concentration can be seen even without a block in protein synthesis.

Oligopeptides of varying length and composition can be transported by the same carrier system. Its non-specificity permits the carriage of other compounds covalently attached to an oligopeptide. Subsequent hydrolysis in the cell yields the passenger compound and the component amino acids, any one of which can meet the needs of auxotrophic mutants that have lost the corresponding transport system.

Most **phosphorylated compounds** (e.g., nucleotides) are excluded from bacteria. However, glycerol-P and glucose-P induce systems that transport the intact molecule.

Inhibitors that act inside a cell presumably enter via a hospitable system that ordinarily transports a normal metabolite (e.g., sulfonamides compete with p-aminobenzoate for entry; Ch. 7). Moreover, **competition** for a shared transport system is an occasional mechanism of **growth inhibition**. For example, arginine blocks entry of lysine, thus inhibiting growth of lysine auxotrophs (in response to added lysine) but not growth of the wild type (which makes its own lysine). Such interference with cellular transport may be significant in human diseases in which metabolites (such as phenylalanine) accumulate.

Inorganic ions (e.g., K+, Mg++, PO$_4^{3-}$) have a constant intracellular level in normal growth, even when the external concentration is very low. This homeostasis is maintained by membrane transport systems (the actual intracellular level depending in part on the concentration of counterions, e.g., nucleic acids). Thus mutations that impair transport of a specific ion (K+, Mg++) markedly increase the extracellular concentration required for growth. Fe ion is transported after chelation by various *siderochromes* (Ch. 5, Inorganic requirements); the complexes enter through specific, mutable transport systems.

NONSPECIFIC ENTRY

Cryptic mutants, which have an intracellular enzyme for a particular metabolite but no corresponding transport system, hydrolyze the substrate at a very low rate. Moreover, this rate is directly proportional to extracellular concentration over a broad range, rather than saturable (Fig. 6-26). Like the

Fig. 6-26. Hydrolysis of o-nitrophenyl-β-D-galactoside (ONPG) by *E. coli* cells. Upper curve: induced wild type; lower curve: induced cryptic mutant. Ordinates on left apply to upper curve, and ordinates on right (with expanded scale) to lower curve. [After Herzenberg, L. A. *Biochim Biophys Acta* 31:525 (1959).]

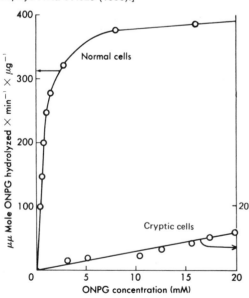

exit process discussed above, this nonspecific entry may represent true diffusion through pores (or through solution in the membrane), or transport through low affinity for systems designed to transport other compounds, or both. With compounds that induce at intracellular sites nonspecific entry appears to be essential for initiating the induction.

Thus following **initial** induction of the β-galactoside system by MTG a low external concentration will suffice for **maintenance** of induction because the inducer can now be concentrated (Ch. 13, Fig. 13-8). On the other hand, some inducers of transport systems (citrate, glucose-6-P) may somehow act at the membrane, since they are effective with cells in which their endogenous metabolism, without induction, far exceeds the rate of initial entry from outside.

Because specific transport systems are saturable, increases in the external concentration of a permeant above saturation levels will not further raise the intracellular concentration. Hence spheroplasts can be protected by hypertonic solutions even of compounds (e.g., sugars, KCl) that can enter the cell.

MECHANISM OF ACTIVE TRANSPORT

Direct study of the molecular mechanism of active transport proved difficult, in part because the characteristic asymmetrical (vectorial) spatial orientation of the process is lost when the components of its machinery are isolated from the cell. However, several useful approaches have now developed.

ISOLATION OF BINDING PROTEINS

A specific binding protein of a transport system was first isolated by Fox and Kennedy from the membrane of *E. coli* after selective labeling with N-ethylmaleimide, an irreversible -SH agent that poisons β-galactoside transport. The procedure used is outlined in Figure 6-27. Induced cells yielded a species of labeled membrane protein that was absent from uninduced cells, or from mutants lacking β-galactoside transport; and the protein was found to be temperature-sensitive in mutants with a *ts* transport system.

In this procedure the -SH reagent unfortunately inactivates the protein that it identifies. However, a similar purification procedure without that reagent can now be used to recover active β-galactoside-binding protein. The MW of the binding protein is 30,000; a fully induced cell has about 8000 molecules in its membrane (ca. 0.35% of the total protein of the cell). This large value, for a single species of transport protein, suggests that the total of all such proteins may constitute much of the cytoplasmic membrane.

The specific "binding proteins" obtained from cells by osmotic shock as noted above (see Periplasmic proteins), may well have been artificially released from a transport system in the membrane. They are observed for certain permeants in *E. coli* (e.g., galactose, the aliphatic amino acid group, sulfate), but not for others (e.g., β-galactosides), though the mechanisms of transport seem otherwise identical.

CARRIER MODEL FOR ACTIVE TRANSPORT

Active transport somehow converts chemical energy into osmotic work: i.e., there is net transport of a permeant into a cell despite the presence of a higher concentration inside. By analogy with the familiar enzymatic linkage of endergonic and exergonic reactions in biosynthesis, it seemed possible that permeants might similarly undergo covalent modification in a cyclic transformation. However, such a mechanism could not account for the active transport of inorganic ions, which do not form covalent bonds. Alternatively, the energy-linked steps might alter the structure of a **carrier protein** rather than that of the permeant. This mechanism was supported by the observed noncovalent interaction of a permeant with its binding protein.

In this model, schematized in Figure 6-28, the carrier protein has a conformation with a **high affinity** for its permeant when facing the outside medium. When it then rotates (or diffuses) in the membrane, so that it faces the inside, some energy-linked reaction changes its conformation in a way that **lowers** its affinity for the permeant (i.e., increases the K_m); the permeant can then be unloaded despite a high intracellular concentration. On the way back to face the outside the protein would somehow snap back into its initial, high-affinity conformation. Equilibrium is reached when the concentration inside causes a rate of reloading of the carrier (in its high K_m state), plus a rate of nonspecific exit, equal to the rate of loading of the carrier facing the outside solution, plus the nonspecific entry.

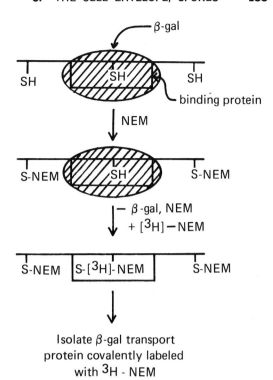

Isolate β-gal transport protein covalently labeled with ^{3}H - NEM

Fig. 6-27. Design of experiment for selectively labeling binding protein of β-galactoside transport system. With a high concentration of a β-galactoside present, to protect (by noncovalent binding) any —SH group on the active site of its transport system, the many other —SH groups (in intact cells or in membrane fragments) were largely inactivated by reaction with N-ethylmaleimide (NEM). The cells or particles were then transferred to a solution without β-galactoside, and the uncovered group was reacted with radioactive NEM. The membrane proteins were solubilized by complexing with a detergent (which covers lipophilic surfaces with a hydrophilic layer) and then were separated by chromatography. [Based on Fox, C. F. and Kennedy, E. P. *Proc Nat Acad Sci USA 54*:891 (1965).]

We have seen that mutations in a binding protein affect both active and facilitated transport. However, other mutations, not affecting the binding protein, "uncouple" β-galactoside transport (i.e., impair only active transport). Hence active transport must involve additional, undefined protein components.

Kinetic studies support the conclusion that energy is used in active transport. Thus when *E. coli* cells are "preloaded" with a β-galactoside and transferred to permeant-free medium (with an energy source still present) they lose the internal pool only slowly (as expected from its low affinity for the

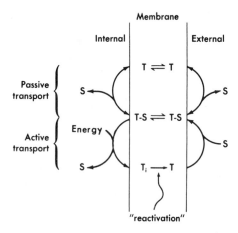

Fig. 6-28. Model for specific transport. T = specific transport protein; S = its substrate (permeant). In passive specific transport (facilitated diffusion), without the application of energy, T moves back and forth reversibly, either unloaded (T) or loaded (T-S); the direction of net flow of S depends on whether the internal or external concentration of S is greater. When energy is applied, to bring about active transport, T is converted on the inner side of the membrane to a relatively "inactivated" form (T$_I$), with a much lower affinity for S. Hence T-S, by conversion to T$_I$-S, can unload S within the cell even when the internal concentration is high; and T$_I$ will return unloaded until that concentration becomes high enough to satisfy the high Michaelis constant of T$_I$ (or to block its formation from T-S). During this cycle T$_I$, in a spontaneous, exergonic reaction, becomes T on the outer aspect of the membrane, and can thus again pick up external S at a low concentration.

Although the arrows suggest diffusion of T from one side of the membrane to the other the movement may be a hinge-like conformational change in a protein, fixed in the membrane, that allows the binding site to face either side.

carrier). However, if the supply of energy is cut off (e.g., by azide) the accumulated compound is transported outward more rapidly, with the same relation of rate to concentration as in its **inward flux** (with or without an energy supply). Thus energy seems to affect only the exit process.

Membrane Vesicles. Kaback found that vesicles formed from membrane fragments of *E. coli,* though lacking cytoplasmic enzymes, retain systems capable of active transport of amino acids and sugars (as well as flavoprotein oxidases that permit them to derive the required energy from D-lactate or succinate). Since the vesicles lack demonstrable adenine nucleotides, and added ATP is ineffective, electron transport appears to be directly coupled

in some way, without ATP, to permeant transport. On the other hand, fermentative cells, deriving their energy from glycolysis, also carry out active transport. Moreover, membrane fragments of bacterial and animal cells exhibit a K+-dependent ATPase activity, which is thought to be involved in K+ transport. It therefore seems possible that intact cells may derive energy for active transport either directly from electron transport or from ATP.

THE PHOSPHOTRANSFERASE SYSTEM

Bacteria have evolved an additional transport mechanism, in which a sugar undergoes the first step in its metabolism, phosphorylation, in the course of entering the cell. This **group translocation** process, discovered by Roseman, resembles active transport, but there is a fundamental difference: the compound released to the cytoplasm is not the same as that taken up from the medium. *E. coli* uses this system to transport several related compounds: glucose and mannose, their reduction products (sorbitol and mannitol), and fructose.

In the phosphotransferase system three components (Fig. 6-29) have been separated. Cytoplasmic enzyme I transfers phosphate from P-enolpyruvate to a small membrane protein (HPr), of MW 9000, which transfers the ~P across the membrane. On the outer surface various P-lipid-dependent membrane enzymes (called enzyme II), each specific for a particular sugar or polyol, transfer the phosphate from HPr-P to that compound. The phosphorylated compound is then released as such to the cytoplasm, where its subsequent dephosphorylation can lead to the same overall result as active transport.

A mutant defective in any particular enzyme II cannot take up the corresponding substrate, whereas mutants defective in enzyme I or in HPr are incompetent for the whole class of compounds that are transported in this way. Indeed, this pleiotropic effect of a single mutation provided the key to the recognition of this novel transport mechanism.

Organisms differ in their choice of the two mechanisms. For example, lactose is taken up by a carrier system in *E. coli* but by a phosphotransferase in *Staphylococcus*. Phosphotransferases are **not** found in certain obligate aerobes (e.g., *Pseudomonas*), which lack (Fig. 3-3) the PEP-generating glycolytic pathway.

Fig. 6-29. The phosphotransferase transport system.

$$\text{P-enolpyruvate} + \text{HPr} \underset{\text{(Cytoplasm)}}{\overset{\text{Enzyme I}}{\rightleftharpoons}} \text{Pyruvate} + \text{P-HPr} \quad \text{(Reaction 1)}$$

$$\text{P-HPr} + \text{Sugar} \xrightarrow[\text{(Membrane)}]{\text{Enzyme II}} \text{Sugar-P} + \text{HPr} \quad \text{(Reaction 2)}$$

$$\underset{\text{(outside)}}{\text{P-enolpyruvate} + \text{Sugar}} \longrightarrow \underset{\text{(inside)}}{\text{Sugar-P} + \text{Pyruvate}} \quad \text{(Over-all reaction)}$$

CHEMOTAXIS AND CHEMORECEPTORS

It has been known since 1900 that motile bacteria exhibit chemotaxis: oriented movement in a gradient of food concentration. Many sugars and amino acids serve as **attractants,** while some compounds (e.g., phenol, acid, base) serve as **repellents** (negative chemotaxis). This phenomenon was long thought to result directly from an unequal supply of energy to different regions of a cell. However, Adler found that chemotaxis can also be evoked by certain non-metabolizable analogs, which cannot supply energy. Since the effective analogs are all substrates for transport systems, whether of the carrier or the phosphotransferase type, it appears that transport mechanisms share common chemoreceptors with the signaling mechanisms of chemotaxis. Indeed, they appear to depend on the same specific binding protein: the two systems that respond to galactose exhibit a similar set of affinities for various analogs, and their activities are affected in parallel by mutations that lower (or eliminate) the affinity of the binding protein. However, chemotaxis and transport must also have unshared components, for either response may be lost selectively by mutation.

Chemotactic receptors must be able to transmit information to the flagella by some mechanism, still unknown, which can be considered a primitive precursor of the nervous system. (A hint is provided by the finding that 3′,5′-cyclic GMP stimulates flagellar activity.) How flagella sense a gradient is also uncertain. It seems unlikely that a cell only 2 to 3 μm long can recognize the infinitesimal difference in concentration at its two ends in the usual gradient. Tracings of patterns of movement suggest an alternative mechanism: cells move in a straight line for a time (Ch. 2), then "tumble" aimlessly, then advance in a new direction; and those linear excursions that advance into a gradient of positive chemotaxis last longer.

SPORES

Under conditions of inadequate nutrition specialized cells, called **spores** (Gr., "seed") or **endospores,** are formed within certain gram-positive cells: aerobic rods (bacilli), anaerobic rods (clostridia), and a few cocci (sporosarcinae). The surrounding mother cell is called a sporangium. Like the seeds of higher plants or the cysts of protozoa, these spores are **cryptobiotic** (i.e., they have no metabolic activity), and they are much more **resistant** than the parental, vegetative cells to the lethal effect of heat, drying, freezing, toxic chemicals, and radiation. Electron microscopic and biochemical studies have revealed much of the structural basis for the remarkable properties of these cells. Moreover, the regulatory changes involved in their formation are now of wide interest, as a unicellular model for cellular differentiation.

Resistance to heat is the property that has received most attention, because of its importance in sterilization.* However, the main ecological role of spores is probably survival in the dry state (or in a non-nutrient medium). For though extreme dehydration (lyophilization; Ch. 64) stabilizes viable vegetative cells ordinary drying kills them rapidly, whereas dried spores survive quantitatively and die very slowly on storage. In fact, viable spores have been recovered from sealed soil samples stored at room temperature for 50 years.

In contrast to bacteria fungi give rise to **exospores,** by budding or septation of their long cells (hyphae; Ch. 43); these are not especially resistant to heat. However, exospores are resistant to drying and they are readily detached from the syncytial, rooted fungal colony: hence they may serve primarily to promote spread.

FORMATION AND STRUCTURE

Spores are unusually dehydrated, highly refractile cells. They do not take ordinary stains (Gram, methylene blue). In the **light microscope** the first visible stage in sporulation is the formation of an area of slightly increased refractility, the **forespore,** at one end of a cell. The refractility gradually increases until the mature spore is formed in about 6 hours. At a later stage the spores are found partly or completely freed of the sporangial wall, which is digested by a lytic enzyme. In a well-sporulating culture most cells form a spore.

Spores are usually smooth-walled and ovoid, but in some species they are spherical or have characteristic ridges. In bacilli the spores usually fit within the normal cell diameter, but in the slender clostridia they cause a bulge, which may be either terminal ("drumstick") or more central (Fig. 6-30).

The **electron microscope** shows that spore formation in bacilli begins with migration of a condensed nucleoid toward one end of the cell. Ingrowth of a mesosome, followed by two layers of membrane, forms a **spore diaphragm** or septum, which separates the nucleoid and surrounding cytoplasm from the sporangium. The diaphragm then grows (at its periphery) toward the tip of the cell until the entire forespore is engulfed by a double membrane, which is attached to the tip of the cell and surrounded by sporangial cytoplasm (Fig. 6-31).

---

* Indeed, as was noted in Chapter 1, the controversy over spontaneous generation was prolonged by the failure of boiling to kill spores present in the culture media.

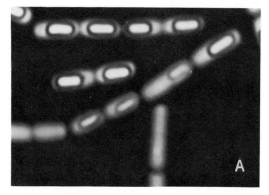

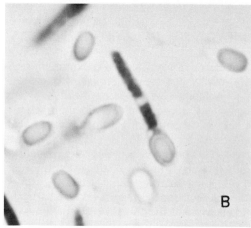

Fig. 6-30. Visualization of spores: light microscope. **A.** *Bacillus cereus:* elongated subterminal spores, nigrosin stain. **B.** *Clostridium pectinovorum:* large, terminal spores, and spores freed from parent cell (sporangium); cells stained with I_2 (which stains granulose). ×3600. (Courtesy of C. Robinow.)

The specialized spore integument is thus laid down by membranes that are initially continuous with the mother cell membrane but become differentiated in composition and function. During forespore formation DNA synthesis ceases, as does net RNA and protein synthesis, but there is extensive phospholipid synthesis and protein turnover.

The facing surfaces of the membranes surrounding the protoplast (core) correspond to the wall-synthesizing surface of the parental cell membrane, and in the maturation of the spore a large amount of material is laid down between the two layers. The resulting thick envelope (integument) eventually occupies as much as half the cell volume. Several layers can be distinguished (Fig. 6-32).

1) The innermost layer is the thin **spore**

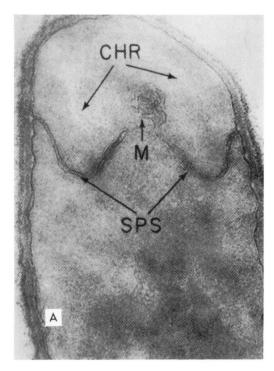 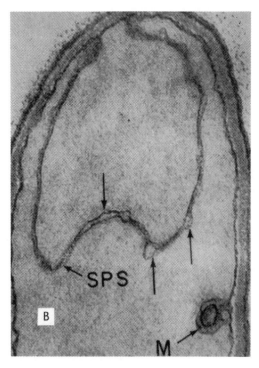

Fig. 6-31. Early stage in sporulation. **A.** The protoplasm at one end of the cell, containing a chromosome (CHR), is cut off from the rest of the cell by a transverse spore septum (SPS), formed by ingrowth of a double membrane and mesosome (M) from the protoplasmic membrane of the other cell. **B.** The periphery of this double membrane moves toward the tip of the cell, ultimately enclosing the whole forespore. [From Ohye, D. F., and Murrell, W. G. *J Cell Biol 14*:111 (1962).]

wall or membrane. It is much thinner than the vegetative cell envelope and is usually difficult to resolve into a membrane and a wall component. 2) Next is the thickest layer, the **cortex,** which contains a concentric laminated structure with clear areas between the layers. 3) Outside the cortex is the **coat,** which is the most densely stained structure of the spore in electron micrographs. 4) Spores of some species are further loosely shrouded in a delicate **exosporium.**

The **spore wall** and the **cortex** each contain **peptidoglycan,** but different kinds. The spore wall has the same type of cross-link in its peptidoglycan as the vegetative cell wall (to which it ultimately gives rise on germination), and it can be stained by antibody to the latter. The cortex, however, does not react with that antibody. Moreover, its peptidoglycan has a different type of cross-link and a much looser structure. Thus in *B. subtilis* only 6% of the muramic residues in the cortex are cross-linked; some are substituted with L-Ala in-

stead of a tetrapeptide, and in others the customary acetyl substituent on the NH_2 group has been displaced by a lactam ring, formed with the lactyl group on the same muramate.

Muramic lactam

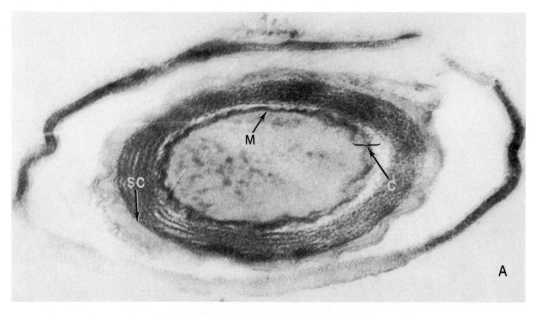

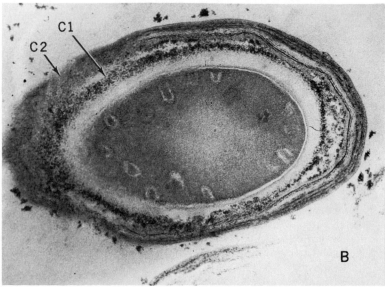

Fig. 6-32. Electron micrographs of sectioned spores. **A.** Fixed to bring out the laminated character of the thick cortex (C) surrounding the spore membrane (M); the spore coat (SC) is artificially separated from the cortex. **B.** Fixed to distinguish the thin inner coat (C1) and the thicker outer coat (C2). The spore membrane is also well preserved, but the cortex has been largely dissolved. (**A** courtesy of C. Robinow. **B** courtesy of D. F. Ohye and W. G. Murrell.)

The loose cortical peptidoglycan is extremely sensitive to lysozyme (if the barrier presented by the coat is breached), and its autolysis plays a key role in germination (see below).

The **coat** is made of an antigenically novel, keratin-like protein, which constitutes as much as 80% of the total protein of a spore. This protein cannot be solubilized unless its numerous S-S bonds are reduced. Through kinetic studies on the parallel appearance of various properties and structures during spore maturation, and through studies of spores with a defective coat or cortex, it has been shown that **the impervious protein coat is responsible for the resistance of spores to chemicals.**

The moderate increase in resistance to killing by ultraviolet or ionizing radiation may also depend on the coat. This structure probably impedes entry of radiochemical products formed in the medium; it also contains S-S groups, which can combine with free radicals (Ch. 64).

The **exosporium** is fundamentally a lipoprotein membrane, with 20% carbohydrate, and it appears to be formed as an enveloping extension of the outer forespore membrane. It is not essential for survival, but it may play a role in laying down the coat.

Small molecules. A striking feature of spores is their huge content of Ca^{++}, for which active transport units appear in the membrane of the mother cell early in sporulation. Normally the Ca^{++} is accompanied by a roughly equivalent amount of **dipicolinic acid,** which can chelate Ca^{++}; dipicolinate is almost unique to bacterial spores and may constitute as much as 15% of their weight. A mutant blocked in dipicolinate formation cannot complete spore formation unless this constituent is supplied.

Dipicolinic acid

(Pyridine-2,6-dicarboxylic acid; for biosynthesis see bacterial pathway to lysine, Ch. 4)

Although the low density of the cortex in electron micrographs has suggested a large content of small molecules (presumably largely dipicolinate), which would be lost during fixation, more direct studies indicate that most of the dipicolinate is located in the core.

These studies have compared the attenuation of the soft β-emission of ^3H-labeled substances in the intact spore and after its disruption, when the emission no longer has to pass through the surrounding cell layers before being counted. ^3H-dipicolinate in the spore suffers about as much attenuation as ^3H-uracil (known to be in the core), while more peripherally located materials (^3H-diaminopimelate in the cortex, ^3H-lysine in the coat) are more attenuated.

The mechanism for accomplishing the large thermodynamic work of eliminating the water from the maturing spore is a mystery, but the cortex seems to play a key role. Thus a block in cortex formation (by mutation or penicillin) prevents effective dehydration, though the spore develops a normal coat; and conversely, a block in coat formation by chloramphenicol can permit formation of a refractile (but unstable) spore, with a cortex. The **contractile cortex theory** suggests that water may be expelled by contraction of the loose polyanionic cortical peptidoglycan, resulting from ionic cross-linking by Ca^{++}.

COMPOSITION OF CORE

The spore contains much less material than the corresponding vegetative cell (Table 6-4), but it must contain everything necessary for forming a complete, growing cell in a medium adequate for germination. Indeed, chemical analyses verify the presence of enough DNA for a complete genome (3×10^9 daltons), together with small amounts of all the stable components of the protein-synthesizing machinery (including ribosomes, tRNAs, and accessory factors and enzymes; Ch. 12). It is not certain whether messenger RNA is also present or appears only after germination.

A wide variety of catabolic and biosynthetic enzymes have also been identified. Energy is derived by glycolysis and by a simplified, soluble flavoprotein oxidative pathway (Ch. 4); the membrane-bound cytochromes are virtually absent. Energy for initiating germination is stored not as ATP but as the much more stable **3-P-glycerate,** which is readily converted to the $\sim$ P donor P-enolpyruvate. Most free amino acids are virtually absent, suggesting that germination initially involves protein turnover.

TABLE 6-4. Composition of *B. megaterium* Spores and Stationary Cells

Constituent*	Cells	Spores
Protein	910	330
Soluble protein	680	165
DNA	91	40
Ribosomal RNA	380	140
Transfer RNA	95	35
Acid-soluble P	500	52

* Protein is expressed as gm/10^{15} cells; nucleic acid and acid-soluble P as moles P/10^{15} cells.

After Kornberg, A. Spudich, J. A., Nelson, D. L., and Deutscher, M. *Annu Rev Biochem 37:*51 (1968).

Proteins and Heat Resistance. Various enzymes in spores are much more thermostable than the corresponding enzymes in vegetative cells. However, after thorough purification many corresponding spore and vegetative enzymes have turned out to be indistinguishable. It thus appears that the striking thermostability of spores depends on their internal environment rather than on intrinsic properties of their proteins though at least one enzyme in a bacillus (aldolase) is cleaved by a protease to yield a smaller and more stable spore enzyme.

Dehydration and ionic conditions are undoubtedly major factors in stabilizing spore proteins. Ca dipicolinate evidently plays a large role, by some unknown mechanism, for its content markedly influences heat resistance. (Fig. 6-33).

REGULATORY CHANGES IN SPORULATION

The commitment to sporulate is usually triggered by exhaustion of the C or N source. Indeed, some organisms exhibit **endotrophic** sporulation, i.e., in distilled water, in which all components must be derived from materials already present in the mother cell. Some organisms sporulate when they exhaust glucose and shift to utilizing the accumulated acids. This response to a poorer C source has suggested that the mechanism may involve release from catabolite repression (Ch. 13), which is known to regulate simultaneously a whole class of genes; but the connection has not been firmly established.

Whatever the metabolic signal, it must result in a broad shift in the pattern of gene expression. A number of spore-specific genes, which are inactive (repressed; Ch. 13) in the vegetative cell, must become active, while many other genes become inactive. Losick has revealed a major mechanism in this shift: **changes in the structure of RNA polymerase** (Ch. 11) that alter its specificity in selecting regions of DNA for transcription into RNA. Moreover, this step is essential, since sporulation is prevented by a mutation in the polymerase that prevents its change in specificity. This major advance is further discussed in Chapter 13. The changes in the polymerase appear to involve a protease, which may play a crucial role in triggering sporulation. Other proteases play an additional role, causing a protein turnover of 20% per hour (compared with 5% in nonsporulating stationary cells).

Unfortunately, it is not possible to distinguish the metabolic activities of the two compartments of the sporulating cell, for forespores have not been separated physically from sporangia. Radioautography indicates that both genomes are being transcribed; and it seems probable that the spore wall and cortex are built from precursors on opposite sides of the double membrane.

An altered RNA polymerase is probably only one of the regulatory mechanisms in sporulation. Even in the much simpler (and somewhat parallel) process of phage synthesis genes are expressed in several successive sets, a product of one set releasing the next (Ch. 45). Sporulation has not been nearly as thoroughly studied. However, from hybridization studies with pulse-labeled RNA (Ch. 11), and from the kinetics of appearance and labeling of various enzymes, it is clear that the pattern of activation of spore-specific genes changes at various stages of sporulation; moreover, some of the "vegetative" genes remain active.

Genetic studies have revealed a variety of loci at which mutations destroy the ability to form spores. Some of these **asporogenous** mutants are blocked in early steps (including protease formation), whose absence prevents any detectable formation of spore components. Others (morphological mutants)

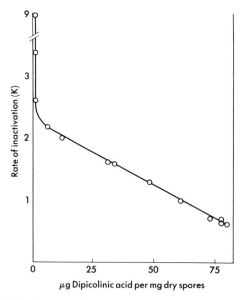

Fig. 6-33. Heat resistance of spores of *B. cereus* with different dipicolinic acid content, resulting from growth in media of different composition. K is an index of the rate of exponential loss of viability at 80°. [From Church, B. D., and Halvorson, H. *Nature 183*:124 (1959).]

initiate sporulation but are blocked at various stages in its completion, hence forming immature or abnormal spores. Over 20 spore-specific genes, scattered on the chromosome, have been defined; probably hundreds exist.

Asporogenous mutants have generally lost the ability to synthesize those **antibiotics** and **toxins** whose formation is characteristically associated with sporulation. However, the association of these activities does not define cause and effect, and the suggestion of a regulatory role for antibiotics in sporulation (Ch. 7) remains speculative.

GERMINATION

The overall process of converting a spore into a vegetative cell is often called germination. Three stages can be distinguished: **activation, germination proper** (initiation), and **outgrowth.**

Activation. Though some bacterial spores will germinate spontaneously in a favorable medium, others (especially if freshly formed) remain dormant unless they are activated by some traumatic agent, such as heat, low pH, or an SH compound. **Aging,** with its multiple,

undefined consequences, is probably the most important natural cause. Activation presumably damages the impermeable coat, since grinding with glass powder is also effective.

Dormancy (i.e., unresponsiveness to a germination medium) has the same biological function in spores as in plant seeds. In many wild-type higher plants the seeds will not germinate until an outer coat has been damaged by some agent provided in the ecology of that plant; abrasion, chemical or bacterial attack, heat, freezing, light, maceration. Germination is thereby spread in time and space, promoting survival of the species by preventing uniform germination in response to conditions that are only temporarily favorable. Domesticated plants, in contrast, have been selected for uniformity of germination.

Germination. Unlike activation, germination requires water and a triggering **germination agent.** Various species respond to various metabolites (e.g., alanine, dipicolinate) or inorganic ions (e.g., Mn^{++}), which penetrate the damaged coat and evidently activate a spore-lytic enzyme. The resulting hydrolysis of the cortical peptidoglycan (Fig. 6-34)

Fig. 6-34. Electron micrograph of germinating spore of *B. megaterium.* The cortex has lost its laminated appearance, and the coats have begun to disintegrate. Note deposition of vegetative cell wall (CW) outside membrane. Appearance of cytoplasm already closely resembles that of vegetative cell, with mesosomes, many ribosomes, and well-demarcated nuclear body. (Courtesy of C. Robinow and J. Marak.)

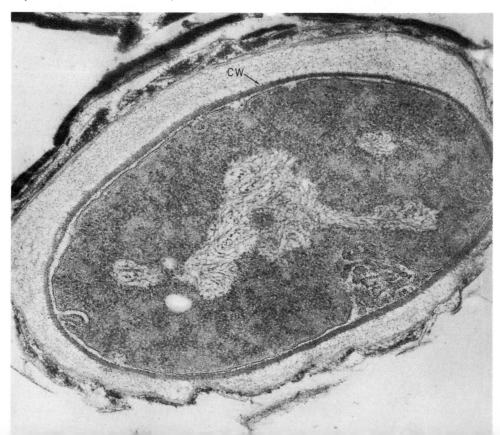

proceeds to completion within minutes or even seconds: up to 30% of the dry weight of the cell is released, mostly as peptidoglycan fragments ("spore peptide")* and Ca dipicolinate. With the breakdown of the cortical barrier the cell rapidly takes up water, accompanied by K^+ and Mg^{++}, and so the cell loses its refractility (the usual test for germination), along with its resistance to heat and staining.

The **loose cross-linking** of the cortical peptidoglycan undoubtedly promotes its rapid digestion during germination. It has been suggested that the peptidoglycan is protected from hydrolysis by bound

---

* It was in "spore peptide" released from germinating spores that muramic acid was first found.

Ca^{++} ions and becomes susceptible when these ions are removed. However, the state of the sporelytic enzyme is also important: during germination it shifts from a particle-bound to a soluble state.

Outgrowth. In a nutrient medium germination leads to immediate outgrowth. However, in a starvation medium, or in the presence of an inhibitor of protein synthesis, germination still goes to completion and the cell becomes rehydrated, but there is very little metabolic activity. Outgrowth is a gradual resumption of vegetative growth (Fig. 6-34).

Protein synthesis increases progressively as its initially scanty machinery is expanded by protein and RNA synthesis; the spore wall becomes a thicker vegetative cell wall; and after about an hour

Fig. 6-35. Sequence of germination and outgrowth in *B. megaterium,* observed with light microscope. **A.** 5 minutes: all cells highly refractile. **B.** 50 minutes: most cells have lost refractility and begun to grow out of spore integument. **C.** 135 minutes: elongated cells still attached to refractile spore coat. **D.** 165 minutes: cells longer, some freed from spore coat. ×3600, reduced. (Courtesy of C. Robinow.)

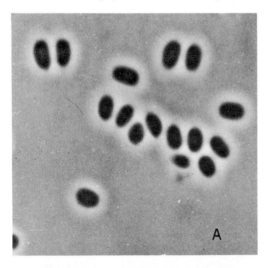
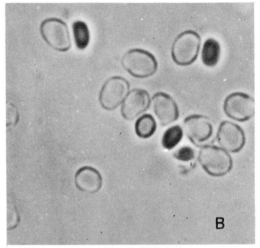
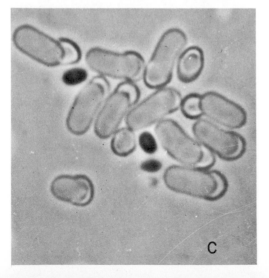
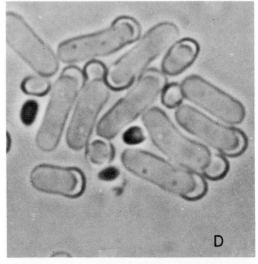

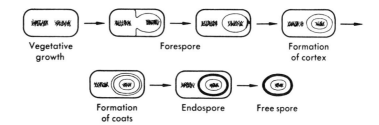

Fig. 6-36. Stages in sporulation.

DNA synthesis begins and the cell, twice its initial volume, begins to burst out of the spore coat (Fig. 6-35). (In the slender clostridia, however, such **"emergence"** from the coat may result from germination, even before outgrowth.) Since the spore, like the sporulating cell, possesses the spore-specific RNA polymerase, vegetative RNA polymerase must be formed early.

SUMMARY

We have seen that bacterial endospores are differentiated cells formed within a vegetative cell; they encase a genome in an insulating, dehydrated vehicle that makes the cell ametabolic and resistant to various lethal agents, but permits subsequent germination in an appropriate medium. Spores are formed by the invagination of a double layer of cell membrane, which closes off to surround a chromosome and a small amount of cytoplasm. A thin spore wall, and a thicker cortex with a much looser peptidoglycan, are synthesized between the two layers; outside the cortex is a protein coat, rich in disulfide crosslinks. The stages of sporulation are presented diagrammatically in Figure 6-36.

The characteristic features of spore physiology are beginning to be understood. The keratin-like properties of the coat account for the resistance to staining and to attack by deleterious chemicals, while the dehydration and the presence of a large amount of Ca dipicolinate contribute to the heat resistance. We do not know the mechanism by which these cells become essentially completely dehydrated, and therefore highly refractile, but the cortex appears to play a major role: in germination, following activation by mechanical or chemical damage to the surface coat, the attack of a lytic enzyme on the peptidoglycan of the cortex permits uptake of water and loss of Ca dipicolinate.

Sporulation involves an extensive shift in the pattern of gene transcription, brought about at least in part by a change in the specificity of RNA polymerase. This change probably results from the attack of a specific protease on the polymerase. Other protease(s) are responsible for the extensive protein turnover that accompanies sporulation. The sequential regulatory processes in this microbial differentiation are being further analyzed through the use of mutants blocked at various stages in sporulation.

SELECTED REFERENCES

Cell Envelope: Books and Review Articles

ADLER, J. Chemoreceptors in bacteria. *Science 166*: 1588 (1969).

ARCHIBALD, A. R., BADDILEY, J., and BLUMSON, N. L. The teichoic acids. *Adv Enzymol 30*:223 (1967).

COHEN, G. N., and MONOD, J. Bacterial permeases. *Bacteriol Rev 21*:169 (1957).

CRONAN, J. E., JR., and VAGELOS, P. R. Metabolism and function of the membrane phospholipids of *Escherichia coli. Biochim Biophys Acta 265*:25 (1972).

DEMAIN, A. L., and BIRNBAUM, J. Alteration of permeability for the release of metabolites from microbial cell. *Curr Top Microbiol Immunol 46*:1 (1968).

GHUYSEN, J. M. Use of bacteriolytic enzymes in determination of wall structure, and their role in cell metabolism. *Bacteriol Rev 32*:425 (1968).

GLAUERT, A. M., and THORNLEY, M. J. The topography of the bacterial cell wall. *Annu Rev Microbiol* 23:159 (1969).

GOLDFINE, H. Comparative aspects of bacterial lipids. *Adv Microbial Physiol* 8:1 (1972).

HAROLD, F. M. Antimicrobial agents and membrane function. *Adv Microbial Physiol* 4:46 (1970).

HAROLD, F. M. Conservation and transformation of energy by bacterial membranes. *Bacteriol Rev* 36:172 (1972).

HEPPEL, L. A. The effect of osmotic shock on release of bacterial proteins and on active transport. *J Gen Physiol* 54:953 (1969).

KABACK, H. R. Transport across isolated bacterial cytoplasmic membranes. *Biochim Biophys Acta* 265:367 (1972).

KENNEDY, E. P. The Lactose Permease System of *Escherichia coli*. In *The Lactose Operon*, p. 49. (J. R. Beckwith and D. Zipser, eds.) Cold Spring Harbor Laboratory, Cold Spring Harbor, N.Y., 1970.

LEIVE, L. (ed.). *Membranes and Walls of Bacteria.* Dekker, New York, 1973.

LENNARZ, W. J., and SCHER, M. G. Metabolism and function of polyisoprenol sugar intermediates in membrane-associated reactions. *Biochim Biophys Acta* 265:417 (1972).

LIN, E. C. C. The genetics of bacterial transport systems. *Annu Rev Genet* 4:225 (1970).

LÜDERITZ, O. Recent results on the biochemistry of the cell wall lipopolysaccharides of *Salmonella* bacteria. *Angew Chem* (Eng) 9:649 (1970).

OSBORN, M. J. Structure and biosynthesis of the bacterial cell wall. *Annu Rev Biochem* 38:501 (1969).

PARDEE, A. B. Membrane transport proteins. *Science* 162:632 (1968).

ROSEMAN, S. The transport of carbohydrates by a bacterial phosphotransferase reaction. *J Gen Physiol* 54:1388 (1969).

ROTHFIELD, L. I., ed. *Structure and Function of Biological Membranes.* Academic Press, New York, 1971.

ROTHFIELD, L. I., and ROMEO, D. Role of lipids in the biosynthesis of the bacterial cell envelope. *Bacteriol Rev* 35:14 (1971).

RYTER, A. Structure and functions of mesosomes of gram-positive bacteria. *Curr Top Microbiol Immunol* 49:151 (1969).

SALTON, M. R. J. The Bacterial Membrane. In *Biomembranes,* vol. I. (L. A. Manson, ed.). Plenum Press, New York, 1971.

SCHLEIFER, K. H., and KANDLER, O. Peptidoglycan types of bacterial cell walls and their taxonomic implications. *Bacteriol Rev* 36:407 (1972).

STROMINGER, J. L. Penicillin-sensitive enzymatic reactions in bacterial cell wall synthesis. *Harvey Lect* 64:179 (1968–69).

STROMINGER, J. L., and GHUYSEN, J. M. Mechanisms of enzymatic bacteriolysis. *Science* 156:213 (1967).

VAN ITERSON, W. The Bacterial Surface. In *Handbook of Molecular Cytology,* p. 174. (A. Lima-de-Faria, ed.). North Holland, London, 1969.

Cell Envelope: Specific Articles

BRAUN, V., and WOLFF, H. The murein-lipoprotein linkage in the cell wall of *Escherichia coli*. *Eur J Biochem* 14:387 (1970).

HAZELBAUER, G. L., and ADLER, J. Role of the galactose-binding protein in chemotaxis of *Escherichia coli* toward galactose. *Nature New Biol* 230:101 (1971).

OSBORN, M. J., GANDER, J. E., and PARISI, E. Mechanism of assembly of the outer membrane in *Salmonella typhimurium*. *J Biol Chem* 247:3973 (1972).

WINKLER, H. H., and WILSON, T. H. The role of energy coupling in the transport of β-galactosides by *E. coli*. *J Biol Chem* 241:2200 (1966).

Spores: Books and Review Articles

BALASSA, G. The genetic control of spore formation in bacilli. *Curr Top Microbiol Immunol* 56:99 (1971). A critical, balanced review of sporulation, as well as an analysis of the results obtained with mutants.

GOULD, G. W., and HURST, A. (eds.). *The Bacterial Spore.* Academic Press, New York, 1969.

KEILIN, D. The problem of anabiosis or latent life: History and current concept. *Proc Roy Soc Lond* B150:149 (1959).

SUSSMAN, A. S., and HALVORSON, H. O. *Spores: Their Dormancy and Germination.* Harper & Row, New York, 1966.

TIPPER, D. J., and GAUTHIER, J. J. Structure of the Bacterial Endospores. In *Spores V*, p. 3. (H. O. Halvorson, R. Hanson, and L. L. Campbell, eds.). American Society of Microbiologists, Washington, D.C., 1972.

chapter 7

CHEMOTHERAPY

The Development of Chemotherapy   148
Effects of Drugs on Growth and Viability   149
Mechanisms of Bactericidal Action   151
THE ACTIONS OF METABOLITE ANALOGS   151
Competitive Inhibition   151
Noncompetitive Reversal of Growth Inhibition   152
ANTIMETABOLITES   153
Biosynthetic Incorporation of Antimetabolites   153
Additional Actions of Antimetabolites   153

SPECIFIC AGENTS   154
SULFONAMIDES AND RELATED AGENTS   154
PENICILLIN AND RELATED ANTIBIOTICS   155
Production and Chemistry   155
Semisynthetic Penicillins   156
Cephalosporins   157
Antibiotics with a Similar Spectrum   158
OTHER   ANTIBIOTICS   158
Aminoglycosides   158
Isoniazid   159
Broad-spectrum Antibiotics   159
Synthetic Compounds Used for Urinary Antibacterial Therapy   160
PHYSIOLOGY AND ECOLOGY OF ANTIBIOTIC PRODUCTION   160
Biosynthesis   160
Ecological Role   161

DRUG RESISTANCE   161
PHYSIOLOGICAL (PHENOTYPIC) EFFECTS OF MUTATIONS
    TO DRUG RESISTANCE   161
RESISTANCE GENES IN PLASMIDS   162
Penicillinase   162
Enzymes Directed by Resistance Factors (RF)   163
GENERAL FEATURES OF DRUG RESISTANCE   163
Induced Phenotypic Resistance   163
Prevention of Resistance   164
Synergism and Antagonism   164
Genetic Dominance of Sensitivity and Resistance   165

CODA   165

Chemotherapeutic agents are defined as **chemicals that can interfere directly with the proliferation of microorganisms, at concentrations that are tolerated by the host.** Accordingly, their essential feature is **selective toxicity.** Some of these drugs are **bacteristatic,** the inhibition of growth being **reversed** when the drugs are removed. Others are **bactericidal,** exerting an **irreversible,** lethal effect.*

This chapter will consider general features of antimicrobial action in vitro, as well as certain specific actions. Others are taken up elsewhere (interference with wall synthesis and membrane function in Ch. 6, nucleic acid synthesis in Ch. 11, protein synthesis in Ch. 12, and viral replication in Ch. 49). Chemotherapeutic action in the host will be considered in Chapter 23, and disinfection (nonselective antimicrobial action) in Chapter 64.

THE DEVELOPMENT OF CHEMOTHERAPY

Origins. The Peruvian Indians used cinchona bark (containing quinine) centuries ago to treat malaria, without knowing that it acted directly on a parasite. The idea of such direct action was forcefully advanced by Paul Ehrlich, in Germany, at a time when organic chemistry was blossoming. While a medical student in 1870 he introduced dyes that are still used in histology for the selective staining of basophilic and acidophilic cell components. Becoming involved later in the exciting development of antitoxins, he formulated the essentially correct "side-chain" theory to account for the extraordinarily specific interactions of antibodies with antigens. In 1904, drawing on these experiences, he began to

seek synthetic chemicals that would exhibit a greater affinity for parasites than for host cells (*nihil agit nisi fixatur*); and for this selective action he coined the word "chemotherapy." But though Ehrlich did discover dyes that were useful against trypanosomes, and arsenicals that were useful against spirochetes, he was disappointed at not finding any "magic bullets" against bacteria (other than spirochetes).

Following Ehrlich's death in 1915 the medical world returned to the conviction that chemotherapy was an impractical dream. According to the prevailing view, the struggle between host and parasite was too complex to permit such a direct attack, and one could better seek to stimulate host defenses. This vitalistic attitude hindered further progress for 20 years, but it did have an experimental basis, since the drugs then known to exhibit antimicrobial activity in vivo were inactive in vitro. We now know why: these drugs were precursors that became active only after metabolic conversion by the host. For example, Ehrlich's antisyphilitic arsphenamine was active only in vivo; but many years later this arseno compound ($RC_6H_4As=AsC_6H_4R$) was found to be oxidized in the body to the corresponding arsenoxide ($RC_6H_4As=O$), which is active in vitro as well as in vivo.

Sulfonamides. Antibacterial chemotherapy was launched at the I. G. Farben industry in Germany, where in 1935 Domagk developed a dye, Prontosil, that dramatically cured streptococcal infections. The dye proved inactive in vitro, but a year later Tréfouel, in France, showed that patients received it excreted a simpler, colorless product, **sulfanilamide,** which was active in vitro.

This compound was then used directly as a drug and was soon succeeded by more potent derivatives; the class is known in medicine as the **sulfonamides.** Sulfanilamide unequivocally reestablished Ehrlich's principle of direct chemotherapeutic action on the parasite. Since then thousands of antimicrobial compounds have been discovered by screening for

---

* Though the traditional words are "bacteriostatic" and "bactericidal," the continued use of different stems in these parallel words hardly seems desirable.

activity in vitro. Of these, however, only a few dozen have been sufficiently selective to be useful.

$$NH_2$$

$$SO_2NH_2$$

Sulfanilamide

Antibiotics. The success of sulfonamides renewed interest in **antibiotics,** which are defined as **antimicrobial agents of microbial origin.**[*] Fleming reported in 1929 that a contaminating colony of the mold *Penicillium notatum* lysed adjacent colonies of staphylococci (Fig. 7-1); but the lytic agent seemed too unstable to be useful. However, 10 years later Chain at Oxford showed that after purification the active material, called penicillin, was stable, and it proved remarkably effective in certain infections.

The success of penicillin encouraged the systematic search for additional antibiotics. In 1944

---

[*] Inhibition of some microbes by others was first recorded in 1877 by Pasteur, who observed sterilization of anthrax bacilli in a contaminated culture. In the next half-century a number of similar accidental observations were reported, together with a few abortive attempts at putting the inhibitory substances to therapeutic use.

Fig. 7-1. The discovery of penicillin. Note lysis of colonies of *Staphylococcus aureus* surrounding a contaminating colony of *Penicillium notatum.* [From Fleming, A. *Br J Exp Pathol 10:*226 (1929).]

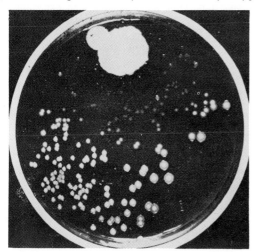

Waksman and colleagues, in a laboratory devoted to soil microbiology, discovered streptomycin (from *Streptomyces griseus*), which extended chemotherapy to the tubercle bacillus and to many gram-negative organisms.[†] The further search was then largely taken over by pharmaceutical and fermentation industries, whose massive effort has yielded many valuable products. At the same time a new, high-pressure pattern of drug promotion has emerged. One factor has been the large financial consequence of having one antimicrobial drug displace another in the market: in 1971 the wholesale purchase of antibiotics for medical purposes in the United States amounted to 600 million dollars.

The term chemotherapeutic is sometimes used to distinguish synthetic compounds from antibiotics, but Ehrlich's original definition is more useful: those antimicrobial compounds that are nontoxic enough to be useful in therapy. Some drugs discovered as antibiotics are now produced by chemical synthesis.

EFFECTS OF DRUGS ON GROWTH AND VIABILITY

The action of any new antimicrobial agent is first studied with growing cultures. Such studies yield several kinds of information that are of value in guiding and understanding therapeutic use. These include the concentrations required to inhibit various organisms (antimicrobial spectrum), effects of environmental conditions on this sensitivity, the kinetics of inhibition, the presence of a bactericidal or only a bacteristatic action, the presence or absence of bacterial lysis, and the metabolic requirements for bactericidal action.

1) **Sensitivity** may be determined by the **tube dilution method,** in which identical inocula are incubated in tubes of medium containing different concentrations of the drug. The endpoint is the lowest concentration that prevents the development of turbidity on incubation. In the more convenient **agar diffusion method** a sample of drug is deposited on solid medium heavily seeded with bacteria. When the bacterial lawn grows out a clear

---

[†] This group had earlier isolated **actinomycin,** the first of many antibiotics obtained from streptomycetes. It was soon discarded because of its toxicity, but in a cancer chemotherapy screening program nearly two decades later it proved useful against certain tumors. Actinomycin D is also valuable in cell physiology for stopping RNA synthesis (Ch. 11).

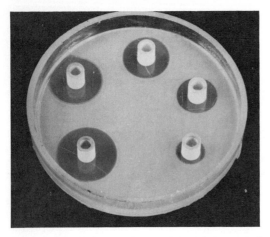

Fig. 7-2. Plate assay for an antibiotic. Plate seeded on the surface with the standard test strain of *S. aureus* (Oxford strain). To the open cylinders were added equal amounts of penicillin solutions containing 4, 2, 1, 0.5, and 0.25 U/ml, and the plates were incubated overnight. Today filter paper discs are generally used instead of cylinders. (From Chain, E., and Florey, H. *Endeavour.* Jan. 1944, p. 3.)

zone may surround the drug sample (Fig. 7-2).

The diameter of this zone can be used either to measure the sensitivity of the organism, with a known drug sample, or to assay the concentration of drug, with an organism of known sensitivity. The diameter is influenced by the growth rate and the density of the inoculum, since it reflects a race between multiplication of the organism and diffusion of the drug.

Stable paper test discs, containing a known amount of drug, are widely used for determining the **sensitivity spectrum** of strains freshly isolated from patients. If the dose is properly chosen the appearance of a zone of inhibition will indicate that the organism is inhibited by the drug at levels attainable in the patient. The test is used to recognize those agents that are sufficiently active; from this set the one to be used is then selected on the basis of various pharmacological considerations, and should **not** be simply the drug yielding the widest zone of inhibition.

2) **The kinetics of inhibition, and the presence of lysis,** may be observed by **turbidimetric** measurements following addition of drug to a growing culture. Several patterns are illustrated in Figure 7-3A: thus sulfonamide inhibition of growth is delayed for several generations; chloramphenicol and streptomycin cause an almost immediate leveling off of turbidity; and penicillin causes lysis, indicated by a sharp fall in turbidity.

3) **Bactericidal action** is, of course, evident when there is gross lysis (e.g., with penicillin). To quantitate the rate of killing, however, and also to recognize killing when the cells remain unlysed, **viability counts** are required (Fig. 7-3B). For this purpose samples are taken at intervals, diluted appropriately, and plated on an adequate medium. The dilutions must be great enough to eliminate further action of the drug. Indirect indices of cell death, such as inhibition of respiration or of dye reduction, or intracellular staining, are less reliable than viability

Fig. 7-3. Kinetics of antimicrobial action of representative drugs. Drug added at arrow to exponentially growing culture.

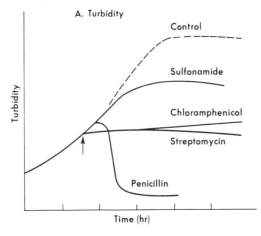

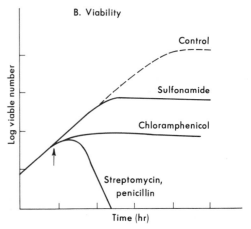

counts, especially for drugs that do not damage the cell membrane.

The distinction between bactericidal and bacteristatic action is not rigorous. At borderline concentrations some ordinarily cidal agents appear to be static; conversely, bacteristasis from any cause is eventually followed by a decrease in viable number. Nevertheless, with one set of agents the viable count remains essentially constant for the first few hours; while with the other set it declines by decimal orders of magnitude. The distinction is important both for understanding mechanism of action and for guiding clinical use (Ch. 23).

MECHANISMS OF BACTERICIDAL ACTION

The difference between an irreversible, bactericidal action and a reversible, static action depends not simply on whether the agent reacts reversibly or irreversibly with its receptor, but on whether an irreplaceable element of the cell is irreversibly damaged. Three general sites of cidal action can be recognized: 1) the **DNA chain** (whose damage and repair will be discussed in Ch. 11); 2) the **cell membrane** (unless the hole can be repaired); and 3) **any enzyme (or other protein) required for protein synthesis.** Thus with most enzymes irreversible inactivation of all the molecules present permits at least a trace of continued protein synthesis, which can regenerate the depleted species; but if all the molecules of any protein required for protein synthesis are irreversibly damaged its regeneration is impossible.

Persisters. When a bactericidal chemotherapeutic agent is added to a growing culture killing is initially more or less exponential but it often levels off after some time: successive samples continue, for hours, to contain a small number of "persisters," which generate colonies when transferred to fresh solid medium. Since the descendants of such cells are no more resistant than the original population the phenomenon is evidently one of **phenotypic** rather than **genotypic resistance.** The effect is difficult to study, since it is not perpetuated in the progeny: presumably some cells, perhaps because of abnormal cell division, have a block in the metabolism required for a bactericidal response, and the

block is spontaneously overcome during subsequent incubation for a viability count.

This phenomenon undoubtedly also occurs in the body, and contributes to the usual failure of brief chemotherapy to be curative. Indeed, persisters are likely to be even more frequent in the body than in the test tube because nutritional conditions will be less uniform (Ch. 23). Fortunately, immune mechanisms complement chemotherapy, for their attack on bacteria depends on surface receptors rather than on metabolic activities; hence persisters, like cells subject to bacteristatic action, are finally cleaned up by immune (and phagocytic) responses.

THE ACTIONS OF METABOLITE ANALOGS

COMPETITIVE INHIBITION

Sulfonamides and PAB. For many decades certain inhibitors have been known to compete with normal substrates for binding to the active site of an enzyme or other protein—for example, carbon monoxide ($C = O$) competes with oxygen ($O = O$) for binding to hemoglobin. However, the principle of competitive inhibition seemed relatively unimportant in biochemistry and pharmacology until it dramatically explained the action of the first useful antibacterial agent, sulfanilamide.

This inhibitor was found to be relatively ineffective in cultures containing yeast extract. Woods, in England, identified the antagonist in the extract as a structural analog of the drug, p-aminobenzoate (PAB, Fig. 7-4); he further showed that the antagonism was competitive. This finding led to the suggestion that 1) PAB must be an **essential metabolite,** i.e., the substrate of an essential (though then unknown) metabolic reaction, and 2) sulfanilamide competitively inhibits this reaction. (The concept of essential metabolite should be distinguished from that of **essential nutrient:** sulfonamides are equally effective against organisms that can synthesize PAB (Fig. 4-17) and against those few that require it for growth.)

PAB was soon identified as a component of folic acid (Fig. 7-4), and the postulated inhibition of a

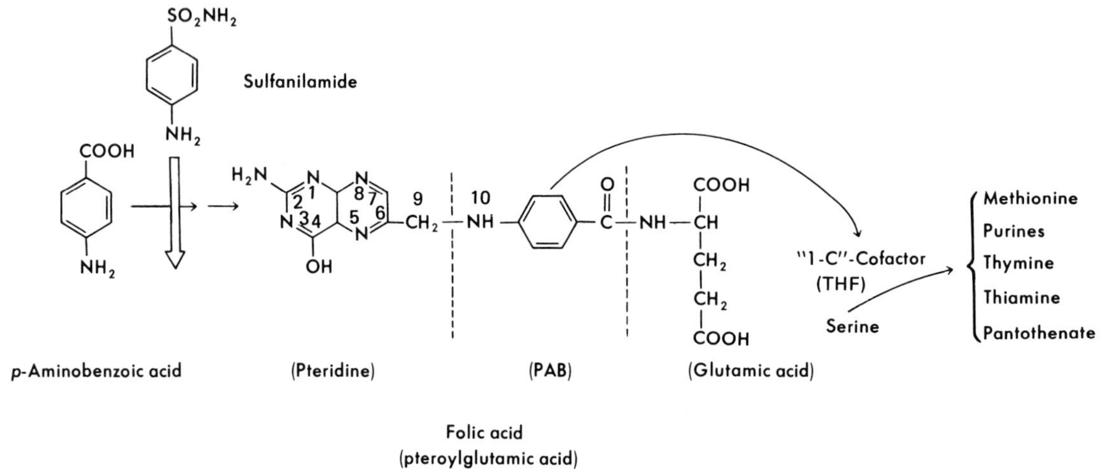

Fig. 7-4. Metabolism interfered with by sulfonamides. A reduction product of folate, tetrahydrofolate (THF), is the cofactor of 1-C transfer (Fig. 4-15).

PAB-incorporating enzyme on the pathway to folate was later demonstrated with extracts. Further studies have shown, however, that the story is not so simple: as with many other analogs, sulfonamides serve as substrates and thus are converted to products that jam later reactions; moreover, sulfonamides also compete with PAB for entry to the cell via a specific transport system (see below, Additional actions).

The function of PAB as a precursor of a coenzyme explains the delayed bacteristatic effect of sulfonamides. As we have seen in nutritional studies with mutants (Fig. 5-5), deprival of a required vitamin (in contrast to deprival of a building block) slows growth only gradually. Similarly, a sulfonamide-induced block leads to a gradual dilution of the folate in the growing cells.

The characteristics of competitive inhibition of an enzyme follow from the Michaelis-Menten application of the mass law to enzyme kinetics. If a substrate (S) and its analog (A) can each form a reversible complex with an enzyme (E), the extent of formation of each complex will be proportional both to the concentration of the corresponding unbound ligand and to its affinity for the enzyme (i.e., the inverse of the dissociation constant, K, of the complex).

$$\frac{(ES)}{(E)} = \frac{1}{K_S}(S) \qquad (1)$$

$$\frac{(EA)}{(E)} = \frac{1}{K_A}(A) \qquad (2)$$

Hence, in a mixture of E with both S and A the ratio of the two complexes will be proportional to the ratio of the concentrations of the two ligands.

$$\frac{(ES)}{(EA)} = \frac{K_A}{K_S} \cdot \frac{(S)}{(A)} \qquad (3)$$

Since the rate of the enzyme reaction is proportional to the concentration of the ES complex, the analog slows the reaction to the extent that it replaces ES by the inactive EA. The **criterion for competitive inhibition** follows from equation (3): at the usual, saturating concentration of substrate the **rate** of the reaction is **constant for a particular analog/substrate ratio.** In other words, with a doubled concentration of the analog one can maintain the same enzyme activity by also doubling the concentration of substrate.

NONCOMPETITIVE REVERSAL OF GROWTH INHIBITION

Inhibition of growth can be reversed not only by the substrate of the blocked enzyme but also by a later product in the same sequence. The latter reversal is noncompetitive, i.e., an adequate quantity of product restores growth, regardless of how tightly the preceding reaction is shut off. Since our body fluids contain folic acid sulfonamide chemotherapy depends on the fortunate fact that most bacterial species are impermeable to it. Moreover, though sulfonamide inhibition is noncompetitively reversed by a mixture of the compounds whose biosynthesis depends on folate (principally methionine, purines, and thymine; (Fig. 7-4), only the amino acids and vitamins in this mixture are present in our body fluids: purines and pyrimidines are made in cells as nucleotides and do not circulate, at least in forms (free bases or nucleosides) that can be utilized by bacteria.* However, autolysis releases these noncompetitive antagonists; **hence sulfonamides are ineffective in sites of extensive tissue destruction,** such as purulent exudates, burns, and wounds.

ANTIMETABOLITES

The explanation of sulfonamide action generated the hope that the hitherto empirical search for new agents could be replaced by the rationally based synthesis of antimetabolites, i.e., analogs of known essential metabolites. Accordingly, thousands of analogs of vitamins, purines, pyrimidines, and amino acids have been synthesized. Many proved to be effective inhibitors of bacterial growth in vitro; but unfortunately they failed to exhibit the selectivity required for chemotherapy, in part because they were modeled on metabolites that are also essential (unlike PAB) for the host.

Antimetabolites have had some success, however, in cancer chemotherapy, and they have been useful in studies in cell physiology. Moreover, the concept of structural analogy has had broad application in pharmacology. For example, many antibiotics have been found to serve as structural analogs of intermediates in macromolecule synthesis.

Most analogs compete reversibly with the substrate, but some react irreversibly with the corresponding enzyme (e.g., penicillin, Ch. 6).

BIOSYNTHETIC INCORPORATION OF ANTIMETABOLITES

The classic interpretation of competitive inhibition assumed that analogs can only **block** the active sites of corresponding enzymes.

---

* Thus mutations of *Salmonella* to various amino acid requirements do not affect virulence in mice; but a requirement for PAB or for purines eliminates virulence unless the required factor is injected along with the organisms.

However, radioactive labeling showed that many analogs can also **deceive** enzymes, i.e., can serve as competitive substrates, which are incorporated into coenzymes, proteins, RNA, or DNA.

Some **amino acid analogs** (e.g., p-fluorophenylalanine) can replace as much as 50% of the corresponding amino acid in the protein being formed, whereas others (e.g., 5-methyltryptophan) inhibit growth without being substantially incorporated. The incorporation may or may not impair the function of the product. Thus, when p-fluorophenylalanine replaces much of the phenylalanine and tyrosine in protein in *Escherichia coli* the altered β-galactosidase formed is still functional but the β-galactoside transport protein (Ch. 6) is not.

Base analog incorporation has been extensively studied in relation to cancer, inhibition of viral synthesis, and mutagenesis (Ch. 11). A few key findings will be summarized here.

To affect growth bases must first be converted to the corresponding nucleotides, by the enzymes (one set for purines and one for pyrimidines) that convert normal bases and nucleosides to nucleotides. However, since bases are synthesized in cells as nucleotides (Ch. 4) these "salvage" enzymes are not essential for growth and mutants that have lost them are viable. Such strains are readily selected because they are completely resistant to purine or to pyrimidine analogs (which are activated by the same enzyme).

Some base analogs are incorporated primarily into DNA and others into RNA. Thus 5-bromouracil substitutes well for thymine and can largely replace it in bacteria or bacteriophage, with little loss of viability but with mutagenic consequences. 5-Fluorouracil, in contrast, resembles uracil more than thymine and is incorporated into RNA but not detectably into DNA. However, it also becomes converted to a deoxyribonucleotide, 5-fluorodeoxyuridylate, which inhibits the conversion of the normal deoxyuridylate, by thymidylate synthetase, to deoxythymidylate (Fig. 4-23). The lack of this compound causes the cell to undergo "thymineless death" (Ch. 13, Control of DNA initiation), but this effect can be prevented by the presence of thymine. Incorporation of 5-fluorouracil into mRNA is then found to produce errors in the translation of information from DNA into protein, resulting both in decreased production of active enzymes and in correction (phenotypic suppression) of many mutations (Ch. 12).

ADDITIONAL ACTIONS OF ANTIMETABOLITES

Specific membrane transport systems (Ch. 6) are an additional site of competition between analogs and the corresponding normal metabolites, with two consequences: metabolites can restore growth not only by the mechanisms discussed earlier but also by blocking the entry of the analog; and analogs, conversely, can inhibit growth by inhibiting entry of an essential nutrient.

In addition, the sensitivity of a cell to an inhibitor may be determined by the affinity of its transport system rather than by the affinity of an intracellular receptor. Thus, the competitive ratio of sulfonamide to PAB is much lower with intact cells than with the extracted enzyme that converts PAB. Furthermore, sulfanilic acid (p-$NH_2 \cdot C_6H_5 \cdot SO_3H$) is less potent than sulfanilamide in inhibiting growth but much more potent in inhibiting the enzyme. Hence even in this classic example the quantitative relations observed with intact cells probably reflect competition at a transport system rather than at an enzyme.

In **pseudo feedback inhibition** an analog mimics the feedback inhibitory action of a biosynthetic endproduct on the initial enzyme of its pathway, thus interfering with the formation, rather than with the function, of the metabolite. This effect will be further described in Chapter 13.

SPECIFIC AGENTS

SULFONAMIDES AND RELATED AGENTS

Since the discovery of sulfanilamide (*p*-aminobenzene sulfonamide) several congeners, known as the sulfonamides, have been synthesized. Some of those in current use are shown in Figure 7-5. Sulfonamides are active in vitro against all bacteria (except the few that require exogenous folic acid), but the level of sensitivity varies widely from one species to another. Since various sulfonamides do not differ in spectrum a single member of the group will suffice in sensitivity tests.

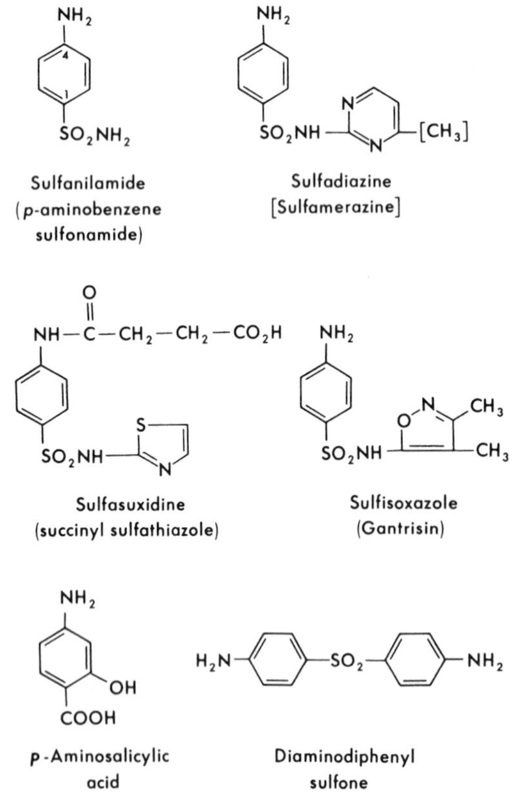

A variety of substitutions on the sulfonamide group (N^1) of sulfanilamide were found to increase activity, but compounds substituted on the amino group (N^4) are inactivated. Some members of the latter group, however, are useful because they are poorly absorbed and are slowly hydrolyzed in the gut to yield an active compound, sulfathiazole.

Among the N^4-substituted sulfonamides potency correlates closely with the pK of the acidic sulfonamide group (which can dissociate, releasing a proton). The more active derivatives (e.g., sulfadiazine) are stronger acids (pK ca. 7) than the parent sulfanilamide (pK 10). However, even more electrophilic substituents, which still further increase the acidity of the sulfonamide group, weaken the drug: maximal potency is found at 50% dissociation (pK=pH of test medium).* With the pK thus setting a maximum to potency in the sulfonamide series, further improvements have been based on other pharmacological properties, such as toxicity, absorption, and excretion.

Bell and Roblin have provided an explanation for the relation of potency to pK, based on the negativity of the sulfone group, serving as an analog for the carboxyl of PAB. The latter group (pK 4.8) is essentially fully ionized in the cell. In sulfonamide analogs the sulfone (-SO_2-) group occupies the

Fig. 7-5. Sulfonamides and other PAB analogs.

Sulfanilamide
(*p*-aminobenzene
sulfonamide)

Sulfadiazine
[Sulfamerazine]

Sulfasuxidine
(succinyl sulfathiazole)

Sulfisoxazole
(Gantrisin)

p-Aminosalicylic
acid

Diaminodiphenyl
sulfone

---

* For this reason the sulfonamides selected for maximal activity at pH 7.4 are ordinarily less potent in the urine (pH ca. 5 to 6) than those with a lower pK, such as sulfacetamide.

corresponding position; and though it cannot ionize, the greater its electronegativity the closer its resemblance to the ionized carboxyl. Increasingly electrophilic substituents on the sulfone increase its electronegativity and hence increase activity as long as they are unionized; but when the substituent becomes so electrophilic that it ionizes it becomes the center of negativity, instead of the sulfone, and the molecule becomes a poor competitor.

Diaminodiphenyl Sulfone. This drug (Fig. 7-5) was synthesized as a congener of sulfanilamide. It does not have a sulfonamide group, however, and unlike the sulfonamides it exhibits marked specificity: it is valuable in the treatment of only one bacterial infection, leprosy. The failure of the leprosy bacillus to grow in vitro has made it impossible to test whether this drug competes with PAB.

p-Aminosalicylic Acid (PAS). Among analogs of PAB created by substitutions on the ring p-aminosalicylate (PAS; Fig. 7-5) is useful against tuberculosis. It penetrates mammalian cells and is bacteristatic. PAS is antagonized competitively by PAB and noncompetitively by the products of 1-carbon metabolism, and bacteria can incorporate it into an analog of folic acid.

Though PAS is effective against the tubercle bacillus its potency against other bacteria is low. The contrast with the antimicrobial spectrum of sulfonamides implies that bacterial species can differ in the structure of the active site in the enzyme that metabolizes PAB. Presumably in the tubercle bacillus this active site has a shape particularly well fitted to PAS, while in other bacteria it is better fitted to sulfonamides.

PENICILLIN AND RELATED ANTIBIOTICS

PRODUCTION AND CHEMISTRY

We have noted above the discovery of penicillin, and in Chapter 6 we considered its mechanism of action at a molecular level. Here we shall consider its production and its modifications.

This valuable antibiotic was discovered in England in 1939, but because of the pressures of World War II on British laboratories it was developed chiefly in the United States, as one of the first large government-supported medical research projects. The procedures devised for the large-scale

production of such a labile "fermentation" product have provided a model for the development of all subsequent antibiotics.

Production was greatly improved by the isolation of high-yield mutants of the original *Penicillium notatum* (obtained by large-scale empirical testing following mutagenic treatment), and also by modifications of the culture medium. Moreover, though molds, as obligate aerobes, are conventionally cultivated as a surface mycelial growth on shallow layers of medium, it proved more economical to produce penicillin in aerated submerged culture. Fermenters have therefore been developed that can maintain good aeration in 50,000-gallon cultures. These advances have increased the yield 5000-fold, compared with the original laboratory cultures. A high-yield strain is a precious commercial secret that is particularly hard to guard, since a few spores, carried out under a worker's fingernails, can start a culture with considerable black-market value.

Structure. The substance originally designated as penicillin was found to be a mixture: the active compounds have in common a binucleate structure, **6-aminopenicillanic acid,** but differ in the acyl side chain (Fig. 7-6). Isotopic studies provide evidence for the biosynthesis of this nucleus by an unusual condensation of L-cysteine and D-valine.

The four-membered (β-lactam) ring of penicillin has a strained configuration, and its CO-N bond is therefore readily hydrolyzed, yielding an inactive product, penicilloic acid (Fig. 7-6). Hence penicillin is unstable in solution and is relatively rapidly destroyed by the acid of the stomach. The lability to acid varies with different penicillins; with G it permits roughly ⅕ of an orally administered dose to be absorbed. The same bond is also hydrolyzed by penicillinase (β-lactamase), an enzyme formed by certain bacteria (see below).

Modified Biosynthesis. It was found that the nature of the side chain in the penicillin can be controlled by varying the acids present in the medium. Thus when **benzylpenicillin** (penicillin G) proved to be the most satisfactory of the original penicillins, an almost pure yield was obtained by providing an excess of the corresponding acyl donor (phenylacetic acid). It is still the most widely used and inexpensive penicillin. Penicillin G was not patented; hence more profitable derivatives are vigorously promoted for general

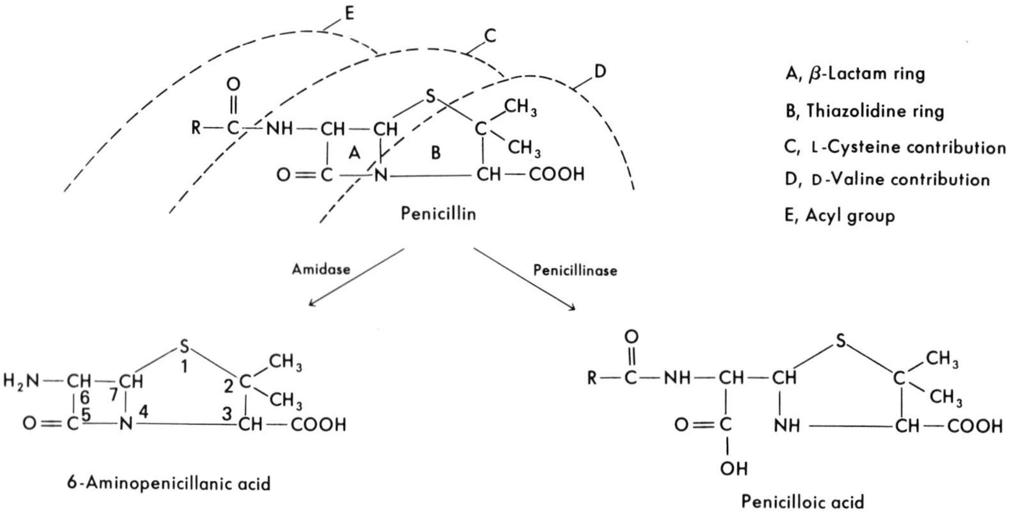

A, β-Lactam ring

B, Thiazolidine ring

C, L-Cysteine contribution

D, D-Valine contribution

E, Acyl group

6-Aminopenicillanic acid

Penicilloic acid

Side chain (R)	Penicillin
$C_6H_5CH_2$ —	Benzyl (G)
$C_6H_5OCH_2$ —	Phenoxymethyl (V)

Fig. 7-6. Structure of some penicillins.

use, though they are better only in special situations (see below).

By adding various other acyl donors to the medium a number of novel penicillins were obtained. Among these penicillin V (phenoxymethyl penicillin, Fig. 7-6) proved unusually resistant to gastric acidity.

SEMISYNTHETIC PENICILLINS

Though penicillin G is an extremely effective and nontoxic drug, it has several limitations: a relatively narrow antimicrobial spectrum (see below), destruction by acid, destruction by penicillinase, and elicitation of allergic responses. Efforts at improvement have therefore continued. A major advance was the discovery, in a British drug firm (Beecham) in 1960, that if the medium lacks any acyl side chain donors the mold produces mostly a precursor, **6-aminopenicillanic acid,** rather than penicillin itself.* Since 6-aminopenicil-

lanic acid can be condensed chemically with acids, the number of possible penicillins is now almost unlimited; several are illustrated in Figure 7-7.

Penicillinase and Drug Resistance. Many staphylococci, and certain other species, form the enzyme penicillinase, whose rapid destruction of the antibiotic makes the organism resistant (see below, Drug resistance). One of the early semisynthetic penicillins, **methicillin** (Fig. 7-7), provided an effective solution to the problem, for it is essentially inert to penicillinase. This resistance is probably promoted by the steric effect of the bulky o-methoxy groups near the site of enzymatic attack. Methicillin is very labile to acid and is only 1/30 as potent as penicillin G (against nonproducers of penicillinase); hence it must be given parenterally and in large doses. However, several later penicillinase-resistant products (oxacillin, cloxacillin, nafcillin) are more resistant to acid and hence can be given orally; they are also more potent but still only about 1/10 as potent as penicillin G.

The penicillinase-resistant penicillins not only fail to be hydrolyzed by the enzyme but they stabilize it in an inactive conformation, as tested by sub-

---

* This discovery arose from the observation of discrepancies between chemical determinations of penicillin, based on reactivity of the β-lactam ring, and microbiological assays, which require the whole molecule.

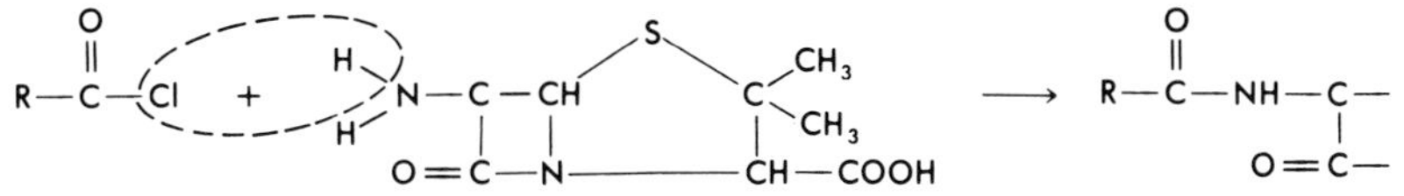

Fig. 7-7. Some semisynthetic penicillins.

The figure shows the reaction:

Acyl chloride + 6-Aminopenicillanic acid → Penicillin (P)

R group	Chemical name	Generic name	Trade name
Class I: Resistant to staphylococcal penicillinase			
(dimethoxyphenyl, OCH₃/OCH₃)	Dimethoxyphenyl P	Methicillin	Celbenin
			Dimocillin
			Staphcillin
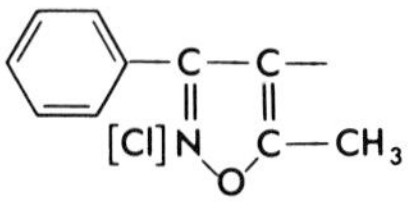 (5-methyl-3-phenyl-4-isoxazolyl, [Cl]N, C—CH₃)	5-Methyl-3-phenyl-4-isoxazolyl P	Oxacillin [Cloxacillin]	Prostaphlin Resistopen [Tegopen]
(2-ethoxy-1-naphthyl, OC₂H₅)	2-Ethoxy-1-naphthyl P	Nafcillin	Unipen
Class II: Broader spectrum			
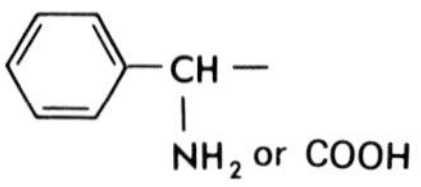 (CH—, NH₂ or COOH)	α-Aminobenzyl P or α-Carboxybenzyl P	Ampicillin Carbenicillin	Penbritin Polycillin

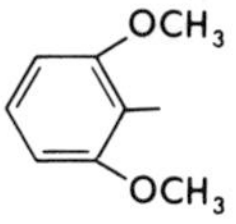

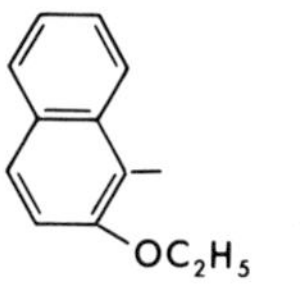

sequent addition of a sensitive penicillin. However, the enzyme has much greater affinity for the sensitive than for the resistant drug; hence on simultaneous addition the latter fails to protect the former from destruction.

Various organisms have yielded mutants with moderately increased resistance to penicillin not involving penicillinase; this form of resistance extends to penicillinase-resistant compounds.

Broadening of Antimicrobial Spectrum. Semisynthetic penicillins have also included products with a broadened antimicrobial spectrum. Thus the presence of an amino group on the side chain makes **ampicillin** (α-aminobenzyl penicillin; Fig. 7-7) much more active than penicillin G against many gram-negative bacilli, perhaps because the positive charge promotes penetration through the negatively charged lipopolysaccharide. Ampicillin is also half as active as penicillin G against gram-positive organisms, and it can be given orally. As might be expected from the lack of a bulky group near the β-lactam ring, ampicillin is sensitive to penicillinase.

CEPHALOSPORINS

Cephalosporin C, an antibiotic similar to the penicillins but much less potent, was isolated in 1952 from a mold of the genus *Cephalosporium*. Ten years later the nucleus (7-aminocephalosporanic acid) was isolated and used as the basis for a series of derivatives, analogous to the semisynthetic penicillins. Among these cephalothin, cephaloridine, and cephaloglycine have proved useful.

Cephalothin
[Cephaloglycine]

These antibiotics have a broad enough spectrum to be useful against some gram-negative bacilli as well as against staphylococci, and they are also resistant to staphylococcal penicillinase, but their most important advantage is that **they do not cause allergic reactions** in **most** patients who are allergic to penicillin (though allergy to cephalosporins can also develop). The absence of allergic cross-reactions is readily understood, for in most patients who react to penicillin the actual allergen is a derivative, formed in the body, with the β-lactam ring opened up (Ch. 20); hence the major immunological determinant is the S-containing ring, which differs in penicillins and cephalosporins.

Cephamycins are a structurally related group of compounds produced by certain streptomycetes.

ANTIBIOTICS WITH A SIMILAR SPECTRUM

We have considered in Chapter 6 various additional antibiotics that interfere, at one level or another, with wall synthesis: bacitracin, vancomycin, and cycloserine. In addition, a group of antibiotics have been developed that resemble penicillin in antimicrobial spectrum but have quite different, bacteristatic actions; these include **macrolides** and **lincomycin** (whose actions are described in Ch. 12) and novobiocin (see below). These alternatives to penicillin were originally most useful for patients allergic to penicillin or those infected with penicillinase-producing organisms; they have therefore been partly displaced by cephalothin and the semisynthetic penicillins.

It has been suggested that such antimicrobials offer some advantages because they do not select, as does penicillin, for wall-less L forms (Ch. 6), which are completely unresponsive to penicillin. However, the role of L forms in disease is quite unsettled (Ch. 40), and in general penicillin is the drug of choice for a sensitive organism: it is less toxic, and

its rapidly bactericidal action provides a more dramatic response than is observed with the alternative, bacteristatic agents.

Macrolide Structure. The most widely used macrolide is **erythromycin** (Fig. 12-26), produced by *Streptomyces erythreus* (a red organism). Other macrolides, including oleandomycin and carbomycin, have a similar mechanism of action on protein synthesis and exhibit cross-resistance. However, a classification based on a large lactone ring is too general, for it includes antibiotics with quite different actions: certain ionophores (Ch. 6) and polyenes (Ch. 43), which affect membrane permeability, and rifampin (Ch. 11), which inhibits initiation of transcription.

Novobiocin. This antibiotic, produced by the actinomycete *Streptomyces niveus,* has the structure shown below. The presence of an aromatic lactone (cyclic ester) causes it to be classed chemically as a coumarin; it also contains a novel sugar and a substituted phenol.

(sugar)   (coumarin)   (substituted phenol)

Its antibacterial spectrum resembles that of penicillin but in addition includes penicillin-resistant staphylococci and some strains of *Proteus* and other gram-negative organisms. Novobiocin is bacteristatic; it inhibits primarily DNA and RNA synthesis in cells and also in extracts, apparently by directly inhibiting the polymerases since it does not complex with DNA. With some strains the cell membrane becomes leaky, and a slow bactericidal action is seen.

OTHER ANTIBIOTICS

AMINOGLYCOSIDES

We have already noted the discovery of **streptomycin,** and we shall consider in Chapter 12 its action on the ribosome. Other clinically useful aminoglycosides, with similar structures (Fig. 12-30), have a similar action (on the small subunit of the ribosome), moderately similar antimicrobial spectrum, and variable cross-resistance (based on mutational alterations in the ribosome). These include **kanamycin, neomycin, paromomycin,** gentamycin,* and **hygromycin.** Here we may note features of their action that are relevant to their chemotherapeutic use.

1) Aminoglycosides are rapidly **bacteri-cidal.** Unlike penicillin, which requires substantial growth before the cell wall is irreversibly damaged, these antibiotics require only a small amount of protein synthesis for killing; this action can be prevented by starvation or by inhibition of protein synthesis (e.g., by a reversible inhibitor; see Fig. 7-9 below).

2) **Penetration** into bacteria, rather than growth rate, generally determines the rate of killing, which increases with drug concentration (unlike penicillin) up to a very high level. The effect in extracts (Ch. 12), in contrast, is saturated at a very low level (one or two molecules per ribosome).

3) **Anaerobiosis** antagonizes aminoglycosides, for unknown reasons: these drugs are not effective with any obligate anaerobes, and with facultative organisms the potency decreases ca. 10 times on shifting from aerobic to anaerobic conditions.

4) **Acidity** and **salts** (especially of polyvalent ions such as Mg^{++} and HPO_4^{--}) strikingly decrease their effectiveness, perhaps by impairing penetration into the bacterial cell.

5) Aminoglycosides are relatively ineffective against **intracellular bacteria** (e.g., mycobacteria). It is not certain whether the drug fails to penetrate or whether the host cell has too high a concentration of antagonistic ions.

Spectinomycin somewhat resembles aminoglycosides in structure but has basic groups only on the inositol ring. It also acts on the small subunit of the ribosome, and it selects for one-step high-level resistance, but it is bacteristatic rather than bactericidal (Ch. 12). Spectinomycin is useful in treating gonorrhea.

ISONIAZID

This compound (Fig. 7-8), long known to organic chemists, was discovered in 1951 to have chemotherapeutic value against tuberculosis. In the postsulfonamide, antibiotic era it represents the only major advance in anti-

---

* The name **gentamicin** was assigned by the discoverers, but it would be burdensome to perpetuate such an aberrant orthography.

Fig. 7-8. Isoniazid (isonicotinic hydrazide; pyridine-4-hydrazoic acid) and two metabolites that it resembles.

bacterial chemotherapy based on screening synthetic compounds. Isoniazid is bactericidal to growing organisms at a remarkably low concentration (< 1 $\mu g/ml$). Its useful action is curiously restricted to *Mycobacterium tuberculosis,* on which it has a dramatic effect. Other agents used in treating tuberculosis will be noted in Chapter 35.

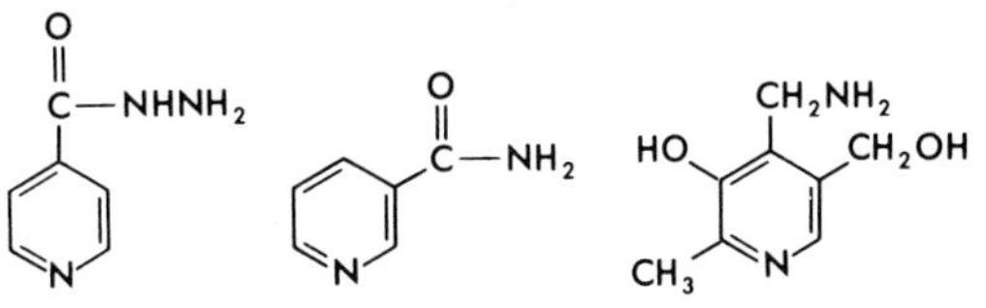

Isoniazid is bactericidal (in contrast to PAS), and it is effective against intracellular as well as extracellular bacteria (in contrast to streptomycin). These properties contribute to its value in tuberculosis. An early effect on mycobacteria is inhibition of the synthesis of mycolic acid, a major constituent of the cell envelope. Resistant mutants are not as serious a problem as with streptomycin: they exhibit only small-step increments of resistance, and they are often less virulent (at least in guinea pigs) than the sensitive parent strains.

Isoniazid resembles nicotinamide in structure (Fig. 7-8), and the extracted enzymes that convert nicotinamide to diphosphopyridine nucleotide (DPN = NAD) can also incorporate isoniazid instead. Isoniazid also resembles pyridoxamine, and it can inhibit a number of enzymes (e.g., transaminases) that require pyridoxal (or pyridoxamine) phosphate as cofactor. It is not certain whether either of these competitions is responsible for the antimicrobial action of the drug. The clinical toxicity of isoniazid, however, resembles pyridoxine (B_6) deficiency and is reported to be antagonized by large doses of this vitamin.

BROAD-SPECTRUM ANTIBIOTICS

The definition of a broad spectrum is relative; the term is conventionally applied to **chloramphenicol** and the **tetracyclines,** which are essentially identical in antimicrobial spectrum. Though differing markedly in structure they are also identical in their over-all mechanism of action, causing a bacteristatic, reversible inhibition of protein synthesis. However, they inhibit the ribosome in different ways, as will be shown in Chapter 12. The key to their usefulness is their selective action on bacterial but not on mammalian ribosomes.

Chloramphenicol (Chloromycetin) is produced by *Streptomyces venezuelae.* It contains a nitro and a dichloroacetyl group (Fig. 12-27), both unusual in a natural product.

The tetracyclines have four fused rings; hence their generic name. **Chlortetracycline** (Aureomycin) is produced by *Streptomyces aureofaciens,* presumably named for the golden color of its colony. Several closely related compounds were discovered as products of other streptomycetes. Formation of the chlorine-free compound tetracycline is favored by a low-chloride fermentation medium, or by mutations that block chlorination.

Other natural and semisynthetic tetracyclines include oxytetracycline, doxycycline, and minocycline. The several tetracyclines have identical antimicrobial spectra and complete cross-resistance. They are all being vigorously promoted on the basis of the patent laws, which regard naturally occurring compounds possessing any chemical differences as independently patentable "compositions of matter." The physician is therefore barraged with claims of superiority based on small differences and selected literature references.

SYNTHETIC COMPOUNDS USED FOR URINARY ANTIBACTERIAL THERAPY

The **nitrofurans** are bacteristatic against a variety of gram-positive and gram-negative bacteria. One of these drugs, **nitrofurantoin** (*Furadantin*), is widely

Nitrofurantoin   Nalidixic acid

Methenamine

used against low-grade, chronic urinary tract infections. It is excreted in the urine in high concentrations, but the concentrations attained in the body fluids are too low to provide any systemic chemotherapy.

Nalidixic acid has a similar use in urinary tract infections caused by gram-negative bacilli. It inhibits DNA synthesis and also causes membrane damage; the combination suggests that it may act on the membrane-attached site of DNA replication (Ch. 10).

Methenamine is not active in vitro, but on reaching the urine, after oral administration it is split, if the urine is acidic, into ammonia and the bactericidal compound formaldehyde; a mixture with mandelic acid (Mandelamine) is often used to promote acidity of the urine. *Proteus* is resistant because it forms urease, which splits urea to carbon dioxide and ammonia and hence makes the urine alkaline.

Antifungal chemotherapy (with polyenes and griseofulvin), and **antiviral chemotherapy,** will be discussed in Chapters 43 and 49.

PHYSIOLOGY AND ECOLOGY OF ANTIBIOTIC PRODUCTION

The polypeptide antibiotics are produced by sporulating bacteria, and the penicillins and cephalosporin by sporulating fungi (which also produce griseofulvin and fusidic acid). All the other major useful groups of antibiotics are produced by **streptomycetes,** a sporulating group of actinomycetes. The prominent role of the streptomycetes is curious and unexplained. In addition to the dozen-odd useful antibiotics noted, this otherwise narrow group of organisms has yielded about 500 distinct antibiotics; few such products have been isolated from other microbial groups.

BIOSYNTHESIS

Isotopic studies, as well as structural relations, support the conclusion that all the major biosynthetic pathways contribute precursors to the various antibiotics, and often a single antibiotic incorporates products of several pathways. The polypeptides (including penicillins and actinomycins) are derived from amino acids, and the extraordinarily varied sugar residues of many antibiotics from the carbon chain of glucose; puromycin is one of many modified nucleosides.

The long chains and rings of macrolides, polyenes, tetracyclines, and portions of other antibiotics are derived from lipid precursors (acetyl CoA and propionyl CoA). These undergo sequential condensation, just as in fatty acid biosynthesis (Ch. 6), but without the reductive steps between successive condensations; the resulting poly-β-ketones, e.g.,

$$RCO\text{-}CH_2CO\text{-}CH_2CO \cdots,$$

can readily condense internally to form the rings, both large and small, that are found in many antibiotics.

ECOLOGICAL ROLE

The production of antibiotics was initially assumed to reflect a fundamental ecological relation (antibiosis) between competing organisms, i.e., the opposite of symbiosis. However, several considerations oppose this view. 1) Antibiotic-producing organisms constitute only a tiny fraction of the microbial population in soil samples, and thus do not appear to have a striking advantage. 2) The strains found in nature excrete only small amounts of an antibiotic; heavy excretion is an artefact dependent on selection of regulatory mutants by man. 3) Antibiotics appear only after growth has ceased, rather than during competition for growth.

Compounds that are formed after growth ceases are called **secondary metabolites.** This is a broad class, in which antibiotics have become prominent only because of their ready detection and commercial value. The fantastic variety in the structures of secondary metabolites suggests an almost playful activity of enzymes, creating novel molecular patterns in cells when their normal metabolites accumulate and cannot engage in the serious business of growth. A fortuitous rather than highly selected origin of antibiotics is further suggested by their limited biosynthetic specificity; many organisms excrete a **group** of closely related compounds rather than a single antibiotic: several penicillins, actinomycins A to D, etc.

It seems more than coincidence, however, that **antibiotics appear only in sporulating organisms;** moreover, they appear **at the time of sporulation,** and their production is eliminated by mutations that block any early stage in this process. Sporulation is associated with synthesis of a new kind of wall (Ch. 6) and degradation of old wall, and walls contain unusual components (D-amino acids, novel sugars; Ch. 6), similar to those in many antibiotics. Hence the altered wall metabolism of sporulating organisms might provide the precursors that are forged into antibiotics. Alternatively, it has been suggested that the inhibitory actions of antibiotics may play a regulatory role in the major biosynthetic shifts accompanying sporulation. At present both these suggestions are speculative.

DRUG RESISTANCE

That microbes can become resistant to a drug during treatment ("drug-fast") was discovered by Ehrlich with protozoa. The rediscovery of this phenomenon with antibacterial drugs led to the fear that in our race with the adapting microbial population we would be required, like the Red Queen, to keep running faster and faster merely to stand still. Indeed, the problem is clinically very important in infections with certain bacteria (especially staphylococci, tubercle bacilli, and enterobacteria), but fortunately not with many others.

The term drug resistance ordinarily refers to **genotypic** changes, which persist during further cultivation in the absence of the drug, and not to a **phenotypic** adaptation (see below). Drug-resistant strains emerge either by infection by a plasmid (i.e., a block of nonchromosomal, accessory genes) or by mutation and selection.

The expression of genotypic drug resistance may involve various phenotypic mechanisms. Indeed, resistance to the same drug may depend on different mechanisms in different strains.

PHYSIOLOGICAL (PHENOTYPIC) EFFECTS OF MUTATIONS TO DRUG RESISTANCE

1) **Increased destruction of the drug** is often the mechanism of plasmid-borne resistance. It is exemplified by penicillinase, which will be discussed below.

2) **Decreased activation of the drug** is seen in mutants resistant to purine or pyrimidine analogs (used in cancer chemotherapy), for these compounds must be converted to nucleotides before they can interfere with essential reactions, and the enzymes involved are not essential for the cell. The antimicrobial agents in current clinical use, however, are all active without further chemical alteration.

3) **Formation of an altered receptor** is an important mechanism. The most striking example is the alteration in a specific **ribosomal protein** (Ch. 12) by one-step mutations

that increase the level of resistance to streptomycin (Str) many hundredfold. Similarly,
altered **enzymes,** resistant to an inhibitor,
have been found in mutants resistant to that
inhibitor, whether a substrate analog (e.g.,
sulfonamides) or an analog of an allosteric
feedback inhibitor (Ch. 12).

Mutations to ribosomal resistance are seen with
some but not all aminoglycosides, though all have
the same general mechanism of action on the
ribosome. One-step high-level resistance is also seen
with spectinomycin and erythromycin, bacteristatic
inhibitors of ribosomal function. A more selective
phenylalanine-activating enzyme is present in a
mutant that can no longer incorporate *p*-fluoro-
phenylalanine into protein. Such mutational
alterations should be valuable in studies aimed at
correlating structure and function.

4) **Decreased permeability** has been demonstrated in mutants resistant to amino acid
analogs; the loss or alteration of a specific
membrane transport protein also impairs
transport of the corresponding normal amino
acid. Permeability barriers are probably also
responsible for a good deal of the "natural
resistance" that limits the antimicrobial spectrum of various agents.

Thus actinomycin inhibits RNA synthesis in many
gram-positive bacilli and cocci, but not in gram-
negative bacilli; yet the extracts of both groups are
equally sensitive. Furthermore, such unresponsive
cells become sensitive following a treatment that increases nonspecific permeability (Ch. 6).

Decreased uptake of the drug by **intact cells** has
been used as an index of decreased permeability
but is not reliable; it could also be due to a reduction in number or in affinity of the binding sites.
Moreover, much of the binding may involve receptors that are irrelevant to drug action. However,
studies on binding to **isolated cell components**
(ribosomes, membranes) have yielded significant
results.

Other theoretically possible mechanisms of genotypic resistance include increased levels of a reversing metabolite or of the sensitive enzyme. These
have not been observed, however, as significant
mechanisms.

RESISTANCE GENES IN PLASMIDS

Fluctuation analysis showed that resistant
strains arise in pure cultures as **mutants,**
selected by the drug (Ch. 8). Such resistance
is specific for a particular drug (or a class of

related drugs). However, after this mechanism was finally accepted it was unexpectedly
found that other resistant strains arise in
nature by a second mechanism, which can
simultaneously produce resistance to **multiple
drugs: infection** with an extrachromosomal
genetic element (plasmid) carrying genes for
resistance. These **resistance transfer factors**
(RTF) are of great clinical and theoretical
importance; their replication and their transfer between cells will be described in Chapter 46.

The wide use of antibiotics in medicine and
agriculture has increased the distribution of
such resistance factors. In one hospital in
Japan the frequency of multiple resistance in
Shigella isolates rose from 0.2% in 1954 to
52% in 1964. In Great Britain it has become
illegal to use in cattle those antibiotics that are
widely used in man.

Transfer factors found in nature (especially
among Enterobacteriaceae) carry a variety of
combinations of genes for resistance to
various antimicrobial agents. These genes
code for enzymes that inactivate the agents;
the most fully studied enzyme of this kind is
penicillinase.

PENICILLINASE

The resistance of certain strains of *Staphylococcus* to penicillin is due to the formation of
an enzyme, penicillinase, that inactivates
penicillin by opening the β-lactam ring (Fig.
7-6). The formation of this enzyme is directed
by a plasmid, which sometimes also carries
genes for resistance to other antibiotics.
Unlike RTF this plasmid does not generate
the machinery for its own transfer, and so it
does not spread around freely in mixed cultures. However, it can be transferred by a
transducing phage (Ch. 46), which is often
present in the same cell. Moreover, the cell
can be "cured" of its resistance by agents that
are known to eliminate plasmids (Ch. 46).

Plasmids have presumably acquired their
genes for various enzymes by genetic exchange with previous hosts in which these
genes were part of the chromosome (Ch. 46).
Indeed, chromosomal genes for penicillinase
production are widely distributed, e.g., among
Enterobacteriaceae and bacilli.

In some organisms penicillinase is formed in
quantity only when **induced** by penicillin; in others

it is **constitutive** (Ch. 3, Phenotypic adaptation). Some organisms release much of the enzyme into the medium and retain the rest in the periplasmic space (including the mesosomes) and attached to the cytoplasmic membrane (Ch. 6).

Because penicillinase clearly was widely distributed even before the medical use of antibiotics, and little or no penicillin is formed in nature (see above, Ecology), the enzyme presumably has some unknown function in the organism in which it is part of the genome. It may have some function in wall morphogenesis or in sporulation, such as cleavage of the kind of bond of which penicillin is an analog (Ch. 6).

Population Density. With penicillinase-producing organisms the **apparent sensitivity** can vary enormously with bacterial **population density,** for it depends on a race between killing of bacteria and destruction of penicillin. Thus individual cells, widely separated on solid medium, show only a moderate increase (ca. 10-fold) in resistance compared with those of non-penicillinase-producing strains, whereas a dense lawn of the same cells can show a 1000-fold increase. Quantitation of the level of this kind of resistance is thus almost meaningless. However, in the usual disc assay, which involves a heavy inoculum, penicillinase-producing staphylococci are recorded as resistant; and clinically they are completely resistant to penicillin G, since staphylococci characteristically produce localized infections with a high bacterial density. Indeed, in mixed infections such organisms may protect sensitive neighbors by destroying the drug.

Sensitive staphylococcal strains fail to yield penicillinase-producing mutants in vitro, for they lack the genetic background necessary for producing the required gene by mutation. Hence penicillinase-producing staphylococci, seen so frequently in hospitals, must reach the patient by contamination from the environment (see Ch. 27, Epidemiological evidence).

ENZYMES DIRECTED BY RESISTANCE FACTORS (RF)

The RF-mediated resistance to various agents was found to differ from that due to penicillinase in being relatively insensitive to population density. Nevertheless, Okamoto and Suzuki showed in Japan that extracts of such strains could inactivate the drugs to which the cells were resistant. A number of such reactions were then defined; these are presented in Table 7-1. In contrast to the hydrolytic exoenzyme penicillinase (Ch. 6), the RF-directed enzymes are intracellular and require intracellular metabolites (ATP, acetyl CoA), and the resulting levels of resistance are modest.

The negligible influence of population density suggests that the level of resistance is determined by the relation between the rate of antibiotic penetration and the rate of its inactivation in the individual cell, whereas penicillin does not require penetration for either its action or its destruction, and resistance depends largely on elimination of the reservoir of drug in the medium. As with penicillinase, the physiological role of these detoxifying enzymes remains unknown.

GENERAL FEATURES OF DRUG RESISTANCE

INDUCED PHENOTYPIC RESISTANCE

In most cases a borderline concentration of a drug causes a partial inhibition that does not change during further growth. However, some kinds of resistance increase adaptively during growth in the

TABLE 7-1.   Enzymes Determined by R Factors

Enzyme	Substrate
β-Lactamase	Various penicillins
Chloramphenicol acetylase	Chloramphenicol
Streptomycin phosphotransferase	Streptomycin
Streptomycin adenylate synthetase	Streptomycin, spectinomycin
Kanamycin phosphotransferase	Kanamycin, neomycin B, paromomycin
Kanamycin acetyltransferase	Kanamycin, neomycin B
Gentamycin adenylate synthetase	Gentamycin, kanamycin

presence of the inhibitor, allowing resumption of growth **(transient bacteristasis).** In contrast to genotypic resistance, such induced phenotypic resistance is lost after several generations of growth without the inhibitor.

Several mechanisms have been recognized.

1) **Ribosomal resistance** to erythromycin can be produced by an alteration in the methylation of ribosomal RNA (Ch. 12), and in some strains **erythromycin** itself **induces** that alteration (though in others it is genetic).

2) In *Proteus* (but not in *E. coli*) the **resistance factor** (RF) can **split** into two autonomously replicating plasmids, separable by virtue of their different density: one carries the genes for conjugation (the RTF, or Δ), and the other carries the genes for resistance (Ch. 46). In the stationary phase the **resistance plasmid increases** several-fold in the **presence of any of the antibiotics** whose inactivation it directs; moreover, reassociation can yield an RTF carrying multiple copies, in tandem, of its r determinants. Thus a substrate can induce its enzyme, in this novel regulatory mechanism, by influencing not only transcription (Ch. 13) but also replication of the corresponding gene.

3) A few antimetabolites exert **pseudofeedback inhibition** (Ch. 13). For example, 2-thiazolealanine, an analog of histidine, inhibits the first enzyme of histidine synthesis, thus lowering the intracellular histidine level and causing immediate bacteristasis. However, this starvation for histidine in turn derepresses the enzymes of the histidine pathway, and so the level of the partly inhibited enzyme rises. With borderline concentrations of inhibitor this process restores growth in a few minutes.

PREVENTION OF RESISTANCE

Since no method is available for effectively decreasing spontaneous mutation rates one cannot prevent **formation** of resistant mutants in growing populations, but one can prevent their **selection.** Where effective resistance develops through the accumulation of small increments, by successive mutations, therapeutic "escape" can be avoided by continuously maintaining drug concentrations high enough to inhibit the first-step mutant. A more general approach, applicable even to one-step mutations to high-level resistance (e.g., with streptomycin), is the simultaneous use of two non-cross-resistant drugs. Such **combination therapy,** originally suggested by Ehrlich, has a simple genetic rationale: if 1 cell in 10^6 mutates to resistance to drug A and 1 in 10^7 to drug B, only 1 in 10^{13} will develop both

mutations. This approach has been especially successful in the treatment of tuberculosis and will be discussed further in Chapters 23 and 35.

Attempts to prevent or cure the infection of bacteria by drug resistance plasmids have not been successful.

SYNERGISM AND ANTAGONISM

In addition to being legitimately used to prevent drug resistance, mixtures of antimicrobial agents are unfortunately often used to cover a broad range of possible agents, instead of making a specific diagnosis (Ch. 23, Combined therapy). Apart from the increased danger of toxic reactions, and the other disadvantages of dealing with an infection of unknown etiology, it should be noted that some mixtures are **antagonistic,** i.e., a drug that requires growth for its bactericidal action (penicillin, streptomycin), given together with a bacteristatic agent (a sulfonamide, tetracycline, or chloramphenicol; Fig. 7-9). These interactions are readily seen in vitro

Fig. 7-9. Rapid protection by chloramphenicol against killing by streptomycin in growth medium. Streptomycin (30 μg/ml) was added to exponentially growing *E. coli* at 0 time. At intervals the viable count was determined, and samples were transferred (arrows) to flasks containing sufficient chloramphenicol to yield a final concentration of 20 μg/ml. These flasks were similarly incubated, and samples were removed for viable counts after 30 and 60 minutes. [From Plotz, P. and Davis, B. D. *J Bacteriol 83:*802 (1962).]

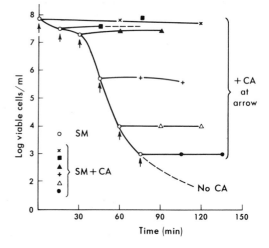

and have also been demonstrated under certain circumstances in experimental animals; and though they are difficult to establish with certainty in patients the extrapolation seems reasonable.

Synergism of bactericidal action can be seen with certain mixtures of two bactericidal drugs, such as penicillin and an aminoglycoside: a low concentration of one can promote the lethal action of the other, in terms of both rate and minimal effective concentration.

Sublethal pretreatment with penicillin increases sensitivity to streptomycin and accelerates its uptake, whereas pretreatment with streptomycin does not influence the subsequent interaction with penicillin. Presumably damage to the wall by penicillin facilitates subsequent entry of streptomycin and/or loss of antagonistic ions.

GENETIC DOMINANCE OF SENSITIVITY AND RESISTANCE

Analysis of the phenotypic mechanisms of resistance has made it possible to understand why some genes for resistance are dominant and others recessive to sensitivity. Thus the production of an altered, resistant enzyme is dominant, for in a heterozygote producing both sensitive and resistant molecules the drug would inhibit only the sensitive fraction. On the other hand, streptomycin-resistant ribosomes are recessive. Study of this phenomenon has revealed a complex competition between sensitive and resistant ribosomes, described in Chapter 12 (Aminoglycosides).

The production of a detoxifying enzyme by RTF is, of course, "dominant" (or, more precisely, epistatic, for though the gene for resistance is added to a sensitive cell it does not compete with a sensitive allele).

CODA

It would be difficult to exaggerate the impact of antibacterial chemotherapy on the practice of medicine: the sulfonamides introduced a second Golden Age in the field of infectious diseases. For nearly two decades the advances were almost exclusively practical. However, with our increasing knowledge of bacterial organization and macromolecular synthesis it has become possible to classify most antimicrobial agents in terms of their mechanisms of action. Conversely, chemotherapy has contributed to cell physiology, in a true dialogue. We have already seen how penicillin has contributed to our understanding of cell wall structure and synthesis. Later chapters will show how other antibiotics have thrown light on nucleic acid and protein synthesis, and how selective inhibition of these reactions has helped to reveal regulatory interactions.

In this chapter we have considered the effects of chemotherapeutic agents on bacterial growth and viability, including the influence of both environmental and genetic factors on the sensitivity of the bacteria; these interactions, observed in vitro, are all pertinent to the chemotherapy of the infected patient (Ch. 23). We have also summarized current knowledge of the mechanism of action of certain chemotherapeutic agents; the actions of others will be considered in later chapters.

SELECTED REFERENCES

Books and Review Articles

ALBERT, A. *Selective Toxicity,* 3rd ed. Wiley, New York, 1965. Chemotherapy considered as part of a broader problem, with emphasis on physiochemical factors. Chapter III provides an excellent historical account.

ANDERSON, E. S. The ecology of transferable drug resistance in the enterobacteria. *Annu Rev Microbiol 22*:131 (1968).

CITRI, N., and POLLOCK, M. P. The biochemistry and function of β-lactamase (penicillinase). *Adv Enzymol 28*:237 (1966).

DAVIES, J. F., and ROWND, R. Transmissible multiple drug resistance in Enterobacteriaceae. *Science 176*:758 (1972).

EHRLICH, P. *Collected Papers,* vol. III, *Chemotherapy.* Pergamon Press, New York, 1960.

FLOREY, H. W., CHAIN, E., HEATLEY, N. G., JENNINGS, M. A., SANDERS, A. G., ABRAHAM, E. P., and FLOREY, M. E. *Antibiotics.* Oxford Univ. Press, London, 1949. An exhaustive and fascinating account of early work.

GOLDBERG, I. H., and FRIEDMAN, P. A. Antibiotics and nucleic acids. *Annu Rev Biochem 40*:775 (1971).

GOTTLIEB, D., and SHAW, P. D. (eds.). *Antibiotics: I, Mechanism of Action; II, Biosynthesis.* Springer, New York, 1967.

HITCHINGS, G. H., and BURCHALL, J. J. Inhibition of folate biosynthesis as a basis for chemotherapy. *Adv Enzymol 27*:417 (1965).

Joint Committee on the Use of Antibiotics in Animal Husbandry and Veterinary Medicine. *Report Command No. 4190.* HMSO, London, 1969.

MEYNELL, G. G. *Drug Resistance and Other Bacterial Plasmids.* Macmillan, London, 1972.

MOYED, H. S. Biochemical mechanisms of drug resistance. *Annu Rev Microbiol 18*:347 (1964).

NEWTON, B. A. The properties and mode of action of the polymyxins. *Bacteriol Rev 20*:14 (1956).

RICHMOND, M. H. The plasmids of *Staphylococcus aureus. Adv Microbial Physiol 2*:43 (1968).

SCHAEFFER, P. Sporulation and the production of antibiotics, exoenzymes, and exotoxins. *Bacteriol Rev 33*:48 (1969).

STOKSTAD, E. L. R. Antibiotics in animal nutrition. *Physiol Rev 34*:25 (1954).

UMEZAWA, H. *Index of Antibiotics from Actinomycetes.* University Park Press, State College, Pa., 1967.

WAKSMAN, S. A., and LECHEVALIER, H. A. *The Actinomycetes,* vol. III, *Antibiotics of Actinomycetes.* Williams & Wilkins, Baltimore, 1962. Useful as a reference on the isolation and the chemistry of several hundred antibiotics.

WATANABE, T. Infectious drug resistance in bacteria. *Curr Top Microbiol Immunol 56*:43 (1971).

WOLSTENHOLME, G. E. W., and O'CONNOR, M. (eds.). *Bacterial Episomes and Plasmids.* (Ciba Conference). Little, Brown, Boston, 1969.

ZÄHNER, H., and MAAS, W. K. *Biology of Antibiotics.* Springer, New York, 1972. A brief, clear presentation, designed for interested medical students. Paperback.

Specific Articles

BELL, P. H., and ROBLIN, R. O., JR. A theory of the relation of structure to activity of sulfanilamide compounds. *J Am Chem Soc 64*:2905 (1942).

BROWN, G. M. Biosynthesis of folic acid. II. Inhibition by sulfonamides. *J Biol Chem 237*:536 (1962).

DUGUID, J. P. The sensitivity of bacteria to the action of penicillin. *Edinburgh Med J 53*:407 (1945).

part II

BACTERIAL AND MOLECULAR GENETICS

BERNARD D. DAVIS
RENATO DULBECCO

chapter

BACTERIAL VARIATION AND POPULATION DYNAMICS

GENOTYPIC AND PHENOTYPIC VARIATION   170
SELECTIVE PRESSURES AND GENETIC ADAPTATION   171
  Dissociation and Phase Variation   171
RANDOM OR DIRECTED MUTATIONS   173
  Fluctuation Analysis   173
  Cytoplasmic Inheritance   174
HAPLOID AND DIPLOID STAGES   175
DETECTION AND SELECTION OF MUTANTS   176
  Phenotypic Lag   177
MUTATION RATES   179
  Determination of Mutation Rate   179

The development of bacterial genetics provided much of the conceptual and technical background for the development of molecular genetics, with which it is now inextricably linked. Moreover, molecular genetics has illuminated innumerable aspects of biology and medicine. Accordingly, we shall review not only bacterial genetics in the next two chapters, but also molecular genetics in the rest of this bloc. Phage genetics, which has also contributed extensively to molecular genetics, will be presented in Chapters 45 and 46.

GENOTYPIC AND PHENOTYPIC VARIATION

The extensive changes observed in repeatedly transferred cultures of bacteria led some authorities in the 1870s (Naegeli, Buchner) to propose the doctrine of **pleomorphism,** which interpreted the wide range of microbes encountered in nature as different stages in the life cycles of only a few species. However, after Koch introduced reliable pure culture technics (Ch. 1) much of this apparent variation disappeared and the opposing doctrine of **monomorphism** prevailed. But its success led to an exaggerated emphasis on microbial stability. Moreover, since no nucleus could be demonstrated in bacteria their inheritance was long believed to reside in some vague plastic properties of the entire cell. There was thus a long delay in recognizing that variation in microbes depends on two processes fundamental to all living organisms: 1) **mutation and selection,** yielding a **genetically** altered population, and 2) **phenotypic variation** (physiological adaptation), involving altera-

tions within the range of potential of a given genotype.*

Several factors contributed to this delay. 1) Rare spontaneous mutants appearing among the enormous numbers of bacterial progeny are often subject to strong but not obvious selective pressures, and the resulting evolution of an altered population seemed too rapid for the mechanisms already demonstrated in higher organisms. 2) Genetic recombination is required for recognition of units of inheritance linked in chromosomes, but it was not discovered in bacteria until the 1940s. 3) Genotypic and phenotypic variation (or adaptation) can often bring about the same phenotypic change, such as capsule formation; the quantitative technics and the concepts necessary to distinguish between the two processes were not developed until late. Yet the criterion is simple: a **phenotypic adaptation** to an environmental change involves essentially **all** the cells in the culture, whereas a **genotypic adaptation** involves a **rare** mutant, which is then selected (i.e., grows faster than the parental strain) because it is better adapted to the new environment. The enumeration of parental and mutant cells, recognized by the **differences** between their **clones** (i.e., progeny derived from a single cell) in the **same** environment, is thus crucial.

Genotypic adaptation is especially simple to follow with mutations that alter colonial morphology (Ch. 5) or alter the ability to form a

---

* In 1900, within a few months after de Vries in Holland discovered mutations in higher plants, his countryman Beijerinck proposed that the same mechanism must be responsible for the heritable changes observed in bacteria. But this idea was premature by nearly half a century.

170

colony on appropriate media. During incubation additional mutants may appear among the cells newly formed on the plate. Such mutant subclones yield **sectored colonies** when they involve R-S (rough-smooth surface) variation, color mutants, or fermentation mutants (whose ability to ferment a sugar may be recognized by the resulting precipitation of a dye; Fig. 5-9).

SELECTIVE PRESSURES AND GENETIC ADAPTATION

Many of the inheritable changes observed in bacteria during cultivation have obvious **adaptive value** (i.e., increased fitness for a new environment). For example, on first isolation a pathogen often grows slowly in a laboratory medium (because it is better adapted to conditions in the animal host) but on repeated transfer it will adapt to faster growth, through selection of rare mutants appearing among the progeny. This adaptation is often accompanied by decreased ability to grow in the animal host, i.e., **loss of virulence (attenuation);** and virulence may be restored by passage of a large inoculum through an animal host, which selects rare virulent mutants. As we shall see in many chapters on pathogenic bacteria and viruses, repeated transfer under unusual conditions is widely used to produce attenuated strains for use as vaccines.

DISSOCIATION AND PHASE VARIATION

With some kinds of variation in bacterial populations the adaptive advantage is not obvious and even seems to be belied by spontaneous reversal during further cultivation. These phenomena were long considered to reflect some kind of life cycle (analogous to sporulation or gametogenesis) and were given special names. In **dissociation** the original, virulent isolate of various pathogens from a patient plates out homogeneously as smooth (S) or mucoid colonies, but after repeated passage in liquid medium plating may yield mostly or entirely rough (R) colonies (Fig. 8-1); and the transition may later be reversed (i.e., R⇌S). In **phase variation** an enteric organism, with flagella of a given surface structure (recognized immunologically), rapidly shifts during cultivation from this antigenic type (phase 1)

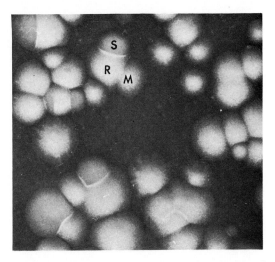

Fig. 8-1. Rough (R) and smooth (S) colonies of *Brucella abortus*. Because of the difference in the reflection of light in the photograph the R colonies appear considerably lighter than the S colonies, as well as more stippled. (Courtesy of W. Braun.)

to another (phase 2); and on further transfers the cultures may shift back and forth between the two "phases."

Analysis of the kinetics of these shifts showed that they are also due to pressures, however subtle, for the **selection of spontaneous inheritable variants.** Moreover, though the frequency of appearance of these variants **per unit time** seems high, **per cell generation** it is low enough to be consistent with a mutational origin. The apparently quixotic reversals in direction of the population shifts have two causes: the mutations involved are readily reversible (for reasons that are not understood), and within a growing culture the composition of the medium undergoes continuous changes, which create **new selective pressures.** These changes become especially rapid as the population becomes dense and the culture approaches the stationary phase: the concentration of nutrients, pO_2, and pH are lowered, and metabolic products accumulate. Hence the patterns of phase variation and dissociation during laboratory transfers depend not only on the medium used but also on such details as the inoculum size and the duration of incubation before transfer.

Thus with a mixture of R and S *Brucella* cells in fresh medium the S cells grow faster. However, they excrete D-alanine, whose accumulation in the grow-

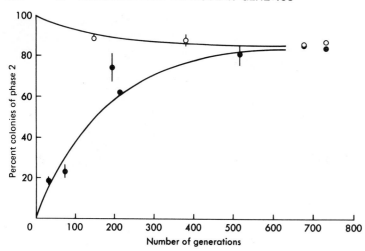

Fig. 8-2. Attainment of equilibrium between two phases of *Salmonella typhimurium* after prolonged exponential growth. Each initial culture contained cells of only phase 1 (●) or 2 (○). Growth was kept exponential in a series of successive subcultures by transferring from each a small inoculum taken before the population density approached saturation. From these kinetics it can be calculated that the mutation rates were 5.2×10^{-3} per division for phase $1 \to 2$ and 8.8×10^{-4} for $2 \to 1$ [After Stocker, B. A. D. *J Hyg 47*:398 (1949).]

ing culture, as well as the decrease in pO_2, cause S cells to grow more slowly than R cells. With phase variation, in which the alternative genotypes are less subject to differential selective pressures, a stable equilibrium can be reached (Fig. 8-2) whose value depends only on the mutation rates in two directions.

Because of occult selection pressures stock cultures are best preserved in the frozen or the dried (lyophilized) state. Storage in a refrigerator, for example, allows slow further growth and introduces new selective pressures.

Periodic Selection. The heterogeneity of a population is increased not only by selection of favorable mutations but also by the progressive accumulation of a variety of "neutral" mutations, i.e., mutations with no effect on the growth rate. However, these may be eliminated from time to time by **periodic, indirect, negative selection;** i.e., by strong positive selection for a new mutant that has emerged from the predominant wild-type population.

For example, in a culture growing in the chemostat (Ch. 5) with limiting tryptophan the propor-

Fig. 8-3. Periodic selection. In the population of a tryptophan auxotroph of *E. coli* growing in a chemostat under tryptophan limitation the fraction of phage T5-resistant mutants increases progressively, since existing mutant clones maintain a normal growth rate and fresh mutations continually add to their number. (The linear nature of the increase will be discussed below; see Mutation rates.) The accumulated mutant population, however, is periodically almost wiped out when the entire population is displaced by the progeny of another kind of mutant, which can grow faster. (T5-resistant mutants were used in this study simply because they grow at a normal rate and are very easy to select, and therefore to count.) [Modified from Novick, A., and Szilard, L. *Proc Natl Acad Sci USA 36*:708 (1950).]

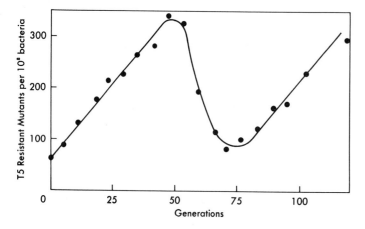

tion of phage-resistant mutants increases for about 30 hours, as expected, but then rapidly declines and slowly rises again (Fig. 8-3). The decline is due to the appearance and rapid outgrowth of a new kind of mutant, with improved ability to utilize tryptophan at a very low concentration. The resulting indirect selection against the phage-resistant cells occurs not because they are resistant but because, on purely numerical grounds, their small population does not include any cells with the mutation for faster growth. In the new population resistant mutants accumulate again, and several such waves of periodic rise and fall are possible.

As we shall see in later sections, population genetics is also pertinent to the properties of pathogenetic bacteria in the host. Not only do infecting populations vary in composition but the initial and the later stages of an infection differ in selective pressures, much like fresh and aging cultures: selective pressures are exerted by the effects of high bacterial density, by host defenses, and by chemotherapeutic agents.

RANDOM OR DIRECTED MUTATIONS

We have noted three features of microbial cultures that contribute to an extraordinarily rapid evolution: the large numbers of individuals, the short generation time, and the strong selective pressures. As a result, what looked superficially like excessive genetic instability has turned out to be simply rapid selection—often literally overnight—of a hereditary variant that appears spontaneously during the multiplication of an originally homogeneous inoculum. These variants were found to have much the same range of frequencies as the familiar spontaneous mutations in high organisms (i.e., 10^{-5} to 10^{-10} per generation); and their appearance was found to be accelerated by agents known to be mutagenic in higher organisms. Hence around 1940 bacteriologists began to refer to inheritable changes in bacteria as mutations.

Bacteria remained, however, the last stronghold of Lamarckism—the doctrine of the inheritance of adaptively acquired characteristics. For unlike higher organisms, whose germ cells are distinct from somatic cells, the single bacterial cell is at the same time soma and gamete and is fully exposed to the environment. Hence an inhibitory

compound that enters it reaches its genome and thus might conceivably direct as well as select for resistant mutations. To be sure, resistance appears in only a tiny fraction of the population; but the drug might conceivably have a directive influence with a low level of effectiveness. The drug could thus be producing the resistant clones that it was being used to detect.

FLUCTUATION ANALYSIS

Definitive evidence on this problem was furnished by a statistical approach called fluctuation analysis, designed by Luria and Delbruck in 1943 to study resistance to bacteriophage and subsequently applied by Demerec to drug resistance (Fig. 8-4). Small inocula, containing no resistant mutants, were incubated in 100 identical tubes of medium. When growth was complete the content of each tube was poured in a plate of medium containing the inhibitor, and after incubation the number of colonies (i.e., resistant mutants) was determined. If the resistant mutants arose only **during** exposure to the drug each plate should have received an identical, mutant-free inoculum, and the postulated interaction of drug and cell should have the same chance of producing resistant mutants in each plate; hence each should yield the same number of resistant colonies, except for a predictable statistical variation. If, on the other hand, the mutations had occurred in the liquid medium, **before** exposure to the drug, the individual tubes would differ from each other in the time of random appearance of the first mutant; hence the size of the resulting clone of resistant cells, formed before cessation of growth in each tube, should fluctuate widely from tube to tube.

Wide fluctuation was indeed observed, a few "jackpot" tubes (those with a very early mutation) having a large number of mutants. In contrast, a control experiment, with multiple samples plated from a single flask, had a much smaller fluctuation, accounted for by the expected statistical distribution about the mean. Drug-resistant mutants were thus conclusively shown to appear **spontaneously,** rather than being directed by the presence of the drug. This proof did much to overcome a widespread anthropomorphic reluctance to

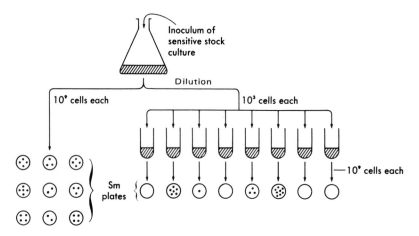

Fig. 8-4. Fluctuation analysis of mutation to streptomycin (Sm) resistance. A small number of sensitive cells were inoculated in a flask containing 100 ml of broth, and also in 100 tubes each containing 1 ml of the same medium. After full growth was reached, 1-ml samples were inoculated in plates of medium containing the drug, and the number of Sm-resistant colonies appearing after overnight incubation was determined. The fluctuation in their number was much greater among the samples that had grown out in separate tubes than among those from the same flask. [Based on Luria, S. E., and Delbruck, M. *Genetics 28*:491 (1943).]

accept the overwhelming role of chance events in genetic adaptation.

In this ingenious experiment the separation of populations in tubes allowed the effect of an early spontaneous mutation to be amplified and thus detected. Lederberg subsequently used separation on solid media to demonstrate the same effect more directly: a heavy inoculum of sensitive bacteria (ca. 10⁴ cells) was grown on plates without drug to form a lawn, which was then replica-plated onto drug-containing medium. The results showed that the lawn contained localized clusters of resistant cells, i.e., clones had been formed **before** exposure to the drug (Fig. 8-5).

It was thus gradually recognized that mutable unit factors, indistinguishable from the genes of higher organisms, govern inheritance in bacteria. Nevertheless, bacterial genes still could not be studied in depth because they could not be manipulated by recombination. Between 1944 and 1951 three novel mechanisms for transferring genetic material were discovered in various species: transformation, conjugation, and transduction. These will be considered in the next chapter.

CYTOPLASMIC INHERITANCE

Although the central role of chromosomal DNA in microbial heredity has been firmly established, as we shall see in the next chapters, there are a few well-documented examples of nonchromosomal (cytoplasmic) mutations (using the term mutation broadly to denote any abrupt hereditary change). Thus acriflavine can "cure" bacteria of self-reproducing cytoplasmic genetic units called plasmids (Ch. 46). Similarly, in eukaryotes acriflavine can destroy the capacity of yeasts to form mitochondria, and hence to respire (Ephrussi); while streptomycin can cure algae of the ability to make chloroplasts, and hence to grow photosynthetically. These mutations in eukaryotes do not show Mendelian segregation in genetic crosses, but rather exhibit "maternal" inheritance, i.e., the progeny of a zygote all inherit the pattern of the cell that provides the cytoplasm.

Though all these examples appear to involve elimination of DNA-containing cytoplasmic units, DNA does not appear to be involved in certain "inheritable" changes in the cell envelope. Thus penicillin converts bacteria quantitatively to wall-deficient, spherical L forms by interfering with wall synthesis (Ch. 6); and though these L forms will

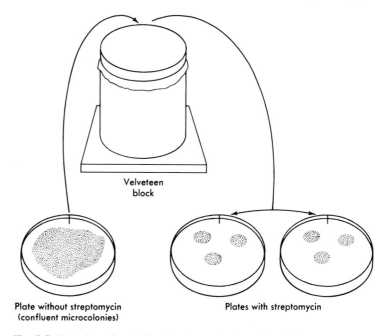

Fig. 8-5. Use of replica plating to demonstrate undirected, spontaneous appearance of streptomycin-resistant mutants. About 10^5 sensitive cells were spread on a plate of drug-free solid medium and allowed to reach full growth (10^{10}-10^{11} cells). Sterile velveteen, covering the flat end of a cylindrical block, was pressed lightly on this continuous heavy lawn ("master plate") and was then pressed successively on two plates of medium containing streptomycin, at a concentration that killed sensitive cells. A few colonies of resistant cells appeared on each plate, usually in coincident positions; and cells harvested from the corresponding positions on the master plate yielded a much larger proportion of resistant colonies than cells harvested from other parts of the plate. Evidently resistant clones, arising in the absence of drug on the master plate, were the source of most of the resistant colonies on the replica plates.

Plate without streptomycin
(confluent microcolonies)

Plates with streptomycin

Velveteen
block

often resume the normal bacterial form when grown in the absence of the drug, as expected, in some species they are stable and grow indefinitely without reversion. Moreover, a high proportion (over 25%) of the cells of one *Salmonella* species are converted into stable L forms by penicillin. Hence the drug cannot simply be selecting spontaneous mutants. Instead, it apparently prevents the cell from retaining a wall material that is needed as a primer before more can be synthesized.

The wall-less L forms produced by penicillin are resistant to penicillin; hence this special case formally represents a **directed, adaptive change in inheritance.** However, such an irreversible loss of cytoplasmic components bears no resemblance, in mechanism or in scope, to the processes of genotypic and phenotypic adaptation. Nevertheless, extragenic inheritance of the pattern of organization of a cell envelope may be a significant

mechanism elsewhere in biology, for in certain large single-celled protozoa, such as *Paramecium* or *Acetabularia,* surgical alterations in the cell cortex may be perpetuated in the progeny.

HAPLOID AND DIPLOID STAGES

Bacteria are almost always haploid, i.e., their chromosomes all have the same genetic structure, whereas most eukaryotic cells are diploid, i.e., have paired homologous chromosomes which may carry different alleles (heterozygosity). To be sure, more than one copy of a chromosome is often present in a bacterium, because chromosomal replication precedes cell division (Ch. 13); hence a mutation in a gene in one chromosome can make the cell temporarily heterozygous. However, at cell division the mutant and wild-type chromosomes are segregated into different progeny cells, in contrast to the persistence of homologous pairs in the progeny of diploid cells.

In the life cycle of higher organisms haploid cells appear only briefly, as gametes. Conversely, the problem of dominance appears in bacteria only under the special circumstances that yield partial diploids (Ch. 9).

All bacterial species studied thus far carry all their genetic loci in a single linkage group. The inference of a single chromosome per genome has been confirmed by direct observation (Ch. 10).

Fungi, in contrast, have typical eukaryotic nuclei, with several different chromosomes enclosed wthin a membrane. In many of these organisms classic sexual reproduction has been observed, with a regular cycle of alteration between diploid and haploid stages. In some species it is one stage, and in others the other, that multiplies vegetatively. The genetic behavior of these organisms will be described in Chapter 43.

DETECTION AND SELECTION OF MUTANTS

The isolation and the detection of various kinds of mutants are essential for genetic studies, which are always based on observations of phenotypic differences produced by alternative (allelic) forms of a gene. Moreover, mutants are also valuable for studying such problems of cell physiology as metabolic pathways and metabolic regulation. We shall discuss here the use of selective methods in the isolation of mutants; the mutagenic agents that can be used to increase their production will be discussed in Chapter 11.

Wild Type and Mutant Alleles. It is customary to refer to strains found in nature, or to certain standard strains, as the wild type, and then to distinguish as mutants the altered strains derived from them, usually in the laboratory. But though the concept of a wild type is readily applicable to those many genes for which a particular allele is highly predominant in the species as found in nature, other genes are **polymorphic,** i.e., they have no single wild-type allele but rather two or more prevalent alleles. Examples are genes determining surface antigens in bacteria or blood group substances in man.

Different mutations, interacting with different environments, create a wide range of selective pressures. Thus genes for **surface antigens** are often polymorphic (see Dissociation and phase variation, above), because in the presence of antibodies to the initial antigen mutations to another antigen may promote survival. Mutations to **drug resistance** obviously increase fitness for a drug-containing environment, though presumably at the expense of at least slight loss of fitness for the earlier environment (since the organism had evolved a drug-sensitive wild type). In contrast, mutations to **auxotrophy** (defective synthesis of an essential component) generally depend on the investigator for their survival: the organism is disadvantaged in many environments, compared with its parent or its **prototrophic** revertants (i.e., strains with the nutritional properties of the wild type).

Biochemical Markers. As long as the detectable genetic "markers" were restricted to those mutations with visible morphological effects the study of microbial genetics was seriously impaired. The introduction of **biochemical markers,** with their much greater variety, revolutionized the field.

To be sure, some geneticists had recognized early that a simple enzymatic defect probably underlay certain visible mutations in higher organisms, including color in plants, eye color in *Drosophila,* and albinism in man (Garrod). However, Beadle and Tatum were the first to demonstrate, in 1941, the possibility of systematically isolating microbial mutants (of the mold *Neurospora*) with various specific enzymatic defects. These studies led to the **one gene–one enzyme** hypothesis, which established proteins as concrete intermediates between the genes and characters of formal genetics. In addition, these auxotrophic mutants yielded an unexpected dividend by facilitating the analysis of many metabolic pathways (Ch. 4).

Screening and Scoring. The enormous size of microbial populations proved to be a great asset for genetic studies, but only because it is also possible to **select** certain kinds of rare genotypes with great efficiency. Thus even 1 lactose-positive mutant in 10^9 lactose-negative cells can be **screened** by inoculation on a medium in which lactose is the sole carbon source. On the other hand, one cannot similarly **screen** for mutations from positive to negative; though they can be readily detected **(scored)** on an appropriate plate containing a dye that will stain only fermentation-positive colonies (Fig. 8-5).

Quantitative selection is possible for several classes of mutants: fermentation-positive (or similarly able to use a new source of N, S, etc.), prototrophic, drug-resistant, and phage-resistant. In contrast, mutants altered in color, cell morphology, colonial morphology, or excretion of various metabolites are readily scored but are not readily selected.

The metabolic activities of the background population limit its permissible density in the scoring and the selection of mutants. Thus with strains that cannot grow on a given medium an excessively dense inoculum can nevertheless consume the carbon source and thus prevent outgrowth of the mutants (e.g., prototrophs) to be selected. Moreover, high density promotes **cross-feeding** of auxotrophic cells by other cells that excrete the required metabolite. Indeed, when cells of two nutritionally complementary mutants are mixed on a plate of minimal medium they may give the illusion of undergoing genetic recombination because the number of prototrophs increases. However, the increase may be due to spontaneous reversion, at a normal rate, in an expanded background population.

Replica plating (Fig. 8-5) is useful in scoring clones, in a mixed population, for their ability to grow in different media (Fig. 8-6). Thus a plate containing 100 colonies, replica plated onto 10 different selective media, can provide the same information that is obtained by 1000 repetitive transfers of cells from each colony to each medium.

Selection of Auxotrophs. Auxotrophic mutants have been especially valuable in studies of biosynthesis (as already noted) and also of genetic fine structure (Ch. 10). They are conveniently **enriched** by the use of penicillin, which kills only growing cells: in a minimal medium, in which auxotrophs cannot grow, this agent will selectively kill the parental cells, which can grow (Fig. 8-7). However, this method does not select quantitatively for auxotrophs because they undergo residual growth (and hence killing), as a result of incomplete ("leaky") blocks, stored metabolites, and cross-feeding by the cells lysed by penicillin.

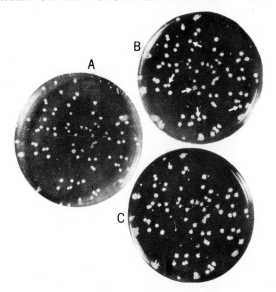

Fig. 8-6. Detection of auxotrophic mutants by replica plating. The master plate **(A)**, containing enriched medium, was replicated by a velvet press (Fig. 8-5) onto a plate of enriched medium **(B)** and one of minimal medium **(C)**. Arrows indicate colonies of auxotrophic mutants, which grow on B but not on C. [From Lederberg, J., and Lederberg, E. M. *J Bacteriol 63*:399 (1952).]

Conditionally Lethal Mutations. Some mutants are auxotrophic at 35–40° but not at 20°. These strains form a **temperature-sensitive** enzyme, which is rapidly denatured, even at ordinary growth temperatures. Indeed, this finding provided the first direct evidence that in the 1 gene–1 enzyme relation the gene does not simply regulate the synthesis of an enzyme but actually determines its structure.

Temperature sensitivity is one type of a broader group of conditionally lethal mutations, i.e., mutations whose effects are lethal (or at least prevent growth) under one set of conditions but not under another. Their development has permitted the isolation of mutations in a great variety of genes whose products cannot be supplied exogenously. Such mutations have been especially valuable in mapping viral genomes and will be discussed further in Chapter 45.

PHENOTYPIC LAG

After a mutation has occurred it is still not phenotypically expressed until the cells have undergone some growth. Thus if a suspension

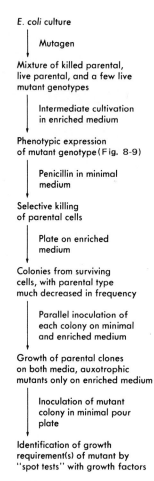

Fig. 8-7. Penicillin method for selecting auxotrophic mutants of bacteria (Davis; Lederberg). Similar results may be obtained with other agents that also kill only growing cells: e.g., metabolite analogs such as 8-azaguanine (Ch. 10) or radioactive metabolites that are allowed to disintegrate during prolonged storage after their incorporation.

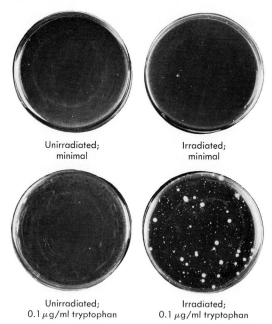

Fig. 8-8. Lag in phenotypic expression of induced mutations. Part of a culture of a tryptophan auxotroph was irradiated with ultraviolet to 0.1% survival, and equal numbers of cells, taken before and after irradiation, were plated on minimal medium to select for prototrophic mutants. Identical inocula were also plated on medium supplemented with 0.1 μg/ml of tryptophan, which permitted a few generations of growth of the large number of auxotrophic cells plated. It is seen that many back-mutants were induced, but very few of these initiated colony formation unless provided with a trace of the required growth factor. (B. D. Davis.)

of auxotrophic cells is mutagenized the prototrophic reversions that have been induced yield few colonies unless a trace of the previously required growth factor is provided in the selective minimal medium (Fig. 8-8). Similarly, in the selection of auxotrophs intermediate cultivation after mutagenesis is essential to allow phenotypic expression before exposure to penicillin.

This delay in phenotypic expression has two main components (Fig. 8-9). 1) **Nuclear segregation,** requiring cell division, is neces-

sary when a mutation yields a recessive allele (e.g., a noncompetent biosynthetic gene), whose presence would be masked by a wild-type gene in a companion chromosome. 2) There is a **phenomic lag** before any change in the genome is reflected in the phenome (defined as everything in the cell other than the genome). Thus for expression of an auxotrophic mutation the enzyme molecules formed by the gene before mutation must be diluted out by further growth.

The effectiveness of a mutagenic treatment is also influenced by the enzymatic repair of genetic damage, whose extent depends in part on the lag between mutagenesis and resumption of growth. This topic will be discussed in Chapter 11.

The quantitation of mutation or recombination presents a dilemma: when selecting for resistance to

Fig. 8-9. Delay in phenotypic expression of recessive mutations. o = wild-type allele; ● = mutant allele; x = product of the wild-type allele.

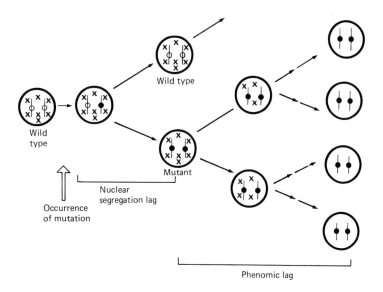

Wild type

Wild type

Mutant

Nuclear segregation lag

Occurrence of mutation

Phenomic lag

a bactericidal drug one must first allow enough growth for phenotypic expression; yet the treated cells must be separated on plates before they begin to multiply. The problem can be solved by plating the treated cells in the depths of growth medium, incubating for several hours to allow phenotypic expression, then adding a layer of medium containing the drug, which diffuses into the medium below and allows selective outgrowth of the resistant clones.

MUTATION RATES

Even low mutation rates can be measured with precision by using classes of bacterial mutations (e.g., to resistance or to protrophy) that are easy to select quantitatively from large populations. This development has revealed that the "spontaneous" mutation rate rises with temperature, and also varies in some degree with almost any change in the culture medium. In addition, many weakly mutagenic substances have been detected, expanding enormously the class of recognized mutagens. These even include some normal components of our diet (e.g., caffeine; see Ch. 11).

Definition of Mutation Rate. In growth at different rates the spontaneous mutation rate (i.e., the probability of appearance of a given type of mutant) remains relatively constant

per cell division rather than per cell per unit time. Accordingly, it is customary to define mutation rate, α, as

$$\alpha = m/d,$$

where m is the number of **mutations** and d the number of cell **divisions.**

If one grows a population from a small inoculum the value of d essentially equals the final number of cells present, since the initial number is negligible and each division produces one additional cell. The value of m, on the other hand, cannot be determined simply by screening for the total number of **mutants** of a given kind, i.e., the **mutant frequency** in that population. That number represents not only the mutants produced in the most recent generation but also the accumulated progeny of mutations that occurred in earlier generations; hence the mutant frequency ordinarily becomes higher the later a culture is harvested. The mutation rate, in contrast, is constant for a given class of mutations under constant conditions.

DETERMINATION OF MUTATION RATE

1) The most precise and reliable method for determining the true mutation rate is to measure the slope of the increase in mutant frequency with continued growth. Especially smooth curves are obtained (Fig. 8-3) in the chemostat, which

provides prolonged steady-state growth with a constant population size. This method is applicable only when the mutant multiplies at the same rate as the wild type.

2) Another method utilizes mutation on solid medium, in which the progeny of an early mutant can be held together in a colony. For example, 10^5 streptomycin-sensitive cells may be spread on a porous membrane on nonselective medium and allowed to grow to 10^9 cells (determined by enumerating the cells washed off a plate grown in parallel). When the membrane is then transferred to a drug-containing plate the progeny of each mutation to resistance, whether a single cell or a clone, will give rise to one colony.

3) A widely used statistical method is based on the Poisson distribution of random, chance mutations in a series of identically inoculated and incubated tubes of culture medium. The probability that X mutations occurs in a tube ($P_{(x)}$) depends on m, the average number of mutations per tube, averaged over all the tubes, and is:

$$P_{(x)} = (m^x/x!)e^{-m}$$

where e is 2.718, the base of natural logarithms. The value of m cannot be measured directly, since, as we have seen, the number of mutants in any tube does not measure the number of mutational events in that tube, owing to multiplication of mutants. However, the number of tubes containing **no** mutants corresponds to the Poisson prediction, which reduces in this case to:

$$P_{(o)} = e^{-m}$$
$$\text{or} \quad \ln P_{(o)} = -m$$

Hence from a determination of the proportion of mutant-free tubes, $P_{(o)}$, one can calculate the average mutation frequency per tube, m. For example, an average of 1 mutation per tube yields $P_{(o)} = e^{-1} = 1/e = 0.37$.*

---

* For a more extensive discussion of the use of the Poisson distribution see Appendix, Chapter 45.

SELECTED REFERENCES

Books and Review Articles

ADELBERG, E. A. (ed.). *Papers on Bacterial Genetics.* Little, Brown, Boston, 1966. A collection of reprints of outstanding papers, with a valuable historical introduction and bibliography.

BEADLE, G. W. Genetics and metabolism in *Neurospora. Physiol Rev 25*:643 (1945).

SAGER, R. On non-chromosomal heredity in microorganisms. In *Function and Structure in Microorganisms,* p. 324. (15th Symposium, Society of General Microbiologists; M. R. Pollock and M. H. Richmond, eds.), Cambridge Univ. Press, London, 1965.

Specific Articles

LANDMAN, O. E., and HALLE, S. Enzymically and physically induced inheritance changes in *Bacillus subtilis. J Mol Biol 7*:721 (1963).

LEA, D. E., and COULSON, C. A. The distribution of the mutants in bacterial populations. *J Genetics 49*:264 (1949). In Adelberg collection.

LEDERBERG, J., and IINO, T. Phase variation in *Salmonella. Genetics 41*:743 (1956).

LEDERBERG, J., and LEDERBERG, E. M. Replica plating and indirect selection of bacterial mutants *J Bacteriol 63*:399 (1952).

LURIA, S., and DELBRUCK, M. Mutations of bacteria from virus sensitivity to virus resistance. *Genetics 28*:491 (1943). In Adelberg collection.

NOVICK, A. Experiments with the chemostat on spontaneous mutations of bacteria. *Proc Natl Acad Sci USA 36*:708 (1950).

chapter

GENE TRANSFER IN BACTERIA

TRANSFORMATION   182
    Discovery   182
    Identification of the Transforming Factor   183
    Mechanism of Transformation   184
    Entry of DNA   185
    Significance of Transformation   186

CONJUGATION   187
    Discovery   187
    Resemblance to Zygote Formation   187
  POLARITY IN BACTERIAL CROSSES   188
    The F Agent   188
    High-Frequency (Hfr) Recombination   189
  KINETICS OF HFR × F— MATING   190
    Interrupted Mating   190
  STRUCTURE OF THE CHROMOSOME AND THE F AGENT   191
    The Cyclic Chromosome   191
    F-Duction   192
    Distribution of Conjugation Factors   194
  PHYSIOLOGY OF CONJUGATION   195
    Bridge Formation; F Pili   195
    Transfer of DNA   196

TRANSDUCTION   197
    Mapping by Transduction   197

Though mutations are the fundamental source of genetic variation, the additional mechanism of genetic recombination enormously accelerates and expands the process of biological diversification and evolution. It is hardly surprising that so powerful a device should have evolved early, at the stage of single-celled organisms. Indeed, in some fungi sexual reproduction, the classic mechanism for gene exchange, has long been recognized as an occasional alternative to clonal, vegetative reproduction.* In bacteria, however, for a long time only the latter could be demonstrated. Moreover, variation in these organisms was supposed to rest on some vague plasticity. But when the evidence for distinct, mutable genes in bacteria became overwhelming (Ch. 8) the search for their recombination was renewed in a more sophisticated way.

At the level of gene structure and function the results unified bacterial variation with classic genetics: inheritance depends on units

linked in a chromosomal chain. However, at the level of gene transfer bacterial heredity yielded several surprises, which have made sex seem something of a luxury in evolution. Instead of the expected zygotes, bacteria form **merozygotes** (Gr. *meros,* part): partial diploids, containing the entire genome **(endogenote)** of a recipient cell plus a genetic fragment **(exogenote)** transferred from a donor cell. The transfer can be accomplished by three distinct mechanisms: cell conjugation **(mating),** viral infection **(transduction),** or uptake of naked DNA **(transformation).**

If the development of science were strictly logical, conjugation would surely have been discovered before transformation, for naked DNA is very susceptible to mechanical and enzymatic damage, and so large a molecule would not be expected to penetrate into the cell. Nevertheless, transformation was discovered first. Moreover, the discovery was made with the pneumococcus, though its ready autolysis and fastidious nutritional requirements make it an exceptionally difficult experimental subject. The key was Avery's lifelong devotion to the organism that was the leading cause of death, in Western countries, until the advent of chemotherapy. Like Mendel's discovery of units of inheritance in 1865, transformation was so novel that its significance was not widely appreciated for some time; the discovery was not rewarded by a Nobel Prize, though it founded molecular genetics.

---

* The essential features of true sexual reproduction are the formation of haploid gametes from diploid parents by the process of meiosis, and the subsequent fusion of a male and a female gamete to form a diploid zygote. Meiosis is important in the genetics of fungi; it is reviewed in Chapter 43.

TRANSFORMATION

DISCOVERY

The analysis of bacterial transformation not only revealed the possibility of gene transfer in bacteria but also identified DNA as the material basis of heredity. Hence it seems worthwhile to recall the circumstances leading to the fortuitous observation of the phenomenon in 1928 by Griffith, a health officer

in England concerned with the epidemiology of pneumococcal pneumonia.

The work of Avery and Goebel (Ch. 25) had shown that all virulent pneumococci form smooth (S) colonies, owing to the presence of a carbohydrate capsule of one or another antigenic type. In cultures such a strain often yields avirulent variants, which have lost the ability to form a capsule and hence form rough (R) colonies. Conversely, when a large number of cells of some R strains are inoculated in a mouse, reversions to the parental, virulent S type appear and kill the animal.

The mutational origin of these shifts (Ch. 8, Dissociation) was not yet recognized when Griffith undertook his experiments. He postulated that R cells retain traces of S antigen, and that the antigen released by the disintegration of many R cells can be accumulated by a surviving R cell and somehow convert it into an S cell. To test this hypothesis he injected mice with a small number of live R cells, with or without heat-killed S cells as added sources of antigen. The killed S cells did indeed promote conversion of R cells to S cells of the same type. Moreover, the process could not simply be promotion of reversion, for R cells derived from one S type could be **transformed** into another S type (Fig. 9-1). Griffith did not realize the implication of this observation: that some chemical substance was endowing a cell with a specific new heritable property and hence must be replicated in subsequent generations.

IDENTIFICATION OF THE TRANSFORMING FACTOR

Deeper analysis became possible when transformation was extended from mice to the test tube—first with heat-killed S cells and then with extracts of frozen and thawed cells. (These studies were facilitated by using antiserum against R cells, instead of the mouse, as a selective agent: R cells are agglutinated by the antiserum and sediment, while any S cells that appear grow diffusely in the medium.) The implications of transformation for cell heredity, however, were apparently still not generally recognized by geneticists or biochemists, perhaps because the genetic marker used, bacterial virulence, was so foreign to their fields. Hence only a single group of medical bacteriologists continued to work on the phenomenon. After a decade, in 1944, Avery, MacLeod, and McCarty, at the Rockefeller Institute, succeeded in isolating the active substance and identifying it as deoxyribonucleic acid (DNA; Fig. 9-2).

It is difficult today to appreciate how revolutionary was this identification. Though cytologists

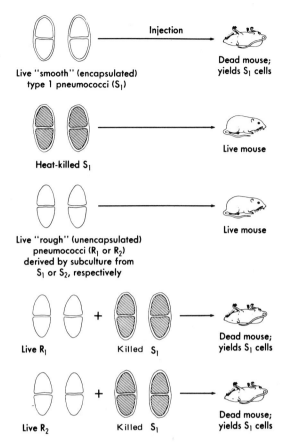

Fig. 9-1. The Griffith experiment. R cells not only were apparently reverted to S cells of their parental type by heat-killed S cells of the same type, but also were transformed to a different S type by heat-killed cells of that type.

had recognized that DNA, as well as protein, is present in the chromosomes of animal and plant cells, genetic specificity was universally assumed, for several reasons, to reside in the protein. Thus sequences of only four different bases seemed inadequate for the enormous variety of specificity required; these bases were present in roughly equal proportions in the early samples, suggesting a monotonous repeating tetranucleotide sequence and hence a structural role; and DNA was not even recognized as a macromolecule until late. Indeed, some skeptics maintained for years that the transforming activity must reside in traces of contaminating protein. However, the activity was destroyed by a trace of highly purified DNase, but not by proteases; and Hotchkiss later purified transforming DNA to less than 0.02% protein.

Even accepting DNA as the active principle, it took some time for investigators to realize that they

Fig. 9-2. Transformation of R to S pneumococci by DNA from S cells. Left (1): Colonies on blood agar of an R variant derived from type 2 pneumococci. Right (2): Colonies from cells of the same strain that had grown in the presence of DNA from type 3 pneumococcus, plus antiserum to R cells (see text). (Type 3 forms especially glistening mucoid colonies.) [From Avery, O. T., MacLeod, C. M., and McCarty, M. *J Exp Med 79*:137 (1944).]

now had in their hands the stuff of heredity, rather than a mutagenic or an inducing agent. This conclusion was verified a few years later when Hershey and Chase, infecting bacteria with radioactively labeled bacteriophage, showed that its DNA penetrated while most of the protein remained outside (Ch. 45).

Transformation has been accomplished with only a few bacterial species, including *Hemophilus influenzae, Neisseria, Streptococcus, Staphylococcus, Bacillus subtilis,* the plant pathogen *Xanthomonas phaseoli,* the nitrogen-fixing plant symbiont *Rhizobium,* and *Acinetobacter.* With *Escherichia coli* or other Enterobacteriaceae, in which so many well-studied mutants are available, transformation unfortunately has not generally been successful; low frequencies have been achieved with spheroplasts, or with simultaneous infection with a phage, or with strains lacking certain DNases, or following treatment of the cells with Ca Cl$_2$.

MECHANISM OF TRANSFORMATION

At first it was thought that transformation involves addition of a molecule of DNA, carrying perhaps one gene, to the genome of a cell. We now know, however, that in the usual preparation of transforming DNA the bacterial chromosome is randomly fragmented into pieces averaging roughly 1/200 of its length (Ch. 10); and the introduced exogenote undergoes recombination with the chromosome, replacing a homologous segment by means of a double crossover (whose mechanism is described in Ch. 11). Such gene exchange was revealed by the demonstration of **reciprocal** transformation: type 3 DNA can transform type 2 cells (as well as R cells) into type 3; and these can be transformed back into type 2 by type 2 DNA.

Quantitation. Capsule production was an unsatisfactory marker for detailed study of transformation: its enzymatic basis remained entirely unknown for many years; and precise quantitative studies became possible only when "transformation" was extended to other, sharply selectable genes, such as those conferring drug resistance or fermentative ability. (Auxotrophic mutants have not been useful in studies with the pneumococcus because of its fastidious growth requirements.)

With such markers the number of transformants at low DNA/cell ratios is found to be strictly proportional to the amount of DNA; hence the **effectiveness** of DNA preparations can be compared. Conversely, at high DNA/cell ratios, saturating the cells, the number of transformants obtained with different sets of cells provides an index of their state of **competence** to accept DNA (Fig. 9-3).

Linkage. Because transforming DNA consists of relatively small fragments of the chromosome, most markers are unlinked: that is, usually the DNA from an A+B+ donor, like a mixture of DNA from an A+B− and DNA from an A−B+ donor, transforms both markers in an A−B− cell with only a very low frequency. (The value approximates the product of the frequencies of single transformation to A+ and to B+, as expected if the two markers are on different fragments of DNA which can independently enter and transform the same competent cell.) However, the genes for streptomycin resistance (*str*) and mannitol fermentation (*mtl*) are close enough to yield **joint** transformation in the pneumococcus. Thus DNA from an *str*R *mtl*+ donor, applied to an *str*S *mtl*− recipient,* yielded double transformants with a frequency (0.1%) well above the product (0.006%) of the frequencies of the single transformations. It thus became possible to transform for loss of a function as well as for gain: for though one cannot select directly for such a defective gene, it could be brought in as an unselected marker linked to the selected *str*R gene. Thus DNA from an *str*R *mtl*− donor, added to an *str*S *mtl*+ recipient, yielded *str*R *mtl*− (as well as *str*R *mtl*+) recombinants.

ENTRY OF DNA

Transformation is produced by fragments of DNA of MW 300,000 to 10^7 or more. The frequency of linkage increases when the fragments are large, i.e., in gently prepared DNA. The uptake of such large "molecules" was most unexpected, since the osmotic barrier of the cells will exclude even a small sugar unless a specific transport system is present. However, unlike studies on transport of metabolites, which are concerned with the entry of thousands of molecules per cell, transformation recognizes the entry of a single long-chain molecule.

DNA entry appears to involve special sites, since with *Hemophilus* cells the maximal uptake of radioactive DNA is only 10 fragments per cell, suggesting the presence of about 10 entry sites. The sites of entry appear to recog-

---

* In bacterial genetics superscripts R and S denote resistance and sensitivity, respectively, to an inhibitor or to a phage. The superscripts + and − denote ability and inability, respectively, to synthesize a required metabolite (prototrophy vs. auxotrophy) or to use a carbon source (fermentation-positive and -negative). In classic genetics, in contrast, + means **absence** of the deviation from the wild type, usually visible, for which the mutant is named. Thus *cin* = cinnabar eye color, while *cin*+ = **not** cinnabar.

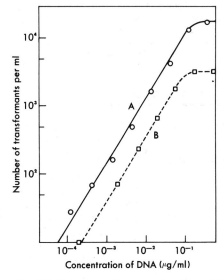

Fig. 9-3. Transformation as a function of DNA concentration. **A.** Transforming DNA alone. **B.** Same DNA mixed with three equivalents of nontransforming DNA. A similar difference would be seen between two preparations of transforming DNA of different effectiveness. (After Hotchkiss, R. D. In *The Chemical Basis of Heredity*. Johns Hopkins, Baltimore, 1957.)

nize the external surface of any double-stranded DNA; for while DNA from distant species cannot transform or be integrated into the chromosome, it enters the cell as well as does homologous DNA, and it can competitively inhibit penetration of transforming DNA (Table 9-1). Single-stranded DNA, however, has very little capacity to penetrate or to inhibit transformation.

In the transformed pneumococcus one strand of entering DNA is rapidly hydrolyzed, whether during or after entry, and the other somehow participates in recombination. Thus immediately after penetration into this organism there is an **eclipse** period of about 5 minutes, during which markers in the transforming DNA cannot be extracted from the recipient cells in a form that is active in transformation. In *Hemophilus*, in contrast, there is no eclipse.

Competence. The uptake of DNA requires cellular energy, and in synchronized cultures the proportion of competent cells varies regularly during the cycle of cell division. These findings, and radioautography with tritiated DNA, suggest that DNA enters in zones of wall synthesis. Entry, however, is also strongly affected by environmental factors that af-

TABLE 9-1. Inhibition by Nontransforming DNA of Both the Uptake and the Action of Transforming DNA

Ratio of competing to transforming DNA	Percentage inhibition	
	^{32}P-DNA uptake	Transformation
0	0	0
3	44	42
10	73	71
30	87	86

Hemophilus influenzae *cells were exposed, under the usual conditions of transformation, to* ^{32}P-labeled transforming DNA from a streptomycin-resistant mutant of the same organism, plus varying amounts of unlabeled DNA from another source. After several minutes the cells were treated with DNase to remove adsorbed DNA that had not penetrated, and the cells were analyzed for radioactivity and for the number of transformants to streptomycin resistance.

From Schaeffer, P. In Biological Replication of Macromolecules. *Academic Press, New York, 1958.*

fect the cell surface. Thus in nonsynchronized cultures competence rises and then falls during the growth cycle; and during the brief optimal phase the culture supernatant contains a protein **competence factor,** of MW 5,000 to 10,000, whose addition increases the competence of cells from other phases. DNA uptake can also be promoted by serum albumin with the pneumococcus, and by protamine with *E. coli* spheroplasts.

With improved technics it has become possible to transform about 5% of the cells for a given marker, compared with Avery's fraction of 10^{-4}. When pains are taken to keep the DNA relatively undegraded, it is highly efficient in transformation: nearly one transformant for a given marker can be produced per genome-equivalent of DNA taken up by the recipient culture.

Integration. Integration of transforming DNA involves recombination with the bacterial chromosome. Thus in the pneumococcus the introduced DNA, which is single-stranded during the eclipse period, develops several new properties at essentially the same time: 1) recovery of its transforming ability, implying double-strandedness; 2) genetic link-

age to host genes, implying insertion into the chromosome; and 3) initiation of replication in synchrony with the chromosome. Furthermore, after transforming DNA labeled with heavy isotopes joins unlabeled recipient DNA, fragments of variable intermediate density can be recovered.

SIGNIFICANCE OF TRANSFORMATION

Transformation may well be a significant mechanism of genetic recombination in nature: it has been achieved with live cultures mixed in the test tube and also in the mouse peritoneal cavity. Such recombination may have epidemiological significance, for with pairs of pneumococcal strains of low virulence the mouse can select for recombinants with increased virulence.*

Transformation can be used for short-distance genetic mapping by conventional measurements of linkage. In addition, since replication of a gene in a cell doubles the content of the corresponding transforming DNA, transformation can be used for large-scale mapping by determining the sequence in which the successive regions of a chromosome are replicated (Ch. 10) in a synchronized culture.

The main promise of bacterial transformation lies in the possibility of testing the intracellular activity of DNA that has been synthesized or modified in vitro. In addition, since the efficiency of integration depends on the degree of genetic homology, interspecific transformation of various genes provides an index of their evolutionary divergence.

The ability of transforming DNA to enter bacterial cells is paralleled by the ability of the DNA extracted from viruses of bacteria, animals, and plants to infect, though inefficiently, cells (or spheroplasts) that are normally hosts for the corresponding viruses (Ch. 45).

---

* The production of different degrees of virulence by recombination is not surprising, for virulence is a quantitative, polygenic character, i.e., one influenced by many genes. In the pneumococcus mutations can affect virulence by altering the quantity of capsule produced, as well as by other, less well-defined effects.

CONJUGATION

DISCOVERY

The discovery of auxotrophic mutations of bacteria in the early 1940s provided a powerful tool for renewing the search for mating, since the efficient selection of their wild-type alleles should make it possible to detect recombinants even if these were rare. Furthermore, the discovery of pneumococcal transformation provided a stimulus to renew the search for a more conventional mechanism of gene transfer. Under these circumstances Joshua Lederberg quit medical school, became a graduate student of Tatum, and made the dramatic discovery, in 1946, of a process in bacteria resembling sex.

The frequency of recombinant formation between auxotrophs was initially very low, comparable to the frequency of their reversion. The two were distinguished by the ingenious use of **double auxotrophs,** A^-B^-: these would not revert at a detectable rate to A^+B^+ since two independent mutations, each of low frequency, would be required. Thus when 10^8 cells of each of two double auxotrophs of E. coli strain K_{12}* were mixed and plated on minimal medium a few prototrophic colonies appeared, whereas even 10^9 cells of either parent alone yielded none (Fig. 9-4).

Recombination requires cell contact, unlike transformation. Thus culture filtrates of either parent did not transform cells of the other. There might conceivably still be a very labile transforming factor. However, when fluid was rapidly forced back and forth between growing cultures of the two parental strains, separated in a U-tube by a sintered glass disc (permeable to macromolecules but not to cells), no recombinants appeared.

RESEMBLANCE TO ZYGOTE FORMATION

Linkage. To map the genome, mutants of E. coli K_{12} with a variety of auxotrophic markers were isolated and the linkage of their markers was studied by crossing different

---

* The use of this strain, which had been carried for many years as a stock laboratory strain at Yale, was most fortunate, for the large majority of subsequently tested E. coli strains proved to be infertile.

strains: i.e., for each mating pair recombinants were **selected** for certain markers from each parent, and, among these recombinants the frequency of each **unselected** marker was determined. After several false starts the results finally indicated that the units of inheritance in E. coli are all physically connected with each other, in a single chromosome. Moreover, recombinants were clearly haploid products of genetic exchange, for auxotrophic and prototrophic alleles were equally readily expressed and the recombinants were stable. Diploid products of genetic addition, in contrast, would be expected to exhibit dominance and to segregate occasional haploid progeny with different genotypes.

A typical cross is depicted in Table 9-2. Among the recombinants selected for the T^+ marker from one parent, and for M^+ from the other, unselected marker L is encountered more frequently in the allele provided by the former parent; hence L is considered more closely linked to T than to M, according to the rules of classic genetics.

Because of the low frequency of mating, its immediate products could not be studied directly. Since the progeny observed many generations later were haploid, and reflected linkage between all markers, it seemed that mating had produced diploid cells, which quickly yielded stable haploid segregants. Yet the assumption that two haploid bac-

Fig. 9-4. Diagrammatic representation of the initial experiment of Lederberg and Tatum. The mutants, obtained by irradiation and selection, were cultivated in a medium that included growth factors A, B, C, and D; the test for recombination, and the control tests for reversion, were carried out in a minimal medium lacking these factors.

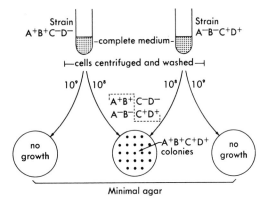

TABLE 9-2. Relative Frequency of Various Recombinants

Plates supplemented with	Selected markers	I B⁻M⁻T⁺L⁺ × II B⁺M⁺T⁻L⁻			
		Unselected alleles found			
		Type	No.	Type	No.
Biotin (B)	$M^+T^+L^+$	B^-	10	B^+	60
Threonine (T)	$B^+M^+L^+$	T^-	9	T^+	37
Leucine (L)	$B^+M^+T^+$	L^-	5	L^+	51

Conclusions:

1) Unselected B is more often B+ (from parent II) than B− (from parent I); therefore B linked with M (the selected marker from parent II);

2) Similarly T+L+ more frequent than T+L− or T−L+; therefore T and L more closely linked with each other than with B or M.

After Lederberg, J. Genetics 32:505 (1947).

terial cells fuse to form a diploid zygote created problems, for with the crossing of more and more pairs it became increasingly difficult to translate the linkage data into a consistent map: a given pair of markers showed very different degrees of linkage in different crosses. Nevertheless, the unification of bacterial variation with genetics had been such a triumph that valiant efforts were made to force *E. coli,* by means of *ad hoc* assumptions, into a Procrustean bed of classic genetics. The first step toward the correct solution was the discovery of two mating types.

POLARITY IN BACTERIAL CROSSES

Heterothallic species of fungi have two mating types (plus and minus, or male and female), and only gametes from strains of opposite type can fuse with each other. **Homothallic** species, in contrast, can form zygotes from gametes of a single strain. Since all the *E. coli* mutants derived from the K₁₂ wild type were interfertile, it was naturally assumed that bacteria are homothallic. However, in a classic example of serendipity, while studying the kinetics of mating, Hayes in London found this assumption to be false.

Hayes followed the formation of recombinants in a cross, in liquid medium, between a *str*ᴿ (resistant) auxotrophic and a *str*ˢ (sensitive) prototrophic parent. In a control test the presence of the lethal antibiotic streptomycin **during** mating prevented the formation of recombinants when a particular parent was

*str*ᴿ and the other *str*ˢ; but unexpectedly it did not do so when the *str* alleles were reversed in the two parents. Hence one of the parents could apparently serve as **donor** of genetic material even though killed by streptomycin, while the other parent appeared to serve as **recipient,** having to provide not only genetic material but also viable cells.

THE F AGENT

The inference of two mating types was confirmed by the finding that some stock cultures had lost their fertility with certain complementary stocks but not with others. Further studies, with clones freshly derived from single cells of various mutant stocks, led to the recognition that cells of *E. coli* K₁₂ exist in either of two inheritable mating types, called F⁺ and F⁻. F⁻ × F⁻ crosses are uniformly sterile, while F⁺ × F⁻ crosses are fertile, with a relatively low frequency (10^{-5} to 10^{-6} recombinants per cell pair). The F⁺ cells serve as **genetic donors (males),** since they can function even when killed (by agents that do not lyse the cell, such as streptomycin or ultraviolet irradiation). The F⁻ cells serve as **recipients (females),** which must be viable. The process is evidently **not symmetrical cell fusion;** rather, genetic material is transferred from male to female through some kind of **conjugation** bridge.

The donor property of F⁺ cells is due to a **sex factor** (fertility factor), the F agent. It was discovered because its presence can lead not only to chromosomal transfer but also, much more often, to transfer of itself. Thus when F$^+$ and F$^-$ cells, carrying different genetic markers, are grown together (at population densities sufficient for frequent cell contact), within an hour up to 70% of the originally F$^-$ cells become F$^+$. This spread of the F agent requires cell contact, just like the rare production of recombinants (which are also generally F$^+$).

The F agent is easily lost from an F$^+$ cell, especially during prolonged incubation in the stationary phase or on growth at an elevated temperature (42°) and a low population density (to prevent reinfection). The cell then becomes F$^-$. In nature loss of the agent must occur readily, compared with its spread, since F$^-$ strains are much more prevalent than F$^+$. The structure of the F agent will be discussed below.

Asymmetrical Genetic Contributions. The anomalies of the genetic linkage map in *E. coli* were further clarified by the discovery that the two parents make unequal **genetic** (as well as unequal **cytoplasmic**) contributions. Specifically, in the recombinants the unselected markers are derived much more frequently from the F$^-$ than from the F$^+$ parent. This finding suggested that the transient heterozygotes were diploid for only part of the bacterial chromosome (**merozygotes**). This inference was soon verified by the more direct studies described below (under Kinetics).

HIGH-FREQUENCY (HFR) RECOMBINATION

The mechanism of conjugation became much clearer when Cavalli by chance isolated from an F$^+$ strain a subclone with a 1000-fold increase in its rate of recombination with F$^-$ strains. Such strains, called Hfr (for "high frequency of recombination"), are evidently derived from F$^+$ by a reversible change in the state of the F agent, for F$^-$ cells (which lack the agent) never mutate to Hfr, and Hfr can revert to F$^+$. The shift in state from F$^+$ to Hfr has a further striking consequence: the **F agent becomes essentially nontransmissible** (i.e., recombinants from Hfr × F$^-$ crosses are ordinarily F$^-$). As we shall see in the next sections, the mechanism responsible for these associated changes was brilliantly analyzed by Jacob and Wollman, at the Pasteur Institute in Paris, who showed that in Hfr strains the F agent has become **integrated** in the chromosome in some way. In that position it still leads to conjugation, which now transfers the F-attached chromosome instead of the F agent (Fig. 9-5).

Role of Hfr in the Fertility of F$^+$ Strains. The low-frequency fertility of F$^+$ cultures could now be explained by their low-frequency mutations to Hfr. These mutants were demonstrated by **indirect selection:** when a lawn of F$^+$ cells was replica-plated (Ch. 8, Fig. 8-5) onto a thin lawn of a complementary F$^-$ strain, on a medium that would select for recombinants, a few colonies appeared; and from the corresponding regions on the stored F$^+$ plate one could often obtain Hfr clones. Thus in part, at least, crosses of F$^+$ **cultures** involve crosses of rare, unrecognized Hfr **cells.**

We have noted that the recombinants are generally F$^-$ when derived from Hfr × F$^-$ crosses, but F$^+$ when from F$^+$ × F$^-$ crosses; yet both crosses depend on Hfr cells. The explanation is that F$^+$ × F$^-$ crosses must be performed at high cell density, and contain many more F$^+$ than Hfr cells. Hence any merozygotes formed have a high probability of secondary contact with F$^+$ cells, resulting in multiple conjugation.

Fig. 9-5. Transitions between F$^-$, F$^+$, and Hfr cells. The reasons for invoking cyclic structures arose later, and will be presented below.

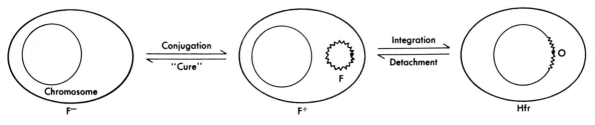

KINETICS OF HFR × F⁻ MATING

As long as bacterial conjugation was a rare event its study had to rely on progeny many generations removed from the actual conjugants. With the discovery of Hfr mutants, however, gene transfer could be studied directly. The mating process could then be analyzed in terms of its component steps: cell pairing (bridge formation), chromosome transfer, and integration.

INTERRUPTED MATING

Jacob and Wollman studied the kinetics of mating in a cross in which various readily selectable alleles were transferred from an *str*ˢ to an *str*ᴿ strain containing the complementary alleles. By mixing with a 20-fold excess of F⁻ cells each Hfr cell was assured opportunity for effective contact. Samples were taken at intervals, diluted, and plated on various selective media. The results showed that **effective contact,** firm enough to withstand the manipulations of dilution and plat-

ing, began at once; and all possible pairs had formed by 50 minutes at the population density (10^8/ml) employed. Moreover, the **level of the plateau** varied: the markers could be arranged in a sequence, with a **gradient** of decreasing frequency of transmission.

The existence of such a gradient suggested an orderly sequence of entry, with two features: all the cells of the Hfr strain must start chromosomal transfer at the same locus (the **origin),** and the probability of entry of a gene must decrease with increasing distance from this origin. These inferences were directly confirmed by a bold experimental procedure: the artificial interruption of mating (i.e., of transfer) by mechanical agitation (by means of a Waring Blendor, rapid pipetting back and forth, or strong vibration). As Figure 9-6 shows, entry started at about 8 minutes for genes very close to the origin; the other markers start entry later. Moreover, the increasing **time of initial entry** of various markers corresponds to their order in the **gradient of decreasing transmission.** Markers near the origin (T⁺L⁺) are seen to have yielded a maximum of 20% recombinants (per donor cell).

Mapping by Interrupted Mating. The time of **initial** entry of a gene represents the time re-

Fig. 9-6. Kinetics of conjugation, studied by interrupting further chromosomal transfer, as well as further pair formation, at various times. Cross: HfrH *str*ˢ *thr*+ *gal*+ × F⁻ *str*ᴿ *thr*− *leu*− *gal*−. Exponential broth cultures were mixed at time 0 (10^7 Hfr and $2 × 10^8$ F⁻ cells per milliliter) and aerated in broth. At intervals samples were diluted, agitated briefly in a Waring Blendor, and plated 1) on glucose minimal medium plus streptomycin to select *thr*+ *leu*+ *str*ᴿ recombinants; and 2) on galactose minimal medium plus threonine, leucine, and streptomycin, to select *gal*+ *str*ᴿ recombinants. [After Wollman, E. L., Jacob, F., and Hayes, W. *Cold Spring Harbor Symp Quant Biol* 21:141 (1956).]

Fig. 9-7. Gradient of transfer of markers from *E. coli* K₁₂ strain HfrH. The ordinate represents the **total entry** of each marker (fraction of input Hfr, with excess of F−), ranging from 20% to ca. 0.1%. The abscissa represents the **time of initial entry** for each marker; the distal end of the chromosome requires 90 minutes at 37°. [After Wood, T. H. *J Bacteriol* 96:2077 (1968).]

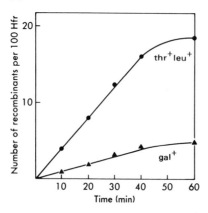

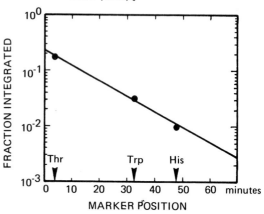

quired to form the earliest mating pairs (which is the same for all genes), plus the time to transfer the required length of chromosome. The differences between genes thus provide a direct **physical** measurement, in units of time, of **map distances** on a chromosome. Previously these could only be inferred, in all genetic systems, from the relative frequency of crossover between genes in recombination tests. The **sequence** and the **map distances** inferred from time of initial entry correspond to those obtained from linkage studies. Moreover, the frequency of marker transfer shows a simple relation to map distance (Fig. 9-7), which can be used for mapping.

The segment of chromosome transferred in 1 minute (ca. 1%) corresponds to about 15 μ of double-stranded DNA, or 5×10^4 nucleotides per strand (Ch. 10). This genetic length corresponds to about 20 recombination units

(i.e., 20% crossing over). Mapping by interrupted mating is thus ideal for long distances in the chromosome (> 1 minute) and complements the use of transduction (see below) for short distances.

STRUCTURE OF THE CHROMOSOME AND THE F AGENT

THE CYCLIC CHROMOSOME.

Independent Hfr mutants selected from the same F+ parent were found to **differ in the sequence of genes injected.** However, the various sequences bear an orderly relation to each other, as shown in Figure 9-8. This relation suggested that the E. coli **chromosome is a ring,** which can be interrupted at various points (though not necessarily at random) by **integration of the F agent;** integration deter-

Fig. 9-8. Formation of different Hfr strains by integration of F agent at different locations on bacterial chromosome. The site of initial entry (origin) is depicted as an arrowhead in the middle of F.

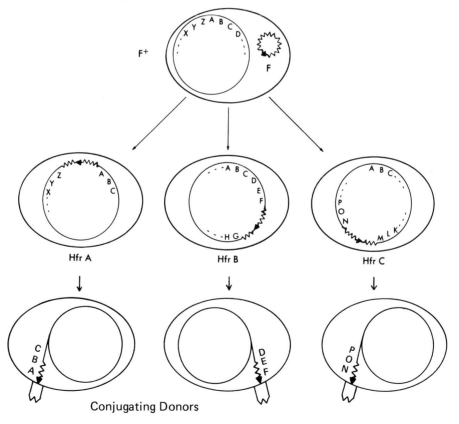

Conjugating Donors

mines both the origin and the direction of transfer of the chromosome. This formal evidence for a cyclic (often called "circular") chromosome was eventually confirmed by radioautography following gentle lysis (Ch. 10).

In Figure 9-8 the autonomous F agent is also represented as a ring, i.e., a small, supernumerary chromosome. Indeed, characteristic double-stranded, covalently closed rings (Ch. 10) have been recovered from F+ but not from F− cells (Fig. 9-9). There is about one per chromosome, and the MW (6×10^7) is 2% that of the chromosome. Chapter 46 will show that such a ring can fuse with the bacterial chromosome by a single genetic cross-over. This model unifies the behavior of F+ and Hfr strains: in either the F agent causes the formation of a conjugation bridge and can enter it, starting at a characteristic site in F.

In F+ strains the entire F agent usually enters, because its short DNA chain is rarely broken during conjugation. In Hfr strains the bacterial chromosome may be regarded as a huge insertion into the F agent, which is usually broken before complete transfer. If part of the agent enters first, while the rest remains at the distal terminus, we can readily

Fig. 9-9. Electron micrograph of F agent recovered from F+ cells. The length is 32 μm. (Courtesy of S. R. Palchaudhuri, A. J. Mazaitis, W. K. Maas, and A. K. Kleinschmidt.)

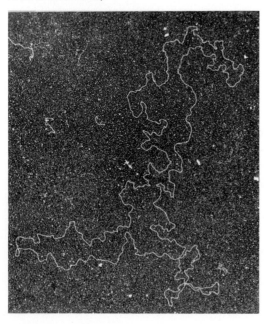

understand why the products of most Hfr crosses, lacking the distal portion, are F− (as noted above). Indeed, those rare products that have incorporated the most distal markers are Hfr.

Once an F agent has been integrated it is replicated as part of the chromosome: its own mechanism of initiation of replication is evidently repressed, as is superinfection by a second F agent. Moreover, though acridine dyes selectively block replication of the F agent (by an unknown mechanism), and thus "cure" growing F+ cultures of their maleness (see Ch. 46), they do not cure, or inhibit growth of, Hfr cultures (Ch. 46).

The Genetic Map of E. coli. Once the several unique features of bacterial conjugation were elucidated it finally became possible to construct a consistent, detailed chromosome map. This is shown in Figure 9-10.

Indeed, with all the complications that are hidden in the system it is remarkable that the early crosses gave sufficiently reproducible linkage data to be at all encouraging. On the other hand, since a stock F+ culture may have accumulated a mixture of different Hfr types we can understand how it could donate all its markers at a similar frequency, thus mimicking the production of classic zygotes by cell fusion.

F-DUCTION

Hybrid F Agents. One item yet remained in the treasure-chest of genetic novelties in *E. coli:* abnormal excision of the F agent (in the reversion of Hfr to F+), resulting in inclusion of an adjacent segment of host chromosome (Fig. 9-11). Strains carrying such a hybrid agent exhibit an **intermediate donor** property, resembling both F+ and Hfr. That is, they engage in conjugation, but with a frequency somewhat lower than that of F+; and most of the cells transfer the whole F agent (making the recipient F+), along with a fixed segment of the host chromosome (which is smaller than that transferred by Hfr). This transfer is called **F-duction** because it resembles the **transduction** of a limited segment of host chromosome by a bacteriophage (see below). In addition, a fraction of the cells at any time behave in conjugation like a typical Hfr, because the **extensive homology** of the hybrid agent with a segment of the host chromosome promotes its frequent reversible integration, by genetic crossing-over (Ch. 46).

In a complementary kind of abnormal excision part of the sex factor is left in the bacterial chromo-

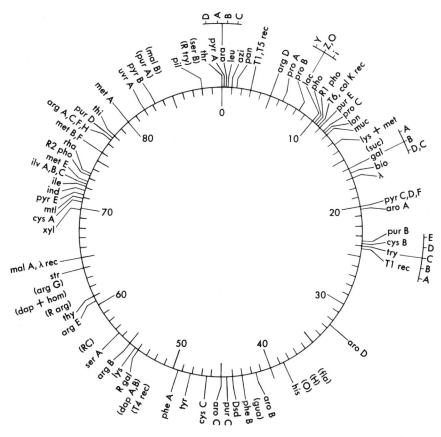

Fig. 9-10. Map of chromosome of *E. coli*. The numbers represent map distances in minutes required for transfer under the usual conditions (broth, 37°); zero is close to the origin of the first Hfr strain isolated, HfrH, (designated for Hayes). [From Taylor, A. L., and Thoman, M. S. *Genetics 50*:659 (1964); consult for abbreviations.] A later, more detailed map is presented for *E. coli* in Taylor, A. L., and Trotter, C. D. *Bacteriol Rev 36*:504 (1972), and for *Salmonella* in Sanderson, K. E. *Bacteriol Rev 36*:558 (1972).

some at the former site of Hfr integration. This region is then the site of preferential integration of a subsequently infecting F agent. These strains are called *sfa* (= sex factor-attracting).

F′ Agents. Jacob and Adelberg developed an ingenious method for isolating hybrid F agents containing known markers; these are called **F′ (F-prime) agents,** or **F-genotes.** The method was based on two expectations: a detaching F-agent might occasionally incorporate an adjacent segment at the distal terminus of the Hfr chromosome; and the F′ carrier of these genes, being much shorter than the bacterial chromosome, would transfer them early and efficiently. Accordingly, recombinants were selected (by interrupted

mating) for the early transfer of a locus known to be near the **distal** terminus of an Hfr chromosome (and hence normally transferred late and infrequently).

For example, when a donor with terminal *lac+* is crossed with a *lac−* recipient, and the mating is interrupted early, *lac+* recombinants are occasionally obtained. They have not been formed by normal chromosomal transfer, for they have not received various less distal Hfr markers. Rather, a hybrid F′ unit (F-*lac+*) has been formed in the donor, and its transfer to the *lac−* cell has made the latter a **heterogenote** (i.e., diploid and heterozygous for some region of the chromosome), of the type *lac−*/F-*lac+*. Thus these cells, unlike ordinary recombinants, 1) segregate stable *lac−* haploids, with a frequency of about 10^{-3} per cell division,

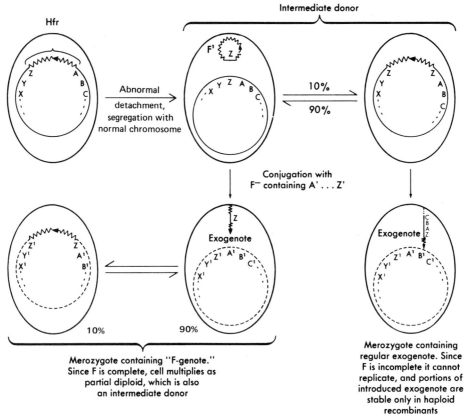

Fig. 9-11. Formation of intermediate donor by incorporation of portion of chromosome into detached F agent.

and 2) are intermediate donors, converting *lac−* females to *lac+* heterogenotes which can in turn pass on the F'.

A simpler and more versatile method is based on the use of *recA−* recipients. These lack an enzyme required for recombination and integration (Ch. 11); hence they cannot form recombinants by the usual mechanism of gene exchange. However, since a complete F' can replicate in a cell without being integrated it can form a stable heterogenote in *recA−* recipients. Hence an F' carrying genes near **either** side of the origin of an Hfr can be readily isolated, by mating that Hfr with a *recA−* recipient defective for the desired gene. The resulting recombinants have been formed by rare excision of an F' from the Hfr chromosome, followed by its transfer to the F− cell.

Many types of F' (F-*gal,* F-proline, etc.) have been obtained. The size can vary widely, as shown by genetic and by electron microscopic observations; some include as much as ⅓ of a bacterial chromosome. They yield relatively stable hetero-

zygotes, which have proved valuable in analyzing dominance relations in various mechanisms of gene regulation (Ch. 13).

DISTRIBUTION OF CONJUGATION FACTORS

Conjugation undoubtedly contributes heavily to bacterial variation in nature. Its importance in the spread of multiple drug resistance, for example, will be considered in Chapter 46.

Most strains of *E. coli* are F− and can accept the F agent (and genes transferred by it). However, the level of fertility is often low. Moreover, because of restrictions in the compatibility of the DNA of different strains transfers **between** strains are often less effective than within a strain (Ch. 46).

Genes have been transferred by the F factor from *E. coli* to other Enterobacteriaceae, such as

Salmonella and *Shigella*. Such interspecific transfer is considerably more effective with F′ than with Hfr donors because the former do not require integration. Indeed, F′ can even be transferred from *E. coli* to organisms with a grossly different DNA base composition, such as *Proteus, Pasteurella,* and *Vibrio*.

Other sex factors, differing in various ways from the F agent, have been found in various strains of *E. coli,* in *Pseudomonas aeruginosa,* and in the actinomycete *Streptomyces coelicolor*. The factors of *E. coli* include resistance transfer factors and certain colicinogens, which are generally much less efficient than F in mobilizing the host chromosome. These factors, and the mechanism of integration and detachment, will be further discussed in Chapter 46, where sex factors are considered as members of the broader class of **episomes:** agents that can replicate either autonomously (as plasmids) or as part of a chromosome.

PHYSIOLOGY OF CONJUGATION

BRIDGE FORMATION; F PILI

The cell pairing that initiates conjugation must involve mutual recognition of complementary macromolecules on the surface of the male and the female cell. The isolation of bacteriophages that can adsorb only to male cells of *E. coli* (Ch. 45) confirmed this differentiation and also revealed the mating structure. Crawford and Gesteland showed that male cells adsorb such phages onto two or three rod-like "F pili", up to several μm in length (Fig. 9-12). These are not found in female cells, and the adsorption does not occur with any of the several hundred other pili found on both male and female cells (Ch. 2).

F pili are clearly necessary for conjugation: fertility is lost when they are removed by mechanical or cultural means, and it returns immediately on their regrowth. Moreover, the initial connection by a pilus does not appear to be followed by a more intimate contact (even though cells with apposed walls, often seen under the microscope, were long assumed to be mating). Thus no energy supply is required for effective pairing; motile males can be seen to tow nonmotile females, held at the distance of an (invisible) F pilus; and such pairs, separated by micromanipulation, can transfer genes without closer contact. The

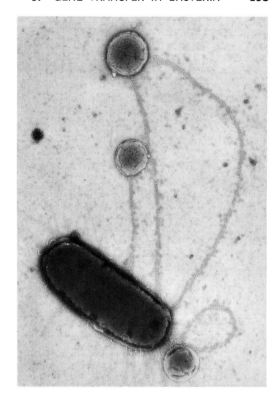

Fig. 9-12. Attachment of an F+ *E. coli* cell, by means of F pili, to three F− minicells. The few long F pili are covered with F-specific MS2 phage particles. These phages are icosahedral (Ch. 44); other, filamentous male-specific phages have been found to adsorb only to the tips of F pili. [See Curtiss, R., III, Caro, L. G., Allison, D. P., and Stallions, D. R. *J Bacteriol 100*:1091 (1969).]

structure and dimensions of the pilus, in contrast to those of a cytoplasmic bridge, can explain certain features of mating: its ready spontaneous interruption, the failure of a bristling array of other pili to prevent effective contact, and the non-transfer of radioactively labeled cytoplasmic components.

Electron micrographs of cross sections suggest that the F pilus is tubular. Moreover, its diameter (ca. 70A) is similar to that of the tails of certain phages, through which the phage DNA is known to be injected (Ch. 45); and male-specific phage MS2 (Fig. 9-12), whose RNA presumably enters male cells through the F pilus, can also infect female cells during conjugation (in a sort of venereal disease of bacteria).

As we shall see in Chapter 46, sex factors closely resemble viruses in their mechanism of replication, and the pilus that they induce parallels, in structure and function, the coat formed by a virus. The F pilus is an aggregate of a single small protein, F **pilin** (MW ca. 12,000). Because sex pili are so thin their staining with specific antibody can be readily recognized in electron micrographs.

The receptor for pili in F⁻ cells has not been identified. It is also present in male cells, but the frequency of DNA transfer in F+ × F+ crosses is very low. However, male cells often lose their F pili in the stationary phase and then behave temporarily as F⁻ phenocopies, able to conjugate with other males. F pili are normally found free in the medium, and when removed mechanically from growing cells they are replaced within 5 minutes. Hence they may well be constantly shed and replaced.

TRANSFER OF DNA

The mechanism of transfer of DNA in conjugation could not be analyzed until certain key features of chromosome replication were understood (Chs. 10 and 13). Thus the replication of any autonomous, self-replicating bloc of DNA (replicon), such as a chromosome or a sex factor, is regulated by control over its initiation at a specific "replicator" site. The growing point, once initiated, moves continuously along the DNA molecule, synthesizing a new, complementary strand on

each old, template strand; and the chromosome is attached to the plasma membrane, which separates the daughter chromosomes at cell division.

The similarity of this process to the continuous chromosome **transfer** in conjugation led Jacob and Brenner to suggest a direct connection between the two processes. According to their model the F agent attaches to the cell membrane near a conjugation bridge, whose formation it induces. In ordinary cell division the agent replicates and migrates with the growing membrane like an ordinary chromosome. Cell pairing, however, in some unknown way triggers an extra initiation of replication, which then proceeds asymmetrically: one of the original strands goes into the conjugation bridge and the other remains in the cell and is replicated (Fig. 9-13). The energy of replication, derived from the pyrophosphorolysis of deoxyribonucleoside triphosphates (Ch. 10), may provide the energy required to drive the transfer. This model fits the finding that normal replication of the chromosome, which appears to proceed in both directions from the initiation point, takes half as long (ca. 45 minutes; Ch. 10) as transfer of the whole chromosome in conjugation.

This model implies a requirement for DNA synthesis in the male, but not in the female, during conjugation. Supporting evidence was obtained in experiments in which DNA synthesis was blocked in one partner (by a temperature-sensitive mutation, addition of an inhibitor, or thymine deprival) while the other partner lacked or was resistant to the blocking mechanism. However, contradictory results were also reported. In a more decisive experiment Vapnek and Rupp separated the F agent from other DNA in donor and recipient cells by virtue of its special physical properties as a closed ring (Ch. 10), separated its two strands by virtue of their difference in density, and showed by labeling that the original partner of the transferred strand, remaining in the male, became a double strand during conjugation.

This model also explains how the F agent, which ordinarily multiplies in synchrony with cell division, can produce the extra copies needed to spread to F⁻ cells in a mixed culture. We can also understand why transfer is not blocked when the male is killed by streptomycin, for this antibiotic permits continued energy production and DNA synthesis long after killing (Ch. 12).

Fig. 9-13. Model for the mechanism of DNA transfer during conjugation. Attachment of the tip of a pilus to an F⁻ cell somehow activates a machinery (F rep), attached to the cell membrane at the base of the pilus, which begins asymmetrical DNA replication at the initiator site of an F agent and directs one product into the conjugation bridge (now identified as an F pilus). [Modified from Jacob, F., Brenner, S., and Cuzin, F. *Cold Spring Harbor Symp Quant Biol* 28:329 (1963).]

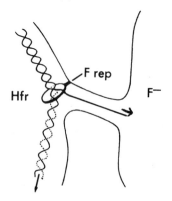

Hfr   F rep   F⁻

Single-strand Transfer. The DNA transferred in conjugation can be recovered as a single strand immediately after transfer to minicells (Ch. 6), which lack other DNA. This strand is later replicated; it forms a closed ring when it includes the entire F agent. Moreover, when prelabeled DNA participates in conjugation, and the progeny DNA is recovered from donors and from recipients and separated into its strands, it can be shown that the same strand of the F agent is always transferred, in the $5' \to 3'$ direction.

TRANSDUCTION

Transduction is a process of bacterial gene transfer mediated by bacteriophage particles. Some bacteriophages mediate **generalized** transduction, which may transfer any bacterial genes (or extrachromosomal elements such as the F agent). Others mediate **specialized** transduction, in which a particular phage strain can transfer only certain genes. Since an understanding of these processes requires a knowledge of bacteriophage physiology and genetics, detailed consideration will be deferred to Chapter 46. We shall consider here only the use of generalized transduction as a means for obtaining desired recombinants and for mapping the bacterial chromosome.

In **generalized transduction** the donor cell is infected with a temperate phage, which ordinarily does not lyse the cell. However, sometimes it does cause lysis, and in this process the assembling phage coat occasionally encloses a fragment of host chromosome, as outlined in Figure 9-14. When such a transducing particle encounters a susceptible recipient cell it injects the DNA and leaves the phage coat outside, just as in normal phage infection (Ch. 45).

The transducing particles in a lysate are always accompanied by a large excess of normal phage particles (Fig. 9-14), but because these are temperate their superinfection of a transduced cell usually does not cause lysis.

Generalized transduction is akin to transformation in transferring only small fragments of bacterial DNA. However, their sizes are much more uniform, being determined by the capacity of the phage coat. After the DNA penetrates the cell the intracellular events are much the same as in transformation. Transduction is more widely used than transformation: it is available in a greater variety of species, and it is simpler and more reproducible (since the DNA is protected from mechanical and enzymatic damage).

MAPPING BY TRANSDUCTION

Mapping by transduction is based on the assumption that the population of phage particles includes an assortment of segments, of uniform length, derived at random from the donor chromosome (Fig. 9-14). Each bacterial gene should then be represented with more or less equal frequency (depending on the state of replication of the chromosome). The distance between two mutational sites

Fig. 9-14. Formation of transducing phage particles by phage-infected bacterium. The entry of the phage genome leads to its own replication and also to formation of all the other components that are finally assembled into the mature phage (Ch. 45). In this assembly the coat occasionally encloses a fragment of bacterial DNA instead of the usual phage DNA, thus forming a transducing particle.

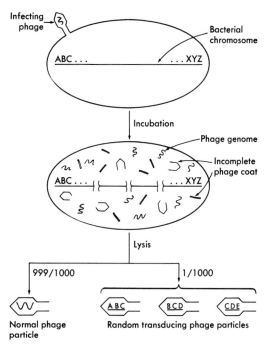

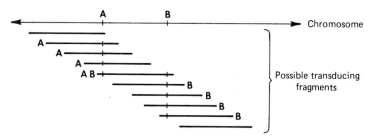

Fig. 9-15. Relation of distance between two markers to the frequency of their joint transduction. The various segments indicate possible fragments of cellular DNA of the proper length to be transferred in phage particles. The probability of including any marker in a fragment is proportional to its length, but the probability of including two markers is proportional to the difference between that length and the intermarker distance. This is one reason why the frequency of cotransduction falls off sharply with increasing distance between A and B. In addition, cotransduction requires two crossovers outside the AB region, one on either side. With those fragments that contain both A and B a greater distance between them decreases the probability of the required crossovers and also increases the probability of an undesired crossover within the AB region.

that can **cotransduce** in the same particle can be estimated in two ways: from the frequency of **joint transduction** of distinct genes, or from the frequency of **recombination** between sites within a gene.

Joint Transduction. In classic genetic mapping the frequency of recombination between two markers is assumed to be directly proportional to the distance between them. In mapping by transduction, however, the frequency of cotransduction of two markers depends not only on the probability of recombination between them but also on additional features, outlined in Figure 9-15. Accordingly, this frequency falls off very rapidly with increasing distance and reaches zero at the length of the transducing fragment (1 to 2% of the chromosome length).

Sequences and map distances can be determined in **two-factor crosses,** selecting for one marker and comparing the frequency of cotransduction of various others, and in **three-factor crosses,** identifying the "inside" member of a set as the unselected one whose cotransduction is increased by selection for **both** the others.

In the example provided in Table 9-3 the donor was wild-type *E. coli* and the recipient was a mutant negative for leucine synthesis (*leu−*), threonine synthesis (*thr−*), and arabinose utilization (*ara−*). Transductants were selected separately for the wild-type allele of each marker. For example, *leu+* was selected by plating on a medium lacking leucine but containing threonine and a carbon source other than arabinose. The selected colonies were then scored for each of the unselected markers. It is seen that *thr* is relatively distant from *ara* (6.7 and 4.3% linkage when one or the other was selected) and even more distant from *leu* (*4.1* and *1.9%* linkage), whereas *leu* and *ara* are closely linked (55 and 72% linkage). These results suggest the sequence diagrammed at the top of the table.

This sequence was verified by selection of different pairs in a three-factor cross. As Table 9-3 shows, selection for the two extremes (*thr+* and *leu+*) yielded a very high frequency of unselected *ara+* (80%), as would be expected if *ara* were the middle one of the three loci ("inside marker"). (The value was not 100% because an extra pair of crossovers occasionally occurred, from *thr+* to *ara−* and back to *leu+*.) In contrast, selection for the much closer *ara+-leu+* pair (not shown) yielded a larger number of transductants, but with a lower frequency of incorporation of the unselected, "outside" *thr+* marker, similar to that obtained with selection for *ara+* alone.

Frequency of transductional recombination, between similar mutations in donor and recipient, has been extremely valuable in **fine-structure genetics,** i.e., mapping of sites at very close intervals (Ch. 11). In this procedure the strains tested all have the same phenotypic deficiency, due to different mutations in the same gene (or in adjacent

TABLE 9-3.   Joint Transduction in *E. coli* by Phage P1

	thr⁺		ara⁺	leu⁺	
Donor: Wild type *Recipient:* Multiple mutant					(Fragment) (Chromosome)
	thr⁻		ara⁻	leu⁻	

Selected marker	Number of transductants per phage P1 plated	Percentage of selected colonies that also have the following unselected marker		
		leu⁺	*thr⁺*	*ara⁺*
Two-factor				
thr⁺	2.5×10^{-5}	4.1	—	6.7
leu⁺	5.0×10^{-5}	—	1.9	55.4
ara⁺	3.5×10^{-5}	72.6	4.3	—
Three-factor				
thr⁺ leu⁺	1.0×10^{-7}	—	—	80.0

The dashed line in the diagram depicts the double crossover necessary to exchange the chromosomal segment containing all three markers. Crossovers in various other regions would result in transduction of one or two markers. In transduction, unlike conjugation, the donor need not be "counterselected" since it is introduced as a lysate rather than as intact cells.

After Gross, J., and Englesberg, E. Virology 9:314 (1959).

genes of similar function). One adds phage obtained from various donors to the same recipient and compares the frequency of wild-type recombinants. Since these have arisen by a crossover **between** the sites of mutation in donor and recipient, their frequency increases

with the distance between these sites (Fig. 9-16).

Transductional recombination does not distinguish whether two nearby mutations are in the same gene or in two adjacent genes that may have a similar phenotypic effect

Fig. 9-16. Recombination between a transducing and a recipient marker in same gene. **A.** A *trp⁻* recipient is transduced with phage from a wild-type donor. Of the set of transducing fragments (Fig. 9-15) carrying a normal replacement for the defective *trp₁* site one typical member is shown. Given a crossover (1) to the left of the *trp₁* site, the second crossover (2) required for recombination may occur anywhere to the right of that site, up to the end of the fragment. **B.** The same experiment is performed with transducing phage from another *trp⁻* mutant (*trp₂*). To produce a prototrophic recombinant the second crossover must now occur between *trp₁* and *trp₂*; otherwise the correction of one defect in the cell will be accompanied by the introduction of another.

With various donors, the shorter the distance between their *trp⁻* site and *trp₁* the lower the frequency of prototrophic recombinants. In "self-transduction" of *trp₁* × *trp₁* none are obtained.

The sequence of *trp₁* and *trp₂* relative to an outside marker (in another gene) can be determined unequivocally by a three-factor cross with two different alleles of the marker in the two parents. In a reciprocal transduction, with either strain serving as donor in the *trp₁* × *trp₂* cross, one donor will bring in a nearby unselected marker to the right of the cross and the other will bring in one to the left.

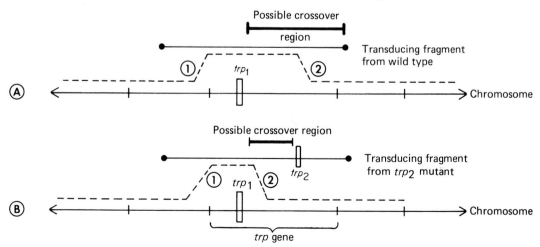

(e.g., requirement for the same endproduct). Chapter 46 (Fig. 46-17) will show how this distinction can be made by the use of a special form of transduction (abortive trans-duction), which leads to gene transfer without recombination and thus can show functional complementarity between two sets of genes.

SELECTED REFERENCES

Books

ADELBERG, E. A. (ed.). *Papers on Bacterial Genetics.* Little, Brown, Boston, 1966. A paperback reprinting of key articles, with a useful historical introduction and bibliography.

HAYES, W. *The Genetics of Bacteria and Their Viruses,* 2nd ed. Wiley, New York, 1968.

JACOB, F., and WOLLMAN, E. L. *Sexuality and the Genetics of Bacteria.* Academic Press, New York, 1961.

Specific Articles

Transformation

GRIFFITH, F. The significance of pneumococcal types. *J Hyg 27*:113 (1928).

HOTCHKISS, R. D. Gene, transforming principle, and DNA. In *Phage and the Origins of Molecular Biology,* p. 180. J. Cairns, G. S. Stent, and J. D. Watson, eds.) Cold Spring Harbor Lab., Cold Spring Harbor, N.Y., 1966.

HOTCHKISS, R. D., and GABOR, M. Bacterial transformation. *Annu Rev Genet 4*:193 (1970).

SCHAEFFER, P. Transformation. In *The Bacteria,* vol. V. (I. C. Gunsalus and R. Y. Stainier, eds.) Academic Press, New York, 1964.

TOMASZ, A. Some aspects of the competent state in genetic transformation. *Annu Rev Genet 3*:217 (1969).

Conjugation

BRINTON, C. C., JR. Contributions of pili to the specificity of the cell surface, and a unitary hypothesis of conjugal infectious heredity. In *Specificity of Cell Surfaces.* (B. D. Davis and L. Warren, eds.) Prentice-Hall, Englewood Cliffs, N. J., 1966.

CAVALLI-SFORZA, L. L., LEDERBERG, J., and LEDERBERG, E. M. An infective factor controlling sex compatibility in *Bacterium coli. J Gen Microbiol 8*:89 (1953).

COHEN, A., FISHER, W. D., CURTISS, R., III, and ADLER, H. I. DNA isolated from *Escherichia coli* minicells mated with F+ cells. *Proc Natl Acad Sci USA 61*:61 (1968).

HAYES, W. The mechanism of genetic recombination in *Escherichia coli. Cold Spring Harbor Symp Quant Biol 18*:75 (1953). In Adelberg collection.

IHLER, G., and RUPP, W. Strand-specific transfer of donor DNA during conjugation in *E. coli. Proc Nat Acad Sci USA 63*:138 (1969).

JACOB, F., BRENNER, S., and CUZIN, F. On the regulation of DNA synthesis in bacteria. *Cold Spring Harbor Symp Quant Biol 28*:329 (1963).

LEDERBERG, J. Gene recombination and linked segregation in *Escherichia coli. Genetics 32*:505 (1947). In Adelberg collection.

LEDERBERG, J. Aberrant heterozygotes in *Escherichia coli. Proc Natl Acad Sci USA 35*:178 (1949).

LOW, B. Formation of merodiploids in matings with a class of Rec− recipient strains of *Escherichia coli* K12. *Proc Natl Acad Sci USA 60*:160 (1968).

OU, J., and ANDERSON, T. F. Role of pili in bacterial conjugation. *J Bacteriol 102*:648 (1970).

VAPNEK, D., and RUPP, W. D. Asymmetric segregation of the complementary sex-factor DNA strands during conjugation in *Escherichia coli. J Mol Biol 53*:287 (1970).

WOLLMAN, E. L., JACOB, F., and HAYES, W. Conjugation and genetic recombination in *Escherichia coli. Cold Spring Harbor Symp Quant Biol 21*:141 (1956). In Adelberg collection.

Transduction

HARTMAN, P. E., LOPER, J. C., and SERMAN, D. Fine structure mapping by complete transduction between histidine-requiring *Salmonella* mutants. *J Gen Microbiol 22*:323 (1960).

chapter 10

STRUCTURE AND REPLICATION
OF NUCLEIC ACIDS

PROPERTIES OF NUCLEIC ACIDS   203

MOLECULAR ORGANIZATION OF NUCLEIC ACIDS   203
Primary Structure   203
Secondary Structure   204

PHYSICAL PROPERTIES OF DNA MOLECULES   205

Size and Configuration   205
Denaturation: Melting Temperature (Tm)   208
Renaturation   209
Cyclic DNA   211

ORGANIZATION OF SEQUENCES IN NUCLEIC ACIDS   212

NUCLEOTIDE SEQUENCES   212
The Analytical Approach   212
Homology: Hybridization   212
Distribution of Sequences over the Genome   212
Study of DNA Sequences by Renaturation Kinetics   216
Evolutionary Considerations   218

DNA REPLICATION   219

SEMICONSERVATIVE REPLICATION   219
ORGANIZATION OF THE REPLICATING CHROMOSOME   219
Sequential Replication: Unidirectional Type   219
Symmetry of Replication   221
Units of Replication   221
Divergent Replication   222
Role of Membrane in Replication   223
MECHANISMS OF REPLICATION   225
Enzymes Involved in Replication   225
The Replicating DNA   228

We have seen that DNA carries the **genetic information** of a cell, i.e., the instructions for its own replication and also for the structure of the other macromolecules, whose activities are in turn responsible for the total structure, function, and growth of the cell. In most viruses the genetic information similarly resides in the DNA, while in others it is in RNA (Ch. 45). In this and the following chapters we shall examine the main molecular aspects of genetic information: 1) how it is encoded in nucleic acid and other macromolecules, 2) how it is transmitted from one macromolecule to another, and 3) how its expression is regulated.

The molecular events connected with storage and transmission of information are governed by two general principles. 1) The informational molecules are **linear** and have periodic backbones (the phosphate-sugar chain in DNA, the peptides in proteins) which allow stepwise synthesis and formation of regular secondary structures. The information itself is carried, in vast amounts, in an aperiodic sequence of side chains (bases in nucleic acids, amino acid side chains in proteins).* 2) Nucleic acid and proteins are synthesized on a **template,** which serves neither as a catalyst nor as a substrate, but as a source of information determining the sequences being synthesized. Thus, a DNA strand serves as template for the complementary strand in **replication,** or for an RNA strand (the messenger RNA) in **transcription,** while the latter in turn serves as template for a polypeptide chain in **translation.**

The idea of a template found its first sig-

nificant support when Watson and Crick, in 1953, proposed the double-stranded structure of DNA: the complementarity of the two strands immediately suggested that each serves as a direct template for the synthesis of the other. This prediction was confirmed in 1958, when A. Kornberg discovered a DNA polymerase, in extracts of *E. coli,* that requires a template DNA and faithfully replicates its composition.

The study of nucleic acid structure and formation, coordinated with bacterial genetics, gave rise to the interdisciplinary field of molecular genetics (sometimes called molecular biology). When this field was extended from gene replication to gene expression, it was found to have converged with the pursuit of the mechanism of protein synthesis by biochemists. The rapid solution of many fundamental problems resulted from a dialogue between the two groups—even though molecular biologists were accused of "practicing biochemistry without a license."

Protein Synthesis. The critical step was the development of a system for synthesizing proteins in vitro, achieved in 1954 by Zamecnik with extracts of mammalian cells and subsequently extended to bacteria. The involvement of an RNA template was suggested by the requirement for RNA but not DNA. Two classes of RNA were separated: ribonucleoprotein particles (now called **ribosomes**), which were pelleted at 100,000 g, and supernatant, smaller "soluble" RNA molecules (the **transfer RNA [tRNA]**), to which, as Hoagland showed, specific amino acids were covalently attached by specific activating enzymes. On the basis of studies of the regulation of gene expression (Ch. 13) Jacob and Monod, in 1961, discovered an unstable additional fraction of cellular RNA, the **messenger RNA (mRNA),** into which the information for protein sequences is **transcribed** from DNA. Biochemical studies with synthetic messengers then rapidly deciphered the **genetic code,** which governs the **trans-**

---

* Chapter 13 will consider the processing of another kind of information that is also important for the cell: the concentrations of its metabolites.

lation of the nucleotide sequence of mRNA into a polypeptide sequence.

As background for this and the following chapter, we may further note that the bacterial ribosome contains many different proteins and three molecules of **ribosomal RNA (rRNA),** designated from their sedimentation constants as 23S, 16S, and 5S.

As this particle moves along the mRNA, each coding unit it encounters (**codon**) directs the binding of the corresponding **aminoacyl-tRNA (aa-tRNA),** whose amino acid is then incorporated into the growing polypeptide chain. These developments will be described in Chapter 12.

PROPERTIES OF NUCLEIC ACIDS

MOLECULAR ORGANIZATION OF NUCLEIC ACIDS

Extraction. Nucleic acids are extracted with the aid of substances that denature the proteins with which they are associated, e.g., sodium dodecyl sulfate or water-saturated phenol. The conditions for extraction (temperature, ionic strength, and special additives) depend on the type of nucleic acid (RNA or DNA) and on the nature of the contaminants (e.g., whether containing lipids or not). Analyses are facilitated by the fact that RNA is completely hydrolyzed by concentrations of alkali that do not hydrolyze DNA.

PRIMARY STRUCTURE

All nucleic acids contain four main bases: guanine, adenine, cytosine, and either thymidine in DNA or uracil in RNA, commonly symbolized as G, A, C, T, and U. The DNA of certain bacteriophages contains 5-hydroxymethylcytosine instead of cytosine, or 5-hydroxymethyluracil instead of thymine. Each nucleic acid strand has a **polarity,** because phosphodiester bonds connect the 3′ position* of one nucleotide residue to the 5′ position of the next, and polynucleotide chains terminate with a free 3′ position at one end (the 3′ end) and a free 5′ position on the other end (the 5′ end) (Fig. 10-1).

The free rotation of the phosphodiester bonds between adjacent nucleosides tends to make the polynucleotide chains very flexible, but this tendency is compensated for by the reciprocal attraction of the planar aromatic rings of the bases (especially purines), which tend to **stack** neatly on top of each other, stiffening the chain. Nucleic acids have multi-

ple negative charges owing to the primary phosphoryl groups; in vivo these are neutralized by inorganic cations (especially Mg^{++}) and by basic organic molecules, such as polyamines or, in cells of higher organisms, histones.

Various **modifications** are found in nucleic bases occupying specific positions in polynucleotide chains.

Most common in DNA are **methylations** that yield 5-methylcytosine and 6-methylaminopurine (methyladenine). In addition, in some bacteriophages the DNA is glucosylated on the hydroxyl group of hydroxymethylcytosine or hydroxymethyluracil. Transfer RNAs contain many methylated derivatives, including 5-methyluracil (ribothymidine), 5,6-dihydrouracil, and pseudouridine (with the ribose attached to ring C-5 rather than N-1).

*Primed numbers in nucleotides refer to positions on the ribose or deoxyribose residue, and other numbers to positions in the purine or pyrimidine rings.

Fig. 10-1. Diagram of double-stranded DNA. The two strands are connected by hydrogen bonds (vertical parallel lines) between complementary bases. Each strand has a polarity determined by the direction of the phosphodiester bonds. The 5′ → 3′ direction of each strand is indicated by an arrow. The two strands constituting the same molecule have opposite polarities.

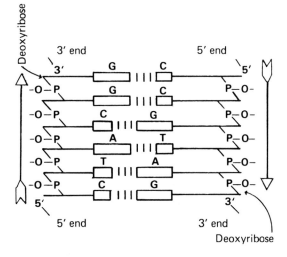

Ribosomal RNA contains 2-methyl substituted ribose as well as methylated bases. Modifications are carried out by specific enzymes after polymerization; at least eight different tRNA methylases have been identified. Different microbial species and strains have modifying enzymes that recognize different nucleic acid sequences. Another kind of modification is the **cleavage** of some precursor molecules to produce mature molecules (see Chs. 45, 47, and 48).

Modifications do not affect the genetic information. In DNA they can function as **species-specific markers,** which allow specific nucleases of various cells to distinguish between self and non-self DNA; only the latter is destroyed (see Ch. 45, Host-induced modification). In tRNA modifications affect functional efficiency (see Ch. 12, tRNA).

SECONDARY STRUCTURE

Chargaff showed that in DNA the proportion of A equals that of T and the proportion of G equals that of C. This finding, and X-ray crystallographic studies of Franklin and Wilkins, led Watson and Crick to recognize in 1953 that the structure of the usual DNA is a double-stranded helix with a diameter of about 20 A. A turn of the helix encompasses about 10 nucleotide pairs, but this number varies slightly, depending on conditions, with important consequences for cyclic DNAs (see below). In the helix, the two strands are complementary and antiparallel, i.e., with inverse polarity (Fig. 10-1). **Complementarity** is determined by the steric relations of the bases: A of one strand is always paired with T of the other (by two hydrogen bonds), as is G with C (by three hydrogen bonds; Fig. 10-2). The **base ratio** of a given DNA is thus $(A + T)/(G + C)$. The same relations apply to **double-stranded RNA,** with U substituting for T. In **single-stranded DNA and RNA,** in contrast, the proportions of the four bases can vary independently.

In helical nucleic acids the genetic information, i.e., the sequence of the bases, is inside, near the axis of the helix and cannot be read unless the strands separate. The information is **redundant** since it is duplicated in each strand. This feature is essential not only for replication, where each strand determines its complement, but also for recombination and repair of damage, where one strand is partly demolished and is then rebuilt as a complement of the intact strand (see Ch. 11).

Fig. 10-2. Pairing of adenine with thymine (2 hydrogen bonds) and of guanine with cytosine (3 hydrogen bonds) in double-stranded DNA. The arrows indicate bonds to 1′-C of deoxyribose (dR) in the sugar-phosphate backbone of DNA (Fig. 10-1). Similar pairing occurs in helical RNA with U substituting for T and ribose for deoxyribose.

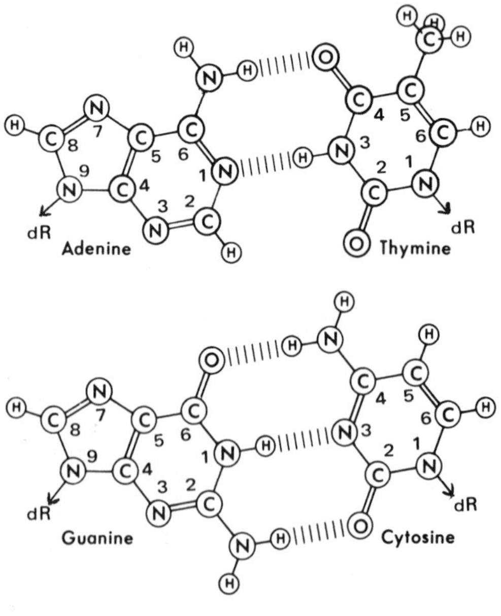

The hydrogen bonds between complementary bases and the stacking of the bases in each strand make helical molecules **very rigid,** as shown by the high viscosity of their solutions and by electron micrographs. The resulting chemical stability is of great importance for ensuring the constancy of genetic information. Yet these molecules undergo considerable **dynamic changes** (especially near single-strand breaks), which are of great significance in replication, transcription, and recombination. Thus they form short unpaired segments that move rapidly along the molecule ("breathing"); they can exchange a strand with an identical strand of another molecule; and they can undergo a certain amount of reversible deformation by changing the tilt of the base pairs relative to the helix axis.

In contrast, molecules of **single-stranded nucleic acids are very flexible** and in solution form random coils; but these tend to give rise to internal base pairs and are therefore **folded** to some degree (Fig. 10-3). The proportion of paired bases (also called

Fig. 10-3. Schematic representation of single-stranded nucleic acids in solution. **a.** Unfolded random coil present either at high temperature or in the presence of denaturing agents. **b.** Folded random coil present at low temperature in salt solutions; parallel lines represent hydrogen-bonded bases. Short-range folds, i.e., between adjacent sequences (A) are thought to be much more common than long-range folds (B). [Modified from Studier, F. W. *J Mol Biol* 41:189 (1969).]

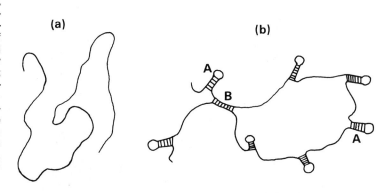

the **helical content**) varies with the composition of the nucleic acid and the nature of the solvent; if it is high, as in tRNA (see Ch. 12), a definite three-dimensional structure may result. Differences in secondary structure of diverse RNAs are probably important for their specific interactions with certain proteins, as in the complexing of tRNA with activating enzyme and with ribosomes, and in the assembly of rRNA and proteins to form ribosomes (Ch. 12).

PHYSICAL PROPERTIES OF DNA MOLECULES

SIZE AND CONFIGURATION

DNA molecules from bacterial or other cells are so long (millimeters or centimeters) that they are very fragile; even pipetting causes extensive shearing, and so most preparations consist of fragments. Accordingly, the size of cellular DNA can be determined only by methods that measure its length without purification and with a minimum of handling, e.g., electron microscopy and radioautography.

Electron Microscopy. Elegant electron micrographs of DNA are obtained by Kleinschmidt's method, in which the molecules are spread on a solution of a basic protein and then are collected on a membrane. The basic molecules absorb to the nucleic acid, increasing its thickness and therefore its visibility. This method reveals the length and shape of the nucleic acid.

Radioautography. Tritium-labeled thymidine of high specific activity is incorporated into replicating DNA, and the molecules of the DNA, extracted by mild methods, are collected on membrane filters or glass surfaces. These are overlaid with a photographic emulsion and kept in the dark for several weeks or months, during which β-radiations from disintegrating tritium atoms cause the formation of silver grains in the adjacent emulsion. After photographic development and fixation, the length and over-all shape of the DNA molecules can be determined. Moreover the molecular weight can be calculated, because the nucleotide pairs are 3.4 A apart along the helix axis.

With smaller nucleic acids, size and configuration can also be determined from migration velocity in sedimentation or in electrophoresis.

Sedimentation. The sedimentation coefficient ($s°_{20,w}$) is based on the velocity of sedimentation in a centrifugal field in water at 20° and at very low nucleic acid concentration. Absolute values are determined in the analytical ultracentrifuge. Relative values are often obtained in the preparative ultracentrifuge by **zonal centrifugation** through a **density gradient,** commonly of sucrose or CsCl, formed in a centrifuge tube by the appropriate mixing of two solutions of different specific gravity (Fig. 10-4). This technic is also invaluable for separating macromolecules in a mixture.

Sedimentation coefficients of macromolecules and viruses are usually given in Svedberg units (S). The sedimentation coefficient of nucleic acids is related to the molecular weight by empirical relations, which differ for single-stranded and double-stranded molecules. For single-stranded molecules the S values are strongly affected by the degree of helical content, which varies with ionic strength and temperature; therefore, molecular weights are best determined from S values obtained in solutions that prevent formation of secondary structures, such as

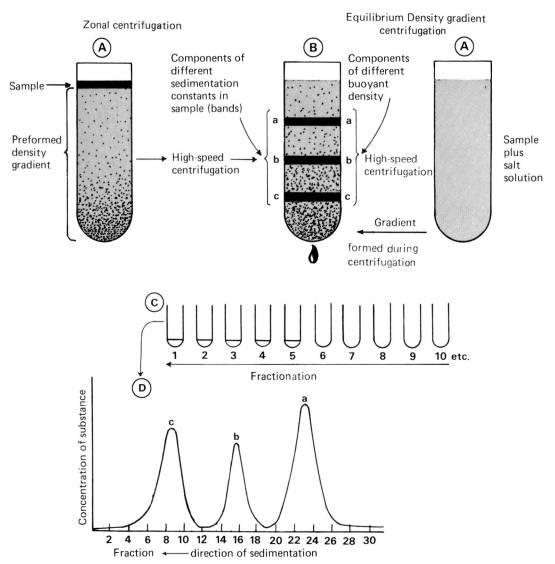

Fig. 10-4. Zonal and equilibrium density gradient centrifugation. In **zonal centrifugation** a linear density gradient is prepared with an inert solute in a plastic centrifuge tube, as indicated in **A**; its purpose is simply to prevent convection. The sample is layered at the top of the gradient. After centrifugation the various components of the sample have moved different distances, depending on their sedimentation coefficients, and thus form **bands** as shown in **B**. Components are separated on the basis of differences in their sedimentation coefficient which depend on particle size, shape, and density.

In **equilibrium density gradient centrifugation** the sample is mixed in the centrifuge tube with a solution of a salt of high molecular weight (e.g., CsCl or Cs_2SO_4) to obtain a mixture of uniform specific gravity similar to that of the nucleic acid **(A')**. Centrifugation causes the salt to form a concentration—and therefore density—gradient; each component of the sample collects in a band at a level where its density equals that of the gradient **(B)**. The **band width** is inversely proportional to the square root of the **molecular weight,** because larger molecules have less tendency to diffuse away from the band against the pull of the gravitational field.

In both technics the bottom of the tube is punctured with a fine needle and fractions are collected, in the form of drops, into a series of tubes, as shown in **C**; during this operation the bands maintain their relative positions because the density gradient prevents mixing. Different bands are therefore collected in different groups of tubes. When their contents are analyzed a diagram similar to that of **D** is obtained.

TABLE 10-1. Lengths of Different Kinds of Nucleic Acid Molecules

	No. of nucleotides per strand	Length (μ)
Transfer RNAs	70–80	ca. 0.03
5S ribosomal RNA	120	ca. 0.04
16S ribosomal RNA	1600	0.53
23S ribosomal RNA	3100	1.0
Phage MS2 RNA	4000	1.3
Polyoma DNA (cyclic)	5000	1.7
Phage λ DNA	5×10^4	17.0
Phage T2 DNA	2×10^5	68.0
E. coli DNA	4.5×10^6	1,530 (1.5 mm)
Mammalian, largest strand observed	5×10^6	1,800 (1.8 mm)
(*Drosophila* total haploid genome)	2×10^8	68,000 (68 mm)
(Human total haploid genome)	3×10^9	10^6 (1 meter)

low concentrations of formaldehyde or a high concentration of dimethylsulfoxide.

Polyacrylamide Gel Electrophoresis. This technic is especially suitable for single-stranded nucleic acids. The sample is applied to one end of a glass cylinder containing polymerized acrylamide; then a constant voltage is applied across the length of the cylinder, with the anode at the sample end. The negatively charged RNA molecules move toward the cathode, the larger molecules moving more slowly. At the end of the run, RNA molecules of different sizes are distributed as thin bands at various distances from the origin. The bands may be visualized by staining or by radioautography (when the RNA has been formed from radioactive precursors) or by analysis of thin slices. The rates of migration of the RNAs are inversely proportional to the logarithms of their molecular weights and also depend on the pore size of the gel (which is determined by the concentration of acrylamide used).

Such determinations of molecular size, combined with measurements of the amount of nucleic acid per cell or particle, show that bacterial and many viral **genomes** consist of a **single molecule of nucleic acid,** sometimes fragmented (see Ch. 44). These molecules must be tightly coiled because in their extended form they are extraordinarily long in comparison with the dimensions of the microbe to which they belong. For instance, the DNA of *E. coli* (MW 3×10^9 daltons) is

TABLE 10-2. Properties of Some Widely Used Nucleases

Enzyme*	Substrate†	Point of attack	Products after exhaustive digestion
DNases—endonucleases			
Pancreatic DNase I	ds or ss		5′-Nucleoside monophosphates and oligonucleotides with 5′-P and 3′-OH
Restriction nucleases	ds	Specific sequences	Fragments terminated with characteristic sequences
DNases—exonucleases			
E. coli exonuclease I	ss	3′-OH end	5′-Nucleoside monophosphates
E. coli exonuclease III	ds	3′-OH end	5′-Nucleoside monophosphates and single strands of high molecular weight
Snake venom exonuclease	ds or ss	3′-OH end	5′-Nucleoside monophosphates
Spleen exonuclease	ds or ss	5′-P end	3′-Nucleoside monophosphates
RNases—endonucleases			
Pancreatic RNase A	ss	3′-End of pyrimidines	3′-Pyrimidine nucleoside monophosphates, purine oligonucleotides terminated by a 3′-pyrimidine nucleoside monophosphate
RNase T1	ss	3′-End of guanines	Oligonucleotides terminated by a 3′-guanosine phosphate
RNase H (cellular)	DNA-RNA hybrids		5′-Nucleoside monophosphates and oligonucleotides
RNases—exonucleases Snake venom exonuclease Spleen exonuclease		Same as for DNase action	

* Endonucleases hydrolyze phosphodiester bonds within a polynucleotide chain; exonucleases hydrolyze strands from one end.

† ds = Double-stranded nucleic acid; ss = single-stranded nucleic acid.

about 400 times longer than the long axis of the cell, and that of bacteriophage T2 (MW 1.2×10^8 daltons) 500 times longer than the entire viral particle (virion). RNA molecules of all kinds are much smaller. Among the well-characterized ones the largest, from a virus, has a molecular weight of about 10^7 daltons (Table 10-1).

Several other properties of nucleic acids are of particular biological interest for understanding the structural basis of their activities and as tools for purifying and fractionating nucleic acids.

Buoyant Density. Buoyant density is determined by **equilibrium density gradient centrifugation** in either the analytical or the preparative ultracentrifuge (Fig. 10-4). Homogeneous nucleic acid accumulates as a symmetrical **band** whose width is related to the molecular weight; since smaller molecules diffuse more rapidly, their bands are wider.

Several characteristics of nucleic acids determine their buoyant density. Thus at neutral pH 1) RNA has a higher density than DNA of the same strandedness; 2) single-stranded DNA, which is less hydrated, is denser than double-stranded DNA of the same average base composition; and 3) the density of double-stranded DNA increases linearly

with its proportion of G plus C. Buoyant density is very useful for determining properties of DNA available only in small quantities. However, the presence of substitutions on the bases, especially glucosylation, which is found in some viruses (Chs. 44 and 45), alters the relation between base ratio and density.

Chromatographic Properties. DNAs can be fractionated on various columns. Methylated serum albumin adsorbed to diatomaceous earth is polycationic and retains the nucleic acids (polyanions due to their phosphate groups) which can be eluted separately according to number of strands, size, and base ratio. Hydroxyapatite selectively retains helical nucleic acid since the phosphates are more accessible; and benzoylated-naphthoylated DEAE (N,N-diethylamino-ethyl-) cellulose selectively retains single-stranded nucleic acids whose aromatic rings are more accessible for interaction with the aromatic substituents of the adsorbent.

Enzymatic Digestion. Nucleases are widely used for identifying nucleic acids, for determining strandedness and details of structure, and for sequencing. The properties of some of the most used enzymes are given in Table 10-2.

DENATURATION: MELTING TEMPERATURE (Tm)

Transient breakage of individual bonds between the bases occurs continuously in DNA at physiological temperatures but it does not weaken the helix, which is held by a large number of intact bonds. Only at high temperatures are enough bonds disrupted simultaneously to cause the collapse or **melting** of the helical structure. A sudden **helix-coil** transition results. This **denaturation,** and its reversal, have provided the powerful analytical tool of nucleic acid hybridization (see below), which has been of immense importance.

Denaturation increases the optical density of nucleic acids at 260 nm,* the peak of their absorption spectrum (OD_{260}). The extent of this **hyperchromic shift** is proportional to the change in the helical content; a complete helix-coil transition causes an increase of ca. 40%. A plot of OD_{260} as a function of temperature yields an S-shaped curve (Fig. 10-5). The slope is steep for helical nucleic

Fig. 10-5. Melting curves of two nucleic acids of the same length but of different structure. Double-stranded DNA of bacteriophage ϕX174 gives a sharp transition (curve A). Single-stranded DNA of the same phage gives a very flat curve (B). Tm = melting temperature, i.e., the temperature at which 50% of the hyperchromic effect of heating has appeared. For sample A, Tm is 79°. [From Chamberlin, M. and Berg, P. *Cold Spring Harbor Symp Quant Biol 28*:67 (1968).]

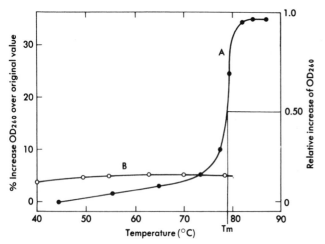

---

* 1 nm (nanometer) = 10 A.

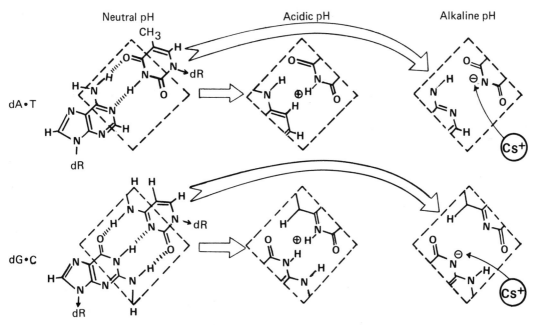

Fig. 10-6. Effect of pH on DNA base pairs; only the parts involved in interpair bonding are represented. Both acidic (<3) and alkaline pHs (>12) cause ionization (of different kinds at the two pHs) of ring nitrogens, which by abolishing the H bonds and creating a steric hindrance disrupt the DNA helix. At alkaline pH the negatively charged thymine and guanine bind Cs+ which therefore increases the buoyant density; hence equilibrium sedimentation in alkaline CsCl yields two bands with DNAs in which one strand is rich in T + G. Short parallel lines = hydrogen bonds. [Modified from McConnel, B., and von Hippel, P. *J Mol Biol* 50:297 (1970).]

acids, in which the shift from helix to random coil occurs rapidly in the whole molecule; the slope is shallow and variable for single-stranded nucleic acids, where individual helical segments melt independently at different temperatures, depending on their length and composition. Denaturation also increases the buoyant density of nucleic acid and markedly decreases its viscosity.

The **melting temperature (Tm)** is defined as that at which 50% of the maximum increase in OD_{260} occurs. For **double-stranded DNA, the Tm is a strict function of the base ratio,** since the triple hydrogen bonds of GC require a higher temperature for rupture than the double hydrogen bonds of AT. Small cyclic DNA molecules, however, behave atypically (see below).

Denaturation can also be brought about at low temperature by extremely low or high pHs, which disrupt hydrogen bonds (Fig. 10-6); indeed, sedimentation in an alkali sucrose gradient (pH 12.5) is a standard procedure for isolating single strands of DNA. Low temperature denaturation is also caused by substances (such as formamide, dimethyl-sulfoxide, or certain salts, e.g., sodium perchlorate or trifluoroacetate) that interact with the bases, disrupting the bonds between them.

RENATURATION

Heat-denatured DNA tends to re-form double-stranded helices **(renaturation)** on cooling, especially when it is made up of relatively small, homogeneous molecules, as is viral DNA. Renaturation is maximized by **annealing,** i.e., maintaining the denatured DNA at a temperature somewhat below Tm, to melt out the folds, and at a high ionic strength (e.g., 0.3 M NaCl) to decrease electrostatic repulsion between the strands; once **nuclea-**

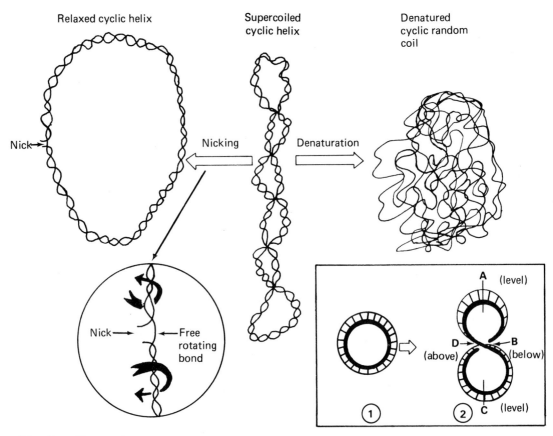

Fig. 10-7. Characteristics of cyclic helical DNA. Cyclic DNA molecules in solution usually have fewer helical turns per unit length than linear molecules under the same conditions. For instance, linear DNA of the length of polyoma DNA (see Ch. 63) would have 500 helical turns in solution, but native polyoma DNA has 485 built-in turns. Whereas in linear DNA the number of helical turns per unit length can easily change by rotation of one strand around the other, in cyclic molecules this rotation is prevented by the covalent continuity of the two strands. However, cyclic DNAs can change the required helical turns per unit length by developing **superhelical turns** (or **twists**). The inset shows that a twist corresponds to a change of a helical turn: when the duplex without helical turns (1) is twisted once (2), the thin strand goes around the thick strand full circle; it is level with the thick strand at point A, it goes below it at B, it is level again at C, and goes above the thick strand at D. At A the two strands are level again. Polyoma DNA compensates for its deficiency of helical turns by acquiring 15 twists.

The introduction of a single-strand break **(nick)** in a cyclic molecule restores free rotation of one strand in respect to the other. The result is disappearance of the superhelical turns. Cyclic molecules **denature** when the bonds stabilizing the helix completely collapse; although the two strands retain the same number of helical turns (which is a topological invariant), they are locked out of register in the random coil. The entangled strands cannot renature when the denaturing conditions are removed.

tion is initiated between two short complementary sequences, the helical region extends rapidly in a zipper-like fashion by a process of one-dimensional crystallization. Renaturation of heat-denatured DNA is minimized by rapid cooling **(quenching),** which permits intrastrand folding to be stabilized before complementary strands have time to pair. Cyclic structures, and the covalent cross-links between complementary strands produced by some cytocidal agents (see Ch. 11), prevent the melted strands from separating completely and thus promote renaturation.

Renaturation can also be promoted, at low temperature, by formamide or dimethylsulfoxide, which melt out the folds. In nature, renaturation at physiological temperature (e.g., in recombination) is probably facilitated by special molecules. In fact, a gene of bacteriophage T4 specifies a **DNA-unfolding protein** which binds to single-stranded DNA preventing it from folding and increasing its renaturation rate at 25° by 100- to 1000-fold.

CYCLIC DNA

Covalently closed, ring-shaped polynucleotide chains are the genetic material of some viruses and episomes, as well as of mitochondria and plastids of higher organisms. Most of them are double-stranded, but a few are single-stranded (as in bacteriophage ϕX174, Ch. 45). In cyclic helical DNA, the two strands cannot unwind unless one is broken, and the number of helical turns per molecule is fixed at synthesis. This results in several special properties of importance for these molecules.

First, under the artificial conditions of analysis, small cyclic molecules are twisted **(superhelical)** because they require for stability a higher than the built-in number of helical turns (Fig. 10-7). Each twist compensates for the deficiency of one helical turn. The twisted molecules are more compact and sediment faster than linear DNA of the same length, affording a method for separating the two forms. If even a single phosphodiester bond is broken in one of its strands **(nick)**, a cyclic molecule loses its superhelical twists **(relaxes)** because the free ends of the broken strand can then rotate around the intact strand, equalizing the number of helical turns to that required for stability. In large cyclic molecules accidental nicking produced by physical or chemical means usually prevents recognition of the superhelix.

Another property of cyclic molecules is that the degree of twisting varies with the composition of the solution. Thus, at an alkaline pH where the required number and the built-in number of helical turns coincide, the molecules become untwisted without strand breakage. Such **relaxed rings** become again twisted upon neutralization. Twisting is also strongly affected by **acridine dyes** (e.g., ethidium bromide). In linear DNA the planar molecules of the dye intercalate between adjacent nucleotide pairs (Fig. 10-8) causing the helix to unwind until a maximum of one dye molecule is intercalated for every two base pairs. This lowers the buoyant density of the DNA in CsCl. Helical cyclic molecules accept much less dye per unit length because they cannot unwind as extensively and so they have a higher buoyant density than linear DNA in the presence of the dye. This difference affords another useful method for separating cyclic from linear DNA.

Finally, cyclic molecules do not denature at the temperature expected from their base ratio because the hindered unwinding prevents extension of the initially melted areas. At much higher temperatures there is a sudden collapse of all the interbase bonds, generating a compact coil with the two strands still irregularly wound around each other, which sediments much faster than the native molecule (see Fig. 10-7).

In replication cyclic DNAs occasionally produce larger rings containing multiple units **(oligomers)** or forms in which two or more rings of unit length are interlocked like links in a chain **(catenates)**. These forms may conceivably arise either as anomalies of replication or by recombination.

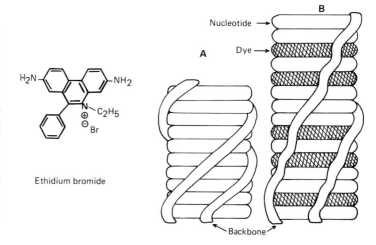

Fig. 10-8. Sketches representing the DNA helix seen from the side **(A)** and the helix of DNA containing ethidium bromide **(B)**. Both the nucleotides (white bands) and the intercalated dye (dark bands) are seen edgewise. The phosphate-deoxyribose backbones appear as smooth coils. The intercalation increases the length of the DNA and the stability of the helix; the pitch of the helix and the number of helical turns per nucleotide are decreased. [Modified from Lerman, L. S. *J Mol Biol* 3:18 (1961).]

Ethidium bromide

ORGANIZATION OF SEQUENCES IN NUCLEIC ACIDS

NUCLEOTIDE SEQUENCES

The biological function of macromolecules depends on minute details of their structure. In DNA the details of particular interest center on the sequence of the bases. These sequences specify the three-dimensional structure of proteins and of stable RNAs; determine where replication, transcription, and translation begin or end; allow DNA molecules to align at recombination and to separate at mitosis; and control their interaction with regulatory molecules.

Nucleotide sequences are studied in small part by the rewarding but laborious process of direct determination and for the most part by the much less informative but simpler approach of nucleic acid hybridization.

THE ANALYTICAL APPROACH

Sequencing of Nucleic Acids. Substantial progress has been made in sequencing nucleic acids. Overlapping fragments are first prepared using two nucleases of different specificities, such as A and T1 for RNA (see Table 10-2). The overlaps allow the sequence of the whole molecule to be reconstructed after determining the sequences of the fragments. In this way, several tRNAs, 5S ribosomal RNA, and fragments of 16S and 23S ribosomal RNA and various viral RNAs have been sequenced.

The many important conclusions deriving from the analytical approach will be reviewed in the appropriate sections of this book.

HOMOLOGY: HYBRIDIZATION

The degree of similarity in the sequences of different nucleic acids can be determined by measurement of hybridization in vitro. A nucleic acid of unknown sequence and a reference nucleic acid, denatured and annealed together, form hybrid helices if they have homologous regions (Fig. 10-9). Hybrid helices are formed by even short sequences of complementary bases (between 30 and 50), allowing the recognition of **partial homology.** For more complete characterization, the hybrid helices can be purified by a combination of various procedures: 1) enzymatic digestion, using nucleases specific for single-stranded DNA; 2) equilibrium density gradient centrifugation, since denatured DNA has a higher buoyant density than the corresponding helical DNA; and 3) chromatography through hydroxyapatite columns, which retain only helical DNA at 60° (in 0.14 M phosphate buffer, pH 6.8) and then release it at 90°.

The helices can be tested for accuracy of pairing by measuring their Tm. If the Tm is the same as that of native DNA, the sequences are accurately matched, while a decrease of 1° corresponds to approximately 1.5% of unpaired bases.

Hybridization between an unknown labeled RNA and a reference DNA has been very important in studies of transcription. In a widely used procedure the denatured reference DNA is dried on nitrocellulose filters, which are then immersed in the RNA solution under annealing conditions. When the filters are washed they retain only RNA that has hybridized with the DNA.

Measurement of the Degree of Homology. The proportion of sequences shared by two nucleic acids with **complementary** sequences (as two DNAs, or a DNA and an RNA) is determined by **saturation** experiments, which measure the maximum proportion of DNA hybridized by one that can form helices with an excess of the other nucleic acid (Fig. 10-10 I). The degree of homology between nucleic acids containing stretches of **identical** sequences (such as different RNAs) is determined by **competition** experiments, which measure the proportion of DNA hybridized by one RNA in the presence of an excess of the other RNA (Fig. 10-10 II). These widely used technics have cast much light on the organization of nucleic acid sequences in bacteria and viruses.

DISTRIBUTION OF SEQUENCES OVER THE GENOME

These studies require a specialized methodology, some examples of which are given below.

Separation of the Strands of Helical DNA. Complementary strands of a DNA can be separated by equilibrium centrifugation if they differ in average

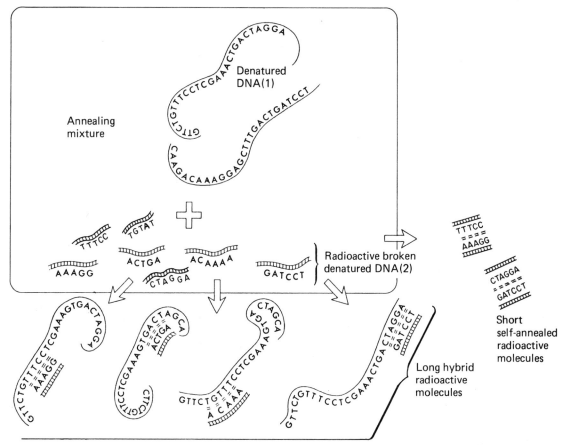

Fig. 10-9. DNA-DNA hybridization experiment. Reference DNA (1) is nonradioactive and in long molecules; unknown DNA (2) is radioactive and broken in small pieces. They are denatured by heat and annealed together for several hours at 65° in 0.3 M NaCl. The radioactive hybrid segments are part of long single-stranded molecules and can be separated from self-annealed radioactive DNA by their faster sedimentation in a neutral sucrose gradient.

base composition. Strands differing in the proportions of G and T are separated in alkaline CsCl gradients, where Cs++ binds to the two bases (see Fig. 10-6). A more general method uses neutral CsCl containing a synthetic homopolyribonucleotide which binds to runs of the complementary base increasing the buoyant density of the DNA. For instance, polyguanylic acid increases the density of strands containing runs of cytosine.

Isolation and Synthesis of Genes. Pure gene DNA is required for determining the base sequences that determine various genetic functions. Such preparations are also a step toward **gene therapy,** which aims at replacing or supplementing defective genes in man. Two main approaches are available for preparing genes. One, applicable to bacteria, is to use noncomplementary DNA strands from two transducing phages (see Ch. 46) which contain complementary sequences of bacterial genes, for instance, of the *E. coli* lactose operon (Fig. 10-11).

After annealing, only the bacterial sequences form a helix, which can be isolated in pure form after enzymatic degradation of the unhybridized phage DNA. The other, applicable to any gene whose mRNA can be isolated, is to make a DNA copy of the RNA using the reverse transcriptase (see Ch. 11).

Small genes of known sequences can also be synthesized. Thus, brilliant success has been achieved by Khorana in synthesizing the gene for a tRNA, using a combination of chemical and enzymatic methods similar to those he used in preparing repeating deoxyribopolynucleotides for deciphering the genetic code (see Ch. 12).

A general feature of nucleic acid sequences is that in all organisms their base ratios are not uniform over the genome; thus, when the DNA is broken into fragments of 2000 to 3000 nucleotides, these range from about 35

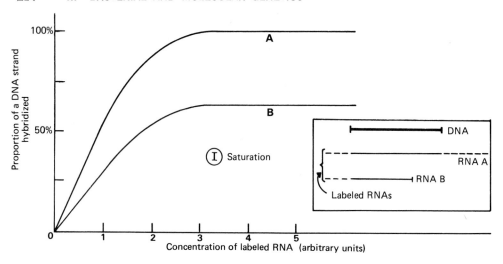

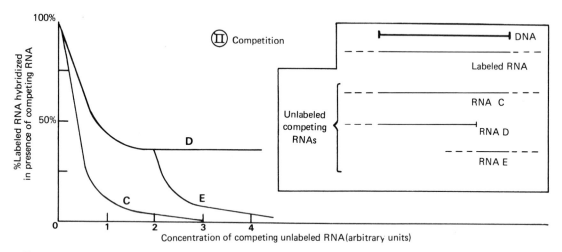

Fig. 10-10. Results of hybridization experiments between denatured DNA immobilized on nitrocellulose membrane filters and labeled RNA. **I.** In the **saturation experiment** each curve results from a number of determinations with the same amount of DNA on a filter and increasing concentrations of RNA in the hybridization mixture. In each case, hybridization is carried out until the number of counts hybridized approaches a maximum (usually hours). The filters are then treated with RNase A, specific for single-stranded RNA (see Table 10-2) to remove the unhybridized RNA. The radioactivity is then measured and the proportion of DNA hybridized is calculated from the bound radioactivity and the known specific activity of the RNA. The horizontal part of the curves gives the saturation level of hybridization. Curves A and B obtained with the same DNA and two different RNAs show that RNA A can saturate completely one of the DNA strands (i.e., the maximum, since only one DNA strand is transcribed into RNA), whereas RNA B saturates only 60% of a strand. Thus, RNA A transcribes all sequences of the DNA, and RNA B 60% of them (see inset). **II.** In the **competition experiment,** constant amounts of DNA on filters are hybridized, first to increasing amounts of an unlabeled unknown RNA and then to saturating amounts of a labeled known RNA. If both RNAs share sequences with the DNA, the unlabeled RNA will compete with the hybridization of the labeled one. The amount of labeled RNA hybridized without competing RNA is taken as 100%. Curve C shows complete competition caused by an RNA identical to the labeled RNA (or perhaps containing additional sequences; see inset). Curve D shows the result obtained using a partially homologous competing RNA, which allows 40% of hybridization by the labeled RNA; the competing RNA, therefore, has 60% of the sequences of the labeled RNA. In Curve E, the addition to competing RNA D of another competing RNA containing the sequences missing in RNA D again completely abolishes hybridization by the labeled RNA.

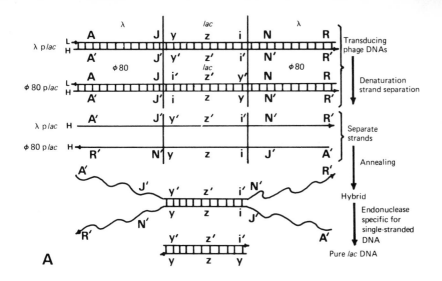

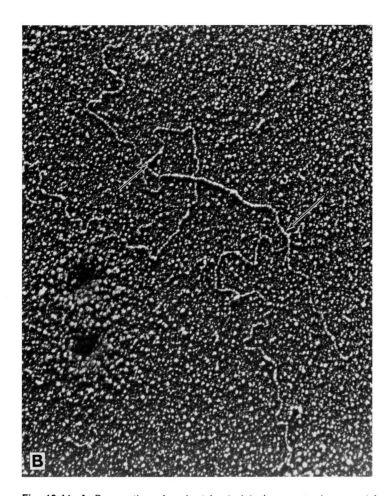

Fig. 10-11. **A.** Preparation of a short bacterial chromosomal segment in pure state. The DNAs of derivatives of the related phages λ and φ80 carrying sequences of the *E. coli lac* operon in opposite directions (see Ch. 46, Specialized transduction) were denatured and were separated into heavy (H) and light (L) strands by equilibrium centrifugation in CsCl in the presence of a synthetic polynucleotide. The H strands of the two phages, carrying complementary *lac* sequences but essentially identical phage sequences, were annealed together, yielding molecules hybrid only for the *lac* sequences. The product was treated with a nuclease specific for single-stranded DNA to destroy the unhybridized segments, yielding the chromosomal segments in pure state. **B.** Electron microphotograph, obtained by the Kleinschmidt technic, of a molecule resulting from annealing. It shows a rigid and thick double-stranded segment (the helical *lac* DNA, between the two arrows) with four thin flexible tails (the unhybridized phage strands). [From Shapiro, J., MacHattie, L., Eron, L., Ihler, G., Ippen, K., and Beckwith, J. *Nature* 224:768 (1969).]

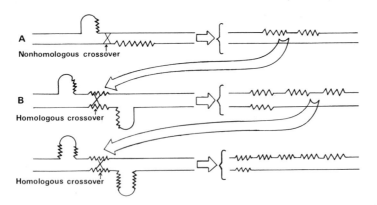

Fig. 10-12. Origin of gene clusters. The clusters are thought to arise by a rare initial nonhomologous recombination (**A**; see Ch. 11, Genetic recombination) and to increase in size by subsequent unequal homologous recombinations (**B**), which are more frequent. The latter process can also lead to elimination of some of the replicas of the gene. Whether the clusters grow or shrink during evolution depends on their selective value.

to 55% GC in *E. coli,* and similarly in phage λ and in man. Moreover, heterogeneity on a smaller scale is revealed by 1) the interaction of synthetic ribopolymers which bind to runs of a given base and 2) local melting, recognizable by electron microscopy, of AT-rich segments at temperatures below the Tm of the whole DNA.

An important feature is the **repetition and clustering of similar genes.** This form of differentiation along the chromosome is seen, for instance, in ribosomal RNAs which, being manufactured in the cell at much higher rate than most gene products, require multiple genes. Saturation hybridization experiments of cellular DNA with either of the main ribosomal RNAs reveal that the total genome contains about 7 of these genes in bacteria, 14 in yeast, and 400 to 700 in amphibians. Genetic evidence shows that these genes are clustered together. A finer differentiation is indicated by the regular alternation of genes for 16S and 23S RNA. There are about 30,000 genes for 5S RNA in the genome of a higher organism, and they are also clustered together.

Gene clusters have probably evolved by duplications through unequal crossing over, according to the scheme of Figure 10-12. This is supported by sequence analyses show-

ing that *E. coli* genes for 5S RNA are constituted by two quite similar portions, clearly arisen by duplication (Fig. 10-13). Many examples of repetition of very similar sequences also result from the study of proteins (e.g., the immunoglobulins).

STUDY OF DNA SEQUENCES BY RENATURATION KINETICS

The rate of renaturation is proportional to, among other things, the square of the DNA concentration (second order kinetics), since the formation of a helix requires the encounter of two complementary strands. The rate is also inversely proportional to the **complexity** of the DNA, i.e., to the length of the **unique** sequences (i.e., containing no repetitions; Fig. 10-14). This is because the encounter of two complementary sequences in perfect register is more likely to happen when the nonrepetitive sequences are short and therefore are present in higher concentration for a given total **DNA concentration.** Britten has shown that the rate of renaturation can be used to determine the length of unique sequences and the number of their repetitions, without determining the sequences themselves. By this method hardly any repeat sequences are demonstrable in viral and bac-

Fig. 10-13. Sequence of *E. coli* 5S ribosomal RNA, showing the large homologous segments of the two halves (boxes). Dashes indicate gaps introduced in the sequence to maximize homology. Two short sequences of the two ends are also equal (underlined). The numbers indicate the order of nucleotides from the 5′ end. [From Brownlee, G. G., Sanger, F. and Carrell, B. G. *J Mol Biol 34*:379 (1968).]

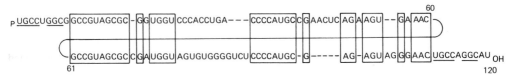

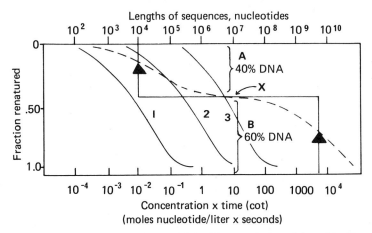

Fig. 10-14. Renaturation kinetics of double-stranded nucleic acids. The nucleic acids were fragmented to a constant length of about 500 nucleotides, denatured, and reannealed, using different concentrations or reannealing times. The proportions renatured were determined by chromatography through hydroxyapatite columns, which selectivity retain helical nucleic acids. The curves give the proportion of renatured nucleic acid versus the product: concentration of denatured nucleotides $\times$ time (Cot).

The solid curves were obtained with nucleic acids that do not contain appreciable proportions of extensively repeated sequences: 1 = double-stranded RNA of phage MS2; 2 = DNA of phage T4; 3 = *E. coli* DNA. These curves have similar shapes, reflecting the second order kinetics of renaturation, and are displaced in proportion to the molecular length. The lengths, determined from the Cot value at 50% renaturation, are similar to those determined by other means (see Table 10-1). Therefore, the Cot value of 50% renaturation can be used to determine the **length of unique sequences** in any double-stranded nucleic acid. The dashed line of calf thymus DNA splits up into two curves separated by the inflexion at X. Curve A, consisting of 40% of the DNA, contains sequences of the length of 10^4 nucleotides, as shown by the position of its midpoint (arrow); curve B, consisting of 60% of the DNA, contains sequences of the length of 3×10^9 nucleotides (arrow). Since there are 6×10^9 nucleotide pairs in a calf thymus cell, component B must be composed of independent sequences. In contrast, component A, which corresponds to 2.4×10^9 (i.e., 40% of 6×10^9) nucleotide pairs per cell, contains highly repeated sequences ($2.4 \times 10^9/10^4 = 2.4 \times 10^5$ per cell). [Modified from Britten, J., and Kohne, D. E. *Science 161:*579 (1968).]

terial DNAs but many in eukaryotic cell DNAs. In the mouse, for instance, 10% of the whole genome contains about 10^6 repeat sequences, each 300 to 400 nucleotides long. About 30% corresponds to about 100 sequences of 10^5 nucleotides; 60% has essentially unique sequences. The number of repeating units in each class is probably even larger (and their size smaller), because any mismatching of bases due to mutations tends to reduce the rate of reassociation.

The electron microscope has provided direct visual evidence for repetitive DNA in amphibian cells. On renaturing partially or totally denatured DNA fragments Thomas obtained a high proportion of circular structures (Fig. 10-15), which could have been produced only if the DNA contains many repeat sequences.

Whereas unique sequences probably correspond to genes for structural or regulatory proteins, the significance of highly repetitive DNA is obscure. It seems that most of it may not carry genetic information, since it is present largely in heterochromatin (condensed nonfunctional DNA), and only some of it is transcribed into messenger RNA. Since it is present in all eukaryotes but not

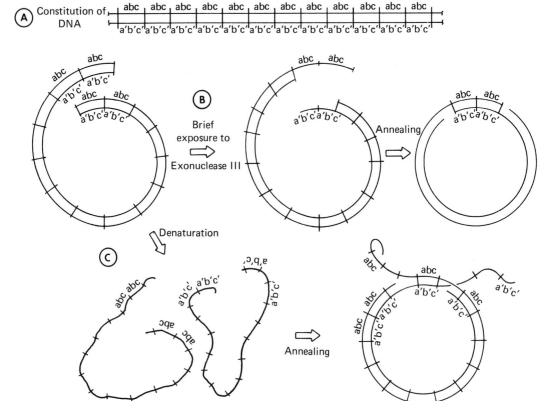

Fig. 10-15. Circularization of repeat sequence in amphibian DNA. A fraction of the DNA is supposed to have the constitution indicated in **A**, with many repeating identical sequences. Two procedures (**B** and **C**) were used to produce partially or completely single-stranded molecules, which were then self-annealed. Both procedures generated circular molecules recognizable by electron microscopy, of contour length between 0.2 and 5.0μ, in up to ⅓ of the DNA. [Modified from Thomas, C. A., Jr., Hannkalo, B. A., Misra, D. N. and Lee, C. W. *J Mol Biol 51:*621 (1970).]

in lower organisms, it may function in chromosome mechanics in mitosis or meiosis. Other roles have also been suggested (see Ch. 47).

EVOLUTIONARY CONSIDERATIONS

Extensive duplication of certain genes in clusters, duplication of sequences within a gene, and reiteration of short sequences show that tandem reduplication of the genetic material (see Fig. 10-12) is a basic evolutionary mechanism. The fate of the duplicated DNA depends on mutations, which cause it to diverge, and on selection, which determines the acceptable amount and kind of divergence. **Divergence of DNA sequences** accounts for taxonomic differentiation; in fact, in both bacteria and higher organisms, the degree of

DNA homology between species or strains parallels other criteria for taxonomic relatedness. Divergence involves a very large number of nucleotide changes, since DNA homology (based mostly on highly and moderately repetitive sequences) is essentially absent between not too distant species, such as primates and the mouse. Such extensive divergence is in part due to special features of the genetic code, such as its degeneracy (see Ch. 12).

Sequences vary widely in their rate of divergence. Thus, ribosomal RNAs and probably also many of the unique sequences have diverged relatively little. Probably divergence has been slow here because the ability of ribosomes and of many essential proteins to change is severely limited by functional requirements (for instance, cytochromes are almost identical in all living cells). In contrast, the divergence of highly repetitive DNA sequences in eukaryotes must have been very rapid since they are different in even closely related species.

The uniformity of certain highly reiterated sequences (such as those of genes for ribosomal RNA) **within an organism** is most remarkable. This uniformity implies the existence of an effective mechanism for counteracting the effect of muta-tions, which would tend to cause divergence of the replicate genes within the same cell. Callan has proposed that a **master gene** periodically pairs with the identical **slave genes** and thus corrects their mutations. A molecular mechanism that could produce such a correction will be discussed later (see Ch. 11, Error correction).

DNA REPLICATION

The Watson-Crick structure of DNA suggests a model for its reproduction called **semicon-servative,** whereby each strand (i.e., half molecule) serves as template for, and combines with, a new strand (Fig. 10-16), thereby being conserved in the new double helix. By taking advantage of the wonderful symmetry built into the DNA structure, this model accounts in a natural way for the transmission of genetic information from parent to progeny.

SEMICONSERVATIVE REPLICATION

The semiconservative model of DNA replication was verified by Meselson and Stahl in a classic experiment presented in Figure 10-17. The most notable feature is the total conversion of the DNA, after one generation, into a form of hybrid density, i.e., containing half old and half new DNA. Since the density remained unchanged after the DNA was fragmented into short pieces, the old and new strands were complementary rather than joined end to end.

ORGANIZATION OF THE REPLICATING CHROMOSOME

SEQUENTIAL REPLICATION; UNIDIRECTIONAL TYPE

Replication of DNA occurs at a **growing point** which moves linearly from an **origin** to a **terminus** either in one direction **(unidirectional replication)** or in both directions **(divergent replication),** depending on the organism, and, for the same organism, on growth conditions. In bacteria unidirectional replication is most frequently observed, as shown by genetic and chemical evidence.

Fig. 10-16. The replication of DNA. The box indicates the growing fork. (Modified from Watson, J. D. *Molecular Biology of the Gene.* Benjamin, New York, 1970.)

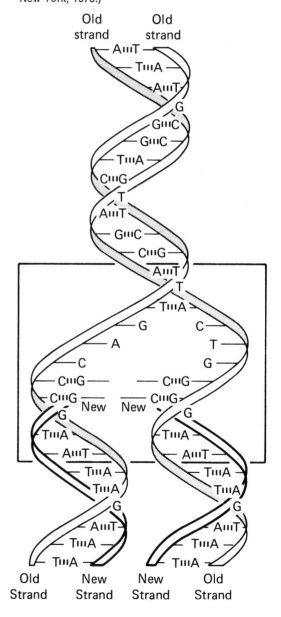

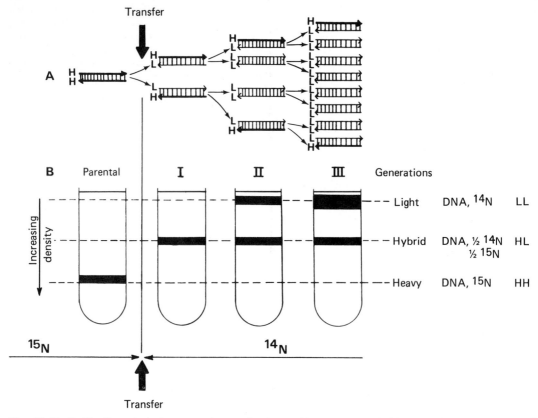

Fig. 10-17. Verification of semiconservative replication in Meselson and Stahl's experiment. A culture of *E. coli* is grown with $^{15}NH_3$ (heavy isotope) and then transferred to $^{14}NH_3$ (light isotope). The expected changes of the constitution of the DNA are shown in **A**, where the heavy parental DNA strands (H) are represented by heavy lines and the light newly synthesized DNA strands (L) by thin lines. The results obtained from DNA extracted from samples taken at various times after transfer and banded by equilibrium density gradient centrifugation in CsCl are shown in **B**. The results confirmed the predictions: before transfer the DNA is heavy; one generation after transfer it is hybrid (one strand heavy, the other light); and at subsequent generations a constant amount of DNA per culture is hybrid and an increasing proportion is light.

Genetic Evidence. Genetic evidence for sequential unidirectional replication has been obtained with synchronously growing cultures (Ch. 5) using two different methods. In one method, different *B. subtilis* genes were identified by transformation in the DNA extracted from the cells; upon addition of a DNA density label, the various genes shifted their density in succession, revealing the order of replication. Similar experiments are carried out with transduction in *E. coli,* which does not undergo transformation. In another method *E. coli* cultures were briefly exposed, at various times after resumption of DNA synthesis, to the mutagen nitrosoguanidine, which is especially effective on replicating

genes, and the mutation frequencies of various genes were determined. The genes showed peaks of mutagenesis at characteristic times, again indicating the order of replication (Fig. 10-18).

Experiments of both kinds have shown that in several *E. coli* strains replication begins at about 60 minutes on the standard genetic map (see Fig. 9-8) and proceeds clockwise; but some strains have a different origin and direction. Integrated episomes only rarely affect the origin or direction of synthesis, though they have their own origin when not integrated, and they determine a new origin for the Hfr chromosome when it engages in conjugation (see Ch. 9).

Fig. 10-18. Origin and direction of bacterial DNA replication determined by nitrosoguanidine mutagenesis. Mutagenesis peaks are sequentially induced in genes of an *E. coli* TAU-bar culture, synchronized by amino acid starvation followed by resumption of growth and replication at time zero. The time sequence is indicated by the oblique lines which identify two successive cycles of synchronized replication beginning before arg and terminating after tyr. This method determines not only the order of replication, but also the distances between genes, assuming a constant replication rate. The order of the genes and the origin of replication are indicated on the right, superimposed onto the standard K_{12} genetic map (see Ch. 9). The origin is at about 55 minutes and direction is clockwise. [Modified from Cerdá-Olmedo, E. and Hanawalt, P. C. *Cold Spring Harbor Symp Quant Biol 33*:599 (1968).]

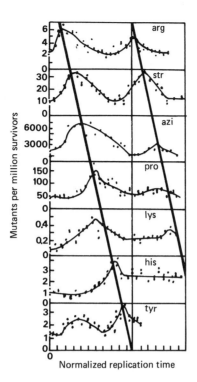

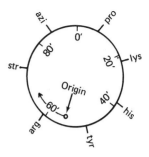

Chemical Evidence. Cairns examined by radioautography the intact chromosome of *E. coli* cells that were labeled with ^{3}H-thymine and then disrupted as gently as possible by lysing protoplasts with detergents. The results clearly showed, as was suggested earlier by the variation of Hfr strains in conjugation (Ch. 9), that the *E. coli* chromosome is cyclic throughout replication (Fig. 10-19) and that the ring is doubled between the origin and the growing point. When the growing point finally reaches the terminus next to the origin the ring has been duplicated and the daughter molecules separate.

SYMMETRY OF REPLICATION

Both the scheme of semiconservative replication given above and that derived from the interpretation of Cairns' radioautographs are symmetrical because the two strands have equal roles. A different, **asymmetrical,** type of replication of cyclic molecules has been proposed by Gilbert and Dressler. This process, called the **rolling circle** (Fig. 10-20), occurs in the replication of the cyclic DNA of bacteriophage ϕX174 (Ch. 45). The possibility

of rolling circle replication has been tested in other cases, using some sharply different features of the symmetrical and asymmetrical models (Fig. 10-21). With several DNAs (bacteria, initial replication of phage λ, and polyoma virus), replication follows the symmetrical model, which appears to be the general mechanism. The asymmetrical model is mostly used in the conversion of cyclic into linear molecules, as in the replication of some viruses (Ch. 45, Phage DNA replication).

UNITS OF REPLICATION

The above results show that DNA replication begins at special DNA sites which are identical in related bacterial strains. Topographically, therefore, it is possible to distinguish between **chain initiation** and **elongation.** The two functions appear to require different proteins because some *E. coli* mutants with a temperature-sensitive block in DNA synthesis cannot initiate replication at high temperatures but can complete already initiated chains, while others cannot complete chain growth when the temperature is raised. The proteins involved appear to be specific for

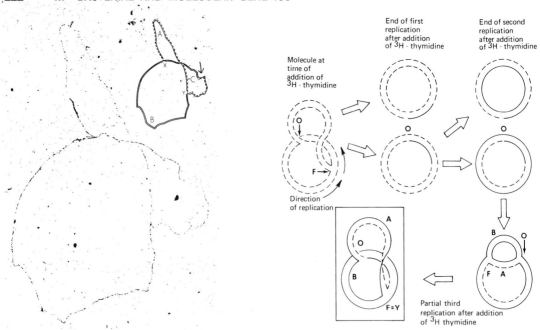

Fig. 10-19. I. Radioautograph of an *E. coli* chromosome after about 1.8 generations of incorporation of ³H-thymine. As indicated in the box, part of the DNA with a low grain density has label in one strand, while the rest, with twice the grain density, has label in both strands. Dashed lines = unlabeled strand; solid line = labeled strand; O = origin; Y = growing fork; A and B = the two parts of the chromosome already replicated; C = unreplicated part; arrow = position of the growing point at the moment of initial addition of ³H-thymine, part way through preceding replication cycle. **II.** Diagram explaining the different labeling of the various parts in relation to the previous history of the chromosome. The boxed figure corresponds to the radioautograph. [From Cairns, J. *Cold Spring Harbor Symp Quant Biol* 28:43 (1963).]

bacterial DNA, since at least some of the mutations do not prevent synthesis of bacteriophage DNA at high temperature in the same cell. Similarly, some temperature-sensitive mutations of the F episome can prevent its replication at 40° without affecting the replication of cellular DNA in the same cell. These findings define, in both anatomical and functional terms, well-defined units of DNA replication (**"replicons"**), each with a unique origin (the **"replicator"**) and termination.

DIVERGENT REPLICATION

The clearest evidence for divergent replication, with a replicator centrally located in the replicon, has been obtained in some bacteriophages and in mammalian cells.

In λ and in T4 DNA, two divergent replicating forks are recognizable by electron microscopy (see Ch. 46). The evidence for divergent replication in mammalian cells is based on the slow equilibration of the thymidine pool. Thus, when ³H-thymidine in the medium is replaced with unlabeled thymidine, labeled residues continue to enter the DNA at a decreasing rate: hence the radioautographs show series of grains with progressively longer spacings, which indicate the direction of synthesis. Chain growth in adjacent segments of DNA is divergent.

Characteristics of Replicons. Replicons vary in number and size in different organisms. Thus, in *E. coli* there is one replicon per cell; it contains 4.5×10^6 nucleotide pairs and is replicated at the rate of 3.5×10^5 nucleotides per minute under optimal conditions, requiring about 30 minutes for complete replication. Mammalian cells, which contain about 1000 times more DNA, probably contain 30,000 to 40,000 replicons, with about 10^5 nucleotide pairs each, and a replication rate of about 3000 nucleotides per minute at 37°. In mammalian cells many replicons are linearly arranged in the same DNA molecule, and their radioautographs after ³H-thymidine incorporation show that they are not replicated at the same time. Thus, the time required for DNA doubling depends not only on the rate of DNA chain elongation but also on the number of forks per DNA molecule in bacteria and the number of active replicons in animal cells.

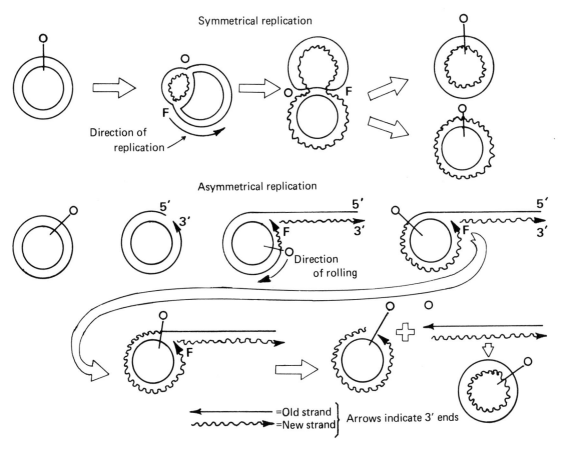

Fig. 10-20. Symmetrical and asymmetrical replication of cyclic DNA. In the symmetrical replication the newly synthesized strands are not covalently connected to the preexisting strands. The product is two identical cyclic molecules. In the asymmetrical replication, one of the old strands is broken at the origin and its 3′ end is continued by new synthesis, while an unconnected second new strand is synthesized as the complement of the 5′-ended old strand, which in the process is detached from its old complement. The elongation of the 3′-ended strand causes the circle to roll if the fork is held fixed. When the origin has completed a revolution, the extended strand is set free by another single-strand break at the origin, releasing a cyclic and a linear molecule. The latter may be subsequently cyclized.

ROLE OF MEMBRANE IN REPLICATION

Electron microscopy shows that the bacterial DNA is always attached to the cell membrane or its mesosome (Ch. 6). The attachment probably occurs at the growing point because after a short exposure to a radioactive precursor the DNA just synthesized is found preferentially associated with the membrane fraction when the cell is lysed. Furthermore, the membrane fraction also contains DNA polymerase III (see below), which appears to be directly involved in replication.

The cell membrane also determines the **segregation** of the daughter molecules to different cells at the end of replication. Electron microscopy has revealed that splitting of the mesosome by localized growth of the membrane causes the segregation. Some plasmids (independently replicating DNA molecules within the same cell; see Ch. 46) segregate together with the chromosome, suggesting that they are attached to the same mesosome. The cell membrane, therefore, has a dual role in DNA replication and segregation. In the latter capacity it resembles the spindle that segregates the chromosomes of higher cells at mitosis.

Fig. 10-21. Test of the symmetry of DNA replication in germinating *B. subtilis* spores carrying the three genetic markers *met, leu,* and *ade* at the indicated positions. After synchronization by thymine starvation, the density label 5-bromodeoxyuridine (BUDR) was added to the medium and then a pulse of radioactive ³H-BUDR was given, as indicated at the top. Incubation with unlabeled BUDR was continued until most of the chromosomes had two forks (as shown below). The diagram reproduces the expected constitution of the replicating molecules under either symmetrical or asymmetrical replication. For simpicity, the molecules are represented as linear. stretching from origin (O) to terminus (T) although they are actually cyclic. At the end of the experiment the extracted and fragmented DNA was banded at equilibrium in CsCl to separate DNA with BUDR in both strands (HH, of highest density) from DNA with BUDR in one strand (HL, of intermediate density) and from DNA without BUDR (LL, of lowest density). The DNA of the various bands was also tested for the three markers by transformation. The transformation tests showed that 86% of the chromosomes had two forks because the HH band contained 86% of the amount of *ade* marker present in the HL band; but neither fork had reached the *leu* marker, since both *leu* and *met* were all in LL DNA. Under the asymmetrical model at least 50% of the label in the chromosomes with two forks (circled) must be in HL DNA, whereas under the symmetrical model it should all be HH. The results showed that 86% of total label, therefore 100% of that in chromosomes with two forks, was HH, thus ruling out the asymmetrical model. [Data from Quinn, W. G., and Sueoka, N. *Proc Natl Acad Sci USA* 67:717 (1970).]

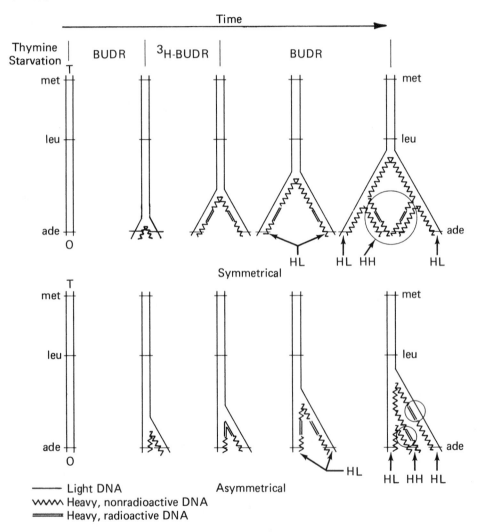

MECHANISMS OF REPLICATION

Initiation and chain growth must be considered separately.

Initiation requires protein as well as RNA synthesis, because it is blocked by inhibitors of either synthesis. The RNA appears to be transiently incorporated in the DNA chain: thus in incomplete replication of phage M13 (see Ch. 45, Phage with single-stranded DNA) in vitro, the new DNA is covalently linked to a ribonucleotide. In normal replication this RNA link must then be removed because none is present in the final product. The participation of RNA may be required in initiation because DNA synthesizing enzymes (see below) can only add nucleotides to a preexisting primer whereas RNA polymerase can initiate without primer (see Ch. 11, Transcription).

Hence the **initiation complex** appears to include the DNA template, new proteins, RNA polymerase, and ribonucleoside triphosphates to generate a temporary primer for DNA synthesis.

Important in **chain growth** are the nature of the synthesizing enzymes and the structural changes of the DNA.

The discovery by A. Kornberg of a DNA polymerase seemed to solve the first problem, but later studies showed that replication is carried out by other enzymes discovered more recently. As the properties of these enzymes were studied, it became apparent that they could not account by themselves for semiconservative replication, which requires the essentially simultaneous replication of both antiparallel strands. Since the enzymes synthesize DNA in $5' \rightarrow 3'$ direction, they can account for the replication of one strand only. After searching in vain for an enzyme able to synthesize DNA in the opposite direction, the solution came from the study of the structures present at the growing point, which is selectively labeled after a short pulse of ³H-thymidine.

The main contribution was Okazaki's finding that a large proportion of the label is in **short segments** immediately after synthesis. This suggested that replication proceeds through the synthesis and end-to-end joining of short segments. At least two enzymatic activities are therefore required: one to synthesize the segments, and ligase to join them.

By this procedure, either strand can be replicated by synthesizing DNA in constant $5' \rightarrow 3'$ direction since the direction of synthesis of the individual segments is independent of the over-all direction in which the strand grows.

Another difficult question was **how DNA unwinds** at replication: unwinding would require rotation of the molecules, raising formidable hydrodynamic problems because of their great length. The problem was made even more serious by the discovery of cyclic DNA, which, if intact, cannot unwind. These difficulties were solved by 1) the realization that replicating DNA contains single-stranded nicks which act as swivels, allowing local unwinding without rotation of the whole molecule, and 2) the later discovery of a "swivel" enzyme.

ENZYMES INVOLVED IN REPLICATION

Polymerases. Three DNA polymerases (I, II, and III) have been isolated from E. coli; enzymes with similar properties exist in other microorganisms and in animal cells. In vitro these enzymes catalyze the reaction:

$$n(dATP, dGTP, dTTP, dCTP) \xrightarrow[\text{DNA template}]{\text{Enzyme}}$$
deoxynucleoside triphosphates

$$(dAMP, dGMP, dTMP, dCMP)n + n(P.P)$$
polynucleotide   pyrophosphates

The isolation of **DNA polymerase I** by A. Kornberg made DNA replication accessible to biochemical study and was thus a most important development, though not the replicating enzyme.

DNA polymerase I is an unusually large single polypeptide (MW 109,000) with two distinct enzymatic activities in the same molecule (polymerase and 5' exonuclease) which can be separated by proteolysis. Polymerization requires a template primer, i.e., a double-stranded molecule, with a single-stranded extension of the 5' end of the strand (Fig. 10-22A). The unpaired terminal segment serves as **template;** the other strand, with a 3'-OH end, is the **primer,** to which the new DNA is attached by a 3'-5' phosphodiester bond. A suitable template-primer molecule can be prepared from helical DNA by partial digestion with exonuclease III (Fig. 10-22B); from single-stranded DNA (Fig.

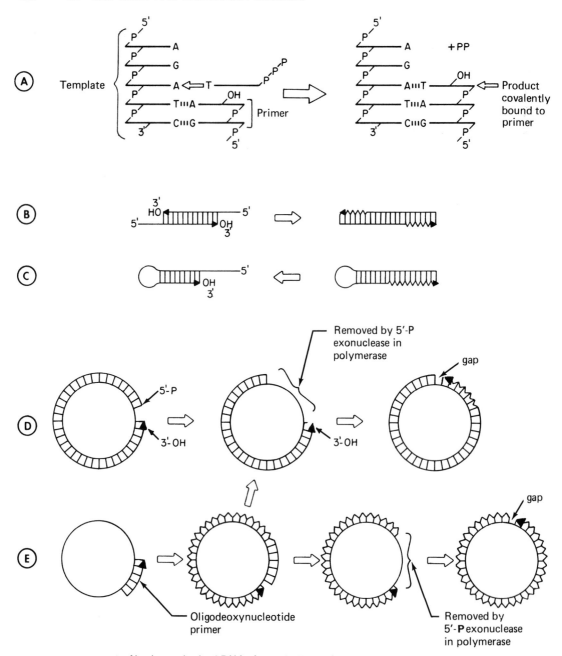

www▶ = Newly synthesized DNA. Arrow indicates 3' end

Fig. 10-22. DNA synthesis by DNA polymerase I. **A.** General structure of a suitable template-primer molecule and the reaction step. **B–E.** Suitable template-primer combinations. The residual gaps in **D** and **E** can be sealed by ligase to yield a continuous strand. [Modified from Kornberg, A. *Science* *163*:1410 (1969).]

10-22C) by hairpin folding; and from nicked DNA by taking advantage of the exonucleolytic activity of the polymerase toward 5′ P-terminated strands (Fig. 10-22D). Intact double-stranded DNA, such as viral cyclic DNA, does not support synthesis; neither does cyclic single-stranded DNA unless oligodeoxynucleotides are provided which pair with accidentally complementary sequences of the DNA, forming the required structure (Fig. 10-22E).

During in vitro synthesis, using double-stranded DNA as template, maximal activity is obtained by introducing many nicks in the template, but although the yield may be equivalent to many times the amount of template provided, the product is abnormal, in the form of multibranched molecules; and when transforming DNA is the template, the product lacks biological activity. However, synthesis on a single-stranded ring template with an imperfectly paired oligodeoxynucleotide can go around the circle until it reaches the 5′ P end of the oligonucleotide, which is exonucleolytically digested by the polymerase and replaced by new synthesis (Fig. 10-22E). By this method, synthesis of biologically active ØX174 DNA has been accomplished in a tour de force with DNA polymerase I and ligase. Hence the enzyme yields a **product that is faithfully complementary to the template.**

That DNA polymerase I is not the normal replicating enzyme became clear when Cairns isolated an *E. coli* mutant (pol A⁻) which lacks this enzyme but multiplies normally. This mutant has markedly increased sensitivity to ultraviolet irradiation, suggesting that the **main role of DNA polymerase I is repair synthesis** (see Ch. 11).

Studies with pol A cells have led to the identification and purification of two other **DNA polymerases (II and III)** which had gone undetected before because they are present in much smaller amounts than polymerase I.

T. Kornberg showed that **DNA polymerase III** probably carries out replicative synthesis because it is thermosensitive in *E. coli* strains in which DNA replication is made temperature-dependent by mutations. Whereas all mutations tested affected polymerase III, none affected the other two polymerases. Moreover, DNA replication in toluene-treated cells (to make them more permeable to external substances) is inhibited by sulfhydyl-blocking reagents, which inactivate polymerases II and III (but not I) and not by antibodies inhibiting polymerase I. The replicative role of polymerase III is also supported by other

observations: lack of 5-endonuclease activity, which although important in repair (see Ch. 11), is not required in replication; its association with the membrane fraction bound to the replicating DNA, and its similarity to a polymerase specified by phage T4 whose elimination by mutation blocks replication of the phage DNA.

DNA polymerase II and III require a template primer like that required by polymerase I, and the product is covalently linked to the primer.

Polynucleotide Ligases. These enzymes, which join polynucleotide chains, are present in all cells and some are specified by certain phages; their role in DNA replication is shown by mutations that abolish both ligase activity and DNA replication. *E. coli* ligase restores a phosphodiester bond break in single-stranded DNA, between a 3′-OH and a 5′-P end (such as produced by pancreatic DNase). The restoration is perfect, as shown by return of biological function in transforming DNA. The reaction requires nicotinamide-adenine dinucleotide (NAD) as an unusual adenyl donor; it proceeds in three steps:

1) NAD + enzyme → enzyme-AMP + NMN (nicotinamide mononucleotide)

2) enzyme-AMP + nicked DNA → enzyme + nicked DNA-AMP

3) nicked DNA-AMP → AMP + joined DNA

The ligase specified by phage T4 and those of animal cells require ATP instead of NAD as adenyl donor, but the reactions are otherwise identical.

Other Enzymes. Some **endonuclease** may participate in DNA replication because mutations in a gene of phage T7 specifying an endonuclease inhibit replication.

A specific **swivel enzyme** may be responsible for unwinding the helix during replication, since an activity converting cyclic helical DNA from the twisted to the relaxed state (see above) without permanent nicking is present both in bacteria and in the nuclei of animal cells. **DNA-unfolding proteins,** such as that specified by phage T4 (see above, DNA denaturation) and required for its replication may be needed to keep locally denatured (single-stranded) DNA from folding at random.

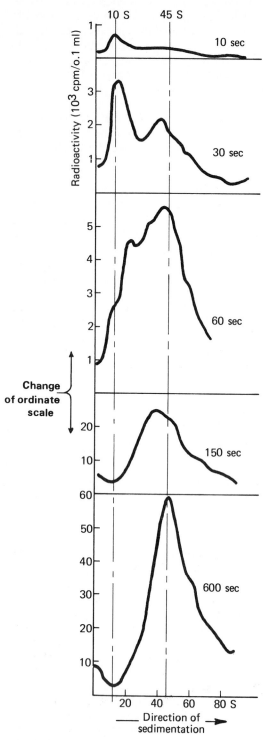

Fig. 10-23. Sedimentation in alkaline sucrose gradient of *E. coli* DNA labeled by progressively longer pulses of ³H-thymidine. The length of the pulse, at

THE REPLICATING DNA

The essential events of DNA replication take place at or near the **replicating fork,** where the unreplicated molecule and the two daughter molecules meet (see Fig. 10-16). The newly synthesized DNA is recognized because it is radioactively labeled after very brief exposure of cells (5 to 10 seconds at 37°) to ³H-thymidine. In many bacteria and phages, this DNA contains single-stranded segments **(Okazaki segments)** a few thousand nucleotides long, which are recognized by their sedimentation value (10S) in alkali gradients (Fig. 10-23). If the short labeling is followed by a chase without label, the label originally incorporated becomes covalently bound to the larger DNA. The Okazaki segments are evidently synthesized in the 5′ → 3′ direction, since the ³H-thymidine incorporated during a short pulse into the terminal portion of the fragments is rapidly released by exonuclease I, which attacks single-stranded DNA from the 3′-OH end. Incomplete Okazaki segments contain a short polyribonucleotide, apparently the primer, which is removed before synthesis is completed, probably by RNase H which hydrolyses RNA in RNA-DNA hybrids (see Table 10-2). The segments are connected by ligase since they accumulate unjoined in ligase-deficient mutants. These findings support the replication scheme of Figure 10-24.

Discontinuous synthesis of Okazaki segments occurs in the 5′-ended growing DNA chain; the 3′-ended chain may be synthesized continuously (i.e., without segments). In fact, segments formed in *B. subtilis* (whose DNA replicates unidirectionally) hybridize to only one of the two strands. In DNAs with bi-

20°, is indicated near each curve. After very short pulses (top) a large proportion of the label has a sedimentation constant of 10S (Okazaki segments). The amount of radioactivity in that fraction increases until 30 seconds, then remains essentially constant (considering the change of scale), as would be expected of an intermediate. In contrast, the 45S peak, which under the conditions of extraction contains the bulk of the DNA, continues to increase. The label, therefore, flows from the TTP pool into the Okazaki segments and from them into long DNA. [Modified from Okazaki, R., Okazaki, T., Sakabe, K., Sugimoto, K., Kainuma, R., Sugino, A., and Iwatsuki, N. *Cold Spring Harbor Symp Quant Biol 33*:129 (1968).]

Fig. 10-24. General characteristics of synthesis at a DNA fork. Synthesis always occurs in 5′ → 3′ direction. The 5′-ended growing strand (circled, left branch) is replicated by discontinuous synthesis, whereas the 3′-ended growing strand (circled, right branch) is elongated by continuous synthesis. Gaps are later closed by ligase.

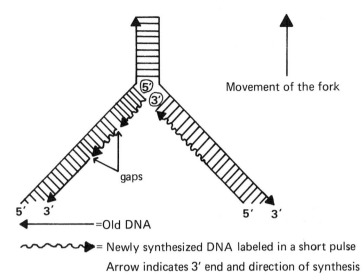

Movement of the fork

gaps

5′ 3′

5′ 3′

◄━━━━━━━ =Old DNA

〜〜〜▶ = Newly synthesized DNA labeled in a short pulse

Arrow indicates 3′ end and direction of synthesis

directional replication (such as in phage T4) the segments, as expected, hybridize with both strands.

The Problem of Unwinding. How nature has solved the unwinding problem is shown by the behavior of cyclic DNAs, which replicate while maintaining a cyclic configuration and therefore must unwind at swivel points.

Cyclic DNAs, such as those of papova viruses (Ch. 63) and of mitochondria of animal cells, have intact parental DNA strands during replication (Fig. 10-25); the unwinding is probably carried out by a swivel

Fig. 10-25. Replication of cyclic SV40 (see Ch. 63) DNA without permanent breaks in the parental strands, suggesting the participation of a "swivel" enzyme. The molecule seen in the electron micrograph (Kleinschmidt's technic) is about half replicated; its constitution is shown at right, where continuous lines represent the parental strands and dashed lines the progeny strands. The continuity of the parental strands is shown by the tight supercoiling of the unreplicated part: had a strand been nicked, the whole molecule would have relaxed. [From Sebring, E. D., Kelly, T. J., Jr., Thoren, M. M., and Salzman, N. P. *J Virol 8*:478 (1971).]

enzyme, which from time to time breaks one of the strands, allowing unwinding, and immediately reseals it. It is likely that a similar mechanism, in conjunction with single strand breaks, operates in the replication of the long nuclear DNA: swivels near the growing fork would allow unwinding of short DNA stretches without any mechanical problem; and as the growing fork progresses, new swivel points are probably created ahead.

SELECTED REFERENCES

Books and Review Articles

GILHAM, P. T. RNA sequence analysis. *Annu Rev Biochem 39*:227 (1970).

MCCARTHY, B. J. Arrangement of base sequences in deoxyribonucleic acid. *Bacteriol Rev 31*:215 (1967).

MCCARTHY, B. J., and CHURCH, R. B. The specificity of molecular hybridization reactions. *Annu Rev Biochem 39*:131 (1970).

SANGER, F., and BROWNLEE, G. G. A two-dimensional fractionation method for radioactive nucleotides. *Methods Enzymol. 12* (part A): 361 (1967).

SPIEGELMAN, S. S. Hybrid nucleic acids. *Sci Am 210*:48 (1964).

SZYBALSKI, W., and SZYBALSKI, E. H. Equilibrium density gradient centrifugation. *Progr Nucleic Acid Res Mol Biol 2*:311 (1971).

WATSON, J. D. *Molecular Biology of the Gene,* 2nd ed. Benjamin, New York, 1970.

Specific Articles

AGARWAL, K. L., BUCHI, H., CARUTHERS, M. H., GUPTA, N., KHORANA, H. G., KLEPPE, K., KIMAR, A., OHTSUKA, E., RAJBHANDARY, U. L., VAN DE SANDE, J. H., SGARAMELLA, V., WEBER, H., and YAMADA, T. Total synthesis of the gene for an alanine transfer ribonucleic acid from yeast. *Nature 227*:27 (1970).

BAUER, W., and VINOGRAD, J. Interaction of closed circular DNA with intercalative dyes. I. The superhelix density of SV40 DNA in the presence and absence of dyes; II. The free energy of superhelix formation in SV40 DNA. *J Mol Biol 33*:141 (1968); *47*:419 (1970).

BIRNSTIEL, M., SPIERS, J., PURDOM, I., JONES, K., and LOENING, U. E. Properties and composition of the isolated ribosomal DNA satellite of *Xenopus laevis. Nature 219*:454 (1968).

CALLAN, H. G. The organization of genetic units in chromosomes. *J Cell Sci 2*:1 (1967).

GEFTER, M. L., HIROTA, Y., KORNBERG, T., WECHSLER, J. A., and BARNOUX, C. Analysis of DNA polymerases II and III in mutants of *E. coli* thermosensitive for DNA synthesis. *Proc Nat Acad Sci USA 68:*3150 (1971).

LARSEN, C. J., LEBOWITZ, P., WEISSMAN, S. M., and DUBUY, B. Studies of the primary structure of low molecular weight ribonucleic acid other than tRNA. *Cold Spring Harbor Symp Quant Biol 35:*35 (1970).

MESELSON, M., and STAHL, F. W. The replication of DNA in *Escherichia coli. Proc Natl Acad Sci USA 44*:671 (1958).

MOSES, R. E., and RICHARDSON, C. C. Replication and repair of DNA in cells of *Escherichia coli* treated with toluene. *Proc Natl Acad Sci USA 67*:674 (1970).

OLIVEIRA, B. M., HALL, Z. W., ANRAKU, Y., CHIEN, J. R., and LEHMAN, I. R. On the mechanism of the polynucleotide joining reaction. *Cold Spring Harbor Symp Quant Biol 33:*27 (1968).

SCHILDKRAUT, C. L., MARMUR, J., and DOTY, P. The formation of the hybrid DNA molecules and their use in studies of DNA homologies. *J Mol Biol 3*:595 (1961).

SOUTHERN, E. M. Base sequence and evolution of guinea pig α-satellite DNA. *Nature 227*:794 (1970).

WETMUR, J. G., and DAVIDSON, N. Kinetics of renaturation of DNA. *J Mol Biol 31*:349 (1968).

WICKNER, W., BRUTLAG, D., SCHEKMAN, R., and KORNBERG, A. RNA synthesis initiates in vitro conversion of Ml3 DNA to its replicative form. *Proc Natl Acad Sci USA 69*:965 (1972).

WOLF, B., NEWMAN, A., and GLASER, D. A. On the origin and direction of replication of the *Escherichia coli* K12 chromosome. *J Mol Biol 32:*611 (1968).

chapter 11

ORGANIZATION, ALTERATION, AND EXPRESSION OF THE GENETIC INFORMATION

ORGANIZATION OF GENETIC INFORMATION 233

UNITS OF INFORMATION 233
  Fine-Structure Genetics 233
  Codons 235

RADIATION DAMAGES AND THEIR REPAIR 237

RADIATION DAMAGES 237
  Damage by Ultraviolet Light 237
  Damage by Ionizing Radiation and by Radioactive Decay 237

REPAIR OF DAMAGE TO DNA 238

Photoreactivation 238
Excision Repair 238
Postreplication (Recombinational) Repair 240
Factors Determining Sensitivity to Radiation 240
Repair of UV Damage in Higher Organisms 240
Error Correction 241

GENETIC RECOMBINATION 241

MOLECULAR MECHANISMS OF RECOMBINATION 241
  Crossing Over 242
  Gene Conversion 246
GENETIC MAPPING 246

MUTATIONS 252

GENERAL PROPERTIES 252
  Consequences of Mutations for Protein Structure 253
INDUCTION OF MUTATIONS 253
  Action of Chemical Mutagens 254
  Point Mutations 255
INDUCTION OF MUTATIONS BY RADIATION 259
  Directed Mutagenesis 259
  Mutagenic vs. Lethal Action of Mutagenic Agents 259
CHEMICAL MUTAGENESIS IN HIGHER ORGANISMS 259

SPONTANEOUS MUTATIONS 259

    Mutations Affecting Mutability 260

    Site Influence on the Frequency of Spontaneous and
        Induced Mutations 261

    Evolutionary Consequences of Mutations Affecting
        Mutation Rates 261

TRANSCRIPTION OF DNA 262

    RNA Polymerase and Sigma Factor 263

    Initiation 263

    Chain Elongation 265

    Termination 265

    Antibiotics Affecting Transcription 266

   REVERSE TRANSCRIPTION 267

APPENDIX 268

    QUANTITATIVE ASPECTS OF KILLING BY IRRADIATION 268

ORGANIZATION OF GENETIC INFORMATION

UNITS OF INFORMATION (FIG. 11-1)

Early in the development of genetics, it was recognized that mutations affecting the same hereditary trait were usually located in the same position (**locus**) on a chromosome, distinct from loci whose mutations affected other traits. The positions of different loci (genetic linkage) were determined by the frequency of recombination between them in a genetic cross. These positions were confirmed most directly by cytological observations in insects and later by interrupted mating experiments in bacteria (Ch. 9). Genetic loci, thus recognized by their mutations and located by recombination tests, were identified as the **units of genetic function** and were called **genes.** A molecular basis for gene function was suggested by the generalization of Beadle and Tatum: "one gene–one enzyme" (see Ch. 8).

A specific form of a gene, as noted in Chapter 8, is called an **allele: wild-type** allele refers to the prevalent nucleotide sequence, whereas **mutant** allele refers to a sequence which has been modified by mutation. The phenotypic recognition of mutant alleles is fundamental to all genetic studies, but the definition of a given allele as wild type is somewhat arbitrary: alleles are not fundamentally different from each other because they have all evolved by successive mutations; and more and more genes are being recognized as **polymorphic,** i.e., having more than one allele widely distributed in nature.

FINE-STRUCTURE GENETICS

Genes were at first treated as formal points connected by a line—the chromosome. Later the gene developed dimensions, when recombination was observed not only between genes but also, in rare instances, between mutational sites belonging to the same gene. However, because of the low resolving power of recombinational studies in higher organisms, such intragenic recombination was detected with only a few loci. Microbial genetics had a tremendous impact on the theory of the gene by making it possible, through the use of huge populations and strongly selective techniques, to detect recombinations at far lower frequencies, by several decimal orders of magnitude. This striking increase in resolving power revealed that intragenic recombination is a general phenomenon. Moreover, Benzer showed, with the r_{II} locus of bacteriophage T4, that **within a gene** mutations can occur at **many different sites,** distinguishable by recombination. **Fine-structure** genetic analysis was thus begun.

Complementation. Fine-structure studies in viruses and bacteria showed that the chromosome can be mapped as an extremely large number of mutable sites distinguishable by crossing over. Adjacent genes form a quasi-continuum, in which a gene is defined as the set of those sites that all affect the same function (Fig. 11-1). The function, in turn, is defined phenotypically, e.g., the synthesis of a certain enzyme or a regulatory protein or an untranslated RNA. A more general way for distinguishing units of genetic function is the **complementation test,** which is applicable even where the gene product is unknown. In this test, two mutant genomes of similar phenotype are brought into the same cell. When the two mutations eliminate different gene products, the wild-type allele of each

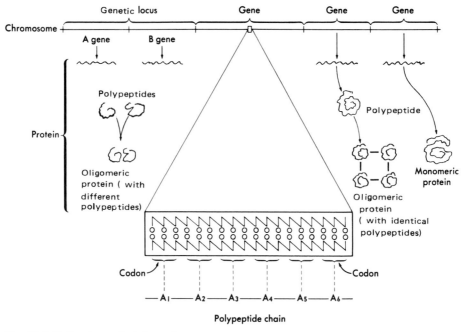

Fig. 11-1. Units of genetic information.

gene still functions, and so the heterozygous cell is prototrophic **(intergenic complementation).** In contrast, mutants that are altered at different sites within the same gene generally do not complement each other. (Exceptions will be discussed in Chapter 12 under Intragenic complementation.)

The usefulness of the complementation test is shown by Benzer's analysis of the phage r_{II} locus. This locus specifies a single phage property, but complementation tests separate the r_{II} mutants into two clearly distinct classes, A and B (Fig. 11-2), among which any mutant complements members of the other class but not of its own. Moreover, the two sets of mutations occupy two adjacent, nonoverlapping parts of the chromosome. It was concluded that this locus is made up of two genes (which for some time were called cistrons).* This result was obtained without

* The term was derived from the ability of two mutations in different units of function to complement each other (by virtue of the presence of their wild-type alleles) when on different chromosomes **(trans),** but not when on the same chromosome **(cis).** This term has persisted mainly in the definition of mRNAs (see Ch. 13), which are called mono- or polycistronic depending on whether they transcribe a single or several genes.

knowing the proteins specified by the two genes. However, in later studies of the tryptophan synthetase locus of *E. coli* by Yanofski two adjacent genes revealed by complementation could each be shown to form a distinct polypeptide. The earlier generalization about gene function then became, more precisely, "one gene–one polypeptide chain."

This definition connects the informational units in the chromosome to the informational units in the protein and deserves special attention on several points. 1) An enzyme is not an informational unit because it may contain polypeptide chains specified by distinct genes. 2) The mRNA intermediates between genes and polypeptides may transcribe several adjacent genes without affecting the gene-polypeptide relation. 3) Some genes specify stable RNA (rRNA or tRNA) rather than polypeptides as their product. 4) With certain animal viruses (see Ch. 48), the polypeptide chain is secondarily cleaved into fragments with different functions; then the part of the genome corresponding to each fragment is a gene.

Locus and Gene. The term **locus** is used to define a chromosome segment that contains the sites of a cluster of mutations affecting a

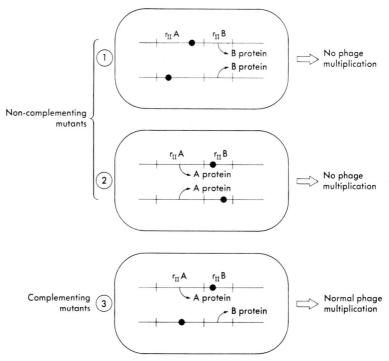

Fig. 11-2. Complementation test showing that the r_{II} locus of bacteriophage T4 contains two genes. *E. coli* K_{12} (λ) cells, which will not support multiplication of an r_{II} mutant, were infected simultaneously by two r_{II} mutants. The mutations were localized, by recombination tests, both in the A segment **(1)**, both in the B segment **(2)**, or one in each segment **(3)**. In **1** only B protein is presumably made, and in **2** only A protein, since in neither is there viral multiplication (except for a small proportion of cells in which wild-type phage has been produced by genetic recombination). In **3**, however, viral multiplication takes place in all cells, so presumably both A and B proteins are made. The A and B segments, therefore, are two separate genes.

given trait, but that does not (or has not yet proved to) consist of a single gene. We shall therefore speak of an r_{II} or a tryptophan synthetase locus, each constituted of two genes.

Loci concerned with related functions are often clustered, and the cluster will appear to be a single locus until its phenotypic effects can be analyzed with sufficient refinement. For example, rough mutants of *Salmonella,* which have lost the ability to make the O polysaccharide of the normal cell surface, all map in a large O locus, but it has recently been shown that this locus specifies a dozen enzymes, which are all concerned with steps in the synthesis of the O polysaccharide (Ch. 6). The O region of the chromosome is therefore now known to consist of a number of loci. The loci concerned with the surface antigens of mammalian cells (Ch. 21) undoubtedly have a similar complexity.

CODONS

A gene contains numerous smaller units, the codons, each made up of three bases; the codons follow each other without any intercalated base as comma. The functional separation between codons **(reading frame)** is thus established at the beginning of the translation and is maintained by reading sequentially groups of exactly three bases.

These properties of the code (triplet, commaless) were discovered through the ingenious use of acridine-induced mutations in bac-

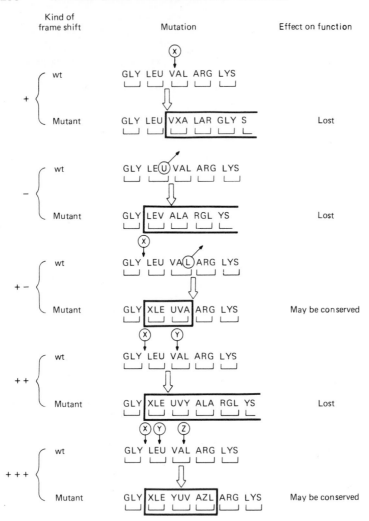

Kind of frame shift | Mutation | Effect on function

Fig. 11-3. Effect of frame shifts on a sequence of bases which are read three at a time, from the left. The sequences are arbitrary, but in order to emphasize the consequences of frame shifts the three letters in each codon (underlined) are the first letters of amino acids (glycine, leucine, valine, arginine, lysine). Plus frame shift means the insertion, and minus the deletion, of a nucleotide. The scheme shows that two mutations of opposite signs can correct each other (except for the region between them), two of the same sign cannot, but three of the same sign can. The heavy lines indicate the abnormal region. If these stretches are short, the function of the gene may be conserved.

teriophages (Fig. 11-3), which are mostly either deletions or additions of one nucleotide (see below, Mutations). The resulting shift of the reading frame causes formation of jumbled protein distal to the shift (see Fig. 12-8), but a second acridine mutation a short distance away sometimes corrects the frame shift and restores the gene function (which can sometimes tolerate a short jumbled segment between the two mutations). Pairs of mutations that correct each other are given opposite signs (+ and −, where + means addition of a nucleotide). Two mutations of the same sign do not correct each other, but **three** can, because after either removal or addition of three nucleotides, the original reading frame is restored. The triplet nature of the code has also been clearly demonstrated by biochemical studies in vitro (see Ch. 12).

The information contained in the genome has been likened to a book containing paragraphs (loci), sentences (genes), words (codons), and letters (bases). Whereas the gene is the **unit of genetic function,** the codon is the **unit of translation;** the base is the site of point mutations and thus the **unit of mutation.**

RADIATION DAMAGES AND THEIR REPAIR

Ultraviolet (UV) radiation in the absorption spectrum of nucleic acids (below 3000 A) and ionizing radiations (X-rays, γ-rays) cause both mutagenic and lethal changes in nucleic acids. The study of these effects, carried out at first to understand how the photons act on the nucleic acids, led to the discovery of elaborate mechanisms for repairing the damages. These processes have turned out to be related to the basic processes of error correction and recombination, which help explain the remarkable stability of the gene throughout the lifetime of a cell. Knowledge of the production, consequences, and repair of the damages is of practical importance, especially since radiations, from industrial and military sources, become more and more abundant.

RADIATION DAMAGES

DAMAGE BY ULTRAVIOLET LIGHT

The main consequence of UV action is the formation of a **dimer** from adjacent pyrimidines on the same strand, which become connected by a cyclobutane ring after activation of the 5:6 double bond (Fig. 11-4). Less frequently this bond is hydrated to form a dihydropyrimidine, or forms an adduct with another pyrimidine (connected by a single bond), or cross-links are formed between paired strands. The photoproducts are chemically stable and are readily recovered by chromatography of hydrolysates of the altered nucleic acid.

These changes in DNA are all lethal if not repaired. Best understood is the role of dimers, which block both replication and transcription. Replication is only transiently halted and soon proceeds beyond a dimer, leaving a gap in the newly synthesized strand. The effect on transcription and therefore on function of individual genes can be determined by measuring the efficiency of complementation of UV-irradiated genes. For example, in *E. coli* K_{12} (λ), an r_{II} mutant of phage T4 does not multiply, but it does when accompanied, in the same cell, by a phage genome carrying an intact $r_{II}{}^+$ (wild-type)

gene (even though other parts of the second genome may be damaged) (see Ch. 45). This system thus provides a test for the integrity of the r_{II} gene. As Figure 11-5 shows, and as would be expected, the **functional survival of the gene** is much more UV-resistant than is phage viability, which requires integrity of the whole genome.

DAMAGE BY IONIZING RADIATIONS AND BY RADIOACTIVE DECAY

An X-ray photon, which has about 10^4 times as much energy as a UV photon, causes ejection of an electron from the target atom, and that electron causes several dozen ionizations in molecules along its trajectory **(ion cluster)**. In a cell the ionizations occur principally in water and result in the formation of highly reactive, short-lived radicals and protons. These products cause numerous alterations in the bases and single-stranded breaks, i.e., breakage of phosphodiester bonds, which are detected by decreased sedimentation rates in **alkaline** sucrose gradients. The clustering of the ions sometimes causes breaks in both strands close enough to cause a **double-strand**

Fig. 11-4. Thymidine dimer resulting from UV irradiation of DNA. Thymine-cytosine and cytosine-cytosine dimers are also formed.

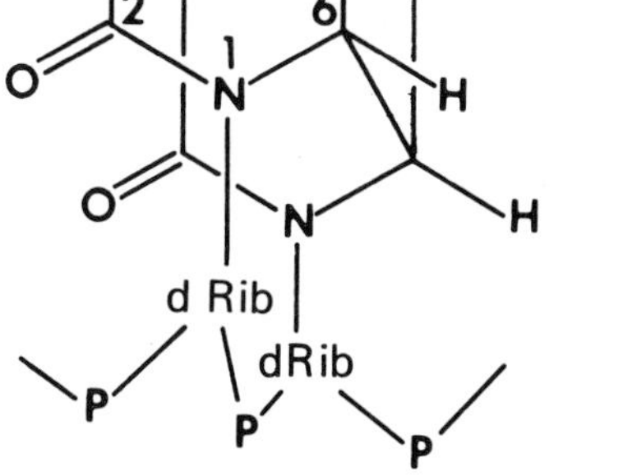

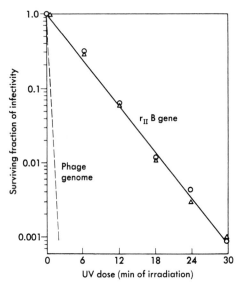

Fig. 11-5. Kinetics of UV damage to a gene and to a genome. Functional survival of the $r_{II}B$ gene compared with the entire genome of T4 bacteriophage particles irradiated with UV light. Survival of the genome was measured from the fraction of K_{12} (λ) cells yielding phage after single infection by irradiated T4r_{II}B+ particles (dashed line). Survival of the $r_{II}B$ gene was measured by simultaneously infecting the same cells with unirradiated $r_{II}B$ mutant phage, which cannot multiply in this host unless complemented with an undamaged $r_{II}B$+ allele. [Data from Krieg, D. *Virology* 8:80 (1959).]

break (detected by reduced sedimentation rate in **neutral** sucrose gradients). Whereas most single-strand breaks are repaired, double-strand breaks are lethal.

The **decay of radioactive atoms** of various kinds incorporated in DNA has different effects, depending on the location of the atom. ^{32}P decay to ^{32}S breaks the phosphodiester bond. Morover, decay releases an electron of such energy that the whole atom **recoils** (as in the firing of a gun); this recoil may cause a break in the complementary strand, producing a double-strand break. As with X-rays, breaks in one strand of a helix, but not in both, may be repaired.

REPAIR OF DAMAGE TO DNA

The effects of UV irradiation are alleviated by three mechanisms of universal occurrence:

photoreactivation, excision repair, and post-replication repair.

PHOTOREACTIVATION

This phenomenon was discovered through the curious observation that bacteria exposed to UV irradiation yield highly variable viability counts, depending on how long the inoculated plates remain on the laboratory bench before being placed in a dark incubator. The cause is an enzyme that combines specifically with pyrimidine dimers: this enzyme remains inactive in the dark, on irradiation with light of the long UV or short visible region it cleaves the dimer, restoring the original pyrimidine residues. Photoreactivation occurs with UV-damaged RNA as well as DNA. UV-irradiated virus particles (which do not contain the photoreactivating enzyme) show photoreactivation if they are allowed to infect cells.

Photoreactivation increases, sometimes enormously, the viable fraction of UV-treated bacteria: after maximal photoreactivation other, less frequent damages still remain. The net effect is as though the UV dose had been strongly reduced (but the pattern of damage is different).

EXCISION REPAIR

This type of repair, also known as **dark reactivation,** is observed if a UV-irradiated suspension of bacteria is stored for a few hours in the cold, or in an inadequate medium, before being allowed to resume growth; the fraction of colony-forming cells increases appreciably during this preincubation. Studies on the state of the DNA after irradiation, and of the effects of mutations in repair enzymes, show that repair occurs in two main steps (Fig. 11-6). First, a single-stranded oligonucleotide segment containing the pyrimidine dimer is degraded and its components appear in the acid-soluble intracellular fraction and in the medium. The gap is then reconstructed. In the degradation phase, a **dimer-specific endonuclease** hydrolyzes a phosphodiester bond adjacent to the dimer, and then a **dimer-specific exonuclease** excises the dimer and a number of neighboring nucleotides. Both enzymes have been purified from *Micrococcus luteus.* In the reconstruc-

Fig. 11-6. Steps in excision repair.

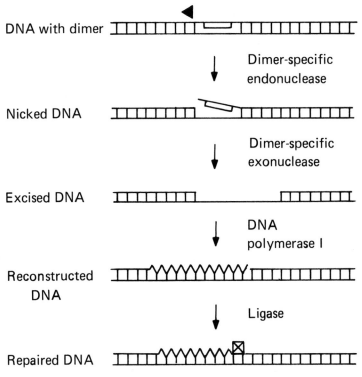

tion, a DNA polymerase (probably polymerase I) rebuilds the degraded segment as a complement of the intact strand, and then ligase reestablishes continuity of the chain. Enzymatic excision of dimers is hindered by several chemicals, including caffeine, whose mutagenic effect may thereby be explained.

Several mutations in *E. coli* affect the excision repair process. For instance, mutations in genes *uvr A, B,* and *C* prevent dimer ex-

Fig. 11-7. Repair of pyrimidine dimers by recombination. When DNA-containing dimers (PD) replicate, the new strands (wavy lines) have corresponding gaps. Multiple crossovers (Co) between the two daughter helices restore an intact molecule. Evidence for this mechanism was obtained with *E. coli* cells by growing in heavy isotopes (^{13}C, ^{15}N), which were irradiated and transferred to light isotopes (^{12}C, ^{14}N) in the presence of ^{3}H-thymidine. Shortly after irradiation the label was in short strands of light density; after repair it was in long strands of density intermediate between light and heavy. Sonication to break the strands produced short pieces, some of light, some of heavy density. [From Rupp, W. D., *et al. J Mol Biol* 61:25 (1971).]

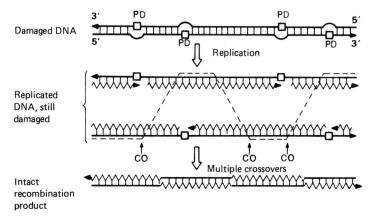

cision. Pol A⁻ mutants, lacking polymerase I activity, and ligase-deficient mutants cannot carry out reconstruction after excision; hence they show extensive DNA degradation after UV irradiation, but not if the cells fail to excise (i.e., are also *uvr*⁻).

POSTREPLICATION (RECOMBINATIONAL) REPAIR

In mutant bacteria incapable of excision repair or photoreactivation the presence of even hundreds of pyrimidine dimers in the DNA nevertheless does not prevent its replication. Shortly after synthesis the new DNA strands in the irradiated bacteria are in relatively short pieces (but longer than Okazaki segments), which sediment slowly in an alkali gradient. Upon incubation of the cells for an hour or so, the pieces become longer until they reach the normal length. The new DNA strands are at first fragmentary because DNA

Fig. 11-8. UV inactivation of the single-stranded and double-stranded DNA of bacteriophage φX174 (see Ch. 43). Each DNA in solution was exposed to increasing doses of UV light and then was assayed for infectivity in the dark on protoplasts of two *E. coli* strains, one capable and the other incapable of carrying out repair. The proportion of the residual titer is plotted versus the dose of irradiation. [Data from Yarus, M., and Sinsheimer, R. L. *J Mol Biol* 8:614 (1964).]

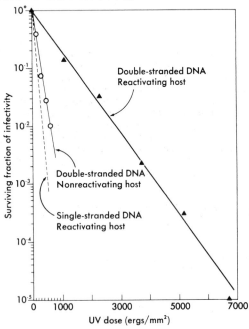

replication of each strand stops at a dimer and resumes beyond it, leaving a single-stranded gap of a few hundred nucleotides (perhaps a whole Okazaki segment is missing). The subsequent disappearance of the gaps is attributed to crossovers between the two daughter molecules produced in replication, whose gaps do not usually coincide (Fig. 11-7), for postreplication repair is eliminated by *rec A* mutations which abolish crossing over.

FACTORS DETERMINING SENSITIVITY TO RADIATION

Sensitivity to the lethal effect of radiation is measured by the slope of the survival curve (see below, Appendix). Organisms with comparable amounts of DNA and similar base ratios may display different sensitivities if they differ in repair efficiency. Thus photoreactivation eliminates over 90% of the effect of UV irradiation, but is absent in the dark or in *phr* mutants. Another factor is strandedness: single-stranded DNA cannot be repaired by dimer excision or by recombination, and is very sensitive to UV in the dark, whereas double-stranded DNA is highly resistant, with as many as 2500 dimers accumulated per lethal hit (Fig. 11-8). UV sensitivity in the dark is also increased by mutations, noted above, that prevent dimer excision or subsequent reconstruction. However, these mutants can still repair dimers by postreplication repair. The most sensitive are *uvr*⁻ (*rec A*⁻) double mutants, which lack both dark repair mechanisms. For them essentially every dimer is lethal in the dark.

Sensitivity may also be increased manyfold in the wild type by the incorporation of **5-bromouracil** instead of thymine. The presence of this analog in the DNA does not interfere with normal replication (see Mutagenesis, below) but it does impair response to the enzymes of photoreactivation and of dark repair. 5-bromouracil-containing DNA undergoes single-strand breaks if exposed to UV light of long, near visible, wave length.

REPAIR OF UV DAMAGE IN HIGHER ORGANISMS

The significance of repair of UV damage in man is dramatically demonstrated by the finding that patients with a fatal hereditary disease **xeroderma pigmentosum** lack the

Fig. 11-9. Removal of a cross-link produced by a bifunctional alkyl residue (A) between two guanine residues (G) in opposite strands. The process consists of a sequence of degradation and resynthesis, as in repair of UV damage. Wavy lines = new synthesis.

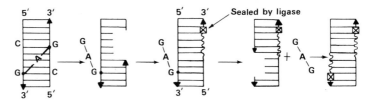

dimer-specific endonuclease. As a result of the failure to repair UV damage, the skin of these individuals develops severe burns after minor exposures to sunlight, and frequent skin cancers usually lead to early death.

ERROR CORRECTION

Excision applies not only to various kinds of radiation damage, but also to various kinds of chemical damage and to replication errors. For example, it can remove cross-links between two guanines in the opposite strands produced by **bifunctional alkylating agents** (e.g., nitrogen mustard, mitomycin C). The cross-link is recognizable because it prevents the complete separation of the strands during heat denaturation and thus leads to immediate renaturation on cooling. Repair of this damage probably proceeds as in Figure 11-9. In animal cells the excision mechanism repairs damages produced by certain carcinogenic agents (such as N-acetyl-2-acetylaminofluorene), since repair fails in xeroderma pigmentosum cells.

These findings indicate that helical DNA segments distorted for a variety of reasons can be equally repaired. Especially important are distortions due to **heterozygosity in hybrid DNA,** whose correction produces **gene conversion** (see below).

GENETIC RECOMBINATION

We have seen in Chapter 9 that genetic recombination is recognized by analysis of the distribution of markers (identifiable alleles) among the progeny derived from a cross between two parental genomes. We shall examine here the molecular events underlying this redistribution.

Early genetic studies with eukaryotes indicated that the pairing of homologous chromosomes **(synapsis)** must be extremely precise, since a cross between two different mutants regularly yielded the wild type and the reciprocal recombinant (double mutant) with no other genetic changes in either. Studies in fine-structure genetics localized the site of recombination with the greatest possible precision, between **two adjacent nucleotides.** Thus, if one of the glycines in tryptophan synthetase is replaced by glutamic acid, arginine, or valine, crosses between pairs of these mutants yield a very low frequency of recombinants in which the glycine has been restored (Fig. 11-10). These results would have appeared extraordinary before the genetic code was discovered, but they are now easily understood as the consequence of rare recombination **within a codon.**

MOLECULAR MECHANISMS OF RECOMBINATION

Phages and fungi have been particularly useful in elucidating molecular aspects of recombination. Phages offer a unique opportunity for correlating the properties of a DNA molecule with its genetic behavior because the DNA is small enough to be isolated without breaks. It is very difficult, however, to study a **single** act of phage recombination (see Ch. 45). In fungi, in contrast, the products of a single recombination can be regularly examined through tetrad analysis (see Ch. 43), but the DNA molecules cannot be studied physically or chemically. Results obtained with these two systems have finally been merged into a model that is generally consistent, although not complete in many details.

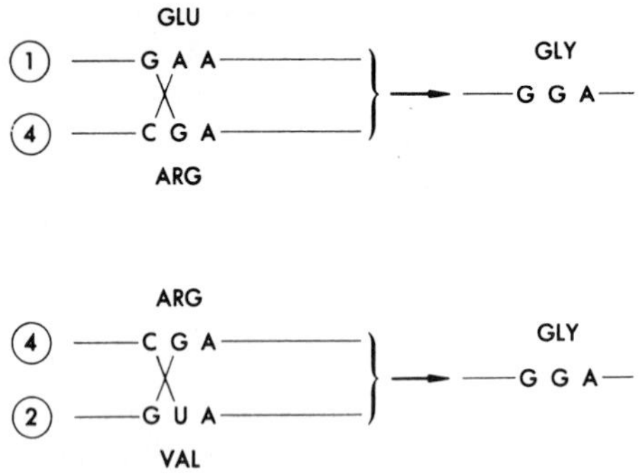

Fig. 11-10. Recombination within a codon. The crosses involved three mutants of the A protein of tryptophan synthetase (identified by numbers on the left) with different amino acids replacing the glycine. The nucleotide composition of the codons is inferred from the amino acid changes in the mutants and the restoration of glycine in the recombinants (see Ch. 12, Genetic code). As would be expected from the known codons, the cross of mutants 1 and 2 failed to restore glycine. Recombination is indicated as a reciprocal crossover but may have occurred by gene conversion (see below). [Data from Yanofsky, C. *Cold Spring Harbor Symp Quant Biol* 28:581 (1963).]

CROSSING OVER

This process is studied by the recombination of **distant markers,** such as those contained in nonadjacent genes, and occurs by **breakage and reunion of DNA molecules.** The evidence derives from a classic experiment of Meselson and Weigle with isotopically and genetically labeled bacteriophage λ (Fig. 11-11). In this experiment the infecting particles of the two **parental** strains, carrying different genetic markers, had DNA labeled with the heavy isotopes ^{13}C and ^{15}N, while the **new** DNA formed during replication of the phage was light, containing only the normal isotopes ^{12}C and ^{14}N. The progeny phage particles had different densities, depending on the DNA they contained, and accordingly formed different bands in density gradient equilibrium centrifugation in CsCl (bands 1 and 3 in Fig. 11-11). Thus particles with nonreplicated DNA were the heaviest and formed a separate band (band 1), from which they could be isolated. Some of these particles were found genetically to be recombinant and therefore must have arisen by union of fragments of unreplicated DNA molecules present in the infecting particles. These special recombi-

nants had another important property—they were **slightly lighter** than the original parental particles (band 2), indicating that a small proportion of the heavy parental DNA had been replaced by new synthesis. This finding suggests that limited DNA synthesis of repair type is required to seal the fragments together.

This conclusion appears to be valid for all organisms, because in eukaryotes recombination occurs at the pachytene stage of meiosis, **after** the DNA has replicated, and it is also accompanied by a small amount of repair synthesis.

Molecular Mechanisms. Studies with bacteriophages led to the following picture of crossing over (Fig. 11-12). A combination of endo- and exonucleases exposes the base sequences in one strand. Since many DNA molecules undergo such changes at the same time in the same cell, some single-stranded segments can pair with a complementary segment on a different molecule. The resulting **joint molecules** contain unpaired tails, which are degraded, and single-stranded segments, which are soon reconstructed by DNA polymerase (probably polymerase I). Finally, the nicks are closed by ligase, yielding **integrated molecules.**

There is clear experimental evidence for most of these steps. Thus, powerful nucleases appear in cells infected by phage T4, which has an extremely active recombination system (see Ch. 45). Joint molecules accumulate during replication of phage mutants with an inactive DNA polymerase or ligase gene; they can be recognized because alkali denaturation yields single strands shorter than the regular strands.

Crossing over in bacteriophage is **nonreciprocal** since only one recombinant is formed; but the model is readily adaptable to **reciprocal** crossing over (i.e., producing two reciprocal recombinants), as observed in bacteria and in eukaryotes (Fig. 11-13). The molecular mechanisms appear to be similar.

In bacteria crossing over probably involves an ATP-dependent exonuclease active on double-stranded DNA; this enzyme is synthesized under the control of two genes (*rec B* and *rec C,* specifying different subunits) whose mutations strongly

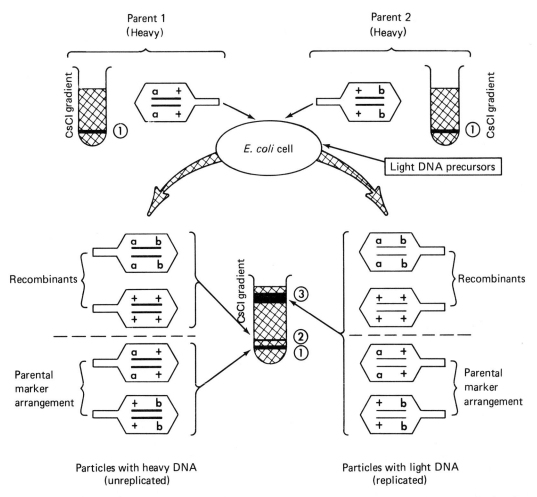

Fig. 11-11. Scheme of the Meselson and Weigle experiment showing recombinant formation by breakage and reunion of molecules of phage λ DNA. The differences between particle densities in CsCl density gradients have been exaggerated for clarity. [From Meselson, M., and Weigle, J. *Proc Natl Acad Sci USA* 47:857 (1961).]

reduce recombination frequency. Another gene, *rec A*, appears to control the action of the exonuclease; its mutations completely abolish recombination. In eukaryotes the activities of endonuclease and ligase are increased in the pachytene stage of meiosis, when crossing over occurs.

Several special forms of recombination occur by different mechanisms. 1) In the rare integration of episomes in aberrant positions (see Ch. 46), and in the formation of deletions in bacteria (see below), exchanges occur at low rates between non-homologous segments, perhaps using accidental microhomology. This **nonhomologous recombination** is not affected by *rec* mutations and therefore involves different enzymes. 2) The integration of prophage (Ch. 46) is produced by phage enzymes

that recognize specific DNA sites rather than homology **(site-specific recombination)**; this process is also not affected by *rec* mutations. 3) In eukaryotic cells, enzymes recognizing specific sites in DNA may cause alignment and pairing of chromosomes before recombination: since certain mutations lower the frequency of recombination of small fractions of the genome, each cell may have many such enzymes recognizing different sites.

Hybrid DNA in Recombination. According to the above model, a recombinant molecule should have a hybrid DNA segment, with one strand from each of the two parental molecules (Figs. 11-12 and 11-13). Indeed,

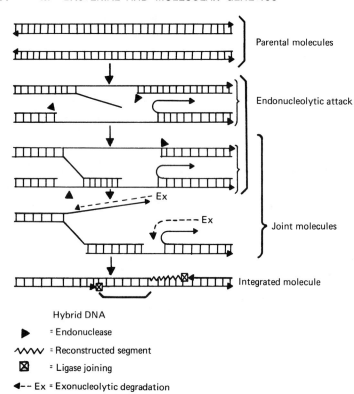

Fig. 11-12. Model of non-reciprocal crossing over in phage without DNA replication, only repair synthesis. The hybrid DNA corresponding to the region of initial pairing of the two strands contains information (base sequences) derived from different parents.

Parental molecules

Endonucleolytic attack

Joint molecules

Integrated molecule

Hybrid DNA

▶   = Endonuclease

ᨓ   = Reconstructed segment

⊠   = Ligase joining

◀-- Ex = Exonucleolytic degradation

Fig. 11-13. Model of reciprocal crossing over. The initial stage is similar to that of the previous model (Fig. 11-12), except that both DNA molecules participate similarly. This step leads to a symmetrical exchange of one strand (I). A second symmetrical exchange between the other two strands (II) produces a crossing over with two recombinant molecules. These are symmetrical except at the point of exchange, where both contain hybrid DNA (HD). The boxed diagram illustrates the case in which the second exchange occurs between the strands involved in the first exchange (III). In this case there is no crossing over, but the two molecules contain hybrid DNA. [From Holliday, R. *Genet Res* (*Camb*) 5:282 (1964).]

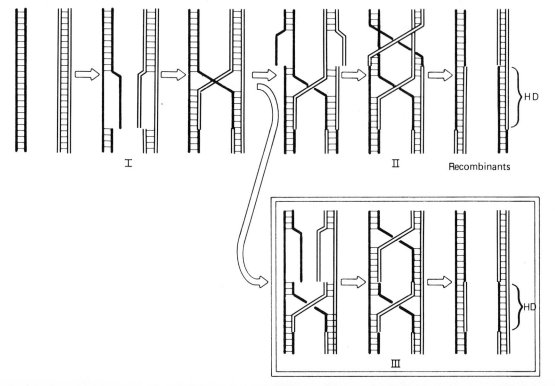

I

II

Recombinants

HD

III

HD

in cells mixedly infected by a phage mutant and wild type, 1% of the progeny phage are **heterozygous** for the mutation: i.e., on subsequent replication, half of the progeny of such a particle are mutant and half are wild type. There is a profound difference, however, from diploid heterozygosity as found in eukaryotic cells (Fig. 11-14): phage DNA heterozygosity is **intramolecular** and is therefore lost upon replication, whereas heterozygosity of diploid cells is **intermolecular** and therefore is maintained indefinitely, by mitosis, through cell duplication. In crossing over the hybrid segment probably occurs at the junction of

the DNA fragments, because in recombination of phage DNAs differing at three not closely linked markers (e.g., ABC, abc), molecules heterozygous at the middle marker (B/b) are usually recombinant for the outside markers (Ac or aC). The probability of heterozygosity is just about the same for any marker. From this probability it can be estimated that the hybrid segment is short (about 1% of the length of the DNA—about 2000 base pairs—in phage T4); hence molecules can be heterozygous for several markers only when these are closely linked.

Formation of the hybrid DNA in crossing

Fig. 11-14. Phage and cell heterozygosity.

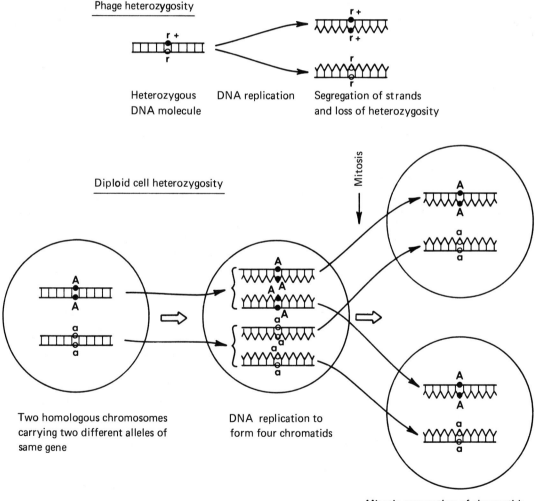

Phage heterozygosity

Heterozygous   DNA replication   Segregation of strands
DNA molecule   and loss of heterozygosity

Diploid cell heterozygosity

Mitosis

Two homologous chromosomes   DNA replication to
carrying two different alleles of   form four chromatids
same gene

Mitotic segregation of chromatids
to two daughter cells;
heterozygosity persists

over is probably common to all organisms, although not directly demonstrable with longer chromosomes. In eukaryotes, its existence is indicated by the aberrant marker ratios in the progeny of certain crosses. The analysis of this complication, described in the next section, revealed a second important mechanism of recombination involving only one DNA strand rather than a true crossover.

GENE CONVERSION

With phages or cells distinguished by several markers, several recombinations may be observed in the same DNA molecule. If these occurred independently, the frequency of multiple recombinants should equal, as a first approximation, the product of the frequencies of each recombination by itself. With distant markers this is observed. With closely linked markers, however, multiple recombinations are considerably **more** frequent than expected. This phenomenon is called **high negative interference.** (In classic genetics, interference was defined as a deficit of multiple crossovers: presumably a synapsis mechanically inhibited the formation of a second synapsis nearby.) These multiple recombinations take place in hybrid DNA because they occur between closely linked markers for which phage DNA molecules are heterozygous. The recombination cannot be due to conventional crossovers, because distant markers flanking the group of close markers often do not show enhanced recombination.

Further studies with phage could not reveal the mechanisms because the products of a single act of recombination could not be identified. The phenomenon was clarified by studies in fungi, in which the four (or eight) spores in each ascus define the four products resulting from recombination among four chromatids (each a helical DNA molecule) in one meiosis (Fig. 11-15). In a cross of mutant (m) and wild type (wt), the ratio of mutant to wild type in each ascus is normally 2:2, but occasionally it is 3:1 or 1:3. This exceptional result, called **gene conversion,** implies that a part of one chromatid, instead of being **exchanged,** becomes **changed** to match a chromatid deriving from the other parent.

Gene conversion is believed to be produced by **sequence correction** in heterozygous hybrid DNA, following the model of error correction discussed above (Fig. 11-16). At a heterozygous site the two strands cannot regularly pair with each other and a "bulge" appears, which is excised and replaced by new DNA, copying the intact strand. The base sequence characteristic of one allele is thus eliminated. Asymmetrical bulging (e.g., produced by insertions or deletions) may lead to preferential elimination of one allele, and thus to preferential conversion in the direction of the mutant in some cases and in the direction of the wild type in others. Correction initiated by a mismatched site may extend to other mismatched sites in nearby hybrid DNA, causing simultaneous conversion of two or more markers.

One gene conversion is formally equivalent to two crossovers, since it replaces one allele by another without recombination of flanking markers. It is responsible for almost all recombinations between very closely linked markers in fungi and possibly in phage and higher organisms as well. Gene conversion is not important, however, in classic recombination between more distant markers (which is due to crossing over), except insofar as it throws light on the underlying molecular mechanism.

Gene Conversion Without Crossover. Studies in fungi have shown that crossing over creates favorable conditions, but is not required, for gene conversion. The explanation would appear to be that a crossover can begin and not go to completion (as shown in Fig. 11-13, III).

In some fungi the frequency of gene conversion decreases with the distance of a marker from one end (or sometimes from either end) of the gene **(conversion polarity).** The intergenic divider may therefore contain special sequences recognized by enzymes important in recombination and perhaps also in chromosome pairing; for if the single-strand breaks that initiate formation of hybrid DNA occur preferentially at this region, the nearby markers would be included more often in hybrid DNA.

GENETIC MAPPING

Genetic mapping aims at determining the **order** of markers in DNA and the **distances**

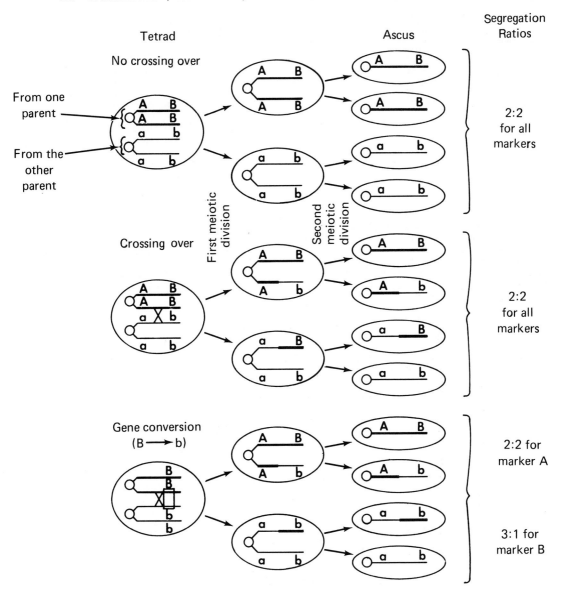

Fig. 11-15. Chromatid segregation during meiosis in ascomycetes, showing the composition of asci and the effect of gene conversion. The boxed area indicates where gene conversion occurs. It is represented as associated with crossing over, but the association is not required (see Fig. 11-13). Asci are shown with four spores, but often they contain eight, each spore being doubled by a subsequent mitotic division. The events presumably occurring in the boxed area are shown in Figure 11-16.

between them. It can be achieved by recombinational, topological, and physical methods.

Recombination frequency is the classic method, intermarker distances being calculated from the proportion of recombinants between markers. The standard procedure is a three-factor cross with pairs of double or single mutants, as shown in Figure 11-17. By crossing various overlapping sets of three markers in this way all the available markers in a chromosome can be ordered in a unique sequence, as shown in Figure 45-29. The various markers are then assigned to the same or to different genes by means of com-

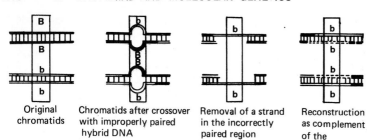

Fig. 11-16. Hypothetical molecular events in gene conversion, representing the segments of only two of the four participating chromatids (i.e., half tetrad) of Figure 11-15 (boxed area). Incorrect pairing in hybrid DNA is thought to activate the error-correction mechanism to form a correctly paired DNA, thus eliminating the mutant (or wild-type) information in one of the strands.

Original chromatids | Chromatids after crossover with improperly paired hybrid DNA | Removal of a strand in the incorrectly paired region | Reconstruction as complement of the other strand

Fig. 11-17. Three-factor cross. A, B, and C indicate the mutant alleles; + indicates the corresponding wild-type alleles. The test consists in measuring the proportion of wild-type genomes in the progeny of the cross. The centrally located marker is identified by comparing the recombination frequencies in double-mutant × single-mutant crosses (AB × C; AC × B; A × BC: column I), with those of single-mutant crosses (column II). In the example shown, only one ((1) = AC × B) of the double-mutant × single-mutant crosses gives many fewer recombinants than crosses between single mutants, indicating that B is located between A and C.

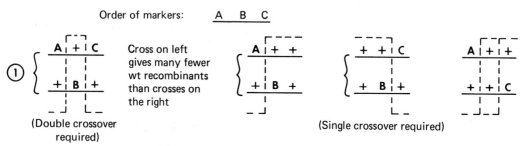

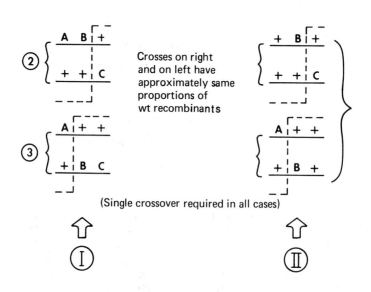

plementation tests, and a **genetic map** is thus established.

Distances are evaluated from the proportion of recombinants between fairly close markers, for with more distant markers multiple crossovers tend to decrease the recombination values; instead of remaining additive, the values asymptotically approach a maximum, with increasing distance, of 50%. For instance if we take into account only single and double crossovers, the recombination frequencies expected between three markers, with the order A, B, and C, are given by

$$P_{ac} = P_{ab} + P_{bc} - 2P_{ab} \cdot P_{bc}$$

where P_{ac}, P_{ab}, and P_{bc} are the proportions of the respective recombinants.

Recombination mapping is generally valid down to fairly short distances. However, it is less accurate for very close markers (e.g., within a gene) because of the role of gene conversion, whose frequency does not depend simply on distance (Fig. 11-18).

Deletion mapping is based on the inability of a deletion, in a pairwise cross, to produce wild-type recombinants with an overlapping deletion or with any point mutation within the deleted region. Since the crosses give a simple yes or no answer, the order of the deletion ends, and their relation to point mutations can be defined unambiguously (see Fig. 11-19). This approach gives marker order over any distance, whether minute or large; it is not affected by gene conversions and it is applicable to many biological systems. Deletion mapping, however, does not provide distances. These can sometimes be obtained by physical mapping of the product as in Figure 11-18.

Transduction mapping takes advantage of the transfer by generalized transducing phages of DNA segments of **nearly constant length** (see Ch. 46), taken at random from the DNA of the donor cell. In abortive transduction (Ch. 9), the probability of cotransduction of two genes depends only on the ratio of their physical distances to the known length of the transduced DNA fragments; hence the results afford an unambiguous way not only to order markers, but also to determine their

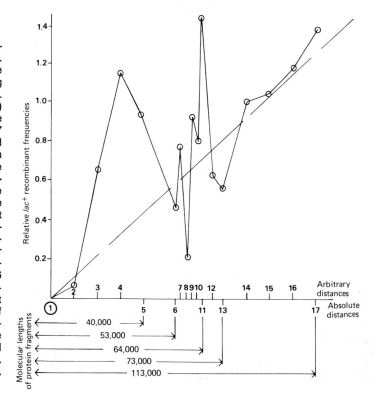

Fig. 11-18. Relative recombination frequencies in the *E. coli* β-galactosidase gene when various chain-terminating (nonsense) mutants (see Ch. 12 for terminator mutations) are crossed to mutant 1. The **order** of the markers 1 to 17 was independently determined by deletion mapping, which affords an unbiased order (see below). Moreover, the **distances** between some of the markers and marker 1 are given by the molecular weight of the protein fragments produced by the mutants. The distances between the other markers are arbitrary. By conventional mapping, mutants 6 to 10 would incorrectly be considered closer to 1 than mutant 4. The classic expectation of linear dependence of recombination frequency on distance (dashed line) is thus verified for only some of the markers. [Modified from Norkin, L. C. *J Mol Biol* 51:633 (1970).]

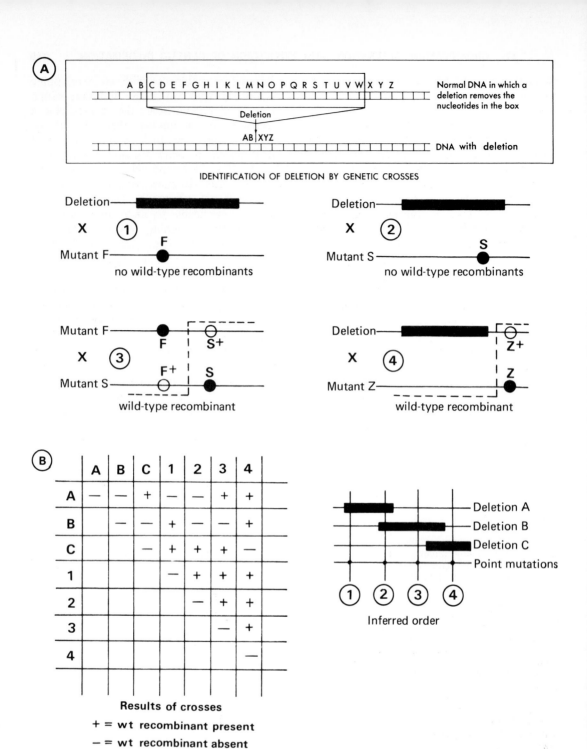

Fig. 11-19. A. Properties of deletions. Deletions (heavy lines) are identified by genetic crosses with selected point mutants. Cross 1 between the deletion mutant and point mutant F and cross 2 between the deletion mutant and point mutant S yield no wild-type recombinants; however, cross 3 between the two point mutants yields wild-type recombinants, showing that mutations F and S occur at different sites. The deletion mutant gives wild-type recombinants in cross 4 with point mutant Z, whose site is outside the deletion. **B.** Principles of deletion mapping. Three deletions and four point mutations are ordered using presence or absence of wild-type recombinants in pairwise crosses. The results of the crosses given in the checkerboard determine the order indicated at right.

distances. In stable transduction, the distances are less reliable because the factors that complicate recombination similarly influence integration of the transduced segment.

Physical mapping is the most direct. Protein mapping is useful for sites within a gene; it can be carried out by measuring the length of the polypeptide chains resulting from the introduction of terminator (nonsense) mutations (see Fig. 11-18). **Nucleic acid mapping** is best performed by hybridizing DNA from strains differing by two or more deletions or insertions. As Figure 11-20 shows, where two strands lack homology, the unpaired DNA forms a "bush" and the length of the paired region between bushes can be unambiguously determined. An absolute **physical map** of the λ genome obtained from a series of deletions fits the genetic map quite well.

These methods have been invaluable because of their directness, but one is limited to genes whose product can be identified even when incomplete, and the other to short DNAs that can be obtained pure.

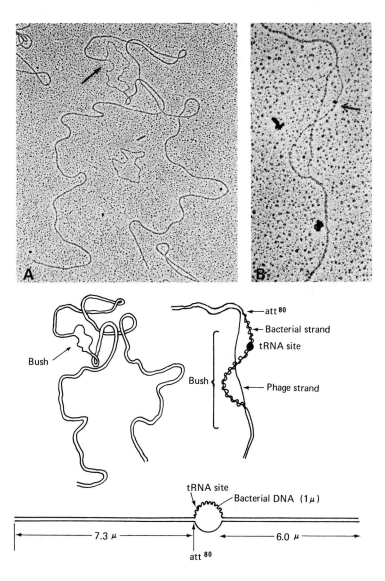

Fig. 11-20. Physical mapping of deletions in the DNA of a transducing derivative of phage φ80 (see Ch. 46), which has exchanged 0.7 μ of its DNA for 1 μ of *E. coli* DNA containing two tRNA genes. **A.** Heteroduplex DNA was made by hybridizing complementary DNA strands deriving from the mutant and from the wild-type φ80. The left end of the exchanged piece corresponds to the φ80 attachment site (att⁸⁰) which is therefore precisely mapped. The DNA was prepared under conditions that prevented random folding of the single-stranded loops. **B.** The bacterial tRNA site is recognizable by hybridization with the RNA after labeling it with the iron-containing (and electron-opaque) protein ferritin. (Courtesy of M. Wu and N. Davidson.)

Bush

att 80
Bacterial strand
tRNA site
Bush
Phage strand

tRNA site
Bacterial DNA (1μ)
7.3 μ
6.0 μ
att 80

MUTATIONS

GENERAL PROPERTIES

The term **mutation** applies broadly to all heritable changes in the genetic material other than incorporation of foreign genetic material. They are recognizable through their effect on the phenotype. However, certain mutations have no phenotypic consequences (are **silent**): some do not cause amino acid replacement because the new codon specifies the same amino acid; others cause an amino acid replacement that does not alter the function of the protein; and mutations may be compensated for by suppression (see Ch. 12). Because of silent mutations, over-all measurements of mutation rates give minimal values.

Mutational changes in the genetic material (i.e., DNA, or RNA in some viruses) include nucleotide **replacements, deletions,** and **insertions** (Table 11-1). Two kinds of replacements are produced by different mechanisms (Fig. 11-21): **transitions,** in which a purine is replaced by a purine and a pyrimidine by a pyrimidine (e.g., AT → GC), and **transversions,** in which a purine is replaced by a pyrimidine and vice versa (e.g., AT → CG).

Deletions and insertions are generated by different mechanisms and have different properties, depending on the size. Microdeletions and microinsertions, usually of a single nucleotide, generate shifts of the reading frame and, as we have seen above (Units of information; Fig. 11-3), can be compensated for by another frame shift, opposite in direction, a short distance away. The regular (longer) deletions and insertions involve sequences of many nucleotides (up to thousands).

Nucleotide replacements, microdeletions, and insertions (of any length) behave in recombination as **point mutations,** i.e., give wild-type recombinants with all mutations except those at the same site. Deletions, in contrast, fail to give wild-type recombinants with point mutations at more than one site (Fig. 11-19). Point mutations undergo **true reversion** through appropriate replacement, dele-

TABLE 11-1.   Main Properties of Different Kinds of Mutations

Nature of nucleic acid change	Effect on coding properties	Recombinational behavior	Production of reversions	Consequences for protein		Other properties
				Structure	Function	
Nucleotide replacement	Missense	Point mutation	Yes	Amino acid substitution	1) None 2) Temperature-sensitive 3) Lost	CRM may be present
	Nonsense	Point mutation	Yes	Premature termination of polypeptide chain	Usually lost	Extragenic suppression
Microdeletion; microinsertion		Point mutation	Yes	Frame shift	Usually lost	Intragenic suppression
Insertion		Point mutation	Yes	Altered	Usually lost	May introduce terminator codons
Deletion		Segment	No	Altered	Usually lost	
Silent				1) No amino acid substitution 2) Amino acid substitution with little effect on over-all structure	Conserved	

tion, or insertions. However, **deletions do not revert;** and because of their genetic stability they are especially useful for metabolic and genetic studies where reversions would interfere; they are also extremely useful in genetic mapping (see above).

The above properties of various mutations afford criteria for their recognition. Most difficult to recognize are insertions, which are readily distinguished from point mutations only if the inserted DNA has special properties (e.g., contains recognizable genes).

CONSEQUENCES OF MUTATIONS FOR PROTEIN STRUCTURE (TABLE 11-1)

Among the replacements, **missense mutations** cause one amino acid to substitute for another. The protein may remain functional (though often with quantitative alterations) if the amino acid substitution does not markedly affect its tertiary structure or its active site. Some altered enzymes exhibit increased sensitivity to heat denaturation, and if this effect is detectable in the whole organism, the mutation is called **temperature-sensitive.** In others the function is lost, but the mutant protein still remains recognizable immunologically as a **cross-reacting material** (CRM). **Nonsense** or **terminator mutations** create a codon that prematurely terminates the growing peptide chain and almost always destroys the function of the protein (see Ch. 12).

Microinsertions and microdeletions cause a shift of the **reading frame** and therefore the production of a jumbled protein with loss of function (see Fig. 11-3). Larger deletions destroy function except in rare cases when they remove small unessential parts of proteins. Deletions at the boundary of two genes may cause the two polypeptide chains to be synthesized as a single chain, with partial retention of one or both functions. **Insertions** usually destroy the function of the protein; moreover, they may introduce terminator codons which affect the expression of distal genes.

INDUCTION OF MUTATIONS

The induction of mutations by a **mutagenic agent** was first shown in 1927 by Muller, applying X-rays to the fruitfly *Drosophila*. With

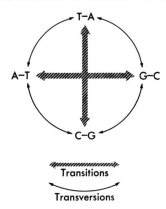

Fig. 11-21. Classes of nucleotide replacement. When a mutation substitutes one nucleotide for another in a strand, at the next replication the new nucleotide is paired to its regular partner. The resulting pair is represented. Example: if in the upper TA pair, T is replaced by C, the result is the lower CG pair. Note that each base pair can undergo one kind of transition and two kinds of transversions.

bacteria, which did not require the penetrating power of X-rays, UV irradiation was found to be equally effective and more convenient. Powerful chemical mutagens were subsequently discovered, and knowledge of their effects on nucleic acids has thrown considerable light on the mechanisms of mutation. In addition, the possibility of selecting rare mutants from large populations of bacteria has shown that many chemicals (e.g., formaldehyde, caffeine, Mn^{++}) and even elevated temperature are slightly mutagenic. Bacteria thus provide sensitive test systems for detecting potential environmental mutagens for man.

Reversion and Suppression. Some chemical mutagens can induce reversion of the mutations they have produced. As is shown below, this property is very useful for understanding the mechanism of mutagenesis. Such **genotypic** or **true reversion,** however, must be clearly distinguished from **suppression**—a change at a different site in the genome that phenotypically corrects the mutation. The distinction is made by crossing an apparent revertant to the wild type: as Figure 11-22 indicates, only suppressed mutants can yield recombinants in which the original mutated gene has been segregated from its suppressor and therefore again produces the mutant

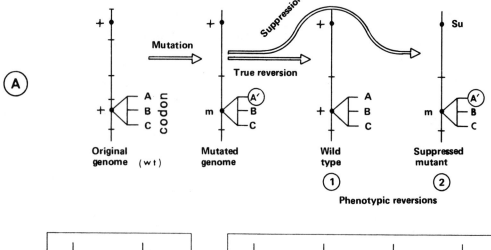

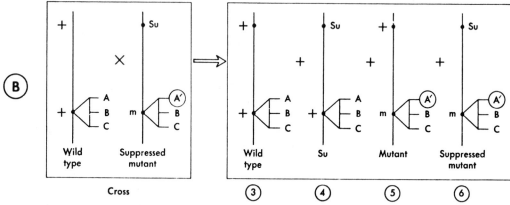

Fig. 11-22. Phenotypic reversions. **A.** Phenotype reversions caused either by true reversion of a mutation (as in 1) or by suppression (as in 2). **B.** A genetic cross between a suppressed mutant (with wild-type phenotype) and a true wild type yields four types of progeny, in one of which (5) the primary mutation is expressed. (The recombinant carrying only the suppressor mutation [4] may or may not be phenotypically distinguishable from the wild type.) A true reversion (1) would not produce mutant progeny in a similar cross. + = wild-type allele; Su = suppressor mutation; m = suppressible mutation; ABC = wild-type codon; A'BC = mutated codon.

phenotype. The mechanisms of suppression will be considered in the next chapter.

ACTION OF CHEMICAL MUTAGENS

The chemical mutagens fall into several main groups, with different actions: 1) base analogs, 2) deaminating agents, 3) alkylating agents, and 4) acridine derivatives. Additional agents, of less significance for present purposes, include Mn^{++} and formaldehyde. The base analogs, the acridines, and Mn^{++} require replication for their action because they cause

mutations only in the product of replication. The other mutagens, in contrast, are chemically reactive and cause mutations even in nonreplicating template, including transforming DNA or phage DNA in vitro. A very reactive, strongly mutagenic alkylating agent, nitrosoguanidine, seems an exception because it acts only in cells, but apparently the true mutagen is a metabolic derivative. As already shown in Chapter 10 (DNA replication), nitrosoguanidine produces mutations in closely linked clusters at the growing fork of replicating DNA.

We shall now consider these agents in terms of the types of mutations they induce.

POINT MUTATIONS

Transitions (Fig. 11-21). Substitutions of AT for GC, or vice versa, are produced in DNA when cells or viruses are grown in the presence of certain base analogs, such as **5-bromouracil** (5-BU) (often employed as its deoxynucleoside 5-bromodeoxyuridine, 5-BUDR). In the cells these are converted to the corresponding nucleoside triphosphates and thus can be incorporated into new DNA in place of a normal base. Indeed, 5-BU so closely resembles thymidine (T) that its substitution for as much as 90% of the T in bacteriophages is compatible with subsequent normal replication. However, T itself occasionally undergoes a transient internal rearrangement **(tautomerization)** from the keto to the enol state, in which it pairs with G instead of A; and the enol form is much more frequent (though still rare) with 5-BU (Fig. 11-23).

The tautomerization of 5-BU **after** incorporation in DNA in place of T can lead to a **replication error,** causing the positioning of a G instead of A in the new strand; tautomeriza-

tion **before** incorporation can lead to an **incorporation error** in which BUDR triphosphate is recognized as though it were dCTP. Because of this double action 5-BU can induce reversion of the mutations that it has produced (Fig. 11-24). Hence mutations of unknown origin can be defined operationally as transitions if 5-BU induces them to undergo true reversion.

Incorporation errors yield a mutant clone within a generation after exposure to the analog, but replication errors may be delayed. Indeed, a strand of DNA carrying the analog exhibits genetic instability, since a replication error may occur and yield a mutant clone after any number of generations.

2-Aminopurine, a purine analog, causes transitions about as effectively as 5-BU, its tautomerization causing it to be read as either A or G.

Several **chemically reactive mutagens** induce more selective transitions. **Nitrous acid** (Fig. 11-25) oxidatively deaminates the amino-substituted bases, thus converting A to hypoxanthine (which resembles G), and C to U. Nitrous acid can also cause interstrand cross-links probably by creating aldehyde groups, which then react with amino groups on the opposite strand. **Hydroxylamine** specifically converts cytosine to a derivative that

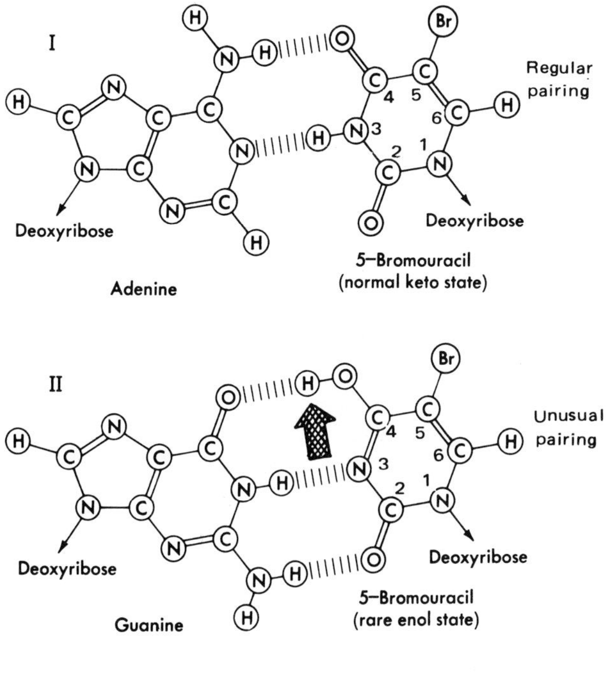

Fig. 11-23. Regular and unusual base pairing of 5-BU. **I.** Regular base pairing (in the common keto form) with adenine. **II.** Base pairing (in rare enol form) with guanine. The heavy arrow in II indicates the displacement of the proton in the tautomerization of 5-BU.

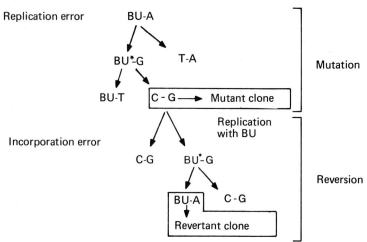

Fig. 11-24. Induction of mutations and their reversions by 5-BU. The sequence can be inverted, a replication error correcting a previous incorporation error. BU* = transient enol form.

pairs with A, producing a GC → AT transition, which is not reversed by hydroxylamine. Monofunctional **alkylating agents,** such as ethyl ethanesulfonate (EES) or nitrosoguanidine (see above), also produce transitions, primarily through alkylation of G at the 7N position, which increases the frequency of an N1 ionized form (Fig. 11-26) that pairs with T instead of C. These observed chemical interactions are evidently responsible for most

Fig. 11-25. The oxidative deamination of DNA by nitrous acid and its effects on subsequent base pairing. **A.** Adenine is deaminated to hypoxanthine, which pairs with cytosine instead of thymine. **B.** Cytosine is deaminated to uracil, which pairs with adenine instead of guanine.

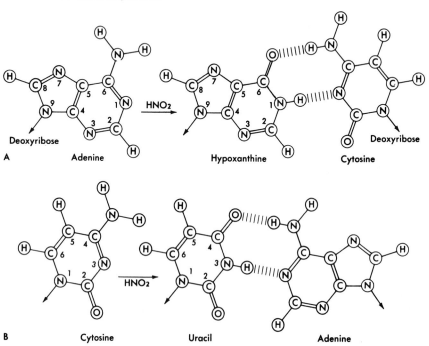

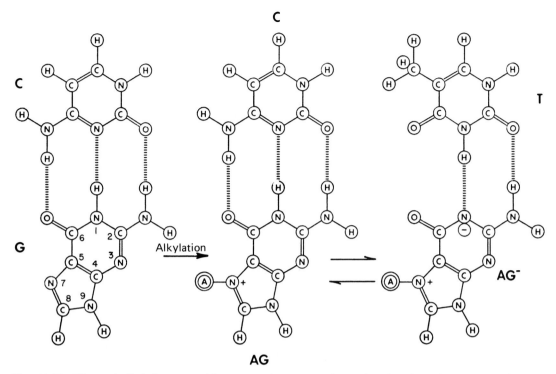

Fig. 11-26. Effect of alkylation on pairing properties of guanine. Alkylation by ethyl ethanesulfonate (EES) produces 7N-alkylguanine (AG), which has normal pairing but also has a higher frequency of ionization at N1. N1-ionized guanine (AG⁻) pairs with thymine instead of cytosine; A = alkyl group.

of the mutagenic effects of these agents with DNA in vivo, since the amino acid substitutions are those predicted from the chemical changes according to the genetic code (see Ch. 12).

It seems likely that transitions occur most frequently as secondary consequences of base mispairing, i.e., when the resulting distortion of the helix elicits DNA repair. If the strand with the altered base is conserved, a transition results; in the opposite case the transient base change goes unnoticed.

Transversions. These are recognized operationally as point mutations that are not reverted by the agents that induce transitions or those that induce frame shifts. Though frequent among spontaneous mutants, they are not produced by most mutagens. However, their frequency is increased by certain mutations in DNA polymerase (see below, Mutations affecting mutability) and also by Mn^{++} (which increases the effect of the mutant polymerase). Transversions thus appear to

depend not on a chemical change in the DNA but on the conformation of the polymerase.

Reading-frame Shifts. These mutations, already discussed above (codons), are induced by **acridine** derivatives. Usually only one base is removed or inserted (though additions of up to four bases have been observed). Unknown mutations can be identified as frame shifts if they can be reversed by acridine (often by a second frame shift, Fig. 11-3), but not by mutagens of the preceding groups.

Acridines shift the reading frame by intercalating between successive base pairs in DNA (see Fig. 10-8).

These mutations evidently require single-strand breaks, for they preferentially affect replicating DNA, and their frequency is increased when the closing of a gap is slowed by mutational impairment of DNA ligase. They also seem to require the presence of short runs of the same base pair, which promote **illegitimate** (i.e., shifted) **pairing** of a free end and the uninterrupted strand (Fig. 11-27). As is

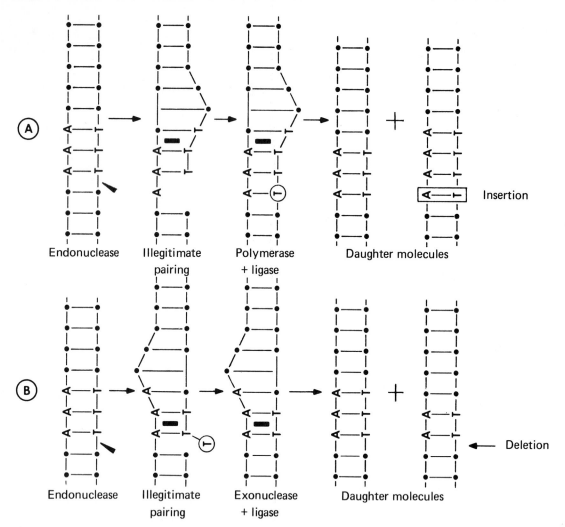

Fig. 11-27. Possible mechanism of frame shift by acridines at sites of single-base reiterations near a single-strand break. The intercalated acridine molecules would stabilize the illegitimate pairing long enough to allow repair, leading to base insertion (**A**) or deletion (**B**).

suggested in the model of this figure, the intercalated acridine may act by stabilizing the temporary illegitimate pairing until the gap is closed.

The considerable recombinational activity explains why certain phages (e.g., coliphage T4) have a high frequency of frame shifts. In bacteria, frame shifts are induced only by special alkylating acridine derivatives (e.g., ICR 191). The species specificity suggests differences either in cell permeability to acridines or in interaction of acridines with part of the machinery of replication in addition to the DNA.

Epoxides of carcinogenic polycyclic hydrocarbons (which represent the active forms of the carcinogens) cause frame shifts in bacteria, suggesting a correlation between mutagenic and carcinogenic activity of these compounds.

Deletions. Deletions in DNA are induced by agents that can cause cross-links between the complementary strands: nitrous acid, bifunctional alkylating agents, and irradiation. Apparently a segment of DNA around the cross-link is not replicated, while the segments on either side replicate and join with each other. This joining may well involve the usual process of genetic recombination but on the basis of local microhomology; the result would be excision of a defective loop of DNA.

Many synthetic cross-linking agents are available. In addition, the antibiotic **mitomycin** is converted metabolically, in cells, into such an agent.

INDUCTION OF MUTATIONS BY RADIATION

Of the two main mechanisms of **UV-light**-induced pyrimidine dimer—excision and post-replication repair by recombination—the latter is the main source of UV-induced mutation. Thus mutants that have lost the capacity for postreplication repair (*rec A, exr*) show no induction of mutations, although they are more sensitive to the lethal action of UV. Conversely, mutant strains unable to excise dimers (*uvr*) show a marked increase both in induction of mutation and lethality; and caffeine, which interferes with dimer excision, increases both effects in wild-type cells. The mutations induced by UV light are mostly base substitutions with fewer frame shifts and occasional deletions.

Mutations induced by X-rays or ^{32}P decay also appear to result from imperfection of recombinational repair, since they are rare or absent in recombination-deficient strains.

DIRECTED MUTAGENESIS

A goal of chemical mutagenesis has been to induce mutations selectively in a given gene, but this goal cannot be attained because the agents act randomly over the whole genome. However, an ingenious approach, based on transduction (Ch. 9), has been worked out (Fig. 11-28) whereby temperature-sensitive mutations can be introduced in small segments (about 1%) of the bacterial genome at any location. Since a very high density of mutations per unit length of DNA can be induced, this approach permits saturation of the cellular genome with induced mutations.

MUTAGENIC VS. LETHAL ACTION OF MUTAGENIC AGENTS

At the concentration required for mutagenesis most agents also cause extensive killing of microorganisms by producing changes in nucleic acid that upset its replication. Thus UV radiation produces photohydrations and intra-strand pyrimidine dimers which are lethal if not repaired; bifunctional alkylating agents produce cross-links between strands; nitrous acid cross-links and deaminates G to xanthine, which cannot be read. However, the ratio of mutagenic to lethal action varies widely. Essentially devoid of killing (except for lethal mutations) are certain base analogs (e.g., 5-BU) and monofunctional alkylating agents (e.g., EES) which cause transitions.

CHEMICAL MUTAGENESIS IN HIGHER ORGANISMS

Most work on mutagenesis in higher organisms has been carried out with ionizing radiations because of their ability to penetrate. Some mutagenic chemicals, such as alkylating agents, are also active, but others may not reach the DNA in germ cells. However, mutagens may also be carcinogens and induce cancers by causing somatic cell mutations or by activating latent viruses. Thus in the 12 years following the atomic bomb explosion in Hiroshima, the incidence of leukemia among the heavily irradiated survivors was about 50 times higher than in a comparable nonirradiated population. The widespread use of chemicals with potential mutagenic action in the human environment, and the possible relation between mutagenesis and carcinogenesis, are serious problems for human health. As a practical matter, alkylating agents should be handled as carefully as radioactive chemicals, though they are not yet subject to close legal restrictions.

SPONTANEOUS MUTATIONS

These represent a very heterogeneous class, the proportions varying in different systems. For instance, frame shifts constitute about 80% of all spontaneous mutations in phage T4 but only 15 to 20% in bacteria and fungi, probably reflecting the differences in recombination frequencies per unit DNA length (the role of recombination in frame shifts has been discussed above). Moreover, transversions, which are rare among induced mutations, constitute more than 80% of spontaneous reversions of amber mutations in phage T4

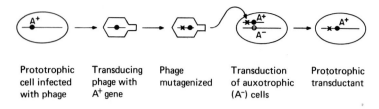

| Prototrophic cell infected with phage | Transducing phage with A+ gene | Phage mutagenized | Transduction of auxotrophic (A⁻) cells | Prototrophic transductant |

x = Induced temperature–sensitive mutation

Fig. 11-28. Localized mutagenesis, allowing the isolation of temperature-sensitive mutants in the neighborhood of any selected gene A.
Transducing phage (see Ch. 9) carrying the wild-type A+ allele is
obtained as rare particles in the lysate of A+ cells infected by the
phage. The lysate is exposed to intense mutagenic treatment and is
then used to infect A− cells. Cells receiving DNA containing the A+
gene and transduced to prototrophy are selectively isolated and tested
for the presence of temperature sensitivity of growth. If so, they contain temperature-sensitive mutations in the transduced DNA, close to
gene A. Any region of the genome adjacent to an essential gene can
be saturated with induced mutations. [From Hong, J. S., and Ames,
B. N. *Proc Natl Acad Sci USA 68:*3158 (1971).]

and almost 50% of the mutations in electrophoretically altered human hemoglobins (see
Fig. 12-9). However, in the latter group the
selection requiring an amino acid replacement
with a different charge discriminates against
transitions, which tend to replace one amino
acid with a chemically similar one (see Ch. 12,
Evolution of the code).

The discovery of base pairing in DNA
suggested immediately that spontaneous mutations could be explained by keto-enol tautomerization, which would give rise to
transitions (Fig. 11-23). However, the frequencies just described suggest that spontaneous mutations do not arise primarily from
tautomerizations; neither do they fit the
pattern to be expected if they were mostly
induced by natural background irradiation.
Rather, they appear to arise primarily from
enzymatic imperfections during DNA replication or recombination: the next section will
describe the role of DNA polymerase in the
production of transversions; and the frequent
spontaneous frame shifts may similarly be
due to errors in recombination.

MUTATIONS AFFECTING MUTABILITY

Mutator mutations in the gene for DNA polymerase in phage T4 and a mutator mutation
in a gene of unknown function in *E. coli*
increase the spontaneous mutation rate by

inducing nucleotide substitutions. The phage
mutants tested cause a high proportion of
transversions of the GC → TA type, while the
opposite type predominates in *E. coli*.

The prevalence of one type of substitution suggests that during the replication of the phage DNA
the mutant polymerase favors the incorporation of
a certain base, perhaps forming an unusual base
pair with the template. This is possible because
DNA bases can form 29 different pairs, of which
only 2 are normally utilized in DNA. Similar errors are probably produced (or permitted) less
frequently by the normal enzyme, contributing to
the normal rate of spontaneous mutations.

The same T4 gene is also subject to **antimutator**
mutations, which decrease the rate of both spontaneous and induced transitions. The altered polymerase appears to discriminate against bases in the
tautomeric form that causes transitions.

Spontaneous deletions are caused by two mechanisms. Deletions of one group are uniformly distributed over the genome and occur with similar
low frequency in different *E. coli* strains. They are
probably the result of defective repair, since their
frequency is increased by mutations that impair the
activity of DNA polymerase I. Presumably they
arise by a mechanism similar to that postulated for
induced deletions, around an excised and not properly reconstructed segment of one strand. Deletions
of another class occur in "hot spots" and only in
certain *E. coli* strains, probably as the result of
nicking by a mutated nuclease, which recognizes
certain DNA sequences normally respected. The
frequency of these deletions is affected by mutations

in a locus of unrecognized function, perhaps specifying the nuclease.

SITE INFLUENCE ON THE FREQUENCY OF SPONTANEOUS AND INDUCED MUTATIONS

Figure 11-29 shows that the point mutation frequencies among the various sites in a gene are far from random, varying from 10^{-3} to about 10^{-8} (i.e., one mutation in 10^8 new phages or one error in 10^8 replications of a nucleotide). Hence the frequency of mutation at a nucleotide pair must be strongly influenced by factors in the neighborhood. In fact, in the triplet UAA of phage T4 mutations affecting the middle A are about 20 times more frequent than in the triplet UAG, whereas those affecting U occur at a similar rate in the two triplets. Apparently the difference in the third nucleotide affects the mutability of the middle A. The extreme hot spots of Figure 11-29 may be a special property of phage whose high recombination rate provides increasing opportunity for frame shifts.

EVOLUTIONARY CONSEQUENCES OF MUTATIONS AFFECTING MUTATION RATES

The nonrandom effect of mutator polymerase on the GC proportion could explain the large variations observed among different organisms. Indeed, the experimental introduction of a mutator gene with TA → GC bias in *E. coli* resulted in a 0.5% increase in GC in about 1500 generations. In order to be tolerated, these substitutions must occur where they have acceptable consequences for the proteins, i.e., mostly in the third bases of codons (see Ch. 12, the genetic code).

These observations raise the question of the evolutionary consequences of high mutation rates. The rates in various organisms (Table 11-2) are correlated with the size of the genome, so that the **rate per genome, rather than per nucleotide,** is relatively constant. This result suggests that mutation rates balanced at this level have a selective value.

An advantage of higher mutation rates has been in fact demonstrated by chemostat experiments

Fig. 11-29. Sites of occurrence of mutations in a segment of the $r_{II}B$ gene of bacteriophage T4. Note the very uneven distributions of mutational sites and also the sites (hot spots) with very large numbers of recurrences, given in circles. Small circles on the base line in the map of spontaneous mutations indicate sites at which no spontaneous, but only induced, mutations occurred. [Data from Benzer, S. *Proc Natl Acad Sci USA* 47:403 (1961).]

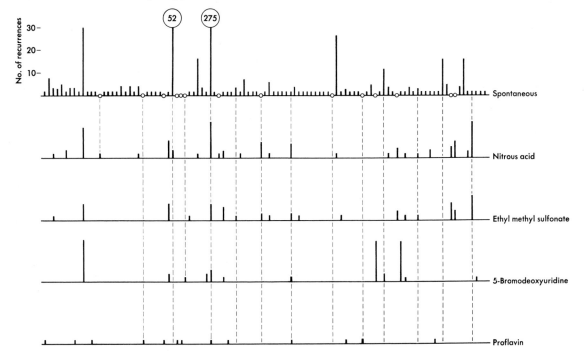

TABLE 11-2.   Spontaneous Mutation Rates in Different Organisms

Organism	Base pairs per genome	per base pair Mutation rate replication	Total mutation rate per genome
Bacteriophage	4.8×10^4	2.0×10^{-8}	1.2×10^{-3}
Bacteriophage T4	1.8×10^5	1.7×10^{-8}	3.0×10^{-3}
Salmonella typhimurium	4.5×10^6	2.0×10^{-10}	0.9×10^{-3}
Escherichia coli	4.5×10^6	2.0×10^{-10}	0.9×10^{-3}
Neurospora crassa	4.5×10^7	0.7×10^{-11}	2.9×10^{-4}
Drosophila melanogaster	2.0×10^8	7.0×10^{-11}	1.4×10^{-2}
E. coli with a mutator mutation	2.0×10^6 AT pairs	3.5×10^{-6} (per AT pair)	

Data from Drake, J. W., *Nature 221*:1132 (1969).

(see Ch. 5) in which an *E. coli* mutator strain outgrew the wild type, and by the lack of natural selection for antimutator mutants, which are so easily obtained in the laboratory. At the same time, mutator polymerases may have some unfavorable feature, because in heterozygous diploid *E. coli* cells, the corresponding wild-type allele is dominant.

TRANSCRIPTION OF DNA

The transfer of genetic information from DNA to RNA was inferred from the function of mRNA and from the requirement of DNA for RNA synthesis in vitro (see below); it was later demonstrated directly by the formation of a **hybrid RNA-DNA helix** when a template and its transcript are annealed together (see Ch. 10, Hybridization). The term **transcription** emphasizes retention of the language of base pairs.

The powerful technique of hybridization has made it possible to identify and quantitate the template for a given kind of RNA, and vice versa. For example, it showed that *E. coli* DNA provides the template for synthesizing not only mRNA but also stable RNA (Fig. 11-30); and saturation of the system with different kinds of RNA revealed that only 0.8% of the genome is transcribed into rRNA and 0.05% into tRNA. Similarly, it is possible

Fig. 11-30. Hybridization of rRNA with homologous DNA. A small amount of radioactive *E. coli* rRNA was melted and annealed with a large amount of *E. coli* DNA and also with the unrelated DNA of bacteriophage T5. Equilibrium centrifugation in a CsCl density gradient separated hybridized RNA, which bands close to DNA, from unhybridized RNA, which has a much higher buoyant density and thus appears in earlier fractions (collected from the bottom of the tube; see Fig. 10-4). The results show excellent hybridization with *E. coli* DNA, but none with T5 DNA. [Modified from Spiegelman, S. *Sci Am 210*:48 (1964).]

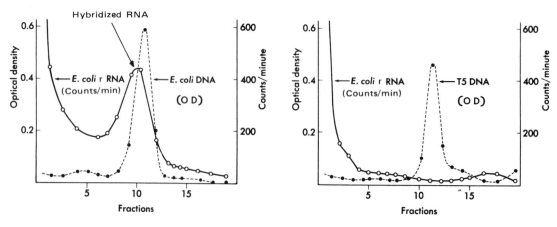

to follow the kinetics of formation and destruction of specific mRNAs, corresponding to specific bacterial genes, by hybridizing the total cellular RNA with the DNA of specialized transducing phages (Ch. 46) carrying those genes. Physiological transcription (in the cell and with unbroken DNA in vitro) is generally **asymmetrical,** i.e., the RNA transcribes only one DNA strand. (However, both strands of mitochondrial DNA are transcribed at first and one transcript is then destroyed.) Studies with DNAs whose strands can be separated show that tRNA transcribes some segments from one strand and other segments of the same molecule from the other strand. Owing to the opposite polarity of the DNA strands, transcription on opposite strands proceeds in opposite senses.

RNA POLYMERASE AND SIGMA FACTOR

Transcription can be achieved with suitable enzymes extracted from cells. The product is similar to native RNA, for when its formation is coupled with protein synthesis small amounts of functional enzymes can be obtained, such as *E. coli* β-galactosidase. Hence in vitro studies of the mechanism of transcription are clearly relevant to the process in the cell.

RNA is synthesized by **DNA-dependent RNA polymerase,** also called **transcriptase,** which catalyzes the reaction:

$$n(\text{ATP, GTP, UTP, CTP}) \xrightarrow[\text{Enzyme}]{\text{DNA template}}$$

Nucleoside triphosphates

$$(\text{AMP, GMP, UMP, CMP})n + n(\text{P.P})$$

Polynucleotide   Pyrophosphate

After a short labeling pulse, digestion with exonucleases of suitable specificity (see Table 10-2) shows that, as in DNA synthesis, the chains grow at 3′ end (see Fig. 10-22).

The "core" RNA polymerase of *E. coli* (MW ca. 400,000) is composed of two α-chains (MW 41,000), one β-chain (MW 155,000), and one chain whose similar size (MW 165,000) led it to be designated as β′. The normal function of the enzyme requires its interaction with a protein, called sigma (σ) factor (MW, 86,000), that promotes its attachment to ɪspecific initiation sites, but does not

remain on the enzyme during chain extension (Fig. 11-31).

The σ-factor was discovered by Burgess and Travers through the observation that chromatography on phosphocellulose, as a purification step, yielded RNA polymerase that was still active but exhibited altered relative activity with different DNA templates (i.e., a different specificity; hence σ). This effect was traced to separation of the core enzyme from its complex with the previously unrecognized σ-factor; moreover, readdition of this factor restored the original specificity.

During chain extension the σ-factor is released and made available for another initiation (Figs. 11-31 and 11-32). Its binding to the core polymerase probably causes an allosteric transition which is then maintained by other components of the initiation complex, allowing σ-factor to be released.

The σ-factor has provided the key to certain dramatic shifts in specificity of transcription in sporulation (Ch. 6) and in phage infection (Chs. 45 and 46). As will be described in greater detail in the relevant chapters, two changes are responsible: a different σ-factor is produced, and a core subunit is altered in a way that increases its relative affinity for the new compared with the old specifier. In addition, the formation of rRNA may be regulated by a factor (ψ) that apparently interacts with RNA polymerase, in contrast to the factors (repressors) that selectively regulate the synthesis of various mRNAs by interacting with specific sites on the DNA (Ch. 13). Factors that modify the specificity of RNA polymerase have great potential importance for differentiation, since they influence the selective transcription of large blocs of genes.

The function of transcriptase does not require the complexity found in the bacterial enzyme. Phage T7 specifies a transcriptase that is a single polypeptide of MW 100,000, while *Neurospora* mitochondria have one of MW 64,000.

The interaction of transcriptase with DNA involves three main steps: initiation, chain elongation, and termination.

INITIATION

The triphosphate of the initial, 5′-terminal residue of RNA is retained. The use of sub-

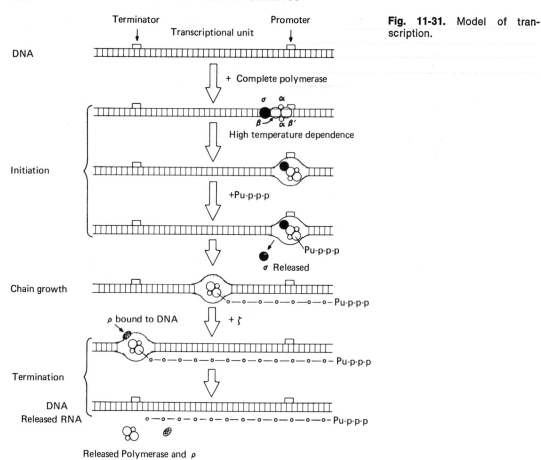

Fig. 11-31. Model of transcription.

strates with ^{32}P in the terminal (γ) position, followed by alkaline hydrolysis of the chain and identification of the 5′-terminal base (B) recovered as PPP-B-P, showed that this base is always a purine.

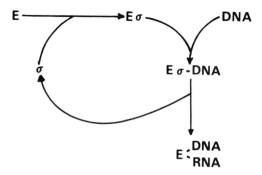

Fig. 11-32. The recycling of RNA polymerase core enzyme and of σ-factor.

Physiological initiation requires the complete enzyme (with σ-factor), the proper nucleoside triphosphate, and a special DNA sequence called a **promoter.** The promoters are apparently sequences rich in pyrimidines on the transcribed strands: they are identified by mutations that prevent physiological initiation (Ch. 13) in vivo and in vitro.

In initiation (Fig. 11-31) RNA polymerase first binds loosely with σ-factor to a promoter. A second step, requiring σ-factor, makes the binding much tighter; this step is highly dependent on temperature and probably involves a local melting of the DNA helix. The initial nucleoside triphosphate is then bound and the σ-factor is released, to engage in another cycle (Fig. 11-32). Transcription can reinitiate many times at the same promoter before the first complete molecule is released (Fig. 11-33).

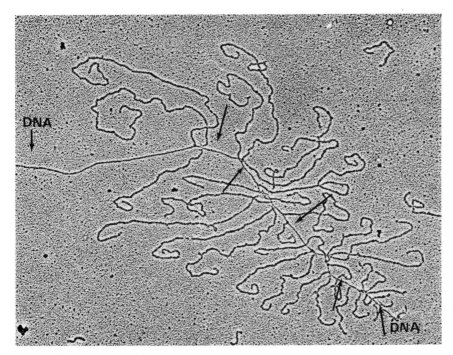

Fig. 11-33. Transcription of the DNA of bacteriophage T7 (arrow) in vitro by *E. coli* DNA polymerase. The transcriptase recognizes a promoter at the right end of the DNA (in the figure); and transcribes about 15% of the DNA. The remainder (not shown) is not transcribed owing to absence of the required T7-specified transcriptase. The RNA molecules are held in stretched state by addition of the antifolding protein specified by phage T4 (see Ch. 10). The shortest are near the promoter and grow in length as they approach the terminator at the left. This picture shows that transcription can reinitiate many times before the first complete molecule is released. (Courtesy of H. Delius and N. Axelrod.)

CHAIN ELONGATION

In this phase, which occupies most of the period of enzyme activity, mononucleotides complementary to the template are incorporated. The growing end of the RNA is attached to the DNA noncovalently by the enzyme and perhaps by a short stretch of RNA-DNA base pairs in a region of local DNA strand separation (Fig. 11-31). Thus if RNA is labeled by a brief radioactive pulse the most recent synthesized molecules are found connected to the DNA, and the complex is broken by mild heating or by proteolysis. Hence as synthesis proceeds the double helix apparently opens ahead of the region of growth and closes behind it. The DNA thus does not unwind as a whole, but simultaneously unwinds and rewinds. At the end of transcription the template DNA is unaffected: thus cyclic polyoma virus (Ch. 63) DNA retains its original physical and biological properties.

TERMINATION

Transcription of phage DNA in vitro by purified polymerase (including σ-factor) was found to yield RNA with increased average chain length, compared with a crude extract. Normal values were restored by a protein, called the rho (ρ) factor (MW 200,000 and composed of four identical polypeptide chains) which does not affect the number of initiations (measured as incorporation of γ-^{32}P-GTP). Purified ρ-factor binds to DNA, causing release (hence ρ) at normal termination signals, since the chains made in its presence in vitro are indistinguishable from those synthesized in vivo under comparable conditions.

ANTIBIOTICS AFFECTING TRANSCRIPTION

Initiation is blocked by the naturally occurring **rifamycins** as well as the more effective semisynthetic derivative **rifampin** and the **streptovaricins.** These all have an aromatic chromophore spanned by a long aliphatic bridge (Fig. 11-34). These antibiotics inhibit initiation by bacterial RNA polymerase and the small RNA polymerase of *Neurospora* mitochondria, but not the cytoplasmic enzyme of eukaryotes.

They all bind to the same polymerase subunit, since mutants selected for resistance to

Fig. 11-34. Antibiotics that interfere with initiation of transcription (rifamycin and derivatives, streptovaricin) or chain extension (streptolydigin, actinomycin D.) Sar = sarcosine (N-Me-glycine); Me-Val = N-Me-L-valine.

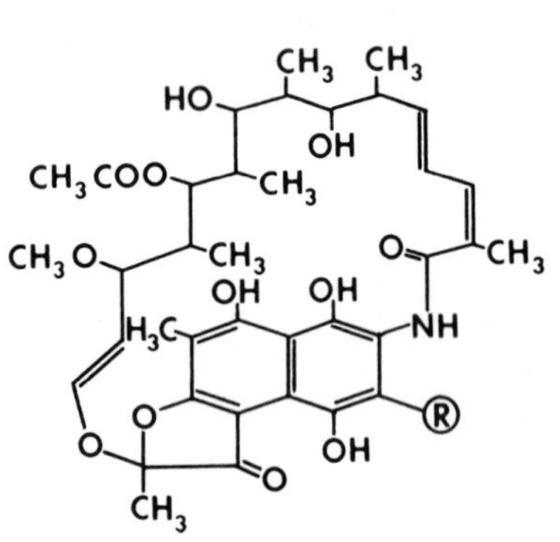

R=-H   Rifamycin

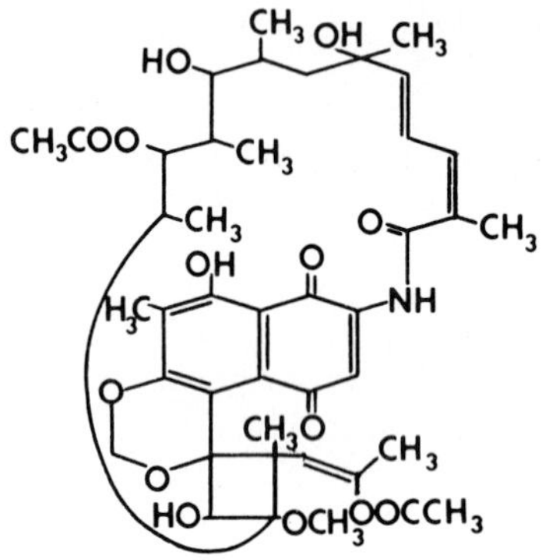

Streptovaricin A

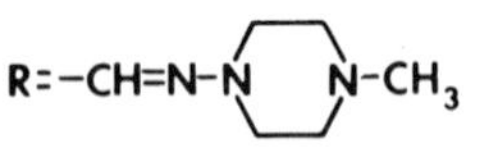

R=-CH=N-N   N-CH₃   Rifampin

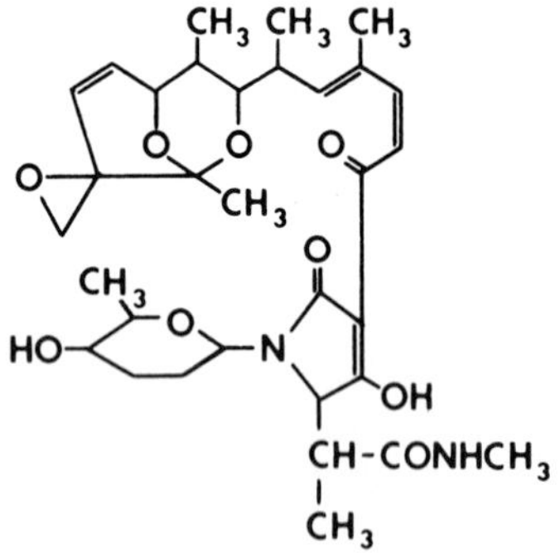

Streptolydigin

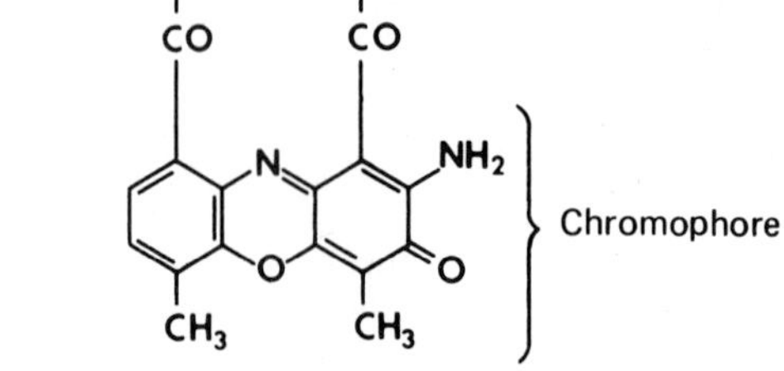

Actinomycin D

one antibiotic are also resistant to the others. The binding site is on the β-subunit: rifampin binds tightly and specifically, in a 1:1 molar ratio, to this isolated subunit; and resistant mutants have an altered subunit, with decreased affinity for the inhibitor. These antibiotics do not prevent binding of the enzyme to DNA but they prevent completion of the process of initiation. However, once that complex is completed **these antibiotics do not intefere with chain extension.**

Chain extension can be blocked in two ways. **Streptolydigin** (Fig. 11-34) does so by binding to the polymerase; and resistant mutants have an altered polymerase. In contrast, **actinomycin D** binds to helical DNA at GC pairs; the chromophore (Fig. 11-33) intercalates into the helix, as demonstrated earlier for acridines, and the attached cyclic peptides appear to bind to the external surface of the DNA. Like all antibiotics that bind to DNA, actinomycin inhibits both transcription and replication, but the former is much more sensitive. This difference suggests that the unwinding of the helix in replication, behind a swivel point, can dislodge actinomycin, whereas the local, transient unwinding in transcription is probably less forceful and can therefore be easily blocked by the reinforcing effect of the antibiotic on the helix.

Other antibiotics that bind to DNA and inhibit transcription and replication, with various relative sensitivity, include nogalomycin, daunomycin, chromomycin, mithramycin, olivomycin, anthiramycin, phleomycin, and bleomycin. Many of these have intercalating polycyclic chromophores with attached sugars.

The use of rifampin to permit RNA chain initiation, and actinomycin to block all RNA synthesis, has proved most valuable in studying the kinetics of RNA synthesis and breakdown. With many intact organisms, however, including *E. coli,* the effective use of some of these inhibitors requires the elimination of a permeability barrier, either by chemical treatment or by mutation.

Animal cells have several RNA polymerases, which function in different locations (see Ch. 47). They are insensitive to rifamycins, but one of them

is inhibited by a fungal cyclic peptide α-amanitin. The selectivity of rifamycins has made them useful in antibacterial chemotherapy (especially against tuberculosis). Agents that act on DNA, of course, act equally on the DNA of animal and bacterial cells; actinomycin has thus been useful in tumor chemotherapy but not in antimicrobial chemotherapy.

REVERSE TRANSCRIPTION

The enzyme **RNA-dependent DNA polymerase** (or **reverse transcriptase**) present in some viruses (leukoviruses, see Ch. 63) inverts the normal flow of genetic information by synthesizing DNA on an RNA template. The DNA appears to be an intermediate for the replication of the viral RNA. Besides being present in viruses, such enzyme may be present in cells, where it has been implicated in **gene amplification,** e.g., in the accumulation of many copies of the DNA genes for rRNAs in nuclei of amphibian oocytes. The enzyme produces gene amplification in vitro: the DNA of the globin gene has been synthesized from the mRNA extracted from reticulocytes.

The reverse transcriptase contains two subunits one of which is an RNase specific for DNA–RNA hybrids (RNase H). The enzyme differs in its action from regular DNA polymerases (see Ch. 10) for its ability to recognize native RNAs as template. However, it is similar to *E. coli* polymerase I in many other properties (see Fig. 10-22): 1) it requires a template-primer combination; in contrast to the *E. coli* enzyme reverse transcriptase can use either RNA or DNA as template; 2) the product is covalently bound to the primer; and 3) synthesis occurs in 5′ → 3′ direction. With viral RNA as template the primer is probably supplied by a hairpin folding (as in Fig. 10-22C); with globin mRNA a short polythymidylic acid is used to pair (as in Fig. 10-22E) with a polyadenylic acid tail at the end of the RNA (see Ch. 47). The enzyme produces first a DNA-RNA hybrid, and later, perhaps after hydrolyzing the RNA strand with its RNase, single- and double-stranded DNA. The activity is inhibited by rifampin derivatives.

APPENDIX

QUANTITATIVE ASPECTS OF KILLING BY IRRADIATION

We shall consider here certain mathematical relations between the dose of radiation and its lethal effect which will be useful later for understanding problems involving uses of radiation in research and for sterilization.

The inactivation or death of a microorganism is defined as the loss of its ability to initiate a clone; this effect is the consequence of a certain number of chemical **events,** each consisting, for instance, in the unrepaired change of a chemical group or the **breaking of a chemical bond.** In a population, those individuals that have experienced enough of these events are inactivated. The relation between the dose of radiation and the proportion of surviving organisms can be calculated as follows.

It is assumed that the events occur at random and independently in the susceptible chemical groups in various individual organisms, with a probability (p) per group, proportional to the dose.

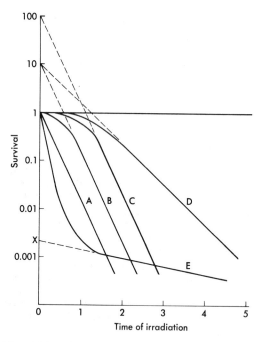

Fig. 11-35. Survival curves of microorganisms irradiated by UV light. A, a single-hit curve; B, a 10-hit curve; C, a 100-hit curve, all with the same target size. D, a 10-hit curve for organisms with a smaller target size than that of B. E, a multi-component curve produced by a populaton containing organisms with different target sizes. The dose rate is assumed to be constant.

If there are n susceptible groups per organism, there will be, **on the average,** pn inactivating events per organism. If a single such event is sufficient for inactivation, the proportion of surviving organisms (S) which have experienced no such event is, from the Poisson distribution (Ch. 44, Appendix), $S = e^{-pn}$. In turn, $p = kD = krt$, where k is a constant that measures the probability of unrepaired damage of the chemical group, D is the dose of radiation, r is the dose rate (i.e., the dose of radiation per unit time), and t is the time. Thus the basic equation of inactivation is $S = e^{-krtn}$. The equation is used in its logarithmic form: $\log S = -krtn \log e = -Krtn,$ where $K = k \log e$.

When the inactivation of bacteria or viruses by UV light, X-rays, or ^{32}P decay is followed by plotting the surviving fraction (S) versus time on semilogarithmic paper, the **survival curve** is often seen to be a **straight line passing through the origin:** curve A in Figure 11-35. Such a curve is called a **single-hit** curve, since it is generated by a process in which a single **event** in an organism destroys its viability.

The slope of a one-hit survival curve, $-Krn$, is the basis for radiation analysis of **target size;** or as a first approximation this slope is proportional to the size of the genome (e.g., n) if the proportion of repaired damages is constant. If K is the same for two different organisms, the ratio of the values of $Kn,$ or **relative target size,** can be directly determined from the slopes of their survival curves. However, differences in efficiency of repair influence the value of K and hence the observed target size.

Other important types of survival curves, called **multiple-hit curves,** are represented by curves B, C, and D in Figure 11-35; they have a shoulder near the origin before becoming linear, because several events (hits) must accumulate in a viable unit before it is inactivated. To determine the number of events required for inactivation, the straight part of the survival curve is extrapolated back to meet the ordinate axis. The position of the intersection measures the number of hits in logarithmic units. Curve B corresponds to 10 hits and curve C to 100 hits; curve D also corresponds to 10 hits but has a different slope. The slope of the straight part of a multiple-hit curve has the same meaning as for a single-hit curve, i.e., it is proportional to the target size of the organism. Thus, curves A, B, and C reflect the inactivation of organisms with the same target size.

Multiple-hit curves are obtained whenever one susceptible component can replace another in the reproduction of the unit whose survival is being measured. For instance, when bacteria are infected by several UV-irradiated virus particles, curves of this type describe the survival of the ability to yield active virus (Ch. 45, Multiplicity reactivation). Similarly, in clumps of bacteria, the survival of a single organism is sufficient to maintain the colony-forming ability of a clump. Since the straight part of the curve is reached when the surviving clumps have only one surviving organism, its slope measures the inactivation (and the target size) of the last surviving individual in the clump. Curve A would then describe the inactivation of single organisms, curve B clumps of 10 organisms, and curve C clumps of 100 organisms.

Another mechanism generating multiple-hit curves is a decrease of repair efficiency at high radiation doses. Thus postreplication repair becomes inefficient when the distances between dimers are short and recombination is infrequent.

A third type of survival curve is **multicomponent** (E in Fig. 11-35), generated by a population composed of organisms with different target sizes. Those with the larger target, i.e., more sensitive, are inactivated first with a steep slope, while at the end only the most resistant ones survive, with a lower slope for their survival curve. The proportion of this group can be estimated by extrapolating back the final straight part of the curve to the ordinate axis (about 2×10^{-3} in Fig. 11-35). Curves of this type are commonly obtained when viruses are exposed to chemical inactivating agents (Ch. 45), where the differences in sensitivity depend primarily on differences in penetration. Recognition of these curves is important in the preparation of safe vaccines.

SELECTED REFERENCES

Books and Review Articles

DRAKE, J. W. *The Molecular Basis of Mutation.* Holden-Day, San Francisco, 1970.

FREESE, E. Molecular mechanisms of mutation. *Chem. Mutagens 1:*1 (1971).

MILLER, R. W. Delayed radiation effects in atomic-bomb survivors. *Science 166:*569 (1969).

MOLDAVE, K., and GROSSMAN, L. Nucleic acid, part 7. In *Methods in Enzymology,* vol. 21. Academic Press, New York, 1971.

SETLOW, R. B. Cyclobutane-type pyrimidine dimers in polynucleotides. *Science 153:*379 (1966).

Transcription of genetic material. *Cold Spring Harbor Symp Quant Biol 35* (1970).

TRAVERS, A. Control of transcription in bacteria. *Nature New Biol 229:*60 (1971).

WITKIN, E. M. The role of DNA repair and recombination in mutagenesis. In *Proceedings, XII International Congress on Genetics,* vol. 3, p. 225, 1969.

ZILLIG, W., ZECHEL, K., RABUSSAY, D., SCHACHMER, M., SETHI, V. S., PALM, P., HEIL, A., and SEIFERT, W. On the role of different subunits of DNA-dependent RNA polymerase from *E. coli* in the transcription process. *Cold Spring Harbor Symp Quant Biol 35:*47 (1970).

Specific Articles

ANRAKU, N., and TOMIZAWA, J. Molecular mechanisms of genetic recombination of bacteriophage. *J Mol Biol 12:*805 (1965).

BRENNER, S., BARNETT, L., CRICK, F. H. C., and

ORGEL, A. The theory of mutagenesis. *J Mol Biol 3:*121 (1961).

BURGESS, R. R., TRAVERS, A. A., DUNN, J. J., and BAUTZ, E. K. F. Factor stimulating transcription by RNA polymerase. *Nature 221:*43 (1969).

CLEAVER, J. E. Defective repair replication of DNA in xeroderma pigmentosum. *Nature 218:*652 (1968).

COUKELL, M. B., and YANOFSKY, C. Influence of chromosome structure on the frequency of *tonB trp* deletions in *Escherichia coli. J Bacterial 105:*864 (1971).

COX, E. C., and YANOFSKY, C. Altered base ratios in the DNA of an *Escherichia coli* mutator strain. *Proc Nat Acad Sci USA 58:*1895 (1967).

DAVIS, R. W., and DAVIDSON, N. Electron-microscopic visualization of deletion mutations. *Proc Nat Acad Sci USA 60:*243 (1968).

DI MAURO, E., SNYDER, L., MARINO, P., LAMBERTI, A., COPPO, A., and TOCCHINI-VALENTINI, G. P. Rifampicin sensitivity of the components of DNA-dependent RNA polymerase. *Nature 222:*533 (1969).

FRANKLIN, N. C. Extraordinary recombinational events in *Escherichia coli.* Their independence of the *rec+* function. *Genetics 55:*699 (1967).

HOWARD-FLANDERS, P., BOYCE, R. P., and THERIOT, L. Three loci in *Escherichia coli.* K-12 that control the excision of pyrimidine dimers and certain other mutagen products from DNA. *Genetics 53:*1119 (1966).

HOWELL, S. H., and STERN, H. The appearance of DNA breakage and repair activities in the synchronous meiotic cycle of *Lilium. J Mol Biol 55:*357 (1951).

OKADA, Y., AMAGASE, S., and TSUGITA, A. Frameshift mutation in the lysozyme gene of bacteriophage T4: Demonstration of the insertion of five bases, and a summary of in vivo codons and lysozyme activities. *J Mol Biol 54:*219 (1970).

OKADA, Y., STREISINGER, G., OWEN, J. (E.), NEWTON, J., TSUGITA, A., and INOUYE, M. Molecular basis of a mutational hot spot in the lysozyme gene of bacteriophage T4. *Nature 236:*338 (1972).

ROBERTS, J. W. Termination factor for RNA synthesis. *Nature 224:*1168 (1969).

RUPP, W. D., and HOWARD-FLANDERS, P. Discontinuities in the DNA synthesized in an excision-defective strain of *Escherichia coli* following ultraviolet irradiation. *J Mol Biol 31:*291 (1968).

SETLOW, R. B., REGAN, J .D., GERMAN, J., and CARRIER, W. L. Evidence that xeroderma pigmentosum cells do not perform the first step in the repair of ultraviolet damage to their DNA. *Proc Nat Acad Sci USA 64:*3 (1969).

TOWN, C. D., SMITH, K. C., and KAPLAN, H. S. DNA polymerase required for rapid repair of X-ray-induced DNA strand breaks in vivo. *Science 172:*851 (1971).

WRIGHT, M., and BUTTIN, G. The isolation and characterization from *Escherichia coli* of an adenosine triphosphate-dependent deoxyribonuclease directed by *rec* B, C genes. *J Biol Chem 246:*6543 (1971).

chapter 12

PROTEIN SYNTHESIS

MACHINERY OF PROTEIN SYNTHESIS   272

TRANSFER RNA   272
Charging of tRNA   272
Structure of tRNAs   272
GENETIC CODE   275
Degenerate Triplet Code   275
Direction of Translation   279
Evolution of the Code   279
RIBOSOMES   281
Structure   281
Assembly   283
POLYPEPTIDE CHAIN ELONGATION   284
Ribosomal Microcycle   284
Chain-elongation Factors   284
INITIATION AND TERMINATION   286
The Initiating tRNA   286
Role of Ribosomal Subunits   287
Initiation Factors   288
Chain Termination; Release Factors   290
Intergenic Dividers   290
SUMMARY OF THE CYCLES IN PROTEIN SYNTHESIS   291

MODIFICATIONS OF PROTEIN SYNTHESIS   292

GENETIC SUPPRESSION   292
Genotypic Suppression   292
Mechanism of Codon-specific Translational Suppression 293
Efficiency of Suppression   294
Genotypic Ribosomal Ambiguity   295
Phenotypic Suppression   296
INHIBITORS OF PROTEIN SYNTHESIS   296
Classification   296
Antibiotics Acting on Recognition   297
Antibiotics Acting on Peptidyl Transfer   298
Antibiotics Acting on Translocation   299
Streptomycin and Related Aminoglycosides   300
Other Actions on Protein Synthesis   302
Summary   303

THREE-DIMENSIONAL PROTEIN STRUCTURE   304

GENETIC DETERMINATION OF PROTEIN STRUCTURE   304
Specification of Higher Order Structures   304
Structural Features Related to Function   305
Intragenic Complementation   305
Summary: Significance of Quaternary Structure   306
EVOLUTION OF PROTEINS   307

> *Art and Science cannot exist but in minutely organized Particulars*
> *And not in generalizing Demonstrations of the Rational Power.*
> WILLIAM BLAKE (JERUSALEM)

Some years ago it emerged that biochemists studying the mechanism of polypeptide formation and geneticists studying the mechanism of gene expression were really involved in the same problem. The convergence of these approaches contributed to the rapid unraveling of the complex process of protein synthesis, in which a specific transfer RNA (tRNA) serves as the "adapter," on a ribosome, between a codon in a messenger RNA (mRNA) molecule (Ch. 10) and the cognate amino acid. We shall review this material briefly and then consider related topics that are of particular importance for genetics and for microbiology: the genetic code for translating a language of 4 bases into one of 20 amino acids; the suppression of mutational errors by errors in translation; the effect of various antibiotics on protein synthesis; and the conversion of the one-dimensional product into a three-dimensional, functional protein. The discovery of mRNA will be discussed in the next chapter because it is so intimately linked with the analysis of gene regulation.

MACHINERY OF PROTEIN SYNTHESIS

TRANSFER RNA

CHARGING OF tRNA

In translation each **trinucleotide codon** pairs with a complementary **anticodon** sequence in a corresponding tRNA. Each tRNA is charged with the proper amino acid by a specific aminoacyl-tRNA (aa-tRNA) synthetase (ligase):

$$\text{Amino acid} + \text{ATP} + \text{Enz}$$
$$\downarrow$$
$$\text{5'-Aminoacyl} - \text{AMP} - \text{Enz} + \text{PP}_i$$
$$| \text{ tRNA}$$
$$\downarrow$$
$$\text{Aminoacyl} - \text{tRNA} + \text{AMP} + \text{Enz.}$$

The product has a rather high-energy ester linkage, between the amino acid and the terminal 3'-OH of the tRNA, which is subsequently used in peptide bond formation. Accurate translation requires a high level of fidelity in all three specific reactions: enzyme-amino acid, enzyme-tRNA, and codon-anticodon. The conventional designation of a specific (e.g., **valine**) uncharged tRNA is tRNA^{Val}, and the corresponding aminoacylated tRNA is **valyl**-tRNA or Val-tRNA.

The synthetases ordinarily have an affinity appropriate for the amino acid concentration maintained in the cell by endogenous synthesis. However, a mutation reducing that affinity can make a strain "**pseudoauxotrophic**": growth requires addition of the corresponding amino acid, though its biosynthetic pathway is unimpaired.

STRUCTURE OF tRNAs

The tRNAs are small (75 to 85 nucleotides) and are easily purified by chromatography. Their similarity of function is reflected in a striking similarity of general structure, which permits cocrystallization of different tRNAs. Sequencing, first achieved by Holley, revealed several regions of potential base pairing, suggesting a "cloverleaf" structure with four arms and three loops, plus a variable extra arm (Fig. 12-1); this secondary structure has been confirmed by chemical and physical studies.

Fig. 12-1. Generalized clover-leaf model of tRNA. I–IV = unpaired regions (loops); a–e = base-paired regions; solid small circles with centered dots = base pairs; R = purine; Y = pyrimidine; T = ribothymidine; ψ = pseudouridine; arrow = 5′ → 3′ direction. Letters indicate nucleotides common to all sequences (but often with substituents, not indicated). Circled nucleotides, joined by light lines, are known to be paired or adjacent in the tertiary structure; these contacts suggest the folding depicted in Figure 12-2. [From Levitt, M. *Nature* 224: 759 (1969).]

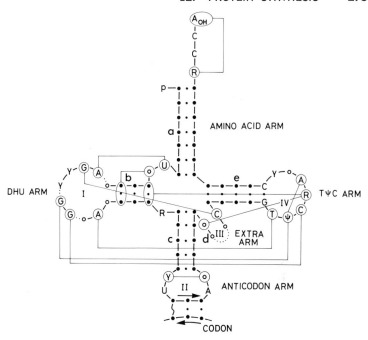

Loop I (starting from the 5′ end) is called the **dihydroU (DHU)** loop (or arm) because in most tRNAs it contains this residue. The **anticodon** loop (II) was identified by the presence of the predicted anticodon sequence in various specific tRNAs; moreover, many mutant tRNAs of altered specificity have a nucleotide substitution in that sequence (see Genotypic suppression, below). Arm III (the "extra arm") is highly variable in length. Loop IV (**pseudoU** loop) always contains a GTψC sequence, which may play a role in binding to the ribosome since an accessible complementary sequence has been identified in ribosomal RNA.

The several base-paired arms, though constant in location, vary widely in sequence. However, the 3′ terminus that accepts the amino acid is regularly -C-C-A, which is presumably recognized, along with the aminoacyl ester, by the ribosomal enzyme that incorporates the various amino acids (regardless of side chain) into the growing polypeptide.

tRNAs also have extensive tertiary structure, much like proteins. Hydrodynamic properties and X-ray crystallography reveal a long, narrow molecule (ca. 25 × 35 × 90 A), which fits the requirement, noted below, for the close parallel binding of two tRNAs on a ribosome. The specific tertiary structure has been inferred primarily from determining which residues are accessible or inaccessible to various chemical reagents. Thus the DHU loop is so tucked in that one of its bases can form a covalent bond, on ultraviolet irradiation, with a base at the junction between arms a and b (Fig. 12-1). A neighbor in the same loop evidently pairs with a

base at the 3′ end of the extra arm, since mutation in one of these bases releases the other for chemical substitution. These and other interactions suggest the model presented in Figure 12-2B in which the four arms are stacked in such a way as to lie almost on a single axis. By high resolution X-ray crystallography Rich has demonstrated a sharp bend (Fig. 12-2C), indicating that at least in one conformation the molecule has an L-shape.

Aminoacylation, which adds a positively charged amino group, significantly alters the shape of tRNA. This difference presumably contributes to the binding of aa-tRNA and release of free tRNA by the ribosome.

The anticodon is always flanked at its 5′ end by two successive pyrimidines, which provide a flexible region with little stacking, while at the 3′ end two purines provide a rigid region with strong stacking. Accordingly, as Figure 12-3 shows, the regular helical pairing of anticodon to codon allows a certain degree of ambiguity ("wobble") at the 5′ end but not at the 3′ end of the anticodon (see Genetic code, below).

Specific codon recognition can be achieved with tRNA fragments containing only the anticodon loop, and enough of the stem to close it. Recognition of the correct synthetase appears to involve primarily the stem near the CCA end (a in Fig. 12-1), together with features of tertiary structure; this recognition is not altered by profound changes in the loops.

tRNAs are synthesized as longer chains (precursor tRNA), which then undergo enzymatic modification: about 40 nucleotides at the 5′ end and 3

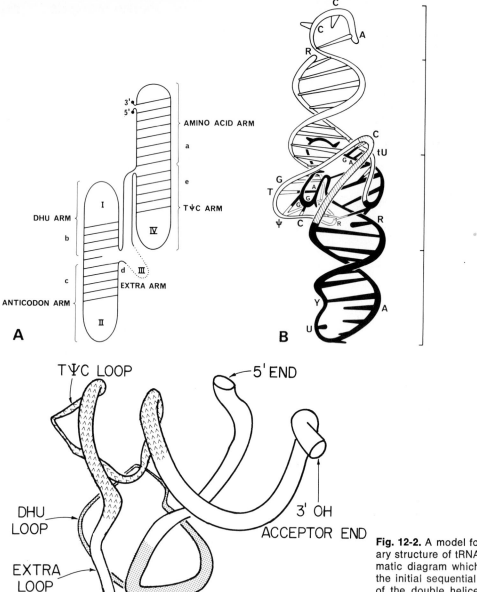

Fig. 12-2. A model for the tertiary structure of tRNA. **A.** Schematic diagram which indicates the initial sequential alignment of the double helices of arms a and e, and the similar alignment of arms b and c. **B.** A less schematic diagram, based on a model with permissible bond distances and angles. The structure in scheme A is further folded, to provide the contacts indicated as light lines in Figure 12-1. The four helical arms are almost completely aligned along a single axis. [A and B from Levitt, M. *Nature* 224:759 (1969).] **C.** Three-dimensional structure of yeast phenylalanine tRNA inferred from X-ray crystallography. [Kim, S. H., et al., *Science* 179:285 (1973)].

AMINO ACID ARM

a

e

TΨC ARM

DHU ARM

I

IV

b

c

d III

EXTRA ARM

ANTICODON ARM

II

A

B

TΨC LOOP

5' END

DHU LOOP

3' OH

ACCEPTOR END

EXTRA LOOP

ANTICODON LOOP

C

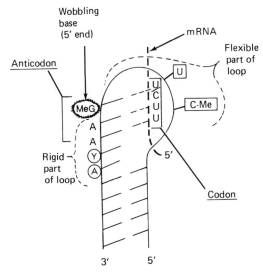

Fig. 12-3. Schematic diagram of the anticodon loop in phenylalanine tRNA. Y is an alkylated purine. The 3' end of th anticodon is rigid, owing to the strong stacking of the two adjacent purines (circled). In contrast, the 5' end is flexible, since the two adjacent pyrimidines (boxed) have only weak stacking. In the diagram this short flexible segment is stretched out as a crude way of indicating that it does not prevent the codon from pairing with the anticodon to extend the double helix. However, the end of this helix is distorted by the subsequent flexible segment, and so the 5'-end base of the anticodon can wobble, resulting in decreased specificity of pairing. In this case the 5'-end MeG can pair, as indicated, with either U or C. [Modified from Fuller, W., and Hodgson, A. *Nature 215*:277 (1967), and Ghosh, K., and Ghosh, H. P. *Biochem Biophys Res Commun 40*:135 (1970).]

at the other are removed by specific nucleases, and up to 20% of the bases are modified (mostly by methylation).

tRNA can be visualized as an evolutionary product that has retained the base pairing of a nucleic acid while gaining some of the properties characteristic of proteins. These include the presence of a large variety of residues (modified bases), which may promote the formation of a specific tertiary structure. Indeed, mutational replacement of a single residue in tRNA, as in proteins, can yield a temperature-sensitive product.

Degeneracy of tRNAs. Fractionation by column chromatography often yields several

distinct tRNAs that transfer the same amino acid (isoaccepting tRNAs); some recognize the same and some recognize different codons.

The tRNAs of mammalian cells are generally similar to those of bacteria. However, differentiation and malignant transformation sometimes change the number and the chromatographic mobilities of isoaccepting tRNAs in different tissues; the main effect is a decrease in methylation, due to inhibition of tRNA methylases. These alterations may conceivably influence the selective translation of different genes.

GENETIC CODE

DEGENERATE TRIPLET CODE

Since 4 kinds of bases must code for 20 kinds of amino acids more than 1 base must be used to specify an amino acid. As 2 would produce only 16 (i.e., 4^2) different combinations, the minimum is 3, although it produces more combinations than are necessary ($4^3 = 64$). The excess is used by assigning more than one codon to the same amino acid. The code is thus **degenerate;** but it is not **ambiguous,** because no codon normally specifies more than one amino acid. Its general nature as a triplet code was established by ingenious experiments of Crick *et al.* with multiple frame-shift mutations in bacteriophage, as discussed in Chapter 10 (Units of genetic information).

In Vitro Evidence. In 1961 Nirenberg and Matthaei discovered that polyU, a synthetic homopolymer of uridylic acid, directs the formation of polyphenylalanine: hence UpUpU (designated also as UUU) is the coding triplet for phenylalanine. Following this fortunate observation much of the code was rapidly cracked by in vitro studies with synthetic polynucleotides of known composition and random sequence. But though these studies revealed the **composition** of many codons they did not reveal the **sequences.** These became known when Nirenberg and Leder found that isolated trinucleotides, of known sequence, can code for binding of corresponding aa-tRNAs to ribosomes. However, some of the results were ambiguous since additional factors influence the binding. A definitive approach came from the use of polyribonucleotide messengers consisting of

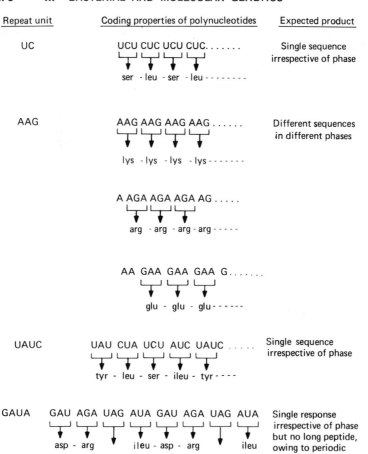

Fig. 12-4. Use of repeat polynucleotides for determining the genetic code. The polynucleotides were used as messengers in vitro in conjunction with a protein-synthesizing system from *E. coli*. The polypeptide chains produced were isolated and analyzed, and their composition defined the coding properties of the contributing triplets. [Adapted from Khorana, G. *Harvey Lect* 62:79 (1968).]

First letter of triplet (5' end)	Second letter of triplet				Third letter of triplet (3' end)
	U	C	A	G	
U	Phe⌉ Phe⌋ Leu Leu	Ser⌉ Ser⌋ Ser Ser⌋	Tyr⌉ Tyr⌋ Ochre Amber	Cys⌉ Cys⌋ Opal Try⌉	U C A G
C	Leu⌉ Leu⌋ Leu Leu	Pro Pro Pro Pro	His⌉ His⌋ Gln Gln	Arg⌉ Arg⌋ Arg⌋⌉ Arg	U C A G
A	Ileu⌉⌉ Ileu⌋ Ileu⌋ Met	Thr⌉ Thr Thr⌋ Thr⌋	Asn Asn Lys⌉ Lys⌋	Ser⌉ Ser⌋ Arg⌉ Arg⌋	U C A G
G	Val Val Val Val⌋ or Met	Ala⌉ Ala Ala⌋ Ala⌉	Asp Asp Glu⌉ Glu⌋	Gly⌉ Gly⌋ Gly⌉ Gly⌋	U C A G

Fig. 12-5. The genetic code. Glu, Asp = glutamic and aspartic acid; Gln, Asn = glutamine and asparagine; ochre, amber, opal = terminator codons. Shaded triplets code for polar amino acids. This diagram also summarizes the pattern of degeneracy (Figs. 12-6 and 12-7). Brackets at the right of amino acids indicate codons known to be recognized by the same tRNA, in which wobbling of the 5'-end base of the anticodon leads to ambiguous reading of the 3' end of the codon. Some codons are recognized, as indicated by more than one anticodon.

specific **repeating** sequences. Khorana prepared these valuable reagents by chemically synthesizing the complementary oligodeoxynucleotides, enzymatically converting them into longer chains of repeating units, and then transcribing these with RNA polymerase. These polynucleotides generated polypeptides with repeat sequences of amino acids, as illustrated in Figure 12-4. Such studies rapidly deciphered the entire genetic code, shown in Figure 12-5.

Origin of Degeneracy. The degeneracy of the code results from a mixture of **tRNA degeneracy** (multiple tRNAs) and **anticodon degeneracy** (multiple readings by a given tRNA). The latter depends on the loose positioning of the base at the 5′ end of the anticodon (Fig. 12-3), which allows nonstandard pairing. Indeed, there are 29 theoretically possible ways in which pairs of normal bases with either two or three hydro-

Fig. 12-6. Rules for anticodon degeneracy determined by wobble of the 5′-end base of the anticodon corresponding to the 3′ end of the codon. Arrows indicate possible stable pairing, from anticodon to codon. I = inosinic acid (hypoxanthine nucleotide). [Adapted from Crick, F. H. C. *J Mol Biol* 19:548 (1966).]

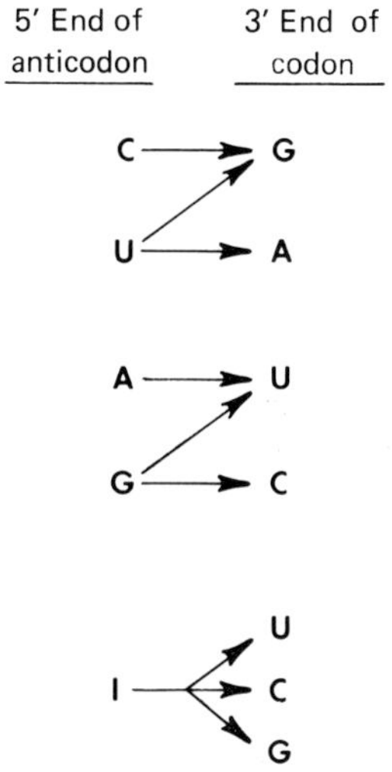

5′ End of anticodon	3′ End of codon

```
C ――→ G
       ↗
U ――――→ A


A ――→ U
       ↗
G ――――→ C


         U
       ↗
I ――←――→ C
       ↘
         G
```

gen bonds can be formed, and from a consideration of their angles and distances Crick accounted for the known patterns of degeneracy by predicting the specific **wobbles** depicted in Figure 12-6. These have been confirmed by the sequencing of many tRNAs. The mixture of the two kinds of degeneracy is illustrated in Figure 12-7.

In Vivo Evidence. The code that was elucidated in vitro could be shown to apply also in the cell. Thus when two frame-shift mutations of opposite sign were introduced by acridine (Ch. 11) into a gene (for T4 phage lysozyme or for *E. coli* tryptophan synthetase) the amino acid substitutions between the two mutations were found to be precisely those predicted (Fig. 12-8).

Evidence is also provided by the amino acid substitutions that are found in nature in abnormal human hemoglobins (Fig. 12-9). These substitutions are almost all consistent with a single base change, which would be expected to be the most frequent mutation yielding an altered product. Further confirmation was obtained with nitrous acid mutagenesis, which produced only those amino acid substitutions, in tobacco mosaic virus and *E. coli* tryptophan synthetase, that correspond to the expected specific nucleotide replacements induced by this agent (A → G and C → U; see Ch. 11, Mutagenesis).

Punctuation. Since the end of one codon is not separated by a special signal from the beginning of the next, punctuation is required to ensure accurate positioning (reading frame) at the beginning of translation. Chain synthesis must also be terminated at a specific position. AUG and GUG serve for initiation, and UAG, UAA, or UGA for termination.

The terminating triplet sequences were identified by producing nonsense mutations, which interrupt chain completion, in well-studied proteins, such as *E. coli* alkaline phosphatase. A variety of revertants were then isolated, in which insertion of an amino acid in the initially mutant site allowed chain completion; the protein was then purified from each revertant and the inserted amino acid was determined. Figure 12-10 shows these insertions, which could all be derived by single base substitutions only from UAG. This terminator was named **amber** (translated

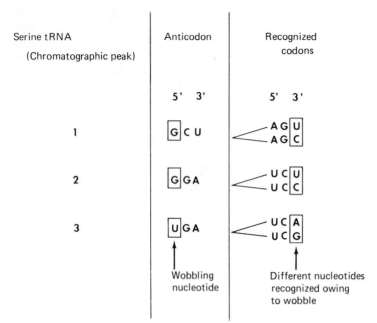

Serine tRNA

(Chromatographic peak)

Anticodon

Recognized codons

Fig. 12-7. The origin of degeneracy of the genetic code is illustrated with the serine codons. Part of the degeneracy derives from the existence of three tRNAs, each with a different anticodon. The other part derives from 5'-end wobbling of the anticodon, which results in 3'-end degeneracy in the codon.

Fig. 12-8. Consequences of shift of the reading frame in *E. coli* tryptophan synthetase. I. The normal correspondence between mRNA and polypeptide chains; the vertical lines indicate the reading frame. II. Acridine-induced **deletion** of a nucleotide, indicated as a circled minus sign, shifts the reading frame and causes production of a jumbled, inactive protein (indicated by heavy box, open on right). III. A second mutation, causing the **insertion** of a nucleotide at the circled plus sign, restores the reading frame beyond that point. The result is a polypeptide chain with only a short jumbled segment, compatible with function. [Data from Brammar, W. J., Berger, H., and Yanofsky, C. *Proc Natl Acad Sci USA* 58:1499 (1967).]

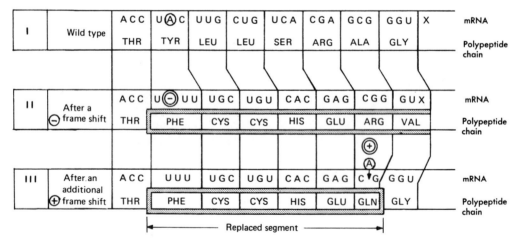

Fig. 12-9. Amino acid replacements in abnormal human hemoglobins caused by spontaneous mutations. Each replacement can be explained by a change of a single nucleotide in the assigned codons.

Original amino acid	Assigned codon	Amino acid in the mutant	Assigned codon	Replacement
Lysine	AAA ⟶ Glutamic Acid		GAA	A ⟶ G
Glutamic Acid	GAA ⟶ Glutamine		CAA	G ⟶ C
Glycine	GGU ⟶ Aspartic Acid		GAU	G ⟶ A
Histidine	CAU ⟶ Tyrosine		UAU	C ⟶ U
Asparagine	AAU ⟶ Lysine		AAA	U ⟶ A
Glutamic Acid	GAA ⟶ Valine		GUA	A ⟶ U
Glutamic Acid	GAA ⟶ Lysine		AAA	G ⟶ A
Glutamic Acid	GAA ⟶ Glycine		GGA	A ⟶ G
Valine	GUA ⟶ Glutamic Acid		GAA	U ⟶ A

from the German name of a contributor, Bernstein).

Similar analyses, and conversions of one terminator to another by a mutagen specific for an A-to-G replacement, identified the other terminator codons as UAA (**ochre**) and UGA (**opal**). These deductions were later confirmed by binding studies: these three trinucleotides do not bind any normal aa-tRNA, but they do bind a termination factor (see below).

Fig. 12-10. Identification of an amber nonsense triplet. The diagram summarizes the amino acid substitutions found in various revertants of an amber mutant, in a position originally occupied by a tryptophan residue. The standard codons for the substituted amino acids are listed: the underlined codons are those related to UAG by a single base change. UAG is the only triplet that has this relation to at least one codon for each of the substituted amino acids. [From Garen, A. *Science 160:* 149 (1968).]

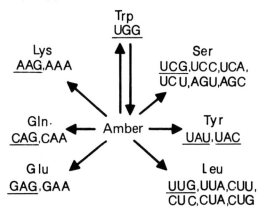

Trp
UGG

Lys
AAG,AAA

Ser
UCG,UCC,UCA,
UCU,AGU,AGC

Gln·
CAG,CAA

Amber

Tyr
UAU,UAC

Glu
GAG,GAA

Leu
UUG,UUA,CUU,
CUC,CUA,CUG

DIRECTION OF TRANSLATION

Pulse labeling showed that the polypeptide chain starts at its N terminus and adds residues at its C terminus (Fig. 12-11). With this knowledge, the translation of known oligonucleotides could establish the direction of reading of mRNA. For example, AAAUUU $(5' \rightarrow 3')$ must be translated $5' \rightarrow 3'$, since it codes in vitro for Lys-Phe ($NH_2 \rightarrow COOH$) and not for Phe-Lys. This direction also fits the constitution of the jumbled polypeptide segment synthesized between two frame shifts (Fig. 12-8).

As we shall see in Chapter 13 (Fig. 13-14A), in bacteria translation and transcription occur simultaneously on the same mRNA, in the same direction, and at the same rate.

EVOLUTION OF THE CODE

Universality. The same code is found in man, *E. coli,* and a plant virus; hence all present terrestrial life may have had a common (monophyletic) origin. Moreover, the code could not change once life had achieved a certain degree of complexity, for a mutation that shifted the meaning of any codon would alter the synthesis of practically every protein. However, organisms vary in their use of the different codons. Thus in different bacteria the proportion of G + C in DNA ranges from about 30% to 70%. These differences fall within the limits of degeneracy of the

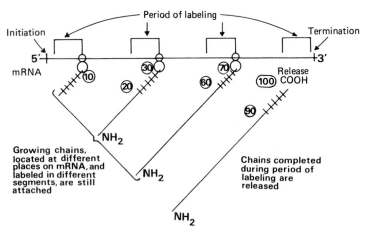

Fig. 12-11. Direction of synthesis of polypeptide. Reticulocytes synthesizing predominantly one protein (hemoglobin) were briefly pulsed with a radioactive acino acid. Those globin chains that were completed and released during the period of labeling were isolated, partly hydrolyzed, and "fingerprinted" by paper chromatography. Only fragments derived from the COOH-terminal region were found to be radioactive. Segments of the same length are labeled in chains not yet completed, but these chains remained attached to ribosomes and were thus eliminated before analysis. Incorporation at the COOH end was also demonstrated in pulse-labeled bacteria by attacking the total protein with carboxypeptidase, which sequentially hydrolyzes proteins from this end. [After Dintzis, H. M. *Proc Natl Acad Sci USA* 47:247 (1961) and Goldstein, A., Goldstein, D. B., and Lowney, L. I. *J Mol Biol* 9:213 (1964).]

code; they may be accounted for by selection for the use of triplets with C or G rather than U or A (or vice versa) at the less specific 3′ end of codons (see Ch. 11, Mutations affecting mutation frequency).*

In contrast to the code itself, the specificity of the **machinery** for its translation is not universal. Thus *E. coli* enzymes can charge some tRNAs of yeast but not others. Obviously the enzyme-recognizing region of a tRNA molecule can change in ways that are balanced by changes in the enzyme. Ribosomes of various organisms also differ in their specificity for tRNAs of various origins.

Evolutionary Advantages. Though the term "degeneracy" suggests a sloppy consequence of the evolution of 64 codons to specify 20

* DNA rich in GC would have codons primarily of the type XXG or XXC; that poor in GC would have codons of the type XXA or XXU. With equal nucleotide frequencies in the first two positions, in codons of the first type the proportion of G + C would be $(\frac{1}{2} + \frac{1}{2} + 1) /3 = \frac{2}{3}$; in those of the second type it would be $(\frac{1}{2} + \frac{1}{2} + 0) /3 = \frac{1}{3}$.

amino acids plus some punctuation, it is easy to see an evolutionary benefit in this pattern. A new gene arises by duplication of a pre-existing gene, followed by selection for a series of missense mutations that yield an increasingly useful protein. With a nondegenerate code of 64 codons for 20 units most nucleotide replacements would give rise to a nonsense triplet, which would halt further improvement of the affected protein; with a degenerate code most mutations yield an altered protein.

The present code minimizes the consequences of mutation and of incorrect base pairing: the ratio of missense to nonsense mutations is high; in most codons a transition $(AT \rightleftharpoons GC)$ in the 3′ position would not change the amino acid specified (yielding nearly a 2½-letter code); and most single-nucleotide shifts lead to a codon for a similar amino acid (e.g., polar or nonpolar; see shading in Fig. 12-5). It thus seems that if terrestrial life had to start again it might well develop nearly the same genetic code.

RIBOSOMES

STRUCTURE

An *E. coli* cell contains up to 15,000 ribosomes, the number depending on the growth rate (Ch. 13). Their integrity in extracts, and their function, are very sensitive to ionic conditions. When the cells are disrupted by grinding with an abrasive (e.g., alumina), in the presence of suitable concentrations of Mg^{++} (2-20 mM) and K^{+} (50-100 mM), the lysate contains mostly single ribosomes, of sedimentation constant 70S, MW 2.6 × 10^6, and diameter ca. 180 A. However, if growing *E. coli* cells are disrupted more gently (e.g., by enzymatic digestion of the cell wall) ca. 80% of the ribosomal particles are recovered as **polysomes** (Fig. 12-13). In these the ribosomes are connected by a strand of mRNA, which is very sensitive to cleavage by RNase.

At low Mg^{++} concentrations (0.1 to 1.0 mM) or high K^{+} the ribosome dissociates into a **large (50S)** and a **small (30S) subunit.** Electron micrographs with negative stain reveal a groove between the subunits of a ribosome and also show that each subunit has an irregular shape, quite unlike the symmetry of viruses (Fig. 12-12). The composition of the subunits, each with a large RNA molecule (23S or 16S) and many protein molecules, is given in Figure 12-13.

Gently prepared lysates of growing cells also contain ca. 10% of the ribosomal particles as single 70S ribosomes and 10% as "native subunits" (see below). The 70S ribosomes are of two kinds: complexed (with peptidyl-tRNA and mRNA) and free (i.e., lacking these ligands). They are readily distinguished because the free ribosomes are easier to dissociate into subunits (e.g., by lowering the Mg^{++} concentration).

The ribosomal proteins and RNA are for

Fig. 12-12. Electron micrographs of negatively stained polysomes and ribosomes. **A.** Rat liver polysomes, with arrows showing mRNA strands connecting ribosomes. **B.** Various views of *E. coli* 70S ribosomes, showing irregular shape, densely stained cavity, and grooves. **C.** Various views of *E. coli* 50S subunit. **D.** Photographs of plasticene model based on views in **C.** The small subunit presumably fits into the hollow formed by the main body and the projecting "nose." **E.** "Mother and Child," sculpture by Walter Hannula. [**A** from Nonomura, Y., Blobel, G., and Sabatini, D. *J Mol Biol 60*:303 (1971). **B** Courtesy of Y. Nonomura and D. Sabatini. **C** and **D** from Lubin, M. *Proc Natl Acad Sci USA 61*:1454 (1968). **E** Reproduced with permission of Alva Museum Replicas, Inc., Long Island City, N.Y.]

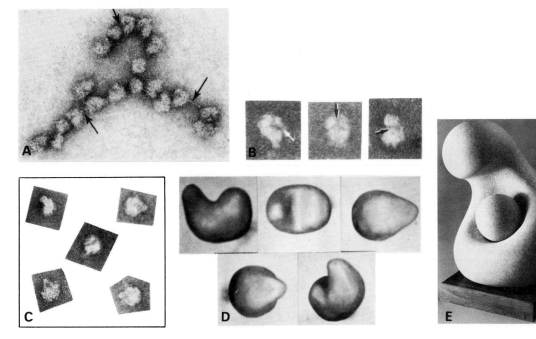

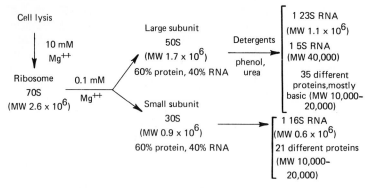

Fig. 12-13. Composition of bacterial ribosomes.

the most part tightly associated with each other, and rather drastic means are required for their dissociation and solubilization (phenol-H_2O to precipitate the protein and recover the RNA in the aqueous phase; 2 M LiCl-urea, or acetic acid, for the reverse). Gel electrophoresis and chromatography have revealed that the small bacterial subunit contains 21 different proteins and the large subunit 34. Most of these proteins have a low MW (10,000 to 20,000). Wittmann has isolated all the proteins from *E. coli* ribosomes and prepared antibodies to most; except for L7 and L12 (which differ only in acetylation), each ribosomal protein is antigenically unique. These antibodies are proving useful in blocking the binding of various ligands to the ribosome and thus identifying the proteins that provide various binding sites. Another approach to mapping the **topography** of the ribosome is the use of cross-linking reagents to form covalent bridges between neighboring proteins.

The proteins of the large subunit (designated L1 to L34) are each present as one molecule per ribosome. Among the proteins of the small subunit (S1 to S21) some are present in a 1:1 molar ratio but others are "fractional" (i.e., recovered as 0.3 to 0.7 molecules per subunit. It is not certain whether this result reflects functional heterogeneity of this class of subunits or artificial partial loss of some components during purification. In fact, the small subunit is known to be looser than the large one, to have a less uniform shape in electron micrographs, and to be more easily inactivated on exposure to abnormal ionic conditions.

Ribosomal RNA (rRNA), which constitutes the bulk of cellular RNA, does not have the base ratios characteristic of double-stranded nucleic acid, and the base composition may be very different from that of the DNA of the same cell (Table 12-1). Nevertheless, this RNA does have a high degree of secondary structure, for denaturation produces a large hyperchromic shift (Ch. 10). Unlike the RNA in viruses, which is

TABLE 12-1.   Nucleotide Composition of Ribosomal RNA

Organism	rRNA from large subunit (mole %)				DNA (mole %)
	C	A	U*	G	GC
Bacillus subtilis	22.8	26.4	20.7	30.0	42
Escherichia coli	21.0	25.5	21.0	32.5	50
Pseudomonas aeruginosa	22.1	26.3	21.5	30.0	67
Saccharomyces cerevisiae	19.2	26.4	26.0	28.4	36
Man	32.2	15.9	16.8	35.1	40

* Including pseudoU.

Based on Attardi, G., and Amaldi, F. *Annu Rev Biochem 39*:183 (1970).

entirely surrounded by protein, that in ribosomes is interdigitated with the proteins: both are susceptible to enzymatic attack and chemical derivatization, and all the proteins of the small subunit have at least part of their surface accessible for binding antibodies.

rRNA may well function in tRNA binding, for 5S RNA contains a loop with a sequence complementary to one regularly present in loop IV (Fig. 12-1) of tRNAs.

The ribosomes of **eukaryotic cell cytoplasm** are larger than those of bacteria (80S, made of a 60S and a 40S subunit). They have somewhat larger RNA molecules (28S and 18S, as well as 5S) and an even greater increase in protein content. The ribosomes in chloroplasts, however, are as small as those of bacteria, while those of animal cell **mitochondria** are even smaller; both types resemble bacterial ribosomes in their patterns of antibiotic sensitivity. It has accordingly been suggested that mitochondria have evolved from cytoplasmic symbionts, though selective pressure for a small ribosome in mitochondria also seems plausible.

ASSEMBLY

In 1969 Nomura assembled active small (30S) subunits of bacterial ribosomes from their separated RNA and proteins; assembly of large subunits (50S) followed. This formation of an active particle from 58 inert component molecules is a milestone in explaining the function of a cell in terms of the organized interactions of its parts.

Success depended on two special features: a salt concentration high enough to prevent nonspecific aggregation, but without preventing specific assembly, and a rather high temperature (ca. 40°). At 0°, which was previously used to minimize protein denaturation and enzymatic degradation, improper contacts or folds are evidently "frozen," as in the renaturation of DNA.

Assembly in vitro provides a useful tool for identifying the proteins coded for by various "ribosomal" genes. For example, when various single proteins from streptomycin-resistant (str^r) ribosomes (see below, Aminoglycosides) were used to replace the corresponding protein in a mixture from sensitive (str^s) ribosomes, only protein P4 (= S12) was found to confer resistance on the reassembled ribosome. The genes for many ribosomal proteins are clustered (at ca. 64 minutes) on the *E. coli* map (Fig. 9-9); they may well be transcribed as a unit (see Operons, Ch. 13), since a block in translation of one stops the translation of all those on one side of it (see Polarity, Ch. 13).

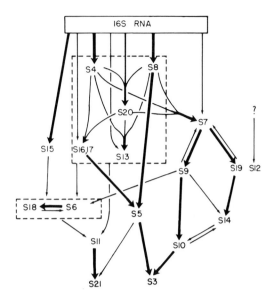

Fig. 12-14. Assembly map of small subunit of *E. coli* ribosome. Arrows indicate the effect of one protein on the binding of another; thick arrows represent a strong effect and thin arrows a weak effect. Boxes in dashed lines enclose a group of proteins with an effect not yet traced to specific proteins. [After Nomura, M. *Science* 179:864 (1973).]

Assembly evidently involves an orderly sequence, since cold-sensitive subunit-assembly-deficient (*sad*) mutants have been isolated, which form active ribosomes at 40° but accumulate specific incomplete precursor particles at 20°. Analysis of possible sequences of assembly in vitro has given rise to **assembly mapping,** which sheds light on this process. It also sheds light on the **topographical relations** of the ribosomal components, which is a major problem since the lack of repeating structures makes the ribosome much more complex than a virus. It has been found that some proteins can bind independently to an rRNA and protect specific sequences from attack by nuclease; but some proteins require the preceding binding of certain other proteins (Fig. 12-14). Moreover, the proteins that bind first (i.e., directly to rRNA) bind to the region (toward the 5' end) that is synthesized first, suggesting that ribosomal proteins rapidly assemble in the cell on still nascent RNA. The mechanism and regulation of ribosome formation in the cell will be further considered in Chapter 13.

Evolution. It is not surprising that the ribosome, with its intricate interdependence of many components, has changed very slowly in evolution. Thus bacteria with gross differences in average DNA composition retain strikingly similar ribosomal genes, as shown by their similar rRNA composition (Table 13-1), by efficient hybridization between rRNA and DNA from different organisms, by more extensive integration of ribosomal genes than of other genes in interspecific gene transfer, and by the assembly of active ribosomes from a mixture of rRNA from one species and ribosomal proteins from another.

POLYPEPTIDE CHAIN ELONGATION

Protein synthesis involves a **macrocycle** of ribosomal attachment to mRNA and subsequent release, with many intervening **microcycles** of amino acid addition.

RIBOSOMAL MICROCYCLE

We have already seen (Fig. 12-11) that a polypeptide chain grows at its carboxyl end. The major features of the process of chain elongation were worked out with polyU-coded synthesis of polyphenylalanine. Gilbert showed that the growing chain is attached by ester linkage to a tRNA (peptidyl-tRNA, pp-tRNA), like the amino acid in aa-tRNA. Moreover, both aa-tRNA and pp-tRNA are bound to the ribosome noncovalently, since they can be released by altered ionic conditions. These findings suggested a cycle in which pp-tRNA alternately occupies two adjacent sites on the ribosome: the **P** (peptidyl or donor) and the **A** (aminoacyl or acceptor) site.

The ribosomal microcycle has three major steps (Fig. 12-15):

1) **Recognition:** a trinucleotide codon in the mRNA, residing in the **recognition region** of the A site, specifies the binding of a particular aa-tRNA by base pairing with its complementary anticodon (c in Fig. 12-15).

2) **Peptidyl transfer:** the nascent polypeptide replaces its ester bond to tRNA by a peptide bond with the α-amino group of the aa-tRNA in the A site (c in Fig. 12-15); the peptide is thus transferred from the P to the A site and becomes one residue longer.

3) **Translocation:** the pp-tRNA moves back from the A to the P site, and the mRNA moves with it. The free tRNA (b) left by the preceding peptidyl transfer is thus displaced, and the next untranslated codon (d) is brought into the A site. The ribosome is now ready for step (1) again. Releasability of the polypeptide by puromycin, a small analog of aa-tRNA (see Fig. 12-25 below), is an operational test for the presence of pp-tRNA in the P rather than the A site.

Though this model is undoubtedly oversimplified, its general features are well established. The two subunits have different roles, which have been revealed in part by studies under conditions that exaggerate binding affinities (e.g., high Mg++ concentrations, addition of alcohol). Thus mRNA can bind to the small subunit (though contact also with the large subunit is not excluded). With aa-tRNA both subunits evidently contribute binding energy: at high Mg++ specific binding (directed by a codon) can be demonstrated with the small subunit, and nonspecific binding (without mRNA) with the large subunit. The peptidyl transfer reaction can be carried out by the large subunit alone under certain conditions, as shown by transfer to puromycin.

CHAIN-ELONGATION FACTORS

Chain elongation requires not only ribosomes, Mg++, K+, GTP, and aa-tRNAs, but also additional components of the supernatant (i.e., the "soluble" portion of the cell lysate, from which the ribosomal particles have been sedimented). Lipmann showed that three factors are involved, called T_u and T_s (for unstable and stable transfer factors) and G (for GTPase). These elongation factors are now designated as EFT_u, EFT_s, and EFG (though the initial distinction between T and G functions does not fit present knowledge); they have been crystallized.

The T factors function in **recognition.** EFT_u forms a complex with aa-tRNA and GTP, which is the form in which the aa-tRNA initially binds to the A site of the ribosome. The GTP is then hydrolyzed and $EFT_u \cdot GDP$ is released. The GDP is displaced from this complex by EFT_s (Fig. 12-16), allowing the regeneration of an $EFT_u \cdot aa$-tRNA $\cdot$ GTP complex.

A nonhydrolyzable analog of GTP, GDP-CH_2-P (also called GMPPCP), can also promote the initial binding of an $EFT_u \cdot aa$-tRNA complex, but it cannot provide phosphate bond energy and peptidyl transfer does not follow. The energy is not required directly for peptide bond formation, however, because even in the absence of GTP this bond can be formed with puromycin (see Fig. 12-25

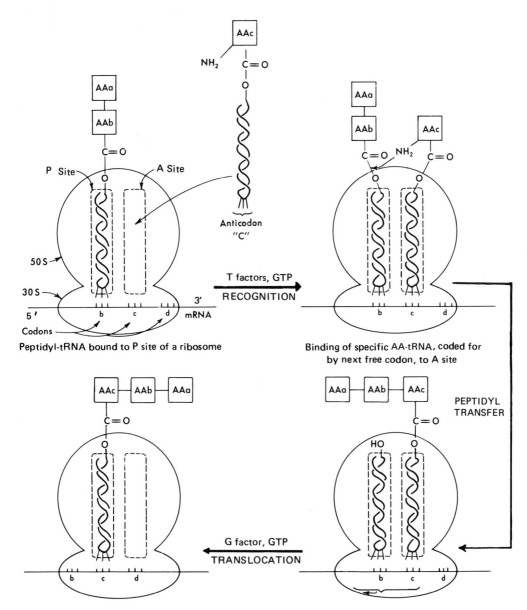

Fig. 12-15. Schematic representation of the addition of an amino acid to a growing polypeptide chain.

below). Hence GTP hydrolysis evidently provides energy to "lock" the aa-tRNA in a position where it can accept peptidyl transfer.

This two-stage recognition sequence permits delicacy to be combined with strength. Thus the initial interaction with aa-tRNA must have barely enough energy for binding, since lack of pairing of even one nucleotide of a triplet will lead to rejection; but before accepting transfer of the expensive polypeptide chain the aa-tRNA is firmly "locked in" by an enzymatic reaction deriving energy from GTP hydrolysis.

Factor EFG functions in **translocation:**

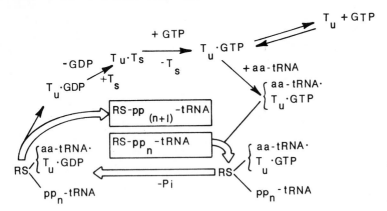

Fig. 12-16. T-factor cycle. One $EFT_u \cdot EFT_s$ complex (abbreviated in diagram as $T_u \cdot T_s$) must be available for each incorporation of an amino acid. RS = ribosome.

without it a ribosome can form a dipeptide but not a longer peptide. A complex of this factor and GTP binds to the ribosome (on its large subunit), where the GTP is hydrolyzed in the course of translocation; EFG and GDP are then released.

The binding of EFG is specifically blocked by antibodies to only one species of ribosomal protein (the essentially identical L7 and L12). EFT_u and EFG cannot bind to the ribosomes simultaneously: fixation of the latter (by fusidic acid: see below) prevents binding of the former. It is not certain whether the sites overlap or whether the conformations required are mutually exclusive.

Translocation is a very complex reaction: Chemical energy from GTP is converted into mechanical energy, i.e., into conformational changes that move pp-tRNA and mRNA extensively on the surface of the ribosome while keeping them firmly bound. A two-step model has been suggested, in which the tRNA retains a firm grip on each subunit in turn while moving relative to the other.

Eukaryotic ribosomes similarly require a recognition factor (transferase I) and a translocation factor (transferase II). Evolution has handled the machinery of protein synthesis conservatively: ribosomes from prokaryotes and from eukaryotic cytoplasm, despite a substantial difference in size, can function with some tRNAs and even with the translocation factor from either class of organisms.

INITIATION AND TERMINATION

As we have noted above, mRNA must carry the information for initiating translation at the proper position and with the proper reading frame: hence a special initiating tRNA seemed likely. However, a direct search was unsuccessful. Indeed, all the special components of initiation (ribosomal subunits, three protein initiation factors, and a special tRNA) were discovered more or less accidentally and assigned a function afterward. The crucial step was the discovery, by Marcker and Sanger, that hydrolysis of tRNA carrying labeled methionine yields not only methionine but also N-formylmethionine, which was soon found to participate in the complex process of initiation.

Ironically, a major reason for the delay in our understanding of initiation was the use of convenient **synthetic** messengers; for while these could shed much light on the genetic code and the ribosomal microcycle they bypassed the process of physiological initiation. Progress depended on the introduction of the RNA of small RNA viruses (Ch. 44) as a stable **natural** messenger.

THE INITIATING tRNA

The bacterial cell has two tRNAs for Met, designated as $tRNA_F$ and $tRNA_M$; a specific enzyme attaches a formyl group (HCOO-) only to the Met in $Met-tRNA_F$, with formyltetrahydrofolate as donor. The resulting N-formyl-$Met-tRNA_F$ ($fMet-tRNA_F$) serves as the sole initiating tRNA in bacteria, fixing the reading frame at AUG (or GUG). Thus without $fMet-tRNA_F$ the repeating copolymer $(AUG)_n$ codes for three different amino acid homoploymers, each depending on a different starting frame (see Fig. 12-4); while with $fMet-tRNA_F$ present only polymethionine is made.

Though an initiating AUG codon is recognized only by $fMET-tRNA_F$ AUG çodons **within** a gene are recognized only by $Met-tRNA_M$. It may be that secondary structure makes intragenic AUG codons inaccessible to initiation, and the advancing ribosome opens up that structure. Thus in phage RNA the AUG at the beginning of a gene is present in a sequence of seven unpaired nucleotides at the end of a hairpin loop, while the same sequence also occurs within the gene but in a paired helix, where it does not initiate. Moreover, initiation at internal sites becomes possible when the secondary structure is opened by treatment with formaldehyde.

The special function of fMet-tRNA$_F$ seems to depend on both the structure of its tRNA and its formylation (which simulates the peptidyl bond in pp-tRNA). Thus neither unformylated Met-tRNA$_F$ nor chemically formylated Met-tRNA$_M$ can form an initiation complex; conversely, unformylated Met-tRNA$_F$ (unlike Met-tRNA$_M$) cannot complex with elongation factor T$_u$ and hence cannot incorporate Met in an internal position in a polypeptide chain. The two tRNAs have the same anticodons and no striking differences in sequence, but tRNA$_F$ has a shorter amino acid stem (by 1 nucleotide pair) than other tRNAs; its striking specificity may be due to this feature or to some aspect of tertiary structure.

An initiation site sequenced in viral RNA has been identified as AUG. In synthetic messengers fMet-tRNA$_F$ can also read GUG, which normally codes for valine (Fig. 12-5), but the role of this codon in initiation is not established.

Eukaryotes have a similar system of initiation, with an initiating and a noninitiating Met-tRNA species, but the initiating species is unformylated in the cytoplasm of mammalian cells (though it is formylated in yeast). This initiation without formylation emphasizes the unique structure of the initiating tRNA. The two mammalian species are called Met-tRNA$_F$. and Met-tRNA$_M$. because, curiously, the former can be formylated by the *E. coli* transformylase, and it can then initiate with *E. coli* ribosomes.

Enzymatic Modification of the fMet-Polypeptide. Before completion the polypeptide loses its formyl group. Moreover, though half the proteins found in *E. coli* have N-terminal methionine the rest have also lost this residue. The enzymes involved in these cleavages evidently recognize relatively long polypeptides, since in vitro no cleavage occurs with a phage protein halted as a hexapeptide on the ribosome.

ROLE OF RIBOSOMAL SUBUNITS

Even under ionic conditions that do not cause dissociation of ribosomes into subunits a

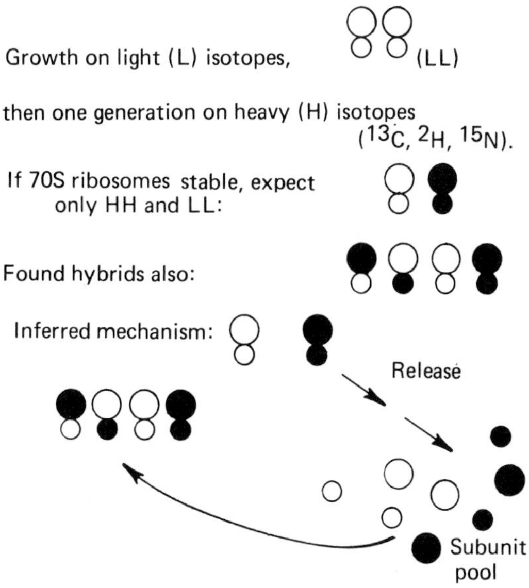

Fig. 12-17. Exchange of subunits by density-labeled ribosomes in *E. coli.* The "light" ribosomes sediment in a sucrose gradient at 70S and the "heavy" ribosomes at ca. 76S. "Hybrid" (HL) ribosomes (one heavy, one light subunit) were detected as a large peak of intermediate S value; they could also be distinguished from HH and LL ribosomes by equilibrium density gradient centrifugation. [After Kaempfer, R., Meselson, M., and Raskas, H. *J Mol Biol 31*:277(1968).]

small proportion (ca. 10%) of the particles in cell lysates are always present as subunits. The significance of these "native" subunits remained obscure for a decade, until Kaempfer and Meselson showed, by the use of heavy isotopes (Fig. 12-17), that ribosomes exchange subunits during protein synthesis. Hence they evidently go through a pool of subunits at some stage in their macrocycle. Nomura defined the role of subunits more specifically by showing that the small subunits can form complexes with the initiating fMet-tRNA$_F$ and natural messenger (phage RNA), even more effectively than 70S ribosomes.

Work with purified components established the following sequence (see Fig. 12-19):

1) The free 70S "runoff" ribosome released from a polysome is dissociated into two subunits before initiation.

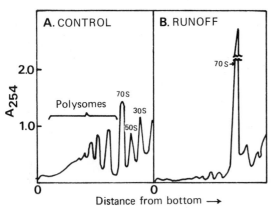

Fig. 12-18. Production of 70S ribosomes by runoff in *E. coli*. Control growing cells **(A)** were rapidly chilled by pouring the culture onto ice. "Runoff" cells **(B)** were prepared by starvation for the carbon source for 5 minutes before chilling. (A smaller sample was analyzed because of the high 70S peak.) The cells were gently disrupted by treatment with lysozyme (to lyse the wall) and deoxycholate (to lyse the membrane). The lysates were analyzed in a sucrose gradient (see Ch. 10, Zonal centrifugation). Similar results were obtained when net runoff was promoted by starvation for a required amino acid or a required nucleic acid base, or by treatment with puromycin to release ribosomes prematurely, or by treatment with actinomycin to cause depletion of mRNA. [After Subramanian, A. R., Davis, B. D., and Beller, R. J. *Cold Spring Harbor Symp Quant Biol 34*:233 (1969).]

2) mRNA and fMet-tRNA$_F$ bind to the small subunit to form a first-stage "30S" initiation complex.

3) A large subunit joins this complex to yield a second-stage, completed "70S" initiation complex.

4) The completed complex transfers the fMet to the next aa-tRNA, establishing the first peptide bond and yielding a peptide-bearing "polysomal" ribosome.

The subunit mechanism of initiation permits the ribosome to be bound to the mRNA in a way that ensures firm attachment and yet permits movement along the chain.

A rather long piece of the mRNA is "buried" in the ribosome: when polysomes are treated with a trace of an RNase that readily attacks exposed single-stranded RNA a segment of 30 nucleotides is protected by each ribosome. With ribosomes blocked at initiation of a viral gene these segments are so uniform that they can be sequenced.

Mechanism of Ribosome Dissociation. The products of polysome runoff can be caused to accumulate in cells through the use of antibiotics or metabolite deprival to accelerate ribosome release or impair reinitiation. These products are then found as free 70S ribosomes; the level of the subunits remains unchanged (Fig. 12-18). Moreover, these accumulated ribosomes have been released from polysomes without dissociation and reassociation, since experiments with mixed "heavy" and "light" polysomes in vitro, similar to that of Fig. 12-17, showed that the ribosomes do not exchange subunits during runoff.

The conversion of runoff ribosomes into stable subunits could be shown to depend on complexing with a **dissociation factor,** later identified with initiation factor IF$_3$ (see below). This protein causes dissociation of free (but not of complexed) ribosomes, in a stoichiometric reaction. The limited supply of the factor thus **regulates the level of subunits** in the cell.

The dissociation reaction might involve a **direct** interaction of the factor with the ribosome, actively displacing the large from the small subunit; or the factor might interact only with a trace of 30S subunits in spontaneous equilibrium with the ribosomes, thereby shifting the equilibrium. These alternative molecular mechanisms are closely analogous to the two models (shift of equilibrium or induced fit) proposed for the action of allosteric effectors (Ch. 13, Mechanism of allosteric interactions):a conformational change is expressed in one case as dissociation into subunits and in the other as an alteration in enzyme activity.

INITIATION FACTORS

The requirement for special initiation factors was recognized when Ochoa discovered that bacterial ribosomes purified by washing with 1 M NH$_4$Cl had lost their activity with viral mRNA as messenger (which initiates physiologically) but remained active with polyU (which is used at Mg^{++} concentrations high enough to permit nonspecific initiation without AUG or fMet-tRNA$_F$). Moreover, the proteins in the wash fluid could restore full activity with viral mRNA. Chromatography yielded three required factors: IF$_1$, IF$_2$, and IF$_3$.

IF$_3$ apparently acts first: as noted above, it

causes free ribosomes to dissociate into sub-units and then remains complexed with the small subunit (see Fig. 12-19), which also binds IF_1 and IF_2. Addition of mRNA, fMet-tRNA$_F$, and GTP yields a **30S initiation complex,** which is joined by the large subunit to form a **70S initiation complex.** Hydrolysis of GTP then places the fMet-tRNA in the puromycin-reactive P site and releases initiation factors, just as translocation does with pp-tRNA and EFG.

Just as a nonhydrolyzable analog of GTP permits the EFT$_u$·aa-tRNA complex to be bound in the A site but does not lock it into a reactive position (see above), so fMet-tRNA can be bound with the analog but will not then react with puromycin.

The process of recognition of fMet-tRNA apparently does not involve the A site, since it is not affected by antibiotics that block recognition in the

A site or those that block translocation from the A to the P site. Moreover, GTP hydrolysis during initiation does not shift the position of the messenger on the ribosome, since the portion of AUG (U)$_n$ protected from RNase action is unchanged.

IF_1 and IF_2 function in initiation much as EFT$_s$ and EFT$_u$ do in chain elongation (Fig. 12-16): i.e., IF_2 can form a complex with fMet-tRNA in solution (as well as on the initiating 30S particle); and after GTP hydrolysis on the 70S complex IF_1 is required for the release of IF_2 and GDP. Both these factors are required for binding of fMet-tRNA with tri-nucleotide Ap Up G, as well as with natural mRNA, but IF_3 is required only with the latter: it converts ribosomes to the subunits required for physiological initiation, and it also appears to select the initiating region on the mRNA. Two "interference factors" (factor i), which complex with IF_3, also influence the specificity of initiation: infection with phage T4 alters their concentrations and thereby promotes initiation on phage rather than on host mRNA.

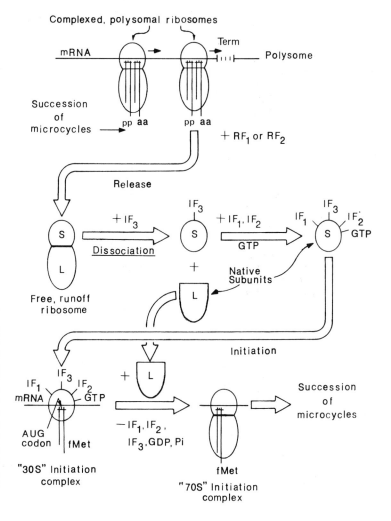

Fig. 12-19. Ribosome-polysome macrocycle. S = small subunit; L = large subunit; T = termination triplet. IF_3 stabilizes S in a conformation that does not couple with L. At some stages in the formation of the 70S initiation complex factors IF_1, IF_2, and IF_3 are released, and they then reattach to a small subunit from a free, runoff ribosome. [After Subramanian, A. R., and Davis, B. D. *Proc Natl Acad Sci USA* *61*:761 (1968).]

Initiation in eukaryotic systems also involves ribosomal subunits and requires initiation factors.

Initiation-factor Cycle and Ribosome Macrocycle.

Unlike the elongation factors, which are present mostly in the supernatant fraction of the cell lysate, the initation factors are detectable only in the native small subunits; they are not found in polysomal or free ribosomes. This special location showed that these factors cycle off the ribosomal particles during completion of initiation. Indeed, evidence for a cycle of attachment and detachment is essential for distinguishing a true factor from an easily eluted ribosomal protein.

Figure 12-19 diagrams the ribosome-polysome macrocycle, interlocking with the initiation-factor cycle.

CHAIN TERMINATION; RELEASE FACTORS

The nonsense triplets cause not only chain termination but also rapid release; hence they must recognize a special release factor. Capecchi isolated this factor by using a highly purified system (containing all the other factors required for protein synthesis) to translate a phage RNA with an early termi-

nation mutation in one of its genes: fractions from a cell extract could then be assayed for their activity in releasing the resulting short polypeptide from the ribosomes. (In a simpler assay, developed later, fMet is released from fMet-tRNA bound in the presence of ApUpG plus a termination trinucleotide.) The release factor (RF) turned out to be a protein without any RNA, although it reads a codon.

Indeed, there are two release factors: RF1 recognizes either UAG or UAA, and RF2 either UAA or UGA. These factors promote hydrolysis of the bond between polypeptide and tRNA, i.e., peptidyl transfer to water. This reaction appears to involve ribosomal peptidyl transferase (see above), since inhibitors of transferase also block release (Fig. 12-20).

An additional factor from the supernate (ribosome release factor = RRF) is required to release the ribosome from the mRNA (and free its tRNA) after polypeptide release.

INTERGENIC DIVIDERS

Deletions that remove intergenic punctuation, without shifting the reading frame, cause the production of "fused" polypeptides. However, there is more to punctuation than an initiator and a terminator codon, at least in phage, in which correlation of protein and RNA sequences has revealed an apparently untranslated sequence of 36 nucleotides between two successive genes. This intergenic divider (Fig. 12-21) contains two successive terminator codons at the end of the first gene. This "double stop" may protect against the effect of mutations and translational errors that might wreck two proteins by skipping a stop.

The function of the rest of the intergenic segment is not known. Since the length of the segment is about the diameter of a ribosome it may protect initiation at a gene from being blocked sterically by the flow of ribosomes up to the end of the preceding gene. An even longer untranslated segment, however, is present at the beginning of the phage RNA.

This mechanism of spacing between genes is clearly sensible for RNA phages, in which different genes in the same RNA are independently translated. With cellular mRNA, however, the genes of an operon are translated coordinately, and it is not known whether or not there are similar spacers, and whether the ribosome is released at the end of each gene or at the end of the messenger.

Fig. 12-20. Effect of inhibitors of peptidyl transferase on polypeptide release. After fMet-tRNA was bound to *E. coli* ribosomes by incubation with initiator trinucleotide AUG various concentrations of an antibiotic were added. Peptidyl transferase activity was measured as the initial rate of formation of fMet-puromycin, on incubation with puromycin. Release was measured as the initial rate of formation of free fMet, on incubation with a release factor and terminator trinucleotide UAG. [From Vogel, Z., Zamir, A., and Elson, D. *Biochemistry 8*:5161 (1969).]

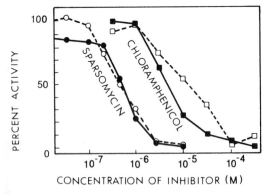

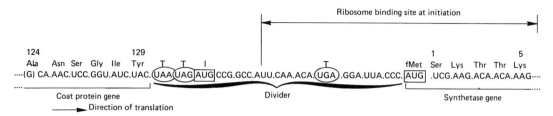

Fig. 12-21. Nucleotide sequence at the junction of two genes (an intergenic divider) of the RNA of phage R17, a polygenic messenger. Numbers above amino acids represent positions in the known COOH-terminal sequence of the coat protein and the NH₂-terminal sequence of the synthetase, which permit identification of the corresponding codons in this large oligonucleotide. The intervening sequence (the "divider") contains not only the expected terminator codon (T) at the end of the coat protein gene, and an initiator codon (I) at the beginning of the synthetase gene, but a second T immediately after the first, and an additional I and T, of unknown function, within the divider. The ribsosome binding site is the segment protected from RNase degradation in an initiation complex. [From Nichols, J. L. *Nature* 225:147 (1970).]

SUMMARY OF THE CYCLES IN PROTEIN SYNTHESIS

The ribosome goes through both a macrocycle and successive microcycles. In the **macrocycle** the free 70S ribosome released from mRNA is converted into a stable pair of subunits by complexing of the small subunit with initation factor IF₃. This subunit complexes with factors IF₁ and IF₂ and then forms a "30S" initiation complex by binding mRNA (phasing the initiator codon AUG in the recognition position), the initiating tRNA (fMet-tRNA_F), and GTP. The large subunit is then added to yield a "70S" initiation complex. GTP hydrolysis places the fMet-tRNA_F in a reactive position, the initiation factors are released to recycle, and chain elongation begins. The limited supply of IF₃ regulates the level of subunits and hence the potential rate of initiation.

In the **microcycles** of chain elongation a triplet codon in the A (aminoacyl) site of the ribosome codes for binding a cognate aa-tRNA ("recognition"). This step involves chain-elongation factors EFT_u and EFT_s and the hydrolysis of GTP. Polypeptide (or fMet), attached by an ester bond to tRNA in the P (peptidyl) site, is transferred to form a peptide bond with the aminoacyl residue of the aa-tRNA, thus yielding a longer peptidyl-tRNA, in the A site. The peptidyl-tRNA is then moved back from the A to the P site, and the mRNA is moved in parallel by the length of one codon. This translocation step involves factor EFG and hydrolysis of another GTP molecule. The ribosome is now ready for another microcycle. After many such cycles the ribosome reaches a terminator codon (UAG, UAA, or UGA), at the end of a gene, which calls for a supernatant factor (RF1 or RF2) that releases the polypeptide.

The distribution of the various factors fits their proposed functions. Each incorporation of an amino acid requires a set of chain-elongation factors, and these are present in the supernatant in abundance (around 1 of each per ribosome, constituting in all 5% of the soluble protein). The other factors, however, are used only once in the translation of each gene, and there is only about 1 molecule of RF1 or RF2 per 50 ribosomes and perhaps 1 molecule of each initiation factor per 10 ribosomes.

Protein synthesis is expensive. The peptide is a low-energy bond, and in the nonribosomal synthesis of cell wall polypeptides (Ch. 6) its formation requires expenditure of only one ~P. Yet in protein synthesis it costs four ~P bonds: two in the amino-acylation of tRNA, one in recognition, and one in translocation. Moreover, mRNA turnover (Ch. 13) requires further substantial expenditure, as does the supply of all the stable components of the machinery. This enormous price is evidently required for combining unlimited freedom of sequence with high fidelity and with reliable retention of the growing chain until its completion.

The selection pressure for accuracy must be great: with a polypeptide of 200 residues 1 error in 2000 would damage 10% of the product. Evolution therefore had to develop mechanisms that could bind mRNA and the growing peptide securely, yet allow rapid movement and dependable discrimination among similar aa-tRNAs. The two-stage binding of aa-tRNA or fMet-tRNA can thus be understood in terms of a lock-and-key analogy: it takes little energy to engage the correct key reversibly in its slot but much more energy to turn it against a spring.

A landmark in the study of protein synthesis was the formation of an active enzyme in vitro. First accomplished by Gesteland in 1967 with a lysozyme coded for by phage DNA, and by Zubay, the procedure has been successfully extended to many bacterial enzymes by using transducing phage DNA (Ch. 46) as a rich source of specific bacterial genes.

Several technics are beginning to shed light on the topographical relations of the various ribosomal proteins, and various regions of their RNA's, to each other and to the binding sites of the several ribosomal ligands. The successive binding and release of these ligands must place the ribosome in a variety of conformations, whose definition is a major challenge for the future.

MODIFICATIONS OF PROTEIN SYNTHESIS

GENETIC SUPPRESSION

GENOTYPIC SUPPRESSION

As Figure 11-22 showed, the phenotypic effect of a mutation can be reversed not only by a true reversion, which restores the original gene, but also (and more often) by a mutation elsewhere. In classic genetics such "suppression" of a mutation could be recognized when the second mutation mapped in a second, nonadjacent gene. Studies in microbial genetics have revealed several molecular mechanisms for such **extragenic** suppression. They have also revealed additional kinds of suppression: **intragenic** suppression, by mutations in the same gene as the primary mutation; and **phenotypic** suppression, by environmental factors.

The following mechanisms of genotypic suppression have been identified.

Intragenic suppressors:
1) **Intracodon** suppressors replace the mutant codon by one that differs from the original codon but nevertheless codes for an amino acid that restores protein function.
2) **Reading-frame mutations** add a shift in the frame opposite in direction to one already introduced in the gene (i.e., addition of a base following deletion of another, or vice versa). This compensatory shift restores normal reading except for the segment between the two mutations (Fig. 12-8).
3) An **amino acid substitution** in the same polypeptide, at some distance from the primary mutation, can sometimes restore a folding required for function. Thus tryptophan synthetase A of *E. coli* is in-

activated by a primary mutation (Gly → Glu) in a certain position, or by a second mutation 36 amino acid residues away (Tyr → Cys): but with **both** mutations the protein formed is active.

Extragenic suppressors:
4) **Codon-specific translational** (or **informational**) **suppressors** cause errors in the translation of a codon, which sometimes correct or alter the original mutational error and yield a functional protein. They often have **pleiotropic effects,** i.e., influence more than one phenotypic trait, both by suppressing mutations in more than one gene in the same strain and by causing mistranslation of normal genes. Bacterial strains carrying a translational suppressor mutation are designed as su^+ and those lacking it as su^- (in contrast to the usual convention of designating wild-type alleles as $+$).
5) **Generalized translational suppressors,** which are not codon-specific, will be discussed in a later section (Genotypic ribosomal ambiguity).
6) **Metabolic suppressors** of various types have also been observed. These gene-specific mutations may supply an enzyme that carries out or bypasses a blocked reaction, or they may restore activity to a mutant enzyme by altering the concentration of a cofactor or inhibitor. They have been useful primarily in the study of metabolism and will not be considered further here.
7) **A polarity suppressor** (*SuA*) will be discussed in Chapter 13 (Polarity mutations).

Translational suppressor mutations have made profound contributions to molecular genetics. They make it possible to recover and maintain mutations, otherwise lethal, in essential genes (Ch. 11, Conditional lethal mutants); they have distinguished the three termination codons; they have revealed the biological importance of imperfect fidelity in translation; and they have shown that the machinery of protein synthesis can influence this fidelity.

MECHANISM OF CODON-SPECIFIC TRANSLATIONAL SUPPRESSION

Some suppressors are **missense suppressors,** causing the substitution of one amino acid for another. The vast majority, however, are **nonsense suppressors,** which can insert a particular amino acid at the site of a particular **nonsense** codon. Closely related are **frame-shift** suppressors, which restore the correct reading frame at or near a site where it has been shifted by insertion (or deletion) of a nucleotide.

The three nonsense (or terminator) mutations (see Genetic code, above) are distinguished by their response to different sets of suppressor (su^+) mutations. The mechanism of this suppression has been analyzed with **amber mutations** in the gene for a known protein, such as alkaline phosphatases or phage coat protein (see Ch. 45).* When the cells also contain an **amber suppressor** the mutant gene produces both a protein fragment (terminated by the mutation) and a protein of normal length. It is thus clear that the suppressor causes occasional but not regular insertion of an amino acid at the site of the amber codon (Fig. 12-22).

Analysis of such products showed that **various su^+ mutations,** located in different parts of the bacterial chromosome, cause the insertion of **different amino acids.** Moreover, these amino acids each have a codon that differs from the amber codon in only one nucleotide (just as in the reversions of a nonsense codon; Fig. 12-10). These results suggested that each su^+ mutation might have changed a **single base** in some tRNA, so that it would now read the amber codon and hence

---

* Nonsense mutations are conveniently studied in phage. The mutant phages are carried in "permissive" su^+ host cells, while infection of nonpermissive su^- hosts reveals the effect of mutation.

Fig. 12-22. Genetic suppression of nonsense mutation. In the translation of an amber mutant gene (in cell or in extracts) an su^+ tRNA reads the amber codon (UAG) in competition with an R (release) protein. A particular suppressor causes incorporation of a particular amino acid at the amber site, corresponding to the su tRNA present; the various amber su tRNAs are derived from those tRNAs in which a replacement of one nucleotide can yield the required anticodon (CUA), which reads UAG in antipolar fashion. The **efficiency** of suppression is the fraction of the readings yielding complete protein. The rate of cell growth restored by an su^+ mutation depends not only on its efficiency but also on the frequency of harmful interference of the su tRNA with normal termination elsewhere in the genome.

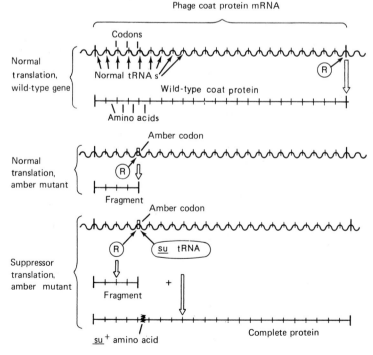

could allow continuation of the polypeptide. This hypothesis was readily confirmed by studies of protein synthesis in vitro.

Thus when phage RNA with an amber mutation was used as a messenger premature termination was observed with extracts of an *su*− strain, but extracts of an *su*+ strain could suppress this effect. Fractionation of the *su*+ extract showed that **the active component was a tRNA.** Similarly, in binding studies with ribosomes trinucleotide UpApG is inactive in normal extracts but it coded with this *su*+ extract for the binding of a seryl-tRNA (which had read UCG rather than UAG before it was changed by the *su* mutation). These properties of the *su*+ tRNA could be used to guide its isolation.

Structural Changes in Suppressor tRNAs. The expected base substitution in an anticodon has been demonstrated in some suppressor tRNAs. However, substitutions in other locations have also been found.

For example, a mutation in the DHU loop (Fig. 12-1) of a tryptophan tRNA affects the tertiary structure in a way that increases wobble, so that the unchanged anticodon CCA can now pair not only with UGG (*trp*) but also with UGA* (nonsense). Moreover, reversions from *su*+ to *su*− have re-

---

* By convention both codon and anticodon are designated starting with the 5′ end. Since they pair in antipolar fashion one member must be read backward to reveal the matching (Fig. 12-7).

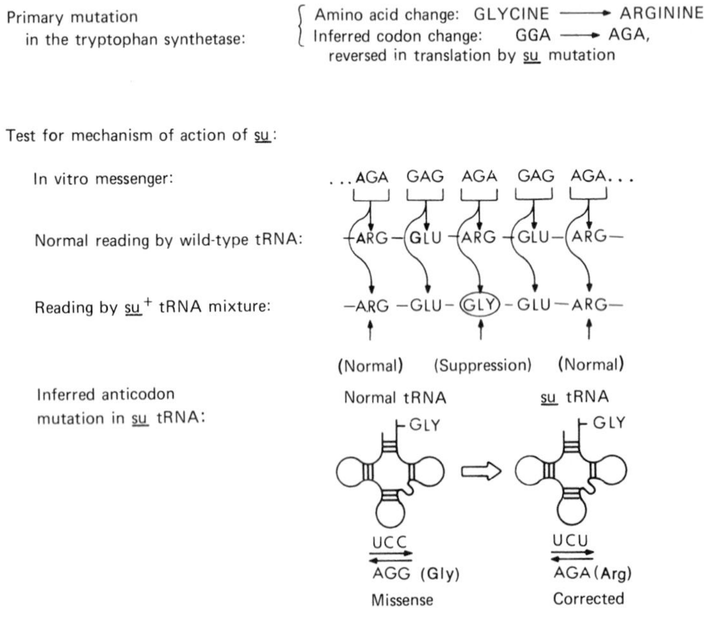

vealed various alterations that inactivated a tRNA. In particular, a mutation that eliminated the complementarity of one pair of bases in an arm yielded a **temperature-sensitive tRNA,** emphasizing the importance of tertiary structure in tRNAs just as in proteins.

The alterations produced by *su*+ mutations provide a powerful tool for correlating structure and function in tRNAs. These mutations also make it possible to map (and thus to manipulate the quantity of) the genes for the various tRNAs. These genes do not appear to be strongly clustered.

In missense suppression the mechanism (Fig. 12-23) is similar to that of nonsense suppression. **Frame-shift suppressors** have also been observed, and in one case, where a slip in replication in a sequence of three Cs has inserted an extra nucleotide in the mRNA, the suppressor tRNA reads the **four nucleotides** as CCC.

EFFICIENCY OF SUPPRESSION

Every nonsense suppressor has a characteristic efficiency (i.e., the fraction of the mutant chains that are completed). The value depends on the concentration of the *su* tRNA in the cell, its affinity for its activating enzyme, and the ability of the aa-tRNA to compete with release factors (which read the same codon) on the ribosome.

Fig. 12-23. Evidence for altered codon recognition by a missense suppressor tRNA. Unfractionated tRNA was obtained from an *E. coli* strain carrying a mutation (*su*+) that suppressed a Gly → Arg mutation in the tryptophan synthetase gene. This tRNA was tested for altered codon recognition with a repeating AG polymer, together with ribosomes and protein factors from wild-type *E. coli*. In addition to the expected incorporation of Glu (GAG) and Arg (AGA), some Gly was incorporated by a suppressor tRNA$_{Gly}$ that recognized AGA instead of GGA, the normal Gly codon. [Data from Carbon, J., Berg, P., and Yanofsky, C. *Proc Natl Acad Sci USA* 56:764 (1966).]

In cells synthesizing predominantly a single protein (e.g., phage coat protein) efficiency can be precisely measured, as the ratio of mutant fragment to completed protein. With mutant enzymes restoration of activity also provides a useful measure of efficiency, though it is not a direct measure of frequency of chain completion. Suppressors with even very low efficiency (2–5%) can be isolated, since the regulatory response to a block in a biosynthetic sequence (derepression, Ch. 13) leads to maximal gene activity, and so even a low frequency of correction of the mutational error can restore detectable growth.

Su+ mutations themselves usually slow down growth: the cell may be deprived of a valuable tRNA; and if the codon being suppressed is also normally present elsewhere in the genome its misreading there results in an error rather than a correction. Hence, paradoxically, the more efficient the suppressor of a widely distributed codon (above a certain level) the more deleterious its effect on growth. Conversely, if the suppressed codon is absent or rare in the normal genome the corrections outweigh the errors and a stronger suppressor will support faster growth.

Amber suppressors (anticodon *CUA*) recognize only amber mutations, while ochre suppressors (anticodon *UUA*) suppress amber as well as ochre. (This result would be predicted, since anticodon U wobbles but C does not (Fig. 12-2).) However, the two groups are easily distinguished, for amber suppressors are highly efficient (30 to 75% restoration) and do not themselves slow growth, whereas the known ochre suppressors are all weak (4 to 12% efficiency) and do slow growth. These ochre suppressors are mutated in a **minor tRNA,** comprising only ca. 10% of the tRNA for the same amino acid. Since there is no theoretical reason to doubt that mutations to strong suppression occur with ochre just as frequently as with amber, their failure to be recovered suggests that strong ochre suppressors would be lethal, presumably by interfering too often with normal termination. It therefore seems likely that **the ochre codon (UAA) is a frequent normal terminator,** and perhaps even the only one (but see Fig. 12-21).

UGA suppressors recognize the tryptophan codon, UGG, as well as UGA; hence only one kind, which inserts Trp, has been encountered. Growth is rapid.

Missense suppressors are weaker than amber but stronger than ochre suppressors; hence the frequent substitutions that they must cause elsewhere in the genome must usually be less damaging than failure to terminate. These suppressors are isolated much less frequently than nonsense suppressors, though they are theoretically possible in great variety. Their rarity is readily understood: unlike nonsense mutations, which prevent formation of an active enzyme wherever they occur, missense mutations have a markedly deleterious effect only in certain critical locations in a polypeptide chain; and in these locations most shifts to other amino acids, by suppression, will also be deleterious. The most frequently encountered missense suppressors replace a larger amino acid by glycine, whose unobtrusive presence is apparently more easily tolerated than most other substitutions.

tRNA Gene Duplication. The altered specificity of a tRNA is compatible with viability only when the cell also possesses an unmutated gene for the same or for an isoaccepting tRNA; otherwise its original codon could no longer be translated. Some codons seem to be served by only one gene for tRNA, since *su* mutations in these tRNAs are not found unless the cells become diploid for the appropriate region of the genome. Such organisms can be obtained not only by gene transfer but also by rearrangements within a cell that yield tandem duplication (see Fig. 10-12).

GENOTYPIC RIBOSOMAL AMBIGUITY

The fidelity of translation can also be influenced by some mutations that alter the ribosome. Thus mutations to streptomycin (Str) resistance (*str^r*), altering protein S12, **restrict** both the misreading response to the drug (see below, Phenotypic suppression) and amber (*su*) suppression. Conversely, selection for loss of this restriction has yielded a class of mutants (altered in another protein, S4) with **increased ribosomal ambiguity** (*ram*). These mutations overcome the "restricting" effect of an *str^r* mutation, moreover, when the latter is removed they cause suppression by themselves, and they also increase amber suppression. The pattern of ambiguity induced by *ram* suppresses a wide variety of mutations, including nonsense and frame shifts, in contrast to the codon-specific suppression observed with *su* mutations in tRNA.

Since *str^r* mutations restrict and *ram* mutations increase amber (*su*) suppression, these genetic distortions of the ribosome evidently influence its relative affinity for the release factor (RF) and the *su* tRNA, which are competing for a termination codon.

Normal Ambiguity. Clearly a low frequency of error is inevitable in transcription and translation in the cell, just as in DNA replication. But while the consequences of a replication error are amplified in a clone of mutant progeny, a rare substitution in a particular protein is hard to recognize. Thus incompletely blocked ("leaky") auxotrophic mutants might be forming a mutant protein of low activity, or they might be reflecting the background level of ambiguity, i.e., the formation of a small amount of fully active, normal protein along with the predominant inactive mutant protein.

The increased accuracy seen in str^r mutants indicates that nature has not selected, in the wild type, for minimal ambiguity. There might be a biological value to low-level ambiguity; or excessive precision might involve, as in any manufacture, too high a cost (e.g., in catalytic rate). Str^r mutants often grow a bit more slowly than the wild type.

PHENOTYPIC SUPPRESSION

Certain substances can increase translational ambiguity by being incorporated into mRNA, and others by decreasing the accuracy of codon recognition on the ribosome. These effects are not codon-specific.

Ambiguous Messengers. 5-Fluorouracil (Fig. 12-24) is incorporated extensively into RNA in place of U but then is sometimes misread as C; hence in RNA viruses it is mutagenic. In mRNA this analog causes phenotypic suppression of certain mutations: for example, with an amber mutation (UAG) it has been shown to restore function by causing occasional insertion of glutamine (codon CAG).

Stimulation of Ribosomal Ambiguity. Gorini observed that streptomycin (Str) and related aminoglycoside antibiotics distort the recognition region of the ribosome (see below, Aminoglycosides) and thus can cause phenotypic suppression at sublethal concentrations: in certain auxotrophs of *E. coli* they restore the synthesis of a small amount of the mutant enzyme in active form and hence allow slow growth without the required amino acid. Most of the responsive auxotrophs have nonsense mutations, which Str evidently causes to be read incorrectly by a normal tRNA. Moreover, in polypeptide synthesis in vitro, with synthetic polynucleotides as messengers, Str causes extensive incorporation of certain incorrect amino acids, each corresponding to the misreading of a single base on the codon (see below, Aminoglycosides).

Other agents that can increase ribosomal ambiguity in vitro include elevated Mg^{++}, nonoptimal pH, elevated temperature, and various organic solvents. Indeed, 5% methanol can even cause suppression in cells and thus support the growth of some auxotrophs. It is clear that environmentally induced alterations in the conformation of the ribosome can have marked effect on the noise level in the process of information transfer.

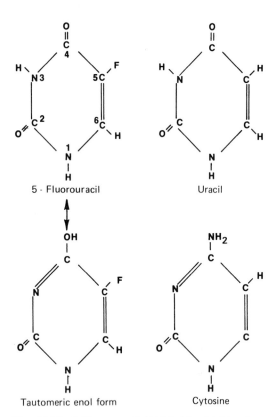

Fig. 12-24. 5-Fluorouracil (5-FU). The highly electronegative fluoro group makes the adjacent 4-carbonyl more negative and hence increases tautomerization to the enol form, which is an analog of cytosine rather than uracil.

INHIBITORS OF PROTEIN SYNTHESIS

The majority of antibiotics have turned out to be inhibitors of protein synthesis. They have been exceedingly useful in demonstrating the role of protein synthesis in various cell functions and also in analyzing ribosomal action, and are also clinically useful chemotherapeutic agents (Ch. 7).

CLASSIFICATION

In analyzing the action of an antibiotic on the ribosome the first step is to determine whether it interacts with the **small or the large subunit.** This can be done in several ways. A few agents (e.g., streptomycin) select for mutants with resistant ribosomes, which are no longer inhibited by the antibiotic in vitro; the site of action can then be localized by separating the subunits from wild-type and from mutant ribosomes, reconstituting them into hybrid ribosomes, and determining which subunit carries resistance. In addition, many antibiotics can be shown (with the aid of radioactive labeling) to bind selectively to one subunit. Finally, an antibiotic may inhibit a reaction that occurs on a subunit: at elevated Mg^{++} small subunits can specifically bind an aa-tRNA in the presence of the appropriate codon, while in the presence of 10 to 20% ethanol large subunits can bind fMet-tRNA and carry out its release reaction with puromycin (see below). Indeed, under the latter conditions binding does not require the whole tRNA molecule: in the *"fragment reaction"* it is sufficient for the fMet to be attached to a short terminal oligonucleotide (e.g., a hexanucleotide) of its tRNA.

Ribosomal inhibitors are further classified in terms of interference with recognition, peptidyl transfer, or translocation (Fig. 12-15), or with initiation. However, this classification often simplifies the picture excessively. The ribosome is not a group of discrete enzymes: it is a single body of conformationally interacting macromolecules, and distortion in one region may influence more than one function. Moreover, in the course of its cycle the ribosome has many different conformations, and these need not interact identically with an antibiotic. Specifically, many antibiotics bind to free or initiating ribosomes but bind less firmly, or not at all, to already engaged polysomal ribosomes (whose conformational flexibility is restricted by the bound pp-tRNA and mRNA). Among these initiation-specific antibiotics some (e.g., kasugamycin) block the initiation process itself (i.e., inhibit complexing with fMet-tRNA and mRNA), while others (streptomycin, erythromycin, spectinomycin) allow formation of the initiation complex but block subsequent chain elongation at various stages.

ANTIBIOTICS ACTING ON RECOGNITION

Puromycin blocks protein synthesis on either bacterial or eukaryotic ribosomes, but not by inhibiting any reaction: instead it causes **premature release** of the polypeptide chain. This release occurs because puromycin can bind to the 50S moiety and replace aa-tRNA as an acceptor of the nascent polypeptide from the P site, but its small area of contact does not then provide enough binding energy to retain the polypeptide (or fMet) on the ribosome. The release can be demonstrated in several ways: the loss of nascent polypeptide from the ribosomes, the accompanying separation of free ribosomes from mRNA (seen as breakdown of polysomes), and the identification of free peptidyl-puromycin.

The structural resemblance of puromycin to tyrosyl-adenosine (Fig. 12-25) suggested that it func-

Fig. 12-25. Puromycin and its metabolic analog, the aminoacyl end of tRNA. [After Yarmolinsky, M. B., and de la Haba, G. L. *Proc Natl Acad Sci USA 45*:1721 (1959).]

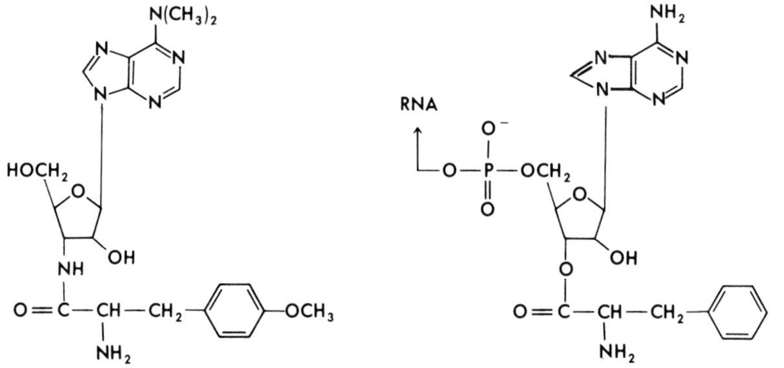

Puromycin

Termination of phenylalanyl tRNA

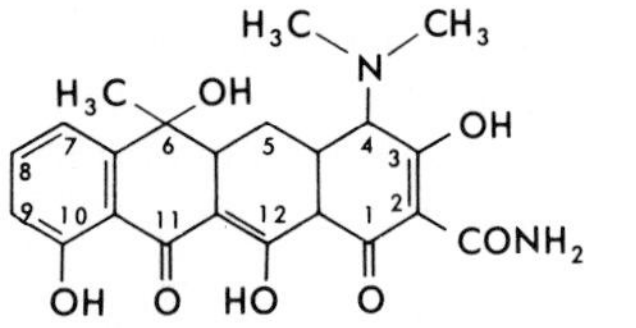

Fig. 12-26. Tetracycline. Chlortetracycline (Aureomycin) is 7-chlorotetracycline; oxytetracycline (Terramycin) is 5-hydroxytetracycline; demethylchlortetracycline (Declomycin) is chlortetracycline without the 6-methyl group.

The puromycin reaction has been very useful, as noted earlier in this chapter. For example, its release of fMet in the "fragment reaction" on isolated 50S subunits has shown that peptidyl transferase is part of the large subunit, and that its action does not require GTP or soluble factors.

Tetracycline (Fig. 12-26) binds to the small subunit of bacterial ribosomes; it blocks the binding of aa-tRNA to the A site. Thus with free ribosomes the binding of aa-tRNA, under conditions permitting both sites to be occupied, is halved by tetracycline. Moreover, the residual bound aa-tRNA, and similarly the pp-tRNA in ribosomes inhibited by tetracycline, are in the P site (i.e., reactive with puromycin).

Streptomycin also interferes with recognition, but in a complex manner that will be discussed below.

tions simply as an analog of the terminus of tyrosyl-tRNA (i.e., tyrosyl-A). However, synthetic analogs with various other amino acids side chains are inactive, and so the phenyl ring may play an essential role.

On 70S ribosomes the puromycin reaction is blocked by inhibitors of peptidyl transfer, and also, indirectly, by inhibitors of translocation, since these keep the pp-tRNA from reaching the P site. However, the binding of fMet-tRNA (or its terminal fragment) in the P site does not involve translocation; hence compounds that inhibit release of the fMet, without blocking its binding, can be considered inhibitors of peptidyl transfer.

ANTIBIOTICS ACTING ON PEPTIDYL TRANSFER

Chloramphenicol (Fig. 12-27) was shown early to bind to the large subunit of bacterial

Fig. 12-27. Antibiotics that interfere with peptidyl transfer.

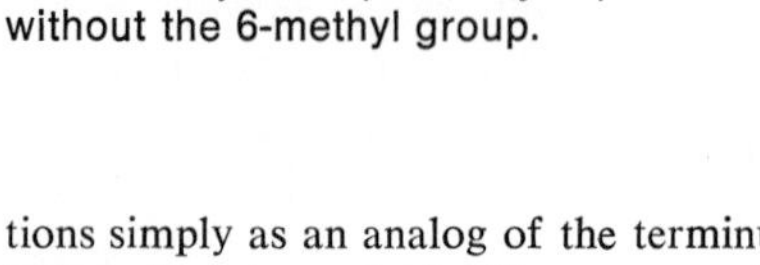

Chloramphenicol

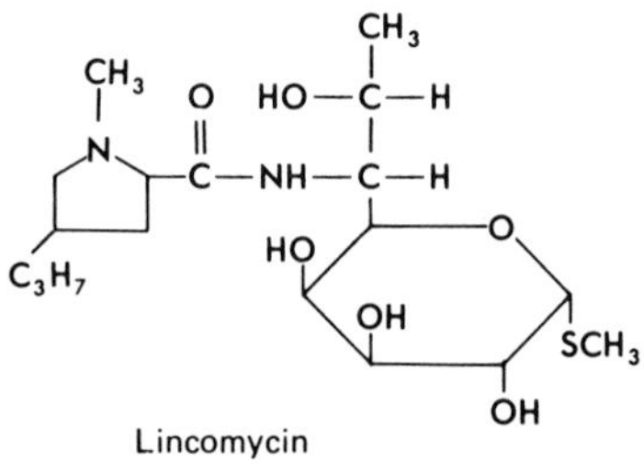

Lincomycin

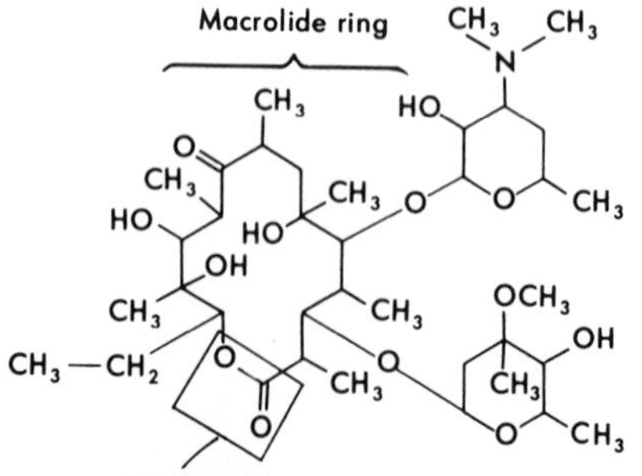

Erythromycin A

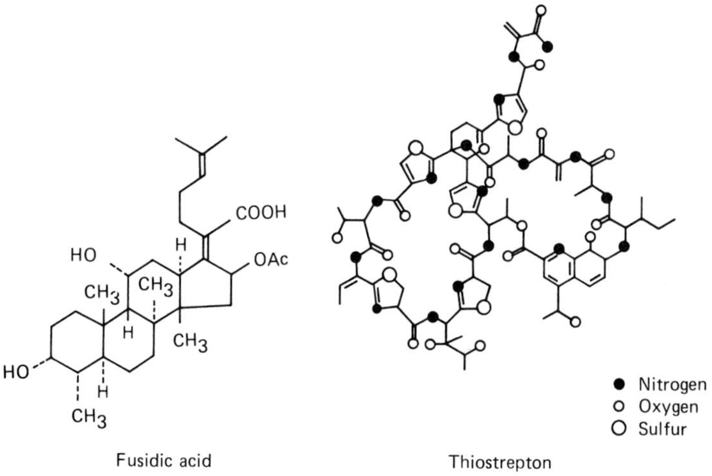

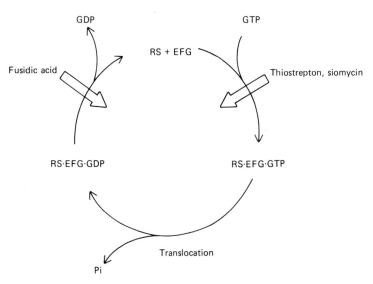

Fig. 12-28. Antibiotics that interfere with translocation.

Fusidic acid

Thiostrepton

● Nitrogen
○ Oxygen
◎ Sulfur

ribosomes. It blocks the puromycin reaction with pp-tRNA on the ribosome and also the "fragment reaction" with fMet-oligonucleotide on the large subunit. Chloramphenicol is thus the prototype inhibitor of the peptidyl transferase reaction. **Lincomycin** (Fig. 12-27) also appears to block this reaction, and it competitively inhibits the binding of chloramphenicol.

Sparsomycin, a small antibiotic (C_{13}), blocks peptidyl transfer at still another stage (on both eukaryotic and bacterial ribosomes), for it stabilizes the complexing of an fMet-tRNA fragment with the large subunit, whereas chloramphenicol and lincomycin have the opposite effect. Moreover, sparsomycin does not compete with the binding of chloramphenicol.

Despite the competition between chloramphenicol and lincomycin the large structural differences suggest that they bind to different sites, which presumably overlap or influence each other's conforma-

tion. Morever, the ribosomes of gram-positive and those of gram-negative cells are similarly responsive to chloramphenicol but the latter require a much higher concentration of lincomycin.

It should be emphasized that a block in the peptidyl transfer reaction does not imply action on the catalytic center: distortion of the complex ribosomal surface might keep either the donor or the acceptor group away from that center. Moreover, even though the transfer reaction must involve a catalytic center, "the enzyme" peptidyl transferase may turn out to be a region of the ribosome rather than a protein with definable boundaries.

Inhibitors of peptidyl transfer have been useful in studying the mechanism of polypeptide release, as was illustrated earlier (Fig. 12-20).

ANTIBIOTICS ACTING ON TRANSLOCATION

Fusidic acid, a steroid (Fig. 12-28), inhibits translocation (in both prokaryotic and

Fig. 12-29. Antibiotics and translocation. RS = ribosome.

GDP

GTP

RS + EFG

Fusidic acid

Thiostrepton, siomycin

RS·EFG·GDP

RS·EFG·GTP

Translocation

Pi

eukaryotic systems) by preventing the release of the EFG (or transferase II) GDP complex from the ribosome after GTP hydrolysis (Fig. 12-29). Synthesis is thus halted after one round of translocation. Bacterial mutants resistant to fusidic acid are altered in EFG rather than in the ribosome.

By fixing the EFG·GDP complex on the ribosome fusidic acid secondarily excludes the binding of the EFT·aa-tRNA·GTP complex in the A site; hence even though it interacts with the translocation factor it fixes pp-tRNA in the puromycin-reactive P site. Interaction between the binding sites for EFT_u and EFG is further shown by the finding that binding of both factors is blocked by **thiostrepton** (Fig. 12-28) or **siomycin.** These structurally related antibiotics bind irreversibly to the large subunit.

Erythromycin cannot be definitely categorized in terms of a block in either peptidyl transfer or translocation. It inhibits binding of chloramphenicol, but unlike the latter it does not act on preformed polysomes, nor does it block release of fMet by puromycin. It does inhibit release of certain polypeptides by puromycin, the results varying with length and composition. Some forms of erythromycin **resistance** are unusual in that the ribosome is altered not in one of its proteins but in the methylation of a base in the 23S RNA; others are altered in a protein of the large subunit.

STREPTOMYCIN AND RELATED AMINOGLYCOSIDES

The aminoglycoside group of antibiotics, which include a number of useful chemotherapeutic agents (Ch. 7, 23), contain an inositol ring substituted with 1 or 2 amino or guanidino groups and with various sugars (including 1 or 2 aminosugars). Members of this group of polycationic compounds act similarly but differ from each other in potency, antimicrobial spectrum, and cross-resistance. The irreversibility of their binding to the ribosome under physiological conditions can account for their bactericidal action. The action of aminoglycosides has been studied extensively, mostly with streptomycin (Str; Fig. 12-30).

Resistance and Dependence. One-step mutations can give rise to a large increment in resistance (several hundredfold), which is due to an alteration in the **small subunit** that

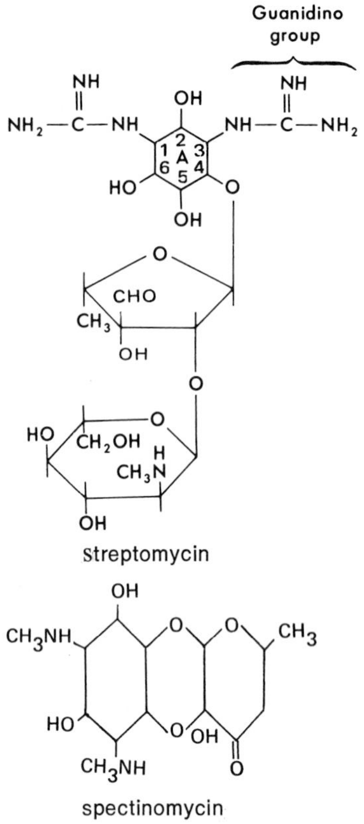

Fig. 12-30. Streptomycin and spectinomycin. The inositol with two guanidino substituents is called streptidine; in other aminoglycosides the inositol carries two amino instead of guanidino substituents (streptamine) or lacks the 2-OH group (2-deoxystreptidine or 2-deoxystreptamine). Aminoglycosides also vary in the sugars attached to position 4, or 4 and 6, of the inositol ring. A cationic group on the inositol and at least one on a sugar are essential for the characteristic bactericidal activity of aminoglycosides.

markedly reduces its affinity for Str. Many resistant isolates further exhibit Str **dependence,** i.e., they require the antibiotic for growth (Fig. 12-31). The two classes (str^r, str^d) are mutated in a small region of the same gene, which codes for protein S12 (formerly P10). (Lower level resistance to various aminoglycosides may be produced by mutations that alter the permeability of the cell, or by introducing a resistance factor (episome) carrying a gene for inactivation of the antibiotic (Ch. 7).)

Although the very similar action of the various aminoglycosides suggests that they bind to closely

Fig. 12-31. Mutations to altered response to streptomycin. Three strains of *E. coli* have been streaked in parallel, for equal distances, at right angles to a strip of filter paper containing streptomycin. **Top,** a resistant mutant; **middle,** the sensitive wild type; **bottom,** a dependent mutant. (B. D. Davis.)

related sites, the *str*r or *str*d mutations have little effect on the interaction of the ribosome with most aminoglycosides other than Str. Moreover, high level resistance and dependence are not prominent with aminoglycosides other than Str.

Effect on Recognition. With synthetic messengers Str partly inhibits aa-tRNA binding and peptide synthesis. In addition, it causes

misreading (decreased accuracy of recognition), as shown in both binding and synthesis. Str does not impair the reaction of pp-tRNA with puromycin, which suggests that it distorts codon-anticodon interaction. This finding has provided a plausible explanation for **Str dependence,** in which absence of Str blocks protein synthesis: presumably *str*d mutations distort the ribosome in a way that prevents recognition, and this effect can be compensated for by an opposite distorting effect of the antibiotic on the recognition region (Fig. 12-32).

Str dependence provided perhaps the first demonstration that a drug can act on a macromolecule not only by blocking an active site on its surface but also by altering its conformation. The effect of the *str*d mutation can also be **genotypically suppressed** by mutational alterations in proteins S4 (*ram*) or S5.

Lethal Action. Though the misreading effect has cast much light on the interaction of Str with the ribosome, the resulting production of abnormal protein does not explain its lethal action.

Thus when all protein formation is stopped by adding puromycin killing by Str still continues. Conversely, in a *str*r strain misreading by Str can be markedly increased by a *ram* mutation (see Phenotypic suppression, above) without causing killing.

The key to the lethal action of Str was provided by shifting investigation from syn-

Fig. 12-32. Schematic diagram to show how the same distorting effect of streptomycin is believed to impair the activity of sensitive ribosomes but reactivate dependent ribosomes.

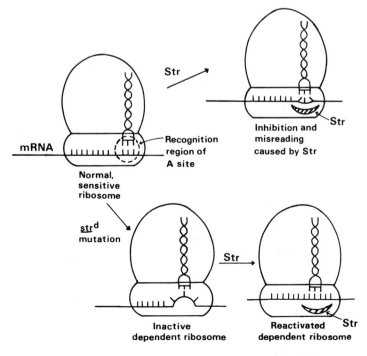

thetic to natural messengers. The results showed that **Str permits formation of an initiation complex** but then blocks its further activity. Moreover, the Str-blocked complex is **unstable:** the fMet-tRNA is gradually released and the ribosomes drop off the messenger (Fig. 12-33). Finally, though the released Str ribosome can no longer synthesize protein it is not inert: it can reinitiate (but at a reduced rate because Str impairs its dissociation into subunits). Hence the Str-killed cell continues to maintain a certain level of polysomes, with rapid turnover of their mRNA despite the lack of protein synthesis. Str is thus seen to cause a **cyclic polysomal blockade,** with repeated reinitiation, inhibition, and release.

The half-life of the block (5 minutes) is short enough to provide a supply of released Str-ribosomes that can block new mRNA, while it is long enough (compared with a normal initiation rate of *ca.* 1 ribosome per site per second) to be effectively permanent. Hence the cyclic blockade of initiation sites by Str-ribosomes can explain why **sensitivity to Str is dominant in** *str^r/str^s* **heterozygotes,** despite the presence of resistant as well as sensitive ribosomes.

Str interacts not only with free or initiating ribosomes but also with complexed ribosomes(i.e., preformed polysomes). However, the effect of the latter interaction is less drastic, slowing rather than halting chain elongation. These two different actions in vitro explain how Str can cause in cells either phenotypic suppression (described in an earlier section) or irreversible inhibition of protein synthesis, depending on the concentration. Moreover, its less drastic interaction with polysomal ribosomes is presumably reversible under physiological conditions, which would explain why killing by Str is prevented by agents that fix ribosomes in polysomes (e.g., chloramphenicol).

Str illustrates strikingly the ability of an antibiotic to have pleiotropic effects on the ribosome, and also to interact with it differently in different parts of its cycle. The lethal action of Str is also accompanied by early damage to the cell membrane (without gross lysis); the mechanism is unknown.

Spectinomycin resembles aminoglycosides in possessing an inositol ring with two basic substituents (Fig. 12-30), but it lacks additional basic substituents. It also similarly selects for 1-step highly resistant mutants with an altered protein in the small subunit (S5). However, it cannot be considered a member of the aminoglycoside group, for its action is only bacteristatic: the selective inhibition of initiating ribosomes observed in vitro is evidently reversible.

OTHER ACTIONS ON PROTEIN SYNTHESIS

Chain termination requires translocation of pp-tRNA into the P site, recognition of an **R** factor, and participation of peptidyl transferase. It is therefore blocked by antibiotics that interfere with any of these steps.

Initiation is inhibited by agents that interfere with synthesis of the formyl group (e.g., trimethoprim, hydroxylamine). It also appears to be inhibited selectively by **edeine** (a polypeptide), **kasugamycin,**

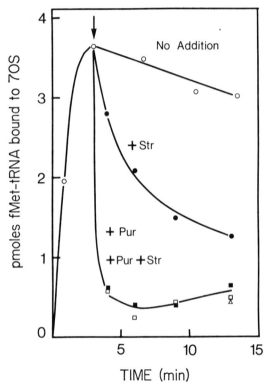

Fig. 12-33. Breakdown of initiation complexes, formed by incubating 70S free ribosomes, initiation factors, ApUpG, and radioactive fMet-tRNA. On passage of samples through a nitrocellulose filter, which adsorbs protein and ribosomes, only the radioactivity complexed with ribosomes is retained. It is seen that Str (added at the arrow) causes gradual breakdown of the complex, while puromycin (Pur), alone or in the presence of Str, causes very rapid breakdown. In parallel experiments, in which an incomplete initiation complex was formed (lacking the 50S subunit, or with GMPPCP instead of the hydrolyzable GTP), Str or puromycin failed to cause breakdown. [From Modolell, J., and Davis, B. D. *Proc Natl Acad Sci USA* 67:1148 (1970).]

and **aurintricarboxylate** (a synthetic dye). **Pactamycin** selectively blocks addition of the large subunit in formation of the initiation complex in animal systems, but in *E. coli* lysates it blocks preformed polysomes at an almost equally low concentration.

Amino acid activation is a minor area of antibiotic action. **Borrelidin** inhibits threonine-tRNA ligase of *E. coli* and yeast. Some synthetic analogs of amino acids become attached to the corresponding tRNAs and are then incorporated; others block tRNA charging.

Antibiotics that interfere with **transcription,** and thus block protein synthesis indirectly, are discussed in Chapter 11.

SUMMARY

Table 12-2 summarizes the actions of various antibiotics. Inhibitors are available for each major ribosomal function. The different ways of blocking translocation and peptidyl transfer should contribute to the further resolution of these complex reactions.

Some antibiotics act, as noted in Table 12-2, on eukaryotic as well as on prokaryotic ribosomes. Others are specific for **eukaryotic** systems: **cycloheximide** (a glutarimide) blocks eukaryotic ribosomes, and diphtheria toxin (Ch. 24) inactivates transferase II.

TABLE 12-2. Antibiotics Inhibiting Protein Synthesis

Antibiotic	Site of action			Altered in resistant mutants	Specific Effects
	Cell type	Subunit	Step		
Puromycin	Eu, Pro	L	R, P	———	Releases peptidyl-puromycin
Tetracyclines	Pro	S	R	———	Blocks binding in A site
Streptomycin, other aminoglycosides	Pro	S	R	Protein S12	Irreversible; blocks recognition, causing ribosome release; distorts recognition, causing misreading
Chloramphenicol, lincomycin	Pro	L	P	———	Blocks fragment binding to P site, puromycin reaction with pp-tRNA
Sparsomycin	Eu, Pro	L	P	———	Promotes fragment binding in unreactive position; blocks puromycin reaction
Erythromycin	Pro	L	P/T	23S RNA	Blocks P site, inhibiting both peptidyl transfer and translocation
Siomycin, thiostrepton	Pro	L	T, R	———	Irreversible; blocks binding of EFG + GTP, and of EFT·aa—tRNA·GTP
Fusidic acid	Eu, Pro	L	T	EFG	Blocks release of EFG and GDP
Rifamycins	Pro	Transcription		β-subunit	Blocks initiation
Streptolydigin	Pro	Transcription		Polymerase	Inhibits extension
Actinomycin D	Eu, Pro	Transcription		———	Binds to DNA

Eu, Pro = eukaryotic, prokaryotic cells; L, S = large, small subunit; R = recognition; P = peptidyl transfer; T = translocation.

THREE-DIMENSIONAL PROTEIN STRUCTURE

GENETIC DETERMINATION OF PROTEIN STRUCTURE

We have considered in some detail how a gene determines the synthesis of a corresponding polypeptide chain. However, gene activity is usually measured not in terms of amount of polypeptide but in terms of enzyme activity. Moreover, the physiological significance of mutations, and their evolutionary selection, depend largely on their effect on enzyme activity. Accordingly, we must ask how polypeptide sequences are related to three-dimensional structure and to function.

Sequence analysis, X-ray crystallography, optical probes of conformation, and studies of denaturation and renaturation have given considerable insight into these relations. Two very important principles have emerged, though less dramatically than the principle of base pairing. 1) Information encoded in a one-dimensional sequence can specify a complex three-dimensional protein. 2) The dynamic function of catalytic and regulatory proteins depends not on a fixed three-dimensional structure but on highly specific variations of this structure in the fourth dimension, time.

From the structure of tRNA (Figs. 12-1 and 12-2) we have already seen that weak interactions between various residues within the chain of a macromolecule can lead to specific folding. Proteins which probably followed nucleic acids in evolution, have made possible an infinitely greater range of three-dimensional structures, for their greater variety of residues can interact in many ways and over a wide range of angles.

Hierarchy of Structures. Four levels of structure can be distinguished in proteins. **Primary** structure refers to the sequence of amino acid residues. **Secondary** structure, when present, refers to a helical or other orderly arrangement in a segment of the primary chain. Additional folds form the more or less globular **tertiary** structure of the **monomer** (folded chain). Finally, many proteins form a **quaternary** structure by assembling identical or different **monomers** to yield an **oligomer**. The tertiary and quaternary structures are formed by hydrogen, hydrophobic, ionic, and van der Waals bonds, and may be stabilized by covalent S-S bonds. As we shall see in the chapters on the structure of antibodies and of viruses, quaternary structures can usually be dissociated without seriously altering the component monomers.

In soluble proteins the residues exposed on the surface are mostly polar, while the core contains a higher proportion of nonpolar amino acids. However, proteins that are incorporated in membranes have nonpolar surface regions in contact with lipids (Ch. 6).

SPECIFICATION OF HIGHER ORDER STRUCTURES

The formation of these structures by spontaneous folding reflects the selection for a structure with minimal free energy. Thus when small proteins, such as ribonuclease, are denatured (converting the folded polypeptide chain into a random coil) subsequent renaturation spontaneously restores the native conformation. However, with large protein monomers (such as serum albumin) renaturation often yields many different conformations of almost equivalent energy. We must therefore ask how the unique conformation of the native protein is selected.

Two factors have been recognized as promoting native protein conformation.

1) **Progressive folding during chain synthesis** may yield relations that are difficult to achieve by renaturation of the longer, completed chain. Thus in the cell β-galactosidase monomers must be almost completely folded before they leave the ribosome, for while still attached they can associate with already released, inactive monomers to constitute **ribosome-bound active enzyme.** However, with some enzymes late parts of a chain also play an essential role.

2) **Binding with a natural ligand** (i.e., a small molecule that interacts physiologically with the protein) may **direct or select** for the native conformation of the released polypeptide. Thus when denatured bovine albumin is renatured in the presence of long-chain fatty acids (which are a natural ligand) the restoration of native molecules is increased. Similar observations have been made with denatured enzymes in the presence of their substrates or their inhibitory effectors (Ch. 13). Hence in the cell a new protein may well be present in various "precursor" conformations until

brought into the "native" state by interaction with specific ligands. The ligand is not required for, but it does increase, the stability of the native conformation that it has induced.

Some functional proteins are monomeric but most are oligomeric. **Oligomer assembly** may occur easily in vitro, but it is often accelerated and stabilized, like protein folding, by specific ligands. Thus aggregation of components of a viral coat (capsomers) may require the viral nucleic acid (see Ch. 44, Virus capsid); and assembly of monomers of the long bacterial flagellum is initiated by a membrane-bound structure that forms its base (Ch. 2).

Enzymatic Modification. Some proteins (e.g., the head protein of phage T4) require elimination of a segment by a specific endoprotease before they can aggregate. Enzymatic cleavage of a chain may also generate two chains that remain connected by the already developed tertiary structure, as in insulin. Finally, certain residues in a finished protein may subsequently undergo various enzymatic modifications, such as phosphorylation, methylation, hydroxylation, adenylylation, or addition of oligosaccharides.

STRUCTURAL FEATURES RELATED TO FUNCTION

Flexibility is particularly important for both catalytic activity and its regulation in enzymes: that is, substrates and regulatory effectors, in forming the maximal number of weak bonds with a protein, cause it to rearrange preexisting weak bonds and thus to alter its conformation. Such changes may be recognized by X-ray crystallography, by altered accessibility of side chains to specific chemical reagents, and by changes in spectral or antigenic properties. The **central core,** the largest part of the molecule, not only is responsible for its general shape but also contributes to its flexibility. The size of cores seems to be limited: a very large chain, such as those in β-galactosidase (MW 135,000) or immunoglobulins (Ch. 16), forms two or more connected **globular domains.**

The phenotypic effects of mutations in different parts of a gene depend on the relations between protein structure and function. Mutations in the core may change its ability to undergo the necessary conformational changes during function. Mutations or chemical treatments that alter a **ligand-binding site** may impair its ability to interact appropriately with its specific ligand: substrate, prosthetic group (such as heme), coenzyme, or regulatory effector. (The **substrate site** contains essential chemical groups (hydroxyl, sulfhydryl) involved in specific catalysis.) In contrast, in those peripheral parts of the enzyme not involved in specific interactions or in core flexibility mutations and chemical changes (including proteolysis) may have little or no effect on activity (**silent** mutations).

Catalytic activity almost certainly involves mutual conformational stresses between the protein and its substrate, which strain susceptible bonds in the substrates and also bring these bonds close to the reacting groups of the enzyme. The reaction converts the substrate to the product, which has a different shape and is therefore released; the enzyme then returns to its pristine conformation. Evidence for this mechanism has been provided mainly by X-ray diffraction. Thus lysozyme possesses a surface cavity that can be shown by models to fit perfectly the substrate with a slight strain in the bond under attack. Moreover, induction of conformational changes by ligands (including unreactive, stably bound analogs of substrates) has been demonstrated.

The importance of conformation is vividly illustrated by a **paradoxical effect of antibodies.** Though antibody to β-galactosidase inactivates the normal enzyme it **restores activity** to the inactive enzyme produced by certain mutants. Moreover, in normal β-galactosidase the monomer is totally inactive until aggregated with identical molecules.

Conformational changes are the basis of **allosteric regulation** of enzyme activity, which will be discussed in Chapter 13.

INTRAGENIC COMPLEMENTATION

Conformational changes also account for the unexpected phenomenon of intragenic complementation. Thus in a heterozygote a recessive, defective gene does not ordinarily cause a metabolic block if an allele yielding a normal product is also present, because each gene directs formation of a corresponding free polypeptide. Hence as we have already seen (Ch. 10, Units of genetic information), when two genomes (or their relevant portions) mutated in different genes are placed in the same cell normal function is restored without genetic recombination. This intergenic complementation would not be expected, however, when the two genomes are mutated in the same gene. Indeed, absence of complementation was demonstrated for many allelic pairs of mutations, and complementation thus provided an intellectually satisfying operational basis for defining the gene or cistron.

However, in 1957 Fincham and Giles discovered that certain pairs of allelic mutations exhibit some degree of **intragenic (interallelic) complementation.** This effect was puzzling until the enzymes involved were found to contain multiple copies of a monomer. Intragenic complementation was then produced in vitro: with dimeric alkaline phosphatase (or with tetrameric β-galactosidase) pairs of mutants that displayed complementation in vivo yielded extracts that also became active when mixed: hence the two mutant enzymes must have formed hybrid molecules by exchange of subunits. Intragenic complementation in the cell thus results from **formation of functional hybrid molecules from different mutant monomers.**

This restoration of function is restricted to monomers whose abnormality in tertiary structure is small enough to be corrected by the mutual stresses generated by monomer interaction (Fig. 12-34). However, as a rule the correction is imperfect and **the level of enzyme activity is low.** Since **intergenic complementation,** in contrast, **yields a normal level,** the complementation test for defining a gene is still useful.

Intragenic complementation usually involves missense mutations. However, it can also occur between deletion mutants in large genes and thus is not completely restricted to oligomeric proteins. For example, the first ⅔ and the last ⅓ of the β-galactosidase chain each form a globular domain; and in a heterozygote that synthesizes both these fragments (because it has genes with complementary deletions) their association yields a stable, functional molecule.

SUMMARY: SIGNIFICANCE OF QUATERNARY STRUCTURE

Quaternary structure, involving like and sometimes unlike monomers, is essential for the function of most proteins. This pattern has

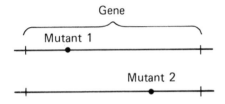

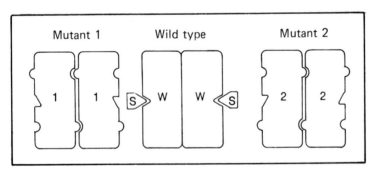

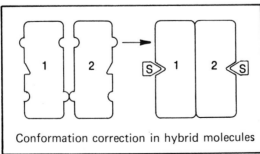

Conformation correction in hybrid molecules

Fig. 12-34. Intragenic complementation in a gene specifying a dimeric protein. Neither mutant alone produces functional dimers. The hybrid dimers, however, are functional because they mutually correct each other.

many consequences. Inactive monomers can interact to create an active oligomeric enzyme; the addition of a particular monomer may alter the substrate specificity of another monomer; intragenic complementation can result from interaction of monomers carrying different defects; separation of catalytic and regulatory sites on different monomers may allow more effective regulation; effectors can regulate activity by stabilizing the monomeric or the oligomeric state, as well as by altering the conformation of the oligomer; cooperative responses of subunits permit a small change in effector concentration to have a large regulatory effect (Ch. 13); oligomers may be more stable to physical and chemical agents than the constituent monomers; oligomers may act as multienzyme complexes, in which metabolically linked reactions are brought together physically in a large particle without requiring an excessively large core; and making the machine out of parts, rather than as a single casting, reduces the cost of an error and hence the cost of evolutionary experimentation.

In summary, in oligomers the monomers do not simply aggregate but strongly influence each other's conformation, yielding a protein with important new features. The same principle is extended in the assembly of proteins and RNA into ribosomes. The evolution of quaternary structure has thus provided advantages in terms of both functional flexibility and economy.

EVOLUTION OF PROTEINS

We have already seen that DNA hybridization can be used to quantitate evolutionary relatedness (Ch. 10). In higher organisms, however, much of the DNA consists of highly reiterated sequences, of unknown function. Alternatively, evolutionary changes can be estimated by comparing the products of a single homologous gene in different organisms; and this approach, initiated with determinations of immunological cross-reactivity of homologous proteins, has been immensely improved with determinations of sequence. Such studies, however, have other limitations: they test only a small sample of the genome, and they miss mutations to synonymous codons (about ¼ of all possible base substitutions). Moreover, bacteria interchange genes promiscuously, and so the comparison of single gene products may be less reliable than the results of DNA hybridization.

The **phylogenetic tree** inferred from one extensively studied protein, cytochrome c (Fig. 12-35), closely parallels that inferred by earlier biologists from the total phenotypic differences between species; and consistent results have been obtained with several other proteins. Among different proteins the rate of evolution varies widely, ranging from histone IV, with only 2 of its 101 residues differing between peas and cattle, to a short fibrinopeptide released on the activation of fibrinogen, with 10 of its 16 residues differing between rabbits and cattle. This variation depends primarily not on differences in mutation rate but on differences in the fraction of possible amino acid substitutions that are compatible with satisfactory function. Thus globular proteins evolve much more slowly than the terminal fibrinopeptide, whose structural requirements are obviously much less stringent. Identification of the invariant residues in such evolutionary studies contributes, like laboratory mutations, to the correlation of structure and function.

The relative constancy of the rate of evolution in a given protein has suggested that most of the observed substitutions are not selected for but are effectively **neutral mutations,** which become fixed in a new species through the statistical chance of having appeared in its progenitor (genetic drift). Of course, cumulative combinations of such changes within a polypeptide may then have selective value, which can be further increased by quaternary combinations of polypeptides (as in the α- and β-chains of hemoglobin, obviously derived from a common progenitor). As with nucleic acids, divergence is much wider in microbes, with their long history and short generation time, than in higher organisms (see Fig. 12-34). Thus in two strains of B. subtilis a protease (subtilisin) with essentially the same specificity differs in 83 of 275 residues.

Though a non-Darwinian evolution thus appears to be prominent at the level of macromolecular sequence, the more important selective, Darwinian evolution depends on those rarer mutations that produce proteins with useful novel properties. Protein sequences support the basic mechanism of innovation already indicated from work on nucleic acids: duplication of DNA segments, followed by mutational modification of the redundant genes. Thus some large polypeptides (such as immunoglobulin H chains) contain two or more segments with much

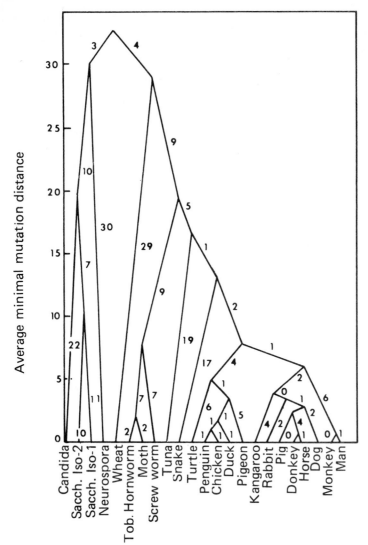

Fig. 12-35. Phylogenetic tree derived from amino acid sequences of cytochrome c of different species. The total of the numbers along the shortest path between any two species ("mutation distance") represents the minimal number of nucleotide substitutions required to account for the observed differences in their sequences. The nodes, representing hypothetical common progenitors, connect those species, or subsets of species, which are closer in sequence to each other than to any other members of the total set. The long evolution and rapid reproduction of microbial species has generated large mutation distances, as illustrated by the large differences between yeasts (the four species at the left) compared with the differences between animals. [From Margoliash, E., Fitch, W. M., and Dickerson, R. E. *Brookhaven Symp* 21:259 (1968). See also Fitch, W. M., and Margoliash, E. *Science* 155:279 (1967).]

homology, suggesting derivation by tandem gene duplication followed by fusion and cumulative mutation. Indeed, the fusion of two adjacent genes has been accomplished experimentally, by deletion of the intervening punctuation; and the resulting large polypeptide has retained both of the original enzymatic activities.

Of all the effects that mutation can have on an enzyme a change in specificity is clearly especially important for evolution. It is striking that this kind of change does not require a long sequence of mutations. Thus wild-type *Aerobacter* can grow on ribitol but not on its isomer, xylitol; but selection with xylitol as sole carbon source can yield a one-step mutant whose ribitol dehydrogenase has gained the ability to act on xylitol.

SELECTED REFERENCES

Books and Review Articles

ANFINSEN, C. B., JR. (ed.). *Aspects of Protein Biosynthesis,* Parts A and B. Academic Press, New York, 1970. An excellent set of chapters, ranging from transcription to the tertiary structure of proteins.

CAIRNS, J. (ed.). *The Mechanism of Protein Synthesis. Cold Spring Harbor Symp Quant Biol 34* (1969). Extensive collection of papers including excellent summary by Lengyel.

CHAMBERS, R. W. On the recognition of tRNA by its aminoacyl-tRNA ligase. *Progr Nucleic Acid Res Mol Biol 11*:489 (1971).

CRAMER, F. Three-dimensional structure of tRNA. *Progr Nucleic Acid Res Mol Biol 11*:391 (1971).

DAVIES, J. Errors in translation. *Progr Mol Subcell Biol 1*:47 (1969).

DAVIES, J., and NOMURA, M. The genetics of bacterial ribosomes. *Annu Rev Genetics 6:* 203 (1972).

DAVIS, B. D. Role of subunits in the ribosome cycle. *Nature 231*:153 (1971).

FITCH, W. M., and MARGOLIASH, E. The usefulness of amino acid and nucleotide sequences in evolutionary studies. *Evol Biol 4*:67 (1971).

GEIDUSCHEK, E. P., and HASELKORN, R. Messenger RNA. *Annu Rev Biochem 38*:647 (1969).

GORINI, L. Informational suppression. *Annu Rev Genet 4*:107 (1970).

GORINI, L., and DAVIES, J. The effect of streptomycin on ribosomal function. *Curr Topics Microbiol Immunol 44*:100 (1968).

KURLAND, C. G. Structure and function of the bacterial ribosome. *Annu Rev Biochem 41*:337 (1972).

LENGYEL, P., and SOLL, D. Mechanism of protein synthesis. *Bacteriol Rev 33*:264 (1969).

LIPMANN, F. Polypeptide chain elongation in protein biosynthesis. *Science 164*:1024 (1969).

NOMURA, M. Bacterial ribosome. *Bacteriol Rev 34*: 228 (1970). Especially informative on ribosome reconstruction.

PESTKA, S. Inhibitors of ribosome functions. *Annu Rev Microbiol 25*:487 (1971).

SMITH, J. D. Genetics of Transfer RNA. *Annu Rev Genetics 6:* 235 (1972).

WOESE, C. R. *The Genetic Code.* Harper & Row, New York, 1967.

YCAS, M. *The Biological Code.* Wiley, New York, 1969.

Specific Articles

ABELSON, J. N., et al., Mutant tyrosine transfer ribonucleic acids. *J Mol Biol 47*:15 (1970).

BODLEY, J. W., ZIEVE, F. J., and LIN, L. The hydrolysis of a single round of guanosine triphosphate in the presence of fusidic acid. *J Biol Chem 245*: 5662 (1970).

HELSER, T. L., DAVIES, J. E., and DAHLBERG, J. E. Mechanism of kasugamycin resistance in *Escherichia coli. Nature New Biology 235*:6 (1972).

INFANTE, A. A., and BAIERLEIN, R. Pressure-induced dissociation of sedimenting ribosomes: effect on sedimentation pattern. *Proc Natl Acad Sci USA 68*:1780 (1971).

LEE-HUANG, S., and OCHOA, S. Messenger discriminating species of initiation factor F_3. *Nature New Biology 234*:236 (1971).

MODOLELL, J., CABRER, B., PARMEGGIANI, A., and VAZQUEZ, D. Inhibition by siomycin and thiostrepton of both aminoacyl-tRNA and factor G binding to ribosomes. *Proc Natl Acad Sci USA 68*: 1796 (1971).

MONRO, R. E. Catalysis of peptide bond formation by 50S ribosomal subunits from *Escherichia coli. J Mol Biol 26*:147 (1967).

MONRO, R. E., CELMA, M. L., and VAZQUEZ, D. Action of sparsomycin on ribosome-catalyzed peptidyl transfer. *Nature 222*:356 (1969).

OLEINICK, N. L., and CORCORAN, J. W. Two types of binding of erythromycin to ribosomes from antibiotic sensitive and resistant *B. subtilis. J Biol Chem 244*:727 (1969).

OZAKI, M., MIZUSHIMA, S., and NOMURA, M. Identification and functional characterization of the protein controlled by the streptomycin-resistant locus in *E. coli. Nature 222*:333 (1969).

STÖFFLER, G., and WITTMANN, H. G. Sequence differences of *Escherichia coli* 30S ribosomal proteins as determined by immunochemical methods. *Proc Natl Acad Sci USA 68*:2283 (1971).

SUBRAMANIAN, A. R., and DAVIS, B. D. Rapid exchange of subunits between free ribosomes in extracts of *Escherichia coli. Proc Natl Acad Sci USA 68*: 2453 (1971).

SUBRAMANIAN, A. R., and DAVIS, B. D. Release of 70S ribosomes from polysomes in *Escherichia coli. J Mol Biol 74*:45 (1973).

TOCCHINI-VALENTINI, G. P., and MATTOCIA, E. A mutant of *E. coli* with an altered supernatant factor. *Proc Natl Acad Sci USA 61*:146 (1968). On a fusidic-resistant G factor.

WALLACE, B. J. and DAVIS, B. D. Cyclic blockade of initiation sites by streptomycin-damaged ribosomes in *Escherichia coli:* an explanation for dominance of sensitivity. *J Mol Biol 75*:377 (1973).

WALLACE, B. J., TAI, P.-C., HERZOG, E. L., and DAVIS, B. D. Partial inhibition of polysomal ribosomes of *Escherichia coli* by streptomycin. *Proc Natl Acad Sci USA 70*:1234 (1973).

WITTMANN, H. G., et al. Correlation of 30S ribosomal proteins of *Escherichia coli* isolated in different laboratories. *Mol General Genet 111*:327 (1971).

YOURNO, J., and KOHNO, T. Externally suppressible proline quadruplet. *Science 175*:650 (1972).

ZIMMERMAN, R. A., GARVIN, R. T., and GORINI, L. Alteration of a 30S ribosomal protein accompanying the *ram* mutation in *Escherichia coli. Proc Natl Acad Sci USA 68*:2263 (1971).

chapter 13

METABOLIC REGULATION

ALLOSTERIC REGULATION OF ENZYME ACTION   313

Endproduct (Feedback) Inhibition   313
Enzyme Activation   314
Branched Pathways   314
Molecular Basis: Allosteric Transitions   317
Mechanism of Allosteric Interactions   318
Physiological Significance of the Cooperative Response   319

REGULATION OF THE SYNTHESIS OF SPECIFIC PROTEINS   319

OPERON REGULATION   320
Enzyme Induction   320
Enzyme Repression   321
Regulator Genes   322
Operons and the Operator   323
MESSENGER RNA   325
Discovery of a Rapidly Turning Over Messenger   325
Simultaneous Synthesis, Translation, and Breakdown of
mRNA   327
REPRESSOR ACTION   330
Isolation of Repressors   330
Effect of Repressor Binding   331
OTHER MECHANISMS REGULATING GENE EXPRESSION   332
Promoter Variation   332
Alterations of RNA Polymerase: Phage Infection and
Sporulation   332
Catabolite Repression: Cyclic AMP Activation   332
Positive Regulation by an R Gene Product   334
Regulation at the Level of Translation   334
Polarity Mutations   335

REGULATION OF STABLE RNA SYNTHESIS   336

Relation of Cell Composition to Growth Rate   336
Response to Shifts of Growth Rate and Protein Synthesis   337
Relation to Protein Synthesis: "Relaxed" Mutants   337
Ribosome Biosynthesis   338
tRNA Synthesis   339

REGULATION OF DNA SYNTHESIS   339

    Growing Points   339
    Control of Initiation   340
    Relation between DNA Replication and Cell Division   340

OTHER TYPES OF REGULATION   343

  SYNTHESIS OF OTHER COMPONENTS   343
  MACROMOLECULE TURNOVER   343
    Protein Breakdown   343
    Nucleic Acid Breakdown   343

FUNCTIONS OF VARIOUS REGULATORY MECHANISMS   344

    Repression vs. Inhibition   344
    Responses to Endogenous Stimuli   344
    Pseudofeedback Inhibition by Metabolite Analogs   346
    Teleonomic Value of Various Regulatory Mechanisms   346
  SIGNIFICANCE FOR HIGHER ORGANISMS   347
    Differentiation   347
    Cybernetics and Evolution   347

In earlier chapters we have considered how a cell converts various foodstuffs into building blocks and polymerizes these into specific macromolecules. The charting of this elaborate pattern was the original goal of the study of intermediate metabolism. However, the qualitative arrows of the conventional metabolic map, with its enzymes, intermediates, and products, ignore an important quantitative dimension: the relative distribution of the metabolic flow among these compounds. Organisms evidently found great evolutionary advantage in being able to vary this distribution according to their needs under different circumstances, since even the simple bacterial cell possesses several elaborate mechanisms for this purpose.

These mechanisms of **intracellular** regulation are more universal than the neural and hormonal interactions **between** cells in higher organisms, but because their effects can be detected only by chemical analysis they were discovered much later. Bacteria were prominent in this development: they can adapt to a wider variety of environments than the cells of higher organisms, and they readily yield mutants altered in regulation, which have been most valuable in sorting out the mechanisms. Several fundamental features of intracellular metabolic regulation have emerged. 1) Specific macromolecules (DNA, RNA, and protein) are synthesized by an enzyme or catalytic particle that attaches to an initiation site and then moves along the template (processive synthesis), and the overall rate is ordinarily governed by the **frequency of initiation** and not by the rate of chain growth. 2) The **signal** for this regulation is the concentration of some small molecule in the cell. 3) This molecule interacts with the initiation site indirectly, by reversibly altering the conformation (and hence activity) of a **regulatory protein (allostery).** 4) The biosynthesis of small molecules, via enzymatic sequences, is governed by an analogous mechanism, which regulates the activity of the first (i.e., initiating) enzyme of each pathway. 5) Linkage of genes in chromosomes, known to promote their equipartition in cell division, is also used for the coordinate regulation of functionally related genes.

The discovery of these principles has had a large impact on many areas of biology. It has encouraged vigorous exploration of the role of variable gene expression in differentiation in higher organism; allosteric proteins undoubtedly detect many pharmacological, hormonal, and sensory stimuli, as well as metabolite levels; and the alterations of allosteric regulation in human disease are only beginning to be explored. The analysis of intracellular regulatory mechanisms has also encouraged biochemistry to broaden its biological base by focusing on questions of teleonomy (Ch. 1) as well as of mechanism (on "why" as well as "how"); indeed, the regulatory responses of an enzyme are now recognized as valid criteria for defining its function in the cell.

ALLOSTERIC REGULATION OF ENZYME ACTION

ENDPRODUCT (FEEDBACK) INHIBITION

Two observations independently suggested that endproducts of metabolic pathways somehow influence their own biosynthesis. First, when *E. coli* is making all its amino acids from radioactive glucose the addition of an unlabeled amino acid rapidly and al-

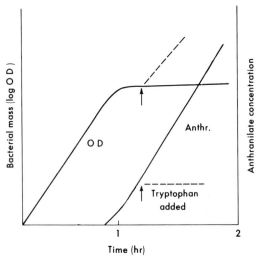

Fig. 13-1. Feedback inhibition by an endproduct. Effect of tryptophan on accumulation of anthranilate by cells of a tryptophan auxotroph blocked immediately after anthranilate (Ch. 4, Aromatic biosynthesis). The culture starts accumulating anthranilate (measured fluorometrically) in the medium as growth ceases for lack of tryptophan (solid lines). Dashed lines show effect of a small addition of tryptophan (1 μg/per milliliter) in restoring growth and blocking further accumulation. A similar, immediate inhibition of anthranilate synthetase by tryptophan can be demonstrated in vitro.

most completely cuts off synthesis of that amino acid: i.e., further protein synthesis utilizes the exogenous amino acid almost exclusively. (Because of active transport (Ch. 6) even a very low concentration in the medium has a strong sparing effect.) Second, auxotrophic mutants accumulate precursors (Ch. 4) only after the cells have exhausted the endproduct of the blocked pathway, rather than during growth; and readdition of the endproduct inhibits further accumulation within a few seconds (Fig. 13-1).*

Though it was already known that endproducts can **repress formation** of their own biosynthetic enzymes (see enzyme repression, below), this mechanism would only gradually

---

* Accumulations in auxotrophic mutants do not inhibit growth, because they occur only after growth has ceased. However, blocks in central or catabolic pathways may lead to growth inhibition—for example, when galactose or mannitol is fed to a mutant that can convert it only as far as the corresponding phosphate.

affect the rate of formation of their products and hence could not explain the immediate responses observed. Umbarger then reasoned that since isoleucine blocks precursor accumulation by mutants blocked at any step in its biosynthetic pathway it must inhibit the **first enzyme** of that pathway, threonine deaminase; and in 1957 he demonstrated this inhibition in a bacterial extract. The effect involves a direct interaction (without lag) between the endproduct and the enzyme, rather than formation of a regulatory derivative of the endproduct. Moreover, the interaction is highly specific: isoleucine does not inhibit any later enzyme of the pathway, and its precursors are not inhibitory (unless they can be converted to isoleucine). Such **endproduct inhibition (feedback inhibition)** has now been observed with the initial enzyme of many biosynthetic pathways.

ENZYME ACTIVATION

Regulation of enzyme activity involves not only negative but also, in some pathways, **positive effectors.** Thus ATP stimulates the activity of aspartate transcarbamylase and also antagonizes its inhibition by CTP (see Fig. 13-6, below). In this way the parallel purine and pyrimidine pathways (yielding ATP and CTP) are cross-linked by a positive response, as well as being subject to separate negative feedback. Similarly, valine stimulates the first enzyme of isoleucine synthesis. In general, however, in biosynthesis feedback inhibition seems to be the main mechanism of regulation, while stimulation is more prominent in amphibolic and catabolic pathways.

For example, AMP, formed by the utilization of ATP in biosynthesis, stimulates the action of muscle phosphorylase, which initiates glycogen breakdown and thus provides a source of energy for regenerating ATP. This positive allosteric effect contributes to regulating the energy supply.

BRANCHED PATHWAYS

Branched pathways present a special problem in regulation, for if feedback from one branch blocked the common pathway any parallel branch would be starved (pseudoauxotrophy). One way of solving the problem is the formation of **more than one enzyme (isozymes)** for the same reaction. For example,

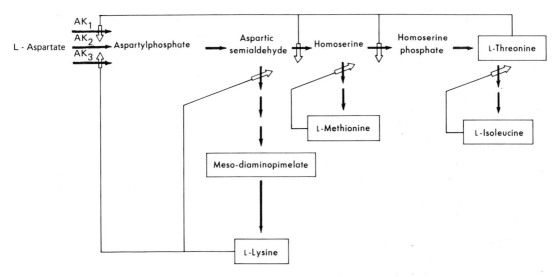

Fig. 13-2. Sites of endproduct inhibition in *E. coli* in the family of amino acids derived from aspartate. Thick lines = biosynthetic pathway; thin lines = inhibition; AK_1, AK_2, AK_3 = isozymic aspartokinases. AK_1 and AK_3 are subject to endproduct repression (see below) as well as inhibition; AK_2 is not inhibitable, but it is repressed by methionine.

the biosynthesis of the aspartate family involves a common pathway that later branches to threonine and isoleucine, to methionine, and to diaminopimelate and lysine (Fig. 13-2). *E. coli* makes three species of the first enzyme of this common pathway (aspartokinase); these all contribute to a common pool of the precursor, aspartyl phosphate, but are subject to inhibition by different endproducts. In addition, at each subsequent **fork** in the pathway the **initial enzyme of each branch** is also subject to inhibition by the appropriate endproduct (Fig. 13-2). Complete regulation of flow is thus provided by the presence of a valve at each branch point, just as in a proper hydraulic system.

The formation of regulatory isozymes was discovered with acetolactate synthetase (Fig. 13-3),

Fig. 13-3. Anabolic and catabolic metabolism of acetolactate. The enzymes of the catabolic pathway (including acetolactate synthetase B) are induced in large amount by fermentative conditions but are not influenced by valine, which inhibits (and also represses, see below) synthetase A but not B.

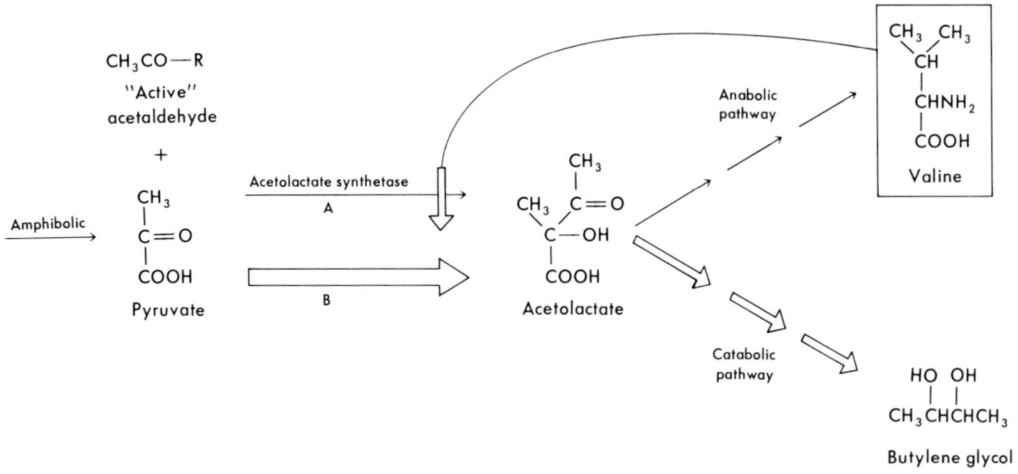

which contributes to both an anabolic pathway (the biosynthesis of valine) and a catabolic pathway (the butylene glycol fermentation of *Aerobacter;* Ch. 3). In aerobic growth the cell maintains only the level of this enzyme required for valine biosynthesis. Under anaerobic, acidic conditions, however, the level of activity was found to increase manyfold, through the induced synthesis of an additional enzyme that catalyzes the same reaction. This "catabolic" enzyme differs from the "anabolic" enzyme in ways that fit their functions: the former has a much lower pH optimum, and only the latter is inhibited by valine. In this way a **common pool** of acetolactate can be used under appropriate conditions for either biosynthesis or fermentation or both, without mutual interference.

Alternative Mechanisms. Other ways have been found, as illustrated in Figure 13-4, for regulating a branched pathway without interference between its branches. 1) In **concerted** (or multivalent) feedback inhibition a single enzyme responds to a mixture of two or more endproducts but not to individual ones (at relevant concentrations). In a related mechanism methionine and its precursor S-adenosyl methionine are **synergistic** in their action on the first enzyme of their pathway. 2) In **sequential** feedback inhibition endproducts govern the activity of the branches; the resulting accumulation of a branch-point intermediate then inhibits the first enzyme of the common pathway. 3) **Cumulative** endproduct inhibition is seen with glutamine synthetase (in *E. coli*), which plays a key role in forming nitrogenous compounds from carbohydrates: six endproducts each contribute only partial inhibition, even at high concentration, and their effects are additive. Different bacterial groups may employ different patterns to regulate the same pathway, as has been shown for the aspartate pathway (Table 13-1) and the aromatic pathway.

In some pathways in the mold *Neurospora* the problem of branched pathways is solved by **compartmentation,** i.e., the intermediates in a sequence

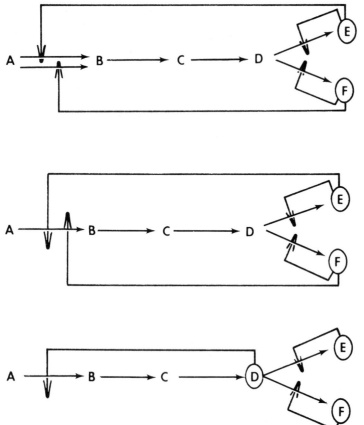

Fig. 13-4. Alternative patterns of regulation of branched pathways.

TABLE 13-1. Patterns of Control of Aspartokinase Activity in Some Bacterial Species

Species	Mode of regulation	Feedback inhibitors
Escherichia coli *Salmonella typhimurium* *Klebsiella aerogenes*	Isoenzymes	Lysine (enzyme 3) Threonine (enzyme 1) None (enzyme 2)
Pseudomonas aeruginosa *Rhodopseudomonas capsulata* *Brevibacterium flavum*	Concerted feedback inhibition	Lysine plus threonine
Rhodopseudomonas spheroides	Sequential feedback inhibition	Aspartic semialdehyde
Bacillus subtilis *Bacillus stearothermophilus*	Isoenzymes and concerted feedback inhibition	*Meso*-diaminopimelate (enzyme 1) Lysine plus threonine (enzyme 2)

remain enzyme-bound rather than enter a common pool. In mammalian cells (Ch. 47) compartmentation extends to membrane-bound organelles; isozymes are also prominent, and their different catalytic properties, often involving mixtures of different catalytic subunits in the same enzyme, are adapted to a variety of physiological situations.

Alteration of enzyme activity by reversible chemical reactions (e.g., phosphorylation, adenylylation), and by rapid breakdown of unused enzymes, has so far been more prominent in some animal systems than in bacteria. However, the former is seen in bacterial glutamine synthetase and RNA polymerase.

MOLECULAR BASIS: ALLOSTERIC TRANSITIONS

The feedback inhibition of most enzymes by the endproduct is **competitively antagonized** by the substrate, even though the two compounds usually differ markedly in structure

(e.g., aspartate and CTP) and hence must be **bound to different sites.** Competition could be readily explained if these sites overlapped. However, as Monod and Changeux recognized, kinetic considerations (see below) indicate that the two ligands have opposing **conformational effects** on the enzyme (Ch. 12). That is, an **inhibitory effector,** complexing with an **effector site,** induces or stabilizes an inactive conformation, which has less affinity for the substrate; conversely, the **substrate,** complexing with a **catalytic site,** induces or stabilizes an active conformation, which has less affinity for the effector (Fig. 13-5). (Both reactions are rapidly reversible.) Enzymes with such alternative stable conformations, promoted by the binding of quite different molecules, were designated as allosteric (Gr. "other shape"). This model also readily explains the action of positive

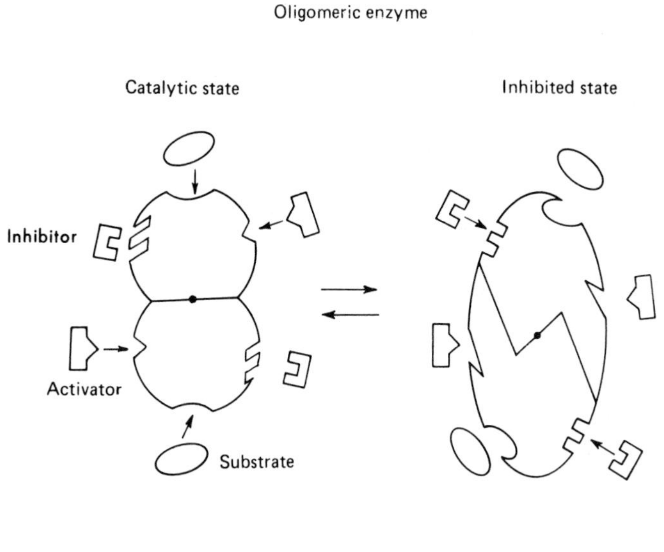

Oligomeric enzyme

Catalytic state   Inhibited state

Fig. 13-5. Regulatory changes in an allosteric model that assumes spontaneous transitions in tertiary structure, which can be stabilized by binding various ligands (as indicated by arrow). The inhibitory effector stabilizes the inactive state of the enzyme; the substrate (and for some enzymes the stimulatory effector) stabilizes the active state. (After Changeux, J.-P. *Sci Amer* April, 1965, p. 36.)

Inhibitor

Activator

Substrate

effectors, whether they act by exerting a specific conformational effect or by displacing a negative effector (without exerting its effect).

The evolution of proteins with allosteric interactions between structurally unrelated ligands, binding to entirely different sites, may seem rather elaborate, but it has provided maximal regulatory flexibility. Thus regulatory effectors might conceivably act more simply, by binding directly to the catalytic site or to an overlapping site, but there would then be structural constraints on the range of possible effectors for a given enzyme.

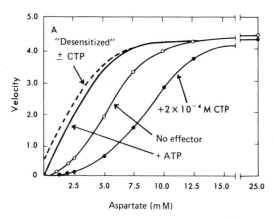

Fig. 13-6. Kinetics of an allosteric enzyme, aspartate transcarbamylase from *E. coli*. This initial enzyme of the pathway to pyrimidine nucleotides converts aspartate plus carbamyl phosphate to carbamyl aspartate. In these experiments enzyme activity in an extract is measured at various concentrations of aspartate, with the other substrate constant.

Solid line = native enzyme. The sigmoid middle curve, obtained without effectors, indicates cooperative interaction of multiple substrate molecules with an enzyme molecule. The endproduct of the reaction, CTP, causes feedback inhibition, which is overcome competitively by increased concentrations of aspartate (lower curve). Conversely, ATP is stimulatory (top curve); and since it was added at a high enough concentration to stabilize all the enzyme molecules in a fully active conformation the kinetics become hyperbolic, as with ordinary enzymes (i.e., binding of the first molecule of substrate by such a stabilized enzyme does not influence the binding of additional molecules). The effect of ATP can be antagonized by CTP, and vice versa.

Dashed line = enzyme "desensitized" by heating for 4 minutes at 60° or by treatment with 10^{-6} M Hg^{++}, which separates the catalytic from the regulatory subunits. The curve is hyperbolic; in addition, inhibitability by CTP has been lost. [After Gerhart, J. C., and Pardee, A. B. *J Biol Chem* **237**:891 (1962).]

MECHANISM OF ALLOSTERIC INTERACTIONS

This topic has become a major one in protein chemistry and can be reviewed here only very briefly. A special feature of many allosteric enzymes derives from the presence of **multiple catalytic sites** and **multiple effector sites,** on symmetrically arranged polypeptide subunits.

Usually a catalytic site and an effector site are paired on the same polypeptide. However, the first enzyme of pyrimidine biosynthesis, aspartate transcarbamylase (see Fig. 13-6), has separate catalytic and regulatory subunits (six of each, arranged in threefold radial symmetry), which can each bind the appropriate ligand.

The interactions of the subunits in allosteric proteins illustrate the importance of quaternary structure (Ch. 12). Because of these interactions many allosteric enzymes do not exhibit simple, first-order mass-law kinetics. Rather, the variation of activity with **substrate concentration,** and the variation of inhibition with **effector concentration,** are both **sigmoidal** (Fig. 13-6), with a steep rise in activity over a narrow concentration range. The profound physiological significance of these kinetics will be discussed below.

The molecular basis for the sigmoidal kinetics is that the enzyme acts to a certain extent as a unit, with **cooperative** subunit interactions: a change in the tertiary structure of one monomer influences the other monomers and hence the quaternary structure. Thus the binding of one ligand molecule increases the rate of binding of additional molecules to similar sites on the same enzyme. Such cooperative interactions can also explain the concerted inhibition (see above) of some branch-point enzymes by a mixture of endproducts.

It is not certain whether the ligands simply **stabilize** transitions of an allosteric enzyme that occur spontaneously, through random thermokinetic motions within a broad range of possible conformations, or whether the spontaneous range is too narrow and the ligand **induces** the required strain in the molecule (induced fit, see Ch. 12, Structural features related to function).

Experiments and calculations show that the most stable states of the whole enzyme are those in which all identical monomers are in the same conformation, and the enzyme is symmetrical. Monod, Wyman, and Changeux have emphasized the **concerted nature** of the transition between fully active and inactive states, and have interpreted the enzyme kinetics in terms of the proportions of the enzyme molecules in the two states. But though allosteric enzymes do tend to switch from a state fully favorable for binding the substrate to one fully favorable for binding the inhibitor (or vice versa), Koshland has shown that at subsaturating ligand concentrations the **intermediate steps,** with an asymmetrical conformation of the enzyme, are sometimes stable enough to be recognizable; moreover, a high concentration of both ligands may strain some enzymes enough to bind both. It thus appears that effectors (and substrates) are not restricted to **stabilizing** a conformation but may also use their binding energy to **induce** a conformation (induced fit), as noted in the last section of Chapter 12 (Structural features related to function).

Whatever the detailed mechanism, it is clear that an excess of either substrate or effector can fix an allosteric enzyme in one or the other stable conformation, eliminating looser transitional conformations. This conclusion, suggested initially by kinetic studies, is now supported by additional lines of evidence. Thus **either** ligand can **decrease the accessibility** of the enzyme to alteration by proteases, denaturing agents, or chemical reagents that attack specific substituents. Moreover, after partial denaturation of an enzyme either ligand will promote restoration of the native state (by "induced fit"): hence with such denatured enzymes, which lag in their catalytic response to substrates, a **negative** effector can paradoxically **stimulate** the initial activity by promoting the transition from a denatured, inactive form to a native but inhibited form, which can be readily activated by substrate.

Competition between substrate and effector, as we have seen, derives from a conformational interaction in which the binding of either **decreases the affinity** for the other: i.e., the effector increases the K_m of the enzyme. With some enzymes, however, such as the first enzyme of histidine biosynthesis in *E. coli,* binding of the effector **decreases the turnover number** (V_{max}) instead of increasing the K_m. As predicted, this kind of conformative interaction is not associated with competition between substrate and effector.

Desensitization. The **connection** between the catalytic and the effector site is evidently **more labile** than is either site, for mild denaturation, by heat or by chemicals (e.g., $Hg++$, high pH), destroys the allosteric interaction between these sites without destroying either the catalytic activity or the ability to bind the effector. Desensitizing treatments also eliminate cooperation between catalytic sites, yielding classic, first-order Michaelis-Menten kinetics, i.e., a hyperbolic curve compared with the sigmoid curve for the native enzyme (Fig. 13-6). Desensitization sometimes involves **separation** of the catalytic and the regulatory subunits, but when the two kinds of sites reside in the same subunit desensitization results from a conformational change.

PHYSIOLOGICAL SIGNIFICANCE OF THE COOPERATIVE RESPONSE

The sigmoid allosteric curves of Figure 13-6 resemble the oxygen dissociation curve of hemoglobin, which permits a large fraction of the bound O_2 to be transferred between lungs and tissues with only a moderate pO_2 difference. In allosteric enzymes, similarly, the higher-order kinetics cause a large response of enzyme activity to small changes in concentration of substrate (or of regulatory endproduct). Allosteric enzymes are thus designed, like hemoglobin, for efficient **homeostasis,** i.e., for maintaining the concentration of their ligands in the organism within a narrow range. Moreover, the molecular mechanism similarly involves a multimeric protein: when hemoglobin is dissociated into its monomers the equilibrium shifts from sigmoidal to first-order, just like the kinetics of the separated catalytic subunit of aspartate transcarbamylase.

The marvelous homeostatic properties of hemoglobin thus derive from a property of proteins that had evolved already in bacteria. With hemoglobin the allosteric transition has been demonstrated in detail, by X-ray crystallography, as a large difference in conformation between the oxygenated and the reduced forms.

REGULATION OF THE SYNTHESIS OF SPECIFIC PROTEINS

We have already seen (Ch. 3) that variations in the growth medium can markedly alter the enzymatic composition of bacteria. Further analysis showed that some compounds induce, and others repress, the formation of various enzymes specifically involved in their metabolism.

These apparently opposite actions were eventually found to have much the same mechanisms, affecting in each case a small group of genes. After discussing their classical mechanism we shall consider additional mechanisms, discovered later, which act on much larger classes of genes.

OPERON REGULATION

ENZYME INDUCTION

The enzymes specifically required for the utilization of a particular foodstuff usually are found in a bacterial cell only when it is grown on that compound. This response to the environment was designated decades ago as **adaptive enzyme formation.**

Monod, at the Pasteur Institute, studied this process in detail with the enzyme (β-galactosidase), which hydrolyzes **lactose** (glucose-4-β-D-galactoside; see Table 13-2), a key sugar in the identification of E. coli (Ch. 29). Induction of this enzyme involves a mechanism separate from its actual activity, for some β-galactosides (e.g., phenyl-β-D-galactoside) are good substrates but poor inducers, while the opposite is true of a sulfur analog, methyl-β-D-thiogalactoside (Table 13-2). With the focus of interest shifting from adaptive function to mechanism the response was renamed **induced enzyme formation.**

This shift of emphasis was associated with an important refinement in experimental design. In previous studies of the kinetics of induction the addition of a natural substrate could influence enzyme synthesis not only directly, by interacting with the regulatory system, but also indirectly, by providing an increased supply of food. With **nonmetabolizable inducers,** however, induction occurred under **gratuitous conditions,** which eliminated this secondary effect. Another improvement was to plot enzyme content not as a function of **time** (during which the mass of synthesizing protoplasm is increasing) but as a function of **total protein** (or of total bacterial mass). This **differential rate of enzyme synthesis** (dz/dB) proved useful in comparing the efficiency of induction under various conditions. In particular, it was found that within a few minutes of addition of inducer (at a concentration having maximal effect) the new enzyme becomes a **constant fraction** of the total new protein (Fig. 13-7); moreover, **all** the cells respond immediately.

Studies with radioactive amino acids showed that the **induced enzyme protein is formed entirely de novo** during induction,

TABLE 13-2. Induction of β-Galactosidase by Various Compounds

Compound	Substituent	Relative induction*	Relative hydrolysis rate†	Relative affinity‡
β-D-galactosides	Glucose (product: lactose)	17	30	14
	Phenyl	15	100	100
β-D-thiogalactosides	Isopropyl	100	0	140
	Methyl	78	0	7
	Phenyl	0	0	100

* **Relative induction** is the maximal enzyme concentration attained, after growth with a saturating concentration of inducer, relative to the value obtained with isopropyl-β-thiogalactoside (designated as 100).

† The **hydrolysis rate,** obtained at a saturating concentration, is given relative to that of phenyl-β-galactoside (100).

‡ The **affinity** is given relative to that of phenyl-β-galactoside (100); for substrates it is determined from the concentration giving half-maximal hydrolysis, and for nonsubstrates it is determined from the inhibitory effect of a given concentration on the hydrolysis of a substrate of known affinity.

From F. Jacob and J. Monod. J Mol Biol 3:318 (1961).

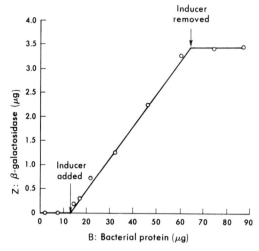

Fig. 13-7. Kinetics of induced enzyme synthesis in a growing culture of *E. coli.* Accumulation of β-galactosidase is plotted as a function of total protein synthesis (dz/dB = differential rate of synthesis). At a saturating concentration of inducer the value for this enzyme, as for other enzymes under conditions of maximal induction, turned out to be surprisingly high (ca. 5% of total protein). [From Jacob, F., and Monod, J. *J Mol Biol 3*:318 (1961).]

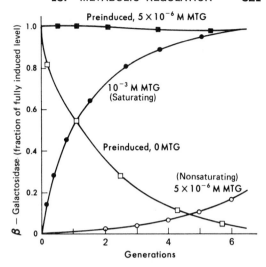

Fig. 13-8. Slow induction and full maintenance of β-galactosidase by a low concentration of inducer. At zero time methyl thiogalactoside (MTG), a gratuitous inducer, was added at the indicated concentration to both a preinduced and an uninduced culture of *E. coli,* in steady-state growth in the chemostat (Ch. 5), and the amount of β-galactosidase per cell was determined at intervals. It is seen that uninduced cells (circles) responded rapidly to a saturating concentration of inducer, and very slowly to a nonsaturating concentration. However, with preinduced cells (squares) the same low concentration can maintain the induced state, though in the absence of inducer the enzyme is no longer formed and its concentration is exponentially diluted by cell growth and division.

The response of a **cell** to inducer is all-or-none. The slow response of the **culture** to low concentrations represents a slow increase in the fraction of induced cells; each cell that begins to respond forms transport units that ensure its further rapid induction. [Adapted from Novick, A., and Weiner, M. *Proc Natl Acad Sci USA 43*:553 (1957).]

i.e., the inducer stimulates expression of a specific gene, rather than activating a pre-existing zymogen. Enzyme induction was thus no longer a special feature of bacterial adaptation but became a key to a very general problem: gene regulation.

Persistent Induction: Inducible Transport Systems. Though the differential synthesis of induced enzymes rapidly becomes constant after addition of excess inducer (Fig. 13-7), at **nonsaturating inducer concentrations** the response increases gradually over a long period of time (Fig. 13-8) and involves only a fraction of the cell population. The explanation is the induction of an **active transport system** (Ch. 6), which concentrates the inducer in the cell. Thus a low concentration of inducer only occasionally causes formation of transport molecules in a cell; but once that process is initiated it becomes **autocatalytic,** and the cell soon becomes fully induced. Moreover, because of the active transport of the inducer in already induced cells, the fully induced state can be **maintained** by a concentration of inducer too low to **initiate** a uniform response in uninduced cells (Fig. 13-8). This persistent **positive feedback** superficially resembles differentiation in higher organisms: preinduced and nonpreinduced cells of the same genotype can be maintained for

many generations with different phenotypes in a medium with a low concentration of inducer.

ENZYME REPRESSION

Enzymes not requiring induction (e.g., biosynthetic enzymes) were originally designated as **constitutive,** and were thought to be synthesized in an invariant manner. However, addition of an endproduct (e.g., an amino acid) to the medium was later found to diminish or abolish formation of the enzymes of its own biosynthetic pathway. (As with feedback inhibition, intermediates in the pathway are effective only when they can be converted to the endproduct.) This regulatory mechanism has become known as **endproduct repression,** the endproduct being the **corepressor.**

When the level of an **amino acid** in a cell is abnormally lowered (e.g., by limiting its supply to an auxotroph in a chemostat) the enzymes of its biosynthetic pathway (and usually its aminoacyl-tRNA synthetase) are **derepressed**, i.e., their concentrations rise above the usual levels. In at least some of these pathways the repressive endproduct appears to be not the amino acid itself but a later derivative on the path to protein synthesis. Thus when charging of valyl-tRNA synthetase and the corresponding tRNA are partly blocked, by raising the temperature of a mutant in which the enzyme is temperature-sensitive, the enzymes of valine biosynthesis are derepressed even in the presence of excess valine. Moreover, the enzymes of histidine biosynthesis are derepressed by mutations in several different genes that affect a tRNA, i.e., decrease its concentration, or decrease its charging, or modify its structure (U → pseudo U, Ch. 10).

Repression has been observed in various kinds of metabolic pathways. For example, alkaline phosphatase is formed in *E. coli* cells when one of its substrates (e.g., glycerophosphate) is used as an obligatory source of phosphate, but not when even a trace of inorganic phosphate (i.e., the endproduct) is present. Similarly, in the shift of facultative organisms from aerobic to anaerobic conditions (and vice versa) many enzymes are induced and others repressed. Since most enzymes are either inducible or repressible the term **constitutive** is now used primarily to characterize mutations that eliminate these regulatory responses.

Branched Pathways. In branched (multifunctional) pathways regulation of enzyme formation requires special arrangements, just as was noted above for endproduct inhibition, to avoid interference between the branches. Two mechanism have been observed. 1) The common pathway may have parallel, isofunctional enzymes, catalyzing the same reaction but each subject to regulation of its synthesis (and often its activity, as noted above), by a different endproduct. For example, carbamyl phosphate, a precursor of arginine and of pyrimidine nucleotides (Ch. 4), is formed in *E. coli* by two kinds of carbamyl phos-

phokinase, each repressible by one of the endproducts. 2) The pathway to isoleucine, leucine, and valine in *E. coli* involves **multivalent repression:** all three products must be present to achieve repression.

The same pathway may be governed by different mechanisms in different organisms. For example, the branched aromatic pathway has parallel initial enzymes in enteric bacteria but multivalent repression in gram-positive bacilli. Evolution has evidently selected for greater variety in regulatory mechanisms than in pathways. The selection presumably takes into account the genetic background, the efficiency of the regulatory mechanism, its expense to the cell, and the frequency of environmental fluctuation in the organism's ecological niche.

Repression is not confined to biosynthetic endproducts: in fact, many inducible enzymes are also repressed by glucose, as shown in the early observation of "diauxie" (Fig. 3-7). The mechanism of this "glucose effect" will be discussed below (see Catabolite repression).

We shall now discuss the common mechanism underlying induction and endproduct repression.

REGULATOR GENES

Studies of the induction of β-galactosidase revealed the existence of a **regulatory** gene (*lac*R = *i*)* as well as a **structural** gene (*z*) for this enzyme. That *z* directly controls the structure of the enzyme is clear, for some *z⁻* mutants form an **altered protein,** recognizable in some cases as an immunologically **cross-reactive material (CRM;** see Ch. 11, Mutations) and in others as a temperature-sensitive enzyme. Mutations in *R,* in contrast, affect only the **quantity** and not the structure of the enzyme. Thus *R⁻* mutants are **constitutive** for the corresponding enzymes:† with ordinarily inducible enzymes they eliminate the requirement for inducer, and with repressible enzymes they eliminate repression by the endproduct.

---

* The regulator locus for β-galactosidase was discovered and designated as *i* (for induction) before it was known to direct the manufacture of a repressor. Regulator loci are now denoted by the symbol *R.*

† Colonies of constitutive *lac* mutants, grown without an inducer, can be recognized by spraying with a chromogenic substrate (*o*-nitrophenyl-β-galactoside), whose hydrolysis yields the yellow product *o*-nitrophenol.

Fig. 13-9. Dominance of R^+lac (inducible) over R^-lac (constitutive) allele. The dashed arrows indicate that the product of the R gene influences the activity of a z gene in a separate chromosome (*trans* effect) as well as in its own chromosome (*cis* effect).

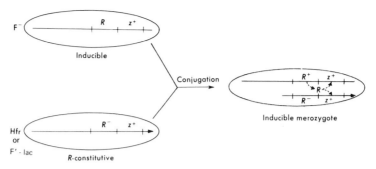

R^+ **Dominance; Trans Effect.** To test the dominance of constitutivity or inducibility a merogenote carrying the genes for *lac* constitutivity (R^-z^+) was introduced into an R^+z^+ cell (inducible) or an R^+z^- cell (*lac⁻*). The resulting merozygote (R^+z^+/R^-z^+ or R^+z^-/R^-z^+) was inducible (see Table 13-3, below). It follows that inducibility (R^+) is dominant over constitutivity (R^-). Since the R^+ gene thus influences a z^+ gene even on a different chromosome it must give rise to a **diffusible cytoplasmic product** (Fig. 13-9), which is designated as the **regulator, repressor,** or **aporepressor.**[*]

Repressors and Effectors. These properties of R mutants led Jacob and Monod to propose a mechanism of **negative operon control,** in which **repression is fundamental;** induction would be a release of repression. According to this model an R-gene product (repressor) can be rapidly and reversibly shifted between an active and an inactive conformation by complexing with a specific small-molecule **effector.** In **inducible** systems the repressor is **active** in its natural conformation, and the inducer makes it inactive, while in **endproduct repression** the repressor is **inactive** until combined with a corepressor.

Mutations can alter the activity of the repressor in various ways. Thus $lacR^-$ (constitutive) alleles make an ineffective repressor (or none), while R^S (superrepressed) mutants make a repressor that remains active

even in the presence of the inducer. The structure and molecular action of repressors will be considered in a later section.

OPERONS AND THE OPERATOR

The evidence for a cytoplasmic repressor led to several important predictions. 1) Such a specific repressor must act on a corresponding specific, genetically determined receptor or **operator.** 2) The genes of a given metabolic pathway are often adjacent to each other in bacteria, and so a single operator might well be able to regulate the whole cluster. Indeed, as had already been observed, in a histidine auxotroph grown with different degrees of histidine limitation the levels of **all the enzymes of histidine biosynthesis vary in constant proportions** (Fig. 13-10); and **coordinate induction** of multiple proteins was similarly observed in several catabolic systems, including *lac*. 3) The postulated operator should be subject to mutational alterations that would destroy its response to the repressor, rendering the cell constitutive. 4) If the operator is physically linked to the sequence of genes that it regulates such **operator-constitutive** (O^c) mutations should be **dominant,** in contrast to the recessive R^- constitutivity; and they should be *cis*-dominant (i.e., affecting only the z gene on the same chromosome), in contrast to the *cis-trans* dominance of R^+.

These predictions were confirmed by the isolation of *cis*-dominant constitutive *lac* mutants (using the selective conditions described in Figure 13-11). Thus heterozygote $O^cz^+R^+/O^+z^-R^+$ is constitutive, but $O^cz^-R^+/O^+z^+R^+$ is inducible (Fig. 13-11; see also Table 13-3).

---

[*] Later work revealed rare dominant R-constitutive (R^{-D}) mutants. The basis is the multimeric structure of the repressor protein, which causes most molecules in a heterozygote to be a hybrid of wild-type and mutant subunits; with certain mutants these hybrids are inactive.

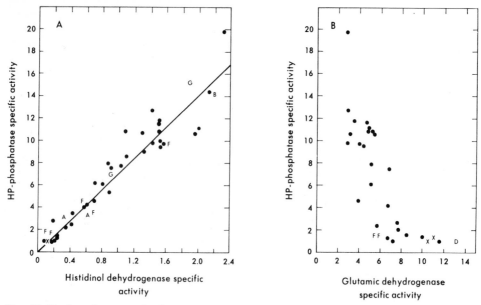

Fig. 13-10. Coordinate synthesis of enzymes of a biosynthetic pathway. Cells of several different histidine auxotrophs were harvested in different stages of derepression, caused by growth on a limited supply of histidine, and the levels of several enzymes were determined, including histidinol phosphate (HP) phosphatase and histidinol dehydrogenase. Throughout the range of derepression these enzymes (and others of the histidine pathway) retained a constant ratio of concentrations. Glutamic dehydrogenase, an enzyme not on the histidine pathway, did not vary in parallel. [From Ames, B., and Garry, B. *Proc Natl Acad Sci USA* 45:1453 (1959).]

Fig. 13-11. Dominance of operator constitutivity. Repressor (R) attaches to the wild-type operator (O+) (blocking expression of the *cis* z+ gene) but not to the O^C operator. In isolating O^C mutants merodiploid cells were used in order to avoid the more frequent (R−) constitutive mutants. Thus in an $R+O+z+/R+O+z+$ cell mutation of either $R+$ gene to $R−$ would usually be recessive and hence not expressed, whereas mutation of O+ to O^C would be dominant and hence would be expressed.

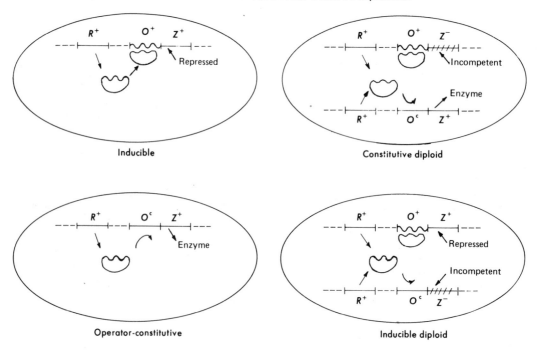

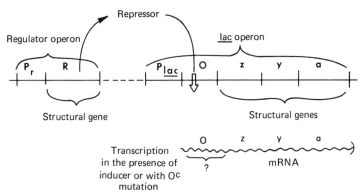

Fig. 13-12. Map of the *lac* operon and its regulator operon. Structural genes *z, y,* and *a* code respectively for β-galactosidase, the β-galactoside transport protein (Ch. 6), and β-galactoside transacetylase (an enzyme of unknown function in the cell's economy). Transcription is initiated at the promoter (P), and its rate of extension to the sequence of structural genes is regulated by the intervening operator (O) locus. The promoter for the *lac* operon (P_{lac}) has been mapped from P-negative mutations, which inactivate the whole operon. The regulator operon is a primitive operon: it has a single structural gene (for a repressor protein) and must have its own promoter, but it does not appear to have an operator.

O^c *lac* mutants are constitutive not only for β-galactosidase but also for two additional proteins, whose genes map, together with *lac,* in a sequence adjacent to the operator (Fig. 13-12). On the other side of the operator is the **promoter** (P), which binds RNA polymerase (Ch. 11). Such a group of linked genes and regulatory elements, which functions as a **unit of transcription,** is called an **operon.** However, the **regulator (R) gene,** which synthesizes a soluble repressor, is not part of the operon; in *lac* it appears to be next to the operon, but in many systems the two are separated by unrelated genes. The effects of mutations in P, O, and R are illustrated in Table 13-3.

Not all biosynthetic pathways have their genes linked in a single operon. However, in some pathways with several separate operons (e.g., arginine) these still constitute a functional unit (**regulon**), since they are regulated by the same *R* gene. One arginine operon has a central O locus, which regulates a promoter on each side.

Before discussing the interaction between repressor and operon we must consider some properties of mRNA. We shall first review its remarkable discovery, which arose as part of the operon story; its interaction with the ribosome has already been reviewed in Chapter 12.

MESSENGER RNA

DISCOVERY OF A RAPIDLY TURNING OVER MESSENGER

When the operon model of regulation was proposed in 1961 the process of information transfer from DNA to protein was still obscure. Protein could be synthesized in vitro without DNA, and so there must be intermediate templates; but chemical analyses revealed no well-defined RNA fraction except tRNAs and rRNAs. It was therefore assumed that various templates were built into various ribosomes. However, it became difficult to reconcile this view with the extremely rapid response of β-galactosidase synthesis to addition or removal of an inducer, especially since Davern and Meselson had already shown that ribosomes are stable in growing cells: subunits formed during growth with heavy isotopes remain fully "heavy" during subsequent growth in a medium containing only light isotopes. Activation of a preexisting gene product might explain the response to inducer, but it was excluded when Pardee, Jacob, and Monod demonstrated a similar rapid response to the introduction of a z^+ gene into R^-z^- cells. It was therefore concluded that the intermediate template of

TABLE 13-3. Effects of Regulatory Mutations on β-Galactosidase Synthesis

Genotype	Approximate enzyme level	
	Uninduced	Induced
I. R mutants		
R+Z+ (wild type)	1	1000
R+Z+/F'—R+Z+	2	2000
R−Z+	1000	1000
R−Z+/F'—R+Z−	1	1000
R^{s}Z+	1	**1**
R^{s}Z+/F'—R+Z−	1	1
II. O mutants		
R+O+Z+ (wild type)	1	1000
R+O^{c}Z+	500	1000
R−O^{c}Z+	1000	1000
R+O^{c}Z+/F'—R+O+Z−	500	1000
R+O^{c}Z−/F'—R+O+Z+	1	1000
III. P. mutants		
R+O+P+Z+ (wild type)	1	1000
R+O+P−Z+	0.01	10
R+O+P−Z+/F'—R+O+P+Z−	0.01	10
R+O+P−Z−/F'—R+O+P+Z+	1	1000

F' is an episome carrying the designated genes in a merozygote.

R− fails to make repressor; Z− fails to make β-galactosidase. R^s makes a "superrepressor" that fails to be altered by the inducer. O^c is a constitutive operator (seen to be still slightly responsive to repressor). P− is a nearly inactive promoter; in its presence the level of synthesis is very low, but the ratio with and without inducer (in an R+O+ strain) is unchanged. The "basal level" of enzyme in uninduced cells will be discussed below (Effect of repressor binding).

The data show that R+ is *cis*- and *trans*-dominant to R−, and R^s is *cis*- and *trans*-dominant to R+. O and P, in any allele, are *cis*-dominant.

protein synthesis cannot be a permanent part of the ribosome: it must be a separate "messenger" RNA molecule, that turns over very rapidly in the cell.

Chemical Evidence. The predicted novel RNA fraction was soon demonstrated with the help of bacteriophage, which blocks host RNA and protein synthesis as well as causing phage protein synthesis (Ch. 45). Indeed, earlier experiments had shown that though such infected cells carry out no net RNA synthesis they incorporate radioactive phosphate into RNA; hence they are synthesizing an RNA that turns over rapidly. Moreover, the base composition of this labeled material was found to resemble that of phage DNA, quite unlike host DNA.

The messenger concept provided a theoretical explanation for this mysterious labile RNA. Brenner, Jacob, and Meselson then showed that in **phage-infected cells,** synthesizing phage components from radioactive precursors, the ribosomes (formed before infection) become complexed with **newly formed RNA** as well as with newly formed polypeptide; and further synthesis with unlabeled precursors "chases" both these labeled chains from the ribosomes within a few minutes. Hence the ribosome complexes transiently with new mRNA while forming new nascent polypeptide. Finally, when Spiegelman developed the hybridization technic (Ch. 10) he demonstrated directly that this phage-induced RNA is specifically complementary to phage DNA.

In **uninfected cells** labeling of mRNA is harder to detect because the stable forms of RNA accumulate and soon represent most of the labeled product. However, at any moment

more than half the RNA being synthesized in steady-state growth in bacteria is mRNA. Accordingly, this RNA can be heavily labeled (compared with the stable RNA) by exposing cells to radioactive precursor (phosphate or base) for a brief period (< 1 minute)—the shorter the better. With this **pulse-labeling** technic growing cells were shown to have a rapidly labeled, labile RNA fraction transiently complexed with the ribosomes (Fig. 13-13). Moreover, this RNA has a base composition close to that of the cellular DNA,* and it can be specifically hybridized with that DNA. (Unlike mRNA, rRNA, constituting the bulk of cellular RNA, deviates markedly from this average composition, Ch. 12).

---

* The DNA-like composition of mRNA has helped in its recognition but is not logically necessary, since mRNA is transcribed from only one strand in any region of DNA. In fact, in the DNA of certain bacteriophages the two strands differ significantly in composition. However, in the much longer chromosome of bacteria the strands have about the same average composition.

Studies with Specific mRNAs in Vitro. The most direct evidence for mRNA has come from the use of **transducing phages** (Ch. 46) carrying specific bacterial genes. These have been used in two ways to study the relation between mRNA synthesis and enzyme synthesis. First, hybridization with the DNA of such phages (see Fig. 13-15, below) can be used to measure the mRNA of the genes that they carry; and following the addition or removal of the inducer or repressor this concentration in cells has been found to parallel closely the rate of synthesis of the corresponding enzymes. Second, DNA enriched for particular genes can be used for the synthesis of specific proteins in vitro (see Table 13-4, below), and the rate of this synthesis is also found to be proportional, under varying conditions of induction, to the level of the corresponding mRNA.

SIMULTANEOUS SYNTHESIS, TRANSLATION, AND BREAKDOWN OF mRNA

We saw in Chapter 12 that transcription and translation take place in the same direction

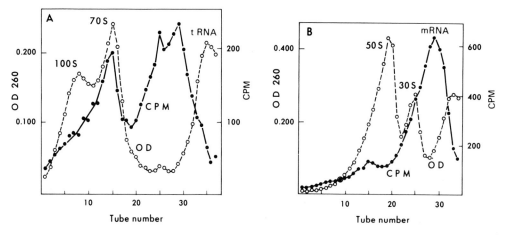

Fig. 13-13. Zonal sedimentation of pulse-labeled RNA. RNA was labeled in *E. coli* by growth for 20 seconds with ^{14}C-uracil, and further metabolism was stopped rapidly by pouring on ice and adding azide. The cells were lysed by grinding with alumina, the lysates were sedimented in a sucrose gradient with Mg++-Tris buffer; and fractions were collected through the bottom of the tube and analyzed for RNA (optical density at 260 nm) and radioactivity (CPM). **A.** Extracted and sedimented in 10 mM Mg++, which preserves the integrity of the ribosomes and their association with mRNA. The pulse-labeled RNA (closed circles) sediments partly with the 70S (ribosome) and 100S (disome) peaks, and partly as a broad slower peak of free mRNA and nascent rRNA. **B.** Sedimented in 0.1 mM Mg++, which dissociates the ribosomes into their 50S and 30S subunits and releases the labeled RNA, now all seen as a broad, slower peak. The lysis by grinding in this early work produced much fragmentation; gentler methods later yielded much longer polysomes, binding a larger fraction of the pulse-labeled RNA. [From Gros, F., Hiatt, H., Gilbert, W., Risebrough, R. W., and Watson, J. D. *Nature 190*:581 (1961).]

$(5' \rightarrow 3')$. Electron micrographs of gently prepared bacterial lysates (Fig. 13-14A) have shown that they occur simultaneously on the same RNA chain; and since the ribosomes on these growing polysomes are seen to be pressing close to the attachment to DNA it is evident that translation keeps pace with transcription. Moreover, the progressive increase in polysome length with distance from an evident site of initiation indicates that many chains are growing simultaneously on the same operon (as with phage, Fig. 11-32).

The same conclusions were reached from hybridization studies with the DNA of selected parts of the tryptophan operon. As Figure 13-15 shows, when transcription of this operon is abruptly initiated many copies of mRNA from the first gene can be demon-strated before any mRNA from the last gene appears. Moreover, a brief pulse of de-repression is seen to cause a wave of transcription followed by degradation, with the early part of the messenger (at the 5' end) disappearing before transcription reaches the last part. Hence **mRNA breakdown** proceeds **in the same direction as synthesis,** i.e., from the 5' to the 3' end.

Molecules of mRNA are thus subject to simultaneous growth, translation, and degradation. The rate of these processes (and the simultaneous translation) in *E. coli* at 37° is ca. 45 nucleotides ($= 15$ amino acids per ribosome) per second. (Protein synthesis in vitro, in much more dilute solutions, reaches only ca. 1/10 this rate.)

The observed kinetics suggest that most polysomes are attached to DNA, even though electron

Fig. 13-14. Electron micrographs of active operons recovered in gentle lysate of *E. coli.* The rectilinear thin fibers can be destroyed by treatment with DNase, while the attached chains can be destroyed by RNase. (They are also released by protease, which suggests that RNA polymerase is essential for their attachment.) **A.** Formation of messenger. The growing mRNA chains are essentially fully loaded with ribosomes, and they exhibit an irregular gradient of increasing length along the gene. Most regions of DNA are free of such appended polysomes, which is consistent with other evidence that only a small fraction of the genome is being transcribed at any time. An RNA polymerase molecule (MW 4×10^5) is visible at the presumptive site of initiation (arrow) and at the site of attachment of most chains. **B.** Formation of rRNA. Since ribosomes do not attach to this RNA it is seen as coils, probably with aggregated protein, rather than as extended chains. The products exhibit two adjacent gradients of size, whose lengths of DNA correspond reasonably to those expected for the formation of 16S and 23S RNA. This operon for rRNA is much more crowded with growing chains than is the typical mRNA-synthesizing operon shown in **A.** [From Miller, O. L., Jr., Hamkalo, B. A., and Thomas, C. A., Jr. *Science 169*:392 (1970).]

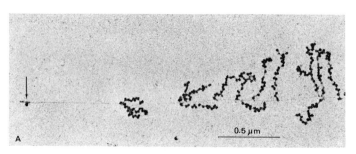

Fig. 13-15. Sequential synthesis and degradation of the mRNA of the tryptophan operon in *E. coli*. Cells were derepressed at 0 time to initiate transcription, with ³H-uridine present to label the RNA formed. After 1.5 minutes tryptophan was added to restore repression, i.e., to prevent new initiations. At various times RNA was extracted and portions were hybridized to the DNA of different transducing φ80 phages carrying selected parts of the operon, as indicated in the upper part of the figure. The curves show that such a pulse of derepression causes a wave of transcription that moves from one end of the operon to the other in about 6 minutes. [After Morse, D. S., Mosteller, R. D., and Yanofsky, C. *Cold Spring Harbor Symp Quant Biol* 34:725 (1969).]

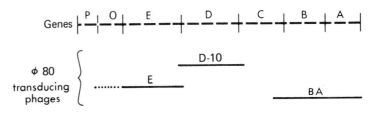

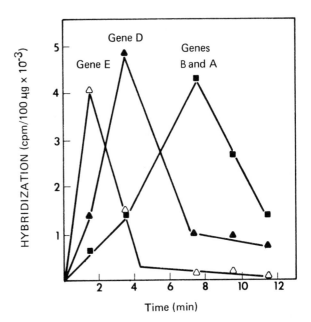

micrographs (Ch. 2) give the impression of a sharp border between the ribosome-free nuclear body and the ribosome-packed cytoplasm. When transcription reaches the end of the operon the polysome presumably is released and continues to be translated until it is completely degraded.

The kinetics of breakdown of mRNA can be studied by pulse-labeling the mRNA in growing bacteria and then abruptly stopping its synthesis by the addition of actinomycin. The labile fraction of the RNA decays with a half life of about 2 minutes, at 37°. After a pulse of 45 seconds, this fraction (which includes incomplete rRNA and tRNA) is well over 50% of the labeled RNAs, while in steady-state labeling it is only ca. 5%. More specific measurements of mRNA of the *lac* operon, using hybridization, have shown that after repression the decay of the mRNA is also strictly exponential (Fig. 13-16), suggesting random initiation of breakdown

(starting at the 5′ end). Moreover, after the onset of repression the capacity for residual β-galactosidase synthesis declines in strict proportion to the amount of its residual mRNA. This close coordination of physical and functional loss suggests that the degradative enzyme normally moves closely behind or with the last ribosome.

This conclusion fits other evidence that mRNA breakdown occurs only in regions that are not protected by close packing of ribosomes: when polysomes are "frozen" in the cell (e.g., by chloramphenicol) the mRNA is much more stable; conversely, removing ribosomes has the opposite effect, as will be described below (see Polarity mutations). The other kinds of RNA in the cell are evidently stabilized in other ways: secondary structure, methylation, and bound proteins (in ribosomes).

From measurements of the amount of *trp* mRNA in a cell (determined by hybridization), compared with the rate of synthesis of the corresponding enzymes, this mRNA appears to be translated ca. 30

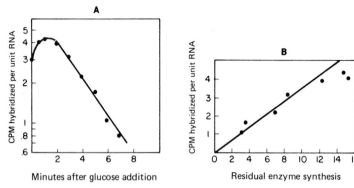

Fig. 13-16. Kinetics of decay of *E. coli lac* mRNA at 37°. Cells were grown under inducing conditions with ³H-uridine long enough for steady-state labeling of the mRNA. The *lac* operon was then repressed by adding glucose (see below, Catabolite repression), which blocks further initiation of transcription of this operon. **A.** Cells were harvested at intervals and lysed, and the *lac* mRNA was measured by hybridization to the DNA of a transducing phage (see Ch. 46) carrying the *z* gene of the *lac* operon. It is seen that the *lac* mRNA level is maintained for about 2 minutes (completion of initiated chains) and then decays exponentially (half life 2.2 minutes). **B.** Following repression the declining further synthesis of β-galactosidase was measured at short intervals. The residual enzyme synthesis at a given time is the amount of enzyme still synthesized between that time and the cessation of further synthesis. This measure of mRNA activity is seen to be proportional, at various times, to the amount of mRNA measured by hybridization. [From Adesnik, M., and Levinthal, C. *Cold Spring Harbor Symp Quant Biol 35*:451 (1970.)]

times before destruction. This value should equal the average number of ribosomes on that polysome in midphase, with one end of the mRNA growing and the other being degraded. However, it is not certain that the average interval between the initiation of synthesis and that of decay is the same for all mRNAs.

The enzyme(s) responsible for mRNA breakdown has not been identified. Polynucleotide phosphorylase would be more economical than RNase since it converts RNA to nucleoside diphosphates rather than monophosphates and thus would halve the cost of the ultimate conversion to the triphosphates used for RNA synthesis.

$$(BRP)n + nH_2O \xrightarrow{\text{RNase}} n(BRP)$$

$$(BRP)n + nP_i \xrightarrow[\text{phosphorylase}]{\text{Polynucleotide}} n(BRPP)$$

However, the phosphorylase attacks RNA from the 3′ end, whereas mRNA is destroyed over-all from the 5′ end; hence the phosphorylase would have to work on fragments snipped off near the 5′ end by an endonuclease. This mechanism is only speculative.

mRNA in mammalian cells is more stable than that in bacteria. This and other differences from bacterial mRNA will be discussed in Chapter 47.

REPRESSOR ACTION

The kinetics of induction and repression, which parallel those of mRNA synthesis and breakdown, suggested from the start that a repressor (R-gene product) regulates transcription. However, the repressor could conceivably yield the same results by influencing the stability of mRNA, perhaps via an effect on its translation. A decisive solution was obtained only after the repressor was isolated, which made possible direct studies of its action in vitro.

ISOLATION OF REPRESSORS

Two repressors (*lac* and phage λ) were isolated from *E. coli* in 1966; though they recognize specific DNA sequences they turned out to be proteins rather than nucleic acids. These isolations required considerable ingenuity, because the amount of a repressor normally present is small, and there was no direct assay for its acivity.

The lac repressor was isolated by Gilbert

and Muller-Hill by assaying fractions of the cell extract for their capacity to bind an inducer, IPTG (Table 13-1). The final product was a tetrameric protein of MW 160,000, constituting only ca. 0.002% of the normal cell protein (ca. 10 molecules per cell).

The concentration in the cell has now been raised 1000-fold by two devices: selection for mutations that increase the quantity of the *R* gene product, and multiple infection with phage carrying this mutant operon (see Ch. 46, Transducing phages). Moreover, purification is now greatly facilitated by affinity chromatography: passage through a column of agarose covalently linked to the inducer, which retains the binding protein.

Ability of a protein to bind IPTG suggests but does not prove that it is the *lac* repressor. Proof was provided by certain mutations: R^- cells lack the binding protein, and cells with various other *R* mutations (increased or decreased response to IPTG; temperature sensitivity) yield a protein with appropriate changes. In addition, *lac* DNA (in a transducing phage) specifically binds the protein and O^c mutations weaken the binding, as would be expected if the repressor directly blocks transcription of the operon.

Binding to DNA. The repressor binds to double-stranded but not to single-stranded DNA; hence it evidently reads features of the bases that are accessible in the double helix. Its affinity for the operator is extremely high (dissociation constant ca. 10^{-12} M), which results in infrequent spontaneous dissociation and reassociation; hence very few molecules of repressor per cell suffice for regulation.

Repressors, like regulatory branch-point enzymes, are allosteric proteins, whose conformation is altered by the appropriate effector molecule. Thus binding of IPTG to the *lac* repressor markedly decreases its affinity for DNA. Moreover, this inducer does not simply shift an equilibrium but actively causes a conformational change, for it accelerates the release of already bound repressor. This property explains the striking speed of induction in the cell.

The repressor binds a DNA segment of ca. 20 nucleotide pairs, which it protects from DNase action. Since it is a tetramer four molecules of inducer can contribute binding energy to counteract the binding to DNA. The first 50 amino acids of the repressor (at the NH_2 end) apparently form a pro-

trusion that fits a groove in the DNA: various mutations in this short region destroy the binding to the operator but not to the inducer.

Phage λ repressor (Ch. 46) was isolated by Ptashne without benefit of even a binding assay, by isolating the protein in a chromatographic peak that was present in cells infected with this phage but not in those infected with a mutant lacking the repressor gene. Repressors of a number of operons have now been isolated. Some enzymes may act as repressors: the first enzyme of histidine biosynthesis, which responds allosterically to histidine, binds to histidine operon DNA.

With the identification of repressors the initial distinction between a structural gene and a regulatory gene has become a distinction between genes that produce **an enzymatic or a regulatory protein.** However, neither the P nor the O locus of an operon seems to yield a protein. Since the O locus is distal to the promoter it would be expected to be transcribed, but sequence studies indicate that mRNA synthesis may begin only after the polymerase has traversed the operator and reached the first gene for an enzyme.

EFFECT OF REPRESSOR BINDING

Effect in Cells. When a deletion eliminates the *lac* promoter and operator the *lac* operon is fused with the nearby *trp* operon and the fused pair is transcribed as a unit, subject to *trp* regulation. However, if a deletion leaves the *lac* operator intact its repression can prevent this extension of transcription from the *trp* to the *lac* genes. The lac repressor thus appears to block transcription by preventing the unwinding of DNA. With λ, in contrast, binding of the repressor to its operator appears to block binding of the polymerase to the adjacent promoter (Ch. 46).

Since the binding of repressor is reversible an operator is transiently free from time to time even under repressive conditions, permitting initiation of a round of synthesis. Moreover, a newly formed operator requires some time to contact a repressor molecule. Hence uninduced cells have a **basal level** of enzyme, ca. 1/1000 of the fully induced level (Table 13-3). The small number of repressor molecules per cell explains the transient **"transduction escape synthesis"** seen on infecting uninduced cells with a high multiplicity of transducing phage: when the number of *gal* operons thus introduced exceeds the limited supply of *gal* repressor some operons must remain unrepressed.

Effect on in vitro Transcription. Assay of mRNA by hybridization showed that the *lac* repressor blocks specific transcription of *lac* genes in a transducing phage DNA in vitro, even when not coupled with translation. This finding provided the final, most decisive evidence that repressor affects transcription directly, rather than via an effect on translation. Such repression in vitro, coupled to translation (see Table 13-4 below), now provides a convenient assay for specific repressors of various operons (for which transducing phages are available).

OTHER MECHANISMS REGULATING GENE EXPRESSION

We have seen that the operator provides fine control of the expression of small groups of genes. Several additional mechanisms have since been discovered; some provide a coarser control and some influence large groups of genes.

PROMOTER VARIATION

Operons vary widely in their maximal rate of transcription, in the absence of repression; hence their promoters presumably differ in affinity for RNA polymerase (Ch. 11) and thus in frequency of initiation. Indeed, **promoter mutations,** which increase or decrease the maximal rate of transcription of an operon, have been isolated. In addition, inefficient **secondary promoters** are found in the middle of some operons, and they may be created by mutation; they permit weak expression of distal genes despite repression of the proximal genes.

As long as repressors were the only known mechanism of gene regulation, the question of what in turn **regulates repressor formation** was disturbing. For repressors clearly cannot each constitute as much of the total protein (5%) as unrepressed β-galactosidase; neither can their operons each be controlled by the product of another operon, in infinite serial regression. Promoter variation solves this problem: it provides a primitive, inflexible method of regulating those operons with a low output (e.g., repressor proteins, enzymes of vitamin synthesis), where finer regulation might not contribute enough economy to justify the extra R and O loci.

ALTERATIONS OF RNA POLYMERASE: PHAGE INFECTION AND SPORULATION

Since σ specifies the sites of initiation of transcription (Ch. 11) its discovery introduced a new class of possible regulatory elements. Indeed, alteration in σ (and in other aspects of RNA polymerase) are observed under circumstances that drastically and irreversibly reprogram the transcription in a cell: sporulation and phage infection. But σ does not appear to play a role in the reversible shifts in transcription seen in vegetatively growing cells.

In **sporulation** the cell becomes irreversibly committed to radical changes in structure and composition (Ch. 6). The key to the long-standing problem of the mechanism of this differentiation came from an unexpected source: cells of *Bacillus subtilis* that had started to sporulate were found to have lost the ability to support the multiplication of a phage (ϕe). The known mechanism of phage regulation suggested an alteration in transcription, and this notion was reinforced when Losick found that certain mutations in RNA polymerase (Ch. 11), selected because they make the enzyme resistant to rifampin, also prevent sporulation. He then showed that in sporulating cells the RNA polymerase is altered in several ways: the σ-factor is lost, a β-subunit of MW 155,000 has been cleaved to (or replaced by) one of the MW 110,000, and smaller subunits have been added. The altered polymerase not only lacks but can no longer bind σ, which was shown in vitro to be necessary for transcribing phage ϕe and also for transcribing at least one major class of host genes, those producing rRNA. Presumably this change in the polymerase also turns off other vegetative operons in the cell's genome and turns on sporulation-specific operons (see Ch. 6).

Chapter 45 will describe changes in RNA polymerase seen on **phage infection;** synthesis of a new, small polymerase by phage T3, and modifications of *E. coli* polymerase (adenylylation, amino acid addition) by T4.

CATABOLITE REPRESSION: CYCLIC AMP ACTIVATION

The phenomenon of diauxie (Ch. 3) showed that the presence of glucose prevents various other carbon sources, which support slower growth, from inducing the corresponding enzymes. This device guarantees preferential utilization of the most profitable energy sources.

More detailed kinetic studies, with cells induced to form β-galactosidase during growth on succinate or glycerol, showed that addition of glucose causes initial, transient complete repression of further synthesis of the induced enzyme, followed after 20 to 30 minutes by permanent, partial repression (Fig. 13-17). Constitutive mutants, which do not depend on an inducer, also exhibit catabolite repression. The effect is exerted by **any carbon source that supports more rapid growth** than the natural inducer, and so it seemed to resemble endproduct repression: the corepressor was presumably some product **(catabolite)** whose level rises with an increased rate of catabolism.

Catabolite repression has been a frequent source of indirect effects, often unrecognized, in studies of regulation. Thus at borderline concentrations inhibitors of protein synthesis preferentially inhibit formation of induced enzymes, and such effects were long attributed to preferential action on the translation of particular messengers. We now know that the mechanism is indirect: when a major outlet for energy is curtailed even a poor source (such as succinate) will exert the same effect as glucose.

Mechanism. Though the "glucose effect" was observed very early, its mechanism remained a mystery for decades. The actual signal turned out to be not the postulated repressive catabolite but rather adenosine-3′,5′-monophosphate (cyclic AMP, cAMP). This compound was already known to regulate many metabolic processes in mammalian cells.

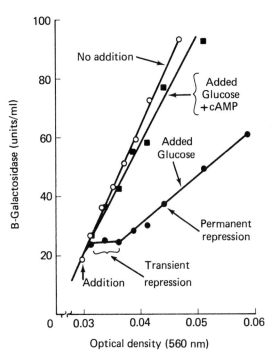

Cyclic AMP

Fig. 13-17. Catabolite repression of β-galactosidase synthesis and its reversal by cyclic AMP. Glucose, with or without cAMP, was added to a culture of *E. coli* growing on succinate in the presence of an inducer of the enzyme, IPTG. The cAMP is seen to overcome both the transient (complete) and the permanent (partial) repression by glucose. The brief lag before repression reflects the completion and the translation of already initiated messenger. [After Pastan, I., and Perlman, R. *Science* 169:339 (1970).]

Sutherland showed that it is present in *E. coli,* and that growth on glucose lowers its level; and Pastan and Perlman later found that added cAMP specifically overcomes catabolite repression of various inducible enzymes (Fig. 13-17). Finally, mutations blocking its synthesis (adenyl cyclase-deficient) were found to have lost simultaneously the ability to grow on (and be induced by) many different carbon sources, and addition of cAMP corrected all these pleiotropic defects.

The cAMP system acts **positively** on transcription, rather than negatively like repressors. Thus a **cAMP receptor** (or catabolite gene activator) protein (CRP or CAP), which binds cAMP, has been isolated from *E. coli;* and mutants with a defective CRP have lost the ability to grow on a variety of carbon sources. Moreover, the transcription of *lac* DNA in vitro requires both CRP and cAMP, as can be conveniently shown by its coupling with translation (Table 13-4).

CRP evidently **interacts with the promoter** rather than the operator, since mutations that eliminate catabolite repression of the *lac* system have been mapped in the promoter locus. Moreover, the cAMP·CRP complex can be shown in vitro to bind to the promoter DNA. Apparently with certain promoters this binding is necessary for the binding of the initiating RNA polymerase.

POSITIVE REGULATION BY AN R GENE PRODUCT

The original model of negative regulation, by a repressive R-gene product, has been extended to many operons, but it is not universal: positive regulation (i.e., activation by the R-gene product, just as by cAMP·CRP)

has been observed with the arabinose (*ara*) and maltose (*mal*) operons, and also with phage λ lysogenization (Ch. 46).

The *ara* system has a regulator locus (C) that influences the *ara* operon in a *trans* chromosome as well as in the *cis* chromosome; hence it must form a cytoplasmic product, like the usual R loci. However, instead of having the usual choice of a neutral or a repressive state one form of this product is required to activate the operon, and its alternative state is repressive; the inducer promotes a shift to the activator conformation. Some mutations in the C-gene product make it a permanent repressor and others make it a constitutive activator.

REGULATION AT THE LEVEL OF TRANSLATION

This mechanism has not been found in growing bacteria, where such a limitation of mRNA utilization would be less economical than limitation of mRNA synthesis. However, when *E. coli* is infected with certain phages their takeover of biosynthesis is hastened by a shift in the proportion of two interference factors that interact with initiation factor IF$_3$ and influence the relative frequency of initiation on phage-coded and on host messengers.

TABLE 13-4. Effect of Cyclic AMP and its Receptor (CRP) on DNA-directed in Vitro Synthesis of β-Galactosidase

Source of bacterial extract	Cyclic AMP (5×10^{-4} M)	CRP protein	β-Galactosidase (relative values)
CRP⁻ strain	−	−	1
CRP⁻ strain	+	−	1
CRP⁻ strain	−	+	1
CRP⁻ strain	+	+	5
CRP⁺ strain	−	−	1
CRP⁺ strain	+	−	20
CRP⁺ strain	−	+	1
CRP⁺ strain	+	+	24
CRP⁺ strain without IPTG	+	+	1

The incubation system contained DNA from a transducing phage (φ80*dlac*) carrying the *lac* operon, as well as IPTG (Table 13-1) to overcome repression, a crude bacterial extract (s-30), and all the additional components required for transcribing the DNA and then translating the product. The enzyme formed was assayed after incubation for 1 hour, which yielded maximal formation.

The control mixtures, lacking some component required for normal initiation, are seen to have about 1/20 the full activity. This background synthesis represents imperfect regulation in the in vitro system, i.e., false initiation distal to the regulatory loci.

Modified from G. Zubay, D. Schwartz, and J. Beckwith. *Proc Natl Acad Sci USA* 66:104 (1970).

POLARITY MUTATIONS

Nonsense (terminator) mutations not only block expression of the gene in which they occur but often decrease the expression of distal ("downstream") genes of the operon, without affecting that of proximal ones. The degree of inhibition increases with the length of the "dead space" between the mutation and the beginning of the next gene (Fig. 13-18). This "polarity" effect evidently depends on failure to translate the dead space, for it disappears when a suppressor mutation (Ch. 12) restores translation of the nonsense codon.

It seemed initially that the block in translation in these cells must somehow cause a block in transcription, for hybridization studies showed a decreased content of the mRNA of the gene distal to the mutation. However, pulse-labeling studies on the *trp* operon, similar to those of Figure 13-15, have shown that the mRNA distal to a polarity mutation is **hyperlabile.**

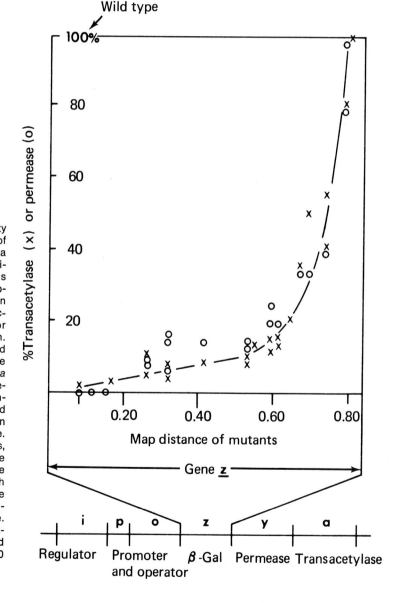

Fig. 13-18. Gradient of polarity effect in relation to position of mutation. *E. coli* strains with a nonsense (translation-terminating) mutation at various positions in the β-galactosidase gene (z) were grown under conditions of full induction and were then assayed for the products of the *lac* operon. None of the mutants yielded any z gene product. The amounts of the y and the a gene products are seen to decrease, in parallel, with increasing length of the "dead space" between the mutation and the start of the next gene. Studies with genetic deletions, on either side of the nonsense mutation, have shown that the important variable is the length of this dead space and not the length of the translated fragment of the mutant gene. [From Newton, W. A., Beckwith, J. R., Zipser, D., and Brenner, S. *J Mol Biol 14*:290 (1965).]

In addition, by selecting for activity of the distal genes in polar mutants, Morse has isolated mutations that suppress the polarity effect and has shown that these *su A* strains are defective in a novel endonuclease **(endonuclease A)**. Evidently this endonuclease initiates premature degradation in the mRNA of the dead space, which lacks protection by ribosomes; and once initiated, the degradation progresses through the entire distal mRNA. The longer the dead space the sooner this process is likely to be initiated.

As observed with nonsense mutations the polarity effect appears to be a laboratory artefact rather than a component of normal operon regulation.

However, the underlying mechanism may serve to scavenge normally encountered idle mRNA, and to help shift ribosomes to new mRNAs in cells adapting to a new medium. For example, when deprival of a required amino acid halts ribosomes at an empty codon they are slowly released from polysomes; the resulting "empty" regions of mRNA are rapidly broken down, and *su A* mutations slow this breakdown even though they do not influence normal mRNA degradation.

Polarity mutations provide an additional criterion for defining an operon and for sequencing its genes, and have thus revealed an operon for ribosomal proteins (see below, Ribosome biosynthesis).

REGULATION OF STABLE RNA SYNTHESIS

RELATION OF CELL COMPOSITION TO GROWTH RATE

It has long been known that among the various cells of an animal the RNA content parallels the rate of protein synthesis. In bacteria, similarly, as growth slows on approach to the stationary phase the RNA concentration per unit mass falls. To analyze the mechanisms regulating such **macromolecular changes** Maaloe compared cells in **steady-state, balanced growth** (i.e., exponential growth, with unchanging composition) at various rates, and also during **shifts** from one growth rate to another. The growth rate was varied by varying the carbon source supplied; similar results have been obtained by controlling the rate in the chemostat (Ch. 5), with a constant carbon source.

As Table 13-5 shows, variations in the rate of balanced growth, over a 10-fold range, have little effect on the ratio of protein to DNA in the cells. The ratio of rRNA to DNA, however, is essentially proportional to the rate of growth, and hence of protein synthesis. It thus appears that the **cell forms ribosomes only in the amount needed** for the growth rate supported by its current environment.

The obvious inference is that in steady-state growth ribosomes are being used at or near full capacity, and with constant efficiency. This equally efficient use at all growth rates is possible because the rate of protein synthesis depends on the frequency of initiation (Ch. 12): as later work showed, both in fast and in slow steady-state growth the bulk of the ribosomes are located in polysomes, and they move along the mRNA at the same rate.

TABLE 13-5. RNA Distribution and Protein Synthesis in *S. typhimurium* at Various Growth Rates

Carbon source	Growth rate (generations/ hr)	DNA (μg/mg bact. dry wt.)	rRNA/ DNA	tRNA/ DNA	Protein/ DNA	Protein synthesis per hr	
						Per unit RNA	Per unit rRNA
Broth	2.4	30	8.3	2.0	22	3.7	4.5
Glucose	1.2	35	3.9	2.4	21	2.8	4.6
Glycerol	0.6	37	2.4	2.4	21	1.8	3.6
Glutamate	0.2	40	0.9	2.1	21	1.0	3.3

The growth rate was controlled by providing carbon sources that differ in their maximum rate of utilization. The cells were harvested at low densities, below 0.15 mg dry weight per milliliter. Above this value alterations of the medium by the metabolizing cells began to cause unbalanced growth, resulting in progressive changes in cell composition. Conventionally plotted growth is much less sensitive to such changes and appears to remain exponential until about half-way to the saturation level of 1 to 2 mg per milliliter.

From N. O. Kjeldgaard. Dynamics of Bacterial Growth. Thesis, Univ. Copenhagen, 1963.

The ribosome concentration may set the limit to the maximal growth rate of bacteria (a doubling time of ca. 20 minutes at 37°). Thus electron micrographs show that in cells growing at this rate the average distance between ribosomes in the "cytoplasmic" region is approximately the diameter of a ribosome; further crowding might limit the supply of substrates to the ribosomes. Such cells contain about 25,000 ribosomes per genome, and about 15 molecules of tRNA per ribosome. In cells growing 1/10 as fast the values are 2500 and around 150, respectively.

Since the rRNA/DNA ratio is proportional to the growth rate, and the rate of DNA synthesis is proportional to the growth rate, the rate of rRNA synthesis is proportional to the square of the growth rate.

RESPONSE TO SHIFTS OF GROWTH RATE AND PROTEIN SYNTHESIS

The responses to an altered growth rate were further studied by observing the effect of an abrupt shift in medium on macromolecule synthesis. (Since the bulk of bacterial RNA is rRNA, net RNA synthesis is a convenient index of ribosome synthesis.) The results (Fig. 13-19) confirmed the inference that ribosomes are working at or near capacity in balanced growth. Thus in a **"shift up"** to a richer medium, with an increase in the supply of building blocks and energy, the **rate of net RNA synthesis** increases **abruptly,** while the

rate of protein synthesis increases only **gradually,** as the cells build up their ribosome concentration: the ratio of the two components (and rates) become constant only when the cells reach the ribosome concentration required for steady-state growth at the rate supported by the new medium. An even more dramatic response is seen in a **"shift down"** to a poorer medium, where the cell initially has more ribosomes than it needs. Protein synthesis slows, because of the diminished supply of building blocks and energy; but **net RNA synthesis and ribosomal protein synthesis ceases abruptly,** and it resumes only when further growth has diluted the ribosomes to about the level apropriate for the new growth rate. (The synthesis of ribosomes is coordinated, by an unknown mechanism, with that of polyamines, which are largely bound to ribosomes in cells.)

RELATION TO PROTEIN SYNTHESIS: "RELAXED" MUTANTS

Further work showed that net RNA synthesis also stops immediately when charging of any tRNA is blocked, either by depriving an auxotroph of the required amino acid or by a *ts* mutation in a charging enzyme. Hence the mechanism regulating ribosome synthesis evidently senses some shift in the activity of the protein-synthesizing machinery. However, the

Fig. 13-19. Response of DNA, net RNA, and protein synthesis to shift up and shift down of the medium. The curves are idealized, and the amount of each component per milliliter of culture is normalized to 1.0 at the time of the initial shift. The shift down depicted does not involve a diauxic lag. When there is such a lag, as in adapting to a new carbon source, the absence of an energy supply temporarily halts net synthesis of all three classes until turnover of mRNA and protein (see below) has built up the required new enzymes.

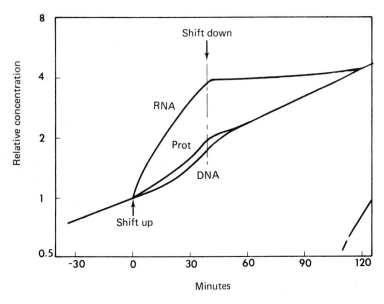

connection has proved elusive, and several initially plausible hypotheses have fallen.

For example, all transcription was once thought to depend on simultaneous translation, but we now know that rRNA is not translated at all and that mRNA can be formed without translation (e.g., in chloramphenicol-inhibited cells and in the polarity mutants described above.) Alternatively, the inhibition of net RNA synthesis by a block in tRNA charging, but not by a block at a later stage in protein synthesis (e.g., by chloramphenicol), suggested uncharged tRNAs as regulatory elements. However, rRNA synthesis is also inhibited by a block in formylation of Met-tRNA, which stops protein synthesis but should not cause accumulation of any uncharged tRNA; hence some other aspect of protein synthesis is evidently involved.

Relaxed Mutants. Another approach has been provided by mutants that have lost the response of their rRNA regulation to amino acid deprival. These strains are designated as having **relaxed control** of rRNA synthesis (RC^{rel} or rel^-), in contrast to the wild-type **stringent control** (RC^{str} or rel^+). After many years of search for a metabolic difference between stringent and relaxed cells it was found that only the former rapidly accumulate a novel nucleotide, 3',5'-**guanosine tetraphosphate (ppGpp,** also called MS-I). Moreover, the level of ppGpp varies inversely with the growth rate in steady-state growth of stringent cells. In extracts its synthesis, by transfer of pp from ATP to GDP, requires "idling" ribosomes, mRNA, uncharged tRNA, factor EF-G, and a protein factor extractable from ribosomes of stringent (but not those of relaxed) cells. ppGpp is probably the long-sought signal.

Total vs. Net RNA Synthesis. The model of coordinate regulation of all RNA synthesis originally seemed plausible because of the assumption that the incorporation of labeled uracil reflects *total* RNA synthesis (including turnover). However, unlike amino acids, free bases are not true intermediates but must be converted into the nucleotide intermediates by "salvage" pathways (Ch. 4), and these pathways are inactive when there is no drain on the nucleotide pool. Hence the incorporation of ^{14}C-uracil measures only *net* RNA synthesis; it fails to detect mRNA recycling in the absence of such net synthesis. (The classical labeling of phage mRNA in infected cells obscured this principle, for though the infection stopped net RNA synthesis the rapid synthesis of phage DNA drew on the ribonucleotide pool for its precursors.) It is now clear that the regulation of total RNA synthesis represents the sum of many individual processes, which each regulate the transcription of an operon.

RIBOSOME BIOSYNTHESIS

Ribosomal RNA. It would obviously be desirable for the cell to synthesize equivalent amounts of the 3 rRNA species. This goal is achieved by **linkage of the rRNA genes in an operon.**

Because these genes do not form protein this operon cannot be demonstrated in the usual way. However, when initiation of transcription is blocked with rifampin, and radioactive precursor of RNA is immediately added, the rRNA molecules being completed show a gradient of increasing specific activity, in the sequence 16S-23S-5S. Since the stages of transcription of the various copies of the operon are randomly distributed at the time of addition of label this gradient indicates that some chains of 5S RNA are being synthesized after all synthesis of 16S RNA has ceased.

Hybridization studies showed that an *E. coli* genome contains about 6 copies of the rRNA genes (Ch. 11). Since these few genes (ca. 1% of the genome) carry out roughly **half the total RNA synthesis** of the steady-state cell they are much busier than even unrepressed genes for mRNA (Fig. 13-14B); hence their promoters, interacting with yet unknown regulatory elements, must have a high affinity for RNA polymerase.

Maturation. In eukaryotic cells the initial product of transcription of rRNA genes (45S RNA) remains intact during transport to its site of cleavage and ribosomal assembly (Ch. 47), but in bacteria it is apparently cleaved too early to be detected (see Fig. 13-14). The immediate products are three species of **precursor rRNA** molecules, which accumulate under conditions permitting rRNA synthesis without ribosomal protein synthesis (e.g., chloramphenicol treatment). These precursors each have an extra segment at each end, amounting to ca. 10% of the total. In the course of ribosome assembly in the cell mature rRNAs are formed by elimination of the terminal segments and methylation of various bases. The function of the extra terminal segments is not known; they are not required for assembly (Ch. 12), but they may make it more efficient.

Ribosomal Proteins. The pool of ribosomal proteins is quite small, turning over in 1 minute of steady-state growth. rRNA appears to aggregate with these proteins and fold up as it is being transcribed. This conclusion is suggested by electron micrographs (Fig. 13-14B); moreover, the portion of 16S RNA that is synthesized first (the 5′ end) can be shown to bind the proteins that enter first (Fig. 12-14) into subunit assembly in vitro. In the absence of ribosomal protein synthesis the precursor rRNA molecules that accumulate do not remain free but aggregate nonspecifically with basic proteins.

The mechanism of coordination of ribosomal protein and RNA synthesis is not known but clearly is not based on a common operon: the genes for the former are clustered in *E. coli* at 63 to 64 minutes on the 90-minute map, while those for rRNA (located by synchronized chromosome replication followed by hybridization) are at about 73 minutes.

The ribosomal proteins may well all be formed by one large operon, for by the ingenious use of polarity, Nomura has shown that the genes for certain proteins of both subunits, as well as for elongation factor G, are located in the same operon. Thus in heterozygotes, containing both a drug-resistant and a sensitive allele of the genes for these proteins, a terminator mutation in any of these genes changes the ratio of resistant to sensitive ribosomes for the genes on one side but not on the other side of the mutation.

tRNA SYNTHESIS

This process is clearly not regulated in strict coordination with rRNA synthesis: as Table 13-5 showed, the tRNA concentration remains essentially constant at various growth rates, at about twice the DNA and 1/10 the protein concentration. This constancy fits the constant efficiency of ribosomal function in steady-state growth: for whether the ribosomal population is sparse or crowded, the frequency of contact of any ribosome with tRNA depends on the tRNA concentration. Hybridization studies indicate that the sum of the tRNA loci constitutes ca. 0.1% of the *E. coli* genome, which is sufficient for 1 gene for each of the approximately 50 species of tRNA formed.

REGULATION OF DNA SYNTHESIS

GROWING POINTS

Chapter 10 summarized the evidence that the bacterial chromosome is a single, cyclic molecule, which is replicated by its continuous movement through a replicating machinery (growing point). In a nonsynchronized culture of *E. coli* growing on glucose (doubling time 50 minutes at 37°) Maaløe found, by radioautography with ³H-thymidine, that about 90% of the cells are engaged in DNA synthesis at any moment: i.e., about 45 minutes are required between initiation and completion of replication. In slower growth the gap between rounds of replication is greater, but the rate of incorporation of thymidine per actively replicating chromosome is the same. Thus with DNA, as with other specific macromolecules, the rate of steady-state synthesis depends on the **frequency of initiation.**

In rich media, with doubling times as short as 20 minutes, the total DNA must be duplicated in a correspondingly shorter time. The mechanism, as Sueoka showed, is the development of **additional growing points** (branching) before completion of the first round of replication (Fig. 13-20): each growing point still moves at its inexorable rate. This mechanism was revealed by using transformation (in *B. subtilis*) to measure the relative frequency of genes located near the origin or near the termination of the sequence of replication: with a doubling time of 50 minutes the ratio is about 2:1, but in fast growth it approaches 4:1.

The conventional model of the growing point as a fork moving along a linear chromosome is very schematic: for the chromosome is actually cyclic, and in replicating it threads through the membrane-bound replication machinery. Moreover, it can do so from both sides simultaneously: for measurements of the relative frequency of various genes of *E. coli,* determined by transduction and also by hybridization of integrated phage genomes (Ch. 46), have shown that under some circumstances replication proceeds **in both directions** from the point of initiation (73 minutes on the *E. coli* chromosome). However, experiments conducted in other studies have demonstrated unidirectional replication: it seems probable that both modes occur. It is of interest that in conjugation 90 minutes is required for transfer of a complete chromosome (Ch. 9), which must proceed in a single direction.

CONTROL OF INITIATION

Maaloe also showed that initiation depends on formation of a "replication protein," which can act only once (perhaps becoming a permanent part of the membrane attachment site of a new chromosome).* Thus when protein synthesis is blocked (e.g., by deprival of a required amino acid) any replication of DNA that is already under way is completed but no new rounds are initiated: the number of cells engaged in replication falls linearly for about 45 minutes. Analysis of this regulatory mechanism was advanced further by studies with temperature-sensitive mutants (summarized in Chapter 10: Units of replication), which led Jacob and Brenner to infer a functional unit of DNA replication, the **replicon.** Its regulation involves two matched genetic elements, much like R locus and operator: an **initiator** locus produces a specific cytoplasmic initiator, which initiates a round of replication at a corresponding **replicator** locus. **Different replicators** (e.g., on bacterial chromosome, phage genome, or F episome) **respond to different initiators.**

The initiator appears to be synthesized at a rate proportional to the rate of protein synthesis. Thus as we have seen, initiations in a cell are spaced by a period corresponding to the doubling time. Moreover, if protein synthesis is interrupted the chromosomes are all completed, but on resumption of growth the resumption of DNA replication is not synchronous: protein synthesis has to catch up with the accumulated excess of DNA. In contrast, with agents that stop or slow DNA replication while allowing protein synthesis (thymine starvation, mitomycin, BUDR) the initiator continues to accumulate, new forks are formed, and when DNA

---

* The studies that revealed this important feature of replication made ingenious use of **"thymineless death,"** i.e., the rapid death of cells of a thymine auxotroph when deprived of thymine in an otherwise adequate medium. Maaloe found that this effect could be used to measure the fraction of the cells engaged at any time in DNA synthesis, since only such cells die under these circumstances. The mechanism is not certain, but attempted DNA synthesis in the absence of thymine has been shown to cause progressive degradation of DNA behind the blocked growing point, conceivably due to a misused repair mechanism (Ch. 11) or to the exonuclease activity of DNA polymerase. In addition, an increased mutation rate, and induction of prophages, have been observed.

synthesis is restored there is a burst of synchronized synthesis.

Control of Chromosome Replication by an Integrated Episome. Integrated episomes, such as an F agent (see Ch. 9 and 46), usually do not initiate replication of the cell chromosome; their replicon becomes inactive after integration. However, in bacteria with a *ts* mutation affecting chromosomal initiation the ability to replicate the chromosome at the nonpermissive temperature may be restored after integration of an episome. The initiator locus is evidently reactivated, since initiation is inhibited, as with the detached episome, by acridines (see Ch. 9).

RELATION BETWEEN DNA REPLICATION AND CELL DIVISION

Cell division depends on DNA replication. Thus interference with DNA synthesis (by an inhibitor, by UV irradiation, or by a temperature shift with a *ts* mutant) prevents septum formation while allowing further growth. Indeed, UV irradiation has long been known to cause filamentous growth, i.e., increase in cellular mass without cell division. More specifically, cell division is triggered by **completion of replication** of the chromosome. Thus, in a synchronized culture, with one growing point per chromosome, inhibition of DNA synthesis before completion of chromosome replication prevents initiation of cell division; while acceleration of initiation of DNA replication has no effect on cell division until the new rounds reach completion.

Since the chromosome is attached to the membrane at the growing point, a second attachment site, requiring additional proteins, must be formed on completion of replication. Septum formation is initiated in the region between the attachment sites, ensuring segregation of the two chromosomes to the two daughter cells.

The growth behavior of a culture is defined by three time periods: I, the time for accumulating enough initiator in a cell to cause initiation of chromosome replication (i.e., form a growing point); C, the time required for an initiation site to traverse the length of the chromosome; and D, the time between completion of replication and completion of cell division (i.e., physical cell separation). In *E. coli* period I is a variable equal to τ, the doubling time, while C and D are constants (about 45 and 25 minutes, respectively, at 37°) over a wide range of growth rates (Fig. 13-21).

Period D can be split into several components. 1) Completion of a membranous septum, which splits

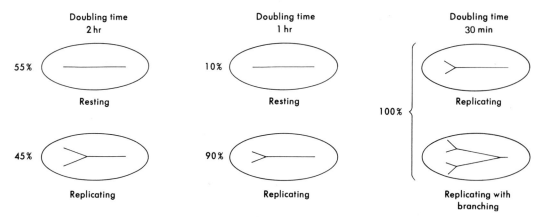

Fig. 13-20. States of chromosomal replication in *E. coli* cultures growing at different rates. Numbers denote per cent of the population in each state. For convenience the cyclic structure of the chromosome is ignored.

Fig. 13-21. DNA replication and cell division at various rates of initiator synthesis. For convenience the cyclic chromosome is diagrammed as though it were linear. I = doubling time = time required for initiating a new round of replication after last initiation. C = time required for travel of a growing point through the length of a chromosome. At completion of a chromosome pair (time X) the attachment site is doubled, the two chromosomes begin to separate, and a septum is initiated (dashed line) between them. D = time between X and physical separation of the daughter cells.

When the growth rate is slow (e.g., I = C + D) the chromosome duplicates with a single fork, there is a gap without DNA synthesis before cell division, and the daughter cells start with one unbranched chromosome. When I = C the gap is eliminated: there is still only a single fork but a new fork is formed just when the old one ends, and the daughter cells have a branched chromosome. During faster growth (I < C) there is multiforked replication and the daughter cells have two branched chromosomes. [Modified from Helmstetter, C., Cooper, S., Pierucci, O., and Revelas, E. *Cold Spring Harbor Symp Quant Biol* **33**:809 (1968).]

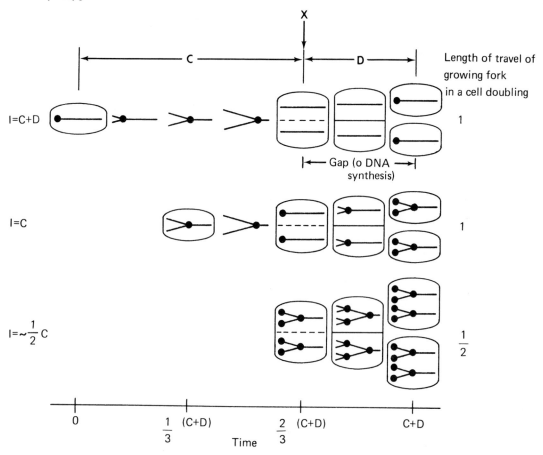

the cell protoplast into two compartments **(physiological division)**, can be determined operationally in synchronized cultures as the time when one half of a cell survives phage infection of the other half, yielding a two-hit curve. 2) Formation of a **rigid cross-wall** is identified operationally by the kinetics of killing by sonication (i.e., both halves have to be disrupted before viability is destroyed). 3) Finally, there is the classic **physical division** into two independent, clonable cells. In *E. coli* 10 minutes of D are required for forming a weak septum, 8 more for completing a strong cross-wall, and 7 more for scission of the wall.

In a rich medium with multiforked growth the doubling time, which is less than C, corresponds to the time for one growing fork to reach the position of the next fork (Fig. 13-21); this is also the time between two successive chromosome completions. Period D creates a constant phase difference between the rhythm of chromosome completion and that of cell division. A gap in DNA synthesis appears only when $\tau > C$.

DNA Synthesis in a Nongrowing Cell.

In the transition from exponential to stationary phase the size and composition of the cells may change even after cessation of over-all growth. Specifically, when growing cells are abruptly made stationary by transfer to a medium lacking a nitrogen or a carbon source, or lacking a required amino acid or nucleic acid precursor, **the DNA and the cell number increase** by 40 to 75%, while the optical density and the protein and RNA content remain essentially constant (Fig. 13-22).

Each cell evidently completes its current round of DNA replication, for when such stationary cultures are inoculated into fresh medium they initially exhibit synchronized growth. Because of this delayed division assays for viable cell number may give erroneously high values if there is a substantial delay between dilution in buffer and subsequent plating.

This completion of a chromosome during starvation depends on turnover of other, less essential cell components (see below) to supply the required material and energy. Indeed, the observed large increase of DNA is economically possible because DNA constitutes only ca. 5% of the cell mass.

Fig. 13-22. Persistence of DNA synthesis and cell division in starved cells. A tryptophan auxotroph of *S. typhimurium* was deprived of the required amino acid in an otherwise rich medium, and was supplied with it again after 150 minutes. Note the "overshoot" of DNA synthesis and cell division after the deprival, and the subsequent delayed response after the restoration. [From Schaechter, M. *Cold Spring Harbor Symp Quant Biol 26*:53 (1961).]

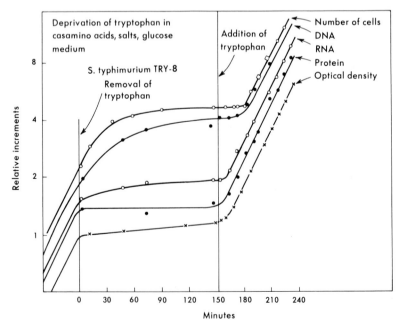

OTHER TYPES OF REGULATION

SYNTHESIS OF OTHER COMPONENTS

Cell Envelope. Regulation of the formation of cell wall and membrane is not well understood. As we saw in Chapter 6, in some species a block in basal wall synthesis (e.g., by penicillin) causes extensive intracellular accumulation of nucleotide-linked precursors of peptidoglycan and of teichoic acid, which implies lack of tight feedback inhibition. Cells starved for various constituents continue to lay down wall, in part because of the completion of cell division described above. Another adaptive change in the stationary phase increases the stability of the already present lipids by converting the double bond in unsaturated fatty acids into a **cyclopropane ring** (Ch. 6).

Storage Materials. Selective starvation for nitrogen, phosphorus, or sulfur often directs metabolism of the carbon source into storage materials, up to 25% of the dry weight; these include a **glycogen-like polysaccharide** and **poly-β-hydroxybutyrate** (Ch. 4). Nitrogen starvation also leads to **polymetaphosphate** accumulation in many organisms, up to 1 to 2% of the dry weight. Its formation and depolymerization, under different cultural conditions, have been shown to involve induction of appropriate enzymes.

MACROMOLECULE TURNOVER

PROTEIN BREAKDOWN

Bacterial metabolism is normally linked to growth and not to maintenance; hence protein turnover is not very prominent relative to synthesis. For example, when an inducer is removed the enzyme molecules already induced persist during further growth, and the concentration per cell decreases primarily by exponential dilution. However, slow turnover of protein also occurs: not only are radioactive amino acids slowly incorporated in nongrowing cells but previously incorporated labeled amino acids are released in parallel. (This release can be measured, despite the recycling of turnover, by incubating cells with an excess of the same unlabeled compound, since added amino acids exchange freely with the intracellular pool.) Two different kinds

of protein breakdown can be distinguished in bacteria: steady-state and starvation-induced.

Breakdown during steady-state growth occurs in *E. coli* at an average of ca. 1% of the total protein per hour at 37°. Although it is not known which proteins are catabolized, pulse-labeling shows that a fraction of newly synthesized protein breaks down much more rapidly; moreover, growing cells can selectively degrade abnormal proteins (e.g., containing amino acid analogs or errors in sequence, or released prematurely by a termination mutation or by puromycin). Their common fate suggests that these several classes (and probably some normal proteins) share a conformational feature (perhaps a loose loop, or a free end, or a high rate of spontaneous denaturation) that promotes initiation of breakdown by a protease.

Starvation-induced breakdown becomes an important mechanism during adaptation to a nutritional limitation. Thus when *E. coli* is starved for a carbon source, a nitrogen source, or a required amino acid the rate of protein breakdown jumps to 5 to 8% per hour, accompanied by an equal rate of resynthesis. (There are indications that certain normally stable proteins are preferentially degraded by this mechanism.) This turnover clearly helps to adapt to oscillations in nutrition, since it provides amino acids during a shift from a repressed to a derepressed state.

Different enzymes are evidently responsible for the two classes of breakdown, for specific inhibitors of "serine proteases" (e.g., *p*-toluenesulfonylfluoride), at concentrations that do not interfere with growth, block the starvation-induced breakdown but not the background turnover.

NUCLEIC ACID BREAKDOWN

RNases. Starvation for Mg++ or phosphate activates an adaptive mechanism that causes loss of **ribosomes,** without loss of viability. The enzymatic basis for this net RNA breakdown, like that for mRNA turnover, is not known. *E. coli* has several **ribonucleases,** as well as polynucleotide phosphorylase. RNase I, an endonuclease, is probably used to hydrolyze extracellular RNA: it is found in the periplasmic space, and it is inhibited by Mg++ at the levels present in the cell. RNase II, a K+-requiring 3′-exonuclease, probably functions in some aspects of turnover. Other RNases appear to perform specific cleavages in connection with RNA maturation and phage infection.

DNA does not break down in living cells, except for the limited excision in repair and in recombination (Ch. 11). This stability is hardly surprising, since conservation and replication of the genetic material may be considered the ultimate objective of all metabolic activity.

FUNCTIONS OF VARIOUS REGULATORY MECHANISMS

REPRESSION VS. INHIBITION

Repression has often been viewed as a "course" and sluggish control, and inhibition as a "fine" and immediate control, over the synthesis of metabolic building blocks. However, it seems clear that repression is concerned primarily with economy in macromolecule synthesis, and endproduct inhibition with economy in metabolite synthesis. Thus mutants that have lost the feedback inhibition but not the repression of a pathway excrete its endproduct, while loss of repression does not cause excretion as long as normal inhibition is retained.

In the several mechanisms that regulate gene expression in bacteria (repression, activation by cAMP, promoter variation, RNA polymerase modification) *transcription is central*. By blocking both mRNA and protein synthesis, rather than the latter alone, these mechanisms maximize economy in macromolecule synthesis. Regulation at the level of translation has been observed only in the wholesale shift of macromolecule synthesis that follows phage infection. The flexibility of regulation is increased by having mechanisms available that can regulate the same genes both in small, linked groups and in larger, scattered sets.

RESPONSES TO ENDOGENOUS STIMULI

Feedback inhibition and repression were both discovered as mechanisms that spare endogenous synthesis in response to **exogenous** supplies. However, their effects on the **domestic** economy of the cell may be even more important.

Thus **feedback regulation of metabolite synthesis** explains the ability of bacteria to form their building blocks in the correct proportions (i.e., without excretion) under varying conditions of pH and temperature, which affect different enzymes quite differently. Similarly, an auxotroph deprived of the required amino acid ceases to make the other 19 amino acids, though it has the necessary enzymes and fuel. **Mutants resistant to feedback** by an endproduct most sharply demonstrate the significance of the normal mechanism: they develop endogenously an elevated level of the compound and therefore excrete it, as noted above. These mutants are readily recognized as **feeder colonies** when plated on media heavily seeded with an auxotroph requiring the excreted product. (Indeed, such loss of feedback control underlies the empirical selection of mutants for the commercial production of various microbial products). Since elimination of repressions by mutations in O or R loci, marked by elevating the level of the corresponding enzymes (derepression).

Endogenous regulation by endproduct repression is equally impressive. In steady-state growth most operons are obviously active at only a fraction of their capacity, this reserve capacity for enzyme formation is put to good use in certain adaptations. Thus in wild-type *E. coli* the enzymes of arginine biosynthesis are a constant fraction of the protein synthesized in minimal medium, but in the presence of arginine they are no longer formed and after several generations the levels become negligible. Hence when the exogenous arginine is removed the cells cannot form the compound and growth temporarily ceases. Within 10 minutes, however, growth becomes normal again. During that interval **total** protein synthesis, at the expense of protein turnover, is slow; but the enzymes of the now derepressed arginine pathway exhibit striking **preferential synthesis** (Fig. 13-23). Their rate of synthesis returns to normal when they reach normal levels (and can thus maintain a normal intracellular arginine level in the growing cells). However, an elevated (derepressed) level of biosynthetic enzymes is attained if the intracellular level of the corresponding corepressor

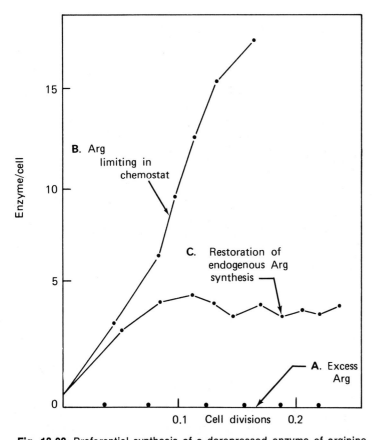

Fig. 13-23. Preferential synthesis of a derepressed enzyme of arginine biosynthesis, ornithine transcarbamylase. **A.** Arg⁻ His⁻ mutant, grown with **excess Arg** and His and transferred to a chemostat with limiting His and excess Arg. **Complete repression** of the enzyme continues. **B.** The same cells, in a chemostat with **limiting Arg** and excess His, show **completely derepressed** synthesis of the Arg enzyme: i.e., because of continued Arg limitation this enzyme continues indefinitely to be synthesized up to the capacity of the gene. Only the early part of the curve of exponential increase, which finally levels off at 45 arbitrary units per cell, is shown. Thus, 50% of the final level is reached after one generation, 75% after two, etc., just as with a fully induced enzyme. **C.** Arg+ His⁻ mutant, **grown with excess Arg** and His and transferred to chemostat with limiting His and **no Arg.** The enzyme is initially synthesized at the fully derepressed rate, relative to total protein synthesis, as in **B.** However, within ca. 0.05 generations the restoration of Arg synthesis in this Arg+ strain causes formation of the enzyme to level off at the steady-state, **partly repressed** rate (maintaining 2.5 units per cell) characteristic of the wild type in minimal medium.

We thus see that normal synthesis uses only about 1/20 of the capacity of the gene, while the excess capacity is used for rapid adjustment to deprival of exogenous arginine. [From Gorini, L., and Maas, W. K. *Biochim Biophys Acta* 25:208 (1957).]

is kept low, for example in an incompletely blocked mutant ("leaky auxotroph") growing in minimal medium, or in an auxotroph provided with a slowly utilizable source of the required endproduct or growing in the chemostat with the endproduct limiting.

Those gene products that are required only in small amounts are often regulated, inflexibly, by their low-activity promoter. There is then no reserve capacity to derepress. Thus when a *lacR+* gene is introduced into an *R−* cell its product (the *lac* repressor) accumulates slowly, and so the cell is not converted from constitutive to inducible until after 60 to 90 minutes.

Gene Dosage Effects. Snythesis of a biosynthetic enzyme in bacteria does not normally vary with the number of copies of the corresponding gene per cell (e.g., in merodiploids, or in the abrupt doubling in synchronized DNA replication), since the rate is regulated by endogenous feedback. A fully induced or derepressed gene, however, is expressed to capacity, and the rate of enzyme synthesis per cell is then proportional to the number of copies present (Table 13-3). In man, curiously, biosynthetic enzymes do not seem to be regulated by such endogenous feedback, for "carriers" heterozygous for an enzymatic defect (e.g., galactosemia) produce only half as much of the enzyme as the normal homozygotes.

PSEUDOFEEDBACK INHIBITION BY METABOLITE ANALOGS

Regulatory mechanisms have obviously had to be selected in evolution for great accuracy in discriminating between a particular normal metabolite and others with a similar structure (e.g., leucine vs. isoleucine). However, **synthetic analogs, which have not been exerting evolutionary selective pressure** for specificity, can often **deceive regulatory mechanisms,** just as they can deceive enzymes. For example, 5-methyltryptophan inhibits growth by mimicking the feedback effect of tryptophan on the first enzyme of its pathway, thus preventing further synthesis of tryptophan. Added tryptophan restores growth noncompetitively.

Mutants resistant to such pseudofeedback inhibition are readily isolated, and their altered enzyme is often resistant to the normal stimulus as well. Hence analogs provide a means for selecting **feedback-resistant mutants,** whose excretion of the endproduct has been noted above.

TELEONOMIC VALUE OF VARIOUS REGULATORY MECHANISMS

The developments described in this chapter show that even the primitive bacterial cell has evolved an intricate set of feedback circuits in its metabolism. These mechanisms contribute to faster growth and faster adaptation in several ways. They promote **efficient conversion of foodstuffs:** steady-state growth would be slower if metabolic flow were wasted on synthesis of unused products. Moreover, in **adaptation to enrichment** many feedback responses spare the machinery, material, and energy for making components that are available at less expense from the medium. (For example, *E. coli* grows three times faster in broth than on glucose, not because its manufacture of amino acids is rate-limiting, or because broth provides multiple energy sources, but primarily because the sparing of small molecule synthesis permits more biosynthetic effort to be channeled into polymerization.) Finally, in **adaptation to exhaustion of a nutrient** several features of metabolic regulation promote rapid restoration of the missing, repressed enzymes. Thus normal mRNA turnover permits new mRNA to be formed; protein turnover primes the pump (i.e., supplies small amounts of the endproduct required for initiating synthesis of the missing enzymes); and as we have seen above, the reserve capacity of the operon permits preferential formation of the mRNA for the missing enzymes. Without these mechanisms the cell might never recover in such a transition, despite its genetic capacity to grow on minimal medium.

A close link between RNA and protein synthesis also has evident value in preventing wasteful accumulation of either class without the other. A block in RNA synthesis stops protein synthesis directly, for lack of mRNA, but protein synthesis influences RNA synthesis by more elaborate mechanisms. The synthesis of ribosomes in the cell is thereby adjusted to yield precisely the level required for a given rate of steady-state growth. Inhibition of protein synthesis also influences, though it does not completely block, mRNA synthesis: the large decrease in energy expenditure leads to catabolite repression, and the accumulating metabolites repress transcription of their operons.

In contrast to RNA and protein synthesis, chromosome replication proceeds to completion even under conditions of extreme starvation. This continued synthesis, at the expense of turnover of other constituents, has evident teleonomic value. Thus since a growing point in DNA, with single-stranded segments, would be less stable than the one-dimensional crystal of the double helix, the completion of replication of the chromosome in nongrowing cells evidently promotes its subsequent prolonged survival. Completion of a chromosome pair also triggers cell division, which may further promote survival since the zone of septum formation appears to be a region of weakness in the cell wall.

SIGNIFICANCE FOR HIGHER ORGANISMS

DIFFERENTIATION

The systems of gene regulation revealed in bacteria have provided a model for the even more complex regulatory processes of differentiation in animals to which the technics of molecular genetics are being rapidly extended. We now know that the various differentiated cells each contain the same complete set of genes, of which only a small fraction are active in any one type of cell. The mechanisms responsible for triggering and maintaining this selective gene activation constitute an area called **epigenetics.** The remainder of developmental biology may reasonably be grouped as **morphogenesis,** since at all levels of organization, from the polypeptide leaving the ribosome to the aggregation of cells into an organ, a single kind of process—the mutual attraction of complementary molecular surfaces through weak chemical forces—shapes the proteins and the products of their catalytic activity into an organism.

Gene regulation and morphogenesis can both be studied not only in bacteria but even in the simple model of viral multiplication: different groups of genes are activated at different times, and the resulting products spontaneously aggregate to form a mature virus (Ch. 45). These systems, however, do not provide an adequate model for one striking characteristic of epigenetic changes: their self-perpetuating nature, inherited from one generation of cells to the next within the differentiated tissue. Thus the regulatory mechanisms seen in bacteria are homeostatic: their **negative-feedback loops** promote **constancy** of the cell in different environments. In contrast, differentiated cells, by definition, **maintain their differences in an identical environment.** This process suggests the presence of **positive-feedback loops,** in which activation of a group of genes results in formation of a product that ensures perpetuation of its activity.

CYBERNETICS AND EVOLUTION

The regulatory devices discovered in bacteria have extended the concept of **molecular information transfer** beyond the sequence of a nucleic acid template: the level of a regulatory metabolite also provides information to the proper receptor. Moreover, the concept of **cybernetics** (self-regulating feedback loops), developed for mechanical and electronic devices, has turned out to be equally applicable to living organisms.

This extension may have important implications for our intellectual climate. The theory of natural selection had a forceful impact on the public in the nineteenth century, and superficial analogy with Darwin's principle of "survival of the fittest" was often used to justify unrestrained competition in human social organization. Today the growing recognition of the importance of self-regulatory mechanisms in biology provides an alternative analogy, which emphasizes the contribution of cooperation, at a variety of levels, to evolutionary survival.

SELECTED REFERENCES

Books and Review Articles

ATKINSON, D. E. Regulation of enzyme function. *Annu Rev Microbiol* 23:47 (1969).

BAUTZ, E. K. F. Regulation of RNA synthesis. *Progr Nucleic Acid Res Mol Biol* 12:129 (1972).

Cold Spring Harbor Symposia: Cellular Regulatory Mechanisms, vol. 26 (1961); Synthesis and Structure of Macromolecules, vol. 28 (1963); Replication of DNA in Microorganisms, vol. 33 (1968); Transcription of Genetic Material, vol. 35 (1970).

GEIDUSCHEK, E. P., and HASELKORN, R. Messenger RNA. *Annu Rev Biochem 38*:647 (1969).

GERHART, J. C. A discussion of the regulatory properties of aspartate transcarbamylase from *Escherichia coli. Curr Top Cell Regulation 2*:275 (1970).

GILBERT, W., and MULLER-HILL, B. The Lactose Repressor. In *The Lactose Operon*, p. 93. (J. R. Beckwith and D. Zipser, eds.) Cold Spring Harbor Laboratory, Cold Spring Harbor, N.Y., 1970. This volume contains many additional relevant articles.

HELMSTETTER, C. E. Sequence of bacterial reproduction. *Annu Rev Microbiol 23*:223 (1969).

KOSHLAND, E. E., JR. Conformational aspects of enzyme regulation. *Curr Top Cell Regulation 1*:1 (1969).

LARK, K. G. Initiation and control of DNA synthesis. *Annu Rev Biochem 38*:569 (1969).

LEWIN, B. M. *The Molecular Basis of Gene Expression*. Wiley, New York, 1970.

LEWIS, J. A., and AMES, B. N. The percentage of tRNA^{His} charged in vivo and its relation to the repression of the histidine operon. *J Mol Biol 66*: 131 (1972).

LOSICK, R. The question of gene regulation in sporulating bacteria. *Symp Soc Develop Biol.* In press.

NEIDHARDT, F. C. Effects of environment on the composition of bacterial cells. *Annu Rev Microbiol 17*:61 (1963).

PAIGEN, K., and WILLIAMS, B. Catabolite repression and other control mechanisms in carbohydrate utilization. *Adv Microbial Physiol 4*:252 (1970).

PASTAN, I., and PERLMAN, R. Cyclic adenosine monophosphate in bacteria. *Science 169*:339 (1970).

PITTARD, J., and GIBSON, F. The regulation of biosynthesis of aromatic amino acids and vitamins. *Curr Top Cell Regulation 2*:29 (1970).

SANWAL, B. D. Allosteric controls of amphibolic pathways in bacteria. *Bacteriol Rev 34*:20 (1970).

STADTMAN, E. R. The role of multiple molecular forms of glutamine synthesis in the regulation of glutamine metabolism in *Escherichia coli. Harvey Lect 65*:97 (1969–1970).

UMBARGER, H. E. Regulation of amino acid metabolism. *Annu Rev Biochem 38*:323 (1969).

Specific Articles

DE CROMBRUGGHE, B., CHEN, B., GOTTESMAN, M., PASTAN, I., VARMUS, H. E., EMMER, O., and PERLMAN, R. L. Regulation of *lac* mRNA synthesis in a soluble cell-free system. *Nature New Biol 230*: 37 (1971).

GERHART, J. C., and SCHACHMAN, H. K. Distinct subunits for the regulation and catalytic activity of aspartate transcarbamylase. *Biochemistry 4*:1054 (1965).

GOLDBERG, A. L. Effects of protease inhibitors on protein breakdown and enzyme induction in starving *E. coli. Nature New Biol 234*:51 (1971).

HASELTINE, W. A., BLOCK, R., GILBERT, W., and WEBER, K. MSI and MSII made on ribosome in idling step of protein synthesis. *Nature 238*:381 (1972).

JACOB, F., and MONOD, J. Genetic regulatory mechanisms in the synthesis of protein. *J Mol Biol 3*: 318 (1961).

LAZZARINI, R. A., CASHEL, M., and GALLANT, J. On the regulation of guanosine tetraphosphate levels in stringent and relaxed strains of *Escherichia coli. J Biol Chem 246*:4381 (1971).

LOSICK, R., SHORENSTEIN, R. G., and SONENSHEIN, A. L. Structural alteration of RNA polymerase during sporulation. *Nature 227*:910 (1970).

MONOD, J., CHANGEUX, J.-P., and JACOB, F. Allosteric proteins and cellular control systems. *J Mol Biol 6*:306 (1963).

MORSE, D. E., and PRIMAKOFF, P. Relief of polarity in *E. coli* by "suA." *Nature 226*:28 (1970).

MOYED, H. S. False feedback inhibition: Inhibition of tryptophan biosynthesis by 5-methyltryptophan. *J Biol Chem 235*:1098 (1960).

NOMURA, M., and ENGBACK, F. Expression of ribosomal protein genes as analyzed by bacteriophage Mu-induced mutations. *Proc Natl Acad Sci USA 69*:1526 (1972).

PARDEE, A. B., JACOB, F., and MONOD, J. The genetic control and cytoplasmic expression of "inducibility" in the synthesis of β-galactosidase by *E. coli. J Mol Biol 1*:165 (1959).

ROSENBUSCH, J. P., and WEBER, K. Subunit structure of aspartate transcarbamylase from *Escherichia coli. J Biol Chem 246*:1644 (1971).

TRAVERS, A. Control of transcription in bacteria. *Nature New Biol 229*:69 (1971).

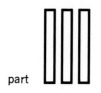

part III

IMMUNOLOGY

HERMAN N. EISEN

chapter

INTRODUCTION TO IMMUNE RESPONSES

THE ORIGINS OF IMMUNOLOGY   352
DEFINITIONS   353
ANTIGENIC DETERMINANTS   354
CELLULAR IMMUNITY   355
IMMUNOLOGY AND IMMUNITY   356

Immune responses are remarkably versatile adaptive processes in which animals form specifically reactive proteins and cells in response to an immense variety of organic molecules and macromolecules. This ability seems to have been acquired relatively late in evolution since these responses are encountered only in vertebrates, for whose survival they are of vast importance: they constitute the principal means of defense not only against infection by pathogenic microorganisms and viruses but probably also against host cells that undergo transformation into cancer cells.

THE ORIGINS OF IMMUNOLOGY

From almost the earliest written observations of mankind it seems to have been suspected that persons who recover from certain diseases become incapable of suffering the same disease again: they become immune. Thucydides, for example, pointed out 2500 years ago, in a remarkable description of an epidemic in Athens (of what might have been typhus fever or plague), that fear of contagion led to neglect of the sick and dying, but whatever attention they received was "tended by the pitying care of those who had recovered, because they . . . were themselves free of apprehension. For no one was ever attacked a second time, or with a fatal result." This awareness led to deliberate attempts, beginning in the Middle Ages, to induce immunity by inoculating well persons with material scraped from skin lesions of persons suffering from smallpox. The procedure was hazardous, but in the late eighteenth century a safe, related procedure was established by the English physician Jenner.

Jenner shared the widely held impression that those individuals who had had cowpox (a benign disease acquired from cows infected with what seemed to be a mild form of smallpox) were spared in subsequent smallpox epidemics. To test this belief he inoculated a boy with pus from a lesion of a dairymaid who had cowpox; some weeks later reinoculation with infectious pus, from a patient in the active stage of smallpox, failed to cause illness. Repetition of the experiment many times led to Jenner's classic report, followed by widespread adoption of this procedure and confirmation of his conclusion that **vaccination** (L. *vacca* = cow) leads to immunity against smallpox.

It is remarkable that Jenner's work was not extended until nearly 100 years later, when a fortuitous observation led Pasteur to recognize and exploit the general principles underlying vaccination. While studying chicken cholera Pasteur happened to use an old culture of the causative agent (*Pasteurella aviseptica*) to inoculate some chickens, who failed to become ill. When the same animals were reinoculated with a fresh culture, which was known to be virulent, they again failed to become ill. The finding that aged cultures lose virulence but retain the capacity to induce immunity against cholera may have prompted Pasteur's epigram that "chance favors only the prepared mind." His observations were soon applied to many other infectious diseases.

A variety of procedures have been used to destroy the viability or attenuate the virulence of pathogenic organisms for purposes of vaccination; examples are aging of cultures or passing the microorganisms through "unnatural hosts" (e.g., the agent of rabies through the rabbit). The latter effect provides a plausible explanation for Jenner's earlier success: passage of smallpox virus through cows had probably selected for a variant virus that multiplied unusually well in cows but poorly in humans, while retaining ability to induce immunity against the virus that is virulent in man.

Induction of immunity to infectious diseases does not always require inoculation of the causative microorganism. Following the demonstration of a powerful toxin in culture filtrates of diphtheria bacilli in 1888, von Behring showed that nonlethal doses of the bacteria-free filtrates could induce immunity to diphtheria. These results were then generalized when Ehrlich and Calmette similarly established immunity to toxins of nonmicrobial origin, e.g., snake venoms and ricin from castor beans.

The basis for these immune responses was revealed in 1890, when von Behring and Kitasato demonstrated that induced immunity to tetanus was due to the appearance in the serum of a capacity to neutralize the toxin; this activity was so stable it could be transferred to normal animals by infusions of blood or serum. Moreover, Ehrlich showed that protection against the toxic effects of ricin on red cells in vitro involved the combination of specifically reactive components of the sera with the toxin; a similar combination presumably accounted for the effects of immune sera on infectious agents. These observations opened the way to analyses of substances responsible for immunity and to the practical treatment of many infectious diseases by infusions of serum from immune animals.

Within the ensuing 10 years most of the now known serological* reactions were discovered. For example:

1) **Bacteriolysis:** Cholera vibrio disintegrated when incubated with serum from animals that had previously been inoculated with this organism (Pfeiffer and Issaeff, 1894).

2) **Precipitation:** Cell-free culture filtrates of plague bacilli formed a precipitate on being added to serum from animals previously inoculated with cultures of plague bacilli (Kraus, 1897).

3) **Agglutination:** Bacterial cells suspended in serum from an animal previously injected with these bacteria underwent clumping (Gruber and Durham, 1898).

These reactions were all specific: the immune serum reacted only with the substance inoculated for induction of the immune response, or, as we shall see later, with substances of similar chemical structure. By about 1900 it was realized that immune responses can also be elicited by nontoxic agents, such as proteins of milk or egg white injected into rabbits.

DEFINITIONS

The inoculated agents, and the substances whose appearance in serum they evoked, were called antigens (Ags) and antibodies (Abs), respectively.

An **antigen** has two properties: **immunogenicity,**† i.e., the capacity to stimulate the formation of the corresponding Abs; and the **ability to react** specifically with these Abs. The distinction is important, because substances known as **haptens** (described more fully below) are not immunogenic but react specifically with the appropriate Abs. "Specific" means here that the Ag will react in a highly selective fashion with the corresponding Ab and not with a multitude of other Abs evoked by other Ags.

The term **antibody** refers to the proteins that are formed in response to an Ag and react specifically with that Ag. All Abs belong to a special group of serum proteins, the **immunoglobulins,** whose properties will be considered in Chapter 16.

Though the definition states that Abs are formed in response to Ag, sera may contain immunoglobulins that react specifically with certain Ags to which the individuals concerned have had no **known** exposure. These immunoglobulins are called **natural antibodies.** When present in serum they are usually in low titer, but they may exercise a significant role in conferring resistance to certain infections. It is not clear whether natural Abs are formed without an immunogenic stimulus or in response to unknown exposure to naturally occurring Ags, e.g., in inapparent infections or in food.

In contrast to the restricted group of proteins that possess Ab activity, an enormous variety of macromolecules can behave as Ags —virtually all proteins, many polysaccharides, nucleoproteins, lipoproteins, numerous synthetic polypeptides, and many small mole-

---

* The specific reactions of immune sera are generally referred to as serological reactions.

† The term immunogen is often used for the substance that stimulates the formation of the corresponding Abs.

cules if they are suitably linked to proteins or to synthetic polypeptides.

It is important to recognize the operational nature of the definition of Ag. For example, when rabbit serum albumin (RSA) is isolated from a rabbit and then injected back into the same animal, Abs specific for RSA are **not** formed. Yet the same preparation of RSA injected into virtually any other species of vertebrate may evoke copious amounts of anti-RSA Abs. Moreover, the formation of these Abs depends not only on the injection of RSA into an appropriate species, but also on the other conditions employed: viz., the quantity of RSA injected and the route and frequency of injection. It is clear, therefore, that **immunogenicity is not an inherent property of a macromolecule,** as is, for example, its molecular weight or absorption spectrum; immunogenicity is operationally dependent on the biological system and conditions employed. One cardinal condition is that the putative immunogen be somehow recognized as alien (i.e., not self) by the responding organism.

ANTIGENIC DETERMINANTS

The reaction between an Ag and the corresponding Ab in an immune serum involves an actual combination of the two. We shall consider in Chapter 15 the nature of this combination, which is the fundamental reaction common to most immunological phenomena. For the present, however, it is useful to distinguish between the Ag molecule as a whole and its **antigenic determinants,** i.e., those restricted portions of the Ag molecule that determine the specificity of Ab-Ag reactions.

Attempts to evaluate the size and conformation of an antigenic determinant involve indirect approximations (Ch. 16, Size of active sites of antibodies) which indicate that these determinants are much smaller than a typical macromolecule, being equivalent in volume to, perhaps, five or six amino acid residues.

The great diversity of antigenic substances was first emphasized by Obermayer and Pick (1903), who attached NO_2 groups to rabbit serum proteins, injected these proteins into rabbits, and found that the serum obtained reacted with the nitrated proteins of rabbit, horse, or chicken sera, but not with the corresponding unmodified proteins. They inferred, therefore, that the Abs formed were capable of specifically recognizing the nitro groups or other uniquely altered structures in the nitrated proteins.

Such responses to chemically defined substituents of modified proteins were vigorously developed by Landsteiner to explore the chemical basis of antigenic specificity. Diazonium salts were used to couple a wide variety of aromatic amines to proteins, as indicated by the representative reactions in Figure 14-1.

Rabbits injected with p-azobenzenearsonate globulin form Abs that react with this protein and with other proteins containing p-azobenzenearsonate substitutuents, but not with the latter proteins in unsubstituted form. Landsteiner and Lampl showed in 1920 that p-aminobenzenearsonate alone can competitively inhibit the reaction with p-azobenzenearsonate proteins, whereas other aromatic amines (with a few important exceptions) do not.

Though benzenearsonate combines in a highly selective manner with the Abs formed

Fig. 14-1. Attachment of an aromatic amine to a protein via azo linkage (-N=N-) formed by reaction with diazonium ions.

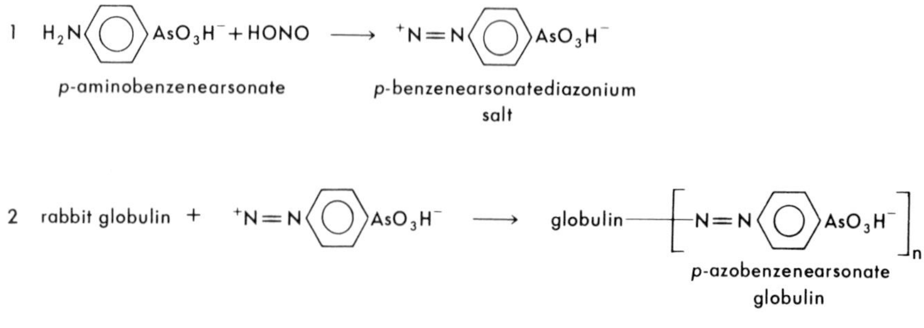

in response to injection of the corresponding azoprotein (as further described in Ch. 15), injections of benzenearsonate itself do not evoke the formation of Abs. Substances of this type are defined as **haptens: they are not immunogenic but they react selectively with Abs of the appropriate specificity.** In the example cited the hapten is a small molecule; it should be noted that the definition does not say anything about molecular weight; even some macromolecules can function as haptens.

While the formal difference between Ag and hapten is clear, in practice it may be difficult to decide whether a substance is weakly immunogenic or completely non-immunogenic. Generally, however, small molecules (MW <1000) are not immunogenic, unless covalently linked to proteins in vitro or in vivo (Fig. 14-2; for other examples see Ch. 17, Contact skin sensitivity). Certain macromolecules also become immunogenic only when bound by proteins: however, covalent bonds are unnecessary here, probably because the cumulative effect of many weak noncovalent ones between large interacting molecules (e.g., between DNA and protein) can establish stable (nondissociating) complexes.

The diazo reaction introduces azo groups as substituents in tyrosine, tryptophan, histidine, and lysine residues (Fig. 14-2). Other methods for coupling haptens to proteins will be referred to in subsequent chapters; some are even more useful because they provide a more stable linkage and substitute fewer kinds of amino acid residues.

Proteins with substituents covalently linked to their side chains are referred to as **conjugated proteins;** and the substituents, sometimes including the amino acid residues to which they are linked (when these are known), are called **haptenic groups.** These distinctions are shown in Figure 14-3. While a haptenic group is thus part of an antigenic determinant, it is not yet clear just how much of the complete antigenic determinant it represents.

CELLULAR IMMUNITY

Coincident with early studies on the role of serum Abs, Metchnikoff's observation of phagocytosis suggested that ingestion and destruction of microbes by polymorphonuclear leukocytes and macrophages are responsible for resistance to most infectious agents. The resulting controversy between advocates of Abs and of cells was reconciled by the finding, in 1903, that coating of particles with Abs (opsonins; Gr., "to prepare food") increased their susceptibility to phagocytosis.

Fig. 14-2. Some azo substituents on tyrosine, lysine, and histidine residues of a representative azoprotein.

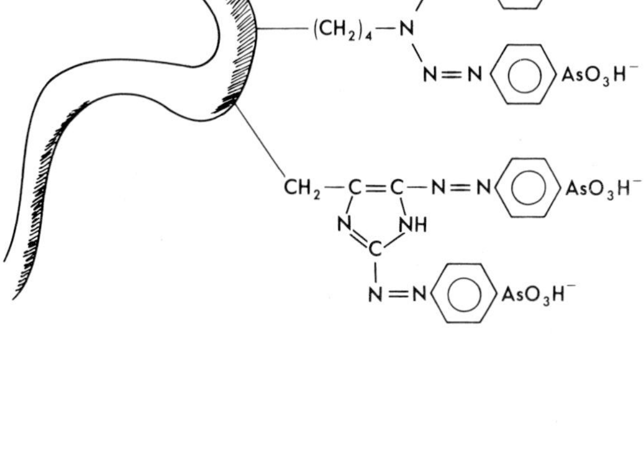

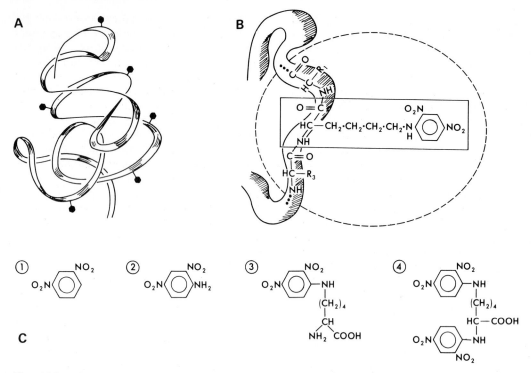

Fig. 14-3. Distinctions between conjugated proteins, antigenic determinants, haptenic groups, and haptens.

   A. Conjugated protein with substituents represented as solid hexagons.

   B. A representative **haptenic group:** a 2,4-dinitrophenyl (Dnp) group substituted in the ε-NH₂ group of a lysine residue. The haptenic group is outlined by solid line, whereas the **antigenic determinant** might be visualized as the area outlined by the broken line. Amino acid residues contributing to the antigenic determinant need not be the nearest covalently linked neighbors of the ε-Dnp-lysine residue, as shown; they could be parts of distant segments of the polypeptide chain looped back to come into contiguity with the Dnp-lysyl residue.

   C. Some **haptens** that correspond to the haptenic group in **B.** 1) m-dinitrobenzene; 2) 2,4-dinitroaniline; 3) ε-Dnp-lysine; 4) α,ε-bis-Dnp-lysine. With respect to antibodies specific for the Dnp group, haptens 1, 2, and 3 are univalent haptens (one combining group per molecule), and hapten 4 is bivalent.

However, the Ab-cell controversy reappeared in different form in the 1920s as a result of studies of allergy (hypersensitivity)—a state induced by an Ag, in which a subsequent response to the Ag causes inflammation or even acute shock and death. Though many allergic states can be transferred by the Abs in serum, Landsteiner and Chase showed (1942) that transfer in others can be accomplished only with living leukocytes (later shown to be lymphocytes).

Various other immune responses were subsequently also found to be mediated by living lymphocytes rather than by Ab molecules, e.g., immunity to tubercle bacilli and many other infectious agents, the capacity for accelerated rejection of allografts (i.e., grafted cells from genetically different individuals of the same species), and resistance to many experimental cancers. In all these **cell-mediated immune reactions** the Ag apparently combines specifically with particular lymphocytes. Thus mediators of immune responses can be either freely diffusible Ab molecules **(humoral immunity)** or specifically reactive lymphocytes **(cell-mediated immunity):** whether the Ag-binding receptors on these cells are surface immunoglobulins (cell-bound Abs) or a different type of molecule is still uncertain.

IMMUNOLOGY AND IMMUNITY

Though immune responses were originally of interest because of their role in **immunity** to infection, it has long been evident that these responses can be induced against an almost limitless variety of substances, of which microbial Ags are a small minority: hence immunity represents only a limited facet of a more general adaptive response. Coincident with this expanded view of immunology, diverse pathological effects have been traced to immune responses to noninfectious and nontoxic Ags (e.g., allografts, red blood cells, pollens, drugs), and even to an individual's own constitutents (autoimmunity); hence immunological considerations have now come to pervade almost all segments of modern medicine. In addition, interest has grown in 1) the Ab-Ag reaction as a model for specific noncovalent interactions in general (e.g., those between viruses and their host cell membranes, or those causing cell aggregation in such phenomena as fertilization and morpho-genesis); 2) Ab formation as a model for cellular differentiation; and 3) the use of Ab molecules as highly specific and sensitive analytical reagents for exploring the structure of complex macromolecules (e.g., enzymes, blood group glycopeptides, cell surface macromolecules) and for measuring trace amounts of many physiologically important substances (e.g., hormones, vasoactive peptides, cyclic AMP, prostaglandins).

While there is thus much more to immunology than immunity in the literal sense, the historical role of infectious diseases has left an indelible imprint on nomenclature, which is not always appropriate: for example, induction of Ab formation and of specific cellular responses is referred to as immunization even when infectious agents are not involved. **Vaccination** is the term reserved for immunization in which a suspension of infec-

TABLE 14-1. Comparison of Enzymes and Antibodies

Property	Enzymes	Antibodies
Phylogenetic distribution	Ubiquitous; made by all cells	A late evolutionary acquisition; made only in vertebrates (and in certain cells of the lymphatic system)
Structure	Proteins with variable chemical and physical properties; an enzyme of a given specificity and from any particular organism is homogeneous; many have been crystallized	A group of closely related proteins having a common multichain structure with the chains held together by —SS— bonds. Molecules of a given specificity are heterogeneous in structure and function
Constitutive	Yes	"Natural" antibodies?
Inducible	Often	Yes
Function	Specific reversible binding of ligands* with breaking and forming covalent bonds	Specific reversible binding of ligands* without breaking or forming covalent bonds
Reaction with ligands*	Wide range of affinities; populations of enzyme molecules of a given specificity are uniform in affinity for their ligand	Wide range of affinities; but populations of antibody molecules of the same specificity are usually heterogeneous in affinity for their ligand
Affinity	Usually measured kinetically	Usually measured with reactants at equilibrium (because the reactions are so fast)
Number of specific ligand-binding sites per molecule	Different in different enzymes, depending on number of polypeptide chains per molecule; usually one site per chain	2 per molecule of the most prevalent type (MW ~ 150,000); each site is formed by a pair of chains (a light plus a heavy chain)
Inducers	Primarily small molecules	Usually macromolecules, especially proteins and conjugated proteins

* Ligand = substrate or coenzyme in case of enzymes, and antigen or hapten in case of antibodies.

tious agents (or some part of them), called a **vaccine,** is given to establish resistance to an infectious disease.

In the chapters that follow we shall consider the molecular before the cellular aspects of immune responses: not only are structure and functions of Ab molecules better understood than those of the relevant cells, but the properties of Abs and their interactions with Ags provide the conceptual framework in which immune cellular reactions are viewed.

It would be logical to consider first the structure of Abs and then, perhaps, how these molecules are made and function. However, the present view of Ab structure is extensively based on the use of Ab molecules themselves as analytical reagents. We shall therefore begin with a consideration of specific Ab-Ag reactions. Some properties that will be emphasized are summarized in Table 14-1, where Abs are compared with enzymes.

SELECTED REFERENCES

Books and Review Articles

AMOS, B. (ed.). *Progress in Immunology.* Proceedings of the First International Congress of Immunology. Academic Press, New York, 1971.

ARRHENIUS, S. *Immunochemistry.* Macmillan, New York, 1907.

BURNET, F. M. *Cellular Immunology.* Cambridge Univ. Press, London, 1970.

EHRLICH, P. *Studies in Immunity.* Wiley, New York, 1910.

HEIDELBERGER, M. *Lectures in Immunochemistry.* Academic Press, New York, 1956.

LANDSTEINER, K. *The Specificity of Serological Reactions,* rev. ed. Harvard Univ. Press, Cambridge, Mass., 1945; reprinted by Dover Publications, New York, 1962 (paperback).

TOPLEY, W. W. C., and WILSON, G. S. *The Principles of Bacteriology and Immunity,* 2nd ed. Williams & Wilkins, Baltimore, 1936.

ZINSSER, H., ENDERS, J. F., and FOTHERGILL, L. D. *Immunity: Principles and Applications in Medicine and Public Health,* 5th ed. Macmillan, New York, 1939.

chapter 15

ANTIBODY-ANTIGEN REACTIONS

REACTIONS WITH SIMPLE HAPTENS 360
 VALENCE AND AFFINITY OF ANTIBODIES 360
   Equilibrium Dialysis 361
   Effects of Temperature, pH, Ionic Strength; Rates of Reaction 364
 SPECIFICITY 366
   Specificity and Affinity 370
   Carrier Specificity 370

REACTIONS WITH SOLUBLE MACROMOLECULES 370
 PRECIPITIN REACTION IN LIQUID MEDIA 370
   The Quantitative Precipitin Reaction 371
   Lattice Theory 373
   Valence and Complexity of Protein Antigens 375
   Nonspecific Factors in the Precipitin Reaction 377
   Nonprecipitating Antibodies 377
   Reversibility of Precipitation 379
   Applications of the Precipitin Reaction 379
   Hapten Inhibition of Precipitation 380
   Cross-Reactions 381
   Avidity 384
   The Flocculation Reaction 385
 PRECIPITIN REACTION IN GELS 386
 IMMUNOELECTROPHORESIS 389

REACTIONS WITH PARTICULATE ANTIGENS 391
   The Agglutination Reaction 391
   Blocking or Incomplete Antibodies 392
   Passive Agglutination 393
   Differences in Sensitivity of Precipitation,
     Agglutination, and Other Reactions 393

RADIOIMMUNOASSAYS 395

FLUORESCENCE AND IMMUNE REACTIONS 397
   Immunofluorescence 397
   Fluorescence Quenching 398
   Fluorescence Polarization 399
   Fluorescence Enhancement 399

DIAGNOSTIC APPLICATIONS OF SEROLOGICAL TESTS 399

THE UNITARIAN HYPOTHESIS 400

APPENDIX 401
   Purification of Antibodies 401
   Ferritin-Labeled Antibodies 401
   Intrinsic and Actual Affinities 402

The combination of antibody (Ab) with antigen (Ag) underlies most immunological phenomena. However, most Ags in wide use are macromolecules, especially proteins, and even when their covalent structure is completely established we hardly ever know the identity and conformation of their functional groups or even the number and variety per molecule. We shall therefore first consider specific Ab reactions with simple haptens, on which most of our understanding of the Ab-Ag reaction is based. Subsequently the more complicated reactions with macromolecular Ags will be taken up. Since the distinction between haptens and Ags, based on immunogenicity, is largely irrelevant for the present discussion (Ch. 14, Definitions), we shall frequently use the generic term **ligand** to include both.

REACTIONS WITH SIMPLE HAPTENS

Abs specific for simple haptens are usually obtained with immunogens prepared by attaching the hapten covalently to a protein. In addition, Abs formed in response to high molecular weight polysaccharides may react with small oligosaccharides corresponding to short sequences in the immunogen, e.g., the cellobiuronic acid of type 3 pneumococcus (below, Fig. 15-18). Since the Abs that react with haptens are easily isolated (Appendix, this chapter), the formation of specific Ab-ligand complexes can be examined in detail with relatively simple systems.

VALENCE AND AFFINITY OF ANTIBODIES

In the simplest reaction a small **univalent ligand,** with one combining group per molecule, binds reversibly to a specific site on the Ab. At low concentrations of unbound ligand only a small proportion of the Ab's combining sites are occupied by ligand molecules; and as the ligand concentration increases the number of occupied Ab sites rises until all are filled. At saturation the number of univalent ligand molecules bound per Ab molecule is the **antibody valence** (or n, see below).

If we assume that the binding sites on Ab molecules are equivalent and act independently of each other the **intrinsic binding reaction** can be represented as

$$S + L \underset{k'}{\overset{k}{\rightleftharpoons}} SL \qquad (1)$$

where S is a representative binding site on Ab, L is ligand, and k and k' are the rate constants for association and dissociation, respectively. The ratio of these rates (k/k') is the **intrinsic association constant** K, which measures the tendency of site and ligand to form a stable complex; i.e., **K represents the intrinsic affinity of the representative Ab binding site** for the ligand, or

$$K = \frac{k}{k'} = \frac{[SL]}{[S][L]} \qquad (2)$$

the terms in brackets referring to the concentrations, at equilibrium, of the occupied Ab sites (SL), vacant Ab sites (S), and free (unbound) ligand molecules (L). If the total concentration of binding sites ($S + SL$) and of ligand molecules ($L + SL$) are known, a measurement at equilibrium of any one term on the right side of equation (2) can lead to the association constant. It is usually most convenient to measure the free ligand concen-

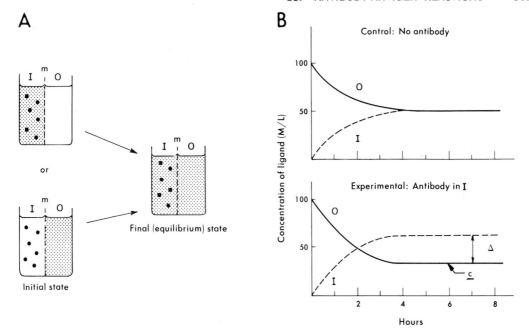

Fig. 15-1. Equilibrium dialysis. **A.** Small univalent haptens (small dots) can diffuse freely between the compartments (I, O), but Ab molecules (large dots) cannot. At equilibrium the greater concentration of hapten in I is due to binding by Ab. **B.** Change in hapten concentration with time. Equilibrium is reached in about 4 hours (the time varies with temperature, volumes of the compartments, nature and surface area of the membrane (m), etc.). In **B** the hapten was initially in compartment O; c is the concentration of free (unbound) hapten at equilbrium in I and O; Δ is the concentration of Ab-bound ligand in compartment I at equilibrium.

tration, and several methods are available to distinguish free (L) from bound (SL) ligand. The most generally applicable is equilibrium dialysis (below); it is also the most satisfactory method because it avoids perturbing the equilibrium, unlike some more rapid methods that physically separate free and Ab-bound ligands (e.g., precipitation of Ab with high salt concentration, below).*

EQUILIBRIUM DIALYSIS

The principles are shown in Figure 15-1. A solution containing Ab molecules specific for a simple haptenic group, such as 2,4-dinitrophenyl, is placed in a compartment (I, for

---

* Hapten inhibition of precipitation, described later in this chapter, is widely used to obtain relative values of association constants for a series of homologous haptens in relation to a reference hapten and a reference Ag. Other methods are also useful with special systems, e.g., ultracentrifugation, fluorescence quenching (see below).

"inside") which is separated by a membrane (m) from another compartment (O, for "outside") that contains a solution of an appropriate univalent ligand (e.g., 2,4-dinitroaniline; Ch. 14, Fig. 14-3). The membranes used are impermeable to Abs, but freely permeable to small molecules (MW <1000). If the concentration of ligand is measured periodically, it is observed to decline in O and to rise in I until equilibrium is reached; thereafter the concentrations in the two compartments remain unchanged. If compartment I had simply contained the solvent, or a protein that was incapable of binding, the ligand's concentration would ultimately become the same in both compartments. But if I contains Ab molecules that can bind dinitroaniline, the final (equilibrium) concentration of total ligand in I will exceed its concentration in O.† The difference represents

---

† Even without specific binding by Ab, charged ligands could be unevenly distributed across the membrane at equilibrium because of the net charge

ligand molecules bound to Ab molecules. Because the reaction is **readily reversible** the same final state is attained regardless of whether the Ab and ligand are placed initially in the same or in adjacent compartments (Fig. 15-1A).

By dividing the numerator [SL] and the denominator [S] by the Ab concentration, equation (2) may be expressed as:

$$K = \frac{r}{(n-r)c},\qquad(3)$$

or, more conveniently, as:

$$\frac{r}{c} = Kn - Kr,\qquad(4)$$

where, at equilibrium, r represents the number of ligand molecules bound per Ab molecule at c free concentration of ligand, and n is the maximum number of ligand molecules that can be bound per Ab molecule (i.e., the Ab valence). A set of values for r and c is obtained by examining a series of dialysis chambers at equilibrium, each with the same amount of Ab but a different amount of ligand. The derivation of equation (4) is outlined in the Appendix, this chapter.

Two features of equation (4) are notable. 1) A plot of r/c vs. r should give a straight line of slope $-K$, provided the original assumption was correct that all Ab sites are identical and independent. As we shall see, linearity or nonlinearity of this plot provides information on the uniformity of the ligand-binding sites in the sample of Ab molecules. 2) When the concentration of unbound ligand (c) becomes very large r/c approaches 0 and r approaches n; i.e., the number of ligand molecules bound per Ab molecule approaches the number of ligand-binding sites, or Ab valence.

Antibody Valence. Figure 15-2 shows representative data for the binding of univalent ligands by Abs of the most prevalent type (MW 150,000, IgG class; see Ch. 16). The limiting value for r is 2 as c approaches

infinity. That is, there are two binding sites per Ab molecule of this molecular weight. Bivalence of Abs has been repeatedly observed with many systems in which the ligands are small dialyzable molecules. Even in the more complex reactions with macromolecular Ags the same stoichiometry has been observed (Fig. 15-11).

Heterogeneity with Respect to Affinity. The relation between r/c and r is **not** linear for most Ab-ligand systems (Fig. 15-2, A–C). According to the assumptions made above, nonlinearity means that the Ab's binding sites are not uniform or not independent. However, bivalent Ab molecules can be cleaved by proteolytic enzymes into univalent fragments (Ch. 16), which exhibit similar nonlinearity in the reaction with their ligands. Hence the nonlinearity cannot be due to interactions between sites, e.g., to the modification of a vacant binding site by occupancy of the other site on the same molecule. Rather, as we shall see, several independent lines of evidence demonstrate that Ab molecules of any particular specificity are usually highly diversified with respect to affinity for the ligand.

However, occasional animals form Abs (particularly to certain Ags, such as polysaccharides of streptococcal cell walls and pneumococcal capsules) that have greatly restricted heterogeneity. In their uniform affinity for ligands these **homogeneous Abs** resemble enzymes and the immunoglobulins produced by plasma cell tumors (the neoplastic counterparts of the cells that normally synthesize Abs; see Ch. 16, Myeloma proteins; Fig. 15-2 D–F).

For an Ab-ligand pair with diverse binding constants (nonlinearity in r/c vs. r) it is useful to determine an average value, K_0, which is defined by the free ligand concentration required for occupancy of half the Ab-binding sites. Thus, substitution of $r = n/2$ in equation (4) leads to

$$K_0 = \frac{1}{c}.\qquad(5)$$

K_0 is designated the **average intrinsic association constant;** it is a measure of **average affinity,** and is usually referred to simply as "the affinity." Its unit is the reciprocal of con-

---

on the protein. This inequality, or Donnan effect, is avoided by carrying out the dialysis in the presence of a sufficiently high salt concentration, e.g., 0.15 M NaCl. Donnan effects are irrelevant with uncharged ligands.

Fig. 15-2. Specific binding of ligands plotted according to equation (4). For all Ab-ligand systems **(A–E)** the extrapolation indicates two ligand-binding sites per Ab molecule (MW 150,000, IgG class; see Ch. 16). In **B** affinity is higher at 7.1° than at 25°. In **C** two purified anti-2,4-dinitrophenyl (anti-Dnp) Abs differ about 30-fold in affinity for dinitroaniline, and even larger variations occur (Ch. 17, Changes in affinity). In **E** the anti-Dnp protein was produced by a plasma cell tumor (Ch. 16, Multiple myeloma). Non-linearity, showing heterogeneity with respect to affinities of binding sites, is pronounced in **B** and **C,** but slight in **A;** linearity, showing uniformity of binding sites, is evident with the antipolysaccharide Abs **(D),** the anti-Dnp myeloma protein **(E),** and the enzyme muscle phosphorylase a, which has four binding sites per molecule (MW 495,000) for Amp **(F).** [Data based on: **(A)** Eisen, H. N., and Karush, F., *J Am Chem Soc* 71:363 (1949); **(B)** Karush, F., *J Am Chem Soc* 79:3380 (1957); **(C)** Eisen, H. N., and Siskind, G., *Biochemistry* 3:996 (1964); **(D)** Pappenheimer, A. M., Redd, W. R., and Brown, R., *J Immunol* 100:1237 (1968); **(E)** Eisen, H. N., Little, J. R., Osterland, C. K., and Simms, E. S., *Cold Spring Harbor Symp Quant Biol* 32:75 (1967); **(F)** Madsen, N. B., and Cori, C. F., *J Biol Chem* 224:899 (1957).]

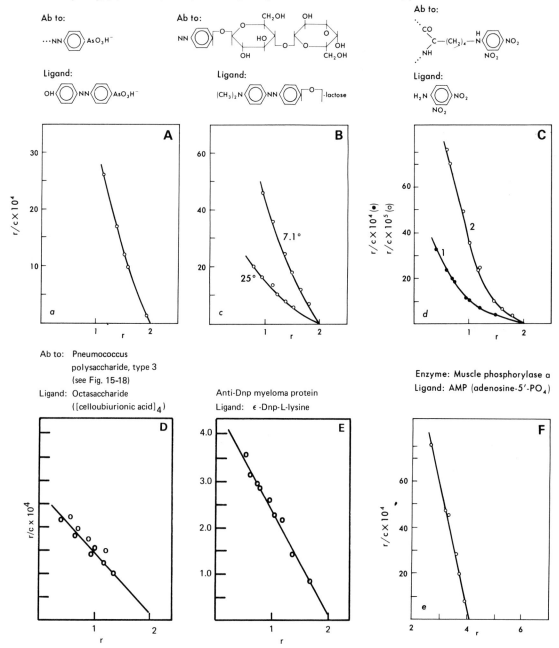

centration, i.e., liters mole^{-1} when c is in moles liter^{-1}. The higher the affinity of a given population of Ab molecules for ligand, the lower the concentration of free ligand need be for the binding sites to become occupied to any specified extent. (The analogy to the equilibrium constant for ionization reactions should be evident; by convention, however, the ionization constant is expressed as the dissociation constant [reciprocal of the association constant], which for the reaction $HA \rightleftharpoons H^+ + A^-$ is the H^+ concentration at which half of A binds a hydrogen ion.)

The heterogeneity of Abs with respect to affinity is often evaluated by the Sips distribution function,

which is similar to the normal distribution function commonly used in statistics. The Sips function leads to the explicit statement:

$$\log \frac{r}{n-r} = a \log K + a \log c \qquad (6)$$

where r, n, K, and c have the same meanings as before, and a is an index of the dispersion of equilibrium constants about the average constant, K_0. The term a is similar to the standard deviation in the more familiar normal distribution.

As is shown in Figure 15-3, a plot of $\log (r/n—r)$ vs. $\log c$ is indeed a straight line, and a, the index of heterogeneity with respect to affinity, is obtained as its slope. Values for a range from 1 to 0. When $a = 1.0$, all sites are identical with respect to K, and equation (6) is then equivalent to equations (3) and (4); the smaller the a value, the greater the degree of heterogeneity.

Some representative average association constants are shown in Table 15-1. With different conditions of immunization substantial differences are observed among Ab molecules of the same specificity (Ch. 16); for example, different populations of anti-Dnp antibodies differ as much as 100,000-fold in affinity for ε-Dnp-lysine.

Thus the term antibody, which is used in the singular, is in most instances a collective noun referring to a population or set of molecules defined by the capacity to bind a given functional group. With simple ligands it is relatively easy to specify the functional group and thereby the corresponding set of Ab molecules. For example, the set that binds lactose is the "antilactose antibody." With protein Ags, however, the functional groups (antigenic determinants) have only rarely been identified; but it is reasonable to assume that each of the groups on a protein also specifies a set of heterogeneous Ab molecules.

Fig. 15-3. Contrast between uniform binding of substrate by enzyme (AMP by phosphorylase) and heterogeneous binding of a ligand by Ab (dinitroaniline by anti-Dnp). Data are from Figure 15-2 and plotted according to equation (6). The slopes (a) are 1.0 for the enzyme-substrate pair and 0.5 for the Ab-ligand pair. (At $a = 0.5$ about 20% of the protein's binding sites fall outside the range $K_0/40$ to $40 K_0$.)

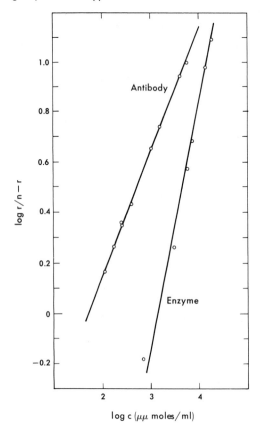

log c ($\mu\mu$ moles/ml)

EFFECTS OF TEMPERATURE, pH, IONIC STRENGTH; RATES OF REACTION

Because Abs are relatively stable proteins, their reactions can be studied over a wide range of conditions. The resulting changes in association constants provide some insight into the forces that stabilize the Ab-ligand complex. For example, the binding of ionic ligands, such as p-aminobenzoate by antibenzoate, decreases as pH is dropped from 7 to 4 and as salt concentration is raised from

TABLE 15-1. Average Intrinsic Association Constants for Representative Antibody-Ligand Interactions

Antibody specific for	Ligand	Average intrinsic association constants (K_0) in liters mole^{-1}
p-Azobenzenearsonate	OH⟨○⟩NN⟨○⟩AsO$_3$H$^-$	3.5×10^5
p-Azobenzoate	I⟨○⟩COO$^-$	3.8×10^4
ε-2,4-Dinitrophenyl-lysyl	$^-$OOC(NH$_2$)CH—(CH$_2$)$_4$NH⟨○⟩—NO$_2$ (NO$_2$)	1×10^7
p-Azophenyl-β-lactoside	(CH$_3$)$_2$N⟨○⟩NN⟨○⟩—O-lactose	1.6×10^5
Mono-Dnp-ribonuclease	Mono-Dnp-ribonuclease	1×10^6

Based on references given in Karush, F., *Adv Immunol 2*:1 (1962); Singer, S. J., in *The Proteins,* vol. 3, p. 269, Academic Press, N.Y., 1965; Eisen, H. N., and Pearce, J. H., *Annu Rev Microbiol 16*:101 (1962).

0.1 to 1.0 M NaCl. But similar changes do not affect the binding of nonionic ligands, such as 2,4-dinitroaniline by anti-Dnp Abs. This means that the COO$^-$ group of benzoate probably interacts with a positively charged group of the Ab combining site, but ionic interactions are not important for the binding of the neutral Dnp group.

Temperature variations have, broadly speaking, two kinds of effects. With some systems **increasing** temperature **decreases** the association constant; i.e., binding is exothermal. In others, association constants are unaltered between 4 and 40°. But no Ab–simple ligand systems have increasing affinity with increasing temperature. Nevertheless, the general practice of incubating mixtures of antisera and Ags at 37° is often helpful in speeding up some of the secondary reactions of Ab-Ag complexes, e.g., precipitation, agglutination, and complement fixation (see below and Ch. 18).

Thermodynamics. The formation of the Ab-ligand complex results in a change in free energy, ΔF, which is exponentially related to the average association constant by

$$\Delta F = RT \ln K_0, \qquad (7)$$

where R is the gas constant (1.987 cal/mole-deg.), T the absolute temperature, and $\ln K_0$ the natural logarithm of the average intrinsic association constant. $\Delta F°$, the standard free energy change, is the gain or loss of free energy in calories, as 1 mole of Ab sites and 1 mole of free ligand combine to form 1 mole of bound ligand; when the units of K_0 are

liters per mole (i.e., reactant concentrations are in moles per liter) ΔF in equation (7) is $\Delta F°$.

$\Delta F°$ values for various Ab-hapten pairs range from about −6000 to −11,000 calories (per mole hapten bound), corresponding to association constants of 10^5 to 10^9 liters mole^{-1} at 30° (Table 15-1).

It is sometimes useful to determine whether the free energy change comes about from a change in heat content (enthalpy) or in the entropy of the system. This determination is based on

$$\Delta F° = \Delta H° - T\Delta S° \qquad (8)$$

where $\Delta H°$ is the change in enthalpy (measured in calories), T is absolute temperature, and $\Delta S°$ is the entropy change. $\Delta H°$ is determined experimentally in sensitive calorimeters or by measuring the average intrinsic association constant (K_0) at two or more temperatures:

$$\Delta H° = \frac{R \ln \dfrac{K_2}{K_1}}{\dfrac{1}{T_1} - \dfrac{1}{T_2}}, \qquad (9)$$

where K_1 and K_2 are average association constants (K_0) at temperatures T_1 and T_2. $\Delta H°$ values range from 0 (no change in affinity with temperature), in which case the driving force for complex formation is the $T\Delta S$ term of equation (8), to −30,000 cal/mole ligand bound, in which case the decrease in heat content drives the reaction. The formation of apolar or hydrophobic bonds is essentially athermal ($\Delta H° \cong 0$), whereas the formation of hydrogen bonds is exothermal ($\Delta H° \cong -1000$ cal/hydrogen bond).

Rates of Reaction. Abs combine with ligands so rapidly that rates can be measured only with special technics. In one approach ligand is added at extremely low concentration (e.g., 10^{-10} M). This slows the rate to where changes can be measured at ca. 5-min intervals, but unusually sensitive methods are then required for determining the changing, low ligand concentrations. One effective procedure exploits the finding that haptens, such as Dnp, can be coupled to bacteriophage, which remain infectious and can be counted by plaque technics at concentrations as low as 10^{-17} M (a virion is taken as equivalent to a single macromolecule). The rate of decreasing infectivity following addition of Abs to the haptenic group measures the forward (association) rate, and reappearance of infectivity after adding an excess of the simple hapten measures the backward (dissociation) rate. In a second approach Ab and hapten are present initially at concentrations that present no unusual analytical problems (10^{-4} to 10^{-5} M), but the resulting extremely rapid reaction can be followed only in special instruments for measuring changes at microsecond (10^{-6}) intervals.

The forward (association) reaction for binding small haptens is one of the fastest biochemical reactions known: the rate constant (k in equations 1 and 2) is only about 10 times less than the theoretical limit of 10^9 liters/mole/second for diffusion-limited reactions. This implies that a high proportion of collisions between ligand and Ab are fruitful (i.e., they lead to specific binding), perhaps because of a "cage effect" of the water solvent, which holds collided ligand and Ab molecules together until the ligand slips into the protein's binding site. Forward rate constants for binding protein Ags are about 10 to 100 times lower than for small haptens.

Forward rate constants differ only slightly for Ab-hapten pairs that differ enormously in intrinsic affinity (Table 15-1); hence the differences in affinity are due almost entirely to wide variations in reverse (dissociation) rate constants. The activation energy of the forward reaction (measured by effects of temperature on association rate constants) is low, suggesting that the Ab's combining site is rigid. In contrast to enzymatic reactions, where binding of substrate is usually complicated by sequential conformational modifications of the enzyme, the combination of Ab and hapten appears to be a straightforward and uncomplicated bimolecular reaction.

Phage Neutralization as Assay for Antibody. The rate of neutralization of phage has been developed into perhaps the most sensitive general assay method for measuring Abs because, as is noted above, the infectivity of phage with covalently attached ligands can be blocked ("neutralized") by Abs to the attached ligand. The initial rate of neutralization is proportional to Ab concentration: it measures the forward (association) reaction which, as we have just emphasized, is essentially independent of the

Ab's affinity for the ligand. Because phage with a wide variety of attached ligands, including proteins, can remain infectious for bacteria, and can be measured at exceedingly low concentrations (ca. 10^{-17} M, above), the assays based on neutralization are widely applicable and extremely sensitive: for example, they have been used to measure Ab production by single cells (Ch. 17) and trace levels in serum of the Abs that cause human hypersensitivity to penicillin (Ch. 19; see Table 15-5, below). Unbound ligand competes with the ligand attached to phage and can inhibit the neutralization of phage infectivity caused by a standard amount of a standard antiserum to the attached ligand. From the extent of inhibition it is possible to measure trace amounts of unbound ligand in diverse biological fluids (serum, lymph, urine) following the same principle and with about the same sensitivity as in radioimmunoassays (e.g., see Fig. 15-33, below).

SPECIFICITY

Immune reactions are highly specific: a given population of Ab molecules will usually have different affinities for ligands whose structures differ only in the most minute detail. The basis for this discrimination lies in the forces responsible for the stability of the Ab-ligand complex. Insight into the nature of these forces has been acquired by comparing the binding of structurally related ligands.

The early classic studies by Landsteiner were elegant and illuminating. They were based on ability of antisera to a conjugated protein, such as sulfanilate-azoglobulin, to form specific insoluble complexes (precipitates) with other proteins conjugated with the same azo substituent, e.g., sulfanilate azoalbumin, but not with the albumin itself. Thus the reaction was specific for the sulfanilate azo group. The Abs' reactivity with diverse groups could be evaluated by comparing the precipitating effectiveness of various conjugates, each substituted with a different azo group. Some representative results are shown in Figures 15-4 and 15-5.

Some more recent examples of the dependence of affinity on the structure of ligands are given in Figures 15-6 and 15-7. From these and many other examples the generalizations discussed below have been drawn.

1) The ligands bound most strongly by a given set of Ab molecules are those that most closely simulate the structure of the deter-

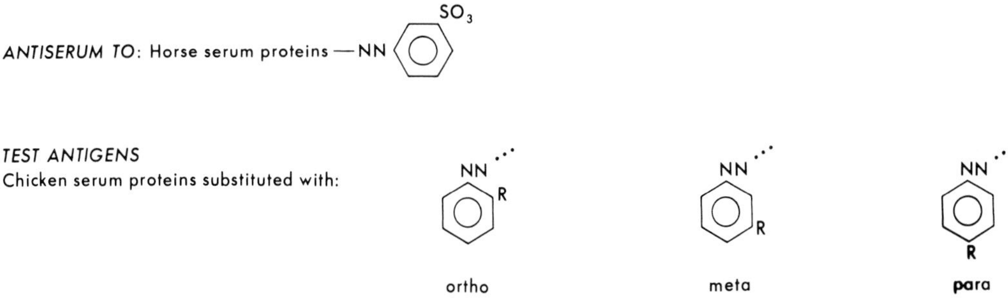

ANTISERUM TO: Horse serum proteins —NN ⟨◯⟩ SO₃

TEST ANTIGENS
Chicken serum proteins substituted with:

	ortho	meta	para
$R = SO_3^-$	+±	**++**	±
$R = AsO_3H^-$	0	+	0
$R = COO^-$	0	±	0

Fig. 15-4. Prominent effect of position and nature of acidic substituents of haptenic groups on the reaction between Abs to *m*-azobenzenesulfonate and various test Ags. R in the test Ag refers to the acidic substituents SO_3^-, AsO_3H^-, and COO^-. The homologous reaction is most intense (largest amount of precipitation) and is shown in heavy type. [From Landsteiner, K., and van der Scheer, J. *J Exp Med 63*:325 (1936).]

minant groups of the immunogen. This generalization is part of the broader rule that Abs react more effectively with the Ag that stimulated their formation than with other Ags; within this context the former is generally designated **homologous antigen** and the latter **heterologous antigens.** Similarly, haptens that resemble most closely the haptenic groups of the immunogen are the **homologous haptens** (Ch. 14, Fig. 14-3).

2) Those structural elements of the determinant group that project distally from the central mass of the immunizing Ag are **immunodominant,** i.e., they are especially influential in determining the Ab's specificity. Thus, Abs to *p*-azophenyl-*β*-lactoside and to 2,4-dinitrophenyl bind the **terminal residues** almost as well as the larger haptenic structures of which these residues constitute the end groups: for example, compare lactose with a phenyl-*β*-lactoside (Fig. 15-6), and dinitroaniline with ε-Dnp-lysine (Fig. 15-7). A particularly striking example is provided by Abs to the human blood group substances: anti-A is specific for terminal N-acetyl galactosamine residues, and anti-B is specific for

Fig. 15-5. Effect of nature and position of uncharged substituents of haptenic groups on the reactions between Abs to the *p*-azotoluidine group and various test Ags. The homologous reaction is shown in heavy type. [From Landsteiner, K., and van der Scheer, J. *J Exp Med 45*:1045 (1927).]

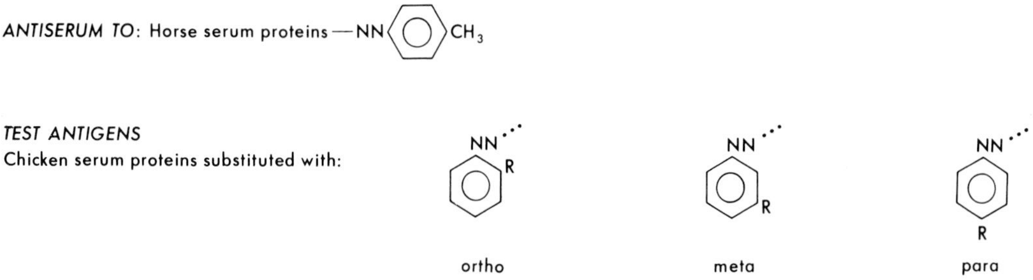

ANTISERUM TO: Horse serum proteins —NN⟨◯⟩CH₃

TEST ANTIGENS
Chicken serum proteins substituted with:

	ortho	meta	para
$R = CH_3$	+±	+±	**++**
$R = Cl$	+	+±	++
$R = NO_2$	±	+	+±

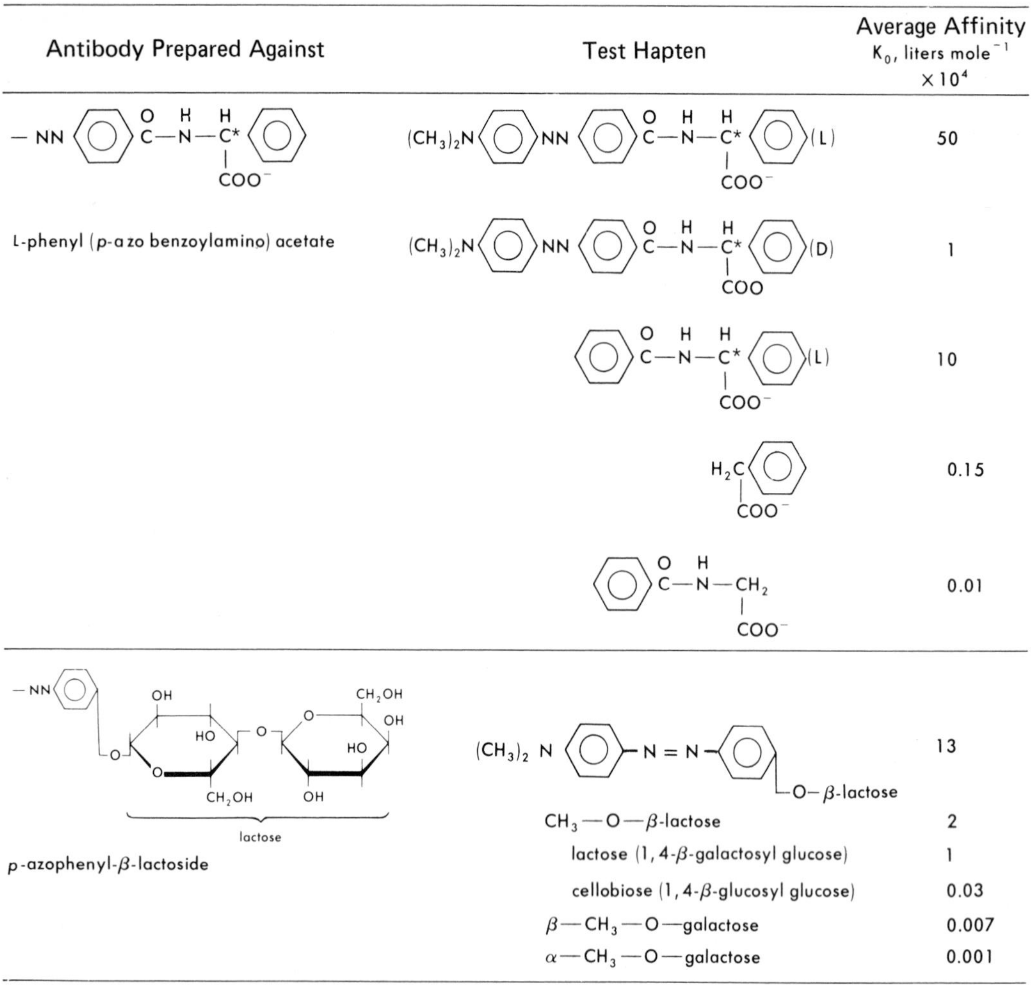

Antibody Prepared Against	Test Hapten	Average Affinity K_0, liters mole^{-1} $\times 10^4$
L-phenyl (*p*-azo benzoylamino) acetate	(structure, L)	50
	(structure, D)	1
	(structure, L)	10
	(structure)	0.15
	(structure)	0.01
p-azophenyl-β-lactoside (lactose structure)	(CH$_3$)$_2$ N ⬡ — N = N — ⬡ —O—β-lactose	13
	CH$_3$—O—β-lactose	2
	lactose (1,4-β-galactosyl glucose)	1
	cellobiose (1,4-β-glucosyl glucose)	0.03
	β—CH$_3$—O—galactose	0.007
	α—CH$_3$—O—galactose	0.001

Fig. 15-6. Specificity of Ab-hapten reactions: Dependence of affinity on structure of the hapten. [From Karush, F. *J Am Chem Soc 78*:5519 (1956); *79*:3380 (1957).]

terminal galactose residues (Ch. 21, ABO system).

However, nonterminal residues also contribute to specific binding, sometimes decisively. For example, in the cell wall lipopolysaccharide that determines the serological specificity of various groups of *Salmonella* the sugars that react specifically with Abs to group E organisms are nonterminal mannosyl galactose residues (Fig. 15-20).

3) The **resolving powers** of Abs are, in general, comparable to those of enzymes. Some Abs readily distinguish between two molecules that differ only in the configuration about one carbon (e.g., glucose and galac-

tose, or D- and L-tartrate; see also Fig. 15-6).

4) The specific binding of a ligand by an Ab molecule may be regarded as a competitive **partition** of ligand between water and Ab-binding sites, which are relatively hydrophobic. Hence **ligands that are sparingly soluble in water,** such as dinitrophenyl haptens, tend to form particularly **stable complexes** with Ab, whereas ligands that are highly soluble in water, such as sugars and organic ions (e.g., benzonate), tend to form more dissociable complexes.

Many observations on the interactions between Abs and their ligands have made clear that the strength of the over-all bond between

Antibody Prepared Against	Test Hapten	Average Affinity K_0, liters mole^{-1} $\times 10^5$
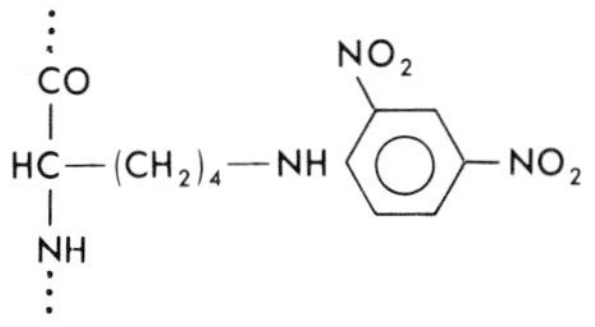 2,4-dinitrophenyl-L-lysyl group of Dnp protein	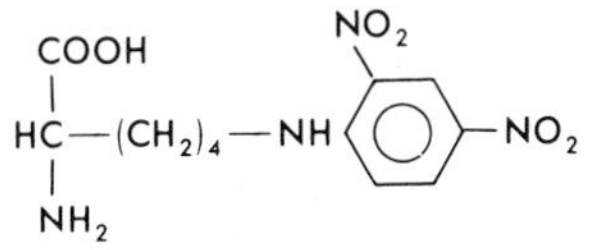 ϵ-Dnp-L-lysine	200
	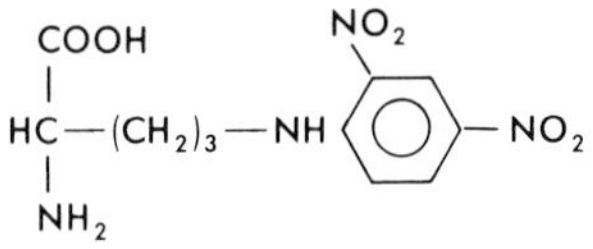 δ-Dnp-L-ornithine	80
	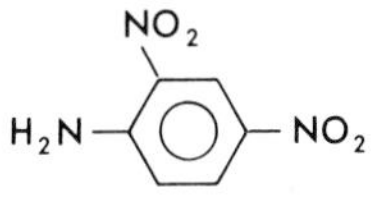 2,4-dinitroaniline	20
	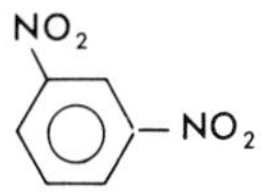 m-dinitrobenzene	8
	 p-mononitroaniline	0.5

Fig. 15-7. Specificity of Ab-hapten reactions: Dependence of affinity on structure of the hapten. The haptens that approximate the haptenic group of the immunogen are bound more strongly. [From Eisen, H. N., and Siskind, G. W. *Biochemistry* 3:996 (1964).]

an Ab and a ligand reflects the sum of many constituent noncovalent interactions among atomic groups of the ligand and side chains of amino acid residues in the binding site of Ab. The greater the number and strength of the constituent interactions the more stable (i.e., the less dissociable) is the Ab-ligand complex. It is apparent intuitively that the number of bonds formed is greater the more closely the three-dimensional surface of the ligand matches, in a complementary sense, the three-dimensional contour of the Ab site. However, binding strength depends not only on geometrical complementarity but on chemical features of the paired groups: for example, it is greater when the groups attract, as when an anionic group of the ligand is contiguous to a cationic group of the Ab, or a hydrogen-bond acceptor of the ligand is adjacent to a hydrogen-bond donor of the Ab.

Virtually all the known noncovalent bonds appear to participate in various Ab-ligand interactions: ionic bonds, hydrogen bonds, apolar (hydrophobic) bonds, charge-transfer bonds. The strength of these bonds depends on distance between the interacting groups; for some bonds their strength is inversely proportional to distance to the sixth or seventh power. Hence the stability of immune complexes is critically dependent on the closeness of approach of ligand groups to Ab groups. Bulky substituents on ligands can hinder close approach and thereby diminish the strength of binding (**steric hindrance**; for example, see Fig. 15-16).

The binding sites of Ab molecules probably exist on the surface as shallow depressions, rather than as deep clefts, since they are accessible to determinant groups on macromolecules, including those on cell surfaces.

SPECIFICITY AND AFFINITY

As noted above, the specificity of an Ab population refers to its capacity to discriminate between ligands of similar structure by combining with them to detectably different extents: **the greater the difference in affinity for two closely related structures, the more specific the antibody.** Long before affinity was measurable this viewpoint was recognized by Landsteiner, who defined specificity as "the disproportional action of a number of related agents on a variety of related substrates"; by "related agents" he meant Abs, and by "related substrates" he meant haptens and Ags.

Sera differ greatly in specificity: for example, antiserum to *m*-azobenzenesulfonate is highly specific (it distinguishes sharply among *p*-, *o*-, and *m*-azobenzenesulfonates), whereas antiserum to *p*-azotoluidine is poorly specific (it does not differentiate methyl, chloro, or nitro in the para positions; Figs. 15-4 and 15-5). This difference in specificity probably arises because of the following considerations: 1) uncharged toluidine is more apolar than charged benzenesulfonate; 2) Abs to apolar groups tend to have higher affinity than Abs to polar groups for their respective ligands (see item 4 under Specificity, above); 3) high-affinity Abs usually appear to be less specific than low-affinity Abs, probably because Abs with high affinity for the homologous ligand often give easily detectable reactions with related ligands (for which their affinity may be very much lower), but Abs whose affinity for the homologous ligand is so low as just to permit a detectable reaction are not likely to react perceptibly with any other ligand. The high specificity (resolving power) of low-affinity Abs and the low affinity of Abs for carbohydrates could account for the exquisite specificity of some anticarbohydrate antisera and their resulting practical value in diagnostic typing of blood groups (Ch. 21), salmonellae (Chs. 6, 29), etc.

CARRIER SPECIFICITY

Some of the Abs made against a hapten-protein conjugate seem to react exclusively with the haptenic group, for they combine no better with the immunogen than with conjugates of the same hapten with unrelated proteins. However, many other Abs exhibit "carrier specificity": for maximal reactivity (highest affinity) they require not only the haptenic group and the amino acid residue to which it is attached but also (in varying degree) neighboring, unsubstituted residues of the immunogen. Other carrier-specific Abs react only with the protein moiety of the conjugate. Carrier specificity is particularly significant in some of the reactions of Ags with specific receptors on cell surfaces, as we shall see in connection with Ab formation (Ch. 17) and cell-mediated immunity (Ch. 20).

REACTIONS WITH SOLUBLE MACROMOLECULES

All complexes formed by Abs and small univalent ligands, considered in previous sections, are soluble. With macromolecular Ags, however, the complexes frequently become insoluble and precipitate from solution. Though the Abs responsible for this **precipitin reaction** used to be regarded as members of a unique class, called "precipitins," it is now clear that most Abs are capable of precipitating with their Ags (for exceptions see Blocking Abs and the Unitarian hypothesis, below).

PRECIPITIN REACTION IN LIQUID MEDIA

Since its discovery in 1897 the precipitin reaction has been used extensively as a qualitative or semiquantitative assay for estimating Ab titers in sera. Attempts to measure precipitates quantitatively were of limited value until Heidelberger and Avery discovered, in 1923, that an important Ag of the pneumococcus, the capsule, was a polysaccharide. This discovery had several important consequences: 1) it established that some macromolecules besides proteins could be immunogenic; 2) the structural basis for the specificity of natural Ags could be explored because the antigenic determinants of polysaccharides, unlike those of proteins, are not markedly influenced by the macromolecule's conformation; hence small oligosaccharides, derived from polysaccharides, could be used as simple haptens; 3) the Abs precipitated from serum could be identified as proteins; and 4) quantitative procedures for measuring proteins in general could be applied to the analysis of precipitated Abs since the included Ag did not interfere.

THE QUANTITATIVE PRECIPITIN REACTION

For purposes of illustration consider an antiserum prepared by immunizing a rabbit with type 3 pneumococci. When the purified capsular polysaccharide is added to the antiserum a precipitate appears. The reaction is specific: it does not occur with serum obtained before immunization or from rabbits immunized with other Ags. Analysis of the precipitate after thorough washing reveals only protein and the type 3 polysaccharide. Moreover, when the precipitated protein is separated from the polysaccharide (Appendix, this chapter) it can be precipitated completely and specifically by the type 3 polysaccharide. Thus, all the precipitated protein is evidently Ab, which can be measured with precision by a variety of quantitative procedures, e.g., by Kjeldahl analysis for nitrogen. (Trace amounts of some other proteins, called complement, are also precipitated; these are considered in Ch. 18.)

As is shown in Figure 15-8 and in Table 15-2, the amount of protein precipitated in a series of tubes, each with the same volume of antiserum, increases with the amount of polysaccharide added up to a maximum, beyond which larger amounts of the Ag lead to progressively less precipitation. The precipitation of a **maximum** amount of Ab by an **optimal** amount of Ag may appear inconsistent with the binding reaction discussed earlier, in which the number of Ab sites occupied by ligand increased monotonically without going through a maximum. This apparent discrepancy is due to special features of precipitation, which are discussed below (Lattice theory).

When the Ag is a protein instead of a polysaccharide the precipitated Ag protein must be deducted from the total precipitated protein. One must therefore know the Ag content of the precipitate. Fortunately, in certain regions of the precipitin curve (Ab-excess and equivalence zones in Fig. 15-8) the precipitated Ag is essentially equivalent to the total amount of Ag added, as had been inferred many years ago from the absence of Ag (in qualitative precipitin tests) in the corresponding supernatants. Subsequently a direct demonstration was provided by the use of Ags labeled with radioactive iodine or with intensely colored substituents.

Fig. 15-8. Precipitin curve for a monospecific system: One Ag and the corresponding Abs.

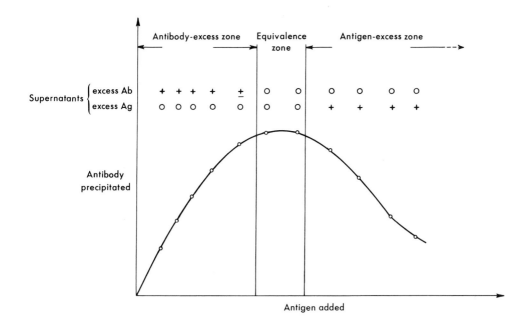

Zones of the Precipitin Curve. Useful information can be obtained from an examination of the supernatants by qualitative tests to detect unreacted Ab and unreacted Ag. For this purpose each supernatant is divided into aliquots, to one of which is added a small amount of fresh Ag (to detect excess Ab), and to the second a small amount of fresh antiserum (to detect excess Ag). If the Ag is homogeneous (i.e., consists of a single uniform group of molecules), or if the Ag preparation is heterogeneous but the antiserum is capable of reacting with only one of the components, then **none** of the supernatants contains **both** unreacted Abs and unreacted Ag in detectable amounts. Instead, the residual soluble reactants are distributed as shown in the precipitin curve of Figure 15-8: on the ascending limb, or **Ab-excess zone,** the supernatants contain free Ab; on the descending limb, or **Ag-excess zone,** the supernatants contain free Ag. In the region of maximum precipitation, the **equivalence zone** or **equivalence point,** the supernatants are usually devoid of both detectable Ab and detectable Ag, and the amount of Ab in the corresponding precipitate is taken to represent the weight of Ab in the volume of serum tested. Sometimes, as in Tables 15-2 and 15-3, the maximum amount of Ab is precipitated when there is a slight excess of free Ag in the supernatant; this is commented on below (see Nonprecipitating antibodies).

Up to this point we have been considering a **monospecific** system, i.e., one in which only one Ag and the corresponding Abs form the precipitates. However, most Ags, including those that satisfy the usual physical and chemical criteria of purity, are actually contaminated by small amounts of unrelated Ags, which may provoke independent immune responses. The precipitin reaction between such a contaminated Ag and its antiserum is thus usually the sum of two or more independent precipitin reactions, each monospecific.

This complex (but commonplace) situation is shown schematically in Figure 15-9. Contrary to what was observed with a monospecific system, some supernatants contain **both** unreacted Abs and unreacted Ag, because the Ag-excess zone of one system overlaps the Ab-excess zone of another. Qualitative testing of supernatants in the precipitin reaction thus provides a simple means for detecting the existence of multiple systems. However, as we shall see later the precipitin reaction in agar gel provides a more sensitive way to detect multispecificity, and can also indicate how many monospecific systems are present in a given pair of reactants.

Antibody/Antigen Ratios in the Precipitin Reaction. With most monospecific systems the Ab/Ag ratio in precipitates varies nearly linearly over the Ab-excess zone with the

TABLE 15-2.   Precipitin Reaction with a Polysaccharide as Antigen

Tube No.	S3 added (mg)	Total protein (or antibody) precipitated (mg)	Supernatant test
1	0.02	1.82	Excess Ab
2	0.06	4.79	Excess Ab
3	0.08	5.41	Excess Ab
4	0.10	5.79	Excess Ab
5	0.15	6.13	No Ab, no S3
6	**0.20**	**6.23**	**Slight excess S3**
7	0.50	5.87	Excess S3
8	1.00	3.76	Excess S3
9	2.00	2.10	Excess S3

The antigen (S3) is purified capsular polysaccharide of type 3 pneumococcus. Each tube contained 0.7 ml of antiserum obtained by injecting rabbits repeatedly with formalin-killed, encapsulated type 3 pneumococci. The supernatant of tube 6, which had the maximum amount of precipitated Ab, contained a slight excess of Ag; this is often observed and reflects the presence of some nonprecipitable or poorly precipitable Ab (see text).

Based on Heidelberger, M., and Kendall, F. E. *J Exp Med* 65:647 (1937).

TABLE 15-3.   Precipitin Reaction with a Protein as Antigen

Tube No.	EAc added (mg)	Total protein precipitated (mg)	Antibody precipitated, by difference (mg)	Supernatant test	Ab/Ag in precipitates Weight ratio	Ab/Ag in precipitates Mole ratio
1	0.057	0.975	0.918	Excess Ab	16.1	4.0
2	0.250	3.29	3.04	Excess Ab	12.1	3.0
3	0.312	3.95	3.64	Excess Ab	11.7	2.9
4	0.463	4.96	4.50	No Ab, no EAc	9.7	2.4
5	**0.513**	**5.19**	**4.68**	**No Ab, trace EAc**	**9.1**	**2.3**
6	0.562	5.16	(4.60)	Excess EAc	(8.2)	(2.1)
7	0.775	4.56	(3.79)	Excess EAc	(4.9)	(1.2)
8	1.22	2.58	—	Excess EAc	—	—
9	3.06	0.262	—	Excess EAc	—	—

Each tube contained 1.0 ml of antiserum obtained by injecting rabbits repeatedly with alum-precipitated crystallized chicken ovalbumin (EAc).

Ab content of precipitates in tubes 6–9 could not be determined by difference because too much EAc remained in the supernatants. The latter was measured independently in the supernatants of tubes 6 and 7, allowing an estimate to be made of EAc and Ab in the corresponding precipitates (values in parentheses).

Mole ratio Ab/Ag was estimated by assuming molecular weights for EAc and Ab of 40,000 and 160,000, respectively.

Based on Heidelberger, M., and Kendall, F. E. *J Exp Med 62*:697 (1935).

amount of Ag added (Fig. 15-10). With Ab in large excess the mole ratio greatly exceeds 1.0, showing that many Ab molecules can combine simultaneously with one molecule of Ag; i.e., **Ag is multivalent.** In the Ag-excess zone the role ratio of Ab/Ag tends to plateau with a limiting value of slightly over 1.0 (Table 15-3, Fig. 15-10). These variations

are explicable in terms of the lattice theory (below).

LATTICE THEORY

Multivalency of Ags suggested to Marrack, and to Heidelberger and Kendall, that precipitation could be a consequence of the growth

Fig. 15-9. Precipitin curve for a multispecific system. The precipitation observed (—•—) is the sum of two or more precipitin reactions (– – –). The significant difference from the monospecific system shown in Figure 15-8 is that some supernatants have **both** excess Ag and excess Abs (indicated by pluses in heavy type).

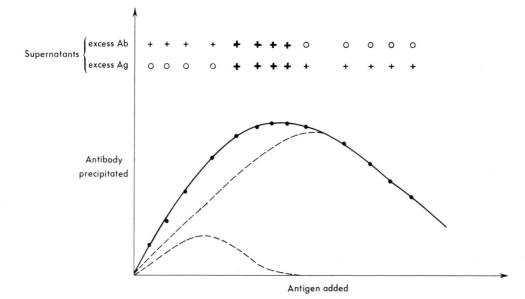

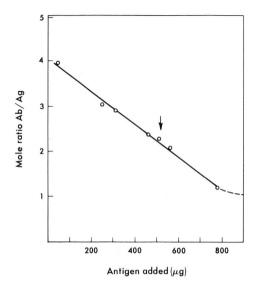

Fig. 15-10. Continuous decline in Ab/Ag ratio of precipitates with increasing amount of Ag added to a fixed volume of antiserum. Chicken ovalbumin (EAc) is the Ag and the serum is rabbit anti-EAc. The limiting mole ratio and slope vary about 30% in different sera. The arrow marks the equivalence zone. (Data are those of Table 15-3.)

of Ab-Ag aggregates in such a way that, broadly speaking, each Ag molecule is linked

to more than one Ab molecule and, in turn, each Ab molecule is linked to more than one Ag molecule. When the aggregates exceed some critical volume they settle out of solution spontaneously (because sedimentation rate is proportional to $V(\rho-\rho_0)g$, where V is the volume of a particle, ρ and ρ_0 are the densities of the particle and solvent, respectively, and g is the gravitational field). The assumption that Abs are multivalent was subsequently validated by equilibrium dialysis with univalent haptens (Fig. 15-2).

As is shown in Figure 15-11, the suggested **alternation** of multivalent Ag and Ab molecules (i.e., the **lattice theory**) accounts for the wide and continuous variations in Ab/Ag ratios of specific precipitates (Fig. 15-10), which for some time previously had defied rational explanation and seemed to be in conflict with the law of fixed proportions in formation of chemical compounds. With systems in which the Ag is distinctively labeled, and thus can be measured directly, it can be shown that precipitates formed in the presence of excess Ag have Ab/Ag ratios that approach 1.0 as a limiting value. This suggests that the least complicated precipitating complex is a large linear aggregate with alter-

Fig. 15-11. Hypothetical structure of immune precipitates and soluble complexes according to the lattice theory. Numbers refer to mole ratios of Ab to Ag. Dotted lines with precipitates are intended to indicate that the complexes continue to extend as shown. The precipitates may be visualized as those found in the Ab-excess zone **(A)**, the equivalence zone **(B)**, and the Ag-excess zone **(C)**. The soluble complexes correspond to those in supernatants in moderate **(D)**, far **(E)**, or extreme **(F)** Ag excess. Black circles, Ag molecules; open ellipses, Ab molecules.

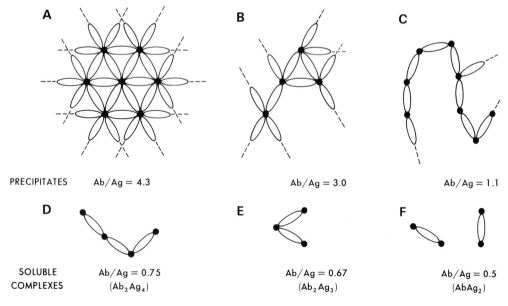

nating Ab and Ag molecules (. . . Ab·Ag·Ab· Ag·Ab·Ag . . .). Moreover, in the region of Ag-excess complexes of even lower Ab/Ag ratios are found, but they remain in the supernatant because they are small (Fig. 15-11D–F); they account for the "descending limb" of the precipitin curve. The soluble complexes have been demonstrated by ultracentrifugation and electrophoresis, and their mole ratios vary considerably; e.g., 0.75 (Ab$_3$Ag$_4$) in slight Ag excess and 0.67 (Ab$_2$Ag$_3$) in substantial Ag excess. In extreme excess the ratio approaches 0.5 (AbAg$_2$), as expected from the bivalence of Ab molecules.

VALENCE AND COMPLEXITY OF PROTEIN ANTIGENS

The limiting mole ratio of Ab/Ag in extreme Ab excess is often taken as a measure of the Ag molecule's valence (e.g., 4 in Fig. 15-10). Since Ab molecules are bivalent, however, the actual number of binding sites on the Ag can approach twice the limiting mole ratio. And the limiting ratio provides only a **minimal** estimate of the Ag valence: a larger number of reactive sites could exist but fail to be expressed, either because the spatial limitations at the surface preclude the packing of more Ab molecules about one Ag molecule, or else because the particular antiserum used lacks Abs for some potentially functional groups.

An Ag molecule of high molecular weight should be able to bind more Ab molecules simultaneously than an Ag of low molecular weight; as Table 15-4 shows, this relation is, in general, found. It should be especially noticed that all Ags capable of giving a precipitin reaction have a valence of at least two. Even some small bivalent haptens form specific precipitates; for example, some Abs to the 2,4-dinitrophenyl (Dnp) group are specifically precipitated by α,ε-bis-Dnp-lysine (Ch. 14, Fig. 14-3; Fig. 15-13).

The multivalency of many **polysaccharides** is readily understandable, since they are made up of repeating residues (e.g., Fig. 15-18). With **proteins,** however, the chemical basis for their multivalency is less obvious. The covalent structures of many proteins (e.g., ribonuclease, myoglobin) make clear that groups of amino acid residues almost never recur as repetitive sequences in a given polypeptide chain. Hence each antigenic determinant should occur once per chain, and two or more times only in those molecules with multiple copies of a particular chain. Nevertheless, protein molecules made up of a single chain behave as though multivalent. This is because there are a variety of determinants per chain and a particular determinant need occur only once per molecule of immunogen in order to stimulate the formation of the corresponding Abs.

The consequences can be illustrated with bovine pancreatic ribonuclease, with one

TABLE 15-4. Correlation Between Molecular Weight and Valence of Antigens

Antigen	Molecular weight	Approximate mole ratio Ab/Ag of precipitates in extreme antibody excess*
Bovine pancreatic ribonuclease	13,600	3
Chicken ovalbumin	42,000	5
Horse serum albumin	69,000	6
Human γ-globulin	160,000	7
Horse apoferritin	465,000	26
Thyroglobulin	700,000	40
Tomato bushy stunt virus	8,000,000	90
Tobacco mosaic virus	40,000,000	650

* The mole ratios are representative, and tend to be higher with antisera obtained late in the course of immunization. Since the Ab molecules involved are at least bivalent (cf. IgG and IgM Abs in Ch. 16), the Ag valences must be somewhat higher than the ratios listed.

Based on Kabat, E. A. In *Kabat and Mayer's Experimental Immunochemistry,* 2nd ed. Thomas, Springfield, Ill., 1961.

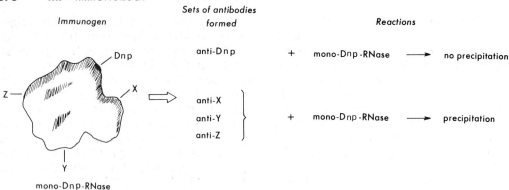

Fig. 15-12. Antigen valence illustrated with mono-Dnp-RNase, which induces the formation of anti-Dnp and other Abs, arbitrarily called anti-X, anti-Y, and anti-Z. The immunogen is **univalent** with respect to the anti-Dnp Abs, with which it forms soluble complexes. The same immunogen, however, is **multivalent** with respect to the mixture of diverse sets of Abs (anti-X, anti-Y, etc.) that are formed against its various nonhaptenic determinants (X, Y, Z). [Based on Eisen, H. N., Simms, E. S., Little, J. R., and Steiner, L. A. *Fed Proc 23*:559 (1964).]

chain per molecule, carrying one Dnp group (mono-Dnp-RNase). In the antiserum made against this immunogen some Abs are anti-Dnp and others are specific for different determinant groups (X, Y, Z in Fig. 15-12). Mono-Dnp-RNase behaves as a multivalent molecule with this antiserum, giving a classic precipitin reaction. But if the anti-Dnp molecules are isolated from the antiserum and then mixed with mono-Dnp-RNase only soluble complexes are formed, because the same Ag is univalent with respect to this particular set of Ab molecules. These observations emphasize the operational nature of the definition of valence: **a given molecule of antigen can be univalent with respect to some antibody molecules and multivalent with respect to others.**

These observations also reinforce the view that a typical globular protein is a mosaic within one molecule of many determinants, and that an antiserum to a protein consists of many sets of Ab molecules, each with the capacity to react with a single kind of functional group. Thus, proteolytic cleavage of bovine serum albumin (BSA, MW 70,000) yields several large fragments, each of which gives a precipitin reaction with different sets of Abs in antiserum prepared against the whole BSA molecule.

The minimum number of sets of Abs in an antiserum to a globular protein is indicated by the limiting mole ratio of Ab/Ag in the far Ab-excess region of its precipitin curve.

Since each set is probably made up of Ab molecules of different affinities and even of somewhat different structures (see Ch. 16), terms such as "the anti-BSA Ab" or "tetanus antitoxin" are deceptively simple.

From the argument that a protein Ag is a constellation of determinants, it follows that **the precipitin reaction usually involves cooperation between different Abs.** This conclusion is supported by the observations in Figure 15-13: the small ligand R-X, in which the functional groups R and X each occurs once per molecule, does not precipitate with an anti-R serum or with an anti-X serum, but does precipitate with a mixture of the two. These considerations lead to the schematic view of the precipitin reaction shown in Figure 15-14.

Homospecificity of Antibodies. Since there are diverse functional groups in most Ags it may be asked whether a bivalent Ab molecule can have binding sites that are specific for two different ligand groups. All existing evidence indicates that this does not occur: **bivalent Abs are homospecific.** For example, in antisera prepared against dinitrophenylated bovine γ-globulin (Dnp-BγG) one could imagine Ab molecules with one binding site specific for Dnp and one specific for a nonhaptenic group of BγG. However, removal of all the Abs that can react with BγG does not reduce the capacity of the antiserum to bind Dnp ligands. This finding is in accord with

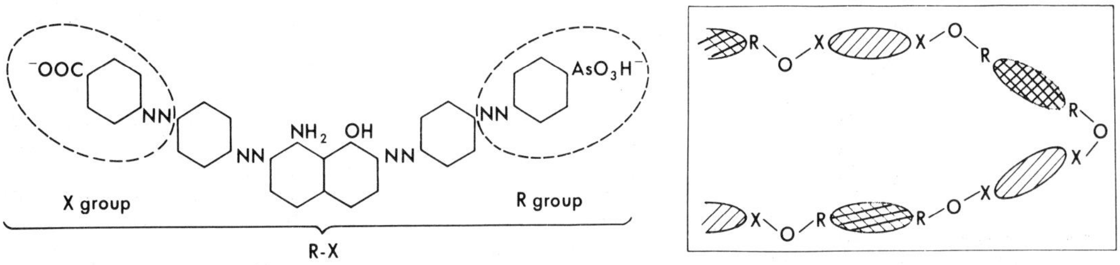

$$R\text{-}X + \text{anti-R serum} \longrightarrow \pm \quad \text{precipitation}$$

$$R\text{-}X + \text{anti-X serum} \longrightarrow \pm \quad \text{precipitation}$$

$$R\text{-}X + \text{anti-R} + \text{anti-X sera} \longrightarrow + + + + \quad \text{precipitation}$$

Fig. 15-13. Cooperation between Abs of different specificities (anti-R and anti-X) in the precipitin reaction with a synthetic ligand, R-X. (The small amount of precipitate ($\pm$) formed by R-X with anti-R alone or with anti-X alone is probably due to some aggregation of R-X.) The inset shows a hypothetical segment of the precipitate with alternation of anti-R and anti-X molecules in a linear aggregate. [Based on Pauling, L., Pressman, D., and Campbell, D. H. *J Am Chem Soc 66*:330 *(1944).*]

and explained by the symmetrical structure of Ab molecules (Ch. 16).

NONSPECIFIC FACTORS IN THE PRECIPITIN REACTION

The precipitin reaction involves two distinct stages: rapid formation of soluble Ab-Ag complexes and **slow** aggregation of these complexes to form visible precipitates. Thus, by measuring free Ag concentrations it has been found that the specific interactions are completed within a few minutes, but precipitate formation usually requires several days to reach completion.

While the lattice theory is clearly relevant for the first stage, an older view, that of Bordet, contains elements of interest for the second. This view suggests that the close packing of Ab molecules when bound to an Ag molecule provides opportunities for neighboring Ab molecules to react with each other, mostly by way of ionic bonds between oppositely charged groups (Fig. 15-11). As a consequence the complexes, which are usually predominantly made up of Ab molecules, become relatively hydrophobic and so tend progressively to associate with each other and to become increasingly insoluble. This view is supported by the effects of ionic strength on precipitation (less precipitation at low salt concentrations) and by observations on chemically modified Abs. For instance, when the negative charge on Ab molecules is increased by acetylation of free amino groups the ability to pre-

cipitate can be lost without impairing the Ab's ability to form a specific, soluble complex with Ag.

NONPRECIPITATING ANTIBODIES

Nonprecipitating Abs were first recognized because of discrepancies between the amounts of Ab precipitated after the addition of Ag by two different procedures. In **procedure A** each of a series of tubes, containing the same amount of antiserum, receives a different amount of Ag systematically increasing as in Table 15-2. By analysis of the precipitates the quantity of Ag necessary to precipitate the maximal amount of Ab is established. In **procedure B** about 1/10 the equivalent amount of Ag is added repeatedly to a single volume of antiserum, from which the precipitate formed after each addition is removed before the next addition. The sum of all the Ab precipitated is usually considerably less than that precipitated at the equivalence point in procedure A. The difference represents **nonprecipitating antibodies,** i.e., molecules that cannot alone precipitate with Ag but can become incorporated into Ab-Ag precipitates of the same specificity.

Under the influence of the lattice theory, **nonprecipitating Ab molecules** were assumed to have a single combining site. However, careful studies have never demonstrated intact Ab molecules to be truly univalent, and the inability of some Abs to precipitate with Ags arises from other mechanisms. For example, some Abs have too low an affinity for Ag. Hence the Ag must be added at relatively high concentrations in order to occupy a significant proportion of Ab combining sites; because

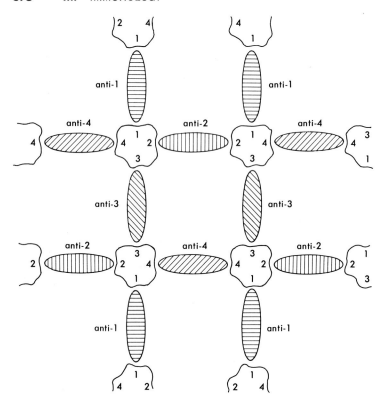

Fig. 15-14. Schematic illustration of the diversity of Abs formed against a pure Ag and their cooperation in the precipitin reaction with that Ag. Further complexities arise because each set of Abs (anti-1, anti-2, etc.) is probably heterogeneous with respect to affinity for the corresponding antigenic determinant; in addition Abs of the same specificity may differ considerably in structure (Ch. 16). Antigenic determinants are labeled 1,2,3,4.

this level of Ag is likely to be in excess, relative to Ab, only small soluble Ab-Ag complexes are formed. Another mechanism arises from monogamous bivalency (below).

Monogamous Bivalency. Some bivalent Ab molecules with high affinity for Ag do not form specific precipitates because they preferentially combine with two determinant groups of a **single** Ag particle, forming cyclic complexes (Ab:Ag) rather than cross-linked ones (Ag·Ab·Ag; Fig. 15-15). This type of binding, called "monogamous bivalency" by Karush, requires that a given antigenic group occur repetitively on the Ag surface. Besides hapten-protein conjugates, polysaccharides, and multichain proteins, this commonly occurs with surface determinants of bacteria, red cells, and viruses. The commonplace and important "blocking Abs" (which do not agglutinate bacteria, red cells, etc., but combine specifically with them) could also owe their behavior to monogamous bivalency (see below, Incomplete Abs).

Abs in general would be expected to engage preferentially in monogamous bivalent binding (because it is energetically advantageous to form binary Ab:Ag rather than ternary Ag·Ab·Ag complexes). However, most seem unable to do so, and act instead to cross-link Ags, perhaps because insufficient flexibility prevents the two combining sites in most Ab molecules from being adjusted to fit neighboring antigenic sites on a given Ag particle (Fig. 15-15; see Ch. 16, Overall structure of Abs).

For Ab-ligand pairs that can form monogamous complexes the equilibrium constant with multivalent ligands can be many orders of magnitude higher than the same Ab's intrinsic affinity for the corresponding univalent hapten. For instance, the reaction between anti-Dnp Abs and Dnp bacteriophage, with many Dnp groups per phage particle, has an association constant about 100,000-fold greater than the reaction between the same Abs and univalent Dnp-lysine (e.g., 10^{12} vs. 10^7 liters/mole): evidently the two bonds involved in the monogamous complex greatly decrease the probability that the Ab:Ag pair will dissociate. Because similar cyclic complexes are probably formed with unmodified

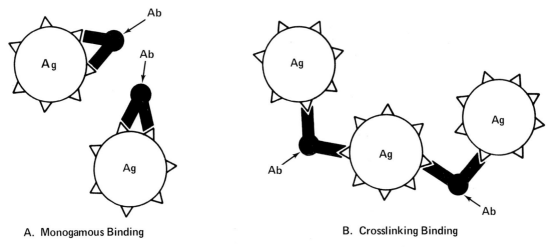

A. Monogamous Binding   B. Crosslinking Binding

Fig. 15-15. Monogamous bivalent binding, leading to cyclic Ab:Ag complexes, is contrasted with conventional binding, which leads to cross-linking of Ag particles (. . . Ag · Ab · Ag . . .).

viruses and bacteria, whose surfaces often have identical, repeating antigenic groups (e.g., various structural subunits of wall or envelope), some Abs can probably combine with and neutralize pathogenic microbes under in vivo conditions where the free Ab concentration is exceedingly low (e.g., 10^{-12} M).

REVERSIBILITY OF PRECIPITATION

After a precipitate has formed, its complexes can dissociate and reequilibrate with a fresh charge of Ag. When the latter is in sufficient excess small soluble complexes are formed and the precipitate dissolves. This reversibility provides the basis for many of the procedures used to isolate purified Abs (see Appendix, this chapter). In practice, however, it may be difficult to observe dissociation: hence the formation of Ab-Ag aggregates was formerly held to be irreversible, as in a particularly striking example described by Danysz.

The **Danysz phenomenon** is well illustrated with the diptheria toxin-antitoxin system. When an equivalent amount of toxin is added to an antitoxin serum the residual toxicity depends upon how the reagents are mixed. If the toxin is added to the antiserum all at once the mixture is nontoxic; but if the same quantity of toxin is divided into two or more portions and added at intervals of about 30 minutes the mixture is toxic. This phenomenon was interpreted to mean that the first addition of toxin led to the formation of irreversible complexes with a high Ab/Ag ratio, leaving insufficient free Ab to neutralize all the toxin subsequently added.

However, all Ab-Ag reactions are reversible in principle; those that appear irreversible probably have unusually slow dissociation rates, requiring days or even months, rather than minutes, to reestablish equilibrium when the system is perturbed. **Apparent irreversibility** is especially striking with many Ab-virus complexes: some are so stable that they do not perceptibly dissociate even when a mixture of virus and antiserum is diluted many thousandfold, with a corresponding reduction in the concentrations of free virus and free Ab (Ch. 50). The extraordinary stability of such complexes, reflecting unusually high equilibrium constants, probably derives from monogamous multivalent binding (above). Even ordinary single-bonded complexes of Ab with small univalent haptens can appear to be irreversible if the intrinsic association constant is sufficiently high ($> 10^8$ liters/mole); e.g., the hapten cannot be separated from the Ab by dialysis. Nevertheless, reversibility can always be demonstrated by "exchange": addition of more ligand, in great excess, will replace the molecules that appeared to be irreversibly bound.

APPLICATIONS OF THE PRECIPITIN REACTION

Measurement of Antibody Concentration. The quantitative precipitin reaction is performed as described above: various amounts of Ag are added to a constant volume of antiserum, and the resulting specific precipitates are washed and analyzed for total protein: the results, plotted as shown in Figure 15-8, measure the amount of precipitable Ab.

The precipitin assay is highly reproducible and precise. However, it requires the analysis of multiple precipitates, and its accuracy (i.e., its relation to "true" values) is limited because a proportion of the Abs, varying from serum to serum, may not bind to the Ag or precipitate under the conditions of the assay (see Nonprecipitating antibodies, above). Despite these reservations the quantitative precipitin reaction remains the most reliable general method for measuring Ab concentrations.

Measurement of Antigen Concentration. Once a precipitin curve has been constructed for a given antiserum and its Ag, the serum can be used to measure the concentration of that Ag in unknown solutions. The assay is carried out in the Ab-excess region; and it is only necessary to establish that the antiserum is free of extraneous Abs that could precipitate other Ags in the test solution. Antisera are stable on prolonged storage, and the precipitin reaction can measure small quantities (microgram range) of Ags in complex solutions; e.g., rabbit antisera have been used to measure immunoglobulins in human spinal fluid. However, more sensitive assays are often preferred (e.g., Radioimmunoassays, see below and Table 15-5).

HAPTEN INHIBITION OF PRECIPITATION

Qualitative Hapten Inhibition. Long before its quantitative features were characterized by Heidelberger and his associates the precipitin reaction was widely used as a visual, qualitative test to detect Ab-Ag reactions. It was, in fact, in just this simple fashion that Landsteiner had exploited it, by means of hapten inhibition, in his classic investigations of immunological specificity. In general terms, in the precipitin system he used the antiserum was prepared against one conjugated protein, which may be designated X-azoprotein A, and the precipitating agent was another conjugate with the same azo substituent attached to a different protein, X-azoprotein B. A and B were chosen so that they did not react with other's antisera (e.g., horse and chicken serum

Fig. 15-16. Competitive inhibition of the precipitin reaction by a univalent ligand. Ag, multivalent ligand; Ab, antibody; H, univalent ligand.

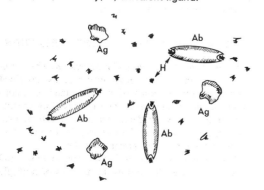

proteins, Figs. 15-4 and 15-5), and X-azoprotein B was prepared with many X-azo groups per molecule, so that it formed precipitating complexes with anti-X Abs. In contrast, simple haptens with one X group per molecule formed soluble complexes with anti-X and could competitively inhibit the precipitin reaction (Fig. 15-16). The greater the Ab's affinity for the hapten, relative to its affinity for the precipitating azoprotein, the more effectively was its precipitation inhibited.

Quantitative Hapten Inhibition. By combining hapten inhibition with quantitative measurements of precipitates, Pauling and Pressman were able to obtain more insight into the specificity of Ab reactions than had been possible with the qualitative method. In their assay analyses are carried out by adding equal volumes of an anti-X serum to a series of tubes with varying amounts of univalent X haptens. After a few minutes a multivalent precipitating agent (e.g., an X-azoprotein) is added in that amount which, in the absence of hapten, would give roughly maximal precipitation of anti-X Abs. The amount of Ab precipitated decreases as the concentration of added hapten increases (Fig. 15-17). The binding sites in a heterogeneous population of anti-X molecules differ in capacity to discriminate between the univalent and the multivalent ligand, and the relative effectiveness of univalent ligands is characterized by specifying the concentration required to inhibit precipitation by 50%. The concentration required of the reference hapten for 50% inhibition, divided by the concentration required of a second hapten, provides an index of the Ab's affinity for the second hapten, relative to the reference hapten. Some representative results are shown in Figure 15-17.

Hapten inhibition of precipitation has been used effectively to identify determinant groups of complex Ags, such as proteins and blood group substances (Ch. 21). The reactions of globular protein Ags are usually inhibitable by some of the large fragments (MW $\gg$ 1000) derived from the Ag by proteolytic enzymes, but not by small peptides or by the denatured protein. Such results suggest that **most determinant groups of proteins are "conformational,"** representing a particular three-dimensional spatial arrangement of a cluster of amino acids, rather than simply their sequence. However, a few protein-antiprotein reactions are inhibited specifically to some extent by small dialyzable peptides (derived from the Ag); these determinants appear to be **"sequential,"** their reactivity being manifested whether the sequence is part of the intact protein or a small oligopeptide. Such determinants are conspicuous in fibrous proteins (e.g., silk fibroin, synthetic polypeptides), but are evident only in exceptional globular proteins (e.g., coat protein of tobacco mosaic virus).

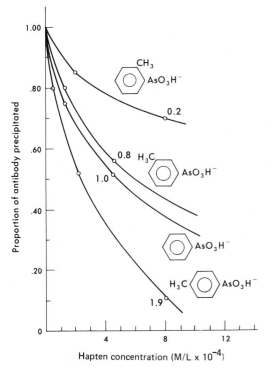

Fig. 15-17. Hapten inhibition of the precipitin reaction, illustrated with various univalent benzenearsonates. The number on each curve gives the affinity for the corresponding hapten **relative** to benzenearsonate ($= 1.0$). The antiserum was prepared against *p*-azobenzenearsonate; hence the hapten with -CH_3 in the para position is bound best, i.e., it is most inhibitory ($K' = 1.9$). With the methyl group in the meta or ortho position steric hindrance reduces affinity for Ab. [From Pressman, D. *Adv Biol Med Phys 3*:99 (1953).]

CROSS-REACTIONS

Besides reacting with its immunogen, an antiserum usually reacts with certain other Ags (heterologous Ags) that are sufficiently similar to the immunogen; some cross-reactions are illustrated in Figures 15-18 and 15-19.

Cross-Reactions Due to Impurities in Antigens. Most immunogens are complex mixtures of many different kinds of antigenic molecules. This is obviously true when the immunogen is a cell. It is also usually true, though less obvious, with purified proteins, since these are nearly always contaminated with other immunogenic proteins. Even at trace levels (e.g., 1%) the contaminants can often

provoke the formation of detectable amounts of Ab. Hence, antisera prepared against these mixtures consist of several Ab populations, each reactive with one Ag.

If, for example, crystallized chicken ovalbumin (EAc) contaminated with trace amounts of chicken serum proteins were used as immunogen, the anti-EAc serum would probably contain low levels of Abs to the contaminants and would probably form precipitates (i.e., cross-react) with some chicken serum proteins. Supernatant tests would probably reveal that the precipitin reaction was not monospecific; and this would be even easier to establish by reactions in agar gel (see below, Fig. 15-27).

Cross-Reactions Due to Common or Similar Functional Groups in Different Antigens. We have already discussed some of the evidence that a single protein contains many different antigenic determinants, each of which can potentially evoke the formation of a corresponding set of Ab molecules. If two different molecules happen to have one or more groups in common they usually cross-react.

This type of cross-reaction is frequently observed with polysaccharides and provides the basis for classifying many groups of closely related bacteria. The 500 or more varieties of salmonellae, for example, have been arranged into several serological groups, each identified by mutual cross-reactions: an antiserum to any strain of a particular group reacts with the other strains of that group. The common antigenic determinants responsible for the group-specific cross-reactions have been shown to be short sequences of particular sugar residues. For example, the determinant that defines group O *Salmonella* has as its principle residue colitose (3,6-dideoxy-L-galactose), attached as a terminal sugar on a branch of the cell wall lipopolysaccharide. Other terminally attached dideoxyhexoses are responsible for the cross-reactions that characterize other groups of salmonellae. The determinants responsible for cross-reactions need not, however, be terminal residues. The strains of group E *Salmonella,* for example, cross-react because their cell wall lipopolysaccharides all possess nonterminal mannosyl-rhamnose residues as repeating units (Fig. 15-20).

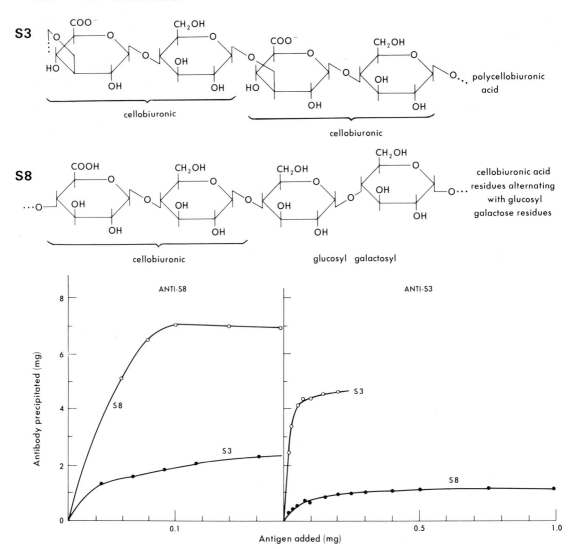

Fig. 15-18. Cross-reactions between type 3 and type 8 capsular polysaccharides of pneumococci. **Left:** Horse antiserum to type 8 pneumococcus, reacted with purified S8 and S3 polysaccharide. **Right:** Horse antiserum to type 3 pneumococcus, reacted with S3 and S8 polysaccharide. [Based on Heidelberger, M., Kabat, E. A., and Shrivastava, D. L., *J Exp Med 65*:487 (1937); and Heidelberger, M., Kabat, E. A., and Mayer, M., *J Exp Med 75*:35 (1942).]

Another example is provided by the pneumococcus. The capsular polysaccharide of type 3 is a linear polymer of repeating cellobiuronic acid residues (β-1,4-glucuronidoglucose), while in that of type 8 these residues alternate with glucosyl-galactose residues. Hence antisera to either Ag cross-react extensively with the other (Fig. 15-18).

A similar principle accounts for the cross-reactions that are commonplace with conjugated proteins; for example, antisera to 2,4-dinitrophenyl (Dnp)-bovine γ-globulin react with Dnp-human serum albumin because Dnp groups are present in both conjugates.

Cross-reactivity does not require that a functional group of the heterologous ligand be **identical** with the corresponding group of the immunogen; it need only be sufficiently **similar.** For example, Abs to *m*-azobenzenesulfonate cross-react with *m*-azobenzenearsonate (Fig. 15-4), and Abs to the 2,4-dinitrophenyl-lysyl group cross-react with

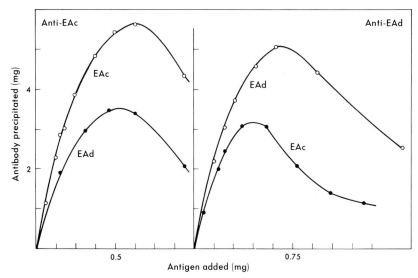

Fig. 15-19. Cross-reactions between chicken and duck ovalbumin, EAc and EAd, respectively. **Left:** Rabbit antiserum to EAc, reacted with EAc and EAd. **Right:** Rabbit antiserum to EAd, reacted with EAd and EAc. [Based on Osler, A. G., and Heidelberger, M. *J Immunol 60*:327 (1948).]

Fig. 15-20. Portion of the cell wall lipopolysaccharide of three strains of salmonellae. The strains are assigned to the same group (O) because of their serological cross-reaction, due to their common mannosyl rhamnose residues (boxed in). They are also distinguishable serologically because each has some unique structural feature, e.g., the terminal glucose (G) residue in *S. minneapolis,* and the α or β glycosidic bond linking galactose (Gal) to mannose (M). [Based on Robbins, P. W., and Uchida, T. *Fed Proc 21*:702 (1962).]

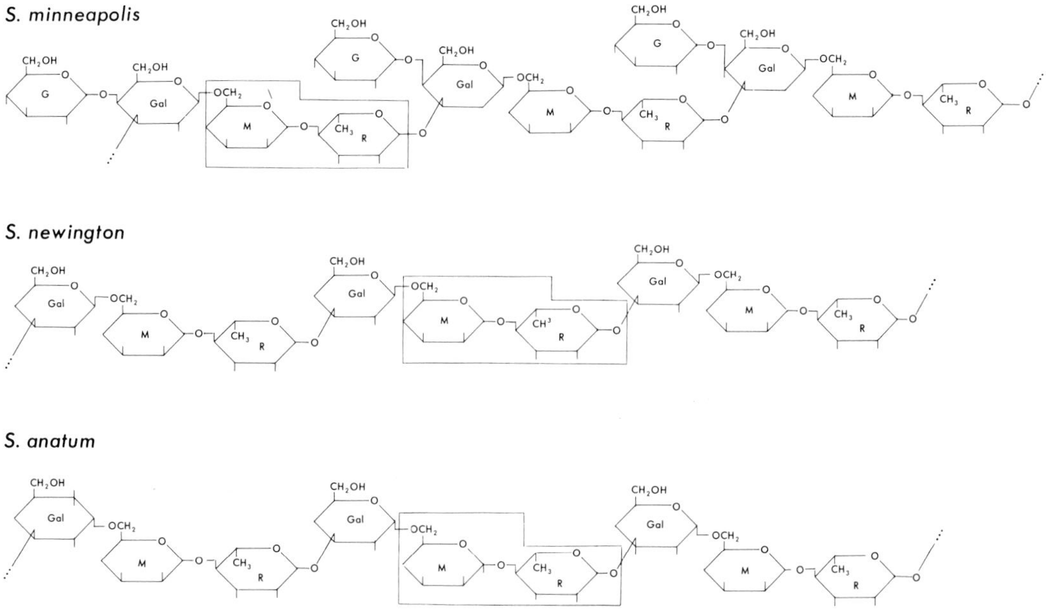

S. minneapolis

S. newington

S. anatum

2,4,6-trinitrobenzene. All Abs exhibit some cross-reactions: i.e., they bind some Ags with determinant groups that are not identical with those in the immunogen.

General Characteristics of Cross-Reactions. The following generalizations are drawn from the study of many cross-reactions.

1) An antiserum precipitates **more copiously** with its immunogen than with cross-reacting Ags (Figs. 15-18 and 15-19), because a heterologous ligand usually reacts only with some proportion of the total Ab to the immunogen. In addition, purified Abs of a given specificity will almost always have **greater affinity** for the homologous ligand than for cross-reacting ligands (Figs. 15-6 and 15-7).

2) The mutual cross-reactions between a pair of Ags are usually not quantitatively equivalent; Abs to the first Ag may react more extensively with the second Ag than Abs to the second react with the first (Fig. 15-18).

3) Different antisera to a given immunogen are likely to vary in the extent of their cross-reactions with diverse heterologous Ags, even when the antisera are obtained from the same animal (at different times after immunization, Ch. 17).

4) Cross-reacting ligands tend to be bound more strongly by Abs with high affinity than by those with low affinity for the homologous ligand. Thus, fewer cross-reactions are exhibited by low- than by high-affinity Abs (see Specificity and affinity, above). Since polysaccharide-antipolysaccharide systems are, in general, characterized by low affinities (see above), this rule may account for the extraordinary specificity of the sera used in typing bacteria and red blood cells according to the saccharides on their surfaces (Ch. 21).

Comparison of Various Types of Cross-Reactions. The cross-reactions discussed above may be contrasted as follows. 1) The presence of antigenic impurities is trivial conceptually but of considerable importance technically in practical applications of serological reactions. 2) The presence of **identical** functional groups in different Ags is particularly relevant for those groups that are relatively insensitive to the fine structure of macromolecules, such as lactose, 2,4-dinitrophenyl, etc. Groups of this type are usually

found in polysaccharides, nucleic acids, and conjugated proteins. 3) Cross-reactions based on **similarity** rather than identity of determinants are probably of particular importance among proteins, whose determinant groups are mainly conformational (see Hapten inhibition, above) and not likely to be precisely duplicated in different proteins. While the second and third types of cross-reactions may be distinguished formally, it is not usually possible to differentiate between them experimentally.

Removal of Cross-Reacting Antibodies (Adsorption and Absorption). It is usually necessary to remove certain cross-reacting Abs before an antiserum is sufficiently **monospecific** for use as an analytical reagent, e.g., in typing bacteria, measuring enzyme levels in bacterial extracts, etc. Removal is accomplished simply by allowing the antiserum to react with the appropriate cross-reacting Ags. Large complexes (e.g., precipitates or Abs bound to cells) are easily removed along with the cross-reacting Abs. When, however, the complexes are soluble and difficult to remove, cross-reacting ligand may be added in large excess to saturate the cross-reacting Abs, eliminating them functionally without excluding them physically. An effective alternative is to pass the serum over a column of agarose beads to which the cross-reacting Ags are covalently attached, removing the corresponding Abs. We shall use the terms **adsorption** and adsorbed serum when Abs are removed by specific binding on the surface of **particulate** Ags (e.g., red cells, bacteria, agarose beads coated with Ag), and **absorption** and absorbed serum when Abs are neutralized by reaction with **soluble** Ags.

AVIDITY

Early in this chapter we saw that an Ab's affinity for univalent ligands is measured by the intrinsic association constant, K, which is the property of a representative **single** site (see also Appendix, below). However, Ab molecules and most Ag particles are multivalent, and their tendency to pair depends not only on intrinsic affinities (per site), but also upon the **number** of sites involved per reactant. For instance, the equilibrium constant in the formation of multivalent monogamous Ab:Ag complexes (above) can exceed by perhaps 100,000-fold the intrinsic equilibrium constant of the Ab for the homologous univalent ligand. In most reactions the number of functional groups on Ag particles is usually obscure and different ones are likely to be engaged by different sets of Abs (see Fig. 15-12); in addition, the pairing of Ab and Ag can also be influenced by nonspecific factors involved in aggregation or close-packing of macromolecules.

Because of these complexities the term **avidity** is used to denote the over-all tendency of Abs to combine with Ag particles, reserving the term **affinity** for the intrinsic association constant that characterizes the binding of a univalent ligand (see Appendix, below).

Differences in the avidity of different antisera for the same Ag can sometimes be demonstrated simply by diluting antiserum-Ag mixtures. The concentrations of free Ab and free Ag are reduced, causing dissociation of the immune complexes: **dissociation is less evident with more avid antisera.** This procedure is particularly useful when the unbound Ag can be measured with high sensitivity (as toxin or infectious virus). Variations in the shape of precipitin curves also reveal differences in avidity. For example, in Figure 15-21 antiserum A is more avid than B, which is more avid than C, etc.

Antisera can differ in avidity because their Abs differ in intrinsic affinity for a given determinant (Fig. 15-21); or they might differ in the number of sets of Abs for diverse determinants of the Ag, or in the proportion of Ab molecules that can engage in monogamous binding. Abs with more available combining sites also have a greater tendency to bind ligands than Abs with fewer available sites, even when both Abs have the same intrinsic association constant (see Appendix, especially equation 16); hence sera can differ in avidity if they have different proportions of Abs with 2 or 10 combining sites per molecule (see Ch. 16, IgG and IgM). Avidity differences among antisera are commonplace and easily recognized, but the basis for the differences is usually obscure.

THE FLOCCULATION REACTION

The precipitin reactions of certain antisera differ from the classic precipitin reaction in that insoluble aggregates are not observed until the amount of Ag added exceeds some relatively large value (cf. Figs. 15-18, 15-19, and 15-22). In these anomalous **flocculation** reactions precipitation is inhibited by extreme Ab excess as well as by Ag excess: precipitation is observed only over a narrow range of Ab/Ag ratios.

Flocculation reactions are regularly given by certain horse antisera, particularly those prepared against diphtheria toxin and some streptococcal toxins; but horse antisera to most other proteins and to polysaccharides give the usual precipitin re-

Fig. 15-21. Precipitin curves showing differences in avidity of four antisera for the same Ag. The sera were prepared against 2,4-dinitrophenyl (Dnp)-bovine γ-globulin and tested with human serum albumin substituted with about 30 Dnp groups per molecule (Dnp-HSA). The order of avidity of the sera is A>B>C>D. A similar order in affinity for ε-Dnp-lysine was found for anti-Dnp molecules isolated from each of the sera: K_o in liters mole^{-1} were: (A) $> 10^8$, (B) 1×10^7, (C) 5×10^6, and (D) 1×10^6. (From Steiner, L. A,. and Eisen, H. N. In *Immunological Diseases*. M. Samter, ed. Little, Brown, Boston, 1965.)

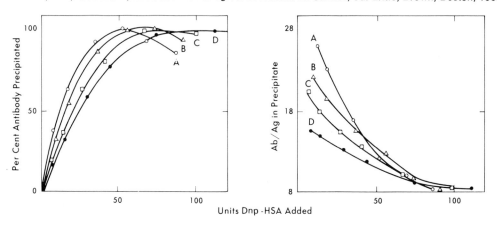

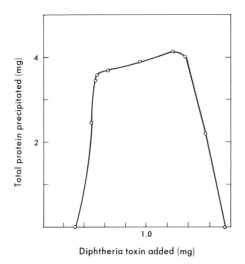

Fig. 15-22. Precipitin reaction of the flocculation type. The indicated amount of toxin was added to one ml of horse antiserum to diphtheria toxin. [Based on Pappenheimer, A. M., Jr., and Robinson, E. S. *J Immunol* 32:291 (1937).]

actions. The explanation for the special features of the flocculation reaction must lie in the properties of the Abs involved, rather than the Ags, because the same Ags (and all others tested) give only typical precipitin reactions with rabbit antisera. Some human antisera to thyroglobulin give flocculation reactions.

The basis for the flocculation reaction has not been clarified. Possibly the involved Abs are unusually soluble or the special antisera that give this reaction contain some high-affinity, nonprecipitating Abs (see Monogamous binding, above), and it is only after they are saturated that the remaining conventional Abs can react and precipitate with additional Ag.

PRECIPITIN REACTION IN GELS

When Abs and Ag are introduced into different regions of an agar gel they diffuse freely toward each other and form readily visible opaque bands of precipitate at the junction of their diffusion fronts. Simple and ingenious applications of this principle provide powerful methods for analyzing the multiplicity of Ab-Ag reactions within a system. The most widely used methods are illustrated in Figure 15-23 and are described below.

Single Diffusion in One Dimension. This procedure, developed by Oudin in France, is generally performed by placing a solution of Ag over an antiserum that has been incorporated in a column of agar gel. By diffusion, a concentration gradient of Ag advances down the agar column (provided the concentration of Ag in the upper reservoir is high relative to the concentration of Ab in the agar phase), and a precipitate forms in the agar at the advancing diffusion front. This precipitate extends upward to that level in the gradient at which Ag excess is sufficient to prevent precipitation. With continuing diffusion of Ag from the reservoir, the leading edge of the precipitate advances downward. The trailing edge of the precipitate likewise advances, since the additional Ag migrating into the

Fig. 15-23. Some arrangements for gel diffusion precipitin reactions. **A.** Single diffusion in one dimension. **B.** Double diffusion in one dimension (Preer method). **C.** Double diffusion in two dimensions. Arrows show direction of diffusion. Stippled areas are opaque precipitation bands.

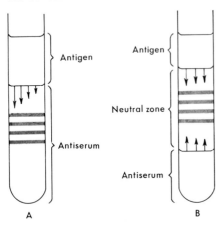

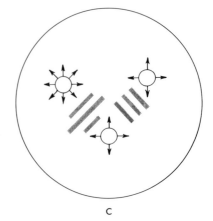

region of the specific precipitate dissolves it by forming soluble complexes, as expected from the precipitin reaction in liquid. Thus, a **band** of precipitate migrates down the column of agar. The distance traveled is proportional to the square root of time, in accord with the laws of diffusion.

The rate of migration depends on the diffusion coefficient of the Ag (which varies with molecular weight and shape) and on its **concentration** in the upper reservoir (Fig. 15-23A). Accordingly, when several Ags diffuse into an antiserum that can react with each of them, several bands of precipitation are observed, and each migrates at a distinctive rate. It is possible, though, that different Ab-Ag systems will form overlapping bands that migrate at indistinguishable rates; hence the number of bands observed represents the **minimum** number of Ab-Ag systems in the substances being analyzed.

The rate of band migration varies with the concentration of Ag introduced, and the optical density of a precipitate (which can be measured by suitable photometers) depends on the concentration of Ab in the antiserum. Hence it is possible, with the use of appropriate standards (solutions of known Ag concentration and antisera of known Ab concentration), to measure the concentration of Ag in unknown solutions and the concentration of precipitating Abs in antisera.

The arrangement shown in Figure 15-23A can be reversed by incorporating Ag in the agar column below and overlaying it with antiserum. If the concentration of Abs greatly exceeds the concentration of relevant Ags, precipitation takes place at the interface and migrates into the lower gel phase. This arrangement can distinguish between precipitin and flocculation reactions (Fig. 15-24).

Double Diffusion in One Dimension. In this procedure the antiserum in agar is overlaid by a column of clear agar which in turn is overlaid by Ag, added either as liquid or incorporated into agar. Ag and Ab molecules diffuse across the respective interfaces into the clear neutral zone and advance toward each other. At that junction of their diffusion fronts where Ab and Ag are in equivalent proportions a precipitation band forms, and it increases in density with time as more Ab and Ag molecules continue to enter this zone in optimal (i.e., equivalent) proportions (Fig. 15-23B). If the Ag and Ab are added to their respective reservoirs in proportions that correspond to the equivalence zone in solution, the precipitation band has maximal sharpness and is

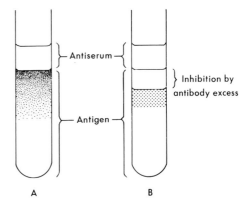

Fig. 15-24. Gel diffusion illustrating differences between the usual type of precipitin reaction **(A)** and the flocculation reaction **(B)**. Precipitation is inhibited by excess Ab in the flocculation reaction but not in the precipitin reaction. Compare Ab-excess zones of Figures 15-17, 15-18, and 15-21.

stationary. If, however, either is added in relative excess, the band migrates slowly away from the excess.

This method can detect as little as 10 μg Ab per milliliter. It is particularly valuable for determining the number of Ab-Ag systems in complex reagents. For example, with highly purified diphtheria toxin and a human antiserum to diphtheria toxin as many as six bands of precipitation have been observed. These same reactants in liquid solution would very likely have displayed a precipitin curve with a single zone of maximum precipitation; with careful supernatant analysis it might have been possible to recognize that more than one Ab-Ag system was involved, but not how many.

Double Diffusion in Two Dimensions. Additional insight can be obtained by the simple but elegant procedure, developed mainly by Ouchterlony in Sweden, of placing Ag and Ab solutions in separate wells cut in an agar plate (Fig. 15-23C). A large number of geometric arrangements are possible; some of the simpler ones are shown in Figures 15-25 to 15-27. The reactants diffuse from the wells, and precipitation bands form where they meet at equivalent proportions. If the concentration of Ab introduced is in relative excess over Ag, the band forms closer to the Ag well; and the converse occurs if the Ag is introduced in relative excess.

The arrangements shown in Figures 15-25 to 15-27 are particularly useful for comparing Ags for the presence of identical or cross-reacting components. If a solution of Ag is

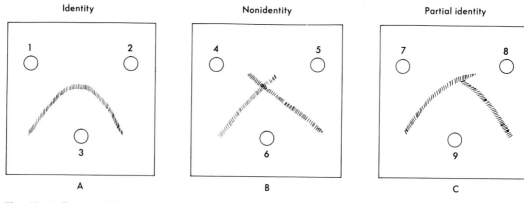

Fig. 15-25. Double diffusion precipitin reactions in agar gel illustrating reactions of identity, nonidentity, and partial identity. In **A** the same Ag was placed in wells 1 and 2, and the antiserum was in well 3. In **B** different Ags were placed in wells 4 and 5, and antisera to both were placed in well 6. In **C** an Ag and its antiserum were placed in wells 7 and 9, respectively, and a cross-reacting Ag was placed in well 8.

placed in two adjacent wells and the corresponding Ab is placed in the center well, the two precipitin bands eventually join at their contiguous ends and fuse (Fig. 15-25A). This pattern, known as the **reaction of identity,** is seen whenever indistinguishable Ab-Ag systems react in adjacent fields. If, on the other hand, unrelated Ags are placed in adjacent wells and diffuse toward a central well that contains Abs for each, the two precipitin bands form independently of each other and cross (**reaction of nonidentity,** Fig. 15-25B). If, however, the Ag in one of the wells and the antiserum in the central well constitute a homologous pair, and the Ag in an adjoining well is a cross-reacting Ag, the precipitation bands fuse, but in addition form a spur-like projection that extends toward the cross-reacting Ag (**reaction of partial identity** or **cross-reaction,** Fig. 15-25C). From what is known of precipitin reactions in liquid the spur can be readily interpreted: it represents the reaction between homologous Ag and those Ab molecules that do **not** combine with the cross-reacting Ag and hence diffuse past its precipitation band. Since these non-cross-reacting Abs represent only a fraction of the total Ab involved in the homologous precipitin reaction (see Figs. 15-18 and 15-19, for example), the spur is usually less dense than the band from which it projects and tends to have increased curvature toward the antiserum well (Fig. 15-25C).

In the event that neither of the Ags is homologous with respect to the antiserum (i.e., neither is the immunogen) a pattern of partial identity might be observed, with the spur projecting toward the less reactive Ag; or there might be partial fusion with two crossing spurs, indicating that some Abs react only with one of the cross-reacting Ags and

Fig. 15-26. The use of purified Ags to identify components of a complex series of precipitin bands. The center well (AS) contains rabbit antiserum prepared against unfractionated human serum. Wells a and c contain human serum; well b has purified human serum albumin (HSA); and well d has a purified human immunoglobulin (HIg). Thus, bands 4 and 6 correspond to HIg and HSA, respectively.

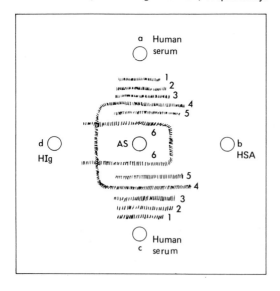

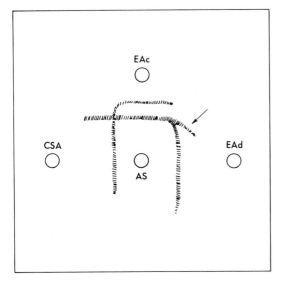

Fig. 15-27. An example of the use of gel diffusion to discriminate between two types of cross-reactions. In liquid medium an antiserum (AS) prepared against crystallized chicken ovalbumin (EAc) precipitates copiously with EAc and less well with crystallized duck ovalbumin (EAd; see Fig. 15-19); it also precipitates slightly with crystallized chicken serum albumin (CSA). From the gel precipitin pattern shown it is concluded that 1) EAc is contaminated with CSA, hence the antiserum contains some antibodies to CSA; 2) CSA and the main component of EAc are unrelated antigenically (reaction of nonidentity); and 3) some antigenic determinants of EAd are similar to or identical with some determinants of EAc (reaction of partial identity). The Abs that cause the spur (arrow) are those that in the quantitative precipitin reaction of Figure 15-19 precipitate with EAc but not with EAd.

some only with the other. Unless inspected carefully for attenuation and curvature of the spurs, double spur formation can be mistakenly interpreted to mean that the adjacent systems are unrelated.

As in diffusion in one dimension, if one well contains a mixture of different Ags and the facing well contains Abs to several of them, the number of bands formed between the wells represents the minimum number of Ab-Ag systems involved. These bands may be identified if the appropriate pure Ags are placed in adjacent wells, as illustrated in Figure 15-26.

Double diffusion in two dimensions provides a simple means for evaluating, to a limited extent, the basis for cross-reactions observed in liquid media. Thus, of the several classes of cross-reactions discussed earlier, that which arises from a common impurity can usually be recognized unambiguously, as is shown in Figure 15-27. On the other hand, when two purified Ags, such as chicken and duck ovalbumins (Fig. 15-19), give rise to a cross-reaction it is not possible by gel diffusion analysis to decide whether their common determinant groups are identical or only similar; in both cases partial fusion would be observed, with projection of the spur toward the cross-reacting Ag (Fig. 15-27).

Because rates of diffusion vary inversely with molecular weight, the curvature of the precipitin band also provides a clue to the molecular weight of the Ag (provided Ag and Ab are present in roughly equivalent amounts). If the Ag and Ab have about the same molecular weight the precipitation band appears as a straight line; if the Ag has a higher molecular weight the band is concave toward the Ag source; and if the Ag is of lower molecular weight the band is concave toward the Ab reservoir.

Radial Immunodiffusion. This procedure is carried out in a layer of agar (e.g., on a glass slide) containing a uniformly dispersed monospecific antiserum. Diffusion of Ag from a well cut in the agar leads to the formation of a ring of precipitation, whose diameter is proportional to the initial concentration of the Ag (Fig. 15-28). From the results obtained with known concentrations a calibration curve is constructed, permitting quantitation of the concentration of the Ag. This simple procedure has been widely adopted for measurement of many Ags (e.g., immunoglobulins in diverse biological fluids).

IMMUNOELECTROPHORESIS

By combining electrophoresis with precipitation in agar, Grabar and Williams developed a simple but exceedingly powerful method for identifying Ags in complex mixtures (Fig. 15-29). The mixture of Ags is introduced into a small well in agar that has been cast on a plate, say an ordinary microscope slide. By applying an electric field across the plate for 1 to 2 hours the proteins are made to migrate, each according to its own electrophoretic mobility. The electric gradient is then discontinued, and antiserum is introduced into a trough whose long axis

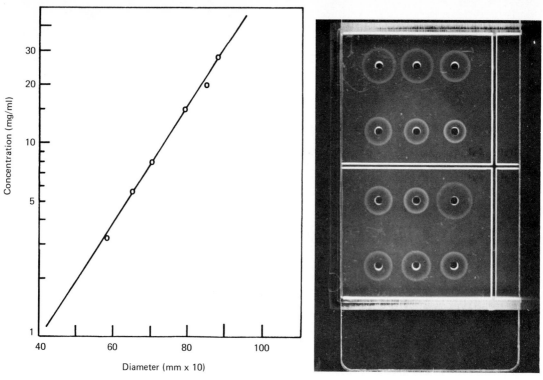

Fig. 15-28. Radial immunodiffusion measurement of Ag concentrations. A standard preparation of Ag, human immunoglobulin G (HIgG), was added at six different concentrations to wells in agar containing goat antiserum to HIgG (upper panel). The diameters of the resulting circular precipitates are plotted on the left. The lower panel (right) shows six human sera with different concentrations of HIgG. (Courtesy of Dr. C. Kirk Osterland.)

Fig. 15-29. Immunoelectrophoresis. **A.** A thin layer of agar gel (ca. 1 to 2 mm) covers a glass slide, and a small well near the center (marked origin) receives a solution containing various Ags. After electrophoresis of the Ags the current is discontinued and antiserum is added to the trough. Precipitation bands form as in double diffusion in two dimensions. The apex of each precipitin band corresponds to the center of the corresponding Ag. **B.** Human serum, placed at the origin, was analyzed with an antiserum prepared against unfractionated human serum. (Courtesy of Dr. Curtis Williams.)

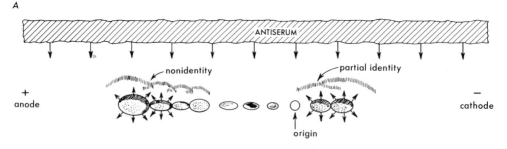

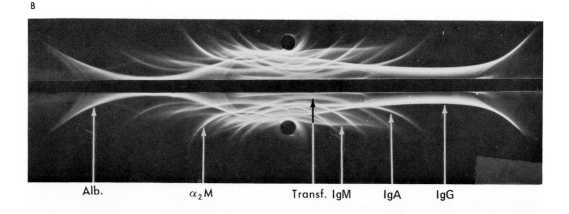

parallels the axis of electrophoretic migration. The Abs and Ags now diffuse toward each other, and precipitin bands form at the intersection of their diffusion fronts. The principles involved in the precipitation stage are those described above: the familiar reactions of identity, nonidentity, and partial identity may be seen.

Immunoelectrophoresis has revealed as many as 30 different Ags in human serum. Many applications of this method appear in subsequent chapters.

REACTIONS WITH PARTICULATE ANTIGENS

THE AGGLUTINATION REACTION

Bacterial and other cells in suspension are usually clumped (agglutinated) when mixed with antisera prepared against them. The principles involved are fundamentally the same as those described above for reactions with soluble Ags. Nevertheless, agglutination requires special consideration, for it is widely used as a simple and rapid method for identifying various bacteria, fungi, and types of erythrocytes; and, conversely, with the use of known cells it provides a simple test to detect and roughly quantitate Abs in sera.

Mechanisms of Agglutination. Agglutination is carried out in physiological salt solution (0.15 M NaCl). The ionic strength is important, for at neutral pH bacteria ordinarily bear a net negative surface charge which must be adequately damped by counter-ions before cells can approach each other closely enough for Ab molecules to form specific bridges between them. Hence even with Abs bound specifically to bacterial cell surfaces agglutination may not occur if the salt concentration is too low (e.g., $< 10^{-3}$ M NaCl). Conversely, the addition of sufficient salt can lead to agglutination even in the absence of Abs.

When a mixture of readily distinguishable particles, such as nucleated avian erythrocytes and nonnucleated mammalian erythrocytes, is added to a mixture of their respective antisera, each clump that forms consists of cells of one or the other type (Ch. 16, Fig. 16-10). Thus, as expected from the lattice theory (above), **each cell-Ab system agglutinates independently of the others in the same mixture.** The basic similarity between agglutination and precipitation is also brought out by the quantitative agglutination reaction, which measures the amount of Ab adsorbed by bacteria; e.g., the maximum amount taken up by encapsulated pneumococci is identical with the maximum amount precipitated from the same volume of serum by the cells' isolated, soluble capsular polysaccharide (see Unitarian hypothesis, below).

Titration of Sera. The agglutination reaction is widely used as a semiquantitative assay. A given volume of a cell suspension is added to a series of tubes, each with the same volume of antiserum at a different dilution, usually increasing in twofold steps. The reaction is speeded up by shaking and warming to 37° (sometimes to 56°); then, after the cells have been allowed to settle, or have been centrifuged lightly, the clumping is detected by direct inspection. The relative strength of an antiserum is expressed as the reciprocal of the highest dilution that causes agglutination. If, for example, a 1:512 dilution gives perceptible agglutination but a 1:1024 dilution does not, the titer is 512.

Agglutination titers are not precise ($\pm 100\%$), but they are easily obtained and provide valid indications of the **relative** Ab concentrations of various sera with respect to a particular strain of bacteria. Hence agglutination titrations are of immense practical value in following changes in Ab titer during the course of acute bacterial infections (see below, Diagnostic application of serological tests).

Agglutination titers obtained with **different** sera and their respective Ags are, however, not necessarily comparable: for example, an antiserum to type 1 pneumococcus with 1.5 mg of Ab per milliliter agglutinated these organisms at a dilution of 1:800, while an antiserum to type 1R pneumococcus with 9.6 mg of Ab per milliliter agglutinated the latter organisms to a titer of only 1:80. Apparently the number and distribution of determinant groups on the bacterial cell surface can markedly influence titer.

Agglutination reactions are, in principle, neither more nor less specific than other serological reactions. However, they present difficulties in actual practice because a cell surface possesses a great diversity of antigenic determinants, of which some will usually cross-react with different but related cells. Hence in order to achieve a high level of specificity it is nearly always necessary to adsorb antisera with sufficient amounts of appropriate cross-reacting cells.

Surface vs. Internal Antigens of Cells. When a bacterial, fungal, or alien animal cell is introduced as an "immunogen" it is broken up in the host animal and many of its surface and internal components (of cytoplasm and nucleus) are immunogenic. However, the Abs that cause agglutination are those specific for surface determinants, which are often called **agglutinogens** (i.e., they induce formation of **agglutinins,** the agglutinating Abs). Surface Ags are much more potent immunogenically when administered as part of a morphologically intact cell that when given in purified form, possibly because in the former condition they are more likely to facilitate interactions among precursors of Ab-forming cells (Ch. 17, Immunogenicity).

Prozone. By analogy with the precipitin reaction it would be expected that agglutinating activity would decrease progressively with dilution of an antiserum. Curiously, however, some sera give effective agglutination reactions only when diluted several hundred- or thousandfold: when undiluted or slightly diluted they do not visibly react with the Ag. The latter region of the titration is called the **prozone,** as in tubes 1 to 3 below.

Tube No.	1	2	3	4	5	6	7	8	9
Serum dilution	1:8	1:16	1:32	1:64	1:128	1:256	1:512	1:1024	1:2048
Clumping	0	0	0	+	+	+	+	+	0

By means of labeled Abs, or the antiglobulin test described below, it can be shown that un-agglutinated cells in the prozone actually have Abs adsorbed on their surface. Indeed, it might be expected, on statistical grounds, that when Ab molecules are in great excess relative to the number of functional groups on the cells, the simultaneous attachment of both sites of individual bivalent Ab molecules to different cells would be improbable. Nevertheless, the prozone phenomenon is not due simply to Ab excess, but often involves special blocking or incomplete Abs.

BLOCKING OR INCOMPLETE ANTIBODIES

In certain sera with a pronounced prozone a portion of the total Ab appears to react with Ag particles in an anomalous manner: the bound Ab not only fails to elicit agglutination but actively inhibits it, as shown by subsequently mixing the particles with antiserum at a dilution that would otherwise evoke a brisk clumping reaction. These inhibitory Ab molecules are referred to as **blocking** or **incomplete** Abs, and they are particularly evident in certain human antierythrocyte sera (anti-Rh; see Ch. 21). Some sera, in fact, contain only blocking Abs; their presence is revealed by the ability to inhibit specifically a standard agglutinating mixture.

Blocking Abs were once thought to be univalent. However, they must be bivalent, because they can agglutinate the cells to which they are adsorbed if the reaction is carried out under special conditions, e.g., in the presence of serum albumin at high concentration (instead of in conventional salt solutions) or if the red cells are first treated with a proteolytic enzyme (trypsin or ficin). It is possible that blocking Abs can engage in monogamous binding (to repeating antigenic groups of the cell surface), like some nonprecipitating Abs in the precipitin reaction. (Some entirely different Abs are also called "blocking Abs" for a different reason: they seem to protect the cells on which they are adsorbed against destructive attack by specifically reactive lymphocytes; see Chs. 20, 21, Enhancement in tumor immunity and in allograft reaction.)

The Antiglobulin Test. An ingenious method for detecting incomplete or blocking Abs,

Fig. 15-30. The antiglobulin (Coombs) test for incomplete Abs. Cells coated with specifically bound, incomplete (non-agglutinating) Abs are clumped by other Ab molecules (from another species), which react specifically with antigenic determinants of the incomplete Abs.

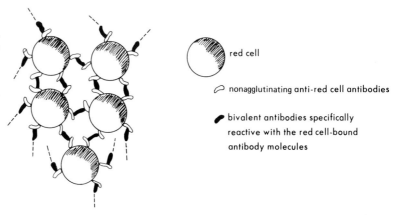

○ red cell

◠ nonagglutinating anti-red cell antibodies

➤ bivalent antibodies specifically reactive with the red cell-bound antibody molecules

called the Coombs or antiglobulin test, is of considerable importance in the recognition of certain hemolytic diseases. The test exploits the ability of Ab molecules to participate simultaneously in binding to an Ag on the red cell surface and in complexing with Ab to itself (Fig. 15-30). As we shall see subsequently (Ch. 16), Abs are highly immunogenic in a foreign species.

For example, rabbits injected with human immunoglobulins (HIg) form anti-HIg that reacts with most HIg molecules, regardless of their specificities as Abs (Ch. 16). Hence erythrocytes with incomplete human Abs bound specifically to their surface can be agglutinated by rabbit anti-HIg serum. Since an Ab molecule can function both as an Ab and as a ligand in two independent and simultaneous reactions, the ligand-binding sites of Abs must be different from, or represent only a fraction of, their numerous and diverse antigenic determinant groups (Ch. 16).

PASSIVE AGGLUTINATION

In the agglutination reactions described above the functional ligand groups are components of cell wall or cell surface membrane. It has also been possible to extend the agglutination reaction to a wide variety of soluble Ags by attaching them to the surface of particles. In such **passive agglutination** reactions the particles most widely used are erythrocytes **(passive hemagglutination),** a synthetic polymer such as polystyrene, or a mineral colloid such as bentonite. Adsorption ordinarily depends on noncovalent bonds and is generally achieved by simply mixing the particles with the Ags. Thus erythrocytes readily adsorb many polysaccharides. For the attachment of proteins, however, it is usually necessary first to treat the cells with **tannic acid** or chromic chloride (whose mechanisms of action are not established). Covalent linkage of proteins to the red cell surface has also been achieved by the use of bifunctional cross-linking reagents (Fig. 15-31).

As in conventional agglutination tests, passive agglutination is highly sensitive (see below), but its precision is low (at best ± 100%); and Ab levels are measured only in relative terms, not in weight units.

Inhibition of passive agglutination provides a sensitive assay for Ags, and it has been widely used for this purpose to measure certain hormones in plasma and urine (e.g., urinary chorionic gonadotropin in a rapid test for pregnancy). For example, polystyrene particles or tanned red cells with an adsorbed purified protein or polypeptide hormone can be agglutinated by the corresponding antiserum; and added free ligand specifically inhibits agglutination by competing for Ab. Thus, by comparison with standard Ag solutions, the concentration of Ag in biological fluids, e.g., blood or urine, can be evaluated (see Radioimmunoassays, below).

DIFFERENCES IN SENSITIVITY OF PRECIPITATION, AGGLUTINATION, AND OTHER REACTIONS

The methods used for measuring Ab levels differ considerably in the minimal amounts of Ab that can be detected and measured. Of the two considered so far in this chapter, aggluti-

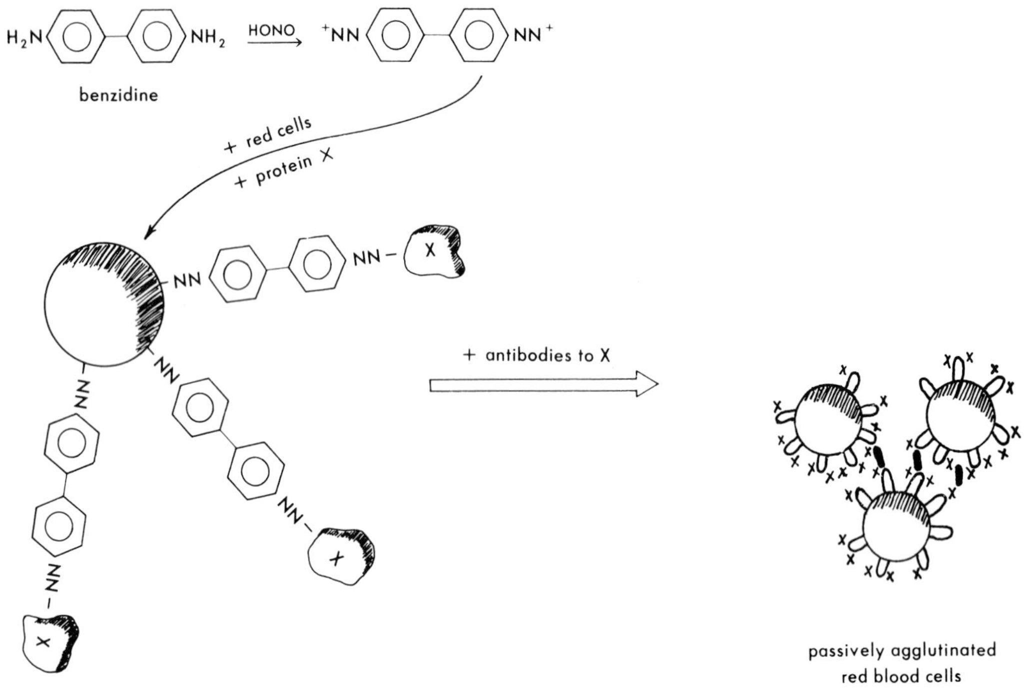

Fig. 15-31. Passive hemagglutination. Bis-diazotized benzidine is used to attach protein X to the red cell surface. Red cells coated with X are specifically clumped by anti-X Abs. The antiglobulin test shown in Figure 15-30 may be regarded as another form of passive hemagglutination, the Ag attached to the red cell surface being an adsorbed, noncovalently bound, incomplete Ab molecule. Besides benzidine, a variety of other bifunctional reagents have been used, e.g., toluene-2,4-diisocyanate (Fig. 15-37).

nation is considerably more sensitive than precipitation: the relatively voluminous particle, coated with a thin layer of Ag, serves to **amplify** the reaction.

Thus, precipitin reactions in liquid media or gels are usually not observed when antisera are diluted more than ten- to fiftyfold, while many antisera to bacteria or red blood cells retain agglutinating activity after being diluted many thousandfold. Passive agglutination, especially passive hemagglutination, is particularly sensitive: some mouse antisera to hemocyanin have titers of almost 10^5.

The difference in sensitivity is clearly illustrated when agglutination and precipitation involve the same Ag. In one instance, for example, an antiserum to hen ovalbumin (EA) lost ability to precipitate EA when diluted 1:5, but at 1:10,000 dilution it could still agglutinate collodion particles coated with EA. These large differences are probably related to the difference in the number of particles needed for a visible reaction; e.g., about 10^7 bacterial cells (corresponding to perhaps 10^9 surface Ag mole-

cules), but about 6×10^{12} molecules of soluble Ag (1 μg of a protein of MW 100,000). The sensitivities of some routine assay methods for Abs are compared in Table 15-5.

TABLE 15-5. Minimal Concentrations of Antibody Detectable by Various Quantitative and Semi-Quantitative Methods
(Approximate Values)

Method	μg antibody/ml
Precipitin reactions in liquid media	20
Precipitin reactions in agar gel	60
Bacterial agglutination	0.1
Passive hemagglutination	0.01
Complement fixation*	1
Passive cutaneous anaphylaxis†	0.02
Phage neutralization‡	0.001–0.0001

* See Chapter 18.
† See Chapter 19.
‡ See Rates of reaction, above, this chapter.

RADIOIMMUNOASSAYS

The most versatile and sensitive of the various assays that use Ags or haptens with radioactive labels was introduced by Berson and Yalow for measurement of serum insulin concentrations. The method, which can measure trace concentrations of any substance that can serve as Ag or hapten (Table 15-6), is based on competition for Ab between a standard radioactive indicator ligand (*L) and its unlabeled counterpart (L) at unknown concentration in the test sample: the higher the concentration of unlabeled competitor the lower the ratio of bound (B) to unbound or free (F) indicator (B/F). Specific precipitates do not form even if the ligand is multivalent, because Ags and ligands are used at exceedingly low concentrations. Nevertheless, bound and free ligands are readily separated (and counted) by using either special technics for particular ligands (e.g., chromatography, electrophesis, adsorption of certain free ligands on talc) or the general method in which soluble Ab-*L complexes are specifically precipitated with antisera to the Ab moiety of the immune complex (Fig. 15-32).

The double Ab approach is feasible because, as noted before, the Abs of one species (donor) are usually immunogenic in other species: the resulting antisera (anti-antibodies) react with essentially all Abs of the donor species, regardless of their ligand-binding specificities (e.g., rabbit antihuman immunoglobulins in the antiglobulin test,

above; see also Ch. 16). The concentration of unlabeled ligand in the test sample is determined by comparison with the inhibitory effect of known standards (Fig. 15-33).

In the double Ab reaction the first Ab (antiligand) is added at a level that enhances competition between labeled and unlabeled ligand, i.e., the Ab/L ratio corresponds to the Ag-excess side of the equivalence zone in a liquid precipitin reaction (see Fig. 15-8). The second Ab (anti-Ab) is introduced in excess to ensure complete precipitation of the first Ab and its complexes with ligand (i.e., Ab-excess side of the equivalence zone in Fig. 15-8). The method is exceedingly discriminating because it is possible to choose antisera with great specificity for the test ligand. Its sensitivity is limited only by the amount of radioactivity that can be introduced into the indicator ligand: with carrier-free radioactive iodine as extrinsic label it is possible to detect ligands at almost the picogram level (e.g., at ca. 10^{-10} gm). However, the method is sometimes limited by difficulty in determining whether the inhibitor in the unknown is identical with the radioactive indicator or only similar enough to cross-react.

Ammonium Sulfate Precipitation. Abs and soluble Ag-ligand complexes are precipitated by ammonium sulfate at high concentration (40% of saturation), but many unbound ligands are soluble at this ionic strength. Hence addition of the salt at this concentration to a mixture of Ab and radioactive ligand can provide a measure of bound ligand if the specific

TABLE 15-6. Substances Measured by Radioimmunoassays: A Partial List

Insulin	Testosterone
Growth hormone	Estradiol
Parathyroid hormone	Aldosterone
Adrenocorticotropic hormone	Intrinsic factor (see Ch. 20)
Glucagon	Digitalis
Secretin	Cyclic AMP (cAMP), cGMP, cIMP, cUMP
Vasopressin	Prostaglandins
Bradykinin (see Ch. 19)	Australia antigen (see Ch. 16)
Thyroglobulin	Carcinoembryonic antigen (see Ch. 21)
Gastrin	Human IgE (see Ch. 19)
Calcitonin	Morphine

Each substance corresponds to ligand X in Figure 15-32. The same principle has been applied in nonimmune assays, in which Abs are replaced by particular proteins; for example, vitamin B_{12} is measured by its binding to intrinsic factor.

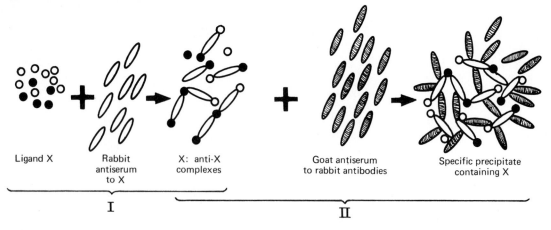

Ligand X   Rabbit   X: anti-X   Goat antiserum   Specific precipitate
            antiserum   complexes   to rabbit antibodies   containing X
            to X

I   II

Fig. 15-32. Radioimmunoassay. The Ab-ligand complexes formed in step I are separated from unbound ligand by specific precipitation in step II with goat antiserum prepared against rabbit Abs. The effect of competition between labeled (●) and unlabeled (○) ligand molecules is illustrated in Figure 15-33.

binding reaction does not depend on ionic interactions (i.e., is not affected by high ionic strength). This approach has been used in the **Farr technic** to measure Ab concentrations in serum: by titrating the serum sample with increasing amounts of labeled ligand (or testing the ligand with increasing dilutions of the antiserum) the **Ag-binding capacity** of the serum can be established: 1 nmole of bound ligand is equivalent to 75 μg of Ab (i.e., one occupied site in Ab corresponds to half the molecular weight of 150,000, see Fig. 15-2). This approach has also been used with radioactive univalent ligands to estimate intrinsic association constants. Even though the high salt concentration might modify the over-all Ab structure (e.g., cause the molecule to become highly insoluble) the ligand-binding site seems to be largely unperturbed: under these conditions the affinity for at least some ligands has been close to that measured in solution by equilibrium dialysis (see above).

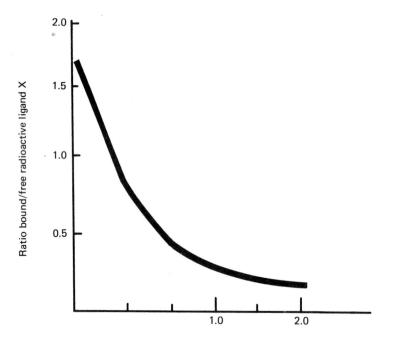

Fig. 15-33. Calibration curve for a radioimmunoassay. A fixed amount of radioiodine-labeled ligand X competes with various amounts of unlabeled ligand X for a limiting amount of anti-X Abs.

Concentration of unlabeled ligand X (ng/ml).

FLUORESCENCE AND IMMUNE REACTIONS

The fluorescent* properties of Ab molecules and of certain organic residues that can be attached to them provide the basis for a number of analytical methods. The most important of these is called **immunofluorescence.** Introduced by Coons, this method is widely used for rapid identification of bacteria in infected materials and for the identification

and localization of cellular Ags (and Abs, Ch. 17).

IMMUNOFLUORESCENCE

Of the several reagents that introduce fluorescent groups into proteins, the most widely used with Abs is fluorescein isothiocyanate (Fig. 15-34). Abs substituted with one or two fluorescein residues per molecule are intensely fluorescent but retain their specific reactivity; e.g., they can specifically bind to bacteria in smears and tissue sections. After a specimen on a slide has been covered for several minutes with a solution of fluorescein-labeled Abs (usually in the form of the globulin fraction of an antiserum), the slide is rinsed to remove

---

* When molecules absorb light they subsequently dispose of their increased energy by various means, one of which is the emission of light of longer wave length. When the emission is of short duration (10^{-8} to 10^{-9} seconds for return of the excited molecule to the ground state) the process is called **fluorescence.** (When the emission is of long duration the process is called phosphorescence.)

Fig. 15-34. Immunofluorescence. The fluorescent isothiocyanates **(1)** form thioureas, substituted with the fluorescent group and ε-NH_2 groups of lysine residues of Abs or other proteins **(2)**. In the **direct** staining reaction **(3)** the labeled Abs are specific for the Ag of interest. In the **indirect** reaction **(4)** the labeled Abs are specific for the Abs of another species (e.g., goat anti-rabbit immunoglobulin; see Ch. 16).

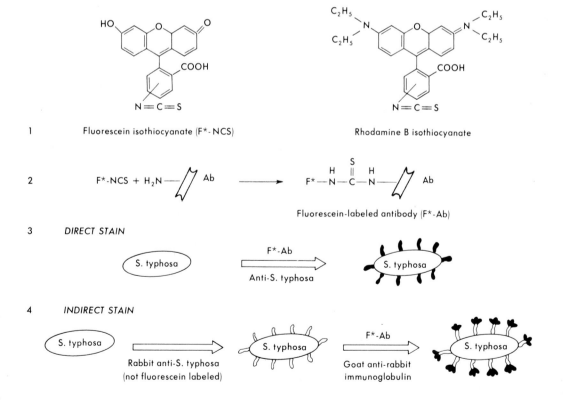

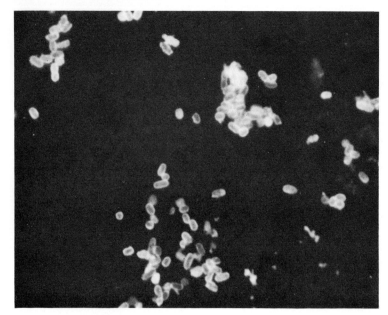

Fig. 15-35. Immunofluorescence staining. Stained with a fluorescein-labeled globulin fraction of a specific antiserum, *E. coli* 0127:B8 is clearly visible in a fecal smear from a patient with infantile diarrhea. Many other types of bacteria are abundant in the smear but are not stained. Ca. × 1000. (Photograph supplied by National Communicable Disease Center, Atlanta.)

unbound fluorescent protein and then is examined in the light microscope, with suitable light source and filters to provide the proper incident light. Since the emitted light (about 530 nm) is of longer wave length than the background incident light (about 490 nm), the Ab-coated cell stands out as a sharply visible yellow-green mass (Fig. 15-35).

As with all immunological reactions it is necessary to establish specificity with suitable controls. Thus staining should be blocked by preliminary treatment of the smear or tissue section with unlabeled Abs (to saturate antigenic sites) and by addition of excess soluble Ag (to saturate the labeled Abs).

Fluorescein-labeled Abs can be used in both direct and indirect reaction sequences, outlined in Figure 15-34, for the detection and localization of a wide variety of Ags. The indirect reaction both amplifies the response and permits one labeled preparation to be used as a stain for a wide variety of serologically specific reactions.

Different chromophores can be introduced into Abs with other reagents, e.g., orange-red fluorescence with rhodamine B isothiocyanate (Fig. 15-34). This modification provides opportunities for applying immunofluorescence to a number of special problems, e.g., to the possible coexistence of different Abs in the same cell (Ch. 17, Antibody formation).

In all these applications a recurrent problem arises from the nonspecific staining of tissues, especially by labeled proteins with more than two or three fluorescein residues per molecule. To minimize this difficulty ion-exchange chromatography and adsorption with acetone-dried tissue powders are used to remove the more highly substituted proteins. In addition, the use of high-titer antisera (e.g., > 2 mg of Ab per milliliter) reduces nonspecific staining by making it possible to use dilute solutions of labeled protein.

FLUORESCENCE QUENCHING

Like other proteins, unlabeled Ab molecules fluoresce in the upper ultraviolet region by virtue of their aromatic amino acid residues. Ab fluorescence is excited maximally at 290 nm and the emitted light has maximal intensity at about 345 nm (characteristic of the absorption and emission spectrum of tryptophan). When the binding sites in the Ab are specifically occupied by certain kinds of haptens (or Ags) the energy absorbed by the irradiated tryptophan residues is transferred to the bound ligand, which emits light at some still longer wave length or does not fluoresce at all. The net result is quenching or damping of the Ab's characteristic fluorescence. Thus the binding sites of a preparation of purified Abs can be titrated as shown in Figure 15-36; and the titrations can be used to calculate average intrinsic association constants, which agree with those determined by equilibrium dialysis. Quenching is particularly effective with those ligands whose absorption spectrum overlaps the protein's

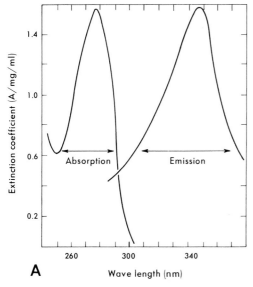
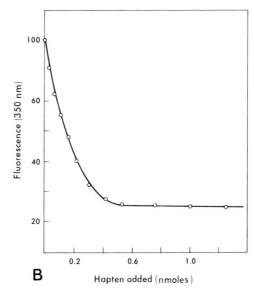

Fig. 15-36. Fluorescence quenching. **A.** Absorption and emission spectra of a purified Ab, specific for the 2,4-dinitrophenyl group. The height of the emission band maximum is set, by convention, equal to that of the absorption maximum, though only about 20% of the energy absorbed by Abs appears as fluorescence emission. **B.** Decline in fluorescence of the Ab as it binds ε-Dnp-lysine; about 75% of the fluorescence is quenched when the Ab's ligand-binding sites are saturated. [From Velick, S. F., Parker, C. W., and Eisen, H. N. *Proc Natl Acad Sci USA 46*:1470 (1960).]

emission spectrum; but the overlap need not be extensive.

The fluorescence quenching method is not as generally useful as equilibrium dialysis, since many haptens, such as simple sugars, do not quench. However, the method offers a number of advantages: it is rapid, requires only small amounts of Ab, and can be carried out with ligands that are too large to be dialyzable (for example, cytochrome c quenches Ab to cytochrome c).

FLUORESCENCE POLARIZATION

When fluorescent molecules are excited by a beam of polarized light the extent of polarization of the emitted light varies with the molecule's size: emission from small molecules, with much rotational motion, exhibits little polarization. If, however, the fluorescent molecule is bound to a relatively large particle, such as an Ab molecule, its rotational movement is restricted and the polarization of its fluorescence is increased. Hence the proportion of

ligand molecules bound by Abs can be determined from the polarization of fluorescent emission. This method is potentially useful with those small Ags and haptens whose fluorescence spectrum (natural or introduced by fluorescent substituents) is distinctly different from that of Abs (see Fluorescence quenching, above).

FLUORESCENCE ENHANCEMENT

Certain small organic molecules, such an anilino-naphthalenesulfonates, have strikingly different fluorescence spectra when they exist as free molecules in water and when bound in the active sites of Abs, which have a much lower dielectric constant than water. It is thus possible to measure the specific binding of such ligands to the corresponding Abs by following the appearance of fluorescence at the appropriate wave length. The method is exquisitely sensitive and can be used with Abs in complex media, such as serum.

DIAGNOSTIC APPLICATIONS OF SEROLOGICAL TESTS

The etiological diagnosis of infectious diseases is usually established both by isolating the putative causative microorganism and by

demonstrating Abs to it. Several principles are involved in the diagnostic use of serological tests.

Changing Titer. The Ab formation initiated by an infectious agent may continue for months or even years after the clinically apparent infection has subsided. Hence the presence of an elevated titer of Abs to a given microbe indicates only that infection (or vaccination) has occurred at some time in the past. In order to establish that an acute illness is due to a particular agent it is desirable to show a **change** in titer to that agent during the illness, i.e., the absence (or low level) of detectable Abs during the first week, their appearance and progressive increase in titer during the second and subsequent weeks, and perhaps their eventual decline. For practical purposes a pair of serum samples is compared, one drawn in the acute phase of the illness, the other during convalescence. If only a single serum sample is available, from late in the illness, a high titer is sometimes accepted as provisional diagnostic evidence; the critical level depends on the disease and the assay, and it is necessary to be sure that the high titer is not the result of prior vaccination.

Identification of Etiological Agent. Even when no likely agent has been isolated, a provisional etiological diagnosis can be established by testing the serum with Ags from microbial strains under suspicion: the Ag that elicits the highest titer reaction is assumed to be from the etiological agent. This procedure can be misleading: thus other Ags, not tested, might have elicited a still higher titer. Moreover, some curious cross-reactions between very different microorganisms may also serve to mislead: in a classic example persons with rickettsial infections form Abs that react in high titer with certain *Proteus* strains (Ch. 38, Weil-Felix reaction).

Time-Course of Immune Response in Infectious Disease. In the course of a given infectious disease the appearance and the persistence of Abs to the etiological agent follow a different time course when measured with different antigenic preparations and different assays. In brucellosis, for example, agglutination titers appear early in infection and persist at low levels for years afterward, whereas precipitin levels appear later and disappear much sooner; and in many rickettsial and viral diseases the Ab titers measured by complement fixation (Ch. 18) appear later and subside sooner than those measured by neutralization of infectivity (Ch. 50).

A number of possibilities can account for these variations. 1) The many Ags in a microbe differ in amount, immunogenicity, and stability in the animal host. 2) There are structural and functional differences in the Abs specific for a given antigenic determinant, and different types of Abs are formed at different stages of the immune response, e.g., IgM Abs (Ch. 16) are formed early in the response and are especially effective in the agglutination assay (Ch. 17, Antibody formation). 3) The various serological methods differ in their sensitivity (Table 15-5), and a method that requires less Ab will, other things being equal, detect Ab earlier and more persistently. Though the basis for the differences between different assays is not always clear, it is important for practical purposes to be aware that such differences exist.

THE UNITARIAN HYPOTHESIS

In the last decade of the nineteenth century immunologists were confronted in rapid succession by discoveries of a bewildering variety of immune phenomena. Particularly impressive, no doubt, was the extraordinary diversity of activities displayed by a given antiserum. A serum prepared, for example, against the cholera vibrio could 1) protect guinea pigs against an otherwise lethal infection with virulent cholera vibrios, 2) agglutinate a suspension of these organisms, 3) form a precipitate with a filtrate of the broth in which they had grown, 4) specifically enhance their phagocytosis by leukocytes, etc. This wide range of phenomena gave rise to the prevailing **pluralistic** view, according to which each of these diverse activities was ascribed to a qualitatively different molecular form, named agglutinins, precipitins, opsonins (promoters of phagocytosis), protective antibodies, etc.

With the improved methods of analysis developed in the 1930s, however, this view became clearly untenable. For example, Heidelberger found that all the Ab activities of antiserum to type 3 pneumococci were removed by precipitation with the organism's purified capsular polysaccharide; and the protein subsequently isolated from the precipitate duplicated the diverse activities of the antiserum. Moreover, this capsular polysaccharide is a polymer of cellobiuronic acid (Fig. 15-18); and Avery and

Goebel found that a cross-reacting antiserum, prepared against a conjugated protein substituted with the *p*-azobenzyl ether of cellobiuronic acid, not only precipitated the type 3 polysaccharide and agglutinated type 3 cells, but also protected mice against lethal infection with these organisms.

From these and similar observations there emerged the **unitarian hypothesis,** according to which a given population of Abs will, on uniting with the corresponding ligand, produce any of the diverse consequences of Ab-Ag combination, depending on the state of the ligand and the milieu in which combination takes place: if the ligand is soluble and multivalent, precipitation; if the ligand is a natural constituent of a particle's surface or artificially attached to the surface, agglutination; if the ligand is part of the surface of a virulent bacterium, protection against infection.

These contrasting views may be epitomized by saying that with respect to capacity to elicit the total range of in vivo and in vitro manifestations of Ab-Ag complex formation, the early pluralistic view maintained that a given Ab molecule is **unipotent,** whereas the unitarian view held that the Ab molecule is **totipotent.** While it has been clear since the 1940s that the pluralistic view is not valid, it has also become clear that the unitarian view is a gross oversimplification, e.g., certain Abs do not precipitate their Ags. In subsequent chapters we shall see more striking evidence that among the Ab molecules specific for a given ligand group, some are able to elicit certain reactions (e.g., Anaphylaxis, Ch. 19) and others are not. Though antibodies are multipotent to an impressive degree, they are definitely not totipotent.

APPENDIX

PURIFICATION OF ANTIBODIES

Methods of purifying Abs are based on the dissociability of Ab-ligand complexes. Two stages are usually involved. 1) Abs are precipitated from serum with soluble Ags or adsorbed by insoluble antigenic materials; the latter are often prepared by coupling small haptenic groups or soluble proteins to an insoluble matrix, such as agarose. 2) After extraneous serum is washed away Abs are eluted from the insoluble complexes by specific or nonspecific procedures.

Specific Procedures. With aggregates whose stability depends largely on specific ionic interactions, such as those involving types 3 and 8 pneumococcal polysaccharides (Fig. 15-18), strong salt solutions (e.g., 1.8 M NaCl) elute purified Abs effectively.

When the specific antigenic determinants are simple haptenic groups, such as 2,4-dinitrophenyl, small univalent haptens that encompass the crucial part of the determinant (e.g., 2,4-dinitrophenol) are useful for competitive displacement from the precipitating Ag or adsorbent, yielding soluble Ab-hapten complexes. Depending upon the properties of the Ag, the adsorbent, and the hapten, diverse procedures are then used to isolate the soluble Ab-hapten complexes and finally to separate the hapten from the Ab (e.g., ion-exchange resins, dialysis, gel filtration).

When small univalent haptens are employed for specific elution of Ags it is desirable to use those haptens that are both 1) **weakly bound** by the Ab and 2) **highly soluble.** Highly concentrated solutions of hapten can then be used to elute the Ab in high yield, and the weakly bound hapten is easily sep-

arated from the soluble hapten-Ab complex, e.g., by dialysis or gel filtration.

Nonspecific Procedures. For the isolation of Abs to protein Ags it is usually necessary to expose specific aggregates to conditions that cause reversible denaturation of the Ab, allowing it to dissociate from the Ag. Organic acids at pH 2 to 3 are often effective; various procedures are then used to separate the denatured Ab and Ag, depending upon the properties of the Ag. Since Abs usually regain their native structure on being restored to physiological conditions, neutralization of the Ag-free material yields active Ab, usually without excessive losses due to persistent denaturation.

Yield and Purity. Though Abs can be isolated from serum in high yield (50 to 90%) and with high purity (> 90% of the recovered Abs usually react specifically with Ag) the purified molecules are usually heterogeneous with respect to affinity (see Fig. 15-2) and in many of the physical and chemical properties described in Chapter 16.

FERRITIN-LABELED ANTIBODIES

As noted earlier, fluorescein-labeled Abs provide specific stains for the detection and localization of Ags in the light microscope. For use in the electron microscope Abs must be rendered much more highly electron-scattering than proteins in general, and this can be accomplished by attaching a molecule of ferritin to an Ab molecule. Ferritin, a protein molecule of about the same size as an Ab molecule, has high electron-scattering capacity because of its uniquely high iron content (about

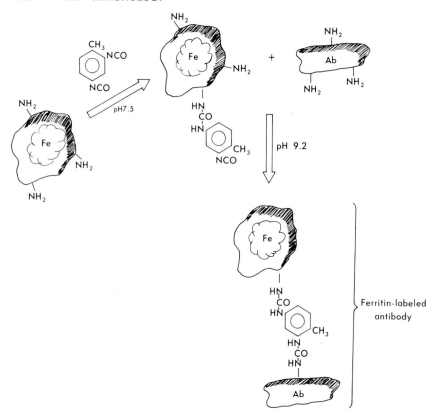

Fig. 15-37. Preparation of ferritin-labeled Ab by reaction with toluene-2,4-diisocyanate. Fe, ferritin. [Based on Singer, S. J., and Schick, A. F. *J Biophys Biochem Cytol* 9:519 (1961).]

20%). Ferritin-Ab conjugates were introduced by Singer, who used a bifunctional reagent to form stable covalent links between the two proteins, as is shown in Figure 15-37.

Ferritin and Hybrid Antibodies. Unmodified ferritin is also useful in conjunction with bivalent "hybrid" Ab molecules in which one combining site is obtained from anti-ferritin Abs and the other site from Abs to an Ag of interest, e.g., a particular cell surface macromolecule. The hybrid Ab can thus link a ferritin molecule (noncovalently) to the Ag target. The preparation of hybrid Abs is described in Chapter 16 and their use with ferritin to stain particular immunoglobulins on mast cells is illustrated in Figure 19-4; the use of hybrid Abs to identify cell surface molecules by light microscopy is given in Chapter 17 (indirect rosettes, Fig. 17-3).

INTRINSIC AND ACTUAL AFFINITIES

The successive steps in the binding of a univalent ligand (L) by a multivalent antibody (B) with n combining sites can be represented by the reactions on the left and their equilibrium constants on the right.

$$B + L \rightleftarrows BL \qquad k_1 = \frac{[BL]}{[B][L]}$$

$$BL + L \rightleftarrows BL_2 \qquad k_2 = \frac{[BL_2]}{[BL][L]}$$

$$\cdots \qquad \cdots$$

$$BL_{i-1} + L \rightleftarrows BL_i \qquad k_i = \frac{[BL_i]}{[BL_{i-1}][L]} \tag{10}$$

$$\cdots \qquad \cdots$$

$$BL_{n-1} + L \rightleftarrows BL_n \qquad k_n = \frac{[BL_n]}{[BL_{n-1}][L]}$$

We wish to find the general expression for the ratio of all bound ligand molecules (L_b) to all Ab molecules (B_t) at equilibrium.

$$L_b = BL + 2BL_2 + 3BL_3 \cdots + iBL_i \cdots + nBL_n, \text{ or}$$
$$= B[k_1L + 2k_1k_2(L)^2 \cdots + ik_1k_2 \cdots k_i(L)^i \cdots + nk_1k_2 \cdots k_n(L)^n] \tag{11}$$

and

$$B_t = B + BL + BL_2 + BL_3 \cdots + BL_i \cdots + BL_n, \text{ or}$$
$$= B[1 + k_1 L + k_1 k_2 (L)^2 \cdots + k_1 k_2 \cdots k_i (L)^i \tag{12}$$
$$\cdots + k_1 k_2 \cdots k_n (L)^n]$$

Assume all Ab combining sites are equivalent and independent. Then for the representative step in which the ith site becomes occupied (third step in equation 10), the concentration of vacant Ab sites (S) is:

$$[S] = (n-(i-1)) [BL_{i-1}] \tag{13}$$

and the concentration of occupied sites (SL) is:

$$[SL] = i [BL_i] \tag{14}$$

When equations (13) and (14) are combined, the equilibrium constant for the ith reaction becomes

$$k_i = \frac{n-i+1}{i} \cdot \frac{[SL]}{[S][L]} \tag{15}$$

$[SL]/[S][L]$ is defined as K, the **intrinsic association constant**, or the **intrinsic affinity**, for the general reaction in which a representative site binds a ligand molecule $(S + L \rightleftharpoons SL)$; i.e.,

$$k_i = \frac{n-i+1}{i} K \tag{16}$$

The constants for the individual steps in equation (10) can thus be expressed in terms of the intrinsic constant [e.g., $k_1 = nK$; $k_2 = (n-1)/2\ K$; $k_n = K/n$], and equations (11) and (12) can be reduced with the aid of the binomial theorem to*:

$$L_b = nK[B][L](1 + K[L])^{n-1} \tag{17}$$

and

$$B_t = [B](1 + K[L])^n \tag{18}$$

or,

$$\frac{L_b}{B_t} = \frac{nK[L]}{1 + K[L]}$$

Since L_b/B_t is r (moles ligand bound per mole Ab)

and $[L]$ is c (equilibrium concentration of free ligand) we have

$$r = \frac{nKc}{1 + Kc} \tag{19}$$

which is the same as

$$\frac{r}{c} = Kn - Kr \quad \text{(equation 4)}$$

Intrinsic affinity, K, provides a convenient and rigorous basis for analyzing the function of Ab combining sites and for comparing sites of different Abs with each other and with those of other proteins. However, the multiplicity of binding sites on Ab molecules and on most Ag particles means that for biologically significant Ab-Ag reactions in vivo the actual equilibrium constants can differ greatly from the intrinsic constants. Because most microbes have many identical envelope subunits on their surface, their complexes with Abs probably conform in most instances to Karush's concept of the monogamous complex (above); the equilibrium constants for formation of these complexes can exceed by perhaps 10^5 the intrinsic affinity of their Ab component. The great difference represents a form of statistical cooperativity between sites on the same molecule which is widespread in nature; for example, the bond between a pair of complementary nucleotides (e.g., A-T) is weak but between two polynucleotide strands with many complementary pairs it is enormously greater (Ch. 10).

Even without monogamous binding, actual and intrinsic binding constants are not necessarily the same. For instance, in a reaction that is limited, say by a low level of Ag, to only one site of an Ab molecule that has 2 or 10 sites (IgG vs. IgM, see Ch. 16), the observed equilibrium constant will be 2 and 10 times the respective intrinsic constants (see equation 16). Other things being equal, Ab molecules with more combining sites will form more stable complexes with ligands.

---

* The derivation is due to Dr. B. Altschuler, New York University.

SELECTED REFERENCES

Books and Review Articles

KABAT, E. A. *Kabat and Mayer's Experimental Immunochemistry*, 2nd ed. Thomas, Springfield, Ill., 1961.

KARUSH, F. Immunologic specificity and molecular structure. *Adv Immunol* 2:1 (1962).

LANDSTEINER, K. *The Specificity of Serological Reactions*, rev. ed. Harvard Univ. Press, Cambridge,

Mass., 1945. Reprinted by Dover, New York, 1962.

PRESSMAN, D., and GROSSBERG, A. L. *The Structural Basis of Antibody Specificity*. Benjamin, New York, 1968.

WILLIAMS, C. A., and CHASE, M. W. (eds.). *Methods in Immunology and Immunochemistry*, vols I, II, III. Academic Press, New York, 1967, 1968, 1971.

Specific Articles

ARNON, R. A selective fractionation of anti-lysozyme antibodies of different determinant specificities. *Eur J Biochem 5*:583 (1968).

BERSON, S. A., and YALOW, R. S. *Radioimmunoassay: A status Report in Immunobiology.* (R. A. Good and D. W. Fisher, eds.) Sinauer, Stamford, 1971.

EISEN, H. N., and SISKIND, G. W. Variation in affinities of antibodies during the immune response. *Biochemistry 3*:966 (1964).

HAIMOVICH, J., SELA, M., DEWDNEY, J. M., and BATCHELOR, F. R. Anti-penicilloyl antibodies: Detection with penicilloylated bacteriophage and isolation with a specific immunoadsorbent. *Nature 214*:1369 (1967).

HORNICK, C. L., and KARUSH, F. The interaction of hapten-coupled bacteriophage ϕX174 with anti-hapten antibody. *Isr J Med Sci 5*:163 (1969).

KLINMAN, N. R., LONG, C. A., and KARUSH, F. The role of antibody bivalence in the neutralization of bacteriophage. *J Immunol 99*:1128 (1967).

MÄKELÄ, O. Assay of anti-hapten antibody with the aid of hapten-coupled bacteriophage. *Immunol 10*:81 (1966).

PARKER, C. W., GOTT, S. M., and JOHNSON, M. C. The antibody response to a 2,4-dinitrophenyl peptide. *Biochemistry 5*:2314 (1966).

SELA, M., and HAIMOVICH, J. Detection of Proteins with Chemically Modified Bacteriophages. In *Protides of the Biological Fluids.* (H. Peeters, ed.) Pergamon Press, Elmsford, N.Y., 1970.

SELA, M., SCHECHTER, B., SCHECHTER, I., and BOREK, F. Antibodies to sequential and conformational determinants. *Cold Spring Harbor Symp Quant Biol 32*:537 (1967).

chapter 16

ANTIBODY STRUCTURE: THE IMMUNOGLOBULINS

CLASSES OF HEAVY CHAINS AND TYPES OF LIGHT
CHAINS: ISOTYPES   407
IgG IMMUNOGLOBULINS   408
Chains   409
Fragmentation with Enzymes   409
Over-all Structure   411
Disulfide Bridges   412
IgG Subclasses   414
OTHER IMMUNOGLOBULIN CLASSES   416
IgM   416
IgA   418
IgD   418
IgE   418
GENETIC VARIANTS: ALLOTYPES   419
Human Allotypes   419
Rabbit Allotypes   423
Mouse Allotypes   423
UNIQUE DETERMINANTS OF INDIVIDUAL IMMUNOGLOBULINS:
IDIOTYPES   423
HOMOGENEOUS IMMUNOGLOBULINS   425
Myeloma and Bence Jones Proteins   425
Comparison Between "Pathological" and Normal
Immunoglobulins   425
Homogeneous Antibodies   426
AMINO ACID SEQUENCES: THE BASIS FOR IMMUNOGLOBULIN
DIVERSITY   427
General Features of Light and Heavy Chain Sequences:
Domains   428
Variable (V) and Constant (C) Segments   429
V Subgroups and Hypervariable Regions   431
The Hinge   433
GENETIC BASIS OF IMMUNOGLOBULIN STRUCTURE   436
Two Genes–One Chain   436
Translocation Hypothesis   436
Origin of Diversity   438
EVOLUTION OF IMMUNOGLOBULINS   440
COMBINING SITES: THE BASIS FOR ANTIBODY
SPECIFICITY   442
Composition of Combining Sites   443
Size of Combining Sites   444
Reconstruction of Combining Sites   445
APPENDIX   446

Though antibodies (Abs) were found early to be a heterogeneous group of proteins (sedimentation constants 7S and 19S), for many years they were regarded simply as "γ-globulins," the class of plasma proteins with least electrophoretic mobility. Thus Tiselius and Kabat showed that the γ-globulin level was increased by intensive immunization and was lowered when Abs were removed from serum by specific precipitation with the corresponding antigen (Ag) (Fig. 16-1). Moreover, purified Abs to various Ags, isolated from specific precipitates, seemed to be indistinguishable from γ-globulins, not only chemically (electrophoretic mobility, solubility, amino acid composition), but also in their reactivity as Ags: for instance, when goats were immunized with γ-globulins from a nonimmune rabbit the resulting antiserum reacted not only with the immunogen but with rabbit Abs of diverse specificities. Some of the electrophoretically faster moving serum proteins also exhibit Ab activity; and many γ-globulins lack activity, probably because they are specific for unidentified ligands. Accordingly, **all proteins that function as antibodies or that have antigenic determinants in common with antibodies are now called immunoglobulins (abbreviated Igs).**

All Igs have a similar structural organization, but they are an immensely diversified family that can be arranged into groups and subgroups on the basis of variations in antigenic properties and amino acid sequences. In this chapter we emphasize the relations between these structural features and the two kinds of functions that are characteristic of every Ab molecule: 1) **specific** binding of one or a few of an almost limitless variety of ligands, and 2) regardless of this specificity, participation in a limited number of **general or effector reactions,** e.g., complement fixa-

tion (Ch. 18), allergic responses (Ch. 19). As we shall see, the two functions are carried out by different parts of the Ab molecule.

Three categories of antigenic determinants are used to classify Igs. 1) Those that differentiate among the main Ig classes are the same in all normal individuals of a given species: they are called isotypic determinants or **isotypes** (Gr. *iso* = same). Various isotypes are associated with different effector functions (see Appendix, Table 16-5), but they are essentially unrelated to ligand-binding activity; thus Abs of diverse Ig classes can be specific for the same antigen (Ag). **Heterologous antisera,** raised in one species against another species' Igs, are used to recognize isotypes (see below).

2) Other Ig determinants reflect regular small differences between individuals of the same species in the amino acid sequences of their otherwise similar Igs. The differences are specified by allelic genes and are called **allotypes** (Gr. *allos* = other). These markers are not associated with particular ligand-binding or effector functions, but they are of great importance in probing the genetic basis for Ab structure and biosynthesis. Allotypes are detected with homologous or **alloantisera,** which are formed by one individual against Igs with different allotypes from another individual of the same species.

3) Antigenic determinants of a third kind, **idiotypes** (Gr. *idios* = individual), are unique to the Ig molecules produced by a given clone of Ig-producing cells (Ch. 17). These determinants are probably located in or close to the specific ligand-binding sites of Abs, and they are detected with various antisera; e.g., those produced by one individual against an Ab from another individual of the same species and with the same allotypes.

406

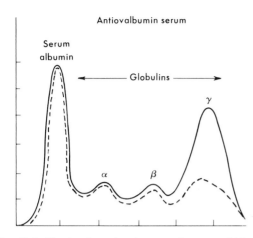

Fig. 16-1. Electrophoresis of serum from a rabbit intensively immunized with ovalbumin, before (——) and after (– – –) removal of Ab molecules by specific precipitation with ovalbumin. [From Tiselius, A., and Kabat, E. A. *J Exp Med* 69:119 (1939).]

CLASSES OF HEAVY CHAINS AND TYPES OF LIGHT CHAINS: ISOTYPES

Heavy-chain Isotypes. Human Igs are divided into five principal classes on the basis of chemical and isotypic properties. The three classes shown in Figure 16-2 are called IgG, IgA, and IgM.* The other two, IgD and IgE,

are present at such low concentrations in normal serum that their detection requires more sensitive methods, e.g., radioimmunoassay for IgE (Chs. 15 and 19).

As will be described later, Ig molecules are made up of small (light) and large (heavy) polypeptide chains. **Each of the five classes has similar sets of light chains but an antigenically distinctive set of heavy chains, which is named with the corresponding Greek letter (γ chains in IgG, μ in IgM, α in IgA, δ in IgD, ε in IgE).** Thus heterologous antisera prepared against human IgG contain Abs specific for heavy chains and others specific for light chains. After absorption with the heavy chains of IgG these antisera, specific for light chains, give reactions of identity with Igs of all classes (Fig. 16-3). In contrast, if the anti-IgG is absorbed with light chains (isolated from any class) the serum becomes monospecific for heavy chains of IgG; i.e., it reacts with Igs only of this class (Fig. 16-3).

Kappa and Lambda Light-chain Isotypes. There are also two main types of light chains,

---

* The main classes are also sometimes referred to as γG, γA, γM, γD, γE. Up to 1965 various other designations were used: γ, γ_2, γ_{ss}, 7Sγ for IgG; $\beta_2 A$, $\gamma_1 A$ for IgA; $\beta_2 M$, $\gamma_1 M$, 19Sγ for IgM.

Fig. 16-2. Immunoelectrophoresis of human serum showing IgG, IgA, and IgM. Serum in the center wells (c) was subjected to electrophoresis in agar in 0.1 M barbital, pH 8.6. The rabbit antiserum in the long troughs, parallel to axis of migration of the human proteins, had been prepared against unfractionated human serum. (Courtesy of Dr. C. Kirk Osterland; see also Fig. 13-29, Ch. 13.)

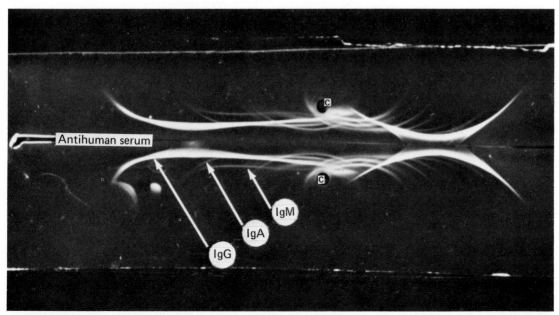

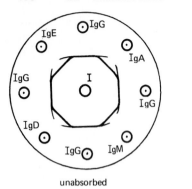 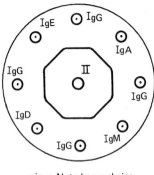 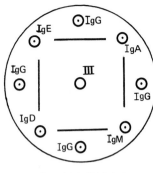

unabsorbed   minus Ab to heavy chains   minus Ab to light chains

Fig. 16-3. Gel diffusion precipitin reactions of human IgG, IgM, IgA, IgD and IgE with rabbit antiserum to human IgG. The antiserum (center wells) was used without absorption (I) or after absorption with heavy (II) or light chains (III) from IgG.

but since they are both present in all Ig classes (Figs. 16-3 and 16-4), they were long indistinguishable. Their distinction became possible only after it was recognized that the **myeloma proteins** secreted by myeloma tumors are Igs. Each of these tumors is a clone of neoplastic cells of the type that normally makes Abs (Ch. 17, Plasma cells); and each clone usually produces a distinctive and homogeneous myeloma protein. These "pathological" proteins have the same basic structure as normal Igs; some of them bind Ags specifically in the same way as ordinary Ab molecules. Some myeloma tumors also secrete homogeneous light chains, called **Bence Jones proteins** (p. 425), and heterologous antisera to individual Bence Jones proteins distinguish two main types of light chains, **kappa (κ) and lambda (λ).** κ and λ chains differ extensively in amino acid sequences (see below), and **one type or the other is present in every Ig molecule, regardless of its heavy-chain class** (Fig. 16-4). About 60% of human Igs have κ chains and 40% have λ chains; the ratio varies somewhat in different Ig classes (Appendix, Table 16-5).

Igs in other vertebrate species also have κ and λ chains, as defined by homologous amino acid sequences (see below). However, some species have predominantly one type (ca. 95% of all light chains in mice are κ and almost 100% in horses and in birds are λ), while others (e.g., guinea pigs) have nearly equal amounts.

Fig. 16-4. Two types of light chains. Antisera to Bence Jones proteins of kappa (κ) and lambda (λ) types were mixed in the center well; κ and λ Bence Jones protein were placed in the lateral wells and normal human γ-globulin in the well at the top.

In this chapter we shall first consider the main classes of Igs, then genetic variants, and next individual Igs, whose amino acid sequences will be shown to account for both the unique and common properties of all Igs. The final sections consider the genetic basis for diversity among Igs and the structural basis for the specificity of their combining sites.

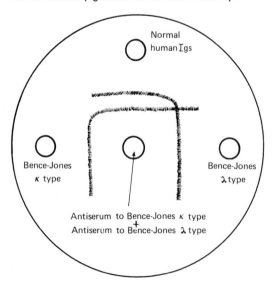

IgG IMMUNOGLOBULINS

These proteins used to be referred to as 7S γ-globulins because their sedimentation coefficient ($S^{\circ}_{20,w}$) in neutral, dilute salt solution is 6.5 to 6.6 Svedberg units; their molecular weight is 150,000. They constitute over 85% of the Igs in the sera of most normal and hyperimmune individuals; hence they are responsible for many phenomena previously observed with crude γ-globulin fractions.

CHAINS

Multiple Chains. The IgG proteins have 20 to 25 disulfide (S-S) bonds per molecule and the subunit structure was not established until these were split. In 1959 Edelman discovered that reduction of the S-S bonds in human IgG led to a drop in molecular weight, suggesting that multiple chains are linked in the intact molecule by these bonds. The products were soluble only in special solvents (e.g., 8 M urea), probably because many intrachain S-S bonds were also cleaved, and this causes polypeptide chains, in general, to become grossly denatured. Nevertheless, electrophoresis in 8 M urea revealed two components.

Shortly thereafter Fleischman, Pain, and Porter showed that the covalent bridges between chains could be selectively ruptured by scission of only a few S-S bonds (about four). The chains then came apart when the reduced molecule was exposed to an organic acid (acetic, propionic), which broke noncovalent hydrophobic bonds and imposed many positive charges on the chains, leading to their mutual repulsion. By gel filtration two components were then recovered with essentially 100% yield (Fig. 16-5A); the MW of the heavier was about 50,000 and of the lighter was about 25,000. Since recovery was quantitative the arithmetic was simple: on the assumption that each of the fractions was made up of a single kind of chain, **the original molecule of 150,000 daltons consisted of two heavy chains (2 × 50,000) plus two light chains (2 × 25,000).**

Though antigenic and amino acid sequence analyses have revealed a great variety of heavy chains and of light chains, it has been clearly established that **any particular Ig molecule has identical heavy chains and identical light chains** (see Rabbit allotypes, below).

FRAGMENTATION WITH ENZYMES

Early attempts were made to analyze the architecture of Ab molecules by enzymatic frag-

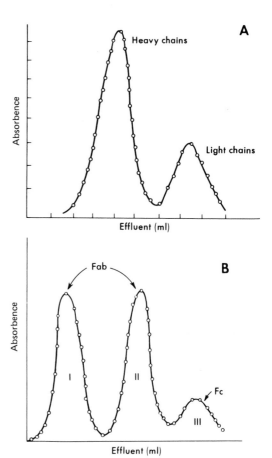

Fig. 16-5. A. Separation of heavy and light chains of IgG. The protein was reduced and alkylated (···S-S··· → ···SH; + ICH$_2$·COOH → ··· S·CH$_2$·COOH), and subjected to gel filtration in an organic acid (1 M propionic). **B.** Separation of Fab and Fc fragments of IgG. The protein was digested with papain (in presence of cysteine); after dialysis (during which some of the Fc crystallized) the digest was chromatographed on carboxymethylcellulose. [**A** based on Fleischman J. B., Pain, R. H., and Porter, R. R. *Arch Biochem Suppl 1*:1974 (1962). **B** based on Porter, R. R. *Biochem J 73*:119 (1959).]

mentation, but this approach contributed little until certain S-S bonds were also cleaved.

Papain Digestion. In 1959 Porter showed that digestion of rabbit IgG with papain, in the presence of cysteine, decreased the sedimentation coefficient from 7S to 3.5S, with loss of only about 10% of the protein as small dialyzable peptides. Ion-exchange chromatography separated the digest into three fractions.

It was subsequently recognized that fractions I and II (Fig. 16-5B) are the same except for small differences in charge, corresponding to differences among the IgG molecules from which the fragments were derived (see Charge heterogeneity, below). When obtained from various purified Abs, these fragments retained the specific combining sites of the parent molecule but were **univalent,** i.e., contained a single ligand-binding site per fragment; and their total yield accounted fully for the ligand-binding activity of the original molecules. These fragments are now called **Fab (antigen-binding fragments).**

The third fragment (III) did not combine with Ags, and it appeared to be uniform in all rabbit IgG molecules, regardless of their specificity as Abs. Moreover, in contrast to intact Ab molecules, this fragment was easily crystallized; hence it is called **Fc (crystalliza-ble fragment).**

Of the total digest prepared with papain, about ⅔ is Fab (MW ca. 45,000) and ⅓ is Fc (MW 50,000). Hence **an intact, bivalent IgG molecule is made up of two univalent Fab fragments joined through peptide bonds to one Fc fragment.**

Cysteine was added with papain because the activity of the enzyme depends on the reduced state of its SH groups. Subsequently, however, it became clear that the cysteine was performing an additional, critical role, for it also reduced some S-S bonds in the immunoglobulin substrate. Thus fragmentation of IgG molecules into Fab and Fc actually requires two independent reactions, which can be carried out concomitantly or in sequence: 1) hydrolysis of a few peptide bonds (two are sufficient), and 2) reduction of a critical S-S bond (see Pepsin, below).

Other Proteases. Other proteolytic enzymes also split Igs, and **pepsin** has been especially useful: in the absence of cysteine it cleaves from IgG Abs a 5S bivalent fragment, MW ca. 100,000, which can be split by reduction of one S-S bond to yield univalent fragments (Fig. 16-6). The latter are indistinguishable

Fig. 16-6. Schematic diagram of the four-chain structure of IgG, showing some interchain S-S bonds and regions susceptible to proteolytic cleavage (papain, pepsin at arrows). The piece of heavy chain within the Fab fragment is the **Fd piece.** Dots represent noncovalent interchain bonds. The spherical Ag represents an antigen molecule in a ligand-binding site. [Four-chain model from Fleischman, J. B., Porter, R. R., and Press, E. M. *Biochem J* 88:220 (1963). Y-shape at top based on Figs. 16-7 and 16-8. Scheme for pepsin digestion based on Nisonoff, A., Wissler, F., Lipman, L., and Woernley, D. *Arch Biochem* 89:230 (1960).]

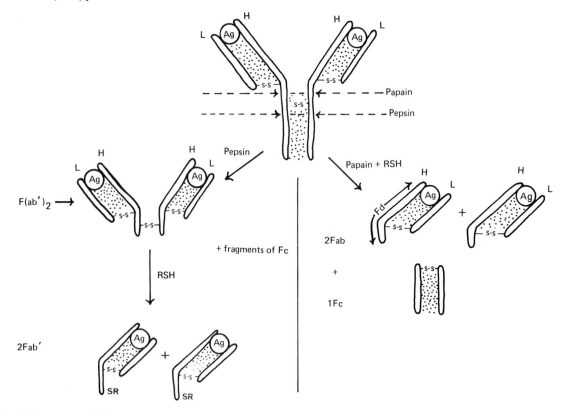

from the Fab fragments prepared with papain with respect to sedimentation constant (3.5S), ligand-binding activity, and antigenicity; but the pepsin fragment has a molecular weight about 10% higher, and it contains some covalently linked carbohydrate. To indicate these small differences the univalent fragment obtained with pepsin is called **Fab′**; the corresponding bivalent fragment is **F(ab′)₂**. The Fc fragment is not recovered in peptic digests, but is fragmented.

The variety of fragments obtained with various proteases from Igs of many species fit into a coherent pattern: **all IgG Abs have two compact globular domains (corresponding to Fab fragments) joined to a third compact domain (corresponding to the Fc fragment) by connecting regions that are highly susceptible to attack by proteases** (Fig. 16-6).

OVER-ALL STRUCTURE

Relations Between Chains and Fragments. The immunogenicity of the proteolytic fragments made it possible to match fragments with dissociated chains. Thus, goat antisera to the isolated Fc fragment of rabbit IgG form specific precipitates with heavy chains, as well as with Fc, but not with light chains. In contrast, antisera to Fab fragments react with both light and heavy chains (Table 16-1). These findings led Porter to formulate the schematic structure for IgG shown in Figure 16-6, which fits all the information now available.

Shape. A variety of analytical approaches (based on hydrodynamic and fluorescence properties, electron microscopy, and X-ray diffraction) suggest that IgG molecules are **Y-shaped,** with a "hinge" at about the middle of the heavy chain, connecting the two Fab domains with the Fc domain (Figs. 16-7, and 16-8). There may also be **segmental flexibil-ity:** by movement about the hinge, the angle between Fab segments might vary as much as 0 to 180°. But it is not yet clear whether each molecule can assume such a wide variety of angles.

Charge Heterogeneity. IgG migrates in zone electrophoresis as a broad protein band (see Figs. 16-2 and 16-16) because the individual proteins that make up the class vary in net charge (isoelectric points probably range from about 5.0 to 7.5). If IgG is arbitrarily separated by electrophoresis into rapidly and slowly migrating fractions, the heavy chains isolated from each fraction show corresponding differences in mobility. Moreover, each still migrates as a diffuse zone; thus the heavy chains are also inhomogeneous.

The light chains are also heterogeneous. They are small enough to penetrate easily into polyacrylamide gels, where they can be resolved by electrophoresis into many (at least 10) distinct bands, which seem to differ in unit electric charge (Fig. 16-18). Since this heterogeneity is observed in denaturing ("unfolding") solvents, such as 8 M urea, it is due to diversity in covalent structure, and not simply to folding or aggregation (see Amino acid sequences, below).

Carbohydrate Content. Igs are glycoproteins with oligosaccharides covalently linked to the Fc domain of heavy chains. In some IgG molecules carbohydrate is also present in Fab, i.e., on the light chain or the N-terminal half of the heavy chains (Fd piece, Fig. 16-6).

Each of the oligosaccharide substituents has two branches on a central stem, for which, so far, three different sequences have been established (Fig. 16-9). At its base, the stem is linked to an asparagine residue of the heavy chain through a covalent bond with N-acetyl glucosamine. Though a common sequence is present in all branches (galactose $\beta1 \rightarrow 6$ N-acetyl glucosamine $\beta1 \rightarrow 2$ mannose), sialic acid or additional galactose residues are present at some branch termini; and the fucose content of the oligosaccharides is also variable. This "microheterogeneity" has been evident in all samples analyzed, including monoclonal IgG myeloma proteins; the ragged ends probably arise from degradation by glycosidases that act on these proteins as they circulate in blood.

Carbohydrate is less abundant in IgG (3% by weight) than in other Ig classes (Appendix, Table 16-5). Besides having oligosaccharides like the one

TABLE 16-1. Goat Antibodies to Rabbit Fab and Fc Fragments: Precipitation Reaction with Chains Isolated from Rabbit IgG

	Antisera to	
	Fab	Fc
Heavy chains	+	+
Light chains	+	−

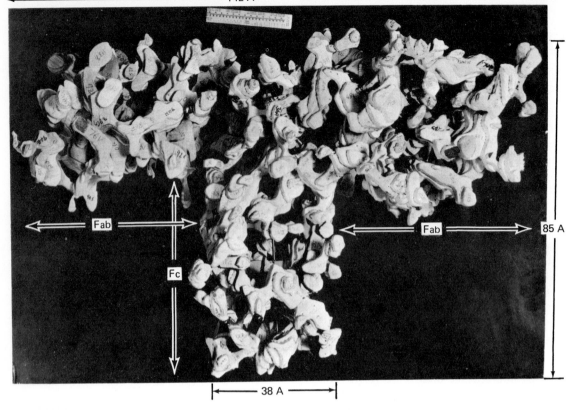

Fig. 16-7. Balsa wood model of an IgG molecule based on a three-dimensional electron density map at 6-A resolution from X-ray diffraction data. The dyad axis relating the halves of the molecule lies in the plane of the paper. The crystallized protein was a human myeloma protein of the IgG-1 subclass (see below). [From Sarma, V. R., Silverton, E. W., Davies, D. R., and Terry, W. D. *J Biol Chem 246*:3753 (1971).]

at the extreme left of Figure 16-9, other Ig heavy chains have additional oligosaccharides, which are distinctive for various classes: α chains (of IgA) have oligosaccharides linked via O- glycosidic bonds to serine OH groups; μ chains (of IgM) have mannose rich groups; ε chains (of IgE) have some that are rich in mannose and others that contain additional galactose residues.

The function of the carbohydrate is unknown: the saccharides could provide binding sites for selective attachment of different Igs to specific receptors on various cells (e.g., IgE on mast cells; see Ch. 19), or the addition of sugar residues during biosynthesis could play a role in the transfer of Ig chains into membranous vesicles, from which they are secreted at the cell surface (Ch. 17).

DISULFIDE BRIDGES

Stability. Proteins of the IgG class are unusually stable. In serum or as purified Abs they can remain unaltered for years at 0°. Moreover, after exposure to denaturing conditions (e.g., 70°, pH 11, pH 2, 8 M urea) they

largely recover native structure when returned to dilute salt solution at physiological conditions of pH and temperature. A major factor is the presence of a large number of disulfide (S-S) bonds (20 to 25 per molecule). Three to seven of these S-S bonds link the four chains together; the remainder are distributed within the chains, stabilizing their respective conformations (Figs. 16-6 and 16-20).

In order of increasing resistance to reduction the S-S bonds fall into groups that link: 1) the heavy chains, 2) light to heavy chains, and 3) cysteine residues within chains (intrachain). Hence, under mildest conditions the IgG molecule is split into symmetrical halves, each with one heavy plus one light chain. The halves separate when, at low pH, they acquire large positive charges; they reassociate to form native molecules when returned to neutral pH. Even if their SH groups are blocked chemically, e.g., by treatment with iodoacetic

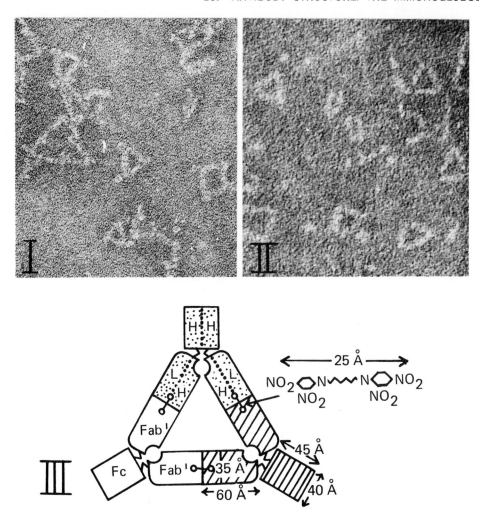

Fig. 16-8. Three IgG anti-Dnp antibody molecules joined in a triangular complex by a ligand with two Dnp groups per molecule [Dnp-NH-(Ch$_2$)$_8$-NH-Dnp; see **III**]. The ligand is too small to be resolved. The complex in **I** forms the basis for the diagram in **III**. The picture in **II** was obtained after treatment of the complex in **I** with pepsin, removing the corner projections, which correspond to Fc. **I** and **II**, Electron micrograph, ×500,000. [From Valentine, R. C., and Green, N. M. *J Mol Biol* 27:615 (1967).]

acid, the halves can still associate (though without S-S bridges) to form molecules that are essentially indistinguishable from the original in sedimentation properties and antigenicity.

Hybrid Antibodies. When half-molecules (H·L) from purified Abs of a particular specificity are mixed with an excess of half-molecules from nonspecific IgG and allowed to reassociate, most of the Ab behaves in precipitin reactions as though univalent, i.e., as

though formed from one specific plus one nonspecific half-molecule. Thus these "hybrid antibodies" specifically inhibit, rather than contribute to, the corresponding precipitin reaction.

Hybrids can be prepared more easily from the F(ab')$_2$ fragments obtained by pepsin digestion: mild reduction yields Fab' fragments, each with one ligand-binding site and one free SH group. If these fragments are prepared from two purified Abs of different specificity, mixed, and allowed to reoxidize,

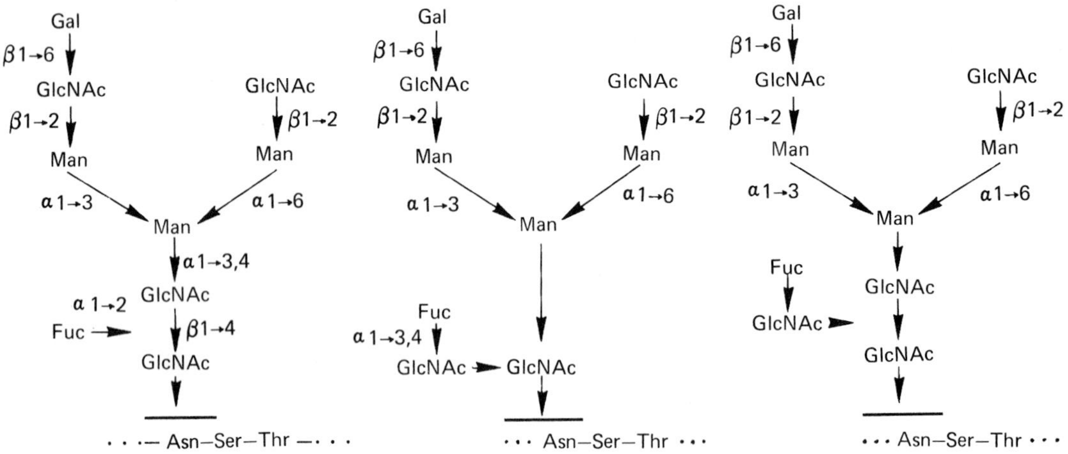

Fig. 16-9. Glycopeptides from three IgG myeloma proteins, showing branched oligosaccharides. The basal N-acetyl glucosamine (GlcNAc) is attached to an asparagine (Asn) residue of the heavy chain. An asparagine or aspartic acid residue in the sequence ···Asx-X-Thr··· forms the point of attachment of oligosaccharide in many glycoproteins. Sialic acid groups are attached to the free ends of some branches, but are not shown. Each of the oligosaccharides shown (with 2 sialic acid end groups the MW would be ca. 2500), occurring once per heavy chain, could account for the carbohydrate content of a typical IgG molecule; with 3 to 4 times more carbohydrate on other Ig classes, additional oligosaccharide moieties must be present on IgM, IgA, IgE (see Fig. 16-12). [From Kornfeld, R., Keller, J., Baenziger, and Kornfeld, S. *J Biol Chem* 246:3259 (1971).]

hybrid bivalents appear with a different ligand-binding specificity at each of the two sites (Fig. 16-10). The large proportion of hybrids indicates that the recombination of Fab' fragments is random.

Symmetry. In contrast to the pairs formed in the test tube, **natural IgGs are symmetrical: in any particular molecule the two light chains are identical, as are the two heavy chains, and the two ligand-binding sites are also identical** (Ch. 15). As we shall see later, the symmetry comes about because a given cell makes, at any one time, only one type of light chain and one type of heavy chain (Ch. 17, One cell–one Ig rule).

IgG SUBCLASSES

Careful antigenic analysis with heterologous antisera has revealed that proteins of the IgG class can be differentiated into four subclasses (IgG-1 through IgG-4), each with a distinctive heavy chain, called γ1, γ2, γ3, γ4.

Antisera vary greatly in capacity to differentiate among the four subclasses. Rabbit antisera to the total IgG fraction of human serum do not distinguish among them, while monkey antisera to the same mixture of immunogens can distinguish three of the four. The sharpest serological distinctions are made with antisera against the individual, homogeneous Igs produced by different clones of neoplastic plasma cells (see Myeloma and Bence Jones proteins, below). Adsorption of such an antiserum by appropriate myeloma proteins of other IgG subclasses make it monospecific for the subclass of its immunogen, i.e., it reacts only with Igs whose heavy chains belong to the same subclass (Fig. 16-11).

With the aid of monospecific antisera, the subclasses have been detected in all normal human sera: IgG-1, -2, -3, and -4 make up about 70, 19, 8, and 3%, respectively, of the IgG proteins. Their characteristic determinants (Isotypes, above) are localized in their Fc segments. The subclasses also differ in their general effector functions and in several other properties (Appendix, Table 16-5). Fundamental differences in their amino acid sequences and patterns of interchain S-S bonding are described later (Amino acid sequences and Fig. 16-20, below).

Four subclasses of IgG have also been identified in the mouse, the only other species in which many myeloma proteins are available for the preparation of subclass-specific antisera. But fewer subclasses have been recognized in other species (e.g., two in the guinea pig and in the rat), perhaps because homogeneous Igs have not yet been obtained from them and the only IgG subclasses recognized are

Fig. 16-10. Formation of hybrid bivalent Ab fragments. **Top.** F(ab')$_2$ fragments prepared from anti-O and anti-● were reduced to Fab' fragments, then mixed and allowed to reoxidize. For precipitation of hybrid F(ab')$_2$ fragments it was necessary to have both Ag O and Ag ● present (Ch. 15, Fig. 15-13). **Bottom.** Dual specificity of hybrid Ab fragments revealed by mixed hemagglutination. The Fab' fragments were obtained from rabbit Abs specific for chicken ovalbumin (Ea) or for bovine γ-globulin (BγG). Human red cells (small, spherical, nonnucleated) were coated with Ea, and duck red cells (large, oval, nucleated) were coated with BγG. In **a** reoxidized F(ab')$_2$ fragments prepared exclusively from anti-Ea clumped only the human cells. In **b** the reoxidized fragments prepared exclusively from anti-BγG clumped only the duck cells. In **c** a mixture of the two F(ab')$_2$ fragments of **a** and **b** formed separate clumps of human and of duck cells. In **d** hybrid F(ab')$_2$ fragments formed mixed clumps, with both human and duck cells. [From Fudenberg, H., Drews, G., and Nisonoff, A. *J Exp Med 119*:151 (1964).]

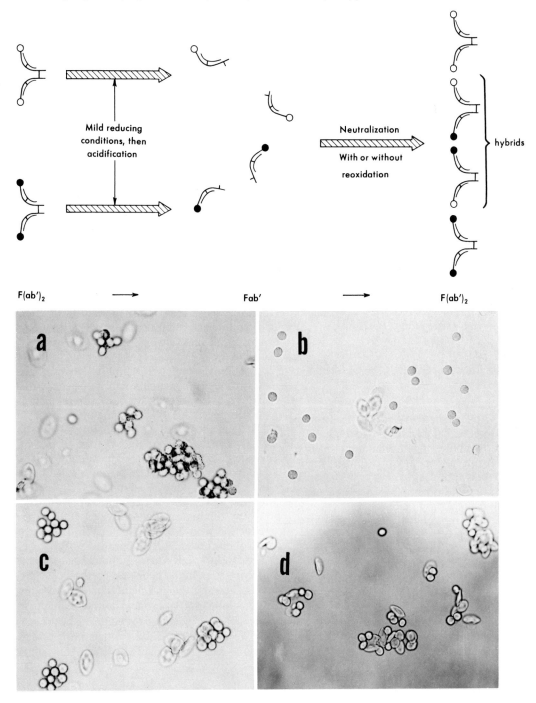

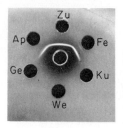 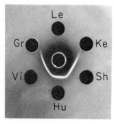

Fig. 16-11. IgG subclasses. Twelve IgG myeloma proteins (each from a human with a myeloma tumor) were tested with monkey antiserum to the protein from patient Zu, after the antiserum had been absorbed with Igs of other subclasses. Five additional proteins were precipitated (Ap, Fe, Vi, Hu, Sh); hence they belong to the same subclass as Zu (IgG-3). The six nonreacting IgGs belong to other subclasses. [From Grey, H. M., and Kunkel, H. G. *J Exp Med 120*:253 (1964).]

those that happen to be relatively abundant and to differ substantially in electrophoretic mobility (Fig. 19-6, Ch. 19).

Antigenic differences among subclasses are less pronounced than among classes. For example, after Abs to light chains are removed, antisera prepared against IgG-1 cross-react with IgG-2, IgG-3, IgG-4 but not with Igs of the other classes (IgM, IgA, etc.). Amino acid sequence data (below) confirm these relations by revealing greater homologies among heavy chains of the γ subclasses than among those of diverse classes (Fig. 16-23).

OTHER IMMUNOGLOBULIN CLASSES

IgM

The first Abs studied in the ultracentrifuge (Heidelberger and Pedersen) had high sedimentation values (predominantly 19S). Obtained from horses immunized with pneumococci, they appeared at first to be curiosities, because most other Abs (made in other species to other Ags) subsequently proved to be 7S IgG molecules. However, hemagglutination and other sensitive assays then revealed that 19S Abs are formed in almost every immune response, though nearly always early and only at low levels; they are usually soon overshadowed by larger amounts of IgG Abs to the same Ag (Ch. 17).

Now called IgM (M, macroglobulin), its molecular weight of about 900,000 is accounted for by five tetrameric subunits linked

through S-S bonds between cysteine residues near the hinge region of the heavy chains (Fig. 16-12A). Under mild reducing conditions, which cleave only interchain S-S bonds, IgM molecules dissociate into 7S subunits which yield (after treatment with acetic or propionic acid) light and heavy chains in the same proportion as from IgG molecules (Fig. 16-5A). However, the IgM heavy chains (μ chains) are ca. 20% longer and have more oligosaccharide branches and a higher carbohydrate content (ca. 10%); their molecular weight is accordingly higher (ca. 70,000 compared with 50,000 for the γ chains of IgG).

IgM is split by papain at 37° into a variety of heterogeneous fragments, unlike IgG which yields well-defined Fab and Fc fragments under these conditions. However, Fab and Fc can be derived from the IgM molecule under special conditions (trypsin at 56 to 60°). The Fab fragment resembles that of IgG (one light chain plus one Fd piece), but the Fc is a decamer, with 10 heavy-chain fragments joined by S-S bonds (Fig. 16-12A). The presence of 10 Fab domains and a correspondingly large number of combining sites per molecule probably accounts for the relatively high avidity of IgM antibodies for Ags (Ch. 15). The high avidity is notable especially because IgM Abs generally have much lower intrinsic affinities (per site) than IgG Abs of the same specificity.

The 7S subunits (IgM$_s$) derived by mild reduction do not give agglutination reactions even though their combining sites seem to be unaltered (as shown by equilibrium dialysis measurements of ligand binding; Ch. 15). The dependence of agglutinating activity on more than two sites could be due to low intrinsic affinity (per site) or to unknown steric factors that interfere with effective function of both sites on the 7S subunit. Whatever the reason, IgM can be distinguished from IgG Abs by the loss of agglutinating activity after reduction with 2-mercaptoethanol. The test is useful but not infallible, because Abs of other classes (IgA, IgE) can also lose agglutinating activity after mild reduction (below).

Besides heavy and light chains, IgM immunoglobulins have one additional polypeptide, called J, per 10 light chains. The **J chain** (MW ca. 20,000) differs antigenically and in amino acid composition from the other

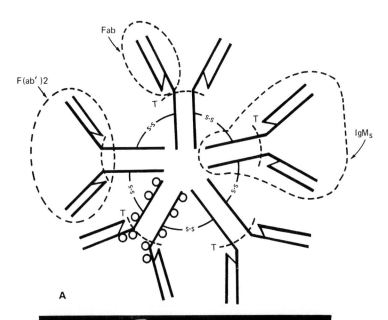

A

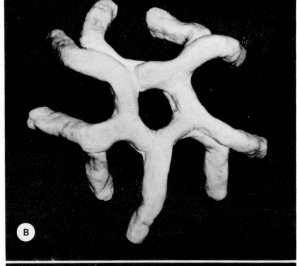

B

C

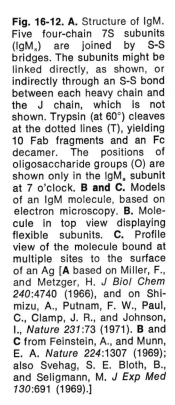

Fig. 16-12. A. Structure of IgM. Five four-chain 7S subunits (IgM$_s$) are joined by S-S bridges. The subunits might be linked directly, as shown, or indirectly through an S-S bond between each heavy chain and the J chain, which is not shown. Trypsin (at 60°) cleaves at the dotted lines (T), yielding 10 Fab fragments and an Fc decamer. The positions of oligosaccharide groups (O) are shown only in the IgM$_s$ subunit at 7 o'clock. **B and C.** Models of an IgM molecule, based on electron microscopy. **B.** Molecule in top view displaying flexible subunits. **C.** Profile view of the molecule bound at multiple sites to the surface of an Ag [**A** based on Miller, F., and Metzger, H. *J Biol Chem 240*:4740 (1966), and on Shimizu, A., Putnam, F. W., Paul, C., Clamp, J. R., and Johnson, I., *Nature 231*:73 (1971). **B** and **C** from Feinstein, A., and Munn, E. A. *Nature 224*:1307 (1969); also Svehag, S. E. Bloth, B., and Seligmann, M. *J Exp Med 130*:691 (1969).]

chains. Its function is not known; it could be the glue that stabilizes multimeric Igs (see also IgA, below).

IgA

Immunoelectrophoresis of human serum led to the identification of IgA (Fig. 16-2), whose basic structural unit corresponds to the four-chain molecule shown in Figure 16-6. Unlike IgM, which is a pentamer, and IgGs, which are uniformly monomers, IgA occurs in a variety of polymeric forms: monomer (7S), dimer (9S), trimer (11S), and even higher multimers (ca. 17S). The various multimers are evidently S-S linked monomers, because mild reduction is necessary (and sufficient) for their dissociation into 7S molecules. As with IgM, the reduced 7S molecules are defective in cross-linking activity though they retain two combining sites with unaltered affinity for small ligands.

IgA is normally present at about ⅕ the concentration of IgG in human serum (Table 16-5). But in man and in some other mammalian species **IgA is the principal Ig in exocrine secretions (milk, respiratory and intestinal mucin, saliva, tears).** The cells that produce IgA are concentrated in the subepithelial tissue of exocrine glands and respond to Ags that enter locally (e.g., by ingestion, inhalation, via the conjunctivae, etc.). The secreted IgA Abs seem to be important in protecting mucosal surfaces from invasion by pathogenic microbes. About as much IgA as IgG is synthesized per day in man, but loss in exocrine secretions probably accounts for its relatively low level in serum.

Human serum IgA occurs mostly as the monomer, while exocrine IgA is largely a dimer. After cleavage of S-S bonds, all forms of IgA yield heavy and light chains in the same proportions as IgG and IgM. Multimeric IgA also yields one **J chain** per four light chains. Exocrine IgA has, in addition, a fourth polypeptide chain, **secretory piece** (MW ca. 60,000), whose function is not known; it may be responsible for the unusual resistance of exocrine IgA to proteolysis, and hence contribute to its persistence in intestinal and other secretions.

As in the IgG class, antisera to human myeloma proteins of the IgA class differentiate subclasses, IgA-1 and IgA-2, with anti-genically distinctive α chains (α1, α2). About 11% of normal IgA and of IgA myeloma proteins belong to the IgA-2 subclass (Appendix, Table 16-5).

IgD

These Igs were discovered not as Abs but as rare myeloma proteins that failed to react with specific antisera to IgG, IgM, and IgA. These myeloma proteins, and the antigenically similar normal immunoglobulins, are called IgD. They are found in normal sera at about 1/50 the level of IgG (Appendix, Table 16-5).

IgE

Though present in normal human sera at only about 1/25,000 the level of IgG, the IgE proteins are responsible for severe, acute, and occasionally fatal allergic reactions (Ch. 19). IgE was discovered by the Ishizakas, who suspected its existence because the Abs that mediate acute allergic skin reactions in man are inactivated by rabbit antisera to crude mixtures of human Igs, but not by monospecific antisera for IgG, IgM, IgA, IgD (see Ch. 19, Reagins).

Like μ chains of IgM, ε chains of IgE are longer than γ and α chains (of IgGs and IgAs) by about 100 amino acid residues, suggesting that they have an additional domain (Amino acid sequences, below). Though other Igs are remarkably heat-stable, when IgE is heated for 4 hours at 56° it loses the ability to sensitize skin, but retains the capacity to bind Ags. Both activities are lost on reduction and alkylation of S-S bonds (Ch. 19).

Several other mammalian species have Igs that resemble IgE; i.e., they are also present in serum in trace amounts, are labile to heat and S-S splitting reagents, and have skin-sensitizing activity (Ch. 19, Homocytotropic Abs and reagins). Abs with these properties from several species (man, rat, rabbit, monkey) also cross-react antigenically.

The cells that produce IgE are relatively abundant in mucosa of the respiratory and intestinal tracts, and Abs of the IgE class are also found in exocrine secretions. Though present here also only in trace amounts, they probably play an important role in localized allergic reactions, e.g., hay fever, asthma.

Some properties of various classes and subclasses of human Igs are summarized in the Appendix (Table 16-5).

GENETIC VARIANTS: ALLOTYPES

Though animals only rarely form Abs to components of their own tissues, some substances derived from one individual are regularly immunogenic in certain other individuals of the same species (**isoantigens** or **alloantigens**). Oudin showed that some Igs behave in this manner and introduced the term **allotypes** for their various forms (see above). Other alloantigens, on cell surfaces, determine host responses to blood transfusions and organ transplants; they are discussed at length in Chapter 21.

HUMAN ALLOTYPES

A chance observation led to the discovery of genetic variants of human Igs. Serum from persons with rheumatoid arthritis often contains **rheumatoid factors,** which are Igs (generally IgM) that react with pooled human IgG; for example, they specifically agglutinate human red blood cells (RBC) coated with human anti-RBC Abs of the incomplete type (Ch. 15, Blocking Abs; Fig. 19-20; Ch. 21, Rh system). In studying these reactions Grubb observed that some patients' rheumatoid factors reacted only with cells that had been coated with Abs from certain persons. Moreover, any particular agglutination reaction could be specifically inhibited by serum only from particular individuals, whose Igs presumably carried the same antigenic determinants as the Ab molecules adherent to the red cells and therefore competed for the same rheumatoid factor (Fig. 16-13).

Some normal sera also contain Igs that agglutinate particular Ab-coated RBC. Called SNaggs (for serum normal agglutinators), they discriminate more sharply among related IgG variants than do most rheumatoid serum agglutinators (Raggs), but both are valuable reagents for typing human Igs.

At least some of the agglutinating activities in normal sera are probably due to maternal-fetal incompatibilities. In about 10% of human pregnancies mothers have Ig allotypes their babies lack. With transfer of maternal Ig across the placenta about half of these babies are stimulated to make antiallotype Abs. Detected with the sensitive hemagglutination assay (Fig. 16-13), these Abs generally reach peak titers at about 6 months of age (after the large amount of maternal Ig present in the infant's blood at birth has been eliminated; see Ch. 17, Ontogeny). Antiallotypes are also detected in about 1% of normal adults. Antiallotypes are also frequently formed in persons who receive multiple blood transfusions (for chronic anemias, open-heart surgery, etc.), because donors often have allotypes foreign to recipients.*

Through the reactions of Raggs and SNaggs with sera from thousands of persons, many human allotypes have been identified and classified into several groups, of which the principal ones are *Gm* (for gamma) and *Inv* (for inhibitor of the reaction with serum from patient V). These groups are specified by independently segregating genes and they are located on different chains. **The Gm markers are on γ (heavy) chains, and thus found only in IgG; the Inv markers are on light chains of κ type, and occur in all Ig classes.**

Gm Allotypes. The nomenclature of Gm allotypes has grown increasingly cumbersome as more varieties have been recognized. Hence a numerical system, resembling those used for Rh blood factors and *Salmonella* Ags (Chs. 21 and 29), has recently been adopted. Agglutinating sera of a given specificity are designated by number. An Ig that reacts, or fails to react, with a particular agglutinator receives the corresponding positive or negative number; for instance, an Ig with allotype Gm(1,-4) reacts with anti-Gm(1) but not with anti-Gm(4) (the absence of other numbers in this example would mean that the Ig was not tested with other agglutinators). As is conventional with other systems, alleles are italicized, their products are not: e.g., the allele *Gm⁵* specifies a γ chain of allotype Gm(5).

Over 20 Gm allotypic determinants have

---

* Most human Abs to IgG allotypes appear to have low affinity, and transfusions of blood with foreign allotypes of this class rarely cause allergic reactions (Chs. 19 and 21), even in recipients with high titers of the corresponding antiallotype. Severe reactions have occurred, however, when the discordant allotype is IgA, some of which is polymeric and may form more stable complexes with its antiallotype (Ch. 15, Multivalent binding).

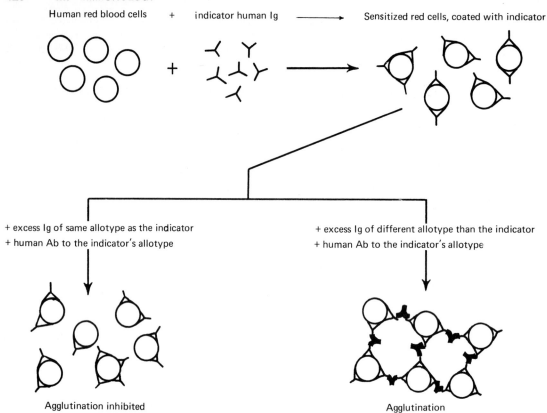

Fig. 16-13. Agglutination-inhibition assay for human allotypes. In the assay for Gm and Inv allotypes the indicator Ig is nonagglutinating Ab specifically bound to red cells (anti-Rh on Rh+ red cells; see Ch. 21). In the assay for some other allotypes the indicator Ig is adsorbed nonspecifically on CrCl$_3$-treated red cells. [Based on Grubb, R., and Laurell, A. B. *Acta Pathol Microbiol Scand 39*:195,390 (1956).]

been recognized. Each IgG subclass has its own group of allotypes (Table 16-2), specified by codominant* alleles at a complex autosomal (i.e., not sex-linked) locus. Hence a given allotypic determinant may be present on all or on approximately half the molecules of its subclass, depending on whether the individual is homozygous or heterozygous. Moreover, allotypic markers on myeloma proteins have shown that the chain specified by one allele can have several allotypic determinants, as expected from the multiplicity of antigenic determinants on protein chains in general (Ch. 15). Thus some IgG-1 molecules are Gm (1, 17): their γ1 chains have the Gm(1) marker on Fc and the Gm(17)

marker on the Fd piece (Table 16-2). Some γ1 chains in certain ethnic groups have three markers: Gm(1), Gm(4), and Gm(22) (Table 16-3 and Fig. 16-14).

The inheritance of allotypes follows mendelian rules. For example, a *Gm*[17] homozygous mother [whose γ1 chains and IgG-1 molecules are all Gm(17)] and a *Gm*[22] homozygous father [whose γ1 chains and IgG-1 molecules are all Gm(22)] invariably have heterozygous children, *Gm*[17]/*Gm*[22], with each marker on about half their IgG-1 molecules. The Gm allotypes are of considerable value in population genetics and in anthropology, and in cases of disputed paternity they are accepted as legal evidence in some countries (Norway, France).

The alleles for Gm markers are linked in complexes that differ conspicuously in diverse racial groups; for example, a common

---

* Alleles in a heterozygous diploid individual are codominant when both are expressed more or less equally.

TABLE 16-2.   Allelic Groups of Gm Allotypic Markers on Heavy Chains of IgG Subclasses

Subclass	Heavy chain	Gm allotype	Molecular local-ization*	Comments
IgG-1	γ1	1 2 17	Fc Fc Fd	Probably represent one allele; they are found on the same chains, but (2) is detected less often than (1). Present in ca. 60% of Western Europeans (and their descendents) and in ca. 100% of all others.
		3 4 22	Fd Fd Fc	Probably represent one allele; (3) and (4) are detected only when the γ1 chain is associated with light chain, and (3) is detected less often than (4). Present in >90% of Caucasians and Chinese, but in only 13% of Japanese.
IgG-2	γ2	23	Fc	Present in ca. 50% of Caucasians, 90% of Chinese, ca. 0% of Negroes. An antiserum is lacking for Gm (−23), the allelic product.
IgG-3	γ3	5 6 10 11 13 14	Fc Fc Fc Fc Fc Fc	A family of γ3 determinants present in ca. 100% of Negroes, 90% of Europeans and Chinese, and absent in Australian aborigines. When (6), (10), (11), (13), (14) are detected, they are nearly always associated with Gm(5).
		21	Fc	Allelic with the Gm(5) family.
IgG-4	γ4	4a 4b	? ?	Agglutinators of (4a) react with all IgG-1 and IgG-3 molecules. Agglutinators of (4b) react with all IgG-2 molecules.

Based on summary by Grubb, R. *The Genetic Markers of Human Immunoglobulins,* Springer, New York, 1970. Data for IgG-4 from Kunkel, H. G., Joslin, F. G., Penn, G. M., and Natvig, J. B. *J Exp Med 132*:508 (1970).

* See Figure 16-31 for amino acid substitutions associated with some of these markers.

heritable complex in caucasoids is 22,4,23,5 (Table 16-3). Each of these complexes behaves as a stable heritable unit. For example, a person who is heterozygous for γ1 and γ3 genes (with, say, the following allotypes—γ1: 1, 17, 4, 22; γ3: 21, 5) would have inherited one set of genes from one parent (γ1: 1, 17; γ3: 21) and the second set from the other parent (γ1: 4, 22; γ3: 5).

Surveys of large populations have shown that the crossover frequency between genes for γ1, γ2, and γ3 is extremely low; indeed, these genes may be adjacent, and probably represent the most closely linked human genes recognized so far.* However, a few pedigrees

have provided evidence for recombination, yielding the order of genes shown in Figure 16-14. An ancient crossover event probably accounts for the combination of markers on some γ1 chains in mongoloid populations (Fig. 16-14).

Some relations among allotypes suggest that genes for certain γ subclasses might have arisen in evolution by duplication of preexisting alleles of an older subclass (see below, Evolution of immunoglobulins). Thus some heavy-chain markers that are allelic in one γ subclass are uniformly present (i.e., nonallelic) in certain other γ subclasses: e.g., one allelic variant of γ4—Gm(4a)—occurs on all γ1 and γ3 chains and another γ4 variant—Gm(4b)—occurs on all γ2 chains. Hence an ancient γ4a allele might have given rise by duplication to genes that evolved into those of present-day γ1 and γ3 and

* Rare individuals have normal levels of IgG but fail to react with any of the known Gm agglutinators, suggesting that there are additional Gm alleles. However, in some Gm-less persons the γ chains were "hybrids," with the Fd piece of γ3 type, and Fc of γ1 type. This unusual chain probably arose from crossover between γ3 and γ1 genes: it is Gm-less

because Gm markers are not known for the Fd of γ3 and are lacking in the Fc of some γ1 chains. (This γ chain resembles the "hybrid Lepore type" hemoglobin chain, which is half β chain and half δ chain.)

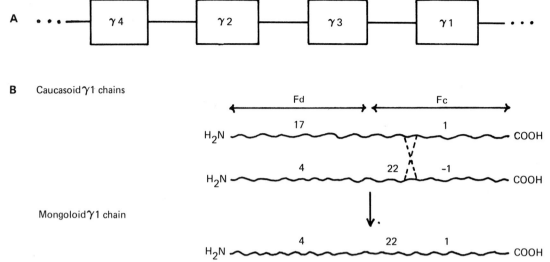

Fig. 16-14. A. Order of linked genes for γ chains of human IgG. The position for γ4 is still tentative. **B.** Alignment of allotypic markers in common γ1 chains. Recombination (dashed lines) between alleles for the caucasoid chains could account for the markers in a mongoloid γ1 chain. [Based on Natvig, J. B., Kunkel, H. G., and Litwin, S. D. *Cold Spring Harbor Symp Quant Biol 32*:173 (1967).]

an ancient γ4b allele might have similarly given rise to γ2; through subsequent mutation-selection the duplicated products would have eventually evolved their own allelic variants while retaining the imprint of the γ4 allele from which they originated.

Amino acid residues associated with Gm and other allotypic markers are given later (see Fig. 16-24).

TABLE 16-3. Common Gene Complexes in Different Human Populations*

Racial group	IgG-1	IgG-2	IgG-3
Caucasoid	····1, 17	·····−23	····21····
	····22, 4	····· 23	···· 5····
Negroid	····1, 17	·····−23	···· 5····
Mongoloid	····1, 17	·····−23	···· 5····
	····1, 17	·····−23	····21····
	···1, 22, 4	····· 23	···· 5····

* The gene complexes are aligned horizontally; see Figure 16-14 for the order of linked genes. Data for the racial distribution of IgG-4 allotypes are not yet available.

Based on Natvig, J. B., Kunkel, H. G., and Litwin, S. D., *Cold Spring Harbor Symp Quant Biol 32*:173 (1967); and on Muir, W. A., and Steinberg, A. G., *Semin Hematol 4*:156 (1967).

Am Allotypes. Genetic variants of IgA were discovered through reactions to blood transfusions (footnote, p. 419, and Ch. 21): recipients who lack an IgA variant can suffer severe reactions when transfused more than once with blood that contains it. (The first transfusion induces formation of the anti-allotype, which reacts with that allotype in subsequent transfusions.) With these human antiallotypes and the hemagglutination assay described above (Fig. 16-13), two heritable variants of α2 chains have been detected in IgA-2, the minor subclass of IgA; by analogy with Gm allotypes of IgG they are called Am(1) and Am(2).

Inv Allotypes. Two allelic variants of κ light chains, differing in a single amino acid (at position 191; see sequences below) are distinguished by agglutinating sera. Kappa chains with leu-191 are called Inv(1,2) and those with val-191 are Inv(3).

The remarkably discriminating agglutinators (Fig. 16-13) for these allotypes thus distinguish between light chains on the basis of a single -CH$_2$- group (in a leucine vs. a valine residue). Moreover two kinds of antisera to Inv(1,2) have been recognized: anti-Inv(1) and anti-Inv(2). They might recognize subtle modifications of the allotypic marker due to pairing of kappa with different heavy chains (see Light chain–heavy chain interaction, below).

Rare individuals are Inv(-1,-2,-3), suggesting additional Inv alleles. The geographical and ethnic distribution is as uneven for Inv as for Gm alleles (Table 16-2): e.g., Inv(1,2) is present in 10 to 20% of Europeans and in >90% of Venezuelan Indians.

Allotypic variants of human λ light chains have not yet been definitely identified.

RABBIT ALLOTYPES

Careful experiments by Oudin in France revealed genetic variants of rabbit Ig chains at the same time that human allotypes were disclosed in Sweden by Grubb's chance observation. Though the principles are the same in man and in rabbit, studies of the experimental animal contributed additional important insights.

Rabbit allotypes were first recognized by precipitin reactions with antisera from rabbits immunized with Abs of other rabbits.* Examination of rabbit populations with a variety of these antisera eventually disclosed three distinct sets of allotypes, each associated with a different Ig chain: three *a* allotypes (a1, a2, a3) on heavy chains, three *b* allotypes (b4, b5, b6) on κ chains (which make up about 80% of rabbit light chains), and two *c* allotypes (c7, c21) on λ chains. Each of these sets is specified by an autosomal locus with codominant alleles. Studies of pedigrees have shown that **the three loci (for heavy, κ, and λ chains) are completely unlinked: they segregate independently, as though located on different chromosomes.**

Rabbit allotypes played a key role in establishing the concept that each Ig chain is coded for by two genes (below), and they were also instrumental in establishing the

identity of heavy chains and the identity of light chains in any given molecule, even when synthesized in a heterozygous animal. For example, successive additions of anti-b4 and anti-b5 precipitate 50 and 30%, respectively, of the Ig molecules in serum from a heterozygous b4,b5 rabbit. When the order of additions is reversed (anti-b5 then anti-b4) the amounts precipitated by each antiallotype are unaltered. Thus anti-b4 does not precipitate molecules with the b-5 marker and anti-b5 does not precipitate those carrying b-4 (Fig. 16-15). Similar findings with pairs of antisera to *a* allotypes and with anti-κ and anti-λ, reinforced by amino acid sequence analyses (below), established that **despite the great diversity of Ig chains synthesized in an individual, any particular Ig molecule has identical heavy chains and identical light chains.** The basis for this principle is that Ig-producing cells synthesize, at any one time, only one kind of heavy chain and one kind of light chain; the chains become irreversibly associated in the cell before they are secreted (Ch. 17).

MOUSE ALLOTYPES

Recognition of allotypes in mice has been simplified by the existence of inbred strains, whose individual members are isogenic (i.e., have essentially the same genome) and are homozygous at virtually all loci. Accordingly, allotype-specific antisera are easily produced by immunizing mice of one strain with Igs from another. A pair of allotypes for each of three IgG subclasses, and three allelic variants of α chains (IgA) have been identified.

Not a single recombinant has been detected in about 2000 progeny of crosses and F_2 back-crosses between inbred strains with different allotypes for heavy chains. Thus genes for γ and α chains appear to be tightly linked, as with the various γs in man (see Fig. 16-14). Allotypes for mouse light chains have not been identified.

---

* Unlike the sensitive hemagglutination assay for human allotypes (Fig. 16-13), precipitin reactions require high titers of Ab (Table 15-5, Ch. 15). Since soluble Igs are poor immunogens, antiallotype sera are produced by immunizing rabbits with other rabbits' aggregated Abs (cross-linked by an Ag), usually incorporated in Freund's adjuvant; the adjuvant is too irritating for use in humans (Ch. 17, Adjuvants).

UNIQUE DETERMINANTS OF INDIVIDUAL IMMUNOGLOBULINS: IDIOTYPES

Unlike particular isotypes and allotypes, which are found on many Igs, a given idiotype is essentially restricted to a particular Ab in a

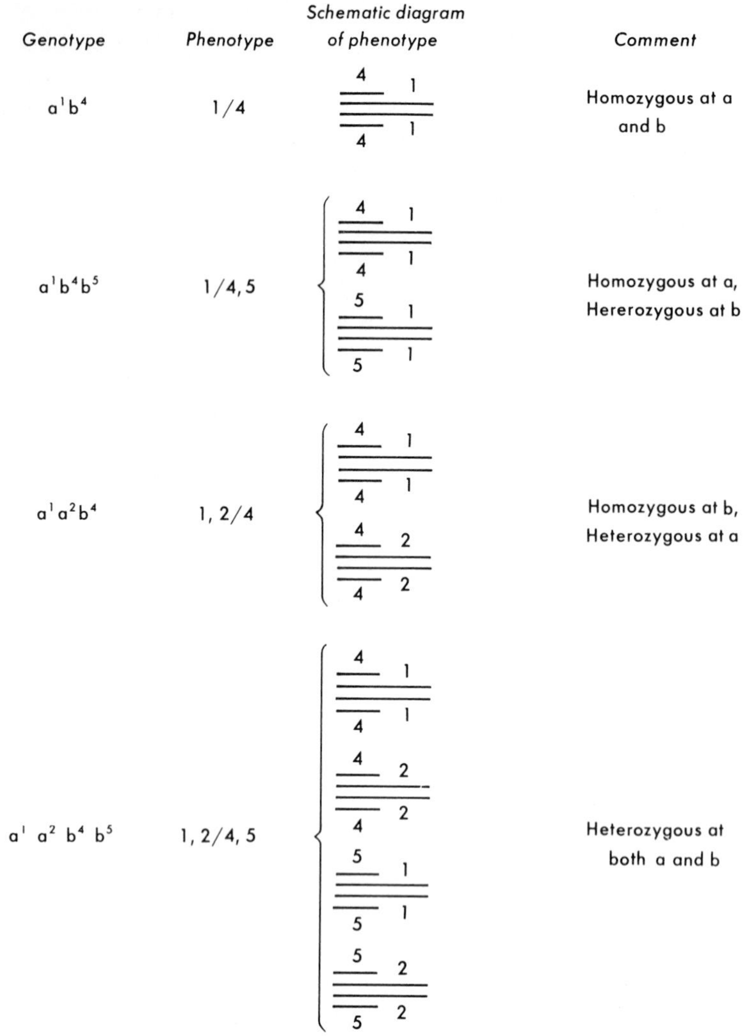

Genotype	Phenotype	Schematic diagram of phenotype	Comment
$a^1 b^4$	1/4		Homozygous at a and b
$a^1 b^4 b^5$	1/4,5		Homozygous at a, Hererozygous at b
$a^1 a^2 b^4$	1, 2/4		Homozygous at b, Heterozygous at a
$a^1 a^2 b^4 b^5$	1, 2/4, 5		Heterozygous at both a and b

Fig. 16-15. Some allotypes of rabbit IgG. [Based on Dray, S., and Nisonoff, A. *Proc Soc Exp Biol Med 113*:20 (1963).]

particular individual.* Idiotypes were discovered by Oudin, who injected anti-*Salmonella* Abs from one rabbit (donor) into recipient rabbits with the same allotypes as the donor. Some recipients formed Abs (antiidiotypes) that reacted specifically with the immunogen but not with other Igs from the donor or to a significant extent with anti-*Salmonella* Abs from other rabbits. It was found later, moreover, that each idiotype-specific antiserum recognizes only a particular subset of Ab molecules in the heterogeneous population of Abs used as immunogen. Idiotypes on some relatively homogeneous human Abs (e.g., to dextrans) have also been identified with rabbit antisera formed against the purified Abs from one person and then absorbed with all the other Igs from the same individual.

Idiotypes are confined to Fab fragments, where they appear to be localized at the ligand-binding sites. Thus, haptens can block competitively the reaction of an antihapten Ab with its antiidiotype. Idiotypes are also

---

* With certain inbred (genetically uniform) strains of mice identical or similar idiotypes are present in some of the Abs (of a given specificity) formed by various members of the strain. With outbred animals, in contrast, there are virtually no idiotypic cross-reactions among Abs of the same specificity from different individuals.

dependent on the singular conformation of appropriately combined light and heavy chains: when the isolated inactive chains reassociate, the antiidiotype usually reacts only with those correctly reconstituted molecules that recover the original Ag-binding activity.

Antigenic properties and amino acid sequences of myeloma proteins, considered in the following sections, reinforce the view that idiotypes are associated with those amino acid residues that are responsible for the uniqueness of each Ab.

HOMOGENEOUS IMMUNOGLOBULINS

MYELOMA AND BENCE JONES PROTEINS

Certain human diseases provide anomalies that are crucial for understanding normal structures and mechanisms, just as microbial mutants play a vital role in elucidating the physiology of wild-type organisms. Long apparent for the endocrine and nervous systems, this generalization has come more recently to apply to the immune response, especially through studies of the Igs secreted by the neoplastic plasma cells in patients with **multiple myeloma.** These **myeloma proteins** have been indispensable in revealing fundamental patterns of amino acid sequences and in the discovery of various light chain types, IgE, IgD, and the subclasses of IgG and IgA. A similar proliferative disease of lymphoid cells that produce IgM is called **Waldenström's macroglobulinemia.** Because the tumor in each patient appears to be a single clone, the homogeneous Ig it secretes is often called a **monoclonal** or **M protein.**

Similarly homogeneous Igs are also produced in occasional older individuals without obvious neoplastic or other proliferative disorders of lymphoid cells (ca. 2% of persons over 70). Mouse plasma cell tumors that secrete M proteins (IgG, IgA, IgM) are of great experimental value, as they can be readily induced in certain inbred strains (BALB/c, NZB) by intraperitoneal injections of mineral oil or implants of lucite shavings.

Many individuals with plasma cell tumors also excrete in the urine a peculiar **Bence Jones** protein, named after the physician who first studied it in the mid-nineteenth century. This protein has unusual thermosolubility properties (in man, but not in mice): it precipitates on being heated to 45 to 60°, redissolves on boiling, and precipitates again on cooling. In contrast, serum Igs, like most other proteins, are coagulated irreversibly on boiling. These urinary proteins came under intensive study after Edelman and Gally and Putnam *et al.* showed that the Bence Jones protein excreted by a patient is nearly always identical with the light chain of his myeloma protein.

Myeloma and Bence Jones proteins are particularly valuable for antigenic and chemical analyses: not only is each homogeneous, but many are also available in large quantities. For example, a patient may have serum levels of a myeloma protein as high as 10 gm/100 ml, and may excrete up to 2 gm of a Bence Jones protein per day in his urine.*

Studies with labeled amino acids show that Bence Jones proteins are synthesized independently of myeloma proteins, rather than derived from them by cleavage. With mouse myeloma tumors, cells that secrete only light chains (Bence Jones protein) arise as frequent variants (possibly mutants) from tumor cells that produce complete myeloma proteins (Ch. 17).

COMPARISON BETWEEN "PATHOLOGICAL" AND NORMAL IMMUNOGLOBULINS

The fundamental identity of pathological and normal Igs has been established not only by their structural similarity but by the finding that some myeloma proteins bind ligands specifically, in the same way as authentic Abs (e.g., at one site per Fab fragment). Moreover, in response to certain immunization procedures occasional animals form Abs that appear to be as homogeneous as myeloma proteins (see below). It is now clear that **normal Igs and conventional Abs are mixtures of many individual proteins, each resembling a myeloma protein in its homogeneity and individuality.**

Homogeneity. The contrast between normal Igs and myeloma proteins can be epitomized by considering an individual whose myeloma

---

* The production of normal Igs and Abs is correspondingly poor, and these patients are sometimes detected clinically because they suffer from repeated bacterial infections.

tumor produces an IgG-1 protein. His normal IgG-1 will have κ and λ chains (Fig. 16-4) and, if he is heterozygous at loci for heavy and light chains, it will contain many allotypes—e.g., Gm(1), Gm(22), Inv(1,2), Inv(3). In contrast, the heavy chains of his myeloma protein will be **either** Gm(1) or Gm(22), its light chains will be **either** κ or λ, and if κ **either** Inv(1,2) or Inv(3). The myeloma protein's light chains would, like the Bence Jones protein he might excrete, appear in acrylamide gel disc electrophoresis as one

Fig. 16-16. Electrophoresis of human sera on a cellulose acetate film. **A.** Normal serum. **B.** Serum with elevated, heterogeneous Igs (polyclonal hypergammaglobulinemia). **C–E.** Sera from three patients with multiple myeloma. Each myeloma protein (arrow) is a compact band (monoclonal Ig); note their different mobilities. **F, G.** Sera from children with virtual absence of immunoglobulins (agammaglobulinemia, see Ch. 17). Serum albumin is the compact dark band at left. (Courtesy of Dr. C. Kirk Osterland.)

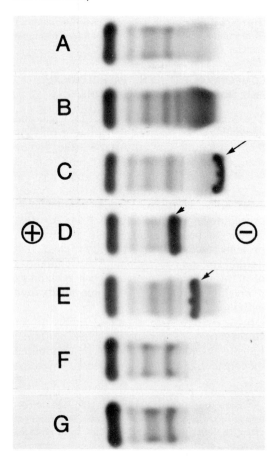

or two bands (monomer and dimer forms of the same chain), not the many bands observed with the light chains of his normal IgG-1 (Fig. 16-18). In addition, with conventional zone electrophoresis the normal Ig would form a broad ("polyclonal") band, whereas the myeloma protein would probably migrate as a compact ("monoclonal") peak (Fig. 16-16). The singularity of the amino acid sequences of a myeloma protein's heavy and light chains (below) also shows that in a given four-chained Ig molecule both light chains are identical, as are both heavy chains.

Though homogeneous in amino acid sequence some myeloma proteins are polydisperse in charge and in sedimentation velocity; this is especially common with IgA myeloma proteins, in which dimers, trimers, and higher multimers of the 7S monomer are formed by the same tumor. Electrophoretic inhomogeneity probably also arises from erratic losses of amide groups from glutamine and asparagine residues, perhaps from the action of plasma amidases.

Individuality. Like other Igs, myeloma proteins have isotypic antigenic markers, corresponding to the class (or subclass) of their heavy chain and the type of their light chain, and also allotypic markers, in accord with the genotype of the person or mouse in whom the tumor arose. Moreover, antisera to individual myeloma proteins reveal that each also has individual (idiotypic) antigenic peculiarities, recognized as spurs in gel diffusion tests. These idiotypic reactions can be blocked competitively, like the idiotypic reactions of conventional Abs, by ligands that are bound specifically by the myeloma protein.

When a pool of normal Igs is used to absorb an antiserum to a myeloma protein, the Abs to its isotypic and allotypic determinants are readily removed, but antiidiotypes are removed, if at all, by enormous amounts (e.g., 100 mg/ml antiserum). Hence the unique determinants in the ligand-binding regions of many individual myeloma proteins, like those of conventional Abs, are each represented at exceedingly low levels (if at all) in normal serum Igs.

HOMOGENEOUS ANTIBODIES

The homogeneity and high serum levels of myeloma proteins have been matched by Abs to bacterial capsular polysaccharides produced by animals after intensive immunization with streptococcal or pneumococcal vaccines (Figs. 16-17 and 16-18). In many of these animals serum Ab levels of 20 to 50 mg/ml are attained, and the Abs are of **restricted heterogeneity,** i.e., they are separable into two or three distinctive, homogeneous Abs.

A rare animal may produce a massive amount of a single Ab, as homogeneous as a myeloma protein, but unlike actual myeloma tumors, production depends on continued administration of immunogen. A clinical analogy is seen in occasional elderly persons with a benign form of Waldenström's macroglobulinemia who produce, without known immunization, two to four monoclonal IgM proteins, some of which react specifically with bacterial polysaccharides. It is possible that these individuals are subjected to prolonged antigenic stimulation by one of their own cryptic microorganisms or perhaps by cross-reacting "self" Ags (see Ch. 17, Tolerance; Ch. 20, Autoimmune disease). Ags that ordinarily elicit highly heterogeneous Abs (such as Dnp-proteins in the production of anti-Dnp Abs) appear to elicit Abs of restricted heterogeneity in animals of certain highly inbred strains, suggesting that genetic control, not just the nature of the Ag, determines **clonal dominance,** the ability of one or

a few clones of Ab-forming cells to overgrow others (Ch. 17, Clonal selection).

AMINO ACID SEQUENCES: THE BASIS FOR IMMUNOGLOBULIN DIVERSITY

Since Abs are highly diverse with respect to ligand-binding activities their combining sites must have diverse shapes. How can this variability be reconciled with the limitations imposed by the over-all structural uniformity described above? The cogency of this question is emphasized by the contrast with enzymes of diverse specificities, which are conspicuously dissimilar in structure (Table 16-4). A general answer emerged from analyses of amino acid sequences, which also established a chemical basis for the distinctions among isotypes, allotypes, and idiotypes, and brought into focus fundamental genetic issues underlying the diversity of Abs and their evolutionary development.

Most preparations of purified Abs are mixtures of many kinds of Ig molecules. Accordingly, the most extensive sequences have been established for myeloma and Bence

Fig. 16-17. Homogeneous antibody (arrow) raised in a rabbit. Serum samples were analyzed by electrophoresis on a cellulose acetate membrane. **A.** Before immunization. **B.** After 16 injections (4 weeks) of streptococcal vaccine (group C); the sample had about 40 mg Ab per milliliter serum. **C.** After absorption of the sample in **B** with purified envelope carbohydrate of group C streptococci. Compact, dark bands of serum albumin are at the right. [From Eichmann, K., Lackland, A., Hood, L., and Krause, R. M. *J Exp Med* **131**:207 (1970).]

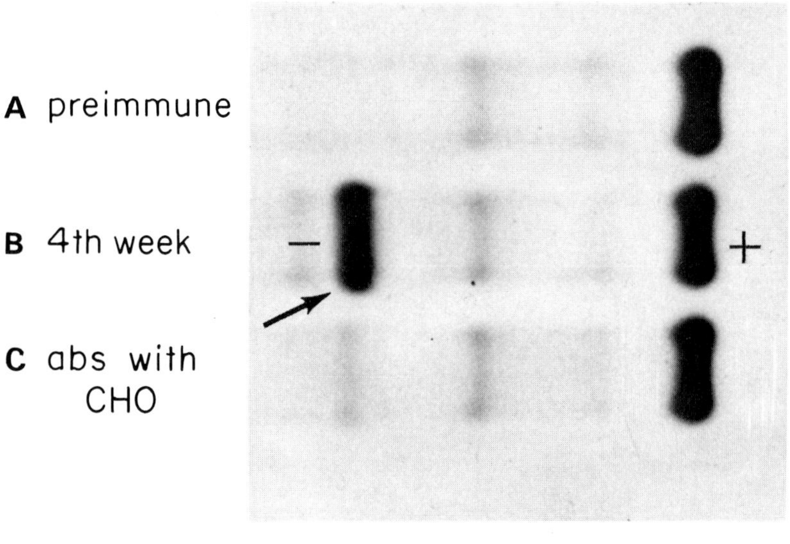

TABLE 16-4. Contrast Between Structural Diversity of Enzymes and Gross Structural Uniformity of Antibodies

Enzymes	Molecular weight	No. of chains	IgG antibodies specific for	Molecular weight	No. of chains
E. coli alkaline phosphatase	80,000	2	Diphtheria toxin	150,000	4
Pancreatic RNase	13,700	1	Bacteriophage T2	150,000	4
Phosphorylase a	495,000	4	Lysozyme	150,000	4
Yeast enolase	67,000	1	Azophenyl-β-lactoside	150,000	4
Glutamic acid dehydrogenase	336,000	6	2,4-Dinitrophenyl	150,000	4

Jones proteins, which are homogeneous. Similar (but less extensive) sequences have been determined for Igs from normal serum, and especially Abs of restricted heterogeneity [elicited by certain bacterial polysaccharides in outbred rabbits (genetically highly diversified) and by hapten-protein conjugates in inbred strains of guinea pigs].

GENERAL FEATURES OF LIGHT AND HEAVY CHAIN SEQUENCES: DOMAINS

Light chains vary slightly in length (211 to 217 residues). Heavy chains are twice as long: γ and α chains have about 450 residues, while μ and ε chains have about 550 residues. All chains consist of linearly repeating, similar

Fig. 16-18. Light chains from purified rabbit antibodies to pneumococcal polysaccharide (type 8) examined by polyacrylamide disc electrophoresis. **A.** From a rabbit with heterogeneous (polyclonal) Abs. **B–D.** From a rabbit after the first **(B)**, second **(C)**, and third **(D)** courses of immunization. Virtually homogeneous (monoclonal) Ab was obtained after the second course **(C)**. Bence Jones proteins and light chains from a myeloma protein migrate as one band (or sometimes as two due to presence of a dimeric form of light chain). [From Pincus, J. H., Jaton, J.-C., Bloch, K. J., and Haber, E. *J Immunol 104*:1149 (1970).]

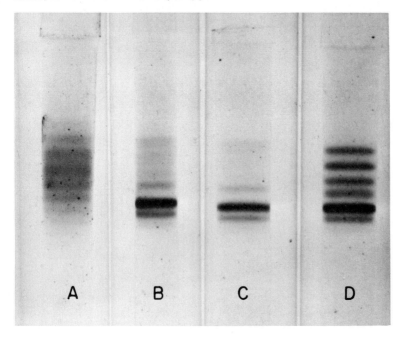

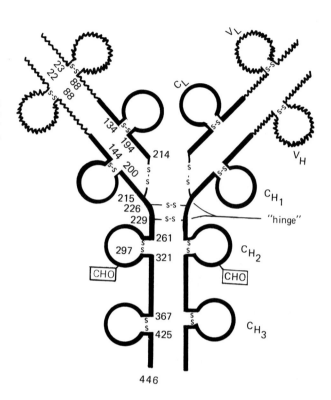

Fig. 16-19. Linear periodicity in amino acid sequences suggests that light (L) and heavy (H) chains have repeating domains, each with about 110 amino acid residues and an approximately 60-membered S-S bonded loop. Domains with variable sequences are represented by jagged lines (V_L, V_H); those with invariant sequences in a given class of H or type of L are represented by smooth lines (C_L, C_{H1}, C_{H2}, C_{H3}; see Fig. 16-21). Numbered positions refer to cysteinyl residues that form S-S bonds, or to the point of attachment of an oligosaccharide (CHO). For other arrangements of interchain S-S bonds see Figure 16-20. [Based on Edelman, G. M. *Biochemistry* 9:3197 (1970).]

(but not identical) segments of about 110 residues: in each segment an S-S bond establishes an approximately 60-membered loop. Similarities in sequence suggest that each segment may be folded into a compact globular **domain,** stabilized by the S-S bond (Fig. 16-19). Adjacent domains are evidently linked by less tightly folded regions: thus pepsin can cut isolated light chains into halves, just as it cleaves between two heavy chain domains in intact Ig molecules in the production of Fab′ fragments (Fig. 16-5). Light chains contain two such domains, and heavy chains contain four (γ and α chains) or five (μ and ε chains). (Occasional domains have additional, smaller S-S bonded loops.)

Other cysteine residues form the interchain S-S bonds that link chains within molecules. The interchain bonds differ in number and position in different classes and subclasses of heavy chains (Fig. 16-20) and types of light

chains; e.g., the light-heavy interchain S-S bond involves a cysteine that is C-terminal in κ chains and penultimate in λ chains (see Fig. 16-23).

VARIABLE (V) AND CONSTANT (C) SEGMENTS

The first two Ig chains that were sequenced (human κ chains by Hilschmann and Craig and by Putnam *et al.*) revealed a remarkable pattern that has since been found consistently in all others. The two chains had different amino acid residues at many positions; but, strikingly, the differences were all clustered in the **amino-terminal half** of the chain, now called the variable or V segment or V_κ (Fig. 16-21). In the carboxyl half of the chain, called the constant or C segment (or C_κ), the sequences were identical except at position 191, where leucine and valine occur as

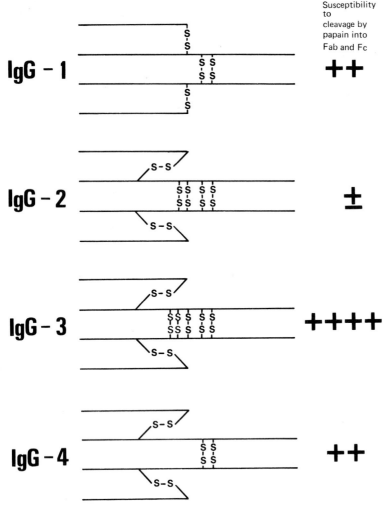

Susceptibility to cleavage by papain into Fab and Fc

Fig. 16-20. Interchain S-S bridges in different IgG subclasses. The variations contrast with the virtual constancy of intrachain S-S bonds (Fig. 16-19). The heavy-heavy interchain bridges are in the hinge region, and their variations may be related to other differences among IgG subclasses, e.g., in susceptibility to papain digestion (Table 16-2). The heavy-chain cysteinyl residue forming the heavy-light S-S bridge in IgG-2,-3,-4 is about 100 residues closer to the amino terminus (left) than in IgG-1. [Based on data from Frangione, B., Milstein, C., and Pink, J. R. L. *Nature 211*:145 (1969).]

heritable variants (see Inv allotypes, above). A similar pattern is evident among human λ light chains, which also differ at many positions in the amino-terminal domain (V_λ), but have essentially the same sequence in the carboxy-terminal domain (C_λ).

Alternatives are found at two positions in the C_λ domain of λ chains: 153, ser-gly, and 190, lys-arg. Antisera can differentiate between λ chains with arg-190 and those with lys-190 (called Oz+ and Oz−, respectively, after the patient whose Bence Jones protein elicited the distinguishing antiserum). All of the variants at positions 153 and 190 have been found, in various combinations, among the peptides in proteolytic digests of light chains from each of the 30 persons tested. Hence the C_λ variants are probably isotypes (i.e., present in all individuals), rather than allotypes like the Inv variants of C_κ.*

Heavy chains also differ extensively in sequence in the amino-terminal domain (V_H), but within a class or subclass the sequence in the remainder of the chain is identical (C_H segment), except for a few substitutions

_____

* If the C_λ variants were allotypes the homozygous state (e.g., corresponding exclusively to λ_{arg} or to λ_{lys}) ought to have been observed even in a sample as small as 30 individuals.

that correspond to allotypic markers. The V_H segments are slightly longer than those of V_κ and V_λ (ca. 115 as compared with ca. 107 residues).

Though many positions in V segments are variable many others are invariant (or show little variation) within a given type (V_κ, V_λ, or V_H; see Fig. 16-21). For instance, the amino terminus of most V_λ segments is a cyclized glutamine residue (pyrrolidone carboxyl) without a free NH$_2$ group, whereas all V_κ sequences begin with a free amino group (usually of glu or asp). **The V_κ sequences are always associated with C_κ, and V_λ with C_λ. In contrast, the V_H sequences are not restricted to particular classes or subclasses of heavy chains.** The significance of these strict V-C associations is considered later in this chapter.

The over-all characteristics of amino acid sequences of Igs, derived initially from partial sequences of many chains, has been firmly established by Edelman *et al.,* who determined the complete sequences of both light and heavy chains in a myeloma protein (Fig. 16-22). Sequences that distinguish diverse chains and allotypic variants are illustrated in Figures 16-23 and 16-24.

V SUBGROUPS AND HYPERVARIABLE REGIONS

Comparison of many chains has revealed that sequence variability in V segments is dis-

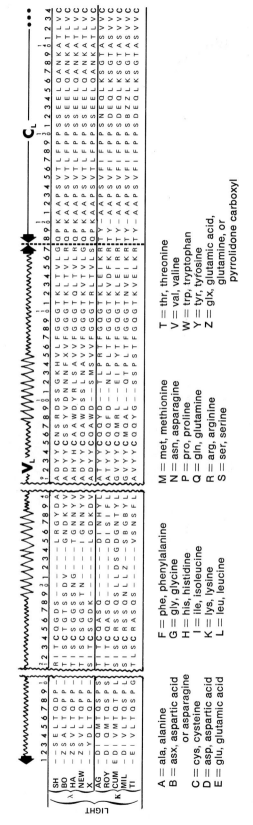

Fig. 16-21. Variable (V) and constant (C) segments illustrated by selected amino acid sequences from some human light chains (named SH, etc., for the myeloma patient from whom the protein was derived). Sequences are aligned for maximal homology. At some positions in V segments the residues seem to be invariant in all light chains, including those from mice (e.g., Thr-6, Tyr-93, Gln-7, Phe-107, Gly-108, Gly-110, Thr-111). Amplitude of the jagged line above represents extent of variability. (Based on references in Dayhoff, M. ed. *Atlas of Amino Acid Sequence and Structure,* vol. 5. National Biomedical Research Foundation, Silver Spring, Md., 1972.)

A = ala, alanine
B = asx, aspartic acid or asparagine
C = cys, cysteine
D = asp, aspartic acid
E = glu, glutamic acid
F = phe, phenylalanine
G = gly, glycine
H = his, histidine
I = ile, isoleucine
K = lys, lysine
L = leu, leucine
M = met, methionine
N = asn, asparagine
P = pro, proline
Q = gln, glutamine
R = arg, arginine
S = ser, serine
T = thr, threonine
V = val, valine
W = trp, tryptophan
Y = tyr, tyrosine
Z = glx, glutamic acid, glutamine, or pyrrolidone carboxyl

HEAVY CHAIN (1–446)

```
1                                                                              20
PCA-VAL-GLN-LEU-VAL-GLN-SER-GLY-ALA-GLU-VAL-LYS-LYS-PRO-GLY-SER-SER-VAL-LYS-VAL-
                                                                               40
SER-CYS-LYS-ALA-SER-GLY-GLY-THR-PHE-SER-ARG-SER-ALA-ILE-ILE-TRP-VAL-ARG-GLN-ALA-
                                                                               60
PRO-GLY-GLN-GLY-LEU-GLU-TRP-MET-GLY-GLY-ILE-VAL-PRO-MET-PHE-GLY-PRO-PRO-ASN-TYR-
                                                                               80
ALA-GLN-LYS-PHE-GLN-GLY-ARG-VAL-THR-ILE-THR-ALA-ASP-GLU-SER-THR-ASN-THR-ALA-TYR-
                                                                              100
MET-GLU-LEU-SER-SER-LEU-ARG-SER-GLU-ASP-THR-ALA-PHE-TYR-PHE-CYS-ALA-GLY-GLY-TYR-
                                                                              120
GLY-ILE-TYR-SER-PRO-GLU-GLU-TYR-ASN-GLY-GLY-LEU-VAL-THR-VAL-SER-SER-ALA-SER-THR-
                                                                              140
LYS-GLY-PRO-SER-VAL-PHE-PRO-LEU-ALA-PRO-SER-SER-LYS-SER-THR-SER-GLY-GLY-THR-ALA-
                                                                              160
ALA-LEU-GLY-CYS-LEU-VAL-LYS-ASP-TYR-PHE-PRO-GLU-PRO-VAL-THR-VAL-SER-TRP-ASN-SER-
                                                                              180
GLY-ALA-LEU-THR-SER-GLY-VAL-HIS-THR-PHE-PRO-ALA-VAL-LEU-GLN-SER-SER-GLY-LEU-TYR-
                                                                              200
SER-LEU-SER-SER-VAL-VAL-THR-VAL-PRO-SER-SER-SER-LEU-GLY-THR-GLN-THR-TYR-ILE-CYS-
                                                                              220
ASN-VAL-ASN-HIS-LYS-PRO-SER-ASN-THR-LYS-VAL-ASP-LYS-ARG-VAL-GLU-PRO-LYS-SER-CYS-
                                      H                        H              240
ASP-LYS-THR-HIS-THR-CYS-PRO-PRO-CYS-PRO-ALA-PRO-GLU-LEU-LEU-GLY-GLY-PRO-SER-VAL-
                                                                              260
PHE-LEU-PHE-PRO-PRO-LYS-PRO-LYS-ASP-THR-LEU-MET-ILE-SER-ARG-THR-PRO-GLU-VAL-THR-
                                                                              280
CYS-VAL-VAL-VAL-ASP-VAL-SER-HIS-GLU-ASP-PRO-GLU-VAL-LYS-PHE-ASN-TRP-TYR-VAL-ASP-
                                               CHO                            300
GLY-VAL-GLU-VAL-HIS-ASN-ALA-LYS-THR-LYS-PRO-ARG-GLU-GLU-GLN-TYR-ASN-SER-THR-TYR-
                                                                              320
ARG-VAL-VAL-SER-VAL-LEU-THR-VAL-LEU-HIS-GLN-ASN-TRP-LEU-ASP-GLY-LYS-GLU-TYR-LYS-
                                                                              340
CYS-LYS-VAL-SER-ASN-LYS-ALA-LEU-PRO-ALA-PRO-ILE-GLU-LYS-THR-ILE-SER-LYS-ALA-LYS-
                                                                              360
GLY-GLN-PRO-ARG-GLU-PRO-GLN-VAL-TYR-THR-LEU-PRO-PRO-SER-ARG-GLU-GLU-MET-THR-LYS-
                                                                              380
ASN-GLN-VAL-SER-LEU-THR-CYS-LEU-VAL-LYS-GLY-PHE-TYR-PRO-SER-ASP-ILE-ALA-VAL-GLU-
                                                                              400
TRP-GLU-SER-ASN-GLY-GLN-PRO-GLU-ASN-TYR-LYS-THR-THR-PRO-PRO-VAL-LEU-ASP-SER-ASP-
                                                                              420
GLY-SER-PHE-PHE-LEU-TYR-SER-LYS-LEU-THR-VAL-ASP-LYS-SER-ARG-TRP-GLN-GLN-GLY-ASN-
                                                                              440
ASN-VAL-PHE-SER-CYS-SER-VAL-MET-HIS-GLU-ALA-LEU-HIS-ASN-HIS-TYR-THR-GLN-LYS-SER-
                446
LEU-SER-LEU-SER-PRO-GLY
```

Fig. 16-22. Complete amino acid sequence of the light (χ) and heavy (γ1) chains of a human IgG1 myeloma protein (EU). Intrachain SS bonds are indicated by shaded bands and the interchain SS bridge between heavy (H) and light (L) chains by arrows. CHO is an oligosaccharide attached to the Asx residue (aspartic acid or asparagine) at position 297 (see Fig. 14-9). [Based on Edelman, G. M. *Biochemistry* 9:3197 (1970).]

LIGHT CHAIN (1–214)

```
1                                                                              20
ASP-ILE-GLN-MET-THR-GLN-SER-PRO-SER-THR-LEU-SER-ALA-SER-VAL-GLY-ASP-ARG-VAL-THR-
                                                                               40
ILE-THR-CYS-ARG-ALA-SER-GLN-SER-ILE-ASN-THR-TRP-LEU-ALA-TRP-TYR-GLN-GLN-LYS-PRO-
                                                                               60
GLY-LYS-ALA-PRO-LYS-LEU-LEU-MET-TYR-LYS-ALA-SER-SER-LEU-GLU-SER-GLY-VAL-PRO-SER-
                                                                               80
ARG-PHE-ILE-GLY-SER-GLY-SER-GLY-THR-GLU-PHE-THR-LEU-THR-ILE-SER-SER-LEU-GLN-PRO-
                                                                              100
ASP-ASP-PHE-ALA-THR-TYR-TYR-CYS-GLN-GLN-TYR-ASN-SER-ASP-SER-LYS-MET-PHE-GLY-GLN-
                                                                              120
GLY-THR-LYS-VAL-GLU-VAL-LYS-GLY-THR-VAL-ALA-ALA-PRO-SER-VAL-PHE-ILE-PHE-PRO-PRO-
                                                                              140
SER-ASP-GLU-GLN-LEU-LYS-SER-GLY-THR-ALA-SER-VAL-VAL-CYS-LEU-LEU-ASN-ASN-PHE-TYR-
                                                                              160
PRO-ARG-GLU-ALA-LYS-VAL-GLN-TRP-LYS-VAL-ASP-ASN-ALA-LEU-GLN-SER-GLY-ASN-SER-GLN-
                                                                              180
GLU-SER-VAL-THR-GLU-GLN-ASP-SER-LYS-ASP-SER-THR-TYR-SER-LEU-SER-SER-THR-LEU-THR-
                                                                              200
LEU-SER-LYS-ALA-ASP-TYR-GLU-LYS-HIS-LYS-VAL-TYR-ALA-CYS-GLU-VAL-THR-HIS-GLN-GLY-
                214
LEU-SER-SER-PRO-VAL-THR-LYS-SER-PHE-ASN-ARG-GLY-GLU-CYS ──→ H-220
```

tributed in a remarkable pattern, with two important features. First, those of a given type (V$_\kappa$, V$_\lambda$, V$_H$) can be arranged into sets, or **subgroups,** each of which is characterized by distinctive amino acids at certain positions, dispersed at unequal intervals throughout almost the entire length of the V segment. As is shown in Figure 16-25, for example, human V$_\kappa$ sequences fall into three subgroups; five subgroups have similarly been recognized in human V$_\lambda$ and three in human V$_H$. V segments that belong to the same subgroup differ in about 10 to 15 of their approximately 110 residues; those from different subgroups of the same type (e.g., V_{κ_I} vs. $V_{\kappa_{II}}$) differ in about 25 to 35 residues. V segments of a given subgroup can be arranged in a genealogical order, as though the structural genes coding for them were derived sequentially by a series of single-step mutations (Fig. 16-26). Though divisions into subgroups are somewhat arbitrary, these sets are of considerable importance for theories concerned with the genetic basis for the diversity of Igs (see below).

The second important feature is that variability is especially pronounced at a few restricted regions, of which there are three in light chains (24-34, 50-55, 89-97), and four in heavy chains at approximately homologous positions (31-37, 50-65, 86-91, 101-109) (Fig. 16-27). When V segments of various chains are aligned for comparison it is necessary to introduce gaps (or insertions) into some of them in order to achieve maximal sequence homology, and the gaps also tend to cluster in the hypervariable regions. Chemical studies of the combining sites suggest that they are formed by the hypervariable regions (Affinity-labeling, below).

THE HINGE

Near the middle of heavy chains a special stretch of about 15 residues includes all the cysteines that form the heavy-heavy interchain S-S bonds. This region probably corresponds to the "hinge" in the Y-shaped molecule (Figs. 16-6 to 16-8). Amino acid **sequences at the hinge are distinctive for each class and subclass of heavy chains** (Fig. 16-28); **they are not homologous with other parts of light or heavy chains.** Diversity of sequences in the hinge region may be associated with some of the different biological properties of different Ig classes, e.g., in natural half life (see Appendix, Table 16-5).

COOH ENDS →

Human γ1 ...	Z	K	S	L	S	L	S	P	G
Human γ2 ...	Z	K	S	L	S	L	S	P	G
Human γ3 ...	Z	K	S	L	S	L	S	P	G
Human γ4 ...	Z	K	S	L	S	L	S	L	G
Human α ...	M	A	E	V	D	G	T	C	Y
Human μ ...	M	S	B	T	A	G	T	C	Y
Human κ ...	K	S	F	N	R	G	E	C	
Human λ ...	K	T	V	A	P	T	E	C	S
Mouse α ...	M	S	E	G	D	G	T	C	Y
Mouse κ ...	K	S	F	N	R	N	E	C	
Mouse λ1 ...	K	S	L	S	R	A	D	C	S
Mouse λ2 ...	K	S	L	S	P	A	E	C	L

Fig. 16-23. Representative C-terminal nonapeptides illustrating differences between heavy-chain classes and subclasses, and between κ and λ light chains. The cysteine that is C terminal in κ and penultimate in λ chains contributes to the S-S bonds that link light and heavy chains (see Fig. 16-22). For abbreviations, see legend, Figure 16-21. (Based largely on references in Dayhoff, M., ed., *Atlas of Amino Acid Sequence and Structure*, vol. 5. National Biomedical Research Foundation, Silver Spring, Md., 1972.)

Fig. 16-24. Amino acid substitutions associated with some allelic allotypes. Each substitution can be accounted for by a single base difference in the corresponding codon (e.g., Phe → Tyr is UU_C^U → UA_C^U). (Based on references in Grubb, R. *The Genetic Markers of Human Immunoglobulins*, Ch. 3. Springer, New York, 1970.)

Allotypes	Amino acid substitutions	Position of substitution	Domain
Gm(1) Gm(-1)	··· Arg–**Asp**–Glu–**Leu**–Thr ··· ··· Arg–**Glu**–Glu–**Met**–Thr ···	356,358	C_{H3} of γ1
Gm(4) Gm(-4)	··· Asp–Lys–**Arg**–Val–Glu ··· ··· Asp–Lys–**Lys**–Val–Glu ···	214	C_{H1} of γ1
Gm(5) Gm(21)	··· Arg–**Phe**–Thr–Gln–Lys ··· ··· Arg–**Tyr**–Thr–Gln–Lys ···	Ca. 436	C_{H3} of γ3
Inv(1, 2) Inv(3)	··· Lys–**Leu**–Tyr–Ala–Cys ··· ··· Lys–**Val**–Tyr–Ala–Cys ···	191	C_κ

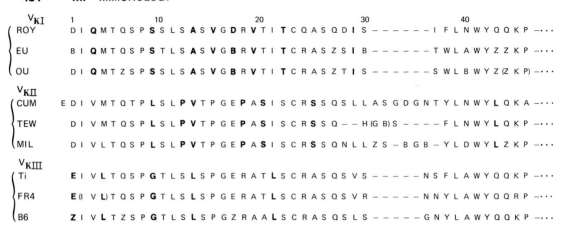

Fig. 16-25. V_K subgroups. Amino acid sequences at N-terminal region of some human κ chains demonstrate three subgroups. Subgroup-specific residues are in boldface. Alignment maximizes homology; dashes (gaps) represent deletions (or insertions). Residues in parentheses are known by peptide composition, and not actually sequenced. See Figure 16-21 for abbreviations. [Data are from several laboratories, reviewed by Hood, L., and Prahl, J. *Adv Immunol 14*:291 (1971).]

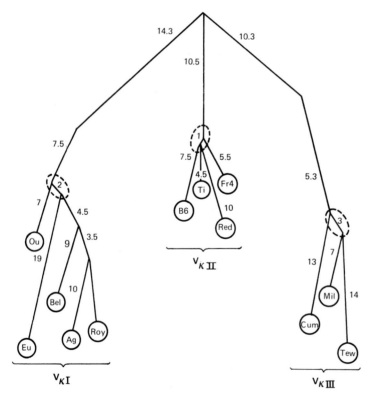

Fig. 16-26. A genealogical tree for human V_κ domains, based on residue-by-residue comparison of several κ chains. The numbers alongside each branch represent the average number of nucleotide differences required after each nodal point to code for the observed amino acid differences. Divergences of uncertain order are enclosed in a dashed line. Similar genealogies are characteristic of families of proteins that arise in evolution by successive mutations (Ch. 12). The circled name or number identifies the sequenced Bence Jones protein. [Based on Smith, G. P., Hood, L., and Fitch, W. M. *Annu Rev Biochem 40*:969 (1971).]

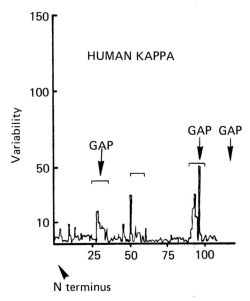

Fig. 16-27. Hypervariable regions in V segments of human κ chains. Variability (*v*) is defined as the number (*n*) of different amino acids at a given position in the chains under comparison divided by the frequency (*f*) of the most common amino acid at that position ($v = n/f$). Variability can range from 1.0 (no variation) to 400. Regions with greatest variability are 89 to 97 in V_L and 95 to 102 V_H domains. [From Kabat, E. A., and Wu, T. T. *Ann NY Acad Sci 190*:382 (1971).]

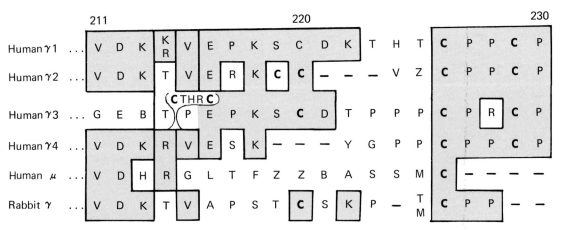

Fig. 16-28. Hinge region of some heavy chains. Shaded residues are identical with γ1. Cys in boldface form interchain heavy-heavy S-S bonds (see Fig. 16-20); cys not in boldface (220 in human γ1) forms interchain heavy-light S-S bonds. The alternatives at 214 in human γ1 and at 225 in rabbit γ are associated with allotypes (Fig. 16-26). The sequences are aligned with an insertion in γ3 and gaps in other chains in order to maximize homology with γ1. Other than the differences shown, the various human γ chains (subclasses) seem to be identical at almost all other positions in their respective C regions (e.g., Fig. 16-23). For abbreviations, see legend, Figure 16-21. (Based on references in Dayhoff, M., ed., *Atlas of Amino Acid Sequence and Structure*, vol. 5. National Biomedical Research Foundation, Silver Spring, Md. 1972.)

In γ chains the hinge region contains an unusual abundance of proline and hydrophilic residues (lysine, aspartic), which promote the exposure of this region, without tight folding, to solvent. This feature is probably related to the susceptibility of this region to attack by proteases, and to the apparent flexibility of the Fab arms of the Y-shaped molecule. The hinge region of μ chains is not rich in proline; instead it has an attached oligosaccharide, which might similarly promote exposure to solvent.

GENETIC BASIS OF IMMUNOGLOBULIN STRUCTURE

TWO GENES—ONE CHAIN

The finding that chains of the same type have identical sequences in the C domain but different sequences in the V domain seemed difficult to reconcile with the doctrine that every polypeptide chain is coded for by one gene. The mendelian inheritance of C-region allotypes implies that each C sequence is coded for by a **single** structural gene. But there must be **multiple** V genes, for each V subgroup must be specified by at least one gene. Otherwise, many parallel and independent somatic mutations would be required to generate the many members of each subgroup in each individual, and this seems most unlikely. The generally accepted solution (based on an early suggestion by Dreyer and Bennett) is that any one of many V genes can be associated with a single C gene in the synthesis of an Ig chain: i.e., **each immunoglobulin chain seems to be coded for by two genes, one for the V segment and one for the C segment.**

The following observations support this hypothesis:

1) In addition to the a allotypes (Rabbit allotypes, above) the γ chains of rabbit IgG contain genetic markers *d11* and *d12*, determined by alleles at the d locus. Markers a and d are linked but can separate in out-breeding populations (by recombination?), and mutations in a and d alter different parts of the same γ chain. The residues associated with various a markers seem to be localized in the V_H domain and appear to occupy several (ca. 15) widely dispersed positions near the amino terminus, whereas the single residue associated with the d

marker (a methionine for *d11* and a threonine for *d12*) is located near the middle of the chain (position 225; Fig. 16-28).* Thus one gene seems to code for the amino terminal section (about 115 residues) and the other gene for the remainder of the chain; and somehow they become associated in specifying a complete chain. The finding that the **same** a allotypic markers, located in the N-terminal section, can occur on various classes of heavy chains (γ, μ, α, ε; the Todd phenomenon) also supports the two genes–one chain hypothesis, and, indeed, first prompted its serious consideration.

2) About 1% of patients with multiple myeloma form two distinct myeloma proteins. In one patient with such a "biclonal" tumor the IgM and IgG myeloma proteins had the same idiotype, which became understandable when it was later found that the μ and γ chains of the two proteins had the same V_H sequence and were associated with the same light chain.

TRANSLOCATION HYPOTHESIS

If it is true that two genes specify one polypeptide chain, how do they cooperate? Do the genes fuse, or are their separate products joined in cells at the level of mRNA or polypeptide? Suggestive evidence for fusion at the DNA level is provided by rare patients who produce unusually short heavy chains ("heavy-chain disease"). Secreted in urine, these chains (γ and α) lack some of the carboxyl (internal) end of the V_H segment and some of the adjacent amino end of the C_H segment. They can be understood as the product of an aberrant recombination between a V and a C gene with a large internal deletion (for which there is ample genetic precedent). The alternatives would not only require a novel mechanism for joining mRNA or polypeptide molecules, but would require that the mechanism be able to function without the ends that are normally ligated.

The fusion of V and C genes is also con-

---

* The association of several different amino acid residues with each a allotype is an exception to the rule that allelic polypeptide chains generally differ from each other in only one or two residues per 100; e.g., normal and sickle cell α hemoglobin chains, Inv allotypes of human κ light chains (above).

sistent with evidence that in Ab-forming cells (pulse-labeled with radioactive amino acids) the nascent Ig chains elongate continuously from the amino terminus, without interruption, as do proteins in general. In addition, the RNA for an Ig light chain has been found to be large enough to code for the complete chain.

As noted before, there are stringent limitations to the association between V and C segments: i.e., V_κ sequences are always associated with C_κ, and V_λ sequences with C_λ. V_H sequences, however, are not differentiated among the various C_H classes (C_γ, C_μ, C_α, etc.). A basis for this rule has been suggested by the finding that the genes for κ light chains, those for λ light chains, and the set of closely linked genes for the diverse classes of heavy chains segregate independently and hence are probably on separate chromosomes. **This separateness could account for the V-C restrictions if only V and C genes on the same chromosome can fuse to code for a given chain** (Fig. 16-29A,B).

Supporting evidence for this view is provided by a study of two pairs of allelic markers in different parts of the same heavy (γ) chains of rabbit IgG: *a1* and *a3* in the V_H segment, *d11* and *d12* in the $C\gamma$ segment. When homozygous rabbits were crossed (a1,d11 $\times$ a3,d12) the doubly heterozygous progeny had the expected variety of markers, but their individual γ chains were exclusively of parental type: i.e., **either** a1,d11 or a3,d12. No IgG molecules with a mixture of parental allotypes were detected. (As noted above, however, recombination between the *a* and *d* markers must occur occasionally, because various combinations are found in large outbreeding populations; e.g., a1,d12 and a3,d11.)

One scheme for joining separate V and C genes is shown in Figure 16-29B, which depicts a chromosome with a linear array of

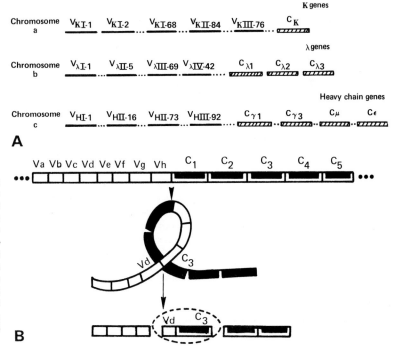

Fig. 16-29. A. Studies with allotypic markers show that the three families of V and C genes (κ, λ, heavy) segregate independently, as though located on different chromosomes (a, b, c). **B.** Scheme for translocation of a V to a C gene, forming a complete structural gene ($V_d \cdot C_3$) for an Ig chain. [**B** derived from Dreyer, W. J., and Bennett, J. C. *Proc Nat Acad Sci USA* 54:864 (1965).]

linked V_H and C_H genes. One of many V_H genes is presumed to move next to and join a C_H gene further along on the same chromosome by crossover at a region of homology, as in the integration of a temperate phage into the bacterial chromosome. [This crossover might recognize nucleotide sequences at positions corresponding to the ends that become ligated: thus the N-terminal 3 to 4 amino acid residues of C_{HI} segments are the same in γ, μ and α chains.]

ORIGIN OF DIVERSITY

In any particular species the number of different C segments is limited: in man there are 4 or 5 for C_L and about 10 for C_H, and where allotypic variants of these sequences are known they are inherited as uncomplicated mendelian alleles. It appears therefore that each C segment is coded for by a single structural gene, and transmitted from generation to generation in the germ line. With V segments, however, the number of distinctive amino acid sequences is enormously greater: any individual must be able to synthesize thousands of different V_L and V_H sequences in producing an immense variety of Abs. Figure 16-30 summarizes the main theories that try to account for the origin of this great multitude of V sequences.

Germ Line Theory. According to this theory germ cells carry structural genes for all the V_L and V_H sequences an individual can synthesize. These genes would have arisen during evolution through conventional mechanisms for gene duplication, mutation, and selection. The germ line theory not only provides an explanation within the known framework of molecular genetics but also readily accounts for the genealogical relation among V sequences (Fig. 16-26) and their orderly, though arbitrary, arrangements into sub-

Fig. 16-30. Theories of origin of diversity of immunoglobulin V genes. (Based on Edelman, G. M. In Neurosciences, Second Study Program, F. O. Schmitt (ed.), Rockefeller Univ. Press, N.Y., 1970.)

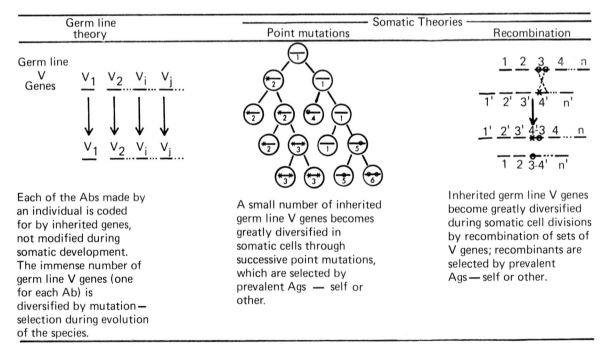

Germ line theory	Somatic Theories	
	Point mutations	Recombination

Each of the Abs made by an individual is coded for by inherited genes, not modified during somatic development. The immense number of germ line V genes (one for each Ab) is diversified by mutation—selection during evolution of the species.

A small number of inherited germ line V genes becomes greatly diversified in somatic cells through successive point mutations, which are selected by prevalent Ags — self or other.

Inherited germ line V genes become greatly diversified during somatic cell divisions by recombination of sets of V genes; recombinants are selected by prevalent Ags — self or other.

groups. Moreover, the required number of V genes could easily be accommodated in the vertebrate genome.

A typical diploid mammalian cell has about 10^{-11} g DNA, which could code for about 4×10^9 amino acid residues in proteins.* Hence as many as 1×10^4 V genes (each coding for 110 residues) would occupy only about 0.025% of the genome (e.g., $1 \times 10^4 \times 110/4 \times 10^9$). If the number of V_L and V_H genes were approximately equal and their products could pair at random, and if each paired V_L-V_H sequence established a unique, specific combining site, this number of genes could generate about 1×10^8 different Abs—or perhaps 1×10^7 if only 10% of the V_L-V_H pairs resulted in stable Ig molecules (see Reconstruction of combining sites, below). This number is probably sufficient to account for the apparently limitless diversity of antigenic determinants to which an individual can make an immune response.

There are, however, difficulties with the germ line theory. Foremost, perhaps, is the implication that the preservation and diversification of thousands of germ line V genes throughout evolution is governed, like that of other genes, by selection for survival value and loss by genetic drift: it is difficult to see how this mechanism would provide Abs to Ags that probably never existed in the past history of a species, such as those of recently synthesized organic chemicals (e.g., benzenearsonate). The preservation of so many unused genes during eons of evolution seems impossible unless a given Ig can react specifically with many structurally dissimilar ligands, and such strange cross-reactions have not been generally found. However, some do occur: e.g., between anti-Dnp Abs and a 1,4-naphthoquinone (menadione) that resembles the vitamin K-like molecules produced by common intestinal bacteria.

Another difficulty is that a few positions have the same residues in all V sequences of a given type in a particular species (species-specific residues) or in a given allotype. It is difficult to see how such invariant species-specific and allotype-specific codons can be present in thousands of independent germ line V genes, for this development would seem to have required a highly unlikely series of parallel, independent mutations plus exceedingly strong selective pressure for these residues. Moreover, if the same allotype-specific residues are present in thou-

sands of V sequences, recombination among the multitude of similar alleles in heterozygous individuals must somehow be prevented. Otherwise these allotypes would not be recognizable as distinct mendelian markers.

Though severe, these difficulties may not be insurmountable. Hood has suggested that sets of V genes might undergo rapid expansion and contraction in evolution through preferential duplication of particular genes in a set, and that this process could lead to the appearance of species-specific or allotype-specific residues. Moreover, a precedent exists for the absence of recombination among genes of certain chromosomes, e.g., in male *Drosophila*.

Somatic Mutation Theory. This theory maintains that a small number of germ line genes become highly diversified through mutation in somatic cells, yielding differentiated clones of immunologically competent cells (Ch. 17, Lymphocytes) that differ in V genes. Though the number of V genes per somatic cell is presumed to be as small as in germ cells, their variety in the individual animal could be immense if there were a sufficient number of different cells (lymphocytes).

The severest problem faced by this theory is that of accounting for the accumulation of a vast number of different sequences during the period between fertilization of an egg and acquisition of immunological competence by the embryo (Ch. 17, Ontogeny). Special mechanisms have been proposed for increasing mutation rates in cells destined to make Abs, but they seem unnecessary: even with ordinary mutation rates (ca. 10^{-7}/base pair/cell division) the mutation rate per V gene might well be about one per 10^5 cell divisions, since there are ca. 330 nucleotide base pairs per V gene and perhaps 10% of the mutations would be compatible with a functional Ig chain. It is likely that many (e.g., 1000) immunologically competent cells with V gene mutations could arise each day among the special sets of rapidly dividing lymphocytes of even so small an animal as the mouse (e.g., the ca. 1×10^8 cells of the thymus; see Ch. 17). However, to account for the observed diversity of V sequences the cells with mutations would have to be selected for, relative to the vastly greater number of cells with unmutated germ line V genes.

The effects of an immunogen on the corresponding Ag-sensitive cell (Ch. 17) suggest that if a cell expressing a particular mutant V gene should happen to encounter the appropriate Ag the cell would be stimulated to proliferate (positive selection). The need for a large number and variety of selective Ags during ontogeny is bypassed in Jerne's proposal

---

* On the basis of a triplet code and a weight of 600 daltons for each nucleotide pair, 1×10^{-11} g DNA/$3 \times 600 = 6 \times 10^{-15}$ moles amino acid equivalents; and $6 \times 10^{-15} \times 6 \times 10^{23}$ (Avogadro's number) $= 36 \times 10^8$ amino acid residues in proteins.

for **negative selection:** germ line V genes are assumed to code for V domains that are specific for certain self-Ags (histocompatibility Ags on cell surfaces, Ch. 21), leading to suppression of the cells that express these anti-self genes, as part of the normal mechanism for maintaining self-tolerance (Ch. 17). Hence cells with V gene mutations coding for Igs that are specific for *anything* other than self-Ags would be preferentially preserved. The resulting selection would thus operate in favor of mutations in hypervariable regions (which determine Ab specificity, see below), without requiring the fortuitous presence of the corresponding Ags. Random mutations in C genes and in relatively invariant regions of V genes probably also occur but would not accumulate, either because they are selected against (if, for instance, they are incompatible with a functional Ig molecule) or because they are neutral and not subject to any selective pressure.

Somatic Recombination of Germ Line Genes. According to this theory, diversity arises in somatic cells through recombination among a limited number of germ line V genes for each subgroup. If there are *a* different codons at each of *n* positions, unlimited recombination could yield a^n sequences. If, for example, there were several germ line V genes for a given subgroup with 12 variable positions and 4 different codons represented at each of these positions, unlimited recombination could result in 4^{12}, or about 10^7, different sequences in this V subgroup. The germ line V genes are assumed to have been selected in evolution for ability to generate highly diverse somatic recombinants rather than for ability to determine V regions of a particular specificity. Thus this theory avoids the necessity of either preserving through eons of evolution a large number of germ line genes for Ags that have yet to come into existence (like the germ line theory) or of generating through successive point mutations a large repertoire of useful mutations during embryonic development (like somatic mutation theory).

However, recombination requires pairing of homologous chromosomes, which is an exceedingly rare event in somatic cell mitosis (unlike its regular occurrence during meiosis in formation of gametes). So far, comparison of the known V sequences has not yielded evidence for recombination. However, the number of sequences analyzed is still relatively small, and recombination would be difficult to detect if the number of recombining genes were large (e.g., 100).

Other, more unconventional suggestions include the proposal that episomes coding for hypervariable regions might be inserted into V genes in chromosomal DNA. If the variability were present in the episomes its origin would still have to be explained; if it were generated during the insertion process (e.g., by incision and repair errors) it is not clear why those chains that differ more in the hypervariable regions also differ more in the less variable parts of the V segments. Another proposal, without known precedents in other genetic structures, suggests that the vertebrate chromosome contains parallel, interconnecting networks of DNA strands that are transcribed by a polymerase that can switch back and forth among the strands, generating a large number of different mRNAs for V segments.

A firm decision between germ line and somatic theories cannot be made on available evidence. But preliminary results from hybridization of radioactive mRNA for an Ig light chain to homologous DNA suggest that the mouse genome probably has between 40 and 500 genes for light chains; the smaller number, if valid, would certainly rule out the germ line theory.

EVOLUTION OF IMMUNOGLOBULINS

Homologies Among Chains. Because serological analyses are sensitive to small antigenic variations they are particularly valuable for revealing differences among Ig chains. Amino acid sequences, in contrast, are especially useful for revealing similarities among chains, including those that do not cross-react at all serologically. The relatedness of two chains is apparent from the frequency with which corresponding positions are occupied by the same amino acid. A more searching comparison determines the minimum number of mutations necessary to change from one chain to the other. Both approaches give essentially the same result, and for both it is usually necessary to introduce occasional gaps in one or another chain, corresponding to genetic deletions, to bring out the full extent of their homology.

The sequences shown in Figures 16-21, 16-23, and 16-28 illustrate variations in homology, and the matrix of Figure 16-31 summarizes homology relations among a number of C segments. Extensive analyses of this kind lead to the following generalizations:

1) Homology is greater between the C_κ sequences of different species (man and mouse), and between

Fig. 16-31. Matrix comparison of frequencies of identical amino acid residues in homologous positions of the constant (C) regions of various Ig chains, expressed as identities per 100 residues (%). The most striking homologies (>60%) are encircled. Because there are 20 amino acids, 5% identity is the background level expected when randomly chosen chains are compared. (In fact, when the first 110 residues of the human hemoglobin α chain are compared with the above Ig chains identities range from 4 to 8%.) (Based on sequences and references in Dayhoff, M. D., ed. *Atlas of Protein Sequence and Structure*, vol. 5. National Biomedical Research Foundation, Silver Spring, Md., 1972.)

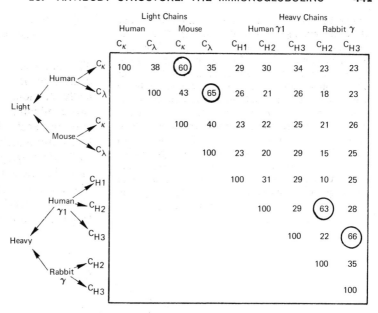

| | Light Chains | | | | Heavy Chains | | | | |
| | Human | | Mouse | | Human γ1 | | | Rabbit γ | |
	C_κ	C_λ	C_κ	C_λ	C_{H1}	C_{H2}	C_{H3}	C_{H2}	C_{H3}
Human C_κ	100	38	(60)	35	29	30	34	23	23
Human C_λ		100	43	(65)	26	21	26	18	23
Mouse C_κ			100	40	23	22	25	21	26
Mouse C_λ				100	23	20	29	15	25
Human γ1 C_{H1}					100	31	29	10	25
Human γ1 C_{H2}						100	29	(63)	28
Human γ1 C_{H3}							100	22	(66)
Rabbit γ C_{H2}								100	35
Rabbit γ C_{H3}									100

the C_λ sequences of different species, than between C_κ and C_λ of the same species.

2) Successive domains in the constant part of heavy chains of a particular class (C_{H1}, C_{H2}, C_{H3} of γ chains) have extensive homology with each other, with the corresponding domains of other species (cf. human and rabbit γ), and also with C_κ and C_λ sequences.

3) Hardly any homologies are detectable between V and C sequences, even when those of the same chain are compared.

4) In spite of variations among different molecules, there is more homology **within** each of the V_κ, and V_λ, and the V_H sets than among these sets (Fig. 16-21). Nevertheless, discernible low-level homology among all V sequences sets them apart from all the C sequences.

5) The V_H and V_L sequences of the same molecule may have no more homology than those from different molecules.

Order of Evolution. From these observations it has been suggested that structural genes for present-day Igs evolved from a primordial gene coding for a polypeptide with the general characteristics of V and C segments, viz., about 110 amino acids with two cysteine residues that form an approximately 60-membered S-S loop. The small degree of homology between V and C sequences suggests that as one of the earliest events duplication of the primordial gene probably gave rise to the primitive precursors of V and C genes. Genes for constant segments of diverse heavy chains could have arisen from unequal ("illegitimate") crossover, yielding duplicated C genes (Fig. 16-32). The light chains in immuno-

globulins of sharks (among the most primitive existing vertebrates) appear to be only of the κ type. This finding and the extensive homology between avian and mammalian λ chains, suggest that C_κ is the more primitive C_L gene and that it probably gave rise by duplication to the C_λ gene some time before birds and mammals diverged, about 250 million years ago. Present day μ chains are assumed to correspond most closely to primordial heavy

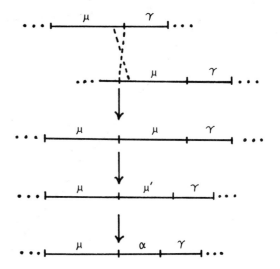

Fig. 16-32. Illegitimate crossing over could generate additional genes, whose occasional missense mutations could be selected for, leading eventually to the appearance of new Ig classes or subclasses (e.g., $\mu \to \alpha$).

chains, because sharks and teleost fish have only one class, resembling μ of higher forms. Chains resembling γ are present only in amphibia and higher vertebrates; hence C_γ may have arisen from C_μ (with a deletion). Recognized clearly so far only in mammals and in birds, α chains probably arose by a still later duplication, most likely of the μ gene, since μ and α are more alike than either is like γ, at least near the C-terminus (see Fig. 16-23). This order of evolution (Fig. 16-33) is in accord with the successive appearance of Igs during embryonic development and infancy (IgM first, then IgG, finally IgA), and with the principle that ontogeny recapitulates phylogeny. Ontogeny and phylogeny of the immune response are considered further in Chapter 17.

COMBINING SITES: THE BASIS FOR ANTIBODY SPECIFICITY

The biological activity of an Ab molecule centers on its ability to bind ligands specifically. As noted before, binding studies reveal two combining sites per four-chain molecule, one in each Fab region (see Ch. 15, Equilibrium dialysis, and above, this chapter), while Fc fragments have other sites that engage in general effector reactions (Table 16-5; also see Ch. 18, Complement Fixation; Ch. 19, Allergic Responses). After an Fab fragment is completely unfolded (e.g., in 7 M

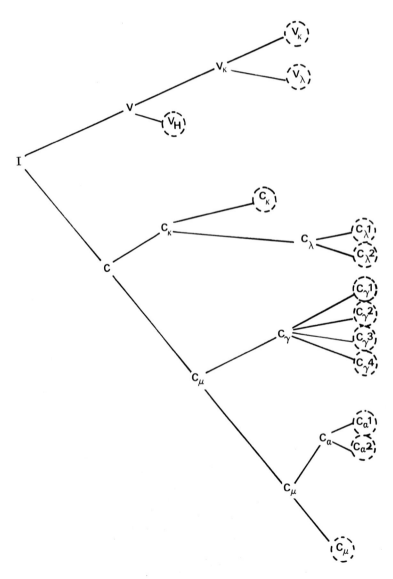

Fig. 16-33. Possible order of evolution of contemporary Ig genes (in dashed circles) from a primitive precursor (I) over approximately 4×10^8 years. δ and ε are not included as too little is known of their structure.

guanidine-HCl) and has had all its S-S bonds cleaved by reduction, it can spontaneously regain some of its original activity following air oxidation (2SH→S-S) and annealing (refolding) in dilute solution under physiological conditions of pH and ionic strength. Hence, just as with enzymes, **the amino acid sequences of Ig chains determine the shape, composition, and specificity of the Ab molecule's combining site.**

COMPOSITION OF COMBINING SITES

A number of elegant methods have been devised to identify the amino acid residues that form the active site. In the widely used method of **affinity-labeling** chemically reactive small molecules serve as univalent haptens. In reacting with the corresponding Ab, these reagents are first bound specifically in the active site by noncovalent bonds, and then form stable covalent bonds with susceptible amino acids in or close to the site. For example, Abs to the 2,4-dinitrophenyl group bind *m*-nitrobenzenediazonium salts, which can form azo derivatives of tyrosine, histidine, or lysine residues. The derivatives are stable during separation of the chains and their cleavage into peptides, making it possible to establish the position of the labeled residue

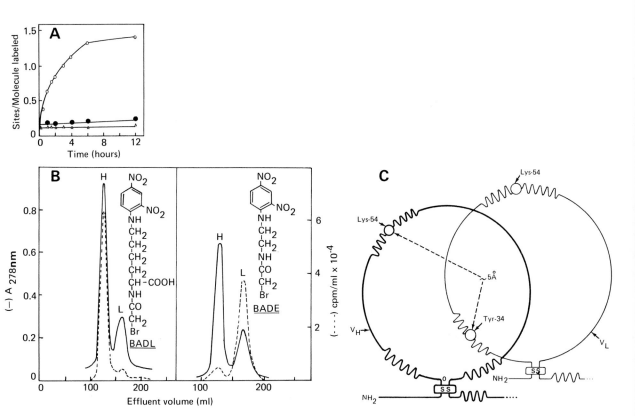

Fig. 16-34. Affinity-labeling of a mouse myeloma protein with antihapten activity (anti-Dnp). **A.** Rate of labeling the combining sites of the myeloma protein (O) with a bromoacetyl-Dnp reagent (BADE; structure shown in right panel of **B**). The presence of excess ε-Dnp-L-lysine specifically blocked the binding and reaction of BADE (●) ("hapten protection"). In a control test BADE did not label nonspecific mouse Ig (△). **B.** Separation of heavy (H) and light (L) chains of the myeloma protein after reaction with 14C-BADL, which labeled only the H chain, or with 14C-BADE, which labeled only the L chain. **C.** Localization of labeled residues in hypervariable regions (jagged lines) of V_L (light lines) and V_H (heavy lines) domains. A bifunctional Dnp reagent with bromoacetyl groups at two positions (ca. 5 A apart), one resembling its position in BADL and the other its position in BADE, labeled **both** Lys-54 on V_H and Tyr-34 on V_L, cross-linking H and L chains (dashed lines). A second myeloma protein with anti-Dnp activity was specifically labeled by BADE in Lys-54 of V_L domain. [Based largely on Haimovich, J., Eisen, H. N., Hurwitz, E., and Givol, D. *Biochemistry 11*:2389 (1972).]

in a chain. The specificity of the labeling reaction can be vertified by **hapten protection,** in which an excess of a conventional hapten (which cannot form a covalent bond) specifically blocks the covalent reaction (Fig. 16-34A).

Because of their homogeneity, myeloma proteins with combining sites of known specificity are especially useful for affinity-labeling, furnishing labeled peptides in high yield. From studies of these proteins and of conventional, heterogeneous Abs it is clear that amino acid residues of **both** heavy and light chains of a single Ig can be specifically labeled, and that **the labeled residues fall within the hypervariable regions of the** V_L **and** V_H **segments** (Fig. 16-34). However, the residues that become labeled are sometimes common to many chains (conserved amino acids) and are not unique to Abs of a particular specificity. Hence the labeled residue is useful in locating regions involved in the combining site, but it is not necessarily of singular importance in establishing the unique structure of that site.

The close approximation of hypervariable regions of both V_L and V_H in the combining site is emphasized by the specific reaction of a bifunctional affinity-labeling reagent, with which it has been possible to label covalently **two** amino acid residues in the same site, one in a hypervariable part of V_H and the other in a hypervariable part of V_L. From the dimensions of the reagent it is clear that substituted residues in hypervariable V_L and V_H are within 8 A in the intact molecule (Fig. 16-34).

Comparison of sequences of selected Igs provides independent evidence that hypervariable regions determine the specificity of combining sites. Thus several Abs with the same ligand-binding specificity have had virtually identical residues in hypervariable regions of their V_L and V_H segments (Fig. 16-35); moreover, some of the identical residues are rarely present in the corresponding regions of chains from randomly selected Igs.

SIZE OF COMBINING SITES

The dimensions of the combining sites have been estimated primarily from binding studies (Ch. 15). Abs to dextran (poly-D-glucose) bind with increasing strength glucose oligosaccharides of increasing size up to a maximum of about six or seven residues (Fig. 16-36). Similar results have been obtained with Abs to polythymidylic acid or to polyalanine, where binding also appears to be maximal with oligomers of four, five, or six residues. If the combining sites for these ligands were shaped as an invagination that encompassed an extended four- to six-membered oligomer, about 20 of the approximately 220 amino acid residues of the V_L and the V_H segments of an antibody molecule might make contact with the ligand. (A similar number of contact residues have been demonstrated by X-ray diffraction to constitute the active site of

Fig. 16-35. Evidence that variable residues, especially in hypervariable regions, are associated with specificity of combining sites. Sequences from N terminal to position 40 for light (κ) chains of three human monoclonal antibodies (Dav, Fin, Lay; specific for IgG), are compared with Bence Jones κ chains of V_{K1} subgroup, of which protein AG is the prototype. Listed above the AG sequence are the substitutions, and their approximate frequencies, in other light chains of V_{K1} subgroup. The identity of Dav and Fin is impressive but Lay, with the same specificity, is somewhat different, showing that various sequences are compatible with a given specificity. Variability is pronounced at the hypervariable region (30 to 32 and 34), but it is certainly not limited to that region. See Fig. 16-21 for abbreviations. [Based on Capra, J. D., Kehoe, J. M., Winchester, R. J., and Kunkel, H. G. *Ann NY Acad Sci 190*:371 (1971); AG is due to Titani, K., Whitley, E., Avogardo, L., and Putnam, F. W. *Science 149*:1090 (1965); for sources on substitutions above AG see refs. in Wu, T. T. and Kabat, E. A. *J Exp Med 132*:211 (1970).]

							I.16 K.16 I.16 T.16 K.16 S.16 W.16 S.20		
	V.05 L.04 L.12	T.04 T.14	V.10 L.05 R.05	L.05 A.05 L.07 A.07	R.30		S.16 V.07 N.50 H.16 Y.33 A.20		G.30
AG	D I Q M T Q S P S S L S A S V G D R V T			I T C Q A S Q B I			S B F L N W Y Q Q K P		
Dav	_____ T V _____			D			N S W _ I _____ Y __		
Fin	_____ T V _____			D			N S W _ I _____ Y __		
Lay	_____ V _____						N A Y _____		
	1	10		20			30		40

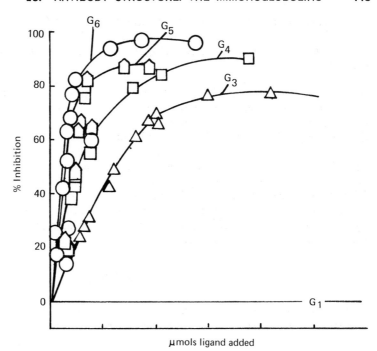

Fig. 16-36. Estimate of size of Ab combining sites. Inhibition of the dextran:antidextran precipitin reaction with a homologous series of polyglucose (G_n) oligosaccharides of increasing size: G_3 (△), G_4 (□), G_5 (□), G_6 (o). Isomaltoheptaose (G_7) was no more inhibitory than isomaltohexaose (G_6) with four of five antidextran sera, but inhibited marginally more with a fifth antiserum. [From Kabat, E. A. *J Immunol* 77:377 (1956); *84*:82 (1960).]

lysozyme, an enzyme that binds a hexasaccharide.) Additional variable residues of V segments probably also play a vital role in determining specificity by modulating the three-dimensional positions of the "contact" amino acids.

Among the Ab molecules of a given specificity some exhibit maximal binding with smaller ligands than others. This means that active sites vary in size or shape (or both) in different Ab molecules. These variations are consistent with the diversity of affinities observed among Abs of a given specificity (Ch. 15).

RECONSTRUCTION OF COMBINING SITES

Light Chain–Heavy Chain Interaction. Though interchain S-S bonds stabilize the native Ig molecule, they are not indispensable for the characteristic structure. Thus when heavy and light chains are separated, and the SH groups that constituted their S-S links are chemically blocked (e.g., with iodoacetate), they can reassociate through noncovalent bonds, spontaneously forming four-chain molecules that closely resemble native 7S Abs in physical, chemical, and antigenic properties; they even yield Fab fragments when

treated with papain. Chains derived from different Igs, and even from different species, also pair and form 7S molecules (e.g., human light with rabbit heavy chains). It is significant nonetheless that when a mixture of light chains from diverse Igs compete for a limiting amount of heavy chain from one Ig, the homologous pairs (derived from the same molecule) usually recombine more readily with each other than with heterologous chains (from different Igs).

The mutual affinity of all light and heavy chains probably derives from invariant sequences in C domains while the preferential binding of homologous chains probably is due to additional interaction between some residues in V_H and V_L domains. (Indeed an Ig cleaved by pepsin between V_L and C_L and between V_H and C_H domains yielded a stable fragment (called Fv) that consisted essentially of V_L plus V_H linked entirely by noncovalent bonds; this fragment retained all the ligand-binding activity of the intact Ig from which it was derived.) It is nevertheless possible that many light chains might pair equally well with a given heavy chain, and that many heavy chains might pair equally well with a particular light chain. Hence the number of different Ig molecules that could be assembled from *l* light and *h* heavy chains could be about (0.1) (*l* × *h*), if 10% of all chain combinations formed stable molecules (Origin of diversity, above).

Activity of Recombined Chains. The heavy chains isolated from an Ab can bind the corresponding ligand, though much less strongly than the original molecule. The isolated light chains bind far less well. Nevertheless, the light chains must make a profound and specific contribution to the combining site, because 7S molecules reconstituted from the chains of a given Ab have much greater affinity for ligand than those formed from the same heavy chain and other light chains.

The 7S molecules reconstituted from the chains of a homogeneous Ab (e.g., a myeloma protein with active sites) are fully as active as the native molecule. However, when recombination is carried out with chains from conventional Ab preparations, which are nearly always heterogeneous populations of molecules, the reconstituted molecules have far less activity for the test Ag than the native ones, probably because many mismatched heavy-light pairs are formed and lack appreciable affinity for the original ligand. It seems likely, though, that many of the mismatched pairs possess their own unique combining sites, specific for other (unknown) ligands.

APPENDIX

TABLE 16-5. Some Properties of Classes and Subclasses of Human Immunoglobulins

Property	IgG				IgA		IgM	IgD	IgE
	IgG-1	IgG-2	IgG-3	IgG-4	IgA-1	IgA-2			
Sedimentation coefficient (S)	7	7	7	7	7–13	7–13	18–32	7	8
Molecular weight ($\times 10^{-3}$)	150	150	150	150	150–600	150–600	900	?	190
Heavy chains	$\gamma 1$	$\gamma 2$	$\gamma 3$	$\gamma 4$	$\alpha 1$	$\alpha 2$	μ	δ	ε
Light chains: κ/λ ratio	2.4	1.1	1.4	8.0	1.4	1.6	3.2	0.3	?
Carbohydrate (%, approx.)	3	3	3	3	7	7	12	13	11
Average concentration in normal serum (mg/ml)	8	4	1	0.4	3.5	0.4	1	0.03	0.0001
Half life in serum (days, in vivo)	23	23	8	23	6	(6?)	5	3	2.5
Heavy chain allotypes	Gm	Gm	Gm	Gm		Am			
Earliest Ab usually detected in primary immune responses*							+		
Most abundant Ab in most late immune responses*	←		+	→					
Conspicuous in mucinous exocrine secretions					+	+			+
Transmitted across placenta*	+	±	+	+	0	0	0	?	0
Effector Functions:									
Active in complement fixation†	++	+	++	0‡	0‡	0‡	++	?	0‡
Sensitizes human mast cells for anaphylaxis (homocytotropic)§	0	0	0	0	0	0	0	0	+
Sensitizes guinea pig mast cells for passive anaphylaxis (heterocytotropic)§	+	0	+	+	0	0	0	?	0
Binds to macrophages‖	++	+	++	±			0		

* Chapter 17.
† Chapter 18.
‡ Can activate complement via bypass mechanism (Ch. 18).
§ Chapter 19; IgE molecules are reagins and responsible for atopic allergy.
‖ "Cytophilic" Abs (Ch. 20).

SELECTED REFERENCES

Books and Review Articles

DORRINGTON, K. J., and TANFORD, C. Molecular size and conformation of immunoglobulins. *Adv Immunol 12*:333 (1970).

EDELMAN, G. M., and GALL, W. E. The antibody problem. *Annu Rev Biochem 38*:415 (1969).

GREEN, N. M. Electron microscopy of the immunoglobulins. *Adv Immunol 11*:1 (1969).

GRUBB, R. *The Genetic Markers of Human Immunoglobulins.* Springer, New York, 1970.

KOCHWA, S., and KUNKEL, H. G. (eds.). Immunoglobulins. *Ann NY Acad Sci 190*:1 (1971).

KRAUSE, R. M. The search for antibodies with molecular uniformity. *Adv Immunol 12*:1 (1970).

LENNOX, E. S., and COHN, M. Immunoglobulins. *Annu Rev Biochem 36*:365 (1967).

METZGER, H. Structure and function of γM macroglobulins (IgM). *Adv Immunol 12*:57 (1970).

NATVIG, J. B., and KUNKEL, H. G. Human Immunoglobulins: Classes, subclasses, genetic variants, and idiotypes. *Adv Immunol 16*:1 (1973).

OUDIN, J. Genetic regulation of immunoglobulin synthesis. *J Cell Physiol 67 (Suppl. 1)*:77 (1966).

POTTER, M. Immunoglobulin producing tumors and myeloma proteins of mice. *Physiol Rev 52*:631 (1972).

SMITH, G. P., HOOD, L., and FITCH, W. M. Antibody diversity. *Annu Rev Biochem 40*:969 (1971).

TOMASI, T. B., and BIENENSTOCK, J. Secretory immunoglobulins (IgA). *Adv Immunol 9*:2 (1968).

Two genes–One polypeptide chain: A symposium. *Fed Proc 31*:177 (1972).

Specific Articles

BRIDGES, S. H., and LITTLE, J. R. Recovery of binding activity in reconstituted mouse myeloma proteins. *Biochemistry 10*:2525 (1971).

BRIENT, B. W., and NISONOFF, A. Inhibition by specific haptens of the reaction of anti-hapten antibody and anti-idiotypic antibody. *J Exp Med 132*:951 (1970).

HAIMOVICH, J., EISEN, H. N., HURWITZ, E., and GIVOL, D. Localization of affinity-labeled residues on the heavy and light chains of two myeloma proteins with anti-hapten activity. *Biochemistry 11*:2389 (1972).

HOOD, L., and TALMAGE, D. Mechanism of antibody diversity: Germ-line basis for variability. *Science 168*:325 (1970).

KINDT, T. J., MANDY, W. J., and TODD, C. W. Association of allotypic specificities of group *a* with allotypic specificities A11 and A12 in rabbit immunoglobulins. *Biochemistry 9*:2028 (1970).

JERNE, N. K. The somatic generation of immune recognition. *Eur J Immunol 1*:1 (1971).

MORRISON, S. L., and KOSHLAND, M. E. Characterization of the J chain from polymeric immunoglobulins. *Proc Nat Acad Sci USA 69*:124 (1972).

OUDIN, J., and MICHEL, M. Idiotypy of rabbit antibodies: I and II. *J Exp Med 130*:595, 619 (1969).

TOSI, S. L., MAGE, R. G., and DUBISKI, S. Distribution of allotypic specificities among IgG molecules. *J Immunol 104*:641 (1970).

WANG, A. C., WILSON, S. K., HOPPER, J. E., FUDENBERG, H. H., and NISONOFF, A. Evidence for the control of synthesis of the variable regions of the heavy chains of IgG and IgM by the same gene. *Proc Nat Acad Sci USA 66*:337 (1970).

WOFSY, L., METZGER, H., and SINGER, S. J. Affinity-labeling: A general method for labeling the active sites of antibody and enzyme molecules. *Biochemistry 1*:1031 (1962).

WU, T. T., and KABAT, E. A. An analysis of the sequences of the variable regions of Bence Jones proteins and myeloma light chains and their implications for antibody complementarity. *J Exp Med 132*:211 (1970).

chapter 17

ANTIBODY FORMATION

CELLULAR BASIS FOR ANTIBODY FORMATION 450
    Method for Studying Antibody Formation at the Cellular Level 451
  LYMPHOCYTES 456
    B and T Cells 456
  MACROPHAGES 463
  CELL COOPERATION IN INDUCTION OF
      ANTIBODY FORMATION 464
  ANTIGEN-BINDING RECEPTORS ON LYMPHOCYTES 467
  BLAST TRANSFORMATION 468
  PLASMA CELLS 469
    One Cell—One Antibody Rule 470
    Synthesis, Assembly, and Secretion of Immunoglobulins 471
  LYMPHATIC ORGANS 473
    Primary Lymphatic Organs 473
    Secondary Lymphatic Structures 475
  GENETIC CONTROL OF ANTIBODY FORMATION 477

ANTIBODY FORMATION IN THE WHOLE ANIMAL 480
  ADMINISTRATION OF ANTIGEN 480
    Route 480
    Adjuvants 481
    Dose 481
  THE PRIMARY RESPONSE 482
    Changes with Time 484
  THE SECONDARY RESPONSE 486
    Cross-Stimulation 488
  FATE OF INJECTED ANTIGEN 489
    Effect of Antibodies 490
    Antigenic Promotion and Competition 490
  VACCINATION AGAINST MICROBIAL ANTIGENS 490

INTERFERENCE WITH ANTIBODY FORMATION 492
  IMMUNOLOGICAL TOLERANCE 492
    Conditions Promoting Tolerance 493
    B Cells vs. T Cells 494
  ANTIBODIES AS IMMUNOSUPPRESSIVE AGENTS 496
    Suppression by Antibody as Antiimmunogen 496
    Suppression by Antibody as Antireceptor 497
    Suppression by Antilymphocyte Serum (ALS) 498
  NONSPECIFIC IMMUNOSUPPRESSION 499

PHYLOGENY AND ONTOGENY OF ANTIBODY FORMATION 502
  PHYLOGENY 502
  ONTOGENY 503

IMMUNODEFICIENCY DISEASES 506

CELLULAR BASIS FOR ANTIBODY FORMATION

This chapter is concerned with the sequence of events that begins with entry of antigen (Ag) into the body and culminates in the appearance in serum of the corresponding antibodies (Abs). In the past 70 years, the general framework for viewing this process has undergone several radical transitions, which have assigned different roles to the immunogen.* According to **selective** theories the immunogen stimulates certain cells to make greater amounts of the Ab molecules that they already make at a low level in advance of immunization. **Instructive** theories, in contrast, assume that the immunogen is an obligatory participant in the formation of the corresponding Ab, providing essential information for determining its structure.

The first detailed theory, proposed by Ehrlich in 1900, was selective. Cell surfaces were supposedly covered by a variety of Ab-like receptors; in combining with matching ones the immunogen was supposed somehow to stimulate selectively their further synthesis and secretion by the cells. This theory was abandoned by the 1920s when recognition of the enormous variety of immunogenic substances, including newly synthesized organic chemicals, suggested that the diversity of Ags is almost limitless: it seemed beyond belief that the corresponding Abs could all preexist.

In its place an instructive "antigen-template" theory (suggested by Pauling, Breinl and Haurowitz, and by others) came to be widely favored in the 1940s and 1950s. According to this view the specificity of an

Ab molecule was determined not by its amino acid sequence but by the process of molding the nascent molecule around the immunogenic determinant; the immunogen would act as a template at the site of protein synthesis, creating a complementary conformation at the combining sites of the Ab. However, this model became untenable when the specificity of the Ab molecule was shown to be determined by its amino acid sequence (Ch. 16), for alteration of the corresponding messenger RNA sequence could hardly have been immediately acquired from the encounter with Ag. Identification of the cells forming a specific Ab also contributed: with highly radioactive Ag it could be shown that these cells contained no detectable Ag (less than 10 molecules per cell).

Clonal Selection. Even before the antigen-template theory had to be discarded Burnet proposed a **clonal selection** theory, based on the success of mutation-selection in explaining microbial adaptations. According to this theory an immunologically responsive cell (lymphocyte, see below) can respond to only one Ag (or a few related ones), and this capacity is somehow acquired before the Ag is encountered (see Ch. 16, Origin of diversity). Accordingly, an individual's lymphocytes are an immensely diversified pool of cells—some capable of responding to one Ag, others to a second, etc. When an immunogen penetrates the body it binds to an Ab-like receptor on the surface of the corresponding lymphocyte, which is then somehow stimulated to proliferate and generate a clone of differentiated cells. Some of these (plasma cells, see below) secrete Abs, while others circulate through blood, lymph, and tissues as

---

* In this and in subsequent chapters "immunogen" is used as a synonym for "antigen" when emphasizing its function in inducing immune responses (Ch. 14).

an expanded reservoir of Ag-sensitive "memory" cells. The same immunogen encountering these cells months or years later evokes a more rapid and copious secondary response.

Older theories are contrasted with clonal selection in Figure 17-1. Several lines of evidence support the clonal hypothesis. 1) A given Ag is bound specifically by only a small proportion of an individual's lymphocytes, whose stimulation results in production of the corresponding Abs; inactivation of these cells eliminates ability to form these Abs. 2) Immunoglobulin (Ig)-forming cells make only one kind of Ig molecule, at least at one time; and in an animal stimulated with two Ags, or even with an Ag that has two distinguishable determinants (A and B) some cells make anti-A, others make anti-B, and none make both.

It has been known since the 1950s that the **principal Ab-producing cells are plasma cells** (see below). However these cells become prominent in tissues only several days after antigenic stimulation, and little was known of the early cellular events until several major developments in the past decade. Gowans' use of purified cell populations established that the **Ag-sensitive cells** with which the immunogen reacts initially are **small lymphocytes,** and these cells were later recognized (largely from immunological defects following extirpation of thymus or bursa of Fabricius) as being of two types: **T cells,** which are derived from the thymus, and **B cells,** which are thymus-independent and are derived from the bursa of Fabricius in birds and possibly from bone marrow in mammals. As we shall see, the full immunogenic activity of many Ags depends upon cooperation between B and T cells and on ancillary support from large phagocytic cells: motile **macrophages** and sessile **reticular** cells. Other, "T-independent" immunogens (especially those of high molecular weight with repeating antigenic determinants) can stimulate B cells without any assistance from T cells and macrophages. The mystery surrounding the earliest events has also begun to lessen with the demonstration that specific receptors on Ag-sensitive B lymphocytes are Igs, linking the structural characteristics of Ab molecules (Ch. 16) and the principles of the Ab-Ag reaction (Ch. 15) with the initiation of cellular events in the induction of Ab formation.

Before describing the cells involved in Ab formation (B and T lymphocytes, plasma cells, and macrophages), we review briefly the main methods used to study the consequences of Ag reaction with these cells.

METHODS FOR STUDYING ANTIBODY FORMATION AT THE CELLULAR LEVEL

Immunofluorescence, introduced by Coons (see Ch. 15, Fig. 15-34), can be used to identify individual cells that contain Abs. **Killed cells,** in which the cell interior is accessible, are stained by the "sandwich" technic, which amplifies the method's sensitivity: the cells (usually in a tissue section) are incubated with multivalent antigen X and then with fluorescein-labeled Ab against X. The cells that contain anti-X bind the Ag, and each molecule of bound Ag can then bind many molecules of fluorescent anti-X (Fig. 17-2).

With **living cells,** whose surface membrane is not penetrated by extracellular Ab molecules, the membrane-bound Ab molecules that serve as Ag-binding receptors can be detected on the cell surface with fluorescent anti-Igs, whose specificity also helps identify the receptor's heavy-chain class, light-chain type, and allotype (Fig. 17-2).

Ag-binding receptors on lymphoid cells can also be detected by **direct rosette formation:** red cells (or Ag-coated red cells) bind to lymphocytes or plasma cells whose membrane contains the corresponding receptor (Fig. 17-3). **Indirect rosettes** are formed with the aid of bridging "hybrid" Ab molecules (Ch. 16): one site binds to the cell-bound Ig and the other to an Ag on the erythrocyte.

Ab production by individual cells can be detected in small **microdrops** (ca. 10^{-6} ml) of dilute cell suspensions; some drops contain a single cell. Ab produced in such microdrops (kept under mineral oil to prevent evaporation) can be measured by sensitive technics, such as specific immobilization of motile *Salmonella* or neutralization of bacteriophages (see Ch. 15, Rates of reaction).

The **hemolytic plaque assay,** introduced by Jerne, is a particularly useful method for counting Ab-forming cells. In a typical test spleen cells from a mouse immunized with sheep erythrocytes (SRBC) are plated in agar with the SRBC. During the following in-

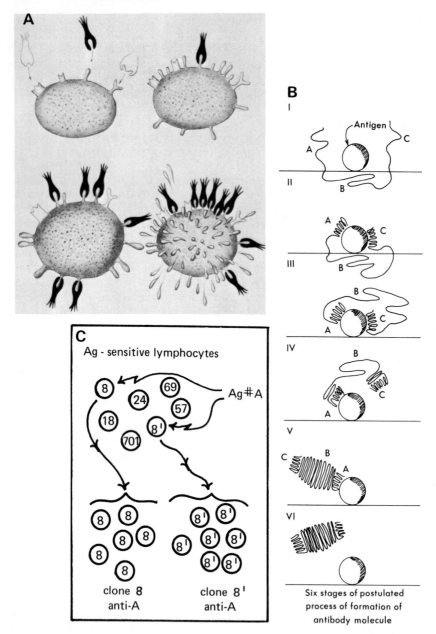

Fig. 17-1. Comparison of selective and instructive theories of Ab formation. **A.** Ehrlich's selection hypothesis, showing multiplication and shedding of a cell receptor following its specific combination with an immunogen; each cell was assumed to be able to combine with any Ag. **B.** Antigen-template theory: immunogen reacts with the nascent Ig chain, A-B-C, causing it to acquire sites that are complementary to determinants of the immunogen. **C.** Burnet's clonal selection hypothesis resembles **A,** but with the important difference that the population of reactive cells is diversified and each cell can respond to only one or a few immunogens. [**A** from Ehrlich, P. *Proc R Soc Lond, Biol 66*:424 (1900). **B** from Pauling, L. *J Am Chem Soc 62*:2643 (1940). **C** based on Burnet, F. M. *The Clonal Selection Theory of Acquired Immunity.* Vanderbilt Univ. Press, Nashville, 1959.]

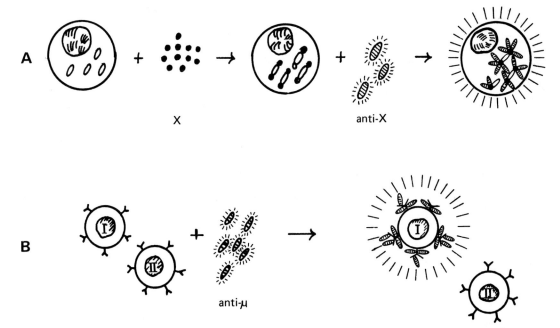

Fig. 17-2. Identification by immunofluorescence of Ab-producing cells **(A)** and Ag-sensitive cells with Igs on the surface membrane **(B)**. In **A** cells (e.g., plasma cells) are killed, rendering Ab at sites of synthesis inside the cell accessible to the test reagents (sandwich technic): multivalent ligand X and fluorescein-labeled Ab (anti-X) (see Fig. 17-24B). In **B** the cells (e.g., small lymphocytes) are living and only their surface Ig molecules are accessible for reaction with labeled Ab. Assume that the surface Igs are IgM in cell I and IgG in cell II; with fluorescein-labeled anti-μ only cell I would be stained, and with labeled anti-γ only II would be stained (for aggregation of surface Igs see Fig. 17-10).

cubation some of the lymphoid cells synthesize and secrete Abs that lyse surrounding SRBC when complement is added (Ch. 18). This leaves clear plaques, resembling those produced by lytic phage on a lawn of susceptible bacteria, with the Ab-secreting cell in the center of each plaque (Fig. 17-4). By using red cells with various covalently attached haptens and Ags this method can be extended to detect cells forming a wide variety of Abs.

The unmodified ("direct") plaque technic counts cells that produce IgM Abs, but not those that produce IgG Abs. The reason is that lysis of an Ab-coated red cell by complement requires only a single IgM molecule on the erythrocyte's surface, while a pair of adjacent IgG molecules is necessary (Ch. 18): with relatively few Ab molecules secreted by a single cell, the close packing of bound Ab molecules is likely to be infrequent, especially if spacing of the red cell's antigenic determinants are unfavorable. However, cells that secrete Abs of the IgG class can be counted by the "indirect" technic: addition of an antiserum to IgG results in clusters of anti-IgG

Abs on each molecule of bound anti-red-cell Ab, and the red cell can then be lysed by complement (Fig. 17-4). If the added antiserum is specific for a particular allotype it is possible to recognize cells that produce that allotype; antiserum to IgA, which does not fix complement (Ch. 18), similarly allows "indirect" enumeration of cells that produce Abs of the IgA class.

Transplantation is extensively used to analyze the types of cells needed for Ab production. In transfers of living lymphoid cells* from one animal to another **(adoptive immunity)** the recipient is usually first heavily irradiated (e.g., 750 rad) and sometimes the thymus gland is also removed; by eliminating essentially all the recipient's functional lymphoid tissue these procedures convert it into a

---

* The term "lymphoid cells" is used to emphasize that the cell suspensions ordinarily isolated from lymph node, spleen, blood, etc., are highly heterogeneous mixtures of lymphocytes (themselves highly diversified), macrophages, rare plasma cells, and some granulocytes and erythrocytes.

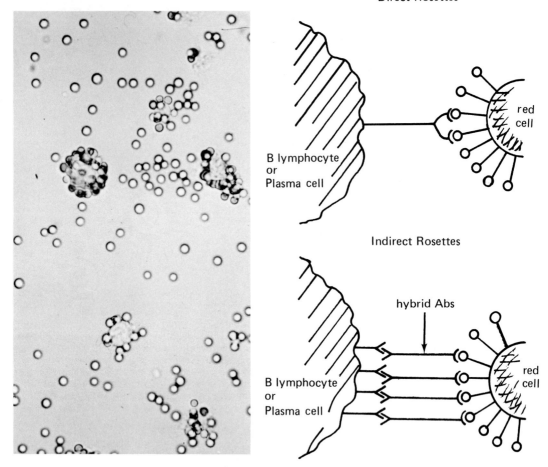

Fig. 17-3. Identification of surface Igs by rosette formation. Rosettes formed by the indirect method (using hybrid Abs with one combining site for the B cell's surface Ig and one for a ligand on red cells) are more stable than those formed by the direct method, because many more bonds can be formed between B cells and red cells. The photograph shows direct rosettes formed with trinitrophenyl (Tnp)-coated red cells and a myeloma cell whose surface Ig has anti-Tnp combining sites. (Courtesy of Drs. K. Hannestad, M. S. Kao, and R. G. Lynch.)

"walking tissue culture." In addition, donor and recipient are generally chosen from the same inbred strain to avoid complications that would otherwise arise from an immune attack by the host against histocompatibility Ags on the surface of the donors cells or vice versa (Ch. 21).

The principles governing survival or rejection of transplanted cells are considered in detail in Chapter 21. It is useful to realize here that histocompatibility Ags are **alloantigens** (Gr. *allos* = different): like immunoglobulin allotypes (Ch. 16), they segregate genetically and various ones are thus present

in some and absent in other members of a given species. An animal almost invariably destroys transferred cells carrying histocompatibility Ags that it lacks. However, members of a highly inbred strain are essentially identical genetically **(syngeneic)** and antigenically: they accept grafted cells freely from each other, while rejecting genetically different (allogeneic) cells from other inbred strains of the same species.

Cell Separations. Lymphoid cells differ in various physical properties and, most importantly, in the specificity of their Ag-binding surface receptors. **Velocity sedimentation** at unit gravity (1 g) sepa-

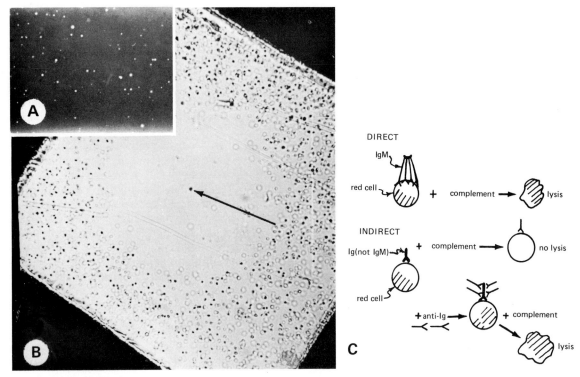

Fig. 17-4. Hemolytic plaque assay of Ab-producing cells. **A.** Multiple plaques in a petri dish. ca. ×15. **B.** A single plaque with its central Ab-secreting cell (arrow). Though these cells are only rarely observed in mitosis, ca. 40% of those stimulated by Ag appear to have arisen from a recent mitotic division (based on ^{3}H-thymidine content in DNA). Some representative plaque-forming cells are shown in Figure 17-11 ca. ×100 **C.** Contrast between **direct plaques**, which reveal cells that secrete Abs of IgM class, and **indirect plaques** which require appropriate anti-Igs to obtain complement-dependent lysis of the red cells. [**A** courtesy of Dr. L. Claflin; **B** from Harris, T. N., Hummeler, K., and Harris, S. *J Exp Med* *123*:161 (1966).]

rates cells of different size (large cells sediment faster than small ones). **Density-gradient centrifugation** separates cells of different densities; the centrifugation is carried out in a medium whose different densities are either distributed continuously in a gradient or discontinuously in multiple layers.

Macrophages adhere to glass, and passage through a column of fine glass beads removes these cells from populations of lymphoid cells. B cells also tend to adhere to glass and nylon surfaces, but are more efficiently removed on glass beads that are coated with anti-Igs. In a form of "affinity chromatography," columns of glass or polyacrylamide beads coated with Ag retain cells with the corresponding Ag-binding surface receptors, and a few of the retained cells can even be recovered (though with difficulty) by elution with Ag at high concentration.

In an automatic **"cell-sorting" machine** a stream of microdrops containing single cells is monitored for particular cell-surface markers (by binding fluorescein-labeled Abs); microdrops with fluores-

cent living cells are collected, yielding particular subsets from heterogeneous cell populations.

Induction of Ab Formation in Cell Culture. Incubation with Ag induces cells in cultured fragments of spleen to produce Ab. However, dispersed cell suspensions are not consistently induced unless they are allowed to reaggregate into small clusters. The importance of cell aggregation in culture probably reflects the need for cooperative interactions among several cell types (two kinds of lymphocytes, macrophages; see below). In contrast to bacterial cells, which usually function well as solitary individuals, many functions of lymphoid and other animal cells require cell-cell interactions (as though animal cells, like animals, have societal needs). With sheep red cells as Ag, the incidence of plaque-forming cells induced in short-term primary cultures of dispersed spleen cells (5 to 7 days) is close to that found in the spleen of an immunized animal (e.g., $500/10^6$ nucleated spleen cells).

Living vs. Dead Cells. Lymphoid cells (especially lymphocytes, see below) are fragile. Many are killed inadvertently during isolation, or deliberately by cytotoxic immune reactions of their surface Ags (Ch. 18; Ch. 20, Cytotoxic assays). Permeability to anionic dyes (such as trypan blue) is commonly used to distinguish dead from living cells: the dyes penetrate into and stain dead cells, while living cells remain unstained.

Living and dead cells can also be separated by centrifugation through a dense solution of polyglycol and sodium triiodobenzoate: dead cells are denser and sediment to the bottom while live cells accumulate at the surface.

LYMPHOCYTES

B AND T CELLS

Lymphocytes are relatively small (5 to 15 μm diameter), round, nondescript cells that are ubiquitous in blood, lymph, and connective tissues. As is described below, the two funda-mentally different kinds, B and T cells, differ in origin, in surface macromolecules, in circulation patterns, and above all, in the mode and consequences of their interaction with Ags.

The induction of Ab formation by many immunogens requires specific interaction with both B and T cells, with the T cells somehow regulating the proliferation and differentiation of B cells into Ab-secreting plasma cells. Thus, in working with recipient mice that were both thymectomized and irradiated (to destroy virtually all lymphocytes), Claman *et al.* found that the amount of Ab formed against sheep red cells in recipients of bone marrow and thymus cells exceeded by far the sum of the amounts formed in recipients of either one alone (Table 17-1): **the cooperation of both populations was evidently necessary to restore ability to make an optimal Ab response.**

Subsequent studies showed that the active cells in bone marrow and thymus are precursors of B and T cells, respectively (see below), and that the **Ab-secreting cells are derived from the marrow and not from the thymus cells.** For instance, when marrow and thymus cells were obtained from donors that differed in Ig allotype but were otherwise syngeneic (see Ch. 21, Congenic strains) the

TABLE 17-1. Cooperation Between Thymus and Bone Marrow Cells in Induction of Antibody Synthesis*

	Recipients	
Donor cells	% Spleen fragments secreting antibody	Serum antibody activity†
None	0.8	None detected
Bone marrow (1 × 10⁷)	1.3	None detected
Thymus (5 × 10⁷)	12.3	0.2
Bone marrow (1 × 10⁷) + Thymus (5 × 10⁷)	53.7	2.1

* X-irradiated (800 r) mice were injected with the indicated numbers of thymus or bone marrow cells (or both, or neither) from normal donors, and the recipients were then given sheep red cells (SRBC) as Ag. Eleven days later Ab production was determined by measuring the serum Ab titer and the percentage of spleen fragments that secreted hemolytic anti-SRBC Abs in culture.

† Log_2 hemolysin titer.

Based on Claman, H. N., Chaperon, E. A., and Triplett, R. F. *J Immunol* 97:828 (1966).

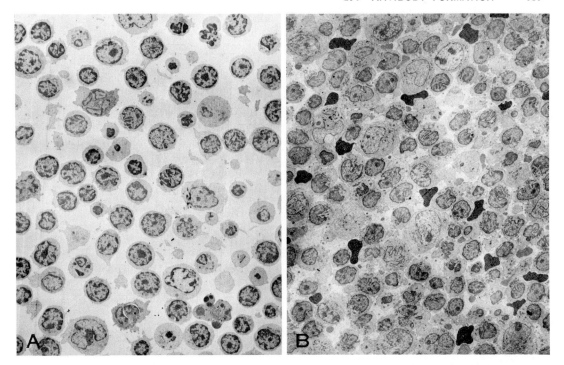

Fig. 17-5. Heterogeneity of lymphocytes in lymph from thoracic duct of rat **(A)** and human **(B)**. Most of the cells are of T lineage (Table 17-2). Some have well developed rough endoplasmic reticulum; others lack these organelles. Dark, irregular forms in **B** are erythrocytes. **A,** ×2400; **B,** ×1200. [From Zucker-Franklin, D. *Seminars in Hematology 6*:4–27 *(1969)* **(A)**; *J Ultrastruct Res 9*:325–339 *(1963)* **(B)**.]

allotype of the Ab produced was always that of the bone marrow donor.

Appearance and Generation Times. B and T lymphocytes look alike and both are motile, nonphagocytic cells of varying size (Fig. 17-5). In the smallest ones (5 to 8 μm diameter) mitochondria are scanty, ribosomes are single, and no endoplasmic reticulum is discernible; in the light microscope the cytoplasm forms a barely perceptible rim around the dense nucleus (Fig. 17-6). Specific binding of Ag by membrane-bound receptors on the cell surface can stimulate transformation of small lymphocytes into larger ones (up to 15 μm diameter), whose more abundant cytoplasm contains endoplasmic reticulum and a prominent Golgi apparatus, and is richer in mitochondria and in polysomes (Fig. 17-6). In accord with the appearance of a secretory system (endoplasmic reticulum, Golgi apparatus) some of the large lymphocytes (of B type) secrete Abs (see Fig. 17-11). The large cells also divide more

rapidly (generation time ca. 6 to 48 hours), and some of those of B lineage differentiate into mature plasma cells, the most active of all lymphoid cells in synthesis and secretion of Igs (Figs. 17-6 and 7-11). Many large lymphocytes also revert back eventually into small ones, which probably function as "memory" cells (see Secondary response, below). Small lymphocytes rarely divide unless stimulated by Ag (for important exceptions in the thymus and bursa of Fabricius, see below).

Even after 40 days' continuous infusion of ^{3}H-thymidine into mice and rats the DNA remains unlabeled in most of the small lymphocytes in blood.

Another approach has shown that these cells also survive for long periods in man without dividing. After intensive X-ray therapy chromosomes in many lymphocytes become badly damaged; the cells can survive but they are unable to complete a mitotic division. When blood lymphocytes are subsequently treated with certain mitosis-stimulating plant proteins (mitogens, see below) and forced to

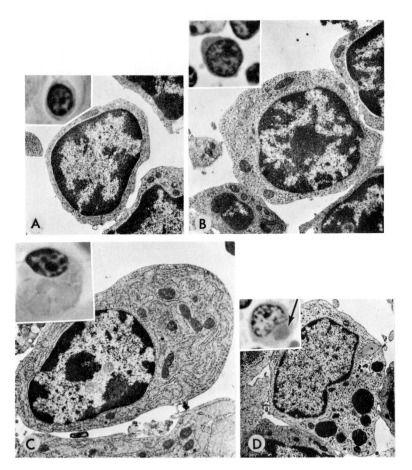

Fig. 17-6. Principal cells involved in antibody formation. **A,** Small lymphocyte; **B.** large ("transitional") lymphocyte; **C,** plasma cell; **D,** macrophage. The macrophage contains a phagocyted red cell (arrow, inset **D**), corresponding to the larger of the dark cytoplasmic inclusions in the associated electron micrograph. The plasma cell (inset, **C**) has its secretory vesicles distended with immunoglobulin. Electron micrographs are: **A,** ×7500; **B,** ×6600; **C,** ×6600; **D,** ×5500. Light microscope photographs of the corresponding cells (upper left insets) stained with toluidine blue. Magnifications are approximately: **A,** ×4000; **B,** ×3000; **C,** ×3000; **D,** ×2000. (Courtesy of Dr. R. G. Lynch.)

undergo mitosis, cytological examination reveals some cells with such pronounced chromosomal aberrations that they are unlikely to have divided once since incurring their X-ray damage. From the decreasing incidence of such cells with time after irradiation it has been estimated that the mean generation time of circulating small lymphocytes (B and T cells) in man is about 5 years; some small lymphocytes seem to survive 10 years without division, which might account for the long persistence of immunological memory (below).

Differences Between B and T Cell Surfaces. B cells have abundant surface membrane-bound Ig molecules (ca. 100,000/cell), which are restricted in a given cell to molecules of a particular allotype and isotype, and probably also restricted to one idiotype, accounting for the cell's selective response to a given Ag (see One cell–one antibody rule, below). T cells in contrast, seem to have far fewer surface Igs (perhaps 1000-fold less per T cell than B cell);

TABLE 17-2. Comparison of Mouse B and T Lymphocytes

Properties	B cells	T cells
Differentiation (from uncommitted Ag-insensitive "stem" cells to Ag-sensitive cells) in:	Bursa of Fabricius (in birds) or as yet unknown equivalent in mammals	Thymus
Ag-binding receptors on the cell surface:	Abundant Igs, (restricted to 1 isotype, 1 allotype, 1 idiotype per cell)	Nature of specific receptors is uncertain; Igs are sparse
Cell surface antigens:*		
θ	−	+
TL	−	+
Ly	−	+
PC	+ (plasma cells)	−
H-2 transplantation Ags	+	+
Approximate frequency (%) in:		
Blood	15	85
Lymph (thoracic duct)	10	90
Lymph node	15	85
Spleen	35	65
Bone marrow	Abundant	Few
Thymus	Rare	Abundant
Functions		
secretion of antibody molecules	Yes (large lymphocytes and plasma cells)	No
Helper function (react with "carrier" moieties of the immunogen)	No	Yes
Effector cell for cell-mediated immunity	No	Yes
Distribution in lymph nodes and spleen:	Clustered in follicles around germinal centers	In interfollicular areas
Susceptibility to inactivation by:		
X-irradiation	++++	+
Corticosteroids	++	+
Antilymphocytic serum (ALS)	+	++++

* θ occurs at high levels in thymus and in brain. Two allotypes are known: θ-AKR (in AKR and a few other inbred mouse strains), and θ-C3H (in C3H, BALB/c, and most other mouse strains).

TL is present on normal thymus cells of only some mouse strains (TL+), but is present on leukemic lymphocytes of TL+ and TL− strains.

Ly is present on thymus cells and circulating lymphocytes, but absent from all nonlymphoid cells. There are two loci: Ly-A and a second one with linked Ly-B and Ly-C; two alleles are known at each.

PC is present on plasma cells (including myeloma cells).

H-2 histocompatibility Ags (see Ch. 21). B and T cells also differ in ability to adsorb Igs: B cells, but not T cells, bind Ab·Ag·complement (C) complexes through surface sites that are specific for activated third component of C (C3, see Ch. 18). Other sites on B cells bind aggregated Igs (cross-linked, for instance, by Ag), probably through specific reaction with the Fc domains of the aggregated molecules.

though predominantly IgM molecules, it is not certain that they are restricted in idiotype.

B and T cells also differ in other surface macromolecules. In mice, where they are best characterized, T cells have a variety of heritable alloantigens (θ, TL, Ly; Table 17-2) that are lacking in B cells. Like other **alloantigens** (see Ch. 16, Ig allotypes, and especially Ch. 21), these differ in various inbred mouse strains and are thus recognizable by **alloantisera,** raised in a strain that lacks the alloantigen by injecting appropriate cells from mice that possess it.

The most important of the T cell alloantigens is θ, which is defined by antisera raised in AKR mice against T cells from C3H mice. After absorption with various C3H cells (other than T cells), to remove Abs to histocompatibility Ags, the resulting antiserum reacts specifically with T cells of C3H and of other strains of mice with the same θ allotype. In conjunction with complement (Ch. 18), anti-θ destroys T cells selectively in a simple procedure that is widely used to determine whether various properties of a population of lymphocytes are due to its T cells or to its B cells. Though useful, this approach is not foolproof because circulating T cells (in blood, lymph, lymph nodes, spleen) have less θ on their surface than those in the thymus; and some are not killed by anti-θ (plus complement) but can be eliminated by antisera to other distinctive surface antigens (Table 17-2).

B and T cells also differ in the areas they occupy in lymphatic organs (see Fig. 17-8), and in susceptibility to inactivation by certain immunosuppressive agents (see Table 17-2 and Suppression by antilymphocytic serum, below).

Origin of B and T Cells. Transplantation experiments in chickens have shown that B cells arise from migrant bone marrow stem cells (primitive precursors of hematopoietic and lymphoid cells) that lodge in the bursa of Fabricius (see below) where they begin to synthesize Igs (Fig. 17-7). In mammals, which lack a bursa, it is not certain where the B cell precursors become committed to synthesize a particular Ig: this could happen in the bone marrow itself or in lymph nodes or in gut-associated lymphoid structures, such as appendix or tonsils.

T lymphocytes also originate from bone marrow stem cells, which migrate to the thymus where they divide rapidly (generation time about 8 hours). Most of the rapidly dividing cells die without leaving the thymus; the survivors, mature T cells, differ from entering stem cells in several important properties that are acquired in the thymus, or possibly in other tissues under the aegis of a thymus hormone (see below). 1) They develop characteristic surface Ags (Table 17-2); theta (θ) is particularly useful for detecting and tracing these cells (see above). 2) They can react specifically with one or a few Ags, i.e., they become Ag-sensitive ("immunologically committed") lymphocytes. 3) Besides acquiring the capacity to regulate B cell responses to Ags, specifically reactive T lymphocytes can become specific effector cells for cell-mediated immune responses: they can destroy tumor cells, cause rejection of allografts, and promote the differentiation of resting macrophages into highly bactericidal cells capable of destroying microbial pathogens (Chs. 20 and 21). T cells are heterogeneous and it is likely that T cell effectors of cell-mediated immune reactions (Ch. 20) differ from those that regulate the differentiation of B cells into Ab-secreting cells (see below).

Circulating T cells also differ from thymus cells in having more histocompatibility Ags. About 25 to 50 times fewer circulating T cells than thymus cells are necessary to support cell-mediated immune responses (e.g., Ch. 21, Graft-vs.-host reaction), perhaps because mature T cells in the thymus (ca. 5% of the total) are diluted by the more abundant immature ones.

Circulation of B and T Cells. Lymphocytes circulate through blood, lymph, and peripheral lymphatic tissues (lymph nodes, spleen, Peyer's patches of intestinal wall, etc.), where the T cells tend to congregate in special interfollicular T-cell areas while B cells cluster in adjacent follicles (see Figs. 17-8, 17-15, 17-16). Lymphocytes subsequently leave these areas via efferent lymph channels that eventually fuse to form the thoracic duct, through which lymphocytes return to blood and recirculate. **The majority of circulating lymphocytes are T cells** (Table 17-2), except in rare disorders affecting the thymus and T cells (see Immunodeficiency diseases, below) and in some individuals with chronic lymphocytic leukemia.

Separation of B and T Cells. Separation of these cells is necessary for evaluating their respective

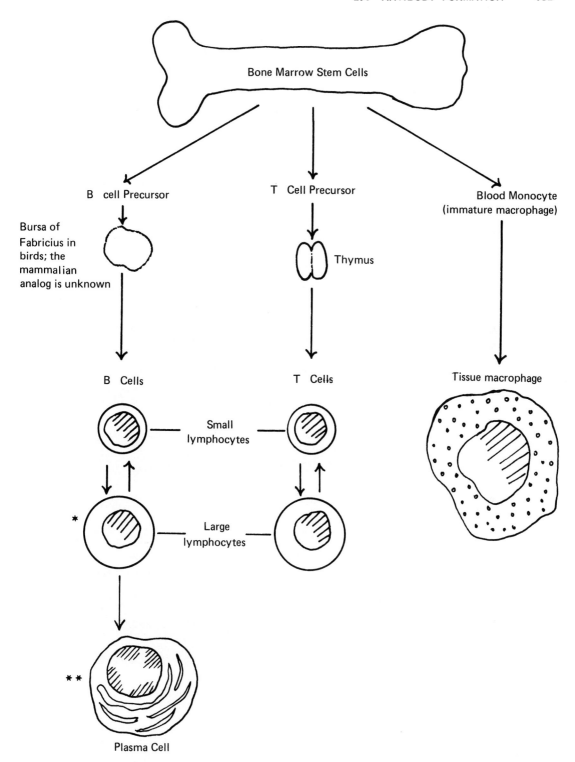

Fig. 17-7. Maturation pathways of the principal cells in the immune response. Antibody molecules are secreted by large lymphocytes (*) and especially by plasma cells (**) in the B-cell lineage. Lymphocytes of B and T lineages are morphologically indistinguishable.

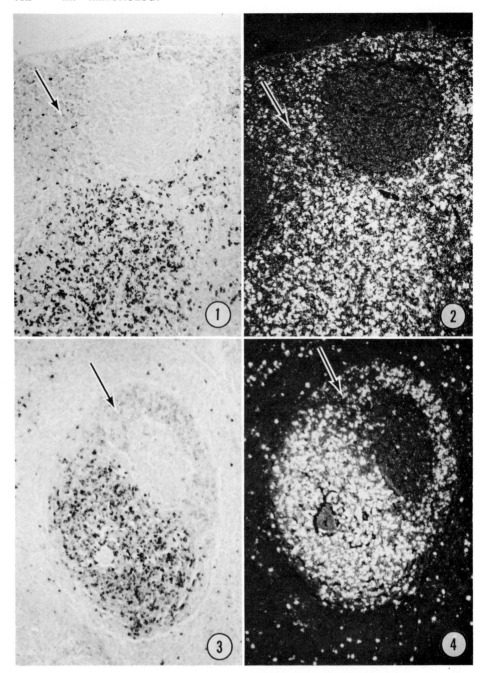

Fig. 17-8. B and T zones in secondary lymphatic organs. The T cells are highly radioactive (because of ³H-ribonucleoside incorporation into RNA) and are concentrated in interfollicular and deep cortical areas of lymph nodes **(1,2)** and in periarteriolar areas of white pulp of spleen **(3,4).** B cells (arrows) are clustered around germinal centers of lymph node **(1,2)** and spleen white pulp **(3,4).** (For reasons that are not clear the B cells incorporate much less ³H-ribonucleoside into RNA and are less radioactive than the T cells.) See also Figures 17-15 and 17-16. Bright- and dark-field views of the same field of a lymph node **(1,2)** and of spleen **(3,4).** [From Howard, J., Hunt, S., and Gowans, J. *J Exp Med 135*:200 (1972).]

properties. The following are some experimental procedures.

1) Thymus cells are injected into heavily irradiated recipients (ca. 800 r). If an Ag is administered at the same time, or is present as an alloantigen on the recipient's own cells (e.g., a histocompatibility Ag lacking in the donor), the corresponding thymus cells proliferate, yielding an enriched population of Ag-stimulated or "educated" T cells.

2) Repeated passage of heterogeneous lymphoid cell populations over glass beads coated with anti-Igs removes B cells (and macrophages), yielding a T-cell suspension.

3) Treatment of heterogeneous populations with anti-θ sera and complement destroys almost all T cells, leaving lymphocytes of predominantly B type.

4) Virtually all lymphocytes of mutant "nude" mice with defective development of the thymus are B cells (see Thymus, below).

MACROPHAGES

These large, highly phagocytic cells (15 to 20 μm diameter) do not form Abs. However,

the following indirect evidence suggests that they have an important accessory role, cooperating with T cells in aiding the response of B cells to many immunogens.

1) Ags, especially in particulate form, are conspicuously taken up by macrophages.

2) Trace amounts of Ag bound to the macrophage surface are far more potent immunogenically than the same amount of free, unbound Ag.

3) Macrophages adhere to glass and plastic surfaces; and in spleen cell cultures some Ags can induce B lymphocytes to differentiate into Ab-secreting plasma cells only in the presence of both adherent cells and T cells (see Cell cooperation, below, and Fig. 17-9).

The action of macrophages on Ags has long seemed paradoxical. The cell's sticky surface binds Ags, which are then endocytized and degraded (see below). Cleavage of covalent bonds of soluble Ags usually reduces or destroys immunogenicity (see Ch. 15, Conformational antigenic determinants). Yet Ags associated with macrophages are more im-

Fig. 17-9. Evidence that induction of Ab synthesis by certain (T-dependent) Ags requires three cells. **A.** Induction of Ab formation to sheep erythrocytes by cultured mouse spleen cells, which were separated into adherent (predominantly macrophages) and nonadherent (lymphocyte) populations. By adding one cell population in excess and titrating with graded amounts of the other, the dependence of induction on the number of titrated cells indicated (from slopes, b) that one adherent and two nonadherent cells were needed to induce the appearance of one Ab-secreting cell. With unseparated spleen cells, the slope was 2.6, or close to 3. Induction was measured by the number of plaque-forming cells (PFC) that appeared. **B.** Some hypothetical inductive complexes. Evidence in **A** favors a 3-cell model (B-T-macrophage). [**A** from Mosier, D. E. and Coppelson, L. W. *Proc Natl Acad Sci USA* **61**:542 (1968).]

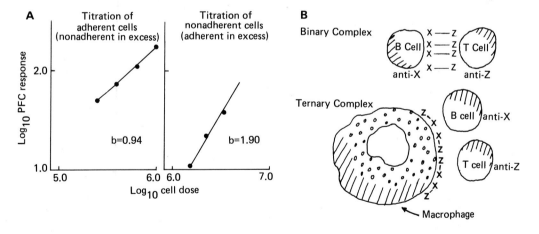

munogenic than unbound Ag. It was thought that an intermediate level of degradation ("processing") might be the solution, but it now appears that while the internalized Ag is indeed destroyed, the unmodified Ag on the sticky surface is exceedingly immunogenic: removing it with trypsin or blocking it with Abs eliminates the immunogenicity of macrophage-associated Ag. However, with complex Ags, such as bacterial cells, phagocytosis and digestion by macrophages might contribute to immunogenicity by releasing internal molecules that would not otherwise become free to react with lymphocytes.

Macrophages wander through tissues. Together with reticular cells that are immobilized on reticulum fibers and other phagocytic cells in the endothelial lining of vascular, sinus-like spaces (of spleen, lymph nodes, liver, etc.), they constitute the scavenger **reticuloendothelial system** which traps, ingests, and degrades foreign particulate matter (bacteria, many viruses, aggregated macromolecules). The degradative activity is due to effective phagocytosis plus abundant cytoplasmic vesicles **(lysosomes)** that are filled with various hydrolytic enzymes: proteases, RNase, DNase, lipase, esterases, lysozyme, phosphatases, etc. When coated with opsonizing Abs and fragments of activated complement (C3b and C5b, see Ch. 18), foreign particles adhere to the sticky cell membrane and are then ingested into a phagocytic vacuole **(phagosome)** that fuses with lysosomes to form the **phagolysosome,** in which degradative enzymes attack the ingested material.

"Cytophilic" antibodies stick to the macrophage surface and help bind the corresponding Ags to the cell; for example, red cells form rosettes around macrophages that have been incubated in antiserum to red cells. The use of myeloma proteins established that **only certain Ig classes are cytophilic** (e.g., human IgG-1 and IgG-3; see Ch. 16), doubtless through binding sites in the Fc domain. Though these Abs dissociate from macrophages at 37°, and thus have questionable activity in vivo, they seem to promote destruction of some target tumor cells in vitro (Ch. 20, cytotoxic mechanisms).

Macrophages can divide, but they probably survive for only a short time in acute inflammatory lesions (due to infection, irritants, or adjuvant mixtures with Ag), where they arise by differentiation of monocytes carried to the lesion via blood. The blood monocytes (which derive from "promonocytes" in bone marrow) are less active as phagocytes and have fewer lysosomes (Figs. 17-6 and 17-7). The life span of mature tissue macrophages is not known, but they can survive many weeks in tissue culture. In some chronic inflammatory lesions macrophages differentiate further into epithelioid cells and into multinucleated giant cells (Chs. 22 and 35).

An additional role of macrophages in cell-mediated immunity will be discussed in Chapter 20: Ags react specifically with T lymphocytes and cause release of factors that stimulate macrophages to differentiate into "angry" cells, which have greatly increased numbers of lysosomes and are more actively phagocytic.

CELL COOPERATION IN INDUCTION OF ANTIBODY FORMATION

As noted above, both thymus and bone marrow cells are needed for optimal restoration of responsiveness in thymectomized-irradiated mice. Cooperation between cell populations is also evident from the "carrier effect," which provides hints about how the cells interact.

Carrier Effect. When an animal is injected once ("primed") with a hapten-protein conjugate, such as Dnp-bovine γ-globulin (Dnp-BγG), and then given the same immunogen many weeks later, anti-Dnp Abs are formed rapidly and copiously (Secondary response, below). However, if the second injection is made with Dnp attached to certain other carrier proteins, say ovalbumin (EA), a substantial antihapten response is usually not elicited unless the animal has also been previously primed with EA itself. This suggests that recognition of both hapten and carrier contributes to the secondary response.

The same result can be obtained by adoptive transfer: an irradiated recipient of spleen cells from a mouse primed with hapten-protein I does not make significant amounts of antihapten Ab in response to hapten-protein II, unless it is also given "helper" spleen cells from a second mouse primed with protein II. The helpers are T cells: their effect is eliminated by treatment with antisera to θ (plus complement).

The requirement for carrier recognition seems, however, not to be absolute: it can be bypassed by giving a large amount of hapten-protein II or by preparing it from special proteins (e.g., hemocyanin). Other measures that bypass the carrier requirement are described below (see Allogeneic effect, also soluble factors released by T cells; Ch. 20).

The distinction between hapten and carrier determinants also applies to natural immunogens (proteins, red cells), some of whose determinants (hapten-like) bind the elicited Abs (formed by B cells), while others (carrier-like) react with helper T cells in the induction of the B cells. For instance, a brisk secondary response in formation of non-cross-reacting Abs to both *a* and *b* subunits of tetrameric lactic acid dehydrogenase (porcine) could be elicited in rabbits primed with the a_2b_2 form of the enzyme and then boosted with a_2b_2, but no Abs to either subunit were formed in animals boosted with b_4, as though b were a hapten-like subunit requiring the carrier effect of the a subunit. Though distinguishable functionally, the structural distinctions between haptenic and carrier determinants of natural Ags are completely obscure.

T-cell Function. How do T cells promote induction of B cells? According to Mitchison's **focusing hypothesis,** simultaneous binding of an Ag by B and T cells provides additional bonds, which stabilize its binding by membrane receptors of the B cell (Fig. 17-9). Stabilization would be especially helpful for those Ags that have only a single copy of a given determinant per molecule, such as serum albumin and other proteins with one chain per molecule (Ch. 15). In these Ags, moreover, different determinants would have to react with the T and B partners, requiring different specificities of the cooperating cells. This prediction is in accord with the carrier effect, and also with the rule that **T-dependent immunogens have more than one kind of determinant per particle:** the haptenic or hapten-like determinants react with B cells and the carrier determinants with T cells ("associative recognition"). **Thus haptens are not immunogenic unless attached to a carrier;** and proteins are especially effective as carriers (and as immunogens) because they have a great variety of potential carrier determinants per particle, facilitating the engagement of many helper T cells.

T cells seem to exercise their effects both through contact with B cells and the release of diffusible factors that act at short range on nearby B cells. Thus T cells can aid B cells when they react at the same time, and in the same locale, with **separate** immunogenic particles, and even when the cooperating cells are separated by an artificial membrane that is permeable to macromolecules but not to cells. Nevertheless, the highest Ab yields are obtained when the cooperating B and T cells are specific for determinants on the **same** immunogenic particle. Indeed, there seem to be two diffusible factors: a low-molecular-weight (dialyzable) nonspecific substance that enhances the response of B cells of any specificity, and a high-molecular-weight (nondialyzable) specific factor that augments only those B cells that can react specifically with the same immunogen that activated the T cell's release of the factor. It has been suggested that the specific factor might be an Ig of a special class (8S IgM) that

moves from the carrier-activated T cell to the surface of a macrophage where its specific binding of multiple Ag units would facilitate their binding to and activation of B cells (see Fig. 17-9).

The importance of contact between cooperating cells is suggested by an **"allogeneic effect."** In this complex reaction the carrier requirement for eliciting antihapten secondary responses (Carrier effect, above) can be overcome in animals primed with hapten on protein I if, at the time they receive the hapten on a second, non-cross-reacting protein (II), they are also injected with allogeneic lymphoid cells from a donor whose incompatibility in transplantation Ags is such that the injected cells attack the host's cells (including his specifically primed, Ag-binding B cells). This "graft-vs.-host" reaction almost certainly requires cell-cell contact (see Chs. 20 and 21). The allogeneic effect is not obtained with the reverse incompatibility,* i.e., when the primed host's T cells attack the unprimed, allogeneic donor's cells, which argues against massive release of diffusible and stable "activating" factors as the primary basis for T-B cooperation.

Though T cells exercise an important regulatory role on responses of B cells to T-dependent immunogens it is important to note that these Ags can induce B cells, in absence of T cells, to make Abs of the IgM class, and to differentiate into memory cells, whose response to a subsequent introduction of the immunogen elicits the augmented secondary response (see below). However, the further maturation of the primary response, usually characterized by production of IgG Abs (see IgM–IgG switch and Secondary response, below), seems not to occur unless T cells are engaged (and also macrophages, below).

Macrophage Function. With Ags whose induction of Ab synthesis in primary spleen cell culture requires accessory T cells, there is a further requirement for adherent cells, prob-

---

* Unidirectionality in the attack of host against donor or vice versa is arranged by using animals of inbred strains and their hybrids; i.e., injection of parental lymphoid cells (AA or BB) into F1 hybrid hosts (AB) causes the graft-vs.-host reaction, and injection of lymphoid cells from hybrids into a host of parental type elicits the more commonplace host-vs.-graft response (Ch. 21). The difference comes about because T cells from one individual react against essentially all cells from genetically different (allogeneic) individuals with alien surface histocompatibility Ags (see Chs. 20 and 21, and above, discussion of transplantation as a method for studying Ab formation).

ably macrophages, suggesting that **the inductive complex has three components (ternary complex), perhaps with B and T lymphocytes binding to Ag on the sticky surface of a macrophage** (Fig. 17-9). While each of the cooperating B and T cells reacts specifically with an antigenic determinant of the immunogen, **the macrophage acts nonspecifically:** it is just as effective if drawn from a tolerant or genetically unresponsive donor as from a normal one (see below, Immunological tolerance and Genetic control of Ab formation). Macrophages and T cells are effective even when irradiated, indicating that induction does not require their proliferation, whereas multiplication of B cells is the essence of induction: their irradiation eliminates the response (Table 17-3 and Immunosuppression, below).

The B and T cells that react specifically with the immunogen must be rare, and so their simultaneous reaction in vivo with a given Ag particle would be highly improbable. However, a successive reaction would be more probable if Ag were concentrated on the nonspecifically acting macrophages, and the probability would be further increased if the macrophages were strategically stationed at critical positions along pathways followed by circulating lymphocytes (e.g., sinus spaces of lymph nodes and spleen), where large numbers of B and T cells could continuously pass over their surfaces.

For experimental analysis macrophages are often obtained from peritoneal exudates elicited by intraperitoneal injection of irritants (e.g., thioglycolic acid). However, under natural conditions in vivo it is likely that an equivalent function is also (and perhaps largely) performed by the sticky **reticular cells** that abound on the reticulum fiber meshwork permeating lymph nodes and spleen (see below).

T-independent Antigens. The focusing hypothesis is consistent with the properties of certain Ags that can induce differentiation of B cells without any assistance from T cells and macrophages. These T-independent Ags are large polymers, with many copies of the same determinant per particle (e.g., polyfructose [levan], pneumococcal polysaccharides, polyvinylpyrrolidone, some viruses, polymerized flagellin). Repetitive copies of the determinant on single Ag molecules probably facilitates formation of stable multivalent cross-links with membrane receptors on B cells, bypassing the need for ancillary binding to helper T cells and macrophages.*

Indeed, T cells may even play a **suppressive** role with T-independent Ags: their selective elimination

---

* T-independent Ags seem also to be unusually resistant to degradation by host enzymes, but it is not clear how this contributes to their independence of T cells and macrophages.

TABLE 17-3. Roles of B and T Cells and of Macrophages in Induction of Antibody Synthesis

Property	B cells	T cells	Macrophages
Role	Differentiate into Ab-secreting cells	Accessory cells	Accessory cells
Specificity*	Restricted	Restricted	Unrestricted
Memory†	Yes	Yes	No
Proliferation stimulated by Ag-binding	Yes	Yes	No
Inactive in unresponsive individuals‡	Yes (sometimes)	Yes (sometimes)	No
Helping function	No	Release stimulators for B cells§	Bind Ag on surface facilitating B-cell binding to Ag

* Variety of antigenic determinants bound per cell.

† That is, priming by an Ag increases the number of cells able to respond later to that Ag.

‡ Genetically unresponsive or made tolerant (see below).

§ See Chapter 20 for many other factors released by Ag-activated T cells.

by thymectomy or antilymphocytic serum (see below) causes an increase in the amount of Ab formed in response to pneumococcal polysaccharide and to polyvinylpyrrolidone (see Antigenic competition, below, for other evidence that Ag-activated T cells can sometimes suppress Ag-stimulation of B cells.)

ANTIGEN-BINDING RECEPTORS ON LYMPHOCYTES

These receptors have been demonstrated in various ways. For instance, in "hot suicide" experiments highly radioactive Ag (50 μCi/μg) can eliminate specifically the corresponding Ab-forming B cells and helper T cells. In another approach, red cells (used as Ags or coated with extrinsic Ags) form rosettes around the corresponding lymphocytes (Fig. 17-3). Hemolytic plaque assays show that many of the rosette-forming lymphoid cells secrete Ab and are thus of B lineage. The ability of T cells to form rosettes is disputed, perhaps because the density or affinity of their surface Ag-binding-receptors is too low to permit consistent formation of stable rosettes.

B-cell Receptors. Membrane-bound receptors are abundant on the surface of B lymphocytes (ca. 100,000/cell has been estimated with labeled ligands). Binding of Ags by B cells is blocked by antisera to Igs, but not by antisera to other surface macromolecules, such as histocompatibility Ags. Moreover, immunofluorescence staining of living cells with antisera to various Ig chains shows that the Ig on the surface of a given B cell is usually restricted to one class and allotype of heavy chain, and to one type and allotype of light chain (see One cell–one Ab rule, below). Hence each B cell has a particular Ig as receptor, possibly modified for purposes of attachment.

The proportion of B cells with κ or λ light chains in their surface receptors corresponds roughly to the frequency of these chains in serum Igs. However, in some species (e.g., man, mouse) the heavy chain distribution is discordant: of the approximately 15 to 30% of peripheral blood lymphocytes that have surface Igs (B cells), about 50% have μ, 35% have γ, and 10% have α chains. (The corresponding values for serum IgM, IgG, and IgA are about 5 to 10%, 80 to 90%, and

5 to 10%, respectively; see Appendix, Ch. 16.) Moreover, labeled Igs extracted from the B cell membrane are predominantly monomeric IgM$_s$ (8S), not the pentameric 19S molecules that account for most serum IgM (Ch. 16).*

IgM-IgG Switch. Indirect evidence hints that the Ig class of surface receptors on a B cell may change during an immune response. For instance, when a primary response to sheep erythrocytes is induced in cultured mouse spleen cells in the presence of anti-μ Abs during the initial 48 hours, the formation of Abs of all classes is blocked; the presence of anti-γ has little effect during the initial induction period, but if it is added 1 to 2 days later it blocks just the formation of Abs of the IgG class, suggesting that receptors on virgin B lymphocytes are IgM and that after stimulation by Ag some B cells come to have surface receptors of the IgG class. (For other evidence that individual cells can undergo the IgM-IgG switch without stimulation by Ag, see Ontogeny, below, and biclonal myeloma tumors in Ch. 16.) A switch could also account for effects in intact animals: anti-γ blocks only about 20% of the rosettes formed by spleen cells isolated 3 days after immunization with sheep red cells but about 40% of those that can form on day 8.

Cap Formation. When the living B cell reacts with fluorescein-labeled anti-Ig the cell's surface is initially stained diffusely; but after a few minutes the fluorescent material aggregates into **patches,** which then coalesce into a **polar cap** (Fig. 17-10). With further incubation the cap is shed or ingested by the cell and degraded into fragments. The cells thus lose their receptors and cannot be stained again until 6 to 8 hours later, when the surface Igs are regenerated. A similar time is required for regeneration after stripping surface Igs with trypsin; and this time is concordant with evidence for a slow, dynamic turnover of ^{125}I-labeled surface Igs.

Patch formation seems to represent a two-dimensional microprecipitin reaction, cross-linking receptor Ig molecules that are normally diffusible in the cell's fluid membrane: it does not, for instance, occur if the fluorescent reagent is prepared from the univalent Fab fragment of anti-Ig. Patches are also formed when B cells bind a multivalent Ag

---

* In some species (guinea pig) the predominant Ig on B cells seems to be IgG of the γ2 subclass (Ch. 19, Fig. 19-6).

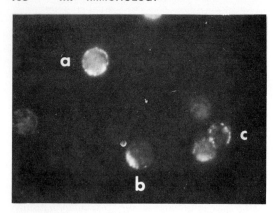

Fig. 17-10. Aggregated Igs in surface membrane of lymphocytes. Living white blood cells (of rabbit) were stained with fluorescein-labeled Abs to a rabbit Ig allotype. The speckled appearance of cells *a* and *c* represents patches of membrane Igs cross-linked by the anti-Ig; the patches are accumulated into a polar cap in cell *b*. The reaction was at 4° and the cells were examined at room temperature. At 37° most of the stained cells would have caps as in *c*. (Courtesy of Dr. Joseph M. Davie.)

Quantitative adsorption of trace amounts of radio-iodine-labeled anti-Igs suggests that the average T cell has on its surface the equivalent of about 100 molecules of Ig per cell, or about 0.1% of the amount present on the average B lymphocyte. Whether this small amount is functionally significant is not known. Some of the difficulties in evaluating surface Igs of T cells could derive from the use of antisera to class-specific (isotypic) determinants (Ch. 16), which could be buried in the T cell membrane or sterically hindered by other surface macromolecules; and since Ag binding seems perhaps to be blocked more by antisera to light chains than by those to the known heavy chains (including μ) the T-cell receptors might belong to an as yet uncharacterized class (IgT). The receptors could also lack domains with isotypic determinants and consist of just the V_H plus V_L regions required to constitute the Ag-binding site of an Ig (Ch. 16). The most intriguing possibility is that a novel family of diversified proteins, not Igs at all, makes up the Ag-binding receptors on T cells (see discussion of Ir genes below under Genetic control of antibody formation).

and it is possible that cross-linking of surface receptor molecules (by multivalent Ag or aggregated univalent Ag and even experimentally by bivalent anti-Ig) is a crucial triggering event that sets off cell proliferation and differentiation (Blast transformation, below).

Polar caps seem to arise from patches that are swept together in the moving membrane of motile lymphocytes. Thus conditions that block motility inhibit cap formation (without blocking aggregation into patches); e.g., lowering temperature to 4°, blocking ATP generation (with azide or cyanide), or adding colcemid, an agent that disaggregates cytoplasmic microtubules and disorganizes the cell's locomotory system. Patches can also be formed on the surface of various nonlymphoid cells by fluorescein-labeled Abs to their surface histocompatibility Ags; coalescence of patches into a polar cap occurs in motile fibroblasts but not in nonmotile epithelial cells.

T-cell Receptors. The nature of Ag-binding receptors on T cells is uncertain. The immunological activities of these cells (e.g., in graft-vs.-host reactions, see Ch. 21) are not consistently blocked by antisera to Igs; and these cells have virtually no surface Ig detectable by immunofluorescence (< 1% of the cells are stained).

BLAST TRANSFORMATION

Incubation with Ag causes small lymphocytes with the corresponding surface receptors to undergo "blast transformation": the cells enlarge; the nucleolus swells; polysomes, rough endoplasmic reticulum, and microtubules develop; and rates of macromolecule synthesis increase markedly (Ch. 20, Fig. 20-9). Similar changes are initiated by the binding of anti-Ig Abs to surface receptors of B cells, and of certain mitogenic plant proteins to the surface of B and T cells. The changes suggest that a membrane perturbation brought about by cross-linking appropriate surface macromolecules stimulates lymphocytes to divide.

The **plant mitogens** are specific for various sugars of different surface glycoproteins (not Igs), which seem to be approximately equally represented in B and T lymphocytes. However, the two cell types respond differently; soluble **phytohemagglutinin (PHA)** and **concanavalin A** stimulate transformation in T cells only, but when these proteins are insolubilized by attachment to large particles (Sephadex beads) they also stimulate B cells, which can even start secreting Igs. Another protein, **pokeweed mitogen,** stimulates transformation of B and T cells. The lack of a response of blood lymphocytes to soluble

PHA is used clinically to reveal T-cell deficiencies in rare developmental anomalies (see below, Immunodeficiency diseases) and in certain neoplastic diseases (e.g., Hodgkin's disease).

Lipopolysaccharides (endotoxin) of gram-negative bacteria are mitogenic for B cells. Unlike the plant mitogens, which react with cell surface glycoproteins, the lipid moiety of endotoxin **(lipid A,** see Chs. 6, 29) probably reacts with lipids of the lymphocyte membrane.

PLASMA CELLS

Attention focused on plasma cells as the principal Ab-producing cells because their frequency in a tissue reflects the level of Ab synthesis:

1) As lymphoid tissues become hyperplastic in response to immunogenic stimulation, plasma cells become conspicuous. They are especially numerous when Ab production is at its height.

2) In intensively immunized animals the amount of Ab extractable from various tissues correlates roughly with plasma cell content.

3) Plasma cells in tissues are unusually prominent in those diseases in which serum Igs are markedly elevated: e.g., in multiple myeloma (which is a plasma cell tumor), various chronic infectious diseases, cirrhosis of liver, and certain diseases of unknown origin, such as Boeck's sarcoid.

Ag-stimulated small B lymphocytes differentiate into plasma cells, which synthesize and secrete Abs more actively than any other cell type (Fig. 17-11). "Immature plasma cells"

Fig. 17-11. Electron micrographs of representative Ab-secreting cells (hemolytic plaque-forming cells, PFC, as in Fig. 17-4). Cells in **A,B,C** are from lymph nodes and spleen; the one in **D** is characteristic of PFC in efferent lymph emerging from an antigenically stimulated lymph node. Most PFC from within lymph nodes and spleen are plasma cells **(C)**, except during the first few days after immunization when lymphocytes **(A)** and transitional cells **(B)** constitute the majority. [From F. G. Gudat, T. N. Harris, S. Harris, and K. Hummeler, *J Exp Med 134*:1155 (1971) **[A,B]** *J Exp Med 132*:448 (1970) **[C]**; K. Hummeler, T. N. Harris, S. Harris, and M. B. Farber, *J Exp Med 135*:491 (1972) **[D].]**

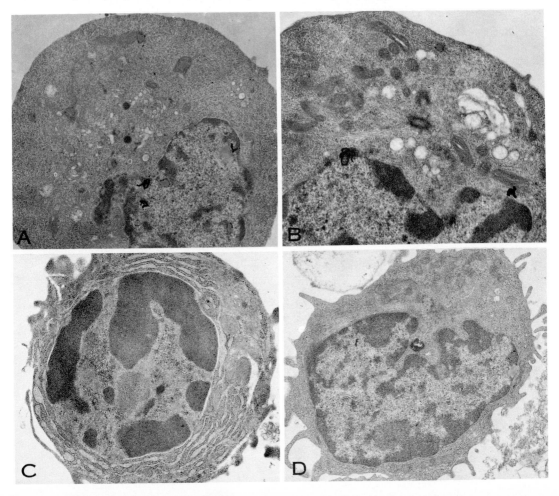

and "large lymphocytes" are arbitrary designations for transitional stages between the small B lymphocyte and the fully differentiated plasma cell. In its fully differentiated form the plasma cell is anatomically distinctive: the nucleus is eccentric and contains coarse, radially arranged chromatin (cartwheel), the cytoplasm has a conspicuous Golgi apparatus, and abundant rough and smooth endoplasmic reticulum is usually packed into thin onionskin-like lamellae or sometimes distended with Igs (as shown by immunofluorescence; Figs. 17-6 and 11).

The mature plasma cell is believed to survive only a few weeks and to die after a few (or no) cell divisions; in contrast, neoplastic plasma cells (myeloma cells), like other neoplastic differentiated cells, are capable of an indefinite number of cell divisions. Synthesis of Igs decreases during the mitotic stage of the cell cycle in myeloma cells.

ONE CELL–ONE ANTIBODY RULE

The Igs in individual plasma cells appear to be uniform in staining reactions with fluorescent Abs to diverse Ig chains: **a given cell has only one class and one allotype of heavy chain and one type and allotype of light chain.** The Abs used as reagents in establishing this one cell–one Ig rule are usually specific for the invariant parts of Ig chains, but the rule also applies to the variable regions, which determine the Ig's ligand-binding site: in animals immunized with two Ags or one Ag with two distinguishable determinants (e.g., arsanilate and polyalanyl groups) some plasma cells make Abs to one, others make Abs to the second, and none make Abs to both.* Hence the rule is **one plasma cell–one Ab.** The uniformity of the heavy chains and of the light chains produced in each Ig-synthe-

---

*One search for double-producing plasma cells was carried out by an interesting modification of the hemolytic plaque assay with a mixture of rabbit and camel red cells, each coated with a different Ag: a clear plaque meant that the central Ab-forming cell secreted Abs to both Ags, while turbid plaques meant Ab to only one of the two Ags. Over 40,000 plaques were examined from animals immunized with two different Ags or one Ag with two determinants: no clear plaques were found above the low number observed as background with cells from unimmunized animals.

sizing cell accounts for the **homospecific symmetry** of the Ab molecule (i.e., the same specificity at each of its combining sites), and it explains the absence of natural "hybrids," with different heavy or different light chains in the same molecule.

Cross-reactions show that different cells producing Abs to the same immunogen can differ in their respective Abs' active sites: by testing individual cells in microdrops with different phages (T2 and T4) labeled with cross-reacting haptens (mono- or dinitroiodophenyl, NIP or NNP), some cells from an animal immunized with NIP-BγG were found to neutralize NIP-phage but not NNP-phage, and other cells neutralized both. In addition, individual cells from an animal immunized with a Dnp-protein differ in the concentration of Dnp-hapten required to block their formation of hemolytic plaques in vitro, showing that the Abs secreted by different cells can differ in affinity for the hapten. **Thus the great diversity of Igs in an individual, and the often immense heterogeneity of the Abs he forms against a given Ag, arise from corresponding diversity among Ab-producing plasma cells.**

Small B cells also conform to the one cell–one Ab rule. As noted above, immunofluorescence shows that the surface Igs on a given cell have one isotypic and allotypic form of heavy chain and of light chain (B-cell receptors, above), and reactions with antiidiotypic antisera suggest that there is also one idiotype per cell (see Idiotype suppression, below). Thus the Ig molecules produced by an individual cell of B type (small lymphocyte or plasma cell) are probably as uniform in primary structure as the myeloma protein produced by a myeloma tumor (each of which is a clone of neoplastic plasma cells).

All these findings emphasize that synthesis of Ig molecules in cells of B lineage is subject to the principle of **allelic exclusion:** the cell expresses only one of its several alleles for light chains and only one of its many alleles for heavy chains† (for possible exceptions see

---

† A somewhat similar situation occurs in mammalian females, where random inactivation of one of the two X-chromosomes per cell during embryogenesis causes **mosaicism:** one X is active in some cells and the other X in other cells (the Lyon effect). But structural genes for Igs are not X-linked, and allelic exclusion among autosomal genes appears to be unique for Igs. It does not, for instance, occur

IgM-IgG shift, below). It also secretes Ig molecules or produces Ig membrane receptors that in each cell are restricted to a particular Ag-binding specificity and idiotype. Though all cells express only a small portion of their genes, the degree of restriction seems to be extreme in regard to Ig genes; but it is not clear just how or when the selective expression of particular Ig alleles is initiated in differentiation of stem cells into B cells. Nevertheless, **the restricted specificity of single lymphocytes and the resulting diversity of the mass of lymphocytes in a given individual strongly support the clonal selection hypothesis.**

The **frequency** of a given Ig-producing plasma cell in lymphoid tissues is roughly proportional to the level of the corresponding protein in serum. For instance, there are about 3 to 6 plasma cells that produce δ chains per 1000 that produce γ chains, in accord with the much lower level of IgD than of IgG in serum. And cells that form α chains constitute the vast majority of Ig-forming cells in intestinal and respiratory tracts, and in a number of secretory glands, in accord with the preponderance of IgA over other Igs in mucinous secretions.

In contrast to fully differentiated plasma cells, most of the small Ag-sensitive B lymphocytes that represent the beginning of the differentiation pathway seem to produce only IgMs (see B cell receptors, above). The difference suggests that during the differentiation triggered by Ag some cells switch from production of IgM to IgG or IgA or another class (see below).

IgM-IgG Shift. Rare individual cells in microdrops seem to secrete anti-*Salmonella* Abs of both IgM and IgG classes; and immunofluorescence reveals rare lymphocytes with IgM on the surface and IgG internally. These "double-producers" could correspond to cells in the process of shifting from synthesis of IgM to IgG Abs.* This shift, if it occurs consistently following immunization, could contribute to the shift in preponderance of serum Abs from IgM to IgG that occurs early in most immune responses (see discussion of Change with time after immunization, below). A shift from IgM to IgG production can apparently occur independently of Ag stimulation during embryogenesis of B cells (see Ontogeny, below).

Two considerations can reconcile the double-producers with the one cell–one Ab rule: 1) The rule refers to the activity of cells at a given time: if a cell stops making IgM molecules and starts making IgG molecules, it could for some hours contain and possibly even secrete both Igs without actually synthesizing both of them at the same time. 2) The two Igs of double-producers probably have the same idiotype and presumably therefore the same Ag-binding specificity; i.e., they are likely to differ in just the isotype of the heavy chain.

The two genes–one chain hypothesis (Ch. 16) suggests how the shift could occur with minimal perturbation: in making an IgM Ab the cell expresses genes for V_L, C_L, V_H, and $C\mu$ domains, and in shifting to production of IgG only the gene for $C\mu$ would have to be shut off and the one for $C\gamma$ turned on. Both the IgM and the IgG molecules would thus have the same V_L and V_H and, therefore, the same idiotype and same combining-site specificity. The plausibility of this sequence is supported by several observations. For instance, the IgM and the IgG myeloma proteins made by one **biclonal human myeloma tumor** seem to be identical (by peptide map, idiotypic determinants, partial amino acid sequences) except for C_H, which was μ in one protein and γ in the other (Ch. 16).

As noted above, in the absence of T cells the T-dependent Ags can stimulate B cells to make Abs of the IgM class, but the copious production of IgG molecules to these Ags requires the assistance of T cells and macrophages. The IgG response could either reflect a shift within individual cells (from IgM to IgG production) or a requirement by B cells of the IgG class for some product of stimulated T cells.

---

with hemoglobin (Hgb): in a heterozygote with normal and sickle cell alleles each red cell has both normal and abnormal Hgb molecules.

* Some cultures of permanent lymphoid cell lines consistently produce two kinds of Igs (IgM and IgG, or IgG and IgA). However, these cells are usually aneuploid (abnormal number of chromosomes) and may have many anomalous regulatory mechanisms; it is possible, nevertheless, that a very small proportion of normal lymphoid cells also can simultaneously secrete Igs of two classes.

SYNTHESIS, ASSEMBLY, AND SECRETION OF IMMUNOGLOBULINS

In plasma cells pulse-labeled with radioactive amino acids it appears that Igs, like secreted proteins in general, are synthesized on polyribosomes attached to the rough endoplasmic reticulum, rather than on unbound polyribosomes. Though the cells promptly incorporate radioactive amino acids, the newly synthesized Ig molecules are secreted only after a

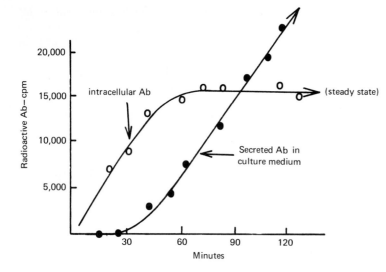

Fig. 17-12. Lag in secretion of antibodies. Labeled amino acid was added at zero time to a suspension of lymph node cells from an immunized rabbit, and cells and supernatant were then assayed at intervals for newly synthesized Abs. [Based on Helmreich, E., Kern, M., and Eisen, H. N. *J Biol Chem 236*:464 (1961).]

20 to 30 minute lag (Fig. 17-12), during which they presumably enter and pass through cisternae of endoplasmic reticulum, traverse the Golgi apparatus, and then move through secretory vesicles toward the cell surface, from which Ig molecules are shed as the vesicles fuse with the surface membrane. During this migration the completed chains are assembled into molecules, the interchain S-S bonds are formed, and sugars are added successively by hexosetransferases to form the oligosaccharide groups of complete Ig molecules (Fig. 17-13).

The order of assembly of chains has been studied primarily in myeloma tumors, in which various patterns have been deduced from the variety of in-

complete molecules found intracellularly and from pulse-chase experiments. Thus, when cells are harvested at various times after a 30-second pulse of radioactive amino acid the labeled nascent light (L) or heavy (H) chain can be followed as it joins with other chains to establish complete four-chain Ig molecules ($H_2 \cdot L_2$). In some tumors complete molecules are made by joining two H · L half-molecules; in others, H-chain dimers (H_2) add one L chain at a time; in still other tumors both patterns are observed. Regardless of the particular pattern, assembly and secretion are orderly: interchain S-S bonds form slowly (2 to 20 minutes after the constituent chains are completed); the rate of secretion equals the rate of synthesis and, accordingly, the intracellular level of Ig molecules is constant; the more recently synthesized molecules leave the cell only after pre-

Fig. 17-13. Scheme for assembly, intracellular transport and secretion of Igs. RER = rough endoplasmic reticulum, i.e., with bound polyribosomes. [Based on Uhr, J. W. *Cellular Immunol 1*:228 (1970).]

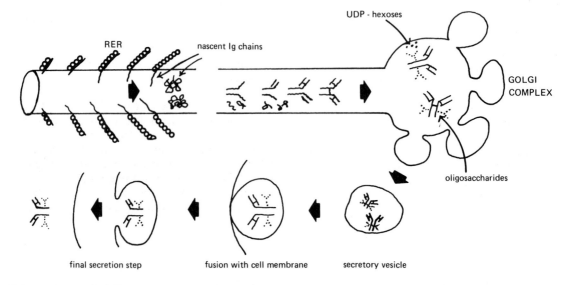

viously synthesized ones have been secreted (Fig. 17-12).

In myloma cells that make IgA or IgM molecules the intracellular Ig seems not to progress beyond the four-chain monomer, though polymers are found in the culture medium. The monomers probably associate with each other and with J chains (which are made in the same cells) as Ig molecules are secreted (for J chains, see Ch. 16).

The order of incorporation of radioactive sugar precursors (hexoses and hexosamines) is in accord with the sequence of sugar residues in Ig oligosaccharides (Ch. 16). Glucosamine, which links oligosaccharides to Ig chains, is incorporated first, and followed by mannose, galactose, and later by sialic acid groups. Fucose appears to become attached last, as Ig molecules exit from the cell.

The addition of sugars may not be necessary for secretion, as has been suggested, because some myelomas secrete only light chains (Bence Jones protein), many of which lack carbohydrate.

LYMPHATIC ORGANS

Besides circulating through blood, lymph, and tissue spaces, lymphocytes are aggregated into primary and secondary lymphatic structures, where different stages in their differentiation are carried out. In the primary organs (thymus, bursa of Fabricius or its analog in nonavian species) lymphocytes become committed to react specifically with particular Ags, and in secondary lymphatic organs the committed cells react with Ags, which stimulate their terminal differentiation: B cells into Ab-secreting plasma cells, T cells into effectors of cell-mediated immunity (Ch. 20), and both into their respective memory cells, which probably also look like small lymphocytes.

PRIMARY LYMPHATIC ORGANS

Thymus. A role for the thymus in the immune response was long suspected (because it is essentially a mass of lymphocytes) but largely discounted, because decades of study showed that this organ did not form Abs, trap Ags, or contain more than a rare plasma cell. Beginning, however, in 1960 with Jacques Miller's observations in England, mice that were thymectomized **at birth** were found to develop a coherent variety of defects, which

illuminate the contribution of the thymus and define properties of T cells.

Neonatally thymectomized mice have a drastically reduced number of blood lymphocytes, depleted T-cell areas in lymph nodes and spleen (Fig. 17-8), and reduced ability to reject allografts (Chs. 20 and 21); moreover, in response to many Ags they produce few if any Abs, and what is produced seems to be largely limited to the IgM class. They can, however, form Abs normally in response to other, T-independent Ags (e.g., highly polymerized proteins, viruses, polysaccharides; Table 17-4).

The effects of thymectomy are confirmed in mice born without a thymus. The defect is inherited as a recessive autosomal trait linked to hairlessness: the resulting **nude, thymus-less** mutants do not usually survive more than a few months, probably because of extreme susceptibility to infections.[*] Practically all their lymphocytes have readily demonstrable surface Igs (i.e., they are B cells). Their serum IgM levels are normal but Igs of other classes are depressed, and most of the Abs they make in response to experimental immunization are IgM, suggesting that the switch from IgM to IgG production is largely blocked. Perhaps because some T cells migrate out of the thymus before birth, neonatally thymectomized normal mice seem not to be as deficient in T cells or in T cell functions as the nude mutants.

The thymus is a "lymphoepithelial" mixture of lymphoid and epithelial cells. It develops in the embryo from branchial pouches of the pharynx: epithelial buds grow out, pinch off, and migrate (in higher vertebrates) to the midline of the upper thorax where they become infiltrated with lymphocytes. The transfer of cells from diverse tissues in mice with the T6 chromosomal marker to syngeneic mice lacking the marker has shown that thymus lymphocytes derive from migratory blood-forming (hematopoietic) stem cells (originating in yolk sac or liver of the fetus and in bone marrow of the adult). A high incidence of mitotic figures in the thymus, and rapid incorporation of radioactive thymidine into DNA, show that the immigrant cells divide rapidly; but only a small proportion of their progeny survive and be-

---

[*] These mice are, accordingly, produced in special colonies maintained for high frequency of the defective chromosome.

TABLE 17-4. Effects of Extirpation of Thymus or Bursa of Fabricius*

Property	Thymectomy	Bursectomy
Cellular changes		
Circulating lymphocytes	↓↓	0
Spleen		
Germinal centers	0(↓)	↓↓
B zones†	0	↓↓
T zones†	↓↓	0
Plasma cells	0(↓)	↓↓
Functional changes		
Serum Ig level	0(↓)	↓↓
Antibody formation	0(↓)	↓↓
Cell-mediated immunity		
(e.g., rejection of skin		
allografts)	↓↓	0

  * Composite effects of bursectomy and thymectomy in chicks at hatching and of thymectomy in newborn mice.

  † Lymph nodes in birds are poorly organized clusters of lymphocytes alongside lymphatic channels; their B and T zones are readily discernible in the spleen.

  0 = No change; ↓↓ = marked decrease; ↓ = modest decrease; 0(↓) = variable decrease, probably secondary to role of T cells as helpers in induction of Ab synthesis.

come mature T cells. About 90% of thymus lymphocytes are readily destroyed by corticosteroids. The resistant 10%, located mostly in the medulla (Fig. 17-14), have the characteristic surface Ags of mature T cells (Table 17-2), which they also resemble in having marked reactivity with foreign (allogeneic) cells (Ch. 21, Graft-vs.-host reaction).

In most species the thymus is fully developed at birth. It atrophies gradually with time ("age involution"), or in occasional rapid bursts after severe stress ("accidental involution" after severe infection, starvation, trauma), probably in response to high levels of adrenal corticosteroids (Immunosuppressive agents, below). Removal of the thymus from an adult usually has no obvious effects, because peripheral tissues are already fully populated with a great many diversified T cells. However, the adult thymus is functional: its cells serve as competent T cells and T-cell precursors in adoptive transfer, and thymectomy delays recovery of immunological activity after whole-body X-irradiation (see Fig. 17-29).

Thymic Hormone. It is possible that a hormone elaborated by thymus epithelial cells promotes differentiation of precursor lymphocytes into T cells.

Thus the cellular and functional deficiencies of neonatally thymectomized mice can be largely corrected with thymic grafts enclosed in chambers whose porous membrane passes macromolecules but not cells. Moreover, a glycoprotein ("thymosin") in calf thymus extracts seems to repair some defects in thymectomized mice (e.g., the ability to reject allografts).

Bursa of Fabricius. One of the earliest clues to the distinction between B and T cells derived from Glick's serendipitous discovery that chicks deprived at birth of the bursa of Fabricius are unable to produce Abs. Resembling the thymus anatomically, the bursa also develops by lymphocytic infiltration of an epithelial outpouching from the intestine, but in the hind gut (cloaca) instead of the foregut. Lymph follicles of the bursa, unlike those of thymus, are packed with plasma cells. The bursa also undergoes atrophy at puberty (4 months of age in chickens), probably as a result of rising levels of steroid sex hormones. In fact, the bursa can be eliminated ("hormonally extirpated") by simply dipping embryonated eggs for a few minutes in a solution of testosterone.

Extirpation has shown that the bursa and thymus regulate complementary functions

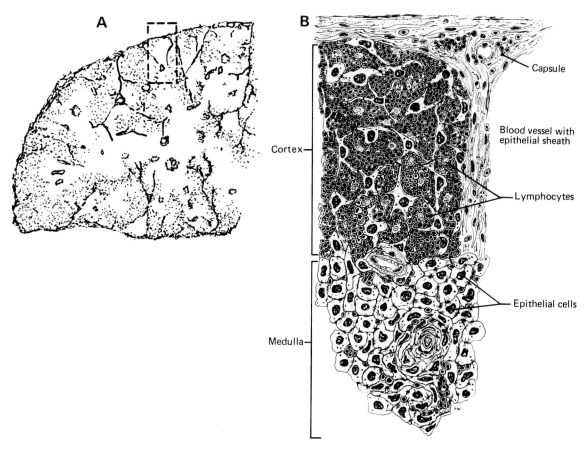

Fig. 17-14. Thymus. **A.** The darkly stippled cortex consists of a dense mantle of T lymphocytes surrounding the lightly stippled medulla, which is made up of epithelial cells plus some lymphocytes (more mature T cells). **B.** Portion of thymus lobule (area enclosed in **A**) in some detail. (Based on Weiss, L. *The Cells and Tissues of the Immune System*. Prentice-Hall, Englewood Cliffs, N.J., 1972.)

(Table 17-4). In contrast to thymus-less animals, neonatally bursectomized chicks have normal cell-mediated immune responses (e.g., they reject skin allografts, Ch. 21) and they have normal levels of lymphocytes in blood and also in T zones of lymph nodes and spleen. However, they are devoid of plasma cells, and their severe deficit in serum Ig levels and in Ab-forming ability is greater the earlier in life the bursa is removed (Ontogeny, below).

Mammals lack a bursa of Fabricius and their structure with equivalent function has not been identified. Possibilities include the bone marrow itself, lymph nodes (which are rudimentary in chickens), various gut-associated lymphoepithelial structures (appendix, tonsils), and the lymphocyte-infiltrated villous epithelium that stretches the length of the intestinal tract.

SECONDARY LYMPHATIC STRUCTURES

The spleen, many lymph nodes, and smaller, less organized clusters of lymphoid cells in many tissues and organs constitute the secondary structures in which committed lymphocytes undergo terminal differentiation in response to Ag stimulation. These structures are highly effective Ag-trapping filters in which lymphocytes pack the interstices of a dense three-dimensional fibrous network, with many sticky reticular cells attached to reticulum fibers. Injected lymphocytes accumulate in spleen and lymph nodes, as though expressing the **homing** instinct that normally causes circulating lymphocytes to leave blood vessels in these structures (via postcapillary venules or sinusoids), and to reside for a time in appropriate areas (B and T zones, Fig. 17-8) before recirculating.

Lymph Nodes. These ovoid structures are widely distributed in extremities, trunk, mediastinum, omentum, etc., where they generally are situated at major junctions of the network of lymphatic channels, which pass tree-like from twigs in the more superficial tissues to a large central collecting trunk, **the thoracic duct;** this vessel pours lymph and lymphocytes into the great vein that returns blood to the heart. Ags deposited in tissues are carried via lymphatic channels through successive lymph nodes, in which they are likely to be trapped by reticular cells and macrophages. (Brain, spinal cord, and eye lack lymphatic drainage, and Ags carried away in blood from these areas are likely to be trapped in the spleen or distant lymph nodes.)

At the center of the subcapsular B-cell nodules the **germinal center** is made up of a mass of rapidly dividing large B lymphocytes (generation time ca.

6 hours; Fig. 17-15). Macrophages with ingested pyknotic nuclei and DNA debris are also abundant, as in many other areas with rapidly dividing cells (as though many of the rapidly formed progeny are defective and are disposed of by phagocytosis). **The size of germinal centers is proportional to the intensity of antigenic stimulation:** they are greatly enlarged in secondary responses (below) and are essentially absent in germ-free mice. These centers become similarly hyperplastic in other secondary lymphatic structures when intensely engaged in producing Abs.

Spleen. This large encapsulated vascular organ (about 200 gm in adult man) traps Ags carried in blood. Like other lymphatic organs, its cortex has densely packed lymphocytes (with germinal centers), and its loose medulla has wide vascular spaces with a variety of lymphoid and other cells (Fig. 17-16). In the spleen, however, the medulla (called "red pulp") encloses cortical areas (called "white pulp"), rather than the reverse, as in other lymphatic structures. Cortical areas surround small central arteries with leaky walls, through which Ags are probably driven by the arterial blood pressure. Separate T and B zones are also discernible in the white pulp (Fig. 17-8); the T area is relatively smaller in spleen than in lymph nodes,

Fig. 17-15. Mammalian lymph node. **A.** Schematic view showing afferent (AL) and efferent (EL) lymph channels and the subcapsular sinus (SS) with criss-crossing reticulum fibers that extend throughout the node. Reticular macrophages, attached to the fibers, trap Ags, especially when in the form of Ab-Ag complexes. GC = germinal center; MS = medullary sinus. **B.** Human lymph node (×45). The "secondary nodule" points to a germinal center with its surrounding mantle of B lymphocytes. The "tertiary cortex" is a mass of T cells. (**A** courtesy of Dr. S. Clark, Jr. **B** from Weiss, L. *The Cells and Tissues of the Immune System.* Prentice-Hall, Englewood Cliffs, N.J., 1972.)

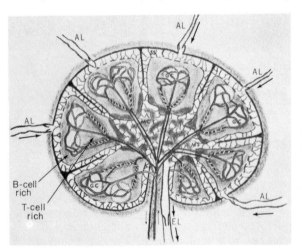

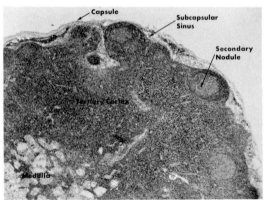

A   **B**

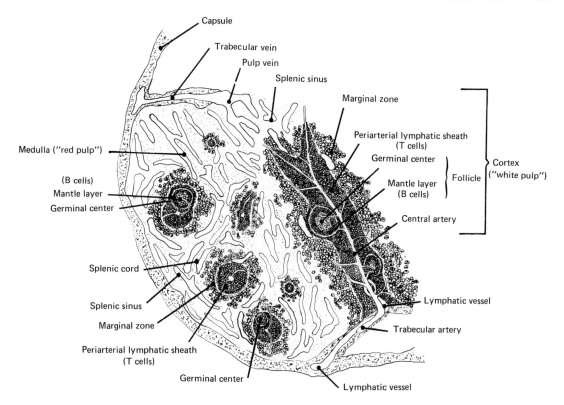

Fig. 17-16. Schematic view of a representative area of the spleen showing masses of lymphoid cells (white pulp) distributed around the central arteries. T cells make up the periarterial lymphatic sheath; B cells form the mantle layer around germinal centers. See also Figure 17-8. (From Weiss, L. *The Cells and Tissues of the Immune System.* Prentice-Hall, Englewood Cliffs, N.J., 1972.)

and after intense antigenic stimulation T cells are virtually replaced by sheets of plasma cells (in contrast to lymph node T areas, where plasma cells are almost never seen).

Gut-associated Lymphoid Cells. Lymphocytes and plasma cells (predominantly with IgA) are spread throughout the inner layers of the intestinal wall as isolated cells or as small cell clusters. Larger clusters form distinct follicles with germinal centers; some follicles become confluent and interdigitated with overlying epithelium, forming small **lymphoepithelial** structures which (unlike thymus and bursa) remain associated with the intestinal wall throughout life. The main ones in man are 1) tonsils (in the pharyngeal wall), 2) appendix (at junction of small and large intestine), and 3) Peyer's patches (oblong lymphoid aggregates found mostly in the wall of the terminal part of the small intestine). As in other secondary lymphatic structures, follicular (B) areas around germinal centers are well developed and interfollicular (T) areas are atrophic in thymus-less individuals.

GENETIC CONTROL OF ANTIBODY FORMATION

Selective breeding can generate strains of animals able to make high or low titers of Abs to complex Ags (foreign red cells, salmonellae, etc.). However, insight into genetic control began to accumulate rapidly only after systematic use was made of 1) inbred mice and guinea pigs, and 2) Ags of limited immunogenicity, such as synthetic polypeptides, unusually small doses of conventional proteins, and alloantigens.

Ability of mice to make high Ab titers to many of these Ags is determined by a dominant, autosomal (not sex-linked) locus, Ir-1: e.g., mice of the C57Bl strain make about 10 times higher titers than CBA mice to the tyrosine-glutamate (T,G) determinant of (T,G)–A—L (Fig. 17-17); and all F_1 hybrids

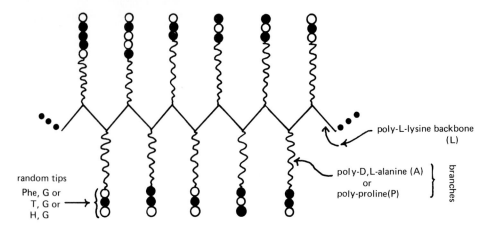

random tips
Phe, G or
T, G or
H, G

poly-L-lysine backbone (L)

poly-D,L-alanine (A) or poly-proline(P) } branches

Immunogen:	Phe, G-Pro--L		T, G-Pro--L	T, G-A--L	Phe, G-A--L	H, G-A--L
Abs specific for:	Phe, G	Pro, L	Pro, L	T, G	Phe, G	H, G
Mouse strains						
High responders	DBA	SJL	SJL	C57Bl	DBA	C3H, CBA
Low responders	SJL	DBA	DBA	SJL, C3H, CBA	SJL	C57Bl, SJL
Ir gene linked to:	H-2	not H-2	not H-2	H-2	H-2	H-2

Fig. 17-17. Genetic control of Ab formation to synthetic branched polypeptides. The most useful ones have a poly-L-lysine (L) backbone and poly-D,L-alanine (A) or polyproline (P) branches, with tips carrying short random amino acid sequences of tyrosine and glutamate (T,G) or histidine and glutamate (H,G) or phenlyalanine and glutamate (Phe,G). The corresponding polymers are designated (T,G)-A—L, (H,G)-A—L, and (Phe,G)-A—L. The antigenic determinants are usually the short terminal sequences (Phe,G or T,G or H,G); but in a few mouse strains the branched backbone (poly-pro-poly-lysine) is immunogenic. (Based on Sela, M. *Harvey Lect,* 1973, in press.)

(i.e., C57 × CBA) are high responders, as are ca. 50% of the progeny of the back-cross between hybrids and the low-responding homozygous parental strain (F$_1$ × CBA). Parallel effects have been obtained in other strains with this and other Ags: for instance, with another branched polypeptide, (H,G)–A—L, the high responders are CBA and the low responders are C57Bl mice (Fig. 17-17).

The ability of a mouse strain to respond to certain Ags was found by McDevitt to be associated with the strain's major histocompatibility Ags, which are located on cell surface membranes: both phenotypes are consistently carried together in the progeny of diverse matings, establishing genetic linkage of Ir-1 with the H-2 locus, which specifies the most potent histocompatibility Ags in

mice (Ch. 21). This unexpected association turned out to be remarkably useful, because independent genetic analyses provided rare mice that represented recombinants between the two ends of the H-2 locus: and by selective breeding the recombinant chromosome could be introduced in homozygous form into mice whose immune responses to branched polypeptides then made it possible to map the Ir-1 locus close to the center of the H-2 locus (Fig. 17-18). A few other Ir genes, segregating from Ir-1, specify ability to make high Ab titers to different Ags; some are linked to other loci (not H-2) (Fig. 17-17).

Similar relations were found independently by Benacerraf *et al.* in inbred guinea pigs, in whom the additional ability to exhibit delayed-type hypersensitivity skin responses provided

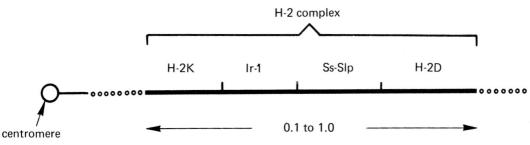

Fig. 17-18. Map of a mouse chromosome (IXth linkage group) showing the immune response-1 (Ir-1) locus and the complex locus for the most potent histocompatibility Ags (H-2). *Ir-1* falls between *H-2K* and *H-2D*, the two genes that define the limits of the *H-2* complex (Ch. 21). *Ss-Slp* governs the quantitative level of a serum protein, Ss, and a cross-reacting protein whose production is limited to males (Slp, sex-limited protein). Products of H-2 and of Ss-Slp are probably entirely unrelated in function. The recombination frequency between the ends of *H-2D* and *H-2K* is between 0.1 and 1.0%, which means that the *H-2* complex might include ca. 1000 genes. By convention, the markers are aligned from left to right starting with the centromere. [From Klein, J. and Shreffler, D. C. *Transplant Rev* 6:3 (1971).]

additional information about genetic control of cell-mediated immunity (Ch. 20). Guinea pigs of one inbred strain (2) are high responders to poly-L-lysine (PLL) carrying Dnp or other haptenic groups: they develop cell-mediated hypersensitivity to the immunogen (Ch. 20) and high Ab titers to the haptenic group. Those of another strain (13) are low responders: they develop no hypersensitivity to the immunogen and little if any antihapten Ab. As with purebred strain 2, all F_1 hybrids (2 × 13) are high responders, and so also are ca. 50% of the progeny of the back-cross of hybrids to the low-responder strain (F_1 × 13). Hence the response to hapten-PLL also depends on a dominant autosomal locus, the PLL gene, which, like Ir-1 of mice, has also turned out to be closely linked to the main locus for guinea pig histocompatibility Ags.

Guinea pigs that lack the PLL$^+$ gene can, nevertheless, make the same anti-hapten Abs as responders if the hapten-PLL Ag (Dnp-PLL) is administered as a complex with an immunogenic carrier, such as bovine serum albumin (BSA). Similarly, mice that ordinarily make little response to (T,G)–A—L can make a high Ab response if this Ag is administered as a complex with an immunogenic carrier (e.g., methylated BSA). Hence the Ir genes do not specify the V-region amino acid sequences that determine Ab specificity (Ch. 16), but rather some element that seems to be concerned with recognition of the carrier moiety of the immunogen (e.g., PLL). From what is known about carrier recognition (see above), products of the Ir-1 gene should be expressed on T cells.

The following circumstantial evidence supports this view.

1) T lymphocytes are the effector cells for cell-mediated immunity (such as delayed-type skin reactions, Ch. 20), and this type of reaction to Dnp-PLL is entirely lacking in PLL$^-$ guinea pigs, even when they are induced to make anti-Dnp Abs in response to Dnp-PLL in complex with an immunogenic carrier.

2) T cells seem to regulate the production of the 7S, rather than the 19S, Abs made in response to T-dependent Ags; correspondingly, mice that are genetic nonresponders to a given Ag are blocked in the formation of the appropriate 7S Abs, but not in the production of the 19S Abs.

Adoptive transfers suggest, however, that while some Ir genes are expressed on T cells, others seem to be manifested on B cells. Thus titrations of cells from donors carrying Ir+ or Ir− genes have been carried out in irradiated, syngeneic recipients by transfer of an excess number of T cells plus varying numbers of B cells, or the reverse (excess B and varying number of T cells): the proportion of recipients making Abs to the subsequently injected Ag indicates the frequency of Ag-sensitive cells in the transferred thymus and bone marrow populations.* With some Ags the responder and nonresponder strains differ in the frequency of Ag-sensitive cells

---

* By application of the Poisson distribution: the frequency of Ag-sensitive cells in a given number of T or B cells (the complementary cell population being given in excess) is based on the proportion of recipients who make no Ab response to an optimal dose of Ag (see Ch. 18, One-hit theory in complement lysis).

in the thymus, not in the bone marrow, while with other Ags the strains differ in marrow cells, not in thymus. While the Ir-1 gene of the H-2 locus seems to be fully expressed on T cells, it is possible that other Ir genes, at other loci, are expressed on B cells.

The Ir genes seem also to control immune responses to many complex Ags, such as red cells, serum proteins (including Igs). With serum proteins as Ags, very small doses elicit responses in certain inbred mouse strains, but not in others, and the response is again genetically controlled by single autosomal dominant loci, which in a number of instances have been linked to the H-2 locus. Evidently with sufficiently low doses of immunogen only the dominant determinant is effective, and complex Ags can approach the simplicity of single haptenic groups.

The linkage of Ir-1 to the H-2 locus suggests that the product of the Ir-1 gene might be simply another histocompatibility Ag that happens, perhaps because of proximity, to modify the specificity or affinity of the cell-bound Ig that serves as Ag-binding receptor. If so, various Ir-1 genes would be expected to be scattered throughout the H-2 locus. However, the recombinant chromosomes so far analyzed show the Ir-1 gene(s) to be consistently clustered in a discrete region next to but distinct from the left-hand end of the H-2 locus (Fig. 17-18). It seems possible, therefore, that the product of Ir-1 genes are the T-cell Ag-binding receptors themselves (or an essential part of them): if so, the receptors would have to be something other than Igs, for Ir-1 is not a structural Ig gene: it is not linked to the locus for heavy Ig chains and responder and nonresponder strains form indistinguishable Abs when the discriminating Ag is attached to a suitable carrier (see above).

The products of the Ir-1 locus could be sufficiently diversified to account for the great range of specificities of T-cell receptors. The recombination frequency between the two ends of the H-2 locus suggests that the area in between, including Ir-1, could accommodate perhaps 1000 genes; moreover, diversity could be amplified by the mechanisms envisioned to help account for Ig diversification—e.g., mutations in dividing stem cells or assembly of various combinations of subunits (Ch. 16, Origin of Ig diversity).

Whatever their product turns out to be, the Ir genes probably have an important role in many diseases. Susceptibility to various leukemogenic viruses, to lymphocytic choriomeningitis virus, to mouse smallpox infection, and perhaps even to some autoimmune diseases in certain hybrid mice seems to be linked to the H-2 locus and many of these phenotypes could be under control of immune responses and thereby of Ir genes.

ANTIBODY FORMATION IN THE WHOLE ANIMAL

The rates of formation, steady-state levels, and types of Abs formed vary widely with conditions of immunization, which are reviewed in this section.

ADMINISTRATION OF ANTIGEN

ROUTE

Natural Immunization. Lymphatic tissues are probably bombarded almost constantly with Ags from transiently invasive or indigenous microbes (normal flora of skin, intestines, etc.). The resulting stimulation is probably responsible for the familiar histological appearance of lymph nodes and spleen, the normal concentration of Igs in serum (ca. 15 mg/ml), and **natural antibodies**—those Igs that react or cross-react (one cannot be sure which) with Ags that have not been known to serve as immunogens in the individual under test. Animals reared under germ-free conditions synthesize Igs at ca. 1/500 the normal rate, have exceedingly low serum levels (especially IgG), and have markedly hypoplastic secondary lymphatic tissues. Nonmicrobial immunogens can also enter the body naturally by inhalation (e.g., plant pollens), ingestion (e.g., foods, drugs), and penetration of skin (e.g., catechols of poison ivy plants).

Deliberate Immunization. For this purpose immunogens are usually injected into skin (intradermally or subcutaneously) or muscle, depending upon the volume injected and the irritancy of the immunogen. Intraperitoneal and intravenous injections are also used in experimental work, especially with particulate Ags. Usually, regardless of route, the Ag eventually becomes distributed widely throughout the body via lymphatic and vascular channels.

Because most Ags are degraded in the intestines, **feeding** is effective only under special circumstances; e.g., with attenuated poliomyelitis vaccine (Ch. 55), which can invade the intestinal wall, and allergic responses to food are often due to earlier antigenic stimulation by ingestion of the same substances. **Inhalation** can also be used: e.g., aerosol administration of attenuated strains of *Pasteurella tularensis*. Preferential synthesis of IgA Abs can occur when immunogens are introduced into the respiratory and intestinal tracts, many of whose abundant plasma cells are committed to produce Igs of this class. Direct **application to the skin** is effective with certain penetrating substances of low molecular weight, which tend especially to induce cell-mediated immune responses (Ch. 20, Contact skin sensitivity).

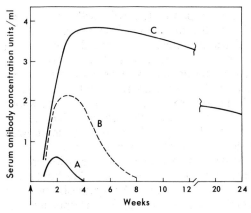

Fig. 17-19. Influence of adjuvants. Schematic view of amounts of Ab produced by rabbits in response to one injection (arrow) of a soluble protein, such as Bovine γ-globulin in dilute salt solution (A), adsorbed on precipitated alum (B), or incorporated in a water-in-oil emulsion containing mycobacteria (Freund's complete adjuvant, C).

ADJUVANTS

The immunogenic potency of soluble proteins is enhanced if they persist in tissues; for example, repeated small injections of diphtheria toxoid evoke a greater Ab response than the same total amount of toxoid given as a single injection. Accordingly, a widely used procedure involves administration of inorganic gels (e.g., alum, aluminum hydroxide, or aluminum phosphate) with adsorbed immunogens that are released slowly for a prolonged period. These gels are called **adjuvants,** a term which applies broadly to any substance that increases the response to the immunogen with which it is injected.

The most effective adjuvants are the water-in-oil emulsions developed by Freund, particularly those in which living or dead mycobacteria are suspended. After a single subcutaneous or intramuscular injection (e.g., 0.5 ml in a rabbit) droplets of emulsion metastasize widely from the site of injection; Ab formation can be detected as early as 4 or 5 days later, and it may continue for 8 or 9 months or longer (Fig. 17-19). The intense, chronic inflammation around the deposits of emulsion precludes their use in man. However, emulsions without mycobacteria ("incomplete" Freund's adjuvant) are less irritating and have been used clinically; their enhancing effect is less than that of "complete" Freund's adjuvant. Some other adjuvants are bacterial endotoxin; large

polymeric anions (certain synthetic polyribonucleotides [poly AU], dextran sulfate, but not dextran itself); and *Bordetella pertussis*.

Most adjuvants promote the maintenance of low, effective Ag levels in tissues and provoke inflammation, in which accumulated macrophages could bind Ag for reaction with B and T cells. Some adjuvants (e.g., mycobacteria) aid responses only to T-dependent Ags: perhaps by acting as independent immunogens or as carriers for Ags that they adsorb, these adjuvants cause proliferation of T cells (evident as hyperplasia in T zones of regional lymph nodes) and might also increase their output of B-cell-stimulating factors (Antigenic promotion, see below).

DOSE

The smallest effective dose of an immunogen depends on the method of administration. Thus 100 μg of the human serum albumin will usually not evoke detectable Ab formation when injected intravenously in a rabbit, whereas the same dose is highly effective if injected subcutaneously in Freund's adjuvant. With the adjuvant as little as 1 to 10 μg might represent the threshold, which is usually much lower in animals that have been previously stimulated by the same immunogen (Secondary response, below). Moreover, a subthreshold dose can sometimes "prime" animals for a subsequent, augmented secondary response (below) without actually eliciting the formation of detectable Ab.

Once the threshold is exceeded, increasing doses lead to increasing, but generally less than proportionate, responses (which can be measured over an approximately 10 million-fold range: from ca. 0.001 μg Ab/ml, about the lowest measurable concentration, to ca. 50 mg/ml in animals intensively immunized with the most potent immunogens).

Doses that are excessive (or, with certain Ags, that are just below the threshold) not only fail to stimulate Ab synthesis, but can establish a state of specific unresponsiveness (Immunological tolerance, below).

THE PRIMARY RESPONSE

The initial exposure to most Ags evokes a smaller and somewhat different Ab response than subsequent exposures. The differences are of cardinal importance in resistance to microbial infections.

The first introduction of immunogen, the **primary stimulus,** evokes the **primary response** in which Abs are first detectable after 1 to 30 days or more (Fig. 17-20). The lag varies with dose, route of injection, the particulate or soluble nature of the immunogen, the type of adjuvant used, and with the sensitivity of the assay. It is common for Abs to be detected 3 to 4 days after injection of foreign erythrocytes, 5 to 7 days after soluble proteins, and 10 to 14 days after bacterial cells.

Special immunogens are needed to follow early rates of Ab formation. Bacteriophages are particularly useful: exceedingly small quantities are immunogenic, avoiding significant masking of newly formed Ab by persistent Ag. In addition, neutralization of phage infectivity (for *E. coli*) can measure extremely low concentrations of Ab, including molecules of low affinity (see Ch. 15, Rates of reaction and Table 15-5). As is shown in Figure 17-21, the phage evokes detectable Abs as early as 1 day after an intravenous injection. The activity in serum increases thereafter for the next 4 to 5 days at an exponential rate (i.e., the rate of increase at any time is proportional to the level of activity at that time), and the doubling time for serum Ab is as short as 6 hours.

The time required to attain maximal Ab levels, and the duration of the peak titer, vary with different immunogens and methods of immunization. The peak is often reached 4 to 5 days after an injection of red cells, and 9 to 10 days after soluble protein is common;

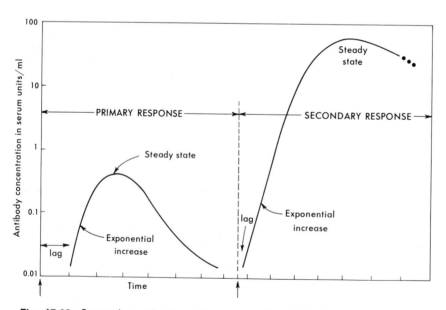

Fig. 17-20. Comparison of serum Ab concentrations following the first (priming) and second injections of the immunogen (arrows). Note the logarithmic scale for Ab concentration. Time units are left unspecified to indicate the great variability encountered with different immunogens under different conditions.

diphtheria toxoid in man may require as long as 3 months. The serum concentration of Abs to soluble proteins may begin to decline within 1 or 2 days after reaching the maximum, but after the same immunogen is given in Freund's adjuvant it may remain elevated for many months (Fig. 17-19).

Antibody Turnover and Distribution. The level of Ab in serum reflects the balance between rates of synthesis and degradation. When the rates are equal, the serum Ab concentration is constant (steady state). The rate of synthesis depends upon the number of Ab-producing cells, which varies enormously with conditions of immunization. However, Ab is degraded at a rate (expressed as half life or $t_{1/2}$) which is determined by its Ig class (Table 17-5): **IgM and IgA are normally broken down much more rapidly than IgG molecules.**

However, infusions of trace amounts of ^{125}I-labeled IgG into individuals with widely varying IgG levels (agammaglobulinemia or multiple myeloma) revealed an inverse relation between the half-time for IgG degradation and the total concentration of this Ig class: at high and low levels of IgG the $t_{1/2}$ was about 11 and 70 days, respectively, as compared with 23 days at normal levels. Injected Fab fragments and light chains disappear rapidly ($t_{1/2} < 1$ day), but the Fc fragment has the same half life as intact IgG. By analogy with some other serum pro-

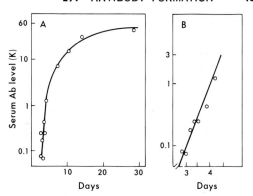

Fig. 17-21. A. The primary response in a guinea pig injected at zero time with 6×10^8 bacteriophage ϕX174 (about 0.01 μg protein). **B.** The first 4.5 days in **A** is repeated with an expanded scale to demonstrate that the increase is exponential. [From Uhr, J. W., Finkelstein, M. S., and Baumann, J. W. *J Exp Med 115*:655 (1962).]

teins, it is possible that shortening of oligosaccharide branches, by random removal of terminal sialic acid or other residues by blood glycosidases, makes Ig molecules susceptible to uptake and degradation in the liver.

The actual serum concentration of Ig depends also upon the volume in which the molecules are distributed. The total mass of IgG is about the same in blood and in extravascular fluids. About 25% of the plasma IgG diffuses out of the vascular compartment each day, and an equal amount is returned via lymphatic channels.

TABLE 17-5. Degradation Rates and Some Other Metabolic Properties of Human Immunoglobulins

Properties	IgG*	IgA	IgM	IgD	IgE
Serum concentration (mg/ml) (average, normal individuals)	12.1	2.5	0.93	0.023	0.0003
Half life (days)†	23	5.8	5.1	2.8	2.5
Rate of synthesis (mg/kg body weight per day)	33	24	6.7	0.4	0.016
Catabolic rate (% of intravascular pool broken down per day)	6.7	25	18	37	89

* IgG-1, IgG-2, and IgG-4 have the same half life (ca. 23 days), but that of IgG-3 is shorter ($t_{1/2}$ = 8 to 9 days). The half life of IgG in some other species is (in days): rabbit (6), rat (7), guinea pig (7 to 9).

† Half life ($t_{1/2}$) is the time required for the concentration (at any particular moment) to drop to ½ the value; it is related to the first-order rate constant for degradation, k (in days^{-1}), by: $t_{1/2}$ (in days) = 0.693/k.

Based on Waldmann, T. A., Strober, W., and Blaese, R. M. in *Immunoglobulins* (E. Merler, ed.). National Academy of Sciences, Washington, D.C., 1970.

CHANGES WITH TIME

The Abs made at various times after immunization differ in type of Ig chain and in affinity for Ag.

Changes in Ig Class. Most immunogens elicit detectable Abs of the IgM class before those of the IgG class (Fig. 17-22).*

---

* Human IgM and IgG Abs can usually be distinguished by treating dilute serum with a reducing agent, such as mercaptoethanol: IgM molecules dissociate into 7S subunits, which lose ability to cross-link Ag particles (as in hemagglutination), though the subunits still bind Ags. IgG molecules, in contrast, retain full activity even when all of their interchain S-S bonds are cleaved (Chs. 15 and 16). The inability of reduced IgM molecules to cross-link Ags may reflect the low intrinsic affinities of most IgM Abs and a correspondingly great need for cooperative multivalent binding (Ch. 15, Monogamous binding; Avidity; Appendix).

It is possible that this sequence is related to the IgM-IgG switch that appears to take place in some B cells following Ag stimulation (see above). Nevertheless, it is not clear whether IgM Abs are always produced in significantly greater bulk before the corresponding IgG molecules, or whether IgM Abs are simply more readily detected because of the greater sensitivity of conventional assays for more avid Abs: with 10 combining sites per IgM molecule, and 2 per IgG, the IgM Abs tend to be more avid (Ch. 15, Avidity; Appendix). Indeed, special analyses suggest that low-avidity IgG (see below) can be produced as early as the more readily detectable IgM molecules.

If the immunogen is administered in complete Freund's adjuvant both IgM and IgG Abs can continue to be synthesized for many months. Otherwise, synthesis of IgM usually ceases to be easily detectable after 1 to 2 weeks, while IgG titers continue to rise (Fig. 17-22). The synthesis of IgM Abs can be arrested by passive transfer of IgG molecules of the same specificity **(feedback inhibition);** this lowers the level of residual immunogen and suggests that Ag levels have to be relatively high to elicit synthesis of IgM Abs, which would be in ac-

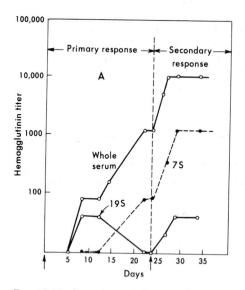

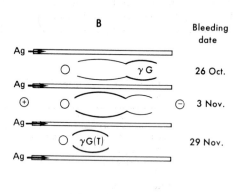

Fig. 17-22. Sequential changes in class of Ab made after immunization. **A.** A rabbit was immunized with diphtheria toxoid as shown (arrows). Antitoxin was measured by passive hemagglutination, using sucrose density gradient centrifugation to separate IgM (19S) and IgG (7S) Abs. **B.** A horse was immunized with tetanus toxoid. Antisera were subjected to electrophoresis, then Ag (toxoid) was added to the long troughs to precipitate antitoxin (Fig. 15-29, Ch. 15). The earliest precipitating antitoxin belonged to one IgG subclass; the more anionic antitoxin that appeared later, IgG (T), belonged to another. [**A** from Bauer, D. C., Mathies, M. J., and Stavitsky, A. B. *J Exp Med 117*:889 (1963). **B** from Raynaud, M. in *Mechanisms of Hypersensitivity*, Shaffer, J. H., LiGrippo, G. A., and Chase, M. W., eds. Little, Brown, Boston, 1959.]

cord with the generally low intrinsic affinities of these molecules for their ligands (see discussion of Changes in affinity, below).

Some thymus-independent Ags seem to elicit the formation of IgM Abs almost exclusively. For instance, the Abs made in horses to pneumococcal capsular polysaccharides, and in rabbits to the Forssman antigen of red cell stroma (Ch. 18) and to *Salmonella* lipopolysaccharide (endotoxin), are predominantly and persistently IgM molecules.

Other Ig chains are also occasionally used preferentially in the formation of Abs to certain other Ags. For instance, in human Abs ("cold agglutinins") to certain red cell Ags and in high-affinity guinea pig Abs to the Dnp group, the light chains seem always to be of κ type; and in horse antitoxin formed late in immunization (Fig. 17-22) and in human Abs to teichoic acid, the heavy chains are of a particular γ subclass. The basis for these restrictions is obscure.

Increase in Affinity. The antihapten Abs formed initially after injection of a hapten-protein immunogen usually have low affinity for the hapten, while the average affinity of the Abs formed later is much higher. These changes are especially pronounced with IgG molecules, and they take place more rapidly after a small than after a large dose of immunogen (Table 17-6). They probably derive from the following: 1) the level of immunogen decreases with time, and 2) the Ag-binding sites of receptor-Igs on an Ag-sensitive B cell are probably the same as on the Ab molecules secreted by the stimulated cell's differentiated progeny (plasma cells). Hence

the concentration of Ag required to stimulate a B cell should be inversely related to the affinity of the Ab molecules secreted by the resulting clone: e.g., high Ag levels generally elicit production of low-affinity Abs.

In previously unimmunized animals most of the B cells that bind a given immunogen are likely to have low affinity for it. The relatively high level of Ag prevailing soon after the immunogen is injected can stimulate many of these cells, leading to the initial production of low affinity Abs. As the Ag level declines, increasingly selective stimulation and expansion of high-affinity Ab-producing clones probably occurs, leading to a rise in average intrinsic affinity of serum Abs. Thus with time after immunization the formation of hemolytic plaques by isolated Ab-secreting cells can be specifically inhibited by increasingly low concentrations of haptens. As expected, prolonged high levels of Ag delay the onset of selective production of high-affinity Abs (Table 17-6).

An immunogen generally has diverse antigenic determinants, and the corresponding Ab molecules can appear after different lag periods: hence the variety of Ab specificities tends to increase with time. This increase and the late appearance of high-affinity Ab molecules may explain why a given Ag behaves as though it has more binding sites per molecule when it reacts with a late antiserum than with an early one (Fig. 17-23). The Ab-Ag complexes formed in late antisera are correspondingly less dissociable: **avidity of antisera increases with time.**

TABLE 17-6. Sequential Changes in Affinity of the Anti-Dnp Antibodies (IgG Immunoglobulins) Made with Increasing Time After Immunization*

Group	Ag injected per rabbit (mg)	Average intrinsic association constants for binding ε-Dnp-L-lysine at		
		2 weeks	5 weeks	8 weeks
I	5	0.86	14	120
II	250	0.18	0.13	0.15

* Immunogen was 2,4-dinitrophenyl bovine γ-globulin. Values are averages for five animals per group, and are given in liters mole^{-1} $\times$ 10^{-6} (30°).

From Eisen, H. N., and Siskind, G. W. *Biochemistry 3*:996 (1964).

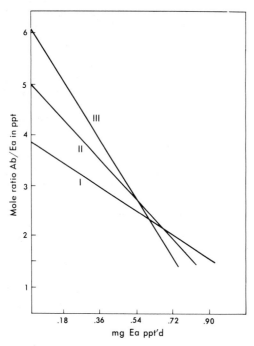

Fig. 17-23. Sequential changes in antisera reflected in the increasing number of functional groups detected per molecule of Ag. Rabbits were immunized with several courses of injections of hen's ovalbumin (Ea). Precipitin reactions with sera obtained after courses I, II, and III show, from the limiting mole ratio in extreme Ab excess (intercept on ordinate), that with time the serum Abs react with an increasing number of sites per Ea molecule (see Ch. 15., Avidity). [Based on Heidelberger, M. and Kendall, F. *J Exp Med* 62:697 (1935).]

Cross-reactivity also increases with time, because Abs with high affinity for the immunogen usually cross-react with related ligands more extensively than low-affinity Abs (Ch. 15), and perhaps also because a greater variety of Ab populations appear.

THE SECONDARY RESPONSE

After Ab levels in the primary response have declined, even to the point of being no longer detectable, a subsequent encounter with the same Ag usually evokes an enhanced **secondary (anamnestic, memory) response** (Gr. *anamnesis* = recall). By comparison with the primary response it is characterized by 1) a

lower threshold dose of immunogen,* 2) a shorter lag phase, 3) a higher rate and longer persistence of Ab synthesis, and 4) higher titers (Fig. 17-24A). In addition, the Abs formed in the secondary response are overwhelmingly of the IgG class, and they have from the outset about the same high affinity for the immunogen as those synthesized at the end of the response to the primary stimulus (Table 17-7). **The capacity for a secondary response can persist for many years and provide long-lasting immunity against infection: long after Abs cease to be detectable, the immunogen can evoke unusually prompt synthesis of relatively large amounts of highly efficient Abs.** The primary response is less protective, because Abs appear more slowly and their combining power is relatively poor, at least initially.

The doubling time of serum Abs in the secondary response appears to be about the same as in the primary response, but the rate of Ab production is much greater (Figs. 17-21 and 17-24A); the difference evidently depends upon the increased number of Ag-sensitive ("memory") cells at the time of the secondary stimulus (Fig. 17-24B). If administered 2 to 3 days after the primary stimulus, agents that prevent cell proliferation (X-rays, 6-mercaptopurine) can block development of the capacity to give a secondary response, showing that cell division is necessary for production of memory cells.

Memory resides in B and in T cells. The carrier effect is not always evident in secondary antihapten responses. Thus, 1 and 2 years after rabbits are primed with Dnp on BγG a vigorous secondary response with prompt production of high-affinity anti-Dnp molecules can be elicited with Dnp conjugated onto hemocyanin (which does not cross-react with BγG): memory is evidently represented in the B cells that promptly produce high-affinity antihapten Abs (see Table 17-7). However, other secondary responses, elicited with primed, cultured spleen cells (B plus T cells) can be blocked with anti-θ sera, indicating a need for T cells; and the greater helping effect of T cells from primed than from unprimed donors in adoptive secondary responses shows that

---

* For instance, the minute amount of diphtheria toxoid injected intradermally in a man in the Schick test (0.003 μg) is usually enough to cause a secondary response in the production of antitoxin, though it is far too little to cause a primary response.

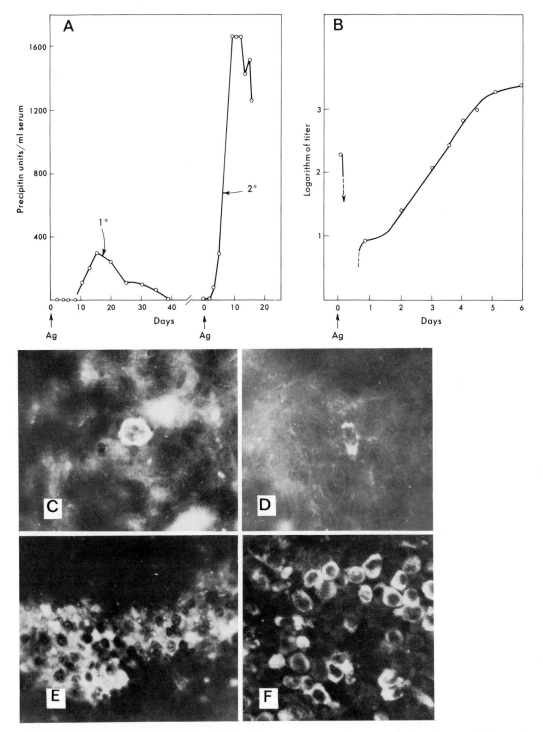

Fig. 17-24. Secondary response in rabbits. **A.** Precipitin titers in a rabbit injected once (1° = primary response), and in another rabbit injected with the same Ag about 4 weeks after a series of priming injections (2° = secondary response). **B.** A rabbit previously immunized with staphylococcal toxoid was given a second large dose intravenously. Note the initial drop in titer (negative phase) followed by the exponential increase; the doubling time is ca. 9 hrs, about the same as in the primary (and secondary) responses to bacteriophage ϕ×174 (see Fig. 17-21). [A based on Dean, H. R., and Webb, R. A. *J Path Bact* 31:89 (1928). B based on Burnet, F. M. Monograph No. 1, Walter and Eliza Hall Institute, 1941.] Greater proliferation of B cells in secondary than in primary response. Following footpad injections with diphtheria toxoid or bovine serum albumin sections of regional lymph nodes were treated with the Ag and then with the corresponding fluorescein-labeled Abs. **C** and **D** show rare Ab-producing cells 4 days after the first injection; **E** and **F** show clusters of Ab-producing cells 4 days after the second injection (see Fig. 17-2). [From Leduc, E. H., Coons, A. H. and Connolly, J. M. *J Exp Med* 102:61 (1955).]

TABLE 17-7. Increased Affinity of the Anti-Dnp Antibodies Formed in the Secondary Response

Group	Rabbit No.	Average intrinsic association constant for binding 2,4-dinitrotoluene (liter mole^{-1} $\times$ 10^{-6})
I. Bled 14 days after primary stimulus	1	0.14
	2	.03
	3	.20
	4	.15
II. Bled 8 days after secondary stimulus	5	34.
	6	84.
	7	>500.

Rabbits were injected with 1 mg Dnp-hemocyanin in Freund's adjuvant and anti-Dnp Abs were isolated 14 days later (group I, primary stimulus). Seven months later animals with less than 100 μg anti-Dnp Ab per milliliter of serum were reinjected with 1 mg of the same immunogen: 7 to 8 days later the level of anti-Dnp Abs had increased to 2 to 6 mg/ml, and affinity for ε-Dnp-L-lysine, the haptenic group of the immunogen, was too high to measure. Hence affinities were all measured with 2,4-dinitrotoluene, a less strongly bound analog. A secondary response to the Dnp group has been elicited in rabbits as long as 2 years after the primary injection.

From Eisen, H. N. *Cancer Res 22*:2005 (1966).

memory can also reside in "educated" T cells (Carrier effect, above).

Indeed, memory T cells are probably more important for most secondary responses than memory B cells. This would account for the lack of secondary responses to T-independent Ags, such as pneumococcal polysaccharides, and the strong secondary responses elicited by most proteins, which are usually T-dependent: tetanus toxoid, for instance, can evoke a powerful secondary response in man 10 to 20 years after primary immunization.

The magnitude of secondary responses also depends on other variables, such as the interval between injections. If it is too short, a substantial level of Ab is likely to be still present from the primary injection: hence immune complexes can form, leading to accelerated degradation of Ag, less stimulation of new Ab production, and even a transient decrease in the serum concentration of free Ab (so-called "negative phase," Fig. 17-24A). With too long an interval the secondary response can also be reduced, probably because Ag-sensitive cells are not infinitely durable. Indeed, it has been estimated from successive transfers of Ag-activated Ab-secreting cells through a series of X-irradiated, syngeneic recipients that a clone of B cells undergoes "**senescence**" (i.e., stops responding to Ag) after about 80 cell divisions. However, with an interval of many months or even years between mitoses of nonactivated lymphocytes, immunological memory can persist for practically a lifetime after the primary

stimulus with some Ags (see Appearance and generation times of lymphocytes, above).

Enhanced synthesis of Abs of the IgM class is normally inconspicuous in the secondary response, but it can become evident when IgG molecules are absent, as in certain severe immunodeficiency states. Increased affinity of IgM Abs in the secondary response is slight and difficult to demonstrate.

CROSS-STIMULATION

A secondary response can sometimes be elicited with an immunogen that is not quite identical to the primary Ag. Most of the Abs made will then react more strongly with the first than with the second immunogen. This phenomenon, called "**original antigenic sin**," was initially recognized in epidemiological studies with cross-reacting strains of influenza virus. An example is shown in Table 17-8, where the secondary response evoked with 2,4,6-trinitrophenyl (Tnp)-proteins in rabbits that had been primed many months before with 2,4-dinitrophenyl (Dnp)-proteins consisted primarily of Abs with the physicochemical and ligand-binding properties of anti-Dnp, rather than of anti-Tnp, molecules: the effect is probably due to specific cross-stimulation by Tnp-proteins of an en-

TABLE 17-8. Cross-Stimulation (Original Antigenic Sin) Illustrated in the Secondary Response to 2,4-dinitrophenyl (Dnp)-protein and to 2,4,6-trinitrophenyl (Tnp)-protein

Rabbit No.	Primary stimulus	Secondary stimulus	Affinity of antibodies formed 7–8 days after secondary stimulus		Ratio of affinities DNT/TNT
			For 2,4-dinitrotoluene (DNT) (liters mole$^{-1} \times 10^{-7}$)	For 2,4,6-trinitrotoluene (TNT) (liters mole$^{-1} \times 10^{-7}$)	
1	Dnp-BγG	Tnp-BγG	25.	0.49	50.
2	Dnp-BγG	Tnp-BγG	17.	.85	20.
3	Dnp-BγG	Tnp-BγG	34.	.84	40.
4	Tnp-BγG	Dnp-BγG	0.20	0.30	0.7
5	Tnp-BγG	Dnp-BγG	0.62	1.0	0.6

The primary stimulus was 1 mg Dnp-bovine γ-globulin Dnp-BγG or 1 mg Tnp-BγG; 8 months later animals with no detectable Ab in serum were reinjected with the immunogen shown, and they produced Abs in abundance. In control rabbits given Dnp-BγG in both injections the ratio of affinities for DNT/TNT ranged from 2 to 100 (i.e., they formed anti-Dnp Abs). In other controls given Tnp-BγG in both injections this ratio ranged from 0.2 to 0.9 (i.e.,they made anti-Tnp molecules). Most of the Abs produced within 1 week of the second stimulus had binding properties that corresponded to the primary rather than the secondary immunogen.

Based on Eisen, H. N., Little, J. R., Steiner, L. A., and Simms, E. S. *Isr J Med Sci 5*:338 (1969).

larged population of anti-Dnp memory B cells remaining after the primary response to the Dnp immunogen.

This principle has been exploited in "serological archaeology," testing human sera during an influenza epidemic with diverse strains of the virus. A given patient's serum tends to react less strongly with the strain that causes his current illness than with a strain that caused his first attack of influenza in some previous epidemic. From the study of sera from very elderly patients it has thus been possible to identify strains that probably caused major epidemics in the past, e.g., in 1918, long before the influenza virus was discovered.

Nonspecific Anamnestic Response. Secondary responses can sometimes be triggered by stimuli that appear to be unrelated to the priming immunogen. For instance, patients with typhus fever, due to *Rickettsia,* sometimes exhibit a rise in Ab titer to unrelated bacteria against which they were previously immunized. These nonspecific responses are probably due to stimulation of one Ag's memory B cells by diffusible factors released by nearby T cells that are activated by the unrelated immunogen (see Cell cooperation, above, and Antigenic promotion, below). A plant (pokeweed) mitogen that causes blast transformation of B cells can also cause a

slight secondary response in antihapten Abs in animals previously primed against the hapten (see also Antigenic promotion, below).

FATE OF INJECTED ANTIGEN

Following intravenous injection of a soluble Ag the decline in its concentration in serum exhibits three sharply distinguishable phases (Fig. 17-25): 1) a brief **equilibration phase** due to rapid diffusion into extravascular space, 2) slow **metabolic decay** during which the Ag is degraded, 3) rapid **immune elimination,** which identifies the onset of Ab formation; during this phase the Ag exists as soluble Ab-Ag complexes, which are taken up and degraded by macrophages. Free Ab appears at the end of the immune elimination stage (Fig. 17-25).

Extensively phagocytized particulate Ags, such as bacteria and red cells, do not diffuse into extravascular spaces, and hence do not exhibit the initial equilibrium phase of rapid decrease in serum concentration after intravenous injection. Trace antigenic fragments can persist in lymphoid tissues, long after the Ag is no longer detectable in blood. Ags that readily establish tolerance (below) tend to be unusually persistent (e.g., pneumococcal poly-

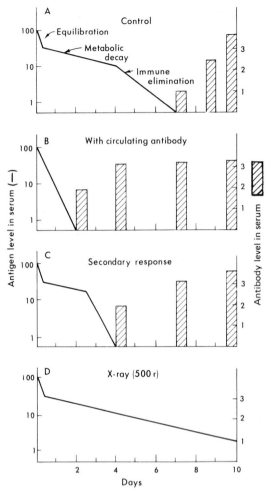

Fig. 17-25. Accelerated elimination of Ag as an index of Ab formation in rabbits injected intravenously at zero time with 75 mg ^{131}I-labeled BγG. The vertical bars represent serum Ab levels. **A.** Control animals, not previously immunized with BγG. **B.** Accelerated elimination begins at time zero when BγG is injected into animals that still have circulating anti-BγG from an earlier immunization; and it begins after an abbreviated lag in previously immunized rabbits (without residual Abs) that make a secondary response **C.** X-irradiation (500 r) before administration of BγG blocks Ab formation and accelerated elimination **D.** [Based on Talmage, D. W., Dixon, F. J., Bukantz, S. C. and Dammin, G. J. *J Immunol* 67:243 (1951).]

saccharides can be detected for about a year after their injection in mice).

EFFECT OF ANTIBODIES

High levels of Ab, whether due to active immunization or to injected antisera, can intercept Ags and block their immunogenicity (Suppression by Ab as antiimmunogen below). However, small quantities of Abs are actually enhancing, as though small Ab-Ag complexes can be more immunogenic than free Ag, perhaps because some complexes (such as bimolecular Ab · Ag) stick more readily on macrophages. For instance, newborn piglets, who have no Igs, form Abs in high titer in response to an Ag administered with small amounts of the corresponding Abs, but not to the Ag alone. In fact before diphtheria toxoid was available humans were immunized against diphtheria with nontoxic mixtures of toxin and antitoxin. But unless very carefully selected doses are given the simultaneous administration of Ab is more likely to block than to enhance immunogenicity (Antibodies as immunosuppressive agents, below).

ANTIGENIC PROMOTION AND COMPETITION

The response to immunogen X can sometimes be enhanced by simultaneous administration of immunogen Y, especially if the responding cell population has been previously primed with Y. This **antigenic promotion** of the response to X could be due to interaction of Y with the corresponding T cells, which release factors that stimulate neighboring B cells with specifically bound X.

Antigenic competition can also occur: the response to one Ag may be reduced if an unrelated, non-cross-reacting Ag is injected at the same time. For instance, poly-L-alanyl-protein readily elicits Abs to the L-polypeptide, but on coimmunization with poly-D-alanyl-protein only the latter elicits Abs to its polypeptide **intermolecular competition).** Moreover, different determinants on the same molecule can also compete **(intramolecular competition):** thus a protein with both D- and L-polyalanyl peptides substituted on the same molecule evokes synthesis only of Abs to the D-polyalanyl groups. Presumably the available receptors for the two determinants differ in affinity, and one determinant becomes dominant because the corresponding cells bind the limited supply of Ag. Intermolecular competition is more obscure: it could involve a suppressive effect of T cells on B cells, rather than the usual augmentation (see, for instance, Allotype suppression, below). Thus the phenomenon was not obtained in thymus-deprived mice unless they were given thymus cells, and the extent to which competition was restored was proportional to the number of injected T cells.

VACCINATION AGAINST MICROBIAL ANTIGENS

The list of procedures that increase resistance to infectious agents (Table 17-9)

TABLE 17-9. Vaccines Preventing Infectious Disease in Man

Disease	Immunogen
Diphtheria	Purified diphtheria toxoid
Tetanus	Purified tetanus toxoid
Smallpox	Infectious (attenuated) virus
Yellow fever	Infectious (attenuated) virus
Measles	Infectious (attenuated) virus
Mumps	Infectious (attenuated) virus
Rubella	Infectious (attenuated) virus
Polio	Infectious (attenuated) virus or inactivated virus
Influenza	Inactivated virus
Rabies	Inactivated virus
Typhus fever	Killed rickettsiae (*Rickettsia prowazeki*)
Typhoid and paratyphoid fever	Killed bacteria (*Salmonella typhi, S. schottmülleri,* and *S. paratyphi*)
Pertussis	Killed bacteria (*Bordetella pertussis*)
Cholera	Crude fraction of cholera vibrios
Plague	Crude fraction of plague bacilli
Tuberculosis	Infectious (attenuated) mycobacteria (bacille Calmette Guérin or BCG)
Meningitis	Purified polysaccharide from *Neisseria meningitidis*
Pneumonia	Purified polysaccharide from *Diplococcus pneumoniae*

justifies Edsall's statement: "Never in the history of human progress has a better and cheaper method of preventing illness been developed than immunization at its best." Unfortunately, the development of the best procedure for a given microbe is a laborious and almost entirely empirical process. Useful generalizations are meager; some of them follow.

Number of Injections. Multiple injections of immunogen (commonly at 1- to 6-month intervals) are usually necessary to establish long-lasting ability to give an effective secondary response, either to natural infection or to a prophylactic ("booster") injection.

Soluble vs. Insoluble Antigens. Immunogens that are aggregated, or are adsorbed on alum or other gels, are usually more effective than soluble immunogens. The increased multivalency of aggregated Ags probably enhances ability to cross-link receptors on B cells; and the slow desorption of Ags from gels maintains low concentration of Ag in tissues for long periods. Freund's complete adjuvant is too irritating for use in man, and the clinical safety of the incomplete adjuvant (without mycobacteria) is still under consideration.

Systemic vs. Local Immunization. The choice of site for injection of Ag is usually determined by convenience, because the ensuing immunity is generally due to systemically disseminated Ig molecules (or to T lymphocytes in cell-mediated immunity, Ch. 20). However, preferential activation of local groups of Ag-sensitive cells can sometimes be valuable, by concentrating the immune response at sites where target microbes invade; e.g., by achieving high IgA Ab levels in secretions of respiratory or intestinal tracts for agents that primarily infect these regions. This may, in part, account for the effectiveness of the attenuated polio vaccine (below), which is administered orally, and for the contention that influenza vaccine elicits more protection when given by aerosol inhalation than by injection into skin or muscle. Trachoma vaccines instilled in the conjunctival sac may similarly generate more protection against trachoma eye infections than injected vaccine.

Though current immunization practices stimulate immune responses effectively, it is still not possible to stimulate them selectively: e.g., to elicit Ab formation without cell-mediated immunity (Ch. 20) or vice versa, or to avoid producing certain types of Abs, such as those of the IgE class that can cause serious allergic reactions (Ch. 19).

INTERFERENCE WITH ANTIBODY FORMATION

IMMUNOLOGICAL TOLERANCE

The ability of an individual's immune response to distinguish between his own and foreign Ags has been recognized at least since 1900, when Ehrlich's doctrine of *horror autotoxicus* maintained that one can form Abs to almost anything except components of his own tissues. The medical importance of this principle, now called "self-tolerance," is apparent from the serious consequences of its occasional breakdown, which can result in autoimmune diseases (Ch. 20). Its key role in the biology of immune responses is also evident, for central to these processes are the regulatory mechanisms that prevent formation of Abs to self-Ags while permitting it to an almost unlimited range of foreign Ags.

Analysis of mechanisms responsible for self-tolerance became possible with the finding that under certain experimental conditions a foreign Ag may be treated in the same way as a self-Ag: i.e., it not only fails to act as immunogen but acts as **tolerogen**, establishing a

specific **unresponsive state.** In this state the animal fails to form Abs (or to develop cell-mediated immunity) even when the Ag is given under what would otherwise be immunogenic conditions. The alternative responses suggest that the tolerogen somehow causes the Ag-sensitive lymphocytes, or their receptors, to be physically eliminated or functionally blocked: the immunogen can then no longer stimulate cell proliferation and differentiation.

Pseudotolerance. When Ags are present at moderately high levels they can both stimulate Ab production and thoroughly neutralize or mask the Ab molecules that are formed (for example, see 10 μg Ag in Fig. 17-26). If tolerance were generally due to this "treadmill" mechanism, the continuing synthesis and elimination of Ab should be accompanied by accelerated elimination of labeled Ag. However, this is not ordinarily observed. Instead, truly tolerant animals not only lack circulating Abs but also lack the cells that synthesize Abs,

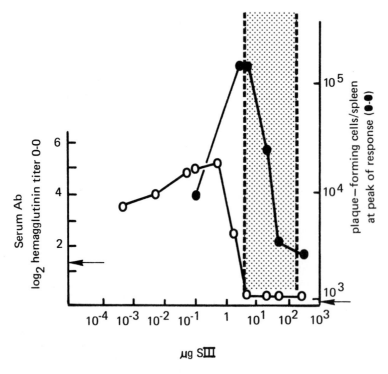

Fig. 17-26. Pseudotolerance; discordance between serum Ab titers and Ab-producing cells in tissues. Mice were injected with various doses of pneumococcal polysaccharide type 3 (SIII) and tested 10 days later for serum Ab titers and for the frequency of Ab-secreting (plaque-forming) cells in the spleen (see Fig. 17-4 for assay). At doses between 10 and 100 μg of SIII per mouse the absence of serum anti-SIII is not due to true tolerance: the Ab produced was neutralized by persistent nondegradable polysaccharide (cross-hatched area). At 250 mg per mouse (not shown) true tolerance was approached: the incidence of plaque-forming spleen cells fell below even the low background level in unimmunized controls (←). [Based on Howard, J. G., Christie, G. H., and Courtenay, B. M. *Proc R Soc Lond, Biol* 178:417 (1971).]

as shown by the hemolytic plaque assay with isolated cells or by immunofluorescence staining for intracellular Ab. **Hence tolerance of an immunogen signifies the absence of synthesis of the corresponding Abs.**

CONDITIONS PROMOTING TOLERANCE

The alternative effects of **dosage** were clearly demonstrated in the 1940s by Felton *et al.:* mice injected with 0.01 to 1.0 μg of a pneumococcal polysaccharide, say of type 2, became resistant to infection with type 2 pneumococci and produced Ab to its polysaccharide; but if given 1000 μg of the same substance they failed to become resistant or to form detectable Abs. This **immune paralysis or unresponsiveness** was persistent and specific: though no response to type 2 could be elicited, the mice reacted normally to immunogenic doses of other Ags including other capsular polysaccharides.

Route of administration is another determinant: soluble Ags tend to be immunogenic when injected with adjuvants into the tissues, but to be tolerogenic when given alone intravenously (perhaps because, like a large dose in tissues, this creates a higher circulating concentration).

Another factor with many Ags is their **physical state:** with BγG, for instance, **aggregated molecules tend to be immunogenic and monomers to be tolerogenic.** Specific unresponsiveness to BγG is induced not only by doses that exceed the immunogenic level **(high-zone tolerance),** as is true with virtually all Ags,* but also by doses just below the threshold for initiating Ab formation **(low-zone tolerance;** Fig. 17-27). Since many Ag preparations contain both immunogenic aggregates and tolerogenic monomers, which probably compete for the same cell receptors, low-zone tolerance would be produced if the tolerogen were effective at lower concentrations than the immunogen.

Diversity and Specificity. Tolerance can be established for any of the substances that induce Ab formation: proteins, polysaccharides, synthetic polypeptides, haptenic groups, etc. Because of their greater immunogenicity, it is more difficult to establish tolerance with particulate Ags (viruses, bacteria, aggregated proteins), with many repeating copies of a given determinant per particle, than with monomeric molecules, such as unaggregated proteins.

As with other immune responses, tolerance is specifically directed to particular determinants. For instance, rabbits made unrespon-

---

* Some Ags cannot establish unresponsiveness because of toxic side effects of high doses.

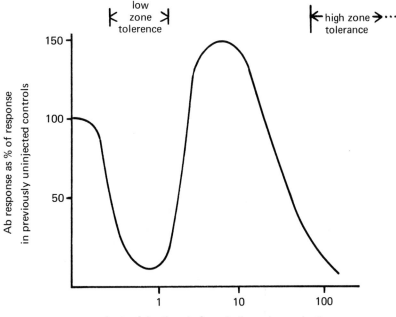

Fig. 17-27. High- and low-zone tolerance. Effect of daily injections of rats with different amounts of flagellin from *Salmonella Adelaide* on Ab response to a later challenge with 10 μg of flagellin. Ab levels are given as percent of values in control rats (no pretreatment). Low-zone tolerance is probably due to an effect on T cells. High-zone tolerance could be due to an effect on T or B cells or on both. Few Ags elicit the low-zone effect, but most elicit the high-zone effect. The responses that exceed 100% are due to the priming effects of the initial injections. [Based on Allison, A. C. *Clinical Immunobiology* *1*:113 (1972) and derived from data of G. R. Shellam and G. J. V. Nossal.]

sive to the Fc fragment of human IgG can still respond to immunization with intact IgG, but the resulting Abs react only with Fab domains.

Newborn vs. Adult. After Burnet's speculation that unresponsiveness results from introduction of any Ag during fetal life, Billingham, Brent, and Medawar showed in 1953 that newborn mice can be made permanently tolerant of cells from genetically different mice (Ch. 21). It then seemed likely that some special features of the fetal lymphatic system rendered it uniquely susceptible to induction of tolerance. In fact, however, adults can also be made specifically unresponsive, particularly if immunosuppressive measures are applied at the same time as high doses of the Ag: it appears that **once Ab formation has been initiated, tolerance becomes more difficult to establish.** One reason may be this: it probably takes several days for a large quantity of injected Ag to achieve a uniformly high tolerogenic level in tissue fluids, and during this period the levels in certain tissues are transiently immunogenic, initiating Ab formation: by forming immune complexes the newly synthesized Abs accelerate degradation of the Ag (Fig. 17-25) and arrest (or reverse) the induction of tolerance. If, however, adults are first depleted of Ag-sensitive lymphocytes by nonspecific immunosuppressive measures (X-irradiation, etc.; see below) their lymphoid system comes temporarily to resemble the newborn's immature system, and they can be more easily rendered tolerant.

The easy establishment of tolerance in the fetal and neonatal period is obviously advantageous, permitting the gradual build up of self-Ags to the levels required to perpetuate tolerance during adult life.

Maintenance. Tolerance is unstable: for instance, some weeks after high-zone tolerance is induced a short burst of Ab synthesis can occur as the declining Ag reaches an immunogenic level. Persistence of Ag is thus necessary, though less is needed to maintain than to establish tolerance. Since it is usually not clear whether tolerance is due to loss or dormancy of Ag-sensitive cells it is generally not known whether loss of tolerance to a particular Ag is due to emergence of new Ag-sensitive cells or to reactivation of suppressed cells (see below).

B CELLS VS. T CELLS

Tolerance to T-dependent Ags can be due to unresponsiveness of either T or B cells. Thus an Ag fails to stimulate Ab formation in irradiated mice given bone marrow and thymus cells if either set of cells is drawn from a donor previously rendered unresponsive to that Ag. However, when prospective donors are treated with various doses of Ag at varying times before their B and T cells are separately transferred, it is apparent that **T cells become unresponsive much more readily than B cells,** i.e., the effect is obtained with lower doses of Ag and the cells remain unreactive longer (Fig. 17-28). Hence T cells are probably largely responsible for natural tolerance to most self-antigens.

This view is supported by the following relations among cross-reacting Ags. Animals tolerant of antigen I will often make Abs that react specifically with I following immunization with structurally similar Ags, such as chemically modified I (e.g., Dnp-I or X-azo-I) or with native cross-reacting I' (e.g., human serum albumin if I was bovine serum albumin): under these conditions the anti-I Abs seem to be limited to those determinants of I that are shared (or cross-react) with I'. These results suggest that anti-I B cells are still present in animals tolerant of I, and prevented from responding only by the absence of specific helper T cells: with new carrier groups on the modified or cross-reacting immunogen other helper T cells can become engaged, permitting stimulation of the persistent anti-I B cells. Similar mechanisms might be involved in some autoimmune responses (Ch. 20): self-constituents modified by, say, infection might initiate Ab formation against the unmodified, native substance.

The specificity of the resulting Ab molecules probably depends on the tolerizing dose of I: with a relatively small dose few anti-I B cells would have been eliminated and the shared anti-I Abs elicited by I' would probably correspond closely to those that would have been raised against I itself in a normal animal; but with a large tolerizing dose, many anti-I B cells would have been suppressed and the Abs that react with I, made in response to the I', would probably differ substantially from those normally induced by I.

Is there a tolerant cell? Are Ag-sensitive lymphocytes destroyed by the tolerogen or do

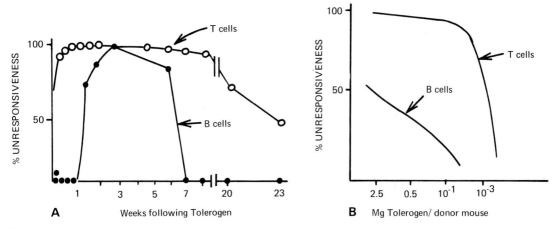

Fig. 17-28. Induction and persistence of tolerance in B- and T-cell populations. Thymus (T) and bone marrow (B) cells were removed at various times from mice rendered tolerant with various amounts of BγG and tested, with complementary cells from normal donors, for ability to cooperate in Ab formation when transferred to irradiated, syngeneic mice. Results are as per cent of control values (untreated donors). In **A** tolerance was induced with 2.5 mg of BγG. In **B** the cells were removed 15 to 20 days after the doses of Ag shown (abscissa). Tolerance appeared sooner, lasted longer, and was established more solidly, and with lower Ag doses, in T than in B cells. [From Chiller, J. M., Habicht, G. S., and Weigle, W. O. *Science 171*:813 (1971).]

they persist as "tolerant" cells? There is evidence for both. When animals tolerant of a T-dependent Ag are thymectomized their tolerance is more persistent (Fig. 17-29): this suggests that recovery of responsiveness depends upon emergence in the thymus (or under its influence) of new T cells that are sensitive to the Ag.

The stability of tolerized B cells has been followed in animals made tolerant of T-independent Ags: spleen cells taken at various times after induction of tolerance were trans-

Fig. 17-29. Increased persistence of tolerance following thymectomy. Mice were first made tolerant by multiple injections of bovine serum albumin (BSA); then half were thymectomized. Loss of tolerance (= recovery of responsiveness) was assayed by response to a challenging injection of BSA in Freund's adjuvant. Symbols are averages for groups of five to eight mice with standard deviation: x = tolerant, thymectomized; O = tolerant, not thymectomized. Control mice (not tolerant and not thymectomized) had a titer of about 10. Nonthymectomized mice recovered responsiveness by 12 weeks; the thymectomized ones had not recovered by 24 weeks. [From Taylor, R. B. *Immunology 7*:595 (1964).]

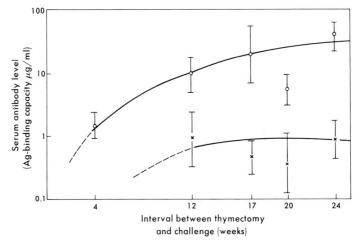

ferred to irradiated recipients, which were then stimulated with Ag. Two weeks after a tolerizing dose of pneumonococcal polysaccharide the transferred cells could already form Abs to this Ag in the surrogate host, but cells taken up to 6 months after a tolerizing dose of levan (polyfructose) were still unresponsive. It is likely that B cells are reversibly suppressed by relatively small tolerizing doses, but are eliminated in the long-lived tolerance established by relatively large doses: tolerance probably endures until the Ag is dissipated and bone marrow stem cells generate new B cells with the appropriate specificity.

Cellular Mechanisms. It seems likely that an Ag-sensitive lymphocyte responds in all-or-none fashion (proliferation and differentiation, or tolerance) when Ag binds to its surface receptors, and that the graded responses observed in the whole animal are determined by the proportions of cells making one or the other response. For instance, with a high but not completely tolerizing dose of Dnp-protein the subsequently formed anti-Dnp Abs have unusually low affinity for Dnp; evidently the B cells with high-affinity surface receptors are preferentially tolerized.

With B cells, whose abundant Ag-binding receptors are Igs, the immunogenic pathway (cell division with differentiation toward plasma cells) seems to be triggered by cross-linking of the receptors by multivalent Ag. Hence aggregated proteins with many copies of the same determinant per particle are highly immunogenic, whereas the same Ag in disaggregated form, with only one or a few copies of a determinant per molecule, is likely to be tolerogenic: it binds without stimulating, and competitively blocks binding of the immunogenic aggregates. Univalent haptens and high levels of multivalent Ag (see Ch. 15, Ag excess in the precipitin reaction) can also be tolerogenic because they do not engage in, but block, cross-linking; their effectiveness probably depends greatly on their persistence and on whether they only temporarily suppress or somehow eliminate the cells to which they bind. Ags also exercise alternative effects on T cells; but the detailed mechanisms are totally obscure, because the nature of the T-cell surface receptor is still unclear.

Possible Additional Mechanisms. Tolerance to many self-Ags may well involve much more than a preemptive interaction that blocks the cell's Ag-driven immune differentiation. Some evidence hints at antagonistic, rather than cooperative, relations between B and T cells under some circumstances. In mice, for instance, thymectomy increases the formation of Ab to polyvinylpyrrolidone (PVP), a T-independent antigen, suggesting that certain **"suppressor" T cells** might specifically block the activity of anti-PVP B cells.

In contrast, the characteristics of **tetraparental (allophenic) mice** suggest that Abs can block T-cell activities (Ch. 20, Cell-mediated immunity). These mice are produced by mixing the cells of early embryos (at about the eight-cell blastula stage) from two genetically different sets of parents, and implanting the single combined blastula in a foster mother's uterus. The progeny are **mosaics,** with cells of genetically different origin contributing to essentially all parts of the body (skin, liver, lymphatic system, etc.). Though these remarkable animals grow and develop normally their lymphocytes can destroy cultured fibroblasts from either set of parents, but the reaction can be specifically blocked with the tetraparental animals' serum (which presumably contains some kind of blocking Abs or Ab-Ag complexes). Hence the "self-tolerance" in these mice seems to be due to opposing immune responses, i.e., each set of cells can make immune responses to the other set, with Abs (or Ab-Ag complexes) blocking potentially destructive cellular immune reactions by killer T lymphocytes (Ch. 20).

Whether it is common in natural self-tolerance for suppressor T cells to block Ab formation to self-antigens, or for blocking Abs to interfere with cell-mediated reactions against other self-antigens is still not clear. Nevertheless, it seems likely that a variety of regulatory mechanisms, with diverse fail-safe features, must have developed in evolution to guard against destructive antiself immune reactions; for without effective mechanisms for self-tolerance the potentially damaging effects of the immune response would probably have outweighed its protective benefits.

ANTIBODIES AS IMMUNOSUPPRESSIVE AGENTS

Depending upon their specificities, some Abs can block the formation of other Abs in various ways: by hindering the reaction of immunogen with appropriate cell surface receptors, or by inactivating lymphocytes, especially T cells (antilymphocytic serum).

SUPPRESSION BY ANTIBODY AS ANTIIMMUNOGEN

As noted before, if Abs are added to the immunogen before its injection, or are present in the animal as a result of previous immunization, the induction of Ab formation is likely to be blocked; inhibition is particularly pronounced when Ab is in relative excess, for the resulting Ab-Ag complexes are then degraded rapidly* (Fate of injected antigen, above).

The inhibition by excess Ab is important clinically. For instance, severe hemolytic disease of the newborn, due to maternal Abs against fetal Rh red cells, has been greatly reduced in incidence by routine injections of anti-Rh Abs into Rh⁻ mothers at the time they deliver Rh⁺ babies: the baby's red cells, entering the maternal circulation in profusion as the placenta separates, are eliminated by the anti-Rh Abs. This prevents the mother from becoming immunized and reduces the risk of an anamnestic anti-Rh response during a subsequent pregnancy with another Rh⁺ baby (Ch. 21, Rh). Another example arises from placental transfer of maternal Abs to measles, polio, etc.: these can block induction of Ab synthesis by the corresponding immunogens in the young infant. Hence active immunization of the newborn is postponed until 6 to 9 months of age, by which time all maternal Abs have been eliminated (see Fig. 17-37).

SUPPRESSION BY ANTIBODY AS ANTIRECEPTOR

Under some circumstances anti-Igs suppress the production of Igs either by eliminating B lymphocytes with the target Ig as surface receptor or by blocking their stimulation by Ag. Depending upon its specificity and the conditions under which it is administered, the attacking anti-Ig specifically suppresses the formation of Igs of the corresponding class (isotype), or allotype, or idiotype (Ch. 16, classification of antigenic determinants of Igs). These effects emphasize the importance of surface Igs for the differentiation of B cells, and perhaps even for the survival of these cells. They also suggest how immunization procedures might eventually be developed to avoid coincidental production of Abs of undesirable type (e.g., IgE Abs that cause severe allergic reactions, see Ch. 19).

Isotype Suppression. Injections of newborn mice with antisera to μ chains leads to absence of serum IgM

---

* Ab-Ag complexes may have an additional, long-range suppressive effect on the corresponding Ag-sensitive lymphocytes, for such cells seem to remain unresponsive when transferred to irradiated recipients.

in the growing animal and decreases in the levels of IgG and IgA. Similar injections of anti-α or anti-γ antisera block only the production of Igs with the corresponding heavy chains (IgA and IgG, respectively). The effects persist for a few weeks, and recovery gradually ensues. Similar suppression cannot be established in adults, probably because their high levels of serum Igs neutralize the injected Abs. The broader effect of anti-μ than of Ab to other heavy chains fits additional evidence that cells with IgM on their surface are precursors of those with cell-bound IgG or IgA (Ontogeny, below).

Isotype suppression is also evident in primary spleen cell cultures, in which the addition of both sheep red blood cells (SRBC) and certain anti-Igs block production of various classes of Abs to SRBC: antisera to μ chains inhibit formation of Abs of all classes, while antisera to γ1 or to γ2 chains block only the formation of those anti-SRBC of the corresponding class. These effects (and others, see IgM-IgG shift, above) suggest that the initially reactive (virgin) Ag-sensitive cells have only IgM on the surface; once stimulated by the Ag, some of these cells apparently shift to produce surface Ig receptors with the same class of heavy chain as in the Ab molecules subsequently secreted by these cells (or their progeny).

Allotype Suppression. The first discovered case of suppression by anti-Ig was found by Dray among hybrid offspring of crosses between rabbits homozygous for different Ig allotypes (see Ch. 16, rabbit *a* and *b* allotypes of heavy chains and of κ light chains, respectively). If the newborn receives antiserum to the paternal allotype, it produces hardly any Ig of that allotype for many months; however, a compensating overproduction of the maternal allotype maintains total Ig at a normal level (Fig. 17-30).

Antiserum to the mother's allotype is not effective in the newborn, whose high levels of maternal Ig, acquired transplacentally or by suckling, neutralize the injected Ab. Antiserum to the father's allotype is suppressive only if given before a neutralizing level of this allotype has been actively produced (i.e., before the third week after birth in rabbits).

More extreme conditions are necessary to suppress both maternal and paternal allotypes; when this happens the corresponding isotype is eliminated. Thus when early rabbit embryos (8- to 16-cell stage), derived from parents homozygous for the same κ-chain allotype (b5), were implanted in the uterus of a foster mother who was injected with large amounts of anti-b5 antiserum, the resulting newborn rabbits did not form κ-containing Igs: their serum Ig levels were normal but all the light chains were λ type, which is ordinarily present in only ca. 10% of rabbit Ig molecules. In the fully suppressed rabbits it was not until 6 months after birth that blood lymphocytes appeared with the

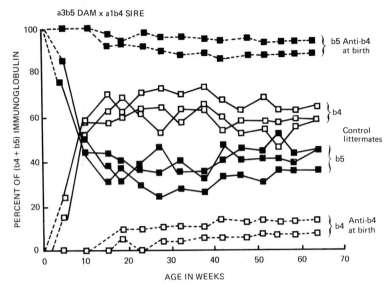

Fig. 17-30. Allotype suppression in rabbits. The parents were homozygous for different light-chain allotypes (father, b4/b4; mother, b5/b5). At birth half the offspring (b4/b5) were given 2.2 mg anti-b4 Abs (– – –); control littermates did not receive Abs (——). For over 1 year, the target allotype (b4) was virtually lacking in the treated animals' Igs while its allelic partner (b5) was overproduced; the level of total Ig in serum was normal as were the heavy chain allotypes. The changes in controls at 5 to 10 weeks are due to synthesis of the paternal allotype (b4, □) and loss of passively acquired maternal Igs (b5, ■). Relative concentrations of b4 and b5 in controls are not unusual for products of codominant alleles. Levels of individual allotypes are expressed as per cent of the sum of both allotypes (b4 + b5). [From Mage, R. G. *Cold Spring Harbor Symp Quant Biol 32*:203 (1967).]

suppressed κ (b5) chains on their surface, and the corresponding serum Igs first appeared much later, perhaps because the additional time was needed for the newly emerging cells to express a sufficiently broad repertoire of V_κ regions to be stimulated by prevalent Ags.

Since the suppressing Ab molecules are catabolized with a half life of about 6 days (Table 17-5), the chronicity of allotype suppression in rabbits (up to 3 years) is puzzling. Allotype suppression with the expected transitory existence can be elicited in most hybrid mice. However, with crosses between certain inbred strains antiallotypes evoke long-lasting suppression which appears, from adoptive transfer studies, to be due to T cells that specifically inhibit just those B cells with the target allotype. Thus cotransfer to irradiated, syngeneic recipients of spleen cells (not serum) from chronically suppressed and from normal (unsuppressed) donors resulted in suppression of the previously unsuppressed normal cells; and the block could be alleviated by prior treatment of the transferred population of suppressing cells with anti-θ serum and complement, destroying their T cells. It is possible that Ab-Ag complexes (antiallotype plus allotype) stimulate some T cells to make a long

lasting cell-mediated immune response (Ch. 20) against the corresponding allotype.

Idiotype Suppression. An animal fails to produce Ab of a particular idiotype if it has been previously immunized (actively or passively) against that idiotype. In a study in inbred mice, for example, rabbit antiserum against the idiotype of one mouse's Abs to the phenylarsonate (R) group was injected into other mice, which were then immunized with R-azo-proteins: none of the resulting anti-R Abs reacted with the rabbit antiidiotype (which could, however, react with a large proportion of the anti-R Abs formed in the control mice that had been injected only with the R-azo-protein immunogen). Evidently the antiidiotype had suppressed a major anti-R clone, by eliminating it or by blocking its reaction with the immunogen; other anti-R clones were unaffected.

SUPPRESSION BY ANTILYMPHOCYTE SERUM (ALS)

Antilymphocyte heteroantisera are usually prepared in horse or rabbit against another species' thymus cells, white blood cells, or cultured lymphocytes. These antisera were developed primarily to suppress cell-mediated immune reactions and to promote acceptance of allografts (Ch. 21). Usually administered as the IgG fraction (antilymphocyte globulin, or ALG), the Abs seem to cause preferential inactivation of the long-lived pool of recirculating lymphocytes, which are primarily T cells; hence ALG dramatically prolongs survival of allografts and reduces other manifestations of cell-mediated immunity (Ch. 20). However, ALG can also reduce Ab formation in primary responses to many Ags (probably T-dependent ones), if given just before the Ag; if given a few hours later, or if given before a second ("booster") injection, Ab formation is not blocked. Because ALG hardly suppresses secondary Ab responses (which are also more resistant to X-ray and other nonspecific suppressants, see below), but can apparently greatly depress secondary cell-mediated immune responses, it is used clinically to prevent immune rejection of allografts (Ch. 21). ALG loses its effectiveness when the Fc domain is removed, suggesting that complement fixation may be involved (Ch. 18).

Clinical use of ALG is somewhat limited by its immunogenicity. It does not suppress immune reactions to itself, and about half of humans injected with horse ALS (to human cells) form Abs to the horse Igs and develop "immune-complex" disease (see Ch. 19). As is described below, the suppressive effect of ALG on T cells greatly increases susceptibility to certain infections and to tumors (Complications of immunosuppression, below, and chs. 20 and 21, Immune responses to tumors).

NONSPECIFIC IMMUNOSUPPRESSION

Whole-body X-irradiation and a variety of cytotoxic drugs (Fig. 17-31) can prevent the initiation of Ab formation; but once Ab synthesis is under way they cannot interrupt it, unless given in doses that are severely cytotoxic for cells in general. All these agents are **much more effective in blocking the primary than the secondary response:** either memory cells are more resistant than virgin Ag-sensitive cells, or the increased number of cells that can respond increases the probability that

some will initiate Ab formation before they can be blocked. Some immunosuppressants act primarily as inhibitors of cell division (**antiproliferative,** e.g., the antimetabolites); others act primarily by destroying lymphocytes (**lympholytic,** e.g., 11-oxycorticosteroids); and others are both lympholytic and antiproliferative (X-rays, radiomimetic alkylating agents).

Most of the **antimetabolites** (purine, pyridimine, and folate analogs; Chs. 6 and 11) were developed as byproducts of cancer chemotherapy screening programs. Their principal action is to **block cell proliferation** (Fig. 17-31), and they are especially toxic for rapidly dividing cells: in tumors, bone marrow, intestinal and skin epithelium, as well as lymphocytes that proliferate in response to stimulation by Ag. Accordingly, the margin of clinical safety between therapeutic and toxic doses is small, but for some of these drugs it is great enough for routine clinical use (e.g., azathioprine and cyclophosphamide).

The frequency of successful kidney transplants in man is now largely due to skillful use of immunosuppressive agents: currently favored are combinations of azathioprine (an antimetabolite), prednisone (a corticosteroid), and antilymphocytic serum or its IgG fraction (ALG), which selectively destroys circulating T cells (see below and Chs. 20 and 21).

Lympholytic agents (X-rays, radiomimetic alkylating drugs, corticosteroids) cause prompt and massive destruction of lymphocytes. The chromosomal damage caused by X-rays and alkylating agents also impairs the capacity of surviving lymphocytes to undergo mitosis, blocking their normal response to subsequent antigenic stimulation.

In the following comparison of various agents it is useful to contrast their activity in the three phases of the Ab response: 1) the preinductive phase, before immunogen is administered; 2) the inductive phase, between immunogen administration and rise in titer of the corresponding serum Ab; and 3) the productive phase, when Ab is being synthesized vigorously. The **lympholytic agents are effective immunosuppressors when given in the preinductive phase, and the antiproliferative agents are most effective in the inductive phase.** All known agents appear to be ineffective in the productive phase, unless given in highly toxic doses. Though this division into phases is convenient, it is an oversimplification, for lymphoid cells do not respond syn-

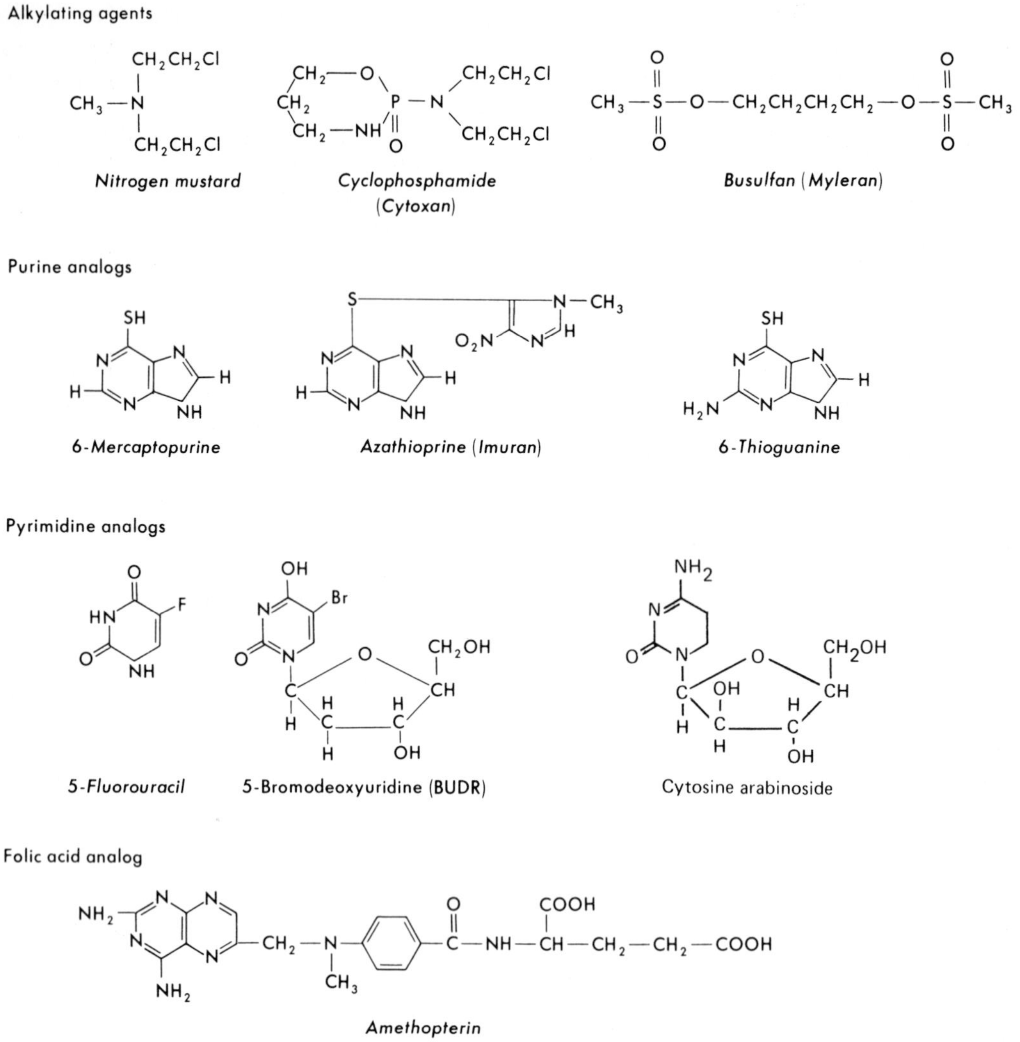

Alkylating agents

Nitrogen mustard   Cyclophosphamide   Busulfan (Myleran)
                            (Cytoxan)

Purine analogs

6-Mercaptopurine   Azathioprine (Imuran)   6-Thioguanine

Pyrimidine analogs

5-Fluorouracil   5-Bromodeoxyuridine (BUDR)   Cytosine arabinoside

Folic acid analog

Amethopterin

Fig. 17-31. Some immunosuppressive drugs. Azathioprine (Imuran) is converted in vivo to 6-mercapto-purine by sulfhydryl or other nucleophilic groups.

chronously: the immunogen can remain active for months, and cells in all three phases may be present simultaneously.

X-rays. Whole-body irradiation with sublethal doses of X-rays (400 to 500 r) tends to suppress the response to most immunogens for many weeks, but usually not permanently. Suppression is greatest when the irradiation is given 24 to 48 hours before Ag, because massive disintegration of lymphocytes occurs promptly after whole-body irradiation. After a large dose (e.g., 500 r) active proliferation of lymphoid cells is resumed after ca. 3 to 4 weeks, and lymph nodes appear normal shortly thereafter. However, damage to chromosomes (the main tar-

gets of X-ray, Ch. 11) can be "stored": i.e., it may not be expressed until the cells attempt to divide, perhaps months or years later (Lifetime of lympho-cytes, above). In the meantime the nondividing cells can apparently function normally.

When immunogen is given **after** massive irradia-tion it may be largely or completely eliminated before the capacity to initiate Ab formation is re-stored. However, if the immunogen is given just **before** irradiation the stimulated cells can apparently continue their differentiation and eventually form Abs, while already differentiated cells continue to synthesize Abs. Indeed, having begun to differenti-ate, the responding cells may even yield more Ab in animals irradiated 1 to 2 days after immunization (Fig. 17-32).

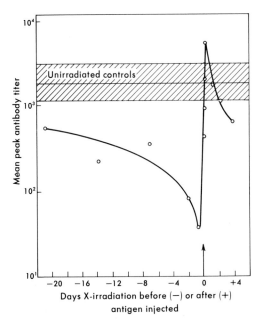

Fig. 17-32. Effect of 500-r X-irradiation of rabbits at various times before and after immunization (arrow) with sheep red cells. [Based on Taliaferro, W. H., and Taliaferro, L. G. *J Infec Dis* 95:134 (1954).]

The **secondary response** is relatively **resistant to irradiation:** the appearance of Abs may be delayed, but peak titers are usually normal. As in the primary response, Ab formation may be increased by irradiation given after the booster injection.

Alkylating Agents. Compounds such as the nitrogen mustards (Fig. 17-31) block cell division by cross-linking strands of DNA (Ch. 11). Often called radiomimetic drugs, their biological effects (such as massive destruction of lymphocytes) resemble those of X-irradiation. However, recovery is more rapid than after X-rays, and these drugs are therefore usually given at frequent intervals (e.g., daily) for sustained immunosuppression.

Corticosteroids. Large doses of 11-oxycorticosteroids cause extensive destruction of small lymphocytes. However, the surviving small T cells appear to be unusually active in some reactions, e.g., graft vs. host (Ch. 20), and even perhaps in cooperation with B cells in induction of Ab formation. Nonetheless, if given just before Ag they inhibit Ab formation in some species (rats, mice, rabbits); but at the doses used clinically a significant suppression has not been observed in man. In therapeutic doses these drugs inhibit inflammation, whether due to allergic reactions (Chs. 19 and 20) or to nonspecific irritants such as turpentine. Accordingly, they are widely used clinically to suppress allergic inflammation, especially of the delayed type.

Antimetabolites. In contrast to X-rays, which are most inhibitory when given just before the immunogen, the antiproliferative metabolite analogs, such as 6-mercaptopurine, usually suppress Ab formation best when administration is begun 2 days afterward, when the Ag-induced proliferation of lymphocytes is particularly active (Selectivity, below). The difference is readily understood: X-rays (and alkylating agents) can damage DNA whether or not it is replicating, while the antimetabolites used as immunosuppressants only damage cells with replicating DNA.

Other Immunosuppressive Drugs. Some antimicrobial drugs also inhibit Ab synthesis. **Chloramphenicol** in relatively high doses inhibits the primary response in intact animals. If added with the immunogen it also inhibits the secondary response in cultures of cells from a previously immunized animal, but after the cells are stimulated to form Abs the addition of chloramphenicol has little effect. Inhibition by chloramphenicol may be related to its ability to inhibit mitochondrial (but not cytoplasmic) ribosomes (Ch. 12).

If **actinomycin D** is added together with the immunogen the formation of Ab is substantially inhibited, because the synthesis of new mRNA is required for initiation of Ab synthesis. However, Ab formation can persist for many days in cell culture in the presence of this drug, suggesting that the corresponding mRNA is relatively stable.

Selectivity. Though immunosuppressants generally affect immune responses as a whole, rather than particular manifestations, some selective suppression is observed. For instance, as noted before, ALS seems to block cell-mediated more than humoral immunity. X-ray, corticosteroids, and antiproliferative drugs suppress IgG more than IgM production, suggesting that fewer cell divisions may be needed to initiate production of Abs of the IgM class.

The most impressive selectivity, however, is brought about by **combined administration of Ag and antiproliferative immunosuppressants** (Fig. 17-33). The clones whose proliferation has been stimulated are selectively eliminated by the drugs, as in the selective destruction of growing bacterial cells by penicillin (Ch. 6). The animal thus becomes specifically unresponsive to the Ag used, and repeated injection of that Ag can then maintain the tolerant state. The clinical utility of this approach is likely to be restricted to those situations in which the relevant Ag is known in advance of an actual response, e.g., in organ transplantation (Ch. 21).

Complications of Immunosuppression. Prolonged immunosuppression is dangerous: not only are the agents highly toxic for various cells, but they can activate latent infections and greatly increase susceptibility to serious infection with the many prevalent fungi, bacteria, and viruses that ordinarily have little pathogenicity (e.g., *Candida, Nocardia,* cytomegalovirus, herpesvirus). In addition, chronically immunosuppressed individuals, just like those with congenital immunodeficiencies (below), have an increased incidence of various cancers, especially lymphomas and reticulum cell sarcomas (see discussion of increased frequency of cancer in chronically suppressed recipients of kidney allografts, Chs. 20 and 21). Suppression of cell-mediated immunity (Ch. 20) rather than of Ab formation is probably responsible, because the most effective immune responses to most tumors seem to be cell-mediated (Chs. 20 and 21).

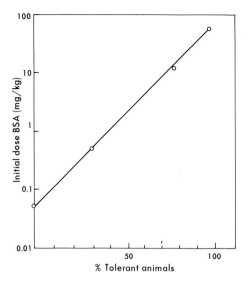

Fig. 17-33. Selective unresponsiveness produced by combined administration of Ag and an antiproliferative immunosuppressive agent. Rabbits were injected with bovine serum albumin (BSA) and 6-mercaptopurine: 1 to 3 months later many were specifically unable to make Abs to BSA, though they could respond to other Ags. The proportion of tolerant animals increased with the dose of BSA given initially with the immunosuppressant. [From Schwartz, R. S. *Progr Allergy* 9:246 (1965).]

PHYLOGENY AND ONTOGENY OF ANTIBODY FORMATION

PHYLOGENY

Invertebrates lack a lymphoid system and do not form recognizable Igs. A primitive form of cellular immunity probably provides their principal defense against invasive parasites (Ch. 20). However, invertebrates of many phyla have soluble, circulating (hemolymph) proteins that agglutinate or lyse various bacteria and vertebrate red cells; and the production of these agglutinins and lysins can be augmented by administration of the target cells. Moreover, treatment of red cells with oyster hemagglutinin enhances their phagocytosis by oyster amebocytes (motile, phagocytic cells that recognize foreign cells, see Ch. 20). Despite these superficial similarities to Ab-enhanced phagocytosis by vertebrate macrophages (Ch. 18) these invertebrate "antisubstances" (e.g., horseshoe crab and oyster hemagglutinins) seem to differ considerably from vertebrate Igs: 1) very little specificity is evident from the range of substances that induce or react with particular invertebrate agglutinins; 2) of high molecular weight (ca. 400,000), they are made of dissociable, apparently identical subunits (MW 22,000); 3) they do not show the electrophoretic heterogeneity expected of proteins with diverse

specificities. It would nonetheless prove interesting if any amino acid homology were found between invertebrate agglutinins or lysins and vertebrate Igs.

In contrast to invertebrates, even the most primitive of present-day vertebrates, the **hagfish** (a cyclostome), has some capacity for immunological memory: it rejects allografts more rapidly after repeated trials (Ch. 21, Second set reaction). While these creatures have been reported to form Abs that agglutinate sheep red cells and that precipitate hemocyanin, they seem unable to make Abs to several other Ags, and their Igs have not so far been characterized. Their lymphoid system is also exceedingly primitive, and a thymus has not been identified. However, the **lamprey,** a slightly more advanced cyclostome, has a better developed lymphoid system, including a rudimentary thymus; it forms Abs to several Ags, and has recognizable Igs (Fig. 17-34).

Beginning with elasmobranchs **(sharks)** and continuing up the phylogenetic scale of complexity, essentially all features of the immune response are recognized in all vertebrates: Ab formation to a broad range of Ags, appearance of plasma cells in response to immunization, allograft rejection, and a fully developed lymphoid system.

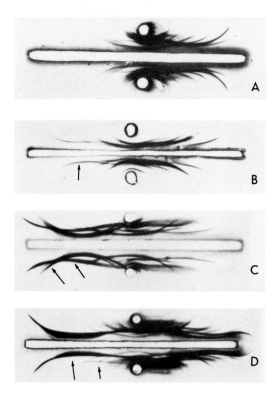

Fig. 17-34. Evolutionary development of immunoglobulins, suggested by electrophoresis of serum from selected vertebrates. Proteins with suggestive electrophoretic characteristics of Igs (arrows) were present in small amount in the lamprey **(B)**, and abundant in the bowfin, a fresh water dogfish **(C)**, and in man **(D)**. Though undetected in the hagfish **(A)**, they must also be produced in this species, which can make Abs to some Ags. [From Papermaster, B. W., Condie, R. M., Finstad, J., and Good, R. A. *J Exp Med 119*:105 (1964).]

Serological analyses indicate that the Igs of lower vertebrates are less diversified than those of mammals. However data on their amino acid sequences are scanty and tentative assignments to particular mammalian classes are based on sedimentation coefficients and electrophoretic mobilities, which are less dependable criteria (e.g., in frogs the most cathodically migrating serum proteins are not Igs).

Fish have two serum Igs (19S and 7S in sharks, 14S and 6S in bony fish), and both seem to belong to the IgM class. From the first N-terminal 10 residues it appears that at least 80% of pooled shark light chains are of κ type, with striking similarity to human κ chains. Several distinct classes of Igs are present in amphibia and in reptiles (IgM-like, IgG-like, and at least one other); the three major mammalian classes, IgG, IgM, and IgA, are present in birds. Considerable insight into evolutionary relations is emerging from amino acid sequence homologies among Ig chains (Ch. 16).

ONTOGENY

Though synthesis of Igs can be detected in mammalian fetuses, protective levels of Abs to common pathogens are not attained until some time after birth. The newborn would thus be extremely vulnerable to many infections were it not for the maternal Abs it receives before or shortly after birth. In some species the fetus receives maternal Igs only from colostrum, which is usually rich in IgA and IgG, while in others the Igs are transferred both in utero and by suckling (Table 17-10).

In man and in higher primates absorption from colostrum is probably of minor importance compared with transfer across the placenta. Maternal IgG, but not IgM, IgA, or IgE, is transmitted freely to the human fetus in utero, suggesting selective transport. Special sites on the Fc domain are evidently required: in rabbits, Fc fragments (of γ chains) are transferred as readily as intact IgG molecules and much more rapidly than Fab fragments.

The order in which Igs are produced in embryogenesis repeats the apparent evolutionary sequence ("ontogeny recapitulates phylogeny"): IgM, then IgG, and finally IgA.

The time course has been especially well studied in the chicken, where B-cell differentiation is initiated in the bursa of Fabricius. IgM-containing cells are evident in the chick embryo's bursa on day 14, and cells with IgG are not detected until 1 week later (day 21, the time of hatching). The spread of Ig-containing cells to peripheral tissues follows the same order: cells with IgM are found in the spleen on day 17 and those with IgG appear there 8 days later (4 days after hatching).

The switch of individual lymphocytes from production of IgM to IgG is suggested by the presence, just before hatching, of many cells in the bursa with both μ and γ chains, and especially from the suppressive effects of anti-μ-chain Abs, whose administration on day 13, followed by removal of the bursa at hatching, yields birds that are completely agammaglobulinemic: they are devoid of IgG and IgA as well as of IgM. If, however, injection of anti-μ and bursectomy are delayed until hatching (i.e., after the switch has taken place) the developing chicks are depleted only of IgM and they have a normal or even an elevated IgG level. (Bursectomy at an intermediate time—after IgM-producing cells have been seeded in peripheral tissues, but before IgG producers have appeared—yields many birds with persistently elevated levels of IgM and virtual absence of IgG.)

TABLE 17-10.   Relation of Placental Structure and Mode of Passive Transfer of Immunoglobulins to the Fetus*

Species	No. of tissue layers between maternal and fetal circulation at term	Placental or amniotic transmission	Importance of transmission via colostrum
Pig	6	—	+++
Ruminants	5	—	+++
Carnivores	3	±	+
Rodents	2	+ (yolk sac)	+
Man	2	+++ (placenta)	—

* In chickens, and presumably in other birds, β-globulins containing Abs are transmitted from hen to ova via follicular epithelium and are stored in the yolk sac, from which the proteins are absorbed into the fetal circulation shortly before hatching.

From Good, R. A., and Papermaster, B. W. *Adv Immunol 4*:1 (1964); based on Vahlquist, B. *Adv Pediat 10*:305 (1958).

The differentiation of B cells from IgM to IgG production in the bursa is independent of Ag (i.e., it is not modified by administration of extrinsic immunogens); this is in sharp contrast to the Ag-driven IgM-IgG switch that seems to take place in peripheral tissues during induction of Ab formation (see above, One cell-one Ab rule, Isotype suppression, and Fig. 17-35).

In human fetuses lymphocytes with IgM or IgG on the cell surface are detected by immunofluorescence by the tenth week of gesta-

Fig. 17-35. Scheme for successive stages in differentiation of B cells. Stem cells (from yolk sac or liver in early and late embryogenesis, respectively, or from bone marrow after birth or hatching) differentiate into IgM-bearing B cells, of which some differentiate further into IgG-bearing, and then into IgA-bearing, B cells, which are then stimulated by an appropriate immunogen to undergo further (2nd stage) differentiation into plasma cells or to proliferate and continue in circulation as Ag-sensitive memory cells. [Based on Cooper, M. D., Lawton, A. R., and Kincaide, P. W. *Clin Exp Immunol 11*:143 (1972).]

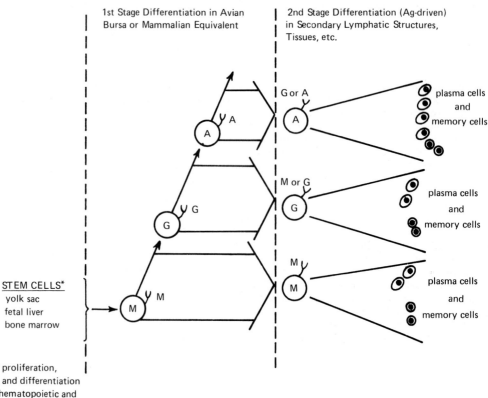

1st Stage Differentiation in Avian Bursa or Mammalian Equivalent

2nd Stage Differentiation (Ag-driven) in Secondary Lymphatic Structures, Tissues, etc.

STEM CELLS*
yolk sac
fetal liver
bone marrow

* Capable of proliferation, self-renewal, and differentiation into diverse hematopoietic and lymphoid cells.

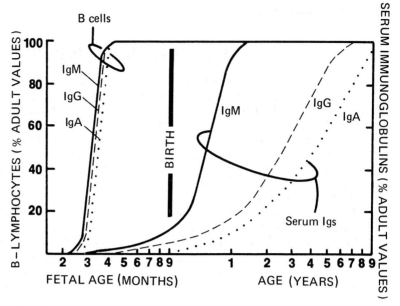

Fig. 17-36. Maturation of B cells and serum Ig levels in man. With both cells and their secreted products the ontogenetic order (IgM, IgG, IgA) recapitulates the presumptive evolutionary sequence of their appearance. [Based on Cooper, M. D., and Lawton, A. R. *Am J Pathol 69*:513 (1972), from data of Lawton, A. R., Self, K. S., Royal, S. A., and Cooper, M. D. *Clin Immunol and Immunopathol 1*:104 (1972)].

tion, and cells with surface IgA are detected later (12 weeks). By the fifteenth week the proportions of fetal spleen and blood cells with each Ig class is essentially the same as in the normal adult. However, active synthesis and secretion of Igs by fetal cells occurs much later, perhaps because restricted diversity of Ag-sensitive cells and of foreign Ags in the sheltered fetus reduces the chances for Ag-triggering of the fetus's lymphocytes. IgM and lesser amounts of IgG normally begin to be synthesized and secreted by fetal spleen cells in about the twentieth week, but production of IgA seems to start only some weeks after birth. However, with severe fetal infection, as in congenital syphilis, Abs (of the IgM class) are formed vigorously and plasma cells are abundant in the infected 6-month fetus (Fig. 17-36).

Because IgG molecules are readily transferred from maternal to fetal blood and the other Igs are not, the newborn infant's blood contains high levels of IgG (with Gm allotype of the mother), traces of IgM (of fetal origin), and essentially no IgA. The newborn's total IgG (and total Ig) declines until ca. 8 to 10 weeks of age, when it starts to rise as the neonate's biosynthesis becomes sufficiently active. Adult serum levels of IgM are reached at about 10 months of age, of IgG at about 4 years, of IgA at about 9 to 10 years (Figs. 17-36 and 17-37), and of IgE at about 10 to 15 years.

Fig. 17-37. Changing plasma level of Igs in human infant during first year. Maternal IgG, which accounts for almost all the infant's Igs at birth, has essentially disappeared by about 6 months. [Based on Gitlin, D. *Pediatrics 34*:198 (1964).]

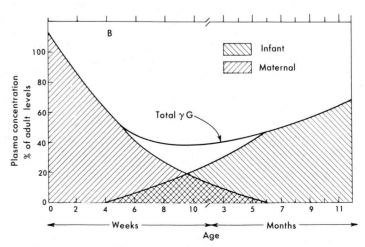

IMMUNODEFICIENCY DISEASES

Many genetic and congenitally acquired defects in ability to form Abs and to develop cell-mediated immunity have been recognized since the 1950s, when it became possible to prolong the survival of affected individuals through the use of antibiotics, Igs, and even bone marrow or thymus grafts from normal donors (Ch. 21, Allografts). These immunodeficient states help illuminate mechanisms in normal immune responses, just as mutations have clarified many other pathways. Classification of many of the clinical syndromes is unclear, but several distinctive clinical patterns are recognized (Table 17-11). Some affect only the B cell system (low levels or absence of one or all Ig classes), others the T cell system (defective cell-mediated immunity, Ch. 20), and some affect both, perhaps through defects in common stem cells. An increased incidence of cancer in these patients underscores the significance of the normal immune system in eliminating potential cancer cells (Chs. 20 and 21).

Suspected subjects with recurrent severe infections, or with close kinship to affected persons, are evaluated by procedures listed in Table 17-12. **In testing suspects for responses to routine immunization procedures it is important to avoid use of live vaccines**—e.g., attenuated viruses (polio, vaccinia, measles, ru-

TABLE 17-11. Primary Immunodeficiency Disorders*

Probable defect in	Disease
I. B cells	Infantile X-linked aggammaglobulinemia Selective Ig deficiency (usually IgA) Transient hypogammaglobulinemia of infancy X-linked Ig deficiency with hyper-IgM
II. T cells	Thymic hypoplasia (pharyngeal pouch syndrome or DiGeorge's syndrome) Episodic lymphopenia with lymphocytotoxin

III. T cells, B cells, and stem cells	Immunodeficiency with ataxia-telangiectasia†
	Immunodeficiency with thrombocytopenia and eczema (Wiskott-Aldrich syndrome)‡
	Immunodeficiency with thymoma
	Immunodeficiency with dwarfism
	Immunodeficiency with generalized hematopoietic hypoplasia
	Severe combined immunodeficiency
	    Autosomal recessive
	    X-linked
	    Sporadic
	Variable immunodeficiency (commonest type, largely unclassified)

* Omits immunodeficiencies secondary to X-irradiation, cytotoxic drugs, lymphomas with replacement of normal by neoplastic lymphoid cells (e.g., multiple myeloma), or excessive loss of Igs through leaky lesions in intestine ("exudative enteropathy") or kidneys (nephritis, nephrosis). Severe, recurrent infections are also seen in rare children with deficiencies in some complement components (C3, C5; see Ch. 18) or in bactericidal activity of granulocytes. Cell-mediated immunity is conspicuously deficient in categories II and III, but normal in I.

† Inherited as autosomal recessive character; 80% lack serum and secretory IgA; IgM and IgG are usually normal.

‡ Inherited as X-linked character. IgM levels are usually low; IgG and IgA are usually normal or elevated. Pronounced inability to make Abs to polysaccharides.

Based on *Bull WHO 45*:125–142 (1971).

TABLE 17-12. Some Tests Used for Clinical Evaluation of Immune Status

Tests for B-cell functions	Serum Ig levels*
	Serum Ab levels†
	Biopsies examined for plasma cells by histological methods and immunofluorescence‡
	Viable blood lymphocytes stained for surface Igs
Tests for T-cell functions	Skin tests for delayed-type hypersensitivity§
	Lymphocyte transformation induced by plant mitogens (phytohemagglutinin or concanavalin A) or by incubation with allogeneic lymphocytes‖
	Release of macrophage-inhibition factor (MIF) upon incubation of lymphocytes with common Ags¶

* Radial immunodiffusion is preferred (Ch. 15). It requires little serum (ca. 10 μl), and it is precise ($\pm$10%), sensitive ($\geq$10 μg/ml), and not too slow (24 hr): with radiolabeled anti-Igs the sensitivity can be increased 100-fold to where, for instance, 0.05 μg/ml of IgE can be detected (Ch. 19). The distribution of Ig levels in the normal population is essentially bell-shaped and lacks the discontinuities that permit sharp definition of an Ig deficiency. By common agreement, some deficient levels are < 2 mg/ml for IgG and absence (< 10 μg/ml) of IgA.

† Commonly measured are "natural" Abs to blood group substances A and B (Ch. 21) or to sheep red cells or to *E. coli*, or induced Abs after immunization with potent, harmless and potentially helpful Ags, such as diphtheria and tetanus toxoids and *Bordetella pertussis* ("triple vaccine"), inactivated **(not attenuated)** polio vaccine, or polysaccharides from pneumococci, or *Hemophilus influenzae,* or *Neisseria meningitidis.*

‡ Plasma cells are sought in biopsies of lymph nodes that drain intracutaneous sites of injection of Ags in preceding footnote, or of the rectal mucosa, whose lamina propria layer normally contains many plasma cells (IgA-containing).

§ The Ags injected intradermally are derived from prevalent microbes: e.g., mumps virus, tuberculin (from *Mycobacterium tuberculosis*), streptokinase-streptodornase from hemolytic streptococci, and culture media supernatants from various fungi (e.g., *Candida, Trichophyton, Coccidioides* [useful in California], or *Histoplasma* [useful in Mississippi Valley]). Skin patch tests are performed after deliberate skin sensitization with 2,4-dinitrochlorobenzene (Ch. 20).

‖ Blast transformation can be evaluated by changes in cell morphology (see Ch. 20, Fig. 20–8) or by radioautography or measurement of incorporation of ^{3}H-thymidine into DNA. For the mixed lymphocyte culture, see Chapter 20.

¶ See Chapter 20.

bella) or bacteria (the bacillus Calmette-Guerin [BCG] variant of the tubercle bacillus) —which can cause overwhelming infections in severely affected persons. By staining blood lymphocytes with fluorescein-labeled, class-specific anti-Igs, it is possible to measure the proportion of cells with various membrane-bound surface Igs, and thus to determine whether defects occur before or after differentiation of stem cells into B cells.

Infantile X-linked Agammaglobulinemia. This X-linked genetic defect occurs in male infants who begin to suffer from recurrent bacterial (pyogenic) infections at ca. 9 to 12 months of age, when maternal Igs received transplacentally have completely disappeared (Fig. 17-37). These patients form exceedingly little if any Ab and plasma cells following deliberate immunization, and their serum Igs of all classes are greatly depressed (IgG to less than 10% and IgA and IgM to ca. 1% or less of the normal level; see Ch. 16, Fig. 16-16 F, G). Correspondingly, their blood lymphocytes lack surface Igs, as though the defect occurred in stem cell → B cell differentiation. However, differentiation into T cells seems to be unaffected, as all their cell-mediated immune responses are normal (Ch. 20). Since they generally recover without difficulty from measles, mumps, and other viral diseases of childhood, it appears that **resistance to many virus infections is due to cell-mediated immunity (T cells) rather than to serum Igs.**

Selective IgA Deficiency. This is the commonest abnormality, occurring in ca. 0.1% of all persons, often without any clinical manifestations. In some families the defect is inherited as an autosomal dominant, and in others as an autosomal recessive. IgA deficiency is also associated with two other autosomal genetic abnormalities—ataxia-telangiectasia and partial deletion of chromosome 18—and it is also frequently associated with congenital infections (rubella, toxoplasmosis), suggesting that in some cases the defect may be acquired. Despite the diversity of associated genetic and congenital lesions, patients with selective IgA deficiency may have a **normal number of blood lymphocytes with surface IgA.** Moreover, the addition of a plant protein (pokeweed mitogen) that stimulates blast transformation of normal B cells can also stimulate this change in IgA-bearing cells from these patients. The defect seems, therefore, to lie in Ag triggering of terminal differentiation of the IgA-bearing B lymphocytes into IgA-secreting plasma cells. Either the cells are unable to respond to cross-linking of their surface IgA molecules by Ags, or the surface Igs are defective (e.g., they might lack good combining sites for Ag, through absence of V_H or V_L or both, or the combining sites of all IgA-bearing cells might be so uniform that the cells are unlikely to encounter an Ag and to be stimulated into secretion of IgA molecules). A block in terminal differentiation of B cells can occur in other Ig deficiencies; for example, rare children who are deficient in serum Igs of all classes but who have a normal number of lymphocytes with surface Igs of diverse classes.

However, **concordant deficiencies** can also occur: absence of serum Igs of a given class and of lymphocytes carrying surface Igs of that class suggests a block in differentiation of stem cells into B cells or of B cells with one type of Ig into B cells with another type (see Ontogeny, above, and Fig. 17-35). For instance, some individuals with "X-linked agammaglobulinemia with hyper IgM" have elevated levels of serum IgM and of lymphocytes carrying IgM but lack IgA and IgG, both as serum molecules and as surface receptors on B cells: the defect here is likely to be in the differentiation of IgM-bearing B cells into IgG- and IgA-bearing cells (see Fig. 17-35).

Thymic Hypoplasia. The most disabling immunodeficiencies are associated with T cell defects, including those due to faulty embryogenesis of the thymus. The fetus with defective development of the third and fourth branchial pouches is born without parathyroid glands and with no (or rudimentary) thymus **(DiGeorge's syndrome);** many are also born with anomalies of the great blood vessels. The disorder is not familial, and is probably due to some teratogenic agent. In some heritable cases of thymic aplasia the parathyroids develop normally (Nezelof syndrome).

If the thymic aplasia is total the infants should resemble mutant, thymus-less ("nude") mice (Thymus, above): all their lymphocytes should be B cells (with surface Igs) but Ab formation in response to T-dependent Ags should be grossly defective. In fact, however, most of the affected infants who survive long enough to be studied probably have less than total aplasia of the thymus, and their serum Igs and ability to produce Abs can be in the normal range. They also have normal frequency of plasma cells and normal germinal centers of lymphatic tissues.

However, T zones in these tissues are underdeveloped and the associated defective cell-mediated immune reactions must be severely disabling: these patients usually succumb to recurrent bacterial, viral, and fungal infections during infancy. Surprisingly, some children who survive longer seem to spontaneously recover some T cell function. In a few patients grafts of fetal thymic tissue (allografts) have been followed by dramatic improvement in clinical status.

Severe Combined Immunodeficiency. This disorder, with defects in B and in T cell function, is probably the most common of the severe heritable deficiencies. It is transmitted in some families as an autosomal recessive character and in others it is X-linked: hence reported cases are predominantly male. Severe recurrent infections due to bacteria, viruses, fungi, and protozoa begin at ca. 3 to 6 months of age and generally end fatally by the second year: death has resulted from generalized chickenpox, measles, vaccinia, and progressive BCG infection (after attempted immunization with BCG).

Serum Igs are exceedingly low and no Ab formation has been detected after trial immunization. Cell mediated immunity is also lacking (Ch. 20): skin allografts are not rejected, contact skin sensitivity to dinitrofluorobenzene cannot be induced, and delayed-type allergic skin reactions are not elicitable, even to candida Ags in infants with candida infections. The bone marrow lacks lymphocytes and plasma cells; the thymus is almost totally aplastic; and it is likely that stem cells do not differentiate into B and T cells.

Early efforts to save these infants with bone marrow transplants from normal donors invariably ended in fatal graft-vs.-host reactions (Chs. 20 and 21). However with sibling donors who have matching major histocompatibility Ags, successful transplants have been carried out, leading to long lasting restoration of B and T cell function (and only minor, transient graft-vs.-host reactions).

Some infants with severe combined immunodeficiency have normal serum Igs and some plasma cells in tissues (sometimes called Nezelof syndrome). Nevertheless they fail to form Abs in response to deliberate immunization and their clinical course is only slightly less severe.

SELECTED REFERENCES

BOOKS AND REVIEW ARTICLES

Antigen-Sensitive Cells. Their Source and Differentiation. Miller, J. F. A. P., Mitchell, G. F., Davies, A. J. S., Claman, H. N., Chaperon, E. A., and Taylor, R. B. *Transplantation Reviews 1*:3 (1969).

BENACERRAF, B., and MCDEVITT, H. O. Histocompatibility-linked immune response genes. *Science 175*: 273 (1972).

COHN, Z. A. The structure and function of monocytes and macrophages. *Adv Immunol 9*:164 (1968).

GOWANS, J. L. Lymphocytes. *Harvey Lect. ser 64,* p. 87 (1970).

GREY, H. M. Phylogeny of immunoglobulins. *Adv Immunol 10*:51 (1969).

KATZ, D. H., and BENACERRAF, B. The regulatory influence of activated T cells on B cell responses to antigen. *Adv Immunol 15*:1 (1972).

MILLER, J. F. A. P., BASTEN, A., SPRENT, J., and CHEERS, C. Interaction between lymphocytes in immune response. *Cell Immunol 2*:469 (1971).

SCHARFF, M. and LASKOV, P. Synthesis and assembly of immunoglobulin polypeptide chains. *Progr Allergy 14*:37 (1972).

SISKIND, G. W., and BENACERRAF, B. Cell selection by antigen in the immune response. *Adv Immunol 10*:1 (1969).

STEINER, L. A., and EISEN, H. N. Variations in the immune response to a simple determinant. *Bacteriol Rev 30*:383 (1966).

STERZL, J., and SILVERSTEIN, A. M. Developmental aspects of immunity. *Adv Immunol 6*:337 (1967).

WEIGLE, W. O. Immunological unresponsiveness. *Adv Immunol 16*:61 (1973).

WEISS, L. *The Cells and Tissues of the Immune System.* Prentice-Hall, Englewood Cliffs, N.J., 1972.

World Health Organization Technical Report No. 448: Factors regulating the immune response, 1969.

SPECIFIC ARTICLES

CRABBE, P. A., CARBONARA, A. O., and HEREMANS, J. F. The normal human intestinal mucosa as a major source of plasma cells containing IgA immunoglobulins. *Lab Invest 14*:235 (1965).

DAVIE, J. M., and PAUL, W. E. Receptors on immuno-competent cells V. Cellular correlates of the "maturation" of the immune response. *J Exp Med 135*:660 (1972).

GASSER, D. L., and SHREFFLER, D. C. Involvement of H-2 locus in a multigenically determined immune response. *Nature (New Biol) 235*:155 (1972).

HENRY, C., KIMURA, J., and WOFSY, L. Cell separation on affinity columns: The isolation of immuno-specific precursor cells from unimmunized mice. *Proc Nat Acad Sci USA 69*:34 (1972).

JULIUS, M. H., MASUDA, T., and HERZENBERG, L. A. Demonstration that antigen-binding cells are precursors of antibody-producing cells after purification with a fluorescence-activated cell sorter. *Proc Nat Acad Sci USA 69*:1934 (1972).

KINCAIDE, P., LAWTON, A., BODEMAN, I., and COOPER, M. Suppression of IgG synthesis as a result of Ab-mediated suppression of IgM synthesis in chickens. *Proc Nat Acad Sci USA 67*:1918 (1970).

MAGE, R. Quantitative studies on the regulation of expression of genes for Ig allotypes in heterozygous rabbits. *Cold Spring Harbor Symp Quant Biol 32*:203 (1967).

MISHELL, R. I., and DUTTON, R. W. Immunization of dissociated spleen cell cultures from normal mice. *J Exp Med 126*:423 (1967).

MITCHISON, N. A. The carrier effect in the secondary response to hapten-protein conjugates: I, II. *Eur J Immunol 1*:10 (1971).

RAJEWSKY, K., SCHIRRMACHER, V., NASS, S., and JERNE, N. K. The requirement of more than one antigenic determinant for immunogenicity. *J Exp Med 129*:1131 (1969).

SHEARER, G. M., MOZES, E., and SELA, M. Contribution of different cell types to the genetic control of immune responses as a function of the chemical nature of the polymeric side chains (poly-L-prolyl and poly-L-alanyl) of synthetic immunogens. *J Exp Med 135*:1009 (1972).

TAKAHASHI, T., OLD, L. J., MCINTYRE, K. R., and BOYSE, E. A. Immunoglobulin and other surface antigens of cells of the immune system. *J Exp Med 134*:815 (1971).

TAYLOR, R. B., DUFFUS, W. P. H., RAFF, M. C., and DEPETRIS, S. Redistribution and pinocytosis of lymphocyte surface Ig molecules induced by anti-Ig antibody. *Nature (New Biol) 233*:225 (1971).

chapter 18

COMPLEMENT

THE COMPLEMENT-FIXATION ASSAY 512
  Measurement of Complement 514
  The Quantitative Complement-Fixation Assay 514
  Competent and Incompetent Ab-Ag Complexes 515
REACTION SEQUENCE IN IMMUNE CYTOLYSIS 515
  The One-Hit Theory 519
ALTERNATIVE PATHWAY FOR COMPLEMENT ACTIVATION 520
OTHER REACTIONS MEDIATED BY COMPLEMENT 521
COMPLEMENT DEFICIENCIES 523
CODA 524

Shortly after the discovery in the 1890s that immunity to diphtheria is due to serum antibodies (Abs) (Ch. 14), a curious observation led to the disclosure of another remarkable group of substances in serum that contribute to host defenses through modifying the behavior of diverse antibody-antigen (Ab-Ag) complexes. Pfeiffer observed that cholera vibrios disintegrated when injected into the peritoneal cavity of a guinea pig that had been previously immunized against the organism. The vibrios were also lysed within a few minutes in vitro when added to serum from the immunized animals. However, if the serum had been previously heated to 56° for a few minutes, or simply allowed to age for a few weeks, it lost its lytic activity, though its Abs were retained; and the addition of fresh **normal** serum to the inactivated antiserum restored its bacteriolytic capacity. Hence lysis required **both** specific Ab and a complementary, labile, nonspecific factor present in normal (as well as in immune) serum. Originally called **alexin** (Gr., "to ward off"), the unstable factor was subsequently named **complement;** it is referred to as **C.**

C is now known to consist of 11 proteins that make up about 10% of the globulins in normal serum of man and other vertebrates. These proteins are not immunoglobulins (Igs), and they are not increased in concentration by immunization. They react with a wide variety of Ab-Ag complexes, and exert their effects primarily on cell membranes, causing lysis of some cells and functional aberrations in others, e.g., degranulation of mast cells with release of histamine (Ch. 17), increased permeability of small blood vessels, directed migration of polymorphonuclear leukocytes, increased phagocytic activity by leukocytes and macrophages, and bacteriolysis.

Of these far-ranging effects, red cell lysis (hemolysis) has been analyzed in greatest de-

tail, because it is so simple to measure; it also provides the basis for the C-fixation assay, an important laboratory procedure for detecting and measuring many different kinds of Ags and Abs.

THE COMPLEMENT-FIXATION ASSAY

In addition to the test Ag and antiserum the assay requires a number of standard reagents (referred to as the immunological zoo): **sheep** red blood cells, **rabbit** Abs to sheep cells (these Abs are also often called **hemolysins),*** and fresh **guinea pig** serum as a source of C. Guinea pig serum is more active than serum from other species; it is used promptly after collection, or it may be stored at $-70°$ or lyophilized.

If sheep erythrocytes in neutral isotonic salt solution are optimally coated with nonagglutinating amounts of Abs to the cells, the addition of C in the presence of adequate concentrations of Mg^{++} and Ca^{++} promptly causes the cells to lyse. The extent of lysis is evaluated qualitatively by inspection, or quantitatively by determining the concentration of supernatant hemoglobin after sedimentation of in-

---

* Rabbits immunized with intact sheep red cells form two kinds of hemolysins: **isophile Abs,** which are species specific (for determinants on red cells of sheep), and **heterophile** Abs, which are specific for the Forssman antigen, a glycolipid found in the red cells and tissue cells of many species in addition to sheep, and even in some bacteria (Ch. 21). In order to obtain more uniformly effective antiserum it is preferable to immunize rabbits with boiled stromata of sheep red cells, which evoke only the anti-Forssman (heterophile) Ab. The latter are almost entirely IgM, whereas the isophile Abs belong to both IgG and IgM classes. As noted later, the IgM are more efficient that the IgG hemolysins.

tact cells and stroma. Ab-coated ("sensitized") erythrocytes thus become indicators to detect active C. In general, when Abs combine with Ags in the presence of C some components of C are bound and inactivated. As a result, C activity, i.e., ability to lyse sensitized red cells, is lost.

C-fixation assays are therefore performed in two stages. **In stage 1** antiserum and Ag are mixed in the presence of a carefully measured amount of C and then incubated, usually for 30 minutes at 37° or overnight at about 4°. If the appropriate Ab-Ag complexes are formed C is inactivated or "fixed." **In stage 2** a suspension of sensitized red cells is added to determine whether active C has survived. **Hemolysis** indicates that C persists and, therefore, that an effective Ab-Ag reaction **has not** occurred in stage 1. Conversely, **absence of hemolysis** indicates that C has been fixed and, therefore, that an Ab-Ag reaction **has** occurred in stage 1 (positive C-fixation reaction). These steps are outlined in Figure 18-1. With a known Ag the assay can be used to detect (and measure) Abs in unknown sera; and with a standard antiserum to a known Ag it can be used to detect that Ag in complex biological materials.

Conditions of Assay. If the amount of C used in the assay is excessive, some active C may persist and cause hemolysis even though an Ab-Ag reaction has taken place in stage 1. On the other hand, if the amount of C added is insufficient, its deterioration may lead to the absence of lysis in stage 2 even if an Ab-Ag reaction has not occurred in stage 1. Accordingly, just the right amount of C is required.

Usually 5 units is an effective compromise in conventional assays (10^8 sensitized cells in a total reaction volume of 1.5 ml). The unit of C is discussed below.

The interpretation of the C-fixation assay depends on the outcome of a number of **controls.** For example, the Ag and the antiserum must be individually tested to ascertain that they are not "anticomplementary," i.e., that each does not inactivate C without the other.* To provide a margin of safety, anticomplementarity is tested with Ags or antiserum at a higher concentration than that used in the assay, and C is reduced to an amount that is just sufficient to lyse the indicator cells (Table 18-1). It is also essential to ascertain that the red cells do not lyse spontaneously, and that C survives stage 1 in the absence of an authentic Ab-Ag reaction.

Dependence on Mass and Ratio of Antibody-Antigen Complexes. The amount of C fixed depends not only on the mass of Ab-Ag complexes formed, but on their Ab/Ag ratio. When the ratio corresponds to that in the Ab-excess or the equivalence regions of the precipitin curve (Ch. 15) C is fixed most effectively.

---

* The most common artefact arises from anticomplementary properties of Ags and antisera. Particularly frequent with Ags prepared from tissue homogenates, this difficulty is especially troublesome with C-fixation assays in the serological diagnosis of diseases due to viruses, rickettsiae, and chlamydia, in which antigenic material is obtained from infected tissue. Undiluted or slightly diluted antisera are also frequently anticomplementary, usually owing to some denatured and aggregated Igs (below).

Fig. 18-1. Complement-fixation assay. Ab = antibody; Ag = the corresponding antigen; C = guinea pig complement; EA = sheep erythrocytes complexed with rabbit Abs to the cells (sensitized red cells) as indicator for active complement. C fixation has occurred in reaction (1) but not in reaction (2) or (3).

STAGE 1	STAGE 2
1. Ab + Ag + C′ ⟶ Ab-Ag-C′	+ EA ⟶ No lysis
2. Ab + C′ ⟶ Ab + C′	+ EA ⟶ Lysis
3. Ag + C′ ⟶ Ag + C′	+ EA ⟶ Lysis

TABLE 18-1.   Complement-Fixation Test for Antibodies in Human Serum to Poliovirus*

Virus dilution	Serum dilution								Control with	
	1:10	1:20	1:40	1:80	1:160	1:320	1:640	1:1280	5 C'H$_{50}$	3 C'H$_{50}$
1:20	0	0	0	½	4	4	4	4	4	4
1:40	0	0	0	0	2	4	4	4	4	4
1:80	0	0	0	0	0	3½	4	4	4	4
1:160	0	0	0	0	0	0	4	4	4	4
1:320	0	0	0	0	0	0	2	4	4	4
1:640	0	0	0	0	0	0	1	4	4	4
1:1280	½	½	½	½	½	½	3	4	4	4
1:3200	4	3	3	3	3	3	4	4	4	4
1:6400	4	4	4	4	4	4	4	4	4	4
Control with										
5 C'H$_{50}$	4	4	4	4	4	4	4	4		
3 C'H$_{50}$	4	4	4	4	4	4	4	4		

* Assay with 5 C'H$_{50}$ units of complement; fixation at 4° for 20 hours before addition of sensitized indicator red cells. 0 = no lysis (positive C fixation); 4 = complete lysis (negative C fixation). The highest dilution of Ag (virus) that gives positive C fixation is 1:1280. A 1:640 dilution of serum gives positive C fixation with dilute virus (1:640), but with more concentrated virus the system is then in Ag excess and it is necessary to use less dilute serum. The Ag alone and the serum alone are not anticomplementary at any dilution examined, even when tested in controls with only 3 C'H$_{50}$ units of complement.

From M. Mayer *et al.*, *J Immunol* 78:435 (1957).

When Ag is present in excess, however, fixation is less; and with Ag in extreme excess C is not fixed at all (see Fig. 18-3). Since information about optimal proportions is not generally available, C-fixation tests are best carried out by means of a "checkerboard titration," in which both Ag and antiserum concentrations are varied as shown in Table 18-1.

MEASUREMENT OF COMPLEMENT

The proportion of sensitized red cells that are lysed increases with the amount of C added (Fig. 18-2). Because 100% lysis is approached asymptotically it is convenient to **define the unit of complement (the C'H$_{50}$ unit) as that amount which lyses 50% of sensitized red cells** under conditions that are arbitrarily standardized with respect to concentration of sensitized cells, concentration and type of sensitizing Ab, ionic strength and pH of the solvent, concentrations of Mg^{++} and Ca^{++}, and temperature.

The dose-response curve of Figure 18-2 follows the von Krogh equation,

$$x = K \left(\frac{y}{1-y} \right)^{1/n}$$

in which x is the amount of C added (i.e., milliliters of guinea pig serum), y is the proportion of cells lysed, and n and K are constants. The curve described by this equation (which was arrived at empirically) is sigmoidal when $1/n < 1$; for fresh normal guinea pig sera $1/n$ is usually about 0.2. In estimating the number of C'H$_{50}$ units per milliliter of guinea pig serum it is convenient to plot log x vs. log $(y/1-y)$; the data fall on a straight line,

$$\log x = \log K + \left(\frac{1}{n} \right) \log \frac{y}{1-y}$$

in which the intercept at 50% lysis ($y = 1-y$; log $y/1-y = 0$) gives the volume of guinea pig serum that corresponds to one C'H$_{50}$ unit.

THE QUANTITATIVE COMPLEMENT-FIXATION ASSAY

The amount of C fixed in a reaction between soluble Ag and Ab can be determined as the difference between the amount added and the amount remaining after the reaction has gone to completion. When the reaction of antiserum with increasing amounts of Ag is followed (Fig. 18-3), the amount of C fixed varies as the amount of precipitate in the precipitin reaction, increasing over the Ab-excess region to a maximum at the equivalence zone and then decreasing in the Ag-excess zone. The C reaction can thus be used as an alternative to the precipitin reaction, e.g., to measure quantitatively the concentrations of Ab or of Ag, or

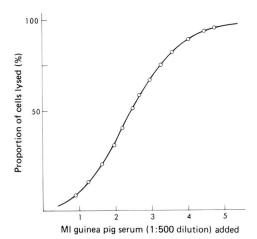

Fig. 18-2. Dose-response curve of immune hemolysis. The curve follows the empirical von Krogh equation (see text).

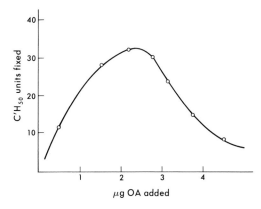

Fig. 18-3. Fixation of C by varying quantities of chicken ovalbumin (OA) and 12.5 μg Ab from rabbit antiserum to OA. Note resemblance to precipitin curves of Chapter 15, with decreasing C fixation in the region of Ag excess. [From Osler, A. G., and Heidelberger, M. *J Immunol 60*:327 (1948).]

to compare closely related Ags for their reactivity with a standard antiserum. C fixation offers the important advantage that it can be used with high precision to measure very small amounts of Ab or Ag, detecting as little as 0.5 μg of Ab. However, it measures the reactivity of only certain classes of Abs (see below), and it measures concentrations in relative rather than absolute weight units.

COMPETENT AND INCOMPETENT AB-AG COMPLEXES

The small complexes formed by Abs and univalent ligands do not fix C. But the aggregates formed with multivalent ligands of diverse chemical nature, size, and charge are effective. The almost limitless variety of effective Ags suggested that C reacts with the Ab moiety of Ab-Ag aggregates. Indeed, when Igs of an appropriate class (see below) are simply aggregated by heat or some other denaturing conditions they fix C in essentially the same way as when specifically cross-linked with multivalent Ags.

Igs from different species and of diverse classes differ greatly in ability to fix guinea pig C (which is used in most assays); e.g., cattle Igs are ineffective and those from some birds are only marginally active. Human antibodies of the IgG-1, IgG-2, IgG-3, and IgM classes are highly competent, while IgG-4, IgA, and IgE react only in a special way (Alternative

pathway for C activation, below). Since Igs of diverse classes have characteristically different heavy chains (Ch. 16), it seems likely that sites in these chains become exposed through conformational changes brought about when Ab molecules are aggregated by Ag or by denaturation.

REACTION SEQUENCE IN IMMUNE CYTOLYSIS

Various procedures destroy serum C activity, which can be restored when certain inactive preparations are recombined: hence it has long been apparent that C consists of more than one substance. From the 4 classic fractions (C'1, C'2, C'3, C'4), defined in Figure 18-4, 11 active proteins have been isolated; their properties are summarized in the same figure.

After one component (C1q) is bound by competent Ab-Ag complexes, the others react in an ordered sequence: in several steps an activated protein cleaves the next reacting member of the series into fragments, of which the largest usually also behaves as an activated proteolytic enzyme, cleaving and thereby activating the next protein in sequence. Some of the smaller proteolytic fragments have "phlogistic" activity (Gr. *phlogistos* = inflammatory): they cause inflammatory tissue changes, such as increased vascular permeability, and

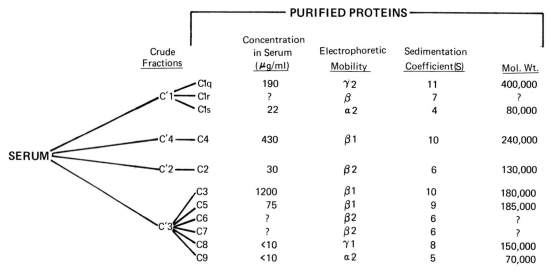

Fig. 18-4. Some properties of purified human C proteins. The four crude fractions (C'1 to C'4) were characterized originally by conditions that inactivated or removed them from fresh serum. C'1 and C'2 are extremely heat-labile: they are inactivated in a few minutes at 56° whereas inactivation of C'3 and C'4 at this temperature requires 20 to 30 minutes. C'1 is precipitated with "euglobulins" by dialysis against H_2O at pH 5, whereas C'2, a "pseudoglobulin," is soluble under these conditions. C'3 is inactivated by zymosan (see Alternate pathway for C activation, below) and C'4 by exposure to ammonia or hydrazine. C'1 to C'4 are numbered in order of discovery, which unfortunately is not the same as the order in which the components react in the C cascade (see Fig. 18-5). [Based in part on *Bull WHO* 39:935 (1968).]

attraction of polymorphonuclear leukocytes (chemotaxis). The reaction sequence is illustrated in Figure 18-5, and individual steps are discussed below. Activated C proteins with enzymatic activity are designated by an overbar (e.g., $\overline{C1}$ is the enzymatically active form of C1).

1. S + A ⇌ SA. Immune cell lysis is initiated by the specific binding of one IgM or one pair of IgG Ab molecules (A) (One-hit theory, below) to an antigenic site (S) on the cell surface, forming a "sensitized" cell. The surface Ag need not be a natural constituent of the cell membrane: in **passive hemolysis** soluble Ags or small haptens are attached artificially to the red cell membrane, and the cells are then sensitized with the corresponding Abs.

The requirement for a pair of adjacent IgG molecules implies that the effective antigenic determinants on the cell surface must be closely spaced (Fig. 18-5). Improper spacing could explain the inability of Abs to many cell surface Ags to cause lysis (for example, Abs to Rh and to some other antigens on red cells fail to sensitize for immune hemolysis; Ch. 21).

2. SA + C1 + Ca++ ⇌ SAC1. With attachment to antigenic sites on the membrane, the Ab molecule probably undergoes a conformational change that

exposes sites in Fc domains of the heavy chain; these sites then interact specifically with C1. C1 exists in serum as an aggregate of three proteins—C1q, C1r, and C1s—which are associated noncovalently (in a mole ratio of 1:2:4) in a complex with Ca++. C1q, the **recognition** unit of the complex, consists of five subunits, each with one binding site for the heavy chains of those Ig classes (e.g., IgG-1, IgG-2, IgG-3, IgM) that can trigger the entire C sequence. In contrast, Abs that belong to Ig classes that do not react with C1q (e.g., IgG-4, IgA, IgE) cannot trigger the early steps in the C cascade; they can, however, activate the later-acting C components via an alternate pathway (below).

Unlike other C proteins, C1q has stable combining sites and requires no activation.* However, in binding to Igs, C1q probably undergoes a conformational change that results in modifying C1r, which in turn, converts C1s to a proteolytically active enzyme, $\overline{C1s}$, whose natural substrates are C4 and C2. (Like trypsin and some other proteases, $\overline{C1s}$ also has amino acid esterase activity, and its

---

* C1q is also unusual in its striking chemical similarity to collagen: i.e., it has a high content of glycine, hydroxyproline, and hydroxylysine, with a galactose-glucose disaccharide attached to the hydroxyl of hydroxylysine; and it can be inactivated by collagenase.

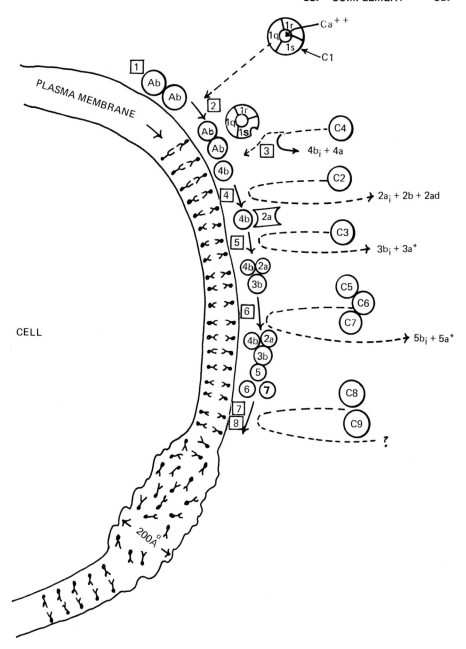

Fig. 18-5. Reaction sequence in lysis of a cell by activated C. The leaky lesions that develop after steps 7 and 8 probably correspond to "holes" (Fig. 18-6) and to local swellings of the membrane (ca. 200 A) seen in electron micrographs. The lesions could represent local perturbations in the membrane's lipid bilayer, due to a detergent-like effect of activated C8 and C9. Three activated components have enzymatic activity: C$\overline{1}$, C$\overline{4b,2a}$, C$\overline{4b,2a,3b}$. They cleave the following substrates: C2 and C4 (C$\overline{1}$), C3 (C$\overline{4b,2a}$) and C5 (C$\overline{4b,2a,3b}$). C5b,6,7, is commonly written with an overbar, as though it were an established enzymatic complex; however, its enzymatic activity has not been demonstrated. The numbers in boxes correspond to the numbered steps in the reaction sequence in text.

catalytic activities are blocked specifically by DFP, diisopropylphosphofluoridate.)

Only the C1 complexes, not the separated subunits, are activated by binding to Ab-Ag complexes. **The stability of the C1 complex depends on calcium ions (Ca++), which are thus required for immune lysis of cells.** The binding of C1q, C1r, C1s (i.e., C1) to Ab-Ag complexes is reversible, and dissociated C1 can transfer to other SA sites. Chelating agents that bind Ca++ cause C1 complexes to come apart and to dissociate from Ab-Ag aggregates.

C1q can also form specific precipitates with soluble Ab-Ag complexes whose Ab moiety belongs to Ig classes that can bind and activate the C1 complex (IgG-1, IgG-2, IgG-3, IgM). This precipitation reaction is useful for detecting soluble immune complexes in diseases where they seem to be responsible for severe inflammatory lesions (e.g., Rheumatoid arthritis, Ch. 19).

3. $SAC\overline{1}$ + C4 → $SAC\overline{1,4b}$ + C4b + C4a. Two fragments result from the cleavage of C4 by the $C\overline{1s}$ moiety of $SAC\overline{1}$: inactive C4a (MW 15,000) and active C4b (MW 230,000). About 10% of C4b is bound by $SAC\overline{1}$, forming the next activated complex, $SAC\overline{1}$,4b; the remaining C4b decays rapidly in solution ($C4b_i$). With the aid of ^{125}I-labeled C4, it has become clear that C4b binds to both the membrane-bound Ab and the membrane itself: some C4b can remain on the membrane after C1 is removed (by removing Ca++ with a chelating agent).

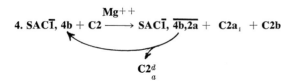

4. $SAC\overline{1}$, 4b + C2 $\xrightarrow{\;Mg++\;}$ $SAC\overline{1}$, $\overline{4b,2a}$ + $C2a_i$ + C2b

$C2_a^d$

In the presence of Mg++ C2 is split by the $C\overline{1s}$ moiety of $SAC\overline{1}$,4b. One fragment, C2a (MW 80,000), binds to the membrane complex, forming the next activated complex, $SAC\overline{1,4b,2a}$: the C4b,2a moiety of this complex is called **C3 convertase**, because it specifically and effectively attacks C3, which reacts next. With the formation of the C3 convertase, $C\overline{1}$ becomes unnecessary: it can be removed (by a chelating agent that withdraws Ca++ without affecting the remaining reactions in the complete sequence).

C2 is adsorbed on cell-bound C4b of $SAC\overline{1}$,4b before it is cleaved by $C\overline{1}$; this minimizes loss of C2a, which is readily inactivated in solution to $C2a_i$. The C2a moiety of $SAC\overline{1,4b,2a}$ is also unstable (half life at 37° is 10 minutes): in dissociating from the complex (as inactive $C2_a^d$), it regenerates $SAC\overline{1}$,4b which can bind and cleave additional C2 to form more $SAC\overline{1,4b,2a}$.

5. $SAC\overline{1,4b,2a}$ + C3 → $SAC\overline{1,4b,2a,3b}$ + $C3b_i$ + C3a In this critical step the reaction is amplified, because each $SAC\overline{1,4b,2a}$ unit cleaves hundreds of C3 molecules into fragments. A small fragment, C3a (MW 7000), has pronounced phlogistic activity (see Anaphylatoxins, below). The large fragment, C3b, decays in solution (to $C3b_i$) or binds to the cell; however only those fragments that are bound in close proximity to $C\overline{4b,2a}$ can join in forming the new enzyme C4b,2a,3b. The other C3b fragments, scattered over the cell membrane, are not involved in the hemolytic reaction but they contribute to immune adherence (below), which increases the cell's susceptibility to phagocytosis; i.e., C3b on the cell membrane is a potent **opsonic** agent (below).

6. $SAC\overline{1,4b,2a,3b}$ + C5·C6·C7 → $SAC\overline{1,4b,2a,3b,5b,6,7}$ +$C5b_i$ + C5a. In this step activated C4b,2a,3b in the SAC1-3b complex cleaves C5. A small fragment, C5a (MW 15,000), has some of the phlogistic activity of C3a (Anaphylatoxins, below); in addition, it is a powerful attractant for polymorphonuclear leukocytes (see Chemotaxis, below; and Arthus reaction, Ch. 19). About 10% of the larger activated fragment, C5b, combines with C6, and the complex then binds C7 to form the ternary complex C5b,6,7 on the cell membrane at or near the $SAC\overline{1-3b}$ complex; the remaining C5b in solution becomes rapidly inactive ($C5b_i$). Membrane-bound C5b,6,7 is the initiator of the hemolytic steps (7 and 8, below); and like C5a, it is chemotactic for polymorphonuclear leukocytes. This complex is unstable in solution, but it is possible that traces bind to non-sensitized "bystander" cells, whose subsequent reaction with C8 and C9 (below) would then lead to lysis.

7. $SAC\overline{1,4b,2a,3b,5b,6,7}$ + C8 → $SAC\overline{1-8}$. Following the binding of C8, cells with $SAC\overline{1-8}$ complexes may develop functional membrane lesions and undergo slow lysis at 37°.

8. $SAC\overline{1,4b,2a,3b,5b,6,7,8}$ + C9 → $SAC\overline{1-9}$. In this terminal step, up to six molecules of C9 are bound per $SAC\overline{1-8}$ complex (though one suffices for full activity) and the rate of cell lysis becomes greatly accelerated: small, intracellular molecules leak out of the cell and extracellular water enters rapidly, causing the cell to swell and to rupture its membrane.

Membrane Lesions. After binding C5-C9, the red cell membrane acquires circular lesions (Fig. 18-6) that could perhaps correspond to leaky foci. While these lesions could be due to a direct enzymatic attack on the membrane by activated C, it seems more likely that they

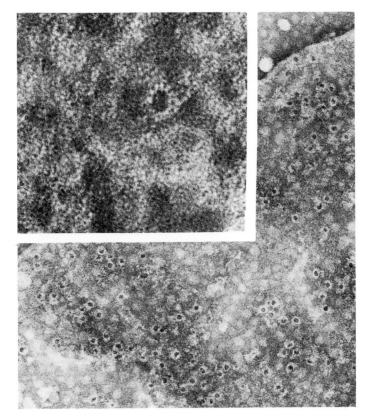

Fig. 18-6. Electron micrograph of the membrane of a sensitized sheep red cell lysed by C. Many defects ("holes") are evident. ×187,000 (reduced). Inset shows a representative lesion at greater magnification (×720,000, reduced). Preparations by R. Dourmashkin. (From Humphrey, J. H., and Dourmashkin, R. R. In *Complement*. Ciba Foundation Symposium, G. E. W. Wolstenholme and J. Knight, eds. Little, Brown, Boston, 1965.)

arise from a detergent-like action of hydrophobic patches present on activated late-acting proteins (C8,C9). Thus, membranes treated with the hydrophobic polyene antibiotic filipin (Ch. 7) develop indistinguishable circular lesions. Moreover, when completely synthetic model membranes (liposomes) with incorporated Ags are lysed by C plus Abs no covalent changes have been found in the lipids or other constituents.

Unlike red cells, nucleated cells are not usually lysed by bound, activated C but their membranes become damaged and leaky: K^+ and small organic molecules (e.g., AMP) escape before macromolecules. A readily detectable change is the entry of ionic dyes that are excluded by intact cells. Hence dyes like trypan blue and eosin are often used to detect C-damaged (i.e., stained) lymphocytes and other nucleated cells.

THE ONE-HIT THEORY

The sigmoidal dose-response curve (Fig. 18-2) originally suggested that lysis depends on the accumulation of many damaged sites (S^*) per cell. However, Mayer and his colleagues have provided evidence that one S^* lesion per red cell is sufficient. Thus with a limiting amount of complement the extent and velocity of hemolysis is independent of the total number of sensitized cells in the reaction mixture. Moreover, with SA, or with SAC$\overline{1}$, or with SAC$\overline{1}$,4, the average number of lytic lesions per cell is linearly related to the concentration of C1, or C4, or C2, respectively. (The average number of lytic lesions per red cell is calculated from the proportion of red cells that are not lysed, by applying the Poisson distribution.† Similarly, the activation and binding of C1 is linearly related to the concentration of IgM antibodies to red cells or to the square of the concentration of the corresponding IgG antibodies (Fig. 18-7). These linear relations indicate that **the S* site is established by the binding of one IgM or one pair of IgG antibody molecules, followed by one molecule each of C1, C4, and C2.** This is sufficient to activate and bind many molecules of C3 and the remaining C components and to cause lysis.

The sigmoidal shape of the hemolytic dose-response curve (Fig. 18-2), which had constituted the principal basis for interpreting immune hemolysis as a multi-hit or cumulative process, has been reconciled with the one-hit theory by considering the properties of SAC$\overline{1}$,4b,2a. This complex undergoes two competing reactions (step 4, above): one hindering hemolysis, with loss of C2a; and the other, by activating C3, producing the next complex in the sequence. The loss of C2a is a unimolecular reaction, with a fixed half life at a given temperature, whereas the rate of reaction with C3 is exponentially dependent on the C3 concentration, and is greatly augmented by high concentrations of this component. Qualitatively, at least, these relations account for the sigmoidal shape of the dose-response curve.

ALTERNATIVE PATHWAY FOR COMPLEMENT ACTIVATION

Additional opportunities for C to participate in immune reactions are provided by an alternative mechanism for inducing phlogistic activities by initiating the C sequence with C3, without triggering the sequence through C1, C4, C2. In this pathway various initiators (see below) apparently react with the properdin system (see below) and bring about the formation of a proteolytically active substance ("C3 activator") that cleaves C3 in the same way as C$\overline{4b}$, 2a (Fig. 18-7). With the forma-

tion of C3b, the remaining C proteins (C5-C9) can probably be sequentially modified as in the classic hemolytic sequence. This **bypass sequence** (C3-C9) is scarcely able to cause cell lysis, however, because C5b,6,7 decays rapidly in solution; nevertheless, the phlogistic activity of the C3-C9 sequence might be as great as in the complete C1-C9 system.

Initiators of the bypass sequence include endotoxin (lipopolysaccharide of cell walls of gram-negative bacteria), zymosan (polysaccharide from yeast cell walls), and aggregated Abs of those Ig classes (e.g., human IgG-4, IgA-1, IgA-2, IgE, and guinea pig $\gamma 1$) that cannot trigger the classic hemolytic sequence because they are incapable of binding C1q. Abs that initiate the bypass sequence also differ from those (IgM, IgG-1, -2, -3) that initiate the complete sequence via C1 in the location of the relevant heavy-chain sites: from the effects of pepsin digestion (Fig. 16-6, Ch. 16), it appears that these sites are located in the Fc domain of the C1q-reactive Abs and in the F(ab')$_2$ region of Abs that initiate the bypass sequence.

The Properdin System. Pillemer *et al.* argued in the 1950s that certain non-Ab, normal serum proteins, which they called the properdin system (Gr. *perdere* = to destroy), were necessary for both degradation of C3 by zymosan and for some antimicrobial activities of normal serum, e.g., killing *Shigella dysenteriae* and neutralizing some viruses. Others dismissed these proteins as natural C-fixing Abs (for zymosan and some bacterial Ags); but 15 years later properdin was purified and shown not to be an Ig (i.e., it does not react with antisera to Igs).

The steps by which aggregated Igs of certain classes and other initiators activate the properdin system are not known. Nonetheless, this system and the early-acting C1, C4, C2 components of the classic C-reaction sequence seem to represent **alternative pathways** for triggering the biologically most effective components of the C system—i.e., C3 and late-acting proteins (Fig. 18-7).

The properdin system is now known to consist of properdin itself and at least three other serum proteins, called A, B, and the 0° factor. All are needed for an as yet undefined reaction sequence, triggered by various initiators (zymosan, certain aggregated Igs), that culminates in cleavage of C3 (Fig. 18-7).

---

† $P(k) = e^{-m} \, m^k/k!$, where $P(k)$ is the proportion of cells with k lytic lesions per cell, m is the average number of lytic lesions per cell, and e is the base of natural logarithms. Unlysed cells have no lytic lesions ($k = 0$); their frequency is e^{-m} (because $0! = 1$), and $m = -2.303 \log_{10} P(0)$. For more on the Poisson distribution, see Ch. 44, Appendix.

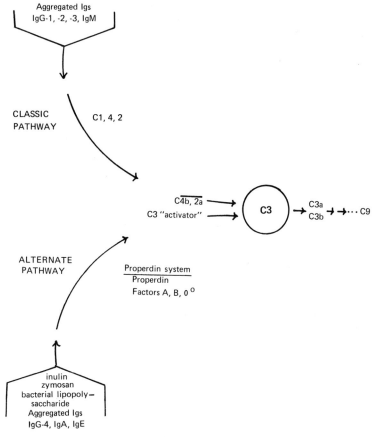

Fig. 18-7. The properdin system as the alternative pathway for activating C3 and late acting components. Different classes of immunoglobulins activate the two pathways. Cobra venom can also bring about cleavage of C3 by activating factor B of the properdin system.

Activation of factor B (also called C3 pro-activator) appears to be directly involved in generating the C3-cleaving activity. The resulting C3b fragments probably account for the properdin system's antimicrobial activities: the fragments are powerful potentiators of phagocytosis (Opsonization, below), and they can probably also trigger the activation of later-acting C components (C5-C9).

Other Proteases. Besides "C3 activator" in the properdin system and $\overline{\text{C4b,2a}}$ in the classic system, C3b and C3a can be formed from C3 by certain other proteases; e.g., those released from lysosomes of degenerating polymorpho-nuclear leukocytes or from some bacteria (β-hemolytic streptococci), and some of those involved in blood clotting (below).

Blood Clotting and C. The activation of C probably always occurs to some extent during blood coagulation, in which plasminogen (a serum proenzyme) is converted to plasmin, a trypsin-like protease. Plasmin (like trypsin) can initiate the C sequence by converting C1 to $\overline{\text{C1}}$, and by cleaving C3 to C3a and C3b. Kallikrein, another serum protein activated during blood clotting (see Kinins, Ch. 19), can also activate C1 to $\overline{\text{C1}}$. A more direct connection between C and clotting has been revealed by the finding that C6 is required for normal clot formation: blood from rabbits with a genetic deficiency of C6 (see below) clots very slowly, and the addition of purified C6 corrects this defect.

OTHER REACTIONS MEDIATED BY COMPLEMENT

Though cell lysis has dominated the study of C, the main physiological effects of activated

C are related to cellular and tissue changes associated with inflammation: e.g., increase in capillary permeability (anaphylatoxin activity), and directed migration (chemotaxis) and enhancement of phagocytic activity of polymorphonuclear leukocytes (Table 18-2). Because these phlogistic activities are generated in the immediate vicinity of the Ab-Ag complexes that activate C they lead to increased local concentrations of immunoglobulins, C, and "activated" phagocytes. The resulting destruction or containment of pathogens is probably more important to host defenses than is cell lysis itself, especially since many bacteria and probably most viruses are not susceptible to the lytic action of C (Bactericidal reactions, below).

Anaphylatoxins. This term is derived from the old observation that guinea pigs can undergo fatal shock, resembling anaphylaxis (Ch. 19), when injected with normal serum that has been incubated briefly with various substances (e.g., inulin, Ab-Ag complexes, talc). The effect was attributed to the appearance in the incubated serum of "anaphylatoxins," which have recently been identified as the small polypeptides cleaved from C3 and C5 during C activation (C3a and C5a).

The effects of C3a and C5a arise largely from their degranulation of mast cells with release of histamine; this causes (among other effects) a marked increase in capillary permeability (Ch. 19). Injection of C3a into human skin promptly elicits a wheal, and the effect can be specifically blocked by antihistamine drugs (Ch. 19).

C3a consists of the 60 N-terminal amino acid residues of C3; C5a is twice as long and probably represents the corresponding segment of C5. Both polypeptides have C-terminal arginine, indicating their origin from a trypsin-like attack of $\overline{C4b,2a}$ on C3 and of $\overline{C4b,2a,3b}$ on C5 (Fig. 18-5). The C-terminal arginine is required for anaphylatoxin activity, which is lost when just this residue is removed by a specific serum enzyme, called **anaphylatoxin inhibitor.**

Though similar in origin, activity, and structure, C3a and C5a seem to act on different mast cell receptors: isolated guinea pig ileum treated repeatedly in vitro with C3a loses its responsiveness to this substance ("tachyphylaxis"), but still responds to C5a, and vice versa. Both peptides are active at extremely low concentrations (1×10^{-8} to 5×10^{-10} M): human C5a is about 4 times more potent than C3a, but human serum contains about 20 times more C3 than C5 (Fig. 18-4).

Chemotaxis. With diffusion chambers divided into compartments by porous membranes that pass macromolecules but not cells, it has been shown that certain C derivatives in one compartment attract polymorphonuclear leukocytes from the adjacent compartment. This

TABLE 18-2. Principal Activities of Activated Complement (C) Proteins and Their Fragments

Activity	C protein or fragment
Anaphylatoxin: Histamine release from mast cells and increased permeability of capillaries	3a; 5a
Chemotaxis: Attract polymorphonuclear leukocytes	5a; $\overline{5b, 6, 7}$; ?3a
Immune adherence and opsonization: Adherence of Ab-Ag-C complexes to leukocytes, platelets, etc., increasing susceptibility to phagocytosis by leukocytes and macrophages	3b; 5b
Membrane damage: Lysis of red cells; leakiness of plasma membrane of nucleated cells; lysis of gram-negative bacteria	8; 9

chemotactic activity is exhibited by C5a, the ternary complex C$\overline{5b,6,7}$ (see step 6, above), and probably also by C3a. This activity is neither stimulated by histamine nor blocked by antihistamine drugs (see Ch. 20, Fig. 20-11).

Immune Adherence and Opsonization. The binding of C3b to Ab-Ag aggregates and to Ab-sensitized cells causes them to adhere to polymorphonuclear leukocytes, macrophages, and certain other cells (e.g., platelets, primate red blood cells). This "immune adherence" is probably responsible for the **opsonic activity** of C3b: i.e., the increased susceptibility to phagocytosis of bacteria, enveloped viruses, and other particles with Ab and C3b on their surface. An undefined serum protein can specifically inactivate and reverse the opsonic activity of soluble or cell-bound C3b.

Bactericidal Reactions. Gram-negative bacteria coated with specific antibody can be lysed by C acting through the same reaction sequence as in red cell lysis. **Gram-positive bacteria and mycobacteria, however, are not susceptible to the lytic action of complement,** and the basis for their resistance is not understood.

A cytotoxic C reaction is used in clinical diagnosis to detect Abs to the treponemes that cause syphilis. Here, however, the reaction leads not to immediate lysis but to loss of motility (*Treponema pallidum* immobilization, TPI test, Ch. 37).

Tissue Destruction. Though activation of C potentiates the defensive function of Abs, it can also cause severe damage to normal tissues. In some allergic disorders (especially those involving small blood vessels and glomeruli) the binding and activation of C by otherwise innocuous Ab-Ag complexes attracts leukocytes, whose subsequent degeneration releases lysosomal enzymes, causing local necrosis of host tissues (Arthus reaction, Ch. 19).

Immune Conglutination. A variety of Ab-Ag-C complexes are specifically precipitated by "immune conglutinin," an Ab (of the IgM class) specific for activated C3 and C4. Present in low titer in most normal sera, the level of immune conglutinin increases after various infections and after immunization with many Ags. This anti-C antibody is directed against hidden determinants in native C3 and C4, which become exposed when these proteins are activated and bound by Ab-Ag complexes. Since a given individual's activated C reacts with his own immune conglutinin, it could be regarded as autoantibody (Ch. 17).

Conglutination. Antibody–red cell–C complexes are agglutinated by a protein, **conglutinin,** which is present in the serum of certain mammals (cattle, other ruminants). The biological significance of conglutinin is not known: its formation is not increased by immunization, and it is not an Ig. It specifically binds some sugar residues of C3 (which is a glycoprotein) in a reaction that requires Ca++.

COMPLEMENT DEFICIENCIES

Since C activation augments the protective role of antibodies against infectious agents, individuals deficient in C proteins should be unusually susceptible to infections. Severe, recurrent pyogenic infections have been observed in rare persons with markedly depressed serum levels of C3 (due to its excessive degradation or genetically determined absence) or with a heritable defect in C5. Surprisingly, infections have not been troublesome in human families and strains of animals with heritable deficiencies of C2 or C5 in man, C4 in guinea pigs, C5 in mice, or C6 in rabbits (Table 18-3), perhaps because a normal level of C3 and the bypass mechanism for its activation are sufficiently protective. C6-deficient rabbits give poor Arthus reactions, supporting other evidence that late-acting C components are necessary for this allergic reaction (Ch. 19).

Hereditary Angioedema. Humans with this disease are deficient not in C but in a carbohydrate-rich glycoprotein that inhibits activated C$\overline{1}$ (i.e., C$\overline{1s}$). This disorder is inherited as an autosomal dominant trait and affected individuals suffer periodically from acute and transitory local accumulations of edema fluid, which can become life-threatening when localized in the larynx, where they obstruct the tracheal airway. Serum obtained from patients during attacks has increased C$\overline{1s}$ activity (reflected in increased amino acid esterase activity) and decreased C4 and C2. Injected into skin, the serum also causes increased permeability of cutaneous blood vessels (hives, Ch. 19). The responsible factor is a small polypeptide, "C-kinin," which is split from C2 by C$\overline{1s}$; its activity resembles that of bradykinin (a nonapeptide that increases capillary permeability, see Ch. 19), but the two are distinguishable by radioimmunoassay (Ch. 15). Acute attacks seem to be triggered by activation of Hageman factor

TABLE 18-3.   Complement-Deficiency States

Primary deficiency	Species	Associated with	
		Increased susceptibility to infection	Comments
C1 inhibitor	Man	No	Become deficient in C1,C4,C2
C2	Man	No	Heritable defect with C2 about 5% of normal serum level
C3	Man	No	Heritable partial defect; two patients had severe recurrent infections
C5	Man	No	Heritable defect revealed by decrease in serum enhancement of phagocytosis in vitro; only 1 in 15 affected members in one family had recurrent infections
C4	Guinea pig	No	
C5	Mouse	No	
C6	Rabbit	No	Blood coagulation defect corrected by purified C6

(the plasma protein that initiates clotting), which causes the activation of two serum proenzymes: plasminogen and prekallikrein (Ch. 19). The resulting proteases, plasmin and kallikrein, activate C1 (and also C3), and thereby the remaining C sequence, resulting in formation of anaphylatoxin and liberation of histamine. Kallikrein, in addition, cleaves bradykinin from a serum protein substrate and thereby also releases histamine (Ch. 19). Administration of ε-aminocaproate, an inhibitor of plasmin activation of C1, reduces the frequency of attacks and antihistamines reduce their intensity.

Cobra Venom Factor. Anomalous immune reactions in C-deficient individuals help clarify the normal function of C. Since animals with genetic or congenital deficiences (above) are not always available, soluble Ab-Ag complexes, formed in vitro, have been injected to deplete normal C levels (with simultaneous administration of antihistamines to prevent anaphylaxis due to production of anaphylatoxin and massive release of histamine; see Aggregate anaphylaxis, Ch. 19). However, the most selective and prolonged depletion (of C3) is brought about with venom from the cobra (*Naja naja* or *N. haja*). Depletion of C3 blocks certain allergic reactions (Arthus) but not others (cytotropic anaphylaxis, delayed-type hypersensitivity; Chs. 19 and 20).

The active component, cobra venom factor or CoF (MW 140,000), forms a complex with a β-globulin in serum (probably factor B of the properdin system). The complex, like C3 "convertase" ($C\overline{4b,2a}$), cleaves C3 into a small (C3a) and a large (C3b) fragment. Intact C3 is thus eliminated and remains completely undetectable for 4 to 96 hours. Other C components are hardly affected.

CODA

The reaction of antibodies with antigens on cell surfaces is often followed by cell lysis and by extensive inflammatory changes in tissues (vasodilation, increased vascular permeability, accumulation and increased phagocytic activity of leukocytes). These changes are due to complement (C), a set of eleven proteins (C1q, C1r, C1s, C2 . . . C9) that normally exist in inactive form in serum. Except for C1q (which initiates the classical sequence of reactions), the other C proteins are converted one after the other from inactive to activated forms, some of which are enzymes whose substrate is the next protein in the sequential chain reaction. An alternative set of reactions, involving another group of four serum proteins, the properdin system, can bypass the early proteins (C1q, C1r, C1s, C2, C4) and trigger the sequence by activating C3, which then fires the other components (C5 to C9). Aggregated immunoglobulins of some classes (e.g., IgG-1, IgG-2, IgG-3, IgM) can activate the entire sequence via C1q; but others (e.g., IgG-4, IgE) seem to activate the sequence via the alternative or bypass (properdin) system.

Many of the inflammatory (phlogistic) effects of the C sequence can be accounted for by individual components or their activated products: C3b on the surface of a cell (or virus?) is a powerful opsonic agent, increasing susceptibility to phagocytosis by polymorphonuclear leukocytes and macrophages; C3a and C5a cause vasodilation and increase capillary permeability (anaphylatoxin activity); C5a, C5b·6·7, and probably C3a, attract leukocytes (chemotactic activity); C8 and especially C9 can cause severe damage to cell membranes (cytotoxic activity), resulting in lysis of some cells.

The cytotoxic effects of activated C are almost entirely confined to cells with antibodies bound to the cell membrane (sensitized cells). Neighboring cells, without bound antibody, are spared because the activated C proteins deteriorate rapidly in solution.

The importance of the C system to host defenses against microbial pathogens is brought out by the increased susceptibility to infections of rare persons with depressed serum levels of C3 and C5. But many humans, mice, and rabbits with heritable deficiencies in C2, C5, or C6 seem not to be especially troubled by infections, perhaps because a normal level of C3 is sufficiently protective. Though largely beneficial, the C system can also cause damage in tissues around otherwise benign Ab-Ag aggregates, by causing local accumulation of leukocytes whose degeneration releases destructive lysosomal enzymes. Thus C amplifies both the protective effects of Abs and their occasional capacity to cause tissue damage (hypersensitivity).

SELECTED REFERENCES

BOOKS AND REVIEW ARTICLES

ALPER, C. A., and ROSEN, F. S. Genetic aspects of the complement system. *Adv Immunol 14*:252 (1971).

GIGLI, I., and AUSTEN, K. F. Phylogeny and function of the complement system. *Annu Rev Microbiol 25*:309 (1971).

KINSKY, S. C. Antibody-complement interaction with lipid model membranes. *Biochim et Biophys Acta 265*:1 (1972).

LEPOW, I. H. Biologically Active Fragments of Complement. In *Progress in Immunology*. (D. B. Amos, ed.) Academic Press, New York, 1972.

MAYER, M. M. Highlights of complement research during the past 25 years. *Immunochemistry 7*:485 (1970).

MÜLLER-EBERHARD, H. J. Biochemistry of Complement. In *Progress in Immunology*. (D. B. Amos, ed.) Academic Press, New York, 1972.

SPECIFIC ARTICLES

ALPER, C. A., ABRAMSON, N., JOHNSTON, R. B., JANDL, J. H., and ROSEN, F. S. Increased susceptibility to infection associated with abnormalities of C-mediated functions of the 3rd component of C (C3). *N Eng J Med 282*:349 (1970).

MAYER, M. M., MILLER, J. M., and SHIN, H. S. A specific method for purification of the 2nd component of guinea pig C and a chemical evaluation of the one-hit theory. *J Immunol 105*:327 (1970).

MILLER, M. E., and NILSSON, U. R. A familial deficiency of the phagocytosis-enhancing activity of serum related to a dysfunction of the 5th component of C. *N Eng J Med 282*:354 (1970).

MÜLLER-EBERHARD, H. J. The molecular basis of the biological activities of C. *Harvey Lect* (1973).

NAFF, G. B., PENSKY, J. and LEPOW, I. H. The macromolecular nature of the 1st component of human C. *J Exp Med 119*:593 (1964).

ROMMEL, F. A., GOLDLUST, M. B., BANCROFT, F. C., MAYER, M. M., and TASHJIAN, A. H. Synthesis of the 9th component of C by a clonal strain of rat hepatoma cells. *J Immunol 105*:396, (1970).

SANDBERG, A. L., OSLER, A. G., SHIN, H. S., and OLIVEIRA, B. Biological activities of guinea pig antibodies. II. Modes of C interaction with γ1 and γ2 immunoglobulins. *J Immunol 104*:329 (1970).

SHIN, H. S., SNYDERMAN, R., FRIEDMAN, E., MELLORS, A. and MAYER, M. M. Chemotactic and anaphylatoxic fragment cleaved from the 5th component of guinea pig C. *Science 162*:361 (1968).

chapter 19

ANTIBODY-MEDIATED (IMMEDIATE-TYPE) HYPERSENSITIVITY

ANAPHYLAXIS   529
   Cutaneous Anaphylaxis in Man   530
   Reagins and Blocking Antibodies   531
   IgE Immunoglobulins   531
   Prolonged Desensitization   535
   Cutaneous Anaphylaxis in the Guinea Pig   536
   Genetic Control of Reagin Production   537
   Generalized Anaphylaxis   538
   Pharmacologically Active Mediators   540
   Anaphylactic Responses in Isolated Tissues   543
   Aggregate and Cytotoxic Anaphylaxis   545
   Complement and Anaphylaxis   546
ARTHUS REACTION   546
SERUM SICKNESS SYNDROME   549
IMMUNE-COMPLEX DISEASES   551

For some years after the discovery of anti- toxins and antimicrobial antibodies (Abs) the immune response appeared to be purely protective. Though it was found soon there- after that the same mechanism could be acti- vated by innocuous substances, such as milk proteins, it probably came as a surprise when Portier and Richet showed, in 1902, that im- mune responses also possess dangerous po- tentialities.

While studying the toxicity of extracts of sea anemones these French investigators ob- served that dogs given a second injection, several weeks after the first, often became acutely ill and died with a few minutes. Richet called this response anaphylaxis (Gr. *ana* = against; *phylaxis* = protection) implying in- correctly that it represented an increase in susceptibility to a toxic substance rather than the expected increase in resistance.* Almost simultaneously, however, observers in the United States and in Germany noted similar responses in guinea pigs to widely spaced in- jections of nontoxic antigens (Ags); and with the increasing use of horse and rabbit antisera to treat various infectious diseases in man, diverse pathological consequences of the im- mune response soon become commonplace.

In an attempt to organize a chaotic set of observations von Pirquet introduced the term allergy (Gr., "altered action") to cover any

altered response to a substance induced by previous exposure to it. Increased resistance, called immunity, and increased susceptibility, called hypersensitivity, were then regarded as opposite forms of allergy. Through usage, however, **"allergy" and "hypersensitivity" have become synonymous: both refer to the altered state, induced by an Ag, in which pathological reactions can be subsequently elicited by that Ag, or by a structurally simi- lar substance.**

In previous chapters the administration of an immunogen to stimulate Ab formation was called immunization. Within the context of the allergic response, however, the immuno- gen or Ag is often referred to as the **allergen** or **sensitizer,** and immunization as sensitiza- tion; and the immunized individual, previ- ously called immune, is called sensitive or hypersensitive or allergic. It will also occa- sionally be useful to emphasize the distinction between the substance used to establish the allergic state ("inducer") and that used to evoke the allergic response ("elicitor").

Two Basic Mechanisms. Allergic responses were originally divided into two classes, immediate and delayed, on the basis of the lag in their appearance—several minutes after the administration of Ag in one, and several hours or even a few days in the other. These terms are still used, but they are now en- dowed with a different meaning. Not only the reactions that appear within minutes, but also some of the more slowly evolving ones, are mediated by freely diffusible Ab mol- ecules. To emphasize this common feature both are now called **immediate type** (indicat- ing that "immediate" is not to be taken literally). In contrast, the **delayed type** are those slowly evolving responses that are medi- ated by specifically reactive ("sensitized") T

---

* Magendie reported about 60 years earlier the sudden death of dogs repeatedly injected with egg albumin, and Flexnor noted shortly afterward that "animals that had withstood one dose of a foreign serum would succumb to a second dose given after the lapse of some days or weeks, even when this dose was not lethal for a control animal." These early discoverers of anaphylaxis were, however, overlooked; as often happens, valid observations were ignored until they could be accommodated within a conceptual framework.

lymphocytes rather than by freely diffusible Ab molecules; hence they are also called **cell-mediated hypersensitivity.** They constitute part of a larger group of reactions, called **cell-mediated immunity,** in which similar mechanisms are also involved in resistance to many infectious agents and to neoplastic cells.

In this chapter we consider the allergic reactions due to soluble Ab molecules, and in the next those due to sensitized lymphocytes. In allergy to drugs and in autoimmunity either reaction can be involved, and so a discussion of these clinically important subjects is postponed to the end of the next chapter.

Antibody-mediated Responses. The most important Ab-mediated responses are grouped in Table 19-1 on the basis of underlying mechanisms. The arrangement reflects the principle that mere **combination of Ab and Ag is seldom damaging unless the immune complexes trigger certain cells to release various mediators,** which serve as the immediate causes of pathological change. However, in special circumstances major aberrations follow the combination of Abs (and comple-

ment) with unusual Ags, without involving additional mediators (e.g., massive destruction of transfused red cells; Ch. 21).

ANAPHYLAXIS

Injection of a soluble Ag into a hypersensitive animal can cause an explosive response within 3 to 4 minutes. If the Ag is injected intravenously the response, called **systemic or generalized anaphylaxis,** can lead to shock, vascular engorgement, and asphyxia due to bronchial constriction; if death does not follow promptly recovery is complete within about 1 hour. If the Ag is injected into the skin the same type of reaction occurs in miniature form at the local site: called **cutaneous anaphylaxis,** it is characterized by transient redness and swelling, with complete return to normal appearance in about 30 minutes. Systemic and localized anaphylaxis occur not only in actively immunized ("sensitized") individuals, but also in those who are **passively sensitized** with certain antisera or purified Abs.

The basic mechanisms have been largely

TABLE 19-1. Antibody-Mediated Allergic Reactions

Protype	Examples	Mechanism	
		Activated cells	Mediators released
I. Anaphylaxis	Anaphylactic shock Wheal-and-erythema responses Hayfever Asthma Hives	Mast cells Basophils* (platelets in some species)	Low moleculer weight, e.g., histamine (see Fig. 19-8)
II. Serum sickness	Arthus reaction Serum sickness syndrome Immune-complex diseases (glomerulonephritis, ? rheumatoid arthritis, etc.)	Neutrophils*	High molecular weight, (lysosomal enzymes)
III. Reactions to transfused blood	Red cell incompatibilities (e.g., maternal-fetal, as in Rh disease; Ch. 21) Autoantibodies to some self-Ags (Ch. 20; e.g., to platelets, or to antihemophilic globulin, causing bleeding and purpura)	None	None

* The principal white blood cells (leukocytes) are polymorphonuclear granulocytes, monocytes (i.e., not fully differentiated macrophages), and lymphocytes. On the basis of affinity for various dyes the granulocytes are classified as neutrophils (>95%), basophils (ca. 1%), or eosinophils (ca. 1%).

illuminated by experimental and clinical studies of passive cutaneous anaphylaxis, which can be elicited simply and safely at multiple sites in the same individual, providing ideal opportunities for controlled observations on the nature of the mediating Abs and Ags.

CUTANEOUS ANAPHYLAXIS IN MAN

The response begins 2 or 3 minutes after Ag is injected into the skin of a sensitive person: itching at the injected site is followed within a few minutes by a pale, elevated, irregular wheal surrounded by a zone of erythema (hive or urticarium). This **wheal-and-ery-thema** response reaches maximal intensity about 10 minutes after the injection, persists for an additional 10 to 20 minutes, and then gradually subsides (Fig. 19-1).

Fig. 19-1. Cutaneous anaphylaxis (wheal-and-erythema response) in man. Fifteen minutes before the photograph was taken the subject was injected intradermally with 0.02 ml containing about 0.1 μg protein extracted from guinea pig hair. Note the irregularly shaped wheal, with striking pseudopodia. The surrounding erythema is not easily visible in the photograph. No reaction is seen at the control site where 0.02 ml of buffer alone was injected.

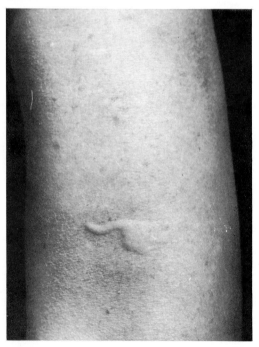

Atopy. A special group of persons, constituting about 10% of the population in the United States, is especially prone to hypersensitive responses of this type. These individuals readily become sensitive "spontaneously" (i.e., without deliberate immunization) to a variety of environmental Ags (often called **allergens),** such as airborne pollens of ragweed, grasses, and trees, and also to fungi, animal danders, house dust, and foods. When they inhale or ingest the appropriate allergen their response is prompt: most frequent and prominent among the manifestations are hayfever, asthma, and hives. The tendency to develop this form of allergy, called **atopy** (Gr., "out of place"), is heritable: it has been recognized in dogs and in cattle, and its genetic regulation is evident in inbred mouse strains (below).

Passive Transfer. Until the development of highly sensitive assays, such as passive hemagglutination (Ch. 15), the sera of atopic persons usually gave no detectable reactions with allergens in vitro. Nevertheless, these sera can, even after extensive dilution (1000-fold or more), sensitize passively the skin of normal persons. Passive sensitization is performed by injecting about 0.05 ml of serum (or serum dilution) from the sensitive donor into the skin (dermal layer) of a nonsensitive recipient. After 1 day, and up to as long as 6 weeks, injection of the corresponding Ag into the same skin site elicits the wheal-and-ery-thema response. To elicit the reaction it is usually necessary to allow a **latent period** of at least 10 to 20 hours after the injection of serum.

This transfer response is called the **Prausnitz-Küstner** or **P-K** reaction after those who first described it.* Patients are commonly tested for wheal-and-erythema responses to

---

* As described in 1921, Küstner was extremely sensitive to certain fish, but his serum gave no detectable reaction with extracts of these fish and did not sensitize guinea pigs for passive anaphylaxis. Prausnitz injected a small amount of Küstner's serum into a normal person's skin, and the injected site was tested 24 hours later with fish extract; the immediate appearance of a wheal-and-erythema response provided the basis for much of the clinical and experimental work on allergy of succeeding decades.

intradermal injections of extracts of plant pollens, fungi, food, animal danders, etc. to identify etiological Ags. P-K tests are sometimes used to avoid direct skin tests on young children or on adults with disseminated skin disease.

REAGINS AND BLOCKING ANTIBODIES

The appearance of atopic allergy depends upon the production of a special kind of Ab. Thus if ragweed extract is injected repeatedly into nonatopic human volunteers antiragweed Abs (predominantly of IgG class) appear in serum and may be detected by conventional assays, such as passive hemagglutination. However, these Abs are incapable of sensitizing human skin for wheal-and-erythema responses; instead, they combine with Ag and specifically **block** its ability to evoke this response in a sensitive person's skin (or in a normal person's skin at a P-K site). These **blocking antibodies** differ substantially from the **skin-sensitizing Abs, called reagins,** that

cause the wheal-and-erythema reaction (Table 19-2). Thus the reagins are heat-labile and do not cross the human placenta, whereas blocking Abs (like IgGs in general) are heat-stable and readily cross the placenta. Most important, the sensitizing Abs persist at passively prepared (P-K) skin sites for up to 6 weeks, whereas blocking Abs diffuse away almost completely within 1 to 2 days. In addition to reagins, serum from atopic individuals usually contains some blocking Abs of the same specificity. Until the blocking Abs diffuse away they tend to competitively inhibit the reaction of injected Ag with reagin at P-K sites; the latent period in the P-K reaction probably also reflects the time required to fix reagin to tissue receptors (see Mast cells, below).

IgE IMMUNOGLOBULINS

Careful studies by the Ishizakas showed that rabbit antisera prepared against a reagin-rich serum fraction could precipitate and remove

TABLE 19-2. Comparison of Human Reagins and Blocking Antibodies To Pollen Antigens*

	Reagins	Blocking antibodies
Immunoglobulin class	IgE	IgG (predominantly)
Activity in Prausnitz-Küstner (P-K) test	Yes†	No (inhibits)
Persistence in human skin (P-K test)	Up to 6 weeks	Up to 2 days
Stability		
To heat (56°, 4 hr)	Labile	Stable
To sulfhydryls (0.1 M 2-mer-captoethanol)	Labile	Stable
Transfer across human placenta	No	Yes
Passive sensitization of guinea pigs for anaphylaxis	No	Yes†
Detection in in vitro assays	Radioimmunoassay	Hemagglutination and others
Molecular weight‡	185,000	150,000
Sedimentation coefficient ($S°_{20, w}$)	8.2	6.6
Carbohydrate‡	12%	3%
No. of amino acid residues per heavy chain‡	ca. 550	ca. 440
Heavy chains	ε	γ (predominantly)
Light chains	κ, λ	κ, λ

* Highly purified protein Ags have been isolated from ragweed and grass pollen extracts by ion-exchange chromatography. Several active fractions have been obtained from each extract, and different fractions are active in different persons. As little as 10^{-4} μg of some ragweed fractions evoke specific wheal-and-erythema responses. Fatal anaphylaxis has occurred on very rare occasions in response to skin tests with small amounts of crude extracts. Indeed, pollen antigen may be as potentially lethal (on a weight basis) for a pollen-sensitive person as botulinus toxin is for humans in general.

† Reagins are homocytotropic; blocking Abs are heterocytotropic (below).

‡ Values for reagins are those for an IgE myeloma protein.

reaginic activity from human atopic sera, even after the rabbit antisera had been completely freed of Abs to the then-known Ig classes by absorption with representative myeloma proteins (IgGs, IgA, IgM, IgD). They concluded that reagins belong to another Ig class, and this was then promptly verified by Johansson and Bennich's independent discovery in Sweden of an unusual human myeloma protein that also did not react with Abs to any of the then-known Ig classes, because of distinctive antigenic features of its heavy chain. An exchange of monospecific antisera between the two groups established that reagins and the rare myeloma protein belong to the same novel class, called IgE (Fig. 19-2).

A second IgE myeloma protein was subsequently identified in the United States, and myeloma cells from the Swedish patient were adapted to tissue culture where they continued for some time to produce their IgE. With these two crucial human proteins as immunogens, large amounts of anti-IgE sera have now

been prepared in rabbits and in goats. After absorption with human κ and λ light chains (to render them monospecific for ε chains) these antisera provide a key reagent for measurement of IgE in human sera and for diagnostic tests of human atopic allergy.

Serum IgE concentrations are measured by a form of radioimmunoassay in which anti-IgE is attached covalently to particles of an inert adsorbent which are then mixed with a standard amount of radioactive *IgE (labeled with ^{125}I) and the human serum to be tested. Unlabeled IgE in the test serum competitively reduces the specific binding of *IgE; hence radioactivity associated with the washed beads decreases in proportion to the serum IgE concentration (Fig. 19-3).

Another radioimmunoassay measures IgE antibodies of a particular specificity, e.g., to dog dander (epithelial scales). Protein extracts of the dander are coupled to Sephadex particles which are then trapped in small cellulose discs. The discs are incubated with about 0.05 ml of a patient's serum, washed, treated with radioactive anti-IgE (labeled with ^{125}I), washed, and counted. A positive test (adherent radioactivity) can detect a few nanograms

Fig. 19-2. Antigenic identity of reagins and an IgE myeloma protein. A reagin-rich fraction of human atopic serum was placed in the center well and antisera for each of the human Ig classes were placed in peripheral wells. Antisera to reagin and to the IgE myeloma gave reactions of identity. [From Ishizaka, K. *Immunoglobulins:* Biologic Aspects and Clinical Uses. E. Merler, ed. *Natl Acad Sciences,* Washington, D.C. (1970).]

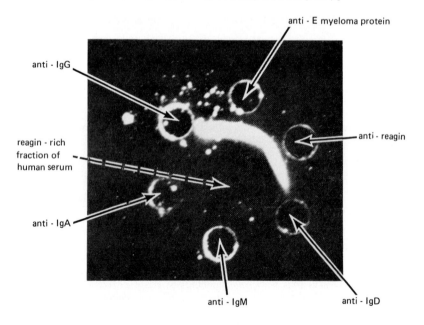

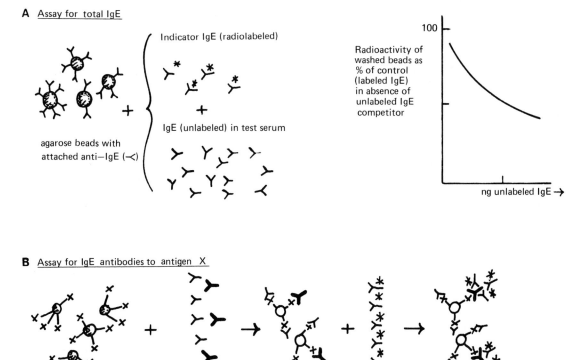

A Assay for total IgE

Indicator IgE (radiolabeled)

IgE (unlabeled) in test serum

agarose beads with attached anti—IgE (—≺)

Radioactivity of washed beads as % of control (labeled IgE) in absence of unlabeled IgE competitor

100

ng unlabeled IgE →

B Assay for IgE antibodies to antigen X

agarose beads with attached antigen X

antiserum with anti-X Abs (some are IgE, boldface)

radiolabeled anti-IgE

Radioactivity on washed beads

ng Ige Abs

Fig. 19-3. Assays for total IgE concentration in serum **(A)** and for IgE antibodies of a particular specificity (anti-X) **(B)**. IgE molecules (in heavy type) function as Ag in the assay at top, and as both Ag and Ab in the one below.

per milliliter of IgE Abs of a particular specificity (Fig. 19-3).

In sera of atopic persons the IgE levels are often three or four times or more above the upper limit found in normal human sera, which is about 350 ng/ml. The extraordinarily low level of these biologically potent molecules (about 30,000-fold lower than the average normal concentration of IgG) ex-

plains why they were previously not detected in atopic sera that were highly active in P-K skin tests.

Chemical Properties of IgE. The IgE molecule has four chains: one pair of heavy ε chains plus one pair of κ or one pair of λ light chains. The ε chains are distinctive: they are longer than γ (approximately 540 and 440 amino acid residues, respectively), as though ε has an additional domain in its constant region (Ch. 16, Amino acid sequences), and they also have about three times more carbohydrate (Table 19-2). The differences account for the higher molecular weight of IgE than of IgG molecules. IgE and other Ig classes are compared further in the Appendix of Chapter 16.

Affinity for Mast Cells. IgE Abs have high affinity for receptors in skin: despite their extremely low concentration in serum they become attached to skin in P-K tests, and they can remain anchored for many weeks. The affinity apparently derives from sites in the Fc domain. Thus the Fc fragment of IgE myeloma protein (but not the Fab fragments) can specifically block P-K reactions, by competitively displacing reaginic Abs from skin receptors.

Immunofluorescence and radioautographs with radioactive IgE show that the tissue receptors are on **mast cells,** which are found in close association with capillaries in con-nective tissues throughout the body. These cells are distinguished by their high content of histamine, which is concentrated in large cytoplasmic granules and is secreted as one of the key triggering events in anaphylaxis (see below, Fig. 19-9 and Chemical mediators). IgE also binds to **basophilic leukocytes** (basophils; Fig. 19-4). These cells make up ca. 1% of all blood leukocytes; they resemble mast cells in appearance and in histamine content and secretion. Other leukocytes do not bind IgE or contain histamine.

Homocytotropic vs. Heterocytotropic Antibodies. The immediate-type reactions mediated by special Abs that attach to mast cells are sometimes called **cytotropic anaphylaxis.** Because human IgE molecules (reagins) bind to human (and monkey) mast cells, but not to those of other species, they are called **homocytotropic antibodies.** Human blocking Abs, in contrast, are **heterocytotropic:** they can bind (fortuitously) to mast cells of some phylogenetically distant species (e.g., guinea

Fig. 19-4. Human IgE on the surface of a human basophil. Washed white blood cells were incubated successively (with intervening washes) with human IgE (a myeloma protein), burro antihuman IgE, hybrid 7S Abs (Ch. 16) in which one combining site was specific for burro Ig and the other for ferritin, and finally with ferritin—the iron-rich, electron-dense particles seen under brackets on the cell surface (see Ferritin-labeled Abs, Ch. 15). Basophil granules (BG is a typical one) resemble the histamine-containing granules of tissue mast cells (Fig. 19-9); basophils are the only blood leukocytes that bind IgE and contain and secrete histamine in response to specific binding of Ag or anti-IgE to the cell surface's IgE. Electron micrograph at ×77,500. [From Sullivan, A. L., Grimley, P. M., and Metzger, H. *J Exp Med 134:* 1403 (1971).]

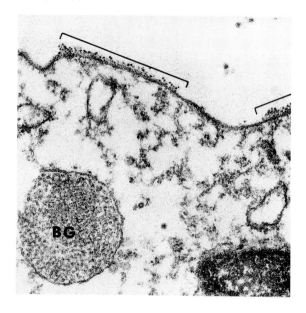
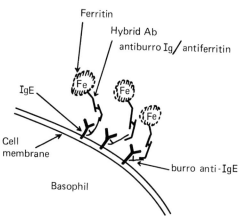

Fig. 19-5. Degranulation of mast cells and basophils and secretion of vasoactive amines by cross-linking IgE on the cell surface. The surface Abs are bound noncovalently via sites in Fc domain.

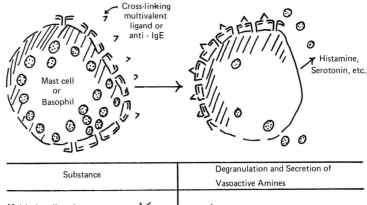

Substance		Degranulation and Secretion of Vasoactive Amines
Multivalent ligand		+
Univalent ligand		− (in excess can inhibit activity of multivalent ligand)
Anti-IgE (bivalent)		+
Anti-IgE F(ab′)$_2$ (bivalent)		+
Anti-IgE Fab′ (univalent)		− (in excess can probably inhibit activity of bivalent Ab or F(ab′)$_2$)

pigs) but not to human cells. In accord with these differences, human homocytotropic Abs can sensitize human (or monkey) skin and other tissues, but not those of guinea pigs, while human heterocytotropic Ags can sensitize guinea pigs but not humans.

Cross-linking of IgE. Aggregation of IgE molecules already bound to the mast cell surface probably initiates anaphylaxis (Fig. 19-5). Thus, when a skin site is passively sensitized with antihapten reagins the corresponding multivalent ligand can elicit a wheal-and-erythema response, but univalent ligands are usually specifically inhibitory, just as they competitively block the precipitin reaction in vitro.*

Other means of aggregating IgE on the cell surface are also effective. For example, the injection of anti-IgE Abs or their bivalent F(ab′)$_2$ fragments (but not their monovalent Fab′ fragments) can provoke wheal-and-erythema responses ("reverse cutaneous anaphylaxis"; Fig. 19-5). An injection of the aggregated Fc fragments of IgE myeloma

* Certain univalent ligands seem to elicit cutaneous anaphylaxis. It is unclear whether these ligands aggregate in tissues, and thus become functionally multivalent, or whether there is a special form of cutaneous anaphylaxis that can be evoked with non-cross-linked Ab-ligand complexes.

proteins (but not the monomeric Fc fragment) can also elicit the response, doubtless because they also bind to mast cells.

PROLONGED DESENSITIZATION

Atopic individuals form blocking Abs as well as reagins, especially after repeated injections of the Ag. Accordingly, such persons are commonly immunized ("desensitized") by repeated injections of small, increasing amounts of allergen, at intervals (e.g., weekly) and in doses that avoid systemic anaphylactic reactions. The level of blocking Abs, but not of reagins, often rises considerably. However, therapeutic benefits are not consistently evident, nor are they regularly correlated with the titers of blocking Abs.

Blocking Abs can be measured in atopic sera after heating (56°, 4 hours) to inactivate reagins; dilutions of the heated serum are then mixed with various amounts of Ag (e.g., pollen extract) and injected into passively sensitized (P-K) skin sites. The titer of blocking Ab is taken as the highest dilution of heated serum that inhibits the P-K reaction with a standard amount of allergen, or as the maximal amount of allergen that can be inhibited.

IgE and Intestinal Parasitism. While IgEs have obvious pathological effects, it is likely that they also have beneficial ones, contributing to their evolutionary development. A hint of benefits is suggested by the exceedingly high serum levels of IgE in persons with chronic parasitic infections: for instance, values of 3000 to 10,000 ng/ml (ca. 30 times average normal levels) are found in Africans and others with chronic intestinal roundworm infestations. With a relative abundance of IgE-producing plasma cells in the normal intestine* it is possible that IgE is especially effective in controlling intestinal parasites. The same relative abundance in the respiratory tract (which is derived embryologically from the fetal gut) probably contributes to the reactions underlying asthma and hayfever. Individuals that lack ability to synthesize IgE should provide opportunities to clarify the physiological role of IgE molecules.

---

*IgE-producing plasma cells are conspicuous in human surgical specimens of tonsils, adenoids, bronchial and intestinal mucosa; they are rare in spleen and lymph nodes. However, even in respiratory and intestinal tracts IgE-producing cells are greatly outnumbered by IgA producers.

CUTANEOUS ANAPHYLAXIS IN THE GUINEA PIG

Cutaneous anaphylaxis can also be elicited in actively or passively sensitized guinea pigs. **Passive cutaneous anaphylaxis (PCA)** has been especially well developed by Z. Ovary into a powerful model system for evaluating abilities of various Abs and Ags to elicit anaphylactic responses. PCA and the human P-K reactions are fundamentally the same; but special measures are taken to increase visibility of the response in animal skin.

In PCA an antiserum (or purified Ab) is injected intradermally. After a latent period of several hours the corresponding Ag is injected intravenously along with a dye, such as Evans blue, that is strongly bound to serum albumin. Hence, as serum proteins rapidly pour into the dermis at the site of the reaction the response appears as an irregular circle of

Fig. 19-6. Passive cutaneous anaphylaxis in the guinea pig. In **A** the guinea pig was injected intradermally at three sites with 0.1 ml containing 1) 100 μg rabbit anti-chicken ovalbumin (Ea), 2) 10 μg anti-Ea, and 3) buffered saline. Four hours later 1.0 ml containing 2 mg Ea and 5 mg Evans blue was injected intravenously; the photo was taken 30 minutes later. Note blueing at 1 and at 2, and absence of blueing at the control site (3). In **B** a similar sequence was followed except that 30 minutes after the intravenous injection of Ea the animal was sacrificed and skinned. The photo was taken of the skin's undersurface. The amount of rabbit anti-Ea injected initially was 100μg (at 4), 10 μg (at 5), 1 μg (at 6), and 0.1 μg(at 7). The control site, which did not turn blue (8), had been injected with buffered saline. Another site (not shown) had been injected with 0.01 μg anti-Ea; it also failed to react.

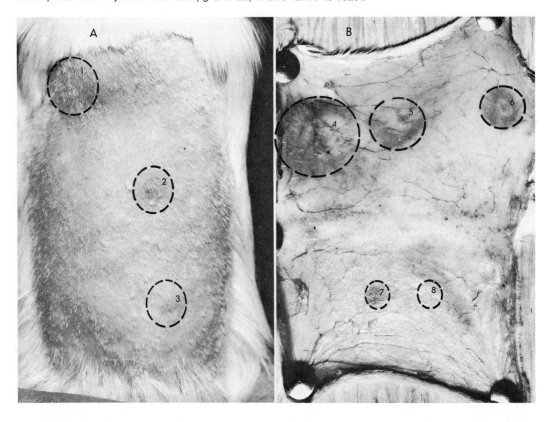

stained skin; the area is an index of the reaction's intensity (Fig. 19-6).

Two kinds of guinea pig Abs produce PCA reactions in guinea pigs: 1) reagins and 2) IgG molecules of the γ1 subclass (Fig. 19-7). Both bind to mast cell surfaces and are **homocytotropic,** i.e., they sensitize guinea pigs but not other species. The reagins resemble human IgE: they are present at trace levels in serum (nanograms per milliliter), they persist at injected skin sites for weeks, and their skin-sensitizing activity is destroyed by sulfhydryl compounds and by heat (56°, 4 hours); they also cross-react with rabbit antisera to human IgE. In contrast, the homocytotropic γ1 subclass of IgG is present at high levels in serum (milligrams per milliliter), is sulfhydryl- and heat-stable, and persists at injected skin sites for only 1 to 2 days.

The other major subclass of guinea pig IgG, called γ2 (Fig. 19-7), is heterocytotropic: molecules of this subclass can sensitize mouse skin, but not guinea pig skin, presumably because they can fortuitously bind to and activate mouse mast cells.

PCA can be used to measure each of the foregoing Abs in a single sample of guinea pig serum. The response in mice measures γ2 Abs (heterocytotropic), whereas PCA in guinea pigs measures the homocytotropic Abs: the Ag is injected into passively sensitized guinea pigs after a latent period of 3 or 4 hours to measure γ1 molecules, and after 2 to 3 days to measure reagins. Alternatively, with PCA in guinea pigs, γ1 molecules alone are measured in a heated test serum and γ1 plus reagins in an unheated sample.

Many other species (rat, rabbit, dog, mouse) also have two classes of homocytotropic Abs (reagins and a heat-stable subclass of IgG, e.g., γ1 in guinea pig and mouse and IgGa in rat); in man, however, only reagins have been identified so far. Abs of the IgM and IgA classes do not sensitize animals of the same or other species for anaphylactic responses.

GENETIC CONTROL OF REAGIN PRODUCTION

Inbred mouse strains differ in ability to produce reagins, indicating that their susceptibility to anaphylaxis, like atopy in man, is heritable. Such strains differ not only in reagin production but in over-all responses to low doses of Ag (0.1 to 1.0 μg/mouse), to which some strains make no Abs while others pro-

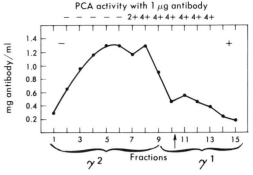

Fig. 19-7. Separation of homocytotropic from other Abs of the same specificity. Purified 7S guinea pig Abs, specific for the 2,4-dinitrophenyl (Dnp) group, were subjected to electrophoresis in starch (arrow marks point of application). Fractions eluted from 0.5-inch cuts of the starch block were tested for ability to mediate PCA in guinea pigs. The more anionic anti-Dnp molecules, called γ1, mediated PCA, whereas the less anionic antibodies, called γ2, did not. With a rabbit antiserum to guinea pig Igs, the γ1 and γ2 fractions are also distinguished antigenically. [Redrawn from Ovary, Z., Benacerraf, B., and Bloch, K. J. *J Exp Med 117*:951 (1963).]

duce both reagins and other Abs. Atopic persons similarly seem to differ from non-atopics primarily in ability to produce Abs (reagins and others) in response to natural exposure to trace amounts of environmental Ags, rather than exclusively in ability to form reagins.

Mouse strains that do not respond to low doses of Ag can produce reagins (and other Abs) in response to high doses (100 μg/mouse). Similarly, **with sufficient immunization nearly all persons can be actively sensitized to give wheal-and-erythema responses,** which can be elicited by the appropriate Ag in over 90% of persons injected with a sufficient quantity of horse serum (see Serum sickness syndrome, below), and in essentially all those recovering from pneumococcal pneumonia. Indeed, certain Ags, like extracts of *Ascaris* (a common nematode parasite in the intestine), seem able even with minimal immunization to sensitize nearly all persons to give wheal-and-erythema reactions.

One mouse strain (SJL) seems unable to form reagins to several Ags while able to produce Abs of other classes. Thus unlike Ir genes, which control ability to produce Abs (or cell-mediated immunity) of a particular specificity (Chs. 17 and 20), the "reagin genes" seem to regulate the ability to synthesize Abs of a particular class (IgE).

In crosses between the low-reagin strain and another strain all progeny formed high reagin titers, showing that reagin production is dominant. However, the progeny of back-crosses between the hybrids and the poor-producing parents (SJL) segregated into good, poor, and intermediate producers, suggesting that more than one gene controls reagin production.

GENERALIZED ANAPHYLAXIS

Systemic anaphylaxis in man is a rare event brought on occasionally in hypersensitive individuals by insect stings (especially bees, wasps, hornets), or by injection of horse serum or (more commonly at present) penicillin. Because of this hazard patients who receive foreign proteins (e.g., horse antitoxin) or pencillin should first be questioned about previous allergic reactions, and sometimes tested for wheal-and-erythema responses: when horse serum is used, greatly diluted serum (1000-fold or more) should be injected, for even minute amounts of undiluted serum (ca. 0.05 ml) can precipitate systemic anaphylaxis. (For penicillin hypersensitivity and skin tests, see Ch. 20.)

Fundamental mechanisms in systemic anaphylaxis are the same as in the cutaneous wheal-and-erythema response, but certain features of the generalized reaction, which has been studied mostly in the guinea pig, are illuminating.

Mode of Administration of Antigen. Anaphylaxis depends not only on the number of Ab-Ag complexes formed in tissues but also on the **rate** at which they form, for the complexes act by causing the release of pharmacologically active mediators that are rapidly degraded (below). Intravenous injection of Ag is therefore especially effective. Inhalation of Ags dispersed in aerosols can also provoke fatal shock, but responses elicited by subcutaneous and intraperitoneal injections come on more slowly and are less often fatal.

Fixation of Antibodies. Less Ab is needed for passive anaphylaxis if a latent period intervenes between injection of antiserum and of Ag: for example, 0.18 mg of anti-egg albu-

min (EA) rendered guinea pigs uniformly susceptible to fatal shock when challenged 48 hours later, whereas 12 mg was required if the Ag was injected immediately after the antiserum. The latent period is needed both for binding of cytotropic Abs to mast cells and for reducing the circulating level of unbound Abs, which lessen shock by competing with cell-bound Ab for the Ag. (Shock elicited by simultaneous injection of large amounts of Ab and Ag probably involves somewhat different mechanisms; see Aggregate anaphylaxis, below.)

The necessity for mast cell binding accounts for the inability of many Abs to mediate passive anaphylaxis. For example, guinea pigs are sensitized only by guinea pig IgE and $\gamma 1$, by human IgG-3, and -4 (but not by IgG-2; see Ch. 16, Appendix), and not by any Abs from chickens, goats, cattle, and horses.

Reverse Passive Anaphylaxis. Passive anaphylaxis can also be evoked by reversing the order of injections if the Ag, which is now injected first, is itself an Ig of the type that is readily bound to guinea pig mast cells (such as rabbit IgG). After a latent period the intravenous injection of antiserum, specific for the Ig used as antigen, can then cause anaphylaxis. This procedure, reverse passive anaphylaxis, is not effective with other Ags because they do not bind to mast cells. Reverse passive cutaneous anaphylaxis can be similarly carried out. It is used occasionally to evaluate an Ig's ability to bind to mast cells: the Ig under test is injected into a normal guinea pig's skin, and then antiserum to the Ig (plus blue dye) is injected intravenously.

Quantities of Antibody and Antigen Required for Anaphylaxis. The levels of Ag required are substantially greater than those necessary for precipitation in vitro: guinea pigs sensitized with 180 μg of anti-egg albumin (EA) require for a uniformly fatal response over 500 μg of EA or ca. 25-fold more than is usually needed for maximal precipitation of this amount of Ab in the EA/anti-EA precipitin reaction. Much of the injected Ag probably never has a chance to react with Abs in vivo, because it is taken up by phagocytic cells or excreted. Ags that form large complexes with circulating, soluble Abs tend to be rapidly phagocytized, and are also not efficient in provoking anaphylaxis. In fact, as suggested above, high levels of circulating Abs may protect against anaphylaxis because they compete with mast-cell-bound Abs for the Ag. Thus when an animal is passively sensitized with a small amount of antiserum and then given a sufficiently large dose of the same antiserum immediately before the Ag, fatal shock can be replaced by mild symptoms.

Acute Densensitization. Because the **speed** of complex formation determines whether anaphylaxis will occur (see above), shock can be prevented by administering Ag slowly. For example, if 100 μg of a particular Ag (injected intravenously) is required to provoke fatal shock, the same quantity given in 10 divided doses at 15-minute intervals would not elicit shock. Moreover, if the full dose were then given all at once shortly after the last small injection shock would probably still not be elicited, presumably because the supply of reactive Ab would have been depleted.

Densensitization by repeated, closely spaced injections of small doses of Ag is often resorted to clinically when it becomes necessary to administer a substance, such as penicillin or horse antiserum, to a person known or suspected to be intensely allergic to it. The procedure is effective but requires great care to avoid anaphylaxis and it has only temporary value. Several weeks afterward hyper-

sensitivity is likely to be fully restored, in contrast to the densensitization based on formation of protective ("blocking") Abs (Prolonged densensitization, above).

Species Variations. Guinea pigs are preferred for the study of anaphylaxis because they react uniformly and intensely. However, anaphylaxis has also been provoked in many other mammals, in fish, and in chickens, and it can probably be elicited in all vertebrates. The pathological manifestations differ in various species (Table 19-3) and even, as suggested above, when the Ag is injected by different routes. In a sensitized guinea pig, for example, intravenous injection leads to respiratory distress due to constriction of bronchi, and at autopsy the lungs appear bloodless and are greatly distended with air; whereas subcutaneous or intraperitoneal administration produces primarily hypotension and hypothermia, and death occurs only after many hours, with engorged blood vessels in abdominal viscera as the main pathological finding. The differences are probably due

TABLE 19-3.   Anaphylaxis in Different Species

Species	Principal site of reaction (shock organ)	Pharmacologically active agents implicated	Principal manifestations
Guinea pig	Lungs (bronchioles)	Histamine Kinins SRS-A	Respiratory distress: bronchiolar constriction, emphysema
Rabbit	Heart Pulmonary blood vessels	Histamine Serotonin Kinins SRS-A	Obstruction of pulmonary capillaries with leukocyte-platelet thrombi; right-sided heart failure; vascular engorgement of liver and intestines
Rat	Intestines	Serotonin Kinins	Circulatory collapse; increased peristalsis; hemorrhages in intestine and lung
Mouse	?	Serotonin Kinins	Respiratory distress; emphysema; right-sided heart failure; hyperemia of intestine
Dog	Hepatic veins	Histamine Kinins ? Serotonin	Hepatic engorgement; hemorrhages in abdominal and thoracic viscera
Man	Lungs (bronchioles) Larynx	Histamine ? Kinins SRS-A	Dyspnea; hypotension; flushing and itching; circulatory collapse; acute emphysema; laryngeal edema; urticaria on recovery

Based mostly on Austen, K. F., and Humphrey, J. H. *Adv Immunol 3*:1 (1963).

mostly to differences in distribution or reactivity of released pharmacologically active mediators (Table 19-3).

PHARMACOLOGICALLY ACTIVE MEDIATORS

Following Dale's observations, in 1911, that injections of histamine duplicate manifestations of anaphylaxis, a number of vasoactive substances have been found to be released from tissues in response to Ab-Ag complexes. The direct action of these substances on blood vessels and smooth muscle accounts for nearly all manifestations of anaphylaxis.

Of the several active substances so far identified, two (histamine and serotonin) preexist in cells and are promptly released by appropriate Ab-Ag complexes, while the kinins and "SRS-A" are produced only after the complexes are formed. The main properties of these mediators are reviewed below (Fig. 19-8).

1) **Histamine,** formed by decarboxylation of L-histidine, is distributed widely in mammalian tissues, particularly in granules of connective tissue mast cells (which are especially abundant near blood vessels) and basophilic leukocytes of blood. As noted before (Affinity for mast cells, above), the mast cells and basophils bind on their surface only those Igs that cause anaphylaxis: human cells bind IgE (via specific sites of the Fc domain) but not other human Igs. (Mast cells can also bind appropriate heterocytotropic Abs.) The addition of Ag to tissues from a sensitized individual leads to discharge of mast cell granules and release of histamine (Fig. 19-9).

The **antihistamines** block the anaphylaxis-like effects of histamine injected in guinea pigs. (These drugs are less effective, however, against true anaphylaxis because they do not antagonize the other pharmacological mediators.) Further evidence for the role of histamine is provided by the decline of histamine in tissues, and its rise in plasma, during

Fig. 19-8. Substances that mediate anaphylaxis. Kinins are given with conventional abbreviations for amino acids with terminal α-NH$_2$ at left.

MEDIATOR	STRUCTURE	SOURCE	PROPERTIES USED FOR IDENTIFICATION
Histamine	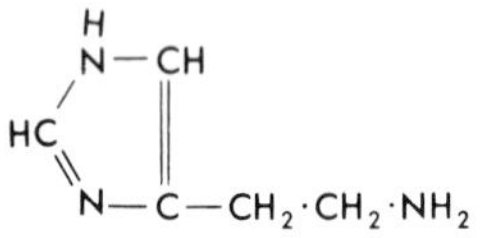 $\begin{array}{c} H \\ N-CH \\ HC \quad \parallel \\ N-C-CH_2 \cdot CH_2 \cdot NH_2 \end{array}$	Mast cells Basophils Platelets Others?	Contracts guinea pig ileum; inhibited by antihistamines
Serotonin	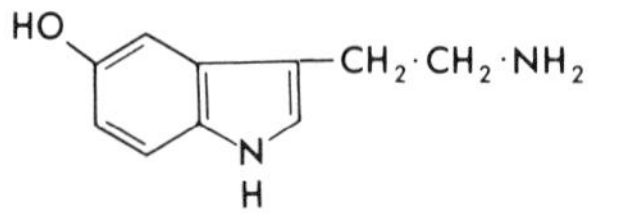 HO— [indole ring] —CH$_2$·CH$_2$·NH$_2$	Enterochromaffin cells Mast cells Platelets	Contracts guinea pig ileum and rat uterus; inhibited by lysergic acid
Kinins			
Bradykinin	Arg·pro·pro·gly·phe·ser·pro·phe·arg	Kininogen (α-globulin, plasma)	Contracts rat uterus; destroyed by chymotrypsin
Lysyl-bradykinin	Lys·arg·pro·pro·gly·phe·ser·pro·phe·arg	Tissues	
Methionyl-lysyl-bradykinin	Met·lys·arg·pro·pro·gly·phe·ser·pro·phe·arg	Tissues	
SRS-A	Acidic lipid, MW $\simeq$ 400	? Mast cells	Contracts human bronchiole; no effect on rat uterus; not inhibited by antihistamines or destroyed by chymotrypsin

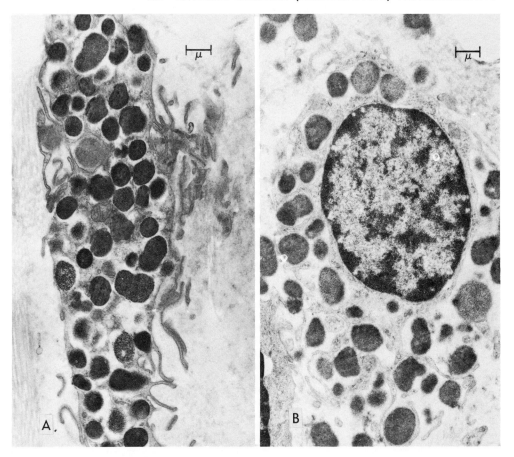

Fig. 19-9. Electron micrographs of mast cells from rat dermis. The intact cell **(A)** contains small, dense granules, each about the size of a mitochondrion. Mitochondria, which are generally scarce in mast cells, are not visible. The nucleus also is not visible in this section. The degranulating cell **(B)** contains larger, paler granules. The release of granules, associated with secretion of histamine, involves fusion of the membrane surrounding each granule with the cell membrane, releasing swollen granules into the extracellular space. ×7000. [Courtesy of S. L. Clark, Jr.; based on Singleton, E. M., and Clark, S. L., Jr. *Lab Invest 14*:1744 (1965).]

anaphylaxis. In addition, sensitized animals that are temporarily depleted of histamine by certain drugs ("histamine liberators") are not susceptible to fatal shock: when their histamine levels are restored, their susceptibility to anaphylaxis returns.

The mast cell granules that contain histamine also contain heparin, and this acidic mucopolysaccharide is responsible for the characteristic metachromatic staining with some basic dyes, e.g., toluidine blue. Heparin is released with histamine, in certain species: dogs undergoing anaphylactic shock have incoagulable blood. However, heparin does not account for the more important manifestations of anaphylaxis.

Species vary widely in their susceptibility to histamine; man and the guinea pig are exquisitely sensitive, while the mouse and rat are insensitive (Table 19-3).

2) **Serotonin** (5-hydroxytryptamine), formed by decarboxylation of L-tryptophan (Fig. 19-8), dilates capillaries, increases capillary permeability, and contracts smooth muscles in susceptible species. It is found mainly in blood platelets, intestinal mucosa, and brain.

Ab-Ag complexes cause release of serotonin in vitro from platelets in most species, and from mast cells in the mouse and rat; but it is not released

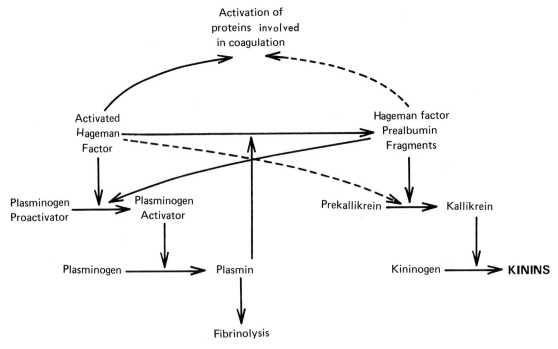

Fig. 19-10. Scheme for production of kinins. The interrelations between blood coagulation and kinin production extends to the complement proteins (C), which can be activated by plasmin and by kallikrein (see Ch. 18, Blood clotting and C). For the structure of kinins, see Figure 19-8. [Based on Kaplan, A. P. and Austen, K. F. *J Exp Med 136*:1378 (1972).]

from the brain during allergic reactions. Its effects are inhibited by lysergic acid diethylamide and by reserpine. Rats and mice are highly susceptible to serotonin, whereas humans and rabbits are highly resistant. Serotonin is probably not important in human anaphylaxis.

3) **Kinins** are basic peptides (Fig. 19-8) formed as endproducts of the sequential action of several plasma proteins (Fig. 19-10). The first, Hageman factor (HF), is activated by various negatively charged substances (footnote, p. 545), including perhaps certain Ab-Ag complexes, initiating the reaction sequence that leads to blood clotting. One component of this sequence, plasmin, cleaves from HF a fragment that converts another proenzyme, prekallikrein, into **kallikrein,** * which cleaves the basic nonapeptide **brady-**

kinin from an α-globulin in plasma (Fig. 19-10).

Tissue kallikreins form the same peptide with a lysine or methionyllysine at the N terminus; aminopeptidases in blood remove the additional N-terminal residues, converting these kinins into bradykinin. Kallikrein also splits (via C1 acting on C4, C2; Ch. 18) another bradykinin-like peptide from complement; called **C-kinin,** the fragment has somewhat similar activities to bradykinin but differs in being susceptible to destruction by trypsin. Several mechanisms prevent dangerous accumulation of kinins, which are powerful hypotensive agents as well as inducers of bronchiolar constriction. Their activities are abolished, for instance, by 1) a plasma carboxypeptidase that removes the C-terminal arginine (see Fig. 19-8), and 2) an inhibitor of activated first component of complement (Ch. 18).

The level of bradykinin in the blood increases during anaphylaxis, causing contraction of smooth muscles, an increase in capillary permeability, and marked vasodilation. Injected into normal animals, these peptides duplicate some of the signs of anaphylaxis.

---

* Kallikreins were discovered by mixing tissue extracts, a source of enzyme, with plasma which contains the substrate. The enzyme was named for its abundance in pancreas (Gr. *kallikreas*-pancreas); the product, bradykinin, was named for the slowness of the contraction it induces in isolated guinea pig ileum.

4) **SRS-A** is released along with histamine when Ag is added to lung fragments of sensitized guinea pigs. Like bradykinin, it causes slow contraction of isolated guinea pig ileum in the presence of antihistamines and is therefore called **SRS-A (slow reacting substance of anaphylaxis).** This substance is not destroyed by proteolytic enzymes (hence is not a kinin), and its action on smooth muscle is not blocked by antagonists of histamine or serotonin (Fig. 19-11). Few chemical properties of SRS-A are known: it is probably an acidic lipid, MW ca. 400.

SRS-A appears to be synthesized as well as released following the immune reaction, for it has been found in tissues of animals only during anaphylaxis. It appears to be important in human allergy: it is released from lung fragments of ragweed-sensitive persons by ragweed pollen extracts in vitro, and it is a powerful constrictor of isolated human bronchioles. Moreover, human allergic bronchospasm is hardly benefited by antihistamines, suggesting that another mediator is a major factor.

Two additional mediators may be involved in anaphylaxis. Ags also stimulate release from lung fragments of allergic individuals of **ECF-A** (eosinophil chemotactic factor of anaphylaxis), which attracts eosinophils, but not other leukocytes, through millipore filters (see Fig. 20-11, Ch .20). This factor could account for the characteristic infiltration of eosinophils at sites of repeated reaginic reactions in atopic individuals (e.g., in nasal and bronchial mucosa of persons with recurrent allergic reactions of the respiratory tract). The role of **prostaglandins** is anaphylaxis is uncertain.

Some of the ubiquitous prostaglandins seem to inhibit Ag-initiated histamine release from sensitized basophils, perhaps through increasing intracellular levels of cAMP (Modulation, below). However, other prostaglandins seem to cause vasodilation and increased permeability of venules. It is possible that some prostaglandins promote allergic inflammation while others are inhibitory.

The manifestations of anaphylaxis vary among species (Table 19-3) because of differences both in the amount of the various mediators that they release and in the responses of their smooth muscles and blood vessels to these substances (Table 19-4).

Because the pharmacologically active mediators are rapidly degraded and excreted they act only transiently. Their failure to accumulate, and their slow resynthesis, account for the efficacy of repeated, closely spaced injections of small doses of Ag in bringing about temporary desensitization (Acute desensitization, above).

ANAPHYLACTIC RESPONSES IN ISOLATED TISSUES

Many organs from sensitized animals respond to Ag in vitro. In the Schultz-Dale reaction the isolated uterus from a sensitized guinea pig contracts promptly when incubated

Fig. 19-11. Assays for histamine and SRS-A. Standards and test samples were added to an isolated strip of guinea pig ileum, whose contractile response was recorded on a moving strip of paper (kymograph). Antihistamines block response to histamine. Methysergide can be added to block serotonin and chymotrypsin to degrade kinins. Note the latent period and slow response to SRS-A and the faster response to histamine. Time scale is about 30 seconds (vertical markers). (Based on Orange, R. P. and Austen, K. F. in *Immunobiology,* [R. A. Good and D. W. Fisher, eds.] Sinaver, Stamford, 1971.)

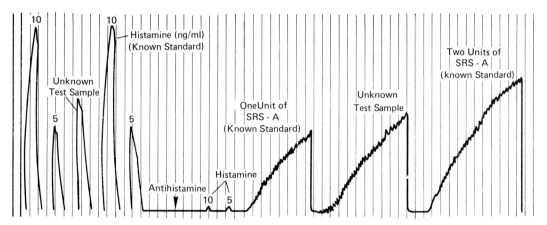

TABLE 19-4. Species Variation in Tissue Levels and Susceptibility to Histamine and Serotonin

Species	Lung content (μg/gm)		Bronchiolar sensitivity (minimal effective dose in μg)	
	Serotonin	Histamine	Serotonin	Histamine
Cat	<0.2	34	0.01	2
Rat	2.3	5	0.01	>5
Dog	<0.1	25	0.05	0.3
Guinea pig	<0.2	5–25	0.4	0.4
Rabbit	2.1	4	>8	0.5
Man	<0.3	2–20	>20	0.2

From various sources summarized in Austen, K. F., and Humphrey, J. H. *Adv Immunol 3*:1 (1963).

with Ag, which doubtless reacts with cytotropic Abs on tissue mast cells and causes release of mediators (Fig. 19-12). Similar reactions are obtained with isolated segments of ileum, gallbladder wall, and sections of arterial wall. These responses can also be elicited with tissues from passively sensitized animals, and with isolated normal tissues that are sensitized simply by incubation with antiserum. Because of the high affinity of reagins for the mast cell surface, the isolated tissues retain reactivity after extensive washing.

In response to Ag, minced fragments of lung from sensitized individuals release measurable amounts of histamine, SRS-A, and ECF-A. The liberation of histamine is complete within 5 minutes, whereas SRS-A is not detected for several minutes and its concentration then rises slowly. In one of the simplest in vitro reactions Ag elicits the release of histamine from washed leukocytes of atopic persons and the degranulation of basophils (demonstrated by staining smeared cells; see Fig 19-9): extremely small amounts of Ag suffice (e.g., 10^{-13} mg/ml of purified ragweed Ag). The degranulation has been used as a diagnostic assay for penicillin allergy (see Ch. 20, Allergy to drugs). Human basophils can also be passively sensitized with atopic sera and with purified human IgE.

The release of vasoactive amines from mast cells appears to be due to secretion rather than to cell lysis. Thus releasing cells remain impermeable to ionic dyes that can penetrate into killed but not into living cells, and the release requires an intact glycolytic pathway and is controlled by intracellular levels of cyclic AMP (Modulation, below). The clinical benefits of some drugs (sodium cromoglycate, diethylcarbamazine) in preventing bronchospasm in asthma are probably due to inhibition of secretion of vasoactive amines: these drugs do not hinder combination of Ab and Ag or the response to histamine.

Modulation. Cyclic AMP (cAMP) inhibits mediator release from Ag-stimulated mast cells: both **the rate and extent of release are enhanced when intracellular cAMP levels are low and are reduced when the levels are**

Fig. 19-12. Smooth muscle contraction in vitro in response to Ag (Schultz-Dale reaction). A uterine horn, excised from a guinea pig 13 days after a sensitizing injection of a horse serum euglobulin, was suspended in Ringer's solution to which various protein fractions from horse serum were added (arrows): at A, 1 mg of pseudoglobulin was added; at B, 10 mg of pseudoglobulin; at C and at D, 10 mg of euglobulin (the immunogen). Following the specific response at C the muscle was almost totally desensitized, because the tissue-bound Abs were saturated with Ag or because the content of vasoactive amines was depleted. Time scale markers at 30-second intervals. R = changes of Ringer's solution. [From Dale, H. H., and Hartley, P. *Biochem J 10*:408 (1916).]

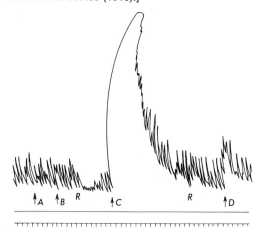

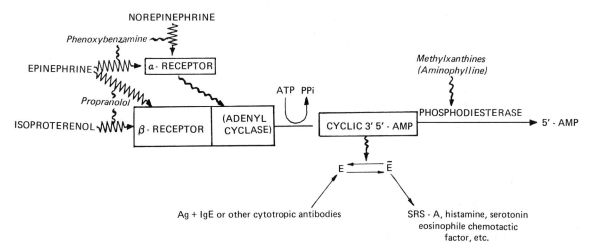

Fig. 19-13. Schematic view of the modulating effect of the cyclic-AMP system on the immunological release of SRS-A and vasoactive amines. Stimulation of α and β receptors on the cell surface membrane supposedly inhibits and enhances, respectively, the enzyme (adenyl cyclase) that generates cAMP (cyclic 3',5'-AMP). E and $\bar{E}$ are inactive and activated forms of a hypothetical proteolytic enzyme, whose existence is suggested because release of mediators requires Ca^{++} and is blocked by diisoprophylphosphofluoridate (which specifically phosphorylates and blocks a group of proteases with esterase activity). Curved lines ($\sim\sim\sim$) refer to inhibition, jagged lines ($\vee\vee\vee\wedge$) to stimulation. (Based on Austen, K. F. *Sixth International Symposium on Immunopathology.* Schwabe, Basel, 1970.)

high. Hence drugs that stimulate adenylcyclase, the enzyme that synthesizes this nucleotide, or that inhibit the phosphodiesterase that degrades it, block the release of histamine and SRS-A (isoproterenol, epinephrine, aminophylline; Fig. 19-13). These drugs have been used clinically for control of anaphylaxis and allergic bronchospasm for many years (i.e., before this mechanism of action was suspected). The modulating effects of cAMP and of cyclic GMP (which blocks the effect of cAMP) suggest a possible explanation for the influence of emotional states on the intensity of acute allergic reactions (asthma, hayfever, hives, perhaps atopic eczema).

AGGREGATE AND CYTOTOXIC ANAPHYLAXIS

All passive cytotropic responses need a latent period for attachment of cytotropic Abs to mast cells. In contrast, there are other forms of passive anaphylaxis that do not involve IgE or other cytotropic Abs and that do not require a latent period. One class, **"aggregate anaphylaxis,"** seems to be caused by relatively large amounts of **soluble** Ab-Ag complexes interacting with complement (C)

(see below). For example, in a normal guinea pig a single intradermal injection of such complexes, prepared by dissolving a specific precipitate in a concentrated solution of Ag, can evoke passive cutaneous anaphylaxis; and fatal shock can follow intravenous injection of antisera that have been incubated for a few minutes with soluble Ag.* Even some heat-aggregated Igs, without any Ag, can elicit cutaneous anaphylaxis, suggesting that the essential role of the Ag is simply to crosslink certain classes of Ab molecules. Effective soluble immune complexes are those that fix C (e.g., with average molar composition of about Ag_3Ab_2); those that do not fix C are ineffective (e.g., the Ag_2Ab complexes that form in extreme Ag excess; Ch. 18).

Another form of acute allergic reaction,

---

* Normal serum becomes similarly toxic after incubation with suspensions of various particles (kaolin, talc, barium sulfate, inulin, agar), which apparently activate Hageman factor, plasmin, kallikreins (Fig. 19-10), and C with formation of anaphylatoxin (Ch. 18). The response to these incubated sera (without Ag) is sometimes called "anaphylactoid" shock.

called **cytotoxic anaphylaxis,** sometimes follows the injection of Abs to natural constituents of cell surfaces (prototype III, Table 19-1). For example, guinea pigs injected with rabbit Abs to the Forssman Ag, a constituent of all guinea pig cells, undergo acute shock. Acute hemolytic **transfusion reactions** in man are also sometimes associated with shock and could be considered a form of cytotoxic anaphylaxis (Ch. 21).

COMPLEMENT AND ANAPHYLAXIS

Aggregated homocytotropic Abs (e.g., human IgE, guinea pig IgE-like reagins and γl) can activate C3 via the alternate pathway that bypasses the early-acting C components (Ch. 18). It is doubtful, nonetheless, that any C activation is necessary in cytotropic responses, for cytotropic cutaneous anaphylaxis is not altered in animals depleted of C3 (by cobra venom, see Ch. 18).

However, C fixation is probably essential for aggregate anaphylaxis; there is a consistent correlation between ability of aggregates to fix C in vitro and ability to evoke this form of anaphylaxis (see above), in which the release of histamine is probably caused by the anaphylatoxin formed in the C reaction sequence (Ch. 18).

ARTHUS REACTION

Shortly after the discovery of anaphylaxis, Arthus, a French physiologist, described a substantially different kind of Ab-dependent allergic reaction. When rabbits were inoculated subcutaneously each week with horse serum there was at first no noticeable response, but after several weeks each injection evoked a localized inflammatory reaction. Similar responses were soon described in man and in many other vertebrates, and were called **Arthus reactions.** These reactions are not limited to the skin: they can take place when Ags are injected into the pericardial sac or synovial joint spaces. **The principal requirement is the formation in tissues of bulky immune aggregates that fix C and attract polymorphonuclear leukocytes: lysosomal enzymes released by the cells cause tissue damage, characteristically with destructive inflam-** mation of small blood vessels ("vasculitis").

Patients with serum sickness or with certain forms of glomerulonephritis (below) develop similar lesions in small blood vessels and in kidney glomeruli, respectively; and those with high serum levels of Abs to the thermophilic *Aspergillus* that thrives in decaying vegetation, or to molds used to produce cheese, develop severe localized lung lesions of Arthus type when they inhale these fungi or fungal spores (farmer's lung or cheesemaker's lung).

The main features are illustrated by the passive cutaneous form of the Arthus reaction, in which an antiserum is first injected intravenously into a nonsensitive recipient and the corresponding Ag is then injected into the skin. Alternatively, in the **reverse passive Arthus reaction** the antiserum is injected in the recipient's skin and the Ag is then injected into the same dermal site or intravenously.

Time Course. After intradermal injection of Ag the Arthus response comes on more slowly than cutaneous anaphylaxis and is much more persistent. Local swelling and erythema appear after 1 to 2 hours, followed by punctate hemorrhages. The changes are maximal in 3 to 4 hours and are usually gone in 10 to 12 hours; but severe reactions, with necrosis at the test site, subside more slowly (cf. tuberculin skin reaction, Ch. 20).

Type and Amount of Antibody. The higher the level of precipitable Abs the more intense and persistent the lesion. Abs of almost any class of Igs, and from almost any species, can mediate the reaction, and even the injection of Ab-Ag complexes formed in the test tube can evoke the response, though with less intensity than when the aggregates form in situ. The passive Arthus reaction requires a large amount of Ab, ca. 10 mg when injected intravenously in a rabbit and ca. 100 μg when injected into the skin. In contrast, a few nanograms of reagin (almost 100,000-fold less) is sufficient for passive cutaneous anaphylaxis in man (Prausnitz-Küstner reaction, above).

Though most Igs appear to be effective, different types account for different features of the Arthus reaction: in guinea pigs, for instance, γ1 Abs induce the edematous changes, and γ2 Abs appear to be responsible for the hemorrhage and necrosis (see Fig. 19-7).

Histopathology. In anaphylaxis the inflammatory changes are limited to vasodilation and exudation of plasma proteins; inflammatory cells are not conspicuous. The Arthus response, however, is characterized by classical inflammation: blood flow through small vessels is markedly retarded; thrombi rich in

platelets and leukocytes form within small blood vessels; erythrocytes escape into the surrounding connective tissue; and after several hours the skin site becomes edematous and heavily infiltrated with polymorphonuclear leukocytes (Fig. 19-14). Finally, localized patches of necrosis appear in walls of affected small blood vessels. As the lesion begins to subside, after 4 to 12 hours, neutrophils become necrotic and are replaced by mononuclear cells and eosinophils. Within a

Fig. 19-14. The passive Arthus reaction in a rat, showing localization of Ag and complement in the wall of an affected blood vessel. The skin site was excised 2 to 3 hours after an intradermal injection of 300 μg of rabbit Abs to bovine serum albumin (anti-BSA) and an intravenous injection of 6 mg of BSA. (The Ab was injected intradermally, and the Ag intravenously, to conserve Abs.) **A.** Note intense polymorphonuclear leukocyte infiltration in and around the wall of a small blood vessel adjacent to skeletal muscle. **B.** The section was stained with fluorescent rabbit Ab to a purified component of rat C (C3; see Ch. 18). **C.** The section was stained with fluorescent anti-BSA to localize the aggregated Ag in the blood vessel wall and in the adjacent perivascular connective tissue. The same result would be obtained by staining the aggregated Ab (rabbit anti-BSA) with fluorescent anti-rabbit Ig. [From Ward, P. A., and Cochrane, C. G. *J Exp Med 121*:215 (1965).]

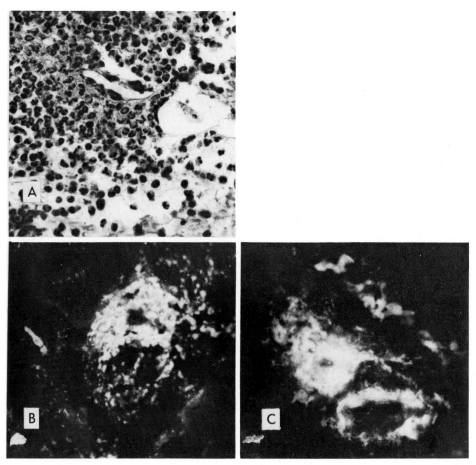

few days, the phagocytized immune complexes are degraded and inflammation disappears.*

The response in the cornea emphasizes the role of blood vessels. The injection of Ag into an immunized rabbit's normal cornea, which is devoid of functional blood vessels, can result in concentric opaque rings of Ab-Ag precipitates, like bands in gel precipitin reactions in vitro (Ch. 15), but little or no inflammation is observed. If, however, functional blood vessels are present (e.g., as a sequel of some earlier trauma to the cornea), then the Ag can elicit an Arthus response in the cornea, as in any other tissue.

Role of Complement and Granulocytes. At the site of the local reaction immunofluorescence reveals Ab-Ag aggregates with C

---

* In gross and microscopic appearance the Arthus reaction resembles the **Shwartzman reaction,** in which hemorrhagic and necrotic inflammatory lesions are evoked with the endotoxin of gram-negative bacteria (Ch. 26). Two skin injections, spaced about 1 day apart, are generally used to elicit the Shwartzman reaction: one injection consists of endotoxin and the other can be endotoxin or any of a wide variety of immunologically unrelated substances, such as agar, starch, or even Ab-Ag precipitates. An immune mechanism does not appear to be a fundamental feature of the Shwartzman phenomenon.

components (C3) localized in blood vessel walls, between endothelial cells and the internal elastic membrane (Fig. 19-14). The aggregates are also evident within granulocytes in perivascular connective tissue. If an animal's C activity has been greatly reduced (e.g., by depleting C3 with cobra venom; Ch. 18), or if its level of circulating polymorphonuclear (PMN) leukocytes has been depressed (e.g., by an anti-PMN serum or by nitrogen mustards), no inflammatory reaction appears, even though the immune complexes form in blood vessel walls.

It has been suggested therefore that the Arthus reaction depends on the following sequence: 1) Ag and Ab diffuse into blood vessel walls, where they combine and form complexes and fix C; 2) the resulting chemotactic factors (C5a and C5b,6,7; see Ch. 18) attract PMN leukocytes, which ingest the aggregates and release lysosomal enzymes; 3) the enzymes cause focal necrosis of the blood vessel wall and the other inflammatory changes. From the release of acid-soluble peptides (from radiolabeled Ags) it is evident that lysosomal enzymes degrade the immune complexes, whose disappearance is associated with subsidence of inflammation.

Increased permeability of the blood vessel endothelium (due to vasoactive amines released from mast cells or basophils or plate-

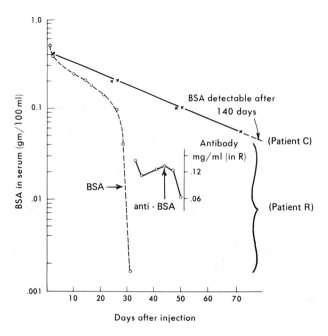

Fig. 19-15. Serum sickness syndrome in man following the injection of 25 gm of BSA at zero time. Patient R had serum sickness from day 24 to 31; patient C did not exhibit serum sickness or form Abs to BSA. [Data of F. E. Kendall; From Seegal, B. *Am J Med 13*:356 (1952).]

lets) seems to aid the penetration of Ab-Ag complexes into blood vessel walls (see Serum sickness syndrome, below). However, antihistamines have little (if any) clinical benefits on Arthus lesions.

SERUM SICKNESS SYNDROME

From about 1900 to 1940 several types of bacterial infection in man were treated routinely by injecting patients with a large volume of antiserum prepared in horses or rabbits. The recipients often developed, from 3 days to 2 or 3 weeks later, a characteristic syndrome called serum sickness. Heterologous antisera are now used much less in medicine (e.g., for tetanus, rabies, and prevention of allograft rejection with horse antisera to human lymphocytes), but the syndrome is also encountered as an allergic reaction to penicillin and other drugs (Ch. 20).

The syndrome includes 1) fever, 2) enlarged lymph nodes and spleen, 3) erythematous and urticarial rashes, and 4) painful joints. The disease usually subsides within a few days. In the few patients who have died at the height of the illness, autopsy has disclosed vascular and perivascular inflammatory lesions like those of the Arthus reaction.

The mechanisms have been analyzed in rabbits and in humans injected with large amounts of purified foreign protein. The opportunity to make detailed observations in man arose in connection with attempts, during World War II, to use bovine serum albumin (BSA) as a plasma expander in the treatment of traumatic shock (Fig. 19-15).

Mechanisms. The illness usually becomes evident 7 to 14 days after the initial injection of Ag. During this interval the Ag level declines, but it is still high enough after Ab production starts to form the small soluble Ab-Ag complexes (in Ag excess) that initiate focal vascular lesions (in coronary arteries, glomeruli, etc.; Fig. 19-16). Serum sickness is thus usually observed only after exceptionally large amounts of foreign protein are injected, e.g., 25 gm BSA in man, or 1 gm in a rabbit. However, in a previously sensitized person, with an accelerated (anamnestic) Ab re-

Fig. 19-16. Representative cardiovascular and renal lesions in experimental serum sickness in the rabbit. The Ag was BSA (see Fig. 19-17). **A.** Medium-sized coronary artery: endothelial cell proliferation, necrosis of media, polymorphonuclear leukocyte infiltration through all layers, and mononuclear cells in the media and adventitia are evident. **C.** An affected glomerulus showing increase in size, proliferation of endothelial and epithelial cells, and obliteration of capillary spaces. **B.** Section through a normal glomerulus of a control rabbit for comparison with **C:** note the much lower density of glomerular cells and patency of capillaries. (From Dixon, F. J. In *Immunological Diseases.* M. Samter, ed. Little, Brown, Boston, 1965.)

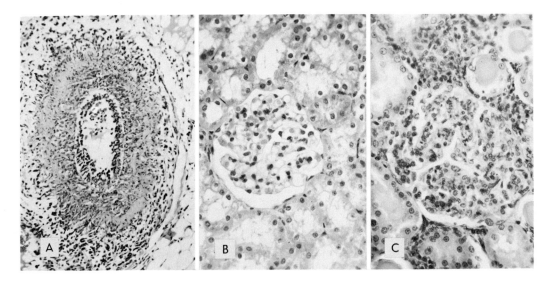

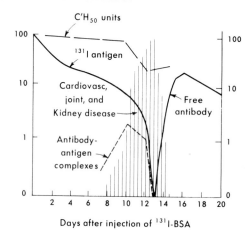

Fig. 19-17. Serum sickness in the rabbit. Changes in serum levels of free Ag (^{131}I-labeled BSA), free Ab (anti-BSA), Ag-Ab complexes, and complement activity ($C'H_{50}$ units, see Ch. 18) following the injection of rabbits at zero time with 250 mg of ^{131}I-BSA per kilogram of body weight. Ordinate (log scale) refers to ^{131}I-BSA in total blood volume, as per cent of amount injected; anti-BSA as micrograms of Ag bound per milliliter of serum; C activity as per cent of normal serum. All animals had cardiovascular, joint, and kidney lesions (Fig. 19-16), shown by shaded area, on day 13. (From Dixon, F. J. In *Immunological Diseases.* M. Samter, ed. Little, Brown, Boston, 1965.)

sponse, the reaction appears earlier and therefore requires much less Ag: e.g., 3 or 4 days after 1 ml of horse serum.

As the manifestations of serum sickness appear, the decline in level of free Ag is accelerated (Figs. 19-15 and 19-17). During this period soluble Ab-Ag complexes can be detected in serum if ^{131}I-labeled BSA (*Ag) is used as Ag (for example, ammonium sulfate at 50% saturation precipitates the *Ag-Ab complexes, but not the free *Ag). As Ab-Ag complexes form they fix C, which is depressed at the height of the illness (Fig. 19-17); and, as in the Arthus reaction, the most abundant C component (C3) can be detected by immunofluorescence in immune aggregates within the focal blood vessel lesions (Fig. 19-14). As the complexes disappear free Abs become detectable and inflammatory lesions regress.

The injection of preformed complexes (made in vitro) also elicits the characteristic lesions in rabbits. The most effective complexes are those prepared in moderate Ag excess (Ag_3Ab_2). Complexes formed at equivalence or in extreme Ag excess (Ag_2Ab) are ineffective: the former are generally particulate and tend to be rapidly cleared from the circulation, and the latter fail to fix C (Ch. 18).

Vasoactive Amines. Increased vascular permeability seems to be necessary for the penetration of immune complexes from plasma through endothelium into the blood vessel wall, and vasoactive amines probably aid at this step.

Thus vascular deposition of injected preformed complexes is diminished in rabbits that are treated with antihistamines and antiserotonins or are depleted of platelets (which are the major circulating reservoir of vasoactive amines in rabbits). The immune complexes probably contain enough cytotropic Abs to activate blood basophils, which are thought to produce (in rabbits) a soluble factor that clumps and lyses platelets, releasing vasoactive amines. These amines could also be released by platelets that are clumped on Ab-Ag-C complexes (immune adherence due to the bound C3b fragment, Ch. 18). The pathogenetic steps are summarized in Figure 19-18.

Relation to Other Allergic Reactions. Serum sickness involves both Arthus and anaphylactic elements (Table 19-1). The focal vascular lesions and the requirement for immune complexes and for C suggest that the syndrome is essentially a disseminated form of the Arthus reaction, with the same injected substance, given in a single large amount, serving first as immunogen and then as reacting Ag. However, the role of homocytotropic Abs seems larger than in the Arthus reaction, since urticarial skin lesions are also prominent in serum sickness. Moreover, after a person has recovered from the disease he will generally give a wheal-and-erythema response to intradermal injection of the responsible Ag. The amines probably contribute also to development of the focal vasculitis, as in the Arthus reaction.

Multiplicity of Antigens. The overlapping kinetics of Ag disappearance and Ab appearance explain some of the confusion that arose in earlier studies of serum sickness due to foreign sera. For example, just before, during, and for some weeks after clinical manifestations had run their course both horse serum proteins and Abs to these proteins were detectable in a given patient's serum. In addition, even after a single injection of horse serum a patient

1. Ab-Ag complexes fix C, causing either release of a leukocyte factor
    that acts on platelets or immune adherence (Ch. 18) that clumps
    and damages platelets.

2. Platelets lyse, releasing vasoactive amines (histamine, serotonin).

3. Permeability of vascular endothelium increases.

4. Ab-Ag complexes penetrate into blood vessel walls or form within the walls,
    fixing C and forming chemotactic factors (C5a, $\overline{C5b, 6, 7}$) for poly-
    morphonuclear leukocytes (neutrophils).

5. Neutrophils penetrate into blood vessel walls, ingesting immune complexes
    and releasing lysosomal enzymes.

6. Lysosomal enzymes damage neighboring cells and connective tissue elements,
    leading to more inflammation.

7. If immune complexes are formed in an acute episode ("one-shot" serum
    sickness) the lesions abate as complexes are degraded.

8. If immune complexes are formed repeatedly (as in persistent viremia,
    malaria, some other forms of human glomerulonephritis) chronic
    inflammatory disease can develop in small blood vessels and kidney
    glomeruli.

Fig. 19-18. Pathogenesis of lesions in and around blood vessels in serum sickness in rabbits.

might suffer a series of attacks of serum sickness, separated by illness-free intervals. These effects were evidently due to the multiplicity of Ags in horse serum; each bout of illness represented the response to a particular Ag and the patient's serum under these circumstances resembled the supernatants of a precipitin reaction performed with horse serum as the Ag (Ch. 15; Fig. 15-9).

IMMUNE-COMPLEX DISEASES

Glomerulonephritis. The pathogenesis of experimental Arthus and serum sickness lesions probably accounts for some forms of glomerulonephritis, a kidney disease in which obstructive inflammatory lesions of glomerular blood vessels can lead to renal failure. Immunofluorescence of biopsies usually reveals lumpy deposits of Ig and C3 (probably Ab-Ag-C complexes) beneath the glomerular endothelium (Fig. 19-19). The deposits resemble those of serum sickness, especially the chronic experimental model of Dixon *et al.*, in which Ag is administered almost daily for many weeks at a rate that approximates Ab synthesis and provides continuous production of immune complexes.

Ags have been identified in human glomerular lesions in special circumstances. *Plasmodium malariae* Ags have been recognized by immunofluorescence in kidneys of patients with the chronic nephritis associated with malaria, and Abs to malarial Ags have been eluted from kidney biopsies (at pH 2 to 3 to dissociate Ab-Ag complexes, see Appendix, Ch. 15).

Similarly, Abs to single- and double-stranded DNA have been eluted from kidney tissue of patients with systemic lupus erythematosus; these patients often have high serum levels of Abs to nucleic acids and develop progressive glomerulonephritis with lumpy glomerular deposits containing Ig, C, and DNA (see Ch. 20, Autoimmune diseases).

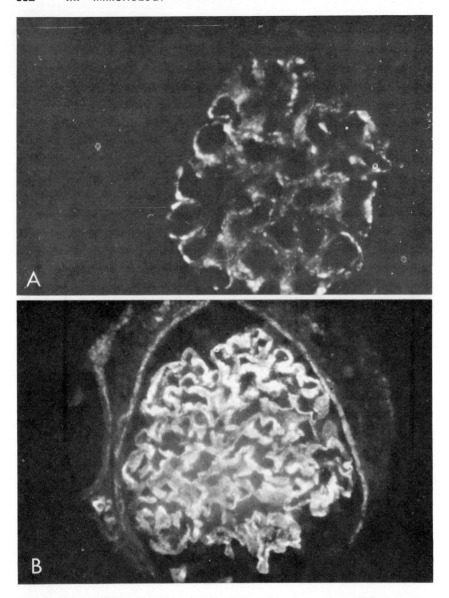

Fig. 19-19. Immune complexes in glomeruli revealed by immunofluorescence. Kidney biopsies from patients with glomerulonephritis were stained with fluorescein-labeled Abs to human Igs. **A** Lumpy deposits due to Abs-Ag-C in glomerulus from a patient with systemic lupus erythematosus. **B** Linear deposits due to Igs attached specifically to the glomerular basement membrane (Goodpasture's syndrome; see Autoimmune diseases, Ch. 20). (Courtesy of Dr. C. Kirk Osterland.)

Most cases of human glomerulonephritis occur as a sequel to infection with β-hemolytic streptococci (especially Type XII, the "nephritogenic" strain; see Ch. 26), but streptococcal Ags have not been consistently detected in the associated glomerular deposits of Ig and C, perhaps because reactive sites are usually covered by antistreptococcal Abs; the specificity of the Abs eluted from affected kidneys has not yet been determined.

Viral Complexes. Observations with mice suggest that chronic viral infection may be a source of immune-complex disease. Animals

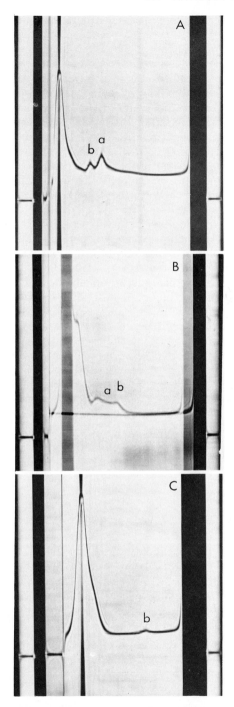

infected at birth with lymphocytic choriomeningitis (LCM) virus become chronic carriers of the virus, but they are not tolerant of it (see Ch. 17, Pseudotolerance). Instead, they produce large amounts of antiviral Abs that do not neutralize infectivity. Virus- antivirus-C complexes in serum are revealed by a reduction in titer of infectious virus when specific precipitates are formed by addition of antisera to mouse Igs or to mouse C components (usually anti-C3; see Ch. 18). Progressive renal disease in these mice is associated with inflammatory vascular lesions and lumpy glomerular deposits containing LCM virus, antiviral Ig, and C.

Immune-complex disease with glomerulonephritis also occurs in mice as a result of neonatal infection with murine leukemia viruses, murine sarcoma virus, Coxsackie B virus, and polyomavirus; and similar lesions seem to be responsible for the high mortality rate in an economically important disease of mink the probably also derives from a neonatal viral infection (Aleutian mink disease).

Antikidney Antibodies. In a rare form of human glomerulonephritis immunofluorescence reveals not lumpy but "linear" glomerular deposits of Ig that follow the basement membrane continuously (Fig. 19-19). The pattern resembles that seen in the experimental nephritis produced with heteroantisera to basement membrane (e.g., the so-called Matsugi nephritis produced in rabbits with duck antisera to rabbit kidney). Some monkeys inoculated with Igs eluted from human kidneys with linear deposits have developed glomerulonephritis, suggesting that the human lesion could be due to autoantibodies to glomerular basement membrane, or to the cross-reacting membranes in lung (see Ch. 20, Goodpasture's disease).

Rheumatoid Arthritis. In this common, chronic inflammatory dissease of joints the fluid in joint (synovial) cavities contains high levels of Igs (much of it synthesized locally in synovial membrane) and C-fixing aggregates of Igs, as well as granulocytes and C components that attract granulocytes (C5a,

Fig. 19-20. Immune complexes in human serum. Sera were diluted with buffered saline and subjected to velocity sedimentation at about 50,000 RPM in the analytical ultracentrifuge. **A** Serum from a patient with rheumatoid arthritis showing at *a* specific complexes (22S) of "rheumatoid factor" of the IgM class with IgG (ligand); the 19S at *b* represents unbound IgM immunoglobulins. **B** Serum from a patient with rheumatoid arthritis showing a less common pattern with polydisperse immune complexes involving "rheumatoid factor" of the IgG class with normal IgG (ligand) (*a*); the normal (unbound) 19S IgM peak is at *b*. **C** Control: serum from a normal person; *b* represents the normal IgM peak. No complexes are evident. (Courtesy of Dr. C. Kirk Osterland.)

C5b,6,7). Hence the joint fluids contain all the ingredients for an Arthus reaction. However, the actual Ags remain uncertain (as in most human diseases that are suspected to arise from immune complexes).

The IgG in joint fluid might conceivably function as Ag for the characteristic **rheumatoid factors** of rheumatoid arthritis, which are IgM and IgG molecules that react specifically with antigenic determinants on Fc domains of various IgGs (see Ch. 16, Human Gm allotypes). The IgG-IgM and the IgG-IgG complexes could then initiate the C-granulocyte-lysosomal enzyme sequence that results in Arthus inflammation (Fig. 19-18). Alternatively, the IgGs could function as Abs that bind special Ags in affected joints, such as DNA (and probably other, unidentified Ags): the rheumatoid factors would then not be essential participants but would have arisen secondarily as Abs to new antigenic sites that appear on conformationally altered IgG molecules in immune complexes. Indeed, IgM molecules that behave like rheumatoid factors are found in diverse situations where immune complexes are present at high levels for protracted periods (e.g., experimental chronic serum sickness, see above).

Evidence for Persistent Soluble Immune Complexes. In rheumatoid arthritis, glomerulonephritis, systemic lupus erythematosus, and other chronic diseases where immune complexes are probably pathogenic their presence is often revealed by the formation of precipitates when sera, or joint fluids in rheumatoid arthritis, are simply stored at 4°. These **cryoprecipitates** contain IgM (probably anti-antibodies), IgG, and sometimes additional components that could represent Ags (e.g., single-stranded DNA in patients who form anti-DNA, such as those with systemic lupus or with rheumatoid arthritis).

The presence of soluble immune complexes can also be revealed by: 1) ultracentrifugation (Fig. 19-20); 2) appearance of breakdown products of C3 (recognized by immunoelectrophoresis with specific antisera); 3) increased levels of "immune conglutinins," which are Abs (of IgM class) to antigenic sites that appear on activated C3 and C4 components of C (Ch. 18); 4) precipitation with C1q, a stable subunit of the first C component, which reacts specifically with soluble immune complexes if the Ab moiety of the complex belongs to certain Ig classes (e.g., IgG-1 and IgM in man; Ch. 18); and 5) certain highly avid monoclonal IgMs with rheumatoid activity (e.g., from patients with Waldenström's macroglobulinemia; see Ch. 16).

Autoimmune diseases will be discussed further in the next chapter.

SELECTED REFERENCES

BOOKS AND REVIEW ARTICLES

AUSTEN, K. F., and BECKER, E. L. (eds.). *Biochemistry of the Acute Allergic Reaction.* Oxford, Blackwell, 1968, 1971.

BECKER, E. L. Nature and classification of immediate-type allergic reactions. *Adv Immunol 13*:267 (1971).

BECKER, E. L., and HENSON, P. M. In vitro studies of immunologically induced secretion of mediators from cells, and related phenomena. *Adv Immunol.* In press.

BENNICH, H., and JOHANSSON, S. G. O. Structure and function of human IgE. *Adv Immunol 13*:1 (1971).

BLOCH, K. J. The anaphylactic antibodies of mammals including man. *Prog Allergy 10*:84–150 (1967).

COCHRANE, C. G. Immunologic tissue injury mediated by neutrophilic leucocytes. *Adv Immunol 9*:97–162 (1968).

COCHRANE, C. G., and KOFFLER, D. Immune complex disease in experimental animals and man. *Adv Immunol 16*:186 (1973).

ISHIZAKA, K., and ISHIZAKA, T. Biologic function of IgE antibodies and mechanisms of reaginic hypersensitivity. *Clin Exp Immunol 6*:25 (1970).

OSLER, A. G., LICHTENSTEIN, L. M., and LEVY, D. A. In vitro studies of human reaginic allergy. *Adv Immunol 8*:183 (1968).

SAMTER, M. (ed.) *Immunological Diseases.* Little, Brown, Boston, ed. 2. 1971.

Symposium: Cellular mechanisms and involvement in acute allergic reactions. *Fed Proc 28*:1702–1735 (1969).

UNANUE, E. R., and DIXON, F. J. Experimental glomerulonephritis: Immunologic events and pathogenetic mechanisms. *Adv Immunol 6*:1 (1967).

VON PIRQUET, C. F., and SCHICK, B. *Serum Sickness.* (Engl. trans. B. Schick.) Williams & Wilkins, Baltimore, 1905.

ZVAIFLER, N. J. The immunopathology of joint inflammation in rheumatoid arthritis. *Adv Immunol 16:*265 (1973).

SPECIFIC ARTICLES

COCHRANE, C. G., Mechanisms involved in the deposition of immune complexes in tissue. *J Exp Med 134:*75S–89S (1971).

HENSON, P. M. Interaction of cells with immune complexes: Adherence, release of constituents, and tissue injury. *J Exp Med 174:*114S–135S (1971).

ISHIZAKA, K., ISHIZAKA, T., and HORNBROOK, M. M. Physico-chemical properties of human reaginic antibody. IV. Presence of a unique immunoglobulin as a carrier of reaginic activity; V Correlation of reaginic activity with gamma E-globulin in antibody. *J Immunol 97:*75–85, 840–853 (1966).

JOHANSSON, S. G., and BENNICH, H. Immunological studies on an atypical (myeloma) immunoglobulin. *Immunology 13:*381–394, 1967.

JOHANSSON, S. G., BENNICH, H., BERG, T., and HÖGMAN, C. Some factors influencing the serum IgE levels in atopic diseases. *Clin Exp Immunol 6:*43 (1970).

KOCHWA, S., TERRY, W. D., CAPRA, J. D., and YANG, M. L. Structural studies of IgE. I. Physicochemical studies of the IgE molecule. *Ann NY Acad Sci 190:*49 (1971).

KOFFLER, D. AGNELLO, V., THOBURN, R., and KUNKEL, H. G. Systemic lupus erythematosus: Prototype of immune complex nephritis in man. *J Exp Med 174:*169S–179S (1971).

LEVINE, B. B. Atopy and mouse models. *Int Arch Allergy 41:*88–92 (1971).

LEVINE, B. B., CHANG, H., and VAZ, N. M. Production of hapten-specific reaginic antibodies in the guinea pig. *J Immunol 106:*29–33 (1971).

MCCLUSKEY, P. T. The value of immunofluorescence in the study of human renal disease. *J Exp Med 184:*242S–255S (1971).

MCPHAUL, J. J., and DIXON, F. J. Characterization of human anti-glomerular basement membrane antibodies eluted from glomerulonephritic kidneys. *J Clin Invest 49:*308 (1970).

ORANGE, R. P., AUSTEN, W. G., and AUSTEN, K. F. Immunological release of histamine and slow-reacting substance of anaphylaxis from human lung. I. Modulation by agents influencing cellular levels of cyclic 3′,5′-adenine monophosphate. *J Exp Med 134:*136S–148S (1971).

WINCHESTER, R. J., KUNKEL, H. G., and AGNELLO, V. Occurrence of γ-gloublin complexes in serum and joint fluid of rheumatoid arthritic patients: Use of monoclonal rheumatoid factors as reagents for their demonstration. *J Exp Med 134:*286S—295S (1971).

chapter 20

CELL-MEDIATED HYPERSENSITIVITY AND IMMUNITY

**GENERAL PROPERTIES OF CELL-MEDIATED
IMMUNE RESPONSES 558**
DELAYED-TYPE HYPERSENSITIVITY 560
Responses to Infectious Agents and to Purified Proteins 560
Responses to Small Molecules: Contact Skin Sensitivity 562
Specificity 564
OTHER CELL-MEDIATED IMMUNE RESPONSES 566
Resistance to Infectious Agents 566
Rejection of Allografts 567
Graft-vs.-Host Reaction 568
Tumor Immunity 568
PHYLOGENY AND ONTOGENY 570

CELLULAR BASIS FOR CELL-MEDIATED IMMUNITY 570
T Lymphocytes as Mediators 570
Macrophages as Nonspecific Accessories 571
Genetic Control 571
Relation to Helper Cells in Antibody Formation 572
Transfer Factor 572

CELL-MEDIATED IMMUNE REACTIONS WITH
CULTURED CELLS 572
Blast Transformation 573
Products of Activated Lymphocytes 575
Cytotoxic Mechanisms 577

INTERFERENCE WITH CELL-MEDIATED IMMUNITY 578
Tolerance: Suppression by Antigen 578
Suppression by Antibodies 579
Nonspecific Suppression 580
Clinical Deficiencies 580

SOME SPECIAL ALLERGIC REACTIONS 581
ALLERGY TO DRUGS 581
Penicillin Allergy 581
Special Tissues 585
AUTOIMMUNE RESPONSES 585
Mechanisms 585
Characteristics of Autoimmune Diseases 586
Some Representative Autoimmune Diseases 588
Pathological Consequences of Autoimmune Reactions 593

ALLERGY AND IMMUNITY 593

GENERAL PROPERTIES OF CELL-MEDIATED IMMUNE RESPONSES

Following the discovery of anaphylaxis a bewildering variety of other allergic responses were recognized. Though classification proved difficult it became clear in the 1920s that anaphylactic, Arthus, and serum sickness reactions differed from a special group, called delayed-type, in which the responses always evolved slowly and passive transfer could not be achieved with antisera. It gradually became apparent much later that these **delayed-type hypersensitive** responses are mediated by specifically reactive lymphocytes, now known to be T cells, rather than by conventional, freely diffusible ("humoral") antibody (Ab) molecules: hence they are now also referred to as **cell-mediated hypersensitivity.**

Delayed-type hypersensitive skin responses have long been used to diagnose infectious diseases and to screen populations for those with previous or current infections (Table 20-1). Moreover, these responses appear to be responsible for a number of autoimmune diseases; and in the form of allergic contact dermatitis they constitute one of the commonest skin diseases of man. However, the practical significance of these responses extends beyond allergic reactions, for similar mechanisms (independence of humoral Abs and dependence upon specifically reactive lymphocytes) underlie many other immune reactions, including resistance to a variety of infectious agents,* rejection of grafted cells from genetically different individuals of the same species (allografts), and resistance to most tumors. Thus, all these responses, including delayed-type hypersensitivity, are now called **cell-mediated immune reactions** (Table 20-2). The principal differences between these responses and those due to conventional Ab molecules (humoral immunity) are summarized in Table 20-3. Though the distinctions are fundamental, many clinical reactions are actually mixtures, with both Ab- and cell-mediated reactions to the same Ag occurring in the same individual and even in the same inflammatory lesion.

Various forms of cell-mediated immunity, best studied in different species, illuminate different aspects of their common mechanisms; and no one manifestation can be used to illustrate all the fundamental features. For instance, the peculiar specificity requirements are best appreciated in skin reactions, which are recognized readily in guinea pigs but only with difficulty in mice, rats, and chickens; while the effects of neonatal thymectomy (and bursectomy in chickens) are effectively studied in the latter species but not in guinea pigs.† Hence an understanding of the basis for the diagram that summarizes the essentials of all of these reactions (Fig. 20-1) requires an appreciation of the key manifestations of diverse cell-mediated reactions.

---

* **Bacterial** or **infectious allergy,** older terms for delayed-type hypersensitivity, are misleading synonyms, for despite the frequent association of these reactions with chronic bacterial infections, microbial antigeus (Ags) can also evoke immediate-type hypersensitivity and nonmicrobial Ags can elicit delayed-type responses.

† Newborn guinea pigs appear to have fully differentiated lymphocytes; hence the effects of thymectomy are probably as difficult to discern in them as in adult mice, rats, etc.

TABLE 20-1. Some Delayed-Type Skin Reactions Used as Diagnostic Tests and for Epidemiological Surveys

Disease	Type of etiological agent	Antigenic preparation used in skin test
Tuberculosis	Bacteria	Tuberculin
Leprosy	Bacteria	Lepromin
Brucellosis	Bacteria	Brucellin
Psittacosis	Bacteria	Heat-killed organisms
Lymphogranuloma venereum	Bacteria	Extract of chorioallantoic membrane of infected chick embryo
Mumps	Virus	Noninfectious virus from yolk sac of infected chick embyro
Coccidioidomycosis	Fungus	Concentrated culture filtrate
Histoplasmosis	Fungus	Concentrated culture filtrate
Blastomycosis	Fungus	Concentrated culture filtrate
Leishmaniasis	Protozoan	Extract of cultured *Leishmania*
Echinococcosis	Helminth	Fluid from hydatid cyst
Contact dermatitis	Simple chemical	Patch tests with simple chemicals

TABLE 20-2. Cell-Mediated Immune Responses

Delayed-type hypersensitivity
Resistance to many infectious agents (especially intracellular parasites)
Resistance to most tumors
Rejection of allografts
Graft-vs.-host reaction
Some drug allergies
Some autoimmune diseases

TABLE 20-3. Basic Differences Between Humoral and Cell-Mediated Allergic Reactions

Property	Humoral	Cell-mediated
Time course in already sensitized individual	Minutes to hours	One or more days
Histology of inflammatory lesion	Edema, polymorphonuclear leukocytes (granulocytes)	Mononuclear cells*
Transfer with	Serum (Abs†)	Lymphoid cells (T lymphocytes†)
Specificity	Small determinants, ranging from a benzenoid molecule to a hexasaccharide or a hexapeptide	Large molecules, usually proteins (especially on cell surfaces)

* A noncommittal term for lymphocytes and macrophages, which cannot be readily distinguished in the light microscope in ordinary histological preparations.

† The active, unit reacting specifically with Ags. Lymphoid cell preparations are heterogeneous cell populations from lymph nodes, spleen, or blood; predominantly lymphocytes, they also include many macrophages (or monocytes, the immature macrophage in blood), some granulocytes, red cells, and probably a rare plasma cell.

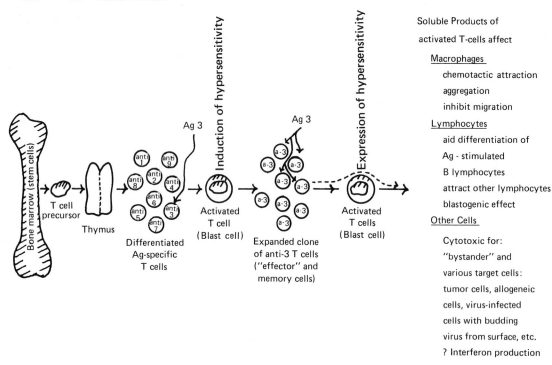

Fig. 20-1. Essentials of cell-mediated immune responses. Precursor cells differentiate into mature Ag-sensitive T cells, which can be specifically activated by the corresponding Ag to 1) proliferate into an expanded clone (includes specific "memory" and "effector" cells; whether they are the same is uncertain), and 2) produce soluble factors that affect a variety of other cells.

DELAYED-TYPE HYPERSENSITIVITY

RESPONSES TO INFECTIOUS AGENTS AND TO PURIFIED PROTEINS

The response to proteins of the tubercle bacillus has been studied longer than other cell-mediated responses because of its suspected role in tuberculosis; it serves as a general model for delayed-type allergic reactions to soluble proteins and to microbial Ags.

Koch observed in 1890 that viable tubercle bacilli inoculated subcutaneously into guinea pigs evoke a much more intense inflammatory reaction in previously infected than in uninfected animals (Koch phenomenon). Moreover, filtrates of cultures of *Mycobacterium tuberculosis,* even after concentration by boiling,* elicited an inflammatory reaction many

hours after injection into tuberculous but not into normal animals. Similar preparations from other bacterial and fungal cultures elicit similar delayed-type responses in those infected with the corresponding organisms (Table 20-1).

Cutaneous Reaction. After 0.1 μg of tuberculin* is injected intradermally into a sensitized individual no change is observed at the inoculated site for at least 10 hours. Erythema and swelling then gradually appear and increase progressively; maximal intensity and size (up to ca. 7 cm diameter) are reached in 24 to 72 hours, and the response then subsides over several days.

In highly sensitive humans 0.02 μg of tuberculin can cause necrosis, ulceration, and scarring at the

---

* After concentration by boiling and removal of debris, the culture filtrate was called tuberculin (now called **old tuberculin** or **OT**). In modern preparations the active proteins are concentrated from autoclaved cultures by precipitation with ammonium

sulfate; the somewhat purified product is **purified protein derivative** or **PPD**. A standard batch, PPD-S, has been designated as the basis for the international unit: 1 tuberculin unit (TU) equals 0.02 μg of total protein (see Ch. 35 for further details).

inoculated site. In contrast, at least 0.5 μg, and usually several times this amount, is required to elicit a faint response in highly sensitized guinea pigs: and even more is necessary in cattle and rabbits. In sensitized rats and mice erythema is not discernible and histological examination is required (see below).

Histologically the response is characterized by massive accumulation of inflammatory cells. Initially (e.g., at 12 hours) granulocytes are present around blood vessels, but at peak intensity the lesions are populated exclusively by mononuclear cells. Electron microscopy and histochemical stains (for lysosomal enzymes) demonstrate that the cells are lymphocytes and macrophages (Fig. 20-2).

Comparison Between Arthus and Delayed-type Skin Reactions. Arthus reactions sometimes simulate delayed-type responses. It should be recalled, however, that the Arthus reaction usually appears ca. 2 hours after Ag is injected into the skin, is maximal at 4 or 5 hours, and subsides by 24 hours (Ch. 19). It is also more boggy than indurated, reflecting a large amount of edema fluid and only a modest accumulation of inflammatory cells, which are mainly polymorphonuclear leukocytes. Delayed-type reactions are not only later in onset and more persistent but they are indurated because of the intense accumulation of mononuclear cells.

A bimodal reaction sometimes follows a single test injection: one maximal at ca. 4 hours (Arthus) and the other at ca. 24 hours (delayed-type). The situation is, however, often more complex, and a severe Arthus reaction can remain conspicuous for 24 hours or more: its distinction from the delayed-type reaction is then hardly possible by either gross or microscopic inspection.

Noncutaneous and Systemic Reactions. Delayed responses also occur in tissues other than skin, e.g., tuberculin can cause severe inflammation and necrosis in the cornea of highly sensitized guinea pigs.

A systemic response, **tuberculin shock**, ensues when a sensitized guinea pig is injected intraperitoneally with a relatively large amount (e.g., 5 mg) of tuberculin. Prostration develops after 3 to 4 hours, body temperature falls ca. 4 or 5°, and death may follow in 5 to 30 hours. A systemic reaction (headache, malaise, prostration, but only rarely fatal) also occurs

Fig. 20-2. Delayed-type allergic reaction in guinea pig skin. The animal was sensitized by injecting 5 μg of hen's egg albumin (HEA) in complete Freund's adjuvant into its toepads. Six days later it was injected intradermally with 5 μg of HEA in saline. The skin site was excised 24 hours later, when induration and redness (which had probably first become evident at about 12 hours) were maximal. The infiltrating mononuclear lymphoid cells appear by electron microscopy and with special stains to be lymphocytes and macrophages. **A**, ×64; **B**, ×355. [From Coe, J. E., and Salvin, S. B. *J Immunol* 93:495 (1964).]

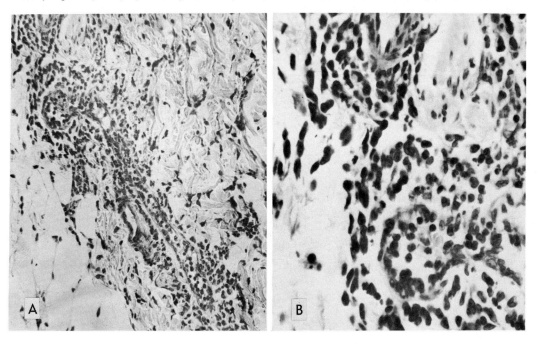

in highly sensitized persons who are injected with an excessive amount of tuberculin or who inhale aerosols of the material in the laboratory.

Tuberculous individuals exposed to large amounts of tuberculin also develop **focal** reactions, with increased inflammation in their lesions in lungs and elsewhere. The reactions are probably due to interaction of the Ag with a high concentration of specifically reactive lymphocytes in the infected lesions: they resemble histologically the responses elicited in the skin with tuberculin. Similar systemic responses have been evoked with Ags of histoplasma, brucella, vaccinia, and pneumococci in persons with delayed-type sensitivity to these organisms.

Sensitization. The time required for induction is about the same as for induction of Ab formation (Ch. 17), i.e., 4 to 14 days after Ag is first administered.

Small quantities of Ag, especially when associated with the surface of living cells, are the most effective immunogens. Sensitization to tuberculin is most effectively induced by infection with virulent tubercle bacilli (or with the attenuated BCG strain). Killed bacilli are less effective, and purified tuberculin is ineffectual unless given in appropriate adjuvants (below). However, declining delayed-type sensitivity seems to be enhanced by repeated intradermal injections of purified tuberculin (in saline), suggesting that it can stimulate secondary responses.

Through efforts to improve the immunogenicity of soluble proteins, Dienes found that when egg albumin was injected into an animal's tuberculous lesions intense, generalized delayed-type hypersensitivity to the Ag was established; this observation led eventually to Freund's development of the complete adjuvant that bears his name (water-in-oil emulsions with suspended tubercle bacilli; Ch. 17). The active component of the tubercle bacillus appears to be a lipid (a mycoside called wax D, Ch. 35), and a water-in-oil emulsion containing the lipid and purified tuberculin can establish delayed-type hypersensitivity to tuberculin.

Complete Freund's adjuvant is also an extremely potent enhancer of Ab formation (Ch. 17), but the dose of incorporated protein is crucial: small doses (e.g., 1 to 50 μg in a guinea pig) produce intense delayed-type hypersensitivity, while milligram amounts tend to induce only vigorous Ab formation, perhaps because high Ab levels seem to block the

induction of cell-mediated immunity to the same Ag. Perhaps for similar reasons, other adjuvants, such as alumina, only aid Ab formation and even appear to selectively inhibit the induction of delayed-type hypersensitivity (Immune deviation, below).

Cutaneous Basophil Hypersensitivity. A slightly different form of delayed-type skin hypersensitivity can be induced by injecting small amounts of protein Ags, or Ab-Ag complexes (in great Ab excess) into the skin. The allergic response subsequently elicited resembles the tuberculin reaction in timing (maximal inflammation at 24 hours) and in susceptibility to transfer with living lymphocytes but not with antiserum. It differs, however, in two respects. 1) It is transient: it can be elicited about 1 week after sensitization is initiated, but not after about 2 to 4 weeks, when Ab formation occurs. 2) The infiltrating cells are lymphocytes and basophil leukocytes, rather than lymphocytes and macrophages; and epidermal necrosis does not occur. Called **cutaneous basophil hypersensitivity** (or the **Jones-Mote reaction** after authors of an early description), this form of delayed-type skin hypersensitivity emphasizes the wide variety of hematopoietic cells that can be affected by Ag-stimulated T lymphocytes: other lymphocytes (Ch. 17 and below), macrophages, basophils.

RESPONSES TO SMALL MOLECULES: CONTACT SKIN SENSITIVITY

Induction of this form of hypersensitivity, like its expression, can be evoked merely by contact of simple chemical sensitizers (MW < ca. 1000) with the intact skin. The ability to respond to this **percutaneous** application is called **contact skin sensitivity;** the ensuing inflammation is **allergic contact dermatitis.** These reactions are responsible for common allergic skin diseases in man (caused by contact with such simple substances as catechols of poison ivy plants and a great variety of synthetic chemicals, drugs, antibiotics, and cosmetics). They also provide investigative advantages because percutaneous tests do not elicit Arthus responses: hence the assay for contact skin sensitivity is an unambiguous test for delayed-type allergy (see above).

Delayed-type skin responses to simple chemicals seemed initially to offer important opportunities to clarify the specificities of cell-mediated reaction, just as simple haptens illuminate the specificities of Ab molecules. However, this reaction depends upon ability of sensitizer and eliciting agent to form covalent bonds with skin protein in vivo (see Fig. 20-4), and the complexity of the derived molecules obscured the unusual specificity requirements of this and other cell-mediated responses (Carrier specificity, below).

Induction. Contact skin sensitivity is usually induced by percutaneous application or intradermal injection of the sensitizer. The skin can be bypassed only by administering the sensitizer in complete Freund's adjuvant. It has been suggested that skin lipids might exert an adjuvant effect comparable to the mycoside of *M. tuberculosis* (above).

Elicitation. Beginning 4 or 5 days after the sensitizing exposure an application of the sensitizer almost anywhere on the skin surface of man or guinea pig elicits delayed inflammation.*

Guinea pigs are tested with a drop of a dilute solution on the skin (after hair has been removed), followed by light massage. Humans, however, are tested with a **patch test:** a 1-cm piece of filter paper soaked with a dilute solution of sensitizer is placed on the skin, covered with tape, and left for 24 hours. The area of contact is examined a few hours after the patch is removed, and the response is scored again the next day.

Since sensitizers are potential irritants (because they react indiscriminately with most proteins; see below), they must be used at concentrations that avoid nonspecific inflammation, which are determined by testing nonsensitized individuals: with many potent sensitizers 0.01 M solutions are suitable.

Most simple sensitizers are relatively hydrophobic and readily penetrate the intact skin. They are usually administered in solvents that are partly non-volatile, such as 1:1 acetone-corn oil, to prevent excessive concentration by evaporation. Hydrophilic sensitizers penetrate the skin less well, and relatively high concentrations are required (e.g., 0.5 M penicillin). These substances probably penetrate human skin via sweat ducts, and their penetration of guinea pig skin, which lacks sweat glands, can be augmented by addition of nonionic detergents to the solvent.

The **time course** of the contact skin response is the same as that of the tuberculin reaction:

erythema and swelling appear at ca. 10 to 12 hours and increase to a maximum at 24 to 48 hours. Unusually intense reactions produce necrosis, and complete recovery of the skin site can take several weeks, even without necrosis.

Histologically, the dermis (deep layer of skin) at the site of contact is invaded by mononuclear cells, as in delayed-type responses to tuberculin (cf. Figs. 20-2 and 20-3). The epidermis (superficial layer of skin), however, looks different: it is hyperplastic and is invaded by mononuclear cells. In addition, **intraepidermal vesicles** regularly form in human (but not in guinea pig) skin; they sometimes coalesce to form large blisters filled with serous fluid, granulocytes, and mononuclear cells (Fig. 20-3).

Reactions with Proteins in Vivo. In accord with the general requirements for immunogenicity (Ch. 17), the actual immunogens are not the simple substances themselves, but the covalent derivatives they form with tissue (skin) proteins in vivo. For instance, among a group of 2,4-dinitrobenzenes those that can form stable, covalent derivatives of protein -SH and $-NH_2$ groups in vivo are sensitizers, while those that cannot form such derivatives are not (Fig. 20-4).†

Persistence. Once established in guinea pig or human, contact skin sensitivity probably persists for years, though it tends to wane. A patch test to evaluate its persistence can boost the level of sensitivity.

The persistence of sensitizers in tissues is evident in the "flare reaction." In this phenomenon a small amount of sensitizer, applied percutaneously as a diagnostic test to the skin of an insensitive person, causes no reaction; but 10 to 20 days later the test site flares up with a typical contact skin reaction. At that time repetition of the test anywhere on the skin elicits the characterisitc allergic response. Evidently enough sensitizer remains at the first site to provide an effective test dose when the subject be-

---

* Allergic contact dermatitis can be elicited on the ears of mice, but it cannot be evoked elsewhere on the skin in this species, or in many others, perhaps, as has been suggested, because of anatomical differences in small cutaneous blood vessels.

---

† Some macromolecules, such as denatured nucleic acids, become immunogenic on forming noncovalent complexes with proteins; the complexes are extremely stable because they involve many noncovalent bonds per interacting molecule. However, the simple inducers of contact sensitivity can form only a few noncovalent bonds per molecule and hence they definitely must form covalent derivatives.

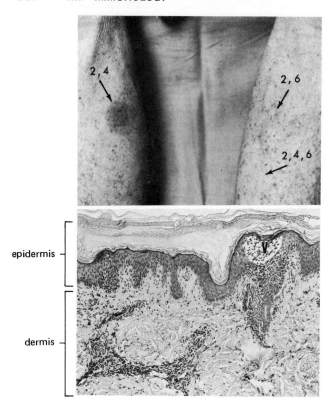

Fig. 20-3. Allergic contact dermatitis in man. The subject was sensitized to 2,4-dinitrofluorobenzene (DNFB) and then tested with 2,4-dinitrochlorobenzene (2,4), 2,6-dinitrochlorobenzene (2,6), and 2,4,6-trinitrochlorobenzene (2,4,6). The positive response was evident at 24 hours and photographed at 72 hours. Specificity is shown by the strong reaction to 2,4-dinitrochlorobenzene, and the absence of reactions to the 2,6 and 2,4,6 analogs. DNFB and all the analogs tested form dinitrophenyl (or trinitrophenyl) derivatives of skin proteins in vivo. Histology of the skin reaction is shown below. Note the characteristic intraepidermal vesicle (V) and the dense infiltration of dermis and epidermis by lymphoid cells. Epidermal cells around vesicles generally have a foamy cytoplasm ("spongiosus").

comes sensitive a few weeks later. The flare reaction thus resembles serum sickness (Ch. 19), though more extended in time and involving cell-mediated, rather than humoral, allergy.

SPECIFICITY

The determinants of specificity are still incompletely defined, but they appear to be larger and more complex than those involved in Ab-mediated reactions. For instance, guinea pigs sensitized with 2,4,6-trinitrophenyl-bovine γ-globulin (Tnp-BγG) respond more intensely to the immunogen than to the unsubstituted protein (BγG), and not at all to Tnp conjugated onto unrelated carriers, such as ovalbumin (Tnp-Ea). Hence the response appears to be specific for the total immunogen (Tnp plus carrier protein) or a large part of it: **delayed-type responses are carrier-specific.**

In Ab-mediated reactions, in contrast, the determinants are both smaller and better defined, in vitro and in vivo: for example, anti-Tnp Abs, produced in response to Tnp-BγG, form precipitates with Tnp-Ea, bind Tnp-lysine, and mediate anaphylactic and Arthus

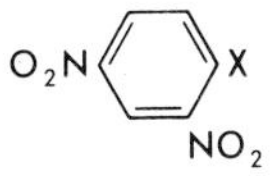

Substituents (X) in C-1 of O_2N benzene ring with X and NO_2	Reaction with protein in vivo		
	with ϵ-NH_2 of lysine residues	with SH of cysteine residues	Ability to induce and elicit contact skin sensitivity
X = —F	+	+	+
—Cl	+	+	+
—Br	+	+	+
—SO_3	−	+	+
—SCN	−	+	+
—SCl	−	+	+
—H	−	−	−
—CH_3	−	−	−
—NH_2	−	−	−

Fig. 20-4. Correlation among C-I substituted 2,4-dinitrobenzenes between ability to form 2,4-dinitrophenylated proteins in vivo and ability to induce and to elicit contact skin sensitivity. (Based on Eisen, H. N. in *Cellular and Humoral Aspects of the Hypersensitive States,* H. S. Lawrence, ed. Hoeber, New York, 1959).

responses with Tnp conjugated onto almost any protein (Table 20-3).

The need for large and complex antigenic units is also suggested by the apparent requirement that agents triggering cell-mediated responses must be immunogenic: for instance, with hapten conjugates of poly-L-lysines of varying lengths, only oligomers with more than seven lysines can both establish and elicit delayed hypersensitivity, whereas smaller nonimmunogenic oligomers react perfectly well with Ab molecules raised against the larger conjugates.

The requirement for large antigenic units helps clarify the specificity of contact skin allergy: for instance, an individual who has been sensitized with Tnp-BγG, and who gives delayed-type responses to an injection of this conjugate, will not respond to a contact test with Tnp-chloride (TNCB or picryl chloride), which readily forms Tnp conjugates with skin proteins in vivo. Conversely, guinea pigs sensitized with TNCB give delayed-type skin reaction to this substance, but not to conjugates of Tnp on BγG or other common proteins. It appears, therefore, that self-proteins of skin, specifically modified by covalent attachment of particular sensitizer groups (such as Tnp) are the actual antigenic units in contact skin responses.

The simple group, however, seems also to be required: for example, guinea pigs with contact skin sensitivity to TNCB do not give contact skin reactions to the corresponding 2,4-dinitrophenyls

(such as the chloro- or fluoro-derivatives, DNCB or DNFB), which readily form Dnp conjugates in vivo, probably by reacting with the same amino and sulfhydryl groups of the same skin proteins as TNCB (Fig. 20-4). Hence, both the specific haptenic and carrier groups seem to be needed (see also Fig. 20-3).

Why are the agents that trigger delayed-type responses larger and more complex than those that react with Abs? One possibility is that the binding sites of receptors on T lymphocytes (see below) are structurally different from binding sites of Abs; another is that triggering the cellular responses requires more than merely binding of Ag: the cell's membrane presumably undergoes a perturbation that results in "activation" and release of diverse products (see below). It is also possible that the triggering of delayed-type responses requires cell-cell cooperation, though the cells are likely to be T lymphocytes with or without macrophages (see below), rather than T and B cells (and macrophages) as in induction of secondary responses in Ab formation, where the need for carrier specificity is also prominent (Ch. 17). Indeed it is possible that large multideterminant ligands and "carrier specificity" are characteristic requirements of all cellular responses to Ags, whether involved in delayed-type hypersensitivity or induction of Ab synthesis, rather than a distinguishing difference between cell-mediated and humoral immunity.

OTHER CELL-MEDIATED IMMUNE RESPONSES

RESISTANCE TO INFECTIOUS AGENTS

Many bacteria survive phagocytosis and even multiply within phagocytic cells (e.g., tubercle bacilli, leprosy bacilli). Clinical and experimental observations have shown, however, that macrophages from individuals with such infections have augmented ability to kill the infecting bacteria and also many other antigenically unrelated ones. Though the antimicrobial activity of these altered macrophages is nonspecific, their conversion into "angry killers" is due to a specific immunological process: when an animal is primed by infection with one organism and then cured and later reinfected with a small number of bacteria of the same type, its macrophages promptly become nonspecific killers, but a similar small number of antigenically unrelated bacteria as a challenge would not have triggered the conversion of macrophages into killer cells. Like other cell-mediated immune reactions the specific ability to activate macrophages can be transferred—not with antisera but with lymphocytes whose specific reaction with Ag releases soluble factors that probably cause differentiation of macrophages into angry killers (below). Modified cells are more phagocytic, and have more lysosomes and lysosomal enzymes than normal macrophages (Figs. 20-5, 20-6).

Fig. 20-5. Comparison of resting **(A,C)** and activated **(B,D)** macrophages. All are peritoneal cells from normal **(A,C)** or infected mice **(B,D)** 14 days after injection of tubercle bacilli (BCG, bacillus Calmette-Guerin). In the phase-contrast photographs **(C,D** ×1900) the activated macrophages **(D)** are larger (the field is almost filled by one-half a cell), more spread out, and have increased content of organelles, especially lysosomes (dense, spherical bodies); the translucent spherical bodies represent ingested culture medium (pinocytotic vesicles). In the electron micrographs **(A,B;** magnification uncertain) the dense spherical lysosomes (L) are abundant in activated cells and rare in resting cells, most of whose organelles are mitochondria (M). [**A,B** from Blanden, R. V., Lefford, M. J., and Mackaness, G. B. *J Exp Med 129*:1079 (1969); **C,D** courtesy of Dr. G. B. Mackaness.]

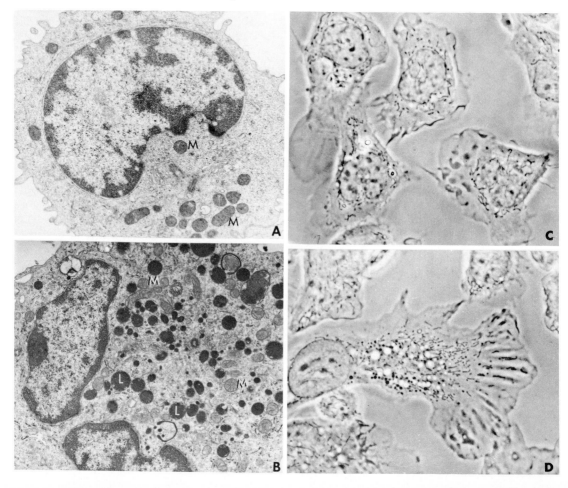

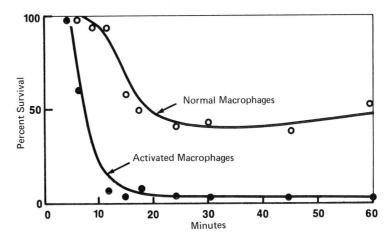

Fig. 20-6. Increased bactericidal activity of activated macrophages. *Salmonella typhimurium* coated (opsonized) with anti-*Salmonella* antibodies are ingested and killed more rapidly and in greater numbers by activated than by normal macrophages. [From Mackaness, G. B. *Hosp Practice,* 73 (1970).]

Besides its protective role in infections with intracellular bacteria (tuberculosis, leprosy, brucellosis), this type of cell-mediated immunity is probably important in many viral, fungal, and protozoan infections. The evidence comes not only from a lack of Ab protection in experimental infections, but from the greater severity of clinical infections in those with diminished cell-mediated immunity due to heritable or acquired defects (Clinical deficiencies, below) or to certain immunosuppressive agents (antilymphocyte serum, below). Activated macrophages are also important in some other cell-mediated immune responses (Accessory cells; Tumor immunity, below).

REJECTION OF ALLOGRAFTS

Cells, tissues, or organs that are transferred from a donor to a genetically different recipient of the same species are called **allografts.** Because of extensive polymorphism of certain surface glycoproteins (Ch. 21) the grafted cells almost invariably contain on their surfaces histocompatibility or transplantation Ags that are lacking on host cells (and vice versa).*

---

* Chapter 21 deals in detail with the principles governing survival of transplanted cells. It is necessary to repeat here (see also Chs. 17 and 21) that cell surface histocompatibility Ags, like Ig allotypes, are genetically segregating alloantigens (Gr., *allos,* other): they are present in some and absent in other members of the same species. Genetically identical (syngeneic) individuals (monozygotic twins or members of highly inbred animal strains) have identical histocompatibility Ags and accept grafted cells freely from each other. Genetically different

The resulting host response leads to destruction of the allograft through a delayed-type hypersensitivity reaction, which depends on specific lymphocytes (below) and seems usually to be independent of Abs.

When one allograft is followed by another from the same donor the second rejection occurs more rapidly than the first (Ch. 21, Second-set reaction). The acceleration is specific: it does not occur with another allogeneic graft, donated by a third party, to the same recipient. As with delayed-type tuberculin hypersensitivity, the site of the allograft rejection is characterized by chronic infiltration with lymphocytes and macrophages, and the capacity for the second-set reaction can be transferred by living lymphocytes, but not by antiserum. Moreover the intensity and speed of graft rejection do not correlate with the level of serum Abs to transplantation Ags; indeed, these Abs are often undetectable.

Under special circumstances, however, certain allografts can be rejected wholly, or partly, as a result of reactions with conventional Abs. In contrast to "solid" tissue and organ grafts, dispersed allogeneic lymphocytes can be lysed by the combined action of Abs to their transplantation (or other surface) Ags, plus complement (C). The rejection of tissue allografts can also sometimes be speeded up by transfer of massive amounts of antiserum from hyperimmune animals, but the histological appearance is then anomalous, with edema, vascular occlusion, and granulocytes, rather the usual mononuclear in-

---

(allogeneic) individuals of the same species reject each other's grafts, owing to the host immune response against histocompatibility Ags on the grafted cells.

filtrate. Abs to vascular endothelium can also accelerate rejection of kidney allografts, probably through occlusion of nutrient blood vessels (Ch. 21, Effect of blood group incompatibilities on kidney allografts).

GRAFT-VS.-HOST REACTION

This form of the allograft reaction occurs when lymphocytes are transferred from an immunologically competent donor (normal adult) to an allogeneic incompetent recipient (e.g., newborn). These reactions have increasing clinical importance because of therapeutic attempts to transfer normal thymus or bone marrow cells to immunodeficient humans [e.g., infants with genetic defects (see Ch. 17 and below), patients with leukemia treated with cytotoxic drugs and whole-body X-irradiation]. The recipient exhibits loss of weight, skin rash (in guinea pigs and man), increase in spleen size, stimulation of macrophage activity (below); chick embryos receiving allogeneic lymphocytes from adult chickens develop characteristic focal lesions on the chorioallantoic membrane. The intesity of the reaction is generally proportional to the number of transplanted lymphocytes, and a sufficient number can cause the recipient's death.

Nodules appear in many organs, but are most numerous in the spleen. Cells from mouse donors with the easily recognized T6 chromosomes show that a typical nodule starts with proliferation and blast transformation of donor lymphocyes (see below and Ch. 17), followed by massive proliferation of host fibroblasts.

TUMOR IMMUNITY

Clinical observations have long hinted at immunity against tumors, e.g., cancers disappear spontaneously from rare patients, and unusually slow-growing tumors are often prominently infiltrated with lymphoid cells. Almost all carefully studied tumor cells are now known to have distinctive surface Ags that are lacking or masked on normal cells (Ch. 21). Considerable evidence shows that these tumor-specific Ags are immunogenic in the **autochthonous host** (the one in whom the tumor arises), and that cell-mediated responses are especially capable of restricting tumor growth.

1) In a classic study by Klein and coworkers, a sarcoma induced in a mouse by methylcholanthrene was excised and carried by successive transplants in syngeneic mice while the surgically cured autochthonous host was immunized with his X-irradiated sarcoma cells.* The immunized animal was then able to reject a graft of his own tumor, but was not resistant to another sarcoma.

2) Under standard tissue culture conditions a patients' tumor cells usually form fewer colonies if incubated with his own lymphocytes than if incubated with lymphocytes from other persons (Hellstrom).

3) Oncogenic polyomavirus causes tumors to appear in adult mice only if the animals' ability to make cell-mediated responses has been eliminated, e.g., by thymectomy and antilymphocyte serum (ALS; see below, Ch. 17, and Table 20-4). Moreover, humans with defective cellular immunity have an abnormally high incidence of various cancers: in chronically immunosuppressed recipients of kidney allografts, malignant lymphomas, particularly reticulum cell sarcomas, appear ca. 4000 times more frequently (ca. 0.5%) than in the population at large (same age and sex distribution).

The role of cell-mediated immunity in resistance to tumors is also evident in the behavior of **transplanted tumors.** Tumor cells usually have the same histocompatibility Ags as normal cells (Ch. 21) of the autochthonous host and hence are rejected (like normal tissue allografts, Ch. 21) when transplantetd to allogeneic recipients. However, tumor grafts can flourish and eventually kill recipients that are syngeneic with the autochthonous donor. The number of cells required for a successful (i.e., lethal) syngeneic graft increases if the recipient has been previously immunized against that tumor (Fig. 20-7).

For instance, 1000 tumor cells might suffice to produce progressive tumors in 50% of unimmunized syngeneic recipients ($TD_{50} = 1000$ cells/recipient), whereas this number might produce only a rare tumor in recipients who were previously immunized with that tumor, or even with its isolated tumor-specific Ag. This resistance can, however, be overcome by a larger inoculum, which might range from perhaps 3 to 1000 times the TD_{50} dose. Tumors vary greatly in virulence: TD_{50} doses can range from 10 to 10^6 cells per unimmunized recipient. This wide variation, and the effect of immunization (Fig. 20-7), suggest that **progressive growth of a tumor (as graft or as indigenous clone) requires a critical mass above which it can grow more rapidly than it can be destroyed by the immunity it elicits.**

---

* X-irradiation (e.g., 750 r) blocks cell division by damaging DNA (Ch. 11) but leaves the cell Ags intact.

TABLE 20-4. Increased Incidence of Tumors (Due to Polyomavirus) in Mice with Impaired Cell-Mediated Immunity

Preliminary treatment of adult mice, CBA strain*	Treatment 7 wk later†	No. of mice	Percent that develop tumors
Normal rabbit γ-globulin (control)	None	24	0
Thymectomy + ALS*	None	14	100
Thymectomy + ALS	Inject lymphoid cells from normal CBA mice	10	90
Thymectomy + ALS	Inject lymphoid cells from CBA mice immunized with polyomavirus	11	0

* Polyomavirus was injected after thymectomy and the first of three biweekly injections of ALS (γ-globulin fraction of rabbit antiserum to mouse lymphocytes).

† Ten days after final injection of ALS.

Based on Allison, A. C. *Proc R Soc Lond* (*Biol*) *177*:23–39 (1971).

The resistance induced by immunization is specific for the tumor used as immunogen, and sometimes for closely related tumors (e.g., those raised in other animals with the same oncogenic virus, Ch. 63). It can usually be transferred with viable lymphocytes but not with serum. Indeed, **passively transferred serum can sometimes decrease the recipient's resistance:** i.e., transfer of antiserum plus tumor cells can result in a higher incidence and earlier appearance of tumors, perhaps because Abs bind to surface Ags on the tumor cells, preventing either induction of cell-mediated immunity or its expression (attachment of specifically reactive lymphocytes: see below and Ch. 21, Enhancement).

Current efforts to increase clinical resistance to established tumors include injections of avirulent tubercle bacilli (BCG strain) into human skin tumors, or painting the tumors with 2,4-dinitrochlorobenzene, a potent skin sensitizer (DNCB, Fig. 20-4). The ensuing delayed-type allergic reaction within the tumor may increase destruction of tumor cells by providing "angry" macrophages (see Resistance to infectious agents, above) or cytotoxic

Fig. 20-7. Dose-response curves for growth of transplanted tumors. Recipients were syngeneic with the autochthonous host. Besides use of X-irradiated cells as immunogen, resistance can sometimes be induced with a subthreshold inoculum (e.g., 100 cells, curve A): before it reaches the critical mass the tumor is eliminated by the immunity it induces. A recipient will often also become resistant to another graft of the same tumor if its first tumor graft is excised after about 1 week's growth.

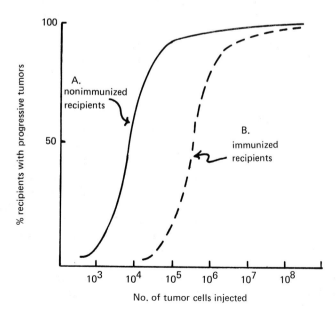

factors (Products of activated lymphocytes, below); or tubercle bacilli and allergic inflammation may have an adjuvant effect on the immunogenicity of the tumor-specific Ags (Ch. 17, Adjuvants).

Surveillance. All of the foregoing suggest that cell-mediated immunity probably functions continuously in normal individuals as a **surveillance** mechanism, destroying newly arising neoplastic clones before they exceed a critical mass. Occasional clones probably develop into clinically significant tumors because 1) their cell surface tumor-specific Ags have exceptionally low immunogenicity, 2) the host may lapse in its ability to make cell-mediated immune responses to Ags in general, 3) tumor variants may emerge with high growth rates or increased resistance to cytotoxic effects of immune reactions (below), and 4) "blocking" Abs to tumor Ags can sometimes prevent attack by specifically reactive lymphocytes (below and Ch. 21, Enhancement). Selective pressure of immune surveillance may account for feeble immunogenicity of distinctive tumor-specific Ags of most well-established tumors in their autochthonous hosts. Immunity to tumors is discussed further (Ch. 21).

PHYLOGENY AND ONTOGENY

As was noted before (Ch. 17) all species of vertebrates have both cell-mediated and humoral immunity at birth or hatching or shortly afterward. Invertebrates do not form Abs (Ch. 17) and their defenses against invasive parasites are probably based on primitive forms of cell-mediated immunity. Metchnikoff's classic description of motile, phagocytic cells (amebocytes) engulfing a penetrating thorn in a sea star indicated that macrophage-like cells of invertebrates could recognize and react against foreign objects; and amebocytes of a marine worm are now known to phagocytize various foreign cells (but not their own). Suggestive evidence for rejection of skin allografts has also been found in earthworms (but not in other invertebrates).

Since ontogeny generally recapitulates phylogeny, the time sequence for acquisition of diverse immune responses during fetal development of vertebrates should clarify evolutionary sequences. However, the evidence is not easily interpreted: fetal lambs become capable of rejecting allografts at 75 days of gestation (halfway through pregnancy), but they make Abs to certain Ags before 60 days and to other Ags only after birth (Ch. 17). Human fetuses also synthesize Igs (Ch. 17), but delayed-type skin reactions seem not to be elicitable until some months after birth, perhaps because of immaturity in accessory cells (below) or in the skin itself, rather than in the immunologically specific cells.

CELLULAR BASIS FOR CELL-MEDIATED IMMUNITY

Conventional Ab molecules are excluded from a role in these reactions because countless attempts to transfer the capacity for delayed-type skin reactions with serum have failed, and the few reported successes have not been corroborated. Moreover, these reactions, and allograft rejection, can be elicited in individuals who are almost devoid of serum Igs: e.g., children with immunodeficiency disorders (X-linked agammaglobulinemia, Ch. 17), mice treated from birth with Abs to μ chains (Ch. 17).

T LYMPHOCYTES AS MEDIATORS

Though suspected from the early 1900s, cells were first shown to mediate delayed-type allergy in 1942, when Chase transferred delayed sensitivity to tuberculin from sensitive to normal guinea pigs with viable lymphoid cells. Since then similar results with many other Ags, and with other forms of cellular immunity,

have established that **transfer by viable cells from lymph nodes, spleen, or peritoneal exudates is an essential characteristic of cell-mediated immune reactions.** However, it is not an exclusive property: similar populations of lymphoid cells from an intensively immunized animal can continue to form Abs in the recipient (or even in tissue culture, Ch. 17) and thus can probably also occasionally transfer immediate-type allergy.

The cell suspensions ordinarily used to transfer sensitivity are usually referred to as "lymphoid cells," to emphasize that they are crude mixtures of lymphocytes (themselves highly heterogeneous), macrophages, and other cells (erythrocytes, granulocytes). After fractionating a mixture by sedimentation Gowans demonstrated that **small lymphocytes** are the active cells in the graft-vs.-host reaction.

Extirpation experiments then showed that **the active small lymphocyte is a T cell:** for example, allograft survival is greatly prolonged in neonatally thymectomized chickens or mice but not in bursectomized chickens (Ch. 17, Table 17-4). Moreover, lymphoid cell populations enriched for T cells (by passage over glass beads to remove B cells, Ch. 17) require fewer cells to transfer cell-mediated reactions, while populations enriched for B cells (by destroying most of the T cells with Abs to the θ surface antigen, Ch. 17) are less active.

The association of T cells with cell-mediated immunity and of B cells with humoral immunity is in accord with zonal effects within the lymph nodes that drain sites of injected Ags: when the stimulus leads to delayed-type hypersensitivity (e.g., painting the skin with a contact sensitizer) the regional lymph node's T zones become greatly hyperplastic, and when the stimulus causes Ab formation with little or no cell-mediated immunity proliferation of lymphocytes is largely confined to B-cell areas (Ch. 17, see germinal follicles in Figs 17-15 and 17-16).

MACROPHAGES AS NONSPECIFIC ACCESSORIES

Macrophages are even more abundant than lymphocytes at sites of delayed-type reactions or allograft rejections, and are required. Thus lethally irradiated animals (900 r) developed cell-mediated responses only when both lymphocytes and bone marrow cells were transferred. Administration of ³H-thymidine to the marrow donor showed that rapidly dividing "promonocytes" in bone marrow (tagged in DNA) are precursors of blood-borne monocytes (immature macrophages, Ch. 17) which infiltrate the developing skin lesion and differentiate there into mature macrophages with abundant lysosomal enzymes (see Fig. 20-5). **The macrophages act nonspecifically, for nonsensitized and sensitized donors provide equally effective marrow cells. The lymphocytes, however, are effective only if obtained from a sensitized donor.**

Relatively few specific T cells are needed to initiate an allergic response. After adoptive sensitization of a normal recipient with labeled lymphocytes (from an actively sensitized donor who had been treated with ³H-thymidine) fewer than 10% of the mononuclear cells in the recipient's skin lesion were labeled, whereas over 90% were labeled when normal recipients were given ³H-thymidine and sensitized lymphocytes were then transferred from an unlabeled donor. Studies with cultured cells (below) show that on reacting with Ag the sensitized lymphocytes release substances that cause the chemotactic accumulation and differentiation of macrophages. It is likely that the accumulated macrophages can behave as the "angry killers" that destroy diverse, antigenically unrelated bacteria in cell-mediated immunity to infections (Figs. 20-5 and 20-6).

GENETIC CONTROL

The ability of guinea pigs to develop delayed-type hypersensitivity to some Ags is governed by a genetic locus that seems to determine specific binding of Ag by lymphocytes. Thus with hapten–poly-L-lysine (PLL) conjugates as Ag animals of an inbred "responder" strain (2) can develop delayed skin sensitivity but those of a "nonresponder" strain (13) cannot. The nonresponders, however, are able to exhibit delayed skin reactions to the PLL conjugates if adoptively sensitized with lymphoid cells from an immunized responder*: hence the nonresponder is defective in the induction of sensitivity, rather than in its expression.

Breeding experiments show that the responder phenotype is governed by a single, dominant autosomal gene: e.g., all F_1 hybrids (2 × 13) are responders (PLL+/PLL−), as are 50% of the progeny of the back-cross between the heterozygotes and nonresponders (PLL−/PLL−). The PLL gene is linked to the locus that specifies potent histocompatibility alloantigens on guinea pig cells.† The PLL gene's product may be a histocompatibility Ag; or it may be the elusive Ag-binding receptor on T cells, for antisera produced in nonresponders to histocompatibility Ags of the responder strain block specific reactions of the responder's T lymphocytes (such as Ag-induced blast transformation), but their B cells, whose receptors are clearly immunoglobulins (Ch. 17), are not affected.

---

* Though the transferred lymphocytes are allografts they can mediate delayed-type hypersensitivity in the recipient for about 1 to 2 weeks; i.e., until they are destroyed by the recipient's immune response to their histocompatibility Ags.

† The corresponding locus in mice, Ir-1, is linked to the locus for the most potent histocompatibility alloantigens in this species (H-2; see Ch. 17, Genetic regulation of Ab formation; also Ch. 21). Ir-1 probably also specifies delayed-type allergic reactions, but those elicited with soluble proteins are difficult to discern in mice.

RELATION TO HELPER CELLS
IN ANTIBODY FORMATION

The T lymphocytes that specify cell-mediated immune responses have many properties in common with those that serve as helper cells in Ab formation:

1) The reactions with Ag appear to be specific for larger determinants ("carrier-specific") than those of conventionalAbs.

2) Both require macrophages as nonspecific accessory cells in carrying out their respective activities (delayed-type allergic reactions and modulation of the B-cell response to immunogen).

3) Both can function without dividing: i.e., X-irradiation does not block their activities, though it inhibits their induced proliferation (Nonspecific suppression, below).

4) Recognition of Ag by both cells might be specified by the same genetic locus (called *Ir-1* in mice and *PLL* in guinea pigs), which could conceivably code for their Ag-binding receptors (see Ch. 17, Genetic control of Ab formation).

However, the effector T cells of cell-mediated immunity react specifically with the concerned Ag, whereas helper function can be less specific. For instance, T cells triggered by diverse Ags can modulate nonspecifically the B cell responses to other Ags (see Ch. 17, Allogeneic effect, and Antigenic promotion and competition).

The ability of Ag-sensitive T lymphocytes to engage in both functions would provide a basis for the old view that delayed-type hypersensitivity is a preliminary stage in the induction of Ab formation: the initial proliferation of T cells in response to an immunogen could contribute to the subsequent differentiation of neighboring B cells. Delayed-type hypersensitivity reactions might thus serve as an "adjuvant" in the formation of Abs to T-dependent Ags. Some conventional adjuvants probably also exercise their effects by attracting (and activating) many T cells: thus Freund's complete adjuvant promotes Ab formation to many T-dependent Ags, but not to T-independent ones, such as pneumococcal polysaccharides (Ch. 17).

TRANSFER FACTOR

Though cell-mediated responses are readily transferred with living lymphocytes, disrupted cells are ineffective in all animal species tested except man, in whom Lawrence found that extracts of blood leukocytes can apparently transfer the cell-mediated responses of the leukocyte donor. Recipients become reactive to Ag 1 to 7 days after receiving the extract and sensitivity can persist for years, even when the extract is prepared from as little as 5 ml of blood (corresponding to about 10^7 total lymphocytes).

The active component, "transfer factor," has been difficult to characterize since it can be assayed only in humans. It is dialyzable, stable on prolonged storage, and has a molecular weight of less than 10,000; it is inactivated by heat (56°, 30 minutes) but not by trypsin, RNase, or DNase. Clinical trials indicate that transfer factor may be effective even in those with deficient cell-mediated immunity (below). For instance, patients with disseminated *Candida* infections (chronic mucocutaneous candidiasis) lack delayed skin responses to *Candida* proteins, but after receiving transfer factor from *Candida*-sensitive donors their previously intractable infections sometimes subside, and some of them manifest delayed-type skin reactions to *Candida*. Similar events have occurred in patients with disseminated vaccinia, as well as in children with congenital cell-mediated immunodeficiencies (Wiskott-Aldrich syndrome, below).

Transfer factor seems to lack antigenicity; it is not inactivated by anti-Igs; and it seems too small to be an informational molecule, such as mRNA. Possibly it is an immensely potent nonspecific adjuvant, enhancing induction of cell-mediated immunity to prevalent Ags or to those injected intradermally to evaluate its effects. Clarification of the nature and mode of action of transfer factor is likely to illuminate fundamental mechanisms of cellular immunity.

CELL-MEDIATED IMMUNE REACTIONS
WITH CULTURED CELLS

The reactions of cultured cells have been intensively explored to develop assays for measurement of cell-mediated immunity and to analyze mechanisms. In contrast to methods for measuring Abs and Ab-mediated reactions, assays for cell-mediated responses are cumbersome and imprecise. Nonetheless, the in vitro analyses clearly emphasize several phases: specific reaction of Ag with sensitized T lymphocytes stimulates their differentiation into blast cells and their secretion of factors that act nonspecifically on a variety of other cells. The products of activated lymphocytes ("PALs" or "lymphokines") have not so far been obtained in sufficient quantity or purity for detailed analysis; hence they are consid-

TABLE 20-5. Products of Activated Lymphocytes

Product	Activity
A. Affects macrophages	
Migration-inhibition factor (MIF)	Inhibits migration of normal macrophages
Aggregation factor (AF)	Agglutinates suspended macrophages
Chemotactic factor (CF)	Causes migration of macrophages toward increasing concentration of the factor
B. Affects lymphocytes	
Blastogenic or mitogenic factor (BF or MF)	Induces normal lymphocytes to differentiate into blast cells (increasing their incorporation of thymidine into DNA)
Potentiating factor (PF)	Enhances blast transformation in Ag-stimulated cultures
Helper factor	Enhances differentiation of Ab-forming cells in culture
C. Affects cultured cells	
Lymphotoxin (LT)	Damages or lyses various cultured cell lines (human Hela, mouse L)
Proliferation inhibitory factor	Blocks proliferation of cultured cells without lysing them
Interferon (?)	Inhibits viral multiplication in cultured cells
D. In vivo effects	
Skin reactive factor	Produces indurated skin lesions in normal guinea pigs
Macrophage disappearance factor	Intraperitoneal injection causes macrophages to adhere to peritoneum

Courtesy of Dr. Barry R. Bloom.

ered below only in terms of their effects on various target cells in vitro (Table 20-5).

BLAST TRANSFORMATION

Transformation Due to Antigen. The addition of Ag to cultured lymphocytes from sensitized individuals causes a small proportion of the cells to differentiate into large, rapidly dividing blast cells (Fig. 20-8). The morphological changes (which include conversion of nuclear chromatin from heterochromatin to euchromatin, and the development of conspicuous nu-

cleolus and polyribosomes; see Ch. 17) are associated with increased synthesis of protein, RNA, and especially of DNA, as though cells were preparing to enter mitosis. The peak of the response occurs ca. 5 days after Ag is added, and the changes are most conveniently revealed by increased incorporation of ^{3}H-thymidine into DNA.

Similar changes are brought about with lymphoid cells from individuals who are primed for Ab production but lack obvious cell-mediated hypersensitivity for the stimulating Ag (e.g., from persons with immediate-type hypersensitivity to plant pollens,

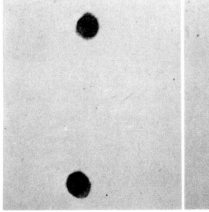
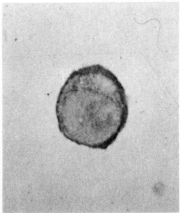

Fig. 20-8. Blast transformation of lymphocytes. Small (resting) lymphocytes at left; the same after stimulation by Ag or a plant mitogen (lectin) at right. (Based on Hirschhorn, K. *Human Transplantation.* Rapaport, F. T., and Dausset, J., eds. Grune and Stratton, New York, 1968.)

penicillins, etc.; Chs. 17 and 19). Nevertheless, the reactions seem to occur more regularly and with greater intensity among cells from those with delayed-type hypersensitivity.

Transformation by Mitogens.   Certain plant proteins that react specifically with various sugars of glycoproteins on the lymphocyte surface can also induce differentiation into blast cells. The changes seem to be the same as those induced by Ag, but the **plant "mitogens"** act nonspecifically: prior sensitization of the donor is unnecessary, and the majority of lymphocytes of a given class react.

B and T cells respond to different mitogens: for example, soluble concanavalin A (Con A) and phytohemagglutinin (PHA) transform T cells, but when these plant proteins are attached to insoluble particles (Sephadex beads) they also stimulate B cells (Fig. 20-9). **Pokeweed mitogen** (at low dose and in insoluble form) stimulates B and T cells. Though not of plant origin, **lipopolysaccharide** of cell walls from Enterobacterioceae transforms B cells exclusively.

The time course of events triggered by PHA spreads out over 4 or 5 days in culture. Within 5 minutes an increased turnover of phosphatidyl inositol of the cell membrane is noted, followed a few minutes later by acetylation of histones; increased synthesis of RNA is measurable at 2 hours but morphological changes are not detected until 20 hours, and increased incorporation of thymidine into DNA is measurable only after 2 to 4 days' incubation with the mitogen (or Ag).

Transformation by Mixed Lymphocyte Culture.   Small lymphocytes can also be transformed by culturing together the cells from genetically different individuals of the same species: because of extensive polymorphism at loci for cell surface proteins, the two cell populations are virtually always different antigenically, and they stimulate each other to undergo blast transformation. If individual X is made tolerant of individual Y, but not vice versa, Y's cells will undergo blast transformation in the mixed culture, but not X's. (The distinction between X and Y can be made in the "one-way reaction" through prior incubation of one or the other set of cells with mitomycin C; the drug blocks DNA synthesis without affecting the surface Ags and one can thus tell which cells stimulate and which respond.)

The results of mixed cultures help predict the histocompatibility "matching" and fate of human allografts (Ch. 21), but the high proportion of reactive cells in nonsensitized individuals (ca. 3%) raises some doubts about the immune mechanisms involved, especially since the proportion is not increased in sensitized individuals and it seems to be even lower with combinations of lymphocytes from different species, where the antigenic incompatibilities must be greater.

Specificity of Blast Transformation. The specificity of T cells has been demonstrated in cultures of cells from individuals who are sensitive to several Ags: if cells are stimulated by one Ag to incorporate bromodeoxyuridine (a light-sensitive analog of thymidine) and then are killed selectively by exposure to visible light, further response to that Ag is eliminated without affecting other cells in the same culture that are sensitive to other Ags.

If a given T cell can react with only one Ag, and the repertoire of Ags that evoke cell-mediated responses is as great as in Ab formation, the cells that respond to a given Ag should represent a very

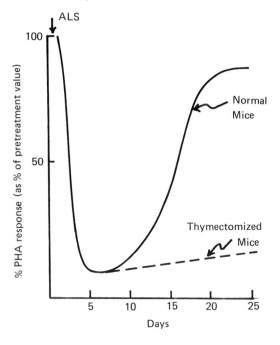

Fig. 20-9. Evidence that lymphocytes responding to phytohemagglutinin (PHA) are predominantly T cells. Injection of antilymphocyte serum (ALS) on day 0 abolished cells that respond by blast transformation to PHA; recovery of responsiveness to PHA was not observed in thymectomized mice, but was almost complete by day 20 in normal mice. [Based on Tursi, A., Greaves, M. F., Torrigiani, G., Playfair, J. H. L., and Roitt, I. M. *Immunology* 17:801 (1969).]

small fraction of an individual's lymphocytes. However, incubation with an Ag for a few days will often convert as many as 1% of an individual's lymphocytes into blast cells, and the proportion can reach 5% or more. These values could reflect death and lysis of nondividing cells in the culture, proliferative increase of Ag-activated cells (whose generation time is probably ca. 8 hours) and, most of all perhaps, recruitment of nonsensitized cells by the "blastogenic" factor released from authentically sensitized cells in response to Ag stimulation (below). The high proportion of blast cells elicited by single Ags has not been accounted for. Whether T cells are as restricted in specificity as B cells (Ch. 17) and Ab molecules is still uncertain.

Virus Plaque Assay for Cell-mediated Immunity. The number of Ag-activated lymphocytes can be counted by a novel plaque assay that is reminiscent of the hemolytic plaque assay for counting Ab-secreting cells (Ch. 17). This assay depends upon the capacity of Ag- (or mitogen)-activated T cells to support the replication of many viruses (mumps, measles, polio, vesicular stomatitis virus, etc.) that fail to multiply in resting lymphocytes. Each activated lymphocyte can then serve as an **infectious center,** producing a distinct cytopathic plaque when plated on a monolayer of virus-susceptible indicator cells (e.g., mouse L cells).

About 1 per 1000 lymphocytes from tuberculin-sensitive guinea pigs forms a plaque in culture after 24 hours' incubation with tuberculin, and the number increases linearly over several days, whether or not antimitotic agents are present (e.g., vinblastine). The infected cells therefore differ from blast cells, which increase exponentially and are prevented from increasing by the presence of antimitotic agents. Since X-irradiated (nondividing) lymphocytes can also mediate delayed-type hypersensitivity in adoptively sensitized animals the effector cell in vivo may be more like the responding lymphocyte in the virus plaque assay than the cell undergoing blast transformation.

The removal of glass-adherent cells abolishes the ability of Ag to establish infectious centers, suggesting that a macrophage is also required. This result fits the observation that the number of plaques formed is proportional to the square of the number of cultured lymphoid cells, indicating that two cells are necessary to form one center. Function of the macrophage in this reaction is obscured, as in blast transformation, where it seems also to be required.

PRODUCTS OF ACTIVATED LYMPHOCYTES

Effects on Macrophages. Ag-stimulated lymphocytes are associated with macrophages not only in many reactions in vivo but also in vitro. It has been known for over 40 years that the addition of Ag inhibits the outward migration of macrophages in cultured lymphoid cells from a donor with delayed-type hypersensitivity to that Ag (Fig. 20-10); in contrast, cultures from donors who lack this form of allergy are unaffected, even if the donor can make Abs to the Ag, or has immediate-type hypersensitivity to it. The response exhibits the "carrier specificity" that characterizes delayed-type skin reactions to hapten-protein conjugates (legend, Fig. 20-10), supporting the belief that it is an authentic cell-mediated immune reaction in culture.

Inhibition of macrophage movement is due to a **migration-inhibition factor** (MIF), which is released when Ag interacts with sensitized T lymphocytes. Thus if only 1% of the cells in a mixed culture are lymphocytes from a sensitized donor, migration of macrophages in the nonsensitive major population can still be inhibited by addition of Ag. Culture supernatants of Ag-stimulated sensitized lymphocytes have a similar effect. Peripheral blood lymphocytes from sensitive humans also release MIF on incubation with Ag.*

The culture supernatants of Ag-activated T cells can also **attract** macrophages: when separated by microporous filters from macrophages, these mobile cells move along the resulting gradient toward the supernatant (Fig. 20-11). But when mixed, the active supernatant blocks motility, apparently without killing the macrophages: net movement ceases because there no longer is a gradient for the cells to follow. In addition, the immobilized macrophages stick to each other, paralleling perhaps the disappearance of these cells from peritoneal cavity (and of monocytes from blood) when tuberculous guinea pigs are injected with large amounts of tuberculin. All three activities—inhibition of migration, chemotactic attraction, aggregation—could be due to the same substance (MIF).

PHA (above) not only stimulates blast transformation of T cells but causes release of MIF. De-

---

* Because of the difficulty in obtaining human macrophages the human culture supernatants are assayed with guinea pig macrophages (obtained from the peritoneal cavity 1 to 2 days after injecting thioglycollate or another irritant).

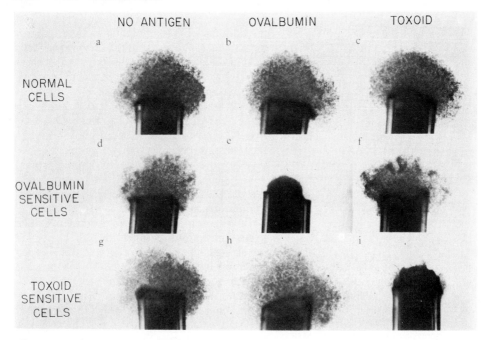

Fig. 20-10. Inhibition of macrophage migration: a response in vitro that parallels delayed-type hypersensitive reactions in vivo. Peritoneal exudate cells (predominantly macrophages and lymphocytes) are cultured in capillary tubes in presence or absence of Ags (ovalbumin, diphtheria toxoid). An Ag blocks outward migration of macrophages from the cell population derived from guinea pigs sensitive to that Ag. The response is also carrier-specific: for example, Dnp-guinea pig albumin (Dnp-GPA) blocks migration of cells from animals sensitized with Dnp-GPA, but not cells from those sensitized with Dnp or another protein, say Dnp-BγG. [From David, J. R., Al-Askari, S., Lawrence, H. S., and Thomas, L. *J Immunol 93*:264 (1964).]

Fig. 20-11. A two-compartment chamber demonstrates chemotaxis: the movement of cells along a concentration gradient of an attracting substance (chemotactic factor). Movement of macrophages from one chamber to and through the pores of a Millipore filter reveal chemotactic activity in culture supernatants of Ag-activated T lymphocytes (left). In the control (right) the uniform mixing of the culture supernatant with macrophages immobilizes the cells, as in the macrophage inhibition assay of Figure 20-10. Culture supernatants of Ag-activated lymphocytes are also chemotactic for normal lymphocytes (probably owing to another attractant).

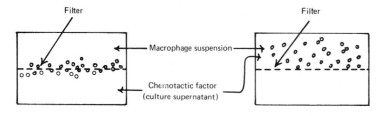

Filter   Filter

Macrophage suspension

Chemotactic factor
(culture supernatant)

spite this aid to purification only trace amounts of impure MIF have been obtained: the active substance appears to be a polydisperse relatively heat-stable glycoprotein (MW between 35,000 and 65,000).

Macrophages modified by MIF are "sticky" (i.e., they are more adherent to glass and to each other) and they exhibit increases in 1) movement of the surface membrane ("ruffled borders"), 2) phagocytic activity, 3) O_2 consumption, 4) lipid biosynthesis (perhaps related to increased activities at the cell membranes), 5) glucose degradation via the hexose monophosphate shunt (related to lipid biosynthesis), and 6) H_2O_2 production. All these changes are consistent with the differentiation of macrophages into "angry killers" of ingested bacteria, but so far increases in lysosomal enzymes and in bactericidal activity have not been definitely established in MIF-modified macrophages.

Effects on Cultured Cell Lines.

Supernatants from activated lymphocytes (stimulated by reaction with Ag, or plant mitogen, or with allogeneic cells in mixed lymphocyte cultures) can destroy or inhibit growth of various cultured cell lines (HeLa cells, L cells, etc.). The active substance, called *lymphotoxin* (LT), could account for necrosis of normal "bystander" cells at sites of delayed-type allergic skin reactions and in the severe inflammatory reaction around some chronic infections (e.g., tuberculosis).

Release of LT is not consistently associated with blast transformation; it precedes the characteristic increase in synthesis of DNA, and it is brought on by some mitogen inducers of blast transformation (e.g., PHA) but not by others (Con A).

Human lymphocytes stimulated by prolonged incubation with Ag (> 3 days) seem also eventually to produce **interferon**, which can block viral infection of other cells.

Effect on Lymphocytes.

Prolonged incubation of sensitized lymphocytes with Ag releases a **"blastogenic"** factor, whose addition to other cultures increases both proliferation of non-sensitized lymphocytes and the Ag-induced transformation of sensitized lymphocytes. Accordingly, the large number of transformed cells that appear with time in an Ag-treated culture probably represents **recruitment of nonsensitized lymphocytes** by the "blastogenic" factor from a much smaller number of authentically sensitized cells, activated specifically by the Ag. Culture supernatants of Ag-activated lymphoid cells are also chemotactic for lymphocytes (Fig. 20-11).

Effect on Skin. Culture supernatants of Ag-treated lymphocytes produce localized inflammation in guinea pig skin. The reaction has little if any erythema, and its course is slightly more rapid than that of a typical delayed-type reaction: the indurated lesion appears at 3 hours, is maximal at 6 to 12 hours, fades at 24 hours, and is prominently infiltrated with mononuclear cells. The skin activity could be due to the combined effects of some of the factors discussed above.

CYTOTOXIC MECHANISMS

A variety of cell-mediated reactions can destroy target cells (i.e., those with the relevant surface Ag) and also bystander cells, which happen to be in the immediate vicinity of a specific reaction between a T lymphocyte and its Ag. One model system indicates that target

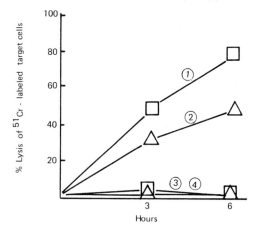

Fig. 20-12. Contact killing of target cells by sensitized lymphocytes. The target cells, cultured mastocytoma cells from C57Bl mice or lymphoma cells from DBA mice, were labeled with ^{51}Cr; released radioactivity measured their lysis. The sensitized lymphocytes were obtained by graft-vs.-host reactions: i.e., the transfer of spleen cells from C57Bl mice to X-irradiated DBA mice (yielding C57Bl anti-DBA cells), or from DBA to X-irradiated C57Bl mice (yielding DBA anti-C57Bl cells).

The cultures contained 1) anti-DBA lymphocytes plus labeled DBA and unlabeled C57Bl target cells; 2) anti-C57Bl lymphocytes plus labeled C57Bl and unlabeled DBA target cells; 3) anti-DBA lymphocytes plus labeled C57Bl and unlabeled DBA target cells; 4) anti-C57Bl lymphocytes, plus labeled DBA and unlabeled C57Bl target cells. [Based on Brunner, K. T., Nordin, A. A., and Cerottini, J. C. *Int Adv Immunol* (1970).]

cells are destroyed by specific contact with T lymphocytes ("contact killing"); another suggests that a toxic substance, released by Ag-activated T cells, lyses target and bystander cells; in a third it appears that normal lymphoid cells destroy target cells that are coated with Ab molecules (Ab-dependent cell mediated cytotoxicity). Each of these mechanisms is defined by a model in vitro system; which ones are important in vivo is still unclear.

Contact Killing. In this system thymus cells are injected into X-irradiated allogeneic recipients, where they home to the spleen and proliferate in a graft-vs.-host reaction (above). When subsequently isolated from the spleen these cells (now "educated" against host alloantigens) lyse cultured cells carrying the host's histocompatibility Ags (target cells). If the active spleen cells are added to a mixture of cells of various origins only the target cells are lysed (Fig. 20-12), suggesting that killing depends upon specific cell–cell contact, rather than on release of a soluble cytotoxic factor. Cytotoxicity is prevented by anti-θ and by ALS (see above, and Ch. 17), but not by anti-Igs, indicating the T lineage of the killer lymphocytes.

A linear relation between the number of added lymphocytes and the proportion of target cells that lyse suggests that the reaction is "one-hit" (i.e., a single sensitized T cell can kill a target cell); moreover as many as 1% of the lymphocytes in a highly sensitized individual appear to be specific killers

for a particular type of target cell. A soluble cytotoxic substance has not been detected in this system (Fig. 20-12), perhaps because too few lymphocytes are used (see below).

Cytotoxicity by Released Lymphotoxin. In this system, incubation of a more concentrated suspension of lymphocytes with their Ag, or with PHA, releases into the medium a toxic substance (**lymphotoxin,** see above), that can damage or lyse a wide variety of cells (e.g., human HeLa cells, mouse L cells, and other cultured cell lines).

Antibody-dependent Cell-mediated Cytotoxicity. This system differs fundamentally from T-cell-mediated immune reactions: target cells with Ab molecules bound specifically to their surface Ags are destroyed by normal lymphoid cells (i.e., from nonsensitized donors). The active cells (from spleen, lymph nodes, blood, but not thymus) are not T cells, and they adhere to glass. They seem to be B lymphocytes, or macrophages, or "A" cells (adherent mononuclear lymphoid cells whose lineage is obscure). Like B lymphocytes (Ch. 17), the cytotoxic cell evidently has surface receptors for the Ab moiety of Ab-Ag complexes, recognizing perhaps an altered conformation in the Fc domain that results from Ab binding to Ags on target cells. Complement is not involved in binding the Ab or in the destruction of the target cell.

INTERFERENCE WITH CELL-MEDIATED IMMUNITY

TOLERANCE: SUPPRESSION BY ANTIGEN

Similar conditions lead to unresponsiveness for cell-mediated reactions and for Ab formation: in general, **an Ag tends to be tolerogenic under conditions where it is not immunogenic** (Ch. 17). Thus unresponsiveness is more readily established in neonatal than in immunologically mature adults. In fact, current interest in tolerance was largely stimulated by Billingham, Brent, and Medawar's dramatic demonstration of allograft tolerance in newborn mice; the injection of living cells from one inbred strain into newborn mice of a second strain established enduring and specific tolerance in the recipients for allografts from the first strain (Ch. 21). As is discussed in Chapters 17 and 21, the inoculated cells can continue to multiply throughout the recipient's lifetime; hence they probably maintain tolerogenic levels of histocompatibility alloantigens.

It is also possible that this induced form of unresponsiveness is actually pseudotolerance, resulting from production of blocking Abs (see Tetraparental mice, below; and Ch. 21, Enhancement).

Variations in **route** of administration are also important. Guinea pigs fed 2,4,6-trinitrochlorobenzene (picryl chloride) became specifically unable to develop contact skin sensitivity to this potent sensitizer, or to form anti-Tnp Abs to picryl proteins, while remaining normally responsive to other Ags. Similarly, unresponsiveness to neoarsphenamine can be established by injecting this substance intravenously, rather than intracutaneously (an immunogenic route). The large doses presumably correspond to high-zone tolerance (Ch. 17); low-zone tolerance in cell-mediated immunity has not yet been demonstrated. Since T cells are more readily and more persistently tolerized than B cells, tolerance should be more easily established for cell-mediated responses than for Ab formation; but so far a difference seems not to have been clearly demonstrated.

Immune Deviation. Administration of Ag under some conditions (adsorbed on alumina gel, or sometimes in solution) seems to induce Ab formation without intervening delayed-type hypersensitivity. When this happens the cell-mediated sensitivity is not likely to be induced by a subsequent injection of the same Ag under what would ordinarily be effective conditions (e.g., incorporation in complete Freund's adjuvant). This effect borders on selective tolerance in respect to cell-mediated immunity and the mechanism is obscure. It may not be due entirely to interception of Ag by the Abs formed as a result of the first inoculum, because the phenomenon has not been regularly reproduced with antisera. If it could be consistently achieved, this "immune deviation" could have considerable clinical value in enhancing survival of allografts (Ch. 21).

Desensitization. In contrast to the establishment of tolerance, where Ag acts as tolerogen *before* it has a chance to function as immunogen, desensitization is established by administering the Ag *after* an individual has already been sensitized to it. Partial desensitization probably occurs whenever a sensitized individual is tested simultaneously at many skin sites with the same Ag: the reaction at each site is usually less intense than after a single test is applied, probably because the number of sensitized T cells is limited. Similarly, in overwhelming infections (such as miliary tuberculosis or disseminated fungal or protozoan infections) delayed-type hypersensitive skin responses can no longer be elicited, probably because the large amount of released Ag saturates sensitized lymphocytes. Comparable events occur in certain industries, where workers with allergic contact dermatitis sometimes lose skin sensitivity after prolonged and intense exposure to the sensitizer. However, deliberate desensitization by repeated administration of Ag or simple sensitizers has been difficult to achieve regularly in patients without frequently precipitating severe allergic reactions. Moreover, any desensitization that is achieved is short-lived (e.g., 5 to 10 days).

It is reported that American Indians formerly chewed poison ivy leaves as prophylaxis or treatment of poison ivy dermatitis. Recently a similar approach, refined by feeding the purified catechol responsible for poison ivy sensitization, has been found to be of dubious value: the presence of the catechol in feces occasionally produces in sensitized individuals an unusually severe perianal contact dermatitis, once referred to as the "emperor of pruritus ani."

SUPPRESSION BY ANTIBODIES

Antibodies as Antiimmunogen. The effects of Abs on cell-mediated immunity are variable: some complexes with Ab reduces the Ag's immunogenicity and others enhance it (see also Ch. 17 for parallel variations in induction of Ab formation). The differences probably derive from variable effects of Abs with different heavy chains, and of Ab·Ag complexes with different mole ratios (below).

In the most provocative effect antisera block cell-mediated rejection of tumor grafts, resulting in **"immune enhancement"** of tumor growth. This effect could arise in two ways: Ab molecules covering tumor-specific antigenic sites on the surface of tumor cells could block induction of cell-mediated immunity **(afferent enhancement)** or block its expression (i.e., attack by sensitized lymphocytes or **efferent enhancement**). The latter mechanism appears to operate in many cancer patients: in tissue culture the patient's tumor cells usually form fewer colonies if incubated with his own lymphocytes than with those from other persons, and his own serum (but not other sera) can specifically block his lymphocytes' inhibitory activity (Ch. 21, Tumor immunity). Abs to histocompatibility Ags may similarly prolong survival of allografts of normal cells (Ch. 21, Enhancement).

Observations on **tetraparental (allophenic) mice** suggest that "blocking" Abs might be similarly involved in natural self-tolerance. These animals are produced by dissociating early embryos (at about the eight-cell stage) from two genetically different sets of parents, mixing the cells, and implanting the single combined blastula in a foster mother's uterus. The progeny are genetic mosaics with cells of both origins contributing to essentially all organs and tissues. The lymphocytes of these remarkable animals **can destroy cultured fibroblasts from either parent.** Nevertheless, the animals grow and develop normally, apparently because their serum contains Abs (or Ab·Ag complexes) that specifically block the cytotoxic reaction. Hence natural tolerance in these genetic mosaics seems to be due not to absence of an immune response but to two canceling ones: serum Abs block the destructive potentialities of cell-mediated immunity. In general, however, injections of antisera into sensitized individuals cannot be depended upon to block delayed-type hypersensitivity reactions to soluble Ags.

Antilymphocyte Sera (ALS). The antisera made in one species against the thymus cells of another are particularly destructive against the latter's recirculating pool of lymphocytes, which are mostly T cells (Ch. 17): studies with ^{51}Cr-labeled cells show, however, that lymphocytes in lymph nodes and in other secondary lymphatic organs are much less affected. The differences could account for ALS's selective depression of cell-mediated immune reactions (i.e., circulating T-cell effectors of cell-mediated responses are abolished, but T cells in lymphatic organs persist and remain able to support Ab production). Treatment of recipients with ALS therefore leads to greatly prolonged survival of allografts. However, Abs are made against the serum proteins of ALS (including the antilymphocyte Abs themselves). The resulting risk of anaphylaxis and the serum sickness syndrome limits the clinical usefulness of these antisera in the prevention or treatment of rejection of allografts (Ch. 21).

Antisera to θ, one of the distinctive surface alloantigens on mouse thymus cells (Ch. 17), selectively inactivate T cells; accordingly these antisera are effective inhibitors of all cell-mediated reactions in appropriate strains of mice, blocking both induction and expression of responses. As noted before, θ is an alloantigen in mice; anti-θ antisera are not yet available in other species.

NONSPECIFIC SUPPRESSION

The cytotoxic agents that block Ab formation by preventing mitotic division (of cells in general), or by causing massive destruction of lymphocytes, were considered in Chapter 17. No new principles are involved in their application to cell-mediated immunity. The agents that block proliferation of lymphocytes (X-rays, alkylating drugs, purine and folate analogs; Ch. 17, Fig. 17-31) block induction, but not expression of cell-mediated immunity: evidently **Ag-sensitive T cells and macrophages can generate inflammatory responses without themselves undergoing cell division.**

The agents most widely used clinically to inhibit cell-mediated allergic responses are the 11-oxycorticosteroids. In therapeutic doses they act as nonspecific suppressors of inflammation, but they probably **do not block in-duction** of the sensitive state. Indeed it is possible that mature T cells are particularly resistant; for the small proportion of mouse thymus cells (ca. 5%) that resist steroid action are both more mature than the others (they have less θ and more histocompatibility Ags) and more effective in carrying out a graft-vs.-host reaction.

The survival of human kidney allografts depends upon skillful use of immunosuppressants. ALS (horse antiserum to human lymphocytes) is usually administered for the initial 2 weeks, while prednisone (a corticosteroid) and imuran (a purine analog; Ch. 17, Fig. 17-31) are usually given continuously. These measures prolong graft survival, but the resulting suppression of all cell-mediated immune responses increases susceptibility to overwhelming infection by prevalent "opportunistic" bacteria and fungi, and increases the risk of cancer (Ch. 17, Immunosuppression in Ab formation; also Ch. 21).

CLINICAL DEFICIENCIES

The fundamental distinctions between cell-mediated and Ab-mediated immunity have come to be appreciated partly from clinical experience with patients having massive defects in one or the other. Defects in B cells, leading to deficiencies in Igs and in Ab formation, were considered in Chapter 17 along with clinical procedures for evaluating suspected immunodeficient states (see especially Tables 17-11 and 17-12). Some T cell deficiencies are considered here.*

Thymic Aplasia. The principal form of this rare disorder, the **DiGeorge syndrome,** arises from defective embryonic development of the third and fourth pharyngeal outpouchings that give rise to the thymus, parathyroid, and thyroid glands (Ch. 17); hence hormonal deficiencies (hypoparathyroidism, rarely also hypothyroidism) are present in addition to profound immune defects. Another form of thymic aplasia, inherited as an autosomal recessive trait, occurs without associated hormonal deficiencies (Nezelof syndrome). Infants with either

---

* As was indicated before (Ch. 17), live (attenuated) vaccines must be avoided in evaluating immune responses of infants with suspected deficiencies: disseminated infections following immunization with BCG (an attenuated strain of tubercle bacilli) and vaccinia have been fatal in severely affected infants.

form suffer from severe recurrent infections; they also lack delayed-type skin hypersensitivity to Ags of prevalent microbes, and fail to develop contact skin sensitivity to potent sensitizers, such as 2,4-dinitrochlorobenzene (DNCB, Fig. 20-4). Their blood lymphocytes are nearly all B cells (with normal distribution of surface of Igs; Ch. 17), with absent T cell function: for instance, they respond poorly to stimulation with phytohemagglutinin (PHA) or to allogeneic cells in mixed lymphocyte cultures (see Ch. 17, Table 17-12; also Ch. 21).

Since many Ags are T-dependent, these infants would also be expected to form Abs poorly in trial immunizations and to have low serum Ig levels (especially IgG). However their serum Igs are usually in the normal range, as is their ability to form Abs to diverse Ags, including proteins that are expected (from responses in mice) to be T-dependent. It is likely that thymic aplasia is not complete in infants who survive beyond 1 or 2 years after birth; indeed, in some who survive longer there seems to occur a spontaneous recovery of some T-cell functions (such as stimulation of thymidine uptake by PHA).

Severe Combined Deficiency Disease. Infants with this disease suffer from crippling absence of both T and B cell functions, probably because hematopoietic stem cells fail to differentiate normally (see Ch. 17, Table 17-11). Most attempts to reconstitute the defective system with transplants of allogeneic bone marrow cells and T lymphocytes (peripheral blood leukocytes) from normal donors have resulted in fatal graft-vs.-host reactions. However, in a few instances where the donor was a sibling with well-matched histocompatibility Ags (see Ch. 21, HL-A matching) the graft-vs.-host reactions were minimal and prolonged recovery of all immune functions has been achieved. With female donor and male recipient the female sex chromatin (Barr body) demonstrated long-term persistence of donor cells in the recipient (see Ch. 21, Chimeras).

Acquired T-Cell Deficiencies. In certain diseases of adults widespread involvement of the reticuloendothelial system with neoplasia (Hodgkin's disease) or with intracellular parasitism (lepromatous type of leprosy or disseminated leishmaniasis) or with chronic granulomas (Boeck's sarcoid) is associated with striking deficiencies in cell-mediated immunity and preservation of Ab-forming ability. Affected individuals do not give delayed-type skin reactions to any Ags and their blood lymphocytes do not respond to PHA. However, their serum Igs are normal (or even markedly elevated) and they form Abs normally in response to immunization, suggesting that some T-cell function is preserved. The reasons for their selective deficiencies are unknown. As noted above some of the deficiencies can be corrected by "transfer factor" from white blood cells of normal individuals (see above).

SOME SPECIAL ALLERGIC REACTIONS

ALLERGY TO DRUGS

Hypersensitivity reactions to drugs, particularly to penicillin, are among the more common allergic disorders in man. Most of the principles involved were noted in earlier sections of this chapter and in Chapter 19. **Sensitizing drugs, like inducers of contact skin sensitivity, must form stable covalent derivatives of proteins in vivo** (see above), which can induce Ab formation and virtually any type of hypersensitivity. In the following section both the immediate- and delayed-type responses are discussed.

Most drugs would be too toxic to be useful if they reacted rapidly with proteins under physiological conditions. If, however, the rate of reaction is sufficiently slow a drug may be tolerably nontoxic and yet may form effective hapten-protein immunogens. For most drugs that cause allergic reaction, however, it is likely that reactive contaminants, or reactive metabolic derivatives formed in vivo, are solely (or additionally) responsible for the formation of immunogenic conjugates. The mechanisms involved are well illustrated with penicillins, which are models for drug allergies in general.

PENICILLIN ALLERGY

Penicillin is probably the least toxic drug in use: 20 gm per day can be given safely for prolonged periods. Nevertheless, up to 10% of persons given penicillin repeatedly can become so sensitized that an injection of 1 mg may elicit fatal anaphylactic shock.*

Immediate-type Hypersensitivity. Penicillin is unstable, and most of its solutions contain at least small amounts of **penicillenate,** a highly reactive

---

* Oral administration is less hazardous because the drug is absorbed more slowly, but it can also lead to severe reactions, such as the serum sickness syndrome (Ch. 19).

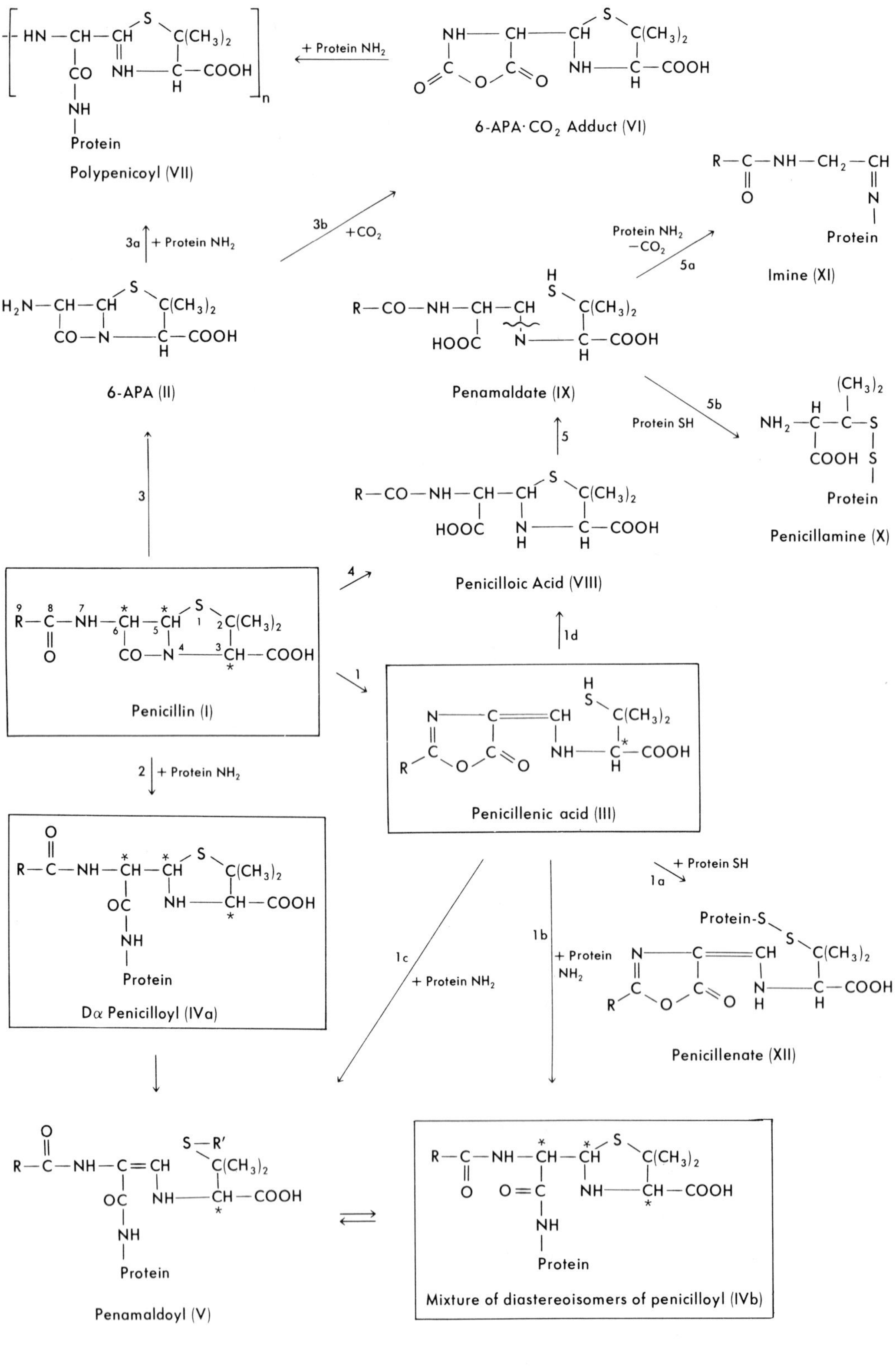
Polypenicoyl (VII)
6-APA·CO_2 Adduct (VI)
+ Protein NH_2
Imine (XI)
3a + Protein NH_2
3b $+CO_2$
Protein NH_2 $-CO_2$
5a
6-APA (II)
Penamaldate (IX)
Protein SH
5b
Penicillamine (X)
5
3
Penicilloic Acid (VIII)
4
Penicillin (I)
1
1d
Penicillenic acid (III)
2 + Protein NH_2
Dα Penicilloyl (IVa)
+ Protein SH
1a
1b + Protein NH_2
1c + Protein NH_2
Penicillenate (XII)
Penamaldoyl (V)
Mixture of diastereoisomers of penicilloyl (IVb)

derivative that forms **penicilloyl** and other substituents of amino and sulfhydryl groups of proteins (Fig. 20-13). Through its strained β-lactam ring penicillin itself can be directly attacked by nucleophilic groups (e.g., ε-NH$_2$ of lysyl residues), leading also to formation of penicilloyl-protein conjugates (though inefficiently at physiological pH). Penicilloic acid, formed from penicillenic acid, also forms a variety of minor determinant groups (Fig. 20-13). Conjugates with different haptenic moieties stimulate the formation of specific non-cross-reacting Abs (Fig. 20-13).

Intradermal injection of small amounts of the protein conjugates (e.g., 0.01 μg) could provide diagnostic test for allergy to penicillin.* However, even small amounts of the conjugates are potentially immunogenic. A safer reagent has been prepared by introducing penicilloyl substituents into poly-D-lysine, with an average of 20 residues per molecule, to form a ligand that is multivalent but not detectably immunogenic. This reagent elicits specific wheal-and-erythema responses as effectively as penicilloyl-proteins, and its specificity can be demonstrated by inhibition with univalent ligands, e.g., ε-penicilloyl-aminocaproate:

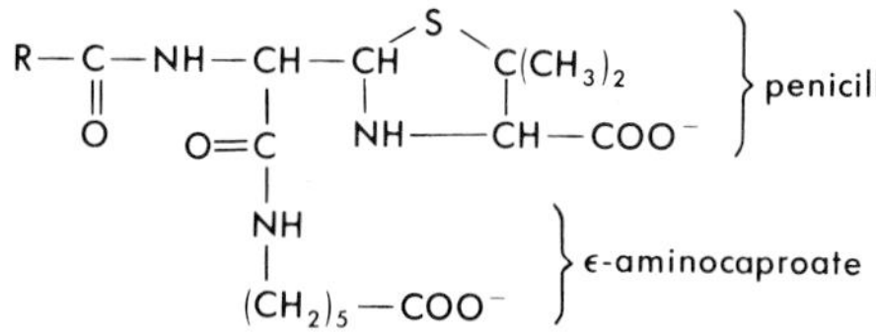

High-molecular-weight contaminants in solutions of penicillin can sometimes also cause anaphylactic reactions. The most significant are **penicilloyl proteins** in which the proteins are probably acquired,

---

† Pencillin itself only infrequently yields wheal-and-erythema responses, even in highly sensitive persons. Positive reactions, when they occur, are probably due to high-molecular-weight contaminants (see below). The unconjugated drug is univalent: if bound by Abs it should inhibit rather than elicit allergic reactions (Ch. 19).

during manufacture, from the fungus (*Penicillium*) or from the *E. coli* amidases that are sometimes added to cleave R-CO side chains in the preparation of 6-aminopenicillanic acid (6-APA) for production of semisynthetic penicillins (Fig. 20-13; also Ch. 7). Removal of the protein (e.g., by proteolysis with enzymes attached covalently to solid Sepharose beads, to avoid introducing another immunogenic impurity) reduces the frequency of reactions to penicillin. However, reactions can still be elicited with other high-molecular-weight contaminants (polymers of 6-APA or of penicillins).

Ligand and Antibody Competition. Many persons with reaginic (IgE) Abs to penicilloyl do not suffer anaphylactic reactions after injections of penicillin, even when the solutions contain penicillenate, a major source of penicilloyl conjugates. One probable reason is that the penicillin molecule itself, and the penicilloic acid formed as its major degradation product (Fig. 20-13), are univalent for antipenicilloyl Abs. Though these cross-reacting ligands are probably bound only weakly, they are present initially in great excess over the penicilloyl proteins that form in vivo, and they are thus likely to act as specific inhibitors of anaphylaxis. However, the univalent small molecules are excreted fairly rapidly, and so they are unlikely to influence the later, serum sickness-like reactions (e.g., rash, fever, joint pains) that often occur 3 to 14 days or more after penicillin is injected into persons with high antipenicilloyl Ab titers. Another reason for the infrequency of anaphylactic reactions is that persons with reaginic antipenicilloyl Abs also usually have non-reaginic Abs (IgGs, IgM) of the same specificity, which probably act as blocking Abs (Ch. 19; Table 20-6).

Hemolytic anemia due to penicillin allergy sometimes occurs in those receiving high doses of peni-

---

Fig. 20-13. Haptenic substituents derived from reactions of penicillin and its degradation products. R differs in different penicillins. (In penicillin G, the most widely used, it is benzyl 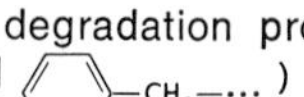) Asterisks designate asymmetrical carbons.

The Abs most often identified in penicillin-treated persons are specific for penicilloyl (IVa,b). Penicilloyl groups formed from penicillenic acid (III) are a mixture of diastereoisómers (IVb); those formed by direct reaction in vitro with penicillin at high pH largely retain the D-α configuration of penicillin itself (IVa). Some human antipenicilloyl Abs react better with D-α-penicilloyl than with the diastereoisomeric mixture. Penicillenate (III) is much more reactive than penicillin at physiological pH values, and it is probably the usual source of the penicilloyl groups formed in vivo. The immune responses induced by penicillin are sometimes specific for other, **"minor-determinant"** groups derived from the drug. 6-APA (II) is 6-aminopenicillanic acid, the intermediate to which many different acyl (R) groups may be attached synthetically to form a variety of penicillins (see Ch. 7). The structures of particular importance (I, III, IVa, and IVb) are enclosed. (From Parker, C. W. in *Immunological Diseases.* M. Samter, ed. Little, Brown, Boston, 1965).

TABLE 20-6.   Antibody-Mediated Reactions to Penicillin

Time of onset after penicillin administration	Clinical findings	Antibodies		Comments
		Class	Specificity	
2–30 min	Diffuse urticaria, hypotension, shock, respiratory obstruction	IgE	Penicilloyl and minor determinants	
3–72 hr	Diffuse urticaria, pruritus; rarely respiratory symptoms	IgE	Penicilloyl	IgG antipenicilloyl rises to high titer and probably acts as blocking Ab, leading to spontaneous cessation of the reaction
		IgM	Penicilloyl	
3 days to several weeks	Urticaria (sometimes recurrent), arthralgias, erythematous eruptions	IgE	Penicilloyl and minor determinants	
		IgM	Penicilloyl	

Derived from Levine, Bernard B. Immunochemical mechanisms of drug allergy in *Textbook of Immunopathology* [P. A. Miescher and H. J. Mueller-Eberhard, eds.], Vol. 1, Grune and Stratton, N.Y., 1968.

cillin for several weeks (as in treatment of subacute bacterial endocarditis): antipenicillin Abs react with penicillin adsorbed on red cells, causing excessive cell destruction and anemia. (Similar events with other, unrecognized exogenous Ags could simulate autoimmune hemolytic anemia; see chronic viral infection and autoimmune disease, below.)

Delayed-type hypersensitivity to penicillin is also common, as allergic contact dermatitis, particularly among handlers of penicillin in bulk (e.g., nurses and personnel in the pharmaceutical industry). One of the determinants that specify these reactions involves D-penicillamine, which elicits patch test reactions as effectively as penicillin in some persons with contact skin sensitivity to the drug. This compound forms stable mixed disulfides with cysteine in vitro, and probably acts in vivo by combining with protein sulfhydryl groups (X in Fig. 20-13).

Diagnosis of Immediate- and Delayed-type Penicillin Allergy. An important difference between immediate-type and delayed-type drug hypersensitivity is brought out by the diagnostic procedures used to detect penicillin allergy. The intradermal injection of penicillin itself is often not reliable for detecting immediate-type sensitivity; and even the available synthetic multivalent ligands are not effective in all sensitive individuals. In contrast, patch tests with penicillin itself are almost infallible for identifying delayed-type allergy to the drug. The differences can probably be explained by the following considerations. 1)

Intradermally injected penicillin diffuses away too rapidly to form many local protein conjugates, whereas penicillin from a patch test percolates into the skin slowly and has more time to form various derivatives that become conjugated with skin proteins in situ. 2) The pronounced carrier specificity of the cell-mediated response precludes competitive inhibition by penicillin and its low-molecular-weight derivatives.

The diversity of haptenic groups introduced into proteins by penicillin is unusual, but the principles are undoubtedly the same for all allergenic drugs. Perhaps the most important are 1) **an allergic reaction to a drug is specific for the haptenic group(s) introduced by the drug, or by its derivatives, into proteins in vivo, rather than for the drug itself;** and 2) **the haptenic derivatives may differ greatly in structure and in chemical properties from the drug itself** (Fig. 20-13).

Variations among Individuals. A striking feature of drug allergies is the extreme range in susceptibility of different individuals. Though an adequate explanation is not available, several possibilities are apparent. 1) Genetic differences could be important if enzymatic reactions are involved in converting a drug into metabolic derivatives that introduce haptenic groups into proteins; 2) the levels of Abs formed to any immunogen vary widely among different individuals, sometimes because of genetic differences (Ch. 17); 3) the tendency to form reaginic Abs is also variable, and is especially pronounced in atopic individuals (Ch. 19). In addition, certain infections could influence the response to concomitantly administered drugs; for example, killed *Bordetella pertussis* (a commonly used adjuvant in mice and rats) seems to selectively favor formation of reaginic Abs, and mycobacterial mycosides and endotoxins of gram-negative bacteria also enhance the immunogenicity of most Ags.

SPECIAL TISSUES

Most simple sensitizers react with proteins indiscriminately. Allergy to certain drugs, however, involves special tissue elements, which implies selective reactions with certain proteins. Some examples are the hemorrhagic manifestations of allergy to quinidine and to Sedormid (2-isopropyl-4-pentenoyl urea). The administration of these drugs to sensitive persons causes **thrombocytopenia**, resulting in bleeding in various tissues, including the skin (purpura).

In vitro, in the presence of the appropriate drug, the serum of a sensitive person can agglutinate platelets, his own or a normal person's, and if C is added the platelets are lysed; but platelet lysis occurs at much lower concentrations of free quinidine in vivo than in vitro. It is possible that the drug, or one of its metabolic derivatives, reacts selectively under physiological conditions with proteins on the platelet surface to form effective haptenic groups.

AUTOIMMUNE RESPONSES

Clinical and experimental observations show that individuals can sometimes respond immunologically to certain of their self-Ags. These important exceptions to the principle of self-tolerance help analyze its fundamental mechanisms, and they are frequently associated with disease. It is often not clear, however, whether these anomalous responses cause, or are the result of, disease (see below); hence it is necessary to emphasize the distinction between an **autoimmune response,** in which an individual makes Abs or becomes allergic to a self-Ag, and an **autoimmune disease,** which is a pathological condition arising from an autoimmune response. **Autoimmune reactions can be both Ab- and cell-mediated.**

MECHANISMS

Autoimmunity could arise, in principle, from the following mechanisms: 1) a change in the distribution of a self-component, allowing access to lymphoid cells that never encountered it previously; 2) a change in the structure of a self-component or introduction of a cross-reacting Ag, either of which could lead to the formation of Abs and appearance of lymphocytes that react (or cross-react) with the native self-component; and 3) emergence of abnormal ("forbidden") clones of lymphocytes. In addition, immune responses to persistent and unrecognized extrinsic Ags (such as those of certain chronic virus infections, see below) can generate chronic allergic disorders that are difficult to distinguish from those due to anti-self immune reactions.

Altered Distribution of Self-Antigens. Many self-Ags (e.g., from eye lens, spermatozoa, brain tissue) have little or no opportunity to establish tolerance because they are normally confined anatomically to sites that prevent their access to lymphoid cells. Abs to such an Ag are readily produced experimentally by removing the Ag and injecting it back into the same animal as though it were a conventional foreign Ag.

Clinical incidents that release such Ags have the same effect: thus Abs to heart muscle appear after myocardial infarction, and Abs to thyroid after the trauma of partial thyroidectomy. In these instances disease gives rise to the autoimmune response, rather than the reverse. Nonetheless, such responses can be visualized as self-perpetuating: once the response is initiated the resulting allergic inflammation in the target organ probably leads to contact of Ag with infiltrating lymphoid cells; hence resulting in further sensitization.

Altered Forms of Self-Antigens. Few self-Ags are totally isolated from lymphoid cells: for example, thyroglobulin (Tg), once thought to be completely confined to thyroid nodules, evidently leaks into lymphatics, for sensitive assays reveal its presence in serum at concentrations (ca. $0.01 \ \mu g/ml$) that could easily establish T-cell tolerance. Hence modified Tg molecules, altered by introducing or exposing new groups with "carrier" function, could engage the corresponding T cells and help stimulate persistent (not tolerized) B cells to produce Abs that react with determinants on native Tg. Thus rabbits immunized with chemically modified rabbit Tg or with cross-reacting hog Tg develop thyroiditis, associated with Abs that react with native rabbit Tg.* A clinical parallel is seen in the encephalitis that occasionally develops in people injected with rabies vaccine, which is a suspension of infected rabbit brain (Allergic encephalomyelitis, below; see Ch. 17, discussion of T and B cells in Immunological tolerance).

Some bacteria carry antigenic determinants that resemble those of the host; e.g., in rheumatic fever some of the Abs produced against group A streptococci seem to cross-react with human heart. Bacterial infections might also facilitate autoimmune responses through the adjuvant effects of certain products (e.g., mycosides, endotoxins).

Forbidden Clones. In developing the clonal selection hypothesis Burnet assumed that lymphocyte clones are somehow eliminated as they arise if they are specific for self-Ags. A breakdown in the hypothetical elimination mechanism, or the appearance of mutant lymphocytes that are resistant to this mechanism, could lead to the emergence of autoreactive "forbidden clones." There is no firm evidence in support of this possibility, but a hint of its validity is provided by clinical findings: when an individual who suffers from one autoimmune disorder is examined closely, signs of other autoimmune responses are likely to be disclosed.

Chronic Virus Infection. Mice infected at birth with certain temperate viruses (e.g., lymphocytic choriomeningitis virus, lactic dehydrogenase virus, and others that also seem not to injure infected cells) become life-long carriers, producing Abs that form virus:Ab complexes without neutralizing the virus' infectivity (see Ch. 19, discussion of Viral complexes under Immune-complex disease). Budding virions on infected cells, or viral Ags adsorbed on erythrocytes or other cells, simulate self-Ags: their combination with antiviral Abs (or perhaps with specific T cells) can give rise to chronic disorders, such as hemolytic anemia, that resemble autoimmune diseases. These circumstances in model animal systems suggest that some of the ostensibly autoimmune human diseases might also be due to unrecognized chronic viral infections (for an anal-

---

* Similar lesions and Abs can even appear when rabbit Tg itself is incorporated into Freund's adjuvant and injected back into rabbits: denaturation during preparation could expose new carrier-like groups.

ogous situation, see Hemolytic anemia due to penicillin, above).

CHARACTERISTICS OF AUTOIMMUNE DISEASES

Antibody- vs. Cell-mediated Reactions.   All the mechanisms that cause allergic reactions to foreign Ags can participate in autoimmune diseases (Table 20-7). 1) Various **autoantibodies** (with or without C) can act directly on cells, e.g., lysing erythrocytes or platelets, or injuring cells of the thyroid gland (thyroiditis). 2) **Autoimmune Ab·Ag aggregates** are illustrated by systemic lupus erythematosus, in which some complexes formed from DNA, anti-DNA, and C become lodged in kidney glomeruli and lead eventually to progressive glomerulonephritis and renal failure (Ch. 19, Immune-complex diseases). Similarly, IgGs that function as autoantigens combine with certain IgM (or IgG) molecules that serve as autoantibodies (rheumatoid factors), forming aggregates that localize around synovial membranes and probably contribute to the joint inflammation of rheumatoid arthritis (Ch. 19). 3) **Autosensitized lymphocytes** have not been shown to cause human disease, but hints are provided by the striking mononuclear infiltrates in many autoimmune disorders (atrophic gastritis in pernicious anemia, thyroiditis, allergic encephalomyelitis, etc.).

Stronger support for cell-mediated mechanisms is provided by much experimental work in which bits of tissue (from thyroid, brain, adrenal, etc.) are removed surgically, emulsified in Freund's adjuvant, and injected back into the same animal. After 1 to 2 weeks the recipient develops both serum Abs and lesions infiltrated with mononuclear cells in the corresponding organ. Moreover, viable lymphocytes from the affected animal, but not serum, will usually transfer the disease to a normal recipient.

Failure to transfer the disorder with serum should not, however, be unduly stressed. Autoantibodies, like other Abs, are likely to be heterogeneous; and with the continuous presence of the target self-antigen those Abs with highest affinity are probably removed preferentially, leaving as free autoantibodies in serum the relatively ineffectual low-affinity molecules.

Localized vs. Disseminated Disease.   Autoimmunity can affect almost every part of the

TABLE 20-7. Some Autoimmune Disorders in Man

Organ or tissue	Disease	Antigen	Detection of antibody*
Thyroid	Hashimoto's thyroiditis (hypothyroidism)	Thyroglobulin	Precipitin; passive hemagglutination; IF on thyroid tissue
		Thyroid cell surface and cytoplasm	IF on thyroid tissue
	Thyrotoxicosis (hyperthyroidism)	Thyroid cell surface	Stimulates mouse thyroid (bioassay)
Gastric mucosa	Pernicious anemia (vitamin B_{12} deficiency)	Intrinsic factor (I)	Blocks I binding of B_{12} or binds to I:B_{12} complex
		Parietal cells	IF on unfixed gastric mucosa; CF with mucosal homogenate
Adrenals	Addison's disease (adrenal insufficency)	Adrenal cell	IF on unfixed adrenals CF
Skin	Pemphigus vulgaris	Epidermal cells	IF on skin sections
	Pemphigoid	Basement membrane between epidermis-dermis	IF on skin sections
Eye	Sympathetic ophthalmia	Uvea	Delayed-type hypersensitive skin reaction to uveal extract
Kidney glomeruli plus lung	Goodpasture's syndrome	Basement membrane	IF on kidney tissue; linear staining of glomeruli (see Ch. 19, Fig. 19-19B)
Red cells	Autoimmune hemolytic anemia	Red cell surface	Coombs' antiglobulin test
Platelets	Idiopathic thrombocyto-penic purpura	Platelet surface	Platelet survival
Skeletal and heart muscle	Myasthenia gravis	Muscle cells and thymus "myoid" cells	IF on muscle biopsies
Brain	? Multiple sclerosis	Brain tissue	Cytotoxicity on cultured cerebellar cells
Spermatozoa	Male infertility (rarely)	Sperm	Agglutination of sperm
Liver (biliary tract)	Primary biliary cirrhosis	Mitochondria (mainly)	IF on diverse cells with abundant mitochondria (e.g., distal tubules of kidney)
Salivary and lacrimal glands	Sjögren's disease	Many: secretory ducts, mitochondria, nuclei, IgG	IF on tissue
Synovial membranes, etc.	Rheumatoid arthritis	Fc domain of IgG	Antiglobulin tests: agglutination of latex particles coated with IgGs, etc.
	Systemic lupus erythematosus (SLE)	Many: DNA, DNA-protein, cardiolipin, IgG, microsomes, etc.	Precipitins, IF, CF, LE cells (see text)

* IF = immunofluorescence staining, usually with fluorescent antihuman Igs (see Ch. 14).
CF = complement fixation (Ch. 18).
Based on Roitt, I. *Essential Immunology*. Blackwell, Oxford, 1971.

body (Table 20-7). Some responses are directed to **organ-specific** Ags and may even be limited to a particular cell type (e.g., parietal cells of gastric mucosa in pernicious anemia). Other responses are directed to widely distributed Ags and are associated with **disseminated** disease (e.g., antinuclear Abs in systemic lupus erythematosus). In still other diseases the responses are **intermediate** between these extremes: for instance, in Goodpasture's disease, characterized by chronic glomerulonephritis and pulmonary hemorrhages, Abs are deposited on basement membrane of kidney glomeruli and lung parenchyma, in accord with a strong cross-reaction between these particular basement membranes, shown by localization in vivo of ^{125}I-labeled heteroantisera (e.g., rabbit antirat glomeruli).

Multiplicity of Responses. An individual who makes one autoimmune response is likely to make others. For instance 10% of persons with autoimmune thyroiditis have pernicious anemia, which is present in only 0.2% of the population at large (of the same age and sex distribution). Thyroid disease is similarly found with excessive frequency in those who suffer from pernicious anemia. Serological evidence for multiple reactions is even more frequent: 30% of those with autoimmune thyroiditis have Abs to parietal cells of gastric mucosa (involved in pernicious anemia), and 50% of those with pernicious anemia have Abs to thyroid, though the two kinds of Abs are entirely non-cross-reacting. Similar associations are found among the disseminated group of autoimmune diseases: persons with systemic lupus erythematosus often have evidence of rheumatoid arthritis, autoimmune hemolytic anemia, or thrombocytopenia.

Genetic Factors. Relatives of patients with autoimmune thyroiditis commonly have antithyroid Abs, and relatives of those with pernicious anemia have a high incidence of Abs to gastric parietal cells. Moreover, some inbred animal strains suffer from a high frequency of autoimmune disorders: all mice of the NZB strain (New Zealand Black) eventually develop autoimmune hemolytic anemia, and most of the hybrids (B×W) made by crossing NZB with another inbred strain (New Zealand White, NZW) develop a syndrome

that is strikingly similar to systemic lupus erythematosus in man.

The genetic factors and the multiplicity of autoimmune disorders in some individuals suggest that the key to autoimmunity may reside in a breakdown of the still unknown central mechanisms that are responsible for self-tolerance (Forbidden clones, above).

Age and Sex. Autoimmune responses are improbable events; their frequency increases with age (Fig. 20-14). They are also more frequent in women than in men, perhaps because different cells in females express different X chromosomes (Lyon effect): the resulting cellular mosaicism might increase opportunities for a breakdown in self-tolerance (see tetraparental mice under Tolerance, above, and Allelic exclusion, footnote p. 470).

SOME REPRESENTATIVE AUTOIMMUNE DISEASES

Acquired Hemolytic Anemia. In this disorder a person forms Abs that react with his own red blood cells. Such Abs are sometimes found in persons with other diseases that might alter self-antigens or introduce cross-reacting ones: e.g., malignant neoplasms of lymphoid tissue, syphilis, mycoplasma

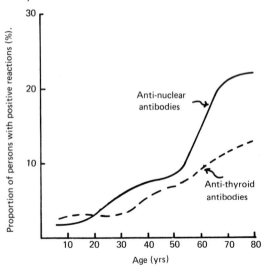

Fig. 20-14. Increasing incidence of autoantibodies with age in general population. Antinuclear Abs were detected by immunofluorescence (e.g., Fig. 20-17) and antithyroid by passive hemagglutination with thyroglobulin on tanned red cells. (From Roitt, I. *Essential Immunology.* Blackwell, Oxford, 1971).

pneumonia, infectious mononucleosis. More often, however, this disorder is idiopathic (i.e., it is unassociated with other diseases or with known exposure to agents toxic to red cells).

Abs to the patient's red cells are demonstrated by two procedures. 1) In the **direct antiglobulin (Coombs) test** the patient's washed red cells are agglutinated by rabbit antiserum to human Igs (Ch. 15). 2) In the **indirect test,** red cells from another person are incubated with the patient's serum, washed, and examined for clumping by the rabbit antiserum.

The anti-red cell Abs have no hemolytic activity in vitro, but they accelerate destruction of red cells in vivo. Thus, transfused normal red cells have a shortened survival in affected patients only if the injected cells have the particular surface Ag that binds the patient's autoantibody, indicating that red cell breakdown is due to the bound Abs. The Ab-coated (opsonized) cells are phagocytized by macrophages and their breakdown occurs especially in the spleen. Therapy includes nonspecific immunosuppressants (Ch. 17) and splenectomy.

Mice of the inbred NZB strain almost invariably develop hemolytic anemia as they age, and viable lymphoid cells (including Ab-forming clones) from older (affected) animals transfer the disease efficiently to young, unaffected mice of the same strain. These mice are chronically infected from birth with C-type viruses (resembling murine leukemia virus; see Ch. 63), and it is possible that what looks like autoimmune disease is really due to a conventional immune reaction to viral Ags absorbed on red cells or present on budding virons of infected cells.

Thrombocytopenic Purpura. In this disease platelets can decline to ca. 1/10 the normal level and bleeding occurs in many organs, including the skin, causing petechial rash and purpura. In the dramatic experiment that first provided evidence of the immune nature of this disease, a human volunteer injected with a patient's plasma suffered a precipitous fall in platelets and extensive bleeding into internal organs and skin (Fig. 20-15). A patient's serum can cause clumping of normal platelets, and lysis if C is present. Infants born to mothers with the disease may have transient thrombocytopenia and bleeding, owing to placental transmission of the maternal Abs.

Allergic Encephalomyelitis. When laboratory animals (e.g., rat, guinea pig, monkey) are injected with suspensions of central nervous system tissue from individuals of the same or other species they develop patchy areas of vasculitis and demyelination in the brain and spinal cord (Fig. 20-16). This response is readily evoked with a single injection of a small amount of brain or spinal cord tissue, even from the same animal, in Freund's adjuvant.

The immune nature of the disease is indicated by

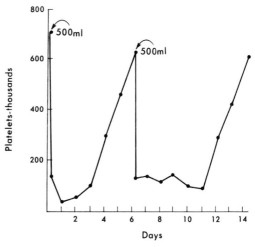

Fig. 20-15. Passive transfer of thrombocytopenic purpura. The drop in platelet count of a normal human volunteer following transfusion (arrows) of blood from a patient with idiopathic thrombocytopenic purpura. Similar results were obtained with Ig fractions from other donors with the disease. Some of the recipients suffered bleeding in internal organs simulating the natural course of the disease. [From Harrington, W. J., Minnich, V., Hollingsworth, J. W., and Moore, C. V. *J Lab Clin Med* 38:1 (1951).]

the following. 1) Lesions appear 9 days or more after the primary injection, but the onset is more rapid in animals that have recovered and are then reinjected (secondary response). 2) The lesions are specific: they follow inoculation only of myelin-containing tissues, and appear only in myelinated tissue, especially white matter of brain. 3) Abs that react with brain tissue can be demonstrated in serum, and lymphoid cells from inoculated animals produce cytopathic effects on myelinated brain tissue and glial cells in culture. 4) The lymphoid cells can also cause specific neural lesions in nonsensitized syngeneic recipients. 5) Intradermal injection of myelinated tissue evokes a delayed skin response in affected animals. Though serum Abs are present they do not seem to play a major role in the disease: the intensity of the lesions does not parallel their levels and serum, unlike lymphoid cells, fails to cause lesions in recipient animals.

The responsible self-antigen is a basic protein, extractable at low pH from myelinated nervous tissue and localized by immunofluorescence in myelinated nerve fibers. It is heterogenetic (i.e., distributed in many species) and organ-specific, being confined to the central nervous system.

The experimental background clarifies the post-vaccination encephalitis that often occurs in humans after vaccination with the standard rabiesvirus, which is a suspension of infected rabbit brain.

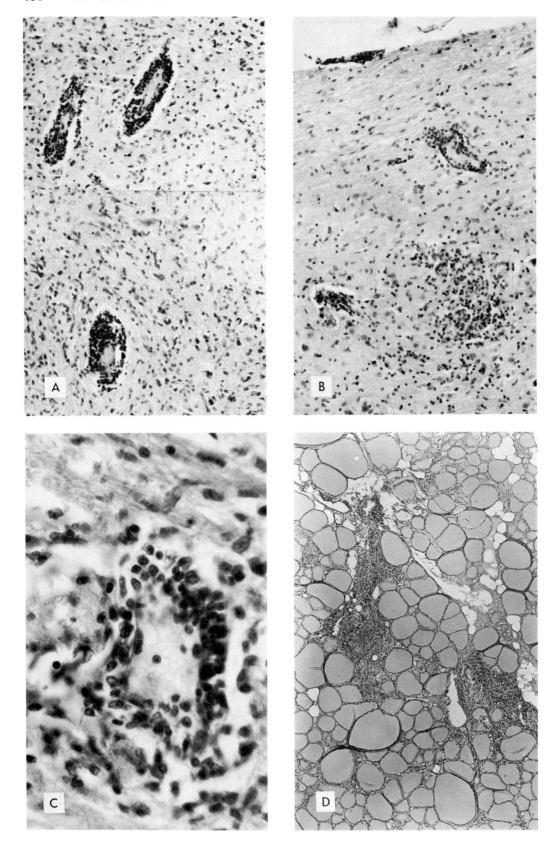

Rabies vaccine from virus grown in tissue culture or in duck embryo tissues that lack myelin (Ch. 58) is now preferred. The demyelinating encephalitides that occasionally follow measles and vaccinia could arise because infection of the nervous system brings the encephalitogenic Ag into contact with inflammatory (lymphoid) cells. A similar mechanism is suspected for multiple sclerosis, a common disabling neurological disorder of unknown etiology with characteristic focal demyelinating lesions.

Thyroiditis. Patients with chronic inflammation of the thyroid (Hashimoto's disease) suffer destruction of secretory cells and loss of thyroid function. Their sera contain some Abs that react in high titer with thyroglobulin (Tg), and others that react with Tg-free particulate fractions of thyroid (? cell membranes; e.g., in immunofluorescence assays as illustrated in Fig. 20-17). The possibility that these Abs cause the disease is supported by observations on an inbred strain of chickens that spontaneously develops antithyroid Abs and thyroiditis (and finally hypothyroidism and obesity): the Abs appear to be causal, since neonatal bursectomy, but not thymectomy, prevents the anomalies. However, experimental thyroiditis induced in mice, rats, etc, is very similar in pattern to experimental allergic encephalomyelitis (above). The suspicion therefore remains that cell-mediated rather than (or in addition to) humoral autoimmunity is involved in the lesions of the human disease.

The interaction of Abs with Ags on cells sometimes causes cell proliferation and differentiation rather than destruction—e.g., when anti-Igs react with B lymphocytes (blast transformation, Ch. 17), or when appropriate Abs react with sea urchin eggs. Similar mechanisms appear to be involved in some patients with hyperplastic and hyperfunctional thyroid glands, who have antoantibodies to the thyroid. If present during pregnancy, these autoimmune Abs (also called **long-acting thyroid stimulators**) can cross the placenta and cause neonatal hyperthyroidism, which subsides within a few weeks of birth (as the maternal IgG is degraded; see Ch. 17, Ontogeny). The autoantibody might bind specifically to the same cell surface receptors as thyroid-stimulating hormone, for the actions of both are potentiated by theophylline, indicating that they act through the adenyl cyclase system (see Ch. 19, Fig. 19-13).

Pernicious Anemia. Defective red cell maturation in this disease is due to lack of vitamin B_{12}, caused by faulty absorption of the ingested vitamin. Affected persons have an atrophic gastritis, and their poor absorption of vitamin B_{12} is due to lack of intrinsic factor (IF), a protein that is necessary for this absorption and is secreted into the stomach by special (parietal) cells of the gastric mucosa. Most patients with the disease have Abs to parietal cells (revealed by immunofluorescence of gastric biopsies) and cell-mediated immunity to these cells could also be present. In addition, autoantibodies to IF itself are also involved, for large amounts of oral B_{12} (which ordinarily cure the anemia) are ineffective if given together with serum from many patients, and the active sera contain Abs to IF. Evidently the inhibitory effect of anti-IF is exercised within the stomach, for absorption of the fed vitamin remains normal in immunized human volunteers even when they develop a high serum titer of these Abs, or intense delayed-type hypersensitivity to IF. The gastric Abs, produced by plasma cells in the gastric lesions, are able to function within the stomach (normally extremely acidic) because loss of cells from the affected mucosa reduces secretion of HCl. Thus, by decreasing production of IF and of HCl, and by secreting Abs to IF the autoimmune atrophic gastritis contributes to (and may even cause) B_{12} deficiency and pernicious anemia.

Systemic Lupus Erythematosus. At various stages in this complex disease patients produce several of an immense variety of autoantibodies to various blood cells, clotting factors, and intracellular components (e.g., mitochondria, nuclear components; Fig. 20-17). The resulting difficulties include hemolytic anemia, leukopenia, thrombocytopenic purpura, and bleeding tendencies. The Abs to some cell com-

Fig. 20-16. Autoimmune allergic reactions. **A–C.** Allergic encephalomyelitis in a rat sensitized adoptively with viable lymph node cells from a donor rat that had been immunized with rat spinal cord (injected in complete Freund's adjuvant). The recipient animal, which had severe ataxia and hind-leg paralysis 5 days after receiving the donor's lymphoid cells, was sacrificed at 7 days. Sections of brain show a focal inflammation with perivascular infiltration by mononuclear cells. H & E stain. **A** and **B**, ×130 (reduced); **C,** ×500 (reduced). **D.** Thyroiditis in a rabbit that had been immunized repeatedly with hog thyroglobulin. A small proportion of the rabbit's Ab (to the hog protein) reacted also with rabbit thyroglobulin. Note intense focal infiltration of the immunized rabbit's thyroid with mononuclear cells. ×60 (reduced). Similar lesions appear in animals injected with their own thyroglobulin or with homogenates of a bit of their own thyroid tissue. [A–C from Patterson, P. Y. *J Exp Med 111*:119 (1960). **D** from Witebsky, E., and Rose, N. R. *J Immunol 83*:41 (1959).]

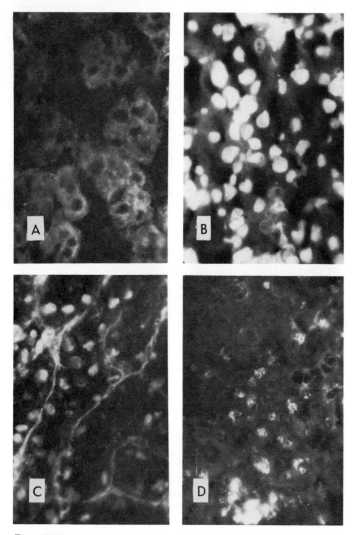

Fig. 20-17. Variety of antitissue Abs in sera of different patients with systemic lupus erythematosus. Mouse kidney sections were incubated with the sera and, after washing, were stained with fluorescein-labeled rabbit antiserum to human 7S Igs. Similar reactions occur with human tissues. **A.** Normal human serum control. **B.** Abs in a lupus serum react uniformly with all nuclei, giving homogeneous nuclear staining. **C.** Abs react with nuclei and with basement membrane of renal tubules. **D.** Abs react with selected parts of nuclei, giving speckled staining. All ×250. (Courtesy of Drs. E. Tan and H. Kunkel.)

ponents (e.g., to DNA) seem to cause difficulty mainly through deposition of Ab · Ag · C complexes in walls of small blood vessels leading to widespread vasculitis (Arthus lesions); deposits in capillaries of renal glomeruli are especially threatening, for they can lead to glomerulonephritis and renal failure (Ch. 19). The production of so many autoantibodies is accompanied by elevated serum Ig levels, and

during the active phases of the disease serum C falls to low levels as Ab · Ag aggregates bind these proteins.

The multiplicity of Abs to self-Ags and the appearance of "lupus erythematosus" (LE) cells are diagnostic. These cells appear simply on incubation of blood or bone marrow: the breakdown of some cells evidently releases DNA which combines with serum anti-DNA, and the resulting Ab · Ag aggregates are taken up by granulocytes, which acquire a characteristic appearance (a large amorphous mass in the center of the cell, with the surrounding multilobed nucleus pushed to the periphery).

PATHOLOGICAL CONSEQUENCES OF AUTOIMMUNE REACTIONS

When autoantibodies were first discussed, at the turn of the century, they were imagined to occur either not at all or with disastrous consequences. Now, with the development of increasingly sensitive and reliable assays for Abs and for hypersensitivity, it is clear that autoimmune responses are not so uncommon. However, their pathological effects are often conjectural, with three possibilities to be considered. 1) The response can be innocuous. For example, though anticardiolipin (i.e., Wasserman Abs) in a person with syphilis (Ch. 37) can react with cardiolipin extracted from his own tissues, these Abs do not appear to be pathogenetic: they may be present at high titer in apparently healthy persons. Presumably cardiolipin is buried within membranes and is thus inaccessible. 2) The response can be secondary to another disease (e.g., antithyroglobulin Abs following the

trauma of partial thyroidectomy); but it may then be responsible for continuing disease (i.e., inflammation in the affected organ introduces new lymphoid cells that increase autosensitivity). 3) The response can be the main causal factor in disease, e.g., in autoimmune hemolytic anemia or thrombocytopenic purpura.

Witebsky's Criteria. The spectrum of allergic effects is so wide—from hemolysis to demyelinating inflammatory lesions—that autoimmune processes are now commonly invoked for many diseases of unknown etiology; and highly sensitive assays often reveal autoantibodies. In an effort to provide guidelines for interpretation Witebsky suggested the following criteria (reminiscent of Koch's postulates for bacterial etiology): an autoimmune response should be considered the cause of a human disease if 1) it is regularly associated with that disease 2) immunization of an experimental animal with Ag from the appropriate tissue causes it to make an immune response, 3) the responding animal develops pathological changes similar to those of the human disease, and 4) the experimental disease can be transferred to a nonimmunized animal by serum or by lymphoid cells.

ALLERGY AND IMMUNITY

Many of the allergic reactions described in this and in the preceding chapter can be elicited with microbial Ags in persons who are suffering, or have recovered, from the associated infectious disease. Two questions arise: 1) Do allergic responses occur naturally during the infectious disease? 2) If they occur, do they enhance or diminish the host defenses?* Broadly speaking, the answers seem to depend on 1) the mass and distribution of microbial Ags, 2) the concentration and types of Abs present, and 3) the intensity and localization of cell-mediated responses. The range of possibilities is large, as is illustrated by the following examples.

1) **No detectable allergic reactions.** In some infectious diseases allergic reactions are

not likely to occur, or to be significant if they do. For example, in diseases caused primarily by exotoxins, such as diphtheria and tetanus, the quantities of toxin involved are too small to elicit an obvious allergic response even if the host were hypersensitive. Moreover, only a minority of persons successfully immunized with the corresponding toxoids exhibit allergic responses to intradermal tests with these proteins.

2) **Coincidental allergic reactions.** In the course of many infections various sterile skin eruptions appear and are probably due to allergic reactions to microbial Ags. For example, transient erythematous rashes and urticaria, resembling cutaneous anaphylaxis, often occur in group A streptococcus infections; erythematous skin nodules (erythema nodosum), resembling the Arthus reaction, occur sometimes in the course of tuberculosis, coccidioidomycosis, and many other infectious diseases;

---

* These questions are also considered elsewhere from different viewpoints (see Ch. 22, and also chapters dealing with particular microorganisms).

and vesicular lesions ("ids"), resembling delayed-type reactions, occur often in fungal infections (dermatophytosis and candidiasis; Ch. 43). Though these reactions are useful for clinical diagnosis there is no evidence that they affect the host's resistance.

3) **Significant allergic reactions.** Mycobacterial diseases regularly produce delayed-type allergy to the bacillary proteins, and the mass of bacteria in tissues is unusually large. Hence it is probable that intense allergic reactions occur regularly at infected foci in these diseases and account for much of the observed inflammatory reaction and tissue necrosis. As noted elsewhere (Chs. 22 and 35), the consequences at these severe reactions depend on their localization: if the reaction occurs in the skin and causes ulceration it can lead to the drainage of virulent microbes from the body, but if it takes place on the surface of the brain or in a bronchus the same response can lead to fatal meningitis or to a spreading pulmonary infection.

The sterile cardiac lesions that sometimes develop after streptococcal infections can also cause severe disease (e.g., rheumatic fever), and seem to be allergic in origin (Ch. 26). Thus the Abs to a cell wall Ag of β-hemolytic streptococci cross-react with an Ag from human cardiac muscle, suggesting that rheumatic carditis may be an autoimmune disease. A similar mechanism may contribute to the glomerulonephritis that frequently follows recovery from infection with the type 12 group A streptococcus (Ch. 26).

The following considerations suggest, however, that allergic responses to microbial Ags generally favor host defenses. 1) A given microbe contains many Ags (probably hundreds with bacteria and fungi), of which only a few are directly concerned with virulence. Thus very few Ab-Ag reactions that occur during an infectious disease are likely to result in specific neutralization of substances that cause toxicity and virulence. Nevertheless, many of the other reactions could well contribute indirectly to the host's resistance by provoking localized allergic responses around the microbial agents, leading to the accumulation of granulocytes, lymphocytes, macrophages, and serum proteins, including Abs and C. 2) Delayed-type responses are associated with "activated" macrophages, which can destroy those bacteria (such as *M. tuberculosis*) that normally proliferate within phagocytic cells (Chs. 22 and 35).

Delayed-type (T-lymphocyte) reactions with cell surface Ags also seem to be more effective than the Ab-C system in destroying target animal cells: such reactions with infected cells that have budding virions on the surface may be a particular effective means for eliminating infectious centers in many virus diseases. Thus children who are unable to produce Abs but who have an intact cell-mediated immunity system recover normally from mumps, measles, and other common childhood viral infections (Ch. 17, Infantile X-linked agammaglobulinemia). The same effect, functioning as a surveillance mechanism, seems to be important in aborting incipient cancer clones via attack on cell surface tumor-specific Ags (see also Ch. 21). The Ag-binding receptors on T lymphocytes might have high affinity for Ag, related perhaps to the large determinant recognized, and this could provide for early elimination of small numbers of target cells. On balance therefore, allergic responses are likely to benefit the host. Perhaps it is for this reason that allergic reactivity has persisted through evolution, being represented in all species of vertebrates, along with Ab formation, from primitive cyclostomes up.

SELECTED REFERENCES

Books and Review Articles

BLOOM, B. R. In vitro approaches to the mechanisms of cell-mediated immune reactions. *Adv Immunol* 13:101 (1971).

BLOOM, B. R., and GLADE, P. (eds.). *In Vitro Methods in Cell-Mediated Immunity.* Academic Press, New York, 1971.

Cell-mediated immune responses. *WHO Tech Rep Ser* No. 423, 1969.

DAVID, J. R. Mediators produced by sensitized lymphocytes. *Fed Proc 30*:1730 (1971).

LAWRENCE, H. S. Transfer factor. *Adv Immunol 11*:195 (1969).

LAWRENCE, H. S., and LANDY, M. (eds.). *Mediators of*

Cellular Immunity. Academic Press, New York, 1970.

MACKANESS, G. B., and BLANDEN, R. V. Cellular immunity. *Progr Allergy 11*:89 (1967).

PERLMANN, P., and HOLM, G. Cytotoxic effects of lymphoid cells in vitro. *Adv Immunol 11*:117 (1969).

Proceedings of Conference on Antilymphocyte Serum. *Fed Proc 29*:101 (1970).

ROITT, I. M., GREAVES, M. F., TORRIGIANI, G., BROSTOFF, J., and PLAYFAIR, J. H. L. The cellular basis of immunological responses (an occasional survey). *Lancet 2*:367 (1969).

SIMONSEN, M. Graft-vs.-host reactions. *Progr Allergy 6*:349 (1962).

SMITH, R. T., and LANDY, M. (eds.). *Immunologic Surveillance.* Academic Press, New York, 1971.

TURK, J. L. *Delayed Hypersensitivity.* North Holland, New York, 1967.

WAKSMAN, B. H. *Atlas of Experimental Immunobiology and Immunopathology.* Yale Univ. Press, New Haven, 1970.

Specific Articles

ALLISON, A. C. New antigens induced by viruses and their biological significance. *Proc R Soc Lond (Biol) 177*:23 (1971).

BLANDEN, R. V., LEFFORD, M. J. and MACKANESS, G. B. Host response to Calmette-Guerin Bacillus (BCG) infection in mice. *J Exp Med 129*:1079 (1969).

BRUNNER, K. T., MAVEL, J., CEROTTINI, J. C., and CHAPUIS, B. Quantitative assay of the lytic action of immune lymphoid cells on ^{51}Cr-labeled allo-geneic target cells in vitro. *Immunology 14*:181 (1968).

DAVIE, J. M., and PAUL, W. E. Receptors on immunospecific cells: Receptor specificity of cells participating in cellular immune responses. *Cell Immunol 1*:104 (1970).

DVORAK, H. F., DVORAK, A. M., SIMPSON, B. A., RICHERSON, H. B., LESKOWITZ, S., and KARNOVSKY, M. J. Cutaneous basophil hypersensitivity: A light and electron microscopic description. *J Exp Med 132*:558 (1970).

GINSBURG, H. Graft-vs-host reaction in tissue culture: Lysis of monolayers of embryo mouse cells from different strains differing in H-2 histocompatibility locus by rat lymphocytes sensitized in vitro. *Immunology 14*:621 (1968).

GOLDSTEIN, P., WIGZELL, H., BLOMGREN, H., and SVEDMYR, E. Cells mediating specific in vitro cytotoxicity: Probable autonomy of thymus-processed lymphocytes (T cells) for the killing of allogeneic target cells.

MCCLUSKEY, R. T., BENACERRAF, B., and MCCLUSKEY, J. W. Studies on specificity of the cellular infiltrate in delayed hypersensitive reactions. *J Immunol 90*:466 (1963).

RUDDLE, N. H., and WAKSMAN, B. H. Cytotoxicity mediated by soluble antigen and lymphocytes in delayed type hypersensitivity. *J Exp Med 128*:1267 (1968).

WAKSMAN, B. H., ARNASON, B. G., and JACKOVIC, B. D. Role of thymus in immune reaction in rats. III. Changes in lymphoid organs of thymectomized rats. *J Exp Med 116*:187 (1962).

WARD, P. A., OFFEN, C. D., and MONTGOMERY, J. R. Chemoattractants of leukocytes with special reference to lymphocytes. *Fed Proc 30*:1721 (1971).

chapter

ALLOANTIGENS ON CELL SURFACES: BLOOD GROUP SUBSTANCES AND HISTOCOMPATIBILITY ANTIGENS

BLOOD GROUP SUBSTANCES   598

THE ABO SYSTEM   598
 Genetic Determination of the Alloantigens   599
 Distribution of Alloantigens: Secretors and Nonsecretors   600
 Chemistry of the A, B, H, and Le Alloantigens   600
 The AB Alloantibodies   603
OTHER RED CELL ALLOANTIGENS   604
 MNSs Blood Groups   605
 Rh Antigens   605
 The Kell and Duffy Blood Groups   607
BLOOD TYPING   608

TRANSPLANTATION IMMUNITY: THE ALLOGRAFT REACTION   610

DEFINITIONS   610
THE ALLOGRAFT REACTION AS AN IMMUNE RESPONSE   610
 The Second-Set Reaction   610
 Transfer of Allograft Immunity   611
 Humoral vs. Cell-Mediated Immunity   611
 Tolerance of Allografts   611
GRAFT-VS.-HOST REACTION   614
HISTOCOMPATIBILITY OR TRANSPLANTATION ANTIGENS   614
HISTOCOMPATIBILITY GENES   614
 Histocompatibility Genes in the Mouse   615
 Histocompatibility Genes in Man   618

IMMUNITY AGAINST TUMORS   620

CODA   623

This chapter deals with the immunology of blood group substances, which are of practical importance in the transfusion of blood, and with histocompatibility antigens (Ags) on diverse cells, which are responsible for the reactions that limit the survival of transplanted organs and tissues. The characteristic immunological effects of all these Ags derive mainly from their position on cell surface membranes. Similar effects are elicited by certain distinctive tumor-specific Ags on surface membranes of tumor cells, which are therefore also considered in this chapter.

The blood group and transplantation Ags are **alloantigens** (Gr. *allo* = other), which are identified operationally as substances from certain individuals that are immunogenic in some other members of the same species, but not in the donor. They may be defined formally as Ags that segregate genetically within a species: they are present in some individuals and absent in others.

An older synonym, isoantigen (Gr. *iso* = same), emphasizes that the variations occur among diverse individuals of the **same** spe-

cies; this term is still widely applied to blood group substances, while alloantigen is the synonym commonly applied to histocompatibility Ags. We shall use alloantigen for both not only for consistency but because it is the differences among these Ags that should be emphasized, rather than the sameness of the species.

Alloantigens were first recognized on erythrocytes over 60 years ago (see below), and those on nucleated cells were later shown to account for the general observation that tissues transplanted from one individual to another are ultimately rejected, unless donor and recipient are genetically identical (**syngeneic,** see p. 610). Though Ags of this type were thought for many years to be limited to cells, several soluble proteins have recently also been recognized as alloantigens (e.g., haptoglobins, transferrins, and some β-lipoproteins). The alloantigenic forms of immunoglobulins (Igs), known as allotypes, have been exploited to great advantage in clarifying the genetics and structure of antibody (Ab) molecules (Ch. 16).

BLOOD GROUP SUBSTANCES

THE ABO SYSTEM

Following Harvey's discovery of the circulation, repeated attempts were made to transfuse blood from one individual to another. Disastrous reactions in the recipients were frequent, however, and were not understood until Landsteiner discovered the alloantigens of human red blood cells in 1900.

Stimulated by observations in the 1890s

that animal species could be distinguished by the reactions of their red cells and serum proteins with specific antisera, Landsteiner sought to determine whether individuals of the **same species** could be distinguished in the same way. When individual samples of serum and erythrocytes from 22 human subjects were mixed in all possible combinations the red cells of some persons were found to be clumped by the sera of certain other indi-

TABLE 21-1. Division of Human Populations into Four Blood Groups on the Basis of Red Cell Agglutination by Normal Human Sera

Serum from group	Red cells from group			
	A	B	O	AB
A	0	+	0	+
B	+	0	0	+
O	+	+	0	+
AB	0	0	0	0

+ = clumping; 0 = no clumping.

viduals. On the basis of these reactions the subjects could be classified into three groups, A, B, and O; and within 1 year a less common fourth group, AB, was also recognized (Table 21-1).

From these results Landsteiner concluded that 1) two different Ags or blood group substances, A and B, are associated with human red cells and one, both, or neither may be present in any given individual's cells; 2) **Abs** for these alloantigens (called **alloantibodies** or isoantibodies) are regularly present in the sera of those individuals who lack the corresponding alloantigen, and never present in the sera of those who possess it. These observations, later confirmed in vast numbers of humans, contributed greatly to the early acceptance of the doctrine of self-tolerance (Ch. 17).

The **blood groups** or types in human populations are named for the red cell Ags: group A has the A alloantigen, group B has the B alloantigen, group O has neither, and in group AB **each** red cell has both A and B. The corresponding serum alloantibodies are anti-A in group B persons and anti-B in group A; group O has both anti-A and anti-B; group AB has neither.

When anti-A sera are adsorbed with cells from certain A persons (A_2), making up about 20% of the A population, they lose their ability to clump these cells but retain their ability to clump A cells from the remaining 80% (A_1). On the other hand, adsorption of these sera with A_1 cells abolishes their capacity to agglutinate all A cells. Thus there are two kinds of A: A_1 adsorbs anti-A completely, and A_2 adsorbs only some of the anti-A. Correspondingly, there are two types of AB cells: A_1B and A_2B.

GENETIC DETERMINATION OF THE ALLOANTIGENS

Family studies established the heritability of blood cell alloantigens. Analysis of large populations led Bernstein to propose that the *ABO* gene has three alleles—*A, B,* and *O,* with *A* and *B* dominant over *O.* Since man is diploid, the two alleles per individual provide the six genotypes and four phenotypes shown in Table 21-2.

Since each allele has the same probability of being inherited, the Bernstein theory accounts for the distribution of blood groups within families. For example, children of an O and an AB parent (*OO* × *AB*) average 50% A (genotype *AO*) and 50% B (genotype *BO*); none are AB or O. The large number of families examined are consistent with the foregoing scheme; the small proportion of inconsistent results (less than 1%) are reasonably ascribed to illegitimacy, technical errors in typing (or possibly very rare mutations).

H Antigen. Though the *A* gene controls the formation of the A substance, and the *B* gene the B substance, the *O* gene is an "amorph," i.e., it does not specify a particular red cell alloantigen. Nevertheless, O cells have a distinctive Ag that has been recognized from their agglutination by some normal animal sera (from eels and cattle), antisera (goat anti-*Shigella*), or plant proteins (extracted from the seeds of *Ulex europeus*). This Ag (as well as the A and B alloantigens) is lacking in the red cells of certain rare humans, first recognized in Bombay; and such **Bombay-type individuals** can form Abs in high titer to the characteristic Ag of O cells, either spontaneously or in response to injec-

TABLE 21-2. Genotypes and Phenotypes in the ABO Blood Group System

Genotype	Phenotype
AA ⎱ AO ⎰	A
BB ⎱ BO ⎰	B
OO	O
AB	AB

tions of O cells. The O cell Ag is present not only in man, but also in a variety of other species; i.e., it is **heterogenetic*** and is therefore called **H substance.** This Ag cannot, however, be considered the product of the *O* gene since it is also detectable, with anti-H sera, on the red cells of persons who **lack** the *O* gene, such as A₂B or homozygous A₂ individuals.

As we shall see below there are good reasons for believing that 1) the gene for H substance is independent of the *ABO* gene, and 2) the H substance provides an obligatory precursor to which the A and B groups are attached. **In O individuals the H substance is exposed and fully expressed.** Partial expression in red cells of other types is revealed by reaction with monospecific anti-H sera, the order of reactivity being $O >> A_2 > A_2B > B$; the sera react weakly with some A_1 and A_1B cells, and, as noted above, not at all with Bombay-type. In Bombay-type persons a defective (or absent) *H* gene apparently leads to the absence of H substance, and thereby of A and B determinants as well (see below).

Another pair of blood group activities, called Lewis a and b (Leᵃ and Leᵇ), are associated with the A, B, H substances. The Leᵃ activity is specified by another gene, *Le,* that segregates from the *ABO* and *H* genes. The Leᵇ activity is not due to a separate allele, but represents instead a novel specificity formed by the combined presence of the individual structures that, by themselves, determine the Leᵃ and H activities (see below).

---

* An Ag is called heterogenetic when it is formed by a variety of phylogenetically unrelated species. The best known examples are the **Forssman** Ags, one of which is of clinical importance in the diagnosis of infectious mononucleosis (Ch. 61). These Ags, defined by their ability to induce rabbits to form hemolytic Abs to sheep red blood cells (sheep hemolysins), are found in many animal and some bacterial species, e.g., in tissues of guinea pig, horse, cat, and chicken, but not rat or rabbit. The anti-Forssman Abs, sometimes called **heterophile** Abs, are of two types, differing in their ability to distinguish human A and AB from O and B cells. Actually, A and B substances are probably also heterogenetic; similar substances are found in a variety of animal and bacterial species.

DISTRIBUTION OF ALLOANTIGENS: SECRETORS AND NONSECRETORS

The A, B, H, and Leᵃ substances are not confined to the red cell membrane. Though absent on connective tissue and muscle cells, they are present as surface components of many epithelial and virtually all endothelial cells (Fig. 21-1), and they are also abundant (in some persons) in many secretions: saliva, gastric juice, pancreatic secretions, sweat, meconium, ovarian cyst mucin, etc.

The presence of A, B, and H in secretions is controlled by alleles *Se* and *se* at a separate genetic locus. About 80% of the all persons are **secretors** (*Se/Se* or *Se/se*): depending on their blood type, their secretions contain water-soluble glycopeptides with A, B, or H activity. The other 20% are **nonsecretors** (*se/se*): their secretions lack A, B, and H, but contain instead a similar glycopeptide with Leᵃ activity. In about 1% of all persons the secretions lack all these activities but contain a related glycopeptide that corresponds to the "core" of the blood group glycopeptides (see below).

Because of their abundance and solubility, blood group substances in secretions have been invaluable for chemical and antigenic analyses. They are serologically indistinguishable from the blood group substances that are isolable in trace amounts, and with great difficulty, from red cell membranes; the few chemical differences are not relevant for their antigenic properties. The purified substances from secretions are glycopeptides, whereas those from red cells are glycopeptides plus glycolipids, with the same oligosaccharides.

A, B, and H antigens when present in saliva are all found to be determinants on the same macromolecule, for all are coprecipitated by a monospecific antiserum to any one.

CHEMISTRY OF THE A, B, H, AND LE ALLOANTIGENS

Regardless of their A, B, H, or Leᵃ activity, the purified water-soluble glycopeptides have a similar over-all structure: they are polydisperse macromolecules (MW 200,000 to 1,000,000) of similar composition (75 to 80% carbohydrate, 15 to 20% protein);

Fig. 21-1. Localization of A and B alloantigens in human tissue by immunofluorescence. The A substance is shown in lymph node (**A**), epidermis (concentrated in stratum corneum) (**B**), Hassall's corpuscle of the thymus (**D**), and the goblet cells of a villus in the small intestine **(F)**. The B substance is shown in squamous epithelium of the tongue (**C**) and in transitional epithelium of renal calyces (**E**). [From Szulman, A. E. *J Exp Med 111*:789 (1960).]

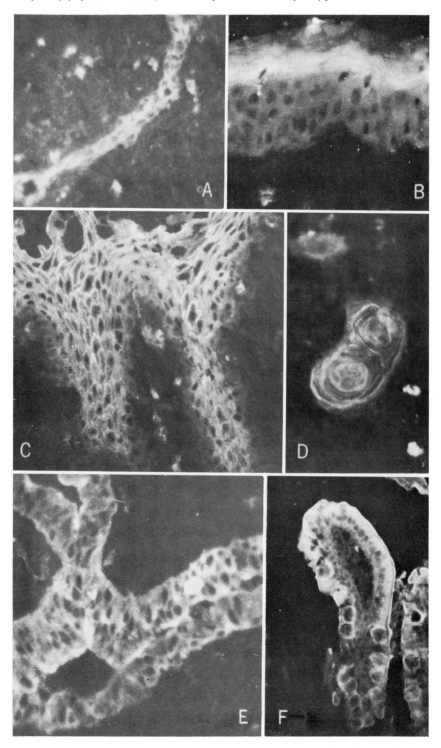

they all consist of multiple heterosaccharide branches attached by glycosidic linkage at their internal, reducing ends to serine or threonine of the polypeptide backbone (see Fig. 21-3, below).

Degradation Products. Stepwise enzymatic removal of sugars from the free, nonreducing ends of branches eliminates some determinant groups and exposes others that were previously detected only feebly or not at all (Fig. 21-2). Thus, when purified A substance is digested with a crude enzyme preparation (from clostridia) the A activity is lost, and H activity increases. Similarly, when B substance is digested with another enzyme (an α-glycosidase from coffee) the B activity is lost and H activity appears. Furthermore, when the H-active glycopeptides derived from these procedures (or isolated directly from O individuals) are treated with still another enzyme preparation, the H activity disappears and the capacity to react with Abs to the Le^a determinant appears. Finally, with further degradation Le^a activity is lost; the remaining glycopeptide cross-reacts with type 14 pneumococcal polysaccharide (Fig. 21-2).

These relations led Watkins and Morgan, and Ceppellini, to postulate that the blood group macromolecule of the ABO and Lewis systems is built from a single large glycopeptide, on which the various sugars that correspond to the Le^a, H, A, and B specificities are added sequentially by enzymes that are specified by the corresponding genes (*A, B, H, Le*). As each sugar is added it introduces a new antigenic specificity, masking the previous one.

This hypothesis has been thoroughly substantiated. Fragmentation of purified blood group glycopeptides (by periodate oxidation,

alkaline borohydride reduction, partial acid hydrolysis) has yielded a large variety of small oligosaccharides from whose properties the structure shown in Figure 21-3 was gradually deduced. Since many of these fragments specifically inhibit various alloantibodies (e.g., in hemagglutination or precipitin assays) specificities can be assigned to particular saccharides. For instance with fucose residues attached ($\alpha 1 \rightarrow 2$) at residues ② and ④ of Figure 21-3 the heterosaccharide has H activity; but if N-acetyl galactosamine is also present ($\alpha 1 \rightarrow 3$) on both ② and ④, A$_1$ activity is established and H activity is negligible, even though fucose is still present on ② and on ④. If galactose ($\alpha 1 \rightarrow 3$) is the additional residue at ② and at ④ the substance has B activity, rather than A, and H activity is also masked.

Structural analysis also accounts for the antigenic difference between A$_1$ and A$_2$. In A$_2$, N-acetyl galactosamine is present only on ④ (branch II), and branch I lacks a terminal substituent on ②; hence A$_2$ also has a good deal of H activity.

Thus the antigenic difference between A and B determinants, of vast clinical importance for blood transfusions, is determined by the presence or absence of acetylated amino groups in the terminal sugars of complex, branched glycopeptides. While these few atoms are the crucial elements, the actual A and B determinants are probably as large as the terminal tetra- or pentasaccharides shown in Figure 21-3.

The Watkins-Morgan hypothesis postulates that the direct products of the *A, B, H,* and *Le* genes are enzymes (glycosyltransferases) that carry out the additions of particular sugars in the biosynthesis of the heterosaccharides, rather than the antigenic sugars themselves.

In support of this idea Ginsburg *et al.* identified in human milk the four postulated enzymes: they transfer the appropriate sugar moieties from nucleotide-activated substrates to a milk tetrasaccharide that resembles the termini of branches I and II of the core heterosaccharide in the blood group glycopeptide (Fig. 21-3). The presence or absence of these glycosyltransferases in different individuals coincides with their A, B, O, Le blood type (Fig. 21-4)

The A and B glycosyltransferases can function only if the core heterosaccharide has

Fig. 21-2. Antigenic determinants of blood group mucopeptides revealed by sequential enzymatic removal of terminal sugar residues from A and B substances. [From Watkins, W. M. *Science* 152:172 (1966).]

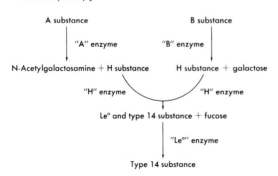

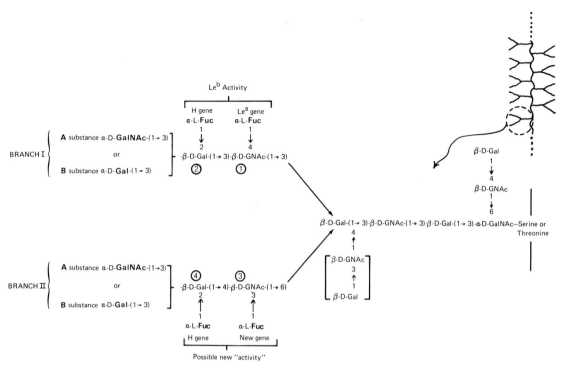

Fig. 21-3. Proposed structure of the ABO megalosaccharide. Branched heterosaccharides project from a polypeptide backbone in secreted water-soluble substances, or are associated with lipid in cell membrane glycolipids. A representative branched structure (enclosed in dotted line) is shown in detail. The residues responsible for A, B, H, Le^a, Le^b activities are in heavy type; they are added by glycosyltransferases to the "core"structure, whose residues are in light type. The core, devoid of the heavy type substituents, reacts with antisera to type 14 pneumococcal capsular polysaccharide; a fragment of the core (branch II) reacts with certain monoclonal human Abs ("cold agglutinins" called anti-I). [Modified from Lloyd, K. O., and Kabat, E. A., *Proc Natl Acad Sci USA* 61:1470 (1968), and Lloyd, K. O., Kabat, E. A., and Liverio, E., *Biochemistry* 9:3414 (1970).]

terminal fucose residues, corresponding to H activity. In the Bombay-type blood mentioned earlier the absence of these fucose residues—owing to an absent or defective *H* gene—prevents addition of the terminal hexoses; the resulting glycopeptide in these persons lack A and B as well as H determinants. The *O* allele evidently produces an inactive enzyme or no protein at all.

From the foregoing results it is clear that both allelic (*A, B*) and nonallelic (*H, Le*) genes cooperate in sequential additions of different sugar determinants on a complex heteropolymer. The process resembles other complex interactions of genes that determine the numerous different specificities of the cell wall antigens (O polysaccharides) in the salmonellae (Ch. 29) and other bacteria.

THE AB ALLOANTIBODIES

The anti-A and anti-B alloantibodies are natural Abs (Ch. 14), probably formed in response to ubiquitous A- and B-like polysaccharides of many intestinal bacteria and of some foods. Additional inconspicuous immunogenic stimulation probably comes from inapparent infections. Because of self-tolerance (Ch. 17), an individual would form only those natural Abs that are specific for the alloantigens he lacks; for example, a person of type A forms anti-B, not anti-A. [Chickens have blood groups, including alloantibodies, that resemble the human AB system; when raised under germ-free conditions chicks form alloAgs (A, B, etc.) but not anti-A and anti-B.]

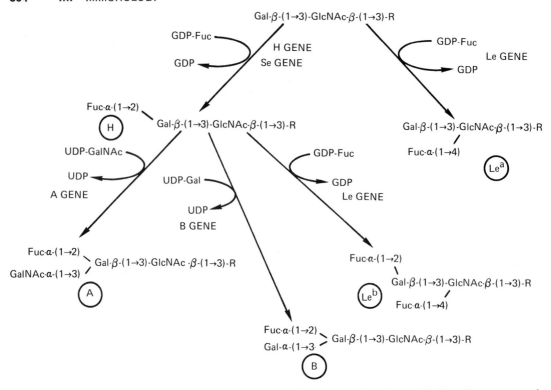

Fig. 21-4. Glycosyltransferases in the synthesis of A, B, H, and Le antigenic specificities. Enzymes specified by *A, B, H,* and *Le* genes transfer hexoses from activated precursors (uridine or guanosine diphosphosugars) to a tetrasaccharide (from human milk) that correponds to branch I of the core heterosaccharide of Figure 21-3. R is lactose. [From Kobata, A., Grollman, E. F., and Ginsburg, V. *Biochem Biophys Res Commun* 32:272 (1968).]

Classes of Immunoglobulin. Natural anti-A and anti-B are predominantly IgM molecules in persons of groups B and A, but they are both IgM and IgG in those of group O. The Abs formed in response to intensive antigenic stimulation by injections of A and B substance are largely of the IgG class.*

Evolution. It has been suggested that some severe infectious diseases could have served as selective agents in the evolution of blood group substances and accounted for the different frequencies of A, B, O blood types in different ethnic groups (Table 21-3). If, for instance, a particular blood group substance were the same as a surface Ag of a microbial pathogen, persons with that alloantigen would be immunologically unresponsive to it and might be more susceptible to infection (Ch. 17, Tolerance), assuming that the corresponding Abs (if formed) would be protective. So far the lack of

---

* Purified A and B substances are sometimes injected deliberately to raise the Ab titer, as in the preparation of typing sera.

correlation between blood type and incidence or severity of certain infectious diseases (e.g., smallpox, whose virion has been thought to contain an A-like substance) does not support this idea, but it nevertheless remains an attractive possibility.

That pathogens might operate as selective agents in another way is suggested by evidence that a blood group Ag of chicken red cells serves as specific receptor for an avian leukosis-sarcoma virus. Similarly, the MN blood group substances (see below) resemble cell surface receptors for influenza virus: isolated MN blocks agglutination of human red cells by the virus (Ch. 56, Influenza virus hemagglutinin).

OTHER RED CELL ALLOANTIGENS

The discovery of the ABO system was facilitated by the presence of anti-A and anti-B Abs in most normal human sera. Since then many other red cell alloantigens have been

TABLE 21-3. Frequency of ABO Blood Types in Various Ethnic Groups

Population	Phenotypes (%)			
	O	A	B	AB
Scotland (Stornoway)	50	32	15	3
Sweden (Uppsala)	37	48	10	6
Switzerland (Berne)	40	47	9	4
Pakistan (N.W. frontier province)	25	33	36	5
India (Hindus in Bombay)	32	29	28	11
United States (Chippewa Indians)	88	12	0	0
Eskimos (Hudson Bay)	54	43	1.5	1.5

Based on Mourant, A. E. *The Distribution of the Human Blood Groups.* Thomas, Springfield, Ill., 1954.

recognized, but different means were required for their detection, because the corresponding natural Abs are not found in normal human sera. One approach involves skillfully adsorbed animal antisera to human red cells. Another depends on the appearance of alloantibodies in persons who have received multiple blood transfusions, and in women who have had multiple pregnancies; for a fetal alloantigen, inherited from the father and absent from the mother, can stimulate maternal Ab formation if it enters the maternal circulation. These several approaches are illustrated with selected blood group systems described below.

MNSs BLOOD GROUPS

MN Antigens. About 25 years after the ABO system was recognized Landsteiner and Levine found that occasional rabbit antisera to human blood would, after adsorption with red cells from certain persons, agglutinate many but not all human erythrocytes, regardless of their ABO type. The adsorbed antisera were thus specific for an additional red cell alloantigen, which they called M. Rabbit antisera subsequently elicited with various samples of M+ cells revealed still another red cell alloantigen, which was called N. All human red cells react with either anti-M or anti-N sera, or with both. From the distribution of the M and N Ags in many human families, it has become clear that these Ags are governed by a pair of allelic genes that segregate inde-

pendently from the ABO locus: irrespective of an individual's ABO type, his cells also have M or N, or both. The *MN* alleles are codominant; hence genotypes *MM, NN,* and *MN* correspond to phenotypes M, N, and MN.

Ss Antigens. Human sera (e.g., from a mother with Abs to her fetus' red cells) later identified another pair of alloantigens, called S and s, that are associated genetically with M and N. S and s are distributed in families as though governed by a pair of codominant allelic genes: i.e., every person has one or the other or both.

S is much more frequently found on M or MN than on N cells. This suggests that the S specificity might be due to a mutation of the *M* gene (and less frequently of *N*) that adds a determinant group (S) without modifying the original determinants: a similar basis for the s specificity would mean that there are four alleles for the *MN* locus (*MS, Ms, NS, Ns*). However, family studies are also consistent with the alternative possibility that these alloantigens represent two separate but tightly linked genes (*M,N* and *S,s*): for example, the child of homozygous MS and Ns parents (*MMSS × NNss*) would be MSNs but would contribute to each of his offspring *either MS or Ns,* as though each pair were inseparable.

It is not possible to make a firm decision between the alternative genetic possibilities. But the nonrandom association of *S* with *M* argues against linkage of separate genes; for even with tightly linked genes, and correspondingly infrequent crossover events, over a long period of evolution genetic equilibrium should have led *S* (and *s*) to be as frequently associated with *N* as with *M*, unless there are unknown selective advantages to the *M-S* combination, or unless contemporary populations resulted from relatively recent interbreeding of different racial groups that originally had considerably different frequencies of these genes.

Rh ANTIGENS

Like ABO, the Rh system was discovered as a byproduct of curiosity about alloantigens and was then found to solve an important clinical problem. In the 1930s, Landsteiner and Wiener, investigating the phylogeny of M antigen, found that rabbit antisera to *Rhesus* monkey red cells (after removal of anti-M) clumped the red cells of about 85% of all persons (Rh+) but not those of the remaining 15% (Rh−). Wiener then showed that some severe reactions to blood transfusions could be explained by Rh incompatibility: donors were Rh+ while recipients were Rh− but had anti-Rh Abs.

At about the same time Levine and Stetson described the case of a mother who, after giving birth to a baby with a hemolytic disease called erythroblastosis fetalis, suffered a severe reaction upon receiving blood from her husband, though both parents had the same ABO type. The newly discovered anti-*Rhesus* serum was subsequently found to clump the father's cells (Rh+), not the mother's (Rh−); and in similar cases erythroblastotic babies were found to be Rh+ with an Rh− mother and an Rh+ father. Evidently fetal red cells, carrying the paternal Ag, can cross the placenta and immunize the mother; maternal Abs can then enter the fetal circulation and react with fetus' red cells, causing massive breakdown of red cells.

Nonagglutinating Rh Antibodies. The foregoing explanation presented one difficulty: most sera from affected Rh− mothers did not clump red cells from the erythroblastotic Rh+ babies or from other Rh+ individuals. It was later discovered, however, that human anti-Rh Abs are often of the "incomplete" type (Ch. 15): they combine specifically with Rh+ cells without causing agglutination. Because these nonagglutinating Abs are of the IgG class, they cross the placenta; and they are ultimately responsible for destruction of the fetal red cells: they probably act as opsonins and promote phagocytosis and degradation by macrophages. The initial puzzle was complicated by the presence in some mothers of anti-Rh Abs of the IgM class: these Abs are readily recognized because they agglutinate Rh+ cells but they do not cross the placenta (Ch. 17, Ontogeny) and therefore do not affect the Rh+ fetus.

The search for anti-Rh Abs in hemolytic disease provided the main impetus for the development of a number of ingenious methods for detecting incomplete Abs (see, for example, the antiglobulin or Coombs test, Ch. 15). Though these Abs were originally considered to be univalent they almost certainly are bivalent since they can clump red cells

under special circumstances, e.g., when 1) reactions are carried out in the presence of a high protein concentration, such as 30% serum albumin, or 2) red cells are first treated with proteolytic enzymes, such as trypsin or ficin. (See Ch. 15, Monogamous binding, for other bivalent Abs that do not cross-link Ag particles.) Nonagglutinating anti-Rh Abs also played a key role in the discovery of diverse alloantigenic forms of human immunoglobulins (Ch. 16, Allotypes; Fig. 16-13).

Frequency of Hemolytic Disease in the Newborn Due to Blood Group Incompatibility. Fetal red cells probably enter the maternal circulation in small numbers in most pregnancies. However, only about 1 in 100 Rh+ babies with Rh− mothers have significant hemolytic disease. Yet the Rh factor is more immunogenic than most other alloantigens and it accounts for the vast majority of fetal deaths due to red cell incompatibilities. [On rare occasions incompatibilities in the ABO and the Duffy and Kell blood group systems (see below) also cause severe hemolytic disease of the newborn.] One reason is that entry of fetal red cells into the maternal circulation is usually slight during early pregnancy, and is only pronounced during expulsion of the placenta at parturition. Hence it is only after the immunogenic stimulus experienced at delivery of her first Rh+ baby that the Rh− mother can make a substantial anamnestic response to the small numbers of Rh+ fetal cells that cross the placenta during subsequent pregnancies. Thus the first Rh+ baby is usually unaffected and the risk of hemolytic disease increases in successive pregnancies.

Prevention of Rh Disease. The incidence of Rh disease is less when mother (Rh−) and baby (Rh+) are also incompatible in ABO type (e.g., an O mother with B baby): rapid removal of fetal cells from maternal circulation by anti-B (or anti-A) evidently reduces the antigenic Rh stimulus (see discussion of Suppression of Ab formation by other Abs acting as antiimmunogens, Ch. 17). This observation led eventually to the development of a simple and effective means for reducing the incidence of erythroblastosis in the successive children born to high-risk parental combinations (Rh+ father, Rh− mother). At each delivery the mother is injected with anti-Rh Abs (γ-globulin fraction of human anti-Rh serum), causing rapid clearance of Rh+ fetal cells from maternal blood: this minimizes immunogenic priming of the mother.

Multiplicity of Rh Antigens. Analyses of many fetal–maternal incompatibilities have yielded a large variety of anti-Rh sera, of which there now seem to be about 27 distinctive types, each corresponding to a particular alloantigenic determinant or "specificity." Family studies have established that these serologically distinctive specificities are all genetically linked. However, since structural information about Rh antigens is lacking the genetic basis for their extreme polymorphism remains controversial. Before so many specificities were identified there was wide acceptance of Fisher's view that three allelic pairs of closely linked genes (*Cc, Dd, Ee*) were responsible. However, the multiplicity of specificities now known can hardly be reconciled with this scheme, but they can be accommodated by Wiener's view that all Rh specificities are due to a single Rh locus with many alleles. Various terminologies for the Rh Ags are still widely used (Table 21-4).

Antigen D. Despite the large number and complexity of the genetically linked Rh Ags their clinical utility is simplified by the outstanding importance of one of them, the factor called D. This alloantigen was responsible for the first case recognized by Levine and Stetson; it accounts for over 90% of all cases of erythroblastosis fetalis. Hence anti-D sera, from transplacentally immunized women, are used routinely to establish Rh type.* A person who is D− is commonly referred to as being Rh−; he will, however, usually possess some of the other antigenic determinants of the Rh locus.

---

* Guinea pig and rabbit antisera to *Rhesus* red cells do not distinguish consistently between D+ and D− red cells in umbilical cord blood, and so they are not used for typing purposes.

TABLE 21-4. Notations for Rh Alloantigens and Genes*

Some Rh alloantigens		Some genes or gene combinations		
Fisher-Race	Wiener	Fisher-Race	Wiener	Commonly used
D	Rh_0	cDe	Rh_0	R_0
C	rh′	CDe	Rh_1	R_1
E	rh″	cDE	Rh_2	R_2
c	hr′	CDE	Rh_z	R_z
e	hr″	cde	rh	r
		Cde	rh′	R′
		cdE	rh″	R″
		CdE	rh_y	R_y

* If a person's red cells react only with anti-C, anti-D, and anti-e sera, the cells are called CDe in the Fisher-Race terminology, and the genotype is assumed to be *CDe/CDe*. In the Wiener terminology the same antisera are called anti-rh′, anti-Rh_0, and anti-hr″, and the red cells that react with all three are called Rh_1, as though a large antigen, Rh_1, is composed of three distinguishable antigenic determinants (rh_1, Rh_0, hr″). The more recent terminology of Rosenfield, Allen, Swisher, and Kochwa (*Transfusion 2*:287, 1962) is purely descriptive: the antisera are numbered and the red cells' antigenic structure is given by listing the sera with which they have been tested, with a minus sign before the number meaning that no reaction was observed. Thus, anti-Rh1 = anti-D; anti-Rh2 = anti-C; anti-Rh3 = anti-E; anti-Rh4 = anti-c; anti-Rh5 = anti-e; anti-Rh6 = anti-ce or anti-f, etc. The CDe (Rh_1) cells receive the following designation: Rh: 1, 2, -3, -4, 5, -6.

THE KELL AND DUFFY BLOOD GROUPS

Besides ABO and Rh, several other Ag systems are important causes of severe transfusion reactions (and rarely of hemolytic disease in the newborn). The most common Ags of the Kell system are called K and k. K, present in about 10% of the population (Table 21-5), is a potent immunogen. Abs to K, formed as a result of pregnancy or transfusion, are usually incomplete and are detected by the Coombs antiglobulin test (Ch. 15). Unlike the red cells coated with Abs to most of the other blood group Ags, cells coated with anti-K are lysed by complement (C; see below).

The most common Ags of the Duffy system are termed Fy^a and Fy^b. Abs to Fy^b, formed as a result of transfusion or pregnancy, are being detected with increasing frequency in transfusion reactions and usually also require the Coombs antiglobulin test for their detection.

TABLE 21-5.   Incidence of Some Red Blood Cell Phenotypes in the United States*

Blood group system	Phenotype	Frequency (%)	
ABO	O	44	
	A ($A_1 + A_2$)	42	
	B	10	
	AB ($A_1B + A_2B$)	4	
MN	M	27	
	N	24	
	MN	50	
Ss†	S	11	
	s	45	
	Ss	44	
P	P_1	80	
	P_2	14	
	p	rare	
Rh‡	DCe (Rh_1, R_1)	54	⎫ 85% react with anti-D
	DCE (Rh_z, R_z)	15	⎬ (= "Rh-positive")
	DcE (Rh_2, R_2)	14	⎪
	Dce (Rh_0, R_0)	2	⎭
	dce (rh, r)	13	⎫
	dCe (rh′, R′)	1.5	⎬ 15% do not react with anti-D
	dcE (rh″, R″)	0.5	⎪ (= "Rh-negative")
	dCE (rh_y, R_y)	rare	⎭
Lutheran	Lu^a	6	
	Lu^b	94	
Kell	K+	6	
	K−	94	
Lewis	Le^a	22	
	Le^b	78	
Duffy	Fy^a	38	
	Fy^b	28	
Kidd	Jk^a	83	
	Jk^b	17	

* Based on *Zinsser's Microbiology,* 13th ed. (D. T. Smith, N. F. Conant, and J. R. Overman, eds.), Appleton, New York, 1964.

† The incidence of S, s, and Ss is from Race, R. and Sanger, R. *Blood Groups in Man,* 2nd ed., Blackwell, Oxford, 1954. Anti-S sera agglutinate 73% of M, 54% of MN, and 32% of N cells.

‡ From Levine, P., Stroup, M. and Pollack, W. In *Bacterial and Mycotic Infections of Man,* 3rd ed. (R. J. Dubos, ed.), Lippincott, Philadelphia, 1958.

BLOOD TYPING

The identification of red cell alloantigens, called blood typing or grouping, is required for blood transfusions. In addition, it is commonly used for genetic analysis in cases of disputed paternity, and for anthropological surveys of human populations. It has even been used in archeological work, since the red cell Ags are extraordinarily stable: they have apparently been identified (by specific inhibi-

tion of hemagglutination reactions) in Egyptian mummies thousands of years old. King Tutankhamen, for instance, has been typed A_2MN.*

Blood typing is performed by agglutination

---

* Blood typing and other serological procedures are widely used in forensic medicine: to distinguish human from animal blood; to identify human blood types in blood stains, semen, or saliva; to distinguish horse meat from beef; etc.

reactions; unknown red cells are typed with known antisera and unknown alloantibodies with red cells of known type. The C-fixation test is not used; red cells coated by Abs to most alloantigens are not lysed by C. This might be due to a peculiar distribution of most alloantigens on the red cell surface, or blood group Abs might often belong to classes of immunoglobulins that do not react effectively with the first component of C (Ch. 18).

Transfusion Reactions. Blood transfusion can lead to serious reactions if massive intravascular clumping and hemolysis of red cells take place.* Such reactions are usually the result of an attack on the donor's red cells by the recipient's alloantibodies. Rarely is the reverse reaction serious, between the donor's Abs and the recipient's cells, since 1) the alloantibody titer in most normal individuals (as donors are supposed to be) is usually low to begin with, and 2) the dilution of donor blood in the recipient (usually about 10-fold) is generally sufficient to lower the titer to an innocuous level. Nevertheless, each lot of blood drawn from a prospective donor is screened against standard samples of red cells with 8 to 10 of the Ags that are most often responsible for transfusion reactions; bloods with a high titer of either agglutinating or incomplete Abs are used for other purposes (for example, the washed red cells might be injected).

The blood considered for a particular transfusion must, of course, have the same major red cell Ags as those in the prospective recipient's blood. In addition, "major" cross-matching must be performed routinely, with the prospective donor's red cells and the recipient's serum; if no agglutination is observed the cells must be washed and tested for adsorbed incomplete Abs by addition of rabbit antiserum to human γ-globulins (Ch. 15, Coombs' test). As a further precaution "minor" cross-matching is similarly carried out, i.e., with the prospective donor's serum and the recipient's red cells.

Universal Donors and Recipients. Under emergency conditions certain persons can donate blood to any recipient, and certain others can accept blood from any donor. The "universal donors" are type O; their red cells cannot react with either anti-A or anti-B in the recipient. However, type O donors are avoided if they have unusually high titers of anti-A and anti-B. The "universal recipients" are type AB; they lack both anti-A and anti-B. As an additional precaution under such circumstances, soluble A and B substances are often added to the transfused blood as specific inhibitors to minimize the possibility of AB alloimmune reactions.†

Uniqueness of the Individual's Red Cell Alloantigen Pattern. The foregoing precautions are not always reliable, however, since there are other red cell alloantigens that can cause serious incompatibilities (Table 21-5). In fact, so many different blood group alloantigens have now been identified (at least 60) that no two individuals are likely to be found with identical combinations, except for monozygotic twins.

In spite of this extreme diversity blood transfusions are extraordinarily successful, even when recipients undergo multiple transfusions from many different donors. Aside from the care exercised in the selection of prospective donors, through scrupulous typing procedures, the infrequency of transfusion reactions appears to be due to the fortunate fact that most red cell alloantigens are only feebly immunogenic. Moreover, transfused red cells survive for only a limited time (an average of perhaps about 3 to 4 weeks), so that even if Ab formation should be stimulated in the recipient the transfused cells are

---

* After multiple transfusions or multiple pregnancies alloantibodies are also sometimes formed to diverse alloantigens on white blood cells and on platelets (HL-A transplantation antigens, below). These Abs sometimes also cause serious transfusion reactions, which are characterized by high fever when leukocytes are involved, or by thrombocytopenic purpura when platelets are involved (Ch. 20; Fig. 20-15). Alloantibodies to Am allotypes of IgA-2 Igs (Ch. 16) can also cause transfusion reactions.

† Except for life-saving circumstances, A and B substances are not added to blood that is to be transfused into women of reproductive age; any enhancement of their immune response to A or B would increase the possibility that the babies they subsequently bear might have hemolytic disease due to A or B incompatibilities.

likely to be few in number by the time his alloantibodies reach an effective level; the destruction of a small number of cells is not likely to be serious. As we shall see below, however, nature is less benevolent in the immune response to the alloantigens of transplanted tissues.

TRANSPLANTATION IMMUNITY: THE ALLOGRAFT REACTION

The immunological mechanisms underlying host reactions to transplanted tissues are difficult to decipher because of the complexity of the substances involved. But these reactions are among the most significant problems in immunology today, both because their suppression could make organ transplantation as feasible in the future as blood transfusions are at present, and because similar mechanisms are probably also involved in immune reactions against cancer cells.

DEFINITIONS

Four terms are used to describe tissue grafts.

1) **Autografts** are transplants from one region to another of the same individual.

2) **Isografts** are transplants from one individual to a genetically identical individual. These are possible only between monozygotic twins or between members of certain lines of mice and other rodents that have been so highly inbred (by brother-sister mating) as to be **syngeneic** or **isogenic,** i.e., genetically identical.*

3) **Allografts** or **homografts** are transplants from one individual to a genetically nonidentical (i.e., allogeneic) individual of the same species.

4) **Heterografts** or **xenografts** (Gr. *xenos* = foreign) are transplants from one species to another.

In these four types of grafts the donors are designated respectively as autologous, isologous, homologous, or heterologous with respect to the recipient.

---

* Syngeneic in the transplantation field refers operationally to the absence of any discernible tissue incompatibility, i.e., to genetic identity with respect to the genes controlling histocompatibility Ags. It approximates the situation in isogenic individuals, who are identical with respect to all their genes.

THE ALLOGRAFT REACTION AS AN IMMUNE RESPONSE

Among vertebrates autografts and isografts are usually enduring, but allografts (and heterografts) are regularly rejected.† Successful transfer of experimental tumors between individuals of an inbred mouse strain, but not between mice of different strains, provided the first indication that rejection of a graft was due to genetic differences between host and donor. It is now clear that grafted cells with genetically determined surface **histocompatibility or transplantation antigens** that are lacking in the recipient elicit an immune response that is directed against these Ags and leads to destruction (rejection) of the graft.

In investigating mechanisms of the allograft reaction skin grafts have been used most extensively, because of technical advantages: they are easy to prepare, their rejection is readily detected, and their median survival times provide an estimate of the intensity of the host's immune response. However, the same principles are involved in grafts of other tissues and cells.

THE SECOND-SET REACTION

When an allograft of skin is placed on a recipient animal, in a bed created by excising a slightly larger piece of skin, the graft at first becomes vascularized and its cells proliferate; but after about 10 days it quite abruptly becomes the seat of intense inflammation, withers, and is sloughed. If a second graft is then made to the same recipient, with another piece of skin from the same donor, it is rejected much more rapidly than the first graft, perhaps in 5 to 6 days. This accelerated re-

---

† Allografts are accepted by most invertebrates, which display few if any adaptive immune responses (see Chs. 17 and 20).

jection, the "second-set" reaction, is specific for a particular donor: if after the accelerated rejection another donor, antigenically different from the first, provides skin grafts to the same recipient, first- and second-set reactions to successive grafts are again seen. Thus, the capacity to reject an allograft (**transplantation immunity**) is acquired by virtue of exposure to the donor's cells; and it is specific for transplantation Ags of that donor. The shorter survival of the second graft results from persistence of the immunity acquired from the first graft or, perhaps, from an anamnestic response.

The second-set reaction to a skin graft can be induced not only by a prior skin graft, but just as well by prior inoculation of various other cells (e.g., spleen cells) from the same donor. In fact, virtually all cells induce transplantation immunity except erythrocytes; some histocompatibility Ags are present on red cells, but apparently not in immunogenic form (H-2 antigens in the mouse, below).

TRANSFER OF ALLOGRAFT IMMUNITY

Adoptive Transfer. The second-set reaction can be transferred "adoptively" from an immunized donor to a nonimmune recipient by viable lymphoid cells (T lymphocytes, Ch. 20). Thus if a mouse of strain X is immunized with a graft from a mouse of strain Y and viable lymphoid cells from the immune X animal are inoculated into a nonimmunized X mouse, the latter can respond as would the donor, giving a specific second-set reaction to a Y graft. [When the donor of sensitized lymphoid cells is syngeneic with the recipient the adoptive immunity can be enduring; when the donor is allogeneic the recipient's adoptive immunity is short-lived (1 to 2 weeks) and is terminated by an allograft reaction against the donor's lymphoid cells.]

HUMORAL VS. CELL-MEDIATED IMMUNITY

The rejection of most allografts appears to be an expression of cell-mediated hypersensitivity because the capacity for accelerated rejection is transferred by viable lymphocytes, not by serum. Moreover, in some species, such as the guinea pig, lymphocytes from a sensitized animal can elicit a delayed cutane-

ous reaction if injected into the skin of the animal to whose tissues they are sensitive. As in other cell-mediated reactions, the site of an allograft undergoing rejection is intensely infiltrated with lymphocytes and macrophages; granulocytes and plasma cells are much less conspicuous (Ch. 20).

Nevertheless, an animal that has rejected an allograft will often have serum Abs for the donor's histocompatibility Ags, some of which are also present on red blood cells; **hemagglutination reactions** have therefore provided a simple but powerful means of characterizing these Ags and for their genetic mapping (see below). However, these serum Abs seem not to be responsible for rejection of skin grafts: they generally appear late in the course of the allograft response, their levels bear no consistent relation to the intensity or rapidity of the rejection reaction, and as noted above they are generally incapable of transferring the ability to mount a typical second-set reaction.

However, some antisera from intensively immunized individuals can interfere specifically with the healing-in of a fresh graft, causing unusually rapid rejection (the "white-graft" reaction). Moreover, allogeneic lymphocytes, unlike "solid" tissue cells, are readily destroyed by Abs to histocompatibility Ags, and complement is required. Broadly speaking, therefore, rejections of allografts, like allergic reactions to most Ags, can be a manifestation of either humoral or cell-mediated immunity; but the cell-mediated reaction is primarily responsible for rejection of skin and other "solid" tissue grafts.

TOLERANCE OF ALLOGRAFTS

Under a number of exceptional circumstances allografts are not rejected.

Privileged Sites. There are a few special ("privileged") sites where allografts may flourish for prolonged periods without inducing immunity, such as the meninges of the brain and the anterior chamber of the eye. Lymphatic drainage is lacking in these sites; hence stimulation of immunologically competent cells of the host's lymphatic tissue is minimal.

Pregnancy. Histocompatibility Ags are formed early in embryonic life, and many of these Ags, inherited from the father, are alien to the mother. Hence in man and other mammals **the intrauterine fetus is actually an allograft.** Its failure to evoke an allograft reaction, even when the mother has been previously immunized deliberately to the father's histocompatibility Ags, is probably explained by the absence (or masking by mucinous secretions) of these Ags in the special fetal cell layer (trophoblast) at the placental interface between the fetus and its maternal host.

Induction of Immunological Tolerance. The induction of immunity to allografts has dose-response features that resemble those of immune responses in general. The routine success of corneal transplants in man (which are always allografts) is due to the small amount of tissue transplanted and the relative avascularity of the transplantation site. At the other extreme, large doses of allogeneic cells can establish tolerance to allografts, especially when introduced around the time of birth (see below).

Current interest in immunological tolerance in general actually began with experiments on allografts, which grew out of a crucial observation on red cell alloantigens in nonidentical cattle twins, which frequently have in utero anastomoses of their placental blood vessels. Owen observed in 1945 that each twin often had two antigenically different kinds of red cells, which persisted as the twins grew to maturity. It was inferred that hematopoietic stem cells from each twin, transferred through the common circulation in utero, had settled in the marrow of the other and then survived in the genetically foreign soil through the animal's lifetime. Each twin thus produced two kinds of red cells, with its own and with its twin's characteristic alloantigens. Such individuals, with mixtures of genetically different cells, are called **chimeras,** after the monster in Greek mythology with a lion's head, a goat's body, and a serpent's tail. (Rare blood group chimeras among nonidentical human twins have also been found.)

Surmising that such chimeras might also be generally tolerant of each other's tissues, Billingham, Brent, and Medawar then demonstrated that they accepted skin grafts from each other without an allograft reaction. By taking advantage of the fact that mice of a given inbred strain are exact genetic replicas, it was subsequently shown that mice of one strain (A) could be rendered permanently tolerant of skin allografts from another strain (CBA) if embryonic or newborn A mice were inoculated with viable cells (e.g., of spleen) from CBA animals. When the inoculated animals matured they accepted allografts permanently from CBA donors, though they rejected grafts from any other strain in a normal manner.

The newborn mouse appears to develop allograft tolerance with ease because its immune apparatus is relatively immature, and it therefore cannot reject the foreign cells by an allograft reaction. Once established, the tolerance persists, probably because the foreign cells continue to proliferate and to maintain a sufficient level of the tolerated transplantation Ags. An individual rendered tolerant in this manner is, therefore, a chimera, carrying allogeneic cells in advance of the allograft that reveals tolerance.* Since tolerance of allografts depends on an adequate level of foreign histocompatibility Ags (i.e., allogeneic cells) it seems to be fundamentally the same as the tolerance that can be established to other Ags, such as globular proteins and polysaccharides (Ch. 17). Another mechanism that might account for the persistence of allogeneic cells in a chimera is discussed below (blocking Abs in enhancement).

The immunological balance in the chimera can be readily tipped, and the tolerance abrogated, by introducing immunologically competent cells that recognize the tolerated tissue as genetically alien. Thus when a mouse of strain X, rendered tolerant as a newborn to tissues of strain Y, carries a successful and enduring Y graft, the graft can be made to undergo prompt rejection by inoculating the tolerant host with viable lymphoid cells from a strain-X animal that had previously been immunized against Y (Fig. 21-5). The rejection can also be elicited by inoculating the tolerant X animal with lymphoid cells from a normal, nonimmune animal of strain X, but

---

* The chimerism can be demonstrated by using spleen or white blood cells from such a tolerant animal (A strain tolerant of CBA) as immunogens to induce in a third strain of mice allograft sensitivity to both A and CBA.

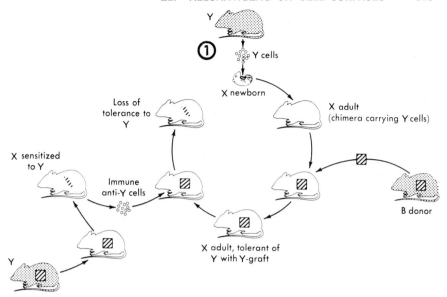

Fig. 21-5. Establishment and loss of tolerance to histocompatibility Ags. Newborn X was injected with Y cells. The subsequent injection of the tolerant X animal (shown carrying a Y skin graft) with lymphoid cells from a Y-sensitive X donor leads to adoptive immunity, with rejection of the previously tolerated graft. The risk of a graft-vs.-host reaction can be eliminated in step 1 (top) if the donor of cells to the newborn is an X × Y hybrid, rather than a purebred Y-strain mouse, as shown.

the effect requires more cells and takes longer to develop.

Adult animals have, with difficulty, also been rendered tolerant of allografts by inoculation of viable cells from prospective donors, but special measures are required: the adult is first converted temporarily to a state of immunological incompetence, resembling that of the newborn period, by intensive X-irradiation or treatment with cytotoxic drugs (Ch. 17), and multiple injections of the allogeneic cells are then usually given. Reactions of allogeneic lymphoid cells against the immunologically incompetent host creates other severe problems (Graft-vs.-host reaction, below and Ch. 20).

The genetic basis for both transplantation tolerance and transplantation immunity is evident with the hybrid offspring (AB) of a cross between inbred parental strains (AA and BB). **The AB hybrids are tolerant of grafts from either parental strain, but mice of either parental strain reject grafts from the hybrids:** a graft is permanently accepted only when essentially **all** its histocompatibility Ags are present in and therefore tolerated by the recipient.

Enhancement. Paradoxically, the survival of certain allografts may be prolonged in adult recipients by previous intensive immunization, either active (with the corresponding allogeneic cells) or passive (with Abs to the histocompatibility Ags of these cells). The mechanism for this graft "enhancement" is still not completely understood; it appears, however, that if Abs bound to cell surface histocompatibility Ags are not cytotoxic they can sometimes specifically protect the target cells against destruction by the "killer" lymphocytes responsible for cell-mediated immunity (Ch. 20, see "efferent enhancement" in discussion of Suppression of cell-mediated immunity by Abs). It is also possible that these Abs block immunogenicity of the graft's histocompatibility Ags, preventing induction of cell-mediated immunity (Ch. 20, "afferent" enhancement). Enhancement by serum Abs is less readily demonstrated with normal tissue allografts than with certain solid tumor allografts (Sarcomas and carcinomas, below).

Enhancement might be responsible for the long-lasting chimerism established in neonatal animals with allogeneic cells (Induction of immunological tolerance, above): the host, rather than being truly tolerant, might make Abs to the allogenic histocompatibility Ags, causing prolonged survival of the donor's cells. Thus, spleen cells from these chimeras appear, in in vitro cytotoxicity tests, to destroy donor cells, and the effect appears to be blocked specifically by the chimera's serum. (Other assays suggest, in contrast, that the chimera is truly tolerant; for example, his lymphocytes seem to be incapable of effecting a graft-vs.-host reaction in an appropriate host: see below.)

GRAFT-VS.-HOST REACTION

In establishing tolerance by injecting a newborn animal with viable allogeneic cells the inoculum is usually prepared as a suspension of spleen cells: when derived from an immunologically mature (adult) animal some T cells in the suspension can react with the alloantigens of the neonatal host.

Evidence of such a reaction was only occasionally observed with the particular donor-recipient strains used in the initial demonstration; but it has since been seen consistently with many other pairs of strains. In these graft-vs.-host reactions the inoculated newborn animal fails to gain weight normally, develops skin lesions and diarrhea, and dies after a few weeks **("runting syndrome").** This symptom complex has also been seen when immunologically competent cells from allogeneic donors are injected into an adult that cannot reject the cells, e.g., one depleted of lymphocytes by X-irradiation or cytotoxic drugs or an adult hybrid (F_1), who is tolerant because one of its parents is syngeneic with the donor (i.e., AA donor cells make a graft-vs.-host reaction against the B alloantigens in an AB recipient). For more on these reactions, see Ch. 20.

HISTOCOMPATIBILITY OR TRANSPLANTATION ANTIGENS

Partially purified histocompatibility Ags (obtained by limited proteolysis of mouse spleen cells) are glycoproteins (10% carbohydrate, MW ca. 35,000 to 50,000). The small amounts available so far have been able to induce allograft hypersensitivity (accel-

erated rejection) and, in larger doses, to establish tolerance with respect to Ab formation. However, the purified Ags have not yet established tolerance of allografts, suggesting that they may have some defective (or missing) antigenic determinants.

The specificity of histocompatibility Ags seems to reside in the protein moiety (removal of most of the carbohydrate leaves the antigenic activity intact). Moreover, the allelic alloantigens of hybrid mice (one inherited from each parent are isolable as separate molecules; and in immunofluorescence with intact cells they appear to form separate aggregates ("patches") with the respective fluorescent Abs, labeled with a red (rhodamine) chromophore for one alloantigen and with a green one (fluorescein) for the other alloantigen (see Ch. 17, Fig. 17-10). These properties are in sharp contrast to the ABO blood group substances, whose alloantigenic determinants (A, B) are saccharides and where the final products of allelic genes (as well as some nonallelic ones) are joined covalently in a single heterosaccharide; moreover, the immediate products of the genes for blood group substances are enzymes that synthesize the antigenic determinants, whereas the immediate products of the genes that determine graft-host histocompatibility are almost certainly the histocompatibility Ags themselves (Histocompatibility genes, below).

HISTOCOMPATIBILITY GENES

Histocompatibility Ags are specified by histocompatibility or H genes.* Because a host immune response to any of a donor's transplantation Ags can lead to rejection of a graft, permanent acceptance (in the absence of chronic immunosuppressive therapy) requires that essentially all of the donor's histocompatibility alleles be present in the recipient. Hence the probability that an allograft will be accepted when donor and recipient are drawn at random from an outbred population, such as man, depends upon 1) the number of H genes or loci in the species, and 2)

---

* H is a convenient abbreviation but the reader must obviously not confuse it with the completely unrelated H (heterogenetic) antigen of blood group substances (earlier, this chapter).

the number of alleles at each locus and their frequencies in the population.

The *H* loci and their alleles have been characterized most extensively in the mouse, because its many inbred strains permit detailed genetic analysis with skin grafts. The *H* genes in this species seem to correspond closely to those in man, in whom methods of analysis are necessarily more limited.

HISTOCOMPATIBILITY GENES IN THE MOUSE

The number of independent *H* loci at which two inbred strains differ can be estimated by mating them (AA × BB), crossing the F_1 progeny (AB × AB), and using animals of the F_2 generation as recipients for skin grafts from the purebred parental strains. As is illustrated in Figure 21-6, the number, *n,* of *H* loci, is provided by the proportion (*x*) of F_2 animals that accept grafts from one of the purebred parental strains: $x = (3/4)^n$. This technic yields minimal values because some *H* genes are linked to one another and behave as single rather than as multiple genes, and some inbred strains have certain *H* alleles in common. In the mouse, where many pairs of strains have been subjected to this test, about 30 *H* loci have been detected.

These loci, and the number of alleles at each, have been characterized by more complex proce-

dures that depend upon selected matings to produce **"congenic" lines*** that differ from a standard inbred strain by a small chromosomal segment containing a single locus (the one selected for; see Fig. 21-7). From the fate of grafts exchanged among various lines, and between these lines and standard inbred strains, it has been possible to establish pedigreed lines that differ only at single *H* loci and to enumerate the alleles at these loci (about 19 alleles at the *H-2* locus, 3 at *H-1,* etc.). Congenic lines are exceedingly valuable for analyzing the properties of various individual genes, not just those involved in transplantation, and a general method for their production is given in Figure 21-7.

H-2 Locus. The transplantation Ags specified by one locus, called *H-2,* are outstanding because of the intensity of the reactions they evoke: allografts exchanged between mice that differ only at this locus are usually rejected in about 11 days (first set), whereas those exchanged between mice that have the same *H-2* alleles but differ at any one of the

---

* Also called congenic "resistant" lines because of the assay originally used to detect alleles in the progeny of successive back-crosses of the type shown in Figure 21-7. A tumor arising in one purebred strain and carrying that strain's histocompatibility Ags was used as a graft: recipients that were **resistant** to the tumor (rejected it as an allograft) lacked one or more of the donor strain's histocompatibility Ags (and alleles).

Fig. 21-6. Estimate of the number of independent histocompatibility genes at which two inbred strains of mice differ. Shaded squares correspond to progeny in the F_2 generation that accept a graft from parent A. (Courtesy of Dr. Ralph J. Graff.)

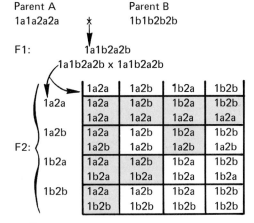

Parental Strains Differ by One H Gene

Parent A   Parent B
 1a1a   ×   1b1b

F1:   1a1b

1a1b × 1a1b

	1a	1b
1a	1a1a	1a1b
1b	1ab	1bb

$(3/4)^1$ or 75% of F2 progeny accept grafts from parent A or parent B

Parental Strains Differ by Two Independent H Genes

Parent A   Parent B
1a1a2a2a   ×   1b1b2b2b

F1:   1a1b2a2b

1a1b2a2b × 1a1b2a2b

	1a2a	1a2b	1b2a	1b2b
1a2a	1a2a	1a2b	1b2a	1b2b
	1a2a	1a2a	1a2a	1a2a
1a2b	1a2a	1a2b	1b2a	1b2b
	1a2b	1a2b	1a2b	1a2b
1b2a	1a2a	1a2b	1b2a	1b2b
	1b2a	1b2a	1b2a	1b2a
1b2b	1a2a	1a2b	1b2a	1b2b
	1b2b	1b2b	1b2b	1b2b

$(3/4)^2$ or 56% of F2 progeny accept grafts from parent A or parent B

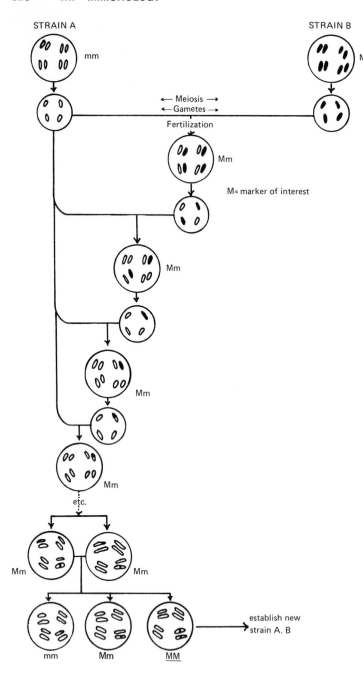

STRAIN A

mm

STRAIN B

MM

← Meiosis →
← Gametes →

Fertilization

Mm

M = marker of interest

Mm

Mm

Mm

Mm

Mm

etc.

Mm

Mm

mm

Mm

MM

establish new
strain A. B

Fig. 21-7. Production of congenic strain A.B. Following an initial cross between an inbred A mouse and an inbred B mouse and back-cross of the F₁ generation to strain A, a series of successive backcrosses to strain A is carried out with progeny that possess the B marker (M) of interest (e.g., their skin grafts are rejected by A-strain mice). Offspring of the tenth backcross generation (N10) are intercrossed, and progeny that are homozygous for the B-strain marker (e.g., they reject skin grafts from A-strain donors) are then inbred by successive brother-sister matings to establish the new congenic A.B strain. This strain is essentially identical with inbred strain A except for a chromosomal segment (e.g., with a histocompatibility gene) of inbred strain B. The principle is a general one: a congenic strain can be established for any dominant gene. (Courtesy of Dr. Ralph J. Graff.)

other *H* loci have median survival times that range from 20 to upward of 200 days.

The H-2 alloantigens are also outstanding because they are present on red blood cells, which can be agglutinated by sera from animals that have been intensively immunized by appropriate skin grafts plus injections of allogeneic cells (especially lymphocytes). As with many hyperimmune sera (Ch. 15), these sera also exhibit complicated cross-reactions: it appears that the Ag associated with each *H-2* allele has both a **unique ("private")** antigenic specificity and a variety of **broader ("public")** specificities, which it shares with certain other histocompatibility Ags, determined by other *H-2* alleles. Whether the shared specificities represent identical or similar structural elements in various H-2 Ags is uncertain and is not likely to be known until the detailed structure of these molecules is established.

With serological analysis to facilitate screening of many progeny from matings between inbred strains, it has been possible to detect rare crossovers within the *H-2* locus, which actually consists of two linked genes, called *H-2D* and *H-2K*. In between these genes there are several others (Fig. 21-8), including a number of so-called *Ir* genes,* which determine ability to make an immune response to various Ags (Ch. 17, Genetic control of Ab formation).

When surface macromolecules of intact

---

*About 1 in 200 progeny are recombinants between D and K. If the crossover frequency between two markers is proportional to the linear distance between them there is room between the D and K ends of the *H-2* locus for up to about 1000 genes.

cells are stained with differently colored fluorescent Abs of various specificities it is clear that the antigenic determinants specified by a *D* allele and those specified by the *K* allele of the same chromosome are located on separate molecules: i.e., anti-D and anti-K form separate aggregates in the same cell's membrane (Ch. 17, see Cap formation and Fig. 17-10). However the multiple determinants associated with any particular allele seem to be present in the same surface molecule. Thus *D* and *K* of a given chromosome (**haplotype;** see the human HL-A locus, below) specify different membrane molecules, each with multiple antigenic determinants (or specificities).

Non H-2 Loci. The other *H* loci determine Ags that are much less immunogenic in allogeneic mice. Not only are survival times much longer when grafts are exchanged between congenic lines that differ only at one *non H-2* locus, but many more inoculated cells are re-

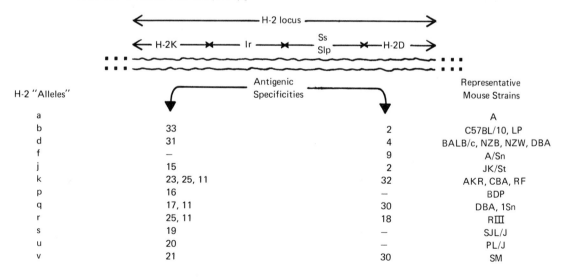

Fig. 21-8. *H-2* locus in the mouse. Different *H-2* "alleles," designated by lower-case letters, correspond to different patterns of antigenic specificity (designated by numbers), which are found in the representative inbred mouse strains listed at the right. Each of the "alleles," representing a combination of linked alleles of two genes (*K* and *D*) in the locus, corresponds to a haplotype in the human *HL-A* locus. Genes *Ss* and *Slp* specify certain serum proteins; the *Ir* (immune respónse) region represents a cluster of genes that govern ability to make immune responses to about 20 different Ags (Ch. 17). Most of the numbered antigenic specificities shown are unique ("private") for a given allele. Some alleles (—) are associated with distinctive patterns of several specificities, each shared with various other alleles. [From Klein, J., and Schreffler, D. C. *Transplant Rev* 6:3 (1971).]

H-2 "Alleles"	Antigenic Specificities		Representative Mouse Strains
a			A
b	33	2	C57BL/10, LP
d	31	4	BALB/c, NZB, NZW, DBA
f	—	9	A/Sn
j	15	2	JK/St
k	23, 25, 11	32	AKR, CBA, RF
p	16	—	BDP
q	17, 11	30	DBA, 1Sn
r	25, 11	18	RIII
s	19	—	SJL/J
u	20	—	PL/J
v	21	30	SM

quired to induce transplantation immunity, and many fewer are needed to establish tolerance, with such strain combinations than with combinations that differ only at *H-2*. Nevertheless, the cumulative effect of many *non H-2* differences can approximate that of an *H-2* incompatibility and can lead to virtually as rapid allograft rejection.

Diverse *H* loci are associated with different genetic markers on various linkage groups (i.e., chromosomes): for example, the *H-1* locus is linked with albinism, and *H-2* with fused tail vertebrae. One of the *non H-2* loci appears to be associated with the Y sex chromosome. It is revealed by exchange of grafts (skin, spleen, thyroid, etc.) between males and females of a given inbred mouse strain: grafts from female to male are accepted permanently, but those from male to female are rejected (after long survival times). Male tolerance of female cells is due to the presence of X chromosomes in both sexes, but some product of the Y chromosome on cell surfaces of the male can apparently function as a weak histocompatibility Ag in otherwise isogenic females. Because this Ag is weak, tolerance of it is readily established; e.g., by transplanting an extra large piece of skin from male to female.

HISTOCOMPATIBILITY GENES IN MAN

When leukocytes from one mouse strain are injected into newborn mice of certain other strains the recipients became tolerant of grafts from the donor strain (Tolerance, above). Mouse histocompatibility Ags seem therefore to be fully expressed on these readily available cells. The analysis of human histocompatibility Ags has accordingly been largely based also on leukocytes.

Antisera to human leukocyte (HL) Ags are derived from three sources: 1) patients who have received multiple blood transfusions; 2) women who, through pregnancy, have been immunized against alien leukocyte alloantigens of their offspring, inherited from the father; 3) human volunteers immunized with leukocytes from selected donors.

These sera agglutinate appropriate leukocytes (leukagglutination) and, together with complement (Ch. 18), they also damage leukocyte membranes in a cytotoxic reaction that is revealed by "dye-exclusion" tests: ionic dyes, such as eosin and trypan blue, penetrate into and stain only those cells with damaged membranes (Ch. 17, Living vs.

dead cells). Monospecific sera, useful for typing, have been identified by screening many human sera against panels of leukocyte samples from many (100 or more) persons.

With standardized typing sera over 20 human leukocyte Ags have been identified. Their distribution in family pedigrees has established that they are determined by a single complex genetic locus that is remarkably like that of *H-2* in the mouse. Called **HL-A** (for human histocompatibility locus A), it consists of two closely linked loci (sometimes called the "first" and "second"), each with a large number of alleles (*HL-A1, HL-A2,* etc.). In a doubly heterozygous individual the cells carry four different HL-A antigens (called a "full house"). Because these Ags are produced by two closely linked genes (like *D* and *K* in the mouse *H-2* locus, see Fig. 21-8), they are associated in pairs or **"haplotypes"** (one pair inherited from each parent, Fig. 21-9).

Mixed Lymphocyte Cultures. Serological typing of HL-A Ags is still limited: monospecific sera for some of them are probably still not available and the results of cytotoxicity tests are not always unambiguous. More important, grafts are sometimes rejected when donor and recipient appear to have identical HL-A haplotypes, suggesting that incompatibility of other transplantation Ags can be significant. Other tests, resembling cross-matching in blood transfusions (above), are therefore used sometimes for further evaluation of HL-A matched pairs. In the mixed lymphocyte culture (MLC) test isolated blood

		Haplotype	Haplotype	
Father		1, 5	9, 12	
Mother		2, 7	10, 13	
Child No.	1	1, 5	2, 7	
	2	1, 5	10, 13	HL-A
	3	1, 5	10, 13	identical sibs
	4	9, 12	2, 7	
	5	9, 12	10, 13	

Fig. 21-9. A hypothetical family pedigree illustrating some features of the *HL-A* locus and its antigens. At gene 1 there are 9 alleles (*HL-A 1, 2, 3, 9, 10, 11, Ba, Li*); at gene 2 there are 14 alleles (*HL-A 5, 7, 8, 12, 13, 4C, BB, FJH, LND, AA, SL, Maki, 407*). Each haplotype consists of one allele (antigen) from gene *1* and one from gene *2*. Every child has one haplotype in common with each parent. With doubly heterozygous parents (as in the example) the probability that two sibs have identical HL-A antigens is 0.25.

lymphocytes from recipient and prospective donor are maintained together for several days in tissue culture. Under these conditions blast transformation occurs if allogeneic cells are present; i.e., the antigenically stimulated lymphocytes enlarge, become basophilic (owing to increase in content of ribosomes), and incorporate ^{3}H-thymidine into DNA at an enhanced rate (Chs. 17 and 20). The test can be made unidirectional by treating one set of cells, say from the prospective donor, with mitomycin C before mixing with the prospective recipient's cells; DNA synthesis is blocked in the treated set and any increase in DNA synthesis in the mixed culture is due to the reaction of some of the recipient's lymphocytes against the treated cells, which must have surface Ags the recipient lacks.

Mixed lymphocyte cultures seem to be responsive primarily to HL-A Ags, for which they provide a sensitive measure of compatibility: e.g., the response is greater with cells from sibs that differ in two rather than in just one haplotype. However, a positive response has been obtained in rare instances with sibs that have two identical haplotypes; and among unrelated individuals with identical HL-A Ags a positive MLC response is seen more often. The lower frequency of discordant results among sibs suggests that other genes (the *MLC* locus), linked to HL-A, contribute to, or even determine, the MLC reaction: because of the linked relation, siblings having the same *HL-A* alleles usually have the same *MLC* locus, but unrelated persons with the same *HL-A* alleles differ more often at the *MLC* locus. These relations might account for the greater success with HL-A identical kidney grafts when donor-recipient pairs are siblings than when they are unrelated (Fig. 21-10).

The rare acceptance of an allograft when related donor-recipient pairs (e.g., uncle-nephew) differ in all HL-A Ags suggests that the Ag specified by the MLC locus might be the primary determinant of allograft immunity.

A positive MLC response has also been obtained with lymphocytes from congenic mouse strains that have identical *H-2D* and *H-2K* loci and differ only in the *Ir-1* locus. The human *MLC* locus might thus correspond to *Ir-1* of mice which, it may be recalled, occupies a position near the center of the *H-2* locus and determines the ability of mice to make immune responses to particular Ags (Ch. 17), perhaps because it codes for surface macromolecules that serve as Ag-binding receptors on T lymphocytes (Ch. 17).

ABO Blood Group Substances and Transplantation Immunity. ABO blood group substances are present on the surface membranes of many types of cells, in addition to erythrocytes, and these substances can also function as transplantation Ags: kidney grafts seem to be subject to especially rapid and violent rejection when donor and recipient belong to different ABO groups. Group O human volunteers injected with AB red cells (or with purified A and B substances; above) reject subsequent skin. grafts from any AB donor in an accelerated manner (second set), but grafts from an O donor are rejected as a "first set." It is not clear if accelerated rejection of AB-incompatible grafts is due to conventional anti-A and anti-B alloantibodies (above) or to cell-mediated reactions.

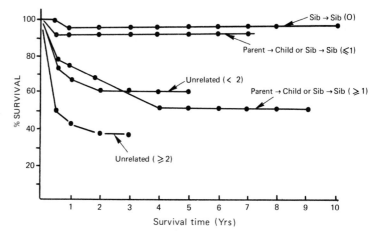

Fig. 21-10. Correlation between survival of human kidney grafts and HL-A compatibility. Number of incompatibilities based on leukocyte typing is given in parentheses. Since these are two linked genes (haplotypes) in the HL-A locus, every parent-child combination and half the sib-sib combinations will have at least one haplotype in common, or a maximum incompatibility of two. Uncertainties arise when fewer than four HL-A antigens are typed in donor and recipient. [From Dausset, J., and Hors, J. *Transplant Proc* 3:1004 (1971).]

If donor and recipient belong to the same ABO group, injected red cells from one donor do not induce accelerated rejection (second set) of a subsequent graft from the same donor: hence red cells lack HL-A Ags, at least in immunogenic form (or amount).

IMMUNITY AGAINST TUMORS

Tumor-specific Antigens. Immune reactions of experimental animals to tumor grafts, and of human subjects to their cancers, depend on **tumor-specific antigens (TSAs)** that are lacking or masked in normal cells. Because TSAs are usually present in the tumor cell's surface membrane, the host can react against his tumor as though it were an allograft. However, an intense host reaction can **select for tumor cell variants** with modified TSAs; this could account for the general finding that TSAs of most well-established tumors have little immunogenic potency in the **autochthonous host** (i.e., the individual in whom the tumor arose).

Tumor Transplantation. The fate of transplanted tumors depends upon the recipient's relation to the autochthonous host. When the latter and the recipient are **syngeneic** the outcome varies with the number of transplanted cells: with more than a critical number the tumor quickly becomes too massive to be destroyed by any first-set immune response that might be made against its TSAs.* However, an effective subthreshold number of cells can elicit a sufficient response to block the graft's growth; and the resistant host will then usually reject a much larger transplant of the same tumor (compare dose-response curves in immunized and unimmunized syngeneic recipients, Ch. 20, Tumor immunity, Fig. 20-7).

Recipients can also become specifically resistant to a tumor after injections of X-irradiated tumor cells (which are incapable of proliferating but have intact Ags) or after total excision of a tumor graft that has grown for 1 to 2 weeks in situ. Whichever method is used in syngeneic animals, the resistance is specific for the particular tumor used as immunogen or, in some instances, other tumors induced by the same oncogenic virus (see below). When the autochthonous host and the recipient are **allogeneic,** transplanted tumor cells are rejected, even in massive numbers, by the recipient's first-set response to the tumor's normal histocompatibility Ags (for rare exceptions see Loss of normal transplantation Ags, below).

Many tumors are weakly immunogenic in the autochthonous host. As noted before (Ch. 20), a mouse immunized with X-irradiated sarcoma cells, derived from its own totally excised tumor, can reject a subsequent graft of that tumor (maintained during the immunization procedure by successive transplants in syngeneic recipients), without becoming resistant to other sarcomas. (For other evidence, see Enhancement, below, and Tumor immunity, Ch. 20.)

Cell-mediated vs. Humoral Immunity. The rejection of solid tumor grafts in syngeneic or allogeneic hosts seems usually (like that of normal tissue allografts) to be due to cell-mediated immunity: resistance can usually be transferred adoptively with viable lymphocytes but not with antisera. In conjunction with C, however, serum alloantibodies against TSAs can destroy neoplastic lymphocytes (from leukemias and lymphosarcomas), just as other humoral Abs (plus C) can destroy allogeneic normal lymphocytes. With most other kinds of tumors, however, Abs to their TSAs are not destructive and some may even protect the tumor against cell-mediated immune reactions (Enhancement, below).

Enhancement. Many grafts of malignant epithelial or connective tissue cells (carcinomas or sarcomas, respectively) grow more rapidly and in a higher proportion of recipients if the recipients are first subjected to prolonged immunization or given antisera from syngeneic animals that were intensively im-

---

* The tumor is then lethal for the recipient. Experimental tumors are maintained routinely by serial transplantation of relatively large numbers of cells into animals that are syngeneic with the autochthonous host.

munized against the tumor. Evidently, TSAs on the tumor cell surface become covered by noncytotoxic Abs, blocking induction of immunity or recognition and attack by specifically reactive killer lymphocytes (Ch. 20, see afferent and efferent enhancement).

It is not clear why Abs to TSAs destroy leukemic lymphocytes but seem harmless or even protective for many carcinomas and sarcomas; possibly the target cells differ in susceptibility of the surface membrane to lytic attack by activated C or in the distribution of their surface TSAs: with widely dispersed TSAs, Abs might be less likely to activate C than if the TSAs were closely packed (Ch. 18). Even with the ordinarily susceptible neoplastic lymphocytes of leukemia some Abs to TSAs actually protect the cells against other, cytotoxic Abs: the difference could arise from differences in ability to activate the C cascade (Ch. 18).

Though enhancement by noncytotoxic Abs seemed initially a laboratory curiosity, there is growing evidence for its clinical importance. For instance, enhancing activity is present in the sera of many individuals with progressive tumors, but lacking in those with regressing tumors (below). As was noted above, enhancing Abs seem also, under certain circumstances, to prolong the survival of normal tissue allografts.

Unique vs. Shared Tumor-specific Antigens. Injection of methylcholanthrene or other chemical carcinogens into isogenic mice raises tumors that are antigenically distinct from one another; little or no cross-immunity has been detected when animals immunized with one tumor are challenged by grafts of other tumors, induced with the same carcinogen. Diverse tumors raised by a given carcinogen in one animal are also antigenically distinguishable by cross-grafting; and there is even evidence for antigenic variation within individual, chemically induced tumors, which could either have a multicellular origin, with different clones growing together in one tumor mass, or a unicellular origin, with the development of antigenic variants at subsequent cell divisions. The antigenic individuality of these tumors suggests that chemical carcinogens activate different host genes in different clones of tumor cells.

In contrast, different tumors induced in mice by a given oncogenic virus have the same TSAs, which differ from the TSAs of tumors induced by other oncogenic viruses. For instance, the histologically diversified tumors produced with polyomavirus in mice (Ch. 63) all have a common TSA, which is even shared (cross-reacts) with tumors induced by this virus in hamsters. This Ag, which induces resistance to tumor grafts, is not part of the virion; whether it is coded for by the virus or by a host gene activated by the virus is not known. In addition to such shared TSAs, some virus-induced tumors also have TSAs that seem to be unique for different tumors raised by the same virus.

Loss of Normal Histocompatibility Antigens. Some tumor cells differ from their normal counterparts not only in having gained TSAs, but also in having lost some normal histocompatibility Ags: in exceedingly rare instances a mouse tumor becomes freely transplantable in diverse allogeneic mice, and some human cancer cells seem also to have less HL-A antigen than normal cells of the same histological type.

A more restricted type of antigenic loss is seen when tumors originating in hybrid strains of mice (AB) develop specific subline variants capable of growing in one or the other of the parent strains (AA or BB). Thus, an AB tumor may develop variants that can grow in the AA parent, indicating an apparent loss of the histocompatibility Ags contributed by the B half of the hybrid's genome. This loss has been postulated to be the result of chromosomal deletion or of recombination in somatic cells (Ch. 43, Parasexual cycle in fungi).

Antigenicity of Human Tumors. Clinical observations have long suggested that a patient's immune response might suppress growth of his cancer cells. Thus tumors with slowed growth are often found to be heavily infiltrated by lymphocytes and plasma cells; on rare occasions tumors even undergo "spontaneous" regression.

Direct evidence for TSAs on human cancer cells has been provided by immunofluorescent staining, and C-dependent cytotoxicity tests show that serum from a patient may contain Abs that react wtih his own tumors cells and sometimes with tumor cells of the same type from other individuals (e.g., malignant lymphocytes of Burkitt's lymphoma, neuroblastoma).

Fetal Antigens. When rabbit antisera prepared against extracts of gastrointestinal tract carcinomas are absorbed with extracts of normal intestinal mucosa, the residual Abs react with an Ag (a glycoprotein with ca. 50% carbohydrate; MW 1×10^5 to 2×10^5) that is present not only in gastrointestinal carcinomas but also in embryonic gastrointestinal mucosa and in liver and pancreas (which are endodermal derivatives of the gut). Hence this **carcinoembryonic antigen (CEA)** appears to be specified by a normal gene that is expressed transiently in endodermal cells during fetal development, and in adult life if these cells undergo malignant change.

Serum from nearly all patients with colon or rectal adenocarcinoma contains CEA, and many also contain Abs to CEA. Both disappear with successful surgical removal of the cancer, and reappearance of CEA can be the first diagnostic clue of the tumor's recurrence. With widespread metastases the CEA concentration rises and anti-CEA disappears (masked by Ag excess). With highly sensitive radioimmunoassays that can detect ca. 1 ng, CEA (or a substance that cross-reacts with it) has also been detected in sera of some persons with carcinoma of lung or breast.

Other fetal Ags are associated with certain other human cancers: α**-fetoprotein** is found in serum of patients with hepatomas or embryonal carcinomas; γ**-fetoprotein,** a serum protein of fetal blood, is found in subjects with various types of cancer. The detection in adult serum of fetal Ags, by sensitive and rapid serological tests, promises to provide screening assays for the early detection of human cancers.

Cell-Mediated Immunity and Enhancement of Human Tumors. As noted before (Ch. 20), the prevalence of cell-mediated immune responses by human hosts to their own cancers has been revealed by the Hellstroms, who showed that under standard tissue culture conditions a patient's tumor cells usually form fewer colonies if incubated with his own lymphocytes than if incubated with lymphocytes from other persons. With certain types of cancer (e.g., neuroblastoma) the inhibitory effects of one patient's lymphocytes are also exerted on tumor cells from other subjects, suggesting that these tumors share common TSAs and might have been induced by a particular oncogenic virus.

Many human cancer cells seem, however, to be protected against the host's killer lymphocytes by the concomitant presence of "enhancing activities" in the patient's serum, which can specifically block his lymphocytes' inhibition of colony formation by his cultured tumor cells. Preliminary findings indicate that the enhancing activity could be due to Abs that bind to TSAs on tumor cells, or to soluble TSAs that block receptors on specifically reactive lymphocytes, or to soluble Ab-TSA complexes that combine with either target cell or killer lymphocyte. A further indication that tumor growth depends upon the net outcome of multiple, antagonistic immune responses is the further suggestion of "deblocking" activities that appear in serum of some mice with regressing tumors: these activities block enhancing activity, and thus reinforce the inhibitory effects of killer lymphocytes.

Immunological Surveillance. Increasing evidence for TSAs on almost every variety of tumor, and for host immune responses to these Ags, has raised hopes that serological or allergic skin tests might be developed for early detection of most human cancers, and that immunization might become practical in the prevention or treatment of certain tumors. It seems quite possible that neoplastic cells arise at frequent intervals in normal individuals but rarely establish cancers because they are rapidly detected and destroyed by host immune responses against their surface membrane TSAs. Indeed, some believe that this monitoring system, called immunological surveillance, may have been more important than protection against infection in promoting the evolution of the immune system.

When the possibility of immunological surveillance was first suggested (by Lewis Thomas) it seemed far-fetched. However, its credibility has grown with evidence for an increased incidence of neoplasms in animals and in humans with diverse spontaneous and induced immunological defects. In one striking episode, for instance, accidental contamination of an animal room with polyomavirus led to the appearance of tumors in all mice with severely defective cell-mediated immune responses (due to thymectomy and antilymphocyte serum), whereas no tumors appeared in any of the normal isogenic mice that shared the same space. Moreover, about 10% of persons with various immunological deficiency diseases (Chs. 17 and 20) develop diverse malignant tumors. And approximately 1% of patients under chronic immunosuppression therapy for maintenance of renal allografts develop a variety of cancers; this incidence is about five-fold higher than in the population at large (of the same age and sex distribution). Reticulum cell sarcoma accounts for almost half of the reported tumors in these chronically immunosuppressed patients, in whom the incidence of this lymphoma is increased about 4000-fold, perhaps because the normal precursors of the neoplastic lymphoid cells are targets for antilymphocytic serum and other immunosuppressive agents (Chs. 17 and 20).

CODA

Many genes, with multiple alleles at each, specify the great variety of macromolecular alloantigens that cover cell surfaces. The antigenic uniqueness of each individual arises from the distinctive combinations of these alloantigens on his cells. Immune responses to these antigens are responsible for both reactions to transfused blood and for rejection of transplanted tissues and organs.

About 60 human red cell alloantigens ("blood group substances") have been identified, largely with the aid of natural antibodies in normal human sera or antibodies formed in response to blood transfusions or to pregnancies in which the fetus carries paternal alloantigens that are foreign to the mother. The immunological specificity of the best characterized substances (with A, B, H, Le^a, and Le^b activities) resides in groups of four or five sugars at the free ends of oligosaccharide chains that branch from a polypeptide or lipoidal backbone. In the biosynthesis of these complex heteropolymers the critical sugar residues are added sequentially to a "core" polymer by enzymes (glycosyltransferases) that are determined by the corresponding blood group alleles.

The alloantigens (histocompatibility or transplantation antigens) responsible for allograft rejection are specified by histocompatibility genes, of which there are about 30 in the mouse with approximately 20 alleles at the most intensively studied locus (*H-2*). Polymorphism is doubtless as great in man and in other vertebrates. The antigenic individuality of a person's red cells may turn out to be no less distinctive that that of his tissue cells, but in blood transfusions the red cells need survive only a few weeks for the procedure to be scored as a success; whereas success of an allograft requires enduring survival, and therefore virtually complete antigenic identity of graft and host. Nevertheless, many human kidney allografts survive for many years. This impressive, but still limited, level of success occurs because 1) only a few of the great number of histocompatibility antigens are potent immunogens, 2) tissue-typing methods have improved to where badly matched donor-recipient pairs can be avoided, and especially because 3) host immune responses can be chronically depressed by skillful administration of immunosuppressive drugs. Though immunosuppressive measures greatly prolong graft survival, they increase susceptibility to overwhelming infection by ubiquitous "opportunistic" bacteria, viruses, and fungi, and they increase the risk of developing cancer, especially malignant lymphomas.

Tumor cells behave as allografts because their surface membranes possess tumor-specific antigens (TSAs) that are lacking or masked in normal cells. Host immune responses to tumors probably select for tumor cell variants with altered TSAs, which could account for the low-grade immunogenicity (in the autochthonous host) of TSAs of well-established tumors. Diverse host immune responses to TSAs can have antagonistic effects: some antibodies to TSAs protect tumor cells from destruction by cell-mediated immune reactions. TSAs are specified by various genes (of oncogenic viruses or host cells) and some are "fetal": they are normally expressed transiently on certain fetal cells during embryonic development, and later in life only on some kinds of tumor cells.

Immunological surveillance by the host immune system probably eliminates many of the potentially neoplastic cells that arise during a normal lifetime; hence individuals with long-term impairment of immune mechanisms (es-pecially of cell-mediated immunity) have an increased probability of developing cancer.

SELECTED REFERENCES

Red Cell Isoantigens

KABAT, E. A. *Blood Group Substances: Their Chemistry and Immunochemistry.* Academic Press, New York, 1956.

KABAT, E. A. In *Blood and Tissue Antigens.* (D. Aminott, ed.) Academic Press, New York, 1970.

KOBATA, A., and GINSBURG, V. Uridine diphosphate N-acetyl-D-galactosamine: D-galactose α-3-N-acetylgalactosaminyltransferase, a product of the gene that determines blood type A in man. *J Biol Chem 245*:1484 (1970).

MOLLISON, P. L. *Blood Transfusions in Clinical Medicine,* 3rd ed. Blackwell, Oxford, 1961.

MORENO, C., LUNDBLAD, A., and KABAT, E. A. A comparative study of the reaction of A_1 and A_2 blood group glycoproteins with human anti-A. *J Exp Med 134*:439 (1971).

MORGAN, W. T. Croonian lecture: A contribution to human biochemical genetics: The chemical basis of blood group specificity. *Proc R Soc Lond (Biol) 151*:308 (1960).

RACE, R. R., and SANGER, R. *Blood Groups in Man,* 5th ed. Thomas, Springfield, Ill., 1968.

WATKINS, W. M. Blood-group substances. *Science 152*:172 (1966).

Transplantation Immunity

BACH, F. T. Transplantation: Pairing of donor and recipient. *Science 168*:1170 (1970).

BILLINGHAM, R., and SILVERS, W. *The Immunobiology of Transplantation.* Prentice-Hall, Englewood Cliffs, N.J., 1970.

DAUSSET, J., and RAPAPORT, F. T. (eds.). *Human Transplantation.* Grune & Stratton, New York, 1968.

KLEIN, J., and SHREFFLER, D. C. The H-2 model for the major histocompatibility systems. *Transplant Rev 6*:3 (1971).

LAFFERTY, K. J., and JONES, M. A. S. Reactions of the graft-vs.-host type. *Aust J Exp Biol Med Sci 47*:17 (1969).

MEDAWAR, P. B. The immunology of transplantation. *Harvey Lect 52*:144 (1956–57).

MEDAWAR, P. B. *The Uniqueness of the Individual.* Basic Books, New York, 1958.

SNELL, G. D. The H-2 locus of the mouse: Comparative genetics and polymorphism. *Folia Biol (Praha) 14*:335 (1968).

Symposium on biological significance of histocompatibility antigens. *Fed Proc 31*:1087 (1972).

Symposium on human histocompatibility antigens. *Fed Proc 29*:2010 (1970).

Tumor Immunity

DYKES, P. W., and KING, J. Progress report: Carcinoembryonic antigen (CEA). *Gut 13*:1000 (1972).

GOLD, P., and FREEDMAN, S. D. Specific carcinoembryonic antigens of the human digestive system. *J Exp Med 122*:467 (1965).

HELLSTRÖM, K. E., and HELLSTRÖM, I. Immunological enhancement as studied by cell culture techniques. *Annu Rev Microbiol 24*:373 (1970).

KALISS, N. Immunological enhancement. *Int Rev Exp Pathol 8*:241 (1969).

KLEIN, G. Tumor immunology in *Clinical Immunobiology.* (Bach, F. H. and Good, R. A., eds.) vol. 1, p. 219, Academic Press, New York, 1972.

KLEIN, E., and COCHRAN, A. J. Immunology and malignant disease. *Haematologica 5*:179 (1971).

SJÖGREN, H. O. Transplantation methods as a tool for detection of tumor-specific antigens. *Progr Exp Tumor Res 6*:289 (1965).

part  IV

BACTERIAL AND MYCOTIC INFECTIONS

Edited by MACLYN McCARTY

chapter **22**

HOST-PARASITE RELATIONS IN BACTERIAL DISEASES*

HISTORY   629

BASIC CONCEPTS AND TERMINOLOGY   630

INFECTION VS. DISEASE   630
  Germ-free Animals   631
PATHOGENICITY AND VIRULENCE   631
IMMUNITY   631
HYPERSENSITIVITY   632
LATENCY   632
COMMUNICABILITY   632

PATHOGENIC PROPERTIES OF BACTERIA   632

INVASIVENESS   632
  Intracellular vs. Extracellular Parasitism   632
  Antiphagocytic Capsules   633
  Adaptation to Microenvironment   634
  Production of Extracellular Enzymes   635
TOXIGENICITY   635
  Exotoxins   637
  Endotoxins   638
VIRULENCE   639
  Measurement   639
  Variation   640

ANTIBACTERIAL DEFENSES OF THE HOST   641

SKIN AND MUCOUS MEMBRANE BARRIERS   641
  Mechanical Factors   641
  Chemical Factors   642
  Microbial Factors   642
PHAGOCYTIC DEFENSE   643
  Mononuclear Phagocytes   643
  Neutrophilic and Heterophilic Leukocytes   644
  Inflammation and the Sequence of Cellular Defense   645
  Inflammation and the Delivery of Phagocytic Cells   645
  Fever   648

* Revised with the generous help of ZANVIL A. COHN, M.D.

    The Phagocytic Process   649
    Postphagocytic Events   650
  HUMORAL FACTORS IN ANTIBACTERIAL DEFENSE   651
  TISSUE STRUCTURAL FACTORS IN ANTIBACTERIAL
     DEFENSE   652
  CHEMICAL FACTORS IN ANTIBACTERIAL DEFENSE   653
  ROLE OF HYPERSENSITIVITY IN BACTERIAL DISEASES   656
    Allergy vs. Immunity   656
    Acquired Antibacterial Cellular Immunity   656
  VARIABILITY OF HOST DEFENSE   657
    Augmentation   657
    Depression   657
    Genetic Defects of Leukocyte Function   659

INAPPARENT BACTERIAL INFECTIONS   660
  DORMANT AND LATENT INFECTIONS   660

COMMUNICABILITY OF BACTERIAL INFECTIONS   662

APPENDIX   663
    Computation of 50% Endpoints (LD_{50} or ID_{50}) by Method of
     Reed and Muench   663

HISTORY

In 1720 Benjamin Marten, a London practitioner, published a remarkable book entitled *A New Theory of Consumptions: More Especially of a Phthisis or Consumption of the Lungs*. In it he wrote "The Original or Essential Cause, then, which some content themselves to call a serious disposition of the Juices, others a Salt Acrimony, others a strange Ferment, others a Malignant Humour, may possibly be some certain Species of **Animalculae** or wonderful minute living creatures that, by their peculiar Shape or disagreeable Parts are inimicable to our Nature; but, however, capable of subsisting in our Juices and Vessels." As to the contagiousness of the malady he stated: "The minute Animals or their Seed . . . are for the most part either conveyed from Parents to their Offspring hereditarily or communicated immediately from Distempered Persons to sound ones who are very conversant with them. . . . It may, therefore, be very likely that by an habitual lying in the same Bed with a consumptive Patient, constantly eating or drinking with him or by very frequent conversing so nearly as to draw in part of the Breath he emits from the Lungs, a Consumption may be caught by a sound Person."

Although based on Leeuwenhoek's discovery of bacteria in the early 1680s, Marten's germ theory of disease was ignored by his contemporaries. Only when the specific microbes causing bacterial diseases were discovered, during the late 1800s did Marten's views on the etiology of contagions become generally accepted.

In 1882 the Russian zoologist Elie Metchnikoff, while studying the behavior of mobile cells in the transparent starfish larva, noted that the introduction of a sharp thorn into the body of the larva caused the mobile cells to surround the foreign body and apparently to attack it (Fig. 22-1). Moreover, these cells could also ingest and destroy foreign particles, including bacteria; and Cohnheim had earlier seen white blood cells migrating from tissue capillaries to form pus at sites of injury.

Metchnikoff concluded that inflammation served as an important defense reaction of the body. "In man," he wrote, "microbes are usually the cause which provokes inflammation; therefore, it is against these intruders that the mobile mesodermic cells have to strive. The mobile cells must destroy the microbes by digesting them and thus bring about a cure. Inflammation, therefore, is a curative reaction in the organism and morbid symptoms are no other than the signs of the struggle between the mesodermic cells and the microbes."

Metchnikoff's theory of the destruction of bacteria by phagocytic cells was vigorously attacked by his medical colleagues. Pathologists had often observed bacteria in the leukocytes of patients who had died of bacterial diseases, but inferred that the cells merely

Fig. 22-1. A. Transverse section of body cavity of starfish larva (*Astropecten*). ect. = ectoderm; end. = endoderm; mes. = mesodermal wandering cells; pl. = "plasmodium" composed of mesodermal cells fused together around the foreign body that Metchnikoff thrust through the ectoderm. **B.** Higher magnification of "plasmodium." f. = foreign spicules; nucl. = nuclei of fused mesodermal cells. (From Adami, J. G. *Inflammation.* Macmillan, London, 1909. Redrawn from Metchnikoff's original illustrations.)

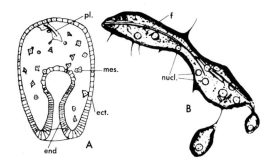

transported the microbes through the tissues. When in 1890 Buchner demonstrated bactericidal substances in blood serum, and Behring and Kitasato announced the discovery of tetanus antitoxin, the view became widely accepted that both natural and acquired resistance to bacterial infections was due exclusively to constituents of the serum. Thus during the closing decades of the nineteenth century a lively controversy arose between the great majority of pathologists and clinicians, who adhered to the doctrine of **humoral immunity,** and the few followers of Metch-

nikoff, who championed the theory of **cellular immunity.**

In 1904 Neufeld and Rempau reported that the serum of animals immunized with killed pneumococci contained antibodies which greatly accelerated in vitro phagocytosis of the same strain of living pneumococci. These phagocytosis-promoting antibodies were termed **opsonins*** (Gr. *opsonein* = to prepare food for). This joint role of antibodies and phagocytes in destroying invading microbes finally reconciled the two opposing theories.

BASIC CONCEPTS AND TERMINOLOGY

Host-parasite relations in infectious diseases involve two major topics: the properties of the offending microbes that enable them to produce disease, and the manner in which the infected host responds to the microbial invasion. To summarize current knowledge in these two broad areas pertaining to bacterial diseases of man, it is necessary first to define a number of terms and concepts.

INFECTION VS. DISEASE

Healthy humans, as well as other animals, are **infected** with a variety of bacteria from infancy through old age (Ch. 42, Distribution). This is not surprising when one considers the ubiquitous distribution of bacteria in the environment. In fact, most of the air we breathe is heavily contaminated. Soon after birth our body surfaces, both external and internal, become colonized with bacteria. These we disseminate as we breathe, speak, cough, sneeze, handle objects, sweep the floor, etc. The surprising fact is that we can live in peaceful coexistence with microorganisms capable of producing overt **disease.** We are able to do so only because our tissues possess efficient natural mechanisms of antibacterial defense which restrict the microbes to areas where they can be tolerated. When the defenses are penetrated and bacteria gain access to tissues not normally infected, disease usually results.

The distinction between **infection** and **disease** is of paramount importance to the

clinician. Confronted with the problem of establishing the etiology of a patient's illness, he must interpret bacteriological findings with great caution. The mere fact that he can culture a given microbe from a patient's body may be totally irrelevant. For example, over 90% of throat cultures taken at random will be positive for a certain kind of streptococcus (viridans). A positive culture is therefore not indicative of disease. When, on the other hand, viridans streptococci are repeatedly cultured from a patient's blood, which normally should be sterile, the diagnosis of subacute bacterial endocarditis can be made with virtual certainty.†

Unlike most diseases, infections may even benefit the host. The presence of relatively avirulent bacteria at a given tissue site often

---

* While the concept of opsonins as a special class of antibodies is no longer considered valid (Ch. 15, Unitarian hypothesis), opsonization of bacteria is a distinct immunological reaction.

† Healthy persons infected with a given organism are often referred to as **carriers,** and the percentage of individuals so infected in a population is known as the **carrier rate.** Not only does the existence of carriers often make it difficult for the physician to reach a definitive bacteriological diagnosis, but the occurrence of **mixed infections,** in which more than one potential pathogen is recovered from a lesion, frequently requires that a distinction be made between primary and **secondary invaders.** The latter decision is usually based on a knowledge of the relative disease-producing potential of the microorganisms in question.

prevents the growth of more virulent species. When the delicate ecological balance thus achieved by bacterial antagonism (p. 642) is upset, serious disease may ensue (e.g., staphylococcal enteritis). In addition, synthesis of specific metabolites essential to the host (e.g., vitamin K) may result from infection of such tissues as those of the gastrointestinal tract (Ch. 42, Actions beneficial to the host).

GERM-FREE ANIMALS

The varied effects of bacterial infection on the body economy of the normal host are clarified by the study of **axenic*** (germ-free) animals reared in a bacteria-free environment. These animals are unusually susceptible to a wide variety of infections. They show retarded development of antibody-forming lymphoid organs and are deficient in immunoglobulins. In addition, the turnover rate of the mucosal epithelial cells in their gastrointestinal tracts is depressed, and their responses to such stresses as injection of endotoxin and acute radiation injury are strikingly modified. Germ-free rats fail to develop dental caries (Ch. 42).

It is clear that life-long bacterial **infection** influences the responsiveness of the animal host to its environment, including its susceptibility to bacterial disease.

PATHOGENICITY AND VIRULENCE

Most bacterial species are incapable of penetrating the natural defenses of the host and are referred to as **nonpathogenic.** Others, which under the right circumstances are clearly **pathogenic,** often coexist with the host in a truce that is only occasionally broken. Still others are so highly pathogenic that they quickly overcome the natural defenses of the tissues and generally produce disease whenever they infect. Pathogenicity or **viru-**

---

* Axenic (from Gr. *xenos* = stranger) animals purposely infected with one or more known bacterial species are referred to as **gnotobiotic,** another adjective of Greek derivation, meaning "of known life." Although axenic animals are by definition free of detectable bacterial infection, they may nevertheless be exposed to bacterial and other antigens by ingesting them in their food.

lence,† then, may vary over a wide range, depending upon the strain of microbe, the strain of host, and the conditions under which the two are brought together.

The phrases "strain of microbe" and "strain of host" are used advisedly, since avirulent variants of pathogenic species of bacteria are not uncommon and host resistance is strain- as well as species-variable. The "conditions" under which the host is exposed to the microbe are also critical. For example, nearly all strains of pneumococci cause disease when injected into mice, whereas natural pneumococcal infections in mice rarely, if ever, occur. Furthermore, the relative virulence of a given strain of pneumococcus for a given strain of mice depends upon the state of the organisms (age of culture), the state of the host (age, nutritional status, freedom from other diseases, etc.), and the site of inoculation. Thus in any laboratory procedure designed to measure virulence the conditions must be precisely defined.

IMMUNITY

The term immunity refers to the relative resistance of the host to a given microbe. It may be of such a degree as to prevent infection altogether, or it may merely combat invasiveness and thereby prevent disease. As already mentioned, both **cellular** and **humoral** factors have been recognized. In addition, there are nutritional **tissue factors** that influence the host's resistance by affecting the growth and survival of the microbe, and the body fluids contain enzymes (e.g., lysozyme) that attack many bacterial species.

Natural immunity depends upon innate properties of the host that render it resistant to a microorganism, whereas **acquired immunity** is due either to the active formation of specific antibodies or specifically reactive lymphocytes by the host or to the passive acquisition of these components from another host. **Active immunity** is induced by exposure to the infectious agent itself or to one of its antigens, usually as a result of natural infection or vaccination. This form of immunity

---

† The term virulence, which is synonymous with pathogenicity, is sometimes reserved to indicate the **degree** of pathogenicity of a given strain of microbe for a given strain of host. The authors prefer not to make this distinction and will use virulence and pathogenicity interchangeably.

may persist for months or even years. **Passive immunity,** on the other hand, is transient (usually lasting only a few weeks); it results from injection of antibodies or, in newborns, from antibodies acquired in utero from the maternal circulation or postnatally from the mother's colostrum. **Adoptive immunity** is a laboratory artifice, achieved by transferring viable lymphocytes from actively immunized to unimmunized individuals; its duration depends on survival of the transferred cells and their progeny (see Allograft reactions, Ch. 21).

HYPERSENSITIVITY

Not all immune reactions are necessarily beneficial. Previous experience with a bacterial antigen may render the tissues of the host **hypersensitive** to further exposures. Subsequent contact with the antigen may then result in a local interaction that damages surrounding cells. Hypersensitivity states* are particularly common in chronic bacterial diseases, such as brucellosis and tuberculosis. In some circumstances hypersensitivity intensifies an illness and in others it hastens recovery.

LATENCY

We have differentiated recognizable infections with and without disease. In addition, some organisms give rise to **latent** or **inapparent** infections, which are persistent but not sufficiently active to be recognized. Latent stages of chronic infectious diseases, e.g., syphilis and tuberculosis, have long been recognized. Persistence of the microorganisms within the tissues of the host accounts for the late recrudescences observed in such diseases and also for prolonged states of immunity.

COMMUNICABILITY

Microbes that produce natural **diseases** are both pathogenic and relatively **communicable.** The latter property relates to the ease with which **infection** is transmitted from one individual to another. A high degree of communicability may be exhibited by nonpathogenic organisms, such as the saprophytes that normally inhabit the skin and mucous membranes, as well as by bacteria that produce disease. Bacterial strains that cause epidemics must be virulent as well as highly communicable.

PATHOGENIC PROPERTIES OF BACTERIA

Bacteria cause disease by two basic mechanisms: the **invasion of tissues** and the **production of toxins.** Whereas invasion leads to demonstrable damage of host cells only in the immediate vicinity of the invasion, soluble toxins transported by lymph and blood may cause cytotoxic effects at sites remote from the original lesion. Some species of bacteria appear to owe their pathogenicity to invasiveness alone (*Diplococcus pneumoniae*), while others are almost solely toxigenic† (*Clostridium tetani*). Many species are both invasive and toxigenic (*Streptococcus pyogenes*).

Rarely are invasiveness and toxigenicity completely separable, since invasiveness may involve factors of short-range toxicity and toxigenicity must require at least some degree

of bacterial multiplication in the tissues.‡ Nevertheless, it is useful to distinguish between the two processes.

INVASIVENESS

INTRACELLULAR VS. EXTRACELLULAR PARASITISM

Pathogenic bacteria may be conveniently divided into two additional categories: those that can survive and even multiply within the phagocytic cells of the host, and those that are promptly destroyed by phagocytes. The factors which lead to ultimate destruction of the former (*Mycobacterium tuberculosis, Brucella abortus, Salmonella typhosa*, etc.)

---

* The distinction between **immediate** and **delayed** types of hypersensitivity is discussed in Chapter 19.

† Synonymous with toxinogenic.

---

‡ Toxic products may also be released from damaged cells.

are only partially understood; undoubtedly, they relate to complex metabolic and enzymatic interactions between the ingested bacterium and the phagocyte. When in balance, these interactions permit a state of intracellular parasitism to persist. When out of balance they lead to destruction of either parasite or host cell, or in certain situations to the death of both. Bacteria that behave as **intracellular parasites** frequently give rise to relatively chronic diseases, such as tuberculosis.

In contrast, those species that are promptly destroyed when phagocytized (*Diplococcus pneumoniae, Klebsiella pneumoniae, Streptococcus pyogenes,* etc.), damage the tissues of the host only so long as they remain outside phagocytic cells. The manner in which they are disposed of following phagocytosis will be discussed below. Since their presence in the tissues usually stimulates the production of opsonizing antibodies, which render them susceptible to phagocytosis, the diseases they produce are likely to be acute and of relatively short duration (e.g., pneumococcal pneumonia). When bacteria behave as **extracellular parasites,** phagocytosis per se is a critical event in disease.

ANTIPHAGOCYTIC CAPSULES

Pathogenic bacteria that behave as extracellular parasites owe their virulence to **antiphagocytic surface components,** often demonstrable as definite **capsules** (Fig. 22-2, A and B). These surface structures, consisting of hydrophilic gels, inhibit ingestion of the organisms by phagocytic cells. The chemical structures of the capsular gels of a number of bacteria have been identified. The capsule of type 3 pneumococcus, for example, is composed of a high-molecular-weight polymer of glucose and glucuronic acid, while that of *Bacillus anthracis* contains a polypeptide of D-glutamic acid. Capsules of some bacteria (e.g., *Bacillus megaterium*) contain both protein and carbohydrate (Fig. 22-2B).

That such capsular structures may indeed interfere with phagocytosis, and thus contribute to virulence, is clearly exemplified by *Diplococcus pneumoniae.* If the behavior of a fully encapsulated S (smooth) strain is compared with that of a nonencapsulated R (rough) strain, the former is found to resist phagocytosis (in the absence of antibody)

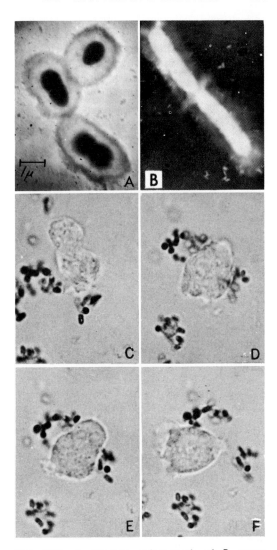

Fig. 22-2. A. Electron micrograph of *D. pneumoniae,* type 1. Capsule has been reacted with type 1 antibody to accentuate its visibility (quellung reaction). ×10,500, enlarged. **B.** Similar photograph of capsule of *B. megaterium* reacted with homologous polysaccharide antiserum. The specific anticarbohydrate antibody increases the visibility of the transverse septa of the capsule, which are composed of polysaccharide, but does not affect the rest of the capsule, which is made up of protein. ×2600, enlarged. **C–F.** Series of photomicrographs taken at 30-second intervals showing failure of granulocyte in hanging drop to phagocytize encapsulated Friedländer's bacilli (*K. pneumoniae*), despite their direct contact with the cell's advancing pseudopod. ×1250, enlarged. [**A** from Mudd, S., Heinmets, F., and Anderson, T. F. *J Exp Med* 78:327 (1943). **B** from Baumann, G., and Tomcsik, J. *Schweis Z Pathol Bakteriol* 21:906 (1958). **C–F** from Smith, M. R., and Wood, W. B. Jr. *J Exp Med* 86:257 (1947).]

and is highly virulent for mice, whereas the latter is readily phagocytized and is virtually avirulent. Enzymatic removal of the capsular gel, as may be readily accomplished in the case of type 3 pneumococci, likewise renders the organisms both nonpathogenic and susceptible to phagocytosis. Similarly, antibody to the capsular polysaccharide combines specifically with the gel and destroys its antiphagocytic properties. Furthermore, during the exponential phase of growth the capsules are largest and the organisms most virulent, and in the stationary phase the capsules become smaller and the organisms are less virulent. This difference must obviously be borne in mind when the virulence of a given strain is measured.

The physicochemical forces at the capsule-phagocyte interface that prevent phagocytosis are poorly understood. Although the cytoplasmic membranes of polymorphonuclear leukocytes are known to be composed primarily of lipoproteins, the chemical groups exposed at their outer surfaces have not been identified. It has been postulated that surface lipids impart to the membrane a hydrophobic character which prevents the pseudopod of the leukocyte from forming an intimate contact with the hydrated capsular gel (Fig. 22-2, C to F). Attempts to relate the antiphagocytic effect of the capsule to electrostatic charges have been unsuccessful.

ADAPTATION TO MICROENVIRONMENT

Since the surfaces of many potentially invasive bacteria are not antiphagocytic, it is evident that factors other than resistance to phagocytosis must be responsible for their virulence. By and large, bacterial species that are easily phagocytized and incapable of intracellular survival are noninvasive (e.g., *Staphylococcus albus*); unless toxigenic they do not cause disease. Other species, though readily phagocytized, resist intracellular destruction and are virulent (e.g., *Salmonella typhi*). The factors that enable one bacterial strain and not another to survive phagocytosis are not yet known, though some of the mechanisms involved in the killing of ingested bacteria have been elucidated (see below, Intracellular killing).

Invasive strains of bacteria, whether resistant to phagocytosis or not, must also be capable of multiplying in the deeper tissues

of the host. Biochemical characterization of the environments provided by various organs of the body is fragmentary; little is known of the metabolic properties that enable invasive strains to thrive in certain tissues.* It is clear, however, that many bacteria are **organotropic:** they are highly selective in regard to the tissues they invade. For example, both meningococcus (*Neisseria meningitidis*) and pneumococcus (*Diplococcus pneumoniae*) are frequently present in the human nasopharynx, yet of the two only pneumococcus invades the lower respiratory tract to cause pneumonia.

A high degree of selectivity is also involved in the habitation of superficial tissues by specific bacteria. Whereas *Neisseria meningitidis* infects only the upper respiratory tract of man, the closely related species *N. gonorrhoeae* characteristically invades the genitourinary tract. Similarly, in healthy individuals staphylococci tend to inhabit the skin and the vestibule of the nose, whereas pneumococci are usually confined to the throat.

Experiments of Buddingh and Goodpasture in chick embryos suggest that the cell type plays a key role in bacterial organotropism. When the chorioallantoic membrane of the 15-day-old embryoned egg is infected with *Bordetella pertussis* (the agent of whooping cough), the microorganisms, in addition to producing a local lesion on the membrane, frequently localize on the ciliated epithelium of the bronchial mucosa. When, on the other hand, the experiments are performed with 12- or 13-day embryos, in which ciliated cells have not yet appeared, infection of the bronchial epithelium fails to occur. *B. pertussis* appears, therefore, to have a special predilection for the environment provided by ciliated epithelial cells of the bronchi, a fact entirely in keeping with the pathology of whooping cough (Ch. 30, Fig. 30-4). Likewise, *Neisseria meningitidis*, which causes acute meningitis in man, localizes in the meningeal tissues of the embryo, and *Streptobacillus moniliformis*, which causes severe polyarthritis, invades the embryonic joints.

The pattern of organotropism may vary with the host species. When virulent bovine tubercle bacilli are injected intravenously (or even intrarenally) into guinea pigs, the kidneys remain free of progressive

---

* For a possible example of a known metabolic determinant, see the relation of erythritol to the pathogenesis of infectious abortion in cattle (Ch. 32, Pathogenicity).

lesions, though the lungs are heavily infected. In contrast, the kidneys of rabbits treated in the same fashion become riddled with tubercles. It has been suggested that the greater susceptibility of the rabbit kidney may be due to the absence of a tuberculocidal polyamine, spermidine, which is present in renal tissues of the guinea pig.

Finally, invasiveness may vary over a wide range in **different parts of the same organ.** Thus in the rabbit kidney the tissues of the medulla are far more susceptible to experimentally induced staphylococcal and *Escherichia coli* diseases than are those of the cortex. A possible explanation is that the inflammatory response (p. 645) in the medullary tissues (probably for anatomical reasons) is definitely slower than in the cortex. The resulting delay in mobilization of phagocytic cells allows the bacteria invading the medulla to establish a firm foothold. Other more subtle factors may also be involved.

PRODUCTION OF EXTRACELLULAR ENZYMES

A number of bacterial species elaborate extracellular enzymes to which their invasive properties* have been attributed. Many grampositive bacteria, for example, produce **hyaluronidase.** Originally referred to as "spreading factor," this enzyme promotes diffusion through connective tissue by depolymerizing hyaluronic acid in the ground substance. This action has been assumed to facilitate bacterial invasion, but there is no conclusive evidence that it does so. In fact, in experimental *Clostridium perfringens* infections the administration of antihyaluronidase antibodies fails to influence the spreading of the lesion.

Equally unconvincing is the evidence concerning the pathogenetic role of staphylococcal coagulases, which clot plasma by a thrombokinase-like action. These enzymes, produced by most virulent strains, have been thought to be responsible for the fibrin barrier that characteristically surrounds and appears to localize acute staphylococcal lesions, and for the formation of antiphagocytic fibrin envelopes about the cocci themselves. However, pathogenic non-coagulase-producing strains have been identified that produce

---

* Bacterial extracellular enzymes known to be cytotoxic are usually classified as toxins and are discussed in the next section.

identical tissue lesions and are no more readily phagocytized. It therefore seems unlikely that coagulases per se account either for staphylococcal virulence or for the tendency of staphylococcal lesions to remain relatively circumscribed.

Streptococci and staphylococci also elaborate enzymes (streptokinase and staphylokinase) that catalyze the lysis of fibrin. Whether fibrinolysis actually influences the invasiveness of the organisms is also uncertain, although it has long been assumed that the relative absence of fibrin in spreading streptococcal lesions may be due to the action of streptokinase. On the other hand, the collagenases of *Clostridium histolyticum* and *C. perfringens,* which break down the collagen framework of muscles (leaving the fibers intact), clearly facilitate extension of gas gangrene due to these organisms (Ch. 34).

TOXIGENICITY

The idea that bacteria may cause disease by producing toxins was first suggested in 1884 by the classic experiments of Loeffler. He noted that guinea pigs injected subcutaneously with diphtheria bacilli developed widespread systemic lesions in which no bacilli could be found. He concluded that the bacilli growing at the site of infection had generated a soluble poison which became widely disseminated by the blood stream. This conclusion was confirmed 5 years later when injection of bacteria-free filtrates was shown by Roux and Yersin to produce an essentially identical disease. Many other species of pathogenic bacteria were later also found to be toxigenic (Table 22-1).

The diffusible toxins produced by certain gram-positive bacteria (and an occasional gram-negative species) are called **exotoxins,** since they are present in the filtrates of growing cultures in which no appreciable autolysis has occurred. All that have been characterized chemically have been found to be proteins. Toxins of a different sort, called **endotoxins,** are produced by many gram-negative bacteria. They are complex lipopolysaccharides of the bacterial cell wall and are released into the surrounding medium only if the organisms become autolyzed or are artificially disrupted

TABLE 22-1.   Exotoxins Produced by Principal Toxigenic Bacteria Pathogenic for Man

Bacterial species	Disease	Toxin	Action	Toxicity per mg, expressed as LD$_{50}$^{kg}*
Clostridium botulinum	Botulism	Six type-specific neurotoxins	Paralytic	1,200,000 (G)
Clostridium tetani	Tetanus	Tetanospasmin	Spastic	1,200,000 (G)
		Tetanolysin	Hemolytic cardiotoxin	
Clostridium perfringens	Gas gangrene	α-Toxin†	Lecithinase: necrotizing, hemolytic	200 (M)
		β-Toxin		
		γ-Toxin		
		δ-Toxin		
		ε-Toxin	Necrotizing	
		η-Toxin		
		θ-Toxin	Hemolytic cardiotoxin	
		ι-Toxin	Necrotizing	
		κ-Toxin	Collagenase	
		λ-Toxin	Proteolytic	
Clostridium septicum	Gas gangrene	α-Toxin	Hemolytic	
Clostridium novyi	Gas gangrene	α-Toxin	Necrotizing	50,000 (M)
		β-Toxin	Lecithinase: necrotizing, hemolytic	
		γ-Toxin	Lecithinase: necrotizing, hemolytic	
		δ-Toxin	Hemolytic	
		ε-Toxin	Lipase: hemolytic	
		ζ-Toxin	Hemolytic	
Corynebacterium diphtheriae	Diphtheria	Diphtheritic toxin	Enzyme altering transferase II	3,500 (G)
Staphylococcus aureus	Pyogenic infections	α-Toxin	Necrotizing, hemolytic, leukocidic	50 (M)
		Enterotoxin	Emetic	
		Leukocidin	Leukocidic	
		β-Toxin	Hemolytic	
		γ-Toxin	Necrotizing, hemolytic	
		δ-Toxin	Hemolytic, leucolytic	
Streptococcus pyogenes	Pyogenic infections and scarlet fever	Streptolysin O	Hemolytic	0.5 (M)
		Streptolysin S	Hemolytic	
		Erythrogenic	Causes scarlet fever rash	
		Streptococcal DPNase	Cardiotoxic	
Pasteurella pestis	Plague	Plague toxin	Necrotizing (?)	25 (M)
Bordetella pertussis	Whooping cough	Whooping cough toxin	Necrotizing	
Shigella dysenteriae	Dysentery	Neurotoxin	Hemorrhagic, paralytic	1,200,000 (R)

* LD$_{50}$^{kg} denotes median lethal dose (LD$_{50}$, p. 639) per kilogram of guinea pig (G), mouse (M) or rabbit (R).

† The designation of toxins by Greek letters is purely arbitrary and is based on the order in which they were identified.

Modified from VAN HEYNINGEN, W. E., In *General Pathology,* p. 754. (H. W. Florey, ed.) Saunders, Philadelphia, 1962.

by mechanical or chemical means. Endotoxins are generally less potent and less specific in their action than exotoxins.

EXOTOXINS

The most thoroughly studied exotoxins are those of *Corynebacterium diphtheriae, Clostridium tetani,* and *clostridium botulinum* (type A). All three have been highly purified and two (tetanus and botulinus) have been crystallized. Relatively heat-labile proteins, they share with the neurotoxin of *Shigella dysenteriae* (Shiga) the distinction of being the most powerful poisons known. From the figures listed in Table 22-1 it will be seen that a milligram of tetanus (or botulinus) toxin is enough to kill more than 1 million guinea pigs. It has been suggested that the somewhat lower potency of diphtheria toxin, as estimated by its median lethal dose (LD_{50}) in guinea pigs, may be due to its affinity for body cells in general, in contrast to the highly selective actions of the tetanus and botulinus toxins on specific cells of the central nervous system. Even so, 1 oz. of diphtheria toxin would be more than enough to kill everyone in the city of New York.

Toxicity estimates of this kind are obviously inexact. Indeed, one of the most striking characteristics of bacterial exotoxins is the degree to which their potencies vary when measured in animal hosts of different species. Men, horses, and guinea pigs, for example, are much more susceptible to diphtheria toxin than mice or rats, and are far more sensitive to tetanus and botulinus toxins than dogs. Rabbits (per unit of body weight) are 10,000 times more susceptible than guinea pigs to the lethal action of shigella neurotoxin. The reasons for these striking differences are not known, though recent studies on the effects of diphtheria toxin on cultured mouse and human cells suggest that species-specific components of the protein-synthesizing system may be involved (Ch. 24, Pathogenicity).

A few exotoxins have been identified as enzymes. It is therefore not surprising that the amino acid composition and the size of toxins are no different from those of enzymes in general.

Exposure of exotoxin to heat, acid, or proteolytic enzymes (e.g., in gastrointestinal tract) causes marked loss of toxicity. (An important exception is the enhancement of potency of type E botulinus toxin by proteolysis,

Ch. 34.) In addition, introduction of substituent groups, e.g., by iodine, ketene, diazonium salts, and formaldehyde eliminates toxicity but leaves much of the original antigenicity: these altered toxins are called **toxoids.** Those prepared from cultures of *Corynebacterium diphtheriae* and *Clostridium tetani* are widely used to induce active immunity in man. The change from toxin to toxoid, which is accomplished industrially by treatment with dilute formaldehyde, also occurs with denaturation; it tends to occur spontaneously even at low temperatures.

The **pharmacological actions of the known bacterial exotoxins** are comparatively slow, some requiring several days. As already mentioned, diphtheria toxin in susceptible hosts inhibits protein synthesis and thus damages a wide variety of cells. The α-toxin of *Clostridium perfringens* is a lecithinase which likewise acts upon many different kinds of cells, including erythrocytes, affecting primarily their membranes. When injected intravenously into laboratory animals it causes hemolysis, but in human infections hemolytic reactions are rarely observed.

The actions of the neurotoxins of botulinus and tetanus are restricted to cells of the central nervous system. Although both seem to depress the formation and/or release of acetylcholine, botulinus toxin affects both pre- and postganglionic synapses of the peripheral autonomic system, as well as cholinergic mechanisms in peripheral motor nerves, whereas tetanus toxin appears to act only on motor cells within the cerebrospinal axis where, according to Eccles, it causes spasm by blocking the function of inhibitory synapses.

The "shiga neurotoxin" of *Shigella dysenteriae* differs from botulinus and tetanus toxins in that it acts only secondarily upon the central nervous system. Its primary action is limited to small blood vessels in the brain and spinal cord.

Not all bacterial exotoxins have been shown to play a role in the production of disease. The pathogenic properties of diphtheria, tetanus, and botulinus toxins, staphylococcal enterotoxin, and the erythrogenic toxins of *Streptococcus pyogenes* have been conclusively established. The evidence is also convincing that the exotoxin of *Bordetella pertussis* is responsible for the deep bronchial wall lesions in whooping cough. Similarly, observations implicating the α-toxin of *Staphylococcus aureus,* the neurotoxin of *Shigella dysenteriae,* the lethal toxin of *Bacillus anthracis,* and the plague and cholera exotoxins are reasonably decisive. The pathogenetic role of the necrotoxic proteases elaborated by certain of the gas gangrene organisms (e.g., *Clostridium histolyticum*) also seems fairly clear. There is yet no conclusive evidence, however, that any of the numerous other exotoxins produced by gram-positive bacilli and cocci are pathogenically active in vivo. These include streptococcal leukotoxin (Ch. 26) and the various leukocidins which tend to damage polymorphonuclear leukocytes in vitro.

ENDOTOXINS

The following properties of bacterial endotoxins distinguish them from exotoxins:

1) They are produced primarily, if not solely,* by gram-negative bacteria.

2) They are complex macromolecules containing lipopolysaccharide.

3) They make up an integral part of the bacterial cell wall and are released only if the integrity of the cell is disturbed.

4) They are relatively heat-stable.

5) They are less potent and less specific than most exotoxins in their cytotoxic actions.

6) Their toxicity appears to reside in the lipid fraction, whereas their specific antigenic determinants reside in polysaccharide.

7) They do not form toxoids.

The bacterial endotoxins are extremely complicated structurally. Originally isolated as phospholipid-polysaccharide-protein complexes identical with the somatic O antigens of bacterial cells, their biological activity has been shown, principally by Westphal and Lüderitz, to reside in a lipopolysaccharide fraction which can be separated by phenol extraction from most of the phospholipid and all of the protein of the original complexes. The structure of the polysaccharide moiety, which contains dideoxyhexoses (Ch. 29, Antigenic structure) not known to occur any-

where else in nature, determines the specific antigenicity of the molecule. Although endotoxins were originally obtained only from smooth strains of gram-negative bacteria possessing demonstrable O antigens, they have also been extracted from rough strains which lack O antigens. Further details of the chemistry of O antigens and of lipid A are discussed in Chapters 6 and 29.

Bacterial endotoxins have numerous **biological effects.** When injected in sufficient amounts they cause within an hour or two irreversible shock, usually accompanied by severe diarrhea. At autopsy a few hemorrhages in the wall of the gastrointestinal tract are often the only discernible lesions. In smaller doses they cause fever, transient leukopenia followed by leukocytosis, hyperglycemia, hemorrhagic necrosis of tumors, abortion, altered resistance to bacterial infections (see below), a wide variety of circulatory disturbances, and vascular hyperreactivity to adrenergic drugs. They are also capable of eliciting the **Shwartzman phenomenon.†**

Animals given repeated injections of an endotoxin become relatively unresponsive to its pyrogenic and other biological effects. This **tolerance** is nonspecific, i.e., it inhibits the response to other endotoxins (and to traumatic shock and radiation damage). Tolerance subsides within a week or two after the last injection; it therefore does not correlate with the much longer persistence of

---

* Pathogenic gram-negative bacteria from which endotoxins have been isolated include *Salmonella typhi, Shigella dysenteriae, Vibrio cholerae, Brucella melitensis, Neisseria gonorrhoeae,* and *Neisseria meningitidis.* Endotoxins have also been obtained from relatively avirulent gram-negative species, e.g., *Escherichia coli.*

† When endotoxins are injected subcutaneously into rabbits, in doses of a few micrograms, a mild inflammatory reaction occurs in the skin. If, 24 hours later, an intravenous injection of the same, or another, endotoxin is given in the same amount, the originally injected skin site becomes hemorrhagic within a few hours. Histologically it is characterized by the presence of leukocyte-platelet thrombi, particularly in venules. Indeed, the reaction, first described by Shwartzman in 1928, will not occur unless a sufficient number of leukocytes are present in the circulation. If both injections are given by vein the animal usually dies within 24 hours after the second, and at autopsy bilateral cortical necrosis of the kidneys is regularly found. Neither this generalized Shwartzman reaction (which is accentuated by cortisone), nor the localized form, has been satisfactorily explained in immunological terms; both appear to be essentially toxic in nature.

active immunity. Tolerance results primarily from an enhanced clearance of endotoxin by cells of the reticuloendothelial system; it can be nullified by "reticuloendothial blockade" with such agents as Thorotrast or India ink. However, there is evidence that the process also involves humoral factors detectable by passive transfer.

The properties of bacterial endotoxins that have aroused the most interest are those relating to pyrogenicity, the ability to cause irreversible shock, and the effect upon nonspecific immunity. 1) The pyrogenic action appears to be indirect and to be mediated through an endogenous pyrogen released from polymorphonuclear leukocytes. 2) Extensive experimental findings indicate that the terminal irreversibility of hemorrhagic shock is due in large part to the absorption of bacterial endotoxins from the bowel. Although this conclusion has been contested on the grounds that irreversible hemorrhagic shock may be produced in axenic animals, quantitative studies of comparative susceptibility have not been performed, and the possibility that the axenic animals may be sensitized to endotoxins contained in their food has not been excluded. 3) How prior treatment with endotoxin influences nonspecific immunity to bacterial infections is only partially understood. Whereas large doses of endotoxin depress both resistance to infection and phagocytosis, small doses enhance them and augment antibody synthesis (Ch. 15). Indeed, if an appropriate dose of endotoxin is given a biphasic response occurs in which resistance and phagocytosis are first depressed and later enhanced.

Despite the many biological effects of bacterial endotoxins, their role in the pathogenesis of bacterial diseases remains poorly defined.

VIRULENCE

MEASUREMENT

The virulence of a bacterial strain, whether due primarily to invasiveness, toxigenicity, or a combination of the two, is ordinarily measured in terms of the median dose that will kill within a stated period of time 50% of the animals inoculated. This number is known as the LD_{50}.* The principal reason for measuring the LD_{50} rather than the minimum lethal dose (MLD), i.e., the smallest dose needed to kill all the animals injected, is revealed by the dose-response curves depicted in Figure 22-3. In graphs A and B the doses of bacteria injected, expressed in logarithms to the base 10, are plotted on the abscissa, and the mortalities observed in separate groups of animals receiving each dose are plotted in percentages on the ordinate. The rates of change in mortality are greatest in the middle portions of the curves (C and D) and are smallest at the extremes. Therefore, it is possible to determine with much greater accuracy the dose which will kill 50% of a group of animals than that which will kill 100%. Furthermore, with a given number of animals the LD_{50} can be measured more precisely when the dose-response curve is relatively steep, as in B, rather than when the slope changes more gradually, as in A.

The sigmoid shape of the dose-response curve is thought to be due both to the heterogeneity of the animal population injected and to the statistical distribution of chance events in the infection. When animals of the same age, weight, sex, and inbred strain are used, the heterogeneity is minimized and dose-response curves are steeper. Since the susceptibilities of different strains, as well as species, of animals for a given organism may vary over a wide range, and since the severity of the disease produced may depend to a large extent on the site of inoculation, these two variables must also be defined. In addition, the state of the culture to be injected must be rigidly controlled because of variations in virulence in the different phases of the growth cycle.

Similar technics may be employed to measure the **LD_{50} of a bacterial toxin** or the **transmissibility** of a bacterial strain to a given animal host. The endpoint in the latter type of test is determined by measuring the dose required to cause a demonstrable infection (usually detected by cultural methods) in 50% of the animals exposed. The dose required, known as the ID_{50} (ID, infectious dose), will depend in part upon the method of exposure, e.g., by aerosol, intravenous injection, intraperitoneal inoculation. There-

---

* LD, lethal dose. The LD_{50} is most readily measured by the Reed-Muench method (see Appendix at end of chapter).

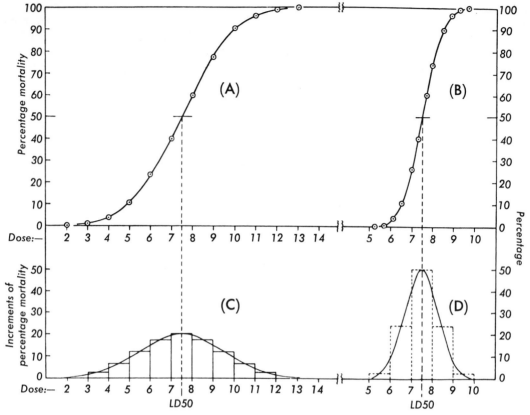

Fig. 22-3. Examples of dose-response curves used to measure bacterial virulence (for explanation see text). Quantitative measurements of the lethal effects of toxins (in contrast to viable organisms) result in much the same kind of dose-response curves. (From *Topley and Wilson's Principles of Bacteriology and Immunology.* [G. S. Wilson and A. A. Miles, eds.] Williams & Wilkins, Baltimore, 1964.)

fore, as in measurements of the LD_{50}, the precise manner in which the infectious agent is introduced into the test animal must be specified.

VARIATION

Variation in virulence may result from changes in the parasite, in the host, or in both. Alterations in the bacterial strain may result either from mutations or from changes in the organism's environment.

Genotypic Variations. The best-studied example of a mutational change affecting bacterial invasiveness is that involving encapsulation of pneumococci. Multiplying populations of smooth (S) fully encapsulated pneumococci give rise spontaneously to un-

encapsulated, one-step, rough (R) mutants. Conversely, one-step R → S mutations occur in populations of rough pneumococci. Which of the two variants (S or R) will eventually predominate in a given population depends upon the conditions under which the organisms multiply (Ch. 8; Ch. 25, Genotypic variations). If, for example, pneumococci originally in the S form are repeatedly subcultured on an artificial medium which selectively favors the growth of the R variant, the unencapsulated rough form will finally overgrow the culture. When, on the other hand, a population containing both variants is injected into a susceptible laboratory animal, the unencapsulated R cells will be rapidly destroyed by the phagocytic defenses of the host, while the S cells will continue to multiply and eventually kill the animal. The pneumococci cultured from the blood at postmortem will, therefore, be of the virulent encapsulated variety.

The virulence of the progeny of a given strain of pneumococcus may thus vary depending upon the selection of mutants. Furthermore, the virulence of each pneumococcal cell depends not only on the presence or absence of a capsule but also on its size. Variants of intermediate (I) virulence may arise which produce only small capsules (Ch. 25, Capsular antigens). Any one of the three genotypes (S, I, or R) may be obtained by transformations performed in vitro. Similar genetic variations of other pathogenic capsule-bearing bacteria (e.g., *Streptococcus pyogenes, Klebsiella pneumoniae, Pasteurella pestis*) affect invasiveness in the same manner.

The toxigenicities of virulent bacterial species are likewise subject to genotypic variation. Of particular interest is the fact that toxin production by diphtheria bacilli depends upon **lysogeny,** i.e., only those bacilli carrying a specific temperate bacteriophage (β-phage) generate diphtheria toxin. Since lysogeny is heritable, its influence on toxigenicity may justifiably be defined as genotypic. Lysogeny also controls the production of the scarlet fever toxin by *Streptococcus pyogenes,* and it seems likely that temperate phages will be found to determine the toxigenicity of other pathogenic bacteria.

Phenotypic Variations. Modifications in invasiveness and toxigenicity may also result from metabolic changes controlled by environmental factors. As already mentioned, pneumococci produce larger capsules and thus are more virulent during the exponential phase than during later stages of the growth cycle. They also form more generous capsular envelopes when cultured in carbohydrate-rich rather than carbohydrate-deficient medium. Similarly, toxigenic strains of *Corynebacterium diphtheriae* and *Clostridium tetani* elaborate their toxins only when cultured in media in which the iron content is critically controlled, and *Bordetella pertussis* thrives in vivo only in the special microenvironment provided by ciliated epithelial cells of the bronchial mucosa. Other examples might be cited, but the foregoing should suffice to indicate how **both genotypic and phenotypic factors influence the virulence of bacteria.**

ANTIBACTERIAL DEFENSES OF THE HOST

The mammalian host becomes **infected** with many of the bacteria that abound in its environment, and it lives in balance with these parasitic intruders by restricting them to relatively superficial sites. The infecting bacteria cause **disease,** however, when they invade deeper tissues. To remain in a state of health, therefore, the host must constantly defend itself against bacterial **invasion.**

SKIN AND MUCOUS MEMBRANE BARRIERS

Mechanical, chemical, and microbial factors all play important parts in limiting the colonization of bacteria on the surfaces of the body.

MECHANICAL FACTORS

The epithelial surfaces of the skin and mucous membranes form physical barriers that are more or less impermeable to bacteria. The impregnability of the skin appears to be greater than that of most mucous membranes: the intact surface of healthy epidermis seems to be rarely, if ever, penetrated by bacteria. Even the widely held opinion that organisms such as *Pasteurella tularensis* and the brucellae may do so is unsubstantiated. When the integrity of the epithelial surface is broken, however, subcutaneous infection often develops. The bacteria most likely to cause such dermal infections are those that normally inhabit the hair follicles and sweat glands of the skin, i.e., staphylococci. When the skin is moist, as in hot humid climates, dermal infections are particularly common.

Many pathogenic bacteria, on the other hand, are able to penetrate mucous membranes, on whose surfaces they often thrive in the moist secretions. The mere ingestion of sufficient numbers of salmonella or tubercle bacilli, for example, will lead to penetration of the gastrointestinal mucosa and invasion of regional lymph nodes. Similar penetrations of the mucosal barrier of the conjunctiva by leptospires, and of the respiratory tract by pneumococci, occur following heavy exposures.

Just how the penetration is achieved is not known. It has been suggested that toxic products of the bacteria may damage the surface epithelial cells, and attempts to demonstrate a microscopic lesion at the site of entrance have met with little success. There is much indirect evidence that prior injury of mucosal cells by viral infections, particularly in the respiratory tract, is frequently involved. Mucosal irritations caused by noxious gases, including certain inhalation anesthetics, or by special climatic conditions, such as cold or low humidity, have also been implicated.

Other mechanical factors that help to protect the epithelial surfaces include 1) the lavaging action of tears in the conjunctival sacs and lachrymal ducts and of saliva in the mouth and pharynx; 2) the trapping effect of the mucus-coated hairs in the anterior vestibule of the nose; 3) the adhesive qualities of the mucus that lines the respiratory and gastrointestinal tracts; 4) the expulsive effects of coughing, sneezing, and ciliary action* in the respiratory tract; 5) the cleansing of the urethra by the flow of urine; and 6) the desquamation of stratified epithelium from the surface of the skin. Were it not for the continuous operation of these essentially mechanical processes, myriads of bacteria would invade the deeper tissues of the body.

CHEMICAL FACTORS

Mechanical factors alone, however, do not account for the remarkable resistance of the skin and mucous membranes to bacterial invasion. The **acidity** of the gastric secretions preserves the usual sterility of the stomach, as evidenced by the frequency with which a heavy colonization of the gastric mucosa accompanies the hypochlorhydria of pernicious anemia. The low pH of the skin (3 to 5),

due in part to acid products of bacterial metabolism, undoubtedly discourages the growth of many of the microorganisms that impinge upon its surface. The acidity of the adult vagina (pH 4.0 to 4.5) caused by the growth of lactobacilli probably serves a similar function. **Unsaturated fatty acids** on the skin kill certain pathogenic bacterial species, e.g., *Streptococcus pyogenes,* but stimulate the growth of others (diphtheroids). The selectivity of these processes may help to account for the fact that so few kinds of bacteria ordinarily inhabit the skin. The principal self-sterilizing factor in semen appears to be spermine;† that in tears, nasal secretions, and saliva is **lysozyme,** an enzyme capable of breaking down the mucopeptide layer in the cell walls of many bacteria. Were it not for antibacterial substances in these body secretions, they would be constantly infected.

Mucous secretions of the respiratory and gastrointestinal tracts also contain antibodies, especially of the IgA class. Their influence on the flora of the throat is indicated by the finding that immunization with pneumococcal type-specific polysaccharides lowers the carrier rates for pneumococci of the corresponding types. Similarly, antibodies to various enteric bacteria may be demonstrated in the feces (coproantibodies), although their protective function remains uncertain.

MICROBIAL FACTORS

Microbial antagonism suppresses the growth of many potentially pathogenic bacteria and fungi at superficial sites, where they might otherwise initiate disease. This important phenomenon is due in some cases to competition for essential nutrients, and in others to the production of substances that suppress competing species (e.g., colicins, Chs. 29 and 46). Its importance in maintaining the proper microbial balance in the mouth and gastrointestinal tract is forcefully illustrated by the *Candida* and *Staphylococcus* infections that

---

* In man, the ciliary "escalator" of the bronchi and trachea keeps the mucus "blanket" moving upward toward the pharynx at a rate of 1 to 3 cm/hour.

† This polyamine, at the concentrations present in seminal fluid, is bactericidal to a wide variety of pathogenic bacteria.

sometimes occur during treatment with broad-spectrum antibiotics.

PHAGOCYTIC DEFENSE

When bacteria or other invading parasites succeed in penetrating the skin or mucous membranes, cellular defense mechanisms are immediately brought into play: local macrophages and blood-borne phagocytic cells accumulate around the invaders and initiate the phagocytic process. The most important phagocytic cells are the mononuclear phagocytes and the polymorphonuclear leukocytes.

MONONUCLEAR PHAGOCYTES

This cell arises from a rapidly dividing precursor in the bone marrow that gives rise to the immature, poorly phagocytic **promonocyte,** which remains capable of active cell division (Fig. 22-4) and is released into the blood stream as a **monocyte.** The monocyte is a somewhat more differentiated cell: it makes up 3 to 7% of the circulating white cells; it is actively phagocytic and bactericidal; but it has a reduced capacity to divide. After circulating briefly in the blood stream (less than 2 days) monocytes emigrate into tissues where they develop into the larger and more actively phagocytic **macrophages**

or tissue histiocytes (Fig. 22-5A). The total body pool of macrophages constitutes the **reticuloendothelial system.**

Although the macrophage retains the ability to divide under certain inflammatory stimuli, the vast majority of tissue and inflammatory macrophages are derived from the blood monocyte. The stimuli that cause the monocyte to increase in size and develop more active pinocytic and phagocytic activity are unknown. These forms of cellular ingestion bring soluble and particulate molecules into the cytoplasm, where they are degraded by a variety of hydrolytic enzymes.

In contrast to the relatively small number of circulating monocytes, the macrophages are numerous. They are widely scattered throughout the connective tissue and are found just outside the basement membrane of most small blood vessels. Larger and more concentrated numbers are present in liver sinusoids (Kupffer cells), spleen, lung (alveolar macrophages), bone marrow, and lymph nodes. In these sites the cells are continually exposed to foreign material and carry out their functions as scavengers. It is believed that tissue macrophages are long-lived cells, retaining the ability to synthesize new digestive enzymes and organelles as a response to repeated bouts of phagocytic activity.

Fig. 22-4. Life history of mononuclear phagocytes illustrating the pools, cell size, and presence of cytoplasmic granules. Nucleoli are prominent in the dividing promonocyte and are progressively lost as maturation occurs.

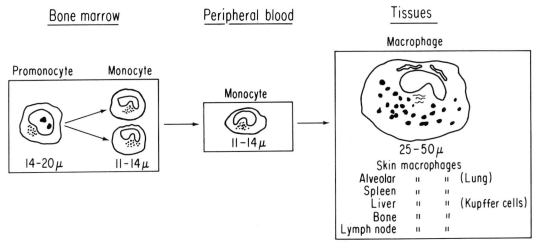

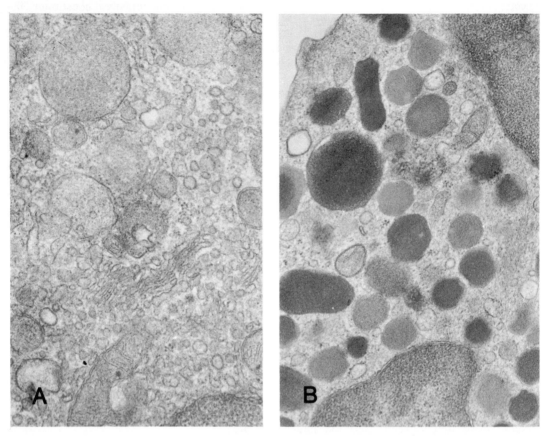

Fig. 22-5. A. Mouse peritoneal macrophage after 1 day of cultivation. Many small electron-lucent pinocytic vesicles are presented in the peripheral cytoplasm. The large dense bodies are secondary lysosomes. ×40,000. **B.** Mature rabbit polymorphonuclear leukocyte. The cytoplasm is filled with two types of granules. The larger and more dense primary or azurophil granule and the smaller, less dense secondary granule. ×30,000. (Courtesy of Dr. Martha E. Fedorko).

NEUTROPHILIC AND HETEROPHILIC LEUKOCYTES

The neutrophils of man and the equivalent heterophils of laboratory rodents are critical for host defense. They arise from a multipotential stem cell in the bone marrow, and through a complex series of divisions and maturational steps they give rise to adult polymorphonuclear granulocytes (Fig. 22-5B). In contrast to mononuclear phagocytes, large numbers of granulocytes are stored in the bone marrow as a reserve pool of fully active phagocytic cells. During differentiation in the marrow, many cytoplasmic granules are formed, the nucleus becomes multilobed, cell division ceases, and mitochondria and endoplasmic reticulum are lost from the cytoplasm. At the same time the cell becomes motile and actively phagocytic. Under normal conditions the adult cell is released into the circulation, where it represents 30 to 70% of the circulating white cells and emigrates out of blood vessels within a day. It represents an expendable, short-lived end cell, produced in huge numbers and containing a full complement of bactericidal agents when released from the marrow. It also contains large stores of glycogen; since most of its metabolic energy is derived from glycolysis, the granulocyte can function efficiently in anaerobic environments.

INFLAMMATION AND THE SEQUENCE OF CELLULAR DEFENSE

The lodgment and multiplication of microbes or the presence of various foreign substances results in an **inflammatory response** characterized by 1) dilatation of surrounding blood vessels; 2) increased vascular permeability, permitting escape of plasma and a few erythrocytes; and 3) diapedesis of monocytes and polymorphonuclear leukocytes. Early in the response, granulocytes predominate in the inflammatory exudate as the result of their greater numbers in the blood stream. In the later phases, after many of the granulocytes have died, monocytes and macrophages become the predominant cell type.

Organisms that escape phagocytosis in the local lesion may be transported by the lymphatics to regional lymph nodes. Here they initially encounter the resident macrophages of the nodal sinusoids which filter particulates from the afferent lymph. If massive seeding of bacteria occurs, there follows acute inflammation of the node **(lymphadenitis)** and an influx of blood granulocytes and monocytes, similar to the local lesion. If the defenses of the regional node are breached, microbes may then be transported from the efferent lymph through the thoracic duct and into the blood stream.

The presence of viable bacteria in blood is termed **bacteremia** and usually indicates an invasive infection with unfavorable prognosis. Microbes are removed from the circulation by **clearance** mechanisms that depend largely on fixed macrophages lining the sinusoids of a number of parenchymatous organs, especially liver and spleen. On occasion, platelets and endothelial cells of capillaries are capable of engulfing certain microbial agents.

The ability of cellular defenses to destroy large numbers of bacteria in the blood stream can be readily demonstrated in laboratory animals by the intravenous injection of a "pulse" of organisms followed by enumeration of the viable bacteria by serial blood cultures (Table 22-2).

INFLAMMATION AND THE DELIVERY OF PHAGOCYTIC CELLS

The process of diapedesis, by which phagocytic cells leave the blood and accumulate at extravascular sites of tissue injury, is described with remarkable clarity in Cohnheim's *Lectures on General Pathology,* published in 1882. Nearly 3 decades earlier both Addison and Waller had noted the adherence of white blood corpuscles to the inner linings of blood vessels in areas of injury and had inferred that the stuck cells somehow emigrated into the tissues to form pus. These early descriptions have been fully confirmed. In addition, electron microscopy and use of the rabbit-ear chamber (a device which permits direct microscopy in vivo at oil-immersion magnifications) have led to further elucidation of the inflammatory process.

When a standardized noxious stimulus,

TABLE 22-2. Comparative Rates of Clearance from Blood Stream (in Normal Rabbits) of Intravenously Injected Pneumococci of Varying Degrees of Virulence

Time after injection (hr)	No. of viable pneumococci per milliliter of of blood		
	Avirulent	Slightly virulent	Highly virulent
0	8,900,000	1,030,000	1,070,000
2	206	20,800	137,000
5	20	340	25,000
24	0	1,300	1,500,000*
48	—	134	Death
96	—	0	—

* The recrudescence of the bacteremia, leading to death, results from failure of the cellular defenses to destroy all the bacteria sequestered in the vascular beds of such organs as liver and spleen, where they eventually establish a firm foothold and reinvade the circulation.

From *Topley and Wilson's Principles of Bacteriology and Immunology,* p. 1180. (G. S. Wilson and A. A. Miles, eds.), Williams & Wilkins, Baltimore, 1964.

such as a sharply localized microburn, is applied to the tissue in a rabbit-ear chamber, there is a brief period of vasoconstriction, and a phase of vasodilatation and increased blood flow ensues. This vasomotor change is followed by increased vascular permeability, intravascular hemoconcentration, stasis, and extravascular edema.* In the venules particu-

* These vascular phenomena account for the classic physical signs of acute inflammation: swelling, redness, heat, and pain.

larly the number of visible leukocytes gradually increases. They can be seen dragging along the endothelium, well outside the axial stream of rapid blood flow. If a single venule in this state is carefully observed, it will be noted that the sticking of the white cells to the endothelium begins on the side of the vessel nearest to the site of the burn (Fig. 22-6A). This "unilateral sticking" is promptly followed by a generalized adherence of leukocytes to the entire lining of the vessel (Fig. 22-6B). Soon thereafter diapedesis is noted

Fig. 22-6. A. Unilateral sticking of leukocytes to endothelium of venule in rabbit-ear chamber 30 minutes after injury. The leukocytes (L) are adhering to the endothelial surface nearest to the zone of injury, which is located below the vessel. ×200. **B.** One hour after injury leukocytic (L) sticking has become generalized. ×400. **C–E.** Diapedesis of leukocyte (L_2) is visible in lower half of photograph. ×350. **F, G.** Extravascular migration of leukocytes. The white cells (a and b) are seen to be migrating toward the stationary cell (c). ×350. [From Allison, F., Smith, M. R., and Wood, W. B., Jr. *J Exp Med 102*:655 (1955).]

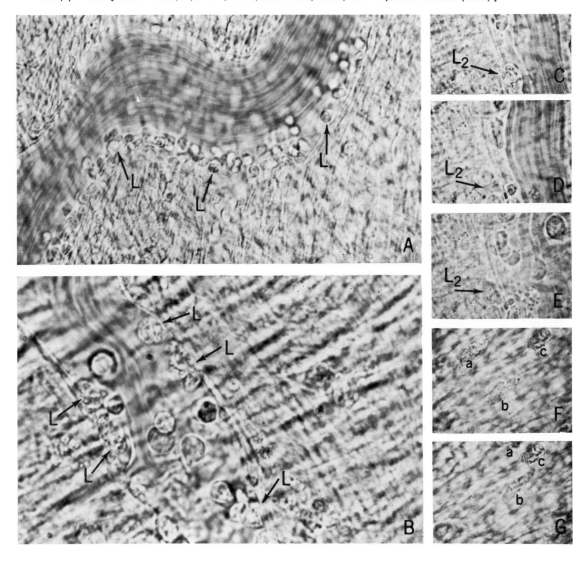

(Fig. 22-6C), and leukocytes that have passed through the vessel wall can be seen moving about in the interstitial spaces of the extravascular tissues (Fig. 22-6D). Electron photomicrographic studies demonstrate that both granulocytes and monocytes pass **between,** rather than through, endothelial cells in leaving the lumen of the venule (Figs. 22-7 and 22-8).*

It is not clear if the localization of granulocytes at the tissue lesion is due to chemotaxis. Continuous observation of individual cells migrating through the extravascular tissue spaces in the rabbit-ear

---

* In contrast, lymphocytes leave the postcapillary venules in lymph nodes by penetrating endothelial cells. The lymphocyte is ingested by the luminal side of the endothelial cell, transported in a phagocytic vacuole through its cytoplasm, and ejected through its parenchymal surface (Fig. 22-8).

chamber has not revealed the undirectional migration expected if positive chemotaxis were involved. The cells frequently change direction and move about at random; yet eventually they become heavily concentrated in and around the site of injury. It has been suggested that the progressive accumulation of leukocytes at the injury site may be due to microenvironmental factors that prevent them from leaving the area of damage once they have entered it.

Although it is difficult to observe chemotaxis of leukocytes and monocytes in the ear chamber, considerable evidence exists that directional movement can take place under in vitro conditions, e.g., to a point source of bacteria. Boyden's quantitative method is based on the use of a vessel with two chambers separated by a Millipore filter: cells placed in one chamber crawl through the filter in response to a variety of stimuli placed in the other chamber. With this approach a number of chemotactic factors for both monocytes and leukocytes have been identified, e.g., peptides derived from some complement components (C3a, C5a; Ch. 16),

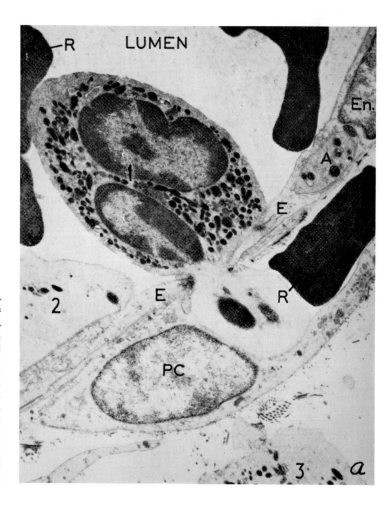

Fig. 22-7. Polymorphonuclear leukocyte (1) passing out of inflamed venule between endothelial cells (E) and coming in contact with periendothelial cell (PC). Red cells (R) are seen within the lumen of the venule and in the periendothelial space. Cell 2 is a granulocyte adhering to the endothelium, and cell 3 is one that has already passed into the connective tissue. ×13,500. [From Marchesi, V. T., and Florey, H. W. Q J Exp Physiol 45:343 (1960).]

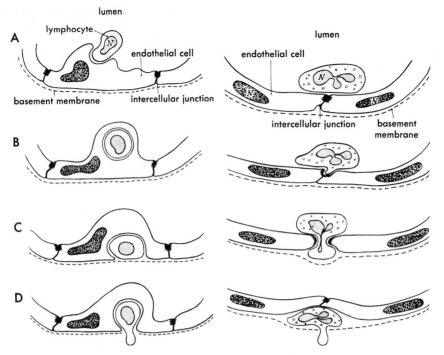

Fig. 22-8. Schematic diagrams contrasting the manner in which lymphocytes leave the postcapillary venules of normal lymph nodes (left) and the diapedesis of granulocytes (and monocytes) in inflammation. At stage C both types of cells, having passed the endothelial barrier, are constrained by the basement membrane and the periendothelial sheath, which they penetrate in stage D (see Fig. 22-7). [From Marchesi, V. T., and Gowans, J. L. *Proc Roy Soc Lond (Biol) 159:*283 (1964).]

a variety of bacterial products, and fractions of leukocytes and other tissue cells. It may be that once cells have migrated into a lesion under in vivo conditions, they are surrounded by chemical stimuli that cancel unidirectional movement and maintain them in the general area.

In the **chronic inflammation** associated with tuberculosis a progressive accumulation of macrophages takes place. Organized into discrete **granulomata** (surrounding bacteria and indigestible bacterial products), these cells become tightly interdigitated; some take on the appearance of **epithelioid cells,** and others fuse to form multinucleated **giant cells.**

Despite the vast amounts of descriptive information available on inflammation, virtually nothing is known about its molecular basis. The sticking of leukocytes to endothelium (Fig. 22-6B), for instance, is still not understood in chemical terms. It is highly probable that specific molecules released from the damaged cells travel, presumably by diffusion, to the nearby vessels, since the early unilateral sticking of leukocytes regularly occurs on the side of the

vessel nearest to the site of injury (Fig. 22-6A). However, the chemical factors that transmit the "message of injury" from the originally damaged tissue cells to the surrounding blood vessels have never been defined.

FEVER

Most forms of fever, including those resulting from bacterial infections, are somehow related to inflammation. There is good evidence that a "messenger molecule," which is detectable in the blood during fever and which acts upon the thermoregulatory centers of the hypothalamus, comes from polymorphonuclear leukocytes participating in the inflammatory response. The release of this **leukocytic pyrogen** can be stimulated in vitro not only by exposure of the cells to endotoxin, to certain viruses, or to bacteria that they can phagocytize, but also to an "activating factor" that is present in exudate fluids.

Although there has been much clinical

speculation about the possible beneficial effects of fever, there is no evidence that it significantly augments antibacterial defense, except possibly in certain experimental pneumococcal infections and in a few specific diseases (e.g., syphilis and gonorrhea) in which the infecting organism is unusually heat-sensitive.

THE PHAGOCYTIC PROCESS

Attachment and Ingestion Phases. Phagocytosis can be separated into two phases. First a bond is formed between the particle and the surface of the phagocyte plasma membrane. This step requires divalent cations, has little

temperature dependency, and involves a variety of membrane receptors. For instance, the **Fc receptor** binds the Fc portion of opsonizing IgG and may play a role in the binding of **cytophilic antibody;** other receptors bind complement components and denatured proteins.

A particle attached to the membrane may initiate the **ingestion** phase: the particle is enveloped by an invagination of the plasma membrane, which pinches off and enters the cytoplasm as a **phagocytic vacuole** (Fig. 22-9, A and B). In general, particles coated or **opsonized** with antibody molecules are more rapidly and efficiently ingested both by granulocytes and by macrophages.

Fig. 22-9. A,B. Invagination of plasma membrane of leukocyte during phagocytosis. ×20,000, reduced. **C,D.** Degranulation of polymorphonuclear leukocytes resulting from phagocytosis of gram-positive bacilli. ×2000, reduced. **E.** Ingested cocci undergoing destruction in phagocytic vacuole of polymorphonuclear leukocyte. Cell walls of three of the organisms are partially or completely eroded. ×30,000, reduced. **F.** Fusion of membranes of leukocytic granules (G) with membrane of phagocytic vacuole (V) resulting in discharge of granular contents into vacuole containing phagocytized pneumococcus (P). ×20,000, reduced. [**A,B** from Goodman, J. R., Moore, R. E., and Baker, R. F. *J Bacteriol 27*:736 (1956). **C,D** courtesy Hirsch, J. G. **E** from Florey, H. W. *General Pathology.* Saunders, Philadelphia, 1962. **F** from Lockwood, W. R., and Allison, F. *Br J Exp Pathol 44*:593 (1963).]

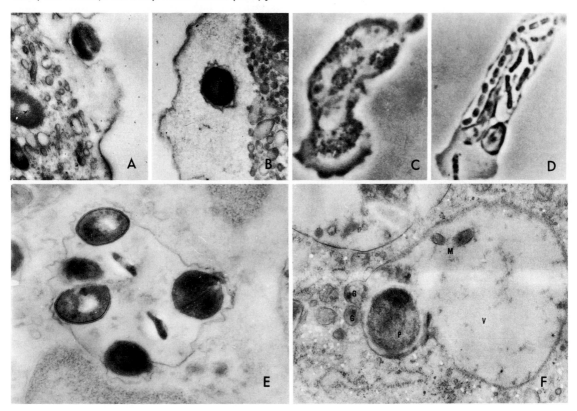

The membrane movements and interactions associated with ingestion depend upon cellular energy sources. In polymorphonuclear leukocytes the glycolytic pathway is of paramount importance, and iodoacetate and fluoride inhibit phagocytosis. Glycolysis depends on either exogenous glucose or the large store of glycogen present in the cytoplasm of leukocytes. Glycolysis may also be an important energy source in macrophages but can be supplemented or overshadowed by respiration and oxidative phosphorylation.

In both macrophages and leukocytes phagocytosis is accompanied by a burst of oxygen consumption, increased glucose utilization and lactate production, and a 10-fold increase in activity of the hexose monophosphate shunt pathway. Increased turnover of phosphatidic acid and other phospholipids has been reported. No net synthesis of lipid occurs in the leukocyte, but in the macrophage net synthesis of plasma membrane components takes place in proportion to the amount of membrane interiorized during the ingestion phase.

POSTPHAGOCYTIC EVENTS

Formation of the Phagolysosome. Cinemicrophotography shows that once the phagocytic vacuole has formed it migrates into the cytoplasm; there it collides with lysosome-like granules which explosively discharge their contents into the bacterium-containing vacuole. Electron microscopy has revealed that the membranes of both granule and vacuole actually fuse (Fig. 22-9F) resulting in a **phagolysosome** or digestive vacuole. The number of granules lost through this **degranulation** is directly related to the number of particles ingested. With very large particles, granules may discharge into an incompletely closed surface vacuole, thereby liberating their contents to the extracellular medium.

Intracellular Killing of Organisms. By 10 to 30 minutes after ingestion many pathogenic and nonpathogenic bacteria are inactivated: they fail to grow if placed on artificial media and often exhibit structural alterations (Fig. 22-9E). It is assumed that killing occurs within the phagolysosome as the result of exposure to contents of lysosomes.

Both lysosomal constituents and metabolic products contribute to the bactericidal activity in the leukocyte; little definitive information exists concerning the macrophage. As a consequence of the metabolic burst following phagocytosis, **lactic acid** is produced and the

pH within the phagocytic vacuole decreases. Rough estimates from indicator dyes suggest that the pH may drop to 4.0, which is a low enough to kill certain pathogens. Another consequence of the metabolic burst is the increased formation of **hydrogen peroxide,** which is also bactericidal; H_2O_2 results from the oxidation of both NADH and NADPH.

In addition to the products of metabolism, a number of components of the lysosome granule are potent antibacterial agents. **Lysozyme** (muramidase), an aminopolysaccharidase, is capable of hydrolyzing the mucopeptide components of susceptible bacteria, rendering them sensitive to osmotic shock. Gram-negative bacteria, which are normally resistant to lysozyme, are rendered susceptible by antibody plus complement. Granules of leukocytes (but not of macrophages) are also rich in a number of highly **cationic proteins** that damage the permeability barrier of both gram-positive and gram-negative bacteria. **Lactoferrin,** an iron-binding protein, has been identified in granules and the apoprotein has bacteristatic activity.

Klebanoff has described an important bactericidal **halogenating** system which utilizes both a granule constituent, **myeloperoxidase,** and a metabolic product, hydrogen peroxide. Working best at a slightly acid pH, peroxidase, peroxide, and halide ions result in the halogenation and killing of bacteria and viruses. The use of radioactive iodine and radioautography in the intact leukocyte indicates that the surface of the microorganism is iodinated (presumably on tyrosine residues) within the phagocytic vacuole. The peroxide may be produced by the leukocyte itself or by the ingested organism.

Many species of bacteria, particularly those that produce chronic diseases, are not killed within phagocytes. Organisms such as the tubercle bacillus may even multiply within the phagolysosome and eventually destroy the cell. Others may multiply only slowly enough to maintain dormant foci persisting for months or years. The latent infections resulting from such indolent states may explain the greatly prolonged immunity that frequently follows tularemia or brucellosis.

Intracellular Digestion. Studies with [14]C- and ^{32}P-labeled bacteria show that dead microbes are rapidly degraded in phagolysosomes to low-molecular-weight components by diverse hydrolytic enzymes (proteases, peptidases, nucleases, phosphatases, lipases, carbohydrases). All these enzymes have maximum activity at the slightly acid pH that prevails in the digestive vacuole. Certain bacterial polysaccharides and complex waxes and lipids are only slowly digested and may be retained within macrophages for weeks or months.

During the maturation of the neutrophilic and heterophilic leukocytes, two types of granules are formed; both can discharge their contents into phagocytic vacuoles. The earliest visible granule has been termed the **azurophil** or **primary** granule; it contains typical lysosomal enzymes. Later in maturation the **secondary** granule becomes the predominant granule of the mature cell: it is smaller, less dense, and contains alkaline phosphatase, lactoferrin, and a portion of the lysozyme.

Both granules are derived from the Golgi apparatus, where newly synthesized enzymes and cationic proteins are packaged within Golgi membranes. The adult leukocyte when released from the marrow contains a full complement of granules, and no further synthesis of antibacterial factors occurs in the tissues.

The formation of macrophage lysosomes occurs by a different mechanism. When released from the marrow the monocyte has a few granules similar in composition to the azurophil granule of the leukocyte. Most of its digestive enzymes are produced at the macrophage stage in the tissues. In response to the pinocytosis or phagocytosis of digestible material, hydrolases are synthesized in the rough endoplasmic reticulum and packaged into tiny Golgi vesicles. These vesicles, which correspond to **primary lysosomes** fuse with the phagocytic vacuole, converting it into a **secondary lysosome** or **digestive body**. Extensive and repeated bouts of enzyme synthesis may take place in response to the endocytic activity of the cell.

Eosinophils. Eosinophilic granulocytes tend to accumulate in exudates caused by allergic reactions, and they are attracted in vitro by antigen-antibody complexes, particularly in the region of antigen excess. Moreover, they ingest such complexes, and in so doing they become degranulated. Since their granules seem to contain substances capable of blocking the action of histamine, serotonin, and bradykinin, all of which are involved in allergic inflammation, it has been proposed that eosinophils protect the tissues of the host not only by phagocytizing and degrading cytotoxic antigen-antibody complexes but also by damping the effects of chemical mediators of the inflammatory response.

HUMORAL FACTORS IN ANTIBACTERIAL DEFENSE

Since antibodies have been described in previous chapters, only points particularly pertinent to host-parasite relations will be considered here. Antibodies to surface and extra-cellular antigens will be given special emphasis because these antigens relate primarily to virulence.

Antitoxins. In diseases such as tetanus or diphtheria, which are due primarily to tissue damage caused by specific exotoxins, it is natural that toxin-neutralizing antibodies should modify the course of the illness. These antitoxins are effective unless administered too late (i.e., after the toxin has become fixed to its target tissues) or in inadequate amounts.

Bacteriolytic Action. Antibodies to cell wall antigens of gram-negative bacilli cause bacteriolysis in the presence of complement.* Highly specific in their action, they affect only bacilli possessing the immunizing antigen itself or one cross-reacting with it. Their role in bacterial diseases, however, remains obscure. In typhoid fever, for example, anti-O antibodies which are capable of lysing *Salmonella typhosa* in the presence of complement are regularly found in the sera of patients during the second week of illness. Since the disease at this stage, if untreated, is usually in full progress, it seems that the mere presence of these circulating antibodies ordinarily is not sufficient to bring the illness under control.

Agglutinating and Opsonizing Actions of Antibody. The significance of agglutinating and opsonizing antibodies is firmly established,

---

* Antibodies to gram-positive bacteria are not bacteriolytic in the presence of complement. In whole blood, however, they may promote the killing of bacteria by leukocytes through their action as opsonins (see below).

particularly in acute bacterial diseases. In pneumococcal pneumonia, for example, agglutination of pneumococci by antibody in the edema-filled alveoli tends to immobilize the invading bacteria and thus stop the spread of the pneumonic lesion. Similarly, antibodies specifically accelerate phagocytosis by combining with antiphagocytic capsular components and rendering them noneffective. By these two actions the antibodies may bring about a termination of the disease. It is this effect that tends to make the disease acute; for if the host can survive long enough to mobilize a sufficient amount of protective antibody, the resulting destruction of the bacteria will end the illness within a few days. Such a result will obviously occur only with extracellular parasites which are readily killed by phagocytes.

Natural Antibodies. When specific antibodies are found in the serum of a host not known to have been previously exposed to the corresponding antigen, they are often termed **natural antibodies,** to distinguish them from **acquired antibodies** (Ch. 14). This distinction, however, is probably artificial in many instances because of the possibility of an unrecognized contact with the corresponding antigen or a closely related one. For example, more than 80% of human subjects over 1 year of age have antibody to type 7 pneumococcus, although the carrier rate for this organism is extremely low (less than 1%). This antibody apears due to exposure to certain streptococci of the common viridans group which possess an antigen that cross-reacts with type 7 pneumococcal polysaccharide. Similarly, natural antibodies to gram-negative bacilli may result from continuous exposure of the host to its own fecal flora. Natural antibodies may also result from contact with foods, plants, and inhalants. In infancy they may be acquired passively both from the maternal circulation before birth and from the colostrum during early suckling.

Significance. The capacity of natural antibodies to prevent bacterial disease is not established. It seems possible that these antibodies account for the great variation in susceptibility of different animal species to experimental pneumococcal disease. Thus, natural antibodies to pneumococci are present in serum of resistant species (pigeons, chickens, dogs, cats) and lacking in serum of susceptible ones (mice, guinea pigs, rabbits).

Studying the unusual susceptibility of mice to *Salmonella typhimurium,* Jenkins and Rowley found 1) that the infecting organism and the tissues of the mouse possess common antigens and 2) that the serum of the mouse contains no natural antibodies to the typhimurium bacilli. It was thus suggested that the absence of protective antibodies could be due to the host's immunological tolerance of a self antigen that happens to be a part of the bacterium and important for its invasiveness. Whether such an explanation can be applied to other examples of natural susceptibility to bacterial diseases remains to be determined. It seems likely that nonhumoral mechanisms of resistance (see below) are also involved.

Other Heat-Labile Humoral Factors. In addition to low-titer specific antibodies, other heat-labile factors in normal plasma may play a role in antibacterial immunity. None of these, however, have been clearly defined. One that has received particular attention is the **nonspecific phagocytosis-promoting factor** (PPF), which stimulates in vitro phagocytosis of a wide variety of bacteria. Recent evidence indicates that it may be all or part of the complement system which becomes fixed to the bacterial surface either spontaneously or by interaction with natural antibodies that by themselves are ineffective as opsonins.

TISSUE STRUCTURAL FACTORS IN ANTIBACTERIAL DEFENSE

Surface Phagocytosis. Many agents of acute bacterial disease such as *Diplococcus pneumoniae, Streptococcus pyogenes,* and *Hemophilus influenzae* possess capsules which protect them from phagocytosis in the absence of antibody; yet there are often no homologous antibodies detectable in the serum before the fifth or sixth day of illness. Survival of the host during the preantibody phase of the infection depends on the cellular defense mechanism known as **surface phagocytosis.**

Phagocytosis of encapsulated bacteria in vivo in the absence of opsonizing antibody was first demonstrated in experimental lesions produced with type 1 pneumococcus and with *Klebsiella pneumoniae.* In both subcutaneous and pulmonary lesions phagocytosis was evident within the first few hours of infection, though no specific antibody could be found either in the blood or in the local tissues. Subsequent in vitro investigations revealed that this form of phagocytosis resulted from a trapping of the encapsulated bacteria by the leukocytes against tissue surfaces (Fig. 22-10, A to E) or between adjacent leukocytes in concentrated exudates (intercellular surface phagocytosis; Fig. 22-10, F to H) or in the interstices of fibrin clots. In vivo studies with the rabbit-ear chamber (Fig. 22-6) have shown that all three of these phagocytic processes occur in the blood stream during experimental bacteremia. In addition they have been observed in monocytic as well as in granulocytic exudates. They become operative almost as soon as healthy phagocytes and encapsulated bacteria come together in the tissues.*

There is, in fact, considerable experimental evidence that the eventual outcome of untreated acute bacterial infections may be settled within the first few hours. If the rate of phagocytosis exceeds the rate of multiplication of the bacteria, relatively prompt recovery is likely to ensue (Fig. 22-11, left). When on the other hand, phagocytosis does not keep pace, the proliferation of microorganisms may prove fatal (Fig. 22-11, right). The early preantibody phase of the illness, therefore, is often critical.

Architecture of Tissues. The efficiency of the preantibody phagocytic defense is greatly influenced by the general architecture of the tissues involved. In potentially "open" cavities (pleural, pericardial, peritoneal, joint, and meningeal), where the leukocytic exudate is initially diluted with large amounts of fluid, surface phagocytosis is much less efficient than in tissues of tightly knit structure, where the exudate tends to be more concentrated and where tissue surfaces are more accessible to the leukocytes. For this reason, striking differences may occur in the severity of acute diseases produced by the same microorganism in different tissues of the body: the case fatality rate in untreated pneumococcal pneumonia is about 30%, whereas in untreated pneumococcal meningitis it is over 99%.

_____

* Very heavily encapsulated bacteria, such as type 3 pneumococci, are especially virulent because they resist surface phagocytosis temporarily until some of the "outer slime layer" of their capsules has been shed.

Necrosis and Abscess Formation. Another condition that profoundly affects antibacterial defense is necrosis. Whether caused by bacterial products or by such factors as trauma or vascular occlusion, necrosis of infected tissue eventually leads to suppuration and abscess formation. In acute abscesses the exudate is predominantly granulocytic; in chronic abscesses (e.g., tuberculous) it is largely monocytic. In the walled-off cavities of both, however, the phagocytic cells are mostly inactive and in various stages of disintegration; furthermore, they provide a medium in which the infecting microorganisms multiply very little but survive for long periods. Although abscesses of this kind sometimes heal spontaneously, they usually require drainage to remove the "dead" exudate and "resting" bacteria.

CHEMICAL FACTORS IN ANTIBACTERIAL DEFENSE

Many substances with antibacterial properties in vitro can be extracted from animal tissues and body fluids, but their significance for antibacterial immunity is not established.

The antimicrobial activities of the polyamine spermine and its oxidation product spermidine have already been mentioned, as have those of lysozyme, phagocytin, the bactericidal histones in leukocytes, and the antibacterial fatty acids present in the skin. Two basic polypeptides—one high in lysine content, the other in arginine have also been recovered from animal tissues. The first inhibits the in vitro growth of _Bacillus anthracis, Staphylococcus aureus, Streptococcus hemolyticus,_ and _Escherichia coli;_ the second is active against _Mycobacterium tuberculosis._ Heat-stable β-lysins which are bactericidal to certain gram-positive organisms (aerobic spore formers and some micrococci) are found only in serum separated from either clotted whole blood or clotted platelet-rich plasma, i.e., their formation results from the clotting of plasma in the presence of platelets. Bactericidal substances have also been extracted from leukocytes (leukins), platelets (plakins), and red cells (hematin and mesohematin). Besides the long-chain fatty acids in the skin, short-chain acids such as lactic acid act primarily by lowering pH. Certain dicarboxylic and keto acids may either stimulate or depress bacterial growth, depending upon their concentrations and the species involved. Many other chemical constituents of tissues and body fluids can undoubtedly influence the growth of pathogenic microbes in vitro.

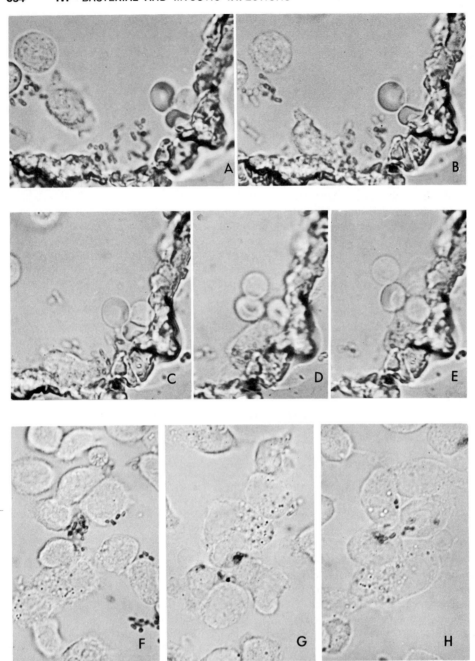

Fig. 22-10. A–E. Surface phagocytosis of encapsulated Friedländer's bacilli suspended with leukocytes in an antibody-free medium and applied to a cut section of normal rat lung. Leukocyte can be seen to trap organisms against alveolar wall in the process of ingesting them. Action photographed took place in a period of 10 minutes. ×1250, enlarged. **F–H.** Intercellular surface phagocytosis of encapsulated Friedländer's bacilli in medium devoid of antibody. Leukocytes in this more concentrated suspension phagocytize organisms by trapping them against the surfaces of adjacent leukocytes. Elapsed time, 30 minutes. ×1250, enlarged. [**A–E** from Smith, M. R., and Wood, W. B., Jr. *J Exp Med 86:*257 (1947). **F–H** from Smith, M. R., and Wood, W. B. Jr. *Science 106:*86 (1947).]

Fig. 22-11. Serial peritoneal smears made from mice infected with avirulent T14 (**left**) and virulent T14/46 streptococci (**right**). Note phagocytosis of T14 organisms by both monocytes and granulocytes in early stages of the infection. Whereas the T14 infection rarely killed a mouse, the T14/46 infection was uniformly fatal. ×1100. [From Foley, M. J., Smith, M. R., and Wood, W. B., Jr. *J Exp Med 110:*603 (1959).]

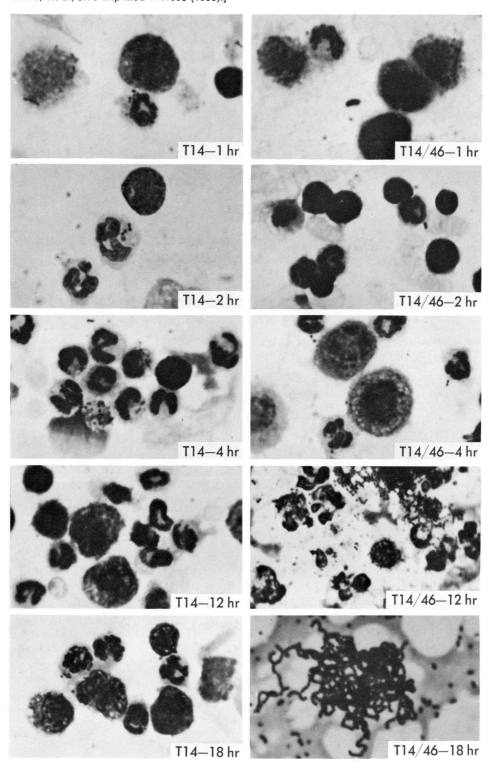

ROLE OF HYPERSENSITIVITY IN BACTERIAL DISEASES

A delayed type of hypersensitivity originally referred to as "bacterial allergy," may develop during the course of bacterial disease. It causes both an accelerated and an exaggerated tissue response to the infecting organism. Like other forms of delayed hypersensitivity, it is not associated with demonstrable serum antibody. However, it can be transferred by lymphocytes with antigen-binding receptors (probably antibodies) in the cell membrane. Combination of microbial antigens with the receptors causes the lymphocytes to release a variety of substances that produce inflammation and necrosis.

Delayed hypersensitivity of this sort was first demonstrated by Koch's classic tuberculosis experiments in 1891. Guinea pigs which had been infected for more than 2 weeks were found to react quite differently from uninfected animals when injected subcutaneously with virulent living tubercle bacilli. In the normal animals the injection resulted, within 10 to 14 days, in 1) for formation of a diffuse nodule, which eventually formed an indolent ulcer, and 2) a simultaneous spread of the infection to the regional lymph nodes. In contrast, the previously infected guinea pigs, a greatly accelerated response, leading to prompt necrosis, occurred within 2 days at the site of injection and extension of the infection to the regional nodes either was delayed or did not occur. Thus the tissues of the previously infected animals reacted more violently to the injected bacteria but tended to localize the infection. Koch demonstrated that the hyperreactive response in the previously infected animals could also be elicited by injections of dead tubercle bacilli or of "tuberculin," a protein fraction which he extracted from the bacilli.

ALLERGY VS. IMMUNITY

Discovery of the Koch phenomenon engendered a prolonged argument about whether the allergic reaction benefits the host. Its tendency to control lymphatic spread of infection appears to be beneficial. But its tendency to cause pulmonary cavitation, erosion of bronchi by caseous lymph nodes, and "internal ulceration" of cerebral tubercles into the subarachnoid space is obviously detrimental. Clearly tuberculin sensitivity either helps or harms the host, depending upon the circumstances. The ulcerating hyperreactive

skin lesions in Koch's guinea pigs represent a very special case, where the primary lesions rupture harmlessly to the exterior and then heal. When ulcerating lesions of this kind rupture **internally,** discharging virulent bacilli into such sites as the leptomeninges, the result may be lethal. Though often associated, allergy and immunity in tuberculosis are certainly not synonymous.

Similar reactions of delayed hypersensitivity are encountered in other bacterial diseases, including brucellosis, leprosy, and glanders, as well as in fungal, viral, and parasitic* diseases.

Delayed skin reactions are also frequently demonstrable during (or following) pneumococcal and streptococcal infections. Streptococcal allergy, which is relatively common, can be readily transferred with lymphoid cells from hyperreactive donors. There is, however, no convincing evidence that delayed hypersensitivity plays a significant role in the pathogenesis of either pneumococcal or pyogenic streptococcal diseases, although allergy to streptococcal erythrogenic toxins may be responsible for the rash in scarlet fever. In staphylococcal disease, on the other hand, study of experimental infections in laboratory animals has revealed that hyperreactivity to antigens, presumably released by the organisms, may be correlated with the amount of tissue necrosis.

ACQUIRED ANTIBACTERIAL CELLULAR IMMUNITY

During chronic infections with "intracellular" parasites, such as tubercle bacilli, *Listeria, Brucella,* and certain *Salmonella,* many laboratory animals develop enhanced cellular immunity at about the time they develop delayed hypersensitivity. If the animal is cured and then reinfected with the same organism, an accelerated enhancement of cellular immunity can be demonstrated. The enhanced immunity can be transferred with lymphocytes (from spleen, thoracic duct), but it is expressed by macrophages of the infected or adoptive host; it is not due to freely circulating antibodies.

Studies by Mackaness with *Listeria monocytogenes* established the main characteristics

---

* According to accepted medical terminology, **parasitic diseases** are illnesses caused by metazoan parasites.

of this important defense mechanism. Macrophages from a normal mouse are destroyed by *Listeria,* which grows rapidly in these phagocytes. In contrast, macrophages from an immune animal efficiently destroy ingested bacilli: the enhanced bactericidal capacity of the macrophage may be associated with the greater number of lysosomes in "immune macrophages." Once modified, the macrophage or effector cell acts nonspecifically: its augmented bactericidal properties are exerted against a variety of antigenically unrelated bacteria (e.g., *Listeria, Salmonella, Brucella*).

However, the **induction** of the activated macrophages **is** immunologically specific, and only the initially infecting organism or antigenically related ones are able to evoke the accelerated secondary response observed with reinfection. It is likely that 1) initial infection induces proliferation of specifically reactive lines of lymphocytes, 2) specific binding of the infecting organisms' antigens to the lymphocyte surface causes release of factors (Ch. 18) that activate macrophages and increase their lysosomal content and phagocytic activity, and 3) once activated, the "angry" macrophages ingest and destroy a wide variety of foreign cells, bacterial or other. Although this form of immunity has been studied only in laboratory rodents, it probably plays an important role in human infections and in the recovery from tuberculosis and brucellosis.

VARIABILITY OF HOST DEFENSE

Resistance to bacterial disease may be either augmented or depressed by a variety of circumstances.

AUGMENTATION

The most effective enhancement procedure available to the physician is **specific immunization** through the use of vaccines or specific antisera. Active immunity produced by vaccination ordinarily lasts for months or years, depending on the immunizing agent; passive immunity resulting from injection of antiserum is rarely demonstrable for more than a few weeks. The limitations of active immunization include the diversity and mutability of antigenic types within a given species, the difficulties involved in preparing nontoxic

stable vaccines containing the critical protective antigens, and the length of time required for the development of active immunity. Currently used immunization procedures will be mentioned in later chapters dealing with specific bacteria.

Some means of enhancing **nonspecific immunity,** such as stimulation of phagocytosis, would obviously be of great value. However, no practical means of achieving this end is known. In appropriate doses bacterial endotoxins enhance the phagocytic capabilities of leukocytes, but they cause a number of unpleasant side reactions, including extreme malaise and fever; in addition, the heightened resistance they induce is relatively short-lived.

Prior mobilization of an inflammatory exudate also increases local nonspecific resistance. For example, if sterile inflammation is artificially induced with a chemical irritant in a pleural or peritoneal cavity, subsequent infection of the same site will be more readily controlled. Although this procedure, like the injection of endotoxin, is obviously impractical, it too illustrates the principle that an upgrading of cellular defense may limit bacterial invasion.

DEPRESSION

A wide variety of conditions may impair antibacterial immunity. **Circulatory disturbances,** both local (causing ischemia or congestion and edma) and general (as in shock), will often interfere with the mobilization and functioning of phagocytic cells, as may **mechanical obstruction** of drainage systems of the body, such as the biliary or urinary tracts. Excessive intake of **ethyl alcohol** has also been shown experimentally to depress the inflammatory response to bacterial infection. **Nutritional deficiency,** as illustrated so often in history by the tragic association of famine and pestilence, may profoundly influence the resistance of large populations to bacterial diseases. Some of the nutritional factors involved are remarkably specific (e.g., the lack of a single amino acid in the diet). Evidence has been advanced by Dubos and Schaedler that the **effect of diet** on immunity may be mediated in part through its action on the bacterial flora of the gastrointestinal tract. As expected from the fact that antibodies are immunoglobulins, individuals with **agammaglobulinemia** or **dysgammaglobulinemia** are particularly prone to bacterial infections. Other **chronic debilitating diseases,** as well as cer-

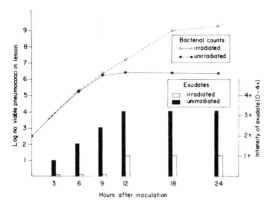

Fig. 22-12. Relation of growth of pneumococci to intensity of exudate in myositis lesions (see Fig. 22-13) of irradiated (650 r) and unirradiated mice. [From Smith, M. R., Fleming, D. O., and Wood, W. B., Jr. *J Immunol* 90:914 (1963).]

tain **acute viral illnesses** (e.g., influenza, Ch. 56), may likewise depress antibacterial immunity. **Hormonal factors** are illustrated by the frequency with which diabetics and patients receiving adrenal steroids are afflicted with both acute and chronic bacterial diseases. How such endocrine imbalances influence the resistance of the host is not altogether clear, although both diabetic acidosis and large doses of cortisone are known to depress inflammation, and high glucose levels in acute exudates have been found to impair phagocytosis. Even less clear is the manner in which **physical fatigue** and overexposure to cold apparently render individuals temporarily susceptible to bacterial illnesses, particularly of the respiratory tract. The latter effect may somehow relate to seasonal variations in host resistance.

Acute Radiation Injury. Compared with other depressants of antibacterial immunity, **acute radiation injury** is perhaps the simplest to understand. Its principal effects on the cellular defenses of the host are illustrated by the following laboratory observations.

When adult white mice are exposed to 650 r of X-irradiation, there occurs an aplasia of

Fig. 22-13. Pneumococcal myositis lesion of unirradiated mouse (**A**) sacrificed at 18 hours after inoculation, compared with that of irradiated mouse (**B**) with severe leukopenia. Note sparcity of granulocytic exudate and relatively unimpeded growth of bacteria in the lesion of the irradiated animal. ×1125. [From Smith, M. R., Fleming, D. O., and Wood, W. B., Jr. *J Immunol* 90:914 (1963).]

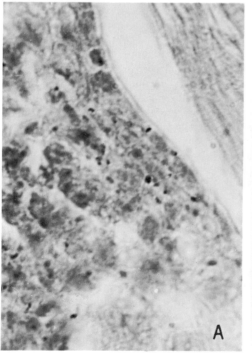

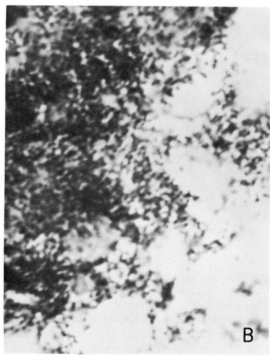

the bone marrow, resulting in profound granulocytopenia within 72 hours. Although other body cells are also affected, those of the marrow are particularly vulnerable because of their high turnover rates. Since the mean half life of granulocytes in the circulation is only 6 to 12 hours,* the fall in peripheral white cell count is precipitous (Fig. 22-14). During the stage of maximum depression of the marrow, normal inhabitants of the nasopharynx and gastrointestinal tract invade the blood stream with great regularity, and the resulting infections are frequently fatal.

If the tissue defenses are challenged with a pathogenic microbe, such as pneumococcus type 1, in normal animals an acute inflammatory response tends to control local multiplication of the organisms. However, the irradiated animals during the leukopenic phase fail to mobilize an inflammatory exudate, and the bacteria quickly overrun the lesion (Figs. 22-12 and 22-13). In fact, the maximum population density of pneumococci reached in the lesions is approximately 1000 times greater in the irradiated mice. Indeed this density is inversely related to the number of leukocytes in the circulation at the time of inoculation. As is shown in Figure 22-14, this correlation obtains both during the early phase of the radiation injury, when the peripheral white cell count is falling, and during recovery, when the count is returning to normal. Depressed mobilization of the exudate is the critical factor:† the leukocytes that are eventually mobilized have normal phagocytic and bactericidal capabilities, and there is no evidence that antibacterial serum factors (e.g., natural antibodies) have been affected.

These results dramatically illustrate the dependence of the host upon its granulocytic

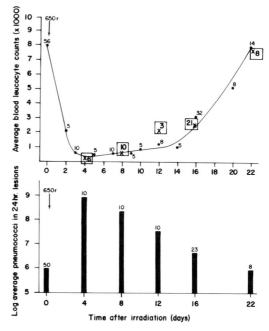

Fig. 22-14. Inverse relation between postirradiation blood leukocyte counts (upper curve) and number of viable pneumococci in 24-hour experimental myositis lesions (bar graph, below) during recovery of mice from acute radiation injury (650 r). Points within squares represent average preinfection leukocyte counts done on animals in which myositis was produced. Other intermediate points indicate average postirradiation counts made on uninfected mice. Numeral above each point refers to number of mice on which blood counts were made. [From Smith, M. R., Fleming, D. O., and Wood, W. B., Jr. *J Immunol* 90:914 (1963).]

defense. Although cells of the reticuloendothelial system may also be damaged by this amount of radiation, the main effect is certainly on the production of granulocytes. Lacking their help, the monocytic cells, which appear to retain much (if not all) of their phagocytic capabilities, cannot by themselves prevent bacterial invasion. This fact is further borne out by the frequency with which bacterial diseases complicate other forms of agranulocytosis, such as those resulting from drug idiosyncrasies or malignancy.

GENETIC DEFECTS OF LEUKOCYTE FUNCTION

Newer methods for studying leukocytes have illuminated the mechanisms underlying cer-

---

* The following facts pertinent to the physiology of granulocytes have been established by radioactive tracer experiments: 1) Maturation from stem cells in the marrow takes 2 to 3 days. 2) Only half of the granulocytes (25×10^9) in the human circulation are detectable in peripheral blood; the rest are marginated on vessel walls. 3) For every circulating granulocyte approximately 100 mature cells are held in the marrow reserve pool. 4) Once a granulocyte enters the tissues or passes into the bowel or respiratory tract, it never returns to the circulation; in contrast to the monocyte, it is an expendable "end cell."

† Its relevance to the bacterial aftereffects of a nuclear weapon disaster is only too obvious.

tain leukocyte defects. The most extensively studied is a fatal disease of childhood—chronic granulomatous disease—genetically transmitted and associated with repeated, widespread infections with *Staphylococcus aureus* and gram-negative organisms. Intraleukocytic bactericidal mechanisms are impaired, although monocytes may also be affected. Leukocytes from such patients phagocytize normally, but fail to kill these bacteria; yet they kill pneumococci and streptococci quite readily. Unlike normal granulocytes, these cells fail to exhibit a postphagocytic burst of oxygen consumption and glucose 1-C oxidation ("shunt" pathway); most important, they fail to produce significant quantities of hydrogen peroxide, which limits the activity of the myeloperoxidase-halogenation system described above. This presumably explains why peroxide-producing bacteria are destroyed by these granulocytes, whereas nonproducing strains remain viable. If these cells are allowed to ingest an inert particle coated with a peroxide-generating enzyme (glucose oxidase), partial correction of both the metabolic and the bactericidal defect is seen. It is likely that other heritable defects associated with either leukocytes or mononuclear phagocytes will be uncovered in the future.

INAPPARENT BACTERIAL INFECTIONS

The balance between bacterial infection and disease is a precarious one. Man, like other animals, is always infected by many species of bacteria, almost any one of which is capable of producing disease under those circumstances discussed above that depress antibacterial defense. These include circulatory disturbances, trauma, alcoholic excesses, nutritional deficiencies, radiation, injury, bone marrow depression due to drugs, and doubtless other as yet undefined factors.

The mere lodgment of bacteria on a body surface does not necessarily lead to infection. Surface bacteria are often washed off or simply fail to survive for any length of time at the site of lodgment. Only if they set up a reasonably permanent residence in the host are they considered to have established an infection. The skin and mucous membranes are the usual sites of such infections. The healthy individual who harbors *Streptococcus pyogenes* in the mucous membranes of his pharynx is referred to as a **streptococcus carrier.** Similarly, a patient recovered from typhoid fever who continues to excrete typhoid bacilli in his feces is called a **typhoid carrier.** Approximately 50% of normal adults are nasal carriers of pathogenic staphylococci.

Since the carrier is in no sense ill, a state of peaceful coexistence has been established between parasite and host. However, this relation is often unstable. An individual may carry a given organism in his throat for a number of weeks, only to be rid of it later. Under epidemic conditions the carrier rate for the causative microbe may approach 100%, even at a time when the morbidity rate is comparatively low. Some forms of bacteria are almost always present, e.g., viridans streptococci in the throat, colon bacilli in the gastrointestinal tract, and staphylococci on the skin; but they can change from time to time in both number and kind.

DORMANT AND LATENT INFECTIONS

Carrier states are often classified as **dormant** infections because the bacteria involved can be detected by appropriate cultural technics. In some infections, however, the microbes cannot be demonstrated by any of the methods presently available. These **latent infections** can be recognized only in retrospect, by the emergence of overt illness. They occur most commonly as "silent" or "quiescent" stages of chronic bacterial diseases, such as tuberculosis and syphilis. The frequency of relapses following the quiescent phases of these diseases clearly indicates that the organisms persist in the tissues, though they can be neither seen nor cultured.

This distinction between dormant and latent infections, based solely on detectability of the microbe, is arbitrary. Nevertheless, it serves a useful purpose, if only to remind us of the limitations of our laboratory methods. That a given microbe will not grow in an artificial culture medium does not indicate that it is nonviable—only that it is not cultivable in that particular medium. If subcultured on a more suitable artificial medium or inoculated into the

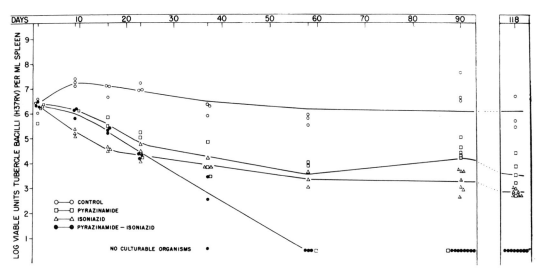

Fig. 22-15. Influence of pyrazinamide and isoniazid, used singly and together, on populations of tubercle bacilli in mouse spleens during 118 days of therapy. The infecting inoculum was 2.0×10^6 culturable units of tubercle bacilli. Only data relating to the spleen are charted, since extensive studies showed the bacterial counts to be consistently highest in this organ. [From McDermott, W. *Public Health Rep* 74:485 (1959).]

tissues of a particularly susceptible animal host, it might multiply and thus be detected. The likelihood of its growing in the new medium will depend upon its metabolic state at the time of transfer. Thus tubercle bacilli that have "slumbered" for long periods in latent tuberculous lesions will sometimes grow out in artificial media only after months, rather than weeks, of incubation. Although it has been suggested that such sluggish organisms may persist in the tissues as L forms, the true basis of their strange behavior is not known.

The failure to see tubercle bacilli in stained sections of a tuberculous lesion is likewise no proof that they are not there. Several thousand bacteria per milliliter must be present if organisms are to be found by ordinary microscopy. Hence the distinction between dormant and latent infections may in reality be only a quantitative one.

The principles involved in inapparent bacterial infections are well illustrated by studies on experimental tuberculosis in mice, in which pyrazinamide and isoniazid converted systemic disease into either a dormant or a latent infection (Fig. 22-15). When either drug was given alone the population of cultivable tubercle bacilli in the tissues steadily fell for some weeks and then stabilized at a low level, where it persisted as a dormant infection for the remaining months of treatment. In contrast, combined therapy for 60 days resulted in the complete elimination of cultivable tubercle bacilli. Following a 90-day treat-

ment-free interval, however, tubercle bacilli again became cultivable from the tissues of over a third of the mice, and when cortisone was given during the posttreatment period the subsequent cultures were positive in virtually all the mice. Thus, it is clear that either dormant or latent infections may be induced experimentally when chronic bacterial diseases are treated with antimicrobial drugs.

Conversely, rats treated with cortisone died with extensive caseonecrotic consolidation of the lungs due to *Corynebacterium pseudotuberculosis muris*. Yet despite repeated attempts it was never possible to demonstrate that the rats were infected with corynebacteria before they received the cortisone.

Immunological Consequences. The possible relation between prolonged immunity and the persistence of bacteria in the tissues has been mentioned above. Skin tests for delayed hypersensitivity (e.g., tuberculin tests) are frequently positive in inapparent infections. Serum antibodies may also be demonstrable. Not all inapparent infections, however, can be recognized by immunological methods: the "antigenic mass" of the few organisms present may be insufficient to sustain an antibody titer detectable by even the most sensitive technics.

COMMUNICABILITY OF BACTERIAL INFECTIONS

Epidemiology, the study of disease in populations, deals with disease in general and therefore is taught separately from microbiology in most schools of medicine and public health. Some basic principles will be briefly reviewed here; points relating to the epidemiology of specific infectious diseases will be discussed in later chapters. In general, factors responsible for communicability from host to host are difficult to study, for the endpoint is the acquisition of an infection, which may be inapparent.

To be **naturally pathogenic** for a given animal species a bacterial strain must be readily transmissible to a susceptible individual; if pathogenic but not transmissible it will cause disease only if artificially inoculated. For example, mice are notoriously susceptible to inoculated pneumococci, but even when dying of infection of the respiratory tract they rarely transmit pneumonia to normal cage-mates. Conversely, bacteria avirulent for a given species (or strain) of host may be highly communicable yet fail to cause overt illness. Clearly, virulence and communicability do not always go together. When bacterial strains possessing both these properties become prevalent in a susceptible population they are likely to produce an epidemic.*

Bacteria may be transmitted from one human subject to another either directly or through some intermediate agent. **Direct transmission** usually results from contact with secretions emitted by an infected person during sneezing, coughing, speaking, or breathing. Very rarely a mother may transmit a systemic bacterial disease, such as typhoid fever, transplacentally to her unborn child. **Indirect transmission** may be through infected food, milk, or water; from contact with infected clothing, eating utensils, or other ob-

jects, collectively referred to as **fomites;** or via infected air **droplets,** their dried-out residues known as **droplet nuclei,** or resuspended particles of **dust.** In the latter category, the sizes of the airborne particles are often critical, since only those of relatively small diameter (ca. 5 μ or less) will penetrate the deepest recesses of the respiratory tract. A few bacterial infections are transmitted indirectly by **arthropod vectors** such as flies or ticks.

Efficient transmission depends upon the following identifiable factors:

1) There must be a ready **source** of the infecting agent. When it is restricted to individuals who are actively ill or recently convalescent isolation measures are effective. However, healthy carriers are responsible for more spread of many bacterial diseases (e.g., meningococcal meningitis) than are ill patients. The natural **reservoir** may also be an animal host, humans being infected only incidentally, either by direct contact or through an intermediate vector. An example is tularemia, which man acquires by direct contact with wild rabbits or via an infected tick or fly.

2) The source must release relatively **large numbers of organisms.** "Open" lesions will disseminate more bacteria than "closed" lesions. Thus bubonic plague, which is transmitted from rat to man by fleas, is characterized by closed infections of lymph nodes and of the blood stream, and in this form it is not transmitted from man to man. Occasionally, however, the lungs become involved, and the open pulmonary lesions then make the disease (pneumonic plague) highly contagious. Similarly, individuals harboring hemolytic streptococci in their noses are more dangerous carriers than those in whom the streptococci are confined to the pharynx; and nursery epidemics of staphylococcal infections have been traced to "cloud babies," who emit veritable clouds of staphylococci into their surroundings.

3) The infecting microbe must be **capable of surviving in transit to a new host**—whether transported by droplet, dust, fomite, food, water, milk, or insect vector. Indeed, mere survival in transit does not guarantee infectiousness. Viable group A streptococci, for example, in floor dust do not infect human

---

* The term **epidemic** is used to refer to a peak in the oscillating incidence of a disease. The increase in incidence needed to justify the use of the term is arbitrary and is usually related to the **endemic** rate of the disease in question and to the degree to which it is feared in the community. The endemic rate refers to the lower, more or less constant incidence in the population. The term **pandemic** is reserved for epidemics that are unusually widespread.

volunteers even when large inocula are instilled into the nasopharynx; direct contact with moist secretions from an infected subject appears to be necessary. Similarly, when suspensions of *Pasteurella tularensis* are disseminated in aerosols, the decay rate of infectiousness exceeds that of viability. The mere cultivation of a microorganism from air or dust, therefore, does not necessarily indicate that it is infectious, for phenotypic changes in an unfavorable environment may depress pathogenicity without destroying viability. Tubercle bacilli in dried droplet nuclei, on the other hand, are infectious.

4) For rapid spread a relatively high proportion of the population must be **susceptible.**

Hence to curtail the spread of a communicable bacterial disease in a given population, it is not necessary to immunize every individual at risk. The statistical principles underlying this important fact are discussed in standard textbooks of epidemiology.

It follows that the **frequency of contacts** between susceptible and infected individuals will be an important factor in determining how fast infection spreads. Crowding may thus constitute a critical term in the cross-infection "equation." This factor undoubtedly contributes to the high rates of respiratory disease during the colder months of the year, when persons are crowded together indoors.

APPENDIX

COMPUTATION OF 50% ENDPOINTS (LD$_{50}$ OR ID$_{50}$) BY METHOD OF REED AND MUENCH*

In biological quantitation, the endpoint is usually taken as the dilution at which a certain proportion of the test animals react or die. A 100% endpoint is still occasionally used, but because it approached asymptotically it is difficult to measure with precision. The most desirable endpoint is one in which half of the animals react and the other half do not. To avoid the costly use of large numbers of animals at many test dilutions, Reed and Muench devised a simple method for estimating 50% endpoints.

---

* From E. H. Lennette. General Principles Underlying Laboratory Diagnosis of Virus and Rickettsial Infections. In *Diagnostic Procedures of Virus and Rickettsial Disease,* p. 45. (E. H. Lennette and N. J. Schmidt, eds.) American Public Health Association, New York, 1964. Also Reed, L. J., and Muench, H. *Am J Hyg 27:*493 (1938).

Their method is applicable primarily to a complete titration series; i.e., the whole reaction range, from 0 to 100% mortality (or infectivity, cytopathic effect, etc), should be represented in the experimental data. However, the method can be utilized even if these conditions are not fulfilled, provided the reactions occur in a uniform manner over the range of dilutions employed. If, however, the results are erratic (e.g., deaths scattered irregularly over a number of dilutions), the endpoint will be inaccurate.

This method for calculating 50% endpoints is illustrated by the example given in Table 22-3.

Accumulated values for the total number of animals that died or survived are obtained by adding in the directions indicated by the arrows in columns c and d. The accumulated mortality ratio (column g) represents the accumulated number of dead animals (column e) over the accumulated total number; for example, at the 10^{-3} dilution, there occurred the equivalent of 5 deaths out of a total of 7 animals.

In the example given in Table 22-3, the mor-

TABLE 22-3. Example of Endpoint Titration

Virus dilution (a)	Mortality ratio (b)	Died (c)	Survived (d)	Accumulated values		Mortality	
				Total dead (e)	Total survived (f)	Ratio (g)	% (h)
10^{-1}	6/6	⬆ 6	0	17	0	17/17	100
10^{-2}	6/6	6	0	11	0	11/11	100
10^{-3}	4/6	4	2	5	2	5/7	71
10^{-4}	1/6	1	5	1	7	1/8	13
10^{-5}	0/6	0	⬇ 6	0	13	0/13	0

tality in the 10^{-3} dilution is above 50%; that in the next lower dilution, 10^{-4}, is considerably below 50%. Therefore, the 50% endpoint lies somewhere between the 10^{-3} and 10^{-4} dilutions of the inoculated

$$\frac{\% \text{ mortality at dilution next above } 50\% - 50\%}{\% \text{ mortality at dilution next above } 50\% - \% \text{ mortality at dilution next below } 50\%} = \text{proportionate distance}$$

or

$$\frac{17-50}{71-13} = \frac{21}{58} = 0.36 \text{ (or 0.4)}$$

Since the distance between any two dilutions is a function of the incremental steps used in preparing the series, e.g., 2-fold, 4-fold, 5-fold, 10-fold, it is necessary to correct (multiply) the proportionate distance by the dilution factor, which is the logarithm of the dilution steps employed. In the

virus. The necessary proportionate distance of the 50% mortality endpoint, which obviously lies between these two dilutions, is obtained from column h as follows:

case of serial 10-fold dilutions, the factor is 1 (log 10 = 1) and so is disregarded; in a 2-fold dilution series, the factor is 0.3 (log of 2); in a 5-fold series, the factor is 0.7 (log of 5), etc. In the procedure which follows, the factor is understood to be negative. Therefore, the negative log of LD_{50} endpoint titer equals the negative log of the dilution above 50% mortality **plus** the proportionate distance factor (calculated above). In the example given, the following values obtain:

negative log of lower dilution (next above 50% mortality) $= -3.0$
proportionate distance (0.4) $\times$ dilution factor (log 10) $= -0.4$

$$\log LD_{50} \text{ titer} = -3.4$$
$$LD_{50} \text{ titer} = 10^{-3.4}$$

SELECTED REFERENCES

Books and Review Articles

ATKINS, E. Pathogenesis of fever. *Physiol Rev 40:*580 (1960).

AUSTEN, K. F., and COHN, Z. A. Contribution of serum and cellular factors in host defense reactions. *N Engl J Med 268:*933, 944, 1056 (1963).

COHN, Z. A. The structure and function of monocytes and macrophages. *Adv Immunol 9:*163 (1963).

HIRSCH, J. G. Phagocytosis. *Annu Rev Microbiol 19:*339 (1965).

KLEBANOFF, S. Intraleukocytic microbicidal defects. *Annu Rev Med 39* (1971).

Microbial Pathogenicity in Man and Animals (22nd Symposium, Society of General Microbiology; H. Smith and J. H. Pearce, eds.). Cambridge Univ. Press, London, 1972.

SPECTOR, W. G., and WILLOUGHBY, D. A. The inflammatory response. *Bacteriol Rev 27:*117 (1963).

WOOD, W. B., JR. Studies on the cellular immunology of acute bacterial infections. *Harvey Lect 47:*72 (1951–1952).

Specific Articles

BAGGIOLINI, M., HIRSCH, J. G., and DEDUVE, C. Resolution of granules from rabbit heterophil leuko-cytes into distinct populations by zonal sedimentation. *J Cell Biol 40:*529 (1969).

BAINTON, D. F., and FARQUHAR, M. G. Origin of granules in polymorphonuclear leukocytes: Two types derived from opposite faces of the Golgi complex in developing granulocytes. *J Cell Biol 28:*277 (1966).

BENNETT, W. E., and COHN, Z. A. The isolation and selected properties of blood monocytes. *J. Exp Med 123:*145 (1966).

BUDDINGH, G. J. Bacterial and mycotic infection in the chick embryo. *Ann N Y Acad Sci 55:*282 (1952).

COHN, Z. A., and BENSON, B. The differentiation of mononuclear phagocytes: Morphology, cytochemistry. *J Exp Med 121:*153 (1965).

COHN, Z. A., FEDORKO, M. E., and HIRSCH, J. G. The in vitro differentiation of mononuclear phagocytes. V. The formation of macrophage lysosomes. *J Exp Med 123:*757 (1966).

COHN, Z. A., and HIRSCH, J. G. The isolation and properties of the specific cytoplasmic granules of rabbit polymorphonuclear leucocytes. *J Exp Med 112:*983 (1960).

DUBOS, R. J., and SCHAEDLER, R .W. The effect of diet on the fecal bacterial flora of mice and on their resistance to infection. *J Exp Med 115:*1161 (1962).

HIRSCH, J. G. Cinemicrophotometric observations on granule lysis in polymorphonuclear leucocytes during phagocytosis. *J Exp Med 116:*827 (1962).

JENKINS, C. R. An antigenic basis for virulence of strains of *S. typhimurium. J Exp Med 115:*731 (1962).

KARNOVSKY, M. J. Metabolic basis of phagocytic activity. *Physiol Rev 42:*143 (1962).

KLEBANOFF, S. J. A peroxidase-mediated antimicrobial system in leukocytes. *J Clin Invest 46:*1078 (1967).

MACKANESS, G. B. The immunological basis of acquired cellular resistance. *J Exp Med 120:*105 (1964).

MACKANESS, G. B. The influence of immunologically committed lymphoid cells on macrophage activity in vivo. *J Exp Med 129:*973 (1969).

QUIE, P. G., WHITE, J. G., HOLMES, B., and GOOD, R. A. *In vitro* bactericidal capacity of human polymorphonuclear leukocytes: Diminished activity in chronic granulomatous disease of childhood. *J Clin Invest 46:*668 (1967).

ROSEN, F. S. The endotoxins of Gram-negative bacteria and host resistance. *N Engl J Med 264:*919 (1961).

SILVERSTEIN, S. Macrophages and viral immunity. *Semin Hematol 7:*185 (1970).

ZUCKER-FRANKLIN, D., and HIRSCH, J. G. Electron microscope studies on the degranulation of rabbit peritoneal leukocytes during phagocytosis. *J Exp Med 120:*569 (1964).

chapter

CHEMOTHERAPY OF BACTERIAL DISEASES*

EXTRACELLULAR AND INTRACELLULAR PARASITISM 668
PREDOMINANTLY EXTRACELLULAR BACTERIA 668
Bacteristatic Drugs 668
Bactericidal Drugs 669
PREDOMINANTLY INTRACELLULAR BACTERIA 670
LIMITATIONS TO EFFECTIVE CHEMOTHERAPY 670
Local Tissue Factors 670
Systemic Host Factors 672
Superinfections and Drug Resistance 672
CLINICAL IMPLICATIONS 673
ANTIBACTERIAL DRUGS IN GENERAL USE 674
The Sulfonamides and Related Drugs 674
Penicillins and Other Drugs That Inhibit Bacterial Cell Wall
Synthesis 674
The Macrolides and Lincomycins 675
Aminoglycosides: Streptomycin, Kanamycin, Gentamycin,
and Neomycin 675
Rifampin 676
Broad-Spectrum Antibiotics: Tetracyclines and
Chloramphenicol 676
Polymyxins 677
SPECIAL ASPECTS OF ANTIMICROBIAL THERAPY 677
Topical Therapy 677
Drug Resistance: Clinical and Epidemiological Significance 677
Combined Therapy 678
Chemoprophylaxis 678

* Revised with the generous help of ROBERT AUSTRIAN, M.D.

The results of antibacterial chemotherapy are determined by a complex set of interactions between drug, host, and offending microbe. These may be diagrammed as follows:

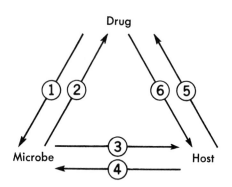

The action of the drug on the microbe (1) has been discussed in Chapter 7, as has its converse (2), illustrated by the inactivation of penicillin by microbial penicillinase. The fate of the drug in the host (5) and its side effects on the host (6) belong to pharmacology and will not be dealt with in detail in this text. The effects of the microbe on the host (3) and the host's defenses against the microbe (4) are extensively treated in several chapters (e.g., 14, 22, and 25). The present one discusses the host factors that influence the curative effects of antibacterial drugs and defines their uses and limitations as chemotherapeutic agents.* The distinction made in Chapter 7 between **bacteristatic** and **bactericidal** drugs will be seen to be of paramount importance.

EXTRACELLULAR AND INTRACELLULAR PARASITISM

The preceding chapter emphasized the distinction between **extracellular** and **intracellular** pathogenic bacteria. The former are harmful to the host only as long as they remain outside phagocytic cells (which promptly kill and destroy them following ingestion), while the latter are capable of surviving and multiplying within phagocytes. Extracellular bacterial parasites tend to produce acute, short-lived diseases (e.g., streptococcal pharyngitis), which often terminate spontaneously as soon as the host has generated enough specific opsonins to bring about destruction of the invading organisms by phagocytosis. In contrast, intracellular bacterial parasites ordinarily give rise to chronic diseases (e.g., tuberculosis), in which neither opsonization nor phagocytosis necessarily constitutes a critical event in the progress of the infection. Since host defenses influence the effectiveness of antimicrobial drugs, the distinction between extracellular and intracellular parasitism is relevant to chemotherapy.

PREDOMINANTLY EXTRACELLULAR BACTERIA

BACTERISTATIC DRUGS

No drug that merely slows down or prevents bacterial multiplication can be curative per se. For a cure most, if not all, of the bacteria invading the tissues must ultimately be destroyed. The role of the host's phagocytic defenses is readily demonstrable in animal models. For example, in experimental pneu-

___
* Antifungal drugs are discussed in Chapter 43 and antiviral agents in Chapter 49.

mococcal pneumonia in rats (Fig. 23-1) sulfonamides suppress the multiplication of the pneumococci in the advancing edema zone of the lesion (Ch. 25, Pathogenesis of pneumococcal pneumonia) and thus stop its spread, but the ultimate destruction of the pneumococci is brought about by the phagocytic cells of the inflammatory exudate (Fig. 23-1C). Thus the cure of the disease results from the **combined actions of the drug and the phagocytes of the host.** Such a dual mechanism appears to operate generally in acute bacterial diseases that are responsive to bacteristatic drugs.

In **suppurative lesions,** on the other hand, assistance by cells is virtually nonoperative, since the phagocytes have become necrotic and nonfunctional (Ch. 22, Necrosis and abscess formation). Hence such lesions respond poorly to bacteristatic drugs, even to those that are not inhibited by products of tissue destruction (Ch. 7, Noncompetitive reversal of growth inhibition).

BACTERICIDAL DRUGS

The relations that obtain with bactericidal drugs are more complicated. As was already emphasized in Chapter 7, bactericidal drugs (with the exception of detergent-like agents such as the polymyxins) are **bactericidal only to growing organisms.** When penicillin, for example, is added to a culture of pneumococci during the exponential phase of growth the organisms are rapidly killed, and the culture is eventually sterilized (Fig. 23-2, curve B). If the drug, on the other hand, is added to the culture in the stationary phase, when multiplication has greatly slowed, no killing occurs (Fig. 23-2, curve C). The same rela-

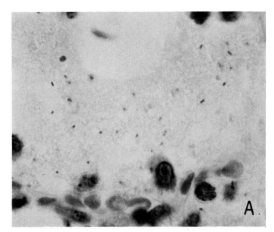
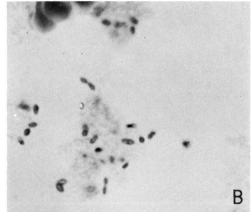
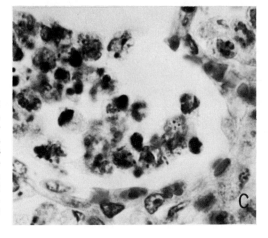

Fig. 23-1. The effects of sulfonamide therapy on experimental pneumococcal pneumonia in rats. **A.** Pneumococci in the edema zone of the pneumonic lesion shortly after the onset of treatment. ×800. **B.** Pneumococci in edema zone 12 hours later showing morphological signs of bacteristasis (pleomorphism and irregular staining). ×1500. **C.** Phagocytosis of pneumococci by leukocytes in alveolar exudate during period of treatment. ×800. [From Wood, W. B., Jr., and Irons, E. N. *J Exp Med 84:*365 (1946).]

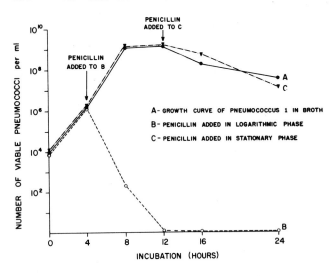

Fig. 23-2. Action of penicillin on type 1 pneumococci in broth culture [From Smith, M. R., and Wood, W. B., Jr. *J Exp Med 103:*487 (1956).]

tion is demonstrable in vivo in rat pneumococcal pneumonia. In the outer edema zone of the lesion, where the organisms multiply rapidly in the untreated animal (Fig. 23-3A), penicillin causes lysis of pneumococci (Fig. 23-3B); in the more central portions of the lesion, where pneumococcal multiplication has slowed, phagocytosis plays the major role in destroying the organisms (Fig. 23-3, C and D).

The part played by phagocytes can be roughly measured by comparing normal with leukopenic animals. The results (Fig. 23-4) indicate that the destruction of the invading bacteria is hastened by the combined bactericidal actions of the drug and the phagocytic cells.

PREDOMINANTLY INTRACELLULAR BACTERIA

With bacteria that can survive for long periods in phagocytic cells, such as tubercle bacilli, *Brucellae,* and typhoid bacilli, the curative effect of chemotherapy is not nearly as rapid or reliable as with extracellular parasites, even when the organisms are highly drug-sensitive in vitro. One factor may be failure of the drug to reach the intracellular bacteria; another may be its relative ineffectiveness in the environment of the cell's cytoplasm (streptomycin; see Ch. 7, Aminoglycosides). Some drugs, however, penetrate animal cells and act upon susceptible microorganisms growing within them. Thus isoniazid can act upon tubercle bacilli within monocytes, and chloramphenicol can slow the growth of *Salmonella typhosa* in cultured mouse fibroblasts. Nevertheless, the intracellular location of the parasites promotes sluggish metabolism and dormancy, as evidenced by the chronic nature of the diseases themselves and the frequent appearance of the carrier state; and dormancy decreases bactericidal drug action. Hence the actual destruction of the organisms must depend primarily upon the cellular defenses of the host. Since these defenses are slow to cure in the absence of treatment, it is hardly surprising that cure is also slow in the presence of drug-induced bacteristasis.

The immune bactericidal processes in monocytic phagocytes, it will be recalled, involve functional as well as anatomical changes in the cytoplasm, which coincide with the onset of delayed hypersensitivity (Ch. 22).

LIMITATIONS TO EFFECTIVE CHEMOTHERAPY

LOCAL TISSUE FACTORS

Abscess Formation and Necrosis. When bacteria reach high enough extracellular population densities in localized tissue sites to kill the surrounding host's cells, *suppuration* results, forming an abscess. Antibacterial chemotherapy, even with potent bactericidal drugs, then becomes relatively ineffective. The rea-

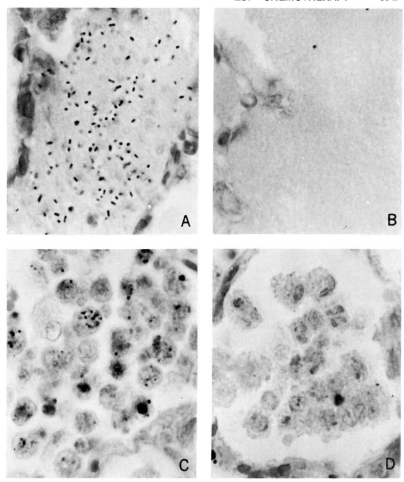

Fig. 23-3. Effect of penicillin therapy on experimental pneumococcal pneumonia. **A.** Pneumococci in edema fluid contained in alveolus at advancing margin of 18-hour lesion (see Ch. 25). **B.** Similar alveolus 12 hours after start of penicillin treatment. Note that all but a few of the pneumococci have already been destroyed by the direct lytic action of the drug. **C.** Alveolus in zone of advanced consolidation (Ch. 25) of same lesion shown in **B.** Note that in this portion of the lesion, where bacterial growth has slowed to a standstill, most of the organisms are being destroyed by phagocytosis rather than lysis. **D.** Similar alveolus to that pictured in **C** 18 hours later. Most of the phagocytized pneumococci have been destroyed by the cells. ×900. [From Smith, M. R., and Wood, W. B., Jr. *J Exp Med 103:*487 (1956).]

son is not failure of the drug to reach the organisms, which is often assumed: subcutaneous pneumococcal abscesses (Fig. 23-5) do not respond to penicillin even when it is injected directly into the abscesses at frequent intervals, and streptomycin has been shown to penetrate caseous tuberculous lesions that it cannot sterilize.

Several factors contribute to the persistence of infection under these conditions. The im-

pairment of circulation in the lesion reduces the flow of antibodies and leukocytes, of oxygen, and of nutrients; it also impedes removal of waste products, including acids (produced by both host cells and bacteria) and specific bacterial toxins. As a result of these deficiencies and accumulations the leukocytes cease to function and eventually die and disintegrate (Fig. 23-5). Moreover, since acidity, lack of oxygen, and lack of nutrients also impair

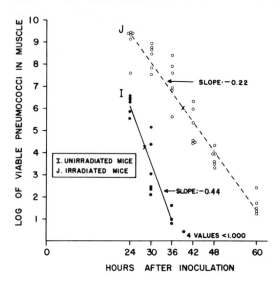

Fig. 23-4. Comparative rates at which pneumococci are killed by penicillin in experimental pneumococcal myositis lesions in unirradated mice and in mice made severely leukopenic by exposure to 650 r of X-ray. Penicillin was administered 24 hours after inoculation. The killing of organisms in the tissues occurred less rapidly in the leukopenic animals. [From Smith, M. R., and Wood, W. B., Jr. *J Exp Med 103:*499 (1956).]

multiplication of the bacteria, they are not killed by penicillin or aminoglycosides. Such abscesses are cured only after drainage, which permits the bacteristatic pus to be replaced by fresh serous exudate, providing nutrients as well as a wave of new leukocytes.

The action of some chemotherapeutic agents is also affected directly. As noted earlier (Ch. 7), anaerobiosis and acidity decrease markedly the potency of aminoglycosides, while certain metabolites released by tissue autolysis antagonize the bacteristatic action of sulfonamides: hence these drugs are ineffective in necrotic lesions.

Foreign Bodies and Obstruction of Excretory Organs. Chemotherapy is also relatively ineffective when there is a foreign body in the lesion, such as splinters of wood, spicules of dead bone (in osteomyelitis), prostheses, or sutures. These provide foci protected from leukocytes, where bacteria may accumulate their products or persist in a quiescent state. Similarly, bacterial infections associated with obstructions of the urinary, biliary, or respiratory tracts (bronchi or paranasal sinuses) tend to persist despite intensive chemotherapy: the increased fluid pressure impairs circulation and lymphatic drainage in the tissues surrounding the obstructed cavity, and so toxic products and eventually quiescent bacteria accumulate within the cavity.

Both these types of lesions are rarely cured by chemotherapy unless the foreign body or obstruction is removed. Surgical intervention alone may permit the natural defenses to eradicate the infection, but chemotherapy is ordinarily continued until the systemic manifestations (fever, leukocytosis, etc.) have subsided.

SYSTEMIC HOST FACTORS

Many disease states that depress natural resistance to bacterial infections also impair response to antimicrobial therapy. Some reduce the number of effective **phagocytic cells** available for mobilization at the infected site (e.g., agranulocytosis, leukemia, radiation injury, overdosage of nitrogen mustards or corticosteroids), while others interfere with **antibody formation** (agammaglobulinemia, multiple myeloma, and chronic lymphatic leukemia). **Uremia** and uncontrolled **diabetes mellitus** also depress resistance of the host and thus affect the response to chemotherapy, although the precise mechanisms are not known. Experiments with alloxan diabetic mice indicate that hyperglycemia (even without acidosis) impairs the phagocytic activity of polymorphonuclear leukocytes.

SUPERINFECTIONS AND DRUG RESISTANCE

One of the most serious and frequent complications of antimicrobial therapy is **superinfection** with organisms resistant to the drug being used, and unfortunately often to many

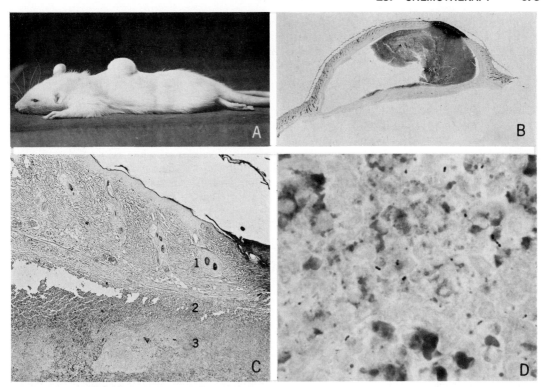

Fig. 23-5. Penicillin treatment of experimental pneumococcal abscesses in rats. **A.** Subcutaneous abscess on back of rat 23 days after inoculation. ×1/3, reduced. **B.** Cross-section of abscess cavity. ×3, reduced. **C.** Wall of abscess showing small vessel (1) in subcutaneous tissue, periphery of abscess cavity (2) where leukocytes, recently arrived from 1 and from other vessels, are still functional, and necrotic pus (3) within core of lesion. ×40, reduced. **D.** Higher magnification of zone 3 in **C,** showing leukocytic debris and unphagocytized pneumococci. ×1200, reduced. [From Smith, M. R., and Wood, W. B., Jr. *J Exp Med 103:*509 (1956).]

other drugs as well. This development may arise from the outgrowth of **indigenous** bacteria or fungi that are normally held in check by the growth of other organisms. For example, staphylococcal enteritis or candidal stomatitis occasionally develops during prolonged treatment with tetracycline (see below). Moreover, in patients with seriously impaired resistance (e.g., cachectic patients, or those with severe burns) penicillin may prevent infection of the local lesion, and has decreased the frequency of terminal bronchopneumonia due to gram-positive organisms, but gram-negative bacillary septicemias frequently occur after 2 to 3 weeks of therapy. *Exogenous* drug-resistant bacteria may also be acquired during therapy, usually from other patients or hospital personnel. Superinfections of either origin may be more serious than the original infection.

CLINICAL IMPLICATIONS

The foregoing considerations suggest the following general principles for the chemotherapy of bacterial diseases.

1) The earlier treatment is begun the more likely it is to be effective. Several advantages are gained: a) organisms that are allowed to reach the stationary growth phase in tissues become refractory to bactericidal drugs; b) some species (e.g., *Staphylococcus aureus*) tend eventually to produce suppurative lesions which may require surgical drainage and may cause irreversible tissue damage in vital organs (e.g., brain); c) drug-resistant mutants are more likely to arise in advanced bacterial lesions, owing simply to the greater numbers of bacteria involved; and d) late lesions are more likely to become superinfected with drug-resistant organisms.

2) In order to eliminate the last infecting bacterial cell and thus prevent relapse therapy must ordinarily be continued for some time after symptoms have subsided. This period is longer when a bacteristatic agent is used, or when treatment is begun late with a bactericidal agent, since most of the invading organisms will have to be eliminated by the relatively slow processes of cellular and humoral defense.

3) In infections of the endocardium, where the cellular defenses of the host are comparatively ineffective, cure almost always requires the use of a bactericidal drug.

4) Chemotherapy can rarely cure estab-lished abscesses, lesions containing foreign bodies, or infections associated with excretory duct obstruction, unless they drain spontaneously or are drained surgically.

5) In the presence of complications that impair natural defenses larger doses of antimicrobial drugs are required.

6) Short-term chemotherapy is generally ineffective in chronic bacterial diseases caused by intracellular parasites. Only if treatment is prolonged, to the point where the relatively ponderous mechanisms of intracellular immunity can effectively intervene, is subsequent relapse likely to be avoided.

ANTIBACTERIAL DRUGS IN GENERAL USE

The presently accepted applications and limitations of the useful antibacterial drugs may be briefly summarized as follows.

THE SULFONAMIDES AND RELATED DRUGS

The sulfonamides have been largely replaced by more effective and less toxic antimicrobial agents, although they are still used widely in the management of infections of the urinary tract and of trachoma. They penetrate the blood-brain barrier rapidly and have been employed extensively in the past in the treatment of meningococcal meningitis, but penicillin is more effective. Prophylactic doses are also effective in preventing reinfection with group A streptococci in patients with rheumatic fever, but penicillin is less likely to select for resistant mutants. Poorly absorbed sulfonamides (sulfasuxidine, phthalylsulfathiazole, sulfaguanidine) are employed to lower the bacterial content of the bowel in preparation for abdominal surgery, but the possible complications may outweigh the advantages.

The principal disadvantages of the sulfonamides are 1) their limited antimicrobial spectrum at attainable drug levels, 2) their purely bacteristatic action, 3) their delayed effect on bacterial growth (Ch. 7, Competitive inhibition), and 4) their tendency to cause such serious side effects as granulocytopenia, aplastic anemia, thrombocytopenia, and periarteritis nodosa.

Two compounds related to the sulfonamides, described in Chapter 7, have a very restricted antimicrobial spectrum: *diaminodiphenylsulfone* is useful only in the treatment of leprosy, and *p-aminosalicylate* in the treatment of tuberculosis. The choice of drugs for the latter disease will be discussed in Chapter 35. The action of the most effective and nontoxic agent, isoniazid, has been discussed in Chapter 7.

PENICILLINS AND OTHER DRUGS THAT INHIBIT BACTERIAL CELL WALL SYNTHESIS

Penicillin G (benzyl penicillin) is effective, at readily attainable blood levels, against **virtually all gram-positive cocci** (pneumococci, streptococci, staphylococci); exceptions are penicillinase-producing strains of staphylococci and some enterococci (e.g., *Streptococcus faecalis*). It is active also against certain **other gram-positive bacteria**, such as *Corynebacterium diphtheriae, Listeria monocytogenes,* and some clostridia; the **gram-negative** neisseriae (gonococci and meningococci), some strains of bacteroides, and *Hemophilus ducreyi* (the cause of chancroid); various **spirochetes** (notably the treponeme of syphilis); and some **actinomycetes.**

In the usual doses **penicillin G is not effective in vivo against mycobacteria, many gram-negative bacilli of the enteric tract, and** *Hemophilus influenzae;* but their insusceptibility is only relative. Indeed, the mechanism of action of penicillin has been demonstrated in part through its effect, at high concentrations, on *Escherichia coli.* In contrast, its ineffectiveness against fungi, protozoa, and viruses is absolute because these organisms lack the penicillin-sensitive reaction (Ch. 6).

Because penicillin can cause a loss of viability by several log units within a fraction of a bacterial generation time, its effect is often more dramatic than that of bacteristatic agents. Furthermore, its action is not antagonized by products of tissue breakdown; indeed, enrichment of the medium accelerates bactericidal action by accelerating bacterial growth. Finally, except for its allergenicity (which may be very serious), it has virtually no toxicity for mammals except in huge doses. Penicillin G can be administered orally, although a substantial fraction is not absorbed; penicillin V (phenoxymethyl penicillin) is better absorbed, but it is also more expensive. Moreover, with penicillin-sensitive organisms penicillin G is considerably more potent than the congeners resistant to penicillinase (e.g., methicillin, oxacillin); hence the latter are valuable primarily against the frequent penicillinase-producing staphylococci, which fortunately are responsive even to penicillins of lower potency. Thus **penicillin G is almost always the drug of choice for the treatment of serious infections caused by susceptible bacteria.** In mixed infections, however, its action on naturally susceptible species may rarely be blocked by the presence of penicillinase-producing staphylococci.

Ampicillin and **carbenicillin** have extended the advantages of penicillin therapy to certain gram-negative bacilli. Ampicillin has proved useful against *Hemophilus influenzae* and certain gram-negative enteric bacilli. Carbenicillin finds its chief application against sensitive strains of *Pseudomonas aeruginosa.* Both these penicillins are inactivated by penicillinase. Ampicillin can be given orally or parenterally, but carbenicillin must be administered by a parenteral route.

The *cephalosporins,* which are closely related to penicillins in structure and action, are described in Chapter 7. Because of their relatively broad spectrum, resistance to staphylococcal penicillinase, and infrequent allergic cross-reactions with penicillin, they have rapidly become very popular. However,

they are considerably less potent than penicillin G against the common penicillin-sensitive organisms, and less effective than ampicillin against *H. influenzae;* but they are the drugs of choice in infections with certain enterobacteria. They are much more expensive than penicillins, and most members of the group must be injected. However, another derivative, cephalexin, can be administered orally.

Vancomycin is useful in treating serious infections caused by some strains of bacteroides and by susceptible gram-positive bacteria in patients allergic to penicillins and cephalosporins. It is moderately toxic (often causing chemical thrombophlebitis) and it must be given intravenously.

THE MACROLIDES AND LINCOMYCINS

The macrolides (erythromycin, triacetyloleandomycin) and lincomycins (lincomycin, clindamycin) are chemically dissimilar groups of antibiotics that inhibit protein synthesis in bacteria (Ch. 12). They are effective against gram-positive bacteria; macrolides also inhibit some gram-negative bacteria (neisseriae, *Hemophilus influenzae*), mycoplasma, and treponemes. Clindamycin is active against many gram-negative anaerobes. These drugs may be given orally, and they are useful at times when allergy prevents the use of penicillin. Resistant mutants may arise during treatment.

AMINOGLYCOSIDES: STREPTOMYCIN, KANAMYCIN, GENTAMYCIN, AND NEOMYCIN

The aminoglycosides have a broad antibacterial spectrum, including many gram-positive as well as most gram-negative organisms. They have been used primarily in the treatment of tuberculosis and of infections with various gram-negative organisms, including *Klebsiella* pneumonia, *Hemophilus influenzae* meningitis, plague, tularemia, *Shigella* dysentery, and systemic or urinary tract infections caused by enteric bacilli.

Because of intracellular residence, salmonellae and brucellae, despite their sensitivity in vitro, can rarely be eradicated by these drugs (Ch. 7, Aminoglycosides; Ch. 32, Pathogenicity). The antagonistic effects of salts, low pH, and anaerobiosis limit the action of aminoglycosides in septic foci and in the urinary tract.

Many bacterial species readily develop, by mutation, high level resistance to streptomycin, usually not extending to other aminoglycosides. Streptomycin dependence (Ch. 7) may also occasionally have clinical significance, particularly in the treatment of tuberculosis. Although such dependent tubercle bacilli have been isolated from treated tuberculous patients, their presence does not necessarily contraindicate further use of the drug, because they may constitute only a minority of the infecting population.

Streptomycin and other aminoglycosides are relatively toxic, often causing damage to the eighth cranial nerve on prolonged administration (e.g., in tuberculosis); renal and peripheral nerve damage are also seen. The risk is increased when drug excretion is impaired by renal insufficiency. Because of this toxicity and the emergence of resistant mutants, aminoglycosides are often given in combination with other drugs (Ch. 7, Combination therapy).

Kamacycin is less effective than streptomycin against *M. tuberculosis* but is more active against gram-positive cocci; moreover, resistant mutants are encountered less frequently. **Gentamycin*** is more toxic than kanamycin or streptomycin and is given in smaller doses. It is effective against gram-positive cocci and many strains of *Pseudomonas,* an activity not manifested by other aminoglycosides. **Neomycin,** because of its marked ototoxicity, should not be given parenterally. It is mostly used topically in various forms of infectious dermatitis, and it is still occasionally employed as an intestinal antiseptic.

RIFAMPIN

Rifampin (rifampicin) selectively inhibits bacterial RNA polymerase (Ch. 11) but not that of mammals. Its principal use appears to be in the combined treatment of tuberculosis. It also inhibits many gram-positive and some gram-negative bacteria, but resistant mutants arise frequently.

---

* For spelling see footnote to Aminoglycosides, Ch. 7.

BROAD-SPECTRUM ANTIBIOTICS: TETRACYCLINES AND CHLORAMPHENICOL

Tetracyclines are bacteristatic inhibitors of protein synthesis (Ch. 12) that are effective against many gram-positive and gram-negative bacteria, though not against Salmonella, Proteus, and Pseudomonas. They are the agents of choice for rickettsiae, chlamydiae, and mycoplasmas, and they may indirectly suppress protozoal infections of mucous surfaces (amebic dysentery, trichomonal vaginitis) by depriving the parasites of their bacterial "fodder." Tetracyclines largely displaced the sulfonamides and streptomycin for use against various gram-positive and gram-negative bacteria; but while they are still widely used for non-hospitalized patients, they are less useful than the bactericidal penicillins in the treatment of life-threatening infections.

The tetracyclines are well absorbed by mouth. They evidently penetrate mammalian cells, since they are effective against some obligate intracellular parasites (e.g., rickettsiae). Tetracyclines rarely give rise to serious toxicity (except for liver damage when given in large doses), but diarrhea, pruritus ani, and oral inflammation are common, and are believed to be largely due to replacement of normal flora by resistant staphylococci and by yeasts (especially *Candida*). Antifungal therapy, however, has been of questionable value. Other serious fungal infections (vaginitis, cystitis, pneumonitis, septicemia) may arise in debilitated patients given tetracyclines, and in treated children *Candida* may cause oral infections (thrush).

Chloramphenicol is a bacteristatic antibiotic with essentially the same antibacterial spectrum as the tetracyclines; in addition, it is moderately effective against salmonellae. It is absorbed more completely from the gut, which may account in part for its less frequent gastrointestinal toxicity. Chloramphenicol, however, is toxic to the bone marrow and often gives rise to some degree of thrombocytopenia, leukopenia, or anemia. Continued administration may then lead to an irreversible, fatal reaction; hence **chloramphenicol should not be administered when an alternative form of therapy is available.** Its use is restricted, therefore, to the treatment of typhoid fever, other salmonelloses, selected cases of meningitis caused by *Hemophilus influenzae* (especially in patients allergic to ampicillin), and infections caused by bacteria resistant to other agents. Periodic hematological examination is imperative.

Because of their wide range of action, broad-spectrum antibiotics have potential value in the treatment of mixed infections (chronic bronchitis, bronchiectasis, peritonitis, abscesses, septic wounds). Unfortunately, physicians often rely on these drugs for "shotgun" therapy in acute febrile illnesses when correct etiological diagnosis (bacteriological or clinical) would allow the use of more effective and less toxic agents.

POLYMYXINS

Polymyxins are antibiotics with a detergent-like action, which destroys the integrity of the membranes (Ch. 6) of gram-negative but not gram-positive bacteria. Because of their neuro- and nephrotoxicity their use is limited to serious infections caused by susceptible organisms. They are among the few drugs effective against *Pseudomonas aeruginosa,* which is a frequent and persistent secondary invader in patients under prolonged chemotherapy. Polymyxins are also used topically.

Antifungal drugs (amphotericin B, nystatin, and grisofulvin) will be discussed in Chapter 43.

SPECIAL ASPECTS OF ANTIMICROBIAL THERAPY

TOPICAL THERAPY

Chemotherapeutic agents are often effective when applied directly to superficial infections, including those of skin, wounds, and eyes. (No benefit can be expected, however, from application of ointments to the superficial drainage of a deep focus.) For such topical application some of the more toxic antibiotics are recommended, including **neomycin, polymyxin, bacitracin,** and **tyrothricin** (a mixture of the polypeptides gramicidin and tyrocidine). A number of other bactericidal compounds (e.g., oxyquinolines, cationic detergents) are also used for therapy of skin infections.

Penicillin and streptomycin should not be used on the skin because application by this route frequently induces allergy, which may then hinder systemic use of the drug in a life-threatening disease. However, to provide high local concentrations at deeper sites of localized infection, these and related drugs can be injected under special circumstances into pleural, pericardial, or joint cavities or into the subarachnoid space.

DRUG RESISTANCE: CLINICAL AND EPIDEMIOLOGICAL SIGNIFICANCE

Resistance to chemotherapeutic agents may be an inherent property of the infecting organism or may result from mutation or

plasmid transfer, followed by selection during therapy. The phenomenon is most pronounced with streptomycin and nalidixic acid because of the occurrence of one-step mutants with extremely high levels of resistance. Mutants with a significant degree of resistance occur frequently during treatment with tetracyclines, macrolides, lincomycins, and sulfonamides, but not with penicillins, cephalosporins, or chloramphenicol,

Bacterial species differ also in their likelihood of yielding mutants resistant to a given drug. Some examples follow.

In those **tuberculous patients** in whom streptomycin is not dramatically effective (e.g., those with large caseous lesions) resistant mutants emerge quite regularly after 1 to 3 months. Although this time may seem long, it should be recalled that the tubercle bacillus has a mean generation time some 30 times longer than that of most bacteria.

The **staphylococcus** is particularly prone to pick up plasmids (Ch. 7) that provide resistance to many antibiotics, including macrolides, tetracyclines, chloramphenicol, and penicillins.

Although no group A β-hemolytic **streptococcus** resistant to penicillin has been isolated from man, strains resistant to sulfonamides, tetracyclines, and macrolides have been recovered. **Pneumococci** resistant to tetracyclines, macrolides, lincomycin, penicillin, or sulfonamides have all been recognized as the cause of human infection. The gonococcus has also developed resistance to various agents.

In hospitals penicillin-resistant organisms are frequently found to be widely distributed not only among patients treated with penicillin but also among the rest of the population. The rapid distribution of these strains is favored by several factors: 1) staphylococci not only are pathogens but also are part of the normal nasal and skin flora; 2) elimination of sensitive organisms in treated patients probably creates an empty ecological niche into which resistant strains can move; 3) staphylococci are hardy enough to be readily transferred from one person to another; and 4) this transmission is facilitated in many hospitals by relaxation in the standards of aseptic medical practice and of environmental sanitation, an unfortunate byproduct of the era of antibiotics.

The widespread use of a drug may thus lead not only to emergence of resistant mutants within individual patients, but also to the spread of resistant strains in response to the ecological alteration produced by the treatment. Such strains are often as virulent as their sensitive parents, and so may persist in the population. However, with species whose wild type is regularly sensitive to a drug the resistant mutants presumably have less evolutionary survival value than the parent; otherwise the resistant strains would be the wild type of that species. Hence if streptomycin were no longer used, streptomycin-resistant tubercle bacilli should in time disappear.

It should be emphasized that much of the present hard core of "natural" drug resistance, as exhibited by many strains of *Staphylococcus, Pseudomonas, Proteus,* and various gramnegative urinary pathogens, does not involve the genetic problem of new mutation to drug resistance: it simply reflects the fact that each agent has limits to its useful antibacterial spectrum, and certain wild-type strains lie outside these limits.

The spread of **resistance transfer factor** (RFT), carrying genes for resistance to several drugs (sulfonamides, chloramphenicol, tetracycline, streptomycin), has created a serious problem in the treatment of *Shigella* dysentery. The nature of this factor is discussed in Chapter 46, and the mechanism of resistance in Chapter 7.

COMBINED THERAPY

Various mixtures of chemotherapeutic agents have been promoted commercially on the basis of alleged synergistic or broadspectrum activity, but they should be used rarely, if ever, in preference to treatment designed for the individual patient. Combinations of antimicrobial drugs may be employed rationally on the following grounds: 1) to prevent the emergence of resistant strains; 2) to take advantage of the synergistic action (Ch. 7) of certain drugs (penicillin and streptomycin in the treatment of certain types of bacterial endocarditis, or gentamycin and carbenicillin in systemic pseudomonas infections); and 3) to provide the most potentially effective therapy in life-threatening emergencies before an etiological diagnosis can be established with certainty.

Combined therapy has been especially successful in the treatment of tuberculosis with isoniazid, streptomycin, and PAS, or isoniazid and rifampin. Nevertheless, multiply resistant tubercle bacilli do emerge occasionally, presumably because of the unequal distribution of the drugs and the difficulty in maintaining an adequate level at all times. Hence combinations of drugs have prolonged the usefulness of individual ones in the treatment of tuberculosis, but they have not completely solved the problem of drug resistance.

Combinations of antimicrobial drugs that are antagonistic rather than synergistic should be avoided. As was noted in Chapter 7, bacteristatic drugs, such as tetracyclines, chloramphenicol, and sulfonamides, interfere with the bactericidal action of drugs that interfere with synthesis of the bacterial cell wall (e.g., penicillins, cephalosporins, and vancomycin).

CHEMOPROPHYLAXIS

The prophylactic use of antimicrobial drugs is likely to be successful only when directed against an organism that rarely gives rise to mutants resistant to the drug employed. It has been used most effectively in preventing: 1) recurrent streptococcal infections in patients with rheumatic fever; 2) subacute bacterial endocarditis in patients with rheumatic or congenital heart disease undergoing dental extractions or other surgical procedures; 3) venereal disease. It has proved effective also in terminating epidemics of meningococcal infection and of shigellosis in closed populations (e.g., schools and military installations). Chemoprophylaxis has been successful in reducing infection with staphylococci following certain surgical procedures, but it has not been uniformly effective in preventing infection with various other organisms; it is not a substitute for good aseptic technic. Isoniazid appears to be useful in preventing tuberculosis in certain high-risk settings, and even more in preventing the disease in those who have become infected (as shown by a positive tuberculin test).

Although chemoprophylaxis is often successful in preventing invasion by specific organisms, it cannot be expected to be effective when susceptible bodily sites are exposed to a wide variety of bacterial species. It has proved unavailing in the prevention of bacterial pneumonia in patients with stroke or with viral lung infection, and it cannot be expected to keep an especially vulnerable region, such as a urinary bladder containing an indwelling catheter, free of bacteria. Indeed, its indiscriminate use may do more harm than good by fostering the development of infections with drug-resistant organisms. For this reason, and because of the hazard of toxic and allergic drug reactions, chemotherapy of common upper respiratory tract disease is not justified. The practice of so treating minor respiratory infections, though unfortunately common, should be condemned.

The chemotherapy of specific diseases is discussed further in the chapters that follow.

SELECTED REFERENCES

Books and Review Articles

GOTTLIEB, D., and SHAW, P. B. (eds.). *Antibiotics. I. Mechanism of Action.* Springer, New York, 1967.

Handbook of Antimicrobial Therapy. *Medical Letter 14,* No. 2 (1972).

MARTIN, W. J. Newer antimicrobial agents having current or potential clinical application. *Med Clin North Am 48:*255 (1964).

STROMINGER, J. L. Penicillin-sensitive enzymatic reactions in bacterial cell wall synthesis. *Harvey Lect 64:*179 (1968–1969).

WEINSTEIN, L. Chemotherapy of microbial disease. In *The Pharmacological Basis of Therapeutics,* 4th ed. (L. S. Goodman and A. Gilman, eds.) Macmillan, New York, 1970.

Specific Articles

EAGLE, H., FLEISCHMAN, R., and MUSSELMAN, A. D. Effect of schedule of administration on the therapeutic efficacy of penicillin. *Am J Med 9:*280 (1950).

LEPPER, M. H., and DOWLING, H. F. Treatment of pneumococcic meningitis with penicillin compared with penicillin plus aureomycin: Studies including observations on an apparent antagonism between penicillin and aureomycin. *Arch Intern Med 88:*489 (1951).

SHOWACRE, J. L., HOPPS, H. E., DUBUY, H. S., and SMADEL, J. E. Effect of antibiotics on intracellular *Salmonella typhosa.* I. Demonstration by phase microscopy of prompt inhibition of intracellular multiplication; II. Elimination of infection by prolonged treatment. *J Immunol 87:*153, 162 (1961).

SMITH, M. R., and WOOD, W. B., JR. An experimental analysis of the curative action of penicillin in acute bacterial infections. I. The relationship of the antimicrobial effect of penicillin; II. The role of phagocytic cells in the process of recovery; III. The effect of suppuration upon the antibacterial action of the drug. *J Exp Med 103:*487, 499, 509 (1956).

chapter 24

CORYNEBACTERIA*

CORYNEBACTERIUM DIPHTHERIAE   682

   History   682
   Morphology   683
   Cultivation   683
   Lysogeny and Toxin Production   684
   Nature and Mode of Action of Toxin   685
   Pathogenicity   686
   Immunity and Epidemiology   688
   Laboratory Diagnosis   689
   Treatment   691
   Prevention   691

OTHER CORYNEBACTERIA (DIPHTHEROIDS)   692

* Revised with the generous help of A. M. PAPPENHEIMER, Jr., Ph.D.

The corynebacteria are gram-positive, rod-like organisms, which often arrange themselves in palisades, possess club-shaped swellings at their poles, and stain irregularly. Taxonomically, they appear to be related to the mycrobacteria (Ch. 35) and nocardia (Ch. 36), since the principal cell wall antigens of all three genera are closely related chemically and serologically. In the species that can cause diphtheria, *Corynebacterium diphtheriae,* lysogenization by a bacteriophage causes synthesis and release of a potent, heat-labile protein toxin.

CORYNEBACTERIUM DIPHTHERIAE

No other bacterial disease of man has been as successfully studied as diphtheria. Its etiology and mode of transmission were established early, and a highly effective method of prevention—immunization with diphtheria toxoid—was subsequently developed. As a result of mass immunization diphtheria has become a rare disease in many countries, including the United States, and toxigenic strains of *C. diphtheriae* have virtually disappeared as common members of the human bacterial flora. Because diphtheria is a prototype of toxigenic disease and its history illustrates the practical benefits that can result from basic research, it will be discussed in some detail.

HISTORY

Records of the existence of diphtheria date back to Hippocrates in the fourth century B.C. Aetius, who lived in the sixth century A.D., wrote that "pestilential lesions of the tonsils . . . occur most frequently in children, but also in adults. . . . Usually in children the evils known as aphthae develop. These are white, like blotches; some are ashen in colour or like eschars from the cautery. The patient suffers from a dryness of the gullet and frequently attacks of choking. . . . A spreading sore supervenes in the region afflicted. . . . In some cases the uvula is eaten up and when the sores have prevailed a long time and deepened, a cicatrix forms over them and the patient's speech becomes rather husky and, in drinking, liquid is diverted upward to the nostrils."

Despite its prevalence diphtheria was not recognized as a specific disease until 1821, when Pierre Bretonneau communicated to the Académie Royale de Médecine his conclusion that the throat distemper was a specific clinical entity which could be differentiated from other afflictions of the throat and which was characterized by the formation of a false membrane* in the respiratory tract. Because of this distinctive feature, Bretonneau called the malady **diphtheritis** (Gr. *diphthera* = skin or membrane), later changed to **diphtheria.** Bretonneau was convinced that diphtheria was a communicable disease, caused by a specific germ and transmitted from person to person, but it was not until 1883 that the causative agent was described by Klebs in stained smears from diphtheritic membranes. That this bacillus was the etiological agent of diphtheria was not proved, however, until a year later, when Loeffler grew the organism on artificial media and produced in guinea pigs a fatal infection which closely simulated the human disease. The organism thereafter

---

* Composed largely of fibrin and trapped leukocytes, rather than true epithelium (Fig. 24-3).

became known as the Klebs-Loeffler (or K-L) bacillus.

When the tissues of the guinea pigs were studied histologically, Loeffler was surprised to find bacilli only in the local lesions at the site of inoculation, though damage was visible in the liver, lungs, kidneys, adrenal glands, and other tissues. He concluded that the bacilli growing at the primary site must have produced a soluble poison which was transported to remote tissues of the body by the blood stream. This conclusion was verified in 1888 by Roux and Yersin, who demonstrated a soluble toxin in the fluid phase of diphtheria bacillus cultures; when the toxin was injected into appropriate laboratory animals, it caused all the systemic manifestations of diphtheria. Two years later, von Behring and Kitasato succeeded in immunizing animals with toxin modified with iodine tritrichloride and demonstrated that the sera of such immunized animals contained an antitoxin capable of protecting other susceptible animals against the disease. On Christmas night 1891, the first diphtheritic child to receive antitoxin was treated in Berlin.

Although the modified toxin was suitable for immunizing animals to obtain antitoxin, it still caused severe local reactions when injected into humans and therefore could not be used as an immunizing agent. In 1909, however, Theobald Smith in America demonstrated that prior neutralization of the toxin with antitoxin greatly reduced the severity of the local reactions without significantly decreasing the immunogenicity of the toxin. Accordingly, **toxin-antitoxin** (TAT) complex was introduced as a practical immunizing agent. Its use was greatly facilitated by Schick's discovery in 1913 of a practical test (see below) for distinguishing nonimmune from immune individuals by their reaction to intradermal injection of a small dose of toxin.

In 1923 Ramon developed an immunizing agent that had definite advantages over the toxin-antitoxin mixtures, which sometimes became toxic through dissociation. This formalin-treated toxin, called toxoid, is noninjurious to tissue cells but still fully immunogenic; it is now used universally for active immunization against diphtheria.

In 1951 Freeman made the remarkable discovery that all toxigenic strains of *C. diphtheriae* are lysogenic, i.e., are infected with a temperate bacteriophage. If such strains lose their specific phage, they cease to produce toxin and thus become relatively avirulent.

MORPHOLOGY

Corynebacteria (Gr. *coryne* = club) tend to be club-shaped when grown on artificial media. They are gram-positive, non-spore-bearing rods, tapered from their septal ends, without flagella or demonstrable capsules. They vary from 2 to 6 μ in length and from 0.5 to 1.0 μ in diameter. Because of the way the individual bacilli divide, they have a tendency in stained smears to form sharp angles with one another, making characteristic figures resembling Chinese letters. When grown on coagulated serum slants rich in phosphate (Loeffler's medium) they contain polymerized polyphosphate granules, called Babes-Ernst bodies, which stain metachromatically with methylene blue or toluidine blue. In some strains well-defined polar bodies are discernible at the ends of each bacillus. These bodies are most frequently seen in organisms growing slowly on suboptimal media; they are less prominent during rapid growth, especially on media containing potassium tellurite (K_2TeO_3, see below). On Loeffler's medium and on blood agar, surface colonies are cream-colored or grayish white; on tellurite agar they are dark gray or black because of intracellular reduction of the tellurite to tellurium.

Three morphologically distinct types of *C. diphtheriae* has been described. The **gravis** strains form large, flat, gray to black colonies, with dull surfaces, on tellurite agar; the colonies of the **mitis** strains are smaller, blacker, and more convex, with glossy surfaces. The colonies of the **intermedius** strains are still smaller and either smooth or rough. The smooth (mitis) strains possess a type-specific thermolabile protein surface antigen which is lacking in rough (gravis) strains. An initially postulated relation between virulence and colony morphology is no longer considered valid. Both smooth and rough strains (Fig. 24-1) may be either toxigenic or nontoxigenic (see below).

In broth rough strains form a pellicle whereas smooth strains grow diffusely.

CULTIVATION

C. diphtheriae is an obligate aerobe. Most strains grow as a waxy pellicle on the surface

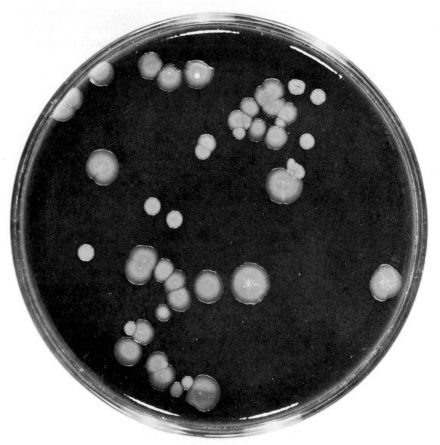

Fig. 24-1. Smooth (small) and rough (large) colonies of C. *diphtheriae* on chocolate agar. [From Barksdale, W. L., Garmise, L., and Rivera, R. *J Bacteriol 81:*531 (1961).]

of liquid media. For primary isolation Loeffler's coagulated serum medium is still useful since it permits growth of diphtheria bacilli with characteristic morphology but fails to support growth of streptococci and pneumococci, commonly present in the throat. Blood or chocolate agar with potassium tellurite, to inhibit the growth of most other bacteria, serves as an even better selective medium for *C. diphtheriae.*

Although the black colonies formed by the diphtheria bacilli are characteristic, other organisms found in the respiratory tract, particularly staphylococci and nonpathogenic corynebacteria (e.g., *C. hofmannii*), may also form black colonies.

C. diphtheriae grows well on relatively simple media containing essential amino acids and an energy source such as glucose or maltose. Most strains are unable to synthesize nicotinic acid, pantothenic acid, or biotin. Under optimal condi-

tions of pH, oxygen supply, and nutrients growth of 20 to 30 gm dry weight per liter may be obtained.

Diphtheria bacilli typically ferment glucose rapidly and maltose slowly, producing acid but no gas. A comparison of the fermentation reactions of the common corynebacteria is shown in Table 24-1.

LYSOGENY AND TOXIN PRODUCTION

Only those strains of *C. diphtheriae* that are lysogenic for β-prophage or a closely related

TABLE 24-1. Fermentation Reactions of Corynebacteria Commonly Cultured from Man

Corynebacteria	Glucose	Maltose	Sucrose
C. diphtheriae	+	+	−(+)
C. xerosis	+	−	+
C. hofmannii	−	−	−

phage, carrying the *tox* gene, produce diphtheria toxin. Since lysogenization by various β-phages carrying a mutated *tox* gene leads to production of antigenically related but nontoxic extracellular proteins, the prophage *tox* gene evidently carries the structural information for the toxin molecule.

Phage multiplication is not a necessary prerequisite for toxin production. Thus in a strain lysogenic for a phage carrying a mutated *tox* gene and superinfected with $β_c{}^{tox+}$ (a clear plaque-forming mutant, analogous to the $λ_c$ mutant of coliphage λ; see Ch. 46), phage multiplication is still repressed; nevertheless, the presence of the $β^{tox+}$ exogenote causes toxin to be synthesized and released together with the nontoxic cross-reacting mutant protein.

Although the structural gene for toxin synthesis is carried by the phage genome, its expression is controlled by the metabolism and physiological state of the host bacteria. The most important factor controlling the yield is the concentration of inorganic iron present in the culture medium. Toxin is synthesized in high yield only by bacteria of abnormally low iron content.

Thus, the classic Park-Williams strain (PW8) of *C. diphtheriae,* isolated in 1898 and still used for commercial production of toxin, possesses the unusual capacity to increase in mass five- to sixfold after depleting its exogenous supply of iron. The differential rate of toxin synthesis may then reach 5% of the total bacterial protein.

The mechanism by which iron controls toxin production has still not been elucidated. Recent evidence suggests that the repressor of the *tox* gene may be an iron-containing bacterial protein. In keeping with such a view, the introduction into *C. diphtheriae* of a high multiplicity of *tox+* phage can overcome the inhibitory effect of iron, presumably by exceeding the number of repressor molecules present.

NATURE AND MODE OF ACTION OF TOXIN

Diphtheria toxin may be isolated from culture filtrates of *C. diphtheriae* as an iron-free, crystalline, heat-labile protein. It is released extracellularly as a single polypeptide chain of approximately 62,000 daltons, with two disulfide bridges. In highly susceptible species, such as guinea pig, rabbit, and man, the pure protein is lethal in doses of 0.1 μg/kg or less. Intradermal injection of only about 10^8 molecules will produce a visible skin reaction. The toxins produced by all naturally occurring strains appear to be immunologically identical. Formol toxoid, prepared by treating toxin with dilute formaldehyde at pH 8 and 37°, is devoid of toxicity but is virtually indistinguishable antigenically from toxin. Immunization with toxoid results in protection against all strains of *C. diphtheriae* capable of causing toxemic diphtheria.

During the past few years the mode of action of diphtheria toxin has become clarified. Strauss and Hendee showed that low concentrations of toxin (10^{-8} M or less) completely block amino acid incorporation by cultured human cells. Subsequently, it was found that suitably activated toxin preparations inhibit polypeptide chain elongation in cell-free extracts by catalyzing inactivation of the translocation factor EF2 (Ch. 12), according to the following reaction:

$$NAD^+ + EF2 \rightleftarrows ADPribosyl\text{-}EF2 + nicotinamide + H^+$$

The ADPribose group is transferred from NAD^+ to become covalently bound in the inactivated factor. The equilibrium of the reaction lies far to the right, but it may be reversed, in inactive extracts of intoxicated cells, by dialysis (to remove NAD) followed by addition of excess nicotinamide and of **an enzymatically active toxin preparation.** Under these conditions, amino acid-incorporating activity may be restored to its normal level. These observations show that toxin inactivates EF2 by ADPribosylation in the living animal as well as in cell-free extracts.

No protein has yet been found in crude mammalian tissue extracts, other than EF2, that will accept the ADPribose group from NAD in the presence of activated diphtheria toxin. Moreover, the reaction is specific for the translocation factor of all species of eukaryotic cells tested, ranging from yeast and protozoa to the higher plants and man; toxin has no effect on the corresponding prokaryotic or mitochondrial factor (i.e., EF-G; see Ch. 12).

The structure of diphtheria toxin is shown diagrammatically in Fig. 24-2. The intact toxin molecule is enzymatically inactive: activation requires reduction of one of the disulfide bridges and hydrolysis of a peptide bond. This change unmasks a 24,000-dalton heat-stable fragment A, on which the enzy-

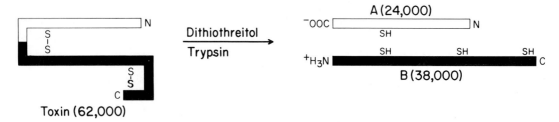

Fig. 24-2. Diagram illustrating "activation" of diphtheria toxin by trypsin in the presence of thiol. All of the transferase II–ADPribosylating activity is present in the N-terminal fragment A. Fragment B is required to enable fragment A to reach the cytoplasm of susceptible animal cells. Mutations of the phage *tox* gene affecting enzyme activity (fragment A) or attachment to the cell (fragment B) may result in a **nontoxic** but serologically related protein.

matic activity is located. Even after it is unmasked fragment A remains bound by weak interactions to a 38,000-dalton, unstable fragment B, which appears to be required for binding to the cell membrane.

Fragments A and B are not separately toxic, but fragment A is active in vitro. Activation of intact toxin in vivo probably occurs at the cell membrane, where fragment A is split off and gains access to the cell cytoplasm. Fortunately for the in vitro analysis, in most preparations of purified toxin part of the molecules had been activated by proteases in the initial crude toxin. A single molecule of fragment A will suffice to convert the entire cell content of EF2 to its inactive derivative within a few hours.

With the realization that the structural gene for toxin synthesis is carried by the β-phage genome, it has become possible to prepare diphtherial strains producing nontoxic mutant proteins of various types. Some are enzymatically inactive because of an altered fragment A but still bind to mammalian cells. Others contain a normal, enzymatically active fragment A, which is unable to gain access to the cell cytoplasm because of an alteration elsewhere in the toxin molecule. When mixtures of these two types of nontoxic mutant molecules are treated with trypsin in the presence of thiols and then allowed to recombine, a fully toxic molecule may be reconstituted.

PATHOGENICITY

Experimental Models. The only known natural reservoirs for toxigenic as well as nontoxigenic *C. diphtheriae* are the upper respiratory tracts of men and horses, and natural infections have been demonstrated only in

man. However, experimental infections have been produced in various laboratory animals.

The closely related *C. ulcerans* and *C. ovis,* frequently found in the nasopharynges of horses and sheep, can be converted to toxigenicity by a *tox+* phage, and can be pathogenic for man, but neither of these strains has been isolated from typical cases of diphtheria.

Although protein synthesis is blocked by toxin in extracts from all eukaryotes, certain animal species are resistant because their cells do not bind toxin and the enzymatically active fragment A cannot gain access to the cell cytoplasm. For this reason rats and mice are more than 1000 times as resistant per unit body weight as susceptible species, such as humans, monkeys, rabbits, guinea pigs, pigeons, and chickens.

Subcutaneous injection of diphtheria toxin into a guinea pig is followed by death within 12 hours to several days, depending upon the dose. The known molecular mechanism of action may explain why even a large dose requires several hours before toxic effects appear. Local swelling and apparent tenderness develop within a few hours, and at autopsy edema and hemorrhage are noted at the site of injection. There is usually marked congestion of the adrenal cortices, and degenerative changes can always be demonstrated in the heart, liver, and kidneys. Sublethal doses of toxin may cause late paralyses similar to those observed in man. A minimal lethal dose (MLD) of diphtheria toxin is defined as the amount that will kill a 250-gm guinea pig on the fourth or fifth day following subcutaneous inoculation.

As was first recognized by Loeffler, when virulent bacilli are injected instead of the toxin the end result is much the same. The systemic lesions which develop are indistinguishable from those caused by the toxin itself, and the injected bacilli remain localized at the original site of inoculation.

Human Disease. Unlike such experimental infections, diphtheria in man usually begins in the upper respiratory tract. When virulent diphtheria bacilli become lodged in the throat of a susceptible individual, they first multiply in the superficial layers of the mucous membrane. There they elaborate toxin, which causes necrosis of neighboring tissue cells and establishes a nidus for further multiplication of the bacteria. The inflammatory response results in the accumulation of a grayish exudate, which eventually forms the characteristic **diphtheritic pseudomembrane** (Fig. 24-3). It usually appears first on the tonsils or posterior pharynx and may then spread either upward into the nasal passages **(nasopharyngeal diphtheria)** or downward into the larynx and trachea **(laryngeal diphtheria).** The nasopharyngeal form of the disease is frequently accompanied by marked prostration and severe toxemia. If recovery ensues, late neurological and cardiac complications are relatively common. Laryngeal diphtheria is particularly hazardous because mechanical obstruction may cause suffocation unless the airway is restored by intubation or tracheotomy.

In contrast to streptococcal pharyngitis (with which it may be easily confused), diphtheria of the upper respiratory tract tends to remain localized. Although enlargement of regional lymph nodes in the neck is common, invasion of other tissues seldom occurs, and bacteremia is not seen. Even the redness and swelling of the mucous membranes in the pharynx are much less pronounced than in streptococccal infections. Fever is usually only moderate.

Proof that the systemic complications of diphtheria in man are due to dissemination of the toxin was provided by a tragic accident in which a large number of children were inadvertently inoculated with unmodified toxin rather than formalin-treated toxoid: 700 became ill with systemic diphtheria and more than 60 died.

Although typical clinically severe diphtheria is

Fig. 24-3. Diphtheria bacilli in pseudomembrane overlying pharyngeal tissue in a fatal case of diphtheria. Note that infection is limited to superficial layers of mucosa. (From Anderson, W. A. D. *Pathology.* Mosby, St. Louis, 1957.)

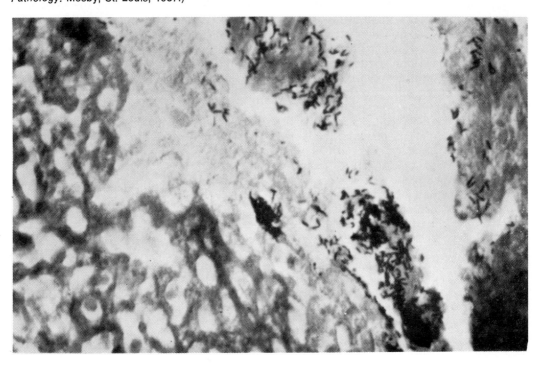

always caused by a toxin-producing organism, mild cases of sore throat with fever and an atypical membrane can be produced by certain *tox⁻* strains. Epidemics of similar mild diphtherial throat infections caused by *tox+* strains have been described among immune populations in persons whose sera already contain circulating antitoxin. In the tropics indolent, ulcerative cutaneous lesions, containing a diphtheritic membrane and *tox+* *C. diphtheriae,* are common.

IMMUNITY AND EPIDEMIOLOGY

Acquired immunity to diphtheria is primarily antitoxic. Newborn infants whose mothers are resistant acquire temporary immunity from transplacental antibodies. Such passive immunity lasts, at the most, only a few months. In areas where diphtheria is not endemic most young children, unless artificially immunized, are highly susceptible. It is probable that immunity can be produced by a mild infection in infants who still retain some circulating maternal antitoxin and in susceptible persons infected with a strain of low toxigenicity.

Persons who have fully recovered from diphtheria may continue to harbor the organisms in the nose or throat for weeks or even months. In the past it was mainly through such healthy **carriers** that the disease was spread and toxigenic bacteria were maintained among a population. Before mass immunization of children, carrier rates for toxigenic *C. diphtheriae* of 5% or higher were observed in New York and London. The advent of universal immunization brought about a dramatic reduction in carrier rate; hence the *tox* gene must have survival value both for a *tox+* phage and for its bacterial host.

Because of the high degree of susceptibility in childhood, artificial immunization at an early age is universally advocated. Formol toxoid, either alum-precipitated or adsorbed on aluminum phosphate gel, is most widely used for this purpose. The final product is so diluted as to contain an appropriate immunizing dose for a child (ca. 10 Lf*) in from 0.5 to 1.0 ml. Two to three doses given 1 month

apart are usually adequate for primary immunization. A booster injection should be given about a year later.

It is common practice in the United States to immunize infants with a combined vaccine containing diphtheria toxoid, tetanus toxoid, and pertussis vaccine. The primary course should be initiated at 3 or 4 months. The immunity ordinarily lasts for several years, although in some cases it may persist for only a few months. In any event, antitoxin levels reached after primary immunization are usually rather low, and in order to ensure continued protection it is advisable to administer several booster injections during childhood, since the likelihood of natural boosters, through chance infection before loss of immunity, has become very small indeed.

Schick test. Whether a given individual is in need of active immunization can be readily determined by means of the Schick test: the intradermal injection of a minute amount of diphtheria toxin. In the absence of circulating antitoxin the injected toxin will cause a local inflammatory response, which will reach a maximum at the end of approximately 5 days. Positive reactions are characterized by local erythema, swelling, and tenderness. If, on the other hand, the blood stream contains a sufficient level of antitoxin, the toxin will be neutralized and no reaction will occur. Thus, a positive Schick test indicates that little or no antitoxin is present in the serum (<0.01 U/ml),† whereas a negative reaction connotes a significant titer (>0.01 U/ml) of circulating antitoxin.

In practice the reading of the Schick test is not so simple, especially among adults and older children and in areas where diphtheria

---

* L stands for the abbreviation of the Latin word *limes* (limit), and 1 Lf is that quantity of toxin or toxoid that flocculates most rapidly when mixed with 1 U of antitoxin (see below).

† Since diphtheria toxin spontaneously changes in time to toxoid, which also combines with antibody, specification and selection of an appropriate unit of antigen cannot be determined on the basis of toxicity alone. Instead one must measure the total antigen by reaction with an arbitrarily standardized antiserum, which, unlike the toxin, is stable. The titers of unknown sera are measured by comparing their ability to neutralize toxin with that of the standard serum. An international standard antiserum is kept at the State Serum Institute in Copenhagen and is available for reference.

is endemic. Many individuals become hypersensitive to the toxin itself or to other antigens in the toxin preparation, as a result of naturally acquired infections or from artificial immunization with toxoid, and react allergically to the intradermally injected toxin. In order to distinguish these delayed hypersensitivity reactions from the primary action of the toxin, a control injection of toxoid (ca. 0.005 Lf) is administered intradermally in the opposite arm. If the individual is immune, but sensitive to one or more antigens in the toxin preparation, he will react to **both** the toxin and the toxoid. The skin reactions, however, usually reach a maximum within 48 to 72 hours and then fade, in contrast to the true **positive Schick reaction,** which persists for many days. This type of response is known as **pseudoreaction.** If, on the other hand, the subject has little or no antitoxin in his serum, but is allergic to an antigen contaminating the toxin, he too will exhibit reactions in both arms, but the reaction at the site of toxin injection will not reach the maximum until 5 days and will persist, whereas that to the toxoid will have subsided by the fifth to seventh day. Such a response is relatively infrequent and is referred to as a **combined reaction.** Individuals exhibiting combined reactions have usually experienced previous exposure to the toxin and almost invariably generate sufficient antitoxin in response to the test itself to become Schick-negative, as indicated by a subsequent test. The various types of responses are summarized in Table 24-2.

Reimmunization of adults and older children should always be approached with caution and should probably be handled on an individual basis to avoid serious local and systemic reactions in persons sensitive to toxoid. It is inadvisable to inject a full immunizing dose of toxoid into anyone who shows a pseudoreaction to the Schick test at 24 or 48 hours; in such cases the test itself will usually have served as a booster.

LABORATORY DIAGNOSIS

A definitive diagnosis of diphtheria can ordinarily be made only by isolating toxigenic diphtheria bacilli from the primary lesion. Exudate from the lesion, taken preferably from the membrane, if present, should be immediately transferred to a Loeffler's slant, a blood agar plate, and tellurite agar. After 24 hours' incubation each culture should be carefully examined, and smears should be made from each type of colony that has grown out on any one of the media. If growth appears on the blood agar plate, but not on the tellurite plate, it may be tentatively concluded that no diphtheria bacilli are present. As a precaution, however, the tellurite plate should be reincubated for an additional 24 hours before being discarded as negative. Smears should immediately be made of any colonies appearing on the tellurite agar and should be stained with methylene blue. The presence of corynebacteria should be readily detected in such smears. Once identified, they should be promptly subcultured on a Loeffler's slant and tested for toxigenicity, either by guinea pig virulence test or by the in vitro gel diffusion method.

The **guinea pig test** is performed by injecting intradermally, into the shaved side of a guinea pig, 0.1 ml of a heavy suspension of the bacilli washed from the Loeffler's slant. After 4 hours, the guinea pig is injected intra-

TABLE 24-2. Reactions to Schick Test

Type of reaction	Skin response				Interpretation
	Toxin		Toxoid		
	36 hr	120 hr	36 hr	120 hr	
Positive	−	+	−	−	Nonimmune; nonsensitive
Negative	−	−	−	−	Immune; nonsensitive
Pseudo	+	−	+	−	Immune; sensitive
Combined	+	+	+	−	Nonimmune; sensitive

peritoneally with 500 U of antitoxin, and 30 minutes later a second sample of the test suspension is injected intradermally on the opposite side. Nonspecific inflammatory reactions may occur at both sites within 24 to 48 hours, but if toxigenic bacilli are present, only the site injected before the antitoxin was administered will progress to form a characteristic necrotic lesion at 48 to 72 hours. Rabbits may be used instead of guinea pigs.

The in vitro **gel diffusion test** is performed by pouring a Petri plate of peptone maltose lactate agar (containing 20% horse serum), into which has been placed a sterile strip of filter paper impregnated with diphtheria antitoxin. After the agar has solidified, a heavy inoculum of the test culture is streaked at right angles across the paper strip, and the plate is incubated for 24 hours. If the inoculated organisms are toxigenic, a visible line of antigen-antibody precipitate will form, radiating from the intersection of the growth and the front of antitoxin diffusing from the paper, as shown in Figure 24-4. The antitoxin used in

Fig. 24-4. Gel diffusion test of toxigenicity. Outer strains are avirulent, whereas those streaked in center of plate are virulent. Note that lines of precipitate generated by the two virulent strains merge to form arcs, indicating that the toxins elaborated are immunologically identical. The serum used had been previously absorbed with nontoxic proteins, obtained from a nontoxigenic culture, and thus behaved as a **monospecific** antiserum. The intersection of the bands with the bacterial growth is removed some distance from the piece of filter paper impregnated with antiserum because horse antitoxin was employed, which inhibits the precipitin reaction in the region of antibody excess. [From King, E. O., Frobisher, M., and Parsons, E. I. *Am J Public Health 39:*1314 (1949).]

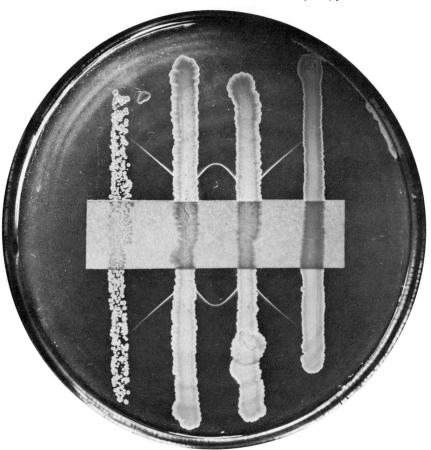

this test must contain no significant amount of antibody against diphtherial proteins other than toxin.

TREATMENT

Once circulating diphtheria toxin has gained entrance to suspectible tissue cells, its noxious action is no longer subject to neutralization by antitoxin. Accordingly, in suspected cases of diphtheria, **antitoxin therapy** must be administered without delay. Indeed, the time factor is so critical (Table 24-3) that the clinician is amply justified in giving antiserum without waiting for the results of the bacteriological tests, provided the circumstances of the infection and the clinical signs of the disease suggest diphtheria. To ensure the maximum immediate therapeutic effect, the antitoxin should be injected intravenously in a single large dose. The amount given should be 100 to 500 U/lb of body weight, depending upon the severity of the disease.* Since the antiserum used is derived from hyperimmunized horses, the usual **skin test for sensitivity to horse serum proteins** should be performed first, and a syringe containing epinephrine should be available for immediate use in the event of an anaphylactic reaction. If the results of the preliminary skin test indicate that

---

* Doses greatly in excess of those theoretically needed to neutralize all the toxin in the patient's tissues are indicated in order 1) to achieve as rapid inactivation of the toxin as possible (including molecules **on** but not yet **in** cells) and 2) to provide sufficient excess of antibody to neutralize toxin subsequently produced during the illness.

TABLE 24-3. Relation of Case Fatality Rate to Time of Treatment with Antitoxin

Day of disease antitoxin administered	Cases treated	Case fatality rate (%)
1	225	0
2	1,441	4.2
3	1,600	11.1
4	1,276	17.3
5 or > 5	1,645	18.7

the patient is sensitive to horse serum, the antitoxin should be given with the greatest possible caution, beginning with a small subcutaneous dose of highly diluted (e.g., 1:10,-000) antiserum, followed by gradually increasing doses administered intramuscularly, and finally intravenously, until the full dose has been injected.

Although diphtheria bacilli are susceptible to the antimicrobial action of penicillin, the tetracyclines, and erythromycin, antibiotic therapy alone should never be relied upon in the treatment of diphtheria. These drugs, however, in conjunction with antitoxin, may hasten the elimination of the causative organisms from the primary lesion.

PREVENTION

The prevention of diphtheria is simpler than that of many infectious diseases because only man appears to be an important reservoir for *C. diphtheriae* and because all strains elaborate the same antigenic type of toxin. For these reasons, it should be possible to eradicate the disease by appropriate methods of immunization, i.e., the widespread use of diphtheria toxoid. In certain areas of the world, including the United States, where children are immunized at an early age and are given booster injections after 1 year and when they enter school, diphtheria has become a rarity. In other countries, however, where diphtheria immunization is rarely practiced, the disease is still prevalent. Its continued spread by droplet infection results from the contact of susceptible individuals, not only with active cases of diphtheria, but also with asymptomatic carriers.

As we have already discussed, carrier rates may be high among urban populations in areas where the disease is endemic. The treatment of carriers presents a troublesome problem. Daily treatment for 1 or 2 weeks with antibiotics such as penicillin, to which diphtheria bacilli are highly sensitive, causes apparent elimination of the organisms from the nose and throat. When treatment is stopped, however, the bacilli often reappear. Fortunately, with the almost complete elimination of the disease itself, very few healthy carriers of virulent diphtheria bacilli remain.

OTHER CORYNEBACTERIA (DIPHTHEROIDS)

Corynebacteria are widely distributed in nature. Many species inhabit the soil, and a number cause disease in animals. *C. ovis,* for example, is responsible for pseudotuberculosis in sheep, horses, and cattle. It resembles the diphtheria bacillus morphologically and produces a potent toxin, which is immunologically unrelated to diphtheria toxin. *C. kutscheri* causes a similar disease in mice and commonly gives rise to latent infections (Ch. 22, Dormant and latent infections).

The two nonpathogenic species of corynebacteria most often found in man are *C. hofmannii* and *C. xerosis.* They grow on tellurite agar and may be easily confused with diphtheria bacilli. *C. hofmannii* is commonly encountered in throat cultures, whereas *C. xerosis* normally inhabits the conjunctival sac. Both are nontoxigenic and may be readily differentiated from *C. diphtheriae* by their fermentation reaction (Table 24-1).

SELECTED REFERENCES

Books and Review Articles

ANDREWES, F. W., BULLOCH, W., DOUGLASS, S. R., FREYER, G., GARDNER, A. D., FILDES, P., LEDINGHAM, J. C. G., and WOLF, C. G. L. *Diphtheria, Its Bacteriology, Pathology and Immunity.* London, H.M.S.O., 1923.

BARKSDALE, W. L. *Corynebacterium diphtheriae* and its relatives. *Bacteriol Rev 34:*378 (1971).

PAPPENHEIMER, A. M., JR. Diphtheria Toxin. In *Microbial Toxins,* vol. IIB. (S. J., Ajl, S. Kadis, and T. C. Montie, eds.) Academic Press, New York, 1973.

WOOD, W. B., JR. *From Miasmas to Molecules.* Columbia Univ. Press, New York, 1961.

Specific Articles

COLLIER, R. J. Effect of dipthheria toxin on protein synthesis. *J Mol Biol 25:*83 (1967).

COLLIER, R. J., and KANDEL, J. Structure and activity of diphtheria toxin. *J Biol Chem 246:*1496, 1504 (1971).

FREEMAN, V. J. Studies on virulence of bacteriophage-infected strains of *Corynebacterium diphtheriae.* J Bacteriol 61:675 (1951).

GILL, D. M., and PAPPENHEIMER, A. M., JR. Structure-activity relationships in diphtheria toxin. *J Biol Chem 246:*1492 (1971).

GILL, D. M., PAPPENHEIMER, A. M., JR., BROWN, R., and KURNICK, J. T. Toxin-stimulated hydrolysis of nicotinamide adenine dinucleotide in mammalian cell extracts. *J Exp Med 129:*1 (1969).

HONJO, T., NISHIZUKA, A., HAYAISHI, O., and KATO, I. Diphtheria toxin-dependent adenosine diphosphate ribosylation of aminoacyl transferase II and inhibition of protein synthesis. *J Biol Chem 243:* 3553 (1968).

MUELLER, J. H. Nutrition of the diphtheria bacillus. *Bacteriol Rev 4:*97 (1940).

PAPPENHEIMER, A. M., JR. The Schick test 1913–1958. *Int Arch Allergy Appl Immunol 12:*35 (1958).

STRAUSS, N., and HENDEE, E. D. The effect of diphtheria toxin on the metabolism of HeLa cells. *J Exp Med 109:*144 (1959).

UCHIDA, T., GILL, D. M., and PAPPENHEIMER, A. M., JR. Mutation in the structural gene for diphtheria toxin carried by beta phage. *Nature New Biol 233:*8 (1971).

UCHIDA, T., PAPPENHEIMER, A. M., JR., and HARPER, A. A. Reconstitution of diphtheria toxin from two nontoxic cross-reading mutant proteins. *Science 175:*901 (1972).

chapter

PNEUMOCOCCI

HISTORY   694
MORPHOLOGY   694
METABOLISM   696
ANTIGENIC STRUCTURE   697
   Capsular Antigens   697
   Somatic Antigens   698
GENOTYPIC VARIATIONS   699
PATHOGENICITY 700
   Host Range   700
   Incidence and Significance of Types   700
   Toxin Production   701
PATHOGENESIS OF PNEUMOCOCCAL PNEUMONIA   701
   Defense Barriers of the Respiratory Tract   701
   Predisposing Factors   701
   Evolution of the Lesion   702
OTHER PNEUMOCOCCAL DISEASES   702
IMMUNITY   704
LABORATORY DIAGNOSIS   704
TREATMENT   705
PREVENTION   705

The pathogenic bacteria to be discussed in this and the following three chapters are often classed together as the **pyogenic cocci.** They include the pneumococci, the streptococci, the staphylococci, and the neisseriae; all but the neisseriae are gram-positive. They are predominantly **invasive** pathogens, which tend to produce acute purulent lesions. Behaving as **extracellular parasites,** they cause tissue damage only as long as they remain outside phagocytic cells; once ingested, they are promptly destroyed. The diseases they cause are **acute,** except when they form abscesses or become lodged on heart valves. They are generally susceptible to antimicrobial drugs, including penicillin, but differ greatly in tendencies to yield drug-resistant mutants. Whereas penicillin-resistant variants of pneumococci are rare and those of group A streptococci virtually nonexistent, resistant mutants are common with staphylococci and group D streptococci and fairly common with viridans streptococci and neisseriae.

HISTORY

The systematic elucidation of the properties of *Diplococcus pneumoniae** (pneumococcus) as an agent of disease has resulted in some of the most important discoveries of biomedical science. First isolated from human saliva in 1881 by Pasteur in France and by Sternberg in the United States, its relation to lobar pneumonia was established a few years later. Recognition of serologically different types in 1910 led to specific antiseria and thus to the first effective treatment for pneumococcal pneumonia. There followed the fundamental observation of Avery, Heidelberger, and

_____

* This term is being replaced in accepted nomenclature by *Streptococcus pneumoniae.*

Goebel on the chemical structure of capsular antigens and their role in bacterial virulence. The discovery that the capsular antigens were carbohydrates had far-reaching effects in immunology, since up to that time it had been believed that all immunogens were proteins. Following Griffith's observation in 1928 that pneumococci of one serological type may be transformed to another type in vivo, Avery, MacLeod, and McCarty discovered that the chemical constituent of the pneumococcal cell responsible for the transforming reaction is DNA (Ch. 9, Transformation). This finding opened the door to molecular genetics and marked the beginning of the current revolution in biology.

Formerly a leading cause of death, pneumococcal pneumonia can now be effectively treated with penicillin and other antimicrobial drugs. Although it is still a common and a serious disease, recovery is usual unless therapy is delayed or the patient is debilitated by a complicating illness.

MORPHOLOGY

Pneumococci in their most typical form are encapsulated, gram-positive, lancet-shaped diplococci. In sputum, pus, serous fluid, and body tissues they may be found in short chains and occasionally as individual cocci. Their tendency to form chains is exaggerated when they are grown in relatively unfavorable media (particularly with a low Mg^{++} concentration) or in the presence of type-specific antibody. Though gram-positive during the exponential phase of growth in artificial media, more and more cells become gram-negative as the culture ages. If incubation is continued the viable count falls and the culture tends to clear. These changes are due to **autolytic enzymes** which first render the cell gram-negative and later bring about lysis. Autolysis is stimulated by surface-active agents, such as bile salts or sodium deoxycholate, and tests for "bile solubility" are useful in identifying pneumococci (p. 705).

The autolytic enzyme activated by deoxycholate is an L-alanine-muramyl amidase (Ch. 4). Its autolytic action is blocked when ethanolamine is substituted for choline in the pneumococcal cell wall (see p. 697). A mutant that lacks the autolytic enzyme activated by deoxycholate (DOC⁻) has also been described.

In liquid cultures most strains of encapsulated pneumococci grow diffusely, tending to sediment only when the medium becomes acid; unencapsulated strains, particularly those that tend to grow in chains (p. 699), exhibit a granular growth, which results in relatively rapid sedimentation. On the surface of solid media (e.g., blood agar plates) the encapsulated organisms form round, glistening, unpigmented colonies with a diameter of

0.5 to 1.5 mm after 24 to 36 hours. In general, the larger the capsule the bigger and more mucoid are the colonies; those of type 3, for example, may reach a diameter of 3 mm. As the mucoid colonies age on blood agar their centers often collapse from autolysis. Colonies growing deep in the agar are ovoid. Both deep and surface colonies become surrounded by a zone of incomplete, α-**hemolysis*** (Ch. 26, History and classification). Hence in gross morphology many pneumococcal colonies on blood agar closely resemble the α-hemolytic colonies of viridans streptococci.

Pneumococcal capsules are most easily demonstrated by suspending the encapsulated organisms in India ink. They also can be seen, and the cells can be typed, by treatment with homologous type-specific antibody (Fig. 25-1), which combines with the capsular

---

* Unless grown anaerobically, in which case β-hemolysis occurs.

Fig. 25-1. A. Quellung reaction of type 3 pneumococcus as seen by light microscopy. ×2000. **B.** Electron micrograph showing type 1 pneumococci reacted with ferritin-labeled antibody; it reveals that the antibody interacts primarily with the surfaces of the capsules. ×21,000. **C.** Electron photomicrograph of pneumococcus type 1 suspended in India ink. Note how closely ink particles abut on cell body (CB) of organism (arrow). ×32,000. **D.** Type 1 antibody has been added to suspension causing capsule (C) to swell and separate India ink particles from body (CB) of cell (arrow). [**B–D** from Baker, R. F., and Loosli, C. G. *Lab Invest 15:*716 (1966).]

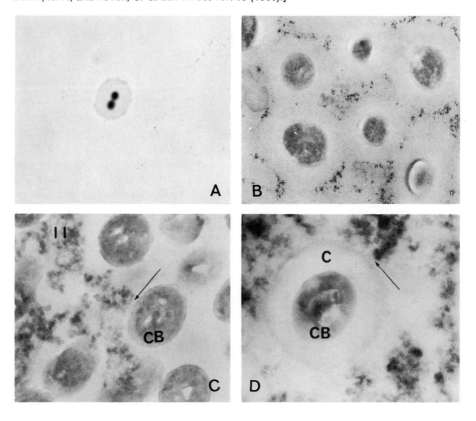

polysacharide and renders it refractile **(quel-lung reaction).**

The word *Quellung* means "swelling" in German. Although there was once much dispute over whether the capsule actually swells when combined with antibody, electron microscopic observations have clearly shown that it does (Fig. 25-1). The quellung reaction, like other polysaccharide precipitin reactions, is inhibited by excess polysaccharide, i.e., in the zone of antigen excess (Ch. 15, Precipitin reaction). Indeed, the reaction may be reversed by the addition of enough homologous polysaccharide or by raising the ionic strength of the preparation to the point where the antigen-antibody complexes dissociate (Ch. 15). A **nonspecific quellung reaction** may also be produced with nonantibody proteins that have the capacity to form strong ionic bonds

with the capsular polysaccharide at an appropriate pH.

In electron photomicrographs pneumococcal capsules can be made out only in hazy outline (Fig. 25-2) unless the cells have been pretreated with homologous antibody, preferably labeled with ferritin (Fig. 25-1). Capsules tend to be largest during the exponential phase of growth when the rate of polysaccharide synthesis is at a maximum (Fig. 25-2). In later phases they become smaller, owing to diffusion of the polysaccharide into the surrounding medium (Fig. 25-2).

The large capsules of pneumococcus type 3, like those of certain group A hemolytic streptococci, stain metachromatically (red) with methylene blue. This property has proved useful in studying the pathogenesis of type 3 pneumococcal infections.

Pneumococcal **L forms** (Ch. 40), deficient in cell wall, have been cultivated on hypertonic agar media containing penicillin.

Fig. 25-2. A. Chart showing relation of growth of type 3 pneumococcus in broth culture to state of capsule and cumulative synthesis of capsular polysaccharide. Electron photomicrographs of type 3 pneumococci from 4-hour (**B**) and 24-hour (**C**) cultures demonstrate loss of outer portion of capsule (slime layer) as culture ages and synthesis of capsular polysaccharide slows (see change in slope of broken curve in chart). ×6700. [From Wood, W. B., Jr., and Smith, M. R. *J Exp Med 90*:85 (1949).]

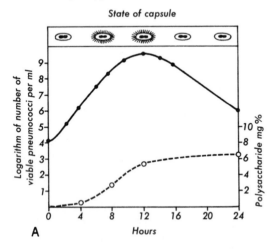

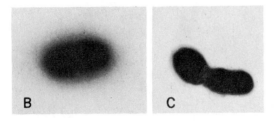

METABOLISM

Pneumococci need a complex medium for growth. Their energy requirements are met primarily by a lactic acid type of fermentation (Ch. 3, Fermentations) and they are classified, along with streptococci, as lactic acid bacteria. Their ability to metabolize inulin is useful in differentiating them from α-hemolytic streptococci. The most satisfactory liquid medium is fresh beef infusion broth, containing 10% serum or blood, and titrated to a final pH of 7.4 to 7.8. No dehydrated medium yet available will support the growth of pneumococci as well as one containing fresh meat infusion. When exposed to air, infusion broth tends to become inhibitory. This difficulty can be overcome by the addition of a reducing agent, such as cysteine or thioglycollate, which also permits a smaller inoculum to initiate growth even in fresh medium.

A synthetic culture medium for pneumococcus has been described by Tomasz. Although not satisfactory for routine use, this medium is useful for recovering macromolecular fractions from pneumococcal cells, since its constituents are all dialyzable and thus can be easily eliminated.

Pneumococci are facultative anaerobes. Since they

produce neither catalase nor peroxidase, the cultures accumulate hydrogen peroxide, often affecting viability. For this reason it is advisable, when storing pneumococcal cultures, to make certain that a source of catalase is present to promote decomposition of the hydrogen peroxide. Red cells serve admirably for this purpose, their presence permitting the organisms to survive in broth cultures at 0 to 4° for long periods. Pneumococci grown in undiluted rabbit blood under petroleum jelly maintain their full virulence for weeks, even for several months, in the refrigerator.

When large yields of pneumococcal cells are needed, the organisms are usually grown in a medium rich in glucose, and the lactic acid is neutralized by continuous or intermittent additions of sodium hydroxide. Glucose stimulates capsule production.

Ethanolamine ($HO\text{-}CH_2\text{-}CH_2\text{-}NH_2$) may be substituted for choline ($HO\text{-}CH_2\text{-}CH_2\text{-}N \equiv [CH_3]_3$) in the synthetic culture medium. The resulting substitution of ethanolamine for choline in the teichoic acid component of the cell wall causes the following abnormalities 1) the cells fail to divide normally as diplococci and they form long chains ("linear clones"), 2) they are resistant to autolysis (p. 695) and do not lyse even when grown in the presence of penicillin, and 3) they lose their ability to undergo genetic transformation. These results indicate that the $-N(CH_3)_3$ moiety of the choline molecule plays a critical role in the pneumococcal cell wall.

ANTIGENIC STRUCTURE

More than 80 serological types of pneumococci have been differentiated by the **immunologically distinct polysaccharides in their capsules.** Three different kinds of somatic antigens* have also been described. A poorly defined **R antigen** (p. 698) and a carbohydrate **C substance** are species-specific. A type-specific protein **(M antigen)** is immunologically independent of the type-specific capsular polysaccharide.

CAPSULAR ANTIGENS

Structure. Pneumococcal capsules are composed of large polysaccharide polymers which form hydrophilic gels on the surface of the

---

* The term **somatic antigen** is used to designate antigenic components of the body (*soma*, Gr.) of a bacterial cell, exclusive of its capsule.

organisms. Although the compositions of many of these capsular carbohydrates are known, the structures of only a few (e.g., types 3, 6, and 8) have been definitely determined. Type 3, for example, consists of repeating cellobiuronic acid units (D-glucuronic acid $\beta 1 \rightarrow 4$ linked to D-glucose) joined by $\beta 1 \rightarrow 3$ glucosidic bonds (Fig. 15-19).

Cross-Reactions. Although pneumococcal capsular polysaccharides exhibit an extraordinary degree of type specificity, some of them cross-react with capsular antigens of other pneumococcal types and of certain other bacterial species (e.g., α-hemolytic and nonhemolytic streptococci, Friedländer's bacilli, and salmonellae). Thus, the type 8 antigen contains D-glucose, D-galactose, and D-glucuronic acid in a sequence that includes cellobiuronic acid; hence, it is not surprising that this polymer should cross-react with the type 3 antigen.

Biosynthesis. The pathway of biosynthesis of type 3 polysaccharide has been elucidated. The principal steps are indicated in Figure 25-3.

Pneumococci that contain the polysaccharide antigens of two serotypes in their capsule are occasionally encountered as products of transformation reactions. The identification of such binary encapsulated strains is illustrated in Figure 25-4.

Pathogenetic Role. The importance of the pneumococcal capsular antigens for pathogenicity can be demonstrated in a number of ways. Thus only smooth encapsulated (S) strains are pathogenic for man and most lab-

Fig. 25-3. Metabolic pathway of type 3 capsular polysaccharide synthesis. UTP = uridine triphosphate; UDPG = uridine diphosphoglucose; UDPGA = uridine diphosphoglucuronic acid; P-P$_i$ = inorganic pyrophosphate. [From Austrian, R., *et al. J Exp Med 110:585* (1959).]

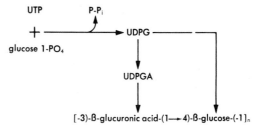

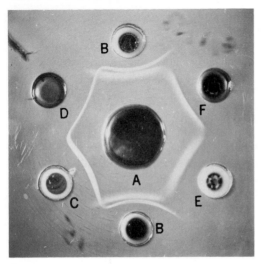

Fig. 25-4. Gel diffusion reactions (Ouchterlony technic) of antisera to type 1 and type 3 pneumococci (in center well, A) with type 1,3 pneumococci (B), type 1 pneumococci (C), type 1 capsular polysaccharide (D), type 3 cells (E), and type 3 polysaccharide (F). [From Austrian, R., and Bernheimer, H. P. *J Exp Med 110:*571 (1959).]

oratory animals (p. 699) and active* or passive immunization against a specific polysaccharide produces a high level of resistance to infection with pneumococci of the homologous type. Moreover, for any given type of pneumococcus virulence appears to be roughly related to capsular size. For example, **intermediate variants** (which produce small capsules) are less virulent than fully encapsulated strains of type 3 but more virulent than rough variants (which produce no capsule at all). Pneumococci of different types with capsules of the same size, on the other hand, may vary widely in virulence (e.g., types 3 and 37). The mechanism by which the capsules impede phagocytosis is not known.

Although the antiphagocytic effect of the pneumococcal capsule may be readily observed both in vitro and in vivo, the physicochemical factors have not been defined that operate at the interface between leukocyte and capsule, and thereby impede the process of ingestion (Ch. 22, Phagocytosis).

---

* The purified polysaccharides, unlike whole pneumococcal cells, are weak immunogens for rabbits. In mice, however, those of many types are highly immunogenic; but when given in doses of more than 1 µg they produce immunological paralysis.

SOMATIC ANTIGENS

In 1930 Tillett, Goebel, and Avery isolated from pneumococcal cells a carbohydrate which, unlike the capsular polysaccharides, contains phosphate residues and is specific for species rather than type. This cell wall antigen, referred to as pneumococcal **C substance,** appears to be analogous to (though antigenically different from) the group-specific C antigens of hemolytic streptococci (Ch. 26, Cellular antigens). As purified from autolysates or deoxycholate lysates of pneumococci, C substance contains galactosamine-6-phosphate as its major constituent, in addition to phosphoryl choline, a diaminotrideoxyhexose, and elements of the cell wall mucopeptide. On the other hand, material obtained by trichloracetic acid extraction also contains substantial amounts of ribitol and glucose, suggesting a teichoic acid polymer. The details of the relations between the two forms of the antigen remain to be clarified.

The C substance, which also forms a portion of the **Forssman antigen** (Ch. 15, Blood group substances) of the pneumococcus, has the interesting property of being precipitated (in the presence of Ca^{++}) by a β-globulin of the serum. This **C-reactive protein is not an antibody;** it is an abnormal protein that is detectable in the blood only during the active phase of certain acute illnesses. A precipitation test designed to determine its concentration in serum is widely used as a measure of "activity" in inflammatory diseases such as rheumatic fever (Ch. 26, Pathogenicity).

The **R antigen,** so named because it was originally extracted from rough (R) pneumococci, is believed to be a protein on or near the surface of the cell.

Sera prepared against rough (R) pneumococci derived from a given smooth (S) type often agglutinate to a higher titer R pneumococci derived from a homologous S type than R strains derived from a heterologous S type. This difference is due to **type-specific somatic M antigens,** which are proteins similar to the type-specific M antigens of group A hemolytic streptococci (Ch. 26, Cellular antigens). Pneumococcal M antigens, however, do not exert a significant antiphagocytic effect; antibodies against them, therefore, are not protective, as are the antibodies to streptococcal M antigens.

Immunization of rabbits with heat-killed R pneumococci causes a slight increase in resistance to infections with pneumococci of any type. This broad immunity is believed to be due to the formation of antibodies to the species-specific somatic antigens: the C polysaccharide and the R antigen.* The degree of immunity thus induced is negligible, however, compared with that mediated by type-specific antibody.

GENOTYPIC VARIATIONS

Cultures of encapsulated pneumococci generate unencapsulated mutants at a low but definite rate. When such cultures are grown in homologous type-specific antiserum, the encapsulated S cells become agglutinated by the anti-S antibody, and the R mutants are thus given a selective advantage.† After a few serial passages in such a medium most of the

---

* Although C polysaccharide is ordinarily confined to the cell wall, an interesting mutant has been isolated which makes a capsule of "C-like" polysaccharide.

† The crowding together of the agglutinated S cells probably places them at a metabolic disadvantage in the medium.

cells recovered will be devoid of capsules and will therefore have lost both their type specificity and their virulence. Also R mutants tend to be selected when S cells are repeatedly subcultured on blood agar.

Conversely, in cultures of R pneumococci occasional cells revert (back-mutate) to S. If anti-R serum is present in the medium the S cells are favored, and on repeated subculture they will eventually replace the R cells. When R pneumococci are injected into a mouse the S mutants are given an even greater selective advantage, since the animal's phagocytes rapidly destroy the unencapsulated cells but not the encapsulated ones, which continue to multiply and eventually kill the host. Thus the tissues of the living mouse provide a highly selective culture medium. This principle underlies the conventional method of maintaining the maximum virulence of pneumococcal cultures by passing them through mice at frequent intervals.

Two kinds of surface colonies may be formed by R pneumococci. The first is the usual small, moderately granular, rough colony already described. The second is a large colony containing long chains of cocci which form intricate filaments responsible for excessive roughness of its surface (Fig. 25-5). Austrian has also described filamentous

Fig. 25-5. A. Colonies of a nonfilamentous (fil−), unencapsulated (S−) variant (usual rough form) of pneumococcus type 2 on blood agar after 24 hours at 36°. ×18. **B.** Colonies of a filamentous (fil+), unencapsulated (S−) variant (very rough form) grown under same conditions. ×18. **C.** Cells from colony of nonfilamentous variant. Gram stain. ×900. **D.** Cells of filamentous variant. Gram stain. ×1100. [From Austrian, R. *J Exp Med 98:*21 (1953).]

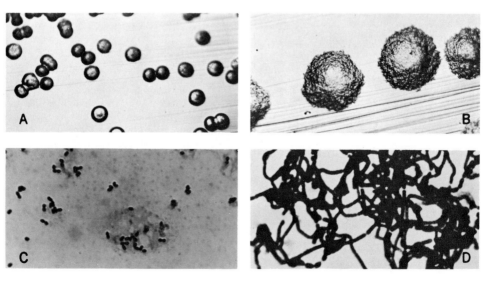

smooth forms and has demonstrated that the chaining trait is a heritable characteristic.

Transformations of pneumococcal types by exogenous DNA have been discussed (Ch. 9, Transformation). How frequently such transformations occur in nature is not known. However, growing populations of pneumococci release transforming DNA into the culture medium, and this release is maximal during the phase of growth when the culture is most responsive to added DNA (Fig. 25–6). Transformations have also been shown to occur in experimental pneumococcal lesions in mice.

Genetic markers which have been successfully transformed in vitro include those relating to 1) the type specificity of capsular antigens; 2) the **amount** of capsular polysaccharide produced (intermediate strains); 3) resistance to antipneumococcal drugs, such as penicillin, streptomycin, sulfonamides, and optochin; 4) the type specificity of somatic M proteins; 5) capacity to produce certain inducible enzymes, e.g., mannitol-phosphate dehydrogenase; and 6) chain formation affecting colonial morphology of R and S mutants.

Pneumococci have also been transformed with DNA prepared from streptococci. This fact, when added to their striking similarities in morphology, metabolism, and antigenic structure, further emphasizes the close relation between streptococci and pneumococci. Indeed, pneumococci are still classified as streptococci (*Streptococcus pneumoniae*) by most British taxonomists.*

PATHOGENICITY

HOST RANGE

In addition to being pathogenic for man, most encapsulated strains of pneumococci will produce experimental disease in mice, rats, rabbits, monkeys, and dogs. Cats and birds are relatively resistant. The reasons for these wide variations in susceptibility are not clear (Ch. 22, Natural antibodies). It is obvious, however, that the results of virulence studies performed in one animal species can be applied to another only with caution. In fact, strain variations in susceptibility within a given animal species may be highly significant. Severe epizootics of pneumococcal pneumonia occasionally occur in monkeys, guinea pigs, and rats.

INCIDENCE AND SIGNIFICANCE OF TYPES

The antigenic types of pneumococci that most commonly caused pneumonia in the 1930s are listed in Table 25-1. More than half the cases of adult lobar pneumonia were due to types 1, 2, and 3. In children, on the other hand, type 14 was especially common and is of particular interest because of the antigenic relation of its capsular polysaccharide to the blood group substances (Ch. 21). This cross-reaction is presumably responsible for the hemolytic

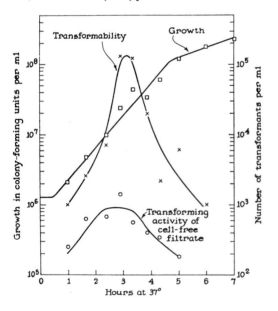

Fig. 25-6. Development of transformability and release of transforming material by pneumococci growing in broth culture. The strain in the culture was sulfonamide-resistant but streptomycin-sensitive. Filtrates prepared from the culture at various intervals were used as donor material to transform a sulfonamide-sensitive strain (**transforming activity**). Transformability of the cultured strain was tested at various stages of the growth phase by adding purified DNA from a streptomycin-resistant culture. The cells from the original culture were exposed to the DNA for 30 minutes and the reaction was terminated by DNase. [From Ottolenghi, E., and Hotchkiss, R. D. *J Exp Med* 116:491 (1962).]

* See *Topley and Wilson's Principles of Bacteriology and Immunity,* p. 726. (G. S. Wilson and A. A. Miles, eds.) Williams & Wilkins, Baltimore, 1964.

TABLE 25-1. Comparative Distributions of Pneumococcal Types in Adults and Children (Under Age 12) with Lobar Pneumonia

Adults		Children	
Type	% of cases	Type	% of cases
1	28.6	14	12.0
3	13.5	1	11.2
2	11.4	6	7.7
5	8.0	19	3.9
8	7.7	5	3.6
7	6.5	4	3.1
4	3.5	3	2.7
6	1.8	7	2.1
All other	19.0	All other	53.7
	100.0		100.0

Data from HEFFRON, R. *Pneumonia, with Special Reference to Pneumococcus Lobar Pneumonia,* pp. 53, 76. Commonwealth Fund, New York, 1939.

anemia sometimes observed in patients recovering from type 14 pneumonia.

In a more recent study of bacteremic pneumococcal infection in selected hospitals throughout the United States, the most common types encountered in adults were 8, 4, 3, 1, 7, and 12, in that order. Although type 2 was a major cause of infection, it has been encountered rarely in this country and in northern Europe in recent years, but remains an important cause of pneumonia in South Africa. In children types 19, 23, 14, 3, 6, and 1 are mainly responsible for otitis media and bacteremic pneumonia.

Type 3 pneumococcus produces the largest capsule and causes the highest case fatality rate in man. Other highly virulent types are 1, 2, 5, 7, and 8.

TOXIN PRODUCTION

Pneumococci produce a hemolytic toxin, called **pneumolysin,** which is related immunologically to the oxygen-labile O hemolysins of hemolytic streptococci, *Clostridium tetani,* and *Clostridium welchii.* They also produce a toxic neuraminidase and release during autolysis a **purpura-producing principle** which causes dermal, as well as internal, hemorrhages when injected into rabbits. Although there is no conclusive evidence that any of these products play a role in the pathogenesis of pneumococcal infections, they may exert short-range toxic effects that have remained undetected.

The type-specific capsular polysaccharides released into the fluid phase by multiplying pneumococci are relatively nontoxic, but they are significant in pathogenesis because they neutralize antibody before the latter has a chance to become bound to the pneumococci invading the tissues.

Since there is no pneumococcal toxin known to be related to pathogenicity, and the virulence of the organism appears to be due primarily to its invasiveness, the cause of death in pneumococcal infections remains a mystery. It is conjectured that, in the terminal stages, a lethal toxin might be generated in vivo, analogous to that described in anthrax (Ch. 33, Pathogenicity).

PATHOGENESIS OF PNEUMOCOCCAL PNEUMONIA

DEFENSE BARRIERS OF THE RESPIRATORY TRACT

Between 40 and 70% of normal human adults carry one or more serological types of pneumococci in their throats, yet epidemics of pneumococcal pneumonia are rare, and morbidity is low. Bacterial antagonism (Ch. 22, Microbial factors), involving primarily α-hemolytic streptococci, tends to limit the growth of pneumococci in the pharynx. The extraordinarily efficient defense barriers of the lower respiratory tract include 1) the epiglottal reflex, which prevents gross aspiration of infected secretions; 2) the sticky mucus to to which airborne organisms adhere on the epithelial lining of the bronchial tree; 3) the cilia of the respiratory epithelium, which keep the infected mucus moving upward into the pharynx (Ch. 22, Mechanical factors); 4) the cough reflex, which aids the cilia in propelling accumulated mucus from the lower tract; 5) the lymphatics draining the terminal bronchi and bronchioles; and 6) the mononuclear "dust cells" (macrophages), which patrol the normal alveoli.

PREDISPOSING FACTORS

Pneumococcal pneumonia develops most often during the course of viral infections of the upper respiratory tract, when the resulting flood of mucous secretion in the nose and pharynx enhances the likelihood of aspiration. Once past the epiglottal barrier, the thin mucus laden with bacteria (including encapsulated pneumococci) is carried by gravity to the farthest reaches of the bronchial tree, where it establishes the initial focus of pulmonary infection. Aspiration is promoted by factors that slow the epiglottal reflex, including chilling of the body, anesthesia, morphine, and alcoholic intoxication.

Aerosol experiments with mice have also revealed that an edematous lung is far more susceptible to pneumococcal infection than the normally dry lung. Pulmonary edema fluid provides a suitable culture medium for aspirated pneumococci and interferes with the phagocytic activities of the tissue macrophages, which constitute the first line of cellular defense in the alveoli. It is not surprising, therefore, that factors which produce either local or generalized pulmonary edema should also predispose to pneumococcal pneumonia. These include inhalation of irritating anesthetics, trauma to the thorax, cardiac failure, influenza virus infections involving the lungs, and pulmonary stasis resulting from prolonged bed rest.

EVOLUTION OF THE LESION

Once infection has become established in a bronchial segment, the pneumonic lesion spreads centrifugally, as depicted in Figure 25-7. In the **edema zone,** at the outer margin of the spreading lesion, the alveoli are filled with acellular serous fluid, which serves as a suitable culture medium for the organisms. Following the outpouring of edema fluid, polymorphonuclear leukocytes and a few erythrocytes begin to accumulate in the infected alveoli and eventually fill them with a densely packed leukocytic exudate **(consolidation).** As they accumulate in sufficient numbers, the leukocytes phagocytize and destroy the infecting pneumococci. When the bacteria have been disposed of, macrophages replace the granulocytes in the exudate, and **resolution** of the lesion ensues. Thus all stages of the inflammatory process are demonstrable simultaneously in the spreading pneumonic lesion. The earliest stage, characterized by increased capillary permeability and the accumulation of serous fluid, is represented by the peripheral edema zone. The late **macrophage reaction,** characteristic of subsiding inflammation, is prominent in the central "burned-out" portion of the lesion.

When alveoli underlying the pleura become involved, **pleurisy** develops and the pleural cavity frequently becomes infected. If unchecked, pleural infections may develop into extensive intrapleural abscesses **(empyema).** The adjacent pericardium may also be affected **(pericarditis).**

Pneumococcal pneumonia is often multilobar. Spread of the infection from one lobe to another results from the flow of infected edema fluid, propelled by the combined effects of coughing and the force of gravity.

Although the majority of pneumococci in the pulmonary lesion are destroyed by phagocytosis, some may be carried by the lymphatics to the regional lymph nodes at the hilus of the lung and thence, via the thoracic duct, to the blood stream. Once bacteremia has developed, organisms may settle on the heart valves or in the meninges or joints. Occasionally, during the bacteremia, pneumococci can be cultured from the urine. In less serious cases the primary pneumonic lesion resolves spontaneously without complications.

The phagocytic mechanisms that operate at the various lines of cellular defense in the lungs, lymph nodes, and blood stream, have been described in Chapter 22. As already pointed out, these mechanisms are relatively inefficient in fluid-filled cavities (subarachnoid space, pleura, pericardium, joints) and are incapable of destroying all the pneumococci after an abscess has formed. Furthermore, many strains of type 3 pneumococcus, which produce usually large capsular envelopes, are highly resistant to phagocytosis in the absence of homologous type-specific antibody. As a result, in acute pneumonic lesions they tend to reach population densities considerably higher than those attained by other types of pneumococci. Their excessive growth in the alveoli occasionally leads to irreversible tissue damage and abscess formation. Type 3 is virtually the only type that causes lung abscesses in man.

OTHER PNEUMOCOCCAL DISEASES

Primary pneumococcal diseases of the upper respiratory tract include sinusitis and otitis media. The latter occurs most commonly in children and may spread to involve the mastoid. Progressive infections of the mastoid or respiratory sinuses sometimes extend directly to the subarachnoid space to cause pneumococcal meningitis. Secondary pneumococcal peritonitis, resulting from transient bacteremia following a primary respiratory infection, occurs most commonly in children with ascites due to nephrosis and in adults with cirrhosis or carcinoma of the liver.

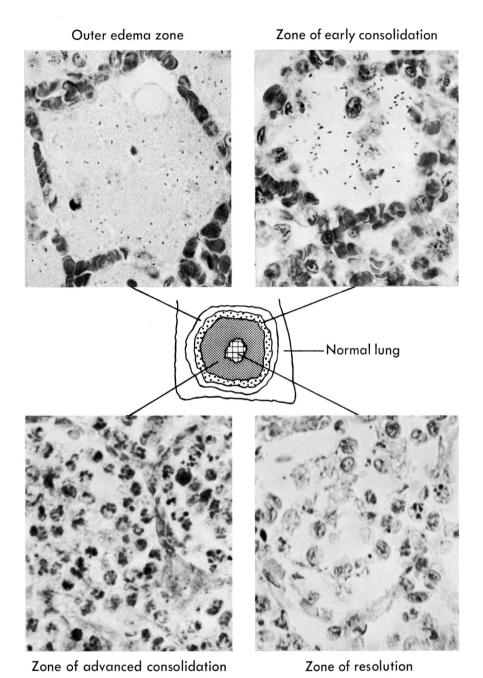

Outer edema zone   Zone of early consolidation

Normal lung

Zone of advanced consolidation   Zone of resolution

Fig. 25-7. Schematic diagram of spreading pneumonic lesion, showing a characteristic microscopic field in each of its four histologically distinguishable zones. [From Wood, W. B., Jr. *Harvey Lect 48:*72 (1951–1952).]

IMMUNITY

Type-specific anticapsular antibody is not usually demonstrable in the serum of patients with pneumococcal pneumonia until the fifth or sixth day of the disease, at which time a spontaneous crisis* is likely to occur (unless, of course, the patient has already been cured by specific treatment). The chances of recovery occurring in untreated cases are approximately 7 out of 10. The spontaneous crisis may even take place in the preantibody stage of the illness, as a result of the effectiveness of the primary cellular defenses that operate in the lung (Ch. 22, Phagocytic defense). Following recovery, anticapsular antibody usually remains detectable in the patient's serum for several months.

Patients recovering from pneumococcal pneumonia often continue to carry the infecting pneumococcus in their upper respiratory tracts for many days or even weeks, despite the presence of circulating homologous antibody. A more or less permanent carrier state may result from chronic pneumococcal sinusitis. Recurrent attacks of pneumococcal disease are usually due to a new serological type of organism, except in cases involving a persistent focus of infection (e.g., bronchiectasis or chronic sinusitis) or in patients with immunological defects.

The presence of protective type-specific anticapsular antibody may be detected by precipitin tests with the homologous polysaccharide or by agglutination, phagocytic, or capsular swelling tests, performed with homologous pneumococcal cells. Circulating antibody may also be detected in the living patient by injecting homologous capsular polysaccharide into the skin; the interaction of the antibody with the injected antigen causes an immediate wheal and erythema reaction at the site of injection (Francis' skin test).

Pneumococcal capsular polysaccharides tend to remain in the tissues of the host for relatively long periods. In experimental pneumonia they have been demonstrated in alveolar macrophages many weeks after the acute disease has subsided. The polysaccharide thus retained at the site of the primary lesion is gradually released and excreted in the urine, where it may be detected by precipitin tests weeks or even months after recovery.

LABORATORY DIAGNOSIS

A tentative diagnosis of pneumococcal pneumonia can be made most rapidly by examining the patient's sputum. Smears made from a fresh sample, raised directly from the bronchial tree, should be stained by the Gram method to distinguish *D. pneumoniae* from *Klebsiella pneumoniae* and staphylococci which also produce acute bacterial pneumonia (Ch. 29, Klebsiella organisms). If typical lancet-shaped, gram-positive diplococci are seen in a smear, a presumptive diagnosis of pneumococcal pneumonia may be made and specific treatment begun. At the same time, the sputum should be cultured on blood agar for final identification (see below). When a specimen of sputum cannot be obtained (e.g., from a small child), a pharyngeal culture is utilized in the same manner.

With typing sera available, pneumococci seen in the sputum may be immediately typed by means of the quellung reaction. (Although this procedure is now rarely carried out in diagnostic laboratories, it was formerly done routinely to enable the physician to treat the patient promptly with the correct type of antiserum.) If no typical gram-positive diplococci are found in the sputum of a patient strongly suspected of having acute bacterial pneumonia, a sample should be emulsified with a small amount of broth in a 1- or 2-ml syringe and injected intraperitoneally into a mouse. When virulent pneumococci are present, the mouse will usually die within 4 days, and the offending pneumococcus can then be recovered from heart blood† or from other infected body fluids.

Since many healthy individuals carry pneumococci in their throats, demonstration of the organism in sputum or throat culture is not necessarily indicative of pneumococcal dis-

---

* The term **crisis** refers to the dramatic defervescence and subsidence of symptoms which characteristically terminate uncomplicated pneumococcal pneumonia.

† Blood drawn from the heart is less often contaminated with extraneous organisms than peritoneal or pleural fluid.

ease. If, on the other hand, pneumococci are cultured from a patient's blood, the diagnosis of a severe acute pneumococcal infection can be made with certainty. Accordingly, in all cases of suspected acute bacterial pneumonia samples of blood obtained by venepuncture (prior to the administration of antimicrobial drugs) should be cultured immediately in both beef infusion broth and thioglycollate medium. (The latter promotes outgrowth of small inocula of pneumococci.) A sample of blood should also be used to make one or more pour plates for bacterial counts.

Serous fluids obtained from pleural, pericardial, peritoneal, or synovial cavities should be cultured in beef infusion broth, in thioglycollate medium, and on blood agar.* At the same time, direct smears of the fluid should be stained by the Gram method. Cultures of spinal fluid are performed in the same manner. In pneumococcal meningitis the spinal fluid frequently contains enough organisms to permit presumptive diagnosis by examining a stained (Gram) smear.

Pneumococci are easily confused with streptococci of the viridans groups (Ch. 26, α-Hemolytic streptococci). Unlike most viridans streptococci, however, pneumococci are generally bile-soluble, virulent for mice, and sensitive to optochin (ethylhydrocupreine). The bile-solubility test is best done with deoxycholate and should be performed on saline suspensions of the bacterial cells, because extraneous proteins from liquid media inhibit the reaction, presumably by binding the deoxycholate. Blood agar plates containing paper discs impregnated with optochin, an antibacterial drug relatively specific for pneumococci, may also be used for presumptive identification.

TREATMENT

Type-specific antiserum and sulfonamides are no longer recommended. The drug generally

---

* Some strains of pneumococcus (<10%) require CO_2 for growth on solid media (incubation in a candle jar).

used is penicillin. Pneumococci significantly resistant to penicillin are rare, and penicillin treatment of uncomplicated pneumococcal pneumonia is almost always effective unless started too late. Its limitations, particularly in the management of established pneumococcal abscesses, such as empyema, have been discussed in Chapter 23. Erythromycin or lincomycin may be used in patients with known hypersensitivity to penicillin, but the pneumococcal isolates should be tested for sensitivity to these drugs. Broad-spectrum antibiotics (e.g., the tetracyclines) may also be effective, although tetracycline-resistant strains of *D. pneumoniae* are becoming common. Chloramphenicol should not be used except to treat pneumococcal meningitis in patients allergic to penicillin.

PREVENTION

Inasmuch as pneumococcal diseases respond promptly to early antimicrobial therapy and afflict less than 1 in 500 persons per year, prophylactic measures are rarely indicated. Occasionally in closed communities, such as military installations and custodial institutions, the pneumococcal carrier rate will become unusually high and an epidemic will result. Under such conditions immunization with pneumococcal polysaccharides of the types prevalent in the carriers may be useful. Indiscriminate treatment of acute respiratory infections with antimicrobial drugs to prevent pneumonia should be discouraged, not only because of the relatively low attack rates, but also because of the danger of drug reactions and the possibility of promoting penicillin resistance of other bacteria being carried by the host.

Ideally, every patient with pneumococcal pneumonia should be isolated. Although isolation rules are often disregarded because of the relatively low cross-infection rates, in crowded hospital wards containing persons with congestive heart failure and other debilitating diseases, it is advisable whenever possible to place patients with pneumococcal pneumonia in separate rooms.

SELECTED REFERENCES

Books and Review Articles

HEFFRON, R. *Pneumonia, with Special Reference to Pneumococcus Lobar Pneumonia.* Commonwealth Fund, New York, 1939.*

WHITE, B. *The Biology of Pneumococcus: The Bacteriological, Biochemical and Immunological Characters and Activities of* Diplococcus pneumoniae. Commonwealth Fund, New York, 1938.*

WOOD, W. B., JR. Pneumococcal Pneumonia. In *Cecil-Loeb Textbook of Medicine.* (P. B. Beeson and W. McDermott, eds.) Saunders, Philadelphia, 1971.

Specific Articles

AUSTRIAN, R. Morphologic variation in pneumococcus. *J Exp Med 98:*21 (1953).

AUSTRIAN, R. The prevalence of pneumococcal types and the continuing importance of pneumococcal infection. *Am J Med Sci 238:*51 (1959.

AUSTRIAN, R., and GOLD, J. Pneumococcal bacteremia with especial reference to bacteremic pneumococcal pneumonia. *Ann Intern Med 60:*759 (1964).

AUSTRIAN, R., and MACLEOD, C. M. A type-specific

---

* Although unrevised, these two monographs remain classics in the field.

protein from pneumococcus. *J Exp Med 89:*439 (1949).

BORNSTEIN, D. L., SCHIFFMAN, G., BERHEIMER, H. P., and AUSTRIAN, R. Capsulation of pneumococcus with soluble C-like (C$_s$) polysaccharide. *J Exp Med 128:*1385 (1968).

HAUSMAN, D., HEATHERBELL, G., START, J., DEVITT, L., and DOUGLAS, R. Increased resistance to penicillin of pneumococci isolated from man. *N Eng J Med 284:*175 (1971).

KELLY, R., and GREIFF, D. Toxicity of pneumococcal neuraminidase. *Infect Immun 2:*115 (1970).

KNECHT, J. C., SCHIFFMAN, G., and AUSTRIAN, R. Some biological properties of pneumococcus type 37 and the chemistry of its capsular polysaccharide. *J Exp Med 132:*475 (1970).

MOSSER, J. L., and TOMASZ, A. Choline-containing teichoic acid as a structural component of pneumoccocal cell wall and its role in sensitivity to lysis by an autolytic enzyme. *J Biol Chem 245:*287 (1970).

OTTOLENGHI, H., and HOTCHKISS, R. D. Release of genetic transforming agent from pneumococcal cultures during growth and disintegration. *J Exp Med 116:*491 (1962).

TOMASZ, A., JAMIESON, J. D., and OTTOLENGHI, E. The fine structure of *Diplococcus pneumoniae. J Cell Biol 22:*453 (1964).

WOOD, W. B., JR., and SMITH, M. R. Host-parasite relationships in experimental pneumonia due to pneumococcus type III. *J Exp Med 92:*85 (1950).

chapter

STREPTOCOCCI

Revised by MACLYN McCARTY, M.D.

HISTORY AND CLASSIFICATION   708

β-HEMOLYTIC STREPTOCOCCI   709

    Morphology   709
    Metabolism   712
    Cellular Antigens   712
    Extracellular Products   717
    Genotypic Variations   720
    Pathogenicity   720
    Immunity   722
    Laboratory Diagnosis   722
    Treatment   723
    Prevention   723
    Epidemiology   724
    β-HEMOLYTIC STREPTOCOCCI OF OTHER GROUPS   724

OTHER STREPTOCOCCI   724

    α-HEMOLYTIC STREPTOCOCCI   724
    NONHEMOLYTIC STREPTOCOCCI   725
    ANAEROBIC STREPTOCOCCI   725

HISTORY AND CLASSIFICATION

Globular microorganisms growing in chains were first described by Billroth in 1874 in purulent exudates from erysipelas lesions and infected wounds. Similar organisms, eventually named streptococci (Gr. *streptos* = winding, twisted) were isolated from the blood in puerperal fever and from the throat in scarlet fever, and in 1882 Fehleisen induced typical erysipelas in human volunteers with streptococci from lesions of patients with the same disease.

At first, each different kind of streptococcal disease was thought to be caused by a specific variety of streptococcus, but it is now known that **a single streptococcal species may be responsible for a variety of diseases.** On the other hand, a number of different kinds of streptococci may be cultured from human patients and animals. In 1903 Schottmüller proposed that the different varieties be classified on the basis of their capacities to hemolyze erythrocytes, and in 1919 Brown introduced the terms **alpha, beta,** and **gamma** to describe the three types of hemolytic reactions observed on blood agar plates.

Primarily through the efforts of Lancefield in the early 1930s, the β-hemolytic streptococci were further differentiated into a number of **immunological groups** designated by the letters A through O. Most strains causing human infections were found to belong to **group A.** That group in turn contains a variety of **antigenic types,** later demonstrated by precipitin tests (Lancefield) and by agglutination reactions (Griffith). The group-specific antigens were identified as carbohydrates and the type-specific antigens as proteins.

More than 55 types of group A β-hemolytic streptococci have been identified.

Hemolytic Classes. β-**Hemolytic streptococci** produce a wide clear zone of complete hemolysis (Fig. 26–1A) in which no red cells are visible on microscopic examination. Two types of β-hemolysin are released. **Streptolysin O** is destroyed by atmospheric oxygen and is therefore demonstrable only in deep colonies. **Streptolysin S** is oxygen-stable and is responsible for surface colony hemolysis. Since most strains produce both S and O hemolysins, they can usually be recognized as β-hemolytic by their surface colonies. To be certain of the hemolytic characteristics of a given strain, however, it may be necessary to examine colonies located beneath the surface of a pour plate. The most common species causing disease in man is *Streptococcus pyogenes,* which belongs to the antigenic group A.

α-Hemolytic streptococcus colonies are surrounded by a narrower zone of hemolysis, with unhemolyzed red cells persistent in an inner zone and complete hemolysis in an outer zone (Fig. 26-1B). (The mechanism responsible for the sparing of some of the red cells is not known.) Green discoloration of the colonies (due to formation of an unidentified reductant of hemoglobin) frequently occurs, depending upon the type of blood in the medium and the duration of incubation. This feature has given rise to the synonym term **viridans group.** *Streptococcus salivarius* is the most commonly encountered species of the α-hemolytic category.

γ-**Streptococci** produce no hemolysis, either on the surface or within the agar. *Streptococcus faecalis* is a typical nonhemolytic species.

708

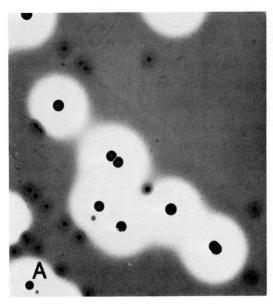 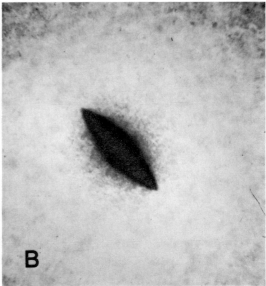

Fig. 26-1. A. Surface colonies of *S. pyogenes* and viridans streptococci on blood agar plate. The wide clear zone of β-**hemolysis** of the former is much more prominent than the latter's narrower dark zone of incomplete α-**hemolysis**. A few small viridans colonies can be seen to have grown within some of the zones of β-**hemolysis**. ×4. **B.** Higher power (×50) view of a deep α-hemolytic colony on blood agar (note oblong shape) showing border of incomplete hemolysis about the colony. No such band of intact red cells is present in the much wider zones of β-hemolysis that surround comparable colonies of β-hemolytic streptococci. (Preparation by Dr. Elaine D. Updyke. Photograph by courtesy of Center for Disease Control, Atlanta, Ga.)

This classification of streptococci based on hemolysis is far from satisfactory, for the following reasons:

1) Many species classified in the β-hemolytic category are, in fact, nonhemolytic, e.g., certain members of the antigenic groups B, C, D, H, K, and O, as well as all of group N (Table 26-1).

2) Most streptococci found in the gastrointestinal tract (enterococci) are nonhemolytic, but some strains produce β-hemolysis (group D).

3) Certain strains of *S. faecalis,* classified as nonhemolytic (γ) after 24 hours' incubation, may show α-hemolysis if incubated for an additional 24 to 48 hours.

4) Anaerobic streptococci, though generally nonhemolytic, are conventionally not classified with the aerobic γ-streptococci, but are considered in a separate category.

Despite its deficiencies, this system of classification has become firmly established.

Since streptococcal disease in man is due primarily to group A organisms which produce β-hemolysis, most of the present chapter will be devoted to β-hemolytic streptococci.

β-HEMOLYTIC STREPTOCOCCI

MORPHOLOGY

Cell Division. β-Hemolytic streptococci (often simply called hemolytic streptococci), like all other streptococci, are gram-positive and characteristically grow in chains; in vivo they commonly occur as diplococci. The length of the chains tends to be inversely related to the adequacy of the culture medium (Ch. 5, Nutrition); in actively spreading lesions within the tissues diplococcal and individual coccal forms are common, whereas in puru-

TABLE 26-1.   Group Classification of β-Hemolytic and Other Immunologically Related Streptococci

Group	Hemolysis*	Usual habitat	Pathogenicity
A	+	Man	Many human diseases
B	+	Cattle	Mastitis
C	±	Many animals	Many animal diseases
		Man	Mild respiratory infections
D	±	Dairy products, intestinal tract of man and animals (enterococci)	Urinary tract and wound infections; endocarditis
E	+	Milk	Unknown
		Swine	Pharyngeal abscesses of swine
F	+	Man	?; Found in respiratory tract
G	+	Man	Mild respiratory infections; rare
		Dogs	Genital tract infections of dogs
H	±	Man	?; Found in respiratory tract
K	±	Man	?; Found in respiratory tract
L	+	Dogs	Genital tract infections of dogs
M	+	Dogs	Genital tract infections of dogs
N	−	Dairy products	None
O	±	Man	Carried in upper respiratory tract of man; endocarditis

* +, all strains hemolytic; ±, some strains hemolytic, others nonhemolytic; −, all strains nonhemolytic.
Modified from MCCARTY, M. Hemolytic streptococci. In *Bacterial and Mycotic Infections of Man.* (R. J. Dubos and J. G. Hirsch, eds.) Lippincott, Philadelphia, 1965.

lent exudates from walled-off lesions and in artificial culture media chain formation is the rule (Fig. 26-2). Prior to division the individual cocci become elongated on the axis of the chain, eventually dividing to form pairs. When the dividing pairs of cocci do not separate chaining results. The bridges between the individual cocci in the chain are composed of cell wall material which has not cleaved (Ch. 6, Cell division). Uncleaved cell walls are particularly striking in mutants with excessive chaining which produce opaque colonies on clear agar (Fig. 26-3). Such variants also exhibit extensive, direct contact between their nucleoids and the cytoplasmic membrane beneath the uncleaved septum; i.e., there is no intervening mesosome.

Factors tending to promote preservation of

Fig. 26-2. A. Group A β-hemolytic streptococci in edema zone (Ch. 25, Evolution of lesion) of experimental pneumonic lesion (rat). Note diplococcal morphology. ×600. **B.** Chain formation characteristic of usual growth of streptococci in artificial media. Smear made from 24-hour culture in serum broth. ×1000. [**A** from Glaser, R. J., and Wood, W. B., Jr. *Arch Pathol 52*:244 (1951).]

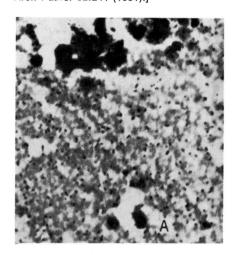
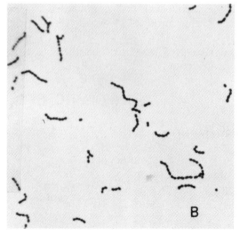

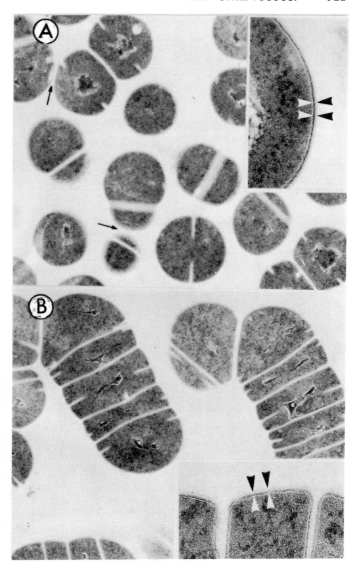

Fig. 26-3. Thin section electron micrographs of parent M⁻ strain of *S. pyogenes* (**A**) and a mutant (**B**) that produces opaque colonies. Whereas the intercellular septa are completely cleaved in the parent strain to form individual cocci and diplococci, most of the mutant's septa remain uncleaved so that the organism grows in long chains. The fine structure of the cell walls in the two strains, however, appears to be identical (see insets, between arrows). [From Swanson, J., and McCarty, M. *J Bacteriol 100*:505 (1969).]

the intercoccal junctions and thus to exaggerate chaining include not only conditions which impair growth (unfavorable medium, cold, antimicrobial agents, etc.), but also the presence of antibodies that react with cell wall antigens. The presence of anti-M antibody, for example, which reacts with the M protein surface antigen (see below) of the streptococcal wall, causes the organisms to grow in long chains in broth culture. Similar results have been obtained with antibodies to another surface protein, the R antigen (Ch. 25, Somatic antigens). Evidently such antibodies may "cover up" substrates that are normally

attacked enzymatically in the cleavage process. Formation of long chains may be used to test for the detection of anti-M antibody (see below).

The synthesis of new cell wall in the equatorial region of each streptococcal cell has been elegantly demonstrated by immunofluorescence (Ch. 6).

Because streptococcal chains are difficult to disrupt without killing the organisms, individual cocci cannot be readily counted by conventional plating methods. It is customary, therefore, to record streptococcal colony counts in **streptococcal units.** These values

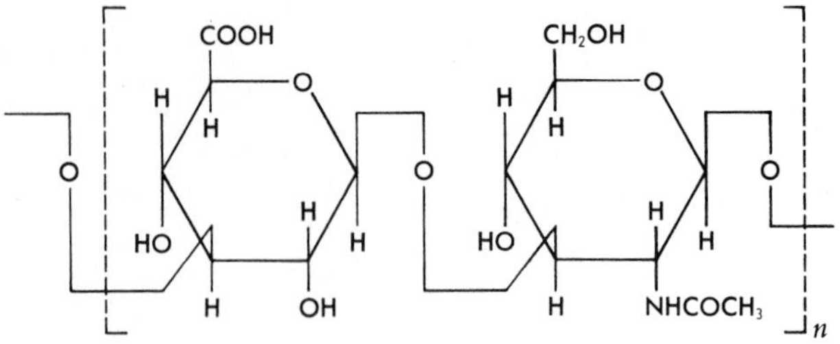

Fig. 26-4. Repeating unit of hyaluronic acid, in which glucuronic acid (left) is linked to N-acetyl glucosamine (right) by a β,1,3-glycosidic bond.

provide only a rough index of the number of cells, since they are obviously influenced by the degree of chaining.

Capsules and Colonial Morphology. Many strains of hemolytic streptococci produce capsules, which in group A are composed of hyaluronic acid (Fig. 26-4). Capsules are demonstrable only in very young (2- to 4-hour) liquid cultures, since the capsular gel tends to diffuse rapidly into the surrounding medium in older cultures.

On blood agar plates group A hemolytic streptococci may form any one of three colony types, designated **mucoid, matt** (Ger. *matt* = dull), and **glossy.** Mucoid colonies are formed by strains which produce large capsules: the abundance of hyaluronic acid gel gives the colony a glistening, watery appearance. The flatter, rougher, matt colonies were originally thought to reflect the production of M protein (whose designation was based on this apparent relation), but they are simply dried out mucoid colonies: as Figure 26-5 shows, the gel becomes dehydrated the surface of the colony shrinks and becomes roughened.

Glossy colonies are smaller; they are formed by cells that do not generate hyaluronate or do not retain it as a capsular gel. Groups F and G include **minute streptococci,** which produce not only smaller cells than other streptococci but also smaller colonies.

L Forms and Protoplasts. L forms (Ch. 40, L forms) of group A streptococci, which lack many cell wall constituents, may be isolated from anaerobic cultures on hypertonic media containing penicillin. Complete removal of the wall to form protoplasts (Ch. 6) may be achieved by treating the cells with a phage-associated mucopeptidase (see below, Group-specific C antigens). Unlike the protoplasts isolated from most other bacterial species,

those of group A streptococci will multiply and produce typical L-form colonies in hypertonic agar medium. During growth they release cell wall M antigen into the medium, as well as hemolysin and deoxyribonuclease (see below). The membranes of protoplasts isolated by osmotic shock possess enzymes capable of synthesizing hyaluronic acid from uridine nucleotide precursors.

METABOLISM

The growth requirements of hemolytic streptococci are very similar to those of pneumococci (Ch. 25, Metabolism). The chemically defined media for both organisms are essentially the same, and both are routinely grown in beef infusion media containing blood or serum. For the isolation of specific antigens and extracellular enzymes, a medium containing only the dialyzable components of the complex meat infusion peptone medium may be used to eliminate macromolecular constituents which might be confused with those produced by the growing organisms. A group of peptides supplied by the complex medium have been shown to be essential for optimal growth, but there is no specific peptide requirement.

Like pneumococci (and lactobacilli; Ch. 42, Distribution), all streptococci are lactic acid bacteria, which derive their energy primarily from the fermentation of sugars, regardless of whether they are growing aerobically or anaerobically. Accumulation of lactic acid in media of high glucose content limits growth unless the pH is corrected.

CELLULAR ANTIGENS

Capsular Hyaluronic Acid. One potential surface antigen of group A hemolytic streptococcal cells, the hyaluronate of the capsule

Colony Forms
Group A

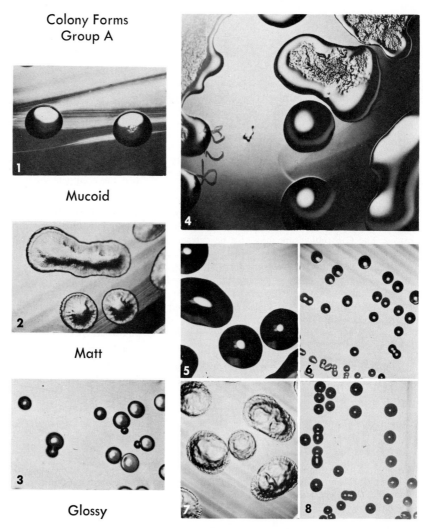

Mucoid

Matt

Glossy

Fig. 26-5. Interrelations among mucoid, matt, and glossy colony forms of group A hemolytic streptococci. **1–3.** The three types of colonies that form on the surfaces of blood agar plates. **4.** Conversion of mucoid to matt form as a result of aging (20 hours) and drying out of colonies. ×6.8. **5.** 19-Hour mucoid colonies of type 17 strain grown on Todd-Hewitt sheep blood agar. **6.** 19-Hour glossy colonies of same strain on same medium containing hyaluronidase. **7.** 24-Hour matt colonies of type 14 strain grown on Todd-Hewitt blood agar. **8.** Glossy colonies of same type 14 strain grown on same medium containing hyaluronidase. **5–8,** ×5.5. Note that presence of hyaluronidase in agar prevents capsule formation and results in formation of glossy rather than mucoid (or matt) colonies. [**1–3** from Lancefield, R. C. *Harvey Lect 36:*251 (1942). **4–8** from Wilson, A. T. *J Exp Med 109:*257 (1959).]

(Figs. 26-4 and 26-8), is **not immunogenic,** presumably because it is chemically indistinguishable from the hyaluronate in the ground substance of connective tissue. Although the hyaluronate of the streptococcal capsules tends, in broth cultures, to diffuse into the medium after a few hours of growth, observations in experimental infections of mice suggest that the capsules may remain intact for a longer time in the living host. The lability of the capsule itself may be related to the elaboration of hyaluronidase by the metabolizing cell (see below, Extracellular products).

Group-specific C Antigens. As already indicated, the separation of β-hemolytic streptococci into immunologically specific groups (A to O) depends upon the presence of group-specific carbohydrate antigens in their cell walls. These **C carbohydrates** may be extracted by a number of technics.

In the one used routinely in grouping streptococci the cells are suspended in dilute hydrochloric acid (pH 2) at 100° for 10 minutes. After neutralization with N/5 sodium hydroxide in M/15 phosphate buffer, the cells are removed by centrifugation; the clear supernatant fluid contains the group-specific antigen. The C carbohydrate may also be extracted by treating the cells with formamide at 150° or autoclaving them at 15 lb pressure for 15 minutes. Other less drastic methods of extraction are based on the use of an enzyme from *Streptomyces albus* or a lysin present in bacteriophage lysates of group C streptococci.* Both dissolve the cell walls by hydrolyzing the mucopeptide and thereby releasing the C carbohydrate.

The group-specific antigen reacts with antisera produced by immunizing rabbits with

---

* The phage-associated lysin acts on group A and E strains as well as on group C.

hemolytic streptococci of the same group. Precipitin reactions performed with appropriate sera permit the grouping of unknown strains. Although most hemolytic streptococci that cause disease in man fall into **group A,** human disease is occasionally due to members of other groups.

The carbohydrate antigen of the cell wall makes up approximately 10% of the dry weight of the organism and, in the case of groups A and C, is composed of rhamnose and hexosamine (Fig. 26–6). Its specific antigenicity depends largely upon the nature of the terminal sugar residue on its oligosaccharide rhamnose side chains. This determinant is N-acetyl glucosamine in the group A antigen and N-acetyl galactosamine in the group C antigen. Closely related strains have also been described (A-variant and C-variant) whose group-specific antigen lacks a terminal hexosamine; the antigenic specificity then appears to reside in the oligosaccharide side chains of rhamnose. Intermediate mutants (A-intermediate and C-intermediate) have been isolated which possess both antigenic determinants, some side chains terminating in a hexosamine and others in rhamnose. These relations are summarized in Figure 26–6.

Fig. 26-6. Compositions and antigenic determinants of groups A, A-intermediate, and A-variant carbohydrates (left) and groups C, C-intermediate, and C-variant carbohydrates (right) of β-hemolytic streptococci. [From Krause, R. M. *Bacterial Rev 27*:369 (1963).]

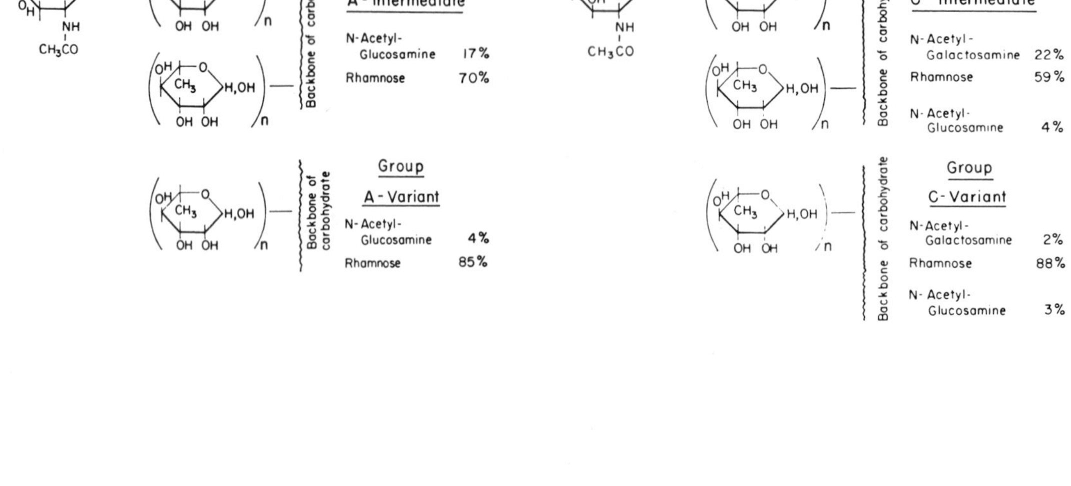

Type-specific M Antigens. Group A hemolytic streptococci can be further broken down into more than 55 immunological types which differ in their cell wall M antigens. M protein is distributed on the surface of the cell in the form of fimbriae (Fig. 26–7), which are lacking on M⁻ strains (Fig. 26-3). The M proteins may be extracted either by the relatively drastic procedure of acid treatment (boiling at pH 2) used to solubilize the C antigen, or by enzymatic lysis of the cell (in the absence of proteolysis) with the phage-associated lysin (see above). Typing is usually done by precipitin tests with the extracted M protein and specific antisera from rabbits immunized with streptococcal strains of each type. The sera are preabsorbed with cells of heterologous types to prevent cross-reactions. Although streptococci may also be typed by agglutination tests, the precipitin method is preferred in most diagnostic laboratories because of the effect of the T antigens (see below) on the agglutination reactions.

The M protein is a readily accessible surface antigen, not blocked by the hyaluronate envelope of encapsulated group A streptococci (Fig. 26-8). Furthermore, anti-M antibody is protective, showing that the M antigen

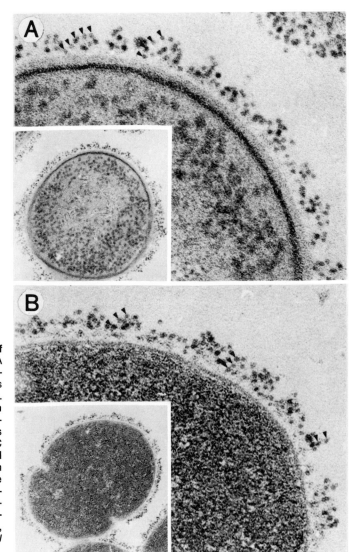

Fig. 26-7. Electron micrographs of fimbriae on surfaces of group A β-hemolytic streptococci. The fimbriae have taken up homologous ferritin-conjugated anti-M antibody. The section of an intact cell shown in **A** is compared in **B** with a nitrous acid extracted cell. This treatment removes most of the C polysaccharide and teichoic acid of the wall leaving the M protein intact. The ferritin particles can be seen to have assumed a linear distribution (arrows) along the surfaces of the fimbriae. ×250,000; insets ×60,000. [From Swanson, J., Hsu, K. C., and Gotschlich, E. C. *J Exp Med 130:*1063 (1969).]

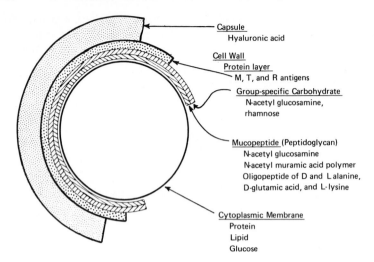

Capsule
Hyaluronic acid

Cell Wall
Protein layer
M, T, and R antigens

Group-specific Carbohydrate
N-acetyl glucosamine,
rhamnose

Mucopeptide (Peptidoglycan)
N-acetyl glucosamine
N-acetyl muramic acid polymer
Oligopeptide of D and L alanine,
D-glutamic acid, and L-lysine

Cytoplasmic Membrane
Protein
Lipid
Glucose

Fig. 26-8. Schematic diagram of capsule, cell wall, and cytoplasmic membrane of group A hemolytic streptococcal cell. [Modified from Krause, R. M. *Bacterial Rev* 27:369 (1963).]

is directly involved in streptococcal virulence; indeed, both the hyaluronate capsule and the M protein are antiphagocytic (see Pathogenicity, below). Finally, the M antigenicity of intact cells may be destroyed by trypsin without affecting their viability or removing the group antigen.

Occasional strains of group A streptococci have been found to contain more than one M antigen, but there is as yet little information on the structure of the individual M proteins.

Other Streptococcal Antigens. Two other kinds of cell wall proteins, which appear to act as surface antigens of group A streptococci, have been identified. Neither seems to influence virulence. The **T antigens** include a number of immunologically distinct proteins, which resist digestion by proteolytic enzymes but are readily destroyed by heat at an acid pH and hence are not present in the usual M-containing acid extract. Their distribution is not related to that of the M antigens.

Two immunologically distinct **R proteins** have thus far been identified: one (designated 3R) is destroyed by either trypsin or pepsin, the other (28R) only by pepsin.

Besides these surface components, four additional kinds of antigens have been identified. A **nucleoprotein fraction** (P antigen) is antigenically similar in hemolytic and nonhemolytic streptococci and in pneumococci; it also cross-reacts with staphylococcal nucleoproteins. The **glycerol teichoic acids,** which make up approximately 1% of the dry weight of the cells, act as the group-specific antigen in groups D and N, but not in other groups. The

mucopeptide itself, which cross-reacts with the structurally similar mucopeptides of many other bacteria, produces many of the same biological reactions as the endotoxins of gram-negative bacteria (Ch. 22, Endotoxin), e.g., fever, dermal and cardiac necrosis, lysis of erythrocytes and platelets, enhancement of nonspecific resistance, etc. (see p. 722, erythema nodosum). The **antigens of the cytoplasmic membranes** have been prepared from osmotically shocked protoplasts. The membranes thus isolated, though free of detectable cell wall antigens, exhibit distinctive antigenic differences when derived from streptococci of different immunological groups.

Apparent Spatial Relations. In summary, virulent group A hemolytic streptococci possess, in addition to their nonimmunogenic antiphagocytic capsules, at least three distinct protein antigens on their cell walls (Figs. 26-8). One of these, the type-specific M protein, is of major importance since it too is antiphagocytic and is therefore directly involved in virulence; the other two (T and R antigens) are unrelated to virulence. The group-specific C antigen is covalently linked to the wall's rigid mucopeptide matrix, which is itself immunogenic. The glycerol teichoic acids appear to reside in the wall but are not bound to this essentially insoluble structure. Other streptococcal antigens, ordinarily not exposed to the surface of the intact cell, include the lipoprotein antigens of the cell membranes and a nucleoprotein (P) antigen, which is presumably located within the cell.

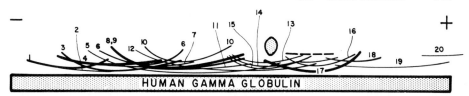

Fig. 26-9. Diagrammatic representation of group A streptococcal extracellular antigens detectable by immunoelectrophoretic analysis performed with crude streptococcal concentrate and 16% solution of pooled human γ-globulin. [From Halbert, S. P., and Keatings, S. L. *J Exp Med 113:*1016 (1961).]

EXTRACELLULAR PRODUCTS

The exceptionally wide variety of diseases in man caused by group A hemolytic streptococci may well be related to the large number of extracellular products they are known to produce. Since new streptococcal toxins and enzymes continue to be discovered, the list in Table 26-2 is surely incomplete. By means of immunoelectrophoresis, for example, there have been found in pooled human γ-globulin 20 different antibodies that react with the extracellular antigens elaborated by the C-203 strain of group A hemolytic streptococcus (Fig. 26-9). This finding suggests that group A streptococci may release as many as 20 extracellular antigens when growing in human tissues. Of those identified thus far, the following appear to be of greatest clinical significance.

Erythrogenic Toxin. The erythrogenic toxin is known to be responsible for the rash in scarlet fever. Strains of group A streptococci that produce this toxin are lysogenic* like toxigenic strains of *Corynebacterium diphtheriae* (Ch. 24, Toxin production). The amount of toxin produced by different lysogenic strains, however, varies widely.

The mode of action of erythrogenic toxin is not clear. When injected into the skin of susceptible children, it causes localized erythematous reactions, which reach a maximum at about 24 hours **(Dick test).** This erythrogenic effect is neutralized by antibody: during convalescence, when the patient's serum

* The finding that toxigenicity was transmissible from one streptococcal strain to another was first reported by Frobisher and Brown in 1927 (*Bull. Johns Hopkins Hosp. 41:*167); at that time, of course, the mechanism involved was not understood.

contains demonstrable antitoxin, the skin test becomes negative; and an injection of homologous antitoxin intradermally at the height of scarlet fever causes a local blanching of the rash **(Schultz-Charlton test).** Accordingly, a positive Dick test is interpreted as indicating absence of circulating antitoxin and thus a state of susceptibility to scarlet fever.

There are at least three immunologically distinct forms of erythrogenic toxin (types A, B, and C) produced by different streptococcal strains. They are presumably proteins. Certain strains of group C and group G hemolytic streptococci, as well as staphylococci, produce erythrogenic toxins closely related to those of group A streptococci.

Streptolysins S and O. The two hemolysins responsible for the zones of hemolysis around streptococcal colonies have already been mentioned. **Streptolysin S** is stable in air and is largely cell-bound; its name derives from the fact that it can be extracted from intact streptococcal cells with serum. This extraction is dependent upon association with serum al-

TABLE 26-2. Some Extracellular Products of Group A β-Hemolytic Streptococci

	Stimulates production of inhibitory antibody
Erythrogenic toxins (A, B, and C)	+
Streptolysin O	+
Streptolysin S	−
Diphosphopyridine nucleotidase	+
Streptokinases (A and B)	+
Deoxyribonucleases (A, B, C, and D)	+
Hyaluronidase	+
Proteinase	+
Amylase	?
Esterase	−

bumin as a macromolecular carrier, and other carriers (e.g., RNA) will similarly form a complex with the hemolysin. No antibody capable of neutralizing the hemolytic action of streptolysin S has been described, but this action is inhibited by serum lipoproteins. Recent evidence indicates that this cell-bound hemolysin is responsible for the leukotoxic action of group A streptococci, manifested by the killing of a proportion of the leukocytes that phagocytize them.

Streptolysin O has been so named because it is reversibly inactivated by atmospheric oxygen. It gives rise to antibodies which neutralize its hemolytic action. Since most strains of group A streptococci produce streptolysin O, patients recovering from streptococcal disease usually have antistreptolysin O antibodies in their sera (see below, Tests for nonspecific antibodies). Like other oxygen-labile bacterial exotoxins (Ch. 22, Exotoxins), the antigenicity of streptolysin O survives its detoxification by oxidation. Although its hemolytic activity is inhibited by cholesterol, the protein-bound cholesterol in normal serum does not have this effect and therefore does not interfere with the measurement of antistreptolysin O antibodies.

The S and O hemolysins can injure the membranes of cells other than erythrocytes. The mechanism of action of streptolysin O is illustrated in Figure 26-10. When added to suspensions of leukocytes in vitro, it causes lysis of the cell's cytoplasmic granules, having first affected and penetrated its outer (plasma) membrane. As a result, the destructive hydrolytic enzymes contained in the granules (Ch. 22, Phagocytosis) are released into the cytoplasm and irreversibly damage the cell. Macrophages are similarly injured, and streptolysin S, though less active, has much the same effect. Hydrolytic enzymes released from the leukocytic lysosomes may well damage other cell structures and thus intensify streptococcal lesions.

When injected intravenously into laboratory animals in sufficient quantities, streptolysin O causes fatal cardiac standstill. A possible relation to the pathogenesis of rheumatic fever has been suggested (see below).

DPNase. Streptococcal cultures contain a diphosphopyridine nucleotidase (DPNase, also called nicotinamide adenine dinucleotidase or NADase) that liberates nicotinamide from DPN. Nephritogenic (type 12) strains are particularly prone to produce this enzyme, but there is no evidence that it plays a role in the pathogenesis of glomerulonephritis (see below). Antibodies that inhibit its action are frequently found in sera of patients convalescing from streptococcal disease.

Streptokinases. In 1939 Tillett and Garner described a substance in streptococcal culture filtrates that promotes the lysis of human blood clots. First termed streptococcal fibrinolysin, it was later shown to catalyze the conversion of plasminogen to plasmin, and so it was renamed **streptokinase.** Two molecular species of streptokinase (A and B), differing in antigenicity and electrophoretic mobility, have been isolated from group A strains. They are immunogenic and induce antistreptokinase antibodies in the course of most diseases caused by group A streptococci. Although their action has often been assumed to prevent the formation of effective fibrin barriers at the periphery of streptococcal lesions, thus permitting the organisms to spread with unusual rapidity, there is no conclusive evidence to support this attractive hypothesis. In fact, the invasiveness of streptococcal lesions appears to be uninfluenced by antistreptokinase antibodies.

Deoxyribonucleases. Group A streptococci also elaborate enzymes that degrade DNA (DNAses). Four immunologically and electrophoretically different types, A, B, C, and D, have been found in streptococcal filtrates. Since these enzymes do not penetrate the plasma membranes of living mammalian cells, they are not cytotoxic. They are capable however, of depolymerizing the highly viscous DNA which accumulates in thick pus as a result of disintegration of polymorphonuclear leukocytes. Enzyme preparations containing both streptokinase and streptococcal deoxyribonuclease (streptodornase) were introduced by Tillett to liquefy purulent exudates (enzymatic débridement) in such diseases as pneumococcal empyema (Ch. 25, Evolution of lesion).

Hyaluronidase is of particular interest because it attacks the polysaccharide gel of the streptococcal capsule and may account, at least in part, for its instability. However, it is difficult to demonstrate hydraluronidase in cultures of encapsulated strains, even after prolonged growth and complete loss of capsules. On the other hand, some streptococcal strains (types 4 and 22) produce large amounts of hyaluronidase from the start of growth and do not form visible capsules in vitro. Since most patients recovering from streptococcal disease have antibodies to hyaluronidase, most strains must elaborate the enzyme in vivo.

Originally called the "spreading factor" be-

Fig. 26-10. Dissolution of leukocytic granules (lysosomes) and eventual destruction of cell resulting from action of streptolysin O. Besides obvious degranulation and formation of cytoplasmic vacuole, cell shows profound nuclear changes, which eventually result in fusion of individual nuclear lobes. First photograph was taken after cell was already partially degranulated and had developed hair-like processes on its membrane. Thereafter pictures were taken at intervals of approximately 1 minute. ×2000. [From Hirsch, J. G., Bernheimer, A. W., and Weissman, G. *J Exp Med 118:*223 (1963).]

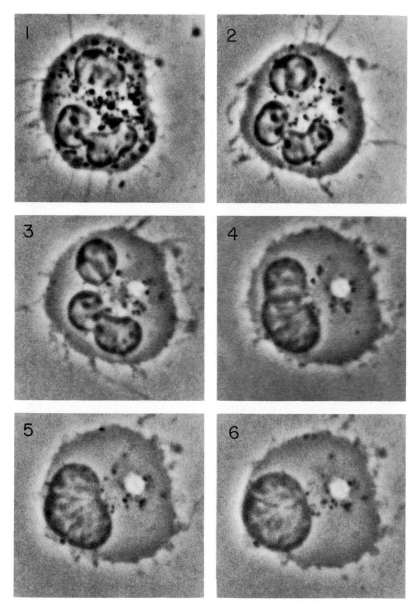

cause of its striking lytic effect on the ground substance of connective tissue, hyaluronidase has long been thought to play a role in the characteristic tendency of streptococci to spread rapidly through mammalian tissues. How important its action really is in this regard has never been determined.

Streptococcal proteinase is capable of destroying another cell factor involved in pathogenesis, the M protein (p. 715). Since this enzyme exhibits a relatively broad specificity it may also affect other extracellular proteins such as the streptolysins and streptokinase. It is released from streptococcal cells only when the pH of the medium is between 5.5 and 6.5. Under these conditions large amounts of the enzyme may appear in culture filtrates, from which it has been obtained in crystalline form. Like many other proteases, it is activated by sulfhydryl compounds and causes necrotic myocardial lesions in laboratory animals when injected intravenously. Whether or not it plays a role in the pathogenesis of poststreptococcal diseases, such as rheumatic fever, is unknown.

GENOTYPIC VARIATIONS

Genetic variations in hemolytic streptococci have been demonstrated to affect numerous traits, including elaboration of hemolysins, colonial morphology (capsule formation), production of M protein, synthesis of other cell wall antigens (including alterations in C polysaccharides, p. 714), and resistance to antimicrobial drugs. The most thoroughly studied variants are those related to colony formation. It is now clear that the transition from mucoid to glossy (p. 712) is analogous to the smooth-rough transition of *Diplococcus pneumoniae:* when repeatedly subcultured on artificial media, mucoid strains tend to become glossy, and on passage through mice glossy mutants revert to the mucoid form.

Under the same selective circumstances analogous changes occur in the production of M protein. Strains producing the M antigen are referred to as M^+ and those producing no M antigen as M^-. Hemolytic streptococci carried in the throat after an attack of streptococcal pharyngitis may eventually lose their M antigens. When such M^- strains are passed through mice, they frequently revert to M^+. Since both formation of capsule and produc-

tion of M protein are related to virulence, variation of either is of pathogenetic significance.

Fortunately, few drug-resistant mutants of group A streptococci have been encountered. Following mass prophylaxis with sulfonamides in military personnel during World War II, sulfonamide-resistant mutants became prevalent, but the introduction of penicillin promptly suppressed their spread. No significant change in the sensitivity of naturally occurring hemolytic streptococci to penicillin has occurred. Even in the laboratory it is extremely difficult to obtain penicillin-resistant streptococci, and those resistant mutants which have been isolated have been found to lack virulence. The absence of virulent penicillin-resistant mutants is of paramount importance for the widespread use of penicillin in the prevention of acute rheumatic fever (see below).

PATHOGENICITY

S. pyogenes causes both **suppurative diseases** and **nonsuppurative sequelae.** The first group includes acute **streptococcal pharyngitis** (with or without scarlet fever) and all its suppurative complications, including cervical adenitis, otitis media, mastoiditis, peritonsillar abscesses, meningitis, peritonitis, and pneumonia. It also includes streptococcal postpartum infections of the uterus **(puerperal sepsis), cellulitis** of the skin, **impetigo, lymphangitis,** and **erysipelas.** The principal diseases of the nonsuppurative category are **acute glomerulonephritis, rheumatic fever,** and **erythema nodosum.**

Suppurative Disease. The pathogenesis of suppurative streptococcal disease is fairly well understood. The factors that determine invasiveness are particularly important. Since hemolytic streptococci ingested by phagocytic cells are almost all killed within minutes, their antiphagocytic properties play a critical role in invasiveness. These in turn depend upon the hyaluronic acid capsule and the M protein. The combined antiphagocytic action of these two factors has been demonstrated both in vivo and in vitro.

The in vivo evidence was obtained in mice and rats infected intraperitoneally with strains of type 14 streptococci that differ in their capacities to produce capsules and M protein. Those strains producing large capsules and generous amounts of M protein were most virulent; those deficient in capsule formation, but producing a full complement of M protein, or vice versa, were of intermediate virulence; while strains deficient in both capsules and M protein were least virulent. These differences were directly correlated with the amount of phagocytosis occurring during the early hours of the infection. Confirmatory evidence has been provided in vitro, as is illustrated in Table 26-3, by using enzymes to deprive cells of these surface factors.

Although the vast majority of streptococci phagocytized are promptly killed, an occasional organism will escape unharmed. One mechanism is **egestion,** in which the engulfed organism is ejected from the cell. This event occurs only rarely in vitro, and there is no evidence that it is frequent enough in vivo to influence the course of the infection. A second escape mechanism results from elaboration of a leukotoxic factor, recently identified as the cell-bound streptolysin S.

The possible effect of streptococcal leukocidins (including streptolysin O) on leukocytes and other cells prior to ingestion has already been discussed.

Scarlet fever occurs as a complication of pyogenic streptococcal disease when the infecting strain produces erythrogenic toxin and the patient (usually a child) is susceptible to

TABLE 26-3. Relative Antiphagocytic Effect of the Hyaluronate Capsule and the M Protein of a Fully Virulent Strain of Group A β-Hemolytic Streptococcus*

Treatment of organism	State of capsule†	Amount of M protein†	% phago-cytosis†
None	+++	+++	3 (± 1.8)
Trypsin	+++	0	49 (± 5.4)
Hyaluronidase	±	+++	41 (± 2.8)
Trypsin and hyaluronidase	±	0	64 (± 0.85)

* Type 14.

† Number of plus signs indicates approximate size of envelope, as visualized in India ink preparations; or amount of M protein demonstrable by quantitative precipitin tests. Figures in parenthesis are standard deviations. The phagocytic tests were performed in the absence of serum (i.e., surface phagocytosis; Ch. 22).

From FOLEY, M. J., and WOOD, W. B., JR. *J Exp Med* 110:617 (1959).

the toxin. It has long been a matter of controversy, however, whether the rash is due to a direct action of the circulating toxin or to a generalized cutaneous hypersensitivity reaction. In favor of the latter possibility is the observation that infants under the age of two rarely have the disease and do not show positive Dick reactions, regardless of the immune state of the mother.

Nonsuppurative Sequelae. Relatively little is known about the pathogenesis of **acute glomerulonephritis** except that it results from infections caused by a limited number of types of group A streptococci. The majority of **nephritogenic strains** belong to type 12; a few have been of types 4, 18, 25, 49 (formerly designated the Red Lake strain), 52, and 55. The manner in which these strains cause acute glomerulonephritis is not fully understood. The characteristic symptoms of hematuria, edema, and hypertension do not appear until about a week after the onset of the acute pyogenic infection, usually of the pharynx or skin. Because of this latent period and because progressive nephritic lesions can be produced experimentally by injection of antibodies to the renal tissues of the host, it has been proposed that glomerulonephritis is an autoimmune disease (Ch. 19, Allergy to self-antigens). In keeping with this hypothesis are the observations that antibodies to human kidney are often present in the sera of patients with acute glomerulonephritis, and that the serum titer of complement frequently falls during an attack. How acute infections caused by specific strains of streptococci might trigger such an autoimmune mechanism is not clear. One possibility is that nephritogenic strains possess antigens that cross-react with and damage the glomerular basement membranes of susceptible individuals. Another is that specific streptococcal antigen-antibody complexes are deposited on the basement membranes. Both hypotheses presume that the resulting inflammatory response is due, at least in part, to the fixation of complement. The presumption is supported by the detection of C3, as well as γ-globulin and streptococcal antigen, in the glomerular lesions.

The pathogenesis of **rheumatic fever** is even more obscure, since it may follow pharyngeal infection with practically any type of group A streptococcus. The latent period between the onset of the acute streptococcal pharyngitis and the symptoms and signs of rheumatic fever is usually 2 or 3 weeks. The consistent finding of antistreptococcal antibodies in the sera of patients with acute rheumatic fever strongly supports the thesis that the disease is due to previous contact with *S. pyogenes*. Moreover, following a streptococcal epidemic, most patients who develop rheumatic fever (roughly 3%) have higher titers of antistreptococcal antibodies in their sera than do those who escape the disease. Nevertheless, it is far from clear how immunological hyperreactivity to streptococcal products could cause the recurring cardiac, joint, and skin lesions that characterize rheumatic fever. Nor is it clear what particular streptococcal products may be involved. M protein, for example, when injected intravenously, becomes deposited beneath the endocardium as well as in the glomeruli; streptococcal protease, injected intravenously, causes subendocardial lesions (as do proteases from other sources); streptolysin O is known to be cardiotoxic, and Halbert has postulated that its slow release from combination with antibody in the plasma is responsible for the characteristic recrudescences of acute rheumatic fever. Moreover, immunological cross-reactions have been described between a streptococcal antigen and tissue antigens of cardiac muscle and between the group A carbohydrate and structural glycoprotein of heart valves. Suggestive as these various findings are, there is still no convincing evidence to support the implication of any one streptococcal product in the pathogenesis of rheumatic fever.

Although **erythema nodosum** is known to occur in association with a variety of diseases (tuberculosis, coccidioidomycosis, sarcoidosis), there is both clinical and experimental evidence that it may also be a poststreptococcal illness. By intradermally injecting suspensions of sonically disintegrated group A streptococcal cell walls, Schwab *et al.* have produced in rabbits chronic remittent skin lesions which resemble erythema nodosum in man. There is evidence that the mucopeptide portion exerts the primary toxic effect. The relation of this model to the human disease remains to be determined, but it has already disclosed the potential cytotoxicity of natural products of the streptococcal cell wall (see mucopeptide, p. 716).

The possibility that **streptococcal L forms** (p. 712) may be involved in the pathogenesis of rheumatic fever, and perhaps even of glomerulonephritis, is currently under investigation. Streptococci in this form might well be difficult to culture and also to visualize in tissue lesions, even by electron microscopy, because of their lack of cell walls. Furthermore, they could survive intensive treatment with penicillin. The cultivation of group A streptococci from the cardiac tissues of an occasional patient who has died of acute rheumatic fever, and the experimental production of latent (Ch. 22, Dormant and latent infections) streptococcal infections in rabbits, has added inconclusive support to this line of reasoning.

IMMUNITY

Of the many varieties of antibodies that are generated in response to acute hemolytic streptococcal disease, only anti-M is known to protect the host against the invasiveness of the organism. The critical role of this antibody is in keeping with the antiphagocytic effect of the antigen. In acute streptococcal disease antibody to the M antigen ordinarily becomes detectable in the serum within a few weeks to several months, and it usually persists for 1 to 2 years; in some individuals it may still be present after 10 to 30 years. Inasmuch as there are more than 55 serological types of group A streptococci, no individual is likely to become immune to group A streptococcal infections in general.

Only a relatively few types of group A streptococci, on the other hand, are nephritogenic; therefore, persistence of anti-M antibodies to one of these types may account for the observation that an initial attack of acute glomerulonephritis greatly decreases the probability of a subsequent attack. No such protective effect, of course, occurs in rheumatic fever, because of the wide variety of streptococcal types that may cause the disease.

Immunity to scarlet fever is associated with the presence of erythrogenic antitoxin in the serum. Since there are at least three immunologic types of erythrogenic toxin (p. 717), occasional second attacks of scarlet fever may be expected.

LABORATORY DIAGNOSIS

Identification of Group A Organisms. The technics used to culture, group, and type hemolytic streptococci have already been described, and the need to employ pour plates to be certain of detecting β-hemolysis has been emphasized (p. 708). A simple method of recognizing *S. pyogenes* (group A) depends on the fact that most group A strains are significantly more sensitive to **bacitracin** than are strains of other groups. An agar plate test using paper discs impregnated with bacitracin may be useful if facilities for serological grouping are not available.

Tests for Anti-M Antibodies. Tests for type-specific streptococcal antibodies in patients' sera based on precipitation, agglutination, or complement-fixation technics are often misleading because of cross-reacting antigens. The most widely used technic, based on the **bactericidal** properties of whole blood, is more specific; as explained elsewhere (Ch. 22, Bactericidal action of antibacterial antibody), it measures indirectly the **opsonizing action** of the homologous anti-M antibody. The **mouse protection test** depends on the same principle.

A microscopic method, based on the tendency of group A streptococci to grow in **long chains** when cultured in the presence of homologous type-specific antibody (p. 711), appears to be almost as sensitive as the bactericidal method. However, it requires the use of strains that produce the right amounts of M antigen and grow in short chains in the absence of antibody.

Tests for Antibodies. Serological tests for streptococcal antibodies to extracellular products are much easier to perform and are therefore more widely used. The **antistreptolysin test** measures antibodies against streptolysin O. Technics designed to measure antibodies against other antigenic products of group A streptococci (e.g., streptokinase, hyaluronidase, DNAse, etc.) may be similarly employed.* The antistreptolysin test, however, is routinely used in most diagnostic laboratories.

TREATMENT

β-Hemolytic streptococci are among the most susceptible of all pathogenic bacteria to the action of antimicrobial drugs. The sulfonamides readily suppress growth, both in vitro

---

* In streptococcal skin infections the test for antibody to streptococcal DNAse B (p. 718) appears to be more reliable than to streptolysin O.

and in vivo, but since they are only bacteriostatic their use will not eliminate the organisms from the upper respiratory tract, nor will it significantly modify the antibody response of the host. Penicillin is bactericidal and hence is far more effective. When used in adequate dosage for a sufficient length of time, it will often rid the pharynx of hemolytic streptococci. Persistence usually indicates a suppurative complication, such as an intratonsillar abscess or purulent sinusitis. When given early in the course of acute streptococcal pharyngitis, penicillin will also depress the patient's antibody response, and such treatment of rheumatic individuals greatly reduces the rate of rheumatic attacks.

Other antibiotics, such as erythromycin, may also be used in the treatment of group A hemolytic streptococcal infections, particularly in patients who have a history of hypersensitivity to penicillin. Many strains of group A streptococci are now resistant to the tetracyclines, which are therefore no longer generally recommended.

PREVENTION

Penicillin is often given continually in small doses to rheumatic patients to **prevent** streptococcal infections and thus reduce the likelihood of recurrences of rheumatic fever. Prophylactic therapy of this kind is possible only because group A hemolytic streptococci do not generate mutants that are significantly resistant to penicillin. Prevention of rheumatic fever with continuous sulfonamide treatment has also been reasonably successful, but because these drugs are not bactericidal and occasionally cause severe reactions (e.g., periarteritis nodosa), penicillin is usually preferred. The responsibility of the physician to eradicate (with penicillin) nephritogenic streptococci from the families of patients with acute glomerulonephritis should be evident from what has previously been said of the pathogenesis of the disease.

The custom in the past of isolating scarlet fever patients but not patients with acute streptococcal pharyngitis was based on the erroneous view that the two diseases are basically dissimilar. From an epidemiological standpoint there is obviously no justification for such a distinction.

As already intimated, immunization of the general

population against group A streptococcal infections is not practical because of the very large number of types. Efforts to prepare suitable M-protein vaccines, however, are being continued in the hope that they will be useful, particularly in military installations, in controlling the spread of individual epidemic strains of streptococci. Active immunization against scarlet fever has been practiced in the past, but is no longer advocated because of the effectiveness of penicillin therapy. Indeed, scarlet fever has become relatively uncommon in the United States, as have streptococcal mastoiditis and puerperal infections, presumably due to the early treatment of suspected streptococcal illnesses with effective antimicrobial drugs.

EPIDEMIOLOGY

The incidence of streptococcal infections varies widely in different geographical areas and appears to be related to climate. Streptococcal diseases are most common in cold, relatively dry areas and occur most often in the winter and spring. Endemic rates are particularly high in the Rocky Mountain states, such as Colorado and Wyoming. Although overt streptococcal **disease** is less prevalent in the southern United States, culture surveys and antibody studies reveal that streptococcal **infections** are not uncommon (Ch. 22, Infection vs disease).

Group A streptococcal **carrier rates** ordinarily run well below 10%. Just before an epidemic, however, they become much higher. Infection is transmitted from the respiratory tract of one person to that of another by relatively intimate contact. The epidemiological studies of Rammelkamp *et al.* have shown that group A streptococci in the air, in dust, on blankets, etc., are far less infectious than the moist secretions ejected from the respiratory tract during speech, coughing, and sneezing. Nasal carriers are known to disseminate many more streptococci into their environments than pharyngeal carriers. The source of puerperal sepsis has often been traced to the upper respiratory tract of the obstetrician or of somebody else in the delivery room.* Milk-borne epidemics of streptococcal disease are now uncommon because of the effectiveness of pasteurization. Isolated outbreaks due to infected food still occur on rare occasions.

β-HEMOLYTIC STREPTOCOCCI OF OTHER GROUPS

Disease in man is only occasionally caused by streptococci of groups B to O (Table 26-1). Though some members of these immunological groups produce no hemolysin, they are all usually classed as β-hemolytic streptococci, for reasons explained earlier in the chapter. Many of them are found primarily in animals, although strains of group C, E, G, H, K, and O are often isolated from the respiratory tract of man, and nonhemolytic strains of group D (*S. faecalis*) are common inhabitants of the human gastrointestinal tract (see below). Group B strains occasionally cause circumscribed epidemics of meningitis in newborn nurseries.

OTHER STREPTOCOCCI

α-HEMOLYTIC STREPTOCOCCI

The α-hemolytic streptococci are often referred to collectively as the **viridans group;**

---

* In the mid-nineteenth century Ignaz Semmelweis in Vienna and Oliver Wendell Holmes in Boston independently concluded that childbed fever was transmitted to patients by the unclean hands and contaminated clothing of the attending obstetrician. Their conclusions, based solely on astute clinical observations, were vehemently rejected by their contemporaries to whom the germ theory of disease was anathema. Modern epidemiological studies, made possible by the Lancefield typing technique, have fully substantiated this unpopular doctrine.

they have never been satisfactorily classified. Antigenic analysis of many strains has led to the recognition of a number of immunological varieties, only a few of which seem to fall into the Lancefield groups (A to O).† One of the most common species is *S. salivarius*.

Viridans streptococci colonize the human upper respiratory tract within the first few hours after birth; rarely does a carefully per-

---

† One set of group F strains, designated "streptococcus MG," produce relatively small colonies and are of clinical interest because they are agglutinated by the sera of patients with mycoplasmal pneumonia.

formed throat culture fail to reveal their presence. They have a very low degree of pathogenicity compared with pneumococci, which also are α-hemolytic and are often cultured from the throat. Unlike pneumococci, however, viridans streptococci are neither bile-soluble nor sensitive to optochin.

The principal significance of α-hemolytic streptococci in clinical medicine relates to **subacute bacterial endocarditis.** This serious illness results from infection of an endocardial surface already damaged by either rheumatic fever or congenital heart disease. Since α-hemolytic streptococci are continually present in the throat and about the teeth, even minor trauma, such as that due to vigorous chewing, may result in their entry into the blood stream. The transient bacteremias that follow dental extraction or tonsillectomy may initiate subacute bacterial endocarditis in patients with abnormal valves. The seriousness of this type of infection is greatly enhanced by the frequency of drug-resistant strains of α-hemolytic streptococci.

In order for viridans streptococci to gain a permanent foothold on the endocardium the organisms must be trapped in a suitable nidus on the endocardial surface, usually provided by a tiny fibrin clot over an area of trauma to the endocardium. Dogs with artificially induced arteriovenous aneurysms in the peripheral circulation regularly develop microscopic clots of this kind on the endocardium. The clot formation results from the severe circulatory disturbance (widened pulse pressure, increased blood flow, etc.) created by the arteriovenous shunt. Similar endocardial lesions have been produced in rats by exposing them to prolonged anoxia, which also causes a hyperkinetic circulatory response. The remarkable frequency of bacterial endocarditis in such animals illustrates the importance of preexisting endocardial damage in the pathogenesis of the disease.

Histological examination of the vegetations on the heart valves, in both naturally acquired and experimentally induced viridans endocarditis, reveals that the organisms grow in large colonies embedded in a fibrinous, relatively acellular exudate. The avascularity of the heart valves accounts for the paucity of phagocytic cells in the lesions and explains why an organism so easily phagocytized in vitro can survive and multiply in the valvular vegetations. As the organisms continue to multiply at the periphery of the lesion, they

break off and are carried away in the blood stream. Quantitative blood cultures, taken at repeated intervals over long periods in patients with subacute bacterial endocarditis, show that the organisms are shed from the vegetation at a surprisingly constant rate. Thrombotic lesions, resulting in the formation of petechiae, commonly develop and constitute a hallmark of the disease.

NONHEMOLYTIC STREPTOCOCCI

The term **nonhemolytic streptococcus** is confusing because it is often used to include any streptococcus that is not β-hemolytic, and because many nonhemolytic species (including a particularly common one, *S. faecalis*) possess the same group-specific cell wall antigens as certain hemolytic streptococci (Table 26-1).

The organisms that fall in the nonhemolytic group are generally of low pathogenicity for man and, like α-hemolytic streptococci, are of concern to physicians primarily as causative agents of subacute bacterial endocarditis. *S. faecalis,* often referred to as **enterococcus** because of the frequency with which it is found in the human gastrointestinal tract,[*] is an exceptionally hardy microorganism, capable of growing under conditions which are lethal to other bacterial species. Its ability to grow in the presence of 0.05% sodium azide, for example, is often utilized in the laboratory to separate it from other streptococci. Enterococci also tend to be relatively resistant to heat (62° for 30 minutes), grow well at temperatures ranging from 10 to 45°, and will multiply in media containing 6.5% sodium chloride. In addition, many strains encountered in clinical practice are highly resistant to antimicrobial drugs, making the treatment of enterococcal endocarditis especially difficult.

ANAEROBIC STREPTOCOCCI

All the varieties of streptococci thus far considered are facultative anaerobes. Obligate

---

[*] Enterococci may also be cultured from the oropharynx.

anaerobic (or microaerophilic) streptococci do exist, however, and may also cause human disease. They are usually nonhemolytic and are smaller than other streptococci. Although a number of different species have been described, they have not been systematically classified.

Because anaerobic streptococci are normal inhabitants of the female genital tract, they occasionally give rise to intrauterine infections. Their virulence for man, as well as for other animals, is low and they tend to multiply only in necrotic or frankly gangrenous lesions. When growing in purulent exudates, they produce a fetid odor; hence their presence in lung abscesses is often suggested by the foul odor of the sputum. Although most anaerobic streptococci are susceptible to the action of antimicrobial drugs, the lesions in which they are found often require surgical drainage as well as chemotherapy.

SELECTED REFERENCES

Books and Review Articles

KUTTNER, A. G., and LANCEFIELD, R. C. Unsolved problems of the nonsuppurative complications of group A streptococcal infections. In *Infectious Agents and Host Reactions.* (S. Mudd, ed.) Saunders, Philadelphia, 1970.

MCCARTY, M. The streptococcal cell wall. *Harvey Lect 65:*73 (1971).

RAMMELKAMP, C. H., JR. Epidemiology of streptococcal infections. *Harvey Lect 51:*113 (1957).

Streptococcal Infections. (M. McCarty, ed.) Columbia Univ. Press, New York, 1954.

The Streptococcus, Rheumatic Fever, and Glomerulonephritis. (J. W. Uhr, ed.) Williams & Wilkins, Baltimore, 1964.

Specific Articles

COBURN, A. F., FRANK, P. F., and NOLAN, J. Studies on the pathogenicity of *Streptococcus pyogenes.* IV. The relation between the capacity to induce fatal respiratory infections in mice and epidemic respiratory diseases in man. *Br J Exp Pathol 38:* 256 (1957).

FREIMER, E. H., and MCCARTY, M. Rheumatic fever. *Sci Am 213:*6 (1965).

GOLDSTEIN, I., HALPERN, B., and ROBERT, L. Immunological relationship between streptococcus A polysaccharide and the structural glycoprotein of heart valve. *Nature 213:*44 (1967).

KAPLAN, M. H., and SVEC, K. H. Immunologic relation of streptococcal and tissue antigens. III. Presence in human sera of streptococcal antibody cross-reactive with heart tissue. Association with streptococcal infection, rheumatic fever, and glomerulonephritis. *J Exp Med 119:*651 (1964).

KRAUSE, R. M. Antigenic and biochemical composition of hemolytic streptococcal cell walls, *Bacteriol Rev 27:*369 (1963).

LANCEFIELD, R. C. Current knowledge of type-specific M antigens of group A streptococci. *J Immunol 89:*307 (1962).

LILLEHEI, C. W., VARGO, J. D., and HAMMERSTROM, R. N. Experimental bacterial endocarditis and proliferative glomerulonephritis. *Dis Chest 24:*421 (1953).

SCHWAB, J. H. Analysis of the experimental lesion in connective tissue produced by a complex of C polysaccharide from group A streptococci. I. *In vivo* reaction between tissue and toxin; II. Influence of age and hypersensitivity. *J Exp Med 116:*17 (1962); *119:*401 (1964).

STOLLERMAN, G. H., SIEGEL, A. C., and JOHNSON, E. E. Evaluation of the "long chain reaction" as a means of detecting type-specific antibody to group A streptococci in human sera. *J Exp Med 110:*887 (1959).

TILLETT, W. S. Studies on the enzymatic lysis of fibrin and inflammatory exudates by products of hemolytic streptococci. *Harvey Lect 45:*149 (1952).

WANNAMAKER, L. W. Differences between streptococcal infections of the throat and of the skin. *N Eng J Med 282:*23 (1970).

ZABRISKIE, J. B. The role of temperate bacteriophage in the production of erythrogenic toxin by group A streptococci. *J Exp Med 119:*761 (1964).

ZABRISKIE, J. B. The relationship of streptococcal cross-reactive antigens to rheumatic fever. *Transplant Proc 1:*968 (1969).

chapter **27**

STAPHYLOCOCCI

History   728
Morphology   728
Metabolism   730
Toxins   730
Extracellular Enzymes   730
Cellular Antigens   731
Typing   732
Genetic Variation   732
Pathogenicity   733
Immunity   736
Laboratory Diagnosis   737
Treatment   737
Prevention   737
Other Micrococci   738

Staphylococci are spherical, gram-positive organisms that grow in clusters (Gr. *staphyle* = bunch of grapes). They cause a wide variety of suppurative diseases in man, including furuncles, carbuncles, osteomyelitis, deep tissue abscesses, wound infections, pneumonia, empyema, pericarditis, endocarditis, meningitis, and purulent arthritis. In addition, certain strains elaborate an enterotoxin that causes food poisoning. Because staphylococci frequently become drug-resistant, they have risen, during the modern era of antibacterial chemotherapy, to a position of special significance in clinical medicine.

HISTORY

Robert Koch (1878) was the first to describe staphylococci in human pus. Two years later Pasteur cultivated the organism in a liquid medium, and in the following year Ogston showed it to be pathogenic for mice and guinea pigs. In 1884 Rosenbach described the two species, *Staphylococcus* (*pyogenes*) *aureus* and *Staphylococcus* (*pyogenes*) *albus,* which are now classified in a separate genus (*Staphylococcus*) of the family Micrococcaceae. Julianelle in the 1930s introduced the first classification of staphylococci based on differences in antigenic structure, and in 1942 Fisk developed the method of typing them with bacteriophages.

Interest in staphylococci as human pathogens was greatly stimulated in 1928 by a tragic accident in Bundaberg, Australia. Of 21 children inoculated with diphtheria toxin-antitoxin from a single rubber-capped vial, which had been kept at room temperature in subtropical heat for several days, 16 became ill in 5 to 7 hours with vomiting, diarrhea, high fever, stupor, cyanosis, and convulsions. Twelve of the children died within 2 days, and those who survived developed staphylococcal abscesses at the site of injection. The results of subsequent studies suggested that the early deaths were due to a soluble toxin elaborated by the staphylococcus cultured from the vial. A staphylococcal toxin which causes food poisoning was later described by Dack and named **enterotoxin.**

MORPHOLOGY

Staphylococci are nonmotile organisms that produce no spores. Their diameters vary from 0.7 to 1.2 μ. The individual cocci of pathogenic strains tend to be slightly smaller than those nonpathogenic strains. All strains are gram-positive but individual cells may become gram-negative. Their characteristic growth in clusters, which results from irregular divisions in two planes perpendicular to one another, is most striking on solid media (Fig. 27-1); in liquid media they often form short chains. Although the latter may be confused with streptococci, differentiation is ordinarily not difficult, since staphylococci rarely form chains containing more than four members, whereas the chains in streptococcal clusters are usually much longer.

On meat digest agar staphylococci produce round, raised, shiny colonies, 1 to 2 mm in diameter, and considerably more opaque than those of streptococci or pneumococci. The two species, *S. aureus* and *S. albus,** are recognized in most instances by the color of the colonies they form (for exceptions see below). Pigment production is optimal when the organisms are grown at 37° on media en-

---

* Although this species is now designated *S. epidermidis* in Bergey's Manual, the older and more familiar name, *S. albus,* will be retained in this text. A third variety originally termed *citreus* because of its lemon yellow color is now classified with the albus species.

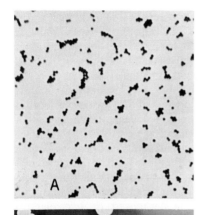

Fig. 27-1. A. *S aureus.* Gram stain. ×1000. **B.** Colonies of *S. aureus* on surface of blood agar plate. Note zones of hemolysis surrounding opaque colonies. ×2. (**A** from *Topley and Wilson's Principles of Bacteriology and Immunity.* [G. S. Wilson and A. A. Miles, eds.] Williams & Wilkins, Baltimore, 1964. **B** from *Zinsser Microbiology.* [D. T. Smith, N. F. Conant, and J. R. Overman, eds.] Appleton, New York, 1964.)

riched with fatty acids (e.g., glycerol mono-acetate). No pigment is produced under anaerobic conditions or in broth cultures. As suggested by their names, the colonies of *S. aureus* are golden yellow, while those of *S. albus* are chalky white. The yellow pigment is composed of two carotenoids, δ-carotene and sarcinaxanthine. Most strains of staphylococcus pathogenic for man form golden colonies which, on blood agar plates, are surrounded by a wide zone of clear hemolysis (Fig. 27-1). In addition, nearly all pathogenic strains elaborate an enzyme known as **coagulase** that clots plasma. Since some pathogenic, coagulase-positive strains fail to produce pig-

ment, it has become customary to classify as *aureus* all strains that elaborate coagulase. *S. albus* does not produce this enzyme and is of relatively low pathogenicity for man except under special circumstances (see below).

Whether or not staphylococci form capsules has long been a matter of controversy. Mucoid variants that produce capsules readily visualized by the India ink technique (Fig. 27-2) are rare in cultures grown on standard media. Yoshida and Ekstedt have shown, however, that they are more common in cultures grown on special media containing peptone, yeast extract, mannitol, lactose, and 3% sodium chloride. Antisera to such encapsulated variants cause a quelling reaction (Fig. 27-2).

A

B

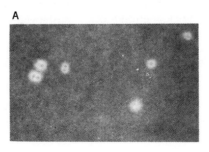

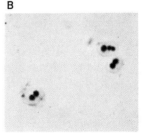

Fig. 27-2. A. Encapsulated cocci from a 4-hour broth culture of *S. aureus,* strain Smith, suspended in India ink. ×930, enlarged. **B.** Staphylococcal quellung reaction. ×1000, enlarged. [**A** from Morse, S. I. *J Exp Med 115:*295 (1962). **B** from Price, K. M. and Kneeland, Y., Jr. *J Bacteriol 67:*472 (1954).]

METABOLISM

Staphylococci are among the hardiest of all non-spore-forming bacteria. They will remain alive for months on the surface of sealed agar plates stored at 4° and may be cultured from samples of dried pus many weeks old. Most strains are relatively heat-resistant, withstanding temperatures as high as 60° for half an hour. Though highly susceptible to the bactericidal action of certain basic dyes, e.g., gentian violet, they are more resistant than most bacteria to disinfectants such as mercuric chloride and phenol.

Typical strains will grow readily in a chemically defined medium containing 14 amino acids, glucose, salts, and two trace growth factors—thiamine and nicotinic acid. On basic meat digest media devoid of blood or serum they grow well over a wide range of pH (4.8 to 9.4). Staphylococci are facultative anaerobes. In aerobic cultures hydrogen peroxide does not accumulate because, unlike pneumococci (Ch. 25, Metabolism), the organisms elaborate catalase. Most pathogenic strains ferment mannitol, in addition to producing yellow pigment, hemolysin, and coagulase. Staphylococci possess lipolytic enzymes which render them resistant to the bactericidal lipids of skin (Ch. 22, Chemical factors).

TOXINS

When grown in artificial media pathogenic staphylococci release a number of different **exotoxins** whose production is stimulated in an atmosphere of 30% carbon dioxide. The toxins most commonly elaborated include four immunologically distinct hemolysins* (Table 27-1), a nonhemolytic leukocidin (Ch. 22, Exotoxins), and four enterotoxins.

The α-hemolysin is a protein injurious to various cells, including rabbit and human leukocytes and tissue culture cells. It is dermonecrotic when injected subcutaneously in rabbits and is lethal to both rabbits and mice in small intravenous doses. In addition it causes aggregation of platelets and spasm of smooth muscle. The β-hemolysin is a

sphingomyelinase responsible for the interesting hot-cold reaction, in which human and sheep erythrocytes in broth or blood agar plates are lysed after incubation at 37° if stored overnight in the refrigerator. The δ-hemolysin is also a phospholipase and is toxic to leukocytes and a wide variety of tissue culture cells. The γ-hemolysin is weaker than the others and has not been well characterized. All the hemolysins appear to act on cell membranes.†

The nonhemolytic Panton-Valentine **leukocidin,** which is produced by most pathogenic staphylococci, is composed of two electrophoretically separable proteins: F (fast) and S (slow). Both components are antigenic, and neutralization of either by antibody nullifies the action of the toxin. Their mode of action, which requires the presence of calcium ions, is thought to involve the same kind of leukocytic degranulation as that caused by streptolysin O and streptolysin S (Ch. 26, Extracellular products).

Staphylococcal **enterotoxins** (types A through D), which are elaborated by about 50% of coagulase-positive strains, are relatively heat-resistant polypeptides of MW ca. 35,000. They cause emesis and/or diarrhea when ingested in microgram amounts.

Production of both α-hemolysin and enterotoxin A is acquired by nontoxigenic strains following lysogenization with appropriate temperate phages from toxigenic strains. The production of other staphylococcal toxins may be similarly controlled by lysogeny.

EXTRACELLULAR ENZYMES

Among the various enzymes released by staphylococci into artificial media, the greatest attention has been given to **coagulase.** Staphylococci appear to be the only bacteria that produce these enzymes that cause citrated (or oxalated) plasma to coagulate. They have a thrombokinase-like action, clotting purified fibrinogen in the presence of a plasma factor CRF (the coagulase-reacting factor) which is believed to be a derivative of prothrombin. Seven immunologically distinct coagulases

---

* A single strain may produce more than one immunological type of hemolysin.

† *S. albus* colonies are often hemolytic owing to production of a fifth hemolysin, designated epsilon (ε).

TABLE 27-1.   Staphylococcal Hemolysins

Serological type	Susceptible erythrocytes	Susceptible leukocytes	Usual source	Animal toxicity
α	Rabbit, sheep, calf	Rabbit, human	Human strains	Dermonecrotic for rabbits; lethal for mice and rabbits; cytopathic for tissue culture cells; aggregates platelets
β	Sheep,* human,* ox*	None	Animal strains	Lethal for rabbits in large doses; aggregates platelets
γ	Rabbit, human, sheep, guinea pig, ox, rat, horse	?	Human strains	Slightly dermonecrotic for rabbits and guinea pigs; lethal for rabbits
δ	Human, rabbit, horse, sheep, rat, guinea pig	Rabbit, guinea pig, human, mouse	Human strains	Edema and induration only in rabbits and guinea pigs; cytopathic for tissue culture cells

* Hot-cold lysis.

have been identified. Although coagulase types correlate with bacteriophage types, they have not proved useful for classification. A positive coagulase test is generally considered the best laboratory evidence that a given strain of staphylococcus is potentially pathogenic for man.

Coagulase-positive strains elaborate a number of other extracellular enzymes. These include staphylokinase (an enzyme which activates the plasma plasminogen system), lipase, hyaluronidase, and DNase. Lipase production may be affected by lysogenic conversion (see below, Relation of virulence to production of extracellular enzymes).

CELLULAR ANTIGENS

Species-specific Polysaccharides A and B. The immunological specificities of *S. aureus* and *S. albus* are determined by phosphorus-containing polysaccharide antigens. **Polysaccharide A** extracted from pathogenic (aureus) strains is a teichoic acid composed of N-acetyl glucosamine residues attached in either α or β linkage to a polyribitol phosphate "backbone." The **B polysaccharide** of *S. albus* is a polyglycerolphosphate teichoic acid with α-linked glucose residues. The ribitol teichoic acid complexes are covalently linked to the muramic acid mucopeptide of the cell wall.

Protein A. Most strains of *S. aureus* possess a surface component known as **protein (or agglutinogen) A.** It is a relatively small basic protein (MW 13,000) which has the unique property of reacting with the Fc fragments of the IgG molecules of most mammalian sera. Because the resulting IgG aggregates fix complement, they cause hypersensitivity reactions in normal rabbits and guinea pigs. They also generate complement-derived chemotactic factor(s) (e.g., C5a) which may account in part for the characteristic purulence of staphylococcal lesions. Protein A is difficult to study as an immunogen because of its tendency to precipitate normal mammalian γ-globulin. It has, however, been shown to have antiphagocytic properties and is released into the medium during growth (see below, Antiphagocytic properties of *S. aureus*).

Capsular Antigens. Of the several capsular antigens formed by mucoid strains of *S. aureus,* two have been chemically characterized. That of the Wiley "wound strain" is a polypeptide made up of glutamic acid, lysine, alanine, and glycine in a ratio of 1:1:2:5; that of the Smith diffuse strain is a polymer of 2-amino-2-deoxy-D-glucuronic acid. The latter is antiphagocytic when on the organism's surface, and in solution it blocks the agglutinating and opsonizing action of anticapsular antibodies.

Other Antigens. Besides **free coagulases,** which cause plasma samples to coagulate, pathogenic staphylococci also form **bound**

coagulase or **clumping factor,** which causes the organisms to clump when incubated in plasma or serum. The substrate for this action is fibrinogen or fibrin monomers. Clumping factor does not require CRF and is immunologically distinct from the free coagulases.

A glycerol teichoic acid antigen and a protein cell wall antigen common to aureus and albus strains have also been described.

TYPING

Serological typing of *S. aureus* has proved to be difficult. Agglutination is hard to interpret because of "false clumping." By special cultural and enzymatic technics, however, over 30 type-specific agglutinogens have been identified.

Most diagnostic laboratories use the less difficult procedure of **bacteriophage typing** of coagulase-positive (*S. aureus*) strains. Coagulase-negative (*S. albus*) strains are rarely sensitive to the typing phages.

Suspensions of the typing phages, which may be divided into five host-range (lytic) groups (Table 27-2), are prepared in the laboratory as lysates of susceptible staphylococcal strains of human origin. Single drops of each suspension are placed on separate squares of an agar surface previously seeded with a heavy culture of the staphylococcal strain to be tested. The pattern of the clear zones of lysis that appear after incubation overnight at 30° permits a reliable enough identification of individual human strains to be of great value in identifying the sources of staphylococcal epidemics. When a given strain is not typable by the routine test dilution

TABLE 27-2. Lytic Groups of Staphylococcal Typing Phages*

Group	Phage no.				
I	29	52	52A	79	80
II	3A	3C	55	71	
III	6	42E	47	53	54
	75	77	83A†	84†	85†
IV	42D				
Not allotted	81	187			

* Recommended for the typing of *S. aureus* of human origin by the International Subcommittee on Phage-Typing of Staphylococci.

† Used only at routine test dilution (RTD). See text.

(RTD) of the recommended phages, more concentrated suspensions (RTD × 100), except in the cases of 83A, and 84, and 85, are employed (Table 27-2).

GENETIC VARIATION

Among the genetic variants of staphylococci, those affecting pigment formation are particularly common. As previously noted, *S. aureus,* when repeatedly cultured on artificial media, may cease to form golden pigment but still retain pathogenicity and ability to produce coagulase. Other mutations have been described that affect pathogenicity, coagulase production, toxin formation, hemolytic properties, antigenicity, bacteriophage susceptibility, drug resistance, and capsule formation (R and S).

A colonial variant, known as the **minute G (gonidial) form,** is occasionally encountered, particularly in cultures containing penicillin or other antibacterial chemicals such as lithium chloride (used to inhibit the growth of coliform bacilli). It is penicillin-resistant, produces no pigment, hemolysin, or coagulase, and forms dwarf colonies which may be so small as to go unrecognized. Although G variants are apparently deficient in cell wall, they appear normal in form and do not require hypertonic medium for growth. When transferred to media devoid of antibacterial agents, they tend to revert to the virulent, large colony form. G variants are sometimes isolated directly from human exudates.

L forms have also been isolated from cultures containing penicillin or lysostaphin (see below). Unlike the G variants, they require hypertonic medium for initial growth and can be cultured anaerobically. It seems likely that L forms are more deficient in cell walls than G forms.*

To the physician, by far the most important staphylococcal variants are those affecting susceptibility to antimicrobial drugs. In no other bacterial infection does drug resistance play such a prominent role. Staphylococcal resistance to penicillin is due to the emergence of strains that produce penicillinase (β-lactamase). The gene for penicillinase is carried by plasmids (Ch. 9, Episomes) which can be transduced by bacteriophages.

---

* Protoplasts totally devoid of cell walls have been prepared with lysozyme from rare strains of *S. aureus* that are susceptible to the enzyme.

Such transduction has been demonstrated not only in culture but also in experimental pyelonephritis. Semisynthetic penicillin derivatives resistant to the action of penicillinase are described elsewhere (Ch. 7, Semisynthetic penicillins). Unfortunately staphylococcal strains resistant to these semisynthetic derivatives also arise, often as a result of lysogenic conversion, which may simultaneously affect the pathogenic properties of the organisms (see below). Therefore, when a new antibiotic is introduced in a hospital for the treatment of staphylococcal disease, its initial effectiveness gradually diminishes as resistant strains emerge among the staphylococci indigenous in the hospital population. The selective pressure exerted by antimicrobial therapy has been discussed earlier (Ch. 23, Clinical and epidemiological significance of drug resistance).

PATHOGENICITY

Nature of Lesions. The hallmark of staphylococcal disease is **suppuration.** Once virulent staphylococci gain a foothold in deeper tissues of the body, their multiplication causes necrosis and eventual **abscess formation.** Much of the localized tissue damage that results is irreversible and therefore leads to permanent scarring. Only in unusually severe infections do the organisms break through the localizing barriers of the lesion(s) and invade the lymphatics and blood stream. If bacteremia becomes established metastatic foci frequently develop.

Constancy of Infection. Man is constantly exposed to staphylococci. The skin and nose of the infant are colonized within a few days of birth and from then on staphylococcal infection persists. *S. albus* is a virtually constant inhabitant of the human skin and mucous membranes. In addition, infection of the skin, nose, oropharynx, and intestinal tract with *S. aureus* is extremely common (Ch. 42). Most human sera contain antibodies to *S. aureus,* and overt staphylococcal disease rarely occurs in adults unless the antibacterial defenses have become depressed.

Factors Predisposing to Disease. Invasive staphylococcal infections most often occur as complications of accidental and operative trauma, of burns and other serious skin lesions, and of such chronic debilitating diseases as cancer, diabetes mellitus, and cirrhosis of the liver. Recurrent superficial staphylococcal infections limited primarily to the skin (folliculitis, acne vulgaris, and hidradenitis) occur particularly during puberty in persons with hyperactive sebaceous glands and in individuals frequently in contact with oil, grease, and other skin irritants.

Human volunteer experiments have shown that even *S. aureus* is relatively avirulent for man compared with many other bacteria; large numbers of coagulase-positive staphylococci injected under the human skin may cause no more than a barely discernible lesion. If, on the other hand, a silk suture contaminated with a relatively small number of staphylococci (<100) of the same strain is placed in the skin of human volunteers, suppurative lesions invariably result. Evidently the resulting foreign body reaction somehow permits the organisms to gain a foothold.

Mice are also normally resistant to subcutaneous injections of staphylococci but susceptible to staphylococcal stitch abscesses. In rabbits other factors that cause tissue damage, such as thermal burns, chemical irritants, or the injection of other bacteria, have been shown to increase the local infectivity of pathogenic staphylococci. The most striking effects result from tissue necrosis; but even when the inciting procedure causes only acute inflammation, some loss of local resistance is demonstrable during the early acellular phase of the inflammatory response (Ch. 22, Inflammation). If, however, the staphylococci are injected into the inflammatory lesions after 2 days (when phagocytic cells have had time to mobilize) local resistance is enhanced. These experimental findings are in keeping with the well-known tendency for human staphylococcal lesions to arise in skin wounds and obstructed hair follicles.

General systemic depressants of antibacterial resistance, such as dietary deficiencies, metabolic disturbances (including diabetes mellitus), injections of bacterial endotoxins, and anaphylactic reactions, have also been shown experimentally to increase susceptibility to staphylococcal disease. It is apparent that *S. aureus,* though not highly virulent for man or most laboratory animals, can produce serious disease when either the local or the general antibacterial defenses are sufficiently depressed. Even *S. albus,* which is much less virulent than *S. aureus,* often infects prosthetic heart valves and indwelling venous catheters and occasionally causes urinary tract infections and subacute bacterial endocarditis.

Antiphagocytic Properties of S. aureus.   Of the various factors that render *S. aureus* more virulent than *S. albus,* resistance to phagocytosis is among the most important. This property long remained unrecognized simply because most human sera, for the reasons already mentioned, contain opsonizing antibodies to *S. aureus.* Phagocytic tests performed with human serum, therefore, failed to reveal the antiphagocytic capacities of the virulent organisms. When such tests were done, however, with rabbit serum which was devoid of these antibodies, *S. aureus* was found to resist phagocytosis, whereas *S. albus* was readily ingested.

Most strains of *S. aureus* that resist phagocytosis in the absence of opsonizing antibody do not possess demonstrable capsules when recovered on standard media. The demonstration of mucoid growth on special media, however, and the conversion of unencapsulated strains to the encapsulated state in experimental infections, suggest that capsule formation may be more common in vivo. This hypothesis is supported by the prevalence of anticapsular antibodies in human sera. Protein A has also been shown to be antiphagocytic. Because it binds to the Fc piece of IgG, the suggestion has been made that it exerts its antiphagocytic effect in the fluid phase by competing with phagocytes for the Fc sites of opsonins (see Ch. 22, Humoral factors in antibacterial defense).

Relation of Virulence to Production of Extracellular Enzymes.   To explain the correlation of staphylococcal virulence and coagulase production it has been suggested 1) that the enzyme promotes the formation of clots which interfere with the functioning of phagocytic cells, and 2) that it causes fibrin to be deposited on the surfaces of the organisms, thus encasing them in antiphagocytic envelopes. However, concerning the first of these possibilities, it has already been emphasized that the presence of fibrin clots in phagocytic systems enhances, rather, than impedes, the phagocytosis of encapsulated organisms, both in vitro and in vivo (Ch. 22, Surface phagocytosis). As regards the second, not only have attempts to demonstrate fibrin capsules been unsuccessful, but coagulase-negative mutants of *S. aureus* have been found to be just as virulent as parent strains that produce the enzyme. There is, in fact, no evidence that coagulase is directly involved in pathogenicity; rather its correlation with virulence appears to be coincidental. There is even doubt that it contributes significantly to the fibrin barrier frequently seen at the periphery of staphylococcal lesions, since coagulase-negative mutants produce lesions of precisely the same character.

The ability of staphylococci to cause boils, on the other hand, seems to be related to lipase production, as indicated by the results of experimental skin infections in which lysogenic conversion of the "Tween reaction" (which measures lipase production) was used to obtain comparable pairs of strains. Clinical studies also suggest that lipase-positive strains tend to cause localized boils and carbuncles, whereas lipase-negative strains, resulting from lysogenic conversion associated with the acquisition of drug resistance (e.g., to methicillin), tend to produce more generalized lesions leading to bacteremia. However, it may be significant that the lipase-negative strains also produce large amounts of hyaluronidase ("spreading factor").

Intracellular Survival.   The invasiveness of *S. aureus* has also been attributed to its ability to survive within phagocytic cells. Rogers and others have shown that a few of the organisms ingested by human leukocytes survive and eventually multiply when the phagocyte disintegrates. Ingested albus staphylococci, on the other hand, are all dead within 20 minutes. It should be emphasized, however, that only a small fraction of the phagocytized aureus organisms are spared. When the leukocytes are disrupted artificially at the end of 90 minutes, for example, less than 5% of the ingested cocci resume multiplication. Aureus staphylococci are more efficiently killed by leukocytes that ingest several cocci than by those that phagocytize only a single organism. This difference is apparently due to the degranulation phenomenon (Ch. 22, Phagocytosis).

α **Toxin.** The role of toxins in the pathogenesis of staphylococcal disease is not clearly defined. The work of Goshi and others suggests that the *α*-hemolytic, dermonecrotic, leukocidal, and lethal effects of staphylococcal toxin are all due to a single component. The common factor separated from staphylococcal filtrates is a protein of MW about 44,000; it is reported to be highly unstable in the purified state, even at $-5°$. Thal has shown that the more stable crude *α*-**toxin** acts selectively on smooth muscle, causing constriction and then paralysis of blood vessels and finally necrosis of smooth muscle cells in the vessel walls. All its effects are blocked by specific antitoxin when given in advance, but not when given after the toxin has reached the tissues. The vasoconstriction it produces may well contribute to the characteristic necrosis of staphylococcal lesions.

The vasomotor action of the toxin also interferes with the mobilization of leukocytic exudates (Ch. 22, Inflammation). Immunization of rabbits with purified *α*-hemolysin toxoid causes the acute inflammatory response in *S. aureus* lesions to be intensified. Since antibodies to the *α*-toxin neutralize its leukocidal as well as its vascular action, the immunity induced may serve to protect exudate leukocytes as well as ensure their mobilization. The role of the leukocidins in general, however, of which the *α*-toxin is only one, has been difficult to evaluate because of their immunological heterogeneity.

Although clinical experience with staphylococcal toxoid has not been encouraging, there is some evidence that antitoxic immunity may affect the progress of experimental staphylococcal disease. The increasing incidence of serious hospital epidemics due to drug-resistant staphylococci (see below) has stimulated a renewed interest in possible methods of immunization.

Delayed Hypersensitivity. The pathogenicity of staphylococci appears to depend not only upon their antiphagocytic properties and toxigenicity, but also upon their tendency to cause delayed hypersensitivity. Repeated staphylococcal skin infections in rabbits result in increased susceptibility to both skin and joint infections with *S. aureus*. The skin lesions in the sensitized animals are far more destructive than in nonsensitized controls (Fig. 27-3). Since the increased susceptibility to infection

Fig. 27-3. Effect of delayed hypersensitivity on severity of dermal staphylococcal lesions in rabbits. **A.** Skin lesion after 48 hours in previously uninoculated rabbit caused by subcutaneous injection of 10^5 *S. aureus* organisms. **B.** Comparable skin lesion in rabbit which had previously received, at other skin sites, four weekly injections of the same strain of staphylococcus. ×40. [From Johnson, J. E., Cluff, L. E., and Goshi, K. *J Exp Med 113:*235 (1961).]

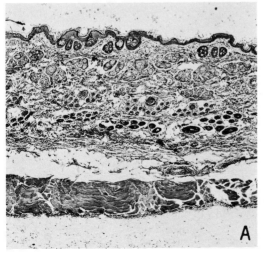 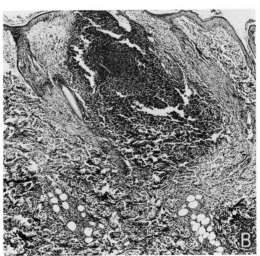

can be passively transferred with lymphoid cells, the hypothesis has been advanced that delayed hypersensitivity may participate in the pathogenesis of staphylococcal disease in much the same way that it does in tuberculosis (Ch. 22, Role of hypersensitivity in bacterial diseases).

Staphylococcal Enteritis. Special mechanisms of pathogenesis operate in staphylococcal diseases of the gastrointestinal tract. Because of their hardiness and their prevalence on the skin and mucous membranes staphylococci are frequently found in the feces. Their proliferation in the gastronintestinal tract, however, is controlled by the antagonistic action of the many other bacterial species present (Ch. 22, Microbial factors). When this balance is upset by a broad-spectrum antibiotic (e.g., a tetracycline), resistant staphylococci may achieve a tremendous population density in the feces. Under such circumstances they sometimes invade the bowel wall and produce **acute staphylococcal enteritis** which may be fatal.

Enterotoxin. The mechanism of **staphylococcal food poisoning** is very different. Characterized by sudden nausea, vomiting, diarrhea, and often shock, occurring within a few hours after ingestion of contaminated food, this acute syndrome is not an infection but a toxemia, due purely to the action of **enterotoxin** released by staphylococci that grew in the food before it was eaten. Of the four known immunological types of enterotoxins (A through D), type B is the most heat-stable. It is also resistant to the proteolytic enzymes of the gastrointestinal tract. The exact manner in which enterotoxins produce emesis is not known.

Combined Virulence Factors. Apart from the action of enterotoxin, the factors that account for the virulence of *S. aureus* include 1) antiphagocytic surface components, which may or may not form a visible capsule; 2) the ability of an occasional coccus to survive phagocytosis; 3) the production of α-toxin, which may promote necrosis, interfere with inflammation, and injure leukocytes; 4) the elaboration of other leukocidins; and 5) the emergence of delayed hypersensitivity, which

enhances tissue necrosis and increases the host's susceptibility to infection. Despite the combined effects of these several factors, the over-all virulence of *S. aureus* for man and most laboratory animals is, as already emphasized, comparatively low; serious staphylococcal disease occurs only when the local or the general antibacterial defenses of the host have been depressed. Although depressions of natural defense are also important in the pathogenesis of other bacterial diseases, their influence in staphylococcal disease is especially critical.

IMMUNITY

Antibacterial immunity to staphylococcal infections has been studied most thoroughly in rabbits inoculated with washed, heat-killed, or formalinized vaccines. When rabbits thus immunized are injected intravenously with staphylococci, their blood streams are cleared more rapidly than those of unimmunized controls, but their tissues (particularly kidneys) are not spared from suppurative lesions.

Rabbits immunized with staphylococcal toxoid, on the other hand, develop **antitoxic** rather than antibacterial immunity. In these rabbits, as already noted, lesions produced by injection of staphylococci are of diminished severity. Toxoid-immunized rabbits also are protected against the lethal action of staphylococcal toxin and therefore outlive controls when given large intravenous doses of living organisms.

In relating these laboratory observations to human staphylococcal disease, it should be stressed that man not only has a high degree of natural resistance to staphylococci, but also has antibodies in his serum acquired as a result of intermittent minor staphylococcal infections of his skin and mucous membranes. Following overt staphylococcal disease, the antibacterial antibody titer of the serum usually rises. The titer of antitoxin may not be affected if the lesions are superficial, whereas in systemic staphylococcal disease it, too, is likely to be increased. For the reasons already emphasized under Pathogenicity the presence of antibodies in the serum does not always protect an individual against staphylococcal disease.

LABORATORY DIAGNOSIS

The finding of gram-positive cocci in stained smears of purulent exudates provides only suggestive diagnostic information, since staphylococci cannot be differentiated from other gram-positive cocci on purely morphological grounds. Exudates therefore should always be cultured on blood agar plates. Although staphylococci will grow on most blood-free basal media, blood agar is recommended because 1) it supports better growth of other pathogenic microbes that may be present in the lesion, 2) it promotes pigment formation by staphylococci, and 3) it permits the direct detection of hemolysins.* Either meat infusion or tryptose phosphate (meat digest) broth may be used as the basal medium. Blood samples should be inoculated not only into broth but also into blood agar pour plates. If the specimen contains large numbers of bacteria of other species which might outgrow staphylococci (as in stool cultures), a selective medium may be used, such as meat infusion agar containing 7.5% sodium chloride, phenolphthalein phosphate agar, or tellurite glycine agar.† When a gram-stained smear of fecal material reveals many granulocytes and a predominance of gram-positive cocci, a presumptive diagnosis may be made of staphylococcal enteritis.

Although most pathogenic strains of staphylococcus form golden pigment on blood agar, particularly when incubated at room temperature for several days, pigment production is not an adequate indicator of pathogenicity because virulent mutants devoid of pigment are frequently encountered, even on primary isolation. Likewise, production of hemolysin, fermentation of mannitol, and ability to liquefy gelatin cannot be relied on, even though most virulent strains share all these characteristics. The property that correlates best with pathogenicity is the elaboration of coagulase. DNase production also correlates well with potential pathogenicity.

The technic for detecting coagulase involves merely the incubation of 0.5 ml of broth culture with an equal volume of citrated rabbit (or human) plasma, previously diluted 1:5 with broth. Two control tubes, one containing a known coagulase-positive culture and the other uninoculated plasma, are included. A coagulase-positive culture will clot the plasma usually after incubation for 3 hours at 37° and almost always after 18 hours. Either citrated or oxalated plasma may be used; but since human plasmas vary in their suitability for detecting coagulase, each lot should first be tested with known positive and negative strains.

Enterotoxin-producing strains were originally identified by injecting culture filtrates intraperitoneally in kittens or intragastrically in monkeys or man. These procedures have now been supplanted by gel diffusion immunoprecipitation technics.

TREATMENT

The therapeutic problem posed by drug-resistant staphylococci is self-evident. Because of the prevalence, particularly in hospital populations, of staphylococci resistant to one or more antimicrobials, it is essential to determine the drug sensitivity of the infecting organism. The technics employed are discussed elsewhere (Ch. 7, Effects on growth and viability). Penicillinase-producing strains are resistant to ordinary penicillins (e.g., benzyl penicillin) but may still be susceptible to semisynthetic penicillin derivatives that are not hydrolyzed by the enzyme. Therefore, until the drug sensitivity is known, the patient should usually be treated with a semisynthetic penicillin derivative such as methicillin.

Because drug resistance may emerge rapidly, chemotherapy of staphylococcal disease should be intensive. Frankly suppurative lesions should always be drained because of the resistance of established abscesses to antimicrobial therapy (Ch. 23, Abscess formation and necrosis).

Lysostaphin is a peptidase that specifically degrades the pentaglycine bridges in the cell wall of *S. aureus*. It is an exoenzyme of a staphylococcal strain known as *S. staphylolyticus*. Its systemic use is limited by its immunogenicity in man and its tendency to select for wall-defective variants which might be resistant to other antibiotics. Topical application is occasionally useful.

PREVENTION

The principal reservoir of staphylococci in nature is man. Cross-infection from one hu-

---

* α-Hemolysin is most readily detected on rabbit blood agar.

† Staphylococci, like corynebacteria, form jet black colonies on tellurite agar (Ch. 24, Metabolism).

man to another is extremely common and may either be airborne or result from direct contact. Indeed, the ubiquity of staphylococci in the population is such that elimination of the human reservoir would appear virtually impossible.

Carrier Rates. By the end of the first 10 days of life 90% of infants have become carriers of *S. aureus*. The carrier rate falls to 20 to 30% by the second year, only to rise again by the fourth or fifth year to a more stable adult rate of about 50%. Some adults (ca. 60%) are intermittent carriers. The factors that control the dynamics of the carrier state remain obscure.

Spread of Disease. How staphylococcal disease is acquired is not always clear. It seems likely that airborne staphylococci may infect exposed burns and open wounds in operating rooms. Under other conditions where trauma is not involved more intimate contact may be necessary for the transmission of significant infection. In hospital nurseries the site of primary colonization in the newborn is the umbilicus, and spread of the infection can usually be shown to result from direct contact with infected attendants. To trace the mode of transmission in any given outbreak, however, is often difficult.

Hospital Epidemics. During recent years an increasing number of hospital epidemics of staphylococcal disease in the United States have been due to group III phage types. The incidence of cases has usually been highest among surgical patients and newborn infants, and the strains isolated have tended to be resistant to the antimicrobial drugs being used in the hospital. Although most of the surgical infections appear to have been acquired in the operating room, the occasional long delay in onset of symptoms suggests that staphylococcal postoperative infections may sometimes be acquired on hospital wards. Attendants known to be nasal carriers of drug-resistant strains should be excluded, if possible, from operating rooms and infant wards because of the high risk of infection in these areas. Patients with open wounds in-fected with staphylococci should be separated, when feasible, from other patients on surgical services. Although none of these methods will eliminate cross-infection, their combined effects, together with strict adherence to accepted hygienic technics of patient care, may significantly lower the incidence of staphylococcal disease.

Prophylactic Procedures. A novel approach to the control of recurrent staphylococcal infections is based on bacterial interference (Ch. 22, Microbial factors). After antibiotic therapy, a relatively avirulent strain of staphylococcus is inoculated locally to prevent subsequent recolonization with a virulent strain. This procedure has been particularly effective in controlling staphylococcal epidemics in newborn nurseries and in managing patients with recurrent furunculosis.

The prophylactic value of immunization with toxoids has never been established.

Food Poisoning. Proper refrigeration of food will prevent staphylococcal food poisoning. In many outbreaks, the incriminated food (often creamed pastry) is found to have been improperly refrigerated and to have been handled by an individual who is a carrier of pathogenic staphylococci or is suffering from an open staphylococcal skin lesion. The heating of contaminated food will often kill the viable organisms without destroying the relatively heat-stable enterotoxin. Although elimination of staphylococcal carriers among food handlers may be difficult, persons with overt staphylococcal disease should certainly not be allowed to prepare or serve food.

OTHER MICROCOCCI

A variety of other gram-positive cocci belong to the family Micrococcaceae. Some species are anaerobic and most are nonpathogenic for man. One species, known as *Micrococcus tetragenus* (or *Gaffkya tetragena*), occasionally causes arthritis, meningitis, pneumonia, soft tissue abscesses, or endocarditis. During multiplication, the individual cocci divide in two planes, forming tetrads. In vivo, the organisms are heavily encapsulated.

SELECTED REFERENCES

Books and Review Articles

COHEN, J. O., ed. *The Staphylococci.* Wiley, New York, 1970.

ELEK, S. D. *Staphylococcus Pyogenes and Its Relation to Disease.* Livingstone, Edinburgh, 1959.

WHIPPLE, H. E., ed. The staphylococci: Ecological perspectives. *Ann NY Acad Sci 128:*1 (1965).

Specific Articles

BERGDOLL, M. S., CHU, F. S., BORGA, C. R., HUANG, I., and WEISS, K. F. The staphylococcal enterotoxins. Jap J Microbiol *11:*358 (1967).

COHN, Z. A., and MORSE, S. I. Interaction between rabbit polymorphonuclear leucocytes and staphylococci. *J Exp Med 110:*419 (1959).

DOSSETT, J. H., KRONVALL, G., WILLIAMS, R. C., and QUIE, P. G. Antiphagocytic effects of staphylococcal Protein A. *J Immunol 103:*1405 (1969).

FORSGREN, A., and SJOQUIST, J. Protein A from *Staphylococcus aureus.* 3. Reaction with rabbit gamma globulin. *J Immunol 99:*19 (1967).

GOSHI, K., CLUFF, L. E., and NORMAN, P. S. Studies on the pathogenesis of staphylococcal infection. V. Purification and characterization of staphylococcal alpha hemolysin; VI. Mechanism of immunity conferred by anti-alpha hemolysin. *Bull Johns Hopkins Hosp 112:*15, 31 (1963).

KOENIG, M. E. Lysostaphin. *J Infect Dis 119:*101 (1969).

JESSEN, O., ROSENDALL, K., BULOW, P., FABER, V., and ERICKSON, K. R. Changing staphylococci and staphylococcal infections. *N Engl J Med 281:*627 (1969).

KOENIG, M. G., MELLY, M. A., and ROGERS, D. E. Factors relating to the virulence of staphylococci. III. Antibacterial versus antitoxic immunity. *J Exp Med 116:*601 (1962).

MELLY, M. A., THOMISON, J. B., and ROGERS, D. E. Fate of staphylococci within human leucocytes. *J Exp Med 112:*1121 (1960).

MORSE, S. I. Isolation and properties of a group antigen of *Staphylococcus albus. J Exp Med 117:*19 (1963).

MORSE, S. I. Isolation and properties of a surface antigen of *Staphylococcus aureus. J Exp Med 115:*295 (1962).

MORSE, S. I. Studies on the chemistry and immunochemistry of cell walls of *Staphylococcus aureus. J Exp Med 116:*229 (1962).

NOVICK, R. P. Staphylococcal plasmids. *J Gen Microbiol 55:*111 (1969).

THAL, A. P., and EGNER, W. The site of action of the staphylococcus alpha toxin. *J Exp Med 113:*67 (1962).

WENTWORTH, B. B. Bacteriophage typing of staphylococci. *Bacteriol Rev 27:*253 (1963).

YOSHIDA, K., and ECKSTEDT, R. D. Relation of mucoid growth of *Staphylococcus aureus* to clumping factor reaction, morphology in serum-soft agar, and virulence. *J Bacteriol 96:*902 (1968).

YOSHIDA, K., TAKAHASHI, M., and TAKEUCHI, Y. Pseudocompact-type growth and conversion of growth types of strains of *Staphylococcus aureus* in vitro and in vivo. *J Bacteriol 100:*162 (1969).

chapter

THE NEISSERIAE

Revised by EMIL C. GOTSCHLICH, M.D.

THE GENUS NEISSERIA   742

  Morphology   742
  Metabolism   743

NEISSERIA MENINGITIDIS   745

  Antigenic Structure   745
  Genetic Variation   746
  Pathogenicity   746
  Immunity   747
  Laboratory Diagnosis   747
  Treatment   748
  Prevention   748

NEISSERIA GONORRHOEAE   749

  Antigenic Structure   749
  Genetic Variation   749
  Pathogenicity   750
  Laboratory Diagnosis   750
  Treatment   750
  Prevention   751

OTHER SPECIES OF NEISSERIA AND RELATED GENERA   751

THE GENUS NEISSERIA

The only known reservoir of the neisseriae is man. The genus includes two gram-negative species of pyogenic cocci that are pathogenic for man: the meningococcus (*Neisseria meningitidis*) and the gonococcus (*Neisseria gonorrhoeae*). It also includes a number of nonpathogenic species that inhabit the upper respiratory tract and therefore may be confused with meningococci.

Epidemic cerebrospinal meningitis, now usually referred to as **meningococcal meningitis,** was not clearly recognized as a contagious disease until early in the nineteenth century. Outbreaks were eventually described on all continents, and the prevalence of the disease gradually became apparent, particularly among military personnel. The causative organism was first isolated by Weichselbaum in 1887 from the spinal fluid of a patient with a purulent form of meningitis. Subsequent studies have revealed that meningococcal meningitis occurs endemically as well as in epidemics, and carrier rates for meningococci in the general population are relatively high and fairly constant. Moreover, a nonmeningeal form of meningococcal disease, known as **meningococcemia,** may result from invasion of the blood stream. As early as 1909 the existence of immunologically specific types of meningococci was recognized, and in 1913 Flexner introduced an effective form of serum therapy. This inconvenient treatment was displaced in 1939 by sulfonamide therapy; and subsequently antibacterial drugs were shown to be effective not only for curing the disease but also as prophylactic agents for terminating epidemics.

Gonorrhea,* the human ailment caused by the gonococcus, was known to be of venereal origin in the thirteenth century.

Gonorrhea was thought to be an early symptom of syphilis by Paracelsus, a teacher of great influence in the 1500s. To distinguish gonorrhea from syphilis the celebrated English physician John Hunter inoculated himself with the purulent exudate from a patient with gonorrhea, but unfortunately, the donor had a double infection. Not until the middle of the nineteenth century were syphilis and gonorrhea clearly differentiated.

N. gonorrhoeae, the causative agent of gonorrhea, was described by Neisser in 1879 and first cultivated by Leistikow and Loeffler in 1882. Treatment was unsatisfactory until the advent of the sulfonamide drugs, but sulfonamide-resistant strains of gonococcus became common within a few years; introduction of penicillin in 1943 restored control of the sulfonamide-resistant strain.† Although penicillin has proved highly effective and has markedly decreased the serious consequences of the disease, its use has not lowered the **incidence** of gonococcal infections in the general population.

MORPHOLOGY

The neisseriae are nonmotile, non-spore-forming, gram-negative cocci which grow in pairs or occasionally in tetrads or clusters. The

---

* The term gonorrhea, meaning "flow of seed," was introduced by Galen (A.D. 130), who had the

mistaken impression that it was due to spermatorrhea.

† Although the neisseriae stain gram-negatively, contain endotoxins, and are lysed by antibody and complement (like most other gram-negative bacteria), they behave pathogenetically like the pyogenic gram-positive cocci and are sensitive to penicillin. Thus they seem, in some respects, to belong in an intermediate category.

individual cocci are small (ca. 0.8 by 0.6 μ) and often assume bizarre shapes as a result of partial autolysis. The ultrastructure of the cytoplasm and of the wall of the meningococcus and the gonococcus are very similar (Fig. 28-1). The cell wall surrounding the cell membrane possesses a dense layer probably consisting of the peptidoglycan and an outer membrane. External to the wall many meningococci have a polysaccharide-containing capsule, but capsules have not been convincingly demonstrated on gonococci.

Gonococci exhibit four colonial forms. When freshly isolated from clinical specimens, they most often grow as small colonies, T1 and T2, which are virulent. Repeated non-selective subculture yields larger colonies, T3 and T4 (Fig. 28-2), which have proved avirulent on inoculation in male volunteers. Pili have been demonstrated on colony types T1 and T2 (Fig. 28-2B, and D) but not on types T3 and T4. Their relation to virulence has not been established. Pili have also been reported to occur in other members of the genus neisseriae, including the meningococcus.

METABOLISM

The principal species included in the genus *Neisseria* are listed in Table 28-1, together with the growth and fermentation character-

Fig. 28-1. **A.** Thin section electron micrograph of a gonococcus. ×160,000. Note the location of (CM) the cell membrane, (DL) the dense layer, which consists at least in part of the peptidoglycan, and (OM) the cell wall membrane containing the endotoxic lipopolysaccharide. **B–D.** Cells labeled with purified antibody against the group A meningococcal polysaccharide conjugated with horseradish peroxidase or ferritin. **B.** Group A meningococci stained with peroxidase conjugated antibody. The polysaccharide forms a capsule around the organisms. ×25,000. **C.** Higher magnification. ×120,000 showing in intensely opaque region immediately external to the cell wall and a less densely staining peripheral zone. **D.** Appearance of the capsule of the group A meningococcus stained with ferritin-conjugated antibody. ×120,000. (Electronmicrographs by Dr. John Swanson.)

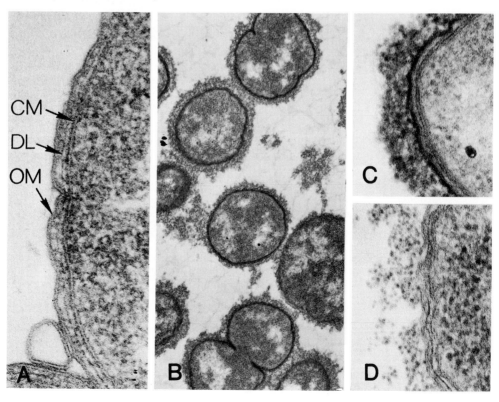

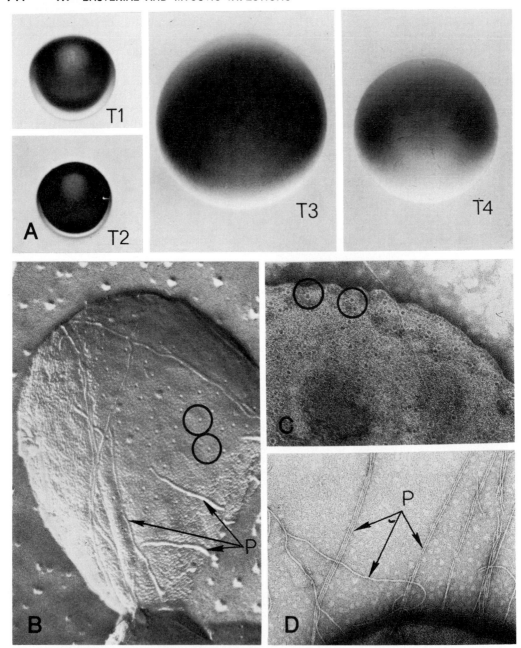

Fig. 28-2 A. Appearance of the four colonial types of the gonococcus as seen through a colony scope. X58. Colonies of types T1 and T2 are small and pigmented dark gold. T2 colonies have a more sharply defined border and are friable when touched with a loop. T3 and T4 colonies are larger. T3 are light brown whereas T4 are colorless and almost transparent. T3 colonies have a granular appearance; the other types appear amorphous. [**A** from Kellogg, D. S., Cohen, I. R., Norins, L. C., Schroeter, A. L., and Reising, G. *J Bacteriol* 96:596 (1968).] **B.** The surface of *N. gonorrhoeae* as seen by freeze etching. X87,000. The circles indicate pits which are characteristically found in this species and meningococci. Pili (P) are seen on the surface of gonococci of colonial types T1 and T2 as well as on meningococci. **C.** The appearance of pits visualized by negative staining with phosphotungstic acid. X100,000. [**C.** from Swanson, J. *J Exp Med* 136:1258 (1972).] **D.** Pili radiating from the surface of a gonococcus (bottom) as seen with negative staining. X60,000. (Electronmicrographs by Dr. John Swanson.)

TABLE 28-1. Principal Differential Characteristics of the Common Species of the Genus *Neisseria*

Organism	Growth on agar devoid of blood	Growth at 22°	Fermentation		
			Glucose	Maltose	Sucrose
N. meningitidis	−	−	+	+	−
N. gonorrhoeae	−	−	+	−	−
N. catarrhalis	+	+	−	−	−
N. sicca	+	+	+	+	+
N. flavescens	+	±	−	−	−

istics that permit their differentiation. The nonpathogenic species grow at 22° and on nutrient agar devoid of blood, whereas the meningococcus and gonococcus do neither; also each species exhibits a characteristic fermentation pattern when cultured in media containing glucose, maltose, or sucrose. However, there are neisseriae indistinguishable from meningococci except that they ferment lactose in addition to glucose and maltose. The fermentation tests are done in semisolid cystine-trypticase agar containing 0.5 to 1.0% of the specific carbohydrate, and only small amounts of acid are generated in the positive reactions. All the species are essentially aerobic but will multiply under microaerophilic conditions.

Meningococci and gonococci are difficult to cultivate, primarily because of their sensitivity to the toxic fatty acids and trace metals contained in peptone and agar. The inhibitory effect of these toxic components may be eliminated by adding to the medium blood, serum, starch, or charcoal, which bind the toxic substances. Blood agar heated at 80 to 90°, to form "chocolate agar," is even more suitable for the growth of gonococci.

Both meningococci and gonococci tend to undergo rapid autolysis. Their autolytic enzymes may be inactivated by heating the culture at 65° for 30 minutes or by adding potassium cyanide or formalin. These methods are used in preparing cell suspensions for serological tests.

The resistance of meningococci and gonococci to extreme physical and chemical conditions is exceptionally low. In most laboratory cultures they die out in a few days. They are particularly susceptible to desiccation and are killed when heated at 55° for 30 minutes. Growth of both species is stimulated by an atmosphere containing 5 to 10% carbon dioxide; for this reason, cultures are usually incubated in a candle jar.

All species of the genus *Neisseria* may be recognized by their ability to oxidize rapidly dimethyl- or tetramethyl-paraphenylene diamine, which causes surface colonies to turn first pink and then black. The usefulness of this test is limited since the nonpathogenic species of the genus also react.

Although the neisseriae share many morphological and metabolic characteristics, their other properties differ enough to warrant individual consideration.

NEISSERIA MENINGITIDIS

ANTIGENIC STRUCTURE

Meningococci have been divided serologically, by agglutination, into groups A, B, C, D, X, Y, and Z. The group A antigen is a capsular polysaccharide consisting of N-acetyl, O-acetyl mannosamine phosphate. The B and C antigens both consist of N-acetyl neuraminic acid (sialic acid), which is partially O-acetylated in the C antigen. In addition, even after removal of the O-acetyl groups, C differs from B in immunological specificity and ease of cleavage by neuraminidases and by acid. The capsular antigens of the other meningococcal groups have not been chemically characterized.

A somatic nucleoprotein fraction (the P antigen), a somatic carbohydrate antigen, and the endotoxin have also been isolated but not chemically defined.

GENETIC VARIATION

When smooth encapsulated strains of meningococci are repeatedly subcultured on artificial media they often yield rough (R) variants which are unencapsulated and relatively avirulent. If mice are infected with a rough variant suspended in mucin, smooth (S) revertents may be recovered from the heart blood.

Various drug-resistant mutants have arisen. Strains highly resistant to sulfonamides are now very common in both military and civilian populations in many parts of the world. From cultures grown in the presence of streptomycin Miller and Bohnhoff isolated **streptomycin-dependent** strains (Ch. 7) whose virulence for mice is enhanced by streptomycin. Penicillin-resistant mutants have been recovered from meningococcal cultures containing increasing concentrations of the drug. When the resistant variants are repeatedly grown in a penicillin-free medium antibiotic-sensitive revertents are at a metabolic advantage and again become predominant.

Genetic transformation by DNA has been observed with the meningococcus: transferred markers include capsule formation and streptomycin resistance. The extracellular DNA that accumulates in the slime of broth cultures is just as active in such transformation as the DNA extracted from intact cells, suggesting that transformation affecting virulence may occur in mixed populations in nature.

PATHOGENICITY

The meningococcus usually inhabits the human nasopharyngeal area without causing any symptoms. This carrier state may last for a few days to months; it is important because it provides the reservoir for the meningococcus and it enhances the immunity of the host. The carrier rate in the general population tends to be appreciable at all times and in military populations may exceed 90%.

When individuals without adequate levels of immunity acquire the meningococcus, usually through contact with a healthy carrier, the resulting nasopharyngeal infection may lead rapidly to bacteremia, which is ordinarily followed by acute purulent meningitis.

The mortality of meningococcal meningitis without treatment is approximately 85% and it can be less than 1% when vigorous antibiotic and supportive therapy is administered early, but it is approximately 15% in the general population.

Infrequently meningococcemia fails to lead to meningitis, but metastatic lesions often develop in the skin, joints, lungs, ears, and adrenal glands. These metastatic lesions are usually thromboembolic and frequently contain cultivable meningococci. Acute adrenal insufficiency (Waterhouse-Friderichsen syndrome) occurs in fulminating cases and is commonly associated with hemorrhagic involvement of both adrenal glands.* Very rarely meningococcemia becomes chronic and lasts for weeks as a fever of unknown etiology.

Laboratory Models. Studies on pathogenesis have been seriously limited by the fact that meningococci are relatively nonpathogenic for most laboratory animals. Meningococcal meningitis has been produced in a variety of laboratory animals by direct injection into the subarachnoid space, but such large numbers of organisms are required that the relevance to pathogenesis in man is questionable. Studies in chick embryos have been more interesting: regardless of the site of inoculation (amniotic sac, body wall, or intravenous), the meningococci eventually localize in the cranial sinuses, lungs, and meninges (Ch. 22). Serial sections of the embryos revealed no evidence of direct extension of the infection from the nasopharynx to the brain, but bacteremia always preceded involvement of the meninges.

On the intraperitoneal inoculation of mice with meningococci suspended in gastric mucin, which protects the organisms from prompt destruction by the peritoneal phagocytes, inocula of as few as 10 meningococci may cause fatal infection. This method is often used to assay antibodies and to study the therapeutic effectiveness of antimicrobial drugs.

---

* Acute adrenal insufficiency of this sort may also be associated with overwhelming bacteremia due to other organisms, such as staphylococci, pneumococci, and streptococci. It may also be produced experimentally in rabbits given intravenous endotoxin within a few hours after stimulation of their adrenal cortices with ACTH, suggesting that the stimulated glands are hypersusceptible to the action of meningococcal endotoxin.

Virulence Factors. The virulence of *N. men-ingitidis* depends in part upon the antiphago-cytic properties of its capsule. As in the case of pneumococci, the nature of the capsular polysaccharide is of importance: most me-ningococcal disease is caused by encapsulated strains of groups A, B, and C, while strains of groups X, Y, and Z, which most probably also possess capsules, are commonly found in carriers but very rarely cause disease. Me-ningococci behave in the animal host as ex-tracellular parasites, i.e., they are unable to survive once ingested by phagocytic cells. Although they are often visible within leuko-cytes (Fig. 28-3), there is no evidence that they can multiply intracellularly. Endotoxins cause extensive vascular damage and are ca-pable of producing both localized and gen-eralized Shwartzman reactions (Ch. 22, Endotoxin) whose pathology resembles that of the thromboembolic lesions of meningococ-cemia. It therefore seems plausible, but has not been proved, that such reactions are involved in the pathogenesis of meningococcal disease.

IMMUNITY

Antibodies are present in the sera of most adults and play a role in preventing meningo-coccal disease. Thus, of 15,000 military re-cruits studied during basic training, 54 contracted group C meningococcal disease, and serum bactericidal activity against this organism was initially present in only 3 of these individuals but in the majority of those who did not contract the disease. Moreover, the newborn very rarely contracts the disease, but with the loss of maternal antibodies sus-ceptibility appears; and children with agam-maglobulinemia are protected by monthly administration of pooled concentrated human γ-globulin. The risk of disease is greatest between the ages of 6 to 24 months, and al-most half the cases occur before 5 years.

Protective antibodies are evoked within a week by the meningococcal carrier state. These antibodies are directed not only against the capsular group-specific polysaccharide but also against other species-specific antigens, including the endotoxin. Antigenic cross-re-activity may explain why the majority of young children do not contract disease on their first contact with a potentially virulent me-ningococcus: they may have immunity result-ing from prior contact with a nonencapsulated or an encapsulated strain with low virulence. In particular lactose-fermenting meningococci of groups B or C are frequent in nasopharyn-geal cultures of children and are avirulent. Moreover, the majority of children and young adults in the United States have antibodies to the group A capsular polysaccharide, al-though group A meningococci have been rarely encountered in the last 15 years; im-munization by some unrelated but cross-re-acting organism must be suspected.

LABORATORY DIAGNOSIS

In cases of suspected meningococcal disease, specimens of blood, spinal fluid, and naso-pharyngeal secretions should be examined for the presence of *N. meningitidis*. Stained smears made from petechial lesions in the skin will occasionally reveal the organism.

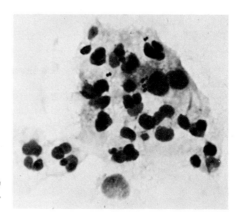

Fig. 28-3. Intracellular meningococci in smear of exudate from spinal fluid of patient with meningococcal meningitis. ×1000.

The accepted **blood culture** procedure is to add 10 ml of blood to 100 ml of a suitable liquid medium (e.g., tryptose phosphate broth) and to spread 0.1 ml of blood on the surface of a blood (or chocolate) agar plate. Both cultures are incubated at 37° in a candle jar and are inspected daily for 7 days before being discarded. When heavy bacteremia is present, diligent search may reveal gram-negative diplococci in routine blood smears.

At the time of lumbar puncture, **spinal fluid** may be permitted to drop directly from the needle onto the surface of blood agar (or chocolate agar) plates. A tube of broth should also be inoculated, and a sample of the spinal fluid itself should be incubated in a candle jar at 37°, along with the broth and agar cultures. Subculture of the incubated spinal fluid will occasionally reveal meningococci when the original drop cultures are negative.

Because meningococci tend to autolyze, smears of spinal fluid that are to be stained for bacteria must be promptly fixed. Likewise, quellung tests should be done immediately. Occasionally, homologous type-specific antibody added to the spinal fluid will give a positive precipitin reaction, even when meningococci cannot be found in stained smears of the centrifugate. Organisms in spinal fluid smears are often more difficult to find than in pneumococcal meningitis.

Nasopharyngeal cultures are best performed with a cotton swab on the end of a bent wire. Care must be taken to avoid touching the tongue, and secretions must be obtained from the posterior nasopharyngeal wall behind the soft palate.* The secretions should be swabbed directly onto a selective medium containing antibiotics to inhibit contaminating flora and streaked with a wire loop. Prompt incubation of the plates in a candle jar is essential. If the cultures cannot be made immediately,

---

* Whereas meningococci usually inhabit only the nasopharyngeal mucosa behind the soft palate, staphylococci reside in the anterior nasopharynx and vestibule of the nose, and pneumococci, β-hemolytic streptococci, viridans streptococci, and influenza bacilli are found most frequently on the mucous membranes of the tonsils and lower pharynx. (An alternative method of obtaining nasopharyngeal cultures, particularly in children, is described in Chapter 30, *B. pertussis,* Laboratory diagnosis.)

the swab should be placed in a transport medium in which meningococci remain viable for a number of hours. Although a positive nasopharyngeal culture does not prove meningococcal disease, meningococcal carriers can be detected only in this way.

Neisseria colonies on agar plates may be tentatively recognized by the indophenol oxidase test. To identify meningococci, fermentation and serological tests must be made. The latter are done with either monotypic or polytypic antisera and bacterial suspensions in saline, containing 0.1% potassium cyanide to inhibit autolysis. Since nonspecific agglutination may be encountered, a control test with normal serum must be included.

TREATMENT

The drug of choice is penicillin. Although penicillin does not penetrate the normal blood-brain barrier, it does so readily when the meninges are acutely inflamed. If a clear history of anaphylactic reactivity to penicillin contraindicates its use, combined therapy with erythromycin and chloramphenicol is recommended. Sulfonamides were widely used for many years: they penetrate the cerebrospinal fluid well, but they are only bacteristatic and resistant strains are now commonly encountered.

The success of antimicrobial treatment is much greater if started early. If signs of adrenal insufficiency develop, therapy with hydrocortisone, pressor amines, and parenteral fluids is required.

PREVENTION

Meningococcal disease occurs most commonly in children between the ages of 6 months and 2 years. It can also at times be a serious problem in military recruits among whom the spread of meningococci is markedly favored. Instances have occurred where the carrier rate has exceeded 90% and the incidence of the disease approached 1% of the recruit population. In the United States the disease used to occur mostly in epidemics (Fig. 28-4), caused predominantly by group A organisms. The last widespread epidemic occurred during the early 1950s. For un-

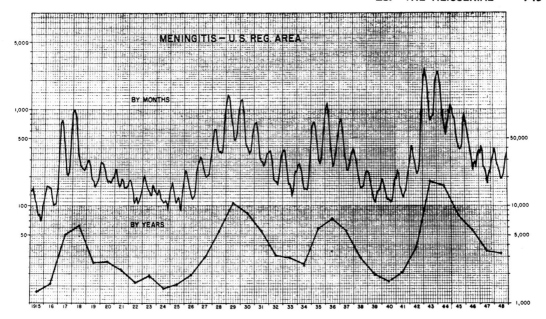

Fig. 28-4. Monthly and yearly variations of meningococcal meningitis attack rates, United States registrations area, 1945–1948. [From Aycock, W. L. and Mueller, J. H. *Bacteriol Rev 14:*115 (1950).]

known reasons, group A meningococci have been exceedingly rare in the United States since that time, and the disease was caused mainly by group B until 1965 and since then by group C. Epidemic group A disease remains a potential threat, since there have been frequent major epidemics caused by this serogroup in certain parts of Africa.

For many years sulfonamide prophylaxis was extremely effective in preventing the disease in both civilian and military populations. The spread of the disease in a military camp could be readily checked by administration of sulfonamides for a few days to all personnel.

However, resistant strains became so common that sulfonamide prophylaxis has become useless. No other antimicrobial has been as successful in chemoprophylaxis.

Recently, the capsular polysaccharide of the group C meningococcus has been isolated in a relatively undegraded form and shown to be an excellent immunogen in human beings, preventing group C meningococcal disease and also the carrier state. Vaccination may thus interfere with transmission of the organism. Polysaccharide vaccines against groups A and B meningococci are not so well developed.

NEISSERIA GONORRHOEAE

ANTIGENIC STRUCTURE

No useful immunological classification of the gonococci exists. Capsular antigen akin to the group-specific meningococcal polysaccharides has not been isolated. The endotoxin of the gonococcus has been investigated, and Maeland has defined two classes of antigenic determinant (a and b). The a determinant is probably on the polysaccharide moiety and

the b determinant on the protein moiety of the ether-extracted endotoxin.

GENETIC VARIATION

Sulfonamide-resistant mutants of gonococci emerge both in vitro and in vivo. Mutants relatively resistant to penicillin have also been isolated from patients. The gonococcus can exhibit several colonial forms (T1, T2, T3,

T4) shown in Figure 28-2A, but the genetic mechanism of the colonial variation is as yet unknown.

PATHOGENICITY

As with meningococcal disease, study of the pathogenesis of gonorrhea has been greatly hampered by the absence of a satisfactory laboratory model. A progressive infection may be established in chick embryos or in the anterior chamber of the rabbit's eye, and in mice a fatal disease may be produced by intraperitoneal injection of gonococci suspended in hog gastric mucin. Manifestly, none of these models simulates the human disease. Recently the disease has been established in chimpanzees by artificial inoculation, and natural transmission by coitus has been observed. In human male volunteers it has been shown that gonococci belonging to types T1 and T2 are able to cause genitourinary disease, whereas types T3 and T4 are not virulent.

Gonococci ordinarily enter the body through the mucous membranes of the genitourinary tract, apparently penetrating between columar epithelial cells; stratified squamous epithelium is relatively resistant to infection. The organism evokes an acute inflammatory response in the subepithelial tissues, giving rise to the purulent urethral or vaginal discharge characteristic of the early stages of the disease. In the male, the acute urethritis may extend to the prostate and epididymis and, if untreated, is often followed by fibrosis and stricture. In the female, the primary infection spreads from the urethra, vagina, and cervix to the adjacent fallopian tubes, where it may become chronic and produce sterility. Bacteremia occurs in fulminating cases, in both men and women, and is occasionally complicated by endocarditis, acute purulent arthritis, or both.

Repeated attacks of gonorrhea are common. Whether this occurs because local immunity cannot be developed or because there are a large number of serologically different strains is not known.

Newborn infants may contract serious gonococcal infections of the eye (**ophthalmia neonatorum**) from passage through an infected birth canal. Unless promptly treated, the disease may result in blindness. Vulvo-

vaginitis in young girls, once thought to be transmitted by bedclothes or towels, is ordinarily due to other species of pyogenic bacteria; when caused by the gonococcus, it has usually been acquired by sexual contact.

LABORATORY DIAGNOSIS

In acute gonococcal disease stained smears of fresh exudate will often reveal the presence of intracellular gram-negative diplococci. This finding, together with a convincing clinical history, may permit the physician to make the provisional diagnosis of acute gonorrhea and to institute specific therapy. Whenever possible, the exudate should be cultured for gonococci before treatment is begun. The diagnosis is definitely established by recovery of typical gram-negative diplococci that ferment glucose but not maltose or sucrose (Table 28-1). Rectal cultures are sometimes positive when urethral or cervical cultures are negative.

In chronic gonorrhea examination of gram-stained material is much less helpful and reliance must be placed on cultural or fluorescent antibody technics: particularly in chronic cervicitis saprophytic species of neisseriae may be present. Because gonococci are not hardy, specimens should be cultured immediately or placed in a special transport medium, e.g., charcoal-absorbed semisolid thioglycollate agar. The medium of choice for the cultivation of gonococci is a selective medium containing antibiotics to inhibit contaminating flora, and incubated in candle jars. After 12 to 16 hours smears of the early growth may be stained with specific flourescent antibody for rapid indentification of *N. gonorhoeae*.

Immunological tests for antigonococcal antibodies in sera of patients have not proved satisfactory.

TREATMENT

The sulfonamide drugs provided the first effective therapy. During World War II, however, sulfonamide-resistant strains of gonococcus were increasingly encountered,* and

---

* Venereal disease, always a major problem of military medicine, tends to be given special attention during periods of intensive mobilization.

sulfonamide therapy became less satisfactory. Fortunately penicillin was developed at this time, and it gradually replaced the sulfonamides in the treatment of gonorrhea. In recent years, strains of gonococcus moderately resistant to penicillin have been encountered, but rarely has the resistance been sufficient to cause a serious therapeutic problem. As a rule, male patients require less intensive treatment than female patients. Only when walled-off abscesses are formed, as in chronic salpingitis, is combined surgical and antibiotic therapy necessary.

PREVENTION

Although present-day methods of treating gonococcal disease are extremely effective, the morbidity rates remain distressingly high. Judging from the human volunteer experiments already described, the incubation period is relatively short, varying from 2 to 7 days. Therefore, if the individual(s) with whom the patient has had sexual contact during the week prior to onset of symptoms can be identified and treated, spread of the disease can be lessened. Chemoprophylactic use of penicillin, within a few hours of exposure, also lowers the incidence of acquired infection. However, in females acute gonorrhea, even at its height, frequently fails to produce recognizable symptoms, so that many infected women go untreated and become chronic carriers. Hence, eradication of the disease by chemotherapy poses an almost unsurmountable problem.

The time-honored method of preventing **ophthalmia neonatorum** (the Credé procedure) consists of instilling drops of 1% silver nitrate solution into the eyes of the infant immediately after birth. Although this method is still practiced, many hospitals now use an ophthalmic ointment containing penicillin.

OTHER SPECIES OF NEISSERIA AND RELATED GENERA

Nonpathogenic species of neisseriae may be mistaken for meningococci or gonococci. *N. catarrhalis* and *N. sicca,* for example, are frequently present in secretions of the normal pharynx, and other nonpathogenic species occasionally inhabit the female genital tract. Despite their lack of general pathogenicity, they occasionally cause bacterial endocarditis.

Other bacterial species that may be confused with the neisseriae include a group of **anaerobic** gram-negative diplococci given the generic name *Veillonella*. They inhabit the mouth and gastrointestinal tract of man and certain animals, and have been found in pus from dental abscesses and in the urine of patients with chronic urinary tract infections. Another group, members of the genus *Mima* (or *Herrelea*), including a species originally named *Bacterium anitratum,* are tiny, gram-negative, aerobic coccobacilli, which may simulate gonococci in morphology and are occasionally cultured from the genitourinary tract (see Ch. 29).

A third group includes the Morax-Axenfeld bacillus (genus *Moraxella*), which is sometimes cultured from patients with conjunctivitis.

SELECTED REFERENCES

Books and Review Articles

SCHERP, H. W. Neisseria and neisserial infections. *Annu Rev Microbiol 9:*319 (1955).

Specific Articles

ARTENSTEIN, M. S., GOLD, R. G., ZIMMERLY, J. C., WYLE, F. A., SCHNEIDER, H., and HARKINS, C. Prevention of meningococcal disease by group C polysaccharide vaccine. *N Engl J Med 282:*417 (1970).

BANG, F. Experimental gonococcus infection of the chick embryo. *J Exp Med 74:*387 (1941).

DEACON, W. E. Fluorescent antibody methods for *Neisseria gonorrheae* identification. *Bull WHO 24:*349 (1961).

GOLDSCHNEIDER, I., GOTSCHLICH, E. C., and ARTENSTEIN, M. S. Human immunity to the meningococcus. I. The role of humoral antibodies; II. Development of natural immunity. *J Exp Med 129:*1307, 1327 (1969).

GOTSCHLICH, E. C., GOLDSCHNEIDER, I., and ARTEN-STEIN, M. S. Human immunity to the meningococcus. IV. Immunogenicity of group A and group C meningococcal polysaccharides in human volunteers. *J Exp Med 129:*1367 (1969).

JEPHCOTT, A. E., REYN, A., and BIRCH-ANDERSEN, A. *Neisseria gonorrhoeae.* III. Demonstration of presumed appendages to cells from different colony types. *Acta Pathol Microbiol Scand B79:* 437 (1971).

KELLOGG, D. S., COHEN, I. R., NORINS, L. C., SCHROE-TER, A. L., and REISING, G. *Neisseria gonorrhoeae.* II. Colonial variation and the pathogenicity during 35 months in vitro. *J Bacteriol 96:*596 (1968).

LIU, T. Y., GOTSCHLICH, E. C., JONSSEN, E. K., and WYSOCI, J. R. Studies on the meningococcal polysaccharides. I. Composition and chemical properties of group A polysaccharide. *J Biol Chem 246:*2849 (1971).

LIU, T. Y., GOTSCHLICH, E. C., DUNNE, F. T., and JONSSEN, E. K. Studies on the meningococcal polysaccharides. II. Composition and chemical properties of group B and group C polysaccharide. *J Biol Chem 246:*4703 (1971).

MAELAND, J. A. Serological cross-reactions of aqueous ether extracted endotoxin from *Neisseria gonorrhoeae* strains. *Acta Pathol Microbiol Scand 77:*505 (1969).

ROBERTS, R. B. The relationship between group A and group C meningococcal polysaccharides and serum opsonins in man. *J Exp Med 131:*499 (1970).

SWANSON, J., and GOLDSCHNEIDER, I. The serum bactericidal system: Ultrastructural changes in *Neisseria meningitidis* exposed to normal rat serum. *J Exp Med 129:*51 (1969).

WISTREICH, G. A., and BAKER, R. F. The presence of fimbriae (pili) in three species of *Neisseria.* *J Gen Microbiol 65:*167 (1971).

THE ENTERIC BACILLI AND SIMILAR GRAM-NEGATIVE BACTERIA

Revised by ALEX C. SONNENWIRTH, Ph.D.

General Considerations   754

ENTEROBACTERIACEAE   756

Metabolic Characteristics   757
Antigenic Structure   759
Endotoxins   765
Phage Typing   765
Colicins   765
Genetic Relations   766
COLIFORM BACILLI   767
Escherichia coli   767
Klebsiella-Enterobacter-Serratia   769
Edwardsiella   771
Citrobacter   771
PROTEUS-PROVIDENCE GROUP   772
SALMONELLAE   772
Arizona Group   776
SHIGELLAE   776

VIBRIOS   779

Cholera Vibrios   779
Other Vibrios   782

AEROMONAS   782

THE NONFERMENTERS   783

PSEUDOMONAS 783
Pseudomonas aeruginosa and Related Forms   783
Pseudomallei Group   784
OTHER NONFERMENTERS   785
Acinetobacter   785
Moraxella   785
Alcaligenes   785
Flavobacterium   786

THE OBLIGATE ANAEROBES   786

BACTEROIDES   786

APPENDIX   788

Frequency of Gram-Negative Rods Isolated from
Human Infections   788

The Enterobacteriaceae, a family of gram-negative rods that can grow in air or in its absence on simple media, are found mostly as normal flora (or pathogens) in the vertebrate gut, though some genera are saprophytes or plant parasites. They are also known as "enteric bacilli," "enterics," or "enterobacteria." These unofficial terms are, however, also applied in medical bacteriology to various superficially similar organisms with quite different metabolism, antigenic structure, pathogenicity, ecology, and evolutionary relations (as indicated by DNA hybridization).*

One of the most significant distinctions in the initial division of these organisms is between those that can **ferment** sugars and those that utilize sugars **oxidatively or not at all** (Table 29-1). The distinction is conveniently based on growth in an aerobic tube and in an anaerobic tube (sealed with sterile mineral oil) of Hugh and Leifson's O-F medium (0.2% peptone, 1.0% glucose, 0.5% NaCl, 0.03% K_2HPO_4, 0.2% agar, and a pH indicator). Fermenters produce acid in both tubes, oxidizers only in the open tube, and nonutilizers of glucose in neither (glucose can ordinarily be utilized if any sugar can).

The **fermenters** include both the many genera ordinarily harmless when present in the gut, as well as the pathogenic genera of Enterobacteriaceae (Table 29-2), *Vibrio,* and *Aeromonas. Yersinia* (formerly included with *Pasteurella*) is physiologically similar but for historical reasons is discussed in Chapter 31.

---

* *Hemophilus, Bordetella, Brucella,* and *Pasteurella* are also gram-negative, aerobic or facultatively anaerobic bacilli of medical importance, but most of them do not grow on simple media and, in general, differ greatly from the organisms described here (see Chs. 30 to 32).

Many **nonfermenters** are glucose oxidizers (*Pseudomonas* and several other genera); a few cannot utilize glucose or other carbohydrates (*Alcaligenes,* some strains of *Acinetobacter;* Table 29-1). The nonfermenters are found primarily free in nature. However, with the advent of antibiotics and the survival of patients with reduced immune responses both nonfermenters and the "nonpathogenic" fermenters are encountered with increasing frequency as agents of opportunistic (but often very serious) infection.

Within the two major groups genera can be differentiated on the basis of various additional characteristics. The **oxidase** reaction tests for the ability of the electron transport chain to convert certain aromatic amines into colored products: this test clearly distinguishes several genera (Table 29-1). The **flagella** (Ch. 2) are peritrichous (or absent) in the Enterobacteriaceae, *Flavobacterium,* and *Alcaligenes,* but polar in *Vibrio, Aeromonas,* and *Pseudomonas.* Other pertinent differential characters include sensivity to bile salts or citrate (used in MacConkey's agar and SS agar, see below), ability to reduce nitrate to nitrite or gas, and various decarboxylase reactions (Table 29-1).

The characterization outlined here, based on phenotypic characteristics, is complemented by studies of DNA (Ch. 2). Table 29-1 shows that phenotypically similar groups are reasonably similar in their DNA base compositions. Moreover, DNA homology is readily demonstrated among various Enterobacteriaceae, but not between these organisms and pseudomonads.

Medical Importance. Accurate identification of the organisms described here is imperative. They differ markedly in susceptibility to various antimicrobial agents and also in virulence, which is important for prognosis

TABLE 29-1. Some Differential Characteristics of Enteric Bacilli (Enterobacteriaceae) and Similar Gram-negative Bacteria

Family, genus, or species	Glucose metabolism	Oxidase	Growth on MacC	Growth on SS	Motility	Nitrate to Nitrite	Nitrate to Gas	Lysine[1]	Arginine[2]	Ornithine[1]	DNA base composition (% G+C)
Enterobacteriaceae	F	−	+	+	+or−	+	−	d[4]	d[4]	d[4]	36–57[7]
Yersinia	F	−	+	+,−	−,+[5]	+	−	−	−	+or−	46–48
Vibrio	F	+	+	−[3]	+	+	−	+	−	+	40–50
Aeromonas hydrophila	F	+	+	+	+	+	−	−	+	−	52–59
Pasteurella	F	+	−[3]	−	−	+	−	−	−	d	35–40
Pseudomonas	Ox	+[6]	+	+	+	−	+or−	− (NC	+ NC	− NC)	57–70
Achromobacter	Ox	−	d	d	+(−)	+or−	−	−	+,−	−	40–70
Acinetobacter											
anitratum (Herellea)	Ox	−	+	−[3]	−	−	−	NC	NC	NC	40–46
A. lwoffi (Mima)	In	−	+	−[3]	−	−	−	NC	NC	NC	40–46
Moraxella	In	+	−or+	−	−	−or+	−				40–46
Flavobacterium	Ox or In	+(−)	+(−)	−	+(−)	−	+or−	NC	NC	NC	32–42
Alcaligenes	In	+	+	(+)−	+	+	+or−	NC	NC	NC	67–70

F = fermentative; Ox = oxidative; In = inactive (in O-F medium, see text); MacC = MacConkey agar; SS = Salmonella-Shigella agar.

+ = positive reaction or growth; − = no reaction or growth; (+) = delayed positive reaction or poor growth; d = different reactions: +, (+), or −; NC = no change; + (−) = majority positive, occasional strain negative.

[1] Decarboxylases; [2] dihydrolase; [3] rare strains grow; [4] various patterns of reactions; [5] motility, when present, demonstrable at 20–25° but not at 37°; [6] except *P. maltophilia;* [7] the range of DNA base composition of practically all Enterobacteriaceae is 50–58 mole % GC, except for *Proteus-Providencia* organisms with a value of 36–53.

Based, in part, on Ewing, W. H., Davis, B. R., and Martin, W. J. *Outline of Methods for the Isolation and Identification of Vibrio cholerae.* Atlanta, National Communicable Disease Center, 1966.

and for recognizing potential danger to contacts. Moreover, identification is essential for epidemiological investigation of the sources of infection.

For example, a recent epidemic of hospital-acquired septicemia (at least 400 cases including 40 or more deaths) was traced to the production of contaminated intravenous solutions; the key was a sharp rise in the isolation of two relatively unusual organisms, *Enterobacter cloacae* and *Erwinia,* in geographically widely dispersed institutions. Had the microbiology laboratory not carried the identification to the species level, the organisms might have easily been confused with common *Klebsiella* and *Enterobacter* and detection of the source of the epidemic might have been long delayed.

Typhoid fever (*Salmonella typhi*), bacillary dysentery (*Shigella*), and cholera (*Vibrio cholerae*) are now largely controlled in the Western world through various public health measures, but they still represent serious and periodically recurring problems throughout the rest of the world. On the other hand, organisms once thought to have little virulence or to be completely nonpathogenic (including members of the normal flora of man and saprophytes) have been increasingly associated with disease in the developed countries. **Endogenous** infections (due to indigenous organisms) and **nosocomial** (hospital-acquired) infections now represent a large proportion of serious bacterial infections in the United States and other Western countries. Thus since the mid-1950s the frequency of staphylococcal, pneumococcal, and streptococcal infections has diminished considerably, while the incidence of infections due to enteric bacteria and other gram-negative rods of low virulence has increased strikingly (e.g., *E. coli, Klebsiella, Enterobacter, Serratia, Proteus, Pseudomonas;* see Appendix). Enteric organisms have always been the most

TABLE 29-2. Principal Divisions and Groups of Enterobacteriaceae

Principal divisions	Groups (genera)	Degree of pathogenicity
Shigella-Escherichia	*Shigella* *Escherichia**	Bacillary dysentery Pathogenic only under special circumstances; certain types produce diarrheal disease
Edwardsiella	*Edwardsiella*	Can cause gastroenteritis and other salmonellosis-like diseases
Salmonella-Arizona-Citrobacter	*Salmonella* *Arizona*	Gastroenteritis, septicemia, enteric fever
Klebsiella-Enterobacter-Serratia‡	*Citrobacter*† *Klebsiella* *Enterobacter*§ *Serratia*	Pathogenic only under special circumstances ("opportunistic," "secondary" pathogens)
Proteus-Providence	*Proteus* *Providencia*	

* *E. coli;* includes alkalescens-dispar organisms, formerly in *Shigella.*

† Formerly *Escherichia freundii;* includes Bethesda-Ballerup "paracolon" organisms.

‡ The genus *Pectobacterium* ("soft-rot coliforms"), not medically significant, is also included in this division.

§ Formerly *Aerobacter.*

Based, in part, on Ewing, W. H., *Revised Definitions for the Family Enterobacteriaceae, Its Tribes and Genera.* Atlanta, National Communicable Diseases Center, 1967.

common agents of urinary tract infections, but they are now also the predominant etiological agents in various endogenous and nosocomial infections (surgical wound infections, hospital-acquired pneumonias, etc.). Treatment of such infections is often made difficult by resistance of the organisms to most antimicrobial agents and by the presence of other serious diseases in the patients.

The term "enteric bacilli" is a misnomer: the facultative aerobes, including the predominant *E. coli,* enterococci (chiefly *Streptococcus faecalis,* Ch. 26), and various lactobacilli account for only 0.1 to 1.0% of the cultivable normal human fecal flora. Various anaerobes—predominantly members of the gram-negative genus *Bacteroides* and of the gram-positive genera *Bifidobacterium* (formerly known as *Lactobacillus bifidus*), *Eubacterium,* anaerobic streptococci, and

clostridia (Ch. 34)—far outnumber all others. The most prevalent species, *Bacteroides fragilis,* averages 10^{10} to 10^{11} per gram of wet feces, contrasted with *E. coli* at 10^7 per gram. *Bacteroides* is essentially harmless but can cause life-threatening disease under certain circumstances; the frequency of recognized *Bacteroides* infections has increased strikingly in recent years. These organisms will also be considered in this chapter.

To complete the introduction of this group of organisms we should note the extraordinary prominence of Enterobacteriaceae, and their phages, in the development of bacterial and molecular genetics (Chs. 9 to 13). This prominence derives partly from their ease of handling and their possession of two major mechanisms of gene transfer, and partly from historical accident.

ENTEROBACTERIACEAE

The gram-negative, nonsporeforming rods included in the family Enterobacteriaceae are relatively small (2 to 3 μ by 0.4 to 0.6 μ).

Shigella and *Klebsiella* are nonmotile, lacking flagella; the rest have peritrichous flagella, but nonmotile variants occur. Capsule production (e.g., by *Klebsiella pneumoniae* and *Enterobacter aerogenes*) gives rise to large mucoid colonies that are easily distinguished from the usual smooth variety.

The principal divisions and groups (genera) (Table 29-3) are distinguished by a variety of biochemical characteristics and are subgrouped by means of further biochemical tests, determination of antigenic structure, or both. Species may be further subdivided by antigenic analysis or by reactions with bacteriophages or bacteriocins.*

METABOLIC CHARACTERISTICS

The Enterobacteriaceae grow readily on ordinary media under aerobic or anaerobic conditions. Most will grow in simple synthetic media, often with a single carbon source. They utilize glucose fermentatively with the formation of acid or of acid and gas, reduce nitrates to nitrites, and give a negative oxidase reaction (Table 29-3). Genera and species can be tentatively identified by their capacity to ferment specific carbohydrates, to utilize certain other substrates (e.g., citrate) as a sole source of carbon, and to give rise to characteristic endproducts (e.g., indole from tryptophan, ammonia from urea, and hydrogen sulfide). Many commonly employed tests are listed in Table 29-3.

Fermentation of lactose is a time-honored differential characteristic in the preliminary examination of suspected cultures. It early became a major criterion because the major enteric pathogens (*Salmonella* and *Shigella*) are lactose-negative. However, a number of other enterobacteria (*Proteus, Providencia, Edwardsiella,* certain types of *Escherichia, Enterobacter,* and *Serratia*) are also lactose-negative (or delayed). Conversely, prompt

---

* The classification of Enterobacteriaceae has been controversial and continues to change. The classification employed here is that of Edwards and Ewing (1962), as revised and expanded by Ewing (1967). Despite its "splitting" tendencies (Classification, Ch. 2) it has been widely adopted because of its usefulness for diagnostic and epidemiological purposes. Table 29-2 shows the taxonomic system (divisions and groups), and Table 29-3 the formal system of nomenclature with specific family, tribe, and genus names. In the text, names of species and genera are printed in italic while those of groups and divisions are in roman type.

lactose fermentation, recognized by the formation of colored colonies on solid media containing lactose and an appropriate indicator (e.g., neutral red), does serve well in delineating the "coliform" organisms (most *Escherichia, Citrobacter, Klebsiella,* and *Enterobacter*); but many *Arizona* strains, resembling salmonellae in pathogenicity, are also lactose-positive. Hence identification requires a combination of tests (Table 29-3).

For many years the term "paracolon" was used to denote organisms that did not ferment lactose promptly and could not be unequivocally identified as members of recognized genera. Since practically all such organisms can now be identified with discrete groups, this wastebasket category is obsolete.

Screening Media. Various ingenious media which reveal a number of differential characteristics in a single culture tube are widely used in the preliminary screening of enteric bacilli. One example, triple-sugar-iron (TSI) agar has the following formula:

Peptone	20 gm
Sodium chloride	5 gm
Lactose	10 gm
Sucrose	10 gm
Glucose	1 gm
Ferrous ammonium sulfate	0.2 gm
Sodium thiosulfate	0.2 gm
Phenol red	0.025 gm
Agar	13 gm
Distilled water	1000 ml

The medium (pH 7.3) is allowed to solidify in tubes so as to form slants, and the inoculum (picked from a single colony on an original isolation plate)† is stabbed through the butt and streaked onto the surface of the slant. TSI agar reveals fermentation of the small amount of glucose or the large amount of lactose or sucrose; in addition, production of H_2S causes formation of black ferrous sulfide, and production of gases (e.g., H_2 and CO_2; Ch. 3) causes bubbles to form in the agar.

The fermentation reactions, which "decolorize"‡ the phenol red indicator, depend upon several fac-

---

† Although it is permissible for diagnostic purposes to "fish" a colony from an original isolation plate, a second subculture on agar should always be made, to be stored for future reference.

‡ Note that acid "decolorizes" phenol red (to a pale yellow) but makes neutral red colored.

TABLE 29-3.   Biochemical Reactions of Enterobacteriaceae, Aeromonas and Vibrio*

TEST or SUBSTRATE	ESCHERICHIEAE Escherichia	Shigella	EDWARDSIELLEAE Edwardsiella	SALMONELLEAE Salmonella	Arizona	Citrobacter	KLEBSIELLEAE Klebsiella	Enterobacter cloacae	aerogenes	hafniae	lique-faciens	Serratia	PROTEEAE Proteus vulgaris	mirabilis	morganii	rettgeri	Providencia	AEROMONAS A. hydrophila	VIBRIO V. cholerae
Oxidase Test	−	−	−	−	−	−	−	−	−	−	−	−	−	−	−	−	−	+	+
Indole Produced	+	− or +	+	−	−	−	− or +	−	−	−	−	−	+	−	+	+	+	+	+
Methyl Red	+	+	+	+	+	+	− or +	−	−	− or +	− or +	− or +	+	+	+	+	+	+	+w
Voges-Proskauer	−	−	−	−	−	−	+	+	+	+ or −	+ or −	+	−	− or +	−	−	−	− or +	− or +
Citrate Utilized	−	−	−	d	+	+	+	+	+	(+) or −	+	+	d	− or +	−	+	+	+ or −	(+) or −
Hydrogen Sulfide	−	−	+	+	+	+	−	−	−	−	−	−	+	+	−	−	−	−	−
Urease	− or +	−	−	−	−	+ or −	+	+ or −	−	−	d	d	+	+	+	+	−	−	−
Motility	+ or −	−	+	+	+	+	−	+	+	+	+	+	+	+	+	+	+	+	+
Gelatin (22°C)	−	−	−	−	(+)	−	−	(+)or−	−	−	+ or −	+or−	+	+	−	−	−	+ or −	(+)
Lysine Decarboxylase	d	−	+	+	+	−	+	−	+	+	+	+	−	−	−	−	−	− or +	+
Arginine Dihydrolase	d	− or +	−	(+)or+	+ or(+)	d	−	− or +	−	−	−	−	−	− or (−)	−	−	−	+ or −	−
Ornithine Decarboxylase	d	d(2)	+	+	+	d	−	+	+	+	+	+	−	+	+	−	−	−	+
Phenylalanine Deaminase	−	−	−	−	−	−	−	−	−	−	−	−	+	+	+	+	+	−	−
Gas from Glucose(1)	+	−(2)	+	+	+	+	+	+	+	+	+	+ or −(3)	+ or (+)	+	+	−	−	+	+

FERMENTATION OF:

TEST or SUBSTRATE	Escherichia	Shigella	Edwardsiella	Salmonella	Arizona	Citrobacter	Klebsiella	cloacae	aerogenes	hafniae	lique-faciens	Serratia	vulgaris	mirabilis	morganii	rettgeri	Providencia	A. hydrophila	V. cholerae
Lactose	+	−(2)	−	−	− or +(+)	+	+	+	+	− or (+)	+	+ or −	+ or −	−	−	− or +	−	+ or −	(+)
Sucrose	d	−	−	−	−	d	+	+	+	d	d	+	+	d	−	d	d	− or +	+
Adonitol	−	−	−	−	−	d	+	− or −	+	−	d	d	−	−	−	d	+(4),−(5)	−	−
Inositol	−	−	−	d	−	−	+	− or +	+	−	d	d	−	d	−	d	+(4),+(5)	−	−
Raffinose	d	d	−	+	−	d	+	+	+	+	+	+	−	−	−	−	−	−	−
Rhamnose	d	d	−	+	+	+	+	+	+	+	−	−	−	−	−	+	−	−	−
Arabinose	+	d	−	+	+	+	+	+	+	+	+	−	−	−	−	− or +	−	+ or −	+
Mannitol	+	+ or −	−	+	+	+	+	+	+	+	+	+	−	−	−	−	−	+	+
Dulcitol	d	d	−	d(2)	−	d	− or +	−	−	d	d	−	d	d	−	− or +	−(4),d(5)	−	−
Salicin	d	−	−	−	−	d	+ or (+)	+ or −	+	d	+	+	d	d	−	d	d(4),d(5)	+	+
Sorbitol	d	−	−	+	−	+	+	+	+	−	+	+	−	−	−	−	−(4),−(5)	d	−

* For physiological characteristics of *Yersinia*, strongly resembling Enterobacteriaceae, see Chapter 31.

+ = 90% or more positive in 1 or 2 days; − = 90% or more negative; d = different biochemical types (+, (+), −); (+) = delayed positive; + or − = majority of cultures positive; − or + = majority negative; w = weakly positive reaction; shaded areas = major differential criteria for identification.

[1] All these organisms ferment glucose with acid formation. Some also form gas (CO_2, H_2) as an end-product; others do not (see text). For diagnostic purposes it is usually sufficient to test gas formation in glucose only.

[2] Certain biotypes of *S. flexneri* produce gas; *S. sonnei* cultures ferment lactose and sucrose slowly and decarboxylate ornithine.

[3] Gas volumes produced by cultures of *Serratia, Proteus,* and *Providencia* are small.

[4] *P. alcalifaciens.*

[5] *P. stuartii.*

Based, in part, on data from Ewing, W. H., *Enterobacteriaceae, Taxonomy and Nomenclature,* Atlanta, National Communicable Disease Center (NCDC), 1966; W. H. Ewing: *Differentiation of Enterobacteriaceae by Biochemical Reactions,* Atlanta, NCDC, 1970; and W. H. Ewing, B. R. Davis, and W. J. Martin: *Outline of Methods for the Isolation and Identification of Vibrio cholerae,* Atlanta, NCDC, 1966.

tors: 1) a delicate balance of the concentrations of the several sugars and the nitrogenous constituents of the medium; 2) inclusion of 10 times as much lactose and sucrose as glucose (see formula); 3) slow diffusion through the agar of the acid products of the fermentations and the alkaline products of nitrogen metabolism; and 4) greater production of acid in the anaerobic environment of the butt (fermentation) compared with the aerobic surface of the slant (respiration).

Thus an organism that ferments glucose only (e.g., *S. typhi*) will produce detectable amounts of acid in the butt of the TSI medium (anaerobic) but not on the surface of the slant (aerobic), where the nitrogenous bases formed will neutralize the less acid endproducts of respiration. Species that ferment the more concentrated lactose (e.g., *E. coli*) or sucrose (*P. vulgaris*), on the other hand, will generate in the butt enough acid to diffuse throughout the medium and thus acidify even the surface (Table 29-3). **Nonfermenters,** incapable of utilizing the sugars fermentatively, will give rise to an alkaline or neutral reaction throughout the medium (slant and butt pink to red).

Although multiple-sugar media of this sort are reasonably satisfactory for preliminary identification of enteric bacilli the fermentation reactions observed should be confirmed by tests with appropriate liquid media, each containing a single sugar as the sole source of carbohydrate. To recognize gas formation (Table 29-3) an inverted small test tube is submerged in the medium to trap a portion of any gas generated. During sterilization in the autoclave the air initially present in the inverted tube is expelled and replaced by medium.

Selective Media. The enteric bacilli are resistant to inhibition by certain bacteristatic dyes (e.g., brilliant green) and surface-active compounds (e.g., bile salts), compared with most gram-positive bacteria. Selective media containing such substances (e.g., MacConkey agar, deoxycholate agar) greatly facilitate the isolation of enteric bacilli from cultures of feces.

Shigellae and salmonellae are less sensitive than the coliform organisms to inhibition by citrate (which may depend on chelation of divalent metal ions). Agar containing both citrate and bile salts, e.g., **SS** (*Salmonella-Shigella*) **agar,** is therefore used for selective cultivation of pathogenic species. However, since some strains of shigellae are inhibited, one cannot rely on such media alone.

ANTIGENIC STRUCTURE

Although the principal varieties of enteric bacilli can be tentatively identified by fermentation and other metabolic reactions in differential media, final identification of individual species is usually based on antigenic structure. Strains with the same antigenic activity may nevertheless exhibit different metabolic reactions (fermentative variants or biotypes).

As is shown diagrammatically in Figure 29-1A, three kinds of surface Ags (**H, O,** and **Vi**)* determine the organism's reactions with specific antisera. If motile species are treated with formalin the labile protein H Ags of their flagella are preserved, and the cells will be agglutinated only by specific antiflagellar (anti-H) antibodies (Abs). The resulting agglutination reaction, for the reasons suggested by Figure 29-1B, will form a light, fluffy precipitate. When reacted only with Abs to their somatic O Ag (i.e., in the absence of anti-H) such bacilli will not agglutinate because their numerous peritrichous flagella keep them apart. When, on the other hand, flagella are absent from a strain or are denatured by heat (100° for 20 minutes), acid, or alcohol, the somatic O Ag at the surfaces of adjacent bacilli can be linked by anti-O Ags, resulting in closely packed, granular clumps.

Certain species have an outermost polysaccharide layer, called a **Vi** Ag. It is usually too thin to be seen as a capsule, and it does not cover the O Ag completely, since it allows adsorption of anti-O Ags. Nevertheless, Vi cells can be agglutinated only by anti-H or

---

* The term **H** (Ger. *Hauch*-breath) was first used to describe the growth of proteus bacilli on the surfaces of moist agar plates: the film produced by the swarming of this highly motile organism resembles the light mist caused by breathing on glass. The designation **O** (Ger. *ohne*-without), first applied to nonswarming (i.e., nonflagellated) forms, is now used as generic term for the lipopolysaccharide somatic Ags of all enteric bacilli and more specifically for their antigenically active polysaccharide components. The term **Vi** designates an additional surface Ag of *S. typhi* originally thought to be primarily responsible for virulence. Its precise relation to pathogenicity is still not clear, and it is a special example of a **K** (Ger. *Kapsel*) or capsular Ag.

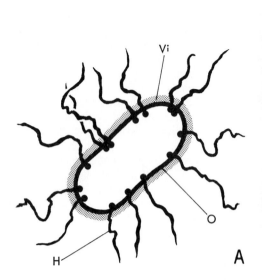
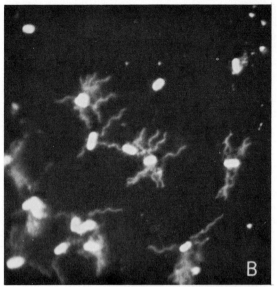

Fig. 29-1. A. Schematic diagram of cellular locations of H, O, and Vi antigens of enteric bacilli. **B.** *S. typhi* stained with fluorescein-labeled immune serum containing anti-H, anti-O, and anti-Vi antibodies. ×3500, reduced. Note that the flagella of different bacilli have in some cases become agglutinated with one another. Note also, however, that the presence of the flagella has prevented the bodies of the bacilli from coming close enough together to agglutinate, despite the presence of the antisomatic (Vi and O) antibodies in the serum. The interflagellar agglutinate is, therefore, understandably less densely packed than if the flagella had been absent and intersomatic agglutination had occurred. [From Thomason, B. M.,Cherry, W. B., and Moody, M. D. *J Bacteriol 74*:525 (1957).]

anti-Vi Ags, and O agglutination becomes possible only if the Vi and H Ags are destroyed (e.g., by boiling for 2 hours). The Vi Ag of *S. typhi* differs from that of certain other salmonellae.

R Antigens. Rough strains of enteric bacilli (R mutants) have none of these three specific surface Ags and tend to agglutinate spontaneously.* The various polysaccharide Ags extractable from their cell walls (R Ags) are also present in smooth (S) cells but are covered by the O Ag chains (Ch. 6). The change from S to R may take place without the loss of flagellar or Vi Ags.

Common Enterobacterial Antigen. In addition to the Ags that differentiate various Enterobacteriaceae from each other, many of these organisms share a common Ag which

can be extracted from the cells, and which is also produced by *Yersinia entercolitica,* an organism physiologically resembling Enterobacteriaceae (Ch. 31). Abs to the common Ag do not precipitate it, and they do not agglutinate cells, but they can be detected by hemagglutination or hemolysis of Ag-coated erythrocytes.

Preparation of Specific Antisera. When animals are immunized with formalin-killed motile strains of enteric bacilli the antisera contain both anti-H and anti-O Ags (as well as anti-R). The Abs to H usually have a higher titer than those to O.† If the immunizing strain also possesses a capsular (Vi) Ag the antiserum will contain, in addition, anti-Vi.

Antisera reacting with individual surface Ags in agglutination tests are prepared by

---

* When suspended in media of proper ionic strength R cells can be prevented from agglutinating spontaneously and thus can be used to test for anti-R Ab.

† This difference may be due in part to the greater immunogenicity of the protein H Ag, as compared with the polysaccharide O Ag, and in part to the greater sensitivity of the H agglutination test.

selective adsorption. For example, anti-H Ab may be removed by adsorption with suspensions of homologous flagella (mechanically removed), or with suspensions of mutants possessing the immunogenic cells' flagella but neither their Vi nor O Ags. Similarly, anti-O or anti-Vi Abs may be selectively removed by appropriate extracts of bacilli.

Kauffmann-White Classification of the Salmonellae. The antigenic complexities of the enteric bacilli have been most thoroughly documented in the genus *Salmonella*. Largely as a result of the systematic studies of Kauffmann in Denmark and White in England more than 1000 varieties have been distinguished on the basis of specific H, O, and Vi Ags, identified by exhaustive cross-adsorption and cross-reaction tests. These tests are illustrated by the following example.

Assume that two strains of *Salmonella,* a and b, have been isolated, each from a separate outbreak of salmonellosis. Antisera prepared against them (anti-a and anti-b) are first reacted with suspensions of each strain (standardized to the same density), and the intensity of the reactions is recorded in roughly quantitative terms (see below). Each serum is then adsorbed with the heterologous strain (i.e., anti-a with b, and vice versa). The adsorbed sera are similarly reacted with the standardized suspensions, and the results are measured in the same manner. If the agglutination reactions observed were as follows:

Antisera	Organisms	
	a	**b**
Anti-a, unadsorbed	4+	2+
Anti-b, unadsorbed	2+	4+
Anti-a, adsorbed with b	2+	0
Anti-b, adsorbed with a	0	2+

it would be concluded that **a** and **b** have unique, as well as shared, antigenic determinants. Using an arbitrary numbering system, **a** might be said to possess determinants (or factors) 1,2, and **b** determinants 2,3. If, on the other hand, strains **a** and **b** were equally effective in removing all the agglutinating Ab from the two antisera, they would be considered antigenically identical. Every new strain isolated would be similarly tested with the existing antisera (anti-a and anti-b,

both adsorbed and unadsorbed) and with antiserum to the new strain itself, before and after adsorption. Whenever a new determinant was detected, it would be given a new number, and according to the Kauffmann-White scheme the organism possessing it would be designated a new "species" (type).

Application of this technic to O, H, and Vi agglutination reactions has resulted in the identification of more than 1000 *Salmonella* "species" (serotypes). Only large *Salmonella* typing centers (Copenhagen, London, Atlanta, etc.) have the collections of specific antisera necessary for such work. In most diagnostic laboratories *Salmonella* strains are merely **grouped** by means of fermentation tests and agglutination reactions performed with group-specific antisera (see below).

O Antigens. Cross-reactions show that each *Salmonella* O Ag has two or three distinguishable determinants, each given a number and each shared with several other O Ags. On the basis of certain strongly reacting major determinants (heavy type in Table 29-4) the salmonellae have been classified into O **groups,** the first 26 designated by letters (A-Z) and subsequent ones by the numbers of their group-determining Ags (50,51,52, etc.); each group has its distinctive major O determinant. In group C (O Ag 6,7), for example, the major determinant is factor 6, and in group E (O Ag 3,10) it is factor 3 (Table 29-4). About 90% of strains isolated from man fall into groups A to E. Additional divisions within a group are made on the basis of minor O determinants.

The members of each O group may be differentiated still further into serotypes ("species") on the basis of their flagellar (H) Ags. A given strain may form at different times either one of two kinds of H Ag. The first, **phase 1 flagellar Ags,** are shared with only a few other species of *Salmonella;* the second, **phase 2 Ags,** are less specific. In the Kauffmann-White classification the former are indicated by small letters and the latter by numbers (Table 29-4). Although the organisms in a given culture may be entirely in one phase (monophasic culture), they frequently can give rise to mutants in the other phase (diphasic culture), especially if the culture is incubated for more than 24 hours. Such phase variation, which is genetically con-

TABLE 29-4. Serological (Kauffman-White) and Chemical Classification of Common Salmonella Species

Species	Group	O antigen*	H antigens Phase 1	H antigens Phase 2	O-specific sugars in determinants
S. paratyphi A	A	(1), **2**, 12	a	—	Mannose, rhamnose, paratose
S. schottmülleri	B	(1), **4**, (5), 12	b	1, 2	Mannose, rhamnose, abequose
S. typhimurium	B	(1), **4**, (5), 12	i	1, 2	Mannose, rhamnose, abequose
S. paratyphi C	C₁	**6**, 7, Vi	c	1, 5	Mannose
S. choleraesuis	C₁	**6**, 7	c	1, 5	Mannose
S. montevideo	C₁	**6**, 7	g,m,s	—	Mannose
S. newport	C₂	**6**, 8	e,h	1, 2	Mannose, rhamnose, abequose
S. typhi	D	**9**, 12, Vi	d	—	Mannose, rhamnose, tyvelose
S. enteritidis	D	(1), **9**, 12	g,m	—	Mannose, rhamnose, tyvelose
S. gallinarum	D	1, **9**, 12	—	—	Mannose, rhamnose, tyvelose
S. anatum	E	**3**, 10	e,h	1, 6	Mannose, rhamnose

* Boldface number signifies major determinant (factor) of group. The determinants represent a subdivision of the O Ags. Parentheses indicate that Ag determinant may be difficult to detect.

Data on O-specific sugars from Lüderitz, O. *Angew Chemie 9*:649 (1970).

trolled (Ch. 8), can also be accentuated by growing the organisms in serum containing Abs to their flagellar Ags, thereby favoring the growth of mutants with the alternate (allelic) Ag which does not react with the antiserum.

Elaboration of the polysaccharide Vi Ag by a species is indicated in the Kauffman-White scheme by the letters Vi, placed by convention after the numbers indicating the individual O factors (e.g., *S. typhi* in Table 29-4). Since the surface coating of Vi prevents agglutination with homologous anti-O Ab (though it permits adsorption of the Ab), it prevents differentiation on the basis of this reaction.

Chemistry of the O Antigenic Determinants. As is true of most proteins, little is known about the chemical determinants of the flagellar (H) Ags. Vi of *S. typhi,* on the other hand, is a simple homopolymer of N-acetyl galactosaminuronic acid: because of its free carboxyl groups the cell surface is highly acidic.

O Antigens. As is described in Chapter 6, the O Ags and their associated endotoxin have the following over-all structure:

O-specific chains, consisting of **repeating** (oligosaccharide) **units,** are attached to a basal **core polysaccharide,** which in turn is attached to **lipid A;** the whole forms a **lipopolysaccharide** (LPS) (Fig. 29-2). The structure of the core (Fig. 6-12) is identical or very similar in all salmonellae but differs in other genera. The serological specificity of each LPS is determined by the O-specific chains, whose many variations form the main basis for classification of *Salmonella.*

A total of 18 different sugars (monosaccharides) have been identified in various *Salmonella* LPS; some species have as many as 9. Among the 5 "basal" sugars of the core L-glycero-D-mannoheptose and 2-keto-3-deoxy-D-mannooctonic acid (ketodeoxyoctonate, KDO) are unique to bacterial LPS (KDO linking the core to lipid A; see Fig. 6-12). The core (or a fraction of it) is responsible for the serological specificity of R (rough) mutants, which are blocked in biosynthesis of the complete O Ag: R_a mutants contain the complete core, while R_b through R_e mutants are deficient in one or more of the basal sugars.

The various O-specific chains contain ordinary sugars (mannose, rhamnose, etc.)

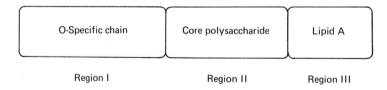

O-Specific chain	Core polysaccharide	Lipid A
Region I	Region II	Region III

Fig. 29-2. Structural diagram of bacterial lipopolysaccharides.

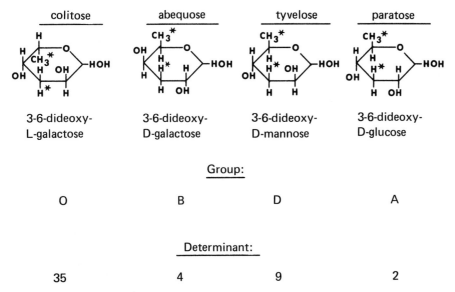

colitose	abequose	tyvelose	paratose
3-6-dideoxy- L-galactose	3-6-dideoxy- D-galactose	3-6-dideoxy- D-mannose	3-6-dideoxy- D-glucose

Group:

| O | B | D | A |

Determinant:

| 35 | 4 | 9 | 2 |

Fig. 29-3. *Salmonella* dideoxyhexoses. The name of each is derived from the species of enteric bacilli from which it was originally isolated, i.e., *E. coli, S. abortus equi, S. typhi,* and *S. paratyphi.* The deoxy positions are indicated by asterisks. [Based on Robbins, P. W., and Uchida, T. *Fed Proc* 21:702 (1962).]

in distinctive combinations; many chains also contain a group of 3,6-dideoxyhexoses that are so far unique to enterobacteria and are named for the organisms from which they were first derived: colitose, abequose, paratose, and tyvelose (Fig. 29-3).* Both ordinary sugars and dideoxyhexoses ("O-specific sugars," Tables 29-4 and 29-5) contribute to the specificity of major, group-defining determinants (see below).

Figure 29-4 illustrates the repeating

---

* Colitose is also found in O Ags of certain *E. coli* strains, including the virulent O-111 and O-55 (see below). The other 3,6-dideoxyhexoses have been found only in relatively virulent *Salmonella* species.

mannose-rhamnose-galactose sequences that constitute the distinctive determinants of the O-specific chains of group E_2 salmonellae (Ag 3,15). The units that compose the chains of several *Salmonella* groups are listed in Table 29-5. Groups A, B, D, and E all contain the repeating mannose-rhamnose-galactose unit but their linkages and substituents differ, determining the immunological specificities of the O Ags (see also Table 29-4).

The structural differences that have been observed in closely related antigens include 1) changes of position of glycosidic linkages, i.e., 1-4 vs. 1-6; 2) altered anomeric configurations, i.e., α vs. β linkage; 3) attachment of additional monosaccharides, such as glucose; 4) presence or absence of acetyl

Fig. 29-4. Structure of the O-specific chains in the lipopolysaccharide of *S. newington* (group E_2). [From Lüderitz, O. *Angew Chem* 9:649 (1970).]

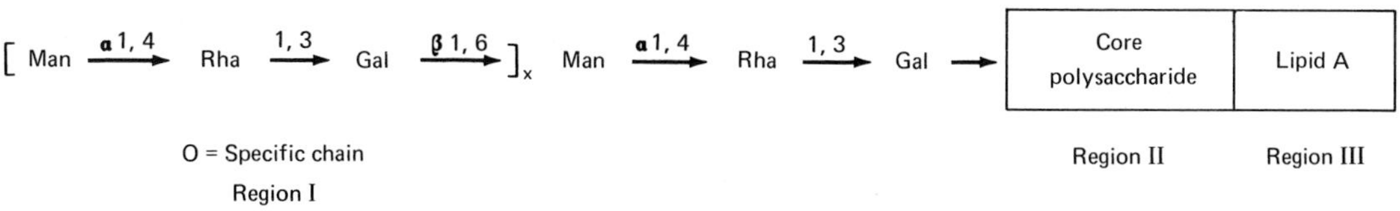

$$\left[\text{Man} \xrightarrow{\alpha 1,4} \text{Rha} \xrightarrow{1,3} \text{Gal} \xrightarrow{\beta 1,6} \right]_x \text{Man} \xrightarrow{\alpha 1,4} \text{Rha} \xrightarrow{1,3} \text{Gal} \longrightarrow$$

| | Core
polysaccharide | Lipid A |

O = Specific chain
Region I

Region II   Region III

TABLE 29-5. Simplified Structures of O Repeat-units of Some Salmonella Serogroups*

Serogroup	Species	O antigen	Structure
A	S. paratyphi A	1,2,12	Par OAc Glc ↓ ↓ ↓ Man→Rha→Gal→
B	S. typhimurium	1,4,5,12	OAc-Abe Glc ↓ ↓ Man————→Rha→Gal→
C_1	S. choleraesuis	6,7	Glc ↙↘ Man→Man→Man→Man→GlcNac
C_2	S. newport	6,8	Abe OAc–Glc ↓ ↓ Rha→Man→Man→Gal→
D_1	S. typhi	9,12	Tyv OAc-Glc ↓ (1-3) α ↓ 2α-Man————→Rha→Gal→
E_1	S. anatum	3,10	OAc ↓ 6α-Man→Rha→Gal→
E_2	S. newington	3,15	6β-Man→Rha→Gal
E_3	S. minneapolis	(3),(15),34	Glc ↓ 6β-Man→Rha→Gal

* Only those type-linkages and anomeric positions of sugars are included which may explain differences between groups.

Par = paratose, OAc = acetyl, Glc = glucose, Abe = abequose, Tyv = tyvelose, Man = mannose, Rha = rhamnose, Gal = galactose.

Modified from Roantree, R. J. The Relationship of Lipopolysaccharide Structure to Bacterial Virulence. In *Microbial Toxins,* vol. 5, p. 18. (S. Kadis, G. Weinbaum, and S. J. Ajl, eds.) Academic Press, New York, 1971.

groups; and 5) deletion of, or substitution for, one of the monosaccharides in the basic trisaccharide units.

Immunological Determinants. The determinants of O Ags are identified from the capacity of fragments of O chains (e.g., di-, tri-, tetrasaccharides) to inhibit serological reactions of monospecific antisera, containing Abs only for a particular determinant. The **immunodominant** sugar of a given determinant accounts for most of the inhibitory activity: e.g., tyvelose for group D factor 9, mannose for group E factor 3. Though the dominant sugar as a monosaccharide at high concentration can specifically inhibit the reactions to some extent, strong inhibition is obtained only when such a sugar is part of a larger fragment whose other sugars, in appropriate linkage

and configuration, define the entire determinant group.

Because the affinity of anti-O Abs for monosaccharides is extremely low, individual sugar residues common to different determinants do not usually account for cross-reactions; for instance, abequose is the dominant sugar in factors 4 and 8, which do not cross-react at all. Different determinants in a given sequence of sugars can correspond to overlapping segments. For example, factors 3 and 10 of Ag 3,10 may be represented as:

$$\text{(O-ac-gal——man——rh)}_n\text{——R}$$

with bracket "3" over man–rh and bracket "10" under gal–man

The structure of antigen (3),(15),34 (Table 29-5) may be written as:

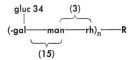

The parentheses around 3 and 15 indicate that the addition of glucose (immunodominant in factor 34) has weakened antigenic determinants 3 and 15 to the point where they are difficult to detect (see footnotes to Table 29-4).

Even though an immunodominant sugar residue can be recognized and a disaccharide can define a determinant, the additional residues of the repeating unit contribute considerably to affinity for binding to the corresponding Abs: i.e., the reaction of antiserum with O Ag is more strongly inhibited by the appropriate tri- or tetrasaccharides than by the component mono- or disaccharides.

ENDOTOXINS

The endotoxic properties of cell walls of gram-negative bacilli, which are due to LPS, have been discussed (Ch. 22, Endotoxins). The extracted LPS of enterobacteria, recovered as a colloidal suspension, may be split by mild acid hydrolysis into lipid A and degraded polysaccharides; the former is mainly responsible for toxicity (Chs. 6 and 22).

Salmonella lipid A consists of highly substituted β-1,6-glucosamine disaccharides, possibly linked by phosphodiester bonds (Fig. 6-11). The amino groups are substituted by β-hydroxymyristic acid and the other available OH groups by other long-chain fatty acids. Lipid A preparations from other enterobacteria have a very similar composition.

The identity of lipid A and endotoxin has been disputed, for by itself lipid A, unlike LPS, is insoluble in water and lacks biological activity. However, the polysaccharide in the intact LPS seems to contribute to the toxicity only by increasing solubility in water. Thus lipid A becomes a highly active endotoxin when solubilized by addition of serum albumin or when dispersed by solution in pyridine before dilution in water. Moreover, Salmonella R_e mutants, whose "LPS" contains only lipid A plus one core sugar, KDO (Figs. 6-11 and 6-12), yield "LPS" with full endotoxic activity.

Endotoxin is generally assumed to play a large role in the vascular, metabolic, and hematological alterations in severe gram-negative infections, but the evidence is indirect: no protective, specific Ab is available, as with exotoxins, and the correlation of symptoms with blood levels of endotoxin has been hampered because only cumbersome and imprecise assays, in whole animals, were used. However, an assay based on the gelation of extracts of blood cells of the horseshoe crab, *Limulus polyphemus,* appears promising.

PHAGE TYPING

Strains of *S. typhi* can be subdivided on the basis of susceptibility to lysis by specific bacteriophages. At least 72 subtypes carrying the same Vi Ag have been distinguished in this way. This information is useful in tracing the sources and progress of epidemics.

All the phages used to type Vi strains of *S. typhi* are descended from a single strain of bacteriophage, some being mutants and others host-modified variants (Ch. 45, Host-controlled modification). The phage susceptibilities of the bacterial strains are determined by lysogeny (Ch. 46) as well as by differences in bacterial genes.

COLICINS

Some enteric bacilli release highly specific proteins called colicins that are members of the broader class of bacteriocins (Ch. 46): molecules that attach to specific receptors on susceptible bacilli, as do bacteriophages, but do not cause bacteriolysis. Various colicins kill sensitive bacteria by different mechanisms: blocking of oxidative phosphorylation, inhibition of protein synthesis, or degradation of DNA.

The production of colicins (colicinogeny) is controlled by plasmids (nonchromosomal blocs of DNA), which may be transmitted to other enteric bacilli by transduction or conjugation (Ch. 46). Responsible for a special form of **bacterial antagonism,** colicinogeny undoubtedly helps to stabilize the microbial population in the adult intestinal tract.

Colicins can also be used for subdividing strains of organisms for epidemiological purposes. In such **colicin typing,** enterobacteria are grouped according to the colicins they produce and those to which they are sensitive. Thus *Shigella sonnei* (see below), which is homogeneous according to phage typing and antigenic analysis, has been divided into 15 types by colicin typing, thus allowing more precise identification in epidemics.

GENETIC RELATIONS

The gastrointestinal tract provides an ideal environment for developing a wide variety of strains of bacteria and bacterial viruses: the population has a high density; it grows as a continuous culture, as in a chemostat; nutritional conditions fluctuate; and genes may be transferred from one strain to another by **transduction, lysogenization,** and **conjugation** (Ch. 9). The result is the accumulation of a wide variety of recombinants with many overlapping patterns of metabolism and antigenicity.

The alteration of somatic O Ags by **lysogenic conversion** is illustrated in Figure 29-5, in which infection by phage ε^{15} provides a new enzyme that causes replacement of factors 3,10 by factors 3,15 in the O Ag, while double infection with phages ε^{15}

and ε^{34} causes still another antigenic change: an enzyme specified by ε^{34} adds a terminal D-glycosyl residue to the determinant 15 resulting from lysogeny with ε^{15}. (These changes in the O Ag require persistent lysogeny with ε^{15} and ε^{34}, indicating that lysogenic conversion, rather than transduction, is involved.)

Another important genetic interaction is the transfer of **multiple drug resistance** by a plasmid (**R factor**) carrying genes for resistance. These plasmids, discovered in shigellae in Japan, are now widespread in salmonellae and in coliforms; they are readily transferred by conjugation, both in vitro and in the intestine, among various enterobacteria. Their genetic properties are discussed in Chapter 46 and their phenotypic effects in Chapter 7.

The medical implications of the **R** factors are obvious. Individuals who harbor the factor in *E. coli* can have it transferred to a pathogen and become the source of a particularly serious epidemic. This process is certain to be accentuated by the selective pressure of antimicrobial therapy. Indeed, R factors have spread remarkably in shigellae and salmonellae. A survey in London (1969) showed that over 50% of healthy individuals harbored R factor-containing enteric bacilli, and often these were the predominant strain of *E. coli*.

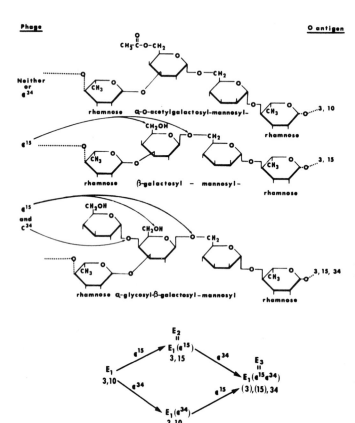

Fig. 29-5. Structures of determinant trisaccharide units of O polysaccharides of group E *Salmonella* (Table 29-5). The changes in structure brought about by lysogenic conversions of determinant 3,10 (*S. anatum,* E_1) to 3,15 (*S. newington,* E_2), and to 3,15,34 (*S. minneapolis,* E_3) by phage ε^{15} and phages ε^{15} plus ε^{34}, respectively (see bottom diagram) are indicated by the curved arrows. [Modified from Robbins, P. W., and Uchida, T. *Fed Proc 21:*702 (1962), and *J Biol Chem* 240:375 (1965).]

Genetic advances have cast new light on the complex interrelations of the enteric bacilli. Above all, they emphasize the extraordinary genetic instability of intestinal bacteria. They show that a shift in antigenic structure may involve only minor modifications of a single kind of macromolecule, readily accomplished by genetic recombination. Likewise, it may be anticipated that the metabolic characteristics of a given immunological type will at times vary. Thus the time-honored lactose fermentation method, universally used to differentiate pathogenic from nonpathogenic enteric bacilli, proved unreliable when a laboratory worker developed typhoid fever due to a lactose-fermenting organism, produced earlier in the laboratory by introduction of a lac-carrying episome (Ch. 46) from *E. coli* to *S. typhi*. In fact, up to 1% of salmonellae in the United States (1960s) were lactose fermenters.

COLIFORM BACILLI

The enteric bacilli included in the "coliform"* group are *E. coli, Klebsiella, Enterobacter* (formerly *Aerobacter*), and several genera previously classified as "paracolon" organisms (see above): *Serratia, Edwardsiella,* and *Citrobacter.*† While in the intestinal tract these organisms generally do not cause disease (except for *Edwardsiella* and some types of *E. coli,* see below). *E. coli* is normally a bowel organism, as are probably some of the *Klebsiella,* whereas the *Enterobacter, Serratia,* and *Citrobacter* groups occur infrequently in the normal intestine and are ordinarily free-living saprophytes. These organisms illustrate the difficulty of determining pathogenicity in absolute terms: usually harmless in their normal habitat, they often cause illness when they reach tissues outside the intestinal tract, especially if the host's defenses are impaired. The

---

* "Coliform" is a general term for fermentative gram-negative rods that inhabit the intestinal tract of man and other animals without causing disease. The term has not been strictly defined: some include in it all the enteric bacteria; others limit it to the lactose-fermenting members.

† Often included among coliforms, *Arizona* organisms are discussed with *Salmonella,* and *Providencia* with *Proteus* (see below).

coliforms (together with Proteus-Providence organisms) are now the predominant etiological agents in various endogenous and nosocomial infections.

They are the most frequent cause of hospital-acquired bacteremia (septicemia), a devastating illness, with *E. coli* and the *Klebsiella-Enterobacter-Serratia* division responsible for about 60% of cases. *E. coli* is the commonest cause of urinary tract infections, often confined to the bladder. *Klebsiella* (usually more virulent than *E. coli*), *Enterobacter,* and *Serratia* are often introduced in the urinary tract by instrumentation. The coliforms are also increasingly involved in pneumonias.

ESCHERICHIA COLI

E. coli is referred to as the "colon bacillus" because it is the predominant facultative species in the large bowel. Its presence in a water supply usually indicates fecal contamination, hence tests for its presence are widely used in public health laboratories (see below).

Cultural and Antigenic Properties. When grown in broth *E. coli* produces a well-dispersed turbidity. Most strains (though not all) have flagella and are motile (Table 29-3). Smooth (S) strains form shiny, convex, colorless colonies but when repeatedly subcultured on artificial media they become rough (R) and form lusterless, granular colonies. Encapsulated variants produce mucoid colonies, particularly when incubated at low temperatures and when grown in media low in nitrogen and phosphorus but high in carbohydrate. Typical *E. coli* colonies are usually easy to recognize by their characteristic appearance on certain differential media: they are lactose fermenters, and on eosin-methylene-blue and Endo agars they have a peculiar metallic sheen. However, some strains ferment lactose late, irregularly, or not at all. On blood agar some strains produce β-hemolysis. *E. coli* produces acid and gas from a large variety of carbohydrates (Table 29-3), but there are strains that ferment glucose with acid but no gas production, such as the alkalescens-dispar organisms formerly classified as *Shigella*. Characteristically, *E. coli* produces indole in media containing tryptophan and is methyl red-positive, but does not produce acetoin or utilize citrate as a sole source of carbon ("IMViC reaction," see *Enterobacter,* below). As mentioned, coliform bacilli are more sensitive than salmonellae or shigellae to inhibition by high concentrations of citrate.

More than 150 different O Ags of *E. coli,* designated O-55, O-111, etc., have been identified by specific agglutination reactions. In addition, approximately 50 H Ags and over 90 different capsular K

Ags have been described. Like the Vi Ags of *Salmonella,* the presence of K Ags as the cell surface often masks the deeper O Ags.

Pathogenicity. The diseases of man most commonly caused by *E. coli* are those of the urinary tract. The organism may conceivably travel from the intestinal tract to the urinary passages and kidneys via hematogenous or lymphatic routes, but most often it follows the "ascending" route from the urethra through the bladder and up the ureters to the kidneys. The normal urinary tract is relatively free of bacteria, but asymptomatic bacilluria is common, particularly in women. When the bacterial count in the urine is greater than 100,000/ml, bacterial **disease** of the urinary tract is usually present. It occurs most commonly in infants (during the diaper period), in pregnant women, and in patients with obstructive lesions of the urinary tract or neurological diseases affecting micturition. Catheterization and other forms of urethral instrumentation are often precipitating factors.

In the kidneys primary lesions tend to occur in the medulla rather than in the cortex. When confined to the bladder (cystitis) the infection is usually well controlled by appropriate antimicrobial therapy. In the kidneys, on the other hand, once suppuration has occurred lesions may continue to progress despite treatment and may eventually lead to scarring and to destruction of tubules and glomeruli. Chronic pyelonephritis of this sort may cause hypertension and frequently terminates in fatal uremia. *E. coli* is often found in peritonitis, appendicitis, and infections of the gallbladder and biliary tract, along with other enteric bacteria. It occurs on the skin of the perineum and genitalia (Ch. 42, Distribution) and frequently infects wounds that become contaminated with urine or feces.

It has been known for 50 years that certain strains of *E. coli* can cause acute diarrhea in man, especially in infants. Such **enteropathogenic *E. coli*** now include 14 distinct antigenic types. Recently, additional strains previously not considered enteropathogenic (some serologically typable, others not) have been implicated in diarrhea, and at least two different pathogenetic mechanisms have been identified. Some *E. coli* strains are invasive (i.e., can penetrate the intestinal epithelium)

and cause a bacillary dysentery-like syndrome ("colitis"), much like virulent *Shigella* strains (see below). Others, isolated from diarrheal illnesses resembling cholera or salmonellosis-like "enteritis," excrete a potent, heat-labile enterotoxin. Injection of the cell-free enterotoxin into an isolated loop of rabbit ileum (a standard laboratory model for diarrhea) causes massive accumulation of fluid similar to that produced by the enterotoxin of the cholera vibrio (see below). Since *E. coli* enterotoxin, like cholera toxin, is active in the upper small bowel but not in the large intestine, toxigenic strains may be carried in the colon without causing symptoms, and disease is produced when the organisms colonize the upper intestine. A plasmid (Ch. 46) is responsible for enterotoxin production in similar strains from swine, and the human disease may well have the same origin; the plasmid is transmissible to other strains and species.

The frequency of dysentery-like and of toxigenic strains of *E. coli* as causes of human diarrhea is not clear, but such organisms play probably a large role in the acute diarrhea of infants, in "traveler's diarrhea," and in "food poisoning" episodes.

E. coli is now the most frequently encountered species in gram-negative sepsis resulting in bacteremia and in severe shock resembling that produced by intravenous injections of endotoxin in laboratory animals. This disease has been occurring with greatly increased frequency in the very young (especially the newborn), in those over age 60, and in patients debilitated by various factors (including corticosteroid therapy, immunosuppressive agents, and leukemia); surgery or instrumentation of the intestinal, biliary, or genitourinary tract may also precipitate such sepsis.

Treatment. Most strains of *E. coli* are sensitive to sulfonamides, ampicillin, cephalosporins, tetracyclines, and carbenicillin. Resistance to any one or all of these agents is frequently encountered (see discussion above on multiple drug resistance). For long-term suppressive therapy of *E. coli* urinary tract infections nitrofurantoin, nalidixic acid, methenamine mandelate, and sulfonamides are used. More toxic agents, such as gentamycin, kanamycin, streptomycin, polymyxin or colistin, and chloramphenicol (now rarely used), should be employed only if all other forms of treatment fail, or in life-threatening systemic infections (sepsis, etc.). Oral colistin, gentamycin, and kanamycin are widely used for enteropathogenic *E. coli* strains in infants, but organisms resistant to these drugs have already been found. Neomycin was widely used in 1960s, but resistant epidemic strains are now widespread. In pediatric institutions constant surveillance of antibiotic sensitivity of enteropathogenic *E. coli* strains, as well as periodic changes of the "routine" antibiotic used, are necessary to minimize the selective pressures resulting in the emergence of highly resistant strains.

KLEBSIELLA-ENTEROBACTER-SERRATIA

The related genera *Klebsiella, Enterobacter* (formerly *Aerobacter*), and *Serratia* are now included in a single division.* The three genera can be readily identified by biochemical tests including decarboxylase reactions (Table 29-3), and most strains of *Klebsiella* and *Serratia* can be typed serologically. These

---

* *Klebsiella pneumoniae* has been considered a significant respiratory pathogen since 1882, but the assessment of *Enterobacter* and *Serratia* in human infections was compromised in the past by discrepancies and confusion in nomenclature and taxonomy. Some of the earlier names of the organisms now included in the two genera are "paracolon" bacteria, *Paracolobactrum aerogenoides, Aerobacter,* and *Hafnia.*

organisms cause serious urinary tract and pulmonary infections in hospitalized patients, and they are second to *E. coli* as causes of gram-negative bacteremia.

The necessity for differentiating *Klebsiella, Enterobacter,* and *Serratia* from each other, and from other enteric bacilli, is underlined by their wide differences in antibiotic susceptibility and in pathogenicity. *Klebsiella* predominates in clinical isolates and as a primary pathogen, and usually produces more severe illness; *Serratia* is usually considered a secondary pathogen.

A number of biochemical tests, used in differentiating these organisms from other *Enterobacteriaceae,* were originally designed to distinguish *Enterobacter* (*Aerobacter*) *aerogenes* from *E. coli* in water supplies. The latter carries out the mixed acid fermentation (Ch. 3, Fermentation) which yields formic, acetic, lactic, and succinic acids, along with ethanol, CO_2, and H_2. The gases are produced from HCOOH by formic hydrogen lyase, but not all enteric bacilli that form this compound synthesize the enzyme. When the enzyme is absent the fermentation occurs without gas production (as among *Shigella* and some species of *Salmonella*), an important diagnostic feature (Table 29-3). *Klebsiella* and *Enterobacter* carry out the **butylene glycol fermentation** (Ch. 3), which forms much less acid, much more ethanol, and large quantities of the neutral endproduct, 2,3-butylene glycol, CH_3-CHOH-CHOH-CH_3 (Table 29-6).

TABLE 29-6. Typical Endproducts Formed in Mixed Acid and Butylene Glycol Fermentations

Product	Mixed acid*	Butylene glycol*
Carbon dioxide, CO_2	88	172
Hydrogen, H_2	75	35
Formic acid, HCOOH	2.5	17
Acetic acid, CH_3COOH	36	0.5
Lactic acid, $CH_3CHOHCOOH$	79	3
Succinic acid, $COOH(CH_2)_2COOH$	11	0
Ethyl alcohol, CH_3CH_2OH	50	70
2,3-Butylene glycol, $CH_3(CHOH)_2CH_3$	0	66
Total acids	129	20

* Moles of product per 100 moles of glucose fermented.
Based on data from Stainier, R. Y., Doudoroff, M., and Adelberg, E. A. *The Microbial World,* p. 582. Prentice Hall, Englewood Cliffs, N.J., 1970.

The **Voges-Proskauer reaction** is used to identify the butylene glycol type of fermentation. At an alkaline pH in the presence of air acetoin (acetyl-methylcarbinol, CH_3-CHOH-CO-CH_3), an intermediate in the formation of butylene glycol, is oxidized to diacetyl (CH_3-CO-CO-CH_3), which in turn reacts with creatine in the peptone medium to form a pink compound. Since certain other bacteria (e.g., lactic acid bacteria) form acetoin without synthesizing butylene glycol, a positive reaction provides only presumptive evidence of butylene glycol fermentation, whereas a negative reaction rules it out.

The **methyl red test** qualitatively distinguishes heavy and light production of acid. Methyl red is yellow above pH 4.5 and is red in more acid media. In glucose-peptone broth cultures incubated for 48 hours only the mixed acid fermentation ordinarily produces enough acid to turn the indicator red. However, some organisms that produce butylene glycol may occasionally form enough acid to change the indicator color; conversely, organisms may carry out the mixed acid fermentation so slowly that the reaction remains negative.

Two other metabolic tests are useful in distinguishing *Enterobacter* from *E. coli:* only the latter produces **indole** in media containing tryptophan, and only the former can grow with **citrate** as sole carbon source (Table 29-3). These four metabolic tests (indole, methyl red, Voges-Proskauer, and citrate) are collectively referred to as the IMViC tests.*

Klebsiellae and many enterobacter strains usually ferment lactose rapidly, whereas the serratiae do not. Differentiation is accomplished by the **motility test** (*Klebsiella* is nonmotile), **decarboxylase tests** with respect to lysine, arginine, and ornithine, urease production, and fermentation of several carbohydrates (Table 29-3).

Klebsiella. *Klebsiella pneumoniae* (Friedländer's bacillus) is the most important human "pathogen" of the *Klebsiella* group. *K. pneumoniae* forms a capsule and hence produces large, moist, often very mucoid colonies. A total of 5 O Ags and 72 capsular (K) polysaccharide Ags have been identified. Type-specific (K) antisera are useful in determining the epidemiology of hospital-acquired *Klebsiella* infections, which represent about ⅔ of all infections due to these organisms.

K. pneumoniae is found in the respiratory tract and the feces of 5 to 10% of healthy subjects and is frequently present as a secondary invader in the lungs of patients with chronic pulmonary disease. It causes approximately 3% of all acute bacterial pneumonias. Its invasive properties, like those of *Pneumococcus,* depend on the antiphagocytic effect of its capsule: unencapsulated (R) strains are avirulent.

Pneumonia caused by *K. pneumoniae,* like that due to *Pneumococcus* type 3, is characterized by the production of thick gelatinous sputum and a high bacterial population density in the edema zones of the active lesions (Fig. 29-6). The destructive action of the unphagocytized organisms on the pulmonary tissue interferes with antimicrobial therapy (Ch. 23, Abscess formation and necrosis) and often results in chronic lung abscesses requiring surgical resection.

Kanamycin, gentamycin, the polymyxins, chloramphenicol, cephalothin, and streptomycin are all commonly used in treatment.

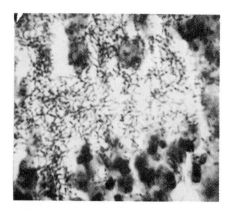

Fig. 29-6. Masses of Friedländer's bacilli (*K. pneumoniae*) in an area of advanced consolidation in a 24-hour experimental pneumonic lesion. Little phagocystosis is seen in such an area, where irreversible tissue damage is likely to occur, leading to the formation of a lung abscess. Gram-Weigert stain. ×850. [From Sale, L., Jr., and Wood, W. B., Jr. *J Exp Med* 86:239 (1947).]

---

* These tests are standard procedures in the study of enteric bacilli. The IMViC formula for *E. coli* is ++−−, for *E. aerogenes* and *E. cloacae* −−++. All 16 possible combinations of the IMViC test have been found.

In urinary tract infections nalidixic acid and nitrofurantoin are effective. The susceptibility of *Klebsiella* to cephalothin sharply distinguishes this organism from *Enterobacter* and *Serratia,* which produce a cephalosporinase. However, klebsiellae strains resistant to cephalothin have appeared.

Klebsiella species have been implicated in chronic inflammatory diseases of the upper respiratory tract: *K. ozenae* in **ozena,** a progressive fetid atrophy of the nasal mucosa; and *K. rhinoscleromatis* in **rhinoscleroma,** a destructive granuloma of the nose and pharynx.

Enterobacter. Organisms in the *Enterobacter* group occur in soil, dairy products, water, and sewage, as well as in the intestinal tract of man and animals. Enterobacters are frequently isolated from sputum (often after antibiotic therapy), urinary tract infections (often hospital-acquired), blood, and wound infections. They are usually considered "secondary" pathogens (i.e., superinfecting an underlying primary infection), or opportunistic (e.g., urinary tract infections following catheterization), or "commensals" (i.e., not causally associated with disease).

Four species of *Enterobacter* are presently distinguished (Table 29-3): *E. cloacae* (formerly *Aerobacter* subgroup A or *A. cloacae*) is the most common form, followed by *E. aerogenes; E. hafniae* (formerly *Hafnia*) and *E. liquefaciens* are less frequent. All are motile and possess ornithine decarboxylase; *E. liquefaciens* produces deoxyribonuclease.

Enterobacters are susceptible to gentamycin, kanamycin, chloramphenicol, the tetracyclines, nalidixic acid, and nitrofurantoin (the latter two being used in urinary tract infections). Many strains are sensitive to polymyxin B or colistin, but *E. liquefaciens* is strikingly resistant to these agents. All enterobacters are resistant to cephalothin and ampicillin.

Serratia. Cultures of *Serratia marcescens* have been used for many years by bacteriologists for demonstration purposes (e.g., to demonstrate bacteremia after dental extraction) because the bright red pigment of some strains is so easily observable.* The organism

---

* Instances dating back to antiquity of the "miraculous" appearance of "blood" on communion wafers, bread, and other foods were likely due to such *Serratia* strains.

has been considered a harmless saprophyte. Since 1960, however, it has been isolated with increasing frequency from man, apparently because of an increase in (often severe) nosocomial infections. On the other hand, the earlier incidence is not known, since many nonpigmented *Serratia* strains were probably included in the past with other slow lactose fermenters as "paracolon" organisms. Generally serratiae are opportunistic pathogens.

Serratiae are motile rods; most strains do not ferment lactose or do so slowly (Table 29-3). Like *E. liquefaciens,* they liquefy gelatin rapidly and produce deoxyribonuclease, but serratiae form little or no gas from fermentable substances; they also fail to ferment arabinose and rhamnose. Although the organisms may produce a red pigment (prodigiosin), more pronounced and brighter when the culture is held at room temperature, at least ¾ of strains isolated at present are nonpigmented. Fifteen O Ags and thirteen H (flagellar) Ags have been identified.

In contrast to *Klebsiella* and *Enterobacter,* almost all strains of *Serratia* are resistant to polymyxin B and colistin. Gentamycin, kanamycin, chloramphenicol, and carbenicillin are useful.

EDWARDSIELLA

The recently (1965) established genus *Edwardsiella* includes a group of motile, H_2S-producing, lactose-negative organisms that resemble salmonellae (see below) in some biochemical features and sometimes in pathogenicity for man. They ferment only glucose and maltose (Table 29-3). *E. tarda* has been isolated from a variety of mammals and reptiles. It is occasionally found in the intestinal tract of man, especially in acute gastroenteritis, and it can produce serious septic infections. However, man is likely only an accidental host. A total of 42 O Ags and 28 H Ags have been identified. Tetracyclines, chloramphenicol, kanamycin, and ampicillin are drugs of choice.

CITROBACTER

The *Citrobacter* group is composed of Enterobacteriaceae previously designated as *Escherichia freundii* and the Bethesda-Ballerup group of "paracolon" organisms (IMViC formula −+−+); the majority of strains form H_2 and ferment lactose, frequently delayed. They are differentiated from salmonellae by their possession of β-galactosidase, lack of lysine decarboxylase, and ability to grow in media containing 1:13,000 KCN (Table 29-3).

The Bethesda-Ballerup portion of the group has been extensively studied, and an antigenic scheme was established, including 32 O groups and 74 H Ags. Some Bethesda strains possess the Vi antigens and thus may agglutinate in polyvalent *Salmonella* antiserum containing Vi Ab. Members of the Bethesda-Ballerup subgroup have been associated with diarrhea in man, but they are not considered true enteric pathogens.

Citrobacter strains occur infrequently in normal feces; they have been recovered from urinary tract infections and various septic processes. Drugs of choice are polymyxin B or colistin, gentamycin, kanamycin, and chloramphenicol; many strains are sensitive to tetracyclines.

PROTEUS-PROVIDENCE GROUP

These lactose-negative, motile bacilli are unusual among Enterobacteriaceae in being able to deaminate phenylalanine* (Table 29-3) and lysine. Rapid and abundant **urease production** distinguishes *Proteus* from *Providencia*.

Proteus. These organisms are commonly found in soil, sewage, and manure. They are found with some frequency in normal human feces, but often in much increased numbers in individuals on antibiotic therapy or during diarrheal diseases due to other organisms. *Proteus* organisms are frequent causes of urinary tract infection (both community and hospital-acquired) and are also involved in other, often serious infections.

P. vulgaris and *P. mirabilis* form a thin spreading growth (swarm) on the surface of moist agar media. To obtain isolated colonies swarming is prevented by cultivation on the relatively dry surface of 5% agar or on ordinary 1 to 2% agar containing chloral hydrate (0.1%). They also produce abundant H_2S and liquefy gelatin; *P. rettgeri* and *P. morganii* do not possess these characteristics. As indicated in Table 29-3, fermentation of most *Proteus* strains is of the mixed acid type. The most commonly encountered species, *P. mirabilis,* does not produce indole from tryptophan, while the others do.

The antigenic structure of *P. vulgaris* is of particular medical interest because strains possessing certain O Ags (OX2, OX19, and OXK) are agglutinated by the sera of patients with various rickettsial diseases (Weil-Felix reaction; Ch. 38, Laboratory diagnosis). These particular O Ags seem to be fortuitously related to antigenic determinants of rickettsial polysaccharides.

Carbenicillin and gentamycin are active against all species, but only *P. mirabilis* is susceptible to ampicillin and cephalothin and often to penicillin G. Kanamycin and chloramphenicol are also used.

Providencia. Organisms in the genus *Providencia* are closely related to *P. morganii* and *P. rettgeri,* and were known formerly as *Proteus inconstans.* They are easily differentiated from *Proteus* by their lack of urease. Members of the group have been isolated from human feces during outbreaks of diarrhea but also in normal individuals. They are primarily associated with urinary tract infections. Providence organisms are highly resistant to antibiotics except for kanamycin, gentamycin, and carbenicillin.

SALMONELLAE

The genus *Salmonella* contains a wide variety of "species" pathogenic for man or animals, and usually for both. They ferment neither lactose nor sucrose and with few exceptions produce abundant H_2S (Table 29-3). Essentially all are motile and decarboxylate lysine and ornithine.† Their complex antigenic structures and genetic relations have already been discussed. Selective media for their isolation from feces contain brilliant green, deoxycholate, selenite, tetrathionate, or citrate to suppress the growth of coliforms. Some of the salmonellae more often encountered in medical practice are listed in Table 29-4.

Originally salmonella were named according to the disease they caused or the animals from which they first isolated. Later, each antigenically distinguishable type was accorded specific rank and was named after the geographical place at which it was first isolated. Recent United States usage recognizes only three species (*S. choleraesuis, S. typhi,* and *S. enteriditis*): the first two do not contain subtypes (serotypes), but *S. enteriditis* contains over 1000 serotypes, each writ-

---

* Phenylpyruvic acid, produced by oxidative deamination of phenylalanine, develops an intense green color on addition of $FeCl_3$.

† *S. typhi* fails to produce gas from carbohydrates and to decarboxylate ornithine. It also produces little or no H_2S in TSI agar.

ten in a nonitalicized form (e.g., *S. enteriditis* serotype paratyphi A, instead of *S. paratyphi* A). However, the classic names will be employed in this section because of their prevalence in the current clinical literature.

Pathogenicity. Three clinically distinguishable forms of salmonellosis occur in man: enteric fever, septicemia, and acute gastroenteritis.

The prototype of the **enteric fevers** is caused by *S. typhi,* acquired by the ingestion of contaminated food or water. **Typhoid fever** usually begins insidiously after an incubation period of 7 to 14 days, with malaise, anorexia, and headache, followed by the onset of fever. The latter often increases in a step-like manner and is accompanied by relative bradycardia. Prostration may be marked, particularly during the first week; and though diarrhea is usually absent, abdominal tenderness and distention are common. Cough and signs of bronchitis may be present; "rose spots," which last for only a few days, may appear on the trunk; and splenomegaly and leukopenia are common. In more severe cases the sensorium is dulled and the patient may become delirious. After the third week the fever usually subsides by gradual "lysis."[*] During the later stages of the disease there may be severe intestinal hemorrhages, or perforation of the bowel causing peritonitis.

Figure 29-7 depicts the kinetics of the bacteremia, shedding of organisms in the feces and urine, and formation of antibody. The ingested organisms multiply in the gastrointestinal tract; some of them enter the intestinal lymphatics, from which they are disseminated throughout the body by the bloodstream and are excreted in the urine. Since the bile is a good culture medium for *S. typhi,* luxuriant growth occurs in the biliary tract[†] and provides a continuous flow of organisms into the small bowel, where they tend to localize in the Peyer's patches. Their ability to persist in the biliary tract may result in a chronic carrier state (see below) in which they continue to be excreted in the feces.

---

[*] In contrast to defervescence by crisis (Ch. 25, Immunity).

[†] Note resistance to bile salts, including deoxycholate.

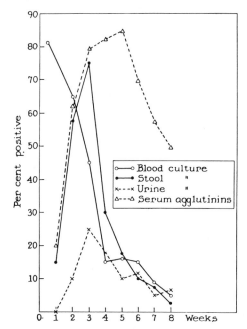

Fig. 29-7. Relative frequencies of positive blood, urine, and stool cultures and serum agglutination tests during the course of typhoid fever. (From Morgan, H. R. In *Bacterial and Mycotic Infections of Man,* p. 381. R. J. Dubos, ed. Lippincott, Philadelphia, 1958.)

In fatal cases of typhoid fever the most prominent lesions found at autopsy are lymphoid hyperplasia (involving the lymph nodes, Peyer's patches, and spleen); focal necrosis of the liver; inflammation of the gallbladder; and patchy inflammatory lesions in the lungs, bone marrow, and periosteum. Intestinal hemorrhages or perforations of the bowel can usually be traced to ulcerations in the Peyer's patches.

In active lesions bacilli are often detectable in the phagocytic mononuclear cells, where they can multiply (as is readily demonstrated in vitro). Intracellular bacilli seem to be protected from the bacteriolytic action of specific Ab, which appears in the blood long before the disease subsides (Fig. 29-7). The studies of Mackaness (Ch. 22, Allergy vs. immunity) suggest that the intracellular killing process is related to the onset of delayed hypersensitivity. Moreover, immunity to experimental *Salmonella* infection may be transferred to a susceptible animal by the injection of cells from an immune donor.

The pathogenesis of typhoid fever is difficult to study because most laboratory animals are resistant. Mice, for example, can be readily infected only by intracerebral injection or by intraperitoneal inoculation with organisms suspended in mucin. Chimpanzees, however, can be infected by the oral route.

Studies with human volunteers show that 10^5 to 10^7 ingested bacilli are required to induce the disease. The experimental human illness has also been useful in assessing the role of endotoxin in pathogenesis; recent findings suggest that circulating endotoxin, which induces tolerance (Ch. 22), does not account for the prolonged febrile phase of typhoid fever.

Enteric fevers caused by other salmonellae ("paratyphoid" fevers) are usually milder and have a shorter incubation period (1 to 10 days). Bacteremia occurs early, fever usually lasts for 1 to 3 weeks, and rose spots are rare. Almost any *Salmonella* may cause enteric fever, but the most common agents in the United States are *S. paratyphi* B (*S. schottmülleri*), and *S. typhimurium*.

Salmonella **septicemias** are characterized by high, remittent fever and bacteremia, ordinarily without apparent involvement of the gastrointestinal tract. Focal suppurative lesions may develop almost anywhere in the body, including the biliary tract, kidneys, heart, spleen, meninges, joints, and lungs. Prolonged septicemia of this type is most commonly caused by *S. choleraesuis*.

In **gastroenteritis** the disease is confined primarily to the gastrointestinal tract: it is the most common clinical manifestation of *Salmonella* infection. Symptoms begin 8 to 48 hours after the consumption of contaminated food, with diarrhea ranging from mild to a fulminant form with sudden and violent onset ("food poisoning"). Headache, chills, and abdominal pain are followed by nausea, vomiting and diarrhea, accompanied by fever lasting from 1 to 4 days. Blood cultures are rarely positive, but the organisms can usually be cultured from the feces. The most common cause of *Salmonella* gastroenteritis in the United States is *S. typhimurium*.

Predisposing factors in *Salmonella* infections include surgical operations on the gastrointestinal tract, treatment with broad-spectrum antibiotics (which disturbs the normal bacterial flora of the bowel), sickle cell anemia, and the acute hemolytic anemia of bartonellosis (Ch. 41). In the hemolytic

phase of the latter disease the incidence of *Salmonella* bacteremia may be as high as 40%.

Salmonellae can be divided into three groups with respect to their host preferences or adaptation. 1) *S. typhi, S. paratyphi* A and C, and *S. sendai* are more or less strictly adapted to man. 2) Others are adapted to particular nonhuman animal hosts. The primary host of *S. choleraesuis* is swine, but it occurs in other animals as well as in man. 3) A far larger number of types produce disease in man and other animals with equal facility. The reasons for these varying interactions are still unknown. There is some evidence that virulent species multiply intracellularly whereas avirulent ones do not, hence the crucial difference may be in the ability of the bacterial cell to withstand intracellular digestion.

Carriers. Following active salmonellosis the organisms occasionally become established in the host and give rise to a carrier state. With typhoid fever approximately 3% of patients become carriers and continue indefinitely to excrete as many as 10^6 to 10^9 *S. typhi* per gram of feces. The source is usually a chronic suppurative focus in the biliary tract. When treated with broad-spectrum antibiotics carriers may come down with active *Salmonella* disease, just as mice are made more susceptible to experimental salmonellosis by treatments that upset the normal balance of their intestinal flora (Ch. 22, Microbial factors).

Immunity. Both anti-O and anti-Vi Abs play a role in immunity to salmonellosis: maximal protection is obtained in chimpanzees only by immunization with both antigens. Anti-H Abs appear to have no protective effect. In experimental salmonellosis in the mouse two mechanisms have been demonstrated: specific O Ab opsonizes *Salmonella* cells, with subsequent engulfing and destruction of a large proportion of these by macrophages; while in "cellular" or "infection" immunity (induced by living vaccines) macrophages become more efficient in killing intracellular bacteria.* In addition, it appears that in man an orally administered, attenuated *S. typhi* strain

---

* This immunity is nonspecific, i.e., immunity against virulent *Salmonella* may be induced by other *Salmonella* species, by living avirulent (rough) strains, or by different intracellular parasites such as *Listeria* (Ch. 22, Acquired antibacterial cellular immunity).

can enhance **local immunity** of the intestinal tract and effectively prevent typhoid bacilli from entering a site in which to multiply.

Laboratory Diagnosis. A diagnosis of *Salmonella* infection is made by isolation of the organism from the blood, feces, urine, or other organs. Isolation from blood or urine is indicative of tissue invasion and ordinarily establishes the diagnosis, but a salmonella organism isolated from the feces is not necessarily the cause of the individual's illness (see Carriers, above). Demonstration of a significant rise in Ab titer to the specific organism isolated from the patient is helpful in confirming the diagnosis.

Blood cultures are particularly important during the first week of illness (Fig. 29-7); occasionally bone marrow cultures are positive when blood cultures are negative. Urine and feces should be cultured repeatedly on differential and selective media. Preliminary incubation of fecal specimens in selenite or tetrathionate broth, which inhibit coliforms but not salmonellae, may facilitate the subsequent isolation of salmonellae on differential solid media. Final identification is based on biochemical reactions (Table 29-3) and agglutination tests with monospecific antisera.

Serological tests for specific agglutinins (Widal test, "febrile agglutination") should be performed on at least two serum specimens: the first obtained as early as possible and the second 7 to 10 days later. Dilutions should be tested with the infecting organism (if available) as well as with standard *Salmonella* O and H Ags.

O agglutinins are ordinarily of greater diagnostic significance, since H agglutinins tend to persist longer following vaccination and occasionally are not produced at all in active infections. Most individuals have Abs to several *Salmonella* serogroups due to inapparent infection or previous immunization, and many Ags are shared by different salmonellae. Hence Widal tests with group Ags, as performed in most clinical laboratories, are often not helpful and occasionally misleading.

Treatment. In *Salmonella* enteric fevers and septicemia chloramphenicol is the drug of choice, despite the danger of serious toxicity; however, chloramphenicol-resistant *Salmonella* strains have appeared. Ampicillin, a much less toxic agent, may be used but is less effective. Although the response is usually rapid, relapses frequently occur unless the patient is treated for at least 2 weeks, since the organisms tend to survive within phagocytic cells.

Antibiotics are not indicated in *Salmonella* gastroenteritis (except in the very young and those over 60), since the disease is brief and limited to the gastrointestinal tract. In addition, the unnecessary use of antibiotics prolongs *Salmonella* excretion, enhances the incidence of the carrier state, and favors the acquisition of antibiotic resistance by the infecting strain.

The bactericidal ampicillin is much more effective than the bacteristatic chloramphenicol in eliminating the carrier state. Prolonged treatment with large doses of ampicillin is effective in 60 to 80% of cases, the organisms being eradicated from the biliary tract. In carriers who have relapses after one or more courses of therapy, cholecystectomy terminates the carrier state in 9 out of 10 cases.

Prevention. Most important in preventing the man-to-man transmission of typhoid fever have been 1) proper sewage disposal, 2) pasteurization of milk, 3) maintenance of unpolluted water supplies, and 4) scrupulous exclusion of chronic carriers as food handlers. Since *S. typhi* infects only man its control is relatively feasible, and the incidence of typhoid fever in the United States has been declining steadily (e.g., from 5593 cases in 1942 to less than 500 in 1967).

In contrast, laboratory-confirmed nontyphoid salmonellosis has increased 20-fold in the same period. Since it is known that in the milder food- and water-borne enteritis epidemics less than 1% of the actual cases are reported, it is estimated that the 24,000 cases of salmonellosis reported in 1970 represent more than 2,000,000 cases per year. Eradication of human disease due to salmonellae that infect animals as well as man is difficult, for elimination of the animal reservoir is impossible. Domestic fowl probably constitute the largest single reservoir, salmonellae having been isolated from 40% of apparently healthy turkeys and many other domestic and wild animals, including pet turtles. Meats and pooled preparations of dried eggs are notorious sources. Practical measures for controlling salmonellosis in animals (only recently begun) and improved technics of food processing and handling should eventually lower the incidence of human infections.

Although typhoid **vaccines** have been employed for many decades, definitive evidence of their effectiveness has become available only recently. Controlled field studies in the 1960s in Yugoslavia, Poland, Russia, and Guyana indicated that an acetone-inactivated dried typhoid vaccine (with the Vi Ag retained) confers protection of up to 90% for 3 years. Conventional heat-killed, formalin-preserved typhoid vaccine (with most of the Vi Ag destroyed) was less effective. Immunization with cell-free extracts containing either O or Vi Ags was virtually ineffective. An orally administered attenuated *S. typhi* strain appears to afford good protection, but additional studies are needed.

Studies in human volunteers challenged with graded doses of *S. typhi* after vaccination showed that vaccine protection depends on the magnitude of the challenging (infectious) dose: protection was 67% against a dose of 10^5 organisms (the approximate infecting dose of bacilli in nature which might result from a water-borne exposure) but zero against 10^7 organisms. Vaccine-induced resistance thus may falter on ingestion of contaminated foods that have accumulated large numbers of organisms upon prolonged incubation.

Routine typhoid immunization is not recommended in the United States but is widely used for military personnel and travelers to parts of the world where typhoid fever is endemic. It is also used in continued household contacts of known typhoid carriers and in community and institutional outbreaks of typhoid fever. The endotoxin present in the vaccine generally causes a mild toxic reaction. The use of triple typhoid vaccine (TAB, containing *S. typhi* and *S. paratyphi* A and B) is no longer recommended, for the effectiveness of the A vaccine has never been experimentally established and recent field trials of the B vaccine failed to show any beneficial effect.

ARIZONA GROUP

The Arizona group (*A. hinshawii:* formerly *Salmonella arizona, Paracolobactrum arizonae*) resemble salmonellae but differ in possessing β-galactosidase, slowly liquefying gelatin (Table 29-3), and utilizing sodium malonate. They have been differentiated into 38 O Ag groups with 332 serotypes; some of the O and H Ags cross-react with those of salmonellae.

Cold-blooded animals, especially snakes, seem to be the natural reservoir, but Arizona organisms have also been isolated from fowl and domestic animals and occur in dried egg powder and other foods. In man they are associated with salmonellosis-like diseases, i.e., gastroenteritis and enteric fevers, with a high incidence (30%) of isolations obtained from patients with bacteremia or localized septic infections. However, this apparent high degree of invasiveness may reflect selective recognition of such cases: many strains ferment lactose rapidly, and while such strains are not ignored when cultured from organs they are likely to go unrecognized in cases of gastroenteritis. Ampicillin, chloramphenicol, and cephalothin are useful in treatment.

SHIGELLAE

The shigellae cause in man a disabling disease known as bacillary dysentery (Gr., "sick gut"). Described in the fourth century B.C., this common illness has been of great military importance through the ages, often rendering whole armies temporarily (but ignominiously) unfit for combat. It spreads rapidly under conditions of overcrowding and poor sanitation, as in disaster areas, prisoner-of-war camps, and overtaxed mental hospitals. In 1896 the Japanese bacteriologist Shiga isolated the first species of this group, now known as *Shigella dysenteriae*.

The shigellae are much less invasive than the salmonellae, rarely causing bacteremia. They produce natural disease in man and higher apes, and inhabit only the intestinal tracts of primates, or rarely of dogs. Other properties that distinguish them from most salmonellae (Table 29-3) include lack of motility, failure to produce gas during fermentation (except for rare strains of *S. flexneri*), and lack of lysine decarboxylase. They possess specific polysaccharide O Ags, but since they are nonmotile they have no H Ags. Certain smooth (S) strains have heat-labile K (envelope) Ags and hence agglutinate more readily in homologous anti-O antiserum after they are heated. Shigellae Ags, like those of salmonellae, may be altered by lysogenic conversion. A system of bacteriophage typing has been developed but not adopted for practical use.

The four species of *Shigella* (Table 29-7) are differentiated from other Enterobacteriaceae by biochemical reactions and from each other by biochemical and antigenic characteristics. All fail to ferment lactose, except *S. sonnei* which does so slowly.

Pathogenicity. All known species of the genus *Shigella* are pathogenic for man.* The lesions produced in the gastrointestinal tract are mostly confined to the terminal ileum and the colon; they consist primarily of mucosal ulcerations, which are covered by a pseudomembrane composed of polymorphonuclear leukocytes, cell debris, and bacteria enmeshed in fibrin. It is believed that the ulcerative lesions arise when the organisms cross the epithelial barrier and enter the lamina propria: local accumulation of metabolic products and release of endotoxin then cause death of the epithelial cells. During recovery the ulcerations fill with granulation tissue and eventually heal, with scar formation in unusually extensive ulcers.

The factors responsible for pathogenicity are not clearly defined except for *S. dysenteriae* type 1 (Shiga bacillus), which produces, in addition to the endotoxin common to all shigellae, a heat-labile exotoxin known as **Shiga neurotoxin** (MW ca. 75,000).

---

* The organisms formerly known as *S. alkalescens* and *S. dispar,* of doubtful pathogenicity in man, are now classified as *E. coli.*

When injected parenterally into rabbits, mice, or guinea pigs, this substance causes paralysis, diarrhea, and death. The same organism also elaborates a potent **enterotoxin** (possibly the same substance), which induces fluid accumulation in ligated segments of rabbit ileum (like *E. coli* and cholera enterotoxin).

Shigella dysentery is characterized by sudden onset of abdominal cramps, diarrhea, and fever, following an incubation period of 1 to 4 days. Both mucus and blood in the feces are common. The loss of water and salts may cause dehydration and electrolyte imbalance, particularly in infants and young children. Specific agglutinins appear in the blood during convalescence, and Abs can often be demonstrated in the feces (coproantibodies). Although circulating Abs seem to have no effect on the course of the disease, the coproantibodies may play a role in the recovery. Stool cultures usually become negative within a week or two of convalescence but may remain positive for a month or more. *Shigella* dysentery varies in severity according to species (in the general order *S. sonnei, S. boydii, S. flexneri,* and *S. dysenteriae*), from a mild intestinal upset with little systemic disturbance to profuse bloody diarrhea with fever and severe prostration. A recent large *S. dysenteriae* type 1 (*S. shigae,* Shiga bacillus) epidemic in Central America, after an absence of several decades, was characterized by fatality rates up to 20% among children and multiple drug resistance of the organism.

Experimental *Shigella* infections can be induced orally in guinea pigs previously starved or pretreated with toxic doses of carbon tetrachloride. The severity of the disease is increased by large doses of opiates, which slow peristalsis.

TABLE 29-7. Simplified Outline of Shigella Classification

Species	Serological subgroup	Serological type(s)*	Mannitol	Ornithine decarboxylase
S. dysenteriae	A	1–10	−	−
S. flexneri	B	1–6	+	−
S. boydii	C	1–15	+	−†
S. sonnei	D	1	+	+

* The serotypes are separately numbered for each species; cross-reactions between those with corresponding numbers of different species are absent or minimal.

† Cultures of *S. boydii* 13 are positive.

+ = 90% or more positive in 1 or 2 days; − = 90% or more negative.

Bacterial antagonism appears to play a major role in natural resistance of the gastrointestinal tract to invasion by *Shigella* organisms. Thus mice pretreated with broad-spectrum antibiotics, which upset the normal flora of the gastrointestinal tract, may be infected with drug-resistant strains. Moreover, germ-free guinea pigs (Ch. 22) are susceptible to orally induced shigellosis and even develop bacteremia; and previous immunization with *Shigella* vaccine does not afford protection. However, if their intestinal tracts are first infected with *E. coli* they become highly resistant.

Immunity. Persons residing in regions where bacillary dysentery of a given type is endemic seem to acquire immunity to the disease, presumably from inapparent infection. Patients with circulating Abs, however, often suffer second attacks of shigellosis when exposed to a large enough reinfecting dose. The ineffectiveness of humoral Abs has been attributed to the superficial nature of the lesions in the intestinal tract. Intestinal immune factors, in contrast, may play an important role, for encouraging results were obtained in recent field trials with live oral vaccines (a *Shigella* mutant–*E. coli* hybrid and a streptomycin-dependent *Shigella* strain) that had lost the ability to penetrate the intestinal epithelial lining.

Laboratory Diagnosis. Although dysentery bacilli are usually excreted in large numbers during the active disease they often remain viable in the feces for only a short time and therefore must be cultured fairly promptly. Indeed, the best method of obtaining specimens for culture is by means of a rectal swab. Specimens obtained from ulcerative intestinal lesions, under direct vision through a sigmoidoscope, are most likely to contain the organisms. Cultures are best made on MacConkey, xylose-lysine-deoxycholate, or eosin-methylene-blue agar,* with final identification of the organism based on biochemical (Table 29-3) and specific agglutination tests.† Identification of *S. dysenteriae* is important, since it causes a particularly severe

---

* Deoxycholate citrate and SS (*Salmonella-Shigella*) agars are often inhibitory.

† Subgroups can be identified in most clinical laboratories, but serotypes can often be determined only in larger laboratories, where adsorbed specific sera are available.

dysentery, thought to be related to its production of exotoxin.

Bacillary dysentery can often be differentiated from amebic dysentery (due to the protozoon *Entamoeba histolytica*) by microscopic examination of the feces: the bacillary lesions are frankly purulent, whereas those caused by amebae are remarkably free of white cells.

Tests for agglutinins are of little immediate diagnostic value because of the brevity of the illness; occasionally the demonstration of a rising titer of serum agglutinins may permit retrospective recognition of the disease.

Treatment. Since the infection, especially that due to *S. sonnei,* is essentially self-limiting and many patients are only mildly ill, opinion is divided on the advisability of antibiotic treatment in mild cases, especially since drug therapy often results in the emergence of multiple drug-resistant strains. For severely ill patients, and for young children and debilitated adults, antibiotic as well as supportive intravenous fluid and electrolyte therapy is needed. Shigellae are usually, but not always, sensitive to ampicillin, tetracyclines, streptomycin, sulfonamides, kanamycin, chloramphenicol, nalidixic acid, and colistin; but resistance develops quickly and it is necessary to determine the sensitivity pattern of the organism isolated from the patient (or at least of the epidemic strain). Presently ampicillin is considered the drug of choice.

S. dysenteriae type 1 strains involved in the recent epidemic in Central America, and also in cases imported into the United States, were resistant to the sulfonamides, streptomycin, the tetracyclines, and chloramphenicol. Large parenteral doses of ampicillin (and in some cases even of penicillin) were effective in curing patients in this epidemic. Controlled studies in Vietnam indicate that oral ampicillin significantly shortens the clinical course of the disease and effectively eliminates shigellae from the intestinal tract within 24 to 48 hours in about 75% of patients.

Prevention. Since the only important source of bacillary dysentery is man and the disease is spread by "food, feces, fingers, and flies," sanitary measures are of the greatest importance. Control is complicated by the high incidence of inapparent infections. Moreover, in contrast to the findings with typhoid in volunteers (see above), the ingestion of as few as 180 viable virulent *Shigella* cells can cause dysentery in man; half of all volunteers became ill after ingestion of 5000 cells. This fact might explain the person-to-person mode of transmission and the recurrence in institutionalized populations.

All patients with clinically recognizable dysentery should be isolated, if possible, until their stool cultures have become negative. Proper chemotherapy significantly shortens the period of infectiousness. Sanitary measures should be directed against infected food handlers, contaminated water and milk supplies, and improper methods of sewage disposal. In widespread epidemics mass chemoprophylaxis may be indicated. Effective control may eventually be achieved through use of oral live vaccines. Since protection afforded by an oral live vaccine for dysentery is serotype-specific, there is a need for a bivalent vaccine in the United States, in which most cases are due to either *S. flexneri* type 2 or *S. sonnei*.

VIBRIOS

CHOLERA VIBRIOS

Cholera, endemic in India for centuries, has from time to time caused devastating epidemics in other parts of the world. Since 1817 seven worldwide pandemics have occurred; the latest of these spread over several continents in the late 1960s. The role of drinking water was established in the London outbreak of 1854, when an anesthetist, John Snow, "with a notebook, a map, and his five senses," proved epidemiologically that "spoiled" water from the Broad Street pump spread the disease. In 1883 the causative agent, the cholera vibrio, was discovered by Robert Koch in epidemics in Egypt and India.

Although Koch demonstrated the organisms in 32 consecutive cases of cholera, and at 64 consecutive autopsies of patients dead of the disease, his conclusions were challenged by other bacteriologists, who found similar but apparently benign vibrios in many other situations. The specificity of the cholera vibrio was established 10 years later, when the cells were found to be rapidly lysed in the peritoneal cavities of previously immunized guinea pigs (Pfeiffer phenomenon). Shortly thereafter Bordet published his classic studies on immune bacteriolysis, showing that Abs and a heat-labile substance, now known as the complement system, were both involved in the process.

Vibrios are curved, motile, gram-negative bacilli, with a single polar flagellum. Of the 34 species described, *V. cholerae* is the major pathogen; some additional species also cause disease in man and animals, while others are found commonly in the normal flora of man.

V. cholerae differs from Enterobacteriaceae in several important respects. 1) In fresh isolates it is shaped like a **comma** rather than a straight rod (Fig. 29-8) (hence it was earlier called *V. comma*), though on repeated subculture it assumes the shape of the enteric bacilli. 2) It is **oxidase-positive** (Table 29-1). 3) It grows luxuriantly in media that are too **alkaline** (pH 9.0 to 9.6) for the growth of most other bacteria, but is sensitive to acid; it is rapidly killed in cultures containing fermentable carbohydrate. (This property is utilized in designing selective media for its primary isolation.) 4) The organism produces a **neuraminidase** which hydrolyzes N-acetylneuraminic acid, a sialic acid found in human plasma, in various mucins, and in surface structures of many mammalian cells (including the specific receptor sites for influenza virus on erythrocytes; Ch. 56, Hemagglutination).

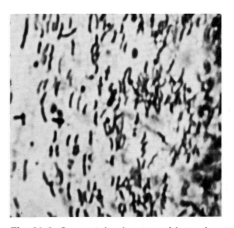

Fig. 29-8. Gram-stained smear of feces from patient with cholera. Predominant comma-shaped organism is *Vibrio cholerae.* ×1200. [From Reimann, H., Chang, G. T. C., Chu, L.-W., Lin, P. Y., and Ou, Y. *Am J Trop Med* 26:631 (1946).]

In many respects, however, *V. cholerae* resembles Enterobacteriaceae (Table 29-3). It ferments lactose slowly (after 3 days or more) but promptly attacks mannose and sucrose, producing acid but no gas; it possesses lysine and ornithine decarboxylases. *V. cholerae* can grow in selective media containing bile salts, bismuth sulfite, or tellurite. Like other motile species of enteric bacilli it produces a variety of O and H Ags, as well as endotoxin. The polysaccharide O determinants of all *V. cholerae* strains belong to the first (group 1) of the six known groups of vibrio O Ags. *V. cholerae* is subdivided into three antigenic types (AB, "Ogawa"; AC, "Inaba"; ABC, "Hikojima"); they share a group-specific O Ag (A) while the secondary Ags (B and C) are type-specific.

V. cholerae elaborates an extremely potent **enterotoxin** that has recently been obtained in highly purified form: it is a heat-labile protein with an MW of about 84,000. Its action upon the epithelial cells of the small bowel is responsible for the massive intestinal fluid and electrolyte losses characteristic of cholera (see Pathogenicity).

El Tor Biotype.   Though most cholera vibrios are not hemolytic, a strain that forms a soluble hemolysin was first isolated in 1906 at the El Tor quarantine station on the Gulf of Suez. This and subsequently isolated similar strains are also unusual in their phage susceptibility. Because the pathogenic properties of these organisms are typical of cholera vibrios, they are no longer considered a separate species, but a biotype of *V. cholerae*. The El Tor vibrios are more resistant to chemical agents, persist longer in man, and survive longer in nature than the classic strain. They are responsible for the latest cholera pandemic that started in Celebes (Indonesia) in 1961 and spread by 1970 to much of Asia, the Middle East, Africa, and parts of Eastern Europe; several cases also occurred in Western Europe.

Pathogenicity.   Naturally acquired cholera has been described only in man. Suckling rabbits are susceptible to oral infection, as are guinea pigs if pretreated by starvation and administration of streptomycin and calcium carbonate. The resulting infection remains localized in the intestine and superficially resembles human cholera. In human volunteers challenges with varying doses of viable *V. cholerae* resulted in a wide gradation of responses,

paralleling those seen in natural infections: 1) no evidence of infection, 2) carrier state, 3) mild diarrhea, and 4) severe watery cholera diarrhea. Gastric acidity seems an important defense against cholera: ingestion of at least 10^8 organisms was necessary to demonstrate some evidence of infection, but neutralizing the acidity of the stomach lowered the number to 10^4 organisms.

Man is usually infected by feces-contaminated water and food. The ingested vibrios multiply throughout the gastrointestinal tract in susceptible individuals, and after 2 to 5 days cause sudden onset of nausea, vomiting, diarrhea, and abdominal cramps. In severe cases the voluminous liquid feces ("rice-water stools") contain mucus, epithelial cells, and large numbers of vibrios (10^6 or more per milliliter). The loss of fluid from the gut may reach 10 to 15 liters per day, leading to extreme dehydration. The liquid feces are virtually free of protein, and contain about the same concentration of sodium, about twice the concentration of bicarbonate, and nearly five times the concentration of potassium as normal plasma. The patient thus develops an extracellular fluid deficit, metabolic acidosis, and hypokalemia. Shock frequently intervenes, and death may occur in a few hours. The mortality rate in hospitalized cases may be as high as 60% without adequate therapy, but is reduced to less than 1% by intravenous replacement of fluid and electrolyte losses (see Treatment). In milder cases symptoms may be minimal. Regardless of severity, the active disease rarely lasts for more than a few days.

Biopsies taken from both the small and the large intestine reveal only hyperemia and signs of superficial inflammation. In postmortem examination the organisms are found in large numbers "adsorbed" to the surface of the mucosa and in the luminal fluid, but in contrast to shigellae (see above) they do not invade the epithelium: the mucosa remains intact. In experimental animals and human volunteers cell-free filtrates of *V. cholerae* cause intestinal secretory changes characteristic of cholera; recently the toxic agent in the filtrate, **cholera enterotoxin** (see above), has been isolated in a highly purified form, active at submicrogram levels. It acts upon the luminal surface of small bowel mucosal cells, reversing the normal direction of ion transport and thus causing accumulation of fluid within the intestinal lumen.

The **action of enterotoxin** seems to be mediated by adenosine-3',5'-cyclic monophosphate (**cyclic AMP, cAMP**). Addition of enterotoxin **or** cAMP to the mucosal surface of isolated ileal mucosa stimulates active secretion of chloride and inhibits absorption of sodium. Similar effects are observed with agents that raise intracellular concentrations of cAMP, i.e., theophylline (an inhibitor of cAMP degradation) or prostaglandins. Moreover, the enterotoxin increases the level of cAMP in mucosal epithelial cells by activating the enzyme adenyl cyclase which converts ATP into cAMP. In canine cholera inhibition of adenyl cyclase (by ethacrynic acid) significantly decreases the fluid and electrolyte loss caused by enterotoxin.

Immunity. Cholera occurs more often when the natural resistance of the host has been impaired; malnutrition seems to be a factor in epidemics. Following recovery specific Abs are demonstrable in both the serum (agglutinating, bactericidal, and toxin-neutralizing) and the feces (coproantibodies). Human volunteers recovered from cholera had Ab to, and a high degree of resistance to rechallenge with, the homologous organism, but not heterologous strains. The immunity induced by vaccination with killed organisms persists for only a few months and is therefore of limited value.

Purified enterotoxin is highly antigenic and toxoid prepared from it protects rabbits and dogs against virulent cholera vibrios, suggesting the potential value of antitoxic immunization in man.

Laboratory Diagnosis. Provisional, rapid recognition of *V. cholerae* can be made by darkfield microscope observation of the characteristic motility of the comma-shaped organisms in a fecal sample. Upon addition of group- and type-specific antisera homologous strains will be clumped and immobilized. In cultures alkaline peptone water (pH 8.4), inoculated with a few drops of feces, allows much more rapid growth of vibrios than of Enterobacteriaceae; smears made after 6 to 8 hours can be stained with fluorescein-labeled specific antisera for quick presumptive identification. The feces should

also be streaked on thiosulfate-citrate-bile-salt-sucrose (TCBS) and taurocholate gelatin agars (TGA): the usual enteric media are suboptimal and often inhibitory for growth of *V. cholerae.** For final identification the organisms in typical colonies are subjected to biochemical (Table 29-3) as well as agglutination tests.

A gram-negative rod (isolated from diarrheal feces) that is oxidase-positive, gives rise to an acid butt–acid slant reaction in TSI agar (see above), and has lysine and ornithine decarboxylase, should be suspected of being a cholera vibrio. Regardless of the outcome, or availability, of agglutination tests, a subculture of such an organism should be submitted promptly to the state public health laboratory for verification. The so-called "noncholera" vibrios, and organisms in the genus *Aeromonas* (see below), may be misidentified by biochemical tests as *V. cholerae*. Demonstration of a rise in serum agglutinins may be of some value in establishing retrospective diagnosis.

Treatment. Prompt and adequate replacement of the lost fluid and electrolytes is mandatory. Response to intravenous isotonic sodium chloride, supplemented with appropriate amounts of sodium bicarbonate (or sodium acetate) and potassium chloride is dramatic; indeed, prompt and adequate intravenous fluid therapy is life-saving in virtually all cholera patients. Tetracycline therapy results in a significant (up to 60%) reduction in gastrointestinal fluid loss, and also eliminates *V. cholerae* from the feces of a large majority of patients. Chloramphenicol and furazolidone also reduce fluid losses, but are somewhat less effective. Antimicrobial therapy is thus helpful but is not a substitute for adequate replacement therapy; even with antibiotic therapy the hospitalized adult patient may still require up to 10 liters of intravenous fluids. Since intraluminal glucose enhances absorption of sodium by the small intestine, maintenance therapy with an orally administered solution of glucose and electrolytes (after initial intravenous rehydration in severely ill patients) can be used where intravenous fluids are limited; oral therapy seems to be adequate in patients with less severe disease.

---

* Imported cholera cases often would not be recognized in the average United States clinical laboratory, because of the inadequacy of the methods employed and the low index of suspicion.

Prevention. Cholera is spread primarily by contaminated water and food, and by contact with infected patients and carriers. Although explosive water-borne outbreaks occasionally occur the disease is most commonly disseminated by convalescent carriers or individuals with inapparent infection, whose fecal matter contaminates water supplies; it is persistently endemic in Calcutta. Mild forms of the disease often go unrecognized, making prevention of new infections difficult. Unlike shigellosis (see above), cholera is not easily spread by person-to-person contact under reasonable conditions of personal hygiene, for a huge number of organisms (10^8 to 10^{10}) are needed to cause illness, as shown in volunteers.

The tendency of cholera to occur in destitute, undernourished persons, living under conditions of poor sanitation, has already been stressed. Improved sanitary control of water and food and improved sewage disposal often stop the spread of the disease. Patients with cholera should be isolated and adequately treated, and their excreta should be disinfected. Contacts should have stool cultures done but need **not** be quarantined. Contacts who are culture-positive or are symptomatic should be treated with 1 gm of tetracycline daily for 5 days. Vaccination with killed cholera vibrios, which reduces the risk to the individual for a few months, is generally used in areas where the disease is endemic, but it plays only a relatively minor role in preventing the spread of cholera across borders; the only effective method for preventing the spread of disease is improvement of basic sanitation.

OTHER VIBRIOS

A group of vibrios very similar biochemically and morphologically to the cholera vibrio, but not agglutinated by cholera polyvalent O antiserum, are commonly referred to as **noncholera vibrios** (NCV); they are agglutinated by their homologous antisera. These vibrios may cause a mild diarrhea or a frank cholera-like disease.

Vibrio parahaemolyticus,* a marine organism, resembles the cholera vibrio in many respects but grows best in the presence of 2 to 6% sodium chloride. It is found with increasing frequency in cases of "food poisoning" in various countries, including the United States, where sea foods are regularly consumed. In Japan it accounts for about half of the cases of bacterial food poisoning. *V. parahaemolyticus* infections of the extremities, eyes, and ears, as well as the bloodstream, have recently been recognized in the United States. These usually occur in persons cut or scratched by the sharp edges of clams or oysters embedded in the sand of marine shore areas. A total of 11 O and 53 K (envelope) Ags are currently recognized.

V. fetus infections occur in cattle, sheep, and goats, resulting in abortions and sterility in infected herds. Human infections due to *V. fetus,* unknown before 1947, are recognized with increasing frequency in infants, pregnant women, and elderly persons. In the great majority of cases the organism is isolated from the bloodstream, but it has also been recovered from the placenta, synovial fluid, and spinal fluid. The organism grows in an atmosphere with about 10% CO_2 or in an anaerobic environment, but not in ordinary air. *V. fetus* reduces nitrates to nitrites and is oxidase- and catalase-positive, but does not ferment carbohydrates; nor does it produce urease, gelatinase, or hydrogen sulfide.

So-called **related vibrios,** morphologically similar to, but serologically distinct from, *V. fetus,* also cause disease in man. Unlike *V. fetus,* they grow better at 42° than at 37°.

AEROMONAS

Organisms of the *Aeromonas* group (aeromonads) are found in natural water sources and soil and are frequent pathogens for cold-blooded marine and fresh-water animals. *A. hydrophila* and *A. (Plesiomonas) shigelloides* ferment glucose and may easily be mistaken for *E. coli* because of their similar reactions in TSI agar and in the indole and urease tests; some strains ferment lactose (Table 29-3). However, they are **oxidase-positive** and possess a single polar flagellum. The combination of sugar fermentations and a positive oxidase test is seen only in aeromonads and vibrios (Table 29-3).

Aeromonads have been associated in man with septicemia, pneumonia, and moderate to severe gastroenteritis. Their recognized incidence in serious human disease has been steadily increasing, and many more cases are probably misdiagnosed as due to coliforms. The organisms have also been isolated from urine, sputum, feces, and bile without evident pathogenic significance.

Aeromonads are resistant to penicillin and ampicillin; most strains are sensitive to gentamycin, the tetracyclines, and colistin.

---

* Initially named *Pasteurella parahaemolytica.*

THE NONFERMENTERS

PSEUDOMONAS

The organisms in the genus *Pseudomonas* are mostly free-living bacteria widely distributed in soil and water, while some are parasites (and pathogens) of plants. Of the many known species, a small (but growing) number is associated with disease, often severe, in man; the most important of these opportunistic pathogens is *P. aeruginosa. P.* (formerly *Actinobacillus*) *pseudomallei,* the agent of melioidosis, and *P. (Actinobacillus) mallei,* causing glanders, are now also classified in this genus.

Most pseudomonads resemble the enteric bacilli, aeromonads, and vibrios morphologically and they grow well on differential enteric media. However, their energy-yielding metabolism is respiratory, not fermentative. *Pseudomonas* organisms are oxidase-positive (except for *P. maltophilia*) and, with rare exceptions, motile by polar flagellation (Table 29-1); they are strict aerobes, except for those species which can use denitrification as means of anaerobic respiration. Some strains produce water-soluble yellow-green fluorescent pigments, while others synthesize, in addition, various type-specific phenazine pigments; many others, however, are non-pigmented.

PSEUDOMONAS AERUGINOSA AND RELATED FORMS

Though *P. aeruginosa* (long known as *Bacillus pyocyaneus*) was frequently isolated from various lesions in man, its presence in human infections was considered insignificant. However, because of the resistance of the organism to many antibiotics it often becomes dominant when the more susceptible bacteria of the normal flora are suppressed. Hence after the introduction of broad-spectrum antibiotics *P. aeruginosa* came to be recognized as a major agent of hospital-acquired infections, especially in persons debilitated by chronic illness and in those treated (or overtreated) with wide-spectrum antibiotics or corticosteroids. Treatment of such opportunistic *P. aeruginosa* infections often fails, and the mortality rate in *Pseudomonas* septicemia may exceed 80%.

This common soil and water organism is a resident of the intestinal tract in only about 10% of healthy individuals, and it is found sporadically in moist areas of the human skin (axilla, groin) and in the saliva. Its nutritional requirements are simple (including use of NH_3 as a source of nitrogen) and it can metabolize a large variety of carbon sources. *P. aeruginosa* can thus multiply in almost any moist environment containing even trace amounts of organic compounds, e.g., eyedrops, weak antiseptic solution, soaps, anesthesia and resuscitation equipment, sinks, fuels, humidifiers, and even stored distilled water.

P. aeruginosa grows readily on standard laboratory media, at temperatures up to 42°. On blood agar strains from clinical material are usually β-hemolytic. Most strains produce a bluish-green phenazine pigment **pyocyanin** (Gr., "blue pus"), as well as **fluorescein,** which is greenish-yellow and fluoresces;[*] the pigments diffuse into and color the medium surrounding the colonies. About 10% of strains do not form pigment. The useful biochemical reactions are shown in Table 29-8.

The organism is oxidase-positive, produces acid oxidatively from glucose,[†] reduces nitrate to gaseous nitrogen, and oxidizes gluconate to 2-keto-3-deoxygluconate (an intermediate in the Entner-Doudoroff pathway[‡]), detectable by appearance of a reducing substance in the medium.

Encapsulated, mucoid strains (resembling *Klebsiella* organisms) are most commonly isolated from the sputum of patients with cystic fibrosis. Individual strains of *P. aeruginosa* can be identified in epidemiological studies by phage typing, as well as by pyocin typing (analogous to colicin typing; see above). At least 13 antigenic groups are known. **Serological typing** may become important, since at-

---

[*] Wood's ultraviolet light can be used for early detection of *P. aeruginosa* infection of burns and wounds.

[†] In TSI medium no change occurs in the slant or butt (Enterobactericeae screening media, see above).

[‡] This pathway for hexose dissimilation is present in the majority of gram-negative bacteria that have over 50% GC in their DNA.

TABLE 29-8. Some Differential Characteristics of Nonfermentative Gram-negative Bacilli

	P. aeruginosa	P. fluorescens	P. pseudomallei	P. mallei	P. alcaligenes	P. acidovorans	P. stutzeri	P. maltophilia	A. anitratus (Herellea)*	A. lwoffi (Mima)†	Flavobacterium spp.	Alcaligenes
Fluorescein	+	+	−	−	−	−	−	−	−	−	−	−
O-F medium	O	O	O	O	N	N	O	O	O	N	O/F	N
MacConkey agar	+	+	+	−	+	+	+	+	+	+/−	+/−	+
Motility	+	+	+	−	+	+	+	+	−	−	−	+/−
Oxidase	+	+	w+	w+	+	+	+	−	−	−	+	+
Growth: 42°	+	−	+	+	+	−	+/−	+/−	+	+/−	−	+/−
Growth: 4°	−	+	−	−	−	−	−	−	−	−	−	?
Gluconate oxidation	+	+/−	−	−	−	−	−	−	−	−	−	−
Glucose (O-F)	+	+	+	+	−	−	−	−	+	−	+/−	−
Gelatinase	+	+	+	−	−	−	−/+	+	−	−/+	+	−
Arginine dihydrolase	+	+	+	+	+	−	−	−	−	−	−	−

* Acid from 10% lactose medium.
† No acid from 10% lactose medium.
O = oxidative, N = nonoxidative, F = fermentative, ? = not determined, +/− = variable, + = positive, − = negative, w = weak.

tempts are being made to employ immunotherapy (see below).

The mechanism of pathogenicity is not clear, but the organism produces endotoxin and a number of extracellular products (lecithinase, collagenase, lipase, hemolysin) that may be of pathogenic significance.

P. aeruginosa causes urinary tract infection, burn and wound infection, septicemia, abscesses, and meningitis. Bronchopneumonia and subacute bacterial endocarditis are also increasing in frequency. For some years the only useful antibiotics were polymyxin B, colistin, and gentamycin, which are more or less toxic. More recently the less toxin carbenicillin has been found valuable, but highly resistant strains have appeared. In life-threatening situations carbenicillin is used in combination with gentamycin. Immunotherapy, both active and passive, has recently been used in burn patients, in whom *P. aeruginosa* infection is often fatal; the results have been encouraging.

Other occasional opportunistic pathogens of the *Pseudomonas* group are *P. fluorescens, P. maltophilia, P. multivorans (cepacia), P. stutzeri, P. acidovorans,* and *P. alcaligenes.* Table 29-8 lists some of their differential characteristics. These organisms are generally more susceptible to antibiotics than is *P. aeruginosa.*

PSEUDOMALLEI GROUP

The pseudomallei group of the genus *Pseudomonas* consists of *P. mallei,* the causative agent of glanders, and *P. pseudomallei,* the agent of melioidosis.

Glanders is a severe infectious disease of horses that can be transmitted to man; it is now extremely rare in the Western world, but still occurs in Asia, Africa, and parts of the Middle East. The disease is characterized by nodular, eventually necrotic involvement of the nasal mucous membranes, lymphatics, lymph nodes, and skin, or by an acute or chronic pneumonitis.

Melioidosis (Lat. *malleus* = severe disease) has long been known as a rare glanders-like disease of man and other animals in Southeast Asia; it has been observed in Americans who have returned from combat in Vietnam. The causative agent, *P. pseudomallei,* is a common free-living inhabitant of soil and water in certain tropical and subtropical regions; man is infected through contamination of skin abrasions and wounds or by inhalation of dust. Direct transmission from man to man or from animals to man does not seem to occur.

Both organisms grow well on standard laboratory media but not on the selective "enteric" media (deoxycholate and SS agars). They are oxidase-positive, but unlike *P. aeruginosa* they do not produce soluble pigment. *P. pesudomallei* colonies on glycerol-nutrient agar become characteristically wrinkled after several days of incubation. *P. mallei*, unlike other pseudomonads, is nonmotile. Other differential characteristics of the two organisms are shown in Table 29-8.

The two organisms are closely related: their DNAs have the same base composition (GC 69%) and hybridize extensively. Complement-fixation and hemagglutination tests with soluble antigens obtained from *P. pseudomallei* cells are of great value in the diagnosis of active melioidosis and the detection of subclinical infections; the latter are widespread in the indigenous Vietnamese population.

The clinical manifestations of melioidosis range from a relatively benign pulmonary infection, often mimicking tuberculosis or mycotic infection, to rapidly fatal septicemia, characterized by multiple abscesses in every organ system except the gastrointestinal tract. Latency and recrudescence are frequent. The disease can appear years after exposure, especially when host resistance is reduced.

Pulmonary melioidosis has been successfully treated with appropriate antibiotics, determined by antibiotic sensitivity testing. Tetracycline is considered the drug of choice; in severe septicemia it is combined with chloramphenicol. Patients require treatment over many months in order to prevent relapse. Gentamycin, the polymyxins, and penicillin and its derivatives are ineffective.

OTHER NONFERMENTERS

ACINETOBACTER

Organisms in the genus *Acinetobacter** are widely distributed in nature (water, soil, milk). They are also carried on the skin and

---

* The chaotic classification and nomenclature of the organisms variously described as *Achromobacter, Moraxella, Mima,* and *Herellea,* has recently been clarified by exhaustive nutritional and genetic studies. The designations used here are gradually appearing in the clinical literature.

in the gastrointestinal, genital, and respiratory tracts of up to 25% of healthy individuals. In the last decade there has been a steady increase in their incidence in a variety of infections, often hospital-acquired (meningitis, septicemia, and wound, genital, and urinary tract infections). Virulence of *Acinetobacter* organisms appears to be low, and they are mainly opportunistic pathogens, but severe primary infections (meningitis, septicemia) do occur in children and young adults in apparently good health.

The organisms are nonmotile, plump, paired gram-negative rods ("diplobacilli"). In the stationary phase they often appear as diplococci, easily mistaken for neisseriae (Ch. 28). They are obligately aerobic but oxidase-negative.

Acinetobacter lwoffi (previously *Mima polymorpha* and *Achromobacter lwoffi*) neither ferments nor oxidizes carbohydrates, while *Acinetobacter anitratus* (previously *Herellea vaginicola* and *Achromobacter anitratus*) utilizes glucose oxidatively and produces acid from 10% (but not from 1%) lactose-containing medium (Table 29-8).

Antibiotics of choice are kanamycin, gentamycin, polymyxin B or colistin, and the tetracyclines. Antibiotic susceptibility tests are required.

MORAXELLA

Moraxellas are similar to *Acinetobacter* (Table 29-8) but are oxidase-positive and highly sensitive to penicillin; they are primary animal parasites, most commonly present on the mucous membranes. Most of these organisms are nutritionally exacting and do not utilize carbohydrates.

M. osloensis and *M. nonliquefaciens* (collectively, *Mima polymorpha* var. *oxidans* in the American clinical literature) are members of the normal flora of man and also are opportunistic pathogens, isolated in pneumonia, septicemia, otitis, urethritis, and rarely, meningitis. Because of their microscopic appearance and positive oxidase reaction they are easily confused with *Neisseria gonorrheae* and *N. meningitidis* (Ch. 28). Their etiological significance in mixed cultures is uncertain. *M. lacunata* (the Morax-Axenfeld bacillus), rarely encountered, causes conjunctivitis and corneal infections.

ALCALIGENES

A. faecalis (Table 29-8) fails to ferment or oxidize any of the usual carbohydrates; hence its oxidation of organic acids or amino acids makes the medium more alkaline. It may occasionally be confused on initial isolation with other non-lactose-fermenters, i.e., *Salmonella* or *Shigella*. It may be encountered in feces or in sputum as a harmless saprophyte, but it has also been associated with serious infections. Since it is hardy, the organism has been involved in contamination of irrigating fluids and intravenous solutions, causing epidemics of urinary tract infections and postoperative septicemia. *A. faecalis* is oxidase-positive and usually, but not always, motile.

These organisms are not uniformly sensitive to any antibiotic; tetracycline and chloramphenicol seem to be the most effective.

FLAVOBACTERIUM

Flavobacteria are widely distributed in soil and water and are encountered as opportunistic pathogens in man. The members of this heterogeneous group are usually oxidative but may be fermentative, form a yellow pigment (hence the name), are oxidase-positive, and usually are nonmotile (Table 29-8).

F. meningosepticum has high virulence for the newborn, especially the premature, in whom it causes epidemics of septicemia and meningitis with a very high fatality rate. While these infections are usually attributed to contaminated hospital equipment and solutions, recent isolation of *F. meningosepticum* from the female genitals suggests this possible source. The organism also occurs in postoperative bacteremia of adults, in whom the illness is much milder. Other flavobacteria are often recovered from the sputum of grossly debilitated patients, but their etiological role is uncertain.

Infants with fatal flavobacterial meningitis have sometimes had temperatures below normal, whereas adults with flavobacterial septicemia usually have high temperatures and recover rapidly. Since many *F. meningosepticum* strains cannot grow at 38° it has been suggested that body temperature is an important factor in the response of infants with flavobacterial meningitis. The organism has an unusual antibiotic sensitivity pattern for a gram-negative bacillus: it is susceptible in vitro to erythromycin, novobiocin, and rifampicin, and to a lesser degree to chloramphenicol and streptomycin and it is resistant to gentamycin and colistin. The infection in infants usually defies antibiotic therapy.

THE OBLIGATE ANAEROBES

BACTEROIDES

The common term "bacteroides" denotes a heterogeneous, large group of nonsporeforming, obligately anaerobic, mostly nonmotile, gram-negative bacilli, which are numerically preponderant in the gastrointestinal tract of man and animals. In the human bowel the gram-negative anaerobes outnumber coliforms at least 100:1. They are also present in the mouth, nasopharynx, oropharynx, vagina, urethra, and on the external genitalia.

These obligate parasites of mucous membranes are separated into the genera *Bacteroides* and *Fusobacterium*.* Originally identified on morphological grounds, they are now differentiated on the basis of metabolic activities. Fusobacteria produce butyric acid as a major product from glucose fermentation, while most bacteroides instead produce combinations of succinic, lactic, acetic, formic, or propionic acids; some produce a mixture of butyric and acetic acids. Patterns of suscepti-

---

* The family Bacteroidaceae also includes the genus *Leptotrichia,* a group of oral organisms.

bility to various antibacterial agents are also used as additional criteria for classification (see below).

Little is known about the **physiological role** of bacteroides. They can deconjugate bile salts, which might alter fat absorption in certain pathological conditions. The organisms may have an important role in our defense against infection: mice become much more susceptible to experimental *Salmonella* infection after the normal *Bacteroides* count in their colon is reduced by treatment with oral streptomycin.

Infections due to *Bacteroides,* once thought to be rare, are now encountered with greatly increased frequency, at least partly because of improvements in anaerobic methodology. Contrary to popular belief, they are at least 10 times as common as infections due to spore-forming anaerobes (Clostridia, Ch. 34). Bacteroides are commensal, noninvasive organisms, but under certain circumstances they can cause severe, often life-threatening infections ("potential pathogens"), mostly in proximity to the mucosal surfaces where they exist normally.

Bacteroides are responsible for about 10%

of gram-negative bacteremias in the United States; they can form abscesses in all regions and organs of the body, and they cause such diverse infections as peritonitis, endocarditis, septic arthritis, uterine sepsis after abortion, and wound infections after bowel surgery or human and animal bites. Some of these infections respond to surgical drainage alone or together with antibacterial therapy, but many result in prolonged hospitalization and even death. Bacteroides are present in pure culture (in about ⅓ of infections) or together with various other anaerobes (particularly cocci) and/or aerobes.

Many bacteroides (as well as other anaerobes) give rise to putrid, foul-smelling discharge and may produce gas in tissue. Presence of either clinical sign strongly suggests anaerobic infection by bacteroides (or clostridia).* The absence of foul odor or gas, however, does not exclude bacteroides infection. Vincent's angina or fusospirochetal diseases, believed by many to be chiefly due to "fusiform" bacilli and spirilla, may be primarily caused by *B. melaninogenicus* (see below). Bacteroides infections are not communicable.

Major predisposing factors are surgical or accidental trauma, edema, anoxia, and tissue destruction (resulting from malignancies or from infection with aerobic organisms). Under such circumstances the anaerobes can invade the tissues and multiply there rapidly. Other predisposing factors are prior antibiotic treatment,† cytotoxic or immunosuppressive therapy, and diabetes.

Some strains produce collagenase or heparinase; all possess endotoxin but there is no evidence for any exotoxin. The organisms are serologically divisible into a number of antigenic groups and types, and Abs are present in healthy individuals; a rise in titer occurs in infections. However, no diagnostically useful serological procedures are yet available.

B. fragilis is the most prevalent species in feces and also in infections. Its growth is stimulated by bile, and it is resistant to penicillin, kanamycin, neomycin, and polymyxin

B. The coccoid *B. melaninogenicus* is recognized by its formation of a brown to black pigment on blood agar. It is inhibited by bile, many strains require hemin and menadione (vitamin K_3) for growth, and it is sensitive to penicillin. The organism is frequently associated, often in mixed infections, with brain and lung abscesses. *B. corrodens,* generally isolated from the oral cavity and from related infections, is sensitive to penicillin, polymyxin B, and colistin. *B. oralis* is rarely isolated from infections.

Fusobacterium fusiforme organisms are characteristically slender rods with sharply tapered ends, while *F. necrophorus* (formerly *Sphaerophorus necrophorus, B. funduliformis*) are pleomorphic, frequently filamentous, and often show free round forms. Both species seem to be more prevalent in the mouth and upper respiratory tract than in the gastrointestinal tract, and both are sensitive to penicillin. *Fusobacterium* infections are rarer than those caused by *Bacteroides*.

Laboratory diagnosis of bacteroides infections depends on careful collection of specimens with as little exposure to oxygen as possible.‡ Because these organisms are prevalent on mucous membranes, special procedures may be required to avoid contamination with this normal flora (transtracheal needle aspiration or direct lung puncture to obtain a reliable sample from pneumonia or lung abscess, uncontaminated by oral flora; urine for anaerobic culture, rarely needed, should be obtained by percutaneous suprapubic bladder puncture). Anaerobic (as well as aerobic) cultivation should be prompt, on prereduced (free of oxygen or of oxidized products) or fresh enriched and selective media, i.e., blood agar, laked (hemolyzed) blood agar with menadione, chopped meat, or blood agar with kanamycin and vancomycin or other appropriate antibiotics for inhibiting aerobes. The **direct Gram stain** of the specimen is of great value: the distinctive morphology of some bacteroides is an early clue to their presence; and seeing organisms that fail to grow aerobically strongly suggests anaerobic infection.

---

* Neither *E. coli* nor *Streptococcus faecalis* produces putrid or fecal odor.

† So-called "preoperative bowel sterilization" with oral, non-absorbable drugs, such as kanamycin or neomycin, temporarily eliminates practically all aerobes and thus may foster invasiveness of the more resistant bacteroides.

---

‡ Since exposure to air for even a few minutes may drastically reduce the numbers of viable anaerobes swabs may be transported to the laboratory in tubes filled with O_2-free CO_2; inoculation should take place within minutes if possible.

The most useful antibiotics are chloramphenicol and clindamycin. About half the *Bacteroides* strains are now resistant to tetracycline, formerly regarded as the drug of choice. Penicillin is the agent of choice in *B. melaninogenicus, F. fusiforme,* and *F. necrophorus* infections and can be used against other strains shown to be sensitive. Kanamycin, gentamycin, streptomycin and cephalosporins are ineffective.

APPENDIX

FREQUENCY OF GRAM-NEGATIVE RODS ISOLATED FROM HUMAN INFECTIONS

The following tabulation lists gram-negative rods isolated from **extraintestinal** specimens during 1970 at the Jewish Hospital of Saint Louis, Missouri (a 530-bed, general, university teaching hospital; no pediatrics). Chapter numbers indicate the location of discussion in this book.

Ch. 29		Ch. 30	
E. coli	2293*	*Hemophilus*	26
Proteus mirabilis	1092	*Bordetella*	0
Klebsiella spp.	1075		
Pseudomonas		Ch. 31	
aeruginosa	602	*Pasteurella multocida*	2
Enterobacter spp.	376	*Yersinia*	0
Citrobacter	250		
Proteus, indole-positive	196	Ch. 32	
Providencia	165		
Bacteroides spp.	149	*Brucella*	0
Serratia marcescens	101		
Pseudomonas, other			
than *aeruginosa*	61		
Acinetobacter			
(*Mima* and *Herellea*)	58		
Salmonella	22 (21)†		
Shigella	20 (20)		
Aeromonas	12 (4)		
Moraxella	2		
Edwardsiella	2 (1)		
Arizona	1 (1)		
Vibrio, "related"	1		

* Number of isolates.
† Numbers in parentheses indicate isolations from feces.

SELECTED REFERENCES

Books and Review Articles

ANDERSON, E. S. The ecology of transferable drug resistance in the enterobacteria. *Annu Rev Microbiol 22*:131 (1968).

BRENNER, D. J., and FALKOW, S. Molecular relationships among members of the Enterobacteriaceae. *Adv Genet 16*:81 (1971).

Committee on Salmonella, National Research Council. *An Evaluation of the Salmonella Problem.* National Academy of Sciences, Washington, D.C., 1969.

FEINGOLD, D. C. Hospital-acquired infections. *N Engl J Med 283*:1384 (1970).

FINEGOLD, S. M., ROSENBLATT, J. E., SUTTER, V. L., and ATTEBERY, H. R. In *Anaerobic Infections.* (Scope monograph.) Upjohn, Kalamazoo, Mich., 1972.

FINLAND, M. The changing ecology of bacterial infections as related to antibacterial therapy. *J Infect Dis 122*:419 (1970).

GORBACH, S. L. Intestinal microflora. *Gastroenterology 60*:1110 (1971).

GRADY, G. F., and KEUSCH, G. T. Pathogenesis of bacterial diarrheas. I, II. *N Eng J Med 285*:831, 891 (1971).

LÜDERITZ, O., WESTPHAL, O., STAUB, A. M., and NIKAIDO, H. Isolation and Chemical and Immunological Characterization of Bacterial Lipopolysaccharides. In *Microbial Toxins,* vol 4, p. 145. (G. Weinbaum, S. Kadis, and S. J. Ajl, eds.) Academic Press, New York, 1971.

MITSUHASHI, S. The R factors. *J Infect Dis 119*:89 (1969).

NETER, E. Endotoxins and the immune response. *Curr Top Microbiol Immunol 47*:82 (1969).

NOMURA, M. Colicins and related bacteriocins. *Annu Rev Microbiol 21*:257 (1967).

ROBBINS, P. W., and WRIGHT, A. Biosynthesis of O Antigens. In *Microbial Toxins,* vol 4, p. 351. (G. Weinbaum, S. Kadis, and S. J. Ajl, eds.) Academic Press, New York, 1971.

SANDERSON, K. E. Genetic homology in the Enterobacteriaceae. *Adv Genet 16*:35 (1971).

SONNENWIRTH, A. C. Gram-negative Bacilli, Vibrios and Spirilla. In *Gradwohl's Clinical Laboratory Methods and Diagnosis,* 7th ed., p. 1269. (S. Frankel, S. Reitman, and A. C. Sonnenwirth, eds.) Mosby, St. Louis, 1970.

STANIER, R. Y., PALLERONI, N. J., and DOUDOROFF, M. The aerobic pseudomonads: A taxonomic study. *J Gen Microbiol 43*:159 (1966).

Specific Articles

CARPENTER, C. C. J. Cholera and other enterotoxin-related diarrheal diseases. *J Infect Dis 126*:551 (1972).

DUPONT, H. L., FORMAL, S. B., HORNICK, R. B., SNYDER, M. J., LIBONATI, J. P., SHEAHAN, D. G., LABREE, E. H., and KALAS, J. P. Pathogenesis of *E. coli* diarrhea. *N Engl J Med 285*:1 (1971).

DUPONT, H. L., HORNICK, R. B., SNYDER, M. J., LIBONATI, J. P., FORMAL, S.B., and GANGAROSA, E. J. Immunity in shigellosis. I, II. *J Infect Dis 125*:5, 12 (1972).

EDMONDSON, E. B., and SANFORD, J. P. The *Klebsiella-Enterobacter* (Aerobacter)-*Serratia* group. A clinical and bacteriological evaluation. *Medicine 46*: 323 (1967).

ETKIN, S., and GORBACH, L. S. Studies on enterotoxin from *Escherichia coli* associated with acute diarrhea in man. *J Lab Clin Med 78*:81 (1971).

FELNER, J. M., and DOWELL, V. R. "Bacteroides" bacteremia. *Am J Med 50*:787 (1971).

FIELD, M. Intestinal secretion: Effect of cyclic AMP and its role in cholera. *N Eng J Med 284*:1137 (1971).

FIELDS, B. N., UWAYDAH, M. M., KUNZ, L. J., and SWARTZ, M. N. The so-called "paracolon" bacteria:

A bacteriologic and clinical reappraisal. *Am J Med 42*:89 (1967).

FINEGOLD, S. M. Gram-negative Anaerobic Rods—Bacteroidaceae. In *Gradwohl's Clinical Laboratory Methods and Diagnosis,* 7th ed., p. 1353. (S. Frankel, S. Reitman, and A. C. Sonnenwirth, eds.) Mosby, St. Louis, 1970.

GALE, D., and SONNENWIRTH, A. C. Frequent isolation of *Serratia marcescens* from patients in a general hospital: Bacteriological and pathogenicity studies of 12 strains. *Arch Intern Med 109*:414 (1962).

GARDNER, P., GRIFFIN, W. B., SWARTZ, M. N., and KUNZ, L. J. Nonfermentative gram-negative bacilli of nosocomial interest. *Am J Med 48*:735 (1970).

HORNICK, R. B., GREISMAN, S. E., WOODWARD, T. E., DUPONT, H. L., DAWKINS, A. T., and SNYDER, M. J. Typhoid fever: Pathogenesis and immunologic control. *N Engl J Med 283*:686 (1970).

HORNICK, R. B., MUSIC, S. I., WENZEL, R., CASH, R., LIBONATI, J. P., SNYDER, M. J., and WOODWARD, T. E. The Broad Street pump revisited: Response of volunteers to ingested cholera vibrios. *Bull NY Acad Med 47*:1181 (1971).

HOWE, C., SAMPATH, A., and SPOTNITZ, M. The pseudomallei group: A review. *J Infect Dis 124*: 598 (1971).

KEUSCH, G. T., and GRADY, G. F. The pathogenesis of *Shigella* diarrhea. 1. Enterotoxin production by *Shigella dysenteriae. J Clin Invest 51*:1212 (1972).

MOORE, W. E. C., CATO, E. P., and HOLDEMAN, L. V. Anaerobic bacteria of the gastrointestinal flora and their occurrence in clinical infections. *J Infect Dis 119*:641 (1969).

PEDERSEN, B. A., MARSO, E., and PICKETT, M. J. Nonfermentative bacilli associated with man. III. Pathogenicity and antibiotic susceptibility. *Am J Clin Pathol 54*:178 (1970).

PIERCE, N. F., GREENOUGH, W. B., and CARPENTER, C. C. J., JR. *Vibrio cholerae* enterotoxin and its mode of action. *Bacteriol Rev 35*:1 (1971).

QUICK, J. D., GOLDBERG, H. S., and SONNENWIRTH, A. C. Human antibody to Bacteroidaceae. *Am J Clin Nutr 25*:1351 (1972).

SONNENWIRTH, A. C. Bacteremia with and without meningitis due to *Yersinia enterocolitica, Edwardsiella tarda, Comamonas terrigena,* and *Pseudomonas maltophilia. Ann NY Acad Sci (Art. 2) 174*:488 (1970).

SONNENWIRTH, A. C., YIN, E. T., SARMIENTO, E. M., and WESSLER, S. Bacteroidaceae endotoxin detection by *Limulus* assay. *Am J Clin Nutr 25*:1452 (1972).

chapter **30**

THE HEMOPHILUS-
BORDETELLA GROUP

Revised by STEPHEN I. MORSE, M.D.

HEMOPHILUS INFLUENZAE   792

　　History   792
　　Morphology   793
　　Metabolism   793
　　Antigenic Structure   794
　　Genetic Variation   794
　　Pathogenesis   795
　　Immunity   795
　　Laboratory Diagnosis   795
　　Treatment   796
　　Prevention   796

OTHER HEMOPHILIC BACILLI PATHOGENIC FOR MAN   797

BORDETELLA PERTUSSIS   797

　　Morphology   797
　　Metabolism   797
　　Variation   797
　　Antigenicity   797
　　Pathogenicity   798
　　Immunity   799
　　Laboratory Diagnosis   799
　　Treatment   799
　　Prevention   799

The organisms of the hemophilus-bordetella group are small, gram-negative, nonmotile, nonsporeforming, aerobic bacilli, which usually require, for primary isolation, enriched media containing blood or its derivatives (Table 30-1). One genus, *Hemophilus,* contains the truly hemophilic species—*H. influenzae. H. parainfluenzae, H. suis. H. aegypticus* (Koch-Weeks bacillus), *H. ducreyi* and *H. hemolyticus*—which need either one or two growth factors (X or V) provided by blood. A second genus, *Bordetella,* consists of *B. pertussis, B. parapertusssis,* and *B. bronchiseptica.*

The most important species of hemophilic bacteria, from a medical standpoint, are *H. influenzae* and *B. pertussis.* The first causes bacterial meningitis in children and the second causes whooping cough. Other species that occasionally produce disease in man include *B. parapertussis* (whooping cough), Koch-Weeks bacillus (conjunctivitis), and the Ducrey bacillus (chancroid).

HEMOPHILUS INFLUENZAE

HISTORY

This prevalent pathogenic species was first isolated by Pfeiffer during the influenza pandemic of 1890. Erroneously thought to be the cause of the disease, it was named the "influenza bacillus" and later designated *H. influenzae.* The primary etiological agent of epidemic influenza is, of course, now known to be the influenza virus (Ch. 56); thus, the precise role played by *H. influenzae* in the pandemics of 1890 and 1918 is not clear. It clearly could be an important secondary invader, since at autopsy it was often the predominant or even the only bacterial species that could be cultured from the lungs; but whether it acted synergistically with the virus to cause particularly malignant lesions remains open to question. In keeping with this possibility was the discovery of Shope, made in 1931, that swine influenza is caused by an influenza virus plus *H. suis,* a bacterial species closely related to *H. influenzae.* When produced experimentally in pigs by inoculation of the respiratory tract, the disease caused by the virus and the bacillus together was found to be more severe than that caused by either agent alone. A similar synergistic effect has been demonstrated in chick embryos infected with the two swine agents or with a human influenza virus strain (C) and *H. influenzae.*

H. influenzae meningitis first described by Slawyk in 1899, is the most common form of bacterial meningitis in children. *H. influenzae* may also cause a fatal epiglottitis (see below). The organism is sometimes isolated from the respiratory tract of children with obstructive bronchiolitis, and from children with otitis media (up to 25%); but its etiological role in those diseases is not certain.

H. influenzae has many properties in common with *Diplococcus pneumoniae.* Both organisms are primarily invasive, gain entrance to the body through the respiratory tract, and have polysaccharide antiphagocytic capsules; in addition, some of the capsular antigens of the two species cross-react. When grown in artificial media both species have a tendency to undergo autolysis and both are bile-soluble. In addition, genetic transformation by DNA occurs with *H. influenzae* as with pneumococcus.

Fig. 30-1. Gram-stained smear of exudate from spinal fluid of patient with *H. influenzae meningitis.* ×1000. (Courtesy of R. Drachman.)

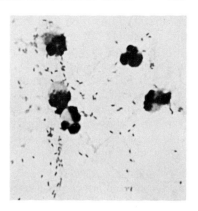

MORPHOLOGY

The outstanding morphological characteristic of *H. influenzae* is its pleomorphism. Although predominantly a small gram-negative cocco-bacillus (1 to 1.5 by 0.3 μ), it often grows in short chains, resembling pneumococci and streptococci, or in long filaments, simulating larger bacilli (Fig. 30-1) Blue-purple bipolar staining is often observed in gram-stained smears, and the organism may be mistaken for *D. pneumoniae* in pathological material.

The organism undergoes profound changes in morphology with aging. Virulent strains have capsules that are readily demonstrable only during the first 6 to 8 hours of incubation in broth cultures. On solid media the presence of encapsulated organisms in 4- to 8-hour colonies causes them to be characteristically iridescent when examined in obliquely transmitted light. After 24 hours of incubation the capsules and iridescence disappear. Signs of

autolysis, including irregular staining and the accumulation of amorphous debris, become evident by 12 hours. Both the autolysis and the capsular destruction apparently result from the action of endogenous enzymes. Immunologically reactive capsular substance is also shed into liquid culture medium.

METABOLISM

Although *H. influenzae* will grow in blood broth and on blood agar, optimal growth occurs in media in which the contents of the red cell have been liberated, either by heat (chocolate agar and Levinthal's medium) or by peptic digestion (Fildes medium). Lysis of the red cells also facilitates the detection of iridescent colonies (see above) on solid media. The optimal pH for growth is 7.6, and multiplication of the organisms in liquid media may be stimulated by aeration. On primary isolation some strains grow best in the presence of 5 to 10% carbon dioxide (candle jar).

The essential growth constituents were first recognized as a heat-stable **X factor** and a heat-labile **V factor.** In 1937 they were both identified by the Lwoffs: X was shown to behave like hematin, and V was found to be replaceable by NAD (DNP), NADP (TPN), or nicotinamide nucleoside. Since the V factor may be rapidly destroyed by enzymes derived from unheated red cells, the use of chocolate agar rather than blood agar is recommended. *H. influenzae* requires both the X and the V factors for aerobic growth, even after having been repeatedly subcultured on artificial media. Under anaerobic conditions, however, it does not require the X factor, which is

TABLE 30-1. Differential Growth Characteristics of the Hemophilus-Bordetella Group

Species	Growth factors		Hemolysis
	X	V	
Genus *Hemophilus*			
H. influenzae	+	+	−
H. parainfluenzae	−	+	−
H. aegypticus	+	+	−
H. ducreyi	+	−	+
H. suis	?	+	−
H. hemolyticus	+	+	+
Genus *Bordetella*			
B. pertussis	−	−	+
B. parapertussis	−	−	±
B. bronchiseptica	−	−	±

involved in respiration. Other essential metabolites required by most strains for optimal
growth include pantothenate, thiamine, uracil,
and cysteine (or glutathione).

The heat-labile V factor may be obtained
from other microorganisms in the immediate
vicinity: this **satellite phenomenon** is illustrated in Figure 30-2. Whether such syntrophy
ever stimulates the growth of *H. influenzae* in
host tissues is not known.

Fermentation reactions of hemophilus organisms
are variable and are of no differential value. Most
strains of *H. influenzae* are capable of converting
tryptophan to indole and of utilizing nitrate as
electron acceptor in the absence of oxygen; these
properties are of diagnostic value.

ANTIGENIC STRUCTURE

As with the pneumococcus, the capsular
carbohydrates of *H. influenzae* evoke protective antibodies. Six types, designated a to f,
have been described. The specific carbohydrates of types a, b, and c are poly-
sugarphosphates, that of type b being a
polyribosephosphate. The individual types are
readily identified by agglutination, precipitation, and quellung tests performed with

specific antisera. Anticapsular antibodies promote phagocytosis and enhance complement-
dependent bactericidal activity of immune
serum.

At least three of the *H. influenzae* capsular antigens are immunologically related to the corresponding antigens of *D. pneumoniae:* type a cross-reacts
with pneumococcus type 6; type b with pneumococcus types 6, 15, 29, and 35; and type c with
pneumococcus type 11.

Much less is known about the somatic antigens of
H. influenzae. A protein (M) antigen common to
all types has been described, as well as an endotoxin
which resembles those from other gram-negative
bacilli.

GENETIC VARIATION

A few noniridescent colonies containing unencapsulated variants of *H. influenzae* are
almost always demonstrable, even in very
early cultures on heavily seeded agar plates.
It is evident, therefore, that the rate of spontaneous S → R mutation is relatively high.
When cultural conditions are suboptimal R
variants will often predominate. Such S → R
shifts have been observed in the upper respiratory tracts of patients recovering from in-
fluenzal meningitis.

Fig. 30-2. Satellite phenomenon: colonies of *H. influenzae* growing only in vicinity of staphylococcal
colonies on agar medium lacking V factor. The agar,
which contained blood autoclaved to destroy the V
factor, was uniformly seeded with a heavy inoculum
of *H. influenzae* and a very light inoculum of *S.
aureus*. (Courtesy of P. H. Hardy and E. E. Nell.)

Transformations mediated by DNA from *H. influenzae* have been extensively studied by Alexander. The traits transferred have included resistance to antimicrobial drugs and the synthesis of specific capsular antigens. Transformations of *H. influenzae* have not yet been shown to occur in vivo, but the demonstration of pneumococcal transformation in host tissues suggests that they may take place.

Transformation reactions are most efficient when the DNA donor and the recipient are of the same species. By this test *H. aegypticus* (Koch-Weeks bacillus) and *H. influenzae* are very closely related.

PATHOGENESIS

Naturally acquired disease due to *H. influenzae* seems to occur only in man, although experimental bronchopneumonia and meningitis have been produced in monkeys and a fatal peritonitis in mice. Diseases due to *H. influenzae* are common in children but rare in adults. The reason is explained under Immunity, below.

Virulence for man is directly related to capsule formation; the organisms produce no demonstrable exotoxin, and there is no conclusive evidence that their endotoxin plays a significant role in pathogenicity. Thus homologous anticapsular antibodies are protective, but antiendotoxic antibodies are not. Virtually all severe infections are caused by type b.

The organism gains entrance to the tissues via the respiratory tract, where it frequently resides without causing trouble. Carrier rates in children may be as high as 30% to 50%, but the organisms are usually unencapsulated. Of the encapsulated strains found, most are type b.

Disease due to *H. influenzae* usually begins as a nasopharyngitis, probably precipitated by a viral infection of the upper respiratory tract. The resulting coryza may be followed by sinusitis or otitis media and may lead to pneumonia; the latter is often complicated by empyema. Bacteremia occurs early in severe cases and frequently results in metastatic involvement of one or more joints or in the development of **acute bacterial meningitis**— the most important clinical entity caused by *H. influenzae*. Indeed, in children *H. influenzae* is the commonest cause of bacterial meningitis, except during epidemics of meningococcal meningitis. The clinical signs are like those of other forms of acute bacterial meningitis. Unless vigorously treated the patient rarely recovers.

A less common but even more serious disease caused by *H. influenzae* type b is epiglottitis and obstructive laryngitis. Its onset is sudden and its course fulminating, often ending fatally within 24 hours. Infection starts in the pharynx and spreads to the epiglottis, which becomes cherry-red and grossly edematous. Laryngeal obstruction ensues. The patient should be immediately hospitalized when the clinical diagnosis is made, for survival may require prompt tracheotomy. Bacteremia is usually a feature of the disease.

IMMUNITY

The incidence of *H. influenzae* meningitis as a function of age is inversely related to the titer of antibody in the blood (Fig. 30-3), whether passively acquired from the mother or actively formed. From the age of 2 months to 3 years antibody levels are minimal; thereafter, antibody increases and the disease becomes much less common.

Like other gram-negative bacilli, *H. influenzae* is susceptible to lysis by antibody and complement. But though the immunological test used in the studies of Figure 30-3 measures primarily bacteriolysis, immunity is not necessarily due to this action. Thus anticapsular antibody, which promotes phagocytosis, is known to be the protective factor in specific antiserum. Furthermore, complement is rarely detectable in the spinal fluids of patients with bacterial meningitis, whereas phagocytosis of influenza bacilli is prominent in the meningeal lesions. Phagocytosis is thus evidently an important defense mechanism in influenza bacillus meningitis, whereas the role of complement remains undefined. The components of the complement system, when present, may conceivably contribute to immunity either by accelerating phagocytosis (Ch. 22) or by participating in a "background" process of immune lysis in vivo, which cannot yet be measured.

LABORATORY DIAGNOSIS

In all cases of suspected bacterial meningitis a sample of blood, as well as of spinal fluid,

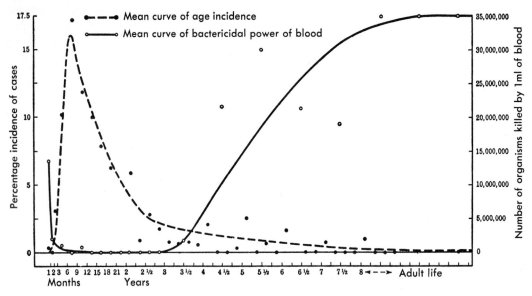

Fig. 30-3. Relation of the age incidence of *H. influenzae* meningitis to bactericidal antibody titers in the blood. [From Fothergill, L. D., and Wright, J. *J Immunol 24:*281 (1933).]

should be cultured. A smear of the spinal fluid, or preferably one made with the sediment from a centrifuged specimen, should be stained by the Gram method. The detection of small, pleomorphic, gram-negative bacilli warrants a provisional diagnosis of influenza bacillus meningitis. If exposure of the organisms in the spinal fluid to specific antiserum results in a positive quellung reaction, the diagnosis is established. The specimen of spinal fluid should be streaked on chocolate agar and incubated in a candle jar. Any organisms suspected of being *H. influenzae* in a blood culture or on a chocolate agar plate should be subjected to the quellung test. If the results are negative, further cultural studies should be performed to determine the requirements of the organism for X and V factors (Table 30-1).* Precipitin tests for free antigen in the spinal fluid may also be performed.

Similar methods are employed in identifying *H. influenzae* in sputum and in purulent exudates.

---

* This test is accomplished with three media: blood (or chocolate) agar contains both X and V factors; autoclaved blood agar has only X factor, the heat-labile V factor having been destroyed; and yeast extract agar provides only V factor.

TREATMENT

Virtually all patients treated early in the course of influenza bacillus meningitis can now be cured. Although type-specific rabbit antiserum is effective, it is no longer used because of its cost and the danger of serum sickness. Ampicillin and chloramphenicol are effective. Some physicians use a sulfonamide in combination with chloramphenicol.

PREVENTION

The incidence of proved *H. influenzae* disease is low, most children acquiring effective natural immunity by age 10. Moreover, the mortality of treated meningitis is less than 10%. Nevertheless, *H. influenzae* meningitis causes 1500 to 2000 deaths per year in the United States, mostly in young children. Moreover, a significant proportion of those who recover have permanent, residual neurological defects; approximately 5% must be institutionalized. For these reasons, the immunizing and protective effects of the polyribosephosphate capsular material from type b, the most frequent pathogen, are now being studied in man. Prophylactic chemotherapy is indicated only in special situations in which young children are at high risk.

OTHER HEMOPHILIC BACILLI PATHOGENIC FOR MAN

H. ducreyi causes chancroid or soft chancre. The paragenital ulcerative lesions produced by this venereal disease lack the firm indurated margins of syphilitic ulcers. Response to sulfonamides and various antibiotics is usually prompt. *H. aegypticus* (the Koch-Weeks bacillus) produces purulent conjunctivitis, and *H. parainfluenzae* is occasionally the cause of bacterial endocarditis. *H. hemolyticus* is nonpathogenic but its β-hemolytic colonies may be easily mistaken for those of *Streptococcus pyogenes* on blood agar, particularly if rabbit blood is employed.

Some differential properties of these organisms are summarized in Table 30-1.

BORDETELLA PERTUSSIS

B. pertussis, the causative agent of whooping cough, was first isolated in 1906 by Bordet and Gengou. It seldom penetrates the mucous membranes of the respiratory tract and consequently is rarely isolated from the blood stream. It does not require either X or V factor for growth.

MORPHOLOGY

Morphologically *B. pertussis* is very similar to *H. influenzae.* It is a small, nonmotile, gram-negative bacillus which forms a capsule when virulent and tends to be pleomorphic, particularly in older cultures.

METABOLISM

An extremely delicate organism, *B. pertussis* survives in vitro for only a few hours in respiratory secretions. The addition to media of blood or blood products, charcoal, starch, or anion-exchange resins is usually necesssary to ensure maximum growth of virulent (phase 1) organisms. These substances probably act by binding fatty acids, which are toxic for the organism. Defined liquid media have yielded adequate growth and preservation of the biological characteristics of some strains. *B. pertussis* does not break down carbohydrates, form indole, or reduce nitrate. Narrow zones of hemolysis surround colonies on blood agar. Bordet-Gengou agar, which contains fresh sheep's blood, is the solid medium most utilized for primary isolation of *B. pertussis* from patients.

VARIATION

When first isolated (phase 1) the organism is a small encapsulated coccobacillus, fully virulent, and possessing filamentous appendages similar to pili. It will not grow on unenriched nutrient agar. Prolonged laboratory passage leads to selection of phase 4 organisms: pleomorphism becomes striking, growth occurs on nutrient agar, and the capsule, virulence, and filaments are lost. The intermediate phases 2 and 3 are not precisely defined. Clearly, this change is equivalent to the classic S → R variation.

ANTIGENICITY

Little definitive information is available concerning the antigens of *B. pertussis,* and the phase 1 antigen which induces specific protection has not been identified. The capsular substance has not been characterized and quellung reactions have not been observed. At least six antigens contribute to bacterial agglutination, but none have been isolated or defined. Bacteria from young cultures will agglutinate erythrocytes from many species, and during growth in liquid medium the hemagglutinin is released into the culture fluid. A heat-labile toxin is produced which induces dermal necrosis in rabbits and is lethal for mice. A heat-stable lipopolysaccharide endotoxin forms part of the cell wall.

Phase 1 cells, as well as cell-free culture supernatant fluids, produce rather unusual biological effects. They induce marked lymphocytosis in experimental animals and cause heightened sensitivity to histamine and serotonin, and anaphylaxis in mice. A single product, not yet isolated in a pure state, is probably responsible for these phenomena.

PATHOGENICITY

Naturally acquired disease due to *B. pertussis* is known only in man. Experimental infections have been produced in various animals (chimpanzees, monkeys, dogs, rabbits, rats, mice, ferrets, and chick embryos).

The remarkable viscerotropic properties of *B. pertussis* that cause it to grow preferentially on ciliated epithelial cells in the bronchi of chick embryos has already been described (Ch. 22, Adaptation to microenvironment). A similar localization is found on the ciliated epithelium of the bronchi and trachea in children dying of whooping cough (Fig. 30-4). In addition, mice are 1000 times more susceptible to *B. pertussis* infection when inoculated intranasally rather than intraperitoneally,

and in intracerebral infection of mice organisms accumulate on the ciliated ependymal cells lining the ventricles.

Although the organism rarely penetrates the mucosa in human disease, subepithelial necrosis and inflammation are characteristic features and are thought to be due to the short-range necrotizing effect of the heat-labile exotoxin. However, the primary determinants of disease production and the mechanisms of immunity have not been delineated.

The disease in children ordinarily begins after an average incubation period of 10 days, with benign nasopharyngeal symptoms, including coryza, sneezing, and a mild cough. After 10 to 14 days the lower respiratory tract becomes involved and paroxysms of severe coughing develop, usually accompanied by characteristic inspiratory "whooping" and often followed by vomiting. Secondary bronchopneumonia is common, and obstruction of the bronchi by mucous plugs may cause severe anoxia, resulting in convulsions. A marked lymphocytosis is often demonstrable

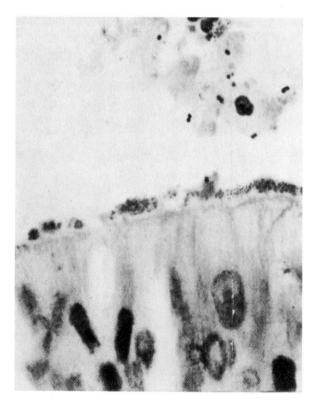

Fig. 30-4. *B. pertussis* lodged among cilia on bronchial epithelium of child dying of whooping cough. ×1000. [From Brown, J. H., and Brenn, L. *Bull Johns Hopkins Hosp 48:*69 (1931).]

during the paroxysmal stage of the disease. In nonfatal cases the paroxysmal stage lasts approximately 2 weeks and is followed by continued cough for another 2 to 3 weeks. Encephalitis occasionally occurs as a late complication. The mortality rate is highest in infants.

Communicability is greatest in the catarrhal stage and the organism is only rarely recovered after the fourth week of illness. A carrier state has not been identified.

IMMUNITY

Naturally acquired immunity against whooping cough is not permanent, but second attacks are usually mild and pass unrecognized; in older persons full-blown whooping cough causes severe distress. The lack of a carrier state probably is of prime importance with respect to natural immunity since a repetitive or continuing antigenic stimulus does not occur. Thus the newborn is highly susceptible to whooping cough, in contrast to *H. influenzae* infection, since maternal serum has little or no protective antibody to the former.

LABORATORY DIAGNOSIS

The time-honored "cough plate" procedure has been supplanted by a nasal swab technic in which a small cotton swab, wrapped on the end of a flexible copper wire, is passed through the nose to the posterior pharyngeal wall, where it is allowed to remain in place while the patient coughs. Upon withdrawal the swab is passed through a drop of penicillin solution, previously placed on the surface of a Bordet-Gengou agar plate, and the drop is then streaked on the plate. Two to three days' incubation at 37° is required for the characteristic small, pearl-white, glistening colonies of *B. pertussis* to form. The growth of contaminants in the inoculum is selectively inhibited by the penicillin. Final identification depends upon agglutination of the organism by pertussis antiserum. A definitive bacteriological diagnosis can be made much more rapidly by the direct staining of nasopharyngeal smears with fluorescein-labeled antibody. Serological tests are of little diagnostic value, because antibodies are usually not detectable until the third week of the disease.

B. parapertussis causes a mild form of whooping cough. Although it shares common antigens with *B. pertussis,* it may be differentiated by fluorescent antibody (or agglutination) tests performed with properly adsorbed antiserum. *B. parapertussis* differs from *B. pertussis* in producing larger colonies, splitting urea, and utilizing citrate as a sole source of carbon. *B. bronchiseptica,* which is associated with bronchopneumonia in rodents and dogs, rarely causes whooping cough in man.

Recently adenoviruses has been isolated from some patients with whooping cough who have no evidence of bordetella infection, and a rise in antibody titer against adenoviruses has also been observed. The significance of these observations is not clear.

TREATMENT

Mild whooping cough requires no specific treatment. The severe form of the disease, particularly in infants, is usually treated with erythromycin, tetracyclines, or chloramphenicol; ampicillin is not effective. The clinical response may not be dramatic, although the drugs rapidly render the patient noninfectious. Secondary pneumonia due to other bacteria may occur and must be treated appropriately. Hyperimmune human γ-globulin is still used occasionally but there are no reliable data on its efficacy.

PREVENTION

Whooping cough occurs throughout the world, and it is estimated that more than 9 persons out of 10 acquire either apparent or inapparent disease, usually in childhood. The disease is highly contagious, and epidemics tend to occur in the United States every 2 to 4 years. As is shown in Table 30-2, the attack rate among nonimmunized children subjected to family exposure of whooping cough is nearly 90%. The degree of protection achieved by active immunization parallels the serum titer of antibody at the time of exposure. When feasible, patients with the disease should be isolated, particularly from young infants, for 4 to 6 weeks or ideally until cultures are negative. Susceptible contacts not previously immunized should receive chemoprophylaxis with erythromycin or a tetracycline. Children under 6 years of age who have previously been vaccinated should receive a booster dose if exposed; erythromycin is often given in addition.

Active immunization of all children at the age of 2 months is strongly advocated. The vaccine consists of either killed phase 1 organisms or a crude extract prepared from them. The immunizing preparation is usually combined with diphtheria and tetanus toxoids. (*B. pertussis* has a marked adjuvant effect in experimental animals and may enhance the immunogenicity of the toxoids.) After the primary course of three injections, at monthly intervals, booster injections are given 1 and 5 years later.

Although the mortality rate for whooping cough in the United States fell markedly in the period 1920 to 1950, its incidence began to fall only in the 1940s, when widespread immunization was begun; the decrease in morbidity attests to the efficacy of the vaccine, as do the data in Table 30-2. "Vaccine-failures" are almost always due to the use of impotent vaccine preparations, although anti-

TABLE 30-2. Incidence of Whooping Cough in Nonimmunized and Immunized Infants Subjected to Intimate Familial Exposure to the Disease

State of immunity	No. of infants	Serum agglutinin titers	% attacked
Unimmunized	438	0	89.7
Immunized	43	<1:20	30.2
	82	1:20	25.6
	91	1:40	18.7
	75	1:80	10.7
	53	1:160	11.3
	62	1:320	0
	48	1:640	0
	39	>1:640	0

From Sako, W. *J Pediatr 30:*29 (1947).

genic shifts of the infecting strain may in some instances be responsible.

Artificial immunization, like natural immunization, is not permanent. There is no protection against *B. parapertussis*.

SELECTED REFERENCES

Books and Review Articles

ALEXANDER, H. E. The hemophilus group. In *Bacterial and Mycotic Infections of Man*. (R. J. Dubos and J. G. Hirsch, eds.) Lippincott, Philadelphia, 1965.

PITTMAN, M. *Bordetella pertussis:* Bacterial and host factors in the pathogenesis and prevention of whooping cough. In *Infectious Agents and Host Reactions* (S. Mudd, ed.) Saunders, Philadelphia, 1970.

TURK, D. C., and MAY, J. R. *Haemophilus Influenzae: Its Clinical Importance*. English Universities Press, London, 1967.

Specific Articles

BUDDINGH, G. J. Experimental combined viral and bacterial infection (influenza C and *Hemophilus influenza,* type B) in embryonated eggs. *J Exp Med 104:*947 (1956).

MICHALKA, J., and GOODGAL, S. H. Genetic and physical map of the chromosome of *Hemophilus influenzae. J Mol Biol 45:*407 (1969).

MORSE, J. H., and MORSE, S. I. Studies on the ultrastructure of *Bordetella pertussis. J Exp Med 131:* 1342 (1970).

MUNOZ, J. J., and BERGMAN, R. K. Histamine-sensitizing factors from microbial agents, with special reference to *Bordetella pertussis. Bacteriol Rev 32:*103 (1968).

WHITE, D. C. Respiratory systems in the hemin-requiring *Hemophilus* species. *J Bacteriol 85:*84 (1963).

ZAMENHOF, S., LEIDY, G., FITZGERALD, P. L., ALEXANDER, H. E., and CHARGAFF, E. Polyribophosphate, the type-specific substance of *Hemophilus influenzae,* type b. *J Biol Chem 203:*695 (1953).

chapter

THE YERSINIAE, FRANCISELLA, AND PASTEURELLA

Revised by MORTON N. SWARTZ, M.D.

YERSINIA PESTIS   802

Morphology   802
Metabolism   803
Antigens   803
Genetic Variation   803
Pathogenicity   803
Immunity   805
Laboratory Diagnosis   805
Treatment   805
Prevention   805

OTHER YERSINIAE   806

FRANCISELLA TULARENSIS   806

Morphology   807
Metabolism   807
Antigens   807
Variation   807
Pathogenicity   807
Immunity   808
Laboratory Diagnosis   808
Treatment   808

PASTEURELLA MULTOCIDA   809

Until very recently the genus *Pasteurella* contained four species of importance in clinical medicine: *P. pestis, P. tularensis, P. multocida,* and *P. pseudotuberculosis.* These organisms produce plague, tularemia, pasteurellosis (hemorrhagic septicemia), and pseudotuberculosis, respectively, primarily in animals but occasionally in humans. The genus has recently undergone reclassification: *P. pestis* and *P. pseudotuberculosis* have been shifted to the genus *Yersinia* (family Enterobacteriaceae, see Ch. 29); *P. tularensis* is now considered a member of the new genus *Francisella* (family Brucellaceae); and the designation *Pasteurella* is now restricted to *P. multocida* and a few other closely related animal pathogens. All these organisms are small, aerobic or facultatively anaerobic, nonsporeforming, gram-negative bacilli.

YERSINIA PESTIS

No infectious disease has created greater havoc in the world than **plague*** The first adequately described pandemic occurred in the sixth century A.D. and is believed to have killed more than 100 million people in its 50-year rampage. In the fourteenth century it again assumed catastrophic proportions and destroyed approximately a quarter of the population of Europe. Because of the severe cyanosis associated with its terminal stages it became known as the **black death.** It spread through Europe into the Middle East, China, and India. Thereafter, with improvement in sanitation and housing, the disease slowly receded in Europe, but serious epidemics continued to occur in other parts of the world. The last pandemic developed in China at the close of the nineteenth century. Outbreaks still occasionally arise in India, China, Vietnam, and Madagascar; sporadic cases are reported in southern Africa, South America, and southwestern United States.

In 1894 Yersin discovered the causative organism of the disease in Hong Kong and named it after his teacher, Pasteur. The role played by fleas in transmitting the infection from rats to man was established in Bombay by the British Plague Research Commission in 1906. More recently, the extensive ecological studies of Meyer in the western United States have led to the discovery of recurrent epizootics of plague among squirrels, prairie dogs, rabbits, and pack rats. This **sylvatic plague** leads to sporadic cases in man and constitutes a potential source of future epidemics.

MORPHOLOGY

The plague bacillus is a gram-negative, nonmotile coccobacillus. When grown under suboptimal conditions, it becomes pleomorphic and may assume long filamentous forms. Its tendency to stain in a bipolar "safety-pin" fashion is best demonstrated by Wayson's stain (methylene blue and carbofuchsin), but can be seen with Gram's stain. Freshly isolated virulent strains, in the smooth (S) phase, produce a generous enveloping slime layer (Fig. 31-1) and form viscous, drop-like colonies which are dark brown when blood is

---

* The term as used here refers to the specific disease caused by *Y. pestis;* it is also often applied generically to any epidemic disease with a high mortality rate.

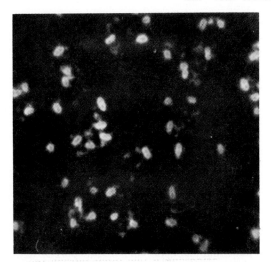

Fig. 31-1. Capsules of smooth phase *Yersinia pestis* stained by the indirect fluorescent antibody technique. ×1200. [From Cavanaugh, D. C., and Randall, R. *J Immunol 83*:348 (1959).]

90,000), which act in combination to render the organism resistant to phagocytosis in the absence of a demonstrable capsule. The V antigen appears to be cell-bound and the W antigen is excreted into the medium. The F1 and the VW antigens are produced during growth at 37° but not at 28°. The so-called murine toxin of *Y. pestis* is intracellular but appears to be independent of endotoxin; it is found as two active proteins: toxin A (MW 240,000) and toxin B (MW 120,000). The former may be a dimer of the latter. The toxin is lethal for the mouse and rat, but other species (rabbit, dog, and monkey) are resistant.

Other ill-defined antigenic components have been extracted from *Y. pestis* (including a lipopolysaccharide endotoxin), but these have not been implicated in pathogenesis.

GENETIC VARIATION

The genetics of virulence of *Y. pestis* have been admirably reviewed by Burrows. Mutants that have lost virulence, for a variety of reasons (see below), are commonly encountered, especially when the organisms are grown under suboptimal conditions. Drug-resistant mutants have been produced in the laboratory, both in vitro and in vivo, but have not created a serious therapeutic problem in natural infections. Nonvirulent variants have been extensively used in the production of living vaccine.

PATHOGENICITY

Plague is a natural disease of both domestic and wild rodents. Rats are the primary reservoir: they usually die acutely, with a high-grade bacteremia but occasionally develop a more chronic form of infection. The disease is transmitted by the bites of **fleas** (e.g., *Xenopsylla cheopis,* the rat flea) which have previously sucked blood from an infected animal. The ingested bacilli proliferate in the intestinal tract of the flea and eventually block the lumen of the proventriculus. The hungry flea, upon biting another rodent, regurgitates into the wound a mixture of plague bacilli and aspirated blood. If its host dies the flea leaves promptly, seeking a replacement. If another rodent is not available it will accept a human host, an accidental intruder in the rat-flea-rat transmission cycle.

A small pustule may be present at the por-

present in the medium. The ability to absorb hemin (or the dye Congo red) from the medium, producing pigmented colonies, is closely correlated with virulence.

METABOLISM

Y. pestis grows well in ordinary peptone broth and agar, both aerobically and anaerobically, particularly in the presence of blood or other tissue fluids. It is nonhemolytic on blood agar medium. Unlike most pathogenic species, *Y. pestis* multiplies rapidly at 28° (temperature optimum), but it then fails to synthesize certain antigens essential for virulence (see below). Glucose and mannitol are fermented, but gas is not produced. Lactose is not attacked. *Y. pestis* may remain viable for weeks in dry sputum or flea feces at room temperature.

ANTIGENS

Y. pestis produces at least 10 different antigenic components recognizable in agar diffusion precipitin reactions.

The capsular or envelope antigen, a heat-labile protein, is often referred to as **fraction 1** (F1). Freshly isolated strains from infected rats or man are all well enveloped. The **VW antigen** system is made up of a protein V (MV 90,000) and a lipoprotein W (MW

tal of entry in the skin but more often there is no discernible lesion. The bacilli introduced by the flea bite enter the dermal lymphatics and are transported to the regional lymph nodes, usually in the groin, where they cause the formation of enlarged, tender buboes. In severe **bubonic plague** the regional lymph nodes fail to filter out all the multiplying bacilli; organisms that gain entrance to the efferent lymphatics disseminate via the circulation (septicemic plague) to the spleen, liver, lungs, and sometimes the meninges. The parenchymatous lesions produced are hemorrhagic; disseminated intravascular coagulation may occur. In the terminal stages bacteremia is often intense.

When metastatic pneumonia develops the sputum may become heavily contaminated, and infection may then be transmitted by way of respiratory droplets. **Pneumonic plague,** particularly under conditions of crowding in cold climates, is relatively contagious, and since the inoculum of virulent bacilli in the infected droplets tends to be large, this form of the disease is extraordinarily malignant. Both the bubonic and the pneumonic forms of the disease can be produced experimentally in rodents and monkeys.

The incubation period of bubonic plague in man varies from 1 to 6 days, depending upon the infecting dose. Onset is usually abrupt with high fever, tachycardia, malaise, and aching of the extremities and back. If the disease progresses to the fulminant bacteremic stage, it causes prostration, shock, and delirium; death usually occurs within 3 to 5 days of the first symptoms. The course of plague pneumonia is even more fulminant; untreated patients rarely survive longer than 3 days. Pulmonary signs may be totally lacking until the final day of illness, making early diagnosis particularly difficult. Late in the disease copious bloody, frothy sputum is produced.

The occurrence of asymptomatic cases is suggested by serological studies on persons living in areas where the disease is endemic and by the finding in Vietnam of a pharyngeal carrier rate of about 10% in family members of plague patients.

Less than 10 organisms of a fully virulent strain of *Y. pestis* injected into a mouse is lethal. The factors responsible for virulence

are complex and only partially understood. Two antiphagocytic components of the cell are definitely involved: the envelope protein (F1 antigen) and the combined VW proteins; for maximal virulence both must be present. Virulent strains all produce VW proteins at 37° but need not produce cell envelopes (F1 antigen; cf. streptococci, Ch. 26, Pathogenicity). An avirulent strain that has been used safely in immunization in man contains the VW and F1 determinants but cannot utilize the iron contained in heme. A bacteriocin, pestocin I, inhibitory for another *Yersinia* species (*Y. pseudotuberculosis*), is found in almost all strains of *Y. pestis* isolated from humans.

The pathogenetic role of the plague toxin is not clear, for there are no highly specific hallmarks of intoxication in the disease. The toxin appears to act primarily on the vascular system, causing hemoconcentration and shock. In favor of a pathogenetic role for the toxin is its lethal effect in very low concentrations ($LD_{50} < 1$ μg protein) on intravenous administration in the mouse. Although all virulent strains produce toxin, not all toxin-producing strains are virulent.

The bacilli contained in the gut of the rat flea possess neither capsular nor VW antigen; consequently they are promptly ingested and destroyed by polymorphonuclear leukocytes. How then does the flea serve as an effective vector? This fascinating riddle was eventually solved by Cavanaugh and Randall, who found that when such bacilli are phagocytized at 37° by monocytes (in contradistinction to granulocytes), they not only survive and multiply intracellularly, but they also emerge as fully virulent organisms possessing both the F1 and VW antiphagocytic factors. Evidently the virulence of the bacilli in the flea is only masked by the low temperature (ca. 25°) at which they have proliferated. The organisms' ability to survive and multiply within the monocytes of the warm-blooded mammalian host permits them to undergo a phenotypic change in which their full virulence is regained. Precisely why they are not killed in monocytes, as they are in polymorphonuclear leukocytes, is not known. The ability to survive and multiply within the monocyte may be the major factor in virulence, rather than resistance to phagocytosis.

IMMUNITY

Recovery from plague appears to confer relatively solid immunity to subsequent infection and rare cases of reinfection have occurred. The antibodies primarily involved are those to the antiphagocytic antigens, F1 and VW complex.* Such antibodies promote phagocytosis and intracellular killing of the organisms by polymorphonuclear leukocytes. Effective vaccines therefore must contain both antiphagocytic antigens; antitoxic sera are not protective.

LABORATORY DIAGNOSIS

Procedures that provide rapid preliminary diagnosis are of paramount importance in the suspect case in view of the swift progression of the untreated disease. Because of the danger of serious laboratory infections, great care must be exercised in handling specimens containing *Y. pestis*. Smears of sputum, or of fluid aspirated from lymph nodes, should be stained by the Gram method and also with either methylene blue or Wayson's reagent to identify bipolar staining. In epidemics of plague pneumonia, fluorescent antibody† is of great value for rapid identification of *Y. pestis* in the sputum (see below).

Cultures may be made in blood broth or on blood agar, but must be handled with caution. Cultured organisms may be identified by biochemical tests, by fluorescent antibody staining with appropriate antisera, or by lysis with specific bacteriophage. Media incorporating antibody to F1 envelope antigen have been used to identify *Y. pestis* colonies in mixed culture.‡ Animal inoculation is sometimes useful when specimens are contaminated with other organisms which may overgrow the cultures; specimens containing *Y. pestis* are usually lethal if injected into mice or guinea pigs, and typical pathological le-

sions are found. Animals so inoculated must be free of ectoparasites and should be kept strictly isolated.

Diagnostic serological tests are of only retrospective value.

TREATMENT

The case fatality rate in untreated bubonic plague is 50 to 75%, and that of plague pneumonia approaches 100%. Fortunately, however, *Y. pestis* is susceptible to the antibacterial actions of streptomycin, chloramphenicol, and the tetracyclines. If instituted early enough antimicrobial therapy dramatically alters the course of pneumonic plague and reduces the mortality of bubonic plague to 1 to 5%. **Time is of the essence,** particularly in pneumonic plague, which can rarely be controlled after 12 to 15 hours of fever. Because a few strains of *Y. pestis* resistant to the highly effective bactericidal drug streptomycin have been noted, treatment involves concurrent use of that agent with either tetracycline or chloramphenicol. Although antiplague serum has shown some therapeutic efficacy, its use has been supplanted by chemotherapy.

PREVENTION

While it is clear that plague is initially transmitted from rodents to man by rodent ectoparasites, epidemiological studies have revealed that man-to-man transmission occurs in epidemics of bubonic plague, the principal vector being the human flea, *Pulex irritans.*

Prevention of the disease is difficult, since elimination of the animal reservoir through rodent control is virtually impossible. Indeed, wholesale poisoning of rats may accentuate an epidemic by forcing infected fleas to leave the dying rats and seek human hosts on which to feed. Insecticides (e.g., DDT) properly directed against human fleas may, however, lower the transmission rate in epidemics. All patients wtih pneumonic plague should be strictly isolated.

Artificial immunization with killed or attenuated vaccines, or with antigenic fractions of the bacilli, appears to provide short-term relative immunity to bubonic plague but not to pneumonic plague, where the inoculum is usually very large. Formalin-treated vaccine

---

* Antibodies to the V protein protect mice against experimental plague infections, but antibodies to the W lipoprotein do not.

† The dominant surface antigen of *Y. pestis* is apparently monotypic.

‡ On antiserum-containing plates F1 antigen (released by treatment of the colonies with chloroform vapor) forms a precipitin ring surrounding each colony.

is recommended for persons entering an area (e.g., Vietnam) where plague is endemic and for laboratory personnel working with *Y. pestis*. Close contacts of patients with pneumonic plague should be treated prophylactically with tetracycline.

To prevent plague from entering uninfected areas, ships from ports known to be infected with the disease have been quarantined and their cargoes fumigated. Large circular shields of metal are also placed around each hawser to the dock to prevent rats from leaving the ship.

OTHER YERSINIAE

Besides *Y. pestis* the genus *Yersinia* consists of two other species, *Y. pseudotuberculosis* and *Y. enterocolitica,* both relatively large gram-negative coccobacilli and both capable of producing disease in man. There is an extensive animal reservoir including mammals (rabbit, pig, cow, mouse, chinchilla) and birds. The organisms may cause extensive epizootic outbreaks characterized by diarrhea, lymphadenopathy with necrosis "pseudotuberculosis"), and septicemia. They may also persist latently in healthy carrier animals. Although the source of human infection is not established, it is likely that infection with these animal pathogens is transmitted through food contaminated by feces or urine. Although originally considered as rare causes of human infection, they have been recognized with increasing frequency in both Europe and the United States (over 1000 cases of *Y. enterocolitica* infection in the period 1966 to 1970).

Clinical infections with these organisms take either of two forms. **Acute mesenteric lymphadenitis** or **enterocolitis** is a relatively benign process which may mimic acute appendicitis or bacillary dysentery. The organism can be isolated from the enlarged mesenteric lymph nodes. Mesenteric lymphadenitis due to yersiniae affects mainly young children, and may be followed by erythema nodosum. The septicemic form, which resembles typhoid fever, occurs particularly in patients with underlying diseases such as cirrhosis of the liver and blood dyscrasias; the mortality may reach 50%. Abscesses may develop in various organs.

A variety of somatic O antigens are present in various strains. Several from *Y. pseudotuberculosis* cross-react with some in group B and group D salmonellae, and antigens present in *Y. enterocolitica* cross-react with ones present in *Brucella abortus.* Some degree of cross-immunity also exists between *Y. pestis* and *Y. pseudotuberculosis*: killed suspensions of *Y. pseudotuberculosis* protect guinea pigs against infection with *Y. pestis.* Agglutinins usually appear by the time of symptomatic disease with *Y. pseudotuberculosis,* but antibodies develop later with *Y. enterocolitica.*

The yersiniae may be misidentified as atypical coliform organisms (Ch. 29). They are oxidase-negative, in contrast to species currently characterized as pasteurellae. Both *Y. enterocolitica* and *Y. pseudotuberculosis* are motile at 22° but not at 37°, which helps to distinguish them from *Y. pestis* and the Enterobacteriaceae. *Y. enterocolitica* and *Y. pseudotuberculosis* can be distinguished from each other by biochemical tests, by susceptibility of *Y. pseudotuberculosis* to specific bacteriophages, by the pathogenicity for the guinea pig of *Y. pseudotuberculosis* (but not of *Y. enterocolitica*), and by agglutination with specific antisera.

Most strains of *Y. enterocolitica* are sensitive to polymyxin. *Y. pseudotuberculosis* is generally sensitive to ampicillin, kanamycin, tetracycline, and chloramphenicol.

FRANCISELLA TULARENSIS

The history of tularemia is less dramatic and more recent than that of plague. While attempting to culture plague bacilli from ground squirrels dying of a plague-like disease in Tulare County, California, in 1912, McCoy and Chapin isolated a new bacterial species which became known as *Bacterium tularense.*

The human illness caused by this organism was described two years later. In an extraordinary series of field, laboratory, and clinical investigations performed in Utah in 1919, Francis conclusively proved that jack rabbits are an important source of human infection and that the disease may be transmitted to man by the bite of a deer fly. The importance of ticks, as both reservoirs and vectors, was subsequently established.

MORPHOLOGY

F. tularensis is a short, nonmotile, unencapsulated, non-spore-forming, gram-negative bacillus. It is markedly pleomorphic in culture, exhibiting bean-shaped, filamentous, coccoid, and bacillary forms. Some of the minute forms resemble spheroplasts and are small enough to pass through Berkefeld filters. Minute coccoid forms are visible in hepatic cells of experimental animals, but identifiable organisms are only rarely observed in tissues of humans dead of the disease. Colonies are slow to grow on primary inoculation, taking 2 to 10 days even on appropriate media. The colonies are minute, transparent, mucoid, and easily emulsified (smooth phase).

METABOLISM

The outstanding growth characteristic of *F. tularensis* is its **requirement for cysteine** (or sulfhydryl compounds) in amounts exceeding those usually present in nutrient media. It grows best on cysteine-glucose-blood agar and on coagulated egg yolk medium, and less well in thioglycollate broth. A satisfactory chemically defined liquid medium has been developed. Multiplication is most rapid at 37°; it is significantly accelerated on blood agar containing both cysteine and thioglycollate.* Though a facultative anaerobe, the organism grows best under aerobic conditions.

Most strains ferment glucose, maltose, and mannose, producing acid without gas. Factors required for growth in synthetic media include 13 amino acids, thiamine, and spermidine.

ANTIGENS

Only a single immunological type of *E. tularensis* has been identified. The immunizing antigens appear to be concentrated in the cell wall. Several kinds of antigens have been extracted: 1) a polysaccharide that causes an

---

* A blood-free tryptose broth medium has also been developed which supports rapid growth; it contains added thiamine, cysteine, glucose, and iron.

immediate wheal and erythema reaction when injected into the skin of patients convalescing from tularemia, 2) a protein antigen that cross-reacts with agglutinating antigens of the genus *Brucella,* and 3) an endotoxin whose role in pathogenesis appears to be similar to that of *S. typhosa* endotoxin in typhoid fever (Ch. 29, Pathogenicity).

VARIATION

Rough (R) mutants of *E. tularensis* are readily recognized by their granular colonial morphology and are generally less virulent and less immunogenic than smooth (S) strains.

PATHOGENICITY

The factors responsible for the pathogenicity of *F. tularensis* are poorly defined. The correlation of virulence with colonial morphology, though not always consistent, suggests that surface components of the bacterial cell may be involved, although no antiphagocytic properties of virulent strains have been demonstrated. No exotoxin has been identified. In general, strains of high virulence for man tend to ferment glycerol and have usually been isolated from tick-borne tularemia in rabbits; isolates of low virulence do not ferment glycerol and have commonly been isolated from water-borne disease of rodents.

The organism behaves primarily as an intracellular parasite, surviving for long periods in monocytes and other body cells. The persistent immune response (see below), and the occasional tendency of the disease to relapse and remain chronic, are probably related to the prolonged intracellular survival of the bacilli.

Human tularemia usually results from direct contact with the tissues of infected rabbits. The organism may gain entrance to the body through an abrasion in the skin or through the mucous membranes of the oropharynx or gastrointestinal tract following ingestion of improperly cooked meat or inhalation of infected aerosols. The infecting dose is only ca. 10 organisms. Tularemia is also transmitted by bites of flies and ticks, important reservoirs and vectors of the disease. *F. tularensis* can be transmitted transovarially by infected female ticks.

At the site of primary lodgment in the skin

or mucous membrane an ulcerating papule often develops. From the original lesion the organisms are carried by the lymphatics to regional lymph nodes, which become enlarged and tender and may suppurate. Further penetration to the blood stream causes transitory bacteremia in the acute phase of the illness, and results in spread to parenchymatous organs, particularly the lungs, liver, and spleen. The characteristic lesions produced in the reticuloendothelial system are granulomatous nodules which may caseate or form small abscesses.

The **ulceroglandular** form of the disease is the most common and results from primary infection of the skin, whereas the **oculoglandular** form is caused by primary involvement of the conjunctivae. The syndrome produced by inhalation of infected droplets or aerosols is referred to as **pneumonic** tularemia; it is apt to occur in laboratory workers. As in plague, pulmonary involvement may also result from hematogenous dissemination from local infection elsewhere. **Typhoidal tularemia** follows ingestion of the organism and resembles typhoid fever, with gastrointestinal manifestations, fever, and toxemia.

The incubation period in tularemia ranges from 3 to 10 days and is followed by the onset of headache, fever, and general malaise. If specific treatment is not instituted the course of the disease is usually protracted, delirium and coma may develop, and the outcome may be fatal. Whereas the case fatality rate in untreated ulceroglandular tularemia is only about 5%, in the typhoidal and pulmonary forms it approaches 30%.

IMMUNITY

Although naturally acquired immunity to tularemia is usually permanent, second attacks of the disease have been described. Agglutinins are usually demonstrable in the serum by the second or third week of illness, and persist for many years after recovery.

Altered cellular reactivity probably plays the major role in immunity. Resistance has been passively transferred by spleen cells from mice that have recovered from infection with an attenuated strain. The tendency of the disease to progress and even to relapse, despite high titers of serum antibodies, is un-

doubtedly due to the ability of the organism to survive within the cells of the host.

LABORATORY DIAGNOSIS

A definitive diagnosis of tularemia cannot be made from the study of exudate smears unless they are stained with specific fluorescent antibody. Cultures made with ordinary bacteriological media are also useless. Only if **special media** (e.g., cysteine-glucose-blood agar) are employed can *F. tularensis* be cultivated. Cultures should be incubated for 3 weeks before being discarded as negative. If an organism grows in the special medium, but fails to grow in ordinary media, it may well be *F. tularensis*. Its identity should be established by staining with fluorescent antibody or by an agglutination test performed with specific antiserum. *F. tularensis* has been isolated from gastric washings, sputum, and tissue specimens by guinea pig inoculation. Great care must be taken to avoid infection of laboratory personnel.

Serological tests are of value in establishing a diagnosis. Agglutinins (and hemagglutinins) appear within 8 to 10 days of the onset of illness and continue to rise for up to 8 weeks. An agglutinin titer may be detectable for years after the disease. The demonstration of a rising titer on serial serological examinations is confirmatory evidence of recent infection. Brucella antibodies cross-react with antigens of *F. tularensis,* but they may be distinguished by comparative agglutination tests performed with antigens of both speices. The delayed hypersensitivity reaction to the tularemia skin test antigen appears to be a sensitive indicator of present or past infection. In tularemia the skin test becomes positive earlier than the agglutination reaction and remains positive for years.

TREATMENT

Because of its bactericidal properties, **streptomycin** is the drug of choice in the treatment of tularemia. The bacteristatic tetracyclines and chloramphenicol are also effective, but relapses tend to occur when treatment is discontinued prematurely. Even when streptomycin is used relapses occasionally occur, probably because of the failure of the drug to affect many of the intracellularly located organisms.

Although human tularemia is usually acquired, in the United States,* through contact with infected rabbits or from the bites of ticks or flies, the disease is also contracted in Russia and other parts of the world† from polluted water supplies contaminated by the carcasses of infected rodents. Gloves should be used in skinning and dressing rabbits, and the meat should always be thoroughly cooked before being eaten. Tularemia has been reported following the bites of cats, dogs, and even snakes, all of which had apparently fed on infected rabbits. An effective attenuated vaccine was developed in the Soviet Union. Although it causes local reactions of moderate severity, on intradermal administration, it affords significant protection against both respiratory and cutaneous challenge. Its use is indicated in laboratory workers and other individuals who are likely to be exposed to *F. tularensis.*

PASTEURELLA MULTOCIDA

P. multocida is the cause of "hemorrhagic septicemia" in a variety of animals and birds; it was described by Pasteur as the cause of fowl cholera. The organism is frequently carried in the respiratory tract of healthy animals such as sheep, dogs, cats, and rats.

When resistance is lowered, as when herds of cattle are shipped, the organisms may become invasive, producing fulminating septicemia or pneumonia ("shipping fever"), which then may spread to other susceptible animals. Killed vaccines are used to protect cattle from shipping fever and to control fowl cholera in areas where the disease is epidemic.

P. multocida is a small, nonmotile, gramnegative coccobacillus. In contrast to yersiniae it is oxidase negative, lacks motility at both 37° and 22°, and cannot grow on MacConkey agar. Small, nonhemolytic, gray colonies are formed on blood agar. Capsular antigens have been found and five serotypes defined on the basis of agglutination reactions. *P. multocida* is usually isolated as smooth or mucoid variants; these exhibit marked mouse pathogenicity in contrast to rough variants.

P. multocida has been recognized more frequently of late as a cause of human disease. Most infections take one of three clinical patterns: local **soft tissue infection,** or even osteomyelitis, following an animal bite (most commonly that of a cat); **systemic infection** such as bacteremia or meningitis, sometimes without any antecedent animal exposure; **respiratory tract infection,** such as sinusitis, empyema, and bronchiectasis.

Most strains of *P. multocida* are sensitive in vitro to penicillin, which is the treatment of choice. This sensitivity may serve in the rapid laboratory differentiation of *P. multocida* from other gram-negative bacilli. Tetracycline is also effective.

SELECTED REFERENCES

Books and Review Articles

FOSHAY, L. Tularemia. *Annu Rev Microbiol* 4:313 (1950).

MONTIE, T. C., and AJL, S. J. Nature and Synthesis of Murine Toxins of *Pasteurella pestis.* In *Microbial Toxins,* vol. III. (T. C. Montie, S. Kadis, and S. J. Ajl, eds.) Academic Press, New York, 1970.

POLLITZER, R. *Plague.* WHO Monogr Ser, No. 22. World Health Organization, Geneva, 1954.

Specific Articles

BALTAZARD, M., BAHMANYAR, M., MOSTACHFI, P., EFTEKHARI, M., and MOFIDI, C. H. Recherches sur la peste en Iran. *Bull WHO* 23:141 (1960).

BURROWS, T. W. Genetics of virulence in bacteria. *Br Med Bull* 18:69 (1962).

CAVANAUGH, D. C., and RANDALL, R. The role of multiplication of *Pasteurella pestis* and mononuclear phagocytes in the pathogenesis of flea-borne plague. *J Immunol* 83:348 (1959).

* A total of 2743 cases were reported to the United States Public Health Service from 1960 to 1969.
† It is of interest that tularemia seems to be primarily, if not exclusively, a disease of the northern hemisphere.

LAWTON, W. D., ERDMAN, R. L., and SURGALŁA, M. J. Biosynthesis and purification of V and W antigen in *Pasteurella pestis*. *J Immunol 91*:179 (1963).

MEYER, K. F. The natural history of plague and psittacosis. *Public Health Rep 72*:705 (1957).

MOLLARET, H. H. *Yersinia enterocolitica* infection: A new problem in pathology. *Ann Biol Clin 30*:1 (1972).

SASLAW, S., EIGELSBACH, H. T., WILSON, H. E., PRIOR, J. A., and CARHART, S. Tularemia vaccine study. *Arch Intern Med 107*:689, 702 (1961).

WEBER, J., FINLAYSON, N. B., and MARK, N. B. D. Mesenteric lymphadenitis and terminal ileitis due to *Yersinia pseudotuberculosis*. *N Engl J Med 283*:172 (1970).

chapter

THE BRUCELLAE

Revised by MORTON N. SWARTZ, M.D.

History   812
Morphology   812
Metabolism   812
Genetic Variation   813
Antigenic Structure   813
Pathogenicity   813
Immunity   815
Laboratory Diagnosis   816
Treatment   816
Prevention   816

The brucellae are nonmotile, short, non-spore-forming gram-negative coccobacilli. The genus contains three principal species (*B. melitensis, B. abortus,* and *B. suis*) pathogenic for man; three other species have been described more recently, of which only *B. canis* appears to have any role in human disease. The original differentiation of the principal species was based on their major animal sources, i.e., goat and sheep for *B. melitensis,* cattle for *B. abortus,* and swine for *B. suis;* this speciation has since been supported by metabolic and antigenic differences. All brucellae are obligate parasites capable of causing acute or chronic illness or inapparent infection; the chronicity depends on the marked capacity for multiplication in phagocytic cells, which is opposed by the development of cellular immunity. In their natural animal reservoirs brucellae show a striking propensity to localize in the pregnant uterus (frequently causing abortion) and in the mammary glands; apparently healthy animals may shed brucellae in their milk for years. The predilection for placental tissues seems to be determined by the presence there of a simple growth factor, erythritol.

Man becomes infected through the ingestion of unpasteurized milk or cheese, or through contact with the tissues of infected animals. Human brucellosis may be an acute or relapsing febrile illness, a chronic illness, or a subclinical infection. Unlike its counterpart in animals it does not tend to localize in the genital tract but rather involves the reticuloendothelial system.

HISTORY

Brucellae were first isolated in 1887 by Bruce from the spleens of British soldiers dying on the island of Malta from a disease known as **Malta fever.** The source of the organism was not discovered until 1904, when it was cultured from milk and urine of apparently healthy goats. When the consumption of raw goat's milk by the soldiers was stopped the incidence of the disease declined sharply. The second organism of the group was isolated in Denmark by Bang in 1897 from cattle suffering from infectious abortion (Bang's disease), and the third was cultured in the United States in 1914 from the fetus of a prematurely delivered sow. In the 1920s Evans recognized that all three organisms were closely related, and they were placed in a separate genus, *Brucella.*

MORPHOLOGY

The *Brucella* organisms are pleomorphic, short, slender coccobacilli. Bipolar staining is sometimes present. Colonies are small, round, convex, smooth, moist-appearing, and translucent. Growth is slow, particularly on initial cultivation, and colonies are not usually visible for 2 days or more. As a rule, *Brucella* isolates recovered from tissues are smooth or S-type colonies, and capsules can be demonstrated with appropriate staining.

METABOLISM

The brucellae are aerobes but can use nitrate anaerobically as an electron acceptor. Their nutritional requirements are relatively complex. They may be cultivated on tryticase soy plain (or blood) agar or in trypticase soy broth. Synthetic media containing a variety of amino acids and vitamins may be used for special purposes. *B. abortus* differs from the other species in requiring, on primary isolation, an atmosphere containing 5 to 10% carbon dioxide. All produce catalase and decompose urea; fermentation of sugars is usually not observed. Various differences, presented in Table 32-1, form the basis for the metabolic differentiation of the four species infecting man. Despite these several differences, the *Brucella* species are often difficult to distinguish. Moreover, stable variant strains (biotypes) often appear as the major strain in a given geographical area.

The genus appears to be quite homogeneous as judged by DNA-DNA homology studies.

Brucellae are killed by pasteurization, but they may survive for many weeks in infected fetal tissues and in soil. In cheese they are killed within a few days by the accumulated lactic acid, but in butter they remain viable for more than a week.

GENETIC VARIATION

When grown on laboratory media the smooth (S) form of *Brucella* isolated from infected tissues tends to be replaced by rough (R) forms, which are less virulent and also exhibit less specific and more nonspecific agglutination. Intermediate (I) and mucoid (M) forms, exhibiting reduced virulence, may also emerge. In 1:1000 acriflavine S-type cells remain evenly suspended, R-type cells clump, and M-type cells form a slimy, thread-like precipitate. Non-S-type colonies do not readily revert either in vitro or in vivo to the fully virulent S type. A strain of the intermediate type, *B. abortus* strain 19, is widely used in a live vaccine for immunization of cattle. The S → R "dissociation" is due to the greater resistance of R cells to alterations of the medium produced by the S cells, including accumulation of D-alanine and lowering of pO_2 (Ch. 8).

ANTIGENIC STRUCTURE

Antisera from animals immunized with a smooth strain agglutinate the three principal *Brucella* species. Two shared determinants have been proposed to account for these cross-reactions. A or abortus antigen is the major surface determinant in both *B. abortus* and *B. suis,* and is a minor determinant (a) in *B. melitensis,* while the M antigen predominates in the latter and is a minor antigen (m) in the others. By adsorbing anti-Am (*abortus* or *suis*) antibodies with a dose of *melitensis* organisms (aM) that will remove all the minor (m) agglutinin but only a small frac-

tion of the major (A) agglutinin, it is possible to prepare monospecific serum to A antigen. Monospecific anti-M serum may be prepared similarly. These sera are useful in diagnosis (Table 32-1).

The failure of monospecific serum to agglutinate the species with the corresponding minor antigen is explained by assuming that the minor antigenic determinants are not identical or are too sparsely distributed on the cell surface to bring about agglutination.

Brucellae do not release exotoxins. Smooth and intermediate (e.g., *B. abortus* strain 19) colonial types contain endotoxin of comparable potency, but R variants and species (*B. canis, B. ovis*) that grow as rough colonies on primary isolation lack endotoxin. Since *B. canis* is pathogenic in dogs and humans, and *B. ovis* in sheep, endotoxin does not seem to be important in virulence, although it may play a role once infection is established.

PATHOGENICITY

The three principal *Brucella* species are pathogenic in a wide range of mammals, although each has a preferred host. Experimental infections are readily produced in guinea pigs, rabbits, mice, and monkeys. In naturally acquired brucellosis the organisms gain entrance to the body via the broken skin, the conjunctivae, or the alimentary tract. Experimental infections have also been induced with aerosols.

At the site of lodgment in the skin or mucous membranes polymorphonuclear cells appear and ingest the organisms, which then multiply within them. The bacteria, mostly intracellular, are then carried via the lymphatics to the regional lymph nodes. There bacteria enter and multiply within mononuclear cells; some host cells die, and the released bacteria and cell contents stimulate local mononuclear cell activation and proliferation. The outcome of this confrontation determines whether the invasive infection is contained. If not, the bacteria, mainly within polymorphonuclear and mononuclear cells, are transported to the blood. Soon the sinusoids of the liver contain large numbers of parasitized leukocytes. Focal aggregations of Kupffer cells, containing large numbers of organisms, develop, and after another few

TABLE 32-1. Differential Characteristics of Brucella Species

Species	CO₂ requirement	H₂S production	Hydrolysis of urea	Growth on dye media		Agglutination in monospecific antisera to:		Lysis by phage	Oxidative metabolic tests			
				Thionine*	Basic fuchsin†	A	M		Glutamic acid	Ornithine	Ribose	Lysine
B. melitensis	−	−	Slow	−	+	−	+	−	+	−	−	−
B. abortus	+‡	+	Slow	−	+	++	−	+	++	−	+	−
B. suis	−	++	Rapid	+	−	++	−	±	±	++	++	++
B. canis	−	−	Rapid	+	±	−	−	−	±	+	+	+

* 1:25,000 thionine.
† 1:000,000 basic fuchsin.
‡ On primary isolation.

days, form typical small granulomas. Similar lesions appear in spleen, bone marrow, and kidney. In certain mammals other than man (cattle, swine, sheep, goats, etc.), brucellae also accumulate in the mammary glands (causing infection of the milk) and in the genital organs, particularly the pregnant uterus, often resulting in abortion.

In bovine infectious abortion the organisms are found mostly in the fetal portion of the placenta, the birth fluids, and the chorion. This remarkable viscerotropism (Ch. 22, Adaptation to microenvironment), according to Smith *et al.,* depends upon the presence of **erythritol,** which may be an important factor in the pathogenesis of infectious abortion. This four-carbon polyhydric alcohol ($HO\ CH_2 \cdot CHOH \cdot CHOH \cdot CH_2OH$) was crystallized from bovine allantoic and amniotic fluids, which had been found to stimulate the growth of *B. abortus.* It is present in appreciable quantities only in the chorion, cotyledons, and fetal fluids of animals prone to infectious abortion (cows, sheep, pigs, and goats). It was not detected in human placentas, which are not especially susceptible to brucellosis.

Classic specific virulence factors, such as exotoxins or antiphagocytic capsular or cell wall constituents, have not been detected among *Brucella* species. Instead, intracellular events largely determine the features and course of the disease: smooth *Brucella* organisms multiply abundantly in nonimmune monocytes, while rough, avirulent organisms do not. A possible virulence factor appears to enhance the intracellular survival. Thus virulent *B. abortus* from cultures of monocytes or from infected bovine placenta survive in mononuclear cells better than the same strain grown on artificial media. Moreover, cell walls from virulent *Brucella* obtained from the bovine placenta, but not from the same organisms grown on artificial media, inhibit the normal intracellular destruction of an avirulent (R) strain by mononuclear cells. It thus appears that the virulence factor may be made only in vivo.

Cellular immunity plays a critical role in this disease. Thus brucellae survive longer in cultured mononuclear cells from nonimmune animals than in analogous cells from actively immunized animals. Mackaness found that the change in the cells coincides with the onset of delayed hypersensitivity (as shown by skin tests). Such cellular immunity is relatively nonspecific and extends to heterologous intracellular bacteria. The probable mechanism is discussed in the chapter on host-parasite relations (Ch. 22, Allergy vs. immunity).

A hypersensitive reaction to suddenly released *Brucella* antigens is presumed to be responsible for the Herxheimer-like reaction (Ch. 37, Treatment) that sometimes follows vigorous chemotherapy, and for the severe systemic as well as local reactions observed within hours after accidental inoculation of previously infected veterinarians with the vaccine strain of *B. abortus.* Bacteremia does not occur despite the profound illness.

The incubation period in human brucellosis is long, often several weeks or even months. The onset of symptoms is usually insidious, with malaise, fever, weakness, myalgia, and sweats. Fever may be remittent, particularly with *B. melitensis* (undulant fever). Vague gastrointestinal and nervous symptoms are common. Bacteremia often is present. The acute illness may be associated with enlarged lymph nodes, spleen, and liver, and with localized vertebral spondylitis. Meningoencephalitis, endocarditis, and interstitial nephritis with focal glomerular lesions occasionally occur in the course of the acute disease. In the later stages of the disease the persistence of vague complaints often suggests a psychoneurosis. The establishment of a definitive diagnosis at this "chronic" stage is especially difficult.

Chronic, recurrent caseating granulomas of liver and spleen of many years' duration have been described with *B. suis,* usually in patients who have not had antibiotic treatment. The lesions usually calcify after some years.

IMMUNITY

Naturally acquired resistance to brucellosis is only relative, since reinfection is common and immunity is not assured by the presence of circulating antibodies, as measured by agglutinin, precipitin, opsonin, or bactericidal tests. Indeed, serum antibodies are usually detectable from the time the first signs and symptoms of the disease develop (see below), and their presence in the circulation does not even prevent bacteremia.

The probable role of delayed hypersensitivity and cellular immunity in the recovery process has already been mentioned.

LABORATORY DIAGNOSIS

The patient with brucellosis commonly presents with an unexplained fever. The diagnosis is usually suggested by the combination of clinical and epidemiological considerations. It may also be suspected when liver biopsy reveals noncaseating granulomas, but these also occur in other infectious and noninfectious processes.

The establishment of a definite diagnosis in man requires **cultivation** of the organism from the blood or from a biopsy of bone marrow, liver, or lymph node; blood should be cultured repeatedly in all suspected cases. Specimens should be incubated in trypticase soy broth under 10% carbon dioxide. At 4- to 5-day intervals, or when visible growth is first noted, a sample is removed for staining and for subculture on blood trypticase soy agar under 10% CO_2. The primary cultures should be incubated for at least 4 weeks before being discarded.

Species are identified on the basis of the tests summarized in Table 32-1. Since the number of viable bacilli present in specimens is usually small, cultivation is often difficult.

Because of the prolonged incubation period in human brucellosis, antibodies are frequently demonstrable in the serum by the time the disease is first recognized. Agglutination tests are performed with phenolized suspensions of heat-killed smooth (S) bacilli. A standard strain of *B. abortus* (No. 456), which is avirulent for man, is generally used for this purpose; it is agglutinated by antibodies to all three principal species. Titers above 1:80 are usually considered indicative of either past or present infection, and a fourfold or greater rise in titer during the course of illness is strong evidence for the diagnosis.

The titer remains elevated (as high as 1:640 to 1:2560) during the active phases of the disease and usually declines as the patient improves. However, high titers may persist for several years without overt disease, and may even be found with no history of clinical disease. As in all bacterial agglutination tests, care must be taken to avoid misinterpretation of the prozone phenomenon (Ch. 15, Prozone). Occasionally, in chronic brucellosis

particularly, the antibodies are of the incomplete, or "blocking" type (Ch. 15, Blocking antibody). They may be detected by using 5% sodium chloride or albumin solution as the diluent in the agglutination test, or by adding the Coombs reagent (anti-human globulin serum) to the test for agglutinating antibodies.

Sera containing antibodies to *Franciscella tularensis, Yersinia enterocolitica, Vibrio cholerae* (due to infection or recent immunization with cholera vaccine), or *V. fetus* can cross-react in the *Brucella* agglutination test. The recent performance of a *Brucella* skin test may cause the appearance of a low titer of agglutinins, and this may be misleading diagnostically.

An intradermal skin test employing suspensions of killed *Brucella* cells or a nucleoprotein fraction (*brucellergen*) has been widely used to demonstrate specific delayed hypersensitivity. Like the tuberculin test, it reveals only prior exposure to the organism, and is of no value in the diagnosis of clinical brucellosis.

TREATMENT

Most strains of *Brucella* are sensitive in vitro to the tetracyclines and streptomycin, but streptomycin does not reach organisms sequestered within mononuclear phagocytes. Tetracycline alone, or in combination with streptomycin in severe infections, is usually effective, often within a few days. However, relapse commonly occurs unless therapy is prolonged for at least 3 to 4 weeks. The difficulty in eradicating the organisms undoubtedly arises because an intracellular location protects them from the action of both drugs and antibodies.

PREVENTION

B. abortus infects cattle almost worldwide, including the United States. *B. suis* infects cattle as well as swine and is presently the most frequently isolated species from cases of human brucellosis in the United States. Goats are important as sources of *B. melitensis* in Mexico and in Mediterranean countries, but this organism is rarely isolated from cases in the United States.

An attenuated live vaccine (*B. abortus,*

strain 19) has been used in calves to decrease the incidence of brucellosis in cattle; it produces a limited infection, followed by reasonable immunity. The identification of infected cattle by an agglutination test (on serum or milk), followed by segregation or slaughter, has also been utilized.

Through the use of public health measures (testing of cows, vaccination of calves, pasteurization of milk products) the incidence of human brucellosis in this country has decreased in the past 25 years from 6300 to 230 cases annually. Most cases now occur in workers in meat-packing plants, in livestock raisers, and in veterinarians. About 10% of cases are attributable to ingestion of raw milk or imported cheeses.

The recent identification of epidemic disease due to *B. canis* in dogs has revealed yet another reservoir for brucellosis. The infection may produce little evidence of illness, even though bacteremia may be demonstrable for months or years. A few cases have already occurred in man.

Although active immunization of humans at high risk is practiced in Russia, public health authorities in the United States have been reluctant to use the vaccines currently available because of their potential pathogenicity for man.

SELECTED REFERENCES

Books and Review Articles

MCCULLOUGH, N. B. Microbial and Host Factors in the Pathogenesis of Brucellosis. In *Infectious Agents and Host Reactions.* (S. Mudd, ed.) Saunders, Philadelphia, 1970.

SPINK, W. W. *The Nature of Brucellosis.* Univ. Minnesota Press, Minneapolis, 1956.

Specific Articles

FITZGEORGE, R. B., and SMITH, H. The chemical basis of the virulence of *Brucella abortus.* VII. The production in vitro of organisms with an enhanced capacity to survive intracellularly. *Br J Exp Pathol 47:*558 (1966).

HALL, W. H. Epidemic brucellosis in beagles. *J Infect Dis 124:*615 (1971).

HOLLAND, J. J., and PICKETT, M. J. A cellular basis of immunity in experimental brucella infection. *J Exp Med 108:*343 (1958).

KEPPIE, J., WILLIAMS, A. E., WITT, K., and SMITH, H. The role of erythritol in the tissue localization of the brucellae. *Br J Exp Pathol 46:*104 (1965).

KEPPIE, J., WITT, K., and SMITH, H. The chemical basis of virulence of *Brucella abortus.* IX. The increased immunogenicity of *Brucella abortus* grown in media which enhance its ability for survival within bovine phagocytes. *Br J Exp Pathol 50:*219 (1969).

SMITH, H., and FITZGEORGE, R. B. The chemical basis of the virulence of *Brucella abortus.* V. The basis of intracellular survival and growth in bovine phagocytes. *Br J Exp Pathol 45:*174 (1964).

SMITH, H., KEPPIE, J., PEARCE, J., and WITT, K. The chemical basis of the virulence of *Brucella abortus. Br J Exp Pathol 42:*631 (1961); *43:*31, 530, 538 (1962).

chapter

AEROBIC
SPORE-FORMING BACILLI

Revised by MORTON N. SWARTZ, M.D.

BACILLUS ANTHRACIS   820

MORPHOLOGY   820

Metabolism   820
Antigenic Structure   821
Genetic Variation   822
Pathogenicity   823
Immunity   825
Laboratory Diagnosis   825
Treatment   826
Prevention   826

OTHER AEROBIC SPORE-FORMING BACILLI   826

The genus *Bacillus* is composed of large gram-positive rods that form spores and grow best under aerobic conditions. Most species are saprophytic and are found on vegetation and in soil, water, and air. The only species that is highly pathogenic for man is *B. anthracis,* which causes **anthrax,** a disease primarily of domestic livestock. Human anthrax is rare in the United States. Although the disease occasionally has been contracted by farmers, veterinarians, and slaughterhouse workers who come into contact with infected livestock (agricultural anthrax), human infection in this country occurs almost exclusively in workers at plants processing imported goat hair, wool, or hides (industrial anthrax).

BACILLUS ANTHRACIS

The anthrax bacillus is unusually large and was the first bacterium shown to cause a disease. As early as 1850 it was seen in the blood of sheep dying of anthrax, and in 1877 Robert Koch grew it in pure culture, demonstrated its ability to form spores, and produced experimental anthrax by its injection into animals.

Four years later, at the celebrated field trial at Pouilly-le-Fort, Pasteur vaccinated 24 sheep, 1 goat, and 6 cows with two injections of a culture of the bacillus attenuated by growth at 42 to 43°. Two weeks later, all the vaccinated animals were injected with a highly virulent culture, along with a similar number of unvaccinated controls. Two days later when Pasteur arrived at the field, he was greeted with acclamation by the crowd that had assembled. Whereas all the vaccinated animals were well, 21 of the 24 unvaccinated sheep and the 1 unvaccinated goat had died, and the 4 unvaccinated cows were obviously ill. During the day, while the crowd watched, the remaining 3 unvaccinated sheep expired. This dramatic demonstration aroused great interest among physicians and provided a potent stimulus to the development of immunology.

MORPHOLOGY

B. anthracis is a large, gram-positive, non-motile, spore-forming rod, 1 to 1.5 μ in width and 4 to 8 μ in length. In smears from infected tissues it appears singly or in short chains, and its capsule is readily demonstrable by Giemsa stain. It does not form spores in the living animal. Spores are formed under conditions unfavorable for continued multiplication of the vegetative form.

The organism grows well on blood agar, where it rarely causes hemolysis (in contrast to its saprophytic relatives). Surface colonies of virulent strains are gray-white, rough, and have a curled edge; when viewed under a hand lens or "colony scope" they usually exhibit the "medusa head" or "curled hair-lock" appearance illustrated in Figure 33-1. If cultivated in the presence of high CO_2 the organisms form capsules and the colonies are smooth and mucoid. At the end of the logarithmic phase of growth. spores begin to appear in the cultures and are numerous after 48 hours of incubation. The oval spores are clearly visible in the centers of the bacilli in specially stained smears (Fig. 33-2).

METABOLISM

The anthrax bacillus is readily cultivated on ordinary nutrient media, and although it grows best aerobically, it will also multiply under strictly anaerobic conditions. Aerobic conditions are required for sporulation, but not for germination.

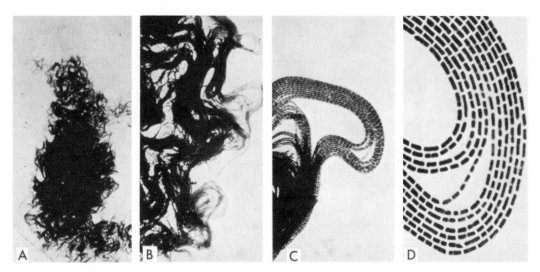

Fig. 33-1. Basis of "medusa head" appearance of surface colonies of virulent strains of *B. anthracis*. Organism grown on nutrient agar and stained with methylene blue. **A.** Whole colony. ×45, reduced. **B—D.** Border of colony. ×145, ×400, and ×1600, all reduced. [From Stein, C. D. *Ann NY Acad Sci* 48:507 (1947).]

The nutritional requirements include thiamine and certain amino acids. Uracil, adenine, guanine, and manganese stimulate the growth of many strains. Glucose, sucrose, maltose, fructose, trehalose, and dextran are fermented in most cultures. Protease, amylase, catalase, lecithinase, and collagenase activities have been demonstrated in cell homogenates or culture filtrates.

Anthrax spores are relatively resistant to heat and chemical disinfectants. They are usually destroyed by boiling for 10 minutes and by dry heat at 140° for 3 hours, but they may survive for 70 hours in 0.1% mercuric chloride. They persist for years in dry earth and may remain viable for months in animal hides.

ANTIGENIC STRUCTURE

Two major groups of antigens are known to be associated with *B. anthracis:* 1) **cellular (somatic) antigens** and 2) **components of the complex exotoxin** elaborated by the organism both in vivo and in vitro.

Cellular Antigens. In addition to a cell wall polysaccharide the organism forms a single antigenic type of capsular polypeptide composed of a γ-polypeptide of D-glutamic acid. The capsule is antiphagocytic and appears to play an important role in pathogenicity, since nonencapsulated (R) mutants are avirulent. However, this role appears to be limited to the establishment of the infection and is not apparent in the terminal phase of the disease, which is much more closely linked to in vivo toxin production and toxemia. Thus, antibodies to the capsule alone are usually not protective. Capsule formation is increased by the presence of both serum and bicarbonate in the medium. The virulence-enhancing effect of the addition of egg yolk to a relatively hypovirulent inoculum of *B. anthracis* is associated with the development of capsules in the inoculated strain. Smooth (S) and rough (R) variants of *B. anthracis* occur and the S→R transition is associated with the selection of mutants which have lost the ability to synthesize capsular polysaccharide. Encapsulated (S) strains may be selected for in cultures of unencapsulated (R) strains by addition to the medium of Wa phage, which attacks only the unencapsulated cells.

The somatic polysaccharide antigen contains equimolar quantities of N-acetyl glucosamine and D-galactose, and cross-reacts with type 14 pneumococcus polysaccharide and with human blood group A substance. In the cell wall it seems to be attached to a peptide containing diaminopimelic acid. The

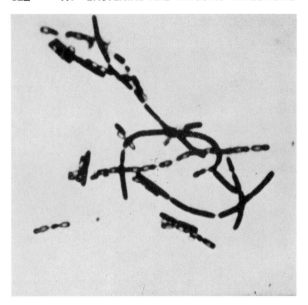

Fig. 33-2. Anthrax bacilli, from 48-hour plate culture, stained with crystal violet. Spores are unstained and clearly visible. ×1200, reduced. (From Burrows, W. *Textbook of Microbiology*. Saunders, Philadelphia, 1963.)

polysaccharide is assumed to play no role in virulence, since antibodies to it are not protective.

Exotoxin Components. Keppie and Smith first demonstrated an exotoxin in *B. anthracis* in 1954. In guinea pigs dying of experimental anthrax a toxic material was found in all infected tissues and exudates but it was most concentrated in edema fluid and plasma.* This crude toxin produces extensive edema when injected subcutaneously in guinea pigs or rabbits, and is lethal when injected intravenously in mice; it is also immunogenic. Subsequently, it was shown that the same toxin is produced in vitro, but it is present in the culture medium only for a short time, when the cell density is about 1×10^8 chains per milliliter. Bicarbonate ion is required early in the growth cycle for production of toxin and possibly for its elaboration into the medium. Bicarbonate also appears to alter cell permeability, thus allowing release of the toxin from the cell. Almost all strains of *B. anthracis* produce toxin, including avirulent (unencapsulated) ones.

Purification has separated the toxin into

three distinct, antigenically active components. All are thermolabile and appear to be proteins or lipoproteins. One component (necessary for the edema-producing activity of the toxin) is designated the **edema factor** (EF) or factor I. A second component, known as the **protective antigen** (PA) or factor II, induces protective antibodies in rabbits and guinea pigs. It can be assayed by its immunogenic activity, complement fixation with appropriate antisera, or immunodiffusion in agar. The third component (essential for the **lethal effect**) is known as the **lethal factor** (LF) or factor III. Upon extensive purification neither EF nor LF retains biological activity individually. However, EF in combination with PA produces edema in guinea pigs; LF in combination with PA is lethal in rats.

GENETIC VARIATION

Mutants derived from wild-type strains of *B. anthracis* exhibit variations in virulence, nutritional requirements, and sensitivity to antimicrobial drugs, bacteriophages, and lysozyme. When repeatedly subcultured on laboratory media at elevated temperatures (42.5°) wild-type strains gradually become avirulent. Indeed, it was by this method that Pasteur prepared his famous attenuated vaccine.

For some time the relation of **capsule forma-**

---

* The discovery of the anthrax toxin illustrates how the existence of a toxin may escape detection because of unfavorable conditions of artificial cultivation.

tion to virulence was confusing for in growth under ordinary laboratory conditions some attenuated strains form capsules and many virulent strains do not, suggesting a negative correlation between virulence and encapsulation. Further studies, however, revealed that virulent strains, which are unencapsulated when cultivated in air, do indeed generate capsular envelopes both in vivo and when grown on appropriate media with added carbon dioxide. In contrast, avirulent mutants, though encapsulated when cultured in the absence of carbon dioxide, are incapable of forming capsules in vivo. Inasmuch as virulence is apparently due to **both capsule formation and toxin production** (see below), not all encapsulated strains are virulent, but only those that produce both capsules and toxin in the infected host are highly pathogenic.

Sporulation and virulence are unrelated, for many nonsporulating mutants are still virulent. Nonproteolytic mutants likewise retain virulence.

PATHOGENICITY

Anthrax is primarily a disease of domesticated and wild animals. Humans become infected only incidentally, when brought into contact with diseased animals, their hides and hair, or their excreta. Many species of mammals, birds, and reptiles acquire the natural disease. Anthrax epizootics still occur in wildlife sanctuaries in Africa. Among laboratory animals, mice, guinea pigs, rabbits, goats, sheep, and monkeys are all highly susceptible.

The LD_{50} for mice, for example, is less than five spores. Rats, cats, dogs, and swine, on the other hand, will usually survive subcutaneous injections of at least a million spores. Susceptibility also varies widely with the site of inoculation.

The most common form of anthrax in man is the cutaneous variety known as **malignant pustule.** The primary lesion usually develops at the site of a minor scratch or abrasion in an exposed area of the face, neck, or upper extremities. It begins as a small, inflamed papule which later becomes vesicular. Eventually the vesicle breaks down and is replaced by a black eschar (Fig. 33-3). A striking "gelatinous" nonpitting edema surrounds the eschar for a considerable distance. At no stage is the lesion particularly painful. In severe cases of cutaneous anthrax the regional lymph nodes become enlarged and tender, and the blood stream is eventually invaded. The systemic form of the disease is frequently fatal.

Another form is inhalation anthrax (**woolsorters disease),** which results most commonly from exposure to spore-bearing dust in industrial plants where animal hair or hides are being handled. It not uncommonly leads to hemorrhagic mediastinitis and to hemorrhagic meningitis. The disease begins abruptly with high fever, dyspnea, and chest pain; it progresses rapidly and is often fatal before treatment can halt the invasive aspect of the infection.

Studies on the pathogenesis of pulmonary anthrax in susceptible laboratory animals have revealed that spores inhaled in aerosols

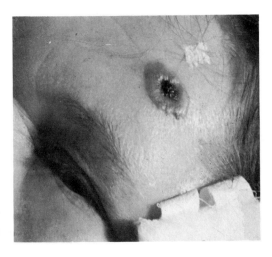

Fig. 33-3. Cutaneous anthrax, 3- to 4-day-old lesion. Note black eschar in center surrounded by rim of edema. [From Gold, H. *Arch Intern Med 96*:387 (1955).]

are phagocytized by alveolar macrophages, which in turn are carried via the pulmonary lymphatics to the regional tracheobronchial lymph nodes. There, the spores germinate, multiply rapidly, and cause an active bacterial infection of the nodes. Although many of the vegetative bacilli are destroyed by the cellular defenses of the lymph nodes, some escape and are carried by efferent lymphatics to the blood stream. Subsequently, they are rapidly "cleared" by the reticuloendothelial system (particularly the spleen), but soon overgrow this defense system and establish a massive, fatal bacteremia.

To produce a fatal pulmonary infection, however, it is necessary to introduce a relatively large number of spores into the respiratory tract. The LD_{50} for the guinea pig, for example, is approximately 20,000 spores. Of great importance also is the manner in which the spores are delivered. Since only particles (droplets) of less than 5 μ in diameter are likely to penetrate to the alveoli, the average particle size in the aerosol is critical: the LD_{50} varies directly with the median size of the particles.

The pathological changes are similar in human cases of inhalation anthrax and in the aerosol-induced disease in laboratory primates. Tracheobronchial and mediastinal lymph nodes are markedly enlarged and hemorrhagic. The striking finding in all cases is an extensive acute hemorrhagic mediastinitis characterized by marked gelatinous edema. Focal pulmonary edema and hemorrhage may be present but there is no pneumonia except in the rare patient with pre-existing pulmonary pathology. Thus, the term "anthrax pneumonia" is less appropriate than "inhalation anthrax."

The intestinal tract, commonly the portal of entry in cattle, is rarely so in man. A few human cases of intestinal anthrax have resulted from the ingestion of poorly cooked meat from infected animals. Abdominal pain, fever, vomiting, diarrhea, and shock are the principal manifestations. The mortality is extremely high and autopsy reveals hemorrhagic inflammation of the small intestine with lymphadenopathy.

Until recent years there was great confusion concerning the pathogenesis of anthrax infections, due primarily to the failure to recognize the toxigenic properties of *B. anthracis*. Discovery of the lethal toxin resulted from a systematic study of the mechanism of death in experimental anthrax. Because of the tremendous number of organisms demonstrable in the blood of animals dying of the infection, it was long assumed that death was due to blockage of the capillaries (log-jam theory). However, this concept was rendered untenable by the observation that, although streptomycin administered a few hours before death promptly controlled the bacteremia, death inevitably occurred if the number of bacteria in the blood had exceeded 3×10^6 per milliliter. Inasmuch as untreated animals usually had 300 times this number of organisms in their blood at the time of death, it was evident that some factors other than the physical presence of the organisms in the circulation was responsible for the fatal outcome. Sterile heparinized plasma, obtained from streptomycin-treated guinea pigs dying of the infection, contains a toxin that causes extensive edema and dermonecrosis when injected into the skin of normal guinea pigs and is lethal when injected intravenously or intraperitoneally.

There is now little doubt that the exotoxin plays a major role in the pathogenesis of the disease. When virulent anthrax bacilli are injected subcutaneously into the susceptible host the encapsulated organisms proliferate freely and appear to resist phagocytosis by the polymorphonuclear leukocytes which accumulate in the lesion. Animal species with leukocytes more sensitive to toxin tend to be more susceptible to the establishment of anthrax than species with highly resistant leukocytes. Thus, anthrax toxin also may play a role in this early stage of infection through a direct harmful effect on phagocytes. In keeping with this conclusion is the protective effect of antitoxic immunity, artificially induced by injections of the protective antigen of the lethal toxin.

The level of lethal toxin in the circulation increases rapidly quite late in the course of the disease, and it closely parallels the concentration of organisms in the blood. Its primary site of action is still unknown. Death from anthrax in humans or experimental animals frequently occurs suddenly and unexpectedly. Cardiac failure, increased vascular permeability, shock, hypoxia, and respiratory failure have all been implicated as the cause. Respiratory failure is regularly seen and may be of cardiopul-

TABLE 33-1.   Relation Between Dose To Establish Anthrax Infection, Number of Organisms per milliliter of Blood at Death, and Susceptibility to Toxin Challenge

Species	Relative resistance to parenteral challenge of spores	Parenteral spore dose to establish anthrax	IV toxin dose (units/kg) causing death	Quantitation of blood at death	
				Bacilli/ml	Toxin units/ml
Mouse	Very susceptible	5	1000	$10^{6.9}$	
Guinea pig	Susceptible	50	1125	$10^{8.3}$	50
Rhesus monkey	Susceptible	3000	2500	$10^{6.8}$	35
Chimpanzee	Susceptible		4000	$10^{8.9}$	110
Rat	Resistant	1×10^6	15	10^{4-6}	15
Dog	Very resistant	50×10^6	60		

Modified from Lincoln, R. E. *et al. Fed Proc 26:*1558 (1967).

monary origin or due to central nervous system depression.

Considerable variation in innate susceptibility to anthrax exists among animal species. Resistance appears to fall into two groups: 1) those resistant to establishment of anthrax, but once infection is established sensitive to the toxin; and 2) those susceptible to establishment of the disease but resistant to the toxin (Table 33-1).

IMMUNITY

Animals surviving naturally acquired anthrax are resistant to reinfection; second attacks in man are likewise extremely rare. Nevertheless, vaccines composed of killed bacilli produce no significant immunity, and anticapsular antibody is protective only in certain species, e.g., mice. Hence the attenuated bacilli of Pasteur's original vaccine were probably capable of producing enough toxin to induce an antitoxic immunity. The Pasteur strain, however, was later found to be sufficiently virulent to cause serious disease in vaccinated animals. Although it was later replaced by a safer non-encapsulated toxigenic strain for use in livestock, no attenuated spore vaccine has ever been considered in the United States to be satisfactory for human use. The best vaccine for man now appears to be an alum-precipitated preparation of the protective antigen of the lethal toxin recovered from culture filtrates. Frequent boosters are necessary to maintain resistance to anthrax challenge.

Acquired immunity to anthrax seems, then, to be due to antibodies to both the thermo-labile toxin and the capsular polypeptide. The relative importance of these two kinds of antibodies appears to vary widely in different species of hosts.

LABORATORY DIAGNOSIS

Anthrax bacilli are readily cultured from skin lesions in the vesicular stage. The organisms may even be seen in stained smears of the blister fluid and may be sometimes tentatively identified by their characteristic morphology. As the lesion ages demonstration of a bacilli becomes increasingly difficult. A specimen should always be obtained for culture prior to the administration of antimicrobial therapy. The organism is encapsulated when grown under CO_2 on a rich medium, and the cells form long chains giving a bamboo-like appearance on Gram stain. Unlike the numerous other saprophytic members of the genus *Bacillus, B. anthracis* is nonhemolytic on sheep blood agar. It is also nonmotile, unlike *B. subtilis* and many strains of *B. cereus.*

In pulmonary anthrax sputum cultures are rarely positive, because the inhaled spores usually do not germinate and multiply until they have reached the mediastinal lymph nodes. Whenever anthrax is suspected the blood should be cultured. Since other aerobic spore-forming bacilli, such as *B. subtilis,* may sometimes be cultured from the skin, respiratory tract, and even the blood, organisms thought to be anthrax bacilli should be tested for pathogenicity by intraperitoneal inoculation of a guinea pig or mouse with a washed culture. The animals succumb in 36 to 48 hours with respiratory failure, and enormous numbers of bacilli can be found in the blood and in smears of the cut surface of the spleen.

Susceptibility to a specific bacteriophage is helpful confirmatory evidence that an isolate is a strain of *B. anthracis*.

TREATMENT

Most strains of *B. anthracis* are sensitive to penicillin, tetracycline, erythromycin, and chloramphenicol. Penicillin G is the drug of choice and tetracycline is an alternative for the patient who is allergic to penicillin. These drugs are usually effective in cutaneous anthrax, although the toxin in the primary lesion often causes the formation of an eschar despite early treatment. In respiratory anthrax chemotherapy is usually ineffective, because the disease is rarely recognized before bacteremia has developed; hence the mortality rate is extremely high. Although antiserum, presumably containing antitoxin, was formerly used in the treatment of human anthrax, it has been superseded by chemotherapy.

PREVENTION

Anthrax still causes heavy loss of livestock, particularly in the Middle East, Africa, and Asia. In the United States the disease is endemic in cattle of Louisiana, Texas, South Dakota, Nebraska, and California. It is transmitted primarily through the ingestion of contaminated forage or the carcasses of infected animals. Spores may remain viable in the soil for many years. Epidemiological evidence indicates that in suitable soils anthrax bacilli can survive ecological competition with other organisms and can maintain an organism → spore → organism cycle for years without infecting livestock. Whether the organism also grows saprophytically is not known.

Animal anthrax can be at least partially controlled by immunization with attenuated vaccines. The disease has been virtually eliminated in South Africa by this means. The carcasses of animals dying of anthrax should be disposed of either by cremation or by deep burial.

Approximately seven cases of human anthrax are reported in the United States each year.* Industrial anthrax is most common among textile workers employed in plants that process goat hair imported from the Middle East.

The control of industrial anthrax poses many practical problems. Economically feasible methods of disinfecting hides without damaging them are extremely difficult to devise. Industrial workers at high risk should probably be immunized with the protective antigen, as should veterinarians practicing in "anthrax districts" and laboratory workers who have contact with *B. anthracis*. It is important to continue to enforce hygienic measures at plants (clothing changes, etc.) to keep incidence of disease at its current low level.

OTHER AEROBIC SPORE-FORMING BACILLI

Many saprophytic species of aerobic spore-forming bacilli are hard to distinguish from *B. anthracis*, except on the basis of pathogenicity. The most commonly encountered are *B. subtilis, B. megaterium,* and *B. cereus.* The DNA base composition indicates that *B. cereus* is most closely related to *B. anthracis. B. subtilis* may occasionally cause human eye infections and it may become an opportunistic invader in the presence of a foreign prosthetic material in the human body. Several outbreaks of food poisoning have been attributed to *B. cereus.*

These essentially nonpathogenic species have been widely used in studies of bacterial physiology because 1) they are easily cultured in minimal media; 2) they can be readily transformed by DNA; 3) they are permeable to actinomycin, which sharply cuts off their RNA synthesis; and 4) they are susceptible to lysozyme. Also of interest is the production of penicillinase by *B. cereus.* A strain of *B. subtilis* produces the polypeptide antibiotic bacitracin.

---

* The disease is much more prevalent in other countries; the number of human cases occurring annually in the world is estimated at 20,000 to 100,000.

SELECTED REFERENCES

Books and Review Articles

LINCOLN, R. E., and FISH, D. C. Anthrax Toxin. In *Microbial Toxins,* vol. III, ch. 9. (T. C. Montie, S. Kadis, and S. J. Ajl, eds.) Academic Press, New York, 1970.

NUNGESTER, W. J. Proceedings of the conference on progress in the understanding of anthrax. *Fed Proc 26:*1491 (1967).

SMITH, H. The use of bacteria grown in vivo for studies on the basis of their pathogenicity. *Annu Rev Microbiol 12:*77 (1958).

Specific Articles

ALBRINK, W. S. Pathogenesis of inhalation anthrax. *Bacteriol Rev 25:*268 (1961).

BERDJIS, C. C., GLEISER, C. A., and HARTMAN, H. A. Experimental parenteral anthrax in *Macaca mulatta. Br J Exp Pathol 44:*101 (1963).

BRACHMAN, P. S. Anthrax. *Ann NY Acad Sci 174:* 577 (1970).

MEYNELL, E., and MEYNELL, G. G. The roles of serum and carbon dioxide in capsule formation of *Bacillus anthracis. J Bacteriol 34:*153 (1964).

SMITH, H., and STANLEY, J. L. The three factors of anthrax toxin: Their immunogenicity and lack of enzymic activity. *J Gen Microbiol 31:*329 (1963).

chapter

34

ANAEROBIC SPORE-FORMING BACILLI

Revised by MORTON N. SWARTZ, M.D.

GENERAL PROPERTIES OF THE CLOSTRIDIA   830

    Morphology   830
    Anaerobiosis   830
    Metabolism and Classification   831

BOTULISM   833

    Botulinum Toxin   833
    Pathogenesis   833
    Immunity   834
    Laboratory Diagnosis   834
    Treatment   834
    Prevention   834

TETANUS   835

    Pathogenesis   835
    Immunity and Antigenic Structure   836
    Laboratory Diagnosis   836
    Treatment   837
    Prevention   837

HISTOTOXIC CLOSTRIDIAL INFECTIONS   838

    Pathogenesis   838
    Laboratory Diagnosis   839
    Treatment   841
    Prevention   841

GENERAL PROPERTIES OF THE CLOSTRIDIA

The anaerobic spore-forming bacilli belong to the genus *Clostridium*. Most species are obligate anaerobes, but some will grow under microaerophilic conditions. Their natural habitat is the soil and the intestinal tracts of animals and man. A few of these saprophytes are also human pathogens under appropriate circumstances, causing these very different diseases: **botulism, tetanus,** and **gas gangrene.** The pathogenicity of these organisms depends on the release of highly destructive enzymes or powerful exotoxins.

MORPHOLOGY

The clostridia are relatively long, narrow, pleomorphic, gram-positive, rod-shaped organisms. Long filamentous forms are common. In 24- to 48-hour cultures many of the bacilli may be gram-negative. All species form spores, but with considerable variation in the required conditions. The highly refractile spores (Fig. 34-1) are oval or spherical and usually wider than the parental cell (L. *clostridium* = spindle); they may be central, terminal, or subterminal in the cell. Most species of clostridia possess peritrichous flagella and are motile (Fig. 34-2). A few produce demonstrable capsules (*C. perfringens,* Fig. 34-3). Colony forms are variable; hemolysis on blood agar is frequent.

ANAEROBIOSIS

The clostridia lack the cytochromes required for electron transport to O_2 (Ch. 3) and the catalases and peroxidases that would destroy hydrogen peroxide; their sensitivity to O_2 may be due to the presence of flavoprotein enzymes that reduce O_2 to this toxic product. Not all clostridia are equally oxygen-sensitive: *C. tetani* requires strict anaerobic conditions; *C. perfringens* is much less fastidious; and *C. histolyticum* produces small but visible colonies on aerobic blood agar plates. Clostridial spores not only are resistant to heat and disinfectants but survive long periods of exposure to air; they germinate only under strongly reducing conditions. Specimens for anaerobic culture should be inoculated immediately upon collection and incubated under anaerobic conditions. These are commonly provided either by an anaerobic jar or by use of semisolid media containing strong reducing agents.

In one relatively efficient type of anaerobic jar the air is evacuated and replaced by hydrogen gas, and the residual oxygen is removed by reaction with the hydrogen, catalyzed by platinized or palladinized asbestos in an electrically heated gauze cage within the cover of the jar. Carbon dioxide (2 to 10%) should be included in the atmosphere since it stimulates growth of many strains of clostridia. The recent development of disposable hydrogen–carbon dioxide generators (Gaspak) permits anaerobic culture in any diagnostic laboratory without the need of vacuum pumps, gas tanks, etc. The cultures are placed in a jar along with a hydrogen generator envelope; water is added to the envelope (containing Mg and $ZnCl_2$), and hydrogen rapidly evolves.

The most commonly used reducing medium, which can be incubated aerobically, contains 0.05% sodium thioglycollate, with 0.07% agar to increase viscosity and thus minimize the spread of oxygen by convection. Most strains of pathogenic clostridia will also grow in deep tubes of fluid media containing fresh animal tissues (e.g., chopped meat) which provide the necessary reducing systems despite autoclaving.

Clostridia are frequently found mixed with enteric

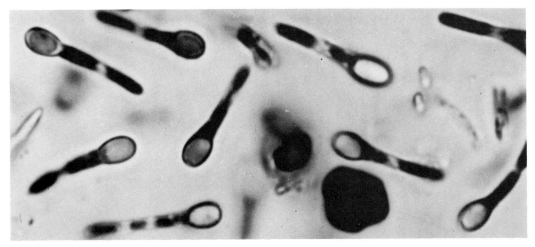

Fig. 34-1. Terminal clostridial endospores in stained wet-mount preparation. ×3600. (Courtesy of C. F. Robinow.)

gram-negative bacilli (e.g., *Proteus*) which may overgrow them in broth and even on anaerobic plates. Since most clostridia sporulate in thioglycollate medium, they may be selectively recovered by "heat shocking" (80°, 30 minutes) a mixed culture before plating.

METABOLISM AND CLASSIFICATION

Most clostridia produce large amounts of gas (mainly CO_2 and H_2) in butyric fermentation (Ch. 3). The fermentation of various sugars is of value in differentiating species. Other biochemical tests include reactions in milk, the liquefaction of gelatin, the reduction of nitrate, and the production of indole from tryptophan (Table 34-1). The characteristic "stormy fermentation" of milk by *C. perfringens* is due to the formation of a clot which becomes torn by the accumulating gas. Some clostridia are predominantly proteolytic and others saccharolytic.

A wide variety of **enzymes** have been identified in the filtrates of clostridial cultures, including collagenase, other proteinases, hyaluronidase, deoxyribonuclease, lecithinase, and neuraminidase. Some of these are known to act as toxins in the animal host; other potent protein exotoxins are **botulinum toxins** and **tetanus toxin.** Many species produce hemolysins. Antibodies to some of these

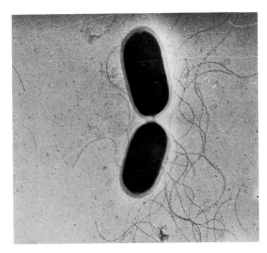

Fig. 34-2. Electron photomicrograph of *C. tetani* showing cell wall and peritrichous flagella. ×11,000, reduced. (Courtesy of Stuart Mudd.)

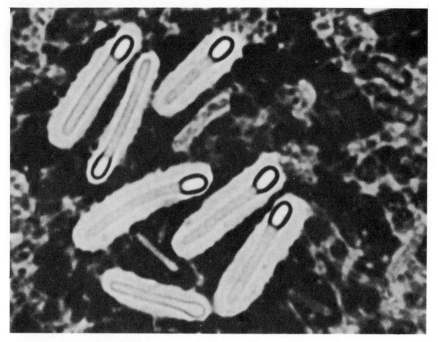

Fig. 34-3. Nigrosin preparation of *C. pectinovorans* showing both capsules and endospores. ×3600. (Courtesy of C. F. Robinow.)

products may be of use in identifying individual clostridial species.

The clostridia can also be classified on the basis of their cellular antigens. As in many aerobic bacteria, the protein flagellar (H) antigens are thermolabile, and the lipopolysaccharide somatic (O) antigens are relatively thermostable. Some species are most conveniently subdivided into immunological types by O antigens, others on the basis of H antigens.

Nontoxigenic mutants of various species have been described.

TABLE 34-1. Biochemical Reactions of Common Species of Pathogenic Clostridia

Disease and species	Fermentation reactions						Gelatin lique- faction	Nitrate reduc- tion	Indole forma- tion
	Milk	Glucose	Maltose	Lactose	Salicin	Sucrose			
Botulism									
C.botulinum	A	+	+	−	+	±	+	−	−
Tetanus									
C. tetani	C	−	−	−	−	−	+	−	+
Gas gangrene									
C. perfringens	ACGS	+	+	+	±	+	+	+	−
C. novyi	CG(D)	+	+	−	−	−	+	−	−
C. septicum	ACG	+	+	+	+	−	+	+	−
C. histolyticum	CD	−	−	−	−	−	+	−	−
C. tertium	ACG	+	+	+	+	+	−	+	−
*C. bifermentans**	CD	+	+	−	+	−	+	−	+
*C. sporogenes**	D	+	+	−	±	−	+	+	−

* Produces no detectable toxin. A = acid; C = clot; D = digestion; G = gas; S = stormy; ( ) = may be absent; ± = variable.

BOTULISM

Human botulism is ordinarily **not an infectious disease** but rather an **intoxication,** resulting from the ingestion of food, containing the toxin; the toxemia is not the result of multiplication of *C. botulinum* in the gastrointestinal tract. It thus resembles staphylococcal food poisoning. A few cases of botulism from wound infections have been reported, but the organism is virtually noninvasive. Ingestion of uncooked sausage (L. *botulus*) was for many years one of the most common causes of the disease.*

BOTULINUM TOXIN

C. botulinum is widely distributed in soil, in the silt of lake and pond bottoms, and on vegetation; hence, the intestinal contents of mammals, birds, and fish may occasionally contain these organisms. The disease results from the ingestion of uncooked foods of animal origin (e.g., sausage, spiced meat, smoked fish), or improperly canned fruits and vegetables, in which contaminating spores of the organism have germinated and produced toxin. Food that is not visibly fermented or spoiled to taste may still contain botulinum toxin.

The spores are relatively heat-resistant, and pressure sterilization is necessary to ensure their destruction. Effective sterilization methods are routine in the canning industry; home-canned and preserved foods have been the source of most outbreaks of botulism in this country. In contrast to the spores, **botulinum toxin is relatively heat-labile,** being completely inactivated at 100° for 10 minutes; hence boiling just prior to ingestion renders home canned vegetables or processed fish safe. Care should be exercised in handling suspect food samples in the laboratory since the toxin may be absorbed from fresh wounds and mucosal surfaces.

The toxin is not inactivated by the acidity of the gastric secretions or by the proteolytic enzymes of the stomach and upper bowel. In fact, crude toxin of *C. botulinum* type E can be potentiated 10- to 1000-fold by partial proteolysis (e.g., limited trypsin treatment). Since activation cannot be shown with toxin-containing serum from patients with botulism, the toxin may have already undergone proteolysis in the gastrointestinal tract.

Six types of *C. botulinum* (A to F) have been recognized; each elaborates an immunologically distinct form of neurotoxin. These are the most powerful toxins known; 1 μg contains 200,000 minimal lethal doses for a mouse, and is nearly a lethal dose for man. Types A, B, E, and F are the principal causes of human illness; types C and D are associated with botulism in fowl and cattle, respectively. The various types tend to be distributed in different kinds of food. Thus, earlier outbreaks in the United States most frequently involved types A or B, in home-canned products; but in the past decade type E, in smoked fish, has been the commonest.

Toxin production is associated with germination of spores and growth of vegetative cells. In cultures of toxigenic *C. botulinum* little toxin appears in the medium until late in growth, when autolysis occurs. The toxigenicity of type C appears to be dependent on the presence of a specific bacteriophage, CEβ: treatment with ultraviolet light or acridine orange, which can "cure" bacteria of plasmids, make the organism nontoxigenic and infection with the phage renders them toxigenic again.

The type A toxin, complexed with a hemagglutinin, has been crystallized as a protein (MW ca. 900,000). Separation from the hemagglutinin can be effected without any loss of activity of the toxin. The purified toxin (MW 150,000) may consist of smaller, toxic subunits.

PATHOGENESIS

The potency of botulinum toxins (Ch. 22, Exotoxins) is undoubtedly related to their affinity for cells of the nervous system, where electron microscopic studies with ferritin-labeled toxin have demonstrated attachment to presynaptic terminals of cholinergic nerves. Moreover, electrophysiological studies have shown that they block release of the transmitter (acetylcholine) of these endings, in which normal vesicles can be seen with the

---

* The celebrated German pathologist Virchow is said to have proposed sausages as weapons when challenged to a duel.

electron microscope.* The molecular nature of the toxin receptor in the presynaptic membrane is not certain but is suggested by the finding that certain gangliosides bind botulinum toxin. This interaction is a function of the number of sialic acid residues in the ganglioside.

After absorption from the gastrointestinal tract the toxin reaches susceptible neurons by way of the blood stream. Symptoms in man usually begin 18 to 36 hours after ingestion.

Weakness, dizziness, and severe dryness of the mouth and pharynx are early symptoms, and nausea and vomiting are common with the type E disease. Neurological manifestations soon develop: blurring of vision, dilation of pupils, inability to swallow, difficulty in speech, urinary retention, generalized weakness of skeletal muscles, and respiratory paralysis.

The fatality rate has been extremely high but has declined to 20 to 35% in this country in the past decade. Fortunately, the disease is much less common in man than in animals. In the United States, 155 cases were reported in the last 8 years.

IMMUNITY

The toxins that cause botulism in man are each specifically neutralized by their antitoxins. Botulinum toxin, as toxoid, is a good antigen: as little as 1 μg will induce high levels of protective antibody in the mouse. Clinical botulism, however, does not induce demonstrable antibody because an amount of toxin sufficient to induce an immune response would be lethal.

LABORATORY DIAGNOSIS

In cases of suspected botulism mice should be injected intraperitoneally with the patient's serum and with aqueous cell-free extracts of

---

* Although the various types of botulinum toxin appear to have the same pharmacological action, they differ greatly in their relative toxicities for different animal species. The ratio of the lethal doses for mice and fowl, for example, are 1:15 for type A, 1:2000 for type C, 1:100,000 for type D, and 1:25 for type E. The differences may well be of epidemiological significance.

the implicated food. Heat-inactivated samples should also be inoculated to serve as controls for the presence of nonspecific toxin substances. Trypsin treatment of a portion of food samples may enhance toxin activity (particularly type E) and make detection easier. If significant amounts of toxin are present the mice will develop paralysis and succumb in 1 to 5 days. The toxin may be typed by appropriate protection tests with specific antisera.

Samples of the suspected food should also be cultured anaerobically, with heat treatment of part of the sample to select for spores. Since only vegetative cells or relatively heat-sensitive spores (type E) may be present, part of the sample should be incubated without heating. If an anaerobic bacillus is recovered from the food, it should be tested for toxin production. The cultural approach is generally less useful than direct tests for the toxin.

TREATMENT

Once toxin has become "fixed" at susceptible nerve endings its harmful action is unaffected by antitoxin (cf. diphtheria; Ch. 24, Treatment). Antitoxin does not appear to significantly alter the clinical course of type A botulism, once neurological symptoms have occurred. However, beneficial results (even after the onset of symptoms) have followed the use of antitoxin in several outbreaks of type E botulism. To neutralize any circulating or "unfixed" toxin, equine polyvalent antiserum (types A, B, and E) should be given intravenously in suspected cases at once, without awaiting the results of laboratory tests. The usual precautions should be taken against hypersensitivity reactions to the foreign serum.

PREVENTION

Other individuals known to have ingested the same food as the patient should also be treated, even if asymptomatic. As already mentioned, boiling of any improperly canned or processed food for at least 10 minutes will destroy botulinum toxin. A pentavalent toxoid evokes a good antibody response but its use in man is justified only in frequently exposed laboratory workers. Toxoids have been used successfully in the immunization of cattle in areas where the disease is endemic.

TETANUS

C. tetani is present in the soil and in the feces of various animals. Only rarely is it found in the feces of man. It produces a potent neurotoxin (tetanospasmin) which, like botulinum toxin, is produced by growing cells and released only on cell lysis. Hence, the disease tetanus appears only when spores of *C. tetani* germinate after gaining access to wounds. The disease is of particular importance in military medicine. Because only **one antigenic type of toxin** is involved, an effective monotypic toxoid could be developed for prophylactic immunization. Less than 200 cases of tetanus occur annually in the United States because of widespread immunization with toxoid. However, the disease remains a major problem worldwide (approximately 350,000 cases a year).

PATHOGENESIS

Infection. The pathogenicity of *C. tetani* stems solely from the effects of its neurotoxin. The introduction and subsequent germination of spores in host tissue are necessary for the development of the disease, but *C. tetani* is not a histotoxic organism. Most cases result from small puncture wounds or lacerations; the infection remains localized to the traumatized tissue at the site of entry, usually with only a minimal inflammatory response. Mixed infection with other organisms may induce more marked inflammation and promote the growth of *C. tetani*. However, in 5 to 10% of cases the initial injury is so trivial as to have been forgotten by the patient and to have left no residue.

Access of tetanus spores to open wounds does not necessarily result in disease: *C. tetani* can frequently be cultured from wounds of patients without tetanus, for, in clean wounds, where the blood supply is good and the oxygen tension remains high, germination will rarely occur. In necrotic and infected wounds, on the other hand, the anaerobic conditions will permit germination. Contaminated puncture wounds, appearing relatively trivial on the surface, may be particularly dangerous, especially when containing a foreign body. Spores may occasionally remain dormant in healed wounds for months (a latent period as long as 10 years has been recorded); trauma to the area may then cause germination and disease.

Tetanus neonatorum results from postnatal infection of the degenerating remnant of the severed umbilical cord.

The importance of tissue necrosis is readily demonstrable in laboratory experiments. Mice inoculated intramuscularly with heavy suspensions of tetanus spores fail to develop the disease unless necrotizing chemicals, such as calcium chloride or lactic acid, are injected with the spores.

Clinical Patterns. The incubation period of tetanus may range from several days to many weeks. An incubation period of less than 4 days is associated with a very high mortality. Although the portal of entry is most commonly a puncture wound or laceration, other foci of infection include burns, skin ulcers, compound fractures, operative wounds, and sites of subcutaneous injection of adulterated narcotics by addicts.

Generalized tetanus, the usual form of the disease, is characterized by severe and often painful spasms and rigidity of voluntary muscles. The usual sequence is local injury followed after some days by mild intermittent muscular contractions near the wound, then trismus ("lockjaw," spasm of the masseter muscles), generalized rigidity, and violent spasms of the trunk and limb muscles. Spasm of the pharyngeal muscles causes difficulty in swallowing. Death ordinarily results from interference with the mechanics of respiration. The patient's sensorium remains clear. Fever is not seen except as a result of increased metabolism (in patients with severe spasms) or of complicating infection.

Local tetanus is a much rarer form of the disease, usually occurring in individuals with partial immunity or as the result of minor wounds containing only a few organisms. It is characterized by localized twitching and spasm in muscles near the wound. It may persist for weeks or months and then subside, or mild trismus may follow. Generalized tetanus can readily be reproduced in experimental animals by intravenous injection of toxin and local tetanus by intramuscular injection.

Toxin Action. The potency of tetanus toxin is similar to that of type A botulinum toxin

(Ch. 22, Exotoxins). The protein can exist as a toxic monomer (MW 68,000) or a nontoxic dimer (still antigenic) of twice that molecular weight. The toxin is produced in vitro in amounts up to 5 to 10% of the bacterial weight, but it serves no known useful function for the bacillus. Toxin is formed during bacterial growth and is released only by lysis.

Animals vary widely in their susceptibility to tetanus toxin: mammals are most sensitive and birds are relatively resistant (e.g., the pigeon is 24,000 times as resistant as the guinea pig). The frog is highly resistant at its natural temperature (10 to 15°) but its susceptibility increases with increasing temperature; at the lower temperature both the fixation of toxin and its action after fixation are blocked.

Upon injection into the central nervous system tetanus toxin becomes rapidly "fixed." Indeed, when mixed with brain emulsion a comparatively large dose can be injected into a susceptible animal without causing damage (Wassermann-Takaki phenomenon). Subcellular fractions of brain rich in synaptosome membranes have a higher toxin-fixing capacity. The substance responsible for this fixation is a **ganglioside** containing stearic acid, sphingosine, glucose, galactose, N-acetyl galactosamine, and N-acetyl neuraminic acid (sialic acid). Fixation depends on the number and positions of the sialic acid residues. Conversion to toxoid destroys reactivity with the ganglioside. It is not clear whether tetanus toxoid affects ganglioside function in the membrane or is localized by ganglioside and then acts enzymatically on a neighboring molecule: it does not alter the ganglioside structure.

The main action of tetanus toxin is on the anterior horn cells of the spinal cord and on the brain stem. Whether it reaches its sites of action via the blood stream or the peripheral nerves has long been a matter of controversy. It now seems clear that tetanus toxin travels to the spinal cord, from the entry wound, in the spaces between the fibers of peripheral nerves. Studies of the movement of other labeled proteins, as well as cord transection experiments, strongly suggest that once toxin has reached the spinal cord it ascends within it to the medulla.

Local tetanus (see above) is apparently due not to the action of the toxin directly on the muscle but to early involvement of anterior horn cells at the level of initial entrance of the toxin. However, when excessive amounts are generated in the tissues enough may reach the blood stream to cause generalized involvement before the signs of local tetanus have developed.

The spasmogenic effect of the toxin has been shown by Eccles to be due to its blocking action at several types of spinal inhibitory synapses, resulting in hyperreflexia and spasm of skeletal muscles. There is evidence that tetanus toxin blocks the release of the inhibitory transmitter (? glycine) from the inhibitory terminals.

In addition to its neurogenic toxin, *C. tetani* produces an oxygen-labile hemolysin, antigenically similar to streptolysin O (Ch. 26, Extracellular products), known as **tetanolysin.** There is no evidence that it plays a significant role in pathogenesis.

IMMUNITY AND ANTIGENIC STRUCTURE

Several serological types of *C. tetani* can be distinguished by their flagellar antigens. All, however, share a common O antigen and produce the same exotoxin.

Immunity to tetanus involves antibody to toxin: antibodies to somatic antigens are not protective. Tetanus toxoid is an excellent immunogen in man, but clinical tetanus does not induce immunity because so little toxin is released. It is therefore extremely important to actively immunize all patients on recovery from tetanus.

LABORATORY DIAGNOSIS

The diagnosis of tetanus is usually based on clinical findings alone. Although attempts should be made to culture *C. tetani* from all suspicious lesions, the use of antitoxin should not be withheld for the 2 to 3 days needed to identify the organism. The tetanus bacillus may be provisionally recognized by its terminal ("tennis racket") spores (Fig. 34-1), the narrow zone of hemolysis on blood agar, and its surface colonies surrounded by an area of swarming (Ch. 29, Proteus group).* Since, however, other gram-positive anaerobic spore-

---

* Staining with fluorescent antibody to the somatic O antigen will probably prove the best method for rapid identification.

forming bacilli may be cultured from infected wounds,* positive identification requires demonstration of production of toxin and its neutralization by specific antitoxin. A mouse protection test is commonly used for this purpose. The organism is cultured from an infected focus in only about 30% of cases; moreover its isolation does not necessarily establish a clinical diagnosis of tetanus.

TREATMENT

Antitoxin should be promptly administered in all cases of suspected tetanus to neutralize circulating, or otherwise accessible, toxin; it is ineffective against toxin already "fixed" in the central nervous system. Tetanus immune globulin (prepared from the plasma of persons who have been hyperimmunized with tetanus toxoid) has recently become available in this country and has supplanted conventional equine antitoxin (introduced by Behring and Kitasato in 1890). The risk of sensitivity reactions with human material is minimal, and antibodies persist longer (half life at 3 to 4 weeks) than with equine antitoxin. A dose of 3000 to 6000 U† should be administered intramuscularly.

In areas where human tetanus immune globulin is not available, equine antitoxin should be administered in a dose of 100,000 U, half given intravenously and half intramuscularly. This amount is far in excess of that needed to neutralize all the preformed toxin present in the body of an adult, as well as that which may be subsequently elaborated. It is essential to test for immediate-type hypersensitivity to horse serum before injecting antitoxin, and to take appropriate precautions if the result is positive (Ch. 20, Serum sickness). Local injections of antitoxin, totaling 10,000 to 20,000 U, may also be made around the suspected primary lesion. Careful surgical débridement of the lesion, and removal of any foreign bodies present, should be carried out only after the antitoxin has been given. Penicillin should also be administered to prevent further toxin formation in the infected tissue, though chemotherapy alone should never be relied upon.

Supportive measures to minimize spasm and aid respiration are of the utmost importance. These measures may be needed for one or more weeks since "fixed" toxin in the central nervous system decays slowly. In severe cases of generalized tetanus the mortality rate is approximately 50% even with the most intensive treatment.

PREVENTION

Because the spores of *C. tetani* are so widely disseminated, the only effective way to control tetanus is by **prophylactic immunization. Tetanus toxoid,** in combination with diphtheria toxoid and pertussis vaccine (DPT), is usually given during the first year of life. Three injections of either fluid or alum-precipitated toxoid should be administered in the initial course of immunization. A single booster injection (DPT) should be given about a year later (Ch. 17, Secondary response) and again on entering elementary school. Subsequent booster immunization with a single dose of adult-type tetanus and diphtheria toxoid (Td) is recommended at 10-year intervals for persons at relatively high risk. Whenever a previously immunized individual sustains a potentially dangerous wound, a booster of toxoid (Td) should be injected. (Booster injections may be effective even after 10 to 20 years, inducing protective levels of antibody within 1 to 2 weeks. Seriously wounded subjects who have **not** previously been immunized, on the other hand, should be given prophylactic human tetanus immune globulin (250 U), since the antibody response to an initial toxoid injection is too slow to be useful.

Intensive programs of prophylactic immunization with toxoid in both civilian and military populations‡ have led to a striking reduction in the incidence of tetanus and have eliminated the danger of inducing hypersensitivity reactions with foreign antiserum.

---

* About 30% of war wounds, for example, become contaminated with Clostridia.

† An American unit of antitoxin is defined as 10 times the smallest amount needed to protect for 96 hours a 350-gm guinea pig inoculated with a standard dose of toxin furnished by the National Institutes of Health.

---

‡ The incidence of tetanus in World War II among troops actively immunized with toxoid was only 1/10 of that in World War I, in which only passive prophylaxis with antitoxin was available.

HISTOTOXIC CLOSTRIDIAL INFECTIONS

A variety of species of the genus *Clostridium,* many of which are toxigenic, are associated with invasive infection in man. Those most commonly encountered are listed under **gas gangrene** in Table 34-1. They are not highly invasive when introduced into healthy tissues; but in the presence of preexisting tissue injury, particularly damaged muscle, they can be responsible for a rapidly progressive, devastating infection characterized by the accumulation of gas and the extensive destruction of muscle and connective tissue. The toxins produced have necrotizing (histolytic), hemolytic, and/ or lethal properties. In addition, enzymes such as collagenase,* proteinase, deoxyribonuclease, and hyaluronidase, elaborated by the growing organisms, cause accumulation of toxic degradation products in the tissues.

C. perfringens (L., "breakthrough"), formerly known as *C. welchii,* is the species most commonly involved, particularly in peacetime. It is normally a commensal in the human intestinal tract and may produce any of the following clinical pictures: 1) wound infection, 2) uterine infection, 3) bacteremia, 4) enteric infection.

PATHOGENESIS

Wound Infection.   There are three types of clostridial wound infection: Wound "contamination," anaerobic cellulitis, and true myonecrosis (gas gangrene). From 80 to 90% of the isolations of *C. perfringens* from hospitalized patients represent simple saprophytic wound contamination, especially of operative wounds, skin ulcers, etc., and usually in association with other organisms. The presence of the organism in this setting does not herald invasive infection.

Anaerobic cellulitis is a clostridial infection that does not involve the muscles and is much less aggressive than gas gangrene. It begins with the introduction of spores into an open wound. Severe wounds, such as those acquired in automobile accidents and military combat, grossly contaminated with soil or fecal mat-

ter and harboring a mixture of organisms, are particularly likely to contain clostridial spores. Germination occurs in devitalized tissue where damage to the blood supply and the presence of foreign material has caused a lowering of the redox potential. The vegetative bacilli multiply, and anaerobic cellulitis usually develops after several days and extends widely, spreading in the fascial planes; the marked gas formation is detected by the resulting crepitus. The infection rarely produces local pain, edema, toxemia, or invasion of healthy muscle.

Gas gangrene is a more serious infection in which toxins produced in ischemic muscle make the adjacent muscle also ischemic and thus allow spread of the anaerobic causative organism; this progressive process rapidly leads to irreversible shock. About 70% of cases are due to *C. perfringens,* and most of the rest to *C. novyi* and *C. septicum.* It is not uncommon, however, to find other aerobic and anaerobic organisms present in cultures of infected material.

The onset is usually sudden, following an incubation period of 6 to 72 hours after injury or abdominal surgery. The involved area is edematous and the skin has a dark yellowish discoloration. Thin dark fluid is exuded, often in tense blebs; aspiration reveals many clostridia but few polymorphonuclear leukocytes. Gas is present in the subcutaneous tissue and muscles.† Exploration of the wound reveals the muscle involvement: swollen, grayish or purple in color, and ischemic.

Uterine Infection.   *C. perfringens* is present in the genital tract of about 5% of women. Following abortion, or uncommonly after prolonged labor, this organism may invade the uterine wall producing extensive necrosis, high fever, and circulatory collapse. Severe bacteremia, unusual in gas gangrene, is characteristic of this process. Severe intravascular hemolysis due to the α-toxin (see below) causes hemoglobinemia, leading to acute renal shutdown.

Bacteremia.   Transient low-grade bacteremia due to *C. perfringens,* or to nonhistotoxic clostridia (e.g., *C. sporogenes, C. tertium*), occasionally occurs in patients with leukemia or gastrointestinal bleeding,

---

* Collagen, unless denatured to gelatin, is not digested by the usual proteases.

---

† The gas that accumulates is predominantly hydrogen, which is less soluble than carbon dioxide (see Ch. 3, Butyric fermentation).

TABLE 34-2. Toxins and Toxigenic Types of *C. perfringens*

	Toxins	Bacterial types				
		A	B	C	D	E
α	(lecithinase)	+++	+++	+++	+++	+++
β	(lethal, necrotizing)	−	+++	+++	−	−
γ	(lethal)	−	++	++	−	−
δ	(lethal, hemolytic)	−	+	++	−	−
ε	(lethal, necrotizing)	−	+++	−	+++	−
η	(lethal)	+	?	?	?	?
θ	(lethal, hemolytic)	+	++	+++	+++	+++
ι	(lethal, necrotizing)	−	−	−	−	+++
κ	(collagenase)	+	+	+++	++	+++
λ	(proteinase)	−	+	−	++	+++
μ	(hyaluronidase)	++	+	+	++	+
v	(deoxyribonuclease)	++	+	++	++	++

+++ = most strains; ++ = some strains; + = a few strains; − = not produced.

or following intraabdominal catastrophe. The clinical picture is like that of bacteremia due to other bacterial species.

Enteric Infections.   *C. perfringens* is a common cause of **food poisoning,** ranking second only to *Staphylococcus aureus* in the frequency of outbreaks in the United States. It produces a self-limited gastroenteritis, lacking constitutional symptoms, about 12 hours following ingestion of heavily contaminated food. The "food poisoning" strains belong to type A (based on the toxin produced, Table 34-2). The incriminated food is usually a meat product that had been stewed or boiled and then held before reheating and serving: heat-resistant spores would survive the cooking and later germinate during cooling in a medium from which oxygen had been driven out.

The mechanism of pathogenicity in *C. perfringens* food poisoning has not been established. Relatively large numbers of viable organisms must be ingested, and they can subsequently be isolated from the feces. An enterotoxin has been found in culture filtrates which produces ileal loop fluid accumulation and diarrhea in rabbits. It is produced only during sporulation, but no role in sporulation has been uncovered. In contrast, α-toxin is produced during logarithmic growth.

A much more serious form of clostridial enteric infection, **enteritis necroticans,** occurred in epidemic form in Germany at the close of World War II and is seen among the natives of New Guinea currently. This acute, severe, inflammatory process in the small bowel is associated with a high mortality. *C. perfringens* type C has been isolated from the lesions. This organism produces large amounts of β-toxin, and during the disease β-antitoxin levels rise.

Toxin Action.   The exotoxins produced by the clostridia that cause gas gangrene are too numerous to discuss.* Those elaborated by *C. perfringens* (Table 34-2) have been most thoroughly studied. One of these, the α-toxin, is a calcium-dependent lecithinase; it is lethal, necrotizing, and hemolytic. These actions, which can all be neutralized by antilecithinase serum, are due primarily to splitting of **lecithin in cell membranes** (Fig. 34-4). (It also catalyzes the hydrolysis of cephalin and sphingomyelin.) Since lecithin is present in the membranes of many different kinds of cells the toxin can cause damage throughout the body, and this can account for intravascular hemolysis in uterine infection and the paucity of leukocytes in the exudate of gas gangrene.

The θ-toxin of *C. perfringens* is also hemolytic and necrotizing but does not hydrolyze lecithin; it is inactive in the presence of cholesterol and it is oxygen-labile. It cross-reacts immunologically with both streptolysin O and tetanolysin. Other species of clostridia elaborate similar exotoxins, each of which is immunologically specific, but their precise roles in pathogenesis remain obscure.

LABORATORY DIAGNOSIS

The bacteriology of gas gangrene is complicated, because most of the infections are mixed.

---

* For detailed review, see MacLennan.

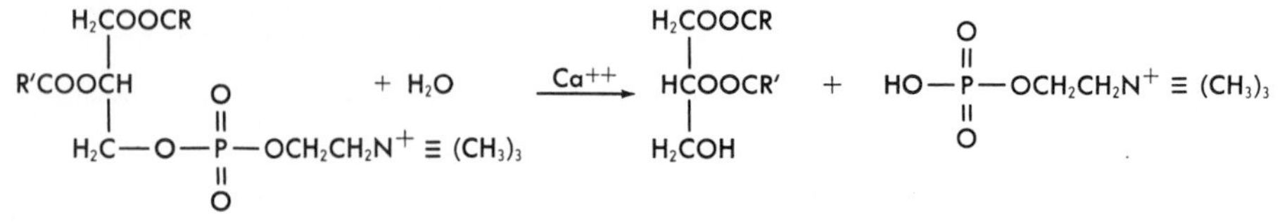

$$
\underset{\text{Lecithin}}{\begin{array}{l} H_2COOCR \\ | \\ R'COOCH \qquad\quad O \\ | \qquad\qquad\quad \| \\ H_2C-O-P-OCH_2CH_2N^+\equiv(CH_3)_3 \\ \qquad\qquad \| \\ \qquad\qquad O \end{array}} + H_2O \xrightarrow{\;Ca^{++}\;} \underset{\text{Diglyceride}}{\begin{array}{l} H_2COOCR \\ | \\ HCOOCR' \\ | \\ H_2COH \end{array}} + \underset{\text{Phosphorylcholine}}{\begin{array}{l} \qquad O \\ \qquad \| \\ HO-P-OCH_2CH_2N^+\equiv(CH_3)_3 \\ \qquad \| \\ \qquad O \end{array}}
$$

Fig. 34-4. Action of *C. perfringens* α-toxin (lecithinase) on lecithin.

Furthermore, the presence of gas in the tissues does not necessarily incriminate clostridia, for klebsiellae, bacteroides, *E. coli,* and anaerobic streptococci may also cause accumulation of enough gas to be detectable by palpation or roentgenography. The finding of many plump gram-positive bacilli but few granulocytes in smears of exudate from crepitant wounds (or from adjacent bullae) or from the uterine cervix suggests clostridial infection; *C. perfringens* rarely forms spores in infected wounds. *C. septicum,* the cause of about 10% of cases of gas gangrene in wartime, may grow in vivo in chains made up of elongated cells.

Samples of the exudate should always be cultivated in thioglycollate medium, as well as in nutrient broth and on blood agar plates, incubated both anaerobically and aerobically. If gram-positive spore-bearing bacilli are found in the thioglycollate growth, or if colonies of such bacilli appear on the blood agar incu-

bated anaerobically but not on that incubated aerobically, the genus *Clostridium* should be suspected.

"Heat shocking" (see above) is helpful when specimens contain other organisms which may outgrow the clostridia. Unfortunately, this procedure cannot be used to isolate *C. perfringens,* which fails to produce spores in most media. It is separated from other organisms by prolonged incubation of the mixed culture in thioglycollate medium containing 0.6% glucose. After 48 to 72 hours of incubation, so many of the contaminating organisms have died out that *C. perfrigens* can be readily identified in subcultures inoculated on anaerobic blood plates.

The identification of individual clostridial species is based primarily on the biochemical tests listed in Table 34-1. *C. perfringens* can be recognized by its capsule formation (Fig. 34-3), lack of motility, and lactose fermentation. Moreover, α-toxin (lecithinase) production causes a visible opacity to form around colonies grown on egg yolk (Nagler) medium (Fig.

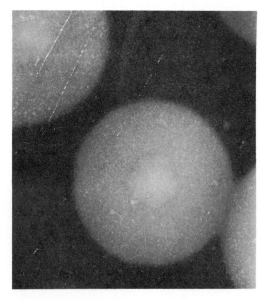

Fig. 34-5. Colonies of *C. perfringens* grown on egg yolk medium. Halo of precipitate about colonies is due to splitting of lecithin in medium by enzymatic action of lecithinase released by organisms. ×5, enlarged. [From McClung, L. S., and Toabe, R. *J Bacteriol 53*:139 (1947).]

34-5),* and type A antitoxin in the agar plate prevents this opacification without inhibiting growth. With other clostridia found in wounds demonstration of histotoxin production in animals, and its neutralization with specific antitoxin, is useful.

TREATMENT

Clostridial wound infection requires surgical debridement, as promptly as possible. It is important to distinguish between anaerobic cellulitis and gas gangrene, since the former requires wide excision and débridement and the latter complete extirpation of involved muscle (usually amputation of a limb). Polyvalent equine antitoxin prepared against toxic filtrates of *C. perfringens, C. novyi, C. septicum* and *C. histolyticum* should be given (after appropriate testing for horse serum sensitivity). Penicillin is also administered to pre-

---

* The opacity is due to the insolubility of the diglyceride formed from egg lecithin (Fig. 34-3).

vent bacteremia and to halt further spread of infection. Hyperbaric oxygen at 3 atm in a compression chamber, to raise the oxygen tension in healthy tissue to a level at which clostridia cannot multiply, may have a place in therapy, especially in gas gangrene of the trunk, where complete débridement may be precluded.

PREVENTION

Prompt surgical débridement of dirty wounds greatly decreases the incidence of anaerobic cellulitis and gas gangrene. Prophylactic antitoxin and antibiotics at the time of injury are widely used, but in the dosage usually employed they probably do not prevent gas gangrene. Toxoids induce protective immunity to experimental infection with several of the histotoxic clostridial species, but a booster dose at the time of wounding would probably not alter the course of the disease with an incubation period of less than 2 days.

SELECTED REFERENCES

Books and Review Articles

BOROFF, D. A., and DAS GUPTA, B. R. Botulinum toxin. In *Microbial Toxins,* vol. IIA, p. 1. (S. Kadis, T. C. Montie, and S. J. Ajl, eds.) Academic Press, New York, 1971.

MACLENNAN, J. D. The histotoxic clostridial infections of amn. *Bacteriol Rev 26:*177 (1962).

NAKAMURA, M., and SCHULZE, J. A. *Clostridium perfringens* food poisoning. *Annu Rev Microbiol 24:*359 (1970).

SMITH, L., and HOLDEMAN, L. V. *The Pathogenic Anaerobic Bacteria.* Thomas, Springfield, Ill., 1968.

VAN HEYNINGEN, W. E., and MELLANBY, J. Tetanus toxin. In *Microbial Toxins,* Vol. IIA, p. 69. (S. Kadis, T. C. Montie, and S. J. Ajl, eds.) Academic Press, New York, 1971.

WRIGHT, G. P. The neurotoxins of *Clostridium botulinum* and *Clostridium tetani. Pharmacol Rev 7:* 413 (1955).

Specific Articles

BROOKS, V. B., CURTIS, D. R., and ECCLES, J. C. The action of tetanus toxin on the inhibition of motoneurones. *J Physiol 135:*655 (1957).

DUNCAN, C. L., STRONG, D. H., and SEBALD, M. Sporulation and enterotoxin production by mutants of *Clostridium perfringens. J Bacteriol 110:*378 (1972).

ECKLUND, M. W., POYSKY, F. T., REED, S. M., and SMITH, C. A. Bacteriophage and the toxigenicity of *Clostridium botulinum* type C. *Science 172:*480 (1971).

KOENIG, M. G., SPICKARD, A., CARDELLA, M. A., and ROGERS, D. E. Clinical and laboratory observations on type E botulism in man. *Medicine 43:*517 (1964).

MILLER, P. A., EATON, M. D., and GRAY, C. T. Formation of tetanus toxin within the bacterial cell. *J Bacteriol 77:*733 (1959).

National Communicable Disease Center, U.S. Department of Health, Education and Welfare report. *Botulism in the United States: Review of cases 1899–1967.*

SUTTON, R. G. A., and HOBBS, B. C. Food poisoning caused by heat-sensitive *Clostridium welchii:* A report of five recent outbreaks. *J Hyg 66:*135 (1968).

VAN HEYNINGEN, W. E., and MILLER, P. A. The fixation of tetanus toxin by ganglioside. *J Gen Microbiol 24:*107 (1961).

chapter 35

MYCOBACTERIA

Revised by EMANUEL WOLINSKY, M.D.

TUBERCULOSIS   844
  PROPERTIES OF MYCOBACTERIUM TUBERCULOSIS   845
    Morphology   845
    Growth   845
    Genetic Variation and Virulence Tests   846
    Lipids   847
    Polysaccharides   848
    Antigenic Structure   848
    Resistance to Chemical and Physical Agents   849
  IMMUNE RESPONSE TO INFECTION   849
    Acquired Immunity   849
    Activated Macrophages and Hypersensitivity   850
  PATHOGENESIS   851
    Reactivation Disease   853
    Variations in Resistance   854
  LABORATORY DIAGNOSIS   856
    Tuberculin Test   856
  THERAPY   857
  EPIDEMIOLOGY   858
  PREVENTION   859
    Chemoprophylaxis   860
    Immunization   860
  OTHER MYCOBACTERIA THAT CAUSE A SIMILAR DISEASE   861
    Mycobacterium bovis   861
    Mycobacterium avium   861
    Mycobacterium ulcerans   861
    "Atypical" Pathogenic Mycobacteria   861

LEPROSY   865
    Mycobacterium leprae   865
    Pathogenesis   866
    Epidemiology   867

OTHER MYCOBACTERIA   867

The mycobacteria are defined on the basis of a distinctive staining property that depends on their lipid-rich cell walls: they are relatively impermeable to various basic dyes, but once stained they retain dyes with tenacity. Specifically, they resist decolorization with acidified organic solvents, and are therefore called **acid-fast.** Mycobacteria range from widespread innocuous inhabitants of soil and water to organisms that are responsible for two devastating human diseases, **tuberculosis** and **leprosy.** In both these chronic diseases, which extend over many decades, intracellular infection, allergy, and cellular resistance play important roles, and slowly evolving granulomatous lesions result in extreme tissue destruction.

Leprosy largely involves the skin, sometimes with shocking disfigurement, whereas tuberculosis is usually confined to internal organs. Moreover, the contagious nature of leprosy was recognized in Biblical times, whereas tuberculosis, though even more contagious, was not recognized as infectious until the last century. Hence the leper, but not the victim of tuberculosis, became a social outcast. Indeed, tuberculosis, though lethal to peasants as well as poets, was romantically regarded in Europe in the eighteenth and nineteenth centuries as enriching the sufferer's shortened life by increasing his esthetic sensitivities.

The leprosy bacillus is strikingly adapted to man: it has been difficult to transfer to any other host, and it has not been grown on artificial culture media. In contrast, the agents of human tuberculosis (*Mycobacterium tuberculosis* and the closely related *M. bovis*) are readily cultivated on simple media and are pathogenic for various lower animals. *M. avium* is the agent of avian tuberculosis and occasionally causes disease in man. Another group, which exhibits little or no pathogenicity for experimental animals, has recently been found to cause disease resembling tuberculosis but generally milder. Though the terminology is not yet settled, in seems better not to apply the term "tuberculosis" to infections with these "atypical" mycobacteria.

TUBERCULOSIS

Pulmonary tuberculosis (consumption, phthisis) has been recognized as a widespread and grave clinical entity for many centuries, and its incidence was probably increased by the social consequences of the Industrial Revolution. However, it was not until the early nineteenth century that some of its diverse manifestations, such as ragged cavities in lungs and small gray nodules sprinkled throughout many organs, were perceived by Laennec to be part of the same process. Its communicable nature was not recognized until 1868, when Villemin produced a similar disease in rabbits by injecting material from tuberculous lesions, both pulmonary and nonpulmonary, of man.

The cause of these lesions was not recognized, however, until Koch discovered the human tubercle bacillus in 1882. His work with this organism represented an impressive fulfillment of those criteria which he emphasized for establishing an organism as the etiological agent of an infectious disease (Ch. 1, Koch's postulates). Since then tuberculosis has been one of the most intensively studied infectious diseases. Not only has it been a major cause of death and prolonged disability, but it often strikes people at the age of greatest vigor and promise.

PROPERTIES OF MYCOBACTERIUM TUBERCULOSIS

The mycobacteria are considered transition forms between eubacteria and actinomycetes (Ch. 36): some of the latter, of the genus *Nocardia,* are weakly acid-fast; while some mycobacteria may exhibit branching. Accordingly, the mycobacteria have been classified in the order Actinomycetales (Table 35-1).

MORPHOLOGY

Acid-Fastness. The microorganism responsible for tuberculosis is difficult to stain and hence to see in tissues. Methods have been devised that promote penetration of the dye. In the widely used Ziehl-Neelsen method the smeared specimen is heated to steaming for 2 to 3 minutes in carbolfuchsin (a mixture of the triphenylmethane dyes rosanilin and pararosanilin in aqueous 5% phenol). Subsequent washing in 95% ethanol–3% HCl ("acid alcohol") decolorizes most bacteria in a few seconds, whereas acid-fast organisms retain the stain much longer. The mechanism

TABLE 35-1. Classification of Mycobacteria

Order:	
	Actinomycetales (four families, including Mycobacteriaceae, and Streptomycetaceae)
Family:	
	Mycobacteriaceae (mycelia absent or rudimentary)
Genus:	
	Mycobacterium (acid-fast rods; nonmotile and nonsporulating)

Principal pathogenic species
  M. tuberculosis (man)
  M. bovis (cattle and man)
  M. avium (birds and swine, rarely man)
  M. microti (field mouse)
  M. leprae (man)
  M. leprae murium (rat)
  M. paratuberculosis (cattle, sheep: Johne's disease)
  M. ulcerans (man)
  M. marinum (fish and man)
  Other mycobacteria (see Table 35-3)

of acid-fastness will be discussed below (under Lipids).

When gram-stained, mycobacteria may appear to to be positive; but they take up the stain weakly and irregularly and without requiring iodine treatment to retain it.

A number of objects other than mycobacteria are acid-fast, including some bacterial and fungal spores, and a substance known as ceroid (which is found in the liver in certain states of nutritional deficiency). These materials, however, do not require the presence of phenol (or aniline) in the staining solution. The acid-fast nocardiae are more easily stained and more easily decolorized than are mycobacteria.

Structure. Tubercle bacilli in the animal are typically slightly bent or curved, slender rods, ca. 2 to 4 μ long and 0.2 to 0.5 μ wide. The rods may be of uniform width but more often appear beaded, with irregularly spaced, unstained vacuoles, or heavily stained knobs. In culture media the cells may vary from coccoid to filamentous. Strains differ in their tendency to grow as discrete rods or as aggregated long strands, called "serpentine cords" (Fig. 35-1).

Electron microscopy reveals a rather thick wall, and large lamellar mesosomes are common. Glycogen granules and polymetaphosphate (volutin) bodies are also seen: the latter stain metachromatically with cationic dyes. These inclusion bodies contribute to the frequently beaded, irregular staining of tubercle bacilli.

The walls of mycobacteria evidently contain the same kind of basal glycopeptide layer as other bacteria, since the cells form spheroplasts when treated with lysozyme in media of high osmolarity. The walls are unusual in their high lipid content (up to 60%) and thus more closely resemble the walls of gram-negative organisms (lipid content ca. 20%) than those of gram-positive organisms (only 1 to 4%). Mycobacteria also resemble gram-negative organisms in containing protein in their wall fraction.

GROWTH

Unlike most other pathogenic bacteria, which are facultative aerobes or anaerobes, the tubercle bacillus is an **obligate aerobe.** It can grow in simple synthetic media, with glycerol or other compounds as sole carbon source and ammonium salts as nitrogen source; asparagine or amino acid mixtures are usually added to promote the initiation and improve the rate of growth.

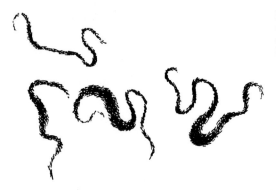

Fig. 35-1. Tubercle bacilli in stained smear from colony. Note parallel growth in "cords." From original drawing of Koch. (Courtesy of Robert Koch-Institut, Berlin.)

Mycobacteria generally show a marked nutritional preference for lipids; egg yolk has been a prominent constituent of the rich media used for diagnostic cultures. Thus, though the tubercle bacillus is very sensitive to inhibition by long-chain **fatty acids,** it is stimulated by them at very low concentrations. A satisfactory concentration is maintained by adding to the medium **serum albumin,** which binds the fatty acids with sufficient affinity to maintain their free concentration at a suitable level for growth.

In ordinary synthetic liquid media the bacilli grow in adherent clumps, which form a surface pellicle. (This "mold-like" property is responsible for the name *Mycobacterium.*) Quantitative studies were facilitated when Dubos introduced the addition of nonionic detergents, such as Tweens,* which cause the bacilli to grow in more dispersed form, though still not as single cells.

Growth of tubercle bacilli in culture media (and in animals) is characteristically slow: **the shortest doubling time observed, in rich media, is ca. 12 hours.** Saprophytic mycobacteria grow more rapidly, but not so rapidly as most other bacteria (which have doubling times as short as 20 minutes).

GENETIC VARIATION AND VIRULENCE TESTS

Human tubercle bacilli freshly isolated from pulmonary lesions produce progressive disease in guinea pigs, which die within 1 to 6 months after infection, depending in part on the size of the inoculum. (Tuberculous skin lesions, called **lupus vulgaris,** often yield less virulent organisms, as do lung lesions from some patients in India and Africa; see below, Atypical pathogenic mycobacteria.) As with other pathogenic bacteria, virulence varies quantitatively: serial passage through artificial culture media selects indirectly for less virulent mutants, while animal passage of large inocula of such strains selects directly for mutants with restored virulence. One attenuated mutant,† carried through several hundred serial cultures on unfavorable (bile-containing) media, is known as **bacille Calmette Guérin** or **BCG.** This bovine strain, which is used to immunize humans against tuberculosis, has remained avirulent for over 40 years.

Most virulent strains produce rough colonies, while many avirulent laboratory strains produce less rough colonies (called smooth, though the cells lack capsules). Thus the correlation between colonial morphology and virulence, though not highly consistent, is the reverse of that seen with many other bacteria, where only the smooth, encapsulated strains are pathogenic.

Virulence Tests. In contrast to pneumococci, virulent strains of tubercle bacilli cannot be reliably distinguished from avirulent strains by cellular or colonial morphology, or by serological tests; and the direct test for virulence in animals is slow and inconvenient. Properties correlated with virulence have therefore been intensively sought. The most significant finding has been that virulent strains grow, on the surface of liquid or on solid media, as inter-

---

* Polyoxyethylene ethers of sorbitol, or of other polyhydric alcohols, esterified to long-chain fatty acids.

† A strain is considered attenuated for a given host if it multiplies in that host to only a limited extent, causing at most minor, transient lesions.

twining **"serpentine cords,"** in which the bacilli aggregate with their long axes parallel (Fig. 35-1), while most avirulent strains grow in a more disordered manner. The correlation, however, is imperfect. Growth in cords can be correlated with the content of a surface lipid ("cord factor") described below. Nonionic detergents (e.g., Tween 80) reduce cord formation (but not virulence), presumably by coating the cell surface. Another property of virulent strains that is lost in many avirulent mutants is binding of the dye **neutral red.**

The failure to find a virulence test that is reliable in both directions is not surprising: since virulence is a multifactorial property, a given bacterial component may be necessary but it cannot be sufficient to make an organism virulent.

LIPIDS

The most striking chemical feature of the mycobacteria is their extraordinarily high lipid content, which amounts to 20 to 40% of the dry weight. Various lipid fractions are defined by the conditions used for their extraction from dried organisms. The striking abundance of lipids in the cell wall (60% of its dry weight) accounts for the hydrophobic character of the organisms, which tend to adhere to each other during growth in aqueous media and to float at the surface unless dispersed with detergents. The lipid-rich wall probably also accounts for some of the other unusual properties of mycobacteria: e.g., relative impermeability to stains, acid-fastness, unusual resistance to killing by acid and alkali, and resistance to the bactericidal action of antibodies plus complement. The huge amount of wall lipid may also contribute to the slowness of growth, both by hindering permeation of nutrients into the cell and by absorbing a large fraction of the cell's biosynthetic capacity. The biological adaptation of the tubercle bacillus thus resembles that of the tortoise—slow but protected by armor.

Among the lipids extracted with neutral organic solvents are **true waxes** (esters of fatty acids with fatty alcohols) and **glycolipids,** also called mycosides (lipid-soluble compounds in which lipid and carbohydrate moieties are covalently linked). The only other bacteria known to contain true waxes

are corynebacteria, some of which are also somewhat acid-fast.

While many different fatty acids are found in the lipids of mycobacteria, one class, the **mycolic acids,** appears to be unique to the cell walls of mycobacteria and to the chemically related nocardiae and corynebacteria. These large saturated, α-branched-chain, β-hydroxylated fatty acids (Fig. 35-2) are found in both waxes and glycolipids.

Cord Factor. The association of virulence with serpentine growth led to the search for a surface component of the bacteria responsible for both properties. An excellent candidate for this role, called "cord factor," was extracted by Bloch from virulent cells by petroleum ether: the extracted cells lose virulence (phenotypically) and disperse in aqueous media. Cord factor has been identified as a mycoside, 6,6′-dimycolyltrehalose (Fig. 35-2).

Several lines of evidence relate cord factors to virulence. 1) This substance is quite toxic: it inhibits migration of normal polymorphonuclear leukocytes in vitro (as do virulent tubercle bacilli), and 10 µg given subcutaneously will kill a mouse. 2) It is more abundant in virulent strains. 3) Its extraction, as noted above, renders cells nonvirulent (though they remain viable). 4) Tubercle bacilli recovered from animals, or from young cultures, are more virulent, and they have a higher content of cord factor, than cells of the same strain from older cultures. Cord factor thus probably contributes substantially to pathogenicity. However, the absence of humoral immunity to the tubercle bacillus (see below) makes it impossible to demonstrate the decisive role of a single component, as has been demonstrated for the toxin of the diphtheria bacillus or the capsular polysaccharide of the pneumococcus. Moreover, other extracted mycobacterial lipids have also been reported to be toxic to macrophages.

A lipid containing a sulfonic acid group (sulfolipid), which has been extracted by hexane from intact organisms, also appears to be more abundant in virulent than in avirulent strains.

Wax D and the Immune Response. Another mycoside of interest is the high-molecular-weight **wax D** (Fig. 35-3), which is not a true wax but contains mycolic acids and a glycopeptide. It is apparently extracted from the basal wall layer, since the peptide contains the characteristic amino acids of that layer, linked to a polymer of hexoses and hexosa-

Mycolic acids

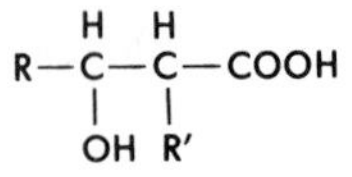

$$R-\underset{\underset{OH}{|}}{\overset{\overset{H}{|}}{C}}-\underset{\underset{R'}{|}}{\overset{\overset{H}{|}}{C}}-COOH$$

Human and bovine tubercle bacilli: $R' = C_{24}H_{49}$

Avian and saprophytic bacilli: $R' = C_{22}H_{45}$

R has about 60 C atoms, 1 or 2 O atoms, and is probably variable in different mycolic acids.

6, 6'-Dimycolyltrehalose (cord factor)

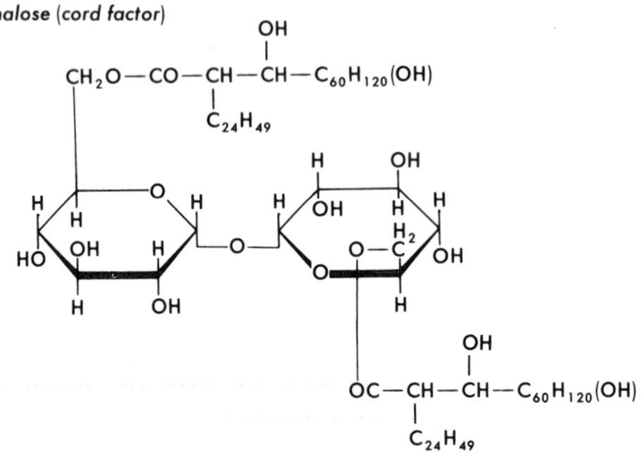

Fig. 35-2. Structures of some distinctive mycobacterial lipids. Trehalose is D-glucose-α-1,1'-D-glucoside.

mines. In a water-in-oil emulsion (Freund's adjuvant) this fraction, like whole tubercle bacilli, **enhances the immunogenicity** of a variety of added antigens (Ch. 17). Moreover, a mixture of wax D and proteins of the tubercle bacillus **induces delayed-type hypersensitivity** to tuberculin, whereas the protein alone is poorly immunogenic.

The **crude phosphatide fraction** has the interesting property of evoking **a cellular response resembling tubercle formation,** including caseation necrosis (see Pathogenesis, below). Rather large amounts must be injected, and comparable fractions from saprophytic mycobacteria are also effective.

Acid-Fastness. After exhaustive extraction of the above-described lipids by organic solvents the residual ghosts remain acid-fast, and they retain **"firmly bound" lipids** largely in the wall fraction. These lipids are removed by hot acid, which also destroys the acid-fastness. Since this property is also lost on sonic disintegration of normal cells it appears to depend on the integrity of the cell wall, including certain lipids.

POLYSACCHARIDES

Polysaccharide is a major component of culture filtrates and extracts of mycobacteria; a major portion exists in chemical union with lipid in the cell wall. The polysaccharide fraction contains glucan, mannan, arabinomannan, and arabinogalactan, and it has been suggested that a mycolic acid–arabinogalactan complex makes up a portion of the cell wall and of the wax D fraction.

ANTIGENIC STRUCTURE

At least 30 different antigens have been recognized among the mycobacteria (Figs. 35-4 and 35-5). Some are species-specific; others are common to many mycobacteria and probably account for the cross-reactivity among mycobacteria, nocardiae, and corynebacteria. Strain- or type-specific antigens are also

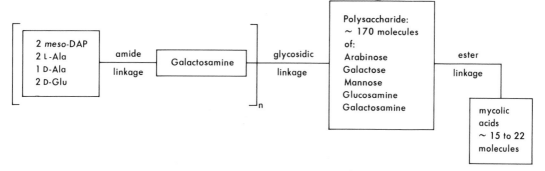

Fig. 35-3. Provisional structure of wax D (a peptidoglycolipid). [From Jollès, P., Samour, D., and Lederer, E. *Arch Biochem Biophys* 98 (Suppl. 1): 283–289 (1962).]

identified by agglutination reactions and by skin testing guinea pigs with partially purified proteins (PPD, below) from diverse mycobacteria.

RESISTANCE TO CHEMICAL AND PHYSICAL AGENTS

Tubercle bacilli are unusually resistant to acid and alkali, and this property is exploited in the diagnostic isolation of the organisms (see below, Laboratory diagnosis). They are also relatively insensitive to cationic detergents, but no more resistant than other nonsporulating bacteria to heat, ultraviolet irradiation, or phenol.

IMMUNE RESPONSE TO INFECTION

ACQUIRED IMMUNITY

The Koch Phenomenon. Shortly after isolating the tubercle bacillus Koch demonstrated that it induces an altered response to superinfection. In the original study bacilli were injected subcutaneously into a guinea pig, and 10 to 14 days later the inoculation site developed a nodule which then became a persistent ulcer. In addition, the bacilli spread to the regional lymph nodes, causing them to enlarge and then become necrotic. When a similar injection was then made at another site the response was faster and more violent, but also more circumscribed: a dusky, indurated lesion appeared in 2 to 3 days and soon ulcerated; but the ulcer healed promptly and the regional lymph nodes remained virtu-

ally free of infection. Koch later showed that a similar second response could be obtained with culture filtrates of the organism (see below, Tuberculin test).

The superinfecting bacilli are evidently better localized than the initial inoculum, and they multiply more slowly in the tissues. The infected animal has thus acquired increased resistance (i.e., partial immunity). But though this immunity may lead to elimination of the bacilli of superinfection, it cannot accomplish the same for the primary lesion, which has developed a denser bacterial population and a different histological pattern. Evidently the level of immunity acquired in the guinea pig is limited, and it can be overcome by local factors in the lesion.

Nonhumoral Immunity. Serum antibodies do not account for this immunity. Thus though the tubercle bacillus enhances antibody formation to a wide variety of immunogens, antibodies to proteins and polysaccharides of the tubercle bacillus itself are generally found in tuberculous individuals only in low titers, and the levels observed have no prognostic value with respect to the course of the disease. Moreover, these antibodies are not bactericidal in vitro, even in the presence of complement, and though they promote phagocytosis of tubercle bacilli in vitro, the organisms multiply within the phagocytes. Finally, increased resistance to infection cannot be transferred passively with serum, but it can with viable lymphoid cells, which transfer delayed-type hypersensitivity to tuberculin.

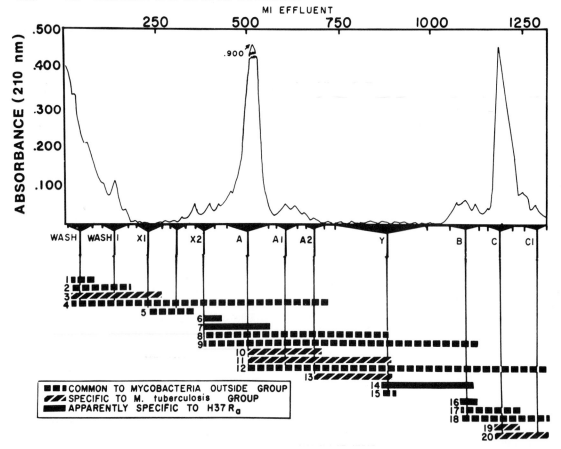

Fig. 35-4. Separation of antigenic fractions by ion-exchange chromatography. Fractions of *M. tuberculosis,* strain H37RA were tested by immunodiffusion against various antimycobacterial antisera. [From Kniker, W. T. *Am Rev Resp Dis 92* (part 2): 19–28 (1965).]

ACTIVATED MACROPHAGES AND HYPERSENSITIVITY

Though macrophages of normal animals permit ingested tubercle bacilli to flourish, as noted above, macrophages from tuberculous animals destroy them more effectively. Some kind of specific cellular immunity in macrophages was therefore long suspected, analogous to the altered mononuclear cells associated with delayed hypersensitivity. However, similar "activated" macrophages can be obtained from animals infected with various other bacteria (such as *Listeria*) that also survive in normal macrophages, and this increased bactericidal capacity is nonspecific, i.e., it extends to a wide variety of microbes

and is associated with an increased number of lysosomes in the "activated" macrophage.

While the enhanced activity of altered macrophages is nonspecific, its **induction** is immunologically specific; Mackaness has shown that lymphocytes modified by specific interaction with the infecting organism can induce the change in the macrophage population (Ch. 22, Acquired bacterial cellular immunity). The activation of macrophages is transient, lasting a few weeks in experimental animals unless antigenic stimulation is repeated or sustained.

Though tuberculin sensitivity is associated with increase in host defenses, it can also lead to harmful reactions. For example, in sensitive individuals the administration of a large

Schematic

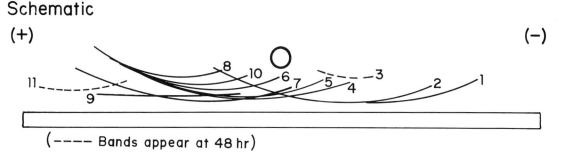

(−−−− Bands appear at 48 hr)

Actual

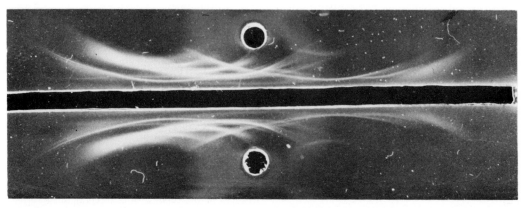

Fig. 35-5. Multiplicity of mycobacterial antigens. A reference culture filtrate of *M. tuberculosis,* strain H37RV, was subjected to electrophoresis and analyzed with a goat antiserum. [From Janicki, B. W., Chaparas, S. D., Daniel, T. M., Kubica, G. P., Wright, G. L. Jr., and Yee, G. S. *Am Rev Resp Dis 104:*602 (1971).]

amount of tuberculin can provoke severe systemic illness, including chills, fever, and increased inflammation around existing lesions.* Moreover, while an allergic ulceration in the skin is evidently beneficial in the Koch phenomenon, leading to drainage of organisms from the body, a similar reaction in the lung (or the meninges) can lead to spread of infection through the bronchial tree (or the subarachnoid space).

In extensive attempts to determine whether immunity to tuberculosis is due to allergic state or develops in spite of this state, earlier workers observed a residual immunity in animals that had been

---

* The disastrous effect of a massive allergic response to tuberculin was tragically revealed when Koch enthusiastically injected tuberculin for therapeutic purposes in tuberculous patients, in the hope of enhancing immunity.

desensitized by the gradual administration of heroic doses of tuberculin. This result, however, does not establish the development of immunity by mechanisms that are independent of hypersensitivity; for desensitization would not immediately eliminate the previously activated macrophages, and their further formation would not necessarily demand the same degree of hypersensitivity required for a skin reaction.

Tuberculin sensitivity is clearly a double-edged sword; it is associated with increased resistance, but under some circumstances the hypersensitive response can also cause a marked exacerbation of symptoms and an increased spread of the bacteria.

PATHOGENESIS

The consequences of inhaling or ingesting tubercle bacilli depend upon both the **viru-**

lence of the organism and the **resistance** of the host (as well as the size and the location of the inoculum). At one extreme, organisms with little virulence for the particular host disappear completely, leaving no anatomical trace behind. At the opposite extreme (e.g., in guinea pigs inoculated with human tubercle bacilli) the bacilli flourish in macrophages as well as extracellularly, are disseminated widely, and produce progressive disease that is fatal within a few months or longer, depending on the inoculum.

Man exhibits a range of responses. The initial infection in a tuberculin-negative individual most often produces a self-limited lesion, but sometimes the disease progresses, presumably because of low resistance or a large inoculum. Because of the delicate balance of resistance, involving local as well as systemic factors, healing and progressing lesions may coexist in the same individual, and the disease often has a chronic, cyclical course (especially without chemotherapy). This pattern stands in marked contrast to the acute course of those infections that give rise to well-defined humoral immunity.

Primary Lesions. After inhalation of tubercle bacilli the initial lesion appears as a characteristic nodule in the lung parenchyma, called a tubercle (L. *tuberculum* = small lump). Similar localized lesions then develop in the draining mediastinal lymph nodes, much as in the experimental Koch phenomenon described above. The pair of lesions is referred to as a **primary complex.** These lesions usually stabilize; and though they are usually not large enough to be seen by X-ray, they appear to retain viable bacilli for the individual's lifetime since reactivity to tuberculin persists. Occasionally, after a highly variable period of dormancy that can last for many decades (Ch. 22), the bacilli resume multiplication and provoke an expanding inflammatory lesion (see below, Reactivation disease).

Secondary Lesions. As a tuberculous lesion expands in a tuberculin-sensitive individual its center undergoes a characteristic **caseation necrosis** (L. *caseus* = cheese), in which the necrotic tissue remains semisolid, with the consistency of cheese. In such areas the enzymes that usually liquefy dead cells are evidently inhibited or scanty. With time caseous

tissues often becomes **fibrosed** and even **calcified** (especially in young children), giving rise to characteristic X-ray findings;* or eventual **softening** of the caseous mass leads to its excavation. When such a lesion expands into a bronchus it discharges necrotic material into the lumen, leaving a cavity in the lung parenchyma if the lesion is large enough. The organisms are thus spread via the bronchi to distant parts of the lung, via droplets and expectorated sputum to other individuals, and via swallowed sputum, which rarely produces lesions of the intestinal tract and the mesenteric lymph nodes.

The necrotic lesions less frequently erode blood vessels (usually pulmonary veins), and when the dissemination is massive enough to overwhelm resistance innumerable metastatic lesions may form in distant organs: bones, spleen, meninges, kidneys, prostate, skin, and so on—although muscles are curiously spared. The disseminated disease is called **miliary tuberculosis** (L. *milium* = millet seed) because of the size of the lesions found in various organs at autopsy. Rarely the metastatic lesions expand while the pulmonary focus remains relatively inapparent; tuberculosis of bone, adrenal, or brain may then be predominant. **Tuberculous meningitis** arises from the rupture of a cerebral tubercle into the subarachnoid space. Tuberculous meningitis and miliary tuberculosis were invariably fatal before chemotherapy.

Histologically the tubercle bacillus evokes two types of reaction. 1) **"Exudative"** lesions are seen in the initial infection, or in the individual in whom the organism proliferates rapidly without encountering much host resistance. Acute or subacute inflammation occurs, with exudation of fluid and accumulation of polymorphonuclear leukocytes around the bacteria. 2) **"Productive"** (granulomatous) lesions form when the individual becomes allergic to tuberculoprotein. The macrophages then undergo a dramatic modification on contact with tubercle bacilli or their products, becoming concentrically arranged in the form of elongated **epithelioid cells,** to form the **tubercles** characteristic of this

---

* Though such pulmonary calcification was long thought to be pathognomonic of healed tuberculosis, epidemiological surveys have shown that in certain geographical regions these lesions are associated far more often with histoplasmin sensitivity (Ch. 43) than with tuberculin hypersensitivity.

disease. In the center of the tubercles some of these cells may fuse to form one or more **giant cells,** with dozens of nuclei arranged at their periphery and viable bacilli often visible in their cytoplasm. Outside the multiple layers of epithelioid cells is a mantle of lymphocytes and proliferating fibroblasts (Fig. 35-6), leading eventually to extensive fibrosis. The subsequent development of caseation necrosis, and its various fates, have been noted above.

In early lesions tubercle bacilli are localized primarily within macrophages, in which they multiply (at least for a time). In more advanced lesions only extracellular bacilli are prominent, probably because the activated macrophages destroy the intracellular bacilli more efficiently. In necrotic lesions the organisms are seen particularly at the periphery; their sparse growth within the caseous material may be due to poor oxygen or food supply.*

Distribution of Lesions. The requirement of the tubercle bacillus for a high O_2 tension may be responsible for the development of pulmonary

---

* *M. tuberculosis* is unusual among obligatory parasites in being able to grow in a very simple medium. This property may promote its growth in the fibrotic, poorly nourished lesions that it causes.

lesions primarily in the most extensive aerated regions: the apices of the upper lobes of man, and the posterior (dorsal) region of the lower lobes of cattle. Thus when rabbits with experimental pulmonary infections were maintained for prolonged periods in an upright position they developed lesions mostly in the upper lobes, rather than in the usual paravertebral portions of the lower lobes. Hence before the advent of chemotherapy pulmonary tuberculosis in man was widely treated by artificial collapse of the lung, which aimed at a decrease in aeration as well as rest from mechanical spread.

REACTIVATION DISEASE

While the concept of reactivation disease is fundamental to our understanding of tuberculosis, it is based essentially on indirect evidence: lifetime persistence of tuberculin hypersensitivity; the indication that this persistence depends on the presence of viable bacilli in the tissues, since it is occasionally eliminated by chemotherapy; and the shifting age distribution of the disease. For as the total incidence of tuberculosis has been declining since 1870, the age of peak mortality has advanced from adolescence to old age (Fig. 35-7). This striking shift of the peak cannot be readily explained in terms of shifts in the

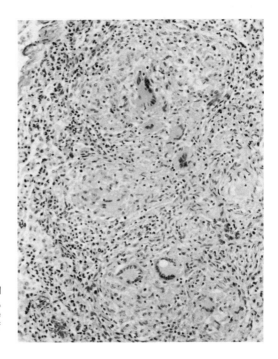

Fig. 35-6. Section of a tubercle, showing several giant cells containing a peripheral ring of nuclei, epithelioid cells, and mononuclear cells toward the periphery of the lesion. ×180, reduced. (Courtesy of B. W. Castleman.)

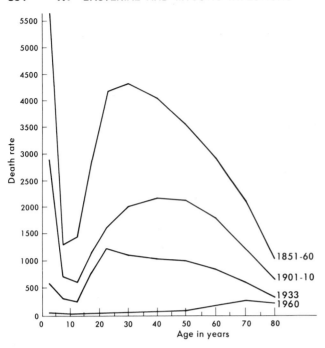

Fig. 35-7. Death rates per million from tuberculosis (all forms) at various ages, in England and Wales. (From Wilson, G. S., and Miles, A. A. *Principles of Bacteriology and Immunity,* 5th ed. Williams & Wilkins, Baltimore, 1964.)

relative exposure of various age groups to exogenous infections. Rather, since the present elderly group had a high risk of acquiring initial infection in an earlier era, and then remained tuberculin-positive, their frequent development of active tuberculosis is best explained as reactivation of dormant foci acquired much earlier in life. The same conclusion applies today even to young adults: in a study of 68,000 U.S. Navy recruits, followed for 4 years in the 1950s, the majority of the cases of active tuberculosis appeared in the 11% of the population who were tuberculin reactors at the start.

It is thus no longer advantageous for a young adult to be tuberculin-positive as a result of earlier inapparent infection: though the hypersensitive state is associated with partial protection against the production of disease by fresh infection, the level of exposure to such infection is now so low that this advantage is outweighed by the possibility of progress or activation of the previous infection. However, individuals with an unusually high degree of exposure to tubercle bacilli, such as sanatorium personnel, are probably better off if they are tuberculin-positive; but this is no longer true even of medical students and nurses, who now have a high probability of graduating without becoming tuberculin-positive.

VARIATIONS IN RESISTANCE

Among persons infected with the tubercle bacillus, as detected by a positive tuberculin test (see below), only a small proportion develop overt disease; and even before the advent of chemotherapy only ca. 10% progressed to fatal disease. The transition from **infection** to mild or to severe **disease** depends strongly on various factors besides the presence of the bacilli; hence in an earlier era, when nearly everyone eventually became tuberculin-positive, tuberculosis could almost be regarded as one of the "endogenous diseases" (Ch. 22).

While many of the observed variations in the response to infection may involve differences in the size, the site, or the virulence of the inoculum, there is also no doubt that resistance in man varies more strikingly with tuberculosis than with most infectious diseases. Thus in a tragic accident in Lübeck, Germany, in 1930, in which 251 children received identical inocula of a virulent strain instead of an attenuated vaccine, 77 developed fatal disease, 127 developed radiological detectable lesions that healed, and 47 showed no signs of disease. Such variations in resistance involve both genetic and nongenetic (physiological) factors.

Genetic Differences. The importance of genetic factors in host resistance has been unequivocally established by Lurie's development of inbred lines of rabbits with high or low tendency to acquire progressive tuberculosis from experimental infection. Multiple factors are evidently involved: some lines are more resistant to **initiation** of disease by small inocula, while in others the resistance primarily influences the **rate of progress** of the disease.

In man it seems likely that there are similar **racial differences** in resistance, associated with different lengths of exposure of the race to the selective pressures of an environment with widespread tuberculosis. Thus in the United States the incidence of tuberculosis, and the ratio of deaths to cases, are especially high among American Indians, Eskimos, and Negroes. These races may have had little or no exposure to tuberculosis until the last two or three centuries, contrasted with descendants of a European population. However, this evidence is only suggestive, because of the concomitant differences in environmental factors that seem to be important (see below). More significant data have been provided by a study in the U.S. Army (from 1922 to 1936), in which the environment was largely equalized: the death/case ratio was four times higher among Negro soldiers than among whites. Evidence for genetic factors in man has also been obtained in a study of twins: when one member of the pair had tuberculosis it was also present in the other member three times more frequently among monozygotic than among dizygotic pairs.

Resistance to tuberculosis has not been correlated with genetically determined variations in any component of the immune response. However, one genetically determined disease, **diabetes,** is known to be associated with decreased resistance.

Since the hereditary component of resistance to tuberculosis is highly polygenic it cannot be as precisely defined or quantitated as monogenic traits. Nevertheless, there is no doubt that tuberculosis has exerted a selective influence on the gene pool of populations where it has been endemic, in the same way that falciparum malaria has selected for the sickle cell trait. Indeed, a few generations of selection were sufficient with rabbits (Lurie's experiments, above) to yield marked variations in heritable resistance.

Physiological Factors. Epidemiological evidence on the frequency of tuberculosis in various populations has long suggested that **malnutrition, overcrowding,** and **stress** decrease resistance to the disease; but this kind of evidence does not firmly establish a causal relation because these conditions are generally also associated with a high rate of infection. However, the ability of one or more of these factors to overcome innate resistance is well illustrated by the experience of inmates of Nazi concentration camps during World War II. Tuberculosis was exceedingly prevalent in this population, which was subjected to extreme stress and prolonged starvation; yet many seemingly moribund persons with far advanced disease exhibited a remarkable recovery when liberated from the camps and renourished. In contrast, before the era of chemotherapy many persons who entered a sanatorium with minimal tuberculosis progressed inexorably to a fatal outcome despite good nutrition and general care.

The most specific and direct evidence on the effect of starvation has been obtained in experimental infections in mice: a reduced protein consumption increased susceptibility, while variations in vitamin and total caloric intake had no demonstrable effect.

The significance of **hormonal** factors is suggested by the striking variations in resistance with **age** and, to a smaller extent, with **sex.** Tuberculosis is apt to be very severe in infants, perhaps because of the immaturity of the immune mechanisms. It then decreases rapidly in both incidence and severity with increasing age, and between 3 and 12 years of age progressive disease is almost unknown, even in children heavily infected by exposure to a tuberculous parent. Susceptibility to the disease increases rapidly at adolescence, and among young adults tuberculosis is more frequent in females. Curiously, in experimental animals no marked influence of age on susceptibility has been noted.

Administration of **cortisone,** in man and in experimental animals, decreases resistance to tuberculosis and tends to mask its symptoms and abolish reactivity to tuberculin. Hence in the therapy of other diseases with this hormone the arousal of dormant tuberculosis is occasionally a serious and readily overlooked complication. Increased secretion of cortisone may also well be involved in the presumptive role of stress, noted above.

Silicosis. Tuberculosis is notoriously frequent among miners and others exposed to dust containing silica. Moreover, in guinea pigs that have inhaled silica suspensions even the avirulent BCG strain produces fatal tuberculosis. The decrease in resistance appears to be due to damage to phagocytic cells: phagocytized silica particles cause rapid disruption of lysosomes, followed by cell lysis; and since the particles are never digested, they can attack cell after cell. Silica is exceedingly insoluble, and it is believed to act by direct contact with lysosomal membranes.

LABORATORY DIAGNOSIS

A provisional diagnosis of tuberculosis is usually made by demonstrating acid-fast bacilli in stained smears of sputum or of gastric washings (containing swallowed sputum). For rapid screening of smears some laboratories employ ultraviolet microscopy following staining with the fluorescent dye mixture rhodamine-auramine.

Whether or not the smear is positive the material should be cultured, for several reasons: 1) cultivation can detect fewer organisms; 2) cultural characteristics distinguish human tubercle bacilli from other acid-fast bacilli, including avirulent contaminants, atypical pathogenic mycobacteria (see below), and *Nocardia* (Ch. 36); 3) tests for drug sensitivities should be initiated before starting chemotherapy. Moreover, during chemotherapy frequent sputum smears and cultures provide a rapid, objective index of the response.

Sputum is prepared for culture by exploiting the unusual resistance of tubercle bacilli to strong alkalis and acids. For example, the material may be shaken at 37° in 0.1 N NaOH. The resulting liquefaction permits concentration of bacteria by centrifugation; and more rapidly growing bacterial contaminants are mostly destroyed by this treatment, whereas a fraction of the mycobacterial cells remain viable (and virulent). The centrifuged sediment should be smeared and stained, as well as cultured.

Solid culture media are preferred for primary isolation; they often contain egg yolk to promote growth of macroscopic colonies from small inocula. However, oleic acid–albumin agar medium (see above, Growth), which is transparent, allows detection of smaller colonies than those required on an opaque egg medium. These media contain malachite green or penicillin at levels that selectively suppress growth of gram-positive, rapidly growing contaminants. A positive culture usually grows out in 2 to 4 weeks.

Part of the sediment from sputum is sometimes inoculated subcutaneously into the groin of a guinea pig. The animal is skin-tested with tuberculin at intervals of several weeks, and when hypersensitivity becomes evident the lesions are examined for bacilli by smear and culture. This test can detect small inocula, but it is slow and expensive. It was long employed to distinguish virulent mycobacteria from contaminants, but it is no longer considered reliable for this purpose (see below, Atypical pathogenic mycobacteria). Modern cultural technics have made guinea pig inoculation unnecessary, except possibly for specimens heavily contaminated with organisms difficult to eliminate, such as *Pseudomonas*.

TUBERCULIN TEST

Significance. Delayed-type hypersensitivity to tuberculin is highly specific for the tubercle bacillus and various closely related mycobacteria. Reactivity appears ca. 1 month after infection in man and persists for many years, often for life; hence the frequency of reactors in the population increases cumulatively with age. **A positive test thus reveals previous infection with various mycobacteria;** it does not establish the presence of active disease.

The persistence of hypersensitivity probably depends on persistence of bacilli in dormant foci, since reactivity occasionally disappears following chemotherapy or surgical removal of localized lung lesions. Moreover, a strong reaction signifies a greater chance of developing the disease than does a weak reaction.

Koch's discovery of the tuberculin reaction in 1890 provided a powerful diagnostic and epidemiological tool, which revealed much more widespread infection than had previously been suspected. Moreover, the importance of the tuberculin test has grown as control measures have aimed increasingly at persons with inapparent infection.

The capacity of the skin to react may be suppressed by advanced age or by terminal or severe acute illness, perhaps because of circulatory changes; by rapidly progressive tuberculosis, perhaps through massive release of antigen; by measles and certain other viral infections or live virus vaccines, through unknown mechanisms; and by cortisone.

Preparation of Tuberculin. Tuberculin as originally described by Koch is known as **Old Tuberculin** (OT). It is prepared by autoclaving or boiling a culture of tubercle bacilli, concentrating it 10-fold on a steam bath, filtering off the debris, and adding glycerol as a preservative. In this impure product the active constituent is a protein remarkable for its heat stability; after being autoclaved it remains soluble and retains specific determinants of the protein in the infecting bacilli. Stock solutions retain full potency for years when stored at 5°.

A slightly more refined tuberculin, called **purified protein derivative** (PPD), is now in use. It is prepared by growing *M. tuberculosis* in a simple synthetic medium, autoclaving, removing debris by filtration, concentrating the filtrate by ultrafiltration, and precipitating several times with 50% saturated ammonium sulfate. The product is mostly a mixture of small proteins (average MW 10,000).

Testing Procedures. Tuberculin hypersensitivity is generally tested for in the United States by **intradermal** injection of 0.1 ml of an appropriate dilution of standardized antigen into the most superficial layers of the skin of the forearm (Mantoux test). The average diameter of **induration** (and not simply erythema) at the injected site is measured at 48 hours, and reactions less than 5 mm in diameter are recorded as doubtful. Intense reactions can cause necrosis and subsequent scarring. Persons infected with cross-reacting "atypical" mycobacteria (see below) frequently exhibit weak reactions (2 to 4 mm in diameter). Ordinarily, however, the population is quite sharply divided into tuberculin-positive or -negative.

In epidemiological work the standard test dose generally used is 5 tuberculin units (TU) of PPD (0.1 μg) in 0.1 ml, corresponding to OT 1:2000. In children suspected of having tuberculosis a fivefold lower concentration is first used ("first strength"), to avoid a severe reaction, and if there is no response within 2 to 3 days the standard strength is used. "Second strength" PPD contains 20 to 50 times the standard dose. It may be used to elicit reactions in those patients whose standard PPD tests are negative or doubtful, where infection with tubercle bacilli or with cross-reacting atypical mycobacteria is suspected (see below).

Other commonly used methods of skin testing involve multiple punctures of the skin by means of various mechanical devices. These methods are slightly less accurate than the Mantoux test and positive or doubtful results should be rechecked by the intradermal technic.

THERAPY

Until specific therapy became available, with the discovery of streptomycin in 1945, the treatment of tuberculosis was limited to rest, good nutrition, and artificial collapse of the lung (see Distribution of lesions, above). Unfortunately, the value of streptomycin was soon found to be restricted by its toxicity to the eighth cranial nerve on prolonged administration, as well as by the frequent emergence of resistant tubercle bacilli during therapy. The real revolution in tuberculosis therapy came with the introduction of isoniazid (INH), which is more effective, remarkably free from toxicity, inexpensive, and easy to take (i.e., small doses, given orally). Additional valuable drugs include *p*-aminosalicylate (PAS), which is also given orally but requires huge doses, and ethambutol, which is tending to replace PAS as a companion to INH in primary treatment. Other useful drugs include kanamycin, capreomycin, and viomycin (by injection), and cycloserine, ethionamide, and pyrazinamide (given orally); they are generally used only in the presence of allergy or resistance to the "primary" drugs. The rapid response to INH and the low toxicity have made ambulatory treatment feasible.

Rifampin, introduced later, is also very effective, relatively nontoxic, and absorbed very well when given by mouth. The combination of rifampin and INH may prove to be the most effective antituberculosis regimen available.

Streptomycin is bactericidal (Ch. 12) but is not effective against organisms in macrophages. **PAS** is only bacteristatic: it interferes, like sulfonamides in many other organisms, with the conversion of *p*-aminobenzoate to folic acid (Ch 7). It is also relatively ineffective against bacteria localized within macrophages, presumably because of the presence of metabolites that reverse its action. **INH** is bactericidal for tubercle bacilli, including those in macrophages; it is much less active against other mycobacteria.

The peculiar lesions in tuberculosis, which protect the organisms from host defenses, also promote a metabolically inactive state, which protects them from the action of bactericidal drugs. Hence even in minimal tuberculosis **chemotherapy must be prolonged** for months to bring about a cure or a long-term arrest. (These two favorable outcomes are distinguished by loss vs. persistence of hypersensitivity to tuberculin.)

Despite the effectiveness of chemotherapy, surgical resection is still occasionally necessary for localized, unresponsive lesions—usually those containing large cavities. Since such lesions chronically shed large numbers of organisms (often drug-resistant), they remain a source of infection to the community.

Drug Resistance and Combination Therapy. The selection of drug-resistant mutants in patients with tuberculosis is favored by several features of the disease: the numerous bacteria in the lesions, especially in the walls of cavities; their further multiplication during the required long periods of therapy; and the limited degree of host resistance, which provides the rare mutant cell with a good opportunity to proliferate before being eliminated by host defenses. Thus, since the mutation of tubercle bacilli to resistance to various drugs is not unusual, specifically resistant strains often appear in patients treated with any one of the drugs. The rationale for preventing the emergence of resistant strains by **combined therapy** with two or more drugs, has been presented in Chapter 7. More than 95% of patients now have their disease arrested by simultaneous treatment with INH and a suitable companion drug, and even among those with advanced bilateral cavitary tuberculosis similar results are being obtained by the monthly **alternation of different therapeutic combinations**—a regimen that is thought to improve the therapeutic response in less tractable cases.

The INH-resistant mutants isolated from treated patients differ from wild-type tubercle bacilli in several additional respects: they lack catalase and peroxidase activity, grow more slowly, and are much less virulent for guinea pigs. However, they retain virulence for man: they may continue to be shed in sputum over months and even years, and they are capable of producing primary tuberculosis in children.

While drug-resistant tubercle bacilli have been transmitted from patients to others, their frequency among freshly discovered cases has remained very low. However, since adult tuberculosis often arises from the reactivation of ancient foci, these cases may have arisen largely from "fossils," sequestered before the period of selection for drug resistance. Accordingly, we cannot predict with certainty the sensitivity patterns of future populations of tubercle bacilli.

EPIDEMIOLOGY

Mechanism of Spread. The human tubercle bacillus is spread principally by droplets and sputum from individuals with open pulmonary lesions. Small droplets produced by coughing are probably the most effective vehicles, since they rapidly dry in the air to yield droplet nuclei, of $<5\ \mu$ diameter, which can reach the alveoli. The organism can survive in moist or dried sputum for up to 6 weeks, but it is killed by a few hours' exposure to direct sunlight.

Distribution. The incidence of tuberculosis may be estimated by tuberculin test surveys as well as from reports of active cases or of deaths. The frequency is higher in impoverished social groups that live under crowded conditions, compared with the more affluent and with those that live in sparsely populated regions. The incidence is also high, however, in certain nonurban groups (American Indians and Eskimos) that are impoverished and probably also have high genetic susceptibility. Socioeconomic conditions undoubtedly affect both host resistance and frequency of infection.

With the progressive decline in prevalence over the past century the peak incidence has shifted from young adulthood, with a preponderance of females, to the age group over 65, with a severalfold preponderance of males.

Stresses responsible for the reactivation of old lesions in this group probably include malnutrition and alcoholism, for the highest incidence of active tuberculosis is now found among elderly, impoverished men living alone in city slums.

The worldwide decline of tuberculosis over the past century probably reflects general social and economic improvement. In addition, since about 1950 death rates have declined especially sharply as a result of effective chemotherapy (Fig. 35-8). The reduction in the number of new cases has thus far been more modest, but since ¾ of these cases in the United States are now believed to arise by activation of dormant infection, the impact of chemotherapy on the frequency of initial infection should increasingly affect the incidence of the disease in the future.

While tuberculosis is no longer the "white plague" of previous centuries, it is still an immense problem. In the world at large it is estimated that ca. 15 million persons have active tuberculosis, and annually 3 million persons die of the disease. Currently in the United States tuberculosis causes ca. 6000 deaths each year, and ca. 37,000 new active cases are recognized. Nevertheless, the tuberculosis problem in this country today is largely one of reactivation, for over 95% of the young population now reach adulthood without infection.*

PREVENTION

A major weapon in tuberculosis control programs in the recent past has been case-finding by mass surveys of apparently well persons (by means of tuberculin tests and chest X-rays), followed by isolation and treatment. However, the yield of new cases in the United States is now too low to justify this program. Instead, X-ray surveys are reserved for high-risk groups such as jail inmates, skid-row

---

* This figure is arrived at by correcting the 10% tuberculin reactors for the estimated frequency of cross-reactions (see Fig. 35-9).

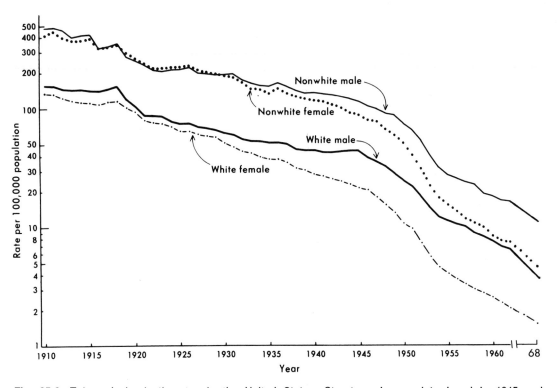

Fig. 35-8. Tuberculosis death rates in the United States. Streptomycin was introduced in 1945 and isoniazid in 1951. (Courtesy of Tuberculosis Program, U.S. Public Health Service, Center for Disease Control.)

males, and the contacts of tuberculin-positive young children encountered in screening programs in the schools.

CHEMOPROPHYLAXIS

The effectiveness and the low toxicity of INH have made possible prophylactic chemotherapy of tuberculosis. This approach was first applied, with striking success, to high-risk groups, such as tuberculin-positive household contacts of cases. Subsequently, in carefully studied populations a 50% decrease in incidence of active disease was achieved by giving a course of treatment to all persons who had recently become tuberculin-positive. It is of interest that despite this protection of the treated population only a small fraction lost their tuberculin hypersensitivity.

Though *M. tuberculosis* has no known reservoir outside man, the standard methods of isolating and treating cases of active tuberculosis would not seem likely to achieve its eradication, for detection of infectious cases is far from complete.

Moreover, since the disease may remain infectious (i.e. sputum-positive) for years while associated with tolerably good health and the ability to earn a livelihood, patients frequently resist isolation and even diagnosis. However, the prophylactic chemotherapy of tuberculin reactors which offers a more painless approach to interrupting the chain of transmission, may well lead to eradication of this pathogen.

IMMUNIZATION

Live vaccines can produce an increase in resistance to tuberculosis, but not complete protection; in experimental animals the effect is judged by the more prolonged course of the disease, and in man by its reduced incidence and case-fatality rate. The most widely used preparation has been **BCG** (see above, Genetic variation), a bovine strain attenuated by many passages on unfavorable media, which was introduced as a vaccine in France in 1923. In the 1940s the **vole bacillus** (or murine tubercle bacillus; see below) was similarly introduced in England: as isolated from field mice, without any "attenuation," it is quite avirulent for man, as well as for guinea pigs and rabbits. Killed vaccines, and various fractions of tubercle bacilli, have given at best only weak and transient protection.

Intradermal inoculation is the route of choice for the live vaccine. **Tuberculin-positive individuals should never be vaccinated:** they may have a severe reaction, with a Koch phenomenon locally and occasionally a flareup of a pulmonary lesion. Moreover, since they have already had an immunogenic exposure, vaccination is unlikely to increase their level of immunity.

The safety and efficacy of BCG have been the subject of bitter dispute over a 40-year period. Fear that the bacilli might revert to virulence has by now been allayed: over the course of the past several decades more than 150 million persons, in various countries, have been injected by the World Health Organization without reactions beyond an occasional small, transient ulcer at the site of inoculation.

Controversy with respect to efficacy arose from several sources: some questionable experimental statistics; probable loss of viability of many lots of vaccine used (before the recent introduction of a stable lyophilized preparation); and the use of populations already naturally immunized by widespread infection with atypical mycobacteria (see below). The most definitive study is that of the British Medical Research Council. Over 50,000 tuberculin-negative children, at ca. age 15, were randomly assorted into three groups, which were respectively immunized with BCG, immunized with vole bacillus, or left unvaccinated. The incidence of tuberculosis in either vaccinated group over the course of the next 9 years was only 20% of that observed in the unvaccinated group. Similar favorable results had been obtained earlier in a smaller controlled study in American Indians, an especially high-risk group.* However, BCG was not accepted by public health authorities in the United States when it might have been useful, and by now the prevalence of the disease here has become too low to justify its widespread use.

Vaccination has a major disadvantage: the resulting development of tuberculin hypersensitivity **destroys the usefulness of the tuberculin test.** Since public health measures in the United States now concentrate on recognizing and treating the recent tuberculin converter, vaccination is recommended only for tuberculin-negative individuals who stand a high risk of infection, such as children of

---

* Much less favorable results were obtained in a carefully controlled study by the U.S. Public Health Service in Alabama and Georgia, but evidence was subsequently obtained that this population had already been largely immunized by natural infection with atypical mycobacteria (see below).

tuberculous parents. Even for them chemo-prophylaxis following tuberculin conversion is considered preferable by many. Vaccination of medical personnel is no longer recommended.

OTHER MYCOBACTERIA THAT CAUSE A SIMILAR DISEASE

MYCOBACTERIUM BOVIS (M. TUBERCULOSIS VAR. BOVIS)

M. bovis causes tuberculosis in cattle and is also highly virulent for man; unpasteurized milk (and occasionally other dairy products) from tuberculous cows have been responsible for much human tuberculosis. The ingested organisms presumably penetrate the mucosa of the oropharynx and intestine (though without apparent damage), giving rise to early lesions in the cervical lymph nodes (scrofula) or the mesenteric nodes. Subsequent dissemination from these sites principally infects bones and joints; such infection of vertebrae was largely responsible for the hunchbacks of a previous generation. When inhaled (e.g., by dairy farmers) the organism can also cause pulmonary tuberculosis of man, indistinguishable from that caused by *M. tuberculosis.*

Tuberculosis due to *M. bovis* has now become very rare in many countries, as a result of the widespread pasteurization of milk and the virtual elimination of tuberculosis in cattle. The conquest of bovine tuberculosis vividly illustrates the effectiveness of public health legislation when the reservoir of a disease can be controlled.

Eradication of bovine tuberculosis was first aimed at in Denmark, through the isolation of tuberculin-positive cattle in separate herds and the separation of their calves at birth; but while this program was successful on individual farms, it required too much cooperation and vigilance to succeed on a national scale. In 1917 the U.S. Department of Agriculture, with widespread support from veterinarians, undertook an audacious program: tuberculin testing of all cattle and **the slaughter of all positive reactors.** As a result the proportion of tuberculin reactors in American cattle has been reduced from 5 to 0.5% (most of which now have no visible lesion at autopsy and may be cross-reactors).

M. bovis is a bit shorter and plumper than *M. tuberculosis,* but the difference is slight. Indeed, for 20 years after discovering the tubercle bacillus

Koch maintained that all mammalian tuberculosis was due to the same organism. By about 1900, however, largely through Theobald Smith's work, *M. bovis* and *M. tuberculosis* were clearly distinguished. For example, *M. bovis* is highly pathogenic in rabbits, whereas *M. tuberculosis* is much less so; and in cultures *M. bovis* tends to grow more slowly (so-called dysgonic growth), and cannot tolerate as high a concentration of glycerol. *M. bovis* also differs from *M. tuberculosis* in being niacin test negative, in not reducing nitrate, and in being resistant to pyrazinamide and susceptible to thiophen-2-carbonic acid hydrazide. Serological tests and skin tests, however, do not differentiate between the two organisms.

MYCOBACTERIUM AVIUM

M. avium causes tuberculosis in chickens, pigeons, and other birds, and often in swine. In a few recorded cases it has caused human tuberculosis. The avian bacillus is readily distinguishable from *M. tuberculosis* and *M. bovis.* The individual cells are smaller; the colonies are smooth; it grows optimally at ca. 41°, a temperature at which human and bovine bacilli will not grow; it is naturally resistant to most of the antituberculosis drugs; and it is pathogenic for chickens and rabbits but not for guinea pigs. Antigenic differences are also readily demonstrated.

MYCOBACTERIUM ULCERANS

This organism, which has been found mainly in Africa and Australia as the cause of a chronic type of cutaneous tuberculosis, may be a temperature-sensitive derivative of a mammalian tubercle bacillus. It cannot grow above 33°, and in mice and rats it causes lesions in only the cooler parts of the body; but it closely resembles *M. tuberculosis* in many other respects, including morphology, growth rate, and immunological specificity. The disease it causes in man, though rare, is of considerable theoretical interest, since the temperature range of the organism provides a simple explanation for its unusual pathogenetic properties.

"ATYPICAL" PATHOGENIC MYCOBACTERIA

In addition to the classic mammalian tubercle bacilli, various other acid-fast bacilli, possessing little or no virulence for guinea pigs or rabbits, have occasionally been cultured from sputum and gastric washings of tuberculous individuals. They were generally found, however, only sporadically and in small numbers; and since virulence for guinea pigs was held sacred since Koch's time in differentiating virulent mycobacteria from contaminating saprophytes, these atypical organisms were disregarded. Nevertheless, such atypical mycobacteria have recently been identified in some patients in large numbers in successive samples of sputum over prolonged periods, and also in the diseased tissues. These findings, together with the absence of typical tubercle bacilli, have eventually established confidence that these **atypical** mycobacteria (formerly also called **unclassified** or **anonymous**) can occasionally cause a disease resembling pulmonary tuberculosis, but generally milder.* Interest in these organisms is further increased by immunological evidence that in certain geographical regions they give rise to very widespread infection, though only rarely to disease.

The term "atypical" is primarily of historical significance. Runyon made a useful provisional grouping according to growth rate and pigment production only in the light (photochromogen) or also in the dark (scotochromogen) (Table 35-2). Subsequent studies have allowed more detailed classification.

M. marinum (synonyms: *M. balnei*, *M. platypoecilus*) causes a tuberculosis-like disease in fish and a chronic skin lesion known as "swimming pool granuloma" in humans. Infection acquired by injury of a hand around a home aquarium or swimming pool can lead to a series of ascending subcutaneous abscesses not unlike the lesions of sporotrichosis (Ch. 43). *M. marinum* grows best at 30 to 33°, only poorly at 37°. Like *M. ulcerans,* it produces lesions in mice only in the cooler parts of the body. It resembles *M. kansasii* (see below) in being photochromogenic, but can be differentiated by its optimum growth temperature and agglutination with specific antisera. The organism is resistant to INH but relatively susceptible to ethambutol, rifampin, and streptomycin.

M. xenopi was first isolated in 1959 from a skin lesion in a South African toad and was associated with human disease in 1965. It has been reported to be the cause of chronic pulmonary disease resembling tuberculosis in adults from England, Europe, Australia, and the United States.

M. kansasii and *M. intracellulare* together account for most of the chronic pulmonary disease in adults attributable to this group of acid-fast bacilli; their incidence varies markedly with geographical location. For example, in Dallas and

Houston the rate is about 10% of newly diagnosed and culture-positive cases of "tuberculosis," and the predominant organism is *M. kansasii;* in Georgia and Florida *M. intracellulare* predominates and the rate is 2 to 4% of cases; and an equal distribution of cases between the two species and a rate of 1 to 2% have been found in Cleveland. Studies from Australia and Japan indicate a predominance of *M. intracellulare.*

M. scrofulaceum is a common cause of lymph node infections in children but only rarely is it associated with pulmonary disease. The rapidly growing species *M. fortuitum* and *M. chelonei* (*abscessus*) may be found in subcutaneous abscesses at the sites of injuries or needle injections, and very rarely in pulmonary disease.

Any of these organisms may play the role of opportunistic pathogens and cause widespread disease in the immune-deficient host. They do not respond as well as ordinary tuberculosis to the presently available drugs because of a greater natural resistance; *M. kansasii* infections are the least troublesome in this respect.

Tables 35-2 and 35-3 summarize some of the important characteristics of these mycobacteria. Determination of serotype by agglutination is especially useful in distinguishing the similar *M. avium* and *M. intracellulare;* for, though *M. intracellulare* is characteristically avirulent for chickens, in contrast with *M. avium*, some strains are intermediate in virulence. Some strains of *M. scrofulaceum* agglutinate as members of "intracellulare" serotypes.

Evaluation of the significance of the isolation of an atypical mycobacterium from a clinical specimen is aided by a species identification; *M. kansasii*, *M. avium-intracellulare*, and *M. scrofulaceum* are potential pathogens, while *M. gordonae*, *M. gastri*, *M. terrae*, and *M. triviale* have not been associated with disease.

Epidemiology. In contrast to ordinary tuberculosis, the atypical mycobacterial infections ordinarily are not communicable, but are acquired by contamination from an environmental source; hence, patient isolation is not justified. In addition to the well-known saprophytic rapidly growing species (see below), many species of Runyon group II and group III mycobacteria may be found in the soil and dust, and some strains belong to the same serotypes as those associated with human disease. *M. kansasii* and *M. xenopi* are not usually found in the soil but they have been isolated from water taps.

Patients with disease due to atypical mycobacteria may or may not respond to a standard tuberculin test. A large-scale epidemio-

---

* It has long been known that organisms with little or no virulence for guinea pigs cause more than half the cases of tuberculosis of the skin.

TABLE 35-2. Summary of Some Characteristics of "Atypical" Mycobacteria That Are Pathogenic for Man or That Resemble Pathogenic Species

Runyon group	Species	25°	37°	42°	Pigment	Catalase 25–37°	68°	Ni-trate	Hydrolysis Tween 80	Reservoir	Helpful	No. of types
I Photo-chromogens	M. kansasii	+	+	−	Yellow*	Strong	+	+	+	? Tap water	+	1
	M. marinum	+	±†	−	Yellow*	Weak	+	−	+	Fish, water	+	1
II Scoto-chromogens	M. scrofula-ceum	+	+	−	Orange	Strong	+	−	−	? Soil	+	3
	M. gordonae	+	+	−	Orange	Strong	+	−	+	Soil	−	?
III Non-chromogens	M. avium-intracellu-lare‡ (Battey bacillus)	±	+	Vari-able	0	Weak	+	−	−	Soil, birds, swine	+	19
	M. terrae, M. triviale	+	+	−	0	Strong	+	+	+	Soil	−§	
	M. gastri	+	+	−	0	Weak	−	−	+	?	+	1
IV Rapid growers	M. fortuitum	+	+	−	0	Strong	+	+	Vari-able	Soil, dust	+	2
	M. chelonei (absces-sus)	+	+	−	0	Strong	+	−	Vari-able	Soil, dust	+	1
	M. xenopi	±	+	+‖	Yellow	Weak	+	−	−	?	±§	?

* Light-dependent.

† Optimum growth at 30–33°.

‡ These two species cannot be differentiated readily except by virulence for chicks and serotype: Serotypes 1, 2, and 3.

§ Do not form smooth suspensions.

‖ Optimum growth at 42–45°.

logical study with various tuberculins, by Palmer of the U.S. Public Health Service, revealed that in general humans infected with various mycobacteria react most intensely to the "tuberculin" from the homologous organism (Fig. 35-9).

This approach revealed a high level of inapparent infection with atypical mycobacteria in certain areas, and therefore a very low disease/infection ratio compared with that of tuberculosis. Indeed, in a nationwide study of over 200,000 U.S. Navy recruits only 8.6% reacted to the tuberculin of M. tuberculosis, while 35% reacted to that of an atypical mycobacterium of group III (the Battey bacil-

lus). Tuberculins of several other atypical mycobacteria, and of known saprophytes, also gave reactions with striking frequency (Table 35-4), and cross-reactions among various strains were frequent. Indeed, comparative studies with several antigens, in different geographical regions, have provided strong evidence that in some regions the bulk of the weak reactions to PPD are cross-reactions. The recognition of such false positives (by testing with an atypical antigen) is especially important if tuberculin reactions are to serve as the basis for instituting chemoprophylaxis. Hypersensitivity to the antigens of various atypical organisms is especially frequent in

TABLE 35-3. Pathogenicity of Mycobacteria

Species	Guinea pig	Rabbit	Mouse	Chick	Man
M. tuberculosis	+	−	+	−	Tuberculosis
M. bovis	+	+	+	−	Tuberculosis
M. avium	−	+	±	+	Tuberculosis (rare)
M. kansasii	−	−	±	−	Tuberculosis-like
M. marinum	−	−	Cool areas	−	Swimming pool granuloma and subcutaneous abscesses
M. scrofulaceum	−	−	−	−	Lymphadenitis in children
M. gordonae	−	−	−	−	No disease
M. intracellulare	−	−	−	±	Tuberculosis-like in adults; lymphadenitis in children
Other group III*	−	−	−	−	No disease
M. xenopi	−	−	−	−	Tuberculosis-like
M. fortuitum	−	−	+	−	Local abscess, rarely tuberculosis-like
M. smegmatis, M. phlei	−	−	−	−	No disease

* *M. terrae, M. triviale, M. gastri.*

the southeastern part of the United States, just as is hypersensitivity to certain fungi, which have a reservoir in soil or water.

Significance. The increasing prominence of the atypical pathogenic mycobacteria is probably relative rather than absolute, since chemotherapy has been decreasing the prevalence of classic tubercle bacilli without exerting a comparable effect on the probable reservoir of the atypical bacteria in nature.

TABLE 35-4. Frequency and Size of Reactions Among U.S. Navy Recruits to 0.1 μG of PPD Antigens Prepared from Various Strains of Mycobacteria

PPD antigen prepared from	No. tested	Per cent	Mean size (mm)
M. tuberculosis	212,462	8.6	10.3
M. fortuitum	3,415	7.7	4.8
Unclassified, group III	3,729	12.0	5.8
M. kansasii	13,913	13.1	6.2
Unclassified, group III	9,473	17.5	7.0
M. smegmatis	14,239	18.3	5.7
M. phlei	15,229	23.1	6.4
Unclassified, group II	10,060	28.4	9.0
M. avium	10,769	30.5	6.7
Unclassified (Battey type)	212,462	35.1	7.7
Unclassified, group III	8,402	39.0	7.2
Unclassified, group II	29,540	48.7	10.3

From EDWARDS, L. B. *Ann NY Acad Sci 106:*32 (1963).

Moreover, because of the need for repeating cultures during chemotherapy the isolated organisms are now being more intensively studied; hence strains with unusual characteristics are more likely to be detected, and are no longer automatically discarded as nonpathogens. The atypical organisms, with their ineradicable reservoir, may soon become the major residue of the problem of mycobacterial infection.

Guinea pigs inoculated with various atypical mycobacteria isolated from man have been found to show some degree of protection against subsequent challenge with *M. tuberculosis.* Hence it seems quite possible that **inapparent infection of man with these organisms may serve as a sort of natural vaccination** against tuberculosis. This background of protection may explain why a large-scale trial of BCG immunization in Georgia gave much less favorable results than a study (cited above) in England, where reactivity to atypical tuberculins is infrequent.

The delayed recognition of the atypical pathogenic mycobacteria illustrates the complexity of the problem of virulence, and the fallibility of laboratory tests for this property. The distinction between pathogens and nonpathogens is evidently not sharp among the mycobacteria. On the one hand, most of the atypical pathogens are probably saprophytes that can become pathogenic under appropriate conditions; and though they are much less virulent than *M. tuberculosis,* they occasionally cause chronic and even life-threatening disease (just as do many saprophytic fungi; see Ch. 43). On the other hand, though the tubercle bacillus is identified with a grave illness, its most frequent fate, after infecting an individual, is to persist for a lifetime without causing disease.

Fig. 35-9. Correlation of sizes of reactions of sanatorium patients, with proved typical or atypical mycobacterial infections, to standard tuberculin (PPD-S) and to comparable preparation (PPD-B) from an atypical bacterium (Battery bacilus) isolated from a patient in the same sanatorium. [From Edwards, L. B., Edwards, P. Q., and Palmer, C. E. *Acta Tuberc Scand Suppl 47*: 77 (1959).]

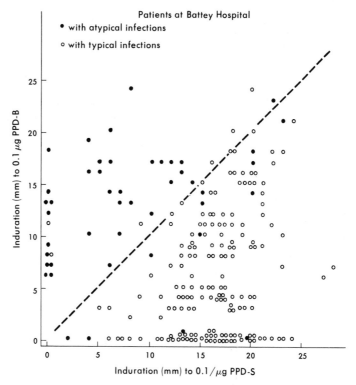

LEPROSY

MYCOBACTERIUM LEPRAE

The organism of human leprosy, discovered by Hansen in Norway in 1879, was the first bacterium shown to be associated with a human disease. Soon after the human tubercle bacillus Hansen's bacillus was also shown to be acid-fast.

Though *M. leprae* grows profusely in lesions it has never been cultivated in the test tube, and until 1959 it had never been convincingly transmitted to laboratory animals. Despite these limitations, Hansen's bacillus has long been conceded to be the etiological agent of human leprosy: 1) the organism is readily demonstrated, often in great numbers, in stained smears of exudates of persons with leprosy, and in tissue sections from lesions, whereas no other organism has been consistently identified in these preparations; and 2) *M. leprae* is virtually indistinguishable in morphology and staining properties from *M.*

tuberculosis, and leprosy has many clinical features in common with tuberculosis. Obviously, however, the failure to cultivate *M. leprae* has hampered its investigation. In particular, since the course of leprosy is notoriously unpredictable, the evaluation of chemotherapy of this disease has often been equivocal.

The human leprosy bacillus is outstanding for its limited host range: it has not even been transmitted to higher primates. In recent years, however, two experimental animal hosts have been found. Chatterjee has shown that when freshly isolated bacilli are serially passed through a particular strain of mice the organism eventually establishes a progressive infection. Even more promising (because it does not require selection of a particularly transmissible mutant) has been the finding by Shepard that *M. leprae* from human lesions will grow very slowly in the **foot pads** of mice; and infections maintained by serial passage remain confined to the foot pad. (This site was chosen because of the preference of the organism for the cooler, cutaneous regions of the human body.)

These two infections in mice make it possible to study experimentally such problems as the susceptibility of the organism to chemotherapeutic agents, properties of bacilli isolated from different forms of the disease (see below), and relations to other mycobacteria. However, the bacilli recovered from the mouse infections have not been propagated in culture media.

Lepromin. Though *M. leprae* cannot be cultivated, a product analogous to tuberculin was first prepared by Mitsuda in 1916 from homogenized leprous tissue, selected to contain abundant quantities of the organism. After intradermal injection many leprous patients give an inflammatory reaction at 24 to 48 hours, resembling the delayed-type response to tuberculin. Active tuberculosis sometimes gives rise to some sensitivity to lepromin.

PATHOGENESIS

M. leprae causes chronic granulomatous lesions closely resembling those of tuberculosis, with epithelioid and giant cells, but without caseation. The organisms in the lesions are predominantly intracellular and can evidently proliferate within macrophages, like tubercle bacilli.

Leprosy is distinguished by its chronic, slow progress and by its mutilating and disfiguring lesions. These may be so distinctive that the diagnosis is apparent at a glance; or the clinical manifestations may be so subtle as to escape detection by any except the most experienced observers armed with a high index of suspicion. The organism has a predilection for skin and for nerve. In the **cutaneous** form of the disease large, firm nodules (lepromas) are distributed widely, and on the face they create a characteristic leonine appearance. In the **neural** form segments of peripheral nerves are involved, more or less at random, leading to localized patches of anesthesia. The loss of sensation in fingers and toes increases the frequency of minor trauma, leading to secondary infections and mutilating injuries. Both forms may be present in the same patient.

Phases of the Disease. In either form of leprosy three phases may be distinguished. 1) In the **lepromatous** or progressive type the lesions contain many "lepra" cells: macrophages with a characteristically foamy cytoplasm, in which acid-fast bacilli are abundant (Fig. 35-10). When these lesions are prominent the lepromin test is usually negative, presumably owing to desensitization by mas-

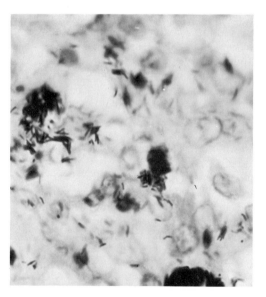

Fig. 35-10. Leprosy bacilli in cells in section of human tissue. Acid-fast stain, methylene blue counterstain, ×1200. (Courtesy of C. H. Binford, Leonard Wood Memorial.)

sive amounts of endogenous lepromin. The disease is then in a progressive phase and the prognosis is poor. 2) In the **tuberculoid** or healing phase of the disease, in contrast, the lesions contain few lepra cells and bacilli, fibrosis is prominent, and the lepromin test is usually positive. 3) In the **intermediate** type of disease bacilli are seen in areas of necrosis but are rare elsewhere, the lepromin test is positive, and the long-range outlook is fair. Shifts of leprosy from one phase to another, with exacerbation and remission of the disease, are common.

Hansen's bacillus may be widely distributed in the tissues of persons with leprosy, including liver and spleen. Nevertheless, no destructive lesions or disturbance of function are observed in these organs. Most deaths in leprous patients are due not to leprosy, per se, but to intercurrent infections with other microorganisms—often tuberculosis. Leprosy itself often causes death through the complication of **amyloidosis,** which is characterized by massive waxy deposits in kidneys, liver, spleen, and other organs. This curious disorder also occurs as a sequel to a variety of other chronic diseases with extensive necrosis and suppuration, or to prolonged, intensive immunization of animals with diverse antigens, especially some bacterial toxins. The waxy deposits contain abundant precipitates of fragments of immunoglobulin light chains.

Diagnosis. Bacteriological diagnosis is accomplished by demonstrating acid-fast bacilli in scrapings from ulcerated lesions, or in fluid expressed from superficial incisions over non-ulcerated lesions. Also useful is the skin test

with lepromin. No useful serological test is available, but, curiously, patients with leprosy frequently have a false positive serological test for syphilis.

Drug therapy with diaminodiphenylsulfone (Ch. 7) is apparently useful, and must be prolonged. Drugs effective against the tubercle bacillus are not strikingly effective in leprosy, but as was noted above, the response is difficult to evaluate.

EPIDEMIOLOGY

Leprosy is apparently transmitted when exudates of mucous membrane lesions and skin ulcers reach skin abrasions; it is not highly contagious. Young children appear to acquire the disease on briefer contact than adults. The incubation period has been difficult to establish but is estimated to range from a few months to 30 years or more. Apparently *M. leprae,* like the tubercle bacillus, can lie dormant in tissues for prolonged periods.

In ancient times leprosy was rampant throughout most of the world, but for unknown reasons it died out in Europe in the sixteenth century and now occurs there only in a few isolated pockets. In the United States leprosy occurs particularly in Texas and Louisiana, but sporadic cases are seen elsewhere. It is estimated that ca. 1000 individuals now have leprosy in the United States, and several hundred are cared for at the national leprosarium in Carville, Louisiana. Recent estimates place the number of lepers in the world at large at ca. 3 million, with the greatest density in central Africa and in parts of the Orient.

OTHER MYCOBACTERIA

In spite of their sluggish growth the mycobacteria are extremely widespread in nature, not only as saprophytes but as parasites throughout the animal kingdom.

The **vole bacillus** (*M. microti*) is found in the field mouse, or vole (*Microtus agrestis*), in which it produces an epizootic chronic disease resembling tuberculosis. The bacillus is indistinguishable from *M. tuberculosis* in morphology, staining, and cultural behavior. In guinea pigs it produces an indolent infection that spontaneously regresses and establishes increased resistance to subsequent infection with virulent tubercle bacilli; it has therefore been used as a vaccine in man (see above, Immunization).

Mycobacterium paratuberculosis causes an often fatal enteritis of cattle and sheep called **Johne's disease,** which is characterized by chronic infiltration of the intestinal mucosa and the mesenteric lymph nodes. Histologically the lesions resemble those caused by tubercle bacilli except that they are less localized and lack caseation. Infected cattle give delayed-type reactions to intradermal injections of culture filtrates (called johnin). The organism does not grow on the media used to cultivate other mycobacteria; it was initially cultivated by the ingenious expedient of enriching the medium with killed tubercle bacilli, and this special requirement can now be satisfied by a novel lipoidal growth factor, mycobactin (an Fe-complexing sideramine), obtained from a mycobacterium.

Mycobacterium leprae murium causes an epizootic chronic disease of wild rats known as rat leprosy; the organism is observed within macrophages in lesions. The disease may be transmitted to mice, guinea pigs, and white rats by inoculation of infected tissue, but the organism, like that of human leprosy, has not been cultivated. Histologically the lesions resemble those of human leprosy, but rats are not susceptible to infection with the human leprosy bacillus.

Mycobacteria of Poikilothermic Vertebrates. Acid-fast bacilli have been isolated from a variety of lesions, some of which resemble tuberculosis, in a number of cold-blooded vetebrates. These bacilli are named according to the species from which they have been isolated. Recent developments have necessitated a reevaluation of some of these names. *M. ranae* (from frogs) is the same as *M. fortuitum; M. chelonei* (from turtles) is the same as *M. abscessus,* a fortuitum-like rapid grower and a potential pathogen; *"M." thamnopheos* (from snakes) is more like a nocardia species than a mycobacterium; *M.*

marinum (from salt water fish) is capable of producing superficial lesions in man (see above).

Saprophytic Mycobacteria. A number of acid-fast bacilli found on plants and in soil and water grow more rapidly in culture than tubercle bacilli and are incapable of producing progressive disease in guinea pigs and rabbits. Their role in nature is primarily concerned with the degradation of lipids. These organisms have not been established as pathogens, but some may be potential parasites. They include *M. phlei,* the timothy hay bacillus, widely found in dust, plants, and soil, and extensively used in metabolic studies on mycobacteria; and *M. smegmatis,* the "butter" bacillus, found in smegma and butter, and also in dust, soil, and water. The latter species includes organisms formerly referred to as *M. butyricum* and *M. lacticola. M. fortuitum* exists as a saprophyte in soil and dust, but it is also a potential human pathogen and it can produce a fatal disease when injected intravenously into mice (see Tables 35-2 and 35-3).

SELECTED REFERENCES

Books and Review Articles

ARNASON, B. G., and WAKSMAN, B. H. Tuberculin sensitivity. *Adv Tuberc Res 13:*1 (1964).

BCG: A discussion of its use and application. *Adv Tuberc Res 8:*1 (1957).

Ciba Foundation Study Group No. 15. *The Pathogenesis of Leprosy.* Little, Brown, Boston, 1963.

Ciba Foundation Symposium on Experimental Tuberculosis. Churchill, London, 1955.

DUBOS, R. J., and DUBOS, J. *The White Plague.* Little, Brown, Boston, 1952.

International Conference on Mycobacterial and Fungal Antigens. *Am Rev Res Dis* (Suppl.) *92* (1965).

KUBICA, G. P., and DYE, W. E. *Laboratory Methods for Clinical and Public Health Mycobacteriology.* USPHS Publication No. 1547. GPO, Washington, D.C., 1967.

LONG, E. R. *The Chemistry and Chemotherapy of Tuberculosis,* 3rd ed. Williams & Wilkins, Baltimore, 1958.

LURIE, M. B. *Resistance to Tuberculosis.* Harvard Univ. Press, Cambridge, Mass., 1964.

NOLL, H. The chemistry of cord factor, a toxic glycolipid of *M. tuberculosis. Adv Tuberc Res 7:*149 (1956).

O'GRADY, F., and RILEY, R. L. Experimental airborne tuberculosis. *Adv Tuberc Res 12:*150 (1963).

RICH, A. R. *The Pathogenesis of Tuberculosis,* 2nd ed. Thomas, Springfield, Ill., 1951.

RUNYON, E. H. Pathogenic mycobacteria. *Adv Tuber Res 14:*235 (1965).

YOUMANS, G. P. The pathogenic "atypical" mycobacteria. *Annu Rev Microbiol 17:*473 (1963).

Specific Articles

AZUMA, I., YAMAMURA, Y., and MISAKI, A. Isolation and characterization of arabinose mycolate from firmly bound lipids of mycobacteria. *J Bacteriol 98:*331 (1969).

BIRNBAUM, S. E., and AFFRONTI, L. F. Mycobacterial polysaccharides. II. Comparison of polysaccharides from strains of four species of mycobacteria. *J Bacteriol 100:*58 (1969).

BLOCH, H. Studies on the virulence of tubercle bacilli: Isolation and biological properties of a constituent of virulent organisms. *J Exp Med 91:*197 (1950).

CHAPMAN, J. S. The ecology of the atypical mycobacteria. *Arch Environ Health 22:*41 (1971).

CHATTERJEE, K. R. Experimental transmission of human leprosy infection to a selected, laboratory-bred hybrid black mouse. *Int J Lepr 26:*195 (1959).

LURIE, M. B. The fate of tubercle bacilli ingested by phagocytes derived from normal and immunized animals. *J Exp Med 75:*247 (1942).

MACKANESS, G. B. The immunological basis of acquired cellular resistance. *J Exp Med 120:*105 (1964).

Medical Research Council. BCG and vole bacillus vaccines in the prevention of tuberculosis in adolescence and early life. Third report. *Br Med J 1:*973 (1963).

MIDDLEBROOK, G., DUBOS, R. J., and PIERCE, C. Virulence and morphology characteristics of mammalian tubercle bacilli. *J Exp Med 86:*175 (1947).

PALMER, C. E., and LONG, M. W. Effects of infection with atypical mycobacteria on BCG vaccination. *Am Rev Resp Dis 94:*553 (1966).

RUNYON, E. H. Identification of mycobacterial pathogens utilizing colony characteristics. *Am J Clin Pathol 54:*578 (1970).

SHEPARD, C. C. The experimental disease that follows the injection of human leprosy bacilli into footpads of mice. *J Exp Med 112:*445 (1960).

chapter

36

ACTINOMYCETES: THE FUNGUS-LIKE BACTERIA

Revised by GEORGE S. KOBAYASHI, Ph.D.

GENERAL CHARACTERISTICS 872
ACTINOMYCES   873
　　Pathogenesis   873
　　Diagnosis   875
　　Epidemiology   876
　　Treatment and Prognosis   876
NOCARDIA   877
　　Pathogenesis   877
　　Epidemiology   878
　　Treatment   879
STREPTOMYCES   879

GENERAL CHARACTERISTICS

The actinomycetes* are gram-positive organisms that tend to grow slowly as branching filaments. In some genera filaments readily segment during growth and yield pleomorphic, club-shaped cells that resemble corynebacteria and mycobacteria; some are even acid-fast. The filamentous growth leads to mycelial colonies (Ch. 43), and some actinomycetes cause chronic subcutaneous granulomatous abscesses. For these reasons the actinomycetes (Gr. *aktino* = ray and *mykes* = mushroom or fungus) were long regarded as fungi. The term ray refers to the characteristic radial arrangement of club-shaped elements formed in infected tissues by microcolonies of the major human pathogen of this group.

Nevertheless, the following more fundamental properties establish the actinomycetes as bacteria.

1) They are prokaryotic (i.e., lack a nuclear membrane).

2) The diameter of their filaments is 1 μ or less, i.e., smaller than yeast cells and much narrower than the characteristic tubular structures (hyphae) of molds (Ch. 43). The filaments of actinomycetes readily segment into bacillary and twig-like forms, with typical dimensions of bacteria.

3) Their cell walls contain muramic acid and diaminopimelic acid (lysine in some species), which are characteristic of bacterial walls (Ch. 6); and they lack chitin and glucans, characteristic of cell walls of molds and yeasts (Ch. 43).

4) Their growth is inhibited by penicillins, tetracyclines, sulfonamides, and other antibacterial drugs, all of which are innocuous for fungi.

5) Their cell membranes do not contain sterols and are therefore insensitive to the polyenes, which are effective fungicidal agents (Ch. 43).

6) Genetic recombination has been observed in one group, *Streptomyces,* and involves a transmissible fertility factor and the formation of merozygotes, rather than the zygote formation seen with fungi.

7) Motile forms possess simple flagella of bacterial type.

8) Anaerobic and chemoautotrophic forms are known.

Together with mycobacteria, the actinomycetes make up the order Actinomycetales (Table 36-1). *Actinomyces* grow as branching filaments when freshly cultivated. However, in older cultures and in infected lesions increased septation leads to spontaneous fragmentation into bacillary and even coccoid elements of irregular diameter, indistinguishable from diphtheroids. Those species of *Nocardia* that are acid-fast resemble mycobacteria. *Streptomyces* species are more fungus-like, forming aerial mycelia and chains of asexual spores. On primary isolation actinomycetes can usually be identified to the genus level on the basis of morphological features alone, but organisms that have been repeatedly transferred in culture may not retain the typical morphological features and therefore require biochemical methods for classification (Table 36-2). Serological methods are also used, but cross-reactions among actinomycetes are extensive.

In the following sections we shall devote particular attention to the more pathogenic genera, the anaerobic *Actinomyces* and the aerobic *Nocardia.* The diseases caused by

---

* In this chapter we follow the practice of referring to all members of the order Actinomycetales, except for the mycobacteria, as actinomycetes (see Table 36-1).

TABLE 36-1. Members of the Order Actinomycetales

Actinomycetes	Family: Mycobacteriaceae
	Genus: *Mycobacterium*
	Family: Actinomycetaceae
	Genus: *Actinomyces*
	Genus: *Arachnia*
	Family: Dermatophilaceae
	Genus: *Dermatophilus*
	Family: Actinoplanaceae
	Genus: *Actinoplanes*
	Genus: *Spirillospora*
	Family: Micromonosporaceae
	Genus: *Micromonosporaceae*
	Genus: *Termoactinomyces*
	Family: Streptomycetaceae
	Genus: *Streptomyces*
	Family: Nocardiaceae
	Genus: *Nocardia*

Based on the classification of H. A. Lechevalier and M. P. Lechevalier, and of L. Pine and L. K. Georg.

these organisms, and also by *Streptomyces,* are collectively called actinomycoses, though the disease caused by *Nocardia* is also called nocardiosis.

Another disease of medical importance, called "farmer's lung," occurs among agricultural workers who have inhaled dust from moldy plant material; it has been traced to allergy to two thermophilic actinomycetes, *Thermopolyspora polyspora* and *Micromonospora vulgaris.*

ACTINOMYCES

Two species of *Actinomyces* have been implicated as the cause of actinomycosis. *A. israelii* is usually responsible for the disease in man and *A. bovis* for the disease in cattle. These organisms are anaerobic or microaerophilic and require rich media (e.g., containing blood or brain-heart infusion); growth is stimulated by carbon dioxide and is poor at temperatures below 37°.

Most strains of *A. israelii* form rough colonies on agar and grow at the bottom of broth tubes as discrete aggregated clumps ("bread crumbs"; Fig. 36-1). A few strains, however, and most strains of *A. bovis,* form smooth colonies and tend to grow more diffusely in broth, and some strains of both species form intermediate colonies. In anaerobic cultures on brain-heart agar macroscopic colonies commonly mature in about 7 days. Gram stains of cultures reveal highly pleomorphic, irregular, club-shaped, gram-positive rods. Occasional cells appear as branched twigs, but long branching filaments are not observed (Fig. 36-1).

PATHOGENESIS

Because most actinomycetes are found in soil, and because human and bovine actinomycosis often affects the jaw and face, this disease was long held to arise from chewing straw and grass. More recent studies, however, have shown that *A. israelii* and *A. bovis* are apparently not present in soil or on vegetation, but are quite regularly isolable from the normal mouth of man and cattle, respectively. Like many respiratory pathogens, they are much more frequently commensal than pathogenic. *A. israelii,* for example, can be cultured from the majority of human tonsils and is nearly always found in scrapings of gums and teeth. Its fastidious growth requirements, noted above, are in accord with this parasitic (or saprophytic) existence.

The conditions that lead the organism to become invasive are not definitely known but may involve trauma (including dental surgery), aspiration of a heavily contaminated tooth or detached bits of dental tartar, or, rarely, human bites.

Actinomycosis is characterized by chronic, destructive abscesses of connective tissues. In one series of patients in the United States about 50% of the infections involved the abdomen, particularly the cecum and appendix, 20% the lungs and chest wall, and 30% the face and neck; but in a larger series in England 60% of all cases involved the cervicofacial area.

Wherever the lesions occur they are basically the same. Abscesses expand into contiguous tissues and eventually form burrowing, tortuous sinuses to the skin surface, where they discharge purulent material. Penetration into mucous membranes is much less frequent. Connective and granulation tissues tend to form a wall around the abscess. Histologically, the lesions are not distinctive except for the presence of the organisms as small colonies, described below ("sulfur granules"; Fig. 36-2).

Actinomycotic lesions of cattle are charac-

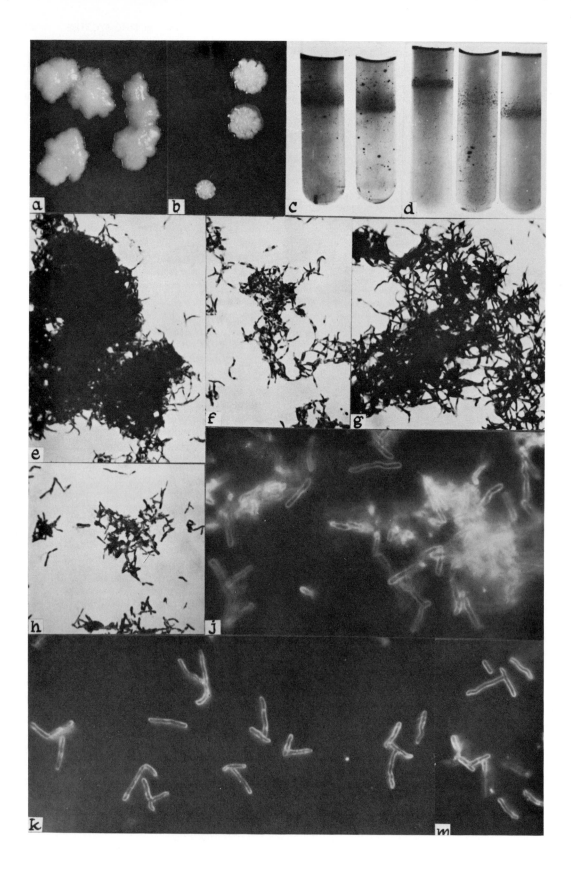

TABLE 36-2. Major Constituents in Cell Wall Preparations of the Most Important Actinomycetes

Cell wall type	Lysine	Ornithine	Aspartic acid	Glycine	L-DAP*	meso-DAP	Arabinose	Galactose	O₂ requirement
I. *Streptomyces*				+	+				
II. *Micromonospora* *Actinoplanes*				+		+			
III. *Streptosporangium* *Dermatophilus* *Thermoactinomyces* *Microbispora* *Actinomadura†*						+			Oxidative metabolism; usually found in soil
IV. *Nocardia* *Mycobacterium‡*						+	+	+	
V. *Actinomyces* (israeli-type)	+	+							Fermentative metabolism; usually associated with animals
VI. *Actinomyces* (bovis-type)	+		+						

*DAP = 2, 6-diaminopimelic acid.

† Formerly *Nocardia* (madurae-type).

‡ See Chapter 35.

All preparations contain major amounts of glucosamine, muramic acid, alanine, and glutamic acid.

Modified according to suggestions of Dr. H. A. Lechevalier from Lechevalier, H. A., and Lechevalier, M. P. *Annu Rev Microbiol 21:*71 (1967).

teristically large, bone-destroying abscesses of the lower jaw; because they appear as disfiguring, swollen nodules the disease in cattle is often referred to as "lumpy jaw." In man, however, lesions of bone are infrequent.

Like most diseases caused by organisms that ordinarily are saprophytes (Ch. 43, Opportunistic fungi) actinomycosis is not transmissible from man to man or from animals to man. It is, in fact, even difficult to establish the infection experimentally in laboratory animals. Repeated inoculation of guinea pigs with *A. israelii* does, however, occasionally establish progressive and fatal infection, while the hamster develops localized abscesses when injected intraperitoneally. The actinomycetes isolated from abscesses are no more effective in establishing experimental infections than those isolated from the normal mouth.*

---

* *Streptobacillus moniliformis,* one of the causes of rat-bite fever (Ch. 41) may be recalled here because it is a normal inhabitant of the mouth of

DIAGNOSIS

When pus from an abscess or infected sputum is examined carefully yellow "sulfur granules," named for their color, are occasionally seen. These small clusters of colonies of actinomycetes range from the barely visible to several millimeters in diameter. Detection of granules is not required to establish a diagnosis of actinomycosis, but their presence facilitates identification of the organism. The granules, made up of one or more colonies embedded in a matrix of calcium phosphate, consist of a central filamentous mycelium surrounded by club-shaped structures in a

---

rats. Moreover, because *S. moniliformis* is also pleomorphic and often grows in filamentous form, it has been called *Actinomyces muris.* The disease that it causes in man, however, is an acute, self-limiting septicemia, with rash and arthritis.

Fig. 36-1. *A. israelii.* **a,b.** Rough colonies grown anaerobically on brain-heart agar. **c,d.** Shake cultures incubated anaerobically; note growth in a layer below the surface. **e–h.** Gram-stained smears, showing masses of filaments. ×750, reduced. **j–m.** Unstained wet films under darkfield illumination; note distinct branching and twig-like forms. ×1200, reduced. [From Rosebury, T., Epps, L. J., and Clark, A. R. *J Infect Dis 74:*131 (1944).]

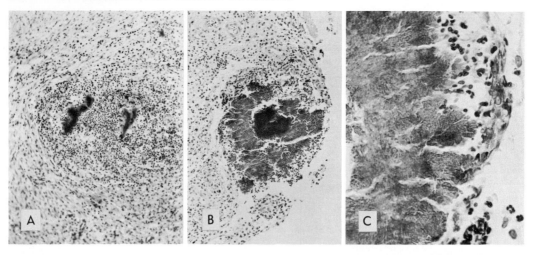

Fig. 36-2. A. Colonies of *A. israelii* (sulfur granules) in a lung abscess. A higher magnification of the periphery of the granule in (**B**) is reproduced in (**C**), showing characteristic radial club-like structures. Hematoxylin-eosin. **A** and **B**, ×130, reduced; **C**, ×600, reduced. (**A** from Emmons, C. W., Binford, C. H., and Utz, F. P. *Medical Mycology.* Lea & Febiger, Philadelphia, 1963. **B** and **C** courtesy of Dr. Alex C. Sonnenwirth.)

characteristic radial arrangement (Fig. 36-2). When granules are crushed and stained the organisms appear as gram-positive bacillary and diphtheroid forms. Branched filaments are not readily discerned, and the club-shaped elements are gram-negative.

Actinomycotic abscesses are almost invariably mixed infections, like many other abscesses; even the washed sulfur granules may contain colonies of associated bacteria. These include fusiform bacilli, anaerobic streptococci, and a tiny anaerobic gram-negative bacillus that bears the formidable name *Bacterium actinomycetemcomitans.* The accompanying bacilli may secrete collagenase and hyaluronidase and thus facilitate extension of the lesion.

Anaerobic diptheroids and a facultative anaerobic actinomycete, *A. naeslundii,* look very much like *A. israelii* and are often found in anaerobic cultures of the oral cavity. *A. naeslundi* may be pathogenic; it has been isolated from a variety of human lesions, and it can cause lesions in mice.

EPIDEMIOLOGY

Actinomycosis is worldwide in distribution but is relatively rare in man. Its incidence is higher in men than in women, and in persons over 20; in older reports it was much more frequent in rural than urban areas. The allegedly greater incidence among farmers is not readily reconcilable with the prevailing view that the disease arises from invasion by an indigenous organism.

Actinomycosis occurs in a variety of wild and domesticated animals besides man and cattle.

TREATMENT AND PROGNOSIS

Actinomyces, like the fungi that cause subcutaneous mycoses (Ch. 43), does not give rise to an effective immune response, and in the absence of therapy these chronic lesions tend to spread slowly but inexorably. Prior to the availability of chemotherapy the prognosis was poor.

A. israelii is sensitive to penicillin, tetracyclines, chloramphenicol, and streptomycin. Penicillin is reported to be most effective clinically. Surgical drainage of abscesses and resection of damaged tissue are important adjuncts to chemotherapy. With combined chemotherapy and surgery the cure rate is now about 90% for cervicofacial actinomycosis, but somewhat less for abdominal and thoracic actinomycosis.

NOCARDIA

In contrast to *Actinomyces,* species of *Nocardia* are aerobic, grow readily over a wide temperature range, grow on relatively simple media (e.g., Sabouraud's glucose agar*), and are inhabitants of soil rather than commensals in animals.

When grown on agar, colonies of *Nocardia* species may be smooth and moist, or rough with a velvety surface due to a rudimentary aerial mycelium. However, when smeared and stained the filaments fragment into bacillary and coccoid bodies. Examination of liquid cultures, especially slide cultures, may be helpful in visualizing branching filaments. Growth in liquid media usually produces a dry, waxy surface pellicle, as with mycobacteria.

Nocardia species are all gram-positive, and the two species most often pathogenic in man, *N. asteroides* and *N. brasiliensis,* are also somewhat acid-fast: they are more easily stained with fuchsin and retain the stain less tenaciously than mycobacteria.[†] Young cultures tend to be more acid-fast than older cultures.

N. asteroides also resembles mycobacteria in being resistant to dilute alkali and to some of the dyes (such as brilliant green) used to inhibit growth of rapidly growing gram-positive bacteria; hence they may grow on the same medium and after the same manipulations used for routine selective isolation of tubercle bacilli from sputum and exudates (Ch. 35). Moreover, the colonies formed by nocardiae resemble those of saprophytic mycobacteria and the unclassified mycobacteria (Ch. 35 and Fig. 36-3). Finally, extensive serological cross-reactions (in agglutination and complement fixation) are observed with

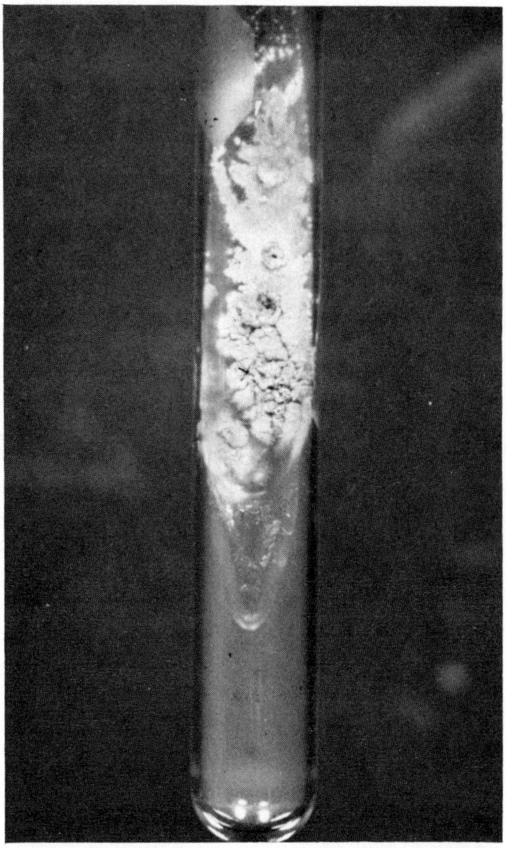

Fig. 36-3. Colonies of *N. asteroides* after 4 weeks on Sabouraud's glucose agar. Note the typical heaped-up irregularly folded appearance. (Courtesy of Dr. Alex C. Sonnenwirth.)

antisera to mycobacteria and nocardiae.[‡] Despite these similarities, however, it is not difficult to distinguish between the two groups: nocardiae grow more rapidly, are less acid-fast, and tend to branch.

PATHOGENESIS

As noted above, *Nocardia* species are regularly isolable from soil, and there are, accordingly, two common modes of establishing infection. Pulmonary nocardiosis arises from

---

* This medium is a particular favorite for the cultivation of fungi (Ch. 43), and its occasional use with *Nocardia* is another vestige of the former view that these organisms are fungi.

† The acid-fast nocardiae are readily stained with basic fuchsin without heating. When subsequently treated for removal of the fuchsin 1% H_2SO_4 without ethanol is used, since acid plus alcohol is too effective as a decolorizing solvent (cf. Ch. 35, Mycobacteria and acid-fastness).

---

‡ *N. asteroides* can also serve as an effective substitute for mycobacteria in enhancing the immunogenicity of diverse antigens in the water-in-oil emulsions described by Freund (Ch. 17).

inhalation of the organisms, while chronic subcutaneous abscesses (mycetomas) arise from contamination of skin wounds, usually on feet and hands of laborers.

Infections with *Nocardia* are hard to establish in laboratory animals. Suspension of the organisms in gastric mucin enhances their pathogenicity, and guinea pigs inoculated with such suspensions regularly develop abscesses and occasionally succumb.

Pulmonary Nocardiosis. In pulmonary nocardiosis, caused by *N. asteroides,* the lesions may be scattered through lung parenchyma, simulating miliary tuberculosis or histoplasmosis, or they may take the form of larger, confluent, partially excavated abscesses that superficially resemble the cavities of chronic pulmonary tuberculosis. The lesions are characterized histologically by suppuration, with granulation and fibrous tissue surrounding the areas of necrosis. Neither the characteristic granulomas of tuberculosis nor the burrowing sinuses of *Actinomyces* abscesses are seen. *N. asteroides* lies scattered through the abscesses in the form of tangled, fine, branching filaments; and aggregation into granules does not occur, in contrast to actinomycotic lesions due to *A. israelii* or *A. bovis* or the mycetomas due to *N. asteroides* and other nocardiae (see below).

In histological sections the bacterial filaments are not seen with hematoxylin-eosin stains; but when visualized with bacterial stains the organisms (which tend to fragment during staining) appear as gram-positive, weakly acid-fast diphtheroid and bacillary forms, which may easily be mistaken for tubercle bacilli.

N. asteroides often spreads from pulmonary lesions by way of the blood stream and establishes metastatic abscesses, usually in subcutaneous tissues and in the central nervous system. Lesions in the brain and the meninges are usually fatal.

N. asteroides is occasionally identified in sputum from patients with chronic pulmonary diseases of unknown etiology. When it is isolated consistently from an individual a presumptive diagnosis of pulmonary nocardiosis is warranted. But isolation in a solitary specimen raises the possibility of its occasional presence as a saprophyte in the upper respiratory tract.

Mycetoma Due to Nocardia. Different species of *Nocardia* are associated with mycetomas in different parts of the world, e.g., *N. brasiliensis* in Mexico. These chronic subcutaneous abscesses are clinically very similar to those due to *Streptomyces* and to various fungi. The abscesses spread locally by direct extension, destroy soft tissues and bone, and form burrowing sinus tracts; colonial aggregates (granules) of the causative microorganism are often present in the abscesses and in their purulent discharge. The detection of granules helps in the isolation and identification of the etiological agent.

A variety of properties distinguish among *N. asteroides, N. brasiliensis,* and some species of *Streptomyces* that also cause mycetomas. These are listed in Table 36-3.

EPIDEMIOLOGY

Though nocardiae are worldwide in distribution, nocardiosis is a rare disease. It seems likely, therefore, that man has a high degree of innate resistance. Indeed, untreated pulmonary nocardiosis is sometimes a self-limited disease, in contrast to actinomycosis due to *Actinomyces.* Generally, however, once nocardiosis becomes evident the disease tends

TABLE 36-3.   Some Aerobic Organisms that Cause Actinomycotic Mycetomas in Man

Species	Microscopic appearance	Hydrolysis of casein	Hydrolysis of starch	Pathogenicity in mouse and guinea pig
Nocardia asteroides	Partially acid-fast	0	0	Abscesses without granules; frequently causes death
Nocardia brasiliensis	Partially acid-fast	+	0	Abscesses with granules caused by some strains
*Streptomyces madurae**	Not acid-fast	+	+	Not pathogenic
*Streptomyces pelletierii**	Not acid-fast	+	0	Not pathogenic

* Sometimes placed in genus *Nocardia* rather than *Streptomyces.* In complement-fixation tests they react with antisera to various mycobacteria, nocardiae, and streptomycetes.

to be progressive and fatal; even with intensive therapy about 50% of patients succumb.

TREATMENT

Various antibacterial drugs are effective in the treatment of experimental nocardiosis, but sulfonamides are reported to be most effective clinically. Drainage of abscesses is an important adjunct. The distinction from fungal mycetomas is especially important because entirely different chemotherapeutic agents are indicated (Ch. 43).

STREPTOMYCES

Streptomyces species are characterized by the stability of their filaments and by the formation of spores on the aerial mycelia, i.e., on the filaments that project above the surface of the culture medium (Ch. 43). The streptomycetes resemble nocardiae more than actinomycetes in growth and appearance, and mutants that lose the ability to form aerial mycelia and spores are difficult to distinguish from nocardiae.

As the distinction between nocardiae and streptomycetes has become more generally appreciated, it has been realized that both cause actinomycotic abscesses (Table 36-3). Since streptomycetes are ubiquitous in soil, infection is attributed to contamination of scratches and penetrating wounds. Mycetomas due to streptomycetes are indistinguishable clinically from those to other actinomycetes.

Until 30 years ago only a small number of soil microbiologists were interested in the streptomycetes. However, since the isolation by Waksman of actinomycin in 1940, and streptomycin in 1943, the streptomycetes have received a phenomenal amount of attention. Innumerable isolates from soil samples, taken from all parts of the world, have been systematically scrutinized for production of antibiotics. They have yielded over 500 different compounds, including most antibiotics of practical value, except penicillin and griseofulvin (Ch. 7).

Streptomycetes and other actinomycetes are also widespread scavengers in soil. By breaking down proteins, cellulose, and other organic matter (including paraffin), they probably contribute as much as all other bacteria and fungi to the fertility of soil and to the geochemical stability of the biosphere.

SELECTED REFERENCES

Books and Review Articles

EMMONS, C. W., BINFORD, C. H., and UTZ, J. P. *Medical Mycology,* 2nd ed., Chs. 8, 9, and 26. Lea & Febiger, Philadelphia, 1970.

GEORG, L. K. Diagnostic Procedures for the Isolation and Identification of the Etiologic Agents of Actinomycosis. In *Proceedings, International Sympsium on Mycoses.* Scientific Publication No. 205, p. 71. Pan American Health Organization, 1970.

GORDON, R. E. Some strains in search of a genus— *Corynebacterium, Mycobacterium, Norcardia* or what? *J Gen Microbiol 43:*329 (1966).

PINE, L., and GEORG, L. K. Reclassification of *Actinomyces propionicus. Int J Systemat Bacteriol 19:* 267 (1969).

ROSEBURY, T. *Microorganisms Indigenous to Man,* Ch. 3. McGraw-Hill, New York, 1962.

WAKSMAN, S. *The Actinomycetes,* vols. 1, 2, and 3 (with H. A. Lechevalier). Williams & Wilkins, Baltimore, 1962.

Specific Articles

COLEMAN, R. M., and GEORG, L. K. Comparative pathogenicity of *Actinomyces naeslundii* and *Actinomyces israelii. Appl Microbiol 18:*427 (1969).

CUMMINS, C. S., and HARRIS, H. Studies on the cell-wall composition and taxonomy of Actinomycetales and related groups. *J Gen Microbiol 18:*173 (1958).

LABERGE, D. E., and STAHMANN, M. A. Antigens from *Thermopolyspora polyspora* involved in Farmer's lung. *Proc Soc Exp Biol Med 121:*463 (1966).

LECHEVALIER, M. P., HORAN, A. C., and LECHEVALIER, H. A. Lipid composition in the classification of nocardiae and mycobacteria. *J Bacteriol 105:*313 (1971).

THE SPIROCHETES

Revised by PAUL H. HARDY, M.D.

GENERAL CHARACTERISTICS 882

SYPHILIS 882
  HISTORY 882
  MORPHOLOGY 883
  METABOLISM 884
  ANTIGENS 885
    Wasserman Antigen 885
    Treponemal Antigens 886
  GENETIC VARIATION 887
  PATHOGENICITY 887
    Syphilis in Man 887
    Experimental Syphilis 888
  IMMUNITY 888
  LABORATORY DIAGNOSIS 889
    Detection of T. pallidum in Lesions 889
    Serological Tests 889
  TREATMENT 890
  PREVENTION 890
  YAWS (FRAMBESIA) 891
  PINTA 891
  OTHER TREPONEMATOSES 892

RELAPSING FEVER 892
  OTHER BORRELIA INFECTIONS OF MAN 893

LEPTOSPIROSIS 894
  Morphology 894
  Cultivation 894
  Antigenicity and Genetic Variation 894
  Pathogenicity 894
  Laboratory Diagnosis 895
  Treatment 895
  Prevention 895

RAT-BITE FEVER (SODOKU) 896

GENERAL CHARACTERISTICS

Spirochetes are motile organisms with a morphology quite different from that of other bacteria. All pathogenic species belong to the family Treponemataceae, which is divided into three genera, *Treponema, Borrelia,* and *Leptospira.* The Treponemataceae are extremely long, flexible, filamentous cells that are usually held in a characteristic spiral, or coiled-spring, shape. Their motility is primarily a rapid rotation around the long axis of the organism, but in a suitable environment they appear to bore their way through the medium. Most spirochetes are not seen in the ordinary light microscope because their transverse diameter is below the resolving power of the instrument. They can be visualized by darkfield illumination* or by staining with special reagents (silver salts or metallic flagellar stains) that are deposited on their surfaces. Spirochetes divide by binary fission (Fig. 37-2).

The genus *Treponema* includes the human pathogens that cause **syphilis** (*T. pallidum*), **yaws** (*T. pertenue*), and **pinta** (*T. carateum*), as well as the agent of rabbit syphilis (*T. cuniculi*). In addition there are a number of ill-defined species among the normal microbial flora in the alimentary tract (e.g., *T. microdentium* and the external genitalia of man and most animals. Unlike the pathogenic species many of these can be cultivated on artificial media.

The genus *Borrelia* is composed primarily of the organisms that cause **relapsing fever** (e.g., *B. recurrentis*). Other species are occasionally found in the mixed flora of ulcerative skin lesions and lung abscesses, but their pathogenic role is not clear. The *borreliae* are the largest Treponemataceae, and are thick enough to be easily identified in Giemsa or Wright-stained blood smears. Their spirals are very coarse and irregular. Unlike treponemes, the borreliae can be kept viable in biological fluids (e.g., blood) for a long time, but their culture in vitro has only recently been reported.

Members of the genus *Leptospira* are the smallest and most tightly coiled spirochetes. They are primarily animal pathogens and man is an accidental host. A variety of wild and domestic animals make up the natural reservoir for pathogenic *Leptospira,* but the organisms can live for varying periods of time in ponds and other bodies of water contaminated with excreta. Such waters may also contain nonpathogenic free-living leptospires, which are distinguishable from the pathogens by cultural and antigenic differences.

SYPHILIS

HISTORY

During the last decade of the fifteenth century syphilis became rampant in Western Europe. The story that it was brought from the West Indies by members of Columbus' crew is probably apocryphal. Records indicate, however, that the disease in Europe, in its acute phase, was extraordinarily malignant for about

---

* Darkfield microscopy requires the use of a special substage condenser which throws the light across rather than directly through the field. Only light that is reflected, refracted, or diffracted by an object in the field reaches the eye. Hence observed objects are light and the background is dark (Fig. 37-1).

Fig. 37-1. *T. pallidum* in a darkfield preparation. ×ca. 2000. (From Pelczar, M. J., Jr., and Reid R. D. *Microbiology.* McGraw-Hill, New York, 1958.)

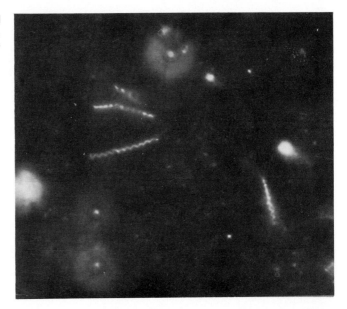

60 years, beginning in the 1490s; thereafter, it became milder, though its late consequences remained serious. It now exists endemically in nearly all parts of the world.

Starting with the poem in which the early epidemiologist Fracastorius (1530) named the disease after a mythical shepherd, syphilis has been the subject of an extensive and colorful literature. The organism causing it was discovered by Schaudinn and Hoffmann in 1905, and the "Wassermann reaction" for detecting antisyphilitic antibodies was described by Wassermann, Neisser, and Bruck in 1906. Widespread treatment of syphilis in the United States with penicillin, beginning in the mid-1940s, led to a decline of more than 85% in the incidence of early cases, but since 1955 newly acquired infections have been increasing again. This resurgence has been attributed to relaxed control measures and increased sexual promiscuity, but a decrease in the indiscriminate use of penicillin for minor ailments is an additional factor.

MORPHOLOGY

T. pallidum, which is morphologically indistinguishable from other pathogenic treponemes, is an extremely slender spiral organism measuring 5 to 20 μ in length and less than 0.2 μ in thickness. It has 4 to 14 spirals that are quite uniform near the center of the cell but frequently increase in periodicity and decrease in amplitude toward the ends, thus giving the cell a tapered appearance. Darkfield illumination is more frequently used than stained preparations for visualizing the organism, because it permits the examination of material containing live spirochetes, and observation of their characteristic motility.

Electron microscopic studies have revealed that all treponemes have essentially the same anatomical structure. The cell body (protoplasmic cylinder) is surrounded by a thin (15 to 20 nm) trilaminar wall that contains muramic acid and is responsible for the basic filamentous shape of the spirochete. Wrapped around the organism from one end to the other are varying numbers of axial fibrils (Fig. 37-3). These organelles originate from knob-like bodies within the cytoplasm near both ends of the cell. Their length varies but is usually sufficient for the fibrils arising from opposite ends of the cell to overlap, and sometimes to extend beyond the distal cell tip. The fibrils create the spiral shape of these organisms, and their contractility is presumably responsible for spirochetal motility.

Both the protoplasmic cylinder and the axial fibrils are encased in a second thin membrane-like covering (Fig. 37-3), normally in close apposition to the inner structures, which is a major osmotic barrier for the cell.

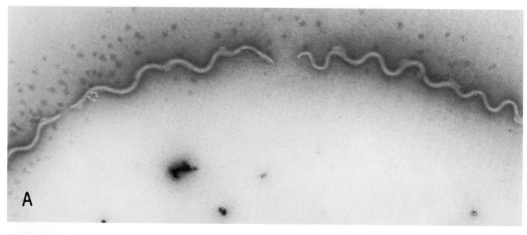

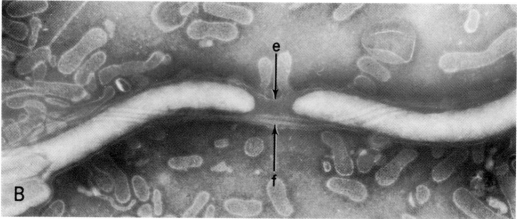

Fig. 37-2. Multiplication of *T. microdentium* by transverse fission. **A.** A chain of multiplying organisms in a pure culture during the logarithmic phase of growth. ×13,000, enlarged. **B.** Daughter cells beginning to separate while still connected by outer envelope (e) and avial filaments (f). See Fig. 37-3. ×64,000, enlarged. [From Listgarten, M. A., and Socransky, S. S. *J Bacteriol 88:*1087 (1964).]

This envelope is quite fragile and easily ruptured, but under hypotonic osmotic shock it will balloon, while the protoplasmic cylinder and fibrils curl inside the resulting sphere. Changes of this type are occasionally seen in old syphilitic lesions, and they are frequent in aging cultures of nonpathogenic treponemes. These spherical bodies were long regarded as the cystic stage of a complex life cycle, but they are largely nonviable and probably represent involutional forms.

In aqueous media young treponemes spin vigorously around their long axis in a seemingly useless type of motion. However, in a more viscous environment or on surfaces they achieve sufficient traction to propel them-

selves; in tissues they exhibit remarkable flexibility as they adapt themselves to the intercellular spaces.

METABOLISM

None of the pathogenic treponemes can be cultured in vitro at present despite numerous reports to the contrary. Attempts to cultivate them in chick embryos and tissue cultures have also failed, although they can be propagated in vivo in laboratory animals (see Pathogenicity). Many nonpathogenic treponemes (e.g., some oral species and the so-called cultivable *T. pallidum* stains such as

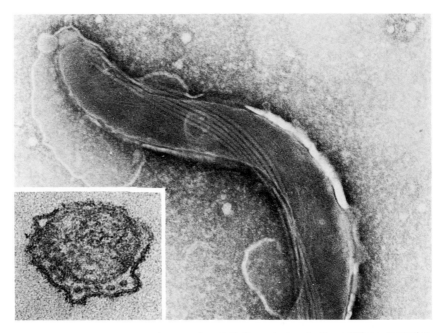

Fig. 37-3. Electron photomicrograph of Reiter strain of *T. pallidum* (negative staining technic) showing fibrils of axial filament arising in end bulb at tip of organism. Note outpouchings of periplastic membrane. ×50,000. Cross-sectional **inset** (thin section) reveals spatial relations between enveloping membrane, fibrils of axial filaments, and central protoplasmic cylinder. ×120,000. [From Ryter, A., and Pillott, J. *Ann Inst Pasteur 104:*496 (1963).]

Reiter, Noguchi, and Kazan), on the other hand, have been cultivated in a variety of artificial media. All are obligate anaerobes and grow very slowly, with average division times ranging from 4 to 18 hours.

Suspensions of *T. pallidum* have remained viable and motile for periods up to 15 days when kept at 35° under anaerobic conditions in a medium containing serum albumin, glucose, carbon dioxide, pyruvate, cysteine, glutathione, and serum ultrafiltrate. Although growth as measured by an increase in cell numbers does not occur, such organisms are killed by penicillin at low concentrations, suggesting some biosynthetic activity. This action requires an incubation period of 5 hours or longer, in keeping with the slow growth rate of treponemes.

The metabolic capabilities of the cultivable treponemes vary, but all have complex nutritional requirements. They utilize glucose or other fermentable carbohydrate as primary energy source, and require multiple amino acids, purines, and pyrimidines as well. Many strains also need bicarbonate and one or more coenzymes, and all require at least one exogenously supplied fatty acid.

For some oral strains a short-chain acid will suffice, but others require one or more 16- to 18-carbon-chain acids, usually supplied as protein-bound lipid in a serum supplement to the growth medium.

When frozen at temperatures below −70° in the presence of glycerol or other cryoprotective agent suspensions of treponemes can be kept viable for years. Of practical importance for the problem of transfusion syphilis is the fact that in blood, serum, or plasma stored at refrigerator temperatures viable *T. pallidum* have not been recovered after 48 hours. On the other hand, organisms may remain alive for as long as 5 days in tissue specimens removed from diseased animals.

ANTIGENS

WASSERMANN ANTIGEN

The first serological test for syphilis was a complement-fixing reaction—the so-called Wassermann test. In its original form this test used as antigen an extract of liver, containing many treponemes, from human fetuses with congenital syphilis. However, the specific ligand involved was later found to be present in alcoholic extracts of normal liver and other mammalian tissues as well. It was subsequently isolated from cardiac muscle and identified as a phospholipid, diphosphatidyl glycerol (Fig. 37-4), which was designated cardiolipin.

Many modifications and variations of the Wassermann test have been devised in hopes of achieving a more specific test. In contrast to the original complement-fixation reaction, most of the later tests are flocculation reactions. These are analogous to precipitins, but they use cardiolipin in an aqueous suspension finely dispersed with the aid of cholesterol and lecithin, instead of a soluble antigen. Each of the various anticardiolipin tests is known by the name of its originator (Wassermann, Kolmer, Eagle, Hinton, Kahn, Kline, etc.), but collectively they are designated serological tests for syphilis (STS) to distinguish them from the antitreponemal tests developed in recent years.

Since cardiolipin is a normal constituent of host tissue, there arose a controversy about whether the primary antigenic stimulus for Wassermann's antibody(anticardiolipin) comes from the invading organism or the host, and accordingly whether the development of this antibody represents an autoimmune response. Wassermann antibody apears in other disorders, notably in lupus erythematosus, and this fact has long been used to support the autoimmune theory.

With respect to the antigenic stimulus for Wassermann's antibody, free cardiolipin is a hapten and must be attached to a suitable carrier to be antigenic. The lipid composition of cultivable treponemes depends to a large extent on the lipids in the growth medium. Hence, it is possible that pathogenic treponemes growing in vivo have access to a great deal of this phospholipid, which could be incorporated into the treponeme: with the microbial cell as a foreign

carrier the bound cardiolipin could serve as an effective immunogenic determinant.

TREPONEMAL ANTIGENS

Two classes of antigens have been recognized in treponemes: those shared by many different spirochetes and those restricted to one or a few species.

Specific treponemal antigens have been detected by a variety of serological procedures. An example is the *T. pallidum* immobilization (TPI) test, a complement-dependent bactericidal reaction performed with treponemes obtained from syphilitic lesions in rabbits. In practice, the treponeme suspension is mixed with a patient's serum and fresh guinea pig serum, and after incubation in an anaerobic environment for 18 hours at 35° the mixture is examined microscopically; in a positive reaction more than half of the organisms are immobilized (i.e., killed). Nonspecific effects are controlled by performing a parallel reaction with heat-inactivated complement, in which the treponemes should remain motile.

The TPI test is highly specific for the detection of syphilis, yaws, and other treponematoses, but it cannot distinguish among these diseases. Evidently the treponemal antigen participating in this reaction is shared by the various pathogenic species but is not present in indigenous organisms. The TPI test becomes positive early in the course of a syphilitic infection. The response roughly parallels the development of superinfection immunity, suggesting that the immobilizing antibodies may be responsible for protection against infection. However, studies in rabbits have shown that vaccination can induce an immune state without producing these antibodies.

The only well-studied treponemal component is a protein that is found in most treponemes, both saprophytic and pathogenic species. This antigen was first obtained from the Reiter treponeme, a spirochete reputed to be a cultivable, nonvirulent variant of *T. pallidum*. In its purest form (extracted from treponemes by cell lysis) it is a macromolecule associated with RNA. This antigen was once widely used in a diagnostic test for syphilis, the Reiter protein complement-fixation (RPCF) test. However, this test is seldom performed at present because it lacks specificity for *T. pallidum*. This antigen, or a very similar protein, is present in many indigenous

Fig. 37-4. Chemical structure of cardiolipin (diphosphatidylglycerol).

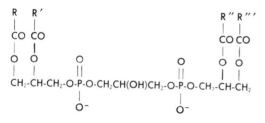

treponemes of the human ailmentary tract; and, as occurs with other indigenous microorganisms, many individuals acquire antibodies to these spirochetes. Though these "natural" antibodies are usually present at very low levels (if at all), they can occasionally yield a reactive RPCF test.

Immunofluorescence and treponeme agglutination have also been used extensively to demonstrate specific antigens. However, cross-reactions occur with both procedures, and the antibodies responsible must be removed by absorption in order to make the reactions specific. Only surface antigens participate in the agglutination reaction, and no cross-reactivity between pathogenic and indigenous treponemes occurs. In immunofluorescent reactions, on the other hand, all antigenic components of the treponemal cell can react. Antibodies to the Reiter protein (see above) are largely responsible for the nonspecific reactions encountered in the fluorescent treponemal antibody (FTA) test used in the diagnosis of syphilis; these antibodies are removed or inactivated in certain modified FTA procedures.

Vaccines. There have been many unsuccessful attempts to make animals immune to syphilis by the administration of treponeme vaccines. Success has been reported recently using as inocula treponeme suspensions freshly harvested from rabbits and rendered noninfectious by γ-irradiation or by in vitro exposure to penicillin. These results suggest that the treponemal antigen inducing the immune state is very labile.

GENETIC VARIATION

Since only in vivo methods are available for studying variations of *T. pallidum,* little is known about its mutations. Tests of cross-immunity and pathogenicity are crude and have yielded only fragmentary information. Rapid passage in rabbits usually results in a detectable increase in virulence for rabbits. Some treponemal strains obtained by primary isolation from patients with yaws produce the typically mild lesions of experimental yaws, but after repeated passage in rabbits cause the more malignant lesions of experimental syphilis. It is possible that such changes involve the in vivo selection of mutants.

No penicillin-resistant variants of *T. pallidum* have been reported.

PATHOGENICITY

SYPHILIS IN MAN

Human syphilis is ordinarily transmitted by sexual contact. In the male, the offending organisms either are present in lesions on the penis or are discharged from deeper genitourinary sites along with the seminal fluid. In women the infectious lesions are most commonly located in the perineal region or on the labia, vaginal wall, or cervix. In approximately 1 out of 10 cases the primary infection is extragenital, usually in or about the mouth.

The organism penetrates mucous membranes but seems to enter the skin only through small breaks. Multiplication at the site of entrance results, within 10 to 60 days, in the formation of a characteristic **primary** inflammatory lesion known as a **chancre,** which begins as a papule and breaks down to form a superficial ulcer with a clean firm base. The predominant inflammatory cells in the lesion are lymphocytes and plasma cells. Although the chancre heals spontaneously, organisms escaping from it at an early stage invade the regional lymph nodes, forming "satellite buboes," and eventually reach the blood stream, where they establish a systemic infection.

Two to twelve weeks after the appearance of the primary lesion, a generalized skin rash usually appears, which also often involves the mucous membranes. During this systemic **secondary** stage of the disease, lesions may develop in the eyes, bones, joints, or central nervous system. All the secondary lesions, particularly those in mucous membranes, contain large numbers of spirochetes, and when located on exposed surfaces are highly infectious. In time they subside spontaneously and once they have healed, the patient is no longer dangerous to others, except for transplacental transmission. Complete healing, however, may be slow, occasionally requiring several years. Rarely, the primary and secondary stages go unrecognized, only to be followed by the late, or **tertiary,** lesions of the disease.

Virtually nothing is known concerning the mechanisms that bring about the destruction of the myriads of treponemes contained in the primary and secondary lesions. Immune immobilization, which is essentially a treponemacidal reaction (with lysis as a subsequent step), or cell-mediated immune responses may play major roles. Electron microscopy has revealed that *T. pallidum* penetrates a variety of host tissue cells, but to what extent phago-cytosis operates as a host defense mechanism is not known.

In approximately half the untreated patients enough treponemes persist in the tissues to give rise to tertiary lesions several years after the primary infection. When present as gummas of the skin or bones these may cause relatively little trouble. Serious manifestations, however, usually result from lesions in the central nervous system, causing **general paresis** or **tabes dorsalis;** in the cardiovascular system, where they affect the aortic valves or cause aortic aneurysms; and in the eyes, where they may produce permanent blindness. Tertiary lesions contain very few organisms but frequently result in necrosis, scar formation, and extensive tissue damage, probably involving a **delayed hypersensitivity response** to products of the small number of organisms that persist.

In about a quarter of untreated cases the tertiary stage is asymptomatic and is recognized only by the persistence of antibodies in the serum; in the remaining 25% the primary and systemic lesions heal so completely that even the STS becomes negative.

Since *T. pallidum* readily passes the placental barrier, a syphilitic mother may transmit the disease to her child, particularly during the secondary stage of her illness. The lesions of **congenital syphilis** resemble, in general, those of acquired syphilis of comparable duration. In infants who fail to survive for more than a few weeks, or are stillborn, the syphilitic process is usually acute and is characterized by massive invasion of nearly all the tissues of the body.

EXPERIMENTAL SYPHILIS

Although the only known natural host of *T. pallidum* is man, experimental infections may be produced in a variety of laboratory animals. The experimental disease in rabbits, monkeys, and chimpanzees simulates the early course of the natural disease in man, but late lesions, resembling those of the tertiary stage of human syphilis, are rarely encountered.

However, viable organisms persist for life in infected rabbits and can be readily isolated from the lymph nodes.

Rabbits may be experimentally infected by inoculation of organisms into the eye, skin, testis, or scrotum. One viable organism injected intratesticularly is said to be infectious, and an average incubation period of 17 days follows the intradermal injection of 500 organisms. Since the time required for the development of a demonstrable lesion at the site of inoculation is shortened by an average decrement of 4 days for each 10-fold increase in the number of spirochetes inoculated, it is estimated that the time required for the organism to divide in vivo is approximately 24 to 36 hours.*

In both experimental and human infections *T. pallidum* tends to proliferate at sites of trauma. In the rabbit treponemes multiply best at peripheral tissue sites in which the temperature is several degrees below that of the internal tissues. Artificial cooling of the skin, for example, may increase strikingly the number of skin lesions that develop in the systemic phase of the infection (Fig. 37-5). Similarly, pretreatment with the antiinflammatory agent cortisone causes the formation of enlarged cutaneous syphilomas, which contain excessive amounts of hyaluronic acid and tremendous numbers of motile treponemes.

IMMUNITY

Resistance to reinfection, as determined both in human volunteers and in experimentally infected rabbits, usually begins about 3 weeks after appearance of the primary lesion. If the disease is eradicated in the early stages by adequate treatment, the host may again become fully susceptible to infection. Conversely, once immunity has developed, the persistence of a latent infection maintains resistance. Late in the course of the disease therapy does not result in loss of resistance to reinfection. The factors responsible for immunity are not clear, and most attempts to

---

* Number of generations involved in 10-fold increase $= \log_2 10 = \log_{10} 10 / \log_{10} 2 = 1/.3 = 3.3$. Hence, 4 days/3.3 $= 1.2$ days $=$ ca. 29 hours.

Fig. 37-5. Effect of temperature on progression of syphilitic lesions as illustrated by their localization in shaved areas of rabbits after intravenous inoculation of *T. pallidum.* The anterior half of the white rabbit's body (**A**) and the posterior half of the black rabbit's (**B**) were shaved lightly (to avoid trauma) on the day of inoculation. Lesions appeared in the shaved (cooler) skin areas of both rabbits on the twenty-second day; 5 days later the rest of each rabbit's body was shaved and no further lesions were found. [From Hollander, D. H., and Turner, T. B. *Am J Syph 38:*489 (1954).]

produce this state in rabbits by the administration of vaccines have failed. However, as noted above, immunity can be produced with freshly harvested treponeme suspensions that are appropriately inactivated.

LABORATORY DIAGNOSIS

DETECTION OF T. PALLIDUM IN LESIONS

In its primary and secondary stages syphilis can often be diagnosed by darkfield examination of fresh exudate fluid obtained from open or from abraded lesions. Since occasionally the exudates from nonsyphilitic lesions may also contain spiral organisms, darkfield observations must be interpreted with care, although an experienced observer can often differentiate *T. pallidum* from other spiral organisms on the basis of its characteristic morphology and motility. Negative darkfield examinations, on the other hand, do not necessarily exclude the diagnosis of syphilis, since

lesions in the later stages of the disease may contain relatively few organisms.*

T. pallidum may also be identified in fluid from active lesions by immunofluorescent staining (Ch. 15) with *T. pallidum* antibodies. Cross-reacting treponemal antibodies must be previously removed by absorption with cultivable treponemes.

SEROLOGICAL TESTS

Tests for either Wassermann or treponemal antibodies are of great value in establishing the diagnosis of syphilis. Flocculation and

---

* It must be remembered that each oil-immersion field in the usual coverslip preparation contains only about 10^{-6} ml of fluid. Accordingly there must be approximately 10^6 organisms per milliliter of fluid under the coverslip for the observer to find an average of one per field. Despite the obvious insensitivity of the method, the spirochetes in early lesions are usually present in sufficient numbers to be detected.

complement-fixation tests for Wassermann antibody are used routinely. However, a wide variety of other diseases, including malaria, infectious mononucleosis, lupus erythematosus, and leprosy, yield a "biological false positive" (BFP) STS reaction; but in general, Wassermann antibody develops a much higher titer in syphilis than in nontreponemal disorders. It is usually first detected 1 to 3 weeks after the primary lesion appears, and reaches a maximum during the secondary stage of infections. Subsequently this antibody may remain at an elevated level in patients who develop late clinical lesions, or it may disappear from the serum of individuals with latent infections. Wassermann antibody is largely a reflection of host tissue involvement in the infectious process, and it usually disappears following treatment.

A variety of tests have been devised to measure treponemal antibodies. The TPI test is the most specific but its technical complexity (see above) limits its performance to a few laboratories where it is used primarily to identify individuals giving BFP reactions for Wassermann antibody. The titer of immobilizing antibody reaches a maximum during or shortly after the secondary stage. Thereafter the titer remains high for a prolonged period.

The FTA test, in one of its variations, is now the most widely performed test for treponemal antibodies. This is an indirect immunofluorescent reaction (Ch. 15) in which *T. pallidum* extracted from animal lesions is exposed first to the patient's serum being tested, and then to fluorescein-labeled antibodies to human immunoglobulins. In its simplest form the FTA test detects cross-reactive and *T. pallidum*-specific antibodies equally well. Several modifications have therefore been devised in an attempt to increase specificity. One of these, the FTA-ABS, uses an extract of the Reiter treponeme to absorb or competitively inhibit cross-reacting antibodies, but this procedure does not achieve immunological specificity.

Antitreponemal antibodies decline more slowly than Wassermann antibody following treatment, and positive reactions may persist for years. Thus, positive serology is not necessarily evidence of continuing disease. Serum antibodies do not penetrate the normal blood-brain barrier, and the presence of Wassermann or treponemal antibodies in the spinal fluid is ordinarily indicative of active central nervous system syphilis.*

TREATMENT

Arsenicals, bismuth, and mercury have been superseded in the treatment of syphilis by penicillin. Its introduction has greatly simplified and improved the treatment: patients in the early stages of the disease may now be treated adequately within 1 to 2 weeks. The fact that short-acting penicillin preparations, even when given as infrequently as once every 24 hours, will usually cure early syphilis,† is doubtless due to the slow multiplication of *T. pallidum* in the tissues. Only rarely is a second course of treatment necessary. Penicillin is less effective in the late stages of the disease and it sometimes fails to eradicate infection when administered to patients with tertiary lesions. Such failures are probably due to the presence of nongrowing treponemes, as there is no evidence that *T. pallidum* ever becomes penicillin-resistant.

The persistence of antibodies in the serum following intensive antisyphilitic therapy does not necessarily indicate the need for further treatment, since titers may decline very slowly after effective control of the disease.

PREVENTION

Many patients with early syphilis are unaware that they have the disease; in women especially the early lesions may be symptomless and unnoticed. In addition, a single promiscuous person may transmit the organism to

---

* The permeability of the barrier is enhanced by the resulting inflammation.

† Within 6 to 10 hours after the initiation of treatment in early syphilis, there often occurs transient fever (2 to 4 hours) and a brief exacerbation of visible lesions (**Herxheimer reaction**), apparently caused by an allergic response to antigens released by the organisms being killed in the tissues. An alternative explanation is that some of the products released have the properties of endotoxins.

numerous sexual partners. Hence control of the spread of syphilis is extremely difficult. When the diagnosis of early syphilis is established in an individual all known contacts should be examined, and they should be treated promptly if signs of the disease (including a positive test for antibody) are present or develop within a few months.

Prophylaxis by thorough cleansing of the genitalia and adjacent areas with soap and water is often ineffective, unless applied early and with great diligence. The prophylactic use of penicillin is also of limited value, because of the low rate at which *T. pallidum* is killed by the drug in vivo. A single oral dose, which may effectively prevent gonorrhea, has little effect on a freshly acquired syphilitic infection; and more prolonged administration is often impractical as a routine prophylactic procedure.

YAWS (FRAMBESIA)

T. pertenue, which causes the tropical disease known as yaws, is virtually indistinguishable from *T. pallidum,* except for the character of the lesions it produces. Both diseases induce the formation of Wassermann and indistinguishable antitreponemal antibodies, and a striking degree of cross-resistance has been demonstrated in both man and laboratory animals. A similar disease has recently been found among apes in central Africa, and the responsible spirochete is indistinguishable from *T. pertenue.*

The **mother yaw,** the primary lesion in the human disease, usually appears 3 to 4 weeks after exposure. It begins as a painless red papule, surrounded by a zone of erythema, and it is often referred to as a **framboise,** or raspberry. Eventually it ulcerates, becomes covered with a dry crust, and heals. Generalized secondary lesions of a similar character make their appearance 6 weeks to 3 months later and commonly occur in successive crops over a period of months or even several years; on the soles of the feet, tender, hyperkeratotic lesions, known as **crab yaws,** often appear. The late, tertiary, lesions are generally restricted to the skin and bones; gummatous nodules and deep chronic ulcerations may disfigure the nose and face and are often disabling. The disease, however, is not as grave as syphilis, since it rarely involves the viscera, and congenital yaws is very uncommon.

Experimental yaws also differs from experimental syphilis. Intratesticular inoculation of *T. pertenue* in the rabbit, for example, regularly results in a granular periorchitis, which is rarely seen following the injection of *T. pallidum.* Similarly, in hamsters *T. pertenue* ordinarily causes a local lesion when inoculated intracutaneously, whereas *T. pallidum* usually does not. Furthermore, the lesions of experimental yaws contain much less hyaluronic acid than do those of experimental syphilis.

Epidemiology. Yaws occurs in the tropics, where the combination of high temperatures and high humidity promotes the persistence of open skin lesions and thus facilitates nonvenereal transmission by direct contact. Children often become afflicted at an early age. In areas of high endemicity more than 75% of the population may acquire the disease before the age of 20. Flies that feed on open lesions become heavily infected with the organism and are thought by some investigators to act as vectors.

Treatment. Yaws responds dramatically to treatment with penicillin, often requiring only a single long-acting injection. However, because it occurs primarily in groups living under poor hygienic conditions and in areas where medical services are limited, the application of this effective control measure is often difficult.

PINTA

Pinta also occurs primarily in the tropics, especially Central and South America. The causative organism, *T. carateum,* is morphologically indistinguishable from *T. pallidum* but differs in being difficult to propagate in laboratory animals. An infection similar to the human disease, though less severe, has been produced in chimpanzees.

The human disease, which may be contracted at any age, is usually transmitted by person-to-person contact, although flies have been implicated as possible vectors. The primary lesion is nonulcerating and is followed within 5 to 18 months by successive crops of flat, erythematous skin lesions which first become hyperpigmented and then, after several years, depigmented and hyperkeratotic. They are most often seen on the hands, feet, and scalp. Late visceral manifestations are extremely rare.

Experimental pinta has been produced in syphilitic patients, and though subjects with pinta regularly develop Wassermann and treponemal antibodies, they occasionally contract syphilis. Penicillin therapy is very effective.

OTHER TREPONEMATOSES

Endemic treponemal diseases occur in many areas of the world where people live under relatively unhygienic conditions. Transmitted by direct contact, these afflictions are often given local names, such as **bejel** in Syria and **siti** in British West Africa. All tend to simulate yaws and are not readily distinguishable from one another.

A venereal form of treponematosis also occurs naturally in **rabbits.** The causative organism, *T. cuniculi,* is morphologically indistinguishable from *T. pallidum.* Its natural occurrence in rabbits may complicate experimental studies performed with treponemes pathogenic for man.

RELAPSING FEVER

Spirochetes of the genus *Borrelia* are morphologically quite different from the treponemes. Not only are they usually longer, but their spirals are more loosely wound and more flexible. Furthermore, they are thick enough to be readily visible when stained with ordinary aniline dyes. They can be propagated in young mice and rats, but host susceptibility varies from one strain to another. Some *Borrelia* strains may also be grown in chick embryos. Cultivation in the test tube has only recently succeeded by growth under microaerophilic conditions in a very complex medium. The organisms have an average division time of 18 hours, and retain animal pathogenicity through numerous subcultures.

In man spirochetes of this genus cause **relapsing fever,** which is transmitted by both ticks and lice from natural reservoir in wild rodents; lice also transmit the organism from man to man. After an incubation period of 3 to 10 days there is a sudden onset of fever, which ordinarily lasts for about 4 days: during this time large numbers of organisms may be demonstrated in the blood and urine, and rarely in the spinal fluid. The fever then declines and the borreliae disappear from the blood. As they decrease in number they become less motile, assume pleomorphic forms, and agglutinate, often in **rosettes.** During the ensuing afebrile period the blood is not infectious, but after 3 to 10 days it teems again with organisms and the fever returns. The ensuing febrile attacks usually number from 3 to 10 and become progressively less severe, until they subside altogether.

Although the case fatality rate is less than 5% when the disease is endemic, it may exceed 50% in severe louse-borne epidemics, presumably owing to increased adaptation to the human host in man-to-man transmission. Autopsy usually reveals miliary necrotic lesions containing large numbers of organisms, particularly in the spleen and liver. Gross hemorrhagic lesions may also be prominent in the gastrointestinal tract and kidneys. Transplacental transmission has been reported.

Experimental infections may be produced in monkeys, mice, rats, and guinea pigs. The experimental disease in monkeys follows the characteristic relapsing course of the illness in man.

Mechanism of Relapse. The most remarkable feature of this disease is its tendency to relapse at regular intervals. The pathogenesis of the sequential relapses is unique because the organisms in each successive attack show antigenic differences. Circulating antibodies specific for the organisms of each wave appear and are responsible for agglutination into rosettes and presumably for subsequent disappearance; the next wave depends on the outgrowth of antigenically distinct mutants for which the host must elaborate new antibodies. The rapidity and regularity of the successive relapses imply that the mutation rates involved are high.

Diagnosis. The diagnosis is usually made by microscopic examination of blood samples obtained during a febrile attack. The organisms can be identified by darkfield microscopy or in stained smears. Because of the many antigenic variants encountered, serological tests are of little diagnostic value. Animal inoculation may be of help when direct microscopic examinations are negative. Infected blood injected into young white rats will usually result in a readily demonstrable spirochetemia within 24 to 72 hours.

Treatment. Relapsing fever responds well to penicillin. Tetracyclines and chloramphenicol are also effective.

Prevention. In the United States sporadic cases are encountered in the South and West. Infected ticks* are plentiful in endemic areas and are thought to be the principal vector, both from animal to animal and from animal to man. Relapsing fever may also be louse-borne, and is not uncommonly associated with typhus epidemics. Tick- and louse-control measures constitute the most effective means of prevention.

---

* Transovarian transmission is known to occur in ticks.

OTHER BORRELIA INFECTIONS OF MAN

Ulcerative lesions of the skin, mucous membranes, and lungs, occurring most commonly in debilitated individuals, often contain borrelia organisms. In **Vincent's angina,** an acute ulcerating disease of the oropharynx, a species designated *B. vincentii* is regularly found in association with a large gram-negative anaerobic bacillus, known as *Bacteroides fusiformis* (Ch. 29, The bacteroides). The etiological significance of the borreliae contained in such lesions remains open to question. The cells vary in length from 4 to 20 μ. The fine structure is basically similar to that of other spirochetes (Fig. 37-6), but the axial filaments are believed to contain only a single fibril. Motility results from rotation of the organism, one or both ends of which are often bent into a hook. Because the tightly coiled spirals are so difficult to recognize, diagnoses of leptospirosis made solely on the basis of direct microscopic examination of the blood are unreliable. Tiny strands extending from the surface of red cells in blood preparations are often mistaken for leptospires.

Fig. 37-6. Electron photomicrograph of chromium-shadowed preparation of a segment of *L. icterohemorrhagiae.* Axial filament is slightly separated from the protoplasmic spiral at a and more closely approximated to it at b. Protoplasmic spiral is somewhat distorted by axial filament at c. Outer sheath is labeled d. ×75,000. [From Simpson, C. F., and White F. H. *J Infect Dis 109*:243 (1961).]

LEPTOSPIROSIS

The first human leptospiral disease to be described was a severe febrile illness characterized by jaundice, hemorrhagic tendencies, and involvement of the kidney. Known as **Weil's disease,** it was shown in 1915 by Inada to be caused by *Leptospira icterohemorrhagiae* (Gr. *leptos* = thin), transmitted to man from infected rats. The organism was shown to cause in guinea pigs the same orange, hemorrhagic lesions that are characteristic of the fatal disease in man.

It has since been learned that many immunologically different strains of leptospiral organisms cause a variety of human illnesses, most of which are not associated with jaundice. Each also seems to have a different natural host. Thus, *L. icterohemorrhagiae* is most commonly found in rats, *L. canicola* in dogs, and *L. pomona* in swine. All the common species are pathogenic for man, with the exception of the saprophytic *L. biflexia,* which inhabits small streams, lakes, and stagnant water. Each species can be identified by specific immunological reactions, but there is considerable antigenic overlap and so absorbed sera must be used for species identification.

MORPHOLOGY

All the leptospires are characterized by extraordinarily fine spirals, wound so tightly as to be barely distinguishable under the darkfield microscope.

CULTIVATION

Leptospires can be readily grown in a variety of artificial media supplemented with 10% heat-inactivated (56°, 30 minutes) rabbit serum. They can also be grown in a synthetic medium containing inorganic salts, a fatty acid, and vitamin B_{12}; growth is stimulated by thiamine. These spirochetes are obligate aerobes and grow best at pH 7.2, at 25 to 30°, and at a slightly increased CO_2 tension. Neither pathogenic nor saprophytic leptospires can synthesize long-chain fatty acids de novo. These compounds (at least 15 carbon atoms long) are essential for growth, being a major energy source, and they must be supplied in the medium in a bound form. On solid media leptospires produce small subsurface colonies that are often difficult to visualize unless stained with oxidase reagent (Ch. 28, Metabolism). Saprophytic leptospires can be differentiated from the pathogenic types by several cultural differences. They produce more oxidase than the pathogens, and their requirements for environmental CO_2 are less. Neither type of leptospire incorporates exogenous pyrimidine; hence cultivation in the presence of 5-fluorouracil permits selective growth of leptospires from material contaminated with other microorganisms.

ANTIGENICITY AND GENETIC VARIATION

All antigenic types of leptospires possess a common somatic antigen (lipopolysaccharide) but vary in their surface or agglutinating antigens. When a given antigenic type is grown in the presence of homologous immune serum antigenic variants frequently appear in the culture; these are analogous to the Borrelia variants that appear in patients with relapsing fever (see above). Differences in surface antigens are most readily demonstrated by agglutination of suspensions of either living or formalinized organisms; complement-fixation and precipitin procedures are also sometimes employed. Mutations affecting virulence, morphology (nonhooked), and colony formation have also been noted. No leptospiral toxins other than a hemolysin have yet been described.

PATHOGENICITY

Lower Animals. Leptospirosis is primarily a zoonosis, wild rodents and domestic animals providing the principal reservoirs. The disease can be produced experimentally in many species of rodents; the young of the guinea pig and the Syrian hamster are particularly susceptible. When injected intramuscularly, subcutaneously, or intraperitoneally with *L. icterohemorrhagiae,* they develop fever within 3 to 5 days, followed by jaundice and hemorrhages into the skin, subcutaneous tissues, and muscles. Death usually occurs in less than 2 weeks. At autopsy the tissues are jaundiced and hemorrhagic, the liver and spleen are both enlarged, and leptospires are readily recoverable from the blood, cerebrospinal fluid, and urine.

In contrast, species that are naturally infected with leptospires (e.g., wild rats) rarely succumb to the infection, which is usually mild, often life-long, and characterized by chronic involvement of the kidneys. The more or less continuous shedding of organisms in the urine is often responsible for transmission of leptospirosis to man. Because the organisms will not survive long in an acid medium, the urine is usually infectious only when alkaline or excreted into alkaline water.

Man. The incubation period in the human disease is from 8 to 12 days. Onset is abrupt, often with chills followed by high fever. Headache, photophobia, and severe muscular pains, particularly in the back and legs, are prominent symptoms. The classical icterohemorrhagic picture is rarely seen: the most constant physical sign is conjunctivitis; albuminuria is common but jaundice occurs in less than 10% of clinically recognizable cases. Lymphocytic meningitis is often present, and when it dominates the clinical findings the illness is often referred to as **swineherd's disease.** The acute illness ordinarily lasts for 3 to 10 days. The mortality rate in clinically recognized cases is approximately 5 to 10%, but serological surveys in heavily endemic areas indicate that subclinical cases are extremely common.

During World War II a hitherto unrecognized form of leptospirosis was encountered among soldiers stationed at Fort Bragg, North Carolina. The disease became known as **pretibial fever** because it was characterized by an erythematous rash most frequently over the shins. An agent originally obtained from the blood of a patient in 1944, and maintained for years by passage in guinea pigs and hamsters on the assumption that it was a virus, was identified nearly a decade later as *L. autumnalis,* a leptospiral species previously isolated in Japan. Subsequently paired acute and convalescent sera, saved from the Fort Bragg epidemic, uniformly showed a rise in titer of specific agglutinins for the same organism. Sporadic cases continue to be reported from Georgia and North Carolina.

Human convalescent sera protect guinea pigs against otherwise fatal inoculations of homologous leptospires. These protective (and agglutinating) antibodies persist in the patient's serum for many years. Phagocytosis probably also plays a part in tissue defense.

LABORATORY DIAGNOSIS

As already indicated, a diagnosis based on direct microscopic examination of blood should be made only by experienced observers. Leptospiremia may be detected by culturing the blood, preferably at 30°, in broth enriched with 10% serum or by inoculating it intraperitoneally into young guinea pigs or hamsters.

Serum antibodies, which usually appear during the second week of illness and reach a maximum titer during the third or fourth week, are not easily identified because of the numerous antigenic types encountered. Agglutination tests, however, may be done with suspensions of killed leptospires, pooled to contain the most common antigenic strains. A broader serological test involves incubation of serum and complement with erythrocytes previously sensitized by genus-specific antigen, extracted from the nonpathogenic *L. biflexa;* when antibody to the genus antigen is present hemolysis results.

TREATMENT

The treatment of leptospirosis is relatively unsatisfactory, even though penicillin, the tetracyclines, and chloramphenicol all inhibit the growth of leptospires in vitro and, if given early enough, control experimental leptospiral infections in animals. The ineffectiveness of therapy in man is due in part to the fact that the disease cannot be recognized until comparatively late in its course. Penicillin appears to be the drug of choice.

PREVENTION

Man usually acquires leptospirosis through contact with water contaminated with the urine of infected animals. Workers in rat-infested slaughterhouses and fish-cleaning establishments, miners, farmers, sewer workers, and swimmers in stagnant ponds and canals run the greatest risk. Since the natural reservoir of leptospirosis in wild animals is far too vast to be attacked directly, and since no satisfactory vaccine is available, preventive measures must be directed at diminishing the chances of contact with contaminated water. The organisms are believed to gain entrance to the body of man through abrasions in the skin and through the mucous membranes of the conjunctiva, nose, and mouth. Because of the acidity of the gastric juice, infection via the intestinal tract is probably rare.

RAT-BITE FEVER (SODOKU)

Although ordinarily not classified as a spirochete, *Spirillum minus* is a short, rigid, spiral organism possessing polar flagella. It is commonly carried by rats and causes one of the two forms of rat-bite fever in man. (The other is due to *Streptobacillus moniliformis,* Ch. 41.) From the primary rat-bite lesion, the organism invades the regional lymph nodes and eventually the blood stream, causing lymphadenitis, skin rash, and relapsing fever.

S. minus has never been cultivated on artificial media; its isolation from human patients depends upon animal inoculation. Guinea pigs and mice are both susceptible and, when infected, harbor in their blood large numbers of organisms, which are often visible in Wright-stained smears. The disease occasionally causes a false positive serological test for syphilis. Penicillin, streptomycin, and the tetracyclines are all effective.

SELECTED REFERENCES

Books and Review Articles

EDWARDS, G. A., and DOMM, B. M. Human leptospirosis. *Medicine 39:*117 (1960).

FELSENFELD, O. Borreliae, human relapsing fever, and parasite-vector-host relationships. *Bacteriol Rev 29:*46 (1965).

GALTON, M. M., MENGES, R. W., SHOTTS, E. B., JR., NAHMIAS, A. J., and HEATH, C. W., JR. *Leptospirosis: Epidemiology, Clinical Manifestations in Man and Animals, and Methods in Laboratory Diagnosis.* PHS Publication 951 GPO, Washington, D.C., 1962.

TURNER, T. B., and HOLLANDER, D. H. *Biology of the Treponematoses.* World Health Organization, Geneva, 1957.

Specific Articles

DEACON, W. E., and HUNTER, E. T. Treponemal antigens as related to identification and serology. *Proc Soc Exp Biol Med 110:*352 (1962).

GJESTLAND, T. The Oslo study of untreated syphilis: An epidemiological investigation of the natural course of untreated syphilitic infection based upon a restudy of the Boeck-Brunsgaard material. *Acta Dermatol 35* (Suppl. 34) (1955).

JEPSEN, O. B., HOUVIND HOUGEN, K., and BIRCH-ANDERSEN, A. Electron microscopy of *Treponema pallidum* Nichols. *Acta Pathol Microbiol Scand 74:*241 (1968).

JOHNSON, R. C., and EGGEBRATEN, L. M. Fatty acid requirements of the Kazan 5 and Reiter strains of *Treponema pallidum. Infect Immun 3:*723 (1971).

KELLY, R. Cultivation of *Borrelia hermsi. Science 173:*443 (1971).

METZGER, M., and SMOGÓR, W. Artificial immunization of rabbits against syphilis. I. Effect of increasing doses of treponemes given by the intramuscular route. *Br J Vener Dis 45:*308 (1969).

MILLER, J. N. Immunity in experimental syphilis. V. The immunogenicity of *Treponema pallidum* attenuated by γ irradiation. *J Immunol 99:*1012 (1967).

NELSON, R. A., and MAYER, M. M. Immobilization of *Treponema pallidum* in vitro by antibody produced in syphilitic infection. *J Exp Med 89:*369 (1949).

RICKETTSIAE

Revised by JAMES W. MOULDER, Ph.D.

GENERAL PROPERTIES   898

MORPHOLOGY   898
CHEMICAL COMPOSITION AND METABOLIC ACTIVITIES   898
PATHOGENICITY   899
CHEMOTHERAPY   900
RELATION TO BACTERIA AND VIRUSES   901
ADAPTATION FOR PARASITISM   901
GENERAL NATURE OF RICKETTSIAL DISEASES   902
LABORATORY DIAGNOSIS OF RICKETTSIAL DISEASES   902
   Immunological Identification   902
   Transmission to Laboratory Animals   903

SPECIFIC DISEASES DUE TO RICKETTSIAE   903

TYPHUS GROUP   903
   Epidemic (Louse-Borne) Typhus Fever   905
   Endemic Murine (Flea-Borne) Typhus   907
SPOTTED FEVER GROUP   908
   Rocky Mountain Spotted Fever   908
   Tick Fevers (Tick Typhus)   909
   Rickettsialpox   909
TSUTSUGAMUSHI DISEASE (SCRUB TYPHUS)   910
Q FEVER   911
TRENCH FEVER   913

GENERAL PROPERTIES

The microorganisms described in this chapter are small, pleomorphic coccobacilli. Most of them are obligate intracellular parasites that can survive only briefly outside animal cells. However, the etiological agent of trench fever (*Rochalimaea quintana*) grows in cell-free media, and the Q fever agent (*Coxiella burneti*) is extraordinarily resistant to heat, drying, and sunlight while passing from host to host.

The natural hosts of rickettsiae include a variety of mammals and arthropods, the latter often serving as both vectors and reservoirs. Except in Q fever, mammals become infected with rickettsiae only through the bites of infected insects (lice and fleas) or arachnids (ticks and mites). With the important exception of epidemic typhus fever, the infection of man is only incidental, and is of no consequence for the survival of the rickettsial species. Once rickettsiae invade man, the diseases they cause vary enormously in severity, from benign self-limited illnesses to some of the most fulminating known. In fact, more human lives may have been lost from rickettsial diseases than from any other form of illness, except malaria.

In the first decades of this century various rickettsiae were isolated and their modes of transmission were determined, but only since about 1940 has it become possible for physicians to cope with rickettsial diseases. The responsible developments include 1) technics for cultivating the organisms, 2) preparation of effective vaccines, 3) use of insecticides on a broad scale to control epidemics, and 4) effective chemotherapy.

MORPHOLOGY

The rickettsiae are small, nonmotile bacteria, appearing as spherical forms about 0.3 μ in diameter, as short rods about 0.3 by 1.0 μ, or sometimes as thin rods about 2 μ long (Fig. 38-1). They occur singly, in pairs, or in strands. Most species are found only in the cytoplasm of host cells, but those of the spotted fever group multiply in nuclei as well as in cytoplasm. Electron microscopy reveals a cell wall, a plasma membrane, granular cytoplasm, and filamentous central bodies, presumed to be nuclei (Fig. 38-2). Rickettsiae stain poorly with the Gram stain but appear to be gram-negative. Special staining procedures are used to demonstrate them in cell smears, e.g., polychromatic stains such as Giemsa's, or basic fuchsin followed by brief acid washing to decolorize host cell constituents (Gimenez's stain, a modified version of the acid-fast stain for mycobacteria).

CHEMICAL COMPOSITION AND METABOLIC ACTIVITIES

Rickettsiae multiply luxuriantly in the yolk sacs of embryonated chick eggs, from which they may be separated in purified form.

The purified organisms resemble ordinary gram-negative bacteria in chemical composition. They have a relatively high content of DNA, probably because RNA and protein leak out of the rickettsial cells during purification. Wall preparations from *Coxiella burneti* contain muramic acid, a cell wall constituent of bacteria (Ch. 6).

In dilute buffered salt solutions, isolated typhus and spotted fever rickettsiae are extremely unstable, losing both metabolic activity and infectivity for animal cells. If, however, the medium is enriched with K^+, serum albumin, and sucrose, the isolated organisms can survive for many hours at 35°. Various rickettsiae fail to attack glucose but do oxidize pyruvate and several substrates of the Krebs

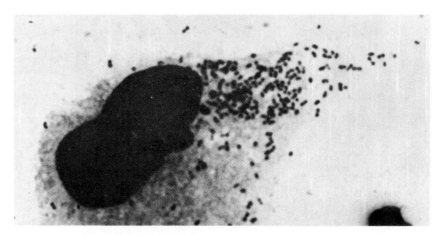

Fig. 38-1. *R. tsutsugamushi* (Karp strain) in Giemsa-stained smear of tunica scraping from an infected guinea pig. ×1500. (Courtesy of N. J. Kramis.)

cycle; extracts contain at least some glycolytic enzymes, most or all of the enzymes of the Krebs cycle, and cytochromes. As expected in organisms with an electron-transport system, cyanide inhibits oxygen consumption, and 2,4-dinitrophenol uncouples oxidative phosphorylation from respiration. It appears that glutamate oxidation, accompanied by oxidative phosphorylation, is probably the principal energy-yielding process in typhus rickettsiae.

Isolated typhus rickettsiae incorporate isotopically labeled amino acids into proteins at very low but significant rates. However, the incorporation is unusual for intact cells in that it requires the addition of **all** amino acids, plus ADP or ATP. Extracts of *C. burneti* can synthesize protein and RNA.

The instability of the isolated typhus rickettsiae probably arises from their tendency to leak essential metabolites, as shown by their **reversible inactivation.** When rickettsiae are kept at 35° in simple salt solution they do not utilize oxygen, and over a few hours they lose infectivity. The loss of activity is probably due to loss of ATP, since it can be prevented by adding ATP or glutamate (from which ATP is generated). Infectivity is also lost when the organisms are maintained at 0° in balanced salt solution, but in this case the loss is restored by addition of DPN and CoA.

It is thus clear that the typhus rickettsiae have many of the metabolic capabilities of bacteria but require an exogenous supply of cofactors to express these capabilities. The response to exogenous cofactors implies an unusual permeability of their limiting cytoplasmic membranes. However, this property may result from damage to the rickettsial cells during isolation, since there is evidence that the permeability properties of rickettsiae growing in their host cells are much like those of ordinary gram-negative bacteria.

PATHOGENICITY

Rickettsiae have a predilection for the endothelial cells of small blood vessels. Hyperplasia of endothelial cells and localized thrombus formation lead to obstruction of blood flow, with escape of red cells into the surrounding tissue. Inflammatory cells also accumulate about affected segments of blood vessels. The angiitis, which is particularly widespread in the small blood vessels of the skin, brain, and myocardium, appears to account for some of the more prominent clinical manifestations, such as the petechial rash, the stupor, and the terminal shock. Rickettsiae may also be seen in macrophages surrounding the vascular lesions.

Large numbers of rickettsiae are most easily grown in the chick embryo yolk sac, but growth requirements are best studied in cell cultures. When isolated scrub typhus rickettsiae are added to cultured mouse lymphoblasts almost all the cells become infected within 1 to 2 hours. Penetration seems to involve

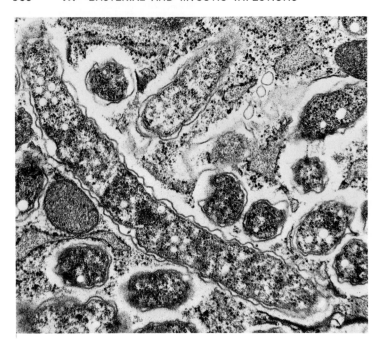

Fig. 38-2. Electron micrograph of a thin section of a monkey kidney cell infected with *R. prowazeki.* One rickettsial cell is sectioned transversely, and several cross-sections are also visible. The rippled cell wall and underlying plasma membrane are easily seen. The cytoplasm contains ribosomes and strands of DNA. ×32,850. [From Anderson, D. R., Hopps, H. E., Barile, M. F., and Bernheim, B. C. *J Bacteriol 90:* 1387 (1965).]

expenditure of energy by the organisms, for if they are first killed by heat or formalin they are not taken up by the cells. Furthermore, penetration is enhanced by L-glutamate and blocked by 2,4-dinitrophenol.

Once inside the cultured cells, rickettsiae multiply by binary fission (Fig. 38-3). They grow best when the infected cells are multiplying rapidly. In cultures with minimal media, where the host cells are not multiplying, the organisms fail to grow. Even in the most favorable cell cultures their generation time, about 18 hours, is longer than that of most bacteria. Cell lysis is the consequence of rickettsial multiplication.

Rickettsial Toxin. When large numbers of rickettsiae are injected intravenously into mice or rats the animals die within 2 to 8 hours. A direct toxic effect is inferred because 1) death occurs so rapidly, 2) UV irradiation can greatly diminish infectivity without reducing toxicity, and 3) the use of antirickettsial drugs does not prevent the rapid deaths. The toxic effect, however, is prevented by antiserum specific for antigens of the rickettsial cell wall. Death is ascribed to damage of endothelial cells, resulting in leakage of plasma, decrease in blood volume, and shock. Although the assumed toxin has never

been isolated, its physiological effects seem to differ from those of endotoxins of Enterobacteriaceae. To exert their toxic effects rickettsiae need not be infectious but they must retain most of their metabolic activities, such as the capacity to carry out oxidative phosphorylation. The reason for this requirement is not known.

Hemolysins. Highly infectious preparations of the rickettsiae that cause typhus fevers hemolyze the red blood cells of sheep and rabbits in vitro, but not those of man. Only in severe experimental infections in rabbits has massive intravascular hemolysis been described.

CHEMOTHERAPY

The first chemotherapeutic agent found to be effective in treating rickettsial infections was *p*-aminobenzoic acid (PAB), given in relatively large doses. Its effect does not appear to be related to its function as a precursor of folic acid, since its inhibition of rickettsial multiplication is overcome by *p*-hydroxybenzoic acid, an intermediate in the synthesis of benzoquinone cofactors in electron transport. However, the tetracyclines are

Fig. 38-3. Time-lapse photographs of *R. rickettsi,* growing in tissue culture of mouse lymphoid cells, showing that multiplication of organisms is by binary fission. Phase-contrast microscopy. × 2000. Time is indicated by figures in upper right corner of each photograph. [From Schaechter, M., Bozeman, F. M., and Smadel, J. E. *Virology 3:*160 (1957).]

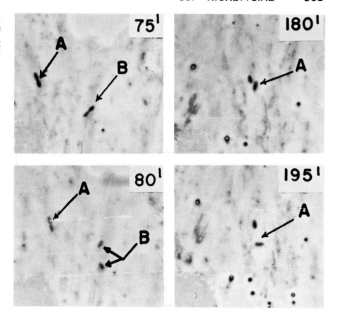

much more effective in inhibiting rickettsial growth and in treating rickettsial diseases. Of the tetracyclines, doxycycline is especially active: a single 50-μg dose effectively treats typhus in man. Chloramphenicol is also effective, but it is not generally used because of the possibility of serious side reactions.

RELATION TO BACTERIA AND VIRUSES

Because rickettsiae depend on the intracellular milieu of animal cells for growth, they were long considered to occupy a special taxonomic niche between bacteria and viruses. The following properties, however, indicate clearly that they are bacteria: 1) they multiply by binary fission; 2) they contain both DNA and RNA; 3) at least one species contains muramic acid; 4) they possess enzymes of the Krebs cycle, of electron transport, and of protein synthesis; 5) their growth is inhibited by a variety of antibacterial agents.

ADAPTATION FOR PARASITISM

Isolated rickettsiae differ strikingly from classic bacteria in the unusual permeability of their cytoplasmic membranes, e.g., to nucleotides. The underlying structural anomaly responsible for this leakiness has not been characterized, and it is not certain that undamaged rickettsiae are similarly leaky. Nevertheless, the existence of this phenomenon provides a potential basis for understanding the peculiar ecological and epidemiological behavior of these organisms. Because their membranes are unusually permeable to ATP, DPN, CoA, and other essential cofactors, typhus rickettsiae may borrow these components from the host cell rather than synthesize them. In acquiring the capacity to accept this kind of intracellular nutritional charity, however, the rickettsiae have lost their ability to be self-sufficient outside host cells.

Accordingly, the survival of most rickettsiae in nature requires that they be transmitted from one host to another under conditions that minimize their exposure to the extracellular world. This requirement is met, in general, by the following modes of transmission: 1) to mammals by direct inoculation into the skin during the feeding of an infected arthropod (louse, flea, mite, or tick); or 2) to arthropods by ingestion of blood from an infected mammal; or 3) from arthropod to anthropod, by means of infected ova. The one known exception is *C. burneti* which is much more stable and is transmitted to man by infected dust or droplets.

GENERAL NATURE OF RICKETTSIAL DISEASES

Rickettsial infections in man usually begin in the vascular system, following the bite of an infected arthropod. The microorganisms proliferate mainly in endothelial cells and become widely disseminated by way of the blood. They establish focal areas of obstruction in small blood vessels due to hyperplasia of infected endothelial cells and to small thrombi. After an incubation period of 1 to 4 weeks the onset of illness is generally abrupt, with headache, chills, and fever, followed by a hemorrhagic rash. Stupor, delirium, and shock occur in severe cases, and patchy gangrene of skin and subcutaneous tissues is often encountered. Case fatality rates range from less than 1% to as high as 90%. Following recovery immunity is usually enduring.

C. burneti, which is generally acquired by inhalation of infected dust particles or droplets, causes pneumonia, though other organs and tissues may become involved.

LABORATORY DIAGNOSIS OF RICKETTSIAL DISEASES

The organism responsible for a particular rickettsial disease is most often identified serologically, i.e., by analysis of antibodies that appear during the course of the illness. Identification is also often established by transmitting the organism to experimental laboratory animals and observing the nature of the disease produced and the immune response.

IMMUNOLOGICAL IDENTIFICATION

As usual in the application of immunological methods to the diagnosis of acute infectious diseases, a serum sample taken at the height of illness is compared with one taken during convalescence. And, as in other acute infectious diseases (Ch. 15), antibodies to the etiological agent appear during the second week of illness and reach maximal titer during, or after, recovery.

In addition, when a particular serum is tested with several antigens, each from a different strain of related organisms, the antigen that reacts most intensely is assumed to represent the etiological agent. Though this is a useful rule in serological diagnosis it suffers from the obvious limitation that still other antigens, not tested, might have given even higher titers.

The cultivation of rickettsiae in yolk sacs of chick embryos has made these organisms available in large numbers and in sufficient purity to provide antigens for detection of antibodies. After ether extraction of lipids from crude suspensions of infected yolk sac, two kinds of antigens are obtained: 1) the soluble **group antigens,** which are released into the environment and show broad specificity characteristic for each of the major rickettsial groups in Table 38-1, and 2) the insoluble **type-specific antigens,** which are presumably associated with the rickettsial cell wall and differentiate species, and strains within a species. Antibodies to both kinds of antigens are routinely detected by complement-fixation reactions.

Other useful immunological tests include agglutination of rickettsial suspensions and immunofluorescence (Fig. 38-4). Sera also may be titrated for ability to neutralize the toxicity of rickettsial suspensions in mice (p. 901), or to protect laboratory animals from experimental infection (protection tests). The protection test is dangerous to perform, however, because of the hazard of accidental human infection, and it is cumbersome because of the difficulty in obtaining standard suspensions of infectious rickettsiae. It is, therefore, not as widely used as in the study of viruses. Neutralization of either toxicity or infectivity depends on antibodies to the type-specific antigens.

One of the most widely used diagnostic tests, the **Weil-Felix reaction,** does not deal with rickettsial antigens at all, but is based on the curious fact that antibodies formed in the course of certain rickettsial diseases react fortuitously with the polysaccharide O antigens of certain strains of the *Proteus* bacillus, called the X strains. Since these strains react only when their O antigens are accessible (i.e., are not covered by flagellar or H antigens) they are referred to as **Proteus-OX strains.** This agglutination reaction originally led to the incorrect suspicion that *Proteus* was the etiological agent of typhus fever, but the test remains a useful and simple diagnostic procedure.

Three strains are used in the Weil-Felix

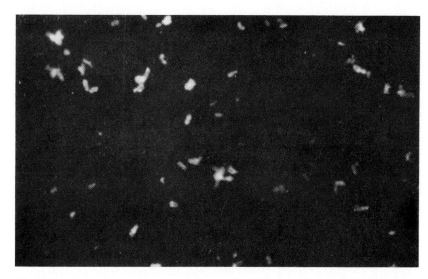

Fig. 38-4. *R. typhi* in smear from infected yolk sac stained with fluorescein-labeled antibody in human convalescent serum. ×650. [From Goldwasser, R. A., and Shepard, C. C. *J Immunol 80:*126 (1958).]

reaction: *Proteus* OX-2, OX-19, and OX-K. The agglutination reactions to all three help to differentiate the various rickettsial diseases (Tables 38-1 and 38-3). In interpreting the results it must be kept in mind that *Proteus* infections are fairly common (particularly in the urinary tract) and that they too may evoke antibodies to the *Proteus*-OX strains.

TRANSMISSION TO LABORATORY ANIMALS

Since rickettsial diseases are characterized by bacteremia during the early febrile phase of illness, intraperitoneal inoculation of a sample of the patient's blood into male guinea pigs or into mice will often provide a means of identifying the causative organism. If the blood is drawn after the first week of illness the inoculations should be made with crushed blood clots from which as much serum as possible has been removed to minimize inhibition of infectivity by antibodies. In the male guinea pigs characteristic scrotal lesions may develop, depending upon the rickettsial species (see below). Although adapted strains of rickettsiae can be propagated readily in yolk sacs of chick embryos, primary isolation of rickettsiae by this means is less reliable.

SPECIFIC DISEASES DUE TO RICKETTSIAE

The differences among the various rickettsial diseases are found in 1) the arthropod vector; 2) the specificities of the antibodies formed; 3) the intracellular localization of the organism, whether only in cytoplasm or also in the nucleus; 4) clinical manifestations, especially the distribution and appearance of the rash; and 5) the behavior of the organism when transmitted from man to experimental animals. On the basis of these parameters, rickettsial diseases fall into several groups, which are compared in Table 38-1. The modes of transmission of some of the major rickettsial diseases are outlined in Table 38-2.

TYPHUS GROUP

R. prowazeki, which causes epidemic typhus, and *R. typhi,* the cause of endemic typhus, have the same group antigen but different type-specific antigens. The diseases they cause differ markedly in epidemiology because of the different vectors involved. However, in-

TABLE 38-1. Rickettsial Diseases of Man

Disease		Natural cycle				Serological diagnosis	
Group and type	Agent	Geographical distribution	Arthropod	Mammal	Transmission to man	Weil-Felix reaction	Complement fixation
Typhus Epidemic	*Rickettsia prowazeki*	Worldwide	Body louse	Man	Infected louse feces into broken skin	Positive OX-19	
Brill's disease	*R. prowazeki*	North America, Europe	Recurrence years after original attack of epidemic typhus			Usually negative	Positive group- and type-specific
Endemic	*R. typhi*	Worldwide	Flea	Rodents	Infected flea feces into broken skin	Positive OX-19	
Spotted fever Rocky Mountain spotted fever	*R. rickettsi*	Western Hemisphere	Ticks	Wild rodents, dogs	Tick bite	Positive OX-19 OX-2	
North Asian tick-borne rickettsiosis	*R. sibiricus*	Siberia, Mongolia	Ticks	Wild rodents	Tick bite	Positive OX-19 OX-2	
Boutonneuse fever	*R. conori*	Africa, Europe, Middle East, India	Ticks	Wild rodents, dogs	Tick bite	Positive OX-19 OX-2	Positive group- and type-specific
Queensland tick typhus	*R. australis*	Australia	Ticks	Marsupials, wild rodents	Tick bite	Positive OX-19 OX-2	
Rickettsialpox	*R. akari*	North America, Europe	Blood-sucking mite	House mouse and other rodents	Mite bite	Negative	
Scrub typhus	*R. tsutsugamushi*	Asia, Australia, Pacific islands	Tromiculid mites	Wild rodents	Mite bite	Positive OX-K	Positive in about 50% of patients
Q fever	*Coxiella burneti*	Worldwide	Ticks	Small mammals, cattle, sheep and goats	Inhalation of dried, infected material	Negative	Positive
Trench fever	*Rochalimaea quintana*	Europe, Africa, North America	Body louse	Man	Infected louse feces into broken skin	Negative	Low titer

Modified from Smadel, J., and Jackson, E. Rickettsial Infections. In *Diagnostic Procedures for Viral and Rickettsial Diseases*, 3rd ed. p. 744. (E. H. Lennette and N. J. Schmidt, eds.) American Public Health Association, New York, 1964.

TABLE 38-2. Modes of Transmission of the Principal Rickettsial Diseases

Disease in man	Etiological agent	Chain of transmission
Epidemic typhus	*R. prowazeki*	. . . Man→Louse→Man→Louse . . .
Endemic typhus	*R. typhi*	. . . Rat→Rat flea→Rat→Rat flea→Rat . . . ↓ Man
Rocky Mountain spotted fever (boutonneuse fever, other spotted fevers)	*R. rickettsi*	. . . Tick→Tick→Tick→Tick . . . ↓ ↓ Dog Man ↓ Tick→Man
Scrub typhus (tsutsugamushi fever)	*R. tsutsugamushi*	. . . Mite→Field mouse→Mite→Field mouse . . . ↓ Man
Rickettsialpox	*R. akari*	. . . Mite→House mouse→Mite→House mouse . . . ↓ Man
Q fever	*C. burneti*	. . . Tick→Small mammal→Tick→Cattle . . . (airborne) ↓ (airborne) Man

Modified from Moulder, J. W. *The Biochemistry of Intracellular Parasitism,* Ch 3. Univ. Chicago Press, Chicago, 1962.

dividual cases of epidemic and endemic typhus can hardly be distinguished clinically, although the epidemic disease tends to be more severe and to have a much higher case fatality rate.

EPIDEMIC (LOUSE-BORNE) TYPHUS FEVER

Transmission. The organism that causes epidemic typhus fever, *R. prowazeki,* is named after Howard Ricketts and Stanislaus von Prowazek, both of whom died of the disease while investigating it. Under natural conditions *R. prowazeki* infects only man and the human body louse, *Pediculus vestimenti,* which nests in clothing and emerges several times a day to take a blood meal from its host's skin. When the ingested blood contains the organism, the cells of the louse's intestinal tract become infected and eventually rupture, discharging the organisms into the feces. While feeding on the skin, the louse defecates, and because its bite causes intense itching, the victim often scratches and rubs the infected feces into his broken skin.

Lice prefer a temperature close to that of normal human skin and tend to leave the bodies of those who have died or who have high fevers. Since lice can neither fly nor jump, and can crawl but a few yards, epidemic typhus flourishes only under conditions of crowding where the infected lice have no difficulty in finding new hosts.

R. prowazeki invades only cells of the louse's intestinal tract. The organism is not passed transovarially, and the infected lice die within 1 to 3 weeks. Nevertheless, the cycle of louse infection is required to maintain the human reservoir, for although other mammalian species can be infected experimentally, no natural reservoir of *R. prowazeki* is known except in man.

The head (scalp) louse also appears to be capable of transmitting epidemic typhus fever but seems to do so less effectively than the body louse.

Pathogenesis. The clinical and pathological manifestations of epidemic typhus can be largely accounted for by the bacteremia and vascular lesions, i.e., occlusion of small blood vessels by swollen endothelial cells and thrombi, perivascular hemorrhages, extravasation of plasma, hemoconcentration, and shock.

From 1 to 2 weeks after a bite from an infected louse, illness begins abruptly with chills, fever, headache, generalized aches and pains, and exhaustion. Within 2 to 3 days body temperature reaches 40° or more, where in untreated patients it tends to remain until death or until recovery ensues in 2 to 3 weeks. A rash usually appears on the trunk by the fourth to seventh day and spreads peripherally to the extremities, sparing the face, palms, and soles. Macular at first, it eventually becomes hemorrhagic. In severe cases, prostration and stupor (*typhus* in Greek means "hazy")* alternate with delirium, and renal insufficiency and gangrene of the fingers, toes, ear lobes, and nose may follow.

Before effective chemotherapy became available the case fatality rate was about 20%, although in some outbreaks it has been recorded as high as 70%.

Recovery from an attack of epidemic typhus gives rise to an enduring immunity which also extends to endemic (murine) typhus (see below).

Diagnosis by laboratory means is based primarily on serological tests. Agglutinins for *Proteus* OX-19 usually appear about 7 days after the onset of the disease, rise to a maximum at about the fifteenth day, and then decline slowly over several months. Complement-fixing antibodies to the group- and type-specific antigens follow the same pattern. Antibodies that protect against experimental animal infections can be detected for many months longer,† and those that react with other rickettsiae, particularly the agent of endemic murine typhus, are often present but in much lower titer. Agglutinins for *Proteus* OX-K (Table 38-3) do not appear.

Treatment. Treatment with tetracyclines or chloramphenicol is highly effective in experimental animal infections with *R. prowazeki* and in other rickettsial infections of man.

TABLE 38-3. Usual Weil-Felix Agglutination Reactions Observed in Rickettsial Diseases

	OX-19	OX-2	OX-K
Epidemic typhus	++++	+	0
Murine typhus	++++	+	0
Spotted fever group	++++	+	0
	+	++++	0
Scrub typhus	0	0	++++
Rickettsialpox	0	0	0
Q fever	0	0	0

From Smadel, J. E., and Jackson, E. R. Rickettsial Infections. In Diagnostic Procedures for Viral and Rickettsial Diseases, 3rd ed., p. 761 (E. H. Lennette and N. J. Schmidt, eds.) American Public Health Association, New York, 1964.

Doxycycline has been very effective in clinical trials.

Epidemiology. Epidemic typhus fever is a disease of human misery. Where humans are crowded together and have few opportunities to bathe or to wash their clothes, body lice flourish, and opportunities for the spread of typhus abound.

Typhus epidemics have repeatedly ravaged armies and civilian populations during and immediately after wars. In Napoleon's march on Moscow in 1812 his army of 500,000 men was reduced in 1 year to less than 200,000, more than half the deaths having been due to typhus. It has been estimated that in Russia, right after World War I (1918 to 1922), 30 million persons had typhus and 3 million died of it.

Prevention and Control. Two powerful means of controlling epidemic typhus have been developed: vaccination and the rapid delousing of human populations.

A number of effective **vaccines** have been developed. The Cox vaccine, for example, is prepared by treating with formalin *R. prowazeki* cultivated in the yolk sac. This vaccine was administered on a large scale to military and civilian populations during World War II, and was highly effective in reducing both the incidence and severity of epidemic typhus. A living vaccine is prepared from the "strain E" variant of *R. prowazeki,* which does not produce serious illness in man; it produces a much longer-lasting immunity. Because egg proteins are present in either vaccine preliminary skin tests with diluted

---

* Until their etiological agents were recognized, typhus and typhoid fever (due to *Salmonella typhosa*) were confused. In German, in fact, *Typhus* refers to typhoid fever, and the disease caused by *R. prowazeki* is called *Flecktyphus*.

† The greater persistence of protective antibodies could be due to a more sensitive assay. Alternatively, a smaller rickettsial mass might be needed to induce these antibodies. This difference could arise if the antigenic determinants were more numerous, stable, or immunogenic (see also Brill's disease, and immune response after prophylactic vaccination).

vaccine should be performed in individuals who have a history of hypersensitivity to eggs, and the full dose should not be administered to those who give a wheal-and-erythema response.

The **elimination of lice** was highly effective in curbing incipient outbreaks of epidemic typhus during World War II. Earlier methods of delousing human populations had been extremely cumbersome: clothes were removed and heated to 120° for 30 minutes, and the unclothed persons were scrubbed with kerosene and soap and their hair was cropped. With the introduction of the insecticide DDT, however, effective delousing became much simpler: solid DDT was blown under sleeves, inside trousers, etc., and huge populations were thus quickly rid of their body lice for prolonged periods.

Brill-Zinsser Disease and the Carrier State. In 1898 sporadic cases of an illness resembling typhus fever were described by Brill in New York City among immigrants from eastern Europe. This disease was first considered to endemic murine typhus (see below); but ultimately the causative agent was isolated (by allowing lice to feed on patients and then on volunteers) and proved to be *R. prowazeki*. However, Brill's disease differs from epidemic typhus fever in a number of respects: 1) louse infestation need not be present; 2) the disease is much milder, and case fatality rates are very low; 3) the characteristic rash is usually absent; 4) antibodies to *R. prowazeki* appear more rapidly (within 3 or 4 days of onset) and reach much higher titers; and 5) most patients have had epidemic typhus many years before. Curiously, persons with Brill's disease do not develop agglutinins for *Proteus* OX-19, and this feature, coupled with absence of a rash, often makes the diagnosis difficult.

Zinsser proposed that this disease represents an exacerbation of a **latent** infection with *R. prowazeki* that persisted following an earlier attack of epidemic typhus. The reasons for the exacerbation are obscure, but once the infection flares up the illness is understandably mild, since the immune response is of the secondary type.* A similar pattern of illness is seen in persons who develop epidemic typhus despite earlier immunization.

Brill's disease also occurs in eastern Europe itself, where typhus has been recurrently epidemic. The epidemiological significance of these sporadic cases is now clear: carriers with latent infection constitute the reservoir of the disease. When louse infestations are prevalent and a carrier develops overt disease, an epidemic of louse-borne typhus is likely to arise. Thus Zinsser remarked that Napoleon's retreat from Moscow was "started by a louse."

ENDEMIC MURINE (FLEA-BORNE) TYPHUS

A relatively mild form of typhus fever occurs endemically in man as a consequence of infection with *R. typhi*. Under natural conditions this organism is common in rats, and is occasionally transmitted to man by rat fleas. The resulting disease tends to be less severe than that caused by *R. prowazeki* but otherwise the two are clinically indistinguishable. The case fatality rate is estimated to be about 5% without chemotherapy, and still less with tetracycline or chloramphenicol treatment.

That *R. typhi* and *R. prowazeki* are antigenically related is indicated by the fact that recovery from either infection leads to enduring immunity to both. In contrast, immunization with either killed organism induces only homologous immunity. This difference is in accord with the observations 1) that natural infection or vaccination with viable microbes usually establishes more solid and more enduring immunity than killed microbes,† and 2) that each immunogen produces a lower titer of antibody to cross-reacting antigens than to itself.

Diagnosis. The agglutination reaction with *Proteus* OX-9 (Table 38-3) and the complement-fixation test with rickettsial antigens are both useful in establishing a diagnosis. In endemic typhus the predominating antibodies are to *R. typhi* rather than to *R. prowazeki* (cf. epidemic typhus). The two organisms

---

* In accord with this interpretation, the antibodies first detected to *R. prowazeki* in classic epidemic typhus are 19S immunoglobulins, whereas those first

formed during Brill's disease are predominantly 7S immunoglobulins. This difference is characteristic of the primary and secondary responses to a variety of immunogens (Ch. 17).

† Presumably because of more prolonged persistence of the immunogens (Ch. 17).

can also be differentiated by the diseases they produce in guinea pigs. Inoculation of blood from patients with epidemic typhus causes a very mild disease, detectable only by a rise in body temperature, whereas blood from patients with endemic typhus produces severe testicular lesions and marked swelling of the scrotum.

Epidemiology. *R. typhi* is transmitted from rat to rat, and occasionally from rat to man, by the rat flea, *Xenophylla cheopis.** The infected flea passes rickettsiae in its feces and infects its host in the same manner as the human body louse (p. 199). However, while *R. prowazeki* causes severe disease in its natural mammalian host (man) and invariably fatal disease in its natural arthropod host (human body louse), *R. typhi* causes only latent infection in both its mammalian (rat) and arthropod (rat flea) hosts. The inapparent diseases in these species constitute effective reservoirs of *R. typhi*.

Control Measures. Where endemic typhus is prevalent the widespread use of DDT has markedly reduced the incidence of fleas in rat populations and of rickettsial infections in rats and humans. Rat populations are also effectively reduced by use of warfarin and other rat poisons. In controlling outbreaks, however, it is important to apply the DDT first, i.e., **before** the rat poisons; otherwise, as the rats die their fleas will depart in droves and infest man.

Geographic Distribution. Endemic typhus is worldwide in distribution. In the Mediterranean area it is called **Toulon typhus,** and elsewhere it is referred to as **Manchurian typhus, Moscow typhus, shop typhus** (Malaya), **tarbadillo** (Mexico), and **red fever** (Congo). In the United States it is most prevalent in the southeastern states; at its peak incidence, from 1930 to 1950, there were about 3000 cases per year. The incidence is now considerably lower.

SPOTTED FEVER GROUP

The rickettsiae that cause the spotted fevers are antigenically distinguishable from those

---

* It is also transmitted among rats by the rat louse, *Polyplax spinulosa*.

of the typhus fevers. In addition, they multiply in both the nucleus and the cytoplasm of susceptible animal cells, whereas typhus and other rickettsiae grow exclusively in the cytoplasm; and they are transmitted by ticks rather than by lice or fleas.

ROCKY MOUNTAIN SPOTTED FEVER

This disease, caused by *R. rickettsi* was first recognized in parts of Idaho and Montana at the turn of the century; it has now been observed in amost every state in the United States and in parts of South America.

Pathogenesis. The onset of the disease in man occurs 1 to 2 weeks after the tick bite and resembles that of typhus, with chills, fever, and prostration, except that the rash begins peripherally on the ankles, wrists, and forehead and then spreads to the trunk, whereas the reverse is characteristic of the typhus fevers. As in typhus and all other rickettsial diseases (except Q fever) the pathological lesions are primarily those of angiitis and intravascular thrombosis, which sometimes lead to necrosis of skin and soft tissues, particularly of the extremities.

The case fatality rates in untreated spotted fevers vary widely. In Montana, in the area of the Bitterroot Valley, they have been reported to be as high as 90%, as compared with 5% in the Snake River Valley of Idaho and 25% in eastern Long Island. Such wide variations suggest that different strains of rickettsiae are involved in different localities. Since the disease responds dramatically to chloramphenicol and tetracyclines, present mortality rates are relatively low when the disease is recognized and treated promptly.

Diagnosis. Serological diagnosis is based on complement-fixation tests performed with crude antigenic fractions isolated from *R. rickettsi* (grown in yolk sac cultures). The group-specific soluble antigen distinguishes rickettsiae of the spotted fever group from those of the typhus and other groups, and the type-specific particulate antigens differentiate the closely related spotted fever organisms.

The Weil-Felix reaction in Rocky Mountain spotted fever is variable: agglutinins may be high for *Proteus* OX-19 and low for OX-2, or high for both, or even low for OX-19 and high for OX-2 (Table 38-3). Occasional patients fail to develop OX agglutinins.

Guinea pigs inoculated intraperitoneally with blood from patients acutely ill with spotted fevers develop fever and a marked scrotal reaction, consisting of a hemorrhagic rash followed by necrosis and ulceration of the skin.

Epidemiology. Spotted fever rickettsiae are primarily parasites of ticks. They do not kill their arthropod hosts and are passed through unending generations of ticks by egg transmission. Men contract spotted fever only when they are bitten by infected ticks, an event totally unnecessary for the survival of spotted fever rickettsiae in nature. The wood tick, *Dermacentor andersoni,* is mainly responsible for transmission of Rocky Mountain spotted fever in the western United States; and the dog tick, *Dermacentor variabilis,* is usually responsible in the eastern United States. Their hatching eggs undergo a series of metamorphoses: to larvae, to nymphs, and finally to adults, over a 2-year period; and the adults can survive for as long as 4 years without feeding. Congenitally acquired rickettsiae often persist through all these stages.

In the infected tick, unlike lice infected with *R. prowazeki,* the rickettsiae are not limited to the cells of the intestinal lining but are widely distributed in many organs, including the **ovaries** and **salivary glands.** The infected ova give rise to infected progeny, thereby perpetuating the infectious cycle; and the rickettsiae in the saliva are transmitted to man by the tick's bite. Thus ticks are not only vectors but also the primary reservoirs of these organisms.

TICK FEVERS (TICK TYPHUS)

A variety of tick-borne rickettsial diseases, similar to Rocky Mountain spotted fever, are widely distributed geographically and are known by different names in different localities. The etiological agents share the same group antigen as *R. rickettsi* but have distinguishing type-specific antigens as revealed in complement-fixation and neutralization tests. An example is **boutonneuse disease,** also known as Marseilles fever. This disease occurs in the Mediteranean area, Africa, and India,

and is caused by *R. conori.* The principal mammalian reservoir of the organism is in dogs, and infection is also transmitted transovarially in ticks. The disease produced in man is much milder than Rocky Mountain spotted fever, the mortality rate being only 1% in untreated patients. A distinctive feature is the small indurated lesion that develops at the site of the tick bite. This lesion, referred to as **tache noire,** becomes necrotic at its center, develops an eschar, and gives rise to enlargement of the regional lymph nodes.

As usual, the immunity established by infection (or vaccination with living organisms) is more solid and broad than that induced with killed organisms. Thus, guinea pigs that recover from infection with *R. conori* are immune both to it and to *R. rickettsi,* and vice versa; whereas vaccination with killed *R. rickettsi* establishes immuniy only to *R. rickettsi.*

Other tick-borne spotted fevers, with similar clinical, pathological, and epidemiological patterns include Siberian tick fever (*R. sibiricus*) and Australian North Queensland tick fever (*R. australis*).

RICKETTSIALPOX

This disease, first recognized in New York City in 1946, has been observed in a number of other cities in northeastern United States, and what is probably the same disease has been described in urban centers within the Soviet Union. The causative agent is *R. akari* (Gr. *akari* = mite). Like the spotted fever rickettsiae, it multiplies in both the nucleus and the cytoplasm of host cells; and its particulate type-specific antigen cross-reacts extensively in complement-fixation assays with spotted fever organisms. However, it is transmitted by a mite rather than a tick, and it does not elicit antibodies to *Proteus*-OX strains.

The house mouse constitutes the natural mammalian reservoir for *R. akari.* A blood-sucking mite transmits the organism among mice and occasionally to man. Since house mice are particularly common near garbage and refuse piles in apartment houses, the disease is characteristically seen among apartment dwellers in urban centers.

About 1 week before the onset of overt disease a distinctive red cutaneous papule and enlarged regional lymph nodes become evident; these correspond to the skin site at which rickettsiae were introduced by a blood-sucking mite. The papule subsequently develops a vesicle, then an eschar (tache noire), and looks somewhat like the site of an inoculation with vaccinia.

As with other rickettsial diseases, the illness is characterized by abrupt onset, fever, headache, and rash. The rash is unusual for a rickettsial disease in that it is vesicular and simulates chickenpox (varicella). The disease is thus called rickettsialpox in the United States and vesicular rickettsiosis in the Soviet Union. The course of the disease usually lasts 1 to 2 weeks, and no fatalities have been reported.

Diagnosis. *R. akari,* transmitted from the patient's blood (drawn during the acute illness), causes a fatal disease in laboratory mice and fever and swelling of the scrotum (without necrosis) in guinea pigs. The organism is also readily grown in the yolk or amniotic sacs of chick embryos. Sera from patients with rickettsialpox fix complement when mixed with suspensions of *R. akari.*

Though the disease is self-limited it may be severe. The response to tetracyclines is dramatic.

TSUTSUGAMUSHI DISEASE (SCRUB TYPHUS)

This disease is prevalent over a wide area in southeast Asia and in the southeastern Pacific region: Japan, Korea Formosa, Indochina, India, New Guinea, and northern Australia. The organism responsible, *R. tsutsugamushi* (Jap. *tsutsuga* = something small and dangerous, plus *mushi* = a mite) is common in mites and in wild rodents. Mites act as both reservoir and vector, transmitting the rickettsiae among rodents and to their own progeny via infected ova. Man becomes an incidental host when infected larvae of various red mites (*Trombicula*) attach to his skin and suck blood.

One to two weeks after being bitten, the human host abruptly develops chills, fever, and intense headache. Within a few days, rash and pneumonitis become evident. Stupor and prostration are often pronounced. At the time of onset, the site of the infected mite's attachment may appear as an indurated, erythe-

matous papule topped by a multiloculated vesicle (which eventually develops an eschar). The regional lymph nodes also become enlarged.

Among the Allied troops in the South Pacific and China-Burma-India theaters during World War II, attack rates in some regions were as high as 900 per 1000 men per year. Scrub typhus also accounted for a portion of the fevers of unknown origin among American troops in Viet Nam. As with spotted fevers, case fatality rates in different localities vary widely without treatment (from 1 to 35%), probably reflecting differences in virulence among diverse strains of *R. tsutsugamushi.* Minor antigenic differences in different strains were also sometimes noted in complement-fixation and neutralization titrations.

The diagnosis of scrub typhus is established most accurately by isolation of *R. tsutsugamushi* from the patient's blood during the acute illness. A variety of laboratory animals are susceptible but white mice are most vulnerable. When inoculated with infected blood they die within 10 to 18 days, and smears of the spleen, stained by the Giemsa method, reveal intracytoplasmic rickettsiae. Most patients also develop elevated titers of agglutinating antibodies for *Proteus* OX-K but not for other strains of *Proteus.*

Patients who recover from scrub typhus are immune for a time, but may suffer second attacks. Studies on human volunteers who had recovered from one attack have shown that immunity to the homologous strain of *R. tsutsugamushi* persists for several years, but that immunity to other strains lasts for only a few months. The diversity of strains greatly complicates the problem of developing an effective vaccine.

That scrub typhus may exist in a latent form, like the carrier state of epidemic typhus, is indicated by the demonstration of viable *R. tsutsugamushi* in lymph node biopsies taken as long as 1 year after recovery. Spontaneous recrudescence of such infections, however, has not been observed, probably because there has been no large-scale migration of infected persons to areas where the disease is absent (cf. Brill-Zinsser's disease above).

Treatment. Tetracyclines are highly effective in controlling the clinical manifestations of scrub typhus. Since this agent is only bacteristatic, ultimate recovery depends upon the patient's immune response.

During World War II the incidence of scrub typhus in military personnel was greatly lessened by clearing away brush, spraying insecticides, and treating clothes and exposed skin with mite repellents such as benzylbenzoate and phthalate esters.

Q FEVER

Q fever was first recognized in Queensland, Australia, by Derrick in 1937. The agent causing it was shown to be a rickettsia by Burnet and Freeman, in Australia, and by Cox, in the United States. The organism is, therefore, referred to as *Coxiella burneti*. It differs from other rickettsiae in several respects: 1) it is unusually stable outside host cells; 2) infection in man is acquired not by the bite of an infected arthropod, but by inhalation of contaminated particles; 3) disease in man is usually characterized by pneumonitis, without a rash; and 4) *C. burneti* does not elicit antibodies for OX strains of *Proteus*. *C. burneti* also differs from other rickettsiae in its antigenic makeup, as shown by complement-fixation tests; and experimental animals rendered immune to coxiella are not resistant to other rickettsiae. The organism of Q fever is therefore placed in a different genus from the other rickettsiae.

Pathogenesis. The disease is acquired by inhaling contaminated dusts and aerosols, particularly in the vicinity of cattle sheds. During parturition of coxiella-infected cattle and sheep, the placenta and amniotic fluid abound in rickettsiae* and contaminate the immediate environment. The infected animals also shed the organisms in their nasal and salivary secretions. The organisms remain viable in wool and in dried secretions and dried excreta for prolonged periods. Man-to-man spread of Q fever is rare, but authenticated instances are on record.

Onset of disease in man is usually abrupt, with chills and fever but without a rash. Pneumonitis is common, and the illness often re-

sembles "atypical pneumonia" (Ch. 40). The uncomplicated disease is rarely fatal in man, but it may be severe enough to require a prolonged convalescence. Subacute bacterial endocarditis caused by *C. burneti* is invariably fatal. According to the few autopsy reports available the organism has been found in a variety of sites: in pulmonary macrophages, in neuroglial cells of the brain, in kidney parenchymal cells, and also outside of cells as, for example, in the lumina of renal tubules.

Diagnosis. The clinical features of Q fever are not specific. The diagnosis may be suspected from a history of exposure to livestock and confirmed 1) by isolation of Q fever organisms from sputum, urine, or blood during the acute illness, or 2) by detecting sequential changes in serum antibody titers.

Guinea pigs and hamsters are preferred for isolation of the organisms, though many laboratory animals are susceptible. Infected guinea pigs become febrile; but several serial passages may be required before the diagnosis can be confirmed by detecting the organisms in stained smears of spleen cells or by development of specific complement-fixing antibodies. The use of hamsters is both simpler and safer: antibodies are detected in the infected animal's serum 4 to 6 weeks after inoculation, by agglutination or complement-fixation reactions with *C. burneti* antigens. The same immunological procedures are used to demonstrate rises of antibody titer in the patients' sera.

Treatment. The disease responds promptly to treatment with tetracyclines, and relapses are rarely observed.

Resistance to Physical and Chemical Agents. *C. burneti* remains viable and virulent for prolonged periods, not only in excreta and secretions, but also in water and in milk (see below). It is relatively resistant to heat, tolerating 70° for several minutes; and it remains viable after 48 hours in 0.5% formaldehyde, and after several days in 0.4% phenol.

Epidemiology. The distribution of infection in mammals and in birds is usually estimated by surveys of coxiella-specific antibodies in sera. In addition a delayed-type allergic skin test, with an antigenic preparation from

---

* Organism counts as high as 10^{10} per gram of placenta have been reported.

C. burneti, is positive in guinea pigs and humans who have recovered from Q fever. Since arthropods do not form antibodies, their infectiousness for susceptible experimental animals must be used to detect the presence of coxiella. Natural infections have been found in 22 species of ticks belonging to 8 genera, in human body lice, and in a large number of wild and domesticated mammals and birds. Hence, there exist diverse natural reservoirs and vectors of *C. burneti.*

The disease has been found in almost every part of the world, including the United States. In Australia, where epidemiological studies have been most intensive the principal natural reservoir appears to be bandicoots (marsupial rats). Ticks act as vectors and also contribute to the reservoir. Cattle and sheep are incidental hosts and are usually infected by tick bites.

The prevalence of infection among cattle is high. On the basis of serum antibody titers it has been estimated, for example, that 10% of cows in the Los Angeles area are infected,

Fig. 38-5. *Rochalimaea quintana* in intestinal lumen of an infected louse. Several of the organisms (R) can be seen in close association with the microvilli which form the striated border of the epithelium (SB). The dense mitochondria (M) in the epithelial cells have a different internal structure from the extracellular microorganisms. A basement lamina (BL) borders the base of the epithelium. ×9500, reduced. [From Ito, S., and Vinson, J. W. *J Bacteriol* 89:481 (1965).]

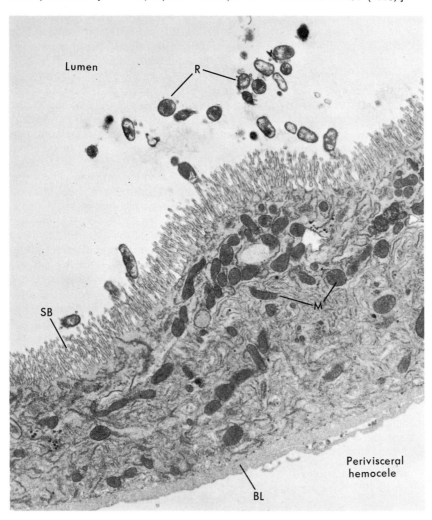

as well as a substantial proportion of sheep. Infected cows shed coxiella in milk, but most of the organisms are destroyed by pasteurization. No major epidemics from ingestion of contaminated milk have been reported. The disease is most common among slaughterhouse employees and farm workers who handle sheep.

Vaccination of humans is effective but is not practical, except for those at high risk, e.g., livestock handlers.

TRENCH FEVER

The disease called trench fever was first recognized during World War I among army troops in northern France, where approximately 1 million cases were reported. The disease disappeared after the war, only to reappear in small outbreaks in Japan and Poland in the 1920s and 1930s, and then in epidemic form on the Russian front during World War II.

Trench fever is characterized by an abrupt onset with chills and fever which tend to subside and recur in repeated cycles of 3 to 5 days' duration. Hence it is often called **5-day fever.** Rash is commonly present during the febrile periods; and the disease usually recurs over a period of 3 to 5 weeks, but is rarely severe. The mechanism responsible for the recurrent cycles is unknown.

The diagnosis has sometimes been established by allowing uninfected lice to feed on persons suspected of having the disease, and then demonstrating "rickettsiae" in the intestines of the lice. The disease has been transmitted to healthy volunteers by blood drawn from febrile patients at the height of their illnesses, and also by allowing lice to feed first on sick individuals and subsequently on healthy volunteers. The causative agent is called *Rochalimaea quintana* (after H. Da Rocha Lima, a pioneer in rickettsial research, and the Latin *quintus,* meaning five). The organism can be cultivated on blood agar. This fact, together with the long-established finding that *R. quintana* normally proliferates within the lumen of the louse gut, rather than within the intestinal epithelial cells (Fig. 38-5), has caused it to be placed in a genus apart from the other rickettsiae.

SELECTED REFERENCES

Books and Review Articles

BREZINA, R. Advances in rickettsial research. *Curr Top Microbiol Immunol 47*:20 (1969).

International symposium on rickettsiae and rickettsial diseases. *Zentralb Bakteriol (Orig) 260*:276 (1968). In English.

LENNETTE, E. H., and SCHMIDT, N. F., eds. *Diagnostic Procedures for Viral and Rickettsial Diseases,* 4th ed. American Public Health Association, New York, 1969.

MOULDER, J. W. *Biochemistry of Intracellular Parasitism.* Univ. Chicago Press, Chicago, 1962.

ORMSBEE, R. A. Rickettsiae (as organisms). *Annu Rev Microbiol 23*:275 (1969).

ROUECHE, B. The Alerting of Mr. Pomerantz. In *The Eleven Blue Men and Other Narratives of Medical Detection.* Little, Brown, Boston, 1954.

ZINSSER, H. *Rats, Lice, and History.* Little, Brown, Boston, 1935.

Specific Articles

GOLDWASSER, R. A., and SHEPARD, C. C. Fluorescent antibody methods in the differentiation of murine and epidemic typhus sera: Specificity changes resulting from previous immunization. *J Immunol 82*:373 (1959).

LENNETTE, E. H., CLARK, W. H., JENSEN, F. W., and TOOMB, C. J. Development and persistence in man of complement-fixing and agglutinating antibodies to *Coxiella burnetii. J Immunol 68*:591 (1962).

PERKINS, H. R., and ALLISON, A. C. Cell wall constituents of rickettsiae. *J Gen Microbiol 30*:469 (1963).

WEISS, E. Some aspects of variations in rickettsial virulence. *Ann NY Acad Sci 88*:1287 (1960).

chapter

CHLAMYDIAE

Revised by ROGER L. NICHOLS, M.D., and
WILLIAM A. BLYTH, Ph.D.

GENERAL CHARACTERISTICS 916

    Taxonomy 916

    Morphology and Developmental Cycle 916

    Chemical and Metabolic Characteristics 918

    Antigens 919

    Immunity 920

    General Aspects of Chlamydial Infections 921

DISEASES DUE TO CHLAMYDIAE 922

    TRACHOMA 922

        History and Epidemiology 922

        Pathogenicity 922

        Antigenic Composition 923

        Immunity 923

        Diagnosis 923

        Treatment 924

    INCLUSION CONJUNCTIVITIS 924

    LYMPHOGRANULOMA VENEREUM (LVG) 925

    ORNITHOSIS 926

GENERAL CHARACTERISTICS

The microorganisms responsible for trachoma, lymphogranuloma venereum, ornithosis, and related diseases are included in a
single genus, *Chlamydia,* earlier called *Bedsonia.* The chlamydiae share a unique developmental cycle, a common morphology, and
a common family antigen (Ag). Because they
are small and reproduce only within host cells
the chlamydiae were long thought, as were
rickettsiae and mycoplasmas, to be viruses.
However, there are profound differences
among these groups of pathogens (Table
39-1), including the diseases they produce
and their management.

Several properties clearly relate chlamydiae
to bacteria: 1) presence of both DNA and
RNA, 2) division by binary fission, 3) cell
walls like those of free-living gram-negative
bacteria, 4) ribosomes similar in size to those
of bacteria, and 5) antibiotic susceptibility.
Chlamydiae are easily visible in the light microscope. The genome is about ⅓ the size of
E. coli. Characteristic inclusions are formed
within the cytoplasm of host cells; when mature these inclusions usually lie close to the
nucleus in typical helmet or mantle configurations.

Many different chlamydial strains have
been distinguished on the basis of antigenic
composition, host range, virulence, and pathogenic effects. However, they can be conveniently divided into two species on the basis
of their susceptibility to sulfonamides and the
type of cytoplasmic inclusion produced in
infected cells (Table 39-2): *Chlamydia trachomatis* (group A) and *C. psittaci* (group
B). There is strong homology of the DNA
within either species but surprisingly little between them, suggesting a long-standing evolutionary separation. Within species there is
considerable antigenic overlap, and a common family ("group") Ag is recognized.

C. trachomatis has a high degree of host
specificity; with the exception of meningopneumonitis, which infects mice, these strains
parasitize only man. In contrast, *C. psittaci*
strains are found in a wide range of animals
(including over 130 species of birds), where
they multiply in cells of the host's placenta,
intestines, joints, lungs, pericardium, conjunctiva, meninges, and reticuloendothelial system. Lesions can be produced in one or more
of these organ systems.

The chlamydiae are important pathogens.
Burnet lists trachoma among the three most
important infectious diseases of man; its
sequelae hamper socioeconomic progress in
some countries. In animals the chlamydiae
are economically important, causing death in
poultry and abortions in meat-producing animals.

MORPHOLOGY AND DEVELOPMENTAL CYCLE

Sir Samuel Bedson, the outstanding early
student of these microorganisms, showed that
their morphology varies during the developmental cycle. His observations have been confirmed and extended in recent years by electron microscopic examination of thin sections
of infected mammalian cells in culture.

The developmental cycle is an orderly alternation between the **elementary body,** specialized for extracellular survival, and the
reticulate body, engaged in intracellular multiplication (Fig. 39-1). The elementary body
is a coccus 300 nm in diameter, while the
spherical or irregular reticulate body ranges

916

TABLE 39-1. Comparison of Chlamydia with Other Pathogenic Agents

Property	Agent				
	Chlamydiae	Bacteria	Rickettsiae	Mycoplasmas	Viruses
Growth outside host cell	0	+	0*	+	0
Independent protein synthesis	+	+	+	+	0
Generation of metabolic energy	0	+	+	+	0
Rigid cell envelope	+ (Reticulate bodies) 0 (Elementary bodies)	+	+	0	Variable
Antibiotic susceptibility	+	+	+	+	0
Mode of reproduction	Fission	Fission	Fission	Fission	Host cell synthesis of subunits; then assembly of virion
Nucleic acids	DNA & RNA	DNA & RNA	DNA & RNA	DNA & RNA	DNA *or* RNA, not both

* Except *R. quintana* (see Ch. 38).

up to 1000 nm (Fig. 39-2). These and the many intermediate forms all stain readily with Giemsa, Macchiavello, Castaneda, and Gimenez stains.

An elementary body is taken into a host cell by phagocytosis (Fig. 39-3). Within 8 hours it reorganizes without loss of identity into an essentially noninfectious reticulate body, remaining within the cytoplasmic vacuole bounded by the host-cell membrane. The reticulate body multiplies by binary fission. Within 8 to 12 hours, 4 to 16 reticulate bodies are recognizable in a vacuole, which lies free in the cytoplasm without physical relation to the nucleus. As increasing numbers of reticu-

late bodies appear their size diminishes, and by 20 hours the small, rigid spheres of elementary bodies appear.

In *C. trachomatis* strains "matrix," a glycogen-containing substance within the inclusion, appears concurrently with early elementary body formation. Its origin and function are unknown. As the transition from reticulate to elementary bodies continues the inclusion assumes a paranuclear position.

The developmental cycle is completed within 48 to 60 hours in humans by reorganization of the essentially noninfectious reticulate bodies into thousands of infectious elementary bodies, which are released by rup-

TABLE 39-2. Division of Chlamydiae into Groups

	Group A	Group B
Property		
Susceptibility to sulfonamides	Sensitive	Resistant
Type of inclusion produced	Rigid	Nonrigid
Presence of glycogen in inclusion	Yes	No
Species	*C. trachomatis*	*C. psittaci*
Diseases induced	Trachoma Inclusion conjunctivitis Lymphogranuloma venereum Mouse pneumonitis	Ornithosis Meningopneumonitis Many clinical and subclinical diseases of mammals and birds

Adapted from Gordon, F. B., and Quan, A. L. *J Infect Dis 115*:186 (1965).

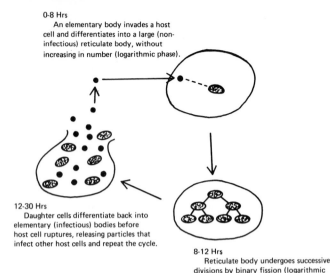

0-8 Hrs
   An elementary body invades a host
cell and differentiates into a large (noninfectious) reticulate body, without
increasing in number (logarithmic phase).

12-30 Hrs
   Daughter cells differentiate back into
elementary (infectious) bodies before
host cell ruptures, releasing particles that
infect other host cells and repeat the cycle.

8-12 Hrs
   Reticulate body undergoes successive
divisions by binary fission (logarithmic
phase).

Fig. 39-1. Schematic representation of infectious cycle of the psittacosis group. The cycle has a superficial resemblance to that of viral infections (including the occurrence of an "eclipse period"), but is fundamentally different.

ture of the host cells. In synchronized growth in cell culture a single developmental cycle can be studied.

CHEMICAL AND METABOLIC CHARACTERISTICS

Chlamydiae are exquisitely adapted to an intracellular environment: they survive phagocytosis; they avoid or stultify the thrust of lysosomal enzyme activity; they inhibit host DNA synthesis; and they are "energy parasites" depending on their hosts for generating metabolic energy in the form of energy-rich substances such as ATP. With the possible exception of a limited ability to make RNA, isolated chlamydiae do not synthesize protein, RNA, or DNA, but within suitable host cells they make all their own macromolecules. However, if the energy-generating systems of their host cells are inhibited, chlamydial multiplication ceases. This suggests that the lack of energy-producing enzyme systems may be what restricts chlamydiae to an intracellular existence. The dependence on host cells is even greater than that of rickettsiae which carry out oxidative phosphorylation, satisfying at least some of their need for ATP; and rickettsiae possess more extensive biosynthetic capabilities (Ch. 38). While the bacterial ancestry of both forms seems certain, chlamydiae represent a more radical evolutionary departure.

Strains of meningopneumonitis, a member

Fig. 39-2. *Chlamydia* grown in cell culture: **a**, elementary body; **b**, similar body beginning to degenerate; **c**, reticulate body. (From Lawn, M. A., Blyth, W. A., and Taverne, J. *J Hyg.* In press.)

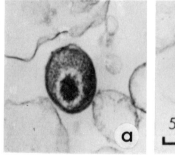
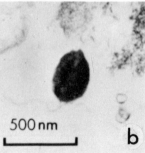
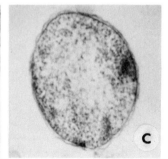

500 nm

Fig. 39-3. Phagocytosis of chlamydial particles by mouse macrophages and by BHK-21 cells. **a.** Chlamydial particles are apparently adsorbed to the cell membrane of a macrophage, which in one instance is indented. **b–c.** Two stages in the phagocytosis of chlamydiae by macrophages. **d.** An elementary body immediately after entry into a BHK-21 cell. **e.** Three organisms in BHK-21 cells 80 minutes after inoculation. The elementary body on the right shows separation between its wall and its cell membrane. The two bodies marked with arrows are assumed to be tangentially sectioned vacuoles containing chlamydiae. All three vacuoles are surrounded by small cytoplasmic vesicles. (From Lawn, A. M., Blyth, W. A., and Taverne, J. *J Hyg.* In press.)

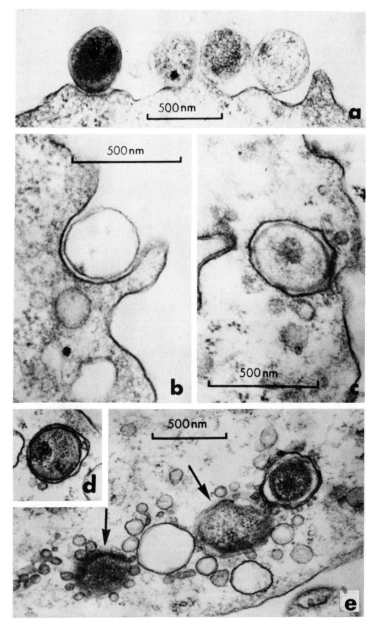

of the *C. psittaci* group, have been most thoroughly studied chemically because they are easily grown in cell culture. Their lipid content is high (40%), and there is a small amount of carbohydrate rich in hexosamine, but the presence of muramic acid is uncertain. Reticulate forms have three to four times more RNA than DNA, while the infectious elementary body has equal amounts.

ANTIGENS

Two classes of Ags can be identified on the basis of antibody (Ab) response to infection or vaccination: **family-specific** and **type-specific.** The family-specific Ag is found in the supernate of lysates of the organism; the type-specific Ag is associated with the cell wall. Thus, as with many pathogenic bacteria and

TABLE 39-3.   Chlamydial Diseases of Man

Disease	Agent	Usual mode of transmission	Geographic distribution
Trachoma	C. trachomatis	Direct contact or transmission on fomites between infected and uninfected individuals	Worldwide: hyperendemic in arid regions of Asia and Africa
Inclusion conjunctivitis	C. trachomatis	In the newborn: from the genital tract of mother In the adult: sexual contact; unchlorinated swimming pools	Worldwide
Lymphogranuloma venereum	C. trachomatis (usually) C. psittaci (rarely)	Sexual intercourse	Worldwide
Ornithosis	C. psittaci	Inhalation of feces from infected birds; contact with infected avian viscera	Worldwide

viruses, the protoplasmic Ags are relatively homogeneous among a family of organisms, whereas the cell wall Ags, concerned with infectivity and virulence, tend to be antigenically diverse.

The family-specific Ag found in all chlamydiae is heat-stable, ether-soluble, resistant to trypsin, and inactivated by periodate; possibly it is a glycolipid. Abs to this Ag can be detected by complement fixation or immunofluorescence; their role in protection is unknown.

Type-specific Ags have been found in isolates from patients with diseases caused by group A chlamydiae: trachoma, inclusion conjunctivitis, and lymphogranuloma venereum. Type-specific Ags have been found in ornithosis and other strains of group B chlamydiae and are correlated with differences in virulence and antibiotic sensitivities. These Ags appeared to be protein, for their reactivity is reduced by trypsin, or by extraction with phenol, but not by periodate. Curiously, species-specific Ags have not been well defined. Abs to type-specific Ags are found in patients or experimental animals.

IMMUNITY

Diseases caused by chlamydiae tend to run chronic and relapsing courses: the same serotype of trachoma can be found in an individual patient for 5 years or longer, which implies continued infection rather than reinfection. Chronicity of infection has led to the suggestion that infection with chlamydiae does not evoke an effective immune response; on the other hand, the host ordinarily restrains and localizes chlamydial disease without serious sequelae.

In animal models infection of the eye can induce quite solid immunity of a few months duration; in pneumonic and other infections in birds and mammals, only partial protection has been observed. Thus an immune response occurs, but it can be overwhelmed by large challenge doses or by adverse conditions.

Abs that combine with cell wall Ags of the parasite appear to prevent penetration of susceptible host cells, arresting the spread of infection. But these Abs do not appear to inactivate parasites already localized within the cells. Individuals free of apparent disease and relatively immune to reinfection may continue to shed virulent chlamydiae, thus serving as healthy carriers. The cellular mechanisms underlying the carrier state are unknown. The roles of circulating Abs, secretory Abs, and the cellular immune reaction in naturally occurring disease are not clear. In passive transfer experiments circulating Ab to guinea pig inclusion conjunctivitis did not

Typical clinical features of disease	Preferred methods of laboratory diagnosis	Preferred chemotherapy
Follicular hypertrophy of conjunctiva, occasionally followed by pannus, conjunctival scarring, and eventual blindness	Isolation, chlamydial inclusions in conjunctival scrapings, or Abs to trachoma in tears by immunofluorescence	Oral sulfonamides
Purulent conjunctivitis usually without pannus or conjunctival scarring	As for trachoma	Newborn: topical tetracycline Adult: oral tetracycline
Granulomatous inflammation of lymph nodes in inguinal and rectoanal regions	Rising titer of Abs to group-specific Ag by immunofluorescence or complement fixation	Oral sulfonamides; oral tetracyclines
Pneumonic consolidation, fever, chills	As for lymphogranuloma venereum	Oral tetracyclines

reach the eye and failed to protect these animals from subsequent challenge; animals with Abs in eye secretions following eye infections were protected. However, this latter group may have been protected by cell-mediated immunity as well.

GENERAL ASPECTS OF CHLAMYDIAL INFECTIONS

Chlamydiae infect a wide range of animal hosts. Group A organisms (except mouse pneumonitis) occur in humans; the primary site is usually the mucous membranes of the eye or the genitourinary tract (Table 39-3). By contrast, organisms of group B rarely infect man but occur widely in birds and mammals: ornithosis is common in feral birds and remains a major economic problem in the domestic fowl industry.

The mechanism by which chlamydiae cause disease or injure cells is unknown. Chlamydial infections of mucous membranes cause damage to tissues deep to the epithelial layer; for example, in trachoma scarring of the tarsal plate occurs frequently. Large intravenous doses of infective elementary bodies frequently induce a fatal toxic reaction within hours in mice, but the role of this phenomenon in natural pathogenesis is not clear.

In general chlamydiae exhibit low pathogenicity except in circumstances that place the host at a disadvantage. Thus apparently healthy birds develop overt signs of ornithosis under crowded conditions, and trachoma increases in severity as well as frequency as personal hygiene and public sanitation deteriorate. (Despite this low pathogenicity, however, less than 20 infective trachoma particles have produced disease in susceptible hosts, including man.)

In the individual case, whether man or animal, chlamydial infections are usually easily controlled, and often cured, by appropriate chemotherapy (sulfonamides for group A; tetracycline for group B). However, epidemiological problems derive from two characteristics: latency of infection and susceptibility to reinfection. Individuals successfully treated in a treatment campaign usually return to an environment in which a few active cases have escaped detection and cause reinfection; the prevalence gradually returns to levels dictated by socioeconomic conditions in human communities and by comparable environmental factors in animal populations. The characteristic of latency may also permit the disease to recur following treatment.

Susceptibility to Chemotherapeutic Drugs. Among the earliest evidence suggesting that chlamydiae were bacteria and not viruses was the demonstration that their susceptibility to chemotherapeutic agents is like that of gram-negative bacteria. The mechanisms of action of antibacterial drugs (sulfonamides, penicillin, etc.) on chlamydiae are the same as on bacteria that live outside host cells, and the minimal inhibitory concentrations are frequently similar. However, some drugs that inhibit chlamydial multiplication penetrate host cells only with difficulty and are therefore required in high concentrations. The drugs act reversibly: they stop multiplication without killing, and an effective host immune response is needed to eradicate the infection.

Chlamydial strains that synthesize their own folic acid are inhibited by sulfonamides, whereas those that require a source of preformed folic acid are not inhibited. In general, *C. trachomatis* strains are sulfonamide-susceptible and *C. psittaci* are resistant, although there are exceptions to both generaliza-

tions. Sulfonamides have been widely used in the treatment of trachoma and lymphogranuloma venereum.

Penicillin interrupts the chlamydial developmental cycle by preventing the reorganization of large cells into small ones. In the presence of penicillin chlamydiae produce very large intracellular spheroplasts which remain viable inside the host cells. Streptomycin is not effective against chlamydial infections, just as it is ineffective against other intracellular bacteria.

Both *C. psittaci* and *C. trachomatis* are susceptible to broad-spectrum antibiotics such as the tetracyclines and chloramphenicol. On a weight basis the tetracyclines are definitely more effective that chloramphenicol. Rifamycin and some of its derivatives, such as rifampicin, can also be used in the treatment of chlamydial infections.

DISEASES DUE TO CHLAMYDIAE

TRACHOMA

HISTORY AND EPIDEMIOLOGY

The agent of trachoma is one of the most successful parasites of man. The Eber's Papyrus, the oldest known medical writing in the Western World, describes the clinical features of trachoma as well as several forms of treatment, at least one of which (copper sulfate) is still in use in certain parts of the world.

In nature **the disease is limited to man,** infecting only epithelial cells of the eye and possibly the nasopharynx; no systemic involvement has been described. Characteristic cytoplasmic inclusions in conjunctival cells were first seen early in this century, but the agent was not grown until 1955, when it was propagated in the yolk sac of chick embryos. Progress in understanding these organisms has since been rapid.

Trachoma is found in nearly every country and **is the greatest single cause of blindness.** It is estimated that over 400 million people have the disease, of whom 6 million are totally blind. Although frequently associated with dry or sandy regions of the world, trachoma is also common in areas of high rainfall such as the equatorial regions of Africa and even in populations living above the Arctic

Circle. The disease flourishes in communities with poor public sanitation and personal hygiene. It is thought to be transmitted primarily within the family by eye-to-finger-to-eye contact or by towels and clothing. Transmission by flies is suspected but has not been established. Subclinical infections are common and serve as important reservoirs.

Many enigmas remain in the epidemiology of trachoma. Why was the disease prevalent among Indians in the reservations of the United States and among whites in the border states (Kentucky, Tennessee, West Virginia), while Negroes in these regions remained relatively unaffected, though Negroes in widely distributed areas of Africa are quite susceptible? Why does the southern portion of India have relatively low rates while in the northern Punjab regions the disease is hyperendemic? Why does onset of the disease occur usually in infants in some countries, in children in others, or in adults in yet others? The three major serotypes of trachoma may be found within inhabitants of the same village or in members of the same family, but almost never in one individual.

PATHOGENICITY

In many communities almost all children are infected in their first few years. Under these holoendemic conditions, the infection becomes chronic in a small percentage of patients; blindness ensues primarily in this group. However, the disease may begin at any age and adult volunteers have developed severe infections after administration of a few egg-infectious doses of trachoma elementary bodies.

Onset of disease is usually abrupt, with inflamed palpebral and bulbar conjunctivae; within a few weeks accumulation of lymphocytes, polymorphonuclear neutrophilis, and macrophages coalesce to form characteristic **follicles** beneath the conjunctival surface. Still later **vascularization** of the cornea begins, usually at the upper limbus, followed by an **infiltration** of the cornea (termed **pannus),** which may produce partial or complete blindness. Scarring of the conjunctiva may cause the eyelids to turn inward so that the lashes scratch the cornea. Distortion of the structures of the external eye also interferes with normal lacrimal flow, growth of lashes, and function of glands; as a result, **bacterial infections** of trachomatous eyes are common. These factors act in concert to destroy vision.

Study of the pathogenesis of trachoma is hampered by the lack of satisfactory animal models. Although Old and New World monkeys can be infected, the disease does not closely resemble the human disease.

ANTIGENIC COMPOSITION

All trachoma strains belong to group A. *C. trachomatis* derived from cases of ocular trachoma from countries around the world have been divided into **three types** on the basis of a mouse toxin protection test. Fluorescent Ab tests confirm these types, which do not differ consistently in virulence.

The nomenclature of typing has not been settled; some laboratories use arabic or roman numerals and others letters to designate types. At least six additional antigenic types have been established for inclusion conjunctivitis, lymphogranuloma venereum, and other group A chlamydiae. Only rarely are the genital strains, whether from inclusion conjunctivitis or lymphogranuloma venereum, analogous antigenically to the trachoma agents.

IMMUNITY

As noted above, trachoma generates partial immunity; and vaccination protects mice against the toxic reaction described above that follows intravenous injection of elementary bodies. In populations where trachoma is holoendemic and not effectively treated, over 90% of patients recover spontaneously without serious sequelae. Experimental evidence suggests that this immunity, although often incomplete, may be effective in diminishing the duration of infection or increasing the dose required for reinfection.

After infection Abs appear in the serum and eye secretions within 2 or 3 weeks. Eye secretions from human patients with active clinical trachoma contain IgG and secretory IgA Abs, which are type-specific for the infecting strain. These Abs, although present in the eyes of chronically infected patients, when mixed with elementary bodies of homologous type neutralized or delayed their capacity to infect the eyes of owl monkeys; but systemic transfer of Abs failed to reach or to protect the eyes from challenge. There is some evidence that a cellular immune response is induced and is implicated in the pathological processes leading to pannus as well as in the immune reactions.

In animal models infection followed by spontaneous cure renders the animal almost totally immune to a later ocular challenge, but this resistance declines after a few months. Animals vaccinated with killed **vaccines** also have temporarily heightened resistance to subsequent challenge, and field trials in humans have significantly diminished clinical attack rates. Moreover, those patients who become infected after vaccination have fewer inclusions in conjunctival scrapings; the transmission rates may thereby be reduced and the epidemic curtailed. However, the short duration of effectiveness severely limits the usefulness of these vaccines, and none are recommended at present.

DIAGNOSIS

Clinical diagnosis of trachoma rests on the finding of characteristic follicles and scars in the conjunctiva and vascularization and infiltration of the cornea. In the typical case diagnosis is easy; in mild cases or after distortion of the anatomy of the external eye the diagnosis of activity may be difficult.

Currently three technics are used for the laboratory diagnosis of trachoma. 1) The agent is readily isolated from conjunctival scrapings of active cases by inoculation in the yolk sac of 7-day chick embryos or in irradiated McCoy cells, bacterial contamination being suppressed by use of streptomycin or other antibiotics to which the agent is not sensitive. 2) Typical cytoplasmic inclusion bodies are found within conjunctival cells stained by fluorescent Ab, Giemsa, or iodine. 3) A recently developed method is a serological test of eye secretions for trachoma-specific Abs, which correlates well with active trachoma as defined by other microbiological tests.

TREATMENT

Antimicrobial therapy is usually curative in the patient with early trachoma. The drug of choice is one of the sulfonamides given for 3 weeks; tetracycline is also effective. Frequent regular application of ophthalmic ointments containing tetracycline has been recommended, but the infection may persist under this regimen. Treatment of chronic trachoma is hampered by anatomical changes in the lacrimal drainage system and by scarring in the eyelids, which predispose to bacterial infections and render clinical cures uncertain.

When trachoma is holoendemic throughout an entire nation, its management poses entirely different problems, as noted above, than does therapy for a single patient. Improvements in standards of living and heightened health consciousness are more important than treatment in reducing the significance of this disease on a public health scale.

INCLUSION CONJUNCTIVITIS

In developed countries inclusion conjunctivitis is the characteristic chlamydial infection of the conjunctiva. Its two synonyms—"inclusion blenorrhea of the newborn" and "swimming pool conjunctivitis"—arise from the distinct epidemiology of the disease in infants and adults, but in both the reservoir of infection is the human genitourinary tract, where the infection is often asymptomatic.

Clinical Description. In the newborn clinical signs appear 5 to 12 days after brith. The conjunctiva becomes inflamed and thickened and there is often copious discharge of pus. Follicles can develop on the conjunctiva, but the disease is usually self-limiting, resolving spontaneously within a few months even without treatment.

In the adult, too, the conjunctivitis resembles the early stages of trachoma, but usually does not progress to a chronic disease; blindness is not a threat. Indeed, in classical descriptions inclusion conjunctivitis, by definition, did not involve the cornea or cause scars on the conjunctiva. However, in an epidemiological situation characteristic of inclusion conjunctivitis in communities without known trachoma, cases clinically indistinguishable from trachoma occasionally occur. Moreover, in some communities where trachoma is endemic, it is often relatively mild; and the agents causing inclusion conjunctivitis are closely related serologically to those of trachoma. Some authorities therefore postulate a continuous spectrum between the two classical diseases (see below, Diagnosis).

Pathogenesis and Transmission. In experimental infections in humans inclusions identical to those of trachoma are found in conjunctival cells, and also in epithelial cells of the human genitourinary tract (see below).

Transmission to the newborn is from the mother's genital tract, and occasionally adults contract the disease from affected babies. Before chlorination became widespread swimming pools were a major source of adult infection, the water being contaminated by genital or eye secretions. More recently, however, the disease in adults is more likely to result from contamination of the eyes by genitourinary exudate borne on fingers or towels.

The reservoir of infection is of course maintained by sexual intercourse. Indeed, in the developed countries nongonococcal (or nonspecific) urethritis and cervicitis are a greater problem than inclusion conjunctivitis, and the organisms of inclusion conjunctivitis can be isolated from a considerable proportion (20 to 40%) of cases. However, the pathogenetic role of these agents is not established. Moreover, chlamydial infection is usually asymptomatic, and nongonococcal urethritis in the male usually causes only a mild purulent discharge whose agent is rarely identified. Women with cervicitis associated with chlamydial infection have a high frequency of positive Papanicolaou cervical smears for carcinoma in situ.

Diagnosis. Diagnosis of inclusion conjunctivitis, like that of trachoma, is not always easy on purely clinical grounds and should be confirmed by finding inclusions or isolating the agent. In nongonococcal urethritis of chlamydial origin infection is usually diagnosed by isolation of the organisms directly in cell cultures.

The serology of inclusion conjunctivitis has been studied very little compared with that of trachoma. Fluorescent Ab tests have recently allowed differentiation of the organisms from the closely related agent of trachoma. This separation may prove to be significant in epidemiology.

Inclusion conjunctivitis is not believed to generate immunity to reinfection. With nongonococcal urethritis repeated attacks are common, but whether each episode implies a new chlamydial serotype is unknown.

Prevention and Treatment. Insofar as inclusion conjunctivitis is a byproduct of venereal disease its prevention is extremely difficult, particularly since genitourinary infection by chlamydia is often asymptomatic. Adequate chlorination of swimming pools and good personal hygiene make direct eye-to-eye spread of little importance.

Both in the newborn and in the adult the eye infection is quickly cured by tetracyclines. By contrast, nongonococcal urethritis is often difficult to cure, but its dependence on chlamydial infection is not certain.

LYMPHOGRANULOMA VENEREUM (LGV)

This disease, first recognized late in the eighteenth century, is caused by group A chlamydiae. Its epidemiology is poorly understood. The disease is not common except in individuals who are sexually highly promiscuous, and then only in certain groups.

Clinical Description. After a latent period of 7 to 12 days the first manifestation of the disease appears at the site of infection, usually as a herpetiform vesicle on the genitals. The vesicle ruptures and then heals without scarring. Since it is painless it may go unnoticed. From 1 week to 2 months later the regional lymph nodes become enlarged and tender and may suppurate. (The enlarged lymph nodes are sometimes referred to as buboes; thus one of the common synonyms of this disease is venereal bubo.) The granulomatous inflammation in the lymph nodes heals with scarring, occasionally obstructing lymphatic channels and thus leading to edema of genital skin. Because lymphatic drainage of the perineum in women is directed to perirectal lymph nodes, perirectal scarring sometimes develops and severe rectal obstruction may be a late manifestation.

Pathogenesis and Transmission. The special characteristics of the infection arise from the severe damage to regional lymph nodes, a site not especially attacked by other chlamydial infections. Most workers set agents that cause LGV apart from the spectrum of strains with varying degrees of predilection for the eye or the genitourinary tract. However, LGV strains may merely be at one end of this spectrum.

Transmission is virtually always by sexual intercourse. Since notification of cases is not required there are no accurate figures on prevalence; indeed, an over-all figure would be meaningless, for like other venereal diseases the incidence of LGV varies with the sexual mores of the group. It has been found frequently in veterans returning from Vietnam.

Diagnosis. Clinical diagnosis is often confirmed by a delayed-type skin reaction to killed organisms grown in yolk sac (the Frei test). Affected individuals often have complement-fixing Abs in their sera. Both these tests, however, give cross-reactions with psittacosis. The skin reactivity can last for many years, probably because of a continuing focus of infection (and possibly a state of "infective immunity"), since it can be eliminated by thorough treatment. It is not known whether the disease confers lasting immunity.

Prevention and Treatment. No effective vaccination exists, but treatment with tetracyclines cures the acute clinical signs. Such treatment has no effect on scarring already present.

ORNITHOSIS

Chlamydia infection of birds was at first thought to be restricted to the parrot family, hence the names **psittacosis** for the disease and *C. psittaci* for the infecting agent. With the realization that a very wide variety of birds harbors the organisms (with or without disease) the term ornithosis became generally accepted, although the term psittacosis is still often applied to the human infection contracted from birds. In birds latent infection is common, but with increased stress, such as overcrowding, epidemics of varying severity occur.

Pathogenesis and Transmission in Birds. Affected birds often show general debility, have discharges from the eyes and nostrils, and suffer diarrhea, but these signs are not specific to ornithosis. Death rates vary widely, sometimes reaching 30% of an affected flock. Chlamydiae can infect practically all the organs and are shed in the feces and discharges. Since the organisms remain infective in dried feces transmission by inhalation of dust is common. Vertical transmission through the egg occurs in some species, but in domestic fowl artifical incubation and brooding usually result in uninfected offspring.

Human Infection. In man, too, ornithosis is frequently asymptomatic or causes only trivial respiratory symptoms. Severe and fatal pneumonia can develop, however, and is a recognized hazard arising from contact with infected birds either as pets, on farms, or in poultry-dressing plants. The incubation period varies from 1 to 3 weeks, after which the disease begins abruptly or insidiously with chills, fever, headache, and an atypical pneumonia. Physical findings may be surprisingly few while X-ray examination may change from day to day. Infected individuals may harbor the organisms (as shown by isolation from sputum) for some years. Transmission from human patients is uncommon but has been observed in hospital personnel. The alleged existence of strains of chlamydiae especially well adapted to man and nonpathogenic for birds has not been confirmed.

Diagnosis. The clinical signs of ornithosis are not sufficiently characteristic to establish diagnosis in either birds or man, and **isolation** of chlamydiae from blood or sputum in eggs or cell culture provides the best evidence of infection. Isolates are identified by use of standard antisera in complement fixation, the fluorescent Ab technic, or neutralization of infectivity in mice.

Infection can also be diagnosed, even in the absence of overt disease, serologicaly if a rising titer of Ab against *Chlamydia* family Ag is demonstrated. Complement fixation and the fluorescent Ab technic are most useful, using Ags grown in yolk sac.

Ornithosis is associated with development of delayed hypersensitivity as well as of serum Ab. Both reactions may wane if the infection is eradicated, but they persist in the carrier state. The protection afforded is characteristically not absolute, and can be overcome by severe challenge.

Prevention and Treatment. Ornithosis cannot be eradicated since avian infection is widespread and does not cause high mortality. Infection can be avoided in flocks of domestic fowl, particularly where birds are kept in enclosed houses throughout their lives. Paradoxically, if infection is then introduced the results are likely to be catastrophic. The use of tetracyclines in the food of turkeys is effective both prophylactically and therapeutically. However, such treatment is likely to produce bacteria resistant to antibiotics useful in man. Vaccines against ornithosis have so far proved of little use.

In humans the pneumonia of ornithosis is cured by therapy with tetracyclines; the greatest problem is early diagnosis.

SELECTED REFERENCES

Books and Review Articles

COLLIER, L. H. The present status of trachoma vaccination studies. *Bull WHO 34*:233 (1966).

Conference on trachoma and allied diseases. *Am J Ophthalmol 63* (Suppl), 1967.

MOULDER, J. W. *The Psittacosis Group as Bacteria.* Wiley, New York, 1964.

NICHOLS, R. L. (ed.). *Trachoma and Related Disorders Caused by Chlamydial Agents.* Cong. Series No. 223. Excerpta Medica, Princeton, N.J., 1971.

STORZ, J. *Chlamydia and Chlamydia-Induced Diseases.* Thomas, Springfield, Ill., 1971.

Specific Articles

BEDSON, S. P., and BLAND, J. O. W. The developmental forms of psittacosis virus. *Br J Exp Pathol 15*:253 (1934).

BELL, S. D., JR., and THEOBALD, B. Differentiation of trachoma strains on the basis of immunization against toxic death of mice. *Ann N Y Acad Sci 98*:337 (1962).

GORDON, F. B., HARPER, I. A., QUAN, A. L., TREHARNE, J. D., DWYER, R. S. C., and GARLAND, I. A. Detection of *Chlamydia* (*Bedsonia*) in certain infections of man. *J Infect Dis 120*:451 (1968).

KINGSBURY, D. T., and WEISS, E. Lack of DNA homology between species of the genus *Chlamydia*. *J Bacteriol 96*:1421 (1968).

MANIRE, B. P., and TAMURA, A. Preparation and chemical composition of the cell walls of mature infectious dense forms of meningopneumonitis organisms. *J Bacteriol 94*:1178 (1967).

MEYER, K. F. The host spectrum of psittacosis-lymphogranuloma venereum (PL) agents. *Am J Opthalmol 63*:1225 (1967).

RICE, C. E. The carbohydrate matrix of the epithelelial cell inclusion in trachoma. *Am J Opthalmol 19*:1 (1936).

WANG, S. P., and GRAYSON, J. T. Classification of TRIC and Related Strains with Micro Immunofluorescence. In *Trachoma and Related Disorders Caused by Chlamydial Agents*. (R. L. Nichols, ed.) *Excerpta Medica,* Princeton, N.J., 1971.

chapter **40**

MYCOPLASMAS AND L FORMS

Revised by LEWIS THOMAS, M.D.

MYCOPLASMAS   930

   History   930
   Morphology and Mode of Replication   930
   Growth and Metabolism   932
   Immunological Classification   934
   Pathogenicity   934

  PRIMARY ATYPICAL PNEUMONIA   938

   Immunity   941
   Laboratory Diagnosis   941
   Treatment   942
   Epidemiology and Prevention   942

  MYCOPLASMAL CONTAMINATION OF TISSUE CULTURES   942

L-PHASE VARIANTS   943

  RELATION TO DISEASE   943

MYCOPLASMAS

Early in the eighteenth century a highly contagious disease of cattle appeared in Europe, causing enormous losses. The characteristic large amounts of serous fluid in the lungs and pleural cavities gave rise to the name **pleuropneumonia.** Injection of a drop of the fluid into the skin of a healthy animal was later found to produce a rapidly progressing edematous lesion; yet no bacteria could be seen. The causative agent was finally cultivated in 1898 on serum-enriched media but was difficult to detect: the colonies were very small, the organisms were tiny, they stained poorly, and they were extremely pleomorphic.

A number of microorganisms with similar morphological and cultural properties have now been isolated. Originally referred to as **pleuropneumonia-like organisms,** or **PPLO,** they have recently been assigned the class name Mollicutes (soft skin), subdivided into a genus requiring sterols, *Mycoplasma,* and a much smaller genus capable of growth without sterols, *Acholeplasma.* Mycoplasmas have been recognized as the cause of disease in a wide variety of animals, insects, and plants, and several nonpathogenic species have been isolated from the mucous membranes of man and other animals, and from soil and sewage. The saprophyte *M. laidlawii,* from sewage, is widely used in metabolic studies.

Mycoplasmas are the smallest known free-living organisms and they differ from bacteria in lacking cell walls. They evidently contain the minimal macromolecular machinery for autonomous reproduction. For this reason, and because their membranes are readily accessible, they have become of great interest to molecular biologists.

In man a mycoplasma causes primary atypical pneumonia, and the so-called **T strains** (which produce especially **tiny** colonies, 15 to 20μ in diameter) may cause nongonococcal urethritis. The role of mycoplasmas in rheumatoid arthritis is controversial.

MORPHOLOGY AND MODE OF REPLICATION

Because of the absence of cell walls, the mycoplasmas do not stain with Gram stain and they are more pleomorphic and plastic (hence the name "plasma") than eubacteria. With Giemsa stain they appear as tiny pleomorphic cocci or short rods, and sometimes as hollow ring forms. Electron photomicrographs reveal a diameter ranging from 150 to 300 mμ, with a limiting membrane 75 to 100 A in thickness, like the cytoplasmic membrane of animal cells (Figs. 40-1 and 40-3).

Some mycoplasmas possess characteristic shapes (which are often distorted during fixation in preparation for electron microscopy): *M. gallisepticum* has a bleb at one region of its periphery; *M. neurolyticum* tends to be long and narrow. Within the cells ribosomes and "nuclear strands" can be identified (Fig. 40-2). There are no mesosomes. In some strains amorphous material on the outer surface of the membrane suggests the existence of a capsule or rudimentary wall (Fig. 40-3).

Replication. In the absence of a rigid cell wall the pattern of replication is quite different from that of typical bacteria, whose division starts with the formation of a well-defined septum. Though the mechanism of division in mycoplasmas has not been unequivocally established, sequential microscopic observations suggest that new elementary particles arise by fragmentation of filamentous cells containing several discrete "DNA com-

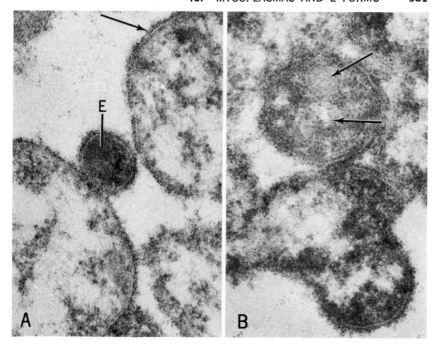

Fig. 40-1. Electron photomicrograph of 48-hour culture of *M. canis*. **A.** Electron-dense "elementary body" (E) and two large cells are seen surrounded by triple-layered limiting membranes. At arrow floccules of dense material are visible on surface of membrane suggesting a rudimentary capsule. **B.** Two "elementary bodies" (arrows) are discernible within cytoplasm of large cell. ×94,000. [From Domermuth, C. H., Nielsen, M. H., Freundt, E. A., and Birch-Andersen, A. *J Bacteriol* *88*:727 (1964).]

Fig. 40-2. *M. hominis*, type 1, 72-hour culture. In cells in **A** ribosomes are clearly visible. In **B** nuclear area (N) of cell is prominent. ×94,000. [From Domermuth, C. H., Nielsen, M. H., Freundt, E. A., and Birch-Andersen, A. *J Bacteriol* *88*:727 (1964).]

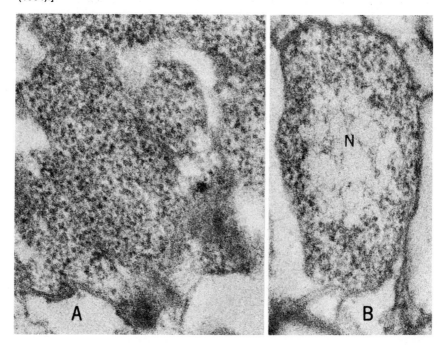

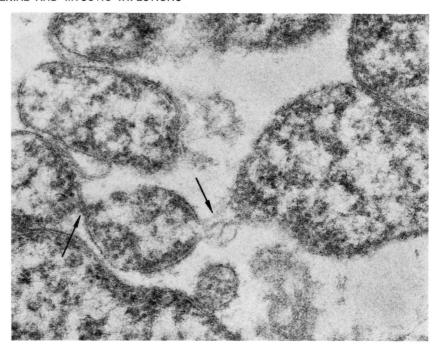

Fig. 40-3. Three connected large cells (arrows) in 48-hour culture of *M. arthritidis*. ×94,000. [From Domermuth, C. H., Nielsen, M. H., Freundt, E. A., and Birch-Andersen, A. *J Bacteriol 88*:727 (1964).]

ponents" (Fig. 40-4). The particles are difficult to measure because of their plasticity, but can pass filters with an average pore diameter of 150 mμ. The elementary particles were formerly thought to arise within large, swollen forms temed "large bodies," but these are probably degenerating or dead cells. The mechanism of reproduction remains controversial.

Colonial Morphology. On solid media mycoplasmas form minute, transparent colonies which are difficult to detect without a hand lens, even when stained. Colonies are best recognized under low-power magnification with a light microscope. Colonial morphology can be visualized under higher magnification by the method of Dienes: blocks of transparent agar on which colonies have grown are covered with coverslips prepared with dried stain, and the colonies are examined with the oil-immersion objective.

Typical colonies range from 10 to 600 μ in diameter and usually appear as tiny round structures with granular surfaces and a dark

central nipple (Fig. 40-5). The latter, which gives the colony a "fried-egg" appearance, is due to a central zone of deep growth beneath the surface of the agar (possibly due to greater ease of division by these soft cells when pinched between agar fibers). Many colonies exhibit only the dark center. The colonies of some human strains cultured aerobically on blood agar are surounded by zones of α- or β-hemolysis.

GROWTH AND METABOLISM

Strains of mycoplasmas vary widely in growth rate. Some produce visible turbidity in broth within 18 hours (yielding up to 5×10^9 colony-forming units) and form sufficient acid to sterilize the cultures if incubated longer than 24 hours. Others will grow only sparsely in broth during a week, and the appearance of colonies on solid media may require several weeks. Many strains can also grow on the chorioallantoic membrane of chick embryos and in tissue cultures. Species vary in their ability to ferment glucose,

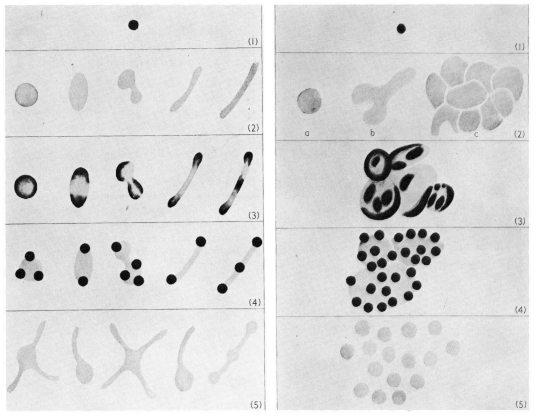

Fig. 40-4. Diagrams of mycoplasmal replication in liquid **(left)** and solid **(right)** media: (1) minimal reproductive unit; (2) pleomorphic mycoplasmal cells; (3) concentrations of opaque cytoplasm at margins of cells; (4) appearance of reproductive granules within cells; (5) new generation of cells arising from granules. (From Klieneberger-Nobel, E. *Pleuropneumonia-like Organisms (PPLO) Mycoplasmataceae.* Academic Press, New York, 1962.)

hydrolyze arginine, or hydrolyze urea; these tests are useful in identifying unknown strains (Table 40-1).

Several phages that produce plaques on lawns of *M. laidawii* have been demonstrated in cultures of unrelated species of mycoplasmas. Two distinct morphological types of mycoplasmal phage have been recognized. DNA obtained from the phage of *M. laidawii* can infect other mycoplasmal strains, causing the formation of virus particles.

Nutrition. Most mycoplasmas require a rich medium for growth, but despite the lack of a cell wall they do not require a medium of very high osmotic pressure. Most species also require a sterol and serum protein.

The most widely used medium contains beef heart infusion, peptone, yeast extract, and 0.5% NaCl, supplemented with 20% horse serum and adjusted to pH 7.6 to 7.8. Most species require nucleic acid precursors and various vitamins. Most species indigenous to man can be cultivated aerobically, but some grow well only in nitrogen with 5 to 10% CO_2 (Table 40-1). For solid medium, 1% agar is used. The high concentrations of serum may produce "pseudocolonies" of lipid or calcium and magnesium soap, sometimes mistaken for mycoplasmal colonies by inexperienced observers.

The requirement for sterols is of major interest, since sterols are not components or nutrients of bacteria or their L forms. Parasitic human mycoplasmal species contain 10 to 20% of their dry weight as total lipid, 50 to 65% of which is nonsaponifiable (hence presumably sterols); nonsaponifiable lipids have not been detected in saprophytic species grown without sterols. However, one of these, *A. laidawii,* is capable of incorporating cholesterol into its membrane when it is made available; the organism then becomes sensitive to polyene antibiotics (Ch. 43, Polyenes). A required component of the horse serum supplement is α-l-lipoprotein, which binds esterified cholesterol.

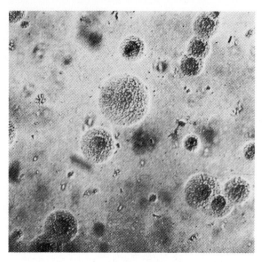

Fig. 40-5. Colonies of human strain of mycoplasma on surface of solid medium. Note fried-egg appearance due to growth penetrating surface of agar in center of each colony. ×100. (From Klieneberger-Nobel, E. *Pleuropneumonia-like Organisms (PPLO) Mycoplasmataceae.* Academic Press, New York, 1962.)

Certain strains of mycoplasmas will grow only in the presence of special additives in the broth, including starch, mucin, and DNA; a mycoplasma causing synovitis in chickens will grow only in broth containing the coenzyme DPN (NAD). A special dialyzable growth factor in yeast extract is required by *M. pneumoniae* and *M. orale* (Table 40-1).

Mycoplasmas are generally susceptible to tetracyclines and to kanamycin, but are resistant to sulfonamides and, lacking a wall, to penicillin. They are uniformly susceptible to gold salts. Thallium acetate is used in media for suppression of gram-negative bacteria and aerobic spore-formers, but it is lethal for the T strains and must be omitted when they are sought. Mycoplasmas are more susceptible than eubacteria to killing by distilled water, physiological saline, and surface-active agents, including soaps.

IMMUNOLOGICAL CLASSIFICATION

Mycoplasmas are classified largely on the basis of immunological reactions, including complement fixation, agglutination, gel diffusion, hemagglutination inhibition, immunofluorescence, and inhibition of growth and certain metabolic activities. Complement fixation has been the most widely exploited but it yields frequent cross-reactions between species; moreover, mycoplasmal antigens are frequently anticomplementary.

The most specific tests are for **growth inhibition** and **metabolic inhibition.** In the former, a mycoplasmal culture inoculated on solid medium exhibits a zone of growth inhibition around a disc of paper impregnated with the corresponding antiserum. This requires relatively high concentrations of antibody, but it is simple and identifies species unequivocally. In the metabolic inhibition test, which is also quite specific and considerably more sensitive, antibody plus complement cause membrane damage and thus prevent acid production from glucose, or ammonia formation from arginine in broth cultures, revealed by appropriate indicators.

The major antigenic determinants in *M. pneumoniae* are contained in lipid extracts of the cell membrane. In addition, a phospholipid resembling cardiolipin, present in *M. pneumoniae* extracts, acts as a hapten; when mixed with antiserum beforehand, it blocks the growth inhibition and metabolic inhibition reactions.

PATHOGENICITY

Animal Infections. Pathogenic mycoplasmas display a high degree of organ and tissue specificity within the infected animals, causing selective lesions in the central nervous system, the joints, pleura, peritoneum, or lungs. They are also extremely species-specific; mycoplasmas that are highly pathogenic for poultry do not affect rodents and those pathogenic for rodents do not infect other laboratory animals. In chronic disease, dense masses of lymphocytes and plasma cells accumulate in the infected tissues; vasculitis, with round cells infiltrating vessel walls, has also been observed in various animal infections.

As Table 40-2 shows, mycoplasmas produce a wide variety of diseases in various animals. These findings point to the great pathogenic potential of these organisms and hence to the usefulness of the animal infections as models for study.

An interesting interplay with certain viruses has been observed in several experimental infections. Nelson found that a nonpathogenic strain of *M. neurolyticum* causes extensive necrosis of the brain in animals simultaneously infected with murine hepatitis virus. Infection of chickens with *M. gallisepticum* combined with the virus of infectious bronchitis produces a lethal disease, unlike the mild illness caused by either agent alone.

TABLE 40-1. Distinguishing Characteristics of Human Mycoplasmal Species

Species	Anerobic	Aerobic	Requirement for yeast extract	Rate	Colonial morphology	Fermentation, glucose	Hemolysis, guinea pig red cells	Hemadsorption	Aerobic reduction of tetrazolium
M. hominis	+	+	0	Rapid		0	Slow	0	0
M. salivarium	+	0	0	Rapid		0	Slow	0	0
M. orale types 1, 2, 3	+	±	+	Moderate		0	Slow	0	0
M. fermentans	+	±	0	Moderate		+	Slow	0	0
T-strain group	+	+	0	Moderate	Small (15–25μ)	0	Slow	0	0
M. pneumoniae	+	+	+	Slow		+	Rapid	+	+

Modified from Hayflick, L., and Chanock, R. M. *Bacteriol Rev 29*:185 (1965).

TABLE 40-2.   Diseases caused by Mycoplasmas

Host	Disease	Mycoplasmal species
Man	Cold hemagglutinin positive: primary atypical pneumonia	*M. pneumoniae*
Cattle	Contagious bovine pleuropneumonia	*M. mycoides* var. *mycoides*
	Mastitis	*M. bovigenitalium*
	Mastitis and arthritis	*M. bovimastitidis*
Sheep and goats	Contagious caprine pleuropneumonia	*M. mycoides* var. *capri*
	Mastitis, contagious agalactia	*M. agalactia*
Swine	Chronic pneumonia or enzootic pneumonia	*M. suipneumoniae* (hyopneumoniae)
	Arthritis and polyserositis	*M. hyorhinis*
Rodents	Rolling disease of mice and rats	*M. neurolyticum*
	Pneumonia of mice and rats; infectious catarrh	*M. pulmonis*
	Arthritis of mice and rats	*M. arthritidis*
Fowl	Chronic respiratory disease, infectious sinusitis, cerebral polyarteritis (turkeys)	*M. gallisepticum*
	Infectious synovitis	*M. synoviae*
	Air-sac disease of turkeys	*M. meleagridis*

Adapted from Hayflick, L. (ed.). *The Mycoplasmatales and the L-phase of Bacteria.* Appleton, New York, 1969.

Human Infections. Despite the long list of mycoplasmal diseases of animals (Table 40-2), no human mycoplasmal disease was identified until 1962, when *M. pneumoniae* was demonstrated as the cause of primary atypical pneumonia, long known as "viral pneumonia" (see below). This organism has also been reported to be associated with various neurological syndromes, including the Guillain-Barré syndrome, meningoencephalitis, and acute cerebellar ataxia, but convincing evidence is lacking.

Various other mycoplasmal species are suspected as the cause of several unrelated diseases. The joint fluid of patients with rheumatoid arthritis was reported to yield mycoplasmas in tissue cultures, but this technique is now known to be unreliable because tissue cultures are commonly contaminated with mycoplasmas. Certain laboratories have reported isolation of mycoplasmas from joint fluid in special broths, but others, using the same methods, have been unable to confirm the findings; the question therefore remains unsettled. The role of T strains and *M. hominis* in genitourinary tract infections also remains to be clarified. Some cases of nongonococcal urethritis probably are due to T strains. However, these microorganisms are ubiquitous in the urethra of normal men and women, and no demonstrable antibody response follows infection; hence a positive culture of T strains does not establish their pathogenic role. The strongest evidence is epidemiological: the demonstration of multiple cases of nongonococcal urethritis in men following coitus with an infected woman and isolation of T-strains from both partners. Following septic abortion *M. hominis* and T strains have been cultured from the maternal blood and the fetal tissues, but their pathogenetic significance is not known.

One possible pathogenic mechanism involves autoimmunity due to mycoplasmal antigens that cross-react with tissue antigens of the host. For example, *M. mycoides* elaborates a galactan that is similar to a carbohydrate in bovine lung, and patients with *M. pneumoniae* infection often form antibodies that react in complement fixation with extracts of normal mammalian lung. Moreover, *M. arthritidis* contains antigens that cross-react with murine tissue components.

Another possible factor in pathogenesis is the capacity of mycoplasmas to become firmly attached to cell surfaces. It is rare to find mycoplasmas inside cells, but the cell surface seems to provide a friendly environment for colonization (Fig. 40-6). Sometimes the organisms are enveloped by cytoplasmic processes (Fig. 40-7), which may protect

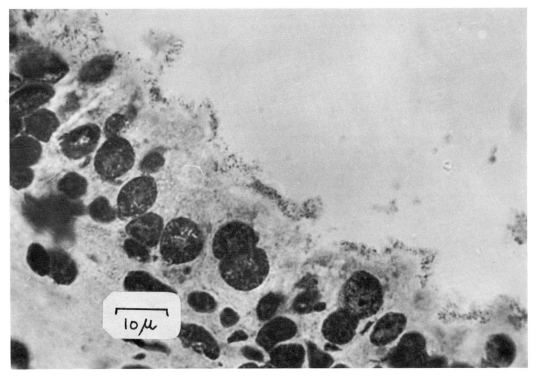

Fig. 40-6. Coccobacillary bodies of *M. pneumoniae* in "mucous blanket" covering bronchial epithelial cells of infected chick embryo. Intensified Giemsa stain. [From Marmion, B. P., and Hers, J. F. *Am Rev Resp Dis 88* (Suppl):198 (1963).]

Fig. 40-7. Monkey kidney cells in tissue culture 5 days after inoculation with *M. pneumoniae*. Colonies of the organism may be seen as dark granular structures in the peripheral cytoplasm of the cell in the center of **A.** No cytopathic effects are noticeable in such cultures. Intensified Giemsa stain. ×200. In **B** can be seen a single monkey kidney cell bearing on its outer membrane a colony of *M. pneumoniae*. ×900. [From Clyde, W. A. Jr. *Am Rev Resp Dis 88* (Suppl):212 (1963).]

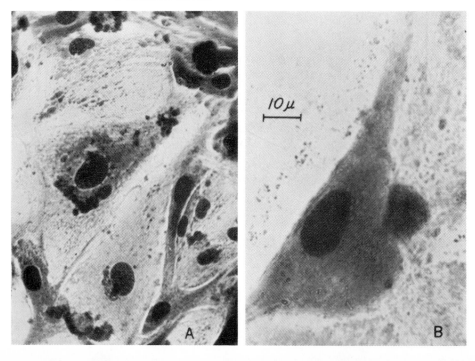

them from the action of antibiotics and antibody. At the same time, coating of the cell surfaces with mycoplasmal antigens may conceivably render them vulnerable to destruction by immunological reactions.

PRIMARY ATYPICAL PNEUMONIA

In the late 1930s a new respiratory ailment was described, occurring especially in institutions and military camps. It was characterized by fever, nonproductive (but often severe) cough, headache, marked prostration, and (later) failure to respond to sulfonamides or penicillin. During the second or third week **cold hemagglutinins*** often appeared in the serum, as well as agglutinins to an α-hemolytic streptococcus, known as **streptococcus MG**

---

* These antibodies agglutinate human group O erythrocytes at 0 to 4° but not at 37°. They are also encountered in trypanosomiasis and malaria and in certain blood dyscrasias and liver disease.

(Ch. 26). Fatalities were rare, but in a few autopsied cases the pneumonia was found to be characterized by bronchiolitis and interstitial mononuclear and granulocytic infiltration.

After many attempts to discover the etiology of this disease, Eaton, Meiklejohn, and van Herrick in 1944 reported recovery of a filtrable agent which could be serially passed in embryonated eggs without producing demonstrable lesions; the products of such blind passage, like the original filtrates, caused pneumonia when inoculated intranasally in hamsters and cotton rats. Graded filtration indicated a diameter of 180 to 250 mμ, and the agent was initially presumed to be a virus.

The low animal virulence and mild cytopathogenicity of the agent prevented confirmation and acceptance for many years, but in 1957 Liu stained specific particles in the lungs of infected (but healthy) chick embryos by indirect immunofluorescence, with serum

Fig. 40-8. Mulberry surface colonies of *M. pneumoniae*. Note absence of peripheral halo (cf. Fig. 40-5). ×600, reduced. [From Morowitz, H. J., and Tourtellotte, M. E. *Sci Am 206*:117 (1962).]

from convalescent patients. This reliable test for infection led to the development of quantitative methods for studying both the etiological agent and its antibodies, including the neutralizing antibodies in convalescent sera. Eventually the agent was grown on tissue cultures; it also was recognized by Giemsa staining, despite the absence of lesions in bronchial sections of chick embryos (Fig. 40-6).

Meanwhile, the infection was shown to respond to tetracyclines (in man, as well as in chick embryos and cotton rats) and to gold salts.* These observations were difficult to reconcile with a viral etiology, and in 1962 Chanock and coworkers succeeded in cultivating the agent on a cell-free agar medium

---

* Mycoplasmas had long been known to be sensitive to gold salts.

containing horse serum and yeast extract.† The minute colonies had a granular surface like a mulberry (Fig. 40-8) and stained with fluorescein-conjugated antibody in convalescent sera (Fig. 40-9).

The cultivated organism, designated *M. pneumoniae*, differed antigenically from other known varieties of human mycoplasmas (Table 40-1); it also grew unusually slowly and fermented glucose. Furthermore, it produced β-hemolysin for guinea pig erythrocytes, in contrast to the α-hemolysin generated by other human species (Fig. 40-10). The zone of hemolysis (in contrast to the colony) is easily seen with the naked eye, thus

---

† Grunholz in Germany had reported in 1950 the cultivation of mycoplasmas from 100 cases of "virus pneumonia" in infants, but unfortunately the report was ignored.

Fig. 40-9. Colonies of *M. pneumoniae* (Eaton agent) stained with fluorescein-labeled acute **(A)** and convalescent **(B)** serum from patient with primary atypical pneumonia. [From Chanock, R. M., Hayflick, L., and Barile, M. F. *Proc Natl Acad Sci USA 48*:41 (1962).]

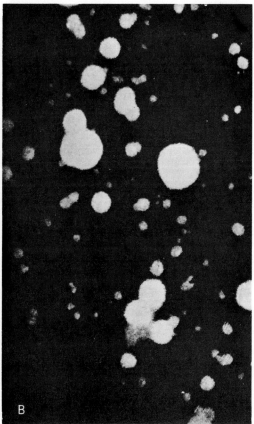

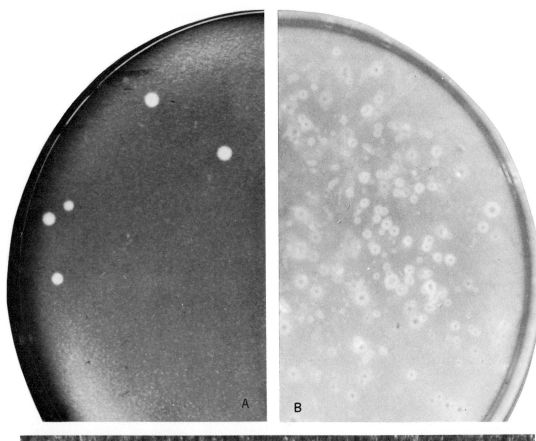

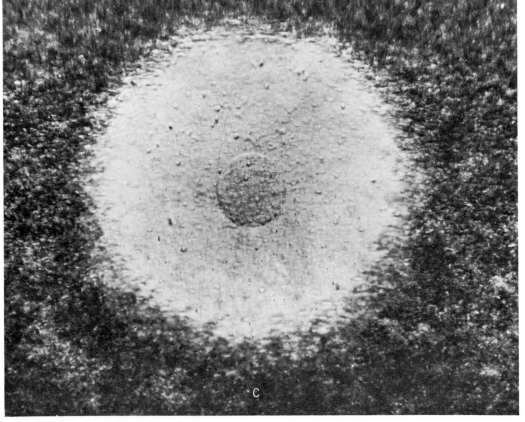

providing the basis for a convenient plaque method of identifying and quantitating *M. pneumoniae*.

Although *M. pneumoniae* is not cytopathic in cultured chick embryo (Fig. 40-6) and monkey kidney (Fig. 40-7) cells, it is cytopathic in human tissue cultures and in naturally infected human lungs. No toxin has been detected. The agent is presumably phagocytized and destroyed by leukocytes, like many bacteria.

The most convincing evidence for the causative role of the newly discovered agent came from a series of experiments in human volunteers, some free of antibodies to *M. pneumoniae* and others possessing naturally acquired antibodies (measured by immunofluorescence). When individuals without such antibodies were infected via the respiratory tract with *M. pneumoniae* grown in tissue culture all 27 acquired antibodies to the infecting organism, 12 acquired cold hemagglutinins, 3 developed pneumonia, and 7 developed respiratory disease without pneumonia. A similar group was infected with the organism propagated in cell-free agar; of 42 individuals, 38 acquired antibodies to *M. pneumoniae*, 3 produced cold hemagglutinins, while 8 developed upper respiratory tract disease with fever; none came down with pneumonia. In contrast, of 25 individuals with antibodies inoculated with the tissue culture organisms, none become ill or developed cold hemagglutinins.

IMMUNITY

M. pneumoniae infection leads to formation of species-specific antibodies to surface antigens of the organism, readily detected by immunofluorescence; it also may elicit cold hemagglutinins and agglutinins for streptococcus MG. Cross-absorption tests reveal that these three antibodies are distinct. The species-specific antibodies are protective, as shown by growth inhibition in vitro and by the human volunteer experiments noted above; they usually become detectable in the serum during the second or third week of the disease. A few second attacks of mycoplasmal pneumonia have been reported.

Cold hemagglutinins and antibodies to streptococcus MG do not develop in all cases of pneumonia due to *M. pneumoniae*. Their response appears to be related to the severity of the disease; it may be due to an autoimmune reaction resulting from the action of the hemolysin on the host erythrocytes. However, there is also evidence that an antigen of *M. pneumoniae* cross-reacts with the I antigen of erythrocytes (Ch. 21). Since DNA hybridization has excluded the possibility that *M. pneumoniae* is the L form of streptococcus MG, the antibodies to this organism may be elicited by a common antigen in the mycoplasma.

LABORATORY DIAGNOSIS

M. pneumoniae can be recovered from sputum or pharyngeal swabbings by direct cultures in liquid or solid medium containing serum and yeast extract, with penicillin and thallium acetate to inhibit contaminating bacteria. In the broth spherical free-floating colonies, 10 to 200 μm in diameter, usually appear only after a week or more of incubation;* they stain intensely with neutral red or tetrazolium and can also been seen with the low-powered lens of a microscope. Colonies on the agar medium are usually stained with methylene blue and azure II, but do not become detectable for 6 to 20 days; they measure up to 150 μm and exhibit a coarse mulberry surface without a peripheral halo (Fig. 40-8). The organism can be presumptively identified by its hemadsorption or β-hemolysis of guinea pig red cells. It is conclusively identified by staining its colonies with homologous fluorescein-labeled antibody (Fig. 40-9) or by the demonstration that specific antibody inhibits its growth on agar or in medium containing tetrazolium (tetrazolium-reduction-inhibition (TRI) test).†

The antibody response to mycoplasmal

---

* *M. pneumoniae* grows slowly: the mean generation time of even well-adapted strains in liquid medium is rarely less than about 6 hours.

† Since *M. pneumoniae* ferments glucose, phenol red may also be used as a growth indicator in liquid media containing 1% glucose.

Fig. 40-10. A. β-Hemolytic plaques in 5-day *M. pneumoniae* culture overlaid with guinea pig erythrocytes in agar and reincubated for 48 hours. **B.** Analogous preparation of *M. hominis* showing α-type hemolytic plaques. **C.** Magnified ($\times$200, reduced) view of β-hemolytic plaque produced by *M. pneumoniae*. The colony is the small central disc. [From Chanock, R. M., Mufson, M. A., Somerson, N., and Couch, R. B. *Am Rev Resp Dis* 88 (Suppl):218 (1963).]

pneumonia is most readily demonstrated by complement-fixation tests on acute and convalescent serum. Although the indirect hemagglutination test (with mycoplasmal antigen attached to red cells) is more sensitive, it is not as useful for diagnostic purposes because high titers of the antibody may be present even in acute phase serum. Cold agglutinins to human O erythrocytes and agglutinins to streptococcus MG may also be measured, although their absence in convalescent sera does not exclude the diagnosis.

TREATMENT

Early trials of the effectiveness of tetracyclines in primary atypical pneumonia were unsatisfactory because lack of specific diagnostic tests probably led to inclusion of many patients with other diseases. However, in a double-blind study on 109 patients with serologically diagnosed mycoplasmal pneumonia, the administration of a tetracycline for 6 days strikingly reduced the duration of fever (Fig. 40-11) and other signs and symptoms.

Despite the proved effectiveness of tetracyclines, the treatment of patients with atypical pneumonia still remains difficult. Because currently available diagnostic tests are slow, the decision regarding initial therapy must be made on clinical and epidemiological grounds. However, a useful method for staining *M. pneumoniae* in sputum with fluorescent antibody has been devised by Hers.

Kanamycin and erythromycin are also effective in experimental infections and erythromycin is being used in treating the disease in man.

EPIDEMIOLOGY AND PREVENTION

Most mycoplasmal pneumonia occurs in children and young adults (under 30), but it may occur at any age (except young infants). In institutions it occasionally occurs in epidemics, which evolve relatively slowly. More often the disease is endemic, the incidence being highly variable and usually reaching a peak during the winter months. Serological studies indicate that only a small fraction (probably less than 5%) of individuals infected with *M. pneumoniae* develop pneumonia, some having only upper respiratory symptoms and others no illness at all.

The incubation period, judging from institutional and family epidemics, as well as from experiments with volunteers, is 9 to 12 days. Infected individuals can shed the organism from their respiratory tracts for at least 4 weeks, despite high titers of protective antibody in their sera. How long they remain infectious is not known. Isolation measures are indicated only in epidemics. No vaccine is available, although trials are in progress.

MYCOPLASMAL CONTAMINATION OF TISSUE CULTURES

Contamination of cell cultures with mycoplasmas is frequent but is often difficult to recognize, either macroscopically or microscopically. To demonstrate such contamination with certainty the offending organism must form characteristic colonies on agar; other methods are much less reliable.

The practice of adding penicillin and streptomycin to cultures to prevent bacterial contamination has

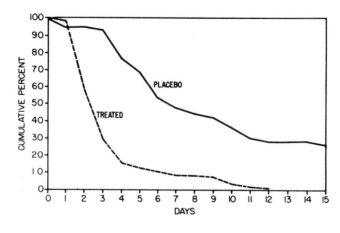

Fig. 40-11. Duration of fever (more than 99°F) in serologically proved cases of mycoplasmal pneumonia treated with either demethychlortetracycline or a placebo. (From Jordan, W. S., and Dingle, J. H. In *Bacterial and Mycotic Infections of Man.* [R. J. Dubos and J. G. Hirsch, eds.] Lippincott, Philadelphia, 1965.)

not only led to relaxation in aseptic technic but has also introduced the possibility of inducing L forms (see below) in media contaminated with airborne bacteria. However, it appears that most contaminants are not L forms but are mycoplasmas that are resistant to streptomycin (as well as to penicillin). Since many of the contaminating species are nonpathogenic but of human origin, and since their presence is a function of the number of culture transfers, most contaminants probably come from the oropharynx of the investigator. But some offending mycoplasmas also come from the animal sera and the trypsin used in the tissue culture; some investigators believe the contaminants originate in the cultured cells.

Prevention of contamination requires faultless aseptic technic. Tetracyclines and kanamycin, which inhibit most mycoplasmas, may be used prophylactically and are sometimes even effective in "curing" contaminated cultures. Decontamination has also been achieved by exposure to high temperatures (that kill the mycoplasmas but spare the animal cells), or by treatment with specific antisera (plus complement).

Virologists must be particularly on guard against mycoplasmal contaminants carried in the cultures or introduced with the inoculum, for mycoplasmas may share with certain viruses some or all of the following properties: filtrability, sensitivity to ether, ability to hemagglutinate and hemadsorb, production of viral interference in vitro, resistance to many antibiotics, inhibition of multiplication by specific antibody, and production of cytopathic effects (CPE).

L-PHASE VARIANTS

Some bacteria (e.g., *Streptobacillus moniliformis*) readily give rise spontaneously to variants that replicate in the form of small filtrable elements with defective or absent cell walls. These organisms, called L forms, can also be formed by many other species when wall synthesis is impaired (e.g., by penicillin); high salt concentrations also favor their development. Their properties are further discussed in Chapter 6.

For many years, L forms were confused with mycoplasmas: the cells are indistinguishable and they form similar "fried-egg" colonies on soft, serum-enriched agar (see above). However, immunological studies and analyses for nucleic acid homology now indicate that the two groups are not taxonomically related; the morphological and colonial similarities are explainable by the absence of cell walls, but there the biological resemblance seems to end. L forms are morphologically equivalent to protoplasts and spheroplasts (Ch. 6), but the term L form is restricted to cells that can multiply. While most protoplasts and spheroplasts formed by the action of cell-wall-degrading enzymes are unable to multply, those of some species (e.g., groups A and D streptococci) grow in appropriate solid media to give typical L-form colonies. Thus, the distinction is not a sharp one. Some L-phase variants (transitional phase variants) readily revert to normal vegetative cells in the host or in favorable media, while others are stable.

RELATION TO DISEASE

The role of L-phase variants in human disease remains unclear, but such cells have been cultivated from cases of pyelonephritis and endocarditis, and have been reported in pulmonary, meningeal, and gastrointestinal infections. Less direct evidence is the finding that cultivation in special hypertonic media (e.g., with 0.6 M sucrose) permits the outgrowth of typical bacteria from tissues that do not yield the organisms in ordinary media. Presumably the hypertonicity is necessary for survival and initial growth of L forms that then revert to eubacteria.

Certain of the agents that cause loss of bacterial walls in vitro may be present in patients. These include penicillin and other antibiotics that impair cell wall synthesis and specific antibody (plus complement and serum lysozyme) and lysosomal hydrolases that degrade intact cell walls. Thus, the milieu exists for the production of L forms in vivo. L forms are less vulnerable to those antibiotics that act primarily on cell wall synthesis, and in addition they may achieve a state of latency by being phagocytized without being killed.

Because of the special properties of L-phase variants, the possibility that they may be implicated in various diseases of obscure etiology has been extensively explored. Unequivocal evidence for their pathogenic role has not been obtained. Similarly, the suggestion that L forms of group A streptococci may be involved in the genesis of rheumatic fever has not received direct support (Ch. 26, Pathogenicity). However, it is clear that L-phase variants can persist for long periods in the tissues of experimentally infected animals, and their main role in infectious disease in man may be as sources of relapse. Whether their multiplication or elaboration of toxic materials may also be important in the pathogenesis of disease must be regarded as an open question.

SELECTED REFERENCES

Books and Review Articles

CLASENER, H. Pathogenicity of the L-phase of bacteria. *Annu Rev Microbiol 26*:55 (1972).

EATON, M. Pleuropneumonia-like organisms and related forms. *Annu Rev Microbiol 19*:379 (1965).

HAYFICK, L. (ed.). *The Mycoplasmatales and the L-Phase of Bacteria.* Appleton, New York, 1969.

HAYFLICK, L., and CHANOCK, R. M. *Mycoplasma* species of man. *Bacteriol Rev 29*:186 (1965).

KLIENEBERGER-NOBEL, E. *Pleuropneumonia-like Organisms (PPLO) Mycoplasmataceae.* Academic Press, New York, 1962.

RAZIN, S. Structure and function in *Mycoplasma. Annu Rev Microbiol 23*:317–356 (1969).

SMITH, P. F. Comparative physiology of pleuropneumonia-like and L-type organisms. *Bacteriol Rev 28*:97 (1964).

SMITH, P. F. *The Biology of Mycoplasmas.* Academic Press, New York, 1971.

STANBRIDGE, E. Mycoplasmas and cell cultures. *Bacteriol Rev 35*:206–227 (1971).

Specific Articles

DOMERMUTH, C. H., NIELSEN, M. H., FREUNDT, E. A., and BIRCH-ANDERSON, A. Ultrastructure of *Mycoplasma* species. *J Bacteriol 88*:727 (1964).

LISS, A., and MANILOFF, J. Transfection mediated by *Mycoplasmatales* viral DNA. *Proc Natl Acad Sci (USA) 69*:3423 (1972).

LIU, C. Studies on primary atypical pneumonia. I. Localization, isolation and cultivation of a virus in chick embryo. *J Exp Med 106*:455 (1957).

RYTER, A., and LANDMAN, O. E. Electron microscopic study of the relationship between mesosome loss and the stable L state (or protoplast state) of *Bacillus subtilis. J Bacteriol 88*:457 (1964).

ZUCKER-FRANKLIN, D., DAVIDSON, M., and THOMAS, L. The interaction of mycoplasmas with mammalian cells. I. HeLa cells, neutrophils, and eosinophils. II. Monocytes and lymphocytes. *J Exp Med 124*: 521, 533 (1966).

OTHER PATHOGENIC BACTERIA

Revised by ALEX C. SONNENWIRTH, Ph.D.

LISTERIA MONOCYTOGENES 946
ERYSIPELOTHRIX RHUSIOPATHIAE 948
STREPTOBACILLUS MONILIFORMIS 948
CALYMMATOBACTERIUM (DONOVANIA) GRANULOMATIS 948
BARTONELLA BACILLIFORMIS 949

LISTERIA MONOCYTOGENES

Listeria monocytogenes infection has been recognized as an important economic problem in domestic and feral animals since 1911, but listeriosis in man, first described in 1929, was considered a rare disease for the next quarter century. However, within the last decade it has been reported in the United States with increasing frequency, now exceeding that of leptospirosis, psittacosis, poliomyelitis, or botulism. It is probable that the disease is actually much more common but not recognized. The organism derives its name from the striking monocytic blood reaction it causes in the infected host.* A similar monocytosis can be produced in mice and rabbits by intravenous injection of a lipid extract of the organism ("monocytosis-producing agent," MPA) or of its component glyceride A.

L. monocytogenes is a small, gram-positive, non-acid-fast, aerobic to microaerophilic, non-spore-forming rod; it exhibits a peculiar end-over-end tumbling motility at 20 to 25°, but often not at 37°.

The organism grows well on blood agar and other general media, but storage of clinical material at 4° for several days before inoculation enhances its isolation rate, perhaps by releasing the organisms from their intracellular location. On blood agar, the colonies usually are surrounded by a narrow band of β-hemolysis, simulating that produced by *Streptococcus pyogenes*. In fact, if not carefully examined in stained smears and tested for motility at 20 to 25°, the organism may be mistaken for a hemolytic streptococcus. Because it often shows palisade formation on microscopic examination and also grows well on potassium tellurite, it is frequently mistaken for a "diphtheroid" (Ch. 24, Other corynebacteria) and discarded as a contaminant. It may also be confused with enterococci because of its growth on bile-esculin agar (a selective medium for enterococci).

L. monocytogenes ferments carbohydrates, producing principally lactic acid without gas (Table 41-1). It elaborates catalase, produces acetoin (acetyl methylcarbinol, Voges-Proskauer test) but not indole, and does not reduce nitrate. Characteristically, instillation of *L. monocytogenes* into the conjunctival sac of a rabbit results in a purulent conjunctivitis in 3 to 6 days, followed by keratitis **(Anton test)**. *Listeria* share general properties with other gram-positive genera. As in corynebacteria and lactobacilli, diaminopimelic acid is present in its cell wall, and the lactic fermentation and base composition (38% GC) suggest a close taxonomic relation to streptococci as well as lactobacilli.

On the basis of somatic (O) and flagellar (H) antigens, 4 main serological groups and 11 serotypes are recognized. Types 4b and 1 predominate among isolates from man and animals, while types 2 and 3 are very rare. **Serologic diagnosis** of listeriosis is highly unreliable because *L. monocytogenes* cross-reacts with a number of common gram-positive bacteria. This is due, at least in part, to its sharing the so-called Rantz antigen (an antigen of undetermined chemical composition) with various streptococci, including enterococci, and with *Staphylococcus aureus*. The organism behaves as an intracellular parasite and has been used in experimental models, particularly by Mackaness, to study the relation of cellular immunity to delayed hypersensitivity (Ch. 22, Acquired antibacterial cellular immunity).

Listeriosis is regarded as a zoonosis.† The astonishingly wide host range includes at least 43 mammals, 22 birds, and ticks, fish,

---

* Because of this property, the organism was once thought to be the causative agent of infectious mononucleosis.

† A disease and infection which is naturally transmitted between vertebrate animals and man.

TABLE 41-1. Listeria and Erysipelothrix: Comparison with Some Common Gram-Positive Bacteria

	Shape*	β-Hemo-lysis	Catalase	Motility	Nitrate reduction	Carbohydrate fermentation†			Kerato-conjunc-tivitis
						Glucose	Salicin	Mannitol	
Listeria monocytogenes	R	+	+	+	−	+	+	−	+
Erysipelothrix rhusiopathiae	R	−	−	−	−	+	−	−	−‡
Streptococcus pyogenes	C	+	−	−	−	+	+	−	−
Streptococcus faecalis	C	−/+	−	−/+	−	+	+	+	−
Corynebacterium	R	−/+	+	−	+/−	+/−	−	+/−	−
Lactobacillus	R	−	−	−	−	+	+/−	+/−	−

* R = rod; C = coccus.

† Acid only; no gas. Some *Lactobacillus* species produce gas.

‡ Conjunctivitis but no keratitis.

Based on Buchner, L., and Schneierson, S. S. *Am J Med 45:*904 (1968).

and crustaceans. The disease in animals is characterized by monocytosis, septicemia, and the formation of multiple focal abscesses in the viscera; meningoencephalitis and involvement of the uterus and fetus have also been described in domestic animals. Apparently healthy enteric carriers are common in many species, including man.

Transmission from animals to man definitely occurs, e.g., from handling of newborn calves, contact with infected dogs, and drinking of infected milk. However in the United States most cases occur among urban residents, with few or no known animal contacts; and extensive distribution of *L. monocytogenes* as free-living organisms on plants and in soil has recently been demonstrated. This habitat likely plays a major role in transmission of the disease.

In man the disease takes various forms (Table 41-2), the most common being a **meningoencephalitis** or **meningitis** which closely resembles other forms of purulent meningitis. In such cases the organism can usually be cultured from the spinal fluid and often from blood and the throat. The mortality rate among untreated patients is approximately 70%. While in adults listeric meningitis, septicemia, or other infections are often superimposed on an underlying primary disorder (neoplastic disease, collagen disease, diabetes, organ transplant), it occurs at least as frequently in the previously normal host.

Listeriosis is often associated with corticosteroid and radiation therapy, which suggests activation of a latent infection.

A less common but even more serious manifestation is **perinatal listeric septicemia** (previously known as **granulomatosis infantiseptica),** resulting from a low-grade uterine infection of the mother, often associated with bacteremia but with few or no symptoms. However, the disease in the newborn has a high mortality rate. It is characterized by numerous foci of necrosis throughout the body, especially in the liver and meninges; the meconium usually contains large numbers of

TABLE 41-2. Clinical Manifestations of Human Listeriosis

Meningoencephalitis, most common in neonates and those over 40 years
Low-grade septicemia in gravida (flu-like), premature or nonviable termination of pregnancy
Septicemia in perinatal period
Infectious mononucleosis-like syndrome
Septicemia in adults
Pneumonia
Endocarditis
Aortic aneurysm (bacterial)
Localized abscesses, external or internal
Papular or pustular cutaneous lesions
Conjunctivitis
Urethritis
Habitual abortion, likely but needs further study

Based on Gray, M. L., and Killinger, A. H. *Bacteriol Rev 30:*309 (1969).

L. monocytogenes. Intrauterine infection can also lead to **abortion,** to **stillbirth,** or to delivery of a healthy child who develops **meningitis** within a few days or weeks after birth.

The organism has also been associated with a large variety of other acute or chronic disorders. Prompt therapy with ampicillin, penicillin, or a tetracycline appreciably lowers the mortality.

ERYSIPELOTHRIX RHUSIOPATHIAE

E. rhusiopathiae (*insidiosa*) is a gram-positive, nonsporulating, microaerophilic bacillus very similar to *L. monocytogenes,* except that it is nonmotile, does not produce catalase, and is usually nonhemolytic or α-hemolytic (Table 41-1). It causes swine erysipelas, a serious economic disease, and in man it produces a disease known as **erysipeloid.** The organism is widespread in nature, occurring in decomposing organic matter and as a parasite in fish, crustacea, rodents, and a number of domestic animals (swine, sheep, turkey, duck). It enters the human skin through minor abrasions, following contact with fish, shellfish, meat, or poultry. Erysipeloid is an occupational hazard of fish and meat handlers, veterinarians, and housewives; epidemics have been reported among crab handlers. The localized cutaneous form of the disease is characterized by a spreading, painful, erythematous skin eruption usually on the fingers and hands; the lesions exhibit sharp margins, extend peripherally, and clear centrally. Less commonly the disease appears in a severe generalized cutaneous form or in a septicemic form (often followed by endocarditis). Penicillin treatment is relatively effective.

STREPTOBACILLUS MONILIFORMIS

S. moniliformis is an aerobic, gram-negative, pleomorphic bacillus, named because of its tendency, in older cultures and on solid media, to form filaments and chains of bacilli with prominent yeast-like swellings (Fig. 41-1). It grows only in media enriched with blood, serum, or ascitic fluid. In young broth cultures and in infected tissues the organisms grow as typical bacilli, and on solid media the

L_1 variant (Ch. 40, L forms) is often demonstrable (Fig. 41-1). Under the crowded conditions in surface colonies the organism has a tendency to form defective cell walls, thus leading to the growth of L forms. All strains appear to be antigenically the same.

S. moniliformis is a normal inhabitant of the upper respiratory tract of wild and laboratory rats (as well as squirrels and weasels). It is the causative agent of one form of human **rat-bite fever,** the other being caused by *Spirillum minus* (Ch. 37, Rat-bite fever). Clinically the two diseases are quite similar, except that arthritis is more common in the *Streptobacillus* variety.* Endocarditis, either acute or subacute, may also occur. The organism can usually be recovered, after 3 to 8 days' incubation, from blood cultures, where it grows on the surface of the sedimented blood cells in the form of "fluff balls." In untreated *Streptobacillus* rat-bite fever the mortality rate is approximately 10%. Penicillin is the agent of choice; streptomycin is also effective.

Streptobacillus disease may also be acquired through chance infection of skin abrasions or by ingestion of contaminated food. When unassociated with a rat bite, the disease is often referred to as **Haverhill fever,** after a milk-borne epidemic that occurred in Haverhill, Massachusetts, in 1929.

CALYMMATOBACTERIUM (DONOVANIA) GRANULOMATIS

Granuloma inguinale is an indolent, granulomatous, ulcerative disease caused by a short, plump, gram-negative bacillus which is antigenically similar to, but not identical with, *Klebsiella pneumoniae* and *K. rhinoscleromatis* (Ch. 29). It is thought to be transmitted during coitus but is not highly communicable. The initial lesion commonly appears on or about the genital organs, beginning as a painless nodule. It soon breaks down, forming a sharply demarcated ulcer, which spreads by direct extension and often destroys large areas of skin in the groins and about the anus and genitalia. Histological examination of the fri-

---

* See experimental model in chick embryo (Ch. 22, Adaptation to microenvironment).

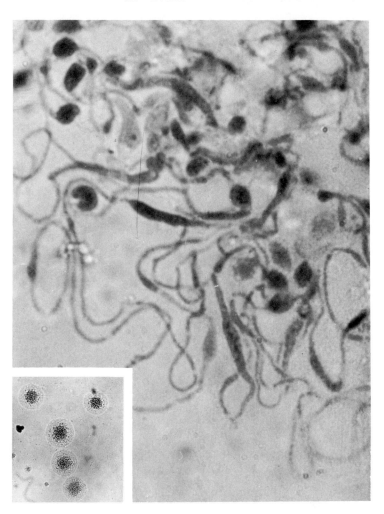

Fig. 41-1. Many irregular filaments of *Streptobacillus moniliformis* with yeast-like swellings growing on solid medium. ×ca. 2520. Inset shows unstained colonies of L₁ variant derived from bacillary form. ×80. (Courtesy of L. Dienes.)

able granulomatous tissue, which forms the base of the coalescing ulcerative lesions, reveals a heavy cellular infiltration with both polymorphonuclear leukocytes and monocytes. In Wright-stained smears of scrapings from the lesions, the encapsulated causative organism *C. granulomatis* can often be seen within the pathognomonic large mononuclear cells (Fig. 41-2).

The coccobacillary microbes, discovered by Donovan in 1905, are referred to as **Donovan bodies;** they were first cultivated in the yolk sacs of chick embryos. Although they cannot be cultured on artificial media when first isolated, they can eventually be adapted to media containing egg yolk and even to beef heart infusion agar.

Antigenic extracts of *C. granulomatis* give positive skin and complement-fixation tests in patients with granuloma inguinale, but these procedures are not routinely used. Diagnosis rests with demonstration of Donovan bodies in stained spreads of granulation tissue or finding of the pathognomonic cell in biopsy material stained with hematoxylin and eosin.

Granuloma inguinale can be treated successfully with tetracyclines, chloramphenicol, or streptomycin; penicillin is not effective.

BARTONELLA BACILLIFORMIS

Inhabitants of the Andes mountains in Peru, Colombia, and Ecuador have long been known to suffer from a severe, febrile, hemolytic anemia **(Oroya fever)** and a relatively

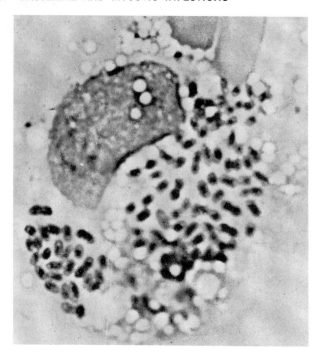

Fig. 41-2. Donovan bodies within large macrophage in film stained with pinacyanole technic. ×ca. 8000. [From Greenblatt, R. B., Dienst, R. B. and West, R. M. *Am J Syph* 35:292 (1951).]

benign nodular eruption **(verruga peruana),** which are two different clinical manifestations of the same, geographically most restricted bacterial disease.* The collective designation of the two syndromes is **Carrion's disease.** The causative agent is *B. bacilliformis,* a small, motile, gram-negative coccobacillus, with flagella at one end. The organism is readily cultured in semisolid nutrient agar containing 10% fresh rabbit serum and 0.5% hemoglobin. It grows best at 28°, the growth first appearing just below the surface of the medium after about 10 days. There is apparently only a single antigenic type.

Bartonellosis is transmitted from man to man by sandflies (*Phlebotomus*) indigenous to the region where the disease is endemic. No other reservoirs of the organism have been found in nature. After an incubation period of 2 to 3 weeks the patient develops intermittent fever and severe constitutional symptoms (including myalgia, nausea, vomiting, diarrhea, and headache), followed by increasingly severe signs and symptoms of **anemia.**

Wright- or Giemsa-stained blood films usually reveal many organisms, either in or on the erythrocytes, and blood cultures are positive. This initial stage of infection is frequently complicated by *Salmonella* superinfection. Among untreated patients the mortality rate averages 40%, salmonellae being responsible for many of the deaths; in those who recover convalescence is slow and blood cultures may remain positive for many months.

The severe anemic stage of the disease may be followed in 2 to 8 weeks by a **cutaneous** stage, characterized by multiple eruptions of hemangioma-like nodules which often ulcerate before healing. This form of the disease also occasionally develops without any obvious signs of a preceding anemia; it causes virtually no systemic manifestations and is not fatal. In this stage the organism is not seen in the blood, endothelial proliferation with new vessel formation being the basic pathological process.

Chemotherapy is effective, even in the severe anemic phase, with penicillin, streptomycin, the tetracyclines, or chloramphenicol. The last should be used when the patient is also suffering from secondary *Salmonella* infection. Control measures, including the use of DDT, are directed against the sandfly.

---

* However, three cases of anemia with *Bartonella*-like bodies were reported recently from Thailand.

SELECTED REFERENCES

BUSCH, L. A. Human listeriosis in the United States. *J Infect Dis 123:*328 (1971).

GRAY, M. L., and KILLINGER, A. H. *Listeria monocytogenes* and listeria infections. *Bacteriol Rev 30:*309 (1966).

GRIECO, M. H., and SHELDON, C. *Erysipelothrix rhusiopathiae. Ann NY Acad Sci 174* (Art. 2):523 (1970).

LAVETTER, A., LEEDOM, J. M., MATHIES, A. W. JR., IVLER, D. and WEHRLE, P. F. Meningitis due to *Listeria monocytogenes:* A review of 25 cases. *N Engl J Med 285:*598 (1971).

MEDOFF, G., KUNZ, L. J., and WEINBERG, A. N. Listeriosis in humans: An evaluation. *J Infect Dis 123:*247 (1971).

PETERS, D., and WIGAND, R. Bartonellaceae. *Bacteriol Rev 19:*150 (1955).

RAJAM, R. V., and RANGIAB, P. N. *Donovanosis (Granuloma Inguinale, Granuloma Venereum).* World Health Organization Monograph Series No. 24. World Health Organization, Geneva, 1954.

ROUGHGARDEN, J. W. Antimicrobial therapy of ratbite fever: A review. *Arch Intern Med 116:*39 (1965).

SCHULTZ, M. G. A history of bartonellosis (Carrión's disease). *Am J Trop Med Hyg 17:*503 (1968).

SONNENWIRTH, A. C. In *Gradwohl's Clinical Laboratory Methods and Diagnosis,* 7th ed., pp. 1202–1209, 1287–1289, 1307–1308. (S. Frankel, S. Reitman, and A. C. Sonnenwirth, eds.) Mosby, St. Louis, 1970.

WEINMAN, D. The Bartonella group. In *Bacterial and Mycotic Infections of Man,* 4th ed., p. 775. (R. J. Dubos and J. G. Hirsch, eds.) Lippincott, Philadelphia, 1965.

chapter **42**

BACTERIA INDIGENOUS TO MAN

Revised by ALEX C. SONNENWIRTH, Ph.D., with
section on oral microbiology by RONALD GIBBONS, Ph.D.,
and SIGMUND SOCRANSKY, D.D.S.

Actions Beneficial to the Host   954
Actions Detrimental to the Host   955
Distribution   955
Endogenous Bacterial Diseases   957
MICROBIOLOGY OF THE ORAL CAVITY   957
Ecological Considerations; Microbial Adhesion   957
Effects of the Oral Flora   959
Nature and Development of Dental Plaque   959
Dental Caries   960
Periodontal Disease   961
Control of Bacterial Plaques   962

Born into an environment laden with microbes, the body of man becomes infected from the moment of birth. Throughout life the skin and mucous membranes, exposed to the outside world, harbor a variety of bacterial* species (called **indigenous** or **autochthonous**), which establish a more or less permanent residence on, and even in, the superficial tissues. Here they usually cause no overt disturbance in health, but are actually **symbiotic** and benefit the host in a number of ways. Some of their actions, however, may not be beneficial, and when the general resistance of the host is sufficiently depressed they may even cause disease.

ACTIONS BENEFICIAL TO THE HOST

The indigenous bacteria play a major role in preventing certain bacterial diseases through effects collectively referred to as **bacterial antagonism** (Ch. 22, Microbial factors). The importance of bacterial antagonism is illustrated by the serious infections that often result from major alterations in the bacterial flora of the gastrointestinal tract. When broad-spectrum antibiotics, such as the tetracyclines, are given in large doses for many days, growth of most of the bacteria that thrive in the normal intestinal tract is suppressed. As a result, antibiotic-resistant strains of potentially pathogenic staphylococci, normally held in check by the antagonistic action of the coliform and other organisms, multiply freely in the bowel and occasionally give rise to a serious, and often fatal, disease—**acute staphylococcal enteritis.** Similarly, the susceptibility of laboratory animals to experimental salmonella and shigella (Ch. 29, Pathogenicity) infections can be

strikingly increased by pretreatment with antimicrobial drugs which modify the gastrointestinal flora. Mice given a single oral dose of streptomycin (50 mg), for example, are about 100,000 times as susceptible to orally induced infection with *Salmonella enteritidis* as untreated mice.

A number of mechanisms are doubtless involved in bacterial antagonism. At least two have been reasonably well defined: 1) competition for nutrients in the environment, and 2) production of inhibitory substances ranging from simple organic acids to the highly specific bactericidins (Ch. 29, Colicins).

The indigenous flora also influences antibody formation, as illustrated by the gross underdevelopment of antibody-synthesizing lymphoid tissues and the low levels of serum γ-globulins in animals reared in a germ-free environment. Such animals are unusually susceptible to primary challenge with pathogenic microbes, even at sites normally free of bacteria (e.g., peritoneal cavity) where bacterial antagonism does not operate. Their lowered resistance is thought to be due in part to the antibody defect† and in part to lack of the poorly understood form of nonspecific immunity that is conferred by endotoxin (Ch. 22, Endotoxins). The presence of undifferentiated cells in their intestinal tracts

---

* Fungi and viruses are discussed in later chapters.

† In the normal host, which has had previous experience with the same (or related) antigens, the antibody response should be of the prompt and intense **secondary** variety (Ch. 17, Secondary response); in the immunologically inexperienced germ-free animal, on the other hand, it should be of the slower and feebler **primary** type (Ch. 17, Primary response). Thus, even if the poorly developed antibody-forming tissues of the germ-free animals are capable of responding, the degree of the response would be expected to be subnormal.

indicates that bacteria in the bowel influence the local process of cellular turnover and maturation.

In addition the resident bacteria synthesize nutrients essential to the welfare of the host. A striking example is the synthesis of vitamin K by the coliform organisms of the gastrointestinal tract. This vitamin, required for the formation of prothrombin in the liver, cannot be synthesized by mammalian cells and therefore must be supplied from an exogenous source, i.e., either by intestinal bacteria in the normal host or by dietary supplement in the germ-free host. Other vitamins including biotin, riboflavin, pantothenate, and pyridoxin are supplied in part by intestinal bacteria.

ACTIONS DETRIMENTAL TO THE HOST

The indigenous bacterial flora, however, does not always benefit the host. **Bacterial synergism** may enable pathogenic species to thrive in tissues which otherwise would not support their growth. For example, the cross-feeding of certain bacteria by specific metabolites excreted by others has been observed in vitro and may well also occur in the body. In another kind of synergism a species that does not alone produce urethritis (*Staphylococcus aureus*) excretes penicillinase, which protects a pathogenic species (*Neisseria gonorrhoeae*) against the action of penicillin.

Endotoxins derived from bacteria normally present in the intestinal tract are thought to cause low-grade toxemia that may be harmful to the host.* Animals of some species reared on diets that suppress endotoxin-producing organisms in the bowel outgrow animals on standard diets. Antibiotics that have a similar effect on the intestinal flora likewise stimulate body growth. Furthermore, it has been suggested that host susceptibility to the harmful effects of endotoxins may be due in part to hypersensitivity to endotoxin antigens derived from the intestinal flora. In keeping with this hypothesis is the observation that germ-free animals, and specially reared pathogen-free animals which have no *Escherichia coli*, *Pseudomonas* sp., or *Proteus* sp. in their gastrointestinal tracts, are extraordinarily resistant to injections of endotoxins.

Under special circumstances the bacteria of the resident flora may themselves be responsible for disease. For example, *Streptococcus mitis* and *S. salivarius,* the most common aerobic species inhabiting the oropharynx, often gain entrance to the blood stream following dental extraction, tonsillectomy, or even vigorous mastication.† If the endocardium of the host has been previously injured by rheumatic fever, the usually harmless streptococci may infect the sites of damage (Ch. 26, α-Hemolytic streptococci) and give rise to **subacute bacterial endocarditis.** Similarly, bacteroides (Ch. 29), which are preponderant in the large intestine, may contribute to serious infections and even produce bacteremia if introduced into the peritoneal cavity or into an open wound. *E. coli* and *Proteus* sp., which are also non-pathogenic under ordinary circumstances, may infect the obstructed urinary tract or open burns of the skin. The role of bacteria in the pathogenesis of dental caries and periodontal disease will be discussed below.

DISTRIBUTION

The various species of bacteria that regularly inhabit the human skin, conjunctivae, nose, pharynx, mouth, lower intestine, external genitalia, and vagina are listed in Table 42-1. Other species, many of them potentially pathogenic, may also be present temporarily as transient visitors (cf. meningococcal carriers, Ch. 28, Prevention).

Although the **skin** is considered as a single site, its bacterial flora varies in different anatomical regions. The flora of the facial area, for example, reflects that of the oropharynx, whereas the perineal flora is influenced by the bacteria of the lower intestinal tract. Although vigorous scrubbing of the skin with soap and water or other disinfectants, as in preparation for surgical operations, will rid it temporarily of most of its surface bacteria, organisms sequestered in hair follicles and sweat glands will soon reestablish the surface infection.

---

* Endotoxemia becomes more severe in the late stages of hemorrhagic shock and contributes to the shock's becoming irreversible.

† The bacteremia is transient and is only demonstrable by serial blood cultures taken at frequent intervals (minutes) following the inciting trauma.

TABLE 42-1.   Bacteria Most Commonly Found on Surfaces of the Human Body

Bacteria	Skin	Conjunctiva	Nose	Pharynx	Mouth	Lower intestine	External genitalia	Vagina
Staphylococci	+	±	+	±	±	±	±	+
Pneumococci		±	±	+	+			
Streptococci								
Viridans	±	±	+	+	+	+	+	+
β-hemolytic				±	±			
faecalis	±			±	+	+	+	+
anaerobic				+	+	+	+	±
Neisseriae		±		+	+	±	+	±
Veillonellae				±	+	±	+	
Lactobacilli				+	+			+
Corynebacteria	+	+	+	+	+	+		+
Clostridia					±	+	±	
Hemophilic bacilli		+	+	+	+			
Enteric bacilli				±	+	+	+	
Bacteroides				+	+	+	+	
Mycobacteria	+		±	±		+	+	
Actinomycetes				+				
Spirochetes				+	+	+	+	
Mycoplasmas				+	+	+	±*	+

+, Common; ±, rare.
* Prevalence related to sexual activity.
Modified from Rosebury, T. *Microorganisms Indigenous to Man.* McGraw-Hill, New York, 1962.

The microbiology of the **oral cavity** will be discussed below. In the **pharynx** viridans streptococci are predominant, while the **nose** is colonized chiefly by staphylococci and corynebacteria ("diphtheroids"). The trachea may contain a few bacteria, but the normal lower respiratory tract is virtually sterile, owing primarily to the efficient escalating action of the mucociliary blanket which lines the bronchi (Ch. 22, Mechanical factors).

In the upper gastrointestinal tract the esophagus contains only the bacteria swallowed with the saliva and food. Because of the high acidity of the gastric juice, few organisms can be cultured from the normal stomach, except immediately after meals or in the presence of gastric disease (e.g., carcinoma). Below the duodenum, the number of resident bacteria increases progressively, reaching a maximum in the colon (ca. 10^{11}/gm of feces); the number found in the small intestine, particularly in the duodenum, jejunum, and upper ileum, is relatively small (10^5 to 10^8/gm). At birth, the entire intestinal tract is sterile, but it soon becomes inhabited by bacteria ingested with food, and the resulting bacterial population tends to be dominated by anaerobic lactobacilli, espe-

cially in breast-fed infants. As the diet is liberalized, coliform bacilli become more prominent and are eventually joined by bacteroides, fusiform bacilli, clostridia, enterococci, and other gram-negative enteric bacilli. In the adult colon bacteroides are usually predominant.*

The **vagina** also becomes infected soon after birth, originally with lactobacilli and later with a variety of cocci and bacilli. During the childbearing years, lactobacilli again predominate and contribute to the maintenance of acidity in the vaginal secretions. After the menopause the lactobacilli become less nu-

---

* Systematic studies of the intestinal flora of mice have revealed similar relations. Lactobacilli and anaerobic streptococci (group N) become established immediately after birth throughout the intestinal tract, whereas bacteroides gain a foothold only after the sixteenth day and are limited to the large bowel. Enterococci and coliform bacilli, after a burst of growth between the tenth and fourteenth days, stabilize at much lower levels (10^3/gm of intestinal contents as compared with 10^9) in the colon. When germ-free mice are infected with each of these bacterial types, either simultaneously or in succession, the same distribution of organisms as is present in normal adult mice promptly develops.

merous and the bacterial flora once more becomes mixed.

The normal **urethra** contains only rare contaminants from the external mucous membranes of the genital organs.

A variety of bacteria may be cultivated from the **conjunctivae,** including diphtheroids (*Corynebacterium xerosis*), neisseriae, *Staphylococcus epidermidis,* and nonhemolytic streptococci. The number of organisms present, however, is usually small.

ENDOGENOUS BACTERIAL DISEASES

A general knowledge of the normal bacterial flora of the human body is essential, not only for interpreting the results of bacteriological findings, but also because of the increasing incidence of endogenous bacterial diseases. Involving bacteria that are indigenous* yet potentially pathogenic, these endogenous diseases are "caused" by factors that lower the resistance of the host, either generally or at specific tissue sites. Such factors include radiation damage, prolonged use of corticosteroid hormones, severe malnutrition, shock, debilitation from other diseases (particularly those involving the bone marrow and/or lymphoid tissues, e.g., leukemia), superinfection and disturbances in bacterial antagonism resulting from antimicrobial therapy, localized obstruction of excretory organs, and predisposing lesions of unrelated etiology (Ch. 22, Depression; Ch. 29). Unlike exogenous bacterial disease, those of endogenous origin have no definable incubation period and cannot be considered communicable in the usual sense.

As the lives of more patients with major illnesses are prolonged by improved methods of treatment, and as more infections caused by virulent exogenous bacteria are controlled by effective antimicrobial drugs, endogenous bacterial diseases have become more common. They now constitute a major proportion of the serious bacterial diseases encountered in clinical practice.

---

* Dormant or latent infections (Ch. 22, Inapparent infections) with exogenous pathogens (e.g., *Mycobacterium tuberculosis*) may also give rise to disease that is in a sense endogenous, but is usually not classified as such.

MICROBIOLOGY OF THE ORAL CAVITY

Recent research, particularly with germ-free animals, has shown that several of the most widespread dental diseases are caused by bacteria. The agents are ubiquitous commensal organisms, whose adaptation to the oral cavity often involves special adhesiveness.

ECOLOGICAL CONSIDERATIONS; MICROBIAL ADHESION

The oral cavity presents a number of surfaces and crypts for the colonization of microorganisms. Since nutrients are scarce, except during meals, microbial growth is generally slow. However, large, dense populations can accumulate on tooth surfaces (dental plaques), on the dorsum of the tongue, and in the gingival crevice area and carious lesions. Saliva contains as much as 10^8 bacteria per milliliter, washed off oral surfaces (primarily the tongue).

The flora varies from site to site within the oral cavity (Table 42-2). Retention in the mouth is achieved either by direct adhesion or by mechanical entrapment in protected crevices. Bacterial adhesion is selective, and it correlates with the indigenous distribution. For example, in the viridans group *Streptococcus salivarius* resides primarily on the tongue, and when exposed experimentally it adheres well to epithelial cell surfaces but not to teeth or developing dental plaques. In contrast, *S. mutans* and *S. sanguis,* which are found mainly on teeth, adhere well to various hard surfaces but not to epithelial surfaces.

Bacterial adhesion would also seem to be important in the colonization of mucosal surfaces elsewhere in the body by such pathogens as the β-hemolytic streptococci and enterobacteria. The M protein surface antigen contributes to the attachment of group A streptococci to epithelial surfaces.

Although adhesion is decisive for colonization of freely exposed oral surfaces, organisms with feeble adherent capacity can proliferate in protective niches of the mouth, including the occlusal fissures of teeth. Thus spirochetes (e.g., *Treponema denticola*) and gram-negative anaerobic rods (e.g., *Bacteroides melaninogenicus*) localize in the

TABLE 42-2.   Composition and Distribution of Oral Flora in the Adult

	Gingival crevice (%)*	Dental plaque (%)	Tongue (%)	Saliva (%)
Gram-positive facultative cocci	29	28	45	46
Streptococcus spp.	27	28	38	41
S. mutans	L†	L-H	L	L
S. sanguis	M	H	M	M
S. salivarius	L	L	H	H
S. mitior	M	H	M	M
Enterococci	M	L	L	L
Staphylococcus salivarius	L	L	H	H
Related nonpathogenic staphylococci	L	L	H	H
Gram-positive anaerobic cocci				
Peptostreptococcus, Peptococcus	7	13	4	13
Gram-negative facultative cocci				
Neisseria spp.	<1	<0	3	1
Gram-negative anaerobic cocci				
Veillonella spp.	11	6	16	16
Gram-positive facultative rods and filaments				
Nocardia, Rothia, Corynebacterium, Bacterionema, Lactobacillus	15	24	13	12
Gram-positive anaerobic rods and filaments				
Actinomyces, Propionibacterium, Leptotrichia	20	18	8	5
Gram-negative facultative rods	1	<1	3	2
Gram-negative anaerobic rods	16	10	8	5
Fusobacterium	2	4	<1	<1
Bacteroides oralis	6	5	5	2
Bacteroides melaninogenicus	5	<1	<1	<1
Vibrio sputorum	4	1	2	2
Spirochetes				
Treponema macrodentium, T. denticola, T. oralia	1	<1	<1	<1

* Approximate percentage of total cultivable flora present in each area.

† Approximate proportions found: L = low, M = moderate, H = high.

gingival crevice area or in periodontal pockets. In these sites nutrition may be a dominant factor. For example, the serum transudate from the gingival crevice provides the serum proteins and hemin required for the growth of the two organisms just named. Diet can also be important: the kind and quantity of dietary carbohydrate have been shown to influence the populations of oral streptococci and lactobacilli.

Acquisition and Nature of the Oral Microflora. At birth the oral cavity is sterile, but microorganisms are immediately introduced through food and other contacts. Initially a large variety of organisms is found, but within days to weeks the favored ones become established. Most of these probably originate from persons since they are not found free-living in nature. One of the earliest settlers is *S. salivarius; S. mutans* and *S. sanguis* do not appear until the teeth erupt. Spirochetes and bacteroides do not regularly colonize the mouth until the onset of puberty. The oral microflora of the adult is quite complex. Gram-positive streptococcal species and filamentous types predominate in most sites (Table 42-2). The filamentous bacteria and the anaerobic gram-positive cocci have been difficult to classify.

The relation between the oral flora and the host is amphibiotic, i.e., under some circumstances the relation approaches symbiosis, and under others the same organisms may cause disease. Many of the organisms closely resemble overt pathogens. For example, the oral α-hemolytic streptococci, staphylococci, and spirochetes have metabolic, antigenic, and morphological relations to well-known pathogens. Perhaps members of our indigenous flora were once pathogens but developed a more stable relation through evolution of both host and parasite.

EFFECTS OF THE ORAL FLORA

The oral flora exerts both beneficial and harmful effects.

Beneficial Effects. Man's indigenous flora contributes to his **nutrition** through the synthesis of vitamins and the digestion of certain foods. The largest populations of bacteria are found in the lower bowel, but absorption of metabolites there is relatively inefficient, and man, unlike rodents, is not coprophagic. Hence vitamin synthesis by the flora in higher portions of the alimentary canal may be more significant. From 1.0 to 2.5 gm of bacteria from the mouth is swallowed daily.

The oral flora may also contribute to **immunity** to various infectious agents, since small numbers of oral organisms entering the blood stream provide a continual antigenic stimulus. The resulting low levels of circulating antibodies may cross-react with certain overtly pathogenic bacteria.

The oral flora may also compete with pathogens. For example, the yeast *Candida albicans* (Ch. 43), a frequent inhabitant of the mouth, is restricted in numbers by indigenous bacteria but often grows out and causes disease when these are suppressed by antibiotics.

Diseases Initiated by the Oral Flora. Oral bacteria also possess a degree of pathogenicity. Thus when plaque or saliva is injected subcutaneously into experimental animals purulent, transmissible abscesses are formed. Oral organisms gaining entrance to tissues in man by various routes may cause not only alveolar abscesses but also more distant abscesses of the lung, brain, and extremities, and surgical wound infection. These infections are usually mixed, with *B. melaninogenicus* playing a dominant role. Other, unmixed infections initiated by oral organisms include candidiasis, actinomycosis, and subacute bacterial endocarditis.

When organisms accumulate sufficiently in the mouth as dental plaque, they may initiate diseases of the teeth and the supporting structures.

NATURE AND DEVELOPMENT OF DENTAL PLAQUE

Dental plaque is a mixed colony of bacteria embedded in an intercellular matrix adhering to the tooth surface (Fig. 42-1). Bacterial cells make up 60 to 70% of the volume, and

Fig. 42-1. Accumulation of bacterial dental plaque 3 days after cleaning. Stained with beta rose.

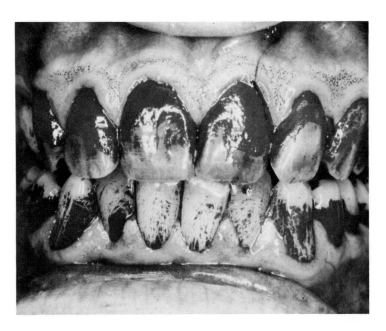

a fully developed plaque may be 300 to 500 cells thick. The matrix is composed of bacterial and salivary polymers and remnants of epithelial cells and leukocytes. The major constituent of enamel, a complex calcium-phosphate salt termed hydroxyapatite, readily adsorbs proteins and other materials; hence a cleaned tooth rapidly adsorbs specific salivary proteins on its surface, which form a thin surface film termed the **enamel pellicle.** Specific surface components of bacteria interact reversibly with glycoproteins of the pellicle. This attachment is made irreversible by excreted polysaccharides or proteins, or salivary glycoproteins, serving as anchoring polymers for various species.

The plaque changes in composition as it grows. Anaerobic bacteria find a suitable environment when sufficient numbers of facultative organisms have accumulated. Moreover, many oral organisms require growth factors produced by neighboring bacteria. For example, *B. melaninogenicus* requires a vitamin K derivative, which is not present in cannulated saliva but is synthesized by neighboring organisms. Similarly, *Treponema macrodentium* requires isobutyric acid and polyamines, which are elaborated by certain plaque bacteria.

The plaque also changes in metabolism, like a colony on agar, as it grows. Organisms in the depths are poorly nourished and are exposed to high concentrations of microbial endproducts; hence they develop thickened cell walls, glycogen accumulation, and mor-phological distortion. Anaerobiosis also develops in the depths and promotes fermentative production of acid.

DENTAL CARIES

The microbial plaque represents a large mass of metabolically active bacteria, which may initiate dental caries (cavities), periodontal disease, or both. Caries—the destruction of enamel, dentin, or cementum—is clearly caused by bacteria. Thus gnotobiotic rats (i.e., raised free of bacteria) fail to develop dental caries, even when the animals are genetically susceptible and are fed cariogenic diets. However, infection with specific types of bacteria can lead to extensive decay. Moreover, experimental caries can be prevented by penicillin. Finally, caries can be transmitted between experimental animals. For example, caging caries-inactive and caries-active hamsters together will spread the infection; and carious lesions can also be initiated in caries-free animals by intraoral inoculation with plaque or with certain pure cultures from carious lesions.

The mechanism of tooth decay appears to be primarily direct demineralization, due to organic acid produced by the overlying bacterial plaque; bacteria also contribute proteases. Only a few bacterial species, belonging mostly to the genera *Streptococcus* and *Lactobacillus,* have significant cariogenic potential. These organisms ferment a variety of carbohydrates, producing and tolerating a

Fig. 42-2. Dental caries in a gnotobiotic rat monoinfected with *S. mutans* for 90 days.

low pH; they can also accumulate on teeth.

Streptococcus mutans (Fig. 42-2), which is highly cariogenic, has a special mechanism of accumulation that may explain the well-known **cariogenicity** of **sucrose** in human diets. When given sucrose this organism synthesizes extracellular polysaccharides similar to dextran, and these are involved in the adhesion of the organism to teeth and other hard surfaces (Fig. 42-3). Substitution of glucose for sucrose in the diet of animals monoinfected with *S. mutans,* or incorporation of dextranase in the diet, prevents plaque formation and dental decay.

The **occlusal pits** and fissures of teeth provide areas where food can become impacted and bacteria can proliferate without requiring marked adhesive abilities. Hence the production of dental caries exhibits less bacterial specificity in these sites than on smooth surfaces.

A third type of decay affects the **root surfaces** of teeth (Fig. 42-4). These are not covered by dental enamel and hence are more susceptible to destruction, under conditions that lead to the development of large subgingival plaques. Filamentous organisms of the *Actinomyces, Nocardia,* and *Rothia* types, in addition to streptococci, have been shown to produce these lesions in experimental animals. Moreover, when the integrity of the enamel on exposed tooth surfaces has been destroyed by acidogenic organisms, these filamentous bacteria may also be significant secondary invaders in carious lesions penetrating into the dentin.

Infected teeth may be a major source of subacute bacterial endocarditis. This disease is often due to *S. mutans* or *S. sanguis,* which are found in much higher proportions on tooth surfaces and in carious lesions than on oral epithelial surfaces and in the intestinal canal.

PERIODONTAL DISEASE

In periodontal disease, the main cause of tooth loss at later ages, the supporting structures of the teeth (i.e., gingiva, periodontal membrane, cementum, and supporting alveolar bone) are infected by organisms residing in the gingival crevice area. In clinical studies a strong correlation has been found between the mass of microorganisms present in this area and the severity of gingivitis and alveolar bone loss. The bacterial etiology of this disease has been established by its transmission in certain strains of experimental animals by a variety of filamentous gram-positive organisms, including *Actinomyces, Rothia, Nocardia,* and *Corynebacterium,* as well as by certain streptococcal species (Fig. 42-4). These organisms all share the ability to form large quantities of subgingival plaque.

Fig. 42-3. Gelatinous plaque-like deposits formed by *S. mutans* when growing on sucrose but not on glucose.

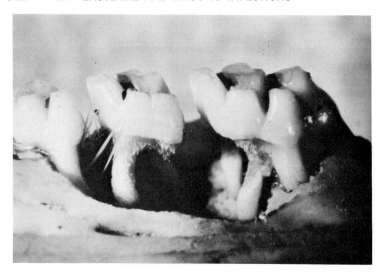

Fig. 42-4. Periodontal destruction and root caries in gnotobiotic rat monoinfected with *Actinomyces naeslundii* for 90 days.

While the mechanisms responsible for the tissue destruction are not clear, several possibilities have been proposed. Bacterial hydrolytic enzymes, endotoxins, and metabolic end-products may lead to the destruction of the intercellular matrix of connective tissue, or may interfere with the anabolism of local tissue cells. Alternatively, an inflammatory response might be initiated directly by bacterial products, or indirectly by an immediate or delayed hypersensitivity response to plaque antigens; the local tissue damage would then be mediated by enzymes present in the host phagocytes or tissue cells. Whatever the mechanisms of tissue destruction, it is clear that removal of microbial deposits in the gingival crevice area prevents or arrests this chronic destructive disease.

CONTROL OF BACTERIAL PLAQUES

Dental caries and periodontal disease could be prevented by control of plaques. Mechanical removal by tooth brushing, dental floss, etc. can be effective, but requires strong motivation. The curtailment of dietary sucrose also reduces decay. However, these measures have thus far proved impractical for the control of dental diseases in the population en masse. Since dental caries and periodontal disease are ubiquitous in civilized populations, and their treatment costs billions of dollars per year and entails considerable discomfort, more effective bacterial control methods are highly desirable. The prophylactic effect of dietary fluoride against caries may involve antibacterial action.

SELECTED REFERENCES

Books and Review Articles

ROSEBURY, T. *Microorganisms Indigenous to Man.* McGraw-Hill, New York, 1962.

Specific Articles

DUBOS, R. T., and SCHAEDLER, R. W. The effect of diet on the fecal bacterial flora of mice and on their resistance to infection. *J Exp Med 115*:1161 (1962).

DUBOS, R., SCHAEDLER, R. W., and COSTELLO, R. Com-positon, alteration and effects of the intestinal flora. *Fed Proc 22*:1322 (1963).

FINEGOLD, S. M. Intestinal bacteria. *Calif Med 110*: 455 (1969).

FINLAND, M. Changing ecology of bacterial infections as related to antibacterial therapy. *J Infect Dis 122*:419 (1970).

MICKELSEN, O. Nutrition: Germfree animal research. *Annu Rev Biochem 31*:515 (1962).

MILLER, C. P., and BOHNHOFF, M. Changes in the mouse's enteric microflora associated with enhanced susceptibility to salmonella infection following treatment with streptomycin. *J Infect Dis 113*:59 (1963).

SCHAEDLER, R. W., DUBOS, R., and COSTELLO, R. The development of the bacterial flora in the gastrointestinal tract of mice. *J Exp Med 122*:59 (1965).

SPRUNT, K., and REDMAN, W. Evidence suggesting importance of role of interbacterial inhibition in maintaining balance of normal flora. *Ann Intern Med 86*:579 (1968).

References on Oral Microbiology

GENCO, R. J., EVANS, R. T., and ELLISON, S. A. Dental research in microbiology with emphasis on periodontal disease. *J Am Dent Assoc 78*:1016–1036 (1969).

GIBBONS, R. J. Significance of the bacterial flora indigenous to man. *Am Inst Oral Biol Symp 26*: 27–36 (1969).

GIBBONS, R. J. Ecology and Cariogenic Potential of Oral Streptococci. In *Streptococci and Streptococcal Diseases.* (L. W. Wannamaker and J. M. Matsen, eds.) Academic Press, New York, 1972.

GOLD, W. Dental caries and periodontal disease considered as infectious diseases. *Adv Appl Microbiol 11*:135–157 (1970).

KEYES, P. H. Research in dental caries. *J Am Dent Assoc 76*:1357–1373 (1968).

SOCRANSKY, S. S. Relationship of bacteria to the etiology of periodontal disease. *J. Dent Res 49*:203–222 (1970).

SOCRANSKY, S. S., and MANGANIELLO, A. D. The oral microbiota of man from birth to senility. *J Periodontol 42*:485–494 (1971).

chapter 43

FUNGI

Revised by GEORGE S. KOBAYASHI, Ph.D.

CHARACTERISTICS OF FUNGI 966
  STRUCTURE AND GROWTH 966
  CELL WALL 967
  METABOLISM 969
  REPRODUCTION 970
    Asexual Reproduction 970
    Sexual Reproduction 972
    Parasexual Cycle (Mitotic Recombination) 974
  TAXONOMY 975
  DIMORPHISM 977
  COMPARISON WITH BACTERIA 977
GENERAL CHARACTERISTICS OF FUNGOUS DISEASES 978
  TYPES OF MYCOSES 978
  OPPORTUNISTIC FUNGI 981
  DIAGNOSIS 981
  CHEMOTHERAPY 983
SOME PATHOGENIC FUNGI AND FUNGOUS DISEASES 984
  SYSTEMIC MYCOSES 984
    Cryptococcus neoformans (Cryptococcosis) 984
    Blastomyces dermatitidis (North American Blastomycosis) 986
    Histoplasma capsulatum (Histoplasmosis) 988
    Coccidioides immitis (Coccidioidomycosis) 991
    Paracoccidioides brasiliensis (South American
      Blastomycosis) 994
  MYCOSES DUE TO OPPORTUNISTIC FUNGI 995
    Candida albicans (Candidiasis) 995
    Aspergillus fumigatus (Aspergillosis) 997
    Phycomycetes (Phycomycoses) 998
  SUBCUTANEOUS MYCOSES 999
    Sporotrichum schenckii (Sporotrichosis) 999
    Chromoblastomycosis 1000
    Maduromycosis 1000
  CUTANEOUS MYCOSES (DERMATOMYCOSES) 1000
    Attack on Keratin 1001
    Epidemiology 1002
    Immunity 1002
    Treatment 1003
    Classification 1003
  SUPERFICIAL MYCOSES 1005

CHARACTERISTICS OF FUNGI

The fungi (L. *fungus* = mushroom) have traditionally been regarded as "plant-like" (see Ch. 1). Most species grow by continuous extension and branching of twig-like structures. In addition, they are mostly immotile and their cell walls resemble those of plants in thickness and, to some extent, in chemical composition and in ultramicroscopic structure.

Fungi grow either as single cells, the **yeasts,** or as multicellular filamentous colonies, the **molds** and **mushrooms.** The multicellular forms have no leaves, stems, or roots, and are thus much less differentiated than higher plants, but they are much more differentiated than bacteria. However, fungi do not possess photosynthetic pigments, and so they are restricted to a saprophytic or parasitic existence. A single uninucleated cell can yield filamentous multinuclear strands, yeast cells, fruiting bodies with diverse spores, and cells that are differentiated sexually (in many species). Moreover, a few species form remarkable traps and snares for capturing various microscopic creatures.

Fungi are abundant in soil, on vegetation, and in bodies of water, where they live largely on decaying leaves or wood. Their ubiquitous airborne spores are frequently troublesome contaminants of cultures of bacteria and mammalian cells. In fact, it was just such a contaminant in a culture of staphylococci that eventually led to the discovery of penicillin (Ch. 7).

Though we shall be concerned mainly with those few fungi that cause diseases of man, other fungi have had an even more adverse effect on human welfare as causes of plant diseases: for example, the potato blight led to death from starvation in Ireland alone of over 1 million persons in the period 1845 to 1860. More recently, the turkey industry of England was threatened by a malady

finally traced to metabolic byproducts of *Aspergillus flavus* which had contaminated the peanut meal used as feed. These **aflatoxins** (named from *A. flavus*) have been isolated and characterized as highly unsaturated molecules with a coumarin nucleus. They are of great interest because of their direct toxicity and long-term carcinogenic effects.

Nevertheless, the over-all effects of fungi on man's condition are probably more benign than malignant. As scavengers in soil they make an essential contribution to the chemical stability of the biosphere (Ch. 1), and their biosynthetic capabilities are used in the industrial production of penicillins, corticosteroids, and numerous organic acids (e.g., citric, oxalic). Moreover, though their role in the production of certain cheeses, bread, and ethanolic beverages, the fungi help provide more than calories to man's food supply.

STRUCTURE AND GROWTH

Molds. The principal element of the growing or vegetative form of a mold is the **hypha** (Gr. *hyphe* = web), a branching tubular structure, about 2 to 10 μ in diameter, i.e., much larger than bacteria. As a colony, or **thallus,** grows, its hyphae form a mass of intertwining strands, called the **mycelium** (Gr. *mykes* = mushroom). Hyphae grow by elongation at their tips (apical growth) and by producing side branches.

Those hyphae that penetrate into the medium, where thy absorb nutrients, are known collectivly as the **vegetative mycelium,** while those that project above the surface of the medium constitue the **aerial mycelium;** since the latter often bear reproductive cells or spores they are also referred to as the **reproductive mycelium.** Most colonies grow at the surface of liquid or solid media as irregular, dry, filamentous mats. Because of the inter-

Fig. 43-1. The coenocytic nature of hyphae. Electron micrograph of a longitudinal section through two cells of *Neurospora crassa* partially separated by a septum (s). Note the streaming of mitochondria (m) through the septal pore (p). Other labeled structures are cell wall (w), outer frayed coat of the cell wall (f), cell membrane (cm), nucleus (N), nucleolus (Nu), nuclear membrane (nm), ribosomal particles (p_i), and endoplasmic reticulum (er). Fixed with O_sO_4 and stained with uranyl nitrate. ×47,000, reduced. [From Shatkin, A. J., and Tatum, E. L. *J Biophys Biochem Cytol* 6:423 (1959).]

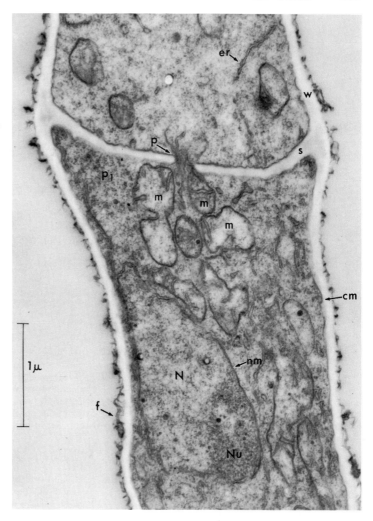

twining of the filamentous hyphae the colonies are much more tenacious than those of bacteria. At the center of mycelial colonies the hyphal cells are often necrotic, owing to deprivation of nutrients and oxygen, and perhaps to accumulation of organic acids.

In most species the hyphae are divided by cross-walls, called **septa** (L. *septum* = hedge, partition; Fig. 43–1). However, the septa have fine, central, pores; hence even septate hyphae are **coenocytic,** i.e., their many nuclei are embedded in a continous mass of cytoplasm.

Yeast. Yeasts are unicellular oval or spherical cells, usually about 3 to 5 μ in diameter. Sometimes yeast cells and their progeny adhere to each other and form chains or "pseudohyphae."

Cytology. Yeasts and molds resemble higher plants and animals in the anatomical complexity of their cells. They are eukaryotic, with several different chromosomes and a well-defined nuclear membrane, and they possess mitochondria and an endoplasmic reticulum (Fig. 43-2). Moreover, their membranes contain sterols, thus resembling higher forms rather than bacteria.

CELL WALL

The cell wall of a fungus, like that of a bacterium, lies immediately external to the limiting cytoplasmic membrane, and in some yeasts it is surrounded by an external capsular polysaccharide (e.g., *Cryptococcus,* p. 984). How-

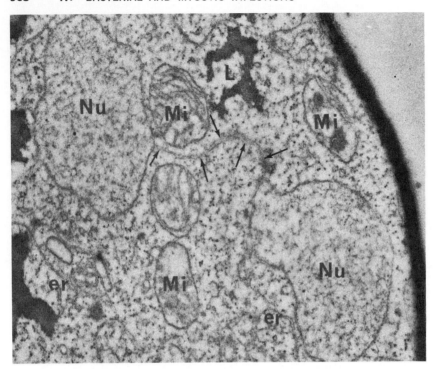

Fig. 43-2. Cytoplasmic organelles in a yeast cell of *Blastomyces dermatitidis,* showing two nuclei (Nu), each surrounded by a nuclear membrane. Also seen are mitochondria (Mi), vesicles and ribosomes of endoplasmic reticulum (er), and lipid droplets (L). Arrows point to membranous connections between nuclei. Osmium-fixed, thin sections. ×35,000, reduced. [From Edwards, G. A., and Edwards, M. E. *Am J Bot 47:*622 (1960).]

ever, unlike bacteria, whose cell walls often contain brick-like structural units (Ch. 2), fungus cell walls appear thatched (Fig. 43-3). Polymers of hexoses and hexosamines provide the main structural wall elements of fungi. In many molds and yeasts the principal structural macromolecule of the wall is **chitin,** which is made up of N-acetyl glucosamine residues. These are linked together by β-1,4-glycosidic bonds, like the glucose residues in cellulose, the main cell wall material in higher plants.

Chitin also makes up the principal structural material of the exoskeleton in crustacea. Hence this substance represents, at a molecular level, an interesting example of convergent evolution, in which distantly related organisms have developed, presumably by an independent evolutionary sequence, similar or identical structures to serve similar needs.

In yeasts extraction of the wall with hot alkali leaves an insoluble **glucan,** made up of β-1,6-linked D-glucose residues, with β-1,3-linked branches arising at frequent intervals. (Glucan resembles cellulose in its insolubility and rigidity.) An additional, soluble polysaccharide is **mannan,** an α-1,6-linked

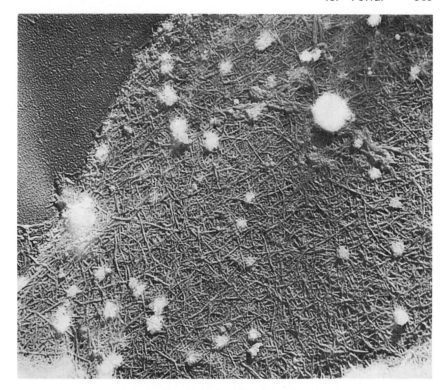

Fig. 43-3. Microfibrillar structure of the wall of a species of phycomycete. Chemical analysis and X-ray diffraction showed the thatched fibrils to be chitin. Electron micrograph shadowed with Pd-Au. ×30,000, reduced. [From Aronson, J. M., and Preston, R. D. *Proc R Soc Lond (Biol) 152*:346 (1969).]

polymer of D-mannose with α-1,2 and α-1,3 branches.

As with bacteria (Ch. 4), the cell wall polysaccharides of fungi are synthesized from various sugar necleotides (UDP-N-acetyl glucosamine, GDP-mannose, etc.). Chitin synthetase is attached to the cell membrane. It requires a primer in vitro, which must contain at least six or seven residues connected as in chitin. A nucleotide antibiotic, polyoxin D, has been found to be a selective inhibitor of chitin synthetase. It inhibits incorporation of ^{14}C-glucosamine into the cell wall of *Neurospora crassa,* which results in an increased accumulation of uridine diphosphate-N-acetyl glucosamine.

The walls of a number of yeasts contain complexes of polysaccharide with proteins rich in cystine residues, and the reversible reduction of -S-S- bonds has been implicated in the formation of buds. In some yeasts lipids, containing phosphorus and nitrogen, are also abundant in the wall (up to 10% of dry weight).

Fungus cell walls can be digested by enzymes contained in the digestive juices of the snail *Helix pomatia*. These juices contain over 30 recognized enzymatic activities, including glucanase, chitinase, and mannanase. Bacteria that produce lytic enzymes for these walls have also been isolated from soil samples, by application of the enrichment technic (Ch. 3, Genotypic adaptation) with purified cell walls as carbon source.

As with bacteria, digestion of the walls of yeast or molds in hypertonic solution yields viable **protoplasts.** Protoplasts are also produced by growth in media that inhibit cell wall synthesis (Fig. 43-4). Yeast protoplasts are deficient in some of the hydrolases of the intact cells, such as invertase and β-fructosidase. These observations suggest that as in bacteria (Ch. 6) secreted enzymes are located in the periplasmic space.

METABOLISM

All fungi are heterotrophic (Ch. 3), requiring organic foodstuffs, and most are obligate

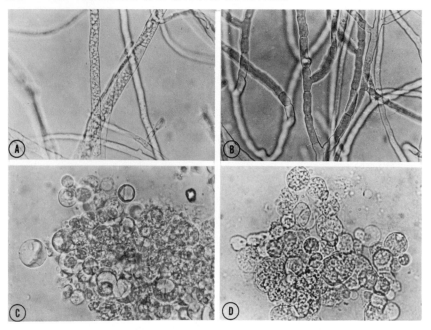

Fig. 43-4. Protoplast formation in hypertonic media by a mutant of *Neurospora crassa* with defective synthesis of cell wall. Cultures were grown in minimal medium with sorbose 0% (**A**), 5% (**B**), or 10% (**C** and **D**). Branching hyphae with septa are evident in **A** and in **B**. Protoplasts with occasional cell wall fragments are present in **C** and **D**. Similar results were obtained with equimolar concentrations of glucose, sucrose, and fructose. [From Hamilton, J. G., and Calvet, J. *J Bacteriol 88:*1084 (1964).]

aerobes. Some, however, are facultative; but none are obligate anaerobes. Except for the absence of autotrophs or obligate anaerobes, the fungi as a group exhibit almost as great a diversity of metabolic capabilities as the bacteria. Many species can grow in minimal media, given an organic carbon source and nitrogen as NH_4^+ or NO_3^-. Thermophilic species can grow at temperatures as high as 50° and above; some can flourish in the high-salt media of cured meats, and others in highly acidic media. Some fungi can hydrolyze such complex organic substances, as wood, bone, tanned leather, chitin, waxes, and even synthetic plastics.

Fungi can be induced by appropriate substrates and analogs to form the corresponding degradative enzymes, and their regulatory mechanisms for controlling enzyme synthesis and activity appear similar to those in bacteria, but differ in some interesting details.

Like bacteria, yeasts show induction, repression, and catabolite repressions; in fact the glucose effect on enzyme induction (Ch. 13) was first observed in

yeast, at the turn of the century. In contrast to bacteria, however, the structural genes for a given metabolic pathway (e.g., histidine synthesis) are scattered over the genome, and operator genes and operons have, accordingly, been difficult to identify. Since fungi are eukaryotic and yet can be handled with almost as much ease as bacteria, they are attractive for extending knowledge of molecular genetics and regulatory mechanisms to higher forms.

REPRODUCTION

In addition to growing by apical extension and branching, fungi reproduce by means of sexual and asexual cycles, and also by a parasexual process. We shall consider asexual reproduction and the parasexual process in particular detail, since the vast majority of fungi that are pathogenic for man lack sexuality.

ASEXUAL REPRODUCTION

The vegetative **growth** of a coenocytic mycelium involves nuclear division without cell division, the classic process of mitosis ensuring transmission of a full complement of chromosomes to each daughter nucleus. The further step of cell division leads to asexual (vegetative) **reproduction,** i.e., the formation of a new clone without involvement of gametes and without nuclear fusion. Three mechanisms are known: 1) sporulation, followed by germination of the spores; 2) budding; and 3) fragmentation of hyphae.

Asexual spores in general are sometimes referred to as **conidia;** more often, however, the term is reserved, as in this chapter, for those asexual spores that form at the tips and the sides of hyphae. Other asexual spores (chlamydospores and arthrospores) develop **within** hyphae. The spores germinate when planted in a congenial medium, i.e., they become enlarged, and, if destined to become a mold, send out one or more germ tubes (Fig. 43-11). The tubes elongate into hyphae and give rise to a new colony.

Chlamydospores, which can be formed by many fungi, are thick-walled and unusually resistant to heat and drying; they are probably formed like bacterial spores, by true endosporulation (Ch. 6), and they similarly promote survival in unfavorable environments. In contrast, **arthrospores,** which form by fragmentation of hyphae, and **conidia,** which form by a process akin to budding, are not unusually resistant; they probably function to promote aerial dissemination.

Spores, which contain one or several nuclei, vary greatly in color, size, and shape. Their morphology and their mode of origin constitute the main basis for classifying fungi that lack sexuality. Some species produce only one kind of spore, and others as many as four different kinds. Various common asexual spores, and their distinctive features, are listed in Table 43-1 and illustrated in Figures 43-5 and 43-6. They should not be confused with sexual spores (see below).

Budding is the prevailing asexual reproductive process in yeast, though some species divide by fission (fission yeasts). Whereas in fission (the usual reproductive process of almost all bacteria) a parent cell divides into two daughter cells of essentially equal size,

Fig. 43-5. Diagram of some representative filamentous fungi and their asexual spores (conidia). (From Ajello, L., Georg, L. K., Kaplan, W., and Kaufman. L. *Laboratory Manual for Medical Mycology* PHS Publ. No. 994. GPO, Washington, D.C., 1963.)

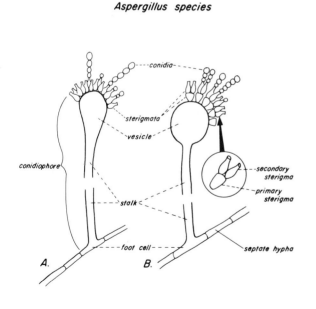

Aspergillus species

A. Conidial head with single row of sterigmata
B. Conidial head with sterigmata in two series.

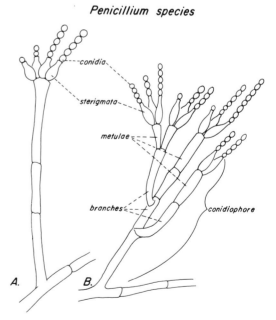

Penicillium species

A. Typical monoverticillate penicillus
B. Typical asymmetric complex penicillus

TABLE 43-1. Asexual Spores Formed by Fungi

Conidia (Gr. *konis*-dust)	This term is used sometimes generically for all asexual spores, or sometimes more specifically for spores borne singly or in clusters along sides or at tips of hyphae or of specialized hyphal branches **(conidiophores).** Highly diversified in shapes, size, color, and septation. Large (usually multinuclear) and small (usually uninuclear) conidia are called **macroconidia** and **microconidia,** respectively.
Arthrospores (Gr. *arthron*-joint; *sporos*-seed)	Cylindrical cells formed by double septation of hyphae. Individual spores are released by fragmentation of hyphae, i.e., by disjunction.
Blastospores (Gr. *blastos*-bud, shoot)	Buds that arise from yeast and yeast-like cells.
Chlamydospores (Gr. *chlamy*-mantle)	Thick-walled, round spores formed from terminal or intercalated hyphal cells.
Sporangiospores (Gr. *angeion*-vessel)	Spores within sac-like structures **(sporangia)** at ends of hyphae or of special hyphal branches **(sporangiophores).** Characteristically formed by species of *Phycomycetes.*

in budding the daughter cell is initially much smaller than the mother cell. As the bud bulges out from the mother cell the nucleus of the latter divides, and one nucleus passes into the bud; cell wall material is then laid down between bud and mother cell, and the bud eventually breaks away (Fig. 43-7). A **birth scar** on the daughter cell's wall, and a **budding scar** on the mother's wall, are visible in electron micrographs (Fig. 43-8). As a result of repeated budding old yeast cells bear many budding scars, but they have only a single birth scar.

Fragments of hyphae (e.g., formed by teasing a mycelium) are also capable of forming new colonies. This capacity is often exploited in the cultivation of fungi, but it is probably not important in nature.

SEXUAL REPRODUCTION

Fungi that carry out sexual reproduction go through the following steps. 1) A haploid nucleus of a donor cell (male) penetrates the cytoplasm of a recipient (female) cell. 2) The male and female nuclei fuse to form a diploid zygotic nucleus. 3) By meiosis the diploid nucleus gives rise to four haploid nuclei, some of which may be genetic recombinants. In most species the haploid condition is the one associated with prolonged vegetative growth, and the diploid state is transient; but in other species, as in higher animals, the opposite is true.

In **homothallic** species the cells of a single colony (arising from a single nucleus) can engage in sexual reproduction. In some homothallic species (hermaphrodites) male and female cells are anatomically differentiated, but in others they are indistinguishable. In **heterothallic** species the cells that engage in sexual reproduction must arise from two different colonies, of opposite mating type. The reproductive cells of heterothallic species may be anatomically distinguishable as male and female **(diecious),** or may not, in which event they are only functionally differentiated into sexually compatible mating types. Among fungi with a sexual stage the anatomy of sex organs and the mating procedures are highly diversified but are characteristic for any particular species; hence they are important for taxonomy.

Sexual Cycle in Neurospora. The sexual process in some fungi has played so vital a role in the development of biochemical genetics, as in the classic studies of Beadle and Tatum in *Neurospora crassa,* that a brief description of a representative cycle is warranted.

Neurospora contains in its nucleus seven different chromosomes, each a single copy (i.e., the vegetative organism is haploid). The haploid state is maintained during mycelial growth and during asexual reproduction (i.e., the formation of conidia). Sexual reproduction occurs when two cells (hyphae or conidia) of different mating type fuse to form a **dikaryotic** cell; the two kinds of haploid nuclei coexist in the same cytoplasm and for a time divide more or less in synchrony. If a cell initiates an ascus, however, two different haploid nuclei fuse and form a diploid nucleus, containing pairs of **homologous** chromosomes (Fig. 43-9).

The diploid cell then initiates the process of

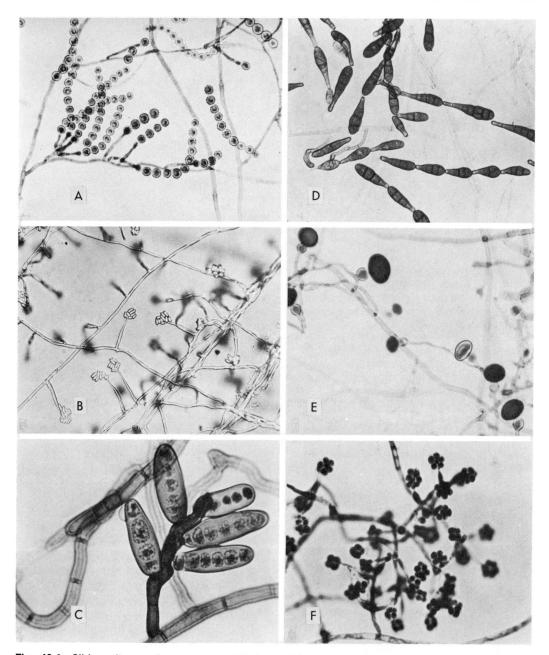

Fig. 43-6. Slide cultures of some representative molds showing diverse forms of asexual spores (conidia), which aid in identifying species, particularly with members of the class Fungi Imperfecti (Deuteromycetes). (From Ajello, L., Georg, L. K., Kaplan, W., and Kaufman, L. *A Manual for Medical Mycology,* PHS Publ. No. 994. GPO, Washington, D.C., 1963.)

meiosis. The homologous chromosomes pair (synapse) with each other (i.e., assume parallel, closely adjacent positions), and each chromosome divides without duplicating its centromere (whose attachment to a spindle fiber subsequently guides its migration to one pole or other during anaphase). Each chromosome thus becomes a pair of identical **chromatids** connected by a centromere (a bivalent), and each pair of homologous bivalents constitutes a **tetrad.** One or more exchanges of parts (crossing

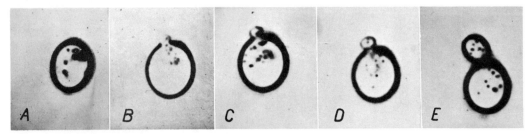

Fig. 43-7. Budding in a yeast cell (*Saccharomyces cerevisiae*). The wall-less bud in **B** was extruded in the 20-second interval between the photos in **A** and **B**. The subsequent photos were taken at approximately 15-minute intervals. The bud in **E** is nearly mature. [From Nickerson, W. J. *Bacterial Rev* 27:305 (1963).]

over) may then occur at random among the four chromatids of the tetrad, resulting in genetic recombination.

Two meiotic divisions follow. In the first there are no divided centromeres to separate (as in mitosis); instead, one member of each pair of homologous chromosomes (in the form of a bivalent) is drawn after its centromere to each daughter nucleus. In the second meiotic division the chromosome is already divided into two chromatids, and only the centromere divides; one product then migrates (as in mitosis) to each pole, drawing its chromosome with it. Thus the four chromatids of each tetrad are distributed to four different cells, each of which ends up with a haploid set of chromosomes. The individual members of each set are thus derived at random by segregation from either parent, and are further scrambled by the genetic recombination occurring at the tetrad stage (Fig. 43-9).

One further feature of this process is that the four haploid cells resulting from the meiosis remain together in the same sac **(ascus)**; and in *Neurospora* (though not in all Ascomycetes) before the ascus is fully matured each cell divides mitotically into two identical spores. In the ascus, which is shaped as a narrow pod, the eight resulting **ascospores** are held in a linear array, whose order reflects the meiotic segregation of their chromosomes.

Analysis of the genetic constitution of all four pairs of spores in the same ascus **(tetrad analysis)** allows the most complete possible description of the genetic events occurring during meiosis: the products of reciprocal recombination can be identified, rather than merely deduced (as with higher organisms) from the statistical distribution of genetic markers among the progeny. Moreover, the haploid nature of the vegetative phase eliminates the complicating effect of dominance on the relation between genotype and phenotype. These organisms have been especially valuable in studying the mechanism of genetic recombination (Ch. 11).

PARASEXUAL CYCLE (MITOTIC RECOMBINATION)

Some fungi go through a cycle that imparts some of the biological advantages of sexuality (i.e., recombination of parental DNA) without involvement of specialized mating types or gametes. This process of parasexuality, first demonstrated by Pontecorvo with *Aspergillus,* involves the following steps (Fig. 43-10). 1) By **hyphal fusion** different haploid nuclei come to coexist in a common cytoplasm. The **heterokaryon** thus formed can be stable, the two sets of nuclei dividing at about the same rate. 2) Rare **nuclear fusion** will yield heterozygous diploid nuclei. These are usually greatly outnumbered by the haploid nuclei, but once formed they tend to divide at about the same rate as the latter; and stable diploid strains may be isolated. 3) Though homolo-

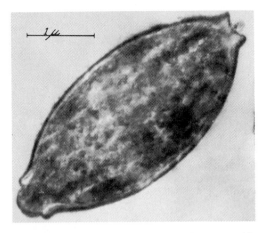

Fig. 43-8. Thin section of an osmium tetroxide-fixed yeast cell (*Saccharomyces cerevisiae*), showing the concave birth scar at one pole and a convex bud scar at the other. ×29,000, reduced. [From Agar, H. D., and Douglas, H. C. *J Bacteriol* 70:247 (1955).]

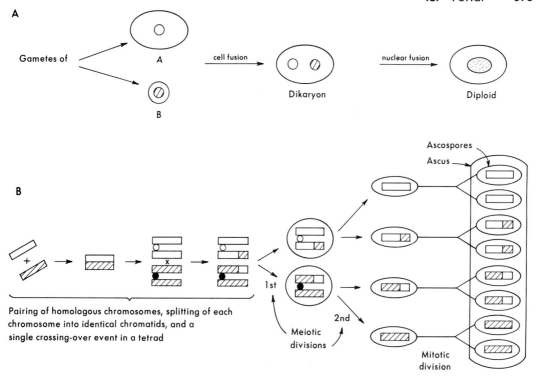

Fig. 43-9. Sexual reproduction in *Neurospora crassa*. **A.** Mating leads to formation of a dikaryon and then eventually, by fusion of nuclei, to a diploid cell. **B.** Meiotic divisions with recombination, followed by mitosis, produces eight haploid ascospores linearly arranged in a narrow pod (ascus).

gous chromosomes are usually arranged independently on the equatorial plate in the mitosis of a diploid cell, rarely (ca. 10^{-4} per mitosis) sufficient **somatic pairing** will occur to permit crossing over, as in meiosis.

The result of such mitotic recombination between heterozygous homologous chromosomes is to make the products **homozygous for genes distal to the exchange point.** Thus two diploid daughter cells with different properties result, each homozygous for some alleles for which the diploid parent cell is heterozygous (Figure 43-10). Further phenotypic changes appear when these new diploid strains yield haploid progeny: as can be seen from Figure 43-10, half of these will be parental in genetic composition and half will be recombinant.

Mitotic recombination has provided unique opportunities for genetic analysis of asexual molds. Its possible application to the study of the genetics of somatic diploid cells, such as human cells in tissue culture, is of great interest, and could contribute to our understanding of many problems of cell physiology and genetics in man. With the limited range of genetic markers available thus far in human cells the recombination rate has been low, but sufficient to be encouraging.

Though recombinants are infrequent in a parasexual cycle, compared with a sexual cycle, the process may be significant for the evolution of asexual fungi.

TAXONOMY

The four major classes of true fungi (division Eumycota) are summarized in Table 43-2.

The Phycomycetes* have nonseptate hyphae; their asexual spores are of various kinds, of which the sporangiospores, contained within sacs (sporangia) formed at the end of specialized stalks (sporan-

---

* The class Phycomycetes is not recognized by some taxonomists, who would separate these organisms into two subdivisions based upon the production of motile or nonmotile spores.

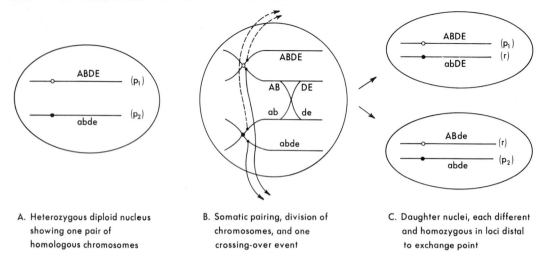

A. Heterozygous diploid nucleus showing one pair of homologous chromosomes

B. Somatic pairing, division of chromosomes, and one crossing-over event

C. Daughter nuclei, each different and homozygous in loci distal to exchange point

Fig. 43-10. Somatic pairing and mitotic recombination in the parasexual cycle of fungi. After fusion of genetically different hyphae to form a heterokaryon, the unlike haploid nuclei occasionally fuse to yield the heterozygous diploid nucleus depicted in **A.** Rare somatic pairing and recombination give rise, through the daughter nuclei depicted in **C**, to partially homozygous lines of diploid cells, as shown. These occasionally also segregate haploid strains, half of which are recombinant (compared with the original haploid parents) in respect to loci dD and eE. p = parental chromosome; r = recombinant chromosome.

giosphores), are unique to this class. Different species have different sexual cycles, and those that live in aquatic environments (Gr. *phyco* = seaweed) have **flagellated gametes.** The flagella resemble in structure cilia of protozoa are higher animals, rather than bacterial flagella (Ch. 2).

The **Ascomycetes** are distinguished from other fungi by the **ascus,** a sac-like structure containing sexual spores **(ascospores).** The ascospores are the endproduct of mating, fusion of male and female nuclei, two meiotic divisions, and usually one final mitotic division, as described above for *Neurospora*. There are thus usually eight ascospores in an ascus. The yeasts are ascomycetes, though they generally do not grow as molds (see Dimorphism, below).

The **Basidiomycetes** are distinguished by their sexual spores, called **basidiospores,** which form on the surface of a specialized structure, the **basidium.**

They include edible mushrooms. Basidiomycetes cause a variety of serious diseases of plants but do not cause infectious diseases of man. However, some species synthesize toxic alkaloids, which may be lethal in man and are often of major pharmacological interest (e.g., ergotamine, muscarine).

The **Deuteromycetes (Fungi Imperfecti)** are particularly important for medicine, as they include the vast majority of human pathogens. Because no sexual phase has been observed, they are often referred to as imperfect fungi. The hyphae are septate, and conidial forms are very similar to those of the ascomycetes; they have therefore long been suspected of being **special ascomycetes,** whose sexual phase is either extremely infrequent or has disap-

TABLE 43-2.   Classes of Fungi

Class	Asexual spores	Sexual spores	Mycelia	Representative genera or groups
Phycomycetes	Endogenous (in sacs)	Anatomy variable	Nonseptate	*Rhizopus, Mucor,* watermolds (aquatic)
Ascomycetes	Exogenous (at ends or sides of hyphae)	Ascospores, within sacs or asci	Septate	*Neurospora, Penicillium, Aspergillus,* true yeasts
Basidiomycetes	Exogenous (at ends or sides of hyphae)	Basidiospores, on surface of basidium	Septate	Mushrooms, rusts, smuts
Deuteromycetes (Fungi Imperfecti)	Exogenous (at ends or sides of hyphae)	Absent	Septate	Most human pathogens

peared in evolution. Indeed, typical ascomycetous sexual stages have recently been observed in several species of this class (Table 43-3). Superficially, the reversible change from sexual to asexual form resembles phase variations in bacteria (Ch. 8), suggesting that the imperfect fungi are mutants in genes that specify sexual development. Indeed, mutations in a gene controlling heterothallism have revealed that the classic "species" *Microsporum gypseum* includes the imperfect stage of two different species of ascomycetes (Table 43-3).

DIMORPHISM

Some species of fungi grow only as molds, and others only as yeasts. Many species, however, can grow in either form, depending on the environment. **This capacity is known as dimorphism.*** It is important clinically, since

TABLE 43-3. Some Pathogenic Imperfect Fungi Discovered to Have a Sexual (Perfect) Stage

Name of imperfect "species"	Perfect form
Microsporum gypseum	*Nannizia incurvata* *N. gypsea*
M. fulva	*N. fulva*
M. nanum	*N. obtusa*
M. cookei	*N. cajetana*
M. vanbreuseghem	*N. grubyia*
Trichophyton ajelloi	*Arthroderma uncinatum*
T. terrestre	*A. quadrifidum*

The "species" listed are dermatophytes: they infect the epidermis, nails, and hair of mammals (p. 1000). Only the asexual (imperfect form is found in infected skin. Conversion to the sexual (perfect) form was facilitated by growth on sterilized soil enriched with keratin (e.g., hair, feathers). The perfect stages were then identified as Ascomycetes by observing production of fruiting bodies when compatible "imperfect" forms were mated (e.g., + and − strains). (A fruiting body contains many asci with their enclosing ascospores.)

---

* The alternatives are not limited to yeast and mold forms. One of the most virulent human pathogens, *Coccidioides immitis,* is dimorphic, growing in infected tissues as spherules (p. 991) and in conventional cultures as a mold. A culture can have a mixture of the two forms (see *Candida albicans,* Fig. 43-23, and *Coccidioides immitis,* below).

most of the more pathogenic fungi (in man) are dimorphic: they usually appear in infected tissues as yeast-like cells, but when cultivated under conventional conditions in vitro they appear as typical molds (Fig. 43-11 and 43-12).

Dimorphism can be experimentally controlled by modifying cultural conditions, a single factor sometimes being decisive. For example, the human pathogen *Blastomyces dermatitidis* grows as a mycelium at 25° but as yeast cells at 37°.

Some fungi produce mycelial forms for growth at surfaces and yeast forms for growth in the depths of a liquid.

This capacity of *Mucor* was observed by Pasteur and provided an early example of microbial adaptation to the environment. Recently the difference in growth pattern has been shown to depend not on anaerobiosis but on pCO_2. As growth switches from mycelial to yeast-type there is no gross change in the over-all growth rate, but the cell wall becomes relatively enriched in mannan-protein complexes, and it composes a much larger fraction of total cell weight (about 30% of the yeast cell but only 10% of the hyphae).

COMPARISON WITH BACTERIA

Fungi resemble bacteria with respect to their role in maintaining the geochemical stability of the biosphere, the methods used for their isolation and cultivation, their capacity to cause infectious diseases, and the many applications of their fermentations to industrial processes. Fungi differ greatly from bacteria, however, in their reproductive processes, their growth characteristics (budding; branching filaments), the greater size of the cell, the greater anatomical complexity of their cytoplasmic architecture, their somewhat less diversified metabolic activities (none are photosynthetic, autotrophic, or obligately anaerobic), the composition and ultrastructure of their cell walls, and the chemotherapeutic agents to which they are susceptible. In addition, dimorphism is not recognized among bacteria but is common among fungi. Finally, as we shall see below, the human diseases caused by fungi are much less common and less varied than those caused by bacteria.

Genetic, biochemical, and antigenic differences have been studied much less among

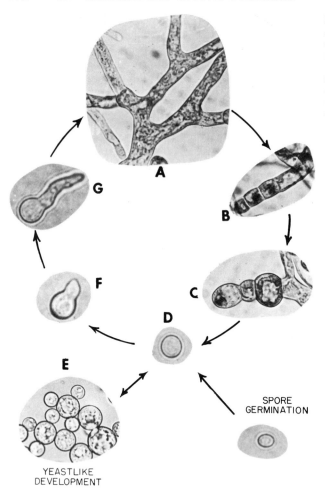

Fig. 43-11. Mold-yeast dimorphism in *Mucor rouxii*. Note absence of septa, typical of phycomycetes, in hyphae of the mold at **A**. Arthrospores are being formed in **B** and **C**. At **D** an isolated arthrospore is shown developing into yeast-like cells (**E**), or into a mold (**A**) by outgrowth of a filamentous tube (**F** and **G**). [From Bartnicki-Garcia, S. *Bacteriol Rev 27*:293 (1963).]

SPORE
GERMINATION

YEASTLIKE
DEVELOPMENT

fungi than among bacteria (Table 43-4). And although modern biochemical genetics had its inception in the classic studies of Beadle and Tatum with the mold *Neurosopra crassa,* only recently have the mating types for two pathogenic fungi for man been discovered, viz., *Blastomyces dermatitidis* and *Histoplasma capsulatum*. Hence the genetics of the medically important fungi remains virtually unexplored.

GENERAL CHARACTERISTICS OF FUNGOUS DISEASES

Because fungi are larger than bacteria they were recognized earlier as agents of disease. However, of the estimated 50,000 to 200,000 species of fungi, only about 100 are known to cause infectious disease in man (mycoses). A few of these, with a special predilection for epidermal structures, seem to depend on parasitic growth in animal tissue, some only in man. For all the other pathogenic species, however, such infection is only incidental and of no ecological importance to the microbe.

TYPES OF MYCOSES

It is useful to divide the mycoses into four groups, differing in the level of the infected tissue. 1) The **systemic or deep mycoses** in-

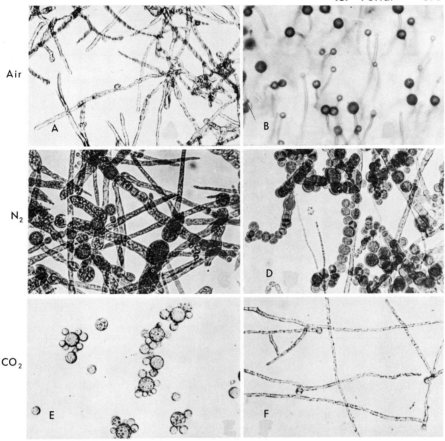

Fig. 43-12. Mold-yeast dimorphism in *Mucor rouxii,* growing in yeast extract-peptone-glucose medium. Air phase for incubation is shown at left. **A.** Filamentous growth in submerged culture. **B.** Surface growth with active spore formation. **C.** Filamentous growth at low concentration of glucose (~ 2%). **D.** Arthrospore formation stimulated by high concentration of glucose (10%). **E.** Yeast-like cells. **F.** Inhibition of yeast-like growth by a chelating agent (diethylene-triamino-pentaacetic acid). All the forms shown can develop under appropriate conditions from a single uninucleated cell (Fig. 43-11). [From Bartnicki-Garcia, S. *Bacteriol Rev 27:*293 (1963).]

volve primarily internal organs and viscera. They are often widely disseminated and involve many different tissues. 2) The **subcutaneous mycoses** involve skin, subcutaneous tissue, fascia, and bone. 3) The **cutaneous mycoses** involve epidermis, hair, and nails. The responsible fungi are known as **dermatophytes** (Gr. *phyton* = plant), and the diseases as **dermatophytoses** or **dermatomycoses.** 4) The **superficial mycoses** involve only hair and the most superficial layer of epidermis.

The natural history of the diseases in these groups will now be summarized, and the following section will describe some representatives of each group.

The **systemic or deep mycoses** are caused by saprophytic fungi in soil, and **inhalation of spores** initiates infection. As will become evident below, these infections resemble clinically the chronic bacterial infections due to mycobacteria (Ch. 35) and to actinomycetes (Ch. 36). The earliest lesions are usually pulmonary, and the initial acute, self-limited pneumonitis is easily overlooked or ascribed to bacteria or viruses. In their subsequent chronic form (which is usually much less frequent) these diseases begin insidiously and progress slowly, and are characterized by suppurative or granulomatous lesions. These sometimes form pulmonary cavities and often

TABLE 43-4.   Contrast Between Fungi and Bacteria

Property	Fungi	Bacteria
Cell volume (μ^3)	Yeast: 20–50 Molds: Not definable because of indefinite size and shape and coenocytic form; but much greater than yeast	1–5
Nucleus	Eukaryotic (well-defined membrane)	Prokaryotic (no membrane)
Cytoplasm	Mitochondria, endoplasmic reticulum	No mitochondria or endoplasmic reticulum
Cytoplasmic membrane	Sterols present	Sterols absent (except for *Mycoplasma* grown on sterols)
Cell wall	Glucans; mannans; chitin, glucan- and mannan-protein complexes No muramic acid peptides, teichoic acids, or diaminopimelic acid	Muramic acid peptides; teichoic acids; some have diaminopimelic acid residues No chitin, glucans, or mannans
Metabolism	Heterotrophic, aerobic, facultative anaerobes; no known autotrophs or obligate anaerobes	Obligate and facultative aerobes and anaerobes; heterotrophic and autotrophic
Sensitivity to chemotherapeutic agents	Sensitive to polyenes and griseofulvin (dermatophytes); not sensitive to sulfonamides, penicillins, tetracyclines, chloramphenicol, streptomycin	Often sensitive to penicillins, tetracyclines, chloramphenicol, streptomycin; not sensitive to griseofulvin or polyenes
Dimorphism	A distinguishing feature of many species	Absent

spread by direct extension, e.g., into contiguous soft tissues such as the pleurae. These fungi are also prone to spread by way of the blood stream, and they can produce metastatic abscesses or granulomas in almost any organ, including the skin. Prior to the development of effective antifungal chemotherapy these disseminated mycotic diseases were almost invariably fatal. They are not contagious.

The **subcutaneous mycoses** are also caused by saprophytes in soil and on vegetation. Infection occurs by **direct implantation of spores or mycelial fragments into wounds** in the skin, commonly in scratches caused by thorns. Hence these diseases tend to be especially prevalent in rural and tropical regions, e.g., in jungle terrain. The diseases begin insidiously, progress slowly, and are characterized by localized subcutaneous abscesses and granulomata that spread by direct extension, often breaking through the skin surface to form chronic, draining, ulcerated, and crusted lesions. Extension may also occur via lymphatics, leading to suppurative, granulomatous lesions in the regional chain of lymph nodes. These diseases are often extremely disfiguring and not infrequently fatal, though dissemination to the viscera is rare.

Those localized subcutaneous abscesses

that are particularly invasive and destructive of soft tissues, fascia, and bone, are known as **mycetomas.** They are characterized by burrowing, tortuous sinus tracts that open onto the skin surface. The purulent discharge and the abscesses frequently contain "granules," which are bits of colonies of the responsible microorganism. Clinically indistinguishable abscesses are also caused by certain bacteria, e.g., by various Actinomycetales (species of *Nocardia* and *Streptomyces,* Ch. 36). The latter are known as **actinomycotic mycetomas** to distinguish them from those due to fungi. (The fungal mycetomas are sometimes referred to as **maduromycoses.**)

The fungi that cause **cutaneous mycoses** have a striking predilection for growth in epidermis, hair, and nails. Only two or three of the pathogenic species have been found in soil, and only one (*Microsporum gypseum*) with any frequency. The many other dermatophytes are found only in mammalian skin; they appear to be obligate parasites of man and other animals, transmitted by direct contact or by bits of infected hair or desquamated epidermal scales. The diseases they cause tend to be chronic, and the inflammatory response, mostly confined to the skin at the site of infection, is not especially destructive.

The dermatophytes acquired by man from contaminated soil (*M. gypseum*) or from animals (e.g., *M. canis* from dogs and cats) tend to produce relatively intense but transitory inflammatory lesions in human skin. But the dermatophytes that are indigenous in man, and apparently obligatory parasites, usually evoke only trivial reactions. If the inflammatory lesion represents an allergic response, it woud appear that man is more tolerant (immunologically) of his indigenous dermatophytes than of alien species.

The fungi that cause **superficial mycoses** are localized along hair shafts and in the more superficial, nonviable, cornified epidermal cells. The pathological lesions are of minor importance.

OPPORTUNISTIC FUNGI

Certain widespread fungal saprophytes almost never establish infections in healthy humans, but can cause serious illness in those with various diseases, such as severe diabetes or malignant tumors of lymphoid tissue (e.g., leukemia, lymphosarcoma). Those individuals who receive extensive treatment with broad-spectrum antibacterial agents (reducing their indigenous bacterial flora), or with immunosuppressive agents are also especially susceptible to infection with these so-called **opportunistic** fungi. The diseases caused by these fungi may be widely disseminated, or they may be localized in the respiratory tract (aspergillosis) or the mucous membranes and skin (candidiasis).

Opportunistic behavior is not, however, an all-or-none phenomenon. Many of the more virulent fungi, which can cause serious systemic diseases in healthy humans, also have an increased tendency to cause progressive infections in humans affected by the conditions mentioned above.

DIAGNOSIS

The fungal origin of a disease is usually first suspected on the basis of its clinical behavior and the appearance of the lesion. As with bacteria that are abundant in man's environment, serological reactions with antigens of the suspected fungus, and delayed allergic skin reactions to intradermal injection of fungal antigens, may provide supporting evidence, but it is not usually conclusive. The most convincing diagnostic evidence is usually provided by **detection of the fungus** in lesions and exudates, by direct microscopic examination and by isolation and cultivation. With some fungous diseases transmission of the infection to experimental animals, by inoculation of tissue suspensions or exudates, facilitates isolation, and the lesions in the test animal may also aid in identification (e.g., *Histoplasma capsulatum* in mice, and *Coccidioides immitis* in mice and guinea pigs; see below). In addition, animal inoculation serves to distinguish between pathogenic and nonpathogenic (saprophytic) fungi, which can be identical in colonial and cellular morphology; the saprophytes are usually innocuous in test animals.

Visualization of Fungi in Tissues. In one of the simplest procedures scrapings of the lesion, or bits of exudate (e.g., sputum, pus), are warmed on a slide in 10% NaOH or KOH (see, for example, Fig. 43-27). Proteins, fats, and many polysaccharides are extensively solubilized (hydrolyzed) and the tissues become optically clear; but the cell walls of most fungi remain largely intact, because of their alkali-resistant glucans (p. 968), and they are thus easily visualized. With some fungi, particularly small yeast cells, visualization of the alkali-resistant cell wall is aided by warming tissue in an equal mixture of 20% NaOH or KOH and a suitable dye.

One of the most widely used staining procedures is based on the periodic acid–Schiff reaction (PAS stain). As is shown in Figure 43-13, periodate cleaves vicinal hydroxyl groups and forms dialdehydes; subsequent reaction of the aldehydes with leucofuchsin (Schiff's reagent) forms colorful quinonoid dyes. Nearly all fungus walls are stained an intense red or magenta by this reaction, because a large number of aldehyde groups is produced on periodate oxidation of their insoluble glucans and mannans (p. 968). Chitin, however, does not stain, owing to the absence of vicinal hydroxyls (p. 968). The reaction is not, of course, specific for fungi, and some tissue polysaccharides (glycogen, hyaluronic acid) are also stained, though not as intensely.

Fungi are all gram-positive, but the Gram

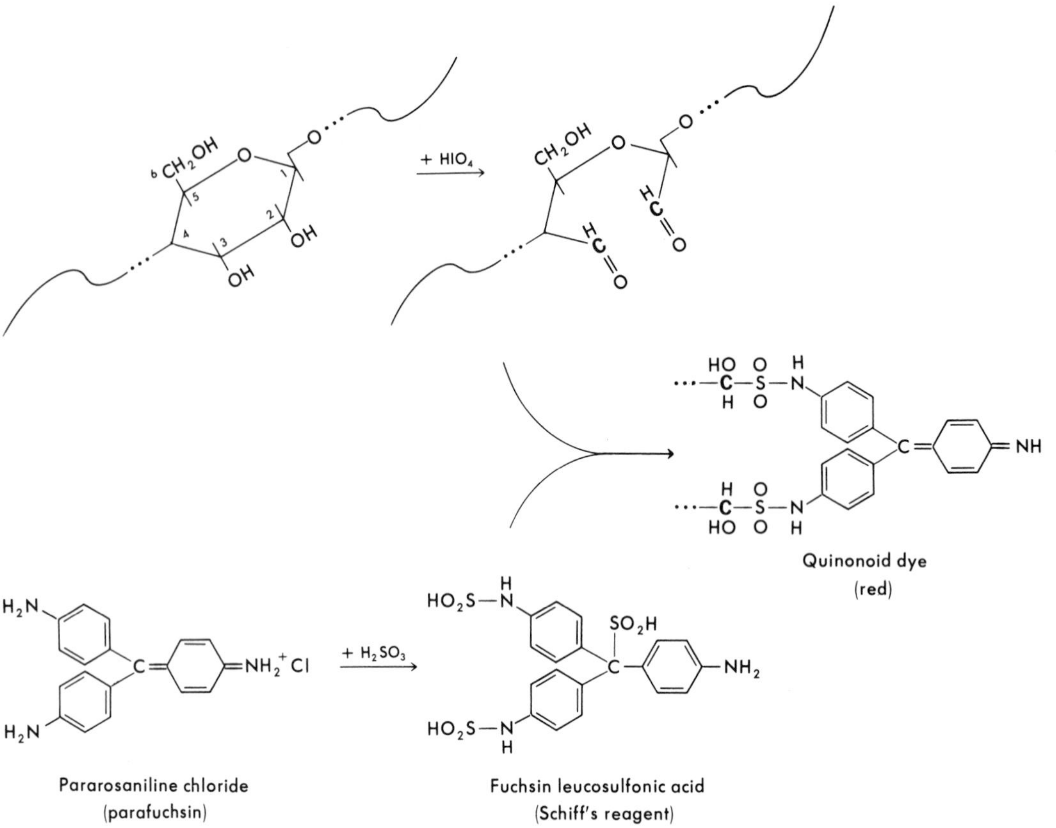

Fig. 43-13. The periodic acid–Schiff (PAS) stain. Fuchsin leucosulfonate (Schiff's reagent) reacts with the aldehyde groups generated by periodate cleavage between vicinal hydroxyls of a polysaccharide, forming a red or magenta quinonoid dye. If the polysaccharide had the sugar residue linked 1,6, rather than 1,4 as shown, a further attack by periodate would have split out C3 as formic acid, and left the two aldehyde groups on C2 and C4.

stain is not particularly useful in recognizing or identifying fungi.

Cultivation of Fungi. Under optimal conditions fungi grow much more slowly than bacteria. Cultures must therefore be maintained for prolonged periods, and it is essential to inhibit growth of bacterial contaminants, e.g., by drugs, or by maintaining low pH and low temperatures (25°), which inhibit bacteria more than fungi. For this reason contaminating fungi often overgrow bacterial cultures stored in the refrigerator.

Sabouraud's agar is the most widely used medium for cultivating fungi. Devised by Sabouraud (a distinguished dermatologist of the nineteenth century) for dermatophytes, it is also useful for virtually all fungi. With glucose and peptone as the sole nutrients, it was originally devised with a pH of 5 to discourage bacterial growth. At present the medium is adjusted to neutral pH (which is more favorable for the growth of most fungi) and the antibacterial agent chloramphenicol (40 μg/ml) is added with cycloheximide (500 μg/ml) to reduce the growth of saprophytic fungi. Cycloheximide, however, inhibits some pathogens, e.g., *Cryptococcus neoformans* and *Candida;* and chloramphenicol inhibits the yeast forms of some dimorphic fungi. For these reasons cultures are often prepared both with and without these drugs.

The anatomy of fungal spores and the manner in which they are produced are important determinative characteristics. Sporulation can be stimulated by cultivation under suboptimal nutritional conditions (e.g., rice medium, cornmeal agar). Mycelia are usually white and fluffy at first, and then become colored as they develop pigmented spores.

Intact colonies may be examined directly under a dissecting microscope, or fragments of colonies may first be teased gently in media containing a dye such as Poirrier's blue in solution with lactic acid and phenol (called **lactophenol cotton blue).** The staining procedures ordinarily used for bacteria are not helpful with fungi as a rule, in part because they fragment structures and perturb anatomical relations of spores.

CHEMOTHERAPY

In view of the substantial physiological differences between fungi and bacteria, it is not surprising that they respond to different things. It is only since 1959 that some effective antifungal chemotherapy has become available with the development of griseofulvin and the polyene agents (Fig. 43-14).

Griseofulvin. Griseofulvin, synthesized by several species of *Penicillium,* has been known for some time to produce hyphal distortions in cultures of dermatophytes and other fungi. However, its use in the treatment of dermatophytoses was delayed for many years because its direct application to infected skin lesions had only a limited effect. When given orally over a period of time, however, it is highly effective in clearing many of these cutaneous infections. The reason for this curious discrepancy appears to be that the ingested drug accumulates in keratinous structures (cornified layer of epidermis, hair, and nails) as they are laid down, and renders them resistant to infection. The keratinous structures already infected at the start of therapy are not sterilized, because griseofulvin fails to attain a sufficiently high concentration in these compact fibrous materials. Hence as older, infected structures are desquamated, with normal epidermal growth, they are replaced with structures that contain a high level of griseofulvin and are resistant to infection. Because the growth of epidermis and hair is relatively slow, treatment must be prolonged (weeks or months), especially with infected nails. Fortunately, griseofulvin has

Griseofulvin

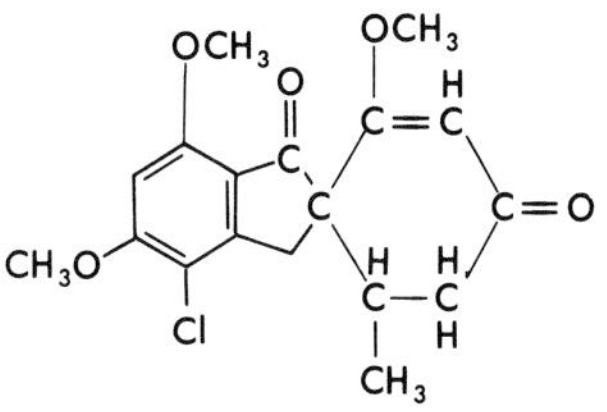

A polyene (pimaricin)

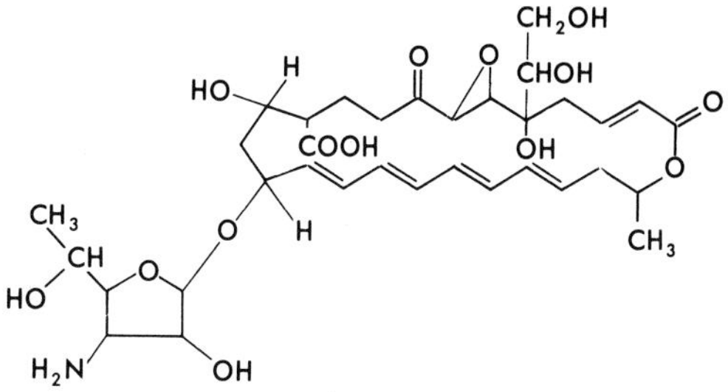

Fig. 43-14. Structures of antifungal agents. Griseofulvin, first isolated from *Penicillium griseofulvum,* is produced by several species of *Penicillium.* Various polyenes are produced by different species of *Streptomyces* (e.g., nystatin by *S. noursei,* pimaricin by *S. natalensis*). Nystatin, and pimaricin (shown above), are tetraenes, i.e., have four alternating double bonds); amphotericin B and candidiacin are heptaenes.

only minimal toxicity, even on prolonged administration. Griseofulvin therapy is ineffective in the deep mycotic infections, but may have value in one of the subcutaneous mycoses sporotrichosis.

Polyenes. In contrast to griseofulvin, the **polyene** antibiotics (e.g., amphotericin, nystatin) are highly effective in the treatment of many systemic mycoses and ineffective in the superficial and cutaneous mycoses. Various *Streptomyces* species produce this group of structurally related complex substances, one of which is shown in Figure 43-14.

At high concentrations, the polyenes are fungicidal when added to growing cultures. At low concentrations, several of the polyene macrolides are fungistatic. (Nystatin was named, however, for its discovery in a New York State laboratory.) These agents cause irreversible membrane damage, as shown by the leakage of small molecules (nucleotides, amino acids, phosphorylated sugars) from the cytoplasm. Moreover, polyenes can be bound not only by intact fungi but by membranes isolated from fungi.

The polyenes act by combining with sterols in the cytoplasmic membrane. Thus various sterols added to the medium form complexes with polyenes, preventing their binding to the membranes and blocking their growth-inhibiting effect.

The cell membranes of bacteria lack sterols, and bacteria are accordingly not susceptible to the polyenes. However, some species of *Mycoplasma* are a notable exception; when these wall-less bacteria are grown in sterol-containing media they incorporate sterols into their cytoplasmic membrane and become sensitive to polyenes; but when grown in sterol-free media they lack sterols and are resistant to these drugs.

Red blood cells contain sterols in their membrane, and polyenes cause hemolysis in vitro. Correspondingly, in treatment with polyenes (especially amphotericin B) one of the toxic side effects often encountered is hemolytic anemia. That the anemia and other cytotoxic effects (e.g., kidney damage) are not more severe and frequent is probably due to the binding of polyenes by serum albumin, which reduces their free concentration in body fluids.

Cycloheximide. This antibiotic inhibits protein synthesis in many yeasts and in animal cells, but not in several species of molds. It is not useful in chemotherapy because of its toxicity for the host, but it is useful diagnostically, inhibiting growth of bacterial contaminants in fungus cultures. It inhibits protein synthesis by blocking the transfer of activated amino acids, on the ribosome, from tRNA to growing polypeptide chains.

SOME PATHOGENIC FUNGI AND FUNGOUS DISEASES

SYSTEMIC MYCOSES

CRYPTOCOCCUS NEOFORMANS (CRYPTOCOCCOSIS)

Fungi of the genus *Cryptococcus* appear as spherical cells that reproduce by budding; they look like true yeasts (ascomycetes), but sexual forms have not been observed. However, under certain growth conditions some isolates of *C. neoformans* produce morphological structures (hyphae with clamp-like connections,

Fig. 43-15. An isolate of *Cryptococcus neoformans* (Coward strain) which produces yeastlike cells (A) and hyphae (B) with clamp-like connections (C). Clamp connections are morphological structures characteristic of basidiomycetes. (Courtesy of Dr. H. Jean Shadomy.)

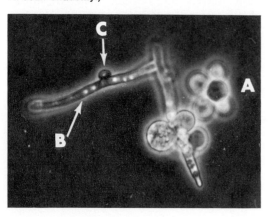

Fig. 43-15) which suggest that this organism be placed in the class Basidiomycetes. In rapidly growing cultures the buds separate from mother cells precociously; hence suspensions often exhibit unusually large variations in cell diameter, from about 4 to 20 μ (Fig. 43-16). **All strains produce capsules,** some much thicker than the enclosed cells. About 12 species have been defined on the basis of antigenic and morphological characteristics; only one, *C. neoformans,* is pathogenic in man. *C. neoformans* is readily differentiated from nonpathogenic species of *Cryptococcus* by its virulence for mice: as few as 1000 cells inoculated intracerebrally cause death in 50% of mice.

Unlike most fungi that cause disseminated infections, *C. neoformans* is **not** dimorphic (Table 43-5): in both infected tissues and cultures it appears as encapsulated yeast cells. It grows readily on Sabouraud's agar at 37° (but not at 40° or higher.) Urease is produced by all species.

The capsule of *C. neoformans,* its most distinctive feature, is easily visualized in India ink suspensions of the cells (Fig. 43-16). It is composed of a polysaccharide containing xylose, mannose, and glucuronic acid, and it cross-reacts with antisera to the capsular polysaccharides of types 2 and 14 pneumococci. Although *C. neoformans* evokes only a feeble immune response in infected humans, hyperimmunized rabbits yield capsule-specific

Fig. 43-16. *Cryptococcus neoformans* suspended in India ink. Note budding, thick capsules, and variations in cell diameter. In fresh spinal fluid, without India ink, the yeast cells are easily mistaken for lymphocytes. ×450, reduced. (Courtesy of Dr. George Kobayashi.)

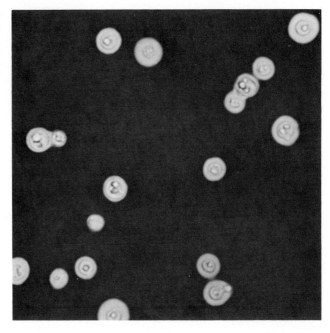

antisera that differentiate three strains, called A, B, and C. When suspended in homologous antiserum the cells undergo a quellung reaction like encapsulated bacteria (Ch. 25).

Strains differ in the size of their capsules and in their virulence for mice, but there is no consistent relation between these variables. The capsule is highly acidic, and quellung is also displayed when the cells react with a cationic polysaccharide. Various cationic dyes, such as toluidine blue, stain the capsule metachromatically.

Pathogenesis. The disease caused by *C. neoformans* is called **cryptococcosis.** Inhalation of yeast cells is assumed to initiate pulmonary infection, with subsequent hematogenous spread to other viscera and the central nervous system. It seems likely that minor infections of the lung are common, transient, and easily overlooked, and that only a very small proportion become disseminated.

In the severe, chronic, and disseminated form of the disease the brain, meninges, lungs, other viscera, skin, and bones are involved to varying extents in different patients. Chronic meningitis is the most frequent and mimics tuberculous meningitis; the lesions may simulate brain abscess or brain tumor. Pulmonary lesions are usually inapparent clinically, but

are almost always found at autopsies of meningitis patients. In a few individuals chronic pneumonitis is the most conspicuous clinical manifestation.

The microorganism appears in tissues as masses of budding encapsulated yeast cells (Fig 43-16). There is often little or no surrounding inflammatory reaction, but sometimes granulomatous lesions are formed, with multinuclear giant cells. The yeast cells may be observed within macrophages, particularly when the periodic acid–Schiff stain is used.

Diagnosis. The diagnosis is usually established either by visualizing *C. neoformans* in spinal fluid (the India ink technic being particularly useful, Fig. 43-16) or by cultivating it from spinal fluid, pus, or other exudates. Cycloheximide is inhibitory and should be omitted from the medium. The only serological test of diagnostic value identifies the soluble capsular polysaccharide in spinal fluid, serum, or urine by means of the precipitin reaction with hyperimmune rabbit serum.

Therapy. Cryptococcosis involving the central nervous system, with or without disseminated visceral lesions, formerly was invariably fatal. With the introduction of polyene anti-

TABLE 43-5. Grouping of Most Frequently Encountered Pathogenic Fungi (for Man) in the United States, with Respect to Tissue Involved and Dimorphism

| Type of mycotic disease | Representative fungus | Morphology in | | Comment |
		Infected tissue	Room temperature culture	
Systemic	Cryptococcus neoformans	Yeast (encapsulated)	Yeast (encapsulated)	No dimorphism
	Coccidioides immitis	Spherules	Mycelia	Dimorphism
	Histoplasma capsulatum	Yeast	Mycelia	Dimorphism
	Blastomyces dermatitidis	Yeast	Mycelia	Dimorphism
Systemic, and particularly opportunistic	Candida (especially C. albicans)	Yeast and hyphae	Yeast and hyphae	Dimorphism
	Aspergillus (most often A. fumigatus)	Mycelia	Mycelia	No dimorphism
	Phycomycetes (Mucor, Rhizopus species)	Mycelia	Mycelia	No dimorphism
Subcutaneous	Sporotrichum schenckii	Yeast	Mycelia	Dimorphism
Cutaneous	Microsporum species	Mycelia	Mycelia	No dimorphism*
	Trichophyton species	Mycelia	Mycelia	No dimorphism*
	Epidermophyton floccosum	Mycelia	Mycelia	No dimorphism*

* The fungi that parasitize epidermis, nails, and hair (dermatophytes) all appear alike in infected skin, but in culture they develop a variety of specialized hyphae and spore structures that differentiate diverse genera and species; in a sense, they do exhibit a certain amount of dimorphism.

biotics, however, it can now often be controlled. Amphotericin B is particularly effective and has cured the infection even when first administered in an extremely ill person.

Epidemiology. Cryptococcosis occurs sporadically and in essentially all parts of the world. The organism has been isolated from soil, particularly when enriched with pigeon droppings. It is also found in pigeon roosts and nests far removed from the soil, e.g., on window ledges and towers of urban buildings. Small outbreaks of acute pulmonary infection have occurred among workers involved in demolishing old buildings.

Since the fungus remains viable in dried material for many months, contaminated materials are a potent source of airborne infection. Pigeon droppings are often highly contaminated: 5×10^7 viable organisms per gram were found in one study. Yet birds are highly resistant to cryptococcal infection, whereas most laboratory animals are quite susceptible.

The explanation may be that *C. neoformans* is unable to grow at the normal body temperature of birds (40 to 42°). In support of this suggestion, mice lethally infected with *C. neoformans* have been observed to survive longer when maintained at 35° than at 25°, and infected chick embryos also survive longer at 40° than at 37°. Hence the *C. neoformans* cells found in pigeon droppings are probably due

not to intestinal infections of pigeons, but to airborne contaminants that find in the droppings a particularly fertile medium.

In the past term **blastomycosis** was used generically for any disease due to infection with yeasts or yeast-like cells; and in the European literature cryptococcosis is still referred to as **European blastomycosis.** In the United States cryptococcosis was formerly called **torulosis** because *C. neoformans* was previously named *Torula histolytica,* partly on the mistaken assumption that the clear mucoid space around the yeast cells in tissues (the cells' capsules) represented liquefaction of host tissue.

BLASTOMYCES DERMATITIDIS (NORTH AMERICAN BLASTOMYCOSIS)*

B. dermatitidis is dimorphic, growing as yeast cells in infected tissues or in cultures at 37°, and as a mold at room temperature (ca. 25°) on conventional media. The mycelia are white at first, darken with age, and have characteristic spherical conidia, about 3 to 5 μ in diameter, borne directly on the sides of hyphae or at the tips of short, slender lateral branches. The conidial walls also darken with age and resemble chlamydospores. Conversion from yeast to mycelial growth, and vice versa, is

---

* South American blastomycosis is due to a different fungus, *Paracoccidioides brasiliensis;* see p. 994.

readily accomplished in vitro simply by altering the temperature. In the sexual stage (called *Ajellomyces dermatitidis*) ascospores develop within asci. The fungus is heterothallic and is classified as an ascomycete.

In sections of infected tissues (and in sputum or in pus expressed from skin lesions) *B. dermatitidis* appears as unusually thick-walled, multinucleated spherical cells, 8 to 15 μ in diameter, without a capsule (Fig. 43-17). Buds, attached to the mother cell by a broad base, are characteristically **unipolar,** i.e., there is not more than one bud per mother cell at any time.

B. dermatitidis is readily isolated on Sabouraud's or an equally simple medium.

Pathogenesis. Infection apparently begins in the lungs and spreads hematogenously to establish focal destructive lesions in bones, skin, prostate, and other viscera; the gastrointestinal tract is spared. The skin lesions are often particularly conspicuous; they seem usually to arise as metastases, from the primary pulmonary lesion, that break through the skin surface and establish spreading, ulcerated, crusted lesions. Skin lesions seem also occasionally to be initiated by direct implantation of spores into broken skin.

The lesions are characterized by granulomatous inflammation, microabscesses, and extensive tissue destruction. The yeast cells are visible within the abscesses and granulomata (Fig. 43-17).

Immune Response. Sera from persons with blastomycosis often give complement-fixation reactions with intact *B. dermatitidis* yeast cells or soluble antigens prepared from them, and patients also often exhibit a delayed-type skin response to **blastomycin,** a crude filtrate of mycelial culture. However, cross-reactions with *Histoplasma capsulatum* or *Coccidioides immitis* diminish the diagnostic value of either reaction.

These reactions may be negative in far-advanced illness. It is not clear whether the reason is 1) neutralization of antibodies by excessive levels of antigens, 2) changes in the nature of the antibodies formed with prolonged antigenic stimulation, or 3) progressive impairment of the host's capacity to respond to immunogens in general.

Diagnosis. The demonstration of nonencapsulated, thick-walled, multinucleate yeast cells in pus, sputum, or tissue sections, and their isolation by culture, establish the diagnosis of blastomycosis. The yeast form of *B. dermatitidis* is inhibited by chloramphenicol and cycloheximide; hence these drugs are omitted from cultures that are to be maintained at 37°. Inoculation of mice intraperi-

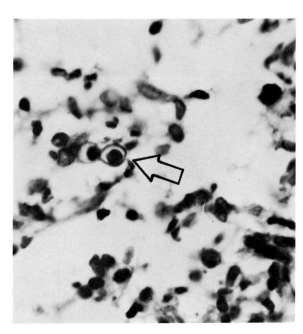

Fig. 43-17. *Blastomyces dermatitidis* in a tissue section; a thick-walled yeast cell (arrow) with bud. The broad connection between mother cell and bud (dumb-bell shape), and the single bud per mother cell, are both characteristic of *B. dermatitidis* (compare with *Paracoccidioides brasiliensis,* Fig. 43-22). Hematoxylin-eosin stain ×450, reduced. (Courtesy of Dr. George Kobayashi.)

toneally with pus or sputum is occasionally helpful; the fungus is usually readily cultivated from the localized abscesses that appear in the peritoneal cavity in 3 or 4 weeks.

Therapy. In cultures of *B. dermatitidis* the polyene amphotericin B is fungicidal at high concentrations. Though this drug is effective in man, hydroxystilbamidine is occasionally preferred because of its lower toxicity.

Epidemiology. The disease caused by *B. dermatitidis* is called North American blastomycosis because it is largely confined to Canada and the United States. It is most often encountered in the Mississippi valley, but occasional human infections have turned up in Mexico and in northern South America. Autochthonous cases have also been encountered in various parts of Africa. Some mycologists therefore suggest that the term North American be dropped and the disease called "blastomycosis."

Infections also occur sporadically in dogs, presumably from inhalation of conidia. *B. dematitidis* is thought to be a soil saprophyte. It has been cultivated in the laboratory on sterilized soil, but the many attempts to isolate it from soil have rarely been successful.

HISTOPLASMA CAPSULATUM (HISTOPLASMOSIS)

H. capsulatum appears in infected tissue as small, oval yeast cells (1 to 3 μ in diameter), usually localized within macrophages and reticuloendothelial cells (Fig. 43-18). Budding is only rarely observed in tissues, since buds separate readily from mother cells. The organism is dimorphic; when cultivated on Sabouraud's or other media at room temperature it forms slowly growing mycelial colonies. These are white at first, become tan with aging, and have fine, silky aerial mycelia. The spores are small or large spherical conidia, with characteristic regularly spaced, spiny projections (Fig. 43-19). The sexual stage of *H. capsulatum* has been designated as *Emmonsiella capsulata* and is classified as an ascomyete.

Hyphae are readily converted to yeast cells in vitro by enriching the medium with blood or yeast extracts, increasing the temperature to 37°, and ensuring the presence of adequate moisture. The conversion becomes apparent macroscopically at the edges of the mycelial mat, where creamy, round, smooth colonies develop. Microscopically, hyphae become constricted at the septa and then fragment into elongated cells, which become oval and multiply by budding. The conversion of mold to yeast also occurs when mycelial fragments are serially passed through animal cell cultures (e.g., HeLa cells) or inoculated into mice.

The reverse conversion (yeast to mold) is brought about simply by decreasing the incubation temperature of yeast cultures from 37 to 25°; germinal hyphae grow out of the yeast cells, elongate, and develop septa.

Pathogenesis. *H. capsulatum* is present in soil, and inhalation of conidia leads to pul-

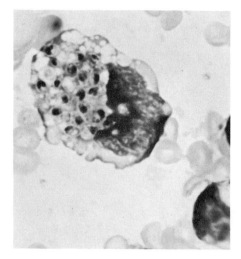

Fig. 43-18. Yeast cells of *Histoplasma capsulatum* in a macrophage; an impression smear of liver from an infected mouse. Wright's stain. ×1250. (Courtesy of Dr. George Kobayashi.)

Fig. 43-19. Characteristic thick-walled spores with spiny projections (tuberculate chlamydospores) in a culture of *Histoplasma capsulatum* at room temperature. ×360. (Courtesy of Dr. George Kobayashi.)

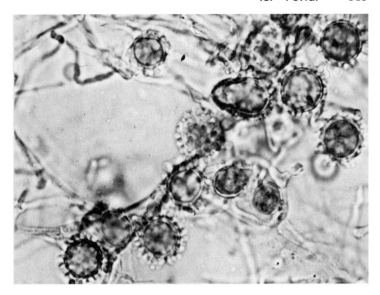

monary infection. Miliary lesions appear throughout the lung parenchyma, and hilar lymph nodes become enlarged. The initial infection is mild. It may pass unnoticed, or may appear as a self-limited respiratory infection. With healing the pulmonary lesions become fibrotic and calcified and give rise to characteristic roentgenographic pattern of healed histoplasmosis, formerly confused with healed primary tuberculosis (see below).

In a very small number of infected individuals the infection becomes progressive and widely disseminated, with lesions in practically all tissues and organs. Fever, wasting, and enlargement of liver, spleen, and lymph nodes occur, and the disease may closely simulate miliary tuberculosis. In fact, this severe disseminated form of histoplasmosis often co-exists in individuals who have tuberculosis or some other severe generalized disease, such as leukemia or Hodgkin's disease. Histoplasmosis is also occasionally seen as a chronic pulmonary disease with cavitation, simulating chronic pulmonary tuberculosis.

The tissue lesions are characterized by granulomatous inflammation (similar to that in tuberculosis), with epithelioid cells, occasional giant cells, and even caseation necrosis. The characteristic features, however, are swollen fixed and wandering macrophages containing many small, oval yeast cells (Fig. 43-18). These are readily visualized in tissue sections or cells smears treated with Wright's, Giemsa's, or the periodic acid–Schiff stain.

Immune Response. Antibodies to *H. capsulatum* are detected by precipitation, agglutination, and complement (C′) fixation. The antigens used are either filtrates of mycelial or yeast-phase cultures (for precipitation and C′ fixation) or heat- or formalin-killed yeast cells (for agglutination and sometimes also for C′ fixation).

Precipitation and agglutination reactions are observed in mild histoplasmosis. With more advanced disease C′-fixing antibodies also appear. The C′-fixation reaction is regarded as having some prognostic significance. Cross-reactions are seen with culture filtrates of *Blastomyces dermatitidis* and *Coccidioides immitis;* and unfortunately the clinical manifestations of these three mycoses may be indistinguishable, particularly in early infections confined to the lungs.

Persons previously infected with *H. capsulatum* regularly give delayed-type skin responses to intradermal injection of **histoplasmin,** a crude, sterile culture filtrate of mycelia grown in synthetic medium.* In both

---

* The same medium is used to cultivate other fungi and also *Mycobaterium tuberculosis* in the preparation of the corresponding culture filtrates as skin test reagents.

appearance and significance the response resembles the tuberculin reaction: it indicates past or present infection with *H. capsulatum.* Cross-reactions also occur with coccidiodin (from *C. immitis*) and with blastomycin (from *B. dermatitidis*).

For both diagnostic and epidemiological purposes it is often desirable to perform repeated skin tests with histoplasmin in the same persons. Fortunately, the response to one test is not invalidated by prior tests, since histoplasmin injected intradermally does not appear to induce delayed-type hypersensitivity; such an injection, however, appears to be able to stimulate a secondary antibody response in some individuals who are hypersensitive, without inducing a primary response.

Diagnosis.   A provisional diagnosis of histoplasmosis is based upon clinical manifestations, serological tests, and a positive skin response to histoplasmin. The latter has, of course, virtually no value in those localities where the fungus is so prevalent that most persons react to histoplasmin (see below). For a definitive diagnosis it is necessary to identify *H. capsulatum* in tissues or exudates and to isolate it by culture. Staining of blood cells, sputum, and tissue biopsies with Wright's or Giemsa's stain often reveals monocytes or macrophages with characteristic intracellular small yeast cells; and the yeast has even been found in urinary sediment of persons with progressive histoplasmosis. Cultivation is readily accomplished on Sabouraud's agar and a variety of other media.

Intraperitoneal inoculation of mice with sputum, to which penicillin, streptomycin, and chloramphenicol are added can detect as few as 10 yeast cells. Autopsy of the mice 2 to 6 weeks later generally reveals the characteristic yeast within reticuloendothelial cells in smears of spleen cells; and cultures of the tissues on Sabouraud's agar at room temperature yields characteristic mycelial growth.

Therapy.   Disseminated histoplasmosis was formerly invariable fatal, but it can be treated effectively with amphotericin B.

Epidemiology.   Histoplasmosis provides a striking example of the power of epidemiological analysis to help in the discovery of previously unrecognized agents of disease—a power that is likely to become increasingly important since the more obvious agents have

already been revealed by more classic methods.

Histoplasmosis first came to light in 1906 when its spherical yeast cells—observed in tissues taken postmortem in Panama—were mistakenly classified (and named) as a protozoon. In the next 40 years a few additional cases were reproted, all fatal, in scattered parts of the world. The discovery of an enormous background of mild or inapparent infection, underlying the rare, fatal disease, was an unexpected byproduct of a survey of tuberculosis in student nurses in different parts of the United States, undertaken by the U.S. Public Health Service in the 1940s. At that time pulmonary calcification was considered to be almost invariably the endproduct of old tuberculosis, and tuberculin-negative persons with such lesions (revealed by X-ray) were assumed to have lost their previous tuberculin hypersensitivity (Ch. 35). However, the nationwide survey revealed that in eastern United States cities few such persons were encountered, while in some midwestern communities they constituted the majority of those with pulmonary calcification.

The striking geographical distribution of the discrepancy between tuberculin sensitivity and pulmonary calcification, and the exclusion of technical artifact by the employment of identical testing materials and personnel in the different localities, clearly implied that some disease other than tuberculosis was producing a large incidence of an indistinguishable pulmonary calcification. Skin testing with culture filtrates of several organisms revealed that histoplasmin sensitivity was much more prevalent in the regions presenting the aberrant calcifications; and the tuberculin-negative persons with these healed lesions were invariable histoplasmin-positive.

In subsequent studies *H. capsulatum* was isolated repeatedly from soil, particularly where contaminated heavily with droppings of chickens, other birds, and bats: not only barnyards, chicken coops, and caves that shelter bats, but also soil beneath trees in which birds nest in urban areas. *H. capsulatum* is not present in fresh bird droppings, which apparently only enrich the soil as a culture medium. Birds do not develop the disease, presumably because *H. capsulatum* does not thrive at their high body temperature.

Mass surveys with chest roentgenograms and histoplasmin skin testing have now revealed that infection is extremely widespread; it is estimated that about 30 million persons in the United States have been infected. The frequency of positive skin tests is particularly high in the Ohio and Mississippi valleys. In some areas about 80% of all adults, and over 97% of those with widespread calcified pulmonary lesions, react positively to histoplasmin. Besides man, a variety of domesticated and wild animals are naturally infected; in one area of Virginia 50% of dogs in a large sample were found at autopsy to have histoplasmosis.

Infection occurs by inhalation of airborne conidia. The yeast cells of *H. capsulatum* in sputum and other exudates are less stable than the conidia, and are readily killed by drying, freezing or heating. Transmission from man to man, and from animal to man, has not been established.

H. capsulatum var duboisii. Histoplasmosis in Africa is caused by a variant of *H. capsulatum,* distinguished by its large size: in infected tissues it appears as large ovoid cells, 10 to 13 μ in diameter. These yeast cells are localized within reticuloendothelial cells, as with the histoplasmosis seen elsewhere. In culture, the mycelium grown from the large variant is identical with that obtained from conventional strains.

COCCIDIOIDES IMMITIS (COCCIDIOIDOMYCOSIS)

The parallel between *H. capsulatum* and *C. immitis* is striking. *C. immitis* was also first observed in postmortem tissue and mistakenly identified, and named, as a protozoon. Though this error was soon corrected when the organism was grown in culture, almost 40 years elapsed before it was also recognized that *C. immitis* could cause both a fatal chronic systemic disease and an acute, benign respiratory infection.

In infected tissues *C. immitis* apppears as spherules, or sometimes as a mixture of spherules and hyphae. The **spherules** are thick-walled structures which may be as small as 5 μ in diameter, but at maturity are usually 20 to 60 μ (Fig. 43-20). They are filled with a few to several hundred globular or irregularly shaped **endospores,** varying from 2 to 5 μ in diameter. When the large spherules rupture the individual endospores are released and in turn develop into spherules: they enlarge, acquire thickened walls, and form their own internal endospores.

Growth of *C. immitis* is not inhibited by chloramphenicol or by cycloheximide, which are usually added to cultures. The organism

Fig. 43-20. Spherules of *Coccidioides immitis* in a tissue of an infected mouse. (Courtesy of Dr. D. Pappagianis.)

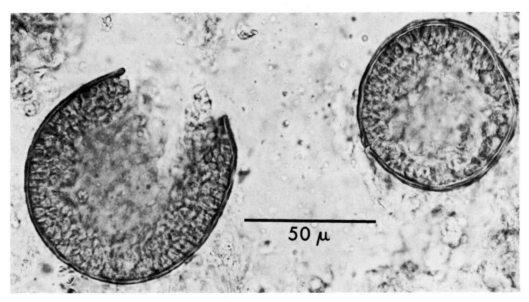

50 μ

is dimorphic: when the spherules are cultivated on Sabouraud's agar or other simple media, even at 37°, they grow out in mycelial form. Growth is rapid, with fluffy white mycelia appearing within about 5 days. A characteristic feature of the hyphae is the cask-shaped **arthrospores,** which alternate with smaller, clear hyphal cells (Fig. 43-21). The hyphae of older cultures fragment easily and release huge numbers of arthrospores; since these are easily airborne and highly infectious, unusual care is required in handling the cultures.

The mycelial (saprophytic) form is readily converted into the spherule (parasitic) form by inoculating mycelial fragments or arthrospores intraperitoneally into mice or guinea pigs. Most mycelia and arthropores degenerate, but several days after inoculation some arthrospores become enlarged, spherical, thick-walled, embryonal spherules: by 1 week, when about 40 μ in diameter, radial partitions appear and subdivide the nascent spherule into cells that subsequently become endospores.

Fig. 43-21. Arthrospores in a culture of *Coccidioides immitis.* The chain-like arrangement of spores, separated by what appear to be empty or vacuolated cells, is characteristic. (Courtesy of Dr. D. Pappagianis.)

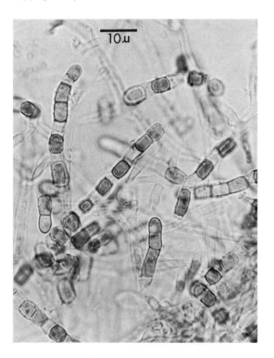

The differentiation of mycelia into spherules also takes place in vitro, but with difficulty. Enrichment of cultures with ascitic fluid, serum, or blood, and incubation at 37° favor the conversion, but it is seldom extensive; mixtures of spherules and hyphae are obtained in such enriched cultures, and some of the developing spherules remain linked together to form hyphal chains.

Pathogenesis.   *C. immitis* grows as a saprophyte in desert soils of southwestern United States and northern Mexico. Infection is established by inhalation of airborne spores, and the disease is known as **coccidioidomycosis.**

During World War II the establishment of military bases in areas of the United States where the disease is endemic provided a unique opportunity to observe the evolution of coccidioidal infections in large numbers of freshly exposed persons, in whom onset of infection could be determined from their immune responses. In about 60% of infected persons disease is not clinically apparent: infection is revealed only by the acquisition of delayed-type hypersensitivity to antigens of *C. immitis.* In about 40%, however, acute pneumonitis develops, often with pleurisy, and various skin eruptions may also occur. For example, erythema multiforme and erythema nodosum are fairly common; these are sterile skin lesions that probably represent allergic responses to fungal antigens (or perhaps fungus-modified tissue antigens). Similar sterile skin eruptions appear, probably for similar reasons, in some other infections, e.g., with *Mycobacterium tuberculosis* (Ch. 35) and β-hemolytic streptococcus (Ch. 26). About 5% of infected persons ultimately develop chronic pulmonary cavitary disease, resembling pulmonary tuberculosis. And in less than 1% of all infected persons dissemination occurs, with infected granulomatous lesions in many organs and tissues, including skin, bone, joints, and meninges.

The inflammatory lesions of acute pneumonitis due to *C. immitis* are histologically like those caused by pyogenic bacteria. But in the chronic pulmonary disease and in the disseminated lesions the inflammation is granulomatous and is characterized by abundant histiocytes, giant cells, and caseation necrosis. Small spherules are found within macrophages or giant cells, and the larger, more mature spherules lie freely in tissue spaces (Fig. 43-20). The spherules are readily visualized with a number of special stains, e.g., the periodic acid–Schiff stain. In the walls of pulmonary cavities hyphae are also seen; thus, **both** forms of this dimorphic fungus (spherules plus hyphae) may be seen in cavitary pulmonary lesions, as in enriched cultures (see above).

Immune Response. **Coccidioidin** is used in precipitin reactions, C′-fixation assays, and skin tests. It is a crude filtrate of a mycelial culture, grown in the same synthetic medium used for the preparation of tuberculin (Ch. 35), histoplasmin, and blastomycin. The skin reaction is of the delayed type, with erythema and induration; those greater than 5 mm in diameter are considered positive. The earliest immune manifestation of infection is a positive skin test; only with more protracted infection do precipitins and C′-fixing antibodies become detectable.

During the first week with overt symptoms 80% of patients in one extensive study had positive coccidioidin skin tests, whereas only 50 % and 10% gave positive precipitin and C′-fixation reactions, respectively. In those with self-limited disease (spontaneously cured), precipitins in serum gradually diminished and were not detectable 4 to 5 months later. Complement-fixing antibodies tended to appear later than precipitins and to persist longer, becoming negative only several years after apparent cure. With active, disseminated disease the titers of complement-fixing antibodies in serum sometimes rose to relatively high levels (e.g., over 1:16) but decreased again as the disease progressed to its terminal stages. Cross-reactions may occur with culture filtrates of B. dermatitidis and H. capsulatum.

A decrease in intensity of the skin response will often, but not invariably, occur in clinically well persons who move away from areas where C. immitis is endemic; their skin reaction may become entirely negative within 12 months. Hypersensitvity may, however, persist indefinitely in others, without apparent disease. This persistence is probably due to survival of viable endospores within healed, walled-off scars: in experimental animals viable organisms have been found in fibrotic and calcified scars as long as 2 to 3 years after infection.

Since coccidioidin is often injected repeatedly in the same individuals during clinical and epidemiological studies the question arises whether skin testing itself can induce coccidioidin hypersensitivity. Fortunately, repeated injections of humans with coccidioidin, in the amounts and concentrations used in routine skin tests, have not induced delayed-type hypersensitivity. Yet this type of response to coccidioidin has been induced in guinea pigs by repeated intracutaneous injections; the discrepancy is curious, since delayed-type hypersensitivity to most antigens is usually induced at least as easily in man as in guinea pigs.

Coccidioidin at high concentration elicits cross-reacting delayed-type skin responses in persons with blastomycosis or with histoplasmosis; conversely, blastomycin and histoplasmin at high concentrations elicit responses in those with coccidioidomycosis. At lower concentrations, however, coccidioidin seems to provide a relatively specific test. Though its value is somewhat limited at present by variations in potency among different lots, the skin tests do provide useful information.

Diagnosis. A provisional diagnosis of coccidioidomycosis is usually based on epidemiological considerations, clinical manifestations, the skin response to coccidioidin, and the detection of antibodies. Definitive diagnosis, however, requires that the C. immitis spherules be identified in sputum, exudates, or tissue sections.

Cultivation of C. immitis in vitro for this purpose is hazardous, because cultures release large numbers of airborne infectious arthrospores; it is best carried out by experienced personnel with access to ventilated hoods. Transmission of C. immitis to laboratory animals is, however, a relatively simple and safe procedure. Sputum or exudate is treated with penicillin, streptomycin, or chloramphenicol, and centrifuged; the sediment is injected either in the testes of guinea pigs or intraperitioneally in mice. If C. immitis is present C′-fixing antibodies may appear in the guinea pig, and mice develop disseminated disease; moreover, the testes and the mouse tissue fluids should contain characteristic spherules with endospores.

Therapy. When confined to the lungs coccidioidomycosis is usually self-limited and heals with scarring. However, the disseminated disease was invariably fatal until the polyene antibiotics became available. Treatment with amphotericin B is now highly encouraging, but failures occur because prolonged therapy is necessary and severe intoxication is common.

Epidemiology. The fungus has a predilection for growth in desert soils, especially after winter and spring rains; and windborne arthrospores readily infect man. Through skin testing of large human populations it has been established that coccidioidomycosis is prevalent in the southwestern United States: central California (especially in the San Joaquin valley), Arizona, New Mexico, western Texas, and southern Utah. In some of these areas 50 to 80% of the population reacts to coccidioidin. The organism has also been found in northern Mexico, parts of Argentina, and Paraguay.

Other mammals are also easily infected, e.g., wild rodents, dogs, and cattle. Histological examination and culture of lungs of trapped wild rodents, and skin testing of domestic cattle, have provided additional important means for identifying regions of prevalence.

PARACOCCIDIOIDES BRASILIENSIS (SOUTH AMERICAN BLASTOMYCOSIS)

The organism that causes South American blastomycosis was first recognized in Brazil, and was thought to be very similar to *C. immitis,* hence its name, *Paracoccidioides brasiliensis*. A number of other designations have been suggested, e.g., that it be classified in the genus *Blastomyces* and named *B. brasiliensis*. The fungus appears in infected tissues

Fig. 43-22. Budding of yeast cells in a culture (37°) of *Paracoccidiodes brasiliensis.* Multiple buds, attached to their mother cell by constricted tubes, are characteristic. Compare with the broad-based, unipolar buds characteristic of *Blastomyces dermatitidis* (Fig. 43-17). ×450, reduced. (Courtesy of Dr. George Kobayashi.)

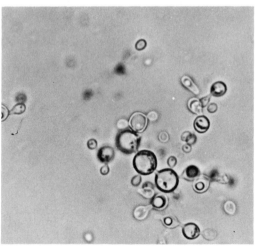

as large, spherical or oval yeast cells, 10 to 30 μ in diameter, and sometimes even 60 μ. Characteristically, multiple buds sprout from a single mother cell and remain attached to it by narrow constricted bands (Fig. 43-22). When the buds are about the same size and all quite small their arrangement around the mother cell is distinctive and this form is often referred to as a **pilot's wheel.** The buds may, however, be equal in size to the mother cell and still remain attached, but because their attachment is inconspicuous they are often referred to as **satellite cells.** Chains of budding cells are also seen.

P. brasiliensis is dimorphic, and in culture at room temperature on Sabouraud's medium it grows as a mycelium with chlamydospores. Conversion to yeast cells is induced by enrichment of the medium (e.g., with brain-heart infusion), adequate moisture, and increase in incubation temperature to 37°.

Pathogenesis.   The disease produced by *P. brasiliensis* is called **paracoccidioidomycosis, paracoccidioidal granuloma,** or **South American blastomycosis.** The earliest lesions arise in the mucous membranes of the mouth or nose and spread by direct extension, e.g., over the mucocutaneous borders to involve the face. Dissemination also occurs, with frequent involvement of lymphoid tissue (including the spleen). The intestinal tract may be involved: the lesions begin in submucosal lymphoid tissue, and may lead to ulceration and even perforation. Subcutaneous abscesses can appear, and by extension to the skin surface they form large, unsightly crusted and ulcerated lesions.

Histologically, skin lesions appear as pyogenic abscesses and granulomatous inflammation with epithelioid cells, giant cells, and necrotic centers. Large spherical or oval yeast cells, with multiple circumferential buds (pilot's wheel or satellite forms), may be observed in routine hematoxylin-eosin stains of tissue sections, but are more clearly brought out with the periodic acid–Schiff reaction. The yeast cells sometimes appear as chain and may be found within giant cells.

Diagnosis.   Detection of *P. brasiliensis* in tissue sections or in exudates, and cultivation as yeast and mycelial forms, establish the diagnosis. Chloramphenicol and cycloheximide are added to Sabouraud's medium and the cultures are maintained as mycelia at room temperature, since these antibiotics inhibit the growth of the yeast cells (but not the mycelia) of this fungus. Transmission to laboratory animals is possible but is not often resorted to as a diagnostic procedure.

Therapy. The disseminated disease is slowly progressive and was formerly invariably fatal. However, amphotericin B arrests the spread of lesions.

Epidemiology. *P. brasiliensis* is probably a soil saprophyte. The disease it causes has been reported mainly in Brazil, and also in most other South American and Central American countries. It is largely a disease of rural areas, where the infection probably arises from the pernicious habit of cleaning teeth with leaves and twigs, and thereby by directly implanting conidia and mycelial fragments in mucous membranes of mouth and gums. Perianal lesions are also frequent, probably because of comparable toilet habits. Thus, of all the principal systemic mycoses, that due to *P. brasiliensis* is notable in that it is **not** respiratory in origin.

MYCOSES DUE TO OPPORTUNISTIC FUNGI

A number of fungi are not pathogenic in healthy humans, but may behave as virulent pathogens in those suffering from a variety of disorders (e.g., malignant lymphomas, severe diabetes), and in those treated intensively with broad-spectrum antibacterial drugs or with immunosuppressive measures. In addition, among the pathogenic fungi discussed above, *C. neoformans, H. capsulatum, B. dermatitidis,* and possibly even *C. immitis* are also somewhat opportunistic, causing progressive infections more frequently under debilitating conditions. The frankly opportunistic fungi are mostly species of *Candida, Aspergillus, Rhizopus,* and *Mucor.*

CANDIDA ALBICANS (CANDIDIASIS)

C. albicans is dimorphic. At the surface of a rich agar medium it grows as oval budding yeast cells, but deeper in the medium hyphae

are found; and **both** forms are characteristically seen in infected tissues and in most cultures. Some hyphae, **pseudohyphae,** have recurring constrictions, as though a chain of sausage-shaped cells were joined end to end (Fig. 43-23).

C. albicans is readily grown on conventional media at room temperature or at 37°. In cultures on agar the early colonies are smooth, creamy, and bacteria-like, but the older, larger colonies appear furrowed and rough. Cultivation on cornmeal agar stimulates the formation of characteristic thick-walled spores (chlamydospores), which distinguish *C. albicans* from other candidae (Fig. 43-23).

On the basis of colonial morphology and serological and fermentation reactions, several medically important species of *Candida* have been distinguished. Of these *C. albicans* is by far the most frequent cause of disease in man (candidiasis); rare species isolated from human lesions include *C. krusei, C. parakrusei,* and *C. parapsilosis.* Because antibodies to *C. albicans* cross-react extensively with the other species of *Candida,* and even with species of *Cryptococcus* and true ascomycetous yeasts (*Saccharomyces*), the tests require hyperimmune rabbit sera absorbed with various cells.

Pathogenesis. *C. albicans* and other species of *Candida* are frequently present on the normal mucous membranes of mouth, vagina, and intestinal tract. When they become invasive, under the special circumstances mentioned above, they establish a variety of acute or chronic, localized or widely disseminated lesions. The following are some examples. 1) **Thrush (oral candidiasis)** consists of discrete or confluent white patches on the mucous membranes of the mouth and pharynx resembling the exudate of diphtheria. The patches are composed of hyphae and yeast cells, and they occur particularly during the first few days of life in the newborn (resulting from infection during birth), and in persons in the terminal stages of a wasting disease, e.g., carcinomatosis. 2) Vaginal mucous membranes are occasionally invaded (**vulvovaginal candidiasis**) during pregnancy and in diabetes. 3) Invasion of bronchial and pulmonary tissues (**bronchopulmonary candidiasis**) is usually secondary to chronic bronchial obstruction with impaired drainage of secretions (e.g., bronchial carcinoma, bronchiectasis). 4) Infections in skin areas that are continu-

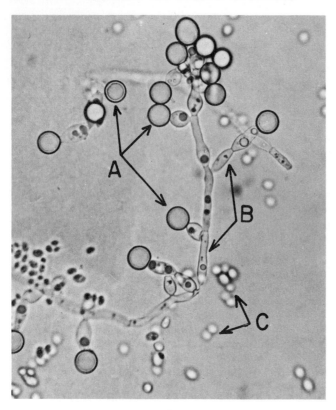

Fig. 43-23. Multiplicity of structural forms in a culture of *Candida albicans*. A = chlamydospores; B = pseudohyphae (elongated yeast cells, linked end to end); C = budding yeast cells (blastospores). ×450, reduced. (Courtesy of Dr. George Kobayashi.)

ously wet and macerated **(intertriginous candidiasis)** are common in the perineum and inframammary folds, and on the hands of those whose occupation requires prolonged immersion in water. 5) **Endocarditis** due to *Candida* is rare, but is occasionally seen in drug addicts, among others; the organisms most often isolated are not *C. albicans,* as in the forms of candidiasis mentioned above, but other species, especially *C. parapsilosis.*

Some individuals with candidiasis develop sterile vesicular or papular skin lesions, called monilids (since the genus *Candida* was formerly called *Monilia* and candidiasis was called moniliasis). These lesions, presumably allergic, resemble the dermatophytids observed in dermatophyte infections (see below).

The clinical impression of heightened susceptibility to candidiasis in diabetes, and during treatment with broad-spectrum antibacterial drugs and corticosteroids, has been supported by experimental observations. For example, mice given cortisone are unusually susceptible to lethal infection with *C. albicans*. And in one study of humans under treat-ment with a tetracycline, over 60% had positive rectal cultures for *C. albicans,* whereas all had had negative cultures prior to therapy. It is not clear whether the reduction in normal intestinal bacterial flora favors growth of *Candida* by decreasing competition for a nutrient or production of inhibitory substances.

Diagnosis. It is not surprising that many normal human sera (15 to 30% in one study) specifically agglutinate cells of *Candida* species, as these organisms are so commonly found on normal mucous membranes; and though agglutinin titers tend to be higher in those with frank candidiasis, serological tests are of little diagnostic value. Similarly, skin tests with aqueous extracts of *C. albicans* **(oidiomycin)** are positive in most normal persons. A presumptive diagnosis of candidiasis is usually made by microscopic demonstration of abundant hyphae and yeast cells in scrapings of lesions, and the diagnosis is supported by isolation of the organism in cultures. However, *Candida* is so ubiquitous that it is often difficult to decide whether or not it is the causative agent. The response to therapy aids in arriving at a decision.

Therapy. Polyene antibiotics are effective: nystatin is generally applied locally to accessible lesions, and amphotericin B is administered orally in the treatment of severe visceral infections.

Epidemiology. Infection of the newborn, by *C. albicans* from the birth canal, is one of the few instances in which a fungus infection is clearly transmitted from one person to another. Usually, however, candidiasis is due to increased susceptibility to a member of the normal human flora (Ch. 42). The underlying mechanisms are obscure.

ASPERGILLUS FUMIGATUS (ASPERGILLOSIS)

Many species of *Aspergillus* have been recognized in nature, but only seven have so far been associated with human disease (**aspergillosis**). Of these, *A. fumigatus* accounts for over 90% of all infections. Most aspergilli, including *A. fumigatus,* are **not dimorphic:** they grow only in mycelial form.

Colonies grow well over a wide temperature range, and *A. fumigatus* can thrive up to 50°. Growth is inhibited by cycloheximide. In culture the mycelia are powdery and have a dark bluish-green cast, hence the name **fumigatus** (L. "smoky"). Conidiophores are up to 500 μ long; each bears a dome-shaped vesicle (ca. 20 to 30 μ in diameter) with a single row of sterigmata (see Fig. 43-5) arranged about the distal half. From these, green conidia (2 to 5 μ in diameter) grow in parallel linear chains (Fig. 43-5), thus accounting for the generic name (L. *aspergillus* = brush).

As noted earlier, most fungi that abound in the environment of birds do not grow well at their high body temperatures (e.g., *Cryptococcus neoformans, Histoplasma capsulatum*). Birds are, however, highly susceptible to infection with various thermophilic species of *Aspergillus* (especially *A. fumigatus*), and commonly suffer fatal aspergillosis. In cattle and sheep abortions can be caused by aspergillosis, but may be due to a toxin produced by the fungus, rather than to overwhelming infection.

Airborne species of aspergilli are ubiquitous, and since these fungi thrive at elevated temperatures, they tend to be particularly abundant in damp, decaying vegetation heated by bacterial fermentations. In compost piles most micro-organisms cease to grow as the temperature rises, but aspergilli flourish under these conditions and can become almost a pure culture.

Pathogenesis. Farmers and others who handle decaying vegetation are often heavily exposed to spores of *Aspergillus.* Asthma and rhinitis due to allergy to antigens of these spores are common; and under predisposing conditions progressive infection can be established. Such conditions include not only those mentioned above (Opportunistic fungi), but also chronic pulmonary diseases with impaired ciliary activity in the bronchi (e.g., bronchiectasis, pulmonary tuberculosis, and bronchial neoplasms).

Most initial infections are pulmonary, following inhalation of spores. These germinate, and hyphae grow and penetrate contiguous tissues by direct invasion (Fig. 43-24). They tend particularly to invade blood vessel walls, producing angiitis and thromboses; severe hemoptysis is conspicuous and life-threatening in pulmonary aspergillosis. Infected emboli may also establish wide-spread metastatic granulomatous lesions in various organs.

Severe local infection can also arise by direct implantation of *Aspergillus* spores in

Fig. 43-24. Hyphae of *Aspergillus fumigatus* in the wall of a pulmonary cavity. Hematoxylin-eosin stain. ×600.

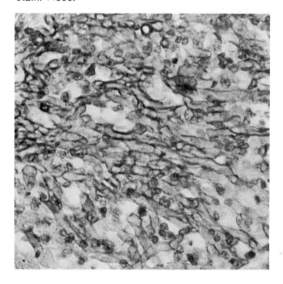

nasal sinuses, with resulting cellulitis of the sinuses and face; in the eye, especially during local treatment with corticosteroids; and in the external ear canal in the presence of concomitant chronic inflammatory disease.

Experimental evidence supports the clinical impression that corticosteroid therapy increases susceptibility to aspergillosis: untreated mice exposed to aerosolized spores developed only mild, transient pneumonitis, but cortisone-treated mice developed fatal pulmonary aspergillosis.

Diagnosis.   Since species of *Aspergillus* are frequent contaminants in cultures, the pathogenic significance of a particular isolate is not easily evaluated. In general, if a species of *Aspergillus* is consistently isolated from a particular patient, in repeated cultures of sputum, exudates, or scrapings of infected tissues, it is presumed to be significant. A definitive diagnosis is established by demonstrating hyphae, which are often abundant, in tissue sections. The hyphae are easily seen on hematoxylin-eosin and on Gram stains: they are 3 to 4 μ in diameter, exhibit dichotomous branching, and are septate (Fig. 43-24). The tissue reaction takes the form of either suppurative nuclear leukocytes, or granulomatous inflammation.

Therapy.   The prognosis in pulmonary aspergillosis is grave, but encouraging therapeutic results have been obtained with amphotericin B.

PHYCOMYCETES (PHYCOMYCOSES)

Seven genera of the class Phycomycetes have been recognized as occasional causes of human disease. *Rhizopus* species are most often involved. The disease produced was formerly referred to as **mycormycosis,** but the generic term **phycomycosis** is more appropriate. *Phycomyces,* like *Aspergillus,* is **not dimorphic:** growth is mycelial in both infected tissues and cultures.

Pathogenesis.   Infection occurs by inhalation of spores, and, rarely, by their traumatic implantation in broken skin or mucous membranes. Severe infections of the central nervous system occur in poorly controlled diabetes. Widely disseminated visceral lesions have been described as complications of severe malnu-

trition, uremia, and hepatic insufficiency, as well as in persons receiving corticosteroids or broad-spectrum antibacterial drugs.

Hyphae are abundant in infected tissues. They grow by direct extension through contiguous tissues and tend to invade blood vessel walls, producing angiitis, thrombi, and ischemic necrosis; in addition, emboli establish metastatic infections in many organs. Acute inflammation is characteristically present at sites of infection.

A useful experimental model for phycomycosis has been developed in rabbits made acutely diabetic with alloxan or treated with corticosteroids. In such animals *Rhizopus* spores introduced into the paranasal sinuses lead to severe nasal, pulmonary, and cerebral lesions, which simulate the disease in diabetic humans.

Diagnosis.   Distinctive hyphae are seen in infected tissues stained with hematoxylin-eosin. In contrast to the hyphae of *Aspergillus,* those of *Phycomyces* are relatively huge (up to 15 μ in diameter) and devoid of septa, though they may have occasional incomplete cross-walls, and they branch haphazardly (Fig. 43-25; see also Fig. 43-11).

Because their walls are rich in chitin and probably correspondingly poor in glucans and mannans, phycomycetes stain poorly with the periodic acid–Schiff stain (Fig. 43-13). Identification of the individual species requires considerable experience. In the following paragraphs the main properties of the more commonly encountered genera are outlined.

Rhizopus.   Species associated with phycomycosis grow rapidly on Sabouraud's medium and form coarse and woolly mycelia, which are initially white but subsequently become peppered with black and brown fruiting structures (sporangia). The hyphae are nonseptate and colorless. Asexual reproductive elements consist of long, unbranched stalks (sporangiophores), which arise directly from a cluster of rhizoids (root-like structures) and are each topped by a spherical sac (sporangium). The latter are dark, and when mature are filled with spores. The species isolated most often from human phycomycosis are *R. oryzae* (most frequently), *R. arrhizus,* and *R. nigricans.*

Mucor.   Cultures of these species grow rapidly on Sabouraud's medium and form fluffy mycelia which are initially white, and subsequently are gray to brown. Hyphae are usually nonseptate, but in old cultures irregular cross-walls are occasionally seen. Sporangiophores, which do not originate from

Fig. 43-25. Phycomycosis (due to *Basidiobolus haptosporus*). The hyphae in the tissue section are unusually wide (about 15 μ in diameter, see arrow), lack septa, and are surrounded by an intense leukocytic infiltrate, including many eosinophils. Hematoxylin-eosin stain. ×450, reduced. (Courtesy of Dr. George Kobayashi.)

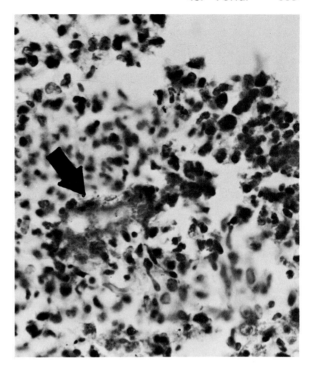

rhizoids, form thick upright tufts. Some are branched, and each terminates in a spherical sporangium. The species most often identified in human infections is *M. corymbifer* (*Absidia corymbifera*).

Phycomycosis is also occasionally due to species of the following genera: *Absidia, Mortierella,* and *Basidiobolus.*

SUBCUTANEOUS MYCOSES

Subcutaneous mycotic infections are usually initiated by penetration of skin with contaminated splinters, thorns, or soil. Once established, these infections tend to remain localized in subcutaneous tissues and to be extremely persistent. The diseases are classified as **sporotrichosis, chromoblastomycosis,** and **maduromycosis.**

SPOROTRICHUM SCHENCKII (SPOROTRICHOSIS)

Sporotrichosis is characterized by an ulcerated lesion at the site of inoculation, followed by multiple nodules and abscesses along the superficial draining lymphatics. Only rarely is there dissemination to the meninges. In infected tissues the organism appears as cigar-

shaped, budding yeast cells (Fig. 43-26); these cells are usually scarce but may sometimes be recognized by the periodic acid–Schiff stain or with fluorescein-labeled antibodies.

S. schenckii is dimorphic. When pus or curettings from skin lesions are cultured on Sabouraud's medium at room temperature (with chloramphenicol and cycloheximide) the fungus grows rapidly in mycelial form as a flat, moist colony. Hyphae are slender (about 2 μ in diameter) and septate, and conidia are individually attached by sterigmata to a common conidiophore (Fig. 43-5). Pigmentation of the conidia accounts for the dark color of the mycelium, the intensity of pigmentation depending on the level of thiamine in the medium. The pigment is allegedly melanin, and tyrosinase has been identified in mycelia.

Identification of yeast cells is also of diagnostic value, since they are usually difficult to visualize in human lesions. Hyphae are converted to yeast forms by cultivation at 37°, at slightly increased pCO_2, in media enriched with proteins, thiamine, and biotin. This change is also brought about by inoculating mycelial fragments into testes of mice; pus withdrawn after 2 to 3 weeks contains abun-

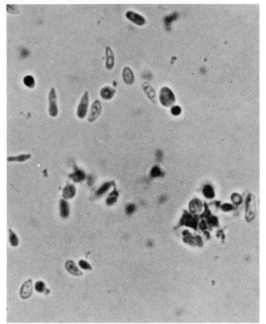

dant characteristic yeast cells. In addition, mouse inoculation differentiates *S. schenckii* from several saprophytes that are morphologically similar but not pathogenic.

S. schenckii has been isolated from soil and plants, and can apparently grow in wood. There is little doubt that penetration of skin by contaminated splinters and thorns is the principal means of introducing infection. The disease is usually sporadic among farmers and gardeners, but a few industrial outbreaks have occurred among workers exposed to batches of heavily infected timbers and plants.

CHROMOBLASTOMYCOSIS

A group of slowly growing, dimorphic fungi produce the subcutaneous mycoses known as chromoblastomycosis. Though the first case of chromoblastomycosis was reported in Boston, the disease occurs primarily in the tropics, with infection arising from penetration of skin by contaminated splinters or soil. The infection is thus seen mostly on the legs of bare-legged laborers, and lesions appear as warty, ulcerating, cauliflower-like growths.

In draining lesions the fungi appear as thick-walled, round cells, about 6 to 10 μ in diameter, whose dark brown color gives the disease its name. The yeast cells apparently multiply by fission rather than by budding. In cultures on Sabouraud's medium at room tem-

perature darkly pigmented mycelial colonies are formed slowly. Several species have been distinguished on the basis of conidial arrangements, but their classification is unsettled. According to one terminology, they are *Fonsecaea pedrosoi, F. compacta, F. dermatitidis, Cladosporium carrionii,* and *Phialophora verrucosa.*

MADUROMYCOSIS

Mycetoma is the generic term for localized destructive granulomatous and suppurative lesions that usually affect the foot or hand, and involve skin, subcutaneous tissues, bone, and fascia (Ch. 36). Mycetomas usually originate with injuries. They are especially prevalent in the tropics.

Lesions are characterized by multiple burrowing sinus tracts that extend through soft tissues and penetrate to the skin surface. The draining pus contains granules (small pieces of colonies of the causative microorganism), which vary in texture, color, and shape, depending on the microorganism, and also vary in size (from ca. 0.1 to 2 mm).

The microorganisms that cause mycetoma are either fungi or actinomycetes (Ch. 36); the fungal mycetomas are often referred to as **maduromycosis.** Diagnosis is based on demonstration of fungus cells directly in KOH-treated granules and pus, and by culture. At least 13 species of fungi have been identified, including *Madurella mycetomi, M. grisea, Allescheria boydii, Phialophora jeanselmei,* and *Aspergillus nidulans.*

Surgical drainage is an important adjunct to therapy. Polyenes, which have not yet been tried extensively, may prove of value, since the causative agents are suceptible in vitro.

CUTANEOUS MYCOSES (DERMATOMYCOSES)

The dermatophytes are fungi that infect only epidermis and its appendages (hair and nails), i.e., structures in which keratin is abundant. The ensuing skin lesions are usually roughly circular, tend to expand equally in all directions, and have raised serpiginous borders. They were therefore thought in ancient times to be due to worms or lice, and they are still called **ringworm** or **tinea** (L., "worm or insect larva"). The names are usually qualified by the area of the skin involved: e.g., ringworm of the scalp **(tina capitis),** of the body **(tinea corporis),** of the groin ("jock itch," **tinea cruris),** and of the feet (athlete's foot, **tinea pedis).**

Dermatophytes are not dimorphic. In infected skin lesions they all look alike with septate hyphae and arthrospores (Figs. 43-27 and 43-28). However, additional differentiated structures appear in cultures and provide the basis for identification.

ATTACK ON KERATIN

The predilection of dermatophytes for epidermis, firmly established by clinical observations, is also demonstrable experimentally. For example, when spores or mycelial fragments are injected intravenously into guinea pigs no lesions develop; but if an area of skin is abraded at the time of the injection dermatophyte infection appears in the scarified skin a few weeks later.

In view of the evident affinity of dermatophytes for keratin-rich tissues, one might expect these fungi to have an unusual capacity to degrade and utilize keratin. In fact, how-

ever, they do so at only a low rate in vitro. Moreover, only a few species of dermatophytes are isolable from soil, where degradation of keratin depends on various other saprophytic fungi and bacteria. Evidently the pathogenicity of dermatophytes depends on more than their ability to attack keratin.

Keratin is a fibrous and very insoluble protein, stabilized by the disulfide groups of frequent cystine residues; and in its native state it is resistant to most proteolytic enzymes. An enzyme preparation of *Streptomyces* cleaves disulfide bonds effectively and digests keratin.* *Microsporum gypseum,* on the other hand, one of the few dermatophytes often isolated from soil and more keratinolytic than the others, does not rupture keratin disulfides and digests this protein to only a limited extent. Nevertheless, many dermatophytes can be cultured on sterile hair and they dissolve localized segments of hair fibers.

Though most dermatophytes are epidermal parasites in nature, most species grow well in simple media, with ammonium salts and glucose as sole sources of nitrogen, carbon, and

---

* This keratinase is used industrially to remove hair from hides in the preparation of leather.

Fig. 43-27. Skin scraping treated with 10% KOH to show hyphae of a dermatophyte among epidermal debris from a human skin lesion. ×450, reduced. (Courtesy of Dr. George Kobayashi.)

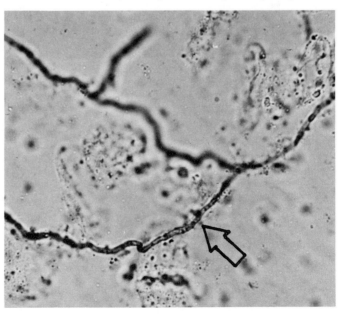

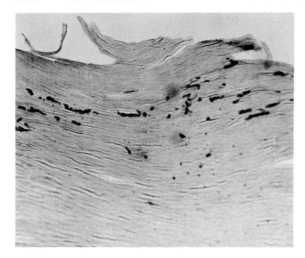

Fig. 43-28. Periodic acid-Schiff stain of skin section of a human lesion showing a dermatophyte in the stratum corneum. See Fig. 43-13. ×100, reduced. (Courtesy of Dr. George Kobayashi.)

energy. Growth is more vigorous, however, in media enriched with proteins or amino acids.

Invasion of Hair. In infection of hairs hyphae first grow from the epidermis into hair follicles, and then into hair shafts. In the **endothrix** type of infection the hyphae then grow only **within** the hair shaft, where they form long, parallel rows of arthrospores (Fig. 43-29). In **ectothrix** infections they grow both within and on the external surface of the hair shaft (Fig. 43-30).

EPIDEMIOLOGY

About 15 species of dermatophytes are found primarily in human skin **(anthropophilic).** Many others are indigenous in domesticated and wild mammals **(zoophilic);** and a few may be free-living saprophytes, since they are isolable from soil **(geophilic),** e.g., *M. gypseum, T. mentagrophytes,* and *T. ajelloi.* Infection is transmitted, though with difficulty (see below), from man to man, or animal to man, or vice versa by direct contact or by contact with infected hairs and epidermal scales (e.g., barber shop clippers, shower room floors, etc.). The reservoir of animal infection is huge: about 30% of dogs and cats in the United States are infected with *M. canis,* a frequent cause of ringworm of the scalp in children.

The incidence of different dermatomycoses varies with age. For example, intertriginous infection of feet (athlete's foot) is common in adults but rare in children, whereas the opposite is true for ringworm of the scalp. Resistance of adults to scalp infection has been linked to the increased secretory activity of sebaceous glands at puberty and the antifungal activity of the C7 to C11 saturated fatty acids in sebum.*

Most dermatophytes have a worldwide distribution, but a few species are restricted geographically. With increased travel in the past 25 years even the localized species are becoming more widely distributed. For example, *T. tonsurans* is endemic in Mexico but has only recently become common in the United States.

IMMUNITY

Human resistance to some dermatophytes is emphasized by the low incidence of conjugal infections; for example, in one study of 60 *T. rubrum*-infected persons, followed for 1 to 20 years, not one spouse acquired active infection. This natural human resistance is probably not due to conventional immune reactions: circulating antibodies are not, as a rule, demonstrable in sera of persons with

---

* Undecylenic acid, an unsaturated C11 fatty acid, is widely used for topical therapy of some dermatomycoses.

Fig. 43-29. Endothrix hair infection with *Trichophyton tonsurans*. Chains of arthrospores are localized within the hair shaft. ×450, reduced. (Courtesy of Dr. George Kobayashi.)

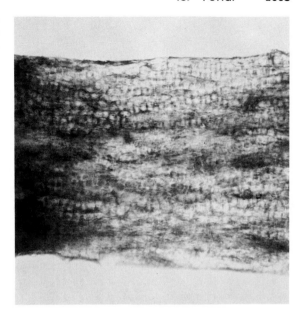

dermatomycosis. However, a fungistatic factor, which seems not to be an immunoglobulin, is demonstrable even in normal serum. This factor may well be responsible for the limited penetration of dermatophyte infections. We do not understand the balance of forces that causes these infections to be so widespread yet difficult to transmit deliberately, and so often self-limited yet difficult to cure.

Hypersensitivity. Persons with dermatophytosis sometimes have another kind of skin lesion, believed to represent an allergic reaction to fungal antigens that spread from the site of infection. These lesions, dermatophytids, are sterile and appear as vesicles symmetrically distributed on the hands.

Hypersensitivity to dermatophytes appears in the course of infection and is usually persistent. Trichophytin, a crude filtrate of broth in which a dermatophyte has been grown, elicits delayed cutaneous responses and sometimes also wheal-and-erythema responses. However, as with tuberculin, these responses do not distinguish between current and prior infection; they have little diagnostic or prognostic value.

TREATMENT

Many infections are eradicated by griseofulvin, but the response varies with the thickness of the keratin and the rate of its replacement.

Infections of scalp and smooth skin are usually cured after several weeks of therapy, but infections of feet, especially of toenails, require many months of continuous treatment. Traditional forms of local therapy with keratinolytic agents are therefore still widely used.

CLASSIFICATION

In identifying specific dermatophytes the form and arrangement of large conidia (macroconidia) is especially important. Additional determinative characteristics are 1) the form and pigmentation of mycelia; 2) the quantity and disposition of small conidia (microconidia); 3) the development of special hyphal structures, e.g., racket-shaped ends of some hyphae (racket mycelia), helically coiled hyphae (spirals), and tightly coiled, twisted hyphae (nodular bodies); 4) the presence of arthrospores and chlamydospores; 5) some physiological characteristics, e.g., growth in culture on sterile hair.

Identification of dermatophytes is hindered by their **pleomorphism**—a word used in mycology in a special sense, to describe the frequent loss of pigmentation and spore formation during laboratory cultivation. The resulting mycelia resist identification. However, transfer to a medium that stimulates sporulation (e.g., potato-glucose agar) reduces the frequency of this conversion. In order to

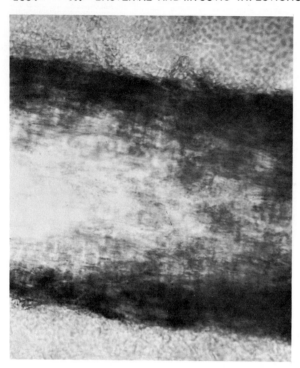

Fig. 43-30. Ectothrix infection of hair by *Microsporum audouini*. Spores are clustered on the surface of the hair shaft. ×450, reduced. (Courtesy of Dr. George Kobayashi.)

maintain sporulating cultures frequent transfers are usually necessary (about once every 10 days), or else storage at −20°.

Genetic crosses have shown that pleomorphic conversion is the result of one or more gene mutations. Thus pleomorphism in dermatophytes resembles phase variation in bacteria (Ch. 8), in which culture conditions select for a mutant strain.

Dermatophytes fall into three genera: in general, *Microsporum* attacks hair and skin but not nails; *Trichophyton* attacks hair, skin, and nails; *Epidermophyton* infects skin and occasionally nails, but not hair. From 20 to 100 species, depending on the classification, cause human infection. A few representative species will be discussed briefly.

Microsporum. Hair infections are of the ectothrix type (Fig. 43-30), with spores packed closely on the external hair surface in a mosaic pattern. In culture the genus is characterized by the production of rough-walled multicellular conidia (Fig. 43-31).

M. audouini is primarily a human pathogen and used to be the most frequent cause of epidemics of ringworm of the scalp in children in the United States. Adults are only rarely infected and animals are highly resistant. When the scalp is irradiated with ultraviolet light (366 nm) infected hairs emit a bright yellow-green fluorescence. This property makes possible the rapid diagnostic screening of large populations of children and also facilitates the selection of infected hairs for culture.

M. canis is primarily a parasite of domesticated and wild animals, and children commonly acquire infections from cats and dogs. An intense but localized inflammatory reaction (called a kerion) develops in the skin and subsides spontaneously after several weeks; it may represent an allergic response. Hairs infected with *M. canis* resemble those infected with *M. audouini* (Fig. 43-30) and also fluoresce in ultraviolet light.

M. gypseum is abundant in soil but is an infrequent cause of human infections. Infected hairs are not fluorescent. Arthrospores appear on the surface or infected hairs and are larger than those of other *Microsporum* species. Sexual forms, with typical asci, have been isolated from cultures (Table 43-3).

Trichophyton. Species of this genus usually produce smooth-walled macronidia in culture, but classification is difficult since spores are often sparse or lacking. A large number of so-called species have been described but are

Fig. 43-31. Macroconidia in a culture of *Microsporum gypseum.* ×450, reduced. (Courtesy of Dr. George Kobayashi.)

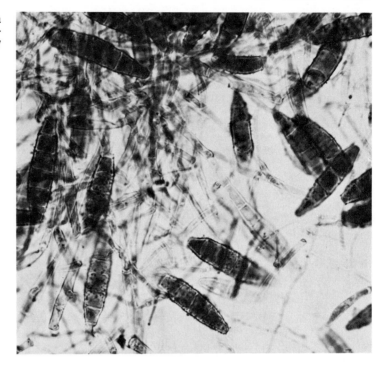

regarded increasingly as variants of 12 or 13 species.

T. schoenleinii is a major cause of **favus,** a severe form of chronic ringworm of the scalp, with destruction of hair follicles and permanent loss of hair. It was the first microbial pathogen of man to be identified (1843), isolated, and established as a pathogen through experimental infection of an animal host. In culture, hyphae are coarse and have knobby, broadened ends and many short lateral branches: these structures, reminiscent of reindeer horns, are called favic chandeliers.

T. violaceum causes endothrix infection in hair, and may also cause favus. It is a common cause of ringworm of the scalp and body in the Mediterranean area.

T. tonsurans causes endothrix infections of hair. With a large influx into the United States of persons from Latin America, ringworm of the scalp is now more frequently due in large cosmopolitan centers to *T. tonsurans* than to *M. audouini.* Early diagnosis of infection is difficult, however, because unlike *M. audouini* infections, hairs infected with *T. tonsurans* do not fluoresce.

T. mentagrophytes and *T. rubrum* are common causes of "athletes foot" and of infections of smooth skin in adult humans. They also infect nails, and *T. mentagrophytes* can cause endothrix infections of hair (scalp and beard).

Epidermophyton. This genus is represented by a single species, *E. floccosum,* and is found only in man. It grows in epidermis (especially in intertriginous areas, as between toes), but hair is not invaded.

SUPERFICIAL MYCOSES

Fungi that produce these infections are limited to invasion of the most superficial layers of skin and hair. Four superficial mycoses are fairly common.

Tinea Versicolor (Pityriasis Versicolor). The disease is common in all parts of the world, but is most frequently seen in warm climates. The infected lesions appear as white or tanned scaly areas on the trunk, and lesions are chronic but asymptomatic. The etiological agent, *Malassezia furfur,* can be readily visualized in clinical specimens (treated with 10% KOH) as clusters of round budding cells (3 to 8 μ in diameter) and mycelial elements. Its taxonomic status is open to question, since attempts to cultivate it were long unsuccessful. However, the cause may be *Pityrosporum furfur,* which has recently been repeatedly isolated from lesions. The yeast-like fungi that constitute this genus require long-chain fatty acids for their growth. *P. ovale,* which morphologically resembles *P. furfur,* is frequently isolated from sebaceous glands and hair follicles but is not pathogenic and is considered to be part of the normal flora of the skin.

A similar disease, **erythrasma,** is caused by a corynebacterium, *C. minutissima.*

Tinea Nigra. These lesions are largely confined to the palms, where they appear as irregular, flat, darkly discolored areas. Infection is particularly prevalent in the tropics, and is rare in the United States. The fungus that causes this disorder is *Cladosporium werneckii.* It appears in scrapings of skin lesions as branched hyphae, and grows slowly as greenish-black colonies on agar, with numerous dark, budding cells.

White Piedra. The fungus that causes this disorder (*Trichosporon cutaneum*) grows on scalp or beard hair. Soft, pale nodules appear on the shafts of hair; they consist of hyphae and oval arthrospores. On agar this fungus grows as soft, creamy colonies that become wrinkled and gray with age, and its septate hyphae fragment easily into arthrospores.

Black Piedra. In this tropical disorder an ascomycete, *Piedraia hortai,* forms hard, dark nodules on the shafts of infected scalp hairs. The nodules contain oval asci with two to eight ascospores. On agar the organism forms greenish-black mycelia that bear chlamydospores.

SELECTED REFERENCES

General Mycology

AINSWORTH, G. C., and SUSSMAN, A. S. (eds.). *The Fungi: An Advanced Treatise,* vols. I, II, III. Academic Press, New York, 1965 (vol. I), 1966 (vol. II), 1968 (vol. III).

HARTWELL, L. H. Biochemical genetics of yeast. *Annu Rev Genet 4:*373 (1970).

KINSKY, S. C. Antibiotic interaction with model membranes. *Annu Rev Pharmacol 10:*119 (1970).

MADELIN, M. F. *The Fungus Spore.* Butterworth, London, 1966.

PONTECORVO, G. The parasexual cycle in fungi. *Annu Rev Microbiol 10:*393 (1956).

RAPER, J. R., and ESSER, K. The fungi. In *The Cell,* vol. VI, p. 139. (J. Brachet and A. E. Mirsky, eds.) Academic Press, New York, 1964.

RAPER, K. B., and FENNELL, D. I. *The Genus Aspergillus.* Williams & Wilkins, Baltimore, 1965.

RAPER, K. B., and THOM, T. *A Manual of the Penicillia.* Hafner, New York, 1968.

ROSE, A. H., and HARRISON, J. S. *The Yeasts,* vols. I, II. Academic Press, New York, 1969 (vol. 1), 1971 (vol. 2).

SERMONTI, G. *Genetics of Antibiotic-Producing Microorganisms.* Wiley, New York. 1969.

Symposium on biochemical bases of morphogenesis in fungi. *Bacteriol Rev 27:*273 (1963).

WEBSTER, J. *Introduction to Fungi.* Cambridge Univ. Press, London, 1970.

WOGAN, G. N. Chemical nature and biological effects of the aflatoxins. *Bacteriol Rev 30:*460 (1966).

Medical Mycology

CAMPBELL, C. C. Serology in the respiratory mycoses. *Sabouraudia 5:*240 (1967).

CONANT, N. F., SMITH, D. T., BAKER, R. D., CALLAWAY, J. L., and MARTIN, D. J. *Manual of Clinical Mycology,* 3rd ed. Saunders, Philadelphia, 1971.

EMMONS, C. W., BINFORD, C. H., and UTZ, J. P. *Medical Mycology,* 2nd ed. Lea & Febiger, Philadelphia, 1970.

HILDICK-SMITH, G., BLANK, H., and SARKANY, I. *Fungus Diseases and Their Treatment.* Little, Brown, Boston, 1964.

LAMPEN, J. O. Amphotericin B and other polyenic antifungal antibiotics. *Am J Clin Pathol 52:*138 (1969).

PEPYS, J., and LONGBOTTOM, L. Immunological methods in mycology. In *Handbook of Experimental Immunology,* p. 813. (D. M. Weir, ed.) Davis, Philadelphia, 1967.

REBELL, G., and TAPLIN, D. *Dermatophytes: Their Recognition and Identification,* 2nd ed. Univ. of Miami Press, Coral Gables, Fla., 1970.

SCHWARZ, J., and BAUM, G. L. Fungus diseases of the lungs. *Semin Roentgenol 5:*3 (1970).

WILSON, J. W., and PLUNKETT, O. A. *The Fungous Diseases of Man.* Univ. of California Press, Berkeley, 1965.

part V

VIROLOGY

RENATO DULBECCO
HAROLD S. GINSBERG

chapter

THE NATURE OF VIRUSES

Distinctive Properties   1010
VIRUSES AS INFECTIOUS AGENTS   1011
Are Viruses Alive?   1014
THE ANALYSIS OF VIRUSES   1014

Preparation of Samples   1014
Electron Microscopic Technics   1015

THE VIRAL PARTICLES   1016

GENERAL MORPHOLOGY 1016
LOCATION OF THE NUCLEIC ACID   1017
THE VIRAL NUCLEIC ACID   1020
Double-stranded Viral DNA   1020
Other Kinds of Nucleic Acids   1023
THE CAPSID   1025
Icosahedral Capsids   1026
Helical Capsids   1028
Theoretical Considerations on the Architecture of the Capsid   1029
THE ENVELOPE   1031
OTHER VIRION COMPONENTS   1033
COMPLEX VIRUSES   1033

ASSAY OF VIRUSES   1033

CHEMICAL AND PHYSICAL DETERMINATIONS   1033
Counts of Physical Particles   1033
Hemagglutination   1034
Assays Based on Antigenic Properties   1036
ASSAYS OF INFECTIVITY   1037
The Plaque Method   1037
The Pock Method   1039
Other Local Lesions   1039
ENDPOINT METHOD   1040
COMPARISON OF DIFFERENT TYPES OF ASSAYS   1041

APPENDIX   1042

QUANTITATIVE ASPECTS OF INFECTION   1042
Distribution of Viral Particles per Cell: Poisson Distribution   1042
Classes of Cells in an Infected Population   1042
MEASUREMENT OF THE INFECTIOUS TITER OF A
        VIRAL SAMPLE   1043
Plaque Method   1043
Endpoint Method   1043
Precision of Various Assay Procedures   1043

Viruses are a unique class of infectious agents. They were originally distinguished because they are especially **small** (hence the original term "filtrable viruses") and because they are **obligatory intracellular parasites.** These properties, however, are shared by some small bacteria, and the truly distinctive features of viruses are now known to lie in their simple organization and composition and their mechanism of replication. A complete viral particle or **virion** may be regarded mainly as a bloc of genetic material surrounded by a coat, which protects it from the environment, serves as a vehicle for its transmission from one host cell to another, and initiates its replication.

DISTINCTIVE PROPERTIES

Composition. The composition of a virus was first determined in 1933 by Schlesinger, who demonstrated that a bacteriophage contained only protein and DNA. A few years later Stanley crystallized tobacco mosaic virus, which was shown to be made up of RNA and protein. Extensive work with many viruses in recent years has confirmed and generalized these findings. It is now accepted that all viruses contain a **single type of nucleic acid,** either DNA or RNA, and a **protein** coat surrounding the nucleic acid. In addition, some viruses contain lipids and carbohydrates. Virions lack constituents fundamental for growth and multiplication, e.g., ribosomes, enzyme systems required for synthesis of nucleic acids and proteins, and systems generating ATP. Virions may contain some **special enzymes (virion enzymes,** Table 44-1) not provided by the host cells. Some of these enzymes are transcriptases required for initiating the viral growth cycle (see Ch. 48). The role of the others is unknown; some may be cellular enzymes adventitiously adsorbed.

Unusual constituents are tRNAs and ribosomes in leukoviruses (Ch. 63); they may be accidentally incorporated during maturation of the viruses (see Ch. 48, Maturation of enveloped viruses).

Size and Structure (Fig. 44-1, Table 44-2). The size of virions was first estimated by W. J. Elford, using filtration through collodion membranes of known pore diameter. Viral sizes are also determined by analytical ultracentrifugation and by electron microscopy. The latter technic seems to provide the most accurate values.

Most viruses are much smaller than bacteria; however, the larger viruses (e.g., vaccinia) are as large as certain small bacteria (e.g., mycoplasma, rickettsia, or chlamydia). The chlamydiae were for a time regarded as large viruses because they also are obligatory intracellular parasites, but their fundamental kinship to bacteria has been established (see Ch. 39).

Virions display a range of highly characteristic morphological types which provide a useful basis for classification. However, the number of distinctive morphological classes is small; hence morphological criteria alone are not adequate.

Mode of Replication. Unlike cells, viruses do not grow in size and then divide because they contain within their coats few or none of the biosynthetic enzymes and other machinery required for their replication. Rather, **viruses multiply by synthesis of their separate components, followed by assembly.** After the protein coat has been shed the viral nucleic acid comes in contact with the appropriate cell machinery, where it specifies the synthesis of proteins required for its own replication. The viral nucleic acid is then itself replicated, the subunits of the viral coat are formed, and these components are finally assembled (see

TABLE 44-1. Virion Enzymes

Enzyme	Virus	Product
DNA transcriptase	Poxvirus (Ch. 54)	Single-stranded RNA
Double-stranded RNA transcriptase	Reovirus (Ch. 60); wound tumor viruses of plants	Single-stranded RNA
Single-stranded RNA transcriptase	Myxovirus (Ch. 56), paramyxovirus (Ch. 57), rhabdovirus (Ch. 58)	Single-stranded RNA complementary to template
RNA-dependent DNA polymerase (reverse transcriptase); DNA-dependent DNA polymerase	Leukovirus (Ch. 63)	DNA-RNA hybrid; single-stranded DNA; double-stranded DNA
Polynucleotide ligase	Leukovirus (Ch. 63)	Closed single-strand breaks in double-stranded DNA
Deoxyribonucleases: Exonuclease, endonuclease	Poxvirus (Ch. 54), leukovirus (Ch. 63)	Oligo- and mononucleotides
Endonuclease	Adenovirus (in the penton, Ch. 52)	Nicks in double-stranded DNA
Nucleoside triphosphate phosphohydrolase	Poxvirus (Ch. 54), reovirus (Ch. 60), frog virus, cytopathic polyhedrosis virus of insects	XTP → XDP
Nucleoside triphosphate phosphotransferase	Many enveloped animal viruses	Phosphate exchange
Protein kinase	Leukovirus (Ch. 63), myxovirus (Ch. 56), paramyxovirus (Ch. 57), herpesvirus (Ch. 53)	Phosphorylated proteins
tRNA aminoacylases	Leukovirus (Ch. 63)	Aminoacylate several tRNAs
Neuraminidase	Myxovirus (Ch. 56), paramyxovirus (Ch. 57)	Splits off N-acetyl neuraminic acid from surface polysaccharides
Endoglycosidase	*E. coli* K bacteriophages	Splits capsular polysaccharides

Chs. 45 and 48). This mode of multiplication accounts for the **obligatory intracellular parasitism.**

Host Range. Viruses multiply only in particular host cells and are accordingly subdivided into three main classes: **animal viruses, bacterial viruses** or **bacteriophages,** and **plant viruses.** Within each class each virus is able to infect only certain species of cells. The host range is determined either by the specificity of attachment of the virions to the cells or by the availability of cellular factors required for viral replication. Specificity of attachment depends on properties of both the virion's coat and specific receptors of the cell surface; these limitations disappear when infection is carried out by the naked viral nucleic acid, because its association with the cells is not restricted by the specificity of the cell receptors. In contrast, limitations at the level of replication still affect the host range for infection by nucleic acid.

VIRUSES AS INFECTIOUS AGENTS

Two properties of viruses explain why they are infectious pathogenic agents: virions produced in one cell can invade other cells and thus cause a spreading infection; and viruses cause important functional alterations of the invaded cells, often resulting in their death.

The role of viruses as infectious agents was recognized long before their true nature was understood. In 1898 Loeffler and Frosch proved that the agent of foot-and-mouth disease in cattle can be transmitted by a cell-free filtrate; cell-free transmission for a plant virus was demonstrated even earlier (Ivanovsky, 1892). These findings paved the way for the recognition of many other viral agents

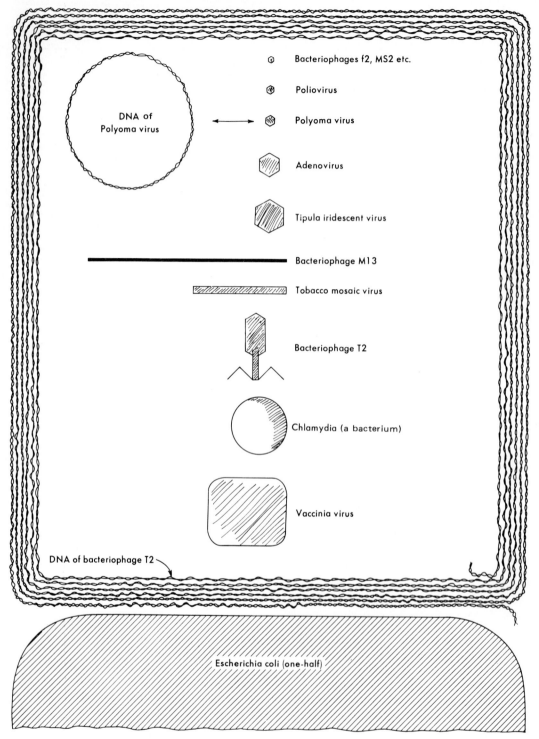

Fig. 44-1. Comparative sizes of virions, their nucleic acids, and bacteria. The profiles, as well as the length of the DNA molecules, are reproduced on the same scale.

TABLE 44-2. Characteristics of Viruses

Morpho-logical class	Nucleic acid*	Example Virus family	Example Virus	Size of capsid (A)	No. of capsomers	Size of virions of enveloped viruses	Special features
Helical capsid							
Naked	DNA	Coliphage fd		50 × 8000			Single-stranded DNA
	RNA	Many plant viruses	Tobacco mosaic	175 × 3000			
			Beet yellow	100 × 8000			
Enveloped	RNA	Myxoviruses	Influenza	90 diameter		900–1000	Fragmented RNA
		Paramyxo-viruses	Newcastle disease	180 diameter		1250–2500 and over	
		Rhabdoviruses	Vesicular stomatitis			680 × 1750	Bullet shape
Icosahedral capsid							
Naked	DNA	Parvoviruses	Adenosatellite	200	12		⎫ Single-stranded cyclic DNA ⎬
			Coliphage φx 174	220	12		⎭
		Papovaviruses	Polyoma	450	72		⎫ Cyclic DNA ⎬
			Papilloma	550	72		⎭
		Adenoviruses		600–900	252		
		Tipula iridescent virus (insects)		1400	1472		
	RNA	Coliphage F2 and others		200–250			
		Picornaviruses	Polio	280	32		
		Many plant viruses	Turnip yellow	280	32		
		Reoviruses		700	92		⎫ Double-stranded RNA ⎬
		Wound tumor virus (plant)					⎭
Enveloped	DNA	Herpesviruses	Herpes simplex	1000	162	1800–2000	
Capsids of binal symmetry (i.e., some components icosa-hedral, others helical)							
Naked	DNA	Large bacterio-phages	T2, T4, T6	Modified icosa-hedral head 950 × 650; helical tail, 170 × 1150			
Complex virions	DNA	Poxviruses	Vaccinia Contagious pustular dermatitis of sheep			2500 × 3000 1600 × 2600	Brick shape

* DNA double-stranded, RNA single-stranded, unless specified in last column.

of infectious disease. Soon the tumor-producing ability of viruses was also indicated by the discovery of the viral transmission of fowl leukosis (Ellerman and Bang, 1908) and of a chicken sarcoma (Rous, 1911). The discovery of bacteriophages, made independently by Twort in England and D'Herelle in France in 1917, was of great significance for the development of virology as a science because it afforded an important model system for investigations of basic virology (Ch. 45).

ARE VIRUSES ALIVE?

Life can be viewed as a complex set of processes resulting from the actuation of the instructions encoded in nucleic acids. In the nucleic acids of living cells these are actuated all the time; in contrast, in a virus they are actuated only when the viral nucleic acid, upon entering a host cell, causes the synthesis of virus-specific proteins. Viruses are thus "alive" when they replicate in the cells they infect. From an epidemiological point of view, viruses are also "alive" since they can cause infection just as well as cellular agents. Outside cells, however, viral particles are metabolically inert and are no more alive than fragments of DNA (e.g., the DNA used in bacterial transformation, Ch. 7).

When Stanley, in 1935, crystallized tobacco mosaic virus, there followed extensive debates on whether such a crystallizable substance was a living being or merely a nucleoprotein molecule. As Pirie pointed out, these discussions showed only that some scientists had a more teleological than operational view of the meaning of the world "life." Just as physicists recognize light either as electromagnetic waves or as particulate photons, depending on the context, so biologists can profitably regard viruses both as exceptionally simple microbes and as exceptionally complex chemicals.

THE ANALYSIS OF VIRUSES

PREPARATION OF SAMPLES

Physical and chemical determinations usually require fairly large amounts of highly purified virus, separated from the constituents of the infected host cells.

The following methods of purification have proved especially valuable.

Differential centrifugation separates particles markedly different in sedimentation constants (dependent upon particle size, shape, and density). Under suitable conditions larger particles are pelleted to the bottom of the centrifuge tube, while most of the smaller ones remain in the supernatant. Low-speed centrifugation is used to remove cell debris from crude virus preparations, and then high-speed centrifugation separates viral particles from smaller molecules.

Zonal sedimentation in a density gradient (Ch. 10) can separate particles even when their sedimentation constants differ only slightly.

Density gradient equilibrium (isopyknic) **centrifugation** (Ch. 10) separates particles according to buoyant density and is used to separate virions of different types from each other and from cellular debris.

Methods that separate particles according to surface properties include: 1) **chromatography on ion exchangers** and **electrophoresis,** which separate particles primarily according to number and distribution of charges on their surface; 2) **chromatography through columns of neutral substances** (calcium phosphate, silica gel, etc.); 3) **partition between two polymer solutions** (such as dextran and polyethylene glycol); and 4) **extraction with a solvent** (such as a fluorocarbon), which removes from the water phase a considerable proportion of cellular degradation products but leaves many viruses unaltered.

How Pure Is a Purified Viral Preparation?   The goal of purification is to increase the ratio of virions to contaminating material, and its results, therefore, can be evaluated by determining the ratio of number of virions to total protein or total nitrogen. This ratio may be hundreds or thousands of times greater in purified than in crude virus. Whereas the relative degree of purification can thus be determined, it is difficult to assess the absolute purity. In fact, whereas the removal of a substance without altering the viral biological properties proves that it is a contaminant, its persistence after extensive purification does not demonstrate that it is a constituent of the virions; it is difficult to remove some contaminants adsorbed to the viral particles. Considerable judgment must therefore be exercised in deciding what may belong to the virions and what may not.

Danger of Contamination with Other Viruses.   In contrast to bacteria, which are propagated in sterile media, viruses are propagated in cells, which often harbor unrecognized viruses. Hence a viral preparation may contain contaminant viruses and they may even be preferentially enriched during viral purification. This possibility can be minimized by ensuring that the particles retained during the various steps of viral propagation and purification have the same physical, immunological, and biological properties.

ELECTRON MICROSCOPIC TECHNICS

The use of X-ray diffraction and electron microscopy has elucidated only recently the architectural properties of viruses, which are of considerable importance for understanding their function and their evolution. Some electron microscopic technics will be briefly reviewed in order to help appreciate the significance of the results obtained.

Shadowcasting (Fig. 44-2). A metal vapor, projected at an angle onto a membrane to which dried viral particles are attached, coats it with an electron-opaque layer of metal. Metal deposition is thick on the side of the particles exposed to the vapor, and strengthens their outlines; it is absent on the opposite side, where a "shadow" forms. The size and shape of the particles can be deduced from a study of the outlined parts and of the shadows.

Negative Staining (Fig. 44-3). The viral particles are mixed with a solution of a salt highly opaque to

Fig. 44-2. Shadowcasting. A sample dried on a film supported by a grid is placed in an evaporator chamber, which is evacuated. Metal atoms projected from a glowing filament impinge at a predetermined angle on the film, where they form a relatively uniform layer of metal. The grid is then examined in the electron microscope. Particles present on the film cause the formation of "shadows," lacking deposited metal. The length and the shape of the shadows provide information on the three-dimensional shape of the particles.

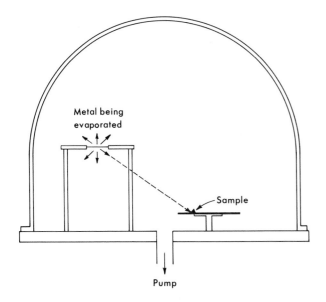

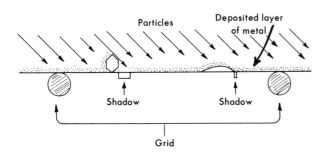

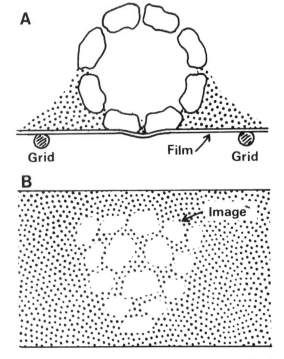

A

Grid   **Film**   Grid

B

Image

Fig. 44-3. Negative staining. The sample is mixed with a solution of a salt having high electron opacity (e.g., sodium phosphotungstate), and a thin layer is spread on the supporting film and dried. The salt solution surrounding the particles penetrates between particles and film and also into indentions in the surface of the particles, as shown in **A.** When the preparation is examined in the electron microscope, surface projections appear as transparent areas on an opaque background, as shown in **B.**

electrons, usually sodium phosphotungstate. The mixture is then spread in a thin layer on a carbon membrane and dried. The parts of the particles that are not penetrated by the salt stand out as electron-lucent areas on an opaque background. This method reveals details of surface structures because the salt penetrates between protruding parts and makes them visible. The side of the particles against the carbon membrane often contributes the predominant detail.

Positive Staining. Certain components of viruses can be stained by salts that become selectively adsorbed. Uranyl acetate stains the viral nucleic acid and other components; antibodies conjugated to an electron-opaque molecule such as ferritin stain the proteins for which they have specificity. Positive staining can be associated with negative staining to improve the resolution.

Thin Sectioning. This is used to study viral particles in cells or in centrifugal pellets.

The different methods enhance different structural details. The size of a virion will be maximal in shadowed preparations, which enhance the contrast at the periphery of the particles; it will be smaller in negatively stained preparations, since the phosphotungstate penetrates the surface details; and it will be even smaller in sections, where the action of the knife tends to collapse the particles. Furthermore, the values obtained by all these methods are less than the size of the particles in water, since drying, as required for electron microscopic examination, causes shrinkage by as much as 30% in linear dimension.

THE VIRAL PARTICLES

GENERAL MORPHOLOGY

Electron microscopy shows that virions belong to several morphological types (Fig. 44-4).

1) Some virions resemble small crystals. Extensive studies, especially by Klug and Caspar, have shown that these virions have an **icosahedral** protein shell (the **capsid**) surrounding the nucleic acid in association with proteins (the **core**). The capsid and the core form the **nucleocapsid.** These virions are called **icosahedral virions.** (The icosahedron is a regular polyhedron with 20 triangular faces and 12 corners.) Examples are picornaviruses (Ch. 55), adenoviruses (Ch. 52),

papovaviruses (Ch. 63), and bacteriophage ϕX174 (Ch. 45; Fig. 44-5A).

2) Some virions form long rods. Their nucleic acid is surrounded by a cylindrical capsid in which a helical structure is revealed by high-resolution electron microscopy. They are called **helical virions.** Examples are tobacco mosaic virus (Fig 44-5B) and bacteriophage M13 (Ch. 45).

3) In virions of more complicated morphology the nucleocapsid—in some cases icosahedral, in others helical—is surrounded by a loose membranous **envelope. Enveloped** virions are roughly spherical but highly pleomorphic (i.e., of varying shapes) because the envelope is not rigid. Examples of **enveloped**

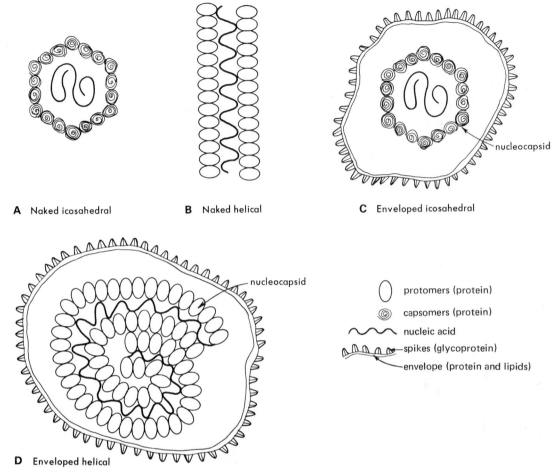

A Naked icosahedral **B** Naked helical **C** Enveloped icosahedral

nucleocapsid

D Enveloped helical

nucleocapsid

protomers (protein)

capsomers (protein)

nucleic acid

spikes (glycoprotein)

envelope (protein and lipids)

Fig. 44-4. Schematic diagram of simple forms of virions and of their components. The naked icosahedral virions resemble small crystals; the naked helical virions resemble rods with a fine regular helical pattern in their surface. The enveloped icosahedral virions are made up of icosahedral nucleocapsids surrounded by the envelope; the enveloped helical virions are helical nucleocapsids bent to form a coarse, often irregular coil, within the envelope.

icosahedral virus are herpesviruses (Ch. 53; Fig. 44-5C). In **enveloped helical** viruses, such as myxoviruses (Ch. 56; Fig. 44-5D), the nucleocapsid is coiled within the envelope.

Virions of more complex structure belong to two groups. Those illustrated by poxviruses (Ch. 54; Fig. 44-5E) do not possess clearly identifiable capsids, but have several coats around the nucleic acid; while certain bacteriophages (Ch. 45; Fig. 44-5F) have a capsid to which additional structures are appended.

The morphological types of the representative viruses are given in Table 44-2.

LOCATION OF THE NUCLEIC ACID

In **icosahedral nucleocapsids** the nucleic acid, together with certain proteins, constitutes a central core within the capsid. Thus, particles without nucleic acid form **empty capsids** (Fig. 44-6), which are present (along with nucleocapsids) in most preparations of icosahedral viruses, both with and without envelopes; since empty capsids have a lower buoyant density they can be separated from the nucleocapsids by equilibrium density gradient centrifugation. Negative staining reveals a similar external configuration in normal vir-

Fig. 44-5. Electron micrographs of representative virions, with negative staining. Markers under each micrograph are 1000 A. **A.** Naked icosahedral: human wart virus (papovavirus, Ch. 63). **B.** Naked helical: a segment of tobacco mosaic virus. **C.** Enveloped icosahedral: herpes simplex virus (herpesvirus, Ch. 53). **D.** Enveloped helical: influenza virus (myxovirus, Ch. 56). **E.** Complex virus: vaccinnia virus (poxvirus, Ch. 54). **F.** Coliphage lambda (Ch. 46). [**A** from Noyes, W. F. *Virology 23*:65 (1964). **B** from Finch, J. T. *J Mol Biol 8*:872 (1964). **C** courtesy of P. Wildy. **D** from Choppin, P. W., and Stockenius, W. *Virology 22*:482 (1964). **E** courtesy of R. W. Horne. **F** courtesy of F. A. Eiserling.]

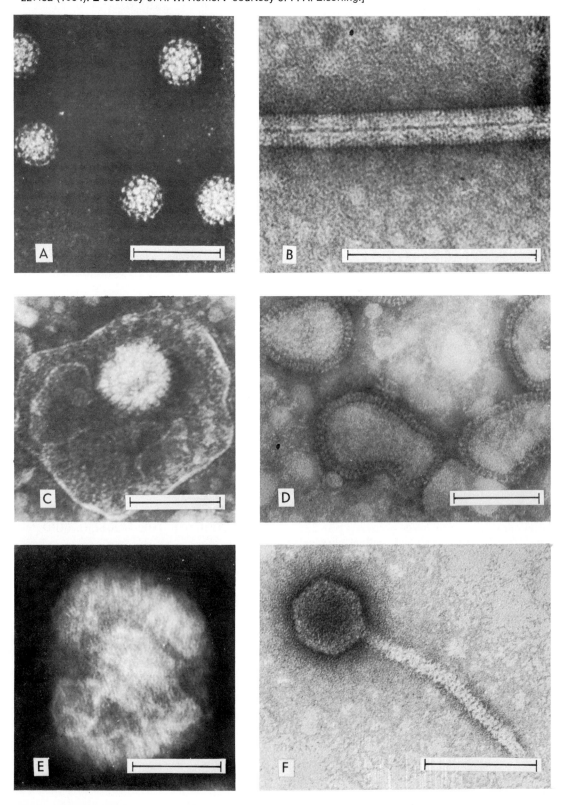

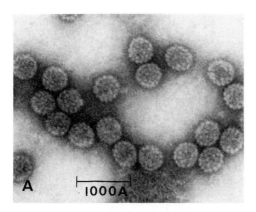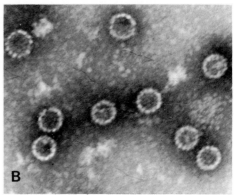

Fig. 44-6. Electron micrographs with negative staining of purified polyoma virus, a papovavirus. The full virions, on the left **(A)**, are not penetrated by the stain and show only the pattern on the surface of the capsid; the empty capsids, on the right **(B)**, are penetrated by the stain.

ions and empty capsids and also fills the hollow center in the empty capsids.

The shape of the nucleic acid in the core is influenced by the symmetry of the capsid. In icosahedral capsids the arrangement of the nucleic acid is not well known; the close packing appears to distort its structure, for in bacteriophage heads 20 to 30% of the DNA bases react chemically as if they were unstacked and probably form bonds with the capsid proteins. In helical nucleocapsids, such as tobacco mosaic virus (TMV),* X-ray diffraction studies show that the RNA is located in a deep helical groove inside the capsid (Fig. 44-7). This location is also confirmed by electron microscopy: after part of the capsid has been stripped by detergents, the RNA protrudes from the center of the shortened rod in the form of RNase-sensitive threads (Fig. 44-8).

The existence of empty capsids shows that the **nucleic acid is not essential for capsid assembly,** although its presence profoundly modifies the structure of the capsid. Thus, in poliovirus (a picornavirus) the association of RNA with the empty capsid causes a break in the polypeptide chain of a capsid protein (see Ch. 48, Maturation of icosahedral viruses); and in TMV empty segments of the capsid consist of stacked discs of protein subunits, whereas association with RNA causes

a slight shift that yields a helix. These changes cause an **increased stability of the nucleocapsid,** which is shown by its greater resistance to disintegration during preparation of speci-

Fig. 44-7. Drawing of a segment of tobacco mosaic virus showing the helical nucleocapsid. In the upper part of the figure two rows of protein monomers have been removed to reveal the RNA. This drawing is based on results of X-ray diffraction studies. [From Klug, A., and Caspar, D. L. D. *Adv Virus Res* 7:225 (1960).]

---

* This virus has been a model for the study of the structure of helical viruses because it can be easily prepared in large amounts and in a highly purified form.

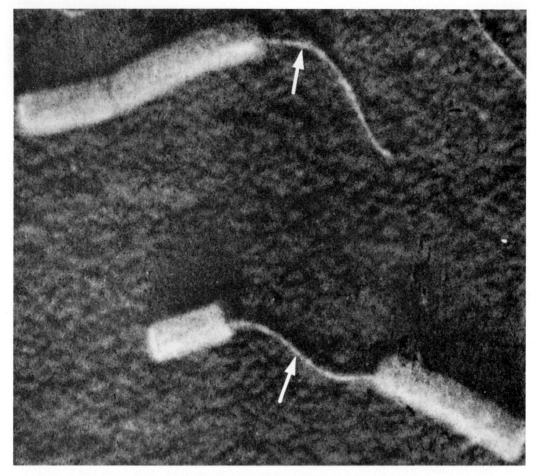

Fig. 44-8. Virions of tobacco mosaic virus in which segments of the capsid have been stripped away by detergent. Threads, indicated by arrows, protrude from the center of the remaining capsid segments; they can be destroyed by RNase, and represent the viral RNA centrally located in the nucleocapsid. [From Corbett, M. K. *Virology 22*:539 (1964).]

mens for electron microscopy (refer to Fig. 44-17).

THE VIRAL NUCLEIC ACID

The proportion of nucleic acid in virions varies from about 1% for influenza virus to about 50% for certain bacteriophages; and the amount of genetic information per virion varies from about 1000 codons in very small viruses to nearly 100,000 in the very large ones. If 300 codons are taken as the size of an average gene, **small viruses contain perhaps three or four genes and large viruses contain several hundred.** The diversity of

virus-specific proteins synthesized in the infected cell varies accordingly.

All the four possible types of nucleic acids in respect to strandedness and composition (single- and double-stranded DNA and RNA) have been found in viruses.

DOUBLE-STRANDED VIRAL DNA

Molecular Weight: Number of Molecules per Virion. The molecular weight of viral DNAs, determined by the methods discussed in Chapter 10, are given in Table 44-3. For the largest DNAs, such as those of the T-even coliphages, the most reliable estimates have been obtained by radioautography (Fig. 44-9) and electron microscopy (Ch. 10). The measured

TABLE 44-3.   Molecular Weights of Viral Nucleic Acids

Type of nucleic acid	Representative virus	Molecular weight (in 10^6 dalton units)	No. of genes*
DNA, double-stranded			
Papovavirus	Polyoma	3	4–5
	Papilloma	6	9
Adenovirus	Types 12, 18	21	30
	Types 2, 4	23	30
Coliphages T3, T7		25	38
Coliphage λ		31	46
Herpesvirus	Herpes simplex	100	150
Coliphages T2, T4, T6		110	165
B. subtilis bacteriophage SP8		130	195
Poxviruses	Vaccinia	160	240
DNA, single-stranded			
Parvoviruses	Adeno-associated†	1.5	4–5
Coliphage φx174		1.7	5
Coliphages fl, M13		2.4	7
RNA, double-stranded			
Reovirus		15‡	22
Rice dwarf virus		15‡	
Cytoplasmic polyhedrosis of silkworms		15‡	
RNA, single-stranded			
Satellite necrosis viruses†		0.4	1–2
Coliphage R17		1.1	3
Togavirus, group A	Sindbis	2	6
Tobacco mosaic virus		2	6
Turnip yellow mosaic virus		2	6
Picornavirus	Polio	2.5	7–8
Myxovirus	Influenza	4‡	12
Paramyxovirus	Newcastle disease	7.5	22
Leukovirus	Rous sarcoma	13‡	40

* Assuming an average gene size corresponding to 1000 nucleotides on a strand.

† These viruses are defective and multiply only in cells infected by a helper virus (adenovirus or tobacco necrosis virus, respectively). They probably specify only their own capsids, perhaps with another small protein.

‡ These virions contain several nucleic acid fragments. The aggregate value is given.

lengths were 52 and 49 μ, respectively, which, assuming an internucleotide distance of 3.46 A along the axis of the helix, correspond to 1.50×10^5 and 1.41×10^5 nucleotide pairs, and molecular weights of 110×10^6 and 103×10^6 daltons, respectively.

In all DNA viruses the molecular weights of the DNA are in good agreement with the average amounts per virion estimated chemically, showing that **virions contain a single DNA molecule.**

Base Composition of Viral DNA. The G+C content of DNA varies considerably in different viruses. It appears meaningful to **compare the viral DNA with that of the host cells,** not only in respect to average base ratios, but also to nearest neighbor frequencies, i.e., the proportion of all possible pairs of adjacent bases, which are more indicative of the sequences. The results show that the largest deviations from the cellular values are found in large viruses which depend least on cellular functions for their multiplication, e.g., T-even bacteriophages (Ch. 45) or herpesvirus (Ch. 53). In contrast, the DNA of small viruses and of those that interact intimately with the host cell, such as temperate phages (Ch. 46) or oncogenic viruses (Ch. 63), are usually quite similar to cellular DNA.

Some viruses contain **abnormal bases.** Thus, **5-hydroxymethylcytosine** (Fig. 44-10) was discovered by Wyatt and Cohen as re-

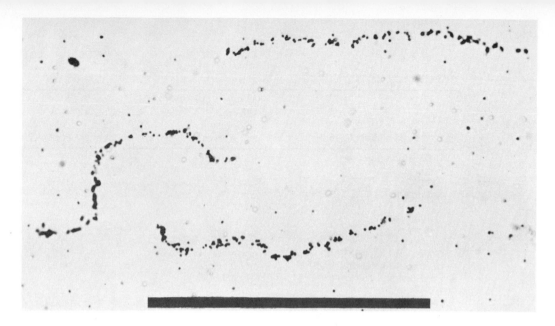

Fig. 44-9. Radioautograph of molecules of DNA of bacteriophage T2 (Ch. 45). A specially prepared glass slide was drawn through a solution containing T2 DNA heavily labeled with ^{3}H-thymidine, mixed with a large excess of unlabeled T2 DNA. This procedure caused some DNA molecules to stick to the glass in an oriented fashion. The preparation was then dried, overlaid with sensitive photographic film, and exposed for 2 months. Electrons released by the decaying ^{3}H atoms initiated formation of silver grains in the emulsion, which were made visible by photographic development. The marker shows 50 μ. (Courtesy of J. Cairns.)

Fig. 44-10. Various forms of glucosylation of 5-hydroxymethylcytosine residues in T-even coliphages. The proportion of the residues with the various types of glucosylation is as follows:

Type of glucosylation	Bacteriophage (%)		
	T2	T4	T6
Unglucosylated	25	0	25
α-Glucosyl	70	70	3
β-Glucosyl	0	30	0
β-Glucosyl-α-glucosyl	5	0	72

[Data from Lehman, I. R., and Pratt, E. A. *J Biol Chem* 235:3254 (1960).]

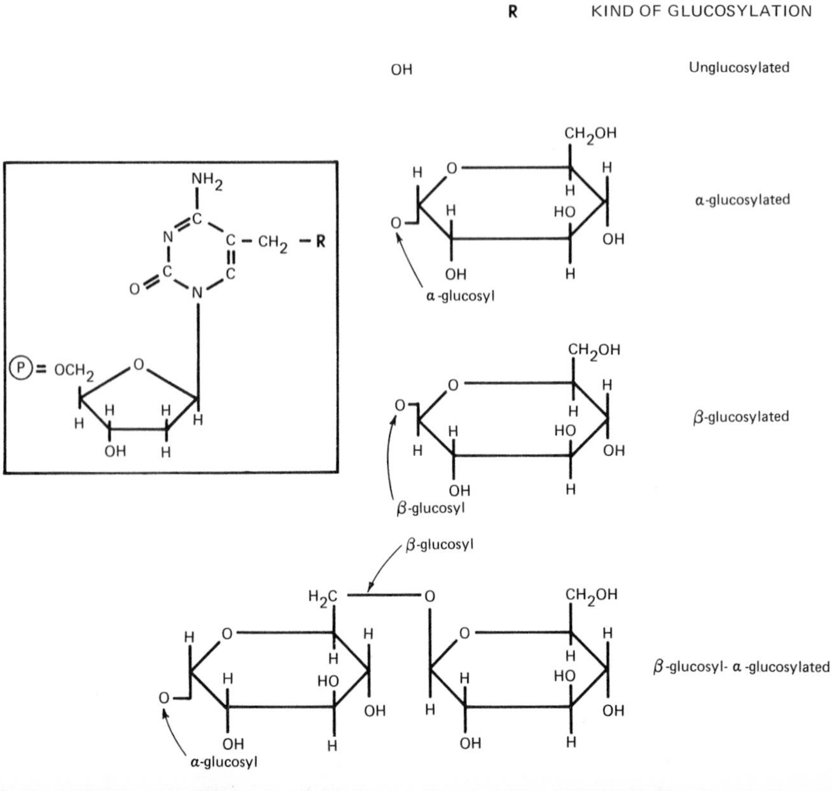

placement for cytosine in the DNA of T-even coliphages (Ch. 45). This important finding made it possible to measure the replication of viral DNA in the presence of host cell DNA, which does not contain this base. The problem of the origin of this base also led to the first demonstration that phage-infected cells synthesize new enzymes, absent in uninfected cells. Certain *Bacillus subtilis* bacteriophages contain **5-hydroxymethyluracil** or **5-dihydroxypentiluracil** instead of thymine (5-methyluracil).

The hydroxymethyl group of 5-hydroxymethylcytosine is glucosylated with a different pattern in different phages (Fig. 44-10). These variations depend on the different glucosylating enzymes made in cells after infection by the corresponding phages (Ch. 42).

Special Configurations of Some Viral DNAs. Small animal viruses and bacteriophages contain **cyclic DNAs** (see Ch. 10 for properties; Fig. 44-11). This finding seemed surprising until the value of the cyclic shape in replication and integration in the host genome was recognized (see Chs. 45, 46, and 63). Its usefulness may also explain why **some linear viral DNAs have structural features that permit them to cyclize** after entering the cells (Figs. 44-12 through 44-14). Thus, individual molecules may cyclize by annealing of **cohesive ends** or by crossing over between **repetitious ends** (terminal duplication); and circles can be generated by recombination of two or more molecules with permuted sequences.

A special feature of T5 DNA is the presence of **single-strand breaks at fixed positions** which may identify DNA parts with special functions during infection (see Ch. 45).

OTHER KINDS OF NUCLEIC ACIDS

Single-stranded DNA. The DNA is single-stranded in very small bacteriophages (e.g., ϕX-174, icosahedral: fl and M13, helical; see Ch. 45) or animal viruses (parvoviruses, see Table 44-2). In the phages, the DNA is cyclic and always of the same strand, called **viral** strand, since molecules from different virions do not form helices on annealing (Ch. 10, Hybridization). In contrast, in the adeno-associated virus (Ch. 52) the two complementary strands exist in different virions.

Double-stranded RNA: Fragmentation. Double-stranded RNA, recognizable for its sharp melting, resistance to RNase, and complementary base ratios, is found in several unrelated icosahedral viruses of animals (reovirus and blue-tongue virus of sheep), of plants (wound tumor virus and rice dwarf virus), and of insects (cytoplasmic polyhedrosis virus of silkworm). A common feature of these viruses is the **fragmentation of the genome;** acrylamide gel electrophoresis of reovirus RNA separates 10 fragments ranging from 0.63×10^6 to 2.6×10^6 daltons, for a total of about 15×10^6 daltons (Fig. 44-15A). Fragmentation is not artifactual since the fragments are present in the intact virions, where their free 3' ends can be labeled by

Fig. 44-11. Electron micrographs demonstrating the cyclic structure of polyoma DNA (a papovavirus) molecules (Kleinschmidt technic). The native molecules **(A)** are twisted (supercoiled, see Ch. 10, Cyclic DNA); the twist is maintained as long as the two strands constituting the molecule are both intact. If one of them is broken the corresponding phosphodiester bond of the other strand acts as a swivel, allowing the DNA molecule to rotate around the helix axis until it forms an untwisted circle **(B)**. Marker, 1 μ. [From Weil, R., and Vinograd, J. *Proc Natl Acad Sci USA* 50:730 (1963).]

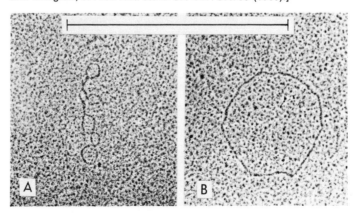

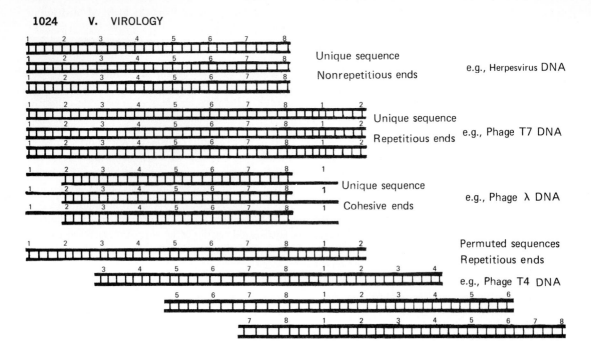

Fig. 44-12. Various types of double-stranded linear viral DNAs. The numbers represent the sequence (e.g., sequence of nucleotides or genes) characteristic of a given virus. Repetitive ends are double-stranded; cohesive ends are also repetitive but single-stranded and complementary. In a DNA with a unique sequence the ends of all molecules are at the same place in the sequence; in those with permuted sequences, the ends of the various molecules are randomly located in the sequence.

suitable reagents. In the virions the fragments may be connected by noncovalent bonds, such as short base-paired segments.

As noted in Chapters 47 and 48, genome fragmentation appears to be a device for avoiding internal initiation of translation in multicistronic mRNAs, which is apparently unfavorable in animal cells. Each fragment specifies a single polypeptide chain.

Single-stranded RNA. Many viruses, either helical or icosahedral, contain single-stranded RNA of total molecular weights between 0.4×10^6 and 13×10^6 daltons (see Table 44-3). Some (e.g., in picornaviruses and paramyxoviruses, Chs. 55 and 57) are in one piece; others are fragmented. The influenza (myxovirus, Ch. 56) virus RNA consists of at least seven different pieces; that of leuko-

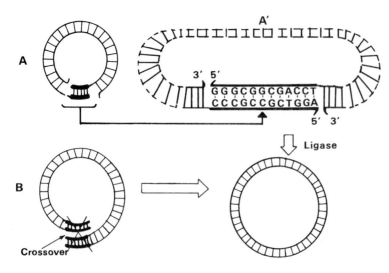

Fig. 44-13. DNA molecules capable of cyclization. **A.** Molecules with cohesive ends cyclize upon annealing. The ends of phage λ have been sequenced and are shown in **A'**. **B.** Molecules with repetitive ends can cyclize by a reciprocal crossover. Repetitive or cohesive ends are indicated by heavy lines. [**A'** modified from Wu, R., and Taylor, E. J. *Mol Biol* 57:491 (1971).]

Fig. 44-14. A. Demonstration of the presence of repetitive ends in DNA. Partial digestion from the 3′-OH ends using *E. coli* exonuclease III (see Table 10-2) produces a molecule that anneals into a ring, recognizable by electron microscopy. **B.** Demonstration of sequence permutations. The DNA is melted and renatured. A proportion of the renatured DNA forms double-stranded rings with tails, recognizable by electron microscopy. Letters indicate sequences; primed letters their complements.

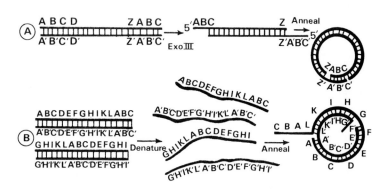

viruses (Ch. 63) has several pieces probably connected by short base-paired segments which melt on mild heating.

Viral RNAs have considerable **secondary structure,** which may have important functional meaning in replication and translation (see Ch. 45, RNA phages). They also have interesting features at their 3′ end. Thus several animal viruses (poliovirus, a picornavirus; Sindbis virus, a togavirus; and leukoviruses, Ch. 63), like the mRNAs of the host cell (see Ch. 47), are terminated by a **polyadenylic acid chain.** Bacterial and plant viruses, like tRNAs, have a CCA end. The similarity to tRNA is strengthened by the ability of turnip yellow mosaic RNA to be charged by valyl-tRNA synthetase (see Ch. 11, tRNA). These findings suggest a relation between some viral RNAs and tRNAs, which may explain the presence of tRNAs in virions of leukoviruses (Ch. 63).

THE CAPSID

The capsid accounts for most of the virion mass, especially in small viruses. In naked virions it protects the nucleic acid from nucleases in biological fluids and promotes the attachment to susceptible cells.

As was pointed out by Crick and Watson, a virus cannot afford too many genes for specifying capsid proteins; hence the capsid must be **formed by the association of many identical protomers** (i.e., it is an oligomeric protein). Thus, poliovirus RNA (MW 2.5 × 10^6) can specify at most 250,000 daltons of different proteins, and some of these must be

used for replication; yet the poliovirus capsid weighs about 6×10^6 daltons.

The prediction has been borne out: capsids of many viruses after dissociation into their component polypeptides were found to con-

Fig. 44-15. Fragmented viral RNAs in acrylamide gel electrophoresis. **A.** Photograph of a stained gel after fractionation of the fragments of labeled reovirus RNA (double-stranded; Ch. 60). At least nine bands can be clearly recognized. **B.** Pattern of radioactivity after fractionation of labeled influenza virus (Ch. 56) RNA; the gel was sectioned and the sections counted for radioactivity. At least five fragments can be distinguished (arrows). [**A** courtesy of W. K. Joklik. **B** modified from Pons, M. W., and Hirst, G. K. *Virology* 35:182 (1968).]

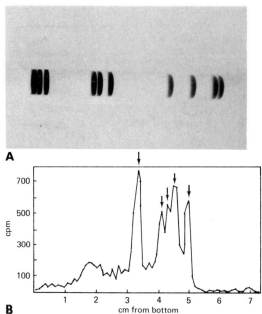

tain a small number of different polypeptide chains separable by acrylamide gel electrophoresis and ranging in molecular weight between 20,000 and 95,000 daltons. Helical capsids usually consist of a single type of polypeptide; icosahedral capsids may have one or several types. The polypeptides of the capsid are **specified by viral genes:** they are characteristic for each virus and absent in uninfected cells, and they are changed in some viral mutants.

The polypeptide chains fold to form the basic structural units of the capsid, the **protomers,** which normally contain a single polypeptide. X-ray diffraction shows that in the capsid the protomers are in constant spatial relation to each other, like molecules in a crystal. The arrangement is different in icosahedral and helical capsids.

ICOSAHEDRAL CAPSIDS

The protomers aggregate in groups of five or six to form substructures known as **capsomers,** which then form the capsid (Fig. 44-16). In electron micrographs capsomers are recognized as regularly spaced rings with a central hole up to 40 A in diameter (but probably not extending through the thickness of the capsid). If the hole is too small, they appear as solid knobs (Fig. 44-17).

The noncovalent bonds between protomers in the capsomers are different from those between capsomers and usually stronger, at least in empty capsids, which often disintegrate into intact capsomers during purification (Fig. 44-18).

The **shape and dimensions of the icosahedron depend on characteristics of its protomers;** the capsid, therefore, is uniquely determined by the properties of its polypeptide chains and ultimately by the viral genes that specify them.

Geometry of the Icosahedral Capsid. All capsids have 12 corners each occupied by a capsomer with five neighbors **(penton),** and 20 triangular faces, each containing the same number of capsomers with six neighbors **(hexon;** Fig. 44-19). Pentons probably contain five protomers, hexons may contain six, although adenovirus hexons appear to have only three. Pentons and hexons are usually made up of different polypeptide chains in animal viruses, but identical ones in plant viruses and in RNA phages. In different viruses the number of hexons per capsid may

Fig. 44-16. Possible constitution of capsomers from protomers. In the icosahedral capsids, the protomers constitute oligomers, of either five or six protomers, called capsomers, which are drawn as seen from the outside of a virion. Every protomer in a capsomer establishes bonds with two neighboring protomers, always through the same chemical groups (diagrammatically indicated as A-a and B-b in the hexon and c-C and d-D in the penton). Each protomer also has other contacts with neighboring capsomers in the complete capsid.

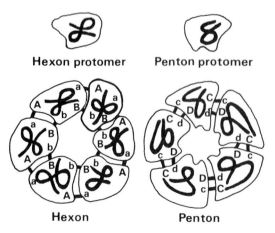

Fig. 44-17. Electron micrograph of GAL virus (chicken adenovirus) by negative staining, showing capsomer structure. The arrowed capsomers are situated on the fivefold axes. Marker, 625 Å. [From Wildly, P., and Watson, J. D. *Cold Spring Harbor Symp Quant Biol* 27:25 (1962).]

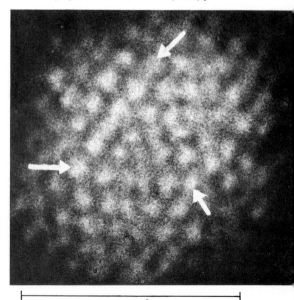

625 Å

Fig. 44-18. Preparation of rabbit papilloma virus (papovavirus) containing mostly empty capsids, i.e., devoid of nucleic acid. Some of the capsids have disintegrated into capsomers during the preparation of the specimen for electron microscopy, each producing a small puddle of capsomers (some of the capsomers of the original capsid have been lost). The angular polygonal shape of the capsomers is evident, but it is not possible to differentiate hexamers from pentamers. Marker, 1000 A. [From Breedis, C., Berwick, L., and Anderson, T. F. *Virology 17*:84 (1962).]

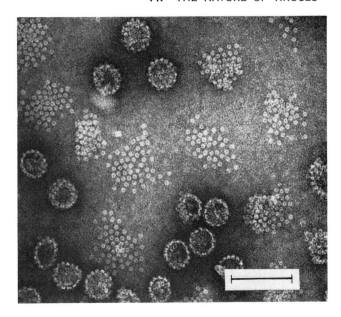

vary, corresponding to one of a set of possible values (see below and Table 44-4).

Symmetry Axes. The basic symmetry of the icosahedron is revealed by the study of certain ideal axes, drawn through its center, whose properties depend on where they cross the surface. Those through the corners are **fivefold axes** of rotational symmetry, as can be seen by viewing the icosahedron along one of these axes as the line of sight (Fig. 44-20): every time the icosahedron is rotated one-fifth of a turn (72°) around the axis, it gives rise to an identical figure. **Threefold**

Fig. 44-19. Assembly of capsomers to form an icosahedral capsid. The capsomers are shown with reference to the edges between triangular faces of the polyhedron. The icosahedron depicted here contains 42 capsomers, of which 12 are pentagonal (P) and 30 hexagonal (H). All are made up of identical protomers. The hypothetical chemical groups involved in the bonds between protomers are indicated by letters (A, B, C, D). Identical intracapsomeric A-D bonds connect protomers of the same capsomer, and identical intercapsomeric B-C bonds connect protomers of different capsomers.

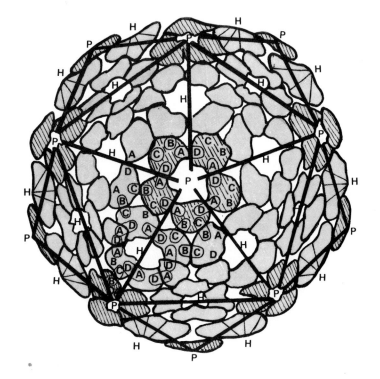

TABLE 44-4. Values of Capsid Parameters and Number of Capsomers Found in Icosahedral Viruses

P*	f*	T*	No. of capsomers	No. of hexons
1	1	1	12	0
	3	9	92	80
	4	16	162	150
	5	25	252	240
3	1	3	32	20
	7	147	1472	1460
	(probably)			
7	1	7	72	60

* For definition of parameters P, f, and T, see text.

axes of rotational symmetry go through the center of each triangular face, and **twofold axes** through the middle of each edge. The icosahedron is thus often referred to as a solid with a **5:3:2 rotational symmetry.**

Two modifications of the icosahedral capsid should be mentioned. One is the capsid of reovirus (Ch. 60) which is made up of two layers of protomers (Fig. 44-21), each satisfying the requirements of an icosahedral capsid. The other is the presence of small protruding fibers at the corners in adenoviruses (Ch. 52) and in bacteriophage φX174 (Ch. 45); they presumably possess fivefold rotational symmetry, as required by their location.

HELICAL CAPSIDS (FIG. 44-22)

The structure of the roughly cylindrical helical capsid is much simpler than that of the icosahedral capsid because the helix has a **single rotational axis,** coincident with the axis of the cylinder. The protomers are thus not grouped in capsomers, but are bound to each other so as to form a ribbon-like structure. This structure folds into a helix because the protomers are thicker at one end than the other. The protomers in successive turns are not exactly aligned, and each protomer establishes bonds with two protomers on each adjacent turn (as bonds between hypothetical groups a and A, and b and B in Figure 44-22). This confers great stability on the structure. The diameter of the helical capsid is determined by characteristics of its protomers, while its **length is determined by the length of the nucleic acid** it encloses.

The capsids of naked helical viruses (e.g., tobacco mosaic virus) are very tight (Fig. 44-5B). In contrast, the capsids of enveloped viruses are very flexible, as they have to coil within the envelope; and the turns of the helices are easily demonstrated in the electron microscope (Fig. 44-23). In these viruses, therefore, the envelope rather than the capsid

Fig. 44-20. Rotational axes in the icosahedron. The edges of the icosahedron, which limit the triangular faces are drawn as heavy lines. The outlines of the capsomers are in thin lines. Pentons are cross-hatched.

A. The icosahedron of Figure 44-19 is seen looking down the center of a pentagonal capsomer which corresponds to the corner of the polyhedron. Rotating the figures by ⅕ of a rotation reproduces the same figure. A fivefold rotational axis, therefore, passes through the center of the pentagonal capsomer.

B. The same icosahedron seen looking down the center of a triangular face. Through this point passes a threefold rotational axis which is situated between three hexagonal capsomers; rotating the figure by ⅓ of a rotation reproduces the same figure.

C. The same icosahedron seen looking down the middle of an edge between two triangular faces. This is an axis of twofold rational symmetry and is situated in the center of a hexagonal capsomer.

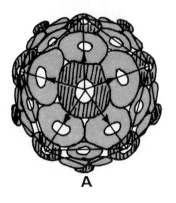
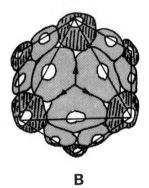
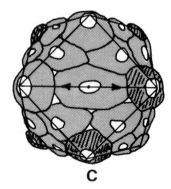

A   **B**   **C**

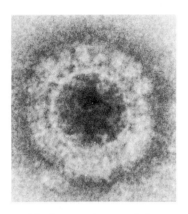

Fig. 44-21. The capsid of a reovirus particle examined with negative staining. The capsid is seen to be composed of two layers: the capsomers are especially evident in the outer layer. The capsid appears to be empty since it is penetrated by the stain. [From Dales, S., and Gomatos, P. J. *Virology* 25:193 (1965).]

may provide the required barrier to nucleases.

Allosteric Transitions of Capsids. The large number of individually weak bonds permits allosteric and other transitions in capsid structure (see Ch. 13, Allosteric transitions). One example is the transition produced by the association with the nucleic acid (see above, Location of nucleic acid); in poliovirus a transition induced by lowering the pH or by interaction with antibodies (see Ch. 50)

markedly alters the exposure of ionized groups to the surface, with resulting changes in electrophoretic mobility. It is likely that allosteric transitions determine also the important functional effects that result from attachment of virions to cells or from heating (see Chs. 48 and 64).

THEORETICAL CONSIDERATIONS ON THE ARCHITECTURE OF THE CAPSID

As already noted, the protomers of a helical or icosahedral capsid are arranged in a regular way dictated, according to the laws of crystallography, by their **uniformity** and their **lack of internal symmetry.** (The latter property follows from the fact that a protomer may be made up of a single polypeptide chain which does not usually have repeating identical amino acid sequences.) These identical protomers can assemble only through identical contacts with their neighbors. How these conditions lead to a **helical** capsid is shown in Figure 44-22. It is not equally obvious why protomers aggregate to form an **icosahedral** capsid with two types of capsomers. This process depends on the laws of crystallography. Consider first the two-dimensional problems of making a ring on a sheet of paper with a number of identical, asymmetrical protomers. Figure 44-24 shows how this can be done with five protomers; the figure obtained has rotational symmetry with a fivefold axis passing through its center. By extending the reasoning to the three-dimensional case, it can be deduced that the capsid must be a solid characterized by rotational symmetry. Such

Fig. 44-22. Constitution of the helical capsid. All protomers are identical and establish bonds with their neighbors. Since each protomer is staggered in respect to its lateral neighbors, it forms bonds with the two of them on each side along the helix axes. This confers considerable stability to the capsid. The protomers assemble first to constitute flat nonhelical discs containing two rings of protomers, with the staggered arrangement found in the finished helix. Under physiological conditions, when the protomers of the discs associate with the RNA they shift slightly to produce a helix. The helix grows in length by the addition and assimilation of discs.

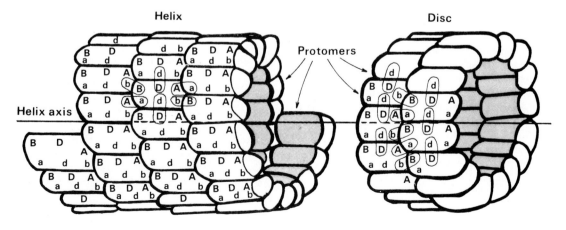

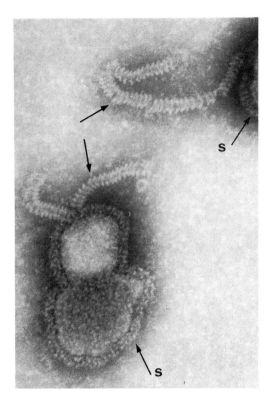

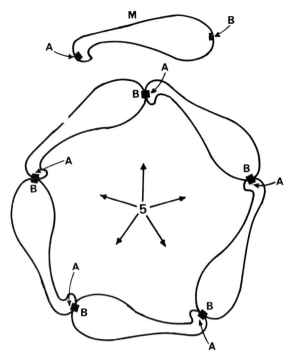

Fig. 44-24. Formation of a closed ring by using five asymmetrical protomers, M, in which group B can form a bond with group A. Since the distance between successive A-B bonds, and the angle of the A-B bonds to the axis of the monomer are constant, a closed ring is formed which has fivefold rotational symmetry around an axis through its center.

Fig. 44-23. The helical capsid of a paramyxovirus with negative staining. Two particles of the simian paramyxovirus SV 5 are seen: from both particles segments of the helical capsid protrude (arrows), probably owing to rupture of the envelope. Note the loose arrangement of the protomers and the hole along the axis of the helix. The envelopes are covered by the characteristic spikes (S). [From Choppin, P. W., and Stockenius, W. *Virology 23*:195 (1964).]

a symmetry is found in three classes of regular solids, the prototypes of which are the tetrahedron, the octahedron, and the icosahedron.

All these solids require **different capsomers** at the corners and in the faces of the polyhedron. The capsomers of the faces would always be hexons, while those of the corners would be made up of five protomers in the icosahedron, of four in the octahedron, and of three in the tetrahedron. **Hexons by themselves can form flat sheets which can roll up to generate open-ended cylinders.** Indeed, such cylinders are often found in preparations of icosahedral viruses and can be attributed to the self-assembly of hexagonal capsomers made in excess during viral multiplication (Fig. 44-25).

The icosahedron has two advantages over the other solids for the formation of a viral shell: 1) the number of protomers is greater and hence, for a

given capsid size, each is smaller, resulting in **economy of genetic information,** and 2) the closer approach to sphericity permits capsomers on corners or faces to fit reciprocally with less strain, thus enhancing the stability of the capsid.

The icosahedron offers even more pronounced advantage in capsids made up of a single protomer type which forms both hexons and corner capsomers (as is frequent in plant viruses). Clearly, the strain will be less in a structure containing pentamers as corner elements, rather than trimers or tetramers.

Number of Capsomers in Icosahedral Capsids. The laws of crystallography permit only certain numbers of capsomers in an icosahedral capsid, given by $10T + 2$, where T (the triangulation number) is Pf^2, with $f = 1$ or 2 or $3 \ldots$ etc., and $P = h^2 + hk + k^2$ (h and k being any pairs of integers without common factors). Identically, $T = (fh)^2 + (fh)(fk) + (fk)^2$, the expression used in Figure 44-26. The minimal permissible number is 12 pentons, all located at the corners of the icosahedron. Higher permissible numbers of capsomers are 32, 42, 72, 92, etc., of which 12 are always section of the coordinates); hence, $T = 25$ and the number of capsomers $= 252$.

THE ENVELOPE

Envelopes are formed by proteins (often with covalently attached carbohydrates = glycoproteins) and lipids. Like the capsid, the envelope contains many copies of a few kinds of polypeptide chains. Usually a protein rich in hydrophobic amino acids joins noncovalently with lipids to form the basic membrane. Some lipids can even be exchanged between the virions and the medium. Glycoproteins are at the outer surface, often as projections called **spikes** (see Fig. 44-23). In myxoviruses (Ch. 56) the spikes, about 100 A long and 70 to 80 A apart, are of two kinds: some bind to red blood cells, conferring on the virions hemagglutinating properties **(hemagglutinins),** * and others have **neuraminidase** activity (see below). The degradation of the spikes by proteolytic enzymes does not affect the integrity of the envelope.

The proteins of the envelope, like those of the capsid, are determined by viral genes. However, the **lipids and the carbohydrate moieties of glycoproteins** depend on the host cell; hence the virion surface may contain polysaccharide-determined cellular antigens. In addition, other cellular antigens, as well as enzymes, may be adventitiously adsorbed onto the viral surface: thus alkaline phosphatase is commonly found in semipurified vaccinia virus preparations but can be removed by further purification.

The liquid state of the lipids prevents the formation of a rigid structure, leading to pronounced **pleomorphism** of the virions. Indeed, enveloped virions containing helical nucleocapsids may assume a bizarre tadpole-like shape when dried for electron microscopy, presumably because one end of the nucleocapsid unravels and pushes the envelope out. This artifact, until recognized, erroneously suggested a similarity of some enveloped viruses to large bacteriophages. The **presence of lipids makes enveloped viruses sensitive**

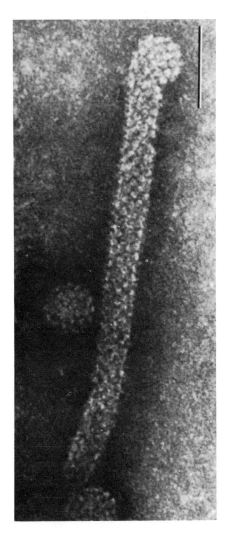

Fig. 44-25. Very long filament of human wart virus hexons (papovavirus) together with two regular virions. The filament, of a diameter close to that of the virions, is made up of hexagonal capsomers. Marker, 1000 A. [From Noyes, W. F. *Virology 23*:65 (1964).]

corner pentons, the others hexons. Only some of the possible values of P and T are found in viruses (Table 44-4). The numbers of protomers in an icosahedral capsid is $5 \times 12 + 6 (10T + 2 - 12) = 60T$, i.e., either 60 or a multiple thereof.

Examples of the arrangements of capsomers in a triangular face of the icosahedron in various types of capsids are given in Figure 44-26. These patterns allow identification of capsid type by electron microscopy. For instance, it can be easily decided that the capsid in Figure 44-17 has $P = 1$ (several hexons in line between two pentons) and $f = 5$ (since the basic vector crosses five times the inter-

---

* These hemagglutinins should not be confused with hemagglutinating antibodies, which are often referred to by the same name.

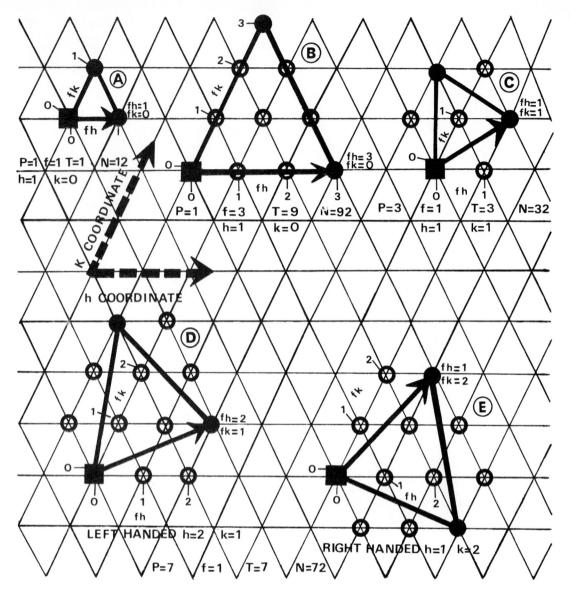

Fig. 44-26. Various forms of icosahedral capsids and their parameters. The triangulation number T is obtained by drawing the basic triangular face of the icosahedron (heavy lines) on a sheet of coordinates (thin lines) forming equilateral unit triangles. Capsomers are located at the intersections of the coordinates; the black circles are pentons, the white ones hexons. One of the sides of the triangular face of the icosahedron, identified by an arrow, is chosen as the vector going from an origin (square) to another penton. The coordinates of the end to the vector in respect to the origin, given near the vector, determine the parameters of the capsid. Capsids with equal values of h and k (hence of P) have equal general shapes; f measures how many times the basic vector crosses an intersection of the coordinates; hence different values of f change the number of capsomers but not the shape (compare **A** and **B**). T measures the number of unit triangles within the triangular face of the icosahedron.

If either h or k is zero ($P = 1$, **A, B**), or if $h = k$ and both are different from zero ($P = 3$, **C**), the capsid is **symmetrical** in respect to the coordinates; when h is different from k and both differ from zero (**D, E**) the capsid is **asymmetrical** and can occur in either a left-handed **(D)** or right-handed **(E)** form. By convention it is left-handed when $h > k$.

On the basis of relations such as those exemplified in this figure, the parameters of a capsid can be determined in electron micrographs of negatively stained virions by identifying two adjacent pentons (as centers of fivefold rotational symmetry) and the neighboring hexons (see Fig. 44-17). Very often considerable technical difficulties are encountered in these determinations.

to disinfection or damage by lipid solvents, such as ether.

OTHER VIRION COMPONENTS

In addition to the coat proteins, virions contain **internal proteins,** generally basic, which are tightly bound to the nucleic acid in the core: in animal viruses some resemble histones and protamines. Small peptides and polyamines (see Ch. 4) such as spermine and spermidine are also present in bacteriophages and some other viruses. These polycationic compounds presumably help the folding of the nucleic acids by linking together different loops. Reovirus virions contain each about 2000 small oligonucleotides of unknown function.

COMPLEX VIRUSES

Poxviruses (Ch. 54). The virions, brick-shaped or ovoid, hold the viral DNA, associated with protein, in a **nucleoid** shaped like a biconcave disc and surrounded by several lipoprotein layers. A layer of coarse fibrils near the outer surface gives the virions a characteristically striated appearance in negatively stained preparations (Fig. 44-5E; see also Ch. 54).

Rhabdoviruses (Ch. 58). This group, which includes rabies virus and vesicular stomatitis virus, is noticeable for its bullet-shaped virions. Negative staining reveals a series of transverse striations on the cylindrical part of the virions, possibly produced by the nucleocapsid wound in the form of a coil under the outer layer of the particle (Fig. 44-27).

Large Bacteriophages (Ch. 45). Some bacteriophages, such as the even-numbered coli-

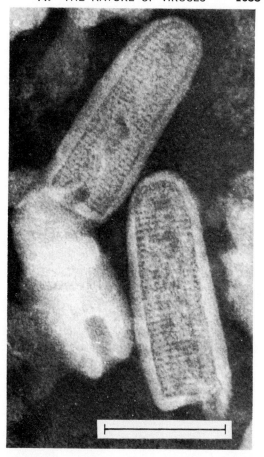

Fig. 44-27. Vesicular stomatitis virions (rhabdovirus) with negative staining. The helical filament, present in a deeper layer, is visible in two particles. Marker, 1000 A. [From Howatson, A. F., and Whitmore, G. F. *Virology 16:*466 (1962).]

phages, T2, T4, and T6, have very complex structures (Fig. 44-5F), including a head and a tail. They are said to have **binal symmetry** because they have components with both icosahedral and helical symmetry within the same virion.

ASSAY OF VIRUSES

The methods used for the assay of viruses reflect their dual nature as both complex chemicals and living microorganisms. Viruses can be assayed either by chemical and physical methods or by the consequences of their interaction with living host cells, i.e., **infectivity.** Assays carried out by different technics can differ vastly in their significance.

CHEMICAL AND PHYSICAL DETERMINATIONS

COUNTS OF PHYSICAL PARTICLES

Virions can be clearly recognized in the electron microscope; if a sample contains only virions of a single type their number can be determined unambiguously. Virions are counted by mixing with known numbers of polystyrene latex particles, spraying droplets of the mixture onto the specimen film, and counting the two types of particles present in the same droplet (Fig. 44-28). This technic does not distinguish between infectious and noninfectious particles.

HEMAGGLUTINATION

Many viruses, from the very small ones such as foot-and-mouth disease virus (picornavirus, Ch. 55) to large ones such as poxviruses (Ch. 54), can agglutinate red blood cells. This important property, discovered independently for influenza virus by Hirst and

Fig. 44-28. Counting of poliovirus particles (picornavirus) mixed with polystyrene latex particles. The mixture was sprayed in droplets on the supporting membrane, dried, and shadowed. The micrograph shows a droplet, whose outline is partly visible (arrows). The small particles are virus, the large ones latex.

There are 220 viral and 17 latex particles in the droplet. Since the latex concentration in the sample was 3.2×10^{10} particles per milliliter, the concentration of viral particles is $220/17 \times 3.2 \times 10^{10} = 4.1 \times 10^{11}$/ml. The precision of the assay based on this one droplet is only about $\pm$ 50% (see Appendix), owing to the small number of latex particles counted. To obtain a greater precision pooled counts from many similar drops would have to be used. (Courtesy of the Virus Laboratory, University of California, Berkeley.)

McClelland and Hare in 1941, affords a simple, rapid method for viral titration. Hemagglutination is usually caused by the virions themselves; in some cases, however, as with poxviruses, it is caused by hemagglutinins produced during viral multiplication, but not by the virions. Breakdown products of virions may also cause hemagglutination, e.g., the hemagglutinins released from myxoviruses by ether treatment.

Although the spectrum of red cell species that are agglutinated and the conditions required differ for different viruses, the phenomenon is basically similar in all cases: a virion or a hemagglutinin attaches simultaneously to two red cells and bridges them, and at sufficiently high viral concentrations multiple bridging yields large aggregates.

Hemagglutination Assay. The formation of aggregates can be detected in a number of ways. The simplest, **the pattern method,** is to leave the suspension of red cells and virus undisturbed in a small test tube for several hours. Nonaggregated cells sediment to the round bottom of the tube and then roll toward the center, where they form a small, sharply outlined, round pellet. Aggregates, however, sediment to the bottom but do not roll; they form a thin film, which has a characteristic serrated edge (Fig. 44-29). The proportion of aggregated cells can be determined more quantitatively by observing the sedimentation in a photoelectric colorimeter, since aggregated cells sediment faster than nonaggregated ones and can be separately measured.

The assay is usually carried out by an endpoint procedure. Serial twofold dilutions of the virus sample are each mixed with a standard suspension of red cells (usually 10^7/ml). The last dilution showing complete hemagglutination is taken as the endpoint. The titer estimated by the pattern method has an inherent imprecision at least as large as the dilution step used (usually twofold); the colorimetric method removes this imprecision. The titer obtained either way is expressed in **hemagglutinating units.**

Hemagglutination is inhibited by antiviral antibodies; hence **hemagglutination-inhibition** provides a convenient basis for measuring antibodies to many viruses (Ch. 50).

Mechanism of Hemagglutination. The phenomenon of hemagglutination throws considerable light on the interaction of myxovirus (Ch. 56) and paramyxovirus (Ch. 57) virions with cell surfaces. It is caused by the

Fig. 44-29. Results of a hemagglutination assay by the pattern method with influenza virus (myxovirus). Two samples, A and B, were diluted serially by using twofold steps; 0.5 ml of each dilution was mixed with an equal volume of a red cell suspension, and each mixture was placed in a cup drilled in a clear plastic plate and left for 30 minutes at room temperature. Each assay was made in duplicate. Sample A causes complete hemagglutination until dilution 320; sample B until dilution 80; in either case the subsequent dilution shows still partial hemagglutination. The hemagglutinating titer of A is 320; that of B, 80.

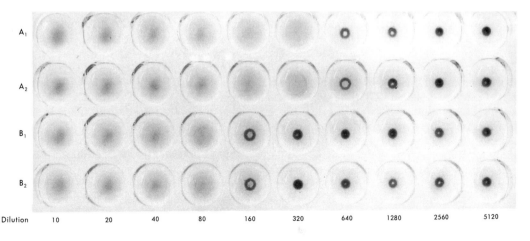

A_1										
A_2										
B_1										
B_2										
Dilution	10	20	40	80	160	320	640	1280	2560	5120

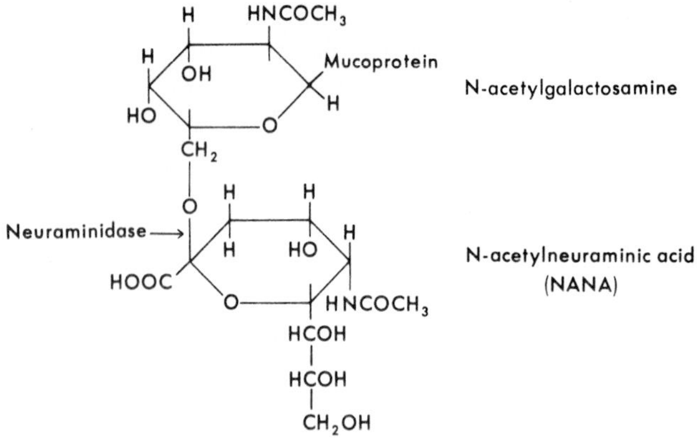

Fig. 44-30. Action of neuraminidase on a serum inhibitor of influenza virus (Ch. 56) hemagglutination. This compound contains N-acetylneuraminic acid (NANA) linked to N-acetylgalactosamine, as part of the mucoprotein; the NANA is released by the enzyme.

attachment of the hemagglutinating spikes of the virion's envelope to receptors of the membrane of the red cell. The receptors are mucoproteins with terminal N-acetylneuraminic acid (NANA) residues (Fig. 44-30). Thus they are inactivated by periodate, which oxidizes glycol groups of sugars, and by neuraminidase, which splits off N-acetylneuraminic acid. Moreover, sialic acid-containing mucoproteins with terminal NANA residues, which are present in biological fluids (e.g., serum, urine, submaxillary gland secretion), bind to the virions and competitively inhibit hemagglutination; and the removal of the terminal NANA by neuraminidase abolishes these activities.

Neuraminidase is obtained from culture filtrates of the cholera vibrio. This enzyme is commonly referred to as **receptor-destroying enzyme** (RDE) because it destroys the receptor activity for myxoviruses (Ch. 56) and paramyxoviruses (Ch. 57) on red blood cells and susceptible host cells.

As noted above, neuraminidases are also present on the surface of myxovirus and paramyxovirus virions and in uninfected cells. By splitting NANA from receptors the viral neuraminidase ultimately dissociates the viruses from red cells at 37°; the virus then spontaneously elutes off the red cells, which disaggregate. At 0°, in contrast, the enzyme is much less active and the virion-red cell union is stable. After the virus has eluted, cells cannot be agglutinated again by a new batch of virus, since they have lost the receptors and cannot regenerate them: **these cells are said to be "stabilized."** The eluted virus, on the contrary, retains all its activities.

Though cells stabilized by a given myxovirus or paramyxovirus are not agglutinable by the same virus, they can sometimes be agglutinated by another virus of these families. Indeed, it is possible to arrange the viruses in a series (called **receptor gradient**) such that any virus will exhaust the receptors for itself and the viruses preceding it in the gradient, but not for those that follow it. This result indicates that viruses differ quantitatively either in their neuraminidases or in their mode of attachment to the receptors.

Heating the virus inactivates its neuraminidase without destroying hemagglutinating activity. This **indicator** virus is useful for studying the union with receptors and mucoproteins without the complication of their enzymatic inactivation.

The nature of the receptors for other hemagglutinating viruses is known only in part. Receptors for polyoma virus (Ch. 63), like those for myxoviruses, are destroyed by neuraminidase, but they differ from myxovirus receptors. Receptors for reoviruses (Ch. 60) probably also contain sugars, because they are inactivated by periodate; they are unaffected by neuraminidase. Receptors for enteroviruses (picornaviruses, Ch. 55) and togaviruses (Ch. 59) appear to involve lipids since they are inactivated by lipid solvents.

ASSAYS BASED ON ANTIGENIC PROPERTIES

Complement fixation or precipitation with antiserum can be used to measure amounts of virus. These methods have low sensitivity and are used only for special purposes (Ch. 50).

ASSAYS OF INFECTIVITY

THE PLAQUE METHOD

This is the fundamental assay method of basic virology, and it is of great importance for diagnostic purposes: it combines simplicity, accuracy, and high reproducibility. First introduced for the study of bacteriophages by D'Herelle, and perfected by the Belgian microbiologist Gratia, this method was a key factor in the spectacular advances of research on phage and later also on animal viruses.

Bacteriophages are assayed in the following way. A phage-containing sample is mixed with a drop of a dense liquid culture of suitable bacteria and a few milliliters of melted soft agar at 44°; the mixture is then poured over the surface of a Petri dish, called a **plate,** containing a layer of hard nutrient agar previously set. The soft agar, before setting, spreads in a thin layer in which bacteriophages and bacteria are evenly distributed. The virions diffuse through the soft agar until each meets and infects a bacterium in which it multiplies; after 20 to 30 minutes the bacterium lyses, releasing several hundred progeny virions. These, in turn, infect neighboring bacteria, which again lyse and release new virus. The uninfected bacteria, in the meantime, grow to form a dense, opaque lawn. After a day's incubation the area stands out as a transparent **plaque** against the dense background (Fig. 44-31A). The soft agar permits the diffusion of phage to nearby cells, but prevents the convection to other regions of the plate; hence secondary centers of infection cannot form.

With **animal viruses** it is possible to use a similar method by replacing bacteria with a suspension of cells cultivated in vitro (Ch. 4). More commonly, however, monolayers of cells growing on a solid support (Ch. 47) are used. The nutrient medium is replaced by a solution containing the viral sample, which is left in contact with the cells for an hour or longer to allow attachment of the virions to the cells. Soft nutrient agar or some other gelling mixture is poured over the cell layer.*

---

* Agar contains sulfated polysaccharides which, by adsorbing some viruses, prevent them from forming plaques. This complication is eliminated either by adding to the agar cationic substances, which

Plaques develop after 1 day to 3 weeks of incubation, depending on the virus (Fig. 44-31B).

Plaques are detected in a variety of ways.

1) The virus often kills the infected cells, i.e., produces a **cytopathic effect;** the plaques are then detected by staining the cell layer either with a dye that stains only the live cells (e.g., neutral red) or with one that stains only the dead cells (e.g., trypan blue).

2) With certain viruses the cells in the plaques are not killed, but acquire the ability to adsorb red blood cells (Ch. 48); the plaques are revealed by **hemadsorption,** i.e., by flooding the cell layer with a suspension of red cells, then removing, by washing, those not attached to infected cells.

3) The infected cells may fuse with neighboring uninfected cells to form **polykaryocytes** (i.e., multinucleated cells), microscopically detectable.

4) Often the cells of the plaques contain large amounts of viral antigens, which can be detected by **immunofluorescence.**

If too many plaques develop on a plate some fuse with others and the counts are too low. The maximum density allowable varies with the size of the plaques and the sharpness of their margins.

The titer of the viral preparation is directly calculated from the number of plaques and the dilution of the sample, as shown in Figure 44-31. As discussed in the Appendix to this chapter, the accuracy of the assay depends on the number of plaques counted. An assay estimated from n plaques will be within $2/\sqrt{n}$ of the true value (e.g., with 100 plaques the true value will be $\pm 20\%$).

The Dose-response Curve of the Plaque Assay. The number of plaques in plates infected with different dilutions of the same viral sample is proportional to the concentration of the virus, i.e., the **dose-response curve is linear** (Fig. 44-32). As shown in the Appendix, the linearity demonstrates that **a single virion is sufficient to infect a cell,** a conclusion fundamental to virology. Moreover, it follows that **the viral population contained in a plaque is the progeny of a single**

---

neutralize the negative charges of the polysaccharides, or by replacing the agar with methylcellulose.

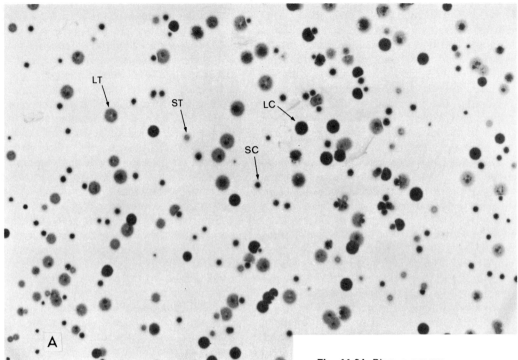

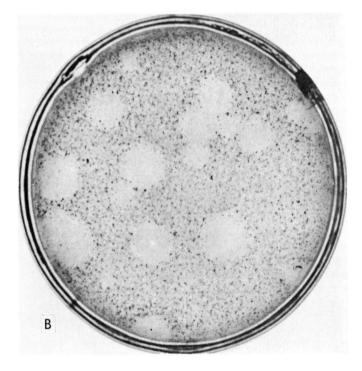

Fig. 44-31. Plaque assays.

A. Phage. The progeny of cells infected by two phage types was diluted by a factor of 10^7; 0.1 ml of the diluted virus was assayed. The plate was counted 18 hours after plating. Four different plaque types, differing in plaque size and turbidity (large clear LC, large turbid LT, small clear SC, and small turbid ST), can be distinguished, showing the great usefulness of plaque formation for genetic work with bacteriophages. (Part of the plate is reproduced.) A total of 407 plaques could be counted on the whole plate. The titer is 4.07×10^{10}/ml in the undiluted sample. The accuracy is $\pm 10\%$.

B. Poliovirus (picornavirus). A sample of poliovirus type 1 was diluted by a factor of 2×10^5 and 0.1 ml was assayed on a monolayer culture of rhesus monkey kidney cells, with an agar overlay containing neutral red. The culture was incubated for 3 days at $37°$ in an atmosphere containing 7% CO_2, which constitutes a buffer with the bicarbonate present in the overlay. Some of the plaques show partial confluence, but they can still be identified as separate plaques. Seventeen plaques can be counted on the photograph. The corresponding titer is 3.4×10^7/ml in the undiluted sample. The accuracy is $\pm 50\%$.

Fig. 44-32. Dose-response curve of the plaque assay. The number of plaques produced by a sample of poliovirus type 1 at various concentrations was plotted versus the relative concentration of the virus (open circles); the accuracy of the assay ($\pm 2\ \sigma$) is indicated for each point. The data are in agreement with a linear dose-response curve which falls between lines 1 and 2, and, therefore, with the notion that a single particle is sufficient to give rise to a plaque. Curves 3 and 4 give the range of data that would be obtained if at least two viral particles were required to initiate a plaque; the deviation is such that the hypothesis is ruled out (see Appendix, this chapter, Dose-response curve of the plaque assay).

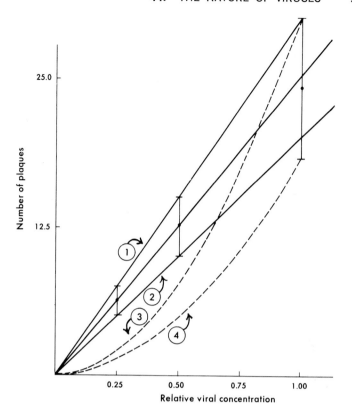

virion, i.e., **a clone,** provided cross-contamination from neighboring plaques is avoided.

Virus isolated from a single plaque thus represents a genetically pure line. Plaques also provide useful genetic markers through their visible characteristics such as size, shape, and turbidity.

THE POCK METHOD

When the chorionic epithelium of the chorio-allantoic membrane of a chick embryo (usually 10 to 12 days of age; see Ch. 48, Fig. 48-1) is infected by certain viruses, characteristic lesions **(pocks)** appear. The counting of these pocks was introduced by Burnet and coworkers for titrating poxviruses (Ch. 54); it has also been applied to other viruses (e.g., herpes simplex, Ch. 53). The formation of a pock, like that of a plaque, begins with the infection of a single cell. Since the chorioallantoic membrane is complex, however, and contains several different cell types and blood vessels, the response to the local infection is also complex; it involves both cell prolifera-

tion and cell death, accompanied by edema and hemorrhage. The pocks appear after 36 to 72 hours as opaque areas, usually white and often hemorrhagic on the transparent membrane (Fig. 44-33).

Pock counting is satisfactory only for viruses that are released from the infected cells too slowly to give rise to secondary infectious centers. This method was important at the time of its introduction; now it is largely superseded by the plaque method. Under optimal conditions the virus derived from a single pock represents a genetically pure line; and viral mutants may be distinguishable by the appearance of the pocks.

OTHER LOCAL LESIONS

Tumor-producing viruses, such as the Rous sarcoma virus (leukoviruses, Ch. 63), can be assayed on monolayer tissue cultures; they produce foci of cells with altered growth properties, derived from one infected cell. Each focus is initiated by a single viral particle.

Many **plant viruses** can be titrated by counting the lesions produced on leaves

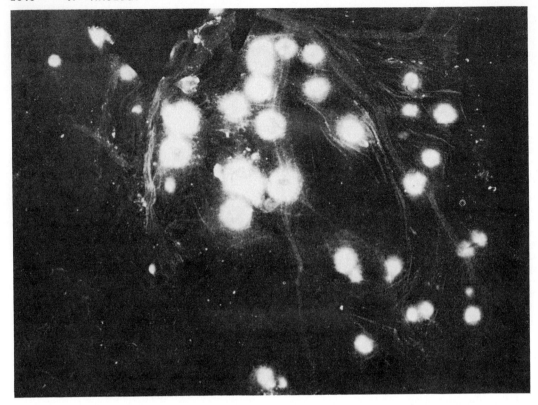

Fig. 44-33. Pocks formed by vaccinia virus (poxvirus) on the chorioallantoic membrane of the chicken embryo. The membrane was removed, washed, and photographed in saline against a dark background. The pocks appear as brilliant white foci, whereas the membrane, which is transparent, is barely visible. [From Coriell, L. L., Blank, H., and Scott, T. F. M. *J Invest Dermatol 11*:313 (1948).]

rubbed with a mixture of virus and an abrasive material. The virus penetrates through ruptures of the cell walls caused by the abrasive, and the progeny spread to neighboring cells, probably through the plasmodesmata (protoplasmic bridges between cells). Recognizable spots, each started by a single virion, are thus produced.

ENDPOINT METHOD

This method was commonly used for assaying animal viruses before the advent of the plaque method. It is still employed for certain diagnostic assays and for special purposes. The virus is serially diluted and a constant volume of each dilution is inoculated into a number of similar **test units,** such as mice, chick embryos, or cell cultures. At each dilution the proportion of infected test

units is scored by 1) death or disease of an animal or embryo, 2) degeneration of a tissue culture, or 3) recognition of progeny virus in vitro (e.g., positive hemagglutination).

Most of the test units develop signs of infection at the lower dilutions of the virus and none at the highest dilutions. A rough idea of the viral titer is given by the intermediate dilutions at which only a fraction of the test units show signs of infection. The transition is not sharp, however, and only by combining the data from several dilutions is it possible to calculate the precise endpoint at which 50% of the test units are infected. At this dilution each sample contains on the average one ID_{50}, i.e., one infectious dose for 50% of the test units. Such interpolation can be carried out in a variety of ways, such as the method of Reed and Muench, which is discussed in Chapter 22, Appendix. Though not mathematically derived, this method yields results in good agreement with more rigorous methods.

In the Reed and Muench method the results obtained at different viral dilutions are pooled; the dilution containing one ID_{50} is obtained by interpolation between the two dilutions that straddle the 50% value of the infectivity index. The interpolation assumes that the infectivity index varies linearly with the log dilution. The accuracy of the method is usually low, since the number of test units used at each dilution is small.

When, for instance, six units are employed, as is common in diagnostic titrations, the estimated titer has a range of uncertainty of at least 36-fold between the minimum and the maximum still compatible with the result obtained (Appendix). Under these conditions, the titration is useful only to ascertain large differences in viral titer: a 50-fold difference is considered significant and is adequate for many routine diagnostic procedures. The precision increases as the square root of the number of test units employed at each dilution.

Viral titers obtained by the endpoint method are expressed in various units, equivalent to the ID_{50}, which describe the criterion used for detecting infection of the test units: LD_{50} (lethal dose) if the criterion is death; PD_{50} (paralysis dose) if the criterion is paralysis; TC_{50} (tissue culture dose) if the criterion is degeneration of a culture. One ID_{50} can be shown to correspond to 0.70 plaque-forming units (Appendix).

COMPARISON OF DIFFERENT TYPES OF ASSAYS

The focal assay methods (plaques and pocks) are most satisfactory for their simplicity, reproducibility, and economy. For example, to match the precision obtained by counting 100 plaques on a single culture one would require more than 100 test units per decimal dilution in an endpoint titration. The precision of any type of assay is adversely affected by variability of the assay units; this can be quite large in the pock assay and even larger in endpoint assays using animals.

The various methods of assay have different sensitivities and measure different properties. Assays based on infectivity are as much as a millionfold more sensitive than those based on chemical and physical properties.

Chemical and physical technics, moreover, titrate not only infectious but also noninfectious virions, such as empty capsids or particles with a damaged nucleic acid. These methods, therefore, can be useful for studies requiring measurement of the total number of viral particles. Hemagglutination can also titrate soluble hemagglutinins (i.e., different from intact virions), such as those obtained from breakdown of the virions or produced during intracellular viral synthesis. Immunological methods also can titrate protein precursors of the virions or viral subunits obtained by disruption of virions.

The ratio of the number of viral particles (determined by electron microscope counts) **to the number of infectious units** measures the **efficiency of infection,** which varies widely among different viruses, and even for the same virus assayed in different hosts. As is shown in Table 44-5, for most viruses the ratio is larger than unity. This result is due in part to the presence of noninfectious particles, and in part to the failure of potentially infectious particles to reproduce. However, even with the highest ratio of particles to infectivity infection is initiated by a single virion, since the dose-response curve remains linear.

TABLE 44-5. Ratio of Viral Particles to Infectious Units

Virus	Ratio
Animal viruses	
Picornaviruses	
Poliovirus	30–1000
Foot-and-mouth disease virus	33–1600
Papovaviruses	
Polyoma virus	38–50
SV 40	100–200
Papilloma virus	ca. 10^4
Reovirus	10
Togaviruses, group A	
Semliki Forest virus	1
Myxoviruses	
Influenza virus	7–10
Herpesviruses	
Herpes simplex virus	10
Poxviruses	1–100
Adenoviruses	10–20
Bacterial viruses	
Coliphage T4	1
Coliphage T7	1.5–4
Plant viruses	
Tobacco mosaic virus	5×10^4–10^6

The ratio of total viral particles to hemagglutinating units is related to the number of red cells that must be aggregated to reveal hemagglutination. Since in the hemagglutination assay by the pattern method each diluted viral sample is mixed with an equal volume of a red cell suspension containing 10^7 cells per milliliter, at the endpoint the virus should theoretically contain about 10^7 hemagglutinating particles per milliliter.* Indeed, experimental determinations with influenza and polyoma viruses give under optimal conditions a value close to 10^7 total viral particles per hemagglutinating unit. But because the ratio of particles to infectious units exceeds 1.0, a hemagglutinating unit corresponds to only about $10^{6.3}$ infectious units of influenza virus and 10^5 of polyoma virus.

---

* If a single particle can form a stable bridge between two red cells, the aggregates at the endpoint with n viral particles would consist of $n + 1$ cells.

APPENDIX

QUANTITATIVE ASPECTS OF INFECTION

DISTRIBUTION OF VIRAL PARTICLES PER CELL: POISSON DISTRIBUTION

In a cell suspension mixed with a viral sample individual cells are infected by different numbers of viral particles, and it is important for several purposes to know the distribution, i.e., the proportions of cells infected by zero, one, two, etc. viral particles.

These proportions depend on the **average number of viral particles per cell,** known as the multiplicity of infection (m). The relevant viral particles are those that initiate infection of a cell; inactive particles or particles that, for whatever reason, never enter a cell are neglected. Hence m is related to the **total** number of viral particles (N) and of cells (C) by the relation $m = aN/C$, where a is the proportion of viral particles that initiates infection.

The proportion $P(k)$ of cells infected by k viral particles is given by the Poisson distribution, assuming that the cells are all identical in their ability to be infected. In practice this is not so because cells vary in size, surface properties, etc., but usually the deviations are small enough to be negligible, at least as a first approximation.

According to the Poisson distribution:

$$P(k) = \frac{e^{-m}m^k}{k!} \tag{1}$$

The value of m can be derived from the known values of N and C if a can be determined; otherwise m can be calculated from the experimentally determinable proportion of uninfected cells, $P(0)$. By making $k = 0$ in equation (1),

$$P(0) = e^{-m}, \text{ and} \tag{2}$$
$$m = -\ln P(0), \tag{3}$$

where ln stands for the natural logarithm.

The use of equations (1), (2), and (3) will now be illustrated with reference to two practical problems.

Problem 1. 10^7 cells are exposed to virus; at the end of the adsorption period there are 10^5 infected cells; what is the multiplicity of infection?

$$P(0) = 0.99, m = -\ln(0.99) = 0.01$$

This problem brings out the point that the multiplicity of infection can assume any value from 0 to ∞. Values smaller than unity indicate that only a fraction of the cells is infected, mostly by single viral particles.

Problem 2. What is the multiplicity of infection required for infecting 95% of the cells of a population?

$$P(0) = 5\% = 0.05, m = -\ln(0.05) = 3$$

The point of this problem is that even at very high multiplicities a certain proportion of the cells remains uninfected. The multiplicity of infection required to reduce the proportion of uninfected cells to a certain value can be calculated from equation (3).

CLASSES OF CELLS IN AN INFECTED POPULATION

It is usually important to determine the proportion of only three classes of cells: uninfected cells ($k = 0$); cells infected by one viral particle (**cells with a single infection,** $k = 1$); and cells infected by more than one particle, irrespective of how many (**cells with a multiple infection,** $k > 1$). The proportions are:

Uninfected cells, $P(0) = e^{-m}$
Cells with single infection, $P(1) = m\,e^{-m}$
Cells with multiple infection, $P(>1) =$
    $1 - e^{-m}(m + 1)$†

---

† This value is obtained by subtracting from unity (the sum of all probabilities for any value of k) the probabilities $P(0)$ and $P(1)$.

Problem 3. To determine the various classes of infected cells if the multiplicity of infection is 10.

$$P(0) = e^{-m} = e^{-10} = 4.5 \times 10^{-5}$$
$$P(1) = 10 \times 4.5 \times 10^{-5} = 4.5 \times 10^{-4}$$
$$P(> 1) = 1 - 4.5 \times 10^{-5}\, 10 + 1 =$$
$$1 - 4.95 \times 10^{-4} = 99.95\%$$

If the cell population includes 10^7 cells these are $4.5 \times 10^{-5} \times 10^7 = 450$ uninfected cells and 4500 cells with single infection; all the others have multiple infection.

Problem 4. To determine the composition of the population of infected cells if the multiplicity of infection is 10^{-3} or 0.001.

$$P(0) = e^{-0.001} = .9990 =$$
$$9.99 \times 10^{-1} = 99.9\%$$
$$P(1) = 0.001 \times e^{-0.001} = 10^{-3} \times 9.99 \times$$
$$10^{-1} = 9.99 \times 10^{-4} = 0.0999\%$$
$$P(> 1) = 1 - 0.9990\,(0.001 + 1) =$$
$$0.000001 = 10^{-6}$$

If the cell population includes 10^7 cells there are $9.99 \times 10^{-4} \times 10^7 = 9900$ cells with single infection, and $10^{-6} \times 10^7 = 10$ cells with multiple infection; most of the cells are uninfected.

MEASUREMENT OF THE INFECTIOUS TITER OF A VIRAL SAMPLE

To measure infectious titer, a viral sample containing an **unknown** number (N) of infectious viral particles is mixed with a known number (C) of cells. N is then calculated from the proportion of cells that remains uninfected according to equation (3): $m = -\ln P(0)$; since, as defined above, $m = a(N/C)$, $N = mC/a = -C \ln P(0)/a$, or

$$aN = -C \ln P(0) \tag{4}$$

Usually the factor a is not determinable and therefore the number (N) of infective viral particles present in the sample to be assayed cannot be calculated. In its place the product aN is obtained and is called the **number of infectious units.**

This is the basis for all measurements of the infectious viral titer. Its application is different in the plaque method and in the endpoint method.

PLAQUE METHOD

In this method m is very small, 10^{-4} or less. This allows the approximation $e^{-m} = 1 - m$; hence $P(0) = 1 - m$. The proportion of infected cells, $P(i)$, is then given by the simple equation, $P(i) = 1 - P(0) = m$. But $P(i)$ equals the ratio of the number of infected cells (C_i) to the total number of cells (C_t); thus

$$\frac{C_i}{C_t} = m = \frac{aN}{C_t},$$

and therefore $C_i = aN$. Since in the plaque method each infected cell originates a plaque, **the number of plaques equals the number of infectious units.** The actual number of cells employed in the assay is irrelevant, **provided it is in large excess** over the number of infectious viral particles, so that m is very small; uncertainties connected with the counting of the cells are therefore eliminated.

The Dose-response Curve of the Plaque Assay. As stated above, the number of plaques that develop on a series of cell cultures infected with different dilutions of the same viral sample is proportional to the concentration of the virus. We shall now show that this linearity proves that **a single infectious viral particle is sufficient to infect a cell.**

Let us assume that more than one particle, say two particles, is required. There would then be two types of uninfected cells: those with no infectious viral particles, and those with just one such particle. According to the Poisson distribution the proportions of cells in these two classes are e^{-m} and me^{-m}, respectively. Thus, under the foregoing assumption, $P(0) = e^{-m}(1 + m)$, which, for very small values of m, approximates to $P(0) = 1 - \frac{1}{2}m^2$. Therefore $P(i) = \frac{1}{2}m^2$, and the dose-response curve would be parabolic rather than linear (Fig. 44-32). If more than two particles were required to infect a cell, the curvature of the dose-response curve would be even more pronounced.

ENDPOINT METHOD

In this method the virus to be assayed is added to a number of test units, each consisting of a large number of cells. A test unit is now equivalent to a single cell of the plaque assay. Therefore m is the multiplicity of infection **of an assay unit, rather than of a cell.**

The virus titer can be calculated from the proportion of noninfected units, $P(0)$, at the endpoint dilution, according to equation (3): $m = \ln P(0)$. If at the endpoint $m = 1$, i.e., there is one infectious unit per test unit, then on the average $P(0) = 0.37$.

The multiplicity corresponding to 50% survival, which is characteristic of the ID_{50} determined by the Reed and Muench method, is calculated from the relation $e^{-ID_{50}} = 0.50$. Therefore $ID_{50} = -\ln(0.50) = 0.70$. One ID_{50} corresponds to 0.70 infectious units.

PRECISION OF VARIOUS ASSAY PROCEDURES

Plaque Method. The statistical precision is measured by the **standard deviation** (σ) of the Poisson distribution, which is equal to the square root of the number of plaques counted. If the number of plaques counted is not too small, 95% of all observations made should fall within two standard deviations from the mean in either direction (i.e., $\pm$ 2 σ). Thus 4 σ is the expected range of variability of the assay. If n replicate assays are made, $\sigma = \sqrt{\bar{x}/n}$, where $\bar{x}$ is the mean value of plaque numbers in the replicate assays. The standard deviation relative to the mean (coefficient of variation) serves as a relative measure of precision: $\sigma/\bar{x}$. This is $\sqrt{x}/x = 1/\sqrt{x}$ for a single assay, and $1/\sqrt{n\,\bar{x}}$ for n replicate assays. The smaller the coefficient of variation, the higher the precision, which therefore increases as the square root of the number of plaques.

Example. If 100 plaques are counted the standard deviation is 10. If the same assay is repeated many times its results will fall between 80 and 120 plaques in 95% of the cases; the coefficient of variation is 1/10. If 400 plaques are counted the coefficient of variation is 1/20.

Reed and Muench Method. An approximate value, empirically determined, of the standard deviation of the titer determined by this method is $\sigma = \sqrt{0.79\ hR/U}$ where h is the logarithm of the dilution factor employed at each step of the serial dilution of the virus, U is the number of test units used at each dilution, and R is the interquartile range, namely the difference between the logarithm of the dilution at which $P(i)$ is 0.25 and 0.75, respectively. σ is expressed in logarithmic units. For the data given in Chapter 22, Appendix, with six assay units (animals) at each dilution, $h = 1.0$ and $R = 1.0$ (both in $\log_{10}$ units) $\sigma = \sqrt{0.79/6} = 0.36$ (in $\log_{10}$ units). The range of variation of the ID_{50} is therefore $\pm$ 0.72 in $\log_{10}$ units, and the highest expected value (within the 95% confidence limits) is 28 times (antilog of 1.44) the lowest value.

SELECTED REFERENCES

Books and Review Articles

CASPAR, D. L. D. Design principles in virus particle construction. In *Viral and Rickettsial Infections in Man,* p. 51. (F. Horsfall and I. Tamm, eds.) Lippincott, Philadelphia, 1965.

KALTER, S. S., and HEBERLING, R. L. Comparative virology of primates. *Bacteriol Rev 35:*310 (1971).

RALPH, R. K. Double-stranded viral RNA. *Adv Virus Res 15:*61 (1969).

SHATKIN, A. J. Replication of reovirus. *Adv Virus Res 14:*63 (1969).

TEMIN, H. M., and BALTIMORE, D. RNA-directed DNA synthesis and RNA tumor viruses. *Adv Virus Res 17:*129 (1972).

Specific Articles

ABODEELY, R. A., PALMER, E., LAWSON, L. A., and RANDALL, C. C. The proteins of enveloped and deenveloped equine abortion (Herpes) virus and the separate envelope. *Virology 44:*146 (1971).

BALTIMORE, D., HUANG, A. S., and STAMPFER, M. Ribonucleic acid synthesis of vesicular stomatitis virus. II. An RNA polymerase in the virion. *Proc Nat Acad Sci USA 66:*572 (1970).

BRENNER, S., and HORNE, R. W. A negative staining method for high resolution electron microscopy of viruses. *Biochim Biophys Acta 34:*103 (1959).

BURLINGHAM, B. T., DOERFLER, W., PETTERSON, O., and PHILIPSON, L. Adenovirus endonuclease: Association with the penton of adenovirus type 2. *J Mol Biol 60:*45 (1971).

CRICK, F. H. C., and WATSON, J. D. Structure of small viruses. *Nature 177:*473 (1956).

DAVIDSON, N., and SZYBALSKI, W. Physical and chemical characteristic of Lambda DNA. In *The Bacteriophage Lambda,* p. 45. (A. D. Hershey, ed.). Cold Spring Harbor, N.Y., Cold Spring Harbor Laboratory, 1971.

DULBECCO, R. Production of plaques in monolayer tissue cultures by single particles of an animal virus. *Proc Nat Acad Sci USA 38:*747 (1952).

GOTTSCHALK, A. The influenza virus neuraminidase. *Nature 181:*377 (1958).

HARRISON, S. C., DAVID, A., JUMBLATT, J., and DARNELL, J. E. Lipid and protein organization in Sindbis virus. *J Mol Biol 60:*523 (1971).

HIRST, G. K. The agglutination of red cells by allantoic fluid of chick embryos infected with influenza virus. *Science 94:*22 (1941).

KATES, T. R., and MCAUSLAN, B. R. Poxvirus DNA-dependent RNA polymerase. *Proc Nat Acad Sci USA 58:*134 (1967).

KLENK, H. D., and CHOPPIN, P. W. Glycosphingolipids of plasma membranes of cultured cells and an enveloped virus (SV5) grown in these cells. *Proc Nat Acad Sci USA 66:*57 (1970).

SCHAFFER, F. L., and SCHWERDT, C. E. Crystallization

of purified MEF-1 poliomyelitis virus particles. *Proc Nat Acad Sci USA 41:*1020 (1955).

SKEHEL, T. T., and SHIELD, G. C. The polypeptide composition of influenza viruses. *Virology 44:*396 (1971).

STANLEY, W. M. Isolation of a crystalline protein possessing the properties of tobacco mosaic virus. *Science 81:*644 (1935).

THOMAS, C. A., JR., KELLY, T., JR., and RHOADES, M. The intracellular forms of T7 and P22 DNA molecules. *Cold Spring Harbor Symp Quant Biol 33:*417 (1968).

MULTIPLICATION AND GENETICS OF BACTERIOPHAGES

MULTIPLICATION OF BACTERIOPHAGES   1049

STRUCTURE OF BACTERIOPHAGES   1049
INFECTION OF HOST CELLS   1051
Adsorption   1051
Separation of Nucleic Acid from Coat   1052
Mechanism of Injection of the Nucleic Acid   1053
Effect of Phage Attachment on Cellular Metabolism   1054
THE VIRAL MULTIPLICATION CYCLE   1054
One-step Multiplication Curve   1054
SYNTHESIS OF VIRAL MACROMOLECULES   1055
Viral Proteins Responsible for Turning Off Host
Macromolecular Synthesis   1056
TRANSCRIPTION OF PHAGE DNA   1057
Characteristics of Phage mRNA   1057
Control of Transcription   1058
Translation of Viral Messengers   1059
VIRAL DNA REPLICATION   1059
Replicative Intermediates   1060
Biochemistry of Replication   1064
ROLE OF DNA MODIFICATIONS: HOST-INDUCED
RESTRICTION AND MODIFICATION   1064
Glucosylation   1065
Methylation   1066
Restricting and Modifying Enzymes   1067
MATURATION   1068
Assembly of the Capsid   1068
Association of DNA and Capsid   1069
Phenotypic Mixing   1071
RELEASE   1071

PHAGE GENETICS   1072

MUTATIONS   1072
Plaque-type Mutants   1073
Conditionally Lethal Mutants   1073
Application of Phage Mutants   1073
GENETIC RECOMBINATION   1073
Special Formal Characteristics of Phage Recombination   1073

Construction of a Genetic Map   1075
Other Characteristics of Bacteriophage Recombination   1077
Molecular and Enzymological Mechanisms of Phage
    Recombination   1077
ORGANIZATION OF THE GENETIC MATERIAL OF PHAGE T4   1078
General Organization of the Map   1079
GENETIC REACTIVATION OF ULTRAVIOLET-INACTIVATED
    PHAGE   1080

MULTIPLICATION OF BACTERIOPHAGES WITH CYCLIC
    SINGLE-STRANDED DNA   1081
Icosahedral Phages   1081
Filamentous Phages   1082

RNA PHAGE   1082
RNA Replication   1082
Uses of RNA Replicase   1083
Regulation of Translation   1084

Model Systems. Viral multiplication is a complex process constituted by many elementary acts; hence the selection of a technically suitable model virus-cell system early in the development of virology has been an important cause of the rapid advance of knowledge in this field. It proved effective in spite of the large variety of viruses and their marked differences in structure and abundance of genetic information, because all viruses are similar in the basic aspects of multiplication. In the 1930s bacteriophages were adopted as model systems by Burnet in Australia and by the Hungarian chemist, Schlesinger, and were later used by a large group of investigators working in close contact, including Delbrück, Luria, Hershey, and S. S. Cohen in the United States and Lwoff in France. These more recent workers concentrated on **the phages of Escherichia coli**, and especially the "T-even" phages (T2, T4, and T6). These bacteriophages were thought to be the simplest possible organisms, but turned out to be among the most complex of all viruses; and their complexity was instrumental for many discoveries. For instance, the presence of 5-hydroxymethylcytosine instead of cytosine in their DNA made it possible to follow the intracellular multiplication of the viral DNA, as noted in the previous chapter; and the enzyme responsible for synthesis of this unusual base gave the first hint that virus-specified enzymes are made in the infected cells. The T phages are also easy to purify, owing to their large size; and they have a high frequency of genetic recombination, which is beneficial for genetic studies.

Several phages containing single-stranded DNA or RNA were discovered later and will be discussed at the end of this chapter. Phage λ, the model for the important phenomenon of lysogeny, will be discussed in Ch. 46.

MULTIPLICATION OF BACTERIOPHAGES

As pointed out by Lwoff, bacteriophages can exist in structurally different states: **extracellular virions, vegetative phage,** and **prophage.** All bacteriophages exist at some stage as vegetative phages during their intracellular multiplication, but not all can become prophages. Those able to become prophages are called **temperate;** those unable to do so are **virulent.** The properties of the extracellular virions were reviewed in the preceding chapter; those of vegetative phage will be studied in this chapter, mostly with the virulent T-even coliphages; and those of prophage will be discussed in the next chapter.

STRUCTURE OF BACTERIOPHAGES

The even-numbered T phages are made up of a head and a tail (Fig. 45-1). The head, which contains the DNA, has the shape of two halves of an icosahedron connected by a short hexagonal prism. In other phages, such as *Salmonella* phage P1 and P2 and other coliphages, the head is strictly icosahedral (Fig. 44-5F). The **tail** varies greatly in dimensions and structure. In its simplest form it is composed of a **tube** (through which the viral DNA passes during cell infection), probably helical in structure. In the T-even coliphages the

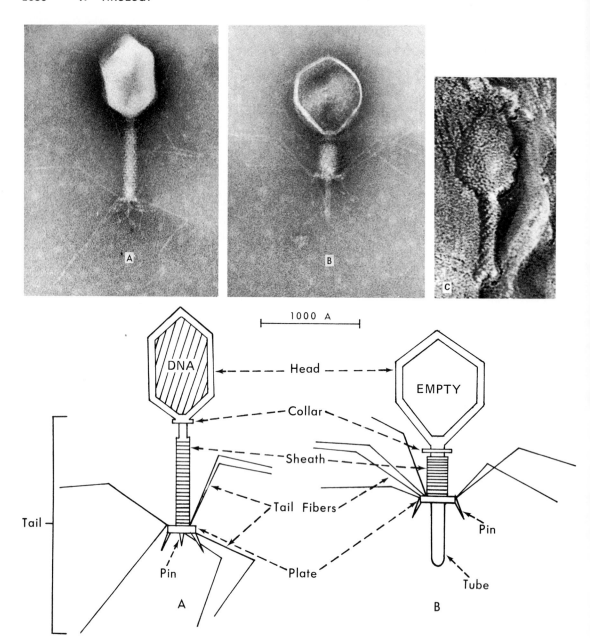

Fig. 45-1. Micrographs of bacteriophage T2. **A** and **B,** negative staining. **A.** Phage before injection, with full head and extended sheath. **B.** Phage after injection, with empty head and contracted sheath. **C,** showing the helical structure of the sheath, was obtained by freeze-etching, i.e., the phage was embedded in ice which was then fractured; the place of fracture was covered with a thin layer of evaporated metal, which was then photographed. (**A** and **B** courtesy of E. Boy de la Tour; **C** courtesy of M. E. Bayer.)

tube is surrounded by a **sheath** formed by parallel rows of protein monomers and connected to a thin disc or **collar** at the head end and to a **plate** at the tip end. The sheath is capable of contraction. The plate of T-even coliphages is hexagonal, has a **pin** at every corner, and is connected to six very long thin **tail fibers** which are the organs of attachment of the bacteriophage to the wall of the host cell. Coliphages T1 and T5 have a sheathless tail terminating in rudimentary fibers; coliphages T3 and T7 have a short stubby tail which terminates in a structure resembling a base plate.

DNA-less capsids **(ghosts)** of T-even phages are obtained by rapidly diluting the phage from a concentrated salt solution into water and thus osmotically bursting the head with release of DNA.

INFECTION OF HOST CELLS

The first step in infection is the attachment of a phage virion to specific **receptors** on a bacterial cell **(adsorption);** then the DNA separates from the viral coat and becomes free in the cell **(release of the nucleic acid).**

ADSORPTION

There is a requirement for host cell metabolism in the adsorption of some phages (e.g. T1) but not of others (e.g., T-even phages) which are adsorbed even at 0°; under certain optimal ionic conditions (e.g., 0.1 N Na+ for phage T2) adsorption becomes irreversible. Adsorption appears to involve the formation of ionic bonds between complementary charges on the attachment sites of the virions and on the cell **receptors.** Thus it is inhibited by low or high pH, when one kind of the interacting group loses ionization.

Receptor. Receptors for different phages of the T series are located in different layers of the cell wall of *E. coli* cells. The deeply located receptors are presumably accessible through holes in the more superficial layers. Purified receptors for phage T5, containing protein and lipopolysaccharide, combine with the tip of the bacteriophage tail in vitro (Fig. 45-2).

Some male-species coliphages adsorb only to the sex pili of F+ cells (see Ch. 9). Phage f2 (RNA-containing) adsorbs laterally on the entire pilus, whereas fd (DNA-containing) adsorbs exclusively on its tip.

Mutations Affecting Adsorption. Adsorption of a bacteriophage can be abolished by bacterial mutations to **bacteriophage resistance** which change the receptors and may also change the antigenic specificity of the cells. For instance, *E. coli* B/2 is a mutant of *E. coli* B resistant to T2. Conversely, if a pure culture of B/2 bacteria is exposed to a large concentration of T2, rare **host-range mutants** (T2h) in the phage population can adsorb to the B/2 cells and initiate normal multiplication.

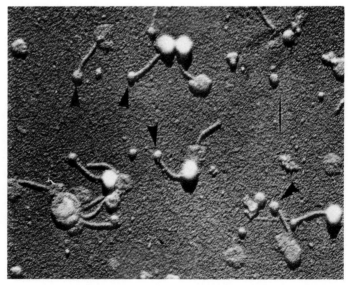

Fig. 45-2. Electron micrograph of bacteriophage T5 mixed with purified receptors. The receptors are in the form of spheres (arrows), about 300 A in diameter, artificially produced upon cell disruption. Some receptors are free and others are adsorbed to the tip of the phage tails. Shadowed preparation. [From Weidel, W., and Kellenberger, E. *Biochim Biophys Acta 17*:1 (1955).]

Similar mutants occur with most phages. Host-range mutants played an important role in the early development of bacteriophage genetics. Their selection illustrates the connection between viral evolution and the evolution of the host cells.

Viral Sites for Adsorption. The T-even coliphages have specialized viral sites for adsorption: the *tail fibers* and the *tail pins*. Thus isolated tail fibers adsorb to the same range of bacteria as the bacteriophage from which they were derived, and antiserum to the fibers inhibits phage adsorption. Electron microscopy shows that the tips of the fibers attach first to the cell surface and are followed by the pins; the adsorbed virion acquires a characteristic position perpendicular to the cell wall with the head sticking out (Fig. 45-3).

SEPARATION OF NUCLEIC ACID FROM COAT

In one of the most significant experiments of modern biology, Hershey and Chase demonstrated in 1952 the separation of the viral nucleic acid from the capsid (Fig. 45-4). They labeled the protein of T2 with ^{35}S or the DNA with ^{32}P, allowed the virus to infect bacteria, and exposed the bacteria to violent agitation in a Waring Blender, which shears the tails of the adsorbed virions. The experiment yielded two results that, at the time, seemed astonishing. 1) With ^{35}S-labeled phage 80% of the label came off; but with ^{32}P-labeled phage, essentially all the label remained with the cells and since it was DNase-resistant it was **within** the cells. 2) The blended bacteria produced progeny phage as if they had not been blended. These results strongly suggested that **phage DNA carries the genetic information of the phage into the cell.**

Ejaculation of the nucleic acid from the coat could also be elicited with wall fragments instead of cells; the viral DNA was then released into the medium, where it was digestible by DNase. This result indicated that injection does not require energy from the cell.

Eclipse. After the DNA is injected, the intact cells can produce plaques. However, disrupted cells do not produce plaques if plated in the

Fig. 45-3. Electron micrograph of particles of phage T5 adsorbed to an *E. coli* cell. The virions attach by the tip of their tail. Note also that the heads of some particles are clear (electron-transparent) having injected their DNA into the cells; others are dark (electron-opaque) and still contain their DNA. [From Anderson, T. F. *Cold Spring Harbor Symp Quant Biol 18*:197 (1953).]

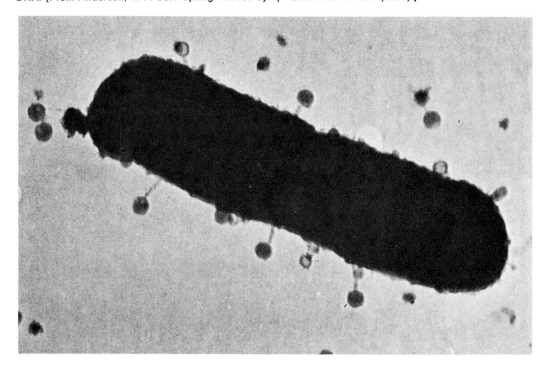

Fig. 45-4. The Hershey and Chase experiment, showing the separation of viral DNA and protein at infections. **A.** The phage protein was labeled by propagation in a medium containing $^{35}SO_4^=$. **B.** Phage DNA was labeled by propagation in medium containing $^{32}PO_4^=$. Phage was adsorbed to host bacteria, and after 10 minutes at 37° the culture was blended. Most of the ^{32}P remained associated with the cells, whereas most of the ^{35}S came off. Labeled components are shadowed.

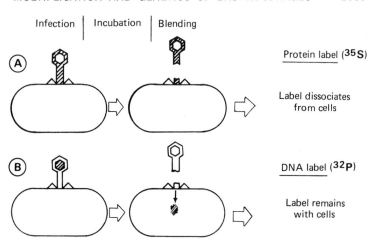

usual way; but the infectivity reappears when progeny virus is formed. The temporary disappearance of infectivity, called eclipse, is due in part to the inability of the naked viral DNA to infect bacteria under ordinary conditions and in part to a transient transformation of the DNA into a noninfectious replicative form (see below, DNA replication).

Infection by Naked Viral Nucleic Acid. Further evidence for the essential role of the nucleic acid is afforded by the infection of bacteria with purified phage DNA. This phenomenon was not readily discovered, because bacterial cells accept infectious viral nucleic acid only under special conditions: 1) as spheroplasts, 2) as intact bacteria in a special competent state in which they can accept transforming bacterial DNA, or 3) when exposed at the same time to other, **helper phage** to which they are susceptible. It is likely

that all three conditions cause openings in the cell wall through which isolated DNA can penetrate.

MECHANISM OF INJECTION OF THE NUCLEIC ACID

Electron microscopic observations have shown that after adsorption of the T-even coliphage the plate is firmly held close to the cell surface by the tail fibers and pins and then **contraction of the tail sheath** pulls the collar and the phage head toward the plate. The tube is then pushed through the cell wall, reaching the outer surface of the plasma membrane without wounding it (Fig. 45-5). Because of this action, as well as its shape, the virion has been likened to a hypodermic syringe, and the release of the nucleic acid is called injection. The injected DNA **spontaneously crosses the plasma membrane** but is thereby exposed to

Fig. 45-5. Contraction of the sheath of T-even bacteriophages during adsorption. Since the plate is held against the surface of the bacterium **(A)** by the tail fibers attached to the cell's receptors, the contraction of the sheath **(B)** pushes the tube of the virion through the cell wall. During contraction there is a rearrangement of the protomers, so that the number of rows is reduced to one-half.

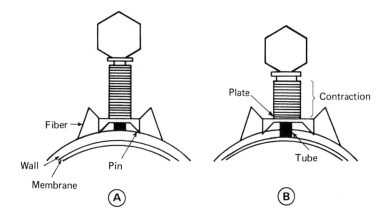

membrane-associated nucleases (see below, Host-induced restriction).

The sheath is comparable to a primitive muscle in becoming shorter and thicker during contraction and in containing ATP which is converted to ADP. Contraction is caused by a change in shape and rearrangement of its protein protomers.

Contraction of the sheath probably favors the release of DNA but is not required for injection, since many phages lack a sheath.

EFFECT OF PHAGE ATTACHMENT ON CELLULAR METABOLISM

The attachment of the phage tail per se, without expression of viral genes (e.g., in the presence of chloramphenicol to prevent new protein synthesis), **alters the permeability of the cell plasma membrane,** which then allows small molecules (e.g., nucleotides) to leak to the medium. Similar alterations are produced by the attachment of **phage ghosts** (see above, Structure of bacteriophages) and of some bacteriocins (see Ch. 46).

These alterations cause profound **metabolic changes.** Cellular DNA and protein synthesis stop almost immediately. The membrane changes and their metabolic effects are **reversible** in cells infected by phage at **low multiplicity;** but in cells infected at **very high multiplicity** or by **ghosts** they lead to cell lysis without viral multiplication (**lysis from without**).

THE VIRAL MULTIPLICATION CYCLE

The process of viral multiplication, initiated by the penetration of the parental nucleic acid, involves many sequential steps, which end in the release of the newly synthesized **progeny virions.** This sequence is called the **multiplication cycle.** Kinetic analysis of the various steps requires large numbers of cells in which infection proceeds **synchronously,** i.e., the cells must be infected simultaneously, and secondary infection by progeny virus must be avoided. The one-step conditions were achieved very simply in the classic studies of Delbrück, by allowing virus to infect cells for a brief time and then preventing further infection either by diluting the virus-cell mixture or by adding phage-specific antiserum. The infected cells are then freed of unadsorbed virus and antiserum by centrifugation.

The multiplicity of infection must be carefully selected according to the demands of the experiment, for if it is too high viral replication may be abnormal, and if it is too low many cells remain uninfected. Since not all virus particles adsorb during the brief initial incubation, the actual multiplicity of infection is usually monitored by determining the proportion of **productively infected cells** (see Ch. 44, Appendix). These cells are identified as **infectious centers,** i.e., cells able to produce a plaque in the regular assay used for the virus.

The viral growth cycle can also be studied in individual infected cells isolated in tubes or in small drops of medium under paraffin oil (**single-burst experiments**).

ONE-STEP MULTIPLICATION CURVE

A one-step multiplication curve describes the production of progeny phage as a function of the time after infection under one-step conditions. The **extracellular** titer is determined by assaying the medium after centrifuging down the cells, and the **total** (extracellular plus intracellular) **titer** by assaying the culture after disrupting the cells. These titers are usually expressed as infectious units per productive cell. Plotting their values versus the time of assay yields multiplication curves such as those of Figure 45-6, in which the following stages can be recognized.

1) **The eclipse period.** Total virus titer is almost nil (only residual from the inoculum). The end of the eclipse period is taken as the time at which an average of one infectious unit has been produced for each productive cell.

2) **The intracellular accumulation period.** Progeny phage accumulate intracellularly but are not yet released into the medium. The end of this period is the time at which one viral infectious unit per productive cell, on the average, has appeared extracellularly. The end of this period also marks the end of **the latent period.**

3) **The rise period.** The extracellular phage titer increases until the end of the multiplication cycle. The average number of infectious units of virus per productive cell at that time represents the **viral yield.**

Progeny phage is released into the medium by the **lysis** of the cells in a burst, which causes a **drop in turbidity** of the culture if

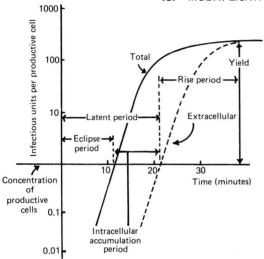

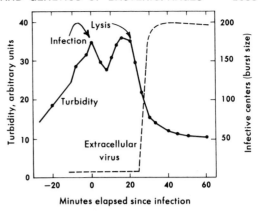

Fig. 45-7. Turbidity of an *E. coli* culture at various times after infection with phage T4r. Multiplicity of infection about 5. The arrow indicates the onset of lysis, which coincides with the end of the latent period. The temporary drop in turbidity immediately following infection is determined by a transient increased permeability of the plasma membrane, which causes leakage of cellular components. [Modified from Doermann, A. H. *J Bacteriol* 55: 257 (1948).]

Fig. 45-6. Diagram of one-step multiplication curve of bacteriophage T2. Bacteria and phage were mixed and adsorption was allowed for 2 minutes; antiphage serum was then added. The bacteria were recovered by centrifugation and were resuspended in a large volume of medium at 37°. A sample was immediately plated to determine the concentration of productive cells. Other samples were taken from time to time and divided into two aliquots: one was shaken with chloroform to disrupt the bacteria and was then assayed (**total virus**); the other was freed of bacteria by centrifugation and the supernatant was assayed (**extracellular** virus). The titers are compared with the concentration of productive cells as 1.0.

most cells are infected (Fig. 45-7). With T-even phages the event can be seen in dark-field microscopy, where the virions are visible. With these phages lysis is delayed by more than an hour if a culture infected with a wild-type strain is heavily reinoculated, before the time of normal lysis, with the same strain (**lysis inhibition**). **Rapid lysis (r) mutants** do not delay lysis under these conditions. The different lysis properties of r⁺ (wild-type) and r phages are reflected in the shape of their plaques: r plaques are about 2 mm in diameter, clear, and with a sharp edge; r⁺ plaques are smaller and surrounded by a halo of increased turbidity. The halo is formed by bacteria infected more than once and therefore, as indicated above, lysis-inhibited (Fig. 45-8). This readily observed plaque difference was a valuable marker in the early development of phage genetics.

Lysis inhibition is very useful in phage research because it **extends the duration of the period of**

viral synthesis. As a consequence, the cells make much more virus; moreover the time-dependent events of viral multiplication can be studied more easily.

SYNTHESIS OF VIRAL MACROMOLECULES

Infection of bacteria by T-even phages causes a **profound rearrangement** of all **macromolecular syntheses.** The first hint of this rearrangement was gained when Hershey measured the synthesis of viral and cellular DNA in the infected cells, taking advantage of the unique presence of 5-hydroxymethylcytosine in the former. He showed that **the synthesis of cellular DNA stops and is replaced by a viral synthesis;** furthermore, the preexisting cellular DNA breaks down and its components are utilized as precursors of viral DNA. It is now clear that within a few minutes after infection the synthesis of all **cellular** DNA, RNA, and protein (i.e., that directed by the cellular genome) ceases. The effect is unrelated to the transient inhibition produced by membrane damage during adsorption; it requires the expression of viral genes. Soon the cellular syntheses are entirely replaced by viral syntheses. This shift represents the fundamental characteristic of viral parasitism: **the substitution of**

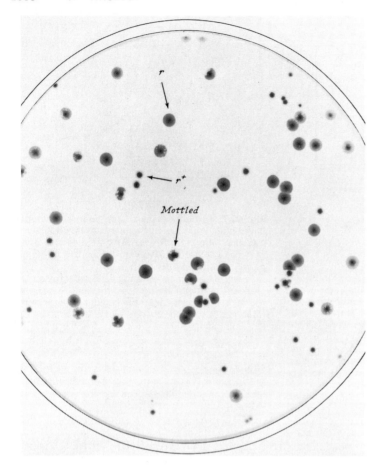

Fig. 45-8. Plaques produced by cells infected by a mixture of T2r and T2r+ phage. Plaques of r type are large and without a halo; those of r+ type (wild-type) are small and surrounded by a halo. Cells infected by both r and r+ phage produce mottled plaques with a sectored halo (dark sectors, r phage; clear sectors, r+ phage). The halo is produced by infected cells with lysis inhibition caused by r+ phage. (From Stent, G. *Molecular Biology of Bacterial Viruses.* Freeman, San Francisco, 1963.)

viral genes for cellular genes in directing the synthesizing machinery of the cell.

This profound reshuffling is not reflected in big changes of **over-all** synthetic rates. RNA synthesis is reduced because rRNAs and most tRNAs are no longer synthesized. Protein synthesis is limited by the available number of ribosomes and thus remains at much the same level as before infection. DNA synthesis resumes after a brief halt at an increased rate, reflecting the rapid multiplication of the viral DNA.

VIRAL PROTEINS RESPONSIBLE FOR TURNING OFF HOST MACROMOLECULAR SYNTHESIS

Following the discovery of deoxycytidylate hydroxymethylase, an intense search for virus-specified enzymes revealed many enzymes that are synthesized only in infected bacteria; they are specified by viral genes because their properties are changed by phage mutations.

These proteins determine the shift either by turning off cellular synthesis or by carrying out viral syntheses.

Turn-off proteins are of three kinds: 1) several viral nucleases are involved in the **degradation of cellular DNA** in cells infected by T-even phage (Fig. 45-9), recognizing cytosine clusters which are absent in phage DNA; 2) within 3 to 4 minutes after infection, when host DNA is not yet degraded, viral proteins (see next section) that change the specificity of RNA polymerase **turn off host RNA synthesis;** 3) **host protein synthesis is inhibited** by changes of tRNAs and of protein-synthesizing enzymes and initiation factors. Thus after T2 infection one of the leucine tRNAs is progressively split into two fragments, causing reduced translation of its codon (CUG), which is abundant in bacterial DNA but not in phage DNA. Later, new tRNAs, tRNA methylases, and aminoacylases appear, either by new synthesis or by modifi-

Fig. 45-9. Four-step degradation of *E. coli* DNA after T4 infection. [Modified from Koerner, J. F. *Annu Rev Biochem* *39*:291 (1970).]

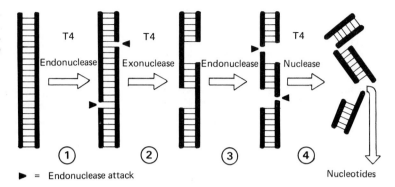

► = Endonuclease attack

cation of cellular components; they presumably favor viral over cellular protein synthesis. The destruction of one of the host initiation factors (see Ch. 12, Protein synthesis) required for the translation of cellular mRNAs restricts ribosomes to the translation of phage mRNA. This event is revealed by the inability of extracts of cells infected by T4 or T5 to translate RNA from an RNA phage (which also requires the missing factor); T4 messengers are translated actively.

TRANSCRIPTION OF PHAGE DNA

The phage DNA is transcribed according to a fixed pattern, which allows the **successive and orderly expression of various sets of viral** genes and consequently coordination of the hundred or so gene functions participating in development of these large viruses. Control is achieved by several means: 1) modification of the *E. coli* transcriptase (see Ch. 11, Transcription); 2) virus-specified synthesis of a few new proteins affecting transcription; 3) coupling of the transcription of certain sets of genes to DNA replication. The molecular bases of this control provide an excellent **model for studying differentiation.**

CHARACTERISTICS OF PHAGE mRNA

Viral messenger RNAs can be identified by hybridization (see Ch. 10) to phage DNA; different classes are distinguished either for their sensitivity to chloramphenicol (an inhibitor of protein synthesis), which inhibits the appearance of some messengers but not of others, or for their different hybridization in competition with RNA extracted at specific times after infecting the cells. The genes transcribed in a given mRNA can be also identified: 1) the mRNA is used in vitro to synthesize functional proteins such as glucosyl transferases, lysozyme, and other phage

Fig. 45-10. Transformation assay for phage T4 mRNA. Separate strands of wild-type T4 DNA are each annealed with mRNA extracted from infected cells. Treatment with a nuclease specific for single-stranded DNA removes the unhybridized segments. The DNA of the genes corresponding to the mRNA is isolated, by denaturing the hybrids, and is then added to cells competent to accept DNA and infected by suitable T4 mutants (transformation). If the DNA contains a gene corresponding to a mutation in the recipient T4 DNA, it may generate a wild-type recombinant, thus identifying the gene. [Modified from Jayaraman, R., and Goldberg, E. B. *Proc Natl Acad Sci USA 64*:198 (1969).]

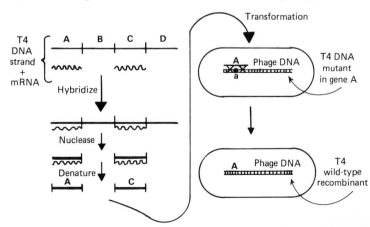

enzymes; 2) the DNA that hybridized with the RNA is isolated and then tested in a phage transformation system (Fig. 45-10). Important for characterizing the specificity of transcribing enzymes is transcription in vitro of phage DNA.

Phage mRNA is usually **polycistronic.** It **turns over** and its degradation products return to the nucleotide pool and some of them are converted into deoxynucleotides and end up in viral DNA. This flow facilitated the recognition of RNA turnover by Volkin and Astrakhan, and therefore the discovery of phage mRNA (Ch. 13). However, because phage transcription does not require as rapid a change of gene action as bacteria, the rate of degradation of phage mRNA is slow. This stability explains the **low polarity of phage nonsense mutations** (see Ch. 13).

CONTROL OF TRANSCRIPTION

The following scheme operates in phage T4 and other large phages (Figs. 45-11, 45-12, and 45-13). 1) When the phage DNA enters the cell it is transcribed by the available host transcriptase, which transcribes sequences containing several **immediate early genes**

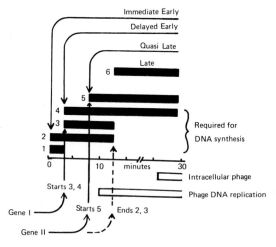

Fig. 45-12. Program of transcription of *B. subtilis* phage SPO1. The black bars (1, 2, 3 . . . 6) indicate the periods during which various classes are transcribed; the open bars indicate the timing of other events. Phage genes I and II affect the beginning of synthesis of certain classes and the end of others, as indicated. Time is from infection. [Modified from Gage, P., and Geiduschek, E. P. *J Mol Biol* 57:279 (1971).]

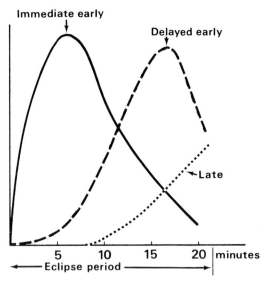

Fig. 45-11. Appearance of different classes of T4 mRNA at 30°. Immediate early mRNA is extracted from cells infected in the presence of chloramphenicol, which blocks delayed early and late RNAs. The other classes of RNA are measured by hybridization with phage DNA in competition with RNA extracted at various times.

scattered throughout the genome. Each transcription is terminated by the host termination factor rho. This transcription, in contrast to those that follow, is not inhibited by chloramphenicol because it involves no new protein. 2) Two new proteins resulting from the previous transcription cause the transcription of **delayed early genes:** an antiterminator interferes with termination of immediate early transcripts, allowing these to be extended to other adjacent genes, and a new σ-like factor causes the host transcriptase to transcribe new sets of genes. Products of early genes shut off cellular macromolecular syntheses and are required for the replication of the viral DNA. 3) Early transcription causes the synthesis of a second viral σ-like factor, which causes the host transcriptase to transcribe a third set of **late genes.** Late transcription, in contrast to the previous ones, requires DNA replication. Apparently this coupling arises because only replicating DNA containing single-strand discontinuities (see Ch. 10, RNA replication) can be transcribed. The main products of late genes are capsid proteins and enzymes for lysing the cells. Additional classes of mRNAs can be recognized (Fig. 45-12). Control is not limited to the

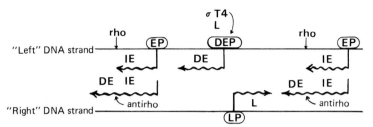

Fig. 45-13. Transcription of different sets of genes on T4 DNA. EP = early promoters; DEP = delayed early promoters; LP = late promoters; IE = immediate early RNA; DE = delayed early RNA; L = late RNA; rho = terminator. Delayed early messengers are transcribed either by interfering with termination of immediate early transcription (antirho) or by initiating at new promoters. Early and late messengers are transcribed on different strands.

beginning of the various transcriptions but also affects their ending (Figs. 45-11 and 45-12) although the mechanisms are less clear. The transcription sequence is under the control of several phage genes and is upset by their mutations.

After infection the host transcriptase undergoes a series of successive modifications (adenylation and phosphorylation) which apparently do not directly affect its initiation specificity, but change its affinity for the various σ-like factors (either of the host or virus-specified). This allows a new factor to be preferentially bound even if those previously used are still present. In addition, the transcriptase binds four small phage-specified proteins, two of which are required for transcription of late genes.

In smaller phages the general organization of transcription is similar, but differs in details. Thus in phages T3 and T7, one of the early genes transcribed by the host polymerase specifies an **entirely new** simple **transcriptase** containing a single polypeptide chain (MW 107,000 daltons), which then transcribes the late genes. In phage T5 there is an additional control mechanism: the phage injects first an 8% segment of its DNA, whose transcription products allow injection of the remainder of the viral DNA.

In many phages, such as T5, T7, or λ, late transcription and DNA replication are not coupled, but DNA replication enhances the amount of late RNA made, presumably by providing more templates for transcription.

The organization of phage transcription is different from that of bacteria. Phage genes of a certain class share a common initiation specificity, but they are not transcribed in the same messager. Hence **a class of coordinately controlled genes is not necessarily an operon** (see Ch. 13).

TRANSLATION OF VIRAL MESSENGERS

The synthesis of phage proteins reflects the temporal sequence of transcription but is also determined in part by additional **independent mechanisms directly affecting translation,** some of which have been noted above (Viral proteins responsible for turning off host macromolecular synthesis).

In cells infected with T-even phage the proteins are synthesized on **preexisting host ribosomes,** as shown in a classic experiment whose details were discussed in Chapter 13 (Messenger RNA). Proteins specified by individual genes are synthesized in **widely different amounts,** probably owing to different efficiencies of initiation of transcription and translation. The most abundant is the main component of the head capsid, a late protein.

VIRAL DNA REPLICATION

Phage DNA replication has been extensively studied because the relatively small size of the molecules and the high rate of synthesis facilitate the observation of DNA in the process of replication **(replicative intermediate).**

These studies had three main results: 1) they allow an evaluation of the biological significance of the two forms of DNA replication (symmetric and asymmetric; Ch. 10, DNA replication) by comparing the properties of the viruses and of their replicative intermediates; 2) they explain how difficulties arising from the presence of unusual bases in the DNA of certain phages have been circumvented; and 3) they clarify the mechanism of recombination.

REPLICATIVE INTERMEDIATES

After the intermediates have been labeled with a short pulse of a radioactive precursor, their structure is inferred from their sedimentation behavior in gradients and sometimes from their shape in electron micrographs (see Ch. 10). Early in infection the replicative intermediates are found associated with **discrete specific sites of the cell membrane.** This association seems essential because a phage cannot replicate in a cell whose sites have been saturated by a related phage, whereas an unrelated phage can. Later in infection replicative intermediates are found free in the cytoplasm.

Different phages may have different replicative intermediates whose nature may change in the course of infection. These differences in the method of replication will be illustrated using phages T4, T7, and λ. The **mature** DNA molecules (i.e., present in the virions) of these viruses are linear but differ in other features, which were reviewed in Chapter 44 (Fig. 44-12). A fourth phage, φX174, whose DNA is cyclic, will be considered at the end of this chapter.

Phage λ, by taking advantage of the presence of complementary (cohesive) ends in the **mature** DNA (i.e., present in the virions) forms a cyclic replicative intermediate. Phages T7 and T4 after an initial round of replication form long linear intermediates containing several genomes covalently linked end to end (concatemers). These replicative intermediates are apparently required because a linear molecule cannot replicate completely.

In the replication of double-stranded DNA, the end of the discontinuously replicated strand (see Ch. 10) remains unreplicated if the synthesis of the last Okazaki segment (which grows backwards) cannot begin at the last nucleotide (Fig. 45-14A). In cyclic molecules this complication does not arise, and in concatemers the internal molecules can be completely replicated. In the first replication of T7 DNA, which occurs bidirectionally from an internal origin, both strands remain completely unreplicated at opposite ends. However, owing to terminal repetition of the molecules, these ends are complementary and can pair with each other generating a dimer (Fig. 45-14B). After the gaps are closed by ligase the replicative defect is eliminated. In successive steps of the same kind long concatemers can be formed.

Concatemer formation may occur by a different mechanism in T4 DNA, whose initial replication appears to be unidirectional. Thus molecules from "petite" mutants, which are ⅔ the normal length, replicate only 50%, as average, in single infection because the origin is located randomly in these permuted molecules (Fig. 45-14C). Normal T4 DNA molecules may replicate completely after becoming cyclic by recombination between the repetitious ends. Cyclization cannot occur in the petite molecules since they lack repetition. However, cyclic T4 molecules have not been seen in electron micrographs: either they break too easily during handling (owing to their large size) or rapidly evolve to concatemers.

Later in infection both λ and T4 DNA replicate bidirectionally, after concatemers are formed. With λ the concatemers are formed by elongation of one strand of the cyclic molecules (Fig. 45-15), according to the rolling circle model (see Chs. 10 and 45, DNA replication). With T4, concatemers are probably generated by recombination (breakage and reunion) of cyclic and linear molecules, because phage with mutations in genes 46 and 47, which lacks recombination (see below, Phage genetics), does not form concatemers.

At the end of the replication concatemers generate mature molecules in different ways (Fig. 45-16). T4 concatemers are cut by a mechanism that **recognizes the length of the mature DNA** but no specific sequences, yielding both permuted sequences and repetitious ends (see below, Maturation). Phage λ concatemers are cut by **nucleases that recognize the specific sequences of the molecular ends,** yielding nonpermuted molecules with cohesive ends.

Origin and Direction of Replication. That replication of phage T4 DNA begins at a constant point in the sequence, as postulated by the replicon theory (see Ch. 10, DNA replication), has been demonstrated using the incomplete replication of petite DNA (Fig. 45-14C). At the end of synthesis genes close to the origin have a greater chance of being replicated than distant genes, resulting in a gradient of gene frequencies. Such a gradient, detected experimentally by transformation (Fig. 45-10), determines the location of the origin and the direction of synthesis as indicated on the genetic map in Figure 45-31. However, it is not known whether concatemers replicate from the same origin: electron microscopy shows that they have several growing forks moving in opposite directions, but their origin is unknown.

A fixed origin and bidirectional replication has

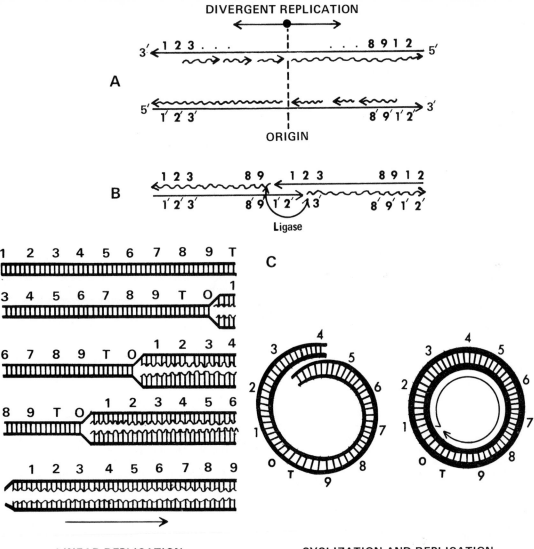

Fig. 45-14. Significance of DNA replicative intermediates. Wavy lines indicate new DNA strands; numbers represent genes (arbitrary). The divergent replication of a linear molecule of T7 DNA (**A**) remains incomplete at the 3′ ends if Okazaki segments cannot initiate at the terminal nucleotides. However, owing to terminal repetitions, the two unreplicated ends can pair (**B**), forming a completely replicated concatemer. The replication of permuted T4 DNA molecules (**C**) lacking terminal repetitions ("petite") from a randomly located origin (O) in the direction indicated by the arrow remains incomplete and never reaches the terminus (T). On the average, only a half molecule is replicated. Genes adjacent to the origin in the direction of replication are replicated most frequently. In contrast, complete molecules can cyclize and then replicate entirely from origin to terminus.

also been demonstrated by electron microscopy for λ DNA (see Ch. 46).

Breakage and Reunion of Concatemers.
Breakage and reunion continues to occur after concatemers are formed, leading to the **dispersion** of the parental DNA into many progeny molecules, each of which contains a small parental segment covalently linked to newly synthesized DNA.

This phenomenon was revealed by the abnormal results of experiments testing whether phage DNA replication is semiconservative (Fig. 45-17). Thus when cells are infected with ^{32}P-labeled phage T4 and then grown in a medium containing 5-bromouracil (5BU), which yields DNA of higher

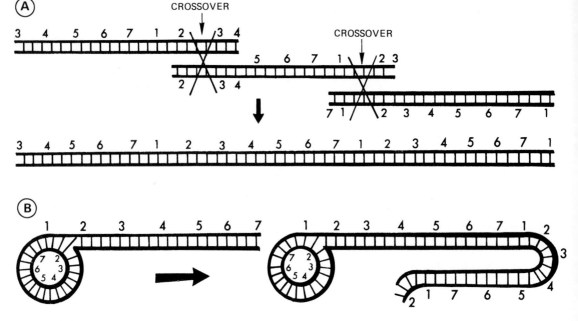

Fig. 45-15. Concatemer formation by recombination **(A)** or rolling circle replication **(B)** (see Ch. 10, DNA replication).

Fig. 45-16. Production of mature viral DNA molecules from concatemers. **A.** Mechanisms generating permuted molecules with repetitious ends, as with phage T4. Constant lengths are cut off the concatemer, irrespective of sequences. **B.** Mechanism generating nonpermuted molecules with cohesive or repetitious ends. Nucleases make staggered cuts on the two strands, recognizing specific sequences, and generate λ-type molecules (with cohesive ends). If the short strand at each end is continued by DNA polymerase as a complement of the longer strand, T7-type molecules with repetitious ends are generated. Black arrowheads indicate endonucleolytic nicking.

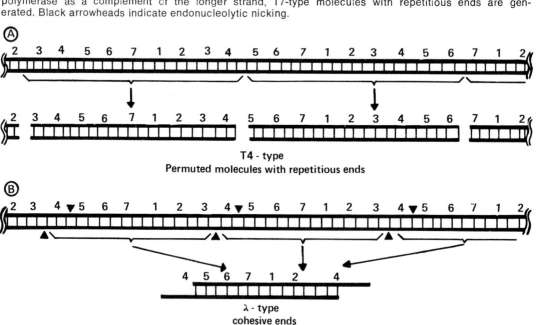

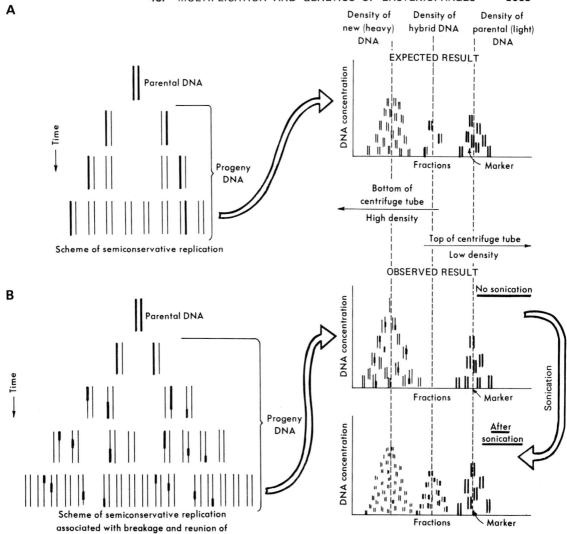

Fig. 45-17. Scheme of replication of T2 DNA. **A.** The results expected on the basis of semiconservative replication. After equilibrium centrifugation in a CsCl density gradient, the parental DNA is expected entirely in a band of hybrid density. **B.** The results expected if semiconservative replication is associated with breakage and reunion. If a progeny molecule has experienced, in its line of descent, many breakage and reunion events, the parental DNA may be present in small segments distributed over a large number of molecules, which therefore would have a density only slightly less than that of new DNA. Sonication, by breaking the molecules into small fragments, allows the identification of the hybrid segments, since their lower density will not be "diluted" by attachment to nonhybrid (all new) segments.

buoyant density, all the ^{32}P would be expected in hybrid molecules of intermediate density, with 5BU in one strand, while the bulk of the DNA would be denser, with both strands containing 5BU (see Ch. 10, DNA replication). However, on equilibrium density centrifugation the **intact progeny phage DNA yielded a single band, at the density of 5BU-DNA,** and it contained all the radioactivity. That this result was due to the dispersion of segments of parental DNA into progeny molecules was shown

by breaking down the DNA by sonication. Banding in CsCl now produced the two bands predicted by the semiconservative model. It could be estimated that the dispersed parental segments were 10% of a whole molecule.

Relation of Breakage and Reunion to Genetic Recombination. In cells infected by two genetically marked phages, the dispersion of the parental DNA must generate recombinant molecules at a frequency dependent on how easily various molecules in the same cell can exchange parts. With phage T4 such recombinants occur at a high frequency; the number of exchanges calculated from genetic evidence (rounds of matings) is close to that deduced from the degree of dispersion, showing that **all replicating DNA in a cell belongs to the same pool.** Moreover, in different phages the degree of dispersion (which is maximal for T-even phages) correlates well with the frequency of genetic recombination between distant markers, showing that **breakage and reunion of DNA molecules is the cause of this recombination.** Although recombination occurs simultaneously with DNA replication, it does not depend on it, since it can take place in thymine-starved cells. Hence the enzymatic mechanisms that produce recombination, although they may have been developed to favor DNA replication, continue to work in its absence.

BIOCHEMISTRY OF REPLICATION

Origin of Components of Viral DNA. The atoms forming the DNA of the progeny virions can derive from the DNA of the infecting virions, from materials present in the cell before infection, and from the culture medium. Suitable labeling experiments show that degradation products of the host DNA contribute about 30% of the precursors for the DNA of phage T2 and 90% of phage T7 (which produces a smaller amount of DNA per cell). The remainder is contributed by precursors assimilated from the medium, except for the small amount of DNA brought in by the infecting virions. The DNA of temperate phages, which do not cause breakdown of the host cell DNA, is synthesized exclusively from exogenous precursors (Ch. 46).

Viral Proteins Participating in DNA Synthesis. DNA replication depends on viral proteins because it is absent in cells infected in the presence of chloramphenicol or when the phage carries a mutation in certain early genes. The functions of many of these genes are now known: some specify polymerizing enzymes (e.g., DNA polymerase, polynucleotide ligase) or proteins required for keeping the DNA in the right configuration (such as the DNA-unfolding protein specified by gene 32; see Ch. 10, DNA denaturation). Other genes specify enzymes for the synthesis of precursors (Fig. 45-18). Especially interesting

among these are enzymes that confer two specific features to the viral DNA: incorporation of 5-hydroxymethylcytosine instead of cytosine, and glucosylation.

Incorporation of 5-hydroxymethylcytosine involves not only the **synthesis of 5-hydroxymethyl-deoxycytidine triphosphate** (dHTP), but also the **dephosphorylation of dCTP and dCDP to dCMP** (Fig. 45-18). The latter step is crucial because **it prevents cytosine incorporation,** which makes the DNA sensitive to nucleases that recognize cytosine clusters and break down cellular DNA. This is shown by the lethal effect of mutations affecting the dephosphorylating enzyme. Dephosphorylation also provides a **major source of precursors** for phage DNA synthesis by converting dCDP, constantly generated from CDP by the ribonucleotide diphosphate reductase, into dCMP, which is the precursor of both TTP and dHTP (Fig. 45-18).

Some *B. subtilis* phages containing 5-hydroxymethyluracil instead of thymine have comparable enzymes for preventing the incorporation of thymine residues in phage DNA.

Phage DNA **glucosylation** is carried out by glucosyl transferases which transfer glucosyl groups from uridine diphosphate glucose to the hydroxymethyl groups of hydroxymethylcytosine residues present in the viral DNA. The **different glucosylation patterns of the DNAs of different T-even phages** (see Fig. 44-10) are determined by the specificities of the corresponding enzymes.

Most of the other T4-specified enzymes involved in DNA synthesis (Fig. 45-18) duplicate enzymes present in uninfected cells and are not essential, since the elimination of any by mutation slows but does not stop viral reproduction. The viral enzyme may be useful to increase the rate of the reaction, as required by the expanded DNA synthesis in infected cells, and possibly to escape regulatory restraints effective on the cellular enzyme. However, two phage-specified enzymes that duplicate cellular enzymes, DNA polymerase and ligase, are essential since their mutations are lethal. The phage ligase requirement seems to derive from extensive phage DNA degradation in the mutants; in fact, if the degradation is prevented (for unknown reasons) by an additional r_{II} mutation in the phage, DNA replication is restored. Host ligase is then used, because a ligase-positive host is required.

ROLE OF DNA MODIFICATIONS: HOST-INDUCED RESTRICTION AND MODIFICATION

The modifications of DNAs, which at first seemed only peculiarities, turn out to have great significance as **species-specific markers,**

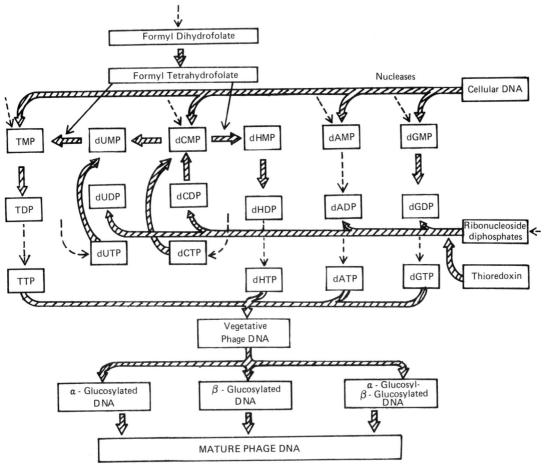

Fig. 45-18. Enzymatic reactions involved in the synthesis of T4 DNA. Heavy arrows indicate enzymes specified by phage genes (whether or not there is a cellular enzyme with the same function). Dashed arrows indicate enzymes of cellular origin.

which determine the fate of a DNA when it enters another organism. The foreign DNA is often broken down while that of the receiving organism remains intact. It is natural that this phenomenon was discovered with phages, which must invade other organisms; but it occurs also in bacteria. Thus recombination between the *E. coli* strains B and K_{12} fails because the DNA of one is destroyed in the other cell.

GLUCOSYLATION (FIG. 45-19A)

The first observation of host-controlled restriction and modification involved glucosylation. With T-even phages the newly formed DNA is rapidly glucosylated. However, in bacteria lacking uridine diphosphate glucose

(UDPG) the progeny viral DNA remains **unglucosylated.** The resulting virions, designated T*, are unable to grow, i.e., are **restricted** in any B strains, because unglucosylated DNA is specifically broken down by nucleases (**restriction enzymes**) present in the plasma membrane of B cells. These enzymes recognize nonglucosylated 5-hydroxymethylcytosine residues in special sequences. T* phage can, however, multiply in *Shigella,* which lacks the restriction enzyme. Moreover, since *Shigella* possesses the glucosylating system, the phage DNA is modified by glucosylation, and therefore the progeny is again unrestricted in *E. coli* B. In a parallel situation, a phage mutant unable to glucosylate (gt⁻) cannot grow in *E. coli;* it grows in *Shigella,* but it is not modified.

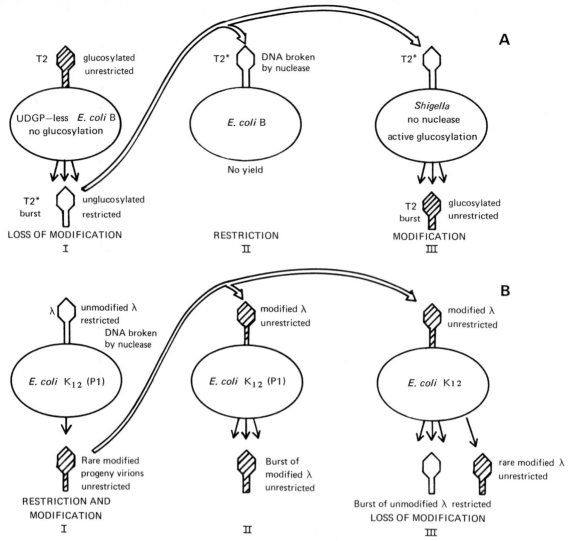

Fig. 45-19. Two examples of host-induced restriction and modification. Shaded outlines indicate modified phage which is unrestricted in the bacterial strains employed. **A.** Restriction and modification occur in different hosts. The DNA present in wild-type T2 is glucosylated, i.e., modified. Unglucosylated phage T* is produced after a growth cycle in a nonglucosylating host (I). T2* is restricted in *E. coli* B (II); it grows in *Shigella*, where it is modified (III) to yield again regular T2. **B.** Restriction and modification in the same host. The usual phage is restricted and at the same time modified (probably by methylation) in *E. coli* K_{12}(P1)(I); the modified phage then grows regularly in the restricting host (II). After one growth cycle in *E. coli* K_{12}, the modification is lost, except for rare virions that inherit a DNA strand from the parent (III).

METHYLATION (FIG. 45-19B)

In the preceding case the restricting and the modifying enzyme are unrelated. In the more general case the two enzymes are closely related and are present in the same cell, constituting a **host-specific (hs) restriction and modification system.** Well studied is the system induced by prophage P1. The plating efficiency of λ is much smaller in P1-lysogenic *E. coli* K_{12} cells (designated as K_{12} [P1]) than

in the corresponding nonlysogenic cells, because the injected λ DNA is rapidly cleaved by a P1-specific nuclease. At the same time, however, a rare molecule will by chance become modified before destruction and is then protected, like host DNA, from the nucleolytic action; the progeny phage, containing modified DNA, then is also unrestricted, i.e., can grow normally in P1-lysogenic cells. This rare successful infection **does not represent selection of a mutant,** for when the progeny is grown through a single cycle in K_{12} not carrying P1 DNA the new progeny viral DNA is not P1-modified and is again restricted in K_{12} (P1), except for those few particles that contain a parental, and therefore modified, DNA strand. (Modification of one strand is sufficient to protect the DNA because if the restriction enzyme attacks the unmodified strand, the single strand breaks, which have a 5'-P and a 3'-OH free end, are sealable by ligase.)

RESTRICTING AND MODIFYING ENZYMES

Several restriction enzymes have been purified. They are unusual **nucleases,** recognizing specific **target** sequences and producing a **double-strand break** for every thousand or more nucleotide pairs. The *E. coli* K_{12} enzyme attacks DNA in a target containing an unmethylated adenine; it has the unusual requirements for S-adenosyl methionine (SAM), apparently because it is related to the corresponding modifying enzyme, a SAM-requiring methylase. The latter enzyme methylates the same adenine to 6-methylaminopurine, destroying the target property. A restriction enzyme from *Hemophilus influenzae,* however, does not require SAM; it recognizes a target of six nucleotides with twofold rotational symmetry (Fig. 45-20A) which it cuts in the middle. The symmetry of the target is explained by a corresponding symmetry of the enzyme, probably related to its ability to break both DNA strands at once (Fig 45-20B).

Genetic Control. The relations between restricting and modifying activities in *hs* bacterial systems are clarified by the study of mutations affecting these properties and their complementations (using F'-merogenotes; see Ch. 9). The results suggest that each strain contains **three linked genes,** *hss, hsr,* and *hsm* (where the endings s, r, and m stand for specificity, restriction, and modification, respectively). The *hss* gene, determining the target specificity of both activities, differs in different systems (such as *E. coli* K and B); *hsr* and *hsm* determine the restriction and the modification properties, respectively. Thus, the wild-type B strain is *hss*B+, *hsr*+, *hsm*+. Hss*B*+ *hsr*−, *hsm*+ is a nonrestricting, modifying mutant, and *hss*− a mutant lacking both activities. The genetic data also suggest that some of the polypeptides specified by these genes are present in both the restricting and the modifying enzymes, explaining the intimate relatedness of the modifying and the restricting function. By ensuring the **identity of modifying and restricting targets,** this relatedness protects the cell's own DNA.

Significance of Restriction and Modification. It seems likely that **restriction is a mechanism of self-recognition** and isolation between species and that **modification is an antirestriction defense** to protect the organism's own DNA and, in phages, to protect them from restricting host enzymes. The **basic modification is methylation.** Glucosylation in T-even phages appears to have superseded an older system based on methylation, which is still recognizable with T* in **Shigella** (P1).

The significance of modification almost certainly extends beyond the realm of restriction. In the same way as the addition of a methyl group in a DNA sequence can prevent the action of a nuclease, other methylations can control the interaction of DNA sequences with other enzymes. It would be surprising if

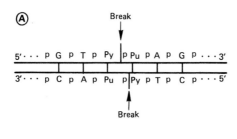

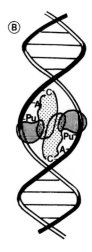

Fig. 45-20. Specificity of a restriction nuclease from *H. influenzae* **(A)** and postulated enzyme structure and mode of action **(B).** In the sequence, Py = a pyrimidine; Pu = the complementary purine. [Modified from Kelly, T. J., and Smith, H. O. *J Mol Biol 51*:393 (1970).]

this mechanism were not utilized in the regulation of DNA replication and function.

MATURATION

Maturation and release are the last two acts of the process of viral multiplication. In **maturation** the various components become assembled to form a complete or **mature** infectious virion; in **release** the mature particles leave the infected cells. The relation between maturation and release is revealed by the one-step multiplication curves (Fig. 45-6): the **total** multiplication curve describes the maturation of the virions; the **extracellular** curve describes their release.

The study of maturation is concerned with the relation between synthesis and assembly and with the mechanism of assembly.

Pools of Precursor Macromolecules. Hershey studied the relation between DNA synthesis and maturation by infecting cells with phage T2 labeled in the DNA. Samples of the mature progeny phage were collected by breaking open the cells at various times after infection. Label was recovered in all samples of progeny phage, including those collected very late; hence the **replicating viral DNA in a cell forms a single pool from which molecules are withdrawn at random** for incorporation into virions. Knowing the total radioactivity introduced into the pool by the parental DNA, the size of the pool at a given time could be calculated from its specific activity (i.e., proportion of labeled DNA) which in turn is equal to the observed specific activity of the DNA in virions matured at that time. At the end of the eclipse period the viral DNA pool was found to be equivalent to about 50 virions; it remains constant afterward when new synthesis equals withdrawal for virion formation. This amount closely agrees with that measurable by chemical analysis of infected cells, suggesting that most of **the viral DNA made is available for maturation.**

Macromolecular Precursors of Virion Proteins. Similar experiments with radioactive pulse-labeling of the viral proteins showed that the capsid precursor proteins are also drawn at random from a common pool during maturation. Most or all gene products must enter the pool since mutations in different phage genes complement efficiently in the same cell.

ASSEMBLY OF THE CAPSID

Experiments determining the time after infection at which a radioactive amino acid labels the progeny phage show that the virion proteins belong to the late class, except three small **internal proteins,** which belong to the immediate early group. Assembly begins with the aggregation of the "soluble" capsid proteins, which is detected by their increased sedimentation rate. At a very early stage components of tail fibers can be recognized by their **serological specificity,** but most partly assembled components can be identified only when they are large enough to have **distinguishing electron microscopic features.**

The assembly of large phages is an interesting model for the morphogenetic processes of higher organisms. Its elucidation has depended almost entirely on the use of **conditional lethal mutations** (Ch. 11), which under nonpermissive conditions block the morphogenetic process at a step requiring the function of the mutated gene. Upon lysis these cells yield partly and often erroneously assembled structures (often recognizable by electron miscroscopy; Fig. 45-21), just as an auxotrophic mutant accumulates the biosynthetic intermediate preceding the blocked reaction. In addition, certain pairs of defective lysates yield **complementation** in vitro, i.e., the accumulated incomplete structures can assemble spontaneously when mixed to form infectious virus. Such complementation studies have shown that the capsid is assembled through three **independent subassembly lines,** which produce the phage head, the fiberless tails, and the tail fibers, respectively (Fig. 45-22). These structures then spontaneously assemble into capsids in vitro and presumably in vivo also.

In contrast to the assemblies of isometric and helical capsids, discussed in the previous chapter, the self-assembly of a single T4 protein usually produces aberrant structures. Thus the protein P23 (i.e., the product of gene 23), which is the main constituent of the head, by itself aggregates randomly into "lumps." Each subassembly requires the **sequential incorporation of the various proteins** in a well-defined order, each protein adding a new detail of form. For instance, in head morphogenesis, with the addition of each new protein the structure more closely resembles a head; and the regular head is formed after the eighth protein is added (Fig. 45-23). Interruption of a sequence by mutation stops all subsequent morphogenetic steps. In contrast to the self-assembly of capsid protomers of

Fig. 45-21. Electron micrographs of polyheads **(A)** and polysheaths **(B,** arrow) present in lysates produced by amber mutants of phage T4. In **B,** one also sees "empty" head membranes (2) and tubes attached to base plates (3). Both micrographs contain also some normal virions, (1) because the cells were simultaneously infected with wild-type phage. The polyheads contain hexagonal capsomers which are no longer recognizable in regular phage, owing to further assembly steps; the polysheath has the diameter of a contracted regular sheath. [**A** courtesy of E. Boy de la Tour. **B** from Boy de la Tour, E. *J Ultrastruct Res* 11:545 (1964).]

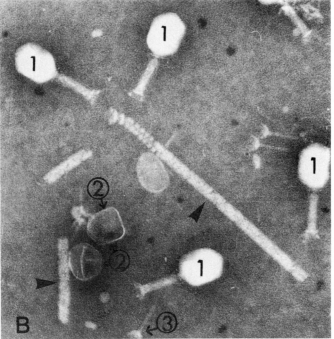

some simple phages, some steps in T-even assembly **require the help of nonviral components.** Thus the base plate is assembled in association with the internal layer of the cell's plasma membrane; and indole is required for the attachment of the tail fibers to the tail. Moreover, the assembly of the head includes irreversible steps of **protein cleavage** (see Fig. 45-22). Four proteins are cleaved after they have been assembled, including the major protein P23 which changes from MW 60,000 to MW 45,000; the released fragments remain associated with the head as **"internal peptides."**

Clearly the complicated stepwise assembly can occur because a substructure present at a given step has suitable surfaces and binding sites for the orderly addition of the next protein. Such order must be the result of a long evolution, during which the ability of the main components to form an isometric capsid by self-assembly was lost in order to allow its orderly interaction with other proteins.

ASSOCIATION OF DNA AND CAPSID

As with other viruses (see Chs. 44 and 48, Maturation), the T-even DNA becomes associated with the capsid after the stage of "empty" head (as defined from electron mi-

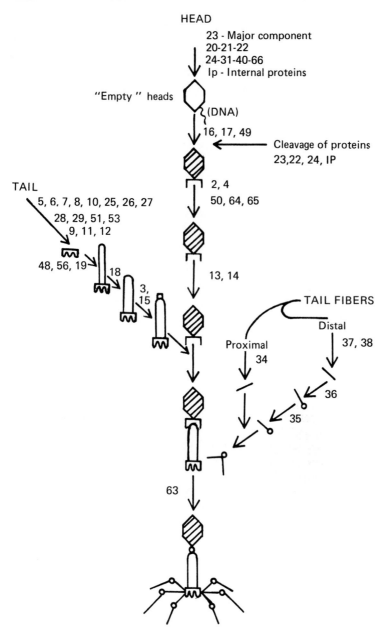

HEAD

23 - Major component
20-21-22
24-31-40-66
Ip - Internal proteins

"Empty" heads

(DNA)

16, 17, 49

Cleavage of proteins
23, 22, 24, IP

Fig. 45-22. The three subassembly lines in the maturation of a T4 virion. The numbers indicate the genes contributing their products to each step. Shaded heads contain a full DNA complement. [Modified from Wood, W. B., Edgar, R. S., King, J., Lielausis, I., and Henninger, M. *Fed Proc* 27:1160 (1968).]

TAIL

5, 6, 7, 8, 10, 25, 26, 27
28, 29, 51, 53
9, 11, 12

48, 56, 19

18

3, 15

2, 4
50, 64, 65

13, 14

TAIL FIBERS

Proximal

34

Distal

37, 38

36

35

63

croscopic images). Thus "empty" heads accumulated at high temperature by a temperature-sensitive mutant (gene 49) rapidly develop into full heads and infectious virions after the temperature is lowered. The DNA probably associates with the "empty" heads before entering the capsid and is easily broken during isolation.

The subsequent folding and tight **packing**

of the DNA within the head apparently occurs when the protein cleavage takes place. Probably a DNA segment binds to a protein molecule of the head capsid, triggering its cleavage, and becomes at the same time locked in. A zipper-like repetition of this event would pack the whole DNA molecule within the head. With the T-even phages, this step appears to **determine the length of the ma-**

Fig. 45-23. Cooperation of various gene products in the formation of the T4 phage head. The numbers in the boxes indicate the gene number (see Fig. 45-31). The gene functions are given near each box. The products indicated by the arrows accumulate when a mutation prevents the function of the next gene in the assembly line. τ-Particles have no corners, only hemispherical caps. [Modified from Laemmli, V. K., Mölbert E., Showe, M., and Kellenberger, E. *J Mol Biol 49*:99 (1970).]

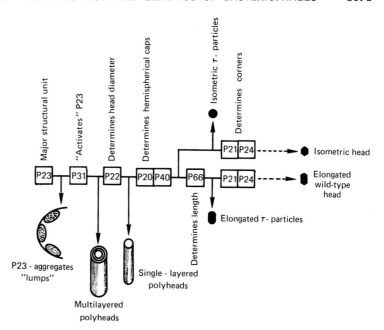

ture DNA as a "**headful,**" i.e., what fits within the head capsid; the excess is presumably cut off by nucleases. Thus if the DNA contains a deletion the repetitious ends are longer; conversely if the head is smaller owing to a petite mutation, the encapsidated DNA is shorter and without repetitious ends.

This method of assembly also explains the encapsidation of random fragments of cellular DNA within phage capsids in generalized transduction (see Ch. 46), which of course requires phages (unlike the T-even phages) that do not destroy host DNA too rapidly.

With phages T7 and λ the capsid does not have a metering function because, as noted above, the ends of the molecules are determined by specific sequences.

PHENOTYPIC MIXING

In bacteria mixedly infected by two related phages (e.g., T2 and T2h) each causes the formation of a complete set of protein protomers. Cells infected by phage T2h+ and T2h thus contain precursor tail fibers of both h+ and h specificity. During assembly either fiber can be used in any virion; hence virions of a given genotype may contain tail fibers of that type, or the other, or a mixture of the two (Fig. 45-24). Virions in which the genotype and the phenotype are discrepant are called **phenotypically mixed.**

Phenotypic mixing may affect many capsid proteins, but the most common example is mixing of the host-range character, which is detected by the two-step procedure outlined in Figure 45-21.

RELEASE

Lysis of cells infected with T-even phage appears to be the result of the contrasting effects of the products of three viral genes. The gene *t* product promotes lysis by altering the plasma membrane of the cells and stopping cellular metabolism; the familiar gene *r* and the similar gene *s* affect the membrane in a way that delays lysis. The final act of lysis is carried out by the product of gene *e,* the lysozyme, which attacks the cell wall after crossing the plasma membrane altered by the action of the other genes. A block in energy metabolism (e.g., addition of cyanide or deprivation of oxygen) accelerates lysis, perhaps by influencing the levels of the products of the various genes.

Lysis by other phages is similar, but with the very small phage φX it does not seem to involve lysozyme. Filamentous phages are released by an entirely different mechanism, without lysis (see below).

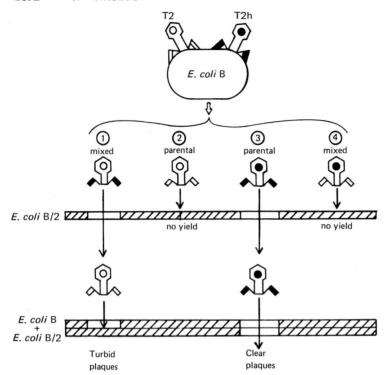

Fig. 45-24. Phenotypic mixing of the h character in bacteriophage T2. *E. coli* B infected by both T2h and T2h+ produces both parental (i.e., either h or h+, both genotypically and phenotypically) and mixed progeny. Phage 1 is genotypically h+, phenotypically h; phage 4 is genotypically h, phenotypically h+, whereas 2 and 3 are parental. After infecting *E. coli* B/2, only phenotypically h phage (1 and 3) produce progeny, which is phenotypically true to the genotype. The genotype is then determined by plating the B/2 yields on a mixture of *E. coli* B + *E. coli* B/2 (mixed indicator strains), where genotypically h+ phage (1) produces turbid plaques (because it does not infect B/2 cells) and genotypically h phage (3, which infects both types of cell) produces clear plaques.

PHAGE GENETICS

Historically, the study of phage genetics provided the main incentive for undertaking the analysis of phage multiplication. The investigators who started the "new wave" of bacteriophage work in the late 1930s recognized that genetics had to turn to simpler organisms in order to determine the molecular bases of the complex formalism previously established. For this purpose, phages offer outstanding advantages: about half the total mass is genetic material; even larger populations are conveniently available than with bacteria; and the number of genes is much smaller than in cells.

The development of phage genetics in the last two decades has been an outstanding intellectual achievement and has contributed heavily to the foundations of molecular genetics.

A highly sophisticated mathematical analysis of phage genetics was built on the basis of technically simple observations, carried out with a few mutant types. In addition, **fine-structure genetics** was launched by Benzer through the study of mutants of a phage gene, even though its protein product was unknown. As we have seen in Chapter 10, the subsequent extension of fine-structure genetics to bacteria led to a profound understanding of the relation between gene and protein. In recent years, similar direct studies have also become possible for phage DNA and proteins. The formal and the molecular aspects of phage genetics have thus become integrated.

The results of general significance have been incorporated in Chapters 10, 11, and 12; the following section will deal with the more special features of phage genetics.

MUTATIONS

The genetics of phage, like that of all organisms, is based on the study of mutants. The types that can be recognized are limited by the haploid nature of phages; hence lethal mutations are unsuitable because the phage carrying them cannot be propagated. The useful mutations are of two main classes.

Mutants producing plaques with changed morphology **(plaque-type mutants)** were extensively used in the initial investigations of bacteriophage genetics, since they could be recognized easily on inspection of the assay plates. Two main drawbacks, however, became evident: some of these mutants do not allow the use of selective technics; and they are localized in a small number of genes, thus leaving much of the chromosome unmapped. These difficulties were overcome by **conditionally lethal mutants.** These mutants, which can be isolated in any gene, have led to the establishment of a fairly complete **phage map** (see below). Moreover, they have been indispensable for **fine-structure** genetics: selection under restrictive conditions of the progeny of genetic crosses between two such mutants makes it possible to detect rare wild-type **recombinants.** Similarly, **complementation** (Ch. 11) is easily detected and the limits of the genes can be established. Suppressor-sensitive mutants do not make the protein of the mutated gene under restrictive conditions and so avoid the complication of intragenic complementation (Ch. 12).

A number of mutant types employed in genetic work with T-even bacteriophages will be briefly examined.

PLAQUE-TYPE MUTANTS

Rapid lysis (r) mutants (Fig. 45-24) have already been described above (Release). Two classes, r_{IIA} and r_{IIB}, are caused by mutations in two adjacent genes of the r_{II} locus. These mutants have been used extensively in genetic work, since they incidentally turned out to be conditionally lethal (see below).

Host-range (h) mutants were also described above (Adsorption). It will be recalled that they can be identified on mixed indicator strains, where they produce **clear** plaques and h^+ (wild-type) phages produce **turbid** plaques.

Minute plaque and **turbid plaque mutants** produce plaques described by their names.

CONDITIONALLY LETHAL MUTANTS

r_{II} **Mutants** multiply in *E coli* B but not in *E. coli* $K_{12}(\lambda)$ (Ch. 46), whereas $r_{II}{}^+$ phages multiply equally well in both hosts.

Benzer used this property to detect minute proportions of r^+ virions in populations of r_{II} phage in his pioneering studies on fine-structure genetics.

Amber or **ochre mutants** are characterized by a nonsense codon (Ch. 12, Genetic code). They can therefore multiply only in bacterial strains carrying a suppressor of that codon (Ch. 12, Suppression), i.e., they are **suppressor-sensitive (sus)** mutants.

Temperature-sensitive mutants (ts) are characterized by a mutation that prevents the expression of its gene at $43°$ but not at $25°$ (see Ch. 11 for mechanism of temperature sensitivity).

APPLICATION OF PHAGE MUTANTS

A useful application in **phage physiology** was to determine the mode of replication of phage DNA. As in bacteria, replication could be **geometric;** alternatively it could occur on a fixed number of effective templates according to a stamping machine model (**linear** replication). The genetic approach could show that the DNA of T-even phages replicates geometrically, whereas that of $\phi X174$ (see below) replicates linearly, although in both cases the tonal amount of viral DNA per cell increases linearly, except at the very beginning. The clue was the **distribution of spontaneous mutants in the yields of individual infected cells,** based on the earlier fluctuation analysis of bacterial mutants (Ch 8). Thus, if DNA replication is exponential, each mutation will produce a clone of mutants whose size is smaller the later the mutation occurs; hence there will be a large spread in the number of mutants in different clones. In contrast, if replication is linear, the size of the mutant clones will be rather uniform (Fig. 45-25). The linear increase of T2 DNA in the infected cells can be explained by the continuous withdrawal of templates for maturation.

GENETIC RECOMBINATION

SPECIAL FORMAL CHARACTERISTICS OF PHAGE RECOMBINATION

Genetic recombination in T-even bacteriophages was discovered by 1946 by Delbrück and by Hershey, who performed genetic crosses by mixedly infecting bacteria with a host-range (h) and a rapid lysis (r) mutant. The yields of these cultures contained four distinct types of particles (Fig. 45-26): the

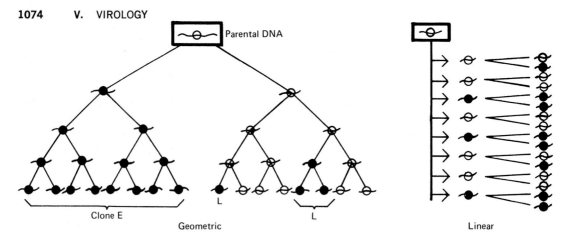

Parental DNA

Clone E

Geometric

L

L

Linear

Fig. 45-25. Clones of mutants in the progeny of individual cells. In the geometric replication of T2 DNA mutations occurring early in the replication of the DNA (E) cause much larger clones than late mutations (L). With ϕX174 replication is partly linear (the stamping machine model); hence the formation of large mutant clones is impossible. White circles = wild-type alleles; black circles = mutant alleles.

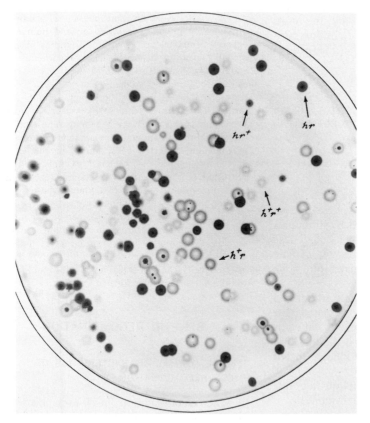

Fig. 45-26. Plaques formed by a mixture of T2 phages carrying mutations at the h and r locus, placed on a mixture of *E. coli* B and *E. coli* B/2 (i.e., resistant to T2 but sensitive to T2h). Phages with the h and those with h+ allele produce, respectively, clear plaques (dark in the photograph) and turbid plaques (gray in the photograph); phages with the r allele are larger than those with the r+ allele. Thus all four possible combinations can be distinguished: T2h+r+ (wild-type), T2hr, T2h+r, and T2hr+. From Stent, G. *Molecular Biology of Bacterial Viruses.* Freeman, San Francisco, 1963.)

the two parental types and two recombinant types with one marker from one parent and the other marker from the other parent.

The study of genetic recombination with bacteriophages turned out to involve complex statistics because **recombination takes place in the vegetative pool with many DNA molecules participating.** Indeed, vegetative phages evidently continue to recombine throughout the period of multiplication, since the **proportion** of recombinants in a mixedly infected cell increases with the time (Fig. 45-27). Furthermore, in triparental crosses (Fig. 45-28) some of the progeny individually incorporate markers from all the parents, a result that requires at least two independent matings. It can be calculated that T-even bacteriophages undergo four or five rounds of mating, on the average, between infection and release. In molecular terms, a round of mating corresponds to the breakage of a DNA molecule and its subsequent reunion to another.

A single mating between two phage DNA molecules, therefore, cannot be studied directly. The consequence of such mating can only be inferred through statistical analysis

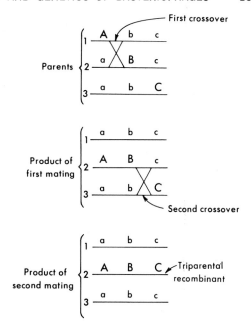

Fig. 45-28. The formation of a triparental recombinant. 1, 2, 3 = three differently marked phages infecting the same cell. The triparental recombinant, which contains alleles from all three parents, requires at least two matings for its formation.

of the consequences of the multiple, successive matings that occur in a phage cross.

CONSTRUCTION OF A GENETIC MAP

The Qualitative Map. The sequence of markers is established for phage in the same way as for higher organisms, i.e., on the basis of the proportion of recombinants (**recombination frequencies**) produced in three-factor crosses (i.e., involving two phages that differ at three markers; (Fig. 45-29) or by deletion mapping, as discussed in Chapter 11 (Recombination).

Statistical Negative Interference. While the order of markers in phage can thus be established in a conventional manner, the **distances** between them do not have the same simple relation to recombination frequency observed in higher organisms, i.e., phage crosses **lack additivity of recombinant frequencies,** even for markers that are not closely linked. This lack of additivity, which is unrelated to the high negative interference discussed in Chap-

Fig. 45-27. Increase in the proportion of recombinants as a function of time after infection. *E. coli* cells were mixedly infected by T2h+r+ and T2hr under conditions of lysis inhibition. At various times after infection a sample was collected, the cells were broken open, and the content was assayed for total viral yield and for the proportion of hr+ and h+r recombinants. The two increase in parallel, showing that recombination continues to occur in the cells while their burst is delayed. [Modified from Levinthal, C. and Visconti, N. *Genetics 38*:500 (1953).]

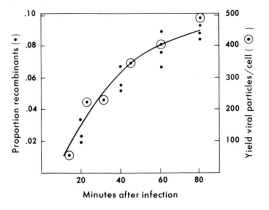

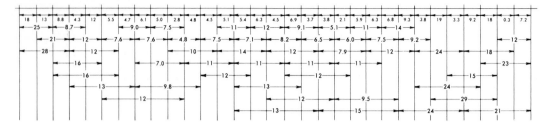

Fig. 45-29. The order of genes in a segment of the T4 genome, determined by three-factor crosses. Each vertical line corresponds to a marker; the observed recombination frequency in each pairwise cross is recorded. Note the consistency of the results: the recombination frequencies between the closest pairs (top row) establish a sequence that yields increasing frequencies, as predicted, in crosses involving more distant pairs. The frequencies, however, are not additive. [Modified from Edgar, R. S., and Lielausis, I. *Genetics* 49:649 (1964).]

ter 11 (Genetic recombination), seems, at first sight, to derive from an excess of double crossovers. A closer study, however, shows that it is produced by the **statistical characteristics of phage replication,** namely, the unequal number of rounds of matings of different progeny DNA molecules before encapsidation.

Purely on statistical grounds, even in identically infected cells, some molecules undergo fewer matings than the average number. Moreover, some cells are infected by a higher proportion of one or the other parental type, or even by one type alone, and the vegetative pool initiated by different parental particles in a cell may not mix well. The presence of molecules with lower than average opportunity for mating causes the **experimentally determined** proportion of recombinants (observed recombinants/ total number of molecules in the yield) to be lower than the **true proportion** of recombinants in the mating fraction of the population (observed recombinants/number of molecules with opportunity for mating). Since the actual fraction of double crossovers is related to the true frequency of recombination, it is always higher than that calculated from the experimental **apparent** proportion of recombinants.

Example. If 50% of the DNA molecules produced have undergone the same number of matings and 50% have not mated at all, the true proportion of recombinants among those that have mated (P_{ac}, etc. where *ab, ac,* and *bc* represent pairs of markers) is twice the apparent proportion (p'_{ac}, etc.). Then the classic relation between single and double crossovers (see Ch. 11, Recombination) **which applies to the true proportions,** must be written for these particular circumstances as

$$2p'_{ac} = 2p'_{ab} + 2p'_{bc} - 2(2p'_{ab}) \cdot (2p'_{bc}), \text{ or}$$
$$p'_{ac} = p'_{ab} + p'_{bc} - 4p'_{ab} \cdot p'_{bc} \qquad (1)$$

rather than

$$p'_{ac} = p'_{ab} + p'_{bc} - 2p'_{ab} \cdot p'_{bc} \qquad (2)$$

as would be written if all molecules had experienced the same number of matings. Comparison of (1) and (2) shows that the actual proportion of double crossovers ($4p'_{ab} \cdot p_{bc}$) is twice the proportion expected from the apparent proportion of recombinants ($2p'_{ab} \cdot p'_{bc}$).

Map Distances. In determining **distances between markers** the statistical negative interference can be corrected by multiplying the observed recombination frequencies by an empirical **mapping function** that restores additivity. The distances between genes can thus be established in **map units** (where 1 unit corresponds to 1% corrected recombination frequency). The length of the map of phage T4 is 2500 map units, each corresponding to about 100 nucleotide pairs. In many phages, however, a map unit corresponds to many more nucleotide pairs; e.g., about 2000 with λ. As noted above (Viral DNA replication), the difference can be attributed to the much higher breaking and rejoining activity during the replication of T4 DNA.

Relative distances calculated from recombination frequencies are in fairly good agreement with **physical distances** determined by electron microscopy of DNA heteroduplexes between the wild-type and various deletion mutants (see Ch. 11, Genetic mapping) or by other indirect procedures. Deviations are usually less than 30% but are more substantial in a few regions of the genome. For example, extremely low frequencies of recombination (10^{-4} to 10^{-8} map units) are observed between certain r_{II} mutants of phage T4, though the average probability of recombination between adjacent nucleotides in T4 is about 10^{-2} units (calculated as the ratio of total length of the map to total number of nucleotide pairs in a viral DNA molecule, i.e., 2,500:150,000). The localized low frequencies evidently result from effects of certain sequences on the molecular mechanisms of recombination, as discussed in Chapter 11.

OTHER CHARACTERISTICS OF BACTERIOPHAGE RECOMBINATION

Recombination Is Nonreciprocal. In the yield of a large population of infected bacteria, a phage cross AB × ab yields equal proportions of the two reciprocal recombinants, Ab and aB. However, **in the yields of individual bacteria one recombinant regularly predominates** especially if the average number of rounds of matings is kept low by prematurely lysing the cells. This result is not due to unequal replication or maturation of the two recombinant types because with phage λ regular nonreciprocal recombination occurs together with a special type of reciprocal recombination within the same cell (see Ch. 46, Lysogeny). A mechanism of nonreciprocal recombination is schematized in Figure 11-12.

Formation of Heterozygous DNA. Cells mixedly infected by two phages carrying allelic markers regularly yield some virions that are heterozygous for one of the markers: with a mixture of r and r+ phage about 1% of the progeny virions is heterozygous and produces characteristic **mottled plaques**, i.e., with alternating r and r+ sectors (see Fig. 45-8).

Extensive studies of this phenomenon demonstrated two classes of heterozygous particles: **internal** and **terminal** (Fig. 45-30). Internal heterozygotes contain heteroduplex DNA, as described in

Chapter 11 (Recombination); **terminal** heterozygotes are peculiar to DNA with **terminal duplications**. Both heterozygotes are formed by recombination. Internal heterozygotes are lost after replication and therefore accumulate when most DNA synthesis, but not recombination, is blocked by FUDR. In contrast, terminal heterozygotes are dissolved by recombination and therefore do not accumulate under FUDR.

High Negative Interference. In addition to the moderate statistical interference discussed above, an apparently large excess of multiple crossover is found in crosses between close markers. The phenomenon is due to gene conversion, which may be especially high with T-even phages owing to the high level recombination and error correction (see Ch. 11, Recombination).

MOLECULAR AND ENZYMOLOGICAL MECHANISMS OF PHAGE RECOMBINATION

The effects of viral mutations on the frequency of recombination in T-even phages were important for formulating the models of recombination discussed in Chapter 11. The results point to the importance of single-strand breaks in DNA and of preventing random folding of single strands. Thus the frequency of recombination is increased if single-strand breaks are generated at higher frequency, as in dCMP hydroxymethylase-deficient mutants, or remain unsealed, as in ligase- or DNA polymerase-deficient mutants. Conversely, recombination fre-

Fig. 45-30. Terminal and internal phage heterozygotes. Two types of internal heterozygotes are found: with crossing over of the external markers A and B (**I**) and without crossing over (**II**), depending on the type of recombination that generated them. The internal but not the terminal heterozygotes are dissolved at replication. Note that recombination generates a molecule whose ends differ from those of the parental molecule because it leads to formation of a concatemer, from which mature molecules are excised at random. Only molecules excised as indicated are terminally heterozygous.

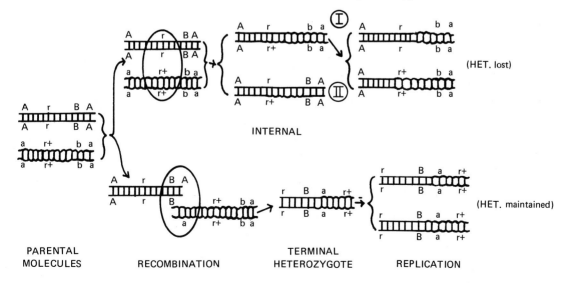

quency is decreased by mutations in gene 46 or 47 which decrease the rate of breakage. The requirement for DNA-unfolding protein is shown by absence of recombination with mutants in gene 32, which specifies such a protein (see Ch. 10, DNA denaturation).

ORGANIZATION OF THE GENETIC MATERIAL OF PHAGE T4

In bacteria the grouping of the genes in the chromosome has important regulatory aspects,

which were discussed in Chapter 13. The question whether the much simpler genetic system of viruses possesses a similar organization can best be approached with bacteriophage T4, whose genetic map of 2500 units is known in considerable detail. This map (Fig. 45-31) presents several important features.

Circularity. If the total length of the map is covered by a series of three-factor crosses between markers at relatively close distances (20 to 30 units), the starting marker, irre-

Fig. 45-31. The genetic map of bacteriophage T4. The numbers refer to genes identified by conditionally lethal mutations, in order of discovery. Genes identified by other means are indicated by letters. The known functions are indicated outside the circle. The pictures inside the circle indicate the recognizable capsid components whose accumulation is caused by mutation of the corresponding gene. Early genes are indicated by a white band, late genes by a black band. Letters H, T, and F indicate that the corresponding genes are involved in head, tail, or fiber synthesis, respectively. The indicated origin and direction of replication apply to the first round and are based on results obtained with petite mutants. [Most of the information from Wood, W. B., Edgar, R. S., King, J., Lielausis, I., and Henninger, H. *Fed Proc* 27:1160 (1968).]

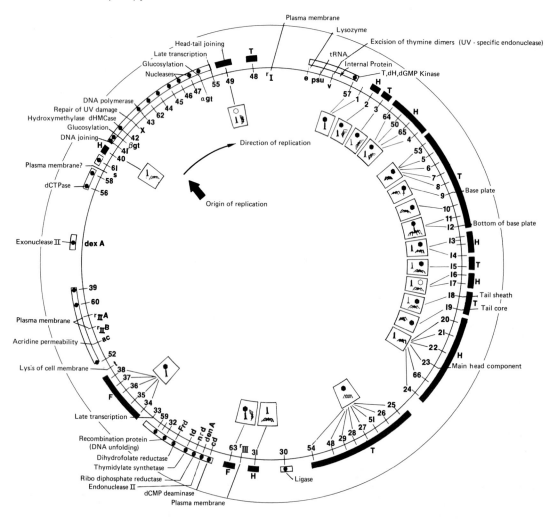

Fig. 45-32. Generation of the circular T4 map. Cutting the concatemer, precursor of mature virions, into equal segments longer than the period of the concatemer generates the indicated collection of permuted, terminally repetitious molecules. They can all be generated from the circle on the right, beginning at various points, as shown by these examples. The genetic map equals that circle.

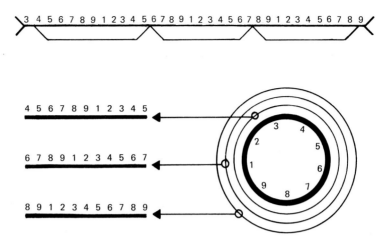

spective of its location, is always closely linked to the terminal marker. Thus the map is circular, despite the physical evidence that the DNA of the virions is not cyclic.

The apparent contradiction is resolved by the structure and mode of formation of the DNA molecules present in virions (see above, DNA replication). Since they are generated from a periodic concatemer by cutting off segments longer than the periodic unit, they can initiate at any gene and have long repetitious ends. Such molecules are said to be **circularly permuted** because they can be ideally derived by copying a circle beginning at various points (Fig. 45-32). The permutation and terminal repetition make any two genes that are adjacent to each other in the concatemer (and in the circle generating the sequences) also adjacent in the mature DNA molecules, and therefore genetically linked. Hence, the genetic map is identical to the theoretical generating circle.

Not all bacteriophages have a circular genetic map. T7 and λ have linear maps (Fig. 45-33) because their DNA molecules are not permuted.

Number of Genes. About 80 genes have been identified by complementation tests in phage T4.

This number probably includes all genes with functions essential for phage multiplication and therefore detectable by conditional lethal mutations. There are reasons, however, for suspecting the **existence of additional, nonessential genes.** First, some genes that are adjacent in the present map give usually large recombination frequencies, suggesting that they are separated by other, yet unknown, genes. Second, T4 DNA, with its 150,000 nucleotide pairs, could contain about 150 genes of 1000 nucleotide pairs, the average gene size.

GENERAL ORGANIZATION OF THE MAP

The genetic map of phage T4 (Fig. 45-31) shows that genes are clustered according to their functions (e.g., synthesis of early enzymes, or of the head, etc.) though with a few exceptions. Similar gene clustering exists in other phages, e.g., T7 (Fig. 45-33) and λ (Fig. 46-4). The clusters recall bacterial operons (see Ch. 13). However, as noted above, the study of transcription shows that they are not operons. The reasons for the arrangement may be related to morphogenesis.

Fig. 45-33. The linear genetic map of phage T7. The white bar indicates the early genes; the black bar the late genes. [Modified from Siegel, R. B., and Summers, W. C. *J Mol Biol* *49*:115 (1970).]

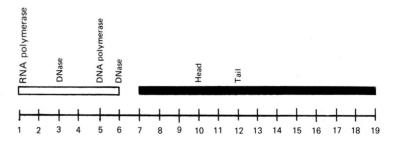

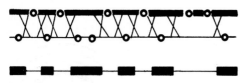

Fig. 45-34. Multiplicity reactivation of UV-inactivated bacteriophage DNA in a cell infected by two phage particles. Although UV damages are present in both DNAs, they do not coincide; hence it is possible, by multiple exchanges, to obtain molecules without damage.

Parental molecules

↓

Molecule without UV damage

GENETIC REACTIVATION OF ULTRAVIOLET-INACTIVATED PHAGE

Radiation damages and mechanisms of repair in phages are similar to those in bacteria (Ch. 11). Large phages (such as the T-even) specify enzymes for dimer excision and post replication repair; however, the smaller phages (e.g., ϕX or λ) rely in large

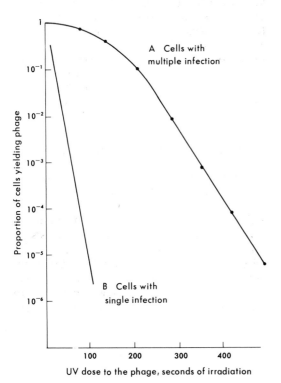

Fig. 45-35. Multiplicity reactivation of UV-irradiated T2. In curve A, bacteria were infected with an average of four T2 phages; the curve shows the fraction of the cells able to yield infectious phage, for different UV doses given to the phage. Curve B shows the results obtained when the cells were infected at low multiplicity (i.e., mostly single infection.) [Modified from Dulbecco, R. *J Bacteriol* 63:199 (1952).]

part or completely on repair mechanisms of the host, which is then said to carry out **host cell reactivation (hcr)**. The survival of these phages after UV irradiation depends on characteristics of the host cell.

In addition to the usual forms of reactivation of UV damages, bacteria infected by **more than a single virion** exhibit **multiplicity reactivation and marker rescue**. Both these mechanisms are based on recombination between **different genomes** in the same cell.

Multiplicity Reactivation. When a cell is infected by irradiated T2 the probability of yielding infectious virus increases disproportionately with the multiplicity of infection. The explanation is that the DNA molecules will have their random UV damages in different locations, and **multiple crossovers,** piecing together intact segments from different molecules, can yield an undamaged molecule (Fig. 45-34).

The effect is especially striking when UV damages are not removed by other repair mechanisms. Conversely, if recombination is prevented by mutations, multiplicity reactivation does not occur.

With the T-even phages the crossovers are produced by enzymes specified by viral genes; hence, for effective reactivation a cell must contain at least one undamaged copy of each of the genes involved in recombination. This requirement determines the shape of the survival curves of multicomplexes as a function of UV dose (Fig. 45-35), which are of the multiple-hit type (Ch. 11, Appendix) because viral multiplication is absent only if all the representatives of a required gene are damaged and have final slopes smaller than those for phage inactivation, since the sum of the required genes does not encompass the whole genome.

In a genetic cross with phage capable of multiplicity reactivation, the UV irradiation of the parents **increases the frequency of recombination,** a useful tool in genetic studies. The explanation is that irradiation forces selection for those DNA molecules that have undergone many crossovers.

Marker Rescue. When a cell is simultaneously infected by a UV-irradiated virion carrying one allele of a particular gene and by one or more undamaged virions carrying another allele, recombination can incorporate the allele of the irradiated phage into the undamaged DNA and hence into the progeny. Effective recombination requires a crossover somewhere between the marker and the nearest UV damage on either side; hence the probability of such marker rescue decreases with the number of damages in the irradiated DNA.

MULTIPLICATION OF BACTERIOPHAGES WITH CYCLIC SINGLE-STRANDED DNA

These very small phages have several unique features. They depend heavily on host functions; they do not inhibit host macromolecule synthesis except for DNA late in infection with icosahedral phages; and formation of single-stranded DNA for the virions requires asymmetric replication.

ICOSAHEDRAL PHAGES

With phage ϕX174, which has been intensively studied by Sinsheimer, replication occurs in three phases (Fig. 45-36). In the **first phase,** the cyclic **viral** strand of DNA is injected into the cell, where it becomes associated with a membrane site from which it is never transferred to the progeny phage. A cell contains only one to four functional sites, depending on its nutritional state. The attached viral DNA is rapidly converted to a double-stranded cyclic **replicative form** (RF) by synthesis of the complementary strand, using a cellular DNA polymerase (since it occurs in the presence of chloramphenicol).

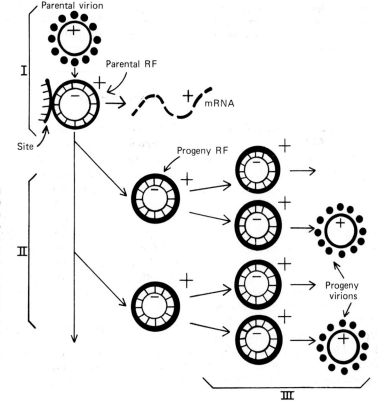

Fig. 45-36. The multiplication of phage ϕX174. Phase I: the viral strand introduced by the infecting virion becomes associated with the membrane site, where it generates the parental RF, which then becomes transcribed into mRNA. Phase II: the parental RF replicates, producing free progeny RFs. Phase III: the progeny RFs generate viral strands which are encapsidated to produce virions. Thick lines = viral or plus (+) strands; thin lines = complementary (−) strands.

In the **second phase** the **parental** RF is transcribed into mRNAs which have the same sequence as the viral (+) strand, and these are transplanted into proteins required for DNA replication and for assembly. The parental RF also repeatedly replicates according to the rolling circle model (see Ch. 10, DNA replication); the daughter molecule containing the parental strand always remains attached to the site, whereas the other is released and remains detached as a **progeny RF,** which does not replicate further in this phase. In the **third phase,** beginning toward the end of the eclipse period, the replication of the parental RF ceases, and the progeny RFs, by rolling circle model, generate viral linear strands. These are closed into circles by the host ligase, encapsidated to form virions, and released by lysis.

Over-all replication follows the **stamping machine model,** with the parental RF as stamp; as noted above (Phage genetics), this method of replication agrees with the observed distribution of mutants in the yields of individual cells.

Nine genes have been identified in ϕX174 by complementation and form a circular map, in the order A through I. All the proteins are synthesized throughout infection, except for the E gene product (required for lysis), which appears only late. Genetic economy is promoted in this tiny phage by giving three capsid genes a second function: with a temperature-sensitive mutation in any of them,

synthesis of viral strands on progeny RFs stops if the temperature is raised. The connection seems to involve binding of capsid protein to the RF at the nick where replication begins (see Ch. 10, DNA replication); there it not only initiates synthesis of a plus strand but also blocks that of the minus strand. Moreover, protein molecules become associated with the virion strand while it grows.

FILAMENTOUS PHAGES

The replication of these helical phages (e.g., fd, M13) resembles that of ϕX174, but there are striking differences. As with animal viruses, the **whole virion,** approximately as thin as the tube of the T-even phage tail, enters the cell; **release occurs without lysis,** the virions being extruded during several hours from the cells, which continue to grow. The infecting DNA is converted to a double-stranded RF, which replicates. However, there is no distinction between parental and progeny RF in the formation of viral strands, and these are released without the help of capsid protein and accumulate in the cell. The capsid protein is incorporated into the cell plasma membrane, increasing its permeability and fragility and making the cell sensitive to virion-specific antiserum. Electron micrographs suggest that mature virus is assembled at the cell and extruded in a single step.

RNA PHAGE

These small icosahedral phages (which include f2, MS2, R17 and Qβ) were isolated by Zinder as male-specific: they attach to F pili of male bacteria, using them as channels for injecting their RNA into the cells. They turned out to contain a small single-stranded RNA, of MW 1.1×10^6 daltons, and to have the simplest genetic organization of all autonomous viruses. Thus all their functions are expressed by only three genes, whose $5' \rightarrow 3'$ sequence is: a gene for maturation (or A) protein, of which there is a single molecule per virion; a gene for the protomer of the viral capsid; and a gene for a special replicase (RNA-dependent RNA polymerase). These phages do not display recombination. The gene order was determined by replicating the phage RNA in vitro into a

series of RNA segments of variable lengths, beginning at the 5' end (see below, Uses of RNA replicase). The genes in each segment were identified by their products after translation in vitro.

The viral RNAs are extremely valuable as messenger for studying protein synthesis in vitro. Much of their base sequence is known, providing important evidence concerning intercistronic punctuation, ribosome-binding sites, and significance of secondary structure (Fig. 45-37) for regulation of translation.

RNA REPLICATION

The replicase of phage Qβ, which has been extensively purified and characterized by Spiegelman, is made up of four subunits of which three are of

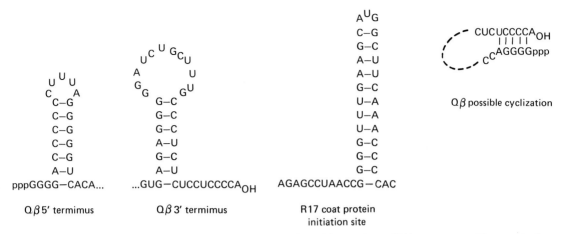

Fig. 45-37. Possible formation of secondary structures inferred from phage RNA sequences. The terminal A at the 3' end is added to the RNA after replication.

cellular origin. Subunit I is an inhibitor of protein synthesis initiation, probably a regulatory subunit of IF$_3$; subunit II is virus-specified; subunits III and IV correspond to the elongation factors of protein synthesis Tu and Ts (see Ch. 12). The association with translation factors is probably needed for regulating the messenger and template functions of the viral RNA. The purified enzyme in addition requires at least one other host protein (Host Factor 1) for recognizing the viral RNA; in the absence of HF1, however, it can recognize the complementary strand, which is synthesized in the course of multiplication. The enzyme is very specific, recognizing its own RNA or of related phages; therefore it does not replicate cellular RNA. Since it is unrelated to the cellular transcription it is rifamycin-resistant. The biochemical features of polymerization are similar to those of the cellular transcriptase.

RNA replication, which is similar to that of some animal viruses (see Ch. 48), involves special intermediates. In cells infected with RNA-labeled virions some of the label is found in molecules completely resistant to RNase, called **replicative form** (RF), and some in molecules partially resistant to the enzyme, called **replicative intermediate** (RI). By many tests (sedimentation, chromatography, thermal denaturation, solubility in concentrated salt solutions, RNase sensitivity), the RF is entirely double-stranded, whereas RI is a double-stranded backbone with one or two single-stranded tails (Fig. 45-38). The pattern of labeling after brief pulses of a radioactive precursor shows that the RF is produced by building a complementary strand on the infecting viral strand; the subsequent

synthesis of a third progeny viral strand on the double-stranded RF converts it into the RI, from which the progeny strands are then released. The synthesis of the viral strands can be semiconservative or conservative in different RIs; the two RI types are distinguishable by the different accessibility of labeled parental RNA to RNase degradation.

A special feature of the synthesis of phage RNA is the **addition of an adenylic acid residue** at the 3' end of both strands after replication. This causes the end of the viral RNA to be CCA, as in tRNAs, a feature that may have special functional significance (see Ch. 44, Viral RNA).

USES OF RNA REPLICASE

The Qβ replicase can synthesize large amounts of infectious RNA in vitro, in contrast to the much more limited success achieved with other DNA or RNA polymerases.

This enzyme finds application in the **synthesis of abnormal RNAs** by using unusual conditions of synthesis (such as unusual nucleoside triphosphates). Using this approach, Spiegelman has studied the **evolution of the RNA as template** in a self-replicating system containing external replicase and precursors. In cells such evolution cannot be studied separately from the evolution of the replicase specified by the RNA, which introduces extremely severe restraints.

Another application is the preparation, by limited synchronous synthesis, of short viral RNA segments, important in sequencing work (see Ch. 10). Synchronization is obtained by preparing a replicating mixture with only one triphosphate, GTP. The replicase molecules bind to the template but syn-

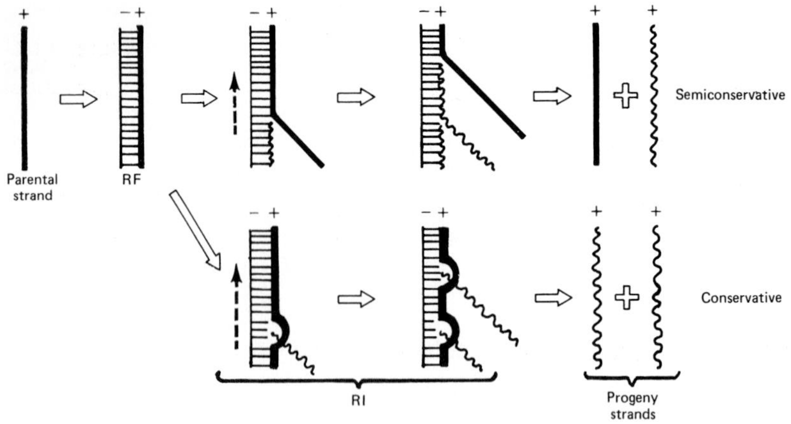

Fig. 45-38. Intermediates in RNA replication. In the semiconservative type of replication the parental strand (straight heavy line) is exposed to RNase attack during replication, whereas it is not in the conservative model. Wavy lines = progeny strands; dashed arrows = direction of replication.

thesis cannot begin for lack of the other triphosphates. When they are later added, synthesis proceeds synchronously on all templates and produces RNA molecules of a fairly uniform length, which depends on the duration of synthesis.

REGULATION OF TRANSLATION

In the infected cells the rates of synthesis of the three viral proteins are adjusted to the requirement of viral multiplication, since capsid protein is made in large excess over either replicase or A protein. In vitro experiments show that translation of the two latter genes is repressed by **changes of the secondary structure of the RNA** after binding capsid protein. The effect is specific, since Qβ protein causes little inhibition with f2 RNA. This observation emphasizes the fundamental role of regulation in even the smallest organisms, as well as the simplicity of the methods by which it can be achieved. It also shows that effective regulation can be obtained at the translational level.

SELECTED REFERENCES

Books and Review Articles

ARCHER, W., and LINN, S. DNA modification and restriction. *Annu Rev Biochem 38*:467 (1969).

BERGQUIST, P. L., and BURNS, D. J. W. The translation of viral messenger RNA in vitro. *Adv Virus Res 15*:159 (1969).

CALENDAR, R. The regulation of phage development. *Annu Rev Microbiol 24*:241 (1970).

DANIEL, V., SARID, S., and LITTAUER, O. Z. Bacteriophage-induced transfer RNA in *Escherichia coli. Science 167*:1682 (1970).

DUCKWORTH, D. H. Biological activity of bacteriophage ghosts and "take-over" of host functions by bacteriophage. *Bacteriol Rev 34*:344 (1970).

HAYES, W. *The Genetics of Bacteria and Their Viruses.* Wiley, New York, 1968.

LURIA, S. E. The recognition of DNA in bacteria. *Sci Am 222*:88 (1970).

MARVIN, D. A., and HOHN, B. Filamentous bacterial viruses. *Bacteriol Rev 33*:172 (1969).

REVEL, H. R., and LURIA, S. E. DNA-glucosylation in T-even phage: Genetic determination and role in phage-host interaction. *Annu Rev Genetics 4*:177 (1970).

STENT, G. *Papers on Bacterial Viruses.* Little, Brown, Boston, 1960.

STENT, G. *Molecular Biology of Bacterial Viruses.* Freeman, San Francisco, 1963.

VALENTINE, R., WARD, R., and STRAND, M. The replication cycle of RNA bacteriophages. *Adv Virus Res 15*:1 (1969).

Specific Articles

ALBERTS, B. H., and FREY, L. T4 bacteriophage gene 32: A structural protein in the replication and recombination of DNA. *Nature 227*:1313 (1970).

AUGUST, J. T. Mechanism of synthesis of bacteriophage RNA. *Nature 222:*121 (1969).

CHAMBERLIN, M., MCGRATH, J., and WASKELL, L. New RNA polymerase from *E. coli* infected with bacteriophage T7. *Nature 228:*227 (1970).

EGGEN, K., and NATHANS, D. Regulation of protein synthesis directed by coliphage MS2 RNA. II. In vitro repression by phage coat protein. *J Mol Biol 39:*293 (1969).

FRANKLIN, R. M. Purification and properties of the replicative intermediate of the RNA bacteriophage RM. *Proc Nat Acad Sci USA 55:*1504 (1966).

GUHA, A., SZYBALSKI, W., SALSER, W., BOLLE, A., GEIDUSCHEK, E. P., and PULITZER, J. F. Controls and polarity of transcription during bacteriophage T4 development. *J Mol Biol 59:*329 (1971).

HUBACEK, J., and GLOVER, S. W. Complementation analysis of temperature-sensitive host specificity mutations in *E. coli. J Mol Biol 50:*111 (1970).

KANO-SUEOKA, T., and SUEOKA, N. Leucine tRNA and cessation of *E. coli* protein syntnesis upon phage T2 injection. *Proc Nat Acad Sci USA 62:*1229 (1969).

LAEMMLI, U. K. Cleavage of structural proteins during the assembly of the head of bacteriophage T4. *Nature 227:*680 (1970).

MARSH, R. C., BRESCHKIN, A. M., and MOSIG, G. Origin and direction of bacteriophage T4 replication. II. A gradient of marker frequencies in partially replicated T4 DNA as assayed by transformation. *J Mol Biol 60:*213 (1971).

MILLS, D. R., PETERSON, R. L., and SPIEGELMAN, S. An extracellular darwinian experiment with a self-duplicating nucleic acid molecule. *Proc Nat Acad Sci USA 58:*217 (1967).

RIVA, S., CASCINO, A., and GEIDUSCHEK, E. P. Uncoupling of late transcription from DNA replication in bacteriophage T4 development. *J Mol Biol 54:*103 (1970).

SCHEKMAN, R. W., IWAKA, M., BROMSTRUP, K., and DENHARDT, D. T. The mechanism of replication of φX174 single-stranded DNA. III. An enzymic study of the structure of the replicative form II DNA. *J Mol Biol 57:*177 (1971).

STREISINGER, G., EMRICH, J., and STAHL, M. M. Chromosome structure in phage T4. III. Terminal redundancy and length determination. *Proc Nat Acad Sci USA 57:*292 (1967).

WATSON, J. D. Origin of concatemeric T7 DNA. *Nature 239:*197 (1972).

LYSOGENY, EPISOMES, AND TRANSDUCING BACTERIOPHAGES

LYSOGENY   1089

NATURE OF LYSOGENY   1089
THE VEGETATIVE CYCLE   1091
  The Genetics of λ   1091
  Transcription   1092
  Replication   1092
THE LYSOGENIC STATE   1094
  Establishment of the Lysogenic State   1094
  Immunity and Repression   1095
  Regulation of Repressor Formation   1096
THE PROPHAGE CYCLE   1099
  Location and State of the Prophage in Lysogenic Cells   1099
  Mechanism of Insertion   1099
  Multiplication of the Prophage   1100
  Defective Prophages   1100
  Excision   1101
MOLECULAR MECHANISM OF PROPHAGE INTEGRATION   1101
  Enzymatic Requirements   1101
  Steric Requirements   1101
CURING OF A LYSOGENIC BACTERIUM   1102
INDUCTION OF A LYSOGENIC CELL   1103
  Mechanism of Induction   1103
EFFECT OF PROPHAGE ON HOST FUNCTIONS   1104
SIGNIFICANCE OF LYSOGENY   1104

EXTRACHROMOSOMAL ELEMENTS: PLASMIDS   1105

  Conjugation   1105
  Plasmid Regulation   1106
  Episomes   1106
RESISTANCE-TRANSFER FACTORS (R FACTORS)   1107
BACTERIOCINOGENS AND BACTERIOCINS   1109
  Mechanism of Action of Bacteriocins   1109
EXTRACHROMOSOMAL ELEMENTS AND THE EVOLUTION
    OF VIRUSES   1110

PHAGES AS TRANSDUCING AGENTS   1111

GENERALIZED TRANSDUCTION   1111
  Distribution and Scope   1111

Mechanism of Generalized Transduction: Role of Defective
    Phage   1112
Abortive Transduction   1112
SPECIALIZED TRANSDUCTION   1113
Low-Frequency Transduction   1113
High-Frequency Transduction   1114
Genetic Changes in Heterogenotes   1116
Uses of Specialized Transduction   1117
Range of Transducing Phages   1117

LYSOGENY

The bacteriophages described in the previous chapter display an extreme form of parasitism, in which the cells are killed at the end of the growth cycle. These viruses are called **virulent.** Contrasted to them are the **temperate** bacteriophages, which produce the phenomenon of **lysogeny:** the persistence of the phage DNA in the host cells for many cell generations, even indefinitely. Under suitable circumstances, the viral DNA in **lysogenic cells** can be induced to multiply like that of a virulent bacteriophage. This form of parasitism favors the persistence and spreading of a virus in a more subtle way than virulence: as Burnet has pointed out in connection with viral diseases in higher organisms, the parasites best adapted to their environment are those that do not rapidly kill their hosts and thus deprive themselves of the opportunity to spread.

Lysogeny has provocative implications for many biological problems. Thus it throws light on the origin of viruses and the evolution of bacteria, and temperate phages provide an important mechanism for gene transfer between bacteria, i.e., transduction. Moreover, the switch between the lysogenic and the vegetative state is useful as a model in the study of differentiation; temperate bacteriophages supply a model for viral oncogenesis (see Ch. 63).

NATURE OF LYSOGENY

Lysogeny characterizes many bacterial strains freshly isolated from their natural environment. Such lysogenic cultures contain a low concentration of bacteriophage, which can be recognized because it lyses certain other related bacterial strains, known as **sensitive** or **indicator** strains.* Lysogeny was recognized in the early twenties; and it was soon realized that lysogenic strains are not simply phage-contaminated bacterial cultures, since the phage could not be eliminated by repeated cloning of the bacteria or by growing the cells in the presence of phage-specific antiserum. Bordet then recognized, in 1925, that the ability to produce phage was a hereditary property of the cells. Nevertheless, disruption of the lysogenic cells does not yield infectious phage; hence the phage must be present in the cells in a noninfectious form.

When a sensitive bacterial strain is infected by a temperate bacteriophage, two alternative responses are seen (Fig. 46-1). Some cells are lysed by phage multiplication and others are lysogenized. Lysogenic strains thus produced are designated by the name of the sensitive strain followed by that of the lysogenizing phage in parentheses, e.g., *Escherichia coli* K_{12} (λ). Because temperate phages lyse only a fraction of the sensitive cells that they infect, they produce **turbid plaques.**

A bacterial strain can be easily recognized as lysogenic by streaking it on a solid medium across a strain sensitive to the phage released; a narrow zone of lysis is seen along the border of the lysogenic strain (Fig. 46-2). Since lysogeny cannot be recognized unless such a sensitive strain is available, many bacterial strains—perhaps most of those known—may be unrecognized as lysogens. Furthermore, many strains are lysogenic for several different phages.

The systems used most in experimental work on lysogeny are λ and related phages

---

* Sensitive strains either are found by chance or, as noted later on, are obtained by "curing" lysogenic strains.

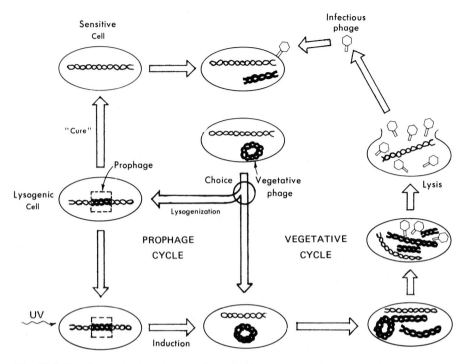

Fig. 46-1. Development of a temperate bacteriophage.

active on *E. coli* K$_{12}$, and phages P1 and P2 with *Shigella dysenteriae* or with several strains of *E. coli*.

Relation to Vegetative Phage Cycle. The ability of lysogenic cultures to produce virus without obvious lysis remained puzzling until Lwoff, in 1950, patiently observing the behavior of single cells in microdroplets, showed that **phage is produced by a small proportion**

of the cells: these lyse and release phage in a burst **(induction),** just like cells infected by phage T4. The other cells of the culture do not give rise to a productive infection and are said to be **immune*** to the released phage. The phage adsorbs to the immune cells and injects its DNA, but **the DNA does not multi-**

---

* This term is totally unrelated to immunity as studied in immunology.

Fig. 46-2. Cross-streaking of lysogenic and sensitive strains of *E. coli* on nutrient agar. **A.** Untreated. **B.** Exposed to a small dose of ultraviolet light, after streaking, to induce the lysogenic cells. In **A** note the narrow bands of lysis of the sensitive strain (vertical streak) flanking the lysogenic strain (horizontal streak). In **B** note that the inducing treatment, by causing cell lysis, markedly reduces the colony density of the lysogenic streak, and the accompanying release of infectious phage causes pronounced lysis at the sensitive strain in the area of crossing.

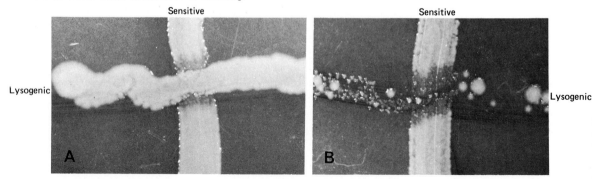

ply and does not cause cell lysis. **Immunity, therefore, is different from resistance,** which, as noted in the previous chapter, prevents adsorption and injection.

These experiments made it clear that the noninfectious form of the phage **is not some kind of incomplete virus,** with an occasional particle becoming completed and "leaking" from the cell. Rather, lysogeny involves a special, stably inherited, noninfectious form of the virus, called **prophage,** whose presence is associated with immunity; and the prophage occasionally shifts abruptly to the **vegetative form** and then reproduces just like a virulent phage. Lwoff further provided a powerful tool for studying lysogeny by showing that the shift from the prophage cycle to the lytic cycle, normally a rare event, could be **induced** in all the cells of a culture by certain environmental influences, such as moderate ultraviolet irradiation (Fig. 46-2).

THE VEGETATIVE CYCLE

The vegetative cycle of temperate phages is similar to that of virulent phages (Ch. 45) but with some modifications. Virions of λ

contain a double-stranded linear DNA with a complementary single-stranded segment 12 nucleotides long at each 5′ end (Figs. 44-13 and 46-3). Under annealing conditions in vitro these "cohesive" ends pair, generating a cyclic molecule with two staggered nicks. Similar cyclization evidently occurs in vivo after infection, with ligase then closing the two nicks, since a labeled infecting phage DNA can be shown to become a covalent closed cyclic molecule. The development of the cohesive ends in evolution can be explained by the requirement for a linear DNA in the virions to allow encapsidation, and for a cyclic form intracellularly during replication (see Ch. 45) and lysogenization (see below).

THE GENETICS OF λ

The genetics of λ is based on several classes of mutants: 1) **conditionally lethal sus** (suppressor sensitive) mutants similar to the nonsense mutants of T4; 2) **temperature-sensitive** mutants, also similar to those found in T4; 3) **plaque-type** mutants, such as minute (*m*), host-range (*h*), and clear-plaque (*c*) mutants; 4) mutants with altered buoyant density, designated as *b* (for buoyancy), which

Fig. 46-3. Different forms of λ DNA. **A.** In the virions the DNA exists as a linear double-stranded molecule with homologous ends (Kaiser-Hogness assay). Apparently the naked DNA joins ends with the helper ends allow the naked phage DNA to be infectious for *E. coli* cells, in the presence of a helper phage with homologous ends (Kaiser-Hogness assay). Apparently the naked DNA joins ends with the helper DNA in the process of penetration, and is then pulled inside the cell. **B.** Under conditions of nucleic acid hybridization (Ch. 10), the linear molecule can reversibly close into a ring by base pairing of the single-stranded ends. **C.** Within the cell, the DNA forms completely covalently closed rings. The rings are not infectious in the Kaiser-Hogness system, but are infectious for spheroplasts. **D.** If the single-stranded ends of **A** are made double-stranded by a polymerase, the molecules can no longer close into a ring. This causes the molecule to lose infectivity, which is restored, **E,** by restoring the single-stranded ends with exonuclease III (see Ch. 10). Lack of infectivity of **D** shows that ring closure is essential for the biological function of the DNA.

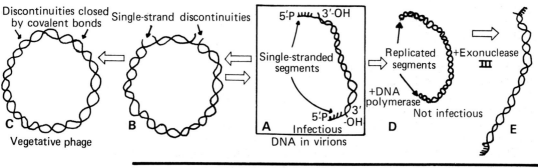

were discovered during purification of the virus by density gradient equilibrium centrifugation; density changes are produced by deletions, which alter the ratio of DNA to protein in virions; and 5) **substitution** mutants in which part of the genome is replaced by bacterial DNA or by DNA of another phage (see Figs. 46-7 and 46-18).

Phage λ undergoes genetic recombination like T-even phages, although at a lower rate. It is produced by the viral *red* (for recombination deficient) locus and gene γ, as well as by the *rec* bacterial system (see Ch. 11, Recombination). This kind of recombination is called **generalized** to distinguish it from a special form occurring in lysogenization (see below).

The λ genetic map determined by various methods (three-factor crosses, deletion mapping, and electron microscopy of heteroduplex DNA using deletions; see Ch. 11, Genetic mapping) is linear, and its ends correspond to the ends of the mature DNA molecules. It is called the **vegetative map** to distinguish it from a permuted map characterizing the prophage (see below).

On the cyclized DNA, on which transcription probably occurs, the genetic information exclusively required for vegetative functions is all contained in a single block of genes (the **vegetative region** of the genome, black band in Fig. 46-4). A second block (the **regulatory region,** white band) also involved in vegetative multiplication, contains mostly genes with control functions; and a third block (the **recombination region**, crosshatched band) is involved in recombination.

TRANSCRIPTION

The extraordinarily detailed study of λ transcription has led to a fairly complete understanding of the switching between vegetative and prophage cycle. This process will be examined in some detail because it is not only the key to lysogenization but also the best model for the switching of genes in differentiation. Aspects important for the vegetative cycle will be reviewed now; those specifically related to lysogenization will be reviewed in a later section.

The general pattern of λ transcription (Fig. 46-4) is similar to that of T4 (Ch. 45). We distinguish immediate early messengers (which appear in the presence of chloramphenicol), delayed early, and late messengers. In vitro experiments suggest that, as with T4, the transition from immediate to delayed early is produced by **interference with termination.** Thus, in vitro, in the presence of rho factor (see Ch. 11, Transcription), two immediate early RNAs can be recognized. One (l_1) is transcribed from the "left" DNA strand, the other (r_1) from the "right" strand. Their promoters, left and right (PL and PR), are identified by mutations that abolish the synthesis of the RNAs. Extension of transcription to include two delayed early RNAs (l_2, r_2) requires the product of gene N (transcribed in l_1) which modifies the host transcriptase in such a way that it does not recognize termination signals. Through its effect on l_2 and r_2 transcription, **gene N controls the expression of most viral functions;** excluded are the immunity region of the genome (see below) and the small regions transcribed into l_1 and r_1 messengers.

Late transcription requires the function of gene Q (whose transcription is under N control), which apparently specifies a σ-like factor (see Ch. 11, Transcription) specific for a promoter just to the right of Q. The factor allows transcription of the "right" DNA strand all the way to gene J and into b_2. The weak polarity of some nonsense mutations (see Ch. 13) suggests that several groups of genes, such as W-B-C, V-G, and H-M, are independently transcribed. Late transcription is strongly increased by DNA replication, but occurs also in its absence.

There are some exceptions to this transcription scheme. Thus, genetic evidence suggests that the N product allows a low rate of independent transcription (i.e., not initiated at PL or PR) of genes O and P (right transcription) and *int* (left transcription). In addition, genes O and P are also weakly expressed in N mutants, suggesting that they follow a weak promoter recognized by the *E. coli* polymerase; this is important for experiments to be discussed below.

REPLICATION

The **replication of λ DNA** utilizes exclusively exogenous precursors, since the bacterial DNA is not destroyed. Initiation depends on the functions of two viral genes (O and P) in conjunction with a host function (since

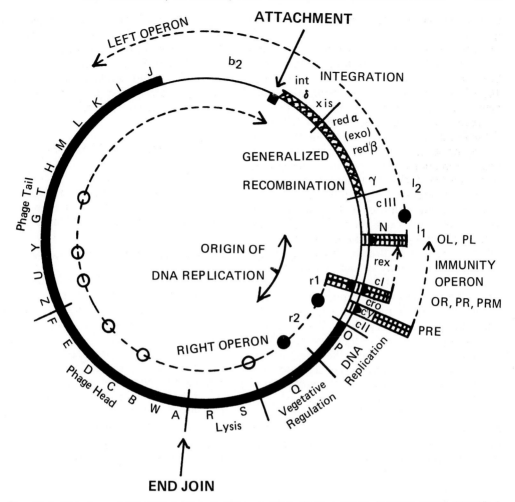

Fig. 46-4. Genetic and functional map of phage λ DNA. Capital letters indicate genes identified by *sus* mutations, which include most of all those required for the vegetative cycle. Genes indicated by black bands perform vegetative functions; those shown as white bands perform general regulatory functions, and those shown as cross-hatched bands perform recombination functions. The DNA is represented in the cyclic configuration in which transcription occurs. In the virions DNA is linear, with ends at end join. The map is linear, i.e., markers close to end join but at opposite sides have the highest recombination frequencies. Transcription is indicated by heavy dashed lines, outside the circle for left transcription, and inside for right transcription. There are three main transcription segments designed as operons: left, right, and immunity operons. Triangles indicate the promoter-operators. PL, OL = left promoter and operator; PR, OR = right promoter and operator; PRE = promoter for repressor establishment; PRM = promoter for repressor maintenance. Circles are points where transcription is controlled either by the N gene product (black circles) or by the Q gene product (white circles). The position of the latter points, except for the first one, is arbitrarily based on polar effects of mutations. The function of the b_2 region is obscure; it is transcribed in both directions by the two convergent operons. (δ indicates a site whose mutations allow λ to form plaques on bacteria lysogenic for the unrelated phage P2.)

it is abolished by mutations in a bacterial gene).

Replication occurs in two stages. At first the cyclic DNA associates with the cell membrane and replicates symmetrically (see Ch. 10, DNA replication). This replication initiates at a site close to gene O, since there is no replication if this region is deleted, even if the O function is supplied by another coinfecting phage. From there, it proceeds in opposite directions and probably terminates when the two forks meet, separating the two daughter molecules (Fig. 46-5). Initiation is coupled to transcription of the initiation region, since it does not occur in suitable mutants lacking r_2 transcription; in this way DNA replication is directly under the control of the phage immunity system (see below).

In the second stage the progeny DNA leaves the membrane and replicates according to the rolling circle model (see Ch. 10, DNA replication), generating long linear concatemers. (If gene N is defective the DNA replicates continuously in cyclic form; see below, Plasmids.) Mature molecules with cohesive ends are formed from the concatemers or from cyclic molecules by pairs of staggered single-strand cuts, 12 nucleotides apart, at the appropriate sites. This step is linked to head assembly, since it requires the functions of several genes which specify head proteins.

Assembly of capsids and virions follows, in somewhat simplified form, the T-even model. The virions are then released through the action of gene S, whose product stops cellular metabolism and weakens the cellular membrane, allowing the lysozyme produced by gene R to lyse the cell wall.

THE LYSOGENIC STATE

This state is determined by the activity of the **regulatory region** of the λ genome, which both generates immunity and causes integration of the phage genome in the cellular DNA.

ESTABLISHMENT OF THE LYSOGENIC STATE

When a sensitive bacterium is infected by a temperate phage the entering DNA has a **"choice"** between vegetative multiplication and lysogenization. The proportion of infected cells that are lysogenized may vary from a few per cent to nearly 100%, depending on both the system and the conditions. Before the choice is made the viral DNA undergoes several rounds of replication, which enhance the chance that a DNA molecule will find the site of integration.

When lysogenization occurs it takes about 2 hours from infection before a stable lysogenic clone is

Fig. 46-5. Scheme of cyclic λ molecules after partial replication. The scheme, derived from electron micrographs, shows two forks, beginning at the same origin (O) each containing a single-stranded arm (ss in the figure). This scheme is compatible with replication proceeding in both directions, one of the strands being replicated discontinuously (see Ch. 10, DNA replication). The single-stranded segments would be those not yet completed by discontinuous replication. When the two replication forks (arrows) meet, the two daughter molecules can separate. However, the separation of the two parental strands (straight lines) requires either single-strand breaks to act as swivel or the participation of a "swivel enzyme" (see Ch. 10, DNA replication). [From data of Inman, R. B., and Schnös, M. *J Mol Biol 56*:319 (1971).]

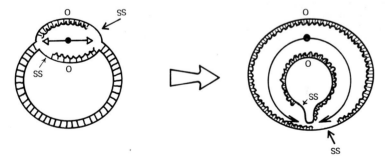

formed. Pedigree analysis shows that during this time the bacterium undergoes several divisions, occasionally producing sensitive as well as lysogenic progeny. Hence there must be a lag between the onset of repression and integration during which cell division, unaccompanied by phage division, may yield a cell without a prophage.

IMMUNITY AND REPRESSION

Immunity, which originally appeared as a curious product of the lysogenic state, emerged as its central feature when it was shown, by Jacob, Wollman, and others, to be produced by **repression of phage genes;** the phenomenon is entirely similar to the repression of bacterial operons described in Chapter 13.

Repression is demonstrated in lysogenic cells by **absence of mRNAs** corresponding to the vegetative and recombination regions of the phage genome. They are replaced by a small mRNA transcribing part of the regulatory region, the **immunity operon** on the "left" DNA strand (see Fig. 46-4). Lysogenic cells contain the **immunity repressor** but no vegetative proteins. On induction the repressor disappears and transcription returns to the vegetative pattern.

The repressor is specified by the cI gene, in the immunity region: cI⁻ (clear*) mutations are repressor-negative and consequently prevent lysogenization but allow vegetative growth.

Temperature-sensitive cI mutants have been especially important in the elucidation of the regulation of repression; they produce a **thermosensitive repressor** which becomes ineffective at high temperature, causing induction of the lysogenic cells.

Immunity to exogenous infection is closely related to repression: 1) it is absent in cells infected by cI⁻ phage, and 2) it breaks down simultaneously with repression when normal lysogenic cells are induced. **Immunity, therefore, is due to repression of exogenous λ DNA.** The repressed DNA cyclizes regularly but remains unexpressed as an **abortive prophage.** It survives for many cell genera-

tions; it is transmitted unilinearly like the genes involved in abortive transduction (see below) and can be recognized by proper genetic tests.

The Repressor. The λ repressor, isolated by Ptashne, is an acidic protein formed by two equal monomers, each with an MW of 30,-000 daltons. In vitro λ repressor binds to λ DNA but not to DNA lacking the immunity region. Therefore, the **binding has the same specificity as repression.** The binding to λ DNA is very strong, since it persists during centrifugation in a sucrose gradient (Fig. 46-6). Binding is prevented if the DNA carries two mutations, called **virulent** (*vir*), close to the right and left promoters (Fig. 46-4).

The λ repressor behaves like that of the *lac* operon (see Ch. 14). The similarity of the two systems is shown by the correspondence of their mutations. Thus i⁻ of the *lac* operon corresponds to cI⁻, and iˢ to *ind⁻*, which makes the prophage not inducible through the synthesis of a superrepressor. The **constitutive operator mutations** of the *lac* operon correspond to the *vir* mutations of λ, which define a left and right operator, OL and OR, associated with the **left and right operon.** Through the left promoter the repressor controls the function of the N gene, and through the right promoter it controls the functions of genes O, P, and Q (which in turn controls all late genes, see above). Therefore, the repressor controls all the phage functions expressed during vegetative multiplication.

Virulent (operator-defective) strains of phage λ cannot lysogenize and hence form clear plaques. But in contrast to cI⁻ mutants, **virulent strains can grow in immune cells** because they are **insensitive to the repressor,** to which cI⁻ strains are fully sensitive.

Specificity of Immunity. Each phage recognizes exclusively its own repressor; different phages, even closely related, form repressors of different specificities. Immunity, therefore, is highly specific. No point mutation in a repressor gene is known to change its specificity into that of a different, even closely related phage. Such a change can be achieved only by exchanging the immunity region (Fig. 46-7); these **heteroimmune** derivatives which

---

* Clear plaques are also produced by mutations in genes cII, cIII, and cY which, as discussed below, participate in lysogenization but not in repressor specification.

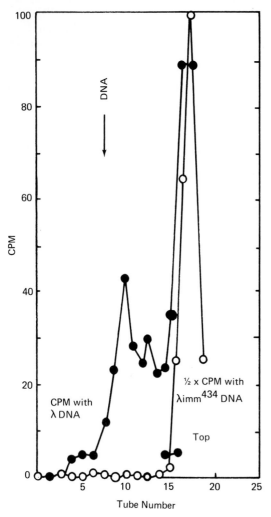

Fig. 46-6. Specific binding of λ repressor to λ DNA. Purified ^{14}C-labeled repressor was incubated with either λ DNA (black circles) or λimm[434] DNA (white circles). λimm[434] contains mostly λ sequences, except for the immunity region, which derives from the related phage 434 (see Fig. 46-7). The mixture was then sedimented in a sucrose gradient. The distribution of radioactivity in fractions collected from the bottom of the tube showed that part of the repressor cosediments with λ DNA (whose position is indicated by the arrow) but not with λimm[434] DNA (sedimenting at the same place on λ DNA) which does not contain the operator recognized by the λ repressor. [Modified from Ptashne, M. *Nature* 214:232 (1967).]

have most genetic properties of a phage, but the immunity of another, have played an important part in unraveling the mechanisms of lysogeny.

REGULATION OF REPRESSOR FORMATION

During the late sixties and early seventies the problem of how lysogenization is produced, maintained, and reversed has attracted a large number of investigators, generating a highly specialized and equally successful field of research. One reason for its popularity is the significance of the problem for gene regulation and differentiation; another is the suitability of lysogeny as a model system, using powerful genetic and molecular tools. The problem has been solved almost completely revealing that the basic facts, although difficult to unravel, are simple. Thus when phage DNA enters a virgin cell, containing no viral gene products, the cooperation of three genes brings forth a rapid accumulation of repressor, while at the same time an antirepressor is also formed. Repressor and antirepressor compete for initiating transcription of the repressor gene. If the repressor wins out all the other genes are shut off and the repressor continues to sustain its own synthesis by a positive-feedback mechanism; if the antirepressor wins out, the repressor gene is shut off and vegetative multiplication proceeds unimpeded.

These studies have dealt mostly with the functional state of the immunity operon, i.e., its rate of transcription. The mRNA of the immunity operon, which constitutes only 10^{-4} of the genome of a lysogenic cell, can be measured by a two-step hybridization procedure (see Ch. 10). The mRNA is first enriched by hybridizing the cellular RNA to the "left" strand of the DNA of a λ plasmid (see below) which contains only a small fraction of the λ genome, including the immunity region; this DNA strand is template for the immunity operon mRNA (see Fig. 46-4). The immunity RNA is then purified by hybridizing the eluted RNA to λimm[434] DNA (see Fig. 46-7) which removes all other RNAs except that corresponding to the immunity region. The repressor labeled in vitro with [135]I is precipitated by specific antibodies. Other useful indicators of the activity of the immunity operon are immunity itself and the resistance of the cells to phage T4 r_{II} (see Ch. 45), which is caused by gene *rex* included in the operon.

These approaches revealed two different methods of repressor control in lysogenic cells and in sensitive cells soon after λ infection (Fig. 46-8).

Infected Cells.   The initial production of repressor after infection requires the activity of three genes— cII, cIII, and cY (see Fig. 46-4)—which have no

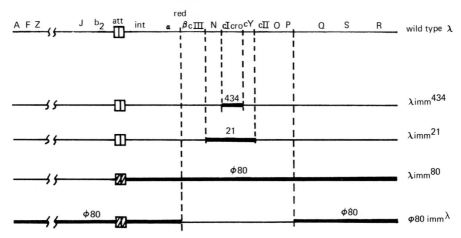

Fig. 46-7. Heteroimmune derivatives of phage λ. They arose from exchanges of λ with the related phages 434, 21, or φ80. Each was obtained by a series of back-crosses to one parent, always using recombinants with the immunity properties of the other parent. In this way, several different immunity regions have been inserted on a more or less complete λ background or, conversely, a λ immunity region was introduced in a φ80 background. Thin lines = λ DNA; heavy lines = substitutions; ☐ = λ attachment sites; ▨ = φ80 attachment sites.

role in lysogenic cells: phages with mutations in any of these genes lysogenize very poorly and form clear plaques. cII and cIII yield *trans*-active diffusible products, like bacterial R genes, whereas cY is only *cis*-active, like a promoter or operator (see Fig. 13-11). cY is therefore considered a **promoter for repressor synthesis establishment** (*pre*), activated by the protein specified by cII and cIII (PA in Fig. 46-8); this interaction allows transcription of the immunity operon in originally repressor-free cells. About 20 minutes after infection repressor synthesis is stopped by the gene *cro* product, whose function will be examined below.

Lysogenic Cells: Positive Feedback. The regulation of maintenance of repressor in lysogenic cells was revealed with prophages in which N⁻–O⁻–P⁻ mutations prevented expression of essentially all genes outside the immunity region, and an additional mutation in cI made the repressor reversibly thermosensitive. At 30° these lysogens are fully repressed and manufacture only RNA of the immunity operon; after the temperature is raised to 42°, to inactivate the repressor, not only do they lose immunity, but, surprisingly, within 5 minutes **stop synthesizing immunity mRNA.** If the temperature is then again lowered, the renaturation of the repressor rapidly restores immunity and also a normal rate of synthesis of immunity mRNA. However, if the heating period lasts long enough to eliminate the preexisting repressor, on returning to 32° the cells fail to restore either immunity mRNA or the repressor (measured serologically). This phenotypic lack of immunity requires the *cro* gene product (see below); in *cro*⁻ cells immunity is slowly restored at 32°. It appears, therefore, that **the immunity operon con-**

tinuously requires functional repressor to activate its own transcription. This activation is accomplished by binding of the repressor, functioning as a positive effector, to a **promoter for repressor maintenance** (*prm*, see Fig. 46-8); deletions have located this binding site at the right operator.

The immunity repressor thus exercises two types of control: a negative control of the left and right operons, by interacting with their operators (OL and OR), and a **positive control** of the immunity operon, by interacting with its maintenance promoter (*prm*), (see Fig. 46-8). The **positive-feedback control** of its own synthesis by the repressor stabilizes the lysogenic state. We may further note that interaction of the repressor with a single region (prm-OR) blocks transcription on one strand and initiates it on the other.

This explanation for the stability of lysogeny has particular significance as a model for the stability of differentiation in higher organisms, where some form of positive feedback must be logically postulated (Ch. 13).

Determination of the Choice Between Lysogenization and Vegetative Multiplication. Gene *cro* plays an important role in the choice that follows infection of sensitive cells. It synthesizes a repressor acting on *prm*, and thus, as noted above, antagonizes the maintenance synthesis of the immunity repressor. Hence *cro*⁻ mutants do not undergo vegetative multiplication but only lysogenize, if all other genes are normal. This vegetative control of the immunity repressor counterbalances the positive control of the immunity repressor by itself, allowing a choice between lysogenization and vegetative multiplication. The balance is influenced by a

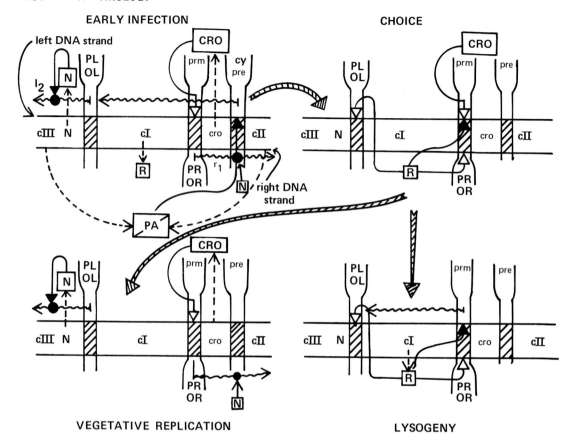

Fig. 46-8. Regulation of repression in phage λ and determination of choice between vegetative replication and lysogeny. In the early period of infection the left and right promoters (PL, PR) become active, allowing transcription (wavy lines) and expression of gene N. The latter in turn allows delayed early transcription, bypassing the terminators (black circles) and thus activating cII and cIII. The cII and cIII products form the positive activator PA which, acting on the promoter for repressor establishment (*pre*), initiates transcription of the immunity operon (positive control, black arrowheads) and synthesis of immunity repressor R. Gene *cro* starts producing the CRO product, which represses the promoter for repressor maintenance (*prm*), but its function is irrelevant at this stage since *prm* is inactive. After about 20 minutes, the various products have accumulated, and new syntheses temporarily cease. The **choice** is determined by the competition between the negative control (white arrowheads) of CRO product and the positive control of immunity repressor R on *prm*. If CRO wins out, *prm* remains inactive and transcription initiated at PL and PR causes vegetative multiplication. If R wins out, it activates *prm* keeping it permanently activated by positive feedback control, and blocks PL and PR.

number of conditions which differentially affect the synthesis or action of the two molecules, such as changes of temperature, ion concentration, or rate of protein synthesis.

We can now understand the necessity for the initial transcription of the immunity operon initiated at *pre,* which builds enough immunity repressor to counterbalance the subsequent *cro* action on *prm*. Without such a **starter mechanism** the immediate early expression of *cro* (which is transcribed in messenger r_1, independent of N product) would commit the system to vegetative multiplication.

An additional effect of the *cro* gene product is to inhibit both left and right transcription initiating at PL and PR; this action can be interpreted also as antilysogenic because it antagonizes the synthesis of *int* product, required for integration of the prophage (see below).

Role of Cyclic AMP. In bacteria cAMP is a signal of hard times, promoting their adaptation to poor food sources when the good ones are not available (see Ch. 13). In infected cells a high cAMP concentration also signals hard times, promoting the energetically economical lysogenization over the expensive lytic multiplication. This is inferred from the reduced frequency of lysogenization of λ or P22 in hosts lacking adenylate cyclase or cAMP-binding protein. The effect is probably on transcription of phage genes controlling lysogenization.

THE PROPHAGE CYCLE

When lysogenization occurs, the vegetative λ DNA, in cyclic form, becomes inserted into the cellular DNA as prophage. Conversely, at induction the prophage becomes excised to generate again cyclic vegetative DNA. The prophage cycle is therefore **an alternative pathway** in the vegetative DNA cycle (see Fig. 46-1).

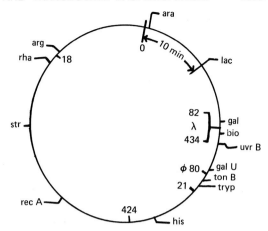

Fig. 46-9. Map of the *E. coli* genome showing the attachment sites for some prophages (inside the circle).

LOCATION AND STATE OF THE PROPHAGE IN LYSOGENIC CELLS

Most prophages are integrated at **fixed locations** on the bacterial chromosome (Fig. 46-9). They are transferred with the Hfr chromosome in bacterial crosses and their sites can be mapped in relation to bacterial genes. In *E. coli,* λ and related phages can settle in a unique site on the host chromosome; P2 can occupy at least nine distinct sites, but two preferentially, and Mu-1 can occupy a large number of sites. However, some prophages, e.g., P1, are separate from the chromosome as plasmids (see below). P1, however, must occasionally interact with the bacterial chromosome because it gives rise to specialized transduction (see below).

In the lysogenic cells the prophage DNA is recognizable by hybridization with radioactively labeled phage DNA. Integrated prophages are connected to the bacterial chromosome by covalent bonds because they remain bound after the DNA is denatured. This is seen, for instance, in zonal centrifugation through an alkaline sucrose gradient, where the viral DNA sediments rapidly, like the cellular DNA.

MECHANISM OF INSERTION

Genetic experiments demonstrate that prophage λ is **linearly inserted** in the bacterial chromosome. Calef first showed that the prophage has a **linear map** by crossing bacteria with genetically marked prophages **(prophage map)**; the order, however, of three genes

(c-mi-h) was different from that of the vegetative map (h-c-mc). The subsequent use of phage containing many markers yielded additional evidence for such a regular permutation, in both λ and other phages.

Campbell explained this permutation by suggesting that in lysogenization the **viral DNA in cyclic form is inserted linearly into the bacterial chromosome by a single reciprocal crossover** which opens the ring at a point different from that where the ends of the mature DNA meet (Fig. 46-10). Subsequent deletion mapping of lysogenic cells (Fig. 46-11) has given conclusive evidence for linear insertion of the λ prophage at a fixed location in lysogenic *E. coli* K_{12}.

Insertion of the prophage in the bacterial DNA thus involves recombination between a specific **phage attachment** site in the vegetative λ DNA and a corresponding **bacterial attachment site.** Insertion creates two recombinant **prophage attachment sites** flanking the prophage (Fig. 46-10).

Experiments with heavy isotopes have shown that the infecting phage DNA can be integrated **without previous replication;** however, since replication of the DNA increases the chance of integration, most prophages derive from replicated DNA.

Fairly frequently, cells are made **doubly lysogenic,** as may be shown by infecting sensitive cells with two λ strains carrying different markers. The two prophages are inserted at the same site in a **tandem linear**

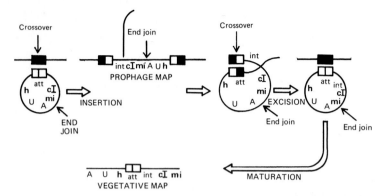

Fig. 46-10. Campbell model for prophage integration explaining the permutation of the vegetative and prophage maps of λ DNA. Both the vegetative and the prophage maps can be derived from the same circle by opening it at different points: end join for vegetative multiplication and maturation, and *att* for prophage insertion. The differences between the two maps is equivalent to shifting the block of markers (int-R) from one end of the map to the other. For terminology, see Figure 46-4. In addition, h = host range mutation; mi = minute plaque mutants; heavy lines = bacterial DNA; thin lines = phage DNA; ■ = bacterial attachment site; ▭ = phage attachment site; ◨ and ◪ = prophage attachment sites resulting from recombination between bacterial and phage sites.

sequence (as shown by transduction). The mechanism of double lysogenization will be examined below.

MULTIPLICATION OF THE PROPHAGE

Whereas the vegetative multiplication of phage DNA is autonomous, i.e., is controlled by phage genes, the **replication of the prophage is regulated by the system that controls replication of the bacterial chromosome.** Thus, certain temperature-sensitive phage mutants cannot multiply vegetatively at 40° but are replicated at this temperature as prophages.

Presumably these mutants are defective, at the nonpermissive temperature, in some element of the phage replicon required for initiating a round of vegetative replication. Genetic evidence shows that the prophage doubles when the growing point of the replicating DNA reaches its integration site.

DEFECTIVE PROPHAGES

Many mutations of the prophage in genes for vegetative functions permit its perpetuation but prevent synthesis of infectious phage. Such **defective prophages** can be recognized,

Fig. 46-11. Evidence for linear insertion of the λ prophage in the bacterial chromosome, and detailed prophage map. Prophage and bacterial genes were mapped by using deletions entering the prophage from either side by taking advantage of the two *chl* (chlorate resistance) genes, A and D. To determine which prophage markers had also been deleted, the cells were induced and then superinfected with λ phage carrying alleles of all the markers; the appearance (or nonappearance) of various prophage markers in the progeny phage indicated whether or not they were present in the partly deleted prophage. The phage gene order corresponds to that of the vegetative map (Fig. 46-4) except for the permutation. Phage DNA = thin line; bacterial DNA = heavy lines; *ara* = arabinose utilization; *gal* = galactose utilization; *blu* = stained blue by iodine; *bio* = biotin synthesis; *uvr* = ultraviolet light resistance. [Data from Adhya, S., Clearly, P., and Campbell, A. *Proc Natl Acad Sci USA* 61:956 (1968).]

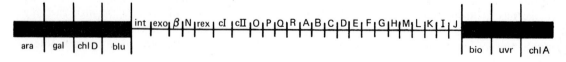

when genetically marked, by inducing the cells (to break down immunity) and then super-infecting with a **helper phage** (usually the homologous wild type) which can comple-ment the missing function of the defective prophage. The progeny then includes both wild-type and mutant virions.

Defective prophages are extremely useful tools for investigation, much like conditionally lethal mutants; they can be propagated as prophages, and the effects of their defect on vegetative multiplication can be observed after induction.

Whereas most known defective prophages contain the immunity system, some **extremely defective prophages** of λ have lost both the immunity genes and genes responsible for detachment; hence these prophages persist but the cells are not immune. These prophages are a useful model for understand-ing the consequence of infection of animal cells with certain tumorigenic DNA-containing viruses (Ch. 63).

EXCISION

Excision of the prophage at induction is ex-plained in the Campbell model as a reversal of integration, i.e., a **reciprocal crossover between the attachment sites at the two ends of a prophage.** The product would be a cyclic phage DNA molecule and an intact bacterial chromosome (Fig. 46-10). Two predictions of this model have been experimentally veri-fied. Excision **without replication** has been demonstrated by isotopic experiments; and the intactness of the bacterial DNA after excision is borne out by the "curing" experi-ments reviewed below.

MOLECULAR MECHANISM OF PROPHAGE INTEGRATION

The Campbell model is formally symmetrical since excision can be regarded as exactly the reverse of insertion. The detailed elucidation of its mechanism, however, has shown that it is **asymmetrical** in respect to both interact-ing sites and enzymes. The asymmetry ensures that either operation is performed only when needed.

ENZYMATIC REQUIREMENTS

Insertion.   The requirement for a special enzyme is evident in experiments with a par-

tially diploid strain of **E. coli** carrying two identical λ attachment sites. Initially λ inte-grates at either site with an equal and high probability. However, after the first prophage has established immunity a new λ infection only rarely causes integration at the un-occupied site. In contrast, if the infecting phage is heteroimmune but has the same attachment site (e.g., λimm⁴³⁴, Fig. 46-7), it again integrates at the free site with high probability. Hence, **insertion requires a product specified by a repressible gene** of the infecting phage. The gene is *int* (Fig. 46-4) because *int⁻* mutants do not integrate.

The *int* product is specific for insertion at an attachment site. Thus, when λimm⁴³⁴ lysogenizes a λ lysogenic cell it always inte-grates at one of the two prophage attachment sites, generating a tandem double prophage. In contrast, when λ rarely lysogenizes a lysogen (*int* function is not expressed) it integrates within the existing prophage (i.e., not at an attachment site), using the bacterial *rec* function, as in any bacterial recombina-tion based on extensive DNA homology. (The *rec* system normally plays a secondary role in integration because it causes much less frequent crossover, per unit length of DNA, than the phage *int* system.)

Excision.   Excision requires another viral function in addition to *int*, **since some mutants integrate but do not excise.** The additional requirement is the function of gene *xis,* ad-jacent to *int*. Thus, excision requires two functions, *int* and *xis*. The *int* product may be the basic recombination enzyme for both insertion and excision, and the *xis* product would then make it to conform to the different site configuration during excision (see below).

STERIC REQUIREMENTS

The basic result is that bacterial, phage, and recombinant attachment sites are not equiva-lent. Rather, the probability of *int* (or *int-xis*) promoted recombination between a pair of sites depends on the *arrangements of half sites* in each of them; certain arrangements are more suitable for *int* alone (i.e., for insertion), others for *int-xis* (i.e., for excision).

The normal process of insertion and ex-cision can then be represented by the equa-tion:

$$\text{P.P}' \; + \; \text{B.B}' \; \underset{\textit{int-xis}}{\overset{\textit{int}}{\rightleftharpoons}} \; \text{B.P}' \; + \; \text{B}'.\text{P}$$

<table>
<tr><td>phage
site</td><td>bacterial
site</td><td>left</td><td>right</td></tr>
<tr><td></td><td></td><td colspan="2">prophage sites</td></tr>
</table>

where P and P′ represent phage half sites, B and B′ bacterial half sites, and the dot the crossover point.

These results are based on studies of both recombination and lysogenization. In fact, *int* and *xis* functions can also cause **vegetative recombination, but limited at attachment sites.** This can be recognized in cells infected with two phages containing recombinant (i.e., half viral, half bacterial) sites such as the transducing phages of Figure 46-18. If these crosses are carried out in the absence of generalized recombination (i.e., in *rec*− bacteria using *red*− phage strains) recombination is still observed between markers on opposite sides of the attachment site; the *int-xis* system is responsible. The recombination frequencies vary markedly depending both on the half-site combination and on the presence or absence of *xis* function (Fig. 46-12). Whereas *int*-mediated recombination recognizes a broad spectrum of sites, that mediated by *int-xis* only occurs between the two sites normally flanking the prophage, as in excision.

The requirement for suitable half-site arrangement in lysogenization is seen in experiments with λdgal (Fig. 46-18). This transducing phage will integrate at high frequency only at the left end of a λimm434 prophage, whereas regular λ will integrate equally well at either end; and λimm434 will integrate well only at the right side of prophage λdgal.

The *int-xis*-promoted recombination is of a very **special kind,** since the precise point where the crossover occurs seems to be determined by the enzyme rather than by DNA homology. The absence of much homology is shown by the rarity of *rec*- and *red*-promoted recombination at the attachment sites, using *int*− phages, and by electron microscopy of DNA heteroduplexes (Ch. 10, Genetic mapping). Thus the homologous segment, if present, extends to no more than 20 nucleotides. Moreover, *int-xis* recombination, in contrast to generalized recombination, is symmetrical (i.e., each event yields two symmetrical recombinant types), even in vegetative crosses.

CURING OF A LYSOGENIC BACTERIUM

Though excision of the prophage is usually followed by phage multiplication and cell lysis, this multipli-

Fig. 46-12. Role of sites and *xis* product in *int*-mediated recombination. Vegetative crosses were made with the phages of Figure 46-18 using *red*− phages and *rec*− *E. coli* to abolish generalized recombination; the phages carried suitable markers on opposite sides of the attachment site. The values given are per cent recombination between attachment sites. The first row gives the results obtained with both *int* and *xis* products present, and the second row shows the effect of eliminating the *xis* product by using a parallel pair of *xis*-phages. The integrative (col. 1) sites of recombination (BB′ × PP′) allow high frequency of recombination with or without *xis*. The excising (col. 2) sites (PB′ × BP′) require the *xis* product for a high frequency. The frequencies given by abnormal combinations vary not only with the kinds of half sites induced, but also with their arrangements (compare cols. 3 and 4). [Data from Echols, H. *J Mol Biol 47*:575 (1970).]

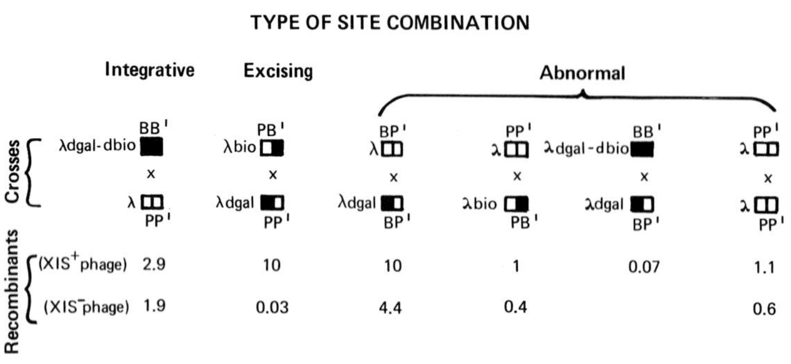

TYPE OF SITE COMBINATION

	Integrative	Excising	Abnormal			
Crosses	BB′ λdgal-dbio ■ × λ □□ PP′	PB′ λbio □■ × λdgal ■□ PP′	BP′ λ □□ × λdgal ■□ BP′	PP′ λ □□ × λbio □■ PB′	BB′ λdgal-dbio ■ × λdgal ■□ BP′	PP′ λ □□ × λ □□ PP′
Recombinants (XIS+ phage)	2.9	10	10	1	0.07	1.1
(XIS− phage)	1.9	0.03	4.4	0.4		0.6

cation can sometimes be prevented: the cell then survives and gives rise to a normal, sensitive clone. The excision is in perfect register with the insertion: thus cells rendered proline-deficient by insertion of prophage within a *pro* gene return to prototrophy after curing. This result confirms Campbell's model for prophage excision.

Curing is obtained by infecting a lysogenic culture with a heteroimmune phage with the same site specificity (e.g., λ lysogens infected with λimm⁴³⁴). The prophage becomes excised by the *int-xis* product of the unrepressed superinfecting phage, and it remains abortive because its own repressor is still present. Cells in which the superinfecting phage establishes its own repression system will then segregate cured daughter cells.

INDUCTION OF A LYSOGENIC CELL

The transition of prophage to vegetative phage represents a **failure of repression:** it can occur either spontaneously or in response to an inducing stimulus, as already indicated. The first sign is **loss of immunity and initiation of synthesis of phage mRNA outside the immunity operon.** Viral DNA and capsid protein are subsequently synthesized.

Spontaneous induction may occur in rare cells as a result of a statistical fluctuation of the interaction among the various regulatory elements discussed above. **Artificial induction** may occur in a large proportion of the cells if external conditions impair either the activity or the synthesis of the repressor.

Many prophages are induced by ultraviolet (UV) light, in doses too small to inactivate the phage but sufficient temporarily to halt cellular DNA synthesis. Other prophages are poorly inducible by UV light but can be induced by a variety of other means that inhibit cellular DNA synthesis, e.g., temporary thymine starvation, X-rays, alkylating agents (including the antibiotic mitomycin), and some carcinogens. Some prophages cannot be induced at all; they exhibit spontaneous induction but at a lower frequency than the inducible phages.

MECHANISM OF INDUCTION

Induction is achieved in three different ways: 1) the repressor is inactivated (e.g., by heat); 2) the prophage is transferred to a repressor-free environment (e.g., by bacterial conjuga-

tion); 3) the action of the repressor is circumvented by providing the prophage with all products required for prophage detachment and vegetative multiplication.

Repressor Inactivation. Heating inactivates the **thermosensitive repressor** produced by some cI mutants. With some mutations inactivation is irreversible, reversible with others. When it is reversible induction requires a longer period of heating, during which protein synthesis must go on in order to allow the buildup of gene products (probably *cro*) able to block the promoter for repressor maintenance; otherwise repressor synthesis is resumed (owing to positive feedback after repressor renaturation) when the temperature is lowered.

The induction of prophage **with a normal cI gene,** e.g., by UV light, is a more complex process. The UV treatment causes the accumulation of an **active inducer,** which prevents the binding of the repressor to the phage operons. In fact, the conjugation of a UV-irradiated nonlysogenic cell to a nonirradiated lysogenic cell causes the latter to be induced, owing to transfer of the active inducer. Since for 20 to 30 minutes after irradiation induction can be reversed by photoreactivation (see Ch. 11), the active inducer is either slow-acting irradiated DNA or a slowly accumulating product of DNA irradiation. The active inducer seems to be an adenine derivative made in normal metabolism, which accumulates after DNA synthesis is inhibited, and inactivates the repressor by binding to it.

This conclusion was suggested by several observations. 1) The inducer causes the irreversible inactivation of a reversible thermosensitive repressor; hence the inducer **binds to the repressor and changes its conformation.** The binding site is probably that altered by the *ind* mutation, which makes the repressor insensitive to UV light. 2) The inducer is probably an **intermediate in the metabolism of adenine** since, under certain conditions, the presence of this nucleoside is required for the thermal induction of the prophage. 3) It is **generated by a cellular pathway,** since heating of λ-lysogenic *E. coli* carrying a temperature-sensitive bacterial mutation affecting DNA synthesis induces the wild-type prophage; adenine facilitates this induction, whereas guanosine or cytidine inhibits it.

Once the repressor is inactivated, the *cro* product blocks the promoter for repressor maintenance, making induction irreversible.

Transfer of the Prophage to a Repressor-free environment. When a lysogenic Hfr cell conjugates with a nonlysogenic F^- cell induction occurs as soon as the prophage is introduced into the nonrepressive F^- cytoplasm **(zygotic induction).** There is no induction, however, if the recipient cell is lysogenic for the same phage.

Circumventing the Repressor. Virulent λ is insensitive to repressor, owing to mutations in both the left and the right operators, and it does not produce repressor, owing to changes in the promoter for repressor maintenance. When this phage infects λ-lysogenic cells it produces all gene products required for prophage excision and vegetative growth. The prophage is therefore induced despite the presence of its own repressor.

EFFECT OF PROPHAGE ON HOST FUNCTIONS

Except for transducing phages (see below), which carry genes known to be derived from a recent bacterial host, most prophages exert no discernible effect on the bacterial phenotype other than immunity to superinfection. Certain prophages, however, change the cell's phenotype, either by expressing new functions **(conversion)** or by modifying quantitatively the expression of adjacent bacterial genes.

In conversion the new functions are viral, since they are abolished by phage mutations. Examples are the change of surface antigens in *Salmonella typhimurium* (which has been carefully analyzed and is discussed in Ch. 29) and the formation of diphtheria toxin by *Corynebacterium diphtheriae* (discussed in Ch. 24). These effects are expressed by both vegetative phage and prophage, but some converting functions are expressed only by the prophage; e.g., the resistance of *E. coli* K_{12} (λ) to T-even phages with an r_{II} mutation (Ch. 45) which is caused by the *rex* gene in the immunity operon.

Prophages often interfere with the regulation of neighboring bacterial genes. The insertion of λ prophage, for instance, increases the repression of the adjacent *gal* operon. However, after the prophage is induced by UV irradiation, there is a burst of enzyme synthesis before the cells lyse. This effect is correlated with local DNA replication initiated within the prophage before excision, which continues toward the left into the *gal* operon. Since the replication of *gal* operator genes is not accompanied by a corresponding increase in the synthesis of *gal* repressor, the new copies of the operon escape repression **(escape synthesis).**

Another interesting effect is displayed by prophage Mu-1, which can be inserted at many sites on the host chromosome. In an *E. coli* culture infected by this prophage many different genes in different cells undergo mutation, i.e., become nonfunctional, in the course of the establishment of lysogenic complexes. The mutations are due to the insertion of the prophage within the gene.

SIGNIFICANCE OF LYSOGENY

Lysogeny indicates a close evolutionary relation between phages and bacteria. Lysogeny, in fact, depends on a remarkable congruence of attachment sites, between the DNA of the phage and that of the host cells; and most phages can lysogenize some bacterial strains. The virulent T coliphages, though long viewed as the prototype phages, are thus exceptional: they represent extreme evolution in the direction of autonomous viral development, leading not only to loss of homology with bacterial DNA, but also to large deviations of biochemical processes.

Prophages give rise to **infectious heredity** in their host cells, i.e., they contribute new genetic characteristics to the lysogenic cells. One way in which this can happen is conversion, in which the prophage introduces new characters that persist as long as the lysogenic state persists. This phenomenon raises the question of the origin of the converting genes that remain active in the prophage, i.e., whether viral or, as in transducing phages, cellular. In fact, converting genes are not essential for phage function, since phage mutations abolishing conversion do not impede either vegetative multiplication or lysogenization of the phage. They may, however, be advantageous in subtle ways: for instance, the conversion of surface antigens by *Salmonella* phage F_{15} prevents further adsorption of the phage, thus eliminating waste of phage on immune cells.

Infectious heredity is also seen with **highly defective prophages** that cannot be induced and do not confer immunity; these cannot be distinguished from segments of cellular genetic material, except by identifying their genes after they are recombined into a regular phage. Through evolutionary changes, these prophages may acquire genetic significance for the cell and thereby become bona fide cellular genes.

EXTRACHROMOSOMAL GENETIC ELEMENTS: PLASMIDS

Bacteria often contain, in addition to a large, circular chromosome, various small, autonomously replicating cyclic DNA molecules, called **plasmids.** Their study has thrown light on the origin of viruses and on the mechanism and ecology of gene transfer by conjugation.

Most plasmids were discovered by virtue of some recognizable function and are therefore also called **extrachromosomal factors.** Thus the **F factor** (the first fertility factor) was recognized because it generates bacterial recombination, the **R factors** because they transfer resistance to antibiotics, and the **bacteriocinogenic factors** (bacteriocinogens) because they synthesize bacterial killing substances (bacteriocins, see below). These diverse groups are now known to share the same basic genetic features, with some variations, and to incorporate in addition variable blocs of optional genetic material, which often affect the host phenotype.

Genes for antibiotic resistance have been most conspicuous among those spread by plasmids, not only because of the selective pressures for their spread but also because widespread tests in medical laboratories have favored their recognition. However, plasmids are important in the transmission of other traits. For example, the **versatility of soil pseudomonas** in attacking various carbon sources has been traced to genes carried on plasmids.

Physical and biological properties distinguish extrachromosomal factors into two main classes: **large** factors, including F and R and certain bacteriocinogens (such as col V and col I), of MW 60×10^6 to 140×10^6, corresponding to 100 to 200 genes; and **small** factors (including many bacteriocinogens), of MW 4×10^6 to 5×10^6 or about 15 genes. The latter class also includes **minicircular DNA,** without known function, of MW ca. 2×10^6. (Similar small circles have also been seen in extracts of animal cells.)

Plasmids share important properties with temperate phages and, indeed, can be generated by some phages. Thus phage P1 appears to lysogenize as a plasmid rather than as a prophage in the chromosome. Moreover, even phage λ can multiply as a plasmid following a mutation (in the N gene), or an extensive deletion, that prevents integration and most vegetative functions but continues to permit restricted replication (see above, Lysogeny).

Replication of plasmids is in various ways correlated to that of the host chromosome. Replication of large plasmids is essentially synchronous with that of the chromosome (perhaps by sharing a membrane site, see Ch. 10) and is blocked by some temperature-sensitive bacterial mutations that prevent host DNA replication. In contrast, replication of the small plasmids is not synchronous and is unaffected by the same mutation, but is blocked by mutations in DNA polymerase I (see Ch. 10), which do not affect large plasmids. Perhaps the use of polymerase I by small plasmids (using a small RNA primer; see Ch. 10) is a primitive method of replication, once used for host DNA.

Most plasmids (except the minicircular DNA) are **transmissible,** i.e., they can cause the formation of a **conjugation** apparatus for their own transfer to another cell. **Nontransmissible** plasmids, in contrast, do not form a conjugation apparatus (but they can be transmitted by transduction). The location of a gene in such a plasmid (e.g., the penicillinase gene in some staphylococci) is generally recognized by two criteria: absence of linkage to chromosomal genes in transduction, and elimination (which works with most but not all plasmids) by acridine "cure" (see Ch. 9).

CONJUGATION

The transmissible plasmids, which can generate a conjugation apparatus, are also called **sex factors.** The mechanism of conjugation has been analyzed most thoroughly with the F factor and has been described in detail in

Chapter 9. In this mechanism the plasmid induced the synthesis of a specific male pilus, which forms a fragile conjugation bridge with recipient cells. Most pili fall into one of two classes, F and I (for colicinogen I, see below), but less frequent classes (N, W, etc.) are being added. Each class is morphologically and antigenically distinct and also adsorbs a different set of male-specific phages (see Ch. 45).

Bridge formation initiates a special form of asymmetrical DNA replication, presumably involving the rolling-circle mechanism (Ch. 10): one strand of the factor is replicated and remains in the donor, while the other strand is transferred without replication across the bridge. The transferred strand is subsequently replicated in the recipient, yielding another closed circle or undergoing recombination with the chromosome.

A sex factor can transfer not only itself but also an additional genome (chromosomal or plasmid) with which it has recombined, as in the mutation of F+ to Hfr (Ch. 9). Most factors, however, are much less efficient than F in mobilizing the bacterial chromosome; and some are sex factors in a limited sense, since they transfer the chromosome with only a very low frequency.

Such low-frequency transfer of the chromosome does not involve demonstrable formation of a stable Hfr derivative with a fixed origin (Ch. 9). It is therefore not certain whether there is unstable integration or whether the apparatus for conjugation, though coded for by the plasmid, may rarely attach to a somewhat favorable sequence in the chromosomal DNA.

A transmissible plasmid must have not only a set of genes that permits and regulates its own replication, but also genes that specify and regulate the machinery of transfer. Complementation of transfer-negative mutations in the F factor has identified 12 genes (*tra*A, B, etc.) involved in this process, of which 7 are concerned with pilus formation. The sequence of these genes has been established by deletion mapping.

PLASMID REGULATION

Extrachromosomal factors, just like prophages, have regulatory components, which limit their multiplication and also other functions. Thus there is a rather constant number of copies of a given plasmid in a given host: *E. coli* can carry one or two copies of F or R, while *Proteus* can carry ten. As was seen in Chapter 10 (DNA replication), this regulation may derive in part from the occupation of discrete **sites on the cell** membrane. Such

sites are not necessarily linked to the sites of chromosome attachment at cell division, for large as well as small plasmids can segregate to chromosomeless minicells in certain mutant *E. coli* strains (see Ch. 6).

Regulation of pilus formation is especially important for transfer. Thus after an R factor has entered a cell it ordinarily exhibits **high frequency transfer** for a few generations, but then a **repressor** accumulates which severely restricts the frequency of pilus formation (and hence of transfer.) The transient high-frequency transfer in newly infected cells permits rapid spread of the factor in a newly infected population; while the subsequent repression may effect some metabolic economy and also protect from attack by male-specific phages.

Many R factors will repress not only their own formation of pili but also that of F pili in an F+ cell that they infect. Derepressed mutants of R factors, altered in a regulatory gene, can be isolated. Indeed, the high-fertility F agent is such a derepressed sex factor, found in nature. Repressive mechanisms are also seen in the ability of many plasmids to exclude entry (surface exclusion) and also to prevent replication (plasmid incompatibility) of a superinfecting related plasmid (and often also of certain phages). Moreover, UV irradiation can derepress bacteriocin formation by bacteriocinogens (see below).

Classification. Plasmids have been classified, and their **evolutionary relations** inferred, on the basis of several properties we have noted above: mutual incompatibility and surface exclusion, genetic complementation, and specificity of pili. A direct and powerful additional tool is heteroduplex electron microscopy (Ch. 10), which determines the extent and relative locations of the regions in which DNA from different sources can hybridize. This approach is particularly useful for plasmids because they are relatively small and carry variable optional material. The results have shown many common regions within the F group or within a group including colicinogen I, but little relation between the two groups.

EPISOMES

Many plasmids can integrate into the host chromosome: hence they can replicate either autonomously or as part of the chromosome. Units capable of shifting between these two modes of replication (including temperate phages) are called episomes. However, this is not an intrinsic property of a plasmid: the frequency of integration varies widely with the plasmid-cell combination, and the same plasmid may be episomal in one host and essentially nonepisomal in another.

Integration of Episomes. In conjugation integrated sex factors determine polarized transfer of the chromosome, with a fixed origin (Ch. 9). The mechanism of integration, as with prophage (Fig. 46-10), is a single reciprocal crossover between two circles of DNA; and detachment involves a similar crossover. The enzymatic mechanism is unknown but probably depends on fortuitous low homology. In contrast, episomes that have incorporated bacterial genes (F′ agents, see Ch. 9) integrate with high frequency at the site of homology determined by those genes (Fig. 46-13). This kind of integration employs the *rec* bacterial system and is rapidly reversible (Ch. 9).

Interspecific Transfer. The F agent can be transferred not only between *E. coli* K$_{12}$ cells but also to other strains of *E. coli* and to *Salmonella, Shigella,* and even *Proteus.* The low GC ratio of the last (38%), compared with 50 to 56% for the factor and for various Enterobacteriaceae, facilitates recognition of the factor as a small "satellite band" of DNA, with a higher density than the chromosomal DNA. Transfer of the agent and of chromosomal genes varies greatly in efficiency with the

host strain, with limits imposed by poor cell pairing, by repression of plasmid replication, by enzymatic destruction of foreign DNA, and by absence of homology with the chromosomes. For example, F is readily accepted by *Shigella* but does not integrate. Conversely, restrictive enzymes in *E. coli* strain B severely limit transfer of F alone but less of the Hfr chromosome from K$_{12}$ (whose size may overwhelm the enzymes).

RESISTANCE-TRANSFER FACTORS (R FACTORS)

A plasmid of unusual medical interest was discovered in 1959 in Japan as a consequence of widespread chemotherapy of bacillary dysentery (Ch. 29, Shigellosis), which favored the selection of resistant *Shigella* strains. Strains resistant to several or all of the four widely used drugs (sulfonamides, streptomycin, chloramphenicol, and tetracyclines) began to appear with high frequency. Moreover, nonpathogenic *E. coli* strains, and pathogenic *Salmonella,* were found with the same property. This unexpected development became understandable when the multiple resistance was found to be transferrable to

Fig. 46-13. Hypothetical explanation for the difference between F and F′ in insertion and in detachment. F synapses on the basis of its own weak homology for the bacterial DNA (a, b). The probability of insertion is very small, owing to the weakness of the interaction; but once it is inserted, the probability of detachment is also small, for the same reason. F′ synapses on the basis of the strong homology of its bacterial segment (c, d, e, f, g, h, i) for the host DNA. Insertion occurs with a high probability because the interaction is strong, and for the same reason detachment also occurs frequently. Heavy lines = DNA of bacterial origin; thin lines = episomal DNA.

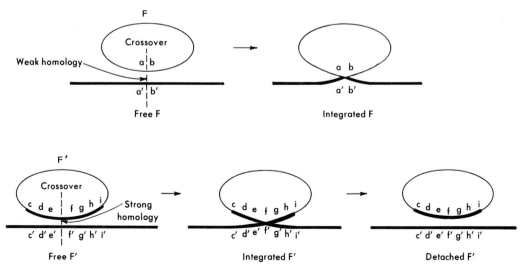

sensitive strains of a variety of Entobacteria-ceae by conjugation in mixed cultures, both in vitro and in the mammalian gastrointestinal tract.

The agents responsible, called R factors, rapidly became extremely widespread: a survey in Japan in 1965 showed their presence in 65% of all dysentery bacilli and 50% of all Enterobacteriaceae isolated from patients. It was not until some years after the initial discovery that such factors were finally sought in Western countries, and they were soon found to be not at all peculiar to the ecology of Japan. This development illustrates the spottiness of our knowledge of microbial distribution. The wide distribution of resistance-bearing plasmids forms a serious obstacle to therapy.

Classification. R factors have beeen extensively investigated in Japan by Watanabe and Mitsuhashi and in England by Anderson and the Meynells. They are heterogeneous not only in the pattern of drug resistance produced but also in several more stable properties. Thus certain strains induce production of sex pili that are similar, antigenically and in phage-attachment specificity, to those induced by the F factor. Moreover, the repressor mechanism that keeps down the frequency of formation of these pili also inhibits formation of F pili when the factor superinfects an F+ strain; hence these are called $fi+$ (or $fin+$) strains. Another family of R factors, though carrying similar sets of resistance genes, have no effect on F fertility and are called $fi-$. Most of these strains induce pili similar to those induced by colicinogenic factor I (see below).

Structure. R factors consist of two distinguishable parts: the basic **transfer factor** (RTF) responsible for conjugation, and a variable **resistance determinant** (R determi-

nant) containing genes for drug resistance (R genes);* the latter can be mapped by their linkage in transduction. In *E. coli* the two parts are usually found as one unit. However, in *Salmonella* and *Proteus* they are often found as separate plasmids (Fig. 46-14), which can be independently transduced and independently inactivated by acridine dyes; in extracts such plasmids can be separated on the basis of differences in density.

Though both plasmids in such a complementary pair are complete replicons, **R-determinant plasmids are not transferred unless they fuse with a transfer factor.** In nature some bacteria carry R determinants alone, and some carry a naked factor without resistance genes. Indeed, the latter factors are widespread in bacteria in nature, but because they do not transfer host genes efficiently they were discovered much later than the rarer F agent. They can be recognized in a *ménage à trois* experiment: in a mixture the transfer factor from one strain will infect a second strain that carries only an R determinant, and the resulting composite plasmid can then confer resistance on a sensitive third strain.

The resistance genes found in various combinations in R factors may have been extracted from the chromosome of various

---

* We are maintaining the original terminology of the Japanese investigators, in which R determinant refers to a part of a complete R factor, whether combined with the transfer factor or segregated. Unfortunately "resistance-transfer factor" has been used by some to denote a transfer factor capable of incorporating an R determinant, and by others to denote a complete factor already carrying the determinant.

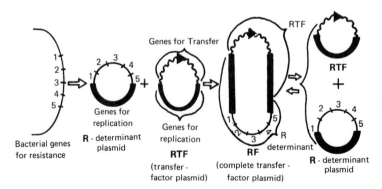

Fig. 46-14. Diagram of hypothetical formation of resistance factors and their observed occasional segregation of an RTF and a resistance-determinant plasmid. These two forms may be found in the same cell or in different cells; when in the same cell they can reversibly associate and dissociate. The plasmid is sometimes symbolized as ▲ for RTF, with various letters following for the specific R determinants.

bacteria one at a time by rare recombination (Fig. 46-14), as in the formation of F′ from the F agent. Indeed, the chloramphenicol-inactivating enzyme formed by an *E. coli* R factor resembles one found in *Proteus.*

Once a suitable R factor is formed it can spread rapidly, owing to its initial high-frequency transfer and to the enormous selective pressure in favor of the cells carrying it. Formation and spread of novel factors appears to go on all the time; for though factors with some R determinants have been isolated from bacteria stored since the preantibiotic era, determinants for an increasing number of drugs have appeared in the flora of patients as additional drugs (e.g., kanamycin, neomycin) have been introduced. R factors have been observed to carry other kinds of genes as well, e.g., genes for resistance to inhibition by various metal ions, for lactose fermentation, and H$_2$S production.

Staphylococcus Penicillinase. The frequent resistance of *Staphylococcus aureus* strains to penicillin is due to formation of penicillinase (β-lactamase) by a plasmid. However, unlike the enteric resistance plasmids, this class is nontransmissible except by transduction. Two groups, α and β, have been distinguished: they probably attach to different membrane sites, for a plasmid can superinfect cells carrying a member of the other group but not of its own group.

BACTERIOCINOGENS AND BACTERIOCINS

In 1925 Gratia in Belgium discovered that filtrates of a particular strain of *E. coli* inhibited the growth of another strain of the same species. The inhibitory substance was later shown to be lethal and was named **colicin.** Some 20 colicins were subsequently recognized and were labeled A to V. Each was specific for a limited variety of strains of several species of Enterobacteriaceae, and they could be distinguished by this host range (antimicrobial spectrum). Similar agents have been isolated since from strains of *Pseudomonas pyocyaneus* (pyocins) and *Bacillus megaterium* (megacins); each acts only on organisms closely related to the one that produces it. The group as a whole is now called **bacteriocins.** Their formation is widespread: 20% of a random group of enteric bacteria yielded colicins against a single test strain of *E. coli.*

The formation of bacteriocins is due to the presence of the corresponding **bacteriocinogen** in the cells. However, the bacteriocinogen is ordinarily repressed, and **most cells carrying it do not contain or produce bacteriocin.** The low level normally found in the cultures is probably produced by a small proportion of the cells, much like the free phage in a lysogenic culture. Furthermore, the cells of some bacteriocinogenic strains can be induced to form bacteriocin, by using the same agents that induce certain lysogenic strains to form infectious phage (e.g., light UV irradiation). The induced cells go on to die, and in at least some systems to lyse. Most bacteriocinogens, therefore, **behave like defective prophages,** which on induction cause the production of only one recognized protein, the bacteriocin.

Most bacteriocinogens induce no pili but may be transferred through the conjugation bridge induced by F or after recombination with F or R factors. Col I, however, induces pili and therefore transfers itself and other colicinogens; it also mediates a low-frequency transfer of the bacterial chromosome, but apparently without integration, because the colicinogen cannot be localized in reference to bacterial genes on the chromosome.

MECHANISM OF ACTION OF BACTERIOCINS

Bacteriocins differ from antibiotic (Ch. 23): they are proteins, have a much narrower antimicrobial spectrum, and are generally much more potent. Some colicins, when added to an excess of sensitive bacteria, kill a number of them that is directly proportional to the amount of colicin, showing that **one molecule is sufficient to kill a bacterium.**

Like phages (Ch. 45), colicins attach to specific cell-wall receptors, and a given colicin will cause the selection of **resistant bacterial mutants** which can no longer adsorb it. Indeed, **some bacteriocins resemble components of the phage tail** in the electron microscope (Fig. 46-15). Furthermore, certain colicins and certain phages (e.g., colicin K and phage T6) appear to share the same receptors, since mutants selected for resistance to either are resistant to the other.

Nomura showed that certain colicins **kill cells without penetrating them,** since their action can be stopped and reversed by trypsin, which destroys the cell-associated colicin. Several colicins (E1, K, A, I) stop all macromolecular syntheses by blocking oxi-

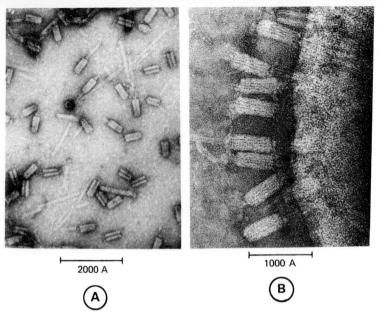

2000 A

A

1000 A

B

Fig. 46-15. Electron micrographs of a *Rhizobium* bacteriocin. **A.** The purified bacteriocin, which is similar to a phage tail, with a tube and a contractile sheath; the two components often separate during preparation; most sheaths are contracted. **B.** The bacteriocin perpendicularly adsorbed to sensitive cells, much as phage tails (see Fig. 45-3). [From Lotz, W., and Mayer, F. *J Virol* 9:160 (1972).]

dative phosphorylation, a function of the bacterial plasma membrane. E2 selectively inhibits DNA synthesis, apparently by dissociating the growing fork from the membrane. E3 stops protein synthesis by altering the 30S components of the ribosomes, whose 16S RNA loses a fragment of about 50 nucleotides from the 3′ end; apparently this colicin penetrates into the cells and interacts with the ribosomes, since the effect can be reproduced in cell-free extracts. The various colicin effects that are mediated through the cell membrane are of considerable interest for contemporary studies of membrane function: these effects are believed to result from allosteric changes of membrane subunits which, by a domino effect, spread from the point of attachment of the colicin to a large part of the cell membrane.

Bacterial mutations affecting the membrane subunits make the cell **tolerant** to the attached colicins, presumably by preventing the allosteric changes or the spreading effect. Also relatively insensitive to a colicin are the strains that produce it, which are unaffected by the levels at which it is ordinarily found in the cultures.

EXTRACHROMOSOMAL ELEMENTS AND THE EVOLUTION OF VIRUSES

The primordial viral DNA must be a replicon in order to replicate autonomously; it was therefore probably derived from the cell's chromosome by excision of a small cyclic plasmid containing the initiation site of the bacterial replicon, and continuing at first to use cellular enzymes for replication. The plasmid could then evolve, by gene duplication and alteration and by incorporating additional host DNA, into transfer factors and into viruses of increasing complexities. The extrachromosomal elements may be stages in this evolution that were frozen because they conferred specific advantages on the host cells, such as recombination (F factors), resistance to inhibitors (R factors), and production of bactericidal substances (bacteriocinogenic factors).

The production by some bacteriocinogens of structures similar to phage tails may imply evolution either from phage to bacteriocinogen or vice versa. It would not be surprising if phage evolution first developed structures for attachment to cells and cell killing, since the resultant selective advantage for the cells harboring the parasite would expand the opportunity for further evolution toward infectivity. It is conceivable, therefore, that viral evolution proceeded from simple plasmids through various forms similar to extrachromosomal factors. The minute plasmids found in cells might be at the initial stage of such evolution.

PHAGES AS TRANSDUCING AGENTS

As was pointed out in Chapter 9, temperate phages can mediate transfer of bacterial genetic material from one cell to another, and this process of **transduction** has been extensively used for mapping the bacterial chromosome. We shall consider here only the virological aspect of transduction.

Two types of transduction can be distinguished: generalized transduction can transfer any bacterial genes, and restricted (specialized) transduction can transfer genes from only a very small region of the host chromosome adjacent to the prophage site.

GENERALIZED TRANSDUCTION

This type of transduction, discovered by Zinder and Lederberg during a search for conjugation in *Salmonella,* is due to the encapsidation of cellular DNA in a phage coat.

The original discovery followed the observation that a cross of strain L22 and strain L2, each carrying a different pair of auxotrophic mutations, produced rare prototrophic recombinants. The process appeared to involve conjugation, since the filtrate of either strain did not transform the other. However, the filtrate of a mixed culture of L22 and L2 caused a fresh culture of L22 also to yield prototrophs. The filtrable agent was resistant to

DNase and was soon recognized to be a phage.

The events occurring in the mixed culture are complex (Fig. 46-16). Strain L22 proved to be lysogenic for a phage, which was named P22; the particles that are occasionally released infect L2 cells. Most of these cells lyse, releasing regular phage particles and also rare particles containing adventitious bits of L2 DNA. Some of the latter particles transfer wild-type alleles of genes for which the host is auxotrophic into L22 cells, and prototrophs are formed by recombination.

The unrecognized lysogenic character of L22 was instrumental in the discovery of transduction. It provided a source of phage to which L2 was sensitive, and the lysogenic L22 cells were immune to the phage released from L2 cells. Hence the L22 cells could accept the L2 DNA present in occasional transducing phage particles without being destroyed by the large excess of regular infectious phage particles.

DISTRIBUTION AND SCOPE

Transduction can apparently occur for any markers of the donor bacterium. However, **only closely linked markers can be cotransduced** by the same phage particle (Ch. 9), because the piece of bacterial DNA carried by a phage must fit inside the phage head and therefore is similar in size to the phage DNA (i.e., ca. 1 to 2% of the bacterial DNA). Transduction is usually carried out

Fig. 46-16. The discovery of transduction in mixtures of lysogenic P22 and sensitive L2 cells.

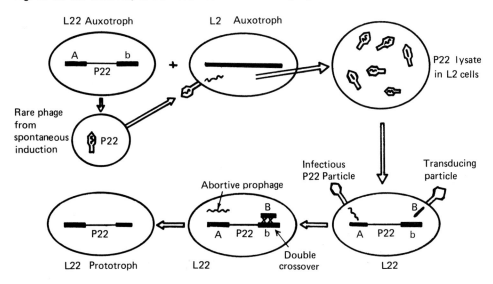

with a high-titer phage preparation obtained from the donor strain **either by lytic infection or by induction of lysogenic cells.**

Phage P1 is widely used in genetic studies of *Salmonella, E. coli,* and *Shigella.* Transduction has also been reported, with different phages, for genera as varied as *Pseudomonas, Staphylococcus, Bacillus,* and *Proteus.* With a recipient culture infected at multiplicity of about 5, the frequency of transduction of a given character ranges from 10^{-5} to 10^{-8} per cell. With *E. coli* the linkage groups obtained by transduction and by conjugation are in gratifying agreement.

MECHANISM OF GENERALIZED TRANSDUCTION: ROLE OF DEFECTIVE PHAGE

Transducing particles are ordinarily defective. This feature was not recognized for some time, for in order to promote a good yield of transductants, the recipient cells were infected at high multiplicity; hence those cells that received a transducing particle also received several infectious particles and thus usually became lysogenic. The study of phage structure, however, led eventually to the realization that infectious DNA cannot coexist with a large piece of foreign DNA within the same phage head. (For instance, P1 DNA, of MW ca. 60×10^6, sometimes transduces the entire prophage λ, of MW 30×10^6, together with adjacent *gal* markers.)

The defectiveness of the transducing particles was demonstrated by Luria by the simple device of using a low over-all multiplicity of infection; then none of the transduced cells were lysogenic. Indeed, the bulk of the transductants contained no detectable phage genes, suggesting that generalized transducing phages might lack phage components.

The composition of the transducing particles has been clarified by physical studies. Thus in equilibrium sedimentation in a CsCl gradient transducing and infectious particles form a common band; hence they have the same DNA/protein ratio and thus **the same amount of DNA** (since the amount of protein per particle does not seem to vary). In a more decisive experiment cells were first grown in a medium containing 5-bromouracil (which makes DNA heavy) and nonradioactive phosphate, then transferred to a medium containing thymine and $^{32}PO_4$, and immediately infected with a virulent mutant of P1 (which stops the synthesis of bacterial

DNA). The transducing particles in the lysate formed a denser band than the other particles; moreover, their DNA was nonradioactive and it could hybridize with cellular DNA. It follows that these transducing particles contain **only cellular DNA, synthesized before infection, and adventitiously enclosed** in the phage head, without any phage DNA. The incorporation of the host DNA in the phage capsid probably occurs by the same mechanism that encapsidates T-even phage DNA (see Ch. 45).

The transducing particles isolated in the above experiment contain a **single DNA piece of rather uniform length** corresponding (in P1) to about 2% of the *E. coli* chromosome. This finding explains why generalized transduction is well suited for genetic mapping.

ABORTIVE TRANSDUCTION (FIG. 46-17)

Some years after the discovery of transduction it was found that the introduced fragment of bacteria DNA (the **exogenote**) does not always undergo either partial insertion (by recombination) or destruction. In a third choice, abortive transformation, the exogenote persists and expresses the function of its genes; but since it is not a complete replicon, it cannot multiply and is transmitted unilinearly, i.e., from a cell to only one of its two daughters. Abortive transduction is easily recognized when a gene of the exogenote codes for an enzyme required for growth on a selective medium, for though the amount is restricted by the unilinear inheritance (Fig. 46-17), it is still sufficient so that abortive transductants for a prototrophic gene can make **microcolonies** on minimal medium, which are detectable with magnification.

The proportion of microcolonies to large, prototrophic colonies unexpectedly revealed that **abortive transduction is several times as frequent as the corresponding complete transduction.** Hence the exogenote that recombines with the bacterial chromosome probably has some special, as yet unidentified, characteristic.

Abortive transduction is a useful tool in **complementation tests** for determining whether two auxotrophic mutations affecting the same character are allelic (Ch. 11). One of the mutants is used as donor and the other as recipient in a transduction experiment; and the cells are plated on minimal medium. Large prototrophic colonies (complete transduction) are formed by recombination between two

Fig. 46-17. Complete and abortive transduction. In complete transduction part of the fragment of bacterial DNA (exogenote) introduced by the transducing phage (thin line) becomes inserted in the DNA of the recipient bacterium (heavy line) and continues to replicate with it. In abortive transduction, the exogenote is not integrated; it neither replicates nor is destroyed and so it is transmitted unilinearly (heavy arrows). A gene contained in the exogenote makes a cytoplasmic product (indicated by x), which is transmitted to both daughter cells. However, in those that do not receive the exogenote in abortive transduction it is progressively diluted out as cell multiplication proceeds.

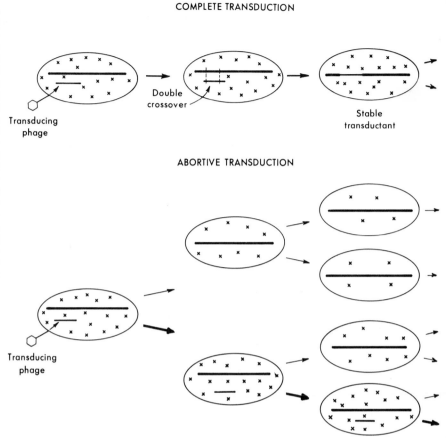

COMPLETE TRANSDUCTION

Transducing phage — Double crossover — Stable transductant

ABORTIVE TRANSDUCTION

Transducing phage

mutations whether or not they are allelic; but microcolonies are formed (abortive transduction) only if the two mutations are not allelic, i.e., if the mutation of the exogenate complements that of the recipient cell (see, however, Ch. 12, Intragenic complementation).

SPECIALIZED TRANSDUCTION

LOW-FREQUENCY TRANSDUCTION

The Lederbergs discovered that phage λ can give rise to transduction in quite a different manner. It transfers only a **restricted group of genes** (the *gal* or *bio* regions) which are located near the prophage (see Fig. 46-11); and it is generated **only on induction of prophage,** and not (in contrast to generalized transduction) in lytic infection. The transducing genes are incorporated into the phage genome by abnormal excision of the prophage (Fig. 46-18).

In a lysate from a normal lysogenic culture only a very small proportion of the virions are transducing, hence the lysate causes **low-frequency transduction** (LFT). Moreover, the **transducing virions differ from each other** in the precise DNA replacements, because they arose by independent recombination events.

Since the excised DNA may differ in length from regular viral DNA, for lack of a precise location of

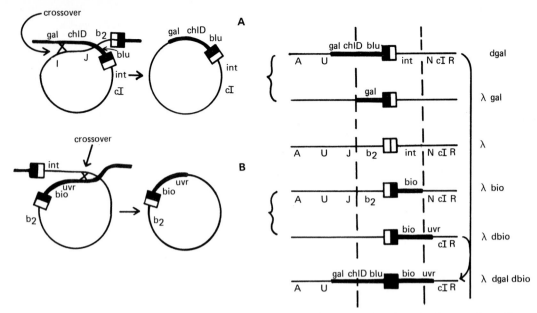

Fig. 46-18. Production of transducing λ derivatives by crossovers outside the attachment sites. Crossovers on the left of the prophage **(A)** generate molecules containing *gal* and other bacterial genes (λdgal) (where d stands for defective), which cannot replicate owing to loss of capsid genes. If the bacterium is deleted between *gal* and the prophage, *gal* genes can be incorporated in replacement of the unessential b₂ region, and the phage is not defective (λgal). Crossovers on the right **(B)** generate molecules containing *bio,* sometimes with other bacterial genes. *Bio* replaces recombination genes, and the phage can replicate (λ*bio*); if the replacement is longer (including N) the phage is defective (λdbio). λdgaldbio, which was useful for studying *int-xis*-promoted vegetative recombination, is obtained by recombination between λdgal and λdbio. Heavy lines = bacterial DNA; thin lines = phage DNA ■ = bacterial half site; ▫ phage half site. See Figure 46-11 for names of bacterial genes.

the excising crossover, and because the ends of λ DNA are produced in an invariant position, transducing particles tend to have **abnormal buoyant density**, which is determined by DNA/protein ratio. The density is usually lower, indicating a shorter DNA. (Longer molecules may also be excised but cannot be packaged in virions.)

HIGH-FREQUENCY TRANSDUCTION (FIG. 46–19)

Defective transducing particles (such as λdgal, λdbio) can replicate in mixed infection with regular λ as **helper** to complement the missing function. By taking advantage of this property, lysates greatly enriched in transducing particles can be produced. Thus in a *gal⁻* λ-sensitive culture exposed to an LFT lysate, a rare cell receiving the DNA of a λdgal⁺ particle, together with that of a regular λ particle, may be lysogenized by both λdgal and λ, forming a double lysogen λdgal−λ. (λdgal alone does not lysogenize because its attachment site is unsuitable for the bacterial site; it is suitable for one of the recombinant attachment sites flanking the prophage.) Such a cell is transformed into a **gal heterogenote,** i.e., carrying two *gal* genes, one the regular bacterial gene, the other in the prophage (indicated as *gal⁻*/λdgal⁺). [In contrast to heterozygous meroploids (Ch. 9), the two allelic genes are on the same chromosome.] By taking advantage of its *gal⁺* phenotype, the heterogenote can be isolated as a pure clone (e.g., by picking a greenish colony on EMB agar).

When the **doubly lysogenic** cells are induced the regular λ genome provides the functions missing in λdgal. The cells then yield a **high-frequency-transduction** (HFT) lysate, in which about half the particles are λdgal and half are λ. This lysate, on lysogenizing sensitive cells, will form new heterogenotes of the same kind, which again will form an HFT lysate on induction.

In contrast to the heterogeneity of the rare particles in LFT lysates, HFT lysates contain a **high proportion** of transducing particles, and these are all **identical** (except for mutations arising during the growth of the clone). Inde-

A. FORMATION OF HFT LYSATE

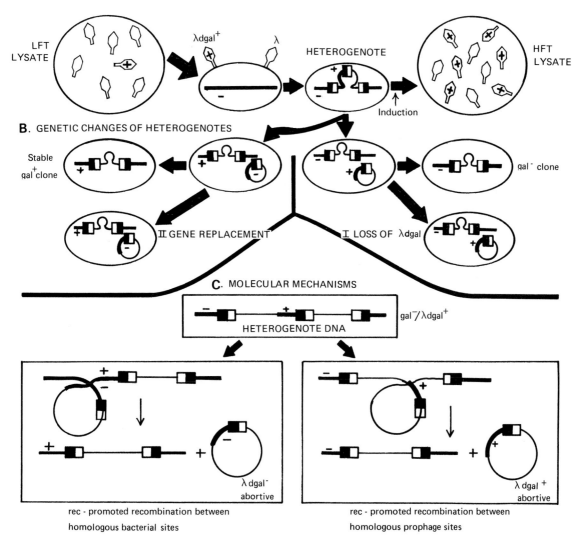

Fig. 46-19. A. Production of HFT lysate. Infection by a λdgal particle present in an LFT lysate causes the formation of a heterogenote, containing both the *gal*− gene preexisting in the bacterium and *gal*+ introduced by λdgal plus regular λ as helper. When the cells of a clone deriving from such a heterogenote are induced, they lyse and produce the HFT lysate. **B.** Occasionally the heterogenote segregates, by loss of the λdgal prophage, a cell with only the original *gal*− gene (I); even more rarely it segregates, by a new exchange, a nonlysogenic cell with the *gal*+ gene of the transducing prophage (II). The probable molecular mechanisms of these segregations are shown in **C.**

pendently derived HFT lysates usually contain transducing particles with different properties.

The properties of phages producing specialized or generalized transduction are given in Table 46-1.

Range of Specialized Transduction. The natural range is limited by the number of genes near the attachment sites of the available temperate phages. In addition to transduction by λ, φ80 is widely used to transduce *E. coli* genes for trytophan synthetase, T1 resistance, and tyrosine tRNA; P(1) for *lac* genes (see below); and P22 in *Salmonella typhimurium* for genes for proline synthesis. However, Beckwith has shown that the range can be artificially enlarged to cover essentially

TABLE 46-1. Comparison of Transducing Phages

	Specialized	Generalized
	Example: *E. coli* K$_{12}$ phage λdgal	Example: *Salmonella* phage P22
Genes transduced	Only gal	Any selectable marker
Localization of prophage	Locus closely linked to gal	Unknown
Source of transducing phage	Induction only	Induction or lytic infection
Capacity of transducing particles to produce infectious viral progeny	Defective (multiply with helper)	No multiplication
Characteristics of the clones of transduced cells	Unstable heterogenotes, segregating stable haploids	Stable haploids
Efficiency of transduction per phage particle	LFT 10^{-6} (from haploid) to HFT 10^{-1} (from heterogenote)	10^{-5} to 10^{-6}

the whole bacterial genome, by obtaining the transducing phage from cells in which an F' episome carrying the desired genes is integrated near the phage-attachment site (Fig. 46-20).

GENETIC CHANGES IN HETEROGENOTES

gal heterogenotes obtained by λdgal transduction show marked segregational instability. About 1 in 10^3 cells lose λdgal, yielding completely stable haploid cells with the old *gal* genes (i.e., those present in the original sensitive cells before their lysogenization by λdgal). More rarely stable bacterial strains segregate with the new *gal* genes (i.e., those present in λdgal). The stability shows that the new genes are no longer associated with λdgal; and the resulting clone can no longer produce an HFT lysate.

These changes probably involve excision by reciprocal crossover between homologous

Fig. 46-20. Preparation of transducing φ80 carrying gene X not normally adjacent to its attachment site. An F' including X (F-X) is first prepared in *E. coli* (see Ch. 9). Rare cells in which the F-X integrates in a gene required for sensitivity to bacteriophage T1 becomes T1-resistant and can be isolated by killing off the other cells with the phage. The cells are lysogenized by φ80, which attaches close to the T1-sensitivity gene; upon induction they yield an LFT lysate with X-transducing particles. [Procedure from Gottesman, S., and Beckwith, J. R. *J Mol Biol* 44:117 (1969).]

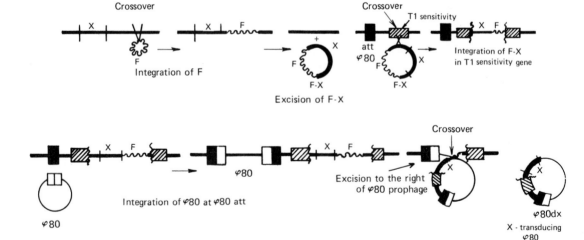

regions of either phage or bacterial regions, followed by loss of an abortive prophage, as shown in Figure 46-19.

Nondefective Transducing Phages. Though most transducing phages are defective, they are plaque forming in some HFT lysates (Fig. 46-18). Examples are λgal (obtained from bacteria with a blu deletion, which puts the *gal* genes immediately adjacent to the prophage), λbio, and φ80 trypt. All these phages have deletions of genes unessential for replication either in b₂ or the recombination region.

Specialized Transduction by a Generalized Transducing Phage. Though phage P1 carries out generalized transduction, as quoted above, it also rarely generates specialized transducing particles carrying *lac* genes, called P1dlac. On infecting a *lac⁻* point mutant of *E. coli,* P1dlac undergoes genetic exchange in the *lac* region, yielding nonlysogenic, stable *lac+* transductants, typical of generalized transduction. In contrast, when these particles infect a cell that has no *lac* region (e.g., *E. coli* mutants with a *lac* deletion, or the *lac⁻* species *Shigella dysenteriae*), they form *lac⁻*/P1dlac+ heterogenotes (i.e., cells that yield HFT lysates for P1dlac and show frequent segregation of *lac⁻* cells). Apparently when the strong homology provided by the *lac* region is absent, the P1 DNA is inserted by a single crossover as in λ lysogeny. However, the *int*-like system is weak in P1, and when the *lac* homology is present it provides the favored site of synapsis and recombination by a double crossover. P1dlac thus resembles F'-lac (see above, Integration of episomes) in **interacting differently with recipients that do or do not possess a lac region.** With transducing λ this effect is not noticeable because *int*-promoted integration is much more frequent than recombination at the *lac* homology.

The nature of the transductants formed thus depends not only on the bacterial genes present in the transducing particles but also on their degree of homology with host genes and on properties of the prophage.

Control of Bacterial Genes Integrated in Phage DNA. These genes often escape normal regulation. If they become disconnected from their operator they become regulated by the phage repressor. Moreover, those remaining connected to their operator may also lose the normal control; **transduction escape synthesis** of their products (in the absence of the regular inducer) is then seen, either when the transducing DNA multiplies after prophage induction or in infection at high multiplicity. One mechanism is transcriptional readthrough from phage to bacterial DNA; another, as in escape synthesis of genes adjacent to an induced prophage (see above), is exhaustion of repressor when its concentration (estimated at about 15 molecules per cell for *lac*) is exceeded by that of the corresponding operators.

USES OF SPECIALIZED TRANSDUCTION

Transducing phages have played an important role in the study of bacterial gene regulation and phage genetics. Thus the increased concentration of the products of the transduced genes in cells containing multiple copies of transducing DNA resulting from multiplication of a transducing phage helps considerably in the purification and characterization of these products (e.g., the *lac* repressor or suppressor tRNAs). The enrichment of the bacterial genes in an HFT lysate helps the study of their sequences and their interactions (e.g., of *lac* operator with repressor, Ch. 13). The different lengths of the phage deletions and the bacterial replacements in different HFT phages constitute a powerful tool for studying the organization of the genetic material, both viral and bacterial.

RANGE OF TRANSDUCING PHAGES

Transducing phages vary widely in defectiveness. Some, as noted, can multiply vegetatively; others are defective to various degrees, since they have lost a variable number of phage markers. However, they produce immunity and, in the presence of helper, lysogenize and give rise to a viral progeny. Particles of P1dlac have been observed with even a wider range of defectiveness; some can lysogenize and can cause lysis on induction, but without yielding infectious particles; some do not cause lysis; some do not cause immunity; and some give rise to prophages that can no longer detach but are recognizable by rescuing their genes by recombination.

The most defective of these specialized transducing particles may be similar in structure to those causing generalized transduction. Thus there is an almost uninterrupted spectrum between DNA molecules with completely viral specificity and those with exclusively bacterial specificity, without a clear demarcation between the two. When much of the DNA is viral the phage DNA tends to interact with the bacterial chromosome at a site specified by the homology between bacterial and viral DNA; when most of the DNA is bacterial the interaction occurs at sites specified by the homology of the transduced bacterial DNA and the DNA present in the cell. The demarcation between "viral" and "bacterial" properties is further confused by the converting phages, which have bacterial functions although they are not known to carry bacterial genes, and by the very similar infectious transducing phages, which carry genes of known bacterial origin. Finally, in abortive transduction a bacterial fragment remains nonreplicating and nonintegrating in the cell.

The distinction between phage genes and bacterial genes is thus blurred, in respect to both function (as in conversion) and heredity (as in defective lysogenization). It is conceivable that much of the heredity of bacteria is of viral origin, since many unknown defective proviruses may exist in nature; and on the other hand, phages may be fragments of bacterial DNA that have acquired the capacity for independent reproduction. Indeed, with a history of mutual interaction of viruses and bacteria in the course of evolution, the question of sharply distinguishing their genes may be meaningless except for recent, observed exchanges.

SELECTED REFERENCES

Books and Review Articles

ECHOLS, H. Lysogeny: Viral repression and site-specific recombination. *Annu Rev Biochem 40:*827 (1971).

GARRO, A. J., and MARMUR, J. Defective bacteriophages. *J Cell Physiol 76:*253 (1970).

HERSHEY, A. D. (ed.). *The Bacteriophage Lambda.* Cold Spring Harbor Laboratory, Cold Spring Harbor, N.Y., 1971.

LWOFF, A. Lysogeny, *Bacteriol Rev 17:*269 (1953).

NOVICK, R. P. Extrachromosomal inheritance in bacteria. *Bacteriol Rev 33:*210 (1969).

WILLETTS, N. The genetics of transmissible plasmids. *Annu Rev Genet 6:*257 (1972).

WOLSTENHOLME, G. E. W., and O'CONNOR, M. (eds.). *Bacterial Episomes and Plasmids* (Ciba Symposium). Churchill, London, 1969.

Specific Articles

BERTANI, G., and SIX, E. Inheritance of prophage P2 in bacterial crosses. *Virology 6:*357 (1958).

BUCHWALD, M., MURIALDO, H., and SIMINOVITCH, L. The morphogenesis of bacteriophage lambda. II. Identification of the principal structural proteins. *Virology 42:*390 (1970).

CAMPBELL, A. Segregants from lysogenic heterogenotes carrying recombinant λ prophages. *Virology 20:*344 (1963).

EISEN, H., BRACHET, P., PEREIRA DASILVA, L., and JACOB, F. Regulation of repressor expression in λ. *Proc Nat Acad Sci USA 66:*855 (1970).

GUERRINI, F. On the asymmetry of integration sites. *J Mol Biol 46:*523 (1969).

HEINNEMANN, S. F., and SPIEGELMAN, W. G. Control of transcription of the repressor gene in bacteriophage lambda. *Proc Nat Acad Sci USA 67:*1122 (1970).

HERSHEY, A. D., BURGI, E., and INGRAHAM, L. Cohesion of DNA molecules isolated from phage lambda. *Proc Nat Acad Sci USA 49:*748 (1963).

HERSKOWITZ, I., and SIGNER, E. R. A site essential for expression of all late genes in bacteriophage λ. *J Mol Biol 47:*545 (1970).

HONG, J. S., SMITH, G. R., and AMES, B. N. Adenosine 3:5'-cyclic monophosphate concentration in the bacterial host regulates the viral decision between lyosgeny and lysis. *Proc Nat Acad Sci USA 68:* 2258 (1971).

IMAE, Y., and FUKASAWA, T. Regional replication of the bacterial chromosome induced by derepression of prophage lambda. *J Mol Biol 54:*585 (1970).

KAISER, A. D., and MATSUDA, T. Specificity in curing by heteroimmune superinfection. *Virology 40:*522 (1970).

KOURILSKY, P., MARCAUD, L., SHELDRICK, P., LUZZATI, D., and GROS, F. Studies on the messenger RNA of bacteriophage. I. Various species synthesized early after induction of the prophage. *Proc Nat Acad Sci USA 61:*1013 (1968).

LWOFF, A., SIMINOVICH, L., and KJELDGAARD, N. Induction de la production de bactériophages chez une bactérie lysogène. *Ann Inst Pasteur 79:*815 (1950).

MORSE, M. L., LEDERBERG, E. M., and LEDERBERG, J. Transduction in *Escherichia coli* K$_{12}$. *Genetics 41:*142 (1956).

PTASHNE, M. The detachment and maturation of conserved lambda prophage DNA. *J Mol Biol 11:*90 (1965).

PTASHNE, M. Phage repressors. In *Strategy of the Viral Genome,* p. 141. (G. E. W. Wolstenholme and M. O'Connor, eds.) Churchill, London, 1971.

REICHARDT, L., and KAISER, A. D. Control of λ repressor synthesis. *Proc Nat Acad Sci USA 68:*2185 (1971).

SCHNÖS, M., and INMAN, R. B. Position of branch points in replicating λ DNA. *J Mol Biol 51:*61 (1970).

SCHWARTZ, M. On the function of the N cistron in phage lambda. *Virology 40:*23 (1970).

SIGNER, E. R. Plasmid formation: A new mode of lysogeny in phage λ. *Nature 223:*158 (1969).

SIGNER, E. R., WEIL, J., and KIMBALL, P. C. Recombination in bacteriophage λ. III. Studies on the nature of the prophage attachment regions. *J Mol Biol 46:*543 (1969).

TAYLOR, A. L. Bacteriophage-induced mutation in *Escherichia coli. Proc Nat Acad Sci USA 50:*1043 (1963).

THOMAS, R. Control of development in temperate bacteriophages. I. Induction of prophage genes following heteroimmune superfection. *J Mol Biol 22:*79 (1966).

WEIL, J. Reciprocal and non-reciprocal recombination in bacteriophage λ. *J Mol Biol 43:*351 (1969).

YARMOLINSKY, M. B., and WEISMEYER, H. Regulation by coliphage lambda of the expression of the capacity to synthesize a sequence of host enzymes. *Proc Nat Acad Sci USA 46:*1626 (1960).

chapter 47

ANIMAL CELL CULTURES

CHARACTERIZATION OF CULTURES OF ANIMAL CELLS   1122
  Transformed Cells   1123
  Storage   1124
  Cloning of Animal Cells   1124
BEHAVIOR OF CELLS IN CULTURE   1125
  Relation of Cells to a Solid Support   1125
CELLULAR GROWTH   1126
  Growth Factors and Inhibitors   1126
  The Cell Growth Cycle   1126
INTERACTION BETWEEN CELLS IN A CULTURE   1127
STRUCTURE OF CELLULAR MEMBRANES   1128
ORGANIZATION AND EXPRESSION OF THE GENETIC
      INFORMATION   1129
  Chromatin   1129
  Transcription   1130
  Protein Synthesis and Breakdown   1133
REGULATION   1133
  Regulation of Gene Expression   1133
  Role of Cyclic AMP   1133
  Regulation of the Cell Growth Cycle   1134
GENETIC STUDIES WITH CULTURED CELLS   1134
  The Karyotype of Cultured Cells   1134
  Genetic Mapping   1134

CHARACTERIZATION OF CULTURES OF ANIMAL CELLS

The development of improved methods for cultivating animal cells in vitro has contributed enormously to the progress of animal virology. In particular, it has provided quantitative technics for studying animal viruses comparable to those used for bacteriophages. In addition, the analysis of gene function, metabolic regulation, and other problems of cell physiology in animal cells have been greatly advanced by their cultivation. Cultured animal cells, however, differ from microorganisms in many respects. This chapter will review the most salient differences.

A major difference is that animal cells are much less stable, both in transferability and in genetic properties, than most bacteria. Thus primary cultures of the separated cells of a tissue may die out on repeated secondary cultivation or may give rise to altered but still diploid cell **strains.** These, in turn, will also eventually die out unless they give rise to more radically altered but permanent cell **lines.** These several types of cultures are all valuable for studies in virology as well as in cell physiology and genetics; their origin and properties will now be described.

Primary and Secondary Cultures. To propagate separated animal cells a tissue fragment is first dispersed into its constituent cells, usually with the aid of trypsin. After removal of the trypsin the suspension is placed in a flat-bottomed **glass or plastic container** (a Petri dish, bottle, or test tube), together with a liquid medium (such as that devised by Eagle) containing required ions at isosmotic concentration, a number of amino acids and vitamins, and an animal serum in a proportion varying from a few per cent to 50%. Bicarbonate is commonly used as a buffer, in equilibrium with CO_2 (about 5%) in the air above the medium. After a variable lag the cells **attach and spread on the bottom of the**

container and then start dividing mitotically, giving rise to a **primary culture.**

In the primary culture the cells retain some of the characteristics of the tissue from which they derived, and are mainly of two types: thin and elongated **(fibroblast-like),** or polygonal and tending to form sheets **(epithelium-like).** In addition, certain cells have a roundish outline and resemble epithelial cells but do not form sheets (**epithelioid** cells). The cells multiply to cover the bottom of the container with a continuous but thin layer, often one cell thick **(monolayer);** if they are fibroblastic they are **regularly oriented** parallel to each other. Primary cell cultures obtained from **cancerous tissues** usually differ from those of normal cells in two ways: they tend to form **thick layers,** and the cells are **randomly oriented.** The cultures are maintained by changing the fluid two or three times a week.

The cells of primary cultures can be detached from the vessel wall by either trypsin or the chelating agent EDTA, and can then be used to initiate new **secondary** cultures. Usually only two subcultures are started from a confluent primary culture of normal cells, because a subdivision into a larger number of subcultures is often unsuccessful. Apparently for survival and multiplication of recently isolated cells the medium must contain adequate concentration of unknown factors produced by the cells themselves (see below). Cancer cells usually withstand a much greater dilution than normal cells.

Cell Strains and Cell Lines. Cells from primary cultures can often be transferred serially a number of times (Fig. 47-1). This process usually causes a selection for some cell type, which becomes predominant. The cells may then continue to multiply at a constant rate over many successive transfers, and the primary culture is said to have originated a **cell strain** (often called a **diploid cell strain)** whose cells appear unaltered in morphological and growth properties. These cells must be transferred at a relatively high cell density to initiate a new culture. Even so, **the transferability of cell strains is limited:** for instance,

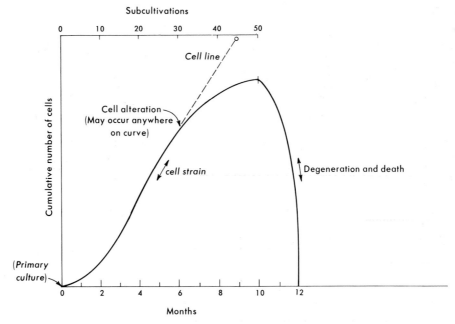

Subcultivations

Fig. 47-1. Diagram representing the multiplication of cultures of cells derived from normal tissues. The primary culture gives rise to a cell **strain,** whose cells grow actively for many cell generations; then growth declines and finally the culture stops growing and dies. During the multiplication of the cell strain altered cells may be produced, which continue to grow indefinitely and originate a cell **line.** The cumulative number of cells is calculated as if all cells derived from the original culture had been kept at every transfer. (Modified from Hayflick, L. *Analytic Cell Culture.* National Cancer Institute Monograph, No. 7, p. 63. 1962.)

with cultures of human cells after about 50 duplications the growth rate declines and the life of the strain comes to an end.

During the multiplication of a cell strain some cells become **altered;** they acquire a different morphology, grow faster, and become able to start a culture from a small number of cells. The clone derived from one such cell, in contrast to the cell strain in which it originated, has **unlimited life** and is designated a **cell line.** Cell lines derived from normal cells have a **low saturation density** under standard conditions (e.g., 10% serum and medium changes two or three times weekly); they grow rapidly to form a monolayer and then slow down considerably.

TRANSFORMED CELLS

During repeated serial transfers cell lines undergo extensive further changes, owing to the **emergence and selection of variants.** Thus, the saturation density increases and the cells overlap each other extensively and become irregularly oriented in respect to each other (transformation, Fig. 47-2). These **transformed cell lines usually are neoplastic, i.e., produce cancer** if they are transplanted into isogenic animals. Lines of transformed cells can also be directly obtained by cultivating neoplastic tissue, or they can be generated by infecting normal primary cultures or cell strains with oncogenic viruses (Ch. 63) or carcinogenic chemicals.

Transformed cultures often produce, by additional changes, cells with low adhesion to the container, which can be propagated in suspension by using a liquid medium poor in divalent ions and stirring constantly. Such **suspension cultures,** such as derivatives of L cells (see Table 48-1), which resemble bacterial cultures, are very useful for experimental virology. Cells of some lines can grow and form **colonies in suspension in semi-solid media,** provided they do not contain sulfated polysaccharides (e.g., methyl cellulose). Some

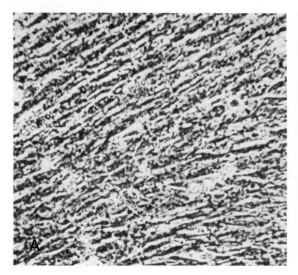 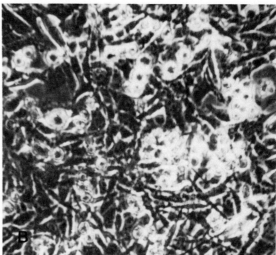

Fig. 47-2 A. An uninfected crowded secondary culture of hamster embryo cells. Note that the cells are arranged in a thin (mostly single) layer with parallel orientation. **B.** A similar culture transformed by polyoma virus (Ch. 63). Note that cells lie on top of each other and are oriented at random.

transformed cells can grow in the presence of sulfated polysaccharides (e.g., in agar); this ability is useful in virological work for isolating pure clones of rare transformants from an excess of untransformed cells.

STORAGE

The tendency of cell lines to change continuously on repeated cultivation is a drawback in experimental work. Cells of constant characteristics can be, however, **maintained in the frozen state.** For this purpose, large batches of cells are mixed with glycerol or dimethylsulfoxide and subdivided in a number of ampules, which are then sealed and frozen. The additives allow the cells to survive freezing. The frozen cells can be maintained in liquid nitrogen for years with unchanged characteristics; when the ampules are thawed most of the cells are viable and can initiate new cultures.

CLONING OF ANIMAL CELLS

Clones can be obtained from most cell strains or lines by transferring cells to a new culture at a very high dilution. For cell lines the proportion of cells that survive and give rise to colonies **(efficiency of plating)** approaches 100%, but for primary cultures or cell strains it is very small. However, Puck showed that the efficiency of plating can be greatly in-creased if they are mixed with a **feeder layer of similar cells made incapable of multiplication by X-irradiation;** these cells are still metabolically active and supply factors that enable the unirradiated cells to survive and multiply. The efficiency of plating can also be increased by Earle's technic of introducing individual cells into very small volumes of medium, as in sealed capillary tubes or in small drops of medium surrounded by paraffin oil. The cell-produced factors thus reach an adequate concentration, permitting the survival and multiplication of the cell without a feeder layer.

Cell Senescence and the Establishment of a Cell Line. Cells obtained from an organism become able to give rise to indefinitely growing cultures only after they have undergone an **"immortality event,"** whose occurrence is revealed by their ability to grow in very sparse cultures and by a high cloning efficiency. Unless this occurs, cultures undergo some kind of cellular senescence and finally die. Cellular senescence is absent in cancer cells.

Senescence appears to depend directly on the number of doublings rather than on astronomical time. It does not depend on the special conditions of growth in vitro; **it also occurs in vivo,** as during serial transplantation of mouse mammary epithelium into mammary fat pads from which the gland was surgically removed.

Cellular senescence is attributed to the **accumulation of unrepaired damage** in cellular constituents. Accumulation of mutations in DNA may lower the growth ability of the cells and finally become lethal. Accidental errors in translation causing changes in proteins that can influence information transfer (such as activating enzymes, DNA- and RNA-polymerases) may also cause progressive deterioration of cellular proteins, even independently of mutation. Indeed, in cultures of the mold *Neurospora* undergoing senescence altered and partly inactive enzymes accumulate in the cells. A plausible interpretation is that damages accumulate if there is **no efficient selection** against the sick cells, as in cultures of cells that have not undergone the immortality event, which must be seeded at relatively high cell density and have their growth soon limited by the density-dependent growth inhibition. The immortality event would prevent senescence by conferring on the cells the ability to initiate growth at low density, thus increasing selection; the effect would be even more pronounced after transformation, which allows the cells to grow to a higher density.

BEHAVIOR OF CELLS IN CULTURE

RELATION OF CELLS TO A SOLID SUPPORT

Electron micrographs show that a cell attaches at a few points to the bottom of the vessel, from which it is elsewhere separated by a layer of medium (Fig. 47-3). The attachment points become evident after mild trypsin treatment: the expanded cytoplasm retracts toward the nucleus but remains connected to the attachment points by thin pseudopodia. Uptake of nutrients seems to occur preferentially at the underface and at the outer margin of the spread cells, where there is also active engulfment of liquid droplets (**pinocytosis**). The droplets are formed by bubbles of the surface membrane, which detach toward the cell interior as small vesicles full of medium.

In a spread cell **microtubules,** which form the cell skeleton, are present in the stretched out parts of the cytoplasm. Contractile **microfilaments,** equivalent to muscle fibers, are located in part at the attachment points in the underface and in part at the upper face where they determine cell mobility.

In dense cultures the cells touch each other mostly at the borders, where the cell membranes have points of close adhesion, called **tight junctions.** The high ion permeabilities at the junctions, detected by a **low electrical resistance** between the cells, equalize the intracellular ionic content and pH throughout the culture.

Cell Locomotion. Active movement is beautifully seen in slow-motion pictures of living cell cultures, using phase contrast microscopy. Fibroblast-like cells have a moving lamella of cytoplasm at the forward edge of the cells in the direction of locomotion. The free edge of the lamella rapidly moves back and forward (**ruffling**), probably by contraction of the microfilaments: at the same time the surface membrane slides toward the center of the cell, as shown by the movement of adhering particles. Locomotion occurs when the ruffling lamella establishes new forward anchoring points and pulls the cell body toward them; at the same time the rear anchoring points vanish. When the ruffling lamella encounters another cell it adheres to its surface and both ruffling and locomotion stop (**contact inhibition of movement**). This effect is pronounced with unaltered or untransformed cells, but is slight or absent with transformed cells, probably because they adhere less to other cells.

The forward edge may also be the **site of membrane synthesis:** in cells infected by Newcastle disease virus (see Ch. 57) viral components recognizable by binding of red cells (hemadsorption, Ch. 48) make their first appearance at the edge and then

Fig. 47-3. Schematic cross-section of a culture on a solid support in liquid medium. Microfilaments in black. Medium, both outside the cells and in pinocytotic vesicles, gray. Arrows indicate the movement of pinocytotic vesicles. [Modified from Brunk, U., Ericcson, J. L. E., Ponten, J., and Westermark, B. *Exp Cell Res* 67:407 (1971).]

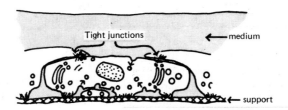

spread on the membrane toward the center of the cell.

CELLULAR GROWTH

GROWTH FACTORS AND INHIBITORS

Most cells cannot multiply unless the cultures contain blood serum as well as all the required amino acids and vitamins. Holley showed that the **serum requirement** of mouse fibroblasts (3T3 line, see Table 48-1) consists of several macromolecular factors separable by chromatography; a factor is needed for cell survival, two for DNA synthesis, and one for cell movement. Some of the factors can be replaced by insulin. Transformed cells require different factors and in smaller amounts. Cells of a few very abnormal lines can totally dispense with serum.

Growth factors are also produced by the cells themselves. Thus the growth of sparse cultures is enhanced by addition of medium obtained from an actively growing culture, where cell-produced factors, probably glycoproteins, accumulate. Especially good producers of growth factors are cell lines that can grow without serum.

More specific growth factors have been identified for differentiated cells of sympathetic ganglia, of the epidermis, and of the liver. It is likely that many other such factors are still unrecognized.

Cell growth rate is also critically dependent on the pH of the medium and on small concentrations of polyanions, such as the glass or plastic on which the cells grow. However, high concentrations of certain polyanions are strongly inhibitory for untransformed cells: for instance, these cells do not grow in agar, which contains sulfated polysaccharides. Transformed cells are less inhibited, and many kinds grow in agar.

Consequence of Serum Exhaustion. The growth effect of serum factors is stoichiometric, i.e., if other requirements are fulfilled the number of cells produced in a culture is proportional to the amount of factor added: on exhaustion of the serum the cells enter a **stationary** phase. Stationary cells show a reduced uptake of glucose, phosphate, and other substances, and a reduced synthesis of ribosomal RNA and protein; they have fewer ribosomes and polysomes and display faster protein degradation. DNA synthesis is either absent or greatly diminished and glycolipids accumulate at the cell surface.

Addition of fresh serum to stationary cultures causes a rapid recovery of uptake rates and of RNA and protein synthesis; about 16 hours later they undergo a synchronous wave of DNA synthesis followed by cell division.

THE CELL GROWTH CYCLE

As with bacteria, the cells of a sparse culture in optimal medium multiply **exponentially** (i.e., with a fixed doubling time, Ch. 5), although individual cells divide at **random** times. The cell growth cycle consists of **four main phases** (Fig. 47-4). DNA synthesis occupies only a fraction of the doubling time, the **synthetic (S) period.** Thus radioautographs of cultures show that a long exposure (e.g., 24 hours) to ³H-thymidine labels most nuclei, but after an exposure of 30 minutes many are still unlabeled. Between the end of the S period and the mitotic **(M) period** there is a G_2 **period** (G for gap) whose length is measured by determining, by radioautography, the time between a brief exposure to ³H-thymidine and the first appearance of label in metaphase chromosomes. After the mitotic phase there is a G_1 **period** before the next S period. Cells of stationary cultures and differentiated

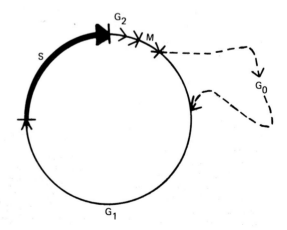

Fig. 47-4. The cell growth cycle.

cells in organs are in a **nongrowth phase designated as G_0** and enter the cycle at G_1 when growth is resumed. For cells with a doubling time of 18 hours typical lengths of the various periods are: $G_1 = 10$ hours; $S = 6$ to 7 hours; $G_2 = 1$ hour; $M = $ about ½ hour.

Synchronization. The cell growth cycle is directly observable in **synchronized cultures.** These can be started with mitotic cells, which are weakly attached to the vessel and hence can be selectively removed by shaking a randomly growing culture.

Random cultures can also be synchronized by inhibiting DNA synthesis, so that all cells become arrested at the end of the G_1 phase. Commonly used is the so-called **double thymidine block,** based on the ability of the nucleoside to inhibit the reduction of ribonucleoside diphosphates to the corresponding deoxynucleoside diphosphates. First, the culture is maintained under thymidine until all cells that were not in S phase reach the end of G_1. Cells that were in S phase remain there during the thymidine block and, after its removal, proceed through the cycle. A second thymidine block after the cells that were in G_1 have progressed beyond S achieves an almost complete synchrony, since there are no cells in S during the second block.

Synchronization by metabolic block may make the cells abnormal because RNA, protein, and lipid syntheses continue in the absence of DNA synthesis, leading to **metabolic unbalance** (as in bacteria, see Ch. 5).

INTERACTION BETWEEN CELLS IN A CULTURE

Animal cells differ from bacteria in their striking ability to interact with each other in various ways. This is a predictable property of cells of multicellular organisms.

Contact Inhibition. In cultures of untransformed cells various cellular activities are inhibited at **high cell densities,** including both cell movement **(contact inhibition of movement)** and cell growth **(density-dependent growth inhibition).** Inhibition of movement occurs when cells are in contact with each other, as already noted. Inhibition of multiplication (Fig. 47-5) derives in part from exhaustion of medium factors but also from cell-to-cell contacts. Thus when continuous perfusion with fresh medium restores multiplication in dense cultures its rate is reduced. Moreover, in **wound experiments,** in which a strip of cells is mechanically removed from a dense culture, cells from the edges penetrate into the wound and within about 10 hours they initiate DNA synthesis and then divide, although the medium is not changed. This type of inhibition, referred to as **contact inhibition of growth** or **topoinhibition** (since it depends on local factors), may result from contacts between the polysaccharide layers that surround the cells. The cellular changes resulting from topoinhibition are similar to those produced by serum exhaustion (Fig. 47-5B), suggesting that an important serum function is to prevent topoinhibition.

The feeder effect works equally well with cells of the same or of different types. Reciprocal inhibition between cell types, however, varies greatly.

Surface Modifications of Cells in Contact. Galactosyl transferases present on the surface of the cells of certain lines can transfer galactosyl residues from uridine diphosphate galactose to suitable acceptors present on the surface of similar cells. With untransformed cells transfer occurs only between cells in contact, the enzymes of one cell apparently attaching the sugar to the receptors on other cells, since transfer is dependent on cell density. With transformed cells the enzymes can transfer to receptors of the same cell since there is no dependence on cell density. The interaction between enzymes and receptors in the untransformed cells may represent a mechanism of intercellular recognition (perhaps involved in topoinhibition), which is lost when transformation occurs.

Exchange of Components Between Cells in Contact. Exchange is not limited to small ions through tight junctions (see above). Thus ^{3}H-choline-labeled membrane phospholipids are transferred to unlabeled cells, as shown by radioautography. Moreover, **metabolic cooperation** is produced by exchange of intermediates between cells, and the effect may be big enough to allow cells with a metabolic defect to grow in mixed cultures with normal cells.

For instance, cells lacking the enzyme hypoxanthine:guanine phosphoribosyl transferase cannot utilize hypoxanthine as a purine source in a medium containing aminopterin (to block the normal purine pathway). However, if they are in contact with normal cells in mixed cultures they incorporate label, as shown by radioautography. The nucleotides formed in the normal cells presumably reach the deficient cells through tight junctions, and perhaps the enzyme itself is transferred.

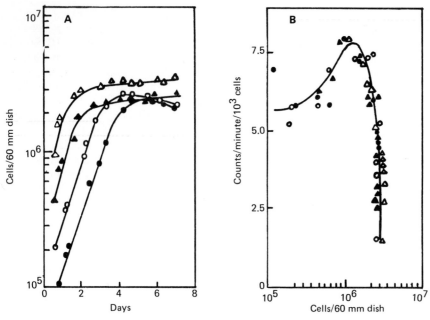

Fig. 47-5 A. Density-dependent growth inhibition in chick embryo cell cultures. Irrespective of initial cell density (within the range shown), cells grow exponentially at constant rate until they form a continuous layer; then growth stops. **B.** The uptake of uridine (measured in brief pulses given at various times) abruptly declines when the cells begin to have extensive contacts. [Modified from Weber, M. J., and Rubin, H., *J Cell Physiol* 77:157 (1971).]

STRUCTURE OF CELLULAR MEMBRANES

Salient Structural Features. Animal cells are surrounded by a **plasma membrane** in contact with the environment, made up of proteins, lipids, and polysaccharides. Various proteins perform structural, transport, and enzymatic functions. Some proteins are embedded in the membrane and are recognizable in electron micrographs of fractured frozen cells where they form, in association with lipids, the so-called **70-A globular particles.** Other proteins are at the outer surface and are recognizable by labeling with reagents that do not penetrate the membrane. External proteins are usually linked to polysaccharides (**glycoproteins**) and are therefore hydrophilic. A prominent sugar is N-acetyl neuraminic acid (see Fig. 44-30). The membrane contains several kinds of lipids: phospholipids, with the ionized groups at the surface; glycolipids, with the sugar moieties at the surface; nonpolar lipids, cholesterol and its esters. Some lipids are firmly bound; others may exchange with the medium.

Membrane proteins undergo turnover with an average half life of about 48 hours. Phospholipids turn over even more rapidly. Furthermore, the membrane displays **continuous internal flow,** owing to the two-dimensional liquid state of the lipids; thus, when two cells with different surface antigens recognizable by fluorescent antisera are fused (see below, Cell hybridization), within an hour the two types of antigen become thoroughly intermixed.

The plasma membrane is continuous with intracellular membranes, such as the **endoplasmic reticulum,** which forms a system of double membranes enclosing narrow channels, stretching from the inner side of the plasma membrane to the outer nuclear membrane. The **rough reticulum** is associated with polysomes and is the site of synthesis of proteins for export, i.e., destined to leave the cell; and the **smooth reticulum,** without polysomes, channels these proteins to the outside.

The endoplasmic reticulum is in turn connected to **mitochondria,** which are the sites of oxidative phosphorylation, and to **lysosomes,** bags full of degradative enzymes whose function is to digest phagocytized material or, when the cell dies, the cellular constituents themselves.

The cell nucleus is enveloped by a **double-nuclear membrane** provided with several thousand **pores** allowing communication between the nuclear and the cytoplasmic compartments. The inner nuclear membrane is connected to the chromosomes, probably at the pores, and it is therefore the counterpart of the plasma membrane of bacteria. The nuclear membrane dissolves at the beginning of mitosis and is re-formed at the end of mitosis by the endoplasmic reticulum which surrounds the chromatin.

Antigenic Structure of the Cell Surface. Many cell specificities are normally determined by surface antigens in animal cells: **blood group** specificity in red cells; **histocompatibility** specificities in tissue cells (see Ch. 21). In addition, new antigens appear in cancer cells (**tumor-specific** antigens) or are induced by viruses, both oncogenic and nononcogenic. All these antigens can be recognized in vitro by various immunological methods: **immunofluorescence** with live cells, which only labels surface antigens (**membrane fluorescence**) since the labeled antibody cannot penetrate inside the cells; **cytotoxic tests,** in which the cells are killed by antibodies in the presence of complement; **colony-reduction tests,** in which the colony-forming efficiency of the cell is reduced by immune lymphocytes; and binding of **ferritin-labeled antibody,** which is recognizable by electron microscopy. Mapping of antigenic sites on the cell surface by electron microscopy reveals that these sites are often in **patches.** Both the blood group and the histocompatibility antigens have been shown to be glycoproteins (see Ch. 21); their sugars are added one at a time by specific glycosyl transferases.

Interaction with Lectins. A number of vegetable proteins, such as wheat germ agglutinin, concanavalin A, and phytohemagglutinin, bind in vitro to carbohydrates of the cell surface; such proteins are called lectins. One consequence is **agglutination** of cell suspensions, since the oligomeric lectins can form bridges between cells. Certain types of transformed cells are more readily agglutinated than their untransformed counterparts because greater patching of lectin binding sites allows more effective bridging. Other effects of lectins include the preferential killing of transformed cells by concanavalin A (see Ch. 63) and stimulation of growth of lymphocytes by phytohemagglutinin.

ORGANIZATION AND EXPRESSION OF THE GENETIC INFORMATION

CHROMATIN

Structure: Histones. In interphase nuclei, the DNA is associated with proteins and some RNA to form complexes collectively known as **chromatin.** Characteristic components are **histones,** small basic proteins rich in lysine and arginine with variable phosphorylation, which are extractable with acid. There seem to be only a few classes of histones, distinguished by their solubility properties and sequences; they are similar in animals and plants, suggesting that they originated early in evolution. Histones are synthesized on polysomes containing a 9S mRNA which differs from other mRNAs in that it lacks a polyA tail (see below). The amount of this RNA that can be hybridized to the cellular DNA shows that there are several hundred histone genes per cell, an example of the repetition of sequence characteristic of animal cells (see Ch. 10). The histones are intimately associated with DNA through multiple ionic and hydrophobic bonds, increasing its stability and raising its Tm (see Ch. 10). In chromatin, the DNA is thicker than free DNA (80 to 120 A diameter instead of 20) and therefore probably supercoiled.

The role of histones and other chromatin proteins seems to be to limit DNA transcription. Thus, the rate of RNA synthesis, using *E. coli* RNA polymerase, is much lower with chromatin as template than with the DNA extracted from it; the readdition of histones to the DNA again limits its transcription. The effect is unspecific because it is independent of the source of DNA or histones. However, the **DNA sites available for transcription in native chromatin are specific,** since RNAs transcribed by

E. coli polymerase on chromatins of different organs of the same animal differ from each other, as shown by hybridization competition experiments (see Ch. 10). Furthermore, activation of cellular functions by hormones is accompanied by an increased template activity of the chromatin. Reconstitution of chromatin from its separate components suggests that transcription specificity is determined by the association of DNA with nonhistone proteins, which then determine the site of attachment of the histones to block transcription.

Functional and Nonfunctional Chromatin.

Active and inactive chromatins have distinguishing physical and cytological properties, as evidenced, for instance, in female cells by the different behavior of the two X chromosomes, of which only one is active. In interphase nuclei one of the chromosomes, presumably the inactive one, is condensed and stainable (the so-called **Barr body)** whereas the other is dispersed. This correlation suggests that inactive chromatin corresponds to the cytologists' **heterochromatin** fraction of certain chromosomes, which remains condensed in interphase. Moreover, in labeling experiments heterochromatin shows no RNA synthesis and **late replication** of its DNA (see below), compared with the active **euchromatin.**

The inactive X chromosome is defined as **facultative** heterochromatin since neither X chromosome is intrinsically heterochromatic, to distinguish it from **constitutive** heterochromatin, which contains the highly repetitive sequence discussed in Chapter 10. Constitutive heterochromatin is found at various places but especially near the **centromeres,** where it can be recognized cytologically by hybridizing the metaphase chromosome with radioactive RNA synthesized in vitro from highly repetitious DNA, using *E. coli* transcriptase. These repeating sequences, apparently interspersed with other DNA, by pairing with each other can form a compact knob for anchoring the spindle fibers; they may also participate in chromosome pairing. Constitutive heterochromatin is also present at the nucleolus organizer, where it insulates the various rRNA genes from each other and from the rest of the genome. Heterochromatin interspersed in the chromosome may function as signal for regulation of gene expression (see below).

DNA Replication. Of the large number of

replicons present in animal cell DNA (see Ch. 10), each appears to replicate at a characteristic time in the S phase. In fact, in a synchronized culture replicons replicating either early or late in one S phase, as shown by the incorporation of 5-bromodeoxyuridine (which allows their identification by the increased buoyant density of the DNA), retain the same timing in the next S phase. The late replication of heterochromatic DNA is shown by radioautographs of mitotic cells after briefly labeling synchronized cultures with ³H-thymidine at various times during the S phase. These regularities point to a **strict temporal organization of DNA replication.**

Histones are synthesized while DNA replicates and immediately move to the nucleus, allowing the incorporation of the newly made DNA into chromatin.

TRANSCRIPTION

In animal cells two main classes of RNA are recognizable in the nucleus after brief pulses with ³H-uridine: the ribosomal RNA precursors (45S and highly homogeneous), together with some of its derivatives (see below), and the heterogeneous nuclear RNA (hnRNA), which has base ratios similar to those of DNA and has S values ranging from 20 to more than 200. The modality of synthesis and fate of the two RNAs are quite different.

The **ribosomal RNA precursor** is synthesized in the nucleolus by an RNA polymerase insensitive to the alkaloid α-amanitin; owing to its high G and C content its synthesis is inhibited by very low concentrations of actinomycin D. Synthesis occurs in the core of the nucleolus, which appears fibrous in electron micrographs. The precursor then undergoes **processing,** i.e., it is fragmented in various pieces to yield three stable rRNAs: 28S, 18S, and 7S (Fig. 47-6). The phenomenon can be studied by briefly labeling the precursor with ³H-uridine, and following the radioactivity in subsequent chases of variable lengths with cold uridine in the presence of actinomycin D (to prevent incorporation of residual labeled nucleotides).

Processing is probably caused by specific enzymes recognizing helical regions of the RNA, because it is inhibited by chemicals intercalating in helical nucleic acids, such as ethidium (see Ch. 10). Processing may also be influenced by the base sequences, since some highly G C-rich segments of the precursor are completely hydrolyzed, and by

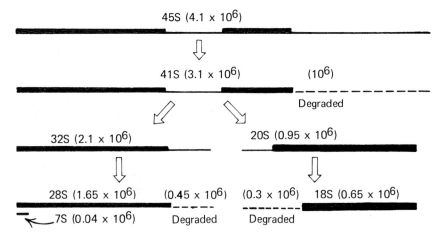

Fig. 47-6. Scheme of processing of the rRNA precursor in animal cells. Heavy lines indicate the sequences corresponding to the mature ribosomal RNAs. Numbers in parentheses indicate molecular weights. [Modified from Weinberg, R. A., and Penman, S. *J Mol Biol 47*:169 (1970).]

presence of methylated bases, because under conditions of undermethylation (caused by methionine starvation) precursors accumulate in the nucleus and no rRNA reaches the cytoplasm. The 28S and 18S rRNAs become rapidly associated with proteins to form precursors of **ribosomal subunits.** The precursor containing the 28S RNA accumulates for some time in the outer, granular part of the nucleolus, whereas the precursor containing the 18S RNA moves rapidly to the cytoplasm. The mature subunits lack some of the proteins present in the precursors, which remain in the nucleus and are recycled from one precursor to another.

The **heterogeneous nuclear RNA (hnRNA)** is synthesized in the nucleoplasm, outside the nucleolus, by a polymerase sensitive to α-amanitin. A large proportion (90% or more) is rapidly degraded in the nucleus, with a half life of about 10 minutes. The degraded parts contain transcripts of highly repetitive DNA, which may have a regulatory role during transcription (see below) but no meaning for protein synthesis. The undegraded parts are probably precursors of cytoplasmic messenger RNA, although the precursor-product relation cannot be clearly established in pulse-chase experiments with ³H-uridine, owing to the rapid degradation of the bulk hnRNA and the small proportion that survives.

Transport of mRNA to the Cytoplasm. The separation of the site of transcription from that of translation requires a transport mechanism with two steps: association with pro-

teins, and possibly the covalent **attachment of polyadenylic acid chains.**

Most cytoplasmic mRNA molecules contain polyA chains, a few hundred nucleotides long, at their 3′ end (an exception is the histone mRNA). These tails are added enzymatically to preexisting RNA molecules and are not translated. Similar chains are present in nuclear hnRNA, but in smaller proportions; hence, it is possible that polyA attachment protects those molecules of hnRNA that are destined to become mRNA. The polyA attachment seems essential because cordycepin (3′-deoxyadenosine), by inhibiting polyA formation, prevents the appearance of cytoplasmic mRNA without interfering with nuclear RNA synthesis. However, polyA is present in non-transported RNAs (of cytoplasmic viruses) and is absent in at least one type of transported mRNA (for histones). The **association of mRNA with proteins during its transport** is indicated by the regular presence of recently arrived cytoplasmic RNA (before it enters polysomes) in particles sedimenting faster than the naked RNA **(informosomes).**

A role of the nucleolus and possibly of ribosomal subunits in mRNA transport is suggested by experiments of Harris which show that the nucleolus is essential for gene expression in hybrid cells formed by the fusion of chicken erythrocytes (which have inactive nuclei and no nucleoli) to mouse cells (see below, Cell hybridization). After cell fusion the erythrocyte nucleus is activated and starts synthesizing RNA, but chicken-specific proteins are not made until it forms nucleoli; furthermore, after the chicken functions are expressed, they can be interrupted by localized ultraviolet irradiation of its nucleolus. The formation of the erythrocyte nucleolus must wait until mouse ribosomal proteins, recognizable by immunofluorescence, enter the erythrocyte nucleus; they are later replaced by chick ribosomal proteins. These results suggest a role of the ribosomal subunits in mRNA transport, coupling transcription with an initial phase of translation; hence there may be considerable similarity to bacteria, where the coupling is complete (see Ch. 12, Translation).

In the cytoplasm ribosomal subunits associate with mRNA to form polysomes, which are found **either free or associated with the rough endoplasmic** reticulum. Membrane-associated polysomes are especially abundant in cells that make protein for export, such as secretory cells. The nascent polypeptide chains are rapidly segregated into the cavities of the endoplasmic reticulum from where they are secreted to the outside. mRNAs of animal cells are rather stable (half life up to several days).

The polysomes containing the 9S histone mRNA are formed during DNA replication; in contrast to other polysomes, their activity rapidly stops if DNA replication is halted by a metabolic inhibitor. Presumably, the activity of these polysomes is controlled by a short-lived factor generated during DNA synthesis, which couples the two functions.

Monocistronic Messengers. In animal cells, in contrast to bacteria, mRNA is at least in the majority, monocistronic (see Ch. 11, Transcription), as shown by labeling all the nascent polypeptides with a radioactive amino acid, and comparing the length of the polysomes (inferred from their RNA content) with the radioactivity. As Figure 47-7 shows, if messengers are monocistronic, the label per unit length should increase linearly with polysome length, as was indeed found in HeLa cells; if the messengers were multicistronic the ratio should level off. Moreover, a monocistronic length has been observed for all

Fig. 47-7. Distinction between mono- and polycistronic translation by comparing polysome label in nascent polypeptide chains with messenger length. The nascent chains, of different lengths, are indicated by parallel thin lines. For each messenger (heavy lines) the total amount of label is given by the area of the triangle covered by the polypeptide chains. It is clear that for multicistronic messengers the area is proportional to L^2 (where L is the messenger length), whereas for monocistronic messengers it is approximately proportional to L (beyond a certain length). The ratio of radioactivity to polysome length is expected to increase proportionally to L for monocistronic messengers, but to reach a constant value for polycistronic messengers. [From data of Kuff, E. L., and Roberts, N. E. *J Mol Biol 26*:211 (1967).]

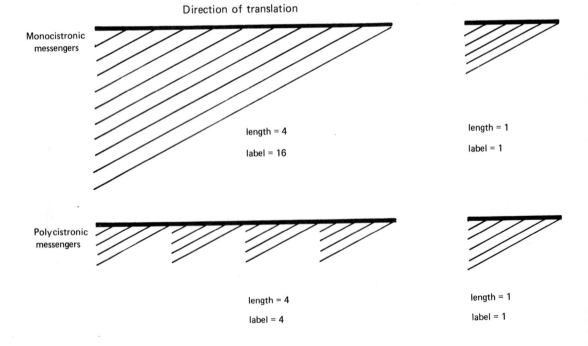

Direction of translation

Monocistronic messengers

length = 4
label = 16

length = 1
label = 1

Polycistronic messengers

length = 4
label = 4

length = 1
label = 1

animal cell messengers with known function so far studied (e.g., those for histones, hemoglobin, γ-globulin, myosin, lens protein). However, extracts of animal cells regularly translate phage mRNA in vitro, although it is multicistronic. As discussed in Chapter 48, the monocistronic property of animal cells may determine important features of organization of genetic material in animal viruses.

PROTEIN SYNTHESIS AND BREAKDOWN

The amino acid precursors for protein synthesis may be derived from the medium or from breakdown of cellular proteins. The average rate of protein breakdown is about 1% per hour, but may be higher for individual proteins. In serum-requiring cells, the rate of protein breakdown is increased by serum depletion.

Proteins responding to regulation often have a very short half life (1 to 2 hours), so that after their synthesis stops their function is rapidly turned off. Examples are ribonucleotide reductase, which is induced at the beginning of the S phase and rapidly decays at its end; RNA polymerase of regenerating liver or of the estrogen-stimulated uterus; and certain catabolic enzymes, such as tyrosine-α-ketoglutarate transaminase of liver cells.

REGULATION

REGULATION OF GENE EXPRESSION

An excellent system for studying regulation of gene expression is the enzyme tyrosine-α-ketoglutarate aminotransferase in hepatoma cells. In this system, as Tompkins showed, **regulation occurs at the level both of transcription and of translation.** Thus glucocorticoid hormones, such as cortisol, induce an increase in enzyme synthesis, and this response requires transcription since it is blocked by actinomycin D. Moreover, once induction has occurred enzyme production is not affected by this inhibitor, showing that the messenger has a long life. On the other hand, insulin also stimulates the synthesis of this enzyme but evidently influences translation, since the response is not affected by actinomycin D.

In embryonic development translational regulation is paramount. Thus in the sea urchin egg

mRNAs made during oogenesis become active after fertilization, during early development. The activation appears to be due to a change in the ribosomes, since a protein extracted from ribosomes of unfertilized eggs inhibits polypeptide synthesis in vitro.

Elements Regulating Transcription. The regulation of genes important in differentiation (which Ephrussi defines as "luxury" genes) appears to rely on special molecular mechanisms not effective in "household" genes, i.e., those that are expressed all the time. The reason is that 5-bromodeoxyuridine (5-BUDR) can block the appearance of differentiated functions without affecting over-all protein synthesis or cell growth rate. Thus 5-BUDR prevents the fusion of myogenic cells into muscle fibers and the synthesis of myosin and actin; and in prolactin-stimulated mammary epithelial cells, it inhibits the synthesis of casein and α-lactalbumin. 5-BUDR also prevents the induction of tyrosine aminotransferase in hepatic cells.

Transcriptional regulation of "luxury" genes may depend on the reiterated sequences present in DNA. Georgiev has proposed that each structural gene is preceded, in the direction of transcription, by a series of sequences recognized by various regulatory molecules. Repetitious DNA sequences at either side of the regulatory sequences would isolate them from each other. The transcription of a given gene could be blocked by a number of regulators, each interacting with a different regulatory sequence, accounting for the intricate gene control during differentiation.

In the absence of repression the whole transcriptional unit would be transcribed into a molecule of hnRNA, from which the regulatory part would be degraded away, leaving the structural message. Evidence for this mechanism is the presence of transcripts of repeating DNA in hnRNA toward the 5′ end of the molecules, and their elimination before the mRNA of the structural gene goes to the cytoplasm.

ROLE OF CYCLIC AMP

The control of DNA synthesis by serum factors or cell-to-cell contact is probably **mediated through the cellular plasma membrane,** since the large serum proteins may be unable to penetrate into the cells. Since the chromosomes are not connected to the plasma membrane, there must be some **vehicle** of communication between them. It now appears that, as in the action of polypeptide hormones and in catabolite repression in bacteria (Ch. 13), one such vehicle is cyclic AMP (3′, 5′-adenosine monophosphate). cAMP is generated from ATP by adenyl cyclase, a ubiquitous enzyme present in the plasma membrane of cells, and it is destroyed by specific diesterases.

In hormone-sensitive cells specific hormone receptors coupled to the cyclase are activated by binding the hormone. The increased concentration of the cAMP in the cells, in turn, activates specific functions which determine the characteristic cell response. For instance, in glucagon stimulation of liver cells cAMP activates a protein kinase present in the cells which converts phosphorylase to the active form, causing the breakdown of the glycogen.

cAMP may regulate the growth of tissue culture cells. In fact, the **concentration of the nucleotide is low in actively growing cells** (as in bacteria, Ch. 13), but much higher in resting cells. Addition of cAMP or its dibutyryl derivative (which penetrates more easily into the cells) slows down multiplication even of transformed cells; and insulin, which stimulates cell growth, depresses the cAMP concentration. The nucleotide also **stimulates differentiation:** under its action, neuroblastoma cells become more like normal neurons, producing long neurites and increased acetyl cholinesterase; and the cells of a fibroblastic sarcoma acquire morphological characteristics of normal fibroblasts. Both these morphological changes appear to be mediated by the formation of microtubules, since they are prevented by mitotic poisons (e.g., vinblastine) which block microtubule formation.

Another cyclic nucleotide, cGMP, also appears to play a regulatory role.

REGULATION OF THE CELL GROWTH CYCLE

The cycle is maintained by a series of catenated events. Studies with metabolic inhibitors show that the $G_1 \rightarrow S$ shift requires previous RNA and protein synthesis in G_1, while completion of S depends on RNA and protein synthesis in the early part of S, and mitosis depends on RNA synthesis during the late S phase. DNA synthesis is controlled by cytoplasmic proteins which can be shown, with isolated nuclei in vitro, to extend the duration of synthesis. DNA replication is also coupled to the synthesis of histones and their phosphorylation, and to that of a number of enzymes required to manufacture DNA precursors, including various nucleoside and nucleotide kinases, deoxycytidylate deaminase, and ribonucleotide reductase.

At mitosis RNA synthesis stops, perhaps owing to the condensed state of the DNA. In addition, the rate of protein synthesis rapidly declines because polysomes disaggregate. At the end of mitosis the reaggregation of preexisting messengers and ribosomes allows protein synthesis to resume even in the presence of actinomycin D.

The various phases of the growth cycle are accompanied by important **changes of the cell surface.** Thus, certain antigens characteristic of infection with oncogenic viruses (see Ch. 63) are present only during a brief period corresponding to the M and early G_1 phases, and lectin binding increases during the same period. These changes are accompanied by variations of the over-all cellular charge as measurable by electrophoresis, and of the composition of plasma membrane, including the proportion of various sugars in polysaccharides and the number of 70-A globules.

GENETIC STUDIES WITH CULTURED CELLS

THE KARYOTYPE OF CULTURED CELLS

The analysis of the chromosomal constitution of tissue culture cells has acquired paramount importance as a tool for studying their genetic properties, especially since it became clear that karyotype anomalies of the body's cells are associated with certain human diseases. In the study of cell cultures the karyotype is an indication of the **degree of abnormality** the cells have attained during their cultivation.

Recent developments of staining technics allow not only the determination of the number of chromosomes in a cell, but also their precise cytological identification. Thus characteristic bands are observed in mitotic chromosomes stained with the fluorescent dye quinacrine mustard and examined under ultraviolet light (Fig. 47-8).

In young cell strains most cells tend to maintain the chromosome number characteristic of the animal, i.e., they are **diploid** ($2n$). The types of chromosomes are also usually normal, and the cells are said to be **euploid.** The cells of older strains and cell lines, in contrast, always contain deviations from the euploid chromosome number and distribution (**aneuploid** cells).

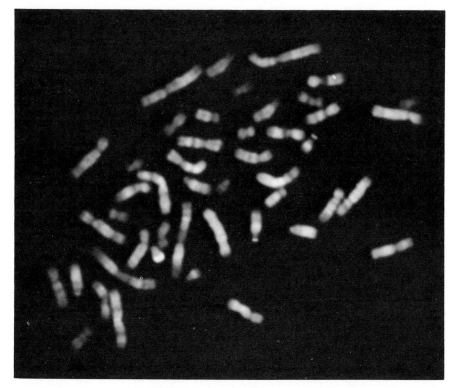

Fig. 47-8. Fluorescence photograph of the metaphase chromosomes of a human cell stained by quinacrine dihydrochloride. (Courtesy of Dr. W. Roy Breg.)

The number of chromosomes may be different from diploid **(heteroploid),** either higher (usually between $3n$ and $4n =$ **hypertriploid)** or lower **(hypodiploid).** In **quasidiploid** cells the number of chromosomes is $2n$ but their distribution is abnormal; for example, a chromosome of one pair may be missing and replaced by an extra chromosome of another pair. In addition, **chromosomal aberrations** (e.g., translocations and deletions) often involve highly characteristic morphological abnormalities in individual chromosomes, which are useful as **markers** for cell identification.

Although in some cell lines most cells are diploid the majority of cell lines are constituted of heteroploid, especially hypertriploid cells. **Individual lines are heterogeneous,** with cells containing different numbers of chromosomes. The most frequent **(modal)** number remains constant if the cells are grown under a constant set of conditions, but changing the conditions often results in selection of a type with a different modal number. Heteroploid cultures show a great deal of variability,

probably accounted for by the great variety of karyotypes they continually produce by unequal segregation of chromosomes at mitosis.

Functional Chromosome Markers. Certain chromosomes can be identified by genes for monomers of oligomeric proteins, which by assembling in various proportions (e.g., AAAA, AAAa, AAaa, etc.) produce oligomers with different electrophoretic mobilities (so-called **isozymes;** Fig. 47-9).

GENETIC MAPPING

Mutations. Genetic mapping in culture offers the opportunity to study the genetic organization in man, where the usual breeding approach is not applicable, and is important for virological studies (e.g., for locating integrated viral genomes, see Ch. 63). Germinal mutations can be employed if they are recognizable in cultures; other mutations are induced in vitro, using suitable mutagenic agents (e.g., nitrosoguanidine or alkylating

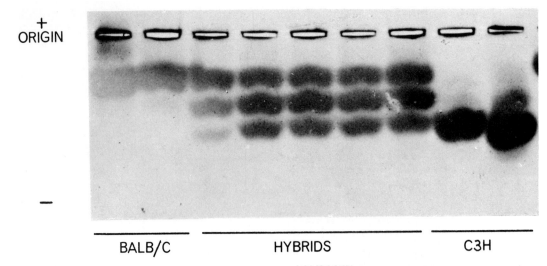

+
ORIGIN

−

BALB/C   HYBRIDS   C3H

Fig. 47-9. Isozyme patterns of the enzyme glucose phosphate isomerase (a dimeric enzyme) in homogenates of cells derived from two inbred mouse strains, BALB/C and C3H, and from hybrid cells obtained by fusing cells of the two strains. The enzymes of the two strains each form a single band with different electrophoretic mobilities, indicating that the enzyme contains a single type of monomer; in addition, the hybrid cells have a third band, produced by enzyme molecules containing a monomer from each parent. The isozyme patterns were recognized by fractionating cell extracts by gel electrophoresis and then treating the gel with a substrate that releases a dye when acted upon by the enzyme. [From Klebe, R. J., Chen, T., and Ruddle, F. H. *J Cell Biol 45*:74 (1970).]

agents, see Ch. 11). Also useful as genetic markers are isozymes.

Cell Hybridization. The relations between genetic markers can be studied in cultures, using somatic hybrid cells obtained by cell fusion. The frequency of spontaneous fusion is low, but can be enhanced by fusing agents, such as Sendai virus (a paramyxovirus, see Ch. 57) inactivated by chemicals or ultraviolet light, which attaches to the cell surface and agglutinates cells by bridging them. Some constituent of the viral envelope breaks the plasma membranes of the agglutinated cells and causes them to fuse into a bi- or multinucleated cell. Harris and Watkins showed that fusion results first in the formation of **heterokaryons,** i.e., without nuclear fusion; if the nuclei enter mitosis at approximately the same time they then also yield a mononucleated **hybrid** cell.

In heterokaryons and in hybrid cells **gene complementation** is observed (see Ch. 12, Complementation) and is used for isolating the hybrid cells from their parents.

For example, to grow in medium containing aminopterin (which blocks the endogenous purine

and pyrimidine pathways), supplemented by thymidine and hypoxanthine, cells must have both thymidine kinase and hypoxanthine:guanine phosphoribosyl transferase. If two parental strains each lack one of these enzymes neither can grow in this medium, but in their hybrids gene complementation allows growth.

Chromosome Elimination. Gradual **chromosome elimination** occurs in hybrid cells during multiplication and is useful for assigning genes to chromosomes and for genetic mapping. The original hybrid cell contains all the chromosomes of both parental cells, but in **heterospecific hybrids,** chromosomes (usually of that parent with longer replication cycle) are subsequently lost. Thus, Weiss and Ephrussi showed that in human-mouse hybrids human chromosomes are lost preferentially. Cytological studies and the behavior of functional markers show that the loss is **random** (within the set from one parent). When a biochemical marker is lost in this process it can be assigned to one of the missing chromosomes.

The disappearance of two biochemical markers at the same time suggests that their genes are linked;

and the frequency (usually low) of independent loss of one of such linked markers owing to translocation to a different chromosome is a measure of their distance. The usefulness of chromosome elimination is enhanced by Pontecorvo's finding that the loss of chromosomes of one set can be enhanced by X-irradiating the parent before fusion, or by growing it with 5-bromodeoxyuridine (which makes the DNA photosensitive and liable to permanent damage on subsequent exposure to blue light).

Cell fusion experiments are useful for studying **regulation of gene expression.** Thus, by fusing a cell performing a given gene function with one lacking it, one can determine whether expression or the lack of it is dom-

inant. The absence of function of "luxury genes" (see above) is usually dominant, suggesting a **negative regulation of the expression of such genes.** The function is sometimes restored in subclones after chromosome loss, probably owing to loss of a negative regulator gene or to a change in the balance between regulator genes of different kinds and between regulator and structural genes. In contrast, "household" genes from both parents continue to be expressed in the hybrid. The normal state is dominant over the neoplastic in hybrids, suggesting that neoplasia is determined by a shift in gene balance to the disadvantage of genes producing negative regulators.

SELECTED REFERENCES

Books and Review Articles

BASERGA, R., and STEIN, G. Nuclear acidic proteins and cell proliferation. *Fed Proc 30:*1752 (1971).

BONNER, J., DAHMUS, M. E., FAMBROUGH, D., HUANG, R. C., MARUSHIGE, K., and TUAN, D. Y. H. The biology of isolated chromatin. *Science 159:*47 (1968).

DARNELL, J. E., JR. Ribonucleic acids from animal cells. *Bacteriol Rev 32:*262 (1968).

HARRIS, H. Cell fusion and the analysis of malignancy: The Croonian lecture. *Proc R Soc Lond (Biol)179:*1 (1971).

MADEN, B. E. H. Ribosome formation in animal cells. *Nature 219:*685 (1968).

MUELLER, G. C. Biochemical events in the animal cell cycle. *Fed Proc 28:*1780 (1969).

PASTAN, I., and PERLMAN, R. L. Cyclic AMP in metabolism. *Nature New Biol 229:*5 (1971).

WESSELLS, N. K., SPOONER, B. S., ASH, J. F., BRADLEY, M. O., LUDUENA, M. A., TAYLOR, E. L., WRENN, J. T., and YAMADA, K. M. Microfilaments in cellular and developmental processes. *Science 171:*135 (1971).

WOLSTENHOLME, G. E. W. (ed.). *Growth Control in Cell Cultures* (A Ciba Symposium). Churchill, London, 1971.

YUNIS, J. J., and YASMINEH, W. G. Heterochromatin, satellite DNA and cell function. *Science 174:*1200 (1971).

Specific Articles

ABERCROMBIE, M. Contact inhibition in tissue culture. *In Vitro 6:*128 (1970).

ABERCROMBIE, M., HEAYSMAN, J. E. M., and PEGRUM, S. M. The locomotion of fibroblasts in culture. III.

Movements of particles on the dorsal surface of the leading lamella. *Exp Cell Res 62:*389 (1970).

CUNNINGHAM, D. D., and PARDEE, A. B. Transport changes rapidly initiated by serum addition to "contact inhibited" 3T3 cells. *Proc Nat Acad Sci USA 64:*1049 (1969).

DARNELL, J. E., PHILIPSON, L., WALL, R., and ADESNIK, M. Polyadenylic acid sequences: Role in conversion of nuclear RNA into messenger RNA. *Science 174:*507 (1971).

GEORGIEV, G. P., RYSKOV, A. P., COUTELLE, C., MANTIEVA, V. L., and AVAKYAN, E. R. On the structure of the transcriptional unit in mammalian cells. *Biochim Biophys Acta 259:*259–283 (1972).

HAKOMORI, S. Cell density-dependent changes of glycolipid concentrations in fibroblasts, and loss of this response in virus-transformed cells. *Proc Natl Acad Sci USA 67:*1741 (1970).

NICOLSON, G. L., and SINGER, S. J. Ferritin-conjugated plant agglutinins as specific saccharide stains for electron microscopy: Application to saccharides bound to cell membranes. *Proc Natl Acad Sci USA 68:*942 (1971).

OTTEN, J., JOHNSON, G. S., and PASTAN, I. Cyclic AMP levels in fibroblasts: Relationship to growth rate and contact inhibition of growth. *Biochem Biophys Res Comm 44:*1192 (1971).

PONTECORVO, G. Induction of directional chromosome elimination in somatic cell hybrids. *Nature 230:*367 (1971).

ROTH, S., and WHITE, D. Intercellular contact and cell-surface galactosyl transferase activity. *Proc Natl Acad Sci U.S.A 69:*485 (1972).

TAYLOR, J. H., MYERS, T. L., and CUNNINGHAM, H. L. Programmed synthesis of deoxyribonucleic acid during the cell cycle. *In Vitro 6:*309 (1971).

MULTIPLICATION AND GENETICS OF ANIMAL VIRUSES

MULTIPLICATION 1140

HOST CELLS FOR VIRAL MULTIPLICATION 1140
  Cell Strains and Cell Lines 1140
PRODUCTIVE INFECTION 1141
  Infection as a Function of the Nucleic Acid 1141
  Release of the Nucleic Acid 1143
ONE-STEP MULTIPLICATION CURVES 1144
EFFECT OF VIRAL INFECTION ON HOST
    MACROMOLECULAR SYNTHESES 1146
SYNTHESIS OF DNA-CONTAINING VIRUSES 1147
  DNA Replication 1149
SYNTHESIS OF RNA VIRUSES 1150
  Viral Proteins 1151
  Replication of Single-stranded RNA 1153
  Site of Replication of Viral RNA (Classes I, II, and III) 1154
  Effect of Inhibitors Acting on Cellular DNA 1154
  Replication of Double-stranded RNA (Class IV Viruses) 1154
  Replication through a DNA-containing Replicative
    Intermediate (Class V Viruses) 1155
MATURATION AND RELEASE OF ANIMAL VIRUSES 1157
  Naked Icosahedral Viruses 1157
  Enveloped Viruses 1158
  Complex Viruses 1161
  Phenotypic Mixing 1163
  Host-induced Modification of the Envelope 1163
ENDOSYMBIOTIC INFECTION 1163
ABORTIVE INFECTION 1163

GENETICS 1165

MUTATIONS 1165
  Mutant Types 1165
  Pleiotropism or Covariation 1166
  Complementation 1167
GENETIC RECOMBINATION 1167
  DNA Viruses 1167
  RNA Viruses 1167
  Genetic Organization 1167
  Biological Significance of Recombination 1168
  Reactivation of UV-irradiated Viruses 1168

MULTIPLICATION

The multiplication of animal viruses follows the pattern of bacteriophage multiplication described in preceding chapters, but with important differences related to the constitution of the viruses and of their host cells. Some animal viruses, for instance, contain a fragmented RNA genome; some contain a transcriptase and other enzymes in the virions, and express the viral genetic information by RNA → DNA transcription. Moreover, animal viruses differ from bacteriophages in their interaction with the surface of the host cells (which do not have rigid walls) and in the mechanism of release of their nucleic acid in the cell.

Like bacteriophages, animal viruses can be differentiated into **virulent** and **moderate;** the latter viruses resemble temperate bacteriophages in their ability to establish stable complexes with the host cells. The differentiation, however, is less sharp, and it is sometimes difficult to decide whether a virus is virulent or moderate.

The main characteristics of animal virus families are given in Table 48-1. Further details are found in later chapters of this book.

HOST CELLS FOR VIRAL MULTIPLICATION

Adult animals or embryos were at first used in experimental or diagnostic work with animal viruses, but except for the chick embryo, they have now been replaced almost completely by cultures of animal cells (Ch. 47).

CELL STRAINS AND CELL LINES

Every type of animal cell culture discussed in Chapter 47 has found application in virology. The choice of species and tissue of origin and type of culture (primary, cell strains, or cell lines) depends on the virus and the experimental objectives. The systems used for the individual viral families will be given in the appropriate chapters. The origin and the characteristics of the most widely used cell lines are summarized in Table 48-2.

In the **production of viral vaccines** for human use, the permanent cell lines are not employed, for they resemble malignant cells in their cultural pattern and their aneuploidy, and so it is feared that viruses propagated in them may acquire genetic determinants of malignancy. On the other hand, the primary cultures or diploid cell strains that are used often contain a variety of latent viruses (Chs. 51 and 63), which may also constitute a health hazard.

The Chick Embryo. Chick embryos have contributed in an important way to the development of virology by conveniently providing a variety of cell types susceptible to many viruses. Various cell types can be reached by inoculating the embryo by different routes (Fig. 48-1). The use of these various cells will be discussed in chapters dealing with individual viruses.

Host Susceptibility. Each animal virus can replicate only in a certain range of **cells.** Nonsusceptible cells are classified as **resistant cells,** in which a block at an early step prevents all expression of viral function, and **nonpermissive cells,** in which a later step is blocked and some viral activities are expressed. In either case, a required cellular function is lacking (e.g., receptors for virus attachment or a factor required for expression of viral genes); a heterokaryon formed

TABLE 48-1. Main Characteristics of Animal Virus Families

Type	Nucleic acid–stranded	Symmetry of nucleocapsid	Naked (N) Enveloped (E)	Diameter of virus (A°)	Viral group (or family)	Specific viruses mentioned in this chapter
RNA	Single-stranded	Icosahedral	N	210–300	Picornaviruses (Ch. 55)	Mengovirus, poliovirus
			E	450	Togaviruses (Ch. 60)	Semliki Forest virus, western equine encephalomyelitis virus
		Helical	E	800–1200	Myxoviruses (Ch. 56)	Influenza virus, fowl plaque virus
				1250–3000	Paramyxoviruses (Ch. 57)	Newcastle disease virus, Sendai virus, measles virus
				700–800 × 1300–2400	Rhabdoviruses (Ch. 59)	
				800–1600	Coronaviruses (Ch. 58)	
		Unknown	E	100–1300	Arenaviruses Ch. 60)	
				1000	Leukoviruses (Ch. 63)	
	Double-stranded	Icosahedral	N	750–800	Reoviruses (Ch. 61)	Reovirus
DNA	Double-stranded	Icosahedral	N	700–900	Adenoviruses (Ch. 52)	
				430–530	Papovaviruses (Ch. 63)	Polyoma virus, SV 40
			E	1800–2000	Herpesviruses (Ch. 53)	Herpes simplex virus, pseudorabies virus, equine abortion virus
		Complex	E*	2000–2500 × 2500–3500	Poxviruses (Ch. 54)	Vaccinia virus
	Single-stranded	Icosahedral	N	180–220	Parvoviruses (Ch. 52)	Adeno-associated virus

* Lipid in outer coat, but not distinct envelope.

by fusing a susceptible and a nonsusceptible cell has the function and is usually susceptible. Resistance is bypassed by infection with naked nucleic acid.

Examples of resistance or nonpermissiveness will be frequently encountered in this chapter and in those dealing with specific viruses.

PRODUCTIVE INFECTION

INFECTION AS A FUNCTION OF THE NUCLEIC ACID

That cells of higher organisms can be infected by naked viral nucleic acid was first shown for tobacco mosaic virus RNA by Gierer and Schramm. This discovery was soon generalized to show that the highly purified RNA or DNA of many animal viruses is infectious and gives rise to synthesis of normal virions. In conjunction with the Hershey and Chase experiment with bacteriophage (Ch. 45, Separation of nucleic acid from coat), this result provided proof for the exclusive genetic role of the viral nucleic acid.

There are several important differences between infections by nucleic acid and by virions.

1) **The efficiency of infection with the nucleic acid is much lower,** by a factor of 10^{-6} to 10^{-8} in ordinary media, showing the important role of the viral coat in infectivity. It increases by several orders of magnitude in hypertonic solutions (e.g., 1 M NaCl) or in the presence of **basic polymers,** such as DEAE-dextran. These additions appear to protect the nucleic acid against nucleases and to increase its uptake by the cells. Even

TABLE 48-2. Cell Lines in Common Use in Virology: Most Are Certified by the Cell Culture Collection Committee and Are Maintained Frozen by the American Type Tissue Culture Collection

Name of cell line	Species of origin	Tissue of origin	Morph-ology*	Ploidy†	Kariology Model no.	Markers
HeLa	Human	Carcinoma, cervix	Epi	Aneu	79	
Detroit-6	Human	Sternal bone marrow	Epi	Aneu	64	
Minnesota-EE	Human	Esophageal epithelium	Epi	Aneu	67	Yes
L-132	Human	Embryonic lung	Epi	Aneu	71	
Intestine 407	Human	Embryonic intestine	Epi	Aneu	76	
Chang liver	Human	Liver	Epi	Aneu	70	
KB	Human	Carcinoma, oral	Epi	Aneu	77	
Detroit-98	Human	Sternal bone marrow	Epi	Aneu	63	
AV_3	Human	Amnion	Epi	Aneu	74	
Hep-2	Human	Carcinoma, larynx	Epi	Aneu	76	Yes
J-111	Human	Peripheral blood‡	Epi	Aneu	111	Yes
WISH	Human	Amnion	Epi	Aneu	74, 75	
$LLC-MK_2$	Monkey§	Kidney	Epi	Aneu	70	
BS-C-1	Monkey‖	Kidney	Epi	Quasidip		
HaK	Syr. hamster#	Kidney	Epi	Aneu	57	
BHK	Syr. hamster					
B14-FAF-G3	Ch. hamster**	Kidney	Fib	Eu		
Don	Ch. hamster**	Peritoneal cells	Fib	Quasidip	22	
CHO	Ch. hamster	Lung	Fib	Eu	22	
		Ovary	Fib	Quasidip	22	
L	Mouse	Connective tissue	Fib	Aneu	78	Yes
NCTC, clone 929	Mouse	Connective tissue	Fib	Aneu	66	Yes
NCTC, clone 2472	Mouse	Connective tissue	Fib	Aneu	52	Yes
NCTC, clone 2555	Mouse	Connective tissue	Fib	Aneu	56	Yes
CCRF S-180 II	Mouse	Sarcoma 180	Fib	Aneu	86	Yes
3T3	Mouse	Connective tissue	Fib	Aneu	75	

* Fib = fibroblast-like; Epi = epithelium-like.

† Aneu = aneuploid; Eu = euploid; quasidip = quasidiploid.

‡ *Monocytic leukemia.*

§ *Macaca mulatta.*

‖ *Cercopithecus aethiops.*

\# *Mesocricetus auratus.*

** *Cricetulus griseus.*

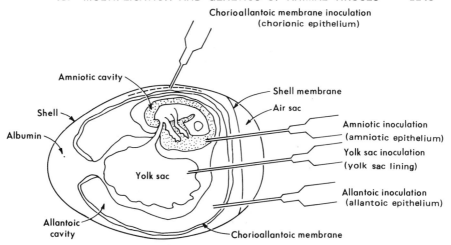

Fig. 48-1. The chicken embryo (10 to 12 days old) and routes of inoculation to reach the various cell types (as indicated). For chorioallantoic membrane inoculation a hole is first drilled through the egg shell and shell membrane; the shell over the air sac is then perforated, causing air to enter between the shell membrane and the chorioallantoic membrane, creating an artificial air sac, where the sample is deposited. The sample comes in contact with the chorionic epithelium. Yolk sac inoculation is usually carried out in younger (6-day-old) embryos, in which the yolk sac is larger.

under the most favorable conditions, however, the efficiency of infection is not more than 1% that of the corresponding virions. This limitation seems to arise from degradation of much of the nucleic acid within the cells.

2) **The host range is much wider with nucleic acids,** which can infect resistant cells. For instance, chicken cells, although resistant to poliovirus, are susceptible to its RNA; but only a single cycle of viral multiplication takes place because the progeny are again virions.

3) **Infectious nucleic acid can be extracted even from heat-inactivated viruses,** in which the protein of the capsid has been denatured; the nucleic acid can withstand much higher temperatures than the protein.

4) **Finally, the infectivity of nucleic acid is unaffected by virus-specific antibodies,** which suggests that this form of a virus could be an effective infectious agent even in the presence of immunity. However, nucleases in body fluids probably greatly limit its role, because a single complete break in a molecule abolishes its infectivity. Indeed, it is not clear whether naked viral nucleic acid plays any role in natural infection. Nevertheless, in the preparation of viral vaccines the ability of nucleic acid infectivity to survive damage to the viral coat must be considered (Ch. 64).

Only small DNA-containing viruses, togaviruses and picornaviruses, yield infectious nucleic acids. Failure in other cases can be ascribed either to the difficulty of extracting a large DNA intact or to **lack of virions' enzymes** (Ch. 44) which are required for initiating the viral growth cycle and are removed during extraction.

RELEASE OF THE NUCLEIC ACID

In ordinary infection entry of the viral nucleic acid into a cell follows the main steps observed with bacteriophages, i.e., adsorption of the virions to the cells, followed by release of the nucleic acid. These steps can be investigated by studying the changes of viral infectivity or the fate of radioactively labeled viral components, or by electron microscopy. The following steps can be identified:

1) **Attachment or adsorption.** The virus becomes attached to the cells, from which it can be **recovered in infectious form** without breaking them. Most animal viruses lack specialized attachment organs and probably have attachment sites distributed over the virions' surface. However, adenoviruses attach through the penton fibers (see Chs. 44 and 52) which if purified still adsorb to susceptible cells.

Adsorption occurs to specific **receptors:** for myxoviruses they are mucoproteins similar to the red cell receptors involved in hemagglutination (Ch. 44), and for poliovirus they are lipoproteins. Whether or not receptors for a certain virus are present on a cell depends on the species and the tissue from which it derives and on its **physiological state.** Thus, cells of monkey kidney and testis, and of human amnion, when first isolated from the organ, lack demonstrable poliovirus receptors but develop them within a few days of cultivation in vitro.

Attachment to receptors appears as a necessary step in infection: thus cells naturally lacking poliovirus receptors or in which receptors for influenza virus are enzymatically destroyed are resistant (see Ch. 51).

As with phages, attachment is favored by optimal **cation concentrations.** Adsorbed viruses can be recovered in infectious forms by various procedures that either destroy the receptors or weaken their bonds to the virions. Poliovirus is removed by detergents, low pH, or high salt concentrations; myxoviruses by neuraminidase; and certain echoviruses by chymotrypsin.

2) **Penetration.** After this step, which rapidly follows adsorption, the attached virus can no longer be recovered from the intact cells. Electron micrographs show that **with enveloped viruses** this step occurs when the virion's envelope fuses with the cellular membrane (Fig. 48-2). With **naked virions** the complete virion penetrates morphologically intact through the cellular plasma membrane into the cytoplasmic matrix.

Viruses of all morphological types can also be taken into the cells by **phagocytosis,**[*] as shown by electron micrographs. It is not clear whether this phenomenon is relevant for penetration, since engulfed virions are still separated from the cytoplasmic matrix by the plasma membrane that surrounds the phagocytic vesicles. Phagocytosis may enhance the rate of penetration by preventing loss of eluted virions.

3) **Uncoating and eclipse.** Uncoating is detected by the accessibility of the viral nucleic acid to nucleases after disruption of the cells. Eclipse is recognized by failure to recover viral infectivity from the disrupted cells (although **nucleic acid** infectivity may be recoverable). For naked viruses the two events probably reflect the same process, i.e., alteration of the capsid. With enveloped viruses eclipse results from the loss of the envelope at penetration, whereas uncoating requires the additional alteration of the capsid.

For some time after eclipse of icosahedral viruses the capsid does not appear morphologically altered (Fig. 48-2). However, invisible structural changes of the capsid evidently occur in poliovirus virions while still attached outside the cells, since about half of those originally adsorbed elute and are unable to reattach; similar alterations are produced in vitro by exposure to fragments of membranes of susceptible cells. The poliovirus changes, which have a high temperature coefficient, are accompanied by the loss of a small superficially situated capsid protein (VP4; see below, Maturation), which is required for adsorption. Later the capsid breakdown becomes morphologically noticeable; it probably results from attack by proteolytic enzymes (Fig. 48-3). The poliovirus evidence suggests that with naked viruses breakdown is initiated by the **interaction of the viruses with cellular constituents** (such as the receptors).

Infection with poxviruses and reoviruses involves some more specialized steps. Poxvirions begin losing their outermost coat immediately after penetration, often in phagocytic vacuoles, releasing complex cores into the cytoplasmic matrix. Some viral messenger RNA is then synthesized by a transcriptase present in the cores (see Ch. 44, Virions' enzymes) and a resulting enzyme completes the uncoating. Reovirions penetrate into lysosomes, where proteolytic enzymes strip the outer capsid, activating transcription of the genome by a transcriptase in the cores.

ONE-STEP MULTIPLICATION CURVES

As for bacteriophage, multiplication curves for animal viruses are obtained under one-step conditions (see Ch. 45). The phage technics are applied to cell suspensions from suspension cultures or obtained after disper-

---

[*] Phagocytosis (i.e., the ingestion of solid particles) is similar to pinocytosis (the ingestion of liquid droplets), which is described in Chapter 47.

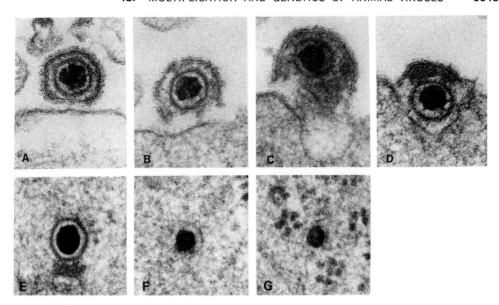

Fig. 48-2. Adsorption **(A)**, penetration **(B–D)**, and digestion of the capsid **(E–G)** of herpes simplex virus on HeLa cells, as deduced from electron micrographs of infected cell sections. The penetration involves local digestion of the viral and cellular membranes **(B, C)** resulting in fusion of the two membranes and release of the nucleocapsid into the cytoplasmic matrix **(D)**. The naked nucleocapsid is intact in **E**, partially digested in **F**, and has disappeared in **G**, leaving a core containing DNA and protein. [From Morgan, C., Rose, H. M., and Mednis, B. *J Virol 2:*507 (1968).]

Fig. 48-3. One-step multiplication curves of two viruses with intracellular accumulation periods of different length. Extracellular virus is measured in the medium surrounding the intact cells; intracellular virus after removal of the medium and disruption of the cells. **A.** Western equine encephalitis virus multiplies in cultures of chick embryo cells with an extremely short accumulation period (about 1 minute). **B.** Poliovirus multiplies in cultures of monkey kidney cells with a long accumulation period (about 3 hours). The relation between the titers of intracellular and extracellular virus in one curve is the inverse of the other. [**A.** From data of Rubin, H., Baluda, M., and Hotchin, J. *J Exp Med 101:*205 (1955). **B.** From data of Reissig, M., Howes, D. W., and Melnick, J. L. *J Exp Med 104:*289 (1956).]

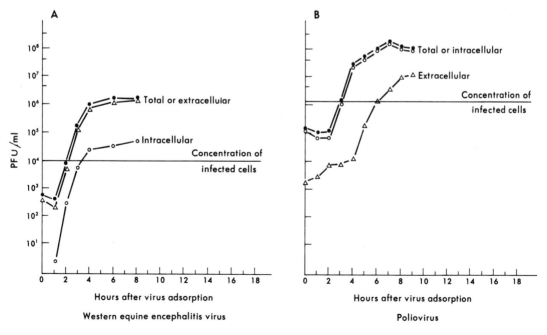

sion of layer cultures by trypsin. For viruses whose cell receptors can be easily destroyed (e.g., myxoviruses, polyoma virus) one-step conditions are possible even without dispersing the cell layers: the infected cultures are washed free of unadsorbed virus and then covered with a nutrient medium containing receptor-destroying enzyme, which prevents adsorption of released virus to the cells. Finally, for viruses that are not readily released from cells (e.g., poxvirus, adenovirus, herpesvirus) the mere washing of the monolayers, after infection, with viral antibody creates approximate one-step conditions, since most of the virus remains within the cells at the end of the growth cycle.

Approximate one-step conditions are also realized by infecting cultures with a large inoculum, which leaves no uninfected cells; the unadsorbed virus is then removed by washing the cells. Since all cells are infected at once, multiplication is necessarily confined to a single cycle. However, high multiplicity of infection may cause abnormal multiplication, and progeny virus may be lost by adsorption to cell debris.

Intracellular Virus.   One-step multiplication curves of animal viruses (Fig. 48-3) show the same stages observed with bacteriophages (Ch. 45). However, the length of the **intracellular accumulation period** varies over a wide range with different viruses; it is very long with some because the progeny virions tend to remain within the cells, and is nonexistent with others which mature and are released in the same act (by acquisition of an envelope at the cell surface). If the accumulation period is very short, the intracellular virus is at any time a small fraction of the total virus, as in Figure 48-3A.

Intracellular virus is measured as **cell-associated virus,** i.e., the virus released by disrupting the cells after they have been washed free of extracellular virus. Cell-associated virus, however, includes both true intracellular virus and extracellular virus that has become secondarily adsorbed to cellular receptors, and its titer is often much higher than that of true intracellular virus (e.g., myxoviruses).

EFFECT OF VIRAL INFECTION ON HOST MACROMOLECULAR SYNTHESES

In these studies viral and cellular nucleic acids are separated by their size, buoyant density,

configuration (e.g., cyclic, see Ch. 10), or hybridization to the nucleic acid present in purified virions. Viral and cellular proteins can be distinguished immunologically or by their different rate of migration in acrylamide gel electrophoresis (see Ch. 10).

As with bacteriophages, **virulent viruses,** either DNA-containing (e.g., adenovirus, vaccinia virus, herpesvirus) or RNA-containing (e.g., poliovirus, Newcastle disease virus, reovirus), turn off cellular macromolecular syntheses (Figs. 48-4 and 48-5) and disaggregate cellular polysomes (Fig. 48-6), favoring a **shift to viral syntheses** (Fig. 48-7). With most viruses the effect does not occur in the presence of inhibitors of protein synthesis (puromycin, cycloheximide), indicating that it is mediated by new proteins, probably virus-specified. The cellular DNA is not usually broken down, but **chromosome breaks** are often observed.

The mechanisms of inhibition of host syntheses are little understood. Cellular DNA synthesis seems to be blocked especially at initiation, since radioautography after ^{3}H-thymidine labeling shows that the replicated segments are of normal length but reduced in number. Inhibition of RNA synthesis often selectively affects the ribosomal RNA; with

Fig. 48-4. Inhibition of cellular DNA synthesis in L cells infected by equine abortion virus in the presence of ^{3}H-thymidine. Viral DNA was separated from cellular DNA because of its higher buoyant density in CsCl. [Modified from O'Callaghan, D. J., Cheevers, W. P., Gentry, G. A., and Randall, C. C. *Virology* 36:104 (1968).]

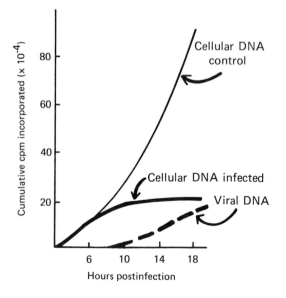

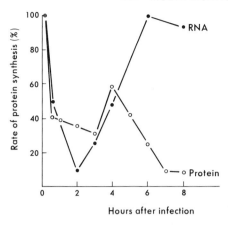

Fig. 48-5. Inhibition of cellular RNA and protein synthesis in L cells infected with mengovirus. The decline in the incorporation of radioactive precursors begins immediately after infection. The resumption of synthesis at about 3 hours is due to synthesis of viral RNA and proteins. [Modified from Franklin, R. M., and Baltimore, D. *Cold Spring Harbor Symp Quant Biol* 27:175 (1962).]

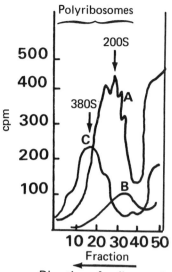

Fig. 48-6. Effect of poliovirus infection on polyribosomes of HeLa cells. Polyribosomes are studied by exposing the cells briefly to radioactive amino acids, then disrupting the cells and sedimenting the cytoplasmic extract through a sucrose gradient. The normal polyribosomes present in uninfected cells (most frequent sedimentation rate 200S; curve A) tend to disappear after infection (curve B), and after viral protein synthesis has begun, they are replaced by polyribosomes around 380S, which contain the viral RNA as messenger (curve C). [Modified from Penman, S., Scherrer, K., Becker, Y., and Darnell, J. E. *Proc Natl Acad Sci USA* 49:654 (1963).]

vaccinia virus it seems to be produced by proteins present in the infecting virions. Inhibition of protein synthesis also seems to occur at initiation: specifically, double-stranded RNA from reovirions or from poliovirus-infected cells inhibits initiation of hemoglobin synthesis in an extract of rabbit reticulocytes.

In contrast to virulent viruses, **moderate** viruses (e.g., polyoma virus) may **stimulate** the synthesis of host DNA, mRNA, and protein. This phenomenon is of considerable interest for viral carcinogenesis and will be discussed in Chapter 63.

SYNTHESIS OF DNA-CONTAINING VIRUSES

The expression and replication of the DNA of animal viruses take place along lines similar to those reviewed in Chapter 45 for bacteriophages.

Transcription.   All viral DNAs are transcribed into more than one RNA molecule: for instance, about 10 are recognized for adenovirus by acrylamide gel electrophoresis. The messengers may, therefore, be monocistronic, like those of animal cells (see Ch. 47 and below).

As with phage, there is a **temporal organiza-** tion. A fraction of the genome is transcribed by **early** messengers and the remainder by **late** messengers. Studies with inhibitors of DNA synthesis (such as arabinosyl cytosine) show that late transcription requires DNA synthesis, whereas early transcription does not. With the small papovaviruses early mRNAs are made continuously throughout infection, allowing transcription of the whole genome in the late period, but with the larger adenoviruses the synthesis of a fraction of early mRNAs stops after onset of DNA replication.

With most viruses early transcription is carried out by a **host** enzyme, since it is not blocked by the addition of inhibitors of protein synthesis at infection. With poxviruses, however, after loss of the outer envelopes a virus-specified transcriptase present in the virion's core transcribes a set of **immediate early** genes (ca. 14% of the genome);

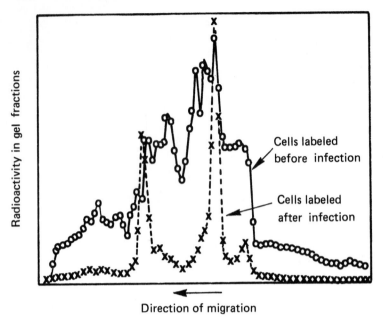

Fig. 48-7. Shift from the synthesis of cellular proteins to those of viral proteins after infection of BHK cells with influenza virus. The two superimposed acrylamide gel electropherograms were obtained from cells labeled with a radioactive amino acid either 1 hour before infection (circles) or between 4 and 7 hours after infection (crosses). The synthesis of viral proteins has completely replaced that of cellular protein. [Modified from Holland, J. J., and Kiehn, E. D. *Science 167*:202 (1970).]

Direction of migration

delayed early genes are transcribed after uncoating is completed.

Most viral DNAs are transcribed in the cellular nucleus (in which they replicate), as shown by cell fractionation after a short pulse of ³H-uridine. Within about 30 minutes after synthesis, the mRNAs are transported to the cytoplasm where they form polysomes; the mechanisms are similar to those used for cellular messengers, including the attachment

of polyadenylic acid tails (see Ch. 47, Transcription).

The DNAs of poxviruses are transcribed and replicated in the cytoplasm. The immediate early RNA is extruded from the cores, also in association with polyA tails, and forms polysomes in the surrounding cytoplasm. A product of this RNA is the enzyme thymidine kinase, whose synthesis is a useful indicator of immediate early transcription (Fig. 48-8).

Fig. 48-8. Evidence that the cessation of synthesis of thymidine kinase in cells infected by vaccinia virus is controlled by the expression of a gene (probably viral). The evidence derives from the effect of actinomycin D and puromycin. In the control cells, i.e., not exposed to any inhibitor, synthesis of the enzyme stops at about 5½ hours after infection; but earlier addition of actinomycin D allows synthesis to continue, showing that the block in this specific synthesis requires new RNA synthesis. Puromycin at 2 hours interrupts all enzyme synthesis, but after its removal at 5½ hours synthesis of thymidine kinase resumes, although it has stopped in the control; this result shows that synthesis of the enzyme stops only after some required protein synthesis. [Modified from McAuslan, B. *Virology 21*:383 (1963).]

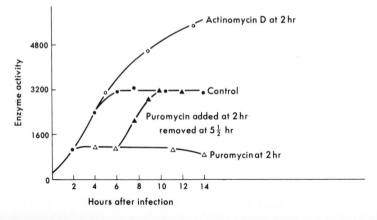

Studying the synthesis of this enzyme it could be shown that immediate transcription ceases after uncoating: thus if the delayed early transcription is inhibited by small doses of actinomycin D, immediate transcription (apparently more resistant to the inhibitor) continues indefinitely. The effect is probably caused by a delayed early gene product because a temporary halt of protein synthesis prolongs the immediate transcription.

Synthesis of Viral Proteins. Acrylamide gel electrophoresis of virus-infected cells reveals the presence of viral proteins, which are synthesized on cytoplasmic polysomes in a temporal sequence corresponding to that of mRNAs. With nuclear viruses structural proteins of the virions, which are specified by late genes, then migrate to the nucleus, whereas enzymes, specified by early genes, tend to remain in the cytoplasm.

The proteins can be identified as viral when they are made after the host protein synthesis is shut off or when they are absent in uninfected cells. Thus, a thymidine kinase appearing in thymidine kinase-less cells after infection with herpesvirus is probably viral.

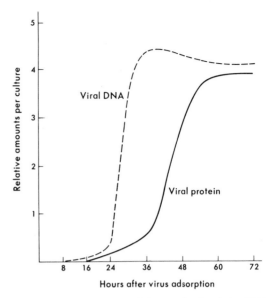

Fig. 48-9. Time course of the synthesis of the DNA and protein of polyoma virus in cultures of mouse kidney cells. Total viral DNA was determined by infectivity after phenol extraction, viral protein by hemagglutination; the amounts are given in arbitrary units. [Data from Dulbecco, R., Hartwell, L. H., and Vogt, M. *Proc Natl Acad Sci USA* 53:403 (1965).]

DNA REPLICATION

DNA replication utilizes precursors derived from the medium, since the cellular DNA is not degraded. Synthesis begins toward the middle of the eclipse period (Fig. 48-9). In contrast to bacteriophage infection, viral DNA remains unused in the infected cells at the end of the multiplication cycle. The difference may reflect the topographical disadvantage of virion assembly in large animal cells.

The replication of the DNA of animal viruses, like that of bacteriophages, depends on viral proteins; it can be prevented by mutations in viral genes or by inhibition of protein synthesis shortly after infection. If the inhibitor is added somewhat later DNA replication occurs but at a lower rate, and beyond the first half of the eclipse period its addition has no effect.

As with bacteriophages, the newly synthesized viral DNA enters a pool. Thus if the infected cells are exposed to a short pulse of a radioactive DNA precursor at any time during the eclipse period the label is distributed among virions finished at any subsequent time.

The mode of replication of the viral DNA is **semiconservative** (see Ch. 10, DNA replication; Fig. 48-10). The **replicative intermediate** is linear for viruses with **linear,** nonpermuted DNA, such as adenoviruses, and has the same length as mature molecules. Replication follows the **symmetrical model** discussed in Chapter 10, since during synthesis the growing strands are shorter than those of the viral DNA and are not linked covalently to parental DNA (see Ch. 10, DNA replication). The replicative intermediate of **cyclic** DNAs (e.g., SV 40) are also cyclic. Sedimentation and electron microscopic studies show that they have continuous parental strands, suggesting that replication is symmetrical and unwinding is carried out by a "swivel enzyme" (see Ch. 10, DNA replication, and Fig. 10-25). Concatemers, common in the replication of phage DNA (see Ch. 45), are not observed with animal viruses. Sedimentation and buoyant density analyses show that the replicative intermediate is associated with large lipid-containing particles derived from membranes.

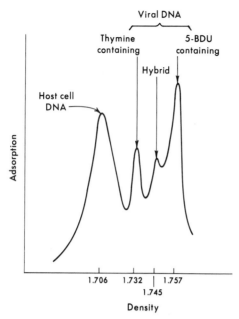

Fig. 48-10. Evidence for semiconservative replication of the DNA of pseudorabies virus. The infected cells were transferred from a medium containing thymidine to one containing the heavy precursor, 5-bromodeoxyuridine (5-BDU), where synthesis was allowed to continue. The DNA later extracted and analyzed by density gradient equilibrium centrifugation shows four density peaks, one of cellular DNA and three of viral DNA; one of the latter has a hybrid density, proving semiconservative replication. [Modified from Kaplan, A. S. *Virology* 24:19 (1964).]

SYNTHESIS OF RNA VIRUSES

We have seen in Chapter 45 that the RNA molecules present in the virions of RNA phages become associated with cellular ribosomes and· function as multicistronic messengers for the synthesis of viral proteins. In animal cells this method of gene expression seems precluded by the difficulty of translating multicistronic messengers (see Ch. 47). This obstacle is overcome in two ways: 1) with some viruses, the viral RNA acts as messenger but is **translated monocistronically into a giant peptide** which is then cleaved to generate distinct viral proteins; 2) in other viruses, **the viral RNA is transcribed into monocistronic mRNAs;** often the genome itself is a collection of separate monocistronic fragments.

RNA-containing animal viruses can be distinguished in five different classes, according to the nature of the RNA in the virions and the relation of this RNA to the viral messenger (Fig. 48-11).

Class I (e.g., picornaviruses). These viruses contain a molecule of single-stranded RNA which **acts as messenger** specifying in vitro the synthesis of capsid protein (recognizable from its peptides). In vivo, the same RNA molecule must also initiate replication in order to explain infection by a single viral particle; since replication requires viral proteins (see below), the messenger function must be expressed first.

Class II (e.g., paramyxoviruses). These viruses have a molecule of single-stranded RNA which cannot serve as messenger, having opposite polarity **(antimessenger).** On this **viral** strand several (five to eight) complementary messenger molecules, apparently monocistronic are synthesized by a **virion transcriptase.**

Class III (e.g., myxoviruses). The single-stranded RNA is in seven or more distinct nonoverlapping pieces. As in the previous case they are of **antimessenger** polarity and are transcribed each into a complementary, apparently monocistronic, messenger by a **virion transcriptase.**

Class IV (e.g., reoviruses). These viruses contain 10 distinct, nonoverlapping segments of **double-stranded RNA;** each is transcribed into a monocistronic mRNA by a **virion transcriptase.**

The function of virions' transcriptases in classes II to IV virions is easily demonstrated in vitro. In vivo the complementary messenger RNAs are evidently made by the same mechanism since cycloheximide does not block their synthesis. In contrast to DNA viruses, there is no differentiation between early and late messengers in any of the preceding classes with the possible exception of reovirus (class IV).

Class V (e.g., leukoviruses). The distinct nonoverlapping segments of single-stranded RNA which constitute the genome probably have **messenger polarity.** Each may be transcribed into DNA by a **reverse transcriptase** present in the virions; mRNA is then probably transcribed on this DNA.

Two main points emerge from this classification: 1) viruses with an antimessenger

Class	Genome constitution and polarity	Transcriptase in virions	Infectivity of RNA	Messenger	Primary gene product

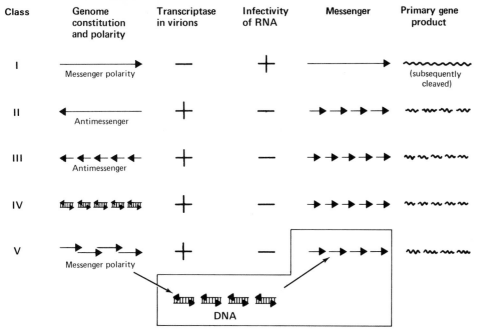

Fig. 48-11. The various classes of RNA viruses. Events enclosed in boxes are hypothetical. The number of multiple genome pieces, messengers, and gene products are only indicative.

strand also have a transcriptase in the virions which synthesizes the complementary messenger; 2) antimessenger viral RNA cannot be infectious when purified and separated from its transcriptase. Hence viruses of class I but not II and III yield infectious RNA. Viruses of class V behave aberrantly because they have a transcriptase in the virions, but the RNA is apparently of messenger polarity. In agreement with the former property, the RNA lacks infectivity. The significance of this aberrant behavior is not clear (see Ch. 63).

VIRAL PROTEINS

These proteins are synthesized in two different patterns which satisfy the monocistronic nature of the messengers.

1) When a virus has several messengers each appears to be translated into a single polypeptide chain, since the number of viral proteins identifiable by acrylamide gel electrophoresis is close to that of the number of messengers.

2) When, as with picornaviruses, the genome serves as a single messenger, a giant multicistronic polypeptide is made which is then cleaved to yield the several viral proteins. This cleavage, called processing, merits special description.

Processing. The mechanism was suggested by Maizel and by Baltimore to solve the paradox that the sum of all the 16 poliovirus proteins observed by electrophoresis corresponded to almost double the coding capacity of the RNA. Short pulse-labeling with radioactive amino acids followed by chases of various lengths showed that the largest peptides are precursors of the smaller ones (Fig. 48-12), and therefore are counted two or more times in the total protein enumeration. Cleavage is carried out by specific enzymes which recognize the secondary and tertiary structure of the protein. Inhibition of cleavage by amino acid analogs incorporated in the precursor peptide or by high temperature allowed detection of the giant polypeptide which is not normally recognized because it undergoes the first cleavage while being synthesized. A second cleavage produces three capsid proteins. A third cleavage of a capsid peptide occurs during maturation (see below).

Altered Processing. The processing pattern is altered by viral temperature-sensitive mutations which cause changes in the secondary and tertiary structure of the precursor peptides. Thus the attenuated L Sc poliovirus strain produces only four large peptides at the nonpermissive temperature (39°); processing is completed, yielding the normal

A   **B**

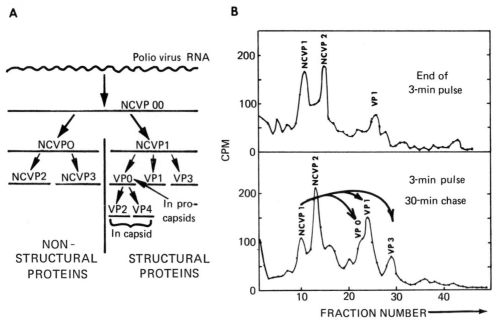

Fig. 48-12. Processing of the poliovirus precursor polypeptide. **A.** General Scheme. NCVP00 is a peptide of about 250,000 daltons observed only when cells are infected in the presence of amino acid analogs, which inhibit protein cleavage. **B.** Evidence for processing of a peptide (NCVP1) during poliovirus synthesis to generate three new peptides, VP0, VP1, and VP3. Infected cultures of HeLa cells were labeled for 3 minutes with a radioactive amino acid during the period of virus synthesis. The extract from a culture at the end of the labeling period was fractionated by acrylamide gel electrophoresis (top). The electropherogram shows a prominent NCVP1 peak, no VP0 or VP3, and a small VP1 peak. After a 30-minute chase with cold amino acids (bottom), NCVP1 is greatly reduced, having generated VP0, VP1, and VP3. VP0 is present in procapsids and is later processed into VP2 + VP4 which are found in the virions. NCVP = noncapsid viral protein; VP = virion protein. [Modified from Jacobson, M. F., Baltimore, D. *Proc Natl Acad Sci USA 61*:77 (1968).]

number of peptides, if the temperature is lowered.

RNA Replicase(s). An enzyme activity similar to that formed in bacteria infected with RNA phages (see Ch. 45) makes its appearance in animal cells infected by viruses with single-stranded RNA. These enzymes are not known to the same extent as the corresponding phage enzymes because they are more difficult to purify; however, they appear to have similar properties. In crude extracts or after partial purification, they incorporate label from ribonucleoside triphosphates into RNAs with the characteristics of those formed in vivo (see below). The incorporation requires all four triphosphates and suitable ionic conditions.

In poliovirus-infected cells **the replicase is unstable** since its level drops rapidly. This finding explains the dependence of viral RNA synthesis on **continuing** protein synthesis: RNA synthesis can be interrupted at any stage (Fig. 48-13) by adding puromycin or the amino acid analog *p*-fluorophenylalanine.

With virions containing a transcriptase, it usually performs a different synthesis than the replicase extracted from the infected cells (e.g., the transcriptase synthesizes messenger strands, the replicase antimessenger strands). However, both enzymes may contain a common subunit whose specificity is modified by interaction with virions' proteins, in the same way as the core replicase of RNA phages changes its specificity after binding a cellular subunit (see Ch. 45).

Regulation of the Formation of Viral Proteins. With reovirus there is evidence for **regulation of synthesis,** since certain peptides (recognizable by electrophoresis) are synthesized at much higher than average rates. With polio-

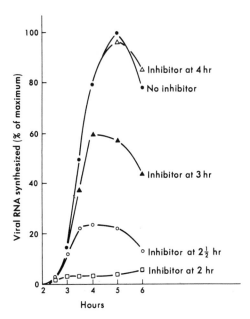

Fig. 48-13. Inhibition of the synthesis of poliovirus RNA by inhibitors of protein synthesis or function. Puromycin or *p*-fluorophenylalanine was added at various times after infection. The synthesis of viral RNA is detected from the incorporation of labeled uridine into RNA in the presence of actinomycin D, which prevents the synthesis of cellular RNA but not of viral RNA. [Modified from Scharff, M. D., Thoren, M. M., McElvain, N. F., and Levintow, L. *Biochem Biophys Res Commun* 10:127 (1963).]

virus regulation is caused by **processing:** proteins resulting from the first cleavage are available earlier in infection than those from late cleavages.

REPLICATION OF SINGLE-STRANDED RNA

Time Course. The synthesis of the RNA of many animal viruses can be studied by blocking the synthesis of cellular RNA by actinomycin D: viral RNA is still synthesized. With **naked icosahedral** viruses its appearance, measured either chemically or by infectious titer, is followed shortly by that of mature virions; for poliovirus the time difference is 30 to 60 minutes, varying with the experimental procedure. The synthesis of the RNA of **enveloped viruses** precedes the appearance of finished virions by a longer time interval (e.g., 2 hours for western equine encephalomyelitis). However, this RNA appears to be assembled early into the nucleocapsid, the extra time probably being spent in adding the envelope.

The fairly close temporal connection between synthesis and assembly may have evolved because of the inherent instability of the naked viral RNA in the cells: in the presence of puromycin, which prevents assembly, the RNA rapidly loses its infectivity.

The replication of single-stranded RNA of animal viruses is similar to that of phage RNA (see Ch. 45). We shall here summarize the findings related to poliovirus (class I) which also apply to other single-stranded RNAs (classes II and III) except for the polarity of the messenger strand in respect to the strand present in virions (see Fig. 48-11). With viruses of class III the genome fragments each have a nucleoside triphosphate at the 5′ end (where synthesis begins); hence they replicate independently.

Replicative Intermediate. Fractionation and characterization of RNA extracted from poliovirus-infected cells according to various procedures (sedimentation, electrophoresis, RNase sensitivity, solubility in 2 M LiCl, and melting profile)* identify three RNA forms whose viral origin is demonstrated by their infectivity: 1) regular viral RNA; 2) a partially double-stranded RNA with single-stranded tails **(replicative intermediate,** RI); and 3) double-stranded RNA (originally called replicative form, RF, although it may not be an intermediate in replication).

In infected cells (or their extracts) a very short pulse of a radioactive precursor labels mostly RI; the label then chases into single-stranded viral strands (Fig. 48-14). This result shows that **RI is the true replicative intermediate.** It contains a complete **complementary** strand and several growing viral strands. Each growing strand is weakly held to the complementary strand, probably by a replicase molecule: if a homogenate of infected cells is treated with a protein denatu-

---

* Single-stranded RNAs migrate faster in sedimentation and electrophoresis because they are less extended; they are insoluble in 2 M LiCl and melt slowly with increasing temperature. Double-stranded RNAs are resistant to RNase and melt suddenly, like DNA (see Ch. 10).

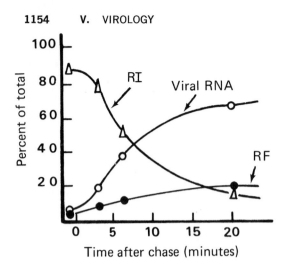

Fig. 48-14. Kinetic evidence that RI is the intermediate generating progeny viral RNA strands in a cell-free system from infected HeLa cells. Label was first incorporated from ³H-uridine triphosphate into the RNAs; at time O the radioactive precursor was replaced with unlabeled uridine triphosphate, beginning the chase. Radioactivity disappearing from RI is found mostly in viral RNA, showing the precursor-product relation of the two molecular types. In contrast, radioactivity slowly accumulates in RF, showing that it, like the viral RNA, is an end product in the synthesis. [Modified from Girard, M. *J Virol* 3:376 (1969).]

rant, the strands separate. It was formerly thought that the complementary strands are held by extensive base pairing, but it is now recognized that pairing is an artefact arising during the classic extraction of the homogenate with phenol, before the anchoring enzyme is denatured and removed (Fig. 48-15).

The RFs may be spent RIs which have finished synthesizing RNA, since they are not kinetic intermediates and they accumulate with time (Figs. 48-14 and 48-15). Poliovirus complementary strands isolated from RFs are not infectious: they are antimessenger and cannot specify the viral replicase required for generating viral strands. **Complementary strands are not found free.** Presumably after they are formed on viral strands, they immediately generate RIs.

SITE OF REPLICATION OF VIRAL RNA (CLASSES I, II, AND III)

The viral RNA replicates in the cytoplasm, as shown by radioautographs of infected cells exposed to a brief pulse of tritiated uridine, in the presence of actinomycin D to inhibit the synthesis of cellular RNA. Labeled RNA is exclusively in the cytoplasm, whereas in uninfected cells (without actinomycin D) it is only in the nucleus (Fig. 48-16).

Cytoplasmic synthesis is also supported by **microsurgical** experiments, in which small bits of cytoplasm are separated from the cell body. If these bits are infected by poliovirus they synthesize RNA, whereas uninfected bits do not. Since the infected fragments also synthesize poliovirus protein, recognizable by immunofluorescence, the RNA they make is viral.

In homogenates of cells infected by poliovirus the RI is found associated with membrane proliferations resembling smooth endoplasmic reticulum (Ch. 47; see Fig. 48-17). The membrane component appears to play an important part since in cells infected by Semliki Forest virus (a togavirus, class I) RI formation is inhibited after treatment with phospholipase C. The infecting viral RNA does not end up in progeny virions and probably remains bound to the membranes.

EFFECT OF INHIBITORS ACTING ON CELLULAR DNA

To test whether cellular DNA has any role in the replication of viral RNA, the effect of various inhibitors on viral replication has been determined. Only viruses of two classes are affected by these inhibitors. Myxoviruses (class III) are inhibited by actinomycin D if it is added to the cells either before infection or shortly afterward; its addition toward the end of the eclipse period has much less effect. The drug inhibits specifically the synthesis of messenger by the virions' transcriptase. Whether this effect results from interference with the function of the virions' transcriptase or with a required cellular function remains to be determined.

Leukoviruses (class V) are inhibited by a larger group of inhibitors, as discussed below.

REPLICATION OF DOUBLE-STRANDED RNA (CLASS IV VIRUSES)

Each piece of reovirus RNA (Ch. 60) is replicated independently since each has a nucleoside diphosphate at the 5′ end (derived from partial dephosphorylation of the original triphosphate). Replication is intimately tied to transcription of the genome by the virions' transcriptase which generates the messenger in the virion cores. Complementary strands

Generation of viral RNA

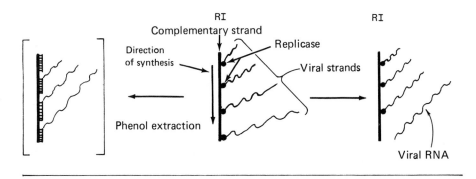

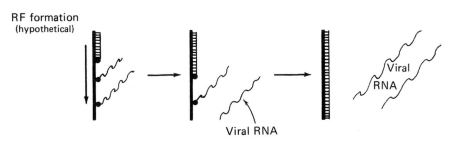

Fig. 48-15. RI and RF in poliovirus RNA replication. During replication progeny viral strands are held to the template (complementary) strand (thick lines) mainly by the replicase; base pairing is probably prevented by some factor bound to the nascent RNA. (Extensive base pairing may occur as an artefact during phenol extraction.) (For two possible structures of the RI, see Fig. 45-38.) RF appears to be formed when a progeny strand becomes base-paired to the template strand, possibly owing to lack of the factor or to alterations of the replicase.

are then made by another enzyme, the replicase. These, paired with their messenger strand templates, form the double-stranded molecules that end up in virions. Hence this replication, unlike that of DNA, is **asymmetrical** and conservative: the progeny is specified only by the minus strand of the virions and the parental DNA does not end up in the progeny (Fig. 48-18). Replication is fundamentally equivalent to that of single-stranded RNA, with the difference that the double-stranded rather than the single-stranded form ends up in virions.

REPLICATION THROUGH A DNA-CONTAINING REPLICATIVE INTERMEDIATE (CLASS V VIRUSES)

The replication of leukoviruses (Ch. 63) is blocked if antinomycin D or an inhibitor of DNA synthesis is added immediately after infection; later in infection viral replication becomes insensitive to inhibitors of DNA synthesis but remains susceptible to actinomycin D. To explain these findings Temin proposed that the viral replicative intermediate is double-stranded DNA, and that progeny viral RNA is obtained from it by regular transcription. Strong support for this theory was afforded by the discovery by Baltimore and Temin of the **reverse transcriptase** present in virions, which synthesizes DNA using the single-stranded viral RNA as template (see Ch. 11, Reverse transcription). This method of replication can fulfill the requirement for monocistronicity since, as with DNA viruses, the viral mRNA may be formed by the regular process of separate trancription of individual genes, using host enzymes.

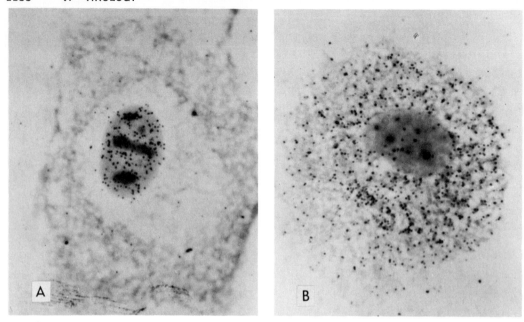

Fig. 48-16. Effect of infection with mengovirus on the distribution of RNA synthesis in the cell. An uninfected L cell **(A)** and one infected 6 hours earlier **(B)** were radioautographed after exposure for 3 minutes to ³H-uridine. The infected cell was also exposed to actinomycin D. In the uninfected cell the silver grains produced by the β-radiation of decaying ³H atoms are concentrated over the nucleus; in the infected cells there are only a few grains over the nucleus, because the synthesis of cellular RNA is inhibited, and many grains over the cytoplasm, due to synthesis of viral RNA. [From Franklin, R. M., and Baltimore, D. *Cold Spring Harbor Symp Quant Biol 27*:175 (1962).]

Fig. 48-17. Electron micrograph of part of the cytoplasm of a HeLa cell infected by poliovirus (picornavirus, Ch. 55), showing a focus of viral reproduction. **Left.** Many viral particles are present in the cytoplasmic matrix (some in small crystals) around or within membrane-bound bodies (B) and vacuoles (Va). **Right.** Presence of empty capsids (arrows). [From Dales, S., Eggers, H. J., Tamm, I., and Palade, G. *Virology 26*:379 (1965).]

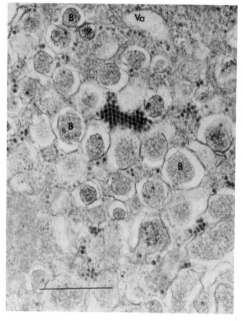

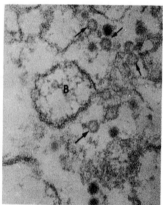

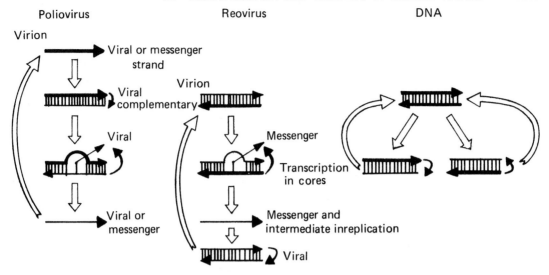

Fig. 48-18. Diagram showing the basic identity of the replication of poliovirus and reovirus RNA; the main difference is the encapsidation of a different form in the virions. In the replication of double-stranded RNA information flows (curved heavy arrows) **from only one strand** of the virion's RNA (the heavy one). This replication therefore is not like that of DNA, where information flows **from both strands.** New strands are shown as thin lines. (The first step in the poliovirus replication is indicated for simplicity as the formation of a double-stranded molecule. In reality such molecule may never exist, because viral strands start forming before the duplex is complete, immediately generating the RI.)

MATURATION AND RELEASE OF ANIMAL VIRUSES

Maturation of naked, enveloped, and complex viruses has different characteristics; it will therefore be useful to consider these several groups of viruses separately.

NAKED ICOSAHEDRAL VIRUSES

Maturation. As with bacteriophages, maturation consists of two main processes: the assembly of the capsid and its association with the nucleic acid. For DNA viruses the two steps are clearly separate, since DNA synthesis precedes the appearance of recognizable capsid components sometimes by several hours. For RNA viruses the two steps proceed almost concurrently.

Pulse-chase experiments with radioactive amino acids show that the newly formed polypeptide chains are rapidly assembled into capsomersized structures which are later assembled into capsids.

Association of the Capsid with the Nucleic Acid. With icosahedral animal viruses, as with bacteriophage (Ch. 45), **empty capsids**

generated during replication (Fig. 48-17) have been shown to be precursors of virions **(procapsids):** labeled amino acids added to infected cells appear first in small proteins, then in capsomer-like components, later in procapsids, and finally in complete virions.

With poliovirus, as with bacteriophages, the association of the nucleic acid with the procapsid is accompanied by peptide **processing,** in which peptide VP0 is split into VP_2 + VP_4 (Fig. 48-12). Presumably an RNA molecule penetrates between the capsomers from the outside, triggering the cleavage reaction. VP_4 fragments may be responsible for locking in the RNA and covering it on the outside, because their loss at uncoating (see above) and during heat inactivation (see Ch. 64) is accompanied by separation of RNA from the capsid.

After they are assembled, icosahedral virions may become concentrated in large numbers at the site of maturation forming the intracellular crystals (Fig. 48-17) frequently observed in thin sections of infected cells.

Release. Naked icosahedral virions are released in different ways, which depend both on the virus and the cell type. Poliovirus, for

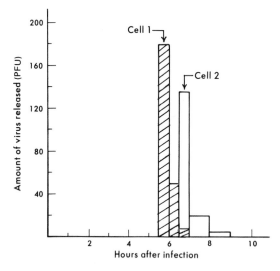

Fig. 48-19. The viral yield from two single monkey kidney cells infected by poliovirus. The cells were obtained by disrupting a monolayer of monkey kidney cells by trypsin; after being infected with poliovirus each was introduced into a separate small drop of medium immersed in paraffin oil. Every ½ hour the medium of each drop was removed, replaced with fresh medium, and assayed for infectivity by plaque assay. Note that with either cell the release was rapid (most virus came out in ½ hour), and note also the difference in latent period. [Data from Lwoff, A., Dulbecco, R., Vogt, M., and Lwoff, M. *Virology 1*:128 (1955).]

instance, is **rapidly released** from HeLa cells, in which it is highly virulent; the intracellular accumulation is then small. The study of single infected cells, contained in small drops of medium under paraffin oil (Fig. 48-19), shows a total yield of about 100 plaque-forming units (more than 10^4 viral particles) released over a period of ½ hour. During release the cells show rupture of surface vacuoles and surface bubbling with detachment of small cytoplasmic blebs. In contrast, virions of DNA viruses that mature in the nucleus do not reach the cell surface as rapidly, and are released when the cells undergo autolysis or in some cases are extruded without lysis. In either case they tend to *accumulate* within the infected cells over a long period.

ENVELOPED VIRUSES

In the maturation of enveloped viruses a capsid must first be assembled around the nucleic acid to form the nucleocapsid, which is then surrounded by the envelope.

Nucleocapsid. The viral proteins are all synthesized on cytoplasmic polysomes and are rapidly assembled into capsid components (recognizable by immunofluorescence or electron microscopy). With paramyxoviruses these components are first detected in a perinuclear zone of the cytoplasm 3 hours after infection and then accumulate there (Fig. 48-20). In contrast, with myxoviruses (e.g., influenza and fowl plague virus) the capsid protein is recognizable first in the nucleus (by 3 hours) and 1 hour later in the cytoplasm (Fig. 48-21).

The role of the nuclear migration of capsid proteins in myxovirus replication is undefined. It is not clear, for instance, whether the nucleocapsids that appear in progeny virions regularly go through the nucleus or remain in the cytoplasm. With measles virus the proportion of capsids found in the nucleus or in the cytoplasm varies in different cell types, and only cytoplasmic accumulation is accompanied by release of infectious virus.

With viruses whose genome is fragmented (class III) the various progeny RNA pieces appear to form a unique nucleocapsid in which they are held together.

Envelope. In all cases, the envelope proteins go directly to the cell membrane. These proteins are virus-specified and are synthesized de novo after infection. In contrast, **the lipids and the carbohydrates are produced by the appropriate enzyme systems of the host cell.** Thus the individual lipids of the viral envelope are very similar to those of the cellular membranes, not only in relative proportions but also in relative specific activities if the virus-producing cells are labeled before infection with a radioactive precursor. Furthermore, if the plasma and the nuclear membrane of a cell differ from each other in composition, the viral envelope will have the constitution of the membrane where its synthesis takes place (e.g., the nuclear membrane for herpesvirus; the plasma membrane for myxo- and paramyxoviruses). The viral membrane is thus formed by aggregation of preexisting cellular lipids with new viral proteins, and the carbohydrates are added to suitable receptor sites on these proteins by cellular glycosyl transferases. The same virus grown in different cells will therefore differ in its lipids and

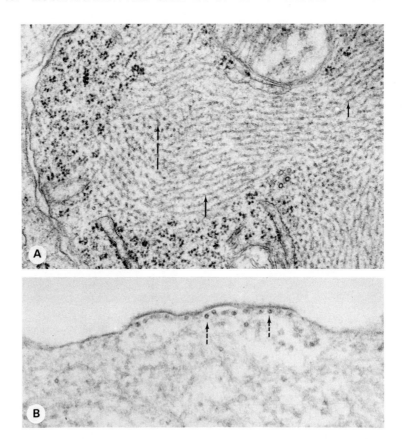

Fig. 48-20. Maturation of enveloped virus with helical nucleocapsid (Sendai virus) as seen in electron micrographs of infected chick embryo cells. **A.** Accumulation of nucleocapsids in the cytoplasm, some cut transversely (dashed arrows), some longitudinally (solid arrows). **B.** Transversely cut nucleocapsids lie under thickened areas of the plasma membrane, covered by spikes, preliminary to budding and virion formation. [From Darlington, R. W., Portner, A., and Kingsbury, D. W. *J Gen Virol* 9:169 (1970).]

carbohydrates, with consequent differences in physical, biological, and antigenic properties (see also Ch. 44). The viral proteins, however, to a certain extent also select the lipids with which they aggregate: thus two strains of influenza virus (A and B) grown in the same cells differ in the proportions of individual phospholipids.

Formation and Release of Virions. Envelopes are formed around the nucleocapsids by **budding of cellular membranes** (Fig. 48-22).

The budding of the virions can be considered the result of an intimate adhesion between the nucleocapsid and the cytoplasmic side of the viral proteins embedded in the cell membrane (Fig. 48-20B), which causes the membrane to curve into a protruding sphere surrounding the nucleocapsid.

The bud detaches from the membrane by a process that can be considered the reverse of penetration. If the budding occurs on the surface membrane, release occurs at the same time. If budding occurs in a vacuole, release requires subsequent fusion of the vacuole with the cell membrane. Either method of release is compatible with cell survival and can be very efficient, allowing a cell to release thousands of viral particles per hour for many hours. Herpesvirus buds out of the nuclear

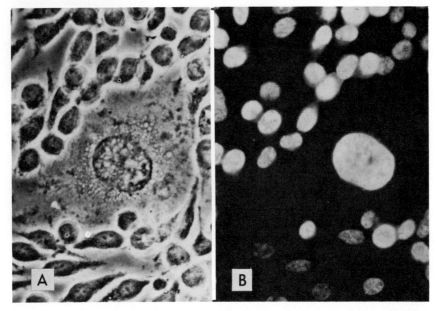

Fig. 48-21. Localization of fowl plague viral antigen in the nucleus 3 hours after infection of a culture of L cells. The cells were fixed and treated with fluorescent antibody to the viral antigen. **A.** Phase contrast micrograph. **B.** Micrograph of the same field in ultraviolet light, where only the antibody bound to the viral antigen is visible. The absence of fluorescence in the cytoplasm is especially evident in the giant cell. [From Franklin, R., and Breitenfeld, P. *Virology 8*:293 (1959).]

membrane into the cytoplasm and reaches the outside through newly made cytoplasmic channels.

With some myxoviruses a deviation from the normal pattern of maturation leads to the formation of infectious cylindrical **filaments** with a diameter similar to that of the spherical particles. Their formation depends on many factors, such as genetic properties of the virus, type of host cell, and environment (e.g., it is greatly enhanced in the presence of vitamin A alcohol or surfactants).

Cellular Alterations Connected to Virus Maturation. With myxo- and paramyxoviruses the budding occurs in membrane regions containing the virus-specified building blocks of the envelope (e.g., the hemagglutinins). These viral constituents become incorporated

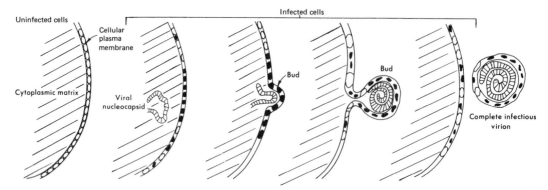

Fig. 48-22. Scheme of the budding of an enveloped virion (myxo- or paramyxovirus). White circles = building blocks of the normal cellular membrane (specified by cellular genes); black circles = viral proteins (specified by viral genes), which become incorporated in the cell membrane **before** budding of viral particles begins.

in the membranes **before budding begins,** conferring on the cell the properties of a giant virion. Thus the infected cells may bind erythrocytes **(hemadsorption,** the equivalent of hemagglutination, Fig. 48-23) or viral antibody, or they may fuse with uninfected cells to form multinucleated syncytia, called **polykaryocytes** (Fig. 48-24). This fusion is equivalent to adsorption of virions to uninfected cells.

Formation of **endogenous polykaryocytes** requires infectious virus, is slow, and is prevented by inhibitors of protein synthesis added together with the virus. In contrast, formation of **exogenous polykaryocytes** by external viruses (especially for the purpose of cell hybridization, see Ch. 47) can be caused by inactivated virus, is rapid, and is not affected by the inhibitors.

COMPLEX VIRUSES

The maturation of **poxviruses** can be inferred by ordering electron microscopic images in a reasonable time sequence (Fig. 48-25). Syn-

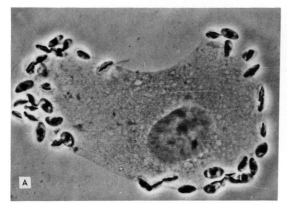

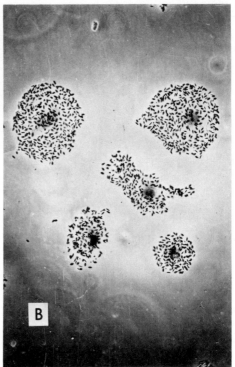

Fig. 48-23. Hemadsorption of HeLa cells infected by Newcastle disease virus. The cells used had been heavily irradiated with X-ray several days before infection; they stopped multiplying but increased in size and became giant cells, facilitating observations. The virus multiplies regularly in these cells. **A.** A cell 5 hours after infection. The ability to adsorb chicken erythrocytes begins to appear at two opposite regions of the cell surface. At these regions new cell membrane appears to be laid down, allowing viral components to become incorporated together with the cellular components. **B.** Nine hours after infection, cells at lower magnification, showing that the entire cellular membrane has developed the capacity for hemadsorption. The erythrocytes are firmly attached to the cells and are not removed by repeated washing. [From Marcus, P. *Cold Spring Harbor Symp Quant Biol 27*:351 (1962).]

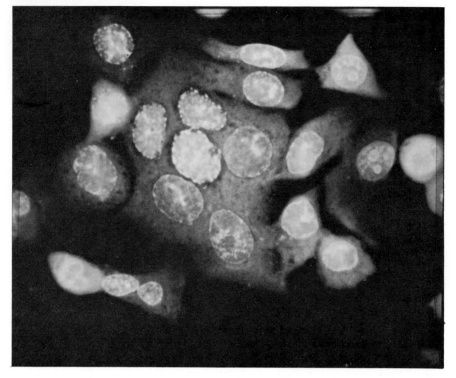

Fig. 48-24. Formation of multinucleated syncytia (polykaryocytes) by Hep-2 cells infected by herpes simplex virus. Five cells have fused completely, and several others partly, into a central mass. The cells were stained with the fluorescent dye acridine orange and photographed in a dark field. [From Roizman, B. *Cold Spring Harbor Symp Quant Biol 27*:327 (1962).]

thesis of the viral constituents takes place in cytoplasmic foci called **factories.** At the beginning of maturation these foci contain fibrils 20 to 25 A in diameter, presumably DNA. A parcel of the fibrous material becomes surrounded by multilayered membranes, which differentiate into two membranes: the inner one contains the characteristic nucleoid, while the external one acquires the characteristic pattern (recognizable with negative staining) of the surface of the virion (Ch. 41). As with bacteriophages and other animal viruses, peptide cleavage during poxvirus maturation appears to perform an important morphogenetic

Fig. 48-25. Diagram of the development of vaccinia virus, reconstructed from electron micrographs of thin sections of infected cells. [Modified from Avakyan, A. A., and Byckovsky, A. F. *J Cell Biol 24*:337 (1965).]

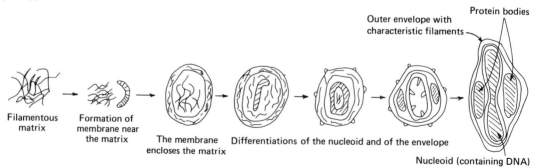

Outer envelope with characteristic filaments

Protein bodies

Filamentous matrix

Formation of membrane near the matrix

The membrane encloses the matrix

Differentiations of the nucleoid and of the envelope

Nucleoid (containing DNA)

Finished virions

role, since it is inhibited together with maturation by the antibiotic rifampin (see Ch. 49). In contrast to simple enveloped viruses, the poxvirus membrane contains newly synthesized lipids, whose composition differs from that of cellular lipids.

The maturation of poxviruses has important differences from the maturation of simpler viruses, since the virion differentiates after the precursors have been enclosed within the primitive membrane. This pattern represents a step toward a cellular organization, where differentiation is all internal. A further step in the same direction can be observed in the development of bedsoniae (Ch. 39). Thus **poxviruses may be transition forms toward bacteria.**

PHENOTYPIC MIXING

In a cell simultaneously infected by certain pairs of related viruses that differ in capsid antigens, such as poliovirus types 1 and 2 (Ch. 55), phenotypic mixing can occur, as with bacteriophage (Ch. 45), because capsids made by building blocks of either type (or both) may enclose the same genome. Indeed, when a part of the capsid has type 1 specificity and another part type 2, the particles are usually neutralized by antiserum to either antigen (see Ch. 50). Hence a mixture of poliovirus of types 1 and 2 yields six classes of virions, with different combinations of genotype (RNA) and phenotype (protein), as shown in Figures 48-26 and 48-27. With adenovirus types differing in the length of the corner fibers (see Ch. 52), phenotypic mixing leads to capsids with random combinations of antigenic types and fiber morphology. Phenotypic mixing affecting envelope glycoproteins can occur even between unrelated viruses, such as rhabdoviruses and paramyxoviruses or leukoviruses.

Hence with animal viruses, as with bacteriophage, building blocks are more or less **randomly assembled from pools.** Moreover, it can be shown that these building blocks are specified not solely by the nucleic acid of the infecting virions, but also by that of the progeny; thus preferential synthesis of one genotype in a mixedly infected cell leads to preferential synthesis of the corresponding proteins.

HOST-INDUCED MODIFICATION OF THE ENVELOPE

Host-induced modification, as of DNA bases found in bacteriophages, has not been observed with animal viruses. However, with enveloped viruses (e.g., myxo- and paramyxoviruses) the host cell influences the **lipids incorporated in the viral envelope.** Thus virions made in different kinds of host cells may differ in such properties as buoyant density, kinetics of heat inactivation, and hemolytic activity. These changes appear after a single cycle of multiplication in a new cell type, and since they affect all the progeny virions, they cannot result from the selection of mutants.

ENDOSYMBIOTIC INFECTION

In certain virus-cell systems the infected cells multiply for many generations, although some or all of them release virus. In such a cell population all or most individuals may be infected (as shown either by release of virus or by the presence of viral nucleic acids or antigens), **even when external reinfection is prevented** by the presence of strong antiviral serum. The cells may be functionally normal: for instance, leukovirus-infected chick embryos develop normally. In some cases, however, the cells are transformed, i.e., neoplastic (see Ch. 63).

Many viruses can cause such endosymbiotic infection both in animal and in tissue cultures, using different mechanisms. In some cases free viral genomes replicate in the cells and virus is produced; but this is possible if viral replication occurs at such a rate as to ensure infection of most daughter cells by random segregation at mitosis without interfering with cellular functions. In other cases the infection is maintained by integrated viral genomes which are regularly transmitted to the daughter cells at mitosis (see Ch. 63).

ABORTIVE INFECTION

The infection consists of an incomplete single cycle in which infectious virus is not produced, although the cells may be killed. Owing to a property either of the cells or of the

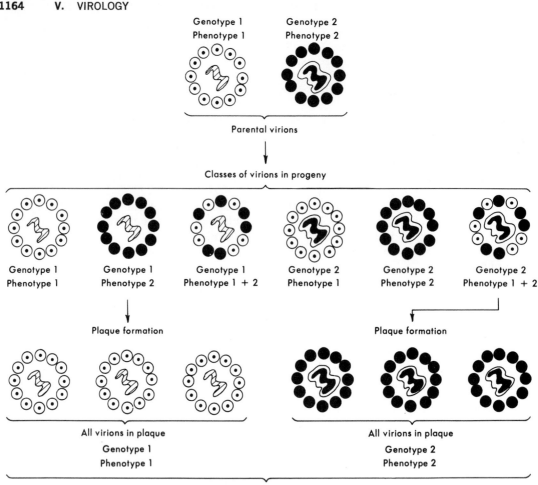

Genotype = the genetic information residing in the nucleic acid

Phenotype = the immunological characteristics residing in the capsid protein

Fig. 48-26. Mechanism underlying phenotypic mixing of the antigenic specificity of poliovirus. Cells mixedly infected by types 1 and 2 produce virions of six genotype and phenotype combinations. Cloning of these by plaque formation yields unmixed virions, with phenotype determined by the genotype of the initiating virion, irrespective of the latter's phenotype.

virus, some step in viral replication is defective.

Cell-dependent Abortive Infection. With certain strains of influenza virus in HeLa (human) cells, with Newcastle disease viruses in L (mouse) cells, or with some strains of herpesvirus in dog cells, nonpermissiveness is a property of the **species** of origin of the cells. In other cases, e.g., with measles virus in nerve cells, it depends on the **tissue** of origin.

The mechanisms are different in different cases. Thus in abortive myxovirus infection (influenza, measles) nucleocapsids accumulate in the nucleus, suggesting that the defect is in a cellular component involved in the assembly in or transport to the cytoplasm. In the herpesvirus case, the proliferations of the nuclear membrane, which normally provide the virion envelopes, are not formed and the nucleocapsid remains unenveloped in the nucleus.

Adenoviruses do not multiply in monkey cells unless the cells also contain an SV 40 genome which acts as **helper,** allowing translation of late adenovirus mRNAs in these cells.

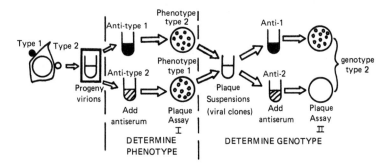

Fig. 48-27. Experimental basis for determining the phenotype and genotype—in respect to capsid antigens—of virions produced in cells mixedly infected by polioviruses, types 1 and 2 (see Ch. 55). Only one genotype determination on virus derived from a single plaque of assay I is shown here; it assigns a type 2 genotype to the progeny virion that generated that viral clone. Normally a number of plaques from either plate of assay I would be tested.

These results show that some steps in viral multiplication are cell-dependent.

Virus-dependent Abortive Infection. This is observed with **defective** viral mutants formed during viral replication. Thus defective myxovirus particles, **lacking the largest RNA piece,** are produced when infection is carried out at a high multiplicity (von Magnus phenomenon). These defective particles, if infecting alone, cause an abortive infection, but replicate if the cells are also infected by infectious particles, which act as helper. In serial passages at high multiplicity the defective particles become enriched until they constitute almost the whole yield, probably because the smaller genome segments have a selective advantage by replicating faster. Production and enrichment of the defective particles are partly cell-dependent.

The small oncogenic DNA-containing viruses (polyoma and SV 40) generate **deletion mutants** which express some gene functions and transform cells to the cancer state but do not produce progeny virions, except in cells coinfected by regular virus as helper.

An extreme case of viral defectiveness is observed with human adeno-associated viruses, which replicate only in cells co-infected by adenovirus; the absolute dependence on helper may be explained by the widespread occurrence of adenovirus infection in man.

GENETICS

The genetics of animal viruses is much less known than that of bacteriophages, largely because technics for selecting and identifying mutants or recombinants are less efficient, and, with many viruses, because of the low frequency of recombination.

guanidine, and fluorouridine for RNA viruses (see Ch. 11). The known mutant phenotypes are numerous and cover a larger range than bacteriophage mutations, because there are more ways for studying their properties: for instance, the various effects on animals.

MUTATIONS

Mutations of animal viruses occur spontaneously and can be induced by various mutagenic treatments, including nitrous acid, bromodeoxyuridine, hydroxylamine, nitroso-

MUTANT TYPES

Only one class of **conditionally lethal** mutations, **temperature-sensitive** (ts), occur in animal viruses; suppressor-sensitive mutations are unknown because no suppressors of nonsense codons (see Ch. 12) have been recognized in animal cells. The ts mutations have been extremely valuable because they occur in many (possibly all) genes, both in DNA-containing and in RNA-containing viruses. In spite of some defects, such as leakiness and high reversion rate, they form the basis for most of our present knowledge of animal virus genetics.

Other kinds of mutants affect special properties. 1) **Host-dependence,** these mutants fail to multiply in certain nonpermissive cell types (see above). Examples are the p mutants of rabbit poxvirus (Ch. 54), restricted in pig kidney cells, and the kb mutants of adenovirus 12 (Ch. 52) in the KB line of human cells. These mutants, like the suppressor-sensitive mutants of phage, can be propagated in permissive cells, are not leaky, and have a low reversion rate; however, they are limited to one or a few genes involved in overcoming the nonpermissiveness of the cells. 2) **Plaque size or type** (Fig. 48-28). These mutants are not as diverse as the corresponding phage mutants because animal virus plaques have less detail (they affect fewer and larger cells). Differences of plaque size may depend on differences either in parameters of the multiplication cycle or in the surface charges of the virions. Small charge differences affect plaque size because agar contains a sulfated polysaccharide which adsorbs the more highly charged virions, especially at certain pHs. 3) **Pock type.** Some poxvirus types produce white smooth pocks on the chorioallantoic membrane of the chick embryo, while others produce red, ulcerated pocks (u character). 4) **Surface properties,** detected by physical methods. 5) **Antigenic properties.** 6) **Hemagglutination** (in myxoviruses). 7) **Resistance to inactivation** by a variety of agents. 8) **Resistance toward or dependence on inhibitory substances during multiplication.** 9) **Pathogenic effect** for organisms. 10) **Host range.** 11) **Functions of certain viral genes in the infected cells,** such as production of thymidine kinase or induction of interferon.

PLEIOTROPISM OR COVARIATION

Mutants selected for a given phenotypic alteration are often found to be changed in other properties. For instance, poliovirus mutants with altered chromatographic properties often, but not always, have decreased neurovirulence for monkeys. This pleiotropism reflects the effect of a single viral protein on several properties of the virus. The connection between the properties is variable: for example, mutations that can modify different charged groups of the viral capsid may have similar effects on chromatographic adsorption but different effects on the more specific adsorption to cells.

Pleiotropism has useful applications. Thus by employing characters that are detectable in vitro it is easy to select attenuated (i.e., nonvirulent) strains for use as live vaccines, reserving the more cumbersome animal testing for final characterization. This approach has been used in the selection of poliovirus vaccines (Ch. 55).

Fig. 48-28. Two plaque-type mutants of fowl plaque virus on a monolayer of chick embryo cells. The wild type produces large round plaques with fuzzy edges; a small plaque mutant produces small plaques with irregularly indented outline and sharp edges. (Courtesy of H. R. Staiger.)

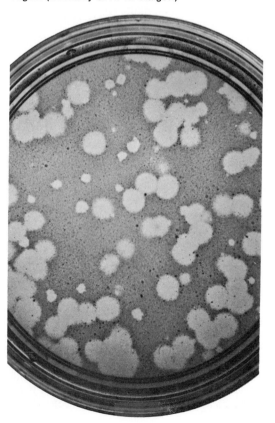

COMPLEMENTATION

With many viruses complementation of ts mutants (see Ch. 11, Mutations) has been useful in determining the functional organization of the viral genome. When two mutants complement each other the yield of the mixed infection at the nonpermissive temperature is greater than the sum of the separate yields at the same multiplicity of infection. A ratio greater than 10 between the two determinations is held significant.

The yield from the mixed infection is always much less than from cells infected with wild-type virus, perhaps because the diffusion of viral proteins in the cells is hampered by the membranes. In many DNA- or RNA-containing viruses several complementation groups have been identified. They probably identify different genes since their mutations usually display different functional alterations.

GENETIC RECOMBINATION

DNA VIRUSES

Among DNA viruses recombination has been studied largely with poxviruses, employing pock type (u) mutants (see above). Strains carrying independent mutations were crossed by mixed infection of tissue culture cells; the yield was then assayed on the chorioallantoic membrane, where the proportion of wild-type recombinants could be determined. In this way, u markers have been arranged in five clusters in a linear order, using the semiquantitative criterion that closely linked markers produce fewer recombinants than distant ones. This relation, however, was obscured by strong selection, during multiplication, for the wild type and for certain mutant genotypes. Recombination of ts mutants occurs in the absence of intracellular selection with a frequency varying between 0.02 and 20% for different pairs of markers. Genetic recombination has also been observed with adenoviruses and herpes simplex virus.

RNA VIRUSES

Burnet first recognized animal virus recombination with influenza virus in 1951. This and other RNA viruses with fragmented genomes (classes III, IV, and V) have much higher recombination frequencies than those with continuous genomes. The ts mutants of influenza virus or reovirus can be subdivided into groups with up to 20% recombination frequency between groups but none within groups. The groups probably correspond to RNA pieces which **assort nearly randomly** at maturation. In contrast, ts mutants of poliovirus, which has a continuous genome, recombine with less than 1% frequency, probably by true recombination (i.e., crossover or gene conversion, see Ch. 11). The low frequencies derive in part from the short length of the genome, in part perhaps from special features of RNA recombination. Most RNA virus recombinants seem to arise at the beginning of replication, probably between nonreplicated molecules.

GENETIC ORGANIZATION

Recombination frequencies between ts mutants of poliovirus are additive, yielding a linear map with a single linkage group. Another way of ordering poliovirus genes is the labeling of various proteins determined after processings, when a radioactive amino acid is added together with the antibiotic pactamycin which inhibits initiation of protein synthesis. Since the whole genome is translated into a single polypeptide chain there will be more label in proteins corresponding to the distal (3' end) part of the RNA, which has the highest chance of being still untranslated when the label is added. The relative labeling of various proteins gives their location in the uncleaved polypeptide chain. The genetic and molecular mappings agree.

The results show that poliovirus contains genes for structural proteins at the right (5' end) part of the RNA, and those for RNA replication in the left part. Thus at nonpermissive temperatures most of the ts right-end mutations allow RNA replication but prevent virion maturation, and at permissive temperature some cause synthesis of thermosensitive virions. Some mutations in the area confer on the virus a cystine requirement (apparently to compensate for alterations of the capsid) or guanidine resistance. Mutations in the left part of the genome prevent RNA replication at a nonpermissive temperature. The poliovirus map, therefore, like that of phages, is organized in discrete functional blocks, further stressing the evolutionary significance of this organization (see Ch. 45, Phage genetics).

Heterozygosity. In the yield of cells mixedly infected by two genetically marked strains of Newcastle disease virus about half the virions are composed of heterozygous particles, which on further multiplication produce virions of the two parental types as well as their own type. These particles differ from phage heterozygotes in two respects: 1) they are **complete** heterozygotes, i.e., for all the markers present; 2) they continue to give rise to a high proportion of heterozygous virions during their multiplication. These two properties have led to the speculation that a high proportion of the

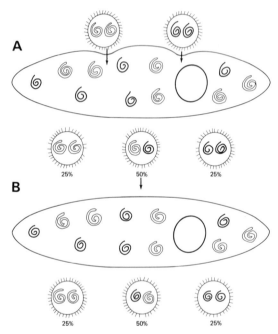

Fig. 48-29. Possible mechanism of persistent heterozygosity with Newcastle disease virus. In **A** a cell is infected by two viral particles of different genotypes; each particle is pictured with two nucleocapsids (i.e., two genomes). Both nucleic acids multiply in the cells, and when the progeny virions are formed most of them contain again two nucleocapsids, selected at random; the heterozygous particles then represent about 50% of the progeny. When a heterozygous particle multiplies **(B)** it again produces two types of nucleocapsids, and if they are again enclosed in randomly selected pairs in progeny particles, the proportion of heterozygous particles in the progeny is again 50%. The observed values are somewhat below 50%, probably because some particles have a single capsid.

Newcastle disease virions are **diploid,** i.e., contain two RNA molecules, and that heterozygous virions have one complete molecule from each parent. Since the virion is surrounded by a floppy envelope, with an outer diameter much larger than that of influenza virus, it could even contain two entire nucleocapsids. When a heterozygous virion replicates it would produce about 50% heterozygous virions if all progeny virions are again diploid, with RNA molecules selected at random (Fig. 48-29).

BIOLOGICAL SIGNIFICANCE OF RECOMBINATION

Cytological studies have suggested that poxviruses multiply in separate factories, each populated with the progeny of only one particle; but to account for recombination, **the factories must mix,** presumably when they grow sufficiently large. Similarly, the occurrence of recombination in RNA viruses, both icosahedral and helical, excludes their replication in entirely separate factories.

With animal viruses, as with phage, recombination may contribute to **genetic variability,** such as the high antigenic variability of influenza virus (Ch. 50).

REACTIVATION OF UV-IRRADIATED VIRUSES

Cross-reactivation. With viruses that give rise to efficient recombination (vaccinia, influenza) markers from a strain inactivated with ultraviolet (UV) light can be rescued by active virus simultaneously infecting the cells. This cross-reactivation could theoretically be generalized for introducing desirable characters into viruses, as in the production of vaccine strains. The survival of a given marker, as a function of the UV dose, is much greater than the survival of the whole virus (Fig. 48-30), as already seen with phages (Ch. 45).

Multiplicity Reactivation. Like bacteriophages, animal viruses capable of recombination also display **multiplicity reactivation** when UV-inactivated: as multiplicity of infection is raised the proportion of cells yielding infectious virus increases excessively.

Both cross- and multiplicity reactivations are much more pronounced for RNA viruses with a fragmented than with a continuous genome, in accord with the different recombination frequencies.

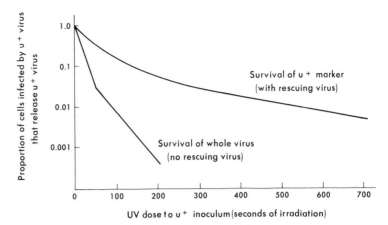

Fig. 48-30. Rescue of u+ markers from UV-irradiated rabbit poxvirus. KB cells were simultaneously infected by variously UV-irradiated u+ virus, at a multiplicity of 0.1 IU per cell (determined before irradiation), and with active u virus at high multiplicity. The infected cells were plated for plaque formation on monolayers of chicken embryo cells, where only the u+ virus forms distinct plaques while u virus forms minute, often invisible plaques). The u+ plaques formed measure the number of infected cells able to release at least one u+ infectious unit, which was used as a measure of the survival of the u+ markers. Parallel cultures, infected only with the irradiated u+ virus, were used to determine the survival of its infectivity in singly infected cells. [Data from Abel, P. *Virology 17:*511 (1962).]

SELECTED REFERENCES

Books and Review Articles

BALTIMORE, D. Expression of animal virus genomes. *Bacteriol Rev 35:*235 (1971).

Basic mechanisms in animal virus biology. *Cold Spring Harbor Symp Quant Biol 27* (1962).

FENNER, F. The genetics of animal viruses. *Annu Rev Microbiol 24:*297 (1970).

JOKLIK, W. F. The molecular biology of reovirus. *J Cell Physiol 76:*289 (1970).

MAIZEL, J. F., JR., SUMMERS, D. F., and SCHARFF, M. D. SDS-acrylamide gel electrophoresis and its application to the proteins of poliovirus- and adenovirus-infected human cells. *J Cell Physiol 76:*273 (1970).

NICHOLS, W. W. Viruses and chromosomal abnormalities. *Ann NY Acad Sci 171:*478 (1970).

SHATKIN, A. Viruses with segmented ribonucleic acid genomes: Multiplication of influenza versus reovirus. *Bacteriol Rev 35:*250 (1971).

Specific Articles

BALTIMORE, D. Structure of the poliovirus replicative intermediate RNA. *J Mol Biol 32:*359 (1968).

BALTIMORE, D., JACOBSON, M. F., ASSO, J., and HUANG, A. S. The formation of poliovirus proteins. *Cold Spring Harbor Symp Quant Biol 34:*741 (1969).

BISHOP, J. M., SUMMERS, D. F., and LEVINTOW, L. Characterization of ribonuclease-resistant RNA from poliovirus-infected HeLa cells. *Proc Nat Acad Sci USA 54:*1273 (1965).

CHOPPIN, P. W., and PONS, M. W. The RNA's of infective and incomplete influenza virions grown in MDBK and HeLa Cells. *Virology 42:*603 (1970).

DALES, S., and SIMINOVITCH, L. Development of vaccinia virus in Earle's L strain cells as examined by electron microscopy. *J Biophys Biochem Cytol 10:*475 (1961).

EHRENFELD, E., and HUNT, T. Double-stranded poliovirus RNA inhibits initiation of protein synthesis by reticulocyte lysates. *Proc Nat Acad Sci USA 68:*1075 (1971).

HOTCHIN, J. E., COHEN, S. M., RUSKA, H., and RUSKA, C. Electron microscopical aspects of hemadsorption in tissue cultures infected with influenza virus. *Virology 6:*689 (1958).

KATES, J., and BEESON, J. Ribonucleic acid synthesis in vaccinia virus. I. The mechanism of synthesis and release of RNA in vaccinia cores. *J Mol Biol 50:*1 (1970).

KRUG, R. M., and FRANKLIN, R. M. Studies on the synthesis of mengovirus ribonucleic acid and protein. *Virology 22:*48 (1964).

MAIZEL, J. V., JR., and SUMMERS, D. F. Evidence for differences in size and composition of the poliovirus-specific polypeptides in infected HeLa cells. *Virology 36:*48 (1968).

MC DONNELL, J. P., and LEVINTOW, L. Kinetics of appearance of the products of poliovirus-induced RNA polymerase. *Virology 42:* 999 (1970).

PFEFFERKORN, E. R., and HUNTER, H. S. The source of the ribonucleic acid and phospholipid of Sindbis virus. *Virology 20:*446 (1963).

ROTT, R., and SCHOLTISSEK, C. Investigations about the formation of incomplete forms of fowl plague virus. *J Gen Microbiol 33:*303 (1963).

SCHONBERG, M., SILVERSTEIN, S. C., LEVIN, D. H., and ACS, G. Asynchronous synthesis of the complementary strands of the reovirus genome. *Proc Nat Acad Sci USA 68:*505 (1971).

SKEHEL, J. J., and JOKLIK, W. K. Studies on the in vitro transcription of reovirus RNA catalyzed by reovirus cores. *Virology 39:*822 (1969).

SPEAR, P. G., and ROIZMAN, B. The proteins specified by herpes simplex virus. I. Time of synthesis, transfer into nuclei, and properties of proteins made in productively infected cells. *Virology 36:*545 (1968).

WECKER, E., and SCHONNE, E. Inhibition of viral RNA synthesis by parafluorophenylalanine. *Proc Nat Acad Sci USA 47:*278 (1961).

ZIMMERMAN, E. F., HEETER, M., and DARNELL, J. E. RNA synthesis in poliovirus-infected cells. *Virology 19:*400 (1963).

chapter **49**

INTERFERENCE WITH VIRAL MULTIPLICATION

GENERAL ASPECTS OF CONTROL OF VIRAL DISEASES
    BY INHIBITION OF REPLICATION   1172

VIRAL INTERFERENCE   1172

DEMONSTRATION OF INTERFERENCE WITH ANIMAL
    VIRUSES   1173

INTERFERON   1173

  Characteristics   1174

  Production of Interferon   1174

  Chemical Induction of Interferon: Double-stranded RNA   1176

  Mechanism of Production   1178

  Regulation of Interferon Production   1178

  Mechanism of Interferon Action   1179

  Protective Role in Viral Infections   1179

  Therapeutic Potential of Double-stranded Polynucleotides   1180

INTRINSIC INTERFERENCE (i.e., NOT MEDIATED BY
    INTERFERON)   1181

  Bacteriophage   1181

  Animal Viruses   1182

SIGNIFICANCE OF VIRAL INTERFERENCE   1183

CHEMICAL INHIBITION OF VIRAL MULTIPLICATION   1183

SYNTHETIC AGENTS   1183

  Selective Agents   1183

  Nonselective Agents Interfering with DNA Synthesis   1185

ANTIBIOTICS   1185

Agents interfering with viral multiplication are useful both for therapeutic and prophylactic purposes and for advancing our understanding of viral biology and of certain aspects of viral diseases.

GENERAL ASPECTS OF CONTROL OF VIRAL DISEASES BY INHIBITION OF REPLICATION

Viral diseases result from a series of growth cycles that kill or alter cells (see Ch. 51). The goal of antiviral treatments is to stop viral replication before it is too extensive, without damaging the surviving cells; for the maximum goal of restoring function to the infected cells is usually unattainable because cellular macromolecules are damaged early in viral infections (see Ch. 58). But even more limited goals present considerable difficulties.

The biology of viruses limits the opportunity to selectively inhibit the parasite without harming the cells. As we have seen in Chapter 23, this is possible in bacteria by taking advantage of many metabolic, structural, and molecular differences that differentiate them from animal cells. Thus sulfanilamide interferes with the function of p-aminobenzoic acid, which is a vitamin in bacterial but not in animal cells; penicillin interferes with the synthesis of the mucopeptide, which has no equivalent in mammalian cells; and strepto-

mycin interacts with molecular features of bacterial ribosomes that are different in animal ribosomes. The dependence of viral multiplication on viral and cellular genes, in contrast, limits the points of differential attack (Fig. 49-1). Even the largest viruses (e.g., poxviruses) contain only about 7% as much genetic information as *E. coli* and should therefore be much less vulnerable.

Another intrinsic limitation of antiviral treatment is that diseases become evident only after extensive viral multiplication and cellular alteration have occurred (see Ch. 51). Therefore the most general approach to control in viral diseases is prophylaxis. Therapy is essentially limited to localized diseases, such as herpetic keratoconjunctivitis (Ch. 53), where the main goal is to prevent the further spreading of the infection, even if it means killing the infected and even some uninfected cells. Their death can be tolerated if the damage is subsequently repaired.

As with bacterial chemotherapy, a third important limitation of antiviral therapy is the emergence of **resistant mutants.** In order to avoid their selection, the same principles valid for bacteria are applicable to viruses: adequate dosage of the drugs, multidrug treatment whenever possible, and avoiding therapy unless clearly indicated. Fortunately, however, genetic resistance to two important antiviral agents (interferon and interferon inducers) does not seem to occur.

VIRAL INTERFERENCE

When viruses of more than one type infect the same cell each may multiply undisturbed by the presence of the others, except for possible recombination or phenotypic mixing. In

certain combinations, however, the multiplication of one type of virus may be inhibited to a varying extent. This inhibition is called **viral interference.**

Fig. 49-1. Points in the growth cycle of animal viruses suitable for differential chemotherapeutic attack. At points A and B only certain viruses use a viral function: A = viruses with a virion's transcriptase, probably virus-specified, use it for partial or complete transcription; B = enveloped virus release depends on incorporation of the viral envelope in the cellular membrane.

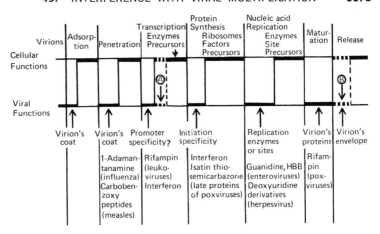

The notion of interference developed first from observations with ring spot virus, which causes characteristic ring-shaped lesions in tobacco plants. The lesions then regress but the virus persists without obvious symptoms; and if the plant is reinoculated with the same virus no new lesions develop. Thus the first infection interferes with the expression of the second infection. Subsequently it was found that in monkeys infection with a neurotropic strain of yellow fever virus (a togavirus), which by itself causes a mild disease, can prevent the usually lethal disease caused by the pantropic strain of the same virus. Protection is not caused by antibody, because the same effect is obtained if an antigenically unrelated togavirus (Rift Valley fever) is used instead of the neurotropic strain of yellow fever. Interference was later found also with bacteriophages, thus opening the way for quantitative studies.

The study of interference with animal viruses took an important turn when Isaacs and Lindenmann, in 1957, discovered that interference can be mediated by a substance produced by virus-infected cells, called **interferon.** Much work since then has shown that interferon accounts for many, but not all, observed instances of viral interference.

DEMONSTRATION OF INTERFERENCE WITH ANIMAL VIRUSES

Interference has been studied especially in the allantoic epithelium of the chick embryo infected with influenza viruses. A typical experiment consists in inoculating the allantoic cavity with influenza A virus, **as interfering virus,** followed 24 hours later by influenza B virus, **as challenge virus:** the multiplication of the second inoculum is partially or totally inhibited. The experiments can be simplified by using inactivated interfering virus which does not multiply: interference can then be determined by measuring the yield of the challenge virus, without the need to distinguish it from the interfering virus.

A pair of viruses that exhibit interference with a certain regimen of infection, as exemplified above, may not demonstrate interference with a different regimen. For instance, if influenza A and B are inoculated **simultaneously** at equal multiplicity in the allantoic cavity they can multiply concurrently in the same cells, as shown by the production of phenotypically mixed progeny particles (Chs. 45 and 48). Even viruses of different families can multiply under proper circumstances in the same cells, as shown in various ways. Thus electron microscopy reveals adenovirus and simian virus 40 virions, which have a different size, in the same nucleus; immunofluorescence demonstrates the unrelated antigens of Newcastle disease virus and parainfluenza virus 3 (both paramyxoviruses) in the same cytoplasm; and the same cell may contain characteristic inclusions of herpes simplex in the nucleus and vaccinia (a poxvirus) in the cytoplasm.

We shall first consider in some detail the role of interferon in viral interference, and shall then consider a heterogeneous group of examples of interference in which interferon plays no demonstrable role.

INTERFERON

Interferon was discovered in the course of studying the effect of influenza virus inactivated by ultraviolet (UV) light on fragments of the chick chorioallantoic membrane, maintained in an artificial medium. The supernatants, although devoid of viral particles, inhibited the multiplication of active influenza virus in fresh fragments; and the inhibitor was found to be a soluble substance. Substances of this class, called interferons, were subsequently shown to be produced by cells infected by almost any animal virus, either DNA- or RNA-containing, in tissue culture or in the animal.

CHARACTERISTICS

Purified interferons from various sources consist of small proteins unusually stable at low pH (they resist a long exposure to pH 2 in the cold) and fairly resistant to heat (mouse interferon loses half its activity after 1 hour at 50°, chick interferon at 70°).

Interferons are **not virus-specific, but cell-specific,** in both their production and their effects; indeed, interferons produced in different species after infection with the same virus differ in antigenic specificity, isolectric point, and molecular weight (e.g., chick interferon, MW 38,000; mouse and human interferon, MW 26,000). Furthermore, a given interferon **inhibits viral multiplication most effectively in cells of the species in which it was produced.** For instance, purified interferon of chick origin is less than 0.1% as effective in mouse cells as in chick cells. (Interferon produced in monkey kidney cells, however, is effective in human as well as in monkey cells.) The basis for species specificity is unknown; it does not depend on a difference in uptake.

Since interferons are produced only in small amounts they are difficult to obtain in pure form, and crude preparations are almost invariably used. These consist of the fluids surrounding the infected cells (e.g., tissue culture fluid, allantoic fluid) or of homogenized tissue suspensions in which viral particles are inactivated at pH 2 and often removed by centrifugation. Any interfering activity of these preparations is attributed to interferon if it is 1) nondialyzable, 2) destroyed by proteolytic enzymes, and 3) cell-specific but not virus-specific.

Interferon is **assayed** by determining its effect on plaque production by a test virus: usually vesicular stomatitis virus (a rhabdovirus), which is very sensitive to interferon and produces plaques on cells of many vertebrate species. Serial dilutions are added to the agar overlay, and the endpoint is a 50% reduction of the number of plaques.

PRODUCTION OF INTERFERON

In order to facilitate understanding of the conditions affecting interferon production as well as of its mode of action it will be useful to consider the over-all scheme of Figure 49-2, which will be justified by the material presented below. Of the three proteins, specified by viral genes, only one, the interferon, has been demonstrated; the presence of the other two is inferred.

Several features of the scheme are noteworthy. 1) The interferon acts both on the cell in which it is produced and on those exposed to it. 2) The antiviral action is not due directly to interferon but to the antiviral protein it elicits. 3) Both interferon production and its antiviral activity require expression of cellular genes, hence are blocked by inhibition of transcription or translation. 4) The rate of interferon production depends on the balance between induction by the inducer and self-induced repression; hence it will be strongly influenced by conditions (such as partial inhibition of protein synthesis) that alter the balance. 5) There is a constant background production of interferon in the absence of inducer, whose presence enhances production by shifting the balance between induction and repression.

Relation to Viral Multiplication. Interferon is produced by cells infected with complete virions, either infectious or inactivated. The amounts produced in different systems differ widely. Viruses that multiply very rapidly and block synthesis of host protein early after infection tend to be poor producers of interferon. Good producers are usually viruses that multiply slowly and do not damage the cells markedly. Notable exceptions are many togaviruses which produce large amounts of interferon even though they multiply rapidly and have a pronounced cytopathic effect. These differences are probably related to a

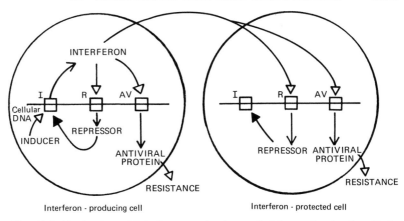

Fig. 49-2. Mechanism of interferon production and of its mode of action. Both interferon production and development of antiviral resistance appear to depend on the interplay between three cellular genes: I = interferon; AV = the antiviral protein, which confers resistance; and R = a repressor of the I gene. Thin lines indicate production of a protein; thick lines indicate its action, with white arrows for induction, black arrows for repression.

different effect on host macromolecule synthesis and to different amounts of inducer (see below).

The relation between viral characteristics and interferon production may be illustrated by considering some specific instances. For example, an attenuated mutant of poliovirus is a much better interferon producer than the wild type, which multiplies better; and the paramyxovirus of Newcastle disease multiplies well in chick embryo cells but produces little interferon, while in human cells it causes a defective infection but produces abundant interferon.

Many viruses with a transcriptase in the virions (Ch. 44, Distinctive properties of viruses) normally produce little interferon but become good producers after inactivation by UV light or heat (e.g., Newcastle disease virus in chick embryo cells; see Fig. 49-7). The possible reason will be discussed below.

Relation to Cells. Although all animal cells appear able to produce interferon, cells of the bone marrow and spleen and macrophages appear to play a special role. Thus lethally X-rayed mice grafted with rat bone marrow cells and then injected with Newcastle disease virus produce only **rat-specific interferon.** Moreover, in a confluent culture of mouse embryo cells added spleen cells from virus-injected mice protect the surrounding cells

from vesicular stomatitis virus. This procedure provides a **plaque assay for interferon-forming cells.** According to this test the proportion of spleen cells releasing detectable interferon increases as a function of the dose of the inducing virus (see Fig. 49-3), like the interferon concentration in the serum of the same animal, to a maximum of approximately 1%.

In vitro studies have shown that interferon is produced by lymphocytes stimulated by mitogens or antigen in the presence of macrophages. Hence interferon production may play a role in the protection afforded by vaccination against viral diseases. With some inducers antilymphocytic serum inhibits interferon production in the animal.

Kinetics of Formation: Refractory Period. Under one-step conditions of viral multiplication the synthesis of interferon begins after viral maturation is initiated; if not interrupted by an early block in synthesis of host macromolecules it continues at the same rate for 20 to 50 hours, then stops (Fig. 49-4). The interferon is mostly released extracellularly. If the cells survive for a longer time, as after infection by UV-inactivated virus, they cannot produce interferon again, in response to reinfection, until after a **refractory period** of at least two cell divisions. Cells continuously stimulated by an inducer stop producing in-

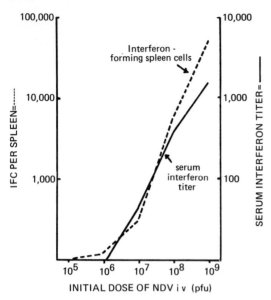

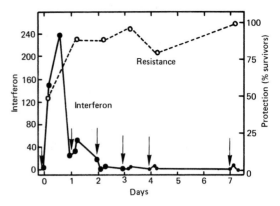

Fig. 49-3. Induction of interferon-producing spleen cells (IFC) in the mouse by administration of Newcastle disease virus (NDV). [Modified from Osborn, J. E., and Walker, D. L. *Proc Natl Acad Sci USA* 62:1038 (1969).]

Fig. 49-5. Refractory period following induction of interferon by double-stranded RNA injected into mice. After the first injection of RNA (arrow) there is a burst of interferon appearance in the serum, followed by the refractory period in which repeated injections do not cause further production. During this period, however, the mice are fully resistant to injection with encephalomyocarditis virus (a picornavirus), suggesting the continued presence of intracellular interferon. [Modified from Sharpe, T. J., Birch, P. J., and Planterose, D. N. *J Gen Virol 12*: 331 (1971).]

terferon but are highly resistant to viral infection (Fig. 49-5). These phenomena can be attributed to an initial burst of interferon synthesis due to low repression, together with

Fig. 49-4. Time course of viral multiplication and interferon synthesis in the allantoic membrane of the chick embryo infected with influenza virus. Owing to the low multiplicity of infection several cycles of viral growth were required before progeny virus could be detected; hence the lag observed is much longer than the regular eclipse period of the virus. Note the considerable delay in the synthesis of interferon. [Modified from Smart, K. M., and Kilbourne, E. D. *J Exp Med 123*:309 (1966).]

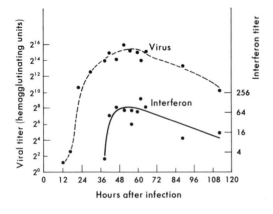

release of preformed interferon. Later, the increased repression decreases the synthesis to a level sufficient for eliciting resistance but not for releasing interferon outside the cell.

Viral strains capable of high interferon production give rise to **autointerference;** in endpoint assays the dilutions containing most virus may be less effective because of the presence of enough interferon to block the further cycles of viral multiplication. Autointerference can also arise by a different mechanism (see below, Interference).

CHEMICAL INDUCTION OF INTERFERON: DOUBLE-STRANDED RNA

The **nature of the stimulus** that induces interferon production has been clarified by the discovery, by the Hilleman group, that **double-stranded RNAs,** such as reovirus RNA and certain synthetic polynucleotides, can induce large production of interferon in many animals. Subsequent work with synthetic models has shown that the activity resides in **polyribonucleotides with a high helical content** such as the double-stranded synthetic polymers formed by one chain of polyribo-inosinic and one of polyribocytidylic acid (poly (I:C)) or by one of polyadenylic and one of polyuridylic acid (poly (A:U)). Single-stranded polynucleotides able to form a stable secondary structure are also effective (e.g., polyguanylic, polyriboinosinic, and polyriboxanthylic acids) as are certain other synthetic anionic polymers. Important for inducing activity are a large molecular weight, a high density of free anionic groups, and resistance to enzymatic degradation. Deoxyribonucleotides and DNA-RNA hybrids are inactive. The interferon released is mostly **produced after induction,** since the amount is considerably decreased by inhibitors of protein synthesis at high concentrations. Some inducers cause mostly **release of preformed interferon,** which occurs very rapidly and is insensitive to inhibition of protein synthesis. Among these are bacteria, rickettsiae, bacterial endotoxin, and phytohemagglutinin.

The activity of synthetic polynucleotides in tissue cultures is greatly enhanced by DEAE-dextran, a cationic polymer that also increases the infectivity of viral nucleic acids (see Ch. 48). Under optimal conditions the DEAE-dextran and the polynucleotide form a compact electroneutral aggregate which is resistant to nucleases and is efficiently taken up by the cells. The aggregate binds to the cell surface, as shown by radioautographs using radioactive poly (I:C); some polynucleotide molecules may enter the cell and interact with cellular components.

Relation Between Viruses and Induction of Interferon by Double-stranded RNA. It is likely that RNA species containing double-stranded segments produced during replication of viral RNAs (RI and especially RF, see Ch. 48) mediate the induction of interferon by most RNA viruses. Viruses containing double-stranded RNA may induce without replication, explaining induction of interferon by fungal extracts containing such viruses. For DNA viruses the inducer may be double-stranded RNA resulting from overlapping

transcription (Fig. 49-6). Thus appreciable amounts of double-stranded viral RNA are extracted from vaccinia-infected cells and induce interferon in tissue cultures.

The role of the RI and RF is suggested by observations with viruses with a virion transcriptase, which synthesizes the messenger strand, complementary to the viral strand, and required for RI and RF formation (see Ch. 48); with increasing UV dose, the interferon-inducing activity decays in parallel to the activity of the transcriptase itself and hence of the ability to form double-stranded RNA (Fig. 49-7). **Small UV doses may enhance the inducing ability** by inactivating viral genes required for the production of viral strands on the replicative intermediate, thus favoring accumulation of RF, which is entirely double-stranded (see Ch. 48). These low doses do not affect the transcriptase, since proteins are much less UV-susceptible than nucleic acids. With viruses that do not use a virion transcriptase UV light does not enhance the inducing ability, apparently because it rapidly inactivates the ability of the viral genome to code for synthesis of new enzymes required for formation of RI and RF.

The ability of some viruses to induce interferon is affected by temperature-sensitive mutations, but no clear pattern for the genetic control of this property has yet emerged.

The ability to synthesize interferon can be considered as a defense mechanism of the cells against viruses, by recognizing the double-stranded RNA, which is a characteristic viral product and is very scarce in normal cells. It was at first thought that the cell was able to recognize foreign nucleic acids on the basis of their sequences. This possibility is now excluded by the studies with synthetic polymers. Moreover, double-stranded RNA extracted from uninfected cells, in which it is present in amounts too low to act as inducer,

Fig. 49-6. Possible origin of double-stranded RNA in vaccinia-infected cells from points where transcriptions of different strands overlap. The complementary RNA segments formed in the overlap region would tend to form a helix if random folding of the RNAs is inhibited in the cells.

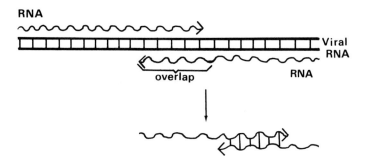

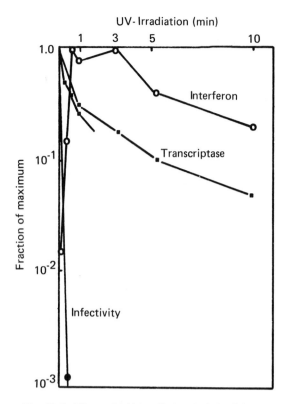

Fig. 49-7. Effect of UV irradiation in infectivity, interferon-inducing ability, and transcriptase activity of Newcastle disease virus (a paramyxovirus). There is a marked enhancement of inducing ability after low UV doses; then it decays with a slope similar to that of transcriptase activity but much smaller than that of infectivity. [Modified from Clavell, L. A., and Bratt, M. A. *J Virol* 8:500 (1971).]

after suitable concentration elicits interferon synthesis in the same cell type from which it was extracted.

MECHANISM OF PRODUCTION

There is strong evidence that interferon production results from the expression of a **cellular gene.** Thus in cells infected by RNA viruses actinomycin D prevents the expression of cellular, but not of viral, genes; and when added to the cells before infection (or before addition of poly (I:C)), the drug prevents the formation of interferon (Fig. 49-8A). However, if the drug is added 2 hours after infection interferon production is normal, suggesting that the required cellular mRNA is synthesized before that time and persists afterward. The virus, therefore, apparently

induces the activation of a cellular gene. The presence of a **specific interferon mRNA** in mouse cells exposed to poly (I:C) has been recently demonstrated by transfer of RNA from these cells into avian or simian cells in which it caused the formation of **mouse-specific** interferon.

The requirement of the expression of viral genes for interferon synthesis affords a tentative explanation for the low "interferogenic" potency of certain viral strains: **viruses that block cellular mRNA or protein synthesis are necessarily poor inducers of interferon production.** Also certain interactions between viruses become understandable. For instance, in tissue culture cells simultaneously infected by a good inducer and a poor inducer interferon synthesis fails, presumably because the poor inducer blocks host RNA and protein synthesis (Fig. 49-8B). Interactions of this type presumably also occur in animals, where they may influence the pathogenesis of viral infections.

REGULATION OF INTERFERON PRODUCTION (SEE FIG. 49-2)

Regulation of interferon synthesis by a labile repressor generated under the influence of interferon is suggested by inhibition of interferon production by **large concentrations of external interferon** after induction by either virus of poly (I:C), and by dramatic enhancement of interferon production by small concentrations of inhibitors of transcription or translation (Table 49–1). The partial inhibition apparently cuts down the concentration of the labile repressor, allowing a much greater synthesis and accumulation of the stable interferon.

Negative regulation by endogenous **intracellular** interferon can explain the **self-limiting character** of interferon synthesis and the refractory period, as discussed above. This regulation may represent an obstacle to the therapeutic use of exogenous interferon in viral diseases.

Low concentrations of external interferon **enhance interferon production (priming)** after induction by an RNA virus. Perhaps the treatment, like small UV doses, inhibits the regular evolution of the viral replicative intermediate (RI, Ch. 48), which depends on viral protein synthesis, favoring accumula-

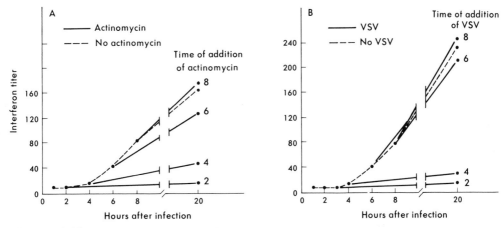

Fig. 49-8. Inhibition of interferon production by actinomycin D or vesicular stomatitis virus (VSV). Ehrlich ascites cells (a transplantable mouse cancer) were infected, at a multiplicity of 5 infectious units per cell, with Newcastle disease virus. Although the virus does not undergo a regular multiplication cyc'e in these cells, it elicits interferon production. Either actinomycin D or vesicular stomatitis virus (which by itself does not elicit interferon production) prevents interferon synthesis if added within 4 hours; later additions have progressively less effect. [Modified from Wagner, R. R., and Huang, A. S. *Virology 28*:1 (1966).]

tion of the double-stranded derivative, RF. Priming may be useful in the large-scale production of interferon in cell cultures.

MECHANISM OF INTERFERON ACTION

That interferon causes antiviral resistance not directly but **by inducing the synthesis of an antiviral protein** is shown by the effect of inhibitors. Thus actinomycin D blocks the effect of interferon when both are added at the

TABLE 49-1. Effect of cycloheximide and actinomycin D on interferon production: Effect of inhibition of RNA and protein synthesis on interferon production in rabbit kidney cell cultures after poly (I:C) induction. The combination of the two inhibitors at suitable times and concentrations brings about dramatic enhancement of interferon production (compare Groups 3 and 1).

Group	Cyclo-heximide, duration of treatment (hr)	Actino-mycin D, hours 3–4	Interferon yield
1	None	None	500
2	0–18	None	1,200
3	0–4	Yes	56,000
4	0–18	Yes	2,000

Data from TAN, Y. H., ARMSTRONG, J. A., and HO, M. *Virology 3:* 503 (1971).

time of infection with a virus insensitive to the drug but sensitive to interferon alone. The cellular gene activated by interferon can be separated from that specifying the interferon by segregation of chromosomes in heterospecific hybrid cells (see Ch. 47): some segregants produce interferon different in specificity from that to which they are sensitive.

After removal of exogenous interferon the inhibition persists for a considerable period, whose length depends on interferon concentration; inhibition may persist even after cell division. The inhibition can be overcome by a large multiplicity of infection, perhaps when it exceeds the number of antiviral molecules.

PROTECTIVE ROLE IN VIRAL INFECTIONS

Experimental evidence on the protective role of endogenous interferon in viral infection comes from studies in both tissue cultures and animals. In **tissue cultures** this role is demonstrated by the establishment of **carrier cultures,** in which most of the cells are made resistant by interferon produced in the culture and only a small proportion are infected at any time (Ch. 51). Interferon production prevents such cultures from being rapidly destroyed by viral multiplication; but the inhibition is also too limited to wipe out the infection.

In **animals** many studies have been concerned with **relating the course** of a viral disease to endogenous **interferon production.** The following observations support a **protective role.** 1) In children receiving live measles vaccine and, at the peak of the circulating interferon titer, live smallpox vaccine, the smallpox vaccination did not "take." 2) In mice recovering from influenza virus infection the titer of interferon was found maximal at the time when the virus titer began to decrease and before additional antibodies could be detected. Furthermore, at this stage the animals exhibited significant protection against the lethal action of a togavirus inoculation intraperitoneally, showing that the interferon concentration was sufficient to prevent a generalized infection. 3) Suckling mice, which are susceptible to coxsackievirus B1 (a picornavirus), produce little interferon in response to this virus; whereas adult mice, which are resistant, produce large amounts.

In other situations, however, any protective role of interferon is masked by other phenomena. 1) Newborn mice produce large amounts of interferon after infection by togaviruses, to which they are very susceptible; whereas adult mice, which are much more resistant to these viruses, produce less interferon when infected. 2) Mice of strains genetically susceptible to West Nile virus (another togavirus) produce much larger amounts of interferon following infection by this virus than do genetically resistant mice. In both cases interferon production appears to reflect the extent of viral multiplication, which is limited by other factors.

These studies suggest that interferon has a major protective role in some, but not all, viral infections. Much depends on the **dynamics of disease,** i.e., the relation between virus titer and interferon titer at various times. Interferon is most effective when present before infection, and when the dose of infecting virus is not too large (as at the beginning of most natural infections). The protection afforded appears to be especially useful because it develops more promptly than antibody production.

Possible Therapeutic Use. Interferon could theoretically be an ideal antiviral agent, since it acts on many different viruses, lacks serious toxic effect on the host cells, has high activity,* and does not cause the selection of interferon-resistant mutants. However, its therapeutic value is limited by various factors: interferon is effective only during relatively short periods, has a reversible action, and has no effect on viral synthesis once initiated in a cell. Moreover, interferon is difficult to produce in large amounts.

Studies on the therapeutic value of exogenous interferon have had some limited success. Influenza in mice was delayed and rendered less severe by injecting large amounts of interferon before infection. Rabbits could be protected against intradermal vaccinia virus infection by interferon prepared in rabbit kidney cells. Some protective effect has also been observed in man: interferon prepared in monkeys inhibited the local development of vaccinia lesions in primary vaccination, inhibited certain stages of vaccinial keratitis, and slowed down the progress of cytomegalic disease (see Ch. 53) in children. A beneficial prophylactic effect has been observed against influenza infection during epidemics.

The therapeutic effect of interferon-producing lymphocytes and macrophages seems promising since they can deliver the interferon at the diseased site, where they accumulate. Mouse peritoneal macrophages induced in vitro to produce interferon markedly inhibited the progress of systemic infection by a togavirus in the mouse.

THERAPEUTIC POTENTIAL OF DOUBLE-STRANDED POLYNUCLEOTIDES

These substances appear much more promising than interferon, because they are easily available and very effective (e.g., 10^{-3} μg of poly (I:C) is sufficient to induce resistance to virus infection in a culture). Like interferon, they do not cause the selection of resistant viral mutants. The polynucleotides are especially useful against **localized disease:** thus herpetic keratoconjunctivitis in rabbits could be controlled by local applications of poly (I:C) even 3 days after infection, when

---

* As little as 4×10^{-3} μg of purified interferon inhibits 100 infective units of eastern equine encephalitis virus (a togavirus) in culture of chick embryo cells; thus on a weight basis the antiviral activity of interferon is greater than the antibacterial activity of effective antibiotics.

the disease was already moderate to severe (Fig. 49-9). For this disease poly (I:C) is even more suitable than the successful arabinosyl adenine or iododeoxyuridine treatment (see below), which is limited by the emergence of resistant viral strains. Intranasal administration of poly (I:C) can protect human volunteers from rhinoviruses, the agents of the common cold, and mice from an otherwise fatal subsequent infection with pneumonia virus of mice (a myxovirus). In **systemic diseases** a preventive effect of the polynucleotide has been experimentally observed against foot-and-mouth disease virus (a picornavirus) and herpesvirus in mice, and against rabies virus in rabbits, if given not later than 24 hours after infection. Effects in man have been limited, however, both for the general reasons discussed above and for more special reasons. Among these are the high toxicity of polynucleotides in animals and therefore the low doses employed in order to avoid toxic complications, as well as the rapid breakdown of poly (I:C) by enzymes in the blood and tissues. The synthesis of chemically modified polynucleotides may overcome these special difficulties.

Resistance to viral infection induced by polynucleotides appears due to interferon production, because large doses of poly (I:C) induce the two phenomena in a parallel way (Fig. 49-10), and resistance induced by poly (I:C) can be duplicated by exogenous interferon. Certain apparent exceptions can be readily explained. Thus development of resistance without production of external interferon at low poly (I:C) concentrations can be attributed to intracellular interferon; and inhibition of resistance but not interferon production by addition of actinomycin D 2 hours after poly (I:C) is explained by inhibition of the synthesis of the antiviral protein.

INTRINSIC INTERFERENCE (i.e., NOT MEDIATED BY INTERFERON)

Inability to detect interferon does not exclude its participation in an instance of viral interference, because the detection methods are insensitive. However, if interference is established **early** in the infectious cycle the participation of interferon can be considered unlikely because its production usually begins later.

Intrinsic interference has been studied

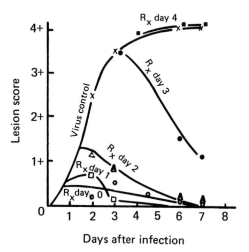

Fig. 49-9. Response of herpetic keratoconjunctivitis to topical treatment with poly (I:C) in rabbits' eyes infected at time zero with 3×10^4 plaque-forming units of herpes simplex virus (see Ch. 53). Rx indicates the day at which treatment begun. In all cases treatment continued for 6 days. The lesions were graded on an empirical scale at the end of the treatment. [From Park, J. H., and Baron, S. *Science* 162:811 (1968).]

with both bacteriophages and animal viruses; several different mechanisms can be distinguished.

BACTERIOPHAGE

Homologous Phages. Within the T-even groups of coliphages homologous phages (such as two mutants of the same strain) can replicate more or less equally in the same cell if both infect it simultaneously. If they infect at different times the phage infecting first multiplies normally and interferes with the multiplication of the phage infecting later; indeed, this challenge phage is almost completely excluded when it infects after the middle of the eclipse period of the interfering phage. Exclusion appears to result from a **change in the bacterial plasma membrane,** which prevents penetration of the DNA of the challenge phage. In fact, radioautographs of cells superinfected with suitably labeled phage show that the DNA remains bound to the cell envelope, where it is then degraded by a membrane-associated nuclease.

If the challenge phage is used at a sufficiently high multiplicity the many DNA

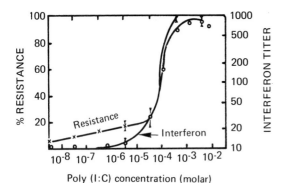

Fig. 49-10. Interferon production and virus resistance in human cells stimulated by poly (I:C). The interferon was measured in the culture medium 12 hours following stimulation. Virus resistance is given by the proportion of cells surviving infection with vesicular stomatitis virus. There is good proportionality between the two effects, except at very low doses of poly (I:C). [Modified from Pitha, P. M., and Carter, W. A. *Virology 45*:777 (1971).]

molecules simultaneously released saturate the nuclease. Some intact molecules can then penetrate into the cells, where they enter the DNA pool of the interfering phage and immediately start replication. Genetic markers of the challenge phage appear in the viral progeny in a proportion that depends on the size of the DNA pool at the time of challenge (Ch. 45).

Heterologous Bacteriophages. The consequences of mixed infections vary with the degree of relatedness of the phages. Simultaneous infection with **related** T-even bacteriophages (e.g., T2 and T4) still allows the production of mixed yields, in which, however, T4 prevails over T2.

Mixed infection with **unrelated** phages usually results in **exclusion;** which phage is excluded depends on several factors. When infection is simultaneous certain phages regularly exclude others: T-even bacteriophages, for instance, exclude T1, T3, and λ. Exclusion can be attributed to profound biological differences between the phages; the T-even phages destroy the host cell DNA, whereas the more readily excluded phages do not destroy it, and even depend on certain of its functions for their multiplication.

Exclusion of the RNA phage F2 (Ch. 45)

by T4 is caused by the inactivation of a translation initiation factor after infection with this phage, which prevents recognition of the F2 RNA (see Ch. 45, Synthesis of viral macromolecules). Furthermore, F2 RNA is rapidly degraded, perhaps by a T4-specified enzyme. In nonsimultaneous infection the challenge phage is regularly excluded if it injects its DNA after the interfering phage has taken control of the whole cell machinery.

ANIMAL VIRUSES

Homologous Interference. When two distinguishable strains of the same virus infect the same cell they must compete for the same precursors, cellular sites, and enzymes. Hence multiplication of one type is often depressed. As with bacteriophages, the effect is most pronounced if one virus has an advantage, either in time or multiplicity. This type of interference takes place with Newcastle disease virus (a paramyxovirus) at **adsorption,** through destruction of cellular receptors; with leukoviruses at **penetration;** and with poliovirus types 1 and 2 as well as other enteroviruses at **replication.**

Defective particles inhibit the replication of the regular particles in the same virus stock **(autointerference).** Thus poliovirus, influenza virus, vesicular stomatitis virus, and Sindbis virus generate at a low frequency particles with a **shorter RNA** than regular particles, and though these virions are not infectious alone they multiply in cells also infected by regular virus as helper (Ch. 48). They have a selective advantage over regular virus, probably because their shorter RNA is replicated more rapidly: hence in serial passages at **high multiplicity** the proportion of regular virus progressively decreases, until defective particles are the main product.

Heterologous Interference. The mechanisms vary in different systems, as shown by two examples. 1) Poliovirus arrests the multiplication of vesicular stomatitis virus (VSV), even when infecting later and at lower multiplicity. Apparently the poliovirus mechanism blocking translation of cellular messengers has the same effect on VSV messengers (see Ch. 48). The interference may be based on recognition of specific sequences in mRNA, since it does not occur with many other viruses tested. 2)

Newcastle disease virus (NDV) fails to multiply in cells previously infected by any of a number of other RNA viruses, e.g., Sindbis virus (a togavirus), owing to interference with NDV RNA replication. NDV RNA may bind to the replicase induced by the interfering virus, forming an inactive complex, because Sindbis mutants deficient in RNA replicase do not interfere. This phenomenon is exploited for assaying noncytopathic viruses on the basis of their antagonism to the cytopathic effect of NDV.

SIGNIFICANCE OF VIRAL INTERFERENCE

Interference, both by interferon and by other mechanisms, is important in several aspects of viral infection. For instance, in human **vaccination** with attenuated poliomyelitis viruses the three strains must be administered in a precise sequence, or at specified concentration ratios, in order to avoid interference of one strain with another. Similarly, the presence of various enteroviruses in the normal intestinal flora may hinder the establishment of infection by the vaccine strains. Viruses already present may also influence the response to an **infecting virus.** For example, dengue virus infection in man is milder in the presence of an attenuated strain of yellow fever virus (both togaviruses). Interference may thus present an obstacle to both infection and vaccination.

Interferon may play an important role in initiating recovery from viral infections, particularly in the light of the uncertain role of humoral antibodies. Thus recovery often occurs before there is a pronounced increase in antibody level; furthermore, recovery usually occurs without delay in agammaglobulinemic individuals (provided cellular immunity is good). Since, however, small amounts of antibodies may be present in both cases there is no strong basis for concluding that recovery may be produced by interference alone.

Finally, in antiviral therapy it seems likely that interferon and interferon-inducers may play an important role, especially in local superficial infection, if the problems related to the large-scale production of the former and toxicity of the latter can be overcome.

CHEMICAL INHIBITION OF VIRAL MULTIPLICATION (FIG. 49-11)

SYNTHETIC AGENTS

SELECTIVE AGENTS (SEE FIG. 49-1)

Adamantanamine. This compound, of peculiar structure (Fig. 49-11), inhibits the multiplication of several viruses including influenza and rubella. It appears to interfere with an early step in virus-cell interaction after adsorption—perhaps the release of the viral nucleic acid in the cells. The compound appears to have some prophylactic value and even marginal therapeutic effect if given early in influenza infections.

Carboxypeptides. Certain carboxy di- and tripeptides (such as carbobenzoxy-D-phenylalanyl-L-phenylalanylnitro-L-arginine) are potent inhibitors of measles (a paramyxovirus) replication in tissue culture cells. They appear to interfere with viral penetration; they also affect the ability of the virus to cause exogenous cell fusion (see Ch. 48) or hemolysis (see Ch. 57), which are probably side effects of the viral function involved in penetration.

Substituted Benzimidazole and Guanidine. A substituted benzimidazole and guanidine each affects only certain picornaviruses. The benzimidazole (2-(α-hydroxybenzyl)-benzimidazole, or HBB) inhibits multiplication of certain enteroviruses in this group; guanidine inhibits multiplication of the same viruses and a few more (Ch. 55). These two compounds interfere with synthesis of viral single-stranded RNA without affecting cellular RNA synthesis. Guanidine apparently interacts with a structural protein that plays a role in RNA replication since some temperature-sensitive mutations in genes for virions' proteins markedly increase guanidine sensitivity.

In spite of their marked effect in vitro, neither HBB nor guanidine has shown promise as a chemotherapeutic agent in animals, owing to the rapid emergence of **resistant** viral mutants. Viruses can also mutate to **depend-**

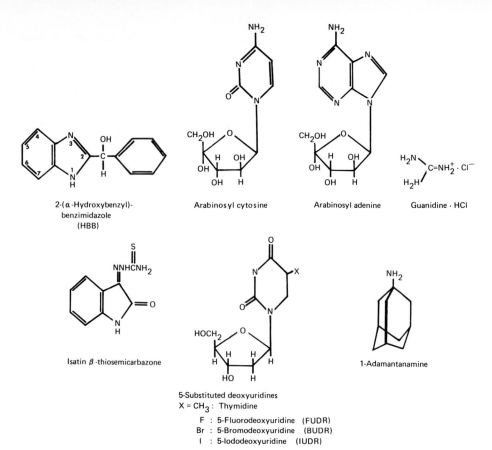

2-(α-Hydroxybenzyl)-
benzimidazole
(HBB)

Arabinosyl cytosine

Arabinosyl adenine

Guanidine · HCl

Isatin β-thiosemicarbazone

1-Adamantanamine

5-Substituted deoxyuridines
X = CH_3 : Thymidine
F : 5-Fluorodeoxyuridine (FUDR)
Br : 5-Bromodeoxyuridine (BUDR)
I : 5-Iododeoxyuridine (IUDR)

Fig. 49-11. Chemical constitution of some antiviral inhibitors.

ence on these drugs; and the mutations to resistance or dependence may affect the response to only one drug or to both (covariation).

With sensitive virus in the presence of guanidine, as with dependent virus in its absence, RNA synthesis is specifically impaired. Hence the drug evidently has the same target when inhibiting a susceptible system or activating a dependent one. A useful model is provided by steptomycin, for which it has been shown that bacterial sensitivity, dependence, and resistance involve alternative configurations of the ribosome (Ch. 12).

Isatin-β-Thiosemicarbazone. This compound inhibits multiplication of poxviruses but not other DNA-containing viruses. It selectively inhibits synthesis of late viral structural proteins, apparently by changing the late viral mRNAs and preventing their formation of polyribosomes. The result is production of nearly spherical defective particles. The β-thiosemicarbazone drugs are of considerable practical use for the control of smallpox.

Indeed, one of the most active derivatives, N-methylisatin-β-thiosemicarbazone, had its initial impressive **prophylactic** success in preventing the spread of smallpox to contacts in an epidemic in Madras, India, in 1963. Oral treatment of household contacts of hospitalized patients was begun 1 day after admission of the patients to the hospital. Among 1101 contacts treated with N-methylisatin-β-thiosemicarbazone and vaccinated 3 mild cases of smallpox occurred; in contrast, among 1126 vaccinated contacts not treated with the drug 78 contracted smallpox and 12 died. The much greater prophylactic effect of the drug can be attributed to its immediate effect on viral multiplication, whereas the effect of vaccination has to wait for the development of antibodies. A number of subsequent studies confirmed its dramatic value.

In contrast to this brilliant **prophylactic** success, N-methylisatin-β-thiosemicarbazone failed as a **therapeutic** agent in patients already suffering from smallpox. This failure is understandable, because the disease appears only after viral multiplication has reached a maximum.

NONSELECTIVE AGENTS INTERFERING WITH DNA SYNTHESIS

Halogenated Derivatives of Deoxyuridine. The hydrogen atom in position 5 of uracil can be substituted by halogens, yielding 5-fluorouracil, 5-bromouracil, and 5-iodouracil. The deoxyribosyl derivatives of these compounds are taken up by cells and phosphorylated by thymidine kinase to the 5'-monophosphates which interfere with normal DNA synthesis in two ways. Thus 5-fluorodeoxyuridine (FUDR) is inhibitory because its 5'-monophosphate inhibits thymidylic acid synthetase (which converts deoxyuridylic to thymidylic acid), stopping DNA synthesis for lack of thymidine triphosphate. 5-bromo- and 5-iododeoxyuridine (BUDR, IUDR) interfere less completely with enzymes synthesizing DNA precursors but also are converted to the triphosphates, and then are incorporated into DNA instead of thymidine; the DNA resulting can continue to replicate, but causes the synthesis of nonfunctional proteins.

Indeed in cells infected by herpesvirus in the presence of IUDR the viral DNA replicates, but viral maturation fails, owing to defects of the capsid protein. If the drug is later removed, virions are formed; they contain DNA replicated in the presence of the drug, in which iodouracil replaces thymine. The inhibition of viral multiplication resulting from this mechanism is especially effective for BUDR and IUDR, but not FUDR. BUDR and IUDR are especially valuable for their inhibitory action on herpes simplex virus and vaccinia virus in surface lesions (e.g., keratitis).

Since these analogs also inhibit DNA replication in uninfected cells they are **most suitable for topical rather than systemic application.** A partly selective effect on infected cells may, however, occur in animals because many uninfected cells are in a resting stage, with a low level of thymidine kinase activity and with no DNA synthesis; they are resistant to the inhibitors. In addition, herpes simplex and vaccinia viruses cause the formation of a thymidine kinase, probably virus-specified (Ch. 48), which phosphorylates both BUDR and IUDR and converts them into their inhibitory monophosphates. A differential effect may therefore be created.

IUDR-resistant herpes simplex strains have been found in nature; they are not deficient in thymidine kinase production, and the mechanism of resistance is unknown.

Arabinosyl cytosine (commonly called cytosine arabinoside) similarly inhibits the multiplication of DNA viruses by interfering with DNA synthesis. Apparently the analog inhibits the reduction of cytidylic to deoxycytidylic acid. It differentiates less between infected and normal cells, perhaps because no virus-specified enzyme phosphorylates arabinosyl cytidine. **Arabinosyl adenine** and **5-iodo-2'-deoxycytidine** have similar effects. Viruses resistant to IUDR are still sensitive to these compounds.

Clinical Application. The analogs interfering with normal DNA synthesis have been successful in controlling herpes simplex keratitis. Arabinosyl cytosine, arabinosyl adenine, and IUDR are mostly used; even after keratitis has appeared, the drugs, dropped directly into the eye every 1 to 2 hours, drastically reduce the period of clinical disease and prevent the formation of corneal scars. Arabinosyl adenine is especially useful in patients with immunodeficiencies or under treatment with immunosuppressive drugs.

The chemotherapeutic value of these drugs is highest in superficial infections, which can be exposed to large doses of the drugs without systemic involvement, for even if DNA is damaged in the neighboring, uninfected superficial cells, almost no damage results because these cells are normally desquamated. The systemic administration of these inhibitors is toxic because it affects DNA synthesis of essential, rapidly dividing cells (e.g., blood cells); nevertheless small doses of arabinosyl cytosine or arabinosyl adenine have proved useful in life-threatening herpes infections.

ANTIBIOTICS

Antiviral activity is displayed by derivatives of rifamycin (e.g., rifampin), and by the related antibiotics tolypomycin and streptovaricin (see Fig. 11-34), against poxviruses and leukoviruses. In bacteria rifampin is known to inhibit initiation by RNA polymerase (Ch. 11); and with leukoviruses these antibiotics inhibit the activity of the reverse transcriptase (RNA-dependent DNA poly-

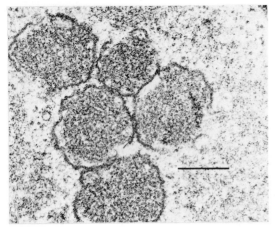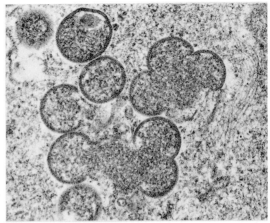

Fig. 49-12. Effect of rifampin on the maturation of vaccinia virus in HeLa cells. **Left.** Electron micrograph from a thin section of infected cells treated for 8 hours with rifampin (100 μg/ml), showing the incomplete and disorganized viral membranes, each surrounding a matrix. **Right.** Thin section of similar cells 10 minutes after removal of rifampin, showing the rapid reorganization of the membranes to a morphology similar to that observed in normal maturation. The bar in the left-hand figure represents 3000 A. (Courtesy of P. M. Grimley.)

merase, Ch. 11). With poxviruses, however, it acts by **interfering with virus maturation:** viral membranes begin to form at the periphery of the filamentous matrix (see Ch. 48, Maturation of complex viruses) but remain incomplete. If rifampin is then removed the membranes close and form virions (Fig. 49-12).

Since this maturation can occur even in the presence of actinomycin D and cycloheximide it can utilize proteins and DNA synthesized earlier in the presence of rifampin. Indeed, this antibiotic blocks the cleavage of several precursor peptides. In addition, though rifampin does not affect transcription by the poxvirus transcriptase in vitro (see Ch. 44, Virions' enzymes), or over-all transcription in infected cells, it prevents the appearance of a particle-associated transcriptase function and of other virions' enzymes. These defects might be the consequence of the maturation defect, if the enzymes require virions for activity or stability.

The mechanism of action of rifampin on poxvirus is unknown. It may conceivably alter the structure of the uncleaved precursor peptides or inactivate the cleaving enzyme, or perhaps block transcription of a few viral genes important in maturation. The activities of the various compounds toward leukovirus, poxvirus, or *E. coli* transcriptase are uncorrelated, suggesting different modes of action. Poxvirus mutants resistant to rifampin and other antiviral rifamycin derivatives are readily isolated.

In vivo these compounds have some local effect: for instance, rifampin inhibits the development of vaccination lesions in man. They have little effect on systemic diseases in animals, perhaps because their high toxicity prevents the use of adequate doses.

SELECTED REFERENCES

Books and Review Articles

DE CLERCQ, E., and MERIGAN, T. C. Current concepts of interferon and interferon induction. *Am Rev Med 21:*17 (1970).

HALL, J. F. Effector mechanisms in immunity. *Lancet 1:*125 (1969).

HILLEMAN, M. R. Toward control of viral infections of man. *Science 164:*506 (1969).

VILCEK, J. *Interferon.* Springer, New York, 1969.

WEHRLI, W., and STAEHELIN, M. Actions of the rifamycins. *Bacteriol Rev 35:*290 (1971).

Specific Articles

ANDERSON, C. W., and EIGNER, J. Breakdown and exclusion of superinfecting T-even bacteriophage in *E. coli. J Virol 8:*869 (1971).

BALTIMORE, D. Inhibition of poliovirus replication by guanidine. In *Medical and Applied Virology,* pp. 340–347. (M. Sanders and E. H. Lennette, eds.) Green, St. Louis, 1968.

BAUER, D. J., ST. VINCENT, L., KEMPE, C. H., and DOWNIE, A. W. Prophylactic treatment of smallpox contacts with N-methylisatin-β-thiosemicarbazone. *Lancet 2:*494 (1963).

CANTELL, K., and TOMMILA, V. Effect of interferon on experimental vaccinia and herpes simplex virus infections in rabbits' eyes. *Lancet 2:*682 (1960).

DE CLERCQ, E., and MERIGAN, T. C. Requirement of a stable secondary structure for the antiviral activity of polynucleotides. *Nature 222:*1148 (1969).

DE MAEYER-GUIGNARD, DE MAEYER, E., and JULLIEN, P. Interferon synthesis in x-irradiated animals. IV. Donor-type serum interferons in rat-to-mouse radiation chimeras injected with Newcastle disease virus. *Proc Nat Acad Sci USA 62:*1038 (1969).

GLASGOW, L. A. Transfer of interferon-producing macrophages: New approach to viral chemotherapy. *Science 170:*854 (1970).

GRIMLEY, P. M., ROSENBLUM, E. N., MIMS, S. J., and MOSS, B. Interruption by rifampin of an early state in vaccinia virus morphogenesis: Accumulation of membranes which are precursors of virus envelopes. *J Virol 6:*519 (1970).

GURGO, C., KAY, R. K., THIRY, L., and GREEN, M. Inhibitors of the RNA and DNA dependent polymerase activities of RNA tumor viruses. *Nature New Biol 229:*111 (1971).

HERTZLER, E. C. Herpetic keratoconjunctivitis: Therapy with synthetic double-stranded RNA. *Science 162:*811 (1968).

HUANG, A. S., and WAGNER, R. R. Defective T particles of vesicular stomatitis virus: Biological role in homologous interference. *Virology 30:* 173 (1966).

ISAACS, A., and LINDENMANN, J. Virus interference. I. The interferon. *Proc R Soc Lond B147:*1258 (1957).

JOKLIK, W. K., and MERIGAN, T. C. Concerning the mechanism of action of interferon. *Proc Nat Acad Sci USA 56:*558 (1966).

KAUFMAN, H. E. Clinical cure of herpes simplex and vaccinia keratitis by 5-Iodo-2'-deoxyuridine. In *Perspectives in Virology III,* p. 90. (M. Pollard, ed.) Hoeber, New York, 1963.

MERIGAN, T. C. Purified interferons: Physical properties and species specificity. *Science 145:*811 (1964).

MERIGAN, T. C. Interferon stimulated by double stranded RNA. *Nature 228:*219 (1970).

OSBORN, J. E., and WALKER, D. L. The role of individual spleen cells in the interferon response of the intact mouse. *Proc Nat Acad Sci USA 62:*1038 (1969).

WOODSON, B., and JOKLIK, W. K. The inhibition of vaccinia virus multiplication by isatin-β-thiosemicarbazone. *Proc Nat Acad Sci USA 54:*946 (1965).

chapter

VIRAL IMMUNOLOGY

HUMORAL IMMUNITY   1190

INTERACTIONS BETWEEN VIRIONS AND ANTIBODIES   1190
  Detection of the Interaction of Viruses with Antibody   1190
NEUTRALIZATION   1191
  Mechanism of Neutralization: Sensitization   1192
  Quantitative Aspects of Neutralization   1193
  Measurement of Neutralizing Antibodies   1197
HEMAGGLUTINATION INHIBITION   1197

DIAGNOSTIC USE OF IMMUNOLOGY   1198

  Types of Virus-specific Antibodies   1198
  Antigenic Classification of Viruses   1198
EVOLUTION OF VIRAL ANTIGENS   1199
METHODS FOR ANTIGENIC ANALYSIS OF VIRUSES   1200
  Neutralization Test   1200
  Hemagglutination-inhibition Test   1201
  Tests on Paired Sera   1201
COMPLICATIONS OF IMMUNOLOGICAL TESTS   1201

CELLULAR IMMUNITY   1201

The immunological properties of viruses are determined by both structural proteins of the virions and other, nonstructural viral proteins made in infected cells (see Chs. 45 and 48).

Humoral antibodies elicited by intact virions are specific for components of the virions' surface, wheras those induced by disrupted virions react with both surface and internal components. In virus-infected animals antibodies are induced not only by complete infectious virions but also by viral products built into the surface of the infected cells or released by dead cells. Thus, animals bearing a tumor induced by an oncogenic DNA virus develop antibodies against the altered cells' nuclear T antigen, a nonvirion protein (see Ch. 63).

Virus-specific proteins in the infected cells' surface membrane can also elicit a **cellular immune response** as though these cells were an allograft. These surface proteins are found after infection not only by surface-budding viruses (see Ch. 48, Maturation) but also by other viruses, e.g., vaccinia, herpes simplex, and oncogenic DNA viruses.

The humoral and cellular immunities elicited by viruses have different significance. Antibodies are useful for identifying, quantitating, and isolating both virions and some of their unassembled components. They also afford criteria for classification and identification of viruses and for serological diagnosis of viral diseases; and as natural or artificially induced immunity they provide the key to protection against many viral infections.

Cellular immunity is important both for pathogenesis of viral diseases (Ch. 51) and for recovery from viral infections, by causing killing of cells with viral antigens on their surface. The effectiveness of this defense shows that killing infected cells is a useful approach to antiviral therapy, provided it is carried out soon enough in development of the disease. This defense accounts for relative infrequency of virus-induced cancer in animals and presumably in man, and for its onset late in life.

HUMORAL IMMUNITY

INTERACTIONS BETWEEN VIRIONS AND ANTIBODIES

While these interactions are based on the same principles as antigen-antibody interactions in general, there are special features to consider, especially in the quantitative aspects of viral neutralization by antibodies.

The interaction of antibodies with virions reflects the presence of different antigenic sites on the virions' surface and the multimeric structure of the coat (Ch. 44). Thus, influenza virions are covered by both hemagglutinins and neuraminidase molecules, and naked icosahedral virions have different proteins in their pentons and hexons. Each component specifies different antibodies whose interactions with the virions have different consequences. Moreover, the multimeric nature of the capsid and the envelope causes the presence of many identical antigenic sites, regularly repeated on their surfaces; as discussed below, these repeating structures interact with antibodies in ways that would be impossible with isolated sites.

DETECTION OF THE INTERACTION OF VIRUSES WITH ANTIBODY

Viral antigens are studied by their reactions with the corresponding antisera, by employing both the usual immunological tests and some special ones, such as hemagglutination inhibition and neutralization. Because neutralization is particularly characteristic of viruses and has widespread application, we shall consider it in detail.

NEUTRALIZATION

Viral neutralization consists in the decrease of infectious titer of a viral preparation following exposure to antibodies. Several features are noteworthy. As with other antigen-antibody complexes, the components can be dissociated and recovered in their original forms. Moreover, the reversibility of the association often decreases with time. Finally, the host cell influences the effectiveness of the interaction: a virion that is neutralized for one kind of host cell may be infectious for another.

Readily Reversible and Stable Virion-Antibody Complexes. These two classes of complexes are distinguished by an experiment of the following kind. Antibody is added to a preparation of influenza virus of known titer at $0°$ to form a neutralization mixture. If a sample is added ½ hour later **without dilution** to the viral assay system (e.g., the allantoic cavity of chicken embryo or cell cultures) a decrease of the viral titer in comparison with untreated virus may be seen; but if the neutralization mixture is **diluted** by a large factor before assay the original titer is restored. Hence the combination between virions and antibody molecules under these conditions is freely reversible, i.e., there is an equilibrium between dissociation and re-formation of the antibody-virion complexes. A decrease of concentration of the reactants diminishes the rate of re-formation of the complexes but not their rate of dissociation.

If, however, the assays are made **several hours after the virus has been mixed** with the antibodies, neutralization persists after dilution. Hence with time the virus-antibody complexes become more stable and are no longer reversed on dilution (at neutral pH and physiological ionic strength).

Results of this type with several viruses show that virions (V) unite with antibody molecules (A) in at least two steps:

1)
$$V + A \rightleftharpoons (VA)r$$

2)
$$(VA)r \rightleftharpoons (VA)$$

$(VA)r$ represents virions complexed with antibodies in a freely reversible way; (VA) represents virions complexed with antibodies in a practically irreversible way, as defined above.

The equilibrium of step 1 is reached very rapidly and shows little temperature dependence, like many antibody-antigen combinations; it merely reflects the attachment of antibody molecules to exposed antigenic sites of the virion. Step 2, however, occurs very slowly at $0°$ but at $37°$ it occurs rapidly, so that within minutes most of the antibody-virion complexes formed are stable.

The reenforcement of the antigen-antibody union with time is not a property of viruses exclusively; it can be observed with ordinary protein antigens (Ch. 15, Danysz phenomenon).

Reactivation of Stably Neutralized Virus. Stabilization of the virus-antibody complex is not associated with a permanent change of either reactant, since the unchanged components can be recovered. Dissociation can be accomplished either by a physicochemical alteration of the antibody-antigen complexes or by competition with a large excess of inactivated virions of similar antigenic specificity. Commonly employed dissociating agents are acid or alkaline pH (Fig. 50-1), sonic vibration, and extraction with a fluorocarbon.

Physicochemical Events Leading to a Stable Association Between Antibody and Virions. Since the combination of antibodies with haptens and soluble antigens is usually readily reversible, except when large complexes are formed (as in the Danysz phenomenon), the firm combination with individual virions probably originates from the multimeric nature of their coat. When a single antibody molecule establishes two specific bonds with two sites on a single virion the stability of the complex increases by many orders of magnitude (Bivalent binding, Ch. 15). For whenever one of the bonds dissociates the other holds, and the dissociated one has time to become reestablished. The specific bonds may

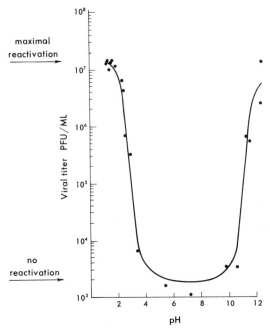

maximal
reactivation →

no
reactivation →

Fig. 50-1. Reactivability of neutralized Newcastle disease virus at different pHs, showing the dissociability of the virus-antibody complexes at acid and alkaline pHs. Virus and antibodies were incubated together until a relative infectivity of about 10^{-4} was obtained (assayed after dilution). Aliquots of the mixture were then diluted 1:100 in cold buffer at various pH values, and after 30 seconds the pH was returned to 7 by dilution into a neutral buffer. The samples were then assayed. [From Granoff, A. *Virology* 25:38 (1965).]

be reenforced by nonspecific bonds between antibody molecules and virions. Although each individual bond may be weak, the collaboration of several may produce considerable stability. In order to dissociate the complex **all** the individual bonds, specific and nonspecific, would have to dissociate by independent activation within a short time, and the probability of such simultaneous dissociation is small. Since the stabilization of the complexes has a high temperature coefficient it probably involves structural modifications of the interacting proteins.

Electron micrographs of influenza virion-antibody complexes afford evidence that an antibody molecule can form two, probably specific bonds, since antibody molecules can be seen bridging two spikes (Fig. 50-2). These bridges may be required for stable neutralization of this virus because mono-

valent antibody fragments produced by papain digestion (Ch. 14), which cannot form bridges, do not form stable complexes, even though they can combine specifically with the virions and cause reversible neutralization. With poliovirus, however, monovalent antibody fragments form stable neutralizing complexes with the virions.

Role of the Host Cells in Neutralization. The host cells play a role in neutralization.

1) **Interaction between neutralized virions and host cells.** Neutralized virions still adsorb to the host cells, provided the number of neutralizing antibody molecules per virion is small; the block in infection appears to be at uncoating. This is best seen with adenovirus virions: although they adsorb through the penton fibers (Ch. 52), antifiber serum does not neutralize, but antihexon serum does. Viruses can also be neutralized after they have attached to the cell, before penetration.

2) **Influence of host cells on the infectivity of antibody-virion complexes.** The same antibody-virus mixture can exhibit different levels of residual infectivity when assayed on different host cells under otherwise identical conditions (Fig. 50-3). Hence neutralization is not the inevitable consequence of certain antibody-virion complexes, but results from a triple interaction of antibody, virion, and host cell. Virions that are scored as neutralized on certain cells, but not on others, will be called **conditionally neutralized.**

MECHANISM OF NEUTRALIZATION:
SENSITIZATION

Neutralization does not require saturation of the surface of the virion with antibody molecules. Thus, cell-adsorbed virions can be neutralized. Moreover, phenotypically mixed virions with two types of capsomers produced by double infection (Chs. 45 and 48), can be neutralized by antisera specific for **either** capsomer. Indeed, kinetic evidence, which will be discussed below, shows that neutralization can be produced by **a single antibody molecule.** The following model suggests how one antibody molecule prevents the whole virion, with its many identical monomers, from infecting the cell. The attachment of a single antibody molecule, especially if it bridges two adjacent subunits, is assumed to cause a

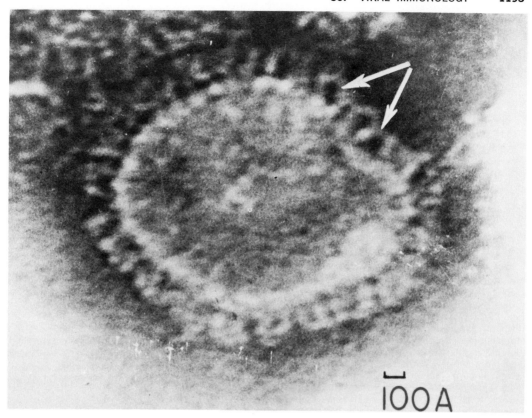

Fig. 50-2. Electron micrograph of an influenza virion with antibody molecules attached to the spikes, some of them forming bridges between two spikes (arrows). [From Lafferty, K. J. *Virology* 21:91 (1963).]

profound local structural alteration which spreads to the whole coat, preventing its regular interaction with cellular constituents involved in uncoating. Such changes are observable in the poliovirus capsid, where exposure to antibody changes both the electrophoretic mobility of the virions and the chemical groups accessible to external reagents. The specific cell dependence of neutralization is understandable in this model, because constituents involved in uncoating in different cells may be affected in different ways by distortions of the viral coat.

Sensitizing Antibodies. The combination of antibody with virion elements does not necessarily cause neutralization. Thus, certain antibody combinations with adenovirus hexons, which probably cause a smaller distortion of the capsid, do not neutralize the virus but do "sensitize" it to neutralization by anti-

bodies to the bound immunoglobulin or sometimes by complement. Sensitization is also caused by early antibodies, or by antibody fragments which bind weakly to the antigen. The anti-Ig or complement may neutralize by causing the initially bound antibody to bind more strongly, with greater distortion of the virion's coat. The antigenic sites of sensitization are different from those involved in neutralization: instead of differing in different adenovirus types they are common to a group.

QUANTITATIVE ASPECTS OF NEUTRALIZATION

Kinetics of Neutralization. Much information on the nature of the antibody-virion interaction leading to neutralization is afforded by studying the kinetics (i.e., the time course) of neutralization. A virus preparation and serum are mixed, and the mixture is then assayed for infectivity at regular intervals. The results are plotted as the logarithm of the fraction of the initial infectious titer that remains un-

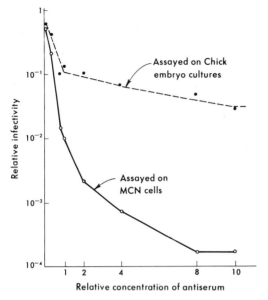

Fig. 50-3. Effect of different host cells on the residual infectivity in neutralizations. Vesicular stomatitis virus was mixed with antiserum at various dilutions at 37°. After 2 hours the mixtures were appropriately diluted and plated for residual infectivity by plaque formation on cultures of either chick embryo cells or MCN cells (a line of cells derived from human leukemic bone marrow). The titer of untreated virus is the same on either cell type. [Modified from Kjellen, L. E., and Schlesinger, R. W. *Virology* 7:236 (1959).]

neutralized **(relative infectivity)** versus the time of sampling.

1. **Relation of the kinetics of neutralization to the stability of the complexes.** The stability of the antibody-virion complexes affects the kinetics of neutralization as revealed by an experiment of the following type. A virus-serum mixture is made and part is immediately diluted (e.g., fivefold); the kinetic curve is determined for the two mixtures by carrying out the assay without further dilution. (It is assumed that the concentration of antibody in the mixture is sufficiently low and will not interfere with the infectivity assay.) In the undiluted and the diluted mixtures the relative concentrations of virus and antibody are equal, but the absolute concentrations differ by a factor of five.

The curve obtained with **readily reversible antibody-virion complexes** begins without a shoulder, then decreases in slope and tends to a horizontal line **(plateau)**, which is reached when the rate of dissociation of virion-antibody complexes equals that of their formation (equilibrium; Fig. 50-4). If the undiluted mixture is diluted fivefold after reaching equilibrium (time A in the figure), its equilibrium shifts to that of the originally diluted mixture.

The curves obtained with the **more stable complexes** also begin without a shoulder and then tend to a plateau. In contrast to the previous case, however, the final relative infectivity is equal in the two mixtures. Since every antibody-virion complex, once formed, persists, the plateau is reached when no more antibodies combine; and the residual infectivity depends on the ratio of antibody concentration to virion concentration, which is the same in the two mixtures. In the diluted mixture, however, the horizontal part of the curve is reached more slowly because collisions between reactants are fewer per unit time. If the mixtures are now diluted their relative infectivity does not change, since the antibody molecules do not dissociate appreciably from the virions.

When the antibody complexes are relatively stable, as is usual in neutralization carried out at 37° with hyperimmune or convalescent serum, the virus-serum mixture can be diluted for assay without altering the results, and so kinetic curves can be obtained for mixtures in which antibody is in large excess over the virions. The curves obtained closely approach straight lines passing through the origin; the slopes are proportional to the concentration of the antibody and to its tendency to combine with the virions and are independent of the virion's concentration (Fig. 50-5). These curves follow the equation:*

$$I = e^{-ktC} \qquad (1)$$

and, taking the logarithm of both sides:

$$\ln I = -ktC$$
$$k = -\frac{\ln I}{tC} \qquad (2)$$

where I is relative infectivity; t, time after mixing the virus and the antibody (in minutes); C, concentration of the serum; and k, a constant proportional to the concentration and the combining power of the antibody in the serum.

2) **The residual relative infectivity.** Kinetic curves of neutralization tend to a plateau of residual infectivity as the time after mixing increases. This plateau is caused by attainment of equilibrium if

---

* This equation is equivalent to that used in Chapter 44, Appendix. ktc is the average number of neutralizing antibody molecules per virion after time t, since molecules of antibody (which is in large excess) continue to attach to the virions at a constant rate. Thus I is the fraction of virions that have **no neutralizing molecule.** As discussed in Chapter 11 Appendix, equation 1 generates single-hit curves and is valid if a single antibody molecule is sufficient to produce neutralization; otherwise the curve would have an initial shoulder (i.e., it would be a multiple-hit curve).

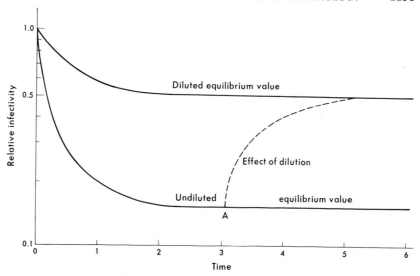

Fig. 50-4. Diagrammatic representation of the time course of neutralization in a virus-antibody mixture with readily reversible antibody-antigen combinations. One mixture is undiluted; the other is diluted fivefold. Note the different equilibrium values reached. If at time A the undiluted mixture is diluted fivefold the viral titer, corrected for dilution, increases, owing to dissociation of virus-antibody complexes, and reaches the same equilibrium value as the mixture originally diluted.

Fig. 50-5. Kinetic curves of neutralization of poliovirus with stable antibody-virion complexes and antibody excess. Note the linearity of the curves in the semilogarithmic plot **A**, with different slopes corresponding to different relative concentration of antibodies (given by the numbers near each line). In **B** the slopes of the curves of **A** are plotted versus the concentration of the antiserum (in relative values), yielding a straight line. The linearity of the two types of curves implies that a single antibody molecule is sufficient to neutralize a virion. [From Dulbecco, R., Vogt, M., and Strickland, A. G. R. *Virology* 2:162 (1956).]

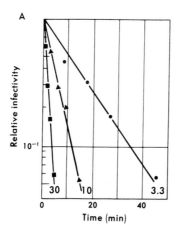

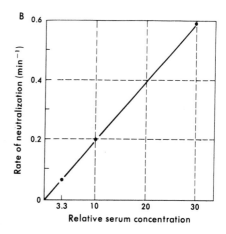

the antibody-virion complexes are readily dissociable, and by exhaustion of free antibody if the complexes are more stable. In either case the residual relative infectivity corresponding to the plateau is a function of the quantities of virus and antibodies in the mixture.

If the **complexes are readily reversible** (and are tested without dilution), the **percentage law** established by Andrews and Elford in 1933 applies: under conditions of antibody excess the proportion (percentage) of virus neutralized by a given antiserum is constant, irrespective of the virus titer. This law can be deduced from the mass law

$$V + A \underset{k_2}{\overset{k_1}{\rightleftharpoons}} \overline{VA}; \quad \frac{k_1}{k_2} = \frac{\overline{(VA)}}{(V)(A)} \; ; \; k(A) = \frac{\overline{(VA)}}{(V)}$$

where V indicates the free virus; A, the free antibody (assumed to have much higher concentration than bound antibody); and $\overline{VA}$, the antibody-virion complexes.

3) The persistent fraction. If the **complexes are stable** the residual relative infectivity, determined after dilution of the virus-serum mixture (Fig. 50-6), theoretically obeys the relations

$$Res = e^{-c(A/V)} \tag{3}$$

where Res is residual infectivity, c is a constant, A denotes the total amount of antibody, and V the total amount of virus.* The problem is quite parallel to that of determining the proportion of uninfected cells at different multiplicities of infection (Ch. 44).

The kinetic curves actually obtained with a large excess of antibodies, however, show an abrupt plateau not justifiable on the basis of the A/V ratio (Fig. 50-7). The fraction of virus that is not neutralized is different in character from the residual infectivity of dissociable antibody-virion complexes at equilibrium, because it is not affected either by dilution or by addition of a fresh charge of antibody. The presence of this **persistent fraction** obscured the study of neutralization of animal viruses for a long time; since it was erroneously attributed to equilibrium, it prevented the recognition of the stable complexes. The persistent fraction was eventually shown to be caused sometimes by the formation of

---

* This equation is derived similarly to equation 1 above, except that the average number of neutralizing antibody molecules per virion is $c(A/V)$. Again this equation requires that a single antibody molecule is sufficient for neutralization; otherwise the curve would have an initial shoulder.

Fig. 50-6. A. Theoretical kinetic curves at different ratios of total antibody (A) to total virus (V) for stable antibody-virion complexes. All the curves tend to plateaus, the relative infectivity of which is a function of the ratio A/V (in arbitrary units). When the residual infectivity values are plotted semilogarithmically versus the A/V ratio they generate a straight line **(B).** This result affords additional evidence that neutralization of a virion requires only the binding of a single antibody molecule.

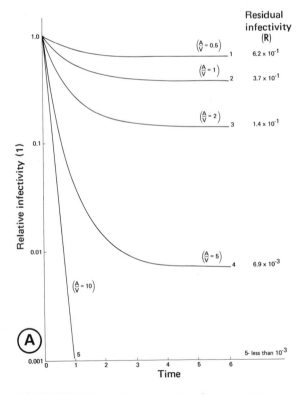

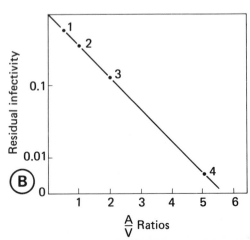

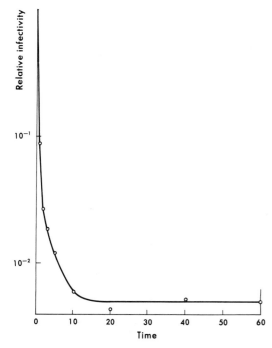

Fig. 50-7. Kinetic curve of neutralization of western equine encephalitis virus (a togavirus), showing the rather abrupt change into a plateau as the time of incubation increases. This plateau is not justified by the A/V ratio employed, which would have allowed a far greater neutralization. The plateau corresponds to the persistent fraction. [From Dulbecco, R., Vogt, M., and Strickland, A. G. R. *Virology* 2:162 (1956).]

aggregates in which some virions remain unneutralized, but usually by interaction of virions with **antibody molecules that do not produce neutralization but prevent neutralization by the conventional interaction.** These inhibitory molecules appear to be of two classes: 1) readily dissociable molecules, whose action is duplicated by univalent antibody fragments; and 2) antibody causing conditional neutralization, which is ineffective with the cells used for determining the residual infectivity. The persistent fraction can be decreased by antibodies to some of these bound molecules.

MEASUREMENT OF NEUTRALIZING ANTIBODIES

This measurement rests on the relations derived above. If the serum-virus mixtures are assayed **without dilution** the relative concentration of antibody is derived from the proportion of neutralized virus, according to the **percentage law.** This method is simple to perform and measures both reversible and stable antibody-virion complexes. The method is unsuitable for accurate measurement, but is adequate and is widely used for diagnostic purposes.

If, alternatively, the serum-virus mixtures are assayed **with dilution** the relative concentration and combining power of the antibody can be calculated either from the **slope of kinetic curves,** according to equations 1 and 2, or from the **residual relative infectivity,** according to equation 3. Although these methods measure mostly stably bound antibody, they yield results of far greater precision than the first method.

The neutralizing power of the serum reflects the degree of protection against a virus. In the early period of immunization, however, such protection is best estimated by carrying out **neutralization in the presence of complement** in order to reveal the weaker antibodies (see above, Sensitization). Through this phenomenon complement appears to be an important factor in early antiviral immunity.

HEMAGGLUTINATION INHIBITION

With virions that agglutinate red cells hemagglutination inhibition can be detected as a decrease of the hemagglutinating titer when the appropriate antibodies are added to the virus before the red cells. The mechanism is basically simple: attachment of antibody molecules to the virions hinders adsorption to the red cells.

Hemagglutination inhibition differs from neutralization in several ways. In the former antibodies interfere with the adsorption of the virions; in neutralization they interfere with cell infection and not necessarily with adsorption. Moreover, there is no evidence for a special role of stable antibody-virion complexes in hemagglutination inhibition, since it is effectively carried out by univalent antibody fragments. Finally, though the number of antibody molecules per virion required to produce hemagglutination inhibition has not been determined, it may have to be high enough to cover all the sites of the virion involved in adsorption.

DIAGNOSTIC USE OF IMMUNOLOGY

Immunological tests can be used with standard antisera to identify and characterize **(type)** a virus isolated from a patient, and with standard antigens to detect antiviral antibodies in the patient's serum. These two applications of viral immunology involve a number of general problems which will be dealt with before considering the diagnostic methodology.

TYPES OF VIRUS-SPECIFIC ANTIBODIES

Viral antigens prepared in different ways elicit the formation of different antibodies. 1) Immunization with **killed virions,** or with virions that cannot multiply in the immunized host, produces antibodies only for **surface components** of the virions; these antibodies have neutralizing and hemagglutination-inhibiting activities against the virions, as well as complement-fixing and precipitating activities with antigens of the viral coat. 2) In contrast, **viruses that multiply in the host** and produce a cytopathic effect in some cells—as in natural infection or in vaccination with "live" vaccines—lead to the formation of antibodies against **all the viral antigens,** including neutralizing or hemagglutination-inhibiting antibodies for surface antigens, and complement-fixing or precipitating antibodies for both surface and internal antigens and antibodies for nonvirion antigens. 3) Immunization with **internal components** of the virions produces complement-fixing and precipitating antibodies active only toward the antigens of these components.

The character of the antibodies recognized in an immune serum depends not only on the method of immunization but also on the **sensitivity of the test method.** Neutralization is the most sensitive, followed by hemagglutination inhibition, complement fixation, and precipitation. These differences parallel the amount of antigen required for each test: for example, with influenza virus about 10^3 viral particles are required for neutralization tests, at least 10^7 for hemagglutination inhibition, and an even greater number for complement fixation. For these reasons, in a given serum the neutralizing antibodies usually have higher titers (hence, are detected more easily) than hemagglutination-inhibiting or complement-fixing antibodies specific for surface components of the virions; precipitating antibodies are even more difficult to detect (Ch. 15, Differences in sensitivity of precipitation, agglutination, and other reactions).

After viral infection the titers of antibodies to different components rise and fall with quite different time courses; the characteristic patterns observed in different infections will be discussed in the chapters on specific viruses.

The antibodies that react in the different tests may not be altogether identical. It seems likely that complement fixation may result from interaction both with the antigens involved in neutralization and with additional surface antigens. Neutralization may be primarily caused by molecules specific for the sites of the virion involved in the release of the viral nucleic acid into the cell; and as noted above, there may be molecules that combine with the surface of the virion without causing neutralization. Neutralization probably requires molecules with higher affinity for virions than do hemagglutination inhibition and especially complement fixation. Only certain classes of immunoglobulins can participate in complement fixation (Ch. 18).

ANTIGENIC CLASSIFICATION OF VIRUSES

Because of their high resolution, immunological methods can differentiate not only between viruses of different families but also between closely related viruses of the same family or subfamily. By these means **family antigens** may be identified; each family or subfamily may be subdivided into **types** (species) on the basis of **type-specific antigens;** some types can even be subdivided into strains (intratypic differentiation). The levels in this classification are obviously somewhat arbitrary. The degree of cross-reaction of an antiserum with related viruses depends on the test method. Usually antibodies detected by neutralization tend to be less cross-reactive, and thus are useful to define the immunological type, whereas those detected by complement fixation tend to be more cross-reactive and are useful to define the family. By proper procedures, however, such as immunization with purified antigens, complement-fixing antibodies as specific as neutralizing antibodies can be prepared. These differences in the specificities of neutralizing and other antibodies are of theoretical interest because they allow some insight into the evolution of viral antigens.

EVOLUTION OF VIRAL ANTIGENS

The great selectivity of neutralization can be understood on the basis of the structure, genetics, and evolution of viruses. Animal viruses that have evolved in the ecology of the mammalian organisms have been opposed by the neutralizing antibodies, which are able to block viral reproduction. Viral evolution must tend to select for mutations changing the antigenic determinants involved in neutralization, since such changes would improve the ability of the mutants to survive in hosts immunized by previous infection with the original type. In contrast, other antigenic sites, detectable by complement fixation but not by neutralization, would tend to remain unchanged, because mutations affecting them would not be selected for. This process would thus force a virus to evolve from a single original type to a variety of types, different in neutralization tests (and sometimes in hemagglutination-inhibition tests), but all retaining some of the original mosaic of antigenic determinants recognizable by complement fixation.

These evolutionary arguments are consistent with the observation that the clearest differentiation of types within a family is present in viruses of rather complex architecture, in which the antigens involved in the attachment of the virion to the cell vary more from one type to another than the other proteins. Thus enveloped viruses have a strain-specific envelope but a cross-reactive internal capsid; adenoviruses have type-specific fibers and family-specific (but also type-specific) capsomers (Ch. 52). Another relevant finding is that the C antigen, which appears in polioviruses after the virions are heated, has cross-reactivity in all three viral types (Ch. 55). The loss of the VP_4 peptide by heating, for example, may reveal antigenic sites normally hidden; these sites would be similar in all polioviruses because they are not affected by selective pressure.

The evolution of antigenic variation appears to have been extensive with some viruses but not with others. The most extensive antigenic variation occurs in influenza virus, a myxovirus (see Chs. 48 and 56), in which the hemagglutinins contain the main antigens responsible for neutralization (see Ch. 44). Experiments in animals show that the high variability of this virus is determined by two factors. One is the fragmented nature of the genome, which allows reassortment of genes derived from different viruses when they infect the same host (see Ch. 48, Viral genetics). Recombination between distant viral strains (e.g., human and equine), which occurs at high frequency, can incorporate antigenic variation arising by mutation in one species into a strain characteristic of another species. Possibly all influenza viruses present in the whole animal kingdom constitute a unique pool for antigenic variation.

The other factor determining high variability is the presence of the virus in cells lining the respiratory tract, where it is exposed only to IgA antibodies, which tend to form reversible complexes with virions because they may be less prone to form bridges between two antigenic sites. Since the probability of neutralization is reduced, the virus can multiply, although at a reduced rate, even in an immune host, producing a large population as a source of antigenic mutants. These, being even less sensitive to the antibody, can then be selected. In contrast, viruses that cause viremia are exposed to the full brunt of the immunological defense; hence even if mutants with a somewhat decreased ability to bind neutralizing antibodies appear they may nevertheless be eliminated.

Also explainable in terms of selective evolution is the frequent **cross-reactivity in neutralization of viruses** of different types within a family, the heterotypic neutralization titer of sera regularly being much lower than the homotypic titer. These cross-reactions can be attributed to a residual antigenic specificity of the primordial type, persisting despite subsequent mutations. A closely related phenomenon is the appearance of so-called **prime strains,** such as influenza A1 or B1, which are poorly neutralized by antiserum specific for the prototype strain, but which induce the formation of antibodies that neutralize well both strains (Ch. 56). It is likely that in prime strains a site involved in neutralization has been modified by mutation, and is still immunogenic but poorly responsive to antibodies. By additional mutations these strains may evolve into a different type.

METHODS FOR ANTIGENIC ANALYSIS OF VIRUSES

Since a number of methods employed in immunological analyses of viruses are similar to those used in other immunological applications, these will not be described here: consideration will be limited to neutralization and hemagglutination inhibition.

NEUTRALIZATION TEST

Neutralization tests are mostly performed for two purposes: 1) to measure the titer of neutralizing antibodies specific for a certain virus, as in serological diagnosis or in typing of viral isolates; and 2) to determine the affinity of different viruses for a standard serum, as in characterizing viral mutants. Both purposes can be fulfilled by the same type of test. Three methods are described below.

Method 1. Determination of the Titer of Neutralizing Antibody by a 50% Endpoint Method.
The titration can be carried out in two ways: **constant virus-varying serum** titration, or **constant serum-varying virus** titration. The first procedure is most commonly used because it has greater accuracy, as will be shown below.

In order to carry out a constant virus-varying serum titration, the serum is inactivated at 56° for 30 minutes to destroy labile substances that have antiviral activity or affect neutralization (see below). The serum is then diluted serially, usually in twofold steps, and each dilution is added to a constant amount of virus (usually between 30 and 100 ID_{50}). The amount of virus is not critical, but it must be adequate to infect all the assay units in the subsequent titration and at the same time small enough to permit detection of a low concentration of antibodies. The mixtures are incubated for a selected time at either 37° or 4°, depending upon the virus employed; then a constant volume of each is assayed for infectivity by inoculation into 5 to 10 units of

a suitable assay system, such as mice, chick embryos, or tissue culture tubes. The neutralization titer of the serum is calculated as the serum dilution at which **50% of the test units are protected** (50% endpoint) by using the method of Reed and Muench (Ch. 44).

The accuracy of the assay can be calculated by the same methods used for viral assays (Ch. 44, Appendix). (Since the serum dilutions are closely spaced the assay is more precise than the corresponding viral assays, which usually employ more widely spaced dilutions.) The constant virus-varying serum titration relies on the constancy of the virus employed; this requirement, however, is not critical, because, according to the percentage law, the proportion of virus neutralized is (at least largely) independent of the titer of the virus. This method of antibody titration is statistically more accurate than the constant serum-varying virus method because in virus-serum mixtures a small change in antibody titer usually produces a much larger change in the infectious titer of the virus. The reason is the exponential relation between residual infectivity and antibody concentration of equation 3 above.

Method 2. Plaque-reduction Test.
In this test an inoculum of about 100 plaque-forming units is incubated with a serial dilution of the serum; each mixture is then added to a monolayer culture, which is overlaid with agar and incubated. The endpoint is an 80% reduction of the number of plaques. The precision of the method depends on the number of plaques at the endpoint (Ch. 44, Appendix).

These two methods are used either qualitatively to demonstrate the presence of virus-specific antibody, or quantitatively to compare antibody titers in different sera.

Method 3. Determination of the Rate of Neutralization by the Plaque Method.
This test measures the slope (k) of the kinetic curves described above. The values of k obtained are extremely reproducible; differences of about 20% are usually significant. Values of k obtained with the same serum provide a sensitive basis for distinguishing viral strains, including laboratory mutants (especially in work with vaccine strains of poliovirus).

In this test a virus sample of known titer is brought to constant temperature in a water bath and mixed with the antiserum. Samples are taken at intervals from the neutralization mixture, diluted, and assayed for plaques. The logarithm of the ratio of the titer of the sample to the original titer (the relative infectivity, I) is plotted versus the time of sampling, yielding a straight line through or near the origin. The slope of that line is k, which is characteristic of the serum and the virus; it is determined from the relation $k = -(\ln I/tC)$, from equation 2 above, where I is the relative infectivity determined after t minutes of incubation of the neutralizing mixture and C is the concentration of the serum.

Example: At a serum dilution of 10^{-3} the relative infectivity is 3×10^{-2} after 10 minutes incubation. Since $\ln 3 \times 10^{-2} = -3.5$, $k = 3.5/(10 \times 10^{-3}) = 350$.

HEMAGGLUTINATION-INHIBITION TEST

For this test a serial twofold dilution of heat-inactivated serum is prepared in saline, and from each dilution 0.25 ml is mixed with 0.25 ml of a viral suspension containing 4 hemagglutinating units (defined in Ch. 44). (If hemagglutination-inhibitory substances are known to be present in the serum or the viral preparation they must be removed in advance; see below.) To each mixture is added 0.25 ml of a 1% erythrocyte suspension. The tubes are shaken and then incubated at the temperature and for the time required for optimal hemagglutination with the virus used. The agglutination pattern is read after incubation as discussed in Chapter 44. The hemagglutination-inhibition titer is the reciprocal of the highest serum dilution that completely prevents hemagglutination. Example:

	Initial serum dilution								HI titer
	1:8	1:16	1:32	1:64	1:128	1:256	1:512	1:1024	
A	0	+	+	+	+	+	+	+	8
B	0	0	0	0	+	+	+	+	64

TESTS ON PAIRED SERA

With a widespread virus, for which a high antibody titer may have resulted from previous exposure, the serological diagnosis of acute infection is based on an increase of the antibody titer in **paired sera,** one obtained soon after onset of disease (acute serum) and the other obtained 1 or 2 weeks later (convalescent serum). This comparison requires considerable precision and is carried out by the plaque-reduction method whenever possible, or by hemagglutination-inhibition or complement-fixation tests. For diagnostic purposes a fourfold increase in the titer is considered evidence of infection.

A summary of the diagnostic serological procedures used for a number of human viruses given in Table 50-1.

COMPLICATIONS OF IMMUNOLOGICAL TESTS

Complications occasionally obscure the significance of immunological tests. Although they concern the specialist, some knowledge of their nature is useful for evaluating the results.

Neutralization Test. Human and animal sera contain **nonspecific viral inhibitors,** especially active against influenza, mumps, herpes simplex, and togaviruses. They are heat-labile and can therefore be eliminated by incubation at 56° for 30 minutes.

Hemagglutination-inhibition Test. Most sera contain **inhibitors of hemagglutination** that are not antibodies; they must be removed for valid tests. The inhibitors for myxoviruses are inactivated by RDE (Ch. 44), trypsin, or periodate; those for togaviruses by extraction with acetone-chloroform.

Complement-fixation Test. The successful performance of this test is frequently hampered by the presence of **anticomplementary substances** in crude tissue suspension used as antigen. This is especially true of the infected brain suspension used with togaviruses; the suspension can be freed of the anticomplementary factors by thorough extraction with acetone in the cold. Extraction with a fluorocarbon has been of value.

CELLULAR IMMUNITY

Cellular immunity against viral antigens present in the cell surface clearly contributes to the **pathogenesis of disease** in animals or man: for instance, in the generation of local lesions by poxviruses and in that of cerebellar lesions in mice after infection with lymphocytic choriomeningitis virus (see Ch. 51). The most direct evidence is the alleviation of these effects by immunosuppressants (such as cyclophosphamide) or antilymphocytic serum.

Cellular immunity also decreases the severity of and plays a role in the recovery from many viral diseases. Thus suppression of cellular immunity markedly increases the severity of infection by poxviruses, herpesvirus, and measles virus. The severity of certain virus diseases in chronically undernourished animals (e.g., coxsackievirus B in

TABLE 50-1. Summary of Serological Tests for Viral Diseases

Virus	Neutralization			Other tests		
	Common virus source	Usual test host	Route of inoculation	Test	Common antigen source	Red cells
Enterovirus Growing in TC	TC fluid (rhesus monkey kidney)	TC (rhesus monkey kidney)		CF HI (some)	TC fluid (rhesus monkey kidney)	Human O
Not growing in TC	Infected mouse torso suspensions	Suckling mice	SC, IP, or IC	CF (some) HI (some)	Mouse torso	Human O
Rhinovirus	TC (diploid human fetal fibroblast strain; monkey kidney)	TC				
Influenza	Chick embryo	Chick embryo	Allantoic sac	HI	Allantoic fluid	Fowl, human 0, guinea pig
	TC (monkey or calf kidney)	TC (monkey or calf kidney)		CF	Allantoic fluid	
Parainfluenza	TC (monkey kidney)	TC (monkey kidney)		CF HI HAI	TC (monkey kidney)	Fowl, human 0, guinea pig
Mumps	Chick embryo (allantoic fluid, membrane)	Chick embryo	Allantoic cavity (production of HA)	CF	Allantoic fluid and membrane	
				HI		Chicken
Respiratory synctial	TC (Hep-2)	TC (Hep-2)		CF	TC (Hep-2)	
Measles	TC (BSC-1)	TC (BSC-1)		CF	TC (BSC-1)	
				HI	TC	Monkey
Rabiesvirus*	Mice TC (hamster kidney, human diploid, chick embryo fibroblasts)	Mice, 4–6 weeks	IC	Fluorescent antibody in brain smears		
				CF	Mouse brain	

* Rabiesvirus is usually identified by a **mouse protection test.** An unknown virus is inoculated intracerebrally both in mice previously vaccinated with an attenuated strain (HEP Flury vaccine virus) and in untreated mice. If the untreated animals die and those vaccinated survive, the identification of rabiesvirus is certain.

Abbreviations: TC, tissue cultures; CAM, chorio allantoic membrane of chick embryo; IC, intracerebral; IP, intraperitoneal; SC, subcutaneous; CF, complement fixation; HI, hemagglutination inhibition; HAI, hemadsorption inhibition; HA, hemagglutinins. Hep-2, BSC-1, HeLa, and KB are cell lines.

TABLE 50-1. Summary of Serological Tests for Viral Diseases (*Continued*)

Virus	Neutralization			Other tests		
	Common virus source	Usual test host	Route of inoculation	Test	Common antigen source	Red cells
Togavirus	Mouse brain TC	Suckling or weanling mice, TC	IC	HI Kaolin-absorbed or acetone extracted sera CF	Mouse brain (fluorocarbon extracted) TC fluids (chick embryo, fibroblasts, hamster kidney, HeLa cells)	Day-old chick or goose cells
Reovirus	TC (rhesus monkey kidney)	TC (rhesus monkey kidney)		HI	TC fluids (rhesus monkey kidney)	Human O, bovine
Herpes simplex	TC or CAM	TC (rabbit kidney) Embryonated chicken eggs	CAM	CF	TC (rabbit kidney) or CAM	
Varicella-herpes zoster	TC (primary human amnion)			CF	TC (primary human amnion)	
Human cytomegalic	TC (human fibroblasts)	TC (human fibroblasts)		CF	TC (human fibroblasts)	
Adenovirus	TC (HeLa or KB)	TC (HeLa or KB)		HI CF	TC (HeLa or KB)	Monkey or rat
Variola; vaccinia	Chicken embryo (CAM) TC	Chicken egg (CAM) TC (monkey kidney, HeLa, KB)		HI CF	Chicken embryo Chicken embryo (CAM)	Chicken cells (selected)

mice; see Ch. 55) appears also to be due to impairment of cellular immunity, because it can be markedly alleviated by injection of immune cells from an immunized syngeneic animal. In man and animals inhibition of cellular immunity is associated with an increased incidence of cancers, both virus-induced and of unknown origin. In man, the effect is seen in individuals who are immunosuppressed in order to avoid rejection of organ transplants: they show both an increased incidence of cancer and of warts (which are virus-induced, Ch. 63).

In experimental animals virus infection at birth (when the animal cannot develop immunity) or neonatal thymectomy shortens the latent period for cancer development and greatly increases its frequency after infection with oncogenic leuko- or papovaviruses (see Ch. 63).

SELECTED REFERENCES

Books and Review Articles

ALMEIDA, J. D., and WATERSON, A. P. The morphology of virus-antibody interaction. *Adv Virus Res* 15:307 (1969).

Diagnostic Procedures for Viral and Rickettsial Diseases. (E. H. Lennette, ed.) American Public Health Association, New York, 1964.

Effect of immunosuppression on viral infections. Immunosuppression: A means to assess the role of the immune response in acute virus infections. Immunology Society Symposium. *Fed Proc 30*:6 (1971).

Specific Articles

DANIELS, C. A., BORSOS, T., RAPP, H. H., SNYDERMAN, R., and NOTKINS, A. L. Neutralization of sensitized virus by purified components of complement. *Proc Nat Acad Sci USA 65*:528 (1970).

DUDLEY, M. A., HENKENS, R. W., and ROWLANDS, D. T., JR. Kinetics of neutralization of bacteriophage f2 by rabbit G-antibodies. *Proc Nat Acad Sci USA 65*:88 (1970).

IWASAKI, T., and OGURA, R. Studies on complement-potentiated neutralizing antibodies (C′-PNAb) induced in rabbits inoculated with Japanese encephalitis virus (JEV). I. The nature of C′-PNAb. *Virology 34*:46 (1968).

JERNE, N. K., and AVEGNO, P. The development of the phage-inactivating properties of serum during the course of specific immunization of an animal: Reversible and irreversible inactivation. *J Immunol* 76:200 (1956).

KJELLEN, L., and PEREIRA, H. G. Role of adenovirus antigens in the induction of virus neutralizing antibody. *J Gen Virol 2*:177 (1968).

LINSCOTT, W. D., and LEVINSON, W. E. Complement components required for virus neutralization by early immunoglobulin antibody. *Proc Nat Acad Sci USA 64*:520 (1969).

MANDEL, B. Reversibility of the reaction between poliovirus and neutralizing antibody of rabbit origin. *Virology 14*:316 (1961).

MCBRIDE, W. D. Antigenic analysis of polioviruses by kinetic studies of serum neutralization. *Virology 7*:45 (1959).

WEBSTER, R. G., CAMPBELL, C. H., and GRANOFF, A. The "in vivo" production of "new" influenza A viruses. I. Genetic recombination between avian and mammalian influenza viruses. *Virology 44*:317 (1971).

chapter 51

PATHOGENESIS OF VIRAL INFECTIONS

CELLULAR AND VIRAL FACTORS IN PATHOGENESIS   1206

VIRAL VIRULENCE   1206
CELL ALTERATIONS PRODUCED BY VIRAL INFECTION   1206
Cytopathic Effect   1207
Development of Inclusion Bodies   1207
Induction of Chromosomal Aberrations   1208
Cell Transformation   1208
CELL SUSCEPTIBILITY   1208
The Role of Cell Receptors   1208
Physiological Factors   1208
Genetic Factors   1209

IMMUNOLOGICAL AND OTHER SYSTEMIC FACTORS   1209

CIRCULATING ANTIBODIES   1209
Protection and Recovery from Infection   1209
Persistence of Antibodies   1210
CELL-MEDIATED IMMUNITY   1211
Immunological Tolerance   1211
NONSPECIFIC SYSTEMIC FACTORS   1211

PATTERNS OF DISEASE   1212

Localized Infections   1212
Disseminated Infections   1212
Inapparent Infections   1214
DISEASE BASED ON VIRUS-INDUCED IMMUNOLOGICAL
RESPONSE   1215
EFFECTS OF VIRUSES ON EMBRYONIC DEVELOPMENT   1216
STEADY-STATE VIRAL INFECTIONS   1216
Latent Infections   1217

SLOW VIRAL INFECTIONS   1219

Viral infection was long assumed to produce only acute clinical disease, but other host responses are being increasingly recognized. These include asymptomatic infections, induction of cancer, and chronic progressive neurological disorders. This chapter will deal with the basic mechanisms involved, leaving to subsequent chapters the analyses of the special effects of various viruses.

The consequences of viral infection depend on a number of factors, including the number of infecting viral particles, their virulence, their path to susceptible cells, their effect on cell function, their speed of multiplication and spread, the host's secondary response to cellular injury (i.e., edema, inflammation), and the host's defenses, both immunological and nonspecific.

CELLULAR AND VIRAL FACTORS IN PATHOGENESIS

The effects of viral infection on cells depend on both the characteristics of the virus (including its virulence) and the susceptibility of the cells.

VIRAL VIRULENCE

Viral virulence, like bacterial virulence (Ch. 22), is under polygenic control. Thus not only can one isolate **attenuated mutants** with reduced virulence, but successive mutations can have an additive effect. In fact, strains of influenza viruses that differ considerably in virulence can yield recombinants of intermediate virulence.

The host reactions to a virus is determined by both viral and host variables. The basis of virulence is poorly understood, but two characteristics are frequently encountered in virulent viruses: they multiply well at the elevated temperatures that arise during illness (i.e., above 39°), and they induce interferon poorly and resist its inhibitory action.

Attenuated mutants are valuable for live virus vaccines. They may be obtained indirectly by **adaptation** or **selection** for other properties. Thus serial propagation of a virus

in a given host tends to select for **adapted variants** with higher virulence for that host but often less virulence for another host. For example, poliovirus attenuated for human neurovirulence was obtained by serial passages in monkey kidney cultures or in brains of cotton rats (Ch. 55, Prevention and control). There are no fixed rules, however, on how adaptation affects attenuation, and this empirical method is often unsuccessful. A more rational approach depends upon the capacity of virulent viruses to multiply at febrile host temperatures: **temperature-sensitive (ts)** mutants, which replicate optimally at 30 to 32° and not at all at 39 to 40°, are usually attenuated.

CELL ALTERATIONS PRODUCED BY VIRAL INFECTION

Cells can respond in four different ways to viral infection: 1) no apparent change, as in some endosymbiotic infections (see below); 2) cytopathic effect and death, as with many virulent viruses; 3) hyperplasia followed by death, as in the pocks of poxviruses on the chorioallantoic membrane of the chicken embryo (Ch. 54); and 4) hyperplasia alone, as in viral transformation to cancer cells (Ch. 63).

CYTOPATHIC EFFECT

The cytopathic effect consists in morphological alterations of the cells, usually resulting in death. There are various types, of which the following are typical examples. With many adenoviruses the cells round up and aggregate in grape-like clusters; with polioviruses they round up and then shrink and lyse, leaving cellular debris; with respiratory syncytial viruses they fuse, producing large syncytia; varicella-zoster virus produces large foci of round cells.

The causes of the various cytopathic effects are not precisely known, but the factors listed below appear to contribute to their development.

1) **Effects on synthesis of cellular macromolecules.** Many virulent viruses cause early depression of cellular syntheses. As noted in Chapter 48, DNA-containing viruses inhibit synthesis of host cell DNA, but not of host cell RNA and protein until late in the multiplication cycle. Many RNA viruses, in contrast, inhibit host cell RNA and protein synthesis early in the multiplication cycle.

2) **Alterations of lysosomes.** These cellular organelles contain hydrolytic enzymes whose release into the cytoplasm after alteration of the lysosomal membrane can cause cell destruction. Two types of lysosome alterations are observed in virus-infected cells. a) A reversible increase in permeability is shown by an increased binding of neutral red, the dye commonly used to stain the live cells in the plaque assay; the cells appear hyperstained and form "red plaques." b) An irreversible damage results in disruption of the organelles and discharge of their enzymes into the cytoplasm. The cells lose their ability to be stained with neutral red and form the usual "white plaques." Protein synthesized late in the viral multiplication cycle, possibly capsid subunits, appears responsible for this profound effect.

3) **Alterations of the cell membranes.** Cells infected by many of the enveloped viruses incorporate viral proteins into cell membranes, preliminary to the formation of the viral envelope (Ch. 48). The cell membrane thus acquires antigenic characteristics of the virions and may, by reacting with virus-specific antibodies and complement or with immune lymphocytes, be a target for **immunological destruction.** The altered cells in some instances (e.g., myxo- and paramyxoviruses) also gain the ability to adsorb red blood cells **(hemadsorption).** Effects on membranes probably also play a large role in altering the cell shape and function. In a striking effect (observed with paramyxoviruses and herpesviruses) the infected cells fuse with adjacent cells and dissolve the common cytoplasmic membranes, forming giant cells **(polykaryocytes).**

4) **Cytopathic effect without synthesis of progeny virus.** A high concentration of virions or of viral coat protein may initiate cytopathic changes without replication of infectious virus (toxic effect). For instance, vaccinia virus at high concentration can rapidly kill cultures of macrophages and other cells; mumps virus lyses erythrocytes and induces syncytial formation in cell cultures by a direct action on plasma membranes; and the cytopathic effect of adenovirus is reproduced by a purified capsid protein, the penton (Ch. 52). Such effects of the viral coat (or some lytic agent associated with it) may be the equivalent of "lysis from without" with bacteriophages (Ch. 45).

Cytopathic effects may also be caused by **incomplete viral synthesis:** for example, in cultured HeLa cells influenza viruses synthesize viral antigens and damage the cells but do not form infectious virions.

Viral toxic effects occur also in animals. In mice, for example, the intravenous injection of a concentrated preparation of influenza, mumps, or vaccinia virus causes death within 24 hours, with hemorrhages and cellular necrosis in various organs; a large intracerebral inoculation of influenza virus produces necrosis of brain cells; and a large intranasal inoculum of Newcastle disease virus produces extensive pneumonia. All these effects are produced with synthesis only of incomplete viral particles or without detectable multiplication of the virus.

DEVELOPMENT OF INCLUSION BODIES

Inclusion bodies, i.e., intracellular masses of new material, are formed by two mechanisms. 1) Some arise as a consequence of the accumulation of virions or unassembled viral subunits (nucleic acid and protein) in the nucleus (e.g., adenovirus), in the cytoplasm (e.g., rabiesvirus, Negri bodies), or in both (e.g., measles). These large accumulations of viral materials appear to disrupt the structure and function of the cells and to cause their death. 2) Other inclusion bodies develop at the sites of earlier viral synthesis but do not contain detectable virions or their components: thus the eosinophilic intranuclear inclusions bodies in cells infected by herpes simplex virus are "scars" left by earlier viral multiplication (Ch. 53).

INDUCTION OF CHROMOSOMAL ABERRATIONS

Cells of **primary cultures** infected by a variety of viruses commonly show chromosomal aberrations, such as breaks or constrictions. Such effects have been reported for measles virus, rubella virus, several herpesviruses, parainfluenza and mumps viruses, adenoviruses, polyoma virus, simian virus 40 (SV 40), and Rous sarcoma virus. Measles virus produces similar chromosomal abnormalities in peripheral leukocytes during **natural infections.** The alterations often appear to be an early expression of the cytopathic effect in cells that will die later. Some of these aberrations, however, have characteristic features; for example, herpes virus induces chromatid breaks only at certain sites of two specific chromosomes. It is also noteworthy that chromatid breaks may continue to occur during the multiplication of cells surviving infection by herpes simplex or polyoma virus, suggesting a persistent infection of the cell clones.

These chromosomal alterations may conceivably result from lysosomal enzymes mobilized in cells, from virus-specified enzymes, or even from interruption of DNA or protein synthesis. The latter possibilities are suggested by the occurrence of chromatid breaks in uninfected cells treated with inhibitors of DNA synthesis (e.g., 5-fluorodeoxyuridine or arabinosyl cytosine) or deprived of essential amino acids.

CELL TRANSFORMATION

A perplexing consequence of infection by certain moderate viruses is the production of tumors or leukemia, discussed in Chapter 63.

Such viruses may have several effects on the cells: stimulation of the synthesis of cellular DNA (e.g., polyoma virus); surface alterations recognizable by the incorporation of new antigenic specificities, distinct from those of virion subunits; chromosomal aberrations; and alterations of the growth properties of the cells so that their division is no longer subject to **contact inhibition** by neighboring cells. This conversion of a normal cultured cell to one resembling a malignant cell has been termed **transformation.** Cells transformed by most DNA-containing viruses (adenoviruses, polyoma, SV 40) do not continue to synthesize infectious virus, but at least a portion of the viral genome persists and continues to function. In contrast, cells transformed by RNA viruses (e.g., avian leukosis, murine leukemia) usually continue to produce virions (Ch. 63).

CELL SUSCEPTIBILITY

THE ROLE OF CELL RECEPTORS

The susceptibility of cells, which plays a critical role in viral infection, is often determined by early steps in the virus-cell interaction, such as the attachment of the virions or the release of their nucleic acids in the cells: as noted in Chapter 48, cells resistant to a virion may be susceptible to its extracted nucleic acid.

With animal viruses, just as was observed earlier with bacteriophages (Ch. 45), resistance of cells is often caused by **failure of adsorption.** Indeed, susceptibility to poliovirus and coxsackieviruses correlates with presence of receptors for viral adsorption on the cell surface: in several animal hosts homogenates of susceptible organs adsorb the virions, whereas homogenates of insusceptible organs do not. In species in which viral susceptibility varies markedly with age the ability of the organs to adsorb the virus often also varies in a parallel way.

The **presence or absence of receptors** for viral adsorption on the cell surface depends on both physiological and genetic factors.

PHYSIOLOGICAL FACTORS

Cultivation in vitro may markedly alter the viral susceptibility of cells from that in the original organ. For instance, polioviruses, which multiply in nervous tissue but not in the kidney of a living monkey, multiply well in cultures derived from the kidneys and receptor activity is present in the homogenates of the cultivated cells but not of the intact kidney. Hence it is possible to propagate many viruses in cells that are readily cultured, without having to use cells that are difficult to grow in vitro.

Marked changes in susceptibility accompany the **maturation** of animals. Many viruses are much more virulent toward newborn than toward adult animals (e.g., coxsackieviruses, herpes simplex virus); while some viruses (e.g., polioviruses, lymphocytic choriomeningitis virus) are more virulent for older animals. With coxsackieviruses and mice the change correlates with receptor activity, although it may also depend on changes in interferon and antibody production (Ch. 49). In contrast, with foot-and-mouth disease virus the change involves rate of viral multiplication rather than adsorption, suggesting that a step following adsorption varies with age. The age-dependence of lymphocytic choriomeningitis

arises by an entirely different mechanism, which will be discussed below.

GENETIC FACTORS

Genetic control of cell susceptibility has been demonstrated in mice with hereditarily altered susceptibility to some togaviruses, to mouse hepatitis virus, or to influenza virus, and in chickens with altered susceptibility to Rous sarcoma virus. In crosses betweeen resistant and susceptible animals the heterozygous first-generation (F_1) progeny are usually of uniform responsiveness, depending on which allele is dominant. (Resistance is dominant with togaviruses, whereas susceptibility is dominant with mouse hepatitis and Rous sarcoma viruses.) Moreover, back-crosses of the F_1 individuals to the parent carrying the recessive allele yield 50% susceptible and 50% resistant animals. These results imply a difference in a single gene or a closely linked cluster. These hereditary differences evidently involve cellular rather than immunological mechanisms, since they can also be demonstrated with macrophages and with cultured cells.

IMMUNOLOGICAL AND OTHER SYSTEMIC FACTORS

CIRCULATING ANTIBODIES

PROTECTION AND RECOVERY FROM INFECTION

Protection. Antibodies in serum and extracellular fluids provide the main protection against viral infection, as shown by both experimental and epidemiological studies. The degree of protection is directly related to the level of neutralizing antibodies in the blood for those infections in which viremia is an essential link in the pathogenesis of the disease, e.g., in measles, poliomyelitis, yellow fever, and smallpox.

IgA antibodies (Ch. 16), which are synthesized by exocrine glands and secreted into the extracellular fluids, are associated with protection of the respiratory and gastrointestinal tracts. Hence, local immunization, particularly with live virus vaccines, may be especially advantageous for prevention of respiratory infections.

A protective role of Abs is also supported by

the prophylactic action of immune serum in **passive immunization.** Inoculation of immune serum or immune γ-globulin before infection, or early in the incubation period, may prevent or modify diseases with viremia and long incubation periods (greater than 12 days), such as measles, infectious hepatitis, poliomyelitis, and mumps. The striking protection of populations by some viral vaccines (Table 51-1) presents additional evidence for the prophylactic function of Abs. The preparation, use, and value of these specific viral vaccines will be discussed in the chapters that follow; the advantages and disadvantages of "killed" and "live" virus vaccines are particularly elaborated for polioviruses in Chapter 55.

The mechanism by which Abs neutralize viruses has been considered in detail in Chapter 50.

Recovery. The role of neutralizing Abs in recovery from viral infection is less clear than is their role in protection. Although Abs gen-

TABLE 51-1. Viral Diseases in Which Immunization Has Been Successful

Disease	Attenuated virus	Inactivated virus
Smallpox	+	−
Yellow fever	+	−
Poliomyelitis	+	+
Measles	+	+*
Influenza	+†	+
Mumps	+	+
Rabies	+	−
Adenovirus infections‡	+	+
Rubella	+	−

  * Less effective.

  † Experimental.

  ‡ Caused by types 3, 4, and 7; vaccines of types 4 and 7 attenuated viruses have been tested and proved effective.

erally develop during recovery, virus may continue to increase in some organs even while Ab is being elaborated. Moreover, in some localized infections with short incubation periods, such as influenza, viral titers may begin to decrease, and recovery to commence, before circulating Abs are detectable. The seemingly normal recovery of patients with agammaglobulinemia provides further evidence that factors other than circulating Abs act to limit the course of viral diseases. Cell-mediated immunity and some nonspecific factors are discussed below.

It is not surprising that neutralizing Abs may not play a major role in recovery from viral infections, for Abs are ineffective against intracellular parasites. Furthermore, viruses can spread to contiguous uninfected cells through cellular processes, and thus remain inaccessible to Abs in the intercellular fluid.

PERSISTENCE OF ANTIBODIES

The time course and the persistence of Ab production vary with the virus, the type of antigenic stimulus, and the type of antibody. For example, 1) neutralizing Abs fall from their maximal level more rapidly, and to a lower titer, following influenza than following poliomyelitis infection; 2) immunity to measles is life-long following infection, whereas following immunization with formalin-inactivated virus it lasts for only a few months (Ch. 57, Measles virus, prevention and con-

trol); 3) after infection complement-fixing Abs generally decrease much sooner than neutralizing Abs.

Long-lasting immunity, with persistence of circulating Abs, results from infection with a number of viruses, especially those causing viremia. It is thus extremely rare for a person to have a second attack of measles, smallpox, yellow fever, or poliomyelitis, to mention only a few examples. In contrast, with most acute localized infections without viremia, particularly respiratory diseases such as influenza, second infections are common. (Adenovirus infections are notable exceptions: they do produce long-lasting immunity, perhaps because they frequently terminate in latent infections of the respiratory tract lymphoid tissue).

Three theories have been proposed to explain lasting immunity to certain viral infections.

1) **Latent infection.** After an acute infection has disappeared a persistent latent infection (see below) could boost immunity by the continued production of small amounts of virus; but the lack of transmission of virus requires explanation. Thus in Panum's classic observations on a measles epidemic in the Faroe Islands, those persons were immune who had been alive during the preceding epidemic, 67 years earlier, whereas the younger islanders were highly susceptible. Hence, not only did immunity persist in the absence of reinfection, but the immune persons had failed to infect their nonimmune contacts during all these years. This failure of viral transmission could be explained by incomplete viral multiplication or the continued neutralization of the virus released in the immune persons.

2) **Repeated infections.** Most diseases to which man maintains lasting immunity (e.g., measles, mumps, chickenpox, and poliomyelitis) are prevalent, and therefore individuals tend to be repeatedly exposed to the causative agents; the resulting inapparent infections probably stimulate Ab production.

3) **Secondary antibody response in disease with a long incubation period.** In viral diseases with a long incubation period and a viremic phase, a prompt and effective secondary Ab response could prevent viremia and dissemination of virus to the target organs. The persistent immunity could thus prevent disease with-

out preventing infection. The plausibility of this mechanism is supported by the observation that in rabies, a disease with an especially long incubation period, even the primary Ab response is rapid enough to make prophylactic treatment by vaccination after exposure worthwhile (Ch. 59, Rabies virus, prevention and control).

Repeated infection is excluded in the example of measles in the Faroe Islands, nor can this mechanism account for the prolonged persistence of circulating Abs against smallpox or yellow fever in previously infected residents of the United States. Moreover, while the capacity to give a secondary response is known to persist for many years, without further antigenic stimuli, its actual duration is not known. Hence a latent infection, which could persist for a lifetime, offers the most reasonable explanation for continuing immunity to many viral infections.

CELL-MEDIATED IMMUNITY

Dysagammaglobulinemias in man have provided the strongest evidence that humoral Abs are not critical in recovery from many viral infections. Patients who lack γ-globulin, but develop delayed-type hypersensitivity normally, recover from viral diseases without difficulty, whereas patients with defective delayed-type hypersensitivity, but normal humoral Abs, show poor recovery from certain viral infections. The most striking example is the severe generalized vaccinia or extensive necrosis of the skin and muscle of the affected extremity **(vaccinia gangrenosa)** that frequently follows smallpox immunization (Ch. 54) of individuals with defective cellular immunity. The complications are unaffected by administration of specific Abs, but they are arrested by local injection of lymphoid cells from recently immunized donors, or by injection of **transfer factor** obtained from such cells (Ch. 20). Development of a delayed-type hypersensitivity reaction to heat-inactivated vaccinia virus accompanies recovery. Moreover, in experimental animals depression of cellular immunity by **antilymphocytic serum** delays recovery from, or increases the severity of, infection with a number of viruses, particularly those pos-

sessing envelopes, such as herpesviruses, togaviruses, lymphocytic choriomeningitis virus, and murine leukemia virus.

IMMUNOLOGICAL TOLERANCE

Some viruses that invade the embryo or newborn animal and reproduce without damaging host cells appear to give rise to immunological tolerance (e.g., visceral lymphomatosis in chicks hatched from eggs laid by infected hens). But tolerance is not a necessary consequence of embryonic infection, and it may even be unusual: it is not seen in humans embryonically infected with rubella (Ch. 57, Pathogenesis) or cytomegalovirus (Ch. 53, Pathogenesis) or in fetal mice infected with influenza virus. And although mice with persistent **lymphocytic choriomeningitis** (LCM) or **murine leukemia** virus infection give birth to offspring who are viremic and lack detectable virus-specific circulating Abs, the newborns do synthesize Abs in abundance. But the Abs are not detectable because they combine with excess virus and are rapidly cleared from the circulation.

Infected tolerant or pseudotolerant animals continuously produce large amounts of virus throughout their life, usually with viremia. This virus is disseminated **vertically** to their offspring, through the ovum, placenta, or milk, and **horizontally** to contacts, through virus shed in excretions and secretions. In such animals infection is asymptomatic for most of their life span, but late in life a chronic disease may develop.

NONSPECIFIC SYSTEMIC FACTORS

Resistance to viral infection depends to some extent on factors that act rather indiscriminately on all or many viruses and are therefore called nonspecific. These include various hormones, temperature, inhibitors other than antibodies, and phagocytes.

Nutrition may also affect the course of viral infections (e.g., the devastating consequences of measles in malnourished children of West Africa); but malnutrition influences so many aspects of the host defenses that specific analysis of cause and effect is difficult. An especially important factor, **interferon,** is formed by infected host cells and interferes with the infection of other cells by many viruses; this agent has been discussed in Chapter 49 and will not be considered further here.

Hormonal Factors. The potential effect of sex hormones on viral infections can be illustrated by several examples: **castration** increases the susceptibility of female monkeys to poliomyelitis, the effect being reversed by administration of estrogens, and **gonadectomy** of young male hamsters during the first week of life increases the incidence of tumors produced by type 12 adenovirus (Ch. 63) as compared to that observed in females. **Pregnancy** increases the severity of several viral diseases: paralytic poliomyelitis is more frequent and extensive; smallpox has a more severe course and abortion is common; complications of influenza, particularly pneumonia, are increased. **Cortisone** enhances the susceptibility of many animals to neurotropic viruses, and of chick embryos to influenza and mumps virus; in man cortisone has caused enlargement and perforation of herpetic corneal ulcers and extensive visceral spread of varicella virus, often terminating in severe pneumonia. These deleterious effects of cortisone appear to be independent of its action on the antibody response, and to result from suppression of inflammatory reactions, cell-mediated immunity, and interferon production (Ch. 48).

Temperature. An increase in the host's temperature may reduce viral replication 1) by suppressing a temperature-sensitive step; 2) by causing a more rapid inactivation of many heat-labile viruses (such as togaviruses, influenza viruses); and 3) by increas-

ing interferon production. Conversely in mice held at 4° rather than 25° after infection with B1 coxsackievirus viral multiplication is excessive in many organs, little interferon is produced, and the mortality is strikingly increased. A rise in body temperature may therefore be an important factor in recovery from viral disease.

Virus-neutralizing Inhibitors. Globulins or lipoproteins present in the sera of uninfected animals can neutralize the infectivity of many enveloped viruses in vitro. They differ from complement in their fractionation properties; furthermore, the lipoproteins are more thermostable. It is assumed that host resistance to infection is enhanced by these inhibitors.

Phagocytosis. This does not appear to be as important a defense mechanism in viral as in bacterial infections. On the contrary, some viruses seem to render polymorphonuclear leukocytes ineffective in combating bacterial infections. For instance, measles virus infects leukocytes, producing chromosomal aberrations, and causes leukopenia; and influenza viruses adsorb to polymorphonuclear leukocytes and reduce their phagocytic function, as discussed in Chapter 22 (Phagocytosis). **Macrophages,** however, appear to be important in viral infections, both on the basis of correlation between host and macrophage susceptibility described above (p. 1209) and because they rapidly take up certain viruses inoculated intravenously, especially the large poxviruses. Such infected macrophages can act as a source of infection for other cells, but with viruses that are unable to multiply in them macrophages appear to play their usual role as scavengers.

PATTERNS OF DISEASE

Viruses cause three basic patterns of infection: **localized, disseminated,** and **inapparent.**

LOCALIZED INFECTIONS

In these infections viral multiplication and cell damage remain localized near the site of entry (e.g., the skin or parts of the respiratory or gastrointestinal tract). The virus may spread from the first infected cells to neighboring cells by diffusion in intercellular spaces and by cell contact, forming a single lesion or a group of lesions as with warts and molluscum contagiosum. In a less strictly localized pattern excretions or secretions within connected cavities transport the virus, causing diffuse involvement of an organ, as with in-

fluenza, the common cold, or viral gastroenteritis. Virus may in time spread to distant sites, but this dissemination is not essential for production of the characteristic illness.

DISSEMINATED INFECTIONS

These infections develop through several sequential steps: 1) the virus undergoes primary multiplication at the site of entry and in regional lymph nodes; 2) progeny virus spreads through the blood stream and lymphatics **(the primary viremia)** to additional organs, where 3) further multiplication takes place; 4) the virus is disseminated to the target organs via a **secondary viremia;** and 5) it multiplies further in the target organs, where it causes cell damage, pathological lesions, and clinical disease. In some instances, such as poliovirus infections (Ch. 55, Pathogenesis), primary and secondary viremias are not distinguishable.

Disseminated infection is illustrated by Fenner's classic investigation of ectromelia (mousepox), summarized in Figure 51-1. Mousepox virus enters through an abrasion of the skin and multiplies locally; from there it spreads rapidly to regional lymph nodes, where it also multiplies. The virus then enters the lymphatics and the blood stream, and this **primary viremia** causes the dissemination of the virus to other susceptible organs, especially the liver and spleen. Viral multiplication results in necrotic lesions in these organs and a more intense **secondary viremia,** which disseminates the virus to the target organ, the skin. There the virus undergoes extensive multiplication, producing papules which eventually ulcerate. With the appearance of the papular rash the asymptomatic period (called

incubation period) terminates and clinical disease begins.

The temporal relation between viral multiplication in the various organs, development of lesions, and formation of antibodies should be noted (Fig. 51-1A). It is particularly striking that **overt disease begins only after virus becomes widely disseminated in the body and has attained maximum titers in the blood and the spleen.**

This model of dissemination is applicable not only to exanthematous diseases, such as smallpox and measles, but also to nonexanthematous diseases, such as poliomyelitis and mumps. Thus the target organ for poliovirus is the central nervous system, and for mumps virus the salivary and other glands.

The dissemination of **neurotropic** viruses to the nervous system may occur by transmission **along nerves** as well as by viremia. For instance, in mice such centripetal transmission of herpes simplex virus after foot-pad inoculation can be followed by assaying segments of

Fig. 51-1. Sequential events in pathogenesis of ectromelia (mousepox) in mice inoculated in foot pad. **A.** Relation between viral multiplication (in foot pad, spleen, blood, and skin), development of primary lesion and rash, and appearance of antibodies (E-AHA). **B.** Diagrammatic representation of the dissemination of virus and pathogenesis of the rash in mousepox. [From Fenner, F. *Lancet* 2:915 (1948).]

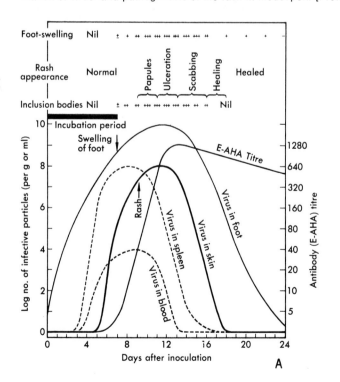

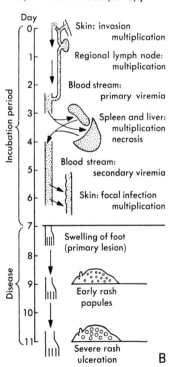

nerves at various times after infection. The virus may conceivably travel either by diffusion within the axons or by multiplication in endoneural cells (Schwann's cells and fibroblasts), in which viral antigens can be localized by immunofluorescence.

For many years, before refined structural studies on virions became possible, animal viruses were classified primarily in terms of their viscerotropism, neurotropism, or dermotropism. However, the **target organ** for a given virus (where the susceptible cells are damaged), and the type of disease produced, bear no relation to the taxonomic position of the virus, as defined in Chapters 44 and 48. In fact, such unrelated viruses as influenza and adenoviruses, may produce diseases that cannot be clinically differentiated; and such related viruses as parainfluenza, mumps, and measles may produce completely different clinical syndromes. The grouping of viruses on the basis of their target organs is presented in Table 51-2.

INAPPARENT INFECTIONS

Transient viral infections without overt disease **(inapparent infections),** though less dramatic

TABLE 51-2.   Grouping of Viruses by Pathogenic Characteristics in Man

Classification according to major target organs	Specific virus	Leisons produced in	
		Target organs indicated	Organ distant from those indicated
Respiratory viruses	Influenza A, B, and C	+	−
	Parainfluenza	+	−
	Respiratory syncytial	+	−
	Measles	+	+
	Mumps	±	+
	Adenoviruses	+	−
	Rhinoviruses	+	−
	Coxsackieviruses (some)	+	+
	Echoviruses (some)	+	+
	Reoviruses	?	?
	Lymphocytic choriomeningitis	+	+
	Coronavirus	+	−
Enteric viruses	Polioviruses	−	+
	Coxsackieviruses	−	+
	Echoviruses	−	+
	Reoviruses	?	?
	Hepatitis virus	+	+
Neurotropic viruses	Polioviruses	+	−
	Coxsackieviruses	+	+
	Echoviruses	+	+
	Rabies	+	−
	Mumps	+	+
	Arboviruses	+	−
	Herpes simplex	+	+
	Virus B	+	−
	Varicella-herpes zoster	+	+
	Lymphocytic choriomeningitis	+	+
Dermotropic viruses	Poxviruses	+	+
	Measles	+	+
	Varicella-herpes zoster	+	+
	Coxsackieviruses	+	+
	Echoviruses	+	+
	Herpes simplex	+	(+)*
	Rubella	+	+
	Human wart	+	−
	Molluscum contagiosum	+	−

* Uncommon.

than the acute infections described above, are very common. Moreover, they have great epidemiological importance, for they represent an often unrecognized source for dissemination of virus, and they confer immunity. For example, for every paralytic case of poliomyelitis in man there were in the United States before immunization 100 to 200 inapparent infections, detected serologically or by viral isolation.

Several factors are involved in the production of inapparent infections.

1) **The nature of the virus.** Moderate viruses or attenuated strains (as in live vaccines) usually cause inapparent infections.

2) The degree of the host's **immunity,** including its ability to produce a prompt secondary antibody response, and the appearance of **viral interference** (Ch. 49). When these defense mechanisms are effective even viruses capable of causing acute disease may remain inapparent.

3) **Failure of the virus to reach the target organ** is also an expression of host defense, but of a more obscure nature. Thus in experimental infections of adult mice with St. Louis or Japanese B encephalitis virus extraneural inoculation causes infection, but only intracerebral inoculation produces disease. A blood-brain "barrier" has been postulated, though its nature is not precisely known. Since passage of viruses from the blood into the brain probably requires active multiplication of the virus in the capillary endothelium, resistance of these cells would impose an effective barrier.

DISEASE BASED ON VIRUS-INDUCED IMMUNOLOGICAL RESPONSE

The immune response, whose protective and ameliorative effects in viral disease have been discussed, can also have an opposite effect, for delayed hypersensitivity appears to play a direct role in the genesis of some viral diseases. For instance, children immunized with **inactivated measles** or **respiratory syncytial** viruses develop unusually severe disease if subsequently infected (Ch. 57), probably owing to an allergic response (of delayed type) to virus or its products. Moreover, formation of the primary lesion induced by

vaccinia virus, at the site of inoculation, is suppressed by agents interfering with the expression of delayed hypersensitivity (Ch. 54, Pathogenesis).

The important role of the immune response in some viral diseases is dramatically demonstrated by studies of lymphocytic choriomeningitis (LCM) virus of mice.* Thus in adults, severe, often fatal, disease follows about a week after intracerebral or intraperitoneal inoculation, but disease fails to develop if the immune response is suppressed (by chemicals, X-irradiation, or antilymphocytic serum). In contrast, after infection in utero or at birth the mice appear normal for 9 to 12 months even though there is widespread viral multiplication, persistent viremia and viruria, and viral antigens in most organs. But circulating Abs are not detectable and specific cell-mediated immunity is depressed. This response was long regarded as the outstanding example of virus-induced immune tolerance (Ch. 17). More careful examination, however, revealed that Abs are produced, but that they are immediately complexed with excess circulating virus or viral antigen and complement. These aggregates are filtered from the circulation, particularly by the renal glomeruli where they accumulate and initiate an inflammatory response, culminating in glomerulonephritis.

Thus in the absence of immunity LCM virus multiplies harmlessly in mice, producing an inapparent infection similar to the endosymbiotic infection it causes in cell cultures (see below). Tissue injury can be initiated in the infected, apparently normal mouse by injection of specific Abs, by establishing parabiosis with an immune mouse, or by transfer of spleen cells from an isologous immune donor. In addition to the glomerulonephritis, necrotic lesions appear in the liver, brain, spleen, and other organs, apparently resulting from reaction of virus-sensitized lymphoid cells with viral antigens present on the surface of many cells. Clearly the brain damage after intracerebral infection of adult animals is generated by a similar mechanism. Hence every aspect of the disease, in both acute and

---

* This virus can also infect man, but it usually produces a mild respiratory infection, only occasionally followed by severe meningitis.

chronic infection, is immunologically mediated.

EFFECTS OF VIRUS ON EMBRYONIC DEVELOPMENT

The variation of susceptibility with age, noted earlier (p. 1209), is especially striking for the embryonic period. Indeed, some viruses that produce mild disease in the adult produce extensive infection and severe malformations in the embryo. The role of viruses in the pathogenesis of congenital anomalies was not recognized until Gregg, in 1941, discovered that rubella virus infection of women during the first 3 months of pregnancy may cause a variety of congenital anomalies (Ch. 57). Of all the viral infections that may occur during the first trimester of pregnancy, **rubella** is the major cause of fetal death and congenital malformations. But a few other viruses have teratogenic effects: **cytomegalovirus** induces a low incidence of microcephaly, motor disability, and chorioretinitis; **Group B coxsackieviruses** are responsible for some congenital heart lesions; and **type 2 herpes simplex virus** may cause microcephaly and other severe central nervous system malformations.

Malformations are readily produced in lower animals by various infections during embryonic development, such as influenza, mumps, and Newcastle disease viruses in chicks, reovirus in mice, and hog cholera virus in swine. Abortion may be produced by Japanese B virus infection in swine and by equine abortion virus (a herpesvirus) in horses, while rat virus (a small DNA virus indigenous to rats) causes the birth of hamsters with malformations of the head resembling those of children with Down's disease (mongolism).

These infections have made it possible to study the pathways of invasion of the embryo. Some leukemia-inducing viruses of mice and chickens reach the embryo through the ovum, whereas LCM virus and rat virus reach it through the placenta. Rat virus injected into pregnant hamsters readily invades the uterus, often killing the fetus.

The passage of a virus across the placenta apparently occurs only when the mother is viremic. Multiplication of the virus in the placenta probably favors transmission, but is not strictly required, since the small coliphage ϕX174, although unable to multiply in animals, is transmitted, but with an extremely low efficiency.

STEADY-STATE VIRAL INFECTIONS

Endosymbiotic Infection. In certain virus-cell systems the infected cells multiply for many generations, although they continue to release virus. These cells generate populations in which all or most individuals are infected (as shown either by release of virus or by the presence of viral nucleic acids or antigens), **even when external reinfection is prevented** by the presence of potent antiviral serum. The cells may be functionally normal: for instance, chick embryos carrying certain tumor viruses (Ch. 63, Avian leukosis viruses) develop normally.

Such endosymbiotic infection is caused by many viruses, both in the animal and in tissue cultures, and by the very small bacteriophage fd in bacteria. Although this type of persistent infection is superficially similar to lysogeny, it is maintained by quite a different mechanism. There is no evidence for association of the viral nucleic acid with the cellular chromosome, or for the operation of a specific regulatory mechanism leading to immunity. The infection persists because 1) the rate of multiplication of virus must be low enough to permit the cells to survive and multiply, and 2) there is a sufficient number of viral nucleic acid molecules or viral particles in each cell to ensure infection of most daughter cells by random segregation at mitosis.

Carrier State. The endosymbiotic infection must be distinguished from the carrier state, often observed in virus-infected cultures of animal cells. This state also causes a persistent viral infection of a cell population, but it is based on the infection of a small proportion of the cells: in these a regular viral multiplication cycle takes place, usually terminating in cell death, but the released virus infects only a small number of the other cells. This limited reinfection depends on special conditions of the culture, e.g., partial resistance of the cells to infection, the presence in the medium of antiviral agents such as weak antibody or interferon, or cell-to-cell transmission of virus without release into the medium. If the continuing reinfection is prevented, as by adding antiviral Abs, the culture is often cured of the infection.

In summary, the differences between endosymbiotic infection and the carrier state are the following:

Endosymbiotic infection

All or most cells infected

Intracellular transmission of infection

Infection cannot be cured by viral antibody

Caused by viruses that can be released without killing cell

Carrier-state infection

Small proportion of cells infected (1% or less)

Usually extracellular, occasionally intracellular, transmission of infection

Infection can often be cured by viral antibody

Caused by any virus

LATENT INFECTIONS

In **persistent inapparent infections,** i.e., **latent infections,** overt disease is not produced but the virus is not eradicated. **This equilibrium between host and parasite** is achieved, as in steady-state infections in vitro, in different ways by different viruses and hosts.

1) In animals, as in cell cultures, some viruses multiply profusely without causing cell damage (i.e., endosymbiotic infections). As described, infection with **LCM virus of mice** may persist for long periods without disease until an adequate immune response has developed. Some slowly developing viral diseases, which have come to be known as "slow viral infections," have this genesis. They will be discussed below.

2) **Herpes simplex virus** has a special pattern of **latency and recurrence.** It usually infects man between 6 and 18 months of age, and the virus persists but cannot be found except during recurrent acute episodes, such as herpes labialis (fever blisters; Ch. 53, Pathogenesis). The **occult** (undetectable) form, which persists in the intervals between recurrent episodes, has not been identified. Since herpes simplex virus contains DNA it may conceivably establish a complex with the host cell genome comparable to **lysogeny** in bacteria. However, there is no direct evidence supporting this hypothesis; instead continued production of virus is suggested by the presence of a higher titer of neutralizing Ab in infected individuals. It therefore seems likely that the virus is maintained, as in virus-carrier cultures, by an infection confined to only a few cells at a time, owing to the presence of Abs, viral interference, or metabolic factors. Because Abs are present most of the extracellular virus would be neutralized and go undetected. Acute episodes may depend on a transient change in the local level or the specificity of Abs, or a change in the susceptibility of the uninfected cells. In man recurrences are provoked by a variety of physical and physiological factors, e.g., fever, intense sunlight, fatigue, menstruation.

In experimental infections of rabbits herpes encephalitis can be reactivated 6 months after an acute encephalitis by inducing anaphylaxis with any antigen or by injecting epinephrine. Similarly, herpes keratitis can be provoked, after an acute corneal ulcer is healed, if the rabbit is made sensitive to horse serum and a corneal Arthus reaction is induced.

3) Latent infections of tissues are sometimes **activated when the cells are cultured.** Thus **adenovirus infections** of man are self-limited (Ch. 52, Pathogenesis), but the virus frequently causes a latent infection of tonsils and adenoids. Though the homogenized tissue fails to yield infectious virus when tested in sensitive cell cultures, cultured fragments of about 85% of "normal" tonsils or adenoids, after a variable time, show characteristic adenovirus-induced cytopathic changes and yield infectious virus.

Apparently so small a proportion of the tonsil and adenoid cells are infected that virus is detected only when additional cells become infected upon in vitro cultivation. Failure to recover infectious virus initially may be attributed to the paucity of virions, to their association with either Ab or receptor material, or to the absence of mature virions. Replication of infectious virus cannot be induced by ultraviolet irradiation, nor are the cells resistant to infection, suggesting that the latent infection is not the result of lysogeny. Similarly, herpes simplex virus may hide in affected tissues after recovery from herpes encephalitis in mice. Virus then reappears when spinal ganglia are cultured in vitro. Cultured monkey kidney cells may also yield unsuspected viruses that cannot be detected in homogenates of the extirpated organs. The oncogenic **simian virus 40** (SV 40, Ch. 63) illustrates the role of host susceptibility in latency. This virus commonly infects **rhesus** and **cynomologus** monkeys in nature, and in cultures of cells from these species infectious virus appears without causing cytopathic changes. In contrast, when the virus is added to cultures of green monkey (Ceropithecus aethiops) kidney cells (which have never been found to carry the virus) extensive cell injury accompanies viral multiplication.

4) Another possibility is the **replication of viral nucleic acid without viral maturation.** Thus the Shope rabbit papilloma virus produces warts equally well in the skin of domestic or wild cottontail rabbits, but infectious virus can usually be detected only in the tumors of the wild animals.* In the warts of wild rabbits virions and viral antigens are recognizable only in the keratinized cells of the outer epidermal layer and not in the growing basal layer. However, when the basal layers of warts from either wild or domestic rabbits are extracted with phenol they both yield infectious viral DNA (Ch. 63).

5) **Swine influenza virus,** a myxovirus related to influenza A virus, illustrates a complex ecological situation in which the virus is latent in two intermediate hosts and requires assistance from a bacterium, *Hemophilus influenzae suis,* to induce the acute respiratory disease in pigs (Fig. 51-2).

Virions in the lung of a sick pig become associated with ova of the lung worm (a common parasite of most pigs). The virions probably penetrate the ova, for infectious virus is not detectable. When the sick pig coughs, the ova, containing **occult virus,** are expelled with pulmonary secretions, swallowed, and eventually passed in the pig's feces.

---

* Since the assay of this virus is extremely insensitive, it is not known whether failure to recover the virus signifies paucity or absence.

Fig. 51-2. Natural history of swine influenza infection.

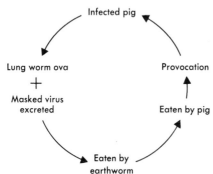

The contaminated feces are then eaten by earthworms, in which the ova develop into larvae which migrate to the salivary glands of the worms. By experimentally maintaining such earthworms and feeding them to pigs Shope has demonstrated that the noninfectious virus is present in the earthworms, where it can remain occult for at least 2 years. To complete the cycle the earthworm is eaten by a pig; the larvae mature further and migrate through the lymph and blood streams to the pig's lungs, where they develop into mature lungworms—but the virus remains noninfectious through this long odyssey. When the parasitized pig is jolted by cold or by inoculation of a suspension of *H. influenzae suis* the occult virus is somehow induced to replicate, and it may initiate an acute disease. The nature of the occult form in the lung worm and earthworm is unknown.

This model for latency of a virus illustrates the complexity of pathogenetic processes. The exact role of the bacterium, however, is still unknown.

Epidemiological Significance of Latent Infections. As these examples indicate, latent viral infections affect the incidence and the pathogenesis of **acute viral diseases** in several ways. 1) In herpes simplex recurrent acute disease is induced in the infected individual when the balance that maintains the latent infection is intermittently distributed. 2) Swine influenza shows how a virus widely seeded in the population may initiate, in response to **environmental** changes, an **explosive epidemic of reactivation,** often occurring in many areas at the same time (Ch. 56, Epidemiology). 3) An occult virus, resuming multiplication, may spread and initiate an epidemic among susceptible contacts. Thus a latent varicella virus, persisting after clinical chickenpox, can be reactivated and produce the different clinical picture of herpes zoster (commonly called shingles); and the patient may then serve as a focus for spread of virus (Ch. 53, Varicella, pathogenesis).

Viral latency can also be seen in the development of certain **chronic diseases.** Two major processes have been observed: 1) Some prolonged viral infections ultimately cause disease through a **humoral and cell-mediated immunological response** as noted above for LCM and probably as the basis of subacute sclerosing panencephalitis (Ch. 57, Measles) and other so-called slow virus infections. 2) Some viruses induce **uncontrolled proliferation of cells,** i.e. **tumors,** which will be discussed in Chapter 63.

SLOW VIRAL INFECTIONS

The name "slow virus" has become associated with those viruses that require prolonged periods of infection (often years) before disease appears. The term, however, is somewhat misleading, for the viral multiplication may not be unusually slow, but rather **the disease process develops over a protracted period.** Table 51-3 lists some chronic degenerative diseases, particularly of the central nervous system, that belong to this group. Consideration is not given here to viral diseases that have a long incubation period but an acute course: e.g., serum hepatitis, rabies. Four chronic neurological diseases of man, **subacute sclerosing panencephalitis, kuru, Creutzfeldt-Jakob disease,** and **progressive leukoencephalopathy** are included; other chronic diseases undoubtedly have similar origins.

Even viruses common in man appear to be responsible for some chronic degenerative diseases. **Subacute sclerosing panencephalitis,** the best substantiated example, follows several years after measles (and perhaps even after measles immunization); a virus identical with or closely related to measles virus appears to be the etiological agent (Ch. 57): the lesions contain specific antigens and huge quantities of nucleocapsids characteristic of paramyxoviruses; and a virus similar to measles virus has been isolated by cocultivation of affected tissue with susceptible cells. Paramyxoviruses appear to be particularly suited to initiate persistent infections: SV 5 (a parainfluenza virus) and mumps virus readily establish endosymbiotic infections in cultured cells (p. 1216). Moreover, accumulations of nucleocapsids resembling paramyxoviruses are seen in lesions of diffuse lupus erythematosus, and a paramyxovirus related to type 1 parainfluenza virus has been isolated from brains of some patients with multiple sclerosis.

Another slow viral disease of man, **kuru,** was first observed in 1957 in the Fore tribe of cannibals living in stone age conditions in New Guinea. This exotic degenerative disease of the cerebellum manifested by ataxia, disturbed balance, clumsy gait, and tremor inexorably progresses to death in 1 to 2 years. The pathological findings resemble those of **scrapie,** a disease of sheep proved to be viral, although the customary inflammatory evidence of an infectious process is absent. This resemblance suggested a viral etiology for kuru, and Gajdushek and Gibbs, using brain material from kuru patients, transmitted the disease serially to chimpanzees. The degenerative process, which had the same clinical and pathological characteristics of kuru in man, appeared 18 to 30 months after the initial inoculation, and after 1 year in subsequent passages. The infectivity of the brain is also shown by the striking decrease in kuru after the tribal chiefs prohibited the custom of women and children consuming the brains of dead kuru victims. The properties of the unique infectious agent, probably a virus, have not yet been described. The evolution of an agent with so specialized a mechanism of transmission is a mystery.

Two other degenerative neurological dis-

TABLE 51-3.   Examples of Slow Viral Infections

Virus	Host	Organ primarily affected
Kuru	Man	Brain
Subacute sclerosing panencephalitis	Man	Brain
Creutzfeldt-Jakob disease	Man	Brain
Progressive multifocal leukoencephalopathy	Man	Brain
Scrapie	Sheep	Brain
Visna	Sheep	Brain
Maedi	Sheep	Lung
Progressive pneumonia	Sheep	Lung
Lymphocytic choriomeningitis	Mouse	Kidney, brain, liver
Aleutian mink disease	Mink	Reticuloendolitical system
Mink encephalopathy	Mink	Brain

eases of man appear associated with unique infectious agents, probably viruses: using brains from patients with the spongiform encephalopathy, Creutzfeldt-Jakob disease (a fatal presenile dementia), a similar disease, has been transmitted to chimpanzees; and papovaviruses (Ch. 63) have been observed in and isolated from brains of patients with progressive leukoencephalopathy (a rare, demyelinating disease); immunological analysis of the virions and chemical analysis of the viral DNA have shown two of these to be closely related to SV 40. Hence, although evidence is sparse, suspicion of the role of viruses in chronic degenerative diseases is now high.

SELECTED REFERENCES

Books and Review Articles

BANG, F. B., and LUTTRELL, C. N. Factors in the pathogenesis of virus diseases. *Adv Virus Res* 8:199 (1961).

Chronic infectious neuropathic agents (CHINA) and other slow virus infections. J. A. Brody, W. Henle, and H. Koprowski, eds.) *Curr Top Microbiol Immunol 40* (1967).

FENNER, F. The pathogenesis of the acute exanthems: An interpretation based on experimental investigations with mousepox (infectious ectromelia of mice). *Lancet 2:*915 (1948).

HOLLAND, J. J. Enterovirus entrance into specific host cells and subsequent alterations of cell protein and nucleic acid synthesis. *Bacteriol Rev 28:*3 (1964).

HORSTMANN, D. M. Viral infections in pregnancy. *Yale J Biol Med 42:*99 (1969).

Latency and Masking in Viral and Rickettsial Infections. (D. L. Walker, R. P. Hanson, and A. S. Evans, eds.) Burgess, Minneapolis, Minn., 1958.

Mechanism of Virus Infection. (W. Smith, ed.) Academic Press, New York, 1963.

MIMS, C. A. Aspects of the pathogenesis of virus diseases. *Bacteriol Rev 28:*30 (1964).

MIMS, C. A. Pathogenesis of rashes in virus diseases. *Bacteriol Rev 30:*739 (1966).

NICHOLS, W. W. Virus-induced chromosome abnormalities. *Annu Rev Microbiol 24:*479 (1970).

PORTER, D. D. A quantitative view of the slow virus landscape. *Progr Med Virol 13:*339 (1971).

Specific Articles

ALLISON, A. C. Lysosomes in virus-infected cells. *Perspect Virol 5:*29 (1967).

BLANDEN, R. V. Mechanisms of recovery from a generalized viral infection. II. Passive transfer of recovery mechanisms with immune lymphoid cells. *J Exp Med 133:*1074 (1971).

BORDEN, E. C., MURPHY, F. A., NATHANSON, N., and MONETH, T. P. C. Effect of antilymphocyte serum on Tacaribe virus infection in infant mice. *Infect Immun 3:*466 (1971).

EVANS, T. N., and BROWN, G. C. Congenital anomalies and virus infection. *Am J Obstet Gynecol 87:*749 (1963).

FULGINITI, V. A., KEMPE, C. H., HATHAWAY, W. E., PEARLMAN, D. S., SIEBER, O. F., JR., ELLER, J. J., JOYNER, J. J., and ROBINSON, A. Progressive vaccinia in immunologically deficient individuals. (D. Bergsiyia and R. A. Good, eds.) In *Birth Defects: Immunologic Deficiency Diseases in Man*, p. 129. The National Foundation, New York, 1968.

GAJDUSEK, D. C., ROGERS, N. G., BOSNIGHT, M., GIBBS, C. J. JR., and ALPERS, M. Transmission experiments with kuru in chimpanzees and the isolation of latent viruses from the explanted tissues of affected animals. *Ann NY Acad Sci 162:*529 (1969).

HOLLAND, J. J. Receptor affinities as major determinants of enterovirus tissue tropisms in human. *Virology 15:*312 (1961).

JOHNSON, R. T. The pathogenesis of herpes virus encephalitis. I. Virus pathways to the nervous system of suckling mice demonstrated by fluorescent antibody staining. *J Exp Med 119:*343 (1964).

OLDSTONE, M. B. A., and DIXON, F. J. Pathogenesis of chronic disease associated with persistent lymphocytic choriomeningitis viral infection. I. Relationship of antibody production to disease in neonatally infected mice. *J Exp Med 129:*483 (1969).

OLDSTONE, M. B. A., and DIXON, F. J. Pathogenesis of chronic disease associated with persistent lymphocytic choriomeningitis viral infection. II. Relationship of the antilymphocytic choriomeningitis immune response to tissue injury in chronic lymphocytic choriomeningitis disease. *J Exp Med 131:*1 (1970).

VOLKERT, M., and LUNDSTEDT, C. Tolerance and immunity to the lymphocytic choriomeningitis virus. *Ann NY Acad Sci.* In press.

chapter

ADENOVIRUSES

GENERAL CHARACTERISTICS   1222
PROPERTIES OF THE VIRUSES   1222
  Morphology   1222
  Physical and Chemical Characteristics   1222
  Stability   1224
  Hemagglutination   1224
  Immunological Characteristics   1225
  Host Range   1227
  Viral Multiplication   1227
PATHOGENESIS   1232
LABORATORY DIAGNOSIS   1234
EPIDEMIOLOGY   1235
PREVENTION AND CONTROL   1235
ADENO-ASSOCIATED VIRUSES   1236
  Parvoviruses   1236

For many years after the isolation of influenza virus in 1933 no new type of respiratory virus was found. In 1953, however, two groups of investigators, searching for the still unknown major causative agents of acute viral respiratory diseases, discovered a new family of viruses (Table 52-1). Rowe and colleagues, using cultures of human adenoids as a potentially favorable host for an elusive respiratory virus, noted cytopathic changes after prolonged incubation of uninoculated as well as inoculated cultures. The pathological alterations were shown to be due to the emergence of previously unidentified agents from latent infections of the adenoid tissues. Hilleman and Werner, studying an epidemic of influenza-like disease in army recruits, isolated from respiratory secretions several similar cytopathic agents in cultures of human tissues. The new agents were called **adenoviruses** to record the original isolation from adenoid tissue.

GENERAL CHARACTERISTICS

It soon became evident that adenoviruses are not the etiological agents for the majority of acute viral respiratory infections. Several characteristics, however, have aroused considerable interest: 1) they are simple DNA-containing viruses (composed of only DNA and protein) that multiply in the cell nucleus; 2) they induce latent infections in tonsils, adenoids, and other lymphoid tissue of man, and they are readily activated; 3) several adenoviruses represent the first examples of common viruses in man that are **oncogenic** for lower animals under special experimental circumstances; 4) they serve as "helpers" for a group of small defective DNA-containing viruses, **adeno-associated viruses** (p. 1236), which cannot replicate in the ab-

sence of adenoviruses; conversely, a number of adenoviruses cannot multiply in primary monkey cells unless the genetically unrelated simian virus 40 (SV 40) is present as a helper (Ch. 48, Abortive infections).

Adenoviruses are widespread in nature. They are defined on the basis of a family cross-reactive antigen and similar chemical and physical characteristics (Table 52-1). The 77 viruses accepted as members of the adenovirus family are distinguished by antibodies to their individual type-specific antigens: 31 are from humans and the rest from various other animals. Comparative studies permit a general classification into several groups (see below).

PROPERTIES OF THE VIRUSES

MORPHOLOGY

Adenovirus particles are 600 to 900 A in diameter according to electron microscopic measurements of both purified preparations and thin sections of infected cells. In thin sections the viral particles have a dense central core, the **nucleoid,** and an outer coat, the **capsid** (Fig. 52-1). Negative staining reveals naked (nonenveloped) isometric particles with capsids composed of 252 capsomers arranged in perfect icosahedral symmetry (Fig. 52-2). There are 240 hexons and 12 corner subunits, termed **pentons.** The hexons are polygonal prisms with a central hole (Fig. 52-3). The pentons, which are more complex, consist of a similar polygonal base with an attached fiber of variable length, depending upon the type (Figs. 52-4 and 52-5).

PHYSICAL AND CHEMICAL CHARACTERISTICS

The virion consists only of DNA and protein (Table 52-2). The DNA is a linear, double-

TABLE 52-1. Characteristics of Adenoviruses

1. Icosahedral symmetry
2. 600 to 900 A in diameter
3. Capsid contains 252 polygonal capsomers and 12 fibers
4. Contain double-stranded DNA
5. Resistant to lipid solvents, i.e., lack lipids
6. Related by family cross-reacting soluble antigens*
7. Multiply in nuclei of cells

* Except for the chicken adenoviruses.

stranded molecule which varies in base composition according to the type. The human adenoviruses, which also transform cells in culture, can be divided into three groups according to their oncogenic propensities and DNA base compositions (Table 52-2). DNA-DNA hybridization substantiates the validity of this classification; viruses within each group share 70 to 100% of their nucleotide sequences (i.e., degree of hybridization), whereas the DNAs of the different groups have much less homology (11-51%). It is not clear whether oncogenicity depends on a viral DNA structure permitting its integration into the host cell's chromosome.

The virion contains six or seven species of proteins; the hexon and fiber proteins have been crystallized. The capsid proteins are relatively unique among animal viruses since, owing to the relatively strong noncovalent bonds between protomers in a capsomer and the weaker bonds between capsomers (Ch. 44), the capsid can be artificially disrupted into intact capsomers (Figs. 52-3 and 52-4). Further dissociation of the hexons into their constituent polypeptide chains, in contrast, requires rigorous denaturing conditions (e.g., 5 to 6 M

Fig. 52-1. A thin section of a crystalline mass of adenovirus particles in a cell infected with type 4. Two types of particles can be seen: dense particles with no discernible internal structure, and less dense particles showing the central body of the viral particle, the nucleoid (arrows), and the outer coat, the capsid. The polygonal shape of adenoviruses is apparent in many particles. Differences in appearance of particles are probably due to the relation of the center of the virion to the plane of section. ×110,000. (Courtesy of C. Morgan and H. M. Rose, Columbia University.)

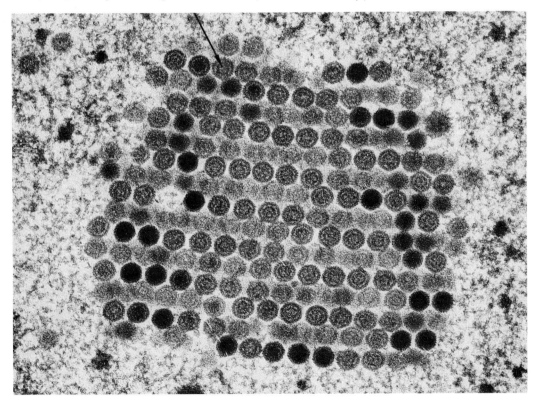

TABLE 52-2. Physical and Chemical Characteristics and Oncogenic Properties of Human Adenoviruses

Group	Types	Oncogenic potential	Viral DNA		
			Percent of virion	Molecular weight	Percent guanine + cytosine
A	12, 18, 31	High	11.6–12.5	$\sim 20 \times 10^6$	48–49
B	3, 7, 11, 14, 16, 21	Low	12.5–13.7	$\sim 23 \times 10^6$	49–52
C	1, 2, 4, 5, 6	None*	12.5–13.7	$\sim 23 \times 10^6$	49–52

* Most of these viruses transform cells in vitro.

Molecular weight of virion $= 8$–145×10^6 daltons. (**Minimum MW** based on sedimentation coefficient, assuming a solid sphere with a partial specific volume of 0.70; maximum **MW** calculated on basis of 700 A and a density of 1.34 gm/cm³).

Protein: Capsid $= \sim 58\%$ of virion; consists of four species; internal proteins $= \sim 30\%$ of virion; consist of two or three species.

guanidine hydrochloride) and a sulfhydryl reagent. The penton base is the least stable of the capsid's morphological subunits.

The virion's core consists of two or three basic internal proteins associated with the DNA. One of the internal proteins is rich in arginine, containing about 20%.

STABILITY

Adenoviruses are relatively stable in homogenates of infected cells; they retain undiminished infectivity for several weeks at 4°, and for months at −25°. Purified virions, however, are relatively unstable under all conditions of storage. The pentons seem more weakly held in the capsid, and may be spontaneously released from purified viral particles resulting in the virions' aggregation or disruption. Adenoviruses are resistant to lipid solvents.

HEMAGGLUTINATION

Human adenoviruses can be divided into three groups (Table 52-3) on the basis of their ability to agglutinate rhesus monkey or rat erythrocytes (these groups are unrelated to the divisions based on viral oncogenicity; Table 52-2). Tips of the penton fibers bind to the red blood cells and cause cross-linking and agglutination by the virions. Dispersed purified fibers and single pentons are monovalent and hence cannot produce a lattice and cause complete hemagglutination unless they are cross-linked by heterologous antibodies from a virus of the same subgroup (i.e., the cross-reacting antibody recognizes the penton base or shaft of the fiber but not the fiber tip). The ag-

glutination is inhibited by type-specific antibodies which combine with the tip of the fiber.

The nature of the receptor on susceptible red cells is not known. In contrast to myxoviruses, the combination of virus with cell receptors is stable, and spontaneous elution of

Fig. 52-2. Electron micrograph of purified particles of type 5 adenovirus embedded in sodium phosphotungstate. The icosahedral symmetry of the virion and subunit structure of the capsid are apparent. Arrow points out a virion's axis of twofold rotational symmetry. Capsomers at apices of triangles are the centers of fivefold symmetry. The capsid parameters are P = 1, f = 5; hence, the capsid consists of 252 capsomers (cf. Ch. 44). ×440,000, reduced.

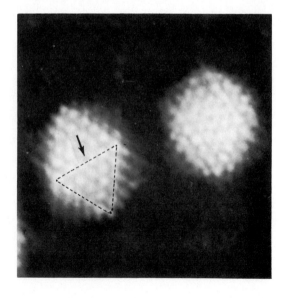

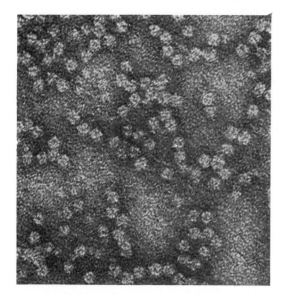

Fig. 52-3. Electron micrograph of purified hexon protein of type 5 adenovirus embedded in sodium silicotungstate. The polygonal shape of the capsomer with the central hole is apparent. ×440,000, reduced. [From Wilcox, W. C., Ginsberg, H. S., and Anderson, T. F. *J Exp Med 118*:307 (1963).]

the hemagglutinin does not occur. Viral antigens, as well as type-specific antibodies, can be titrated by hemagglutination-inhibition procedures.

IMMUNOLOGICAL CHARACTERISTICS

Each adenovirus particle contains at least two morphological subunits that contribute to its immunological reactivity (Table 52-4). The major hexon antigen corresponds to the hollow polygonal capsomers (Fig. 52-4). These contain **family-reactive determinants** which cross-react with a similar antigen in all adenoviruses except the chicken adenoviruses. The hexon capsomers also possess a **type-specific reactive site** (identified by neutralization titrations and cross-adsorption studies).

The complex **penton** (Fig. 52-4) provides minor antigens of the viral particle, and is also found as a family-reactive soluble antigen in infected cells. The purified fiber contains a major type-specific as well as a minor subgroup antigen. Although the fiber is the organ of attachment to the host cell, Abs to the fiber or to the intact penton do not neutralize viral infectivity.

Neutralizing, hemagglutination-inhibiting, and complement-fixing Abs appear about 7 days after onset of illness and attain maximum titers after 2 to 3 weeks. IgA and IgG Abs appear in nasal secretions at the same time or within a week of the detection of serum Abs. Complement-fixing Abs begin to decline 2 to 3 months after infection, but are usually still present 6 to 12 months afterward. Neutraliz-

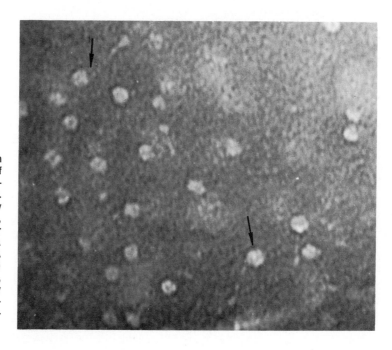

Fig. 52-4. Electron micrograph of the capsid components of purified type 5 adenovirus particles disrupted at pH 10.5. Note the polygonal, hollow capsomers, i.e., the hexons, which are 70–85 A in diameter with a central hole about 25 A across, and the fibers (10–25 A in width and 200 A in length) attached to polygonal bases, i.e., the pentons (arrow). By actual count there are 12 pentons per virion. ×480,000, reduced.

TABLE 52-3. Division of Adenoviruses of Human Origin into Subgroups on Basis of Reaction with Rhesus Monkey or Rat Erythrocytes

Subgroup	Types included in group	Agglutination of RBC	
		Rhesus	Rat
1	3, 7, 11, 14, 16, 21	+	0
2*	8–10, 13, 15, 17, 19, 20, 22–30	+† or 0	+
3*	1, 2, 4–6, 12, 18, 31	0	Partial‡

* Types 8–10, 13, 19, 26, and 27 (group 2) and all group 3 viruses also agglutinate human O RBC.

† Some types (9, 13, and 15) agglutinate rhesus RBC, but to lower titer than rat cells.

‡ Complete hemagglutination occurs when heterologous antibody from same hemagglutination subgroup is added to reaction mixture.

Classification suggested by L. ROSEN. *Virology 5:*574 (1958).

ing and hemagglutination-inhibiting Abs persist longer, decreasing in titer only two- to threefold in 8 to 10 years. Minor rises in heterotypic neutralizing Abs may follow adenovirus infections, especially when heterologous Abs are already present at the time of infection.

Second attacks of disease due to the same type of adenovirus are rare. Such persistent type-specific immunity is unusual among viral respiratory diseases, resistance of relatively short duration being the rule (cf. Influenza and parainfluenza viruses: Immunological characteristics, Chs. 56 and 57).

In addition to their immunological reactivities the penton and its individual components possess striking biological activities. The intact penton reacts with monolayers of cultured cells to cause rounding and clumping, as well as detachment of the cells from glass, and therefore is also termed **toxin** or **cell-detaching** factor. Hydrolysis of the penton's base by trypsin, leaving the fiber intact, destroys the cytopathic effect. The purified fiber, which is present as a soluble protein in infected cells as well as in pentons, can combine with uninfected cells; it blocks their DNA, RNA, and protein biosynthesis, stops their

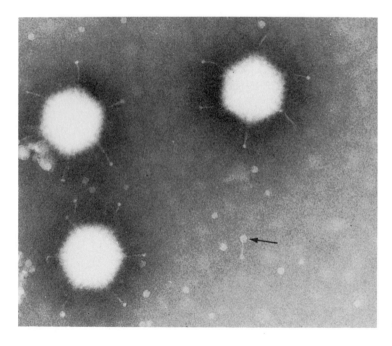

Fig. 52-5. Electron micrograph of purified type 5 adenovirus particles embedded in sodium silicotungstate. Micrographs obtained in areas where the silicotungstate was thin revealed the fiber component of the penton projecting from corners of the virion. Free pentons (arrow) and hexons are also present. ×350,000. [From Valentine, R. C., and Pereira, H. G. *J Mol Biol 13:*13 (1965).]

TABLE 52-4. Biological, Immunological, and Physical Characteristics of Type 5 Adenovirus Proteins

| Property | Hexon protein | Penton proteins | | | Internal proteins |
		Complete*	Base*	Fiber	
Immunological reactivity	Type-specific and family cross-reactive	Family cross-reactive	Family cross-reactive	Type-specific	?
Biological activity	None known	Cytopathic	Cytopathic	Blocks biosynthesis of macromolecules; inhibits viral multiplication	Probably aid in assembly of viral DNA
Hemagglutinin	0	+ Partial	0	+ Partial	0
Morphology	70 A	70 A	70 A	250 × 15 A	———
		250 × 15 A			

* A DNA endonuclease is closely associated with but physically separable from the penton base; its function during infection is unknown.

division, and inhibits their capacity to support multiplication of related or unrelated viruses (Table 52-4).

HOST RANGE

Most adenoviruses of man do not produce recognizable disease in common laboratory animals, but inapparent infections may follow intravenous inoculation into rabbits or intranasal inoculations into hamsters, piglets, guinea pigs, and dogs. In rabbits inoculated intravenously with type 5 strains virus persists for at least 6 months in the spleen, and it emerges when explants of the spleen are cultured in vitro (similar to latent infection of human tonsils and adenoids; see under Pathogenesis). Chick embryos are resistant to all the adenoviruses tested except chicken adenoviruses. At least nine adenoviruses produce tumors when inoculated into newborn hamsters, rats, and mice: the highly oncogenic types 12, 18, and 31 and the weakly oncogenic types 3, 7, 11, 14, 16, and 21 (Ch. 63).

A variety of cultured mammalian cells support multiplication of adenoviruses to high titer and evince characteristic cytopathic changes, including pathognomonic nuclear alterations. Human adenoviruses are propagated readily in continuous epithelium-like cell lines of human origin, such as HeLa cells. Primary cultures of various types of human and animal cells also support viral multiplica-

tion but yield much lower titers, except for primary cultures of human embryonic kidney.

VIRAL MULTIPLICATION

The essential features of multiplication appear to be similar for all adenovirus types. Adsorption to susceptible host cells is relatively slow compared with influenza (Ch. 56, Viral multiplication) and polioviruses (Ch. 55), maximum adsorption requiring several hours with cell cultures. **Penetration** of the viral particle into the cell promptly follows, either by direct entry through the plasma membrane, which appears to re-form without delay (Fig. 52-6), or by a process analogous to phagocytosis. If primary entry is accomplished by engulfment, the virion must still find its way through the vacuole membrane into the cytoplasmic matrix for uncoating.

Uncoating of the viral DNA, which begins immediately after penetration into the cytoplasm, is detected biochemically when the nucleic acid becomes susceptible to DNase hydrolysis and electron microscopically when the virion appears spherical rather than polygonal (Fig. 52-6B). Initially, the pentons and surrounding hexons are displaced, reducing the stability of the capsid. The remainder of the capsid, consisting of hexons, then dissociates leaving the viral DNA and its associated basic internal proteins (i.e., **the viral core).** The core comes into apposition with the nuclear mem-

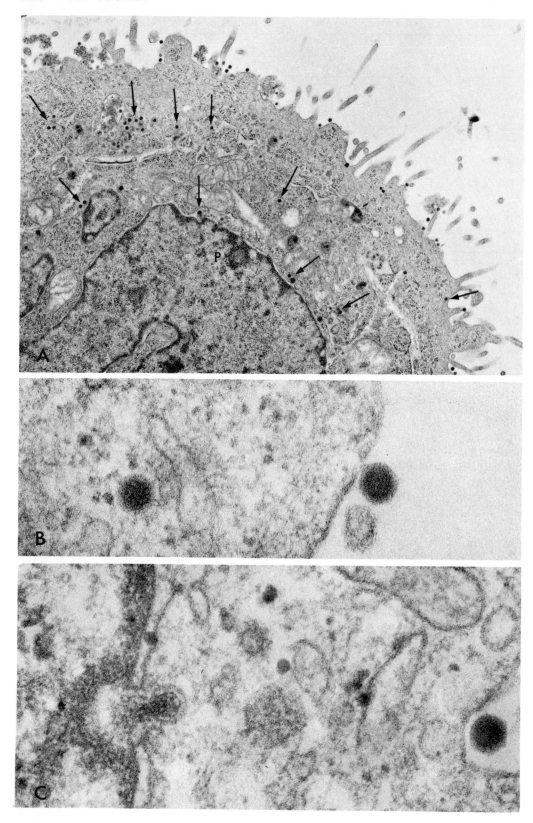

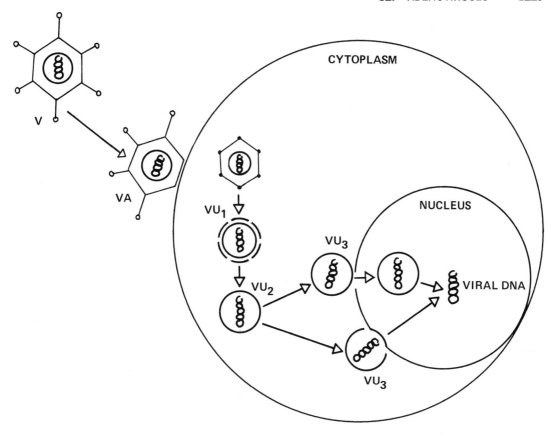

Fig. 52-7. Diagrammatic schemata of uncoating of the adenovirus particle. V = intact virion; VA = attachment of virion to cell, followed by penetration of virion into cell; $VU_1 \rightarrow$ pentons are detached, leaving virion somewhat spherical and the viral DNA susceptible to DNase; VU_2 = capsid disintegrates leaving viral core which migrates to nucleus; VU_3 = either viral DNA is freed from core into nuclear pocket and free DNA enters nucleus, or core enters nucleus through membrane pore and DNA is then dissociated from proteins.

brane, and enters through nuclear pores, or the viral DNA is released into a nuclear pocket (Fig. 52-6C) and gains access to the nucleus, where viral replication takes place (Fig. 52-7). Viral uncoating requires about 2 hours. The characteristics of the multiplication cycle and the biosynthetic events initiated during the long latent period (13 to 17 hours) are diagrammatically summarized in Figure 52-8.

Virus-coded **early messenger RNA,** measured by its specific hybridization with viral DNA, is detected first on polyribosomes as early as 2 to 3 hours after infection. The transcription of early mRNA, which is independent of viral DNA replication, is soon followed by the appearance of several **early proteins** that are identified immunologically and enzymatically (aspartate transcarbamylase, thymidine kinase, deoxycytidylate deaminase, and DNA polymerase). But these enzymes are not unique to virus-infected cells, and it is uncertain whether they are products of the

Fig. 52-6. Adsorption, penetration, and uncoating of adenovirus in HeLa cells. **A.** Numerous particles are adsorbed to the cell surface and are present as free virions in the cytoplasm (arrows). Some partices in phagocytic vacuoles are also noted. A nuclear pocket (P) is also present. ×15,000. **B.** Higher magnification of a polygonal virion adsorbed to the plasma membrane and a virion which has assumed a spherical form in the cytoplasm. ×150,000. **C.** A partially uncoated viral particle releasing core material into a nuclear pocket. For comparison, an unaltered virion is shown on the surface of the cell. ×150,000. [From Morgan, C., Rosenkranz, H. S., and Mednis, B. *J Virol* 4:777 (1969).]

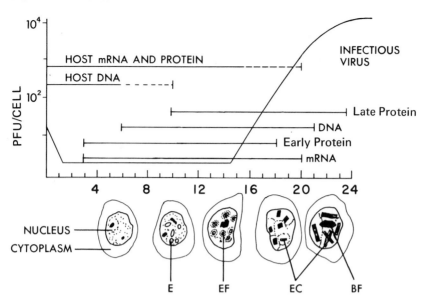

Fig. 52-8. Diagrammatic representation of the sequential events in the biosynthesis of type 5 adeno-virus, its effect on synthesis of host macromolecules, and the concomitant development of nuclear alterations. E = eosinophilic masses; EF = eosinophilic masses with basophilic Feulgen-positive border; EC = eosinophilic crystals; BF = basophilic Feulgen-positive masses. Types 1, 2, and 6 have similar multiplication and nuclear changes.

viral or host cell genome. Inhibition of early mRNA synthesis (by actinomycin D or pyrimidine analogs) or of early proteins (by puromycin or amino acid analogs) prevents viral DNA synthesis.

Replication of **viral DNA** begins in the nucleus 6 to 8 hours after infection, attains its maximum rate of biosynthesis 10 to 12 hours later, and practically ceases by 22 to 24 hours postinfection. Transcription of late **mRNAs** begins shortly after the initiation of DNA replication, for it occurs only on progeny viral DNA molecules. Accordingly, formation of capsid proteins (i.e., **late proteins)** also depends upon prior replication of viral DNA. By hybridization-competition experiments it may be demonstrated that only about 30% of the early mRNAs are transcribed during the period of late mRNA and protein synthesis.

Although viral DNA and mRNAs are synthesized in the nucleus and virions are assembled in the nucleus, viral proteins, like host proteins, are synthesized on polyribosomes in the cytoplasm. After release from the polyribosomes, the polypeptide chains are immediately transported into the nucleus, where like species assemble into the multimeric viral

capsid proteins (Fig. 52-9). Finally, mature particles begin to assemble about 2 to 4 hours after initiation of capsid protein production, possibly the period required to attain component pools of adequate size.

Adenovirus infection produces a profound effect on the physiology of host cells. Production of host DNA stops abruptly 8 to 10 hours after infection, and host protein and RNA biosynthesis ceases 6 to 10 hours later. Division of infected cells accordingly ceases. This inhibition of mitosis is also associated with production of chromosomal aberrations.

Though the replication of adenoviruses proceeds through a well-ordered sequence of events, the process is quite inefficient. Of the viral DNA and proteins made only approximately 10 to 15% is incorporated into virions; the rest remains unassembled and produces the varied **nuclear lesions** that are the hallmark of infection with different types of adenoviruses. Since fibers can block synthesis of macromolecules, and DNA endonuclease activity increases in infected cells, the accumulation of these proteins may also be responsible for some of the biochemical effects noted.

The development of such lesions in cells infected with type 5 adenovirus, and its relation

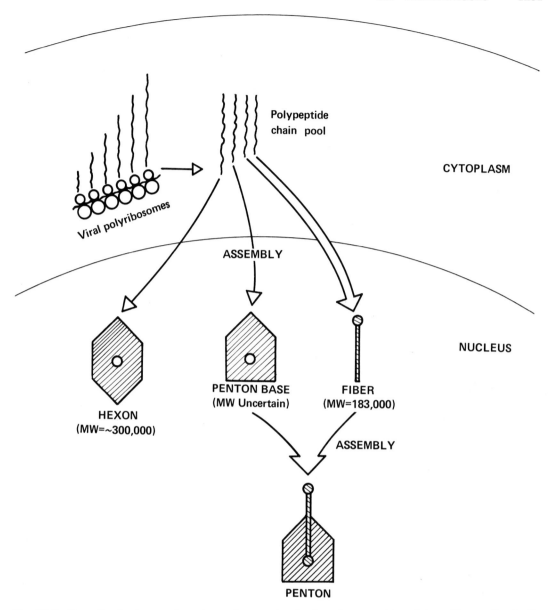

Fig. 52-9. Biosynthesis, transport, and morphogenesis of adenovirus capsid proteins. After synthesis, the released polypeptide chains are rapidly transported into the nuclei. (The molecular weight of the penton base is unknown although it has a sedimentation coefficient of 9S as compared with 12S for hexon, 6S for fiber, and 10.5S for intact penton. [From Velicer, L. F., and Ginsberg, H. S. *J Virol* 5:338 (1970).]

to viral multiplication, are diagrammatically presented in Figure 52-8. The prominent bar-shaped eosinophilic crystals contain the arginine-rich internal viral proteins, but neither viral DNA nor capsid proteins are present. Other common types infecting man produce quite different changes: the infected nucleus has a basophilic inclusion body and large, basophilic crystals composed of viral particles arranged in a crystalline lattice (Figs. 52-10 and 52-11).

Despite these extensive nuclear alterations, newly synthesized viral particles tend to remain in the nucleus; less than 1% of the total virus is in the culture fluid when the maximum viral titer is attained (shown by disrupting

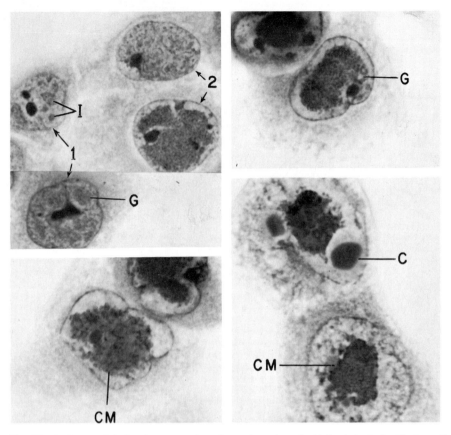

Fig. 52-10. Sequential development of nuclear alterations in HeLa infected with type 7 adenovirus (types 3 and 4 produce similar effects). Earliest changes may be formation of small eosinophilic inclusions (Feulgen-negative) (1) and clusters of granules (G). Clusters of granules gradually become larger and more prominent (Feulgen-positive) and form large central mass (CM). In latter stages nucleus is enlarged and crystalline masses (Feulgen-positive) are apparent (C). ×1050. [From Boyer, G. S., Denny, F. W., Jr., and Ginsberg, H. S. *J Exp Med 110*:327 (1959).]

cells). The infected cells remain intact and metabolically active, despite the extensive cytopathic changes, and even exhibit an increased utilization of glucose and production of organic acids. The culture media therefore become acid, in contrast to the alkalinity that develops in cultures infected with agents that lyse the cells (e.g., polioviruses; Ch. 55, Laboratory diagnosis).

PATHOGENESIS

Although adenoviruses have been detected in several animal species, there are no satisfactory animal models that mimic the infections of man. This lack has impeded progress in the study of pathogenesis, and knowledge of human adenovirus infections is derived primarily from clinical observations and from experiments on volunteers. The diseases listed in Table 52-5 predominantly involve the respiratory tract and the eye. Virus introduced by feeding, or that swallowed in respiratory secretions, multiplies in cells of the gastrointestinal tract and is excreted in the feces, but it does not usually produce gastrointestinal disease. Adenoviruses, most often type 1, 2, 5, or 6, have been isolated occasionally from cases of acute infectious diarrhea and intussusception, but an etiological relation between virus and illness has not been clearly demonstrated. In natural infections and volunteer experiments viremia is not observed, and virus is not commonly transmitted

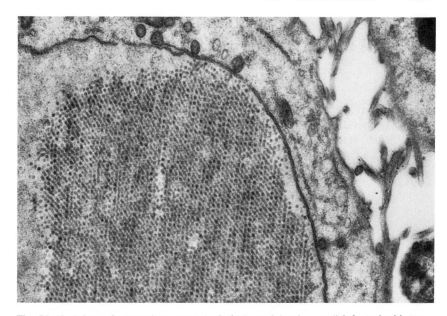

Fig. 52-11. A large intranuclear crystal of viral particles in a cell infected with type 4 adenovirus (corresponds to crystalline structure, C, in Fig. 52-10). Viral particles are also randomly distributed within the nucleus; no virus can be identified in the cytoplasm. Note that the nuclear membrane is intact. ×16,000. [From Morgan, C., Howe, C., Rose, H. M., and Moore, D. H. *J Biophys Biochem Cytol* 2:351 (1956).]

to distant organs (although rare instances of adenovirus meningitis in children have been described).

Adenoviruses usually cause either inapparent infections or self-limited illnesses that are followed by complete recovery and persistent type-specific immunity. Most individuals are infected with one or more adenoviruses before the age of 15. As a corollary, 50 to 80% of tonsils and adenoids, removed surgically, yield an adenovirus when explants are cultured in vitro; the most frequent types are 1, 2, and 5—which are also the types responsible for most infections in young children. Adenoviruses have also been isolated from cultured fragments of mesenteric lymph nodes. It thus appears that following an initial infec-

tion, the virus frequently becomes latent in lymphoid tissues and may persist for long periods. However, recurrent illnesses have not been shown to arise from these latent infections.

It appears likely that latent infections are established so readily because infected cells are not lysed, and large numbers of viral particles remain protected within nuclei of infected cells. In nature the occult virus is confined to relatively few cells, probably by antibodies. Culturing tissues in vitro, however, alters the cellular environment, and possibly dilutes and washes away viral inhibitors, thus permitting the virus to spread to uninfected cells, to multiply more rapidly, and to produce detectable cytopathic changes.

TABLE 52-5. Clinical Syndromes Commonly Caused by Specific Types of Adenoviruses

Disease syndrome	Adenovirus type	
	Most common	Less common
Acute respiratory disease of recruits	4, 7	3, 14
Pharyngoconjunctival fever; pharyngitis	3	5, 7, 21
Conjunctivitis	3, 7	2, 5, 6, 9, 10
Epidemic keratoconjunctivitis	8	7

From a number of fatal cases of nonbacterial pneumonia in infants type 3 or 7 adenovirus has been isolated. (Type 7 has also been associated with a rare case of fatal pneumonia in military personnel.) The pulmonary lesions observed were those of nonbacterial bronchopneumonia, but in addition there were numerous epithelial cells containing central basophilic masses in the nuclei, closely resembling those produced by the same types in cell cultures (Fig. 52-10).

Production of Tumors by Adenoviruses. That certain viruses can induce tumors in animals is clear (Ch. 63). Because attempts to isolate specific viruses from human tumors consistently failed, Trentin and his colleagues suggested that viruses ordinarily causing other diseases in man may, under the proper conditions, initiate tumors. While testing a large number of human viruses they made the startling discovery that type 12 adenovirus induces formation of undifferentiated sarcomas when inoculated into newborn hamsters. It was subsequently shown that several other types (cf. Table 52-2) can also produce similar sarcomas, and that tumors can be initiated in newborn rats and mice. No such tumors were induced by many other viruses tested.

As with tumors produced by polyoma virus and SV 40 (also icosahedral DNA-containing viruses; Ch. 63), infectious virus is not detectable in adenovirus-induced tumors or in cells transformed by virus in vitro. Nevertheless, at least a portion of the viral genome persists in the tumor or transformed cells, perhaps integrated in a host chromosome. Its presence is shown by the following evidence: 1) rapidly labeled RNA species are found which specifically hybridize with denatured DNAs from viruses of the same subgroup but not with other viral DNAs; 2) some of the viral genetic information is transcribed into large RNAs containing viral and cellular RNA sequences in the same polycistronic RNA molecule, which might be expected if viral and cellular DNAs are integrated in the same chromosome; 3) RNA transcribed from viral DNA in vitro specifically hybridizes to DNAs extracted from tumor and transformed cells; 4) the viral mRNA molecules present in transformed cells are homologous to about 50% of the early mRNAs transcribed during productive infection; hence only a fraction of

the viral genome is expressed, explaining why the cells are not killed. Tumor or transformed cells contain a nonvirion protein, termed the **tumor (T) antigen,** and experimental animals with induced tumors develop antibodies specific to this tumor antigen. In cells productively infected with oncogenic adenoviruses, an antigen immunologically similar to the tumor antigen is synthesized early in the multiplication cycle and before the appearance of viral DNA and capsid proteins. Hence, the tumor antigen corresponds to an early viral protein.

Extensive attempts to detect adenovirus DNA, mRNA, or T antigen in human tumors have failed. It seems that adenoviruses, in spite of their marked oncogenic potential in animals, are not oncogenic for man. Probably infection of human cells usually causes cell death, whereas cells of lower animals, being nonpermissive (Ch. 48), survive the viral infection and reveal its oncogenic potential.

LABORATORY DIAGNOSIS

Adenovirus infection can be diagnosed serologically and by isolation of the offending virus from respiratory and ocular secretions and from feces.* For isolation, infected material is inoculated into cultures of continuous lines of human cells (e.g., HeLa or KB) or into human embryo kidney cells. If virus is present cytopathic changes (rounding and clumping of cells) develop after 2 to 14 days, depending upon the quantity of virus in the infected secretions. The virus is identified as an adenovirus by complement-fixation titration with a hyperimmune rabbit or convalescent human serum. This procedure detects the cross-reactive hexon and penton antigens. The specific type of adenovirus can be ascertained most conveniently through hemagglutination-inhibition titrations.† The number of titrations

---

* Although adenoviruses primarily cause diseases of the respiratory tract and eye, virus multiplies in the gastrointestinal tract and can be isolated from feces in high percentage of cases.

† Types 10 and 19 cross-react to such an extent when studied by hemagglutination-inhibition titrations that they cannot be distinguished by this procedure.

required to identify the virus type can be reduced considerably by determining the hemagglutination subgroup (Table 52-3) to which the newly isolated virus belongs. To establish a virus as a new serotype neutralization titration is the method of choice.

The immunological diagnosis of an adenovirus infection is most accurately accomplished by neutralization titrations with acute and convalescent phase sera. Hemagglutination-inhibition assay for antibodies, however, is practically as sensitive, is simpler and less expensive, and is almost as accurate if the sera are reacted with a *Pseudomonas* filtrate to hydrolyze nonspecific inhibitors prior to use. But if a virus is not isolated from the patient all of the common viral types must be used. Complement-fixation titration is the most convenient assay procedure available since any type of adenovirus may be used because all possess the common family cross-reactive antigen (i.e., the hexon protein). Unfortunately, this assay detects less than 50% of the new infections, owing to the constant high level of antibodies in many individuals.

EPIDEMIOLOGY

Man provides the only known reservoir for strains of adenoviruses that infect humans. Person-to-person spread of virus in respiratory and ocular secretions is the most common mode of transmission, though dissemination in swimming pools has also been implicated in epidemics of pharyngoconjunctival fever and conjunctivitis. The spread of **epidemic keratoconjunctivitis** caused by type 8 adenovirus appears to be associated with trauma to the conjunctivae produced with dust and dirt in shipyards and factories, or by improperly sterilized optical instruments. Adenoviruses are commonly present in feces of infected persons, but there is no evidence that disease is transmitted by the fecal-oral route.

Despite the large number and the worldwide distribution of adenoviruses, their clinical importance is largely restricted to epidemics of acute respiratory disease in military recruits and to limited outbreaks among children (except for keratoconjunctivitis caused by type 8). Infections are observed throughout the year, but the greatest incidence and largest epidemics occur in late fall and winter. Types 4, 7, and 3 (in order of decreasing importance) are the viruses most frequently responsible for epidemics of acute respiratory and ocular diseases; types 14 and 21 have also been implicated in some epidemics (Table 52-5). Peculiarly, type 4 adenovirus commonly causes acute respiratory disease (ARD) in military recruits, but rarely produces infections in civilians. This epidemiological phenomenon is without parallel, and the underlying causative factors are unknown.

A relatively high proportion of adults have antibodies to one or more types of adenoviruses (particularly types 1, 2, 3, 5, and 7) indicating previous infections. But epidemiological studies indicate that adenoviruses cause at most 4 to 5% of viral respiratory illnesses in civilians.

PREVENTION AND CONTROL

Isolation of sick persons has little or no effect on the spread of adenoviruses since many healthy carriers exist. Immunization with adequate formalinized and with live virus vaccines has been experimentally successful, as would be expected from the lasting type-specific immunity produced by natural infections. The use of adenovirus vaccine appears to be primarily indicated, however, to protect military recruits, since the incidence of adenovirus disease is usually low in the general population.

Despite some clinical needs and proved effectiveness, vaccines are still generally unavailable for several reasons: 1) adenoviruses multiply poorly in nonhuman and human diploid cells, and therefore viral vaccines of adequate antigenic potency have been difficult to produce consistently (vaccines for human use require that viruses be propagated in selected cells to minimize the risk of transferring latent viruses pathogenic for man); 2) adventitious simian agents (e.g., the highly oncogenic SV 40; Ch. 63) are often present in primary monkey kidney cells, which are generally employed for vaccine production; and 3) some common pathogenic adenoviruses, which must be included in a vaccine (e.g., types 3 and 7) are oncogenic for animals other than man. Although evidence is lacking that any tumors in man are associated with adenoviruses, it is considered unwise by some to immunize with a potentially oncogenic virus, even if the danger is remote, to prevent a self-limiting disease.

Viral capsid proteins, present in abundance as soluble antigens in infected cells, can stimulate production of neutralizing antibodies. Immunization with such antigens may provide suitable vaccines without any of the dangers listed above.

Vaccines for military use should contain at least types 3, 4, and 7 viruses. In closed populations such as chronic disease hospitals or homes for orphans a vaccine containing types 1 to 7 may be useful for infants and young children.

ADENO-ASSOCIATED VIRUSES

Small icosahedral virions, 200 to 250 A in diameter, are often found growing with adenoviruses in cells of man and other animals. Indeed, these **adeno-associated** viruses **(AAVs)** are defective and totally dependent upon the multiplication of the unrelated adenoviruses for their own replication (herpesviruses can assist in partial synthesis of AAV: some viral proteins are made, but infectious virus is not produced). The viral genome consists of a linear molecule of single-stranded DNA with a molecular weight of only 1.4×10^6 daltons. Apparently the quantity of genetic information contained in this small DNA molecule is too limited to code for all required viral functions. The functions that the helper supplies are unknown. It is striking that the complete multiplication of adenovirus is itself inhibited when it offers assistance to the defective AAV (and AAV also reduces adenovirus oncogenicity in hamsters). Although production of infectious AAV in cells coinfected with an adenovirus resembles complementation by two defective, genetically related mutants (Ch. 48), cross-hybridization of DNAs extracted from AAVs and several types of adenoviruses fail to detect complementary regions of their genomes.

Four distinct immunological types of AAVs have been identified as contaminants of human and simian adenoviruses, and AAV antibodies are frequently found in humans and monkeys. About 70 to 80% of infants acquire antibodies to types 1, 2, and 3 viruses within the first decade, and more than 50% of adults maintain detectable antibodies. Viruses can be occasionally isolated from fecal and respiratory specimens during acute adenovirus infections, but not from other illnesses. Thus, AAVs appear to be defective in man as well as in cell cultures.

PARVOVIRUSES

Several other viruses (latent rat viruses, minute mouse virus, and porcine virus) have physical and chemical characteristics similar to AAV although only the AAVs are defective in their natural hosts.* They are called **parvoviruses** (L. *parvo* = small). The latent rat viruses, H_1 and H_3, are particularly intriguing: they are defective in human cells, and adenoviruses, which are in turn inhibited, can serve as their helpers. Moreover, they may serve as models for study of human disease since they can produce dwarfism, mongoloid appearance, and a variety of other congenital lesions when pregnant hamsters and rats are infected; but the suggestion that AAV might have similar effects in man has not been confirmed. The defective AAVs have also been suspected of playing a role in the pathogenesis of "slow" viral infections (Ch. 51), but so far they appear to be only silent, unobstrusive partners of adenovirus infections.

---

* AAVs show an additional feature, not observed in other parvoviruses: the virions can contain either a plus or minus DNA strand, which when extracted from the viral particles form double-stranded molecules unless a reagent that blocks hydrogen bond formation (e.g., formalin) is present.

SELECTED REFERENCES

Books and Review Articles

GINSBERG, H. S., and DINGLE, J. H. The adenovirus group. In *Viral and Rickettsial Infections of Man.* (F. L. Horsfall, Jr., and I. Tamm, eds.) Lippincott, Philadelphia, 1965.

GINSBERG, H. S. The biochemical basis of adenovirus cytopathology. In *Infectious Agents and Host Reactions.* (S. Mudd, ed.) Saunders, Philadelphia, 1970.

GREEN, M. Oncogenic viruses. *Annu Rev Biochem 39:* 701 (1970).

HOGGAN, D. M. Adenovirus associated viruses. *Progr Med Virol 12:*211 (1970).

PHILIPSON, L., and PETTERSSON, U. Structure and function of virion proteins of adenoviruses. *Progr Exp Tumor Res.* In press.

SCHLESINGER, R. W. Adenoviruses: The nature of the virion and controlling factors in productive or abortive infection and tumorigenesis. *Adv Virus Res 14:*1 (1969).

Specific Articles

DALES, S. An electron microscope study of the early association between two mammalian viruses and their host. *J Cell Biol 13:*303 (1962).

HILLEMAN, M. R., and WERNER, J. R. Recovery of a new agent from patients with acute respiratory illness. *Proc Soc Exp Biol Med 85:*183 (1954).

HORNE, R. W., BRENNER, S., WATERSON, A. P., and WILDY, P. The icosahedral form of an adenovirus. *J Mol Biol 1:*84 (1959).

HUEBNER, R. J., ROWE, W. P., and LANE, W. T. Oncogenic effects in hamsters of human adenovirus types 12 and 18. *Proc Nat Acad Sci USA 48:*2051 (1962).

LUCAS, J. J., and GINSBERG, H. S. Synthesis of virus-specific ribonucleic acid in KB cells infected with type 2 adenovirus. *J Virol 8:*203 (1971).

ROWE, W. P., HUEBNER, R. J., GILMORE, L. K., PARROTT, R. H., and WARD, T. G. Isolation of a cytopathogenic agent from human adenoids undergoing spontaneous degeneration in tissue culture. *Proc Soc Exp Biol Med 84:*570 (1953).

TRENTIN, J. J., YABE, Y., and TAYLOR, G. The quest for human cancer viruses. *Science 137:*835 (1962).

WILCOX, W. C., and GINSBERG, H. S. Purification and immunological characterization of types 4 and 5 adenovirus-soluble antigens. *Proc Nat Acad Sci USA 47:*512 (1961).

WILCOX, W. C., GINSBERG, H. S., and ANDERSON, T. F. Structure of type 5 adenovirus. II. Fine structure of virus subunits. Morphologic relationship of structural subunits to virus-specific soluble antigens from infected cells. *J Exp Med 118:*307 (1963).

chapter

53

HERPESVIRUSES

General Characteristics   1240
Chemical and Physical Characteristics   1242

HERPES SIMPLEX VIRUS   1242

PROPERTIES OF THE VIRUS   1242
   Immunological Characteristics   1242
   Host Range   1242
   Viral Multiplication   1242
PATHOGENESIS   1243
LABORATORY DIAGNOSIS   1245
EPIDEMIOLOGY, CONTROL, AND TREATMENT   1245

VIRUS B   1246

**VARICELLA (CHICKENPOX)–HERPES ZOSTER (SHINGLES)
     VIRUS   1246**

PROPERTIES OF THE VIRUS   1246
   Physical and Chemical Characteristics   1246
   Immunological Characteristics   1246
   Host Range   1246
   Viral Multiplication   1246
PATHOGENESIS   1247
LABORATORY DIAGNOSIS   1248
EPIDEMIOLOGY   1248
PREVENTION AND CONTROL   1249

CYTOMEGALOVIRUS (SALIVARY GLAND VIRUS)   1249

PROPERTIES OF THE VIRUS   1249
   Immunological Characteristics   1249
   Host Range   1249
   Viral Multiplication   1250
PATHOGENESIS   1250
LABORATORY DIAGNOSIS   1251
EPIDEMIOLOGY AND CONTROL   1251

EB VIRUS AND INFECTIOUS MONONUCLEOSIS   1252

PROPERTIES OF THE VIRUS 1252
INFECTIOUS MONONUCLEOSIS   1252
   Pathogenesis   1252
   Laboratory Diagnosis   1252
   Epidemiology and Control   1255

Fever blisters (herpes simplex) and chicken-pox (varicella), two of the most common diseases of man, have epidemiological features that puzzled physicians for many years until technics became available for the isolation, identification, and characterization of the responsible viruses. The multiple recurrence of fever blisters in certain individuals was bewildering until about 1950, when Burnet in Australia and Buddingh in the United States showed that herpes simplex virus often becomes latent after initiating a primary infection, usually in children, and is then repeatedly activated by subsequent provocations (Ch. 51, Latent viral infections). A similar mechanism is present in chickenpox (varicella), recurring as herpes zoster (shingles), but the picture was even more obscure because the initial and the subsequent syndromes are very different. Nevertheless, a relation was long suspected, since the sporadic appearance of zoster in an adult was often followed by an outbreak of chickenpox in contacted children. In 1954 Weller demonstrated conclusively, by comparison of the viruses isolated in tissue cultures, that varicella and zoster are caused by the same agent, which closely resembles the virus of herpes simplex.

GENERAL CHARACTERISTICS

During the past decade a number of viruses infecting man and lower animals have been found to have chemical, physical, and biological properties similar to the herpesviruses isolated earlier (Table 53-1); they have therefore been classified in a single family, Herpesviruses (Gr. *herpein* = to creep). Most members of the family have a special affinity for cells of ectodermal origin and tend to produce **latent infections** (Ch. 51). While the mechanism of latency has not been estab-

lished, it is noteworthy that the herpesviruses, like adenoviruses (which also initiate latent infections), are DNA viruses and replicate in the cell nucleus. Herpesviruses also share with another group of DNA viruses, the poxviruses, a tendency to **focal** cytopathogenicity, producing vesicles or pocks both in patients and in egg membranes.

The major herpesviruses that infect man are herpes simplex, varicella–herpes zoster, cytomegalovirus (inclusion or salivary gland virus of man), and EB (Epstein-Barr) virus. A number of viruses infecting lower animals are also members of this family. For example; 1) B virus of monkeys which may infect man accidentally, 2) pseudorabies virus of pigs, 3) virus III of the rabbit, 4) cytomegalovi-

TABLE 53-1. Characteristics of Members of the Herpesvirus Family

Size	1800–2000 A
Symmetry of capsid	Icosahedral
Number of capsomers	162
Lipid envelope	Present
Sensitivity to ether and chloroform	Inactivated
Nucleic acid	Double-stranded DNA*
Site of biosynthesis of viral DNA	Nucleus
Site of assembly of viral particles	Nucleus and cytoplasm
Inclusion bodies	Intranuclear, eosino-philic†
Common family antigen	None

* Molecular weights: herpes simplex viruses approximately 100×10^6 daltons; cytomegalovirus, 64×10^6 daltons. Guanine-cytosine content (moles per cent): 58% cytomegalovirus; 68% herpes simplex type 1; and 70% herpes simplex type 2. DNA-DNA hybridizations indicated that 40 to 46% of the base sequences are homologous in types 1 and 2 herpes simplex viruses.

† Cytomegalovirus-infected cells may also contain basophilic cytoplasmic inclusion bodies.

ruses of animals (inclusion virus of guinea pig, inclusion virus of mouse, and inclusion body rhinitis of pig), and 5) several viruses that may be associated with renal carcinoma of frogs (Ch. 63). These animal viruses appear to have only minor immunological relatedness to herpesviruses from man except for the marked cross-reactivity between herpes simplex virus and B virus. In this chapter only the viruses that infect man will be discussed.

The **morphology** of all herpesviruses is similar. The mature particles range from 1800 A (herpes simplex and varicella-zoster viruses) to 2000 A (cytomegalovirus) in diameter. They consist of 1) a DNA-containing core approximately 760 A in diameter, 2) an icosahedral capsid 950 to 1050 A in diameter, 3) a surrounding granular zone, and 4) an encompassing envelope possessing periodic short projections (Figs. 53-1, 53-4, and 53-6). The capsid is composed of 162 elongated hexagonal prisms (capsomers) 95 by 125 A with a central hole approximately 40 A in diameter. Within a population of virions many particles do not possess envelopes, and some are empty capsids (Fig. 53-1 C and D). The envelope appears to contain both host cell and viral components. Consequently, enveloped herpes simplex virions are aggluti-

Fig. 53-1. The four morphological types of herpes simplex virus particles embedded in phosphotungstate. **A.** Enveloped "full" particle showing the thick envelope surrounding the nucleocapsid. **B.** Enveloped "empty" particle; the capsid does not contain viral DNA and therefore can be penetrated by the phosphotungstate. **C.** Naked "full" particle; the structure of the capsomers is plainly visible. **D.** Naked "empty" particle. All ×200,000. [From Watson, D. H., Russell, W. C., and Wildy, P. *Virology 19*:250 (1963).]

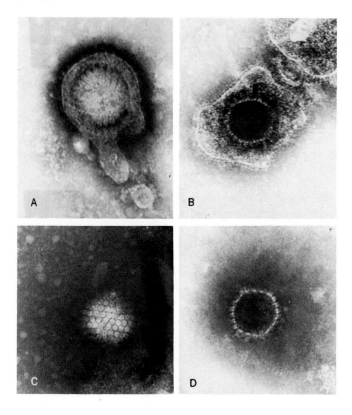

nated by antibodies to uninfected cells, and enveloped virions contain receptors for influenza viruses and will combine with them. The viral envelope is important for adsorption to susceptible host cells.

CHEMICAL AND PHYSICAL CHARACTERISTICS

The chemical composition of purified enveloped viral particles is consonant with their morphology; herpes simplex virions contain 70% protein (consisting of at least 10 distinct species, including 4 envelope glycoproteins), 7% DNA (a linear, double-stranded molecule), 22% lipid (chiefly phospholipid derived from the inner nuclear membrane of the host cell), 2% carbohydrates, and small amounts of the polyamines spermine and spermidine. Other herpesviruses appear to have similar chemical compositions. Some characteristics of a few of the viral DNAs are summarized in Table 53-1.

Like other enveloped viruses, herpesviruses are relatively unstable at room temperature and are readily inactivated by lipid solvents.

HERPES SIMPLEX VIRUS

PROPERTIES OF THE VIRUS

IMMUNOLOGICAL CHARACTERISTICS

Two immunological variants, types 1 and 2, can be distinguished by neutralization titrations, although cross-reactivity between the variants is prominent. Some additional minor antigenic variations have been observed, but do not warrant classification. Infected cells contain, along with virions, soluble antigens, which elicit a delayed-type skin reaction in addition to the usual immunological reactions. Since the virion proteins cannot be isolated in their multimeric forms, unlike adenovirus proteins (Ch. 52), it has not been feasible either to characterize them individually or to compare them with the soluble antigens.

Within 4 to 7 days after a primary infection specific Abs appear and reach maximum titers in about 14 days. Assayed by neutralization or complement-fixation tests, these Abs may drop to undetectable levels after first infections, only to reappear with recurrent episodes. By adulthood Ab titers are generally high and persist indefinitely. Accordingly an increase cannot usually be detected in recurrent disease, although infectious virus can be isolated readily from the lesion.

HOST RANGE

Man is the natural host for herpes simplex virus, but a relatively wide range of animals are also susceptible, including adult and suckling mice, guinea pigs, hamsters, and rabbits. The effects of infection depend upon the route of inoculation. For example, in the rabbit inoculation of the cornea results in keratoconjunctivitis or keratitis, while intracerebral inoculation produces fatal encephalitis. The chick embryo has been a convenient host: production of pocks on the chorioallantoic membrane affords a reproducible method for detection and assay, similar to that employed with poxviruses (Ch. 54, Fig. 54-4).

Many cultured cell types support multiplication of herpes simplex virus, undergo extensive cytopathic changes, and develop intranuclear inclusion bodies. The response of the cells varies with the strain of virus employed. Some strains cause marked clumping of cells, producing pock-like lesions; others produce multinucleated giant cells (polykaryocytes) by fusion of membranes and recruitment of the nuclei of adjoining cells. Chromosomal breaks and aberrations are also observed in infected hamster and human embryonic cells. With suitable cells some strains produce typical plaques.

VIRAL MULTIPLICATION

The viral envelope provides the normal attachment of the virions to susceptible cells. Heparin, dextran sulfate, and other sulfated polysaccharides can react with the envelope and reversibly prevent viral adsorption and infection. Following attachment the viral envelope appears to fuse with the cell's plasma membrane, permitting the viral nucleocapsid to enter directly into the cytoplasm (Ch. 48). Intact virions may also enter in phagocytic vacuoles, from which they are released by similar viral envelope–membrane fusion. In the cytoplasm the capsid separates from the viral core, which enters the nucleus and initiates viral multiplication. The eclipse period is 5 to 6 hours in monolayer cell cultures, and virus increases exponentially until approximately 17 hours after infection, by which time each cell has made 10^4 to 10^5 physical particles, of which 100 to 1000 are infectious. Virions are released by slow leakage from infected cells.

The biochemical events responsible for viral replication are similar to those described for other DNA-containing viruses (Ch. 48). There is an ordered sequential synthesis of virus-induced messenger RNAs (transcribed asymmetrically), enzymes related to DNA synthesis and breakdown (thymidine kinase, DNA polymerase, and a DNA exonuclease have been studied), viral DNA, and viral structural proteins, culminating in the assembly of infectious viral particles; at the same time production of host DNA, RNA, and protein declines and ceases. Only about 25% of the viral DNA and protein made in cell cultures is assembled into virions.

The biosynthetic steps described can be correlated with the development of nuclear inclusion bodies and the formation of viral particles. A basophilic, Feulgen-positive, granular mass develops centrally in the nucleus; it corresponds to the accumulation of newly synthesized viral DNA. Assembly of incomplete viral particles begins within this material (Fig. 53-2). Initially a single coat (the capsid) surrounds the electron-dense DNA-containing nucleoid; the particle is noninfectious and unstable. A second coat (the envelope) appears to be acquired from the inner nuclear membrane as the nuclear membrane reduplicates to permit egress of viral particles into the cytoplasm (Fig. 53-3); some particles, however, may obtain their envelopes from cytoplasmic membranes. The envelope contains virus-specific subunits (particularly glycoproteins) as well as host cell materials. Mature infectious virus is apparently liberated from infected cells slowly either by a process similar to reverse phagocytosis or through membranous tubules that traverse the cytoplasm from nucleus to plasma membrane. Although envelopment and release of viral particles does not occur by budding from the plasma membrane, as with other enveloped

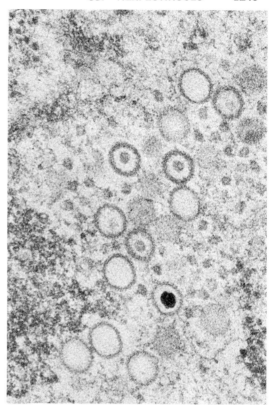

Fig. 53-2. Formation of herpes simplex virus particles within the cell nucleus. Some capsids are in process of assembling, while others are complete. Cores of varying density are forming within the capsids. Indistinct particles probably represent virus sectioned at one margin with loss of density and overlapping structure. ×90,000. (From Nii, S., Morgan, C., and Rose, H. M. *J Virol* 2:517 (1968).]

viruses, the membrane is changed morphologically and it contains virus-specific antigens.

Movement of viral particles from the nucleus into the cytoplasm is accompanied by the transport of soluble antigens into the cytoplasm; concomitantly the originally basophilic intranuclear inclusion body is converted into an eosinophilic, Feulgen-negative mass. Thus, the eosinophilic inclusion body usually observed in infected cells does not contain viral particles or specific viral antigens (detectable by immunofluorescence), but actually is the burnt-out remnant of a viral factory.

PATHOGENESIS

The most striking characteristic of herpes simplex virus infection is its propensity for

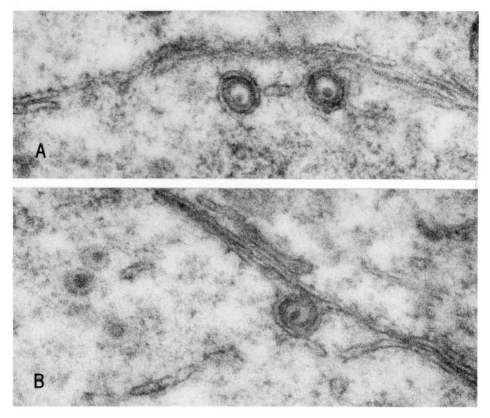

Fig. 53-3. Thin sections of mature particles of herpes simplex virus. Particles show inner core and two or three membranes. In lower picture the nuclear membrane partly surrounds a viral particle (apparently a step in the simultaneous assembly and egress of the complete virions from the nucleus). ×87,000. [From Morgan, C., Rose, H. M., Holden, M., and Jones, E. P. *J Exp Med 110*:643 (1959).]

persisting in a quiescent or latent state in man, with recurrence of activity at irregular intervals. The initial infection occurs through a break in the mucous membranes (e.g., eye, mouth, throat, genitals) or skin, where local multiplication ensues. From this focus virus spreads to regional lymph nodes, where it multiplies further. On occasion virus disseminates, unrestricted by antibody, into the blood and to distant organs.

The initial infection ordinarily occurs in children 6 to 18 months of age; serological surveys have demonstrated that it is most often inapparent. But 10 to 15% of those infected do develop a **primary disease,** usually **herpetic gingivostomatitis,** which is characterized by multiple vesicles in the oral mucous membranes and the mucocutaneous border. Similar lesions are less commonly seen in other regions, including the nostrils, the genitalia or urethra, the cornea, and sites of trauma. More serious rare complications include neonatal generalized infections, meningoencephalitis,* and diffuse skin involvement in children with chronic eczema (Kaposi's varicelliform eruption, eczema herpeticum).

When the initial infection recedes virus persists, producing a **latent infection** at the major site of the primary disease (Ch. 51, Latent viral infections), despite the presence of a high antibody titer. The form in which the virus exists during the occult stage remains unknown; but the balance is readily upset, provoking the second form of herpes simplex, **recurrent disease.** Since neutralizing antibodies are present the virus cannot disseminate, but

---

* Encephalitis occurs more frequently in adults, either as a primary infection or as a flare-up of a latent infection.

it does reach contiguous cells; the current form of herpes simplex therefore remains localized. In a single individual the clinical features are much the same with each episode. For example, if gingivostomatitis was the primary disease, the recurrent form is usually herpes labialis (fever blisters); and if the primary disease was herpetic keratitis, the recurrent disease is keratitis (which may lead to corneal scarring and blindness).

The differences in the pathogenic potentials of types 1 and 2 herpes simplex viruses are noteworthy. Type 1 virus is primarily associated with oral and ocular lesions and is transmitted in oral and respiratory secretions, whereas type 2 is isolated from genital and anal lesions and is passed primarily through genital contact. Type 2 virus appears to have greater neurovirulence in experimental animals. Mothers with genital lesions are the primary source of neonatal herpetic infections with type 2 virus, which is often severe, and often fatal. It is also striking that in the United States more than 80% of women with cervical carcinoma have antibodies to type 2 herpes simplex virus. However, this correlation does not necessarily prove a causal relation but may only be a reflection of the lower socioeconomic status and the possibly more promiscuous sexual activity of many victims of cervical carcinoma (Ch. 63).

Viral multiplication and recurrent disease may be induced by many different factors such as heat, cold, sunlight (ultraviolet light), immunologically unrelated hypersensitivity reactions, pituitary and adrenal hormones, and emotional disturbances. A few of these, such as epinephrine and hypersensitivity reactions, have evoked recurrences in experimental animals (Ch. 51, Latent viral infections).

The presence of intranuclear eosinophilic inclusion bodies in epithelial cells, often giant cells, is characteristic of all the lesions. Electron microscopic examination reveals cells that contain viral particles in various stages of maturation.

LABORATORY DIAGNOSIS

Laboratory procedures are occasionally used to confirm a clinical diagnosis or assist in differential diagnosis (an etiological diagnosis will become more important when effective antiviral chemotherapeutic agents become available). 1) The quickest, simplest, and most economical diagnostic test consists of demonstrating **characteristic multinuclear giant cells,** containing intranuclear eosinophilic inclusion bodies, in scrapings from the base of vesicles. 2) Specific **antigens** may be detected in cells from the lesions by **immunofluorescent technics.** 3) Virus can be **isolated** by inoculating material from lesions (especially vesicle fluid) into susceptible tissue culture cells or newborn mice. 4) An **antibody** rise is detected by neutralization or complement-fixation titrations with sera obtained early in the course of the primary disease and again 14 to 21 days after onset. Patients with recurrent disease have a high initial titer of circulating Abs and often do not show significant increase.

EPIDEMIOLOGY, CONTROL, AND TREATMENT

Close person-to-person contact is the most common mechanism of viral transmission. Secretions from lesions about the mouth (herpes labialis and stomatis) and genitalia are the most frequent sources. Virus is also transmissible on eating and drinking utensils for a brief period after contamination. Occasionally virus is found in the saliva of healthy individuals; the viral shedding may even continue for weeks or months. Rare cases of primary infection have been reported in adults. Approximately 80% of adults have relatively high titers of neutralizing and complement-fixing antibodies, and a substantial fraction of this population is subject to recurrent herpes.

Control is difficult because of the large numbers of persons with inapparent infections and minor recurrent lesions from which virus is shed. It is important whenever possible, however, to prevent contact of infants with persons who have herpetic lesions. A vaccine to prevent the primary infection would not have sufficient clinical application to warrant development, although the required knowledge is available.

Herpetic keratitis, one of the major infectious causes of blindness, has been treated with some success by the repeated local application of 5-iodo-2′-deoxyuridine or cytosine arabinoside (1-β-D-arabinofuranosylcytosine); recurrences may follow cessation of therapy, and drug-resistant herpesviruses may emerge. These analogs and adenine arabinoside (1-β-D-arabinofuranosyladenine) have also been employed intravenously in cases of encephalitis and disseminated herpes simplex infections, but the number of proved infections treated has been too few to permit critical evaluation.

Cortisone, which is commonly employed in the treatment of inflammatory ocular lesions, is **contraindicated** in treating herpes simplex keratitis: it causes greater viral multiplication leading to increased cellular damage, resulting in more extensive lesions and often perforation of the cornea.

VIRUS B

The monkey is the natural host for virus B, which produces latent infections like those of herpes simplex virus in man.

However, in man virus B can also produce acute ascending myelitis and encephalomyelitis. The increased handling of monkeys, and the widespread use of monkey kidney cells in the commercial production of poliovirus vaccines, have augmented the transmission of this virus to man: more than 20 cases have been reported in the past 15 years. Recognition of the danger is critical, since the disease in man is usually fatal.

Virus B closely resembles herpes simplex virus. The two are antigenically related but not identical: antiserum from rabbits immunized with virus B neutralizes herpes simplex virus; however, antibodies to herpes simplex virus neutralize virus B only slightly.

VARICELLA (CHICKENPOX)–HERPES ZOSTER (SHINGLES) VIRUS

The viruses isolated from patients with varicella and from those with herpes zoster (Gr. girdle) are physically and immunologically indistinguishable when tested by all the usual technics (Fig. 53-4). Their identity has been further established by the production of typical varicella in children following inoculation of herpes zoster vesicle fluid.

and continue to increase in titer for about 2 weeks. Titers of complement-fixing Abs decrease over a period of months, but neutralizing Abs persist for many years after the primary varicella infection. In herpes zoster most patients have a relatively high titer of IgG at the onset of the disease, indicating that herpes zoster occurs in partially immune individuals.

PROPERTIES OF THE VIRUS

PHYSICAL AND CHEMICAL CHARACTERISTICS

Varicella-zoster virus is relatively unstable. When virus from tissue cultures is stored even at −40 to −70° infectivity cannot be maintained reliably for longer than 2 months. In vesicle fluid from a patient, however, infectious virus survives many months at −70°, perhaps owing to the high titers of virus and the high concentration of protein.

IMMUNOLOGICAL CHARACTERISTICS

Agar diffusion reveals three separate soluble antigens in both vesicle fluids and tissue culture preparations.

In varicella, Abs usually appear in the serum 4 to 5 days after onset of the exanthem

HOST RANGE

In contrast to herpes simplex virus, varicella virus does not cause reproducible disease in experimental animals or chick embryos, but it can be propagated in cultures of a variety of human and monkey cells. Since little virus is found in the culture fluid, and virus obtained by disrupting cells is unstable, serial propagation is best accomplished by transfer of infected cells. Characteristic cytopathic effects and eosinophilic intranuclear inclusion bodies develop (Fig. 53-5), and metaphase arrest and chromosomal aberrations have been observed.

VIRAL MULTIPLICATION

Multiplication of varicella-zoster virus is confined to the nucleus, and the developmental

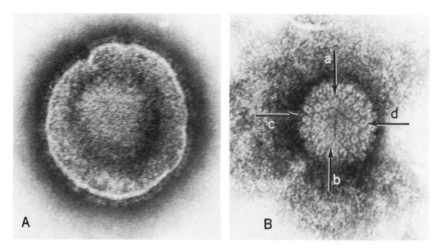

Fig. 53-4. Morphology of varicella–herpes zoster virus particles embedded in sodium phosphotungstate. **A.** Intact virion showing the envelope with surface projections and the centrally placed capsid. **B.** Viral particle in which envelope has ruptured revealing the structure of the capsid more clearly. The arrows point to capsomers situated on axes of fivefold symmetry. Both ×200,000, reduced. [From Almeida, J. D., Howatson, A. F., and Williams, M. G. *Virology 16*:353 (1962).]

stages are similar to those of herpes simplex virus; the biochemical details have not yet been described. Infectious virus after assembly does not emerge readily from infected cells in culture. Natural infections, however, are highly contagious, and the virus is present extracellularly in high titer in vesicle fluid of lesions.

PATHOGENESIS

Varicella (chickenpox) is the **primary disease** produced in a host without immunity; it is usually a mild, self-limited illness of young children. The clinical picture strongly suggests that the virus is spread by respiratory secretions, enters the respiratory tract, multiplies locally and possibly in regional lymph nodes, produces viremia, and is disseminated by the blood to skin and internal organs. The virus prefers ectodermal tissues, particularly in children.

After an incubation period of 14 to 16 days fever occurs, followed within a day by a papular rash of the skin and mucous membranes. The papules rapidly become vesicular and are accompanied by itching. Lesions occur in successive crops, and all stages can be observed simultaneously. In the infrequent adult infections the disease is more severe, often with a diffuse nodular pneumonia; the mortality may be as high as 20%. In children receiving adrenocortical steroids varicella is usually diffuse and harsh.

The vesicles evolve from a ballooning and degeneration of the prickle cells of the skin, along with formation of giant cells with intranuclear eosinophilic inclusion bodies (Fig. 53-5). In disseminated fatal varicella, lesions containing similar giant cells appear in liver, lungs, and nervous tissue. Lesions basically identical with those of herpes zoster may occur in dorsal root ganglia.

Herpes zoster (shingles) is the **recurrent form** of the disease, occurring in adults who were previously infected with the varicella-zoster virus and who possess circulating antibodies. The syndrome develops from an inflammatory involvement of sensory ganglia of spinal or cranial nerves. Herpes zoster usually has a sudden onset of pain and tenderness along the distribution of the affected sensory nerve, accompanied by mild fever and malaise. A vesicular eruption, similar in pathology to varicella (except for its distribution), then occurs in crops along the distribution of the sensory nerve (frequently an intercostal

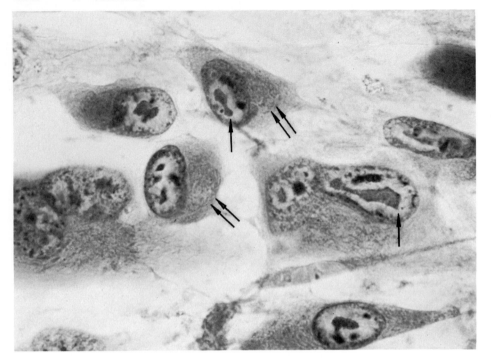

Fig. 53-5. Typical eosinophilic inclusion bodies (arrow) in nuclei of human embryonic cells infected with varicella–herpes zoster virus. Poorly differentiated pale eosinophilic bodies are also present in the paranuclear area of several cells (double arrows). ×1260. [From Weller, T. H., Witton, H. M., and Bell, E. J. *J Exp Med 108*:843 (1958).]

nerve); it is almost always unilateral. The vesicular eruption may last as long as 2 to 4 weeks; the pain may persist for additional weeks or months. If the inflammation spreads into the spinal cord or cranial nerves paralysis results. Meningoencephalitis, which occurs rarely, is usually manifest as an acute illness with severe symptoms (headache, ataxia, coma, convulsions), but most patients recover completely.

Clinically, herpes zoster may be activated by trauma, injection of certain drugs such as arsenic or antimony, tuberculosis, cancer, or leukemia. Moreover, in persons with immunodeficiency states, particularly when induced in the therapy of lymphoproliferative diseases, the activated virus may disseminate causing serious, often fatal, illness. The virus appears to remain latent in ganglionic nerve cells and it seems likely that during activation virus travels along nerve fibers to the skin. It has also been reported that herpes zoster may result from reinfection of a partially immune individual upon exposure to a person with chickenpox, but the evidence is not conclusive.

LABORATORY DIAGNOSIS

A clinical diagnosis of either varicella or herpes zoster seldom offers serious difficulties, but rarely the differentiation between chickenpox and smallpox may require laboratory tests. The two diseases can be distinguished most rapidly and easily by preparing a smear from the base of a vesicle and staining (by the Giemsa method) to detect typical varicella giant cells and cells with characteristic inclusion bodies. Immunofluorescence and electron microscopic examination of vesicular fluid provide rapid confirmation. In addition, virus may be isolated in cultured human or monkey cells and identified by serological technics. Antibody determinations on paired sera from patients may also be useful.

EPIDEMIOLOGY

Varicella virus is usually transmitted in respiratory secretions, producing a highly communicable disease with a high clinical attack rate. Epidemics are common among children, especially in the winter, and second attacks of chickenpox apparently do not occur. Herpes zoster, in contrast, is of low incidence, is not seasonal, and is confined almost entirely to persons over 20 years of age. Moreover, epidemics of herpes zoster never occur, although a person with zoster may be the index case for an outbreak of chickenpox.

PREVENTION AND CONTROL

There are no widely applicable and successful means for preventing and controlling varicella infections, and no vaccine has been developed. In view of the seriousness of the disease in adults, however, one might question the wisdom of attempts to prevent infection in children, unless the preventive procedure can effect as lasting protection as the natural disease. Chickenpox can be modified by administering pooled immunoglobulins to contacts, as in measles (Ch. 57, Prevention and Control). Disseminated disease in the seriously ill (particularly in immunologically suppressed patients) has apparently been successfully treated with cytosine arabinoside or adenine arabinoside.

CYTOMEGALOVIRUS (SALIVARY GLAND VIRUS)

Salivary gland virus disease of newborns is a severe, often fatal illness, usually affecting the salivary glands, brain, kidneys, liver, and lungs. M. G. Smith in 1956 isolated the causative agent. Although the disease is rare, it was subsequently shown that cytomegaloviruses (a term suggested by the large size of the infected cells and their huge intranuclear inclusion bodies) are widespread and commonly produce latent infections which subsequently may be activated by pregnancy, multiple blood transfusions, or an organ transplant in an immunosuppressed recipient. The inclusion of these viruses in the herpesvirus family is based primarily on the morphology of the viral particle (Fig. 53-6), the chemical composition of the virion (Table 53-1), and the characteristics of the intranuclear inclusion body present in infected cells (Figs. 53-7 and 53-8).

PROPERTIES OF THE VIRUS

IMMUNOLOGICAL CHARACTERISTICS

Two distinct antigenic types of cytomegalovirus from man have been identified by neutralization titrations, and there is possibly a third group of viruses that cross-reacts with what may be termed types 1 and 2. Complement-fixation titrations show a spectrum of cross-reactive antigens relating all the human cytomegaloviruses. Investigation of the virion's antigenic structure has been hindered by its low reactivity and poor immunogenicity.

HOST RANGE

Although many species of animals are infected with their own specific cytomegaloviruses, no laboratory animal has proved susceptible to infection with cytomegaloviruses of humans. Virus has been isolated and propagated only in cultured human fibroblasts; despite the prominent involvement of epithelial cells in the disease, cultivated epithelial cells are not infected. Viral multiplication is accompanied

Fig. 53-6. Enveloped full particle of human cytomegalovirus. ×405,000. [From Wright, H. T., Jr., Goodheart, C. R., and Lielausis, A. *Virology* 23:419 (1964).]

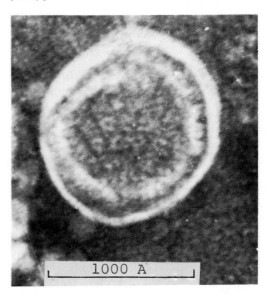

1000 A

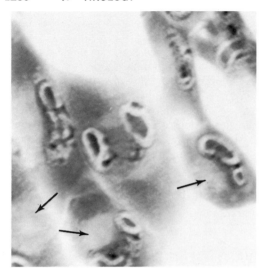

Fig. 53-7. Typical eosinophilic intranuclear inclusion bodies in human uterine cells infected with human cytomegalovirus. A round eosinophilic area adjacent to the nucleus (arrow) is present in many cells. ×515. [From Smith, M. G. *Proc Soc Exp Biol Med* 92:424 (1956).]

first detected 48 to 72 hours after infection with a low virus to cell multiplicity. Under these conditions cytochemical and immunofluorescent technics best reveal the synthesis of viral subunits and the development of the cytologic lesion. De novo biosynthesis of DNA and accumulation of viral antigen are detected initially in the nucleus. Electron microscopic studies suggest that viral particles, like herpes simplex virions, are assembled in the nucleus (Fig. 53-8), attain their envelope at the nuclear membrane, and migrate through reduplications of the nuclear membrane into the cytoplasm in vacuoles. The maturation of viral particles appears to be inefficient, and only rare completely assembled virions can be detected among many incomplete viral particles; hence the yield of infectious virus in tissue cultures is low, and as many as 10^6 particles are needed to initiate infection of a culture.

by development of focal lesions, followed by generalized cytopathic changes including rounding of cells and the appearance of large intranuclear eosinophilic inclusion bodies (Fig. 53-7).

VIRAL MULTIPLICATION

Replication of virus appears to be relatively slow, and newly made infectious particles are

PATHOGENESIS

Cytomegalovirus appears to be transmitted transplacentally from a mother with a latent infection to the fetus. The infection usually remains inapparent in the newborn but the virus may subsequently be activated: protracted periods of viral shedding may follow (although the children develop neutralizing antibodies), and the carrier may serve as a source for dissemination (in different studies 15 to 60% of healthy children were shown to

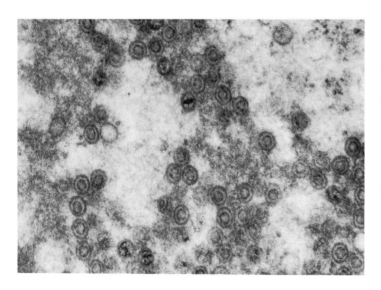

Fig. 53-8. Electron micrograph of a portion of an intranuclear inclusion in a cell infected with human cytomegalovirus. The inclusion is made up of viral particles in various stages of development. The particles are composed of a central core about 400 A in diameter, surrounded by a pale zone and externally by a thin membranous shell. Only a few particles have a dark central core indicating the presence of nucleic acid. ×40,000. [From Becker, P., Melnick, J. L., and Mayor, H. D. *Exp Mol Pathol* 4:11 (1965).]

excrete virus during the first year of life). Patients with neoplastic diseases and recipients of organ transplants, subjected to corticosteroids or other immunosuppressive drugs, are particularly susceptible to viral activation or exogeneous infection, resulting in localized or disseminated disease, or inapparent infection. A syndrome resembling infectious mononucleosis (see below) may also be observed in recipients of multiple tranfusions of blood from latently infected donors (most frequently reported in patients following open heart surgery, probably owing to the large volumes of blood they receive). When manifest illness does rarely occur in newborns and infants up to 4 months of age, it usually has a relentless progression, with hepatic and renal insufficiency, pneumonia, neurological symptoms, and eventual death.

Characteristic cytological lesions are detected in approximately 10% of all stillborns, including those dying from causes other than cytomegalovirus infection. Like adenoviruses, cytomegalovirus can be isolated from explants of adenoids and salivary glands cultured in vitro (Ch. 52, Pathogenesis). Similar findings have been made with cytomegaloviruses of mice, rats, hamsters, and guinea pigs.

The pathological lesion is characterized by necrosis and pathognomonic cellular alterations. The affected cells are greatly increased in size; the nucleus is enlarged and contains a brightly stained eosinophilic inclusion body up to 15 μ diameter (Fig. 53-9). The inclusion body is larger than that produced by any other virus infecting man. Cytoplasmic alterations are also frequently present. The cytoplasm may be swollen and vacuolated and may contain up to 20 minute basophilic and osmophilic structures 2 to 4 μ in diameter. The cytoplasmic bodies contain DNA and polysaccharide, and therefore are positive with Feulgen and PAS (periodic acid–Schiff) stains.

LABORATORY DIAGNOSIS

Infection can be identified by viral isolation, immunological assays, and exfoliative cytological technics. 1) Isolation of virus, in cultures of human embryonic fibroblasts, is the most sensitive method to detect infection in the newborn, but it is also the most expensive and cumbersome. 2) An immunofluorescent

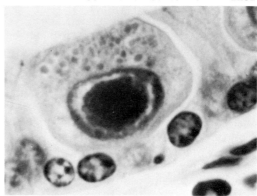

Fig. 53-9. Two duct epithelial cells from a human submaxillary gland showing the typical eosinophilic nuclear and basophilic cytoplasmic inclusions produced by infection with human cytomegalovirus virus. ×1500. [From Nelson J. S., and Wyatt, J. P. *Medicine* 38:223 (1959).]

assay is also suitable for diagnosis in the newborn because one can distinguish between the baby's IgM Abs and the maternal IgG Abs. 3) An indirect hemagglutination test using tanned sheep erythrocytes to adsorb viral antigen is sensitive, capable of detecting IgM and IgG Abs, rapid, and inexpensive. 4) Complement fixation is the least expensive and most convenient method to assay Abs in older children and adults. 5) Identification of characteristic cytomegalic cells with intranuclear and cytoplasmic inclusions (particularly in urinary sediment and bronchial and gastric washings) offers an efficient and inexpensive procedure for making an antemortem diagnosis.

EPIDEMIOLOGY AND CONTROL

Infection with human cytomegalovirus appears to be worldwide and common despite the relative rarity of clinical disease. From 10 to 18% of infants show detectable lesions at autopsy. In adults above 16 years of age typical inclusions in salivary glands are rare; but in a sample taken in the United States, complement-fixing antibodies were found in 53% of the population between 18 and 25 years and in 81% of those over 35 years of age. Transplacental passage and blood transfusion are the most apparent mechanisms for transmitting virus, but person-to-person spread in urine and respiratory secretions seems likely.

Children and adults with immunological deficiencies, naturally or iatrogenically acquired, are particularly susceptible to active disease.

Effective measures for prevention and control cannot be available until more is known of viral characteristics and transmission.

EB VIRUS AND INFECTIOUS MONONUCLEOSIS

In a search for the cause of Burkitt lymphoma, Epstein and Barr observed herpes-like viral particles (termed EB virus; Fig. 53-10) in a small proportion (0.5 to 10%) of the lymphoma cells repetitively cultured. The possible causal relation between the malignancy and the virus will be discussed in Chapter 63 (Burkitt lymphoma). The Henles then discovered that leukocytes from patients with infectious mononucleosis, unlike normal leukocytes, can be serially cultured in vitro, and that a small number of the cultured cells contain EB virus. The EB virus persists in lymphocytes cultured from patients long after the disease, and transfusion of blood from such individuals produces infectious mononucleosis in recipients devoid of Abs. Moreover, patients with infectious mononucleosis develop Abs that react with EB virus-infected cells (demonstrated by the indirect immunofluorescence assay); the Abs persist for years and their presence or absence is correlated with resistance or susceptibility to infectious mononucleosis. Finally, EB virus has been isolated from pharyngeal secretions of patients with infectious mononucleosis. These data imply that the EB virus plays an etiologic role in infectious mononucleosis, although its relation with human malignancies remains uncertain (Ch. 63).

PROPERTIES OF THE VIRUS

Virions purified from cultured lymphoma cells (Fig. 53-11) closely resemble herpesviruses (Fig. 53-1): 1) the virion possesses a large envelope with a diameter of approximately 1100 A; 2) the capsid is of appropriate size, has icosahedral symmetry, and appears to consist of 162 capsomers (Fig. 53-11B); 3) the capsomers are structurally similar to those of herpesviruses; 4) the viral DNA is double-stranded and has a GC content of 59%; 5) the virion consists of the same number and size proteins as herpes simplex virus. Immuno-

logically, however, EB virus and herpesviruses are unrelated. The virus multiplies in cultured lymphoreticular cells, but it is difficult to use cell-free virus for serial passage—possibly because almost all the particles are defective according to electron microscopic examination.

INFECTIOUS MONONUCLEOSIS

PATHOGENESIS

Infectious mononucleosis is an acute infectious disease primarily affecting lymphoid tissue throughout the body. It is characterized by the appearance of enlarged and often tender lymph nodes, enlarged spleen, and abnormal lymphocytes in the blood (from which the disease derives its name). In addition, fever and sore throat are common. Occasionally other diverse manifestations are observed, including mild hepatitis, signs of meningitis or other central nervous system involvement, hematuria, proteinuria, thrombocytopenic purpura, and hemolytic anemia.

Epidemiological and clinical evidence suggest that the virus enters the body in respiratory secretions, multiplies in local lymphatic tissues, and spreads into the blood stream, where it infects and multiplies in leukocytes. Leukocytes can be readily cultured in vitro even after recovery from acute infections, which suggests that latent infections with EB virus are common and may be responsible for the persistence of specific antibodies.

LABORATORY DIAGNOSIS

An increase in titer of Abs to EB virus is required to establish the specific diagnosis. The immunological assays employed are indirect immunofluorescence using a lymphoma-derived cell line containing EB virus, and complement fixation, using a homogenate of an infected continuous cell line. These tech-

Fig. 53-10. Structure of EB virus in cells cultured from Burkitt lymphomas. **A.** Numerous developing immature particles in thin section of a lymphoblast nucleus. ×76,500. **B.** A mature viral particle with envelope, capsid, and nucleoid in cytoplasm. ×42,000. Inset, ×213,500. [From Epstein, M. A., Henle, G., Achong, B. G., and Barr, Y. M. *J Exp Med 121*:761 (1965).]

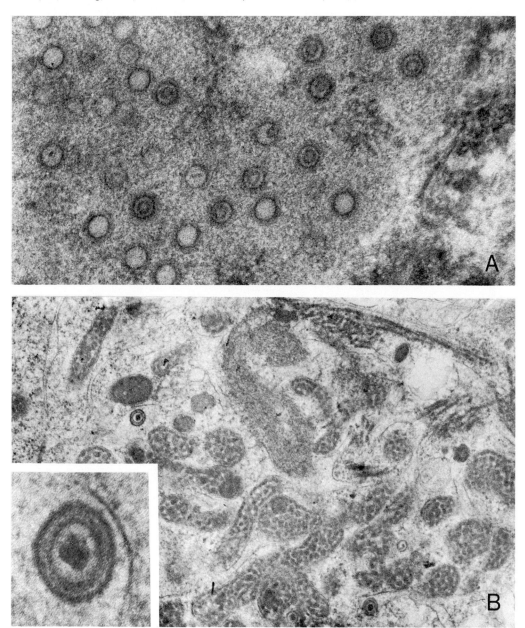

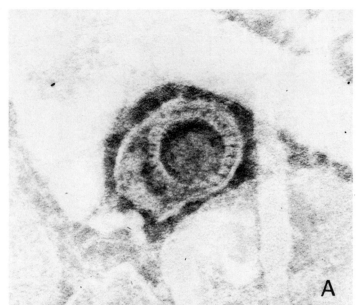

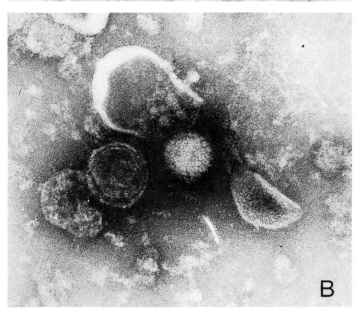

Fig. 53-11. Structure of EB virus obtained from cultured Burkitt lymphoma cells. Electron micrographs are of purified negatively stained virus. **A.** Empty capsid enclosed in envelope. ×200,000. **B.** Viral particle with disrupted envelope. A capsid (about 750 A in diameter) similar to herpesviruses is clearly visible. ×120,000. (Courtesy of Dr. Klaus Hummler.)

nics are specific and sensitive, but they are highly specialized and not yet suitable for the general hospital laboratory. The practical laboratory diagnosis rests upon two unique findings: 1) abnormal, large lymphocytes with deeply basophilic, foamy cytoplasm and fenestrated nuclei appear, often accounting for 50 to 90% of the circulating lymphocytes (initially there is a leukopenia, but by the second week of disease the count may rise to 10,000 to 80,000 cells per cubic millimeter); 2) **heterophil antibodies,** i.e., agglutinins for sheep erythrocytes, develop in 50 to 90% of patients during the course of the disease.

The immunogen eliciting the heterophil Abs is unknown, but it appears to be distinct from the EB virus. Thus, a small percentage of patients develop EB virus Abs but not heterophil Abs; heterophil

Abs can be adsorbed from serum by beef RBC without reducing the Ab titer for EB virus; and heterophil Abs are transient but Abs to EB virus persist for years, perhaps for life.

Heterophil Abs also appear during serum sickness following injection of horse serum, and they may be present in sera from healthy persons. However, differential adsorption of a patient's serum with beef RBC and guinea pig kidney (containing Forssman antigen, Ch. 21, ABO blood system) will distinguish these Abs. Thus, heterophil Abs of infectious mononucleosis are adsorbed by beef RBC but not by guinea pig kidney; serum sickness agglutinins are adsorbed by beef RBC and guinea pig kidney; and normal serum heterophil agglutinins are adsorbed by guinea pig kidney but not usually by beef RBC.

EPIDEMIOLOGY AND CONTROL

Infectious mononucleosis appears to be primarily a disease of relatively affluent teenagers and young adults, such as college students; the peak incidence is from 15 to 20 years of age. The relation to EB virus furnishes some basis for these epidemiological observations: members of lower socioeconomic groups acquire EB virus Abs at an early age, whereas children at higher economic levels are protected from early infection; about 75% of entering college students in the United States are free of detectable Abs.

Successful infection appears to require extensive exposure, or unknown cooperating factors, for multiple cases in families are infrequent. Perhaps the common association of infectious mononucleosis with kissing reflects a need for a large inoculum. The incidence of proved disease is about 15% in the susceptible population of college students. No adequate procedures are presently available to control spread of the virus.

SELECTED REFERENCES

Books and Review Articles

BLANK, H., and RAKE, G. *Viral and Rickettsial Diseases of the Skin, Eye, and Mucous Membrane of Man.* Little, Brown, Boston, 1955.

HANSHAW, J. B. Cytomegaloviruses. *Virol Monog 3:*1 (1968).

KAPLAN, A. S. Herpes simplex and pseudorabies viruses. *Virol Mong 5* (1969).

ROIZMAN, B. Herpesviruses, membranes, and the social behavior of infected cells. In *Viruses Affecting Man and Animals,* p. 37. (M. Sanders, ed.) Warren Green, St. Louis, 1970.

WELLER, T. H. The cytomegaloviruses: Ubiquitous agents with protean clinical manifestations. *N Engl J Med 285:*203, 267 (1971).

Specific Articles

ALMEIDA, J. D., HOWATSON, H. F., and WILLIAMS, M. Morphology of varicella (chickenpox) virus. *Virology 16:*353 (1962).

ANDERSON, S. G., and HAMILTON, J. Epidemiology of primary herpes simplex infection. *Med J Aust 1:*1308 (1949).

BEN-PORAT, T., and KAPLAN, A. S. Phospholipid metabolism of herpesvirus-infected and uninfected rabbit kidney cells. *Virology 45:*252 (1971).

BUDDINGH, G. J., SCHRUM, D. I., LANIER, J. C., and GUIDRY, D. J. Natural history of herpetic infections. *Pediatrics 11:*595 (1953).

COOK, M. L., and STEVENS, J. C. Replication of varicella-zoster virus in cell culture: An ultrastructural study. *J Ultrastruct Res 32:*334 (1971).

EPSTEIN, M., BOAV, Y., and OCHONG, B. Studies with Burkitt's lymphoma. *Wistar Inst Symp Monog 4:*69 (1965).

HENLE, G., HENLE, W., and DIEBL, V. Relation of Burkitt's tumor-associated herpes-type virus to infectious mononucleosis. *Proc Nat Acad Sci USA 59:*94 (1968).

HUMMELER, K., HENLE, G., and HENLE, W. Fine structure of a virus in cultured lymphoblasts from Burkitt's lymphoma. *J Bacteriol 91:*1366 (1966).

KAPLAN, A. S., and BEN-PORAT, T. Synthesis of proteins in cells infected with herpesvirus. VI. Characterization of the proteins of the viral membrane. *Proc Nat Acad Sci USA 66:*799 (1970).

SAWYER, R. N., EVANS, A. S., HIEDERMAN, J. C., and MCCOLLUM, R. W. Prospective studies of a group of Yale University freshmen. I. Occurrence of infectious mononucleosis. *J Infect Dis 123:*263 (1971).

WELLER, T. H., and WHITTON, H. M. The etiologic agents of varicella and herpes zoster: Isolation, propagation, and cultural characteristics in vitro. *J Exp Med 108:*843 (1958).

chapter

POXVIRUSES

GENERAL CHARACTERISTICS   1258

VARIOLA (SMALLPOX) AND VACCINIA   1261

PROPERTIES OF THE VIRUSES   1261
  Morphology   1261
  Chemical and Physical Characteristics   1261
  Antigenic Structure and Immunity   1262
  Host Range   1263
  Viral Multiplication   1264
SMALLPOX: PATHOGENESIS   1265
LABORATORY DIAGNOSIS   1268
EPIDEMIOLOGY   1270
PREVENTION AND CONTROL   1270
  Preparation of Vaccine   1270
  Administration   1271
  Complications   1273
  Control of the Spread of Smallpox   1274
TREATMENT   1274

OTHER POXVIRUSES THAT INFECT MAN   1275

COWPOX   1275
MOLLUSCUM CONTAGIOSUM 1275
MILKER'S NODULES (PARAVACCINIA)   1275
CONTAGIOUS PUSTULAR DERMATITIS (ORF)   1276

The smallpox was always present, filling the churchyards with corpses, tormenting with constant fears all whom it had stricken, leaving on those whose lives it spared the hideous traces of its power, turning the babe into a changeling at which the mother shuddered, and making the eyes and cheeks of the bethrothed maiden objects of horror to the lover.
MACAULAY, T. B. THE HISTORY OF ENGLAND FROM THE ACCESSION
OF JAMES II, VOL. IV

Since the beginning of history smallpox* has left its mark indelibly stamped upon the medical, political, and cultural affairs of man. Records show severe epidemics, century after century, from earliest times. Indeed, by the turn of the eighteenth century the disease had become endemic in the major cities of Europe and the British Isles to the extent that "every tenth person and one tenth of all mankind was killed, crippled or disfigured by smallpox."† Terror of its presence was constant— such that "no man dared to count his children as his own until after they had had the disease."

Today technics are available for prevention and control of smallpox, and the dream of its eradication seems attainable as a result of the World Health Organization's vigorous immunization program. Whereas in 1945 the majority of the world's inhabitants lived in areas in which smallpox was endemic, it is anticipated that in 1972 endemic foci will persist in as few as five countries. Immunization has eliminated smallpox from Western Europe, Russia, China, Japan, North America, and all of Central and South America except for small foci in Brazil. The vaccination of children has been largely responsible for the elimination of smallpox in Western Europe and North America (no documented case has been reported in the United States since 1949). But the majority of the adult population in these areas has regained its susceptibility; and the speed of modern transportation makes introduction of smallpox into a country difficult to detect. Hence constant vigilance and stringent vaccination require-ments are necessary to prevent spread from the few remaining areas where the disease remains endemic. For example, in spite of strict travel regulations smallpox was introduced into New York City from Mexico in 1947, causing great consternation but only a small outbreak of 12 cases and 2 deaths (over 6 million people were vaccinated in about 2 weeks).

The poxviruses are a group of agents that infect both man and lower animals and produce characteristic vesicular skin lesions, often called pocks.

GENERAL CHARACTERISTICS

Poxviruses (Table 54-1) are the largest of animal viruses: they can be seen with phase optics or in stained preparations with the light microscope. The viral particles (originally called elementary bodies) are somewhat rounded, brick-shaped, or ovoid (Figs. 54-1 and 54-2) and have a complex structure consisting of an internal central mass, the nucleoid, surrounded by two membrane layers (Fig. 54-3). The surface is covered with tubules or threads (Fig. 54-2). Poxviruses contain DNA, protein, and lipid. They are relatively resistant to inactivation by common disinfectants, heat, drying, and cold.

All poxviruses studied are related **immunologically** by a common internal antigen extractable from viral particles by 1 N sodium hydroxide. They can be divided into subgroups on the basis of their more specific antigens, their morphology, and their natural hosts:

Group 1 consists of **viruses of mammals:** variola, vaccinia, cowpox, ectromelia, rabbitpox, and monkeypox.

---

* A term initially employed to distinguish the disease from "large pox" or syphilis.

† The Comte de la Condamine, an eighteenth century French mathematician and scientist.

Fig. 54-1. Morphology of vaccinia virus as revealed by electron microscopic examination of shadowed preparation of purified viral particles. Note the central core surrounded by a depression (arrow). ×28,000. [From Sharp, D. G., Taylor, A. R., Hook, A. E., and Beard, J. W. *Proc Soc Exp Biol Med* *61*:259 (1946).]

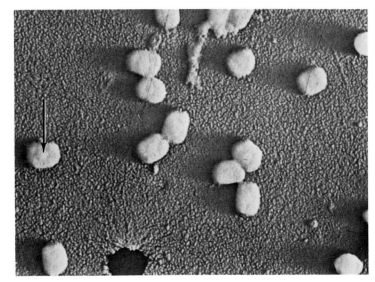

Fig. 54-2. Fine structure of an intact vaccinia virus particle from purified suspension negatively stained with phosphotungstate. The surface is covered with ridges, 60 to 80 A in diameter, which may be long threads of tubules. The virion measures 2500 × 3000A. ×222,500. [From Dales, S. *J Cell Biol* *18*:51 (1963).]

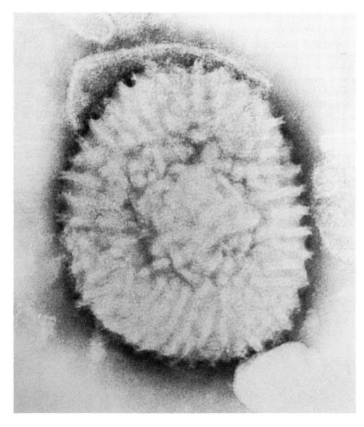

TABLE 54-1. Common Characteristics of Pox-viruses

1. 2500 to 3900 × 2000 to 2600 A
2. Brick-shaped to ovoid
3. Relatively resistant to chemical and physical inactivation
4. Inactivated by chloroform; inactivation by ether variable
5. Contain double-stranded DNA*
6. Virion contains a DNA-dependent RNA polymerase
7. Possesses common family antigen
8. Predilection for epidermal cells
9. Multiply in cytoplasm of cells
10. Produce eosinophilic inclusion bodies

* Vaccinia DNA: MW, 150–160 × 10⁶ daltons; base ratio AT/GC = 1.67. DNAs of cowpox, rabbitpox, and mousepox have similar characteristics. Fowlpox DNA: MW, 200 × 10⁶ daltons; base ratio AT/GC = 1.84.

Group 2 contains the **viruses of birds:** fowlpox, turkeypox, canarypox, and pigeonpox.

Group 3 contains the **tumor-producing viruses:** myxoma and fibroma viruses (Ch. 63).

A fourth group of viruses which resemble

Fig. 54-3. Thin section of an intact mature vaccinia virus particle showing the inner nucleic acid core and surrounding membranes. The elliptical body (EB) on either side of the nucleoid causes a prominent central bulging of the virion. Viral particle is lodged between two cells (arrows). ×120,000. [From Dales, S. *J Cell Biol 18*:51 (1963).]

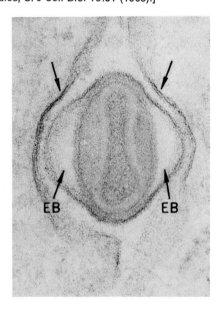

poxviruses, but are not immunologically related to them, will also be described. They are ovoid, somewhat smaller than typical poxviruses, and their surface ridges have a woven pattern. The group contains contagious pustular dermatitis (orf), paravaccinia (milker's nodules), and bovine papular dermatitis viruses.

Some poxviruses, such as those responsible for **molluscum contagiosum** and **Yaba monkey virus** have not been investigated sufficiently to permit immunological classification.

Poxviruses possess the unusual capacity to **reactivate** other members of the family inactivated by heat or 6 M urea* when the infectious and inactivated viruses infect the same cells. This phenomenon is accomplished not by genetic recombination but by a metabolic interaction, the infectious viral particles initiating the enzymatic process of "uncoating" the inactivated virus (Ch. 48). The intact DNA is thus liberated from the denatured viral coat, permitting initiation of viral multiplication. When reactivation is induced by members of the same immunological group **(homoreactivation),** genetic recombination between the two viruses may also occur. When reactivation is accomplished by a virus from a heterologous immunological group **(heteroreactivation),** recombination does not ensue. The extent to which genetic recombination occurs seems to be related to the degree of homology between the viral DNAs and parallels the grouping based on antigenic relations.

Poxviruses vary widely in the extent to which they cause generalized infection, but they exhibit several common pathogenetic characteristics. In the wide variety of animals that are their natural hosts, they have a predilection for infecting **epidermal cells,** in which they produce eosinophilic cytoplasmic inclusion bodies. (Fibroma and myxoma viruses also have a great affinity for subcutaneous connective tissues.) Most poxviruses also multiply readily in epidermal cells of the chorioallantois of chick embryos, where they produce characterisitc **nodular focal lesions,** termed pocks (Fig. 54-4). These lesions re-

* In the reactivable virus the protein capsid has been denatured by such treatment, but the viral DNA is unaffected.

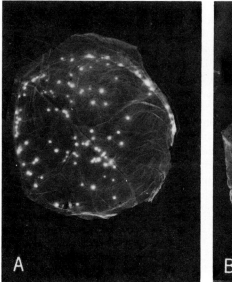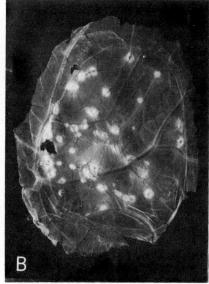

Fig. 54-4. Appearance of pocks produced by variola virus **(A)** and vaccinia virus **(B)** 3 days after viral inoculation of the chorioallantoic membranes of chick embryos. Note the small gray-white pocks produced by variola virus and the large pocks made by vaccinia virus. [From Downie, A. W., and MacDonald, A. *Br Med Bull 9*:191 (1953).]

flect the propensity of poxviruses to cause cellular hyperplasia before cell necrosis. With myxoma, fibroma, and Yaba viruses the hyperplastic aspect of the lesion predominates, and tumors develop.

VARIOLA (SMALLPOX) AND VACCINIA

PROPERTIES OF THE VIRUSES

Smallpox virus is so virulent and contagious that laboratory investigations with it have been understandably limited. The closely related vaccinia virus, on the other hand, is much less dangerous to handle and is one of the most thoroughly investigated of all viruses that infect man. Where comparative data are available the characteristics of the two viruses are similar; accordingly, they will be discussed together.

MORPHOLOGY

Poxviruses have a complex architecture without obvious symmetry. Vaccinia virions are brick-shaped particles with rounded corners; a central dense region with crescentic dense areas on each side is observed (Fig. 54-1).

The viral surface, revealed by negative staining, is composed of unusual thread-like or tubular structures that appear to curve around the particle (Fig. 54-2). Thin sections (Fig. 54-3) disclose a central nucleoid, having a dense core which is usually dumbbell-shaped. The nucleoid is surrounded by two double lipoprotein membranous coats. Between the nucleoid and the viral coat is an ellipsoidal body which causes the central thickening of the virion. The core is composed of thread-like structures which are sensitive to DNase. Viral particles from smallpox crusts and vesicle fluids are morphologically indistinguishable from vaccinia particles.

CHEMICAL AND PHYSICAL CHARACTERISTICS

Vaccinia virus was the first animal virus to be prepared in sufficient quantity and purity to justify detailed chemical and physical measurements. The viral particles are composed of about 3.2% DNA (Table 54-1), 91.8% protein, 5% lipid (including cholesterol, phospholipid, and neutral fat), and 0.2% non-DNA carbohydrate (present in the membrane glycoproteins).*

Stability. Despite the presence of a membrane variola and vaccinia viruses are relatively stable. They are resistant to drying and retain infectivity at moderate temperatures; thus, the exudate and crusts from smallpox patients may yield infectious virus after almost a year at room temperature, and refrigeration is unnecessary for sending specimens to a diagnostic laboratory. Infectivity of viral suspensions is maintained for many months at $4°$, and for years at -20 to $-70°$. Because infectious variola virus can persist on bedclothes, a hazard exists not only for medical personnel but even for laundry workers. The relative resistance of the virus to dilute phenol and other common disinfectants complicates the decontamination of patients' clothing and the sterilization of instruments, furniture, etc. However, variola and vaccinia viruses are inactivated by apolar lipophilic solvents (e.g., chloroform) and by heating at $60°$ for 10 minutes or autoclaving.

ANTIGENIC STRUCTURE AND IMMUNITY

The antigenic structure of poxviruses has been examined largely with vaccinia virus, but smallpox virus is antigenically very similar. In fact, standard vaccinia virus is more closely related antigenically to smallpox than to cowpox virus, from which it was supposedly derived. Viral strains producing severe smallpox (variola major) are indistinguishable immunologically from strains isolated from cases of variola minor (alastrim).

The antigens of poxviruses can be measured by all the usual immunological technics, including hemagglutination inhibition (Ch.

50) and neutralization of infectivity. Viral neutralization is determined by inhibition of cytopathic effects on cells in tissue culture, by prevention of lesions in the rabbit skin, or, most quantitatively, by reduction in number of either pocks on the chorioallantoic membrane of the chick embryo or plaques on a monolayer of cultured cells.

Rivers and his colleagues in the 1930s began dissecting the complex antigenic structure of vaccinia viruses, as had been done earlier with bacteria (particularly the pneumococcus).†

Their pioneering investigations revealed two multicomponent groups of major antigens, consisting of the DNA-containing core (termed the NP or nucleoprotein antigen) and coat material (called the LS antigen because it contained heat labile and stable components). The complex virion contains a minimum of 7 distinct major antigens, revealed by immunodiffusion technics, and 17 identifiable polypeptide chains. Chemical dissection of the virion with trypsin, 2-mercaptoethanol, and a nonionic detergent (e.g., Nonidet P40) permits localization of some of the antigens: 1) two antigens responsible for neutralizing antibodies are present in the surface membrane; 2) one large and three small antigens are present in the core; and 3) one large antigen has been located between the core and the surface structures. The neutralizing antibodies, induced by the two surface antigens, are not family-reactive and only neutralize viruses from the homologous subgroup. A **family antigen,** which is probably a component of the core, can be extracted from virions with weak alkali; it cross-reacts (in complement-fixation and precipitin assays) with a similar antigen from viruses of each of the subgroups as well as from the unclassified poxviruses.

---

* About 0.8% minor constituents including copper (0.06%), biotin, and riboflavin have been detected in highly purified particles, but no metabolic role has been demonstrated, and it is not known whether they are constituents of the particles or host contaminants.

---

† The studies of this group were remarkable for the time. The investigators recognized the viral particle as the infectious unit; they developed precise assay technics; they applied statistical technics to their data; they prepared large quantities of virus in a highly purified state and determined its chemical composition; and they purified antigens from infected cells and studied their immunological, chemical, and physical characteristics. The data they obtained have withstood the test of time and the subsequent development of more elegant technics.

Hemagglutinin. Extracts of variola- and vaccinia-infected cells, but not purified viral particles, hemagglutinate erythrocytes from about 50% of chickens (apparently a genetic trait of the bird). The hemagglutinin is a lipid-containing soluble antigen that is inactivated by lecithinase. Although it is synthesized as a result of the viral infection, it has not been shown to be a structural unit of the viral particle, and Abs that inhibit hemagglutination do not neutralize infectious virus. In contrast to influenza viruses (Ch. 56, Hemagglutination), vaccinia hemagglutinin does not elute spontaneously from the aggregated cells. The hemagglutination-inhibition titration may be used as a convenient and accurate procedure for diagnosis of infection. A new cell surface antigen, which is detected by immunofluorescence shortly after vaccinia infection, is probably responsible for hemadsorption of susceptible chicken erythrocytes to infected cells; the hemadsorption is blocked by virus-specific Abs.

Following infection or immunization Abs develop to each of the viral antigens. The variations observed in their time of appearance and persistence depend on the nature and quantity of the antigen, and not on the titration technic employed. Thus, neutralizing and hemagglutination-inhibiting Abs are first detected about the sixth day after onset of illness in the unvaccinated person, and by the third or fourth day in a previously immunized individual. In contrast, complement-fixing Abs ordinarily appear 2 to 3 days later, and persist less than 2 years. The neutralizing Abs persist for at least 20 years after infection, and hemagglutination-inhibition Abs may also remain many years. In general, Ab titers tend to be higher and persist at high levels longer in those immunized prior to illness. Immunity to smallpox is long-lasting, if not persistent for life, following natural infection. In the rare reinfections that have been reported the disease is usually atypical, very mild, and often without skin rash **(variola sine eruptione)**. If infection occurs in artificially immunized individuals (see below), the clinical disease is also usually milder than in unimmunized neighbors.

The relative importance of **humoral and cellular immunity** remains largely unexplored in man except for cases following immunization with live vaccine in victims of congenital defects of thymus-derived cells (Ch. 51 and see below). Experimentally, however, cell-mediated immunity appears to be critical for successful recovery from infection. For example, 80% of mice who received antithymocyte serum succumbed to infection from a subsequent intravenous inoculation of mousepox virus, whereas only 5% of the control animals died; humoral antibodies and interferon levels were comparable in both groups. Transfer of spleen cells from immunized mice markedly reduced the viral titer although only low levels of specific humoral antibodies were attained and interferon was not detectable. The spleen cell immunity was virus-specific and appeared to be initiated by thymus-derived lymphocytes since the antiviral activity was inhibited if prior to transfer the cells were reacted with antithymocyte or anti-theta serum, and the cells were radiosensitive. Moreover, macrophages from vaccinia-immunized animals do not support multiplication of vaccinia virus although unrelated poxviruses can replicate without restraint.

HOST RANGE

Variola virus has a much more limited host range than does vaccinia or other poxviruses. Monkeys are the only animals, other than man, known to be infected by variola virus under natural conditions; when virus is placed onto scarified skin or inoculated intradermally, local lesions and fever follow.

In rabbits, variola virus can initiate keratitis and local skin lesions; but it cannot be propagated serially. Some strains may be passed intracerebrally in adult mice, with large inocula, but the resulting infections are not consistently lethal. Suckling mice, however, uniformly succumb. On the chorioallantoic membrane of the chick embryo variola virus produces characteristic small, gray-white pocks that can be readily distinguished from the large pocks produced by vaccinia virus (Fig. 54-4). The development of pocks affords a reliable quantitative viral assay. Variola minor (alastrim) viruses have the same host range as those of variola major, but they generally multiply to a lower titer in the chorioallantois of the chick embryo.

Many animals, including the rabbit, calf, and sheep, are susceptible to **vaccinia virus** and have served as convenient hosts for investigation and for production of vaccine. Epidermal cells show the greatest sensitivity, but vaccinia also multiplies readily in the rabbit testis, and a number of strains have been adapted to the rabbit brain (neurovaccinia). In the **chick embryo** the membranes lining the amniotic cavity and the yolk sac support multiplication of vaccinia virus, but about 100 times more virus is necessary to infect them than the chorioallantois. Continued passage of vaccinia on chorioallantoic membranes tends to reduce its virulence for man and rabbits, a property that has been used experimentally for the development of a vaccine.

Cultured cells of many different types, and from a number of animal species, are susceptible to infection by poxviruses. Cells commonly used for propagation of variola virus are human embryonic kidney, monkey kidney, and continuous lines of human epithelium-like cells such as HeLa cells. The slowly developing cytopathic effects become maximal in 5 to 8 days (depending on the size of the viral inoculum). The infected cells contain weakly eosinophilic cytoplasmic inclusion bodies. Infection may be detected even earlier by hemadsorption with susceptible chicken erythrocytes. Variola virus can be assayed in monolayer cultures by counting hyperplastic foci.

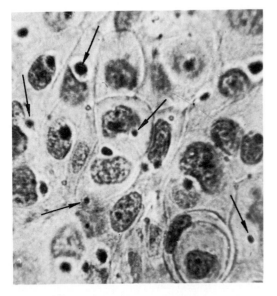

Fig. 54-5. Eosinophilic cytoplasmic inclusion bodies (Guarnieri bodies) in corneal epithelial cells of rabbit eye infected with vaccinia virus. ×900. [From Coriell, L. L., Blank, H., and Scott, T. F. M. *J Invest Dermatol 11*:313 (1948).]

Vaccinia virus propagates more rapidly, to a higher titer, and in a wider variety of animal cells than does variola virus. Cytopathic changes appear 18 to 24 hours after infection, and they are accompanied by earlier development of eosinophilic inclusion bodies similar to the inclusions (Guarnieri bodies) observed in natural and experimental infections (Fig. 54-5). Hemadsorption permits early detection of infection. Virus can be reliably quantitated by plaque assay, with or without an agar overlay,* and by pock counts on the chorioallantoic membrane.

VIRAL MULTIPLICATION

Although poxviruses contain DNA, biosynthesis of the viral components and their assembly into viral particles take place entirely within the cytoplasm of the cell. Virus attaches to the host cell through receptors, whose chemical nature is still undefined, and probably enters the cytoplasm by a process akin to phagocytosis (Ch. 48). After penetration viral DNA is released by a two-stage uncoating process (Ch. 48). Almost immediately after penetration the **first stage** is initiated by preexisting host cell enzymes (i.e., not induced by infection); it results in breakdown of viral phospholipid and part of the protein coat of the viral particle, freeing the nucleoprotein core. The **second stage** commences after a lag of 30 to 60 minutes (dependent upon viral multiplicity) and results in breakdown of the nucleoprotein core to liberate viral DNA. During this lag the DNA within the intact core is protected from DNase activity, but nevertheless a DNA-dependent RNA polymerase present in the core transcribes about 10% of the viral genome after the first stage of uncoating is completed. The resulting mRNA emerges from the core to code for the protein, possibly a proteolytic enzyme, implicated in the final uncoating events. The liberated viral DNA is stable and available to transmit its genetic information for several hours. Cytoplasmic, virus-specific mRNA synthesis is first detected 1 to 2 hours after infection and continues until about 7 hours postinfection.

---

* Because infected cells are not lysed rapidly and most of the newly synthesized virus is not released into the medium, an agar overlay is not necessary to prevent spread of virus and thus permit formation of localized foci.

Synthesis of specific enzymes and a few viral structural proteins begins early in the biosynthetic process, before replication of viral DNA. The products include the second-stage uncoating protein, three proteins associated with the nucleoprotein core, a protein essential for initiation of viral DNA replication, and enzymes related to DNA biosynthesis (thymidine kinase and DNA polymerase) and degradation. Production of these early proteins is blocked 3 to 4 hours after infection by translational control, before the mRNAs have been degraded. Viral DNA begins to be synthesized 2 to 3 hours after infection, and attains its maximum concentration by the time newly made infectious virus is first detected (Fig. 54-6). Late viral proteins are first detected about 4 hours after infection, infectious virus appears about 1 hour later, and both then increase in parallel (Fig. 54-6).

Concomitant with the biosynthesis of cytoplasmic virus-directed mRNA and viral DNA the following effects on host cell macromolecules are noted: production of host proteins stops because transport of host mRNA molecules from the nucleus to the cytoplasm is blocked, and host cell polyribosomes are disrupted (liberating ribosomes for synthesis of viral proteins). Moreover, replication of host DNA ceases, but it is not degraded. Although the host mRNAs and ribosomal RNAs do not leave the nucleus, their synthesis continues unaltered for about 3 hours, after which the rate of host transcription decreases and virtually stops by 7 hours after infection. How these controls of host cell biosynthesis are induced by vaccinia virus infection, and how they relate to cell injury, are still unknown.

The **morphological counterparts** of the foregoing biochemical events have been observed in thin sections of infected cells (Fig. 54-7). As the viral DNA synthesis increases regions of dense fibrous material appear in the cell's cytoplasm. About 3 hours after infection some of the early proteins form trilaminar membrane-like structures, which begin to enclose patches of viral DNA and proceed to form immature particles (Fig. 54-7, A and B). The nucleoid then begins to take shape within the immature particle; an additional membrane encloses the condensing DNA; and finally, the outer coat structures are laid down

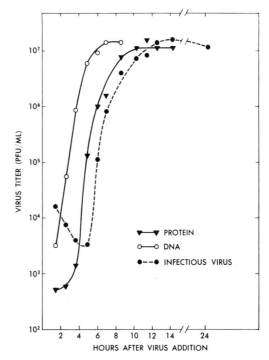

Fig. 54-6. Biosynthetic events in multiplication of vaccinia virus. [Redrawn from Salzman, N. P., Shatkin, A. A., and Sebring, E. D. *Virology* 19:542 (1963).]

on the previously formed membrane completing the assembly of mature virions. These sequential events are summarized in Figure 54-8.

Lysis of infected cells is not a prerequisite for liberation of newly formed virions. The viral particles seem to be released through cell villi by a process resembling phagocytosis in reverse. This is apparently an inefficient process, at least in tissue culture, since less than 10% of the virus formed is released from the infected cells. Radioautographic and immunofluorescent technics reveal that viral materials may also be transmitted directly from cell to cell through villi.

SMALLPOX: PATHOGENESIS

Two basic forms of smallpox are recognized: **variola major,** which has a case fatality rate of approximately 25%, and **variola minor** or **alastrim,** a less virulent form with a mortality rate below 1%. Although a variety of factors

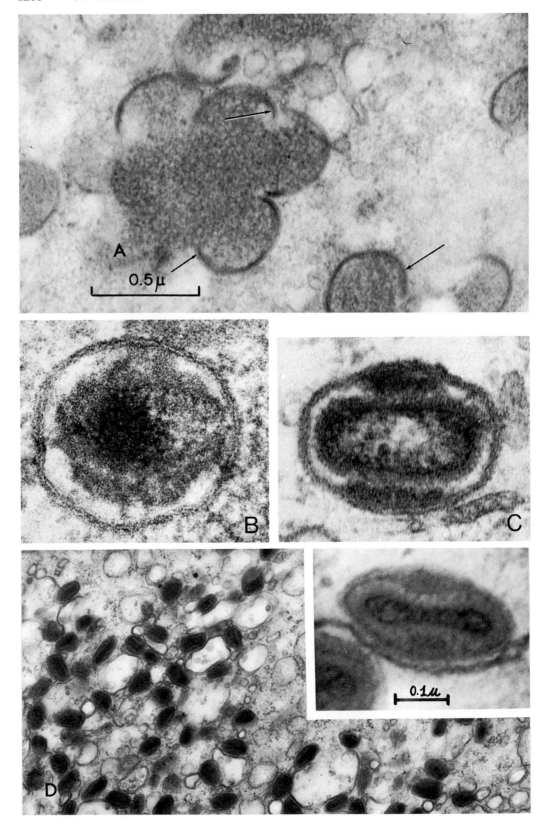

may influence the mortality rate in any epidemic, the evidence is convincing that severe and mild smallpox exist as distinct entities. Nevertheless, it is impossible to distinguish biologically or immunologically the viruses responsible for these two forms of the disease.

Knowledge of the pathogenesis of smallpox is derived from clinical and pathological studies in patients and from detailed laboratory investigations of ectromelia in mice (mousepox; Ch. 51, Patterns of disease) and vaccinia in rabbits. Infection is initiated by the entrance of virus into the upper respiratory tract, where it multiplies first in the mucosa and then in the regional lymph nodes. A transient viremia then disseminates virus to internal organs (e.g., liver, spleen, and lungs). Propagation ensues, yielding a large viral population. A second invasion of the blood stream by virus terminates the incubation period and initiates the toxemic phase of the disease, characterized by prodromal macular rashes and constitutional symptoms: fever, generalized aches and pains, headache, malaise, and prostration. Virus spreads to the skin, where it multiplies in the epidermal cells, and the characteristic skin eruption follows in 3 to 4 days. The rash at onset is macular and progresses through papular and vesicular stages, finally forming pustules in the second week of illness. In severe cases the rash may become hemorrhagic or confluent. The sequence of events in variola minor is similar to that of classic smallpox, but the skin lesions are more superficial, all symptoms are less severe, and the course of the disease is shorter.

The specific lesions begin with a proliferation of the epithelial layer of the skin, followed by its focal degeneration and the formation of multiloculated vesicles. These fill with degenerated epithelial cells and invading polymorphonuclear leukocytes, eventually developing into pustules. Similar lesions occur in the mucous membranes of the mouth, pharynx, larynx, and even the trachea and esophagus; because these lesions are not covered by a horny layer, as in the skin, infectious virus is shed earlier. In the repair stage crusts form, owing to drying of the pustules. In addition to the visible lesions described secondary degenerative lesions develop, similar to those in many other severe febrile infections (e.g., fatty degeneration and focal necrosis of the liver, kidney, adrenals, etc.).

Eosinophilic cytoplasmic inclusion bodies surrounded by a clear halo, termed **Guarnieri bodies,** characteristically develop in cells of the skin and mucous membranes infected with variola and vaccinia viruses (Fig. 54-5). Electron microscopy or staining with flourescein-labeled antibodies reveals that each inclusion body consists of an accumulation of viral particles and viral antigens. Similar masses are observed in cells infected with other poxviruses. More than one may be present in a single infected cell.

Viral multiplication and consequent death and necrosis of invaded cells may not be solely responsible for the lesions of the eruptive phase of smallpox. A hypersensitivity response to the viral antigens may also contribute, for when the antibody response and hypersensitivity are inhibited in infected rabbits by X-irradiation or cytotoxic drugs the characteristic pustules do not form, although viral multiplication is unrestricted. The toxinlike properties of the viral particles may also play a role in the production of cell necrosis. It is clear, however, that pustule formation does **not** result from secondary bacterial in-

Fig. 54-7. Development stages in maturation and assembly of a poxvirus (vaccinia). Electron micrographs of thin sections of L cells taken at increasing intervals following infection. **A.** Dense viral membranes are shown forming within and around clumps of dense fibrillar material in the cytoplasm. Arrows indicate sites where the elaboration of the dense limiting membranes occurred within the aggregations of fibrillar material. ×61,000. **B.** Maturation of a virion within the viroplasmic matrix. The complete trilaminar lipoprotein envelope surrounds the particle. Within is a very dense nucleoid of fibrous material (DNA) surrounded by a less dense homogeneous substance, thought to be material of the lateral bodies and core. ×170,000. **C.** A maturing virion lying in the cytoplasmic martrix of a cell. The core and two lateral bodies are clearly differentiated within the viral envelope. ×170,000. **D.** A large group of mature vaccinia virus particles, smaller and denser than in their formative stages. ×24,000. **Inset** shows internal structure of a single particle at higher magnification. The dense dumbbell-shaped core is surrounded by a zone of lower density. ×150,000. [**A** and **D** from Dales, S., and Siminovich, L. *J Biophys Biochem Cytol* 10:475 (1961). **B** and **C** from Dales, S., and Mosbach, E. H. *Virology* 35:564 (1968).]

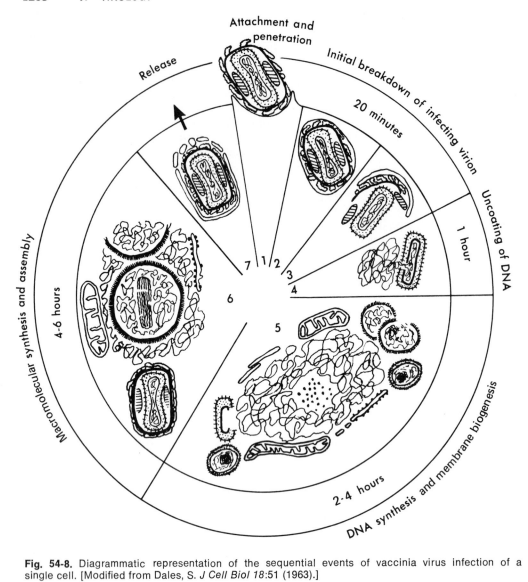

Fig. 54-8. Diagrammatic representation of the sequential events of vaccinia virus infection of a single cell. [Modified from Dales, S. *J Cell Biol 18*:51 (1963).]

fection, since bacterial cultures of the pustule fluid are ordinarily sterile.

LABORATORY DIAGNOSIS

During epidemics, or in countries where small-pox is endemic, most cases can be diagnosed on the basis of the clinical and epidemiological evidence. However, when smallpox appears in countries that are usually free of the disease (where physicians are unfamiliar with the clinical manifestations), or when an atyp-

ical illness occurs in partially immune persons, laboratory procedures are frequently needed to establish the diagnosis. The objectives of laboratory procedures employed for the diagnosis of smallpox and generalized vaccinia are to 1) either identify or isolate the virus, 2) detect viral antigen, or 3) measure a rise in titer of specific Abs. The recommended procedures and the specimens to be collected from the patient depend upon the stage of the suspected disease (Table 54-2). The stability of the infectious virus and its antigens facilitates laboratory diagnosis because materials can be transported directly or mailed without danger of inactivation.

Viral Particles. An early presumptive diagnosis can be made most rapidly by visualizing viral particles (elementary bodies) by electron microscopy or in a Giemsa-stained smear of scrapings from the bases of skin lesions. After crusts form, however, direct observation of viral particles is difficult. A positive smear is decisive, but a negative result requires further tests. The identity of viral particles observed can be confirmed by viral isolation or by immunological technics.

Virus can be isolated most efficiently by inoculating blood or material from skin lesions onto the chorioallantoic membrane of the 12- to 14-day-old chick embryo. With viruses from cases of either variola major or minor, small gray-white pocks appear within 2 to 3 days. These can be distinguished from the large gray-white pocks produced by vaccinia virus (Fig. 54-4). The agent producing the pocks can be specifically identified by 1) observing Guarnieri bodies in histological sections, 2) determining the presence of elementary bodies in impression smears, and 3) identifying viral antigens with a standard antiserum.

Viral Antigens. The detection of soluble viral antigens is probably the most useful, economical, and efficient of the diagnostic procedures. Blood, vesicle fluid, pustule fluid, and saline extracts of skin scrapings or crusts all contain the antigens during the appropriate stage of the disease (Table 54-2). Antigens are most commonly identified by complement-fixation, immunofluorescence, and agar-gel diffusion technics. Care must be taken to use sufficient material: fluid or extracts of crusts from at least six lesions is recommended.

Specific Antibodies. Serological diagnosis is generally reserved for atypical or mild cases occurring in partially immune individuals (e.g., cases of **variola sine eruptione**). A rise in Ab titer can be measured by complement-fixation or hemagglutination-inhibition titrations: a fourfold or greater increase is considered diagnostic. Antibodies measured by hemagglutination inhibition begin to increase about the fifth or sixth day after the onset of illness, and those detected by complement fixation rise about the eighth to tenth day. An Ab increase may be noted several days earlier in persons who have been previously vaccinated. Since Abs measured by

TABLE 54-2. Diagnosis of Variola by Laboratory Tests

Stage of variola	Material tested	Detection of virus			Detection of antibodies‡
		Microscopic examination	Viral isolation*	Antigen detection†	
Preeruptive	Blood		±	±	—
Maculopapular	Blood		±	±	±
	Skin lesions	+	+	+	
	Saliva		+		
Vesicular	Blood		±		+
	Skin lesions	+	+	+	
Pustular	Blood				+
	Pustular fluid	±	+	+	
Crusting	Blood				+
	Crusts	—	+	+	
Time required for completion of test		1 hr	1–3 days	3–24 hr	3 hr to 3 days

* Culture on chorioallantoic membrane of chick embryos or in tissue culture.

† Complement fixation, agar-gel precipitation, or immunofluorescence.

‡ Hemagglutination inhibition, complement fixation, or neutralization; Abs may appear earlier in patients previously vaccinated. Positive indicates a rise in Ab titer.

+, Test usually positive; ±, test may or may not be positive; −, test usually negative. Open spaces likewise indicate test is negative.

Modified from DOWNIE, A. W., and MAC DONALD, A. *Br Med Bull 9*:191 (1953).

neutralization and by hemagglutination-inhibition titrations may persist for years following vaccination, their presence in a single serum specimen, from a previously immunized person, is difficult to interpret. A positive complement-fixing reaction, however, rarely persists longer than 6 to 8 months after vaccination.

EPIDEMIOLOGY

Smallpox is a disease confined to man and spread chiefly by person-to-person contact. Virus is abundant in lesions during the active disease, but it is rarely detected during the incubation period and prodromal stage before the rash appears. Initially the virus is transmitted from the lesions of the upper respiratory tract in droplet secretions or by contamination of drinking or eating utensils; later, when the vesicles or pustules rupture and crusts are formed, the skin lesions also become a source of contagion. Fomites such as bed linens, clothing, utensils, dust, and books are important means of viral dissemination because the virus is resistant to ordinary temperatures and drying. Under unusual circumstances airborne transmission of variola virus can also occur, as has been demonstrated epidemiologically and experimentally. For example, in Meschede, West Germany, smallpox spread from an isolated hospitalized patient to 14 others who were remote from the patient and without a common contact (10 patients were even on other floors but in the same wing as the index case).

Although any person infected with variola virus is potentially contagious, the most dangerous disseminators are persons with unrecognized disease, e.g., the partially immune patient who has relatively few lesions. Such cases, which are easily overlooked or misdiagnosed, have been primarily responsible for introducing smallpox into countries usually free of the disease, such as England and the United States.

PREVENTION AND CONTROL

From the systematic beginnings by Jenner in England at the close of the eighteenth century, artificial immunization has become increasingly effective.* Methods for preparing vaccines have improved; criteria for establishing its potency have been standardized; technics for maintaining the potency have been developed; and recommendations for its use have been established and generally accepted. As a result smallpox is rare in most of the world today; since there are no animal reservoirs, extensive immunization programs could eradicate these remaining foci and eliminate the virus as a threat to man.

PREPARATION OF VACCINE

Cowpox virus was originally utilized for immunization against smallpox. Since its initial use, however, it has been propagated in many different laboratories under diverse conditions, and is now believed, by virtue of its antigenic structure, to have been inadvertently replaced with an attenuated smallpox virus. The vaccinia virus used today is distinctly different from the cowpox virus encountered in nature. There are now a number of different strains of varying virulence (e.g., dermovaccinia and neurovaccinia); the one accepted as the prototype for immunization is a dermal strain of uncertain origin (it infects the skin at the site of inoculation, but under ordinary circumstances does not produce viremia).

Successful immunization requires the use of infectious (attenuated) virus, because of

---

* **Variolation** to protect against smallpox was practiced long before infectious agents and concepts of immunization were clearly recognized. The Chinese powdered old crusts and applied them to the nostrils; Brahmins in India preserved crusts and inoculated them into the skin of the arm or forehead of the unscarred; Persians ingested crusts from patients; and in Turkey fluid from pocks was used for inoculation. It was this latter practice that Lady Mary Wortley Montague (wife of the British ambassador to Turkey) introduced into England in 1718. Crusts and vesicle fluids were selected from patients during epidemics of mild disease, alastrim. The practice spread to the colonies, where it was more widely used than in the British Isles, but never became popular because of the risks involved. Jenner introduced the use of attenuated (cowpox) virus in 1796, prompted by the clinical observation that milkmaids who acquired cowpox usually escaped smallpox, even when the disease was rampant in the community.

the marked lability of the protective antigen. The vaccine most commonly employed is prepared by collecting scrapings of vaccinial lesions on the skin of calves or sheep, treating with 1% phenol to kill contaminating bacteria, and adding 40% glycerol to increase stability of the virus. To maintain infectivity the vaccine must be kept frozen by the manufacturer and refrigerated below 10° after distribution. Since gradual inactivation of virus occurs vaccine should not be kept for more than 3 months.* This requirement presents a serious problem in the tropics or in depressed areas, where refrigeration during transportation and storage is difficult.

Lyophilized vaccines have been prepared and used successfully in some countries, but the success of the standard vaccine has engendered a most conservative attitude toward any change. Effective vaccines can also be made with virus propagated on the chorioallantoic membrane of the chick embryo or in cell culture, but they have not yet been generally accepted. There are suggestions that purified soluble antigens might also be effective.

ADMINISTRATION

Vaccine is administered intradermally, most commonly by gently breaking the epidermis under a drop of vaccine by multiple pressure with the side of a sterile needle. However, inoculation by air jet has been particularly effective for immunization of large numbers; the scratch method and inoculation by needle are also used. Puncture or scarification permits infectious virus to enter the skin, where it multiplies in the deeper layers of the epidermis. The extent of multiplication and spread of virus, and thus the type of reaction that ensues, depend on the state of immunity (and hypersensitivity) of the host. One of three responses is seen (Fig. 54-9): **primary, accelerated** (sometimes referred to as vaccinoid), and **early** or **immediate** (sometimes erroneously termed **immune** reaction).

The **primary response** occurs in an individual who has no effective immunity. A small papule appears on the fourth day after vac-

cination and rapidly progresses to a vesicle and then to a pustule. The reaction reaches its maximum on the eighth to tenth day and is often accompanied by axillary adenitis and fever. Viral spread into the blood cannot be detected under ordinary circumstances, although it clearly does during certain complications (see below). The pustule eventually becomes crusted and may remain 3 to 4 weeks before it separates, leaving a scar.

An **accelerated response** progresses through the same stages as a primary reaction, but does so more rapidly and less intensely. The maximum reaction is reached between the third and seventh day after inoculation. The decreased intensity of the reaction implies that the vaccinated person has some, but not maximum, immunity. The rate of development and the extent of the reaction reflect the balance between the degree of immunity and the extent of hypersensitivity. Increased immunity develops as a consequence of the viral multiplication leading to the reaction.

The **early or immediate response** is first detected as a papule in 8 to 24 hours, attains its maximum by 2 to 3 days, and usually does not progress beyond the vesicle stage, suggesting a high level of immunity. In many cases, however, the early response may blend with the development of an accelerated reaction, indicating inadequate immunity. The early response is **primarily an allergic reaction** of the delayed hypersensitivity type (Ch. 20). It does not require viral multiplication, but is due to the host reaction to viral protein; and, in contrast to the primary and accelerated reactions, the early response can be elicited by noninfectious virus (inactivated by heat at 56° for 30 minutes) as well as by live virus. Hence, **although an immune individual will respond to vaccination with an early response, this reaction does not necessarily indicate immunity;** it only implies previous infection by vaccinia virus. Since hypersensitivity may persist longer than effective immunity the early reaction cannot be interpreted **unless there is evidence that the vaccine contained adequate infectious virus.** If, for example, the vaccine contained only inactivated virus, the early reaction will not provide information on the state of immunity of the host, nor will the procedure induce immunity. One can be certain of the effectiveness of the vaccine, however, if others receiving

---

* After 3 or even 6 months' storage in the refrigerator infectious virus is not totally inactivated, but the standard dose might not contain the very large quantity of infectious, attenuated virus required to infect the vaccinees.

Immediate Response

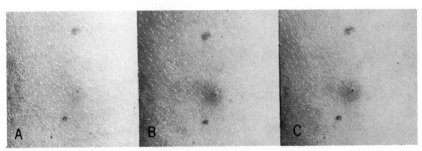

Accelerated Response

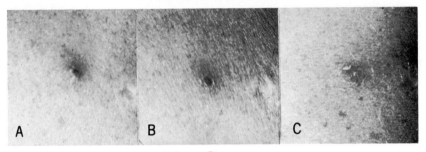

Primary Response

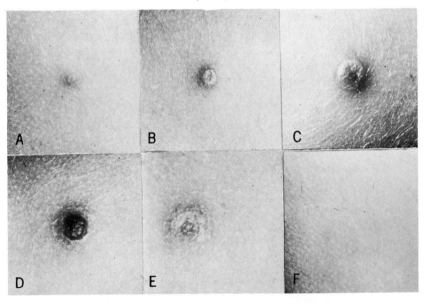

Fig. 54-9. Characteristic cutaneous responses to smallpox vaccine. The vaccine was administered by jet injection. **Immediate response:** A, day 1; B, day 3; C, day 5. **Accelerated response:** A, day 4; B, day 6; C, day 11. **Primary response:** A, day 3, papule; B, day 6, vesicle; C, day 9, pustule with beginning crust formation; D, day 11, crust; E, day 17, crust shed leaving area of desquamation and discoloration; F, day 42, remaining scar. [From Elisberg, B. L., McCown, J. M., and Smadel, J. E. *J Immunol 77*:340 (1956).]

the same material have primary or accelerated reactions. Ignorance of these considerations has led to many erroneous interpretations and tragic failures to immunize persons entering countries in which smallpox is endemic.

Failure to elicit a dermal response is sometimes seen, but **it is never the result of complete immunity;** it simply indicates that the vaccination technic was faulty or the vaccine inadequate. To assure immunity, the individual who has no reaction must be revaccinated until a reaction occurs.

Following the initial vaccination specific Abs, as measured by neutralization or hemagglutination-inhibition titrations, appear in the majority of persons on about the tenth day. Complement-fixing Abs are detected later, and in only about 50% of those immunized successfully. In revaccinated individuals Abs appear earlier and reach higher titers.

Protective immunity develops 7 to 10 days after vaccination, which is rapid enough to be effective in contacts of smallpox cases if administered shortly after exposure (incubation period about 12 days). It has generally been stated that protection lasts for 3 to 7 years and that reimmunization every 3 years will afford uninterrupted immunity. This generalization is not strictly true, however, for cases of mild smallpox have occurred only 1 year after known successful vaccination; and though persistence of neutralizing Abs after the disease is correlated with lasting immunity, smallpox has inexplicably occurred in artificially immunized persons who also maintain neutralizing Abs indefinitely. These facts do not refute the great value of vaccination, but merely emphasize the need for immunization whenever exposure is suspected. Moreover,

persons exposed to high risks of infection* should be revaccinated at least once a year; for while smallpox is usually mild in recently immunized persons, public health considerations make fuller immunity desirable.

COMPLICATIONS

Vaccination is a relatively safe procedure; but complications affecting the skin or central nervous system may occur, especially with initial vaccinations (Table 54-3); occasionally these reactions are fatal. Serious complications may result from widespread secondary implantation of virus on skin diseases such as eczema **(eczema vaccinatum)** or even diaper rash. Viremia frequently occurs in vaccinees who have agammaglobulinemia or dysgammaglobulinemia, and results in progressive and often fatal disease **(generalized vaccinia).** A more frequent and less serious form of generalized vaccinia appears to be related to hypersensitivity (an allergic reaction) since viremia is apparently rare, virus is usually absent from the skin lesions, and vesicles do not form **(erythematous urticarial reactions).** Progressive spread of a primary vaccination response with extensive necrosis of skin and muscle **(vaccinia gangrenosa)** is probably the most alarming complication, and occurs in those rare persons with thymic dysplasias who cannot develop cellular immunity (about 1.5 cases per million primary vaccinees). It is therefore essential that physicians and public health officers be aware of these complica-

---

* Hospital personnel where cases are likely to be admitted or individuals traveling into areas where the disease is endemic.

TABLE 54-3. Incidence of Complications Associated with Smallpox Vaccination in the United States

Complication	Complications per 10^6 primary vaccinations (age at vaccination)					Complications per 10^6 revaccinations, all ages
	<1	1–4	5–19	20+	All ages	
Deaths (from all complications)	5	0.5	0.5	Unknown	1.0	0.1
Postvaccinal encephalitis	6	2	2.5	4	2.9	0.0
Vaccinia gangrenosa	1	0.5	1	7	0.9	0.7
Eczema vaccinatum	14	44	35	30	38	3
Generalized vaccinia	394	233	140	212	242	9
Accidental vaccinia infection	507	577	371	606	529	42

Modified from *Center for Disease Control Morbidity and Mortality Weekly Report 20*:340 (1971).

tions and not attempt to vaccinate persons with skin eruptions, or defects in immunity. Those who have siblings with similar afflictions also should not be vaccinated unless they can be conveniently separated from the family.

Postvaccinal encephalitis, a serious and often fatal form of demyelinating disease, has an incidence of about 2.9 per million primary vaccinations (Table 54-3). Since it seldom occurs in children less than 6 months old and almost never develops after revaccination, if routine immunization is practiced, children should be vaccinated between 3 and 6 months of age whenever possible.

A survey of 7500 women vaccinated during pregnancy revealed no increase in the rate of congenital abnormalities among the offspring. But in 9 cases intrauterine vaccinia infection led to early delivery of a dead or moribund fetus; 6 of the 9 occurred in women vaccinated for the first time. Pregnancy, regardless of trimester, should therefore be considered a contraindication to vaccination, unless there are compelling reasons for immunization.

As the incidence of smallpox plummets in many countries, the risk of a dangerous complication of vaccination increasingly outweighs the advantages of universal immunization. For example, there has not been a single importation of smallpox into the United States since 1949, but each year there is an average of seven deaths resulting from vaccination complications. These alarming statistics have added strength to the critics of compulsory immunization of children in countries that have long been free of endemic smallpox. The U.S. Public Health Service has recognized that the remarkably small risk of smallpox no longer warrants the dangers attendant on universal smallpox vaccination and recommends that routine immunization of children cease. In place of the previous practice a selective immunization of people at special risk is recommended: i.e., all travelers to and from countries in which smallpox is still endemic, and all health services personnel, since they are at greatest risk of contact with the rare cases of imported smallpox.

This recommended radical departure from past practices has not been enthusiastically accepted by all experts. It has been argued that abandonment of childhood immunization will result in the emergence of a susceptible adult population in whom the incidence of immunization complications, particularly postvaccinal encephalitis, is purported to be frighteningly high (10 to 25 times that in young children). It is a segment of this adult population that would require immunization because of occupation or travel needs. It is further feared that a marked alteration of the United States' immunization policy would adversely influence the World Health Organization's successful eradication program.

CONTROL OF THE SPREAD OF SMALLPOX

Control is based upon isolation of the patient, adequate disposal or sterilization of contaminated materials with which infected persons have had contact, and vaccination of all contacts. The essential control measures set forth by the American Public Health Association (1955) have been generally accepted by local public health authorities, the U.S. Public Health Service, and the U.S. Armed Forces. The success of such measures to prevent viral spread depends upon recognition of the disease; this may be more difficult as subclinical or modified disease becomes more frequent owing to immunization.

TREATMENT

Limited but encouraging success in specific chemoprophylaxis and chemotherapy of poxvirus infections has been reported. N-methylisatin-β-thiosemicarbazone (Methisazone), which inhibits multiplication of poxviruses, has proved valuable 1) in the treatment of cases of **vaccinia gangrenosa,** and 2) for prevention of smallpox in initimate contacts of proved cases (Ch. 49). Treatment of smallpox patients with this drug during an epidemic, however, has not been effective. In **vaccinia gangrenosa** local injections of leukocytes or transfer factor from immune persons has effectively halted the necrotic process, and in generalized vaccinia γ-globulin from hyperimmune serum has been employed with variable success. These therapeutic approaches warrant further investigation.

OTHER POXVIRUSES THAT INFECT MAN

Several poxviruses other than variola and vaccinia cause disease in man: cowpox, molluscum contagiosum, contagious dermatitis, and milker's nodules (paravaccinia).

COWPOX

Cowpox is an occupational disease of man acquired from the udders and teats of infected cows. The vesicular inflammatory lesions are usually localized on the fingers, but the virus may be implanted on the face or other parts of the body; the disease is self-limiting.

Cowpox virus has properties similar to those of variola and vaccinia virus, but its antigenic structure differentiates it from the other agents in the subgroup. Although its soluble antigens cross-react extensively with related poxviruses, minor antigenic differences can be demonstrated. The host ranges of cowpox and vaccinia viruses are similar; but the following host reactions to cowpox virus are distinctive: 1) pocks appear more slowly on chorioallantoic membranes; 2) the virus has a tendency to invade mesodermal tissue involving capillary endothelium and thus to produce hemorrhagic ulcers in the pocks; 3) the inclusion bodies are larger and more eosinophilic than classic Guarnieri bodies; and 4) keratitis is produced slowly in rabbits, in comparison with the rapid development effected by vaccinia virus.

MOLLUSCUM CONTAGIOSUM

The molluscum contagiosum virus produces an uncommon skin disease affecting mainly children and young adults. The lesion is a chronic, proliferative process, restricted to the epithelium of the skin of the face, arms, legs, back, buttocks, and genitals. Electron microscopic observations reveal that the molluscum body (a large cytoplasmic inclusion body) is composed of virions indistinguishable from other poxviruses. Mature viral particles develop by a process resembling the formation of vaccinia virions. Molluscum contagiosum has been transmitted experimentally to man; and infections of cultures of HeLa cells and primary human amnion or foreskin cells have been achieved. Although virus could not be serially propagated, viral particles developed, as revealed by electron microscopic examination of thin sections, and cytopathic changes appeared in cultures of infected cells.

The tendency to induce proliferative lesions is even more striking with molluscum contagiosum virus than with other poxviruses. But it should be noted that infected cells do not continue to synthesize DNA, and it is the neighboring uninfected cells whose rate of cell division is stimulated. Although this virus thus appears to provide a link between the common pathogenic viruses of man and tumor-inducing agents (Ch. 63), the mechanism for inducing increased cell proliferation is unknown.

MILKER'S NODULES (PARAVACCINIA)

Jenner recognized the existence of two diseases affecting the udder and teats of cows: classic cowpox and a second condition consisting predominantly of vesicular lesions. The latter disease is also transmitted to man, producing painless smooth or warty "milker's nodules" on the hands and arms. The lesions rarely become pustular. The infected cells contain eosinophilic cytoplasmic inclusion bodies and elementary bodies characteristic of poxvirus infection. Disease is associated with only mild constitutional symptoms and enlargement of regional lymph nodes.

Infection does not confer immunity to either cowpox or vaccinia viruses. Paravaccinia virus cannot be propagated on the chorioallantoic membrane of chick embryos or in laboratory animals usually susceptible to cowpox. The virus was isolated from a milker's nodule of man by inoculating primary fetal bovine kidney tissue cultures. The agent has been serially cultured in these cells, as well as in diploid bovine conjunctival cells and human embryonic fibroblasts. In contrast to many poxviruses, it cannot be propagated in series in continuous human cell lines, such as HeLa cells, and it does not produce pocks on the chorioallantoic membranes of chick embryos. Its cytopathic effects resemble those produced by vaccinia virus, and infected cells prepared with Giemsa stain show metachromatic cytoplasmic inclusion bodies. The virus has stability characteristics similar to those

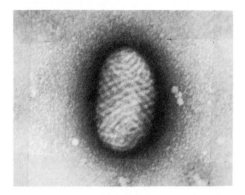

Fig. 54-10. Electron micrographs of viral particle isolated from lesion of milker's nodule. Negatively stained viral particle shows the characteristic morphology of a poxvirus. ×114,000. [From Friedman-Kien, A. E., Rowe, W. P., and Banfield, W. G. *Science 140*:1335 (1963).]

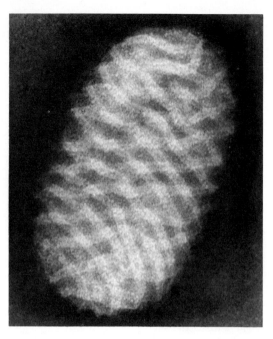

Fig. 54-11. Mature orf virus particle, negatively stained with phosphotungstic acid. The woven pattern of threads or tubules can be clearly seen. The apparent criss-crossing of the tubules results from the visualization of both the front and back faces of the particle. ×600,000, reduced. [From Nagington, J., and Horne, R. W. *Virology 16*:248 (1962).]

of poxviruses. In thin sections the average viral particle measures 1200 by 2800 A and has the typical morphology of a poxvirus. Electron microscopic observations of preparations stained with sodium phosphotungstate (Fig. 54-10) reveal ovoid virions (rather than brick-shaped particles like vaccinia, Fig. 54-1), whose size and surface structures are identical with those of contagious pustular dermatitis (Fig. 54-11) and bovine papular stomatitis viruses.

CONTAGIOUS PUSTULAR DERMATITIS (ORF)

Although a natural affliction of sheep, mainly affecting lambs, contagious pustular dermatitis occurs rarely as an occupational disease of man. Sheep characteristically develop vesicles in the oral mucosa, which become encrusted and heal slowly after several weeks. In man the infection usually causes a single lesion on a finger, beginning as a small, painless vesicle, which becomes pustular, encrusts, and finally heals. Transmission of infection from man to man has not been recorded.

The causative agent, orf virus, can be isolated in various animal cell cultures or on the chick chorioallantoic membrane. It produces characteristic cytopathic changes, but rarely gives rise to eosinophilic inclusion bodies. Electron microscopic study of the viral particles reveals the prominent tubule-like structures characteristic of poxviruses (Fig. 54-11). The virion, in contrast to vaccinia and variola viruses, is ovoid (1600 by 2600 A) and is encircled in a regular pattern by the surface tubules.

SELECTED REFERENCES

Books and Review Articles

DOWNIE, A. W., and KEMPE, C. H. Poxviruses. In *Diagnostic Procedures for Viral and Rickettsial Infections,* 4th ed., p. 281. (E. H. Lennette and N. J. Schmidt, eds.) American Public Health Association, New York, 1969.

JOKLIK, W. K. The poxviruses. *Bacteriol Rev 30:*33 (1966).

LANE, J. M., MILLER, D., and NEFF, J. M. Smallpox and smallpox vaccination policy. *Annu Rev Med 22:*251 (1971).

MCAUSLAN, R. R. The biochemistry of poxvirus replication. In *The Biochemistry of Viruses,* p. 361. (H. B. Levy, ed.) Dekker, New York, 1969.

WILCOX, W. C., and COHEN, G. H. The poxvirus antigens. *Curr Top Microbiol Immunol 47:*1 (1969).

Specific Articles

BANFIELD, W. G., and BRINDLEY, D. C. An electron microscopic study of the epidermal lesion of *Molluscum contagiosum*. *Ann NY Acad Sci 81:*145 (1959).

BLANDEN, R. V. Mechanisms of recovery from a generalized viral infection: Mousepox. II. Passive transfer of recovery mechanisms with immune lymphoid cells. *J Exp Med 133:*1074 (1971).

COUNCILMAN, W. T., MAGRATH, G. B., and BRINCKER-HOFF, W. R. The pathological anatomy and histology of variola. *J Med Res 11:*12 (1904).

FOEGE, W. H., FOSTER, S. O., and GOLDSTEIN, J. A. Current status of global smallpox eradication. *Am J Epidemiol 93:*223 (1971).

FRIEDMAN, R. M., BARON, S., BUCKLER, C. E., and STEINMULLER, R. I. The role of antibody, delayed hypersensitivity, and interferon production in recovery of guinea pigs from primary infection with vaccinia virus. *J Exp Med 116:*347 (1962).

JOKLIK, W. K. The intracellular uncoating of poxvirus DNA. I. The fate of radioactively-labeled rabbitpox virus. II. The molecular basis of the uncoating process. *J Mol Biol 8:*263, 277 (1964).

KATES, J., and BEESON, J. Ribonucleic acid synthesis in vaccinia virus. I. The mechanism of synthesis and release of RNA in vaccinia cores. *J Mol Biol 50:*1 (1970).

KATZ, S. L. The case for continuing "routine" childhood smallpox vaccination in the United States. *Am J Epidemiol 93:*241 (1971).

MARSDEN, J. P. Variola minor: A personal analysis of 13,686 cases. *Bull Hyg 23:*735 (1948).

WOODROOFE, G. M., and FENNER, F. Serological relationships within the poxvirus group: An antigen common to all members of the group. *J Virol 16:*334 (1962).

HISTORY   1280
CLASSIFICATION AND GENERAL CHARACTERISTICS   1281

POLIOVIRUSES   1281

PROPERTIES OF THE VIRUSES   1281
  Morphology   1281
  Physical and Chemical Characteristics   1281
  Immunological Characteristics   1282
  Host Range   1283
  Viral Multiplication   1283
  Genetic Characteristics   1287
PATHOGENESIS   1289
LABORATORY DIAGNOSIS   1292
EPIDEMIOLOGY   1292
PREVENTION AND CONTROL   1293

COXSACKIEVIRUSES   1296

PROPERTIES OF THE VIRUSES   1296
  Physical and Chemical Characteristics   1296
  Immunological Characteristics   1296
  Host Range   1297
  Viral Multiplication   1297
PATHOGENESIS   1298
LABORATORY DIAGNOSIS   1299
EPIDEMIOLOGY AND CONTROL   1299

ECHOVIRUSES   1299

PROPERTIES OF THE VIRUSES   1300
  Physical and Chemical Characteristics   1300
  Hemagglutination   1300
  Immunological Characteristics   1300
  Host Range   1301
  Viral Multiplication   1301
PATHOGENESIS   1301
LABORATORY DIAGNOSIS   1301
EPIDEMIOLOGY AND CONTROL   1302

RHINOVIRUSES   1303

PROPERTIES OF THE VIRUSES   1303
PATHOGENESIS   1304
LABORATORY DIAGNOSIS   1304
EPIDEMIOLOGY   1305
PREVENTION AND CONTROL   1305

Until the present century poliomyelitis (Gr. *poli* = gray and *myelos* = spinal cord) was a disease primarily of infants (hence the name "infantile paralysis"), and this pattern is still seen in communities with primitive sanitation. But with improved sanitation in many countries, in the 75 years prior to widespread immunization in the 1960s, epidemics became increasingly prominent, the age distribution advanced to include young adults, and the disease showed increasing severity as it appeared in older persons. This paradoxical response to improved sanitation was eventually explained by the finding that 1) practically everyone became infected, though the paralytic disease was rare, and 2) the consequences were usually negligible if infection was acquired early in life but might be serious when infection was postponed.

Even in the worst period clinically severe poliomyelitis was not very prevalent. Thus in 1953, 1450 deaths and about 7000 cases with residual paralyses were reported in the United States, compared with about 500 deaths from measles, which was considered hardly more than a nuisance. However, the visibility of the crippled survivors caused even small epidemics to be terrifying. In the United States the problem was dramatized by the presidency of Franklin D. Roosevelt, who was severely handicapped by poliomyelitis acquired as an adult. The generous public contribution of funds to combat polio (The March of Dimes) led to support of research on this problem on an unprecedented scale, resulting within 20 years in essentially complete control by immunization. It is no coincidence that poliomyelitis virus, though one of the most difficult to work with at the start of this program, was eventually studied with such precision that it served as a model for investigation of many other animal viruses.

HISTORY

Progress in this field depended on the development of improved technics for cultivating the virus. Landsteiner* and Popper in 1909 transmitted poliomyelitis to monkeys by intracerebral inoculation of a bacteria-free filtrate of spinal cord from a human patient, and the responsible infectious agent was shown to be a virus. However, the monkey was both expensive and cumbersome for experimentation. For example, as long as a number of monkeys had to be used to titrate antibody in a single serum sample or to test a single specimen for the virus, epidemiological studies could hardly go beyond the demonstration that three antigenically distinct polioviruses existed. Adaptation of polioviruses to the cotton rat by Armstrong in 1939 represented a substantial step forward. The turning point came, however, when Enders, Weller, and Robbins showed in 1949 that polioviruses can be isolated and readily propagated in cultures of **nonneural** human or monkey tissue. The incisive and detailed investigations that followed soon led to control of the disease.

Many related viruses were discovered as accidental byproducts of the intensive pursuit of polioviruses. Thus, before the introduction of tissue cultures, Dalldorf and Sickles discovered a new group of viruses infectious for man, coxsackieviruses,† while attempting unsuccessfully to isolate polioviruses by intracerebral inoculation of newborn mice. Later,

---

* Karl Landsteiner was also distinguished for his profound contributions to the understanding of immunological specificity and for the discovery of human blood groups.

† Named after Coxsackie, New York, the town from which the initial isolates were obtained.

investigations of poliomyelitis wth tissue cultures revealed a third group of viruses in the gastrointestinal tract of man. These were termed Enteric Cytopathic Human Orphan viruses (ECHO viruses) because 1) they were found in the human gastrointestinal tract, 2) they produced cytopathic changes in cell cultures, and 3) they were not clearly associated with disease. **Echoviruses** is the current taxonomical designation.

CLASSIFICATION AND GENERAL CHARACTERISTICS

Polioviruses, coxsackieviruses, and echoviruses are similar in epidemiological pattern; in physical, chemical, and biological characteristics; and in infecting the gastrointestinal tract of man. They were originally given the name **enteroviruses,** but the inadequacy of this

TABLE 55-1. Classification of Picornaviruses Affecting Humans

Group	No. of types
Poliovirus	3
Coxsackievirus	
Coxsackievirus A	23*
Coxsackievirus B	6
Echoviruses	31†
Rhinoviruses	89 (?)

* Type 23 was shown to be identical with echovirus type 9; A23 has been dropped and the number is unused.

† Type 10 has been reclassified as reovirus 1, and type 28 as rhinovirus 1; the numbers are now unused.

term became apparent when some coxsackieviruses and echoviruses were also found to produce acute respiratory infections. In addition a fourth group was discovered, **rhinoviruses,** which have similar chemical and physical characteristics but produce primarily acute respiratory infections. Hence, the new term **picornaviruses** (pico implying small, and RNA, the nucleic acid component) was coined as the family designation, encompassing these four groups of viruses (Table 55-1).

Viruses similar to human picornaviruses have been found in several species of lower animals. These viruses include those of foot-and-mouth disease of cattle, Teschen disease of pigs, and Mengo virus and encephalomyocarditis of mice.

The physical and chemical properties of picornaviruses are summarized in Table 55-2. They are small, contain RNA, and do not contain lipid. Polioviruses, described in detail in the following section, will serve as the prototype of the family.

TABLE 55-2. Summary of Characteristics of Picornaviruses

Size	210–300 A
Morphology	Icosahedral
Capsomers	Probably 32
Nucleic acid	Single-stranded RNA
Reaction to lipid solvents	Resistant
Stability at room temperature	Relatively stable
Stability at pH 3.0	Enteroviruses stable Rhinoviruses labile
Stability at 50°	Enteroviruses relatively labile Rhinoviruses relatively stable

POLIOVIRUSES

PROPERTIES OF THE VIRUSES

MORPHOLOGY

Electron micrographs of purified virus in thin sections of virus-infected cells reveal particles 270 to 300 A in diameter, with a dense core of approximately 160 A. Negative staining shows a capsid with a subunit arrangement consistent with icosahedral symmetry (Fig. 55-1). There appears to be 32 capsomers

per virion. Polioviruses and viruses of the other three picornavirus groups are indistinguishable in size and structure.

PHYSICAL AND CHEMICAL CHARACTERISTICS

Poliovirus was the first virus to be obtained in pure crystalline form (Fig. 55-2). A single molecule of single-stranded RNA constitutes about 30% of the virion; the remainder is protein. Infectious RNA can be extracted

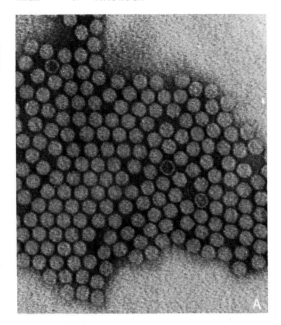

 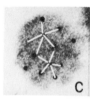

Fig. 55-1. Electron micrograph of a purified preparation of poliovirus negatively stained. **A.** Cubic symmetry of viral particles is evident. ×150,000. **B.** Higher magnification of a viral particle printed in reverse contrast. The capsomers measure approximately 60 A in diameter; their fine structure is not apparent. ×600,000. **C.** Same particle as in **B** marked to display two clear axes of fivefold symmetry (white lines). Mayor, H. D. *Virology 22*:156 (1964).]

from all three sterotypes, and its base composition is identical in each. The physical properties of the purified infectious viruses are also identical (Table 55-3).

Polioviruses are more stable than many viruses (e.g., those with lipid envelopes). Hence their transmission is facilitated because they can remain infectious for relatively long periods in water, milk, and other foods. Polioviruses are readily inactivated by pasteurization (Ch. 64) and many other chemical and physical agents (Ch. 44). MgCl₂ (1 M) appears to stabilize their intercapsomeric bonds and hence markedly increases thermal stability.

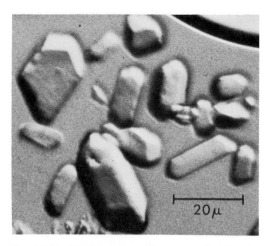

Fig. 55-2. Crystals of purified type 1 poliovirus particles. [From Schaffer, F. L., and Schwerdt, C. E. *Proc Natl Acad Sci USA 41*:1020 (1955).]

IMMUNOLOGICAL CHARACTERISTICS

Polioviruses are of three distinct immunological types, which can be recognized by neutralization, complement-fixation, or gel-diffusion precipitation reactions using type-specific sera.

Crude preparations of each type contain two type-specific antigens detectable by complement fixation or gel diffusion. Each antigen is associated with a different physical form of the viral particle separable by zonal centrifugation or by column chromatography: **the D (dense) antigen** is associated with the fully assembled infectious particles containing RNA, and the **C (coreless) antigen** is associated with noninfectious particles devoid of RNA cores. Empty particles possessing C antigen reactivity appear to be precursors of virions which are not yet completely assembled. The D particle (i.e., the infectious virion) contains four polypeptide chains designated VP1, VP2, VP3, and VP4; the C particle is deficient in VP2 and VP4 but instead contains a large extra polypeptide (VP0) which is the precursor of the two missing proteins. Although C particles do not react in vitro with antibodies to D, injection of C particles into rabbits induces production of antibodies to D as well as to C, and immunization with either type of particle results in the formation of neutralizing antibodies. Various mild denaturing procedures (e.g., exposure to 56°) cause the D particle to lose one protein constituent

TABLE 55-3. Characteristics of Polioviruses

Diameter of virion	270–300 A
Diameter of internal core	160 A
Diameter of capsomer	60 A
Molecular weight of RNA	2.5×10^6 daltons
Base composition (guanine and cytosine)	46 moles per cent
Sedimentation coefficient of virion	157–160 S_{20}
Particle mass of virion	1.1×10^{-17} gm
Molecular weight of virion	6.4–6.8×10^6 daltons

(VP4) and RNA and convert to C antigen reactivity.

There is no single poliovirus group antigen, but antigenic relations between types do exist. The antigenic cross-reactions are particularly prominent when heated virus is employed in complement-fixation titrations. The cross-reactivity, however, can only be demonstrated when sera are obtained from humans who have been infected with more than one type of poliovirus; i.e., after the initial infection the Ab response is strictly type-specific, but upon infection with a second type Abs develop to two or all three of the viruses. Immunological cross-reactivity between types 1 and 2 is also demonstrated in neutralization titrations by cross-absorption experiments and by the development of heterotypic Abs following natural infections or immunization. Slight cross-reactivity can be detected between types 2 and 3 by neutralization, but not between types 1 and 3. The immunological kinship between types 1 and 2 is substantiated epidemiologically: possession of type 2 Abs confers significant protection against the paralytic effects of subsequent type 1 infection.

Antigenic variants of types 1 and 2 viruses have been detected by means of a precise plaque-reduction or complement-fixation technic with cross-absorbed sera. These antigenic differences, however, do not affect the capacity of Abs induced by one strain to protect against infection by all other strains of the same type. Despite these minor intratypic antigenic differences, polioviruses actually show marked antigenic stability, both in nature and in laboratory manipulations.

Neutralizing Abs appear early in the course of poliovirus infection, and they have usually reached a high titer by the time the patient is first seen by a physician (Fig. 55-3); they attain a maximum titer 2 to 6 weeks after onset of disease, decrease to about ¼ that level in 18 to 24 months, and then seem to persist indefinitely. Complement-fixing Abs (anti-D and anti-C) appear during the first 2 to 3 weeks after infection, reach maximum titers in about 2 months, and persist for an average of 2 years. The presence of neutralizing Abs confers clear protection against subsequent infection. Inapparent and nonparalytic infections result in antibody levels as high as those present after severe paralytic disease.

Second attacks of paralytic poliomyelitis are rare and invariably are due to a different viral type from that producing the first illness.

HOST RANGE

Man is the only known natural host for polioviruses. Antibodies are present in some monkeys and chimpanzees studied in captivity, but there is evidence that they acquire the infection only after capture.

Old World monkeys and chimpanzees are susceptible to infection (by the intracerebral, intraspinal, and oral routes) with fresh isolates as well as with laboratory strains. In contrast, nonprimates are relatively insusceptible. However, by serial passage, strains of poliovirus were adapted to cotton rats and mice. Strains of type 2 virus also have been adapted to suckling hamsters and the chick embryo.

An important development was the discovery that these viruses, hitherto considered purely neurotropic, can multiply and produce cytopathology in extra-neural tissues of man cultured in vitro. It then rapidly became evident that many tissues from primates can furnish susceptible cells. Cultures of primate tissues are now widely used for the isolation and identification of polioviruses, as well as for experimental studies. The cytopathic changes produced by polioviruses as well as by other picornaviruses are illustrated in Figure 55-4.

VIRAL MULTIPLICATION

To initiate infection, polioviruses attach rapidly to specific host cell lipoprotein receptors, which are much more prevalent in susceptible than in nonsusceptible tissues. Such **adsorption** to susceptible cells is independent of temperature, but depends upon the concentration of electrolytes such as magnesium, calcium, or sodium. Infectious D particles, but not C particles, can adsorb, suggesting that the protein VP4 is critical for attach-

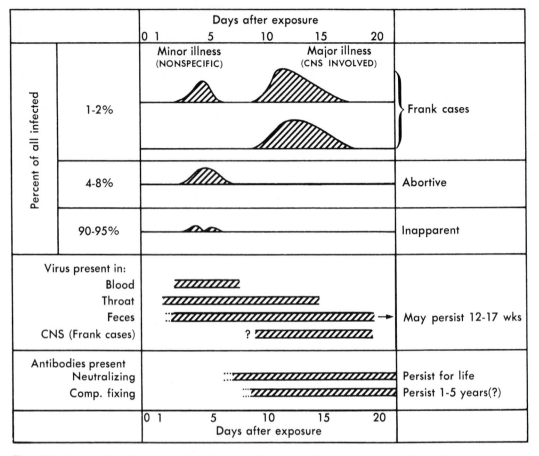

Fig. 55-3. Schematic diagram of the time at which the clinical forms of poliomyelitis appear, correlated with times at which virus is present in various sites and the development and persistence of antibodies. The high incidence of subclinical poliovirus infection is also noted. [From Horstmann, D. M. *Yale J Biol Med 36*:5 (1963).]

ment. Very soon after attachment the viral capsids are altered, probably while still combined to receptors, as the first step in liberation of viral RNA (uncoating). The virions thus become noninfectious, and about 50% of the altered particles elute from host cells.

Penetration of virus into host cells is optimal at about 37°. The process of viral uncoating proceeds rapidly after cell penetration. Uncoating of the viral RNA is completed within 30 to 60 minutes after infection, as measured by susceptibility to RNase.

Replication of infectious virus follows the general pattern of those single-stranded RNA viruses whose genome contains a polyadenylic acid-rich tail and initially serves as a molecule of messenger RNA for synthesis of RNA repli-

case. There then develops the RNA replicative intermediate (Ch. 48; Synthesis of RNA viruses), which serves for synthesis of progeny viral RNA molecules. Replication of viral RNA is totally independent of the biosynthesis of host cell DNA. The production of viral RNA commences soon after viral uncoating is completed, but the initial molecules all become the messengers for very large cytoplasmic polyribosomes. Since internal initiation of protein synthesis does not occur, this messenger of about 6000 nucleotides is translated into a single long polypeptide which is subsequently cleaved enzymatically into the four individual viral capsid proteins as well as into nonvirion proteins. RNA does not appear in viral particles until about 3 hours after infection (Fig. 55-5A), although during

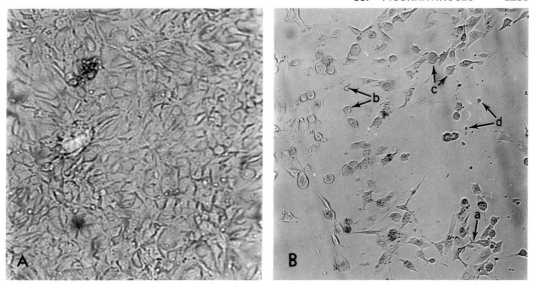

Fig. 55-4. Cytopathic changes in monkey kidney cell cultures infected with type 1 poliovirus. **A.** Monolayer of uninfected cells, unstained. ×200. **B.** Advanced cytopathic changes in infected cultures, unstained. Polioviruses, like most of the other enteroviruses that infect monkey kidney cells, produce marked cell retraction (a), rounding (b), and occasionally ballooning of cells (c), followed by rapid lysis leaving a granular debris (d). ×200. [From Ashkenazi, A., and Melnick, J. L. *Am J Clin Pathol* *38*:209 (1962).]

Fig. 55-5. Biosynthetic events in poliovirus-infected cells. **A.** Time course of 1) synthesis of viral RNA measured by its infectivity and 2) maturation of virions. **B.** Biosynthesis of viral capsid proteins measured by incorporation of ^{14}C-labeled amino acids into antibody-precipitable material. **C.** Rate of total RNA and protein synthesis in poliovirus-infected cells measured by incorporation of ^{14}C-uridine or ^{14}C-L-valine into acid-precipitable material at the indicated times after infection. [**A** from Darnell, J. E., Jr., Levintow, L., Thoren, M. M., and Hooper, J. L. *Virology* *13*:271 (1961). **B** from Scharff, M. D., and Levintow, L. *Virology* *19*:491 (1963). **C** from Zimmerman, E. F., Hector, M., and Darnell, J. E. *Virology* *19*:400 (1963).]

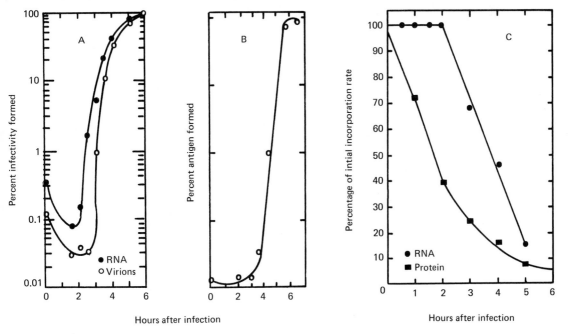

Hours after infection

Hours after infection

the period of virion assembly production of capsid proteins and RNA replication are closely coupled, for newly made viral RNA is incorporated into virions within 5 minutes after synthesis. The final step in morphogenesis appears to be the combination of viral RNA with a shell of viral proteins (termed the procapsid; Ch. 48), during which one of the procapsid proteins (VP0) is cleaved to yield two (VP2 and VP4) of the four capsid structures.

Final assembly of infectious particles is accomplished rapidly, occasionally resulting in small intracytoplasmic crystals composed of virions. Electron microscopic examination reveals that empty shells are also assembled (Fig. 55-6). Approximately 500 virions per cell are produced. Initially, infectious virus is released through vacuoles which fuse with the plasma membrane forming microtubules. After several hours, virions escape in a burst, accompanied by death and lysis of the host cell.

Synthesis of host cell proteins is inhibited very shortly after viral infection (Fig. 55-5C), accompanied by disruption of the host cell ribosomal aggregates (polysomes). The synthesis of normal host cell RNA ceases about 2 hours after infection, shortly after biosynthesis of viral RNA begins in the cytoplasm (Fig. 55-5).

The **cytologic changes** accompanying these biosynthetic events are summarized diagrammatically in Figure 55-7. Intranuclear altera-

Fig. 55-6. Development of poliovirus particles in pieces of cytoplasmic matrix of artificially disrupted cells. Particles in various stages of assembly from empty shells (s) to complete virions (v) can be seen. ×200,000. [From Horne, R. W., and Nagington, J. *J Mol Biol 1*:333 (1959).]

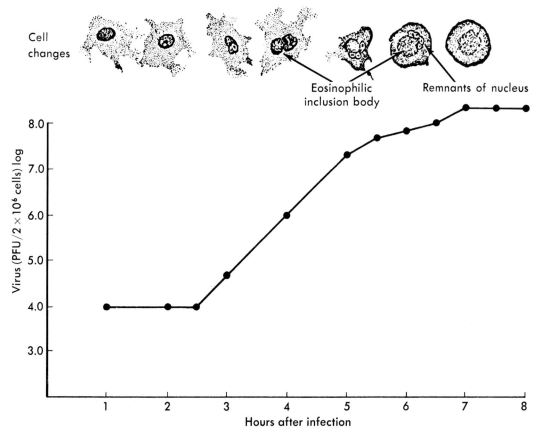

Fig. 55-7. Diagrammatic representation of the multiplication of poliovirus and the accompanying pathological changes in infected cells. A perinuclear cytoplasmic eosinophilic mass (inclusion body) develops as viral multiplication reaches its maximum; the inclusion body impinges on the nucleus, which degenerates as the cell dies. [After Reissig, M., Howes, D. W., and Melnick, J. L. *J Exp Med 104*:289 (1956).]

tions, consisting of rearrangement in chromatin material with condensation at the nuclear membrane, are the first changes detected. One or more small intranuclear eosinophilic inclusion bodies of unknown nature form, and the nucleus becomes distorted and wrinkled and gradually shrinks. These events are probably related to the inhibition of the host cell's protein and nuclear RNA synthesis (Fig. 55-5c). There then develop cytoplasmic alterations consisting of a large eosinophilic mass, which is the site of replication and assembly of viral subunits (Figs. 55-7 and 55-8), and knobs on the cell membrane, which appear to be due to cytoplasmic bubbling associated with release of virus. Finally the nucleus becomes pyknotic, the nuclear chromatin becomes fragmented, and cells become rounded and die.

GENETIC CHARACTERISTICS

The demonstration in 1953 that poliovirus contains only RNA stimulated great interest because 1) it permitted investigation of RNA as genetic material, and 2) knowledge of the inheritable properties of the virus, including attenuated mutants, was required for development of a live virus vaccine.

In general, RNA behaves genetically like DNA. The number of mutant phenotypes observed (Table 55-4) is much larger than the number of viral genes that could be anticipated from the size of the viral RNA. Many pheno-

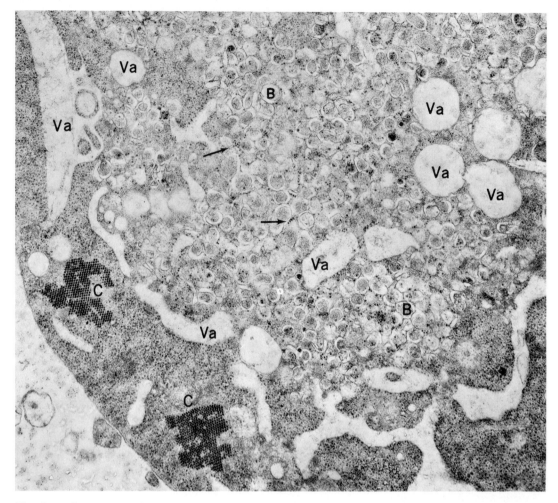

Fig. 55-8. Cytoplasmic changes and assembly of virions in a poliovirus-infected cell. Large numbers of membrane-enclosed pieces of cytoplasm (B) accumulate in the central region of the cell, pushing the nucleus to one side (nucleus is not shown in this picture). Large cytoplasmic vacuoles (Va) also develop. A large number of virions (arrows) are present in the cytoplasmic matrix, both between and within the membrane-enclosed bodies (B). Two large crystals of virions are present (C). ×25,000. [From Dales, S., Eggers, H. J., Tamm, I., and Palade, G. E. *Virology* 26:379 (1965).]

types must therefore arise from mutations of the same gene with different consequences for the corresponding protein. This conclusion is supported by the fact that many mutations affect the viral capsid, which is probably encoded in only four genes.

Early studies revealed that recombination could occur between some of the mutant phenotypes; these data suggested that since the viral genome contains information for no more than 10 proteins, it should be possible to construct a genetic map. With conditionally lethal temperature-sensitive mutants a modest beginning has been made in mapping two genes for replication of viral RNA, one for synthesis of capsid proteins, and two for regulation of cell functions.

These genetic studies have also revealed that many mutations are pleiotropic, i.e., the simultaneous change of two phenotypic traits by the same mutations, although the two phenotypes can also change separately. The *d* and *e* phenotypes (Table 55-4), for example, can arise by the same mutation.

Pleiotropism makes it possible to use mutant phenotypes that are easily detectable in vitro

TABLE 55-4.   Examples of Poliovirus Mutants

Class of mutants	Characteristic	Marker name
Factors that affect cell-virus interaction and therefore viral multiplication	1. Ability to multiply at 40°	rct/40
	*2. Inability to multiply at 40°	rct/40−
	*3. Ability to multiply at 23°	rct/23+
	*4. Heat defectiveness (inability to multiply at 40°, but usual multiplication at 36°)	hd
	5. Plaque size	s
	*6. Resistance to heating in $AlCl_3$	a
	7. Resistance to heat inactivation of virion	t
	8. Inability to grow in MS cells	ms
Variants distinguished by presence or absence of inhibitory substance in media	*1. Sensitivity to agar inhibitor at acid pH	d
	2. Sensitivity to agar inhibitor at neutral pH	m
	3. Cystine inhibition of multiplication	cy^i
	4. Cystine dependence	cy^r
	5. Tryptophan dependence	
	6. Adenine resistance	
	7. Guanidine resistance	g^r
	8. Guanidine dependence	g^d
	9. Hydroxybenzylbenzimidazole (HBB) resistance	HBB^r
	10. Resistance to normal bovine serum inhibitor	bo
	11. Resistance to normal horse serum inhibitor	ho
Mutants whose markers are related to physical characteristics of the virus	*1. Poor elution from Al (OH)₃ gel	$Al(OH)_3$
	*2. Greater adsorbability to DEAE-cellulose	e
Immunological variants	1. Intratypic antigenic variants	

* Phenotypes associated with attenuated viruses.

to select for mutations affecting neurotropism, which could otherwise be studied only by the much more limited and costly approach of inoculating primates. The mutant phenotypes that are frequently associated with attenuation are indicated in Table 55-4; several of them affect the viral capsid, and others multiply preferentially at or below usual body temperatures. These results suggest that neurovirulence depends on the ability of the viral surface to interact with certain cells, and on the ability of the virus to replicate in febrile patients (rct/40 mutants). Furthermore, attenuated viruses induce synthesis of more interferon and are more readily inhibited by it (Ch. 49, Interferon) than are virulent viruses.

PATHOGENESIS

The major sequence of events in the multiplication and spread of polioviruses was revealed by studies in chimpanzees and man as well as in cell cultures. At the cellular level viral multiplication and subsequent cell damage are summarized in Figure 55-7. In man the progression of infection culminating in invasion of the target organs, the brain and spinal cord, is outlined in Figure 55-9.

Infection is initiated by ingestion of virus and its primary multiplication in the oropharyngeal and the intestinal mucosa. It is not known, however, whether virus multiplies in epithelial or lymphoid cells of the alimentary tract. The tonsils and Peyer's patches of the ileum are invaded early in the course of infection, and extensive viral multiplication ensues in these loci, so that as much as 10^7 to 10^8 TCD_{50} of virus per gram of tissue may accumulate. From the primary sites of propagation virus drains into deep cervical and mesenteric lymph nodes, but since its titer there is relatively low, these nodes may not be important sites of viral replication for progressive infections. From the nodes virus drains into the blood, probably resulting in a transient viremia which disseminates virus to other susceptible tissues such as the brown fat (axillary, paravertebral, and suprasternal fat) and the viscera (probably in reticuloendothelial cells). In these extraneural sites virus replicates, and it is continually fed back into the blood stream to establish and maintain a persistent viremic stage. In most

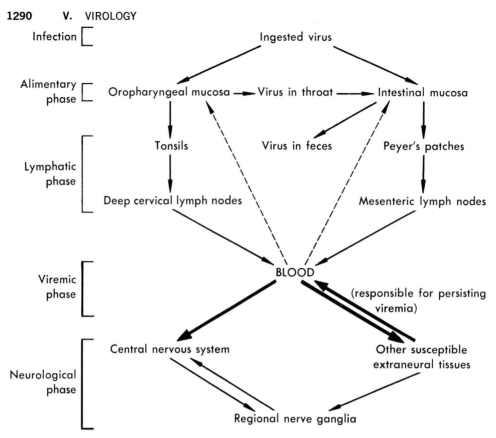

Fig. 55-9. A model for the pathogenesis of poliomyelitis based on a synthesis of data obtained in man and chimpanzees. [Modified from Sabin, A. B. *Science 123*:1151 (1956) and Bodian, D. *Science 122*:105 (1955).]

natural infections, even in nonimmune individuals, only transient viremia occurs; the infection does not progress beyond the lymphatic stage, and clinical disease does not ensue.

If effective viremia is established virus spreads to the central nervous system. But it is still uncertain whether the CNS is invaded through the blood-brain barrier or by transmission of virus along nerve fibers from peripheral ganglia. The fact that viremia is essential in the pathogenesis of poliomyelitis implies that direct invasion of the CNS through capillary walls is the major pathway of penetration. Furthermore, specific antibody in the blood, even with the relatively low levels obtained by passive imunization, effectively halts viral spread and prevents invasion of the brain and spinal cord. The evidence supporting the contention that virus invades the CNS by transmission along nerves is that 1) polioviruses are found in peripheral nerve

ganglia during the progression of infection, and 2) virus can spread along nerve fibers, by means still unknown, in both peripheral nerves and the CNS. It appears from these facts that transmission of virus along nerve fibers from peripheral ganglia may be effective, but it probably is not the primary route of CNS infection.

Poliomyelitis generally conjures up the picture of a severe, crippling, and occasionally fatal paralytic disease. In fact, however, the large majority of infections do not progress beyond lymphatic involvement and never produce the signs or symptoms that have given the disease its notoriety and fearsome image (Fig. 55-3). A moderate number of infections induce transient viremia resulting in a mild febrile disease or so-called "summer grippe." Probably no more than 1 to 2% of infections culminate in an invasion of the CNS and the paralytic clinical syndrome.

The **course of classic paralytic disease,**

summarized in Fig. 55-3, is initiated by a **minor disease,** which is associated with the viremia and characterized by constitutional and respiratory or gastrointestinal signs and symptoms. There follows after 1 to 3 days, or often without any interval, the **major disease,** characterized by headache, fever, muscle stiffness, and paralysis associated with cell destruction in the CNS. **Lesions causing paralysis occur most frequently in the anterior horn cells of the spinal cord (spinal poliomyelitis);** similar lesions may also occur in the medulla and brain stem **(bulbar poliomyelitis)** and in the motor cortex **(encephalitic poliomyelitis)** (Fig. 55-10). Bulbar poliomyelitis is often fatal because of respiratory or cardiac failure; the other forms may result in survival with highly variable patterns of residual paralysis. Paralysis becomes maximal within a few days of onset, and there often follows extensive recovery, which may be aided by physiotherapy. The recovery represents in part compensatory hypertrophy of muscles that have not lost their innervation. In addition, while the virus-infected cells in the lesions are inevitably damaged, neighboring neurons may also contribute to the paralysis through edema and byproducts of necrosis, but paralysis from these latter neurons is reversible. The temporal relation between clinical manifestations, distribution of virus, and the appearance of antibodies is also summarized in Figure 55-3.

The pathogenic scheme described does not consider the role of the host or of the viral type, but the eventual outcome of the complex progression of infection is clearly not dependent upon the antibody level alone. For example, several host factors may alter the course of infection: fatigue, trauma, peripheral injections of drugs and vaccines, tonsillectomy, pregnancy, and age. These factors do not increase the incidence of infection but only the frequency and severity of paralysis. Trauma, including hypodermic injections,

Fig. 55-10. Schematic drawing of a lateral view of the human brain and the midsagittal surface of the brain stem. The usual distribution of lesions is represented by black dots. The spinal lesions usually occur in the anterior horn cells; lesions in the cerebral cortex are largely restricted to the precentral gyrus; those of the cerebellum are largely found in the roof nuclei; lesions of the brain stem centers are widespread. (From Bodian, D. *Papers and Discussions Presented at the First International Poliomyelitis Conference.* Lippincott, Philadelphia, 1949.)

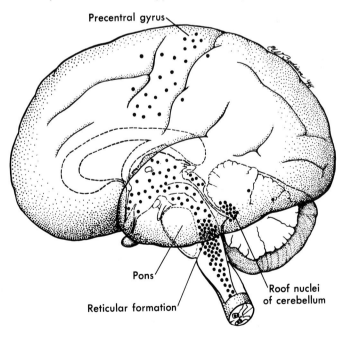

Precentral gyrus

Pons

Reticular formation

Roof nuclei of cerebellum

tends to localize the paralysis to the muscles traumatized. Tonsillectomy, recent or of long standing, markedly enhances the incidence of bulbar poliomyelitis. The mechanisms of action of these localizing host factors are not certain. It is possible that the various physical insults described increase infection of peripheral nerve ganglia or transmission of virus along peripheral nerves associated with the affected area. Some data suggest that the localizing factors increase the permeability of blood vessels in areas of the CNS associated with the traumatized or fatigued peripheral area. The increased severity of paralysis in adults, and even more in pregnant women, may be related to endocrine factors: steroids, for example, greatly enhance the severity of infection in experimental animals.

LABORATORY DIAGNOSIS

Laboratory methods for diagnosis of poliomyelitis are simple, efficient, and readily available, owing to the rapid development of tissue culture methods and immunological technics. Virus is isolated most readily from feces or rectal swabs for about 5 weeks after onset, and from pharyngeal secretions for the first 3 to 5 days of disease. Specimens with added antibiotics are inoculated into several tissue culture tubes. Multiplication of virus is detected by the development of characteristic cytopathic changes (Fig. 55-4B) and by the failure of infected cultures to become acid in 1 to 4 days after inoculation, revealed by the incorporation in the medium of the indicator phenol red (which shifts from red to yellow on acidification). The latter criterion depends on the inability of the dying infected cells to produce organic acids from glucose. Identification of the virus is accomplished by neutralization tests, using a standard serum for each of the three types.

For serological diagnosis Ab levels are compared in sera obtained during the acute phase of the disease and 2 to 3 weeks after onset, using neutralization or complement-fixation titrations. Neutralization titrations are still most widely used. Titration endpoints are based on inhibition of cytopathic effects or on continued acid production; the latter assay is especially convenient, as it can be carried out in disposable plastic trays, and the endpoints

can be determined rapidly without microscopic observations. Complement-fixation tests usually yield less dependable results because of the presence of both D and C antigens in viral preparations and the differences in time of appearance of Abs directed against each.

EPIDEMIOLOGY

Serological surveys show that polioviruses are globally disseminated. In densely populated countries with poor hygienic conditions practically 100% of the population over 5 years of age have Abs to all three types of poliovirus, epidemics do not occur, and paralytic disease is rarely recognized. In countries with improved sanitation, in contrast, the young are shielded from exposure, and prior to the effective use of vaccines, many reached adulthood having escaped infection and therefore without protective Abs. Because the incidence and severity of paralytic disease increase with age, if infection is delayed until susceptibles are above 10 to 15 years, the results are crippling.

But the widespread use of vaccines has strikingly altered the epidemiological picture. Epidemics have been eliminated except in pockets of lower socioeconomic groups where vaccine has not been widespread (although available without cost to the individual). Among the economically disadvantaged living in countries where vaccines are used extensively, small epidemics are again occurring in the very young (reviving the picture of "infantile paralysis") because, unlike events in countries having poor sanitation, polioviruses are not widely disseminated while babies are still protected by high titers of maternal Abs.

It should be reemphasized that the host's reaction to infection depends upon age: i.e., the incidence of disease is highest in childhood, for immunity wanes as maternal Abs are lost; adulthood constitutes the period of greatest susceptibility to extensive paralytic disease.

Poliomyelitis occurs primarily in the summer, similar to the common summer diarrheal diseases, which first suggested a transmission by the fecal route. Indeed, a large amount of virus is excreted in the feces for an average period of 5 weeks after infection, even in the presence of a high titer of circulating Abs (Fig. 55-3). A patient is maximally contagious, however, during the first week of illness, the period when pharyneal excretion of virus also occurs. Multiple modes of infection probably account for the fact that infections can occur in any season of the year.

Person-to-person contact is the primary mode of spread, and transmission within families and schools appears to be the major mechanism of dispersion throughout a community. Flies may occasionally serve as accidental vectors, but they are not an important mode of distribution. Water- and milk-borne epidemics owing to fecal contamination have also been reported. Dissemination of virus is rapid and extensive in nonimmune members of a family or other contact group, but the ratio of paralytic disease to inapparent infections is low—variously reported as 1:87 to 1:200 in temperate zones.

PREVENTION AND CONTROL

Until vaccine became available, in 1954, the only approaches to the control of infection were passive immunization and nonspecific public health measures. To interrupt transmission of the virus patients were isolated, places in which groups of susceptibles might gather (e.g., schools, churches, movies, and swimming pools) were closed, and insecticides were sprayed to reduce the population of flies and other insects suspected of carrying the virus. None of these measures were proved successful in preventing or stopping an epidemic.

The present era of successful control can be attributed to three major discoveries: 1) there are three distinct antigenic types of poliovirus against which protection is required; 2) polioviruses can multiply to high titer in cultures of nonnervous tissues, a finding which afforded a practicable procedure for preparing large quantities of virus free of the nervous tissue capable of inducing demyelinating encephalomyelitis (Ch. 20, Allergy to self-antigens); and 3) viremia is a critical phase in the pathogenesis of poliomyelitis, a discovery which suggested that the infection can be interrupted before the CNS is infected. In addition, the discovery that monkeys, mice, and indeed man could be

protected by passive immunization with immune serum or pooled γ-globulin proved that even low titers of Abs can be effective in preventing paralytic poliomyelitis.

When it became clear that immunization to prevent poliomyelitis was feasible, the development proceeded by two different approaches. One was the preparation of an **inactivated virus vaccine,** based on evidence that poliovirus inactivated with formalin could immunize monkeys. The second was the development of a **live (attenuated) virus vaccine,** modeled on the successful control of smallpox and yellow fever by such vaccines.

Inactivated Virus Vaccine. Salk demonstrated that all three types of polioviruses could be inactivated by 1:4000 formalin, pH 7.0, at 37° in about 1 week, with retention of adequate antigenicity. When purified virus is used, the inactivation follows pseudo first order kinetics (Ch. 49). However, when crude viral preparations are employed virus may aggregate, resulting in a complex inactivation curve: the rate of viral inactivation is not constant, and the tailing-off of the inactivation curve is marked. Failure to recognize this complication led to some serious initial difficulties in vaccine production, exemplified by an incident in which residual infectious virus remained in several lots of commercial vaccine and induced over 100 cases of paralytic poliomyelitis. Fortunately, the errors were soon rectified and a safe, highly effective vaccine was developed.*

Extensive controlled studies showed an effective protection against paralytic poliomyelitis in 70 to 90% of those immunized,

---

* In the presence of 1 M $MgCl_2$ inactivation by formalin shows much less tailing-off, and results in a preparation more suitable for a vaccine. Furthermore, with 1 M $MgCl_2$ inactivation by formalin can be carried out at 50° without loss of immunogenicity, thus fostering inactivation of adventitious viruses which may be present in monkey cell cultures in which the poliovirus is propagated. These viruses, which include SV 40 virus (Ch. 63) are more resistant to formalin inactivation than poliovirus, but are not stabilized by $MgCl_2$ against heat inactivation. Because 1 M $MgCl_2$ also reduces heat inactivation of infectious poliovirus, heating in the presence of $MgCl_2$ may similarly be used with infectious virus preparations to eliminate extraneous viruses such as SV 40.

and subsequent use of vaccine in the general population confirmed its protective ability. Vaccine containing all three poliovirus types is administered in three intramuscular or subcutaneous injections over a 3- to 6-month period. Antibody levels for all three types appear to fall to approximately 20% of their maximum titer within 2 years, and thereafter decline at a slower rate. The actual persistence of Abs is difficult to evaluate, owing to the uncontrolled occurrence of reinfections. Booster injections of vaccine every 2 to 3 years are recommended to raise Ab levels.

Immunization does not prevent reinfection of the alimentary tract unless serum Ab levels are very high (which is unusual except shortly after booster doses). However, infection of the oropharyngeal mucosa and tonsils is generally prevented, and if pharyngeal secretions are a major vehicle for transmission of polioviruses this result may explain the observed decreased incidence of infection since the widespread use of inactivated virus vaccine.

Live (Attenuated) Virus Vaccine. Infection of the alimentary canal with attenuated viruses offered the promise of several hypothetical advantages: 1) long-lasting immunity, similar to that following natural infections; 2) prevention of reinfection of the gastrointestinal tract and therefore elimination of transmission of the virus; and 3) inexpensive mass immunization without the need for sterile equipment.

Three different sets of attenuated viruses have been selected independently by Cox, Koprowski, and Sabin for preparation of vaccines, and all have been successfully used in large field trials. The strains were selected by multiple passage in a foreign host, most frequently tissue culture. The viral strains developed by Sabin were chosen by the Bureau of Biological Standards of the U.S. Public Health Service as the official strains licensed for commercial production of vaccines. These strains lack neurovirulence for susceptible monkeys inoculated both intramuscularly and intracerebrally, but they occasionally cause paralysis following intraspinal inoculation. The types 1 and 2 vaccine strains are genetically stable, probably because they contain several mutations affecting virulence (Table 55-4), although minor increases in neurovirulence may occur following passage in man. Fortunately, the alimentary tract of man does not

offer a selective advantage for mutants with increased neurovirulence. The type 3 vaccine strain, however, reverts more frequently; it is estimated to produce approximately 1 case of paralytic poliomyelitis for every 3×10^6 doses of vaccine. The danger of vaccine-induced disease appears to be greatly increased, however, in children with immune-deficiency diseases. (The genetic markers in mutant strains used for immunization are of particular value in determining whether or not the vaccine is responsible for the rare post-immunization case of disease.)

In most individuals oral administration of a single type in a dose of 10^5 to $10^{5.5}$ TCD_{50} produces infection of the gastrointestinal canal, excretion of virus in high titer for 4 to 5 weeks, and development of Abs to a titer of approximately 1:128 in 3 to 4 weeks. Serological conversion occurs in over 95% of those without Abs to any of the three types at the time vaccine is administered. During the period of relatively high Ab titers reinfection of the alimentary tract by natural infection is prevented. How long reinfection can be inhibited has not been clearly defined.

The persistence of Abs and immunity following natural infections suggested that a similar persistence might follow immunization with infectious attenuated virus. In fact, however, Abs decrease at approximately the same rate as after immunization with inactivated viruses, i.e., a diminution in 2 years to about 20% of the maximum titer. Antibody levels in general are higher, however, following live than following inactivated vaccines.

The optimum schedule for administering the live virus vaccine is still in dispute. Since feeding the three types of viruses simultaneously may result in interference with multiplication of one or more strains (Ch. 49, Viral interference), many prefer giving each virus individually, usually in the order of types 2, 1, and 3 at intervals of 6 to 8 weeks (type 2 has the broadest immunogenic potential, and therefore if given first may permit secondary-type Ab responses to types 1 and 3). Interference may also be minimized by administering all three viruses together in different amounts, with the largest quantity of type 1 and the least of type 2. Three doses of the trivalent vaccine are recommended for maximum Ab response. Preexisting infection of the alimentary canal with other enteroviruses may interfere with successful implantation of the poliovirus vaccine strains. Hence, community immunization programs are usually carried out in the winter or early spring, when infectivity with enteroviruses is less prevalent. Since the greatest incidence of vaccine-induced poliomyelitis is in adult males, immunization is recommended during childhood, preferably from 3 to 18 months of age.

Critique of Poliovirus Vaccines. Each class of the vaccines has advantages and disadvantages.

Inactivated virus vaccine, which is now of high potency and moderate purity, has the distinct **advantage** that it is safe and remarkably effective provided it has been properly employed—for example, in Sweden poliomyelitis has been virtually eliminated by the exclusive use of the inactivated vaccine. It also has the following **disadvantages:** 1) logistic problems of administration by sterile injection to large numbers of people, especially children; 2) greater cost, both for administration

and for several doses of vaccine; 3) requirement for booster immunizations every 2 to 3 years; and 4) failure to eliminate intestinal reinfection and fecal excretion, so that herd immunity may not be produced. This last feature, however, could also be an advantage if immunity is not long-lasting, since natural infection could then proceed at the usual rate, inducing immunity, without producing paralytic disease.

The **advantages** of the **live attenuated virus vaccine** are clear: 1) it is easily administered; 2) it is relatively inexpensive; 3) it results in synthesis and excretion of IgA Abs into the gastrointestinal tract, therefore producing alimentary tract resistance and conferring herd as well as individual immunity; and 4) its effectiveness approaches 100%. It theoretically may produce long-lasting immunity, but controlled evidence is still meager; and we may note that immunization against smallpox by an attenuated virus vaccine does not confer complete, long-lasting immunity (Ch. 54, Prevention and control). The **disadvantages** of

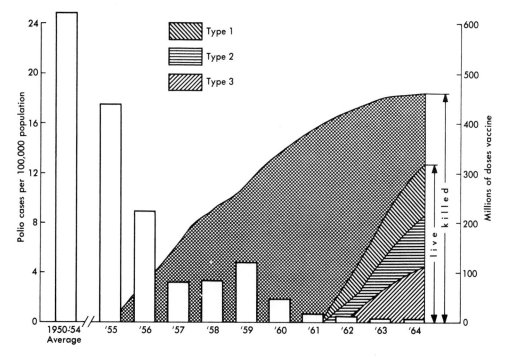

Fig. 55-11. The annual poliomyelitis case rate correlated with the cumulative distribution of vaccine, killed and live, from 1950 to 1964. In the period 1950–54, the average incidence of reported poliomyelitis was 24.8 per 100,000 population; in 1964 it was 0.1 per 100,000 population. (Adapted by the National Foundation from statistics of the Communicable Disease Center, Public Health Service, Atlanta, Ga.)

the attenuated virus vaccine as presently constituted are 1) genetic instability of the type 3 virus employed, and 2) dissemination of virus to unvaccinated contacts. The latter phenomenon might be advantageous by increasing the resistance of members of a group; it is, however, a potential hazard because the transmission is uncontrolled and the viruses may change in virulence.

Despite the safe immunization of millions with live virus vaccine in many countries, its acceptance for general use was slow in the United States, owing to the earlier accident with the inactivated virus vaccine, and to fear of reversion of the attenuated strains to neurovirulence. The initial hesitancy has been overcome, however, and the live virus is now the predominant vaccine used.

Whatever the advantages of either kind of vaccine may be, both have been used in the United States with remarkable effects: in 1955, the year the inactivated virus vaccine was approved for general use, 28,985 cases of poliomyelitis were reported in the United States; in the following year there were 15,140 cases, in 1964 there were only 121 cases (Fig. 55-11), and in 1969 (after the shift to the live vaccine) there were no deaths and a mere 19 cases of paralytic disease were reported.

Because there is no reservoir for polioviruses other than man, it is theoretically possible to eradicate the disease by global immunization. Owing to the practical problems that would be encountered by such a program, however, its fulfillment seems unlikely. Therefore, as sanitation improves and immunization decreases the prevalence of infection as well as disease, more and more people could reach adulthood, when disease is most dangerous, without ever encountering the virus, and therefore without protective antibodies. Hence, the constant threat that virus will be introduced into a poliovirus-free region will probably make vaccination, at least of children, necessary indefinitely. It is depressing that social and economic factors, however, hamper the attainment of even this limited goal: as noted above, many of the economically disadvantaged do not become vaccinated, and as the number of susceptibles increases (for lack of early natural infection) the threat of devastating epidemics once again may arise.

COXSACKIEVIRUSES

The value of the suckling mouse for isolation of viruses was made apparent by the demonstration that yellow fever virus and other togaviruses are more infectious and pathogenic for newborn than for adult mice.

Attempting to apply this unique host to poliomyelitis, Dalldorf and Sickles isolated a new virus from the feces of two children in Coxsackie, New York, in 1948; this new development offered the first major clue that many viruses other than polioviruses infect the intestinal tract of man.

Coxsackieviruses are distinguished from other enteroviruses by their much greater pathogenicity for the suckling than for the adult mouse. They are divided into two groups on the basis of the lesions observed in suckling mice: **group A viruses** produce a diffuse myositis with acute inflammation and necrosis of fibers of voluntary muscles; **group B viruses** evoke focal areas of degeneration in the brain, focal necrosis in skeletal muscle, and inflammatory changes in the dorsal fat pads, pancreas, and occasionally the myocardium.

PROPERTIES OF THE VIRUSES

PHYSICAL AND CHEMICAL CHARACTERISTICS

Those few coxsackieviruses that have been appropriately examined are similar to polioviruses in size, shape, stability, molecular weight, and chemical composition (Tables 55-2 and 55-3). The base composition of the RNA, however, differs from that of poliovirus, especially in the greater proportion of guanine (Table 55-5).

IMMUNOLOGICAL CHARACTERISTICS

Each coxsackievirus (23 group A and 6 group B viruses) is identified by a type-specific

TABLE 55-5. Base Composition of Coxsackievirus and Poliovirus RNA

Base	Virus A9	A10	Polio*
Guanine	1.11	1.13	0.96
Adenine	1.08	1.09	1.14
Cytosine	0.82	0.84	0.88
Uracil	0.99	0.93	1.01

* Average values of the three types of polioviruses. Data from SCHAFFER, F. L., MOORE, H. F., and SCHWERDT, C. E. *Virology 10:*530 (1960).
From MATTERN, C. *Virology 17:*530 (1962).

antigen, measured by neutralization and complement fixation. In addition, all group B and one group A (A9) share a group antigen which is detected by agar-gel diffusion. Cross-reactivities have also been observed between several group A viruses, but no common group A antigen has been found. A virion antigen of a few types causes agglutination of group O human RBC at 37° (maximum titers are obtained with RBC from newborns); the hemagglutinating antigen is type-specific.

Type-specific Abs usually appear in the blood within a week after onset of infection in man, and they attain maximum titer by the third week. Neutralizing Abs persist for at least several years, but complement-fixing Abs decrease rapidly after 2 to 3 months. Resistance to infection, according to epidemiological data, appears to be long-lasting. In patients infected with group B or A9 viruses the Ab directed against the so-called group-specific antigen appears earlier and persists longer than the type-specific Ab (suggesting a secondary response resulting from prior infection by a related virus).

HOST RANGE

Suckling mice inoculated by the intracerebral, intraperitoneal, or subcutaneous route are employed for propagation and isolation of coxsackieviruses. Mice 4 to 5 days old are still susceptible to infection by group A viruses, but group B viruses multiply best in mice 1 day old or less. Even adult mice can be rendered susceptible by the administration of cortisone, X-irradiation, continuous exposure to cold (4°) during the period of infection (group B viruses), or severe malnutrition (group B viruses). Denervation of the limb of an adult mouse also increases susceptibility to group A viruses, with resulting myositis and muscle necrosis limited to the injured extremity.

The striking differences in susceptibility of newborn and adult mice may be partially explained by the finding that only the latter produce interferon when infected by coxsackieviruses. The increased susceptibility produced by cortisone may also be accounted for by this phenomenon; for, as reviewed earlier, cortisone inhibits synthesis of interferon (Ch. 49).

Newborn mice infected with **group A viruses** develop a total flaccid paralysis, resulting from severe and extensive degeneration of skeletal muscles; there are no significant lesions elsewhere. Muscle necrosis may be so extensive that the marked liberation of myoglobulin results in renal lesions similar to those developing in the crush syndrome.

Group B viruses produce quite different manifestations including tremors, spasticity, and spastic paralysis. Degeneration of skeletal muscle is focal and limited. The most prominent pathological lesions are necrosis of brown fat pads, encephalomalacia, pancreatitis, myocarditis, and hepatitis. Adult as well as suckling mice develop pancreatitis, but in adult mice the other lesions do not appear or are so minimal that the mice survive. Necrosis of the myocardium is often noted and is markedly increased by cortisone. Cortisone and pregnancy in adult mice transform an inapparent infection into a fatal one.

Intracerebral inoculation of rhesus monkeys with A7 and A14 viruses produces widespread degeneration of ganglionic cells of the central nervous system followed by flaccid paralysis similar to that caused by poliovirus. (Because of this behavior A7 virus was initially mistaken for a new type of poliovirus.)

Tissue culture technics have become increasingly valuable for study and isolation of coxsackieviruses, and by repeated passage attenuated strains can be obtained. As summarized in Table 55-6, the group B and A9 viruses multiply readily in a number of cell cultures, but most group A viruses do not. The similar tissue culture host range of group B and A9 viruses parallel their antigenic relation (above) and further suggests that these viruses are very closely related, although they have dissimilar pathological reactions in suckling mice.

VIRAL MULTIPLICATION

The multiplication cycle of coxsackieviruses is very similar to that of polioviruses. However, the assembled virions tend to remain

TABLE 55-6. Multiplication of Coxsackieviruses in Tissue Cultures

Viruses	Monkey kidney	HeLa	Human amnion
Group A			
Types 1, 2, 4–6, 19, 22	—	—	—
Type 7	±*	±*	—
Type 9	+	±*	+
All others	±*	±*†	±*
Group B			
Types 1–6	+	+	+

* Not readily isolated in tissue culture, but strains have been adapted to multiply in indicated cells.

† A13, A15, A18, and A21 multiply readily on first passage.

Adapted from WENNER, H. A., and LENAHAN, M. F. *Yale J Biol Med 34:*421 (1961).

TABLE 55-7. Clinical Syndromes Associated with Coxsackieviruses

Clinical syndrome	Group	Type*
Aseptic meningitis	A	**2, 4, 7, 9**
	B	1, **2**, 3, **4**, 5, **6**
Paralytic disease	A	2, **7**, 9
	B	3, 4, 5
Herpangina	A	2, 4, 5, 6, 8, 10
Fever, exanthema	A	2, 4, 9, **16**
	B	4
Acute upper respiratory infection	A	21
	B	2, 3, 4, 5
Epidemic pleurodynia or myalgia	B	**1, 2**, 3, **4, 5**
Myocarditis of the newborn	B	2, 3, 4, 5
Interstitial myocarditis and valvulitis in infants and children	B	2, 3, 4, 5
Pericarditis	B	1, 2, 3, 4, 5
Undifferentiated febrile illness	All	All

* Predominant viruses are printed in boldface type.

within the cell rather than to be immediately released into the culture medium. The cytopathic changes in cells are also similar to those caused by other enteroviruses, but those produced by group A viruses develop much more slowly than those produced by polioviruses.

PATHOGENESIS

Most coxsackievirus infections in man are mild; infections mimicking those of man have not been produced in laboratory animals. Hence, we have very little knowledge of the pathogenesis of human infections or the pathology of the lesions. In biopsies obtained from a few patients with coxsackievirus A infections focal necrosis and myositis were noted, but the lesions were not distinctive. Children who died of **myocarditis of the newborn,** a highly fatal disease caused by group B coxsackieviruses, showed edema, diffuse focal necrosis, and acute inflammation of the myocardium; focal necrosis with inflammatory reaction in the liver, adrenals, pancreas, and skeletal muscle; and occasionally diffuse meningoencephalitis. Group B coxsackieviruses also appear to cause mild interstitial, focal myocarditis and occasionally valvulitis in infants and children.

The coxsackieviruses can produce a remarkable variety of illnesses (Table 55-7), and even the same virus may be responsible for quite divergent types of disease, e.g.,

aseptic meningitis and **herpangina.*** Moreover, a number of group A viruses have not been definitely implicated as causative agents of any human disease. Some viruses in each group are associated with at least one distinctive syndrome, which can usually be diagnosed on clinical grounds alone. Thus, herpangina is caused by group A viruses, and **epidemic pleurodynia†** and myocarditis of the newborn by group B viruses. The other syndromes listed present no clinical features distinctive for coxsackieviruses. Illnesses simulating paralytic poliomyelitis can rarely be induced, particularly by A7 virus. It is curious that only A21 and several group B viruses cause an acute upper respiratory cold-like illness.

---

* **Herpangina** is an acute disease with sudden onset of fever, headache, sore throat, dysphagia, anorexia, and sometimes stiff neck. The diagnosis is dependent upon recognition of the pathognomonic lesions in the throat: at the onset small papules are present, but these soon become circular vesicles that ulcerate.

† Epidemic pleurodynia or epidemic myalgia (Bornholm disease) is an acute febrile disease with sudden onset of pain in the thorax, "a stitch in the side," which is aggravated by deep breathing (simulating pleurisy) and movement. The pain may be chiefly abdominal or associated with other muscle groups, and may be accompanied by muscle tenderness.

Group B viruses may produce myocarditis of newborns in man by intrauterine infection as can certain group A viruses in mice. These findings suggest that these agents may, like rubella virus, be responsible for some cases of congenital heart disease. Indeed, women with coxsackievirus infections during the first trimester of pregnancy have been shown to give birth to newborns with twice the incidence of congenital heart lesions.

LABORATORY DIAGNOSIS

Etiological diagnosis of coxsackievirus infections depends upon isolation of the causative agent from feces, throat secretions, or cerebrospinal fluid by inoculating suckling mice. However, for the initial isolation of group B and A9 viruses (Table 55-6), inoculation of tissue cultures is more suitable. In autopsies of patients with myocarditis and valvulitis the antigen of group B viruses has been demonstrated by immunofluorescence.

A newly isolated virus is grouped as A or B on the basis of the lesions produced in suckling mice. Type identification is considerably more cumbersome, owing to the large number of types. Coxsackieviruses that induce hemagglutination of human group O red blood cells from newborns at 37° (B1, B3, B5, A20, A21, and A24) are rapidly distinguished from other coxsackieviruses, and can be readily identified by hemagglutination-inhibition titrations. With group A viruses, because of the large number of types, identification is initiated by neutralization titrations using pools containing several type-specific sera. Final identification is accomplished with the individual type-specific sera.

Serological diagnosis without viral isolation is not practicable because of the large number of possible antigens (e.g., 29 coxsackieviruses alone) needed for titrations with each patient's serum. However, identification of an isolated virus as the cause of a particular illness requires serological confirmation of infection by neutralization, immunofluorescence, complement-fixation, or hemagglutination-inhibition titrations; this is necessary because many enteroviruses, especially those isolated from feces, appear to be present as harmless inhabitants of the intestinal tract rather than as etiological agents of a current disease.

EPIDEMIOLOGY AND CONTROL

Coxsackieviruses are widely distributed throughout the world, as demonstrated by the occurrence of proved epidemics and by the results of serological surveys. The type prevalent in any locality varies every few years, probably owing to the development of immunity in the population. For example, in 1947–1948 coxsackievirus B1 was predominant in epidemics observed in New York and New England; but by 1951 the B3 virus produced epidemics throughout the world replacing the B1 virus.

Coxsackieviruses are rather highly infectious within a family or the closed population of an institution (about 75% of susceptibles are infected). However, the mechanism of spread may vary with strain of virus and the clinical syndrome. Most clinical infections and epidemics occur in summer and fall, and the viruses are frequently present in the feces, suggesting a fecal-oral spread. Some viruses, however, may be isolated from nasal and pharyngeal secretions and produce acute respiratory disease, suggesting spread by the respiratory route as well.

No effective control measures are yet available. Immunization is not practical because of the large number of viruses that induce disease in man and the relative infrequency of epidemics caused by any single virus.

ECHOVIRUSES

The first echoviruses were accidentally discovered in human feces, unassociated with human disease, during epidemiological studies of poliomyelitis. Viruses are now termed echoviruses if they are found in the gastrointestinal tract, produce cytopathic changes in tissue cultures, do not induce detectable pathological lesions in suckling mice, and have the properties listed in Table 55-2. Most echoviruses, however, are no longer "orphans" in the world of human diseases, but have been associated with one or more clinical syn-

TABLE 55-8.   Clinical Syndromes Associated with Infection by Echoviruses

| Clinical syndrome | Associated echoviruses type | | | |
| | Common | | Uncommon | |
	Epidemic	Endemic	Epidemic	Sporadic
Aseptic meningitis	4, 6, 9, 30		11, 16, 19	1–3, 5, 7, 13–15, 17, 18, 20–22, 25, 31
Neuronal injury				
Paralysis			4, 6, 30	1, 2, 9, 11, 16, 18
Encephalitis				2–4, 6, 7, 9, 18, 19
Rash, fever	4, 9		16	1–7, 14, 18, 19
Acute upper respiratory				
infection		20?		4, 8, 11, 22, 25
Enteritis	6		14, 18	8, 11, 12, 19, 20, 22–24, 32

Modified from WENNER, H. *Ann NY Acad Sci 101*:308 (1962).

dromes ranging from minor acute respiratory diseases to afflictions of the central nervous system (Table 55-8).

Thirty-three viruses were assigned echovirus serotype designations. However, when they were characterized it became apparent that some should be reclassified. Echoviruses 10 and 28 have now been given other names (Table 55-1).

PROPERTIES OF THE VIRUSES

Data on the characteristics of echoviruses are exceedingly fragmentary, and the viruses cannot be adequately compared.

PHYSICAL AND CHEMICAL CHARACTERISTICS

The morphology and general chemical characteristics are similar to those of polioviruses. Infectious RNA has been extracted from a number of echoviruses, but it has not been studied in detail chemically.

Echoviruses are generally heat-stable, but there are marked variations, and some are considerably less stable than polioviruses. But like polioviruses, 1 M $MgCl_2$ stabilizes echoviruses to heat inactivation.

HEMAGGLUTINATION

Of the 31 echoviruses, 12* show **hemagglutinating activity** with human group O erythrocytes. Maximum titers are obtained with

---

* Types 3, 6, 7, 11–13, 19–21, 24, 29, and 30.

RBC from newborn humans (as with some coxsackieviruses), but the optimum temperature for the reaction varies with the virus.[†] The hemagglutinin is an integral part of the viral particle and appears to be a glycoprotein. Some types (3, 11, 12, 20, and 25) elute spontaneously from agglutinated RBC at 37°; but unlike myxoviruses and paramyxoviruses (Chs. 56 and 57) they do not remove the receptors from the cells, which are still agglutinable by the same or other echoviruses.

IMMUNOLOGICAL CHARACTERISTICS

The type designation of each echovirus is dependent upon a specific antigen in the viral capsid, and neutralization titration is the most discriminating method for its identification. There is no group echovirus antigen but hetrotypic cross-reactions occur between a few pairs,[‡] causing major difficulties in the identification of freshly isolated viruses and in the serological diagnosis of infections.

Immunological studies can also be carried out by complement-fixation and hemagglutination-inhibition titrations. Complement-fixation

---

[†] Maximum titers are obtained at 4° for types 3, 11, 13, and 19; and at 37° for types 6, 24, 29, and 30. Hemagglutinin titers for types 7, 12, 20, and 21 are independent of temperature.

[‡] Types 1 and 8 show a major antigenic overlapping by neutralization titrations, and type 12 cross-reacts to a lesser extent with type 29 virus. Antibodies directed against type 23 neutralize type 22 virus, but the reciprocal reaction does not occur. Minor reciprocal cross-neutralization occurs between types 11 and 19, and between types 6 and 30.

TABLE 55-9. Susceptibility of Cell Cultures to Echoviruses

Tissue culture	Echoviruses
Rhesus and cynomologus monkey kidneys	All
Patas monkey kidney	Types 7, 12, 19, 22–25
Human amnion and kidney	All
Continuous human cell lines	Poor until adapted

titrations have the advantage of simplicity but the disadvantage of increased cross-reactivity among echoviruses; it is also difficult to obtain satisfactory antigens for this assay from some isolates.

HOST RANGE

The original notion that echoviruses are not pathogenic for experimental animals has proved to be incorrect for at least 14 of the known viruses.* Intraspinal or intracerebral inoculation of virus into rhesus and cynomologus monkeys initiates viremia, neuronal lesions, and meningitis occasionally associated with detectable muscle weakness. Some strains of types 6 and 9 produce lesions in newborn mice similar to those induced by group B and A coxsackieviruses, respectively.

Cultures of kidney cells from rhesus or cynomologus monkeys are the most suitable for isolation and propagation of all echoviruses (Table 55-9). The final cytopathic changes produced by most echoviruses are similar to those induced by polioviruses and coxsackieviruses (Fig. 55-4B).

VIRAL MULTIPLICATION

Although only fragmentary data are available, the multiplication of those echoviruses studied resembles that of polioviruses.

Echoviruses replicate in the cytoplasm of infected cells. The virions appear to assemble and become oriented in columns supported by a fine filamentous lattice distinct from the endoplasmic reticulum (Fig. 55-12, similar to the assembly process for coxsackieviruses). Crystalline viral arrays may form. Viral particles are subsequently dispersed in the

---

* Types 1–4, 6–9, 13, 14, 16–18, and 20.

cytoplasm and released from the host cell through small rents in the plasma membrane or in cytoplasmic protrusions that are shed from the cell. Eventually infected cells disrupt. Types 22 and 23 echoviruses, in contrast, appear to have a different mode of replication: they cause characteristic nuclear changes, and unlike all other echoviruses they are not inhibited by 2(α-hydroxybenzyl)-benzimidazole or guanidine·HCl (Ch. 49).

PATHOGENESIS

Echoviruses usually enter man by the oral route; a few probably infect through the respiratory tract. The primary cells infected are in the alimentary or respiratory tract, and the majority of infections probably remain limited to them. It is obvious, however, from the clinical manifestations elicited by these agents (Table 55-8), that virus occasionally disseminates beyond the initial organs infected, causing fever, rash, and central nervous system infections. In fact, virus can be isolated from the blood in several of the syndromes listed, and from the cerebrospinal fluid in aseptic meningitis.

The pathology of echovirus infection of man is still unknown, owing to the general mildness of the diseases. Cerebral edema and some focal destructive and infiltrative lesions have been noted in the CNS of the rare fatal case examined, but the pathology is not distinctive. Similar neurological injury has been produced in monkeys and chimpanzees following intracerebral or intraspinal inoculation of the viruses.

LABORATORY DIAGNOSIS

Viral isolation in rhesus monkey kidney cells offers the most sensitive and reliable procedure for diagnosis of an echovirus infection. Feces and throat secretions are the most abundant sources of viruses, but infectious virus persists in feces longer than in any other body excretion or fluid. Use of kidney cell cultures from patas as well as rhesus monkeys are valuable for identification of the specific virus type isolated (Table 55-9). The differ-

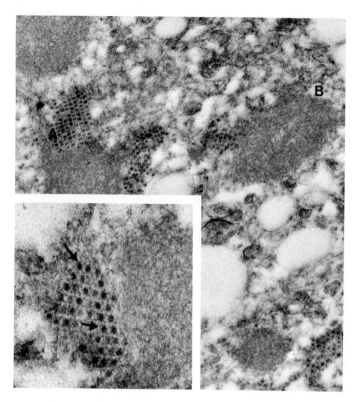

Fig. 55-12. Assembly of type 9 echovirus particles in cytoplasm of infected cells. Viral particles appear to differentiate in columns upon a fine filamentous lattice at cytoplasmic template sites (A). Another crystal-like array of virions associated with finely granular masses is indicated at B. ×50,800. **Inset.** A mass is shown at higher magnification (×112,800). Transected fibrils lie between the particles (arrows). [From Rifkind, R. A., Godman, G. C., Howe, C., Morgan, C., and Rose, H. M. *J Exp Med 114*:1 (1961).]

ential host susceptibility noted, and the limited number of viruses (12) possessing hemagglutinating activity, afford convenient tools for the preliminary grouping of an unknown echovirus and reduce the expense of the immunological identification of a freshly isolated agent. Neutralization titrations provide the final criterion for identification because of their greater specificity.

Diagnosis solely by serological analysis of the patient's paired sera is cumbersome and expensive* and is usually employed only during an epidemic caused by a single virus type.

_____

* Because of the virtual absence of shared antigens it would require that each serum be tested with 31 different viruses.

EPIDEMIOLOGY AND CONTROL

The epidemiological features of echovirus infections resemble those of other enteroviruses, especially coxsackieviruses. But for those echoviruses that multiply in the pharynx as causative organisms of respiratory infections, and particularly for those, such as type 9 echovirus, that produce extensive waves of infection which resemble influenza more closely than poliomyelitis, respiratory secretions may be a more significant route of viral transmission than feces.

Immunization does not appear practicable or warranted because of the large number of viruses and the relative infrequency of epidemics produced by a single agent.

RHINOVIRUSES

Acute afebrile upper respiratory diseases, grouped clinically as the common cold, are the most frequent afflictions of man. Although the diseases are not serious, they cause much discomfort as well as the loss of more than 200 million man-days of work and school each year in the United States alone. Hence, there have been many attempts to discover the etiology of this syndrome. Kruse in 1914 showed that the common cold could be transmitted to man by a filtrable agent, but subsequent studies in every conceivable animal failed to isolate the virus; only man seemed susceptible! From extensive human transmission experiments by Andrewes and his colleagues in England, and by Dowling and Jackson in the United States, the notion emerged that the common cold was caused by a large number of viruses, rather than by a single agent, as had been commonly thought. (They also presented evidence that seems to explode the myth that cold, dampness, and thin clothes provoke the onset of a cold.)

In 1956, using tissue culture methods unsuccessfully employed by many previous investigators, Price, and Pelon and Mogabgab, independently isolated similar viruses from patients with mild upper respiratory infections fitting the description of the common cold.* Subsequently at least 89 immunologically distinct but biologically related viruses have been isolated. Their chemical and physical characteristics led to their classification in a group, designated **rhinoviruses** (Gr. *rhino* = nose), of the family picornaviruses (Table 55-1). Some members of this group multiply in vitro in cultures of either monkey or human cells **(M strains),** but the majority multiply only in human cells **(H strains).**

PROPERTIES OF THE VIRUSES

Only a few rhinoviruses have been investigated in any detail, so it is not yet known how closely the rest are related.

Morphology. The few rhinoviruses examined appear to have icosahedral symmetry and resemble other picornaviruses (Table 55-2). In the subunit structure of the capsid the proteins seem to be somewhat more loosely bonded.

Physical and Chemical Characteristics. Owing to the relatively poor viral yield in cell cultures only a few of the many rhinoviruses have been examined (particularly types 1, 11, and 14). The virion consists of a single molecule of RNA (2.1×10^6 to 2.6×10^6 daltons) and a capsid containing many copies of four different polypeptide chains.

Rhinoviruses are sharply distinguished from other picornaviruses owing to their inactivation at low pH and the maintenance of their infectivity at 50°. When rhinoviruses are held at pH 3 to 5 for 1 hour at 37°, the virions are disrupted, yielding RNA and denatured protein, and more than 99% of their infectivity is lost; the other picornaviruses are not affected by these conditions. Conversely, rhinoviruses are more stable than the other picornaviruses when heated at 50° at neutral pH. Not all rhinoviruses have yet been tested for heat stability, but the H strains appear to be more stable than M strains. Molar magnesium chloride partially stabilizes the M strains.

No hemagglutinating activity has been detected for rhinoviruses, but they have the novel ability to inhibit the spontaneous hemagglutination of trypsin-treated human red blood cells.

Immunological Characteristics. Each of the distinct rhinoviruses identified possesses a type-specific antigen. There is no common group antigen, nor is there consistent evidence for cross-reactivity among members of the group.

Natural infections in man stimulate the production of type-specific neutralizing Abs which confer resistance to reinfection by virus of the same type. The Ab response appears to be greater to the M than to the H strains.

Host Range. Man is the natural host for rhinoviruses. The gibbon monkey is the only known laboratory animal susceptible to infection. After intranasal inoculation, virus multiplies in the nasal and pharyngeal mucosal

---

* The agents were initially classified as echovirus type 28.

cells, type-specific antibodies appear, but disease does not develop.

Cell and organ cultures are the only practical experimental hosts available. The **H viruses** infect and produce cytopathic changes in human embryo kidney cells and in certain human diploid cell lines. The **M viruses** multiply and produce cytopathic changes in primary cultures of rhesus monkey and human embryo kidney cells, human diploid cells, and continuous human cell lines (e.g., KB and HeLa cells). Organ cultures of human embryonic nasal and tracheal epithelium are particularly sensitive hosts for multiplication of rhinoviruses. A few recently isolated rhinoviruses multiplied only in these organ cultures, whereas others could be propagated in human cell cultures after preliminary isolation and several passages in organ cultures. Whether these rhinoviruses with limited host ranges to highly differentiated cells are new types or only uniquely fastidious strains of previously isolated types is unknown.

One may well inquire why rhinoviruses were not isolated sooner, since the diseases they cause are so important and the cells suitable for their isolation and propagation were available. The initial isolations were accomplished with methods employed in earlier unsuccessful studies. The reasons for the ultimate success, where others had failed, are not clear. However, following these first findings came the important discovery of Tyrrell and coworkers that optimal propagation of rhinoviruses, in human embryo or monkey kidney cells, requires special conditions approximating those in the nasal cavities: incubation temperature at 33° and pH of 6.8 to 7.3. The viral multiplication and cytopathology are minimal, or with some rhinoviruses not detectable, if the infected cell cultures are maintained under more usual conditions, i.e., 36° and pH 7.6. With human diploid cell cultures, however, the viruses multiply optimally under ordinary conditions.

The cytopathic changes observed in cell cultures under optimal conditions qualitatively resemble those produced by other picornaviruses, but they are slower to develop. In infected organ cultures ciliary activity diminishes, and superficial epithelial cells begin to shed 18 to 22 hours after infection. Similar to infections in man, only the fully differentiated outer epithelial cells are injured; deeper cells appear unaffected.

Viral Multiplication. The meager data available indicate that the multiplication and biochemical events in the biosynthesis of rhinoviruses do not differ significantly from those of picornaviruses. As for viral isolation, a temperature of 33° is also optimal for maximum viral replication. Supraoptimal temperatures restrict a temperature-sensitive step late in viral multiplication.

PATHOGENESIS

Infection in man appears to be confined to the respiratory tract, and generally fits the syndrome termed the common cold. During the first 2 to 5 days of illness viruses can be isolated from nasopharyngeal secretions but not from other secretions or body fluids. Rhinoviruses have also been associated with some exacerbations of chronic bronchitis and a few cases of bronchopneumonia in children and young adults (i.e., so-called primary atypical pneumonia; cf. Ch. 40). The absence of a satisfactory experimental animal has hindered detailed studies of the pathogenesis of infection.

LABORATORY DIAGNOSIS

The only practical method of establishing the diagnosis is isolation of the etiological agent from nasopharyngeal secretions. Organ cultures of human embryonic nasal or tracheal epithelium are the most sensitive hosts and are required for isolation of some rhinoviruses. Such cultures, however, present more difficulties than cell cultures for routine diagnostic purposes; therefore monolayer cultures of human embryo kidney or human diploid cell stains are the most generally useful for primary isolations. The virus isolated can be typed by neutralization titrations with standard prototype sera. Because of the existence of at least 89 distinct immunological types, the number of titrations required is first narrowed by using pools containing several type-specific sera; final identification is then made with the individual sera present in the effective pool.

Serological diagnosis is not routinely practical because 1) there are no convenient in vitro assays (such as complement fixation) owing to the absence of a family cross-reactive antigen as well as to the small amounts of virus obtained in cell cultures; and 2) neutralization titrations with the patient's sera are not feasible because of the existence of so many distinctive types.

EPIDEMIOLOGY

The epidemiology of specific rhinovirus infections is similar to that of common colds. Seroepidemiological surveys demonstrate that Abs to several prototype viruses are prevalent in many parts of the world and that rhinovirus infections are geographically widespread. Antibodies are found in relatively few infants and children, whereas the majority of adolescents and adults have high titers of Abs to the viruses studied. School children frequently introduce the virus into a family, and virus spreads readily, particularly to those whose nasal secretions lack IgA antibodies. The potential infectivity of rhinoviruses is evident from a study of military recruits during their initial 4 weeks of training: 90% became infected with one or more different viruses and 40% had two or more infections.

The incidence of isolation of rhinoviruses corresponds with the occurrence of minor respiratory infections, with the greatest frequency in fall, winter, and early spring. Rhinoviruses play a significant role as causative agents of common colds, but surely they are not solely responsible for production of these illnesses. For example, in one study of college students over a 2-year period, rhinoviruses were isolated from 24% of patients with common colds, from 2% of those convalescing from the infection, and from 1.6% of well individuals; in a study of military recruits, rhinoviruses were isolated from 31.5% of those with common colds; and investigations carried out in children indicate that rhinoviruses may be isolated from only about 6% of those suffering from common colds. These studies have also demonstrated that inapparent infections occur.

It is common knowledge that the same individual may have repeated episodes of the common cold, even five or six times in a single year. The reason is the prevalence of a large number of immunologically unrelated viruses, including viruses other than rhinoviruses, e.g., A21 coxsackievirus, coronaviruses (Ch. 58), that cause this syndrome. When volunteers were successively infected with four or five different rhinoviruses the infection conferred specific immunity for at least 2 years to homologous but not to heterologous rhinoviruses. The degree of protection appears to depend upon the antibody levels present at the time of reexposure, particularly the level of IgA antibodies in nasal secretions. How long specific immunity persists is unknown.

PREVENTION AND CONTROL

Theoretically it should be possible to prepare a vaccine that could induce immunity for any single virus or for all rhinoviruses. In fact, an inactivated type 13 rhinovirus vaccine has proved effective when administered intranasally to volunteers. However, many types are widespread, the majority of infections are not produced by a few special types, and several may be prevalent concurrently. A vaccine containing 89 different rhinoviruses, therefore, is impractical.

SELECTED REFERENCES: POLIOVIRUSES

Books and Review Articles

BALTIMORE, D. The replication of picornaviruses. In *The Biochemistry of Viruses*, p. 101. (H. B. Levy, ed.) Dekker, New York, 1969.

HOLLAND, J. H. Enterovirus entrance into specific host cells and subsequent alterations of cell protein and nucleic acid synthesis. *Bacteriol Rev 28*:3 (1964).

HORSTMANN, D. M. Epidemiology of poliomyelitis and allied diseases. *Yale J Biol Med 36*:5 (1963).

PAUL, J. R. Poliomyelitis (infantile paralysis). In *Infectious Agents and Host Reactions*, p. 519. (S. Mudd, ed.) Saunders, Philadelphia, 1970.

Symposium: Biology of poliomyelitis. *Ann NY Acad Sci 61*:737 (1955).

Specific Articles

BODIAN, D. Histopathologic basis of clinical findings in poliomyelitis. *Am J Med 6:*563 (1949).

ENDERS, J. F., WELLER, T. H., and ROBBINS, F. C. Cultivation of the Lansing strain of poliomyelitis virus in cultures of various human embryonic tissues. *Science 109:*85 (1945).

GIRARD, M. In vitro synthesis of poliovirus ribonucleic acid: Role of the replicative intermediate. *J Virol 3:*376 (1969).

HORTSMANN, D. M., MCCALLUM, R. W., and MASCOLA, A. D. Viremia in human poliomyelitis. *J Exp Med 99:*355 (1954).

JACOBSEN, M. F., ASSO, J., and BALTIMORE, D. Further evidence on the formation of poliovirus proteins. *J Mol Biol 49:*657 (1970).

JACOBSEN, M. F., and BALTIMORE, D. Polypeptide cleavages in the formation of poliovirus protein. *Proc Nat Acad Sci USA 61:*77 (1968).

SABIN, A. B. Oral poliovirus vaccine: Recent results and recommendations for optimum use. *R Soc Health J 2:*51 (1962).

SALK, J. E. A concept of the mechanism of immunity for preventing poliomyelitis. *Ann NY Acad Sci 61:*1023 (1955).

Special Advisory Committee on Oral Poliovirus Vaccine. Report to the Surgeon General, USPHS. *JAMA 190:*49 (1964).

SELECTED REFERENCES: COXSACKIEVIRUSES

Books and Review Articles

DALLDORF, G. The coxsackie viruses. *Annu Rev Microbiol 9:*277 (1955).

KIBRICK, S. Current status of coxsackie and ECHO viruses in human disease. *Progr Med Virol 6:*27 (1964).

MELNICK, J. L., and WENNER, H. A. The enteroviruses. In *Diagnostic Procedures for Viral and Rickettsial Disease,* 4th ed., p. 529. (E. H. Lennette and N. J. Schmidt, eds.) American Public Health Association, New York, 1969.

Specific Articles

BURCH, G. E., SUN, S., CHU, K., SOHAL, R. S., and COLCOLOUGH, H. L. Interstitial and coxsackievirus B myocarditis in infants and children: A comparative histologic and immunofluorescent study of 50 autopsied hearts. *JAMA 203:*1 (1968).

DALLDORF, G., and SICKLES, G. M. An unidentified, filtrable agent isolated from the feces of children with paralysis. *Science 108:*61 (1948).

PHILIPSON, L., and BERGTSSON, S. Interaction of enteroviruses with receptors from erythrocytes and host cells. *Virology 18:*457 (1962).

SELECTED REFERENCES: ECHOVIRUSES

Books and Review Articles

WENNER, H. A., and BEHBEBONI, A. M. Echoviruses. *Virol Monogr 1:*1 (1968).

Specific Articles

HANZON, V., and PHILIPSON, L. Ultrastructure and spatial arrangement of Echo virus particles in purified preparations. *J Ultrastruct Res 3:*420 (1960).

SELECTED REFERENCES: RHINOVIRUSES

Books and Review Articles

DINGLE, J. H. The curious case of the common cold. *J Immunol 81:*91 (1958).

HILLEMAN, M. R. Present knowledge of the rhinovirus group of viruses. *Curr Top Microbiol Immunol 41:*1 (1967).

KAPIKIAN, A. Z. Rhinoviruses. In *Diagnostic Procedures for Viral and Rickettsial Diseases,* 4th ed., p. 603. (E. H. Lennette and N. J. Schmidt, eds.) American Public Health Association, New York, 1969.

TYRRELL, D. A. J. Rhinoviruses. *Virol Monogr 2:*68 (1968).

Specific Articles

PELON, W., MOGABGAB, W. J., PHILLIPS, I. A., and PIERCE, W. E. A cytopathogenic agent isolated from naval recruits with mild respiratory illnesses. *Proc Soc Exp Biol Med 94:*262 (1957).

PRICE, W. H. The isolation of a new virus associated with respiratory clinical disease in humans. *Proc Nat Acad Sci USA 42:*892 (1956).

chapter **56**

MYXOVIRUSES

HISTORY   1308
CLASSIFICATION   1308
INFLUENZA VIRUSES   1309
PROPERTIES OF THE VIRUSES   1310
  Morphology   1310
  Physical and Chemical Characteristics   1311
  Stability   1314
  Immunological Characteristics   1314
  Genetic Characteristics   1317
  Host Range   1318
  Viral Multiplication   1318
  Toxic Properties   1321
PATHOGENESIS   1322
LABORATORY DIAGNOSIS   1323
EPIDEMIOLOGY   1324
PREVENTION AND CONTROL   1325

HISTORY

In 1918–1919 one of the most devastating plagues in history swept the world, killing approximately 20 million persons and afflicting a huge part of the human population. The underlying disease, **influenza,*** had been known to occur in large epidemics for several centuries. Indeed, the pandemics of 1743 and 1889–1890 were proportionately only slightly less disastrous than that of World War I.

The influenza bacillus (*Hemophilus influenzae,* Ch. 30) was originally thought to be the primary cause of the disease; it was assigned this role by Pfeiffer in the great pandemic of 1889–1890. However, Smith, Andrewes, and Laidlaw in 1933 found that filtered, bacteria-free nasal washings from patients with influenza produced a characteristic febrile illness when inoculated intranasally into ferrets. The viral etiology of influenza was soon confirmed in other laboratories, and it eventually became clear that *H. influenzae* is only one of a number of bacterial pathogens (e.g., *Staphylococcus aureus* and pneumococci) that may cause severe, often fatal secondary pneumonia in patients with influenza.

Following the discovery of the causative agent of influenza, progress in the investigation of the virus and the disease was accelerated by two fortunate findings: 1) influenza viruses can multiply to high titer in the chick embryo, a convenient and inexpensive laboratory animal; and 2) they cause hemagglutina-

---

* Derived from an Italian word reflecting the widespread supposition that the affliction resulted from an evil climatic **influence** by an unhappy conjunction of stars.

tion (HA) of chicken erythrocytes (Ch. 44, Hemagglutination). This reaction, discovered by chance in 1941 by Hirst and also by McClelland and Hare, proved to be of great practical and theoretical importance: 1) it provided a simple method for detecting and quantitating influenza viruses, 2) its specific inhibition by antibodies to the virus provided a highly sensitive **hemagglutination-inhibition** (HI) test for measuring antibody, and 3) its study revealed the mechanism of infection of host cells, since the receptor sites for the virus on the erythrocytes proved to be the same as those on the susceptible host cells. These were shown to be associated with similar mucoproteins, each possessing a terminal N-acetylneuraminic acid (NANA) group attached by a glycosidic bond to an N-acetylgalactosamine residue, which in turn is linked to a protein molecule (Ch. 44, Hemagglutination). As described earlier (Ch. 44), adsorption of virus leads to hydrolysis of the glycosidic bond by a viral enzyme, neuraminidase; the red blood cells thereby become inagglutinable, and the soluble mucoproteins become nonreactive with fresh virus.

CLASSIFICATION

The successful investigations of influenza viruses, and the general availability of tissue cultures, led to the discovery of additional viruses (e.g., parainfluenza viruses). Because these viruses agglutinate erythrocytes and react with similar mucoproteins, they were originally classified together and termed **myxoviruses** (Gr. *myxo* = mucus). When more was learned of their physical and chemi-

TABLE 56-1. Classification of Myxovirus and Paramyxovirus Families

Family	Genus	Species (type)	Subspecies	Subtype*
Myxovirus (Orthomyxovirus)	Myxovirus	Influenza A	Human	$A_0(H_0N_1)$; $A_1(H_1N_1)$; $A_2(H_2N_2)$; A_{HK} (H_3N_2)
			Swine	$(H_{sw}N_1)$
			Horse	
			Avian	
		Influenza B	Human	B†
		Influenza C	Human	
Paramyxovirus	Paramyxovirus	Parainfluenza 1–4		
		Parainfluenza simian 5		
		Mumps		
		Newcastle disease (NDV)		
		Measles		
		Rinderpest		
		Canine distemper		
		Respiratory Syncytial (?)		

* Based upon the immunologically distinct surface antigens, the hemagglutinin (H) and neuraminidase (N), which undergo independent antigenic variation.

† Antigenic variation among strains are known but the information is inadequate to enable division into subtypes.

TABLE 56-2. Characteristics of Myxoviruses and Paramyxoviruses

Characteristics	Myxoviruses	Paramyxoviruses
Particle size	Small (800–1200 A)	Large (1200–2500 A)
Diameter of internal helical core (nucleocapsid)	90 A	180 A
Localization of nucleocapsid	Nucleus	Cytoplasm
Fragmentation of nucleocapsid	+	−
Virion contains RNA polymerase	+	+
Filamentous forms	Common	Unusual
Hemolysin	0	+
Prominent cytoplasmic inclusions	0	+
Syncytial formation	0	+

cal structures it became apparent that major differences among the viruses existed and that they fell into two families, myxoviruses* and paramyxoviruses (Table 56-1), whose distinguishing characteristics are listed in Table 56-2. This chapter will describe myxoviruses (influenza viruses), paramyxoviruses will be discussed in Chapter 57.

INFLUENZA VIRUSES

After the isolation of the causative agent of influenza in 1933 it soon became evident that a complex group of viruses was involved. The agents isolated from humans in England and the United States were found to be similar to, but not identical with, the swine influenza virus isolated by Shope in 1931; and many viral strains isolated proved to be antigenic variants when compared with the initial isolates. In 1940 Francis and Magill, studying patients with influenza in the United States, independently isolated viruses that were im-

* Orthomyxovirus, to contrast with paramyxovirus, is the designation assigned to this family by the International Committee on Viral Nomenclature. Because the term "myxovirus" is so widely used, it will be retained in this edition to take advantage of its familiarity.

munologically distinct from the original strains. The agent isolated in 1933 was termed influenza A virus, and the second discovered was called influenza B virus. A third distinct antigenic type, influenza C virus, was subsequently isolated in 1949. Influenza C virus rarely produces clinical disease and has not been responsible for epidemics. Since the discovery of influenza viruses until the present, new antigenic variants of influenza A and B viruses have continually emerged, highlighted by the appearance in 1947, 1957, and again in 1968, of major variants of influenza A viruses that were only remotely related to the earlier viruses or to each other. The present struggle with this recurring epidemic disease reflects the genetic variability of influenza virus.

PROPERTIES OF THE VIRUSES

MORPHOLOGY

Influenza viruses are somewhat heterogeneous in size and shape, but generally are spherical or ovoid (Fig. 56-1). Filamentous forms occur in recently isolated strains (Fig.

56-2). Influenza A viruses have a mean diameter of 900 to 1000 A while influenza B viruses are somewhat larger, approximately 1100 A in diameter.

Influenza virus particles (like paramyxoviruses) are distinguished by two features: 1) the entire surface is covered by spikes or rods (Fig. 56-1) which are evenly spaced and appear to be arranged in interlocking hexagons so that each rod has six neighbors (type C influenza particles have a somewhat different orientation and the central projections appear absent in many hexagons); and 2) beneath the outer zone is a continuous membrane-like layer (Fig. 56-3). Disruption of the particle by lipid solvents uncovers an inner helical component, the nucleocapsid. Figure 56-3 shows a partially disrupted influenza A particle in which the core is seen as a series of parallel repeating bands. The purified nucleocapsid, shown in Figure 56-4, is composed of several filaments of variable length (averaging 600 A) (the genome of influenza viruses, but not that of paramyxoviruses, is unusual in being composed of several pieces). The elongated nucleocapsid appears to consist of roughly spherical subunits arranged in a double helical chain of two

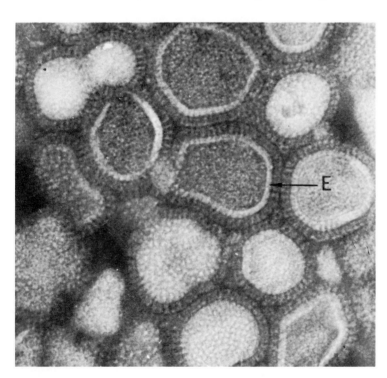

Fig. 56-1. Electron micrograph of intact influenza A virions embedded in phosphotungstate. The virions are of variable shape and size and show evenly spaced, short surface projections covering the entire surface of the particles. E = envelope. ×300,000. [From Hoyle, L., Horne, R. W., and Waterson, A. P. *Virology 13:* 448 (1961).]

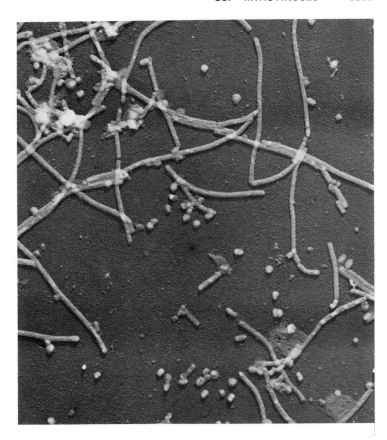

Fig. 56-2. Electron micrograph of influenza A₂(H₂N₂) virus (third passage) showing filamentous as well as a few spherical viral particles. Chromium-shadowed preparation. ×10,400. [From Choppin, P. W., Murphy, J. S., and Tamm, I. *J Exp Med 112*:945 (1960).]

intertwined spirals of the same directional sense (Fig. 56-4, particle designated A).

Upon complete disruption of the virion with sodium dodecyl sulfate (SDS) the **hemagglutinin** and **neuraminidase** subunits can be isolated by electrophoresis or rate zonal centrifugation. Each of these viral surface projections has a unique structure. The hemagglutinin is a rod 140 by 40 A (Fig. 56-5A). The neuraminidase subunit is more complex, with an oblong caput-like structure 85 by 50 A and a centrally attached fiber about 100 A long, which appears to terminate in a knob 40 A in diameter (Fig. 56-5B). The dispersed hemagglutinins are univalent and therefore attach to but do not agglutinate red blood cells. When the SDS is removed the hemagglutinins aggregate into clusters of radiating rods (Fig. 56-5C) which are multivalent and therefore produce hemagglutination. After removal of SDS, neuraminidase subunits aggregate into pinwheel-like structures; but both the single and aggregated neuraminidase units have enzymatic activity.

The filamentous forms of virus (Fig. 56-2), often seen in freshly isolated strains, are very pleomorphic and frequently appear to be composed of spherical subunits; they may be as long as 1 to 2 μ and can be observed by dark-field microscopy. Evenly spaced spikes, similar to those of the more classic spherical particles, project from the surfaces. Filaments, which are apparently infectious, appear to assemble as a result of some defect in the development and emergence of the particles from the cell membrane. The capacity of an influenza virus strain to produce a predominance of filamentous forms is a stable genetic attribute of the strain. However, filamentous forms may also be produced at will in chick embryos in the presence of vitamin A or detergents.

PHYSICAL AND CHEMICAL CHARACTERISTICS

The intact spherical particles contain approximately 0.75 to 0.9% RNA, 75% protein, 6.5% carbohydrate, and 18% lipids. A lower proportion of RNA is found in preparations

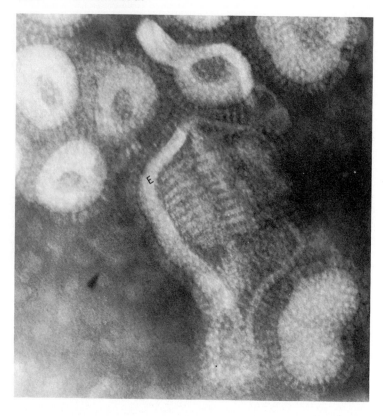

Fig. 56-3. Partially disrupted influenza virus particle. The disrupted outer membrane or envelope (E), which is 60–100 A thick, appears to have collapsed and becomes distorted, revealing the nucleocapsid folded in parallel repeating bands. ×300,000. [From Hoyle, L., Horne, R. W., and Waterson, A. P. *Virology 13*:448 (1961).]

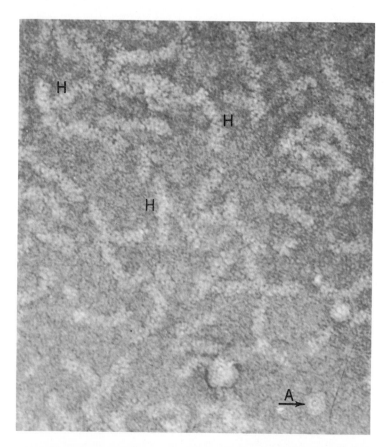

Fig. 56-4. A preparation of highly concentrated nucleocapsid prepared by ether disintegration of purified influenza A virus particles and embedded in phosphotungstate. The helical structure of the nucleocapsid is apparent, particularly in regions marked H. The particle at A is interpreted as part of an elongated structure viewed along the particle axis. ×270,000. [From Hoyle, L., Horne, R. W., and Waterson, A. P. *Virology 13*:448 (1961).]

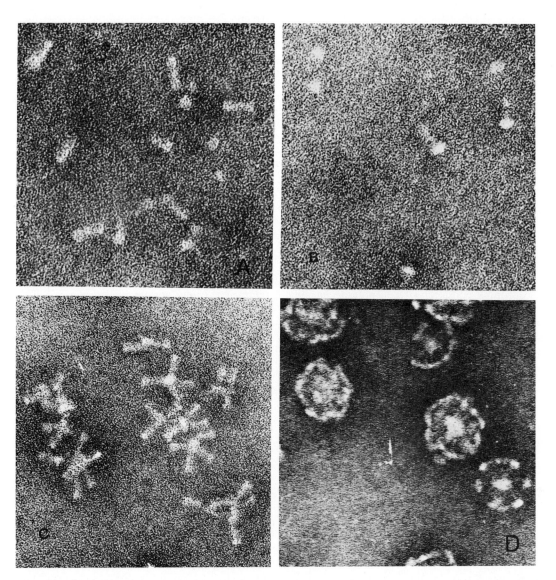

Fig. 56-5. Morphology of the hemagglutinin and neuraminidase subunits of influenza A virus. **A.** Single hemagglutinins appear as thick rods in the presence of sodium dodecyl sulfate (SDS). **B.** Individual neuraminidase subunits dispersed in SDS are seen as oblong structures with a centrally located fiber possessing a terminal knob. **C.** Clusters of hemagglutinins formed by removal of SDS. **D.** Neuraminidase subunits aggregated by the tips of their tails to form pinwheel-like clusters when SDS was removed. ×500,000. [From Laver, W. G., and Valentine, R. C. *Virology* 38:105 (1969).]

containing numerous defective viral particles, or filamentous forms.

Disruption of intact particles with lipid solvents, followed by removal of the envelope components, hemagglutinin and neuraminidase, by reaction with erythrocytes, reveals that the RNA is associated entirely with the inner helical core (the S antigen), which is 5% RNA by weight. In accord with electron micrographs of the nucleocapsid, gentle chemical extraction and physical separation of the nucleocapsid from other viral components yields at least three nucleoprotein pieces. The genome extracted directly from virions appears to consist of at least five or six separate pieces present in three size classes, corresponding to the segments of the nucleocapsid. Each RNA piece has an adenosine triphosphate 5′ terminus, suggesting that the fragmentation of the viral genome reflects the original structure of the nucleocapsid rather than pieces produced artificially by the extraction procedure. This physical structure is consistent with the marked genetic lability and very high recombination rate of influenza viruses (cf. Genetic characteristics, below).

Influenza virus RNAs are single-stranded molecules (a total molecular weight of about 3.9×10^6 daltons, assuming one molecule of each piece per virion). Influenza A and B viruses are probably phylogenetically quite distant for their base compositions differ significantly (i.e., the AU/GC ratio is about 1.25 for type A and 1.42 for type B).

The **hemagglutinin** and **neuraminidase** are glycoproteins containing 4.2% polysaccharide composed of glucosamine, fucose, galactose, and mannose. The lipid of the viral particle is ⅔ phospholipid and ⅓ unesterified cholesterol. The kinds and concentrations of the individual lipids resemble those of the plasma membrane of the host cell in which the virus propagates. Evidence obtained by labeling host cells with ^{32}P before infection and then infecting in the presence of ^{35}P indicates that the lipids incorporated into viral particles, except for phosphatidic acid, are derived from the host cell. At least a portion of the viral polysaccharide is also of host origin. In contrast, the RNA and proteins are specified by the viral genome.

Seven to eight distinct proteins can be separated from completely disrupted viral particles by polyacrylamide gel electrophoresis. A major protein (composed of identical small polypeptides) is associated with the inner surface of the lipid layer and appears to be the major structural protein of the viral envelope. The rod-shaped hemagglutinin contains two species of glycoproteins held together by disulfide bonds and, in accord with electron micrographs (Fig. 55-5), the neuraminidase also consists of two chemically distinct glycosylated polypeptide chains. The nucleocapsid protein is made from identical polypeptides, and there are one or two other minor internal, nonglycosylated polypeptides, one of which may be the virion transcriptase.

Polypeptide maps show little difference between the nucleocapsid proteins from two strains of influenza A, but numerous differences between the hemagglutinins are noted. This result is consistent with evidence that the nucleocapsid proteins of different strains within a given viral type (species) are antigenically similar and therefore have only minimal differences in amino acid sequences, whereas the surface antigens are strain-specific and have many chemical dissimilarities.

STABILITY

The high lipid content makes these viruses susceptible to rapid inactivation by lipid solvents and surface-active reagents. Like most other viruses with lipid envelopes, infectivity of influenza viruses is relatively labile when stored at $-15°$ or $4°$, but infectivity is retained for long periods on storage at $-70°$.

IMMUNOLOGICAL CHARACTERISTICS

Viral Antigens. Influenza viruses possess several antigenic components. 1) The internal or **nucleocapsid antigen,** which corresponds morphologically to the internal helical component (Fig. 55-4), is immunologically identical with a soluble antigen that is present in infected cells. This antigen is assayed by complement-fixation titration. 2) One of the external or surface glycoprotein antigens morphologically corresponds to some of the radiating spikes on the virion's surface, and is biologically and immunologically the same as the **hemagglutinin** (Fig. 55-5A). This antigen is measured immunologically by inhibition of its hemagglutinating activity, neutralization of infectivity, or complement fixation. 3) The other glycoprotein projections on the virion's surface correspond morphologically, immunologically, and biologically to the **neuraminidase** (Fig. 55-5B). Immunological assay is accomplished by inhibition of the enzyme activity. Antibodies specific for neuraminidase do not neutralize infectivity but can inhibit release of budding virus from infected cells and hence diminish the viral yield. 4) The major structural **protein of the viral envelope (M protein)** is measured by complement fixation and immuno-diffusion; its specific antibodies cannot neutralize infectivity nor inhibit hemagglutinin and neuraminidase activities.

Immunological Grouping. On the basis of their nucleocapsid and M protein antigens the large number of influenza viruses are divided into three distinct immunological types (species): A, B, and C. The antigens of each type are unique and do not cross-react with those of the two other types, or with each other.

Within types A and B immunological subtypes are distinguished by major antigenic differences of the external antigens, the hemagglutinin and neuraminidase. In addition, numerous minor antigenic changes of the surface antigens appear to occur within each subtype, especially of the type A viruses. The antigenic drift of the hemagglutinin and neuraminidase proteins, however, is genetically independent. Over the past few decades a new major variant has emerged every few years, but only a single subtype is prevalent at any time.

Antigenic Variation. Influenza A viruses have undergone three major antigenic shifts since 1933. These variations have been detected by immunological studies on sera from persons of different ages and on the viruses isolated (Fig. 55-6). 1) A virus similar to swine influenza virus* presumably accounted for the human infections between 1918 and 1933, since sera collected from persons born during that period contain antibodies to the swine agent. 2) The original human influenza A virus was isolated in 1933 and was responsible for all influenza A infections until at least 1943; moreover, sera collected from persons born during this period have the highest antibody titers to that subtype, A_0 (H_0N_1)†. 3) In 1947 $A_1(H_1N_1)$ viruses emerged and supplanted all prior strains, as indicated by isolations and serum analyses. 4) In 1957 the $A_2(H_2N_2)$ (Asian) influenza

---

* Isolated by Shope in 1931 from pigs with a severe respiratory infection.

† The reader is cautioned to avoid confusion that may result from the terminology currently used: the species is termed type A, and the subtype representing the first human influenza A virus isolated was previously called "A_0". With the recognition that the hemagglutinin and neuraminidase glycoproteins vary independently and determine the antigenic characteristic of a viral strain the subtypes are named accordingly.

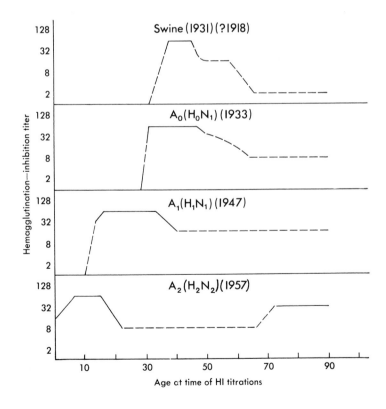

Fig. 56-6. Age distribution of hemagglutination-inhibition (HI) antibodies in the general population to subtypes of influenza A viruses. The highest antibody titers for each virus tested are in those persons who were infected for the first time when that subtype first became prevalent (dates in parenthesis). [Modified from Davenport, F. M., Hennessy, A. V., and Francis, T., Jr. *J Exp Med 98*:641 (1953) with addition of data for A_2 virus.]

viruses became prevalent. 5) In 1968 another relatively large antigenic shift occurred and the A_2 Hong Kong (H_3N_2) virus emerged: the neuraminidase molecules are antigenically similar to the original A_2 (Asian) virus but the hemagglutinin is immunologically unique.

Influenza B viruses have also undergone antigenic variations, but these are neither so extreme nor so frequent as those of A viruses, and immunogenic cross-reactivity occurs among all the subtypes. Hence, classification of influenza B variants into distinct subtypes has not been feasible. The originally isolated B virus was prevalent from 1936 to 1948, and the second variant appeared in 1954. Frequent antigenic shifts then occurred until a new and only distantly related antigenic variant arose in 1962.

The constant antigenic drift of influenza viruses is of considerable practical importance and theoretical interest. Each major shift has found a large proportion of the world population immunologically defenseless against invasion by the newly emerged virus. Furthermore, these major antigenic changes in the surface antigens (particularly the hemagglutinin) rendered ineffective the vaccines then current: the neutralizing antibodies induced by the vaccine did not react with the variant strain. The minor antigenic changes that occur between pandemics more frequently, however, may reduce the effectiveness of the vaccine but do not make it useless. New antigenic variants probably appear by selection of mutants resistant to Abs prevailing in human populations. In fact, new variants have been selected experimentally by passage of viruses in the presence of small amounts of Ab in chick embryos or mice.

Basis for Antigenic Variations. Each new major variant seems to result from adding a new antigenic determinant while retaining some of the previous determinants. Hence, $A_2(H_2N_2)$ virus induces formation of Abs not only against itself but also against the antigens present in $A_1(H_1N_1)$ and $A_0(H_0N_1)$ subtype viruses. However, primary immunization with $A_1(H_1N_1)$ virus will not cause production of neutralizing or hemagglutination-inhibiting Abs that react with $A_2(H_2N_2)$ viruses. Immunological analyses of numerous strains of influenza A virus suggest that the virion contains a mosaic of distinct anti-

genic determinants on the surface projections. The hemagglutinin and neuraminidase subunits are antigenically different from each other, but each consists of several antigenic specificities which cross-react with those of earlier strains. New antigenic specificities are presumably added by minor changes in amino acid sequence or in polysaccharide composition. The complexity of immunological reactivities must reflect, in part at least, the nonconcordant changes of the two independent major antigenic structures, the hemagglutinin and the neuraminidase. With a major antigenic change (a new A subtype) the chemical changes of the hemagglutinin and neuraminidase are sufficiently great to add antigenic reactivities that do not cross-react with prior surface antigens.

Serological responses to successive exposures to influenza viruses, whether by infection or by artificial immunization, indicate that **the antibody response is predominantly directed against the antigens of the viral strain with which one was initially infected,** e.g., if a child were infected first with an $A_0(H_0N_1)$ virus in 1940 and then with an $A_1(H_1N_1)$ in 1947, his antibody response following the second infection would be primarily directed to the $A_0(H_0N_1)$ virus antigens, although he would also develop $A_1(H_1N_1)$ Abs. With advancing age and an increased number of infections the Ab response to infection becomes broader, but the titer of Abs against the antigen of the original infecting virus remains the highest. This phenomenon, termed **"the doctrine of original antigenic sin,"** is reflected in the Ab levels of persons in different age groups (Fig. 56-6); it suggests that influenza virus elicits a primary response to novel determinants of the current infecting strain, but a secondary response and higher Ab titer to the determinants of the original infecting strain.

The prominent antigenic drift, resulting in the appearance of the major antigenic variants described, has led to speculation whether an almost limitless number of major antigenic changes can occur. This question has not been resolved. Before the $A_2(H_2N_2)$ pandemic of 1957 only persons older than 70 years possessed Abs for the new A_2 virus (Fig. 56-6); hence the 1889–1890 influenza pandemic may have been caused by a virus immunologically similar to the A_2 virus of 1957. Persons born after 1890 would not possess the corresponding Abs because the A_2 viral subtype was absent between

1890 and 1957. These observations suggest that influenza viruses have a limited number of possible antigenic variants, and that in 1957 the virus had gone full cycle since 1889–1890. It is noteworthy, in this regard, that the next major variant, which produced extensive influenza epidemics from 1968 to 1970 [A Hong Kong (H_3N_2) virus], was even more closely related immunologically to the virus of the 1889–1890 pandemic than was the 1957 $A_2(H_2N_2)$ variant. But the A Hong Kong (H_3N_2) virus does not appear to be more closely related to the $A_0(H_0N_1)$ subtype than does the $A_2(H_2N_2)$ Japan 1957 strain, and hence the evidence does not completely satisfy the thesis that the ring of immunological variations is closing. Other interpretations are possible. For instance, repeated intranasal infections of guinea pigs with a variety of strains of influenza A viruses, all prevalent before the emergence of $A_2(H_2N_2)$ virus, result in Ab responses with increasingly broad reactivity; eventually Abs that react with the $A_2(H_2N_2)$ influenza virus appear although the animals have not been infected with this virus. Hence, the Abs to $A_2(H_2N_2)$ virus found in 1957 in persons born before 1890 may reflect the broadening specificity of the response to repeated influenza A infections, rather than a response to infection with the A_2 virus in 1889–1890.

Immunity. Either clinically evident or inapparent infection leads to immunity, but the immunity does not appear to be long-lasting. Indeed, reinfections may be caused by variants with minor antigenic differences and have been reported with viruses of antigenic constitution similar to those producing the initial infection 1 to 2 years earlier in institutions (e.g., children's homes) or in military installations. **Immunity is induced by the hemagglutinin on the virion's surface,** since it can be evoked by injection of purified hemagglutinin. Neuraminidase probably also plays a role, for antineuraminidase antibodies effectively reduce viral spread from infected cells and therefore diminish the impact of infection. Thus animals who are challenged with homologous virus after immunization with purified neuraminidase are infected but disease does not develop, and virus is not transmitted to unimmunized cohorts. The immunity is highly type-specific and generally subtype-specific. Antibodies directed against the nucleocapsid antigen, in contrast do not confer immunity.

It follows from these properties of the viral antigens that 1) Abs to influenza A virus do not protect against infection with influenza B viruses, and 2) infection or artificial immuni-

zation with an influenza $A_1(H_1N_1)$ virus affords immunity against infection with other $A_1(H_1N_1)$ viruses but not against influenza $A_2(H_2N_2)$ viruses. It must be emphasized that although immunity to subsequent infection may not be long-lasting, circulating antihemagglutinin and antineuraminidase Abs are present for many years after infection. The role of cell-mediated immunity and its persistence require intensive exploration.

The problems encountered in artificial immunization have two major causes: 1) the marked antigenic variation of the viruses, and 2) restriction of the infections to the respiratory mucous membranes, where antibody concentrations are only approximately 10% of those in the blood. Hence, minor antigenic modifications of the infecting virus permit it to escape neutralization more readily than it could if viremia were an essential part of the infectious process. The situation is analogous to the outgrowth of drug-resistant bacterial mutants when the drug concentration is borderline.

GENETIC CHARACTERISTICS

Influenza viruses, more than any of the other viruses that infect man, show great genetic variability. The mutations capable of producing distinct and almost unrelated antigenic subtypes have already been discussed. Numerous other available genetic markers include alterations in 1) reactions with different species of erythrocytes, 2) avidity for antibody, 3) virulence, 4) reactions with soluble mucoprotein inhibitors, 5) heat resistance, 6) host range, and 7) morphology. Only a few of these mutations have obvious bearing upon the behavior of influenza viruses in nature, but they illustrate the ease with which these agents vary and the types of selective pressures at work.

Variation in the avidity of viruses for specific antibody (Ch. 50) is frequently noted. Thus, viruses isolated in the course of a single epidemic may vary in their susceptibility to neutralization by antisera prepared with homologous virus (isolated during the same epidemic) or with heterologous strains. Strains isolated during the height of epidemics commonly react to high titer only with homologous antibodies. Passage of such strains in the presence of increasing quantities of antibody selects for variants that are inhibited only minimally by either homologous or heterologous antibodies, suggesting that these viruses have been modified so that the usual surface antigens are present in reduced amounts or are partially obscured. Emergence of these variants may permit persistence of virus in the population during interepidemic periods and set the stage for the appearance of new antigenic subtypes. The appearance of strains with minimal cross-reactivity with prior viruses may well be responsible for the initiation of another epidemic.

Numerous mutants with increased virulence for a given host or organ system have been isolated. Conversely, reduced virulence has been correlated with the inability of mutants to multiply effectively at temperatures above 37°. Such temperature-sensitive (*ts*) mutants may prove valuable for live virus vaccines.

Genetic recombination has been extensively studied in influenza viruses because of 1) its special epidemiological and clinical implications for such a variable virus, 2) the unique opportunity it originally afforded to investigate RNA as genetic material, and 3) the numerous markers available.

The first evidence of recombination between animal viruses was obtained by Burnet in 1949, using influenza viruses: infections with mixtures of neurotropic and nonneurotropic strains of different antigenic identity yielded recombinants in which neuropathogenicity from one strain was combined with an antigenic character from the other. Subsequently genetic recombination has been observed with many other naturally occurring strains carrying various markers, and also with conditionally lethal temperature-sensitive mutants. Extraordinarily high recombination frequencies have been reported (up to 50%), but it is likely that these new genotypes are not the consequences of true recombination (Ch. 11) within an RNA molecule but rather that they emerge as the result of independent assortment and segregation of the several segments of the viral genome.

Recombination is detected only between strains that are related antigenically, and not between influenza A and B viruses. When a mixed infection is initiated with high multiplicities of influenza A and B viruses, however, viral particles evolve that have surface antigens of both parent viruses and are therefore neutralizable by Abs to either parent. These doubly neutralizable viruses are not stable on passage and result from **phenotypic mixing** (Ch. 45).

In a process similar to **marker rescue of phage,** recombination may be observed in mixed infection with a virus inactivated by ultraviolet light and an infectious virus with different genetic markers. The recombination frequency is low (10^{-2} to 10^{-3} with closely related strains and 10^{-4} to 10^{-5} with distantly related strains). **Multiplicity reactivation** (Ch. 45, Genetic reactivation of phage inactivated by ultraviolet light) is similarly seen when a viral preparation that is partially inactivated by heat or radiation is inoculated under conditions leading to multiple infection of a single cell; an unexpectedly high yield of infectious virus may result.

HOST RANGE

Strains of human influenza viruses are best propagated experimentally in chick embryos, ferrets, and mice. Wild-type viruses (recently isolated from man) initiate infection with the fewest viral particles, and they multiply to highest titers in the amniotic cavity of chick embryos or the respiratory tract of ferrets. Viruses can be readily adapted to propagation in the allantoic cavity of the chick embryo, as well as in the respiratory tracts of monkeys and many rodents. Chick embryos have proved the most sensitive and the most convenient host. Many strains also multiply readily in cultures of monkey kidney, calf kidney, and chick embryo cells.

VIRAL MULTIPLICATION

Infection is initiated with the attachment of virions to susceptible host cells by reactions between the viral spikes and specific host cell mucoprotein receptors. The host receptors are similar to or identical with those on RBC and with the soluble mucoprotein inhibitors in human and animal secretions (Ch. 44, Hemagglutination). After attachment the viral particles fuse with the cytoplasmic membrane, or they are rapidly engulfed into vacuoles formed by the cell membrane (Fig. 56-7). Infectivity is rapidly lost (**viral eclipse**) as the virions are shorn of their envelope, and the viral nucleocapsid enters directly into the cytoplasm through the rent in the fused virus-cell membranes.

The initial steps in viral replication following entrance of the viral genome into the cell are still unclear. Unlike other RNA-containing viruses (e.g., picornaviruses) ultraviolet ir-

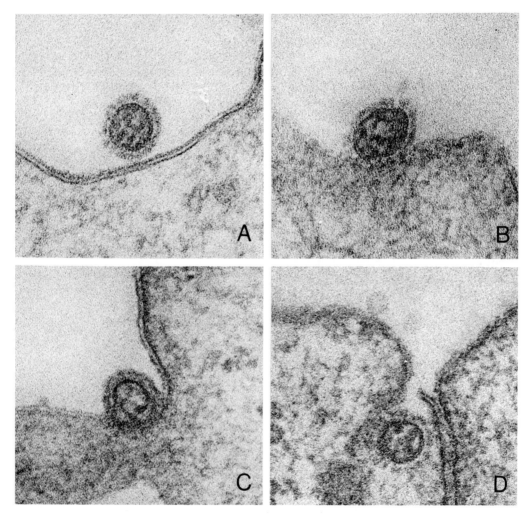

Fig. 56-7. Adsorption and penetration of influenza virus into cells of the chorioallantoic membrane of the chick embryo. **A.** Beginning attachment. **B.** Virion is attached to cell and the viral envelope is fusing with the plasma membrane of the cell. **C.** Fusion of the viral envelope with the plasma membrane is more advanced. **D.** The nucleocapsid has penetrated into the cytoplasm. All ×200,000. [From Morgan, C., and Rose, H. M. *J Virol* 2:925 (1968).]

radiation, actinomycin D, or mitomycin C blocks viral multiplication if administered during the first 2 hours of infection but not thereafter (i.e., before synthesis of viral RNA is established), but chemical inhibitors of DNA biosynthesis (e.g., arabinosyl cytosine, hydroxyurea) do not reduce propagation of infectious virus. Hence, functioning but not replicating host DNA is essential for early events in multiplication of influenza viruses.

Influenza viruses contain an RNA-dependent RNA polymerase **(RNA transcriptase)** within the virion, as do paramyxoviruses (Ch. 57) and rhabdoviruses (Ch. 59), suggesting that the parental viral RNA cannot serve as mRNA (as described for polioviruses, Ch. 55) but that it is complementary to the message and must be transcribed by this enzyme to make functional mRNAs. This possibility is strengthened by the evidence that the virion polymerase has unique requirements for its enzymatic activity (i.e., Mn^+ and NH_4^{++}) which are not shared by the RNA-dependent polymerase **(replicase)** induced in cells infected by influenza viruses. This replicase is responsible for the synthesis of new viral RNA through the usual intermediary replicative forms (Ch. 48). It is uncertain where the viral RNA is replicated. The nucleocapsid protein as shown with fluorescein-labeled or ferritin-labeled antibodies is synthesized in the cytoplasm and is then transported rapidly into the nucleus (Figs. 56-8 and 56-9). The nucleocapsid subsequently moves into the cytoplasm and migrates to the cell membrane. The hemagglutinin and neuraminidase proteins are synthesized in and remain in the cytoplasm (Fig. 56-8).

About 4 hours after infection with influenza A virus discrete patches of the plasma membrane thicken and incorporate hemagglutinin and neuraminidase molecules, which gradually replace the host proteins in these segments (Fig. 56-10). As fragments of the helical nucleocapsid impinge on the altered membrane it buds and forms viral particles which are released as they are completed; virions cannot be detected within the cell.

In the assembly of virions both the collection of the correct set of nucleocapsid fragments and the budding process are imperfectly controlled, accounting for the production of the many noninfectious particles and their morphological heterogeneity. Viral particles are released over many hours, without lysis of the infected cells, but eventually the cells die.*

The sequence of events with influenza B

---

* The mechanism for releasing the budding virions is unclear. It is believed that neuraminidase serves this function, since specific antibody to neuraminidase blocks viral release but does not neutralize infectivity. However, univalent Fab fragments of this antibody do not reduce viral release though they neutralize neuraminidase activity in vitro. The bivalent antibody may thus block viral release by binding virions to the membrane rather than by inhibiting specific enzyme activity.

Fig. 56-8. Multiplication of influenza A virus. A diagrammatic representation of the biosynthesis of virions and viral subunits measured by titrations and immunofluorescence.

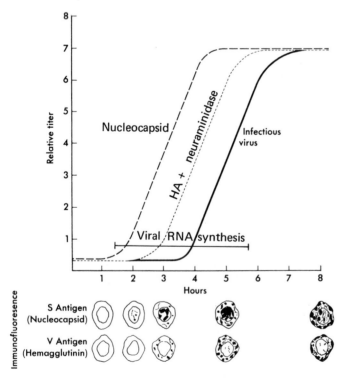

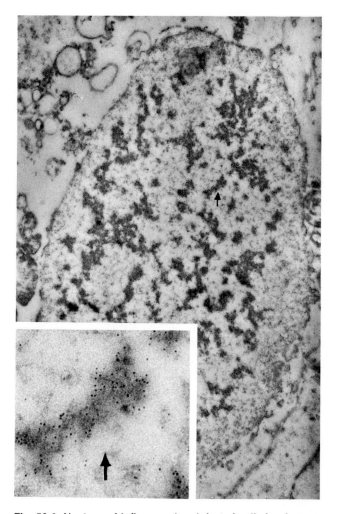

Fig. 56-9. Nucleus of influenza virus-infected cell showing aggregates of dense material labeled with ferritin-conjugated specific antibody (arrow). The chromatin is sparse and the nuclear membranes are disrupted. ×26,000 **Inset.** Higher magnification of the portion of the nucleus marked by arrow. Ferritin-conjugated antibody is present within the aggregates of dense material. The intervening nuclear matrix is nearly devoid of ferritin, i.e., nucleocapsid antigen. ×97,000. [From Morgan, C., Hsu, K. C., Rifkind, R. A., Knox, A., and Rose, H. M. *J Exp Med 114*:833 (1961).]

virus is similar, but the latent period (identical with the eclipse period for viruses that are assembled at the cell membrane) is 1 to 2 hours longer.

TOXIC PROPERTIES

Without production of virions, influenza A and B viruses when added in sufficient amounts can damage cells and induce a pyrogenic effect. After intravenous or intraperitoneal inoculation into mice hepatic necrosis and splenic enlargement occur in 12 to 48 hours; intracerebral inoculation results in brain damage accompanied by convulsions. Fever follows 1 to 2 hours after intravenous inoculation of virus into rabbits, owing to the elaboration of an endogenous pyrogen follow-

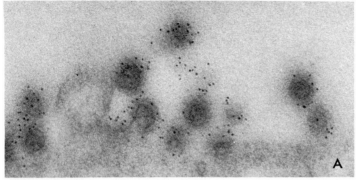

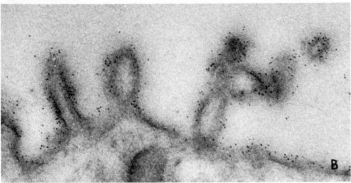

Fig. 56-10. Development of influenza virus particles at surface of an infected cell. Viral particles and components labeled with ferritin-conjugated specific antibody. **A.** In addition to fully formed viral particles, ferritin-antibody complexes have tagged one particle (to the left of center) presumed to be in process of budding and two others (below and to the right) probably at early stages of differentiation. ×140,000. **B.** The surface of an infected cell with several cytoplasmic protrusions to which ferritin-antibody complexes have combined as a result of the virus-specific antigenic change of the surface. No mature viral particles are evident. ×97,000. [From Morgan, C., Hsu, K. C., Rifkind, R. A., Knox, A., and Rose, H. M. *J Exp Med 114*:825 (1961).]

ing reaction of virus with polymorphonuclear leukocytes (Ch. 22, Fever). In vitro this reaction inhibits leukocytes from phagocytizing bacteria or inert particles. The reduction in phagocytic activity is associated with blocking anaerobic glycolysis of the leukocytes (Ch. 22, Phagocytosis).

PATHOGENESIS

Influenza, an acute respiratory disease associated with constitutional symptoms,* results from infection and destruction of cells lining the upper respiratory tract, trachea, and bronchi. Virus enters the nasopharynx and spreads to susceptible cells, whose membranes contain the specific mucoprotein receptors. The virus, however, must first pass through respiratory secretions which contain mucoproteins that can also combine with viral particles. But infection is not blocked because the viral neu-

raminidase hydrolyzes the mucoproteins, rendering them ineffective as inhibitors.†

During acute illness the ciliated epithelial cells of the upper respiratory tract are primarily involved. Viral multiplication is followed by necrosis of infected cells and extensive desquamation of the respiratory epithelium, which is directly responsible for the respiratory signs and symptoms of the acute infection.

Early in the course of the uncomplicated disease constitutional symptoms are more prominent than would be expected from a local infection of the respiratory tract. Viremia, however, is not an essential event in the pathogenesis of influenza infection, and although it has been detected on rare occasions, it cannot be required for producing the constitutional symptoms. These are more likely due

---

* Fever, chills, generalized aching (particularly muscular), headache, prostration, and anorexia.

† The evolutionary selection of viruses containing neuraminidase is understandable. Were influenza virus devoid of its surface hydrolytic enzyme, the secretory mucoprotein would be as effective as antibodies, and infection would be difficult to establish.

to breakdown products of dying cells absorbed into the blood stream. The liberation of endogenous pyrogen from polymorphonuclear leukocytes (see Toxic properties, above) is probably minor in producing fever, since few such cells are found in the upper respiratory tract.

Normally, influenza is a self-limited disease lasting 3 to 7 days. About 10% of patients with clinical influenza have small areas of lobular pulmonary consolidation. Although extensive pneumonia rarely develops, it accounts for most deaths from influenza. Severe influenza virus pneumonias without bacterial invasion occur, but secondary bacterial pneumonias are the major cause of death.

Fatal primary influenza pneumonia occurs most frequently in persons with diminished respiratory function, e.g., patients with mitral stenosis or chronic pulmonary disease and women during the later stages of pregnancy. Viruses isolated from patients dead of primary influenza pneumonia do not appear to be more virulent in experimental animals than those from patients with minor illness.

Fatal nonbacterial influenza pneumonia was observed frequently during the 1918–1919 pandemic, and descriptions of the rare deaths in more recent epidemics are identical. The appearance of the lungs is unforgettable: they are huge, dark red or brightly mottled with subpleural hemorrhages, and distended with edema fluid and blood. An outpouring of the blood and extracellular fluid follows a cut through the lung. Microscopically, marked destruction and often denudation of the tracheal and bronchial epithelium are observed. The bronchioles are dilated and contain cell debris and blood. The alveoli are engorged and distended with hemorrhage and edema fluid, but purulent exudate is absent. Frequently, alveoli are lined by hyaline-like membranes which do not contain fibrin.

The pathogenesis of influenza pneumonia, as observed in ferrets and mice, evolves in two successive stages: 1) necrosis of the bronchial and bronchiolar epithelium, and 2) hemorrhage and edema. The similarities of the ultimate pathological lesions to those in man suggest a comparable sequence of development in the human disease.

Pneumonia during the pandemic of 1918–1919 was predominantly characterized by **secondary bacterial complications.** The most prominent invaders were *Staphylococcus aureus, Hemophilus influenzae,* and β-hemolytic streptococci. In recent epidemics, coagulase-positive *S. aureus* and pneumococci have been most frequent. In addition to the usual severe epithelial injury of influenza, these pneumonias show invasion of the walls of the bronchi and bronchioles by bacteria, destruction of alveolar walls, purulent exudate, abscess formation, and vascular thrombi.

The influence of influenza virus infection on bacterial invasion has been studied experimentally. Introduction of pneumococci or streptococci into the lungs of healthy mice results in little if any inflammatory response. But extensive bacterial pneumonia develops in animals with pulmonary edema induced by influenza virus or irritant chemicals. This effect of influenza viruses is probably promoted by their capacity to inhibit phagocytosis of bacteria by polymorphonuclear leukocytes (Ch. 22, Phagocytosis).

Extrapulmonary lesions have also been observed in fatal cases of influenza pneumonia, including myocarditis; hemorrhage into the adrenals, pancreas, and ovary; and renal tubular degeneration. Viruses have been isolated from such lesions only rarely, however, and the relation of viral infection to the lesions is obscure.

LABORATORY DIAGNOSIS

A presumptive diagnosis of influenza can often be made from clinical and epidemiological considerations. Laboratory confirmation of a clinical diagnosis is generally too costly for individual or sporadic cases, but it is used to establish the presence of the agent in the community, to determine its specific type, and to carry out epidemiological studies. Diagnosis may be established by three general procedures: 1) viral isolation, 2) demonstration of specific Ab increase, and 3) immunofluorescent demonstration of specific antigens in respiratory tract cells present in nasal secretions or sputum.

Viral isolation is most frequently accomplished by inoculating nasal and throat washings or secretions into the amniotic sac of 11- to 13-day-old chick embryos or into monolayers of monkey kidney cell cultures. For general viral isolations use of monkey kidney cells is desirable since many respiratory viruses other than influenza also multiply in these cells. After 2 to 3 days' incubation of embryonated eggs, the presence of virus can be detected in the amniotic fluid by the hemagglutination reaction. Guinea pig or human

RBC are used to assay influenza A and B viruses and chick erythrocytes are employed for influenza C virus. Fresh isolates of influenza virus may not produce cytopathic changes in monkey kidney cells, and the presence of virus is best detected by the **hemadsorption** technic employing guinea pig RBC (Ch. 44). The newly isolated virus from either chick embryos or cell cultures is usually typed by the hemagglutination-inhibition assay, using standard antisera.

Immunological technics are the most frequently used methods for diagnosis of influenza infection. Because the majority of persons already have influenza virus Abs at the time of infection, it is essential to demonstrate an increase in Abs by comparing titers in serum specimens obtained during both the acute and convalescent phases of the disease. **Hemagglutination-inhibition** technics (Ch. 44) are most often employed for this purpose, although they are handicapped by the troublesome presence in serum of nonspecific mucoprotein or protein inhibitors. These may be eliminated by treating the serum after heat inactivation (56° for 30 minutes) with the receptor-destroying enzyme (RDE) from *Vibrio cholerae* (a neuraminidase), with a mixture of trypsin and potassium periodate, or with the adsorbent kaolin.

The **complement-fixation** assay is equally sensitive, and it circumvents the difficulties presented by nonspecific serum inhibitors. With crude preparations which contain the nucleocapsid antigen this test has broad specificity, but if one utilizes the virion's hemagglutinin the assay is just as strain-specific as hemagglutination inhibition. An increase in specific Abs can also be measured by neutralization titrations; but because of its increased expense and the time required, this procedure is employed only for special purposes.

Immunofluorescent technics furnish a method for establishing the diagnosis of influenza while the patient is still acutely ill. Examination of desquamated cells from the nasopharynx after their reaction with fluorescein-conjugated Abs reveals the presence of virus-specific antigens. This technic permits diagnosis of about 75% of the infections detected by isolation of virus or immunological assay.

EPIDEMIOLOGY

Influenza occurs in recurrent epidemics which start abruptly, spread rapidly, and are frequently of worldwide distribution. An influenza A epidemic generally appears every 2 to 4 years and an influenza B epidemic every 3 to 6 years, but the patterns have not been completely predictable. Epidemics of influenza A viruses are usually more widespread and more severe than those of influenza B. Influenza C virus has not caused epidemics, and the majority of infections produced by this agent are inapparent. Although epidemics occur periodically in any given geographical locality, outbreaks occur somewhere every year.

Influenza has the highest incidence in the age group 5 to 9 years. Above 35, the incidence gradually declines with increasing age. The very young and the very old suffer the highest mortality; there are also special groups showing high mortality, such as pregnant women and persons with chronic pulmonary disease or cardiac insufficiency. A striking exception to the usual age-related mortality rate, however, was noted in the severe 1918–1919 pandemic, in which the majority of the 20 million deaths occurred in the young adult group.

Epidemics are common from early fall to late spring. Outbreaks often develop in many places in a country at almost the same time and spread rapidly to neighboring communities and countries; with the common use of air travel intercontinental spread has also become rapid. The rapid dissemination is not entirely due to the speed of modern transportation, however; for this characteristic was also noted when man could travel no faster than the speed of his horse. The pattern of epidemic spread may be related to the occurrence of sporadic cases and the probable seeding of the virus in the population several weeks prior to an explosive outbreak. Even during an epidemic, the ratio of infection to disease, determined by serological and clinical studies, varies from 9:1 to 3:1.

The pathway of widespread dissemination of the virus has now been exemplified by several well-studied episodes, such as the 1957 pandemic caused by a new variant, $A_2(H_2N_2)$ (Asian) subtype, which apparently emerged from central China in February of 1957 (Fig. 56-11). The arrival of the virus in the United States was detected in Naval personnel in Newport, Rhode Island, on June 2 and shortly thereafter in San Diego, California, without traceable connection between the two episodes. The first civilian outbreak was observed at a conference in Davis, California, on June 20, followed by several similar small episodes elsewhere in California. From the conference in Davis the virus was carried directly by some of the more peripatetic members to a meeting of young people in Iowa. From this location the virus was seeded throughout the country. The initial dissemination was followed by small, sporadic outbreaks until September, when epidemics occurred in almost all parts of the country. In similar fashion, the Asian influenza virus spread throughout the world, along paths of travel.

In the summer of 1968, after an appropriate period of antigenic drift (11 years), another influenza A variant appeared in Hong Kong and produced a mild but widespread pandemic whose epidemiological spread was strikingly similar to that of 1957. Although the 1968 Hong Kong virus had antigenic characteristics clearly different from previously isolated $A_2(H_2N_2)$ viruses, the immunological relatedness (owing to cross-reacting neuraminidase molecules) was sufficient so that the new isolates are not considered a new subtype. However, the A Hong Kong (H_3N_2) viruses isolated from epidemics in 1969 and 1970 have showed further antigenic drift, suggesting that a new A subtype may soon emerge. When the postulated virus appears it should answer the question whether only a finite number of immunological variants is possible: if the immunological drift is limited and hence cyclic the next variant should be closely akin to swine $(H_{sw}N_1)$ or $A_0(H_0N_1)$ virus.

Many questions concerning the epidemiology of influenza remain unanswered. Not the least puzzling among the unknowns are the following: 1) Why does the virus not spread rapidly at the time of the initial infections in a community? 2) How does the virus become "masked" or "go underground" in the interval between the seeding and the occurrence of the epidemic? 3) What provocative factors induce the epidemic? 4) Where is the virus during interepidemic intervals?

After seeding, or during the interval between epidemics, virus may conceivably 1) simply be transmitted slowly, producing inapparent infections or sporadic cases; 2) remain latent in the persons previously infected; or 3) reside, active or latent, in an animal reservoir. Influenza virus has rarely been isolated in nonepidemic periods, which speaks against the first possibility. The intriguing ecology of swine influenza virus, which is activated by cold weather in the presence of *H. influenzae suis* (Ch. 51, Latent viral infections) offers one example of

mechanism 2. Finally, human strains of influenza A virus show immunological and genetic relations to influenza viruses of horses, ducks, chickens, and pigs (Table 56-1) which supports mechanism 3 and also suggests that animals may serve as a source of new antigenic variants.

The appearance of influenza viruses after a silent interval no doubt depends frequently on the development of a new antigenic variant, which can escape an immunological barrier existing in nature. However, "old" strains, having only minimal antigenic changes, also initiate epidemics, presumably because Abs in the population fall below the level necessary to prevent infections. The nature of the provoking factors that initiate an epidemic remains a mystery.

PREVENTION AND CONTROL

The high incidence of inapparent infections, the short incubation period, and the high infectivity preclude the successful use of isolation or quarantine procedures to control influenza. Quarantine of travelers entering a country can delay but not prevent entrance of virus. In South Africa in 1957, for example, where ships were quarantined and the passengers and crew forbidden to land, infection did not enter through the ports; but virus finally entered from the north, probably being carried by immigrant laborers traveling overland.

Artificial immunization can prevent influenza to a significant extent (reducing the incidence up to 70%), but not completely. Viruses propagated in chick embryos, partially purified, and inactivated by formalin, can provide a highly effective vaccine if the viruses utilized include a strain closely related immunologically to the currently prevalent strain. This requirement is difficult to satisfy. For example, although influenza vaccines had been shown to be highly effective in 1943 and 1945, the vaccine employed in 1947 was a complete failure because it did not contain the newly emerged antigenic variant, the A_1 (H_1N_1) virus. Hence, influenza must be under constant global surveillance including accurate typing of isolated strains.* Vaccines currently employed contain several strains of influenza A and B viruses in order to cover

---

* The World Health Organization has established centers throughout the world for this purpose.

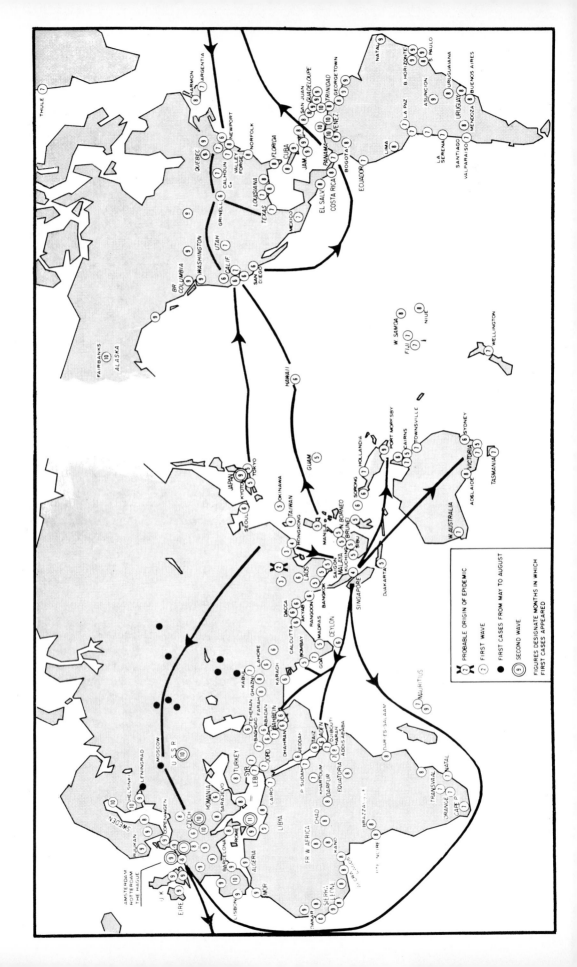

the known antigenic spectrum. New major antigenic variants are added as they appear.

Despite the proved value of the available, inactivated viral vaccines several factors have limited their use and possibly their effectiveness. 1) Pyrogenic reactions, accompanied by constitutional symptoms (not unlike manifestations of mild influenza) and local reactions, have been common, particularly in infants and young children; an incidence of 10 to 20% was not unusual prior to the use of purified vaccines. 2) Secretory IgA Abs in respiratory secretions are probably critical for successful protection, but subcutaneous injection of inactivated virus induces only low levels of such Abs in the respiratory tract. 3) Antibodies begin to decrease about 3 months after immunization and immunity is often lost within 6 months.

Generally, a single subcutaneous injection of 0.1 to 1.0 ml containing at least 300 chick-RBC-agglutinating (CCA) units per inoculum will confer immunity in 2 to 4 weeks. Persons immunized with a new subtype, especially children, show a primary immunological response, while those who have had previous exposure to the antigens in the vaccine exhibit a secondary response. Therefore, if the vaccine contains a new major antigenic variant to which most individuals have no detectable Abs, such as $A_2(H_2N_2)$ in 1957, a second injection is recommended a month after the first.

New methods for preparation and administration of vaccines are now being tested with encouraging results. 1) **Inactivated vaccines,** consisting of purified viruses relatively free of nonviral substances, or vaccines consisting of the hemagglutinin and neuraminidase components of disrupted virions, are being used rather widely. These vaccine formulations permit administration of a greater antigenic mass, therefore effecting greater Abs responses, while they greatly reduce the incidence of major toxic reactions. 2) **Attenuated infectious viruses** (selected by serial passage at low temperatures) have been widely used in the Soviet Union as well as in other countries, and temperature-sensitive mutants are being employed experimentally in the United States; the attenuated viruses have been shown to produce IgA Abs in respiratory secretions. With live vaccines, however, it is impossible to select and test rapidly a suitable isolate of viral subtype that

has recently emerged. 3) **Intranasal aerosol administration** of inactivated viral vaccine induces an adequate response of IgA Abs in nasal secretions as well as of specific IgA, IgM, and IgG Abs in serum, but this method has not yet been adequately evaluated. 4) **Vaccines prepared with adjuvants** (such as light mineral oil, peanut oil, or Arlacel A) result in greater and more persistent Ab response. As yet, adjuvants have not been widely employed because of fear of adverse reactions, including local and generalized reactions to the adjuvant and increased toxicity of the vaccine. The validity of the former concern is uncertain and vaccine in adjuvant may become more generally accepted. Use of purified viral vaccines should largely eliminate the latter fear.

A new approach to immunization has developed from the ability to "tailor-make" antigenic variants of influenza viruses by 1) genetic recombination of a new antigenic variant with an avirulent temperature-sensitive mutant, or with an established strain in order to ensure propagation of the newly emerged subtypes to high titers; and 2) serial passage of current viral strains in the presence of antiserum, in the hope that this selective pressure will cause the emergence in vitro of next year's antigenic variant present as a minor constituent in the viral population.

Although effective vaccines can now be developed, the implications of their widespread use merits consideration. It is undoubtedly desirable to immunize persons at high risk from the severe complications of influenza. The advisability of mass immunization, however, may depend on whether influenza viruses have a virtually unlimited capacity for antigenic mutation, or whether the antigenic variations are very limited in their potential extent. Thus, if infinite antigenic variation is possible, mass immunization programs, which would increase herd immunity, may serve to accelerate the selection and appearance of new antigenic subtypes. But if the potential number of antigenic mutants is limited, mass immunization with a broad antigenic spectrum of viruses would result in a population with broadly reactive antibody levels adequate to provide an effective barrier to the spread of influenza viruses.

Chemotherapy. With the discovery that L-adamantanamine (adamantine) can inhibit an early step in multiplication (uncoating) of some influenza viruses (Ch. 49), hopes were reawakened for the successful chemical control of viral diseases. Adamantanamine has proved modestly successful experimentally in preventing influenza A_2 infections. But the restriction of its effectiveness to influenza A but not influenza B viruses, and its neurological toxic effects (particularly in the aged) limits its clinical usefulness.

Fig. 56-11. Progress of Asian influenza pandemic from its probable origin in central China, February 1957 to January 1958. [From Langmuir, A. D. *Am Rev Resp Dis* 83:1 (1961).]

SELECTED REFERENCES

Books and Review Articles

ANDREWES, C. H. Epidemiology of influenza. *Bull WHO 8:*595 (1953).

ANDREWES, C. H. Factors in virus evolution. *Adv Virus Res 4:*1 (1957).

Cellular Biology of Myxovirus Infections. Ciba Foundation Symposia. (G. E. W. Wolstenholme and J. Knight, eds.) Ciba, Summit, N.J., 1964.

KILBOURNE, E. D. Influenza virus genetics. *Progr Med Virol 5:*79 (1963).

KILBOURNE, E. D., CHOPPIN, P. W., SCHULZE, I. T., SCHOLTISSEK, C., and BUCHER, D. L. Influenza virus polypeptides and antigens—summary of Influenza Workshop I. *J Infect Dis 125:*447 (1972).

KILBOURNE, E. D., COLEMAN, M., CHOPPIN, P. W., DOWDLE, W. R., SCHILD, G. C. and SCHULMAN, J. L. Immunologic methodology in influenza diagnosis and research—summary of Influenza Workshop II. *J Infect Dis 126:*219 (1972).

ROBINSON, R. Q., and DOWDLE, W. R. Influenza viruses. In *Diagnostic Procedures for Virus and Rickettsial Diseases,* 4th ed. (E. H. Lennette and N. J. Schmidt, eds.) American Public Health Association, New York, 1969.

VON MAGNUS, P., CHANOCK, R. M., DAVENPORT, F. M., FARKAS, E., SMORODINCEV, A. H., SABIER, R., and SYRUCEK, L. Influenza. In *Respiratory Viruses.* World Health Organization Technical Report Series, No. 408. WHO, Geneva, 1969.

Specific Articles

CHOW, M., and SIMPSON, R. W. RNA-dependent RNA polymerase activity associated with virions and subviral components of myxoviruses. *Proc Natl Acad Sci USA 68:*752 (1971).

DAVENPORT, F. M., MINUSE, E., HENNESSY, A. V., and FRANCIS, T., JR. Interpretations of influenza antibody patterns of man. *Bull WHO 41:*453 (1969).

DUESBERG, P. H. The RNA's of influenza virus. *Proc Natl Acad Sci USA 59:*930 (1968).

HENLE, W., and LIEF, F. S. The broadening of antibody spectra following multiple exposure to influenza viruses. *Am Rev Resp Dis* (Suppl) *88:*379 (1963).

KATES, M., ALLISON, A. C., TYRELL, D. A. J., and JAMES, A. T. Origin of lipids in influenza virus. *Cold Spring Harbor Symp Quant Biol 27:*293 (1962).

KILBOURNE, E. D. Future influenza vaccines and the use of genetic recombinants. *Bull WHO 41:*643 (1969).

SCHULMAN, J. L., and KILBOURNE, E. D. Independent variation in nature of hemagglutinin and neuraminidase antigens of influenza virus: Distinctiveness of hemagglutinin antigen of Hong Kong/68 virus. *Proc Natl Acad Sci USA 63:*326 (1969).

SCHULZE, I. T. The structure of influenza virus. II. A model based on the morphology and composition of subviral particles. *Virology 47:*181, 1972.

SETO, J. T., and ROTT, R. Functional significance of sialidase during influenza virus multiplication. *Virology 30:*731 (1966).

SIMPSON, R. W., and HIRST, G. K. Temperature-sensitive mutants of influenza A virus: Isolation of mutants and preliminary observations on genetic recombination and complementation. *Virology 35:*41 (1968).

WHITE, D. O., and CHEYNE, I. M. Early events in the eclipse phase of influenza and parainfluenza virus infection. *Virology 29:*49 (1966).

chapter **57**

PARAMYXOVIRUSES AND RUBELLA VIRUS

GENERAL PROPERTIES OF THE VIRUSES   1331

PARAINFLUENZA VIRUSES   1336

PROPERTIES OF THE VIRUSES   1336
  Immunological Characteristics   1336
  Reactions with Erythrocytes   1336
  Host Range   1337
PATHOGENESIS   1337
LABORATORY DIAGNOSIS   1338
EPIDEMIOLOGY   1338
PREVENTION AND CONTROL   1338

MUMPS VIRUS   1339

PROPERTIES OF THE VIRUS   1339
  Immunological Characteristics   1339
  Reactions with Erythrocytes   1339
  Host Range   1339
  Toxicity   1340
PATHOGENESIS   1340
LABORATORY DIAGNOSIS   1341
EPIDEMIOLOGY   1341
PREVENTION AND CONTROL   1342

MEASLES VIRUS   1342

PROPERTIES OF THE VIRUS   1342
  Immunological Characteristics   1342
  Host Range   1343
PATHOGENESIS   1344
LABORATORY DIAGNOSIS   1347
EPIDEMIOLOGY   1347
PREVENTION AND CONTROL   1348

RESPIRATORY SYNCYTIAL VIRUS   1349

  Properties of the Virus   1349
  Pathogenesis   1350
  Laboratory Diagnosis   1352
  Epidemiology   1352
  Prevention and Control   1352

NEWCASTLE DISEASE VIRUS   1353

RUBELLA VIRUS   1353

    Properties of the Virus   1353

    Viral Multiplication   1355

    Pathogenesis   1356

    Laboratory Diagnosis   1357

    Epidemiology   1357

    Prevention and Control   1358

Paramyxoviruses are a relatively homogeneous family on the basis of their chemical, physical, and several biological properties. Pathogenically, however, they differ widely: parainfluenza and respiratory syncytial viruses produce acute respiratory diseases, measles virus causes a generalized exanthematous disease, and mumps virus initiates a systemic disease of which parotitis is a predominant feature.

Rubella virus, the causative agent of German measles, has not been adequately characterized to be assigned an official taxonomic position. It will be discussed in this chapter, however, because of its pathogenic and epidemiological similarities to measles, although it resembles togaviruses (Ch. 60) and has many differences from paramyxoviruses.

GENERAL PROPERTIES OF THE VIRUSES

The characteristics that are similar for all paramyxoviruses will be discussed in this section; the distinctive properties will be described in sections on the individual viruses.

Morphology. The virions are roughly spherical particles of heterogeneous sizes, larger than influenza viruses; the average particle diameter ranges from 1250 to 2500 A for different species, with a range betweeen 1000 and 8000 A. Electron microscopic examination of negatively stained virions discloses that they are similar to myxoviruses. The intact viral particle has a well-defined outer envelope, about 100 A thick, covered with short spikes that are more or less regularly arranged (Fig. 57-1). Disruption of the envelope reveals an inner helical nucleocapsid, which appears to be double-stranded, with a characteristic width of 170 to 180 A, a central hole of 40 to 50 A, and serrations with a regular periodicity of about 50 A (Figs. 57-2 to 57-4). The nucleocapsid is distinctly different from that of influenza viruses (Ch. 56, Table 56-2): its diameter is approximately twice as great; its characteristic periodic serrations are more discrete; and it can be isolated from the virion as a single long helical structure. No filamentous forms have been observed except for occasional ones in preparations of respiratory syncytial virus.

Physical and Chemical Characteristics. Only a few paramyxoviruses have been purified and analyzed (parainfluenza type 1, Newcastle disease virus, and SV 5 or simian parainfluenza virus), but they all have a similar composition, approximately 74% protein, 20% lipid, 6% carbohydrate, and 0.9% RNA. The nucleocapsid contributes 20 to 25% of the mass of the virion; it consists of a single species of protein and one large molecule of single-stranded RNA. The viral envelope has a lipid composition similar to that of the plasma membrane of the host cell; in addition, it contains two glycoproteins, which form the viral hemagglutinin and neuraminidase, and three other distinct proteins associated with the structure of the envelope's membrane. Lysolecithin, probably complexed with envelope proteins, appears to be responsible for the virus-cell fusion and hemolytic activity. Because the viral envelope has a high lipid content, organic solvents or surface-active agents rapidly inactivate the virions by dissolving their envelopes and liberating the envelope proteins and the nucleocapsids from the disrupted particles.

The virions are relatively unstable, losing

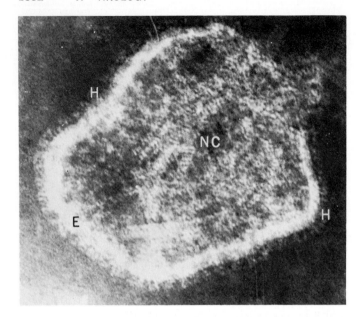

Fig. 57-1. Electron micrograph of type 3 parainfluenza virus embedded in phosphotungstate. The envelope (E), peripheral short projections (H), and internal nucleocapsid (NC) are apparent. The helical nucleocapsid can also be seen escaping from a small break in the envelope. ×210,000. (Courtesy of A. P. Waterson, St. Thomas's Hospital Medical School, London.)

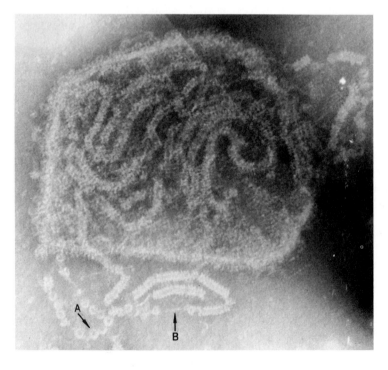

Fig. 57-2. Partially disrupted mumps virus particle showing the envelope and the hollow helical strands forming the nucleocapsid. A number of broken strands have been released revealing periodic structures (A) which make up the helical nucleocapsid. Fine threads connecting separated pieces are visible at B. ×250,000. [From Horne, R. W., and Waterson, A. P. *J Mol Biol* 2:75 (1960).]

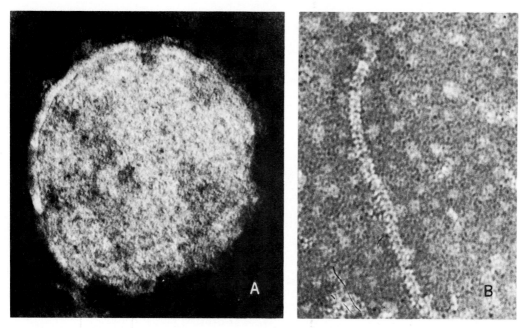

Fig. 57-3. Fine structure of measles virus revealed by negative staining with sodium phosphotungstate. **A.** Particle showing the characteristic envelope and peripheral projections. The nucleocapsid is tightly packed and shows an appearance of concentric rings toward the periphery. ×280,000. **B.** A portion of the helical nucleocapsid released from a disrupted virion. ×240,000. [From Horne, R. W., and Waterson, A. P. *J Mol Biol* 2:75 (1960).]

Fig. 57-4. Strand of helical nucleocapsid of type 1 (Sendai) parainfluenza virus embedded in phosphotungstate. (Examination by tilting through large angles reveals the sense of the helix to be left-handed.) ×200,-000. [From Horne, R. W., and Waterson, A. P. *J Mol Biol* 2:75 (1960).]

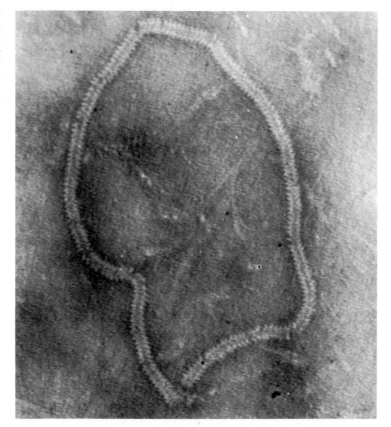

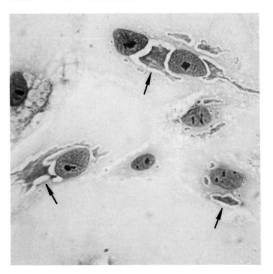

Fig. 57-5. Eosinophilic cytoplasmic inclusions (arrows) in dog kidney cells infected with type 2 parainfluenza virus. Nuclei appear unaffected. ×450. [From Brandt, C. D. *Virology 14*:1 (1961).]

90 to 99% of their infectivity in 2 to 4 hours when suspended in a protein-free medium at room temperature or at 4°.

Viral Multiplication. The multiplication cycles do not appear to differ for each paramyxovirus, and even resemble that of influenza viruses, except for the variations in length of various phases. For example, the latent period is 3 to 5 hours for parainfluenza viruses, 16 to 18 hours for mumps virus, and 9 to 12 hours for measles virus. The basic biosynthetic events (predominantly derived from studies of type 1 parainfluenza and Newcastle disease viruses) are similar to those for other viruses that possess single-stranded RNA, contain an RNA-dependent RNA polymerase within the virion, and are covered by a lipid-containing envelope (Ch. 48, Synthesis of RNA viruses), but there are several differences from myxoviruses: 1) paramyxoviruses multiply without restraint in the presence of actinomycin D; 2) the proteins of the helical nucleocapsid, as well as the hemagglutinin and neuraminidase, are visualized only in the cytoplasm of infected cells (except for measles virus whose nucleocapsid appears in both the nucleus and cytoplasm); 3) excessive numbers of single-stranded RNA molecules complementary to viral RNA are present in infected cells as compared with replicative forms of RNase-resistant (double-stranded) RNA; 4) viral RNA is present as a single, large molecule (about 57S) rather than in pieces; and 5) RNA from purified virions, as well as from nucleocapsid structures isolated from infected cells, self-anneal, to a small degree, demonstrating that in addition to viral strands some complementary strands of RNA (which presumably arise as messenger or replicative RNA) are assembled into viral structures.

After viral RNA is synthesized in the cytoplasm, it is rapidly associated with newly made nucleocapsid protein, but only a relatively small proportion (depending upon the host cell) of the nucleocapsids formed are assembled into virions. The cytoplasmic inclusion bodies (Fig. 57-5), which develop during viral multiplication, are predominantly accumulations of these excess nucleocapsids.

Electron microscopic examination of infected cells reveals the remarkable assembly and maturation of paramyxoviruses at the plasma membrane. Strands of viral nucleocapsid can be seen to impinge upon segments of differentiated cell membranes that are destined to become the viral envelopes (Fig. 57-6A, and B). The altered membrane contains virus-specific proteins but the lipids are those of the host. Intact viral particles are seen only at the cell membrane, where they are assembled and released from the cell by budding. The final assembly of paramyxoviruses is similar to that of influenza viruses (Ch. 56), but it is easier to visualize the paramyxovirus' nucleocapsid and the prominent specific alteration of the plasma membrane (Fig. 57-6).

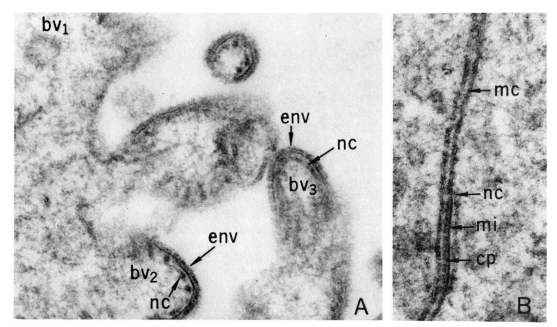

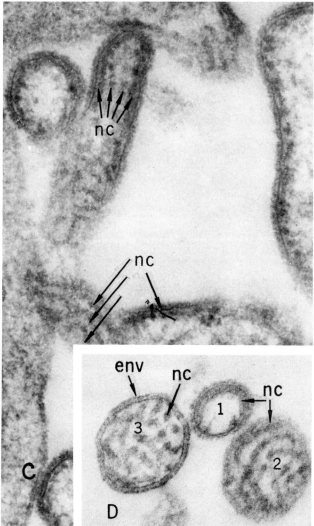

Fig. 57-6. Parainfluenza virions forming at the cell membrane of a chick embryo cell and being released by budding. **A.** Viral buds (bv) at various stages of development. The altered cell membrane forming the viral envelope (env) and the nucleocapsid (nc) beneath the envelope can be clearly distinguished. The nucleocapsid is cut transversely in two of the buds (bv_1 and bv_2) and longitudinally in a third (bv_3). ×105,000. **B.** Differentiation of the viral envelope at the cell membrane. The cell membranes (mc) of two adjacent cells are shown. On contact with the nucleocapsid (nc) the cell membrane differentiates into the internal membrane of the envelope (mi) and the outer layer of short projections (cp). ×105,000. **C.** Arrangement of the nucleocapsid (nc) impinging on the altered cell membrane which is participating in the formation of the viral envelope. Notice the regularity of the arrangement of the nucleocapsid seen in cross-section and in longitudinal views. ×105,000. **D.** Thin section of virions. Note the structure of the envelope covered with short projections (env) and the various arrangements of the nucleocapsids (nc). In particle 1 the nucleocapsid is arranged parallel to the envelope. In particles 2 and 3 the nucleocapsid is arranged more irregularly. ×92,000. [From Berkaloff, A. *J Microsc* 2:633 (1963).]

PARAINFLUENZA VIRUSES

Parainfluenza viruses were recognized in 1957 as important causes of acute respiratory infections of man. There are at least four antigenic types of parainfluenza viruses that infect man and one that infects monkeys.

PROPERTIES OF THE VIRUSES

The major characteristics of parainfluenza viruses infecting man are listed in Table 57-1.

IMMUNOLOGICAL CHARACTERISTICS

Each parainfluenza virus possesses three distinct type-specific antigens: the hemagglutinin and neuraminidase surface antigens and an internal nucleocapsid antigen. No immunological relation between parainfluenza and influenza viruses is detectable. Unlike influenza A and B viruses, progressive major antigenic alterations of parainfluenza viruses have not been detected. Type 4 virus, however, has two subtypes, A and B, in contrast to other members of the group. Although there is no single antigen common to all parainfluenza viruses, immunological relations among the parainfluenza (at least types 1, 2, and 3) and mumps viruses are noted by heterotypic responses to infection in those who have had prior infections with one or more of the other viruses.

The serological response to the initial infection (primary response) in man is as specific as the immunological response in guinea pigs and hamsters. The appearance of heterotypic Abs in man probably results from repeated infections with different members of the group; each additional infection broadens the antigenic response. Predictable cross-reactions do not occur, but heterotypic Ab rises following human infections are sufficiently frequent to permit the following conclusion: 1) these agents probably constitute a group of viruses in which cross-reactive antigens exist, 2) the qualitative and quantitative characteristics of the heterotypic Ab response to infection depend upon prior immunological experience, and 3) immunological diagnosis of infection by a specific virus may be unreliable.

Most adults have a relatively high titer of circulating Abs to all types, but usually lack the critical IgA neutralizing Abs in nasal secretions. Following acute infections the IgA antibodies appear in nasal secretions, but they decrease substantially within 1 to 6 months although specific serum Ab levels remain relatively high. Recurrent infections occur despite the presence of serum neutralizing Abs, but the severity of disease is usually reduced (perhaps owing to a secondary response of IgA antibodies in nasal secretions). The initial infection with a given type is the most severe and usually occurs in children; in adults infections are commonly afebrile and minor.

REACTIONS WITH ERYTHROCYTES

The hemagglutinins and neuraminidases of parainfluenza viruses resemble those of in-

TABLE 57-1.   Properties of Human Parainfluenza Viruses

Size	1500–2500 A
Nucleocapsid	Helical; 180 A in diameter; 1μ in length
Nucleic acid content	RNA; single-stranded; single molecule of 7×10^6 daltons
Virion enzyme	RNA-dependent RNA polymerase
Reaction with lipid solvents	Disrupts
Hemagglutination	Chicken and guinea pig RBC
Hemadsorption	Infected cells adsorb guinea pig RBC
Hemolysin	Guinea pig or chicken RBC
Reaction of receptors to RDE	Receptors destroyed
Cytopathic effect	Syncytial formation; cytoplasmic inclusion bodies
Antigenic analysis	4 types; each shares antigens with other parainfluenza viruses and with mumps virus, but no common antigen
Site of multiplication	Cytoplasm

fluenza viruses in most biological and physical characteristics (Ch. 44, Hemagglutination), but examinations have not been made on the purified structures separated from viral particles. The hemagglutinins appear to be the surface projections on the virion. Maximum hemagglutination titers are obtained with chicken RBC at 4° (types 1 and 2) or with guinea pig RBC at 25° (types 1 and 3).

Unlike influenza viruses, parainfluenza viruses possess a hemolysin which is inhibited by type-specific antisera and is biologically similar to the hemolysins described for mumps and Newcastle disease viruses. Disruption of the viral particle with ether permits solubilization of the hemolysin and its separation from the hemagglutinin; it appears to contain lysolecithin. For hemolytic activity intact mucoprotein receptors on RBC are required. Dialyzable materials in allantoic and tissue culture fluids, possibly calcium, inhibit the hemolytic activity.

HOST RANGE

Primary cultures of cells from monkey and embryonic human kidneys are the hosts of choice for primary isolations, neutralization titrations, and investigations of the biological properties of parainfluenza viruses. Organ cultures of tracheal and nasal epithelium have similarly proved to be sensitive for primary isolations. Types 2 and 3 also multiply well in human continuous epithelium-like cell lines (e.g., HeLa cells). Type 1 virus has been adapted to HeLa and human diploid cells, but neither is satisfactory for initial viral isolation.

In cell cultures cytopathic changes develop very slowly (particularly with type 4); infection is most quickly detected by hemadsorption of guinea pig erythrocytes (Ch. 44). The cytopathic changes that eventually appear consist of stringiness or rounding of cells (types 1 and 4) or formation of large syncytia containing eosinophilic cytoplasmic inclusion bodies (types 2 and 3; Fig. 57-5). Syncytia develop by fusion and dissolution of membranes of infected cells (Ch. 51, Cytopathic effects), perhaps produced by the cell fusion factor or a unique viral hemolysin. This remarkable capacity of parainfluenza viruses (even after ultraviolet or heat inactivation) to induce fusion of cells from the same or differ-

ent animal species has great general utility for studying the genetics of eukaryotic cells as well as for virology (Ch. 47, Cell hybridization).

Intranasal inoculations of as little as 1 to 10 TCD_{50} of virus into hamsters or guinea pigs about 3 months of age result in infections which yield relatively high titers of virus in the lungs within 2 to 3 days, but without pulmonary lesions resulting. The animals develop type-specific Abs and resist intranasal challenge with homologous virus for 1 to 3 months, but then susceptibility returns. This pattern mimics the relation of Abs and susceptibility to recurrent disease in man.

Type 3 virus is highly infectious for cattle. It appears to be harmless, however, except under conditions of stress, such as the herding of cattle together for transportation, when it may induce an acute febrile upper respiratory disease (hence the term "shipping fever").

Parainfluenza viruses have been adapted to propagate in the amniotic sacs of 7- to 9-day-old chick embryos. The virus recovered in the amniotic fluid is then adapted to passage in the allantoic cavity.

PATHOGENESIS

Most parainfluenza virus infections are lacking suitable experimental animals. Infections in volunteers, however, and observations of natural infections, permit a few conclusions concerning pathogenesis. The virus enters by the respiratory route, and in most patients it multiplies and causes inflammation only in the upper segments of the tract. In infants and young children, however, the bronchi, bronchioles, and lungs are occasionally involved. Types 1 and 2 viruses are particularly prone to involve the larynx, trachea, and bronchi of infants, producing laryngotracheobronchitis (croup). As in influenza, viremia is not an essential phase of infection.

Parainfluenza viruses cause a spectrum of illnesses, primarily in infants and young children, ranging from mild upper respiratory infections to croup or pneumonia (Table 57-2). The lack of distinctive clinical features precludes etiological diagnoses on this basis. The occasional infections of adults usually evoke a subclinical illness or a mild "cold." Even in children the majority of infections with parainfluenza viruses appear to be clinically inapparent.

TABLE 57-2.   Clinical Syndromes Associated with Parainfluenza Viruses

Disease	Virus type
Minor upper respiratory disease	1, 3, 4*
Bronchitis	1, 3
Bronchopneumonia	1, 3
Croup	1, 2

  * Clinical disease uncommon.

LABORATORY DIAGNOSIS

An etiological diagnosis of parainfluenza virus infection requires laboratory procedures. Measurement of a **rise** in serum Abs by hemagglutination-inhibition or complement-fixation titration permits diagnosis conveniently and economically. However, immunological technics alone cannot reliably establish the specific type of virus responsible, because of the frequency and the degree of heterotypic Ab responses discussed above.

The specific parainfluenza virus responsible for an infection can be identified by viral isolation. Nasopharyngeal secretions containing antibiotics are added to primary tissue cultures of monkey or embryonic human kidney. (The simian parainfluenza virus SV 5 is a common latent agent in monkey kidney cultures, and its emergence from this tissue must not be confused with its primary isolation from man.) Although cytopathic changes may not be detectable or may develop very slowly, viral infection can be recognized rapidly and conveniently by immunofluorescence or by adsorption of guinea pig erythrocytes to infected cells (Ch. 44, Hemadsorption). Hemadsorption to cells infected with types 1 and 3 viruses can usually be detected within 5 days after inoculation of the patient's secretions, but types 2 and 4 often require 10 days or more. The specific type can be identified by hemadsorption-inhibition technics, utilizing standard sera. Immunofluorescence can yield comparable diagnostic results in only 24 to 48 hours.

EPIDEMIOLOGY

Parainfluenza viruses produce disease throughout the year, but the peak incidence is noted during the "respiratory disease season" (late fall to early spring). Small epidemics occasionally occur with types 1 and 3. Parainfluenza virus infections are primarily childhood diseases: type 3 infections occur earliest and most frequently, so that 50% of children in the United States are infected during the first year of life, and almost all by 6 years of age; 80% of children are infected with types 1 and 2 by 10 years of age. Type 4 viruses appear to induce few clinical illnesses but the infections are common: by 10 years of age 70 to 80% of children have Abs.

Parainfluenza viruses are disseminated in respiratory secretions. Type 3 is the most effective spreader, and during outbreaks in closed populations (e.g., in institutions or hospitals) all children who are free of neutralizing Abs become infected. Under similar circumstances only about 50% of children are infected with type 1 or type 2 virus.

The epidemiological patterns and the clinical manifestations of parainfluenza virus infections, in children and adults, emphasize the protective effect of neutralizing Abs as well as the lack of complete or long-lasting immunity (probably owing to an inadequate level of IgA antibodies in the respiratory secretions). Febrile and severe illness is observed only with the initial infection. Reinfection may be produced by the same virus within as little as 9 to 12 months, but it results in a much milder disease.

PREVENTION AND CONTROL

Reducing the attack rate of respiratory diseases is an important social and economic goal. However, if artificial immunization completely prevented parainfluenza virus infections the incidence of acute respiratory illnesses would probably be reduced only about 15% in children less than 10 years old, and much less in adults. Nevertheless, an effective vaccine, evoking an Ab response in the respiratory tract, would be of value in young children, especially in hospitals and institutions. Live attenuated parainfluenza viruses, in preference to formalin-inactivated viruses, are now being developed for intranasal administration.

MUMPS VIRUS

Mumps has been recognized as a disease since antiquity, but the etiological agent was not isolated and identified as a virus until 1934, when Johnson and Goodpasture produced parotitis in monkeys by inoculating bacteria-free infectious material directly into Stensen's duct. No major progress was made, however, in characterizing the virus and in understanding the pathogenesis and immunological aspects of the disease, until the virus was propagated in the chick embryo and was found (in 1945) to agglutinate chicken erythrocytes, like influenza viruses. The subsequently isolated parainfluenza viruses have been found to have an even closer relation to mumps virus and these agents have been classified together.

PROPERTIES OF THE VIRUS

IMMUNOLOGICAL CHARACTERISTICS

Mumps virus, like influenza and parainfluenza virus, contains the surface hemagglutinin (also termed the **V antigen**), which induces protective Abs; a surface neuraminidase, which although present in small amounts probably plays a biological and immunogenic role similar to its counterpart in myxoviruses; and the internal RNA-protein nucleocapsid, which is immunologically identical with the soluble antigen from cells (called the **S antigen).**

Antigenically mumps virus is distinct from influenza viruses, but cross-reacts significantly with parainfluenza and Newcastle disease viruses. The antigenic relations among these viruses are noted by a heterotypic Ab rise to parainfluenza viruses in sera from mumps patients; the cross-reactivity exists in IgM immunoglobins and signifies recent mumps infection.

Antibodies develop against the S antigen within 7 days, and high titers are attained within 2 weeks after onset of clinical illness. V antibodies appear later (2 to 3 weeks after onset), attain maximum titers in 3 to 4 weeks, and persist longer than the S Abs. V Abs are measured by neutralization, hemagglutination inhibition, or complement fixation with purified hemagglutinin, and appear to reflect the degree of immunity. S Abs, which are conveniently assayed by complement-fixation titrations, do not afford protection against subsequent infection. Immunity develops after subclinical as well as clinical infections and usually persists for many years, although second infections have been reported. No immunological variants have been detected.

Mumps infection induces delayed hypersensitivity which can be observed by a skin test with infectious or inactivated virus. A positive skin reaction correlates roughly with immunity, providing a useful epidemiological tool.

REACTIONS WITH ERYTHROCYTES

The viral particle agglutinates erythrocytes from several animal species (Ch. 44, Hemagglutination). Mumps virus, however, has only weak neuraminidase activity, and therefore hemagglutination is highly susceptible to inhibition by many mucoproteins. The hemagglutinin can also be detected on infected cells in tissue culture by hemadsorption.

Mumps virus, like parainfluenza viruses, hemolyzes susceptible RBC at 37°. This activity and hemagglutination require the same specific mucoprotein receptors on red blood cells. Several viral particles per cell are required to produce hemolysis, and even under optimal conditions only about 50% of the hemoglobin is released. Calcium inhibits hemolysis. The hemolytic agent has not been separated from the virion and chemically identified (as for parainfluenza virus) but appears to be the same as that responsible for the cytolytic property. Viral hemagglutination, hemadsorption, and hemolysis are inhibited by antibodies to the surface (V) antigen.

HOST RANGE

Man is the only natural host for mumps virus, but the virus can infect monkeys and 6- to 8-day-old chick embryos (amniotic cavity or yolk sac); no destruction of cells of the embryos is noted. After adaptation by serial passage in the allantoic sac of the chick embryo, the virus can infect guinea pigs, suckling mice, hamsters, and white rats, but it has lost virulence for man and monkey.

Chick embryo and many types of mammalian cells (including monkey kidney and HeLa cells) support multiplication of mumps virus in vitro. Infected cultures can be recognized by the appearance of viral particles (hemadsorption or agglutination of RBC) and soluble antigen (complement fixation), as well as by the slow development of cytopathic effects (cytolysis and development of giant cells, Fig. 57-7).

TOXICITY

Mumps virus, like influenza viruses, has a lethal, toxin-like action when inoculated intravenously in large amounts into mice, especially newborns. The toxic activity is not dissociable from the viral particle, but unlike influenza virus, infectious virus is not required. The role of this toxic property in the pathogenesis of natural infections is not known.

Fig. 57-7. Eosinophilic cytoplasmic inclusions and giant cell formations in monkey kidney cells infected with mumps virus. Inclusions (arrows) develop in close proximity to the nuclei, which are unaffected. ×450. [From Brandt, C. D. *Virology* *14*:1 (1961).]

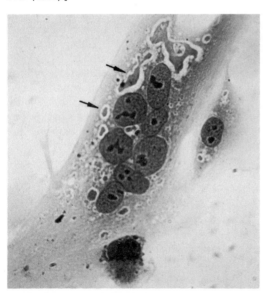

PATHOGENESIS

Mumps virus is transmitted in saliva and respiratory secretions and its portal of entry in man is the respiratory tract. It has been generally taught that virus enters Stensen's duct directly and multiplies initially in the duct and the parotid gland. With the development of a sensitive viral assay, however, evidence suggesting another pathway has been obtained: 1) viremia is noted several days before the development of mumps, and is frequently detected at the time of the onset of clinical symptoms and for 3 to 5 days during the acute illness; 2) virus is present in saliva for 2 to 3 days before and for 5 to 7 days after the beginning of illness; and 3) virus in the urine is common for 10 days or more after the onset.

The pathogenesis of mumps, schematically summarized in Figure 57-8, may be correlated with the recognized clinical course of the disease. Mumps typically has an acute onset of parotitis (with painful swelling of one or both glands) 16 to 18 days after exposure. The incubation period reflects the time required for virus to establish a local infection, spread in the blood to the target organs, and

Fig. 57-8. Schematic representation of the pathogenesis of mumps.

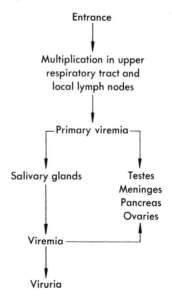

multiply sufficiently to damage cells and induce inflammation. With the onset of disease, as described above, virus can be detected in saliva, blood, and urine, indicating its widespread dissemination. The acute onset of fever and inflammation of salivary glands are often followed in 4 to 7 days by orchitis (in 20 to 35% of males past puberty), meningitis or meningoencephalitis, and occasionally by pancreatitis, ovaritis, or presternal edema. Even more rarely nephritis or paralytic manifestations may appear. The complications noted may develop at the same time as parotitis or even in the absence of salivary gland involvement. These observations support the hypothesis that the virus infects the salivary glands via the blood stream after multiplication in the respiratory tract epithelium and cervical lymph nodes (Fig. 57-8).

The most complete pathological descriptions of mumps virus lesions have been those of affected parotid glands and testes. The parotid lesions consist of interstitial inflammation of the gland and degeneration of the epithelium of the ducts. No characteristic inclusion bodies have been observed in patients' tissues, in contrast to tissue cultures (Fig. 57-7). Testes have diffuse degeneration, particularly of the epithelium of the seminiferous tubules, as well as edema, serofibrinous exudate, marked congestion, and punctuate hemorrhages in the interstitial tissue. Meningoencephalitis or meningitis has been studied pathologically only in rare cases, and proof of etiology has often been lacking. The findings described are typical of postinfectious encephalitis, with perivascular demyelination.

LABORATORY DIAGNOSIS

Diagnoses can usually be made solely by clinical observations. However, to establish the diagnosis of mumps in atypical or subclinical infections procedures similar to those utilized for influenza infection are employed. Virus can be isolated by inoculation of saliva, secretions from the parotid duct, or spinal fluid into the amniotic cavity of 8-day-old chick embryos or appropriate tissue cultures; the latter appear to be more sensitive for primary isolation. Infection can be detected earliest by **immunofluorescence** or hemadsorption technics. Since the virion is unstable at 4° or at room temperature, specimens must either be used immediately or stored at −70°.

Immunological diagnosis is made by demonstrating an increase in Abs after infection, using hemagglutination-inhibition or complement-fixation titrations. Complement fixation with the **soluble (S) antigen** can detect a rise within the first 7 to 10 days after onset of illness. After 14 days serological diagnosis is accomplished most readily with the **viral particles (V antigen)** in hemagglutination-inhibition or complement-fixation titrations. The antigenic cross-reactions between the surface (V) antigens of mumps and parainfluenza viruses complicate immunological diagnoses, making it necessary to test all the related viral antigens; the highest Ab rise occurs in response to antigens of the infecting virus. Where diagnosis is of critical importance the plaque-neutralization assay can detect Ab rises which may not be evident owing to cross-reactive complement-fixing Abs and nonspecific inhibitors of hemagglutination.

Though the skin test may be useful in surveys for estimating immunity, it frequently leads to erroneous conclusions in individual cases and is not reliable for diagnosis. Indeed, because the skin testing itself may induce a rise in Abs and confuse the interpretation of immunological reactions, skin tests should not be used during the course of illness.

EPIDEMIOLOGY

Mumps is predominantly a childhood disease spread by droplets of saliva. Salivary secretions may contain infectious virus as early as 6 days before and as long as 9 days after the appearance of glandular swelling. Virus is also present in the saliva of patients with meningitis or orchitis, even in the absence of clinical involvement of the salivary glands. Although viruria develops, there is no evidence for transmission from this source. Mumps is not nearly so contagious as the childhood exanthematous diseases (e.g., measles and chickenpox), and many children escape infection; hence, disease in adults is not uncommon. Subclinical infections, detectable immunologically, are also much more frequent than in other common childhood diseases.

Mumps appears most commonly during the winter and early spring months, but it is endemic throughout the year. Large epidemics have been observed about every 7 or 8 years.

PREVENTION AND CONTROL

Because subclinical infections are common, control of infection by isolation is not effective, although it is required by law in many localities. The disease can be prevented by immunization. **Infectious attenuated virus,** inoculated subcutaneously, induces an inapparent infection and development of Ab in greater than 95% of antibody-free subjects (children and adults). Infection follows inoculation of the vaccine, but viremia and viruria are not detectable, clinical reactions do not occur, and virus does not spread to exposed contacts. Although the Ab response is not as great as that accompanying natural infections, Ab levels persist for at least 5 years, and the vaccinees are almost uniformly protected upon exposure to mumps infections.

Formalin-inactivated virus also induces Ab production after two subcutaneous injections, and its clinical effectiveness has been demonstrated in controlled studies. However, Abs decline 3 to 6 months after immunization, and neither the effectiveness nor the persistence of immunity appears to be as satisfactory as following the live virus vaccine.

The live virus vaccine is of considerable value for susceptible adults in whom the disease is more severe and the complications more frequent. If widespread immunization of children is to be employed, the vaccine should induce persistent immunity, like that which follows the natural disease. The infectious attenuated virus vaccine may satisfy this requirement, but its use has been too brief and too limited to permit a confident recommendation.

Passive immunization with γ-**globulin from convalescent serum** has been employed to prevent infection after exposure, particularly in men, for whom orchitis is a relatively frequent and severely discomforting complication. Its effectiveness, however, has not been clearly demonstrated. Convalescent whole human serum has also been used for similar purposes, but it is not recommended because of the considerable danger of contamination with hepatitis virus (Ch. 62, Epidemiology).

MEASLES VIRUS

Measles (morbilli)* is one of the most infectious diseases known, and it is almost universally acquired in childhood; fortunately, immunity is essentially permanent. It was first described as an independent clinical entity by Sydenham in the seventeenth century. Although measles was demonstrated to be transmissible in volunteers as early as 1758, and was transferred to monkeys in 1911, it was not until 1954 that the virus, through Enders' persistent and careful search, was isolated reproducibly from patients and was shown to produce cytopathic changes in tissue cultures. This achievement led to a rapid advance in knowledge of the virus and to the development of an effective vaccine.†

PROPERTIES OF THE VIRUS

IMMUNOLOGICAL CHARACTERISTICS

All measles strains studied belong to a single antigenic type. Centrifugation of an extract of virus-infected cells to equilibrium in a cesium chloride density gradient separates, in order of decreasing density, the virions (which in addition to infectivity possess hemagglutinating, hemolytic, and complement-fixing activities), a hemolysin, a soluble complement-fixing antigen, and a hemagglutinin smaller than the virion. The four components are each immunogenic, but their antigenic interrelations have not been defined.

* **Rubeola** is often employed as a synonym; unfortunately, this term has also been used as a synonym for rubella (German measles).

† This brief account illustrates only in part the role of Enders in leading the way to the recent control of two important diseases, poliomyelitis and measles.

Like most myxoviruses and paramyxoviruses, measles virus, or its separated hemagglutinin, agglutinates monkey RBC. In contrast to myxoviruses and other paramyxoviruses, the maximal reaction occurs at 37°. In further contrast to myxoviruses and other paramyxoviruses, measles virus does not elute spontaneously from agglutinated cells, the erythrocyte receptors are not destroyed by *Vibrio cholerae* neuraminidase, and the virions do not contain neuraminidase molecules.

By analogy with influenza, parainfluenza, and mumps viruses, the soluble complement-fixing antigen in cells infected with measles virus is identical with the RNA-containing nucleocapsid of the viral particle, and the hemagglutinin corresponds to the virion's surface projections. The hemolytic factor, which is separable from the viral particle and is soluble in lipid solvents, not only hemolyzes monkey RBC, but also induces fusion of tissue culture cells (giant cell formation), presumably by an effect on the cell membrane.

Circulating Abs are detected 10 to 14 days after infection, i.e., when the rash appears or shortly thereafter, and reach maximal titer by the time the exanthem disappears. Antibody titers (neutralizing, complement-fixing, and hemagglutination-inhibiting) remain high following infection, and immunity often persists for a lifetime (as shown by epidemiological investigations). In monkeys, reinfection can be produced 3 to 6 months after a primary infection, but clinical disease does not ensue.

Measles virus is related to the viruses of canine distemper and rinderpest (of cattle) in antigenic, physical, and biological properties.

HOST RANGE

Measles virus is highly contagious for both humans and monkeys, its known natural hosts. Monkeys in captivity commonly develop spontaneous measles, man probably serving as its source. Because of the resulting immunity only freshly captured and antibody-free monkeys are reliable for testing vaccines or studying pathogenesis.

Although unmodified measles virus from patients replicates poorly in vitro, multiplication has been achieved in a variety of mammalian as well as chick embryo cells. Both primary and continuous mammalian cell cul-

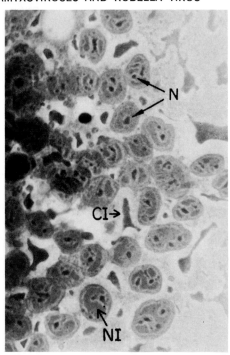

Fig. 57-9. Inclusion bodies of cells infected with measles virus. A large syncytium of cells is illustrated; each large round body is a nucleus. Large eosinophilic cytoplasmic inclusions are indicated by CI, and numerous intranuclear eosinophilic inclusion bodies by NI. The nucleoli (N) are intact. ×750. [From Kallman, F., Adams, J. M., Williams, R. C., and Imagawa, D. I. *J Biophys Biochem Cytol* 6:379 (1959).]

tures are commonly employed.* The development of large syncytial giant cells is generally the major cytopathic effect produced by measles infection of cultured cells (Fig. 57-9). Some strains selected in vitro, however, produce spindle-shaped rather than giant cells. Viruses adapted to propagation in vitro can also multiply in the amniotic sac or in the chorioallantoic membrane of chick embryos, and one strain has been adapted to propagation in brains of newborn mice.

---

* 1) **Primary cultures:** human embryonic kidney; human amnion, monkey or dog kidney; chick embryo cells; and bovine fetal tissue; 2) **continuous epithelium-like cell lines of human origin:** HeLa, KB, Hep-2, amnion, heart, nasal mucosa, bone marrow, and kidney. Viral isolations are best accomplished in primary cultures of human embryonic or monkey kidneys.

Eosinophilic inclusion bodies develop in both the nuclei and the cytoplasm of giant cells (Fig. 57-9). The inclusions are composed of dense, highly ordered arrays of filaments which resemble viral nucleocapsids (Figs. 57-10 and 57-11). Immunofluorescence studies reveal specific viral antigens in both the nuclear and cytoplasmic inclusion bodies.

PATHOGENESIS

Measles is a highly contagious, acute, febrile, exanthematous disease. The pathogenesis in man resembles the general pattern described for smallpox (Ch. 54) and mumps (Fig. 57-8), with local multiplication followed by hematogenous dissemination. Virus transmitted in respiratory secretions enters the upper respira-

Fig. 57-10. Measles virus particles budding from the surface membrane of an infected cell. Viral nucleocapsids are seen within the forming particles; the fuzzy structures on the surfaces of the virions probably correspond to the surface projections seen by negative staining (Fig. 57-3). ×98,000. [From Nakai, T., Shand, F. L., and Howatson, A. F. *Virology 38*:50 (1969).]

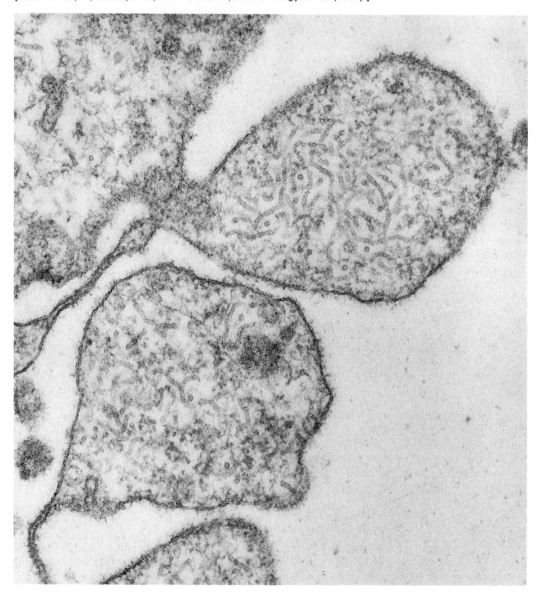

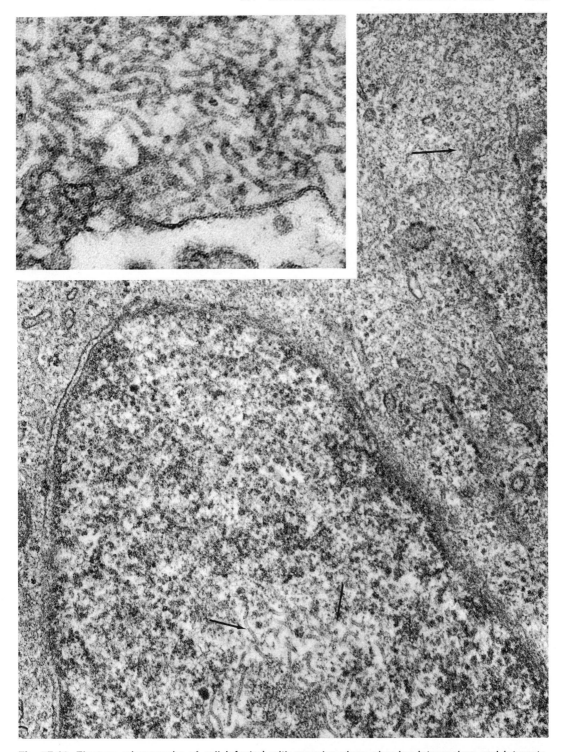

Fig. 57-11. Electron micrographs of cell infected with measles virus, showing intranuclear and intracytoplasmic matrices containing viral nucleocapsids (arrows). ×60,000. **Inset** illustrates a higher magnification of an extensive accumulation of nucleocapsids. Where the tubules are favorably oriented crossstriations of the nucleocapsids can be seen. ×140,000. [From Nakai, T., Shand, F. L., and Howatson, A. F. *Virology* 38:50 (1969).]

tory tract, or perhaps the eye, and multiplies in the epithelium and regional lymphatic tissue. Virus may also disseminate to distant lymphoid tissue by a brief primary viremia. Viral multiplication in the upper respiratory tract and conjunctivae causes, after an incubation period of 10 to 12 days, the prodromal (i.e., prerash) symptoms of coryza, conjunctivitis, dry cough, sore throat, headache, low-grade fever, and Koplik spots (tiny red patches with central white specks on the buccal mucosa in which are noted characteristic giant cells containing viral nucleocapsids). Viremia occurs toward the end of the incubation period, permitting further widespread dissemination of virus to the lymphoid tissue and skin. With the diffuse secondary multiplication of virus the prodromal symptoms are intensified and the typical red, maculopapular rash appears, first on the head and face and then on the body and extremities.

Virus is excreted in the secretions of the respiratory tract and eye, and in urine, during the prodromal phase and for about 2 days after the appearance of the rash. This early shedding of virus, before the disease can be recognized, promotes its rapid epidemic spread. The blood, lymph nodes, spleen, kidney, skin, and lungs also contain detectable virus during this period. Measles virus can multiply in and has been isolated from human leukocytes, suggesting that these cells may play a role in its dissemination in the body and in the pathogenesis of the disease. The leukocytic involvement may also be responsible for the leukopenia observed during the prodromal stage. Measles virus can induce striking aberrations in the chromosomes of leukocytes during the acute disease. Although the chromosomal pulverization produced is probably lethal to the cell, a possible relation between some of the changes and the initiation of leukemia has been suggested.

The **characteristic viremia** in measles, in contrast to the more localized respiratory infections produced by influenza viruses and by many paramyxoviruses, probably accounts for the notably effective immunity conferred by the disease.

Complications. **Bronchopneumonia** and **otitis media,** with or without a bacterial component, are frequent complications of the disease. **Encephalomyelitis** is the most serious compli-

cation, appearing about 5 to 7 days after the rash. Fortunately, its incidence is low in most epidemics (about 1 in 10,000 cases); but in some outbreaks, in particular the widespread infection of infants in Africa, the incidence has been much higher. The mortality rate of encephalomyelitis is about 10% and permanent mental and physical sequelae have been reported in 15 to 65% of survivors.

Measles encephalomyelitis has been attributed to increased virulence of the virus in certain epidemics, but without direct evidence. Indeed, only in very few of the fatal cases tested could measles virus be isolated from the brain or the spinal fluid, which suggested that measles encephalomyelitis may be a hypersensitivity response to either the measles virus or altered host tissue (i.e., an autoimmune phenomenon). However, the pathogenesis remains uncertain.

Subacute Sclerosing Panencephalitis (SSPE). This progressive, degenerative neurological disease of children and adolescents* was unexpectedly discovered to be caused by measles virus (or a virus that is immunologically and biologically very closely related). With this finding it became clear that a single virus could induce both an acute contagious disease and a chronic illness (Ch. 51, Slow viral infections). The following data implicate measles virus in this severe chronic disease. 1) Almost all patients have had measles several years (up to 13) prior to onset of SSPE or have been immunized with live virus vaccine. 2) All have unusually high titers of measles Abs, even in the spinal fluid, and the Ab levels often increase as the disease progresses (IgM as well as IgG viral Abs are present). 3) Affected brain cells have nuclear and cytoplasmic inclusions similar to those seen in measles infections (Fig. 57-9); the inclusions consist of filamentous tubular structures indistinguishable from the nucleocapsids seen in cells infected with measles virus (Fig. 57-11). 4) Immunofluorescence study of the brain lesions reveals antigens that react with Abs to measles but not to distemper virus (a close relative). 5) Measles virus, or a virus almost

---

* Common clinical characteristics are progressive mental and motor deterioration, myoclonic jerks, and electroencephalographic dysrhythmias.

identical with it, has been isolated in ferret brains and by cocultivation of affected brain cells with cells that readily support measles virus multiplication (e.g., African green monkey kidney cells, HeLa cells); and though virus has not been directly isolated from homogenates of brain cells, when the affected cells are cultured in vitro, they show typical inclusion bodies as well as the presence of viral antigen and viral nucleocapsid.

Pathology. The development of very large multinuclear **giant cells** (Warthin-Finkeldey syncytial cells) is the predominant and characteristic feature of the pathology of measles. These distinctive cells are found in nasal secretions during the prodromal stage of the disease, as well as in lymphoid tissue of the gastrointestinal tract, particularly the appendix. Giant cells are also often observed in sputum from patients with bronchopneumonia and may contain eosinophilic nuclear and cytoplasmic inclusions, similar to those seen in infected cell cultures. Moreover, **giant cell pneumonia,** a rare disease of debilitated children, was proved by isolation of virus to be due to measles. The giant cells of measles are presumably produced by the same mechanism that induces syncytial cell formation in infected cell cultures.

In measles encephalomyelitis the brain shows perivascular hemorrhage and lymphocytic infiltration early in the disease; areas of demyelination later appear in the brain and spinal cord.

Brains from patients with subacute sclerosing panencephalitis display a degeneration of cortex and especially the underlying white matter. Characteristically there are seen intranuclear and intracytoplasmic inclusion bodies not unlike those noted in acute measles (Fig. 57-9), perivascular infiltration of plasma cells and lymphocytes, scattered degeneration of nerve cells, hypertrophy of astrocytes, microglial proliferation, and demyelination.

LABORATORY DIAGNOSIS

The epidemiological and clinical features of measles are usually so characteristic that laboratory confirmation of the diagnosis is unnecessary except for investigative purposes.

During the prodromal stage of the disease a rapid and simple presumptive diagnosis can be made by demonstrating characteristic giant cells in smears of the nasal mucosa. Definitive diagnosis can be accomplished by isolation of virus from nasal or pharyngeal secretions, blood, or urine. Serological diagnosis can be made by comparing acute and convalescent sera by hemagglutination-inhibition or complement-fixation titrations, as described for other paramyxoviruses.

EPIDEMIOLOGY

Measles is a highly contagious disease in which virus is spread in respiratory secretions. It is predominantly a childhood affliction which occurs in epidemics during the winter and spring in rural areas, and is more or less endemic in urban areas. In the more highly developed countries epidemics tend to appear in 2- to 3-year cycles as sufficient nonimmune children arise in the population. Where hygienic conditions are poor epidemics occur yearly. But since measles virus does not have a reservoir in nature other than man, and permanent immunity follows infection, persistence of the virus in a community depends upon endemic infections in a continuous supply of susceptible persons. It has been estimated that in urban areas 2500 to 5000 cases per year are required for continued transmission. Hence, it is clear why the disease disappears from small, isolated communities even after exogenous introduction of the virus.

In the United States the highest incidence is in children 5 to 7 years of age, and the disease is relatively mild. The disease is more severe in adults and in young children (who are ordinarily protected to the age of 6 months by transplacental immunity). In communities having primitive and crowded living conditions most cases occur in infants less than 2 years old, and the illnesses tend to be severe and often fatal (e.g., in West Africa). If the virus is introduced into isolated communities, where exposure is rare and measles has not struck in many years, the incidence is very high, and the illness is severe and frequently is fatal to very young children and to the elderly (e.g., Faroe Islands; Ch. 51, Persistence of antibodies).

PREVENTION AND CONTROL

Public health measures alone, such as isolation, have not successfully prevented or even limited measles epidemics, and the injection of pooled human γ-globulin containing specific antibodies* serves only as a temporizing measure. Prior to the development of vaccines for active immunization passive immunization with pooled γ-globulin was used to furnish temporary protection. Because of the long incubation period large doses, even when administered shortly after exposure, prevent the disease, and small doses reduce its severity. This procedure is still effectively employed for susceptibles exposed to measles, particularly adults.

With the successful cultivation of measles virus in tissue culture, vaccines were developed and effective control became possible. Following the methods that had proved so successful in the control of poliomyelitis both **attenuated live virus vaccine** and **formalin-inactivated virus vaccine** were prepared.

The initial "attenuated" virus induced mild measles, with fever in about 80% and modified rash in approximately 50% of the human recipients. Even when γ-globulin, containing measles Abs, was administered at a site different from the vaccine about 20% of the children had fevers ranging between 100 and 103° F and about 8% had mild rashes 6 to 10 days after vaccination. Viruses of greater attenuation were eventually obtained, and they can now be used without an accompanying injection of γ-globulin. Under these conditions, live attenuated virus induces an Ab response in almost 100% of children who previously lacked Abs, but the vaccine still produces fever in 15 to 25% and rash in 10 to 15% of children. However, the virus does not cause neurological complications or spread of virus to susceptibles in the same family.

Immunization with live virus vaccine has been highly effective in large studies, and during epidemics it appears to prevent mea-

sles in more than 95% of children. Where widespread immunization has been employed large epidemics have been eliminated and the incidence of measles has been remarkably reduced. Nevertheless, this vaccine must be utilized with caution until we know that its inoculation, by the unusual subcutaneous route, does not establish endosymbiotic infections that lead to chronic diseases, such as SSPE. Although there has been reported only a single confirmed case of SSPE occurring after immunization (1 year earlier) of a child without a history of clinical measles, this frightening episode must bring a note of restraint.

Immunization with **attenuated** virus vaccines yields Ab titers that are about 10 to 25% of the levels observed after the natural disease. However, the success of the vaccine in field trials suggests that this titer is sufficient to confer resistance. Neutralizing Abs endure without significant decline for at least 2 years after immunization, but decrease two- to three-fold by 5 years; complement-fixing Abs begin to decline in 6 to 8 months. Although effective immunity has been demonstrated at least 8 years after immunization it is still too early to evaluate the total period of adequate protection.

The attenuated virus vaccine has been successfully lyophilized for general distribution. A melange of different attenuated viruses has also been used experimentally to facilitate the multiplicity of immunizations now being recommended: measles virus has been mixed with smallpox virus vaccine, or with attenuated mumps and rubella viruses. It is comforting that the multiple virus formulation has neither increased the reaction rate to measles attenuated virus nor decreased the Ab response to any of the viral immunogens present.

A **formalin-inactivated** virus vaccine has also been developed and shown to elicit Abs in about 75% of the recipients after a course of three intramuscular injections. The general acceptance of this vaccine is unlikely, however, because 1) protection is only temporary, 2) neutralizing Abs as well as complement-fixing Abs begin to decline rapidly within 3 to 6 months, and 3) unanticipated severe disease has been reported in vaccinees who were subsequently infected naturally or reimmunized with live virus vaccine.

---

* Because the majority of adults in most countries have had measles, and because levels of circulating Abs remain high, pooled normal adult γ-globulin is effective in passive immunization.

RESPIRATORY SYNCYTIAL VIRUS

From a chimpanzee with coryzal illness, and from a laboratory worker who had been in contact with the animal, Morris and coworkers isolated a new virus in 1956. The following year Chanock isolated similar viruses from two infants with pneumonia and croup. Subsequent studies indicated that the virus is an important cause of acute respiratory disease in children. Since it characteristically induces formation of large syncytial masses in infected cell cultures, it was named respiratory syncytial (RS) virus.

PROPERTIES OF THE VIRUS

The RS virus appears to be related to measles and parainfluenza viruses; hence it is grouped with paramyxoviruses. But there are several major distinctions from classic paramyxoviruses which suggest that this classification should be only tentative. The properties that distinguish RS virus from other paramyxoviruses are: 1) no hemagglutinin, hemadsorption, hemolytic, and neuraminidase activities are detectable despite the presence of regularly spaced club-like projections on the virion's surface (Fig. 57-12); 2) filamentous virions are present in purified preparations (Fig. 57-12) and in thin sections (Fig. 57-14); the filaments, unlike those of influenza viruses (Ch. 56, Fig. 56-2), are often much narrower than the spherical virions, and they have the appearance of an elongated nucleocapsid rather than a folded helix covered with an envelope; 3) the spherical virions are somewhat less variable in size and are slightly smaller (diameter of 900 to 1300A) than the typical paramyxoviruses; 4) the nucleocapsid appears to have a slightly smaller diameter (130 to 150 A), and the helix has a regular periodicity of about 70 A rather than the 50 A of other paramyxoviruses; 5) the virions and the nucleocapsids are extremely fragile, which makes preservation of infectivity difficult.

Immunological Characteristics. At least three antigenic variants have been noted among the limited number of RS strains studied. The immunological cross-reactivity between variants is too great, however, to warrant division into distinct types, and the antigenic differences noted do not appear to be progressive with successive epidemics. Each variant has a specific surface antigen, which can be detected by plaque-neutralization assays in cell cultures with standard sera from ferrets infected a second time.

Purified virions and extracts of infected cells contain both the specific viral surface antigen and a soluble nucleocapsid antigen, detectable by complement-fixation titrations, which is common to all RS viruses but not to other paramyxoviruses. The soluble antigen is presumably identical with the internal antigen of the viral particle, as demonstrated with influenza viruses.

Immunity is brief following RS virus infections, for although serum Abs persist the titer of IgA antibodies in nasal secretions declines rapidly. Even a level of serum neutralizing Abs as high as 1:256 does not pro-

Fig. 57-12. An intact respiratory syncytial virus particle (arrow) and two filaments negatively stained with phosphotungstate; note regularly arranged club-like peripheral projections. ×110,000. [From Bloth, B., Espmark, A., Norrby, E., and Gard, S. *Arch Gesamte Virusforsch 13*:582 (1963).]

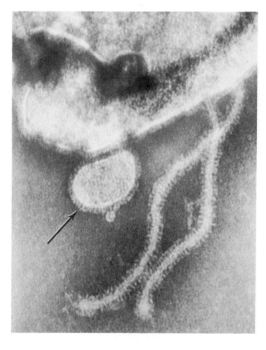

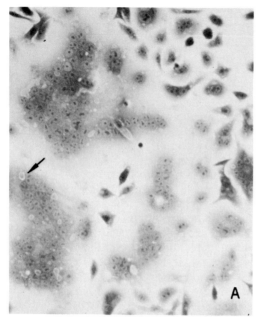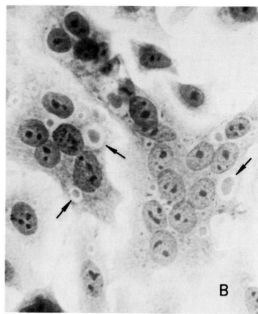

Fig. 57-13. Syncytia formation and cytoplasmic inclusions (arrows) in Hep-2 cells infected with respiratory syncytial virus. Syncytia are characterized by the very large aggregates of intact nuclei and extensive cytoplasmic masses devoid of cell membranes. **A,** × 125; **B,** ×540. [From Bennett, C. R., Jr., and Hamre, D. *J Infect Dis 110*:8 (1962).]

vide adults with effective protection against reinfection, and infants with neutralizing Abs may even develop bronchopneumonia on reinfection.

Host Range. RS virus multiplies and produces cytopathic changes in continuous human epithelium-like cell lines, in human diploid cell strains, and in primary simian and bovine kidney cell cultures. The formation of syncytia and giant cells, as in parainfluenza and measles virus infections, is the predominant and characteristic cytopathology (Fig. 57-13). Prominent eosinophilic cytoplasmic inclusions are commonly found in infected cells, particularly in the syncytia (Fig. 57-13). These inclusion bodies may be considered "scars" of the infection, for they do not contain DNA, RNA, viral particles, or accumulated virus-specific antigens, but consist of densely packed material having a granular or threadlike appearance. About 10 hours after infection, virus-specific antigens, detected by immunofluorescence, appear in the cytoplasm, but not in the nuclei of infected cells. As the quantity of antigen increases it accumulates at the cell membrane. The membrane is thickened where it becomes associated with characteristic filamentous material, and virions assemble and bud from the altered cell membrane (as with other paramyxoviruses; Fig. 57-14).

RS virus has been observed to cause illness only in man and in the chimpanzee. In addition, intranasal inoculation of virus into ferrets, monkeys, and many other mammals produces inapparent infections, from which virus can be recovered.

Unlike most myxoviruses or paramyxoviruses, RS virus has not been propagated in chick embryos, probably owing to its inability to combine with the specific host mucoprotein receptors.

PATHOGENESIS

The pathogenesis of RS virus infections has been largely surmised on the basis of clinical observations. The initial infection results from viral multiplication in epithelial cells of the upper respiratory tract; it most often ends at this stage in adults and older children. In about ½ of infected children less than 8 months old virus spreads into the bronchi,

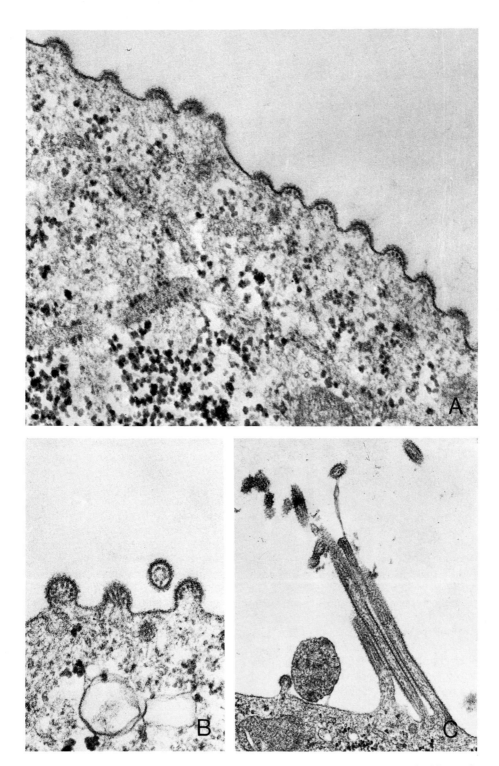

Fig. 57-14. Electron micrographs of thin sections of a cell culture infected with respiratory syncytial virus. Different stages are seen in the morphogenesis of the viral particles at the plasma membrane. **A.** Early stages of budding show thickening of the cell membrane with the appearance of fringe-like projections. Submembranous accumulation of the nucleocapsid is also noted. ×75,000. **B.** Later stages of viral budding and a single free virion are seen. The circular arrangement of the nucleocapsid in the spherical particles suggests on organized packing of this component. ×76,000. **C.** Filamentous forms of viral particles are developing at the cell membrane. The linear arrangement of the nucleocapsid is visible within the filamentous particles. ×42,000. [From Norrby, E., Marusyk, H., and Örvell, C. *J Virol* 6:237 (1970).]

bronchioli, and even into the pulmonary parenchyma. Thus, in children RS virus may cause febrile upper respiratory disease, bronchitis, bronchiolitis, bronchopneumonia, and croup. The dissemination of virus into the lower respiratory tract, with concomitant production of more serious disease, is probably due, at least partially, to the absence of antibodies and lack of a secondary response in the previously uninfected infant. With increasing age, and therefore repeated exposure and greater immunological response to RS virus, infections are likely to be milder and confined to the upper respiratory tract, or they are even more likely to be inapparent. Because of the absence or limited quantity of secretory (IgA) Abs in the respiratory tract, infections occur in adults despite the presence of circulating Abs; most frequently the disease is afebrile and clinically resembles the common cold.

Autopsy examination of several infants dead of RS virus infection has shown severe necrotizing lesions of the epithelium of the bronchi and bronchioles, and interstitial pneumonia consisting of mononuclear infiltration, patchy atelectasis, and emphysema. Cytoplasmic inclusion bodies were noted, but syncytial formation was not observed. The changes described represent one end of the spectrum of disease produced by RS virus.

LABORATORY DIAGNOSIS

A precise diagnosis of RS virus infection requires either isolation of the virus or demonstration of a rise in Abs. Serological diagnosis ordinarily is most reliable and easiest, using complement-fixation (the most convenient and economical) or neutralization titrations. Very young infants, however, may not produce a detectable Ab response, so that under these circumstances a laboratory diagnosis depends upon isolation of virus.

Because RS virus is highly labile, isolation is most efficient when nasal or pharyngeal secretions are inoculated directly into cultures of human continuous cell lines (Hep-2 or KB) or primary monkey kidney cell cultures. Characteristic giant cells develop within 2 to 14 days, but infection can be detected more rapidly by immunofluorescence. This technic can also detect RS viral antigens in exfoliated cells of the respiratory tract, although the sensitivity is approximately ½ that

of viral isolation methods. Newly isolated RS virus can be identified by complement-fixation or neutralization titrations with standard antisera.

Infectivity of RS virus is very unstable. Infectivity is completely destroyed during storage at -15 to $-25°$ for only several days. Viral suspensions can be preserved without complete inactivation by adding protein (5 to 10% normal serum or albumin), freezing rapidly, and maintaining at $-70°$.

EPIDEMIOLOGY

Infections with RS virus appear to have a worldwide distribution and occur in yearly epidemics of varying magnitude. Virus spreads rapidly through the susceptibles in a community, so that epidemics are sharply circumscribed and relatively brief. The outbreaks occur primarily in infants and children between late fall and early spring. Many adults may be infected during the episode, but they have mild disease or inapparent infections. RS infections uniformly occur early in life: approximately ⅓ of infants in the United States develop antibodies in the first year of life, and 95% by 5 years of age.

PREVENTION AND CONTROL

RS virus ranks high on the list of causes of respiratory disease in young children, producing particularly severe disease during the first 6 months of life. Therefore, preventive methods are highly desirable. Isolation or general public health measures are not adequate to control spread of infection; it is difficult to recognize the disease early and inapparent cases are frequent.

Experience with an alum-precipitated, formalin-inactivated vaccine has been discouraging. Though the serum Ab response was good, the clinical response to subsequent RS virus infection was startling and paradoxical: both the incidence and the severity of disease were strikingly increased, particularly in infants. These results resembled the severe reactions in children immunized with inactive measles virus (cf. p. 1348) and then infected with measles virus. It is not certain whether this frightening response is due to a cell-mediated hypersensitivity or to interactions between humoral Abs and virus-infected cells. However, it is clear that inactivated whole virus vaccines containing viruses with lipid envelopes must be used with caution.

Selection of attenuated viral variants, particularly temperature-sensitive mutants, is proceeding for the development of a live RS virus vaccine. But since the disease does not afford effective, lasting immunity, some view the utility of any vaccine with pessimism.

NEWCASTLE DISEASE VIRUS

Newcastle disease virus (NDV) is primarily a respiratory tract pathogen of birds, particularly chickens, but it occasionally produces accidental infections of man. Human infections are almost exclusively confined to poultry workers and laboratory personnel. The disease is characteristically mild and limited to conjunctivitis without corneal involvement. Although predominantly of veterinary interest, this virus merits brief discussion because of the prominent role it has played in the investigations of paramyxoviruses.

NDV possesses the characteristic properties of paramyxoviruses listed in Table 57-1. Morphologically and chemically the virion and nucleocapsid closely resemble those of parainfluenza and mumps viruses. Moreover, NDV's reactions with red blood cells are similar to the reactions of parainfluenza and mumps viruses, and many patients with mumps virus infection develop hemagglutination-inhibiting and complement-fixing antibodies to NDV.

RUBELLA VIRUS

Rubella virus is the etiological agent of German measles. This disease resembles measles but is milder, and it does not have the serious consequences often seen with measles in the very young. Because rubella seems to be such a harmless disease it did not receive much attention earlier, although its probable viral etiology was demonstrated in experiments with human volunteers in 1938. However, interest in rubella was much increased when Gregg, an Australian ophthalmologist, noted in 1941 that women contracting rubella during the first trimester of pregnancy frequently gave birth to babies with congenital defects. Nevertheless, the cultivation of rubella virus was not achieved until 1962 when Parkman, Buescher, and Artenstein detected the virus through its interference with type 11 echovirus in primary grivet monkey kidney cultures, and Weller and Neva demonstrated unique cytopathic changes (Fig. 57-15) in infected primary human amnion cultures.

PROPERTIES OF THE VIRUS

Morphology. The virion is roughly spherical and has a diameter of 500 to 850 A, in both thin sections and negatively stained preparations; it consists of a spherical core of 300 A, covered by a loose envelope. The virions being formed on cell membranes bud into vacuoles and extracellular spaces (Fig. 57-16).

Negative staining technics reveal small spikes projecting 50 to 60 A from the envelopes of most particles. Gentle disruption of the envelope with sodium deoxycholate uncovers an angular core, but no definite symmetry of the nucleocapsid and its subunit structures is discernible.

The morphology and growth characteristics of rubella viruses clearly distinguish it from paramyxoviruses. Indeed the properties described are similar to those of group A togaviruses (Ch. 60); but a final taxonomic assignment must await further characterization of rubella virus and better definition of the togavirus family.

Chemical and Physical Properties. The virion contains one molecule of single-stranded RNA of 38 to 40S. This sedimentation coefficient suggests a molecular weight of 2.5 to 3×10^6 daltons, significantly smaller than the RNAs of paramyxoviruses and togaviruses. Two glycoproteins are present in the envelope, and an arginine-rich protein is associated with the RNA.

Like other enveloped viruses, rubella virus is rapidly inactivated by ether, chloroform, and sodium deoxycholate; relatively labile when stored at 4°; and relatively stable at −60 to −70°.

The hemagglutinin is an integral component of the virion, although it remains biologically

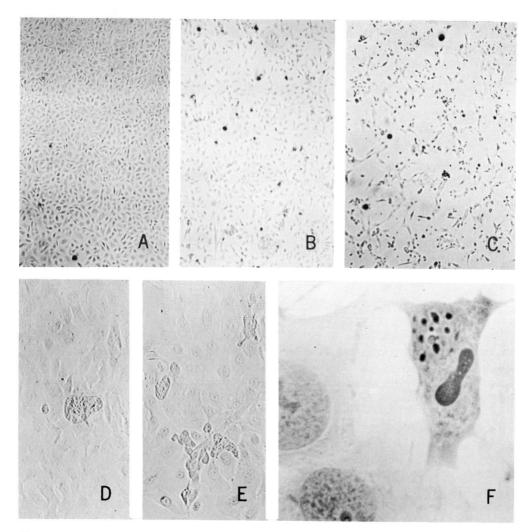

Fig. 57-15. Cytopathic effect of rubella virus in human amnion cultures. **A.** Appearance of normal amnion cell culture as viewed microscopically under low magnification. ×33. **B.** Rubella-infected culture with estimated 20% destruction of cells on fourteenth day after inoculation. ×33. **C.** Rubella-infected culture with 80% cell destruction on twenty-eighth day after inoculation. ×33. **D.** Single affected cell with adjacent uninvolved cells on tenth day after inoculation. ×132. **E.** Scattered infected cells showing ameboid distortion on tenth day after inoculation. ×132. **F.** Infected cell with large eosinophilic cytoplasmic inclusion and basophilic aggregation of nuclear chromatin, as well as portions of two normal cells. H & E stain. ×3500. [From Neva, F. A., Alford, C. A., Jr., and Weller, T. H. *Bacteriol Rev 28*:444 (1964).]

active after gentle disruption of the viral particle. It is most effectively assayed with newborn chick, pigeon, goose, or human group O RBC. The hemagglutinin does not elute spontaneously from the affected erythrocytes, and the receptor-destroying enzyme of *Vibrio cholerae,* i.e., neuraminidase (Ch. 44), does not render RBC inagglutinable (these properties are also similar to those of togaviruses).

Immunological Properties. Only a single antigenic type of rubella virus has been detected. Neutralizing and hemagglutination-inhibiting Abs are commonly present when the rash appears (see below); they attain maximal titer during early convalescence and persist (along with immunity) for many years if not for life. Complement-fixing Abs appear about 6 to 10 days after onset, begin to diminish

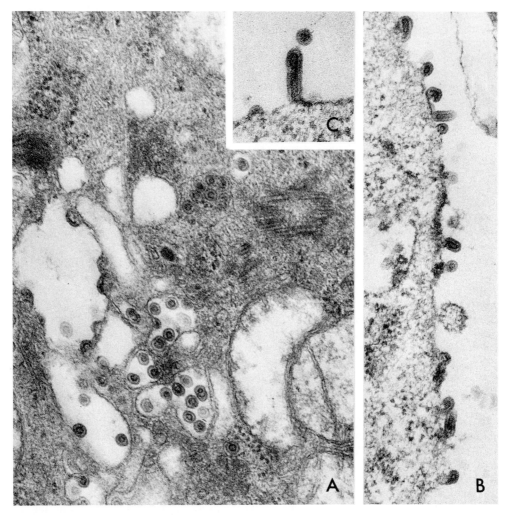

Fig. 57-16. Development of rubella virus in the surface and cytoplasmic membranes of infected cell cultures. **A.** Viral particles are budding from cytoplasmic membranes into vacuoles and into the cytoplasm. Numerous mature virions are present within vacuoles. ×60,000. **B.** Viral particles budding from the surface of an infected cell. ×60,000. **C.** An elongated form in the budding process. ×84,000. [From Oshiro, L. S., Schmidt, N. J., and Lennette, E. H. *J Gen Virol* 5:205 (1969).]

after 4 to 6 months, and disappear after a few years.

VIRAL MULTIPLICATION

Rubella virus can replicate in a variety of primary, continuous, and diploid cell cultures of monkey, rabbit, and human origin. When a large inoculum is employed the eclipse period is 10 to 12 hours. The viral titer reaches a maximum 30 to 40 hours after infection and may remain high for weeks (indeed, carrier cultures are readily established; see Ch. 51). Viral multiplication is confined to the cytoplasm, and its RNA synthesis proceeds through the usual intermediate replicative forms (Ch. 48, Synthesis of RNA viruses). Morphogenesis of the virions occurs at cell membranes: the cell membranes, particularly the plasma membrane, differentiate by incorporating viral proteins, and nucleocapsids bud through the thickened vacuolar and surface membranes to form mature viral particles (Fig. 57-16). Unlike togaviruses (Ch. 60), nucleocapsids do not accumulate in the cytoplasm; and in contrast to classic paramyxoviruses, excess helical nucleocapsids do not form cytoplasmic masses (p. 1334). Because viral components, including the hemagglutinin, are incorporated into the cell surface membrane during budding, the infected cells can be detected by hemadsorption.

In many infected cells (e.g., grivet monkey kidney) an increase in viral titer is associated with increased resistance to infection with some challenge viruses (e.g., picornaviruses, myxoviruses, measles virus), which provides another procedure for detecting infected cells and isolating viruses. The interference is induced by propagation of rubella virus, and is not consistently associated with interferon production.

In most rubella-infected cell cultures cytopathic changes are not detectable, but in primary human cell cultures distinctive cellular alterations do occur. Affected cells appear slowly during 2 to 3 weeks; they are enlarged or rounded, and they often have ameboid pseudopods (Fig. 57-15). Staining reveals a disappearance of the nuclear membrane, prominent clumping of nuclear chromatin, and round or irregular eosinophilic cytoplasmic inclusion bodies (Fig. 57-15). The cytopathic effects are associated with inhibition of biosynthesis of host macromolecules and the inability of infected cells to divide, although the infected cells do not lyse.

In cultures of human diploid cells that are chronically infected (carrier cultures) many chromosomal breaks are evident. Such breaks may have a bearing on the pathogenesis of congenital lesions in the infected fetus since impaired division of affected specialized cells may influence organogenesis (e.g., eye, heart).

PATHOGENESIS

The rash appears 14 to 25 days after infection with rubella virus; the average being 18 days. During this prolonged incubation period the pattern of viral multiplication and dissemination in the body resembles that of smallpox (Ch. 54) and measles (p. 1344). Virus can be isolated from nasopharyngeal secretions (and occasionally from feces and urine) as early as 7 days before and as late as 7 days after the appearance of the exanthem. Respiratory secretions are probably the major vehicle for transmitting the virus.

The disease is initiated by a 1- to 2-day prodromal period of fever, malaise, mild coryza, and prominent cervical and occipital lymphadenopathy. During the prodromal illness, and for 1 to 2 days after the rash appears, virus can be isolated from the blood. The disease is not unlike measles, except that it is milder, is of shorter duration, and has fewer complications.

No characteristic pathological lesions have been described in rubella, except for the serious damage induced in infected fetuses. This damage seems to involve tissues of all germ layers, and results from a combination of rapid death of some cells and persistent viral infection of others. The continued infection in turn frequently induces chromosomal aberrations and finally reduced cell division. The infant infected during the first trimester may be stillborn; if it survives, it may have deafness, cataracts, cardiac abnormalities, microcephaly, motor deficits, or other congenital anomalies in addition to thrombocytopenic purpura, hepatosplenomegaly, icterus, anemia, and low birth weight (the **rubella syndrome).** The greater susceptibility of the early embryo to damage is correlated with the placental transmission of virus: when infection occurs during the first 8 weeks of pregnancy, virus can be isolated as often from the aborted fetus as from the placenta, whereas in the infections after the first trimester viral isolations are less frequent from the fetus than from the placenta.

Persistence. Deformed infants born 6 to 8.5 months after intrauterine infection may still excrete virus in their nasopharyngeal secretions, even in the presence of high titers of neutralizing IgM and IgG antibodies (Fig. 57-17). Indeed, one infant continued to shed virus in the presence of circulating antibodies 2 years after birth, and in another virus was isolated from cataract tissue at 3 years of age. Even the high titers of circulating Abs and the presence of cell-mediated immunity cannot eliminate viral shedding, which persists until the clones of infected cells that are still able to divide eventually disappear. These observations, like those noted in congenital cytomegalovirus infections (Ch. 53) and congenital

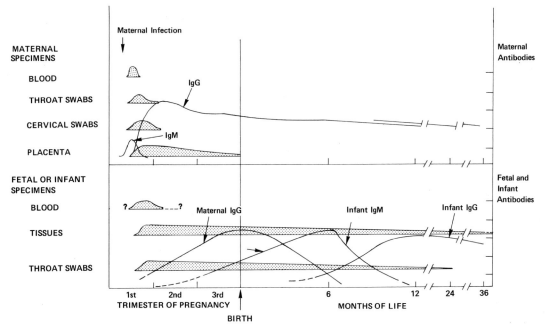

Fig. 57-17. Virological events and antibody response in maternal-fetal rubella infection. [Modified from Meyer, H. M., Parkman, P. D., and Hopps, H. E. *Am J Clin Pathol* 57:803 (1972).]

syphilis (Ch. 37), indicate that intrauterine rubella infection does not induce immunological tolerance to rubella virus antigens.

When a woman has clinical rubella during the first trimester of pregnancy the chance that the baby at birth will have a structural abnormality is approximately 30%. Most of these infants, even those clinically normal, excrete virus at birth.

LABORATORY DIAGNOSIS

Rubella may be confused with measles as well as with infections produced by a number of picornaviruses (echo- and coxsackieviruses; Ch. 55). The diagnosis may be confirmed by inoculating infected materials (usually nasopharyngeal secretions) into susceptible cell cultures. The indirect interference assay, in primary grivet monkey kidney cultures, is the quickest and most sensitive procedure for initial viral isolation. Infection is also detected in cultures of primary human amnion or continuous rabbit cell cultures by cytopathic changes, and in cultures of hamster or grivet monkey kidney cells by hemadsorption. Serological diagnosis can be accomplished by hemagglutination-inhibition, complement-fixation, or neutralization titrations.

EPIDEMIOLOGY

Rubella is a highly contagious disease, spread by nasal secretions. Unlike measles or chickenpox, however, rubella infection is often inapparent, thus fostering viral dissemination and rendering isolation of patients virtually useless. The ratio of inapparent infections to clinical cases is low (approximately 1:1) in children but as high as 9:1 in young adults. Owing to the larger amount of virus present in the nasopharynx, patients are most infectious during the prodromal period; communicability may persist for as long as 7 days after the appearance of the rash. Transmission usually occurs by direct contact with patients or persons with inapparent infections. Children 5 to 14 years of age are the major culprits in disseminating virus. But the apparently normal infant excreting virus acquired in utero is perhaps the most dangerous carrier: its infection is unrecognized and it comes in close contact with nurses, physicians, hospital visitors (including future mothers early in pregnancy), and later with other children at home.

In urban areas of Europe and the Americas, during the winter and spring months, minor outbreaks are noted every 1 or 2 years and major epidemics recur every 6 to 9 years. Infection is almost always followed by permanent protection against clinical disease, although reinfections do occur. Most inapparent infections occur in those in whom immunity has partially waned; the reinfection induces a secondary immune response (IgG antibodies) and probably prevents or reduces the extent of viremia.

PREVENTION AND CONTROL

The extensive epidemic in the United States during 1964 resulted in disabilities of approximately 20,000 infants, causing enormous anguish and a total economic loss of well over 1 billion dollars. To avoid these dire consequences that may result from rubella in the first trimester of pregnancy, prevention of maternal infection is of utmost importance. Isolation procedures, however, are rarely practical, and passive immunization is of questionable value. The development of an effective live attenuated virus vaccine should prevent recurrence of such a devastating experience.

Rubella viruses isolated in African green monkey kidney cells and serially passaged in cultures of duck embryos, rabbit kidneys, dog kidneys, or human embryo diploid cells have been extensively studied to test their attenuation, antigenicity, and effectiveness as vaccines. The different vaccines studied proved to be immunogenic in at least 95% of the recipients, but after immunization antibodies appeared later and did not reach as high levels as following natural infection. Nevertheless, immunization effectively protects the recipients from clinical rubella following exposure and even during epidemics.

Though the vaccines employed are highly protective some drawbacks exist. 1) In 2 to 3 weeks after subcutaneous inoculation small amounts of infectious virus appear in the nasopharynx, but viremia is unusual, and transmission to susceptible contacts has not been observed. 2) At the time of nasopharyn-

geal viral shedding, mild arthralgias and occasional arthritis occur in 1 to 2% of children and in 25 to 40% of adult women; adults also occasionally experience mild rash, fever, and lymphadenopathy (a vaccine made in dog kidney cells produced the highest incidence of reactions, and therefore it has been removed from the market). 3) The relatively low antibody levels attained suggest that the vaccine may not produce long-lasting immunity (unfortunately the vaccines that induce the lowest reaction rates also are the least immunogenic). Despite the marked reduction in incidence of clinical disease, infection frequently is not prevented although exposure may occur only 2 to 3 months after immunization; on the other hand the infection, which may occur in vaccinees while they are still protected from clinical illness, could serve to induce a life-long immunity similar to that following the natural disease.

It should be emphasized that the major goal of immunization is to protect the fetus rather than the vaccines, an objective that is unique. This goal may be accomplished either by immunization of teenage girls who are the prospective mothers or by immunization of young children in order to reduce spread of the virus by its major transmitters, children 1 to 12 years of age. In the United States the latter alternative has been chosen: i.e., routine immunization of all children. But in addition immunization is recommended for all women of childbearing age who are without protective antibodies (birth control must be rigidly practiced for 2 to 3 months following immunization). If vaccination should prove not to effect a durable immunity it may be necessary to recommend a second immunization for all girls when they attain puberty. In Great Britain immunization is given to older persons: all girls between 10 and 14 years, and women of childbearing age who do not have detectable hemagglutination-inhibiting Abs.

If preventive measures fail, however, to avoid the tragedy presented by the birth of a deformed baby, many physicians recommend therapeutic abortion when rubella occurs during the first trimester of pregnancy.

SELECTED REFERENCES: PARAINFLUENZA VIRUSES

Books and Review Articles

CHANOCK, R. M. Parainfluenza viruses. In *Diagnostic Procedures for Virus and Rickettsial Diseases,* 4th ed., p. 434 (E. H. Lennette and N. J. Schmidt, eds.) American Public Health Association, New York, 1969.

KINGSBURY, D. W. Replication and functions of myxovirus ribonucleic acids. *Progr Med Virol* 12:49 (1970).

Specific Articles

BLAIR, C. D., and ROBINSON, W. S. Replication of Sendai virus. II. Steps in virus assembly. *J Virol* 5:639 (1970).

CALIGUIRI, L. A., KLENK, H. D., and CHOPPIN, P. W. The proteins of parainfluenza virus SV5. I. Separation of virion polypeptides by polyacrylamide electrophoresis. *Virology 39:*460 (1969).

CHEN, C., COMPANS, R. W., and CHOPPIN, P. W. Parainfluenza virus surface projections: Glycoproteins with haemagglutinin and neuraminidase activities. *J Gen Virol 11:*53 (1971).

KLENK, H. D., and CHOPPIN, P. W. Plasma membrane lipids and parainfluenza virus assembly. *Virology 40:*939 (1970).

ROBINSON, W. S. Self-annealing of subgroup 2 myxovirus RNAs. *Nature 225:*944 (1970).

SELECTED REFERENCES: MUMPS VIRUS

Books and Review Articles

DEINHARDT, F., and SHRAMEK, G. J. Immunization against mumps. *Progr Med Virol 11:*126 (1969).

HENLE, W. Mumps. In *Diagnostic Procedures for Virus and Rickettsial Diseases,* 4th ed., p. 457. (E. H. Lennette and N. J. Schmidt, eds.) American Public Health Association, New York, 1969.

Specific Articles

LEVITT, L. P., MAHONEY, D. H., CASEY, H. L., and BOND, J. O. Mumps in a general population: A sero-epidemiologic study. *Am J Dis Child 120:*134 (1970).

SELECTED REFERENCES: MEASLES VIRUS

Books and Review Articles

ENDERS, J. F. Measles virus: Historical review, isolation, and behaviour in various systems. *Am J Dis Child 103:*219 (1962).

International Conference on Measles Immunization. *Am J Dis Child 103:*219 (1962).

KATZ, S. L., and ENDERS, J. F. Measles virus. In *Diagnostic Procedures for Virus and Rickettsial Diseases,* 4th ed., p. 504. (E. H. Lennette and N. J. Schmidt, eds.) American Public Health Association, New York, 1969.

KRUGMAN, S. Present status of measles and rubella immunization in the United States: A medical progress report. *J Pediatr 78:*1 (1971).

Specific Articles

BLACK, F. L. Measles endemicity in insular populations: Critical community size and its evolutionary implication. *J Theor Biol 11:*207 (1966).

PANUM, P. L. Observations made during the epidemic of measles on the Faroe Islands in the year 1846. American Publishing Association, New York, 1940. Reprint.

SELECTED REFERENCES: RESPIRATORY SYNCYTIAL VIRUS

Books and Review Articles

BEEM, M., and HAMRE, D. Respiratory syncytial virus. In *Diagnostic Procedures for Virus and Rickettsial Diseases,* 4th ed., p. 491. (E. H. Lennette and N. J. Schmidt, eds.) American Public Health Association, New York, 1969.

Specific Articles

BEEM, M., WRIGHT, F. H., HAMRE, D., EGERER, R., and OEHME, M. Association of the chimpanzee coryza agent with acute respiratory disease in children. *N Engl J Med 263:*523 (1960).

BENNETT, C. R., and HAMRE, D. Growth and serological characteristics of respiratory syncytial virus. *J Infect Dis 110:*8 (1962).

CHANOCK, R. M., KAPIKIAN, A. Z., and MILLS, J. Influence of immunological factors in respiratory syncytial virus disease. *Arch Environ Health 21:*347 (1970).

SELECTED REFERENCES: RUBELLA VIRUS

Books and Review Articles

International Conference on Rubella Immunization. *Am J Dis Children 118:*1 (1969).

NORRBY, E. Rubella virus. *Virology Monographs 7:* 115 (1969).

PLOTKIN, S. A. Rubella virus. In *Diagnostic Procedures for Virus and Rickettsial Diseases,* 4th ed., p. 364. (E. H. Lennette and N. J. Schmidt, eds.). American Public Health Association, New York, 1969.

RAWLS, W. E. Congenital rubella: The significance of virus persistence. *Progr Med Virology 10:*238 (1968).

WELLER, T. H., ALFORD, C. A., JR. and NEVA, F. A. Changing epidemiologic concepts of rubella, with particular reference to unique characteristics of the congenital infection. *Yale J Med Biol 37:*455 (1965).

Specific Articles

DAVIS, W. J., LARSON, H. E., SIMSARIAN, J. P., PARKMAN, P. D., and MEYER, H. M., JR. A study of rubella immunity and resistance to infection. *J Am Med Assoc 215:*600 (1971).

GREGG, N. M. Congenital cataract following German measles in the mother. *Trans Ophth Soc Australia (BMA) 3:*35 (1941).

SEDWICK, W. D., and SOKOL, F. Nucleic acid of rubella virus and its replication in hamster kidney cells. *J Virol 5:*478 (1970).

chapter

CORONAVIRUSES

PROPERTIES OF THE VIRUSES   1362
  Morphology   1362
  Physical and Chemical Characteristics   1362
  Immunological Characteristics   1362
  Host Range   1364
  Viral Multiplication   1364
PATHOGENESIS   1365
LABORATORY DIAGNOSIS   1365
EPIDEMIOLOGY   1365
PREVENTION AND CONTROL   1365

After the discovery of rhinoviruses as major etiological agents of the common cold, more than 50% of the illnesses still could not be associated with known causative agents. A new approach was generated, however, when Tyrrell and Bynoe in 1965 showed that the use of ciliated human embryonic tracheal and nasal organ cultures greatly improved the isolation of respiratory viruses. Not only were known viruses isolated more frequently but also new ones were discovered. The unique properties of the previously unknown viruses (Table 58-1) included their distinctive club-shaped surface projections (Fig. 58-1) which give the appearance of a solar corona to the virion. Hence, the family name **coronaviruses** was proposed to include the initial agents isolated as well as viruses isolated by Hamre and Procknow in 1966 from patients with acute respiratory diseases, using primary monolayer cultures of human embryonic kidney cell, and similar viruses that infect a variety of lower animals.*

PROPERTIES OF THE VIRUSES

MORPHOLOGY

Electron microscopic examinations of negatively stained preparations reveal moderately pleomorphic spherical or elliptical virions. The virion's surface is covered with distinctive pedunculated projections, with narrow bases and club-shaped ends (Fig. 58-1). Thin sections show that an outer double membrane (the envelope) 70 to 80 A thick surrounds a nucleocapsid consisting of an electron-dense shell 90 to 170 A thick and a central zone

---

* Infectious bronchitis virus of chicken; mouse hepatitis virus; transmissible gastroenteritis virus of swine; and a pneumotropic virus of rats.

containing amorphous material of variable density (Fig. 58-2). The nucleocapsid appears to be a loosely wound helix approximately 70 to 90 A in width.

PHYSICAL AND CHEMICAL CHARACTERISTICS

Information on the physical and chemical structure of the virion is fragmentary. The virion contains at least six polypeptides: two glycoproteins are associated with the surface projections; one glycoprotein and a glycolipoprotein which are probably in the envelope; and two major proteins whose location is unknown. The presence of lipid as a major structural component of the virion's envelope is also indicated because lipid solvents inactivate infectivity, and the virions have a low buoyant density. RNA appears to be the nucleic acid constituent of the virion since analogs that inhibit DNA synthesis (e.g., 5-fluoro-2-deoxyuridine) do not reduce viral multiplication, but purified virions have not yet been thoroughly analyzed. Infectivity is inactivated at pHs that deviate as little as 0.5 unit from pH 7.2, which probably reflects the lability of the viral envelope.

IMMUNOLOGICAL CHARACTERISTICS

Coronaviruses isolated from man can be divided into at least three antigenic types by neutralization assays. The strict host-range specificities (see below) suggest that it is probably not a biological accident that the viruses initially isolated in organ cultures are immunologically distinct from those first detected in human embryo kidney cell cultures. The initial isolates of each group serve as the prototype strains for two immunological types (Table 58-2). A third group of six strains obtained in human embryo tracheal

TABLE 58-1. Characteristics of Coronaviruses

Size	800–1600 A
Morphology	Spherical; pleomorphic; pedunculated, club-shaped surface projections
Lipid-containing	Ether or chloroform inactivates
Nucleic acid	RNA
Replication	Buds into cytoplasmic vacuoles

TABLE 58-2. Neutralization Reactions of Representative Coronaviruses

Serum employed	Neutralization titer with virus		
	B814*	229E†	OC 43‡
B814	640	<5	<5
229E	<5	1600	<5
OC 43	<5	160	5120

* Isolated by Tyrrell and Bynoe in human embryo tracheal organ culture.

† Isolated by Hamre and Procknow in human embryo kidney cell culture.

‡ Isolated by McIntosh *et al.* in human embryo tracheal organ culture.

Modified from BRADBURNE, A. F., and TYRRELL, D. H. J. *Progr Med Virol* 13:373 (1971).

organ cultures have a weak cross-reaction with the tissue culture viruses, but they are antigenically unrelated to the other organ culture isolates. The viruses from man appear antigenically unrelated to coronaviruses from other animals in neutralization tests, but complement-fixation assays indicate a partial immunological relatedness of human types with mouse hepatitis virus.

The virion contains at least three distinct antigenic structural units as detected by immunodiffusion studies; probably these same antigens react in complement-fixation assays as well. Two **organ culture** (OC) strains have been adapted to multiplication in the brains of suckling mice, and have attained the capacity to agglutinate erythrocytes from mice, rats, chickens, and humans—perhaps owing to a higher concentration of virus or to the accidental selection of particular mutants. Unlike myxoviruses, coronaviruses do not elute spontaneously from agglutinated RBC, and treatment of susceptible RBC with neuraminidase does not reduce hemagglutination. A rise in

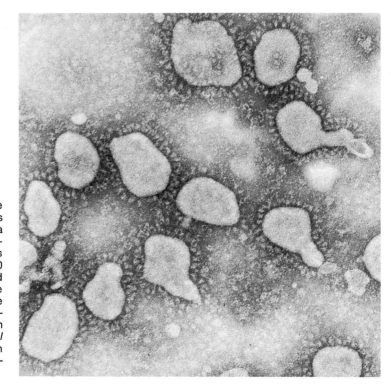

Fig. 58-1. Coronaviruses. The negatively stained particles show the distinctive corona effect produced by the pedunculated surface projections which are approximately 200 A long and have a club-shaped end about 100 A in width. The marked pleomorphism of the virions may be noted. ×144,-000. (From Kapikian, A. Z. In *Diagnostic Procedures for Viral and Rickettsial Infections,* 4th ed. American Public Health Association, New York, 1969.)

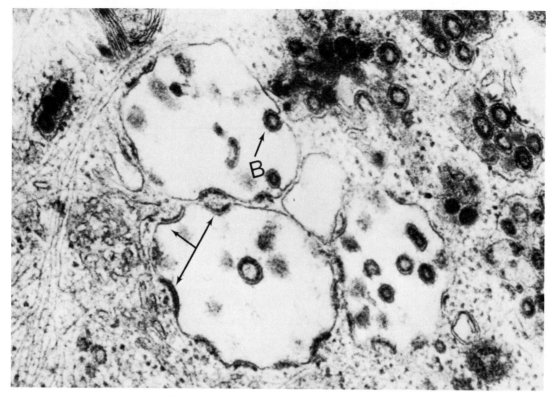

Fig. 58-2. Development of a coronavirus in human diploid cell. Thickened membranes (arrows) indicate sites of early morphogenesis. A particle budding into a vacuole (B) is also present. Several mature particles are present within vacuoles. The mature particles show inner and outer shells with a translucent zone between them. ×50,000. [From Becker, W. B., McIntosh, K., Dees, J. H., and Chanock, R. M. *J Virol 1*:1019 (1967).]

Ab (measured by complement fixation, hemagglutination inhibition, or neutralization) follows infection. But the duration of immunity is still not known.

HOST RANGE

Coronaviruses affecting man appear to be extremely limited in the host cells in which they can multiply. Viruses isolated in organ cultures of ciliated human embryonic tracheal or nasal tissues do not multiply well in monolayer cultures of human diploid or embryonic kidney cells, and vice versa; hence both culture systems must be employed for viral isolations. A continuous human embryonic cell line has shown evidence of susceptibility to both types of viruses, but additional studies are required to determine its general utility. Two of the viruses that had been isolated in organ cultures, the OC strains (Table 58-2), have been

adapted to growth in monkey kidney cells and in the brains of suckling mice, but other laboratory animals have not proved to be susceptible to infection.

VIRAL MULTIPLICATION

Viral replication, as observed with electron microscopic and fluorescent antibody technics, occurs solely in the cytoplasm and in close association with the smooth and rough endoplasmic reticulum and the Golgi apparatus (Fig. 58-2). Nucleocapsid synthesis is not discernible in electron micrographs and these subunits do not appear to accumulate in visible masses. Infrequently, thickened vesicle membranes and budding viral particles are observed; most commonly, however, formed virions are seen within vesicles or extracellularly. Viral multiplication is relatively slow as compared with myxoviruses or togaviruses:

for example, the eclipse period is at least 6 hours and maximum viral yield is not attained until about 24 hours after infection.

PATHOGENESIS

Coronaviruses constitute another group of viruses responsible for common colds and pharyngitis. Patients from whom the viruses were recovered, and volunteers who were inoculated intranasally to establish the etiological relation of coronaviruses to disease, exhibited signs and symptoms of acute upper respiratory infections: coryza, nasal congestion, sneezing, and sore throat were common; less frequent symptoms were headache, cough, muscular or general aches, chills, and fever. Pulmonary involvement has not been noted. In volunteers inoculated intranasally with virus the average incubation period was 3 days and the mean duration of illness was 6 to 7 days. Although coronaviruses have many characteristics that differentiate them from other viruses, coronavirus disease cannot be distinguished clinically from common colds produced by rhinoviruses or even from those occasionally initiated by influenza viruses.

Since the viruses have only been adapted to the brains of suckling mice, in which they cause encephalitis, details of pathogenesis cannot be studied experimentally.

LABORATORY DIAGNOSIS

Coronaviruses are so fastidious in their host requirements that isolation is not practicable for routine diagnosis and is used only for research. In principle, the diagnostic serological procedures of complement-fixation, neutralization, and hemagglutination-inhibition titrations can be used, but complement-fixation titrations are the only practical serological diagnostic procedure: neutralization titrations are difficult and expensive, since they require both organ cultures and cell cultures; neutralizing Abs are present in 50 to 80% of the population, and although reinfection is common, a demonstrable increase in titer is difficult to measure; and hemagglutination has been demonstrated with only two viruses, both belonging to the same immunological type of organ culture viruses that have been adapted to suckling mouse brain. For seroepidemiological surveys, however, neutralization titrations are more informative than complement-fixation assays since complement-fixing Abs decline rapidly after infection.

EPIDEMIOLOGY

Acute upper respiratory infections generally have their greatest incidence in children, particularly those between 4 and 10 years of age. It is therefore especially striking that coronaviruses are more frequently associated with common colds in adults. Virus has been rarely isolated from children, and complement-fixing antibodies are detected in only about 2% of them. Epidemiological studies using neutralization titrations, however, detected antibodies in 20 to 25% of children 5 to 9 years old, but infection (with an increase in titer) was more common in older children and adults. In further contrast to the usual upper respiratory infections, it is a member who is 15 years or older who most frequently introduces infection into a family.

Coronavirus infections occur predominantly during winter and early spring months, and during a particular respiratory disease season only a single immunological type appears to be the causative agent. In a seroepidemiological study covering a 4-year period of acute respiratory infections in Washington, D.C., coronavirus infections occurred in 10 to 24% of adults with upper respiratory illnesses observed during two out of four winters. It is noteworthy that during the periods of prevalent coronavirus infections, rhinovirus infections were uncommon. During epidemics all segments of the population are affected, but virus tends to spread preferentially within families. Unfortunately, because of the technical difficulties comparatively few data on the epidemiologic incidence and spread of coronaviruses are available.

PREVENTION AND CONTROL

While an effective vaccine can probably be developed, additional data on duration of immunity following natural infection, antigenic structure and immunogenic potential of the virions, and incidence of infections are essential before the needs and feasibility of a vaccine are established.

SELECTED REFERENCES

Books and Review Articles

BRADBURNE, A. F., and TYRRELL, D. A. J. Coronaviruses of man. *Progr Med Virol 13:*373 (1971).

KAPIKIAN, A. Z. Coronaviruses. In *Diagnostic Procedures for Viral and Rickettsial Infections.* (E. H. Lennette and N. J. Schmidt, eds.) American Public Health Association, New York, 1969.

Specific Articles

ALMEIDA, J. D., BERRY, D. M., CUNNINGHAM, C. H., HAMRE, D., HOFSTAD, M. S., MALLUCCI, L., MC-INTOSH, K., and TYRRELL, D. A. J. Coronaviruses. *Nature 220:*650 (1968).

HAMRE, D., KINDIG, D. A., and MANN, J. Growth and intracellular development of a new respiratory virus. *J Virol 1:*810 (1967).

HAMRE, D., and PROCKNOW, J. J. A new virus isolated from the human respiratory tract. *Proc Soc Exp Biol Med 121:*190 (1966).

HIERHOLZER, J. C., PALMER, E. L., WHITEFIELD, S. G., KAYE, H. S., and DAWDLE, W. R. Protein composition of coronavirus OC 43. *Virology 48:*516 (1972).

TYRRELL, D. A. J., and BYNOE, M. L. Cultivation of a novel type of common-cold virus in organ cultures. *Br Med J 1:*1467 (1965).

chapter **59**

RHABDOVIRUSES

RABIESVIRUS   1368

    Properties of the Virus   1368
    Pathogenesis   1370
    Laboratory Diagnosis   1371
    Epidemiology   1373
    Prevention and Control   1373

MARBURG VIRUS   1375

Were the basis for evolutionary development not so indelibly imprinted on scientific thought modern virologists might consider the emergence of the striking bullet-like morphology of rhabdoviruses (Gr. *rhabdos* = rod) a reflection of the violence of our times. This unique form, which was first described for vesicular stomatitis virus (VSV) of cattle and horses, is also associated with rabiesvirus (Fig. 59-1) and at least 25 other viruses that are widely distributed in nature, infecting a variety of mammals, fish, insects, and plants.

The known properties of the agents in this group are summarized in Table 59-1. Several rhabdoviruses replicate in arthropods as well as in mammals and hence were previously considered to be arboviruses (e.g., VSV, Hart Park virus, Flanders virus, Kern Canyon virus). Rabiesvirus is the only member of the group known to infect and produce disease in man naturally; it will be considered in detail. The Marburg virus, a simian virus which only accidentally infects man, has some similar features and will be discussed briefly.

RABIESVIRUS

Since history was first recorded, the sound and sight of a vicious rabid (L. *rabidus* = mad) dog have struck terror in those in its vicinity. The change of a docile, friendly animal into a combative beast, often with convulsions, was considered the work of supernatural causes. The infectious nature of rabies was recognized in 1804, but it was Pasteur in the 1880s who suggested that the responsible etiological agent was not a bacterium. He used his knowledge of the properties of infectious agents, and his great intuition, to demonstrate for the first time that the pathogenicity of a virus (before viruses had actually been identified) could be modified by serial passage in an animal other than its natural host. Fifty serial intracerebral passages in rabbits yielded a modified* virus, **fixed virus** (as contrasted with the wild-type or **street virus**), which was used for immunization.

Upon discovery of the filtrable causative agent in 1903, Negri described the presence

of prominent cytoplasmic inclusion bodies **(Negri bodies)** in the nerve cells of infected human beings and animals. Their characteristic appearance and easy recognition made possible the rapid pathological diagnosis of infection.

Although rabies was one of the first diseases of man to be recognized as caused by a virus, its morphological, chemical, and replicative characteristics were unclear until the late 1960s when methods for propagating attenuated viruses in cell cultures in vitro overcame the dangers encountered in handling the virus and the difficulties involved in growing it to high titer.

PROPERTIES OF THE VIRUS

Morphology. The virions, whose average dimensions are 1800 by 750 A, are cylindrical with one rounded and one planer end, resembling a bullet (Fig. 59-1). Regularly spaced projections, with a knob-like structure at the distal end, cover the surface of the virion. Shorter bullet-shaped and cylindrical particles are also occasionally observed in electron mi-

---

* Since even the fixed virus strain is pathogenic for man and other animals, it is misleading to consider it attenuated.

TABLE 59-1. Characteristics of Rhabdoviruses

Morphology	Bullet-shaped; 1300–2400 by 700–800 A
Nucleic acid	Single-stranded RNA; 3.5–4.6×10^6 daltons
Virion enzyme	RNA transcriptase
Nucleocapsid	Helical; 180 A in width
Effect of lipid solvents	Disrupts virions; inactivates infectivity
Maturation	Budding at cytoplasmic membranes
Hosts	Wide variety of mammals, fish, invertebrates, and plants
Common antigens	None

crographs. The flat end of the virion has a depression at the axis of the internal helical nucleocapsid, which is symmetrically wound within the envelope, often giving the appearance of a series of transverse striations (Fig. 59-1A). The purified nucleoprotein is a ribbon-like helical strand, consisting of regular rod-like protein subunits attached to a thread of nucleic acid (Fig. 59-1B).

Physical and Chemical Properties. The viral envelope consists of lipids, a glycoprotein with a molecular weight of approximately 80,000 daltons, and two smaller proteins. The surface projections, i.e., the hemagglutinins, have not been identified with any of the recognized envelope subunits. The nucleocapsid contains 95% protein, consisting of many identical copies of a polypeptide of 69,000 daltons, and a single molecule of single-stranded RNA.

As with most enveloped viruses in the absence of tissue proteins or added protein (normal serum or albumin), infectivity deteriorates rapidly at room or refrigerator temperatures. Inactivation is much slower, however, in crude tissue extracts or in infected tissues stored in neutral glycerol.

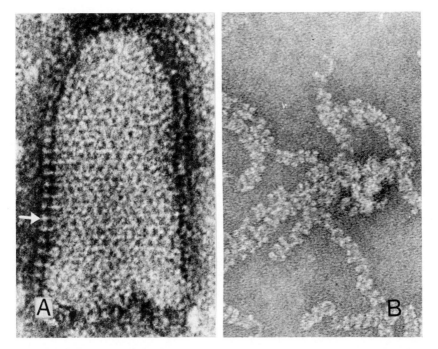

Fig. 59-1. Morphological characteristics of rabiesvirus. **A.** An intact rabiesvirus particle embedded in phosphotungstate and viewed in negative contrast. On the left are well resolved surface projections 60 to 70 A long (arrow). ×400,000. **B.** Helical nucleocapsid isolated from disrupted rabies virions. Note the tightly coiled and partially uncoiled regions of the single-stranded helix. ×212,000. [**A** from Hummeler, K., Koprowski, H., and Wiktor, T. *J Virol 1*:152 (1967). **B** from Sokol, F., Schlumberger, H. A., Wiktor, T. J., Koprowski, H., and Hummler, K. *Virology 38*:651 (1969).]

Infectivity is quite stable in frozen or lyophilized tissue extracts.

Immunological Characteristics. All the rabiesviruses isolated from man and other animals, throughout the world, appear to be of a single immunological type. Selected modified (fixed) and wild-type (street) viruses, prepared by many different methods and propagated in different tissues, are also immunologically similar. Viruses from two cases of human illness (one fatal) in Nigeria are immunologically and biologically similar to the Lagos bat virus and Makola shrew virus, and antigenically related to rabiesvirus.

The virion's surface structures are responsible for the hemagglutination of goose erythrocytes and production of neutralizing as well as hemagglutination-inhibiting Abs. Antibodies to the nucleocapsid, in contrast, are recognized by complement fixation. After infection or immunization there develops a third class of virus-specific Abs, which in the presence of complement lyse infected cells whose plasma membranes have been altered by the virus. These Abs may play a deleterious rather than a protective role during pathogenesis of the disease.

Host Range. Rabiesvirus can infect all mammals so far tested. Among domestic animals dogs, cats, and cattle are particularly susceptible. Skunks, bats, foxes, squirrels, raccoons, coyotes, mongooses, and badgers are the principal wildlife hosts. Birds are also susceptible to infection, but less so than mammals.

To establish laboratory infections hamsters, mice, guinea pigs, and rabbits, in order of decreasing susceptibility, are employed. Intracerebral inoculations are more reliable than subcutaneous or intramuscular inoculations.

Rabiesvirus can be propagated in chick or duck embryos. Attenuated strains developed by multiple passage in embryonated eggs now serve as important sources for vaccines.

Cultures of cells from many different animal species, including reptiles, can also support viral multiplication. Hamster kidney, human diploid, and chick embryo cell cultures maintained at 31 to 33° are most commonly used. Wild virus is propagated with greater difficulty than modified strains. Cytopathic changes are not usually observed in infected cultures, but intracytoplasmic antigen can be detected by immunofluorescence.

Viral Multiplication. Despite the long incubation period in natural infections and in experimental animals, in appropriate cell cultures the characteristics of viral multiplication are not unusual: the eclipse period is 6 to 8 hours, and the initial cycle of multiplication is completed in 18 to 24 hours. Since cytopathic changes are absent or minimal, carrier cultures or endosymbiotic infections (Ch. 51) are established readily.

Viral replication occurs in cytoplasmic "factories," in which prominent matrices of helical nucleocapsids are formed. The virions then assemble by budding from cytoplasmic membranes (and from plasma membranes in many types of cultured cells). In infected cells the masses of nucleocapsids appear as cytoplasmic inclusion bodies which can be identified by specific immunofluorescence.

The molecular details of viral multiplication are still unexplored, but the striking morphological and chemical similarities to VSV suggest a similar mechanism of replication.

The virion RNA of VSV, like paramyxoviruses (Ch. 57), is not infectious, and hybridizes with virus-specific RNAs on polyribosomes (Ch. 48, Class II RNA virus). Hence, the viral RNA cannot serve as a messenger (i.e., it has opposite polarity), and the virion contains a transcriptase to catalyze the synthesis of several species of RNAs which serve as messengers for the RNA replicase and viral structural proteins. Viral RNA replication is accomplished through a characteristic replicative intermediate form (Ch. 48).

PATHOGENESIS

A wound or abrasion of the skin, usually inflicted by a rabid animal, is the major portal of entry into man, virus entering with the animal's saliva. A dense population of infected bats may also create an aerosol of infected secretions, by which virus can obtain entrance into the respiratory tract of man or lower animals. Virus remains localized for periods which appear to vary from days to months, and then apparently progresses along nerves to the central nervous system, where it multiplies and produces severe and often fatal encephalitis. Hematogenous spread of virus to the central nervous system has been claimed but not established. In mammals, the virus must reach the salivary glands if it is to be transmitted, but it is not clear whether it does so via efferent nerves or through blood and lymph vessels.

The **incubation period** is extremely varied, ranging from as short as 6 days to as long as 1 year. It depends primarily upon the size of the viral inoculum and also probably on the length of the neural path from wound to brain. Its length is therefore correlated with the severity and extent of the wound inflicted by the rabid animal, and is less following bites on the face and head; the wounds under these circumstances are also often the most extensive. Illness is ushered in by a **prodromal period,** with complaints including irritability, abnormal sensations about the wound site, and hyperesthesia of the skin. **Clinical disease** becomes apparent with the development of increased muscle tone and difficulty in swallowing, due to painful and spasmodic contractions of the muscles of deglutition when fluid comes in contact with the fauces. Often the mere sight of liquids will induce such contractions; hence the common name, hydrophobia (Gr. fear of water). The final stages of the disease result from the extensive damage of the central nervous system. A fatal outcome has been considered inevitable, but one patient with proved rabies is known to have recovered after being given extensive care for sustaining vital functions. Epidemiological data further suggest that following known exposure only 30 to 50% will develop recognizable rabies: e.g., in an unintentional study about one-half of untreated persons developed clinical rabies and died following severe mutilation by a rabid wolf, which must certainly have effected a viral inoculation, and presumably infection, in all attacked.

Pathologically, rabies is an encephalitis, with neuronal degeneration of the spinal cord and brain. **Negri bodies** within affected neurons are the most characteristic and indeed the only pathognomonic microscopic finding. This cytoplasmic inclusion is a sharply defined spherical or oval, eosinophilic, Feulgen-negative body, 2 to 10 μ in diameter, containing a central mass of basophilic granules (Fig. 59-2). Several may be found in a single cell. Immunofluorescence studies have demon-

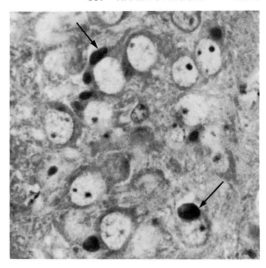

Fig. 59-2. Negri bodies in brain of mouse infected with rabiesvirus. Numerous large dark cytoplasmic inclusion bodies (arrows) are present. Stained by the dinitrofluorobenzene method for protein-bound groups. Note that the matrix of the inclusion body is stained intensely, whereas the internal granules appear as light vacuoles. ×2000, reduced. (Courtesy of Dr. H. Koprowski, Wistar Institute, and Dr. R. Love, Jefferson School of Medicine.)

strated specific viral antigens in the Negri body, and electron micrographs show that it consists of a large matrix of viral nucleocapsids and budding virions which constitute the basophilic granules (Fig. 59-3).

Inclusion bodies are most abundant in Ammon's horn of the hippocampus, but may also be found in large numbers in many other sites in the brain, and in the posterior horn of the spinal cord. In the absence of identifiable Negri bodies, a pathological diagnosis of rabies cannot be made.

LABORATORY DIAGNOSIS

Definitive diagnosis of infections in man, and in suspected animal vectors, depends upon 1) identification of Negri bodies in brain tissue and 2) isolation of virus from brain or saliva. Owing to the relatively slow appearance of circulating Abs after infection, and the need for an immediate definitive diagnosis, serological technics are of little value early in the disease. But serum Abs reach high levels, and if clinical doubts exist during later stages, Ab titrations (neutralization, complement fixation,

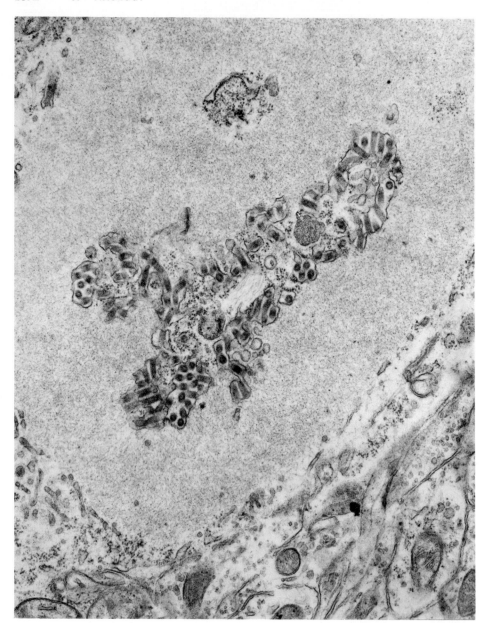

Fig. 59-3. Electron micrograph of a Negri body which contains several inner bodies composed of developing and mature virions. ×19,500. [From Matsumoto, S. *Adv Virus Res 16*:257 (1970).]

immunofluorescence) can establish the diagnosis. Unfortunately, the relatively simple hemagglutination-inhibition assay is of little practical value because normal sera contain high levels of nonspecific inhibitors which cannot be removed readily.

To detect Negri bodies, impression smears are prepared from the region of Ammon's horn and stained.* The distinctive appearance of Negri bodies (stained cherry red with deep blue granules) differentiates them from other inclusion bodies, particularly those produced by distemper virus in dogs (Ch. 57, Fig. 57-9). Fluorescent antibody technics provide reliable confirmation of the specific nature of the inclusion body. The presence of Negri bodies is diagnostic, but failure to detect them does not exclude rabies and should be followed by attempts to isolate the virus.

Virus is preferably isolated by inoculating saliva, salivary gland, or hippocampal brain tissue intracerebrally into infant mice. The mice develop paralysis after an incubation period of 6 to 21 days, depending upon the quantity of virus present. The illness produced in mice is not pathognomonic of rabies, and the virus must be identified by immunological technics or by demonstrating Negri bodies in brain tissue of the inoculated animals.

EPIDEMIOLOGY

Although medical interest in rabies centers upon infection of man, in fact, epidemiologically this is an unimportant dead end to the infectious cycle, since men contract rabies but do not normally transmit the disease. Dogs are the most dangerous source of infection for man, with cats next.

Wild mammals serve as a large and uncontrollable reservoir of **sylvan rabies,** which is a constantly increasing threat to man and domestic animals throughout the world. The most frequent wild sources in North America are skunks, foxes, and raccoons. Moreover, an epizootic in foxes has reintroduced rabies into several countries of Western Europe, which once was relatively free of the disease.

Vampire and insectivorous **bats** are also important reservoirs for rabiesvirus. Theoretically, bats could be one of the most important links in the ecology of rabies: experimentally the virus can remain latent in these animals for long periods; and virus has been detected in the nasal mucosa, brown fat, and salivary glands of apparently healthy bats. Furthermore, the virus can be transmitted to man or other animals from bats through their secretions, without a direct bite, presumably by inhalation of infectious aerosols.

The incidence of rabies in man and dogs has decreased continuously in the United States. But it still remains enzootic in dogs, skunks, bats, and other wild animals and potential danger lurks if current control measures are relaxed. For example, in 1971 in the United States 4392 cases of rabies were identified in animals in every state except Hawaii (an increase of 34%), though there were only 1 to 3 cases of human rabies per year from 1960 to 1971. But these statistics of human rabies should not induce complacency, since a worldwide total of about 1000 fatal cases is reported annually to the World Health Organization, and the actual number must be several times greater.

PREVENTION AND CONTROL

An effective program for rabies control must encompass measures directed toward preventing the disease in animals as well as in man. However, although more cases of human rabies probably arise from animals other than dogs and cats, control measures directed against wildlife are impracticable. In fact, the primary rabies problem in the United States and many other countries is no longer human cases, but rather the emotional stress associated with a suspected exposure and the decision whether or not to undertake immunization with its attendant dangers.

Vaccines. Pasteur's original vaccine was prepared by homogenization of partially dried† spinal cords from rabbits infected with modified (fixed) virus. Daily injections were given for 15 to 20 days with cords desiccated for progressively shorter periods. Frequent difficulties resulted, however, from the inexact

---

* Seller's method, with a stain composed of a mixture of basic fuchsin and methylene blue, is commonly employed.

† Desiccation was employed to control the quantity of infectious agent present since Pasteur had noted earlier that dried cultures of chicken cholera bacteria lost their pathogenicity, but not their immunogenicity, for chickens.

method of viral inactivation and lack of quantitative controls.

A major advance was made with the development of phenol-inactivated vaccines. The vaccine that has been used in the United States (Semple vaccine) consists of a 4 to 5% suspension of nervous tissue (usually rabbit) infected with modified virus and treated with 0.25 to 0.5% phenol. In attempts to reduce the demyelinating complications of immunization, nervous tissues of immature animals are also used: e.g., suckling mouse brain in Latin America and suckling rat brain in Russia. Daily subcutaneous injection for 7 to 14 days is recommended, depending upon the severity and location of the bite. Booster immunizations are given 10 and 20 days after the last daily dose. Neutralizing Abs are usually detectable by 14 days after the first injection.

Despite the widespread acceptance and general use of phenolized vaccine, no valid control study has been carried out to prove its effectiveness. The best data available were gathered from a study that followed 809 persons who had been bitten by animals of proved infectivity: of 581 completely treated, 8% died, whereas among 153 who refused treatment, 50% succumbed from rabies infection. This may be contrasted to a study of persons bitten on the head or neck by a rabid wolf in Iran: 40% of the immunized and 47% of the untreated died from rabies.

Rabiesvirus propagated in embryonated duck eggs and inactivated by β-propiolactone now is the most favored vaccine in the United States, owing to the lower risk of allergic encephalomyelitis (Ch. 20). The course of immunization consists of 14 to 21 daily injections followed by booster inoculations 10 and 21 days later. Although the immunogenic potency of duck embryo vaccine is considerably less than that of the nervous tissue vaccines, it may nevertheless be effective since only 7 cases of rabies have occurred in vaccinees in the United States between 1963 and 1970 (approximately 30,000 persons were immunized each year, and 80 to 90% of these received duck embryo vaccine).

Experimental Vaccines. In recent years several types of infectious vaccines have been prepared from strains of virus adapted to chick embryos. The Flury virus is the most frequently used such strain. The low egg passage (LEP) strain is not pathogenic for dogs when inoculated intramuscularly, although puppies, cats, and cattle may become ill. Vaccination of dogs with the LEP vaccine confers immunity for 3 years. The high egg passage (HEP) strain, which has had more than 180 egg transfers, is no longer virulent. Unfortunately HEP virus has poor immunogenicity in man and is not suitable for use after exposure to a rabid animal.

Since modified rabiesvirus can be propagated in a variety of cell cultures, the virus may be partially purified and greatly concentrated in order to prepare highly immunogenic inactivated vaccines. Vaccines prepared from virus grown in human diploid fibroblasts or baby hamster kidney cell lines have been successfully used in monkeys and rodents; their use in man is to be studied experimentally. Viral components, free of whole virions, are also antigenic and may eventually be useful.

Public Health Measures. Primary control of rabies requires restriction of dogs and other domestic animals, as well as limitation of spread from wildlife to the greatest extent possible. The World Health Organization's excellent report (1966) recommends 1) compulsory prophylactic immunization of dogs with LEP and of cats with HEP Flury vaccines; 2) registration of all dogs, destruction of stray dogs, and isolation and observation of suspect dogs; and 3) attempted control of rabies in wildlife by trapping and other means. The danger of rabies is responsible for current severe restrictions on the transport of dogs across some national boundaries (e.g., England).

Prophylactic Treatment. Despite the low incidence of human rabies in many parts of the world, the question frequently arises as to the course to follow after an animal bite or scratch. Prophylactic immunization cannot be applied routinely because of the relatively high reaction rate noted following its use, particularly with phenolized rabbit vaccine. Demyelinating encephalomyelitis, the major complication, results from the development of hypersensitivity to nervous tissue present in the vaccine. The incidence is variably reported as 1:500 to 1:85,000; the severity of the reactions is likewise varied and is sometimes fatal. The duck vaccine has a great theoretical advantage since it lacks nervous tissue and hence has reduced the danger of neurological complications following immunization. Unfortunately, the avianized vaccine has not completely eliminated the occurrence of encephalomyelitis, and it has not yet been proved effective in man (though it has in dogs).

In deciding the management of a person who has been bitten by an animal suspected of being rabid a number of factors must be considered, and it is not possible to present guidelines for all situations. The chances of demyelinating encephalitis, particularly following use of the phenolized rabbit vaccine containing nervous tissue, may be higher under some circumstances than the risk of contracting rabies. Therefore 1) vaccine should not be given to a person who has had only minimal contact with a questionable source of infection; 2) if exposure to rabies appears definite, treatment with duck vaccine should be instituted promptly; and 3) if a suspected animal can be captured alive it should be observed for at least 10 days for the development of symptoms, since Negri bodies may not be detectable in the brain early in the disease, and the virus can appear in the saliva several days before the onset of symptoms. Suspected bats, however, should be killed and examined at once.

Prophylactic treatment is fundamentally directed toward confining the virus to the site of entry. Local treatment of the wound consists of thorough cleansing (cauterization is no longer suggested) and infiltration with antiserum. The long incubation period usually makes it possible for active immunization to produce an adequate antibody level for this purpose before the virus has reached its target organs in the central nervous system. The combined inoculation of immune serum* and vaccine is the most effective regimen: serum antibodies provide an immediate but temporary barrier to passage of virus, which lasts about 14 days, when antibodies appear from the immunogenic stimulus of vaccine. The combined use of immune serum and vaccine, however, makes it mandatory to give the full course of 14 injections of vaccine in addition to 2 booster injections since serum has a tendency to depress antibody production. Antiserum alone should probably never be used.

Long-term Prophylaxis. Veterinarians, laboratory workers, dog handlers, wildlife workers, and hobbyists (e.g., spelunkers) may be sufficiently exposed to rabies to justify prophylactic immunization. It is recommended that members of this high-risk group receive a course of inactivated duck embryo vaccine, consisting of two subcutaneous injections a month apart followed by a booster injection in several months. Subsequent booster doses should be given every 2 to 3 years thereafter or after a suspected exposure.

---

* Serum is presently prepared by hyperimmunization of horses. However, the allergic reactions (Ch. 19) to the serum (urticaria and serum sickness) are sufficiently frequent (about 15%) to warrant the use of γ-globulin from hyperimmunized humans when adequate supplies are available.

MARBURG VIRUS

In 1967 in Marburg and Frankfurt, Germany, and in Yugoslavia, an acute febrile illness appeared in laboratory workers handling tissues and cell cultures from recently imported African green monkeys. In all, 31 cases occurred, with 7 fatalities. From bloods and organ suspensions from fatal cases, a virus, termed the Marburg virus, was isolated by inoculation of guinea pigs. The virus is inactivated by ether and sodium deoxycholate, and appears to contain RNA since it replicates in the presence of inhibitors of DNA synthesis. Electron microscopic examinations of negatively stained preparations reveal particles that are somewhat rod-shaped, but with a variety of bizarre cylindrical and fishhook-like forms. The virions have a uniform diameter of 750 to 800 A but vary greatly in length from 1,300 to 40,000 A. Most particles have one rounded end

like rhabdoviruses; the other extremity is flat or occupied by a large bleb. Prominent cross-striations and an inner cylindrical structure add to its similarities with rhabdoviruses. Whether to consider it a rhabdovirus is still unsettled.

Marburg virus disease has a sudden onset with high fever, gastrointestinal upset, constitutional symptoms, and marked prostration. The fever is biphasic, the second febrile period being marked by uremia, rash, hemorrhages, and central nervous system involvement. Fatal cases show necrotic foci in many organs, including the brain; the liver and lymphatic tissues are most severely affected.

The disease has not reappeared, but the extensive use of primary monkey cell cultures makes it imperative that physicians and virologists be aware of this simian virus which on occasion also infects man.

SELECTED REFERENCES

Books and Review Articles

DEAN, D. J., EVANS, W. M., and MCCLURE, R. C. Pathogenesis of rabies. *Bull WHO 29:*803 (1963).

Expert Committee Report on Rabies. Fifth Report WHO Technical Report Series No. 321. World Health Organization, Geneva, 1966.

HOWATSON, A. F. Vesicular stomatitis and related viruses. *Adv Virus Res 16:*196 (1970).

MATSUMOTO, S. Rabies virus. *Adv Virus Res 16:*257 (1970).

PLOTKIN, S. A., and CLARK, H. F. Prevention of rabies in man. *J. Infect Dis 123:*227 (1971).

Specific Articles

ALMEIDA, J. D., HOWATSON, A. F., PINTERIC, L., and FENJE, P. Electron microscope observations on rabies virus by negative staining. *Virology 18:*147 (1962).

APPELBAUM, E., GREENBERG, M., and NELSON, J. Neurological complications following antirabies vaccination. *JAMA 151:*188 (1953).

KABAT, E. A., WOLF, A., and BEZER, A. E. The rapid production of acute disseminated encephalomyelitis in rhesus monkeys by injection of heterologous and homologous brain tissue with adjuvants. *J Exp Med 85:*117 (1947).

KISSLING, R. E. Marburg virus. *Ann NY Acad Sci 174:*932 (1970).

PASTEUR, L. Méthode pour prévenir la rage après morsure. *CR Acad Sci (Paris) 101:*765 (1885).

SIKES, R. K. Guidelines for the control of rabies. *Am J Pub Health 60:*1133 (1970).

TIERKEL, E. S., and SIKES, R. K. Preexposure prophylaxis against rabies. *JAMA 201:*911 (1967).

VEERARAGHAVAN, N., and SUBRAHMANYAN, T. P. The value of duck-embryo vaccine and high-egg-passage Flury vaccine in experimental rabies infection in guinea-pigs. *Bull WHO 29:*323 (1963).

chapter **60**

TOGAVIRUSES AND OTHER ARTHROPOD-BORNE VIRUSES; ARENAVIRUSES

History   1378
Immunological Classification   1379

TOGAVIRUSES   1382

PROPERTIES OF THE VIRUSES   1382
Morphology   1382
Physical and Chemical Characteristics   1382
Immunological Characteristics   1383
Host Range   1384
Viral Multiplication   1386
PATHOGENESIS   1387
Group A Togaviruses   1387
Group B Togaviruses   1388
LABORATORY DIAGNOSIS   1388
EPIDEMIOLOGY   1389
Group A Togaviruses   1389
Group B Togaviruses   1390
PREVENTION AND CONTROL   1391

BUNYAMWERA SUPERGROUP VIRUSES   1392

PROPERTIES OF THE VIRUSES   1392
PATHOGENESIS AND EPIDEMIOLOGY   1392

ARENAVIRUSES   1394

PROPERTIES OF THE VIRUSES   1394
PATHOGENESIS AND EPIDEMIOLOGY   1394

GENERAL REMARKS   1396

The arthropod-borne viruses **(arboviruses)** multiply in both vertebrates and arthropods. The former serve as **reservoirs** in the cycle of transmission and the latter mostly as **vectors,** acquiring infection with a blood meal. After the virus is propagated in the arthropod's gut and attains a high titer in its salivary glands, virus is transmitted when a fresh host is bitten. The viruses often cause disease in the vertebrate hosts, but none is evident in the arthropods.

Most of the viruses described in this chapter were classified until recently, on epidemiological grounds, as arboviruses. But increasing knowledge of their chemical and physical characteristics has revealed great heterogeneity among the arthropod-borne viruses. A number of viruses with similar characteristics (Table 60-1), and of great medical significance to man, are now considered to be a single family, **Togavirus** (L. *toga* = coat); groups A and B are distinguished antigenically and classified as genera (Table 60-2). This chapter will also consider various poorly characterized arthropod-borne viruses, important to man (Table 60-2), that resemble togaviruses in containing one molecule of single-stranded RNA and in being inactivated by lipid solvents. In one group **(arenaviruses)** arthropod transmission is not obligatory, and for some members may be absent. Finally, among other viruses carried in arthropods one group (Colorado tick fever, African horse sickness, and blue tongue viruses) resembles reoviruses which have double-stranded RNA, are isometric, and lack an envelope (these will be considered in Ch. 61); while vesicular stomatitis virus is bullet-shaped and classified as a rhabdovirus (Ch. 59).

The frequent association of an enveloped structure with transmission by arthropods may be more than coincidental: enveloped viruses which lose infectivity readily (e.g., on drying, on exposure to bile) are spread either by intimate contact (e.g., myxoviruses) or by insect bite, whereas naked viruses, such as picornaviruses, tend to be more stable and can survive a more circuitous fecal-oral spread.

HISTORY

Yellow fever virus was the first arthropod-borne virus to be discovered, through the work of Major Walter Reed. He headed the U.S. Army Yellow Fever Commission, established in 1901, to try to overcome the disastrous effect of yellow fever on American troops in Cuba during the Spanish-American War. Pursuing the astute observations of a Cuban physician, Carlos Finlay, on the association between yellow fever and mosquitoes, Reed and the members of the Commission demonstrated this transmission of the disease in bold experiments with human volunteers. They also demonstrated the filtrability of the agent. These studies established, for the first time, a virus as an agent of human disease, and an insect as the vector for a virus.

Their discoveries led to the eventual control of yellow fever, which for more than 200 years had intermittently been one of the world's major plagues, and, in fact, was a deciding factor in France's failure to complete the Panama Canal. This disease was by no means purely tropical: for example, an epidemic in the Mississippi Valley in 1878 caused 13,000 deaths, and substantial epidemics occurred in the nineteenth century as far north as Boston.

Because of their vectors, the prevalence of arthropod-borne diseases depends strongly on climatic conditions: they are endemic in areas of tropical rain forests, and epidemics in temperate areas usually appear after heavy rainfall has caused an increase in the vector population. In addition to agents known to

TABLE 60-1. Properties of Togaviruses

Morphology	450–700 A
	Majority 410–500 A
	Icosahedral symmetry; lipid envelope
Nucleic acid	RNA; single-stranded, about 4×10^6 daltons
Effect of lipid solvents	Inactivate
Stability	Unstable without added protein*
Hemagglutination	+ : RBC from newborn chicks or geese
Best animal host	Suckling mouse

* Survive several months at $-20°$, indefinitely at $-70°$ or if lyophilized.

cause human diseases, a large number of additional arboviruses have been isolated from mosquitoes and ticks trapped in forests and also from animals, especially monkeys, caged in the jungle as "sentinels" to permit insects to feed on them. These viruses are not known to cause prominent diseases of man; but they are attracting a good deal of attention because of their threat as the world's expanding and increasingly mobile population impinges progressively on jungles. Over 250 arthropod-borne viruses have now been isolated.

IMMUNOLOGICAL CLASSIFICATION

The presently known viruses isolated from arthropods may be divided into at least 21 antigenically distinct groups; many viruses are still ungrouped. Table 60-2 lists the principal viruses that infect man, the groups to which they have been assigned, and some of their clinical and epidemiological characteristics.

Togaviruses are classified in groups A and B on the basis of hemagglutination-inhibition reactions: members of a group cross-react with each other by this assay, but not with other arboviruses (Table 60-3). Bunyamwera supergroup viruses, in contrast, are classified by complement fixation, which shows greater cross-reactivity among members of the group than does hemagglutination inhibition. Many of the remaining viruses are classified into groups on the basis of the hemagglutination-inhibition reaction. Species within a group are identified by neutralization with standardized antisera; there is less cross-reaction with this test.

The immunological cross-reactions seen within the major arbovirus groups suggest close phylogenetic relations within each group. Indeed, it seems possible that many of these "species" may differ by only one or a few mutations. It thus appears that the current subdivision of arboviruses into more than 250 "species" may be no more valid than the earlier division of *Salmonella* into hundreds of "species" whose antigenic differences are now resolvable into very few alterations in O-polysaccharide biosynthesis (Ch. 29, General properties: antigenic structure). For arboviruses, however, a comparable analysis of the glycoprotein antigens is just beginning; but newer methods for characterizing viral macromolecules are bringing sharper taxonomic criteria into this field.

The immunological cross-reactions among arthropod-borne viruses are of practical as well as theoretical interest. Thus with the viruses that show cross-reactivity in hemagglutination-inhibition or complement-fixation tests, cross-reacting neutralizing Abs are not detectable after primary immunization, but may be evident after repeated immunization. Moreover, infection by one virus may confer a demonstrable increase in resistance to subsequent challenge with another. Epidemiological evidence suggests that such cross-protection may be important in nature; vaccines are being developed to take advantage of this finding.

TABLE 60-2. Classification and Description of Togaviruses and Other Arthropod-Borne Viruses

Family	Group (genus)	Sub-group	Viral species	Vector	Clinical diseases in man	Geographic distribution
Togavirus	A	I	Eastern equine encephalitis (EEE)	Mosquito	Encephalitis	Eastern U.S.A., Canada, Brazil, Cuba, Panama, Philippines, Dominican Republic, Trinidad
			Venezuelan equine encephalitis (VEE)	Mosquito	Encephalitis	Brazil, Colombia, Ecuador, Trinidad, Venezuela, Mexico, U.S.A. (Florida and Texas)
			Western equine encephalitis (WEE)	Mosquito	Encephalitis	Western U.S.A., Canada, Mexico, Argentina, Brazil, British Guiana
			Sindbis	Mosquito	Subclinical	Egypt, India, South Africa, Australia
		II	Chikungunya	Mosquito	Headache, fever, rash, joint and muscle pains	East Africa, South Africa, Southeast Asia
			Semliki Forest	Mosquito	Fever or none	East Africa, West Africa
			Mayora	Mosquito	Headache, fever, joint and muscle pains	Bolivia, Brazil, Colombia, Trinidad
			(13 others named)	Mosquito	Subclinical or none known	
Togavirus	B	I	St. Louis encephalitis	Mosquito	Encephalitis	U.S.A., Trinidad, Panama
			Japanese B encephalitis	Mosquito	Encephalitis	Japan, Guam, Eastern Asian mainland, Malaya, India
			Murray Valley encephalitis	Mosquito	Encephalitis	Australia, New Guinea
			Ilheus	Mosquito	Encephalitis	Brazil, Guatemala, Trinidad, Honduras
			West Nile	Mosquito	Headache, fever, myalgia, rash, lymphadenopathy	Egypt, Israel, India, Uganda, South Africa
		II	Dengue (4 types)	Mosquito	Headache, fever, myalgia, prostration, rash (sometimes hemorrhagic)	Pacific islands, South and Southeast Asia, northern Australia, New Guinea, Greece, Caribbean islands, Nigeria, Central and South America
		III	Yellow fever	Mosquito	Fever, prostration, hepatitis, nephritis	Central and South America, Africa, Trinidad
		IV	Tick-borne group (Russian spring-summer encephalitis group) 9 viruses	Tick	Encephalitis; meningo-encephalitis, hemorrhagic fever	Russian spring-summer encephalitis: U.S.S.R.; Powassan: Canada, U.S.A. Others: Japan, Siberia, Central Europe, U.S.S.R., India, Malaya, Great Britain (Louping ill)
			Bat salivary gland (12 others)	Unrecognized	Encephalitis	California, Texas

Family	Group (genus)	Sub-group	Viral species	Vector	Clinical diseases in man	Geographic distribution
Bunyamwera supergroup	C		Marituba and 12 others	Mosquito	Headache, fever	Brazil (Belém), Panama, Trinidad, Florida
	Bunyamwera		Bunyamwera and 13 others	Mosquito	Headache, fever, myalgia; just fever; or none	Uganda, South Africa, India, Malaya, Colombia, Brazil, Trinidad, West Africa, Finland, U.S.A.
	California group		California encephalitis and 8 others	Mosquito	Encephalitis, or none	U.S.A., Trinidad, Brazil, Canada, Czecho-slovakia, Mozambique
Phlebotomus fever group			3 species	Phlebotomus	Headache, fever, myalgia	Italy, Egypt
Arenaviruses			Junin, Tacaribe, Tamiani, Machupo, Lassa, lymphocytic choriomeningitis, and 5 others	Mosquito, mite, or unrecognized	Headache, fever, myalgia, hemorrhagic signs	South and Central America
	Ungrouped		Silverwater	Tick	None known	Canada
			Rift Valley fever	Mosquito	Headache, fever, myalgia, joint pains, hemorrhagic signs, rash	Africa
			Crimean hemorrhagic fever	Tick	Headache, fever, myalgia, hemorrhagic signs	Southern U.S.S.R.
			36 others	Mosquito	None known	
Others			48 viruses (14 groups of 2–8 viruses)	Mosquito, tick	None known in most	

TABLE 60-3. Results of Hemagglutination-Inhibition Titrations with Group A and B Togaviruses

Immune serum	Group	Virus (antigen)							
		EEE	VEE	WEE	Sindbis	Chikun-gunya	Mayora	Semliki Forest	St. Louis enceph-alitis
EEE	A	**10,240**	80	160	20	40	20	20	<10
VEE	A	160	**640**	80	20	80	80	40	<10
WEE	A	80	160	**10,240**	160	40	80	40	<10
Sindbis	A	80	10	2560	**1280**	10	40	40	<10
Chikungunya	A	20	20	40	<10	**1280+**	80	80	<10
Mayora	A	40	40	320	40	640	**1280+**	1280+	<10
Semliki Forest	A	40	80	160	10	40	320	**2560**	<10
St. Louis encephalitis	B	<10	<10	<10	<10	<10	<10	<10	**2560**

Note cross-reactions among group A viruses, but not between the group A and one group B virus tested. Also note the cross-reactivity among some viruses forming two subgroups: EEE, VEE, WEE, Sindbis; and Chikungunya, Mayora, Semliki Forest.

Modified from CASALS, J. *Ann NY Acad Sci 19:*219 (1957).

TOGAVIRUSES

Among the major togaviruses pathogenic for man (Table 60-2), the group A viruses produce severe encephalitis, particularly western and eastern equine encephalitis viruses (WEE and EEE, respectively), whereas the group B viruses cause more varied illnesses (encephalitis, hemorrhagic diseases, and severe systemic illnesses).

PROPERTIES OF THE VIRUSES

Because of the serious nature of many illnesses caused by togaviruses, particularly encephalitis, and their wide distribution, the viruses have been studied extensively. Most of them, however, are difficult and dangerous to work with, and the physical and chemical characterization is incomplete except for a few less pathogenic viruses that can be readily propagated in cell cultures.

MORPHOLOGY

The group A and B togaviruses have the same general appearance, but group A viruses are significantly larger (Table 60-4). Within each group, however, the sizes of the virions are not identical; but these are enveloped viruses and therefore are susceptible to changes in size and shape according to the conditions employed for preparation of specimens for electron microscopy. The virions are roughly spherical; in thin sections they have an outer membrane (lipoprotein envelope), a core of electron-dense material (ribonucleoprotein), and they tend to pack in crystalline arrays (Figs. 60-1 and 60-2). Negative staining shows an outer membrane covered with fine projections, and a capsid suggestive of icosahedral symmetry (Fig. 60-3). The capsid of Sindbis virus, the only virion adequately visualized, appears to consist of 32 capsomers.

PHYSICAL AND CHEMICAL CHARACTERISTICS

The purified virions analyzed contain one single-stranded infectious RNA molecule per virion (Table 60-1), which amounts to 4 to 8% of the particle weight. The viral RNA, like poliovirus and probably other viral RNAs that serve as messengers, contains a polyadenylic acid-rich sequence. The proportions of protein and lipid reported vary considerably: for example, EEE virus, 54% lipid and 42% protein; Sindbis virus, 28% lipid and 66% protein. The lipids consist of phospholipid, neutral lipids (including cholesterol), and glycolipids with similar proportions in the viruses studied. Three proteins are present in group A viruses: one, high in lysine, is associated with the nucleocapsid; the two others, glycoproteins, are associated with the spikes and the membrane of the envelope.

TABLE 60-4. Sizes of Selected Arthropod-Borne Viruses Determined by Electron Microscopic Measurements

	Group	Virus	Size (A) Purified particles	Size (A) Thin sections
Togavirus	A	EEE	470	550
		WEE	530	450–480
		VEE	650–750	400–450
		Semliki Forest	500	
		Sindbis	600–700	
		Chikungunya	540–580	500–560
Togavirus	B	Japanese B		400
		Yellow fever		380
		St. Louis encephalitis		380
		Dengue type 2	480–500	450–500
		Murray Valley encephalitis	250–275	
		Russian spring-summer encephalitis		250
Bunyamwera supergroup		C (Gumbo Limbo)		1000
		California (5 viruses examined)		980
		Bunyamwera (3 viruses examined)		980
Arenaviruses		11 viruses examined	850–990	500–3000
Ungrouped		Rift Valley Fever	600–750	700
		Eubenangee	1010	

Group B viruses also contain three proteins: a glycoprotein, the predominant species, is present in the spikes of the envelope; a small basic protein is associated with RNA in the ribonucleoprotein core; and the smallest polypeptide is associated with the envelope.

The hemagglutinating component of togaviruses is an immunogenic surface structure of the virion, probably corresponding to the glycoprotein spikes, which agglutinates erythrocytes from newly hatched chicks or adult geese. Maximum hemagglutination is effected within a narrow range of pH and temperatures that is different for group A and B viruses.* Hemadsorption on infected cell cultures is optimum under identical conditions. The virions become firmly bound to the erythrocyte's surface and do not elute spontaneously. Cell lipids inhibit hemagglutination; hence, detection of hemagglutinating activity in cells (particularly brain or spinal cord) requires preliminary extraction with lipid solvents. In fact, treatment of virions with such solvents inactivates infectivity and liberates a viral subunit that agglutinates RBC and usually can induce production of neutralizing antibodies.

---

* Maximum agglutination is attained at pH 6.4 and 37° for group A viruses and pH 6.5 to 7.0 at 4° or 22° for group B viruses.

The infectivity of most togaviruses decreases rapidly at 35 to 37°; infectivity and hemagglutinating activity have maximum stability at about pH 8.5. These activities of group B togaviruses, unlike those of group A, are readily inactivated by small amounts of proteolytic enzymes (e.g., trypsin, chymotrypsin, and papain) and reagents which attack sulfhydryl bonds.

IMMUNOLOGICAL CHARACTERISTICS

The major group A and B togaviruses also fall into immunological subgroups (increased cross-reactivity is noted between members of each subset); group A viruses are divided into two subgroups and those of group B into four subgroups (Tables 60-2 and 60-3). Of practical concern for prolonged immunity and artificial immunization is the finding that viruses isolated at different times (e.g., Murray Valley encephalitis viruses from 1956 and 1969) may be antigenically distinguishable.

Neutralizing Abs appear about 7 days after onset of disease and persist for many years, probably for life. Complement-fixing Abs also rise early but are not detectable after 12 to 14 months. Infection is followed by solid immunity which appears to be correlated with the development and persistence of neutralizing Abs; hemagglutination-inhibiting Abs appear at the same time and are easier to

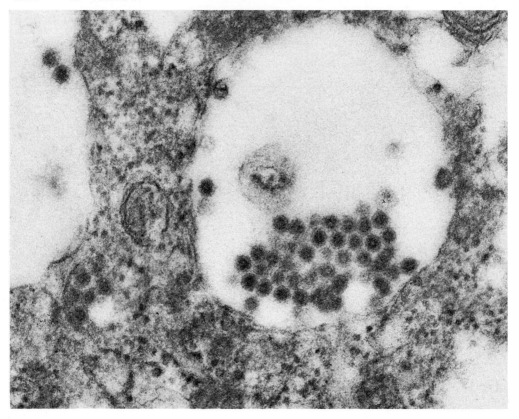

Fig. 60-1. Thin section of mature WEE virus within cytoplasmic vacuole of an infected cell. Viral particle has a dark central nucleocapsid 300 A in diameter and a peripheral membrane about 20 A thick. Note the cubical shape of the virions aggregated in a crystalline-like assay. ×100,000. [From Morgan, C., Howe, C., and Rose, H. M. *J Exp Med 113*:219 (1961).]

assay: Heterologous immunity is also produced within the group experimentally, in accord with the serological cross-reactions.

Most togaviruses maintain their immunogenicity and immunological reactivity following destruction of infectivity by formalin, heat, and β-propiolactone. With the yellow fever and dengue viruses of group B, however, formalin inactivation markedly impairs their capacity to induce formation of neutralizing Abs. Accordingly, preparation of successful vaccines required the development of attenuated viruses (see below). The existence of only a single known immunological type of yellow fever virus accounts in part for the effectiveness of the vaccine.

Cross-reactive hemagglutination-inhibiting Abs develop after infection or artificial immunization; the degree of cross-reactivity increases up to about 1 month after the immunogenic stimulus. For example, an animal infected with EEE virus first develops homologous Abs about 7 days after infection, and 3 to 6 weeks later develops relatively high titers of hemagglutination-inhibiting Abs against all group A viruses. If the same animal is subsequently inoculated with EEE virus **or another group A virus,** a rapid increase of hemagglutination-inhibiting Abs for all group A viruses **(a secondary group response)** will follow; the Ab response is considerably greater than that following a single inoculation of either virus alone. Proposed immunization procedures take advantage of this broadened secondary response.

HOST RANGE

Togaviruses multiply in a wide range of vertebrates and arthropods. The newborn mouse

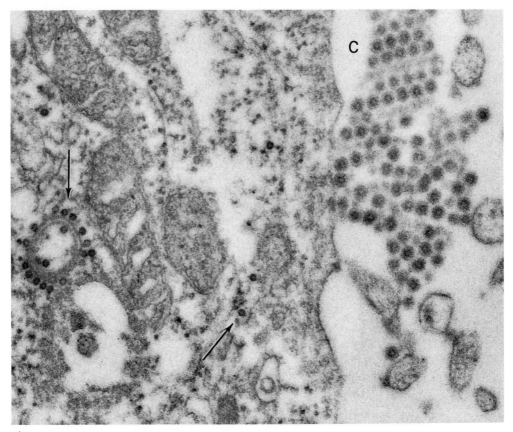

Fig. 60-2. An extracellular crystal (C) composed of WEE virus particles. Dense precursor particles are scattered in the cytoplasm and are present on opposite sides of two concentric lamellae near the left border (arrow). Mature virions are only seen within vacuoles (Fig. 60-1) or outside of cells. ×86,000 [From Morgan, C., Howe, C., and Rose, H. M. *J Exp Med 113*:219 (1961).]

is the laboratory animal of choice: it is highly susceptible to infection by all members of the group, which multiply to high titer in brain (some of the viruses also multiply in muscle), and produce extensive pathological changes. Resistance increases with age: most mice of 3 to 6 months are quite resistant to infection by peripheral routes. Horses are readily and often fatally infected by the group A equine encephalitis viruses (whose initial isolation from horses, during epizootics, gave rise to their names). Monkeys are also useful hosts for studying the pathogenesis of infection, particularly with yellow fever virus.

Wild birds and domestic fowl, particularly newborn, can be infected artificially or by the bite of an infected mosquito. Embryonated chicken eggs are sensitive, convenient hosts for many studies. Mosquitoes (*Culex, Anoph-* *eles,* and *Mansonia*) are the main arthropod hosts in nature for group A viruses.

Many group B viruses infect arthropods other than mosquitoes and ticks, e.g., milkweed bug, black carpet beetle, Indian moth, and common housefly. Some of the viruses can survive in ticks for many months by transovarian transmission (i.e., Russian spring-summer encephalitis viruses) and through periods of molting and metamorphosis, without apparent injury to the host; this survival during periods of poor transmission furnishes a possible mechanism for overwintering.

Most togaviruses can be propagated on a variety of primary and continuous cell cultures including cultures of mosquito cells. They produce cytopathic effects, and infected cells can also be detected by hemadsorption. A sensitive and reproducible plaque assay can be carried out with susceptible cells. Because of their sensitivity and convenience cell cultures are rapidly replacing the suckling mouse for experimental and diagnostic work.

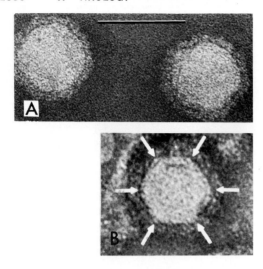

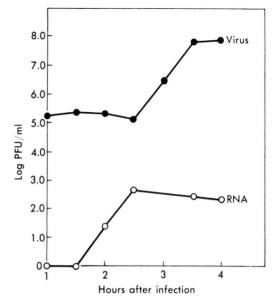

Fig. 60-3. Morphology of a group A togavirus negatively stained. **A.** The spherical particles have a dense nucleocapsid and an envelope covered with fine projections. ×240,000. **B.** In an occasional virion a nucleocapsid is seen with a clear polygonal outline and the semblance of an ordered capsomeric structure. ×360,000. [From Simpson, R. W., and Houser, R. E. *Virology 34*:358 (1968).]

Fig. 60-4. Temporal relation of the biosynthesis of infectious (viral) RNA and infectious WEE virus particles. [From Wecker, E., and Richter, A. *Cold Spring Harbor Symp Quant Biol 27*:137 (1962).]

VIRAL MULTIPLICATION

The multiplication cycles of togaviruses are similar; temporal differences are minor within each subgroup, but relatively large among subgroups. The multiplication cycles for group A viruses are relatively short, as exemplified by WEE virus (Fig. 60-4). In contrast, the multiplication of group B viruses is slower; for example, the latent period of type 2 dengue virus is 10 to 12 hours, and the maximum yield is attained 20 to 30 hours after infection.

After infection virus is adsorbed rapidly by susceptible cells, and eclipse is evident within 1 hour. Biosynthesis of infectious viral RNA and virion proteins, preceded by the production of an unstable RNA polymerase and replicative forms of viral RNA, commences in about the middle of the latent period; the maximum level of viral RNA is attained by the time virions begin to be assembled. Isotopic studies have shown that the phospholipids of the virion's envelope are derived from preexisting cellular phosphatides, but the phospholipid and fatty acid composition of the viral envelope appears to vary with the virus and to depend upon the lipid affinities of

the envelope proteins. The RNA and proteins of the virus are determined by the viral genome and synthesized from new building blocks. Membrane-associated host cell enzymes glycosylate the viral envelope protein, and the characteristics of the carbohydrate moiety of the glycoprotein may depend upon which membranes (plasma, endoplasmic reticulum, or vacuolar) are associated with viral maturation.

After maturation group A virions are rapidly released from the host cell (within 1 minute), suggesting that final assembly of the virion is accomplished at the cell surface and is concomitant with its release (WEE and VEE have been measured). Cells infected with Sindbis virus produce virions at the extraordinary rate of 10^4 virions per cell per hour for approximately 12 hours. Group B viruses, however, are released more slowly; maximum yield appears to be significantly less.

Electron microscopy has contributed additional insight into the development and release of togavirus. RNA replication and the development of the viral nucleocapsids are associated with unique, trilaminar cytoplasmic membranes close to newly formed cytoplasmic vacuoles characteristic of the infections (Figs.

60-2 and 60-5). The nucleocapsids either pass into the lumen of the vacuoles, each acquiring the envelope in the process, or are dispersed in the cytoplasm and subsequently receive their envelope as they are extruded through the cell membrane (Fig. 60-5). The particle is thought to become infectious when it acquires its final coat (i.e., the envelope). Group B viruses characteristically appear to be released in large numbers into prominent cytoplasmic vacuoles, which migrate to the cell surface and disgorge the virions into the extracellular spaces.

PATHOGENESIS

When an infected mosquito bites a prospective host it injects virus from its salivary glands into the blood stream of its victim. Successful infection depends upon the presence of sufficient virus in the mosquito's saliva and the absence of paucity of neutralizing antibody in the host. Details of the pathogenesis of infections in man are largely inferred from experimental studies in animals.

All togavirus infections in man are similar in the initial stages. Virus is removed from the blood by reticuloendothelial cells in which the virus multiplies (mainly in the spleen and lymph nodes). Viremia follows, initiating the **systemic phase** of the clinical disease; finally, virus invades various tissues, including the central nervous system (the encephalitides); the skin and blood vessels (the hemorrhagic fevers); or skin, muscle, and viscera (dengue and yellow fevers). The mechanisms by which encephalitis viruses invade the central nervous system are not yet clearly understood: entrance may be effected directly through the blood-brain barrier or, less likely, by transmission along nerves through the membranes of the nasal cavity.

GROUP A TOGAVIRUSES

Despite similarities in certain pathogenic characteristics, considerable diversity exists in **the diseases** produced by viruses in this group. Two types of clinical syndromes are seen (Table 60-2): 1) EEE, WEE, and VEE viruses produce a **systemic phase** of disease

Fig. 60-5. Characteristic formation of particles of WEE virus at membrane of vacuoles in cytoplasm of infected cell. Extracellular virus is visible at the lower left. One particle seems to be in the process of emerging from an adjacent cell membrane (arrow). ×96,000. [From Morgan, C., Howe, C., and Rose, H. M. *J Exp Med 113*:219 (1961).]

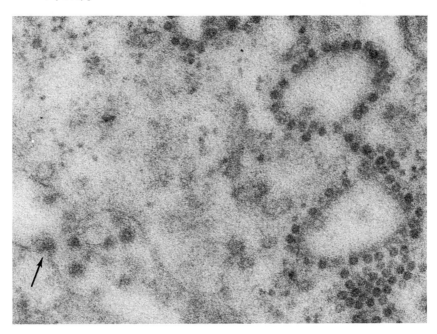

(chills, fever, aches) resulting from the viremia, and an **encephalitic phase** may follow after a variable time; 2) infections by Chikungunya and Mayora viruses are confined to the systemic phase, and the symptoms are solely constitutional.

EEE virus usually produces severe illnesses, with high mortality, and often with severe residual neurological damage among survivors. WEE virus, in contrast, usually causes a less severe disease, and most patients recover completely; it most often appears in infants and children. WEE virus may also frequently initiate abortive disease (fever and headache) or clinically inapparent infections. VEE virus infects horses primarily; transmission to man results usually in a mild disease with variable systemic symptoms and only rarely causes severe encephalitis. In the summer of 1971 a devastating epizootic spread among horses in the southwestern United States, with mild disease occurring in many humans as well. The virus is continuing to disseminate, and additional large outbreaks are expected to occur.

The **pathological** features of infections with group A togaviruses are known in greatest detail for EEE and WEE. Severe infections grossly involve viscera as well as brain and spinal cord. The lesions caused by EEE virus are scattered in the white and gray matter, particularly prominent in the brain stem and basal ganglia; the spinal cord shows milder changes. WEE infections, on the other hand, chiefly affect the brain, producing lymphocytic infiltration of the meninges and lesions of the parenchyma, predominantly in the gray matter. Lesions generally consist of necrosis of neurons, glial infiltration, and perivascular cuffing. Inflammatory reactions in walls of small blood vessels, and thrombi, may occur.

GROUP B TOGAVIRUSES

Group B viruses produce three types of clinical syndromes (Table 60-2): 1) **central nervous system** disease manifested mainly as encephalitis (St. Louis, Japanese B, Murray Valley, Ilheus, and Russian spring-summer viruses); 2) **severe systemic disease involving viscera** such as liver and kidneys (yellow fever virus); and 3) **milder systemic disease** characterized by **severe muscle pains** and **rash,** which may be hemorrhagic (West Nile, den-

gue, and some of the tick-borne viruses). In addition, most of these viruses also produce subclinical or mild infections which can be recognized only by laboratory diagnostic procedures.

The pathogenesis of **yellow fever** differs: after primary multiplication in lymph nodes extensive secondary viral multiplication, with cell destruction, occurs in many viscera, including the liver, spleen, kidneys, bone marrow, and lymph nodes. The organs involved and the severity of the lesions vary with the infecting strain. Clinically recognized yellow fever may be a grave disease, although the mortality is only about 10%.

The pathology of yellow fever is characterized by degenerative lesions of the liver, kidney, and heart, accompanied by hemorrhages and bile staining of tissues. The most distinctive lesions occur in the liver, where a pathognomonic midzonal hyaline necrosis develops with preservation of the basic liver architecture and without inflammatory reaction.

The severity and extent of the encephalitides due to B viruses vary with the etiological agent. For example, St. Louis encephalitis virus produces mild lesions, low mortality, and few residua compared with Japanese B encephalitis virus, which causes a mortality of about 8%, neurological residua in more than 30%, and persistent mental disturbances in about 10% of clinically diagnosed infections. The pathological features of the B encephalitides are indistinguishable from those described for group A viruses.

In uncomplicated dengue fever deaths are rare; hence the pathological lesions are not well known. Biopsies of the characteristic skin lesions reveal endothelial swelling, perivascular edema, and mononuclear infiltration in and about small vessels. In some epidemics, particularly in tropical Asia, **hemorrhagic fever** may be prominent. Hemorrhagic dengue (characterized by high fever, hemorrhagic manifestations, and shock) is most often seen in those sequentially infected with different immunologic types of virus during a limited period (about 2 years), suggesting a hypersensitivity or immune-complex type of disease.

LABORATORY DIAGNOSIS

Togavirus infection is diagnosed by viral isolation or by serological procedures. To isolate a virus from a patient dead of encephalitis, emulsions of the brain and spinal cord are inoculated intraperitoneally and intracerebrally into suckling or newly weaned mice, or into susceptible cell cultures. The viremic phase of the togavirus encephalitides is usually completed by the time a patient seeks medical assistance. Hence, isolation of virus from blood and spinal fluid of patients during life has been accomplished only rarely, and it cannot be considered practicable. With Chikungunya and many group B viruses, however, the responsible virus may be isolated from blood during the initial stages of illness; in yellow fever and dengue fever blood or serum may be a good source of virus during the first 2 to 4 days of disease.

A newly isolated virus is identified by 1) hemagglutination-inhibition titrations with standard antisera, to determine its immunological group; and 2) neutralization titrations to establish its species. Because of the marked cross-reactivity among group B viruses, and particularly within the same subgroup, considerable care is required to identify as agent.

Serological diagnosis is made with patient's sera drawn during the acute illness and during convalescence. Complement-fixation assays are preferred because of their simplicity, speed, inexpensiveness, and comparative lack of cross-reactivity. Antigens are obtained from infected cell cultures or from brains of infected newborn mice extracted with acetone. Complement-fixation titrations are not adequate for epidemiological surveys, however, because complement-fixing antibodies do not persist as long as those measured by hemagglutination-inhibition or neutralization titrations.

EPIDEMIOLOGY

GROUP A TOGAVIRUSES

For most group A viruses man is merely an incidental host. The mosquito is the common arthropod vector. The cycle of infection has been elucidated best for the equine encephalitis viruses, particularly WEE and EEE, and can be simply diagrammed as follows:

Arthropod → Man
⇅
Bird → Horse

Despite the name of the disease, the horse, like man, appears to be a dead end in the chain of infections (Fig. 60-6). In fact, horses are not significant reservoirs in nature, prob-

Fig. 60-6. Epidemiological pattern for WEE virus infections. The chains for rural St. Louis encephalitis are similar, except that horses are inapparent rather than apparent hosts. EEE infections also have a similar summer infection chain, but a few significant differences exist: 1) the identity of the vector infecting man is unknown, 2) domestic birds do not appear to be a significant link in the chain, and 3) it has a bird-to-bird secondary cycle in pheasants whose role is unclear. [From Hess, A. D., and Holden, P. *Ann NY Acad Sci 70*:294 (1958).]

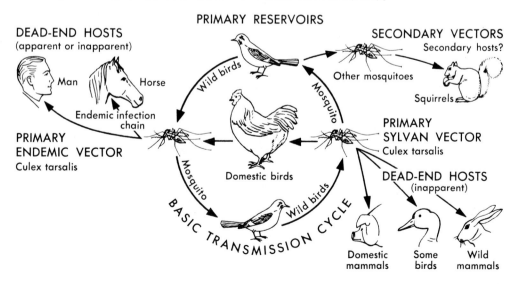

ably because viremia does not usually reach sufficiently high levels to infect mosquitoes with regularity. Birds are the principal natural hosts, but it is not clear which species are the primary reservoirs (Fig. 60-6). Wild snakes and frogs, as well as some rodents, are probably secondary reservoirs for WEE. Birds likewise appear to be the most likely hosts in which viruses can persist in nonepidemic periods and during the seasons in which transmission by mosquitoes is not prominent (overwintering). Hibernating mosquitoes are also a possible reservoir, for infectious virus can persist in them at least 4 months. WEE is generally found in the United States west of the Mississippi River, but increasingly the virus is being isolated along the eastern seaboard. In contrast EEE is confined to the eastern part of the United States and Canada. Both viruses are found in Caribbean islands, Central America, and South America.

The primary vector of WEE virus is the culicine mosquito, *Culex tarsalis.* The primary mosquito vector for EEE is not certain, but *Culiseta melanura* is susceptible to experimental infection and has been found infected in nature. Other mosquitoes also appear to be implicated in the epidemiological cycles involving mammals.

Chikungunya (African, "that which bends up") virus infections may be a notable exception to this general epidemiological pattern. *Anopheles* and *Aedes* mosquitoes are the vectors, but man is the only known vertebrate host.

VEE virus is distributed in the Everglades region of Florida, and the southwestern United States, northern South America, the Amazon Valley, Central America, and southern Mexico. Although horses are invariably infected when human disease appears, the epidemiological pattern of infection has not been elucidated, the reservoir of the virus is still to be found, the primary arthropod vector has not been proved, and its mode of transmission to man is not clear. Similar uncertainties exist in the epidemiology of infections by other group A togaviruses.

GROUP B TOGAVIRUSES

The epidemiological patterns of group B togavirus infections are more varied than those of group A. These patterns will be summarized for a few of the most important diseases.

St. Louis encephalitis* is the major group B togavirus infection in the United States; a severe epidemic occurred in Texas, New Jersey, and six other states in the summer of 1964. (The greatest incidence was in Houston, Texas, where 294 cases with 19 deaths were reported.) The epidemiological pattern is similar to that described for WEE virus infections. Wild birds are the major reservoir of the virus, and *Culex tarsalis* and the *C. pipiens* complex are the most common mosquito vectors. Man is an accidental, dead-end host (Fig. 60-6).

The epidemiology of **Murray Valley encephalitis, Japanese B encephalitis,** and **West Nile fever** is basically the same as that of St. Louis encephalitis except for the species of mosquito vectors and the avian reservoirs. Serological surveys indicate that for each case of clinical disease produced by these viruses several hundred inapparent infections are also induced.

Yellow fever presents another complex ecological situation.† Two distinct epidemiological types exist, **urban** and **jungle yellow fever.** Each has a different cycle, but they may interact. In its simplest form the epidemiological pattern of **urban yellow fever** simply involves man and the domestic mosquito, *Aedes aegypti:*

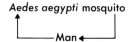

Viremia in man begins 1 or 2 days before, and persists for 2 to 4 days after, the onset of clinical illness. Viremia is greatest during this period and mosquitoes taking a blood meal may be infected. A 10- to 12-day period of viral multiplication in the cells lining the mosquito's intestinal tract is then required **(the extrinsic incubation period)** until sufficient virus accumulates in the salivary glands to permit transmission to man.

Jungle yellow fever is transmitted by various jungle mosquitoes, primarily to monkeys.

---

* The first epidemic due to this virus was recognized in St. Louis, Missouri, in 1933.

† The monograph entitled *Yellow Fever,* ed. by G. K. Strode, reviews in exciting detail many of the important facts personally discovered by the authors.

Man becomes an accidental host when he enters the animals' domain.

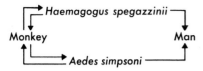

Infection of man may initiate a cycle of urban yellow fever.

Dengue fever resembles yellow fever epidemiologically. There is an urban cycle (man ⇌ *Aedes aegypti* mosquitoes), and probably a jungle cycle with monkey as the mammalian host. This disease is prevalent in the Caribbean islands, as well as more distant subtropical areas (Table 60-2).

The **tick-borne complex** of viral infections introduces two unique features into the epidemiology of togavirus infections: in addition to being transmitted by ticks (*Ixodes*), some of these viruses (e.g., Russian spring-summer encephalitis virus) may also be transmitted to man from the goat by **milk** instead of by an arthropod.

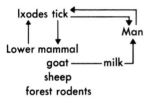

Omsk hemorrhagic fever is transmitted by *Dermacentor marginatus,* probably from muskrats, and Powassan virus (isolated in Canada and the United States) is transmitted by *Ixodes* ticks from small mammals, probably squirrels.

PREVENTION AND CONTROL

Control measures are aimed at 1) preventing transmission of the virus by eradicating, or at least reducing, the population of arthropod vectors, and 2) increasing the host resistance by artificial immunization. The former procedure has been relatively successful. Vaccines have been prepared with formalin-inactivated viruses from nervous tissue of infected animals, chick embryos, and cell cultures, and with attenuated viruses.

Yellow fever has played a special role in the development of concepts and methods for the control of other insect-borne diseases.

Elimination of the vector, *A. aegypti,* proved effective soon after Reed and his colleagues demonstrated the causative agent and the vector requirement. The use of modern insecticides has facilitated this control measure. It should be noted, however, that *A. aegypti* mosquitoes and other possible vectors are still present in many parts of the world, including the southeastern United States. Moreover, mosquito control measures cannot eliminate jungle yellow fever.

As noted above, the loss of immunogenicity by inactivated yellow fever virus made it necessary to seek a vaccine containing infectious virus. Theiler and his coworkers attenuated a mouse-adapted yellow fever virus by serial passage in tissue cultures, first of embryonic mouse tissues, then of embryonic chicks, and finally of embryonic chicks without brain or cord. For the most widely used vaccine the attenuated virus obtained (17D yellow fever virus) is propagated in embryonated chick eggs. The French, chiefly in Africa, utilize a vaccine containing virus attenuated by serial passage in mouse brain. The 17D strain produces many fewer and less severe toxic reactions than the French vaccine. On the other hand, the French vaccine elicits a higher antibody response. The duration of protection afforded by the 17D vaccine is not known, but antibody has been detected at least 6 years after immunization.

Eradication of yellow fever in the United States, and substantial reduction of its incidence in South America and other parts of the world, initially suggested that this disease would be eradicated. However, the reservoir of jungle yellow fever was later discovered, and yellow fever has increased in incidence in parts of Central and South America and has been creeping northward. With the nidus present from which the virus could spread explosively, the danger of epidemics is real.

Formalinized chick embryo EEE and WEE vaccines produce effective antibody responses in horses. But the group A encephalitis virus vaccines have not been utilized in humans except for protection of laboratory workers; their effectiveness in man has not been established. A Japanese B encephalitis vaccine, prepared by formalin inactivation of virus propagated in chick embryos, has been employed with apparent success in children in Japan. However, a similar vaccine was ineffective in U.S. Army personnel stationed in Japan and the Far East. Dengue vaccines containing infectious, attenuated types 1 and 2 viruses seem promising. The benefits to be derived from vaccines containing viruses propagated in cell cultures (inactivated or attenuated) remain to be fully explored.

The marked antigenic cross-reactivity of group B viruses may prove useful for immunization purposes. For example, in experimental studies immunization with an infectious, attenuated virus, such as yellow fever virus, and subsequent injection of one or more inactivated or live attenuated heterologous viruses, resulted in a broad immunological response that protected against a variety of group B viruses.

BUNYAMWERA SUPERGROUP VIRUSES

As noted earlier, arthropod-borne viruses have been classified immunologically on the assumption that antigenic similarities reflect chemical similarities: the togaviruses indicate the validity of this assumption. On the basis of similar immunological evidence and preliminary chemical and morphological data many groups of so-called "arboviruses" have been tentatively gathered into the **Bunyamwera supergroup,** which probably will become recognized as a separate viral family. This large group includes the California encephalitis, Bunyamwera, and C groups, which are the major ones affecting man (Table 60-2). Similar properties have also been described for representative members of the Bwamba, Capim, Guama, Koongal, Patois, Simbu, and Tete virus groups. Together these viruses form the largest set of arthropod-borne viruses (approximately 100 species).

PROPERTIES OF THE VIRUSES

The **development** and **morphology** of all the viruses studied show their marked similarities to each other and their clear distinction from togaviruses. The virions, which are oval particles, first appear within small vesicles or cisternae in the region of the Golgi apparatus (Fig. 60-7). The particles develop by budding into the vacuoles and consist of an electron-dense nucleocapsid core closely bound by an envelope, with spikes protruding from its surface (Fig. 60-7). The nucleocapsid, when released from the virion, consists of single-stranded RNA and a single species of protein arranged in the form of a helix (a striking distinction from the apparent isometric symmetry of togaviruses).

Suckling and newly weaned mice are the laboratory animals of choice for isolation of the viruses. These viruses also multiply and produce cytopathic changes and plaques in a number of cultured cell types; among the most useful are continuous human (e.g., HeLa cells) and baby hamster kidney (BHK 21) cell lines.

Optical hemagglutination titers are obtained with suspensions of RBC from 1-day-old chicks or geese held at room temperature and at pH 6.0 to 6.2. To prepare virus for hemagglutination or complement-fixation assay interfering lipids are extracted from sera or brain tissue with ether or acetone. In contrast to group A and B togaviruses, the cross-reactivity of the Bunyamwera supergroup viruses is maximal in complement-fixation rather than in hemagglutination-inhibition titrations. Hence complement-fixation titrations with one or two standard sera can identify an agent as a group C virus, for example, and specific viral identification can be accomplished by hemagglutination-inhibition and neutralization titrations. With these procedures the Bunyamwera supergroup viruses can be divided into groups (e.g., groups C, California encephalitis, Bunyamwera) and subgroups.

PATHOGENESIS AND EPIDEMIOLOGY

The **California encephalitis viruses,** which were initially isolated from mosquitoes in San Joaquin Valley of California, are widely distributed (Table 60-2). They produce prominent clinical illnesses, manifest by fever, headache, and mild or severe central nervous system involvement, particularly in children. Recovery is usually complete, although mild residua and even rare deaths have been recorded. Clinical disease has been reported from 13 states in all regions of the United States as well as in the countries noted in Table 60-2. The natural reservoir of the virus is unknown, but the agent has been found in the blood of rabbits, squirrels, and field mice in titers adequate to infect mosquitoes. Although California encephalitis viruses have been isolated from *Aedes* and culicine species the specific mosquito responsible for transmission is uncertain.

Among a large number of viruses isolated from experimental (sentinel) monkeys caged

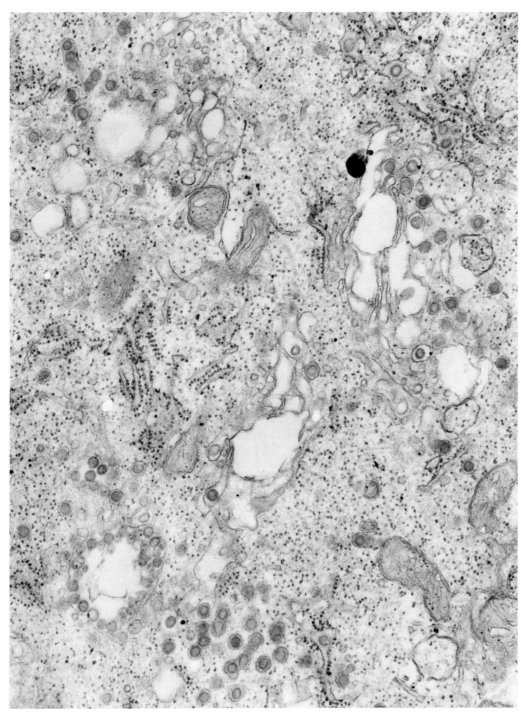

Fig. 60-7. Development of Bunyamwera virus in the cytoplasm of a neuron in a mouse brain. Viral particles are present in the cytoplasm and budding into Golgi vacuoles and cisternae of the endoplasmic reticulum (arrows). The virions have a mean diameter of 900–1000 A, a nucleocapsid core of 600–700 A in diameter, and an envelope of 150–200 A thick. ×39,500. [From Murphy, F. A., Harrison, A. K., and Tzianabos, T. *J Virol* 2:1315 (1968).]

in the forested Belém area of Brazil seven different viruses were recognized as being immunologically related to each other but distinct from togaviruses. Several other immunologically distinguishable but related viruses have been isolated in the Florida Everglades and Central America. These viruses, previously called **group C** arboviruses but now recognized as a division of the Bunyamwera supergroup, produce mild disease in man, consisting of headache and fever: recovery is complete.

The natural reservoir of group C viruses appears to be in monkeys and other forest mammals (e.g., opossums, rats, sloths). The specific mosquito vector has not been definitely established, but culicine and sabethine mosquitoes are likely candidates.

Bunyamwera virus was first isolated from a mixed pool of *Aedes* mosquitoes trapped in Bwamba, Uganda. Fourteen immunologically related but distinct viruses have been recognized, including strains isolated in the United States (Florida, Virginia, Colorado, Illinois, and New Mexico). Disease attributed to the Bunyamwera group is rare and usually mild (Table 60-2).

ARENAVIRUSES

On the basis of morphological observations and subsequent immunological findings the seemingly disparate Tacaribe group of viruses, previously considered to be arboviruses, Lassa virus, and lymphocytic choriomeningitis (LCM) virus are now classified into a single group called **arenaviruses** (L. *arenosus* = sandy), a term derived from the unique electron microscopic appearance of the virions (Fig. 60-8).

Complement fixation reveals the immunological relatedness of the arenaviruses; the Tacaribe viruses, however, are more closely related to each other than to Lassa or LCM viruses. Neutralization titrations reveal the immunological specificity of each arenavirus.

PROPERTIES OF THE VIRUSES

The viral particles are round, oval, or pleomorphic. As viewed in thin sections, the virions consist of a dense well-defined envelope with prominent closely spaced projections and an unstructured interior containing varying numbers of electron-dense granules (Fig. 60-8) which cause the unique pebbly appearance from which the viruses have attained their name. Negative-contrast electron micrography shows spherical or pleomorphic virions with an envelope having pronounced and regularly spaced projections.

Viral particles are formed by budding, chiefly from plasma membranes (Fig. 60-8). At the sites of budding, the host cell membrane becomes thickened, more clearly bilamellar, and covered with projections. The dense particles are present within the budding particles before separation from the infected cell. No apparent damage of the host cell is associated with the viral infection.

The unique electron-dense particles within arenaviruses resemble ribosomes morphologically and have the physical and chemical characteristics of ribosomes. For example, LCM virus contains four species of single-stranded RNAs, two with sedimentation coefficients of 31S and 23S (viral RNAs), and two with the electrophoretic mobility and sedimentation characteristics of the host cell ribosomal subunits (i.e., 28S and 18S). Moreover, like ribosomal RNA, synthesis of the two latter species is inhibited by low doses of actinomycin D (0.15 μg per millileter) without affecting viral replication.

PATHOGENESIS AND EPIDEMIOLOGY

The Tacaribe group of viruses (Tacaribe, Machupo, Junin, Tamiami, and four others) have been principally isolated from bats and cricetid rodents in the Western Hemisphere. But Junin and Machupo viruses have been frequently isolated from cases of Argentinian and Bolivian hemorrhagic fevers, respectively, and appear to be the etiological agents of these severe illnesses. Arenaviruses commonly produce chronic carrier states in their natural hosts, and virus may be isolated from the animals' urine as well as from their blood and internal organs. There is no evidence of viral

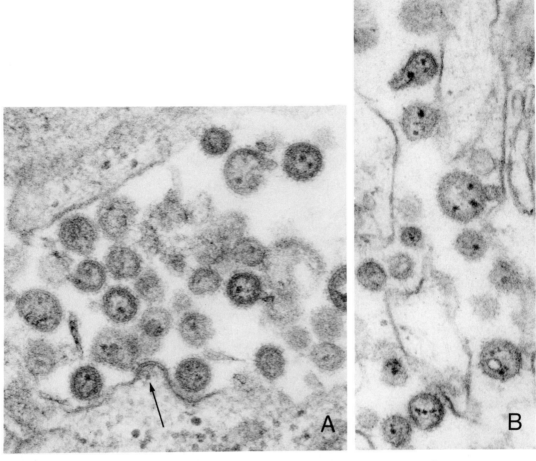

Fig. 60-8. A. Machupo virus (a Tacaribe group virus) particle budding (arrow) from the plasma membrane of an infected cell; the thickened membrane of the budding particle is prominent compared with the neighboring membrane. Many extracellular mature virions (mean diameter of 1100 to 1300 A) are present: their prominent surface projections and internal, ribosome-like particles are readily seen. ×114,000. **B.** Lymphocytic choriomeningitis (LCM) virus particles in an infected culture of mouse macrophages. The morphology is strikingly similar to the Tacaribe group virus in **A.** ×82,000. [From Murphy, F. A., Webb, P. A., Johnson, K. M., and Whitfield, S. G. *J Virol* 4:535 (1969).]

spread from patient to patient, and virus is rarely isolated from arthropods, even during epidemics. Thus, the Tacaribe group viruses, similar to LCM virus of mice, appear to be spread to man in excretions of the naturally infected rodents.

LCM virus infects man rarely, and usually under conditions in which the mouse population is very dense. The disease in man is generally mild, and is manifest most often as a lymphocytic form of meningitis but occasion-

ally as a mild pneumonia. Rarely, LCM virus produces severe and even fatal illnesses associated with hemorrhagic manifestations.

Lassa virus, first isolated in 1969 from an American missionary working in Nigeria, causes a serious febrile illness characterized by severe generalized myalgia, marked malaise, and sore throat accompanied by patchy or ulcerative pharyngeal lesions. The fatal cases also develop myocarditis, pneumonia with pleural effusion, encephalopathy, and hemorrhagic lesions. Virus persists in the blood for 1 to 2 weeks, and during this period it can also be isolated from urine, pleural fluid, and throat washings. The virus is more stable than togaviruses in body fluids, which probably permits its person-to-person contagion and accounts for the hazard it presents for laboratory isolation or study. Arthropods collected in Nigeria, the only known locale of natural infections, have not yielded virus, and insect cell cultures are insusceptible to viral propagation. In mice Lassa virus produces an infection similar to LCM virus, and chronic latent infections can be established. Whether rodents serve as the virus's natural host, and how it is spread to man are unknown.

While arenaviruses were initially grouped together because of immunological relations and similar morphology, they also appear to have somewhat comparable epidemiological, ecological, and pathogenic patterns. Tacaribe group and Lassa viruses (which have been called arboviruses), like LCM virus, do not require arthropods for spread; their natural hosts appear to be rodents in which they often produce chronic infections.

GENERAL REMARKS

Only a few of the known viruses that multiply in arthropods and vertebrates have been discussed in this chapter (Table 60-2). The properties of most of these agents, and even their clinical and epidemiological behavior, are known in only a fragmentary fashion. Some, such as the phlebotomus (sandfly) fever virus, transmitted by the bite of the female sandfly *Phlebotomus papatassi,* assumed importance to the U.S. Armed Forces during World War II, when the disease (which is not serious) appeared in military personnel in the Mediterranean area.

Many of the arthropod-borne viruses, including some that can cause serious disease, also cause a very much larger amount of inapparent infection in endemic areas; hence the native human population carries a high level of immunity, but the insect population is still highly infectious because of the virus reservoir in lower animals. Such diseases could increase dramatically in quantitative significance 1) when ecological alterations cause development of a dense population of infected arthropods next to a nonimmune human population, or 2) when a large, immunologically virgin human population (e.g., military personnel) moves into an endemic area.

Because of their close antigenic relation to human pathogens, those arthropod-borne viruses that have not been associated with human disease cannot be ignored by medical investigators, however esoteric they may seem. Several such viruses have been isolated in the United States or Canada; for example, bat salivary gland virus (group B, U.S.A.); California encephalitis and Trivittatus viruses (California group, U.S.A.); and Silverwater virus (ungrouped, Canada). Since a change in either the host reservoir, the vector, or the genetics of the viral population might permit these agents to infect man, they remain a potential hazard.

The comforting realization that as many as 100 arthropod-borne viruses of seemingly diverse immunological groupings, or even without obvious relatives, may segregate into a major family, the Bunyamwera supergroup, on the basis of physical and chemical characteristics begins to bring order to what previously seemed unmanageable. Thus, from the ecological-epidemiological classification of "arboviruses," which led to assigning the same group name to viruses of widely disparate characteristics, there appears to be emerging several new families (e.g., togaviruses, Bunyamwera supergroup viruses, arenaviruses) and several viruses initially isolated from arthropods have been placed into well-established families (reoviruses, rhabdoviruses, and picornaviruses).

SELECTED REFERENCES

Books and Review Articles

CASALS, J. Arboviruses: Incorporation in a general system of virus classification. In *Comparative Virology*, p. 307. (K. Maramorosch and E. Kurstak, eds.) Academic Press, New York, 1971.

HAMMON, W. MCD. Arboviruses. In *Diagnostic Procedures for Viral and Rickettsial Infections*, p. 227. (E. H. Lennette and N. J. Schmidt, eds.) American Public Health Association, New York, 1969.

HESS, A. D., and HOLDEN, P. The natural history of the arthropod-borne encephalitides in the United States. *Ann NY Acad Sci 70:*294 (1958).

KISSLING, R. E. The arthropod-borne viruses of man and other animals. *Annu Rev Microbiol 14:*261 (1960).

MCALLISTER, R. M. Viral encephalitides. *Annu Rev Med 13:*389 (1962).

PORTERFIELD, J. C. The nature of serological relationships among arthropod-borne viruses. *Adv Virus Res 9:*127 (1962).

Specific Articles

BUCKLEY, S. M. Propagation, cytopathogenicity and hemagglutination-hemadsorption of some arthropod-borne viruses in tissue cultures. *Ann NY Acad Sci 81:*172 (1959).

CASALS, J., and BROWN, L. V. Hemagglutination with arthropod-borne viruses. *J Exp Med 99:*429 (1954).

FRIEDMAN, R. M., LEVY, H. B., and CARTER, W. B. Replication of Semliki forest virus: Three forms of viral RNA produced during infection. *Proc Natl Acad Sci USA 56:*440 (1966).

HOLMES, I. A. Morphological similarity of Bunyamwera supergroup viruses. *Virology 43:*708 (1971).

HURLBUT, H. S., and THOMAS, J. I. The experimental host range of the arthropod-borne animal viruses in arthropods. *Virology 12:*391 (1960).

SHOPE, R. E., and CAUSEY, O. R. Further studies on the serological relationships of group C arthropod-borne viruses and the application of these relationships to rapid identification of types. *Am J Trop Med Hyg 11:*283 (1962).

STOLLAR, V. Studies on the nature of dengue viruses. IV. The structural proteins of type 2 dengue virus. *Virology 39:*426 (1969).

STOLLAR, V., STEVENS, T. M., and SCHLESINGER, W. R. Studies on the nature of dengue viruses. III. RNA synthesis in cells infected with type 2 dengue virus. *Virology 33:*650 (1967).

chapter **61**

REOVIRUSES

PROPERTIES OF THE VIRUSES 1400
    Morphology and Physical Characteristics 1400
    Chemical Characteristics 1400
    Immunological Characteristics 1401
    Host Range 1402
    Viral Multiplication 1403
PATHOGENESIS 1406
LABORATORY DIAGNOSIS 1406
EPIDEMIOLOGY AND CONTROL 1407
OTHER VIRUSES SIMILAR TO REOVIRUSES 1407
    Colorado Tick Fever Virus 1407

The term **reovirus** (**r**espiratory **e**nteric **o**rphan **virus**) refers to a group of RNA viruses that infect both the respiratory and the intestinal tracts, usually without producing disease. Though originally considered members of the echovirus group (and classified as type 10 echovirus) reoviruses are larger and differ in producing characteristic cytoplasmic inclusion bodies (Table 61-1). These inclusion bodies contain specific viral antigens but stain green-yellow with acridine orange, like cellular DNA, rather than red, like the usual single-stranded RNA. This observation led to the discovery that reovirus RNA is a double-stranded molecule with a secondary structure similar to that of DNA and was the first indication that such an unusual nucleic acid exists in nature. Viruses with similar chemical and physical properties are widely disseminated in vertebrates and invertebrates (e.g., blue tongue and cytoplasmic polyhedrosis viruses), as well as in plants (e.g., clover wound tumor and rice dwarf viruses).

PROPERTIES OF THE VIRUSES

MORPHOLOGY AND PHYSICAL CHARACTERISTICS

The virion is icosahedral and possesses a capsid probably composed of 92 hollow capsomers (Fig. 61-1). Free individual capsomers cannot be readily obtained, suggesting that their polypeptide chains are weakly joined. Negatively stained virions and thin sections of viral particles reveal that the core is about 450 A across, contains the viral nucleic acid, and is covered by an icosahedral protein shell (inner capsid) which lies between it and the outer capsid (Fig. 61-1).

CHEMICAL CHARACTERISTICS

Reoviruses are composed of protein and RNA, which makes up about 15% of the weight of the virion. The molecular weight of the RNA is large (16 to 17×10^6 daltons) for a virus, which partly reflects its double-strandedness. The double-stranded nature of the RNA is shown by many properties (Ch. 10, Properties of nucleic acids). Thus, the base ratios are complementary (G = C = 20 moles %, A = U = 30 moles %); thermal denaturation (TM 90 to 95°), exhibits pronounced hyperchromicity; the RNA is resistant to a concentration of ribonuclease (1 μg per milliliter) that completely degrades single-stranded RNA; and electron micrography reveals stiff filaments, like those of DNA.

The reovirus genome has another surprising feature: the RNA extracted from purified virions or infected cells is found in 10 distinct pieces, which are distributed in three size classes (Fig. 61-2). These appear to be distinct components of the virion rather than products of artificial fragmentation. Thus the sizes are extremely reproducible (independent of viral strain, host cell, or extraction procedures); the different pieces do not cross-hybridize; the RNA present within the intact virion contains free 3'-OH terminal cytosines equal in number to extracted RNA; each segment has a 5'-terminal guanosine-5-diphosphate; and electron micrographs of gently disrupted virions demonstrate molecules of viral nucleic acid of similar lengths. The virion also contains about 3.7×10^6 daltons of a heterogeneous collection of small, single-stranded, adenine-rich oligonucleotides, whose function is unknown.

The capsid consists of at least seven species

TABLE 61-1. Characteristics of Reoviruses

1. Diameter of 750–800 A
2. Virion has icosahedral symmetry
3. Outer capsid composed of 92 prismatic, hollow capsomers
4. Possesses an inner capsid
5. Virion contains 10 pieces of double-stranded RNA genome
6. Virion contains an RNA polymerase
7. Virions lack lipid
8. Three immunological types
9. Common soluble family antigens

of polypeptides (Table 61-2) whose molecular weights are also distributed into three size classes. The two largest polypeptides and one of the smallest are associated with the inner capsid. Unfortunately, the structural proteins can be separated only after denaturation of the virion; hence it is impossible to identify the proteins (Table 61-2) associated with specific biological and immunological functions (i.e., hemagglutination, RNA polymerase, nucleoside phosphohydrolase, group antigenicity, and type-specific antigenicity).

IMMUNOLOGICAL CHARACTERISTICS

Three immunological types of reoviruses can be distinguished by neutralization and hemagglutination-inhibition titrations. They are antigenically related, however, by three or four cross-reacting antigens that can be measured by complement-fixation and immunoprecipitin tests. The structural relations of these family antigens to the virion have not been determined, but the type-specific hemagglutinin is evidently a surface component since the intact virion attaches to susceptible erythrocytes.

All three immunological types agglutinate human group O erythrocytes; in addition, type 3 agglutinates ox red blood cells. The reactions of reovirus with susceptible red blood cells differ, however, from those of hemagglutinating picornaviruses and myxoviruses: 1) hemagglutination is not inhibited by reagents that block sulfhydryl groups which inactivate the RBC receptors for echo- and coxsackieviruses; 2) the receptor-destroying enzyme (neuraminidase) for myxovirus receptors does not affect reovirus receptors; 3) reoviruses agglutinate and elute from erythrocytes, but unlike myxoviruses they do not destroy the receptor sites on the red cells. Both trypsin and periodate inactivate erythrocyte receptors for reoviruses, and N-acetyl-D-glucosamine binds to the viral capsid to block hemagglutination, suggesting that the RBC receptor is a glycoprotein. The viral capsid proteins, however, do not contain carbohydrate and hence are not glycoproteins.

In man and other animals type-specific Abs to reoviruses appear 2 to 4 weeks after infec-

TABLE 61-2. Correlation Between Classes of Genomic Segments, Messenger RNAs, and Capsid Polypeptides of Reovirus

Genome				Messenger RNA		Capsid polypeptides			
Size class	Segment	MW daltons $\times 10^{-6}$	Function	Size class	Component	Size class	Component	MW daltons $\times 10^{-4}$	Origin
L	I	2.7	Early	l	l_1	λ	λ_1	15.5	Core
	II	2.6	Late?		l_2		λ_2	14.0	Core
	III	2.5	Late		l_3				
M	IV	1.8	Late	m	m_1	μ	μ_1	8.0	Capsomer
	V	1.7	Late		m_2				
	VI	1.6	Early		m_3		μ_2	7.2	Capsomer
S	VII	1.1	Late	s	s_1	σ	σ_1	4.2	Capsomer
	VIII	0.85	Late		s_2				
	IX	0.76	Early		s_3		σ_2	3.8	Core
	X	0.71	Early		s_4		σ_3	3.4	Capsomer

From GRAHAM, A. F. and MILLWARD, S. In *Nucleic Acid-Protein Interactions and Nucleic Acid Synthesis in Viral Infections*, p. 333. (D. W. Ribbons, J. F. Woessner, and J. Schultz, eds.) North Holland, Amsterdam, 1971.

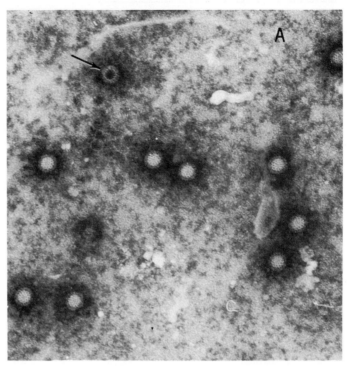

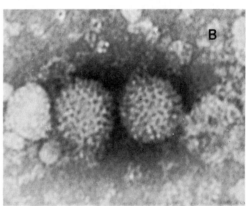

Fig. 61-1. Electron micrograph of type 3 reovirus particles negatively stained with phosphotungstic acid. The virion is 750–800 A in diameter. Cubic symmetry, structure of capsid, and absence of envelope are illustrated by low **(A)** and high **(B)** magnification. Note the empty particle in **A** in which the inner protein coat that covers the core is apparent (arrow). **A**, ×75,000; **B**, ×375,000. [From Gomaîos, P. J., Tamm, I., Dales, S., and Franklin, R. M. *Virology 17*:441 (1962).]

tion. Curiously, heterotypic reovirus Abs appear in the sera of 25% of persons who have primary infections due to type 1 reovirus.

HOST RANGE

Reoviruses appear to be ubiquitous in nature; specific viral inhibitors (presumably antibodies) have been found in the sera of all mammals tested except the whale. Man and many other species, including cattle, mice, and monkeys, are naturally susceptible to reoviruses; all can also be infected experimentally if antibodies are absent at the time of challenge. Newborn mice are particularly vulnerable to experimental infection, which is often fatal; they occasionally develop a chronic illness similar to runt disease (Ch. 21, Graft versus host reaction) when infected with type 3 reovirus.

Cell cultures, including primary cultures of epithelial cells from many animals and various

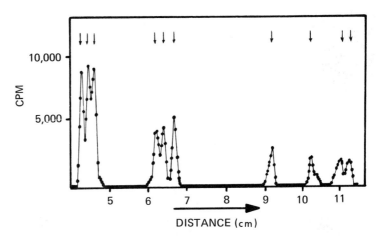

Fig. 61-2. Electrophoretic analysis on polyacrylamide gel of ³H-labeled dsRNA. The ten segments fall into three size classes: L (average MW = 2.7 × 10⁶), M (average MW = 1.6 × 10⁶), and S (average MW = 0.9 × 10⁶). (From Graham, A. F., and Millward, S. In *Nucleic Acid-Protein Interactions and Nucleic Acid Synthesis in Viral Infections,* p. 333. Ribbons, D. W., Woessner, J. F., and Schultz, J., eds. North Holland, Amsterdam, 1971.)

continuous human cell lines, are widely used to isolate and study reoviruses. The infection causes gross cytopathic changes and permits hemadsorption of human group O red blood cells. Distinctive eosinophilic inclusion bodies (Fig. 61-3) are seen in the cytoplasm of infected cells.

VIRAL MULTIPLICATION

Owing to the novel nucleic acid, replication of reoviruses presents some unusual features for an RNA-containing virus. After cell penetration the virions become associated with lysosomes, whose proteolytic enzymes hydrolyze the capsid proteins; the viral core is released but free double-stranded RNA cannot be detected. The eclipse period is long (6 to 9 hours, depending upon viral type and size of inoculum) compared with that of other RNA viruses with icosahedral symmetry (Fig. 61-4 and Ch. 48). Virus then increases exponentially, reaching a maximum titer (from 250 to 2500 plaque-forming units per cell) by approximately 15 hours after infection. Infected cells are not rapidly lysed following viral replication, and release of infectious virus is incomplete.

The propagation of a double-stranded RNA genome is unique. The viral RNA cannot function as messenger, as does poliovirus single-stranded RNA, but like DNA must first be transcribed into mRNAs. Initially (beginning 2 to 4 hours after infection) four segments of the viral genome are primarily transcribed (one large, one medium, and two small single-stranded RNAs are made, corresponding to genomic segments I, VI, IX, and X seen in Fig. 61-2). An RNA polymerase **(transcriptase)** contained within the core of the virion is responsible for this "early" transcription, which is accomplished without synthesis of new proteins or replication of the viral genome. Presumably one or more of the mRNAs formed from the parental genome then functions for the production of a new RNA polymerase, which is transcribing all 10 segments of the viral genome by 6 hours after infection. The late mRNAs are of the same size classes as the pieces of the virion RNA (Fig. 61-2 and Table 61-2), and they hybridize specifically with segments of viral genome of corresponding sizes. Moreover, these mRNAs cannot self-anneal and therefore must be transcribed asymmetrically; and they are all copied at the same rate per unit length. Most of the newly synthesized single-stranded RNA molecules rapidly become associated with polyribosomes and appear to function as monocistronic mRNAs (termed plus strands).

The unusual nature of the viral RNA is further reflected in its replication. Unlike double-stranded DNA molecules (Ch. 48), reovirus double-stranded RNA is replicated **conservatively.** One of the strands is copied in great excess into single strands, termed plus because it does not hybridize with mRNA from polyribosomes. The minus strands are then copied onto these newly formed plus strands, from which they do not separate; hence minus strands cannot be detected as free single-stranded molecules. Replication of the viral RNA requires continuous protein synthesis,

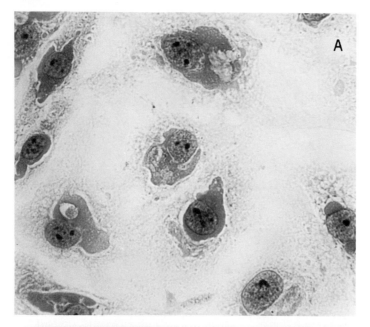

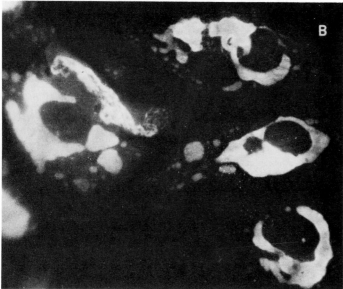

Fig. 61-3. Large eosinophilic cytoplasmic inclusion bodies in monkey kidney cells infected with type 1 reovirus. **A.** H & E stained and viewed with light microscope. ×1000. **B.** Stained with fluorescein-conjugated antibody and viewed with ultraviolet optics. ×1500. [From Rhim, J. S., Jordan, L. E., and Mayor, H. D. *Virology 17*:342 (1962).]

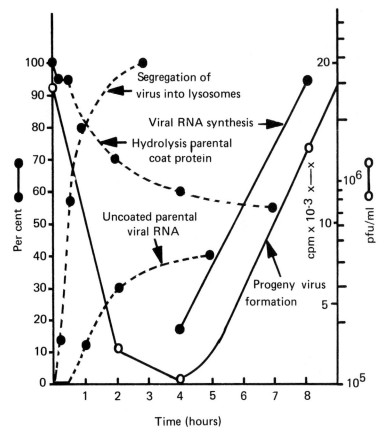

Fig. 61-4. Sequential events in the uncoating and replication of reovirus. [From Silverstein, S. C., and Dales, S. *J Cell Biol 36*:197 (1968).]

apparently for production of the unique virus-encoded replicase and transcriptase molecules. Newly replicated RNA molecules of all 10 size classes are found in infected cells, further strengthening the evidence that the viral genome is indeed segmented. The remarkably high **recombination** rate of **temperature-sensitive mutants** (i.e., 3 to 50%), as with influenza viruses (Ch. 56), further confirms the segmented structure of the viral genome.

About 75% of the newly synthesized single-stranded RNA molecules (plus strands) become associated with cytoplasmic polyribosomes, which serve as the sites for synthesis of viral proteins. All 7 capsid structural proteins can be detected in infected cells as early as 3 hours after infection, but only 6 of these are primary gene products: μ_2 (Table 61-2) is derived from μ_1 by cleavage. In addition, 2 noncapsid viral polypeptides, whose functions are unknown, are also synthesized. Since 10 species of messengers are transcribed, either the 2 additional primary gene products are not translated, or they

are not detectable by the methods currently employed.

The newly synthesized double-stranded RNA (identified by acridine orange staining) accumulates in large masses with excess viral antigens, forming the characteristic cytoplasmic inclusion bodies (Fig. 61-3).

Infectious virions begin to appear 6 to 7 hours after infection (Fig. 61-4), but how the genomic segments (which may be likened to chromosomes) are properly segregated in the appropriate number and linear sequence remains unexplained.

Reovirus infection inhibits biosynthesis of host cell DNA within 6 hours, although host protein synthesis appears unaltered. Along with the inhibition of host DNA synthesis, cell division ceases. In such cells the mitotic index increases more than threefold, but the mitotic sequence is not completed and abnormal mitotic figures form. Excess viral anti-

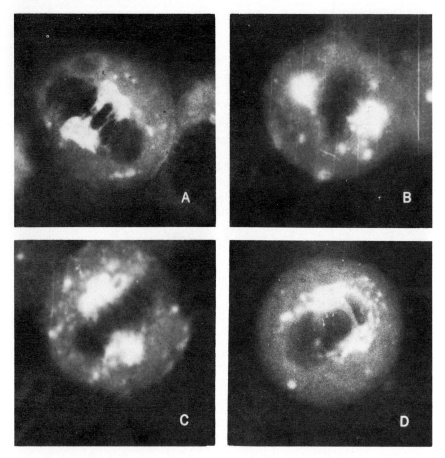

Fig. 61-5. Reovirus-infected cells stained with fluorescein-conjugated antibody. Viral antigen is closely associated with the mitotic spindle of virus-infected cells. Cells were examined by darkfield microscopy with ultraviolet illumination. ×750. [From Spendlove, R. S., Lennette, E. H., and John, A. C. *Immunol 90*:554 (1963).]

gens accumulate and viral particles assemble in close association with the spindle tubules (Figs. 61-5 and 61-6). However, the mitotic spindle is not essential for viral multiplication since viral synthesis proceeds unhindered in nondividing cells arrested in metaphase by colchicine.

PATHOGENESIS

Reoviruses have been isolated from the feces and respiratory secretions of healthy persons, as well as from patients with a variety of clinical illnesses, particularly minor upper respiratory and gastrointestinal diseases. Despite the frequent isolation of these viruses their relation to disease is not yet clear. Human transmission experiments have also failed to clarify the pathogenetic role of reoviruses. In one study, for example, in which young adults were inoculated intranasally with reoviruses (type 1, 2, or 3), mild afebrile respiratory illnesses occurred in approximately ⅓ of the volunteers, but the symptoms were too irregular and the illnesses too mild to establish a causative role of the inoculated viruses.

LABORATORY DIAGNOSIS

Reoviruses may be isolated from throat washings or fecal specimens by means of cell cultures (e.g., human embryonic or monkey kidney), and they are usually identified as belonging to the reovirus group by complement-fixation tests. The specific immunological type

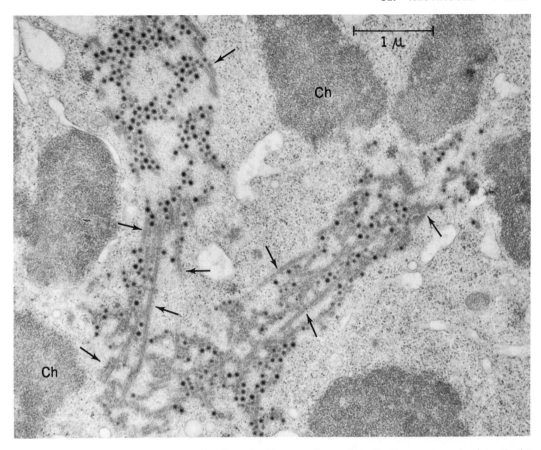

Fig. 61-6. Electron micrograph of cell infected with type 3 reovirus. Section was made through the spindle and chromosomes (Ch) of an infected cell in mitosis showing aggregates of viral particles closely associated with the tubules of the spindle (indicated by arrows). ×15,000. [From Dales, S. *Proc Natl Acad Sci 50*:268 (1963).]

can then be identified by hemagglutination-inhibition or neutralization tests.

To permit recognition of Abs in patients' sera by hemagglutination-inhibition titrations sera should be pretreated with trypsin or periodate, or adsorbed with kaolin, to remove nonspecific mucoprotein inhibitors. Since reoviruses are also frequently isolated from healthy persons, an increase in serum Ab titer must be demonstrated before an illness can be assumed to be due to a reovirus.

EPIDEMIOLOGY AND CONTROL

Though reovirus infections do not seem to be of great clinical importance, further studies are needed to define their pathogenetic potential. Both the respiratory and the gastrointestinal tracts may well be a source of their spread. Unrecognized reovirus infections are common, for approximately 10% of children by 5 years of age and 65% of young adults in the United States have reovirus Abs in their sera. Antibodies are also frequently found in various wild and domestic animals, but it is not known whether the animals serve as a reservoir for human infections.

With the meager data associating these viruses with disease in humans, specific immunization procedures are presently unwarranted.

OTHER VIRUSES SIMILAR TO REOVIRUSES

COLORADO TICK FEVER VIRUS

Colorado tick fever, the only tick-borne viral disease recognized in the United States, occurs predominantly in the Rocky Mountain region. (Powassan virus, a group B togavirus, has been isolated in Canada and is the only other tick-borne virus isolated in the Western Hemisphere.) Colorado tick fever is an acute febrile, nonexanthematous infection characterized by acute onset of fever, chills, headache, and severe pains in the muscles of the back and legs. The course of the disease is short and recovery is complete.

The virions are 750 to 800 A in diameter; the surface consists of regularly spaced, short projections; and the outer capsid appears to be icosahedral. As with reoviruses the virion contains a dense core covered with an inner icosahedral capsid. Colorado tick fever virus not only resembles reovirus morphologically, but its RNA is also double-stranded; whether the RNA is fragmented or a single molecule has not yet been determined. Virions multiply in large numbers free in the cytoplasm and unassociated with cell membranes, unlike groups A and B togaviruses. Only very few viral particles have detectable envelopes or are present within membranous enclosures. The virus multiplies readily in hamsters, suckling and adult mice, and some continuous human cell lines.

Colorado tick fever virus is ether-labile; but unlike togaviruses and viruses of the Bunyamwera supergroup, it is not inactivated by sodium deoxycholate. It is more stable at 4° and room temperature than the togaviruses: it remains infectious for as long as 64 days at room temperature when diluted in 10% rabbit serum in saline; and it is readily isolated from human clotted blood after 3 to 4 days at room temperature. Colorado tick fever virus is immunologically distinct from all other viruses studied. Infection induces long-lasting, probably life-long, immunity.

Dermacentor andersoni is the major vector, and the disease has been reported in the western United States where this tick has been found. The virus has also been isolated from the tick *D. variabilis* collected on Long Island, New York, but no human infections have been reported from this locality. The golden ground squirrel appears to be the major animal reservoir, and the virus has also been isolated from chipmunks, other squirrels, and a deer mouse. Infection of man is only incidental and is a dead end in the chain of transmission.

Prevention of the disease is directed primarily toward avoiding ticks, either by not entering infested areas or by wearing suitable clothing and using arthropod repellents.

SELECTED REFERENCES

Books and Review Articles

GRAHAM, A. F., and MILLWARD, S. Replication of reovirus. In *Nucleic Acid-Protein Interactions and Nucleic Acid Synthesis in Viral Infections,* p. 333. (D. W. Ribbons, J. F. Woessner, and J. Schultz, eds.) North Holland, Amsterdam, 1971.

SHATKIN, A. J. Viruses with segmented RNA genomes: Multiplication of influenza vs. reoviruses. *Bacteriol Rev 35:*250 (1971).

Specific Articles

FIELDS, B. H., and JOKLIK, W. K. Isolation and preliminary genetic and biochemical characterization of temperature-sensitive mutants of reovirus. *Virology 37:*335 (1969).

GOMATOS, P. J., and TAMM, I. The secondary structure of reovirus RNA. *Proc Natl Acad Sci USA 49:*707 (1963).

RHIM, J. S., JORDAN, L. E., and MAYOR, H. D. Cytochemical, fluorescent-antibody and electron microscopic studies on the growth of reovirus (ECHO 10) in tissue culture. *Virology 17:*342 (1962).

SCHONBERG, M., SILVERSTEIN, S. C., LEVIN, D. H., and ACS, G. Asynchronous synthesis of the complementary strands of the reovirus genome. *Proc Natl Acad Sci USA 68:*505 (1971).

SMITH, R. E., ZWEERINK, H. J., and JOKLIK, W. K. Polypeptide components of virions, top component and cores of reovirus type 3. *Virology 39:*791 (1969).

ZWEERINK, H. J., MCDOWELL, M. J., and JOKLIK, W. K. Essential and nonessential noncapsid reovirus proteins. *Virology 45:*716 (1971).

chapter 62

HEPATITIS VIRUSES

PROPERTIES OF THE VIRUSES   1410
PATHOGENESIS   1412
EPIDEMIOLOGY   1413
TREATMENT AND CONTROL   1414

Hepatitis remains one of the few known human viral diseases whose established causative agents have been neither transmitted reproducibly to laboratory animals nor cultivated consistently in tissue cultures. For these reasons, and because the disease tends to be sporadic, its infectious nature was long unrecognized. Until the 1940s the disease was believed to result from obstruction of the common bile duct by a plug of mucus (Virchow) and it was known as **acute catarrhal jaundice.** In 1942 Voeght first transmitted the disease by feeding a patient's duodenal contents to volunteers. Subsequently it was found that the etiological agents are filtrable and that the disease may be transmitted in two ways: through the intestinal-oral route **(infectious hepatitis)** or by injection of infected blood or its products **(serum hepatitis).** As will be noted below, however, these differences in transmission are not absolute, since the virus of infectious hepatitis can also produce disease when inoculated parenterally, and the virus of serum hepatitis can be transmitted orally.

PROPERTIES OF THE VIRUSES

Since human hepatitis viruses have been tested and propagated only in man, few of their characteristics are known (Table 62-1). The infectious particles are relatively **small,** filtration through calibrated gradocol membranes indicating that serum hepatitis virus has a diameter of approximately 260 A. They are also **unusually stable,** their infectivity being retained after treatment with ether, chlorine (1 part per million), merthiolate (1:2000), or nitrogen mustard (0.5 mg per milliliter); after storage at room temperature for 6 months or at -10 to $-20°$ for years; and after heating at $60°$ for 4 hours. Infectivity is destroyed by heating at $100°$ for 30 minutes or by treatment with tricresol (0.2%), β-propiolactone (4 mg per milliliter), or ethylene oxide (liquid or gas). They clearly are not enveloped viruses, and they are more stable than any other virus known to infect man.

The agent causing infectious hepatitis is easily transmitted to human volunteers by both the oral and the parenteral routes, and it appears in both the blood and the feces. The agent of serum hepatitis, in contrast, can be transmitted regularly only by parenteral inoculation, though epidemiological evidence suggests that it is also disseminated by personal contact. Moreover, ingestion of infected plasma has induced hepatitis.

Cross-immunity has not been detected between infectious and serum hepatitis viruses, but unfortunately this result was not decisive: cross-protection was tested with only a few strains of viruses and in only a few volunteers.

Because of these apparent differences the viruses are considered sufficiently distinct to justify separate names: **IH (or A)** and **SH (or B)** virus, for infectious and serum hepatitis, respectively. Unhappily, antigenic properties cannot be used to define their relation, for though extensive efforts have been made to isolate and propagate hepatitis viruses in a large variety of laboratory animals and mammalian cell cultures, the few successes reported have not been confirmed. Among these, particular attention must be given to the reports that IH(A) virus was serially transmitted in marmosets, and that SH(B) virus can be serially passed in rhesus monkeys and chimpanzees.

Hepatitis-associated Antigen. Additional evidence for the existence of two distinct hepatitis viruses evolved from the discovery of the **hepatitis-associated antigen (HAA),** also called the **Australian, SH or HB (hepatitis B) anti-**

gen, which appears only in patients with serum hepatitis.

This antigen was first detected by Blumberg in the serum of an Australian aborigine. In a search for new serum isoantigens he happened to employ test sera from two hemophiliacs who had received multiple blood transfusions. These sera, which contained antibodies to the so-called Australian antigen, were then found to react with sera from a variety of patients who had received multiple transfusions or had been resident in institutions (e.g., for mental defectives or for lepers) in which the inmates had a high incidence of hepatitis. Detection of Australian antigen was then recognized as signaling the presence of active or inactive serum hepatitis.

Discovery of the novel antigen has permitted the diagnostic identification of serum (B) hepatitis, but it has not yet resulted in the isolation and continuous propagation of the causative agent. The **properties of HAA,** however, have some similarities to viruses. Electron micrographs of HAA purified from serum by isopyknic centrifugation (Fig. 62-1) reveal many spherical (probably icosahedral) particles with an average diameter of 200 A

and occasional filamentous structures of variable lengths and a diameter of 200 A. Rare particles are larger spheres (420 A in diameter) with cores of 280 A. The particles appear to be composed of subunits (Fig. 62-1B). Although organic solvents do not inactivate viral infectivity, ether significantly reduces the diameter of the HAA spherical particles, converts the filamentous particles to small spheres, and increases the buoyant density of HAA. These data imply that HAA contains lipid, and suggest that the antigen is not identical with the infectious viral particles.

Purified HAA particles contain two major and one minor proteins, but neither DNA nor RNA has been detected. Immunological analyses of particles from many different sera reveal that HAA is antigenically complex: all particles reveal one common antigenic determinant (a); but additional determinants identify three immunological subgroups of particles (D, Y, and W). Although HAA morphologically resembles viral particles, the absence of nucleic acid and the lipid content

Fig. 62-1. Particles of hepatitis-associated antigen negatively stained. **A.** The majority of particles are about 200 A, but occasional large spheres (arrow) or filamentous particles (double arrow) are seen. ×200,000. **B.** A few particles suggest the presence of inner structures, and one particle appears to be "empty." ×310,000. [**A** courtesy of Dr. Manfred F. Bayer, Institute for Cancer Research, Philadelphia. **B** from Bayer, M. E., Blumberg, B. S., and Werner, B. *Nature 218*:5146 (1968).]

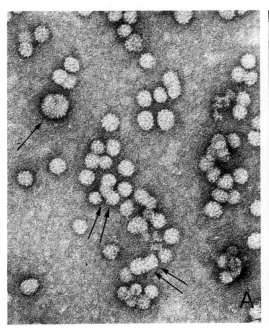
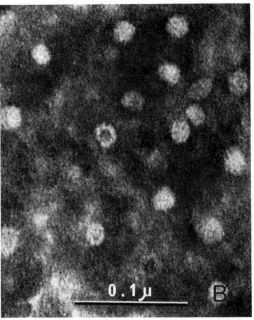

make it an unlikely candidate for the infectious virion of serum hepatitis. The HAA does appear, however, to be formed in response to a specific viral infection; perhaps it consists of aggregates of viral proteins, or even altered host proteins, associated with cell membranes.

The detection and assay of HAA depends upon reactions with Abs in sera from persons who have received multiple transfusions of blood containing HAA, or in sera from rabbits immunized with purified HAA. Complement fixation, immunoelectrophoresis (the technic commonly used, counterelectrophoresis, consists of electrophoresing the antigen and antibodies toward each other to hasten the reaction and increase the sensitivity), gel immunodiffusion, and radioimmunoprecipitation are commonly employed.

We cannot be sure whether the hepatitis viruses are separate viral "species," which presumably separated in evolution long ago, or whether they are recently differentiated variants from a single virus, selected by repeated gastrointestinal or hematogenous passage in man. The latter derivation provides a plausible origin for the B virus of serum hepatitis: the use of needles, and the transfusion of blood or blood products, are historically very recent developments; and unless serum hepatitis (B) virus is transmitted by personal contact or by insects more extensively than is presently apparent, it is difficult to comprehend how it could have arisen and survived without the modern physician and dentist as its vectors.

Although viral hepatitis is known to occur in a variety of animals (dogs, mice, ducks, and turkeys), none of the agents involved appear to be related to human IH (A) or SH (B) viruses. Indeed, mouse hepatitis virus has been sufficiently characterized to distinguish it clearly from either IH (A) or SH (B) virus and to classify it as a coronavirus (Ch. 58), and canine hepatitis virus is an adenovirus (Ch. 52).

PATHOGENESIS

The response of man to infection with hepatitis viruses ranges from inapparent infection and nonicteric hepatitis to severe jaundice, liver degeneration, and death; the recognized disease is often debilitating and convalescence is prolonged. Though minor clinical differences in infectious (A) and serum (B) hepatitis have been described, the distinction must be based at present on 1) the length of the incubation period, which is usually 15 to 40 days in infectious hepatitis and 60 to 160 days in serum hepatitis (Table 62-1), and 2) the presence of HAA in blood and liver cells of patients with serum (B) hepatitis.

Infectious hepatitis (A) virus usually enters by the oral route and multiplies in the epithelium of the gastrointestinal tract; viremia eventually occurs, and virus spreads to cells of the liver, kidneys, and spleen. Virus can be detected in duodenal contents and feces, and also in blood and urine, during the preicteric and the initial portion of the icteric phases, and may persist in the blood and rarely in the feces for many months.

In **serum hepatitis** the B virus is usually present in the blood during the preicteric and icteric stages of the disease and for months or even years afterward. HAA appears in the blood during the latter half of the incubation period, 5 to 8 weeks after infection, and usually disappears before acute symptoms subside or liver function returns to normal. However, in 10 to 20% of adults and about 35% of children HAA persists for prolonged periods. Apparently infection with small viral doses most frequently results in mild or nonicteric hepatitis and persistence of HAA. HAA has also been identified in feces, urine, and bile. As was already noted, however, serum hepatitis (B) virus has not been detected in the feces.

Both infections primarily affect the liver, which gives rise to the signs and symptoms observed. When jaundice occurs it is usually preceded by anorexia, malaise, nausea, diarrhea, abdominal discomfort, fever, and chilliness. This preicteric phase may last from 2 days to 3 weeks. The icteric stage of infectious hepatitis usually has an abrupt onset with a sharp rise in temperature, whereas that of serum hepatitis characteristically appears more insidiously and with less fever. Nevertheless, serum hepatitis is ordinarily a more severe disease, the fatality rate sometimes being as high as 50%, whereas in infectious hepatitis fatalities rarely exceed 1 in 100 (although convalescence is often prolonged). These differences, however, may not reflect the characteristics of the viruses, for persons receiving blood or plasma transfusions are usually ill at the time of inoculation.

Liver biopsies obtained during the course of the illness have revealed early cloudy swelling and fatty metamorphosis at the time of initiation of

TABLE 62-1.   Characteristics of Infectious and Serum Hepatitis

Feature	Infectious (A) hepatitis	Serum (B) hepatitis
Communicability	Contagious	Contagious (minimal)
Characteristic incubation periods*	15–40 days	60–160 days
Onset	Acute	Insidious
Fever, over 38°	Common	Uncommon
Seasonal incidence	Autumn and winter	All year round
Age incidence	Commonest in children and young adults	All ages; commonest in adults
Host range	Man	Man
Size of virus	Not known	About 260 AΔ
Virus in feces	Incubation period and acute phase	Not demonstrated
Virus in blood	Incubation period and acute phase	Incubation period and acute phase
Duration of carrier state		
Blood	Unknown (at least 8 months)	As long as 5 years
Feces	Weeks to months	Not demonstrated
Appearance of hepatitis-associated antigen (HAA)	Absent	30–50 days after infection
Detection of HAA		Blood (less often in feces, urine, and bile)
Duration of HAA		60 days to years
Immunity		
Homologous	Present	Uncertain
Heterologous	None apparent	None apparent
Prophylactic value of γ-globulin	Good	Experimentally successful†

* Considerable overlapping in the incubation periods has been noted in volunteers as well as in patients during epidemics, i.e., as long as 85 days for infectious hepatitis and as short as 20 days for serum hepatitis.

Δ Determined by filtration through calibrated membranes.

† Two injections used, the first shortly after transfusion and the second 1 month later.

clinical symptoms and diffuse parenchymal destruction by the time jaundice has developed. No inclusion bodies are seen in affected cells; intracellular HAA, however, can be detected by immunofluorescence early in the course of serum (B) hepatitis, and occasionally 200 to 300 A virus-like particles (possibly the structures containing HAA) are seen in liver biopsies. The degeneration of cells is not localized in any one part of the liver lobule, in contrast to yellow fever (Ch. 60, Group B togaviruses, pathogenesis) or chemical hepatitis. With recovery marked regeneration of hepatic cells occurs; scar tissue (cirrhosis) develops only after extensive or long-standing destruction of cells. If the infection is fatal, the liver parenchyma is often almost completely destroyed (acute yellow atrophy).

EPIDEMIOLOGY

It is unfortunate that epidemiological names have been employed to designate the two forms of hepatitis, since both viruses can be transmitted by either the oral or the parenteral route. Predominantly, however, virus **A**, the virus of infectious hepatitis (the short incubation disease), is spread by ingestion (particularly in epidemics) and virus B, the serum hepatitis agent, which is more often responsible for sporadic disease, is disseminated by parenteral inoculation.

Infectious hepatitis mimics poliomyelitis in many of its epidemiological features. When environmental factors favor widespread intestinal-oral transmission the disease is endemic, and infection (usually mild or inapparent) occurs in the very young. Under these conditions the disease in adults is uncommon and epidemics are rare. On the other hand, when sanitary conditions are good spread of the virus is restricted, and adulthood is frequently attained without immunity. In such nonimmune populations, especially in military groups, epidemics are likely, and the source of virus can usually be traced to contamination of water or food by human carriers. Because **subclinical infections,** particularly in children, often predominate, secondary person-to-person spread and contamination of food and drink are common. Such transmission is particularly favored owing to the unusual stability of hepatitis viruses and their unique resistance to ordinary concentrations of disinfectants, such as chlorine in water. Not only contaminated water but also the shellfish that breed in it (and concentrate the virus) may be sources of infection: for example, raw oysters and clams obtained from polluted waters have been the origin of numerous epidemics throughout the world.

Subhuman primates, particularly the chimpanzee, are the only known nonhuman hosts of hepatitis virus. Hepatitis has developed among handlers 3 to 6 weeks after the animals arrived in the United States from Africa. Apparently the young animals were infected by man after capture; they excreted virus in their feces, and some showed clinical manifestations of disease.

Infected blood and its products are major sources of hepatitis in the United States. Since unrecognizable chronic viremia frequently follows hepatitis, both serum and infectious, there exists a great danger of spreading infection by transfusions of infected blood, plasma, or convalescent serum; by utilization of contaminated fibrinogen; and by use of inadequately sterilized syringes, needles, or instruments (medical and dental) containing traces of contaminated blood. The last group includes common stylets for blood counts or syringes for drawing blood, needles used in tattooing, and communal equipment used for self-inoculation by narcotic addicts.* Injection of as little as 0.0000001 to 0.000001 ml of contaminated blood may transmit infection to a susceptible recipient.

Despite the clear association of HAA with serum hepatitis, the antigen can be detected in only approximately 25% of cases of posttransfusion hepatitis. It thus appears that the virus of so-called infectious (A) hepatitis, and perhaps other still unknown viruses, may be responsible for many of the cases diagnosed as serum hepatitis.

Although serum hepatitis is usually a

---

* Overt hepatitis follows about 1% of blood transfusions in the United States. On the assumption that even more inapparent infections are transmitted to recipients, it has been estimated that 2 to 3% of the adult population carries a hepatitis virus.

sporadic disease it may occur in epidemics when many samples of serum or plasma are pooled and distributed widely. For example, in the early 1940s more than 28,000 cases of serum hepatitis in American military personnel resulted from the use of yellow fever vaccine containing contaminated human serum to stabilize the live virus. The use of hemodialysis units presents a more modern hazard: patients and staff alike are repeatedly exposed to contaminated blood; hepatitis frequently follows, and HAA is present in the majority of cases.

TREATMENT AND CONTROL

No specific therapy is available for either infectious or serum hepatitis.

In all proved and suspected cases of viral hepatitis great care should be taken in the disposal of feces and of all syringes, needles, plastic tubing, and other equipment used for blood sampling and parenteral therapy. Whenever possible, disposable equipment (including needles and plastic syringes) should be routinely used in hospital and office practice. A syringe, once used, should not be reused with a fresh needle, even merely to obtain a blood specimen. Nondisposable equipment should be autoclaved at 15-lb pressure (121°), boiled in water for at least 20 minutes, or heated to 180° for 1 hour in a sterilizing oven. Subjects giving a history of jaundice or with detectable HAA in their blood should not be used as blood donors (blood HAA has been reported in about 1% of blood donors in New York and 0.2 to 1.2% of different groups of blood donors in Tokyo; the highest carrier rate occurs in commercial donors, particularly drug addicts). It should be noted, however, that the complement-fixation assay for HAA will detect only 40 to 80% of the chronic carriers of a hepatitis virus. Because minute amounts of contaminated plasma can initiate infection, the practice of pooling plasmas should be avoided. If this recommendation is followed plasma from an infected individual, used unwittingly, will infect only one person rather than many.

Infectious (A) hepatitis may be prevented by passive immunization. Pooled human γ-globulin* reduces the incidence of icteric disease (but not of infection) when given early in the incubation period. The recommended dose is 0.06 to 0.15 ml per pound of body weight, administered by intramuscular injection as soon after exposure as possible. Inconsistent results have been reported concerning prophylaxis of **serum (B) hepatitis** by passive immunization. But the addition of 10 ml of 6% γ-globulin to each unit of blood transfused appears to reduce hepatitis significantly.

---

* Infection with hepatitis is so widespread that sera of most adults contain Abs, and a relatively high titer is present in concentrated γ-globulin pools. Hepatitis virus has not been detected in these pools, having been previously precipitated with the fibrinogen in the usual cold ethanol fractionation of plasma.

The discovery of HAA, however, may make active immunization possible. When serum containing the antigen was heated at 98° for 1 minute the infectivity but not the antigenic reactivity was destroyed, and children immunized with this material became resistant to a subsequent challenge.

The continued prevalence of hepatitis (e.g., in the United States 58,000 cases were reported in 1969), and its serious consequences, enhance the challenge to cultivate these recalcitrant viruses. Despite the discovery of HAA, it is unlikely that the viruses causing these diseases will be unambiguously distinguished until they can be propagated in an experimental animal or a cell culture. It appears equally improbable that the diseases will be controlled until this difficult challenge is met.

SELECTED REFERENCES

Books and Review Articles

KOFF, R. S., and ISSELBACHER, K. J. Changing concepts in the epidemiology of viral hepatitis. *N Engl J Med 278:*1371 (1968).

KRUGMAN, S., and GILES, J. P. Viral hepatitis: New light on an old disease. *JAMA 212:*1019 (1970).

LEBOUVIER, G. L., and MCCOLLUM, R. W. Australia (hepatitis-associated) antigen: Physicochemical and immunological characteristics. *Adv Virus Res 16:*357 (1970).

SHULMAN, N. R. Hepatitis-associated antigen *Am J Med 49:*669 (1970).

Specific Articles

GROB, P. J., and JEMELKA, H. Fecal SH (Australia) antigen in acute hepatitis. *Lancet 1:*206 (1971).

HOLMES, A. W., WOLFE, L., DEINHARDT, F., and CONRAD, M. E. Transmission of human hepatitis to marmosets: Further coded studies. *J Infect Dis 124:*520 (1971).

LEBOUVIER, G. L. The heterogeneity of Australia antigen. *J Infect Dis 123:*671 (1971).

Unity of Oncogenic Viruses   1418

DNA-CONTAINING VIRUSES   1419

PAPOVAVIRUSES: POLYOMAVIRUSES   1419
  Properties of the Virions   1420
  Effects in Animals: Pathogenesis   1420
  Viral Multiplication in Tissue Cultures   1420
  Cell Transformation   1422
  Mechanism of Transformation   1424
  Viruses with Hybrid DNA-Pseudovirions   1428
  Comparison of Animal Cell Transformation with
    Bacterial Lysogenization   1428
  Role of Cellular Genes in Transformation   1429
PAPOVAVIRUSES: PAPILLOMAVIRUSES   1429
  Viral Multiplication and Pathogenesis   1429
OTHER DNA VIRUSES   1430

RNA-CONTAINING LEUKOVIRUSES   1430

LEUKOSIS VIRUSES   1430
  Properties of the Viruses   1430
  Effects of Infection in Tissue Cultures   1432
  Effects of Infection in Animals: Gross Leukemia Virus   1432
  Other Leukemia Viruses   1433
  Latency of Leukosis Viruses   1434
  Significance of Leukovirus Latency   1435
SARCOMA VIRUSES   1436
  Nature and Origin of Sarcoma Viruses   1436
  Defectiveness of Sarcoma Viruses   1437
  Mode of Action of Sarcoma Viruses   1440
MOUSE MAMMARY TUMOR VIRUS   1441
  Properties of the Virus   1441
  Pathogenesis   1442

**IMMUNOLOGICAL RELATIONS OF VIRUS-TRANSFORMED
    CELLS TO ANIMAL HOST   1445**

ROLE OF VIRUSES IN NATURALLY OCCURRING CANCERS   1446

The first tumor-producing (oncogenic) virus was discovered in 1908 by Ellerman and Bang, who demonstrated that seemingly spontaneous leukemias of chickens could be transmitted to other chickens by cell-free filtrates. Later (1911) Rous found that a chicken sarcoma, a solid tumor, can be similarly transmitted. The occurrence of virus-induced tumors in chickens was then considered by many as a biological curiosity, either not "true" cancers or perhaps a peculiarity of the species. These notions, however, were shaken by the discoveries of the viral induction of cutaneous fibroma and papilloma of wild rabbits (Shope, 1932) and of the renal adenocarcinoma of the frog (Lucké, 1934).

The discovery of virus-induced tumors in mice provided a particularly suitable system for experimental work. In 1936 Bittner demonstrated that a spontaneously occurring mouse adenocarcinoma is caused by a virus transmitted from mother to progeny through the milk, and in 1951 Gross discovered that a lymphatic leukemia of mice is virus-induced. Subsequently many other virus-induced murine leukemias were discovered. These studies revealed an important characteristic of viral oncogenesis—that the viral etiology of a cancer can easily go unrecognized for several reasons: 1) cancer can be caused by ubiquitous viruses, which may easily be considered as innocuous bystanders; 2) with some viruses most viral particles can infect cells without inducing cancer; 3) the disease may not develop until long after infection; and 4) the cancers do not seem contagious, either because the efficiency of cancer production is low or because the method of transmission of the virus from one individual to another is inapparent (e.g., through the embryo or the milk).

The murine cancer viruses noted above were found to contain RNA. Further impetus to investigations of tumor viruses arose from the more recent discovery that several DNA-containing viruses also cause cancer in mice and other rodents. Thus polyoma virus was isolated from leukemic mice in 1959, long after extracts of various organs had been shown to induce leukemia and parotid tumors in mice. Simian virus 40 (SV 40) was later discovered as a passenger virus in cultures of rhesus monkey kidney cells. Even more attention was aroused when **human** adenoviruses were shown to induce cancer in hamsters, mice, and rats.

By this time the study of bacterial lysogeny had clearly shown that the genetic material of viruses can become permanently integrated with that of the host as a new set of genes of the cell. Among the profound implications of this work was its possible relation to the induction of cancer by viruses. Furthermore, the demonstration of viral carcinogenesis in many animal species suggested that it might also occur in man. These realizations led to an explosive development of interest in the subject.

Soon the oncogenic effect of several viruses was demonstrated also in tissue cultures in the form of readily producible **cell transformation** (Ch. 47). Studies of viral oncogenesis in this model system led shortly to a shattering conclusion: a virus that has induced cancer is often no longer recognizable in the culture by its best known properties, i.e., infectivity and antigenicity. Traces of the virus could, however, be found in the form of viral DNA, RNA, and new antigens (Ags), in ways reminiscent of lysogeny. It thus became clear that time-tested technics and approaches for the identification of viral agents of disease may not be adequate in the search for viral agents of human cancer.

UNITY OF ONCOGENIC VIRUSES

The discovery of many new oncogenic viruses in the sixties revealed a puzzling distribution:

TABLE 63-1.   Distribution of Oncogenic Viruses among Animal Virus Families

Nucleic acid in virions	Viral group (or family)	Oncogenic viruses
RNA	Picornaviruses	None
	Togaviruses	None
	Myxoviruses	None
	Paramyxoviruses	None
	Rhabdoviruses	None
	Coronaviruses	None
	Arenaviruses	None
	Leukoviruses	Leukosis viruses, sarcoma viruses
	Reoviruses	None
DNA	Adenoviruses	Many types
	Papovaviruses	Polyoma virus, SV 40, papilloma virus
	Herpesviruses	Virus of neurolymphomatosis of chicken (Marek's disease)
		Luckés virus of frog renal adenocarcinoma
		Herpes simplex virus type 2 (transformation in vitro after UV irradiation)
		Possibly Epstein-Barr virus (Burkitt lymphoma)
	Poxviruses	Fibroma virus
	Parvoviruses	None

oncogenic members were found in most classes of DNA-containing viruses, but in only one family of RNA-containing viruses, the leukoviruses (Table 63-1). The replication of leukoviruses also displayed a sensitivity, peculiar for RNA viruses, to agents that interfere with DNA replication or transcription. To explain this property Temin proposed that oncogenic RNA viruses, unlike other RNA viruses, replicate through a DNA-containing intermediate, and eventually he and Baltimore each discovered the predicted **reverse transcriptase** (RNA-dependent DNA polymerase, see Ch. 11) in the virions. This discovery brought unity into the field of oncogenic viruses, suggesting that **oncogenesis is an attribute of viral DNA.**

These discoveries, and the problems they raise, will be analyzed below by examining the characteristics of cell transformation induced by several viruses, either in tissue culture or in the animal (i.e., cancer or leukemia). Each system considered conveys a different message and allows the separation of the complicated problem of viral carcinogenesis into distinct and more easily analyzable facets. The experimental findings to be discussed clearly have strong implications for the possible role of viruses in human cancer; for though none of these findings were obtained in man, some were obtained in cultures of human cells, and some were obtained with viruses that can infect man.

DNA-CONTAINING VIRUSES

These viruses are outstanding models for understanding the mechanisms of viral carcinogenesis. Adenoviruses and herpesviruses are reviewed in Chapters 52 and 53. Viruses of the papova group, which have been used most extensively in experimental work, are reviewed here in this section.

PAPOVAVIRUSES: POLYOMAVIRUSES

A virus isolated by Stewart and Eddy was found to produce in mice neoplasias of different types when injected into newborn mice; it was therefore named polyoma virus (agent of many tumors). The virus is widespread in mouse populations, both wild and in the laboratory; it is normally transmitted to animals after birth, through excretions and secretions. The structurally similar SV 40 was discovered by Sweet and Hilleman as an agent that multiplies silently in rhesus monkey kidney cultures (used for propagating poliovirus) but produces cytopathic changes in similar cultures from African green monkeys (*Cercopithecus aethiops*). It was later shown to produce sarcomas after injection into newborn hamsters. The polyomavirus group includes polyoma virus and simian virus 40 (SV 40).

PROPERTIES OF THE VIRIONS

Virions of both polyoma virus and SV 40 are small naked icosahedrons, with a diameter of 450 A and with 72 capsomers (Ch. 44); they contain a cyclic double-stranded DNA molecule (MW 3.5×10^6) which is infectious when freed of capsid protein. Both viruses are very resistant to thermal inactivation and to inactivation by formalin; hence SV 40 (from rhesus kidney cultures) survived in some early batches of formalin-killed poliovirus vaccine.

Despite these similarities between the two viruses, their DNAs have somewhat different base ratios and do not show homology in hybridization experiments. Moreover, the viruses each occur in a single antigenic type, immunologically unrelated to each other. Polyoma virus agglutinates guinea pig red cells by attaching to receptors that are destroyed by neuraminidase, which also hydrolyzes receptors for myxoviruses (see Ch. 44, Hemagglutination).

EFFECTS IN ANIMALS: PATHOGENESIS

Polyoma virus inoculated subcutaneously into newborn animals of many species (e.g., mice, Syrian hamsters, rabbits, guinea pigs, ferrets, and Chinese hamsters) induces tumors, after a latent period of several months. These tumors include adenomas of the parotid, which are prominent in mice; sarcomas, localized at various sites, which often grow enormously but without invading the surrounding tissues; and hemangioendotheliomas of many organs, which may cause massive hemorrhages. Since all these tumors can be produced by virus that has been cloned by isolation from a tissue cul-

ture plaque, their variety seems to depend on the nature of the target cell. Polyoma virus can also cause a **runting disease,** with destruction of lymphoid and parenchymatous cells.

In contrast to the injected newborn, mice infected with polyoma virus through natural contacts do not produce polyoma-induced tumors (as judged from the lack of specific Ags), probably because the animals are already immunologically competent when infected (see below, Immunological relations of virus-transformed cells to animal host).

SV 40 has a less varied effect: it produces pleomorphic undifferentiated sarcomas several months after subcutaneous inoculation into newborn hamsters, and papillary ependymomas after intracerebral inoculation.

VIRAL MULTIPLICATION IN TISSUE CULTURES

Polyoma virus causes a **productive** infection, with production of infectious virus, and plaque formation in cultures of mouse embryo or kidney cells, which are therefore **permissive** (see Ch. 48). In contrast, it causes **transformation** in cultures of **nonpermissive** hamster and rat cells (Table 63-2). SV 40 causes productive infection in cultures of African green monkey kidneys, and it transforms nonpermissive cells of many animals: hamsters, mice, rabbits, cattle, pigs, and man. Productive infection and transformation are not always mutually exclusive in a given cell type: polyoma virus, for instance, causes the productive infection of a small proportion of cells in hamster embryo cultures. If productive infection can be prevented (see below), this virus may also transform permissive mouse cells and SV 40 permissive African green monkey cells.

Productive Infection. As with most DNA viruses, viral synthesis and assembly occur in the cell's nucleus, where viral capsid Ag is detected by immunofluorescence (Fig. 63-1) and virions by electron microscopy: each nucleus can produce up to 10^7 viral particles. The virus either remains in the nucleus or moves to the cytoplasm after rupture of the nuclear membrane. As with other naked viruses, release depends upon **disintegration of the cells.**

The characteristics of the multiplication cycle vary with the state of the host cells. In actively growing cells infected at high multiplicity the synthesis of viral DNA begins at 10 to 12 hours and

TABLE 63-2. Some Permissive and Nonpermissive Cells Used with Polyoma Virus and SV 40

Permissive	Polyoma virus	Secondary cultures of mouse embryo cells Primary cultures of mouse kidney cells 3T3 cells (subcutaneous tissue)
	SV 40	Primary cultures of African green monkey kidney cells BSC-1 (African green monkey kidney cells) CV-1 (African green monkey kidney cells) VERO (African green monkey kidney cells)
Nonpermissive	Polyoma virus	Secondary cultures of hamster embryo cells Secondary cultures of rat embryo cells BHK (baby hamster kidney cells)
	SV 40	Secondary cultures of hamster, rat, or mouse embryo cells 3T3 cells (mouse subcutaneous tissue)

The virus multiplies in the permissive cells producing infectious progeny and killing the cells. In nonpermissive cells it produces an abortive infection usually without cell killing, sometimes resulting in transformation.

is rapidly followed by that of capsids and virions, whereas in resting cells (such as crowded cultures) viral DNA synthesis is delayed and capsid synthesis follows after an additional lag.

In a mouse embryo culture the cells may vary widely in their susceptibility to productive infection by polyoma virus, as shown by the shape of curves relating the proportion of infected cells to input multiplicity (Fig. 63-2). Since the more sensitive cells are killed by the virus the resistant cells are thereby enriched, yielding a carrier culture (Ch. 51). Cultures derived from the surviving cells contain again a mixture of susceptible and resistant cells,

Fig. 63-1. Fluorescence photomicrograph of cultured mouse kidney cells productively infected by polyoma virus. The accumulation of capsid antigen in the nuclei is revealed by its combination with fluorescein-conjugated antibodies, which emit green fluorescence under ultraviolet light.

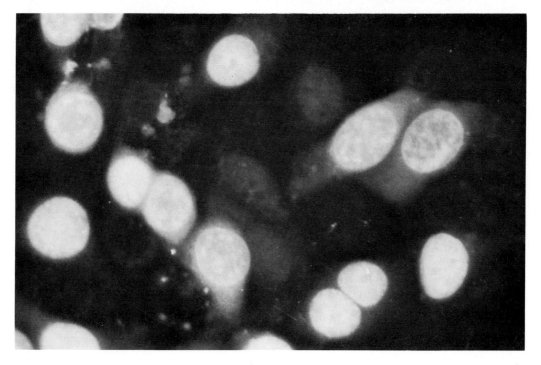

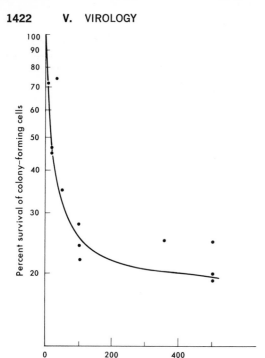

Fig. 63-2. Relation between multiplicity of infection (MOI) with polyoma virus and proportion of mouse embryo cells surviving (i.e., able to form a colony). Were all cells of equal susceptibility a straight line should be obtained in the semilogarithmic plot (Ch. 44, Appendix). The flattening out of the curve discloses heterogeneity of susceptibility, with a very large difference between the most and the least susceptible cells. [From Weisberg, R. A. *Virology* 21:658 (1963).]

showing that resistance is physiological rather than genetic.

Effect of Productive Infection on Cellular Syntheses. In cultures of crowded, resting cells infection stimulates the synthesis of cellular DNA to a level comparable to that found in uninfected growing cultures, and also stimulates formation of enzymes involved in DNA synthesis: thymidine kinase, deoxycytidylate deaminase, and DNA polymerase. Thus productive infection with these viruses, in contrast to virulent viruses, can apparently derepress several cellular genes involved in the synthesis of cellular DNA.

The viral function that stimulates cellular DNA synthesis in crowded cultures may also be involved in transformation, for a salient feature of transformed cells is loss of normal regulatory mechanisms which inhibit DNA

synthesis in uninfected crowded cultures (see Ch. 47).

T Antigen. In cells productively infected by polyoma virus or SV 40 a new Ag appears in the nucleus several hours before the capsid Ag (Fig. 63-3). Since the new Ag is also present in transformed and in tumor cells, it is referred to as the T (tumor) Ag. It does not cross-react with the viral capsid. Antibodies (Abs) to the T Ag (but not to capsid Ag) are present in the sera of hamsters bearing a large transplanted tumor induced by the same virus, probably through immunization by disintegrated cells; these Abs led to the discovery of the Ag in Huebner's laboratory. Since T Ag is formed in the presence of arabinosyl cytosine, which inhibits the synthesis of viral DNA and capsid, it is apparently an "early" viral protein (Ch. 48).

CELL TRANSFORMATION

Polyoma virus and SV 40 cause the formation of proliferative foci of transformed cells in a nonpermissive system (Table 63-2). The fraction of cells transformed is proportional to the titer of the infecting virus (in plaque-forming units determined in permissive cultures);

Fig. 63-3. Time course of the synthesis of T antigen and viral capsid antigen in cells of BSC-1 line (*Cercopithecus* monkey kidney) infected by SV 40. The synthesis of the T antigen precedes that of the capsid antigen. Both antigens were measured by complement fixation with appropriate sera. [Modified from Hoggan, D., Rowe, W. P., Black, P. H., and Huebner, R. J. *Proc Natl Acad Sci USA* 53:12 (1965).]

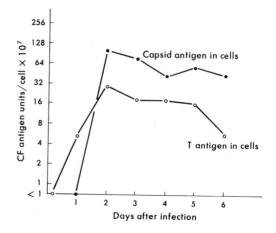

hence transformation is caused by a single virion (Ch. 44, Appendix). The efficiency of such a **stable transformation** is much lower than the plaque-forming efficiency of the virus (with permissive cells): the ratio varies between 10^{-3} and 10^{-5}, depending on the cells employed. The ratio is much higher in infection with viral DNA; hence the rarity of stable transformation with virus is evidently due to its poor ability to infect the nonpermissive cells. However, the virus also causes a **transient transformation** in permissive cultures, which affects all infected cells before they are killed by the infection; and in nonpermissive cells a stable transformation is much less frequent than **abortive transformation,** from which the cells recover and return to normality after a few generations.

The transformed cells are distinguished by the properties listed in Table 63-3. It is likely that many of the changes are interrelated. Thus lack of certain glycolipids in transformed cells probably derives from low activity of the corresponding glycosyl transferases; and the morphological differences (Fig. 63-4) probably depend on the concentration of cAMP, through its effect on the aggregation of the microtubules, which form the cell skeleton (see Ch. 47).

The differences of Table 63-3, observed in rather dense cultures of normal or transformed cells, tend to vanish in very sparse cultures, in which the properties of normal cells become much more like those assigned to transformed cells. This observation suggests that **the virus primarily affects the growth regulation of the cells,** in response to environmental conditions that normally limit growth.

Significance of Surface Changes. The surface characteristics of transformed cells are especially important for both the mechanism and the consequences of transformation. Since, as seen in Chapter 47, the surface is deeply involved in the regulation of cell growth, its changes may be responsible for the regulatory alterations of the cells, through malfunction either of the receptors for environmental regu-

TABLE 63-3. Distinguishing Characteristics of Normal and Transformed Cells

Characteristics	Normal cells	Transformed cells
Cultural behavior (see Ch. 47)		
Thickness	Single layer (though it may ultimately form several layers)	Multilayered
Cell orientation	Regular (parallel if cells are fusiform)	Random
Topoinhibition	High	Low
Serum requirement	High	Low
Efficiency of clone formation	High	Usually low
Cell morphology	Little indentation in the outline	High indentation
Growth in suspension in agar	No	Yes in some cases
Cell surface		
Glycolipids	Normal	Some absent
Glycosyl transferase activities	Normal	Some reduced
Agglutination by lectins	Low	High
Transport activity (e.g., glucose, phosphate, nucleosides)	Regulated by serum or hormones	Some not regulated, permanently active
Antigens	Normal	Some new
Biochemical characteristics		
Intracellular concentration of cAMP	High	Low
Acid production (aerobic glycolysis)	Low	High
Others		
Production of tumor by 10^6 cells inoculated subcutaneously in isogenic animals	No	Yes
Formation of chromosomal abnormalities	No	Yes

These differences are recognizable when **rather dense** cell cultures are compared and apply especially to fibroblastic cultures. Cells transformed by different viruses may differ from each other in some details.

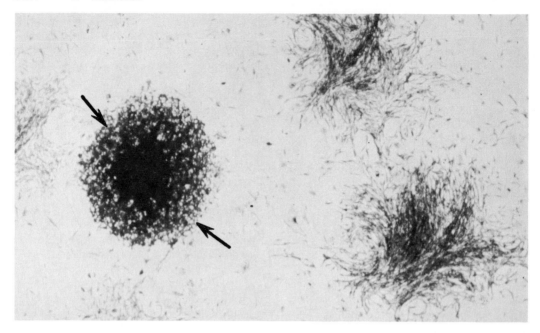

Fig. 63-4. Colonies of the BHK line (hamster kidney) infected by polyoma virus. One colony is transformed (arrows) and is recognizable by its considerable thickness and the random orientation of its cells. The untransformed colonies are thin and contain cells that tend to orient parallel to each other. (Courtesy of M. Stoker.)

latory signals or of the transmission of regulatory signals to the internal machinery. The other important role of the surface characteristics is to determine the fate of the transformed cells in the animal, as will be discussed below (Immunological relations of virus-infected cells to animal host).

The antigenic changes of the cell surface are due both to virus-specific Ags and to unmasking or derepression of cellular Ags. The increased agglutination by lectins (bivalent carbohydrate-binding proteins, see Ch. 47) seems to depend on the clustering of the carbohydrate receptors (which is recognizable by electron microscopy) without increased binding. Apparently clusters of lectin molecules bind the transformed cells to each other more strongly than individual lectins bind normal cells. Agglutination of untransformed cells by lectins is markedly increased by a mild trypsin treatment, showing that the features causing enhanced agglutination are potentially present, but masked, in these cells.

MECHANISM OF TRANSFORMATION

A partial elucidation has been achieved by combining molecular and biochemical studies with the isolation of temperature-sensitive mutants.

Role of Viral Genes. Complementation studies with **temperature-sensitive mutants** of polyoma virus identify four or possibly five genes, two or three early and two late (for capsid proteins). Acrylamide gel electrophoresis of extracts of cells productively infected by polyoma virus or SV 40 also reveals five new proteins, whose aggregate molecular weight closely approaches the coding capacity of the whole viral genome (Fig. 63-5). Hence the whole genome may consist of five genes.

Studies with mutants show that only the function of an early gene, *ts3* (from the name given a mutant) is **permanently** required for transformation, because cells transformed by the mutant retain a fully transformed phenotype at low temperature, when the gene is active but acquire, even after many growth generations, a nearly normal phenotype at high temperature, when the *ts3* gene is inactive (Fig. 63-6). Both the topoinhibition (density-dependent growth inhibition, see Ch. 47) and the agglutinability of the transformed cells by lectins are controlled by the incubation temperature. In lytic infection the *ts3* mutant controls the transient transformation, evidenced by the ability of the cells to replicate their

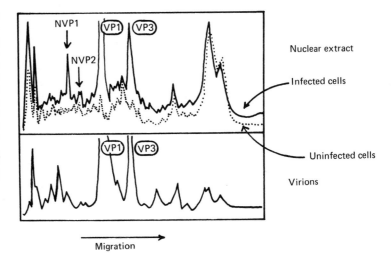

Fig. 63-5. New proteins in SV 40-infected cells. In the nuclear extract four new proteins (absent in extracts of un-infected cells) are recognizable: two (VP1, VP3) are also present in virions, two (NVP1 and NVP2) are not. A fifth new protein, recognizable in extracts of whole cells, is probably cytoplasmic. [Modified from Walter, G., Roblin, R., and Dulbecco, R. *Proc Natl Acad Sci USA* 69:921 (1972).]

Fig. 63-6. Consequences of the *tsa* and *ts3* mutation in polyoma virus on stable transformation of BHK cells. Cross-hatched, heavy outlines represent transformed colonies; outlines with thin, wavy lines represent untransformed colonies. The transformation stage gives the outcome of the interaction of the virus with the cells. The transfer stage shows the result of a temperature shift on colonies resulting from the transformation stage. The wild-type virus produces transformed colonies at either temperature, which are stable in further temperature shifts. The *tsa* mutant produces transformed colonies only at 32°; these are stable to subsequent transfer to 39°. The *ts3* mutant produces transformed colonies at 32°, but these revert to nearly normal when transferred to 39°; on the other hand colonies seemingly untransformed at 39° display transformation when subsequently transferred to 32°. Hence the function of the *tsa* gene is required transiently for transformation; that of the *ts3* continuously.

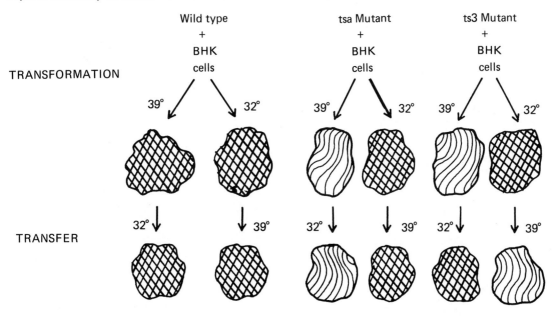

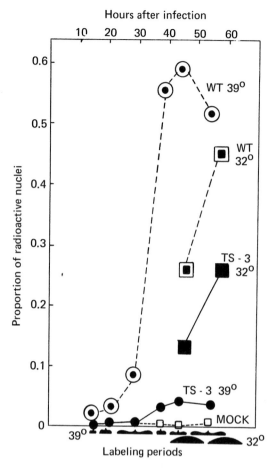

Fig. 63-7. Effect of the *ts3* mutation on the induction of cellular DNA synthesis in dense cultures of 3T3 cells. There is an almost complete inhibition at high temperature. Cellular and viral DNA were not separated; however, with the wild-type virus at either temperature or with the *ts3* mutant at low temperature at least half of the DNA made is cellular [Modified from Dulbecco, R., and Eckhart, W. *Proc Natl Acad Sci USA 67*:1775 (1970).]

DNA in dense cultures (Fig. 63-7). Hence a single viral protein controls both surface properties and growth properties of the cells, clearly showing their interrelation.

The function of another early gene, *tsa* (from the name of another mutant), is required **transiently:** cells are not transformed by the mutant at high temperature, but if transformed at low temperature they remain transformed when shifted to high temperature. The requirement is probably for integration into the cellular DNA (see below).

The *ts3* gene probably reflects the dependence of polyoma virus (and other papovaviruses) on cellular machinery for replicating its DNA. In animals the virus multiplies in cells not actively synthesizing DNA (e.g., in the salivary glands and kidneys), in which many enzymes for DNA replication and for supplying precursors have low activity. The function of the *ts3* gene boosts their activities on which the virus depends.

Integration of the Viral DNA into the Cellular DNA. Cells transformed by SV 40 contain a few viral genomes per cell as shown by hybridization with viral RNA made in vitro using viral DNA as template. During fractionation of the cells the viral DNA always follows the cellular DNA, even when completely denatured to single strands—i.e., in sedimentation in an alkaline sucrose gradient (see Ch. 10, DNA denaturation; Fig. 63-8). Hence the viral DNA is integrated, i.e., covalently bound to the cellular DNA as a **provirus,** equivalent to the prophage of lysogenic cells (Ch. 46). Integration can be considered a general property of polyoma viruses, since the cellular DNA contains viral sequences even during productive infection at high multiplicity.

The integration apparently occurs at several sites in the cellular DNA. They probably correspond to single-strand breaks, because the probability of transformation is markedly enhanced by creating many such breaks by X-raying the cells before infection. The integrated genome appears to be intact. This conclusion is supported by experiments with SV-3T3 cells, i.e., 3T3 cells (nonpermissive for SV 40) transformed by SV 40. When these cells are fused with permissive cells to provide conditions for the full expression of the viral genome, the heterokaryons yield infectious virus. Hence, as for prophage (see Ch. 46), integration and excision may occur by a reciprocal crossover, which allows conservation of the viral genetic information in its entirety.

With polyoma virus excision, like integration, may require the function of the *tsa* gene: thus polyoma-permissive 3T3 cells transformed by the *tsa* mutant of polyoma virus (*tsa*-3T3)* are stable at high temperature, but when the temperature is lowered, causing the appearance of active *tsa* product, the cells release virus and are killed. Hence for its role in integration and excision, the *tsa* gene bears some resemblance to the *int-xis* locus of phage λ (Ch. 46). However, it is also required for the replication of the viral DNA in productive infection.

---

* These cells, although permissive to the virus, are transformed because at high temperature the *tsa* mutant fails to replicate and to kill the cells.

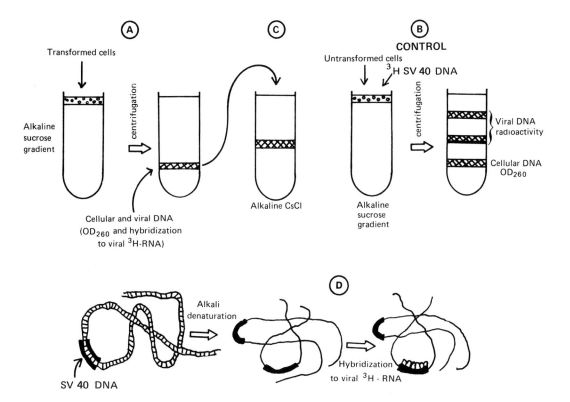

Fig. 63-8. Evidence for integration of SV 40 DNA in the DNA of transformed SV 3T3 cells: The main experiment **(A)** consisted of placing transformed cells on top of a preformed alkaline (pH 12.5) sucrose gradient, where they lysed; the gradient was then centrifuged to sediment the denatured cellular DNA toward the bottom of the tube. This technic, by avoiding the usual extraction of the cellular DNA, minimizes its breakage, so that (as shown in **B)** it sediments much faster than superhelical SV 40 DNA, in spite of its fast sedimentation after denaturation (see Ch. 10, Cyclic DNA). The band of cellular DNA contained all sequences hybridizable to viral RNA (made in vitro with *E. coli* RNA polymerase on SV 40 DNA); hence these sequences were covalently bound to the cellular DNA **(C):** that free DNA was recovered from the gradient was shown by sedimenting it at equilibrium in CsCl (see Ch. 10), where it had the density characteristic of cellular DNA. The molecular species involved in the experiment are shown in **(D)**, where viral sequences are represented by heavy lines and viral RNA (complementary only to one DNA strand) by wavy lines.

Significance of Integration in Transformation.

Integration may be essential only for stabilizing transformation by polyomaviruses because the *tsa* mutant, which apparently does not integrate, produces both the transient transformation of permissive cells and the abortive transformation of nonpermissive cells. It thus seems that whether the viral genome is integrated or free the expression of its genes results in the characteristic cellular changes of transformation.

Conditions Allowing a Stable Transformed State: Consequences of Integration.

Transformation reflects a limited effect of a potentially cytocidal virus: either because the cells are nonpermissive or because the virus is defective viral genes lethal for the cells are not expressed and the cells survive.

In at least some cells nonpermissiveness seems to depend on inability of cellular enzymes to transcribe all the viral genome. For instance, hybridization studies show that in the SV 40-transformed SV-3T3 cells only 40% of the integrated SV 40 viral genome is transcribed and no capsid Ag is made. Both viral and cellular sequences are transcribed in the same messenger molecules, which are longer than the viral genome. Transcription is probably limited by characteristics of the cellular transcriptase since a heterokaryon formed by fusion of SV-3T3 with permissive cells yields virus. Since transcription thus involves cellular initiation or termination signals, viral genomes integrated in an untranscribed part of the cellular genome (e.g. in heterochromatin, see Ch. 47) may remain unexpressed.

Transformation of permissive cells is made possible by mutations (such as the *tsa* mutation) which at high temperature block the expression of late viral genes lethal for the cells; deletions of these genes are also effective.

VIRUSES WITH HYBRID DNA-PSEUDOVIRIONS

The mechanisms involved in the integration of viral DNA into cellular DNA may also be responsible for other recombinations: the insertion of viral DNA into the DNAs of other viruses, and that of cellular DNA into viral DNA.

Adeno-SV 40 Hybrid Viruses. SV 40 acts as helper for adenovirus replication in green monkey kidney cells (see Ch. 48, Abortive infection). In such mixedly infected cells occasionally part of the SV 40 genome becomes incorporated within adenovirus DNA, which in turn may or may not be partially deleted. The hybrid molecules are then enclosed in an adenovirus capsid. Viruses with a nondefective adenovirus genome contain only small parts of the SV 40 DNA, and are useful for mapping SV 40 genes by using a combination of genetic and physical methods (see Ch. 11, Genetic mapping). Hybrid viruses with defective adenovirus genome require regular adenovirus as helper. Cells transformed by some of these hybrids are morphologically indistinguishable from cells transformed by SV 40 alone, suggesting that the helper function of SV 40 for adenovirus is expressed by viral genes important for transformation.

Incorporation of cellular DNA (recognizable by molecular hybridization to cellular DNA) into covalently closed cyclic viral DNA occurs in cells productively infected at high multiplicity by SV 40 or polyomavirus, or after inducing *tsa*-3T3 cells by lowering the temperature. This phenomenon resembles the formation of **specialized** transducing phage (see Ch. 46).

Pseudovirions. A phenomenon akin to **generalized** transduction (see Ch. 46) is also observed: the encapsidation of cellular DNA fragments, without viral sequences, in viral capsids (pseudovirions). The DNA is recognizable because it is linear and hybridizes to cellular but not viral DNA. Pseudovirions may eventually be useful for transducing genes from one animal cell to another; however, the frequency for any gene will undoubtedly be much lower than in bacterial transduction, and possibly not observable, owing to the much smaller proportion of an animal cell genome that can be accommodated in a polyoma capsid.

COMPARISON OF ANIMAL CELL TRANSFORMATION WITH BACTERIAL LYSOGENIZATION (CH. 46)

In both these processes the viruses can elicit either a nonproductive transformation or a productive lytic infection, the viral DNA is integrated in the cellular DNA, and integration and excision of the viral DNA are likely to be symmetrical events. Moreover, the cellular changes induced by oncogenic viruses can be considered akin to lysogenic conversion. However, no viral repressor system has been identified in the animal system, whose stability appears to depend on defects in the cell's ability to express all the genes of the integrated viral DNA.

We have already seen that integration of tumor viruses is promoted by agents (e.g., X-rays) that cause nicks in cellular DNA. Conversely, such agents elicit production of small amounts of infectious virus in some types of transformed cells. However, in accord with the absence of a repression mechanism, regular induction similar to that of prophage does not occur. Indeed, the discovery of integrated DNA in the animal system would have been made much more difficult were free viral DNA present.

ROLE OF CELLULAR GENES IN TRANSFORMATION

In addition to regulating transcription of the provirus, cellular genes perform other functions in transformation. Thus transformed SV-3T3 cells yield mutants in which the transformed phenotype is either less pronounced or temperature-sensitive, although the provirus itself (tested after induction by cell fusion) is not mutated. That changes in some of the components of the cellular system of growth regulation may prevent the effect of the virus is understandable, since the transforming viral protein must interact with cellular proteins. The partial phenotypic reversions are often accompanied by gross changes in the number of chromosomes, suggesting that viral transformation results from the combination of a viral function with an appropriate **balance of cellular genes.** The required gene balance seems different for different viruses, since SV-3T3 cells with a temperature-sensitive phenotype reacquire a temperature-independent state of transformation upon infection with an RNA-containing sarcoma virus (see below) but not with SV 40.

PAPOVAVIRUSES: PAPILLOMAVIRUSES

The papilloma virions are structurally similar to polyoma virions, but somewhat larger (550 A); they also contain a cyclic double-stranded DNA (MW ca. 5×10^6). Different members of the group produce benign warts, called papillomas, in several mammals including man. In the rabbit the tumors regularly become malignant if they persist for a sufficiently long time. The first of these viruses was discovered by Shope, who showed that extracts of warts of wild cottontail rabbits (*Sylvilagus floridanus*) produce warts when inoculated into the skin of either wild or domestic rabbits (*Oryctolagus cuniculus*). Papillomaviruses are readily recovered from extracts of human or other warts.

VIRAL MULTIPLICATION AND PATHOGENESIS

The papillomaviruses of man, rabbit, and cattle induce papillomas in the skin and some mucous membranes. Bioassay of the virus, based on wart formation in the skin, has extremely low efficiency. The ratio of physical particles to infectious units ranges between 10^5 and 10^8.

In tissue cultures these viruses do not appear to propagate. Nonproductive infection, however, may occur: human skin cells can apparently be transformed in culture by the human wart virus, and fetal calf skin cells by the bovine virus.

Skin papillomas begin as a proliferation of the dermal connective tissue, followed by proliferation and hyperkeratinization of the epidermis. In warts of wild cottontail rabbits the nuclei of the keratohyaline and keratinized layer contain viral capsid Ag (recognizable by immunofluorescence) and viral particles (recognizable by electron microscopy and infectivity). The proliferating connective tissue and basal epidermis contain neither Ag nor virions (Fig. 63-9), but they do contain viral DNA, which appears responsible for the stimulus to proliferate.

The papillomas of domestic rabbits contain much less Ag and infectious virus than those of wild rabbits. In contrast to other viruses, the phenol extracted DNA is often more infectious than the corresponding virions (see Ch. 48), suggesting that these papillomas may contain free viral DNA not coated by a capsid. However, the low sensitivity of the assay makes it impossible to test this hypothesis conclusively.

Rabbit papillomas, especially those of domestic rabbits, often progress to cancer several months after their inception. The production of virus or viral DNA is then markedly reduced or even undetectable. Thus one such cancer (Vx7) after prolonged serial transplantation has retained the ability to produce these products in very small amounts and in very few cells; another cancer (Vx2) has lost it completely. The continued production of virus by the Vx7 cancer could be due to the persistence of the viral genome in an integrated state, with occasional excision and maturation, or to a virus carrier state (Ch. 51).

Regression of Papillomas. Papillomas, including human warts, often regress spontaneously. If a rabbit has several separate papillomas they all regress simultaneously, suggesting an im-

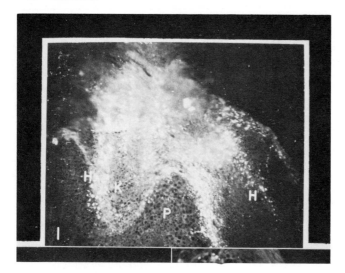

Fig. 63-9. Fluorescence photomicrograph of a frozen section of wild rabbit papilloma stained with fluorescent antiviral antibodies. Capsid antigen is present in nuclei, and therefore appears as small, discrete, bright areas; and it is restricted to the keratohyaline (H) and keratinized (K) layers of the epidermis. There is no capsid antigen in the cells of the proliferating basal layer (P). [From Noyes, W. F., and Mellors, R. C. *J Exp Med 104*:555 (1957).]

munological mechanism (see below). In contrast, the derived carcinomas in rabbits usually do not regress and can be serially transplanted. Presumably their high growth potential can overcome the allograft reaction (Ch. 21).

OTHER DNA VIRUSES

The oncogenic effects of adenoviruses and herpesviruses have been reviewed in Chapters 52 and 53. It is not known whether these viruses, which have a linear and much longer DNA, transform by the same mechanism as polyomaviruses. Thus in lytic infection adenoviruses can induce the replication of the cellular DNA under certain conditions but more often they inhibit it; and this is the rule with herpesviruses.

A possible explanation is that the larger viruses contain genes for the control of cellular growth in addition to other genes more directly involved in viral DNA replication. Some of these genes can shut off host macromolecule synthesis, hiding the effect of the transforming genes. The action of the transforming genes can be detected only when the genes for host shut-off are not expressed, for instance in nonpermissive cells or when the virus genome is partly inactivated (e.g., by UV light).

The larger viruses have linear DNA. Adenovirus DNA has homologous ends and may cyclize in the cells, allowing the integration of the whole genome. DNAs that cannot cyclize (herpesvirus DNA?) can be integrated only in part, by two crossovers. Alternatively the whole DNA may persist as a plasmid (see Ch. 46).

RNA-CONTAINING LEUKOVIRUSES

LEUKOSIS VIRUSES

These viruses have been recognized in many animal species and are probably universal in vertebrates. Originally discovered in chickens from their ability to induce leukemias or solid tumors, they were later recognized in mice, rats, cats, and other species, including primates.

PROPERTIES OF THE VIRUSES

The virions are enveloped, ether-sensitive, and quite unstable outside the cell. In electron

micrographs they appear spherical to pleomorphic, about 1000 A in diameter, and covered with spikes (Fig. 63-10). In sections they have a bull's-eye-like structure, with an electron-dense, probably helical, nucleoid; they are called **C-type particles** (Fig. 63-11). The extracted RNA sediments like single-stranded molecule of MW 1.3×10^7, but regions of base pairing apparently hold together four subunits of approximately equal size since denaturation decreases the sedimentation coefficient to that corresponding to an MW of 3.5×10^6.

Fig. 63-10. Electron micrograph of Rauscher leukemia virus with negative staining. Note the considerable pleomorphism and the presence of tail-like structures in some particles. The tailed forms appear to result from sharp changes of the osmotic pressure of the medium. An internal component is discernible in some particles. Marker = 1000 A. (Courtesy of R. F. Ziegel.)

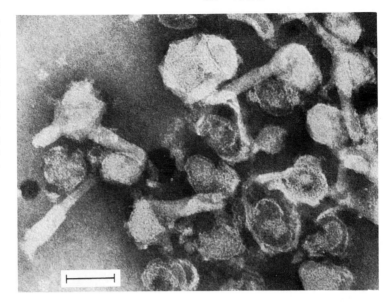

Like cellular messengers (Ch. 47), each RNA subunit has a polyadenylic acid tail at the 3′ end. The virions also contain the reverse transcriptase (Ch. 11) and various other enzymes (see Ch. 44, Virions' enzymes), tRNAs of various specificity, ribosomes, and fragments of cellular DNA.

Immunological Properties. Leukoviruses contain in their virions two main Ags. An **envelope Ag**, detected by neutralization, identifies the various strains. An **internal or group-specific (gs) Ag**, which is common to all leukoviruses of a given animal species, is not involved in immunity but is important in classifying new isolates; its presence in a cell may be the only sign of the existence of a leukovirus genome (possibly defective).

The gs Ags purified from disrupted virions are exquisitely selective reagents. In mammalian leukoviruses two specificities can be distinguished in the same polypeptide chain; **gs1** is common to all viruses of a given species and does not cross-react with viruses of other species, while **gs3** is common to all mammalian (but not avian) viruses.

Multiplication. The presence of a DNA intermediate was first suggested by the effect of inhibitors. In cell cultures 5-bromodeoxyuridine

Fig. 63-11. Thin sections of leukovirus (C-type) particles from cells producing a mouse leukosis virus. **A** and **B**, two phases of budding of the virions at the cell surface; **C**, a detached immature virion, with an electron-lucent nucleoid; **D**, two mature particles with dense nucleoid. (Courtesy of L. Dmochowski.)

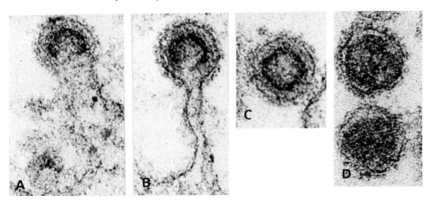

(BUDR), and inhibitors of DNA synthesis such as fluorodeoxyuridine or arabinosyl cytosine, inhibit viral multiplication only if added at the time of infection, apparently by interfering with the replication of the DNA intermediate, but are ineffective later on. In contrast, actinomycin D inhibits multiplication throughout, probably by blocking transcription of the DNA intermediate into viral RNA. Since this RNA is messenger-like, it is probably generated by normal transcription. This multiplication scheme is supported by several kinds of more direct evidence: the reverse transcriptase (Ch. 11) is present in the virions; unlike other viruses there is no evidence for direct RNA replication; and infected cells contain viral DNA (able to hybridize with the viral RNA) which, after extraction, can initiate cell infection.

After a long latent period, complete virus is produced by budding at the cellular surface or in intracellular vacuoles. Virus release then continues indefinitely without apparent harm to the cells.

The segmentation of the genome allows **frequent recombination** in cells infected by two viruses with recognizable markers (see Ch. 48, Recombination of RNA viruses).

EFFECTS OF INFECTION IN TISSUE CULTURES

A striking characteristic of these viruses is their endosymbiotic relation to their host cells (Ch. 48): virus multiplies extensively in many kinds of cells **without apparent effect on their form or function.** Hence most leukosis viruses do not cause transformation in vitro. The avian myeloblastosis virus, however, induces differentiation of chick embryo yolk-sac cells into transformed myeloblasts, which are recognizable by their characteristic morphology and growth ability.

A useful property of murine and feline leukosis viruses is to induce the fusion of XC cells (a line of transformed rat cells) upon cocultivation. The syncytia are easily identified and they can be used for assay purpose (see below).

EFFECTS OF INFECTION IN ANIMALS: GROSS LEUKEMIA VIRUS

Different leukosis viruses have somewhat different effects in animals. We will consider in detail the Gross virus, the first murine leukemia virus to be discovered, which has been extensively studied.

Inbred mouse strains have been especially useful in these studies because their various uniform genetic backgrounds offer different opportunities for virus establishment and tumorigenesis. In **high-leukemia strains** (e.g., AKR and C-58) spontaneous leukemia, usually lymphoid, occurs by 9 months in 85% or more of the individuals. **Low-leukemia strains** (e.g., C3H) have a low spontaneous frequency (0.5% or less). A virus had long been suspected in AKR mouse leukemia, but attempts at cell-free transmission (usually to adult mice of other strains) failed until Gross showed that only very young, especially newborn, mice are susceptible. The reason, discussed in greater detail below, is their immunological immaturity; when the animals later become immunologically competent, the Abs formed against the virus cannot eliminate infected cells (even though they can prevent the formation of new ones).

Newborn C3H mice inoculated with cell-free extracts from organs of leukemic AKR mice develop leukemia only rarely, or only after many months. However, by serial passage of the virus in C3H mice, selecting for early appearance of leukemia, a viral strain able to multiply to a higher titer (called passage A) could be isolated. When extracts from animals infected by this strain are inoculated into newborn or suckling C3H mice, they produce leukemia in 95 to 100% of the animals, after 2.5 to 3 months. Passage A virus produces lymphatic leukemia also in rats.

Transmission of the Gross virus from mother to progeny in AKR mice **must occur congenitally,** for it occurs regularly, even if the newborn mice are isolated from the mother and suckled on females of a low-leukemia strain. Indeed, when embryos or even fertilized ova from AKR females are implanted in the uteri of females of a low-leukemia strain they still develop into mice with a high incidence of leukemia.

In contrast, C3H females infected with the same virus appear to transmit it to their progeny **through the milk,** rather than congenitally, since the incidence is low if the offspring are fostered on uninfected females. Transmission through the milk can be attributed to the much more intense viremia present in infected C3H than in AKR females.

When passage A virus is inoculated into

newborn C3H mice the virus is eclipsed for about a week, and infectivity then reappears and persists for the animal's life. In both the preleukemic and leukemic stages infectious virus is found in many organs, and electron microscopy shows viral particles budding out of many kinds of cells, including those of the thymus, lymph nodes, spleen, bone marrow, the epithelium of mammary and salivary glands, and megakaryocytes. Most of the infected cells do not undergo any noticeable alteration, and leukemia arises only when special **target cells** are transformed by the virus.

Pathogenesis. The **thymus** is the **target organ** for this virus. Not only are the earliest lesions found in this organ, but its removal at birth prevents the development of leukemia. This effect can be reversed by the subsequent grafting of an isologous thymus. The first noticeable changes in the thymus consist in accumulation of many medium-sized and large lymphocytes in the medulla; then a thymic lymphosarcoma develops. Later on, infiltrations of lymphoid cells appear in many organs, and the peripheral blood acquires the characteristic picture of lymphoid leukemia.

The mechanism of action of the thymus was clarified by experiments in which AKR thymus was grafted onto low-leukemia hybrid mice, having AKR as one of the parents, e.g. (AKR × C57Bl)F₁.* Though lymphoid tumors often developed in the hybrid **host** rather than from cells of the grafted thymus, most of them arose from cells of the graft: they contained C57Bl Ags, which could be detected by an allograft reaction in an indicator strain of inbred mice that lacked these Ags. Apparently the grafted thymus retains its reticular and epithelial cells but loses its own lymphoid cells, which are replaced by host cells. These are then transformed into leukemic cells in the graft, presumably by interaction with the fixed cells of the grafted high-leukemia thymus. The nature of the interaction has not been clarified but two possibilities may be considered: 1) the thymus cells produce virus, which infects the cells of the host; or 2) the thymus cells produce a humoral factor required for transformation.

Thymectomized mice that do not develop lymphatic leukemia may, later in life, develop **myeloid leukemia.** Extracts obtained from these animals

* That is, the first generation progeny of a mating between AKR and C57Bl mice. The F₁ animals have the histocompatibility antigens of each parent and accept grafts from either parent (Ch. 21).

again produce lymphatic leukemia in C3H mice. (This effect of the virus does not appear in nonthymectomized animals because its incubation period is much longer than that of lymphatic leukemia.) Since it has not been possible to clone the virus, it is not certain whether passage A virus is a multipotent virus, which can produce different types of leukemia by interacting with different types of target cells, or a mixture containing viruses with different specificities.

The thymus plays an additional, less specific role through its function in the development of the lymphoid system. Such a role will be discussed below.

OTHER LEUKEMIA VIRUSES

Several other **murine leukemia viruses** were isolated subsequently to the Gross virus, often from cancer cells of a type that the virus cannot induce. Such **passenger viruses** were probably picked up accidentally during the numerous transplants of the tumor in mice, in which latent leukosis viruses are ubiquitous (see below, Latency).

Of these viruses, **Maloney virus** produces lymphocytic leukemia; **Friend virus** produces a reticulum cell leukemia with great enlargement of spleen and liver and reticulum cell sarcomas, but without involvement of thymus or lymph nodes; **Rauscher virus** is similar to Friend's but has a wide host range. Both Friend and Rauscher viruses grow rapidly when inoculated into adult mice (Fig. 63-12). **Graffi virus** produces a myeloid and reticular leukemia. The various forms of leukemia induced by the different viruses are ascribable to the different target cells they affect.

Feline leukosis viruses produce lymphosarcomas and lymphoblastic leukemia in cats. **Avian leukosis viruses** induce various forms of leukemia (lymphoblastic, erythroblastic, myeloblastic), and some solid tumors, in chickens and other birds. Congenital transmission through the egg (but not the sperm) has been demonstrated for the lymphomatosis virus. Chickens have also yielded many **nonpathogenic viruses** that are physically and biochemically indistinguishable from leukosis viruses and possess the same gs Ag.

Oncogenic Efficiency. Leukosis viruses present a spectrum of oncogenicity, from those that cause leukemias with high efficiency (e.g., Friend, Rauscher, or avian myeloblastosis) to those that are less efficient (e.g., Gross virus) and to some avian viruses that are not demon-

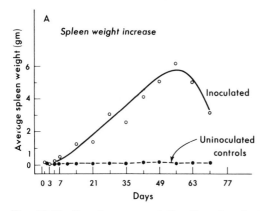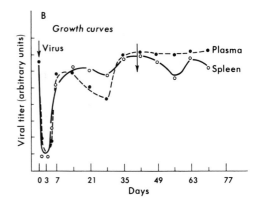

Fig. 63-12. Growth curves of the Rauscher leukemia virus. Young adult BALB/c mice were infected intraperitoneally. **A.** Note the enormous increase in the weight of the spleen. **B.** The inoculated virus is at first eclipsed; then the production of progeny virus begins, and continues at a nearly constant rate for the life of the animals. [Modified from Rauscher, F. J., and Allen, B. V. *J Natl Cancer Inst* 32:269 (1964).]

strably oncogenic. The differences may derive from a requirement for different kinds of gene balance (perhaps produced by cellular mutation), some of which are more rare than others.

Even with the most efficient viruses, the chance that an infected cell becomes transformed is very low: thus infected chick embryos, with most cells producing avian lymphomatosis virus, generate normal adults, which develop lymphomatosis later in life.

Assays. Infectivity assays in animals based on the development of the characteristic leukemias require at least several months and therefore are rarely employed. Friend and Rauscher viruses can be assayed more rapidly on the basis of development of splenomegaly or of proliferative foci in the spleen (Fig. 63-13). Leukosis viruses multiply well in tissue culture but since they do not produce morphological changes in the cell (except for avian myeloblastosis virus), in vitro assay methods are usually indirect.

Thus infectivity assays of murine or feline leukosis viruses are based on the formation of syncytia of cocultivated XC cells. Avian leukosis viruses are conveniently assayed by interference with the development of foci by avian sarcoma viruses (see below). Only the avian myeloblastosis virus can be assayed by formation of foci of transformed myeloblasts.

The most general and convenient assay for all viruses is based on the quantitation of reverse transcriptase present in the virions.

However, crude preparations and cell extracts even more, may be contaminated with cellular DNA polymerases which, although they normally recognize a DNA template, under special circumstances recognize an RNA template. Great caution, therefore, is required in assessing the results.

LATENCY OF LEUKOSIS VIRUSES

Induction. Latency is suggested by absence of virus or gs Ag in cells of adult mice of strains in which embryos or old animals may contain the Ag. Direct evidence has been obtained with cell lines that have been maintained over many cell generations without release of virus from the cells (although the AKR mice from which they were derived regularly develop Gross leukemia at older age). When treated with BUDR or IUDR these cultures produce Gross virus; a less dramatic induction is also obtained with X-rays or carcinogenic chemicals. Similar induction with some cell lines has shown that nonexpressed leukovirus genomes are present in different species, the recovered viruses being either weakly pathogenic or nonpathogenic in animals. In some cases the viral genome is only partially expressed after induction, producing gs Ag but no virus.

Genetic Control of the Expression of Viral Genomes. When chickens containing gs Ag in their cells are crossed with strains that do not contain the Ag, the synthesis of gs Ag segregates as a single dominant mendelian allele which controls the expression of the viral gene specifying the Ag. In mice

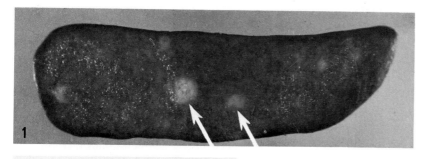

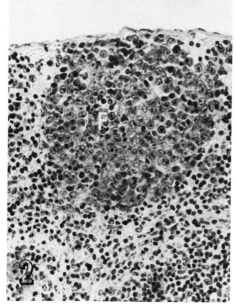

Fig. 63-13. Foci in the spleen of C3H mice inoculated with Friend leukemia virus. 1. Macroscopic appearance of the foci (arrows) in whole spleen. 2. Microscopic appearance at low magnification of a section of spleen with one focus (F). The focus contains large cells, characteristic of Friend's leukemia. [From Axelrod, A. A., and Steeves, R. A. *Virology 24:*513 (1964).]

two cellular genes control the ability of the cells to synthesize complete virus. In addition, several genes linked to genes for histocompatibility Ags (see Ch. 21) determine the susceptibility of an animal to leukemia induction by certain viruses; they probably control cell surface properties important in virus absorption or penetration.

Role of Leukosis Viruses in Neoplasias and Cell Transformation Induced by Chemical or Physical Agents. X-ray treatment induces leukemia in low-leukemia C3H mice, and H. Kaplan showed that filtrates of the leukemic organs contain a virus, similar to the Gross virus, able to induce leukemia in C3H mice. Moreover, many leukosis viruses produce foci of transformed cells in culture in **conjunction with carcinogens** (such as 7,12-dimethyl-benz(a)-anthracene or 3-methylcholanthrene), though neither agent alone can do so. Perhaps mutations in cellular genes induced by the carcinogens provide the genic balance required for transformation by leukosis viruses. Irrespective of their mechanisms, these results have profound implications for the mechanism of oncogenesis by chemicals and radiation, suggesting that it may occur through the activation of latent leukoviruses, which then produce cancer either by independent action or by cooperation with the activating agent. However, just as UV irradiation can kill bacteria both by inducing prophages and by other effects on the DNA, carcinogenesis by a given agent might involve viral activation in some cases and not in others.

SIGNIFICANCE OF LEUKOVIRUS LATENCY

Various interpretations have been offered for the widespread presence of latent leukovirus genomes in cells of mice and other species, and for their interaction with cellular genes. The simplest interpretation is that the DNA genomes are integrated (like those of papovaviruses) as proviruses and are transcribed after an inducing treatment to produce viral RNA; the cellular genes controlling expression of the viral genome may control its transcription, just as cellular genes control the transcription of an SV 40 provirus.

Two other hypotheses consider the latent leukovirus genomes as more deeply adapted to the host genome than simple proviruses. Huebner and Todaro have suggested that these genomes were integrated millions of years ago as **oncogenes** and later evolved to provide normal functions for the organism, presumably in embryogenesis. Derepression of the oncogene late in life would be responsible for seemingly spontaneous cancers. Alternatively, Temin has suggested that cells contain **protoviruses,** true cellular genome segments, which, among other things, specify a reverse transcriptase used by the cells to carry out gene amplification (i.e., the production of many copies of a gene by back-transcribing its messenger) in processes such as differentiation or Ab response. Errors during gene amplification would lead in unspecified ways either to the development of leukovirus genomes or to cancer; the two events would not be necessarily connected to each other.

The formulation of the highly speculative oncogene and protovirus hypotheses is justified by the biology of leukoviruses: the specification of a reverse transcriptase, the formation of a DNA intermediate which may become integrated in the cellular DNA as provirus, the production of viral RNA from this intermediate by the same process of transcription used for the synthesis of cellular mRNAs, and the control of the expression of viral genes by cellular genes. Moreover, in contrast to polyomaviruses, leukovirus multiplication does not kill the cells and does not require the detachment of the provirus.

The suggestion that leukoproviruses, or oncogenes, or protoviruses have a dominant role in spontaneous cancers is based on the relation of leukoviruses to sarcoma viruses (see below). A great difficulty, however, is that most leukoviruses latent in cells are weakly or not at all oncogenic so their activation in a cell is not likely to transform it.

Leukoviruses Producing Slow Infections. Visna virus is recognized as the agent of a slow infection (see Ch. 51) in the central nervous system of sheep; it is morphologically similar to leukosis viruses and contains a reverse transcriptase. Its kinship to leukosis viruses has been finally demonstrated by showing that it too transforms mouse cells in vitro. The possible relation between "slow infection" and oncogenesis will be examined below (Immunological relations of virus-infected cells to animal host).

SARCOMA VIRUSES

These viruses are morphologically and immunologically similar to the C particles of leukosis viruses. However, as infectious agents, they exhibit several important differences: 1) in animals they produce sarcomas rather than leukemias and with a much shorter incubation period; 2) they have a **100% oncogenic efficiency,** since both in the animal and in vitro every infected cell is transformed; 3) they are transmitted from one animal to another by the experimenter* and not congenitally from parent to progeny. However, like leukoviruses, they are transmitted vertically from one transformed cell to its progeny cells both in the animal and in vitro. Since the progeny of a transformed cell remain transformed and are visibly altered, the infection of a cell in a culture causes the formation of a **focus** (Fig. 63-14); hence assay is simple. This property readily distinguishes sarcoma viruses from leukosis viruses.

The Rous sarcoma virus (RSV), isolated from a chicken with a spontaneous sarcoma, was the first to be discovered. It was later developed into a model system for studying viral oncogenesis.

NATURE AND ORIGIN OF SARCOMA VIRUSES

More recently several sarcoma viruses similar to the chicken virus have been isolated from mice heavily inoculated with a leukosis virus, as well as from spontaneous mouse tumors. The mode of isolation of these murine viruses

---

* Transmission from one animal to another in nature does not seem to occur.

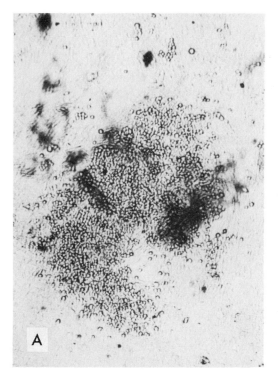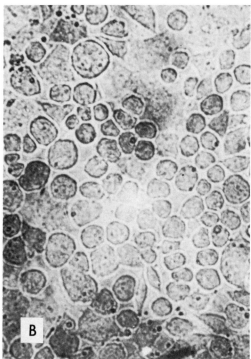

Fig. 63-14. Foci formed by Rous sarcoma virus in cultures of fibroblast-like chick embryo cells. **A.** Photograph of a focus at low magnification. **B.** Photograph at higher magnification, showing the round morphology of the transformed cells. (Courtesy of H. Temin.)

suggests a relation between leukosis and sarcoma viruses. However, the scarcity of the easily detectable sarcomas among the very large number of mice experimentally infected by leukosis virus suggests dependence on some rare event. It is not simply a mutation in a leukosis virus: for within the same species murine leukosis and sarcoma viruses have only limited RNA homology, as shown by hybridizing one RNA with DNA complementary to the other RNA (synthesized in vitro using the reverse transcriptase). Moreover, the subunits of avian sarcoma virus RNA are **longer** than those of corresponding leukosis viruses (Fig. 63-15).

Sarcoma viruses therefore appear to arise when leukosis viruses exchange a considerable proportion of their genetic material for other genetic material **(sarcoma information)** already present in the cell. The infrequency of the event suggests that sarcoma information either is rarely present in cells or is usually not accessible for incorporation, possibly by exchange of parts, into a leukosis virus genome. The

hypothetical exchange is supported by the defectiveness of most sarcoma viruses.

DEFECTIVENESS OF SARCOMA VIRUSES

Rubin and Vogt discovered that stocks of a widely used strain of Rous sarcoma virus selected for high growth potential (Bryan strain) always contain an excess of avian leukosis virus, which was therefore defined as **Rous associated virus** (RAV). Moreover, the envelope of the RSV is the same as that of the RAV in the same stock, a phenomenon explainable as phenotypic mixing (see Ch. 48). Several distinct RAVs were recognized on the basis of the host range (which is determined by the envelope), the immunological specificity of the envelope, and other properties. When such an RSV with a given envelope is grown in cells infected by a different RAV, the progeny RSV has the new envelope. Hence the Bryan strain of RSV has no serological specificity (type) of its own in the envelope, and the various strains obtained us-

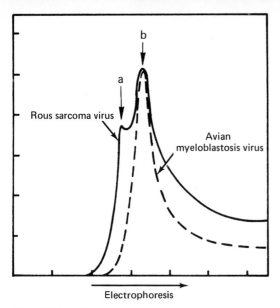

Fig. 63-15. Acrylamide gel electrophoresis of RNA of sarcoma and leukemia viruses after a brief exposure to 100° (to separate the subunits). The sarcoma virus RNA contains large subunits (a) absent in the leukemia virus which only has smaller subunits (b). The small subunits present in the RSV RNA are generated during the serial passage of the virus but are absent in recent clonal isolates.

ing different RAVs are **pseudotypes** (Fig. 63-16). Similarly, the host range changes with the envelope.

The envelope may even be provided by a latent leukosis virus which is activated after infection with the RSV, and thus it may have unusual properties, such as an extremely narrow host range in chicken

(e.g., restricted to chicken of a single genotype). However, such particles are usually infectious for cells of other fowl species, a property that easily permits their assay.

Helper Virus. These observations suggest that the RSV of the Bryan strain is unable to code for a complete envelope of its own, and requires a leukosis virus as helper to provide the missing function. Indeed, viral particles are formed without helper but miss one of the envelope glycoproteins. Defectiveness is common but not universal among sarcoma viruses: the mouse strains are defective, but some chicken strains are not. All avian leukosis viruses can act as helper for defective avian sarcoma viruses, and many mammalian leukosis viruses for mammalian sarcoma virus, but there is no helper effect between avian and mammalian strains. Apparently the specificity of the helper effect is similar to that of the gs3 component of the gs Ag.

When a defective sarcoma virus infects a nonpermissive cell (e.g., RSV infecting mammalian cells), the cells are transformed but no virus is produced; but virus production is elicited by fusing the cells to permissive chicken cells (Fig. 63-17). A murine sarcoma virus with a murine leukosis virus envelope may infect and transform a hamster (or other mammalian) cell; if the cell contains a hamster leukosis virus, or is infected by one, it will produce a new sarcoma virus with the hamster leukosis virus envelope and therefore be able to grow in hamster cells (Fig. 63-18). In this

Fig. 63-16. Formation of pseudotypes of Rous sarcoma virus (RSV). **I.** RSV alone infecting a cell containing no avian leukosis virus produces noninfectious progeny lacking an envelope glycoprotein (although the cells are transformed). **II.** RSV (A) (i.e., with envelope A) and RAV-A produce both RSV and RAV-A, all with envelope A, specified by RAV-A genes. **III.** RSV(A) and RAV-B produce RSV and RAV-B, all with envelope B, specified by RAV-B genes. The envelope of the infecting RSV only determines whether or not it can infect the cell (host range), and has no relation to the envelope of the infectious progeny, which is specified by the helper RAV.

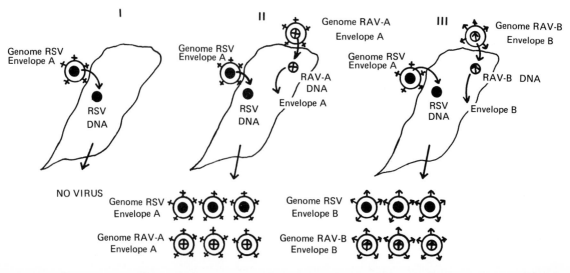

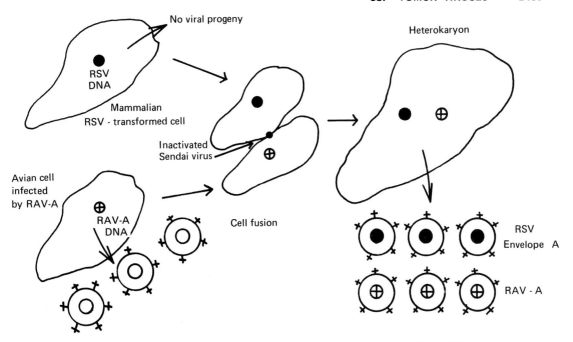

Fig. 63-17. Rescue of a nondefective RSV genome in a nonpermissive mammalian cell by fusion with a permissive chicken cell. Fusion is promoted by inactivated Sendai virus (see Ch. 47).

Fig. 63-18. Adaptation of murine sarcoma virus (MSV) to a new species (hamster) when a rare transformed hamster cell is superinfected by a hamster leukosis virus (HLV). The progeny contain particles with MSV genome and HLV envelope which can multiply in hamster cells and transform them with high frequency.

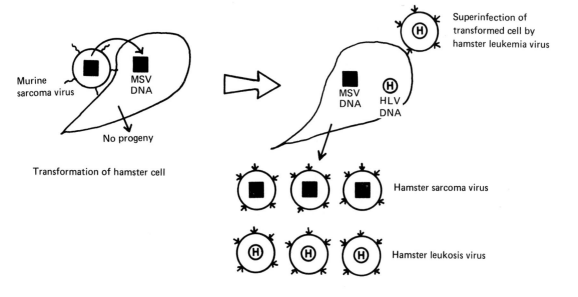

way a **sarcoma virus may be adapted to a new species.**

MODE OF ACTION OF SARCOMA VIRUSES

Sarcoma viruses transform **all** cells they infect in the same way as polyomaviruses cause a transient transformation in all infected permissive cells. It is likely, therefore, that sarcoma viruses also contain information capable of disrupting the growth regulation of the host cells. Moreover, transformation depends on the continuing presence of this product, since certain mutations of Rous sarcoma virus make the transformed state thermosensitive. A direct control of cellular functions is also shown by the occurrence of viral mutants that change the morphology of the transformed cells (Fig. 63-19).

As we have seen, a helper virus may be required for sarcoma virus transmission but not for the expression of the transforming information. In the same way the reverse transcriptase is required for multiplication and the establishment of transformation but not its maintenance, because cells transformed by transcriptase-deficient mutants in the presence of helper (to supply the transcriptase) remain fully transformed after losing the helper. Hence the mere presence and function of viral genes in the cells are sufficient for transformation. These properties of sarcoma virus trans-

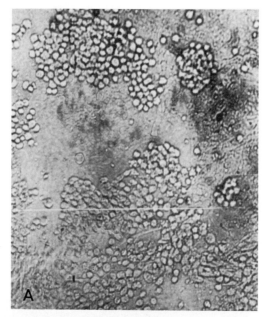

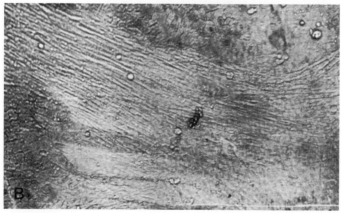

Fig. 63-19. Transformation of cultures of the iris epithelium of the chicken embryo by either wild-type Rous sarcoma virus **(A)** or its morph[f] mutant **(B)**. The cells transformed by the wild-type virus are round, whereas those transformed by morph[f] virus are fusiform and form bundles. The untransformed cells form an epithelial sheet and can be recognized because they contain various amounts of black pigment, whereas both kinds of transformed cells are almost colorless. (Courtesy of B. Ephrussi and H. Temin.)

formation suggest that the mechanism is similar to that of transformation by DNA viruses. However, in contrast to polyomaviruses, the viral function required for transformation is not essential for viral multiplication, since some ts mutants of RSV affect the former but not the latter.

MOUSE MAMMARY TUMOR VIRUS

The study of the induction of the mouse mammary cancer began as a study in mouse genetics. It had been known for a long time that the cancer appears spontaneously in the females of certain mouse strains, and that it is possible to develop inbred lines selected for a very high or for a very low incidence of cancer. The difference between these **high-cancer** and **low-cancer** strains seemed to be purely genetic. However, in 1936 Bittner showed that high-cancer strains can arise from low-cancer strains if the newborn are nursed by females of high-cancer lines. Conversely, newborn mice of high-cancer strains suckled by low-cancer strains had at maturity a low incidence of mammary tumors. This simple experiment revealed that the cancer is induced by a transmissible agent present in the milk **(the milk agent),** later recognized as a virus.

PROPERTIES OF THE VIRUS

The virions resemble those of murine leukemia viruses; they are ether-sensitive, have an envelope with spikes (Fig. 63-20), an RNA of large molecular weight, probably similar to that of leukoviruses, and a reverse transcriptase. Numerous extracellular viral particles or incomplete particles in the process of release from the cells are present in sections of cancer tissues; they are designated as **B-type particles** (Fig. 63-21) and differ from the C-type particles of leukosis viruses by their dense, eccentric core, separated from the outer coat by an electron-lucent space. As with murine leukemias, virus is often found in nonneoplastic cells; for instance, the mammary glands of infected animals contain virus many months before the appearance of cancer.

The **bioassay** of the virus is extremely inefficient. The classic method is based on the incidence of cancer in genetically susceptible, virus-free mice inoculated shortly after birth either by feeding or by intraperitoneal inoculation. This assay requires about 2 years and is purely qualitative, i.e., the proportion of animals developing cancer does not parallel the amount of virus inoculated (and may even bear an inverse relation, apparently owing to inhibiting or interfering agents). A more rapid assay, requiring 4 months, is based on the develop-

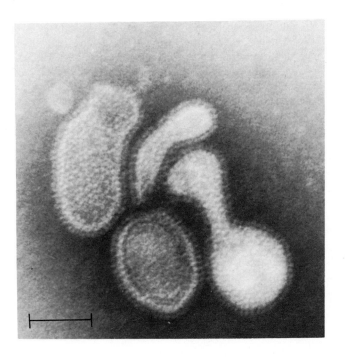

Fig. 63-20. Electron micrograph of the purified mouse mammary tumor virus with negative staining. Note the envelope covered by spikes, which surrounds an internal component. Marker = 1000 A. [From Lyons, M. J., and Moore, D. H. *J Natl Cancer Inst* 35:549 (1965).]

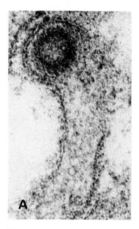
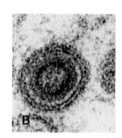
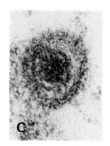

Fig. 63-21. Electron micrographs of thin sections of a mammary tumor producing mouse mammary tumor virus. **A,** the budding of the virions at the cell membrane; **B,** an immature virion with electron-lucent core; **C,** a mature B-type particle with dense, eccentric core. (Courtesy of L. Dmochowski.)

ment of hyperplastic nodules (to be considered below) in females of a virus-free strain exposed to intense hormonal stimulation; unfortunately this method also is qualitative. Since the virions contain a **reverse transcriptase,** the activity of the enzyme is a convenient method of assay, as for leukosis viruses.

PATHOGENESIS

The Bittner virus is usually transmitted from the mother to her sucklings through the milk. In contrast, other strains of mammary tumor virus are transmitted congenitally through the sperm or the egg. Mammary carcinomas, of various histological types, start appearing in susceptible mice infected by Bittner virus at the age of 300 to 350 days.

A key role in the production of the cancer is played in **hyperplastic** alveolar nodules, which are constituted of normal mammary tissue (Fig. 63-22). These nodules contain large amounts of virus and appear to be the sites of cancer development.

Thus, pieces of infected mammary glands transplanted to gland-free mammary fat pads of isologous, virgin, virus-free females undergo neoplastic transformation more frequently if they contain nodules. Moreover, in organ cultures in which part of the mammary gland, without previous dispersion into individual cells, is maintained in vitro, hyperplastic nodules have a tendency to proliferate and to produce additional foci, but without an inva-

sive, malignant character. These results suggest that the virus-transformed cells constitute the hyperplastic nodules, whose progression to cancer requires other factors.

The **susceptibility of the host** to induction of mammary cancer depends on several factors, some genetic and others physiological.

Genetic Factors. Although the high-cancer or low-cancer character of most mouse strains depends primarily on whether the mice are infected by the virus, some strains (e.g., C57Bl) are intrinsically resistant and have a low incidence of cancer even when they are artificially infected by the virus. The results of genetic crosses show that resistance depends on two types of genetic factors: a pair of allelic genes controlling the ability to produce cancer in the presence of viral infection, and probably several other pairs of genes controlling the ability to support viral multiplication.

A cross supporting these conclusions is analyzed in Figure 63-23. Virus-carrying susceptible C3H females were mated to virus-free resistant C-57 males. The F_1 females were back-crossed to C57Bl males, and the females of the resulting back-cross (BC_1) were again back-crossed to C57Bl males, yielding BC_2 progeny. All newborn were nursed by their mothers. The females of the two inbred strains, and of their several generations of hybrids, were scored both for mammary tumors and for the appearance of virus in the milk.

The observed distribution of cancers among the progeny suggests that a single pair of allelic genes

Fig. 63-22. Hyperplastic alveolar nodules in the mammary gland of a multiparous C3H female mouse bearing a mammary tumor. The nodules (some indicated by arrows) are filled with milk. Wholemount of the gland, hematoxylin staining. ×6. [From Nandi, S., Bern, H. A., and Deome, K. B. *J Natl Cancer Inst 24*:883 (1960).]

determines either susceptibility or resistance, susceptibility being dominant. Ability to transmit the virus, in contrast, is not controlled by a single pair of genes, as shown by lack of the expected segregation in the back crosses. Since, however, some nontransmitting animals do appear in the second backcross, and furthermore the ability to transmit the virus is totally lost after a longer series of backcrosses (not shown), this property is apparently subject to polygenic control: only progeny containing the resistance alleles of several genes would lose the ability to transmit the virus.

Physiological Factors. The most important physiological factor in susceptibility is **hormonal** stimulation. Thus after infection such stimulation promotes the production of hyperplastic alveolar nodules in the mammary gland. Moreover the cancer incidence is low in virgin infected females but high in infected females that are force-bred (i.e., made to bear litters in quick succession without nursing them) or given estrogen. It is even more striking that when virus-infected **males** are castrated and are injected with estradiol and deoxycorticosterone over a long period they frequently develop mammary cancer, whereas normal males have no mammary cancer at all.

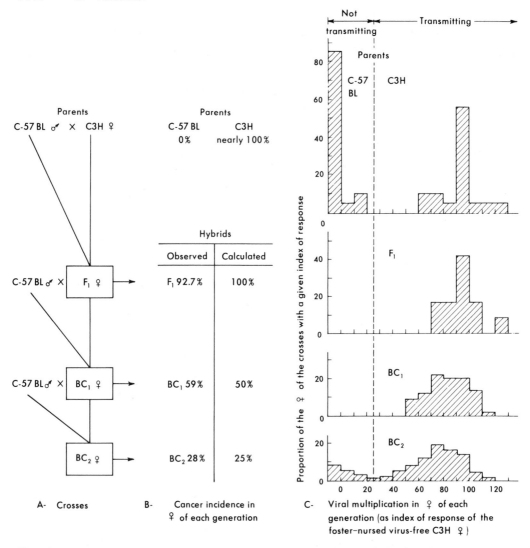

Fig. 63-23. Segregation of the genetic factors controlling resistance to induction of mammary cancer. The crosses are indicated in **A**: the basic cross involves resistant C57Bl males and susceptible, virus-carrying C3H females; two back-crosses follow. In **B** is given the proportion of females in each generation that developed mammary cancer, together with the proportions expected if a single pair of alleles controlled this development, with susceptibility being dominant. The transmission of the virus in the milk of these females was also tested by allowing each animal to suckle a number of newborn, virus-free CH3 females: if these susceptible test animals received virus they later developed mammary cancer, and their age at its appearance was younger the larger the infecting dose. The results are expressed in **C** as an index of response, calculated as the proportion of test animals (for each donor) that developed a tumor, multiplied by the inverse of the average age of its appearance. An index of response above 20 indicates that virus is transmitted to the test females; lower values are not significant and reflect spontaneous cancer incidence in uninfected test mice. For interpretation see text. [Data of Heston, W. E., Vlahakis, G., and Deringer, M. K. *J Natl Cancer Inst* 24:721 (1960).]

IMMUNOLOGICAL RELATIONS OF
VIRUS-TRANSFORMED CELLS TO ANIMAL HOST

Like other viruses, oncogenic viruses elicit the formation of **humoral Abs,** which can be detected by neutralization or complement fixation. With leukoviruses the surface of the infected cells contains viral envelopes in the process of budding (Ch. 48, Maturation of enveloped viruses); binding of Abs to the sites can be shown by cytotoxicity in the presence of complement, or by immunofluorescence.

Ags that appear on the surface of transformed cells have great importance for the development of the neoplasia, for they render the cells foreign to their own body, and the resulting **cell-mediated immune response** tends to destroy the transformed cells. Thus after neonatal thymectomy, or after administration of immunosuppressants, mice develop virus-induced leukemia more frequently and earlier. Conversely, mice immunized with extracts of **tumor** cells and later injected with virus develop leukemia less frequently than controls and after a longer latent period. Moreover, virus inoculation elicits immunity to subsequently grafted transformed cells, presumably because the virus induces the appearance of its specific surface Ag in cells of the host, followed by an immune response.

In cells transformed by leukoviruses the new cell surface Ags correspond to the envelope Ag of the viruses budding at the cell surface. Hence with sarcoma viruses they are usually specific for the helper leukosis virus, and so such tumors lacking helper are less likely to be rejected on transplantation. Cells transformed by the nonenveloped DNA viruses also contain new "transplantation" Ags, but these are unrelated to those of the virions or to the nuclear T Ag. It is not known whether these Ags are virus-specified or are modified host Ags.

A viral origin is suggested by the specificity of the Ags in papovavirus-transformed cells: infection with SV 40 induces resistance to cells transformed by SV 40 but not by polyoma virus, and vice versa (Table 63-4). On the other hand, trypsin treatment of normal cells exposes a new surface Ag cross-reacting with that present on the surface of SV 40-transformed cells; and **cells of many cancers** (not necessarily virus-induced) **contain Ags normally recognizable in the embryo but not in the adult.** Hence even though different viruses cause the appearance of different Ags, they may be cellular Ags made accessible by shortening of the carbohydrate chains at the surface of transformed cells.

Animals infected by leukosis viruses as embryos, or by the mammary tumor viruses at a young age through the milk, may develop a state akin to im-

TABLE 63-4. Specificity of Resistance to Transformed Cells in Papovavirus-Immunized Hosts*

| | No. of cells inoculated per hamster | | | |
| | Polyoma-transformed | | SV 40-transformed | |
Hamsters immunized with	10^3	10^4	10^3	10^4
	Fraction of recipients developing tumors			
Nothing	14/15	15/15	14/15	13/14
Polyomavirus	2/14	5/14	13/14	14/14
SV 40	14/14	13/13	4/13	4/14

* Adult hamsters were immunized by intraperitoneal injection of about 2×10^6 PFU of either polyoma virus or SV 40 and 1 month later they were inoculated subcutaneously with hamster cells transformed by either virus. The fractions give the number of animals later developing tumors, over the number receiving cells. The immunization with either virus is seen to be specific for the cells transformed by that virus.

Data from HABEL, K., and EDDY, B. E. *Proc Soc Exp Biol Med 113*:1 (1963).

munogical tolerance: they contain much virus in their organs and blood and can accept the transplantation of grafted tumors more readily than virus-free animals. However, as with the choriomeningitis virus (Ch. 51), the tolerance is only apparent. Abs are produced but are all bound to excess Ag; the Ag-Ab complexes are eliminated through the kidney, where they can be recognized. **"Slow" Infection.** The presence of viral Ags on the surface of leukosis virus-infected cells is also responsible for a different pathological consequence—the "slow" infection produced in sheep brain by Visna virus. The infection may not be directly harmful but the strong cellular immune response elicited by the viral Ags (which are stronger immunogens than those of the oncogenic leukovirus), produces demyelinization of the nerve cells and finally their destruction.

ROLE OF VIRUSES IN NATURALLY OCCURRING CANCERS (TABLE 63-5)

Animals. The viruses used as model systems, although eminently suited for inquiring into the mechanism of cancer induction, may not be responsible for most natural cancers. Thus the polyomaviruses have not been shown to produce tumors in nature, and the leukoviruses do so much more rarely than in some inbred strains. The high incidences obtained experimentally derive from the adoption of extremely favorable conditions (such as genotype of the host, age of infection, route of inoculation, multiplicity of infection), which do not apply in nature.

Natural cancers seem to be elicited especially by herpesviruses, papillomaviruses, and leukoviruses. The role of **papillomaviruses** has been described above. **Herpesviruses** are responsible for chicken lymphomas (Marek's disease, which is sufficiently frequent to be an economic problem in the poultry industry), for similar lymphomas in monkeys, and for a very widespread renal adenocarcinoma (Lucké) in frogs. The etiological role has been demonstrated in all cases by experimental transmission and in Marek's disease by a considerable decrease of the lymphoma incidence after vaccination. **Leukoviruses** produce a low incidence of leukemias in feral mice, in cats, and in chickens; of the sarcoma viruses only the Rous virus appears to have occurred naturally—all the others were isolated in laboratory animals.

Human Cancers. The search for viruses in human cancers is hindered by the impossibility of testing for the oncogenic activity of a virus by inoculating it into human beings. Hence the search is mainly concentrated on determining which viruses are regularly associated with a given cancer, although the association does not necessarily imply an etiological relation since the virus could be a passenger. Nevertheless, an etiological role can be considered likely if several conditions are satisfied: 1) the

TABLE 63-5. Summary of Oncogenic Viruses

Virus	Transformation in tissue culture	Tumors in lower animals	Implicated in human tumors
Papovaviruses			
Polyomaviruses	+	Sarcomas and other solid tumors	No
Papillomaviruses	+	Papillomas	Yes, warts
Adenoviruses	+	Sarcomas	No
Herpesviruses	+*	Lymphomas Adenocarcinomas	Yes, Burkitt lymphoma, nasopharyngeal carcinoma
Leukoviruses			
Leukosis viruses	−†	Leukemias and some solid tumors	?
Sarcoma viruses	+	Sarcomas	No
Mammary tumor viruses	−	Mammary carcinoma	Possibly

* Only using UV-irradiated virus.
† Except avian myeloblastosis virus.

viral DNA is present and at least partly functional in the cancer cells, 2) the virus can transform human cells in culture, 3) a related virus is oncogenic in experimental animals.

As in lower animals, the search suggests a more important role for DNA viruses than for leukoviruses. *Papillomaviruses* are found in the commonly occurring warts, and their etiological role is well supported. A **herpesvirus,** the Epstein-Barr (EB) virus (see Ch. 53), is implicated in the Burkitt lymphoma (which is similar to lymphomas induced by similar viruses in chickens and monkeys) and in certain forms of nasopharyngeal carcinomas. The tumor cells, although lacking mature virus, contain the viral DNA (recognizable by hybridization with labeled viral RNA), probably in plasmid form. They also contain virus-specific Ags, and upon cultivation in vitro they produce the virus.

That the virus is oncogenic in human cells is suggested by the ability of white blood cells from infectious mononucleosis patients (which contain the same virus), in contrast to those from normal individuals, to give rise to indefinitely growing cultures. The strongest support for the etiological role of the virus is the similarity between the Burkitt lymphoma and the lymphomas produced by a herpesvirus in chicken and monkey. However, the EB virus must act in cooperation with other unknown factors, because the tumors are rare, whereas the virus is ubiquitous. Perhaps mutations in cellular genes must provide the required gene balance for tumor formation.

A Herpesvirus has been also implicated in the etiology of human cervical cancers.

The search for oncogenic human **leukoviruses** has followed the animal model, concentrating on mammary cancers and leukemias. The presence in women's milk of B-type particles possessing a reverse transcriptase has been reported. In view of the complex conditions (genetic, physiological, and infectious) leading to mammary cancer in mice it may be very difficult to establish whether a virus, if present, is the etiological agent of human breast cancer. Epidemiological evidence suggests that virus-containing milk may not be an important virus vehicle, for the incidence of breast cancer in the United States has remained constant during the past 30 years in spite of a large decrease in breast feeding; however, the virus might be transmitted through the germ cells. 70S RNA (characteristic of leukoviruses) is being sought both in the cells of breast cancers and in leukemic cells. However, even if it can be established that the RNA is viral, it may be very difficult to prove that it belongs to the agent of the disease. Indeed, such proof would have been very difficult also in mice had it not been possible to reproduce the disease experimentally by animal inoculation. Nevertheless, these efforts are supported by the discovery of leukoviruses in tumors of two primates, in which the role of the viruses in oncogenesis can be established more easily.

Knowledge derived from the experimental systems suggests ways to determine whether a virus is **not the agent** of a human cancer. The basis is **absence of the viral genome** in the cancer cells, since **in all experimental systems the persistence of the genome has been observed.** The results obtained suggest that neither the very widespread human adenoviruses (although oncogenic in experimental systems, see Ch. 52) nor the papovaviruses, which are frequently recovered from the human brain (where they may produce slow infections, see Ch. 51) are implicated in at least the frequently observed human cancers; these cells do not contain the DNA or the RNA of the viruses, or their specific Ags.

SELECTED REFERENCES

Books and Review Articles

DULBECCO, R. Cell transformation by viruses. *Science 166*:962 (1969).

EMMELOT, P. and BENTVELZEN, P. *The RNA Viruses and Host Genome in Oncogenesis.* North Holland, Amsterdam, 1972.

EPSTEIN, M. A. Aspects of the EB virus. *Adv Cancer Res 13*:383 (1970).

GOOD, R. A. Relations between immunity and malignancy. *Proc Natl Acad Sci USA 69*:1026 (1972).

GREEN, M. Molecular basis for the attack on cancer. *Proc Natl Acad Sci USA 69*:1036 (1972).

GROSS, L. *Oncogenic Viruses.* Pergamon Press, New York, 1961.

KAPLAN, H. S. On the natural history of the murine leukemias: Presidential address. *Cancer Res 27*:1325 (1967).

KIT, S., and DUBBS, D. R. *Enzyme Induction by Viruses.* Karger, New York, 1969.

KLEIN, G. Immunological aspects of Burkitt's lymphoma. *Adv Immunol 14*:187 (1971).

KOLDOVSKY, P. *Tumor Specific Transplantation Antigens.* Springer, New York, 1969.

PASTAN, I. Cyclic AMP. *Sci Am 227*:97 (1972).

PASTERNAK, G. Antigens induced by the mouse leukemia viruses. *Adv Cancer Res 12*:1 (1969).

SJÖGREN, H. O. Transplantation methods as a tool for detection of tumor-specific antigens. *Progr Exp Tumor Res 6*:289 (1965).

TEMIN, H. and BALTIMORE, D. RNA-directed DNA synthesis and RNA tumor viruses. *Adv Virus Res 17*:129 (1972).

Specific Articles: DNA Viruses

BLACK, P., ROWE, W. P., TURNER, H. C., and HUEBNER, R. J. A specific complement fixing antigen present in SV 40 tumor and transformed cells. *Proc Natl Acad Sci USA 50*:1148 (1963).

DULBECCO, R., HARTWELL, L. H., and VOGT, M. Induction of cellular DNA synthesis by polyoma virus. *Proc Natl Acad Sci USA 53*:403 (1965).

DULBECCO, R. and ECKHART, W. Temperature-dependent properties of cells transformed by thermosensitive mutants of polyoma virus. *Proc Natl Acad Sci USA 67*:1775 (1970).

ECKHART, W., DULBECCO, R., and BURGER, M. Temperature-dependent surface changes in cells infected or transformed by a thermosensitive mutant of polyoma virus. *Proc Natl Acad Sci USA 68*:283 (1971).

FRIED, M. Cell transforming ability of a temperature-sensitive mutant of polyoma virus. *Proc Natl Acad Sci USA 53*:486 (1965).

GILDEN, R. V., CARP, R. I., TAGUCHI, F., and DEFENDI, V. The nature and localization of the SV 40-induced complement-fixing antigen. *Proc Natl Acad Sci USA 53*:684 (1965).

HENLE, G., CLIFFORD, P., and SANTESSON, L. EBV DNA in biopsies of Burkitt tumours and anaplastic carcinomas of the nasopharynx. *Nature 228*:1056 (1970).

KLEIN, G. Herpesvirus and oncogenesis. *Proc Natl Acad Sci USA 69*:1056 (1972).

LAVI, S., and WINOCOUR, E. Acquisition of sequences homologous to host deoxyribonucleic acid by closed circular simian virus 40 deoxyribonucleic acid. *J Virol 9*:309 (1972).

LINDBERG, V., and DARNELL, J. E. SV 40-specific RNA in the nucleus and polyribosomes of transformed cells. *Proc Natl Acad Sci USA 65*:1089 (1970).

ODA, K., and DULBECCO, R. Regulation of transcription of the SV 40 DNA in productivity infected and in transformed cells. *Proc Natl Acad Sci USA 60*:525 (1968).

SAMBROOK, J., WESTPHAL, H., SRINIVASAN, P., and DULBECCO, R. The integrated state of viral DNA in SV 40 transformed cells. *Proc Natl Acad Sci USA 60*:1288 (1968).

SHOPE, R. E., Infectious papillomatosis of rabbits. *J Exp Med 58*:607 (1933).

SJÖGREN, H. O. Studies on specific transplantation resistance to polyoma-virus-induced tumors. I, II, III, and IV. *J Natl Cancer Inst 32*:361 (1964).

TODARO, G. T., GREEN, H., and GOLDBERG, B. D. Transformation of properties of an established cell line by SV 40 and polyoma virus. *Proc Natl Acad Sci 51*:66 (1964).

VOGT, M., and DULBECCO, R. Virus-cell interaction with a tumor-producing virus. *Proc Natl Acad Sci USA 46*:365 (1960).

WATKINS, J. F., and DULBECCO, R. Production of SV 40 virus in heterokaryons of transformed and susceptible cells. *Proc Natl Acad Sci USA 58*:1396 (1967).

Specific Articles: Leukoviruses

BALUDA, M. A. Widespread presence in chickens of DNA complementary to the RNA genome of avian leukosis viruses. *Proc Natl Acad Sci USA 69*:576 (1972).

BALUDA, M. A., and GOETZ, I. E. Morphological conversion of cell cultures by avian myeloblastosis virus. *Virology 15*:185 (1961).

GILDEN, R. V., and OROSLAN, S. Group-specific antigens of RNA tumor viruses as markers for subinfective expression of the RNA virus genome. *Proc Natl Sci USA 69*:1021 (1972).

GROSS, L. "Spontaneous" leukemia developing in C_3H mice following inoculation, in infancy, with AK-leukemic extracts or AK-embryos. *Proc Soc Exp Biol Med 76*:27 (1951).

GROSS, L. Effect of thymectomy on development of leukemia in C_3H mice inoculated with leukemic "passage" virus. *Proc Soc Exp Biol Med 100*:325 (1959).

HANAFUSA, H., HANAFUSA, T., and RUBIN, H. The defectiveness of Rous sarcoma virus. *Proc Natl Acad Sci 49*:572 (1963).

KAPLAN, H. S. Influence of thymectomy, splenectomy and gonadectomy on incidence of radiation-induced lymphoid tumors in strain C-57 BL mice. *J Natl Cancer Inst 11*:83 (1950).

LOWY, D. R., ROWE, W. P., TEICH, N., and HARTLEY, J. W. Murine leukemia virus: High frequency activation in vitro by 5-iododeoxyuridine and 5-bromodeoxyuridine. *Science 174*:155 (1971).

MARTIN, G. S. Rous sarcoma virus: A function required for the maintenance of the transformed state. *Nature 227*:1021 (1970).

METCALF, D. Thymus graft and leukemogenesis. *Cancer Res 24*:1952 (1964).

ROUS, P. Transmission of a malignant new growth by means of a cell-free filtrate. *JAMA 56*:198 (1911).

ROWE, W. P., LOWY, D. R., TEICH, N., and HARTLEY, J. W. Some implications of the activation of murine leukemic virus by halogenated pyrimidines. *Proc Natl Acad Sci USA 69*:1033 (1972).

TEMIN, H. The RNA tumor viruses. Background and foreground. *Proc Natl Acad Sci USA 69*:1016 (1972).

TODARO, G. J., and HUEBNER, R. J. The viral oncogene hypothesis: New evidence. *Proc Natl Acad Sci USA 69*:1009 (1972).

WEISS, R. A., FRIIS, R. R., KATZ, E., and VOGT, P. K. Induction of avian tumor viruses in normal cells by physical and chemical carcinogens. *Virology 46*:920 (1971).

Specific Articles: Mouse Mammary Tumor Viruses

BITTNER, J. J. Some possible effects of nursing on the mammary gland tumor incidence in mice. *Science 84*:162 (1936).

LYONS, M. J., and MOORE, D. H. Isolation of the mouse mammary tumor virus: Chemical and morphological studies. *J Natl Cancer Inst 35*:549 (1965).

WEISS, D. W., FAULKIN, L. J., JR., and DEOME, K. B. Acquisition of heightened resistance and susceptibility to spontaneous mouse mammary carcinomas in the original host. *Cancer Res 24*:732 (1964).

chapter 64

STERILIZATION AND DISINFECTION

By BERNARD D. DAVIS and RENATO DULBECCO

Definitions   1452
Criteria of Viability   1452
Exponential Kinetics   1453

PHYSICAL AGENTS   1454
TEMPERATURE   1454
Heat   1454
Freezing   1455
RADIATION   1456
Ultraviolet Radiation   1456
Photodynamic Sensitization (Photooxidation)   1457
MECHANICAL AGENTS   1457
Ultrasonic and Sonic Waves   1457
Filtration   1457

CHEMICAL AGENTS   1457
DETERMINATION OF DISINFECTANT POTENCY   1457
SPECIFIC CHEMICAL AGENTS   1458

INACTIVATION OF VIRUSES   1462
The Degree of Inactivation   1462
Inactivating Agents; Mechanisms of Action   1463

Technics for sterilizing materials—i.e., for freeing them of contaminating viable organisms—were developed as a prerequisite for the preparation of pure cultures in the laboratory and then were rapidly adapted, in medicine, surgery, and public health, to prevent the spread of infectious disease (Ch. 1). This chapter will consider first the action of various lethal agents on bacteria and then special features of their action on viruses.

History. The early arts of civilization included practical means of preventing putrefaction and decay, long before the role of microorganisms in these processes was appreciated. Perishable foods were preserved by drying, by salting, and by acid-producing fermentations (Ch. 3). Embalming was practiced in ancient Egypt, but the essential oils used were probably less important than the dry climate. As was noted in Chapter 1, the canning of food was introduced 50 years before Pasteur's researches gave it a rational basis. Finally, chlorinated lime (calcium hypochlorite) and carbolic acid (phenol) were introduced in the early nineteenth century to deodorize sewage and garbage (and subsequently wounds), even before their germicidal action was recognized.

DEFINITIONS

Sterilization denotes the use of either physical or chemical agents to eliminate **all** viable microbes from a material, while **disinfection** generally refers to the use of germicidal chemical agents to destroy the potential **infectivity** of a material (which need not imply elimination of all viable microbes). **Sanitizing** refers to procedures used to lower the bacterial content of utensils used for food, without necessarily sterilizing them. In contrast to chemotherapeutic agents, disinfectants must be effective against all kinds of microbes, must be relatively insensitive to their metabolic state, and need not be harmless to host cells.

Antisepsis usually refers to the topical application of chemicals to a body surface to kill or inhibit pathogenic microbes. Disinfectants are widely used for skin antisepsis in preparation for surgery, but for prophylactic application to open wounds, or for topical application to superficial infections, they have been largely replaced by various antibiotics, which are painless and less damaging to the tissues.

CRITERIA OF VIABILITY

Though some chemotherapeutic agents are bactericidal while others are only bacteristatic (Ch. 7), an effective disinfectant must be bactericidal. The fundamental criterion of bactericidal action is loss of the **ability of the organism to propagate indefinitely** when placed in a suitable environment; nonviable, "killed" cells may or may not exhibit changes in such properties as morphology, staining, motility, and enzymatic activities.

For example, respiration ceases and normally excluded dyes may penetrate, and phage multiplication is blocked, in cells killed by heat but not in cells killed by ultraviolet (UV) radiation. Similarly, irradiated spores can often germinate but may then exhibit little or no division.

Even direct tests for viability may give quite different counts, with damaged cells, in different media. The repair of certain kinds of damage is evidently influenced by several factors, including osmotic tonicity and nutritional richness. A striking example is Hg^{++}: spores "sterilized" by this agent can be **"resurrected"** by washing a solution of H_2S, which displaces Hg^{++} from its inhibitory complexes with SH groups in cell constituents.

The problem of defining viability assumes practical importance in the preparation of vaccines, which are often sterilized as gently as possible in order to retain maximal immunogenicity. Though their sterility is ordinarily tested in artificial culture medium, what is important is their ability to initiate infection in the animal body.

It should be emphasized that sterilization is not identical with **destruction** of bacteria or their products, though the terms are often loosely interchanged. For example, in preparing solutions for intravenous administration it is not sufficient to ensure sterility; it is also necessary to minimize previous bacterial contamination, since **pyrogenic** bacterial products (endotoxins, Ch. 29), which can survive autoclaving or filtration, may subsequently produce a febrile, toxic response. Hence in the preparation of biologicals and parenteral fluids the water and the reagents used must satisfy criteria of purity quite different from those required for analytical chemical work.

Differential Susceptibility. Susceptibility of cells to disinfectants or to heat varies with their physiological state. The cells in a young culture are generally more susceptible than those in a stationary-phase culture, in which significant changes in the cell wall and membrane have been noted (Ch. 6).

EXPONENTIAL KINETICS

Sterilization of bacteria by many agents exhibits the kinetics of a first-order reaction, in which the logarithm of the number of survivors decreases as a linear function of time of exposure:

$$\ln n = \ln n_o - kt.$$

Such kinetics, observed for interactions with UV or ionizing radiation, are discussed in the Appendix of Chapter 11. What is perhaps surprising is that similar kinetics are often observed for the action of heat (Fig. 64-1) or of various chemical disinfectants, such as phenol. Hence, even though heat and phenol no doubt damage a variety of protein molecules, in a cumulative manner, the lethal

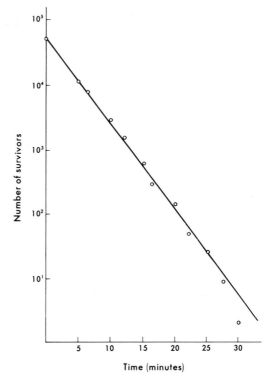

Fig. 64-1. Exponential killing by heat of spores of a thermophilic bacillus. Number of survivors plotted on a logarithmic scale against time of exposure to 120°. [After Williams, C. C., Merrill, C. M., and Cameron, E. J. *Food Res* 2:369 (1937).]

event in a given cell cannot be the denaturation of the last of many molecules of a given essential species of protein, for this process would give a multi-hit curve (see Fig. 64-8 below). The one-hit curve suggests instead damage to one or another single indispensable molecule, perhaps triggering irreversible damage to the cell membrane or irreversibly impairing DNA replication.

Heterogeneity of the population of cells, with respect to either composition or physiological state, will of course distort first-order kinetics in the direction of multicomponent kinetics (Ch. 11; see also Fig. 64-8). Hence it is not surprising that the best exponential killing curves have been obtained with spores, which are more uniform than vegetative cells.

PHYSICAL AGENTS

TEMPERATURE

HEAT

Heat is generally preferred for sterilizing all materials except those that it would damage. The process is rapid, all organisms are susceptible, and the agent penetrates clumps and reaches surfaces that might be protected from a chemical disinfectant. Fungi, most viruses, and the vegetative cells of various pathogenic bacteria are sterilized within a few minutes at 50 to 70°, and the spores of various pathogens at 100°. Consequently, it has been a common practice to sterilize syringes, needles, and instruments for minor surgery by heating for 10 to 15 minutes in boiling water, or even better in a boiling dilute solution of alkali (e.g., washing soda).

The spores of some saprophytes, however, can survive boiling for hours. Since absolute sterility is essential for culture media, and for the instruments used in major surgical procedures, it has become standard practice to sterilize such materials in an autoclave at a temperature of 121° C (250° F) for 15 to 20 minutes. This temperature is attained by steam at a pressure of 15 lb per square inch (psi) in excess of atmospheric pressure, at altitudes near sea level; at high altitudes higher pressures are necessary (e.g., 3 lb psi higher in Denver, at an altitude of 5000 ft). The rapid action of steam depends in part on the large latent heat of water (540 cal/gm): cold objects are thus rapidly heated by condensation on their surface.

In using an autoclave it is important that flowing steam be allowed to displace the air before building up pressure, for in steam mixed with air the temperature is determined by the partial pressure of the water vapor. Thus if the air at 1 atm (15 lb psi) remains in the chamber and steam is added to provide an additional gauge pressure of 1 atm the average temperature will be only 100° (that of steam at 1 atm). Moreover, heating will be uneven since the air will tend to remain at the bottom of the chamber.

Vessels must be loosely plugged or capped and not completely filled with liquid, in order to permit free ebullition of the dissolved air during heating and free boiling of the superheated liquid when the steam pressure is lowered. With bulky porous objects (e.g., bundles of surgical dressings), or with large volumes of liquid, increased time must be allowed to ensure heating throughout. In modern autoclaves the steam pressure may be maintained in an outer jacket while the central chamber is decompressed, so that condensation water is rapidly evaporated.

Pasteurization, introduced to sterilize wine, is now used primarily for milk. It consists of heating at 62° for 30 minutes, or in "flash" pasteurization at a higher temperature for a fraction of a minute. Pasteurization is effective because the common milkborne pathogens (tubercle bacillus, *Salmonella, Streptococcus,* and *Brucella*) do not form spores and are reliably sterilized by this procedure; in addition, the total bacterial count is generally reduced by 97 to 99%.

Kinetics. The sensitivity of an organism to heat is often expressed in practical work as the **thermal death point:** the lowest temperature at which 10 minutes' exposure of a given volume of a broth culture results in sterilization. The value is about 55° for *E. coli,* 60° for the tubercule bacillus, and 120° for the most resistant spores.

For precise studies this qualitative endpoint has been replaced by quantitative determination of the numbers of survivors at different times. Since killing by heat turns out to have simple exponential kinetics the rate of killing can be expressed in terms of the **rate constant** k (see above) in the exponential decay curve (Fig. 64-1). Another convenient index is the "decimal reduction time," D (the time required for a 10-fold reduction of viability), which is inversely proportional to the rate of killing. The logarithm of D varies linearly with temperature (Fig. 64-2); and from the slope of the curve it can be seen that the rate of killing (of the spores studied in this figure) increased about 10-fold with a rise of 10°. From D it is easy to calculate the time required to sterilize (within statistical variation) a sample of a given size: for example, a suspension containing 10^5 cells would require about $5 \times D$ minutes (which would reduce the viable number to 10^{-5} times its initial value; i.e., to 1 cell).

The **mechanism** of sterilization by heat involves **protein denaturation** (though the "melting" of membrane lipids may also be important): the temperature range of sterilization is one in which many proteins are denatured, with a high temperature coefficient. Moreover, both processes require a higher temperature when the material is thoroughly dried, or when the water activity of the medium is reduced by the presence of a high concentration of a neutral substance such as glycerol or glucose.

Moist Heat and Dry Heat. The role of water in heat denaturation of proteins is illustrated by the usefulness of steam in pressing woolen fabrics (i.e., in shifting the multiple weak bonds between fibrous molecules of keratin). The native conformation of a protein is stabilized in part by hydrogen bonds (especially between $>C=O$ and $HN<$ groups), which are more readily broken if they can be replaced with hydrogen bonds to water molecules (Fig. 64-3). Accordingly, bacteria and viruses, like isolated enzymes, require a higher temperature for irreversible damage in the dry state: reliable sterilization of spores by dry heat requires 160° for 1 to 2 hours. Moreover, hot air penetrates porous materials much more slowly than does condensing steam, so that after an hour in an oven at 160° the center of a large package of surgical dressings may not even have reached 100°. Dry ovens are ordinarily used for sterilizing only glassware and metal objects. Intense dry heat is used in flaming contaminated surfaces in the process of aseptic transfer* and in disposing of infectious material by incineration.

FREEZING

When a suspension of bacteria is frozen the crystallization of the water results in the formation of tiny pockets of concentrated solutions of salts, which do not themselves crystallize unless the temperature is lowered below the eutectic point (ca. −20° for NaCl); at this temperature the solution becomes saturated and the salt also crystallizes. The localized

---

* In the early days of developing a satisfactory ritual for aseptic transfer Pasteur recommended quick flaming of the bacteriologist's hands!

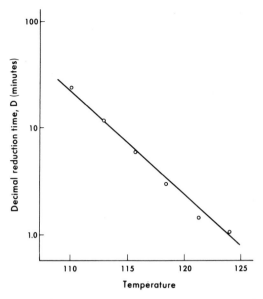

Fig. 64-2. Rate of killing of spores of a thermophilic bacillus at various temperatures, indicated in terms of the time required for a 10-fold reduction in viability. (From Schmidt, C. F. In *Antiseptics, Disinfectants, Fungicides, and Sterilization.* C. F. Reddish, ed. Lea & Febiger, Philadelphia, 1957, p. 831.)

high concentrations of salt, and possibly the ice crystals, damage the bacteria, as shown by increased sensitivity to lysozyme. Only some of the cells, however, are killed, but repeated cycles of freezing and thawing result in progressive decrease in the viable count.

Preservation. Once frozen, the surviving cells retain their viability indefinitely if the temperature is kept below the eutectic points of the various salts present; a household freezing unit (∼ −10°) is not cold enough, but a satisfactory temperature is provided by solid CO_2 (−78°) or liquid N_2 (−180°). Freezing is therefore a useful means of preserving viable cultures.

In preserving bacteria, viruses, or animal cells by freezing it is helpful to add a relatively high concentration of glycerol, dimethylsulfoxide, or protein.

Fig. 64-3. Role of water in promoting denaturation of protein by heat, by facilitating disruption of intramolecular hydrogen bonds between peptide groups.

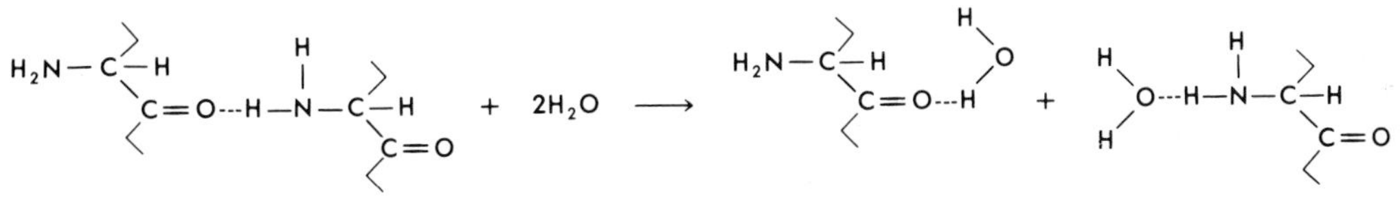

These agents promote amorphous, vitreous solidification on cooling, instead of crystallization thus avoiding local high concentrations of salt. Similarly, protein-rich materials (milk, serum) are added in the preservation of bacteria by **lyophilization** (desiccation from the frozen state).

RADIATION

ULTRAVIOLET RADIATION

With radiation of decreasing wave length sterilization of bacteria first becomes appreciable at 330 nm and then increases rapidly (Fig. 64-4). The sterilizing effect of sunlight is due mainly to its content of UV light (300–400 nm). Most of the UV light reaching the earth from the sun, and all of that below 290 nm, is screened out by the ozone in the outer regions of the atmosphere; otherwise organisms could not survive on the earth's surface.

Photochemistry. The energy of UV light is absorbed in quanta by molecules of an appropriate structure. The absorbing molecule is thereby activated, resulting either in increased interatomic vibration or in excitation of an electron to a higher energy level. The activated molecule may undergo rupture of a chemical bond and formation of new bonds; or it may transfer most of its extra energy by collision to an adjacent molecule, which can then similarly undergo a chemical reaction; or the energy may be entirely dissipated by collision as increased translational energy (heat).

UV absorption by bacteria is due chiefly to the purines and pyrimidines of nucleic acids, with an average maximum at 260 nm, and less to proteins, in which the aromatic rings of tryptophan, tyrosine, and phenylalanine absorb more moderately with an average maximum at 280 nm. The sterilization action spectrum (i.e., the efficiency of sterilization by radiation of various wave lengths; Fig. 64-4) parallels the absorption spectrum of the bacteria, suggesting that absorption either by nucleic acid or by protein can have a lethal effect.

The mechanism of killing by UV light has been discussed in Chapter 11 (Radiation damages), where it was noted that lethal mutations make only a small contribution; the major mechanism is alterations in the DNA that block its replication. Since most of these alterations can be repaired, by several different mechanisms, the quantum efficiency of UV sterilization is ordinarily very low.

The contribution of damage to other parts of the cell is negligible. Thus in isolating phage λ repressor Ptashne used *E. coli* so heavily irradiated that essentially none of the DNA was capable of being transcribed. Despite the overkill the cells could effectively transcribe and translate phage DNA subsequently introduced by infection.

Practical Uses. Inexpensive low-pressure mercury-vapor lamps, emitting 90% at 254 nm, are widely used to decrease airborne infection, e.g., in places of public crowding, barracks, hospital wards, surgical operating rooms, and rooms containing experimental animals. The effectiveness of UV treatment of air in public places seems uncertain, but it has been convincingly demonstrated in hospital wards and in animal houses, where the infected individuals cannot make close contact with other individuals.

In laboratory areas used for bacterial transfers UV lamps are similarly useful to decrease contamination of cultures and infection of workers. It is important to protect the eyes by eyeglasses, since excessive exposure of the cornea to UV causes severe irritation, with a latent period of about 12 hours. (Glass is opaque to UV.)

In preparing killed bacterial vaccines UV irradia-

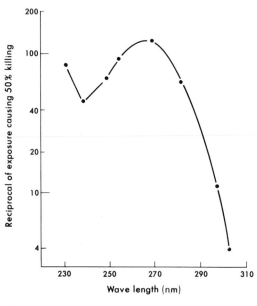

Fig. 64-4. Action spectrum of UV killing of *E. coli.* (Ordinate: reciprocal of the incident energy required for 50% killing.) [After Gates, F. L. *J Gen Physiol 14*:31 (1930).]

tion has a theoretical advantage, since the genome is much more sensitive to UV damage than are the surface antigens. Nevertheless, this approach has not been particularly successful. One problem is the difficulty of avoiding clumps, in whose centers virulent organisms remain unexposed. Moreover, tissue extracts are quite opaque to UV light.

Ionizing Radiations. The lethal and mutagenic actions of X-ray and other ionizing radiations have been discussed in Chapter 11. Intense sources of radioactivity can be used to sterilize hospital goods, foods, etc. With foods, however, the large doses required (millions of rads) often have undesirable effects on flavor.

PHOTODYNAMIC SENSITIZATION (PHOTOOXIDATION)

In strong **visible light** certain fluorescent dyes (e.g., methylene blue, rose bengal, eosin) sterilize bacteria and viruses (and also lyse red blood cells and denature proteins). These dyes retain an absorbed quantum for a comparatively long time (10^{-6} to 10^{-8} seconds), during which the energy sometimes is transferred to another molecule instead of being emitted as fluorescence. This transfer leads to oxidation of certain residues in proteins (especially histidine and tryptophan) and in nucleic acids.

Even in the absence of added dyes, intense visible light is capable of killing bacteria, presumably via physiologically occurring photosensitizing substances, such as **riboflavin** and **porphyrins**. It is therefore inadvisable to expose bacterial cultures to direct sunlight, even when protected by UV-absorbing glass. For example, BCG vaccine in glass ampules can lose all viability and effectiveness on exposure to bright sunlight in outdoor field stations.

MECHANICAL AGENTS

ULTRASONIC AND SONIC WAVES

In the supersonic (ultrasonic) range, with a frequency of 15,000 to several hundred thousand per second, sound waves denature proteins, disperse a variety of materials, and sterilize and disintegrate bacteria. (Audible sonic waves of sufficient intensity are also bactericidal.) The effect has not been of practical value as a means of sterilization but it is useful for disrupting cells for experimental purposes ("sonication").

FILTRATION

Bacteria-free filtrates may be obtained by the use of filters with a maximum pore size not exceeding 1 nm. This procedure is used for solutions that cannot tolerate sterilization by heat (e.g., sera, and media containing proteins or labile metabolites). The Seitz filter (of asbestos) and the Berkefeld filter (of diatomaceous earth) are quite adsorptive. More satisfactory filters are made of unglazed porcelain (e.g., Selas filters), sintered (= fritted) glass (e.g., Corning UF), or nitrocellulose (e.g., Millipore).

The last type (membrane filters) can also be used to recover bacteria quantitatively for chemical or microbiological analysis. Moreover, bacterial colonies can be grown on such a filter resting on a solid nutrient medium, and transfer of the filter permits convenient exposure of the colonies to a succession of different media.

CHEMICAL AGENTS

At sufficiently high concentrations many chemicals, including nutrients (such as O_2 and fatty acids), are bacteristatic and even bactericidal. The term disinfectant is restricted to substances that are rapidly bactericidal at low concentrations. In contrast to lethal radiations (which damage the DNA) and to most bactericidal chemotherapeutic agents (which interact irreversibly with various active metabolic systems) most disinfectants act either by dissolving lipids from the cell membrane (detergents, lipid solvents) or by damaging proteins (denaturants, oxidants, alkylating agents, and sulfhydryl reagents).

The rate of killing by disinfectants in-

creases with concentration and with temperature. Anionic compounds are more active at low pH and cationic compounds at high pH. This effect results from the greater penetration of the undissociated form of the inhibitor, and possibly also from the increase in opposite charges in cell constituents.

DETERMINATION OF DISINFECTANT POTENCY

Phenol Coefficient. Ever since Lister began spraying surgical operating rooms with phenol this compound has been considered the standard disinfectant, though it is required in a higher concentration than almost any other: 0.9% to sterilize a suspension of *Salmonella typhosa* under ordinary conditions in 10 minutes. The phenol coefficient of a compound is the ratio of the minimal sterilizing concentration of phenol (under standard conditions) to that of the compound. In the official test of the U.S. government a broth culture is diluted 1:10 with various concentrations of the test compound; the endpoint is the lowest concentration that yields, after incubation for 10 minutes at 20°, sterile loopful samples. The germicide is generally recommended for use at five times this concentration. Two organisms are ordinarily used: *S. typhosa* (as a representative enteric pathogen) and *Staphylococcus aureus* (the major source of wound infection).

The phenol coefficient provides a reasonable index for comparing various phenol derivatives, which exhibit similar kinetics and mode of action; but it is less satisfactory for other agents, which may differ in their concentration-action curves and in their susceptibility to neutralization by the environment. Thus the concentrations of a disinfectant, *c*, required to sterilize a bacterial population in varying time, *t*, generally correspond to a curve which may be fitted by the equation ($c^n t = k$); and while phenol has the remarkably high concentration coefficient (*n*) of 5 to 6, oxidants such as hypochlorite have a value of about 1. Hence the five-fold increase above the endpoint concentration provides a much wider margin of safety for phenols than for oxidants.

Problems in Evaluating Disinfectants. Some agents (e.g., mercurials, detergents) may adhere to the bacteria and thus exert a bacteristatic action, in the subinoculated samples, that mimics bactericidal action. To test such materials for bactericidal action it is important to include a **neutralizing compound** in the test medium.

In addition, the effectiveness of a disinfecting procedure often depends strongly on the "cleanness" of the material. For example, both gases and organic liquids may fail to reach bacteria encased in crystals deposited by the drying of an aqueous solution: hence porous surfaces (fibers), which take up the drying solution, may be easier to sterilize than nonporous surfaces (metal, glass). Moreover, the presence of a large amount of organic matter (e.g., in excreta or discarded cultures) rapidly neutralizes the action of many types of agents: reactive chemicals (such as oxidants) are chemically altered, and many compounds are adsorbed (especially by serum albumin in samples containing blood).

Because of these considerations, different kinds of disinfectants are used for different purposes, such as skin antisepsis, sanitizing food containers, or rendering discarded cultures harmless. Efficacy must finally be tested under the conditions of use.

As with killing by heat (Fig. 64-1), the sensitivity of an organism to a disinfectant can be expressed more precisely as the slope of the semilogarithmic curve for killing as a function of time, rather than as an endpoint for complete sterilization. However, the curves for chemical disinfection are often imperfectly exponential, with the physiologically more resistant members of the population surviving longer than would be predicted by extrapolation. Hence the endpoint remains of practical value. The statistical problem of defining complete sterilization will be further considered below in connection with the preparation of viral vaccines.

SPECIFIC CHEMICAL AGENTS

Acids and Alkalis. Strongly acid and alkaline solutions are actively bactericidal. However, mycobacteria are relatively resistant, it being common practice to liquefy sputum by exposure for 30 minutes to 1 N NaOH or H_2SO_4. This procedure depends on the survival of a fraction of the population, rather than on complete resistance of the bacteria. Gram-positive staphylococci and streptococci also frequently survive.

Weak organic acids exert a greater effect than can be accounted for by the pH: the presence of highly permeable undissociated molecules promotes penetration of the acid into the cells, and the increasing activity with chain length suggests a direct action of the organic compound itself. (Long-chain fatty acids will be considered below under Surface-active agents.) Lactic acid is the natural preservative of many fermentation products; and salts of propionic acid (CH_3CH_2COOH) are now frequently added to bread and other foods to retard mold growth.

Salts. Pickling in brine, or treatment with solid NaCl, has been used for many centuries as a means of preserving perishable meats and fish. Bacteria vary widely in susceptibility.

Though physiological saline (0.9% NaCl) is widely used as a diluent for bacteria, it is not very suitable: a balanced salt solution, containing Mg^{++} and buffer, permits much better survival. Strains vary widely in their ability to survive in distilled water. Some of the lethal action, however, is due to traces of heavy metal ions, which are more bactericidal in the absence of competitive ions.

Heavy Metals. The various metallic ions can be arranged in a series of decreasing antibacterial activity; Hg^{++} and Ag^+, at the head of the list, are effective at less than 1 part per million (ppm). This potency, however, does not reflect any remarkable effect of comparatively few ions on the cell, for cells take up relatively large numbers of these ions from very dilute solutions. Thus bacteria, trypanosomes, or yeast killed by Ag^+ contain 10^5 to 10^7 Ag^+ ions per cell. The concentration required for killing is therefore markedly affected by the inoculum size.

As was noted above, the antibacterial action of Hg^{++} can be readily reversed by sulfhydryl compounds (whose affinity for Hg^{++} gave rise to the term "mercaptan"). Similar inhibition and reversal can be observed with those enzymes whose activity involves an SH group.

Mercuric chloride, once popular as a disinfectant, is now obsolete. However, various organic mercury compounds (e.g., Merthiolate, Mercurochrome, Metaphen), in which one of the valences of Hg is covalently combined, are used as relatively nonirritating antiseptics, and also as preservatives for sera and vaccines.

Silver has long been used in various forms as a mild antiseptic. Before penicillin became available silver nitrate was used topically for the prophylaxis of neonatal gonococcal ophthalmitis. Organic compounds of arsenic, bismuth, and antimony have been used in the chemotherapy of syphilis and of certain protozoal diseases. Gold has also been used chemotherapeutically but with questionable value. Copper salts have great importance as fungicides in agriculture but not in medicine.

Inorganic Anions. Inorganic anions are much less toxic than some of the cations. Boric acid has found wide use as a mild antiseptic.

Halogens. Iodine combines irreversibly with proteins (e.g., by iodinating tyrosine residues), and it is an oxidant. **Tincture of iodine** (a 2 to 7% solution of I_2 in aqueous alchohol containing KI) is a rapidly acting bactericide. It is a reliable antiseptic for skin and for minor wounds, but it has a painful and destructive effect on exposed tissue. I_2 complexes spontaneously with detergents to form **iodophors,** which provide a readily available reservoir of bound I_2 in equilibrium with free I_2 at an effective but nonirritating concentration.

Chlorine was the antiseptic introduced (as chlorinated lime) by O. W. Holmes in Boston in 1835, and by Semmelweis in Vienna in 1847, to prevent transmission of puerperal sepsis by the physician's hands. Chlorine combines with water to form hypochlorous acid (HOCl), a strong oxidizing agent:

$$Cl_2 + H_2O \rightleftharpoons HCl + HOCl$$

or

$$Cl_2 + 2NaOH \rightleftharpoons NaCl + NaOCl + H_2O$$

Hypochlorite solutions (200 ppm Cl_2) are used to sanitize clean surfaces in the food and the dairy industries and in restaurants; and Cl_2 gas, added at 1 to 3 ppm, is widely used to disinfect water supplies and swimming pools. Chlorine is a reliable, rapidly acting disinfectant for such "clean" materials, but it is less satisfactory for materials containing organic matter that can react rapidly with the Cl_2. The "chlorine demand" of a water supply increases with its content of organic matter, and chlorination must be titrated to a definite level of free Cl_2. This reactivity is a virtue in the sanitizing of food utensils: residual traces of chlorine will be rapidly destroyed on subsequent contact with food, leaving no flavor or odor.

Other Oxidants. Hydrogen peroxide (H_2O_2), in a 3% solution, was once widely used as an antiseptic, but it cannot be strongly recommended. Bacteria vary widely in their susceptibility, since some species possess catalase. **Potassium permanganate** ($KMnO_4$) is of value as a urethral antiseptic in concentrations around 1/1000. **Peracetic acid** ($CH_3CO\text{-}O\text{-}OH$), a strong oxidizing agent, is used as a vapor for the sterilization of chambers for germ-free animals; its reliable disinfection compensates for the inconveniently long flushing required because of its toxicity for animals. These compounds, as well as the halogens, presumably act by oxidizing SH and S-S groups of enzymes and membrane components.

Alkylating Agents. Formaldehyde and ethylene oxide replace the labile H atoms on $-NH_2$ and $-OH$ groups, which are abundant in proteins and nucleic acids, and also on $-COOH$ and $-SH$ groups of proteins (Fig. 64-5). The reactions of formaldehyde are in part reversible, but the high-energy epoxide bridge of ethylene oxide leads to irreversible reactions. These alkylating agents, in contrast to other disinfectants, are nearly as active against spores as against vegetative bacterial cells, presumably because they can penetrate easily (being small and uncharged) and do not require H_2O for their action.

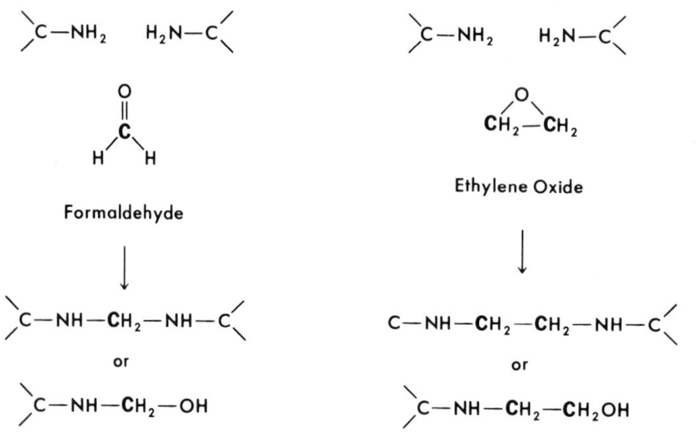

Fig. 64-5. Reactions of formaldehyde and ethylene oxide with amino groups. Similar condensations may take place with other nucleophilic groups. Bridges may be formed between groups on the same molecule or on different molecules.

Formaldehyde is a gas, usually marketed as a 37% aqueous solution (formalin). It has long been used at ca. 0.1% in preparing vaccines: i.e., sterilizing bacteria, or inactivating toxins or viruses, without destroying their antigenicity.

Formaldehyde may be used as a gas for sterilizing dry surfaces. However, it is extensively absorbed on surfaces as a reversible polymer (paraformaldehyde), whose slow subsequent depolymerization provides an irritating residue.

Ethylene oxide, a highly water-soluble gas, has proved to be the most reliable substance available for gaseous disinfection of dry surfaces. However, its action is slower than that of steam, and its use is more expensive and presents some hazard of residual toxicity (vesicant action). Indeed, the potential hazards of mutagenicity and carcinogenicity for man deserve careful investigation, since formaldehyde and ethylene oxide, like other alkylating agents (Ch. 11), have been shown to be mutagenic to bacteria, plant seeds, and *Drosophila.*

Ethylene oxide is widely used to sterilize heat-sensitive objects: plasticware; surgical equipment; hospital bedding; and books, leather, etc. handled by patients. Ethylene oxide is explosive, but this hazard is eliminated by using a mixture with 90% CO_2 or a fluorocarbon.

Surface-active Agents. These compounds (surfactants) are generally called synthetic detergents. Such compounds, like fatty acids (soaps), contain both a hydrophobic and a hydrophilic portion; they therefore form micelles (large aggregates) in aqueous solution, in which only the hydrophilic portion is in contact with water; they can similarly form a layer that coats and solubilizes hydrophobic molecules or structures. Anionic detergents are only weakly bactericidal, perhaps because they are repelled by the net negative charge of the bacterial surface. Nonionic detergents are not bactericidal and may even serve as nutrients.

Cationic detergents, however, are active against all kinds of bacteria. The most effective types are the **quaternary** compounds, containing three short-chain alkyl groups as well as a long-chain alkyl group (e.g., benzalkonium chloride, Fig. 64-6). These compounds are widely used for skin antisepsis and for sanitizing food utensils. They act by disrupting the cell membrane, causing release of metabolites; in addition, their detergent action provides the advantage of dissolving lipid films that may protect bacteria, and they leave a tenacious bactericidal surface film on the treated object.

A variety of cationic germicides are on the market because of patent rights. In the absence of adsorbing macromolecules or lipids they may be rapidly bac-

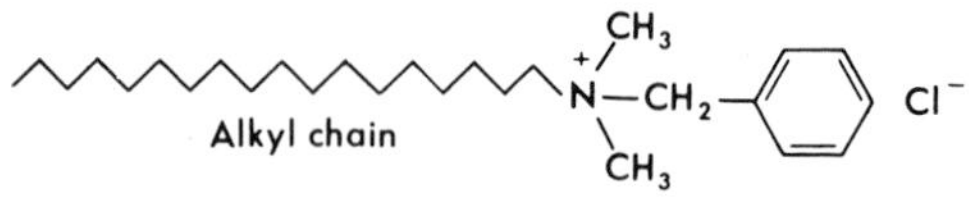

Fig. 64-6. Benzalkonium chloride (benzyldimethyl alkonium chloride; Zephiran). A typical quaternary ammonium detergent; the long-chain alkyl group is a mixture obtained by the reduction and amination of the fatty acids of vegetable or animal fat.

tericidal at concentrations as low as 1 ppm; and unlike many other disinfectants they are not poisonous to man. Their activity is neutralized by soaps and phospholipids, since oppositely charged surfactants precipitate each other. With increasing chain length bactericidal action (and depression of surface tension) goes through a maximum because the tendency to form a surface layer increases but the solubility decreases.

Phenols. Phenol (C_6H_5OH) is both an effective denaturant of proteins and a detergent. Its bactericidal action involves cell lysis.

The antibacterial activity of phenol is increased by halogen or alkyl substituents on the ring, which increase the polarity of the phenolic OH group and also make the rest of the molecule more hydrophobic; the molecule becomes more surface-active and its antibacterial potency may be increased a hundredfold or more. Phenols are more active when mixed with soaps which increase their solubility and promote penetration. However, too high a proportion of soap impairs activity, presumably by dissolving the disinfectant too completely in soap micelles.

With increasing chain length potency first increases and then decreases, presumably because of low solubility. With gram-negative organisms the maximum is reached at a relatively short length (Fig. 64-7); bulkier molecules are excluded, like certain drugs (Ch. 6), by the outer membrane.

A mixture of **tricresol** (mixed *ortho, meta,* and *para*-methylphenol) and soap is a widely used disinfectant for discarded bacteriological materials. Its action is not impaired by the presence of organic matter, because it must be used in a relatively high concentration and it is not extensively destroyed or bound by organic molecules. **Hexylresorcinol** (4-hexyl-1,3-dihydroxybenzene) is used as a skin antiseptic.

Halogenated bis-phenols, such as **hexachlorophene,**

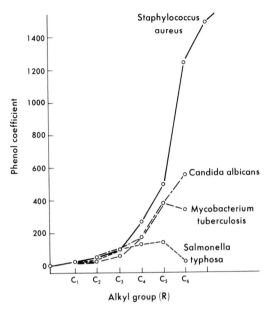

are known to be bacteristatic in very high dilutions, and appear to be less inactivated by soaps or anionic detergents than are ordinary phenols. Hexachlorophene is also not very volatile and lacks the unpleasant odor of many phenols. It is widely used as a skin antiseptic (especially mixed with a detergent, as pHisoHex), and in deodorant soaps (which hinder bacterial decomposition of sweat). However, under conditions permitting its absorption (e.g., skin of premature infants) systemic toxicity may be serious.

Fig. 64-7. Germicidal activity of homologous series of *o*-alkyl-*p*-chlorophenol derivatives against four organisms. The parent compound (*p*-chlorophenol) had a phenol coefficient (see p. 1457) of 4 against all the organisms. Note the strikingly increased activity of the longer-chain (C-6 and C-7) derivatives against the gram-positive *Staphylococcus,* and the opposite response of the gram-negative *Salmonella.* A similar divergence in response at greater chain lengths has also been observed with other homologous series of phenol derivatives. (From Klarmann, E. G., and Wright, E. S. In *Antiseptics, Disinfectants, Fungicides, and Sterilization.* C. F. Reddish, ed. Lea & Febiger, Philadelphia, 1957, p. 506.)

Alkyl esters of **p-hydroxybenzoic acid** are used as a preservative in foods and pharmaceuticals: they act on bacteria much like an alkyl-substituted phenol, but when taken by mouth they are nontoxic because they are rapidly hydrolyzed, yielding the harmless free *p*-hydroxybenzoate.

The **essential oils** of plants, which have been used since antiquity as preservatives and antiseptics, contain a variety of phenolic compounds, including thymol (5-methyl-2-isopropylphenol) and eugenol (4-allyl-2-methoxyphenol); the latter is used in dentistry as an antiseptic in cavities.

Alcohols. The disinfectant action of the aliphatic alcohols increases with chain length up to 8 to 10 carbon atoms, above which the water solubility becomes too low. Although **ethanol** (CH_3CH_2OH) has received widest use, **isopropyl alcohol** (CH_3-CHOHCH₃) has the advantages of being less volatile, slightly more potent, and not subject to legal restrictions as a potential beverage.

The disinfectant action of alcohols, like their denaturing effect on proteins, involves the participation of water. Ethanol is most effective in 50 to 70% aqueous solution: at 100% it is a poor disinfectant, in which anthrax spores have been reported to survive for as much as 50 days; and its bactericidal action is negligible at concentrations below 10 to 20%. Some organic disinfectants (formaldehyde, phenol) are less effective in alcohol than in water because of the lowered affinity of the disinfectant for the bacteria relative to the solvent. On the other hand, alcohol removes lipid layers that may protect skin organisms from some other disinfectants.

Other organic solvents, such as ether, benzene, acetone, or chloroform, also kill bacteria but are not reliable disinfectants. However, the addition of a few drops of toluene or chloroform to saturate aqueous solutions will prevent growth of fungi or bacteria. **Glycerol** is bacteristatic at concentrations exceeding 50% and is used as a preservative for vaccines and other biologicals, since it is not irritating to tissues.

Aerosols. The success of sanitary engineering in controlling waterborne infection has not been paralleled by any comparable control over airborne infection. Lister's practice of spraying operating rooms with phenol was soon abandoned. Propylene glycol ($CH_3CHOHCH_2OH$) and diethylene glycol ($HOCH_2$-$CH_2OCH_2CH_2OH$) reduce the bacterial count in air when dispersed in fine droplets, at concentrations nontoxic to man, but their activity is unfortunately too sensitive to humidity: at high humidities the glycol droplets take up water and become too dilute, while at low humidities the desiccated bacteria no longer attract glycols. Moreover, the glycols do not disinfect surfaces (e.g., bedding, floor dust), from which aerial contamination is renewed.

INACTIVATION OF VIRUSES

Inactivation of viral particles is the permanent loss of infectivity. The purpose may be simply sterilization or it may require retention of other viral properties (e.g., immunogenic specificity in killed vaccines). Accordingly, studies of inactivation are concerned with determining both its **degree** and its **consequences** for the various components of the virions.

THE DEGREE OF INACTIVATION

The exposure of a population of virions to a chemical or physical inactivating agent at a defined concentration for a limited time results in the inactivation of a proportion of the virions; the others retain infectivity. The extent of inactivation (i.e., the proportion of virions that is inactivated) is related to the **dose** (i.e., the product of time × concentration or intensity), as described for the action of radiations in Chapter 11, Appendix. The shape of the survival curves yields information about the mechanism of inactivation; and once the shape is known it is possible to calculate **the dose of the agent required** for a certain degree of inactivation.

What is the required degree of inactivation? As shown in Chapter 11, Appendix, in the usual semilogarithmic plot the survival curves may vary in shape at low doses of the inactivating agents and become straight lines at higher doses. In such a plot a survival equal to zero is never reached (since the logarithm of zero is negative infinity). In practice, the sample exposed to inactivation contains a finite number of virions, and therefore a dose can be reached at which less than a single surviving virion exists on the average in the whole sample. At such doses survival must be interpreted in terms of probability. For instance, if the theoretical survival is ½ a virion, there is a chance of ½ that a virion will survive in the sample; and if many similar samples are inactivated in the same way, half of them will contain an active virion. Although total inactivation thus cannot be reached with certainty it is possible to achieve a safe inactivation. The corresponding degree of theoretical inactivation varies for different purposes; it depends on the acceptable risk of retaining an active virion and on the total amount of the virus to be inactivated.

Example: A vaccine of inactivated virus is prepared, starting with a virus titer of 10^7 infectious units per milliliter, of which 0.1 ml is inoculated per individual; a 1% risk of infectious virus surviving in the inoculum is acceptable. A theoretical survival of 10^{-8} would be satisfactory for a single dose of vaccine. If a million individuals are to be inoculated, the same survival would cause 10^4 individuals to receive an infectious dose, and the risk would be unacceptable. For a million doses, the theoretical survival should become 10^{-14}, which would lower the risk to a 1% chance of one case for the total inoculated population.

Predictions Based on Survival Curves. Such low theoretical survivals cannot be determined experimentally, but can only be inferred by extrapolating the curves observed at much higher survivals. **Extrapolation always involves a risk** because it is never quite sure that the shape of the survival curve in the undetermined segment follows a predictable behavior. Barring unforeseeable events, however, extrapolation is justified if the survival curve is sufficiently well defined in the part that can be determined; faulty extrapolations usually derive from inadequate data in this region. Errors occur especially when the survival curve is either of the **multiple-hit** or the **multicomponent** type (Ch. 11, Appendix) and it is erroneously assumed that the slope of the lower part is known.

The errors that can arise **are of two types,** as shown in Figure 64-8. With a multiple-hit type of survival curve (case A) the dose of inactivating agent required for achieving an acceptable level of infectivity is overestimated, which may decrease the immunogenicity of the vaccine. With a multicomponent type of curve (case B) the dose is underestimated and a dangerously high level of infectious virus could remain. Errors of this type are more common, and their consequences may be more serious.

INACTIVATING AGENTS; MECHANISMS OF ACTION

Some agents are relatively selective, i.e., show a more pronounced action on one viral component than on another. Since this selectivity is important for practical applications, the agents are grouped according to their main target as nucleotropic, proteotropic, lipotropic, and universal (unselective) agents. However, the results vary with the composition of the virus. For instance, a nucleotropic agent will affect almost exclusively nucleic acid when acting on bacteriophages, which contain 50% nucleic acid, but will also damage other components when acting on influenza virus, which contains only 1% nucleic acid. The various inactivating agents are listed below.

Nucleotropic agents: Ultraviolet light of 260 nm wave length; formaldehyde; nitrous acid; hydroxylamine; decay of radioactive elements that are incorporated in nucleic acid, such as ^{32}P or ^{3}H (incorporated as thymidine).

Proteotropic agents: UV light of 235 nm wave length; heat; proteolytic enzymes; mild acid pH; sulfhydryl-containing or sulfhydryl-reactive compounds. Detergents and high concentrations of urea or guanidinium compounds disrupt inter- (and some intra-) capsomeric

Fig. 64-8. Possible errors in predictions based on survival curves. In **A** the survival curve is multiple-hit and in **B** it is multicomponent; in both inadequately determined curves lead to erroneous extrapolations. The estimated dose for acceptable inactivation is too large in **A**; in **B** it is only about 20% of that required, and the corresponding survival is about 10^4 times the acceptable level.

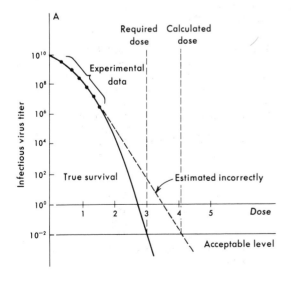
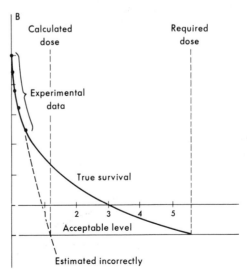

bonds, loosening the capsid of naked viruses.

Lipotropic agents: Lipolytic enzymes, lipid solvents.

Universal agents: X-rays; alkylating agents; photodynamic action.

The properties of some inactivating agents will be briefly analyzed in the following paragraphs.

Heat acts mostly by denaturation of capsid protein, since infectious nucleic acid can be extracted from inactivated virus. Often the effect is quite limited; for instance loss of the small capsid component VP_4 in poliovirus (see Ch. 48) or of a minor component in the capsid of the helical phage f1 (see Ch. 45).

The half life decreases rapidly with increasing temperature. Viruses vary greatly in their sensitivity: the very unstable Rous sarcoma virus has a half life of only 1 to 2 hours at 37°. Viral half lives are usually increased by the presence of various salts (especially Mg^{++}), protein, cystine, or polysulfides, which stabilize the tertiary structure of the protein.

The curves for heat inactivation are usually of the multicomponent type. With poliovirus they clearly differentiate a major rapid inactivation (with a one-hit curve) and a minor slower inactivation. The minor component is probably **free viral RNA**, released from the virions by the heat and able to produce plaques with low efficiency (Ch. 48).

Ultraviolet light at 260 nm acts mainly on nucleic acids, as noted above. This agent is not well suited for viral inactivation because its damage to nucleic

acids can be repaired by a variety of enzymatic and genetic mechanisms (Ch. 11); the inactivation should be measured after allowing maximal reactivation. UV is also absorbed by many substances in biological media, and shadows are projected by opaque particles; hence the risk of virions escaping inactivation is high.

X-rays cause the production of highly reactive and short-lived chemicals within and around the virions; these unstable products interact with various viral components **(direct effect)**. The reactive chemicals also cause changes in components of the medium that produce long-lived inactivating poisons **(indirect effect)**; this effect is minimized by using a medium rich in organic molecules, especially amino acids. The direct effect results in one-hit survival curves; the indirect effect results in multiple-hit curves.

Formaldehyde is widely used for the production of killed vaccines. It reacts mainly with amino groups in proteins and in single-stranded nucleic acids (Fig. 64-5). Thus virions with double-stranded nucleic acid are inactivated mostly by modification of the protein; but in the presence of formaldehyde the helical structure melts at relatively low temperatures, freeing the amino groups. The activation curve is of the multicomponent type, as shown in Figure 64-9; hence extrapolation to a given degree of inactivation is difficult, as discussed above.

Ethylene oxide (Fig. 64-5) is an effective agent for inactivating all viruses, with one-hit kinetics. Other alkylating agents, nitrous acid, and hydroxylamine act on nucleic acids as described in Chapter 11 (Action of chemical mutagens): the former two also act similarly on proteins.

Lipid Solvents. Ether (usually added at 20% to a suspension of virions) readily inactivates enveloped but not naked viruses. As we have seen in many preceding chapters, this simple test is widely used for distinguishing the two groups.

Enzymes. Most viruses are resistant to the action of trypsin, but some are sensitive (e.g., some arboviruses and, under special conditions, some myxoviruses). Viruses are more readily inactivated by proteolytic enzymes of broader specificity, such as pronase. Enveloped viruses may be inactivated by some lipolytic enzymes: some arboviruses are inactivated by phospholipase A, and influenza virus treated with phospholipase C becomes susceptible to inactivation by a protease.

Photodynamic Inactivation. The presence of certain dyes (acridine orange, proflavin, methylene blue, neutral red) can render virions susceptible to inactivation by exposure to visible light, as described above for bacteria. Most viruses become photosensitive simply on being mixed with the dye, but others

Fig. 64-9. Time course of the inactivation of type 2 poliovirus by formaldehyde, showing the marked multicomponent character of the survival curve. (Modified from Gard, S. In *The Nature of Viruses.* CIBA Foundation Symposium; G. E. W. Wolstenholme and E. C. P. Millar, eds. Little, Brown, Boston, 1957.)

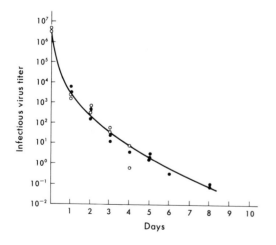

(e.g., poliovirus) require multiplication in its presence, which encloses dye within the capsid. The latter remain sensitive in the absence of external dye, but lose photosensitivity on infecting a cell and releasing the nucleic acid.

SELECTED REFERENCES

Books and Review Articles

ALBERT, A. *Selective Toxicity,* 3d ed. Wiley, New York, 1965.

BRUCH, C. W. Gaseous sterilization. *Annu Rev Microbiol 15:*245 (1961).

CHICK, H., and BROWNING, C. H. The theory of disinfection. *Med Res Counc Syst Bacteriol* (London) *1:*179 (1930).

HUGO, W. B. (ed.). *Inhibition and Destruction of the Bacterial Cell.* Academic Press, New York, 1971.

LAWRENCE, C. A., and BLOCK, S. S. (eds.). *Disinfection, Sterilization, and Preservation.* Lea & Febiger, Philadelphia, 1968.

LEA, D. E. *Actions of Radiations on Living Cells,* 2nd ed. Cambridge Univ. Press, London, 1955.

PHILLIPS, C. B., and WARSHOWSKY, B. Chemical disinfectants. *Annu Rev Microbiol 12:*525 (1958).

WELLS, W. F. *Air-borne Contagion and Air Hygiene.* Harvard Univ. Press, Cambridge, Mass., 1955.

Specific Articles

FREESE, E., SHEU, C. W., and GALLIERS, E. Function of lipophilic acids as antimicrobial food additives. *Nature 241:*321 (1973).

HOTCHKISS, R. D. The nature of the bactericidal action of surface active agents. *Ann NY Acad Sci 46:*479 (1946).

INDEX

A

Abbreviations, in metabolic equations, 41n, 60n
AB hybrids, 613
ABO blood group system. See under *Blood group.*
ABO compatibility, kidney grafts and, 619
Abortion(s)
due to aspergillosis, 997
due to brucellosis, 815
due to listeriosis, 947
due to mycoplasmas, 936
due to *Vibrio fetus,* 782
due to viruses, 1216
therapeutic, in rubella, 1358
Abscess(es). See also under specific organ, e.g., *lung, abscesses in.*
antibacterial, defense inhibited by, 654
bacterial, 980
chemotherapy limited by, 671–673, *674*
due to oral flora, 751, 959
fungal, 980
in South American blastomycosis, 994
micrococcal, 738
mycobacterial, 862, 864t
nocardial, 878
phagocytosis of bacteria and, 654
pneumococcal, *674*
Pseudomonas aeruginosa in, 784
staphylococcal, 733
streptococcal, 720, 723
Absidia corymbifera, 999
Acetabularia, adaptive inheritance in, 175
Acetobacter
commercial production of ascorbic acid and, 50
ethanol oxidation by, 49
pH range of, 93
taxonomic characteristics, 34t
Acetoin, produced by fermentation, 45–46

Acetolactate
bacterial metabolism of, 315, *315*
in fermentations, 42
Acetone, produced by fermentation, 46, 46n
Acholeplasma, 930
Achromobacter. See *Acinetobacter.*
Acid(s), in sterilization, 1458
Acidosis, diabetic, 64
Acinetobacter, 785
transformation in, 184
Acridine (derivatives)
DNA affected by, 211
mutagenic effects of, 254, 257, *258*
in bacteriophages, 235
Acriflavine, 174
Actin(s), 128t
Actinomadura, 875t
Actinomyces, 873–877
cell wall constituents, 875t
in oral cavity, 958t
morphology, 872
periodontal disease due to, 961
taxonomic characteristics, 35t
Actinomyces bovis, 873
Actinomyces israelii, 873, 875, 876, *876*
Actinomyces muris, 875. See also *Streptobacillus moniliformis.*
Actinomyces naeslundii
dental diseases due to, 961, 962, *962*
resemblance to *A. israelii,* 876
Actinomycetaceae, 873t
Actinomycetales, classification, 345, 873t
Actinomycetes, 871–879
cell wall preparations from, constituents of, 875t
characteristics, 872–873
establishments bacteria, 872
Actinomycin(s)
antitumor action of, 149n
bacterial resistance to, 162
discovery, 149n
DNA transcription blocked by, *266, 267,* 1133
immunosuppression by, 501
interferon production affected by, 1178, 1179t, *1179*

Actinomycin(s) *(continued)*
isolation, 879
protein synthesis inhibited by, 303t
RNA synthesis inhibited by, 149n
Actinomycosis
abscesses in, 876, *876*
bacteria accompanying, 876
definition, 873
diagnosis, 875–876
due to oral flora, 959
epidemiology, 876
in cattle, 873
pathogenesis, 873–875
prognosis, 876
species in, 873
sulfur granules in, 875, *876*
treatment, 876
Actinoplanaceae, 873t
Actinoplanes, 873t, 875t
Adamantanamine (therapy)
in influenza, 1183, 1327
in rubella, 1183
structure, *1184*
viruses inhibited by, 1183
Addison's disease, 587t
Adenine, bacterial biosynthesis of, 71
Adenine arabinoside therapy, in herpetic encephalitis, 1246
Adeno-associated viruses, 1236
Adenoma(s), parotid, 1420
Adenosine-3', 5'-monophosphate. See *cAMP.*
Adenosine triphosphate
as link between catabolism and anabolism, 40
bacterial production of, types of, 40
in bacterial fermentation and respiration, 42
Adenosyl methionine, in bacterial methylations, 72
Adenovirus(es), 1221–1237
adeno-associated virus inhibition of, 1236
antibody neutralization of, 1192
antibody sensitization of, 1193
biosynthesis, *1230, 1231*
capsid proteins of, 1230, *1231*

Adenovirus(es) *(continued)*
  cell-dependent abortive infection
      by, 1164
  cell-detaching factor in, 1226
  cell penetration by, 1227, *1229*
  cell transformation due to, 1208
  characteristics, 1222, 1223t
      biologic, 1227t
      chemical, 1222–1224, 1224t
      immunologic, 1225–1227,
          1227t
      physical, 1222–1224, 1224t,
          1227t
  chromosomal aberrations due to,
      1208
  classification according to hema-
      gglutination, 1226t
  clinical syndromes due to, 1232,
      1233t
  cytopathic effect of, 1207
  DNA of, 1222
  family-reactive determinants of,
      1225
  genetic recombination in, 1167
  hemagglutination by, 1224–1225
  history, 1221
  host range of, 1227
  infections by, control of,
      1235–1236
      epidemiology, 1235
      immunity to, 1210
      inclusion body formation due
          to, 1208
      laboratory diagnosis, 1234
      latent, 1217, 1233
      pathogenesis, 1232–1234
      prevention, 1235–1236
      serologic tests, 1203t
  in whooping cough, 799
  kb, mutants of, 1166
  morphology, 1222, *1223, 1226*
  multiplication of, 1227–1232,
      1229
  nuclear lesions due to, 1230,
      1230, 1232
  oncogenicity of, 1224t, 1236,
      1430
  parvovirus inhibition of, 1236
  pentons of, 1225, *1225, 1226*
  phenotypic mixing in, 1163
  properties of, 1222–1232
  proteins of, 1223
  stability of, 1224
  SV40 hybridization with, 1428
  tumor antigen of, 1234
  tumors due to, 1212, 1234
  uncoating of, 1227, *1229*
  viral core of, 1227, *1229*
Adenylate biosynthesis, bacterial,
    81, 82
Adjuvants, 481, *481*
Adrenal insufficiency, 746
Adrenal steroid. See *Corticosteroid.*
Aedes, as vector for diseases, 1390

Aerobacter. See *Enterobacter.*
Aerobactin, 92
Aerobes, obligate, 48
Aeromonads, 782
  biochemical reactions, 758t
  taxonomic characteristics, 34t
Aeromonas hydrophila, 782
Aeromonas shigelloides, 782
Aerosols, antibacterial, 1462
Aerum, adsorbed, 384
Affinity-labeling, 443
Aflatoxin, 966
Agammaglobulinemia
  bacterial infection in, 658
  electrophoretic results in, *426*
  infantile X-linked, 507–508
  meningococcal infection in, 747
Agar. See also *Gel.*
  blood, heated, 94
    to assess bacterial hemolysis, 94
  chocolate, 94
  deoxycholate, in differentiating
      Enterobacteriaceae, 759
  in culture media. See under
      Culture medium.
  MacConkey, in differentiating En-
      terobacteriaceae, 759
  Sabouraud's. See *Sabouraud's
      agar.*
  SS, in differentiating Enterobac-
      teriaceae, 759
  triple-sugar-iron, 757
Agar diffusion method, to deter-
    mine bacterial sensitivity
    to drugs, 149
Agarose, 101
Age factors
  in autoimmune disease, 588, *588*
  in tuberculosis, 853, *854,* 855,
      857, 858, 859
Agglutination
  as serologic reaction, history, 353
  by antibodies, 652–653
  mechanisms, 391
  of bacteria, 391
  of erythrocytes, 391
    by sera, human blood groups
        based on, 599t
  passive, 393
  sensitivity of, 394
Agglutination-inhibition assay, for
    human allotypes, *420*
Agglutination reaction, 391–394
  in measuring bacterial adsorption
      of antibody, 391
  prozone in, 392
  sensitivity of, 393–394
  serum titration by, 391–392
  with surface versus internal anti-
      gens of cells, 392
Agglutination test, *Brucella,* 816
Agglutinator(s)
  rheumatoid serum (Ragg), 419
  serum normal (SNagg), 419

Agglutinin(s), 392
Agglutinogen(s), 392
Agranulocytosis, bacterial disease
    and, 660
Agrobacterium tumefaciens
  DNA base composition in, 37t
  nuclear body of, *32*
Ajellomyces dermatitidis. See *Blas-
    tomyces dermatitidis.*
Alanine, bacterial biosynthesis of,
    72, 73n
Alanine-muramyl amidase, 695
Alastrim, 1262, 1263
Albomycin, 92
Albumin
  bovine serum, proteolytic cleav-
      age of, 376
  in bacterial growth, 94
  in culture media, 94
Alcaligenes, 785–786
Alcaligenes faecalis, 785
  DNA base composition in, 37t
  pH range of, 93
Alcohol(s). See also specific alco-
    hol, e.g., *Ethanol.*
  disinfectant action of,
      1461–1462
  intake of, inflammatory response
      to bacterial infection in-
      hibited by, 658
  produced by fermentation, 47
Alcoholism, tuberculosis and, 859
Aldehydes, fermentation and, 41,
    46
Aldolase, bacterial, 67
Aleutian mink disease, 553
Alexin. See *Complement.*
Alga(e)
  autotrophic metabolism by, 53t
  blue-green. See *Cyanophyceae.*
  chloroplast formation in, strepto-
      mycin prevention of, 174
  evolutionary aspects, 55
  nitrogen fixation by, 56
  photosynthesis in, 54, 55
  taxonomic aspects, 14
Alkali(s), in sterilization, 1458
Alkaline phosphatase
  amino acids in, *279*
  in *Escherichia coli,* 124, 322
Alkylating agents
  disinfectant properties of, 1459,
      1460, *1460*
  DNA damage due to, excision
      repair of, 241, *241*
  immunosuppression by, 501
  mutagenic effects of, 254, 256,
      257, 258, 259
  virus inactivation by, 1464
Allele(s). See also *Gene.*
  bacterial, wild type and mutant,
      176
  codominant, 420n
  for Gm allotypes, 420

Allele(s) (continued)
identifiable. See Marker.
intergenic complementation of, 234
mutant, 176, 233
wild-type, 176, 233
Allergen(s), 530
Allergic reaction(s), 581–593. See also specific allergic reaction, e.g., Hay fever.
antibody-mediated, 529, 529t
at foci of bacterial infection, 594
cell-mediated, 11-oxycorticosteroid inhibition of, 580
cell-mediated versus humoral, 559t
due to IgE, 418
infectious diseases with, 593
serum sickness and, 550
to allotypes, 419n
to drugs. See under Drug.
to microbial antigens, 593, 594
to virus, 1215
with exotoxins, 593
Allergy(ies). See also Hypersensitivity.
atopic. See Atopy.
bacterial, 657. See also Hypersensitivity, delayed-type.
definition, 256, 528
history, 528
immunity versus, in tuberculosis, 657
infectious. See Hypersensitivity, delayed-type.
inflammation in, corticosteroid therapy in, 501
SRS-A in, 543
streptococcal, 657
to drugs. See under Drug.
to food, 481
to mycobacterial proteins, 594
to penicillin, See under Penicillin.
to quinidine, 585
to Sedormid. 585
Allesceria boydii, 1000
Alloantibody(ies)
AB, 603–604
definition, 599
to leukocytic alloantigens, after transfusions and pregnancies, 609n
Alloantigen(s)
blood, transfusion reactions due to, 607
blood cell. See Blood cell alloantigen.
definition, 419, 454, 598
leukocytic, alloantibodies to, after transfusions and pregnancies, 609n
on cell surfaces, 597–624
on T lymphocyte surface, 460
Rh, notations for, 607t

Alloantigen(s) (continued)
theta, 460
Alloantiserum(a) 406, 460
Allogeneic, definition, 567n
Allogeneic cells, injection to induce allograft tolerance, 612, 613
Allograft(s)
definition, 567n, 610
fetus as, 612
from one sex to another, 618
immunosuppressants in, 580, 623
kidney, 580, 619, 619, 623
privileged sites for, 611
rejection of, 567–568. See also Graft-versus-host reaction.
capacity for. See Immunity, transplantation.
humoral versus cell-mediated immunity in, 611
second-set, 567
tolerance of, 611–614
antilymphocyte serum in, 499, 580
enhancement of, 613–614
immune deviation in, 579
immunologic, induction of, 612–613
in chimeras, 612
in newborn mice, 578
intersexual, 618
lack of adaptive immune response and, 610n
mixed lymphocyte culture (MLC) test of, 618–619
Allograft reaction(s), 610–620
as immune response, 610–614
white-graft, 611
Allosteric transitions, of viral capsids, 1029
Allostery, 313
Allotype(s), 419–423
agglutination-inhibition assay for, 420
allergic reactions to, 419n
Am, 422
amino acid substitutions in, 433
definition, 406, 419
Gm, 419–422
alleles for, 420, 421t
in immunoglobulin G, 419
nomenclature of, 419
practical use of, 420
human, 419-423
IgG, antibodies to, 419n
inheritance of, 420
Inv, 422–423
mouse, 423
rabbit, 423, 424
suppression of, 497–498, 498
α-Amanitin, 267
Amber, 277, 279
Amber suppressor, 293, 295

Amebocytes, phagocytic ability, 570
Amidase, bacterial, 114
Amine(s)
aromatic, attachment to protein, 354, 354
vasoactive, in anaphylaxis, 544
Amino acid(s). See also specific amino acid, e.g., Histidine.
addition to polypeptide chain, 285
aliphatic, bacterial biosynthesis of, 72–73
analogs of, bacterial resistance to, 162
azo substituents on, 355, 355
bacterial. See under Bacterium.
bioassays for, 90
fermentative production of, 46
in alkaline phosphatase, 279
mammalian biosynthesis of, 87, 87t
replacement in abnormal human hemoglobin, 277, 279
sequences of, in antibodies, 427–436
in bacterial homologous proteins, 38
in Bence Jones proteins, 428
in myeloma proteins, 425, 428, 431, 432
in polypeptides, determination of, 276, 277
substitutions of, in allotypes, 433
Aminobenzoate, 91t, 151, 152
Aminocaproate, plasmin activation of complement inhibited by, 524
Aminoglycoside(s), 158–159, 676–677. See also Antibiotic and specific aminoglycoside, e.g., Streptomycin.
action of, 159
bacterial resistance to, 162
bactericidal activity of, 300
penicillin synergistic action with, 165
protein synthesis inhibited by, 300–302, 303t
ribosomal effects of, 296
structure, 300
toxicity of, 677
6-Aminopenicillanic acid, 156, 156
2-Aminopurine, 255
Aminosalicylate (therapy), 154, 155, 675
Aminosugars, bacterial, 69
Ammonia, bacterial use of, 83
Ammonium ion, bacterial requirement, 92
Ammonium sulfate, in radioimmunoassay, 395–396

AMP, cyclic. See *cAMP.*
Amphibians
  DNA sequences in, 217, *218*
  genes in genome, 216
  immunoglobulins in, 503
Amphibolic pathway(s), 63–67
  conversion to biosynthetic pathways, 63
  definition, 63
Amphotericin B (therapy). See also *Polyene.*
  in aspergillosis, 998
  in coccidioidomycosis, 993
  in cryptococcosis, 986
  in histoplasmosis, 990
  in North American blastomycosis, 988
  in South American blastomycosis, 995
  in systemic mycoses, *983*
  structure, *983*
Ampicillin (therapy), 157, 676
  effects on *Escherichia coli,* 117, 120
  in *Arizona* infection, 776
  in *Edwardsiella* infection, 771
  in *Escherichia coli* infection, 768
  in *Hemophilus influenzae* infection, 796
  in listerosis, 948
  in *Proteus mirabilis* infection, 772
  in pseudotuberculosis, 806
  in salmonellosis, 775
  in shigellosis, 778
  structure, *157*
Amputation, in gas gangrene, 841
Amylase, bacterial, 95
Amyloidosis, in leprosy, 867
Anabolic pathway(s), 63–67
  relation to amphibolic pathways, *64*
Anabolism, 40. See also *Biosynthesis* and *Metabolism.*
Anaerobe(s),
  aerotolerant, 48
  culture of, 91
    hydrogen-carbon dioxide generators for, 830
  effects of oxygen on, 48
  obligate, 786–788
    definition, 15, 48
    pyruvate formation in, 67
    discovery of, 5
  superoxide effects on, 49
Anaerobiosis, 91
Anaphylactic shock, due to penicillin, 581
Anaphylactoid shock, 545n
Anaphylatoxin(s)
  antihistamines and, 522
  complement and, 522, 522t
  formation of, 545n
  inhibitor of, 522
Anaphylaxis, 529–546

Anaphylaxis *(continued)*
  aggregate, 545–546
  antibody and antigen required for, 538–539
  basophil degranulation in, 544
  bradykinin in, 542
  causes, 538
  C-kinin in, 542
  complement in, 545, 546
  cutaneous, definition, 529
    in guinea pig, 536–537, *536, 537*
    in man, 530–531, *530*
    ligands in, 535n
    passive (PCA), 536, *536*
    reverse, 535, 538
  cytotoxic, 545–546
  cytotropic, 534
  drugs inhibiting, 545
  due to antilymphocyte serum, 580
  due to contaminants of penicillin solutions, 583
  due to immunoglobulins, 545
  due to soluble antibody-antigen complexes, 545
  ECF-A in, 543, 544
  generalized. See *Anaphylaxis, systemic.*
  heparin release in, 541
  histamine in, 540, *540, 541,* 544
  history, 528, 528n
  IgE in, 535
  in isolated tissues, 543–545
  initiation of, 535
  kallikrein in, 542, *542*
  kinins in, *540,* 542, *542*
  manifestations, 539t, 543
  mast cell degranulation in, 540, *541*
  mast cells in, mediator release from, cAMP inhibition of, 544, *545*
  mediators of, pharmacologically active, 540–543, *540*
  passive, mast cell binding of antibodies and, 538
    reverse, 538
  prostaglandins in, 543
  serotonin and, 541–543, *540*
  serum sickness and, 550
  SRS-A in, 543, 544
  systemic, 538–540
    antibody fixation and, 538
    antigen administration and, 538, 539
    definition, 529
    species variations in, 539–540, 539t
    transfusion reactions and, 546
    vasoactive amines in, 544
Anemia
  due to polyene therapy, 984
  hemolytic, acquired, 588–589

Anemia, hemolytic, acquired *(continued)*
    autoimmune, 587t–588
    Coombs test in, 589
    due to penicillin, 583–584
    due to viral infections, 586
    in pneumonia, 701
    in Carrion's disease, 951
    in infectious mononucleosis, 1252
    pernicious, autoimmune aspects, 587t, 588
    Salmonella infections with, 774
Aneuploid cells, 1134
Aneurysms, aortic, syphilitic, 888
Angiitis
  due to rickettsiae, 899
  in Rocky Mountain spotted fever, 908
Angina, Vincent's, 893
Angioedema, hereditary, 523–524
Animal(s)
  axenic, 632
  glycogen formation in, 83
  maturation of, viral susceptibility and, 1209
  mycoplasma infections of, 934
  naturally occurring cancer in, viruses in, 1446
Animal cell(s)
  Barr body in, 1130
  cell membrane of, continuous internal flow in, 1128
    DNA synthesis mediated through, 1133
    70-A globular particles of, 1128
  mitochondria of, oxidative phosphorylation at, 1129
  centromeres of, 1130
  chromatin in, 1129–1130
  chromosome pairing in, 1130
  cloning of, 1124–1125
  cytoplasm of, mRNA transport to, 1131–1132
  DNA replication in, 1130
  DNA synthesis in, plasma membrane in, 1133
  DNA transcription in, 1130–1133
    elements regulating, 1133
    histone limitation of, 1129
  endoplasmic reticulum of, polysomes associated with, 1132
  euchromatin of, 1130
  gene expression in, regulation of, 1133
  genetic information in, 1129–1133
  growth cycle of, surface changes during, 1134
  heterochromatin in, 1130
  histone synthesis in, 1130
  homogenates of, isozyme patterns of glucose phosphate iso-

Animal cell (s), homogenates of
(*continued*)
merase in, *1136*
informosomes in, 1131
mitochondria of, ribosomes of,
283
mRNA in, monocistronic,
1132—1133, *1132*
transport of, 1131
nuclear membrane of, 1129
nucleolus of, 1131
plasma membrane of, 1128
polysomes in, 1132
preservation of, by freezing, 1455
protein metabolism in, 1133
RNA polymerase(s) in, α-amanitin
inhibition of, 267
RNA synthesis in, histone limitation of, 1129
rRNA precursors in, processing
of, 1130, *1131*
synthesis of, 1130
surface of, antigenic structure of,
1129
galactosyl transferase on, 1127
interaction with lectins, 1129
suspensions of, agglutination of,
1129
viral infection of, RNA repli-
case(s) in, 1152
Animal cell culture(s), 1121—1137
behavior of cells in, 1125—1126
cell differentiation in, 1134
cell fusion experiments in, 1137
cell growth in, 1126—1127
cAMP in, 1134
cycle of, 1126—1127, *1126*
regulation of, 1134
density-dependent inhibition of,
1127, *1128*
rate of, 1126
topoinhibition of, 1127
cell hybridization in, 1136
cell interaction in, 1127—1128
cell line(s) in, 1122—1123, *1123*
definition, 1123, *1123*
establishment of, 1124—1125
transformed, 1123
cell locomotion in, 1125—1126
cell membranes in, 1128—1129
synthesis of, site of, 1125
cell microtubules in, 1125
cell senescence in, 1124—1125
cells in contact in, 1127
cell strain(s) in, 1122—1123,
1123
multiplication of, cell alteration
in, 1123, *1123*
transferability of, 1122
cell synchronization in, 1127
characterization of, 1122—1125
cloning of cells in, 1124—1125
contact inhibition of cells in,
1127, *1128*

Animal cell culture(s) (*continued*)
DNA replication in, histone
synthesis and, 1134
DNA synthesis in, 1134
epithelioid cells in, 1122
exchange of components between
cells in, 1127
feeder layer in, 1124
fibroblast-like, 1122
gene expression in, regulation of,
1137
genetic mapping in, 1135—1137
genetic studies with, 1134—1137
growth of, 1126
cAMP in, 1134
inhibitors of, 1126
heterokaryon formation in, 1136
hybrid cells in, chromosome elimination in, 1136
immortality event in, 1124
karyotypes of cells in,
1134—1135
metabolic cooperation between
cells in, 1127
metabolic unbalance in, due to
synchronization, 1127
metabolism in, regulation of,
cAMP in, 1133—1134
microfilaments in, 1125, *1125*
movement of, contact inhibition
of, 1125
multiplication of, *1123*
mutations in, 1135—1136
pinocytotic vesicles in, 1125,
1125
plating efficiency in, 1124
primary, 1122, *1123*
protein synthesis in, 1134
RNA synthesis in, 1134
ruffling of, 1125
secondary, 1122, *1123, 1124*
serum exhaustion in, 1126
solid support in, relation of cells
to, 1125—1126, *1125*
stationary cells in, 1126
storage, 1124
surface modification of cells in,
1127
suspension, 1123
synchronized, 1127
tight junctions in, 1125
transfer of cells from, 1122, *1123*
transformed cells in, 1123—1124,
1124
trypsin in, 1122
variants in, 1123
wound experiments with, 1127
Animal virus(es)
adsorbed, removal of, 1144
assay of, endpoint method in,
1040
attachment to cell, 1143
autointerference by, 1182
building blocks assembled from

Animal virus(es), building blocks
(*continued*)
pools, 1163
cell-associated, one-step multiplication curves in, 1146
cell infected with, hemadsorption
of, 1161, *1161*
complex, maturation of,
1161—1163
defective particles produced by,
1182
diameters of, 1141t
DNA-containing, DNA synthesis
in, 1147—1150
genetic recombinations in, 1167
DNA transcription in,
1147—1149
syntheses in, 1147—1150
DNA replication in, 1149, *1149,
1150*
eclipse of, 1144
enveloped, budding of, 1159,
1159, 1160
envelope synthesis in,
1158—1159
formation and release of virions,
1159—1160
host modification of, 1163
maturation of, 1158—1159
nucleocapsid formation in,
1158, *1159*
factories in, 1162
family characteristics, 1141t
genetic organization in,
1167—1168
genetic recombination in,
1167—1169
biologic significance of, 1168
genetics of, 1165—1168
genetic variability in, 1168
growth cycle of, points suitable
for chemotherapeutic
attack, *1173*
heat-inactivated, infectious nucleic acid from, 1143
heterologous, viral interference
by, 1182—1183
heterozygosity in, 1168, *1168*
homologous, intrinsic interference
by, 1182
host range of, 1143
infection by, abortive,
1163—1165
cell-dependent, 1164—1165
virus-dependent, 1165
as function of nucleic acid,
1141—1143
by nucleic acid versus virion,
1141
effect on host macromolecular
syntheses, 1146—1147,
1146—1148
endosymbiotic, 1163
productive, 1141—1144

Animal virus(es), infection by
(*continued*)
polykaryocyte formation in,
1161, *1162*
infectivity of, assay of, 1037,
1038
intrinsic interference in,
1182—1183
maturation of, 1157—1163
cellular alterations in,
1160—1161
multiplication of, 1140—1165
host susceptibility and,
1140—1141, *1143*
mutants of, complementation of,
1167
defective, abortive infection due
to, 1165
host-dependent, 1166
in vaccines, 1166
pleiotropism in, 1166
types of, 1165—1166, *1166*
mutations of, 1165—1167
induction of, 1165
naked icosahedral, generation of
empty capsids in, *1156,*
1157
maturation of, *1156,* 1157
procapsids of, 1157
release of, 1157—1158, *1158*
nucleic acids of, infection as
function of, 1141—1143
release into host cell,
1143—1144
virus-specific antibodies ineffective against, 1143
nucleocapsid symmetry in, 1141t
one-step multiplication curves in,
1144—1146, *1145*
in intracellular viruses, 1146
penetration of host cell by, 1144
phagocytosis of, 1144
phenotypic mixing in, 1163,
1164, 1165
protein synthesis in, 1151—1153
release of, 1157—1163
replicative intermediate of, 1149
reverse transcriptase in, 1155
RNA-containing, classes of, 1150
genetic recombinations in, 1167
RNA of, single-stranded, physical characteristics, 1153n
RNA replication in, double-
stranded, 1154—1155,
1157
single-stranded, 1153—1154,
1154—1156
replicative intermediates in,
1153—1154, *1154, 1155*
through a DNA-containing
replicative intermediate,
1155
syntheses in, 1150—1157
viral protein synthesis in,
1151—1153

Animal virus(es), RNA containing,
viral protein
synthesis (*continued*)
regulation of, 1152—1153
RNA replicase(s) of, 1152
RNA synthesis dependent on,
1152, *1153*
RNA of, single-stranded versus
double-stranded, 1153n
study of, chick embryos in, routes
of inoculation of, *1143*
ultraviolet-irradiated, reactivation
of, 1163, *1169*
uncoating of, 1144
viral interference with, 1173
viral protein synthesis in host cell
by, 1149
virulence of, host cellular macromolecular syntheses inhi-
bited by, 1146, *1146,*
1147
virulent versus moderate, 1140
Anion(s), inorganic, antiseptic properties of, 1459
Anionic dyes, to distinguish living
from dead lymphoid
cells, 456
Anomaly, congenital, See *Congeni-*
tal anomaly.
Anopheles, as virus vector, 1385,
1390
Anthiramycin, 267
Anthranilate, bacterial synthesis of,
77, 78, *78*
Anthrax, 820. See also *Bacillus*
anthracis.
cutaneous, 823, *823*
hemorrhagic mediastinitis in, 823,
824
history, 820
immunity to, 825
incidence, 826, 826n
inhalation, 823
in laboratory animals, 823, 825t
intestinal, 824
laboratory diagnosis, 825—826
mortality in, 824
pathogenesis, 823—825
prevention, 826
pulmonary, 824
spores in, 824, 825t
susceptibility to, 825, 825t
systemic, 823
toxin in, 824, 825t
treatment, 826
vaccine for, 825
Anthrax bacillus. See *Bacillus an-*
thracis.
Anthropologic surveys, blood typing in, 608
Anthropology, Gm allotypes in,
420
Antiallotypes, 419
Antibacterial. See *Antimicrobial.*

Antibacterial defenses, of host. See
under *Host.*
Antibacterial drugs. See *Drug, anti-*
bacterial.
Antibiotic(s) (therapy). See also
Drug; Chemotherapy;
and specific antibiotic,
e.g., *Rifamycin.*
aminoglycoside. See *Aminogly-*
coside.
as secondary metabolites, 161
bacterial cell wall synthesis affected by, 120—121
bacterial resistance to, 117, 1105
bacterial sporulation and, 143
biosynthesis of, 50, 85—86, *86,*
160—161
broadening spectrum of, 157
broad-spectrum, 159—160,
677—678
candidiasis due to, 996
infections due to, 954
mycotic, 981, 995
Pseudomonas aeruginosa infection due to, 783
salmonellosis due to, 774
shigellosis due to, 778
development of, 149
ecologic role, 161
from fungi, 966
from streptomycetes, 879
in diet, to suppress endotoxin-
producing bacteria, 955
in diphtheria, 691
in leukoviral diseases, 1185
in poxviral diseases, 1185
in viral diseases, 1185—1186
nucleotides as constituents, 80
oxidative phosphorylation inhibited by, 128
plate assay for, *150*
polyene. See *Polyene.*
polypeptide, production of, 160
polypeptide chain elongation inhibited by, 302
production of, by streptomycetes,
160
ecology of, 160—161
physiology of, 160—161
sporulating organisms and, 161
protein synthesis inhibited by.
See under *Protein,*
synthesis of.
resistance to, genes in, 1105
Antibody(ies). See also *Alloanti-*
body; Antiserum; Immu-
noglobulin; Isoantibody;
and under various
microbes.
absorption of, 384
action of, side-chain theory of,
148
adsorption of, 384
affinity for haptens, haptenic

Antibody(ies), affinity for
(*continued*)
structure and, *368, 369*
homogeneity of, 362, *363, 364*
affinity-labeling of, 443, *443*
affinity of, 360–366, 370
actual, 402–403
in hapten binding, 385
intrinsic, 402–403
agglutinating, 652–653. See also
Agglutinin.
amino acid sequences in,
427–436
anticomplement, 523
antiflagellar, 759
anti-Forssman, 512n, 600n
antigen interaction with. See
Antibody-antigen reaction.
antigen precipitated by. See *Precipitin reaction.*
antihapten, changes in affinity of,
485, 485t
antikidney, *552,* 553
anti-Kk, 607
antipenicilloyl, 583, *583*
anti-R, 760
antiragweed, 531
anti-Rh, to prevent erythroblastosis fetalis, 497
antivirus. See under *Virus.*
as antiimunogen, 496–497,
579–580
as antireceptor, 497–499
assay for, phage neutralization as,
366
precipitin reaction in, 370
avidity of, 384–385
bacterial adsorption of, 391
binding by, 364
sites of, 369
in hapten interreactions, 362,
363, 364
intrinsic affinity of, 360
blocking. See *Antibody, incomplete.*
bound to cell surface histocompatibility antigens, 613
brain tissue reacted with, 589
carrier-specific, 370
cell-bound. See *Immunoglobulin, surface.*
cells producing, 453, *453, 455*
combination with pathogenic
microbes, 379
combining sites of, avidity and,
385
composition of, 443–444
size of, 444–445
concentrations of, bacteriophages
to measure, 482, *483*
measurement of, radioimmunoassay in, 396
using precipitin reaction, *371,*
379–380

Antibody(ies) (*continued*)
minimal, 394t
cross-reacting, removal of, 384
cross-reactivity of, time factors in,
486
cytophilic, 464
cytotropic, 550
definition, 353, 364
distribution, 483
ferritin-labeled, 401–402, *402*
fixation of, systemic anaphylaxis
and, 538
fluorescein-labeled, 397, 398, *398*
fluorescence of, quenching of,
398–399, *399*
formation of, 449–509. See also
Immune response.
allogeneic effect in, 465
analysis of, transplantation of
lymphoid cells in, 453
antigen-template theory of, *452*
as model for cellular differentiation, 357
associative recognition in, 465
at cellular level, study of,
immunofluorescence in,
451, 453
methods for, 451–456
by lymphocytes. See under
Lymphocyte.
by plasma cells, 469–473
carrier effect in, 464–465
cells required for, *458, 463*
dendritic, 466
cellular basis for, 450–456
clonal selection theory of,
450–451, *452*
focusing hypothesis of, 465
genetic control of, 477–480,
478
study of, adoptive transfers in,
479
graft-versus-host reaction in,
465, 465n
hapten versus carrier determinants in, 464
helper cells in, relation of T cells
to, 572
indigenous bacteria and, 954
induction of, cell cooperation
in, 464
in cell cultures, 455–456
inhibition of, by disease states,
673
initiation of, immunologic tolerance after, 494
in secondary response, 486
interference with, 492–502
in whole animal, 480–491
macrophages in, 463–464,
465–466
mechanism of, instructive versus
selective theories on,
450, *452*
ontogeny of, 503–505, *505*

Antibody(ies), formation of
(*continued*)
phylogeny of, 502, *503*
plaque-forming cells in, 453,
455, 469
plasma cells in, 469–473
rate of, 483
secondary response in, 464
selective breeding and, 477
study of, chromosome map in,
479
suppression of. See *Immunosuppression.*
T cells in, 465
tolerance of immunogen and,
493
hapten reactions with. See under
Hapten.
heart muscle, 585
heterocytotropic, 534
heterogeneous, 427
heterophile, 512n, 600n
in infectious mononucleosis,
1254
high- versus low-affinity, in crossreactions, 384
homocytotropic, 534, 537
homogeneous, 426–427, *427,
428*
homospecificity of, 376–377
humoral, elicited by viruses, 1190
oncogenic, 1445
hybrid, 413–414, *415*
ferritin and, 402
Ig. See *Immunoglobulin.*
immunosuppressive, 496–499
in anaphylaxis, 538–539
in Arthus reaction, 546–547
in atopy, 531
in autoimmune disorders, 587t
incomplete, 392–393
antiglobulin test for, 392–393,
393, 394
in atopy, 531, 531t
in serum, 535
monogamous bivalency in, 378
valence of, 392
in feces. See *Coproantibody.*
in mucous secretions of respiratory and gastrointestinal
tracts, 643
in precipitin reaction, 376, 377,
378
intrinsic binding reaction of, 360
in viral infections, 1209–1211
isophile, 512n
ligand of, 357t
mast cell binding of, passive
anaphylaxis and, 538
maternal, in newborn, 503, 504t
molecules bound to, fluorescence
of, 399
molecule(s) of, as analytical reagents for exploring structure of complex macromolecules, 357

Antibody(ies), molecules of
  (continued)
  homospecific symmetry of, 470
  monogamous bivalency in,
      378–379, *379*
  natural, artificiality of concept,
      653
    definition, 252, 480, 653
    formation of, 353
    in prevention of bacterial dis-
      ease, 653
  net electrical charge of, 485n
  nonprecipitating, 377–379
  opsonizing actions of, 652–653.
      See also *Opsonin.*
  paradoxic effect of, 305
  phagocytosis-promoting. See *Op-
      sonin.*
  polypeptide chains of, combining
      sites of, 445–446, *445*
    heavy chain interaction with
      light chain, 445–446
    recombination of, activity after,
      446
  production of, in congenital
      syphilis, 505
  in microdrops, 451
  properties of, 357t
  proteins of, 425
  purified, 401
    isolation of, 379
  reaginic, formation of killed *Bor-
      detella pertussis* and, 585
  Rh. See *Rh antibody.*
  resolving powers of, 368
  19S. See *Immunoglobulin M.*
  salmonellae bound by, 368, *383*
  serum levels of, after first and
      second immunogen injec-
      tions, *482*
  maximal, time required for, 482
  skin-sensitizing. See *Reagin.*
  specificity of, antigenic immuno-
      dominant structural ele-
      ments and, 367
    carrier, 370
    combining sites and, 442–446
    definition, 370
    in hapten–antibody reactions,
      366–370, *367*
  structural uniformity of, 428t
  susceptibility to phagocytosis
      and. See *Opsonin.*
  to azobenzenearsonate globulin,
      354
  to bacterial capsules, 635
  to blood group substances, reac-
      tions of, 367
  to cell wall antigens, bacteriolytic
      action of, 652
  to *Histoplasma capsulatum*, 989
  to IgG allotypes, 419n
  to microbes, elevated titer of, as
      diagnostic tool, 400
  to *Mycoplasma pneumoniae*, 936,
      941

Antibody(ies) *(continued)*
  to poliovirus, complement-
      fixation test for, 514t
  to ribosomal proteins, 282
  to tetanus toxin, 836
  to T-dependent antigens, de-
      layed-type hypersensiti-
      vity and, 572
  to thyroglobulin, 585
  totipotency versus unipotency of,
      401
  toxin-neutralizing. See *Antitoxin.*
  turnover of, 483
  valence of, 360–366
    definition, 360
  virion interaction with,
      1190–1191
  virus-specific, types of, 1198
    viral nucleic acids unaffected
      by, 1143
  Wassermann, 886
Antibody–antigen complex(es)
  autoimmune, in systemic lupus
      erythematosus, 586
  competent versus incompetent, in
      complement-fixation
      assay, 515
  cross-linking in, *379*
  cyclic, 378, *379*
  dissociation of, 378, 385
  eosinophil ingestion of, 652
  in rheumatoid arthritis, 586
  mass of, in complement-fixation
      assay, 513–514
  number of, in complex reagents,
      387
  precipitation of. See *Precipitin
      reaction.*
  ratio of, in complement-fixation
      assay, 513–514
  reactions with soluble macromole-
      cular antigens, 370–391
  soluble, anaphylaxis due to, 545
  stability of, 379
  structure of, solubility and, 377
  suppressive effects of, 497, 497n
  vasoactive substance released
      from tissues by, 540
Antibody–antigen reaction(s),
      359–403
  agglutination. See *Agglutination
      reaction.*
  as model for noncovalent inter-
      actions, 357
  cell differentiation and
      proliferation due to, 591
  cross-reactions, 381–384
    bacterial classification using,
      381
    between pneumococcal poly-
      saccharides, 382, *382*
    due to antigen impurities, 381,
      389
    due to functional groups in
      different antigens,

Antibody–antigen reaction(s),
      cross-reactions
      (continued)
      381–384, *383*
    general characteristics, 384
    group-specific, 381, *383*
    low- versus high-affinity anti-
      bodies in, 384
    misdiagnosis due to, 400
    type(s) of, comparison of, 384
      gel diffusion to discriminate
      between, 389, *389*
    with conjugated proteins, 382
  reversibility of, 379
  unitarian hypothesis of, 400–401
  with particulate antigens,
      391–395
Antibody–ferritin conjugates, 402
Antibody–hapten complex. See
      *Hapten–antibody com-
      plex.*
Antibody–ligand complex. See
      *Hapten–antibody com-
      plex.*
Antibody–virion complex. See
      *Virion–antibody com-
      plex.*
Anticardiolipin, 886
Anticodon
  codon interaction with, 294n
    streptomycin distortion of, 301,
      301
  definition, 272
  degeneracy of, 277, *277*
  in charging of tRNA, 272
  wobbling of, *276*, 277, *277*
Anticomplement antibody, 523
Anticomplementary substances,
      from complement-
      fixation test in viruses,
      1201
Antigen(s). See also *Allergen; Allo-
      antigen; Immunogen;
      Self-antigen;* and specific
      antigen, e.g., *PPD anti-
      gen.*
  A, 600. See also *Blood cell
      alloantigen.*
  abortus, 813
  administration of, 480–482
    adjuvants in, 481, *481*
    anaphylaxis and, 538, 539
    antiproliferative immunosup-
      pressant with, 501, *502*
    dosage in, 481–482
    in desensitization, 539
    primary response to, 482–486
      definition, 482
    route of, 480–481
  aggregated versus disaggregated,
      496
  antibodies elicited by, 427
  anticomplementary properties of,
      513n
  antigenic determinants on. See
      Antigenic determinant.

Antigen(s) *(continued)*
application to skin, 481
assay for, 393
B, 600. See also *Blood cell alloantigen.*
Bacillus anthracis, 822
bacterial teichoic acids as, 114
blast transformation due to, 573–574, *573*
blood cell. See *Blood cell alloantigen.*
bound by receptors on lymphocytes, 467–468
capsular, 694
carcinoembryonic, 622
cell surface, cell transformation and, 1445
  in agglutination reactions, 392
  structure of, 1129
  testing for, 1129
cellular, identification and localization of, 397
  in agglutination reactions, 392
cell wall, antibodies to, 652
chlamydial, 919–920
clostridial, 832
Coccidioides immitis, delayed-type hypersensitivity to, 992
common enterobacterial, 760
comparison of, for identical or cross-reacting components, precipitin reaction for, 387, *388, 389*
concentration of, measurement of radial immunodiffusion in, *390*
  using precipitin reaction, 380, 387, 389, *390*
cross-reacting, immunogens versus, in precipitin reactions, 384
definition, 10, 353
degradation of, immune complexes in, 494
densensitization by, 579
detection of, immunofluorescence in, 398
diverse, enhancing immunogenicity of, *Nocardia asteroides* in, 877n
diversity of, 354
Duffy, 607
environmental, 530
erythrocyte, blood groups named for, 599
  cross reaction with O antigens, 117
ethnic distribution of, 13
fetal, 622
Forssman. See *Forssman antigen.*
Francisella tularensis, 807
functional groups of, cross-reactions due to, 381–384, *383*

Antigen(s) *(continued)*
H, 600. See also *Blood cell alloantigen.*
definition, 759n
  in Enterobacteriaceae, 759, *760*
hapten versus, 355
heterogenetic, 600n
heterologous, 367
histocompatibility. See *Histocompatibility antigen.*
HL-A, 618, *618*, 619
homologous, 367
human leukocyte (HLA), 618
hyperreactivity to, in staphylococcal disease, 657
immunodominant structural elements of, 367
immunogenicity of, enhancement by mycosides and endotoxins, 585
immunologic effect of, 493
impurities in, cross-reactions due to, 381, *389*
in anaphylaxis, amounts required for, 538–539
in autoimmune disorders, 587t
in complex mixtures, immunoelectrophoretic identification of, 389–391, *390*
incubation with small lymphocytes, blast formation due to, 468–469
injected, antigenic competition with, 490
  decline in concentration, 489, *490*
  effect of antibodies on, 490
  fate of, 488–490
  response to, antigenic promotion of, 490
in precipitin reactions, valence of, 375
in serum sickness, 550–551
intestinal degradation of, 481
in vaccination, 491
K, 759n
Kell, 607
Kk, 607
Lewis, 600. See also *Blood cell alloantigen.*
localization of, immunofluorescence in, 398
lymph nodal trapping of, 475, 476, *476*
lymphocytes sensitive to, tolerogen effects on, 495
lymphocytes stimulated by. See *Lymphocyte, activated.*
macromolecular, 353
  soluble, in antibody–antigen reactions, 370–391
macrophage action on, 463
microbial, allergic responses to, 593, 594
vaccination against, 490–491

Antigen(s) *(continued)*
MN, 605
molecular weight of, determination of, precipitin reaction in, 389
  valence and, 375, 375t
multivalency of, 373, 376, *376*
Mycoplasma arthritidis, 9 36
O. See *O antigen*
O cell, 600
  on virus surface, 1031
organ-specific, 588
particulate, antibody–antigen reactions with, 391–395
  tolerance and, 493
persistence of, to maintain tolerance, 494
Plasmodium malariae, 551
polysaccharide,· multivalence of, 375, 382
  precipitin reaction with, 372t
precipitated. See *Precipitin reaction.*
precipitin reaction reversed by, 379
properties of, 353
protein, complexity of, 375–377, *376*
  determinant groups of, 380
  precipitin reaction with, 373t
  valence of, 375–377
Rantz, 946
Rh. See *Rh antigen.*
S, of mumps virus, 1339
Ss, 605
somatic, 697
specificity of, 354
splenic trapping of, 475, 476
suppression of. See *Tolerance.*
surface, identification of, *454*
  variety of, 13
T-dependent, formation of antibodies to, delayed-type hypersensitivity and, 572
  tolerance to, 494
testing serum with, to identify etiologic agent in disease, 400
T-independent, 466–467
  IgM antibodies induced by, 484
  tolerance to. See *Tolerance.*
tolerogenic, 492, 493
transplantation. See *Histocompatibility antigen.*
treponemal, 886–887
tumor-specific, 568, 620, 1129
types, 353
univalency of, 376, *376*
V, of mumps virus, 1339
valence of, 375, *375*, 375t
Vi. See *Vi antigen.*
VW, 803, 804
Wassermann, 885–886
Antigen–antibody complex. See *Antibody-antigen complex.*

Antigen-binding fragments (Fab), 410
  from IgM, 416, 417
  idiotypes and, 424
Antigen D, 607
Antigenic determinant(s)
  definition, 354–355, *356*
  in classification of immunoglobu-
    lins, 406
  of blood group mucopeptides, 602, *602*
Antigen-sensitive cells. See *Memory cell.*
Antigen-template theory, of anti-body formation, *452*
Antiglobulin test, 392–393, *393, 394*
  definition, 609
  in hemolytic anemia, 589
  to prevent transfusion reactions, 609
Antihistamine(s)
  anaphylatoxins and, 522
  histamine blocked by, 540, *543,* 545
  in serum sickness, 550
A n t i i m m u n o g e n, 496–497, 579–580
Anti-Kk antibodies, 607
Antilymphocyte globulin, 499
Antilymphocyte serum(a)
  allograft survival and, 499, 580
  anaphylaxis due to, 580
  cell-mediated immune reactions depressed by, 580
  in kidney transplants, 499, 580
  recovery from viral diseases and, 1211
  serum sickness due to, 580
  suppression by, 498–499, 501
Anti-M antibodies, streptococcal, 722
Antimetabolite(s). See *Chemothera-peutic agent, antimeta-bolite.*
Antimicrobial agents and therapy. See also *Chemotherapy.*
  bioassay for, agar culture media for, 102
  antagonistic, 679
  special aspects of, 678–680
  topical, 678
Antimony compounds, in syphilis, 1459
Antimycin, 128
Antiproliferative agents, in immuno-suppression, 499
A n t i r e c e p t o r(s), antibodies as, 497–499
Antisepsis
  definition, 1452
  oxidants in, 1459
Antiseptic(s). See also *Disinfectant and Sterilization.*

Antiseptic(s) *(continued)*
  chlorine as, 1459
  essential oils as, 1461
  eugenol as, 1461
  hydrogen peroxide as, 1459
  iodophors as, 1459
  inorganic anions as, 1459
  mercurial, 1459
  potassium permanganate as, 1459
  skin, hexylresorcinol as, 1461
  thymol as, 1461
Antiserotonins, 550
Antiserum(a). See also *Antibody.*
  anticomplementary properties of, 513n
  avidity of, dissociation of com-plexes and, 385
    precipitin curve and, 385, *385*
    time factors in, 486
  cell-mediated rejection of tumor grafts blocked by, 579
  heterologous, definition, 406
    used to recognize isotypes, 405
  homologous, 406, 460
  injection of, immunity from, 658
  precipitin curve for, antigen con-centration measured by, 380
  reaction with cross-reacting anti-gen to remove cross-reacting antibodies, 384
  sequential changes in, after immu-nization, *486*
  to mu chains, inhibition of antibody formation by, 497
  to human leukocyte antigens, 618
  to Θ, 580
Antistreptococcal antibodies, in rheumatic fever, 722
Antistreptokinase antibodies, 718
Antistreptolysin O antibodies, 718
Antistreptolysin test, 723
Antitoxin(s)
  as part of host antibacterial defense, 652
  tetanus, 837, 837n
Anton test, 946
Appendicitis, due to *E. coli,* 768
Appendix lymphoid cells, 477
Arabinose operons, in bacterial metabolism, 334
Arabinosyl cytosine, viral DNA synthesis inhibited by, 1185
Arachnia, 873t
Arbovirus(es), 1377–1397. See also specific arbovirus, e.g., *Togavirus.*
  classification, 1379, 1380t–1381t
  group C, 1394
  history, 1378–1379
  sizes of, 1383t
Archeology, blood typing in, 608

Arenavirus(es), 1394–1396
  carriers of, 1394
  classification, 1381t
  clinical diseases due to, 1381t
  geographic distribution, 1381t
  infections due to, epidemiology, 1394–1396
    pathogenesis, 1394–1396
  members of, 1394
  properties of, 1394
Arginase, 70
Arginine biosynthesis in *E. coli,* 344, *345*
Arginine pathway, in bacteria, 69, 70, *70*
Arizona hinshawii, 776
Aromatic compounds, bacterial synthesis of, 75–79, *77,* 782
Arsenic therapy, in syphilis, 890, 1459
Artery, isolated, Schultz-Dale reac-tion in, 544
Arthritis
  due to *Actinomyces muris,* 875
  due to *Bacteroides,* 787
  due to gonorrhea, 750
  due to *Streptobacillus monili-formis,* 635
  from rat-bite fever, 948
  micrococcal, 738
  rheumatoid. See *Rheumatoid arthritis.*
  rubella, 1358
Arthroderma spp., 977t
Arthropod(s)
  as vectors for arboviruses, 1377
  bacteria transmitted by, 663
  rickettsiae transmitted by, 901
Arthropod-borne viruses. See *Arbo-virus.*
Arthrospores, 971, 972t, 978t
Arthus lesions, in systemic lupus erythematosus, 592
Arthus reaction, 546–549
  antibodies in, 546–547
  complement in, 546, *547,* 548–549
  corneal, 548
  definition, 546
  delayed-type skin reaction versus, 561
  disorders associated with, 546
  granulocytes in, 548–549
  histopathology, 547–548, *547*
  passive, *547*
  physiologic effects of, 546
  polymorphonuclear leukocytes in, 546
  pulmonary, 546
  reverse passive, 546
  sequence of, 548
  serum sickness and, 550

Arthus reaction *(continued)*
similarity to Shwartzman reaction, 548n
time course of, 546
Ascomycetes, 976, 976t. See also specific yeast, e.g., *Aspergillus*.
asci of, gene conversion in, *247*
Ascorbic acid, *Acetobacter* production of, 50
Ascospore(s), 974, 976
Ascus(i), *247, 974*
Aseptic transfer, 1455
Asparagine, bacterial synthesis of, 71, *71*
Aspartate, bacterial synthesis of, 71–72, *71, 315, 315*
Aspartate transcarbamylase, bacterial activity of, 318, *318*
Aspartokinase, 318t
Aspergillosis, 997–998. See also *Aspergillus fumigatus.*
abortions due to, 997
Arthus reaction in, 546
asthma due to, 997
diagnosis, 998
in birds, 997
opportunistic aspects, 981
pathogenesis, 997–998
pulmonary diseases due to, 997, *997*
rhinitis due to, 997
susceptibility to corticosteroid therapy and, 998
therapy, 998
Aspergillus, 976t
industrial citric acid produced by, 49
parasexuality in, 974
spores of, local infections due to, 997
Aspergillus flavus, 966
Aspergillus fumigatus, 986t, 997–998. See also *Aspergillosis.*
birds infected by, 997
hyphae of, *997*
Aspergillus nidulans, 1000
Aspergillus oryzae, 95
Assay. See specific assay, e.g., *Complement-fixation assay.*
Asthma
bronchospasm in, drug prevention of, 544
due to aspergillosis, 997
IgE produced by plasma cells in, 536
Ataxia telangiectasia, 508
Athlete's foot, 1002, 1005
Atopy. See also *Hypersensitivity.*
antibody development in, 531, 531t

Atopy *(continued)*
passive transfer in, 530–531
reagins in, 531, 531t
Attractants, chemotactic, 137
Aureomycin. See *Chlortetracycline.*
Aurintricarboxylate, 303
Australian antigen, 1410, *1411*
Autoantibodies, 586
Autoclave(s), 5, 1454
Autograft(s), 610
Autoimmune disease(s)
age factors, 588, *588*
antibodies in, 587t
antigens in, 587t
autoimmune response versus, 585
autosensitized lymphocytes in, 586
characteristics, 586–588
genetic factors, 588
human, 587t
due to unrecognized virus infection, 586
humoral versus cell-mediated reactions in, 586
localized versus disseminated, 586–588
sex factors, 588
tissues affected, 587t
Autoimmune response(s), 585–593
as cause of disease, Witebsky's criteria, 593
autoimmune diseases versus, 585
causes, 585
due to altered distribution of self-antigens, 585
due to altered forms of self-antigens, 585–586
due to forbidden clones, 586
mechanisms, 585–586
multiplicity of, 588
pathologic consequences, 593
Autoimmunity, 357
Autolysins, bacterial, 113
Autolysis
of cell wall, in bacterial cell division, 128
of dead bacteria, 114
Autotrophy, 54
Auxotroph(s)
leaky, 346
polyamine, isolation of, 79n
tryptophan, *99,* 100
Auxotrophy
bacterial mutations to, 176
in bacterial genetics, notation of, 185n
Avian leukosis virus
as helper virus, 1438, *1438*
assays of, 1434
cell transformation due to, 1208
leukemia and tumors due to, 1433
Avian myeloblastosis virus, 1432, 1434

Avian sarcoma virus
avian leukosis virus as helper virus for, 1438
subunits of, 1437, *1438*
Azathioprine, 499, 500
Azobacter, taxonomic characteristics, 34t
Azobacter agile, 37t
Azobacter vinelandii, 37t
Azobacteriaceae, classification, 34t
Azo groups, on amino acids, 355, *355*
Azurophil, 652

B

Bacillaceae, classification, 34t
Bacille Calmette Guérin (BCG), 846
in tuberculosis prevention, 860
safety and efficacy of, 861
vaccine of, sunlight destruction of, 1457
Bacillus(i), 24, 24n, *24*
acetoin fermentation in, *45*
anthrax. See *Bacillus anthracis.*
Battey. See *Mycobacterium avium* and *M. intracellulare.*
calcium ion required by, 92
coliform. See *Coliform bacillus.*
colon. See *Escherichia coli.*
enteric, 753–789. See also *Enterobacteriaceae.*
filamentous, 24, *24.* See also *Vibrio* and *Spirillum.*
flagellum(a) of, *28, 29*
Friedländer's. See *Klebsiella pneumoniae.*
fusiform, 24, *24,* 956
glutamine synthesis in, 69n
gram-negative, isolation from human infections, frequency of, 788
nonfermentative, differential characteristics of, 784t
Hansen's. See *Mycobacterium leprae.*
hay. See *Bacillus subtilis.*
Morax Axenfeld, 785
pili of, 30, *30*
Schaudinn's. See *Treponema pallidum.*
Shiga. See *Shigella dysenteriae.*
spores of, 138
thermophilic, *93*
spores of, heat killing of, *1453*
timothy hay, 868
transamination in, 73n
tubercle. See *Mycobacterium tuberculosis.*
vole, 860
Bacillus, 819–827
alanine biosynthesis in, 73n

Bacillus (continued)
capsular polypeptides in, 85
formate fermentation in, 46
penicillinase formation in, 123, *123*
taxonomic characteristics, 34t
transduction in, 1112
Bacillus anthracis, 820–826. See also *Anthrax.*
antigens of, 821–822
capsule of, 85, 634, 822
discovery, 8
DNA base composition in, 37t
edema factor of, 822
exotoxin of, 639, 822, 824
genetic variation in, 822–823
lethal factor of, 822
metabolism, 820–821
morphology, 820, *821, 822*
pathogenicity, 823–825
spores, 8, *8,* 823, 824
virulence, 821, 822, 824
Bacillus brevis, 85
Bacillus cereus
Bacillus anthracis versus, 825, 826
food poisoning due to, 826
nuclear bodies in, *31*
penicillinase produced by, 826
spores, *138*
Bacillus laterosporus, 37t
Bacillus licheniformis, 37t, *123*
Bacillus macerans, 37t
Bacillus megaterium, 826
capsule of, 634, *634*
DNA base composition in, 37t
growth temperature for, 93
megacins from, 1109
penicillin effects on, *115*
plasmolysis of, *25*
purified wall preparation from, *109*
polymyxin effects on, *127*
spores of, 141t, *143, 144*
Bacillus polymyxa, 126
Bacillus pyocyaneus. See *Pseudomonas aeruginosa.*
Bacillus stearothermophilus, 37t, 317t
Bacillus subtilis
aspartokinase inhibition in, 317t
Bacillus anthracis versus, 825
bacitracin produced by, 826
bacteriophages of, 1023, *1058*
DNA base composition in, 37t
DNA replication in, 220, *224,* 228
eye infections due to, 826
life cycle, 5
proteases in, 307
rRNA in, 282t
spores of, 139, *224*
sporulation in, 332
structure, *26*
subtilisin from, 95

Bacillus subtilis (continued)
teichuronic acid produced by, 115
transformation in, 184
Bacillus typhosus. See *Salmonella typhi.*
Bacitracin
bacterial cell wall synthesis inhibited by, 121
in identifying *Streptococcus pyogenes,* 723
produced by *Bacillus subtilis,* 826
structure, *121*
topical use of, 678
Bacteremia
Bacteroides, 787
clostridial, 838–839
death due to, 646t
definition, 646
due to *Escherichia coli,* 768
due to *Klebsiella* and *Serratia,* 769
hospital-acquired, 767
postoperative, due to *Flavobacterium meningosepticum,* 786
Bacterial antagonism
by bacteria indigenous to skin, 954
colicinogeny and, 765
mechanisms, 954
to prevent staphylococcal infections, 738
Bactericide(s). See under *Drug.*
Bacteriocin(s), 1109–1110, *1110*
Bacteriocinogen(s), 1105, 1109–1110
Bacteriology
cytochemical studies in, 24
darkfield microscopy in, 23
electron microscopy in, 23
in genetics, 11
medical, Golden Era of, 9
radioautographic studies in, 24
Bacteriolysis, 353, 1054, 1071
Bacterionema, 958t
Bacteriophage(s), 1047–1085. See also *Virus* and under specific bacterium attacked, e.g., *Bacillus subtilis, bacteriophage of.*
adsorption to bacteria by, 1051–1052, *1052.*
bacterial metabolic changes due to, 1054
as model systems in virology, 1049
bacterial lysis by, 1054, 1071
inhibition of, 1055, 1056
bacterial macromolecular synthesis altered by, 1055, 1056
bacterial mRNA and, 326
bacterial resistance to, mutations

Bacteriophage(s), bacterial resistance to *(continued)*
to, 1051
bacterial RNA polymerase altered by, 332
bacteriocin and, 1109, *1110,* 1110
CEB, toxigenicity of *Clostridium botulinum,* and 833
concatemer formation in, 1060, *1062*
coupled to hapten, to measure rate of hapten–antibody formation, 366
crossing over in, 242, *244*
defective, in transduction, 1112–1113
definition, 9
discovery of, 9, 1014
DNA of, 203, 208, 211, 327n, *1022*
bacterial genes in control of, 1117
cyclic, 1023
heterozygous, 245, *245*
phage genetic information carried into bacterium by, 1052, *1053*
replication of, determination, 1073, *1074*
statistical characteristics of, in genetic recombination, 1076
single-stranded, 1023
synthesis of, in bacterium, 1055
temporary disappearance on infectivity, 1052–1053
transcription of, 1057–1059, *1058*
control of, 1058–1059
effects of restriction enzymes, 1065
environment for, gastrointestinal, 766
evolution of, 1104, 1110
filamentous, replication of, 1082
gene(s) of, conversion of, 246
repression of, lysogeny and, 1095–1099
genetic recombination in, 241, 1073–1078
characteristics, 1073–1075, *1074, 1075*
heterozygous DNA formation in, 1077, *1077*
high negative interference in, 1077
molecular and enzymologic mechanisms in, 1077–1078
nonreciprocality of, 1077
study of, genetic map in, 1075–1077
map distances in, 1076–1077

Bacteriophage(s), genetic
recombination in, study
of genetic map in
(*continued*)
qualitative, 1075, *1076*
statistical negative interference
in, 1075–1076
genetics of, 1072–1081
study of, advantages of, 1072
head of, 1049, *1050*
helical, replication of, 1082
helper, 1053
heterologous, intrinsic interference in, 1182
homologous, intrinsic interference
in, 1181–1182
icosahedral, multiplication of,
1081–1082, *1081*
inactivation of, 1463
in bacterial gene transfer. See
*Bacterium, transduction
in.*
infection of bacteria by,
1051–1054
bacterial DNA transcription in,
263
infectivity of, assay of, plaque
method in, 1037, *1038*
eclipse of, 1052–1053
inhibition of bacterial DNA
synthesis by, 1055
internal proteins in, 1033
intrinsic interference in,
1181–1182
lambda, different forms of DNA
in, *1091*
DNA replication in,
1092–1094, *1094*
mode of, determination of,
1073, *1074*
DNA transcription in, 1092,
1093
functional map of, 1093
genetic map of, 1092, *1093*
genetics of, 1091–1092
multiplication of, 1105
repressor in, 1095, *1096*
regulation of formation,
1096–1099
transduction with, 1113, *1114*
large, complex structure of, 1033
lysogenic state of, 1094–1099
establishment of, 1094–1095
mRNA of, characteristics,
1057–1058
transformation assay for, *1057*,
1058
multiplication of, 1049–1072
with cyclic single-stranded
DNA, 1081–1082
mutagenicity of 5-bromouracil in,
255
mutants of, amber, 1073
conditionally lethal, 1073
host-range, 1051, 1073

Bacteriophage(s), mutants of
(*continued*)
ochre, 1073
plaque-type, 1073
rapid-lysis, 1055, 1073
suppressor-sensitive, 1073
temperature-sensitive, 1073
to determine mode of phage
DNA replication, 1073,
1074
mutations in, 1072–1073
acridine-induced, 235
nonsense, 293n
low polarity of, 1058
rates of, 262t
temperature-sensitive, *260*
transversions due to, 260
mycoplasmal, 933
neutralization of, in assay for
antibody, 366
nucleic acid of, capsid separated
from, 1052–1053, *1053*
injection into bacterium,
1053–1054, *1053*
naked, 1053
P1, in genetic studies of *Escherichia coli, Salmonella,* and
Shigella, 1112
lysogenization of, 1105
particles of, formation of, *197*
phenotypic mixing of, in bacteria,
1071, *1072*
polyheads of, *1069*
polysheaths of, *1069*
prophage form of. See *Prophage.*
reading frame shifts in, 258
receptors of, 1051, *1051*
release from bacterium, 1071
replicate intermediates of, 1060
RNA, 1082–1084
RNA replicase in, 1083–1084
RNA replication in,
1082–1083, *1084*
structure, 1082, *1083*
translation regulation in, 1084
Salmonella P1 and P2, 1049
Salmonella typhi typed by, 765
structure, 1049–1051, *1050*
different states of, 1049
study of, lysis inhibition in, 1055,
1056
study of genetics in, specialized
transduction in, 1117
syntheses by, bacterial syntheses
replaced by, 1055
T2, DNA replication in, *1063*
one-step multiplication curve in,
1055
T4, antimutator mutations in,
260
DNA molecules formed in, *1062*
DNA polymerase in, 260
genes of, 1079
mutations in, 260
genetic map of, *1078,* 1079,
1079

Bacteriophage(s), T4
(*continued*)
genetic material of, circularity
of, 1078–1079
organization of, 1078–1080
mRNA of, transformation assay
for, *1057*
mutations in, frequency of, site
influence on, 261, *261*
spontaneous reversions in, 259
rII locus of, genes of, 235, *235*
UV damage to, 237, *238*
T5, adsorbed to *Escherichia coli,*
1052
T7, DNA transcription in, *265*
genetic map of, 1079, *1079*
transcriptase in, 263
tail of, 1049
tail sheath contraction in, 1053,
1053, 1054
T-even, with DNA-less capsids,
1051
temperate, 1089
definition, 1049
development of, *1090*
plasmid generation by, 1105
turbid plaques produced by,
1089
vegetative cycle of, 1091–1094
to measure antibody concentrations, 482, *483*
transducing, 1111–1118
comparison of, 1116t
defectiveness of, 1117
generalized, specialized transduction by, 1117
localized mutagenesis and, *260*
nondefective, *1114,* 1117
range of, 1117–1118
triparental recombinant, 1075,
1075
ultraviolet-inactivated, genetic
reactivation of,
1080–1081
host reactivation of, 1080
multiplicity reactivation of,
1080, 1080–1081
vegetative form versus prophage
form, 1091
viral interferance with, 1173
virulent, 1049
Bacteriostatic drugs. See under
Drug.
Bacterium(a). See also *Microbe* and
specific bacterium, e.g.,
Bacillus megaterium.
abscesses due to, 980
acetolactate metabolism in, 315,
315
actinomycetes as, 872
adaptability of, 313
feedback mechanisms and, 346
invasiveness and, 635–636
adenine synthesis in, 71

Bacterium(a) *(continued)*
  adenylate synthesis in, *81,* 82
  adhesion to surfaces, 85
  aerobic, obligate, 48
  aerobic metabolism in, industrial products from, 49
  age of, 100n
  agglutination of, 391
  alanine synthesis by, 72
  alleles in, 176
  amino acids in, 28n
    analogs of, activity of, 162
      functional impairment and, 153
    levels of, 322
    requirement of, 90
    synthesis of, 69–79
      D-amino acids, 79
      aliphatic, 72–73
      amino acid inhibition of, 314
      breakdown of nitrogenous compounds and, 69
      carbon skeleton formation in, 69
      CO_2 in, 85
      commercial, 50, 133
      DAP in, 69
      enzymes in, 69
      sulfur reduction in, 69
    transport of, 136
  aminosugar synthesis in, 69
  ammonia used by, 83
  amphibolic pathways in. See *Amphibolic pathway.*
  anabolic pathways in. See *Anabolic pathway.*
  anaerobic, aerotolerant, 48
    effects of oxygen on, 48
    effects of superoxide on, 49
    obligate, 48
  anthranilate synthesis by, 77, 78, *78*
  antibiotic effects on. See under *Antibiotic; Chemotherapy;* and specific drug, e.g., *Streptomycin.*
  antibiotic synthesis by, 85–86, *86*
  antigens of, surface, 176
  antimetabolite action in, THF function and, 74
  arginine synthesis in, 69, 70, *70,* 344, *345*
  aromatic compound synthesis in, 75–79, *77, 78*
  as food for protozoa, 114
  asparagine synthesis in, 71, *71*
  aspartate synthesis in, 71–72, *71,* 315
    allosteric inhibition of, 315, *315*
  aspartate transcarbamylase activity in, 318, *318*
  aspartokinase activity in, *317*
  attenuation of, discovery of, 9

Bacterium(a) *(continued)*
  autochthonous, 954
  autolysins of, 113, 114
  autolysis in, 114, 153
  auxotrophic, cross-feeding of, 177
    in discovery of conjugation, 187
  bacteriocin action on, 1109–1110, *1110*
  bacteriophages and. See *Bacteriophage.*
  benzoquinone synthesis in, 77
  binary fission in, 128
  biosynthesis in, 59–88
    antimetabolite incorporation in, 153
    flow of, redox cofactors in, 68–69
      unidirectional promotion of, 67–69
    from 2-C compounds, 64–67
    ion formation in, 86–87
    oxaloacetate in, 63, *64*
    reversibility of reactions in, 67–68
    small-molecule, 86–87
    succinate in, 49
  biosynthetic pathway in, analysis of, 61–62
    conversion of amphibolic pathways to, 63
    intermediates in, catabolic, 63, *64*
      criteria for, 62–63
    overlapping of degradative pathways with, 68
    precursors in, 62
    ribose-5-P in, 68
    salvage, 63
    study of, 61, *61, 62*
    succinate in, 63, *64*
    uniformity of, 61
  calcium required by, 92
  cAMP in, effects on β-galactosidase synthesis, *333,* 334t
    in reversal of catabolite repression of enzymes, 332–334, *333*
    lysogeny and, 1098–1099
  capsules of, 25–26
    antibodies to, 635
    antiphagocytic, 634–635, *634*
    dextrans in, 85
    environmental factors and, 103
    levans in, 85
    polysaccharides in, 26, 83
    staining of, 25
  carbamyl phosphate synthesis in, 69
  carbon dioxide in, autotrophic assimilation of, 55, *55*
    heterotrophic fixation of, 86
    requirement of, 91–92
  catabolic pathways in, 63–67
    ATP in, 40

Bacterium(a), catabolic pathways in *(continued)*
  definition, 63
  overlapping of biosynthetic pathways with, 68
  catechols in, 92
  cell envelope of, 107–137, 145–146
    in cell division, 128–130
    metabolism regulation in, 343
  cell membrane of, DNA replication and, 223
  cell sap of, 32–33
  cell wall of, 22, 107–121
    enzymes in, 112
    equatorial growth of, *129*
    gram-positive versus gram-negative, 107–108
    granules on, 27
    growth of, 128–129
    in cell division, 128–129
    layers of, 107, *108, 109*
    lipid A synthesis in, 119
    lipopolysaccharides of, polysaccharide core of, 118, *118*
    synthesis of, 119–120, *119,* 129
    morphology, 24–25, *25,* 107
    outer membrane of, 107
    outer surface of, *108*
    peptidoglycans in, 24, 107n, 108, *109,* 110–115
      enzymes attacking, 113–114
      protected by outer membrane of gram-negative organisms, 117
      structure, 110–111, *110*
      synthesis of, 111–113, *112, 113,* 129
      variations in, 111, 111t
    plasmolysis of, for study, 24, *25*
    polysaccharides synthesized in, 83
    separation of, 128
    side chains from, 119, *119*
    storage materials in, 343
    synthesis of, antibiotic action on, 120–121
      drugs inhibiting, 675–676
      enzymes of, 122
      in cell division, mutations in, 129–130
      penicillin interference with, 111, *112,* 113, *115,* 174, 175
    teichoic acids in, *110,* 114–115
  chemical composition, methods of lysis in, 23
  chemoautotrophy in, 53–55
  chemotaxis and chemoreceptors in, 137
  chemotherapy against. See under *Chemotherapy.*

Bacterium(a) *(continued)*
  choline required by, 90
  chorismic acid synthesis by, 77, 78
  chromosome(s) of, 31
    cyclic, 191–192
    F agent on, 191, *191*
    mutations of, 293
    prophage effects on. See under *Prophage.*
    replication of, 340
    single per genome, 176
    site of attachment, 123
    units of inheritance in, 187
  citrulline synthesized by, *70*
  classification, 33–38
    Cohn's, 9
    properties used in, 33
    using antibody–antigen cross reactions, 381
  cobalt requirement of, 92
  coenzyme A synthesis by, 73, 75
  cold shock in, 93
  colicic, 956
  colony(ies) of, crowding of, 103–104
    feeder, 344
    morphology, 102–104
    mutant subclones in, 103
    papillae in, 103, *103*
    phases of, equilibrium between, 172, *172*
      variation in, 171–173
    sectored, 171
    size, 103
    surface texture of, 102–103, *102*
  complement lysis of, 522t, 523
  conjugation in, 187–197
    bridge formation in, 188, 195–196, *196*
    chromosome map of, 192, *193*
    discovery of, 187, *187*
    distribution of, factors in, 194–195
    DNA transfer in, 196–197, *196*
    F agent(s) in, 188–189, 196, *196*
      hybrid, 192–193, *194*
    F-duction in, 192–194, *194*
    F-mating in, 190–191
    F pili in, 195–196, *195*, *196*
    F-prime agents in, 193–194
    frequency of, 187, 188t
    high-frequency recombination in, 189, *189*
    kinetics of, 190–191
    interrupted mating in, 190–191, *190*
    physiology, 195–197
    polarity in, 188–189, *189*
    prophage induction due to, 1103
    resemblance to zygote formation, 187–188

Bacterium(a), conjugation in *(continued)*
    unequal genetic contributions in, 189
  conjunctival, 957
  cross-feeding in, 61, *61*
  crossing over in, 242
  culture contamination by, cycloheximide inhibition of, 984
  culture media for. See under *Culture medium.*
  culture(s) of. See also *Culture.*
    fuel consumption in, 53
    growth of, chemostat in, 99–100, *99*
      curves of, *99*
      in liquid media, 95–101
      periods of, 340
      satellite, on agar culture media, 102
      synchronous versus asynchronous, *98*
    pure, in identifying bacteria, 8
    isolation of, 101
    selective inhibition in, 104
    stock, 102
      preservation of, 172
  cysteine synthesis by, 74, *74*
  cytochrome(s) of, 122
    in oxidative phosphorylation, 51
  cytochrome oxidase in, 51
  cytology of, limited by light microscope, 22
  cytoplasmic inheritance in, 174–175
  cytoplasmic membrane of, 22, 107, *109*, 121–128
    antibiotic effects on, 126–128, *127*
    composition, 121
    enzyme(s) of, 122
    enzyme secretion across, 124
    function, 121–122
    in auxotrophs, 126
    lipids in, 107, 126
    lipopolysaccharide synthesis on, 119
    liproproteins in, 107
    organization of, 121–122
    permeability of, 126
      ionophores and, 127–128
    stretching of, due to osmolarity, 132
    structure, *122*
  cytoplasm of, granular inclusions of, 31
    organization of, 32–33
  daughter cell formation in, 114
  dead, autolysis of, 114
  death of, thymineless, 153
  degradative pathway(s) in. See *Bacterium, catabolic pathways*

Bacterium(a) *(continued)*
    denitrifying, 56
    autotrophic metabolism by, 53t
    deoxyribose residue synthesis by, 81
    detergent breakdown by, 56n
    diaminopimelate (DAP) in, conversion to lysine, 69
      in amino acid synthesis, 69
      synthesis, 72, *72*
    digestion of, enzymatic, for study, 23
      intracellular, 651–652
    dipicolinate synthesis by, 72, *72*
    diploid stages in, 175–176
    disease and infection due to, 631, 642
      agammaglobulinemia and, 658
      agranulocytosis and, 660
      bone marrow depression and, 660
      chemotherapy of. See under *Chemotherapy.*
      chronic, silent stage of, 661
      communicability of, 663–664
      cure of, 669
      dormant, 661–662
      fatigue and, 659
      hormonal factors, 659
      host-parasite interaction in. See under *Host-parasite interaction.*
      hypersensitivity in, 657–658
      inapparent, 661–662
      immunologic consequences of, 662
      initial versus later stages of, 173
      in trachoma, 923
      latent, 661–662
      lymphadenitis in, 646
      mixed, penicillin resistance in, 163
      overexposure to cold and, 659
      protection against, by nonbacterial enzymes, 114
      susceptibility to, 632
        nutritional deficiency and, 658
    disruption of, for study, 23
    dissociation of, 171–173
    division in, cell envelope in, 128–130
      cell wall growth in, *129*
      cross-wall formation in, 342
      DNA replication and, 340–342, *341*
      in starved cell, *342*
      mesosomes in, 123
      physical, 342
      physiologic, 342
      temperature-sensitive alterations in, 130
      UV interference with, 340
    DNA in. See under *DNA.*
    drug resistance by. See under *Drug*

Bacterium(a), drug resistance by
(*continued*)
resistance.
electron transport systems in, 24
mesosomes in, 123
encapsulated phagocytosis of, 653
endogenote of, 182
endogenous, 957
endonuclease A in, 336
endospores of, 137
endotoxin(s) of, 639–640, 955.
See also *Bacterium, exotoxin of;* and *Bacterium, toxin of.*
antigen immunogenicity enhanced by, 585
biologic effects, 639
complement activation by, 520
definition, 636
diets to suppress, 955
exotoxins versus, 639
leukocytic phagocytosis enhanced by, 658
nonspecific immunity and, 640
Shwartzman phenomenon elicited by, 639
species producing, 639n
structure, 639
subcutaneous injection of, 639n
tolerance to, 639
energy production in, 39–58. See also *Fermentation* and *Bacterium, respiration in.*
energy ratio in, 52
energy supply of, regulation of, 314
enterobactin in, 92
enterochelin in, 92
environment for, gastrointestinal, 766
enzyme(s) of. See also specific enzyme, e.g., *Lecithinase.*
allosteric, activation of, 314
catalytic sites in, 318
definition, 317
effector sites on, 318
interaction of, 318–319, *318*
kinetics of, *318*
mechanism of, 318–319, *318*
physiologic significance, 319
altered, drug resistance and, 162
basal level of, 331
catabolite repression of, cAMP reversal of, 332–334, *333*
catalytic site on, 317, *317*
constitutive, 321, 322
constitutive versus adaptive, 57
corepressors of, 322, 323
cross-reactive material produced by, 322
cryptic, 130
dimer-specific, in repair, 238
directed by resistance factors, 163, 163t

Bacterium(a) enzymes of
(*continued*)
dominance in, 323, *323*
drug resistance due to, 163, 163t
effector site on, 317, *317,* 318
endproduct repression of, 314, 323
definition, 322
regulator genes in, 322–323
extracellular, 95
cytotoxic. See *Bacterium, toxin(s) of.*
invasiveness and, 636
feedback inhibition of, *318*
substrate competitive antagonism of, 317
formation of, adaptive, 56
reserve capacity for, 344, *345*
inducible, 323
induction of, 320–321
coordinate, 323, *324*
feedback, 321
persistent, 321
regulator genes in, 322–323
relative, 320t
inorganic ion requirement of, 92
lytic, for fungus digestion, 969
periplasmic, 95
production of, gene dosage effects on, 346
pseudofeedback inhibition by metabolite analogs, 436
structural genes in, 322
regulation of, allosteric, 313–319
cis-trans dominance in, 323, *323, 324,* 326t
constitutivity versus inducibility in, 323, *323*
operator in, 323–325, *324*
operons in, 325, *325*
promotor in, 325
regulator genes in, 325
regulatory, 318
repression of, 321–322
by catabolites, 332–334
in branched pathways, 322
multivalent, 322
required for protein synthesis, drug action on, 151
respiratory, 51
restriction, effects on bacteriophages, 1065
reversible use of, in biosynthesis, 67
sporular, 141, 142
sulfonamide inhibition of, 154
superrepressors of, 323, 326t
synthesis of, differential rate of, 320, *321*
preferential, 344, *345*
temperature-sensitive, 322
to break down nonpenetrating nutrients, 94–95
trace element requirement of, 92

Bacterium(a) enzymes of
(*continued*)
transcription of, repressor effects on, 331
esophageal, 956
essential metabolite in, 151
essential nutrients of, 151
entry of, analog inhibition of, 154
etiologic role of, 8
evolution of, bacteriophage and, 1104
lysogeny and, 1089
rapid, 173
exoenzymes of, 124
unreleased, in cytoplasmic membrane, 122, 124
exogenous, 674
exosporium of, 141
exotoxins of, 637t
definition, 636
endotoxins versus, 639
loss of toxicity in, 638
pharmacologic actions, 638
extracellular, chemotherapy against. See under *Chemotherapy.*
extrachromosomal factors in. See *Plasmid.*
facultative, 48
F agent of, *192*
hybrid, in conjugation, 192–193, *194*
integration with chromosome, 192
structure, 191–195
fatty acid(s) in, requirement of, 90
saturated versus unsaturated, 125
synthesis, 124–125, *125*
carbon dioxide in, 86, 91
fatty acid synthetase in, 125
fermentation by. See under *Fermentation.*
fermentative, active transport in, 136
ferredoxins of, 43
in fermentation, 46
in nitrogen cycle, 56
in nitrogen fixation, 43
in photosynthesis, 43, 54
in pyruvate formation, 66
in respiration, 43
metabolic flow directed by, 68
filament formation in, 128, 130
filtration of, sterilization by, 1457
fimbriae of. See *Bacterium, pilus of.*
flagellum(a) of, *30*
basal body of, 28, *29*
chemotaxis and, 137
darkfield microscopic visualization of, 26
flagellin in, 28

Bacterium(a) flagellum of
(*continued*)
growth of, 29
regenerability of, 26
sheath around, 29, *29*
staining of, 26, *28*
structure, 26—30
fine, 28—29, *29*
types, 26, *28*
flavodoxins of, 43
folate synthesis by, 77
foodstuffs converted by, 346
forespore of, 138
forms of, 24, *24*
F pilin in, 196
fractionation of, DNAase in, 24
for study, 23
fragile, 130
fungi versus, 977—978, 980t
fungus-like. See *Actinomycetes*.
β-galactosidase in, 326t, 333, *333*,
334t
β-galactoside in, 320, 320t, *321*
gene(s) in, clusters of, 216, *216*
drug-resistance and, 162—163
in genome, 216
isolation of, 213
polymorphic, 176
regulation of, as model for
regulation in higher ani-
mals, 347
morphogenesis and, 347
repression versus inhibition in,
344
study of, specialized transduction in, 1117
regulator loci of, 322n
regulatory versus structural, 331
spore-specific, 142, 143
synthesis of, 213
transfer of, 181—200
bacteriophages in. See *Bacterium, transduction in.*
by transformation, 182—186
discovery of, 182—183, *183*
DNA concentration and,
184, *185*
identification of transforming factor in, 183—184,
184
mechanism of, 184—185
species found in, 184
methods of, 182
virus modification and restriction controlled by, 1067
genetic adaptation by, loss of
virulence in, 171
selective pressures and,
171—173
transduction of, pseudovirions
in, 1428
genetic code in, 279
genetic donors in, 188
genetic elements in, extrachromosomal. See *Plasmid*.

Bacterium(a) (*continued*)
genetics of, biochemical markers
in, 176
genetic recipients in, ·188
genome of, 216
single chromosome in, 176
genotypic variations of, virulence
and, 641—642
germination of, causes, 143
glucose in, conversion to galactose, 83, *84*
metabolism of, *47*
oxidation of, 50
selectively labeled, in study of
biosynthetic pathways,
61, *62*
glutamate in; derivatives of,
synthesis of, *70*
synthesis of, 62, 69—70, *70*
commercial, 50
glutamine synthesis by, 69
glycine synthesis by, 73
glycogen storage by, 85
glycolysis in, ion formation in, 87
respiration versus, 52, 52t
glycolytic pathway in, 67
amphibolic aspects of, 63, *64*
flow between TCA pathway
and, 67, *67*
unity of biochemistry and, 60
gram-negative, definition, 22
drug resistance by, 117
outer layer of, 116—120
anchoring lipoproteins of,
116—117
function of, 117
structure, 116—117, *116*
lipopolysaccharide,
117—118
penicillin resistance by, 120
resistance to detergents, 117
spheroplast formation from,
123
surface of, 27
gram-positive, becoming negative
in old cultures, 23
classification, 34t
definition, 22
grinding of, for study, 23
growing point in, 339
growth of, 52—53, 95—104. See
also *Culture*.
ambient water and, 94
base analog effects on, 153
biotin in, *99*
chemotherapeutic effects on,
149—151
cofactors in, 99
colony size and, 103
diauxic, 57, *57*
drug inhibition of, noncompetitive reversal of, 152—153
energy production and, 53
inhibition of, 133

Bacterium(a), growth of,
inhibition of
(*continued*)
by amino acid analogs, 153
by 5-methyltryptophan, 346
by sulfanilimides, 154
due to reduced carbon dioxide
tension, 91
minimal temperature for, 92
on fat, 64
organic factors in, 90
pH and, 93
physical requirements, 92—94
polyamines in, 79
rate of, cell composition and,
336—337, 336t
DNA synthesis and, 337, *337*
protein synthesis and, 337,
337
ribosomes and, 336, 337
synchronized, 100—101, *100*
temperature and, 92—93, *93*
tryptophan in, *99*
guanosine tetraphosphate in, 338
guanylate synthesis by, *81*, 82
halophilic, 93—94
lack of cell wall in, 110n
potassium in, 94
haploid stages in, 175—176
hemin required by, 90
hemolysis by, identification of,
103
histidine synthesis by, 69, 74—75,
76, 78, 322, 323, *324*,
331
antimetabolite inhibition of,
164
homeostasis in, 347
homoserine synthesis in, *71*
host defenses against. See under
Host.
hyaluronidase in, 95, 636
hydrogen, autotrophic metabolism by, 53t
identification of, culture media
in. See also *Culture
medium*.
differential, 103
selective, 56
solid, 102
immunofluorescence in, 397
pure cultures and, 8
imidazole ring synthesis in, 75
in circulation. See also *Bacteremia*.
destruction of, 646
in compost, 93
indigenous, definition, 954. See
also *Host, bacteria indigenous to*.
indigenous to man, 953—963
infected by UV-irradiated virus
particles, 268
inflammation and, 646—649
inflammatory response to. See

Bacterium(a), inflammatory response to *(continued)*
Inflammatory response.
inheritance in, adaptive, 171, 175
cytoplasmic, 174–175
inorganic requirements of, 90–92
inositol required by, 90
in saline solutions, 1458
in saliva, 957
intestinal, 956
vitamins synthesized by, 955
invasiveness of, 633–636
adaptation to microenvironment and, 635–636
extracellular enzymes and, 636
hyaluronidase and, 636
mutational changes affecting, 641
variability in different parts of same organ, 636
iron, autotrophic metabolism by, 53t
iron-chelating compounds produced by, 92
iron-chelating compounds required by, 90, 92
iron required by, 92
iron transport in, 92
isolation of, in solid culture media, 104
isoleucine synthesis by, 71, *71, 73, 74*
isozymes of, 314
allosteric regulation in, 315, *315*
killing of, See also *Drug, bactericidal.*
by silver ions, 1459
by UV, 1456
lactate oxidation in, 68
lactic. See *Lactic bacteria.*
LD_{50} of, 640
leucine synthesis by, 73, *74*
L forms in, 130
lipase dissolution of cell membrane in, for study, 24
lipid(s) in, complex, 125
composition of, 125–126
structure, 124–125
synthesis, 124–126
enzymes in, 122
in auxotrophs, 126
lipoamino acids in, 125
lipoic acid in, 42
lipopolysaccharides of, 762, *762*
lithotrophic, 54
luminescent, 52
lysine auxotrophic, 133
lysine synthesis by, 50, 72, *72*
lysis of, by drugs, 150, *150*
special methods for study, 23
lysogenic. See also *Lysogeny.*
curing of, 1102–1103
double, 1099, 1114
induction of, 1103–1104

Bacterium(a), lysogenic, induction of *(continued)*
artificial, 1103
mechanism of, 1103–1104
spontaneous, 1103
zygotic, 1104
detection of, 1089, *1090*
prophage in. See under *Prophage.*
macromolecule turnover in, 343–344
macrophage killing of, immunologic aspects, 566, *566, 567*
magnesium starvation in, 343
manganese ion requirement by, 92
mass of, 96
maximal growth temperature of, 92
membrane transport in, 130–137
membrane transport systems in, 130–137
active, 130
carrier model for, 135–136, *136*
definition, 131n
mechanism of, 134–136
binding proteins in, 134–135, *135*
membrane vesicles in, 136
amino acid transport in, 133, 136
antimetabolite effects on, 153–154
exit process in, 132–133
facilitated, 131–132
β-galactoside transport by, 132, 133t, 134, *135*
inducers of, 134
inducible, 131
inhibitors transported by, 133
interference with, 134
inorganic ions required by, 92
inorganic ion transport by, 134
oligopeptide transport by, 133
passive, 131–132
model for, *136*
periplasmic membrane in, 124
phosphotransferase, 136, *137*
sugar transport through, 132, *136*
temperature-sensitive, 134
membrane vesicles in, 136
merozygote of, 182
mesophilic, 93
mesosomes of, 122–123, *123*
definition, 31, 122
metabolic flow in, cytochromes in, 68
DPN in, 68–69
ferredoxin in, 68
flavins in, 68
lipoic acid in, 68

Bacterium(a), metabolic flow in *(continued)*
redox cofactors in, 68–69
reversibility of, 67–68
TPN in, 68–69
unidirectional, 67–69
metabolism in, aerobic and anaerobic, 48–52
alternative, substrates for, 56–57
arabinose operons in, 334
autotrophic, 53–56, 53t
carbon dioxide assimilation in, 55, *55*
types of, 54
branched pathways in, allosteric regulation of, 314–317, *316*
endogenous stimuli in, 344–346
energetics of, 52t
gene expression and, mechanisms regulating, 332–336
polarity mutations and, 335–336, *335*
genotypic adaptation in, 56
inhibition of, endproduct (feedback), 313–314, *314, 316, 316*
methionine in, 72
phenotypic adaptation in, 56–57
regulation of, 311–348
at level of translation, 334
by R-gene product, 334
feedback, resistance to, 344
mechanisms in, evolutionary aspects, 322
functions of, 344–347
teleonomic value of, 346–347
significance for higher organisms, 347
types of, 434–444
regulatory proteins in, 313
repression versus inhibition in, 344
understanding of, study of mutants in, 176
metabolite(s) of, analog pseudo-feedback inhibition of, 154
antimetabolite competition with, 153–154
commercial production of, 133
essential, 151
pool of, 33
synthesis of, feedback regulation of, 344
toxicity of, 133
methane, autotrophic metabolism by, 53t
methionine synthesis by, 71, *71, 75*

Bacterium(a) *(continued)*
methylations in, 72
mevalonic acid required by, 90
microcapsule of, 26
microhomology of, 38
mineralization by, 56
minicell formation in, 130, *131*
molybdenum ion requirement by, 92
monomorphism versus pleomorphism in, 170
motility of, 26, 29–30
cell wall and, 107n
vigor of, 26
mucous membrane barriers to, 642–644
murein sacculus of, 111
mutant(s) of, asporogenous, 142, 143
auxotrophic, 176
for genetic studies, 177, *178*
conditionally lethal, 177
detection of, 176–179
drug resistance by, 674, 678
feedback-resistant, 344, 346
in study of biosynthetic pathways, 61, *61*
in study of intracellular processes, 11
metabolism regulation in, 344
operator-constitutive, 323
plaque-type, 1073
resistance to fusidic acid, 300
reversibility of, 171
selection of, 171, 176–179
spontaneous inheritable, 171
streptomycin-resistant, 186t
study of, 176
temperature-sensitive, 92, 177
types, 177
mutation(s) in. See also *Mutation.*
drug resistance and, 161, 176
phenotypic, 161–162
spontaneousness of, 173
fluctuation analysis of, 173–174, *174*
invasiveness changed by, 641
neutral, 172
nonchromosomal, 174
nonsense, 335, *335*
periodic selection of, 172–173, *172*
definition, 172
phenomic lag in, 178
phenotypic lag in, 177–179, *178, 179*
polarity in, 335–336, *335*
quantification of, 178
random versus directed, 173–175
rate(s) of, 179–180
definition, 179
determination of, 179–180
temperature and, 179

Bacterium(a), mutations in *(continued)*
spontaneous, 173, 174, *175,* 259
to auxotrophy, 176
translational suppressor, 292
UV in, 253
mycobactin required by, 90
naphthoquinone synthesis by, 77
nasal, 956
nervous system in, 137
nicotinamide synthesis by, 71, 79
nitrate reduction by, 49
molybdenum ion in, 92
nitrogen-fixing, 53t, 55–56
molybdenum ion in, 92
nonpathogenic, 632
nonspecific entry of compounds into, 134
nuclear body of, 31–32, *31, 32*
nucleases of, 95, 124
nucleosides used by, 80
nucleotide function in, 80
nucleotide sugars isolated from, 83
nucleotide synthesis by, 80–83, *81, 82*
ribose-5-P in, 81
nutrients of, essential, 151
nonpenetrating, enzymatic breakdown of, 94–95
nutrition of, 90–95, 104
inorganic ions in, 92
O antigens of. See *O antigen.*
opsonization of, 631n. See also *Opsonin*
oral, 956
adhesion of, 957
diet and, 958
Leeuwenhoek's, *3*
oral *Candida albicans* suppressed by, 959
organic matter converted to CO_2 by, 56
organotropic, 635
ornithine synthesis by, 69, *70*
osmolarity of, 133
osmotic barrier of, 121
osmotic shock in, 124
oxygen requirements of, 90
pantetheine synthesis by, 75
pantoate synthesis by, 73, 74, 75
parasitism by, 633–634
Pasteur effect in, 52–53
pathogenic. See also *Pathogen.*
definition, 632
heat sterilization of, 1454
pathogenic agents compared to, 917t
pathogenicity, 632, 633–642
teichoic acids and, 115
penicillin resistance by. See under *Penicillin* and *Penicillinase.*
penicillinase in, 124

Bacterium(a) *(continued)*
pentose synthesis by, *80,* 81
peptidoglycan hydrolases of, 113
periplasmic space of, 124
periplasmic vesicles of, 123, *123,* 124
permeability barrier(s) in, 130
to drugs, 162
permeases of, 131n
persisters after bactericidal chemotherapy, 151
pesticide breakdown by, 56n
petroleum formation by, 56
phage action against. See under *Bacteriophage.*
phagocytosis of. See also under *Host, antibacterial defenses of;* and *Phagocyte.*
capsular prevention of, 634, *634*
pharyngeal, 956
phenylalanine synthesis by, 77, 78
phosphatide synthesis by, 125, *126*
phosphorylation in, 136
in photosynthesis, 54, *54*
naphthoquinones in, *50*
oxidative, antibiotic inhibition of, 128
cytochromes in, 51
in respiration, 50–51, *50*
photodynamic sensitization of, 1457
photophosphorylation in, 54, *54*
photosynthesis by, 14n, 53–55
chlorophyll in, 54, *54*
DPNH in, 69
plant photosynthesis versus, 55
photosynthesizing, autotrophic metabolism by, 53t
electron transport in, 123
ferredoxin in, 66
phototaxis by, 55
physiologic adaptation of, 170–171
physiology, 19–166
pilus(i) of, 30–31, *30*
F, in conjugation, 195–196, *195, 196*
plasma membrane of, *108*
permeability of, phage alteration of, 1054
respiration and, 51
plasmids of. See *Plasmid.*
polyamines required by, 79, 90
polyamine synthesis by, 70, 79
ribosome synthesis and, 337
polymetaphosphate granules in, 31, 343
polypeptide synthesis by, 85–86
polyribosomes of, 31
polysaccharide synthesis by, 83–85

Bacterium(a) *(continued)*
  population density of, resistance
    to penicillin and, 163
  population dynamics of, bacterial
    variation and, 169—180
  porphyrin required by, 90
  potassium required by, 92
  preservation of, by freezing, 1455
    by lyophilization, 1456
  prokaryotic aspects, 22
  proline synthesis by, 69, *70, 70*
  promoter(s) in, 332
  prophage effects on, 1104. See
    also *Prophage*.
  proteins in, allosteric, 313
    amino acids in, 28n
      taxonomic aspects, 38
    breakdown of, 343
    cationic, in phagocytosis, 651
    periplasmic, 123—124
    ribosomal, synthesis of,
      338—399
    synthesis of, cell composition
      and, 336—337
      iron ion required for, 92
      methionine in, 72
      potassium ion required for, 92
      rate of, 313
      regulation of, 319—336
        operons in, 320—325
      ribosomes and, 336
      RNA content and, 336, 336t
      RNA synthesis and, 337, *337*
      tetrahydrofolate in, 75
  protoplast of, 25
  prototype of, *23*
  pseudoauxotrophy in, 314
  psychrophilic, 93
  purine ring atoms in, 71, 81, *81*
  purine synthesis by, 69, 75, *75*
    81—82, *81*
    CO_2 in, *81*, 86, 87
  purine used by, 80
  purple, autotrophic metabolism
    by, 53t
  putrescine synthesis by, 79
  pyrimidine ring formed in, 71
  pyrimidines used by, 80
  pyrimidine synthesis by, 69, *82*
    CO_2 in, *82*, 86, 87
  pyrogenic, indotoxins produced
    by, contamination by,
    1453
  pyruvate carboxylation in, *67*, 86
  pyruvate synthesis by, 72—73
  quinon synthesis by, 77, 78
  reading frame shifts in, 258
  recombination in, F factor and,
    1105
  repressors in, 330—332
    binding of, 331—332
      by DNA, 331
    formation of, control of, 332
    isolation of, 330—331

Bacterium(a) *(continued)*
  respiration in, 49—53
    anaerobic, 49
    bioluminescence in, 51
    catalase in, 51
    coenzymes in, 42—43
    cytochrome(s) in, 42, 49, 50, 51
    cytochrome oxidase in, 51
    diphosphopyridine nucleotide
      in, 42
    electron transport in, 50—52
    energetics in, 52—53
    enzymes in, 51
    flavoproteins in, 49, 51—52
    glycolysis versus, 52, 52t
    growth and, 52—53
    lipoic acid in, 42
    luciferase in, 52
    luciferin in, 51
    naphthoquinones in, 42, 51
    oxidative phosphorylation in,
      50—51, *50*
    oxygenases in, 49
    peroxidases in, 51
    plasma membrane in, 51
    pyruvate in, 50
    substrate assimilation in, 52
    terminal, 50
    tetrahydrofolate in, 42
    thiamine pyrophosphate in, 42
    ubiquinones in, 42, 51
  ribose-5-P synthesis by, 68, *80*
  ribosomes of, 283
    artificial assembly of, 283—284,
      283
    chloramphenicol effects on,
      298, 299
    composition of, 203, 281, *282*
    definition, 31
    drugs affecting, 160
    growth and, 336, 337
    in drug resistance, 161, 162,
      164
    magnesium requirement by, 92
    mRNA synthesis and, 326, *327*
    polyamines bound to, 79
    potassium requirement of, 92
    protein synthesis and, 337
    synthesis of, 338—339
      polyamine synthesis and, 337
      RNA synthesis and, 337
  rickettsiae versus, 901
  scoring of, for genetic study, *175,*
    177
  screening of, for genetic study,
    176
  sensitivity to detergents, 94
  sensitivity to drugs, 149, 150, *150*
  sensitivity to heavy metal ions, 94
  septum formation in, 128
  serine synthesis by, 73—74, *73,*
    78
  sex factors in, 188
  shape(s) of, 107, 128

Bacterium(a), shapes of
    (continued)
  aberrant, 130
  sideramines in, 92
  siderochrome(s) of, 92
    chelation by, 134
  sideromycins produced by, 92
  skin barriers to, 642—644
  slime, motility of, 29
  sonication of, for study, 23
  species of, as cluster of biotypes,
    36
    definition, 35—38
      problems in, 35
    indistinctiveness of, 36
  spermidine synthesis by, *70,* 72,
    79
  spermine synthesis by, *70,* 79
  spheroplast(s) of, 25
    definition, 25, 108, 128
    division in, 128
    formation of, 108
  sporangium of, 137, *138*
  spore(s) of, 137—145, 146
    activation of, 143
    aging of, 143
    calcium ion content, 141, 142,
      144
    chemical resistance of, 141
    coat of, 139, 140, *140*
    core of, 141—143
    cortex of, 139
    cryptobiotic aspects, 137
    cytochromes in, 141
    definition, 137
    diaphragm of, 138
    dipicolinate in, 141, 142, *142,*
      144
    dormancy of, 143
    energy sources in, 141
    enzymes in, 141, 142
    exosporium of, 139
    formation of, 138—141
    germination of, 143—145, *143,*
      144
      definition, 143
    heat killing of, *1453*
    heat-resistant, thermal death
      point of, 1454
    integument of, 138, *140*
    longevity of, 138
    mercury ion sterilization of,
      1452
    outgrowth of, 144—145, *144*
      definition, 144
    peptide in, 144
    peptidoglycans in, 139
    resistance of, 5, 137
      to drying, 138
      to heat, 137, 142, *142*
      to ionizing radiation, 141
      to sterilization, 8
      to UV, 141
    RNA polymerase in, 142

Bacterium(a), spores of
(continued)
septum of, 138
small molecules in, 141
sterilization of, by alkalyting
agents, 1459
kinetics of, 1453, 1453
structure, 138–141
viability of, 138
water elimination from, 141
sporulating, antibiotics produced
by, 143, 160, 161
sporulation in, DNA transcription
in, 263
endotrophic, 142
prevention, 142
regulatory changes in, 142–143
RNA polymerase in, 332
toxin formation and, 143
staining of, 22–23
negative, in electron microscopy, 23
of capsules, 25
of acid-fast bacteria, 845
of flagella, 26, 28
starch hydrolysis by, 95
starvation in, 343
cell division in, 342
cell wall storage materials in,
343
DNA synthesis in, 342
survival in, 126
sterilization of, acids in, 1458
alkalis in, 1458
alkalyting agents in, 1459
by freezing, 1455
by phenol, 1453
by radiation, 1456–1457
ultraviolet, 1456, 1456
by sonic and ultrasonic waves,
1457
by visible light, 1457
kinetics of, 1453, 1453
photodynamic sensitization in,
1457
storage of materials by, 52
strains of, indicator, 1089
lysogenic, 1089
pathogenicity, 663
sensitive, 1089
sex factor-attracting, 193
wild type, 176
structure, 22–33, 38, 191–195
chemical characterization, 36
fine, 24–43, 26
gross, 24
immunologic characterization,
36
study of, 22–24
variability of, 36
subclones, mutant, identification
of, 103
substrate assimilation by, 52
succinate metabolism in, 63, 65
sugar(s) in, conversion to other
sugars, 83

Bacterium(a), sugar(s) in
(continued)
synthesis of, 83
transport of, 136
sulfide formation by, 69
sulfonamide effects on, 151–152,
152
sulfonamide ineffectiveness
against, 153
sulfur, autotrophic metabolism
by, 53t
sulfur-oxidizing, pH and, 93
surface molecules of, host-parasite
relationship and, 36
surface topography of, 25, 27
susceptibility to, 664
symbiotic, 954
synergism of, by indigenous bacterial flora, 955
syntheses within, replaced by
bacteriophage syntheses,
1055
synthetics broken down by, 56n
syntrophism in, 61, 61
taxonomy of, 14
based on DNA homology,
36–38
numerical, 36
tetrahydrofolate in, 42, 74, 75
thermophilic, 92, 93
thiamine synthesis by, 75
threonine synthesis by, 71, 71
toxigenicity of, 636–640
history, 636
toxin(s) of, LD$_{50}$ of, 640
lethal effects of, measurement
of, dose-response curve,
641
trace elements required by, 92
tracheal, 956
transduction escape synthesis in,
331
transduction in, 197–199,
197–199
by bacteriophages. See Bacteriophage, transducing.
definition, 197
generalized versus specialized,
197
joint, 198, 198
mapping by, 197–198
species occurring in, 1112
three-factor crosses in, 198
two-factor crosses in, 198
transformation versus, 197
transformation in, competence in,
185–186
DNA entry into cell, 185–186
eclipse period in, 185
nontransforming, DNA in, inhibition of transforming
DNA by, 185, 186t
significance, 186
virulence and, 186

Bacterium(a) (continued)
transmission of, 663–664
transport in, active, 131n
transport systems in, inducible,
321
membrane. See Bacterium,
membrane transport
systems in.
transport of inorganic ions by,
134
tricarboxylic acid (TCA) cycle
(pathway) in, amphibolic
aspects, 63
CoA in, 68
flow between glycolytic pathway and, 67, 67
glyoxylate cycle (bypass) in,
64–67, 66
triphosphopyridine nucleotide
(TPN) in, 42
metabolic flow directed by,
68–69
tryptophan auxotrophic, anthranilate inhibition of,
314
tryptophan synthesis by, 69, 74,
77, 78, 78
typing of, colicin in, 765
tyrosine synthesis by, 77, 78
urea-splitting, 93
urethral, 957
urinary infections due to, 160
UV absorption by, 1456
vaginal, 956
valine synthesis by, 73, 74, 316,
322
variation in, conjugation and, 194
genotypic, 170–171
phase, 171–173
phenotypic, 170–171
population dynamics and,
169–180
viability of, chemotherapeutic effects on, 149–151
quantification of, 150, 150
criteria of, 1452–1453
virulence of, 640–642
antiphagocytic surface components and, 634, 634
capsules and, 25
colony surface texture and, 102,
103
loss of, 171
lysogeny and, 642
measurement of, 640–641
dose-response curve in, 640,
641
LD$_{50}$ in, 640
transformation and, 186
variation(s) in, 641–642
phenotypic, 642
viruses infecting. See Bacteriophage.
viruses versus, 1118

Bacterium(a) *(continued)*
  vitamins required by, 90
  vitamins synthesized by, 77
    B$_{12}$, 92
    K, 955
  volume of, 32
  zinc required by, 92
Bacterium Actinomycetemcomitans, 876
Bacterium anitratum, 751
Bacterium tularense. See *Francisella tularensis*.
Bacterium typhosum. See *Salmonella typhi*.
Bacteroidaceae, 786n
Bacteroide(s), 786—788
  disease and infection due to, 786
    laboratory diagnosis, 787
    treatment, 788
  drug resistance by, 788
  in oral cavity, 958
  intestinal, 956
  physiologic role of, 786
Bacteroides corrodens, 787
Bacteroides fragilis, 787
Bacteroides fusiformis, 893
Bacteroides insolitus, 37t
Bacteroides melaninogenicus, 787
  in dental plaque, 960
  infections due to, treatment, 788
  in oral cavity, 957, 958t
    disease due to, 959
  Vincent's angina due to, 787
Bacteroides oralis, 787
  in oral cavity, 958t
Balfour Declaration, 46n
Bang's disease, 812. See also *Brucellosis*.
Barr body, 1130
Bartonella, taxonomic characteristics, 35t
Bartonella bacilliformis, 949—951.
    See also *Carrion's disease*.
Bartonellaceae, classification, 35t
Bartonellosis. See *Carrion's disease*.
Base analog(s). See also specific
    analog, e.g., *5-Bromouracil*.
  bacterial growth affected by, 153
  mutagenic action of, 254
Basidiobolus, 999, *999*
Basidiomycetes, 976, 976t
Basidiospore(s), 976
Basidium, 976
Basophil(s)
  degranulation of, in anaphylaxis, 544
    by IgE, *535*
    penicillin allergy and, 544
  histamine secretion by, 534, *534*
  IgE bound to, 534, *534*
Bat(s), rabies in, 1370, 1373
Bat salivary gland virus, 1396
B cells. See *Lymphocyte, B*.

BCG. See *Bacille Calmette Guerin*.
Bedsonia. See *Chlamydia*.
Beer
  diseases of, 6
  production of, *Saccharymyces cerevisiae* in, 95
Bee stings, anaphylaxis due to, 538
Bejel, 892
Bence-Jones proteins, 408, 425
  amino acid sequences in, 428
  definition, 425
  myeloma synthesis of, *472*
  structure, 408, *408*
Bentonite, in agglutination, 393
Benzalkonium chloride, antibacterial action of, 1460, *1460*
Benzimidazole, antiviral action of, 1183
Benzoquinones, bacterial synthesis of, 77
Benzylpenicillin, 155, 156
Biliary tract
  diseases of, due to *E. coli*, 768
  Salmonella typhi in, 773
Bioassays
  agar culture media in, 102
  in study of bacterial growth factors, 90, 98
Biochemical markers, in bacterial genetics, 176
Biochemistry
  immunologic methods in, 10
  unity of, microbial composition and, 60—61
Biology
  unity of, at molecular level, 6
  in fermentation, 47
Bioluminescence, in bacterial respiration, 51
Biosphere, chemical stability of, fungi in, 966
Biosynthesis(es). See also *Anabolism* and *Metabolism*.
  bacterial. See under *Bacterium*.
  mammalian, economy in, 87
Biosynthetic reactions. See *Anabolism* and *Metabolism*.
Biotin
  discovery, 91t
  in bacterial fermentation and respiration, 42
  in bacterial growth, *99*
Bird(s). See *Avian* and *Fowl*.
Bismuth
  in growth of pathogenic *Salmonella* and *Shigella*,
  organic compounds of, in syphilis, 1459
Bis-phenols, halogenated, antibacterial action of, 1461
Bittner virus. See *Mouse mammary tumor virus*.

Black carpet beetle, togaviruses carried by, 1385
Black death. See *Plague*.
Blastogenic factor, 575, 577
Blastomyces brasiliensis. See *Paracoccidioides brasiliensis*.
Blastomyces dermatitidis, 986—988, 986t, *987*. See also *Blastomycosis, North American*.
  dimorphism in, 977
  mating types in, 978
  organelles in, *968*
Blastomycin
  Histoplasma capsulatum, cross-reaction with, 990
  skin response to, 987
Blastomycosis
  European. See *Cryptococcosis*.
  North American, 986—988. See also *Blastomyces dermatitidis*.
    diagnosis, 987-988
    epidemiology, 988
    immune response in, 987
    pathogenesis, 987
    therapy, 988
  South American, 994—995, *994*.
    See also *Paracoccidioides brasiliensis*.
    diagnosis, 994
    epidemiology, 995
    therapy, 995
Blastospore(s), 972t
Blast transformation
  due to antigen, 573—574, *573*
  due to Enterobacteriaceae cell wall lipopolysaccharide, 574
  due to incubation of antigen with small lymphocytes, 468—469
  due to mitogens, 574, *574*
  in cell-mediated immunity, 573—575, *573*
  in immediate-type hypersensitivity, 573
  in mixed lymphocyte culture, 574
  lymphotoxin release and, 577
  specificity of, 574—575
Blenorrhea, inclusion, of newborn. See *Conjunctivitis, inclusion*.
Bleomycin, DNA transcription blocked by, 267
Blindness
  due to ophthalmia neonatorum, 750
  due to syphilis, 888
  due to trachoma, 922
Blood
  Bombay-type, 599, 603
  circulating. See *Circulation*.
  clotting of, complement and, 521

Blood, clotting of
  (continued)
    kinin production and, *542*
    heart, use of, advantage of, 704
    passage of virus to brain from, 1215
    transfusion of. See *Transfusion.*
    typing of, 608–610
      anti-D sera in, 607
      in anthropologic surveys, 608
      in archeology, 608
      in disputed paternity, 608
      in forensic medicine, 608
      to prevent transfusion reactions, 609
    universal donor and recipient of, 609
Blood agar. See *Agar, blood.*
Blood cell(s)
  O, antigen of, 600
  red. See *Erythrocyte.*
  white. See *Leukocyte.*
Blood cell alloantigen(s). See also *Blood group substance.*
  chemistry of, 600–603
  distribution, 600, *601*
  genetic determination of, 599–600
  identification of. See *Blood, typing of.*
  individual's pattern of, uniqueness of, 609–610
  localization in human tissue, *601*
  non-ABO, 604–608
  secretors and nonsecretors of, 600
  synthesis of, glycosyltransferases and, 602, *604*
Blood group(s)
  ABO, 598–604
    frequency of, in ethnic groups, 604, 605t
    genotypes in, 599t
    phenotypes in, 599t
    based on erythrocyte agglutination by human sera, 599t
  Duffy, 607
  history, 598
  incompatibility of, hemolytic disease of newborn due to, 606
  Kell, 607
  Lewis activity in, 600
  MNSs, 605
  mucopeptides of, antigenic determinants of, 602, *602*
  named for erythrocyte antigens, 599
  RH, anti-D sera to establish, 607
  specificity of, erythrocyte surface antigens and, 1129
Blood group substance(s), 590–610. See also *Blood cell alloantigen.*
  ABO, as histocompatibility antigens, 619

Blood group substance(s), ABO
  (continued)
    transplantation immunity and, 619–620
    antibody to, reactions of, 367
    antigenic differences between, 602
    degradation products of, 602
    evolution of, 604
    structure, 602, *603*
Bloodstream. See *Circulation.*
Blood vessel(s)
  small, occluded in typhus, 905
    rickettsial attack on, 899, 902
Blue tongue virus, 1400
Body surface, topical chemotherapy to kill or inhibit pathogens on. See *Antisepsis.*
Boeck's sarcoid, 581
Boil(s)
  recurrent, prevention of, 738
  staphylococcal, 734
Bombay-type blood, 599, 603
Bone(s), tuberculosis of, 861
Bone marrow
  depression of, bacterial disease and, 660
  interferon production by, 1175
  stem cells of, B and T lymphocytes from, 460
Bordetella, taxonomic characteristics, 34t
Bordetella bronchiseptica, 799
  growth characteristics, 793t
Bordetella parapertussis, 799
  growth characteristics, 793t
Bordetella pertussis, 797–800. See also *Whooping cough.*
  antigenicity in, 797, 798
  culturing of, 797, 799
  exotoxin of, 637t, 639
  growth characteristics, 793t
  history, 796
  killed, formation of reaginic antibodies and, 585
  metabolism, 797
  morphology, 797
  on ciliated bronchial epithelium, 798, *798*
  organotropicity of, 635
  pathogenicity, 798–799
  variation in, 642, 797
  virulence, 642
Boric acid, antiseptic properties, 1459
Bornholm disease, 1298, 1298n
Borrelia(e)
  general characteristics, 882
  infections due to, 893
  mistaken for leptospires, 893
  taxonomic characteristics, 35t
Borrelia recurrentis. See *Relapsing fever.*

Borrelia vincentii, 893
Borrelidin, 303
Botulism, 833–834. See also *Clostridium botulinum.*
  immunity to, 834
  laboratory diagnosis, 834
  mortality rate, 834
  pathogenesis, 833–834
  prevention, 834
  symptoms, 834
  toxin in, 831, 833
  treatment, 834
Boutonneuse fever, 904t, 909
Bovine papular dermatitis virus, 1260
Bovine serum albumin, proteolytic cleavage of, 376
Bowel sterilization, 787n
Bradykinin, 542, 542n
Brain
  passage of virus from blood to, 1215
  small blood vessels of, rickettsial effects on, 899
  tissue of, antibody reaction with, 589
Breeding, selective, antibody formation and, 477
Brill-Zinsser disease, 904t, 907
5-Bromodeoxyuridin, animal cell DNA transcription blocked by, 1133
5-Bromouracil. See also *Base analog.*
  bacterial incorporation of, 153
  DNA radiation sensitivity and, 240
  herpesvirus DNA synthesis inhibited by, 1185
  mutagenic effects of, 255, *255*, *256*
Bronchial carcinoma, bronchopulmonary candidiasis with, 995
Bronchiectasis
  bronchopulmonary candidiasis with, 995
  due to *Pasteurella multocida* infection, 809
  pneumococcal, 704
Bronchiolar constriction, due to kinins, 542
Bronchiolitis, due to respiratory syncytial virus, 1352
Bronchitis
  due to respiratory syncytial virus, 1352
  due to rhinoviruses, 1304
  infectious, virus in, *Mycoplasma gallisepticum* interaction with, 934
Bronchopneumonia
  due to measles, 1346, 1347
  due to respiratory syncytial virus,

Bronchopneumonia, due to respiratory syncytial virus *(continued)* 1350, 1352
  due to rhinoviruses, 1304
  with whooping cough, 798
Bronchospasm, chemotherapy in, 544, 545
Broth, penassay, 95t
Brucella(e), 811–817. See also *Brucellosis.*
  acquired cellular immunity to, 657, 658
  antigenic structure, 813
  antigens of, hypersensitive reactions to, 815
  characteristics, differential, 814t
    general, 812
    taxonomic, 34t
  cultivation of, 816
  genetic variation in, 813
  history, 812
  metabolism, 812–813
  morphology, 812
  outer layer of, fatty acids in, *117*
  pasteurization of, 813, 1454
  pathogenicity, 813–815
  pH and, 93
  species identification, 814t, 816
  virulence, 815
Brucella abortus, 812
  abortion in cattle due to, 815
  characteristics, 814t
  DNA base composition in, 37t
  immunization against, 813
  metabolism, 812
  rough and smooth colonies of, 171, *171*
  virulence, 815
Brucella canis, 812, 817
  characteristics, 814t
Brucellaceae, 802
  classification, 34t
Brucella melitensis, 812
  characteristics, 814t
  undulant fever due to, 815
Brucella ovis, 813
Brucella suis, 812
  cattle infected by, prevention, 816
  characteristics, 814t
  hepatic and splenic granulomas due to, 815
Brucellergen, 816
Brucellosis. See also *Brucella(e).*
  due to *Brucella melitensis,* 815
  hypersensitivity in, 657
  immunity to, 815–816
    cellular, 815
  incubation period in, 815
  laboratory diagnosis, 816
  prevention, 816–817
  symptoms, 815
  treatment, 816

Bubo(es)
  in bubonic plague, 804
  in lymphogranuloma venereum, 925
  satellite, in syphilis, 887
Budding, in yeast, 128, 971
Bunyamwera viruses, 1392–1394
  classification, 1381t
    immunologic, 1379
  clinical diseases due to, 1381t
  development, 1392, *1393*
  disease and infection due to, epidemiology, 1392–1394
    pathogenesis, 1392–1394
  geographic distribution, 1381t
  morphology, 1392
  properties of, 1392
  vectors of, 1381t
Burkitt lymphoma, 1447
  EB virus in, *1253, 1254*
Burns, *Pseudomonas aeruginosa* infection in, 784
Bursa of Fabricius, 474–475. See also *Lymphatic organs.*
  B cell production of Ig in, shift from IgM to IgG, 504
  hormonally extirpated, 474
Bursectomy, immunologic effects of, 474t, 475
Butanediol, produced by fermentation, 46
Butter, mycobacteria in, 868
Butyric acid, produced by fermentation, 46
Bwamba virus, 1392
Bystander cell(s), destruction of, in vicinity of immune reaction, 577

C

C3 activator, 520, 521
Caffeine, mutagenic effect of, 239, 259
Calcium (ion)
  bacterial requirement of, 92
  in bacterial spores, 141, 142, 144
  in immune cytolysis by complement, 518
Calcium hypochlorite, to deodorize garbage and sewage, 1452
California encephalitis, 1392
California encephalitis virus, 1396
Calymmatobacterium granulomatis, 948–949
cAMP
  in animal cell cultures. See under *Animal cell culture.*
  in bacteria. See under *Bacterium.*
  mediator release from mast cells inhibited by, in anaphylaxis, 544

Canarypox, 1260
Cancer. See also *Leukemia* and *Tumor.*
  antimetabolites in, 153
  cells of, tumor-specific antigens on, 621
  cytomegaloviral disease and, 1249, 1251
  detection of, by immunologic testing, 622
  fetoproteins and, 622
  immunodeficiency diseases and, 506
  immunologic defects with, 568, 569t, 622
  immunosuppression and, 502, 623, 1202
  oncogenes in, 1436
  progression of papilloma to, 1429
  viruses in, 1429, 1446–1447
Candida, cycloheximide inhibition of, 982
Candida albicans, 986t, 995–997, *996.* See also *Candidiasis.*
  endocarditis due to, 996
  in oral cavity, bacterial suppression of, 959
  pseudohyphae of, 995, *996*
  structural forms, *996*
Candida krusei, 995
Candida parakrusei, 995
Candida parapsilosis, 995, 996
Candidiacin, 983. See also *Polyene.*
Candidiasis, 995–997. See also *Candida albicans.*
  bronchopulmonary, 995
  causes, 996
  corticosteroid enhancement of, 996
  diagnosis, 996–997
  epidemiology, 997
  intertriginous, 996
  opportunistic aspects, 981
  oral, 959, 995
  pathogenesis, 995–996
  susceptibility to, in diabetes mellitus, 996
  therapy, 997
  transfer factor in, 572
  vulvovaginal, 995
Capillary permeability, complement and, 522, 522t
Capim virus, 1392
Capreomycin therapy, in tuberculosis, 857
Capsid, viral. See under *Virus* and under specific virus.
Capsomer, viral. See under *Virus.*
Carbamyl phosphate, bacterial synthesis of, 69
Carbenicillin (therapy), 676
  in *E. Coli* infection, 768
  in *Proteus* infection, 772
  in *Providencia* infection, 772

Carbenicillin (therapy) *(continued)*
in *Pseudomonas aeruginosa* infection, 784
in *Serratia* infection, 771
Carbohydrate, in IgG, 411
Carbolic acid. See *Phenol.*
Carbomycin, 158
Carbon dioxide
bacterial use of. See under *Bacterium.*
gonococcal requirement of, 91
meningococcal requirement of, 91
plant, assimilation of, 69
requirement by growing cells, 92
Carboxypeptide therapy, in measles, 1183
Carboxyphosphate anhydride, in energy-producing bacterial fermentations, 41, 42
Carcinoembryonic antigen, 622
Carcinogen(s), leukemia induced by, leukosis viruses in, 1435
Carcinogenesis, mutagenesis and, 259
Carcinoma(tosis)
bronchial, bronchopulmonary candidiasis with, 995
gastrointestinal, 622
protected by antibodies to tumor-specific antigens, 621
thrush with, 995
Cardiac. See also *Heart.*
Cardiac standstill, due to streptolysin O, 718
Cardiolipin
chemical structure, *886*
in bacterial cytoplasmic membrane, 125
in syphilis, 886
Cardiovascular lesions, in serum sickness, 549, *549,* 551
Carditis, rheumatic, 594
Caries, dental. See *Dental caries.*
Carrier(s) and carrier rate(s). See also under specific disease and organism, e.g., *Typhoid, carrier of.*
Carrion's disease, 950. See also *Bartonella bacilliformis.*
anemia in, 951
complicated by salmonellosis, 951
cutaneous stage, 951
symptoms, 950
Casein hydrolysate, in culture media, 94
Castration, poliomyelitis and, 1212
Cat(s). See also *Feline.*
bite of, *Pasteurella multocida* infection due to, 809
Catabolic pathway, in bacteria. See under *Bacterium.*
Catabolism. See also *Metabolism.*

Catabolism *(continued)*
definition, 40
Catalase, in bacterial respiration, 51
Catalysis, protein flexibility and, 305
Catechols, bacterial, 92
Cattle. See also *Bovine.*
abortions in, due to aspergillosis, 997
due to brucellosis, 815
prevention, 813, 816, 817
due to *Vibrio fetus,* 782
actinomycosis in, 873
anthrax in, 8
prevention, 826
Coxiella burneti infection of, 913
foot-and-mouth disease of, 9
Johne's disease in, 867
lumpy jaw in, 875
mycoplasma diseases in, 930, 936t
pneumonia in, 809, 930
tuberculosis in, 861
vesicular stomatitis virus of, 1368, 1370
Cell(s). See also specific kind of cell, e.g., *Warthin-Finkeldey cell.*
changes due to viruses. See under *Virus.*
cultured, cell-mediated immune reactions with, 572–578
differentiation of, antibody formation as model for, 357
due to interaction of antibodies and antigens, 591
diploid. See *Diploid cell.*
distinguishing viruses from, 10
genes of. See under *Gene.*
growth of, CO_2 requirement in, 92
regulation of, polyoma viral effects on, 1423
hyperplasia of, due to poxviruses, 1261
proliferation of, due to interaction of antibodies and antigens, 591
surface alloantigens on, 597–624
variety of, 13
transformed. See also *Cell transformation.*
characteristics, 1423, 1423t
destruction of, by cell-mediated immune response, 1445
in animal cell cultures, 1123–1124, *1124*
Cell culture(s)
animal cell. See *Animal cell culture.*
induction of antibody formation in, 455–456
susceptibility to echoviruses, 1301t

Cell envelope, 22
bacterial. See under *Bacterium.*
Cell line(s)
animal. See also under *Animal cell culture.*
heterogeneous, 1135
used in virology, 1142t
Cell-mediated reaction(s). See *Immune response, cell-mediated.*
Cell physiology
antimetabolites in, 153
chemotherapy and, 165
study of, bacterial mutants in, 176
Cell-sorting machine, 455
Cell strain, animal. See also under *Animal cell culture.*
definition, 1122
Cell transformation. See also under *Cell, transformed.*
carcinogen-induced, leukosis viruses in, 1435
cell characteristics in, 1423, 1423t
cell surface antigens and, 1445
definition, 1208
due to viruses. See under *Virus.*
in animal cells, lysogenization compared to, 1428–1429
Cell type, bacterial organotropism and, 635
Cellular immunity. See *Immunity, cell-mediated.*
Cellulase, bacterial, 95
Cellulitis
clostridial, 838, 841
due to aspergillosis, 998
streptococcal, 720
Cellulose, *Aerobacter* synthesis of, 85
Cell wall
bacterial. See under *Bacterium.*
spirochetal, 107n
Central nervous system
nocardiosis of, 878
syphilis of, 890
Centrifugation
of lymphoid cells, 455
to purify virus samples, 1014
Centromere, 1130
Cephalexin, 676
Cephaloglycine, 157
Cephaloridine, 157
Cephalosporin (therapy), 157–158, 676. See also *Antibiotic.*
biosynthesis, 160
in *E. coli* infection, 768
penicillinase ineffectiveness against, 158
Cephalosporium, 157
Cephalothin therapy
in *Arizona* infection, 776

Cephalothin therapy *(continued)*
in *Klebsiella pneumoniae*
infection, 770, 771
in *Proteus mirabilis* infection, 772
Cephamycins, 158
Cerebellar ataxia, due to *Mycoplasma pneumoniae,* 936
Ceroid, staining of, 845
Cervix
carcinoma of, herpes simplex
virus and, 1245
streptococcal adenitis of, 720
Cervicitis
due to chlamydiae, 924
due to gonococcal infection, 750
Chancre
soft, 797
syphilitic, 887
Chancroid, 797
Cheese fermentation, 44, 45, 46
Cheesemaker's lung, 546
Chemical(s)
allergic contact dermatitis due to,
562
destruction of potential infectivity by. See *Disin-
fection.*
sterilization by, 1457—1462
Chemical bond(s), energy of, in
fermentation, 41
Chemical industry, aromatic products of, 78
Chemoautotrophy
bacterial, 53—55
definition, 54
Chemoprophylaxis, 680
Chemoreceptors, bacterial, 137
Chemostat, in bacterial growth,
99—100, *99*
Chemosynthesis. See *Chemoautotrophy.*
Chemotaxis
bacterial, 137
complement and, 522—523, 522t
negative, 137
Chemotherapeutic agent(s)
definition, 148, 149
bactericidal, definition, 148
bacteristatic, definition, 148
Chemotherapy. See also *Antibiotic;
Drug;* specific chemotherapeutic agent, e.g., *Rufa-
mycin;* and under specific
disease, e.g., *Variola,
treatment.*
antibacterial, 147—166, 668—680
against extracellular bacteria,
669—671
bactericidal drugs in, 670—671
bacteristatic drugs in, 669—670
against intracellular bacteria,
671
bacterial membrane transport
systems affected by,

Chemotherapy, antibacterial
(continued)
153—154
brief, failure of, 151
development of, 148—149
effects on growth and viability
of bacteria, 149—151
general principles, 674—675
immune mechanisms with, 151
interactions of, 669
metabolite analogs in, 151—153,
152
pseudofeedback inhibition by,
164
antimetabolite, 153—154
biosynthetic incorporation of
drug in, 153
in cancer, 153
in cell physiology, 153
in immunosuppression, 499,
501
THF function and, 74
antimicrobial, antagonistic, 679
bioassay for, 102
special aspects, 678—680
topical, 678
antiviral, limitations of, 1172
cell physiology aided by, 165
combined, 164, 679
diseases inhibiting, 673
early, 674
in endocardial infections, 675
in influenza, 1327
limitations of, 671—674
abscesses and, 671—673, *674*
foreign bodies and, 673
local tissue factors in, 671—673
necrosis and, 671—673
obstructed excretory organs
and, 673
systemic host factors in, 673
selective toxicity in, 148
time required for, 675
to control infectious diseases, 11
Chicken(s). See also *Fowl.*
bursectomized, 475
embryos of, in animal virology,
1140, *1143*
immunoglobulin production in,
evolutionary sequence of,
503
infectious bronchitis virus of,
1362n
Marek's disease in, 1446
nonpathogenic viruses of, 1433
synovitis in, due to mycoplasmas,
934
systemic anaphylaxis in, 539
teratogenic viruses in, 1216
tuberculosis in, 861
Chickenpox. See *Varicella.*
Chikungunya encephalitis
epidemiology, 1390
laboratory diagnosis, 1389

Chikungunya encephalitis
(continued)
pathogenesis, 1388
Chikungunya virus, hemagglutination-inhibition titrations
in, 1382t
Childbed fever, 720, 724
Chimera, 612
Chitin, 968
Chitinase, bacterial, 95
Chitin synthetase, in fungi, 969
Chlamydia(e), 915—927
antibody response to, 919
antigens of, 919—920
characteristics, chemical,
918—919
general, 916—922
metabolic, 918—919
classification, 35t
developmental cycle in, 916—918
disease and infection due to,
agents for, 920t
carrier state in, 920
chemotherapy, 921—922, 921t
chronicity of, 920
clinical features, 921t
general aspects, 921—922
geographic distribution, 920t
immunity to, 920—921
laboratory diagnosis, 921t
transmission, 920t
elementary versus reticulate body
of, 916, *918*
evolution of, 916
group A. See *Chlamydia trachomatis.*
group B. See *Chlamydia psittaci.*
groups of, 917t
host specificity of, 916
human diseases due to, 920t
morphology, 916—918, *918*
nongonococcal urethritis due to,
924, 925
pathogenic agents compared to,
917t
staining of, 917
taxonomy, 35t, 916
Chlamydia psittaci, 926. See also
*Lymphogranuloma vene-
reum* and *Ornithosis.*
Chlamydia trachomatis. See also
*Conjunctivitis, inclusion;
Lymphogranuloma vene-
reum;* and *Trachoma.*
antigenic types, 923
sulfonamide therapy in, 922
Chlamydospore(s), 971, 972t
definition, 972t
of *Candida albicans,* 995, *996*
Chloramphenicol (therapy),
677—678
antibiotic action of, 159
bacterial growth inhibited by,
determination of, 150,
150

Chloramphenicol (therapy)
(*continued*)
bacterial RNA synthesis inhibited by, 338
bacterial viability affected by, 150, *150*
biosynthesis of, 160
immunosuppression by, 501
in actinomycosis, 876
in *Alcaligenes faecalis* infection, 786
in anthrax, 826
in *Arizona* infection, 776
in *Bacteroides* infection, 788
in chlamydial infections, 922
in cholera, 781
in *Citrobacter* infection, 772
in *Edwardsiella* infection, 771
in *Enterobacter* infection, 771
in *Flavobacterium* infection, 786
in granuloma inguinale, 949
in *Hemophilus influenzae* infection, 796
in *Klebsiella pneumoniae* infection, 770
in leptospirosis, 895
in melioidosis, 785
in *Neisseria meningitidis* infection, 748
in plague, 805
in pneumococcal infection, 705
in *Proteus mirabilis* infection, 772
in pseudotuberculosis, 806
in relapsing fever, 893
in rickettsial disease, 901
in salmonelloses, 671, 775
in *Serratia* infection, 771
in shigellosis, 778
in tularemia, 808
in typhus, 906
in whooping cough, 799
protein synthesis inhibited by, 298–299, 303t
resistance to, by bacteria, 163t, 679
by *Shigella*, 1107
side effects, 678
streptomycin antagonism with, 164, *164*
Chloramphenicol acetylase, in drug resistance, 163t
Chlorine, antiseptic properties of, 1459
Chloroform, bactericidal action of, 1462
Chloromycetin. See *Chloramphenicol.*
Chlorophyll, in bacterial photosynthesis, 54, *54*, 55
Chloroplasts, ribosomes in, 283
Chlortetracycline. See also *Tetracycline.*
antibiotic action of, 160

Chlortetracycline (*continued*)
structure, *298*
Cholera. See also *Vibrio cholerae.*
control of, 755
coproantibodies in, 781
enterotoxin in, 780
fowl, 809. See also *Pasteurella multocida.*
history, 353, 779
immunity to, 353, 781
laboratory diagnosis, 781
laboratory inadequacies in, 781n
mortality rate, 780
prevention, 782
symptoms, 780
treatment, 781
Choline, bacterial requirement of, 90
Choriomeningitis, lymphocytic. See *Lymphocytic Choriomeningitis.*
Chorioretinitis, due to cytomegalovirus, 126
Chorismic acid, bacterial synthesis of, 77, 78
Chromatin, 1129–1130
Chromatography
in purification of virus samples, 1014
of nucleic acids, 212
paper, growth factor positions in, 102
to separate lymphoid cells, 455
Chromoblastomycosis, 1000
Chromomycin, DNA transcription blocked by, 267
Chromosome(s)
aberrations in, due to viral infections, 1208
bacterial. See *Bacterium, chromosome of.*
DNA-replicating, 219–223
elimination from hybrid cells, in animal cell cultures, 1136
heterochromatin fraction of, 1130
homologous, pairing of, 241
locus of. See *Locus.*
mapping of, 233
in study of antibody formation, *479*
mitotic, stained with quinacrine, 1134, *1135*
pairing of, in animal cells, 1130
used as markers, 1135, *1136*
Chromosome 18, partial deletion of, IgA deficiency and, 508
C′H$_{50}$ unit, 514
Chymotrypsin degradation, *543*
Circulation
bacteria in. See also *Bacteremia.*
destruction of, 646

Circulation (*continued*)
clearance mechanisms in, 646, 646t
disturbances of, phagocytosis of bacteria blocked by, 658
viruses disseminated by, 1212
Cirrhosis, biliary, 587t
Cistron, 234n, 306
Citrate
fermentation of, 68
in citrus fruits, 75n
iron chelation by, 92
Citrate test, in Enterobacteriaceae, 770
Citric acid, industrial, produced by *Aspergillus*, 49
Citrobacter, 771–772
Citrulline, bacterial synthesis, *70*
Citrus fruits, citrate in, 75n
C-kinin, 523
in anaphylaxis, 542
Cladosporium carrionii, 1000
Cladosporium werneckii, 1006
Classification. See under specific group classified, e.g., *Bacterium, classification.*
Clathrates, 127
Clindamycin (therapy), 676
in *Bacteroides* infection, 788
Clonal dominance, 427
Clonal selection hypothesis, of lymphocyte antibody production, 471
Clone(s)
definition, 101n
forbidden, development of autoimmune responses due to, 586
isolation of, 102
scoring of, 175, 177, *177*
Clostridium(a), 829–841
anaerobiosis in, 830–831
antigens of, 832
antitoxin against, 841
cellulitis due to, 838
classification, 34t, 831–832
collagenases produced by, 636
effects of oxygen on, 49
endospores of, *831*
enzymes of, 831, 838
fermentation by, *44*, 831
food poisoning due to, 839
gas formation by, 45
general properties, 830–832
histotoxic infections due to, 838–841. See also *Gas gangrene.*
pathogenesis, 838–839
prevention, 841
treatment, 841
laboratory diagnosis, 839–841
intestinal, 956
invasiveness of, 636
metabolism, 831–832

Clostridium(a) *(continued)*
  morphology, 830, *831, 832*
  pathogenic, biochemical reactions
    of, 832t
  spores of, 138, *138*
    resistance of, 830
  toxins produced by, 637t, 638,
    639, 831, 838, 839,
    839t, *840*
Clostridium aceticum, 53t
Clostridium acetobutylicum, 91t
Clostridium bifermentans, 37t,
    832t
Clostridium botulinum. See also
    Botulism.
  biochemical reactions, 832t
  heat-resistant spores of, 833
  hemolysin produced by, 836
  toxigenicity of, 831, 833, 834n
  types, 833
Clostridium histolyticum
  biochemical reactions, 832t
  oxygen sensitivity of, 830
  toxic filtrates of, antitoxin
    against, 841
Clostridium kluyveri, 37t
Clostridium nigrificans, 37t
Clostridium novyi
  biochemical reactions, 832t
  exotoxins of, 637t
  gas gangrene due to, 838
  toxic filtrates of, antitoxin
    against, 841
Clostridium pectinovorans, 832
Clostridium pectinovorum, 138
Clostridium perfringens
  bacteremia due to, 838
  biochemical reactions, 832t
  capsules of, 830
  colonies of, *840*
  DNA base composition of, 37t
  enteritis necroticans due to, 839
  food poisoning due to, 839
  gas gangrene due to, 838, 840
  infections due to, diagnosis, 840
  in wound infections, 838
  milk fermentation by, 831
  oxygen sensitivity of, 830
  pathogenicity, 839
  toxic filtrates of, antitoxin
    against, 841
  toxigenic types, 839t
  toxins produced by, 637t, 839,
    839t, *840*
  uterine infection due to, 838
Clostridium septicum
  biochemical reactions, 832t
  gas gangrene due to, 838, 840
  toxic filtrates of, antitoxin
    against, 841
Clostridium sporogenes, 832t, 838
Clostridium tertium, 832t, 838
Clostridium tetani. See also *Tet-
    anus.*

Clostridium tetani *(continued)*
  antigenic structure, 836
  biochemical reactions, 832t
  culture of, 91
  DNA base composition in, 37t
  in feces, 835
  morphology, *831*
  oxygen sensitivity of, 830
  pathogenicity, 835
  phenotypic variation in, 642
  serologic types, 836
  toxin of, 831
  virulence of, 642
Clostridium welchii. See *Clos-
    tridium perfringens.*
Clover wound tumor virus, 1400
Cloxacillin, penicillinase ineffectiv-
    eness against, 156
Coagulase
  bacterial, 95
  staphylococcal, 636, 729, 730,
    734
    detection of, 737
Cobalt ion, bacterial requirement
    of, 92
Cobamide, in bacterial fermenta-
    tion and respiration, 42
Cobra venom, complement de-
    pleted by, 524
Coccidioides immitis, 986t,
    991–994. See also *Coc-
    cidioidomycosis.*
  antigens of, hypersensitivity to,
    992
  arthrospores of, 992, *992*
  cultivation of, hazards of, 993
  dimorphism in, 977n
  endospores of, 991
  growth, 992
  immune response to, 993
  spherules of, 991, *991, 993*
Coccidioidin, *Histoplasma capsu-
    latum* cross-reaction
    with, 990
Coccidioidin skin test, 993
Coccidioidomycosis. See also *Coc-
    cidioides immitis.*
  diagnosis, 981, 993
    coccidioidin skin test in, 993
  epidemiology, 993–994
  pathogenesis, 992–993
  therapy, 993
Coccobacillus(i), 24, *24*
Coccus(i), 24
  pyogenic, 693–752. See also
    specific pyogenic cocci,
    e.g., *Streptococcus.*
Codon(s), 235–236
  anticodon pairing with, 294n
    streptomycin distortion of, 301,
    301
  base of, 236
    triplet composition with, 235
  definition, 203, 235, 236

Codon(s) *(continued)*
  mutant, suppression of, 292
  nonsense, mutation suppression
    and, 293
  ochre, 295
  ribosomal recognition of, 284
  recombinations within, 241, *242*
  terminator, 277, *279*
  variety of organismal uses of, 279
Coenzyme(s)
  in bacterial fermentation, 42–43
  in culture media, 94
Coenzyme A, bacterial. See under
    Bacterium.
Cofactors, bacterial. See under
    Bacterium.
Cold
  common. See *Common cold.*
  overexposure to, disease and, 659
Cold shock, in bacteria, 93
Colicin(s), 1109
  bacterial typing using, 765
  definition, 765
  Enterobacteriaceae typing by,
    765–766
  production of, 765
Colicinogen(s), 1109
  in *E. coli,* 195
Colicinogeny, bacterial antagonism
    and, 765
Coliform bacillus(i), 767–772. See
    also specific bacillus, e.g.,
    Escherichia.
  definition, 767n
  genera included in, 767
  identification of, 757
  intestinal, 956
  pathogenicity, 767
Coliphage(s)
  structure, 1049
  T, evolution of, 1104
  T-even, DNA of, 5-hydroxy-
    methylcytosine in, 1021,
    1022
Coliphage lambda, *1018*
Colistin therapy
  in *Acinetobacter* infection, 785
  in *Aeromonas* infection, 782
  in *Bacteroides corrodens* infec-
    tion, 787
  in *Citrobacter* infection, 772
  in *E. coli* infection, 768, 769
  in *Enterobacter* infection, 771
  in *Pseudomonas aeruginosa* infec-
    tion, 784
  in shigellosis, 778
Colitis, due to *E. coli,* 768
Colitose, in *E. coli* antigens, 763n
Collagen, similarity to complement
    C1q component, 516n
Collagenase(s), clostridial, 636
Colon, bacteria in, 956
Colony(ies), bacterial. See *Bac-
    terium, colonies of.*

Colorado tick fever, 1408
Colorado tick fever virus, 1407–1408
Common cold
  due to coronavirus, 1365
  due to rhinoviruses. See *Rhinovirus.*
  parainfluenza virus in, 1337
Complement (C), 511–525
  activation of, 520–521, *521*
  anaphylatoxins and, 522, 522t
  anaphylaxis and, 545, 546
  antibody against, 523
  bactericidal action of, 522t, 523
  blood clotting and, 521
  capillary permeability and, 522, 522t
  chemotaxis and, 522–523, 522t
  cobra venom depletion of, 524
  conglutination and, 523
  C1q component of, 516n
  deficiency(ies) of, 523–524, 524t
  definition, 512
  enzymatic activity of, *517*
  fixation of, 515
    in Arthus reaction, 546, *547*
  fractions of, *516, 517*
  fragments of, inflammatory tissue changes due to, 515
  hemolysis due to, dose-response curve of, *515*, 520
    membrane lesions in, 518–519, *519*
  one-hit theory of, 519–520
  hereditary angioedema and, 523–524
  history, 512
  immune adherence of, 522t, 523
  immune conglutination and, 523
  immune cytolysis by, calcium ion required for, 518
    C3 convertase in, 518
    reaction sequence in, 515–520, *517*
  in Arthus reaction, 548–549
  inflammatory tissue changes and, 518, 520, 522
  in glomerulonephritis, 551, *552*
  in rheumatoid arthritis, 553
  in serum sickness, 550, *550*
  in *Treponema pallidum* immobilization test, 523
  in viral complexes, 553
  mast cells affected by, 522
  measurement of, 514
  opsonic activity of, 522t, 523
  plasmin activation of, 524
  proteins of, properties of, 515, *516*
  tissue destruction by, 523
  unit of, 514
  wheal formation and, 522
Complementation, intragenic, 305–306

Complementation test
  to distinguish genes, 233–234, *235*
  to show genes of rII locus of bacteriophage T4, 235, *235*
Complement-fixation assay, 512–515, *513*
  antibody-antigen complexes in, 515
    mass of, 513–514
    ratio of, 513–514
  conditions of, 513
  controls in, 513
  for antibodies to poliovirus, 514t
  hemolysis in, 513
  quantitative, 514–515, *515*
Compost heaps, bacteria in, 93
Concanavalin. See also *Mitogen.*
Concanavalin A, blast formation and, 468
Concentration, active, 131n
Congenic lines, 615, 615n
Congenital anomaly(ies)
  due to cytomegalovirus, 1216
  due to parvoviruses, 1236
  due to rubella virus, 1216, 1356, 1357, *1357*
    prevention of, 1358
Conglutination, 523
Conglutinin, 523
Conidiophore(s), 972t
Conidium, 971, *971*, 972t, *973*. See also *Fungus, spores of, asexual.*
Conjugation, bacterial. See under *Bacterium.*
Conjunctiva(e), bacteria in, 957
Conjunctivitis. See also *Keratoconjunctivitis.*
  adenoviral, 1235
  due to *Hemophilus aegypticus*, 797
  due to *Moraxella lacunata*, 785
  inclusion, 924–925. See also *Chlamydia trachomatis.*
  in leptospirosis, 895
  in trachoma, 923
  Morax-Axenfeld, 751
  swimming pool, 924–925. See also *Chlamydia trachomatis.*
Constitution, genetic, in infectious disease, 11
Consumption. See *Tuberculosis.*
Contact skin sensitivity. See under *Sensitivity.*
Convertase, C3, in immune cytolysis by complement, 518
Coombs test. See *Antiglobulin test.*
Copper salts, fungicidal properties of, 1459
Coproantibody(ies)
  antibacterial action of, 643

Coproantibody(ies) *(continued)*
  definition, 777
  in cholera, 781
  in *Shigella*, 777
Cord factor, 847
Cornea
  Arthus reaction in, 548
  herpetic ulcers in, 1212
  infections of, due to *Moraxella lacunata*, 785
  transplant of, 612
  UV irritation of, 1456
  vascularization of, in trachoma, 923
Coronavirus(es), 1361–1366
  characteristics, 1363t
    chemical and physical, 1362
    immunologic, 1362–1364
  history, 1362
  host range, 1364
  infection due to, control of, 1364–1366
    epidemiology, 1365
    laboratory diagnosis, 1365
    pathogenesis, 1365
    prevention, 1365–1366
  morphogenesis, *1364*
  morphology, 1362, *1363, 1364*
  multiplication of, 1363, 1364–1365, *1364*
  naming of, 1362
  neutralization reactions of, 1363t
  properties of, 1362–1365
Corticosteroid (therapy)
  aspergillosis enhanced by, 998
  candidiasis due to, 996
  commercial synthesis of, microbes in, 50
  E. Coli infection with, 768
  immunosuppression by, 501
  listeriosis due to, 947
  Pseudomonas aeruginosa infection due to, 783
  thymic effects of, 474
  yeast synthesis of, 125
Cortisone (therapy)
  candidiasis due to, 996
  coxsackieviral infection enhanced by, 1297
  decrease in resistance to tuberculosis and, 855
  effects on *Corynebacterium pseudotuberculosis muris*, 662
  inflammation and, 659
  in herpes simplex keratitis, 1246
  susceptibility to viruses and, 1212
  tuberculin test inhibition by, 857
Corynebacteriaceae, classification, 34t
Corynebacterium(a), 681–692
  anaerobic similarity to *Actinomyces israelii*, 876
  cross-reactivity with mycobacteria

Corynebacterium(a), cross-
reactivity with
(continued)
and nocardiae, 848
cultures from man, fermentation
reactions of, 684t
general description, 682
in oral cavity, 958t
mycolic acids in, 847
nasal, 956
periodontal disease due to, 961
taxonomic characteristics, 34t
waxes in, 847
Corynebacterium acnes, 37t
Corynebacterium diphtheriae. See
also *Diphtheria.*
cultivation of, 683–684
DNA base composition in, 37t
growth of, tellurite in, 104
lysogeny in, 684–685
morphology, 683
nicotinic acid produced by, 91t
pathogenicity, 686–688
animal experiments in,
686–687
in human disease, 687
strains, 683–685
toxigenicity of, gel diffusion test
of, 690, *690*
toxins produced by, 58–59, 635t,
683, 684–685
accidental inoculation with, 687
activation of, 685, *686*
iron ion in, 92
minimum lethal dose of, 686
mode of action, 685–686
structure, 685, *686*
virulence, 642
Corynebacterium hofmanii, 684t,
692
Corynebacterium kutscheri, 692
Corynebacterium minutissima,
1006
Corynebacterium ovis, 686, 692
*Corynebacterium pseudotu-
berculosis muris,* 662
Corynebacterium ulcerans, 686
Corynebacterium xerosis, 684t, 692
conjunctival effects of, 957
Coughing, antibacterial action of,
643
Coulter counter, 96
Cowpox, 7, 10, 352, 1275
Cowpox virus, 1275. See also
Poxvirus.
in immunization against smallpox,
1270
Coxiella(e), taxonomic charac-
teristics, 35t
Coxiella burneti. See also *Q fever.*
cattle and sheep infected by, 913
DNA base composition of, 37t
endocarditis due to, 911
geographic distribution, 912

Coxiella burneti (continued)
metabolism in, 899
natural reservoirs of, 912, 913
pneumonia due to, 902
resistance to heat, 898
resistance to physical and chemical agents, 911–912
transmission of, 901
wall composition of, 898
Coxsackievirus(es), 1296–1299.
See also *Picornavirus.*
cell susceptibility to, 1208
characteristics, chemical and
physical, 1296
immunologic, 1296–1297
clinical syndromes due to, 1298,
1298t
congenital heart defect due to,
1216
discovery of, 1280
disease and infection due to,
control of, 1299
enhanced by cortisone and pregnancy, 1297
epidemiology, 1299
laboratory diagnosis, 1299
manifestations, 1297
malnutrition and, 1202
pathogenesis, 1298–1299
resistance to, 1297
groups of, 1296
host range of, 1297
interferon production with, 1180,
1297
multiplication of, 1297–1298,
1298t
host temperature and, 1212
myocarditis due to, 1298
naming of, 1280n
properties of, 1296–1298
RNA of, base composition of,
1297t
susceptibility to, 1209
Cox vaccine, for typhus, 906
Crede procedure, 751
Creutzfeld-Jakob disease, 1220
Cross-feeding. See *Syntrophism.*
Crossing over, 242–246
by breakage and reunion, 242,
243
gene conversion and, 246
in bacteria, 242
in bacteriophages, 242
in eukaryotes, 243
in hybrid DNA, 245
molecular mechanisms, 242–246
nonreciprocal, 242, *244*
reciprocal, 242, *244*
study of, 242
Cross-reacting material (CRM), 253
produced by bacterial enzymes,
322
Croup, due to respiratory syncytial
virus, 1349, 1352

Crustacean exoskeleton, 968
Cryophile(s), bacterial, 93
Cryoprecipitates, in immune-
complex diseases, 554
Cryptobiosis, in bacterial spores,
137
Cryptococcosis. See also *Cryptococcus neoformans.*
diagnosis, 985
epidemiology, 986
pathogenesis, 985
therapy, 985–986
Cryptococcus neoformans,
984–986, 986t. See also
Cryptococcosis.
capsule of, 984, *985*
cycloheximide inhibition of, 982
general characteristics, 984
morphology, 984, *984, 985*
Crystallizable fragment (Fc), 410
from IgM, 416, *417*
Culex, as togavirus vector, 1385
Culex pipiens, 1390
Culex tarsalis, 1390
Culiseta melanura, 1390
Culture(s)
aeration of, 90
animal cell. See *Animal cell
culture.*
enrichment, organism identification by, 56
increasing air-liquid interface in,
90
lymphocyte, mixed, HL-A antigens and, 619
monolayer, of host cells, development of, 10
of bacteria. See under *Bacterium.*
pure, limiting dilutions to obtain,
8
viable, preservation by freezing,
1455
Culture medium(a)
agar, 101
blood, 103
special uses of, 102
agarose in, 101
albumin in, 94
bacteriologic, composition of, 95t
casein hydrolysate in, 94
coenzymes in, 94
differential, in bacterial identification, 103
enrichment of, 94
indicator dyes in, 103
liquid, bacterial growth in,
95–101
adaptive chemical changes in,
96
bioassays of, 98
cell mass determination in, 96
cell number and, 96
cryptic, 96
doubling time of, 97

Culture medium(a), liquid
(*continued*)
exponential kinetics in,
96–98, *98*
growth cycle in, 96
inhibition by trace contaminants, 96
methods of measurement,
95–96
phases of, 96, *97*
substrate concentration and,
98–99
total cell count of, 96
turbidity in, 96
meat digest, 94
meat infusion, 94
pH of, buffering and, 93
selective, microbe identification
with, 56, 104
serum in, 94
silica gel in, 101
solid, access to oxygen in, 104
bacterial growth on, 101–104
in identifying microorganisms,
102
introduction of, 8
isolation of bacteria with, 104
uses of, 101–102
solidifying agents in, 101
vitamin(s) in, 94
vitamin B$_{12}$ assay, for *L. Leichmannii*, composition of,
95t
yeast extract added to, 94
Cyanophyceae, 14
classification, 14n
electron transport in, 123
motility, 29
Cybernetics, application to living
organisms, 347
Cyclohexamide
interferon production affected
by, 1179t
toxicity of, 984
Cyclophosphamide, immuno-
suppressive action of,
499
Cycloserine (therapy)
bacterial cell wall synthesis inhibited by, 120
structure, *120*
in tuberculosis, 857
Cysteine, bacterial synthesis of, 74,
74
Cystic fibrosis, 783
Cystine, 91
Cystitis, due to *E. coli*, 768
Cytochemical studies, in bacteriology, 24
Cytochrome(s)
b, 50
bacterial. See under *Bacterium*.
c, amino acid sequences of phylogenetic tree derived from,
307, 308

Cytochrome(s), c (*continued*)
definition, 50
o, 51
Cytochrome oxidase, bacterial, 51
Cytolysis, immune. See under *Complement*.
Cytomegalic disease
activation of, 1249
control of, 1252
eosinophilic inclusion bodies in,
1250, *1250*
epidemiology, 1251–1252
laboratory diagnosis, 1251
mitigated by interferon, 1180
pathogenesis, 1250–1251
serologic tests for, 1203t
Cytomegalovirus(es), 1249–1252
cellular effects of, 1251, *1251*
host range, 1249–1250
immunologic characteristics, 1249
morphology, 1249, *1249*
multiplication in, 1249, 1250,
1250
properties of, 1249–1250
teratogenic effects of, 1216
Cytoplasmic membrane, bacterial.
See under *Bacterium*.
Cytoplasmic polyhedrosis virus,
1400
Cytosine arabinoside therapy, in
herpetic keratitis, 1245
Cytotoxicity, cell-mediated, antibody-dependent, 578

D

Dairy products, tuberculosis from,
861
Danysz phenomenon, 379
Daunomycin, DNA transcription
blocked by, 267
DDT
in delousing, to prevent epidemic
typhus, 907
to control rat fleas in endemic
typhus, 908
to control sandfly in Carrion's
disease, 951
Deaminating agents, mutagenic
action of, 254
Death, black. See *Plague*.
Death rate. See under specific
disease, e.g., *Tetanus,
death rate of*.
Decarboxylase test(s), in Enterobacteriaceae, 770
Decimal reduction time, in heat
sterilization, 1454, *1455*
Declomycin therapy, *942*
Degradative pathway, in bacteria.

Degradative pathway
(*continued*)
See under *Bacterium*.
Demethylchlortetracycline, *298,*
942
Dendritic cells. See *Macrophage,
dendritic*.
Dengue
epidemiology, 1391
pathogenesis, 1387, 1388
prevention and control, 1391
yellow fever with, 1183
immunologic characteristics, 1384
multiplication of, 1386
Denitrifiers, 49
Dental. See also *Tooth*.
Dental caries, 960–961, *960–962*
definition, 960
fluoride therapy in, 962
of root, 961, *962*
penicillin prevention of, 960
sucrose and, 85, 961, *961*
Dental plaque(s)
bacterial, control of, 962
bacteroides melaninogenicus in,
960
dental caries due to, 960–961,
960–962
periodontal disease due to, 960
subgingival, 961
Treponema macrodentium in, 960
Deoxycholate agar, in differentiating Entero-
bacteriaceae, 759
Deoxyribonuclease(s). See *DNase*.
Deoxyribonucleic acid. See *DNA*.
Deoxyribose residues, bacterial
synthesis of, 81
2-Deoxystreptamine, *300*
2-Deoxystreptidine, *300*
Deoxyuridine derivatives, viral
DNA synthesis inhibited
by, 1185
Depsipeptide, 128
Dermacentor. See also *Tick*.
Dermacentor andersoni, 1408
in Rocky Mountain spotted fever,
909
Dermacentor marginatus, 1391
Dermacentor variabilis, 1408
in Rocky Mountain spotted fever,
909
Dermatitis
allergic contact, 562, *564*
definition, 562
desensitization in, 579
due to penicillin, 584
elicitation of, 563, 563n
contagious pustular, of sheep. See
Orf.
poison ivy, 562
desensitization in, 579
Dermatomycosis(es), 980–981,
986t, 1000–1004. See

Dermatomycosis(es), *(continued)*
also *Dermatophyte.*
classification, 1003–1004
definition, 979
epidemiology, 1002
immunity to, 1002–1003
nomenclature, 1001
treatment, 1003
undecylenic acid therapy in, 1002n
Dermatophilaceae, 873t
Dermatophilus, 873t
Dermatophyte(s), 980. See also *Dermatomycosis.*
anthropophilic, 1002
cutaneous keratin attacked by, 1001–1002
definition, 979
geophilic, 1002
hyphae of, *1001*
hypersensitivity to, 1003
in stratum corneum, *1002*
pleomorphism of, 1003
predilection for epidermis, 1001
zoophilic, 1002
Dermatophytid(s), 1003
Dermatophytosis. See *Dermatomycosis.*
Desensitization, 579. See also *Immunization.*
acute, 539
antigen administration in, 539
prolonged, 535–536
Desulfovibrio
autotrophic metabolism by, 53t
sulfide formation by, 69
taxonomic characteristics, 34t
Desulfovibrio desulfuricans, 37t
Detergent(s)
antibacterial activity of, 94, 1460–1461, *1460*
bacterial breakdown of, 56n
bacterial resistance to, 117
iodine complexed with, 1459
phenol as, 1461
topical use of, 678
viruses inactivated by, 1463
Deuteromycetes, 976–977, 976t
identification, 973t
pathogenic, sexual stage in, 977t
Dextran(s), in bacterial capsules, 85
Diabetes mellitus
acidosis in, 64
bacterial diseases with, 659
bacteroides infection with, 787
candidiasis and, 995, 996
chemotherapy inhibited by, 673
mycoses with, 995
phycomycotic infections of central nervous system in, 998
Diagnostic tests, delayed-type skin reactions used in, 558, 559t

Diaminodiphenyl sulfone (therapy)
antibacterial activity, 155
in leprosy, 675, 867
structure, *154*
Diaminopimelate (DAP)
bacterial. See under *Bacterium.*
plant synthesis of, 72
Diapedesis, of phagocytes, 646, *647, 649*
Diarrhea
adenoviral, 1232
due to *Citrobacter,* 772
due to endotoxins, 639
due to *E. coli,* 768
due to noncholera vibrios, 782
due to *Proteus* infection, 772
due to *Vibrio cholerae,* 780
due to salmonellae, 774
traveler's, 768
Diazo reaction, in coupling haptens to proteins, 355, *355*
Dick test, 717
Diet
immunity and, 658
oral bacteria and, 958
to suppress endotoxin-producing bacteria, 955
DiGeorge's syndrome, 508, 580
Digestion, enzymatic, of bacteria, 23
Dilutions, limiting, to obtain pure cultures, 8
Diphenicillin, *157*
Diphosphopyridine nucleotide. See *DPN.*
Diphtheria. See also *Corynebacterium diphtheriae.*
antitoxin in, 691, 691t
development of, 683
carriers of, 688
treatment of, 691
case fatality rate in, 691, 691t
epidemiology, 688–689
false membrane in, 682, *687*
history, 353, 682–683
immunity to, 688–689
temporary, 688
immunization against, 353
international standard antiserum for, 688
laboratory diagnosis, 689–691
laryngeal, 687
naming of, 682
nasopharyngeal, 687
prevention, 691
pseudomembrane in, 687, *687*
reimmunization to, 689
Schick test for, 688–689
streptococcal pharyngitis versus, 687
systemic complications, 687
toxin–antitoxin (TAT) complex in, 683
Danysz phenomenon in, 379

Diphtheria *(continued)*
toxoid for, development of, 683
dosage of, 481
formol, 685, 688
immunoglobulin production induced by, *484*
time needed to reach peak levels, 483
treatment, 691
vaccination for, 683
Diphtheroid(s), 692. See also *Corynebacterium.*
Dipicolinate, bacterial. See under *Bacterium.*
Diplococcus(i), 24, *24*
taxonomic characteristics, 34t
Diplococcus pneumoniae. See also *Pneumonia.*
antigens of, 794
capsule of, *634*
DNA base composition in, 37t
Hemophilus influenzae mistaken for, 793
Klebsiella pneumoniae versus, 704
organotropicity of, 635
tetracycline-resistant, 705
Diploid cell(s)
definition, 1134
heterozygous DNA in, 245, *245*
Disease(s). See also *Infection* and specific disease.
antibacterial immunity depressed by, 659
antimicrobial therapy inhibited by, 673
bacterial, prevention of, natural antibodies in, 653
chronicity of, intracellular parasitism and, 634
communicability of, 633
communicable, noncontagious, 7
concept of, impact of microbiology on, 11
contagious, 7
endemic rate of, 663n
germ theory of. See *Germ theory.*
infection versus, 631–632
infectious, immunity to. See *Immunity.*
death due to, 13. See also under specific disease.
latent, 633
parasitic, 657, 657n
Disinfectant(s). See also *Antiseptic* and *Sterilization.*
characteristics, 1452
definition, 1457
evaluation of, problems in, 1458
potency of, 1457–1458
rate of killing by, 1457
susceptibility to, microbial, 1453
uses of, 1458
Disinfection, 1451–1465. See also *Sterilization.*

Disinfection *(continued)*
  alcohols in, 1461—1462
  alkylating agents in, 1459—1460, *1460*
  definition, 1452
  tricresol with soap in, 1461
  viral sensitivity to, 1033
Distemper, canine, 1343
DNA (deoxyribonucleic acid)
  acridine dyes and, 211
  bacterial, attached to cell membrane, 223
  base analog incorporation into, 153
  breakdown of, 344
  composition of, 327n
    taxonomy based on, 36, 37t
  degradation of, by bacteriophage nuclease, 1056, *1057*
  deletions in, 243
  hybridization experiments with, *214*, 216
  in gene transfer by transformation, 183—184, *184*
  in transformation. See under *Bacterium, transformation in.*
  radiation-damaged, excision repair of, 238—240, *239*
  recombinational repair of, *239*, 240
  repair of, DNA polymerase in, 239
  replication of, 221, *222*
    cell division and, 340—342, *341*
    cell membrane in, 223
    origin and direction of, 220, *221*
    replicon in, 340
    unidirectional versus bidirectional, 339
  sequences of, study of, by renaturation kinetics, 216—218, *217*
  synthesis, frequency of initiation and, 339
    growing point in, 339
    growth rate and, 337, *337*
    initiation of, 340
    in nongrowing cell, 342
    in starved cell, *342*
    phage inhibition of, 1055
    regulation of, 339—342
    replication protein in, 340
  repressors bound to, 331
  transfer of, conjugation, 196—197, *196*
  unwinding of, prevention of, 331
  bacteriophage, 203, 208, 211, 327n, *1022*
    different forms of, *1091*

DNA (deoxyribonucleic acid), bacteriophage *(continued)*
    genetic information of phage carried into bacterium by, 1052, *1053*
    5-hydroxymethylcytosine in, 1049
    breaking of, reunion of. See *Crossing over.*
    configuration, 205—208
    crossing over of. See *Crossing over.*
    cyclic, *210*, 211, *211*
      replication of, 229—230, *229*
    degradation of. See *DNase.*
    deletions of, spontaneous, 260
    double-stranded, structure, *203*, 204, *204*
    double-stranded versus single-stranded, 240, *240*
    doubling of, rate of, 222
    electron microscopy of, 205
    error correction in, gene conversion due to, 246, *248*
    fragility of, 205
    genetic recombination(s) in. See *Genetic recombination.*
    helical, separation of strands, 212—213
    heterozygous, 245, *245*
    homology of, bacterial taxonomy based on, 36—38
      in Enterobacteriaceae, 754
    hybrid, *213*
      crossing over in, 245
      recombination in, 243—246, *244*
      multiple, 246
    methylation of, 203
    minicircular, 1105. See also *Plasmid.*
    molecular weight, 205
    molecules of, integrated and joint, 242
    nick in, *210*, 211
    of *E. coli,* 207
    pH effects on, 211
    physical properties, 205—211
    radiation damage to, repair of, 238—241
      excision, 238—240, *239*
      postreplication, *239*, 240
      recombinational, *239*, 240
    radiation sensitivity in, *240*, 240
    radioautography of, 205
    recombination of, frequency of, genetic mapping and, 247, *248*, *249*
      nonhomologous, 243
      site-specific, 243
    relaxation of, 211
    repair of, polymerase in, 227
    radiation effects on, 269

DNA (deoxyribonucleic acid) *(continued)*
  replicating chromosome and, organization of, 219—223
  replication of, 219—230, *219*
    at replicating fork, *219*, 228, *229*
    chain growth in, 225
    chain initiation versus elongation in, 221
    divergent, 222
    definition, 219
    DNA unwinding in, 225, 229—230, *229*
    end-to-end joining of short segments in, 225
    enzyme(s) in, 225—227, *226*
      imperfections of, spontaneous mutations due to, 260
    errors of, excision repair of, 241, *241*
    growing point in, 219
    histone synthesis and, 1134
    in animal cells, 1130, 1134
    initiation of, 225
    mechanisms, 225—230
    Okazaki segments in, 225, 228, *228*
    replicator in, 222
    replicon in, 222
    rolling circle, 221, *223*
    semiconservative, 219, *220*
    sequential, unidirectional, 219—221
      chemical evidence of, 221
      genetic evidence of, 220
    symmetry of, 221, *223*
    test of, *224*
    unfolding proteins in, 211, 227
    unidirectional, 219
    units of, 221—222
  sedimentation of, 211
  sequence(s) of, correction of, gene conversion due to, 246
    divergence of, taxonomic differentiation and, 218
    in amphibians, 217, *218*
  single-stranded, structure, 204
  size, 205—208
  species-specific markers in, 204
  superhelical turns (twists) in, *210*, 211
  synthesis of, in animal cell cultures, 1134
    on RNA template, in reverse transcription, 267
    repair type, 242
  template of, in RNA synthesis, 262
  transcription of, 262—267
    actinomycin D blocking of, 1133
    antibiotic effects on, 266—267, *266*

DNA (deoxyribonucleic acid),
  transcription of
  (continued)
chain elongation in, 264, 265
definition, 262
hybridization of, DNA and
  RNA in, 262, *262*
initiation of, 263—264, *264,
  265*
  antibiotic blocking of, 266,
    266
in phage infection, 263
in sporulation, 263
model of, *264*
promotor in, 264, *265*
reinitiation of, 264, *265*
reverse, 267
rho factor in, 265
RNA polymerase in, 263
sigma factor in, 263, *264*
termination of, 265, *265*
transcriptase in, 263, *265*
unwinding in, 265
transfer of genetic information
  from, to RNA, 262
unfolded by protein, 211, 227
UV damage to, *237*
viral, 211
  base composition of,
    1021—1023
  abnormal, 1021
  association of capsid and,
    1069—1071
  comparison to host cell DNA,
    1021
  components of, origin of, 1064
  configurations of, special, 1023
  cyclic, 1023, *1023—1025*
  double-stranded, 1020—1023,
    1024, 1025
  glucosylation in host, 1065,
    1066, 1067
  host-induced restriction and
    modification of,
    1064—1068, *1066*
    enzymes in, 1067—1068, *1067*
    significance of, 1067—1068
  methylation of, in host,
    1066—1067, *1066,* 1068
  modifications of, 1064—1068
  molecular weight of,
    1020—1021
  renaturation of, 209
  replication of, 221
    biochemistry of, 1064
    breakage and reunion of con-
      catemers in, 1061—1063,
      1063
    double-stranded, 1060, *1061*
    genetic recombination in,
      1063—1064
    origin and direction of,
      1060—1061

DNA (deoxyribonucleic acid),
  viral, replication of
  (continued)
  pools of macromolecules for,
    1068
  replicate intermediate in, 1060
sequences of, 216, 217, *218*
single-stranded, 1023
synthesis of, arabinosyl cytosine
  inhibition of, 1185
  enzymes in, 1064, *1065*
  from concatemers, 1061, *1062*
  interference with, by halo-
    genated deoxyuridine de-
    rivates, 1185
  proteins in, 1064
DNA ligase, mutational impairment
  of, 257
DNA polymerase
  bacterial, in DNA repair, 239
  in DNA repair synthesis, 227, 239
  in DNA replication, 225—227,
    226
  in phage T4, gene for, mutations
    in, 260
  RNA-dependent, in reverse tran-
    scription, 267
DNA replicase, bacterial, 122
DNA—RNA hybrids
  in reverse transcription, 267
  in transcription, 262, *262*
DNase(s)
  in fractionation of bacteria, 24
  staphylococcal, 718, 731
Dog(s)
  Leptospira canicola infection in,
    894
  rabies in, 1373
  Rickettsia conori carried by, 909
Donor, universal, 609
Donovan bodies, 949, *950*
Dose
  infectious (ID$_{50}$), 640
    computation of, 664—665
  lethal, measurement by Reed—
    Muench method, 640
  minimum. See LD$_{50}$.
Dose-response curve, in measurement of bacterial viru-
  lence, 640, *641*
Doubling time, of bacterial growth,
  97, *98*
Doxycycline (therapy), 160. See
  also *Tetracycline.*
  in epidemic typhus, 906
  in rickettsial diseases, 901
DPN (diphosphopyridine nucleotide)
  bacterial. See under *Bacterium.*
  discovery, 91t
  required by *Hemophilus,* 94
DPNase, streptococcal, 718
Droplet nuclei, 663

Drosophila melanogaster, mutations
  in, 253, 262t
Drug(s). See also *Antiseptic; Chemotherapeutic agent;* and
  specific type of drug,
  e.g., *Antibiotic.*
  allergy(ies) to, 581—585
    genetic factors in, 584
    principles of, 584
    special tissues affected by, 585
    variations among individuals,
      584—585
  antagonistic, 164—165, 679
  antibacterial, 675—678
    competitive inhibition by,
      151—152, *152*
    in urinary infections, 160
  antimicrobial action of, deter-
    mination of, 150
  bacterial cell wall synthesis in-
    hibited by, 675—676
  bacterial sensitivity to, 149
  bacterial viability affected by,
    quantification of, 150,
    150
  bactericidal, 151
    action of, criterion of, 1452
    mechanisms of, 151
    of aerosols, 1462
    against extracellular parasites,
      670—671
    chloroform as, 1462
    halogens as, 1459
    organic solvents as, 1462
    phagocytes with, 671, *673*
    phenols as, 1461, *1461*
    surfactants as, 1460—1461,
      1460
    synergistic, 165
  bacteristatic, 151
    hexachlorophene as, 1461
    against bacterial extracellular
      parasites, 669—670
    in suppurative lesions, 670
  germicidal, 1460
  inhibition by, kinetics of, study
    of, 150
  resistance to, 161—165. See also
    under specific drug, e.g.,
    Penicillin, resistance to.
  bacterial, altered enzymes and,
    162
    by exogenous bacteria, 674
    decreased activation of drug
      in, 161
    development of strains in, 161
    drug destruction in, 161
    genetic dominance in, 165
    mutation and, 173, 176, 674
    permeability barriers in, 162
    phenotypic effects of mu-
      tations to, 151, 161—162
    plasmids in, 162—163, 164

Drugs, resistance to, bacterial, altered enzymes and (*continued*)
R factor and, 1109
ribosomal causes of, 161, 162, 164
by *Aeromonas*, 782
by *Bacteroides*, 787, 788
by *E. coli*, 768, 1107
by *Enterobacter*, 771
by gonococci, 679
by herpesviruses, 1246
by *Klebsiella*, 771
by *Neisseria meningitidis*, 746
by pneumococci, 679
by *Proteus*, 679
by *Pseudomonas*, 679
by *Salmonella*, 1107
by *Serratia*, 771
by *Shigella*, 679, 778, 1107
by staphylococci, 679, 954
by streptococci, 679
clinical significance, 678–679
due to enzymes directed by resistance factors, 163, 163t
epidemiologic significance, 678–679
general features, 163–165
induced phenotypic, 163–164
in tuberculosis, 678, 858
multiple, R factor transfer of, in Enterobacteriaceae, 766
penicillinase and, 162–163
prevention, 164
resistance-factor-mediated, 163, 163t
resistance transfer factors in, 162
superinfections and, 673–674
Drying, resistance to, by spores, 138
Dust, in transmission of bacteria, 663
Dust cells, 701
Dwarfism, parvoviral, 1236
Dyes, fluorescent, in photodynamic sensitization of bacteria, 1457
in photodynamic inactivation of viruses, 1464
Dysentery
amebic versus bacillary, 778
bacillary. See also *Shigellosis.*
control of, 755
pathogenesis, 777–778
Shigella species in, 777
symptoms, 777
Shigella. See *Dysentery, bacillary.*
Dysgammaglobulinemia, bacterial infection and, 658

E

Eagle test, 886
Eastern equine encephalitis
epidemiology, 1389, *1389*
pathogenesis, 1388
pathologic features, 1388
prevention and control, 1391
Eastern equine encephalitis virus
immunologic characteristics, 1384
interferon inhibition of, 1180n
Eaton agent. See *Mycoplasma pneumoniae.*
Eberthella. See *Salmonella.*
EB virus, 1252–1255. See also *Burkitt lymphoma* and *Mononucleosis, infectious.*
morphology, 1252, *1253, 1254*
properties of, 1252
ECF-A, in anaphylaxis, 543, 544
Echovirus(es), 1299–1303. See also *Picornavirus.*
assembly of particles in, 1301, *1302*
cell culture susceptibility to, 1301t
characteristics, chemical and physical, 1300
immunologic, 1300–1301
control of, 1302
cytopathic changes due to, 1301
discovery of, 1299
disease and infection due to, 1300t
epidemiology, 1302
laboratory diagnosis, 1301–1302
pathogenesis, 1301
hemagglutinating action by, 1300
host range of, 1301
multiplication of, 1301
naming of, 1281
properties of, 1300–1301
type 10. See *Reovirus.*
Ectothrix infection, 1002
Ectromelia, 1213, *1213*
Ectromelia virus, 1258
Eczema herpeticum, 1244
Eczema vaccinatum, 1273
Edeine, protein synthesis inhibited by, 302
Edema
angioneurotic, 523–524
presternal, due to mumps, 1341
pulmonary, pneumonia and, 702
Edema factor, due to *Bacillus anthracis*, 822
Edwardsiella tarda, 771
EEE virus, 1382

Effector cells, *560*
EFG factor, in protein synthesis, 284
EFT factor, in protein synthesis, 284, *286*
Elderly, Waldenström's macroglobulinemia in, 427
Electron microscopy
antibodies ferritin-labeled for, 401, *402*
in bacteriology, 23
in virology, 1015–1016
of DNA, 205
Electron transport, in bacterial respiration, 24, 50–52
Electrophoresis
in purification of virus samples, 1014
of nucleic acids, 207–208
Elicitor, 528
Embalming, 1452
Embden–Meyerhof pathway, in bacterial fermentation, 43, 46
Embryonic development
translational regulation in, 1133
viral effects on, 1216
Emmonsiella capsulata. See *Histoplasma capsulatum.*
Emphysema, due to respiratory virus, 1352
Empyema, in pneumonia, 702
Enamel pellicle, 960
Encephalitis. See also specific type, e.g., *Ilheus encephalitis.*
after measles, 591
herpetic, 1242, 1244n
treatment, 1246
in whooping cough, 799
postvaccinal, 589, 1274
rabies, 586, 1371
togaviral, 1382, 1387
Encephalomyelitis
allergic, 589–591, *591*
due to rabies vaccine, 1374, 1375
due to virus B, 1246
measles, 1346, 1347
Encephalomyocarditis, in mice, 1281
Endocarditis
bacterial, chemoprophylaxis in, 680
chemotherapy in, 675
diagnosis, 631
due to *Bacteroides*, 787
due to *Brucella*, 815
due to *Candida*, 996
due to *Coxiella burneti*, 911
due to *Erysipelothrix rhusiopathiae*, 948
due to *Hemophilus parainfluenzae*, 797
due to infected teeth, 961
due to *Neisseria*, gonococcal, 750

Endocarditis, due to *Neisseria* (continued)
  nongonococcal, 751
  due to oral flora, 959
  due to *Pseudomonas*, 784
  due to rat-bite fever, 948
  enterococcal, 725
  L-phase variants in, 943
  micrococcal, 738
  streptococcal, 725, 955
Endogenote, bacterial, 182
Endonuclease, in DNA replication, 227
Endopeptidase(s), bacterial, cell wall attacked by, 114
Endoplasmic reticulum, polysomes associated with, 1132
Endospore(s)
  bacterial, 137
  of *Bacillus subtilis*, 5
Endothrix infection, 1002
Endotoxemia, with hemorrhagic shock, 955n
Endotoxin(s). See also *Exotoxin; Lipopolysaccharide;* and *Toxin.*
  bacterial. See under *Bacterium.*
  meningococcal, 746, 747
  of Enterobacteriaceae, 765
  of salmonellae, 765
Endpoint method, of virus assay, 1040−1041
Energy
  bacterial production of. See under *Bacterium.*
  in fermentation, 41−42
  protein synthesis requirements of, 291
Enniatins, 128t
Enol phosphates, in metabolic reactions, 41
Enteric cytopathic human orphan virus(es). See *Echovirus.*
Enteric fever
  due to *Arizona* infection, 776
  in salmonellosis, 773, 774, 775
Enteric infections, clostridial, 839
Enteritis
  due to *E. coli*, 768
  Salmonella, 775
  staphylococcal, 954
    diagnosis, 737
    pathogenesis, 736
Enteritis necroticans, 839
Enterobacter, 769−771
  acetoin fermentation by, 45
  aerobactin in, 92
  biosyntheses in, 62, 72, 85
  differentiating biochemical tests in, 769
  DNA base composition in, 37t
  drug resistance by, 771
  E. coli versus, 46
  fermentation by, *44*, 48, 316

Enterobacter (continued)
  growth on ribitol versus xylitol, 308
  Klebsiella and *Serratia* versus, 769
  polysaccharides of, 83
  taxonomic characteristics, 34t
Enterobacter aerogenes, 46, 771
Enterobacter cloacae, 755, 771
Enterobacter hafniae, 771
Enterobacter liquifaciens, 771
Enterobacteriaceae. See also *Bacillus, enteric.*
  adhesion to body surfaces, 957
  antiflagellar antibodies in, 759
  antigenic structures of, 759−765
  antigen(s) in, common enterobacterial, 760
    O, 117, 759, 760
      lysogenic conversion of, 766, *766*
    R, 760
    surface, 759, 760
      preparation of antisera to, 760−761
  biochemical reactions, 758t
  biochemical tests of, differential, 769, 770
  capsule production by, 756
  cell division in, 129
  cell wall lipopolysaccharide from, blast transformation due to, 574
  characteristics, differential, 755t
    metabolic, 757−759
  classification, 34t, 754, 757n
  colicin typing of, 765−766
  colony variation in, 103
  core of, 118
  differentiation of, 771, 772, 777
    selective media for, 759
  disease and infection due to, recent increase in, 755
  drug resistance by, 766
  endotoxins of, 765
  fermentation by, acetoin, 46
    acid-producing, 45
    formic, 45
    glucose, 759
    mixed, *44, 45*
  fermentation reaction in, 757
  flagella in, tests for, 759, *760*
  general considerations, 754−756
  genes transferred from *E. coli* to, 194
  genetic instability of, 767
  genetic interrelations in, 766−767
  glutamine synthesis in, 69n
  groups and divisions of, 756t
  identification of, screening media in, 757−759
  lipid A in, 765
  medical importance, 754−756
  metabolism in, glucose, 755t
  methionine synthesis in, 72

Enterobacteriaceae (continued)
  motility of, 756
  outer layer of, fatty acids in, *117*
  pathogenicity, 756
  penicillinase production by, 162
  phosphatides in, 125
  resistance transfer factors in, 162
  R factor in, 766, 1108
  species of, indistinctiveness of, 36
  virulence, 103
Enterobactin, 92
Enterochelin, 92
Enterococcus(i). See also *Streptococcus faecalis.*
  intestinal, 956, 956n
Enterocolitis, due to *Yersinia* infections, 806
Enterotoxin, 728. See also *Toxin.*
Enterovirus(es). See *Picornavirus.*
Entner−Doudoroff pathway, in bacterial fermentation, 46
Enzyme(s). See also under organism producing enzyme, e.g., *Clostridium, enzymes of.*
  antibodies compared to, 357t
  bacteria attacked by, to supply food for protozoa, 114
  bacterial. See under *Bacterium.*
  bacterial cell wall peptidoglycans attacked by, 113−114
  bacterial infection prevented by, 114
  catalytic activity regulated by, protein flexibility and, 305
  cryptic, 130
  dimer-specific, lack of, xeroderma pigmentosum and, 240
  fMet-polypeptide modification by, 287
  fractionation of, in study of biosynthetic pathways, 61
  IgG fragmentation by, 409−411, *409*
  in animal cell lysosomes, 1129
  in biosynthesis of mammalian amino acids, 87t
  in breakdown of organic matter in soil, 114
  in crossing over, 242
  in DNA replication, 225−227, *226*
  in immunity, 10
  in nucleic acid modifications, 204
  in processing of animal cell rRNA, 1130
  in site-specific DNA recombination, 243
  in viral DNA synthesis, 1064, *1065*
  in vitro synthesis of, 292
  ligand of, 357t

Enzyme(s) *(continued)*
 nonpenetrating nutrient breakdown by, 94—95
 properties of, 357t
 receptor-destroying, 1036
 regulation of, allosteric, protein
  structure and, 305
 structural diversity of, 428t
 swivel, in DNA replication, 227
 virion, 1010, 1011t
 virus inactivation by, 1464
Eosinophil(s)
 antibody—antigen complex ingestion by, 652
 at sites of reaginic reactions, 543
 host tissue protected by, 652
Eoinophil chemotactic factor, 543
Ependymoma, papillary, due to
  SV40, 1420
Epidemic, 663, 663n. See also
  Pandemic.
Epidemiologic surveys, delayed-
  type skin reactions used
  in, 559t
Epidemiology
 definition, 663
 principles, 663
 science of, founding of, 7
Epidermis, dermatophytes and,
  1001
Epidermophyton, 986t, 1004, 1005
Epigenetics, 347
Epiglottal reflex, 701, 702
Epiglottitis, 795
Episome(s), 1106—1107
 definition, 195, 1106
 integrated, control of bacterial
  chromosome replication
  by, 340
 integration of, 1107, *1107*
  in aberrant position, 243
 interspecific transfer of, 1107
Epstein—Barr virus, 1447
Equilibrium dialysis, in distinguishing bound from free
  haptens, 361—364, *361*
Equine abortion virus, *1146,* 1216
Equine encephalitis virus(es). See
  also specific virus, e.g.,
  *Western equine enceph-
  alitis virus.*
 host range of, 1385
Erwinia
 DNA base composition in, 37t
 septicemia due to, 755
 taxonomic characteristics, 34t
Erysipelas
 streptococcal, 720
 swine, 948
Erysipeloid, 948
Erysipelothrix, taxonomic characteristics, 34t
Erysipelothrix rhusiopathiae, 947t,
  948

Erythema, in group A streptococcal
  infections, 593
Erythema multiforme, in coccidioidomycosis, 992
Erythema nodosum
 in coccidioidomycosis, 992
 poststreptococcal, 722
 tuberculosis with, 593
Erythrasma, 1006
Erythritol, in abortion due to
  brucellae, 815
Erythroblastosis fetalis
 antigen D and, 607
 due to blood group incompatibility, frequency of, 606
 Duffy and Kell antigens in, 607
 prevention, 606, 609n
  with anti-Rh antibodies, 497
 Rh factors in, 606
Erythrocyte(s)
 agglutination of. See *Hemagglutination.*
 binding of, by virus-infected cells.
  See under *Hemadsorp-
  tion.*
 CDe, 607t
 lysis of. See *Hemolysis.*
 membrane transport systems in,
  132
 mumps virus reactions with, 1339
 parainfluenza virus reactions with,
  1336—1337
 phenotypes of, incidence of, in
  United States, 608t
 rosette formation around lymphocytes by, 467
 rosette formation around macrophage by, 464
 S and O hemolysin damage of,
  718
 stabilized, by viruses, 1036
 surface of, covalent linkage of
  proteins to, 393
Erythromycin (therapy), 676
 in anthrax, 826
 in diphtheria, 691
 in flavobacterial infections, 786
 in *Neisseria meningitidis* infection, 748
 in pneumococcal infections, 705
 in primary atypical pneumonia,
  942
 in streptococcal infections, 723
 in whooping cough, 799, 800
 protein synthesis inhibited by,
  297, 300
 resistance to, 162, 164, 300, 303t
Escherichia. See also *Coliform bacillus.*
 taxonomic characteristics, 34t
Escherichia coli, 767—769
 adaptive response to β-galactosides, 131
 Aerobacter versus, 46

Escherichia coli (continued)
 alkaline phosphatase in, 124, 322
 amidases of, in synthetic penicillin production, 583
 amino acids in, fluorophenylalanine replacement of,
  153
 synthesis of, 313
 ampicillin effects on, 117, 120
 anaerobic respiration in, 49
 antibiotic effects on, 267
 antigenic properties, 767—768
 arginine synthesis by, 344, *345*
 aspartate synthesis by, 315, *315*
 aspartate transcarbamylase in,
  318
 aspartokinase in, 317t
 bacteremia due to, 768
 bacteriophage of, adsorbed to,
  1052
  phenotypic mixing of, *1072*
  receptors of, 1051
  restriction of, 1065, *1066*
  sites of attachment, *130*
 biosyntheses in, 49
  of TCA intermediates, *66*
  phage infection and, 334
 borrelidin action in, 303
 cAMP in, 333
  receptor protein for, 334
 carbamyl phosphokinase in, 322
 cell wall of, areas of adhesion
   between inner membrane
   and, *130*
  lipoproteins in, 115
  outer surface of, *108*
  peptidoglycans in, 111
   partial dissolution of, *109*
  separation of cytoplasmic membrane from, *109*
  synthesis of, penicillin inhibition of, 120
 chloramphenicol and streptomycin antagonism in,
   164
 chromosome of, 191
  replication of, *341*
  units of inheritance in, 187
 cold shock in, 93
 colicin(s) in, 1109
 colicinogens in, 195
 cultures of, 767—768
  chromosomal replication in, *341*
  media for, 95t, 99, 346
  protein synthesis in, *100*
 cystine required by, 91
 deoxyribonucleotide kinases in,
  80
 deoxyribose residues in, 81
 diaminopimelate synthesis in, 72
 disease due to, treatment,
  768—769
 division in, 342
 DNA of, 207

Escherichia coli, DNA of
    (continued)
    composition of, 37t, *217*
    degradation by T4 bacter-
        iophage nuclease, *1057*
    hybridization of RNA with, *262*
    radiation-damaged, recombina-
        tional repair of, *239,* 240
    replication of, 220, 221, 222,
        222, 228
    spontaneous deletions in, 260
    synthesis of, 339
    transcription of, 267
  DNA polymerase in, 227
  drug resistance in, 768, 1107
  electron transport system in, 49
  enteric fermentation by, 48
  enterobactin in, 92
  enteropathogenic, 768
  enterotoxin produced by, 768
  enzyme synthesis in, *321*
  episome transfer in, 1107
  F agent in, 194
  fatty acid synthesis in, 125, 125n,
    126
  flagellum of, 29, *29,* 30
  flow between glycolytic and TCA
    pathways in, *67*
  galactosidase formation in, 57
  galactoside hydrolysis in, *134*
  galactoside transport in, 123, 132,
    135
  gene(s) in, frequency of, 339
    mutator, 261
    regulation of, 57
    rRNA, 338
    transfer to other species, 194
  genetic map of, 192, *193, 249*
    showing attachment sites of
        prophage, 1099, *1099*
  genetic studies of, phage P1 in,
    1112
  glucose metabolism in, 46
  glutamine synthetase action in,
    316
  growth in, *66,* 99
    diauxic, 57, *57*
    pH and, *93*
  hexose monophosphate shunt in,
    46
  in food poisoning, 768
  joint transduction in, 198, *199*
  ketodeoxygluconate pathway in,
    46, *47*
  lack of aggregation by, 96
  lactose operon of, gene prepara-
    tions from, 213, *215*
  lactose transport in, 136
  lysine synthesis in, *72*
  methionine synthesis in, *72*
  mRNA in, growth, translation,
    and degradation of, 328,
    328–330
  mutants of, DNA replication in,
    221

Escherichia coli, mutants of
    (continued)
    excision repair of damaged DNA
        in, 239
    glycerol deprivation in, 126
    lipid synthesis in, 126
    peptide stimulation of, 94n
    polyamines required by, 79
  mutation(s) in, mutator, 260
    nonsense, *335*
    rates of, 262t
    reactions to streptomycin, *301*
  naphthoquinones in, 51
  nucleic acid sequences in, 216
  nucleoside diphosphates in, 81
  O antigens of, colitose in, 763n
  outer layer of, 116
  pathogenicity, 768
  penicillin effects on, 120
  phosphatide synthesis in, *126*
  phosphorylation in, 136
  piliated, *30*
  plasma membrane of, *108*
  plasmids in, 1106
  polynucleotide ligase in, 227
  prophage integration with, 1101
  prophage Mu-1, infection of,
    1104
  protein breakdown in, 343
  protein synthesis in, 289
  putrescine synthesis in, 79n
  repressors isolated from, 330
  resistance transfer factors in, 195
  R factor in, 766, 1108
  ribonucleases in, 343
  ribose-t-P synthesis in, 81
  ribosomes in, 281, *281*
    artificial assembly of, 283, *283*
    proteins of, synthesis of, 339
    subunits of, *288*
  RNA polymerase in, 263, 1130
  RNA synthesis in, 262, 338
  RNA transcription by, 1130
  rRNA in, composition of, *216,*
    282t
  shikimic acid utilized by, 75
  shock due to, 768
  strain K$_{12}$, conjugation dis-
    covered in, 187
    two mating types of, 188
  surface of, *108*
  temperature effects on, *93*
  thermal death point of, 1454
  thioredoxin in, 81
  toxigenic strains of, 768
  transduction in, 1112, 1115
  transport systems in, 135
  tRNA suppressors in, 294, *294*
  tryptophan auxotroph of, peri-
    odic selection of, *172*
  tryptophan operon in, mRNA of,
    329
  tryptophan synthetase in, locus
    of, 234
    mutation of, 292, *294*

Escherichia coli, tryptophan
    synthetase in *(continued)*
    shift of reading frame in, 277,
        278
  ubiquinones in, 51
  UV sterilization of, 1456, *1456*
  virus modification and restriction
    in, 1066, *1066*
    enzymes in, 1067
Escherichia freudii, 771–772
Esophagus, bacteria in, 956
Ethambutol therapy
  in mycobacterial infection, 862
  in tuberculosis, 857
Ethanol
  aerobic bacterial oxidation of, 49
  disinfectant action of, 1461
Ether, virus inactivation by, 1464
Ethionamide therapy, in tubercu-
    losis, 857
Ethnic groups, ABO blood groups
    in, 604, 605t
Ethylene oxide
  disadvantages, 1460
  disinfectant properties, 1459,
    1460, *1460*
  virus inactivation by, 1464
Ethyl ethanesulfonate, mutagenic
    effects of, 256, *257*
Eubacteria, gram-positive, classifi-
    cation, 34t
Eubacterial cell, prototype, *23*
Eubacteriales, 26
Euchromatin, 1130
Eugenol, 1461
Eukaryotes, 14–15
  crossing over in, 243
  definition, 14
Eukaryotic cell, ribosomes of, 283,
    286
Eumycetes, 14
Euploid cells, 1134
Evolution
  biological, microbiology and,
    12–16
  cell-mediated immunity and, 570
  chemical, 15
  genetic code and, 279–280
  human, microbes and, 12–13
  microbial, 12
  mutations and, 261–262, 307
  of blood group substances, 604
  of immunoglobulins. See under
    Immunoglobulin.
  precellular, origin of life and,
    15–16
  proteins in, 291, 307–308, *308*
  self-regulatory mechanisms and,
    347
  tandem reduplication of genetic
    material and, *216,*
    218–219
  teleonomy and, 13
  tRNA and, 275

Evolution *(continued)*
  viruses and, 10
Excretory organ obstruction, chemotherapy limited by, 673
Exocrine secretions, immunoglobulins in, 418
Exoenzymes, bacterial, nonpenetrating nutrients broken down by, 95
Exospores, fungal, 138
Exosporium, bacterial, 141
Exotoxin(s). See also *Endotoxin* and *Toxin.*
  bacterial, 637t
  definition, 636
  lack of allergic response with, 593
Extrachromosomal factor(s). See *Plasmid.*
Eye infection(s)
  adenoviral, 1234n
  due to *Bacillus subtilis,* 826
Eyelid scarring, in trachoma, 924

F

Fab. See *Antigen-binding fragment.*
Fab′, 411
F(ab′)₂, 411
Face, aspergillosis of, 998
Factories, viral, 1162
Facultative organisms, 91, 103
F agent, bacterial. See under *Bacterium.*
Famine, pestilence and, 658
Farmer's lung, 546, 873
Farr technic, in radioimmunology, 396
Fat, bacterial growth on, 64
Fatigue, 659
Fatty acid(s)
  bacterial. See under *Bacterium.*
  bactericidal action of, 1460
  in outer layer of enterobacteria, 117
  in yeasts, 125n
Fatty acid synthetase, bacterial, 125
Fatty compounds, bacterial fermentation of, 56
Favus, 1105
Fc. See *Crystallizable fragment.*
Fd piece, *410,* 411
F-duction, in bacterial conjugation, 192–194, *194*
Feces
  antibodies in. See *Coproantibody.*
  Clostridium tetani in, 835
Feline leukosis virus(es), 1433, 1434
Fermentation(s)
  bacterial, 40–48
    acetoin, 45–46

Fermentation(s), bacterial *(continued)*
  adaptive value of, 47–48
  alcoholic, 44–45, *44*
  aldehyde in, 41
  anaerobic conditions required for, 41
  butanediol, 45–46, *44*
  butylene glycol, 45–46
    endproducts in, 769t
    in *Klebsiella* and *Enterobacter,* 769
  butyric, *44*
  butyric–butylic, 46
  citrate, 68
  coenzymes in, 42–43
  cytochromes in, 42
  diphosphopyridine nucleotide in, 42, 43, *43*
  Embden–Meyerhof pathway in, *43,* 46
  energy source in, 41–42, *43*
  Entner–Doudoroff pathway in, 46
  fatty acids in, 48
  ferredoxins in, 43, 46
  flavodoxins in, 43
  formic, 45
  gas formation in, 45
  genetics and, notation of, 185n
  glucose in, 48
  glycolytic pathway in, 41, 43, *43, 48*
  heat dissipation in, 52
  heterolactic, 44, 46, *47*
  hexose monophosphate shunt in, 46
  high-energy thioesters in, 42
  homolactic, 44
  in intermediate metabolism, 41
  ketodeoxygluconate pathway in, 46, *47*
  ketone conversion to carboxyl in, 42
  lactic, 5, 6, 43–44, *44*
  lipoic acid in, 42
  methane, 45
  mixed, *44,* 46
  naphthoquinone in, 42
  of fatty compounds, 56
  oxaloacetate, 68
  Pasteur effect in, 52–53
  phosphogluconate pathways in, 46
  phosphorylation in, 41
  propionic, 44–45, *44*
  pyruvate in, 43–47, *43, 44*
  respiration versus, 40
  source of energy in, metabolic reactions in, 41
  substrate assimilation in, 52
  sugars in, 48
  tetrahydrofolate in, 42
  thermodynamic aspects, 41
  thiamine pyrophosphate (TPP) in, 42

Fermentation(s) bacterial *(continued)*
  tricarboxylic cycle in, 48
  ubiquinones in, 42
  clostridial, 831
  definition, 5n, 40, 40n, 41
  enteric, *E. coli* in, 48
  environmental adaptation of microbe and, 48
  ethanol, 42, *47*
  in classifying Enterobacteriaceae, 754
  industrial, 6
  in Enterobacteriaceae. See under *Enterobacteriaceae.*
  microbes in, 5–6
  of grape juice, 48
  of milk, 44, 48, 831
  of starch. See under *Starch.*
  of vegetation, 48
  respiration versus, 5, 40
  selective cultivation and, 6
  types, 5
  unity of biology in, 47
Fermentation plates, 103
Ferredoxins, bacterial. See under *Bacterium.*
Ferritin-antibody conjugates, 402
Fetal antigens, 622
Fetoproteins, cancer and, 622
Fetus
  as allograft, 612
  Ig production in, 504, *505*
  rubella damage to. See under *Congenital anomaly.*
  syphilis in, 888
Fever. See also specific type, e.g., *Yellow fever.*
  antibacterial action of, 649–650
  inflammation and, 649
Fever blisters. See *Herpes simplex.*
F factor, 1105. See also *Plasmid.*
  interspecific transfer of, 1107
Fibrinolysin, bacterial, 95
Fibroma virus, 1260, 1261
Filtration, sterilization by, 1457
Fimbria(e), bacterial. See *Bacterium, pilus(i) of.*
Fish
  immunoglobulins in, 442, *503*
  mycobacteria in. See *Mycobacterium marinum.*
  systemic anaphylaxis in, 539
Fission, transverse, in *Treponema microdentium,* 884
Five-day fever, 913. See also *Rochalimaea quintana.*
Fixed virus, in rabies, 1368
Flagellin, 28
  immune response to, *493*
Flagellum(a)
  bacterial. See under *Bacterium.*
  in classification of Enterobacteriaceae, 754

Flanders virus, 1368
Flare reaction, 563
Flask, Pasteur's swan-neck, 4, *4*
Flavin(s), bacterial, 43, 68
Flavobacterium meningosepticum, 786
Flavodoxins, bacterial, 43
Flavoproteins, bacterial, 49, 51–52
Flea(s)
  human, 805
  in plague, 802, 805
  rat. See *Xenopsylla cheopis.*
  rickettsiae transmitted by, 898, 901
Flocculation reaction, 385–386, *386*
  precipitin reaction versus, 387, *387*
Fluorescein
  antibodies labeled with, *397,* 398, *398*
  in immunofluorescence, 397, *397, 398*
  produced by *Pseudomonas aeruginosa,* 783
Fluorescence, 397n
Fluorescence polarization, 399
Fluorescence quenching, 398–399, *399*
Fluorescent treponemal antibody (FTA) test, 887, 890
Fluoride, antibacterial action in dental caries, 962
Fluorophenylalanine, bacterial metabolism impaired by, 153
5-Fluorouracil
  bacterial incorporation of, 153
  incorporation into RNA, 296, *296*
  mutagenic effects in viruses, 296
  mutation suppression by, 296
Flury vaccine, 1374
Flury virus, 1374
fMet-polypeptide, enzymatic modification of, 287
Folic acid (folate)
  p-aminobenzoate as component of, 151, *152*
  bacterial synthesis of, 77
  discovery of, 91t
Folliculitis, staphylococcal, 733
Fomites, 663
Fonsecaea species, 1000
Food(s)
  allergic response to, 481
  microbial production of, 50
  preservation of, alkyl esters of *p*-hydroxybenzoic acid in, 1461
    history, 1452
    salts in, 1458
  radiation sterilization of, 1457
  refrigerated, spoiling of, 93

Food industry, fungi in, 966
Food poisoning
  clostridial, 839. See also *Botulism.*
  due to *Bacillus cereus,* 826
  due to *E. coli,* 768
  due to *Vibrio parahaemolyticus,* 782
  Salmonella, 774
  staphylococcal, 736, 738
Foot-and-mouth disease
  discovery of virus for, 9
  double-stranded polynucleotide therapy in, 1181
Foot-and-mouth disease virus, 1281
  susceptibility to, 1209
Foreign substance(s)
  chemotherapy limited by, 673
  nonmicrobial, immune responses to, 10
Forespore, bacterial, 138
Formaldehyde
  disinfectant properties of, 1459, 1460, *1460*
  mutagenic action of, 254
  side effects, 1460
  viruses inactivated by, 1463, 1464, *1464*
Formate, in bacterial fermentations, 42, 45
Forssman antigen
  antibodies against, 512n
  definition, 600n
Fort Bragg fever, 895
Fowl. See also specific type, e.g., *Chicken.*
  aspergillosis of, 997
  disease(s) of. See specific disease, e.g., *Ornithosis.*
    mycoplasma, 936t
  immunoglobulins in, 441, 503
  Newcastle disease virus in. See *Newcastle disease virus.*
  poxviruses of, 1260
  tuberculosis in, 844, 861
Fowl plague virus
  antigen of, *1160*
  nucleocapsid formation in, 1158, *1160*
  plaque-type mutants of, *1166*
Fowl pox, 1260
Fox(es), rabies in, 1373
Fragment(s)
  antigen-binding. See *Antigenbinding fragments.*
  crystallization. See *Crystallizable fragment.*
Fragment reaction, 297, 298
Frambesia. See *Yaws.*
Framboise, 891
Frame-shift suppressors, 294
Francisella tularensis, 806–809. See also *Tularemia.*
  antigens of, 807

Francisella tularensis (continued)
  metabolism, 807
  morphology, 807
  pathogenicity, 807–808
  transmission of, 664
  variation in, 807
  virulence, 807
Francis skin test, 704
Freeze-etching, in electron microscopy, 23
Freezing
  preservation of viable culture by, 1455
  sterilization by, 1455–1456
Frei test, 925
Freund's adjuvant, 562
Friend virus, 1433, 1434, *1435*
Frog(s)
  mycobacteria in, 868
  renal adenocarcinoma in, 1446
Fructose, in bacterial fermentation, 48
Fruitfly. See *Drosophila.*
FTA test, 887, 890
Fungicides, 1459
Fungi Imperfecti. See *Deuteromycetes.*
Fungus(i), 965–1006. See also specific type, e.g., *Mold.*
  abscesses due to, 980
  antibiotics from, 966
  antimycin effects on, 128
  bacteria versus, 977–978, 980t
  benefits from, 966
  cell wall of, 967–969, *969*
    peptidoglycans in, 111
  characteristics, 966–978
  chitin synthetase in, 969
  classes of, 975–977, 976t
  culturing of, 982–983
    cycloheximide inhibition of bacterial contamination in, 984
    Sabouraud's agar in, 877, 982–983
  dimorphism in, 977, 978t
  definition, 977
  disease and infection due to. See *Mycosis.*
  distribution, 966
  effects of polyene antibiotics, 127
  enzymes of, nutrients broken down by, 95
  exospores of, 138
  filamentous, *971*
  first recognized etiologic role, 8
  flagellated gametes in, 976
  gene conversion in, 246, *247*
  genetic recombinations in, 241
  granulomas due to, 980
  growth, 966–967
  heterokaryon formation in, 974
  heterothallic, mating types in, 188

Fungus(i), heterothallic *(continued)*
  sexual reproduction in, 972
  homothallic, sexual reproduction
    in, 972
  hyphal fusion in, 974
  in food industry, 966
  iron-chelating compounds pro-
    duced by, 92
  lysine synthesis by, 72, *73*
  metabolism, 969–970
  mutations in, 259
  mycelium(a) in, aerial, 966
    coenocytic, 970
    definition, 966
    reproductive, 966
    vegetative, 966
  nicotinamide synthesis by, 79
  nuclear fusion in, 974
  nuclei of, 176
  opportunistic, 981, 986t
    mycoses due to, 981, 995–999
  pathogenic, 984–1006. See also
    Mycosis.
    classification according to tissue
      involved and dimor-
      phism, 986
  plant diseases due to, 966
  reproduction, 970–974
    asexual, 970–972
    mitotic recombination in,
      974–975, *976*
    parasexual cycle in, 974–975,
      976
    sexual, 176, 182, 972–974, *975*
      steps in, 972
  snail digestion of, 969
  somatic pairing in, 975, *976*
  spores of, 971
    asexual, 971, *971*, 972t. See
      also *Arthrospore; Coni-
      dium;* and *Chlamydio-
      spore.*
    implantation in wounds, 980
    inhalation of, 979
  sporulation in, antibiotic produc-
    tion and, 160
    cultural stimulation of, 982
  sterilization of, 1454
  structure, 966–967
  taxonomy, 14, 975–977
  thermophilic, 970
  visualization of, in tissue,
    981–982
    lactophenol cotton blue dye in,
      983
    periodic acid-Schiff stain in,
      981, *982*
Furadantin. See *Nitrofurantoin.*
Furazolidone therapy, in cholera,
  781
Furunculosis. See *Boil.*
Fusel oil, 46
Fusidic acid
  bacterial synthesis of, 160

Fusidic acid *(continued)*
  protein synthesis inhibited by,
    299–300, *299,* 303t
Fusobacterium, 786
  in oral cavity, 958t
Fusobacterium fusiforme, 787
  infection due to, treatment, 788
Fusobacterium necrophorus, 787,
  788

G

Gaffkya tetragena, 738
Galactose, bacterial conversion of
  glucose to, 83, *84*
β-Galactosidase, in *E. coli,* 57
β-Galactoside(s)
  bacterial. See under *Bacterium.*
  E. coli adaptive response to, 131
  transport in *E. coli,* 135
Galactosyl transferase, on surface
  of animal cells, 1127
Gallbladder, isolated, Schultz-Dale
  reaction in, 544
Gammaglobulin(s), 7S. See *Immunoglobulin G.*
Gammaglobulin therapy
  in vaccinia, 1274
  in rabies, 1375
Gangrene
  amino acid fermentation in, 46
  gas, clostridial, 832t, 838
    prevention and treatment, 841
    fermentation and, 45
    laboratory diagnosis, 839
Gantrisin, *154*
Garbage deodorization, by phenol,
  1452
Gas
  intestinal, fermentation and, 45
  marsh, 49
Gas gangrene. See under *Gangrene.*
Gaspak generators, 830
Gastritis, atrophic, 591
Gastroenteritis
  clostridial, 839
  due to *Aeromonas,* 782
  due to *Arizona,* 776
  due to *Edwardsiella tarda,* 771
  Salmonella, 773, 774
    treatment, 775
Gastrointestinal tract
  as environment for bacteria and
    bacteriophages, 766
  bacterial resistance to shigellae in,
    778
  carcinoma of, carcinoembryonic
    antigen in, 622
  indigenous flora of, beneficial
    effects of, 954
  mucous secretions in, antibac-
    terial action of, 643

Gastrointestinal tract *(continued)*
  reoviral infection in, 1406
Gel(s). See also *Agar.*
  precipitin reactions in. See under
    Precipitin reaction.
Gene(s), 234–235
  adjacent, fusion of, 308
  allelic. See *Allele.*
  bacterial. See under *Bacterium.*
  cellular, in transformation due to
    viruses, 1429
  clustering of, 216
  codons of. See *Codon.*
  conversion of, 246, *247*
  defective, replacement of, 213
  definition, 233, 236
    operational basis for, 306
  distinguishing between, comple-
    mentation test in,
    233–234
  for immunoglobulin polypeptide
    chains. See under *Immu-
    noglobulin.*
  expression of, in animal cells,
    1133
  H. See *Histocompatibility gene.*
  histocompatibility. See *Histocom-
    patibility gene.*
  Ir, 480
  isolation of, 213–216
  master correction of mutation by,
    219
  mutations within, sites for, 233
  polymorphic, 233
  reading frame of. See *Reading
    frame.*
  regulation of, study of, 57
  repetition of, 216
  Rh, notations for, 607t
  stability of, 237
  synthesis of, 213–216
  theory of, microbial genetics and,
    233
  tRNA, duplication of, 295
  UV damage to, *238*
  viral, 1020, 1021t
Gene complementation. See also
    *Intragenic complementa-
    tion.*
  in heterokaryons and hybrid cells,
    1136
Generation, spontaneous, 3–5
Generation time, of bacterial
    growth, 97, *98*
Gene therapy, 213
Genetic analysis, blood typing in,
  608
Genetic code
  bacterial, 279
  definition, 202
  degeneracy in, *276*
    origin of, 277, *278*
  determination of, in vivo evidence
    in, 278

Genetic code, determination of
(*continued*)
synthetic polynucleotides in,
275, *276*
evolution of, 279–280
in protein synthesis. See under
Protein, synthesis of.
minimizing of mutational consequences and, 280
punctuation in, 277–279
universality of, 279–280
Genetic constitution, in infectious
disease, 11
Genetic elements, extrachromosomal. See *Plasmid.*
Genetic factor(s)
in autoimmune disease, 588
in cell susceptibility to viruses,
1209
in drug allergies, 584
in resistance to tuberculosis, 855
Genetic function, unit of. See
Gene.
Genetic heterogeneity, in man, 13
Genetic information. See also *Molecule, informational.*
in animal cells, 1129–1133
in genome, 236
in helical nucleic acids, 204
molecular aspects, 201–230
organization of, 233–236
protein synthesis and, 202–203
suppressors, 292
synthesis of. See *Nucleic acid,
synthesis of.*
transfer from DNA to RNA, 262
units of, 233–236, *234*
Genetic linkage, 233
Genetic mapping, 246–251
deletion, *249,* 250
in animal cell cultures,
1135–1137
physical, 251, *251*
recombination frequency in,
247–249, *248, 249*
transduction, 249–250, *251*
tandem reduplication of, evolutionary aspects, 218–219
Genetic recombination, 241–251
by breakage and reunion. See
Crossing over.
in fungi, 241
in phages, 241
molecular mechanisms of,
241–246
nonhomologous, 243
site of, 241
within codon, 241, *242*
Genetic suppression, 296
Genetics
bacterial, 11
notations of, 185n
bacteriophage, 1072–1081
fine-structure, 233–235

Genetics (*continued*)
interference in, 246
molecular, development of,
11–12
study of, yeasts in, 15
of animal viruses, 1165–1168
population, Gm allotypes in, 420
Genitals, edematous skin of, 925
Genitourinary infections, 936
Genome
fragmentation in viruses, 1023
information in, 236
UV damage to, *238*
Gentamycin (therapy), 676–677.
See also *Aminoglycoside.*
bacterial resistance to, 163t
in *Acinetobacter* infection, 785
in *Aeromonas* infection, 782
in *Citrobacter* infection, 772
in *E. coli* infection, 768, 769
in *Enterobacter* infection, 771
in *Klebsiella pneumoniae* infection, 770
in *Proteus* infection, 772
in *Providencia* infection, 772
in *Pseudomonas aeruginosa* infection, 784
in *Serratia* infection, 771
Gentamycin adenylate synthetase,
163t
Geochemical cycles, soil microbiology and, 6
German measles. See *Rubella.*
Germicide(s). See *Chemotherapy*
and *Drug.*
Germination agent, bacterial spore
germination and, 143
Germ line theory, of diversity of
immunoglobulin V genes,
438–439
Germ theory
of disease, 6–8
epidemiologic evidence of, 7
Giant cells, in tuberculosis, 649
Gingivitis, 961
Gingivostomatitis, 1244, 1245
Glanders, 784
hypersensitivity in, 657
Glass, protection against UV by,
1456
Globulin(s)
p-azobenzenearsonate, antibodies
to, 354
antilymphocyte, 499
gamma. See *Gammaglobulin.*
immune, tetanus, 837
viral infectivity neutralized by,
1212
Glomerulonephritis, 551–552, *552*
Arthus reaction in, 546
complement in, 551, *552*
cryoprecipitates in, 554
immune-complex disease with,
553

Glomerulonephritis (*continued*)
in lymphocytic choriomeningitis,
1215
in systemic lupus erythematosus,
551, *552, 592*
Plasmodium malariae antigens in,
551
prevention, 723
streptococcal, 552, 721, 722
immunologic aspects, 594
Glucan, in yeast cell wall, 968
Glucocorticoid hormones, in enzyme synthesis, 1133
Glucose metabolism
bacterial. See under *Bacterium.*
in classification of Enterobacteriaceae, 754, 755t
Glucose phosphate isomerase, in
animal cell homogenates,
isozyme patterns of,
1136
Glutamate, bacterial synthesis of.
See under *Bacterium.*
Glutamine, bacterial synthesis of,
69
Glycerol, antibacterial action of,
1462
Glycine, bacterial synthesis of, 73
Glycogen
animal synthesis of, 83
bacterial storage of, 85
Glycolysis, in phagocytic leukocytes, 651
Glycolytic pathway, in bacteria.
See under *Bacterium.*
Glycopeptide. See *Peptidoglycan.*
Glycoproteins, in viral envelope,
1031
Glycosidases, bacteria attacked by,
113
Glyoxylate cycle, in bacteria. See
under *Bacterium.*
Goats, mycoplasma diseases in,
936t
Gold salts (therapy), 1459
in mycoplasma infections, 934
in pneumonia, 939
Gonadectomy, adenoviral tumor
formation and, 1212
Gonococcus. See *Neisseria gonorrhoeae.*
Gonorrhea. See also *Neisseria gonorrhoeae.*
history, 742
incubation period, 751
ophthalmia neonatorum due to,
750, 751
prevention, 751
recurrent, 750
salpingitis due to, 751
treatment, *124–125,* 750–751
spectinomycin in, 159
urethritis in, 750
Goodpasture's syndrome, *552*

Goodpasture's syndrome *(continued)*
  immunologic aspects, 587t, 588
Graffi virus, 1433
Graft. See also *Transplant* and specific type, e.g., *Allograft.*
Graft-versus-host reaction, 568, 614. See also *Allograft, rejection of.*
  antibody formation and, 465n
  due to transplants of allogeneic lymphoid cells, 581
  due to treatment of combined immunodeficiency, 509
  small lymphocytes in, 570
Gramicidin(s), ionophoric action of, 128t
Gramicin S, 85
Gram-negative organisms, 22
Gram-positive organisms, 22
Gram stain, for bacteria, 22
Granules, leukocytic, 652
Granulocyte(s). See also *Leukocyte, polymorphonuclear.*
  half-life of, 660
  in Arthus reaction, 548—549
  physiology, 660n
  radiation damage to, bacterial disease due to, 660
Granuloma(s)
  fungal, 980
  hepatic, due to *Brucella suis,* 815
  in tuberculosis, 649
  paracoccidioidal. See *Blastomycosis, South American.*
  splenic, due to *Brucella suis,* 815
  Staphylococcus aureus infection and, 661
  swimming pool, 862, 864t
Granuloma inguinale, 948—949
Granulomatosis infantiseptica, 947
Grippe, summer, 1290. See also *Poliomyelitis.*
Griseofulvin (therapy)
  in dermatomycosis, 1003
  in mycoses, 983, *983*
  synthesis, 160, 983, *983*
Gross leukemia virus
  infection due to, in animals, 1432—1433
  latency of, 1434
  pathogenesis, 1433
  infectivity of, age of mouse and, 1432
  transmission of, 1432
Growing point, 219
Growth, bacterial. See under *Bacterium.*
Growth factors
  organic, in bacterial nutrition, 90
  positions on paper chromatograms, agar culture media to determine, 102

Guama virus, 1392
Guanidine, substituted, inhibition of enterovirus multiplication by, 1183
Guanosine tetraphosphate, bacterial, 338
Guanylate, bacterial synthesis of, *81, 82*
Guarnieri bodies, 1267
Guillain—Barre syndrome, 936
Guinea pig(s)
  cutaneous anaphylaxis in, 536—537, *536, 537*
  immune responses in, 558n
  saprophytic mycobacteria in, 868
  systemic anaphylaxis in, 539, 539t

H

Hafnia, 771
Hageman factor
  in hereditary angioedema, 523
  in kinin production, 542
Hagfish, immunologic memory in, 502, *503*
Hair
  dermatophytic invasion of, 1002
  fungal infections of, 1005
  Microsporum infections of, 1004, *1004*
  Trichophyton tonsurans infection of, *1003*
Half-life, biologic, 483t
Halogens, bactericidal action of, 1459
Halophiles, bacterial, 93—94
Hamster(s), virus-induced teratogenic effects in, 1216
Hamster leukosis virus, 1438, *1439*
Hansen's disease. See *Leprosy.*
H antigen, 599—600
Haplotype, 618, *618*
Hapten(s), 356
  affinity of antibodies for, 362, *363, 364*
  antibody reaction with, 360—370. See also *Hapten—antibody complex.*
    antibody binding sites in, 362, *363, 364*
    bound versus free hapten in, 361
    equilibrium dialysis to distinguish, 361—364, *361*
    by multivalent antibody, equations for, 402—403
    hapten structure and, *368*
    hapten water solubility and, 368
    intrinsic association constants for, 365t
    rates of, 364—366

Hapten(s), antibody reaction with, rates of *(continued)*
    measurement of, 366
    pH and, 364
    temperature and, 365
    specificity of, 366—370, *367—369*
    steric hindrance of, 369, *380*
  coupled to bacteriophage, to measure rate of hapten—antibody formation, 366
  coupled to proteins, 355, *355*
  definition, 355
  homologous, 367
  immunogenicity, 465
  precipitin reaction inhibited by, 380—381, *380, 381*
  structure of, antibody affinity and, *368, 369*
  terminal residues of, antibody binding of, 367, *368, 369*
Hapten—antibody complex(es). See also *Hapten, antibody reaction with.*
  chemical bonding in, 369
  equilibrium constant of, 378
  stability, 368
  thermodynamics, 365
Haptenic groups, 355, *356*
Hart Park virus, 1368
Hashimoto's disease, 587t, 591, *591*
Haverhill fever, 948
Hay fever. See also *Allergic reaction.*
  IgE produced by plasma cells in, 536
HB antigen, 1410, *1411*
Heart. See also *Cardiac* and *Myocardium.*
  congenital defects due to coxsackievirus, 1216, 1299
  lesions of, due to streptococcal infections, allergic aspects of, 594
  muscle of, antibodies to, 585
Heat
  killing spores of thermophilic bacillus by, *1453*
  resistance to, by bacterial spores, 137, 142, *142*
  sterilization by, 1454—1455
  viruses inactivated by, 1464
Heavy-chain disease, 436
Heavy metals, antibacterial activity of, 1458—1459
Helix pomatia, 969
Helper cells, in antibody formation, relation of T cells to, 572
Hemadsorption
  definition, 1161
  due to viruses, 1161, *1161,* 1207
Hemagglutinating units, viral, 1042
Hemagglutination, 391
  by adenoviruses, 1224—1225
  by variola-vaccinia viruses, 1263

Hemagglutination *(continued)*
by viruses. See under *Virus.*
due to echoviruses, 1300
histocompatibility antigens characterized by, 611
passive, 393, *394*, 394
Hemagglutination assay, of viruses,
1035, *1035*
Hemagglutination-inhibition test
in antigenic analysis of viruses.
See under *Virus.*
in influenza, 1308, 1324
Hemagglutinin(s)
cold, 938n
influenza viral. See under *Influenza virus.*
invertebrate, 502
myxoviral, 1035
parainfluenza viral, 1336
virion, 1031
Hemangioendothelioma, 1420
Hematin, 654
Hemin, bacterial requirement of,
90, 94
Hemoglobin
homeostatic properties of, 319
human, abnormal, amino acid
replacements in, 277,
279
electrophoretically altered,
spontaneous reversions
in, 260
Hemolymph, 502
Hemolysin(s). See also specific
hemolysin, e.g., *Strep-
tolysin.*
parainfluenza viral, 1337
produced by *Clostridium tetani,*
836
produced by *Vibrio cholerae,* 780
rabbit, 512n
rickettsial, 900
staphylococcal, 730, 731t
Hemolysis
activated complement in,
515–520, *517*
bacterial, assessment of, 94
by autoantibodies, 586
by *Listeria monocytogenes,* 946
due to complement. See under
Complement.
immune, dose-response curve of,
514, *515*
in complement-fixation assay,
513
Hemolytic anemia. See *Anemia,
hemolytic.*
Hemolytic disease(s)
antiglobulin test in, *393*
of newborn. See *Erythroblastosis
fetalis.*
Hemolytic plaque assay, of antibody-producing cells,
453, *454*

Hemophilus
DPN produced by, 91t
DPN required by, 94
hemin required by, 94
taxonomic characteristics, 34t
transformation in, 185
Hemophilus aegypticus, 797
growth characteristics, 793t
relation of *H. influenzae* to, 795
Hemophilus—Bordetella group,
791–800
differential growth characteristics,
793t
Hemophilus ducreyi, 797
growth characteristics, 793t
Hemophilus hemolyticus, 797
growth characteristics, 793t
Hemophilus influenzae, 792–796
antigenic structure, 794
culturing of, 793
disease and infection due to,
carrier rates in, 795
prevention, 797
treatment, 796
DNA base composition in, 37t
genetic variations in, 794–795
growth characteristics, 793t
history, 792
immunity to, 795, *796*
immunologic relation to *Diplococcus pneumoniae,* 794
in nasopharynx, 748n
laboratory diagnosis, 795–796
meningitis due to, 793
metabolism, 793–794
mistaken for *Diplococcus pneumoniae,* 793
morphology, 793
pathogenesis, 795
pneumonia due to, 1323
relation of *H. aegypticus* to, 795
satellite phenomenon in, 794, *794*
synergism with influenza virus,
792
transformation in, 184, 186t
virulence, 795
virus-restricting enzyme in, 1067,
1067
Hemophilus parainfluenzae, 797
growth characteristics, 793t
Hemophilus suis, 792, 1218
growth characteristics, 793t
Heparin, in anaphylaxis, 541
Hepatitis
animals subject to, 1412
carriers of, 1413, 1414
characteristics, 1413t
control of, 1414–1415
due to blood transfusion, 1414
epidemiology, 1413–1414
from contaminated hypodermic
needle, 1414
from shellfish, 1414
immunization against, 1209, 1414

Hepatitis *(continued)*
in coxsackieviral infection, 1297
infectious, definition, 1410
infectious versus serum, 1412
mortality rate, 1412
murine, *Mycoplasma neurolyticum* infection with,
934
pathogenesis, 1412–1413
serum, definition, 1410
epidemics, 1414
hepatitis-associated antigen in,
1414
passive immunization in, 1415
signs and symptoms, 1412
subclinical infections, 1413
treatment, 1414–1415
with infectious mononucleosis,
1252
Hepatitis-associated antigen,
1410–1412, *1411*
in hepatitis, 1412, 1413, 1414
in treatment of serum hepatitis,
1415
Hepatitis virus(es), 1409–1415
properties of, 1410–1412
stability of, 1410, 1414
Hepatoma cells, 1133
Herellea, 751. See also *Acinetobacter.*
Herpangina, 1298, 1298t, 1298n
Herpes labialis, 1245
Herpes simplex
history, 1240
recurrent aspects of, 1244, 1245
serologic tests for, 1203t
treatment, 1246
Herpes simplex virus, *1018,*
1242–1246
adsorption to host cell, *1145*
chemical prevention of, 1242
capsid of, digestion of, *1145*
cervical carcinoma and, 1245
diseases and infection due to,
1242
antibodies in, 1217
control of, 1245–1246
epidemiology, 1245–1246
inclusion body formation in,
1208
laboratory diagnosis, 1245
latency of, 1217, 1244
recurrence, 1217
treatment, 1245–1246
dissemination of, 1213
DNA inhibition in, 1185
genetic recombination in, 1167
host cell penetration by, *1145*
host range of, 1242
immunologic characteristics, 1242
mature particles of, *1244*
morphologic types, *1241*
multiplication of, 1242–1243
particles formed in cell nucleus

Herpes simplex virus, particles
  (continued)
    by, 1243, *1243*
  pathogenicity, 1243–1245
  polykaryocyte formation due to,
    1162
  properties of, 1242–1243
  susceptibility to, 1209
  teratogenic effects of, 1216
Herpesvirus(es), 1239–1255. See
    also specific herpesvirus,
    e.g., *Virus B.*
  budding of, 1159
  characteristics, 1240–1242,
    1240t
    chemical and physical, 1242
  chromosomal aberrations due to,
    1208
  disease and infection due to, 1240
    cell-dependent abortive, 1164
    double-stranded polynucleotide
      therapy in, 1181
    latency of, 1240
    recovery from, antilymphocytic
      serum and, 1211
    suppression of cellular immu-
      nity and, 1202
  DNA inhibition in, 1185
  drug resistance by, 1246
  history, 1240
  in synthesis of adeno-associated
    virus, 1236
  Marek's disease due to, 1446
  morphology, 1240t, 1241
  oncogenic effects of, 1430, 1446,
    1447
  pathogenicity, focal, 1240
Herpes zoster. See also *Varicella*
    and *Varicella-zoster virus.*
  activation of, 1248
  epidemiology, 1249
  IgG titers in, 1246
  laboratory diagnosis, 1248
    serologic tests for, 1203t
  pathogenesis, 1247–1248
Herxheimer reaction, 890
Heterochromatin, 1130
Heterogenote
  definition, 193
  gal, definition, 1114
  genetic changes in, 1116–1117
Heterograft, 610
Heterokaryon formation, in animal
    cell cultures, 1136
Heterolactic pathway, in bacterial
    fermentation, *47*
Heteroploid cells, 1135
Heteropolymer, 83
Heterotroph, 86
Hexachlorophene, 1461
Hexose monophosphate shunt, 46
Hexylresorcinol, 1461
Hinton test, 886
Hiroshima survivors, leukemia in,
    259

Histamine(s)
  antihistamine blocking of, *543*
  assays for, *543*
  basophilic secretion of, 534, *534*
  in anaphylaxis, 540, *540, 541,*
    544
  mast cell release of, 540, *541*
  produced by fermentation, 46
  susceptibility to, 541, 544t
  tissue levels of, 544t
Histidine. See also *Amino acid.*
  in bacteria. See under *Bacterium.*
Histidinol, bacterial synthesis of,
    75, 76
Histocompatibility antigen(s), 614
  ABO blood group substances as,
    619
  cell surface, antibodies bound to,
    tolerance of allograft en-
    hanced by, 613
  characterization of, by hemagglu-
    tination reactions, 611
  H-2, 615–617
    specificities of, 616, *617*
  non-H-2, 617–618
  normal, loss of, by tumor cells,
    621
  of tumor cells, 568
  specificity of, 614
  tolerance to, 612, *613*
Histocompatibility gene(s),
    614–620
  HL-A locus of, 618, *618*
  in man, 618–620
  in mouse, 615–618
    H-2 locus of, 615–617, *617*
    non-H-2 loci of, 617–618
    number differing in two inbred
      strains, 615, *615*
  study of, congenic lines in, 615
Histone(s)
  in animal cells, 1129, 1130, 1134
  leukocytic, antimicrobial activity
    of, 654
Histoplasma capsulatum, 986t,
    988–990. See also *Histo-
    plasmosis.*
  antibodies to, 989
  blastomycin cross-reaction with,
    990
  coccidioidin cross-reaction with,
    990
  in macrophage, 988, *988,* 989
  mating types in, 978
  spores, 988, *989*
Histoplasma capsulatum var.
    duboisii, 991
Histoplasmin
  in histoplasmosis, 989
  sensitivity to, pulmonary calci-
    fication and, 852
  skin tests with, 989, 990
Histoplasmosis, 988–990. See also
    Histoplasma capsulatum.

Histoplasmosis *(continued)*
  diagnosis, 981, 990
  diseases associated with, 989
  epidemiology, 990–991
  histoplasmin in, 989
  histoplasmin skin testing in, 990
  immune response in, 989–990
  in Africa, 991
  pathogenesis, 988–989
  simulated by nocardiosis, 878
  therapy, 990
  tuberculosis versus, 989
HL-A antigens, 618, *618,* 619
HL-A compatibility, kidney graft
    survival and, 619, *619*
Hodgkin's disease, 581
Hog cholera virus, 1216
Homograft. See *Allograft.*
Homoserine, in bacteria, *71*
Hormonal disorders, bacterial dis-
    ease and, 659
Hormone(s). See also under gland
    eliciting hormone and
    specific hormone, e.g.,
    Thymosin.
  in mouse mammary tumors, 1443
  in resistance to tuberculosis, 855
  in urine and plasma, measurement
    of, 393
  in viral infection, 1212
Hornet stings, anaphylaxis due to,
    538
Horror autotoxicus, 492
Horse(s)
  abortion in, virus-induced, 1216
  serum of, anaphylaxis due to,
    538, 539
  vesicular stomatitis virus of, 1368,
    1370
Hospital(s)
  bacteremia acquired in, 767
  staphylococcus epidemics in, 738
Host
  antibacterial defenses of,
    642–661
    antibodies in, bacteriolytic ac-
      tion of, 652
      in gastrointestinal and respi-
        ratory mucous secretions,
        643
    natural, 653
    antitoxins in, 652
    augmentation of, 658
    body fluid lysozyme in, 643
    cellular, 646
    chemical factors in, 643, 654
    coughing and sneezing in, 643
    depression of, 658–660
      staphylococcal disease due to,
        733, 734
    epithelial desquamation in, 643
    fever in, 649–650
    gastric acidity in, 643
    humoral factors in, 652–653

Host, antibacterial defenses of
(continued)
inflammation in, 646
intracellular killing of bacteria
in, 650, 651
leukocytes in, 645
mechanical factors in, 642–644
microbial factors in, 643–644
mucous membranes in,
642–644
mucus in, 643
nasal hairs in, 643
opsonization in, 652–653
phagocytosis in, 644–652, 650
events of, 651–652
phagolysosome formation in,
651
respiratory cilia in, 643, 643n
seminal fluid spermine in, 643,
643n
skin in, 642–644
pH of, 643
unsaturated fatty acids in, 643
tears in, 643
tissue structural factors in,
653–654
urethral flow of urine in, 643
vaginal acidity in, 643
variability of, 658–661
bacteria indigenous to, antibody
formation and, 954
beneficial actions of, 954–955
detrimental actions of, 955
temperature of, viral replication
and, 1212
tissues of, eosinophil protection
of, 652
Host-parasite interaction(s)
bacterial surface macromolecules
and, 36
evolutionary, 12
in bacterial diseases, 628–680
basic concepts, 631–633
history, 630–631, 630
terminology, 631–633
Host response. See also Immune
response.
discovery of, 10–11
Housefly, togaviruses carried by,
1385
Human. See Man.
Humoral factors
heat-labile, 653
in antibacterial defenses,
652–653
Humoral reactions, in autoimmune
diseases, 586
Hunchbacks, tuberculous, 861
Hyaluronidase
bacterial. See under Bacterium.
staphylococcal, 731, 734
streptococcal, 718
Hybrid(s) AB, in allograft tolerance, 613
Hybrid cell, in animal cell cultures,
1136

Hybridization, of DNA and RNA,
262, 262
Hydrogen peroxide
antiseptic properties of, 1459
in phagocytosis of bacteria, 651
Hydrolase(s), bacterial, 95
5-Hydromethylcytosine, in bacteriophage DNA, 1049
Hydrophobia. See Rabies.
Hydroxybenzoic acid esters, in
food preservation, 1461
Hydroxylamine
mutagenic effects of, 255
virus inactivation by, 1463, 1464
5-Hydroxymethylcytosine, T-even
coliphage DNA, 1021,
1022
Hydroxystilbamidine therapy, in
North American blasto-
mycosis, 988
5-Hydroxytryptamine. See Serotonin.
Hygiene, 10
Hygromycin. See also Aminoglycoside.
Hypergammaglobulinemia, 426
Hypersensitivity. See also Allergy;
Sensitivity; and specific
type of hypersensitivity,
e.g., Atopy.
antibody-mediated, 527–555
specificity of, 564
cell-mediated, 529, 558. See also
Immunity, cell-mediated.
cutaneous basophil, 562
definition, 528, 633
delayed-type, 560–566. See also
Immunity, cell-mediated.
allograft rejection by, 567–568
antigenic units in, 565
carrier groups in, 564, 565
carrier specificity in, 575
cutaneous, 560–561, 561
Arthus reaction versus, 561
in various species, 561
definition, 528, 558
formation of antibodies to T-
dependent antigens and,
572
genetic control of, 571
in mumps, 1339
noncutaneous, 561–562
recovery from viral diseases and,
1211
responses to infectious agents,
560–562
responses to purified protein,
560–562
responses to small molecules,
562–564
sensitization in, 562
specificity of, 564–566
staphylococcal, 735–736, 735
systemic, 561–562

Hypersensitivity, delayed-type
(continued)
T cells in, 565
to Coccidioides immitis, 992
to penicillin, 584
viral disease enhanced by, 1215
immediate-type, 527–555,
581–583
blast transformation in, 573
definition, 528
in bacterial diseases, 657–658
in fungal diseases, 657
in measles, 1346
in smallpox, 1267
in viral diseases, 657
to Brucella antigens, 815
Hyperthyroidism, 587t, 591
Hypertriploid cells, 1135
Hypha(e), 966, 967, 967
Hypochlorite solutions, in sanitization, 1459
Hypochlorous acid, as oxidizing
agent, 1459
Hypodermic needles, hepatitis
from, 1414
Hypodiploid cells, 1135
Hypotension, due to kinins, 542
Hypothyroidism, 587t

I

ID$_{50}$. See Dose, infectious.
Idiotype(s), 423–425
definition, 406
Fab fragments and, 424
suppression of, 498
Ids, in fungal infections, 594
Ig. See Immunoglobulin.
Ileum, isolated, Schultz–Dale reaction in, 544
Ilheus encephalitis, 1388
Imidazole ring, bacterial synthesis
of, 75
Immune complex(es)
in glomerulonephritis, 551–552,
552
in serum, in rheumatoid arthritis,
553
soluble, detection of, 554
Immune complex disease(s), 499,
551–554. See also spe-
cific disease, e.g., Glo-
merulonephritis.
due to viral infection, 552–553
Immune cytolysis. See under Complement.
Immune deviation, 579
Immune mechanisms, comple-
menting chemotherapy,
in bacterial disorders, 151
Immune paralysis, 493
Immune response(s), 351–358. See
also Host response and
Antibody, formation of.

Immune response(s) *(continued)*
  adaptive, lack of, allograft acceptance and, 610n
  allograft reaction as, 610–614
  anamnestic. See *Immune response, secondary.*
  antiself, 496
  bursa of Fabricius in, 474
  cell-mediated, 559t. See also *Immunity, cell-mediated.*
    cytotoxic mechanisms in, 577–578
    disorders included in, 558
    essentials of, *560*
    general properties, 558–570
    in autoimmune diseases, 586
    resistance to infectious agents and, 566–567, *566, 567*
    transformed cells destroyed by, 1445
    with cultured cells, 572–578
  definition, 352
  evolution of, infectious disease and, 10
  gut-associated lymphoid cells in, 476–477
  history, 528, 528n
  in guinea pigs, 558n
  in infectious disease, 400
  in sharks, 502
  lymphocytes and, 356
  pathologic effects of, 357
  penicillin-induced, *583*
  primary versus secondary, B-cell proliferation in, *487*
  nonspecific immunosuppression of, 499
  secondary, 486–489
    capacity for, duration of, 486
    cross-stimulation in, 488–489, 489t
    interval between injections and, 488
    memory cells in, 486–488
    nonspecific, 489
    resistance to irradiation, 501
    Schick test as, 486n
  shift from preponderance of serum IgM to IgG in, 471
  thymus in, 473
  to flagellin, *493*
  to nonmicrobial foreign substances, 10
  to nontoxic agents, 353
  variety of substances acted against, 357
  virus-induced, disease based on, 1215–1216
Immune serum, prophylactic action against viral disease, 1209
Immune unresponsiveness, 493
Immunity. See also under specific disease, organism, or toxin, e.g., *Variola, immunity to.*

Immunity *(continued)*
  acquired, 632
  adoptive, 453, 633
  against tumors. See *Tumor, immunity against.*
  allergy versus, in tuberculosis, 657
  allograft, transfer of, 611
  antibacterial, depression by debilitating diseases, 659
  antibody-cell controversy in, 356
  cell-mediated, 355–356. See also *Hypersensitivity, delayed-type;* and *Immune response, cell-mediated.*
    against viral antigens, 1201–1202
    antibacterial, acquired, 657–658
    antibody suppression of, 579–580
    blast transformation in. See *Blast transformation.*
    cellular basis for, 570–578
    defective, tumors and, 568, 569t
    definition, 356, 529
    evolutionary aspects, 570
    humoral immunity versus, 611, 631
    inhibition of, 578–581
      by antilymphocyte sera, 580
      cancer and, 1202
      in viral infections, 1211
    macrophages in, nonspecificity of, 571
    nonspecific suppression of, 580
    ontogeny and, 570
    phases of, 572
    phylogeny and, 570
    resistance to infections and, 567
    T cells in, 479
    to viral infections, 1211
    transfer factor in, 572
    transfer of sensitivity by viable cells, 570
    tumor enhancement by, 622
    tumor inhibition by, 622
    virus plaque assay for, 575
  definition, 528, 632
  diet and, 658
  duration of, 486
  enzymes in, 10
  from injection of antiserum, 658
  humoral, against viruses, 1190–1198
    B cells and, 571
    cell-mediated immunity versus, 611, 631
    definition, 356
    allograft rejection and, 611
  immunology and, 356–358
  in viral infections, 1210–1211
  latent infection and, 651, 662
  lysogenic, 1095–1096
    definition, 1090, 1091
    specificity of, 1095–1096

Immunity *(continued)*
  natural, 632
  nonspecific, enhancement of, 658
  nutritional tissue factors in, 632
  oral flora and, 959
  passive, 633
  phagocytic cells in, 10
  through vaccination, 658. See also *Vaccination.*
  to botulism, 834
  to brucellosis, 815
  to bubonic plague, 805
  to chlamydial infection, 920–921
  to cholera, 781
  to dermatomycoses, 1002–1003
  to *Mycoplasma pneumoniae,* 941
  to plague, 805
  to salmonellosis, 774–775
  to shigellosis, 778
  to tetanus, 836
  to trachoma, 923
  to tularemia, 808
  to viruses, nucleic acid infectivity in, 1143
  transplantation, 610–620
    ABO blood group substances and, 619–620
    definition, 611
    genetic basis, 613
    tumor, 568–570
Immunization. See also *Desensitization; Vaccination;* and under specific disease, e.g., *Diphtheria, immunization against.*
  affinity of antibodies after, 485–486, 485t
  against tumors, 568, *569,* 620–621, 622
  artificial, in plague, 805
  as augmentation to host antibacterial defenses, 658
  definition, 357–528
  deliberate, 480–481
  immunoglobulins made after, 484–486, *484*
  natural, 480
  of newborn, timing of, 497
  systemic versus local, 491
Immunodeficiency, combined, severe, 508–509
Immunodeficiency disease(s), 506–509
  cancer and, 506
  classification, 506t
  live vaccines contraindicated in, 506
  patient evaluation in, 506, 507t
Immunodiffusion, radial, 389, *390*
Immunoelectrophoresis, 389–391, *390*
  of human serum, *390, 391*
Immunofluorescence, 397-398
  fluorescein isothiocyanate in, 397, *397, 398*

Immunofluorescence *(continued)*
in identifying antibody-producing cells, *453*
in identifying bacteria, 397
in plaque assay of viral infectivity, 1037
in study of antibody formation at cellular level, 451, *453*
membrane, 1129
problems in, 398
Immunogen(s). See also *Antigen.*
antigenic determinants of, 485
definition, 353n, 450
first versus second injections of, serum antibody concentrations after, *482*
formed from tissue protein, in contact skin sensitivity, 563, 563n, *565*
inhalation of, 480
particles of, determinants of, 465
primary stimulus, 482
T-independent, 451
tolerance of, lack of antibody synthesis and, 493
Immunogenicity
definition, 353
macrophages and, 464
of macromolecules, 354
Immunoglobulin(s) (Ig), 405–447. See also *Antibody.*
affinity of, sequential changes in, after immunization, 485–486, 485t
allotypic. See *Allotype.*
amphibian, 503
anaphylaxis caused by, 545
assembly of, 471–473, *472*
pulse-chase experiments in, 472
attached to lymphocyte surface, 485, 467, *468*
avian, 441, 503
avidity of, 484
cells producing, specific immunoglobulin produced by, frequency of occurrence in various sites, 471
classes of, 407
AB alloantibodies and, 604
biologic properties of hinge region of polypeptide chains and, 433
in penicillin reactions, 584t
in Rh antibodies, 606
properties of, 446t
sequential changes in, 484–486, *484*
classification, 406
complement fixation by, 515
cytophilic, 464
deficiency in thymic aplasia, 581
definition, 406
degradation of, 483, 483t
determinants of, 423–425

Immunoglobulin(s) *(continued)*
disulfide bonding in, 429, *430,* *433*
diversity of, 470
basis for, 427–436
elevated levels in disease, plasma cells and, 469
evolution of, 421, 438, 440–442, *503*
order of, 441–442, *442*
fish, 442, 502, 503
fragments of, *415*
genes of, 438
genetic variants, 419–423
homogeneous, 425–427
human, 515
idiotypic. See *Idiotype.*
in complement activation, 520
incorporation of sugar precursors, 473
induced by diphtheria toxoid, *484*
induced by tetanus toxoid, *484*
in reptiles, 503
intracellular transport of, *472*
invertebrate antisubstances versus, 502
lack of. See *Agamma-globulinemia.*
maternal–fetal incompatibilities in, 419
metabolic properties of, 483t
monoclonal, 426
myeloma proteins as, 408, 425, *426*
nomenclature, 407, 407n
pathologic versus normal, 425–426, *426*
polypeptide chains of, amino acid sequences of, 428–429, *429*
amino-terminal part of, 429
constant (C) segments of, 429–431, *431*
domains of, 428–429, *429*
genes for, 436
fusion of, 436–437, *437*
heavy versus light, 428, *429,* *433*
hinge region in, 433–436, *435*
susceptibility to proteases, 436
homologies in, 440–441, *441*
hypervariable regions of, 431–433, *435*
combining site specificity and, 444, *444*
variable (V) segments of, 429–431, *431*
subgroups of, 431–433, *434*
precipitin reactions of, *408*
proteins of, similarity to myeloma protein, 425
rabbit, genetic variants of, 423,

Immunoglobulin(s), rabbit *(continued)* *424*
secretion of, 471–473, *472*
lag between synthesis and, 472, *472*
rate of, 472
shark, 441, 442
specific types of, proportional amounts of, in serum, 471
structure, 407, 408
genetic basis of, 436–440
subclasses of, properties of, 446t
suppression of, with antisera to mu chains, 497
surface, definition, 356
detection by rosette formation, 451, *454*
synovial fluid, in rheumatoid arthritis, 553
synthesis of, 471–473
B cells in, *504*
in animals reared in germ-free conditions, 480
in bursa, shift from IgM to IgG, 504
in human fetus, 504, *505*
rate of, 472
V genes of, diversity of, germ line theory of, 438–439
origin of, 438–440, *438*
somatic mutation theory of, 439–440
somatic recombination of germ line genes and, 440
positive versus negative selection of, 439
Immunoglobulin(s) A, 418
genetic variants of, 422
in exocrine secretions, 418
in immunoelectrophoresis of human serum, *407*
in protection against viral infection, 1209
J chain in, 418
mucosa protected by, 418
resistance to proteolysis, 418
secretory piece in, 418
selective deficiency of, 508
structure, 418
viral antigenic variability and, 1199
Immunoglobulin(s) D, as myeloma proteins, 418
Immunoglobulin(s) E, 418–419, 531–535. See also *Reagin.*
allergic reactions due to, 418
antigenic identity of, *532*
assays for, *533*
bound to basophils, 534, *534*
bound to mast cells, cross-linking of, 535, *535*

Immunoglobulin(s) E *(continued)*
chemical properties, 533–534
degranulation of mast cells and basophils by, *535*
in anaphylaxis, 535
in asthma and hayfever, 536
in exocrine secretions, 418
in mucosae, 418
intestinal parasitism and, 535–536
mast cell affinity of, 534
plasma cell production of, 536, *536*
radioimmunoassay of, 532
serum, 532–533, *533*
Immunoglobulin G, 408–416
carbohydrate in, 411–412
charge heterogeneity of, 411
disulfide bridges in, 412–414, *430*
electrophoresis of, *407*, 411, *426*
enzymatic fragmentation of, 409–411, *409*
gamma chains of, *422*
genetic variants of. See *Allotype, Gm.*
Gm allotypes in. See *Allotype, Gm.*
IgM distinguished from, 416, 484n
in guinea pig cutaneous anaphylaxis, 537, *537*
in herpes zoster, 1246
in immunoelectrophoresis of human serum, *407*
in racial populations, 422t
in rheumatoid arthritis, *553*, 554
physical characteristics, 408
polypeptide chains in, 409
relation between chains and fragments, 411
sedimentation coefficient, 408
shape, 411, *412, 413*
stability of, 412–413
structure, 410, *410*, 411–412
subclasses, 414–416, *416*
differences in, *430*
antigenic, 416
switch from IgM, lymphocytic antigen-binding receptors and, 467
symmetry, 414
synthesis of, 471
Immunoglobulin(s) M, 416–418. See also *Macroglobulin.*
cells producing, counts of, 453
fragments of, 416, *417*
IgG distinguished from, 416, 484n
induced by thymus-independent antigens, 484
in immunoelectrophoresis of human serum, *407*
in rheumatoid arthritis, *553*, 554

Immunoglobulin(s) M *(continued)*
in Waldenström's macroglobulinemia, 425
J chain in, 416
structure, 416, *417*
switch to immunoglobulin G, 467
synthesis of, feedback inhibition of, 484
Immunologic defects, neoplasia with, 622
Immunologic pseudotolerance, 492–493, *492*
Immunologic self-tolerance
definition, 492
in tetraparental mice, 496
lack of. See *Autoimmune disease.*
mechanisms in, 492
Immunologic state, unresponsive, 492
Immunologic surveillance, of tumors, 622–623
Immunologic tests, to identify O antigen structures, 118
Immunologic tolerance, 492–496, 578–579. See also *Immunosuppression.*
after initiation of antibody formation, 494
B versus T cells in, 494–496, *495*
cell mechanisms in, 496
conditions promoting, 493–494
diversity of, 493–494
due to viral infection of embryos, 1211
high-zone versus low-zone, 493, *493*
immune deviation and, 579
increase in, after thymectomy, 495, *495*
in newborn versus adult, 494
maintenance of, 494
natural, in tetraparental mice, 579
reversal of, 494
route of antigen administration and, 578
specificity of, 493–494
to histocompatibility antigens, 612, *613*
to self-antigens, 496
to viral infections, 1211
Immunology, 349–624
definition of terms, 353–354
development of, 10–11
history, 352–353
immunity and, 356–358
in control of infectious diseases, 10–11
methods in, 10
scope of, 10
Immunosuppressant(s), *500*
antibodies as, 496–499
antiproliferative, antigen administered with, 501, *502*
mycotic infections due to, 981

Immunosuppressant(s) *(continued)*
selective effects of, 501–502, *502*
side effects, 580
therapeutic versus toxic doses of, 499
Immunosuppression. See also *Immunologic tolerance.*
allotype, 497–498, *498*
antimetabolites in, 499, 501
antiproliferative, 499
bacteroides infection due to, 787
by actinomycin D, 501
by alkylating agents, 501
by antilymphocyte serum, 498–499
by chloramphenicol, 501
cancer and, 1202
cytomegalic disease and, 1249, 1251
complications, 502
E. coli infection with, 768
idiotype, 498
in kidney transplants, 499, 580
cancer due to, 623
isotype, 497
lympholytic, 499
nonspecific, 499–502
opportunistic mycoses due to, 995
warts and, 1202
X-rays in, 500–501
Impetigo, streptococcal, 720
Inclusion bodies
due to measles virus, *1343*, 1344
due to paramyxoviral infection, 1334, *1334*
due to poliovirus, 1287, *1287*
due to poxviruses, 1260
due to vaccinia virus, 1264, *1264*
due to variola virus, 1264
due to viral infections, 1207–1208
in cowpox, 1275
in mumps, 1340, *1340*
in paravaccinia, 1275
in rubella, 1356
Incubation period, in viral diseases, 1213
Indian moth, togaviruses carried by, 1385
Indicator dyes, in culture media for bacteria, 103
Indole, produced by fermentation, 46
Indole test, in Enterobacteriaceae, 770
Indophenol-oxidase test, of *Neisseria* colonies, 748
Inducer, 528
Industry, products for, from aerobic microbial metabolism, 49
Infantile paralysis. See *Poliomyelitis.*

Infecting agent(s), reservoir of, 663
Infection(s), 6. See also under *Disease* and specific organism.
  agents of, recognition of, 8—9
  animal. See *Animal, infection of.*
  bacterial. See under *Bacterium.*
  benefits from, 631
  disease versus, 631—632
  dormant, 661
  due to broad-spectrum antibiotics, 954
  due to combined immunodeficiency, 508
  due to complement deficiency(ies), 523
  due to DiGeorge syndrome, 581
  due to Nezelof syndrome, 581
  due to prolonged immunosuppression, 502
  endogenous, 755
  inapparent. See *Infection, latent.*
  laboratory. See under *Laboratory animal.*
  latent, definition, 633, 661
    immunity and, 651
    phagocytosis in, 651
  local, due to aspergillus spores, 997
  mixed, 631n
  nosocomial, 755
  slow, due to viruses, 1446
  susceptibility to, thymus and, 473
  transmission of, 7
Infectious agents, resistance to, 566—567, 566, 567
Infectious bronchitis virus, of chickens, 1362n
Infectious disease(s). See also specific disease.
  allergic reactions with, 593
  control of, chemotherapy in, 11
    environmental sanitation in, 10
    immunology in, 10—11
    personal hygiene in, 10
    vaccination in, 10
  genetic constitution and, 11
  immune response in, 10
    time-course of, diagnostic application of, 400
  in man, vaccines preventing, 491t
  death due to, 13
  multifactorial causation in, 11
  physiologic state and, 11
  selective pressure from, evolutionary effects of, 13
Infectivity. See also under specific infecting organism.
  bacteriophage, eclipse of, 1052—1053
  potential, destruction by germicidal chemical agents. See *Disinfection.*
Infertility, male, autoimmune aspects, 587t

Inflammation and inflammatory response
  allergic, corticosteroids in, 501
  antibacterial action of, 646—649
  characteristics, 646
  complement in, 515, 518, 520, 522
  cortisone injections and, 659
  fever and, 649
  in Arthus reaction. See *Arthus reaction.*
  in bacterial infection, allergic reactions in, 594
    inhibition by ethyl alcohol, 658
  monocytes in, 648, 649
  11-oxycorticosteroid suppression of, 580
  physical signs, 647n
Influenza
  adamantanamine therapy in, 1183
  age factors in, 1315, 1315, 1316, 1324
  antibody production in, persistence of, 1210
  Asian, 1315, 1325, 1327
  chemotherapy in, 1327
  control of, 1325—1327
  epidemics of, 1318, 1324
  epidemiology, 1324—1325
  hemagglutination-inhibition test in, 1308, 1324
  history, 1308, 1309, 1316—1317
  Hong Kong, 1316, 1317, 1325
  immunity to, 1317
  immunization against, 1325
  interferon effects on, 1180
  Japan, 1317
  laboratory diagnosis, 1323—1324
  naming of, 1308n
  pandemics, 792
  pathogenesis, 1322—1323
  pneumonia due to, 1308, 1323
  pregnancy and, 1212
  prevention, 1325
  recovery from, antibodies in, 1210
  self-limiting aspects, 1323
  serologic tests for, 1202t
  susceptibility to, 1212
  swine, 792
  symptoms, 1322, 1322n
  vaccination in, new methods, 1327
  vaccine(s) for, aerosol inhalation of, 491
    effectiveness of, viral antigenic variation and, 1316
    limitations of, 1327
    live virus, influenza virus mutants in, 1318
    new, 1327
    production of, 1325
    reactions to, 1327
Influenza virus(es), 1018, 1309—1327. See also

Influenza virus(es) *(continued)*
    Myxovirus.
  adsorption to leukocytes, 1212
  antibodies to, 1192, 1193, 1316, 1317
  antigen(s) of, 1314
  antigenic types, 1310
  antigenic variation in, 1199, 1315—1316
    basis for, 1316—1317
    confusion in terminology, 1315n
    number possible, 1316
    vaccine effectiveness and, 1316
  autointerference by, 1182
  characteristics, chemical, 1311—1314
    genetic, 1317—1318
    immunologic, 1314—1317
    physical, 1311—1314
  cell-dependent abortive infection by, 1164
  doctrine of original antigen sin in, 1316
  eclipse in, 1318
  fragmented RNA in, 1025
  genetic recombination in, 1167, 1318
  hemagglutination by, serum inhibition of, action of neuraminidase on, 1036
  hemagglutinins of, 1311, 1313, 1314
    antigenic variation and, 1316
    immunity and, 1317
    immunologic function of, 1314
    subunits of, 1311, 1313, 1314
    synthesis, 1320, 1320
  host cell protein synthesis inhibited by, 1148
  host range of, 1318
  immunization against, problems in, 1317
  immunologic activity of, 1314
  immunologic grouping of, 1313, 1315
  immunologic tests of, 1198
  inactivation of, host temperature and, 1212
  infectivity of, 1314, 1322, 1322n
  interferon production with, 1180
  morphology, 1310—1311, 1310—1313
  M protein of, 1313, 1314
  multiplication of, 1318—1321, 1319, 1320, 1322
  multiplicity reactivation in, 1318
  mutants of, for live-virus vaccines, 1318
  neuraminidase of, 1311, 1313, 1314
    antigenic variation and, 1316
    immunity and, 1317
    immunologic function of, 1314
    infectivity and, 1322, 1322n

Influenza virus(es), neuraminidase of
    (continued)
    in viral multiplication, 1320,
        1320n
    subunits of, 1311, *1313*, 1314
    synthesis, 1320, *1320*
nucleocapsid formation in, 1158
prime strains of, 1199
properties of, 1310—1322
protein synthesis in, 1159
RNA transcriptase in, 1319
stability, 1314
susceptibility to, 1209
susceptibility to antiserum neu-
    tralization of, 1317
swine. See *Swine influenza virus.*
synergism with *Hemophilus influ-
    enzae,* 792
teratogenic effects of, 1216
toxic   properties,   1207,
    1321—1322
UV-irradiated,   reactivation   of,
    1168
viral interference with, 1173
virulence, 1206
Information, genetic. See *Genetic
    information.*
Informosomes, 1131
Inhalation
    of immunogens, 480
    of vaccines, 481
Inheritance
    bacterial. See under *Bacterium.*
    in protozoa, 175
Injections, interval between, sec-
    ondary response affected
    by, 488
Injury, polymorphonuclear leuko-
    cytes in, 646—649, *647,
    649*
Inositol, bacterial requirement of,
    90
Insecticides, in plague, 805
Insect stings. See also under specific
    insect.
    rickettsiae transmitted by, 898
    systemic anaphylaxis due to, 538
Insulin
    metabolism by pneumococci, 696
    serum, radioimmunoassay of, 395
Interallelic complementation, pro-
    tein structure and, 306,
    306
Interference, in genetics, 246
Interferon, 1173—1181. See also
    under specific virus eli-
    citing interferon, e.g.,
    *Coxsackievirus, inter-
    feron production with.*
    antiviral protein production in-
        duced by, 1179
    assay of, 1174
    cell specificity of, 1174
    characteristics, 1174

Interferon *(continued)*
    definition, 1173, 1174
    in vaccination, 1175
    lymphocyte production by, 577
    mode of action, 1175n
        mechanism of, 1179
    mouse-specific, 1178
    preformed, release of, 1177
    production of, 1174—1176
        cells in, 1175
            plaque assay for, 1175
        chemical   induction   of,
            1176—1178
        course of viral disease and, 1180
        effects of actinomycin D and
            cycloheximide on, 1178,
            1179t
        effects of UV irradiation on,
            1177, *1178*
        genes involved in, 1174, *1175,*
            1178
        induced by double-stranded
            polynucleotides,
            1176—1178, 1181, *1182*
        relation between viruses and,
            1177—1178, *1177*
        inhibition by actinomycin D,
            1178, *1179*
        kinetics of, 1175—1176
        mechanism of, *1175,* 1178
        priming of, 1178
        refractory   period   in,
            1175—1176, *1176*
        regulation of, *1175,* 1178—1179
        relation to cells, 1175
        self-limiting character of, 1178
        temperature-sensitive viral mu-
            tants and, 1177
        time course for, 1175, *1176*
        UV irradiation and, 1177, *1178*
        viral transcriptase and, 1177,
            1178
        virus   multiplication   and,
            1174—1175, *1176*
    protective role in viral infections,
        1179—1180
    recovery   from   viral   infections
        initiated by, 1183
    therapeutic use, 1180
    viral autointerference and, 1176
Intergenic complementation, pro-
    tein   structure   and,
    305—306
Intergenic dividers, 290, *291*
Intestinal tract
    bacteria in, 956
    lymphoid cells in, 476, 477
    parasitism in, IgE and, 535—536
    specific   immunoglobulins   pro-
        duced in, 471
    ulceration due to South American
        blastomycosis, 994
Intracellular processes, study of,
    bacterial mutants in, 11

Intragenic   complementation,
    305—306
Intrauterine infections. See under
    Uterus.
Invaders, primary versus secondary,
    631n
Invertebrate(s). See also specific
    invertebrate.
    antisubstances of, vertebrate im-
        munoglobulins   versus,
        502
    hemagglutinins in, 502
    hemolymph in, 502
    lack of lymphoid system in, 502
Iodine,   bactericidal   properties,
    1459
5-Iodo-2′ deoxyuridine therapy, in
    herpetic keratitis, 1245
Iodophor(s), 1459
Iodouracil,   herpes   virus   DNA
    synthesis inhibited by,
    1185
Ion(s)
    bacterial cytoplasmic membrane
        permeability to, iono-
        phores and, 127
    heavy metal, bacterial sensitivity
        to, 94
    inorganic, in bacteria. See under
        Bacterium.
Ion cluster, 237
Ionophore(s)
    antibiotic, classes of, 128t
    permeability of bacterial cyto-
        plasmic   membrane   af-
        fected by, 127—128
    Ir-1, 477
Iron (ion)
    chelation of, 92
    in bacteria. See under *Bacterium.*
Iron-chelating compounds
    fungal production of, 92
    in bacteria. See under *Bacterium.*
Irradiation. See *Radiation.*
Isatin-β-thiosemicarbazone,   viral
    multiplication   inhibited
    by, 1184, *1184*
Isoantigen(s). See also *Alloantigen.*
    definition, 598
Isoenzyme. See *Isozyme.*
Isogenic, definition, 610
Isograft(s), 610
Isoleucine, bacterial synthesis, 71,
    71, 73, 74
Isoniazid (therapy), 159
    conversion of systemic tuber-
        culosis to latent infec-
        tion, 662, *662*
    in mycobacterial infections, 858
    in tuberculosis, 671, 857, 858,
        859
    prophylactic use of, 860
    structure, *159*
Isopropanol, produced by fermen-

Isopropanol *(continued)*
tation, 46
Isotype(s)
definition, 406
heavy-chain, 407
light-chain, 407–408
radioactive, in study of bacterial biosynthetic pathways, 61
suppression of, 497
Isozyme(s), 1135, *1136.* See also *Enzyme.*
bacterial, 314, 315, *315*

J

Japanese B encephalitis
epidemiology, 1390
inapparent infection in, 1215
pathogenesis, 1388
prevention and control, 1391
Japanese B virus, abortion in swine due to, 1216
Jaundice, catarrhal, acute. See *Hepatitis.*
Jaw, lumpy, in cattle, 875
Johne's disease, 867
Johnin, 867
Joints, tuberculous, 861
Jones-Mote reaction, 562
Junin virus, 1394

K

Kahn test, 886
Kallikrein
action of, 524
complement activation by, 521
discovery of, 542n
in anaphylaxis, 542, *542*
Kanamycin (therapy), 676–677. See also *Aminoglycoside.*
bacterial resistance to, 163t
in *Acinetobacter* infection, 785
in *Citrobacter* infection, 772
in *E. coli* infection, 768, 769
in *Edwardsiella* infection, 771
in *Enterobacter* infection, 771
in *Klebsiella pneumoniae* infection, 770
in mycoplasma infection, 934
in primary atypical pneumonia, 942
in *Proteus mirabilis* infection, 772
in *Providencia* infection, 772
in pseudotuberculosis, 806
in *Serratia* infection, 771
in shigellosis, 778
in tuberculosis, 857
to prevent tissue culture contamination, 943

Kanamycin acetyltransferase, 163t
Kanamycin phosphotransferase, 163t
Kaposi's varicelliform eruption, 1244
Kappa polypeptide chain(s), 408. See also *Polypeptide chain, light.*
allelic variants of, 422–423
structure, 408, *408*
Karyotype(s), of cultured animal cells, 1134–1135
Kasugamycin, 297, 302
Keratin
characteristics, 1001
cutaneous, attacked by dermatophytes, 1001–1002
enzymatic digestion of, 1001
Keratinase, 1001, 1001n
Keratitis
cowpox, 1275
herpetic, 1242, 1245
treatment, 1185, 1245, 1246
vaccinial, 1180
variola, 1263
Keratoconjunctivitis
adenoviral, 1235
herpetic, 1242
double-stranded polynucleotide therapy in, 1180, *1181*
Kerion, 1004
Kern Canyon virus, 1368
Ketodeoxygluconate pathway, in bacterial fermentation, 46, *47*
Ketoglutarate, in bacterial biosynthetic pathways, 63, *64*
Ketone(s), in bacterial metabolism, 42
Kidney(s)
antibodies against, 552, *553*
disease of, due to *E. coli,* 768
lymphocytic choriomeningitis virus and, 553
lesions of, in coxsackievirus infection, 1297
in serum sickness, 549, *549*
Kidney grafts
ABO compatibility and, 619
immunosuppressive agents in, 499
Kinases, deoxyribonucleotide, in *E. coli,* 80
Kingdoms, taxonomic, 13–14
Kinin(s)
accumulation of, prevention of, 542
blood coagulation and, *542*
chymotrypsin degradation of, *543*
in anaphylaxis, 540, 542, *542*
physiologic effects of, 542
production of, *542*
structure, *540*
KK antigens, 607
Klebsiella, 769–771
differentiating biochemical tests

Klebsiella, differentiating *(continued)*
of, 769
DNA base composition of, 37t
drug resistance by, 771
Enterobacter and *Serratia* versus, 769
pathogenicity, 769
Klebsiella aerogenes, 317t
Klebsiella ozenae, 771
Klebsiella pneumoniae, 769n, 770, 770
capsule of, *634*
Diplococcus pneumoniae versus, 704
genetic variations of, invasiveness and, 642
phagocytosis of, 653, *655*
Klebsiella rhinoscleromatis, 771
Kline test, 886
Koch phenomenon, 560, 657, 849, 860
Koch's postulates, 9
Koch-Weeks bacillus. See *Hemophilus aegypticus.*
Kolmer test, 886
Koongal virus, 1392
Koplik spots, 1346
Kuru, 1219

L

Laboratory animal(s)
anthrax in, 823, 825t
Coccidioides immitis infection in, 993
Yersinia pestis infection in, 805
Lacrimal drainage system, effects of trachoma, 924
β-Lactamase, in drug resistance, 163t
Lactic acid
as preservative, 1458
in phagocytosis of bacteria, 651
oxidation in bacteria, 68
production of, in fermentation, 5
Lactic bacteria
capsule polysaccharides in, 85
classification, 37
growth of, self-inhibition of, 93
lipoic acid produced by, 91t
pyridoxine produced by, 91t
riboflavin produced by, 91t
Lactic dehydrogenase, bacterial, 43
Lactic oxidase, bacteria, 68
Lactobacillaceae, classification, 34t
Lactobacillus(i)
cariogenic potentiality of, 960
in fermentation, 5
in oral cavity, 958t
intestinal, 956, 956n
in vagina, 956
taxonomic characteristics, 34t

Lactobacillus(i) *(continued)*
  deoxyribose residues in, 81
  lactic fermentation in, *44*
  nucleoside triphosphates in, 81
Lactobacillus acidophilus
  DNA base composition in, 37t
  mevalonic acid required by, 125
Lactobacillus bifidus, 37, 37t
Lactobacillus casei
  folic acid produced by, 91t
  homolactic fermentation in, 44
Lactobacillus delbrueckii, 49
Lactobacillus lactis, 91t
Lactobacillus leichmannii, 90, 91t,
    95t
Lactoferrin, 651
Lactophenol cotton blue, to visu-
    alize fungi, 983
Lactose
  fermentation of, in classification
    of Enterobacteriaceae,
    757
  in bacterial fermentation, 48
Lagos bat virus, 1370
Lambda polypeptide chains, 408,
    408. See also *Polypeptide
    chain, light.*
Lamp(s), mercury vapor, in UV
    sterilization, 1456
Lamprey lymphoid system, 502,
    503
Laryngitis, obstructive, due to *He-
    mophilus influenzae,* 795
Laryngotracheobronchitis, para-
    influenza, 1337
Lassa virus, 1394, 1395
Latent rat viruses, 1236
LD$_{50}$
  computation of, 664–665
  definition, 640
Leather, molds on, 94
Lecithinase
  bacterial, 95
  clostridial, 839, 839t, *840*
Lectin(s)
  blast transformation due to, *573*
  definition, 1129
  interaction with animal cell sur-
    face, 1129
Leishmaniasis, 581
Lens(es), glass, molds on, 94
Lepra cells, 866, *866*
Leproma(s), 866
Lepromin, 866
Lepromin test, 867
Leprosarium, national, 867
Leprosy, 865–867. See also *Myco-
    bacterium leprae.*
  amyloidosis in, 867
  cause of death in, 867
  cutaneous, 866
  distribution, 867
  epidemiology, 867
  history, 865, 867

Leprosy *(continued)*
  hypersensitivity in, 657
  incubation period, 867
  lepra cells in, 866, *866*
  lepromin test in, 867
  neural, 866
  pathogenesis, 866–867
  phases of, 866–867
  rat, 868
  T cell deficiencies due to, 581
  treatment, 867
    diaminodiphenyl sulfone
      therapy in, 155
  tuberculosis with, 867
Leptospira
  antigenicity in, 894
  borreliae mistaken for, 893
  cultivation, 894
  general characteristics, 882
  genetic variation, 894
  morphology, *188,* 894
  pathogenic versus saprophytic,
    894
  taxonomic characteristics, 35t
Leptospira autumnalis, 895
Leptospira biflexia, 894, 895
  DNA base composition in, 37t
Leptospira canicola, 894
Leptospira hemorrhagiae, 894
Leptospira icterohemorrhagiae,
    893, 894. See also *Weil's
    disease.*
Leptospira pomona, 894
  DNA base composition in, 37t
Leptospirosis, 894–896
  conjunctivitis in, 895
  experimental, 894
  laboratory diagnosis, 895
  meningitis in, 895
  pathogenicity, 894–895
  in man, 895
  prevention, 895–896
  treatment, 895
Leptotrichia, 786n
  in oral cavity, 958t
Lesion(s). See also under specific
    disease.
  closed versus open, 663
  suppurative. See *Suppurative dis-
    ease.*
Lethal factor, produced by *Bacillus
    anthracis,* 822
Leucine, bacterial synthesis of, 73,
    74
Leuconostoc
  capsule polysaccharides in, 85
  taxonomic characteristics, 34t
Leuconostoc mesenteroides
  DNA base composition in, 37t
  heterolactic fermentation by, 46
Leukemia
  due to viruses, 1208
    feline leukosis, 1433
    Friend, 1433

Leukemia, due to viruses *(continued)*
    Graffi, 1433
    induction of, 1434–1435
    leukosis, 1435
    measles, 1346
  E. coli infection with, 768
  graft-versus-host reaction in, 568
  in birds, 1433
  induced by carcinogens, 1435
  in Hiroshima survivors, 259
  in mice, due to virus, 1433. See
    also *Mouse leukemia
    virus.*
  reticulum cell, 1433
Leukemia virus, murine. See *Murine
    leukemia virus.*
Leukins, 654
Leukocidin(s)
  staphylococcal, 730, 635
  streptococcal, 718
Leukocyte(s). See also specific
    type, e.g., *Lymphocyte.*
  alloantigens of, alloantibodies to,
    after transfusions and
    pregnancies, 609n
  antigens of, antisera to, 618
  basophilic. See *Basophil.*
  chromosomes of, aberrations due
    to measles virus, 1346
  counts of, inverse proportion to
    viable pneumococci, 660,
    660
  genetic defects of, 660–661
  influenza virus adsorption to,
    1212
  lysosomes of, 652
  measles virus infection of, 1212,
    1346
  neutrophilic, antibacterial action
    of, 645
  polymorphonuclear, antibacterial
    action of, 645, *645*
    azurophil in, 652
    complement action on, 522,
      522t
    digestion of bacteria by,
      651–652
    granules in, 652
    in Arthus reaction, 546
    in injury, 646–649, *647–649*
    phagocytic vacuole in, 650, *650*
    phagocytosis of bacteria by,
      cationic proteins in, 651
      endotoxin enhancement of,
        658
      glycolysis in, 651
      lactoferrin in, 651
      lysozyme in, 651
      myeloperoxidase in, 651
      of pneumococci, 670, *670*
    pyrogen released by, 649
  resistant to anthrax toxin, 824
  streptolysin damage of, 718, *719*
Leukoencephalopathy, progressive
    papovavirus in, 1220

Leukoprovirus(es), 1436
Leukosis virus(es), 1430–1436. See also specific virus, e.g., *Rauscher leukemia virus.*
as helper virus, for Rous sarcoma virus, 1438–1440
assays of, 1433–1434
avian. See *Avian leukosis virus.*
feline, 1433, 1434
general properties, 1430–1431
infection due to, in animals, 1432–1433
in leukemia and cell transformation induced by carcinogens, 1435
latency of, 1434–1435
leukemia induction by, genetic control of, 1434–1435
morphology, 1430, *1431*
oncogenic efficiency of, 1433–1434
relation to sarcoma viruses, 1437
RNA subunits of, sarcoma virus RNA subunits versus, 1437, *1438*
sarcoma viruses versus, 1436
Leukovirus(es)
cells infected by, in tissue cultures, 1432
cell transformation due to, 1445
C-type particles in, 1430, *1431*
disease and infection due to, antibiotic therapy in, 1185
slow, 1436
humoral antibodies elicited by, 1445
immunologic properties, 1431
in human cancer, 1447
latency of, 1435–1436
leukoproviruses and, 1436
multiplication of, 1431–1432, 1436
natural cancers due to, 1446
oncogenes in, 1436
phenotypic mixing in, 1163
protoviruses of, 1436
proviruses of, 1436
replication of, 1155
ribosomes in, 1010
RNA-containing, 1430–1443. See also specific virus, e.g., *Mouse mammary tumor virus.*
RNA synthesis in, 1150
tRNA in, 1010
viral interference by, 1182
Levan(s), in bacterial capsules, 85
Lewis antigen, 600. See also *Blood cell alloantigen.*
Lf, 688
L form(s)
definition, 943
due to penicillin, 174, 175

L form(s) *(continued)*
in rheumatic fever, 944
production of, milieu for, 943
Lice. See *Louse.*
Life
definition, 15
origin of, experimental approaches to, 16
precellular evolution and, 15–16
unit of, minimal, 15
Ligand(s), 357t, 360
Ligand–antibody complex. See *Hapten–antibody complex.*
Ligase(s), polynucleotide, in DNA replication, 227
Light
ultraviolet. See *Ultraviolet radiation.*
visible, bacterial sterilization by, 1457
Lignin, aromatic components of, 78
Lime, chlorinated, to deodorize garbage and sewage, 1452
Lincomycin (therapy), 158, 676. See also *Antibiotic.*
in pneumococcal infection, 705
protein synthesis inhibited by, 299, 303t
resistance to, 678
Lipase(s)
bacterial, 95
staphylococcal, 731, 734
to dissolve membrane in study of bacteria, 24
Lipid(s)
in bacteria. See under *Bacterium.*
in mycobacteria, 847–848, *848*
in viral envelopes, 1031
Lipid A, 119, 765
Lipoamino acids, bacterial, 125
Lipoic acid
bacterial. See under *Bacterium.*
discovery of, 91t
Lipopolysaccharide(s) (LPS). See also *Endotoxin.*
bacterial. See under *Bacterium.*
cell wall of Enterobacteriaceae, blast transformation due to, 574
Lipoprotein(s)
bacterial. See under *Bacterium.*
in *E. coli* cell wall, 115
viral infectivity neutralized by, 1212
Lipotropic agents, viruses inactivated by, 1463
Listeria, taxonomic characteristics, 34t
Listeria monocytogenes, 946–948. See also *Listeriosis.*
acquired cellular immunity to, 658

Listeria monocytogenes (continued)
animal species infected with, 946
comparison to other gram-positive bacteria, 947t
general characteristics, 946
hemolysis due to, 946
Listeriosis. See also *Listeria monocytogenes.*
Anton test in, 946
clinical manifestations, 947t
history, 946
meningitis in, 947
treatment, 948
Lithotroph(s), 53t, 54
Liver
acute yellow atrophy of, 1413
arginase in, 70
clearance mechanisms in, 646
granulomas of, due to *Brucella suis,* 815
polysomes in, *281*
urea formation in, 70
L layer, of bacterial cell wall, 107
Lockjaw. See *Tetanus.*
Locus(i), 234–235
definition, 234
genic, 233
positions of, 233
Long-acting thyroid stimulators (LATS), 591
Louse (lice)
elimination of, to prevent epidemic typhus, 907
relapsing fever spread by, 892, 893
rickettsiae transmitted by, 898, 901, 905, 908
Rochalimaea quintana in, *912, 913*
L-phase variants, 943–944
Luciferase, bacterial, 52
Luciferin, bacterial oxidation of, 51
Lumpy jaw, in cattle, 875
Lung(s). See also *Respiratory tract.*
abscesses in, due to actinomycosis, *876*
in pneumococcal pneumonia, 702
in streptococcal infection, 726
artificial collapse in tuberculosis, 853, 857
calcifications in, due to healed tuberculosis, 852
cheesemaker's, 546
dust cells in, 701
edema of, pneumonia and, 702
farmer's, 546, 873
isolated, anaphylactic response in, 544
Lupus erythematosus
paramyxovirus in, 1219
systemic. See *Systemic lupus erythematosus.*
Wassermann antibody in, 886

Lupus vulgaris, 140. See also *Tuberculosis.*
Lymphadenitis
due to *Mycobacterium scrofulaceum,* 862, 864t
in bacterial infection, 646
in rat-bite fever, 896
mesenteric, due to *Yersinia* infection, 806
Lymphangitis, streptococcal, 720
Lymphatic organs, 473–477. See also specific organ, e.g., *Thymus.*
primary, 473–475
secondary, 475–477
Lymph node(s), 476. See also *Lymphatic organs.*
antigens trapped in, 475, 476, *476*
dendritic macrophages in, 476, *476*
germinal center of, 476
lymphocytes in, 475
mammalian, *476*
tuberculous lesions in. See *Scrofula.*
Lymphocyte(s). See also *Leukocyte* and *Lymphoid cell.*
activated, blastogenic factor produced by, 577
lymphotoxin produced by, 577
products of, 572, 573t, 575–577
effects on cultured cell lines, 577
effects on macrophages, 575–577, *576*
effects on skin, 577
skin-reactive factor produced by, 577
allogeneic, destruction by antibodies to histocompatibility antigens, 611
rejection of, 567. See also *Graft-versus-host reaction.*
antibody formation by, 456–463
clonal selection hypothesis of, 471
cooperation between B and T cells in, 456, 456t, 581
interaction with macrophages in, 466, 466t
restricted specificity of, 471
antigen-binding receptors on, 467–468
antigenic suppression of, 490
antigen-sensitive, all-or-none response of, 496
tolerogen effects on, 495
antigen-stimulated. See *Lymphocyte, activated.*
autosensitized, in autoimmune disorders, 586

Lymphocyte(s) *(continued)*
B (bursa), *469*
antibody formation by, allelic exclusion in, 470
antigen-binding receptors on, 467
appearance of, 457–458
cap formation on, 467–468, *468*
comparison of T cells to, 459t
definition, 451
defects of, 508
T cells in, 572
distinguishing T cells from, 474
generation time of, 457–458
humoral immunity and, 571
immunoglobulin G production by, 471
immunoglobulin production in, stages of, *504*
in immunodeficiency diseases, 506t
mitogen effects on, 574
proliferation of, in primary versus secondary response, *487*
senescence in, 488
separation of T cells from, 460–463
surface of, 458–460
change of immunoglobulin class on, 467
immunoglobulins bound to, 458
T cell antagonistic action with, 496
tolerized, stability of, 495
blast transformation of. See *Blast transformation.*
circulation of, 460
differentiation into plasma cells, 457, 469, 470
erythrocytic rosette formation with, 467
from thoracic duct lymph, *457*
homing instinct in, 476
immunoglobulin synthesis by. See also under specific immunoglobulin.
double, 471
immunosuppressive agents and. See under *Immunosuppression.*
in immune response, 356
in lymph nodes, 475
in spleen, 475, 476, *477*
interferon produced by, 577, 1175
in tolerance, 494–496, *495*
large. See *Plasma cell, immature.*
leukemic, destroyed by antibodies to tumor-specific antigens, 621
location in lymphatic organs, 460, *462*

Lymphocyte(s) *(continued)*
macrophage activation by, 471
macrophage interaction with, 463
maturation pathways of, *461*
memory in, 486–487
mixed culture of, blast transformation by, 574
to determine allograft tolerance, 618
nonsensitized, proliferation of, due to blastogenic factor, 577
origin of, 460
plant mitogen effects on, 468, *574*
sensitized, blastogenic factor released by, 577
lymphotoxin from, 578
target cell killed by, 577, *577*, 578
small, antigen sensitivity of, 451
appearance of, 457, *458*
as T cell, 570
division of, 457
incubation with antigen, blast formation from, 468–469
in graft-versus-host reaction, 570
surface membrane of, aggregated immunoglobulins on, 467, *468*
T (thymus), antibodies produced by, 465, 479
relation to helper cells in, 572
antigen-binding receptors on, 468
antigen interaction with, migration-inhibition factor released by, 575
antigen-sensitive, formation of, *560*
appearance of, 457–458
comparison of B cells to, 459t
definition, 451
defects of, 508
deficiencies of, 580
acquired, 581
distinguishing B cells from, 474
derivation of, 473
destroyed by theta alloantigen, 460
development of, 460
function of, absence of, 581
generation time of, 457–458
histocompatibility antigens in, 460
inactivation by antisera to theta, 580
in allotype suppression, 498
in B cell differentiation, 572
in cell-mediated immunity, 479, 570–571
in delayed hypersensitivity, 565

Lymphocyte(s), thymus (continued)
  in immunodeficiency diseases, 506t
  in thymus, 474, 475
  in tolerance to self-antigens, 494
  in transfer of second-set reaction, 611
  memory in, 488
  mitogen effects on, 574, 574
  separation of B cells from, 460–463
  small lymphocyte as, 570
  specificity of, in blast transformation, 574
  suppressor, 496
  surface of, 458–460
    alloantigens on, 460
    T-independent antigens and, 466
    with IgM on surface and internal IgG, 471
  X-ray effects on, 500, 501
Lymphocytic choriomeningitis, 553
  glomerulonephritis in, 1215
  immune response and, 1215
  immunologic tolerance to, 1211
  kidney disease and, 553
  pathogenesis, 1395
  recovery from, antilymphocytic serum and, 1211
Lymphocytic choriomeningitis virus
  morphology, 1395
  properties of, 1394
  susceptibility to, 1209
Lymphogranuloma venereum, 925. See also Chlamydia.
Lymphoid cell(s). See also specific type of cell, e.g., Plasma cell.
  anionic dye reactions with, 456
  appendiceal, 477
  definition, 453n
  density-gradient centrifugation of, 455
  fragility of, 456
  gut-associated. See also Lymphatic organ.
    in immune response, 476–477
  in Peyer's patches, 477
  interactions between, 455
  intestinal, 476, 477
  living versus dead, 456
  macrophages separated from, 455
  preparations of, components and heterogeneity of, 559t
  rosette formation on, surface immunoglobulins detected by, 451, 454
  sensitivity transferred by, 570
  separation of, 454–455
    affinity chromatography in, 455
    cell-sorting machine in, 455
  tonsillar, 477

Lymphoid cell(s) (continued)
  transplantation of, in analysis of antibody formation, 453
Lymphoid tissue hyperplasia, plasma cells and, 469
Lympholysis, 499
Lympholytic agents, in immunosuppression, 499
Lymphoma(tosis)
  Burkitt. See Burkitt lymphoma.
  malignant, mycoses with, 995
Lymphosarcomas, in cats, due to feline leukosis virus, 1433
Lymphotoxin, 577, 578
Lyon effect, 470n, 588
Lyophilization, 1456
Lysine. See also Amino acid.
  bacterial. See under Bacterium.
  beta, antimicrobial activity of, 654
  commercial synthesis of, 50
  fungal biosynthesis of, 72, 73
Lysogen, double, formation of, 1114
Lysogeny, 1089–1105. See also Bacterium, lysogenic.
  animal cell transformation compared to, 1428–1429
  bacterial cAMP and, 1098–1099
  bacterial virulence and, 642
  bacteriophage gene repression and, 1095–1096
  definition, 1089
  DNA transcription in, 1098
  doubly lysogenic bacteria in, 1099
  immunity in, 1095–1096
    specificity of, 1095–1096, 1097
  nature of, 1089–1091
  prophage in. See Prophage.
  repressor formation in, lambda, 1095, 1096, 1096–1097, 1098
  regulation of, 1096–1099, 1098
  significance, 1104–1105
  stability, 1097
  vegetative cycle in, 1091–1094
  vegetative multiplication versus, determination of choice between, 1097–1098, 1098
Lysolecithin, paramyxoviral, 1331
Lysosome(s)
  leukocytic, 652
  macrophage, 464, 652
  silica particles and, 856
  viral effects on, 1207
Lysostaphin, 737
Lysozyme(s)
  animal, antibacterial action by, 114, 643, 651
  bacterial, 95

Lysozyme(s) (continued)
  substrate site for, 305

M

MacConkey agar, 759
Machupo virus, 1395
Macroconidium, 972t
Macroglobulin(s). See Immunoglobulin M.
Macrolides, 676. See also Antibiotic and specific drug, e.g., Erythromycin.
  resistance to, 678, 679
  structure, 158
Macromolecule(s)
  biosynthesis of, 11
  exploring structure of, 357
  immunogenicity of, 354
Macrophage(s). See also Phagocyte.
  action on antigens, 463
  activated, nonspecificity of, 850
    resting versus, 566–567, 567
  antibody formation and, 458, 463–466
    B and T cell interaction in, 466, 466t
  bacteria killed by, immunologic aspects, 566, 566, 567
  chemotactic movement of, 567
  chlamydial particles phagocytosed by, 919
  dendritic, immunogenic activity and, 451
    in antibody formation, 466
    in lymph nodes, 476, 476
    in spleen, 476
  development, 644
  differentiation of, 464
  division of, 464
  effects of products from activated lymphocytes on, 575–577, 576
  enzymes of, 464
  erythrocytic rosette formation around, 464
  for experimental analysis, sources of, 466
  Histoplasma capsulatum in, 988, 988, 989
  immune, in acquired antibacterial cellular immunity, 658
    in Listeria monocytogenes, 658
  in cell-mediated immunity, 571
  in lymph nodes, 476
  interferon produced by, 1175
  in tuberculosis, 649, 850
  life span, 464
  lymphocyte activation of, 571
  lysosomes in, 464
  migration-inhibition factor and, 575, 576

Macrophage(s) (*continued*)
  modified by MIF, 577
  motile, immunogenic activity and, 451
  phagocytic, antibacterial action of, 644, *645*
  phagolysosome in, 464
  phagosome in, 464
  separation from other lymphoid cells, 455
  streptolysin damage of, 718
  surface of, cytophilic antibodies on, 464
  tumor cell destruction by, 570
  viral infection and, 1212
Macrophage inhibition assay, *576*
Madurella grisa, 1000
Madurella mycetomi, 1000
Maduromycosis(es), 1000. See also *Mycetoma*.
Magnesium(ion)
  bacterial requirement of, 92
  in protein synthesis, 284
  ribosomal ambiguity due to, 296
Makola shrew virus, 1370
Malassezia furfur, 1005
Malate
  in apples, 75n
  in glycolytic pathway, 67
Malignant pustule, 823, *823*
Malnutrition. See also *Starvation* and specific type, e.g., *Protein, deficiency of*.
  tuberculosis and, 855, 859
Maloney virus, 1433
Malta fever, 812. See also *Brucellosis*.
Mammal(s)
  biosynthesis in, amino acid, 87, 87t
    economy of, 87
  poxviruses of, 1258
Mammalian cell(s)
  DNA replication in, 222
  tRNA of, 275
Mammalian leukosis virus(es), 1438
Mammalian sarcoma virus, 1438
Mammary tumor
  human, 1447
  hyperplastic alveolar nodules and, 1442, *1443*
Man
  autoimmune disorders in, 587t
  bacteria indigenous to, 953–963
    distribution of, 955–957
  evolution of, microbes in, 12–13
  genetic heterogeneity in, 13
  molecular individuality in, 13
  rRNA in, composition of, 282t
  surface bacteria on, 956t
  virus-induced cancers in, 1446–1447
Manganese (ion)
  bacterial requirement of, 92

Manganese (ion) (*continued*)
  mutagenic action of, 254, 257
Mannan, in yeast cell wall, 968
Mansonia, as togavirus vector, 1385
Mantoux test, 857
Mapping
  genetic. See *Genetic mapping*.
  nucleic acid, 251
  protein, 251
Marburg virus, 1375
Marek's disease, 1446
Marker(s)
  biochemical, in bacterial genetics, 176
  definition, 241
Marseilles fever, 904t, 909
Marsh gas, 49
Masseter muscles, spasm of. See *Tetanus*.
Mast cell(s)
  complement action on, 522
  degranulation of, by IgE, *535*
  IgE affinity for, 534
  IgE bound to, cross-linking of, 535, *535*
  in anaphylaxis. See under *Anaphylaxis*.
Mastoiditis
  pneumococcal, 702
  streptococcal, 720, 724
Matsugi nephritis, 553
Mayaro virus
  hemagglutination-inhibition titrations in, 1382t
  infection due to, pathogenesis, 1388
Measles
  antibody production in, persistence of, 1210
  antibody protection in, 1209
  carboxypeptide inhibition of, 1183
  complications, 1346
  control of, 1348
  encephalitis after, 591
  epidemiology, 1347
  Faroe Island epidemic of, 1210, 1211
  German. See *Rubella*.
  history, 1342
  hypersensitivity response in, 1346
  immunity to, 1210
  laboratory diagnosis, 1347
  pathogenesis, 1344–1347
  pathology, 1347
  prevention, 1348
  sclerosing panencephalitis after, 1219
  serologic tests for, 1202t
  suppression of cellular immunity and, 1202
  symptoms, 1346
  vaccine for, 1342
Measles virus, 1342–1349. See also

Measles virus, (*continued*)
  Paramyxovirus.
  budding of, *1344*
  cell-dependent abortive infection by, 1164
  chromosomal aberration due to, 1208
  dissemination of, 1213
  giant cell pneumonia due to, 1347
  host range, 1343–1344
  immunologic characteristics, 1342–1343
  inclusion bodies due to, *1343, 1344*
  in infected cells, *1344, 1345*
  latent period of, 1334
  leukemia and, 1346
  leukocytes infected by, 1212
  morphology, *1333*
  multiplication of, 1346
  properties of, 1342–1344
  sclerosing panencephalitis due to, 1346
  syncytial giant cells formed by, 1343, *1343*
Meat tenderization, bacterial hydrolases in, 95
Mediastinitis, in anthrax, 823, 824
Medicine
  forensic, blood typing in, 608n
  microbiology and, history of, 6–10
Megacins, 1109
Melioidosis, 784, 785
Memory cells, 451, 457
  formation of, *560*
  in secondary response, 486–488
Mengo virus, 1281
  effect on host cell RNA synthesis, 1147, *1156*
  inhibition of host cell protein synthesis by, 1147
Meninges, nocardiosis of, 878
Meningitis
  Acinetobacter, 785
  adenoviral, 1232
  allergic reaction with, 594
  coxsackieviral, 1298, 1298t
  cryptococcal, 985
  due to anthrax, 823
  due to *Flavobacterium meningosepticum*, 786
  due to *Hemophilus influenzae*, 795
    exudate from spinal fluid in, *793*
    laboratory diagnosis, 795
    mortality in, 796
    prevention, 796
    treatment, 796
  due to lymphocytic choriomeningitis virus, 1215n
  due to *Pasteurella multocida*, 809
  echoviral, 1301

Meningitis *(continued)*
  in infectious mononucleosis, 1252
  in leptospirosis, 895
  listeric, 947
  meningococcal, *747*
    death rate in, 746
    epidemics of, 748, *749*
    history, 742
    in laboratory animals, 746
    sulfonamide therapy in, 675
  micrococcal, 738
  mumps, 1341
  pneumococcal, 702
    laboratory diagnosis, 705
    mortality rate, 654
    treatment, 705
  Pseudomonas, 784
  streptococcal, 724
  tuberculous, 852
Meningococcemia, 746
Meningococcus. See *Neisseria meningitidis.*
Meningoencephalitis
  coxsackieviral, 1298
  due to brucellosis, 815
  due to listeriosis, 947
  due to *Mycoplasma pneumoniae*, 936
  herpetic, 1244, 1248
  mumps, 1341
6-Mercaptopurine, in immunosuppression, 501
Mercuric chloride, 1459
Mercurochrome, 1459
Mercury (ions)
  antibacterial activity of, 1459
  spore sterilization by, 1452
Mercury vapor lamps, in UV sterilization, 1456
Merozygote, bacterial, 182
Merthiolate, 1459
Mesohematin, bactericidal activity of, 654
Mesophiles, bacterial, *93*
Mesosome(s)
  bacterial. See *Bacterium, mesosome of.*
  of *Bacillus subtilis,* 26
Metabolic products, abbreviations of, 60n
Metabolic reactions, in energy production from bacterial fermentations, 41
Metabolic suppressors, genetic, 292
Metabolism, 60. See also *Anabolism; Biosynthesis; Catabolism; Respiration;* and under specific organisms, e.g., *Thiobacillus, metabolism.*
  autotrophic, 53
  bacterial. See under *Bacterium.*
  fermentative. See *Fermentation.*
  heterotrophic, 53

Metabolism *(continued)*
  microbial, study of, founding of, 5
  mutational suppressors and, 292
Metabolite(s)
  analogs of, as antibacterial drugs, 151–153, *152*
  bacterial. See under *Bacterium.*
  secondary, 161
Metals, heavy, antibacterial action of, 1458–1459
Methane, produced by fermentation, 45
Methanol, ribosomal ambiguity due to, 296
Methenamine therapy
  in bacterial urinary infections, 160
  in *E. coli* infection, 768
  Proteus resistance to, 160
Methicillin, 156, *157*
Methionine
  bacterial synthesis of, 71, *71, 75*
  discovery of, 90
Methisazone therapy, 1274
Methylase(s), tRNA, 204
Methylations, bacterial, 72
Methyl red test, in Enterobacteriaceae, 770
Methyltryptophan, bacterial growth inhibited by, 346
Methylsergide, serotonin blocked by, *543*
Mevalonic acid, bacterial requirement of, 90, 125
Microaerophilic organisms, 91
Microbacterium smegmatis, 37t
Microbacter microti, 867
Microbe(s). See also specific type of microbe, e.g., *Fungus.*
  adaptation to environment, fermentation and, 48
  adventitious versus pathogenic, 9
  aerobic metabolism in, commercial products from, 49
  antibiotic production by, commercial, 50
  antibodies to. See under *Antibody.*
  antigens from. See under *Antigen.*
  composition of, unity of biochemistry and, 60–61
  death of, 268
  definition, 2
  differential susceptibility to disinfectants, 1453
  essential nutrients of, 60
  evolutionary aspects, 12–13
  facultative, 48
  food produced by, 50
  geochemical role of, 6
  identification of, enrichment culture in, 56
  in fermentations, 5–6

Microbe(s) *(continued)*
  in steroid synthesis, commercial, 50
  in vitamin synthesis, commercial, 50
  irradiation of, survival curve in. See *Survival curve.*
  target size in, 268, *268*
  pathogenic, antibody combination with, 379
    identification of, immunologic methods in, 10
  taxonomy, 13–16
  viable, elimination of. See *Sterilization.*
Microbial antagonism, 643. See also *Bacterial antagonism.*
Microbiologic specimens, contamination of, 2
Microbiology
  applied fields of, 5
  biologic evolution and, 12–16
  evolution of, 2–12
  impact on concept of disease, 11
  medicine and, history of, 6–10
  of oral cavity, 957–962
  soil, geochemical cycles and, 6
Microbispora, 875t
Microcapsule, bacterial, 26
Microcephaly, 1216
Micrococcaceae, 738
  classification, 34t
Micrococcus, taxonomic characteristics, 34t
Micrococcus denitrificans, 57
Micrococcus luteus, 238
Micrococcus lysodeikticus
  cell wall structure in, 111
  DNA base composition of, 37t
Micrococcus tetragenus, 738
Microconidium, 972t
Microdrops, antibody detected in, 451
Microhomology, bacterial, 38
Micromonospora, 873t, 875t
Micromonospora vulgaris, 873
Micromonosporaceae, 873t
Microorganism. See *Microbe.*
Microscope, early, 2, *3*
Microscopy
  darkfield, 882
    definition, 23
    in bacteriology, 23
    of bacterial flagella, 26
    to detect *Treponema pallidum,* 883, 889
  electron, 23
  first observations, 2–3
  light, limits of, 22
Microsporum, 986t, 1004
  diseases due to, 1004
  sexual stage in, 977t
Microsporum audouini, 1004, *1004*
Microsporum canis

Microsporum canis (continued)
  disease due to, 981
  scalp infections from, 1002, 1004
Microsporum cookei, 977t
Microsporum fulva, 977t
Microsporum gypseum, 977, 977t
  geophilic aspects, 1002
  hair infected by, 1004
  keratinolysis by, 1001
  macroconidia of, *1005*
  mycosis due to, 980
Microsporum nanum, 977t
Microsporum vanbreuseghem, 977t
Migration-inhibition factor, 575, *576*
Milk
  fermentation of, 44, 48, 831
  pasteurization of, 1454
Milk agent. See *Mouse mammary tumor virus.*
Milker's nodules. See *Paravaccinia.*
Milkweed bug, togaviruses carried by, 1385
Mima, 751. See also *Acinetobacter.*
Mima polymorpha. See *Acinetobacter lwoffi.*
Mineralization, 56
Minimum lethal dose (MLD), of diphtheria toxin, 686
Minocycline, 160. See also *Tetracycline.*
Minute mouse virus, 1236
Missense suppressors, 295
Mite(s). See also specific mite, e.g., *Trombicula.*
  rickettsiae transmitted by, 898, 901
  rickettsialpox transmitted by, 909
  Rickettsia Tsutsugamushi transmitted by, 910
Mite repellents, 911
Mithramycin, 267
Mitochondrion(a), in animal cells, oxidative phosphorylation at, 1129
Mitogen(s)
  blast transformation due to, *573*, *574*, *574*
  definition, 457
  plant, effects on lymphocytes, 468
  pokeweed, blast formation and, 468, 574
  secondary response and, 489
Mitomycin, 259
MLC test, of allograft tolerance, 618–619
Mold(s). See also *Fungus.*
  cell wall of, 968
  colonies of, 967
  cytology, 967
  growth, 966–967
    pH and, 93
  protoplasts from, 969

Mold(s) *(continued)*
  red bread. See *Neurospora crassa.*
  spores of, 973t
  starch hydrolysis by, 95
  structure, 966–967
  vitamins required by, 90
Molecule(s)
  individuality of, in man, 13
  informational, 202. See also *Genetic information.*
  small, bacterial synthesis of, 86–87
Mollicutes, 930
Molluscum body, 1275
Molluscum contagiosum, 1275
Molluscum contagiosum virus, 1275. See also *Poxvirus.*
Molybdenum (ion), bacterial requirement of, 56, 92
Monensin, 128t
Monkeypox, 1258
Monilia. See *Candida.*
Moniliasis. See *Candidiasis.*
Monilid(s), in candidiasis, 996
Monocyte(s)
  antibacterial action of, 644
  granules of, 652
  in inflammation, 648, *649*
  phagocytosis by, 464
Monocytosis-producing agent, 946
Mononuclear cells. See *Lymphocyte* and *Macrophage.*
Mononucleosis
  infectious, 1252–1255
    control of, 1255
    epidemiology, 1255
    heterophil antibodies in, 1254
    laboratory diagnosis, 1252–1255
    pathogenesis, 1252
Morax-Axenfeld bacillus, 751, 785
Moraxella, 785. See also *Acinetobacter.*
Moraxella lacunata, 751, 785
Moraxella nonliquefaciens, 785
Moraxella osloensis, 785
Morbilli. See *Measles.*
Morphogenesis, gene regulation and, 347
Mortality rate. See under specific disease, e.g., *Measles, death rate in.*
Mortierella, 999
Mosaic code, 7
Mosaicism, 470n
  in tetraparental mice, 496
Mosquito(es), togaviral infections spread by, 1385, 1389
Motility test, in Enterobacteriaceae, 770
Mouse (mice)
  allophenic. See *Mouse, tetraparental.*
  histocompatibility genes in. See

Mouse (Mice), histocompatibility *(continued)*
    under *Histocompatibility gene.*
  house, as reservoir for *Rickettsia akari*, 909
  strains of, high-cancer versus low-cancer, 1441
    high-leukemia versus low-leukemia, 1432
    inbred, to study leukemia viruses, 1432
  teratogenic viruses in, 1216
  tetraparental, 496
    natural tolerance in, 579
Mouse hepatitis virus, 1362n
  susceptibility to, 1209
Mouse leukemia virus(es), 1432, 1433
Mouse mammary tumor, in castrated males, 1443
Mouse mammary tumor virus, 1441–1443
  bioassay of, 1441
  B-type particles of, 1441, *1442*
  cell transformation due to, 1446
  history of, 1441
  infection due to, pathogenesis, 1442–1443
  morphology, 1441, *1441*
  reverse transcriptase of, 1442
  susceptibility of host to, genetic factors, 1442–1443, *1444*
  physiologic factors, 1443
Mousepox. See *Ectromelia.*
Mouse protection test, to identify streptococcal anti-M antibodies, 723
M protein(s)
  definition, 425
  streptococcal proteinase destruction of, 720
mRNA. See *RNA, messenger.*
Mucopeptide(s). See *Peptidoglycan.*
Mucopolysaccharidases, bacterial, 95
Mucor, 976t, 986t, 998–999
  dimorphism in, 977
Mucor rouxii, *978*, *979*
Mucosa(e)
  IgA protection of, 418
  IgE in, 418
  irritation of, bacterial invasion and, 643
Mucus, antibacterial action of, 643
Multiple myeloma, 436
Multiple sclerosis
  autoimmune aspects, 587t
  paramyxovirus in, 1219
Mumps
  control of, 1342
  delayed hypersensitivity in, 1339
  epidemiology, 1341–1342

Mumps *(continued)*
history, 1339
immunization against, 1209
laboratory diagnosis, 1341
pathogenesis, 1340–1341, *1340*
prevention, 1342
serologic tests for, 1202t
Mumps virus, 1339–1342. See also
Paramyxovirus.
antigens of, 1341
chromosomal aberrations due to,
1208
dissemination of, 1213
endosymbiotic infections due to,
1219
erythrocytic reactions with, 1339
host range of, 1339–1340
immunologic characteristics, 1339
inclusion bodies with, 1340, *1340*
latent period for, 1334
morphology, *1332*
properties of, 1339–1340
susceptibility to, cortisone and,
1212
teratogenic effects, 1216
toxicity of, 1207, 1340
Muramic acid, 144n
Muramidase. See *Lysozyme.*
Muran, in kidney allograft, 580
Murein. See *Peptidoglycan.*
Murein sacculus, bacterial, 111
Murine leukemia, immunologic tolerance to, 1211
Murine leukemia virus(es). See also
specific virus, e.g., *Gross
leukemia virus.*
cell transformation due to, 1208
glomerulonephritis due to, 553
infection due to, recovery from,
antilymphocytic serum
and, 1211
Murine leukosis virus(es), assays of,
1434
Murine sarcoma virus
glomerulonephritis due to, 553
hamster leukosis virus as helper
virus for, 1438, *1439*
Murray Valley encephalitis, 1388,
1390
Murray Valley encephalitis virus,
1383
Mushroom. See *Fungus.*
Mutability, mutations affecting,
260–261
Mutagen(s)
action on animal viruses, 1165
chemical, 253–255. See also
specific mutagen, e.g.,
Nitrous acid.
action of, 254–255
Mutagenesis, 259, *260*
Mutant(s). See also under specific
organism or type of or-
ganism, e.g., *Bacterium,
mutant of.*

Mutant(s) *(continued)*
double, 241
incompletely blocked, 296
leaky, 296
thymus-less, 473
Mutation(s), 252–262. See under
specific organism or type
of organism, e.g., *Bacteriophage, mutation of.*
amber, suppression of, 293
antimutator, 260
consequences of, genetic code
minimization of, 280
correction of, by master gene,
219
definition, 252
deletions in, 252, 258–259
due to chemical mutagens,
254–255
due to replication error, 255, *256*
due to tautomerization of 5-
bromouracil, 255, *255*
effects of, mechanism counteracting, 219
phenotypic, protein structure
and, 305
frequency of, site influence on,
261, *261*
general properties, 252–253
in animal cell cultures,
1135–1136
induction of, 253–259
transversions in, 259
insertion in, 252, 253
intergenic dividers and, 290
microdeletion in, 252, 252t
microinsertion in, 252, 252t
missense, definition, 253
suppression of, by tRNA, 294,
294
mutability affected by, 260–261
mutator, 260
nonsense, 293
definition, 253
suppression of, genetic, *293*
nucleotide replacement in, 252,
252t, *253*
of tryptophan synthetase, 292
pairs of, correcting each other,
236, *236*
point, 255–259
definition, 252
due to errors in replication, 255,
256
reversions in, 252
protein structure affected by,
252t, 253
radiation-induced, 259
rate(s) of, definition, 179
determination of, 179–180
mutations affecting, evolutionary consequences of,
261–262
reading-frame, 257–258, *258*
suppression of, 292

Mutation(s) *(continued)*
reversion(s) of, 253–254
genotypic, 253
phenotypic, *254*
true, 253
ribosomal, 295
silent, 252, 252t
spontaneous, 259–262
abnormal human hemoglobins
due to, 277, *279*
keto-enol tautomerization in,
260
rates of, 261, 262t
transversions in, 259
suppression of, 253–254
amber suppressor in, 293
codon-specific translational,
292, 293–295
definition, 253
extragenic, 292
genetic, definition, 292–296
efficiency of, 294–295
genotypic, 292–294
intragenic, 292
missense, definition, 293
efficiency of, 295
nonsense, 293
phenotypic, 296
reading-frame, 293
tRNA in, 294
temperature-sensitive, 253
terminator. See *Mutation, nonsense.*
transitions in, 252, 255–257
transversions in, 252, 257
definition, 257
unit of, 236
within gene, sites for, 233
Myalgia, epidemic, 1298, 1298n
Myasthenia gravis, 587t
Mycelium(a). See under *Fungus.*
Mycetoma(s), 1000. See also *Actinomycosis* and *Maduro-
mycosis.*
actinomycotic, 980
definition, 980
due to nocardiae, 878
due to streptomycetes, 879
organisms causing, 878t
Mycobacteriaceae, 873t
Mycobacterium(a), 843–869
acid-fast properties, 844
antigenic structure, 848–849,
850, 851
atypical, pathogenic, 862–865
characteristics, 863t
epidemiology, 862
significance, 864
tuberculin test for, 863, *865*
cell wall constituents, 875t
chemoprophylaxis, 864
classification, 845, 845t
cord factor in, 847
cross-reacting atypical, tuberculin
reaction in, 857

Mycobacterium(a) *(continued)*
cross-reactivity with nocardiae and corynebacteria, 848
fermentation of fatty compounds by, 56
infection due to, isoniazid therapy in, 858
  testing for, 856
in poikilothermic vertebrates, 868
lipids in, *142,* 847–848, *848*
mycobactin required by, 92
mycolic acids in, 847
non-tuberculosis, 861–865, 867–868. See also particular species, e.g., *Mycobacterium avium.*
pathogenicity, 864t
pathogenic species of, 845t
polysaccharides in, 848
PPD antigens prepared from, 864t
proteins of, allergy to, 594
resistance to sterilization, 1458
saprophytic, 868
sensitivity to detergents and heavy metal ions, 94
staining of, 844
taxonomic characteristics, 35t
virulent, 847
  detection of, 856
waxes in, 847
Mycobacterium abscessus, 862, 863t, 868
Mycobacterium avium, 844, 861, 863t, 864t
Mycobacterium balnei, 862
Mycobacterium bovis, 861
  culturing of, 844
  pathogenicity, 864t
Mycobacterium butyrica, 868
Mycobacterium chelonei, 862, 863t, 868
Mycobacterium fortuitum, 862, 863t, 864t, 868
Mycobacterium gastri, 862, 863t, 864t
Mycobacterium gordonae, 862, 863t, 864t
Mycobacterium intracellulare, 862, 863t, 864t
Mycobacterium kansasii, 862, 863t, 864t
Mycobacterium lacticol, 868
Mycobacterium leprae, 865–866. See also *Leprosy.*
  dormancy of, 867
  experimental study of, 866
  host range, 865
Mycobacterium leprae murium, 868
Mycobacterium marinum, 157t, 158t, 862, 863t, 864t, 868
Mycobacterium microti, in tuberculosis prevention, 860
Mycobacterium paratuberculosis, 867

Mycobacterium paratuberculosis (continued)
mycobactin required by, 90
Mycobacterium phlei, 868
Mycobacterium platypoecilus, 862
Mycobacterium ranae, 862, 863t, 864t, 868
Mycobacterium scrofulaceum, 862, 863t, 864t
Mycobacterium smegmatis, 868
Mycobacterium terrae, 862, 863t, 864t
Mycobacterium thamnopheos, 868
Mycobacterium triviale, 862, 863t, 864t
Mycobacterium tuberculosis. See also *Tuberculosis.*
  acid-fastness of, 845, 848
  antigens of, *851*
  caseation necrosis due to, 848
  colonies of, 846, *846*
  cultures of, 104
  culturing of, 844, 845, 853, 856
  DNA base composition of, 37t
  drug-resistant strains of, 858
  fatty acids in, 124
  genetic variations in, 846–847
  growth of, 845–846, 853
  histologic reactions to, 852
  host resistance to, 852
  identification of, 8
  immune response to, 848, 849–851
  immunity to, acquired, 143, 849
    activated macrophages in, 850–851
    nonhumoral, 849
  isoniazid action against, 159
  M. bovis versus, 861
  morphology, 845, *846*
  oxygen requirement of, 853
  pasteurization of, 1454
  pathogenicity, 864t
  properties of, 845–849
  range of responses in man, 852
  resistance to chemical and physical agents, 849
  skin eruptions due to, 992
  staining of, 845, 856
  thermal death point of, 1454
  virulence of, 852
    colony surface texture and, 103
    tests of, 846–847
  wax D in, 847–848, *849*
Mycobacterium tuberculosis var. *bovis.* See *Mycobacterium bovis.*
Mycobacterium ulcerans, 861
Mycobacterium xenopi, 862, 863t, 864t
Mycobactin, 867
  bacterial requirement of, 90
  mycobacterial requirement of, 92
Mycolic acids, 847

Mycoplasma(s), 930–943
disease and infection due to, 936t
  history, 930
  in animals, 934
  in man, 936–938
  polyenes in, 984
  virus infections with, 934
division of, 128
effects of polyene antibiotics on, 127
general characteristics, 930
growth, 932–934
  polyamines in, 79
immunologic classification, 934
lack of cell wall in, 110n
metabolism, 932–934
morphology, 930–932, *931, 932*
  colonial, 932, *934*
nutrition, 933–934
pathogenic agents compared to, 917t
pathogenicity, 934–938
replication of, 930–932, *933*
sensitivity to detergents and heavy metal ions, 94
septic abortion and, 936
species infecting man, 935t
sterols in, 121
synovitis in chicks due to, 934
taxonomic characteristics, 35t
tissue culture contamination by, 942–943
T-strain group of, 935t
Mycoplasma agalactia, 936t
Mycoplasma arthritidis
  antigens of, 936
  disease due to, 936t
  morphology, *932*
Mycoplasma bovigenitalium, 936t
Mycoplasma bovimastitidis, 936t
Mycoplasma canis, 931
Mycoplasma fermentans, 935t
Mycoplasma gallisepticum
  disease caused by, 936t
    infectious bronchitis with, 934
  DNA base composition in, 37t
  morphology, 930
Mycoplasma hominis
  alpha-hemolysin produced by, *941*
  distinguishing characteristics, 935t
  in genitourinary infections, 936
  morphology, *931*
  septic abortion and, 936
Mycoplasma hyorhinis, 936t
Mycoplasma laidlawii, 930
  colonies of, phase-produced plaques in, 933
Mycoplasma meleagridis, 936t
Mycoplasma mycoides, 936t
Mycoplasma neurolyticum
  disease caused by, 936t
    murine hepatitis viral infections with, 934

Mycoplasma neurolyticum
(*continued*)
morphology, 930
Mycoplasma orale
distinguishing characteristics, 935t
growth, 934
Mycoplasma pneumoniae. See also *Pneumonia, primary atypical.*
antibodies to, 936—941
antigenic determinants in, 934
beta-hemolysin produced by, 939, *941*
colonies of, *937—939*, 939, 941
disease caused by, 936
distinguishing characteristics, 935t
growth, 934, 941n
identification and quantitation, 941
immunity to, 941
laboratory diagnosis, 941—942
rheumatoid arthritis and, 936
staining of, 942
Mycoplasma pulmonis, 936t
Mycoplasma salivarium, 935t
Mycoplasma suipneumoniae, 936t
Mycoplasma synoviae, 936t
Mycoplasmataceae, 35t
Mycormycosis, 998—999. See also *Phycomycetes* and *Rhizopus.*
Mycoside(s)
definition, 847
mycobacterial, antigen immunogenicity enhanced by, 585
Mycosis(es), 984—1006. See also *Fungus.*
cell-mediated immunity and, 567
chemotherapy, 983—984, *983*
cutaneous. See also *Dermatomycosis.*
definition, 979
deep. See *Mycosis, systemic.*
definition, 978
diagnosis, 981—983
due to opportunistic fungi, 981, 995—999
disorders associated with, 995
general characteristics, 978—984
hypersensitivity in, 657
ids in, 594
pneumonitis due to, 979
subcutaneous, 980, 986t, 999—1000
definition, 979
superficial, 981, 1005—1006
definition, 979
systemic, 979—980, 984—995, 986t. See also specific systemic mycosis, e.g., *Cryptococcosis.*

Mycosis(es), systemic (*continued*)
definition, 978
types, 978—981
Myelitis, due to virus B, 1246
Myeloma
assembly of immunoglobulin chains in, 472
Bence Jones protein synthesis in, *472*
biclonal human, myeloma proteins produced by, 471
light-chain secretion by, 473
multiple, 436
Myeloma cell lifespan, 470
Myeloma proteins, 425
affinity labeling of, *443*, 444
amino acid sequences in, 425, 428, 431, *432*
antigenic markers of, 426
electrophoresis of, *426*
IgD, 418
IgG, glycopeptides from, *418*
immunoglobulin properties of, 408
individuality of, 426
in study of amino acid sequences in polypeptide chains, 425
normal immunoglobulins versus, 425, *426*
produced by biclonal human myeloma, 471
similarity to proteins of antibodies, 425
structure, 426, *432*
Myeloperoxidase, in phagocytosis of bacteria, 651
Myocarditis
due to coxsackievirus, 1297, 1298, 1299
diagnosis, 1299
due to Lassa virus, 1395
in influenza pneumonia, 1323
Myocardium. See also *Heart.*
small blood vessels of, rickettsial effects on, 899
Myxobacteria, motility of, 29
Myxoma virus
cellular hyperplasia due to, 1261
in rabbits, 12
target tissues of, 1260
Myxovirus(es), 1018, 1307—1328. See also specific virus, e.g., *Influenza virus.*
attachment by, 1144
budding in, 1160, *1161*
characteristics, 1309t
classification, 1308—1309, 1309t
envelope spikes of, 1031
hemagglutination by, 1035, 1036
history, 1308
host modification of envelope in, 1163
infection due to, cell-dependent

Myxovirus(es), infection due to (*continued*)
abortive, 1164
virus-dependent abortive, 1165
maturation of, 1160
nucleocapsid formation in, 1158
replication in, 1158
RNA synthesis in, 1150

N

Nafcillin, 156, *157*
Nalidixic acid (therapy)
bacterial cytoplasmic membrane damaged by, 127
in bacterial urinary infection, 160
in *E. coli* infection, 768
in *Enterobacter* infection, 771
in *Klebsiella pneumoniae* infection, 771
in shigellosis, 778
resistance to, 678
Nannizia species, 977t
Naphthoquinone(s), in bacteria. See under *Bacterium.*
Nasopharyngeal cultures, source of organisms for, 748
Nasopharyngitis, due to *Hemophilus influenzae*, 795
Nasopharynx, portions inhabited by various organisms, 748
Necrosis
antibacterial action of, 654
chemotherapy limited by, 671—673
Negri bodies, 1368, 1371, *1371*, *1372*, 1373
Neisseria(e), 741—752
colonies of, 743, *744*
conjunctival, 957
culturing of, 745
in oral cavity, 958t
metabolism, 743—745
morphology, 742—743, *743*, *744*
nonpathogenic, 751
organotropicity of, 635
pH and, 93
sensitivity to detergents and heavy metal ions, 94
species of, 745t
taxonomic characteristics, 34t
transformation in, 184
Neisseria catarrhalis, 745t, 751
DNA base composition in, 37t
Neisseria flavescens, 745t
Neisseria gonorrhoeae, 745t, 749—751
antigenic structure, 749
carbon dioxide requirement of, 91
cervicitis due to, 750

Neisseria gonorrhoeae (continued)
colonies of, 744
culturing of, 750
DNA base composition in, 37t
drug resistance by, 679, 749, 750, 751
endotoxin of, 749
genetic variation, 749–750
in chimpanzees, 750
laboratory diagnosis, 750
pathogenicity, 750
synergistic action with *Staphylococcus aureus*, 955
Neisseria meningitidis, 745–749, 745t
adrenal susceptibility to, 746
antibodies against, 747
antigenic structure, 745
capsules of, 747
carbon dioxide requirement of, 91
carriers of, 746, 748
culturing of, 748
disease and infection due to, adrenal insufficiency in, 746
chemoprophylaxis, 680
immunity to, 747
laboratory diagnosis, 747, 748
prevention, 747, 748–749
treatment, 748
DNA base composition in, 37t
drug resistance by, 746
endotoxins of, 746, 747
genetic variation in, 746
in nasopharynx, 748
intracellular, 747, 747
metastatic lesions due to, 746
organotropicity of, 635
pathogenicity, 746–747
streptomycin-dependent strains, 746
virulence, 747
Waterhouse–Friderichsen syndrome due to, 746
Neisseria sicca, 745t, 751
Neisseriaceae, classification, 34t
Neomycin (therapy), 676–677. See also *Aminoglycoside*.
bacterial resistance to, 163t
in *E. coli* infection, 769
topical, 678
Nephritis
in brucellosis, 815
Matsugi, 553
Nerve(s), viral dissemination along, 1213
Nerve tissue, leprous, 866
Neuraminidase
action on serum inhibitor of influenza virus hemagglutination, 1036
bacterial, 95
from *Cholera vibrio*, 1036

Neuraminidase *(continued)*
influenza viral. See under *Influenza virus.*
in viruses, 1031
parainfluenza viral, 1336
pneumococcal, 701
Neurologic syndromes, due to mycoplasmas, 936
Neurospora, 976t
branched metabolic pathways in, endproduct inhibition of, 316
cell senescence in, 1125
mitochondrial transcriptase in, 263
RNA polymerase of, antibiotic blocking of, 266
Neurospora crassa
cell wall of, 969
hyphae of, *967*
mutant(s) of, in study of biosynthetic pathways, 61
protoplast formation by, *970*
mutation rates in, 262t
sexual cycle in, 972–974, *975*
tetrad analysis in, 974
Neurotoxin, Shiga, 777
Neurovaccinia, 1264
Neutralization test, in antigenic analysis of viruses. See under *Virus.*
Newborn
Flavobacterium meningosepticum infection in, 786
hemolytic disease of. See *Erythroblastosis fetalis.*
immunization of, timing of, 497
inclusion blenorrhea of. See *Conjunctivitis, inclusion.*
maternal antibodies in, 503, 504t
meningitis in, 724
tetanus in, 835
Newcastle disease virus, 1353. See also *Paramyxovirus.*
antibody-neutralized, reactivation of, *1192*
cell-dependent abortive infection by, 1164
interferon production with, 1175, *1178, 1179*
hemadsorption of HeLa cells due to, *1161*
heterozygosity in, 1168, *1168*
pneumonia due to, 1207
teratogenic effects, 1216
viral interference by, 1182
Nezelof syndrome, 581
Nicotinamide
bacterial biosynthesis of, 71, 79
fungal biosynthesis of, 79
Nicotinamide adenine dinucleotidase, streptococcal production of, 718
Nicotinic acid, discovery of, 91t

Nigericin, 128t
Nitrate, bacterial reduction of, 49
Nitrobacter
taxonomic characteristics, 34t
autotrophic metabolism by, 53t
Nitrobacteriaceae, classification, 34t
Nitrofuran (therapy), in urinary infections, 160
Nitrofurantoin (therapy)
in *E. coli* infection, 768
in *Enterobacter* infections, 771
in *Klebsiella pneumoniae* infection, 771
in urinary infection, 160
Nitrogen cycle, bacteria in, 55–56
Nitrogen fixation
bacterial, 55–56
ferredoxins in, 43
Nitrogen mustards, immunosuppression by, 501
Nitrogenous compounds, breakdown of, bacterial amino acid synthesis from, 69
Nitrosoguanidine, mutagenic action of, 220, *221*, 254
Nitrosomonas
autotrophic metabolism by, 53t
taxonomic characteristics, 34t
Nitrous acid
mutagenic effects of, 255, *256*, 258, 259
genetic code information derived from, 277
virus inactivation by, 1463, 1464
Nocardia(e), 877–879
abscesses due to, 980
cell wall constituents, 875t
colonies of, 877
cross-reactivity with corynebacteria and mycobacteria, 848
culturing of, 877, *877*
decolorizing of, 877
general characteristics, 877
in oral cavity, 958t
morphology, 872
mycolic acids in, 847
periodontal disease due to, 961
resemblance to mycobacteria, 877
root caries due to, 961
staining of, 845
streptomycetes, resemblance to, 879
taxonomic characteristics, 35t
Nocardia asteroides
colonies of, *877*
immunogenicity of diverse antigens enhanced by, 877
metastatic abscesses due to, 878
mycetomas due to, 878t
pulmonary nocardiosis due to, 878
resemblance to mycobacteria, 877

Nocardia asteroides (continued)
staining of, 877
Nocardia brasiliensis
mycetomas due to, 878t
staining of, 877
Nocardiaceae, 873t
Nocardiosis
definition, 873
death rate in, 879
epidemiology, 878—879
histoplasmosis simulated by, 878
mycetoma in, 878
pathogenesis, 877—878
pulmonary, 877, 878
treatment, 879
tuberculosis simulated by, 878
Nogalomycin, 267
Nonactin, 128t
Noncovalent interactions, antibody—antigen reactions
as models for, 357
Nonsense suppressors, efficiency of,
294
North Queensland tick fever, 909
Nose
bacteria in, 956
hairs of, antibacterial action of,
643
Nosocomial infections, due to *Serratia marcescens,* 771
Novobiocin (therapy). See also
Antibiotic.
bacterial cytoplasmic membrane
damaged by, 127
in *Flavobacterium* infection, 786
structure, 158
Nuclear body, bacterial, 31—32, *31,
32*
Nuclease(s)
action on nucleic acids, 212
bacterial. See under *Bacterium.*
common, properties of, 207t
in identification of nucleic acids,
208
Nucleolus, in animal cell. See under
Animal cell.
Nucleic acid(s). See also *DNA* and
RNA.
annealing of, 209
bacterial, photoreactivation of,
238
requirement of, 90
bacteriophage, infection by, 1053
bouyant density of, 208, 209,
212
breakdown of, in bacteria,
343—344
break(s) in, double-strand versus
single-strand, 238
chemical stability, 204
chromatography of, 208, 212
competition experiments with,
214
denaturation of, 208—209

Nucleic acid(s) *(continued)*
dynamic changes in, 204
electrophoresis of, 207—208
enzymatic digestion of, 208
extraction of, 203
flexibility of, 203
helical, 204
heterogeneity of, 216
homology of, 212, *213, 214*
hybridization of, 212, *213, 214*
hyperchromic shift in, 208, *208*
lengths of, 207t
main bases of, 203
mapping of, 251
melting temperature(s) of,
208—209, *208*
definition, 209
modifications of, 203
enzymes in, 204
molecular organization of,
203—205
negative charges of, 203
nomenclature, 203n
nuclease action in, 212
nucleation of, 209
nucleotide sequences in, 212—219
base ratios in, 213
distribution over genome,
212—216
organization of, 212—219
study of, 212, *217*
pH effects on, 209, *209*
polarity of, 203
properties of, 203—211
purification of, 212
quenching of, 210
radiation damage of, 237—241
by radioactive decay, 238
by radioactive elements, 238
ionizing, 237—238
UV, 237
renaturation of, 209—211
rigidity of, 204
saturation experiments with, 212,
214
sedimentation coefficient of,
205—207, *206*
single-stranded, flexibility of, 204
structure, *205*
structure, primary, 203—204
secondary, 204—205
synthesis, rate of, 313
template in, 202
viral, 1010, *1012,* 1013t,
1020—1025
infectivity of, 1141
in nucleocapsid, *1020*
location of, 1017—1021
molecular weights of, 1021t
shape of, influenced by capsid
symmetry, 1019
types, 1021t
Nucleocapsid(s), viral. See under
Virus.

Nucleoid, bacterial, 31—32, *31, 32*
Nucleoside(s), bacterial use of, 80
Nucleoside diphosphates, in *E. coli,*
81
Nucleoside triphosphates, in *Lactobacillus,* 81
Nucleotide(s). See also *Purine* and
Pyrimidine.
bacterial. See under *Bacterium.*
in antibiotics, 80
replacement of, in mutation, 252,
252t, *253*
Nucleotropic agents, viruses inactivated by, 1463
Nutrient(s)
bacterial. See under *Bacterium.*
essential, for microbes, 60
Nutrition
host, oral flora and, 959
immunity and, 632
resistance to bacterial disease and,
658
resistance to viral infection and,
1211
Nystatin (therapy). See also *Polyene.*
in candidiasis, 997
in systemic mycoses, 983
structure, *983*

O

O antigen(s)
bacterial, cross reaction with
human red cell antigens,
117
identification of, 118
polysaccharide side chains in,
118
sugars in, 118
blood cell, 600
definition, 759n
in *E. coli,* 763n
in Enterobacteriaceae. See under
Enterobacteriaceae.
in salmonellae. See under *Salmonella.*
structure, 118, 762
O cell antigen, 600
Ochre, 279
suppressors of, 294
Ochre codon, 295
Oidiomycin, 996
Oil(s). See also *Petroleum.*
essential, preservative and antiseptic action of, 1461
Okazaki segments, 225
Oleandomycin, 158
Oligomycin, 128
Oligopeptides, bacterial transport
of, 133
Olivomycin, 267

Omsk fever, 1391
Oncogenes, 1436
One cell—one antibody rule, 470—471
One gene—one enzyme hypothesis, 176, 233
One gene—one polypeptide chain hypothesis, 234
Opal, 279
Operon, bacterial. See under *Bacterium.*
Ophthalmia, sympathetic, autoimmune aspects, 587t
Ophthalmia neonatorum, gonorrheal, 750, 751
Ophthalmic ointment, in trachoma, 924
Ophthalmitis, silver nitrate in, 1459
Opsonin(s)
  antibacterial action of, 355, 631n, 650, 652—653, 652t
  complement and, 522t, 523
  definition, 631
  Rh antibodies as, 606
Oral cavity
  Actinomyces in, 958t
  bacteria in, adhesion of, 957
    diet and, 958
    distribution of, 957
    ecologic considerations of, 957—959
  Bacterionema in, 958t
  bacteroides in, 957, 958, 958t
  Candida albicans in, bacterial suppression of, 959
  flora of, according to site, 957, 958t
    acquisition of, 958—959
    amphibiotic aspects of, 958
    beneficial effects of, 959
    diseases initiated by, 959
    host nutrition and, 959
    immunity and, 959
    nature of, 958—959
    vitamin synthesis by, 959
  Fusobacterium in, 958t
  Lactobacillus in, 958t
  Leptotrichia in, 958t
  microbiology of, 957—962
  Nocardia in, 958t
  Neisseria in, 958t
  Peptococcus in, 958t
  Peptostreptococcus in, 958t
  Propionibacterium in, 958t
  Rothia in, 958t
  spirochetes in, 957, 958, 958t, 959
  staphylococci in, 958t, 959
  streptococci in, 957, 958, 958t, 959
  treponemes in, 958t
  Veillonella in, 958t
  Vibrio sputorum in, 958t
Orchitis, mumps, 1341

Orf virus, 1276, *1276.* See also *Poxvirus.*
Organic acids, in sterilization, 1458
Organic matter, bacterial conversion to atmospheric CO_2. See *Mineralization.*
Organic solvents, bactericidal action of, 1462
Organism(s), definition, 13
Original antigenic sin, 488—489, 489t
Ornithine, bacterial synthesis of, 69, 70
Ornithosis, 921, 926. See also *Chlamydia psittaci.*
Oroya fever, 949
Orthomyxovirus(es). See *Myxovirus.*
Osteomyelitis, due to cat bite, 809
Otitis, due to *Moraxella,* 785
Otitis media
  due to *Hemophilus influenzae,* 795
  due to measles, 1346
  pneumococcal, 702
  streptococcal, 720
Ovalbumin, chicken and duck, cross-reactions between, 383
Ovaritis, mumps, 1341
Overcrowding, tuberculosis and, 855, 858
Oxacillin
  penicillinase ineffectiveness against, 156
  structure, 157
Oxidants, antiseptic, 1459
Oxidase reaction tests, in classification of Enterobacteriaceae, 754
Oxidative phosphorylation. See *Phosphorylation, oxidative.*
Oxidizing agent, hypochlorous acid as, 1459
11-Oxycorticosteroids, cell-mediated allergic responses inhibited by, 580
Oxygen
  bacterial requirements, 90
  effects on anaerobes, 48
Oxygenases, in bacterial respiration, 49
Oxyquinolines, topical use of, 678
Oxytetracycline, 160. See also *Tetracycline.*
  structure, 298
Ozena, 771

P

Pactamycin, 303
Pancreatitis
  in coxsackieviral infection, 1297
  mumps, 1341
Pandemic. See also *Epidemic.*
  definition, 663
Panencephalitis, subacute sclerosing. See *Subacute sclerosing panencephalitis.*
Pannus, 923
Pantetheine, bacterial synthesis, 75
Pantoate, bacterial synthesis, 73, 74, 75
Pantothenic acid, 91t
Papain, immunoglobulins affected by, 409—410, 416
Papilloma(s)
  pathogenesis, 1429—1430, *1430*
  progression to cancer, 1429
  regression of, 1429—1430
Papillomavirus(es), *1027,* 1429—1430
  in human cancer, 1447
  multiplication of, 1429—1430, *1430*
  natural cancers due to, 1446
Papovavirus(es), *1018, 1027,* 1419—1429
  cell transformation due to, 1445, 1445t
  DNA of, cyclic, *1023*
  transcription of, 1147
  in progressive leukoencephalopathy, 1220
  oncogenic, 1429—1430
Para-aminobenzoic acid therapy, in rickettsioses, 900
Para-aminosalicylic acid therapy, in tuberculosis, 857
Paracoccidioides brasiliensis, 994—995, *994.* See also *Blastomycosis, South American.*
Paracoccidioidomycosis. See *Blastomycosis, South American.*
Paracolobactrum arizonae, 776
Paracolon organisms, 757
Parainfluenza, 1336
Parainfluenza virus(es), 1336—1339. See also *Paramyxovirus.*
  chromosomal aberrations due to, 1208
  clinical syndromes of, 1337, 1338t
  cytopathic changes due to, 1337
  disease and infections due to, control of, 1338
    epidemiology, 1338
    laboratory diagnosis, 1338
    latent period for, 1334
    pathogenesis, 1337
    prevention, 1338

Parainfluenza virus(es), disease and infections due to (continued)
  serologic tests for, 1202t
  erythrocytic reactions of, 1336–1337
  hemagglutinins of, 1336
  hemolysin of, 1337
  host range of, 1337
  immunologic characteristics, 1336
  morphology, 1332, 1333
  multiplication of, 1335
  properties of, 1336–1337, 1336
Paralysis
  immune, 493
  infantile. See Poliomyelitis.
Paramecium, adaptive inheritance in, 175
Paramyxovirus(es), 1329–1353, 1359–1360. See also specific virus, e.g., Mumps virus.
  budding in, 1160, 1161
  capsid of, 1030
  characteristics, 1309t
  chemical and physical, 1331–1334
  classification, 1309, 1309t
  general properties, 1331–1336
  hemagglutination of, 1036
  host modification of envelope in, 1163
  inclusion bodies due to, 1334, 1334
  infections due to, persistent, 1219
  in lupus erythematosus lesions, 1219
  in multiple sclerosis, 1219
  morphology, 1331, 1332, 1333
  multiplication of, 1334, 1335
  nucleocapsid formation in, 1158
  phenotypic mixing in, 1163
  RNA synthesis in, 1150
Parasitism
  bacterial, 669
  effect on evolution of host, 12
  intestinal, IgE and, 535–536
  rickettsial adaptation for, 901
Paratyphoid fever, 774
Paravaccinia, 1275–1276
Paravaccinia virus, 1275, 1276. See also Poxvirus.
Paresis, general, 888
Paromomycin. See also Aminoglycoside.
  bacterial resistance to, 163t
Parotid gland(s), adenoma of, due to polyoma virus, 1420
Parotitis. See Mumps.
Particle analyzer, in analyzing bacterial growth, 96
Parvoviruses, 1236
Pasteur effect, in bacteria, 52–53
Pasteurella(e)
  growth of, polyamines in, 79
  taxonomic characteristics, 34t
  transfer of genes from E. coli to, 195

Pasteurella aviseptica, 352
  DNA base composition in, 37t
Pasteurella multocida, 809. See also Cholera, fowl.
Pasteurella haemolytica. See Vibrio parahaemolyticus.
Pasteurella pestis. See also Yersinia pestis.
  DNA base composition in, 37t
  exotoxin of, 637t
  genetic variations of, invasiveness and, 642
Pasteurella pseudotuberculosis. See Yersinia pseudotuberculosis.
Pasteurella tularensis. See Francisella tularensis.
Pasteurization, 1454
  brucellae killed by, 813
  of beer and wine, 6
Patch test, in contact skin sensitivity, 563
Paternity
  disputed, blood typing in, 608
  Gm allotypes in, 420
Pathogen(s)
  as selective agents, in evolution of blood groups, 604
  extracellular enzymes of, 95
  milkborne, sterilization of, 1454
  virulence of. See Virulence.
Pathogenicity. See also under specific organism, e.g., Pneumococcus, pathogenicity.
  definition, 632n
Patois virus, 1392
Pectinase, bacterial, 95
Pediculus vestimenti, Rickettsia prowazeki transmitted by, 905
Pemphigoid, 587t
Pemphigus vulgaris, 587t
Penicillinate, penicilloyl conjugates from, 583
Penicillin(s) (therapy), 675–676. See also Antibiotic.
  allergy to, 549, 581–585
    basophil degranulation and, 544
    contact dermatitis due to, 584
    diagnosis, 584–585
    oral versus injection administration, 581n
  alternatives to, 158
  aminoglycoside synergistic action with, 165
  anaphylaxis due to, 538
    prevention, 539
  anemia due to, 583–584
  antibacterial activity of, 155–157
    determination of, 150, 150
  antibody-mediated reactions to, 584t
  Bacillus megaterium cell wall disorganized by, 115
  bacteria converted to L forms by,

Penicillin(s) (therapy), bacteria converted (continued) 174, 175
  biosynthesis of, 160
    modified, 155–156
  broadening antimicrobial spectrum of, 157
  cell wall synthesis inhibited by, 120
  chemistry of, 155–156
  delayed-type hypersensitivity to, 584
  discovery of, 149, 149
  G, 155, 156
  immune responses induced by, 583
  in actinomycosis, 876
  in anthrax, 826
  in Bacteroides infection, 787, 788
  in chlamydial infection, 922
  in clostridial infections, 841
  in diphtheria, 691
  in erysipeloid, 948
  in Fusobacterium infection, 788
  in glomerulonephritis prevention, 723
  in gonorrhea, 750, 751
  in leptospirosis, 895
  in listeriosis, 948
  in local infections, 678
  in Neisseria meningitidis infection, 748
  in ophthalmia neonatorum prevention, 751
  in Pasteurella multocida infection, 809
  in pinta, 892
  in pneumococcal infection, 670, 671, 671, 672, 672, 674, 694, 705
  in Proteus mirabilis infection, 772
  in rat-bite fever, 896, 948
  in relapsing fever, 893
  in rheumatic fever, 675
  in selecting autotrophic bacterial mutants, 177, 178
  in staphylococcal infection, 737
  in streptococcal infection, 723
  in syphilis, 883, 890, 891
  in tetanus, 837
  in yaws, 891
  penicillinase action on, 155, 156–157
  penicillinase production induced by, 162
  phenoxymethyl, 156, 156
  plate assay for, with Staphylococcus aureus, 150
  production of, 155–156
    by Penicillium, 583
  prophylaxis by, of dental caries, 960
    of gonorrhea, 751
    of rheumatic fever, 723
    of syphilis, 891

Penicillin(s) (therapy) *(continued)*
  reactions of, haptenic substituents
      derived from, 583, *583*
  resistance to, 679
    bacterial, 120, 163t
      in mixed infections, 163
      population density and, 163
    by *Neisseria meningitidis,* 746
    by *Streptococcus,* 720
    by *Staphylococcus,* 162, 732,
        733, 737, 1109
    due to unnecessary prophylaxis
        of pneumococcal infec-
        tion, 705
    in gonorrhea, 751
    non-penicillinase, 157
  semisynthetic, 156–157, 583,
      733, 737
  septicemia due to, 674
  solutions of, contaminants of,
      anaphylactic reactions
      due to, 583
  staphylococcal cell wall synthesis
      affected by, 111, *112*
  streptomycin synergistic action
      with, 165
  structure, 155, *156*
  to prevent tissue culture contami-
      nation, 942
  V, 156
  wheal-and-erythema reactions due
      to, 583n
Penicillinase
  Bacillus formation of, 123, *123,*
      826
  bacterial formation of, 123, *123,*
      124
    penicillin induction of, 162
    plasmids and, 162
  bacterial resistance to penicillin
      and, 155, 156–157
  cephalosporin resistance to, 158
  drug resistance and, 162–163
  drugs not affected by, 156
  methicillin resistance to, 156
  plasmid formation of, 1109
Penicillium, 976t
  penicillin produced by, 149, *149,*
      155, 583
Penicillium griseofulvum, 983, *983*
Penicilloic acid, 583
Penicilloyl, 583, *583*
Pentose(s), bacterial synthesis of,
    80, 81
Pepsin, IgG split by, 410
Peptidase(s), bacterial, 95
Peptide(s). See also *Polypeptide.*
  bacterial spore, 144
  in streptococcal nutrition, 94n
Peptide chain(s), elongation of, in
    protein synthesis. See
    under *Protein, synthesis
    of.*
Peptidoglycan(s)

Peptidoglycan(s) *(continued)*
  bacterial. See under *Bacterium.*
  in fungus cell wall, 111
Peptidoglycan hydrolase(s), bacte-
    rial, 113
Peptidyl transfer, on ribosome, in
    protein synthesis, 284
Peptidyl transferase
  chloramphenicol inhibition of,
      299
  in polypeptide chain termination,
      290, *290*
Peptococcus, in oral cavity, 958t
Peptostreptococcus, in oral cavity,
    958t
Peracetic acid, in sterilization, 1459
Pericarditis, in pneumococcal pneu-
    monia, 702
Periodic acid-Schiff stain, fungus
    visualization and, 981,
    982
Periodontal disease, 961–962
  due to *Actinomyces naeslundii,*
      962
  due to dental plaque, 960
Periorchitis, in yaws, 891
Peritonitis
  due to *Bacteroides,* 787
  due to *E. coli,* 768
  due to salmonellae, 773
  pneumococcal, 702
Peritonsillar abscesses, 720
Permeases, bacterial, 131n
Pernicious anemia, 591
Peroxidases, in bacterial respiration,
    51
Persisters, bacterial, 151
Pesticides, bacterial breakdown of,
    56n
Pestilence, famine and, 658
Pestocin I, 804
Petechiae, rickettsial, 899
Petroleum
  aromatic components of, biosyn-
      thesis of, 78
  formation of, bacteria in, 56
Peyer's patches
  lymphoid cells in, 477
  Salmonella typhi in, 773
Pfeiffer phenomenon, 779
pH, bacterial growth and, 93
Phage. See *Bacteriophage.*
Phagocyte(s). See also *Macrophage.*
  antibacterial action of, 644–652,
      644, 645
  bactericidal drugs with, 671, *673*
  diapedesis of, 646, *647, 649*
  in immunity, 10
Phagocytin, antimicrobic activity
    of, 654
Phagocytosis
  antibacterial, effects of necrosis
      and abscess formation
      on, 654

Phagocytosis, antibacterial
      (continued)
  inhibition by circulatory distur-
      bances, 658
  opsonization and, 355, 650
  tissue architecture and, 654
  by polymorphonuclear leuko-
      cytes, 651
  surface, 653–654, *655, 656*
    intercellular, 654
  susceptibility to, opsonins and,
      355
  virus infectivity and, 1212
Phagocytosis-promoting factor,
      antibacterial action of,
      653
Phagolysosome(s), 464, 651
Phagosome, macrophage, 464
Pharyngitis
  coronaviral, 1365
  streptococcal, 720, 721
    prevention and treatment, 723
    upper respiratory diphtheria ver-
        sus, 687
Pharyngoconjunctival fever, adeno-
    viral, 1235
Pharynx, bacteria in, 956
Phenethyl alcohol, bacterial cyto-
    plasmic membrane dam-
    aged by, 127
Phenol(s)
  bactericidal action of, 8, 1461,
      1461
    kinetics of, 1453
  detergent action of, 1461
  Salmonella typhosa sterilization
      by, 1458
  soaps used with, 1461
  to deodorize sewage and garbage,
      1452
Phenol coefficient
  definition, 1458
  disinfectant potency and,
      1457–1458
Phenomic lag, in bacterial muta-
    tion, 177–179, *178, 179*
Phenotypic suppression, of muta-
    tions, 292
Phenylalanine
  accumulation in man, 134
  bacterial synthesis of, 77, 78
Phialophora species, 1000
Phlebotomus, bartonellosis trans-
    mitted by, 950
Phlebotomus fever virus(es), 1381t,
    1396
Phlebotomus papatassi, 1396
Phleomycin, 267
Phosphate group, abbreviations for,
    41n
Phosphate ion, bacterial require-
    ment of, 92
Phosphatide(s), bacterial synthesis
    of, 125, *126*

Phosphogluconate pathways, in bacterial fermentation, 46

Phospholipase(s)
bacterial, 95
virus inactivation by, 1464

Phosphonomycin, 120, *121*

Phosphorus-32, mutations due to, 259

Phosphorylation
bacterial. See under *Bacterium.*
in fermentation, 41–42
oxidative, at animal cell mitochondria, 1129

Photobacterium, 34t

Photodynamic sensitization
of bacteria, 1457
of viruses, 1457, 1464–1465

Photooxidation, of bacteria and viruses, 1457

Photophosphorylation, in bacteria, 54, *54*

Photoreactivation, of damaged nucleic acids, 238

Photosensitivity, of viruses, 1464

Photosynthesis
algal, 54, 55
bacterial. See under *Bacterium.*
definition, 54
evolution of, 15
ferredoxins in, 43, 66
plant, 54, 55
types of, 14n

Phototaxis, 55

Phthisis. See *Tuberculosis.*

Phthisical diathesis, 11

Phycomyces, 998

Phycomycete(s), 975–976, 976t, 986t, 998–999
wall structure of, *969*

Phycomycosis(es), 998–999. See also *Phycomycetes* and *Rhizopus.*

Phylogeny, influence of taxonomy, 33

Physiologic state, infectious disease and, 11

Physiology
bacterial, 19–165
cell. See *Cell physiology.*
microbial, 11–12

Phytohemagglutinin. See also *Mitogen.*
blast transformation due to, 468, 574, *574*
lymphotoxin release due to, 577
migration-inhibition factor release due to, 575

Picornavirus(es), 1279–1396. See also specific picornavirus, e.g., *Coxsackievirus.*
characteristics, 1281, 1281t
classification, 1281, 1281t
counting of, 1034

Picornavirus(es) *(continued)*
diseases due to, serologic tests for, 1202t
history, 1280–1281
infectious nucleic acids in, 1143
multiplication of, inhibition by substituted benzimidazole and guanidine, 1183
naming of, 1281
plaque assay of, *1038*
protein synthesis in, 1151
RNA replication in, 1150
viral interference by, 1182

Piedra, 1006

Piedraia hortai, 1006

Pigeon(s). See also *Fowl.*
tuberculosis in, 861

Pigeonpox, 1260

Pilin, 30
F, bacterial, 196

Pilot's wheel, 994

Pilus(i), bacterial. See under *Bacterium.*

Pimaricin, 983. See also *Polyene.*

Pinocytosis, 1125

Pinta, 891–892. See also *Treponema carateum.*
experimental, 892

Pityriasis versicolor, 1005–1006

Pityrosporum species, 1005

P-K reaction, 530, 530n

Placenta
syphilis transmitted through, 888
transfer of immunoglobulins to fetus across, 504t
virus multiplication in, 1216
virus passage across, 1216

Plague. See also *Yersinia pestis.*
bubonic, 804
carrier rate in, 804
communicability of, 663
definition, 802n
epidemics of, 7
fleas in, 802, 803, 805
immunity to, 353, 805
laboratory diagnosis, 805
mortality rate, 805
pneumonia in, 804
pneumonic, 804
prevention, 805–806
rats in, 803
septicemic, 804
sylvatic, 802
toxin in, 804
treatment, 805

Plakins, antimicrobial activity of, 654

Plant(s)
CO_2 assimilation in, 69
DAP synthesis in, 72
fungus diseases of, 966
mitogen from. See *Mitogen.*
photosynthesis in, 54, 55

Plant(s) *(continued)*
starch formation in, 83

Plaque, dental, 959–960

Plaque-forming cells, in antibody formation, 453, *455, 469*

Plaque method, of viral assay. See under *Virus.*

Plaque-reduction test, in antigenic analysis of viruses, 1200

Plasma, hormones in, 393

Plasma cell(s), *469.* See also *Lymphoid cell.*
antibody formation by, 451, *458,* 469–473
double antibodies produced by, 470n
one cell—one antibody action of, 470–471
differentiation of lymphocytes into, 457, 469, 470, 496
diversity of, 470
elevated immunoglobulin levels and, 469
idiotype(s) of, 470
immature, *458, 469,* 470
definition, 470
immunoglobulin synthesis by, 471, 536, 536n. See also under specific immunoglobulin.
lifespan, 470
lymphoid tissue hyperplasia and, 469
maturation pathways of, *461*
neoplastic, 470
transitional. See *Plasma cell, immature.*

Plasma membrane(s)
bacterial, *108*
of animal cells, 1128
of *Bacillus subtilis, 26*

Plasmid(s), 1105–1111. *See also F factor* and *R factor.*
acriflavine effects on, 174
classification, 1106
conjunction apparatus generated by, 1105–1106
in drug resistance, 162–163, 164
in spread of genes for antibiotic resistance, 1105
in versatility of soil *Pseudomonas,* 1105
penicillinase formed by, 162, 1109
phage generation of, 1105
pilus formation in, 1106
regulation of, 1106
replication of, 1105
resistance-bearing. See *R factor.*
resistance-determinant, 1108, *1108*
transmissible versus nontransmissible, 1105
types, 1105

Plasmid(s) *(continued)*
  viral evolution and, 1110
Plasmin
  complement activated by, 521, 524
  in kinin production, 542
Plasmodium maleriae, 551
Platelet(s)
  lysis of, due to autoantibodies, 586
  serotonin release from, in anaphylaxis, 541
  vasoactive amines released by, 550
Plating, replica, in scoring clones, *175, 177, 177*
Pleurisy
  in coccidioidomycosis, 992
  in pneumococcal pneumonia, 702
Pleurodynia, epidemic, 1298, 1298n
Pleuropneumonia, in cattle, 930
Pleuropneumonia-like organisms, 930
Pneumococcus(i), 693, 706
  adrenal insufficiency due to, 746
  antibody of, detection of, 704
  antigen(s) of, capsular, 697–698
    M, 698
    R, 698
    somatic, 698–699
  antigenic structure, 697–699
  antigenic types of, 700–701, 701t
  autolytic enzymes from, 695
  capsules of, 696, *696*
    antigens of, 697–698
    demonstration of, 695
    phagocytosis impeded by, 698
    staining of, 696
    structure, 697
  carrier rate of, 705
  cell walls of, 115
    equatorial growth in cell division, *129*
  clearance from circulation, 646t
  colonies of, 699, *699*
    morphologic variation in, *102*
  C substance of, 698
  culturing of, 695, 696, 705
    maintaining maximum virulence of, *699*
  death due to, 701
  disease and infection due to, delayed skin reactions in, 657
    immunity to, 704
    immunization against, 705
    natural antibodies against, 653
    nonpneumonia, 702
    prevention, 705
    recurrence, 704
    treatment, 705
      penicillin therapy in, 672, *674*

Pneumococcus(i) *(continued)*
  drug resistance by, 679
  fermentation by, homolactic, 44
    lactic acid, 696
  genotypic variations, 699–700
    virulence and, 641
  growth of, in irradiated mice, *659, 660*
  hemolysis by, 695
  α-hemolytic streptococci versus, 696
  hosts for, 700
  immunity to, 704
  immunization against, 698, 699
  in nasopharynx, 748
  insulin metabolism by, 696
  interchangeability of serologic types, 694
  inverse proportion to leukocyte blood count, 660, *660*
  metabolism, 696–699
  morphology, 694–696
  mutations of, 186n
  pathogenicity, 115, 700–701
    capsular antigens in, 697–698
  penicillin action on, 670, *671*
  pH and, 93
  phagocytosis of, 653, 670, *670*
    resistance to, 654n, 702
  polysaccharide from, precipitin reaction with, 372t
    types of, cross reactions between, 382, *382*
  quellung reaction of, *695, 696,* 704
  streptococcal relationship to, 700
  toxin production by, 701
  transformation in, 182, *184, 185,* 186
  types, 697, 699–700
    transformations of, 700, *700*
  typing of, 704
  virulence, 186n, 632
    capsule differences and, 698
    colony surface texture and, 103
    genotypic variation and, 641
    phenotypic variation and, 642
Pneumolysin, 701
Pneumonia. See also *Pulmonary disease.*
  adenoviral, 1234
  anthrax, 823
  due to *Aeromonas,* 782
  due to *Coxiella burneti,* 902
  due to *Hemophilus influenzae,* 795, 1323
  due to Lassa virus, 1395
  due to lymphocytic choriomeningitis virus, 1395
  due to Newcastle disease virus, 1207
  due to respiratory syncytial virus, 1349, 1352
  giant cell, 1347

Pneumonia *(continued)*
  herpetic, 1247
  in cattle, 809
  influenza, 1323
  in whooping cough, 799
  Klebsiella pneumoniae in, 770, 770
  metastatic, in plague, 804
  micrococcal, 738
  Moraxella, 785
  mycoplasmal. See *Pneumonia, primary atypical.*
  ornithosis and, 926
  plague, 804
    mortality rate in, 805
  pneumococcal. See under *Diplococcus pneumoniae* and *Pneumococcus.*
    agglutination of pneumococci in, 653
    antigenic types causing, 701t
    crisis in, 704
    empyema in, 702
    Francis' test in, 704
    hemolytic anemia in, 701
    history, 694
    laboratory diagnosis, 704–705
    lesion evolution in, 702, *703*
    lung abscesses in, 702
    mortality rate in, 654
    pathogenesis, 701–702
    pericarditis in, 702
    pleurisy in, 702
    predisposing factors, 701–702
    treatment, 694, 705
      penicillin therapy in, 671, *672*
      sulfonamides in, 670, *670*
  primary atypical, 938–942. See also *Mycoplasma pneumoniae.*
    characteristics, 938
    epidemiology, 942
    incubation period, 942
    prevention, 942
    serum from patient with, *939*
    treatment, 942
  pulmonary edema and, 702
  staphylococcal, 1323
  streptococcal, 1323
  varicella, 1212
  viral. See *Pneumonia, primary atypical.*
  with influenza, 1308
Pneumonitis
  cryptococcal, 985
  fungal, 979
  in coccidioidomycosis, 992
  in glanders, 784
  in Q fever, 911
Pneumotropic virus, of rats, 1362n
Pock(s), produced by poxviruses, 1260, *1261*
Pock method, of assay for virus infectivity, 1039, *1040*

Poison ivy. See *Dermatitis, poison ivy.*
Pokeweed mitogen. See under *Mitogen.*
Polarization, fluorescence, 399
Poliomyelitis. See also *Poliovirus.*
  age factors, 1292
  antibodies in, 1209, 1292
    persistence of, 1210
  bulbar, 1291
    tonsillectomy and, 1212
  case rate, annual, *1295*
  castration and, 1212
  control of, 1293–1296
  encephalitic, 1291
  endocrine factors, 1292
  epidemiology, 1292–1293
  history, 1280
  host factors, 1291
  immunity to, 1210
  immunization in, passive, 1293
  inapparent, 1215
  incidence, *1295*, 1296
  laboratory diagnosis, 1292
  location of lesions, 1291, *1291*
  paralysis in, 1290, 1291
  pathogenesis, 1289–1292, *1290*
  pregnancy and, 1212
  prevention, 1293–1296
  recovery from, 1291
  repeat attack, 1283
  seasonal aspects, 1292
  spinal, 1291
  spread of, 1293
  temporal aspects, *1284*
  vaccination for, necessity for, 1296
    viral interference in, 1183
  vaccines for, 1292
    distribution of, *1295*
    effectiveness of, 491
    inactivated virus, 1293–1294
      advantages and disadvantages, 1295
    live (attenuated) virus, 1293, 1294–1295
      advantages and disadvantages, 1295–1296
  viremia in, 1213
Poliovirus(es), 1281–1296. See also *Picornavirus* and *Poliomyelitis.*
  adsorption to cells, 1283
    capsid changes and, 1144
  antibodies of, appearance of, 1283, 1284
    complement fixation test for, 514t
    neutralization of, 1192, 1193
      kinetic curves of, *1195*
    persistence of, 1283, *1284*
  antigenic analysis of, 1200
  antigenic variants of, 1283

Poliovirus(es) *(continued)*
  antigen(s) of, 1282–1283
    evolutionary aspects, 1199
  assembly of particles in, 1286, *1286, 1288*
  attachment by, 1144
  autointerference by, 1182
  capsid of, 1025, 1144
    allosteric transitions in, 1029
  cell(s) infected by, biosynthetic events in, *1285*
  cell penetration by, 1284
  cell susceptibility to, 1208, 1209
  characteristics, chemical, 1281–1282
    genetic, 1287–1289
    immunologic, 1282–1283
    physical, 1281–1282, 1283t
  counting of, *1034*
  crystals of, 1281, *1282*
  culture of, 1280, 1283
  cytopathic changes due to, 1207, 1283, *1285, 1286, 1287, 1288*
  dissemination of, 1213
  formaldehyde inactivation of, 1464, *1464*
  genetic map of, 1288
  genetic organization in, 1167
  genetic recombination in, 1167
  heat inactivation of, 1464
  HeLa cells infected with, 1147, *1156*
  host range of, 1283
  inclusion body formation with, 1287, *1287*
  laboratory animals infected by, 1283
  maturation of, 1157
  morphology, 1281, *1282*
  multiplication of, 1283–1287, *1287*
    in culture, *1145*
  mutants of, 1166
    interferon production by, 1175
    phenotypic, 1287, 1289t
  phenotypic mixing in, 1163, *1164, 1165*
  photodynamic inactivation by, 1465
  plaque assay of, *1038*
  pleiotropism in, 1288
  polypeptide of, processing of, *1152*
  properties of, 1281–1288
  protein synthesis in, 1151, *1152*
  release of, 1157, *1158*
  replication of, 1284, *1285*
  replicative intermediates in, *1155*
  ribosomes infected by, *1147*
  RNA of, base composition of, 1297t
  RNA replicase in, 1152
  RNA replication in, 1153, *1157*

Poliovirus(es), RNA replication in *(continued)*
  site of, 1154, *1156*
  RNA synthesis in, inhibition of, 1153
  serial propagation of, for vaccines, 1206
  susceptibility to, 1209
  transmission of, 1282
  vaccines for, 1166, 1206
  viral interference by, 1182
Pollen
  immediate-type hypersensitivity due to, blast transformation in, 573
  ragweed, SRS-A release and, 543
Pollution, of streams, 49
Polyadenylic acid, chains of, attachment to animal cell RNA, 1131
Polyamine(s)
  bacterial. See under *Bacterium.*
  cellular function of, 79
  viral function of, 79
Polycyclic hydrocarbons, carcinogenic, epoxides of, reading-frame shifts in bacteria due to, 258
Polyene(s) (therapy). See also specific polyene, e.g., *Nystatin*
  effects on bacterial cytoplasmic membrane, 127
  in candidiasis, 997
  in coccidioidomycosis, 993
  in cryptococcosis, 986
  in *Mycoplasma* infections, 984
  in mycoses, 983–984, *983*
  Streptomyces synthesis of, 983, *983*
Poly-β-hydroxybutyric acid, as bacterial storage material, 52
Polykaryocyte formation, due to viruses, 1161, *1162,*1207
Polymerases, DNA. See *DNA polymerase.*
Polymetaphosphate, bacterial cell wall accumulation of, 343
Polymyxin(s) (therapy), 678
  bacterial cytoplasmic membrane damage due to, 126–127, *127*
  B, in *Acinetobacter* infection, 785
  in *Bacteroides corrodens* infection, 787
  in *Citrobacter* infection, 772
  in *Enterobacter* infection, 771
  in *Pseudomonas aeruginosa* infection, 784
  structure, *127*
  in *E. coli* infection, 768
  in *Klebsiella pneumoniae* infection, 770
  in *Yersinia* infection, 806

Polymyxin(s) (therapy) (*continued*)
produced by *Bacillus polymyxa*,
126
topical use of, 678
toxicity of, 678
Polynucleotide(s)
double-stranded, interferon production induced by,
1181, *1182*
therapeutic potential of,
1180—1181, *1181*
synthetic, for determining genetic code, 275, *276*
Polynucleotide ligases(s), in DNA
replication, 227
Polynucleotide phosphorylase, in
bacterial mRNA break-
down, 330
Polyoma virus(es), *1019*,
1419—1429
animal effects of, 1420
cell transformation due to, 1208,
1422—1428, *1424*, 1445
cell DNA synthesis and, 1422
cell surface changes in,
1423—1424
integration of viral DNA into
cell DNA, 1426, 1428
mechanism, 1424—1428
stability of, 1428
viral genes in, 1414—1426,
1425, 1426
chromosomal aberrations due to,
1208
culturing of, permissive and nonpermissive cells used
with, 1420, 1421t
DNA of, *1023*
glomerulonephritis due to, 553
infection due to, pathogenesis,
1420
productive, 1420—1422, *1422*
effect on cellular synthesis,
1422
members of, 1420
multiplication of, in tissue cultures, 1420—1422, *1421*
proportion of cells surviving,
1421, *1422*
naming of, 1419
runt disease due to, 1420
T antigen production and, 1422,
1422
tumors due to, 1420
virion of, properties of, 1420
virus-dependent abortive infection
by, 1165
Polypeptide(s) (chains)
allelic, amino acid residues of,
436n
amino acid sequences in, *276,
277, 425*
heavy, classes of, 407—408
disorders of, 436

Polypeptide(s) (chains) (*continued*)
in immunoglobulin G, 409, 411
J, 416, 418
light. See also *Kappa polypeptide
chain* and *Lambda poly-
peptide chain.*
in various species, 408
types of, 407—408
of immunoglobulins. See under
Immunoglobulin.
synthesis of, bacterial, 85—86
direction of, 279, *280*
translation between mRNA and,
278, 279
Polyribosome(s)
bacterial, 31
immunoglobulin synthesis and,
471
Polysaccharidases, bacterial, 95
Polysaccharide(s)
bacterial. See under *Bacterium.*
capsular, abnormal, in *Aerobacter, 83*
in *Salmonella* cell wall, *84*
O, 118
synthesis of, genetic aspects,
235
Polysome(s)
in animal cells, 1132
in rat liver, *281*
Polystyrene, in passive agglutination, 393
Population(s)
racial, IgG in, gene complexes in,
422t
susceptibility of, bacterial transmission and, 664
Porphyrin(s)
bacterial requirement of, 90
photosensitization of, in sterilization, 1457
Potassium (ion)
in bacteria. See under *Bacterium.*
in protein synthesis, 284
Potassium permanganate, antiseptic
properties, 1459
Potato blight, 966
Powassan virus, 1408
Poxvirus, *1018*, 1257—1277. See
also specific virus, e.g.,
Vaccinia virus.
antigens of, 1262
as transition forms, 1163
cellular hyperplasia due to, 1261
characteristics, 1258—1261,
1260t
pathogenic, 1260
cultured cells susceptible to, 1264
diseases due to, antibiotic therapy
in, 1185
suppression of cellular immunity and, 1202
DNA transcription in, 1147, 1148
family antigen of, 1262

Poxvirus (*continued*)
genetic recombination in, 1167,
1260
groups of, 1258—1260
inclusion bodies due to, 1260
local lesions due to, cellular
immunity and, 1201
macrophage action on, 1212
maturation of, 1161
morphology, 1033, 1258, *1259,
1260*
multiplication of, 1264
inhibition by isatin-β-thiosemicarbazone, 1184
mutants of, pock-type, 1166
resistant to rifampin, 1186
nucleic acid release by, 1144
of birds, 1260
of mammals, 1258
pock assay of, *1040*
pock production by, 1260, *1261*
rabbit, 1166
reactivation of, 1120, *1169*
rifampin effects on, 1185, *1186*
tumor-producing, 1260
UV-irradiated, reactivation of,
1169
PPD antigens, 159t
PPLO, 930
Prausnitz—Kustner (P—K) reaction,
530, 530n
Precipitation, as serologic reaction,
353
Precipitin(s), 370
Precipitin reaction
antibody—antigen ratios in,
372—373, *374*
antibody concentration measured
by, *371*, 379—380
antigen concentration measured
by, 380, 389, *390*
antigen in, valence of, 375
applications, 379—380
cooperation between different
antibodies in, 376, 377,
378
curve of, 371, *371*
antibody avidity and, 385, *385*
in monospecific system, *371*,
372
in multispecific system, 372,
373
zones of, 371, 372
definition, 370
flocculation, 385—386, *386*
flocculation reaction versus, 387,
387
hapten inhibition of, 380—381,
380, 381
history, 370
in determining molecular weight
of antigen, 389
in gels, 386—389
double diffusion in one dimen-

Precipitin reaction, in gels, double diffusion *(continued)*
　sion in, *386,* 387
　double diffusion in two dimensions in, *386, 387–389, 388, 389*
　of human immunoglobulins, *408*
　single diffusion in one dimension in, *386–387, 386*
　in liquid media, 370–386
　lattice theory of, 373–375, *374*
　nonspecific factors in, 377
　of cross-reaction, 388, *388*
　of identity, 388, *388*
　of nonidentity, 388, *388*
　of partial identity, 388, *388*
　precipitate structure of, 374, *374*
　quantitative, 371–373
　radial immunodiffusion, 389, *390*
　reversibility of, 379
　sensitivity of, 393–394
　stages, 377
　with polysaccharide antigen, 372t
　with protein antigen, 373t
Prednisone, in kidney transplants, 499, 580
Pregnancy
　coxsackieviral infection enhanced by, 1297
　fetus as allograft in, 612
　multiple, formation of antibodies to leukocytic antigens after, 609n
　viral disease enhanced by, 1212
　vulvovaginal candidiasis with, 995
Preservative(s)
　essential oils as, 1461
　lactic acid as, 1458
Pretibial fever, 895
Primary response. See *Immune response, primary.*
Progressive leukoencephalopathy, 1220
Prokaryote(s), 14–15
　definition, 15
Prokaryotic cell, 22, *23*
Promonocyte(s), 644, *644*
Properdin, in complement activation, 520–521, *521*
Properdin system, 520–521, *521*
Prophage(s)
　defective, 1100–1101, 1105
　definition, 1091
　DNA recombination in, 243
　escape synthesis due to, 1104
　excision from bacterial chromosome, 1101
　host functions affected by, 1104
　induction of, 1103
　in lysogenic bacteria, 1099–1101
　　location and state of, 1099, *1099*
　insertion into bacterial chromo-

Prophage(s), insertion *(continued)*
　some, 1099, *1100*
　enzyme requirements in, 1101
　mechanism of, 1099–1100
　integration of, 1101–1102
　multiplication of, in lysogeny, 1100
　replication of, 1100
　transfer to repressor-free environment, 1104
　vegetative form versus, 1091
Prophage map, 1099, *1100*
Propionate, produced by fermentation, 44
Propionibacterium, 958t
Prostaglandins, 543
Protease(s)
　IgG split by, 410
　in *Bacillus subtilis,* 307
　in complement activation, 521
Protein(s)
　allergic reactions to. See under *Hypersensitivity, delayed-type.*
　antibody. See *Immunoglobulin.*
　antibody activity by. See under *Antibody.*
　antigenic. See under *Antigen.*
　attachment of aramatic amine to, 354, *354*
　bacterial. See under *Bacterium.*
　Bence Jones. See *Bence Jones Protein.*
　breakdown in animal cells, 1133
　cAMP receptor, 334
　carrier, 465
　conjugated, 355, *356*
　deficiency of. See also *Malnutrition* and *Starvation.*
　　tuberculosis and, 855
　denaturation of, in heat sterilization, 1454, *1455*
　DNA-unfolding, in DNA replication, 227
　erythrocyte surface covalently linked to, 393
　evolution of, 307–308, *308*
　folding of, 304
　haptens coupled to, 355, *355*
　M. See *M protein.*
　monoclonal. See *M protein.*
　monomeric, 304, 305
　mutations in, evolutionary aspects, 307
　myeloma. See *Myeloma proteins.*
　of complement, 515, *516*
　oligomeric, 304, 305
　ribosomal, antibodies to, 282
　rRNA binding of, 283
　structure, catalytic activity and, 305
　　enzyme activity and, 305
　　flexibility in, 305
　　function and, 305

Protein(s), structure *(continued)*
　genetic determination of, 304–308
　interallelic complementation and, 306, *306*
　intergenic complementation and, 305–306
　intragenic complementation and, 306, *306*
　levels of, 304
　ligand binding and, 304
　phenotypic effects of mutations and, 305
　primary, 304
　quarternary, definition, 304
　　significance of, 306–307
　secondary, 304
　tertiary, 304
　three-dimensional, 304–308
synthesis, 271–309
　antibiotic inhibitors of, 296, 302, 303t
　cycles of, 291–292
　EFG factor in, 284
　EFT factors in, 284, *286*
　energy required for, 291
　evolutionary aspects, 291
　　amino acid sequences in, 275, 277
　genetic code in, 275–280
　　degeneracy of, origin of, 277
　　degenerate triplet, 275–279, *276*
　genetic suppression of mutations in. See under *Mutation.*
　in animal cell(s), 1133
　in animal cell cultures, 1134
　inhibitors of, 296–303. See also specific inhibitor, e.g., *Streptomycin.*
　　acting on peptidyl transfer, 298–299
　　acting on recognition, 297–298
　　acting on translocation, 299–300, *299*
　initiation-factor cycle in, *289,* 290
　initiation of translation in, 286–290
　initiation factors in, 288–290
　in *Salmonella typhimurium,* 336t
　in yeast, 984
　machinery of, 272–292
　magnesium ion in, 284
　modifications of, 292–303
　peptide chain elongation in, 284–286, *285*
　　factors in, 284–286
　　initiation of, 302
　　termination of, 290, 302

Protein(s), synthesis *(continued)*
  peptide chain folding in, 304
  peptidyl transfer in, 290, 298–299
  potassium ion in, 284
  recognition in, streptomycin effects in, 301
  ribosomes in. See under *Ribosome.*
  termination of translation in, 286–290
  translocation in, antibiotics acting on, 299–300, *299*
  tRNA in, 272–275, 286–287
  tissue, interaction with chemicals, to form immunogens, 563, 563n, *565*
  viral, 1033
  in DNA synthesis, 1064
Protein A, 731
Protein–antiprotein reactions, peptide inhibition of, 380
Proteinase, streptococcal, 720
Protein derivative, purified. See *Purified protein derivative.*
Protein mapping, 251
Proteolysis, of bovine serum albumin, 376
Proteus, 772
  drug resistance by, 679
  episome transfer in, 1107
  plasmids in, 1106
  R factor in, 1108
  transduction in, 1112
Proteus inconstans, 772
Proteus mirabilis, 772
Proteus morganii, 772
  DNA base composition in, 37t
Proteus-OX strains, 902
Proteus rettgeri, 772
Proteus vulgaris, 772
  antigenic structure of, 772
  DNA base composition in, 37t
Protomers, viral. See under *Virus.*
Protoplast(s)
  bacterial, 25
  fungal, 969, *969*
Protosil, 148
Protovirus(es), 1436
Protozoa
  enzymes of, bacteria attacked by, 114
  infections due to, cell-mediated immunity and, 567
Protrophy, in bacterial genetics, 185n
Providencia, 772
Prozone, in agglutination reactions, 392
Pruritus ani, emperor of, 579
Pseudoauxotrophy, 314
Pseudohypha(e)
  definition, 967

Pseudohypha(e) *(continued)*
  of *Candida albicans,* 995, *996*
Pseudomonadadeae, classification, 34t
Pseudomonadales, 26
Pseudomonas, 783–785
  carbon sources of, 56
  drug resistance by, 679
  Entner–Doudoroff fermentation in, 46
  psychrophilic, 93
  respiration in, 49
  soil, versatility of, 1105
  taxonomic characteristics, 34t
  transduction in, 1112
Pseudomonas acidovorans, 784
Pseudomonas aeruginosa
  aspartokinase activity in, end-product inhibition of, 317t
  DNA base composition in, 37t
  fluorescein produced by, 783
  infections due to broad-spectrum antibiotics, 783
  pathogenicity, 784
  pyocin in, 783
  rRNA in, 282t
  sex factors in, 195
  typing of, 783
Pseudomonas alcaligenes, 784
Pseudomonas fluorescens, 784
  DNA base composition in, 37t
Pseudomonas mallei, 784–785
Pseudomonas maltophilia, 784
Pseudomonas multivorans, 784
Pseudomonas pseudomallei, 784–785
Pseudomonas pyocyaneus, 1109
Pseudomonas saccharophila, 37t
Pseudomonas stutzeri, 784
Pseudorabies virus, *1150*
Pseudotolerance, immunologic, 492–493, *492*
Pseudotuberculosis. See also *Yersinia pseudotuberculosis.*
  in animals, 692
Pseudovirions, 1428
Psi factor, in rRNA formation, 263
Psittacosis. See *Ornithosis.*
Psychrophiles, bacterial, 93
Puerperal fever, 720, 724
Pulex irritans, 805
Pulmonary. See also *Lung.*
Pulmonary calcification, in tuberculosis, 852
Pulmonary disease. See also specific disease, e.g., *Pneumonia.*
  allergic reaction with, 594
  due to aspergillosis, 997, *997*
  due to *Coccidioides immitis,* 992
Pulmonary vein erosion, in tuberculosis, 852
Puncture. See *Wound, puncture.*
Purified protein derivative (PPD)

Purified protein derivative *(continued)*
  definition, 560n
  preparation of, 857
Purine(s), bacterial. See under *Bacterium.*
Purine ring, 81, *81*
Puromycin
  fragment reaction affected by, 298
  protein synthesis inhibited by, 297, *297*, 303t
  ribosomal effects of, 298
  structure, 297, *297*
Purpura, thrombocytic, 587t
Purpura-producing principle, pneumococcal, 701
Pyelonephritis
  due to *E. coli,* 768
  L-phase variants in, 943
Pyocin(s), 783, 1109
Pyocyanin, 783
Pyrazinamide therapy, in tuberculosis, 662, *662,* 857
Pyridoxal, 91t
Pyrimidine(s), bacterial. See under *Bacterium.*
Pyrogen, leukocytic, 649
Pyruvate
  glycolytic formation of, *43.* See also *Embden–Meyerhof pathway.*
  in energy-producing fermentations, 43–47

Q

Q fever, 911–913. See also *Coxiella burneti.*
  diagnosis, 911
  epidemiology, 912–913
  pathogenesis, 911
  pneumonitis in, 911
  treatment, 911
  vaccination against, 913
Quasidiploid cells, 1135
Quaternary agents, bactericidal action of, 1460, *1460*
Quellung reaction, 696
  with pneumococci, *695,* 696, 704
Quinacrine, 1134, *1135*
Quinidine allergy, 585
Quinine, early use, 148
Quinones, bacterial biosynthesis, 77, 78

R

Rabbit(s)
  allotypes of, 423, *424*

Rabbit(s) *(continued)*
  experimental syphilis in, 888, *889*
  myxoma virus in, 12
  saprophytic mycobacteria in, 868
  syphilis in, 882, 892
  tularemia and, 807, 809
Rabbit-ear chamber, to study inflammation, 646
Rabbitpox, 1258
Rabies
  animals susceptible to, 1373
  control of, 1373–1375
  double-stranded polynucleotide therapy in, 1181
  epidemiology, 1373
  history, 1368
  inclusion body formation in, 1208
  incubation period, 1371
  laboratory diagnosis, 1371–1373
    serologic tests in, 1202t
  pathogenesis, 1370–1371
  prevention, 1373–1375
    long-term, 1375
    vaccination for, 1211, 1373–1374
      development of, 9, 9n
      duck vaccine in, 1375
      encephalitis due to, 586
      encephalomyelitis due to, 1374, 1375
      experimental, 1374
  prodromal period of, 1371
  public health measures in, 1374
  sylvan, 1373
Rabiesvirus, 1368–1375. See also *Rhabdovirus.*
  chemical characteristics, 1369–1370
  fixed, 1370
  host range, 1370
  immunologic characteristics, 1370
  laboratory infections of, 1370
  morphology, 1368–1369, *1369*
  multiplication of, 1370
  Negri bodies elicited by, 1368, 1371, *1371, 1372,* 1373
  physical characteristics, 1369–1370
  properties of, 1368–1370
  street, 1370
Racial factors, in resistance to tuberculosis, 855
Radiation. See also specific radiation, e.g., *X-irradiation.*
  bacterial disease and, 659–660, *659, 660*
  bacterial mutations due to, 259
  bacterial spore resistance to, 141
  DNA damaged by. See under *DNA.*
  dosage of, mathematical relation between lethal effect and, 268

Radiation *(continued)*
  nucleic acids damaged by. See under *Nucleic acid.*
  viruses inactivated by, 1463
Radioautography
  in bacteriology, 24
  of bacterial DNA, 205
Radioimmunoassay(s), 395–397, *396*
  ammonium sulfate precipitation in, 395–396
  calibration curve for, *396*
  Farr technic in, 396
  of serum insulin concentrations, 395
  substances measured by, 395t
Radioisotopes, in study of bacterial biosynthetic pathways, 61
Radiomimetic drugs, immunosuppression by, 501
Radiotherapy, listeriosis due to, 947
Ragg, 419
Rantz antigen, 946
Rat(s). See also *Rodent.*
  in plague, 803
  liver of, polysomes in, *281*
  pneumotropic virus of, 1362n
  Weil's disease spread by, 894
Rat-bite fever, 875, 896
  due to *Streptobacillus moniliformis,* 948
Rat leprosy, 868
Rat virus, teratogenic effects, 1216
Rauscher leukemia virus, 1433, *1434*
  assay of, 1434
  morphology, *1431*
Reading frame
  accurate positioning of, 277
  definition, 235
  mutation of, suppression of, 292
  shift(s) of, *236*
    due to acridine derivatives, 257, *258*
    due to microinsertions and microdeletions, 253
    in bacteria, 258
    in bacteriophages, 258
    in *E. coli* tryptophan synthetase, 277, *278*
    mutation due to, 257–258, *258*
    mutation suppression and, 293
    suppression of, 294
Reagin(s). See also *Immunoglobulin E.*
  antigenic identity of, *532*
  definition, 531
  human, blocking antibodies versus, 531t
  in atopy, 531
  production of, genetic control of, 537–538

Reagin(s) *(continued)*
  sites of reactions, eosinophils at, 543
Receptor-destroying enzyme, 1036
Recipient, universal, in blood transfusion, 609
Recombination, genetic. See *Genetic recombination.*
Rectum
  adenocarcinoma of, carcinoembryonic antigen in, 622
  obstruction of, in lymphogranuloma venereum, 925
Red blood cells. See *Erythrocyte.*
Red fever. See *Typhus fever, endemic.*
Redox cofactors, in directing bacterial metabolic flow, 68–69
Redox potential, in anaerobiosis, 91
Reed–Muench method, to measure lethal dose, 640, 664–665
Regulator loci, of bacterial genes, 322n
Reiter protein, in syphilis, 886
Reiter protein complement-fixation test, 886
Relapsing fever, 892–893. See also *Borrelia recurrentis.*
  death rate in, 892
  diagnosis, 892–893
  experimental, 892
  mechanism of relapse in, 892
  treatment, 893
Reovirus(es), 1399–1408
  capsid of, 1028, *1029*
    polypeptides of, 1401, 1401t
  characteristics, 1401t
    chemical, 1400–1401
    immunologic, 1401–1402
    physical, 1400
  control of, 1407
  cytopathic effects of, 1405, *1406, 1407*
  disease and infection due to, epidemiology, 1407
  inclusion bodies in, 1400, 1403, *1404,* 1405
  laboratory diagnosis, 1406–1407
  pathogenesis, 1406
  serologic tests in, 1203t
  genetic recombination in, 1167
  genomic RNA segments of, 1400, 1401t, *1403*
  hemagglutination by, 1401
  host range of, 1402–1403
  morphology, 1400, *1402*
  multiplication of, 1403–1406
  naming of, 1400

Reovirus(es) *(continued)*
nucleic acid release by, 1144
oligonucleotides in, 1033
properties of, 1400–1406, 1401t
replication of, sequence of events
in, 1405, *1405*
RNA of, double-stranded, propagation of, 1403
messenger, 1401t, 1403
replication of, 1154, *1157*
synthesis of, 1150
teratogenic effects of, 1216
viruses similar to, 1407–1408
Repellents, chemotactic, 137
Replica plating, in scoring clones,
175, 177, 177
Replicator, 222
Replicon(s), 222
Reproduction
asexual, 971
sexual. See *Sexual reproduction.*
Reptile immunoglobulins, 503
Resistance factor (RF), in *Proteus,*
164
Resistance transfer factors (RTF),
162
in *E. coli,* 195
Respiration
anaerobic, 40
bacterial. See *Bacterium, respiration of.*
definition, 40n
fermentation versus, 5, 40
in *Lactobacillus delbrueckii,* 49
terminal, 50
Respiratory disease
adenoviral, 1235
reoviral, 1406
Respiratory enteric orphan virus.
See *Reovirus.*
Respiratory syncytial virus,
1349–1353
budding of, 1350, *1351*
disease and infection due to, 1352
control of, 1352–1353
epidemiology, 1352
laboratory diagnosis, 1352
pathogenesis, 1350–1352
prevention, 1352, 1353
serologic tests for, 1202t
host range of, 1350
immunologic characteristics,
1349–1350
morphogenesis, *1351*
morphology, 1349, *1349, 1350,
1351*
properties of, 1349–1350
syncytial formation in, 1350,
1350
Respiratory tract. See also *Lung.*
cilia of, antibacterial action of,
643, 643n
defense barriers of, against pneumococcal infection, 701

Respiratory tract *(continued)*
immunoglobulins produced in,
471
mucous secretion in, antibodies
in, antibacterial action
of, 643
upper rhinoviral diseases of. See
Rhinovirus.
Reticuloendothelial system
definition, 644
disorders of, T cell deficiencies
due to, 581
scavenger, 464
Reticulum cell sarcoma
due to Friend virus, 1433
in immunosuppressed patients,
623
Reverse transcriptase. See *Transcriptase, reverse.*
R factor(s), 1105, 1107–1109. See
also *Plasmid.*
classification, 1108
drug resistance and, 766, 1109
formation of, *1108*
in Enterobacteriaceae, 766, 1108
spread of, 1109
structure, 1108–1109, *1108*
Rhabdovirus(es), 1367–1376. See
also specific virus, e.g.,
Rabiesvirus.
characteristics, 1369t
phenotypic mixing in, 1163
replication in arthropods, 1368
structure, 1033, *1033*
Rh alloantigen(s), notations for,
607t
Rh antibody(ies), 606
Rh antigen(s), 605–607
multiplicity of, 607
Rh disease. See *Erythroblastosis
fetalis.*
Rheumatic fever
allergic aspects of, 564
carditis in, autoimmune aspects,
594
L forms in, 944
pathogenesis, 721
treatment, 675, 723
Rheumatoid arthritis,
553–554
antibody–antigen complex in,
586
autoimmune aspects, 587t, 588
cryoprecipitates in, 554
immune complexes in, *553*
mycoplasmas and, *936*
rheumatoid factor in, 419, *553,*
554
synovial fluid immunoglobulins
in, 553
Rheumatoid factor(s)
definition, 419
in rheumatoid arthritis, 419, *553,*
554

Rheumatoid serum agglutinator
(Ragg), 419
Rhinitis, 997
Rhinoscleroma, 771
Rhinovirus(es), 1303–1305. See
also *Picornavirus.*
antibodies to, 1303, 1305
chemical characteristics, 1303
cytopathic changes due to, 1304
disease and infection due to,
control of, 1305
double-stranded polynucleotide
therapy in, 1181
epidemiology, 1305
laboratory diagnosis,
1304–1305
pathogenesis, 1304
prevention, 1305
serologic tests for, 1202t
host range of, 1303–1304
immunologic characteristics, 1303
morphology, 1303
multiplication of, 1304
physical characteristics, 1303
propagation of, 1304
properties of, 1303–1304
strains of, 1303
Rhizobiaceae, classification, 34t
Rhizobium
bacteriocin from, *1110*
biotin produced by, 91t
nitrogen fixation by, 56
taxonomic characteristics, 34t
transformation in, 184
Rhizopus, 876t. See also *Phycomycosis.*
species of, 998
Rhodopseudomonas capsulata,
317t
Rhodospirillum rubrum
DNA base composition in, 37t
nitrogen fixation by, 56
Rho factor, 265
Riboflavin
discovery of, 91t
in bacterial fermentation and
respiration, 42
photosensitization of, in sterilization, 1457
Ribose-5-P, bacterial biosynthesis
of. See under *Bacterium.*
Ribosomal ambiguity
genotypic, 295–296
normal, 295–296
stimulation of, 296
streptomycin in, 297
Ribosome(s), 281–284
aminoglycoside effects on, 296
animal cell mitochondrial, 283
antibodies to, 282
bacterial. See *Bacterium, ribosome of.*
chloroplast, 283
codon recognition by, 284

Ribosome(s) *(continued)*
  complexed versus free, 281
  components of, topographic relations of, 283
  definition, 202
  dissociation, 288
  eukaryotic cell, 283
  evolutionary aspects, 283–284
  in *E. coli*, 281, *281*
  inhibitors of, 296–297
  in protein synthesis, 284, 287–288, *288*
    in eukaryotic cells, 286
    inhibitors of, 296–303
    macrocycle of, *289*, 290
    microcycle of, 284
    peptidyl transfer on, 284
    RNA binding in, 284
  leukoviral, 1010
  mutations of, 295
  polyamine bound to, 79
  runoff 70S, 287, *288*
  specificity for tRNAs, 280
  streptomycin effects on, 296, 301, *301*
  structure, 281–283
  subunits of, 287–288, *288*
    regulation of, 288
  topography of, mapping of, 282
  tRNA binding to, 284
Rice dwarf virus, 1400
Ricin, immunity to, 353
Rickettsia(e), 897–913
  adaptation for parasitism, 901
  bacterial aspects, 901
  binary fission, 900, *901*
  characteristics, general, 898–899
    taxonomic, 35t
  chemical composition, 898–899
  culturing of, 899
  disease and infection due to. See *Rickettsiosis.*
  history, 898
  immunologic identification of, 902
  metabolic activities, 898–899
  morphology, 898, *899*, *900*
  natural hosts, 898
  pathogenic agents compared to, 917t
  permeability of cytoplasmic membranes, 901
  predilection for small blood vessels, 899
  relation to bacteria and viruses, 901
  toxin produced by, 900
  transmission of, 898
    modes of, 901
    to laboratory animals, 903
  typhus. See also *Typhus.*
    hemolysins produced by, 900
    metabolism of, 899
    reversible inactivation of, 899

Rickettsia(e) *(continued)*
  use of host cofactors by, 901
  vectors of, 898
Rickettsia australis, 904t, 909
Rickettsia akari, *909.* See also *Rickettsialpox.*
Rickettsia conori, 904t, 909
Rickettsia prowazeki, *900.* See also *Typhus, epidemic.*
  antigens of, 903
  in Brill–Zinsser disease, 907
  relation to *R. Typhi*, 907
  transmission of, 199, 199t, 905, 905t
Rickettsia rickettsi, *901.* See also *Rocky Mountain spotted fever.*
Rickettsia sibiricus, 904t, 909
Rickettsia tsutsugamushi, *899.* See also *Tsutsugamushi fever.*
Rickettsia typhi, *907.* See also *Typhus, endemic.*
  antigens of, 903
  in smear, 903
  relation to *R. prowazeki*, 907
Rickettsiaceae, classification, 35t
Rickettsialpox, 909–910
  diagnosis, 904t, 910
  treatment, 910
Rickettsiosis(es)
  angiitis due to, 899
  chemotherapy, 900–901
  death due to, 898, 900, 902
  differences in, 903
  general characteristics, 902, 904t
  immunity to, 902
  laboratory diagnosis, 902–903
  modes of transmission, 905t
  pathogenesis, 899–900
  shock in, 900
  specific diseases due to, 903–913
  spotted fever group, 908–910
  typhus group, 903–908
  vesicular. See *Rickettsialpox.*
  Weil–Felix reaction in, 904t, 906t
Rifampicin therapy
  in chlamydial infection, 922
  in *Flavobacterium* infection, 786
  in *Mycobacterium marinum* infection, 862
Rifampin, 677
  antiviral action of, 1185
  DNA transcription blocked by, 266
  effects on maturation of vaccinia virus, *1186*
  effects on poxviruses, 1185, *1186*
  vaccination lesions inhibited by, 1186
Rifamycin (therapy)
  antiviral action of, 1185
  in chlamydial infections, 922
  initiation of DNA transcription

Rifamycin (therapy), initiation of *(continued)*
    blocked by, 266, *266*
  protein synthesis inhibited by, 303t
Rift Valley fever virus, 1173
Rinderpest, 1343
Ring spot virus, 1173
Ringworm. See also *Microsporum.*
  of scalp, 1002
Ristocetin, 121
RNA (ribonucleic acid)
  actinomycin D inhibition of, 149n
  bacterial, base analog incorporation into, 153
    breakdown of, 343
  bacteriophage. See under *Bacteriophage.*
  classes of, 202
  double-stranded, interferon production and. See under *Interferon.*
    structure, 204
  5-fluorouracil incorporation into, 296, *296*
  genes of, intergenic dividers of, 290, *291*
  heterogeneous, in animal cells, synthesis of, 1131
  in DNA replication, 225
  in DNA transcription, *264*, 265
  messenger (mRNA), 325–330
    bacterial, transducing bacteriophages and, 327
    bacteriophage and, 326
    breakdown of, 327–330, *329*, *330*
    chemical evidence of, 326–327, *327*
    codons of. See *Codon.*
    composition of, 327n
    definition, 202
    discovery of, 325–327
    hybridization studies with DNA, 328, *329*
    in animal cells. See under *Animal cell.*
    interferon, 1178
    in vitro studies of, 327
    labeling of, 326, *327*
    monocistronic, 234n
    mutations of, 5–fluorouracil suppression of, 296
    polycistronic, 234n
    reading of, direction of, 279, *280*
    synthesis of, 327–330, *328*, *329*. See also *DNA, transcription of.*
    RNA polymerase in, *328*
    translation of, *278*, 279, 327–330, *329*

RNA (ribonucleic acid), messenger
(*continued*)
transport of, in animal cell,
1131–1132
methylations of, 204
ribosomal (rRNA), 282–283
base sequence of, *216*
nucleotide composition of, 282t
precursor of, in animal cells,
1130, *1131*
protein binding to, 283
synthesis of, 338
psi factor in, 263
single-stranded, 204
synthesis of, 327
chain extension in, 263, *264*
chloramphenicol inhibition and,
338
DNA template in, 262
in animal cells, 1129, 1134
protein synthesis and, 337, *337*
ribosome synthesis and, 337
stable, regulation of, 336–339
total versus net, 338
template of, DNA synthesis on, in
reverse transcription, 267
transfer (tRNA), aminoacetyla-
tion of, 273
charging of, 272
anticodon of. See *Anticodon.*
definition, 202
degeneracy of, 275
evolutionary aspects, 275
fragment reaction by, 297
gene(s) of, duplication of, 295
in leukoviruses, 1010
in mutation suppression, 294
in protein synthesis, 272–275,
286–287
missense suppression by, 294,
294
of mammalian cells, 275
polyamines bound to, 79
ribosome specificity for, 280
structure, 272–275, *273*, *274*
synthesis, 339
temperature-sensitive, 294
transfer of genetic information
from DNA to, 262
viral, 208
classes of, *1151*
double-stranded, 1023–1024
propagation of, 1403
fragmentation of, *1025*
secondary structure, 1025
single-stranded, 1024–1025
RNA–DNA helix, 262
RNA polymerase
antibiotic blocking of, 266, *266*
DNA-dependent. See *Transcriptase.*
factors modifying, 263
in DNA transcription, 263
in mRNA synthesis, *328*
sporulation and, 142, 332

RNA replicase(s)
in animal cells infected with virus,
1152
in RNA bacteriophage,
1083–1084
RNases, in bacterial nucleic acid
breakdown, 343
Rochalimaea quintana. See also
Trench fever.
in louse, *912*, 913
media grown in, 898
Rocky Mountain spotted fever,
908–909. See also *Ric-
kettsia rickettsi.*
death rate in, 908
diagnosis, 198t, 908–909
epidemiology, 909
geographic distribution, 198t,
202, 904t, 908
pathogenesis, 908
treatment, 908
Rodent(s). See also specific rodent,
e.g., *Rat.*
as reservoirs for *Borrelia* organisms, 892
as reservoirs for leptospires, 894,
895
in transmission of *Rickettsia
tsutsugamushi*, 910
mycoplasma diseases in, 936t
Rosette formation
erythrocytic, with lymphocytes,
467
in identification of surface antigens, *454*
Rothia, 958t, 961
Roundworms, intestinal, 536
Rous-associated virus, 1437
Rous sarcoma virus
as model for viral oncogenesis,
1436
chromosomal aberrations due to,
1208
defectiveness of, 1437–1440
foci of, *1437*
host range of, 1438
leukosis virus as helper virus for,
1438–1440
mutants of, cell transformation
due to, 1440, *1440*
naturally occurring cancers due
to, 1446
pseudotypes of, 1438, *1438*
Sendai virus as helper virus for,
1439
susceptibility to, 1209
temperature sensitivity of, 1464
RPCF test, 886
rRNA. See *RNA, ribosomal.*
Rubella, 1356
adamantanamine therapy in, 1183
control of, 1358
epidemiology, 1357–1358
fetal damage in, 1216, 1356,

Rubella, fetal damage in
(*continued*)
1357
prevention of, 1358
IgA deficiency and, 508
inclusion bodies in, *1354*, 1356
laboratory diagnosis, 1357
pathogenesis, 1356–1357
prevention, 1358
symptoms, 1356, 1358
vaccines for, 1358
Rubella virus, 1331, 1353–1358
arthritis due to, 1358
chromosomal aberration due to,
1208
cytopathic changes due to, *1354*,
1356
immunologic properties,
1354–1355
morphogenesis, 1355, *1355*
multiplication of, 1355–1356
persistence in newborn,
1356–1357, *1357*
Rubeola, 1342n. See also *Measles*
and *Rubella.*
Runting syndrome, 614, 1420
Russian spring-summer encephalitis
epidemiology, 1391
pathogenesis, 1388
Russian spring-summer encephalitis
virus, in ticks, 1385
Ryter-Kellenberg method, in electron microscopy of bac-
teria, 23

S

Sabouraud's agar
to cultivate fungi, 877, 982–983
to cultivate nocardiae, *877*
Saccharomyces cerevisiae
birth and bud scars in, *974*
budding in, *974*
in beer production, 95
rRNA in, 282t
Saline solutions, bacterial survival
in, 1458
Saliva, bacteria in, 957
Salivary gland virus. See *Cytomegalovirus.*
Salmonella(e), 772–776
acquired cellular immunity to,
657, 658
antibody binding of, 368, *383*
antigenic complexities of, 761
carriers of, 774
cell wall lipopolysaccharides in,
84, 118, *118*
chloramphenicol-resistant, 775
classification of, 761–762, 762t
conversion into L forms by
penicillin, 175

Salmonella(e) *(continued)*
dideoxyhexoses of, *763*
DNA base composition in, 37t
drug resistance by, 1107
endotoxins of, 765
episome transfer in, 1107
food poisoning due to, 774
genetic studies of, phage P1 in, 1112
host preferences of, 774
identification of, 772n
using antibody—antigen cross-reactions, 381, *383*
lactose fermentation by, 767
lipid A of, 117, *117*, 765
lipopolysaccharides of, 762
mutations of, virulence of, 153n
O antigens in, 761, 762—764
structures of, 764t
polysaccharide synthesis in, *84, 235*
outer layer of, 117, *117*
pasteurization and, 1454
pathogenic, growth of, 104
pathogenicity, 773—774
phenol activity against, 1461
R factors in, 776, 1108
shigellae versus, 776
taxonomic characteristics, 34t
transduction in, 1112
transfer of genes from *E. coli* to, 194
Vi antigen in, 762
Salmonella arizona, 776
Salmonella choleraesuis, 774
Salmonella enteritidis
serotypes of, 772
susceptibility to, 954
Salmonella minneapolis, 383
Salmonella newington, lipopolysaccharides of, *119, 763*
Salmonella paratyphi B, 774
Salmonella schottmulleri, 774
Salmonella typhi. See also *Typhoid fever.*
bacteriophage typing of, 765
identification of, 772n
in biliary tract, 773
in Peyer's patches, 773
surface antigens of, *760*
synonyms, 35
Vi antigen of, 762
Salmonella typhimurium
aspartokinase activity in, 317t
colonies of, phases of, equilibrium between, *172*
enteric fever due to, 774
gastroenteritis due to, 774
macrophage killing of, *567*
mutation rates in, 262t
protein synthesis in, 336t
susceptibility of mice to, 653
transduction in, 1115

Salmonella typhosa
in determining phenol coefficient, 1458
infections of, chloramphenicol therapy in, 671
lysing of, by anti-O antibodies, 652
phenol sterilization of, 1458
Salmonellosis
Carrion's disease complicated by, 951
immunity to, 774—775
transfer of, 773
laboratory diagnosis, 775
predisposing factors, 774
prevention, 775—776
treatment, 775
types of, 775
vaccines in, 776
Salpingitis, gonorrheal, 751
Salts, in food preservation, 1458
Sandfly, bartonellosis transmitted by, 950
Sanitation, environmental, in control of infectious diseases, 10
Sanitization
definition, 1452
hypochlorite solutions in, 1459
Sarcina(e), 24, *24*
Sarcina, taxonomic characteristics, 34t
Sarcina flava, 37t
Sarcina lutea, 37t
Sarcoid, Boeck's, 581
Sarcoma(s)
adenoviral, 1234
due to polyoma virus, 1420
due to SV 40, 1420
protected by antibodies to tumor-specific antigens, 621
reticulum cell. See *Reticulum cell sarcoma.*
Sarcoma information, 1437
Sarcoma virus(es), 1436—1441. See also specific virus, e.g., *Rous sarcoma virus.*
adaptation to new species, 1440
cell surface antigens due to, 1445
cell transformation due to, 1440, *1440*
defectiveness of, 1437—1440
foci of, in cultured cells, 1436, *1437*
leukosis viruses versus, 1436, 1437, *1438*
mode of action, 1440—1441, *1440*
nature of, 1436—1437
oncogenic efficiency of, 1436
origin of, 1436—1437
transmission, 1436, *1436*
Satellite cells, 994
Satellite phenomenon in *Hemo-*

Satellite phenomenon *(continued)*
philus influenzae, 130, *130,* 794, *794*
Sauerkraut, bacterial fermentations in, 44
Scalp, fungus infections, 1002
Scarlet fever, 721
Dick test in, 717
erythrogenic toxin in, 717
immunity to, 722
prevention, 723
rash in, 717
Schultz—Charlton test in, 717
Schick test
as secondary response, 486n
development of, 683
in diphtheria, 688—689
reactions to, 689, 689t
Schizomycetes, 14
Schultz—Charlton test, 717
Schultz—Dale reaction
definition, *544*
in isolated organs, 543—544, *544*
Sclerosis, multiple. See *Multiple sclerosis.*
Scrofula, due to *Mycobacterium bovis,* 861
Secondary response. See *Immune response, secondary.*
Second-set reaction, 610—611
transfer of, 611
Secretory piece, 418
Sectioning, thin, in virology, 1016
Sedimentation
velocity, to separate lymphoid cells, 454
zonal, in purification of virus samples, 1014
Sedormid, allergy to, 585
Selective inhibition, in bacteria cultures, 104
Self-antigens
altered distribution of, autoimmune responses due to, 585
altered forms of, autoimmune responses, due to, 585—586
tolerance to, 494, 496
Self-proteins, in contact skin sensitivity, 565
Self-regulatory mechanisms, evolution and, 347
Self-tolerance, 579
Seminal fluid, spermine in, 643, 643n
Semliki Forest virus, 1382t
Semple vaccine, 1374
Sendai virus, *1439*
Sensitivity. See also *Hypersensitivity.*
contact skin, 562—564, *565*
elicitation of, 563
flare reaction in, 563

Sensitivity, contact skin (*continued*)
  histologic aspects, *561, 563, 564*
  immunogen formation in, 563, 563n, *565*
  induction of, 563
  intraepidermal vesicles in, 563, *564*
  patch test in, 563
  persistence of, 563–564
  reaction in, time course of, 563
  reactions with proteins, 563, 563, *565*
  self-proteins in, 565
  transfer of, in cell-mediated immunity, 570
Sensitizer(s), 530
Sepsis
  due to *Edwardsiella tarda,* 771
  uterine, due to *Bacteroides,* 787
Septicemia
  Acinetobacter, 785
  due to *Actinomyces muris,* 875
  due to *Aeromonas,* 782
  due to *Alcaligenes faecalis,* 786
  due to *Flavobacterium meningosepticum,* 786
  due to penicillin therapy, 674
  due to *Yersinia* infection, 806
  hemorrhagic, 809
  Moraxella, 785
  perinatal listeric, 947
  Pseudomonas, 783, 784
  Salmonella, 773, 774, 775
Serine, bacterial synthesis of, 73–74, *73, 78*
Serologic archaeology, 489
Serologic reactions, 353
Serologic test(s)
  diagnostic applications, 399–400
  for syphilis (STS), 886, 889–890
Serologic typing, of *Pseudomonas aeruginosa,* 783
Serotonin
  anaphylaxis and, 541–542, *540*
  methysergide blocking of, *543*
  physiologic effects, 541
  susceptibility to, species variation in, 544t
  tissue levels of, species variation in, 544t
Serratia, 769–771
  differentiating biochemical tests in, 769
  Enterobacter and *Klebsiella* versus, 769
  infections due to, pathogenesis of, 769
  treatment, 771
  taxonomic characteristics, 34t
Serratia marcescens, 771
  DNA base composition in, 37t
Serum(a). See also *Antiserum.*

Serum(a) (*continued*)
  absorbed, 384
  anti-D, to establish Rh type, 607
  antigen-binding capacity of, measurement of, 396
  human immunoelectrophoresis of, *390, 391*
  in culture media, 94
  paired, tests on, in antigenic analysis of viruses, 1201
  specificity of, 370
  titration of, by agglutination reaction, 391–392
  typing, preparation of, 604n
Serum normal agglutinator (SNagg), 419
Serum sickness, 549–551, *550*
  antigens in, multiplicity of, 550–551
  Arthur reaction in, 546
  cardiovascular lesions in, 549, *549, 551*
  complement in, 550, *550*
  cytotropic antibodies in, 550
  due to antilymphocyte serum, 580
  history, 549
  in man, *548*
  mechanisms, 549–550
  pathogenesis, *551*
  physiologic effects of, 549
  relation to other allergic reactions, 550
  renal lesions in, 549, *549*
  vasoactive amines in, 550
Sewage deodorization, 1452
Sex factors
  in autoimmune diseases, 588
  in conjugation, 1105
  in tuberculosis, 855
Sexual reproduction
  essential features, 182n
  in fungi, 176, 182
Shadowcasting
  in bacteriology, 23
  in virology, 1015, *1015*
SH antigen, 1410, *1411*
Shark(s)
  immune response in, 502
  immunoglobulins in, 441, 442
Sheep
  abortions due to aspergillosis, 997
  anthrax in, 8
  contagious pustular dermatitis of, 1276. See also *Orf virus.*
  Coxiella burneti infection of, 913
  Johne's disease in, 867
  mycoplasma disease in, 936t
Shellfish, hepatitis from, 1414
Shiga neurotoxin, 777
Shigella(e), 776–779. See also *Dysentery, bacillary.*
  acid-producing fermentation in, 45

Shigella(e) (*continued*)
  bacterial antagonism against, in gastrointestinal tract, 778
  classification, 777t
  core of, 118
  DNA base composition in, 37t
  drug resistance by, 162, 679, 778, 1107
  episome transfer in, 1107
  genetic studies of, phage P1 in, 1112
  glucosylation of viral DNA in, 1065, *1066,* 1067
  pathogenic, growth of, 104
  pathogenicity, 777–778
  R factors in, 766
  salmonellae versus, 776
  subgroup identification in, 778n
  taxonomic characteristics, 34t
  toxins of, 777
  transduction in, 1112
  transfer of genes from *E. Coli* to, 194
Shigella alkalescens. See *Escherichia coli.*
Shigella boydii, 777
Shigella dispar. See *Escherichia coli.*
Shigella dysenteriae, 776
  dysentery due to, 777
  identification of, 778
  toxins produced by, 637t, 637, 639, 777
Shigella flexneri, 776, 777
Shigella shigae. See *Shigella dysenteriae.*
Shigella sonnei, 777
Shigellosis. See also *Dysentery, bacillary.*
  chemoprophylaxis in, 680
  experimental, 777
  immunity to, 778
  laboratory diagnosis, 778
  prevention, 778–779
  treatment, 778
Shikimate, 75n
Shikimic acid, 75
Shingles. See *Herpes zoster.*
Shipping fever, 809, 1337
Shock
  anaphylactic, due to penicillin, 581
  anaphylactoid, 545n
  cold, in bacteria, 93
  due to *E. coli,* 768
  due to endotoxins, 639, 640
  due to plague toxin, 804
  due to rickettsiae, 899, 900, 902
  hemorrhagic, endotoxemia with, 955n
  in epidemic typhus, 905
  tuberculin, 561
Shope rabbit papilloma virus, 1218
Shwartzman phenomenon
  due to bacterial endotoxins, 639

Shwartzman phenomenon
(*continued*)
similarity to Arthus reaction,
548n
Siberian tick fever, 909
Side-chain theory, Ehrlich's, 148
Sideramine(s), bacterial, 92
Siderochrome(s)
bacterial, 92
bacterial chelation by, 134
Sideromycins, 92
Sigma factor, 263, 264
Silica gel, in culture media, 101
Silica particles, effects on lysosomes, 856
Silicosis, tuberculosis and, 856
Silkworm industry, 8
Silver (ion), antimicrobial action of,
1459
Silver nitrate therapy, 1459
Silverwater virus, 1396
Simbu virus, 1392
Simian virus 5, 1338
endosymbiotic infections due to,
1219
Simian virus 40
adenovirus hybridization with,
1428
animal effects of, 1420
cell transformation due to, 1208,
1422–1428, *1424*, 1445
cell genes in, 1429
cell surface changes in,
1423–1424
integration of viral DNA into
cell DNA in, 1426, *1427*,
1428
mechanism of, 1424–1428
stability of, 1428
viral genes in, 1424–1426, *1425*
chromosomal aberrations due to,
1208
culturing of, permissive and nonpermissive cells used
with, 1420, 1421t
discovery of, 1420
infections due to, latent, 1217
pathogenesis, 1420
multiplication of, in tissue cultures, 1420–1422
provirus formation in, 1426, 1436
T antigen production and, 1422,
1422
virion of, 1420
virus-dependent abortive infection
by, 1165
Sindbis virus
autointerference by, 1182, 1183
hemagglutination-inhibition titrations in, 1382t
morphology, 1382
multiplication in, 1386
protein and lipid proportions in,
1382

Sinus(es), cellulitis of, due to
Aspergillus, 998
Sinusitis
due to *Hemophilus influenzae,*
795
due to *Pasteurella multocida,* 809
pneumococcal, 702, 704
streptococcal, 723
Siomycin, protein synthesis inhibited by, 300, 303t
Siti, 892
Sjogren's disease, 587t
Skin
activated lymphocyte effects on,
577
antigen application to, 481
bacteria on, distribution of, 955
cellulitis of, 720
contact sensitivity of. See under
Sensitivity.
desquamation of stratified epithelium from, antibacterial
action of, 643
eruptions of, due to beta-
hemolytic streptococci,
992
due to *Mycobacterium tuberculosis,* 992
in Carrion's disease, 951
in coccidioidomycosis, 992
fungal infections of. See *Dermatomycosis.*
graft of, second-set reaction to,
611
leprous, 866
necrosis of, in Rocky Mountain
spotted fever, 908
passive sensitization of, 530
pH of, antibacterial effects of,
643
rickettsial effects on small blood
vessels, 899
staphylococcal diseases of, 733
symbiotic bacteria on, 954
trichophyton effects on, 1003
tuberculosis of, 861
unsaturated fatty acids of, antibacterial action of, 643
wounds of, staphylococcal infections in, 733
Skin reaction
delayed-type Arthus reaction versus, 561
in pneumococcal and streptococcal infections, 657
used as diagnostic tests, 558,
559t
to tuberculin, 560–561
in various species, 561
Skin reactive factor, 577
Skin test, *Brucella,* 816
Slow reactive substance–
anaphylaxis
anaphylaxis and, 543, 544

Slow reactive substance–
anaphylaxis (*continued*)
assays for, *543*
in allergy, 543
physiologic effects, 543, *543*
release of, blocked by drugs, 545
cAMP inhibition of, 544, *545*
ragweed pollen and, 543
Smallpox. See *Variola.*
Smegma, mycobacteria in, 868
Smooth muscle contraction, in
Schultz–Dale reaction,
544
SNagg, 419
Snail, fungus digested by, 969
Snakes, mycobacteria in, 868
Sneezing, antibacterial action of,
643
Soap(s)
bactericidal action of, 1460
tricresol with, disinfection by,
1461
used with phenols, 1461
Sodium ion
in halophilic bacteria, 94
potassium ion antagonized by, in
bacteria, 92
Sodoku. See *Rat-bite fever.*
Soil
breakdown of organic matter in,
enzymes in, 114
contaminated, mycosis due to,
980, 981
Solvent(s)
lipid, virus inactivation by, 1464
organic, bactericidal action of,
1462
Somatic mutation theory, of diversity of immunoglobulin
V genes, 439–440
Sonication
definition, 23, 1457
of bacteria, for study, 23
Sonic waves, sterilization by, 1457
Sorbose, produced by *Acetobacter,*
50
Spaerophorus necrophorus. See
*Fusobacterium necro-
phorus.*
Sparsomycin, protein synthesis
inhibited by, 299, 303t
Species
bacterial. See under *Bacterium.*
definition, 33
Spectinomycin, 159
bacterial resistance to, 162, 163t
protein synthesis inhibited by,
297, 302
Spermidine
antimicrobial activity of, 654
bacterial synthesis of, 70, 72, 79
tuberculosis susceptibility and,
636

Spermine
  antimicrobial activity of, 654, 643, *643*
  bacterial synthesis of, *70, 79*
  in seminal fluid, 643, 643n
Spheroplast, bacterial. See *Bacterium, spheroplast of.*
Spirillaceae, classification, 34t
Spirillospora, 873t
Spirillum(i), *24*
Spirillum
  cell wall of, granules on, 27
  taxonomic characteristics, 34t
Spirillum linum, 37t
Spirillum minus, 896
Spirocheta, taxonomic characteristics, 35t
Spirochetales, classification, 35t
Spirochete(s), 881—896
  cell walls of, 107n
  general characteristics, 882
  in oral cavity, 957, 958, 958t, 959
  motility in, 29
Spleen, *477*
  antigens trapped in, 475, 476
  B and T zones in, 476, *477*
  clearance mechanisms in, 646
  dendritic cells in, 476
  Friend virus loci in, 1434, *1435*
  granulomas of, due to *Brucella suis,* 815
  interferon produced by, 1175, *1176*
  lymphocytes in, 475, 476, *477*
Spondylitis, in brucellosis, 815
Spontaneous generation, 3—5
Sporangiophore(s), 972
Sporangiospore(s), 972t
Sporangium
  bacterial, 137, *138*
  fungal, 972t
Spore(s). See also under organism, e.g., *Histoplasma capsulatum, spores of.*
  bacterial. See *Bacterium, spores of.*
  fungal. See *Fungus, spores of.*
  of pathogens, heat sterilization of, 1454
  saprophytic, autoclave sterilization of, 1454
  spontaneous generation and, 4—5
Sporosarcina ureae, 37, 37t
Sporotrichosis, 999—1000
Sporotrichum schenckii, 986t, 999—1000, *1000*
Spots, Koplick, 1346
Spreading factor. See *Hyaluronidase.*
SRS-A. See *Slow reactive substance—anaphylaxis.*
SS agar, 759

Staining
  in bacteriology. See under *Bacterium.*
  in virology. See under *Virology.*
  introduction of, 8
Staphylococcus(i), 24, *24,* 727—739
  abscesses due to, 733
  adrenal insufficiency due to, 746
  boils due to, 734
  capsular antigens of, 731
  cell division in, 128
  cellular antigens of, 731—732
  cell wall of, autolysis in, 128
    synthesis of, 111, *112*
  clumping factor of, 732
  coagulase of, 636, 729, 730, 732, 734
    detection of, 737
  colonies of, 728, *729*
    Hemophilus influenzae satellite phenomenon with, *794*
  culturing of, 737
  delayed hypersensitivity due to, 735—736, *735*
  diseases and infections due to, 728
    carrier rates in, 738
    cephalosporin therapy in, 158
    chemoprophylaxis, 680
    constancy of, 733
    dermal, 642
    due to contamination, 163
    factors predisposing to, 733—734
    hospital epidemics of, 738
    hyperreactivity to antigens in, 657
    immunity to, 736
    immunization against, 735
    laboratory diagnosis, 737
    nature of lesions, 733
    postoperative, 733, 738
    prevention, 737—738
    sites of, 733
    spread of, 738
    suppuration in, 733
    susceptibility to, 733
    treatment, 737
  DNAse produced by, 731
  drug resistance in, 679, 737, 954
  encapsulated, 729, *729*
  enteritis due to, 736, 737
  enterotoxin-producing, 730, 736
    detection of, 737
  extracellular enzymes of, 730—731, 734
  food poisoning due to, 736, 738
  genetic variation in, 732—733
  hemolysins of, 730, 731t
  history, 728
  hyaluronidase of, 731, 734
  in nasopharynx, 748
  in oral cavity, 959

Staphylococcus(i) *(continued)*
  invasiveness of, 636
  lactose transport in, 136
  leukocidins of, 730, 735
  lipase of, 731, 734
  local infectivity of, 733
  metabolism, 730
  morphology, 728—729, *729*
  natural resistance to, 736
  nasal, 956
  oxidase-negative, 51
  pathogenicity, 733—736
  penicillin action in, 111, *112*
  penicillinase-producing, 732, 737
  penicillin-resistant, 156, 162
    novobiocin action on, 158
    sensitive neighbor protected by, in mixed infections, 163
  phenol activity against, *1461*
  resistance to disinfectants, 730
  resistance to sterilization, 1458
  species-specific polysaccharides of, 731
  staphylokinase produced by, 636, 731
  taxonomic characteristics, 34t
  toxins of, 728, 730, 735
  transduction in, 1112
  transformation in, 184
  transmission of, 663
  typing of, 732
  virulence of, 734, 737
Staphylococcus albus, 728
  DNA base composition in, 37t
  immunologic specificities of, 731
  in conjunctivae, 957
  typing of, 732
Staphylococcus aureus, 728, *729*
  antibodies to, 7?3
  antiphagocytic properties, 734
  capsular antigens of, 731
  cell wall of, peptidoglycan structure of, 110, *110, 111, 113, 114*
    synthesis of, *112, 113, 114*
    cycloserine inhibition of, 120
    penicillin action on, 111, *112*
  DNA base composition in, 37t
  drug resistance by, 732
  exotoxin of, 637t, 739
  immunologic specificities of, 731
  in determining phenol coefficient, 1458
  infection due to, chronic granulomatous disease and, 661
  intracellular survival of, 734—735
  L forms of, 732
  minute G form of, 732
  nicotinic acid of, 91t
  penicillin effects on, 111, *112, 149*
    plate assay of, *150*
  penicillin resistance by, 1109

Staphylococcus aureus (continued)
pneumonia due to, secondary to influenza, 1323
protein A of, 731
protoplasts of, 732
serologic typing of, 732, 732t
suppurative lesions due to, 674
synergistic action with *Neisseria gonorrhoeae,* 955
thiamine of, 91t
virulence, 736
Staphylococcus epidermidis. See *Staphylococcus albus.*
Staphylococcus salivarius, 958t
Staphylococcus staphylolyticus, 737
Staphylokinase, 731
staphylococcal invasiveness and, 636
Starch
fermentation of, 95
formation in plants, 83
microbial hydrolysis of, 95
Starvation. See also *Malnutrition.*
bacterial survival in, 126
tuberculosis and, 855
Sterilization, 1451–1465. See also *Disinfection.*
acids and alkalis in, 1458
autoclave, 1454
by chemical agents, 1457–1462
by freezing, 1455–1456
by heat, 1454–1455, *1455*
by phenol, 1453
by physical agents, 1454–1457
by radiation, 1456–1457, *1456*
kinetics of, 1453, *1453*
by sonic and ultrasonic waves, 1457
by UV radiation, 1456–1457
definition, 1452, 1453
history, 1452
in vaccine preparation, 1453
mechanical agents in, 1457
mycobacterial resistance to, 1458
of food, by radiation, 1457
organic acids in, 1458
peracetic acid in, 1459
photodynamic sensitization in, 1457
Steroid(s), cortical. See *Corticosteroid.*
Sterols, in mycoplasmas, 121
Stillbirth
due to smallpox vaccination, 1274
in listeriosis, 947
Stings, insect, anaphylaxis due to, 538
St. Louis encephalitis, 1388, 1390
inapparent infection in, 1215
Stomach acid, antibacterial effects, 643
Strain, attenuated, 846

Stratum corneum, dermatophyte in, *1002*
Streams, pollution of, 49
Street virus, 1368
Streptamine, *300*
Streptidine, *300*
Streptobacillus moniliformis, 875, 948, *949*
L forms from, 943
organotropicity of, 635
rat-bite fever due to, 896
Streptococcal units, 712
Streptococcus(i), 24, *24,* 707–726
adrenal insufficiency due to, 746
alpha-hemolytic, 708, *709,* 724–725
disease due to, diagnosis, 631
endocarditis due to, 725
pneumococci versus, 696
anaerobic, 725–726
beta-hemolytic, 708, *709,* 709–724
adhesion to body surfaces, 957
antibodies of, anti-M, 712, 722, 723
tests for, 723
antigens of, 712–720, 722
group-specific C, 714
M, 715–716, 721, 721t, 722
P, 716
R, 716
T, 716
antiphagocytic properties of, 720, 721, 721t
capsules of, 712, *712*
hyaluronic acid, 712–714, *712,* 720, 721t
cell division of, 709–712
chain formation in, 709, *710,* 711
colonies of, 712, 720
morphology, 712, *713*
culturing of, 712
deoxyribonucleases produced by, 718
DPNase produced by, 718
drug-resistance by, 720
erythrogenic toxin of, 717, 721
extracellular products of, 717–720, 717t
fimbriae of, 715, *715*
genotypic variations in, 720
group A, 716
antigens of, 716, *716, 717*
glycerol teichoic acids of, 716
identification of, 722–723
immunologic types of, 715
leukotoxic action of, 718
mucopeptide of, 716
grouping of, 710t, 714
compositions and antigenic determinants of, 714, *714*
hyaluronidase elaboration by,

Streptococcus(i) *(continued)*
713, *713,* 718
immunity to, 722
L forms of, 712, 722
meningitis in newborn due to, 724
metabolism, 712
morphology, 709–713
M protein of, 721, 721t, 722
nephritogenic, 721
pathogenicity, 720–722
pneumonia due to, 1323
protease produced by, 722
proteinase produced by, 720
protoplasts of, 712
skin eruptions due to, 992
streptokinases produced by, 718
streptolysins produced by, 721, 722
suppurative disease due to, 720–721
typing of, 715
cariogenic potentiality of, 960, *960*
cardiac lesions due to, 594
carriers of, 724
cell division in, 129, *129*
cell growth in, *129*
classification, 708–709
according to hemolysis, 709
colonies of, 708, *709*
disease and infection due to. See also specific disease, e.g., *Scarlet fever.*
allergy in, 594, 657
chemoprophylaxis in, 680
delayed skin reaction in, 657
due to dental extractions, 955
due to tonsillectomy, 955
epidemiology, 724
erythema due to, 593
laboratory diagnosis, 722–723
prevention, 723–724
M-protein vaccine in, 724
treatment, 723
urticaria due to, 593
variety of, 708
drug resistance by, 679, 723
endocarditis due to, 955
gamma, 708
glomerulonephritis due to, 552
hemolytic classes of, 708–709, *709*
history, 708
homolactic fermentation by, 44
in conjunctivae, 957
in nasopharynx, 748, 956
in oral cavity, 958, 959
intestinal, 956n
invasiveness of, 636
lactic fermentation in, *44*
microaerophilic, 726
minute, 712
nonhemolytic, 725

Streptococcus(i) *(continued)*
  pasteurization of, 1454
  peptide stimulation of, 94n
  periodontal disease due to, 961
  phagocytosis of, *656*
  pneumococcal relationship to, 700
  resistance to sterilization, 1458
  root caries due to, 961
  spread of, in mammalian tissues, 720
  taxonomic characteristics, 34t
  transformation in, 184
  toxin produced in scarlet fever, 721
  transmission of, 663, 664
  viridans group of. See *Streptococcus alpha-hemolytic*
Streptococcus cremoris, 44
Streptococcus faecalis, 708, 709, 725
  DNA base composition in, 37t
  folic acid produced by, 91t
Streptococcus MG, 724n, 938
Streptococcus pneumoniae. See *Diplococcus pneumoniae*.
Streptococcus pyogenes
  carriers of, 661
  colonies of, *711*
  confusion with *Hemophilus hemolyticus*, 797
  disease due to, 708, *709*, 720–722
  DNA base composition in, 37t
  exotoxin of, 637t, 638
  genetic variations of, 642
  identification of, 723
  invasiveness of, 642
  rheumatic fever due to, 722
  virulence, 642
Streptococcus salivarius, 724
  infection by, due to dental extraction, 955
  in oral cavity, 957, 958, 958t
Streptococcus sanguis
  endocarditis due to, 961
  in oral cavity, 957, 958, 958t
Streptococcus mitior, 958t
Streptococcus mitis, 955
Streptococcus mutans
  dental caries due to, *960, 961*
  endocarditis due to, 961
  in oral cavity, 957, 958, 958t
Streptodornase, 718
Streptokinase(s), 95
  effects of proteinase on, 720
  production of, 718
  streptococcal invasiveness and, 636
Streptolydigin
  DNA transcription blocked by, *266, 267*
  protein synthesis inhibited by, 303t

Streptolysin(s), 717–718
  effects of proteinase on, 720
Streptolysin O, 708, 718
  cardiac standstill due to, 718
  mechanism of action, *719*
  rheumatic fever and, 722
Streptolysin S, 708, 717, 721
Streptomyces, 873t, 879
  abscesses due to, 980
  antibiotics produced by, 149n, 160, 983, *983*
  cell wall constituents, 875t
  DNA base composition in, 37t
  enzymes from, keratin digested by, 1001
  morphology, 872
  taxonomic characteristics, 35t
Streptomyces aureofaciens, 160
Streptomyces coelicolor, 195
Streptomyces erythreus, 158
Streptomyces griseus, 149
Streptomyces madurae, 878t
Streptomyces natalensis, 983
Streptomyces niveus, 158
Streptomyces noursei, 983
Streptomyces pelletierii, 878t
Streptomyces venezuelae, 160
Streptomycetaceae, 873t
Streptomycin (therapy), 676–677
  bacterial growth inhibited by, 150, *150*
  bacterial resistance to, 163t, 300–301, *301*, 677, 678
  genes for, 185
  mutations to, *174, 175*
  transformation and, 186t
  bacterial viability affected by, 150, *150*
  chloramphenicol antagonism with, 164, *164*
  dependence on, 300–301, *301*
  discovery, 149
  in actinomycosis, 876
  in brucellosis, 816
  in *E. coli* infection, 768
  in *Flavobacterium* infection, 786
  in granuloma inguinale, 949
  in *Klebsiella pneumoniae* infection, 770
  in local infection, 678
  in *Mycobacterium marinum* infection, 862
  in plague, 805
  in rat-bite fever, 896, 948
  in shigellosis, 778
  in tuberculosis, *859*
  in tularemia, 808
  isolation of, 879
  lethal action of, 301–302, *302*
  penicillin synergism with, 165
  prevention of algal chloroplast formation by, 174
  protein synthesis inhibited by, 297, 298, 300–302, 303t

Streptomycin (therapy) *(continued)*
  ribosomal effects of, 296, 297, 301, *301*
  sensitivity to, 302
  Shigella resistance to, 1107
  to prevent tissue culture contamination, 942
  toxicity of, 677
Streptomycin adenylate synthetase, 163t
Streptomycin phosphotransferase, 163t
Streptosporangium, 875t
Streptovaricin(s)
  antiviral action of, 1185
  DNA transcription blocked by, 266, *266*
Stress, tuberculosis and, 855, 856, 859
Subacute sclerosing panencephalitis, 1219, 1346–1347
  clinical characteristics, 1346n
  due to measles virus, 1346
  pathology, 1347
Subclone, 101
Subtilisin, 95
Succinate, in bacterial metabolism. See under *Bacterium*.
Sucrose
  cariogenicity of, 85, 961, *961*
  in bacterial fermentation, 48
Sugar(s). See also *Sucrose*.
  as Ig precursors, incorporation of, 473
  bacterial. See under *Bacterium*.
  in O antigens, 118
Sulfacetamide, 154n
Sulfadiazine, 154, *154*
Sulfamerazine, 154, *154*
Sulfanilamide
  antibacterial activity of, 154
  development of, 148
  structure, *154*
Sulfanilic acid, 154
Sulfasuxidine, *154*
Sulfate ion, bacterial requirement of, 92
Sulfathiazole, succinyl, *154*
Sulfide, bacterial formation of, 69
Sulfisoxazole, *154*
Sulfonamide(s) (therapy), 154–155, 675. See also *Chemotherapy; Drug;* and specific drug, e.g., *Sulfanilamide*.
  activity of, pH and, 154, 154n
  bacterial enzymes inhibited by, 154
  bacterial growth and viability inhibited by, 150, *150*
  competetive inhibition by, 151, *152*
  development of, 148–149
  disadvantages, 675

Sulfonamide(s) (therapy)
(*continued*)
in bacterial transport system, 133
in chancroid and soft chancre, 797
in chlamydial infection, 922
in *E. coli* infection, 768
ineffectiveness at sites of tissue destruction, 153
in *Hemophilus influenzae* infection, 796
inhibition of bacterial THF synthesis by, 74
in meningococcal meningitis, 675, 749
in *Neisseria meningitidis* infection, 748
in nocardiosis, 879
in pneumococcal pneumonia, 670, *670*
in rheumatic fever, 675, 723
in shigellosis, 778
in streptococcal infection, 723
in trachoma, 675, 924
in urinary tract infection, 675
resistance to, 678, 679
by *Neisseria*, 746, 749, 750
by *Shigella*, 1107
by streptococci, 720
structure, *154*
Sulfur granules, in actinomycosis, 875, *876*
Sulfur reduction, in bacterial amino acid synthesis, 69
Summer grippe, 1290. See also *Poliomyelitis.*
Superinfection(s)
drug resistance and, 673–674
due to tetracycline therapy, 674
Superoxide, 49
Suppression, of mutation. See under *Mutation.*
Suppurative disease
chemotherapy in, 670, 671
streptococcal, 720–721
Surfactant(s), bactericidal action of, 1460–1461, *1460*
Surgery
antiseptic, 7
aseptic, 8
staphylococcal infections after, 733, 738
Survival curve
multicomponent, *268*
definition, 269
multiple-hit, *268*
definition, 268
resulting from irradiated microorganisms, 268, *268*
single-hit, *268*
SV 40 virus. See *Simian virus 40.*
Swarmer, 772
Swine
abortion in, virus-induced, 1216

Swine (*continued*)
Leptospira pomona infection in, 894
teratogenic viruses in, 1216
transmissible gastroenteritis virus of, 1362n
Swineherd's disease, 895
Swine influenza, *1218*
Swine influenza virus, 1315, 1315n, 1325
Hemophilus influenzae suis action with, 1218
latent infection due to, 1218
Swivel enzyme, in viral DNA replication, *229*, 230
Synapsis, 241
Syngeneic, definition, 567n, 610, 610n
Synovitis, in chickens, 934
Synthetics, bacterial breakdown of, 56n
Syntrophism
bacterial, 61, *61*
studies of, 102
definition, 61
Syphilis, 882–891. See also *Treponema pallidum.*
antigens in, 885–887
aortic aneurysms due to, 888
blindness due to, 888
cardiolipin in, 886
central nervous system, 890
chancre formation in, 887
congenital, 888
antibody production in, 505
epidemics, 7
experimental, 888, *889*
general paresis in, 888
Herxheimer reaction in, 890
history, 742, 882–883
immunity to, 888–889
laboratory diagnosis, 885, 886, 889–890
darkfield microscopy in, 883
serologic tests in, 886, 887, 889–890
effects of leprosy on, 867
lesions of, temperature effects on, 888, *889*
Treponema pallidum in, 889
pathogenesis, 887–888
placental transmission, 888
prevention, 890–891
rabbit, 882, 892
Reiter protein in, 886
satellite buboes in, 887
stages, 887
tabes dorsalis in, 888
transfusion, 885
treatment, 890
heavy metal compounds in, 1459
vaccines in, 887
Syphiloma(s), 888

Systemic lupus erythematosus
Arthus lesions in, 592
autoimmune aspects, 587t, 588, 591–593, *592*
cryoprecipitates in, 554
due to autoimmune antibody–antigen complexes, 586
glomerulonephritis with, 551, 552, 592

T

Tacaribe virus, 1394, *1395*
Tamiami virus, 1394
T antigen, produced by polyoma virus and SV 40, 1422, *1422*
Tabes dorsalis, 888
Tache noir, 909, 910
Tarbadillo. See *Typhus, endemic.*
Target cell, killed by sensitized lymphocytes, 577, *577*, 578
Tautomerization
keto-enol, in spontaneous mutation, 260
of 5-bromouracil, mutations due to, 255, *255*
Taxonomic differentiation, divergence of DNA sequences and, 218
Taxonomic groups, evolutionary relations of, 14, *14*
Taxonomy
bacterial. See under *Bacterium.*
major divisions of, 13–14
microbial, 13–16
phylogenetic aspects, 33
purposes of, 33
T cells. See *Lymphocyte, T.*
Tears, antibacterial action of, 643
Teichoic acid, bacterial. See under *Bacterium.*
Teichuronic acid, produced by *Bacillus subtilis*, 115
Teleology, 13
Teleonomy, 13
Tellurite, in growth of diphtheria bacilli, 104
Temperature
bacterial growth and, 92–93, *93*
increased, sterilization by, 1454–1456
mutation rate and, 179
syphilitic lesions affected by, 182, *183*, 888, *889*
virus sensitivity to, 1212, 1464
Terramycin, 160
structure, *298*
Teschen disease, 1281
Tetanolysin, 836
Tetanospasmin, 835

Tetanus, 835–837. See also *Clostridium tetani.*
  antitoxin of, unit of, 837n
  clinical patterns, 835
  death rate, 837
  DPT vaccine in, 837
  generalized, 835
  immune globulin in, 837
  immunity to, 353, 836
  immunization against, 837
  incidence, 835, 837n
  laboratory diagnosis, 836–837
  local, 835, 836
  pathogenesis, 835–836
  prevention, 837
  toxin of, 831
    action of, 835–836
    antibody to, 836
    fixing of, 836
    spasmogenic effect of, 836
    susceptibility to, 836
  treatment, 837
Tetanus neonatorum, 835
Tetanus toxoid, *484*
Tete virus, 1392
Tetracycline(s) (therapy), 677–678. See also *Antibiotic* and specific drug, e.g., *Chlortetracycline.*
  antibiotic action of, 159
  candidiasis due to, 996
  disadvantages, 677
  in *Acinetobacter* infection, 785
  in actinomycosis, 876
  in *Aeromonas* infection, 782
  in *Alcaligenes faecalis* infection, 786
  in anthrax, 826
  in brucellosis, 816
  in chlamydial infection, 921, 922
  in cholera, 781, 782
  in *Citrobacter* infection, 772
  in diphtheria, 691
  in *E. coli* infection, 768
  in *Edwardsiella* infection, 771
  in epidemic typhus, 906
  in *Enterobacter* infection, 771
  in granuloma inguinale, 949
  in leptospirosis, 895
  in listeriosis, 948
  in lymphogranuloma venereum, 925
  in melioidosis, 785
  in *Mycoplasma* infection, 934
  in nongonococcal urethritis, 925
  in ornithosis, 926
  in *Pasteurella multocida* infection, 809
  in plague, 805, 806
  in pneumococcal infection, 705
  in primary atypical pneumonia, 939, 942
  in pseudotuberculosis, 806
  in Q fever, 911
  in rat-bite fever, 896

Tetracycline(s) (therapy) (*continued*)
  in relapsing fever, 893
  in rickettsial disease, 900, 910
  in shigellosis, 778
  in trachoma, 924
  in tsutsugamushi fever, 911
  in tularemia, 808
  in whooping cough, 799
  protein synthesis inhibited by, 298, 303t
  resistance to, 678, 679
    by *Shigella,* 1107
    by streptococci, 723
  structure, *298*
  superinfections due to, 674
  to prevent tissue culture contamination, 943
Tetrahydrofolate, in bacteria. See under *Bacterium.*
Tetrahymena geleii, 91t
Tetrazolium-reduction inhibition test, for *Mycoplasma pneumoniae,* 941
Textiles, molds on, 94
Thallus, 966
Thermal death point, 1454
Thermoactinomyces, 873t, 875t
Thermophiles, bacterial, 92, 93
Thermopolyspora polyspora, 873
Θ (theta), antiserum to, 580
  bacterial synthesis of, *75*
  discovery of, 91t
Thiamine pyrophosphate, 42
Thiobacillus
  metabolism, 60
  taxonomic characteristics, 34t
Thiobacteriaceae, classification, 34t
Thioesters, high-energy, in metabolic reactions, 42
Thioglycollate, in culture of anaerobes, 91
Thioredoxin, in *E. coli,* 81
Thiosemicarbazone therapy, in smallpox, 1184
Thiostrepton, protein synthesis inhibited by, *299,* 300, 303t
Thoracic duct, 476
Threonine, bacterial synthesis, 71, *71*
Thrombocytic purpura, 587t
Thrombocytopenia, 585, 588
Thrombocytopenic purpura, 589, *589*
Thrombosis
  due to rickettsiae, 902
  in epidemic typhus, 905
  in Rocky Mountain spotted fever, 908
Thrush, 959, 995
Thymectomy
  effects of, 473, 474t
  susceptibility to infection and,

Thymectomy, susceptibility (*continued*)
  473
  tolerance increase after, 495, *495*
Thymic hypoplasia, 508
Thymol, antiseptic action of, 1461
Thymosin, 474
Thymus, 473–474, *475.* See also *Lymphatic organ.*
  aplasia of, 580–581
  as target organ in Gross leukemia viral infection, 1433
  corticosteroid effects on, 474
  development, 473
  hormones of, 474
  in immune response, 473
  involution of, 474
  structure, 473
  T cells in, 460, 474, *475*
Thymus-less mutants, 473
Thyroglobulin antibodies, 585
Thyroid disease(s), autoimmune aspects, 587, 587t, 588, 591
Tick(s)
  Coxiella burneti carried by, 912, 913
  dog, 1408
  in boutonneuse fever, 909
  in Colorado tick fever, 1408
  in Rocky Mountain spotted fever, 909
  in tularemia, 807
  relapsing fever spread by, 892, 893
  rickettsiae transmitted by, 898, 901
  togaviruses carried by, 1385
  viral encephalitides carried by, 1391
  wood. See *Dermacentor andersoni.*
Tick fever(s), 909
Tinea nigra, 1006
Tinea versicolor, 1005–1006
Tissue(s)
  architecture of, phagocytosis and, 654
  destruction of, due to complement, 523
  sites of, sulfonamide ineffectiveness at, 153
  inflammation of. See *Inflammation.*
  structural factors in antibacterial defense, 653–654
Tissue cultures, mycoplasma contamination of, 942–943
Tobacco mosaic virus, 9, *1018–1020*
  as model for study of helical viruses, 1019n
  infectivity of nucleic acid of, 1141

Todd phenomenon, 436
Togavirus(es), 1382–1392. See also
    Arbovirus.
    characteristics, chemical and
        p h y s i c a l, 1 3 7 9 t,
        1382–1383
    immunologic, 1383–1384
    classification, 1380t
    immunologic, 1379, 1382t
    disease and infection due to,
        1380t
    control of, 1391–1392
    epidemiology, 1389–1391
    l a b o r a t o r y diagnosis,
        1388–1389
    pathogenesis, 1387–1388
    prevention, 1391–1392
    recovery from, antilymphocytic
        serum and, 1211
    serologic tests for, 1203t
    encephalitis due to, 1382
    geographic distribution, 1380t
    hemagglutination-inhibition titra-
        tions in, 1382t
    host range in, 1384–1485
    immunity to, 1384
    inactivation of, host temperature
        and, 1212
    infectious nucleic acids in, 1143
    infectivity of, 1383
    interferon production with, 1174
    morphology, 1379t, 1382, 1383t,
        1384–1386
    multiplication of, 1386–1387,
        1386
    properties of, 1379t, 1382–1387
    susceptibility to, 1209
    vectors of, 1380t, 1385
Tolerance, immunologic. See
    Immunologic   tolerance.
Tolerant cell, 494–496
Tolerogen(s), 492, 493
Tolypomycin, 1185
Tongue, *Streptococcus salivarius*
    on, 957, 958t
Tonsil(s), lymphoid cells in, 477
Tonsillectomy
    bulbar poliomyelitis and, 1292
    streptococcal infections due to,
        955
Tooth (teeth). See also *Dental.*
    disorders of, actinomycosis due
        to, 873
    due to oral flora, 959
    enamel pellicle of, 960
    extraction of, streptococcal infec-
        tion due to, 955
    infected, endocarditis due to, 961
    plaque on, 959–960, *959*
    streptococci on, 957
Torula histolytica. See *Crypto-
    coccus neoformans.*
Torulosis. See *Cryptococcosis.*
Toxin(s). See also *Endotoxin; Exo-*

Toxin(s) *(continued)*
    toxin; and under orga-
        nism producing toxin,
        e.g., *Clostridium, toxins
        produced by.*
    botulinum. See under *Clostridium
        botulinum.*
    formation of, bacterial sporula-
        tion and, 143
    immunity to, 353
Toxoid
    diphtheria. See *Diphtheria, toxoid
        for.*
    diphtheria-tetanus-pertusis, 688,
        800
    staphylococcal, 735
Toxoplasmosis, IgA deficiency and,
    508
Trace elements, bacterial require-
    ment, 92
Trachea, bacteria in, 956
Trachoma, 922–924. See also *Chla-
    mydia trachomatis.*
    blindness due to, 922
    diagnosis, 923–924
    epidemiology, 922
    general features, 921
    history, 922
    immunity to, 923
    pathogenicity, 922–923
    treatment, 924
        sulfonamide therapy in, 675
    vaccine for, 491
Traits, hereditary, selection for,
    12n
Transamination, in *Bacillus,* 73n
Transcriptase(s)
    in DNA transcription, 263, *265*
    reverse, in animal viruses, 1155
        in oncogenic viruses, 1419
        in reverse transcription, 267
    virion, 1010
Transcription, of DNA. See *DNA,
    transcription of.*
Transduction
    abortive, 1112–1113, *1113*
    bacterial. See *Bacterium, trans-
        duction in.*
    bacteriophages in. See also *Bac-
        teriophage, transducing.*
    comparison of, 1116t
    defective, 1112–1113
    definition, 1111
    discovery of, 1111, *1111*
    escape synthesis in, 331, 1117
    generalized, 1111–1112
        distribution and scope of,
            1111–1112
        mechanism, 1112–1113
    genetic mapping and, 249–250,
        251
    high-frequency, 1114–1116,
        1115
    low-frequency, 1113–1114, *1114*

Transduction *(continued)*
    phage particles in, 1112
    specialized, 1113–1118
        by generalized transducing
            phage, 1117
        range of, 1115–1116, *1116*
        uses of, 1117
    types, 1111
Transferase(s)
    in eukaryotic cells, in protein
        synthesis, 286
    peptidyl, in polypeptide chain
        termination, 290, *290*
Transfer factor
    cell-mediated responses trans-
        ferred by, 572
    injection in defective cellular
        immunity in variola and
        vaccinia infections, 1211
Transformation, cell. See *Cell trans-
    formation.*
Transformed cells. See *Cell, trans-
    formed.*
Transfusion(s)
    antiallotype formation due to,
        419
    antibodies to leukocytic antigens
        formed after, 609n
    cytomegaloviral disease from,
        1249, 2151
    hepatitis from, 1414
    reactions due to, 609
        anaphylaxis and, 546
        due to Am allotypes, 422
        due to blood antigen incompati-
            bilities, 607
        due to uniqueness of indivi-
            dual's erythrocytic allo-
            a n t i g e n p a t t e r n,
            609–610
        infrequency of, 609
        prevention of, blood typing in,
            609
    syphilis due to, 885
    universal donors and recipients in,
        609
Translation
    bacterial metabolic regulation at
        level of, 334
    of mRNA, *278, 279,* 327–330,
        329
    suppressors of, codon-specific,
        292, 293–294
    unit of, See *Codon.*
Transmissible gastroenteritis virus,
    of swine, 1362n
Transplant(s)
    corneal, 612
    donors of, designations of, 610
    in same individual, 610
    to another species, 610
    to genetically identical individual,
        610
    to genetically nonidentical indivi-

Transplant(s), to genetically
　　nonidentical *(continued)*
　　dual of same species. See
　　Allograft.
　tolerance of, genetic basis of, 613
　of tumors, 620
Transplantation antigen. See *Histo-
　　compatibility antigen.*
Transport, bacterial. See under
　　Bacterium.
Trench fever, 913. See also *Rocha-
　　limaea quintana.*
Treponema
　antigens of, 886–887
　characteristics, general, 882
　　taxonomic, 35t
　culturing of, 884, 885
　phospholipid incorporation into,
　　886
　Reiter, 886
　storage of suspensions, 885
Treponema carateum, 891. See also
　　Pinta.
Treponema cuniculi, 882, 892
Treponema denticola, 957, 958t
Treponema macrodentium, 958t,
　　960
Treponema microdentium, 884
Treponema oralia, 958t
Treponema pallidum, 883. See also
　　Syphilis.
　cultivable strains of, 885
　detection in syphilitic lesions, 889
　DNA base composition in, 37t
　genetic variation, 887
　metabolism, 884–885
　morphology, 883–884, *885*
Treponema pallidum immobiliza-
　　tion (TPI) test, 886, 890
　complement in, 523
Treponema pertenue, 891. See also
　　Yaws.
Treponemataceae, 882
Treponematosis(es)
　nonsyphilitic, 892
　TPI test in, 886
Triacetyloleandomycin, 676
Tricarboxylic acid (TCA) cycle in
　　bacteria. See under *Bac-
　　terium.*
　in anaerobic growth on glucose, in
　　E. coli, 66
Trichophytin, 1003
Trichophyton, 986t, 1004–1005
　diseases due to, 1004
　sexual stage in, 977t
Trichophyton ajelloi, 977t
　geophilic aspects, 1002
Trichophyton mentagrophytes,
　　1005
　geophilic aspects, 1002
Trichophyton rubrum, 1005
　immunity to, 1002
Trichophyton schoenleinii, 1005

Trichophyton terrestre, 977t
Trichophyton tonsurans, 1005
　distribution, 1002
　hair infection with, *1003*
Trichophyton violaceum, 1005
Trichosporon cutaneum, 1006
Tricresol, disinfectant action of,
　　1461
Trimethoprim, inhibition of bac-
　　terial THF function by,
　　74
Triphosphopyridine nucleotide, in
　　bacteria. See under *Bac-
　　terium.*
Triple-sugar-iron agar, 757
Trismus. See *Tetanus.*
Trivittatus viruses, 1396
tRNA. See *RNA, transfer.*
Trombicula. See also *Mite.*
　Rickettsia tsutsugamushi trans-
　　mitted by, 910
Trypanosome(s), killed by silver
　　ions, 1459
Trypsin, in animal cell culture,
　　1122
Tryptophan. See also *Amino acid.*
　bacterial. See under *Bacterium.*
Tryptophan synthetase
　in *E. coli.* See under *Escherichia
　　coli.*
　mutation of, 292, *294*
　recombination in protein of, *242*
T strains, in genitourinary infec-
　　tions, 936
Tsutsugamushi fever, 910–911. See
　　also *Rickettsia tsutsuga-
　　mushi.*
　attack and death rates, 910
　carrier state in, 910
　diagnosis, 904t, 910
　geographic distribution, 904t, 910
　immunity to, 910
　treatment, 911
Tube dilution method, to deter-
　　mine bacterial sensitivity
　　to drugs, 149
Tubercle formation, in tuberculosis,
　　852, *853*
Tuberculin
　dangers of, 561, 851
　focal reactions to, 562
　noncutaneous reactions to, 561
　old, 560n, 857
　preparation of, 857
　sensitivity to, 562, 657
　　harmful reactions from, 851
　skin reactions to, 560–561
　　in various species, 561
　systemic reaction to, 561
Tuberculin shock, 561
Tuberculin test
　factors suppressing, 857
　in atypical mycobacterial infec-
　　tions, 857, 863, 865

Tuberculin test *(continued)*
　in tuberculosis, 856–857, 859
　positivity of, disadvantage of, 854
　vaccination and, 860
Tuberculin unit, 560n
Tuberculoprotein, allergy to, 852
Tuberculosis. See also *Mycobac-
　　terium tuberculosis.*
　activated macrophages in,
　　850–851
　age factors, 853, *854, 855,* 857,
　　858, 859
　alcoholism and, 859
　artificial lung collapse in, 853,
　　857
　avian, 844
　caseation necrosis in, 852
　causes, 11
　chemoprophylaxis, 680, 860
　chemotherapy, 859, *859*
　　combined, 679
　　isoniazid in, 671, 865
　　resistance to, 678, 858
　　rifamycins in, 267
　chronic inflammation in, macro-
　　phage accumulation in,
　　649
　coccidioidomycosis versus, 992
　cutaneous, 861
　death rates, *854, 859, 859*
　　racial factors in, 854
　distribution, 858
　dormancy in, 661, 662, 852, 856
　　reactivation after, 854, 859
　　testing for, 856
　due to atypical mycobacteria,
　　862, 864t
　due to dairy products, 861
　due to *Mycobacterium avium,*
　　861
　due to *Mycobacterium bovis,* 861
　due to *Mycobacterium ulcerans,*
　　861
　epidemiology, 858–859
　erythema nodosum with, 593
　etiology, 11
　giant cell formation in, 649
　healed, pulmonary calcification
　　in, 852
　histoplasmosis versus, 989
　history, 844
　hypersensitivity in, 657
　immunity to, 851
　　acquired cellular, 657, 658
　　allergy versus, 657
　immunization in, 860–861
　in cattle, 861
　incidence, 858, 859
　　racial factors, 854
　in concentration camps, 855
　laboratory diagnosis, 856–857
　　chest X-rays in, 859
　　Mantoux test in, 857
　　tuberculin test in, 856–857

Tuberculosis *(continued)*
  lepromin sensitivity in, 866
  leprosy with, 867
  lesions of, distribution of, 853
    granulomatous, 649, 852
    types of, 852
  lymph gland, 861
  malnutrition and, 859
  meningitis with, 852
  military, 852
  natural vaccination in, 864
  of bones, 861
  of joints, 861
  organs affected in, 852
  overcrowding and, 855, 858
  pathogenesis, 851–856
  prevention, 859–861
  primary complex in, 852
  reactivation of disease in, 853–854
  resistance to, cortisone and, 855
    variations in, 854–856
      genetic differences in, 855
      hormonal factors in, 855
      physiologic factors in, 855–856
      racial differences in, 855
  sex factors in, 855
  silicosis and, 856
  simulated by nocardiosis, 878
  spread of, 852
    mechanism of, 858
  stress and, 855, 856, 859
  susceptibility to, spermidine and, 636
  systemic, conversion to latent disease, by isoniazid and pyrazinamide, 662, *662*
  therapy, 857–858, *859.* See also *Tuberculosis, chemotherapy.*
  tissue calcification in, 852
  tubercle formation in, 852, *853*
  tuberculin positivity in, 854
  tuberculin test in, 859
Tularemia, 806, 808–809. See also *Francisella tularensis.*
Tumor(s) (cells). See also *Cancer.*
  actinomycin therapy in, 149n
  adenoviral, 1234
    gonadectomy and, 1212
  antigenicity of, 621
  antigen(s) of, adenoviral, 1234
    histocompatibility, 568
      loss of, 621
    surface, 568
  chemically induced, antigenic variation in, 621
  critical mass of, 568, *569*
  defective cellular immunity and, 568, 569t
  due to viruses, 1208
    tumor-specific antigens in, 621
  established, clinical resistance to,

Tumor(s), established *(continued)*
  569
  growth of, immune enhancement of, 579
  immunity against, 568–560, 620–622
    as surveillance mechanism, 570
    cell-mediated versus humoral, 620
  immunization against, 568, *659*, 622
  immunologic surveillance of, 622–623
  macrophage destruction of, 570
  prevention of, immunologic, 622
  resistance to, decrease of, due to passively transferred serum, 569
  spontaneous regression of, 621
  transplantation of, 620
  transplanted, cell-mediated immunity to, 568
    growth of, dose-response curves for, *569*
    enhanced by immunization, 620–621
Tumor-specific antigens, 620, 621
Turbidimetry, 150, *150*
Turkey(s). See also *Fowl.*
  disease due to *Aspergillus flavus*, 966
Turkeypox, 1260
Turtle(s), mycobacteria in, 862, 863t, 868
Tween reaction, 734
Typhoid fever, 773, *773.* See also *Salmonella typhi.*
  carriers of, 661, 774
  control of, 755
  immunization against, 776
  pathogenesis, 773
  prevention, 775
  typhus fever versus, 906
Typhus
  endemic, 907–909. See also *Rickettsia typhi.*
    control measures in, 908
    death rate in, 907
    diagnosis, 904t, 907–908
    epidemiology, 903, 908
    geographic distribution, 904t, 908
    immunity to, 907
    incidence, 908
    transmission of, 904t, 905t, 908
  epidemic, 905–907. See also *Rickettsia prowazeki.*
    Brill-Zinsser disease versus, 907
    carrier state in, 907
    control of, 906–907
    death rate in, 906
    diagnosis, 904t, 906
    epidemiology, 903, 906
    geographic distribution, 904t

Typhus, epidemic *(continued)*
    immunity to, 906, 907
    pathogenesis, 905–906
    prevention, 906–907
    relapsing fever associated with, 893
    treatment, 906
  flea-borne. See *Typhus, endemic.*
  louse-borne. See *Typhus, epidemic.*
  Manchurian. See *Typhus, endemic.*
  Moscow. See *Typhus, endemic.*
  murine. See *Typhus, endemic.*
  scrub. See *Tsutsugamushi fever.*
  shop. See *Typhus, endemic.*
  tick, 909
  Toulon. See *Typhus, endemic.*
  typhoid fever versus, 906
Tyrosine. See also *Amino acid.*
  bacterial synthesis of, 77, 78
Tyrosine-α-ketoglutarate aminotransferase, in hepatoma cells, 1133
Tyrothricin, 678

U

Ubiquinones, bacterial. See under *Bacterium.*
Ulcers, intestinal, in South American blastomycosis, 994
Ultrasound, sterilization by, 1457
  bacteria killed by, 1456, *1456*
Ultraviolet (UV) radiation
  bacterial absorption of, 1456
  bacterial division affected by, 340
  bacterial mutation due to, 253
  bacterial spore resistance to, 141
  bacteriophage T4 damaged by, 237, *238*
  corneal irritation by, 1456
  damage in higher organisms, repair of, 240–241
  energy of, 237
  gene and genome damaged by, *238*
  interferon production and, 1177, *1178*
  mutations due to, 253, 259
  nucleic acids damage by, 237, *237*
  photochemistry of, 1456
  prophage induction by, 1103
  protection against, by glass(es), 1456
  sterilization by, practical uses of, 1456–1457
  viral infectivity and, 1177, *1178*
  viral transcriptase activity and, 1177, *1178*
  viruses inactivated by, 1463

Undecylenic acid therapy, in dermatomycoses, 1002n
Undulant fever. See *Brucellosis.*
Unitarian hypothesis, of antibody–antigen reactions, 400–401
Unity of biology. See under *Biology.*
Uranyl acetate, to stain viruses, 1016
Urea, hepatic formation of, 70
Uremia
  chemotherapy inhibited by, 673
  due to *E. coli,* 768
Urethra, bacteria in, 957
Urethritis
  due to *Moraxella,* 785
  in gonorrheal infections, 750
  nongonococcal, due to chlamydiae, 924, 925
  T strains in, 930, 936
Urinary tract infections
  bacterial, chemotherapy in, 160
  due to *Acinetobacter,* 785
  due to *Alcaligenes faecalis,* 786
  due to *Citrobacter,* 772
  due to *E. coli,* 768
  due to *Enterobacter,* 771
  due to mycoplasmas, 936
  due to *Proteus,* 772
  due to *Pseudomonas aeruginosa,* 784
  due to *Veillonella,* 751
  sulfonamide therapy in, 675
Urine
  antibacterial action of, 643
  hormones in, measurement of, 393
Urticaria, 593
Uterus
  clostridial infection of, 838
  isolated, Schultz–Dale reaction in, 543, *544*
  infections of, streptococcal, 720, 726

V

Vaccination. See also *Immunization* and under specific disease.
  against microbial antigens, 490–491
  attenuation of pathogen virulence in, 352
  definition, 357
  history, 7, 352
  immunity conferred by, 658
  in control of infectious diseases, 10
  interferon in, 1175
  lesions due to, rifampin inhibition

Vaccination, lesions due to (*continued*)
    of, 1186
  limitations, 658
  number of injections in, 491
  soluble versus insoluble antigens in, 491
Vaccine. See also under specific disease, e.g., *Trachoma, vaccine for.*
  definition, 358
  development of, 9
  Flury, 1374
  infectious human diseases prevented by, 491t
  live, contraindicated in immunodeficiency diseases, 506
  live virus, 1206
  M-protein, in streptococcal infections, 724
  preparation of, sterilization in, 1453
  Semple, 1374
  viral, cell strains used in, 1140
  virus mutants in, 1166, 1206
Vaccinia
  diagnosis, laboratory, 1268–1270
    serologic, 1203t, 1269–1270
  due to smallpox vaccination, 1273
  epidemiology, 1270
  mitigated by interferon, 1180
  recovery from, defective cellular immunity and, 1211
  transfer factor in, 572
  treatment, 1274
Vaccinia gangrenosa
  defective cellular immunity and, 1211
  due to smallpox vaccination, 1273
  treatment, 1274
Vaccinia virus, *1018, 1261–1275.* See also *Poxvirus.*
  antigenic structure, 1262–1263
  assembly of, *1267*
  cultured cells susceptible to, 1264
  development, *1162*
  hemagglutination by, 1263
  host cell RNA synthesis inhibited by, 1147
  host cell thymidine kinase synthesis inhibited by, *1148*
  host range in, 1263
  immunity to, 1262
  inclusion bodies due to, 1264, *1264*
  infection of single cell, *1268*
  in immunization against smallpox, 1270
  interferon production with, 1177, *1177*
  maturation of, *1267*
  morphology, *1259,* 1260, 1261

Vaccinia virus (*continued*)
  multiplication of, 1264–1265, *1265*
  pock assay of, *1040*
  pocks produced by, *1261, 1269*
  primary lesion due to, delayed hypersensitivity and, 1215
  properties of, 1261–1265
    chemical, 1261–1262
  rifampin effects on, *1186*
  stability, 1262
  structure, *1267*
  toxic effects of, 1207
  UV-irradiated, reactivation of, 1168
Vagina
  acidity of, antibacterial effects of, 643
  bacteria in, 956
Valine, bacterial synthesis of, 73, 74, 316, 322
Valinomycin, 128
Valvulitis, coxsackieviral, 1298, 1299
Vancomycin, 676
  bacterial cell wall synthesis inhibited by, 121
Varicella. See also *Herpes Zoter* and *Varicella-zoster virus.*
  clinical picture, 1247
  epidemiology, 1248–1249
  immunity to, 1210
  pathogenesis, 1247
  prevention and control, 1249
  serologic tests for, 1203t
Varicella-zoster virus, 1246–1249. See also *Varicella* and *Herpes zoster.*
  characteristics, chemical and physical, 1246
    immunologic, 1246
  cytopathic effect of, 1207
  disease due to, latent aspects, 1248
  host range of, 1246
  inclusion bodies due to, 1246, *1248*
  morphology, *1247*
  multiplication of, 1246–1247
  pathogenicity, 1247–1248
  properties of, 1246–1247
  spread of, cortisone and, 1212
Variola
  antibodies to, 1209
    postvaccination appearance of, 1273
  control of, 1270–1274
  diagnosis, laboratory, 1268–2170, 2169t
    serologic, 1203t, 1269–2170
  epidemiology, 1270
  forms of, 1265
  Guarnieri bodies in, 1267

Variola (*continued*)
history, 1258
hypersensitivity response in, 1267
immunity to, 1210, 1211
cellular versus humoral, 1263
cutaneous responses indicating degree of, 1271
history, 352
immunization in, 1258
immunologic aspects, 1262
mortality rate in, 1265
naming of, 1258
pathogenesis, 1265—1268
pregnancy and, 1212
prevention, 1270—1274
cowpox virus in, 1270
vaccine for, administration of, 1271—2173
allergic reaction to, 1271
cutaneous responses to, 1271, 1272
lyophilized, 1271
preparation of, 1270—1271
vaccinia virus in, 1270
variolation in, 1270n
spread of, 1274
treatment, 1274
thiosemicarbazone therapy in, 1184
vaccination against, 10
complications, 1273—1274, 1273t
introduction of, 7
Variola minor, 1262, 1263
Variola sine eruptione, 1263
diagnosis, 1269
Variolation, 1270n
Variola virus, 1261—1275. See also *Poxvirus*.
antigenic structure, 1262—1263
cultured cells susceptible to, 1264
dissemination of, 1213
hemagglutination by, 1263
host range of, 1263
immunity to, 1262
inclusion bodies due to, 1264
morphology, 1261
multiplication of, 1264—1265
pocks produced by, *1261*, 1269
properties of, 1261—1265
chemical, 1261—1262
stability of, 1262
Vascular changes, in inflammation, 647
Vascular disorders, in phycomycosis, 998
Vasculitis
in Arthus reaction, 546
in systemic lupus erythematosus, 592
Vasoactive substances (amines)
in serum sickness, 550
platelet release of, 550
release from tissues, due to

Vasoactive substances (amines), release from tissues, (*continued*)
antibody—antigen complexes, 540
cAMP inhibition of, 544, *545*
Vaxoconstriction, due to staphylococcal toxin, 735
Veillonella
cell surface of, *27*
in oral cavity, 958t
taxonomic characteristics, 34t
Veneral disease, 750. See also specific disease, e.g., *Gonorrhea*.
chemoprophylaxis, 680
Venezuelan equine encephalitis, 1390
Verruga peruana, 949
Vertebra(e), tuberculous, 861
Vertebrate(s), poikilothermic, mycobacteria in, 868
Vesicular stomatitis virus
autointerference by, 1182
interferon effects on, 1174
interferon production and, 1175
of cattle and horses, 1368, 1370
structure, 1033, *1033*
Viability
microbial, tests for, 1452
of bacteria, criteria of, 1452—1453
Vi antigen(s)
definition, 759n
in Enterobacteriaceae, 759, *760*
in salmonellae, 762
Vibrio(s), *24*, 779—782
biochemical reactons of, 758t
noncholera, 782
taxonomic characteristics, 34t
transfer of genes from *E. coli* to, 195
Vibrio cholerae, 779—782, 779. See also *Cholera*.
DNA base composition in, 37t
El Tor biotype, 780
enterotoxin elaborated by, 780
hemolysin produced by, 780
neuraminidase from, 1036
pathogenicity, 780
pH and, 93
Vibrio comma. See *Vibrio cholerae*.
Vibrio fetus, 782
Vibrio parahaemolyticus, 782
Vibrio sputorum, 958t
Vincent's angina, 787, 893
Vinegar, *Acetobacter* production of, 49
Viomycin therapy, in tuberculosis, 857
Viral interference, 1172—1183
definition, 1172
inapparent infection due to, 1215
interferon in. See *Interferon*.

Viral interference (*continued*)
intrinsic, 1181—1183
noninterferon, 1181—1183
significance, 1183
Viral strand, 1023
Viremia, 1212
Virion(s). See also *Virus*.
antibody interaction with, 1190—1191
definition, 1010
DNA in, 1021
enveloped, 1016, *1017*
enzymes of, 1010, 1011t
helical, 1016, *1017*
hemagglutinating, 1031
icosahedral, 1016, *1017*
killed, immunization with, 1198
maturation of, 1068—1071
morphologic types, 1016, *1017*
phenotypically mixed, 1071
pleomorphism in, 1031
proteins of, internal, 1033
macromolecular precursors of, 1068
restricted, 1065, *1066*
sizes, *1012*
surface components of, antibodies of, 1198
Virion—antibody complexes, 1191
infectivity of, 1192
stability of, kinetics of neutralization and, 1194, *1195*
physiocochemical events leading to, 1191—1192
Virion transcriptase, 1150
Virology, 1007—1447
bacteriophages as model system in, 1049
cell lines used in, 1142t
electron microscopic technics in, 1015—1016
shadowcasting in, 1015, *1015*
staining in, 1015—1016, *1016*
thin sectioning in, 1016
Virulence. See also under specific organisms.
attenuation of, 352
definition, 632n
Virus(es). See also *Virion*; specific viruses; and under organisms attacked by viruses, e.g., *Chicken, nonpathogenic virus of.*
abortion caused by, 1216
adsorption to cells, cell susceptibility and, 1208
allergic response to, 1215
analysis of, 1014—1016
animal. See *Animal virus*.
antibodies elicited by, humoral, 1190
measurement of, 1035
tests of, 1198
antibody neutralization of,

Virus(es), antibody neutralization
    of (continued)
    1181–1197
  cross-reactivity in, 1199
  definition, 1191
  host cells in, 1192
  kinetics, 1193–1196
    curves of, 1195–1197
    virion–antibody complex and,
      1194, 1195
    measurement of antibodies in,
      1197
  mechanism, 1192–1193
  q u a n t i t a t i v e   a s p e c t s,
    1193–1196
  residual infectivity after, 1194,
    1194, 1196
antibody-neutralized, stable, reactivation of, 1191, 1192
antibody sensitization of, 1193
antigen(s) of, antibodies against,
  1198
  cellular   immunity   against,
    1201–1202
  evolution of, 1199–1200
  family, 1198
  type-specific, in classification of
    viruses, 1198
  variation of, evolutionary as-
    pects, 1199
antigenic analysis of, complement-fixation test in,
  1201
  hemagglutination-inhibition test
    in, 1201
  methods, 1200–1201
  neutralization test in, 1200,
    1201
    complications, 1201
  plaque-reduction test in, 1200
  tests on paired sera in, 1201
antigenic properties of, assays
  based on, 1036
arthropod-borne. See Arbovirus.
as infectious agents, 1011–1014
as precellular stage of life, 16
assay of, 1033–1042
  based on antigenic properties,
    1036
  e n d p o i n t   m e t h o d   o f,
    1040–1041
    measurement of infectious
      titer in, 1043
  plaque method in, 1037–1039,
    1038, 1039
    dose-response curve of, 1043
    measurement of infectious
      titer in, 1043
  precision of, 1043–1044
  Reed and Muench method in,
    1044
  types of, comparison of,
    1041–1042
    precision of, 1043–1044

Virus(es) (continued)
  autointerference of, 1176
  bacteria versus, 1118
  bacterial. See Bacteriophage.
  cancer due to. See Cancer.
  capsid(s) of, 1025–1030, 1027
    allosteric transitions of, 1029
    architecture of, theoretic con-
      s i d e r a t i o n s   o f,
      1029–1031
    assembly of, 1068–1069, 1070,
      1071
    definition, 1016
    empty, 1017, 1019, 1019
    formation of, 1025
    h e l i c a l,   1026,   1028–1029,
      1029, 1030
      rotational axis of, 1028
      stability of, 1028, 1029
    icoahedral, 1026–1028, 1026,
      1027, 1029, 1032
      geometry   of,   1026–1027,
        1027
      number of capsomers in,
        1028t, 1030–1031
      shape and dimensions of, 1026
      symmetry of, 1027–1028,
        1028, 1032
    parameters of, 1028t
    protomers of, 1029
    size of, 1013t
    symmetry of, rotational, 1028,
      1028, 1029
    shape of, nucleic acid influ-
      enced by, 1019
  capsomer(s) of, 1013t, 1026,
    1026, 1027
    assembly of, in viral multiplica-
      tion, 1030
    hexagonal, 1030, 1031
  cell hemabsorption due to, 1207
  cell inclusion bodies due to,
    1207–1208
  cell lysis by, 1071
  cell   macromolecular   synthesis
    affected by, 1207
  cell membranes altered by, 1207
  cell receptors for, in viral adsorp-
    tion, 1208
  cell transformation due to, 1208.
    See also under specific
    virus, e.g., Polyoma virus.
  cell genes in, 1429
  immunologic relation of animal
    host to, 1445–1446
  characteristics, 1013t
  chlamydiae mistaken for, 916
  classes of, 1011
  c l a s s i f i c a t i o n,   antigenic,
    1198–1199
  clone of, 1039
  complex, 1018, 1033
  complexes of, complement in,
    553

Virus(es) (continued)
  composition of, 1010
  congenital anomalies and, 1216
  core of, 1016
  counts of, 1033–1034, 1034
  cytopathic effects of, 1207
  definition, 9
  disease and infection due to. See
    also specific disease, e.g.,
    Mumps.
    a n t i b i o t i c   t h e r a p y   i n,
      1185–1186
    antibody production in, persis-
      tence of, 1210–1211
    antibody response in, secon-
      dary, 1210–1211
    autoimmune diseases due to,
      586
    based on virus-induced immuno-
      l o g i c   r e s p o n s e,
      1215–1216
    carrier state in, 1216–1217
    cell   alterations   due   to,
      1206–1208
      in cell membrane, 1207
    cell   hemadsorption   due   to,
      1161, 1161, 1207
    cell   susceptibility   to,
      1208–1209
    chemotherapy in, points in viral
      growth   cycle   attacked
      by, 1172, 1173
    chromosomal aberrations due
      to, 1208
    chronic, 586
    control of, by inhibition of
      replication, 1172
    course of, interferon production
      and, 1180
    disseminated, 1212–1214
    endosymbiotic, 1216
    glomerulonephritis with, 553
    hemolytic anemia due to, 586
    hormonal factors, 1212
    hypersensitivity in, 657
    immune-complex disease due to,
      552–553
    immunity in, 1210–1211
      cell-mediated, 567, 1211
    immunization in, successful,
      1210t
      passive, immune serum in,
        1209
    immunologic   destruction   of
      cells in, 1207
    immunologic tolerance to, 1211
    inapparent, 1214–1215
      definition, 1214
    interferon and, 1179–1180
    latent, 1217–1218
      chronic disease and, 1218
      e p i d e m i o l o g i c   significance,
        1218
    localized, 1212

Virus(es) *(continued)*
  macrophages and, 1212
  mycoplasms infections with, 934
  of embryos, immunologic tolerance due to, 1211
  pathogenesis, 1205–1220
    antibodies in, 1209–1211
    cellular factors in, 1206–1209
    cellular immunity in, 1201
    immunologic factors, 1209–1212
    viral factors, 1206–1209
    patterns of, 1212–1218
  pneumococcal pneumonia and, 701
  pregnancy and, 1212
  protection against, antibodies in, 1209
  quantitative aspects, 1042–1043
  recovery from, antibodies in, 1209–1210
    antilymphocytic serum and, 1211
    delayed-type hypersensitivity and, 1211
    interferon initiation of, 1183
    repeated, 1210
  resistance to, nonspecific systemic factors in, 1211–1212
  nutrition and, 1211
  serologic tests for, 1202t–1203t
  slow, 1219–1220, 1446
    definition, 1217
    examples of, 1219t
  steady-state, 1216–1219
  dissemination of, along nerves, 1213
  by bloodstream, 1212
  distinguishing cells from, 10
  DNA of. See under *DNA.*
  effects of host-specific restriction and modification system, 1066
  effects on macromolecules produced by cell, 1207
  envelope of, *1018,* 1031–1033
    glycoproteins of, 1031
    lipids in, 1031
    spikes covering, *1030,* 1031
  erythrocytes stabilized by, 1036
  evolution of, 1110, 1199
  failure to reach target organ, inapparent infection due to, 1215
  fetal effects of, 1216
  filtrable animal, 9
  first recognition of, 9
  5-fluorouracil mutagenic action on, 296
  GAL, *1026*
  gene(s) in, 1020, 1021t

Virus(es) *(continued)*
  genetic information in, economy of, 1030
  genome fragmentation in, 1023
  grouping of, by pathogenic characteristics, 1214t
  head of, DNA packing in, 1070, *1070*
  heat sterilization of, 1454
  hemagglutinating, receptors of, 1036
  hemagglutination assay of, 1035, *1035*
  hemagglutination by, 1034–1036
    antibody inhibition of, 1035, 1197
    mechanism, 1035–1036
  history, 9–10, 1011
  host range of, 1011
  human wart, *1018*
  hybrid(s) of, 1428
  hybrid DNA-pseudovirions in, 1428
  icosahedral, *1018*
  immunity against, humoral, 1190–1198
    nucleic acid infectivity in, 1143
  inactivation of, 1462–1465
    agents in, 1463–1465
    by alkylating agents, 1464
    by enzymes, 1464
    by ethylene oxide, 1464
    by lipid solvents, 1464
    by UV, 1464
    by X-radiation, 1464
    photodynamic, 1464–1465
    definition, 1462
    degree of, 1462–1463
    dosage of agent required for, 1462
    heat in, 1464
    lipotropic agents in, 1464
    mechanisms, 1463–1465
    nucleotropic agents in, 1463
    photodynamic, 1464–1465
    proteotropic agents in, 1463
    survival curves and, 1463, *1463*
    UV in, 1463
  indicator, 1036
  in evolution, 10
  infectivity of, assays of, 1037–1040
    plaque method of, 1037–1039, *1038, 1039*
    pock method of, 1039, *1040*
    effect of UV irradiation on, 1177, *1178*
    efficiency of, 1041
    index of, 1041
    neutralization by globulins and lipoproteins, 1212
    phagocytosis and, 1212
    relative, 1194
  immunity to, inapparent infec-

Virus(es) immunity to, *(continued)*
    tions and, 1215
  immunologic tests of, complications, 1201
  immunology of, 1189–1204
    diagnostic use of, 1198–1201
    in naturally occurring cancers, 1446–1447
  lysosome altered by, 1207
  macromolecules of, 1055–1057
  maturation of, rifampin interference with, 1186, *1186*
  moderate, inapparent infections due to, 1215
  modification of, host-induced, 1064–1068
  morphologic classes, 1013t
  morphology, 1010
  multiplication cycle of, 1054–1055
    assembly of capsomers in, 1030
    chemical inhibition of, 1183–1186
      by adamantanamine, 1183
      by carboxypeptides, 1183
      by isatin-β-thiosemicarbazone, 1184
      by nonselective agents interfering with DNA synthesis, 1185
      by substituted benzimidazole and guanidine, 1183–1184
      by synthetic agents, 1183–1185
    interference with, 1171–1187
    interferon production and, 1174–1175, *1176*
    maturation and, 1068–1071
    one-step multiplication curve in, 1054–1055, *1055*
    stages in, 1054
  mutants of, attenuated, 1206
    temperature-sensitive, interferon production and, 1177
  mutations in, evolutionary aspects, 1199
  nature of, 1009–1045
  neuraminidase activity by, 1031
  neurotropic, 1213
  neutralized, host cell interaction with, 1192
  nucleases of, degradation of cellular DNA by, 1056, *1057*
  nucleic acids of. See *Nucleic acid.*
  nucleocapsid of, *1017*
    definition, 1016
    helical, *1019*
    icosahedral, 1017
    stability of, 1019
  obligatory intracellular parasitism by, 1010, 1011
  occult, 1217
  oncogenic, 1417–1449. See also

Virus(es), oncogenic *(continued)*
   specific virus, e.g., *Rous sarcoma virus.*
 assay of, 1039
 distribution, 1419t
 DNA-containing, 1419–1430
 etiology, 1418
 history, 1418
 humoral antibodies elicited by, 1445
 reverse transcriptase in, 1419
 summary of, 1446t
 unity of, 1418–1419
 particles of, 1010–1032
   distribution per cell, 1042
   morphology, 1016–1017
   ratio to hemagglutinating units, 1042
   ratio to infectivity units, 1041t
 passage from blood to brain, 1215
 passenger, 1433
 pathogenic agents compared to, 917t
 phenotypic mixing of, 1071, *1072*
 photosensitivity of, 1464
 plant, assay of, 1039
 polyamine function in, 79
 polykaryocyte formation due to, 1207
 preservation of, by freezing, 1455
 properties of, 1010–1011
 protein(s) of, host cell macromolecular synthesis turned off by, 1056–1057
 protein shell of. See *Virus, capsid of.*
 protomers of, 1026, *1026, 1027*
 replication of, host temperature and, 1212
   inhibition of, 1172
   mode of, 1010
 resistance to, host temperature and, 1212
 reverse transcriptase in, 267
 rickettsiae and, 901
 restriction of, host-induced, 1064–1068
 RNA of. See under *RNA.*
 samples of, contamination of, 1014–1015
   infectious titer of, 1043–1044
   preparation of, 1014–1015
   purification of, 1014
 sensitivity to disinfection, 1033
 sensitivity to lipid solvents, 1033
 serial propagation of, for live virus vaccines, 1206
 size, 1010
 slow, 1219–1220, 1219t, 1446
 staining of, negative, 1015–1016, *1016*
   positive, 1016

Virus(es), *(continued)*
 sterilization of, photodynamic sensitization in, 1457
 strains of, attenuated, inapparent infection due to, 1215
   prime, evolutionary aspects, 1199
 study of. See *Virology.*
 surface of, antigens adsorbed to, 1031
 susceptibility to, cortisone and, 1212
 synthesis of, incomplete, cytopathic effects of, 1207
 lysis inhibition and, 1055, *1056*
 target organ for, 1214
 temperature sensitivity of, 1464
 toxic effects of, 1207
 transcriptase activity in, 1177, *1178*
   interferon production and, 1177, *1178*
 tumor-producing. See *Virus, oncogenic.*
 typing of, immunologic tests for, 1198
 viability of, 1014
 virulence of, 1206
Virus B, 1246
Virus plaque assay, for cell–mediated immunity, 575
Visna virus, 1436, 1446
Vitamin(s). See also specific vitamin, e.g., *Folic acid.*
 bacterial requirement of, 90
 bioassays for, 90
 discovery of, 91t
 identification of, 90
 in culture media, 94
 synthesis of, bacterial, 77, 955
   by oral flora, 959
   commercial, microbes in, 50
Vitamin B$_1$. See *Thiamine.*
Vitamin B$_6$, 91t
Vitamin B$_{12}$
 bacterial production of, 92
 deficiency of, pernicious anemia due to, 591
 discovery, 91t
Vitamin K
 bacterial requirement of, 90
 bacterial synthesis of, 955
Voges-Proskauer test, 46
 in Enterobacteriaceae, 770
Vole bacillus, in tuberculosis prevention, 860
Von magnus phenomenon, 1165
Vulvovaginitis, gonorrheal, 750
VW antigen, 803, 804

W

Waldenström's macroglobulinemia, 425, 427
Warfarin, to control rates in endemic typhus, 908
Wart(s)
 benign, due to papillomaviruses, 1429. See also *Papilloma.*
 due to Shope rabbit papilloma virus, 1218
 immunosuppression and, 1202
Warthin-Finkeldey syncytial cells, 1347
Wart virus. See *Papillomavirus.*
Wasp stings, anaphylaxis due to, 538
Wassermann antibody
 complement-fixation tests for, 890
 in lupus erythematosus, 886
 in pinta, 892
 in syphilis, 886
 in yaws, 891
Wassermann antigen, 885–886
Wassermann reaction, 883
Wassermann-Takaki phenomenon, 836
Wassermann test, 885
 modification and variations in, 886
Water, disinfection of, chlorine, 1459
Waterhouse-Friderichsen syndrome, 746
Wayson's stain, for *Yersinia pestis*, 802
Weil-Felix reaction
 in rickettsial diseases, 902, 904t, 906t
 in Rocky Mountain spotted fever, 904t, 909
 Proteus strains used in, 903
Weil's disease, 894. See also *Leptospira icterohemorrhagiae.*
Western equine encephalitis
 epidemiology, 1389, *1389*
 pathogenesis, 1388
 pathologic features, 1388
 prevention and control, 1391
Western equine encephalitis virus, *1384*
 multiplication in, 1386, *1386, 1387*
 in culture, *1145*
West Nile fever
 epidemiology, 1390
 pathogenesis, 1388
West Nile virus, interferon production with, 1180

Wheal-and-erythema response. See
    also *Anaphylaxis, cuta-
    neous.*
  complement and, 522
  due to penicillin, 583n
White-graft reaction, 611
Whooping cough. See also *Borde-
    tella pertussis.*
  adenoviruses in, 799
  bronchopneumonia in, 798
  death rate, 799, 800
  encephalitis in, 799
  immunity to, 799
  incidence, 800, 800t
  laboratory diagnosis, 799
  non-*Pertussis* organisms causing,
    799
  pathology, 635
  pneumonia due to, 799
  prevention, 799–800
  treatment, 799
Widal test, 775
Wine
  diseases of, 6
  fermentation in, 46
Wiskott-Aldrich syndrome, 572
Witebsky's criteria, 593
Woolsorters' disease, 823
Wound(s)
  Bacteroides infection of, 787
  fungal spores in, 980
  puncture, clostridia in, 837, 837n,
    838
  tetanus due to, 835

X

Xanthomonas phaseoli, 184
Xenograft, 610
Xenopsylla cheopis
  in plague, 803
  in transmission of *Rickettsia
    typhi,* 907, 908
Xeroderma pigmentosum, 240
X-irradiation
  Drosophila affected by, 253
  energy of, 237
  immunosuppression by, 500–501
  ion clusters from, 237
  lymphocytes affected by,
    500, *501*
  mutations due to, 259
  virus inactivation by, 1464
X-rays, chest, in tuberculosis pre-
    vention, 859

Y

Yaba virus, 1261
Yaws, 891. See also *Treponema
    pertenue.*

Yaws *(continued)*
  TPI test in, 886
Yeast(s). See also *Fungus.*
  alcoholic fermentation by, 5, 44,
    44
  biotin produced by, 91t
  borrelidin action on, 303
  brewer's. See *Saccharomyces cere-
    visiae.*
  budding in, 128, 971
  cell of, birth and bud scar on,
    972, 974t
  cell wall of, 968
  cycloheximide inhibition of, 984
  cytology, 967, *968*
  enzyme induction and repression
    in, 970
  extract of, added to bacterial
    culture media, 94
  facilitated membrane transport
    systems in, 132
  fatty acid synthesis in, 125n
  fission in, 971
  genes in genome, 216
  genetic aspects, 970
  growth, 967
    pH and, 93
  in study of molecular genetics, 15
  killed by silver ions, 1459
  metabolism, 970
  pantothenic acid produced by,
    91t
  protoplasts from, 969
  pyridoxine produced by, 91t
  pyridoxol produced by, 91t
  respiration of, 174
  starch hydrolysis by, 95
steroid synthesis in, 125
  structure, 967
  thiamine produced by, 91t
  vitamins required by, 90
Yellow fever
  antibody protection in, 1209
  dengue fever coexisting with,
    1183
  epidemiology, 1390
  immunity to, 1210, 1211
  pathogenesis, 1387, 1388
  prevention and control, 1391
Yellow fever virus
  discovery, 9
  history, 1378–1379
  host range of, 1385
  immunologic characteristics, 1384
  viral interference with, 1173
Yersinia(e), 806
Yersinia enterocolitica, 806
  antigen produced by, 760
Yersinia pestis, 802–806. See also
    Plague.
  antigens of, 803
  genetic variation in, 803
  metabolism, 803
  morphology, 802–803, *803*
  pathogenicity, 803–804

Yersinia pestis (continued)
  toxins of, 803
  virulence, 804
Yersinia pseudotuberculosis, 806
  pestocin I inhibition of, 804

Z

Zinc ion, bacterial requirement of,
    92
Zoonosis(es), 946n
Zymogens, bacteria, 95
Zymomonas, taxonomic charac-
    teristics, *34*
Zymosan, in complement activa-
    tion, 520

73 74 75 76 77 78   9 8 7 6 5 4 3